Ecoregions

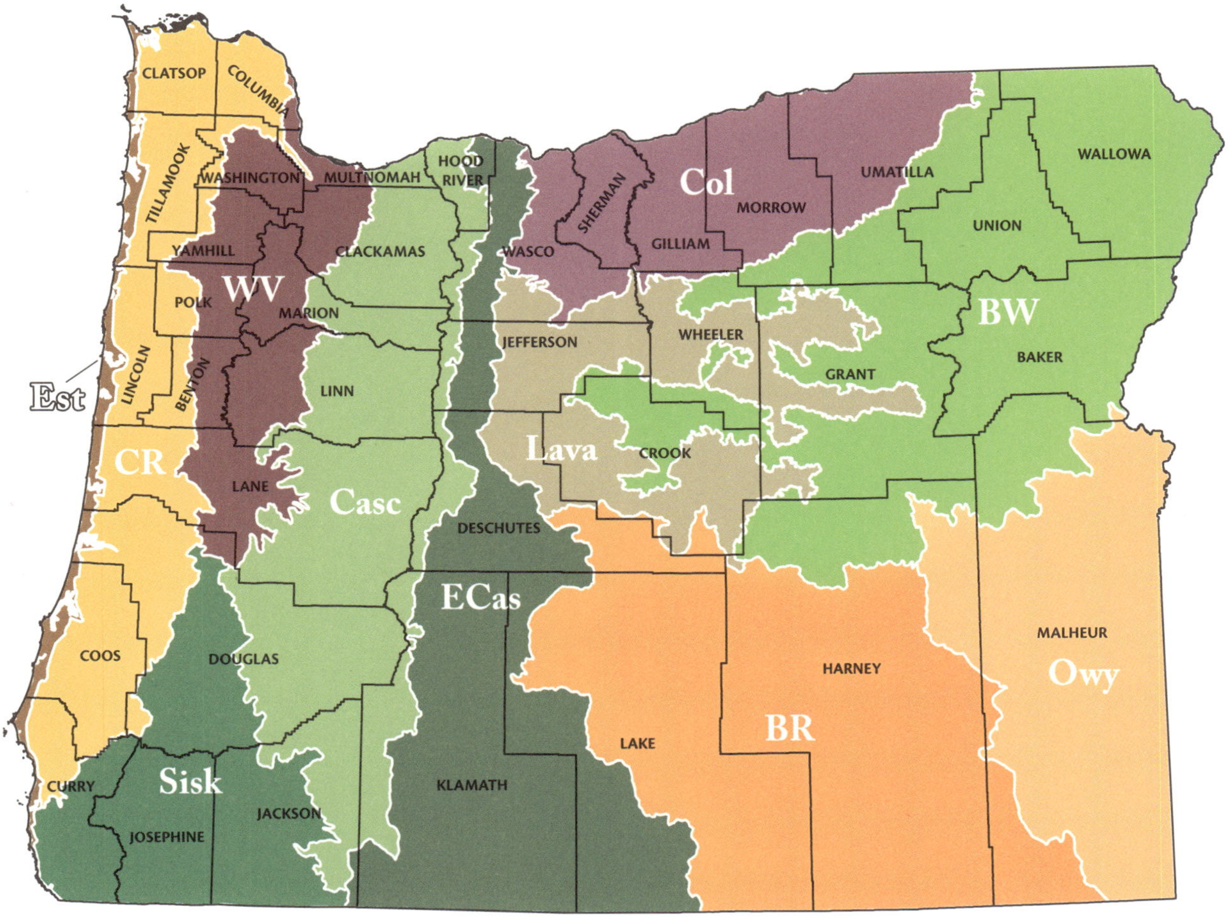

BR	Basin and Range
BW	Blue and Wallowa Mountains
Casc	Cascades
CR	Coast Range
Col	Columbia Basin
ECas	East Cascade Slopes and Foothills
Est	Estuarine Coast
Lava	Lava Plains
Owy	Owyhee Uplands
Sisk	Siskiyous
WV	Willamette Valley

Flora of Oregon

OregonFlora
Dept. of Botany & Plant Pathology
Oregon State University
Corvallis, OR 97331-2902, U.S.A.
www.oregonflora.org

John Myers, principal illustrator
Tanya Harvey, layout and design

Flora of Oregon

Volume 2: Dicots A–F

Stephen C. Meyers, Thea Jaster,
Katie E. Mitchell, Tanya Harvey & Linda K. Hardison, eds.

BRIT
PRESS
Botanical Research Institute of Texas
Fort Worth, Texas
2020

FLORA OF OREGON

Printed in Korea

Please cite this book as follows:

Meyers, S.C., T. Jaster, K.E. Mitchell, T. Harvey & L.K. Hardison, eds. 2020. Flora of Oregon. Volume 2: Dicots A–F. Botanical Research Institute of Texas, Fort Worth, Texas, U.S.A.

Published and distributed by:

Botanical Research Institute of Texas Press
1700 University Drive
Fort Worth, Texas 76107-3400, U.S.A.
817-332-4441
Fax: 817-332-4112
www.britpress.org
shop.brit.org

ISBN: 978-1-889878-61-4

Library of Congress Cataloging-in-Publication Data

Flora of Oregon. Volume 2, Dicots / Stephen C. Meyers, Thea Jaster, Katie E. Mitchell, Tanya Harvey, Linda K. Hardison, editors. — First edition.

volumes cm
Includes bibliographical references and index.
ISBN 978-1-889878-61-4 (alk. paper)
1. Plants—Oregon. 2. Dicots—Oregon. I. Meyers, Stephen C., 1966– editor. II. Title: Dicots A–F.
QK182.F56 2015
581.709795—dc23 2015011057

BRIT
PRESS

MIX
Paper from responsible sources
FSC
www.fsc.org
FSC® C023083

FLORA OF OREGON

is dedicated to

Betty Soreng and Rob J. Soreng

The generosity of the Sorengs reflects two generations of abiding respect and appreciation for the natural resources of Oregon. Their sustained support has allowed OregonFlora to share information with a broad audience. The Sorengs' impact on advancing the understanding of Oregon's botanical resources will serve as an enduring legacy to their scholarship and stewardship of our state's plant life.

The mission of OregonFlora is to engage researchers, plant growers, land managers, and nature enthusiasts by creating the basis of knowledge of Oregon's plants.

DIGITAL RESOURCES

Additional information is presented digitally through the OregonFlora's website, www.oregonflora.org, which is a Symbiota portal. Some of the resources available include:

Taxon Profile Pages

Descriptions, synonymy, images, distribution maps, and external links for every taxon in the state.

Interactive Key

An identification tool based on the plant features recognized by the user. Beginning with a list of species from the surrounding designated area, characters are selected that match the unknown plant to narrow the possibilities.

Mapping Module

Users can draw a shape on the interactive map to learn the plant diversity of the area or enter plant names to view their distribution across Oregon and beyond. The module is supported by a database containing more than 878,000 occurrence records of Oregon's 4,671 terminal taxonomic concepts. Records include *Flora* vouchers (specimens of the OSU Herbarium and other selected records that serve as documentation for all taxa treated in the *Flora of Oregon*), specimens from other Symbiota portal contributors, photo-vouchered observations, and unvouchered observations provided by governmental agencies, organizations, the Native Plant Society of Oregon, and individuals.

Grow Natives

An interactive tool to guide the use of native plant species in gardens and landscapes statewide, offering profile pages, plant collections, and filters to select species with desired features.

Oregon Wildflowers

A mobile app for rapid identification of over 1000 wildflowers, vines, and shrubs statewide.

CONTENTS

OREGONFLORA

Staff

Director
Dr. Linda K. Hardison

Taxonomic Director
Dr. Stephen C. Meyers

Database Managers, Botanists
Thea Jaster
Katie E. Mitchell

Principal Illustrator
John Myers

Graphic Designer, Layout Artist
Tanya Harvey

Advisory Council (2016–2020)

Lynda Boyer, Salem

Jason Bradford, Corvallis

Daniel Luoma, Corvallis

Will McClatchey, Eugene

Joan Seevers, Ashland

Robert Soreng, Washington, DC

Contributors to *Flora of Oregon*, Volume 2

Authors

Ihsan A. Al-Shehbaz
Missouri Botanical Garden, St. Loius, MO

Jason A. Alexander
University of California Berkeley, Berkeley, CA

Geraldine A. Allen
University of Victoria, Victoria, BC

Bruce G. Baldwin
University of California Berkeley, Berkeley, CA

Richard E. Brainerd
Carex Working Group, Corvallis, OR

Steven L. Broich
Corvallis, OR

Gerald D. Carr
Oregon State University, Corvallis, OR

Robert L. Carr
Eastern Washington University, Cheney, WA

Kenton L. Chambers
Oregon State University, Corvallis, OR

Bridget N. Chipman
Corvallis, OR

James I. Cohen
Kettering University, Flint, MI

Ashley Egan
Utah Valley University, Orem, UT

Christian Feuillet
OregonFlora, Corvallis, OR

Mark Fishbein
Oklahoma State University, Stillwater, OK

C. Matt Guilliams
Santa Barbara Botanic Garden, Santa Barbara, CA

Raul Gutierrez
EPA Region 6, New Orleans, LA

Richard R. Halse
Oregon State University, Corvallis, OR

Ronald L. Hartman (deceased)
University of Wyoming, Laramie, WY

Kristen E. Hasenstab-Lehman
Santa Barbara Botanic Garden, Santa Barbara, CA

Thea Jaster
OregonFlora, Corvallis, OR

Ronald B. Kelley
Eastern Oregon University, La Grande, OR

Nurul Ain Khalib
Oregon State University, Corvallis, OR

Donald H. Les
University of Connecticut, Storrs, CT

Melanie A. Link-Pérez
Eastern Kentucky University, Richmond, KY

Luciano Mancera
Texas A&M International University, Laredo, TX

Dominic Maze
City of Portland, Portland, OR

Richard P. McNeill
Rancho Santa Ana Botanic Garden, Claremont, CA

Robert J. Meinke
Oregon State University, Corvallis, OR

Stephen C. Meyers
OregonFlora, Corvallis, OR

Katie E. Mitchell
OregonFlora, Corvallis, OR

Barbara Montoya
Texas A&M International University, Laredo, TX

Abigail J. Moore
University of Oklahoma, Norman, OK

Nancy R. Morin
University of California Herbarium, Berkeley, CA

Mare Nazaire
Rancho Santa Ana Botanic Garden, Claremont, CA

Rory Nichols
Lebanon, OR

Nick Otting
Carex Working Group, Corvallis, OR

Tamra Dickson Prior
Aloha, OR

Richard K. Rabeler
University of Michigan, Ann Arbor, MI

Hamid Razifard
University of Massachusetts Amherst, Amherst, MA

James P. Riser II
Madison, WI

Cindy Roché
Bend, OR

Cameron J. Ross
Oregon State University, Corvallis, OR

Christopher A. Royce
Albany, OR

Erika Saenz
Texas A&M International University, Laredo, TX

Paul Severns
University of Georgia, Athens, GA

Philipp W. Simon
University of Wisconsin-Madison, Madison, WI

Michael G. Simpson
San Diego State University, San Diego, CA

David M. Spooner
University of Wisconsin-Madison, Madison, WI

Scott Sundberg (deceased)
OregonFlora, Corvallis, OR

Debra K. Trock
California Academy of Sciences, San Francisco, CA

Gordon C. Tucker
Eastern Illinois University, Charleston, IL

Michael A. Vincent
Miami University, Oxford, OH

Alan T. Whittemore
U.S. National Arboretum, Washington, DC

Barbara L. Wilson
Carex Working Group, Corvallis, OR

Peter F. Zika
University of Washington, Seattle, WA

Artists

pages

Photographers

figures

ACKNOWLEDGMENTS

The publication of Volume 2 of the *Flora of Oregon* marks the transition into a second quarter-century for our organization. The name change—from Oregon Flora Project to OregonFlora—is a small way to reflect the growth of capacity and the evolution of our program to provide not only comprehensive information about the plants of our state, but also the creation of knowledge around its application and relevancy. This is due in large part to the talented and dedicated staff of OregonFlora: Tanya Harvey, Thea Jaster, Stephen Meyers, Katie Mitchell, and John Myers. Their commitment to this project and the sharing of their enthusiasm with scholars and community members alike expands the meaningfulness of our work. The commitment of several who, in their generosity of spirit, toil alongside our staff with a comparable regularity never ceases to amaze me. Heartfelt thanks go to Kenton Chambers, Christian Feuillet, Dennis Albert, Richard Halse, and Gerald Carr for their contributions of careers' worth of expertise in aid to OregonFlora.

Student involvement provides not only assistance with our many projects, but also an opportunity to mentor young scientists and provide them with skills and perspectives that support their development. Along with numerous students we have employed, funding from the Oregon State University (OSU) Beginning/Continuing Researchers program within the College of Agricultural Sciences, and the URSA Engage program helped support six undergraduates with funding for their OregonFlora research projects.

OregonFlora's relationship with the OSU Herbarium has given us access to the essential materials needed to create this Flora. Thanks go to Aaron Liston, Herbarium Director, for supporting this mutually beneficial partnership. We give our appreciation to the Department of Botany & Plant Pathology at OSU for providing an institutional home for OregonFlora. Special thanks for the people providing supporting services at OSU with whom we regularly interact: the staff at COSINe for hosting our website and providing IT support, Sarah Child for managing our research grants and accounting, and Charlene Wilkinson of the Agricultural Research Foundation for her tireless advocacy and for managing our charitable donations. The Agricultural Research Foundation serves as our fiscal sponsor and facilitates our ability to maintain and interact with donors and foundation-based funders.

The pages that follow list the many benefactors that have financially and programmatically supported OregonFlora. Each one has been integral to the success of our program, and we are honored to consider them part of the OregonFlora team.

The production of Volume 2 of *Flora of Oregon* is possible because of many people's help. The floristic authors and editors shared their expertise and time to contribute to this work. Whether working on a single taxon or an entire family, the authors' effort to prepare a treatment that helps our users readily identify the plants of our state has been what collectively creates a useable, therefore valuable reference. Among the many contributors to this volume we extend additional thanks to Jason Alexander, Ihsan Al-Shehbaz, the Carex Working Group, Kenton Chambers, Christian Feuillet, Richard Halse, Ronald Hartman, Richard Rabeler, and Alan Whittemore for their extensive floristic treatment contributions. We appreciate the reviewers who have critically read the manuscripts and tested keys for Volume 2: Alex Abair, Jerry Black, Christian Feuillet, Sunja Goldenrose, Kirsten Hill, Nathan LeClear, and Kathy Merrifield. James Hall has given his kind permission to again let us feature on the book jacket an illustration by his late wife Bonnie. The front chapters of each volume provide a context that applies the information held within the floristic treatments and are further supported with details gathered in appendices. We are grateful to highlight the expertise and vast knowledge that these authors bring and integrate into a floristic work: Dennis Albert, Melanie Link-Pérez, Rick Martinson, Jeffrey Miller, and Andrew Moldenke. The color photographs add immensely to the content; we appreciate the photographers listed on page x allowing us to use their work and acknowledge Lynda Boyer, Clay Gautier, Tanya Harvey, Kathryn Prive, and Susie Savoie for images of their home gardens. We thank Dennis Albert, Karl Anderson, Dayne Galash, August Jackson, Sheila Klest, George Kral, Richard Little, Troy Maddux, Chris Marshall, Andony Melathopoulos, and Amadeo Ramos for their assistance with and review of the chapters, appendices, glossary, and index.

The digital presentation of OregonFlora information is happening simultaneously with the printing of Volumes of the *Flora*; it provides us with challenging and exciting ways to share and apply floristic data. We have been concurrently working on a significant redesign of the OregonFlora website since 2016. The two formats—digital and paper—are essential to each other, and we express gratitude for those whose support of the website also drives progress on the printed volumes. We thank Ben Brandt, Neil Cobb, Evin Dunn, and Ed Gilbert of Symbiota, along with Lisa Gray, Tom Kerns, Matt Giraud, Arthur Parker, and Roger Leigh for implementing our vision of an engaging and interactive website based on an extensive dataset. Special thanks go to Carl Grimm for securing and rallying the support of Metro and the ACE NW team to develop the *Grow Natives* website tools.

In the years since the publication of Volume 1, the Advisory Council has helped with a sponsorship campaign, strategic planning, benefit concerts, and implementation of new research initiatives for OregonFlora. In addition to the current

Council (listed on page viii) and past Board members (Gail Baker, Jeff Cook, Sam Friedman, Aaron Liston, Carol Savonen, Paul Slichter, and José Vega), thanks go to Margaret Conover, George Kral, Amy Rossman, and Marcia Shaw for their advice and assistance with strategic planning.

It must be mentioned that the final work on this volume has been completed during the global pandemic caused by the novel coronavirus, SARS-CoV-2. The necessary adaptations and the accommodations that have been made to proceed in good health remind me of the value of friendship, consideration for one's fellow humans, and the nourishing presence of our natural world. It is my hope that the efforts of OregonFlora will foster a greater awareness, understanding, and appreciation of the plant diversity that surrounds us, and that we can sustain this perspective moving forward.

I have attempted to accurately acknowledge the many generous contributions made on behalf of this effort; please accept my sincere apologies for any errors or omissions.

Linda Hardison

Perideridia bolanderi & Callophrys gryneus

BENEFACTORS

KEY SUPPORTERS

John and Betty Soreng Environmental Fund of The Oregon Community Foundation

Department of Botany & Plant Pathology, Oregon State University

Oregon/Washington Bureau of Land Management

Native Plant Society of Oregon

Barbara J. Mendius Administrative Trust

SUPPORTING ORGANIZATIONS

Clean Water Services

Metro (Portland OR)

Natural History Museum Fund of Oregon State University (OSU)

Oregon Department of Agriculture Specialty Crops Block Grant Program

Oregon Watershed Enhancement Board

OSU College of Agricultural Sciences Undergraduate Researcher Support Program

OSU Herbarium

OSU Office of Undergraduate Research, Scholarship & the Arts

Tualatin Soil & Water Conservation District

USDA Natural Resources Conservation Service Oregon

West Multnomah Soil & Water Conservation District

Willamette Institute for Biological Control

DONORS

6/1/2015–5/1/2020

Major Donors

Dennis Albert
Karl Anderson
Connie Battaile
Wilbur Bluhm
Kay Butler
Gerald & Barbara Carr
George & Fanny Carroll
Kenton & Henrietta Chambers
Nancy Chapman
Kent Coe
Ellen A. Crumb
Howell Estes
A.J. & Adriana Giustina
Theo Glenn
Goulder Family Foundation
Everett Hansen
Tanya Harvey
Noel & Patricia Holmgren
Lawrence Janeway
Anne Jaster
Albert Johnson
Terry Lettenmaier
Jenny Lippert
Rhoda & Glen Love
Daniel Luoma & Joyce Eberhart
Louise Marquering
Bruce McCune & Patricia Muir
NPSO Emerald Chapter
NPSO Siskiyou Chapter
NPSO Umpqua Valley Chapter
North American Rock Garden Society, Siskiyou Chapter
Richard O'Donnell
Sandy Poinsett
Susan Polchert
Dell Rhodes & Will Bloch
Cindy Roché & Robert Korfhage
Robert A. Ross
Amy Rossman & Christian Feuillet
Charlene Simpson
Paul Slichter
Kerri Stubb
Kent Sundberg & Beth de la Fuente
George Van Vechten
Bruce E. Waugh
Sunia Yang
Desi & Karen Zamudio

Donors

Dana Abel & Charles Quinn
Aislinn Adams
Norma Alberthal
Mary Allardt
Jon Anderson
Joan Arsenault
Yvonne Babb
Betty Bahn
Gail Baker & Clay Gautier
Dan Bates
Carol Baumann
Benton County Master Gardener Association
Benton Soil & Water Conservation District
Julie Biddle & Tom Robertson
Ken & Linda Bierly
Diane Bilderback
Melanie & Kent Bjorge
Lisa Blackburn
Paul Bloch
James & Carol Bluhm
Lynda Boyer
Jason & Kristin Bradford
Kathy Brewer
Paula Brooks
Nancy Broshot
Mary Bushman
Richard Callagan
Pam Camp
Gretchen & Denis Carnaby
Darlene Chirman
Lynda Ciuffetti & Tom Wolpert
Alisa Clark
Jason Clark
Melody Clarkson
Stella Coakley
Glen Cole
Barbara Combs
Jeff Cook
Stanton Cook
Beth Lowe Corbin
Geraldine Custer
Elisabeth Dally
Dancing Oaks Nursery
Sonja DeHerdt
David & Janet Dobak
Kathleen Donham
Lorena Dornfeld
Janet Dorow
Duerksen & Associates Inc.
Sabine Dutoit
Thomas Dyke
Lynn Edwards
Kathleen Ellis
Margaret Ellis
Elizabeth Evans
Jenifer & Roger Ferriel
Robert Fischer
Elizabeth Fisher
James & Eileen Flory
Robert & Elizabeth Frenkel
Sam Friedman
Stuart Garrett
Edwin Gerow & Cheryl Chevis
Sylvia Giustina
Vera Gottlieb
Andrew Gray
Brenda Grewell
Candice Guth
Barbara Halliday
Sue Hamilton
Jeff Handley
Linda Hardie
Jane Hartline
Michael Hartman
Phil Hays
Andrea Heid
Don Heller
Kenneth Himes
James Hoffmann
Beth Horn
Signe Hurd
Hal Hushbeck
Richard & Deanna Iltis
Cheryl Ingersoll
Institute for Applied Ecology
Thea Jaster
Jan Jeddeloh
Judy Jernstedt
Kathryn Jernstedt
Morris & Margaret Johnson
Keith Karoly
Tom Kaye
Doramay Keasbey
Susan Kephart
Marilynne Keyser
Ulo Kiigemagi
Sheila Klest
Ron Klump
Karol Kochsmeier
John Koenig
Rae Kozloff
Bud LaFortune
Lane Community College, Science Division
Lynn Larsen
Richard Larson
Eliza Lindsay
Linn County Master Gardener Assocation
Barney Lipscomb
Aaron & Sara Liston
Richard Little
Jennifer Lorang
Cynthia Lord
Jay & Ann Lunn
Paul MacMillan
Sarah Malaby
Barry Maletzky
Stephen Mark
Craig & Susan Markham
Kathleen Marquardt
Larry Marshik
Rick Martinson
Barbara Massey
Judy Maule
Sara Mauritz
Cheshire Mayrsohn
Dale & Margaret McBride
Cheryl McCaffrey
Wendy McClure
David McLaughlin
Stanley & Margaret Meierhenry
Kathy Merrifield
Greg & Gail Miles
Marilyn Miller
Jeanne & Frank Moore
Molly Morison
Susan Morton
Mountain Rose Herbs
Linda Mullens
Barbara Mumblo
Christopher Mundt
NPSO Corvallis Chapter
NPSO Portland Chapter
Susan Nelson
Gene Newcomb
Julie & Mark Nichols
Pat Nichols

North American Rock Garden Society, SW Oregon Chapter
Kathryn Oliver-Garnett
Bonnie Brunkow Olson
Oregon Dairy Farmers Association
Suzanne Ortiz
John & Helen Ost
Christine Ott-Hopkins
Nick Otting
Patricia Otto
Rob Pabst
Theodore & Laramie Palmer
Betsy Parry
Warren Pavlat
Kathy Pendergrass
Christina Peterson
Itsue & Frederick Pfund
PhotoAssist, Inc.
Allen T. Pierce
Elaine Plaisance
Jim Platt
Ginny & Jim Post
Ray & Peg Prag
Richard Prusz
Carol & Kurt Putnam
Donna Rawson
Phyllis Reynolds
Doug Ripley
Paul & Mary Ann Roberts
Mike Robinson
Wayne Rolle
Bruce & Alice Ronald
Ben Ross
Barbara Roy
Leonard Rydell
Judi Sanders
Susan Saul
Kathleen Sayce
Jim & Marguerite Schellentrager
Patricia Schleuning
Rachel Schwindt
Ernst Schwintzer
Thomas Seddon
Elaine Sedlack
Joan Seevers
Martin Skinner
Alice C. Smith
James D. Smith
Angela Sondenaa
Janelle Sorensen
Darlene Southworth
Julianne Spelletich
Hope Stanton
Deanna Sterett
Peter Stoel & Karen Josephson
William Stone
Greg Sundberg & Susan Sears
Mark Sundberg & Ethel Sennhauser
Ron & Pam Swisher
Janet Throop
Thomas Torak
Julia Unferth
Michael Van Winkle
David Wagner
Jeff Walker
Jay Walters
Megan Ward
Washington Native Plant Society
Ellen Watrous
Nancy Weber
Rick Westcott
Kermit Williams
Mark Griswold Wilson
Mark V. Wilson
Dennis Woodland
Carolyn Wright
Jo Ann Yeager
Karen Yoerger
James Young
Walt Yungen
Gary Zamzow
Judy Zettergren
Don Zobel

SPONSORSHIPS

Donors who have sponsored specific parts of the Flora

Contributions of $135 for a taxonomic treatment, $250 for an illustration, $400 for a front chapter

Volumes 1 & 2

Sarah Canham
treatment of *Astragalus tyghensis*

Kenton & Henrietta Chambers
illustrations of *Microseris borealis* and *M. laciniata* ssp. *laciniata*

Melody Clarkson
in honor of the Flora Fans hiking group
treatment of *Calypso bulbosa*

Dancing Oaks Nursery, Fred Weisensee & Leonard Foltz
illustration of *Asclepias speciosa*

Jan & Dave Dobak
treatments of *Cladothamnus pyroliflorus, Menziesia ferruginea*, and *Moneses uniflora*

Peter Gladhart
in honor of Emily Gladhart, Winter's Hill Estate
illustration of *Anaphalis margaritacea*

Robin Hansen, Hansen Nursery
treatment of *Vaccinium uliginosum*

Jane Hartline, Sauvie Island Natives
illustration of *Symphyotrichum subspicatum*

Humble Roots Nursery
Native Plant Gardening chapter

Leach Garden Friends
in honor of Lilla Leach
illustration of *Kalmiopsis leachiana*

Jennifer Lippert
treatment of Betulaceae (12 taxa)

Aaron & Sara Liston
treatment of *Agrostemma githago*

Daniel Luoma
illustration of *Pleuricospora fimbriolata*

Rick Martinson
Native Plant Gardening chapter

Kristi Mergenthaler
treatment of *Toxicodendron diversilobum*

Morning Glory Café, Claire Aldersong
treatment and illustration of *Calystegia soldanella*

NPSO Cheahmill Chapter
illustration of *Lupinus oerganus*

NPSO Mid-Columbia Chapter
illustration of *Lomatium suksdorfii*

NPSO Portland Chapter
Insect and Native Plant Gardening chapters

NPSO Siskiyou Chapter
treatments of *Calochortus greenei*, *Fritillaria gentneri*, *Quercus sadleriana,* and *Q. kelloggii*, and Insect chapter

NPSO State
illustrations of *Berberis aquifolium, Kalmiopsis fragrans*

Ray and Peg Prag, Forestfarm
Native Plant Gardening and Insect chapters

Mary Ann & Paul Roberts
Insect chapter

Mike Robinson
Native Plant Gardening chapter

Barbara Roy
Native Plant Gardening chapter

Jim & Marguerite Schellentrager
treatment and illustration of *Dudleya farinosa*

Johanna Schmitt & James Dunn
in honor of Gail Baker
illustration of *Streptanthus howellii*

Paul Slichter
treatment and illustration of *Lupinus sericeus*

Kelli Van Norman
treatment of *Arbutus menziesii*

Ellen Watrous
Insect chapter

Bruce Waugh
Native Plant Gardening chapter

Carolyn Wright
treatment of *Astragalus diaphanus*

Karen L. Yoerger
illustration and treatment of *Balsamorhiza sagittata*

Volume 3

Mary Allardt
illustration of *Hypericum anagalloides*

Gail Baker
in honor of Margaret & Donald Baker
illustration of *Rosa*

Gail Baker & Clay Gautier
treatment of *Kopsiopsis hookeri*

Connie Battaile
treatment of *Salix* (39 taxa)

Lisa Blackburn
treatment of *Ceanothus integerrimus*

Stella Coakley
in honor of Kenton & Henrietta Chambers
Climate chapter

Eugene Heritage Rose Group
illustration of *Rosa nutkana*

Kathryn & Pryor Garnett
illustration of *Amelanchier alnifolia*

Hawthorne Auto Clinic, Inc., Elisabeth Dally
Climate chapter

Barbara Mathieson
in honor of Tom Mathieson
Climate chapter

Kristi Mergenthaler
treatment of *Meconella oregana*

Mountain Rose Herbs
treatments and illustrations of *Rosa gymnocarpa* and *Rosa nutkana*

NPSO High Desert Chapter
illustration of *Penstemon peckii*

NPSO Mid-Columbia Chapter
illustration of *Suksdorfia violacea*

NPSO Portland Chapter
treatment and illustration of *Viburnum edule*

NPSO Siskiyou Chapter
treatments of *Lewisia cotyledon*, *Lewisia leeana, Pedicularis howellii*, *Ranunculus austro-oreganus*, and *Sidalcea hickmanii*

Northwest Ecological Research Institute
illustrations of *Eriogonum* (4 species)

Jim & Marguerite Schellentrager
Ethnobotany chapter

Elaine Sedlack
illustration of *Rosa nutkana*

Paul Slichter
treatment and illustration of *Diplacus deschutesensis*

Kent Sundberg
in honor of Norman & Donna Sundberg
treatments of *Darlingtonia californica* and *Valeriana sitchensis*

Mark Sundberg
illustrations of *Anemone deltoidea, Dodecatheon hendersonii,* and *Gentiana sceptrum,* Climate chapter

Bruce Waugh
Ethnobotany and Climate chapters

Karen L. Yoerger
Ethnobotany chapter

INTRODUCTION

SCOPE OF WORK

The *Flora of Oregon* is a three-volume reference publication circumscribing the vascular plants of Oregon—both native and naturalized—that grow without cultivation. It covers Oregon's 254,799 km^2 (98,379 mi^2), bounded on its southern border by the 42nd parallel; its eastern and northern borders predominately by the Snake and Columbia Rivers, respectively; and its western border by the Pacific Ocean. Additional information supporting this publication can be found at the OregonFlora (OF) website (www.oregonflora.org).

Volume 1 includes the ferns and fern allies (pteridophytes), gymnosperms, and monocots. *Volume 2* includes the dicot families A–F. *Volume 3* will cover the remaining families of dicots. The nomenclatural status of dicots referenced in this volume is based on currently existing floristic treatments for *Volume 2* and *Volume 3* of the *Flora of Oregon* or, if not available, on the OregonFlora Checklist version 1.8, dated November 2019.

The *Flora of Oregon* includes both native and exotic (non-native) plant taxa. A *native* plant taxon is defined as a floristic component of Oregon prior to European settlement, or a newly described taxon found in the state. An *exotic* or non-native plant is a floristic component that established in Oregon post-European settlement. The OF recognizes eleven categories of exotic plants and has assigned each non-native taxon to one of these categories. The categories fall into two broad groups: 1) naturalized species (exotic taxa that have persisted in the wild for at least 3–5 years and have spread beyond the area of their original introduction); and 2) non-naturalized species that are persisting or spreading within a humanly altered environment. Examples include a) weeds and b) ephemerals such as waifs (solitary or small groups of exotic plants not documented as having persisted beyond one season) and ballast plants (waifs that have appeared on ship ballast).

Native and naturalized exotic taxa accepted by the OF receive full descriptions. An *accepted taxon* is one with documented occurrences in Oregon and at least one *flora voucher*, that is, a herbarium specimen deposited in the Oregon State University Herbarium (OSC, ORE, and WILLU) or in some cases at another institution. Frequently occurring interspecific hybrids and those that form persistent, self-maintaining populations also receive full descriptions. Rarely occurring hybrids and historic or waif taxa are generally discussed but usually not given full treatments. Waifs discussed in the *Flora of Oregon* include ballast plants, taxa recently escaped from cultivation or otherwise introduced, and taxa from adjacent states that have been reported from a single Oregon location or plant population.

A *terminal taxon* refers to the narrowest taxonomic category recognized by the OregonFlora; a terminal taxon may be a species, subspecies, variety, or forma. An *endemic* is a native terminal taxon whose occurrence has not been documented in adjacent states. At the time of publication of *Volume 2*, there are 4,769 terminal taxa that are presently occurring within the state or that are presumed extirpated. This represents a 9% increase from the 4,378 taxa listed in the initial drafts of the Oregon Flora Project Checklist produced in 1995. The increase is due primarily to the range extension of plants from adjacent states into Oregon, the introduction of exotic species, and taxonomic revisions. *Volume 2* contains full descriptions of 1,668 dicot terminal taxa (Table 1, Figure 1).

Table 1. Floristic summary of the vascular plant taxa of Oregon treated in Volume 2 of the *Flora of Oregon*.

Dicots A–F				
Number of terminal taxa with full floristic treatment				1,668
Total number of terminal taxa (native, naturalized, non-naturalized, sporadic hybrids)				1,874
Number of families				39
Number of genera				436
Number of species rank taxa				1,516
Number of native taxa				1,325
Number of exotic taxa				343
Number of taxa endemic to Oregon				61
Number of taxa in largest families				
family	native	exotic	total	% native
Asteraceae	449	98	547	82
Fabaceae	204	61	265	77
Brassicaceae	147	62	209	70
Apiaceae	108	12	120	90
Caryophyllaceae	72	38	110	66
Boraginaceae	91	16	107	85
Ericaceae	55	3	58	95
Amaranthaceae	42	15	57	74

Before the publication of the *Flora of Oregon*, the most recent comprehensive treatment of the state's vascular plants was Morton Peck's *A Manual of the Higher Plants of Oregon*, which was last updated in 1961.

CONVENTIONS USED

Book Design

Organization

The introductory material is followed by identification keys and detailed descriptions of the families of vascular plants treated within this volume (dicots A–F). The families and content within each family are arranged alphabetically.

Illustrations

Each genus includes an illustration of at least one species, with few exceptions. For genera with more than four species, typically 20–25% of the species are illustrated. In most cases, illustrations represent the plant's habit, flower, and specific details that aid identification. If a taxon is illustrated, the page number for the drawing appears in parentheses to the right of the scientific name in the description.

Maps

The distribution of each terminal taxon within each Oregon county and ecoregion is represented on a map. Each ecoregion in which a taxon occurs is highlighted. The map also shows plant occurrence records reflecting flora vouchers. The identification of the majority of these specimens was confirmed for the OF. Details of each record are available on the OF website.

Measurements

Measurements are given in metric units, with metric conversion scales at the end of each volume. Unless indicated (diameter, width, etc.), measurements refer to length.

Index

The index is a combined alphabetical listing of all common and scientific names of families, genera, species, infraspecific epithets, and synonyms within the floristic treatments.

Keys

Dichotomous keys are provided to identify unknown plants to the level of family, genus, and terminal taxon. A *dichotomous key* is a series of paired, parallel statements. A pair of statements is referred to as a couplet; each statement is referred to as a lead. Of the paired statements, the user selects the statement that most closely matches the characters of the unknown plant and rejects the alternative statement. The selected statement ends either in a scientific name—the identity of the unknown plant—or directs the user to another couplet. The user repeats the selection process until the plant's name is obtained. The user can then compare the plant's features to those in the floristic description to confirm that the identification deduced from keying the specimen is correct.

The characters within each lead are predominately morphological features discernible with the unaided eye or a hand lens. Readily evident features are listed early in the leads. Habitat and distribution information is used when necessary, as are character suites that require multiple phases of development such as mature fruits and flowers. Every opportunity has been taken to develop lead pairs that are parallel; that is, both leads ask questions about the same characters. When a critical character is unique to one of the leads in a couplet, that character is put at the end of the lead and all other elements remain in parallel. In highly diverse taxonomic groups, multiple leads may direct users to the same group in different sections; such sections of a taxonomic group will be designated "in part" in the key. All characters used in the keys appear in the descriptions.

Descriptions

All Descriptions

Descriptions at specific, subspecific, and varietal ranks delimit the features of each taxon as it occurs in Oregon. The content of each description applies to all taxonomic groups hierarchically below the one discussed; that is, the characters used to describe a family are exhibited by all described genera of that family, and a genus description applies to all described species within the genus. The general format described below is followed for all levels of descriptions.

Each treatment begins with the accepted scientific name in boldface type, followed by the common name. If there are multiple common names, the first name listed is designated by the OF as the *primary common name*. The primary common name is preferentially a published regional convention; alternatively it may be a translation of the scientific name or a name provided by the treatment author that describes a distinctive morphological feature. Each description begins with the general habit of the plant and its persistence (annual, biennial, or perennial). Characters generally are listed in the following order: from the ground up; as they appear during the plant's development; from proximal to distal; abaxial before adaxial; and staminate before pistillate. Each major structure or organ is in boldface type and followed by specific details in the following order: number, odor, shape, dimensions, color, texture, bases, margins, central area, tips, surfaces, internal parts, and exudates.

The expected quantity or measurement of a feature is described by a discrete value or range of values. Any extreme values that are uncommonly encountered are placed in parentheses. Two-dimensional measurements are first cited by length, then by width. For example, (5)7–9.5 × 2–3 mm indicates the length is normally 7–9.5 mm but may be as small as 5 mm, while the range of the width is 2–3 mm.

The morphological description is followed by information about range, habitat, and number of subsidiary taxa that are fully treated in the *Flora*. General discussion may follow.

Family Descriptions

Each family treatment begins with the family's scientific name in all caps and boldface type, followed by its common name and acknowledgment of the floristic treatment's contributor. Family names reflect the nomenclature established by the Angiosperm Phylogeny Group IV, or "APG4" (Stevens 2001 onwards) with some exceptions. If members of the family were previously classified within a different family, the original family name is also referenced.

The range and global geographic distribution of the family, its habitat, number of genera worldwide, and the number of genera that are fully treated in the *Flora* are given. When the habitat of family member taxa varies widely, details are given at a lower taxonomic level.

Genus Descriptions

The conventions found in genus descriptions follow those used in family treatments. If a genus is composed of a single species, a genus description is not included, as the details in the species description suffice. If a family is composed of a single genus, a genus description is also not included, as the details in the family description suffice.

Species Descriptions

Each species description begins with the scientific name in boldface type, followed by the publishing author and the common name(s). Below this are listed the synonyms, if any. The morphological description is followed by a paragraph describing the plant's habitat and range. Flowering time and elevation range are based on specimen label data from OSU Herbarium vouchers or information provided by the treatment contributor. Minimum elevation values are rounded down and maximum values are rounded up in 100 m increments; 50 m increments are used if the species is limited to areas with an elevation less than 100 m. The Oregon ecoregions in which the taxon has a documented occurrence are listed; ecoregion abbreviations are defined below. This is followed by information about the species' distribution beyond Oregon, first listing adjacent states and then North American and global occurrences. The species' origin (native or exotic) concludes this section. Taxa designated as endemic are indicated with a ❧ symbol.

When multiple infraspecific taxa are recognized, a key to the subtaxa follows the species description. If a single subspecies or variety occurs in Oregon, it appears in the text as if a full species.

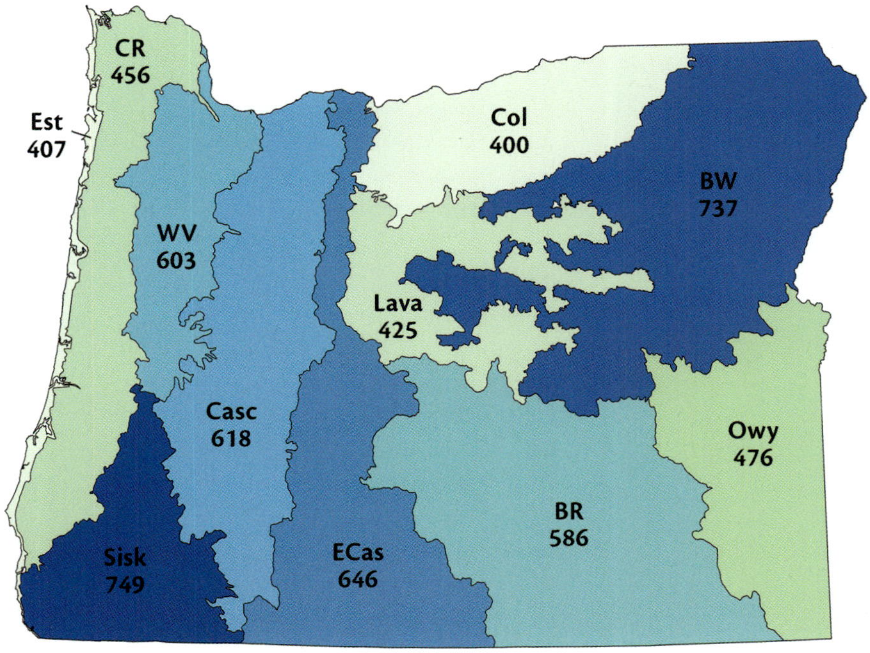

Figure 1. Map depicting the number of Volume 2, fully treated taxa (1,668) that occur in a designated ecoregion.

PHILOSOPHY OF TAXONOMIC APPROACH

The *Flora of Oregon* has been written with the goal of being accessible to a wide array of botanists and plant enthusiasts. The users of this manual include academic botanists, employees of government agencies, ecologists, gardeners, native plant growers, environmental consultants, amateur botanists, and students. To meet the needs of this diverse audience, we have made a concerted effort to present floristic treatments that are scholarly and scientific yet also readable.

The authors and reviewers of the floristic treatments and accompanying chapters range from well-established academic scholars and post-doctoral researchers to graduate students and skilled amateur botanists. Though this diversity has the potential to create problems with consistency across the *Flora*, we feel that these discordances ultimately provide value by presenting varied perspectives.

Broadly speaking, practitioners of taxonomy have always been cast into one of two opposing philosophical camps: splitters or lumpers. In reality, most competent taxonomists fall somewhere in the middle of these two extremes, sometimes splitting taxonomic concepts and sometimes lumping them, according to the logical course suggested by the data.

Some plant groups suffer from a history of inadequate or poor taxonomic work. Compounding this problem, some modern taxonomists have been reluctant to revise the conclusions of the "experts" who preceded them. Seeking to partially rectify this, we have strongly encouraged treatment authors to revise the taxonomy of their groups to reflect their research results, using synonymization, the application of previously described taxonomic concepts, and the description of new taxa.

Some authors recognize taxa that are either cryptic or have few or seemingly vague defining morphological characteristics. Although this can make a plant group appear unduly complex, it is helpful to remember that the characteristics that place a taxon on a distinct evolutionary path are not necessarily evident to human senses.

Whenever possible, we recruit authors with expertise in specific plant groups. Unfortunately, over the past several decades the number of active, university-trained taxonomists has vastly declined. As a result, we have occasionally used authors whose assigned plant group is not the focus of their work. While this unfamiliarity sometimes requires the author to "catch up" with the relevant literature, it also brings the benefit of a fresh perspective on the group. Some authors expressed a desire to devote more time and effort to the study of their assigned plant groups. This was particularly the case with complex groups and questionable taxonomic concepts for which 21st century molecular techniques could help answer longstanding questions. Unfortunately, a dearth of time and financial resources often constrained the authors' admirable enthusiasm. As a result, a hint of some authors' desire is often reflected in the *Flora* by the statement "more study is needed."

OregonFlora has, in part, relied on the eight floras and field guides that cover all or parts of the state as *key references*. Morton Peck's *A Manual of the Higher Plants of Oregon* (1961) is the most recent and complete regional flora. The remaining key references are *Illustrated Flora of the Pacific States* (Abrams and Ferris 1923–1960); *Vascular Plants of the Pacific Northwest* (Hitchcock et al. 1955–1969); *Intermountain Flora: Vascular Plants of the Intermountain West, USA* (Cronquist et al. 1972–2013); *Guide to the Plants of the Wallowa Mountains of Northeastern Oregon* (Mason 1975); *Flora of North America North of Mexico* (Flora of North America Editorial Committee, eds. 1993+); *The Jepson Manual: Higher Plants of California, second edition* (Baldwin et al. 2012); and *Flora of the Pacific Northwest, second edition* (Hitchcock and Cronquist 2018).

ABBREVIATIONS & SYMBOLS

General

2*n*	diploid chromosome number
cm	centimeter
f	forma
ft	feet
in	inch
km	kilometer
m	meter
mi	mile
mm	millimeter
µm	micron or micrometer (0.001 mm)
s.l.	*sensu lato* (in the broad sense)
s.s.	*sensu stricto* (in the narrow sense)
sp.	species (singular)
spp.	species (plural)
ssp.	subspecies (singular)
sspp.	subspecies (plural)
var.	variety

Symbols

	taxon is endemic to Oregon
♀	female
♂	male
°	degrees
×	multiplication; used for range of measurements and to indicate hybrid taxon
×	multiplication or times; used for amount of magnification
=	equal to
<	less than; shorter than
≤	less than or equal to; up to; at most
>	greater than; longer than
≥	greater than or equal to; at least
~	circa; approximately
≈	approximately equal to

Ecoregions

BR	Basin and Range
BW	Blue and Wallowa Mountains
Casc	Cascades
Col	Columbia Basin
CR	Coast Range
ECas	Eastern Cascade Slopes and Foothills
Est	Estuarine Coast
Lava	Lava Plains

Owy	Owyhee Uplands
Sisk	Siskiyou Mountains
WV	Willamette Valley

Institutions

BLM	Bureau of Land Management
CAS	California Academy of Sciences Herbarium. San Francisco, CA
GH	Gray Herbarium, Harvard University. Cambridge, MA
MO	Missouri Botanical Garden Herbarium. St. Louis, MO
NPSO	Native Plant Society of Oregon
ODFW	Oregon Department of Fish and Wildlife
ODSL	Oregon Department of State Lands
OF	OregonFlora
OPRD	Oregon Parks and Recreation Department
ORBIC	Oregon Biodiversity Information Center
ORE	University of Oregon, Oregon State University Herbarium. Corvallis, OR
OSC	Oregon State University Herbarium. Corvallis, OR
OSU	Oregon State University. Corvallis, OR
RNA	Research Natural Area
RSA	Rancho Santa Ana Botanic Garden Herbarium. Claremont, CA
SOC	Southern Oregon University Herbarium. Ashland, OR
TNC	The Nature Conservancy
UO	University of Oregon. Eugene, OR
US	Smithsonian Institution Herbarium. Washington, DC
USFS	United States Forest Service
USFWS	United States Fish and Wildlife Service
UTC	Utah State University Herbarium. Logan, UT
WILLU	Willamette University, Oregon State University Herbarium. Corvallis, OR
WS	Washington State University Herbarium. Pullman, WA
WTU	University of Washington Herbarium. Seattle, WA

States

AK	Alaska
AL	Alabama
AR	Arkansas
AZ	Arizona
CA	California
CO	Colorado
CT	Connecticut
DC	District of Columbia
DE	Delaware
FL	Florida
GA	Georgia
HI	Hawaii
IA	Iowa
ID	Idaho
IL	Illinois
IN	Indiana
KS	Kansas
KY	Kentucky
LA	Louisiana
MA	Massachusetts
MD	Maryland
ME	Maine
MI	Michigan
MN	Minnesota
MN	New Mexico
MO	Missouri
MS	Mississippi
MT	Montana
NC	North Carolina
ND	North Dakota
NE	Nebraska
NH	New Hampshire
NJ	New Jersey
NV	Nevada
NY	New York
OH	Ohio
OK	Oklahoma
PA	Pennsylvania
RI	Rhode Island
SC	South Carolina
SD	South Dakota
TN	Tennessee
TX	Texas
US	United States
UT	Utah
VA	Virginia
VT	Vermont
WA	Washington
WI	Wisconsin
WV	West Virginia
WY	Wyoming

LANDSCAPING WITH NATIVE PLANTS

Richard W. Martinson & Linda K. Hardison

"Native plants just don't work here." While the befuddled customer at the native plant nursery might not have realized its irony, their statement captures what many gardeners and landscapers historically thought about using native plants in managed gardens and landscapes. However, growing native plants is increasingly becoming a passion for many home gardeners and landscapers, and the methods of using native species in these settings vary almost as much as do the environmental conditions across Oregon. This chapter presents a preliminary discussion on gardening with native plants and offers suggestions to planning, designing, and installing a successful native landscape. We provide a basic overview of genetic concerns in propagation practices and plant choice and consider how soil, moisture, and sunlight requirements may differ from non-native cultivars, ways to integrate native species into a landscape, and how using native plants can provide a range of ecosystem services diminished or lost through urban development.

Basic Concepts

A key question in native plant gardening is how to define a native species. All include some frame of reference based on time and space, but a single definition is elusive, and many gardeners have their own description of a native plant. OregonFlora offers this definition:

A native plant is one *"that was an element of the Oregon flora prior to European settlement, a recently described taxon found in Oregon, and/or a taxon disjunct in Oregon if it is considered native in a nearby state; all of which established in the landscape independent of direct or indirect human intervention."*

This definition includes a context of time (prior to European settlement) and space (Oregon). The flora also includes disjunct populations of taxa—those that are widely separated geographically—that are considered native in a nearby state as native to Oregon, unlike most definitions of native plants. The role of humans in the establishment of native plant species warrants consideration: studies in phytogeography have identified extensive effects of human action on plant migration, and in some instances, direct manipulation or movement of plant populations across broad geographic scales (Van Auken 2009; Christy & Alverson 2011).

OregonFlora also provides a distinct definition of a non-native plant: "*a plant from distant parts of North America or from other continents that established in Oregon post-European settlement.*" Examples include naturalized species from beyond Oregon, escaped cultivated plants that naturalize, waifs, and ballast plants. Waifs are small populations of plants persisting for only one season, and ballast plants are defined as waifs growing on ship ballast. However, selecting frames of reference that reflect western European perspectives and social constructs, such as statehood and settlement, leaves out the deep impact of indigenous people on the landscape.

Figure 2. In May, this Willamette Valley garden is overflowing with monkeyflower, popcorn flower, geranium, columbine, and penstemons.

Figure 3. Baby blue eyes is a western Oregon, early-blooming annual that attracts bees like this bicolored striped sweat bee.

It's important to note that the term *native* refers to the presumed origin of a species—where it naturally occurs—and not the source of the plant—such as transplanted from the wild, or where the plant was propagated. This distinction is important in native plant gardening because of the wide degree of variability in how native taxa are defined within the gardening and landscaping industry. The distinction is also important when considering the genetic integrity of plants used in native plant gardening.

Genetics

Humans are always striving to understand their world and in doing so organize and name objects. These efforts are abundantly directed to the natural world, where living organisms are described, named, and classified to reflect differences or similarities between individuals. As our knowledge expands, the groupings and names change to reflect that. Through time, plants have been classified—and accordingly named to reflect relationships—based on their appearance, their physiology, and more recently their genetic information. While DNA studies offer improved ways to identify evolutionary relationships between plant groups, the characters used to classify organisms cannot always be identified in the field. However, some of these characters reflect genetic mechanisms that determine how plants respond to their natural environment (Ungerer et al. 2008). These responses include modifications of diverse traits that increase the ability of native plants to acclimate to geographically and climatically diverse conditions. This is relevant as climate disruption brings changes to the landscape and its plant communities. Genetics can inform us of the extent to which a known species adapts beyond the point at which we identify it as something new.

Native plant taxa evolve and adapt to their environment through genetic changes or heritable traits, thereby helping them survive. These adaptations include strategies that help plants tolerate environmental challenges such as drought, high or low light levels, and soils containing heavy metals or salts. Some adaptations optimize a species for thriving in a "home" environment—one that reflects the physical and biological conditions found where a plant naturally occurs—but contribute to its decreased success in contrasting environments (Anderson et al. 2011). Still other adaptations allow a species to acclimate to environmental conditions across a range of environments. Interestingly, studies have identified areas within a plant's genetic material that change in response to a plant's acclimation to its environment (Excoffier et al. 2009). In native plant gardening, this suggests that the genetics of certain taxa gives those plants the ability to acclimate to changes in environmental conditions and consequently respond well to conditions in urban landscapes. The ability of some native taxa to do so increases their fitness across a heterogeneous landscape (Anderson et al. 2011), but not all genotypes of a given taxon exhibit these adaptive capabilities.

One of the most important keys for success in using natives in a garden or landscape setting is to choose plant material that is genetically adapted to the environmental conditions of the project site. This perspective—creating a landscape or garden that first considers the site and then selecting plants accordingly—turns traditional gardening upside down. Horticultural plant varieties, which are bred to grow in uniform conditions with plenty of water, fertilizer, and potting soil, face fewer limitations on where they can thrive. In contrast, the site is paramount in native plant gardening. Selecting well-adapted species has been shown to markedly increase plant survival and biomass production

Figure 4. Buckwheat, woolly sunflower, and mountain balm are good choices for a warm, sunny, well-drained spot in a garden.

in native landscapes (Martinson et al. 2019) and may help plants acclimate to the unique environmental conditions common in developed settings. Choosing plants grown from seed or cuttings taken from local populations or, at a minimum, originating from the same ecoregion (see below) can also improve the resilience of the planted landscape to stressors such as disturbance or climate change (Kramer et al. 2019).

Figure 5. Paintbrushes are semiparasitic, usually requiring a host species to flourish. This harsh paintbrush (left) was grown by tossing seeds among already established woolly sunflower (right) and Roemer's fescue.

It is interesting to consider how shopping habits may actually impact the genetics of native plant material grown for gardening and landscaping. Consciously or unconsciously, the native plant customer at a garden store or nursery is measuring the attractiveness of the material on display. They typically choose individuals based on their apparent health and vitality and do not buy plants with indications of stress or slower growth rates (Martinson, unpubl. data). Although native species are genetically adapted to conditions of their "home" environment, a subset of phenotypes is particularly well adapted to the altered conditions found in nurseries and urban gardens and consequently have an increased growth rate in these settings. Consumer preference for these bigger, faster-growing individuals may indirectly reduce the genetic diversity and resilience of urban landscapes and increase requirements for inputs such as water and nutrients. This unintentional bias can be overcome by educating both consumers and plant sellers about the importance of offering plant materials that reflect the phenotypic diversity found in nature.

Nativars

In an effort to expand the market share of native plants, some researchers in the native plant industry are looking for ways to enhance or standardize aspects of native plants. The goal of these efforts is to produce plants that are more consistent in traits—often unique ones—such as flower color, bloom time, plant shape and size, or drought resistance, with the rationale that consumers will be more interested in buying plant material that is dependable and consistent; in other words, a known entity.

This raises several interesting issues about both plants and people: Does uniformity increase or lessen a species' aesthetic appeal? How does breeding and propagation of native species to standardize their features impact the genetic variability and diversity? Are nativars the ecological equivalent of their wild-grown parents? Will they maintain resilience to the environment and provide the same ecosystem services? Whether nativars provide ecological functions comparable to plants growing in nature is largely unknown; one aspect—the pollinator value of nativars—is being investigated by researchers (Kramer et al. 2019; Ricker 2019). Clearly more work is needed to understand the potential long-term effects of nativars on biodiversity and the ability of native gardens to become ecologically functional.

Breeding native plant species and propagating selected phenotypes to fulfill certain expectations is motivated by several human perceptions: many gardeners and landscapers want predictability in the plants they are using; growers may value uniformity in their stock as a way to increase the efficiency of their business; merchants can provide customers with a consistent product over time. However, as public awareness of the benefits of planting native species and knowledge of how to successfully use them grows, these motivators may be replaced with the enthusiastic

Figure 6. Native species blend with cultivated plants in this predominantly native landscape in Portland.

embracing of the uniqueness of native plants. This economic and psychological aspect of the native plant horticultural industry is a new area of study, and preliminary results suggest that consistency in plant characteristics is less important to consumers than genetic variability (Martinson, unpubl. data). Gardeners are choosing plants partly because of a personal environmental ethic and partly in response to an understanding of the benefits gained through native plant gardening.

Urban landscapes

Native species can be incorporated into urban gardens and landscapes in many ways and for diverse purposes. They are often used exclusively to create a sense of place or to provide habitat for pollinators or other wildlife. However, natives are also frequently added to a more conventional landscape design to reduce water use in a portion of an ornamental, non-native planting. Recently, interest in landscaping with native plants has increased over concerns of perceived negative environmental impacts associated with conventional landscape practices and as drought becomes more common. Using native plants is often regarded as a fundamental way to reduce resource use—primarily water—in urban landscapes. More than ever, municipalities are encouraging—or in some cases requiring—native plants be used to reduce water use. But in Oregon, landscaping with native plants is viewed by many people as an ethical imperative reflecting a history of environmental consciousness characteristic of the state (Beatley 1995; Wolters & Hubbard 2014). Whatever the reason, the use of native plants in gardens and landscapes is increasing, primarily due to greater availability of plant material and a better understanding of the benefits native plants provide.

Using native plants in gardens was initially popularized in the mid-1970s primarily as a way to increase wildlife habitat in urban environments (Buslaff & Johnson 2004). Common design elements such as lawns or foundation plantings were viewed as "flat, bleak and shaved, [with] shrubs [unable] to display their own shapes..." (Otto, in Buslaff & Johnson 2004). Efforts to introduce "alternative" landscaping methods expanded through the 1970s and early 1980s in response to a growing environmental ethic. A number of publications came from the Environmental Protection Agency, U.S. Fish and Wildlife Service, and the Audubon Society espousing "Naturescaping" as an effort to increase wildlife habitat within urban environments, and reflecting the "back-to-nature" culture of the period (O'Donnell Young 2011). Naturescaping is still emphasized by the Oregon Department of Fish and Wildlife (ODFW) as a landscaping method to increase wildlife habitat, reduce resource use, and reduce noxious species infestation (ODFW 2015), although recent efforts have focused on the concept of ecosystem services.

Plant Choices for Sunny Gardens

The ideal conditions found in an open, sunny site appeal to most gardeners—adequate moisture, abundant organic matter, and unhindered sunshine make it just right for many native species. Diversity can be maximized in these prime conditions, whether following a free-form native prairie look or a more formal planting scheme. Grasses, bulbs, perennials, and annual herbs combine to provide plenty of visual interest through their textures, colors, and forms. Such diversity also benefits a broad array of pollinators, which have coevolved with certain flower morphologies and seek nectar and pollen resources in the warm sunshine from early spring until autumn frosts.

Virginia strawberry, self-heal, woolly sunflower, western buttercup, wild blue flax, and Douglas' aster provide colorful blooms on perennial plants. Bluehead gilia and common madia are annuals that attract numerous insect pollinators, while bright red perennial paintbrushes entice hummingbirds. Graminoids like junegrass, Roemer's fescue, dense sedge, and gray rush provide uniform color and may serve as butterfly caterpillar host plants. Taller plants such as lupines, checkermallows, mule's ears, and larkspur make attractive backdrops for lower-growing choices. Shrubs like tall Oregon-grape, red-flowering currant, mock-orange, and deerbrush bring variety in height to the garden.

Sites east of the Cascades typically enjoy sunnier conditions but also are subject to harsher temperatures and less moisture than west-side gardens experience. Lomatium species, buckwheats, milkweeds, Munro's globemallow, arrowleaf balsamroot, helianthella, purple sage, desert sweet, numerous penstemons, and showy fleabane are among many species available for high desert gardens.

Figure 7. Sunny gardens, such as this one in Bend, can be a riot of color and diversity.

Novel ecosystems

There is a dogged question that arises when discussing native plant gardening: can ecological restoration truly be achieved in urban settings? While native plants are an essential element of ecological restoration, some researchers argue that applying ecological restoration concepts and methods in highly modified urban settings is ineffective (Simmons et al. 2007; Standish et al. 2013) or question if restoration of functional ecosystems in any environment is possible (Perring et al. 2014). A slightly different perspective on the issue may be appropriate for urban landscapes. Creating plant associations that may not have occurred on a site naturally but that are ecologically appropriate for the altered conditions of a developed property can be considered the creation of relevant novel ecosystems. The outcome is a *rehabilitated* urban landscape, and studies show that these carefully designed and constructed native plant landscapes can play a crucial role in preserving ecological functions within urban environments (Simmons et al. 2007).

Figure 8. A rock garden in central Oregon is filled with penstemons, buckwheats, fleabanes, and other natives.

Rehabilitation differs from ecological restoration in that the objectives may not be to reestablish all aspects of the historic plant community but rather to prevent further degradation and to reestablish ecosystem function (Pyke et al. 2002). Nonetheless, neither restoration nor rehabilitation success is guaranteed in highly disturbed and modified environments typical of urban development sites. In these instances, planning and design should be based on reference systems as discussed below (p. 12). The ability of novel ecosystems to become established and persist requires planning that reflects a basic understanding of plants at several levels—interactions within plant associations, the characteristics of each species (germination requirements, phenology, structure, etc.), and site-specific conditions.

Any designed landscape can be thought of as a novel ecosystem. By gardening with native species, we can make conscious efforts to create and support a functional ecological system that will benefit ecosystems as well as appeal to our sense of aesthetics.

Ecosystem services

Ecosystem services are the benefits humans obtain from the diversity of life around us. These occur in four broad categories: regulating, provisioning, supporting, and cultural (Brown et al. 2007; Nelson et al. 2009; Gómez-Baggethun & Barton 2013). Creating healthy plant communities, including landscapes where previously there was no vegetation, reestablishes a number of ecosystem services. Below are some of the services enhanced or regenerated when gardens and landscapes include plant species adapted to a region:

Regulating

- Control floods
- Reduce infestation of diseases and pests
- Moderate the effects of climate change
- Sequester carbon
- Purify water
- Support native pollinators

Provisioning

- Produce food
- Provide fresh water
- Provide fuel and fiber
- Provide medicines

Supporting

- Convert energy (photosynthesis)
- Increase nutrient cycling and mineralization
- Facilitate and support a biodiverse environment

Cultural

- Provide spiritual value
- Provide healing places
- Provide culturally important plant material
- Provide recreation
- Create a sense of place

Gardening with native plants provides benefits beyond the ecosystem services listed above. Native plants can also significantly reduce the maintenance needed in the garden. They can reduce dependency on added water or fertilizer by being physiologically adapted to their environment and by supporting the native soil biota responsible for nutrient cycling. While we are aware that soil is a substrate that supports plants both physically and physiologically, plants—especially native species—are concurrently benefitting the soil and its complex ecosystem. A well-constructed native landscape encourages the growth of a robust population of soil-dwelling bacteria, fungi, microbes, and other organisms essential for healthy plants. These organisms produce exudates that aggregate soil particles, improving the porosity and water-holding capacity of native soils. From creating healthy soils, to supporting native pollinators, to providing ecosystem services that benefit humans physically, socially, and economically, native plant gardens become powerful living systems in our structured environment.

Building Blocks for a Native Plant Garden

When planning a native plant garden or landscape, there are some unique elements to consider that are not normally associated with conventional landscape practices. Native species have evolved under specific environmental conditions and in association with a range of native pollinators and other animals that are dependent on the flowering time and genetic variability of host species. Additionally, a feedback loop develops between native plant communities and soils in which plants affect the creation, structure, and chemistry of soils (pedology); this in turn determines the species composition of the plant community. Here we present four basic concepts that are connected to the relationships between plants, their physical environment, and the diverse living organisms with which they coexist: ecoregions, biodiversity, soils, and subsurface biological communities. Understanding these relationships is critical to ensuring the long-term viability of native plant landscapes, as they are essential elements that impact the design of native plant gardens.

Ecoregions

As anyone who has traveled from the state's coastline to its deserts or from Mt. Hood to the Siskiyous can attest, Oregon is ecologically diverse. Environmental conditions on the west side of the Cascade Mountains reflect a strong marine influence, while the east side of the state lies in the rain shadow of the Cascades and is predominantly a semi-arid cold desert. The forces of geology have also shaped Oregon, creating beaches, broad river valleys, glaciated volcanic peaks, desert plains, and playas. Together, climatic and geophysical conditions influence the plant and animal communities that adapt to a region. Areas with similar environmental characteristics have been classified as ecoregions (Omernik & Griffith 2014) and provide a critical source of information for native plant gardeners

Plant Choices for Shady Gardens

The amount of shade a planting receives can vary from dappled light beneath shrubs to deep shade between tall buildings or under a dense forest canopy. Additionally, shady areas can be dry or wet. These conditions or their likenesses can be readily found in the wild in many different habitats, which serve as good reference environments for plant combinations.

Moist conifer woodlands are an ideal setting for many lush perennials, such as bleeding heart, vanilla leaf, wild ginger, twinflower, false lily-of-the-valley, inside-out flower, darkthroat shooting star, western trillium, and stream violet; these not only provide color spots while flowering but have attractive seasonal (or perennial) greenery.

A dry conifer habitat can support sword fern, toughleaf iris, creeping Oregon-grape, several species of penstemons and lupines, California fescue, woodland strawberry, and Henderson's triteleia. Oak woodland habitats typically provide open shade that supports early-flowering species that go dormant as summer conditions become drier and shadier. These include columbine, crown brodiaea, poison and Menzies' larkspurs, Pacific hound's tongue, Henderson's shooting star, fawn-lilies, and buttercups.

Deep shade in urban areas, such as between buildings or on a north face, can simulate canyons and cliff faces in a variety of habitats. Ground covers, available in varying heights, can be successfully used in these settings: yerba buena, western bunchberry, northern maidenhair fern, largeleaf avens, inside-out flower, oxalis, and kinnikinnick.

Small trees and shrubs for the forest understory include Pacific dogwood, western azalea, Pacific rhododendron, osoberry, huckleberries, currants and gooseberries, and Oregon viburnum.

Figure 9. Tall Oregon-grape has year-round interest: fragrant yellow flowers in spring, clusters of blue berries in the summer, and shiny evergreen leaves that often turn red in the fall and winter.

Figure 10. Desert yellow fleabane thrives in high desert sunny areas or rock gardens and attracts pollinators such as these cedar hairstreaks.

and landscapers. There are four levels of ecoregional classification, each with a more narrowly defined suite of characteristics that ultimately include habitats and plant communities. For example, there are 15 generally accepted Level I ecoregions in North America, three of which are present in Oregon. OregonFlora recognizes eleven Level III ecoregions; these are described in detail in the *Flora of Oregon* Volume 1 (Albert, in Meyers et al. 2015).

How does knowledge of ecoregions help native plant gardeners? A garden or landscape that is designed within an ecoregional framework will contain the individual species, the plant combinations, and the physical elements that *belong* to that region, creating a space that authentically reflects its place on the planet. An ecoregional perspective gives the important big picture; it provides a context to guide the gardener's choices. Additionally, most projects will benefit from additional research into the existing or historic plant community of the project site, as well as an understanding of the microclimates that are present or can be created in a landscape. This means finding an uncultivated, natural ecosystem within the same ecoregion that can serve as a reference. Carefully observing the species found there, how densely they occur, and how they are spatially distributed across the landscape can guide the gardener in creating an effective design focused on native plants.

Biodiversity

Biodiversity is a key component of landscapes that are based on ecological principles. While most people—especially nature enthusiasts—have a sense of the concept, a precise definition of biodiversity is often elusive. The term originated from the Greek word *bios,* meaning life, and *diversity*, which refers to both quantitative and qualitative values: 1) the number of different types of items (richness), 2) the relative abundance of different types of items (evenness), and 3) variety (vaguely defined as the class to which the diversity belongs). Some definitions apply to all aspects of biological diversity, from species richness and ecosystem complexity to genetic variation (Allaby 2010), and do not consider non-living (abiotic) elements in their discussions. Authors in this camp argue that the appropriate term to use when including abiotic components is ecological diversity—not biodiversity. Clearly, how the term *biodiversity* is applied to gardens and landscapes varies by the source of information, the scope of the system being described, and who is defining the term.

The inclusion of human influences in the biodiversity equation is an interesting—and debated—topic. For example, Dasmann (1991) defines biodiversity as "the totality of species, populations, communities, and ecosystems, both wild and domesticated, that constitute the life of any one area or the entire planet… it specifically includes cultural modifications of the natural world." There are two interesting points about this definition: first, Dasmann includes ecosystems in his definition of biodiversity, implying inclusion of abiotic components. Secondly, he argues that the results of human action, while affecting biodiversity, are integral to functional systems (ecosystems). This perspective is especially relevant to gardeners and landscapers, whose efforts are transforming the biodiversity within our managed environments.

A universally accepted definition of biodiversity doesn't seem to exist. Debates continue about abiotic and biotic components and whether human influences play a role in defining biodiversity. In the realm of native plant gardening and landscaping, however, the practical definition and commonly used meaning of the term refers to species richness. This is simply the number of different types of items; i.e., the number of different types of plants in various functional groups, such as grasses, forbs, shrubs, or trees.

Figure 11. Foothill sedge is one of our few upland prairie sedges. It has delicate nodding flower heads and forms large clumps in full sun.

Figure 12. Beneficial actinomycete bacteria are clearly visible as white matter in this shovelful of duff at a central Oregon project site.

Soils

Soils are an immensely important element of gardening; yet like the air we breathe, the substance that literally supports plants and structures often goes unnoticed and underappreciated. Soil is composed of both biotic and abiotic components. Though some definitions refer solely to the abiotic solids that comprise soil, the living (biotic) components are an essential part of the matrix we know as soil. These include bacteria, fungi, nematodes, and protozoa. It also includes the roots of plants living in the soil. Organic material, in the form of intact and decomposing plant and animal matter, contributes both physically and chemically to the soil. However, it is the makeup and proportions of inorganic elements and minerals and the associated physical and chemical characteristics that are largely used to define soil types.

Soil is typically found—as expected—at the earth's surface and is arranged in layers, or horizons. Soil horizons differ from their parent material because they have been altered by climate, relief, and living organisms over time. "Parent material" is the bedrock, or unaltered substrate used in pedogenesis (soil formation) that is lacking animals, roots, or other indicators of biological activity—conditions typically found at soil's lowest depth. The upper limit of soil is defined as the boundary between soil and air, shallow water, live plants, or plant materials that have not begun to decompose. The lower boundary of soil is more difficult to discern and is limited to an arbitrarily set depth of two meters. The US Department of Agriculture began a comprehensive survey of the nation's soils in the late 1800s, creating an invaluable resource that informs both agricultural and urban planning. This work is continued today by the Natural Resources Conservation Service, which provides public access to soil information.

The condition of the soil at a project site is highly relevant: the gardener can consider what soil type is present, how it has been altered or impacted, and what might optimize it for the garden's needs. Impacts to soils are both direct and indirect and can be human-caused and/or environmental in their origins. Direct effects to soils include compaction, chemical alteration, removal, and grading. Indirect effects include loss of subsurface biological communities, heat island effects, changes in hydrologic function, and connection to the region's water cycle. Amendments such as compost can be added to modify the soil structure thereby providing additional organic matter and improving air flow and water movement. Several inches of compost should be simply layered onto the soil surface. This allows its nutrients and compounds to percolate slowly into the soil without disrupting the soil structure.

Subsurface biological communities

A single teaspoon (1 gram) of soil can hold up to one billion bacteria, several yards of fungal hyphae, several thousand protozoa, and hundreds of nematodes (Bandick & Dick 1999). This phenomenal diversity is very important to the function and health of native plant landscapes. Special attention to two of these categories—mycorrhizal fungi and bacteria—can help a native plant-based garden succeed.

Mycorrhizal fungi are a biological component of soil, and our growing knowledge of their importance makes them an exciting and impactful topic in the native plant gardening community. They are soil fungi that form symbiotic relationships between the roots of plants and the surrounding soil. The fungi form hyphae—root-like structures that spread throughout the soil and benefit the plant by

Figure 13. A tiny house in Medford is landscaped with cattails, woolly sunflower, and other natives.

Figure 14. A male Anna's hummingbird nectaring on red-flowering currant. No garden in western Oregon should be without this early blooming shrub.

The many benefits that mycorrhizal fungi provide sound almost too good to be true. However, research supports the observations and provides an excellent example of how science can directly benefit efforts in the home garden. The associations developed between mycorrhizal fungi and plants is an essential—yet often overlooked—part of a created native plant landscape. Fungi can be re-established or introduced into areas where plants had not grown via various commercial inoculants when seeding or planting or through the propagation of community-specific fungi collected from local natural areas.

The second category of beneficial soil-dwelling organisms that are often missing from native plant gardens but that can easily be re-established is bacteria. Several types of bacteria are important in plant community function, but one, the actinomycetes, are commonly found as white filamentous material in leaf mold or duff layers below trees. While some gardeners may mistakenly be tempted to remove these readily visible mats and threads, it is wise to keep them: actinomycetes provide essential ecosystem services through conversion of atmospheric nitrogen into ammonium nitrate, a common fertilizer component. By using duff as a top dressing, or incorporating well-composted material during the landscape

transferring water and nutrients to the plant through their connections with its roots. The fungi receive their nutrition as carbohydrates (sugars made during photosynthesis) from the host plant. Some literature suggests mycorrhizal hyphae are restricted to the shallow soil layers, but some studies have also documented the presence of fungal hyphae in organic layers outside soil profiles (Linderman 1991; Jeyanny et al. 2016) or in deep soil layers and penetrating the bedrock or parent material (van Breemen et al. 2000). Over 90% of all plant families on the planet—from ferns to trees and native species to cultivated ones—form a relationship with mycorrhizal fungi (Wang & Qiu 2006).

How does a plant-fungus symbiosis work? The fungi attach to the roots in two basic ways: by penetrating the root cells (endomycorrhizal fungi) or by colonizing the exterior of root cells (ectomycorrhizal fungi). Endomycorrhizal associations are the most common, being found in most plant families (Linderman 1991; Brundrett 2002); ectomycorrhizal fungi are symbiotic with only about 10% of plant families and are typically found with conifers and some hardwood trees (Burns et al. 2002). With both types of associations, plants exhibit numerous physiological changes after being colonized by mycorrhizal fungi. Some of the documented responses include:

- Increased biomass, resulting in a faster growth rate (Smith & Smith 2012)
- More efficient photosynthesis and respiration, mainly through increased potassium uptake and use in stomatal guard cells (Wright et al. 1998)
- Reduced water use (Young-Cheol et al. 2012)
- Increased disease resistance (Pozo & Concepcion 2007)
- Improved soil airflow and permeability due to increased soil particle aggregation caused by exudates from fungal hyphae (Bearden & Petersen 2000)
- Increased nutrient cycling and mineralization (Marschner 2007)

Figure 15. Western blue flag (*Iris missouriensis*) is a lovely perennial for damp places east of the Cascades.

construction process, many of those bacterial communities can be reintroduced to a site, thereby reestablishing a critically important ecological function of native plant communities.

The material above has highlighted ecological and biological concepts that impact native plant gardening. So how can this information be applied when planning or designing a native plant landscape to ensure success? What are some of the subtle differences between gardening with natives and more conventional landscaping practices?

Interest in native plant landscaping is a relatively new field, and as such, best management practices that are common in conventional horticultural practices are generally lacking. However, a few general methods and approaches discussed here have proven successful when gardening with natives. Knowing that planted landscapes that use native species reflect their place of origin, it follows that the design and planting methods presented will vary based on the location of the project site.

Figure 17. The variable flowers of great polemonium change from light yellow to peach to pale pink-purple as they age. It grows well in part shade.

Creating Your Garden

The design and construction of a successful native plant garden or landscape often begins with a vision. What are the objectives of the effort? Is it to be a low-maintenance landscape or an opportunity to experiment with plant combinations, their placement, and propagation? Becoming aware of what you want to achieve provides a focus for the planted space, making the whole garden much more than the sum of the plants growing there. It becomes an ecosystem that the gardener can create, tend, and enjoy.

Some common objectives of native plant gardens are to reduce resource dependency, reduce maintenance needs, create wildlife or pollinator habitat, or create a sense of place. Often, the vision combines several objectives, so early planning will help determine what actions are needed to satisfy them (Martinson et al. 2017). The objective for the project should also weigh the environmental constraints and opportunities present. The gardener or designer needs to identify existing microclimates (the climate of a small, specific area) or plan where a novel microclimate, such as one associated with a bog or a rocky outcrop, can be constructed. Adding such design elements expands the plant palette that can be used within those specific environments. A thorough understanding of what it takes for each plant species to thrive directs the choice and placement of plants within a designed garden or landscape. The goal is to match the environmental requirements of the plant with the conditions that either exist or that are created in a native plant landscape.

Figure 16. A Willamette Valley garden with camas, largeleaf avens, and fringecup.

However structured or casual the native plant garden is envisioned to be, a good starting point for the gardener or landscaper is to assess the site. What is the nature of the landscape? What existing material features (buildings, fences, pathways) do you hope to preserve or remove? What abiotic features (views, topography, hydrologic patterns, wind patterns, solar aspect) can be highlighted, minimized, or modified? What biological features (soil tilth, wildlife or pet activity, large trees) will be integrated into the garden plan? How a garden fits into the larger environmental context of the neighborhood, city, county, state, or region is also an important consideration when designing a native plant landscape. For example, understanding the symbiosis between native vegetation and migratory pollinator species or the contribution of a native plant garden to wildlife corridors or winter habitat gives your design local or regional importance.

Once you have considered your goals and familiarized yourself with the site, developing a conceptual design is a good next step. This can include a species list and some level of detail in plant placement and density. Conventional designs often use few species with little genetic diversity and are driven by the structural and architectural values present. In contrast, native plant landscapes are often designed to reflect naturally occurring plant associations and try to maximize biodiversity within the physical limitations of a site. An ecologically functional planting is the main goal, with architectural and aesthetic values as secondary concerns. Decisions around plant placement, density, and the distribution of species that can answer the question "How will my choices help provide ecosystem services or ecological function?" will lead to plantings that emphasize the functional relationship between the biotic and abiotic elements of the landscape.

Overall, the design of a native plant garden or landscape is an exercise in ecology. One of the most valuable skills that will help ensure a native garden success is the ability to match species requirements with the environmental conditions of microclimates on the project site. This can be a challenge, predominantly because there aren't many references detailing the adaptability of native species to created environments for most regions of the state. But these considerations, while rarely addressed in conventional practices, are key to successful native plant landscapes and apply to design regardless of location. The ideas presented here reflect the steps taken by professional landscape designers with expertise in native plantings; they will hopefully serve as an inspiration and a road map to anyone interested in integrating native species into their planted landscapes. Tapping the expertise of knowledgeable staff at native plant nurseries, consulting botanical and ecological references (including the *Flora of Oregon* and its online resources), and being an astute observer of your environment will channel enthusiasm for native species and healthy ecosystems into a garden or landscape that truly reflects your place on the planet.

Urban landscape aesthetics

A fundamental difference between native plant landscapes and conventional horticultural plantings is the aesthetic provided by each design style. While conventional horticultural design emphasizes form, color, and texture as primary objectives, native plant landscaping often emphasizes ecological function or habitat value and considers form, color, and texture as secondary objectives. There is aesthetic value—however it is personally defined—in both

Plant Choices for Moist Gardens

While many gardeners lament poorly drained areas in their gardens, wet spots provide an opportunity to grow a different suite of native plants. Wet areas can occur from poor drainage or seepage from higher upslope. Excessive water can also come from runoff of roofs, sidewalks, and other impervious surfaces. Gardeners and landscapers are increasingly constructing rain gardens to handle this runoff, especially in the wetter, western parts of the state.

Rain gardens are created by building a depression that can catch runoff and fill with several inches of rain after a storm. Rather than flowing directly into a storm drain, this water can be contained long enough to soak into the ground slowly. Plants can be placed at various depths in the depression, depending on their moisture tolerance.

Bioswales are similar to rain gardens but are typically long strips following an existing draingage swale. Many urban areas now require the creation of rain gardens or bioswales in new construction. Both help reduce flooding and filter contaminants in the soil that would normally go directly into the sewage system.

For those who don't have problems with excess water but would like to grow moisture-loving plants, it is also possible to build a bog in your garden by digging a depression and lining it with bentonite, landscape fabric, or plastic to retain soil moisture—similar to building a pond, only filling it on top with soil.

Moisture loving trees and shrubs that would do well in these wet gardens include Pacific ninebark, Oregon crabapple, black twinberry, willows, hardhack, Klamath plum, red-osier dogwood, and cluster rose. Various sedges and rushes are found in wet areas, but be careful to choose those that are appropriate for the size of your garden. Camas and monkeyflowers are classic moisture lovers. Large-leaved lupine, western blue flag, skunk cabbage, bleedingheart, sessile trillium, straightbeak buttercup, stream violet, and slender cinquefoil can also provide colorful flowers, while deer fern and maidenhair fern have attractive foliage.

Figure 18. A predominantly native plant rain garden in Eugene.

Figure 19. Coast fawn-lily naturalizes well in western Oregon shady gardens.

approaches. We have a long cultural history of designing landscapes or gardens as outward expressions of our values and status. But balancing a traditional vision with the physiology and functional aspect of native plants can be challenging. Created gardens or landscapes are often constructed in ways that use plant densities or distribution patterns that don't occur in nature. In addition, many common gardening practices often result in conditions that increase nutrient and water availability beyond what is needed by native species. This becomes a lot to juggle: the gardener's desired aesthetics, an ecologically appropriate plant arrangement, and resource (water and nutrient) availability. An unbalanced approach can result in a native plant landscape "going crazy" with excessive production of seedlings or biomass, a general decline in plant health, excessive weed infestation, or wholesale failure if working with xeric species. However, the concepts presented in this chapter can inform maintenance practices that address the unique character of native species and help gardeners achieve a satisfying and productive landscape.

Applications for native plantings

The use of native species in urban environments is booming as individuals, developers, and municipalities recognize the practicality and economic benefits associated with their use. There are native species adapted to almost any of the challenging conditions that exist in developed areas, including poor drainage, reflected heat, heavy foot traffic, and limited light. Beyond their ability to address some of these physical issues, there is significant research documenting the psychological value green spaces provide in urban settings (Dwyer et al. 1992; Bolund & Hunhammar 1999; Shanahan et al. 2015). As city planners embrace the benefits of using native species, it is important that the decision-makers consider the ecoregional perspective we have outlined in this chapter to help inform plant choices and—equally important—maintenance practices.

Native plant landscapes and gardens are also ideal for integrating into green building designs (e.g., LEED, Earth Advantage, Net Zero, Living Futures) and for establishing or conserving natural areas that provide critical habitat, as in the examples below:

- Pollinator gardens
- Bioswales
- Rain gardens
- Green roofs
- Rails to Trails
- Waterfront plantings
- Road medians
- Common areas in urban developments
- Municipal parks and botanical gardens
- Water treatment plants
- Bioremediation projects

Figure 20. A native plant green roof on the MODA building in Bend.

Disturbance

What can be learned from reference communities and historic plant communities provides a starting point but doesn't necessarily limit plant choices for a particular project. This is because conditions within developed sites are often significantly altered through construction activities. Construction of a building on a site, whether it is a house or a multistoried office building, constitutes a severe ecological disturbance. Soils are changed biologically, chemically, and structurally; hydrologic patterns are disrupted or changed; exposure and winds are altered; and most, if not all, existing vegetation is removed. While the resulting conditions change the ability of the site to support the same population of species that existed before construction, they also provide an opportunity to expand species richness and include plants that are appropriate for the pre-construction floral community that may not have been there before development.

Figure 21. A development site in Bend. A *Pinus ponderosa* plant association existed on site prior to development. Construction activities and the multi-story buildings planned for this site will significantly change the biological, structural, and environmental characteristics of the property.

Thoughtful design of a native plant garden or landscape should consider how construction and disturbance change the environmental character of the site. For example, the south side of a new building will be hotter and drier than conditions that previously existed on the property. A plant community designed for those conditions will be markedly different than one planned for the north side of the same building, which will be cooler, shadier, and generally have a higher soil moisture level longer in the growing season. Normally, there are elements of the native flora adapted to the range of conditions that new building construction creates; however, some of those plant communities may not have been present on the site prior to disturbance. Considering the complete diversity that exists within a reference ecosystem can help the gardener successfully respond to the altered conditions of a project site.

Most of the information about native plant landscaping in the state addresses the environment in the Willamette Valley—a region that has wet winters, dry summers, occupies less than 10% of the state's area, and houses approximately 70% of its human residents. Techniques, suggested plant material, and community structure in Willamette Valley gardening guides can provide some background information for gardeners located outside the region. However, clearly not all practices and species will be appropriate for other parts of the state.

The majority of Oregon's acreage lies east of the Cascade Mountains, which is considered an arid or semi-arid environment. As in all environments, the plant communities east of the Cascades have devised adaptive strategies to ensure survival success. At the plant level, vegetation has adapted features to increase water use efficiency, such as

Water and Irrigation

The extreme variation in environmental conditions across the state makes it difficult to describe the basics of native plant gardening in Oregon. Nowhere are generalities more apt to fail than when considering water needs of plants. A common misconception about native plants is that all native species are drought tolerant. The prevalence of this idea may be related to the current concerns over long-term drought in the western United States or may be a result of marketing efforts by retail outlets offering native plants. Though not a wholly accurate claim, it is based on native species' physiological adaptations to their environment. This results in plants of an ecoregion responding optimally to the soil water availability, evapotranspiration, and vapor pressure deficit that are present—features that gardeners and landscapers find so desirable.

Figure 22. The landscaping at a residence in central Oregon is designed to meet the Living Building certification standards. This unique project is based on functional ecology of the sagebrush steppe community, is 90% unirrigated, and is an outstanding example of how a landscape can be designed to meet the dual objectives of beauty and functionality.

leaf color or orientation, pubescence, rooting depth, or dormancy patterns. Community structure—reflected through plant density and distribution, species composition, and species interactions—has adapted in response to environmental conditions such as soil type, hydrologic patterns, and nutrient cycling. Species of all ecoregions across the state exhibit plant/water relations specific to their locale. Understanding that these adaptations exist is appropriately, and often unconsciously, a defining element in native plant landscape design that informs plant selection and maintenance practices.

The development of modern irrigation in the western United States is a fascinating story beginning in the early 1700s, documenting the close relationship between agricultural production and urban irrigation. Throughout that history, the landscaping industry has been very effective in promoting the importance of irrigation for urban landscapes. However, because native species are adapted to local environmental conditions including precipitation levels and seasonal patterns, the irrigation models designed for non-native plantings likely are not optimized for the adaptive physiology these species possess (Mata-González et al. 2005; Mata-González et al. 2014; Martinson et al. 2019). This means many of the installation and maintenance practices common in horticultural plantings are not well suited for the native plant garden.

All plants initially need water to become established in a landscape. To address this need, the native plant gardener may choose to include irrigation systems whose use can, in time, be reduced or discontinued. Alternatively, there are some installation techniques that can be used to help establish native plantings without the use of expensive supplemental irrigation systems. These include:

- adding cornstarch-based polymer products in the planting hole to retain water
- planting slightly below the surface grade (slope) to collect and hold precipitation
- planting in the spring or fall to take advantage of seasonal moisture
- hand watering

Nutrition

As with water needs, native plant species are also well adapted to the nutrition available in their environment. Promoting a healthy soil microbial community is one of the best ways to ensure that your garden or landscape thrives since nutrient uptake is aided by the mycorrhizal associations. Because many created landscapes do not begin with undisturbed "native" soil, adding 2–3 inches of organic matter (compost) to the soil is a recommended first step. Compost, especially if it contains duff or leaf litter from native vegetation, increases the soil microbial populations that are so beneficial to the plants. Compost also helps the soil retain and distribute water, promotes air flow, and aids

Plant Choices for Well-Drained Gardens

Well-drained conditions can be found in garden sites on steep slopes, where there is rocky habitat (natural outcrops or man-made retaining walls), and in the drier eastern and southwestern parts of Oregon. They can also be created by building rock gardens, rock walls, raised beds, or sand beds. Native rock plants tend to be low growing and work well in spots close to buildings or alongside patios, where they can be viewed up close and won't interfere with the view of the rest of the garden. For those on the rainier west side or whose conditions aren't well drained enough for typical rock plants, one can build a mound or raised bed with gravelly soil to get adequate drainage. Sand beds—raised beds containing nothing but gritty sand—can provide habitat for plants that require extremely well-drained conditions.

Some terrific plants for rock gardens and other well-drained sites include mat-forming species such as buckwheats, pussytoes, woolly sunflower, monardella, mat alumroot, and phlox. Small bulbs like brodiaeas, mariposa lilies, and onions look terrific popping up through the mats. Small lomatiums like Hall's lomatium (*L. hallii*) grow naturally in outcrops and provide early blooms. Stonecrops are easy to grow—small cuttings can root into cracks in a rock wall—and their succulent evergreen foliage looks great all winter. Most of our *Iris* species do well in rocky soils. Low-growing penstemons are a must for a rock garden. Eastside gardeners have a number of species of small *Astragalus*, *Erigeron*, and *Townsendia* to choose from. Siskiyou lewisia is a native that is grown in rock gardens around the world. While most people think of rock gardens as needing a lot of sun, many natives grow naturally in shaded rocky areas. In addition to broadleaf stonecrop, try fawn-lilies, columbine, alumroot, brook trillium, or woodland phlox.

Figure 23. Silver bush lupine and Siskiyou iris growing in a private rock garden in western Oregon.

Figure 24. A residence in central Oregon with a mixture of shrubs, grass, and forbs.

in the ability of a plant to access and cycle nutrients (Martinson et al. 2017). There are commercially available mycorrhizal inoculants that can be used when planting; these typically contain numerous commonly found species of soil fungi—both endomycorrhizae and ectomycorrhizae—and therefore can be beneficial in many different environments. Inoculants vary in their concentration, which reflects the recommended application methods, but remember the inoculant must have direct contact with the plant roots to colonize.

Compost and mycorrhizal inoculants are supplements that provide the greatest benefits to native plantings and are aligned with the ecological framework presented in this chapter. Are there ever instances when adding concentrated fertilizers is recommended? The addition of fertilizer when planting or seeding or as a top dressing at the beginning of a perennial's growing season can give plants a boost. However, some (but not all) studies have shown that the addition of chemical fertilizers results in a decrease in soil fungal populations (Linderman & Davis 2004), so choosing organic fertilizers over synthetic products can prevent damage to soil microbial populations. Again, when the native plant garden or landscape effectively reflects its reference environment and ecosystem, adding nutrients that exceed those conditions is not necessarily beneficial; in other words, more is not always better.

Obtaining plants

With a planting design in mind and an awareness of the natural ecosystems it will reflect, the final step to creating your native garden is acquiring plant material. More and more nurseries and garden stores are carrying native plants, but the selection is still limited. Dedicated native plant nurseries are the best resource for a selection of garden-worthy native plants, although it's important to be sure your frame of reference or definition of "native" aligns with that of the seller's. Currently there is no statewide certification program for native plants in Oregon's horticultural industry, and nurseries may have their own guidelines for designating what is native or not. If in doubt, the *Flora of Oregon* volumes as well as the OregonFlora website (https://oregonflora.org) indicate which species are native in Oregon. The website also has an extensive section about gardening with native species with search tools, fact sheets, and a list of known native plant nurseries in the state. See also Appendix 2 (p. 784). It's also important to note that many native plants don't meet the size requirements outlined in the American Standard for Nursery Stock (2014), which can provide a disincentive for the use of native plants in commercial or professionally designed landscapes.

Other great sources of native plants are local plant sales. Organizations such as Master Gardeners and Soil and Water Conservation Districts often hold fundraising sales in spring and fall. Buying native seed is another good way to start your native plant garden: this is sometimes the best way to get annuals established. Some native plant nurseries sell seed, and there are a growing number of mail-order businesses specializing in native plant seed.

Propagating your own plants may be the only way to get native species that are not commonly available through commercial outlets. Plants with adventitious roots (such as penstemons) as well as shrubs and other woody species can readily be propagated from cuttings. Seeds can be collected and grown in pots or sown directly on bare soil. Many gardening books include sections on propagation but rarely address the specific needs of native species.

Figure 25. Deltoid balsamroot (*Balsamorhiza deltoidea*) growing at a native plant nursery in Lane County. Both plants and seeds are sold for restoration and at occasional plant sales to the gardening public.

A basic understanding of propagation techniques will help establish a successful planting, but understanding the reproductive strategies of some unique native species and their response to environmental variables will help in successful propagation of some "specialty" plants. Just remember to never dig a plant from the wild unless rescuing from a construction site, and always obtain permission from the landowner before taking plants, cuttings, or seed.

Maintenance

Any project, whether focused on horticultural varieties or native species, will require a degree of maintenance to get it established. Many native plants form extensive roots during the first season after planting that benefit from reliable water and nutrients. However, the timing and amount of resources a plant should receive can depend upon its reference environment. For example, many species adapted to the forest environment of the Western Cascades generally require more water for establishment than species adapted to conditions in the rain shadow on the east slope of the Cascades. Again, referring to the ecoregional description for your planting area can help inform the maintenance standards needed for a site.

How much maintenance a landscape receives usually reflects the personal decision of the gardener: a native plant landscape can be manicured or reflect varying degrees of entropy often associated with wildlife habitat. When ecological function is the top priority, the impacts maintenance has on the soil biota, biodiversity, plant recruitment and decline, nutrient cycling, and beneficial insects also need to be considered. It can be appropriate and effective to apply standard maintenance practices concerning irrigation, fertilization, weeding, and pest management in conventionally designed plantings that include some native species. However, this comes with the caveat that decisions need to consider the unique physiology of the native plants, how each species responds to a modified and managed environment, and the potential to affect the intended trajectory of the planting. Giving too much supplemental water to some native species can cause them to become overly vigorous or display symptoms (wilting, excessive self-seeding, elongated leaves, chlorosis) similar to drought stress, another environmental extreme (Martinson et al. 2017).

Native plant gardeners may opt for minimal maintenance and allow their planting to better reflect a functional ecosystem. This practice requires patience and a thorough understanding of the intended trajectory of the design. It can also require educating neighbors and friends on the intent of your design and why you chose to create a native plant landscape, which some may view as "wild" or unkempt. Often, constructing the landscape is the easy part of a project; working to gain the acceptance of neighbors and the broader community can be more difficult.

Plant Choices for Container Gardening

If you live in an apartment or condominium or are in a temporary situation, can you still grow native plants? If you don't have a yard, container gardening can be a great answer. Containers can be placed in very small areas or on decks. From traditional terracotta and ceramic pots and window boxes to handmade troughs and found objects, any number of things can be used as a container for growing plants.

Even if you do have garden space, containers can be a great way to provide conditions you might not have in your garden. Extra grit can be added to a soil mix to make it more well draining and better suited for rock plants like small penstemons, buckwheats, or stonecrops. The container can be placed on a dish of water to keep the bottom wet to provide a home for wetland plants like skunk cabbage or monkeyflowers.

If you keep an eye out for roots coming out of the drainage holes at the base of a pot, you can grow otherwise attractive rhizomatous plants like false lily-of-the valley that might be too aggressive for the open garden. Try growing bulbs like lilies, brodiaeas, triteleias, and fritillaries if you have trouble with rodents damaging them in the open garden. They can be moved out of sight and allowed to dry out after they finish blooming, while pots of later blooming plants may be moved to a more prominent spot.

Placing pots up at eye level on a shelf or using hanging pots not only makes them easier to maintain, but you may be able to better see the ground level flowers of wild gingers and other hidden treasures.

Figure 26. Umpqua kalmiopsis is an Oregon endemic that grows on shaded pillar rocks in Douglas County. It roots easily from small cuttings and grows happily in a potting mix with added sand and gravel in a tufa-like handmade container called a trough made from a mix of Portland cement and peat. This gives it the extreme drainage it requires to mimic its natural setting.

Summary

Establishing a native plant garden or landscape is a rewarding endeavor in many ways. It is an act of creativity in which the gardener sets the stage for the growth and evolution of a dynamic ecosystem that will change over time. Native species offer ecological benefits that conventional landscape plants do not. Once established they thrive with minimal maintenance, and they instill a sense of place to the setting. Successfully working with native plants in a garden or landscape setting requires an awareness of the cultural requirements and life cycle of individual species, as well as a broad understanding of the interactions that exist in natural plant communities. Recreating those conditions or establishing a replacement ecosystem appropriate for the specific project site is a key requirement to a successful native plant landscape. This is especially true if the objective is to develop a plant community similar to what existed before disturbance. Designed landscapes can also highlight plant characteristics such as flower time and color, texture, or shape. Happily, these types of designs can satisfy a more traditional aesthetic while still capturing the benefits of native species.

Figure 28. Desert sweet is is a xeric shrub for east-side gardens. It is sweetly fragrant and a magnet for bees.

Design decisions are more likely to become a reality when they are based on a clearly defined objective, whether that is resource conservation, habitat value, minimizing maintenance, satisfying a personal environmental ethic, or simply appreciating the beauty and wonder of plants. Likewise, effective design for a native plant landscape is rooted in an understanding of the ecological and environmental conditions a species occupies in the wild. A successful design will match the needs of a project's plant palette with the environmental conditions—soils, nutrients, water, and light—existing or created on the site. Keen observation of naturally occurring plant associations and using the knowledge gained through that study as a basis for design is a primary component of designing a functional and beautiful native plant garden and is often one of the most enjoyable parts of design!

Increased interest in native plants as key species for pollinator populations and an increased understanding of the benefits of native species to humans has contributed to rapid growth of the native plant industry in Oregon. Our hope is that this trend will continue through education of designers, architects, landscape professionals, and homeowners on the importance and benefits of native plants in created gardens and landscapes.

Figure 27. After hardscaping work, areas of bare soil were seeded with a mixture of 60 species of native wildflowers and grasses, restoring the mountain meadow habitat surrounding this home at 5000′ elevation in the Siskiyou Mountains.

INSECTS AS PLANT TAXONOMISTS

Jeffrey C. Miller, Andy Moldenke, and Dennis A. Albert
with contributions from Tanya Harvey

Introduction

The relationship between plants and insects hinges largely around a rather simple observation: insects have become excellent plant taxonomists, and plants have become excellent insect taxonomists. By evolving together, we see specializations in the shape and function of players on both sides that result in high fidelity relationships. Likewise, what appears to be a generalist relationship between plant and insect still can be limited to certain groups. To understand how these non-human taxonomists have honed their skills, one must consider the physiology, ecology, behavior, and morphology of all the players.

Plants and insects rely upon each other for nutrition, shelter, and reproduction; their natural history is rich in details about their interactions. In this chapter, we present a sampling of key concepts about insect ecology from a botanical perspective. By describing basic as well as unique insect behaviors within the context of the plant communities where such interactions occur, we hope to inspire curiosity and enlightenment about the many different insect activities that can be observed while outdoors. An increased awareness of these fascinating interactions heightens our knowledge and can inspire further protection of plants and the insects associated with them. Through the stories that follow, we address four fundamental questions about plant-insect relationships, focusing especially on bees and butterflies:

- How have plants evolved to attract (in the case of pollination) and deter (in the case of herbivory) insects?
- How do the geographic distributions of plants and insects coincide?
- How common are species-specific relationships, and which plant-insect pairs show such interdependence?
- How do temperature and precipitation affect plant-insect relationships across time and space?

Plant-Insect Pollination Ecology

Because plants are not mobile, they have evolved ways to mate with individuals beyond the immediate area to maintain genetic diversity. Flowers, the often-showy reproductive structures that house the sperm (in pollen) and ovules, are the most apparent evolutionary response to the mating challenge. Bright colors, fragrance, alluring shapes, nectar, and protein-rich pollen entice insects and other animals to visit the reproductive structures of the plant and, in the process, move pollen from the anther of one flower to the stigma of another by hitching a ride on the insects. Cross-pollination, this act of transferring pollen from one plant to another, has led to fascinating variability in both plants and insects. The timing of pollination events—on both a daily and seasonal scale—also has important insights about the ecological consequences of climate change.

Many different types of insects visit flowers: predators, herbivores, fungivores, even aquatic-dwellers. Pollination is not their objective—most insects visit flowers to feed or to encounter mates and promote their own reproduction. The calorie-rich nectar of flowers is an important energy source for flying adult insects, while bees transport the protein-filled pollen back to their nests where it is typically the primary source of nutrition for the complete development of larval offspring. Among insects that do become dusted with pollen during their floral visits are beetles, butterflies, wasps, true flies, and bees, with bees (under most circumstances in Oregon) being the most significant class of pollinator. Though there are far more non-bee species visiting flowers, the bee visitors are much more likely to accomplish pollination due to their relative strength and shape, which allows them to push through partly closed flowers like monkeyflowers, lupines, and gentians (Figure 29). Butterflies usually visit flowers with a shape that allows them to land, such as flat or open flowers and those with multiple small tubes; as they look for nectar they also spread pollen (Figures 39, 59, 60, 71, 90).

Figure 29. Bumble bee (*Bombus*) pushing its way out of the closed flower of King's gentian (*Gentiana sceptrum*).

Figure 30. Goldenrod crab spiders (*Misumena vatia*) hide on flowers like this willow, waiting for unsuspecting pollinators such as this bumble bee (*Bombus* sp.).

Figure 31. This classic cuckoo bee, *Nomada*, has two reddish-colored hyperparasites—in this case *Meloe* blister beetle larvae—tucked between its wings and mid-leg, hitching a ride to the bee's nest.

In certain circumstances, a flower may serve as a death trap to a potential pollinator due to predation by spiders and predaceous insects cryptically poised to capture a visitor (Figure 30). The mobile larvae of a parasitoid (specialized predatory insect) may lie in wait for its host bee to alight on the flower, then climb onto the bee and be transported to the nest and complete its life cycle at the expense of the unsuspecting bee's brood (Figure 31).

The vast majority of flower-visiting insects are associated with flowers only as adults. Additionally, both social and solitary bees and wasps are pollinator groups that are nest-centric, meaning activities throughout their life cycle—including foraging for food—are carried out close to home. Foraging distances vary depending upon the insect size and typically range from ten to several hundred meters, the human equivalent of over three kilometers (2 miles) (Danforth et al. 2019). Of these nest-centric groups, only the bees and pollen wasps (subfamily Masarinae) are dependent upon floral resources for their entire life cycle.

In the case of montane meadows, a large percentage of the flower-visiting insect species are forest inhabitants that emigrate from the forest to the meadow for the purpose of mating. Examples include long-horned beetles (family Cerambycidae) (Figure 32, 45) and hover flies (family Syrphidae) (Figures 35, 41, 57). Once mating is accomplished, the females return to the forest to lay their eggs in a diversity of microhabitats in which the larvae fulfill numerous ecological functions.

In contrast to what is known about butterflies, current knowledge of interactions between other pollinators and plants, and even about pollinators in general is much more limited. Preserved insect collections at universities and museums include many unidentified specimens, and details about the plants they associate with are rare. This lack of documentation makes field observations of pollinators an extremely important resource for scientists to deduce plant preferences, insect behavior, and life histories. Programs such as the Oregon Bee Project (2020) are synthesizing data from personal observations, collections, and scientific literature to better document the bee diversity within our region; as with the floristic work of OregonFlora, it is an ongoing process of discovery and documentation.

Extrapolating from our current knowledge in neighboring California, bee species make up 25–35% of the flower-visiting species in the non-forested meadow, bluff, and grassland communities of Oregon (Moldenke 1971, 1975, 1976; Moldenke & Neff 1974). There are about two dozen species of social bees (the introduced honeybee and native bumble bees with true division of labor between a queen and workers) and several dozen additional species that are social but can behave as solitary bees in extreme environments (sweat bees with the possibility of several females cooperatively joining to form tentative queens and workers). Also present in Oregon are more than 400 species of solitary bees (one egg-laying female with perhaps 1–2 dozen eggs that she provides with a pollen/nectar mix for next year's brood), including an estimated 150 cuckoo bee species (the female must discover the nest of a solitary bee and insert its own egg, resulting in the baby "cuckoo" killing the host larva and growing up on the stored food). Bees of all types are nearly absent in closed-canopy forests due to the need for direct polarized sunlight that they use for body warming and compass bearing to find food and to return to their nest.

Plant-Insect Herbivore Ecology

Approximately half of all insect species must consume plants to complete their development. The Lepidoptera—moths and butterflies—are well studied and familiar to most people and are the focus of our discussion regarding herbivory. Throughout their life history, butterflies interact with plants: adults visit flowers to sip nectar and lay eggs on various plant parts, while the larval form—the caterpillar—is the herbivore. Caterpillar-plant interactions reflect diverse behaviors and taxonomic relationships involving plant chemical signals such as defense mechanisms or feeding stimulants and insect counter-defensive chemistry, as well as patterns in foodplant specialization, biogeography, and population dynamics.

In a coevolutionary arms race, plants have evolved toxic chemicals (allelochemicals) to deter caterpillar feeding. In response, caterpillars have evolved detoxification processes that deactivate or otherwise neutralize the plant's defensive compounds. Some insects, such as the cinnabar moth, have evolved to use the same plant defensive chemicals (alkaloids) as a feeding stimulant (Figure 33). One of the consequences of the allelochemical arms race is the development of caterpillars that are either specialists—feeding only on a single family, genus, or species of plant whose chemicals they have adapted to tolerate, or generalists—feeding on a variety of plant species, perhaps from multiple plant families.

Another consequence of insect herbivory is the evolution of caterpillar appearance. Most caterpillars are more or less green in color and have cryptic markings that provide camouflage (Figures 38, 52). This strategy, surviving by hiding in plain sight (crypsis), is prevalent in species where the caterpillar does not obtain anti-predator protection by storing the normally toxic plant allelochemicals. Conversely, those species with caterpillars that have evolved to usurp allelochemicals to their advantage are typically brightly colored and referred to as aposematic (Figures 33, 55, 63, 79, 84). Typically, toxic caterpillars display black, yellow, white, or orange markings, often in a pattern of alternately colored rings sequentially arranged from head to tail, similar to patterns seen in poisonous snakes. The bright colors warn would-be predators that such caterpillars are poisonous. The colors and supporting poisonous chemicals may be passed on to the adult butterfly, giving it a similar level of anti-predator protection. In general, aposematic species are specialist plant feeders since they require the chemical for their protection, and cryptic species may be either specialists or generalists.

The geographic range of an insect herbivore is limited to the range of its foodplant, up to a point. For example, the distribution of a caterpillar that is highly specific to one species of plant will mirror the distribution of that plant, unless a closely related plant can also serve as its food source. The two plant species together offer a broader area that can support the caterpillar; however, the insect may still specialize on the preferred plant species when both plants are available. Of course, the caterpillar may exhibit a much smaller range than its foodplant based on its intolerance of weather or climate conditions. Overall, insect species that are generalists have the potential to be more widespread than those that are specialists.

Climate and physical environment have a profound influence on insect herbivore-plant interactions. Survival and developmental rates of both caterpillars and their plant hosts are directly correlated to temperature. This means departures or severe fluctuations in normal weather patterns have the potential to disrupt the timing of critical events that must be synchronous for survival, such as egg hatch, bud break, and leaf maturation for caterpillars that prefer fresh, young leaves (Miller 2010). Also, changes in climate patterns can alter the distribution of both plant and insect, resulting in certain species contracting/expanding their range, moving to another geographic location, or dying.

Figure 32. The black red beetle (*Neobellamira delicata*), a member of the flower long-horned beetles subfamily Lepturinae, is a common visitor to flowers, feeding on pollen or nectar. Here it feeds on the flower of heartleaf milkweed.

Figure 33. The non-native cinnabar moth was brought into the country to control the weedy non-native plant, tansy ragwort (*Senecio jacobea*), but their caterpillars will also eat this native arrowleaf groundsel (*S. triangularis*). Both the caterpillars and adult moths show the bright colors found in many toxic insects.

Physiognomy or Structure of Vegetation

The overall shape and structure of a plant community—also called its physiognomy—is one helpful perspective to take when observing an area. Broad physiognomic classes include forests, savannas or woodlands, shrublands, and grasslands. Within each class there are also subclasses—also referred to as "formations" (Beard 1978)—that further describe the area's composition, such as the tree canopy, an understory of smaller trees and shrubs, or herbaceous groundcover. Physiognomy is not strictly defined by the plant species that are found in an area. Rather, it is a more general way of categorizing an area based on the attributes of the vegetation. This is highly useful as we examine some of the behaviors of insects that rely on plants to accommodate their feeding, reproductive, and shelter needs. Physiognomic classes and subclasses, further grouped by plant families, provide our framework for organizing the stories of plant-insect interactions as we would encounter them in nature.

Climatic conditions impact the physiognomic classes that occur across the landscape. The coniferous forests that dominate much of northwestern Oregon extend westward from the crest of the Cascades to the Pacific Ocean and continue north into Washington and south to the Siskiyou Mountains (Albert, in Meyers et al. 2015). Heavy precipitation, fog, and cool temperatures off the Pacific Ocean sustain these vast forests in the Coast Range and Cascades ecoregions. As the predominant physiognomic class of northwestern Oregon, the coniferous forests are characterized by numerous species of large, evergreen trees that can form dense stands whose canopy creates deep shade on the forest floor. Douglas-fir (*Pseudotsuga menziesii*) is a dominant species at lower elevations, while true firs (*Abies*) are dominant at higher elevations.

In southwestern Oregon, climate and geology combine to support a physiognomic class of open oak and conifer woodlands with a diverse shrub and herb understory. The area's warmer, drier, fire-prone conditions and diorite and serpentine bedrock rich in heavy metals define this distinctive biotic region, known for its considerable biodiversity. Woodlands include Jeffrey pine (*Pinus jeffreyi*), members of Fagaceae, California laurel (*Umbellularia californica*), and an understory of light-loving and drought-tolerant species.

East of the Cascade Mountains, desert conditions prevail. Except for a few isolated high elevation areas like the Blue and Wallowa mountains, there is not enough precipitation to support dense forests; instead, open conifer woodlands, grasslands, or sage shrublands dominate the landscape. This high, open condition characterizes several ecoregions: the Columbia Basin, Lava Plains, Eastern Cascade Slopes and Foothills, and the Owyhee Uplands. Open woodlands of ponderosa pine are common on the east slopes of the Cascade Mountains, along with much denser stands of lodgepole pine (*Pinus contorta* var. *latifolia*). The shrublands that extend eastward host sage (*Artemisia*), bitterbrush (*Purshia tridentata*), and other hardy species.

The regions hosting Oregon's three physiognomic classes—the forested northwest, the woodland-dominated southwest, and the grassland and shrublands of the east—contain local variability that results from finer scale climatic and disturbance factors. While most of northwestern Oregon is forested with conifers, there are narrow riparian forests of red alder (*Alnus rubra*) and western redcedar (*Thuja plicata*) along rivers and larger streams. Along the Pacific shoreline, there are open dunes and grassy meadows where winds are too strong for dense forests to establish and survive. Sunny openings in coniferous forests occur on windy ridge tops, very steep slopes, regions where soils are very shallow or wet, or where there has been recent logging or forest fires. A broad expanse of open grassland, prairie, and moist meadows occupy the Willamette Valley between the Coast Range and the Cascades. Prior to European settlement, a combination of Native American management with fire and regular spring flooding of the Willamette River and its many tributaries created a diverse landscape of riparian forests, wet and dry prairies, as well as open, fire-created oak savannas. In the Cascades, extreme cold and wind, combined with shallow soils and exposed bedrock, create montane meadows that support an alpine-type flora (Figure 34).

The open mixed conifer and oak woodlands of the Siskiyou Mountains in southwestern Oregon are also broken up by open rock knobs and rocky outcrops. On thin-soiled slopes or at their bases, open seeps support unique fens with California pitcher plant (*Darlingtonia californica*) and many other rare endemic herbs and shrubs.

Eastern Oregon's open grasslands and shrublands host physiognomic subclasses defined by water: along many streams there are narrow bands of riparian forest dominated by willow (*Salix*), black cottonwood (*Populus trichocarpa*), and quaking aspen (*P. tremuloides*). Basins formed on the basalt plains are occupied by seasonal shallow lakes that dry up in most summers.

Figure 34. Wildflower-filled mountain meadows attract a great variety of pollinators during their relatively short growing season. Here, paintbrush, larkspur, and death camas put on a superb show in the Western Cascades.

Coniferous Forests

Dense upland coniferous forests characterize much of the Coast Range, the west slopes of the Cascade Mountains, and the Blue and Wallowa Mountain ecoregions where moist, cool, and shaded conditions reduce pollinator diversity. Butterflies in these forests are predominately associated with plants in the understory as opposed to the coniferous gymnosperms. However, numerous species of moth caterpillars do feed exclusively on the dominant canopy and are not associated with angiosperms. The specialization of moths on Pinaceae is largely due to the ability of their caterpillars to tolerate terpene compounds that are so common in gymnosperms.

Under the closed canopy, there are very seldom any resident bee species. Except for numerous micro-flies and wasps, there are virtually no pollinators. Nevertheless, there is often a diverse assemblage of herbaceous groundcover species that do bloom in the dark woods. About the only consistent, though very seldom observed, flower visitor is the hover fly, *Platycheirus* (family Syrphidae) (Figure 35). Nearly all the diverse species of forest-dwelling hover flies migrate from the forest canopy, where they mature feeding on aphids and scale insects, to forest openings to mate, then subsequently return to the forest canopy to oviposit (Moldenke pers. obs.). However, several dozen species of *Platycheirus* often do not migrate but drop to the forest floor to search for pollen (necessary for further egg production). These hover flies are not strong fliers and easily escape attention; though infrequent, these generalist flower visitors are significant for the cross-pollination of numerous genera such as *Anemone, Rubus,* strawberries (*Fragaria*), many Saxifragaceae, and beargrass (*Xerophyllum*).

Figure 35. *Platycheirus* hover fly on fringecup (*Tellima grandiflora*). *Platycheirus* is a species-rich genus in Oregon and a common flower visitor in the conifer forest understory.

Araceae

Skunk cabbage (*Lysichiton americanus*) is among the earliest plants to flower within wetland forests of both the Coast Range and the Western Cascades, typically blooming from mid-February on the coast to May in higher elevations. Flowers of the foul-smelling plants are visited by numerous species of fruit flies (*Drosophila*) and Omaliine rove beetles (family Staphylinidae) (Figure 36). These small insects often position themselves in the centers of the numerous flowers borne on the distinctive spadix. Unlike some other species in the family Araceae, western skunk cabbage does not produce its own endogenous heat as an insect attractant in the cold wet springtime.

Asteraceae

Coltsfoot (*Petasites*) is a dioecious plant—one in which an individual plant has flowers of only one sex. Therefore, each head consists of either male or female flowers. Since the species also propagates by runners, an entire stand of coltsfoot is typically only a single sex and can't reproduce without cross-pollination. The sap-feeding beetle *Epuraea*

Figure 36. Omaliine rove beetles on skunk cabbage (*Lysichiton americanus*). The flower-visiting behavior of the peculiar and poorly-understood omaliine rove beetle is unusual within the many tens of thousands of rove beetle species worldwide.

(family Nitidulidae) is frequently encountered nectar-feeding on the heads (Moldenke pers. obs.). Beetle behavior involves long stays upon a single head as they acquire energy-providing nectar; nonetheless they are efficient pollinators since the beetles fly for long distances and likely encounter other coltsfoot plants as they seek sap flows from injured tree branches and trunks, where they oviposit.

Ericaceae

Kinnikinnick (*Arctostaphylos uva-ursi*) and other manzanitas throughout the state have adopted the strategy of flowering very early to ensure pollination success. As the first and often only species in bloom, they minimize the problem of having generalist pollinators distribute their pollen to different species that are synchronously blooming. Their upside-down pendent flowers also preclude the vast majority of potential flower visitors, again permitting the few species capable of visiting the flowers to efficiently cross-pollinate (Figure 37). The most efficient pollinators of all the pendent ericaceous plants are bumble bees (*Bombus*) and a few species of early-flying mason bees (*Osmia*).

Orchidaceae

Under the forest canopy, where most flower-visiting insects are uncommon, many plants are self-compatible, meaning an individual plant's flowers can be fertilized by its own pollen. Some of the chlorophyll-lacking, mycotrophic orchids like the coralroots (*Corallorhiza*) use this strategy. In Oregon, the most abundant flower visitors under dense forest canopy conditions are two distinct groups of hover flies, whose adult diets rely on plant pollen but whose juveniles feed either on yeast in tree wounds or aphids in the tree canopy. Though very few hover flies are furry, it is easy to document that they carry large pollen loads on the surface of various body parts, which allows them to be very efficient pollinators.

Lady slipper orchids (*Cypripedium*) are typically pollinated by male *Andrena* spring bees that haven't yet found their usual flowers of choice. However, like many species of plants, lady slippers tend to attract a number of other nectar-seeking types of insects as well. A flower-visiting insect enters through the large opening in the heel of the slipper; the entrance is shaped like a valve or weir, which promotes entry but excludes the possibility of leaving. There is a unidirectional tunnel leading from the heel to the stigma and later to an anther before an exit hole is accessed. Insects larger than the tunnel diameter usually die imprisoned in the large "room" in the heel. Tiny insects go straight to the exit hole without pollinating. Depending on the species of *Cypripedium*, a subset of male spring bees are probably the most efficient pollinators, since they will automatically brush against the stigma and subsequently against the anther. The frequency of fertilization of *Cypripedium* in Oregon by different flower-visiting insects is poorly studied. *Cypripedium* and other members of the subfamily Cypripedioideae are an evolutionary relict; all other orchids reproduce by encapsulating all the pollen grains they make within a single packet, termed a pollinium, attached to a sticky stalk. While searching for rewards at the flower, a flower visitor might bump into the gluey stalk and it will stick to them. The next orchid flower the insect visits has a stigma with stickier glue that rips the pollinium off the visitor when it comes in contact, accomplishing cross-pollination.

Figure 37. *Nomada* bee on hairy manzanita (*Arctostaphylos columbiana*). With flowers that are difficult for insects to get inside, strong insects, mainly bumble bees, may "cheat" by poking a hole in the side of the flower. Other insects can also use the holes to access nectar and pollen. Unfortunately for the plants, this nectar robbing allows the insects to reap the reward without helping to pollinate the flowers.

Figure 38. The caterpillar of *Lithophane atara*, a noctuid moth, shows the white striping typical of caterpillars that live on conifers.

Pinaceae and Cupressaceae

It would appear that terpenes have restricted the evolution of butterflies colonizing species in the pine family, but the same cannot be said for moths. A few hundred species of moths are specialists on Pinaceae. However, only two species of butterflies in Oregon feed exclusively on Pinaceae.

The cryptic, dark green, white-lined caterpillars of the western pine elfin (*Callophrys eryphon*) are specialists on ponderosa pine and lodgepole/shore pine. The distribution of this butterfly corresponds to the distribution of its foodplants: a thin ribbon along the coast where shore pine (*Pinus contorta* var. *contorta*) resides, and a band across southern Oregon hosting ponderosa pine (*P. ponderosa*) that connects with ponderosa and lodgepole pine (*P. contorta* var. *latifolia*) populations on the eastern slopes of the Cascades.

A green body with white longitudinal lines is a common cryptic pattern that mimics the white stomata lines on conifer needles (Figure 38). It is seen among needle-feeding/foliage-inhabiting species: caterpillars of butterflies and numerous moths, sawflies, and the maggots of true flies. One such species, the pine white (*Neophasia menapia*), is the most abundant butterfly in coniferous forests (Figure 39); nonetheless, its caterpillars are rarely encountered because they occur in the higher reaches of the forest canopy. On occasion, pine white caterpillar populations will explode into epidemic-like numbers causing extensive defoliation.

Typically, populations of the western spruce budworm moth (*Choristoneura freemanii*) are innocuous, but over the course of several years, populations can swell to epidemic proportions, resulting in needle loss and ultimately tree die-offs that, in turn, can facilitate massive forest fires. This occurred in 2003 near Santiam Pass in the central Cascade Mountains. To prevent such outbreaks, programs using aerial spraying of the biological pesticide *Bacillus thuringiensis kurstaki* (Btk) have been implemented over the last four decades. Btk is an effective insect pest management tool for Lepidoptera; however, one of the ecological tradeoffs is that the bacterium is not species-specific. Consequently, caterpillars other than the pest species are negatively affected, as are insectivorous birds that require caterpillars for rearing their young.

Figure 39. Male pine white on Oregon sunshine (*Eriophyllum lanatum*). The appearance of pine whites often signals that summer is on the wane.

How can a native insect species develop populations in such high numbers on the east side of the Cascades but not do so on the west side? One hypothesis is that the presence of endophytic fungi in the needles of foodplants such as Douglas-fir is more widespread and common in the western forests (Carroll 1995; Clark et al. 1989). Endophytic fungi are present in the leaves of nearly all species of plants

Myrmecochory

Some plants have a mutualistic relationship with ants known as myrmecochory, from the Greek word "myrmeco" for "ant" and "chory" for "spread" or "disperse." These plants produce elaiosomes: fleshy appendages attached to their seeds that are rich in lipids and nutrients. Ants take the seeds to their nests where the elaiosomes are fed to their larvae. The seeds are then discarded, often in a midden within their nest, where the conditions are ideal for the germination of the seeds. While the plants undoubtedly expend some energy making elaiosomes, in return, their seeds are protected from herbivory and dispersed by the ants.

Myrmecochory has evolved independently in many different families and occurs in thousands of species throughout the world, predominantly in woodland plants, where extra help dispersing seeds may be an advantage. In Oregon, elaiosomes are found in such species as *Trillium*, violets (*Viola*), inside-out flower (*Vancouveria*), bleedingheart (*Dicentra*), and wild ginger (*Asarum*). Some types of elaisomes are known as arils, and the caruncles of Euphorbiaceae are also a type of elaiosome. The precise name depends on the origin point of the elaiosome.

Figure 40. Ants gathering seeds of *Trillium rivale*. The elaisome is the white attachment on the purple seed.

and have been shown to confer protection from herbivorous insects and other invertebrates via fungal allelochemicals. Additionally, the drier climate of the eastern Cascades is more hospitable to budworm development.

Two native species, the sequoia pitch moth (*Synanthedon sequoiae*) and the Douglas-fir pitch moth (*Synanthedon novaroensis*), are clear-winged moths with similar life histories. Common names for plants and animals can be misleading, as with the assumed foodplant references in the common names of these moths: *S. sequoiae* prefers pines and sequoia trees are not hosts, and *S. novaroensis* is not a specialist on Douglas-fir—it also feeds on pines. However, the "pitch" reference in their common names is deserved based on caterpillar feeding behavior. Caterpillars of pitch moths bore into the heartwood, stimulating an excessive production of pitch in response to the injury. The pitch exudes from the opening of the caterpillar tunnel and drips down the tree trunks. The chemistry of the pitch suggests that the plants are synthesizing defensive compounds to deter caterpillar feeding in addition to producing pitch as a physical consequence of injury.

Figure 42. Johnson's hairstreak is an uncommon lycaenid butterfly found near coniferous forests where its foodplant dwarf mistletoe grows.

Urticaceae

Stinging nettle (*Urtica dioica*) is notorious for the stinging hairs that cover its leaves and stems, a trait shared with the common sheep moth (Figure 62, described on page 42). While the formic acid in these pain-inflicting plant hairs may deter mammalian herbivores, their impact against invertebrate herbivory is minimal, if any at all. Interestingly, caterpillars that utilize nettle are not aposematic but appear to gain protection through the physical presence of the hairs that are so abundant in the feeding microhabitat. There are eight species of caterpillars from the nymphalid butterfly family that specialize on Urticaceae.

Figure 41. Hover flies like this *Chrysotoxum* are also called flower flies because the adults of many species feed mainly on nectar and pollen. For their own protection, they mimic wasps and bees, but they don't sting. The smaller flies on the Columbia lily (*Lilium columbianum*) stamens are *Anthalia* (family Hybotidae), frequent flower visitors.

Viscaceae

Interactions between conifers, the dwarf mistletoes (*Arceuthobium*) that parasitize them, and native gossamer-winged butterflies (family Lycaenidae) that may serve as a biological control agent of the mistletoe exemplify the complex relationships that form in coniferous forests. Johnson's hairstreak (*Callophrys johnsoni*) spends much of its time high in the forest canopy (Figure 42); its caterpillars feed on three species of dwarf mistletoe, each parasitic on a unique species of hemlock or fir. It is unknown if the economic value of timber stands would benefit more from treatment with Btk—which is used to control spruce budworm but also affects nontarget caterpillars such as the Johnson's hairstreak—or from allowing the *Callophrys* caterpillar feeding to naturally control parasitic mistletoe populations.

The pollination of dwarf mistletoes has been studied in detail. All *Arceuthobium* are dioecious, hence cross-pollination must occur to form fertile seeds. However, the flowers are extremely tiny and without showy petals. These innocuous flowers attract tiny fungus gnats (*Bradysia* in the Sciaridae family), which are able to transport the pollen between the separate plants.

Coniferous Forest Openings

There is a rich diversity of plant species found in openings within coniferous forests. Unlike the surrounding shady and rather uniform forest, openings provide a sunny habitat where wildflowers and otherwise suppressed understory plants can flower and thrive. Rocky habitats—cliffs, talus slopes, rocky slopes and ridges, and scree—shape some of the forest openings. Timber harvests and road construction have profoundly impacted Oregon's Coast Range and the Western Cascades. While they have diminished the vast stands of conifers, the disturbed open ground, meadows, and early successional clear-cuts have unintentionally created patches of habitat suitable for other plants as well as their insect pollinators and herbivores.

Asteraceae

A yellow-flowered member of the Aster family, tansy ragwort (*Senecio jacobaea*), is an exotic noxious weed for which insect herbivores from Europe have been intentionally introduced as biological control agents. Among the species introduced into Oregon are an aposematic day-flying arctiid moth, a flea-beetle, and a gall-making fly. The cinnabar moth (*Tyria jacobaeae*) caterpillar is strongly aposematic, displaying alternating orange and black rings the entire length of its body (Figure 33). It feeds predominately on tansy ragwort but also on a non-target species, the native *S. triangularis*. Using sensilla located near its mouthparts, the caterpillars detect alkaloid chemicals in the foliage. The alkaloids serve to stimulate feeding in the caterpillar (a phagostimulant). They also provide toxic protection against vertebrate predators and less successfully so to potential invertebrate predators, notably ants and yellowjacket wasps.

The caterpillars of Hoffmann's checkerspots (*Chlosyne hoffmanni*) are unusual for being gregarious when young and living communally in webby nests in their foodplant, Cascade aster (*Eucephalus ledophyllus*) (Figure 44). The nest protects the tiny caterpillars from predation. After the third instar (stage), they overwinter (diapause) among the dead leaves. The following spring, the caterpillars go their separate ways, with only one or two occupying a plant. This prevents them from overeating the individual aster plants as they grow. In later instars, the caterpillars are protected by their spines and by hiding on the ground when disturbed.

Figure 43. Oreas anglewing

Figure 44. Young Hoffman's checkerspot caterpillars live together in a webby nest on Cascade aster.

Ericaceae and Plantaginaceae

The reproductive effort of *Rhododendron*, as is likely the case with many other genera of Oregon plants, is heavily impacted by the seed-feeding true bug *Kleidocerys*. These bugs suck nutrients from the ovaries and oviposit so that the immatures have access to the ripening and recently shed seeds. *Kleidocerys* is very abundant in Oregon forest openings caused by storms or logging. A similar phenomenon has been observed along roadsides at Metolius Meadows, east of the Cascades summit, with a plant in the plantain family (Plantaginaceae) (Moldenke, unpubl. USFS research). The endangered plant *Penstemon peckii* is limited to three natural populations in central Oregon, each of which has low reproductive success due to different causes. Flowers of the southern population are completely emasculated by thrips, thus have no pollination at all. The northern population is >80% galled by a moth that prevents any flowers from forming. In the Metolius Meadows population, neither the thrips nor the gall-forming moths are present, so seed set is extremely high. However, the seed-bug *Nysius tenellus* completely destroys the developing seeds by feeding on the ovaries. *Nysius* is a relatively small true bug, and without examining the *Penstemon* closely, you would never realize why reproduction was so limited.

Grossulariaceae

An incomplete story exists when we consider the distribution of a butterfly and its foodplant in the oreas anglewing-currant interaction. The genus *Ribes* (currant) has eleven species in Oregon that occur statewide. The oreas anglewing (*Polygonia oreas*) has recently been defined as taxonomically distinct from the transcontinental *P. progne*, its former designation (Figure 43). Both butterfly species are genus specialists on *Ribes*. So why is the oreas anglewing limited to southwestern Oregon when one of its two *Ribes* foodplants (*R. lacustre*) has a broad distribution? The incongruity of the butterfly-foodplant distributions demonstrates that factors other than foodplant availability are defining the range of this butterfly species.

Figure 45. Elderberry long-horned beetles like this species (*Desmocerus aureipennis*) spend their entire lifecycle on elderberries (*Sambucus*). Most long-horned beetle larvae bore into dead wood; these burrow into the live stems. While the adult beetles meet, mate, and forage on the inflorescence, they help move the pollen around, creating a mutualistic relationship with the shrub.

Papaveraceae

Found along the margins of forest meadows and in rocky openings, the distribution of four taxa of the genus *Dicentra* and its genus-specialist feeder the clodius parnassian (*Parnassius clodius*) are examples of closely correlated plant-butterfly interactions. The two subspecies of Pacific bleedingheart (*D. formosa*), steer's head (*D. uniflora*), and dutchman's breeches (*D. cucullaria*) are foodplants for taxonomically separated populations of the clodius parnassian. Although all four species of the native *Dicentra* are documented as caterpillar foodplants, neither the closely related native *Corydalis* nor the exotic ornamental Japanese bleedingheart serve that function. The clodius parnassian caterpillar's velvet gray-black coat marked with two highly contrasting yellow longitudinal dashed lines would be quite visible during the day against the green foliage of the bleedinghearts and is, in fact, a mimicry ruse (Figure 46). The caterpillars are seldom seen because they feed mostly at night when their highly poisonous model, the cyanide-producing millipede *Harpaphe haydeniana*, is active as well.

Polygonaceae

Buckwheats (*Eriogonum*) with large inflorescences are amazing resources for flower-visiting insects throughout Oregon. Plants with large, flat landing platforms composed of upwards-facing tiny flowers attract a wide array of flower visitors, except hummingbirds and sphinx moths. In a research site within the H.J. Andrews Experimental Forest near McKenzie Bridge, Oregon, arrowleaf buckwheat (*E. compositum*) attracted more than 100 different species of insects over 4 years of observation (Jones et al. 2018). The problem that such a flower poses is one of fidelity—getting a generalist insect to make consecutive visits to the same plant species. If it doesn't succeed in programming the visitor to do so within a very short time after the flowers open for the day, all the pollen will have been removed and spread over the same inflorescence or wasted by being scattered over other plant species in the vicinity. Often, the greater the variety of flower visitors that a plant attracts, the greater likelihood that pollination will *not* take place. Such a plant can succeed, though, if it happens to have co-evolved with a specialist bee pollinator. Though there are about 20 known specialist bees associated with *Eriogonum* in California, only the buckwheat black-and-yellow gnat bee (*Perdita claypolei limulata*) is associated with buckwheat and the related small-flowered species, least knotweed (*Polygonum minimum*), at the McKenzie Bridge site.

Rhamnaceae

The plant genus *Ceanothus* is referred to as a cornucopia taxon, a plant group that attracts an extremely high diversity of flower visitors. Even an isolated shrub will support

Figure 46. A clodius parnassian caterpillar on its host plant, bleedingheart.

Carnivory

While insect herbivory is quite common, some plants have turned the tables by adapting to get at least some nourishment from insects, arachnids, and other small organisms. These species are generally found in water or in nitrogen-poor, wet soil such as in bogs. Several carnivorous genera are found in Oregon.

Throughout Oregon, the common bladderwort (*Utricularia vulgaris*; family Lentibulariaceae) can be found in standing water. Bladderworts derive their name from the tiny white "change-purse"-shaped traps that catch protozoa and tiny invertebrates. The traps are under negative pressure; when mechanically triggered by prey, the trap door opens within milliseconds and creates a subsequent current into the trap, pulling water and the prey in to its doom. Many species of bladderworts reproduce in two different ways: with large, colorful, monkeyflower-shaped blossoms designed for potential cross-pollination, and with tiny cleistogamous flowers that set selfed seed under the surface of the water. Flowers of *U. vulgaris* are visually very similar to those of *Erythranthe* (*Mimulus*) *guttatus*, which inhabits similar microhabitats. Both of these unrelated species have an essentially "closed" flower that a bee visitor has to force open.

Also in the bladderwort family, butterworts (*Pinguicula vulgaris* in Oregon) have sticky leaves to trap insects. When an insect contacts the glands on the leaves and struggles to break free, more mucilage is secreted, trapping the insect. Digestive enzymes are released by other glands, breaking down the prey so it can be absorbed by the plant. A particular part of a leaf can only digest once. Sundews (*Drosera*: Droseraceae) operate in a very similar manner, their glistening, gland-covered leaves even partly closing around the victims. Like butterwort, the flowers are held well above the leaves to allow some insects to safely pollinate the plants.

The striking California pitcher plant (*Darlingtonia californica*) of southwestern and coastal Oregon and northwestern California can collect insects in its deep pitchers. Insects that make the mistake of entering these modified leaves are unable to escape because the walls of the pitcher are lubricated and lined with downward-pointing hairs, and the top of the leaf curls over the entrance/exit. Colorless windows appearing to be possible exits confuse the insect until it exhausts itself and falls to the bottom of the pitcher. The same type of cells that the roots use to absorb nutrients are found inside the pitcher.

Figure 47. Dragonfly trapped by great sundew (*Drosera anglica*).

Figure 48. Insect victims of a pitcher plant.

a maelstrom of hyperactive flower visitors of nearly every pollinator guild. The chemical nature of the attraction is not known, but interestingly, even the closely related, tiny green-flowered species of buckthorn (*Rhamnus*) also attract far more insects than would be expected. Throughout the West Coast states, there are 12 species (five in Oregon) of specialist-feeding solitary bees associated with *Ceanothus*.

The foliage of *Ceanothus* species is also well suited to serving as a cornucopia for caterpillars. Nine species of moths and butterflies specialize on the genus *Ceanothus*, with eight species using tobacco brush (*C. velutinus*) in particular (Figure 49). The California tortoiseshell (*Nymphalis californica*) and its two *Ceanothus* foodplants, tobacco brush and deerbrush (*C. integerrimus*), interact in a unique saga beginning with massive migrations of the butterflies into the Cascades (Miller & Hammond 2003). The next generation's caterpillars feed heavily on leaves of tobacco brush and deer brush during the summer, stripping plants bare of their leaves across vast acreages dominated by *Ceanothus*, which is an early pioneer species of burned areas. When caterpillar numbers build to epidemic levels, inevitably a caterpillar-specific disease—a nucleopolyhedrosis virus—is transmitted throughout the population, killing nearly every individual. The plants recover from the defoliation during the following year and, over a few subsequent years, the butterfly population recovers as well. The patchiness of the massive caterpillar outbreaks, which occur in a variety of locations in ensuing years, and the migratory nature of this butterfly are two means for repopulating areas devastated by the virus. In the context of food chain relationships, the caterpillars are "parasites" of the plant, and the virus is a biological control agent of the caterpillars making the virus a "protector" of the plant.

Rosaceae

Figure 50. To bees and many other potential pollinators, the goldfields (*Lasthenia*) blossom that we see as yellow (left) has a dark ultraviolet-absorbing bull's eye in the center (right).

Spring-blooming flowers of the woody Rosaceae are favorites of nearly all flying insects searching for a conspicuous place to mate, more so on native shrubs than on exotic ornamental trees. A large number of spring bees (*Andrena*) are associated with the spring-flowering members of the rose family. The flowers are open-faced and plate-shaped; they exclude no one and hence face a serious challenge training flower visitors to faithfully visit another plant of the same species, especially if there isn't another one nearby. Since an individual shrub usually has numerous flowers, a bee could easily confine its foraging to a single plant. However, foraging bees maximize their net energy returns and concurrently may escape predators by avoiding a pattern of predictable behavior: after visiting several flowers on the same plant, they will suddenly leave their original site and forage elsewhere for another short time. Animal behaviorists term this "leptokurtic" behavior, and the plants benefit from the resultant cross-pollination.

Coastal Bluffs and Meadows

Along the Pacific Coast, strong coastal winds and erosion by storm waves create an environment dominated by herbaceous and shrubby vegetation, where trees are often small and scattered or absent. The cool and wet conditions do not support high butterfly diversity; among those that do inhabit the region are endemics and rare species. A climate with abundant clouds and fog also reduces polarized light that bees need for orientation. This, plus the prevalence of cool, moisture-laden winds favors the most cold-adapted species: the bumble bees. Pollination of coastal plants is additionally accomplished by sweat bees of the genus *Lasioglossum*, as well as fly and beetle species.

Figure 49. The hedgerow hairstreak (*Satyrium saepium*) is one of three species in Oregon whose caterpillars feed exclusively on *Ceanothus*. This female is ovipositing on buckbrush (*Ceanothus cuneatus*).

Asteraceae

A shortage of pollinators may be responsible for evolutionary change in some genera of plants: seaside goldfields (*Lasthenia maritima*) is a genetically self-compatible, largely self-pollinating species that does not require insect pollination and has only a few inconspicuous ray florets. In contrast, slender goldfields (*L. californica* ssp. *californica*) is an obligately outcrossing species found in meadows of southwestern Oregon and California; it has prominent ray florets that attract *Andrena* species specializing on goldfields (Moldenke 1971).

Flowers have several ways of communicating information to insects to make pollination more efficient. Floral guides, also known as nectar guides, honey guides, or guide marks, are patterns on a flower that help guide pollinators to nectar. They can be visible to the human eye, such as the veining patterns on violet or iris flowers or the spots on many monkeyflowers (*Erythranthe*). Unlike humans, both bees and butterflies can see the ultraviolet part of the spectrum (although many bees cannot see the red end), and a number of plant species have guides that can be seen by many insects but are only visible to us under ultraviolet light (Proctor et al. 1996). To humans, the bright yellow flowers of goldfields (*Lasthenia*), like many flowers in the aster family, look uniformly yellow. But to a bee, they resemble a black-eyed Susan with strong UV absorption centrally, including at the base of the ray florets (Figure 50). Strong UV markings occur in members of Ranunculaceae, Brassicaceae, Rosaceae, and a number of other families.

Ericaceae

Two congener species of elfins, the brown elfin (*Callophrys augustinus*) and the hoary elfin (*C. polios*), feed on kinnikinnick (*Arctostaphylos uva-ursi*). The hoary elfin uses kinnikinnick as both a caterpillar species-specific foodplant and a preferred nectar plant for the adult. The brown elfin, in addition to kinnikinnick, also feeds on many other species in the heath family (Ericaceae) such as salal (*Gaultheria*), madrone (*Arbutus*), manzanita (*Arctostaphylos*), rhododendron, and huckleberry (*Vaccinium*) as well as plants from other families, making it a true generalist (Figure 52). In stark contrast, the hoary elfin, a foodplant specialist, is generally less common, and the coastal subspecies, *C. polios maritima*, is listed as imperiled in Oregon.

Figure 52. A brown elfin caterpillar. Most lycaenid caterpillars are rather slug-like, with cryptic coloring to help them hide among foliage or flowers.

Fabaceae

Another endemic coastal butterfly is an uncommon subspecies of the greenish blue (*Icaricia saepiolus littoralis*), a specialist on clovers, particularly springbank clover (*Trifolium wormskioldii*). Remarkably, the caterpillars display dynamically cryptic coloring that can change within hours depending upon which part of the clover it is consuming: caterpillars are greenish when feeding on foliage and stems, and reddish when feeding on flowers (Miller, pers. comm.). Other caterpillars, lycaenids in particular, are known to be able to change their color to match their background.

Violaceae

There are only two Oregon butterfly genera that specialize on violets: the lesser fritillaries, *Boloria*, and the greater fritillaries, *Speyeria*. In fact, they are the only Oregon butterflies to feed on violets. Among these violet-feeders, the most shared foodplant species are *Viola adunca* and the inland species complex of *V. bakeri/praemorsa/vallicola*.

The Oregon silverspot (*Speyeria zerene hippolyta*) is federally listed as a threatened taxon. Once fairly widespread along the coast in pockets of suitable habitat, most populations are now extinct primarily due to loss of habitat. The gray and black spiny caterpillars are specialists on violets that in turn are vulnerable to habitat disturbance and loss. Conservation activities to eliminate exotic grasses, bracken fern, and blackberries that invade coastal meadows help to enhance the habitat needed for both butterfly and its violet host.

Figure 51. Fender's blue on Kincaid's lupine.

Prairies and Moist Meadows

West of the Cascades, most of the grasslands are wet prairie remnants growing in the Willamette Valley, with regions of oak savanna and riparian woodlands. The Willamette Valley, the most densely populated ecoregion in Oregon, is highly modified by agriculture and urban development, which has led to a steep decline in the diversity of pollinator and butterfly species in the region.

Fabaceae

The Fender's blue (*Icaricia icariodes fenderi*) is an endangered subspecies endemic to the Willamette Valley (Figure 51). Believed to have gone extinct prior to World War II, it was rediscovered in 1989 by Paul Hammond. A handful of locations have been found where small populations exist, all in remnant native prairie habitats where its preferred foodplant occurs, the threatened Kincaid's lupine (*Lupinus oreganus*). Presently, *Icaricia icarioides fenderi* is one of three butterflies in Oregon where a subspecies is the taxonomic basis for the federal designation. The limited range of Fender's blue is isolated from the other subspecies of Boisduval's blue (*I. icariodes*), which use many different species of lupine and occur from the Cascades throughout eastern Oregon.

Even though lupine flowers are well known to most hikers, it may surprise them to know that few lupines in North America produce nectar. Most lupines practice deceptive pollination through fragrance and a pollen reward, rather than nectar. The potential exception to this is *Lupinus arbustus* and its near relatives. This is an example of one of the most successful North American plant genera, both in terms of adaptive radiation and abundance, accomplished while rarely offering flower visitors any nectar. It is also characterized by functionally closed flowers, which exclude more than 90% of all the possible insect flower visitor groups, hence resulting in a relatively large pollen resource for the few species that are able to enter (Figure 54).

Figure 54. Bumble bee collecting pollen from velvet lupine (*Lupinus leucophyllus*).

Hydrophyllaceae

A Nature Conservancy Reserve north of Eugene has an enormous population of baby blue eyes (*Nemophila menziesii*), a plant observed as being pollinated by the specialist solitary red-tailed baby-blue-eyes spring bee (*Andrena nigripes*). In a typical phenological sequence, the male bees emerge from last year's nests a few to several days (averaging 9.6 in one study) before the females. The males stake out territories near the flowers where they will eventually mate with the soon-to-emerge females. In some years, the females might emerge after the *Nemophila* has finished blooming; in such cases the females either starve to death or fly off in search of a later-blooming population of *Nemophila* (Moldenke, unpubl. data; Danforth et al. 2019). Prior to habitat loss and savanna fragmentation, microhabitats found in extensive oak woodlands could support populations of *Nemophila* that bloomed over a greater time span, providing late-emerging spring bee females greater chances of survival. This type of weather-induced asynchrony resulting from climate change is the biggest challenge facing plants and their co-evolved specialist pollinators.

Orobanchaceae and Plantaginaceae

Taylor's checkerspot (*Euphydryas editha taylori*) is one of three butterfly species in western Oregon that are designated as federally endangered (Figure 53). This butterfly's native foodplants include harsh paintbrush (*Castilleja hispida*; Orobanchaceae) and golden paintbrush (*C. levisecta*), the latter which itself is a species that is extirpated in Oregon but has been reintroduced into the Willamette Valley. Some populations of the checkerspot now use the weedy non-native English plantain (*Plantago lanceolata*) as a foodplant, perhaps helping to save the butterfly subspecies from extinction. Both genera of foodplants contain iridoid glycosides. Most *Euphydryas* checkerspots use foodplants with iridoid glycosides, which act as a feeding stimulant and are also sequestered in their bodies and make them unpalatable (Figure 55). These chemicals are found in several families that were formerly grouped in Scrophulariaceae, the snapdragon or figwort family, as well as in Caprifoliaceae, the honeysuckle family. So while these checkerspots appear to be taxonomic generalists, using a number of different families, they are physiochemical specialists. Clearly these butterflies are focused on allelochemicals and not present-day plant taxonomy!

Figure 53. Taylor's checkerspot has turned to non-native food sources that contain the same chemicals as its former native caterpillar foodplant that was extirpated from Oregon.

Polemoniaceae

Many species in Polemoniaceae have long, very thin corolla tubes and feeble stems that may not support the weight of a perched bee or butterfly; these flowers are often pollinated by hovering long-tongued bee flies (family Bombyliidae) (Figure 54). Most long-tongued bee flies of the genus *Bombylius* are flower visitors as adults, with larvae that are nest parasites of immature bees. When not nectaring on flowers, female bee flies can be seen hovering and patrolling bare soil in search of solitary bee nests. The female bee fly flicks one of her eggs toward the nest opening. Of the eggs that do not land in the nest, if they persist and hatch, the bee fly larva attempts to find the nest entrance by scent. The larva, being a maggot, has no legs; it moves like an inchworm with a pair of adhesive pads attached to its face and backend. If successful in entering a nest, the bee fly larva attaches to the larval bee and stalls its further development until just before the host pupates, when it drains the host of its internal liquids. When individual flowers are not strong enough to provide a landing platform, a hovering bee fly or the hovering digger bee (*Anthophora urbana*) will be among the primary pollinators.

Montane Meadows

Montane meadows are open, sun-filled habitat found at higher elevations in the Cascades, Blue and Wallowa Mountains, the Siskiyous, and the Coast Range ecoregions. They are characterized by low temperatures, higher moisture from the mountain climate bringing more precipitation and snowpack, a short growing season, high winds, and high levels of solar radiation.

Meadows in the Cascades ecoregion comprise the most species-rich habitat for butterflies in Oregon and are definitely a prime area to visit and explore biodiversity. In contrast, the true alpine zone has the lowest diversity of native bees, and over 90% of them are also found in the Canadian Rockies and northern Canada.

Figure 56. The bee fly (*Bombylius*) is an extremely common and important pollinator of many Oregon species of plants, especially those with very small tubular flowers like this bluefield gilia (*Gilia capitata*). It is capable of hovering flight, often inserting its huge inflexible tongue into the nectaries without ever touching the corolla with its feet.

Figure 55. Several species of checkerspot butterflies in the genus *Euphydryas* feed on paintbrushes, such as this harsh paintbrush (*Castilleja hispida*), as well as penstemons and other genera that contain iridoid glycosides. Both the plant chemicals and their prominent spines protect the caterpillars and allow them to feed out in the open where they are relatively easy to spot.

Many butterfly species are restricted to Cascadian cold, alpine and subalpine habitats, suggesting that climate change poses a threat to their long-term survival. As with many plant species, the crest of the Cascade Mountains is the most important boundary separating Oregon's butterfly species and subspecies into those with either eastern or western distributions.

Apiaceae

Flowers of the carrot family are frequently visited by a minute, shiny, black, non-hairy solitary bee, *Hylaeus*, a genus with 19 species in Oregon. *Hylaeus*, the pollen-regurgitating bee genus, is widely distributed over all continents except Antarctica. Females of *Hylaeus* species behave similarly to the unrelated masarid bee-hornets in that they carry pollen and nectar in their stomachs during foraging trips. Upon returning to their nest, which usually occupies the hollowed-out pithy center of plant stems, they regurgitate their load into an empty nest cell, deposit an egg, and then seal the chamber. *Hylaeus* bees are not generally restricted to feeding on a specific plant species, but their need to regurgitate the pollen apparently often restricts most flower visitation to the carrot family, whose members have very small pollen grains. While most *Hylaeus* are generalist feeders, two species in Oregon are pollinators restricted to alpine and subalpine species of cinquefoils (*Potentilla* and *Drymocallis*; Rosaceae), which also possess unusually small pollen grains.

Figure 57. On the left, each anther of these fresh alpine laurel flowers is still tucked into a pocket in the corolla. These can be seen as bumps on the back of the flower. Some insect, probably a heavier bumble bee, has tripped the stamens of the flower to the hover fly's left, thereby releasing the pollen.

Crassulaceae

An interesting study of caterpillars of the mountain parnassian (*Parnassius smintheus*) feeding on spearleaf stonecrop (*Sedum lanceolatum*) demonstrates how an injured plant's production of defensive phytochemicals causes caterpillars to move to fresh leaves on a different plant that is undamaged by feeding. The caterpillar larvae engage in this energy-expensive and risky behavior to find a new foodplant that is not expressing the defensive chemicals triggered by feeding damage. On a broader scale, the nominate species of mountain parnassian has five foodplant taxa among Oregon's stonecrops. The northeastern Oregon subspecies *magnus* encounters foodplant species *S. lanceolatum*, *S. stenopetalum* ssp. *stenopetalum*, and *Rhodiola integrifolia*. Southwestern Oregon's parnassian subspecies *sternitskyi* uses *S. divergens* and *S. stenopetalum* ssp. *stenopetalum* as foodplants. The mountain parnassian is curiously absent from the Cascades, where its foodplants are nonetheless available.

Ericaceae

The pollen-bearing anthers of newly opened alpine laurel (*Kalmia microphylla*) flowers are tightly held in depressions within the fused petals (Figure 57). When a heavy-bodied insect like a bumble bee alights on the flower, its weight deforms the flower and the anthers are released from the depressions, springing outward, and depositing their pollen through terminal pores onto the bee's body. Terminal pores on anthers are shared by many members of Ericaceae as well as species that are buzz pollinated (Figure 61) but are not often found in other families of flowering plants. The pores function like a saltshaker and illustrate a coevolved adaptation of morphology to achieve pollination in flowers that hang upside-down, such as manzanita (*Arctostaphylos*), madrone (*Arbutus*), and huckleberry (*Vaccinium*). However, this dispersal mechanism is not effective for Ericaceae with erect flowers like *Rhododendron*. Pollen dispersal in *Rhododendron* is facilitated by elongated sticky threads that are present throughout the inside of the anther, which is packed with pollen. By merely touching one pollen grain, a bee can become a vector of dozens of pollen grains.

Melanthiaceae

Flowers of false hellebore (*Veratrum*) species are particularly interesting because they are pollinated primarily by sawflies, a primitive group of herbivores in the Hymenoptera order (Figure 58). Sawflies tend to visit a wide variety of simple dish-shaped flowers, but they are nearly always present on false hellebores. Some of the most abundant sawfly species on false hellebore flowers are those whose larvae also have false hellebore as their foodplant. A number of butterflies also favor their caterpillar foodplants as adult nectar sources. Examples include many pierids such as the margined white (*Pieris marginalis*) and toothwort (*Cardamine*) and other genera of Brassicaceae, monarchs and milkweed, greenish blues and clover, and dotted blues and buckwheat (Figure 59).

Polygonaceae

Buckwheats (*Eriogonum*) are foodplant favorites for many species of butterfly caterpillars, and 21 species or varieties of this plant genus occur in the Siskiyous and 29 in the

Figure 58. Most sawfly larvae are plant eaters, and many of them are specialists. These are on willow. They can be difficult to distinguish without knowledge of their foodplant. Sawfly larvae could be mistaken for those of a moth, but sawflies have more prolegs.

Figure 59. The caterpillars of dotted blues (*Euphilotes enoptes*) use naked buckwheat (*Eriogonum nudum*) as their foodplant. They feed on the flowers, but if the flowers senesce before they pupate, they will eat developing seeds. The adults also frequently nectar on buckwheats.

Figure 61. *Bombus mixtus* buzz pollinating mountain shooting star.

open areas of the eastern Cascades. The caterpillars of several gossamer wing butterflies (family Lycaenidae) of the Siskiyou and Cascade mountains are notably genus-level specialists within *Eriogonum*. Many are dotted blues in the *Euphilotes battoides* complex of species, some of which feed on a single species of *Eriogonum* (Figure 59). Their caterpillars may be tended by ants that defend them against predators.

Figure 60. Growing in the same wetlands as its caterpillar foodplant *Dodecatheon*, bistort (*Bistorta bistortoides*) is a favorite nectar plant of the adult Sierra Nevada blue at the northern end of its range.

The caterpillars range in color from yellow to green or red, depending upon their use of leaves or flowers as their primary food source. A number of lycaenid species are hard to distinguish by morphology, as the species appear to have become isolated by their use of different foodplants. In the Cascades and Siskiyous, the best way to separate the striking Sheridan's green hairstreak (*Callophrys sheridanii*) from the nearly identical bramble green hairstreak (*C. dumetorum*) is to look for nearby buckwheat species (Sheridan's foodplant) or lotuses (*Hosackia* or *Acmispon*, bramble's foodplant).

Primulaceae

The Sierra Nevada blue (*Agriades podarce*) ranges from the northern Sierra Nevada of California into the Cascades of southern Oregon (Figure 60). It spends its life in mountain wetlands, never straying far where its caterpillar foodplants mountain and alpine shooting star (*Dodecatheon jeffreyi* and *D. alpinum*) are found. In 2014, the first population of these pretty blues was found in the Calapooya Mountains in Douglas County by researchers from the Umpqua National Forest and BLM (Harvey, pers. comm.). Subsequent surveying in nearby wetlands quickly turned up eight more populations. Mountain shooting star is abundant throughout wet habitats in the Western Cascades but Sierra Nevada blues could not be found farther north. Many questions come to mind: Have they been in the Calapooyas all this time or is the species moving north? How do they disperse when they stay so close to their foodplant? What factors other than foodplant availability have kept their range limited? Will a warming climate allow them to extend their range further?

Shooting stars (*Dodecatheon*) are very closely related to primroses (*Primula*) and are considered to be in the same genus by some authorities. The main morphological

Figure 62. Sheep moths are large, day-flying moths often mistaken for butterflies. Their coloration is quite variable but always eye-catching.

difference between the two is that *Dodecatheon* flowers have reflexed petals. These distinctive flowers are designed for buzz pollination, a specialized method of pollen collection that bumble bees and some solitary bees do by vibrating the flower anthers at a particular frequency in order to release the pollen onto the bee (Figure 61). Most flowers that are buzz pollinated have pendent flowers with poricidal anthers, so the pollen has to be shaken hard to get it out through pores at the tip (Proctor et al. 1996). The harder the anther is shaken, the more pollen is extracted, so this favors the larger and stronger bumble bees (*Bombus*). The swept-back petals don't interfere with the bee's vibrating. The plant has to offer plenty of pollen to make sure some makes it to other flowers, but it doesn't produce nectar, and by making it difficult for insects other than these bees to access it, the plant increases the likelihood its pollen will be taken to other shooting star flowers. Other flowers that are pollinated this way include many in the tobacco family, Solanaceae, including tomatoes. Many people suggest that using an electric toothbrush to mimic buzz pollination by bees will yield better pollination of home-grown tomatoes!

Ranunculaceae

Monkshood (*Aconitum*) and larkspur (*Delphinium*) share two features: they are abundant in subalpine/alpine communities, and they produce nectar with a sugar chemistry uniquely suited for pollinators at high elevations. Bumble bees are usually the most important pollinators in cold and windy alpine regions, where their large size and dense fur allow them to remain active in very low temperatures, and their ability to pollinate is often limited by the amount of sugar in the nectar. Bumble bees have a narrow straw for a tongue, hence the sugar water they drink cannot be too viscous, a condition associated with high sugar concentrations. The sugars that plants typically produce in nectar are glucose or fructose—monomeric sugars. Some alpine flowers produce sucrose, a disaccharide sugar, or a trisaccharide (as in *Delphinium*), giving the nectar 2–3 times the energy for the same low viscosity.

Rosaceae

One of the very few caterpillars in Oregon that possesses urticating (stinging) hairs is the common sheep moth (*Hemileuca eglanterina*) (Figure 63). The stinging sensation is caused by formic acid, as is found in the nettles plant; it can create a rash as well as intense pain. Common sheep moth is indeed the most common *Hemileuca* occurring in moist habitats of the Pacific Northwest, where they are generalist feeders on flowering trees and shrubs, particularly Rosaceae members such as rose (*Rosa*), hawthorn (*Crataegus*), cherry (*Prunus*), serviceberry (*Amelanchier*), and bitterbrush (*Purshia*). Caterpillars are black with yellowish spines, not an obvious warning for a species that delivers a powerful chemical defense. Ironically, the day-flying adult, which lacks urticating hairs, appears aposematic with its brightly colored orange wings with black markings that are frequently tinged with pink (Figure 62). However, there is no indication that the moth is actually toxic.

Riparian

Riparian habitats occur as narrow bands along rivers and larger streams and can be surrounded by either dense forest or open savannas, woodlands, or grasslands. In western Oregon, the surrounding vegetation is typically dense coniferous forest, or oak savannas in the Willamette Valley. In eastern Oregon, riparian areas are typically surrounded by sage shrublands, grasslands, open bedrock, or volcanic ash deposits.

Figure 63. Sheep moth caterpillars are covered with stinging hairs.

Figure 64. Willows, such as this Sitka willow (*Salix sitchensis*), bloom very early and are a valuable source of nectar for early spring insects including echo azure butterflies (*Celastrina echo*) and many flies.

Asteraceae

The gumweeds *Grindelia integrifolia* and *G. nana* are late summer-blooming species. Their naturally occurring hybrid is serviced by the endangered specialist, the chocolate long-horned bee (*Melissodes pullatella*) (Severns & Moldenke 2010). The bee and *Grindelia* occur in moist areas in the Willamette Valley, where its nests may be submerged, with all their immature bees, for several months of the year. Waterproofing is always a prime necessity for ground-nesting bees, especially in a moist climate, but the ability to persist under standing water is a phenomenal adaptation to western Oregon's wettest habitats.

Primulaceae

Fringed loosestrife (*Lysimachia ciliata*) occurs along the Columbia River primarily east of Portland. Fringed loosestrife produces an oil instead of nectar to attract pollinators. On a worldwide basis, oil-producing flowers have evolved independently over two dozen times. The loosestrife oil-collecting bee (*Macropis nuda*) is a specialist pollinator of the loosestrife genus across North America. The bee uses spongy patches of hair on its legs and feet to absorb the oil for transport to the nest, where it is apparently used for waterproofing the nest, and not exclusively as a food source. Nectar for feeding the larvae must be collected from other species of plants.

Figure 65. Spring bee (*Andrena*) on Mackenzie's willow (*Salix prolixa*).

Salicaceae

The genus *Salix* (willows) is a misunderstood group of species when it comes to pollination. Though the introduced weeping willow and some other species may be partially wind-pollinated, most of the species native to Oregon, specifically Pacific willow (*S. lasiandra*) and Sitka willow (*S. sitchensis*) in the Willamette Valley, are insect-pollinated (Figure 64). Species of willows are characterized by large, heavy, and beautifully ornamented pollen grains, and the early spring flowers are extremely attractive to many pollinators, particularly syrphid hover flies and solitary spring bees (*Andrena*) (Figure 65). About half of the dozens of *Andrena* species that visit *Salix* species are true specialists visiting only willows. The other *Andrena* species are generalists waiting for the rest of the spring flora to bloom.

Figure 66. *Anthidium utahense.* Hornet-mimic (*Anthidium*) solitary bees are excellent mimics of yellow jackets. Their most unique habit is to insulate their individual larval cells with the dense plant hair of mullein plants or cottonwood seeds.

Populus (cottonwoods) is a well-studied genus present in the riparian Basin and Range ecoregion and elsewhere in the state. Its windborne cottony seeds (along with the willows) are a familiar riverside feature. Solitary hornet-mimic bees (*Anthidium*) (Figures 66, 89) use the hairs from *Populus* and *Salix* seeds to insulate their nests. Later in the year, the same phenomenon can be seen in urban backyards as the female bees strip the fuzz from the leaves of mullein (*Verbascum*) and betony (*Stachys*).

Figure 67. The fall webworm feeds on numerous species of deciduous trees. It is one of the few North American moths to become a pest on other continents.

The adult Viceroy (*Limenitis archippus*) is well known as a Batesian mimic of adult monarchs (*Danaus plexippus*)—a non-toxic adult having evolved to look like a toxin-protected adult. However, the viceroy caterpillars, which feed on nontoxic *Salix* and *Populus* plants, do not have protective aposematic coloration because they do not share similar microhabitats with a toxic look-alike. Therefore, no evolutionary drive exists for adopting the warning coloration as occurs in the adult. The viceroy-monarch story demonstrates how co-occurring, free-flying adult butterflies are subject to different agents of natural selection compared to their plant-bound caterpillars.

Silken tents enveloping the branches of trees and shrubs serve as expansive and seasonally expanding domiciles for caterpillars of many moth species. Caterpillars of butterflies also use silk for enhancing their microhabitat but typically do not construct expansive abodes. The fall webworm (*Hyphantria cunea*), a common moth and also a generalist feeder, is prolific in cottonwood trees (Figure 67). The caterpillars construct extensive silken tents, stripping the branches within and around their structures. One natural enemy of this moth, a nucleopolyhedrosis virus, helps keep populations at bay in some years by literally melting mature caterpillars within their tents.

Plant Galls

If you see an unusual growth on a plant, there's a good chance it is formed by an insect: tiny wasps in the family Cynipidae are very common, but also a variety of beetles, flies, moths, and aphids produce plant galls. Certain mites also form galls. Galls typically occur on leaves, twigs, branches, and roots. Females lay an egg in the tissue of a plant, which induces the plant to grow abnormally, forming a gall around the developing larva. Much research has been done on what causes the plant to form a gall—the adult or larval insect produces a chemical that affects the plant's hormones; however, the mechanism is still a mystery.

The immature stage of the insect is largely protected within the gall—although some parasitic wasps are able to breach the gall and parasitize the gall-maker. Other insects, such as caterpillars, may reside within galls but do not prey on the gall-maker. In many galls, the interior contains a cavity lined by a nutrient-rich tissue upon which the larva feeds.

Most gall-making insects are very particular about the species of plants they use, as well as the plant part where they lay their eggs. For instance, a majority of cynipids are associated with oak trees. Galls are also distinctive, enough so to be useful in identification of the species. The California oak gall wasp (*Andricus quercuscalifornicus*) is responsible for the conspicuous "oak apples" so prevalent on Oregon white oak branches.

Galls are often found on willows, asters, goldenrod, and roses. Rosebud- or pinecone-like galls at the tips of willow branches may be home to a willow gall midge (*Rabdophaga*). The magenta, fungus-like growth often seen on Douglas maples (*Acer glabrum*) is caused by the tiny maple erineum mite (*Aceria calaceris*). In general, galls do not affect plants adversely, but they can sometimes radically change the appearance of a plant.

Figure 68. Willow pinecone gall midges (*Rabdophaga*) cause galls in the tips of developing willow branches.

Savannas and Woodlands

Savannas and open woodlands are common on drier slopes of the Willamette Valley and the neighboring lower slopes of the Coast Range and Western Cascades, where Oregon white oak (*Quercus garryana*) is typically dominant. To the south, in the Siskiyou Mountains, mixed woodlands of Jeffrey pine (*Pinus jeffreyi*), Port Orford cedar (*Chamaecyparis lawsoniana*), Oregon white oak, chinquapin (*Chrysolepis*), and tanoak (*Notholithocarpus densiflorus*) dominate much of the landscape, and on the east flanks of the Cascade Mountains, woodlands of ponderosa pine (*Pinus ponderosa*) and lodgepole pine (*Pinus contorta* var. *latifolia*) are extensive. In all these savannas and woodlands, fire has maintained open understory conditions that support a diversity of shrubs, herbs, and graminoids.

Boraginaceae

Cryptantha is a common genus in disturbed and open habitats, with 21 species statewide. Even though the individual flowers are small, collectively they can provide an immense resource. Cryptanthas are most frequently visited, and probably successfully pollinated by the bee fly, *Bombylius*. They are also associated with uncommon specialist solitary bees in the subgenus *Proteriades* (family Megachilidae); our Oregon example is the San Jacinto pre-existing cavity-nesting bee (*Hoplitis jacintana*). These small cylindrical bees, all with partially red abdomens, nest in pithy branches or abandoned beetle burrows in wood, both of which are usually limited resources. A predominant perception in pollination ecology is that bee populations are limited by flower resources; however, *Proteriades* illustrates that bees may also be limited by availability of nesting sites. *Proteriades* is characterized by a very specialized tongue (a character independently evolved several times in different groups of specialist bees) that is covered with backwards-bent stiff hairs that rake the otherwise unavailable pollen out of the tiny corolla tubes of *Cryptantha* (Danforth et al. 2019).

While many species have flowers that fade or change color as they senesce, some also exhibit color changes on part or all of the corolla while it is still fresh as a signal that the flower has already been pollinated. This is seen frequently in flowers of the borage family, many of which have a central ring of crests or scales, called fornices, at the throat of the corolla. A close look at a population of Clearwater cat's eyes (*Cryptantha intermedia*) will show some of the fornices are white, while other, younger ones are yellow. This also can be observed among individual flowers of an inflorescence. Flower color changes occur in many plant families; lupines (Fabaceae) often have a white spot on the banner petals that turns bright purple after pollination, and in the case of mountain owl clover (*Orthocarpus imbricatus*; Orobanchaceae), fresh flowers have a yellow band near their tips that fades to white (Figure 69). Retaining flowers that have been pollinated has been shown to make the inflorescence more attractive to potential pollinators, while advertising the individual flowers that are no longer worth visiting makes pollination more efficient for the insects—a beneficial system for both parties (Weiss 1995).

Figure 69. The upper flowers in the inflorescence of mountain owl-clover (*Orthocarpus imbricatus*) display a yellow band, but the yellow disappears on older, already pollinated flowers, helping potential pollinators focus on the best flowers to visit.

Cyperaceae

Sedge-feeding among caterpillars of butterflies is unusual; one example is the dun skipper (*Euphyes vestris*), not only a

Figure 70. Ants tending a couple of lycaenid butterfly caterpillars while they chew on lupine pods. In addition to protection from predators by the ants, the caterpillars' cryptic coloration makes them difficult to spot on the lupine pods.

Figure 71. Golden hairstreak and tumbling flower beetle on pearly everlasting (*Anaphalis margaritacea*).

Figure 72. Propertius duskywing on its caterpillar foodplant, oak.

sedge-feeder but a sedge specialist. Surprisingly, the foodplants for the dun skipper have not been documented in Oregon although it is widespread across western Oregon. The knowledge of *Carex* as a foodplant comes from records in California and Colorado. Green with white stippling, the cryptic caterpillars live in a tubular nest made of foliage tied together with silk from glands associated with the mouthparts. Many species of skipper caterpillars will make a domicile of some sort from either a single leaf formed into a tent-like roll or several leaves appressed to create a nest like the dun skipper's.

Fabaceae

Ants, lupines, and lycaenid caterpillars exist as a trio of characters in a coevolutionary play. Several lupine species serve as the exclusive foodplant for a number of subspecies of the Boisduval's blue (*Icaricia icarioides*) in the western United States. As the caterpillars feed on the lupine leaves, they are tended by ants, which patrol the lupine foliage and climb onto the caterpillars in an effort to ward off predators looking to eat the caterpillar (Figure 70). The ants are rewarded for their protective service with a droplet of sugar exuded by the caterpillar, in some cases upon command from the ant. In other ant-caterpillar relationships, the ant is a predator of the caterpillars and thus serves to protect the plant from herbivory.

Fagaceae

The abundance and species richness of insects coupled with high plant diversity are hallmarks of oak woodlands found in the foothills of the Coast Range, the Willamette Valley, and the Cascades. Unique features of oak communities help drive increased diversity among insects. Consider that tannins in the mature leaves of oaks can strongly inhibit herbivore digestion and thus negatively affect insect development and survival. To avoid the tannins, much of the insect herbivory on oak leaves happens in early spring before the tannin content rises. Thus, synchrony between insect emergence from overwintering, leaf bud break, and presence of herbivore predators is critical to maintaining Oregon's biodiversity. Climate change threatens to disrupt the synchrony of these interactions to effect an overall reduction in biodiversity.

Notably common in oak woodlands are several western butterfly species that are specialists on oaks and chinquapin: the California sister (*Adelpha californica*), the propertius duskywing (*Erynnis propertius*), and the golden hairstreak (*Habrodais grunus*). The range of foodplant specialization that may occur among the butterflies that are limited to the Fagaceae is well illustrated by comparing *E. propertius* to *H. grunus*.

As a specialist on evergreen oaks, tanoak, and chinquapin, the golden hairstreak (Figure 71), also known as the chinquapin hairstreak, is not associated with deciduous oaks like Oregon white oak. It is not known whether this limitation is based on tannin phytochemistry or the phenology of leaf toughness, both of which are strong determinants of caterpillar feeding behavior. What is known is the northern limits of *H. grunus* are due to range limitations in suitable foodplants. Golden hairstreak adults usually appear in late summer as the chinquapins come into bloom and are rarely seen far from their host trees.

In contrast, the propertius duskywing is found in oak woodlands from southern California to British Columbia feeding on deciduous and evergreen species: Oregon white oak, California black oak (*Q. kelloggii*), canyon live oak (*Q. chrysolepis*), huckleberry oak (*Q. vacciniifolia*), and chinquapin (Figure 72). Although *E. propertius* is generally not rare in Oregon, its distribution is strongly associated with oak woodlands that are themselves in peril. In its northern range, propertius duskywing is listed as a species of special concern in British Columbia.

A magnificent brown, night-flying moth, the polyphemus moth (*Antheraea polyphemus*) is one of over 25 moth species whose caterpillars feed on oaks, though not necessarily exclusively (Figure 74). As one of the largest moths in western North America, this common night-flyer is a generalist feeder and frequently observed near lights in early July.

Chinquapin is one of the few plants visited primarily by long-horned beetles (subfamily Lepturinae of the Cerambycidae); this is true whether the chinquapin is blooming under a partial south-facing canopy or in the open. The lepturine beetles are nearly all diurnal, brightly-colored beetles that feed on the recently dead sapwood of conifer trees. They assemble on the stiff chinquapin inflorescences to mate, and in so doing become covered with pollen and successfully cross-pollinate the shrubs.

Tanoak is a pollination surprise. Most members of the oak family are wind-pollinated; nonetheless, tanoak has become associated with pollen collection by the Kincaid's silk-lining bee (*Colletes kincaidii*) over parts of the bee's range. Normally this solitary bee is a generalist, nectaring on many different plant species (Figure 73). However, in Santa Cruz County, California, they have been observed to be specialized collectors of tanoak pollen. *Notholithocarpus* produces little if any nectar, so the bee must visit California yerba santa (*Eriodictyon californicum*) and other flowers for nectar to feed its young. Even though specialized to the tanoak, *Colletes* do not contribute to its pollination due to the male and female inflorescences occurring separately (monoecious) (Moldenke, unpubl. data). The bee builds nests by tunneling into case-hardened limestone cliffs and backyard cemented stone walls. The female bee uses water from nectar to dissolve the cement crust in the outer hardened limestone layer before burrowing into the more easily dissolved limestone beneath. The bee's reuse of nest burrows for hundreds of generations makes it vulnerable to the cuckoo bee *Epeolus*, which may parasitize more than 70% of the communally concentrated nests (Torchio et al. 1988).

Figure 73. Male *Colletes kincaidii* on daisy (*Leucanthemum vulgare*).

Papaveraceae

California poppy (*Eschscholzia californica*) and the related gold poppy (*E. caespitosa*) have large gorgeous flowers and are a tremendous pollen resource for bees. But no nectar is provided, as no members of the historic poppy family (including Oregon native genera *Argemone, Eschscholzia, Meconella,* and *Platystemon*) produce nectar. This raises the question of how this section of the family lost the ability to produce nectar, and likewise why they have not reacquired it through evolutionary time. *Eschscholzia* is the host for 15 species of specialist-feeding solitary bees representing 7 different phyletic lineages, all of which must forage for nectar on other species of flowers to adequately provide for their young.

Rosaceae

The rose family is not known for having toxins and thus many caterpillars use plants of this group as a foodplant. Although many butterflies nectar on diverse rose family members, there are virtually no caterpillar specialists at

Iridaceae

Iris is a plant genus with copious nectar, but since it takes a very long tongue to sip its nectar, the iris flower doesn't get visited by many kinds of pollinators. In the savanna understory, these rewarding and attractive flowers are often pollinated by the exceedingly long-tongued spider-fly *Eulonchus tristis* (family Acroceridae), whose tongue is much longer than its entire body (Moldenke pers. obs.). Not only is *Eulonchus* a critical pollinator of many species in Oregon, its larvae are also the key parasite and thus population regulator of the omnipresent trapdoor spiders throughout Oregon, the most significant arthropod predators of the forest floor.

Figure 74. Polyphemus moth caterpillars feed on many trees and shrubs.

Figure 75. One of the specialist-feeding *Andrena* bees collecting pollen on large-fruited desert parsley (*Lomatium macrocarpum*).

the species or genus level. One almost-a-specialist species feeds on both chokecherry (*Prunus virginiana*), a native, and green ash (*Fraxinus pennsylvanica*), an exotic of another family, Oleaceae. Prior to the introduction of green ash, the two-tailed swallowtail (*Papilio multicaudata*) had specialized on chokecherry. The ability of caterpillars to use green ash, in a family known to contain allelochemicals, indicates an innate ability to switch hosts, and that the taxa do not necessarily need to be closely related phylogenetically or phytochemically. The distribution of the two-tailed swallowtail is also curious: the butterfly is widespread in western North America but is absent west of the Cascades. Pyle and LaBar (2018) list only *P. virginiana* var. *demissa* as its native foodplant; that it occurs throughout western Oregon suggests it is not a foodplant limitation that defines the range of the swallowtail. The distribution of *P. virginiana* var. *melanocarpa* throughout eastern Oregon suggests it might also be a foodplant, though it has yet to be documented.

Ranunculaceae

Columbine (*Aquilegia formosa*) flowers are favorites of nearly everyone—humans, insects, and hummingbirds alike. The upside-down red flowers have deep nectar tubes, giving a hovering hummingbird an advantage for reaching the nectar rewards. However, the same floral features that make columbines appealing to hummingbirds are what exclude many insects, such that bumble bees are the most physically capable insects to serve as pollinators. The vast majority of pollinator visitations to columbines are performed by bumble bees. The evolutionary dynamic demonstrated is that bumble bees are abundant short-distance vectors of pollen, while hummingbirds might transfer pollen longer distances, providing for an infrequent but grander scale of outcrossing and gene flow.

Viscaceae

Associated with oak woodlands is the great purple hairstreak (*Atlides halesus*). Instead of feeding on oak leaves, the cryptic, green, slug-like caterpillars are specialists on the parasitic mistletoe, *Phoradendron*, especially common on Oregon white oak. The stems and leaves of broad-leafed mistletoes contain toxins that belong to a class of small proteins known as thionins, which are poisonous to animals. In evolutionary terms, it is surprising that the caterpillars of the purple hairstreak do not show colors characteristic of poisonous larvae since they are indeed toxic due to their consumption of thionins. However, the aposematic adults sport brilliant iridescent blue wings with long tails on the hindwings and are marked with bright red patches at the base of the wings. The adult shimmers in the sun, signaling its toxicity while it sips nectar. Oregon records of this showy butterfly indicate a limited distribution along the western boundary of the Willamette Valley and the interior foothills of the Coast Range, the same distribution as its only recorded foodplant in Oregon, oak mistletoe (*Phoradendron villosum*).

Dry Grasslands and Shrublands

Almost all of Oregon's dry grasslands, shrublands, and sage shrublands are located in eastern Oregon, in the rain shadow of the Cascade Mountains. Grasses, sedges, herbs, and shrubs dominate, with scattered groves of juniper.

Windy habitats present a particular challenge to pollination and specifically pollinators. For bees to fly, their body temperatures must approach 95–100°F; their muscle

Figure 76. Indra swallowtail.

Figure 77. The spherical egg of an anise swallowtail, freshly laid on fernleaf biscuitroot (*Lomatium dissectum*).

chemistry is not unlike that of humans. For example, the Rowena Plateau is a windy region, and wind carries away body heat very efficiently. In the morning, a bee must remain in its burrow, warming itself with the heat both the soil surface and its dark face absorb. This initial warm-up takes a good deal longer in the windy Rowena than most other situations, leaving the bee far less time to exploit the floral resources before the afternoon winds build up. These conditions favor bee pollination by the very heavy-bodied bumble bees. In California, windy sites greatly decreased expected bee diversity. One study illustrates this principle: in a 0.5km^2 research site at Point Reyes coastal sage plant community, 172 total bee species were documented, 52 of which are specialists servicing 12 plant genera. In contrast, a comparable North Coast Range site, which lies inland, has at least 377 total bee species, 152 of which are specialists servicing 33 genera of plants (Moldenke & Neff 1974).

Apiaceae

The coevolutionary dynamics of pollination in the genus *Lomatium* (desert parsley or biscuitroot) may provide insights into successful management of endangered plant species. There are three rare species of *Lomatium* west of the Cascades, seven east of the Cascades, and additional locally endemic species east of the Cascades that are not considered rare. Throughout most of the Pacific Coast, *Lomatium* is closely tied to its specialist-pollinating solitary spring bee species, *Andrena pallidiscopa* (Figure 75). *Lomatium* species, with their tiny individual flowers, are not particularly high on the preference list of floral resources on a community-wide basis. Aside from the spring bees, they are visited by super-generalist sweat bees of the genera *Lasioglossum* and *Halictus*. These generalist bees will vector pollen only for very short distances, but the *Andrena* will carefully search out only *Lomatium* to visit over relatively much greater distances. Observations of endangered *Lomatium* by one of the authors has not revealed any of the specialist species at several populations; introducing specialist spring bees from populations where the bees are abundant might be a way for reserve managers to improve the success of its associated rare plant.

Many species of the Oregon carrot or parsley family are fed upon exclusively by the caterpillars of two species of butterflies: the anise swallowtail (*Papilio zelicaon*) and the Indra swallowtail (*Papilio indra*). The Indra swallowtail specializes on one species of *Lomatium*, Gray's biscuitroot (*L. papilioniferum*), and one species of *Cymopterus*, aromatic spring parsley (*C. terebinthinus*) (Figure 76). In contrast, the statewide anise swallowtail has a foodplant list that includes 17 genera and 69 species of Apiaceae (Figure 77).

The caterpillars of swallowtails defend themselves when threatened by nodding their head and arching their thorax to prominently display a viper-like pair of false eyespots. While arching the thorax, the caterpillar also everts an orange-yellow osmeterium, a vesicle located behind the head that resembles a forked tongue. The osmeterium emits a pungent smell coupled with a caustic taste should any potential predator be so bold as to take a bite.

In the Willamette Valley, inflorescences of *Angelica* and wild carrot (*Daucus*) are often found with the small, uniquely-shaped, Mordellidae tumbling flower-beetles. Most species of mordellids are bow-shaped, hunch-backed, with long hind legs and a slender tapered tail (Figures 71, 80). Adults are extremely abundant and can be found walking among flowers within one inflorescence and then buzzing off to an adjacent inflorescence. They are likely important pollinators of many members of the carrot and aster families but are poorly studied. When approached by a person or an insect predator, they kick out with their back legs and tumble head-over-heels off the tiny individual flowers and fall to the ground—hence their common name.

Figure 78. Milkweeds attract numerous insects, including this snowberry clearwing moth (*Hemaris thetis*) nectaring on narrowleaf milkweed (*Ascelpias fascicularis*). As its name implies, its caterpillars feed on snowberries (*Symphoricarpos*).

Apocynaceae

The genus *Asclepias* is similar to plants in the Orchidaceae in that pollination is designed to be a nearly all-or-none phenomenon. Both milkweeds and orchids have many seeds per flower, and thus require an enormous number of pollen grains for successful pollination. Both plant groups present their pollen grains to flower-visiting insects in wrapped units called pollinia. Each pollinium carries hundreds to thousands of pollen grains; in *Asclepias*, the pollinium has two lobes joined with a connector. An insect visitor steps on the flower, and a connector attaches to its foot; it is subsequently removed by fitting into the same structure on another flower. Their nectar-rich flowers attract numerous insects (Figures 32, 78, 91). Specimens in insect collections of one such visitor, pompilid wasps, almost always have the distinctive *Asclepias* pollinia attached to their tarsi, or toes.

The cardiac glycosides found in the milky latex of milkweeds can be highly toxic to a majority of animals but not the caterpillars of *Danaus*, namely the monarch butterfly (*Danaus plexippus*). In Oregon's native flora, three *Asclepias* species are known foodplants for the monarch: showy milkweed (*A. speciosa*), narrowleaf milkweed (*A. fascicularis*), and heartleaf milkweed (*A. cordifolia*) (Figure 79). Young caterpillars are particularly adept at feeding on buds

Figure 79. A late instar monarch caterpillar on heartleaf milkweed.

Figure 80. Numerous insects, including butterflies, moths, bees, beetles, wasps, ants, and grasshoppers, are attracted to rubber rabbitbrush (*Ericameria nauseosa*), a late-blooming shrub that is especially important to insects when very little else is usually in bloom in late summer.

and open flowers early in the spring and then move onto the foliage. In searching for caterpillars in the field, as if alternating rings of bright yellow and black were not obvious enough, a tell-tale sign is the dripping of the milky sap at the feeding scars as well as the frass pellets left behind as the end product of undigested plant tissue. Throughout their development, the caterpillars ingest the cardiac glycosides and sequester the poison internally. The sequestration process carries the poison through pupation and into the adult. Thus, the aposematic coloration in the caterpillar and the adult relay the warning to would-be vertebrate predators that this insect is not to be eaten. However, the same warning is not relayed to potential invertebrate predators, in particular ants and yellow jacket wasps, which pick off many of the small caterpillars. Curiously, the cardiac glycoside-containing chrysalis, which is placed in plain sight on the foodplant, is not aposematic but rather cryptic in its near solid green coloration.

Two other aposematic herbivorous insects that use the toxic nature of the foodplant to their survival advantage are often found on milkweeds: the red milkweed beetle (*Tetraopes femoratus*; family Cerambycidae)—bright red with black spots—and the bright yellow oleander aphid (*Aphis nerii*), relevant here given that cardiac glycosides are also present in oleander.

Asteraceae

The most prominent late summer bee pollinators in central Oregon's Lava Plains ecoregion are the long-horned bees (tribe Eucerini within the family Apidae); only the males have very long antennae. Though there are generalists and several specialists on other types of plants, most of the long-horned bees are in the genus *Melissodes* (18 species likely in the Lava Plains) and are specialized to visit Asteraceae (Figure 82). *Melissodes* belongs to a diverse group of long-tongued bees, some of which are characterized by the specialization of using the tibia and basitarsus of the hind leg to carry pollen. If you observe rather large bees with transverse hair bands on the abdomen and extremely large pollen baskets on the hind legs, it will likely be a *Melissodes*. A similar-sized banded bee that spends a lot of time hovering will be the common male digger-bee, *Anthophora urbana*. Tiny metallic green and black gnat-bees (genus *Perdita*; 2.5–5 mm long) are also abundant specialized pollinators of fall-blooming Asteraceae species.

Flowers of the widespread shrub rubber rabbitbrush (*Ericameria nauseosa*) are one of the only dependable mid- to late-summer nectar sources found throughout the open sagebrush steppe of eastern Oregon (Figure 80). Although the deep corolla tube limits access to nectar for short-tongued insects, they are nevertheless visited by perhaps more than a hundred bee, butterfly, beetle, wasp, and fly species, creating a vast pool of potential pollinators. Since rabbitbrush flowers are the most prevalent nectar source in the landscape at the time, all potential pollinators are facultatively specialized upon them. Most solitary wasps, including species of thread-waisted *Ammophila*, are conspicuously shiny and clearly not very hairy, suggesting that

Figure 81. Dogbane (*Apocynum androsaemifolium*) is an intensely fragrant relative of milkweed. Its sweet scent is a magnet for butterflies, such as this fritillary (*Speyeria* sp., top) and Edith's copper (*Lycaena editha*), and attracts other insects as well. Nectar-seeking insects are especially attracted to sweet scents. Scent enables plants to attract pollinators at a distance and is particularly important in attracting night-flying moths.

Figure 82. Long-horned bee (*Melissodes*) on aster.

Figure 83. Thread-waisted sand wasp (*Ammophila*) on rubber rabbitbrush.

their contribution to pollination is not as great as is that of bees, but they share in the intense competition for nectar as a sugar source (Figure 83). Wasps of the family Philanthidae not only compete for nectar but are also important predators of the ground-nesting solitary bees that visit the flowers. Along with the cadre of native bees and sphecid wasps, the specialized blister-beetle, *Nemognatha* (family Meloidae), is commonly observed on rabbitbrush flowers. The blister-beetle makes its living as a parasite of solitary bees. Unlike nearly all other beetles, the mouthparts of blister-beetles form an elongated straw-like structure that draws nectar up through capillary action into its mouth, making it yet another competitor for nectar.

Sage (*Artemisia*) are not typically fed upon by caterpillars of butterflies, more so by moths. An exception to this observation is tarragon (*Artemisia dracunculus*), which offers anisic aldehyde as a feeding stimulus for the caterpillars of the Oregon swallowtail (*Papilio machaon oregonia*), the official state butterfly of Oregon (Figure 85). Oregon swallowtails, along with other Papilionidae, are victims of specialist parasitoid wasps. Females of the parasitoid genus *Trogus* lay an egg inside a full-grown caterpillar. However, the caterpillar continues to feed on leaves and eventually pupates to form the chrysalis. The maggot-like larva of the parasitoid feeds on internal organs and fat bodies within the chrysalis, killing the butterfly. The wasp then pupates, and the adult wasp emerges out of a hole chewed through the butterfly chrysalis. In the context of food chain relationships, the caterpillars are "parasites" of the plant, and the parasitoid wasp is a biological control agent of the caterpillars making the wasp a "protector" of the plant. In general, plant volatiles emitted from feeding-damaged foliage, caterpillar saliva, and caterpillar frass are known to attract caterpillar natural enemies such as *Trogus* in a coevolutionary relationship involving plants, their herbivores, and predators of the herbivores.

Brassicaceae

The pungent mustard flavor perceived by humans is due to an array of molecules known as glucosinolates; they serve as a phagostimulant for those caterpillars that are mustard-feeders, and also as a feeding deterrent for those that do not feed on the mustard family. There is a continuous coevolutionary adaptation between plants and the insects that eat them to synthesize modified glucosinolates on the one hand and to detoxify those compounds on the other.

The role of Brassicaceae as caterpillar foodplants belongs to the Pieridae, in particular, butterflies of the genera *Anthocharis*, *Euchloe*, *Pieris*, and *Pontia* (Figure 84) (Miller & Hammond 2003). No other butterflies feed on mustard family species, though many moths do. Although statewide distributions exist for some of these butterfly species, their presence in eastern Oregon is most apparent where the Cascades serve as a barrier to their westernmost limits. Mustard-feeding pierids also occur across multiple plant communities, including open coniferous forest, woodlands, savanna, grasslands, and shrublands. The top two hosts of butterflies are the rockcresses *Boechera atrorubens* and *B. pendulocarpa* with four species of pierids known for each. Twenty-five of Oregon's Brassicaceae foodplant species support only one pierid. Conversely, the top generalist mustard feeder, the large marble (*Euchloe ausonides*), uses 16 species of Brassicaeae as foodplants. None of the mustard-feeding pierids are species-specific, although the cabbage white (*Pieris rapae*), a statewide exotic noxious species, lists only two foodplants, both of which are also exotic: nasturtium and broccoli/cabbage cultivars.

Figure 84. Caterpillars of the spring white (*Pontia sisymbrii*) feed primarily on species of rockcress (*Boechera*). They have aposematic coloration to warn predators of the toxic mustard oils in their system.

Caryophyllaceae

Silene is an unusual genus, in that it contains many white, nocturnal-flowering, moth-pollinated species and one red, diurnal, hummingbird-pollinated species (*S. californica*) to the south. Spalding's campion (*Silene spaldingii*), one of Zumwalt Prairie's endangered plant species, is one of the nocturnal-blooming species and is presumably adapted to white-lined sphinx moth (*Hyles lineata*, Figure 86) pollination. Sphinx moth abundance is episodic though, and usually the majority of flowers fail to be pollinated during the night. However, local populations of bumble bees, which can fly early in the morning, have learned to visit the *Silene* flowers as they are wilting. Experiments involving exclusion of bumble bees, achieved with mesh bags placed over the flowers, indicate that many of the flowers are successfully pollinated by the early-morning foraging bumble bees.

Figure 86. White-lined sphinx moth using its extremely long proboscis to nectar on white bog orchid (*Platanthera dilatata*) while in flight.

Fabaceae

The legumes are host to more species of butterflies than any other plant family, fed upon by caterpillars from four families, 13 genera, and 29 species. In particular, species of *Astragalus*, *Lathyrus*, *Lupinus*, *Trifolium*, and *Vicia* are host to many butterflies. Several agricultural crops in the pea family are foodplants for butterfly caterpillars. For example, the Queen Alexandra's sulphur (*Colias alexandra edwardsii*) uses alfalfa as an exotic foodplant and locoweeds (*Astragalus* and *Oxytropis*) as native foodplants. Similarly, a well-known agroecosystem-oriented butterfly is the native and statewide orange sulphur (*Colias eurytheme*), which ventures into the Willamette Valley as a southern immigrant. The multi-state distribution and northern expansion of *C. eurytheme*, also known as the alfalfa butterfly, has been facilitated by the commercial planting of alfalfa and clovers. In native habitats, its caterpillars typically can be found feeding on species of *Lotus*, *Acmispon*, *Thermopsis*, and *Trifolium*.

A commercially important soil-dwelling bee is the alkali bee, *Nomia melanderi* (Figure 87). Alkali bees are recognized by the iridescent opalescent reflections on the surface of the transverse stripes of the abdomen. *Nomia* was the first bee domesticated for the commercial production of alfalfa seed (Danforth et al. 2019). Though the females are attracted to small-flowered legumes in agricultural situations, *Nomia* are most easily observed in the Great Basin on flowers of Cleomaceae. *Nomia melanderi* nests in nature only in regions of alkali deposits on the soil surface. In an effort to enhance pollination of alfalfa, farmers in eastern Oregon, Utah, and Idaho engineer similar soil conditions adjacent to their crop fields for the bees to nest by the hundreds of thousands.

Figure 85. Oregon swallowtail is found mainly in the sagebrush country of the Deschutes, Columbia, and Snake river basins, where its foodplant tarragon is found.

Figure 87. Domesticated alkali bees are used to produce alfalfa seed in the Great Basin and the Owyhee ecoregion, especially around Ontario, Oregon. In this photo, the brown soil is unearthed by the burrow construction and is scattered on the white alkali deposit.

Figure 88. *Dufourea calochorti* is a specialist bee on *Calochortus* species such as this mountain cat's ear (*C. subalpinus*). *Calochortus* are favorites of many pollinators, probably due to the very large and easily located nectaries.

Astragalus, *Lupinus*, *Acmispon*, *Lotus*, and other flowers in the pea family have papilionaceous flowers with petals that are quite difficult to pry apart, making it challenging for most insects to obtain nectar. This exclusion strategy limits flower visitation mostly to the very tough bees, including mason bees (*Osmia*), bumble bees, and hornet-mimic *Anthidium* bees (Figure 89). They also visit and pollinate the closed flowers of the family Plantaginaceae (*Collinsia*) and Phrymaceae (*Erythranthe*, formerly *Mimulus*). These solitary bees carry pollen on scopae, or tufts of hair located on the bee's ventral area of the abdomen. As the bee first enters the flower, the stigma springs up and contacts the bee's pollen-covered belly, followed in quick succession by the stamens that dump new pollen onto the belly as well. Apparently, sufficient bees have not been available during evolutionary time, since these genera of the pea family also demonstrate the evolution of small-flowered, genetically self-compatible species from their large-flowered outcrossing ancestors.

Hydrophyllaceae

Species of *Phacelia* provide a first-rate resource for solitary bees (Figure 89). There are at least 40 specialist species of bees, representing 20 independent lineages, associated with *Phacelia* in California and adjacent regions. In Oregon, *Phacelia* is represented by 15 species in the Owyhee Upland and the Basin and Range ecoregions, whose characteristic clay soils are derived from the weathering of volcanic ash or basalt flows. The soil substrates—ideal for ground-nesting bees—and flower structures associated with *Phacelia* create a combination conducive to the specialization of pollinators. *Phacelia* specialists frequent both the species that require outcrossing, as evidenced by their exserted anthers and stigma, as well as the selfing species whose stamens and adjacent stigma are deep inside the floral tube. The bee specialists on the latter flower shape require special modifications of the tongue guard (rather like a human's cheeks) to rake out the pollen.

Onagraceae

Warner Basin and Hart Mountain are home to two species of evening primrose, *Oenothera cespitosa* and *O. deltoides*, both with large white flowers. Fitting to their size and color, they are pollinated primarily by sphinx moths but also by solitary bees active in the late evening and early dawn. About 50 years ago, Dr. Ned Bohart of Utah State University conducted experiments to investigate the synchrony between the presence of *Oenothera* flowers and bees (Bohart, unpubl. data), since there can be several years between rains and the ensuing production of flowers in this desert environment. He found that greenhouse-raised flowers placed in the desert during drought years would draw solitary bees to the potted flowering plants. When he removed the plants shortly after the bees found them, the bees returned to their nests and hibernated until the following year, demonstrating that the bee's life cycle was flexible and not entirely dictated by the annual schedule of the rain. How flexible other pollinators are in responding to variable precipitation patterns is a critical issue relevant to the dynamics of climate change.

Poaceae

Grasses are one of the top three plant families, along with Fabaceae and Polygonaceae, that serve as foodplants to a high number of butterfly species. In Oregon, 24 species of butterflies utilize 41 species of grasses within 18 genera. Most of the 118 foodplant records for grass-feeding butterflies known from Oregon come from observations made in other western states (Pyle 2002; Warren 2005; James & Nunnallee 2011; Pyle & LaBar 2018). Nonetheless, two groups of butterflies in Oregon are closely associated with grasses in general: the skippers (Hesperiidae) and the satyrines (Nymphalidae).

Figure 89. Mating pair of *Anthidium atrifrons* bees on *Phacelia*. They prefer to harvest pollen and nectar from *Phacelia* and Fabaceae to which they are adapted with their pollen-transport scopa underneath the abdomen.

Figure 90. Mardon skipper is a rare species of the Pacific Coast states. It prefers open grasslands where its caterpillars feed only on several species of grass.

The skipper grass-feeding guild in Oregon comprises 18 species, including one, the dun skipper (*Euphyes vestris*) that specializes only on sedges. Grass-feeding is an evolutionary strategy specialized for survival in a semi-arid environment. Among the grass genera, the highest number of records of foodplant species involve *Poa, Festuca, Bromus,* and *Elymus*. Some skipper species, such as the common roadside skipper (*Amblyscirtes vialis*), are grass-feeding generalists, while others are specialists that feed on only one or a few grasses. The mardon skipper (*Polites mardon*) is federally listed as endangered in Washington and rare and imperiled in Oregon (Figure 90). Four grasses are listed as its foodplants in Oregon: *Festuca idahoensis, F. rubra, F. roemeri*, and *Danthonia californica*. Elsewhere sedges (*Carex*; Cyperaceae) are also listed as foodplants. In the context of conserving this species, Pyle and LaBar (2018) state that habitat loss from meadow overgrowth by exotic grasses, overgrazing by domestic livestock, encroachment by woody species, and forest pest management practices are harmful to the butterfly and its host food plants.

Climate Change

How exactly climate change will impact the relationships between insects and plants is unfolding, resulting in one or a combination of two or three likely outcomes. Fundamentally, a trio of consequences are possible: 1) extinction of local populations, 2) dispersal to a newly suitable biogeograhic location, and 3) adaptation, such as foodplant switching or shifting the timing of life history events (phenology), thus resulting in staying in place..

Figure 91. Will monarchs disappear from the West Coast?

Montane insects may be forced to higher and cooler elevations, leaving alpine species such as some of the *Euphilotes* blues found only on High Cascade peaks with nowhere to go. On the other hand, some species of warmer, drier habitats may benefit from the expected overall warming and drying of the state. The pipevine swallowtail (*Battus philenor*) is found in much of the United States but has only been spotted a handful of times in Oregon. Its foodplants are pipevines (*Aristolochia*), which are not established in Oregon. However, the effects of climate change may well allow both the pipevines and the swallowtail to spread north from California.

One of the most troubling effects of the rapid changing of the climate may be the disturbance of the timing of insect-plant interactions. There are a number of different cues organisms use to set their life cycle "watches." Plant and insect development is affected by photoperiod, moisture, and temperature. Insects are cold-blooded, and temperature literally determines their timing of life history stages (egg to adult) and growth rates within each generation. Specifically, as temperatures change then so does the timing of insect development, resulting in asynchrony between the availabilty of flowers and pollinators as well as leaves and herbivores, namely, caterpillars of butterflies and moths.

Not only synchrony in development of insects and plants will be affected, but so will local dispersal and migration. Monarch butterflies have a particularly difficult time because their migration route takes them through many different ecoregions and habitats. For a successful life cycle, they must have fresh milkweed plants available to lay their eggs on throughout the northward journey of several generations and nectar plants in bloom for the returning generation. If they leave their wintering grounds in California too early, they will have nothing to lay their eggs on. In a drought, there aren't enough nectar plants to sustain them. With increasing catastrophic forest fires in the West and harsher winter storms destroying some of their overwintering grounds, their migration route may even be radically changed. Tagged monarch butterflies that for generations have followed the western migration route are occasionally appearing in the Mexican overwintering grounds of the eastern populations. More may be making that transition as the western population along the West Coast has declined precipitously. The very delicate balance between insects and the plants they depend upon is in grave danger of disruption by our accelerated climate change.

Two grass-feeding skippers that are predominately in California reach their northern distribution limits in the Oregon Siskiyous. The Columbian skipper (*Hesperia columbia*) caterpillars hatch from eggs deposited on their grass foodplants, often the common species *Koeleria macrantha*. The caterpillars of another hesperiid, Lindsey's skipper (*Hesperia lindseyi*) (Figure 93), are forced to search for their grass foodplant (*Festuca* and *Danthonia* are recorded as foodplants in California) after hatching because the female lays her eggs on a non-foodplant such as lichen or lupine, an unusual behavior for butterflies.

In Oregon, the satyrine grass-feeding guild consists of seven species among five genera, including wood nymphs, arctics, ringlets, and alpines. Caterpillars of the satyrines are essentially family-specific feeders, using eleven species among eleven genera throughout western North America, with Colorado showing the most records; only three genera, all with unspecified species, are documented for Oregon. Additional field research by grass taxonomists is needed to identify foodplants of satyrines within Oregon.

Species of the hover fly *Platycheirus* can frequently be observed throughout the state of Oregon, feeding on the pollen of grasses but not pollinating the plants. It is quite unusual to observe insects specifically visiting species of plants that are normally wind-pollinated, such as grasses.

Polemoniaceae

Phlox are common on the dry, rocky soils of the Basin and Range, with six different species occurring on or in the vicinity of Steens Mountain. *Phlox* is an intriguing plant, in part due to its correlation to the migratory behavior of the white-lined sphinx moth, *Hyles lineata*. This moth has a wide North American distribution; endemic populations exist in Oregon. It is a well-known migratory species that overwinters in northern Mexico, where its caterpillars feed on a wide variety of desert annual plants. In the occasional year when there is a huge spring desert bloom, the hawkmoth populations expand dramatically, and the desert soil becomes covered in a "gravel" of caterpillar frass. Only during years of widescale desert blooms, usually associated with El Niño events, do enough moths mature in the southern deserts to eventually migrate up to Oregon. They nectar on the flowers both during day and night and are frequently joined on phlox flowers by the diurnal, and also migratory, painted lady butterfly (*Vanessa cardui*) (Crawford 1981).

Figure 93. Lindsey's skipper, one of the southern Oregon grass-feeding skippers.

Figure 92. This female Leona's little blue was photographed at the only place in the world where you can see them, east of Crater Lake National Park. She is seen here just after laying eggs on her sole host food plant, spurry buckwheat.

Polygonaceae

Buckwheats (*Eriogonum*) are foodplant favorites for 17 species of butterfly caterpillars in Oregon. Many examples of this plant-insect combination illustrate either a species-specific relationship or a general association of the caterpillar with members of this diverse plant genus. The summit blue (*Euphilotes glaucon*) occurs east of a diagonal line extending from Harney County northwest to Wasco County, and feeds only on sulphur buckwheat (*Eriogonum umbellatum*). The tiny Leona's little blue (*Philotiella leona*) is limited to a single population east of Crater Lake and is under consideration for listing as a federally endangered species. Caterpillars exclusively feed on the equally small spurry buckwheat (*Eriogonum spergulinum* var. *reddingianum*) (Figure 92). Habitat loss, notably a lack of bare soil patches and availability of diverse nectar sources surrounding the food plant, are the greatest sources of danger for this imperiled butterfly. Six species of common and easily observed coppers (*Lycaena*) on Steens Mountain rely on a handful of genera from the buckwheat family: *Eriogonum*, docks (*Rumex*), knotweeds (*Polygonum*), and mountain sorrel (*Oxyria*). Mormon metalmark (*Apodemia mormo*), Oregon's only representative of the metalmark family, Riodinidae, feeds solely on species of *Eriogonum* (Figure 94).

Plant-insect interactions: a final perspective

The overarching concept that drives the many interdependent interactions between plants and insects is that of coevolution. The unique components that manifest as the physiology and morphology of an organism have a genetic basis. As they create advantages or liabilities for the species, the resulting behavior or form is subject to natural selection, resulting in the continuation of lineages that are better adapted to their environment. Adaptation may take the form of new features or strategies that other organisms in turn will respond to via modifications of their own; in other words, coevolution.

The entwined natural histories of plants and insects have—and still do—lead to many interesting behaviors and forms for us to contemplate. Coevolution has strongly shaped the **physiological component** of phytochemicals and insect response: plant-generated compounds that can function as feeding stimulants and/or feeding deterrents are countered by the evolution of insect biochemistry that responds to the phytochemicals as beneficial or possibly harmful. Beneficial molecules serve as stimulants for the initiation of feeding or the attraction to a flower. Harmful molecules are detoxified or sequestered to give a protective advantage to the insect.

The imprint of coevolution has a **morphological component** readily seen between pollinators and the flowers they visit: insects develop body shapes, tongue lengths, and pollen-grabbing hairs (Figure 95) that correspond to flower shapes, stamen and stigma placement, and pollen size. For butterfly and moth caterpillars, their coloration strategy—using camouflage (crypsis) or having bold patterns and colors (aposematism)—is a response to host plant colors and forms. The coevolution of such physiological and morphological features within plants and insects can result in very specific and high fidelity interactions. These specialist relationships between pollinator or caterpillar and plant species remind us of the myriad and sometimes fragile connections within ecosystems that become evident with keen observation.

Figure 94. Mormon metalmark nectaring on *Gutierrezia*.

Figure 95. The legs of this *Andrena* bee, here on common madia (*Madia elegans*), are covered with pollen-collecting hairs (scopa).

Within the complex web of plant-insect interactions, players assume the roles of providers, exploiters, and facilitators with each other. Plants serve as **providers** by offering nectar, pollen, leaves, and other plant parts for energy and nutrition as well as structural materials for perches or domiciles. Insects as providers are best illustrated by the fact that they are a primary food source for insectivores such as birds, bats, reptiles, fish, mammals, other insects, and some specialized plants.

The role of plants as **exploiters** is readily seen in the immense diversity of flower shapes and colors—plant life's evolutionary strategy to achieve pollination by convincing insects to visit and disperse their pollen. Insectivorous plants, which often grow in nutrient-poor environments, have adapted to exploit their prey as an alternative source of nutrition. Insects as exploiters—at least from the plant perspective—consist of the millions of species (half of all insects) that consume plant parts.

Finally, plants act as **facilitators** for insects when we view flowers as the stage for insect courtship and mating. The insect facilitators are the pollinators and those herbivores that feed heavily on aggressive dominant plant species. Their herbivory helps reduce the impact of shading, allelopathy, and nutrient/water hoarding that aggressive plant species exhibit; this in turn helps increase plant diversity.

The essays presented in this chapter share some of the fascinating and varied ways in which plants and insects mutually serve as providers, exploiters, and facilitators. Both parties participate in the coevolutionary dance that highlights each as a taxonomist, finely attuned to the life history, location, and characteristics of the other. Awareness of these relationships will hopefully add to the enjoyment and appreciation of our natural world, especially in Oregon.

HERBARIA AND A MODERN FLORA

Melanie Link-Pérez and Linda K. Hardison

There is a place within the Willamette Valley where a person can see and touch every species of flowering plant, conifer, or fern that grows in the state of Oregon, all within an area about the size of a tennis court. Visitors to this special place can marvel at the diversity of plants that grow within the salt marshes or sand dunes along the estuarine coast, examine species that grow only on mountain slopes within the Cascades, or explore the vegetation that characterizes arid sites within the Basin and Range ecoregion—all within a few steps. They can figuratively travel through time and hold plants that are older than the institution that houses them. This treasure trove of Oregon's floristic biodiversity is the Oregon State University Herbarium.

An herbarium is a natural history collection of preserved plants that are pressed flat, dried, and permanently mounted onto archival paper, along with a label bearing information about the specimen. The label includes the name of the plant and the person who collected it, the date of collection, and a description of the habitat and location where the plant was collected. Without the label, the plant specimen has limited or no scientific value. However, with an appropriate label, an herbarium specimen is a time capsule: a preserved piece of the living world from years gone by, as well as a glimpse into the environment it occupied and the person who collected it.

Herbarium specimens are stored within archival folders that are placed in cabinets, where they are arranged in a systematic way to make it easy to access the plants for study. Worldwide, there are approximately 390 million herbarium specimens housed and cared for in just over 3000 active herbaria distributed across 178 countries (Thiers 2020). The specimens within the world's herbaria document four centuries of earth's plant life; they provide physical examples that are the basis of plant nomenclature, record the distributions of species, document the timing of life-history events such as flowering and fruiting, and voucher the compositions of plant communities through space and time. Herbarium specimens are central to the science of botany and its service to humanity.

There are several systems for organizing specimens within an herbarium—including the straightforward method of alphabetizing by plant family—yet virtually all methods are based on the scientific name of the specimens. Unlike arranging a collection by date, collector, or location, organization based on the identity of the object offers the greatest opportunity for accessing and using the information contained within each item. Specimens serve as documented examples of a correctly identified plant, and floras are the guidebooks to their placement within and their relationship to the entirety of an herbarium collection or group of herbarium collections for a specific geographic area, such as a state. Many of the roles an herbarium plays in service to people would be difficult, if not impossible, without a flora. Imagine a farmer visiting the herbarium to determine if a plant discovered in her fields is a noxious weed that could threaten her livelihood; without a flora to approximate an identification of the unknown plant, she would not know where in the collections to begin to look for comparative material.

While an herbarium can be appreciated as a museum of priceless, catalogued archives that tell the history of a region's plant life, it is also an active lending library that supports research well beyond the realm of botany. Pharmaceutics, urban planning, forensics, and art projects are informed by the study of herbarium specimens.

As one might expect, an herbarium is a critical resource in the preparation of a flora. The flora before you and the broader work of OregonFlora are based extensively on the use of herbarium specimens, mostly from OSU's herbarium, which represents the most comprehensive collection of Oregon plants in the world. The key tasks of a flora—to delimit the plants of a given area, define their taxonomic status, describe each one in detail, and provide keys for their identification—can only be accomplished with access to all of the plants being covered, both historically and contemporaneously. Here are key questions that the *Flora of Oregon*—and any flora—seeks to answer:

- What plant taxa (species, subspecies, and varieties; singular = "taxon") are documented as occurring in the state?
- What is the spatial distribution of each taxon? What kind of habitats does a species occupy?
- Is a taxon native or introduced? If introduced, has it become naturalized?
- What variability exists within a taxon?
- How does its occurrence in Oregon relate to its distribution across broader regions?

Each of these questions requires exhaustive study of herbarium collections to accurately answer. The more complete the coverage of the herbarium, in terms of geographic area, species diversity, and time, the more accurate and comprehensive a flora can be.

How to Collect and Preserve Herbarium Specimens

Many people may have preserved an autumnal leaf or a flower from a loved one by pressing the item between the pages of a heavy book, with varied success and often motivated by sentiment. The collecting and pressing of an herbarium specimen, on the other hand, is undertaken with a different motivation altogether: to provide the best representation of the plant as possible. The most useful specimens include both representative vegetative material (leaves, stems, and—for small, herbaceous plants—roots) and reproductive material, such as flowers and fruits for the flowering plants, cones for the cone-bearing gymnosperms, and fronds bearing sporangia for ferns.

Along with the plant, it is important to collect data that will appear on the accompanying specimen label. The label data at a minimum should include the scientific name of the plant, the location from which it was collected, the identity of the person who collected it, and the date of collection. Data can also include a description of the habitat and soil, names of other species associated with the collected plant, details about color or fragrance, notes on population size, information about how common or rare the species appeared to be in the area, whether any insects were observed on the plants, and notes about the use of the plant obtained from conversations with local inhabitants.

After collecting the plant and label data, the specimen is cleaned of debris and arranged with care between sheets of newsprint or blotting paper. All informative structures should be visible: for example, floral parts, leaf arrangement, the tips and bases of leaves, and both sides of leaves. A quantity of plant material that will fit, along with its label, on standard herbarium sheets measuring 11.5 x 16.5 inches (29.2 x 41.9 cm) is selected. The plants and paper are placed for pressing between pieces of corrugated cardboard that are held firmly together between the wooden frame of a plant press with adjustable straps. The cardboards allow air to move freely around the pressed specimens to remove moisture as quickly as possible. Often the plant press is placed near a source of warm, forced air to speed drying, since rapid drying helps retain plant color.

After plants are thoroughly dried, the specimen and its label are mounted on acid-free archival paper (100% cotton cardstock is common). Many types of adhesive have been employed, but in general a good adhesive will not discolor or become brittle over time. Common white glue, slightly diluted, is used in many herbaria. Before being stored in the herbarium, specimens are treated by freezing at temperatures below -18°C for several days to kill insect pests that may be present; most of the common pests that attack herbarium specimens are killed in less than 48 hours at these temperatures. To extend their life, specimens are kept in tightly sealed herbarium cabinets when they are not in use. The tight seals on the cabinets protect the specimens from dust, insect pests, and some humidity.

Handling of specimens is done with care. Specimens must be kept horizontal—they are never flipped from front to back like the pages of a book. The sheet on which the plant is mounted is supported from below or handled at the edges to keep the sheet from bending, which could break the specimen or cause it to become detached from the mounting sheet. When properly cared for, herbarium specimens can retain their usefulness indefinitely.

Figure 96. A collection of *Lupinus albifrons* being laid out for pressing on newsprint before going into the plant press.

The OSU Herbarium was an essential resource when beginning work on the *Flora of Oregon*. To answer the question of which plants occur in the state, specimens were consulted, and a database developed to catalog the information. Specimens also provided information on the nativity of taxa, and whether the non-native species had naturalized. Databasing the specimen locations and assigning mappable coordinates to those lacking latitude/longitude values allowed us to create precise distribution maps for each taxon. This information provided a high quality, historic foundation of plant diversity data, informing us which plant species occurred where through time. Along with field observations and scientific literature, information gleaned from specimens—detailed measurements of plant parts, plant habitats, growth habits, and distribution—was used to prepare the taxonomic treatments that comprise a flora. The preserved specimens provide clear documentation for the claims made in the *Flora of Oregon*; its descriptions synthesize the phenotypic variability all taxa express across time and space.

There are several instances where a flora-writing initiative can strengthen an herbarium's collections.

(1) Augmenting incomplete distribution records. An exhaustive search to document the plant diversity of an area, as preparing taxonomic treatments for a flora requires, must tap into all available sources, including personal and agency herbaria and unvouchered species lists. Whether assembled for research, monitoring projects, or personal enjoyment, species lists can itemize taxa not yet documented by a vouchered specimen, or provide occurrence records from previously undocumented areas. These undocumented leads can be pursued to acquire new specimens for the herbarium.

(2) Resolving inconsistent nomenclature. The names assigned to herbarium specimens reflect the prevailing knowledge and scientific nomenclature of the times. As ongoing research and discovery leads to new information about a species' evolutionary origins or its distribution, that is reflected in its scientific name. One objective of a flora is to establish nomenclature for that point in time when the flora is published. This benchmark guides the annotation of herbarium specimens to reflect a consistent nomenclature for any given taxon and ensures all specimens of a taxonomic concept can be filed together and that superseded names (synonymy) are preserved through annotation labels.

(3) Correcting misidentifications. The scrutiny that authors of a flora's taxonomic treatments give to specimens readily reveals specimens that have either been misidentified or need reclassifying to reflect the taxonomic decisions being presented in the treatment. Such annotation of vast portions of the herbarium, accomplished by experts in various taxonomic groups, provides a level of specimen curation that can rarely be accomplished within the scope of duties of most collections managers. Flora-writing and herbarium management are thus synergistic efforts, with each strengthening and increasing the value of the other.

Herbarium specimens—some of which are centuries old—continue to grow in value and expand our knowledge as the collections are studied and data are added to them. Researchers gain access to the specimens in multiple ways. They can visit the collection housed in an herbarium, examine the specimens, and apply annotation labels that capture the knowledge the researcher brings to substantiate the specimen. If they are not able to visit the collection in person, they can request a loan of specimens be sent to a nearby herbarium to provide temporary access for their study. If high resolution digital images of the specimen are available, researchers can often remotely view and annotate a specimen by submitting their determinations to the herbarium curator via online tools. The physical specimen can then be pulled from the collection, annotated, its database entry updated, and the specimen refiled.

Annotation labels are physically attached to a specimen sheet to document changes in our understanding of a species; this most frequently is seen as a change in the name used to designate the taxonomic concept represented by the plant material. Annotation labels can also note if the specimen contributes to ancillary studies, such as part of an ecological dataset or as source material of a genetic or biochemical analysis. Annotation by researchers is one important way that herbarium specimens are kept up to date.

The digitization of herbarium specimens has revolutionized access to the information contained in herbaria. This occurs in two ways: the capture of an image of the herbarium sheet and the entry of the specimen label information into a database. Transcribing and parsing label information into data fields opens powerful ways to search datasets and to expand one's study to include similarly databased collections throughout the world. When shared with the public, specimen databases and images provide herbarium access to a much wider audience of scientific researchers and the curious public. Providing such access benefits the users as well as the collections: it expands the scientific literacy of the public while also creating opportunities for others to add to the knowledge captured within a specimen record.

Digital presentation of a flora similarly widens the availability of plant data to diverse users. The OregonFlora website serves as a digital version of the three-volume *Flora of Oregon*. There are several benefits to using a digital medium to present a modern flora: information can constantly be updated; material such as color photographs can be presented in quantities and of a quality that would be cost prohibitive in paper form; data is readily shared with other organizations and people; and access through the internet opens the exploration of plant diversity to a larger and more diverse audience. When presented as components of an online flora, an herbarium's digital resources become more than a catalog of images or database of facts; the collections gain context and relevancy to those who may never otherwise encounter such natural history collections.

Printed floras remain an effective method for communicating plant knowledge, most notably through their aids for plant identification. The precise terminology and suites of characters used to create couplets for dichotomous keys make a definitive plant identification feasible. Scientific illustrations, when included, emphasize important features for identification of a species in ways that photographs of individual plants cannot. Many botanists use their printed flora as a journal, noting in the margins when and where a plant was observed, as well as variations not captured within the written description. Both media—printed flora and digital software—have strengths and provide complementary resources about the plants of a region.

Rooted in a tradition hundreds of years old, herbaria remain important contemporary resources for plant-based inquiries. The impact of broad concepts such as coevolution, climate change, species invasions, and extinction can be elucidated by studying the plant life of a region through time. Ecological studies, changes in species distribution, and adaptations in plant structure as a response to environmental change require a dataset that spans significant time periods. Preserved specimens provide tangible documentation that can be compared to the plants found today. Plant molecular and physiological studies are strengthened by referencing a voucher specimen. Methods have recently been developed that allow for non-destructive or minimally destructive sampling of plant material from older specimens (Shepherd 2017), allowing researchers to reach back in time by examining older preserved material. It is a combination of both the physical specimens housed in an herbarium's cabinets and the database that contains their associated label data that make such studies feasible and create a rich repository of information that provides insights into our future natural environment.

Figure 97. A specimen of *Eucephalus gormanii*. The main label indicates the specimen was collected from the wild in August 1933 by Lilla Leach as *Aster gormanii*, her 4413th plant collection. It was accessioned by the University of Oregon Herbarium (now a part of OSU Herbarium) and catalogued with the number 113512. Other labels indicate the specimen has been annotated (re-identified) four times to its current determination of *Eucephalus gormanii*. The first of these annotations was by D.D. Thompson in 1972. The most recent—giving the specimen its currently recognized name—was made by K.L. Chambers in 2008. Its "!" symbol indicates Chambers agreed with the identification on the most recent annotation at that time (*Eucephalus gormanii*, in 2003). The small envelope preserves seeds and plant fragments, and the barcode was added in recent years.

The close relationship between an herbarium and a flora project based on the collections means that both programs are vulnerable to today's common challenges. A lack of financial support for documenting plant diversity threatens the maintenance of both physical collections and the datasets that are developed to create a flora. A decline in the training of plant taxonomists and field botanists leaves the leadership of herbaria and floristic efforts in peril (Belkin 2018; Thaman 2013). It is important to recognize that both the physical collections of an herbarium and the digital resources of a flora require skilled professionals curating the data to reflect the latest research and to reflect consequential nomenclatural changes in the collections themselves.

The contents of an herbarium reveal a picture of a region's plant life through time. When specimens are viewed as the foundation of a flora, the fundamental information each herbarium sheet contains is given context and adds to the rich and complex picture of plant diversity that floras provide us. Both herbaria and floras share a centuries-long history, and both benefit from and are strengthened by each other's presence.

VASCULAR PLANTS OF OREGON

Key to Plant Groups:

1. Plants without seeds, flowers, or cones; plants with an evident alternation of generations . . . . **PTERIDOPHYTES** (Volume 1)

1' Plants with seeds produced by flowers or cones; plants without an evident alternation of generations.

2. Seeds naked, not enclosed in carpels, borne between scales of cone; plants without fruits or typical flowers . . . . **GYMNOSPERMS** (Volume 1)

2' Seeds enclosed in carpels; plants with fruits and typical flowers.

3. Floral parts in 3's or 6's, rarely in 2's or 4's, never in 5's; leaves usually parallel-veined; vascular bundles scattered throughout pith; embryo with 1 cotyledon . . . . . . **MONOCOTS** (Volume 1)

3' Floral parts usually in 4's or 5's; leaves pinnate- or palmate-veined, rarely parallel-veined; vascular bundles arranged concentrically around a central pith; embryo with 2 cotyledons . . . . **DICOTS** (Volumes 2 & 3)

Family Index

Dicots A–F

ACERACEAE

(see Sapindaceae, Volume 3)

ADOXACEAE

(see Viburnaceae, Volume 3)

AIZOACEAE Fig marigold or ice plant family

Stephen C. Meyers

Herbs or **subshrubs** perennial, papillate; rooting at nodes or not. **Stems** prostrate and mat-forming, to erect. **Leaves** cauline, opposite, flat or 3-angled; stipules absent. **Inflorescences** terminal or axillary, flowers solitary; bracts present or absent; bracteoles absent; pedicels present or absent. **Flowers** bisexual, tubular or campanulate; hypanthia present; calyx lobes 4–5, hooded or beaked or not, equal or unequal; petals 0–250; stamens 30–250; pistils 1, ovaries inferior or half-inferior; placentation parietal, styles 0–25, distinct, stigmas 2–25. **Fruits** berries or dehiscent capsules, dehiscence loculicidal or septicidal, usually with multiple valves, with or without membranes covering seeds, or dehiscence circumscissile. **Seeds** many, slightly tuberculate or smooth.

Temperate and tropical regions worldwide. 130 genera; 2 genera treated in *Flora.*

Key to Aizoaceae genera:

1. Petals many; stems rooting at nodes; fruits berries. . . . . **CARPOBROTUS**

1' Petals 0; stems not rooting at nodes; fruits capsules. . . . . **SESUVIUM**

CARPOBROTUS Fig marigold

Herbs or **subshrubs** succulent, glabrous. **Roots** fibrous. **Stems** trailing, mat-forming or rarely erect, 3–25 dm; rooting at nodes. **Leaves** becoming connate basally, straight or curved, 3-angled in cross section, thick, fleshy, margins entire to dentate; sessile. **Inflorescences** terminal or axillary, flowers solitary; bracts absent or 2; pedicel erect. **Flowers** showy, tubular; calyx lobes 4–5, unequal, 2–3 inner smaller than outer, margins membranous; petals distinct; nectaries 0; stamens erect; ovaries inferior, styles 0. **Fruits** berries, fleshy, edible.

Temperate and tropical regions worldwide. 13 species; 1 species treated in *Flora.*

Carpobrotus chilensis (Molina) N.E. Brown (p. 61)

Sea fig, ice plant

Mesembryanthemum chilense Molina

Stems to 1.8 m; bark gray and red, cracked, shiny beneath. **Leaves** 3–7 × 0.5–1 cm, glaucous. **Inflorescences** pedicels 10–45 mm. **Flowers** 3–5 cm in diameter; calyx lobes 10–20 mm; petals 100–140, 10–25 mm, red to purple; stamens 100–250, 3–6 mm, white, anthers yellow; stigmas 8–12(25), 2–3 mm. **Fruits** oval to subglobose, 15–20 mm, green to yellow. **Seeds** 200–350, obovoid, compressed, brown, shiny, slightly tuberculate.

Sandy coastal dunes and bluffs, roadsides. Flowering Jan–Aug. 0–50 m. Est. CA; south to Mexico; South America, South Africa. Exotic.

Originally introduced from South Africa to stabilize dunes and prevent erosion along highways, this species has become an aggressive invader in coastal habitats. Thus far, eradication efforts have been unsuccessful.

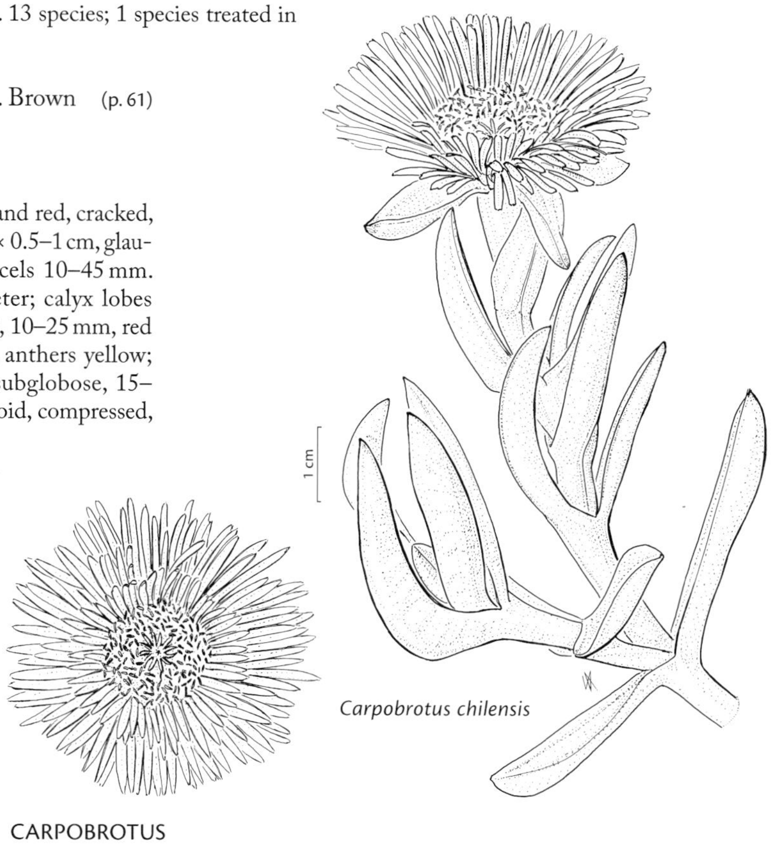

Carpobrotus chilensis

CARPOBROTUS

SESUVIUM Sea-purslane

Herbs succulent, glabrous. **Roots** fibrous. **Stems** prostrate, forming mats, not rooting at nodes. **Leaves** usually flat, margins entire; petiole bases tapered or clasping. **Inflorescences** axillary, flowers solitary; bracts 0 or 2. **Flowers** hypanthia obconic; calyx lobes 5, hooded or beaked near tip; petals absent; stamens connate proximally; ovaries half inferior, styles 2–5. **Fruits** capsules, ovoid to obconic, dehiscence circumscissile.

Temperate and tropical regions worldwide. ~8 species; 1 species treated in *Flora.*

Sesuvium verrucosum Rafinesque (p. 62)

Western sea-purslane, verrucose sea-purslane,

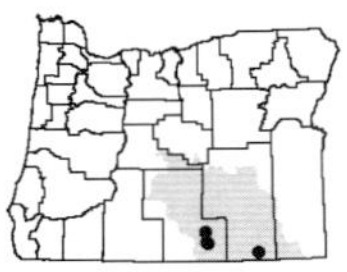

Stems to 1 m, forming mats to ~2 m in diameter. **Leaves** ≤ 4 cm. **Inflorescences** pedicels 0–2 mm. **Flowers** calyx lobes ovate-lanceolate, 2–9 mm, green abaxially, rose or orange adaxially, margins scarious; stamens ~30, filaments reddish; styles 5, stigmas 2–5. **Capsules** ovoid-globose, 4–5 mm. **Seeds** 20–40, ~1 mm, brown to black, shiny, smooth.

Margins of saline or alkaline areas. Flowering Apr–Sep. 50–1400 m. BR. CA, NV; southeast to LA, south to Mexico; South America. Native.

In Oregon, *S. verrucosum* is only known from desert alkali playas (except for one 1902 collection from ballast in Portland). However, in other regions the species can be found in coastal, saline wetlands. Further taxonomic investigation is needed.

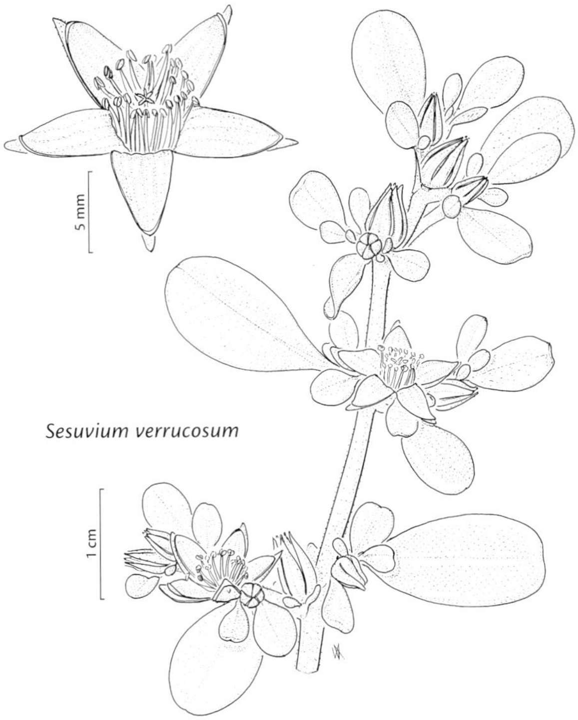

SESUVIUM

AMARANTHACEAE Amaranth family

(also see Sarcobataceae, Volume 3)

Bridget N. Chipman

Herbs, **subshrubs**, or **shrubs** annual or perennial, monoecious or dioecious, glabrous, pubescent, often farinose, sometimes succulent. **Stems** erect, ascending, prostrate or spreading, often much branched, sometimes fleshy and jointed. **Leaves** mostly alternate, sometimes opposite distally, simple, sometimes greatly reduced, margins entire, toothed, or hastate-lobed. **Inflorescences** axillary or terminal, commonly spikes or panicles, or sometimes solitary; with or without bracts. **Flowers** small and inconspicuous, unisexual or bisexual, perianths present or sometimes absent; corollas lacking; filaments free or united at bases; nectaries sometimes present in bisexual flowers; pistils with 1–3(5) carpels, ovaries with 1 cell and 1 ovule, superior or occasionally partially inferior; sometimes subtended by fruiting bracteoles. **Fruits** utricles or achenes. **Seeds** 1, vertical or horizontal in relation to infructescence axis, small, usually lenticular to globose.

Worldwide. ~165 genera; 18 genera treated in *Flora.*

This family contains many plants of economic importance, including beets, chard, and spinach, as well as cereal grains and animal fodder crops.

Beta vulgaris has been occasionally reported as a short-lived waif. It has not, however, become naturalized in our region.

Both the phylogenetic and morphological work support the expansion of Amaranthaceae to include former members of Chenopodiaceae (Stevens 2012). Following a recent molecular study of *Sarcobatus* (Cuénoud et al. 2008), the genus has been assigned to a separate family, Sarcobataceae.

Key to Amaranthaceae genera:

1. Leaves scale-like; stems fleshy and jointed; flowers in depression of rachis; flowers bisexual.
    2. Leaves and branches alternate; shrubs 3–15 dm . . . . **ALLENROLFEA**
    2' Leaves and branches opposite; subshrubs or herbs 0.1–3 dm.
        3. Plants annual herbs, erect, green, becoming red with age . . . . **SALICORNIA**
        3' Plants subshrubs, prostrate to procumbent and mat-forming, grayish green . . . . **SARCOCORNIA**

1' Leaves not scale-like; stems not jointed, rarely fleshy; flowers not sunken in depressions of rachis; flowers unisexual or bisexual.
    4. Plants shrubs, subshrubs, or perennial herbs.
        5. Leaves opposite; plants herbs; perennial, rhizomatous, perianth segments strongly overlapping . . . **NITROPHILA**
        5' Leaves alternate or only opposite on lowermost stems; plants shrubs, subshrubs, or herbs, perennial, not rhizomatous; perianth segments rarely overlapping.
            6. Leaves not fleshy or terete, flat; flowers unisexual.
                7. Leaves densely stellate-tomentose; fruiting bracteoles densely tomentose . . . **KRASCHENINNIKOVIA**
                7' Leaves glabrous, farinose, or scaly-puberulent; fruiting bracteoles glabrous or short-pubescent.
                    8. Fruiting bracteoles samara-like; leaves with rounded buds in axils . . . . **GRAYIA**
                    8' Fruiting bracteoles not samara-like; leaves without rounded buds in axils . . . . . . **ATRIPLEX** (in part)
            6' Leaves fleshy or terete, not flat; flowers bisexual or pistillate.
                9. Plants villous-tomentose; perianth segments with well-developed, prominent, horizontal wings . . . . **NEOKOCHIA**
                9' Plants glabrous to sometimes puberulent; perianth segments without wings . . . . **SUADEA** (in part)
    4' Plants annual herbs.
        10. Leaves flat or occasionally convolute, not subterete or terete.
            11. Flowers unisexual, staminate and pistillate flowers typically distinct.
                12. Perianth segments membranous or scarious; staminate flowers with (2)3–5-parted perianth; pistillate flowers with (1)3–5-parted perianth . . . . **AMARANTHUS**
                12' Perianth segments not scarious; staminate flowers typically with 3–5-parted perianth; pistillate flowers usually without perianth and enclosed by 2 fruiting bracteoles (but with deeply lobed, 5-parted perianths in *A. hortensis*) . . . . **ATRIPLEX** (in part)
            11' Flowers bisexual, occasionally unisexual, if so flowers not obviously distinct.
                13. Perianths 3–5-parted or -lobed.
                    14. Inflorescences densely hairy; leaves linear to lanceolate, margins entire . . . . **BASSIA**
                    14' Inflorescences glabrous, farinose, or rarely hairy; leaves not linear to lanceolate, margins lobed or toothed (but sometimes lance-oblong and/or with entire margins in *Dysphania*).
                        15. Plants with yellow-resinous glandular hairs, not glabrous or farinose; aromatic . . . **DYSPHANIA**
                        15' Plants glabrous or farinose; not aromatic . . . . **CHENOPODIUM**
                13' Perianths of 0–3 segments.
                    16. Fruits 2–4(4.5) mm, plano-convex, usually winged, pericarps smooth; leaves flat or occasionally convolute, not succulent . . . . **CORISPERMUM**
                    16' Fruits 0.5–1.5 mm, subspherical, not winged, pericarps reticulate, papillose or minutely tuberculate; leaves flat and succulent.
                        17. Plants with simple branches or branched from near bases, not forming a rounded plant; leaves 3–40 mm; flowers arranged in axillary clusters . . . . **BLITUM**
                        17' Plants dichotomously branched throughout, forming a rounded plant; leaves 2.5–8.5 mm; flowers usually solitary, rarely in 2–4-flowered clusters . . . . **MICROMONOLEPIS**
        10' Leaves terete to subterete or club-shaped.
            18. Leaves and bracts not tipped with bristles or spines . . . . **SUAEDA** (in part)
            18' Leaves and bracts tipped with rigid bristles or soft spines.
                19. Leaves 0.4–1.7 cm; perianths with wings at tips, not subtended by bracteoles . . . . . . **HALOGETON**
                19' Leaves 1.5–5 cm; perianths with wings abaxially, subtended by 2 fruiting bracteoles . . . . . . **SALSOLA**

ALLENROLFEA Iodine bush

Shrubs deciduous, glabrous; from large taproot. **Stems** erect to decumbent, alternately branched, fleshy, jointed. **Leaves** alternate, scale-like, broadly triangular, bases encircling stem, margins entire, tips acute; sessile. **Inflorescences** cylindrical terminal spikes, flowers in groups of 3 or 5, sunken in rachis in axils of deciduous, peltate, fleshy bracts. **Flowers** bisexual, perianths 4–5-lobed, persistent, stamens 1–2, stigmas 2–3; sessile. **Fruits** compressed utricles, ovoid; pericarp free, membranous. **Seeds** erect, shiny, brown to reddish brown.

Central America, western North America, South America. 3 species; 1 species treated in *Flora*.

This genus is named for English botanist Robert Allen Rolfe.

Similar *Salicornia* is distinguished by its opposite branches and leaves.

Allenrolfea occidentalis (S. Watson) Kuntze (p. 64)

Iodine bush

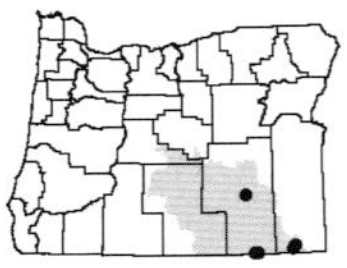

Plants 3–15 dm. **Stems** woody proximally, green, sometimes glaucous; joints (2)3–5(10) mm. **Leaves** blades 2–4 × 2–3 mm. **Inflorescences** 6–25 mm. **Fruits** enclosed by perianth. **Seeds** 0.6–1.2 mm.

Flats, sandy hummocks, moist alkaline soils. Flowering May–Sep. 1000–1600 m. BR. CA, ID, NV; southeast to TX, south to Mexico. Native.

The name iodine bush may be in reference to the iodine-like sap that exudes from crushed leaves.

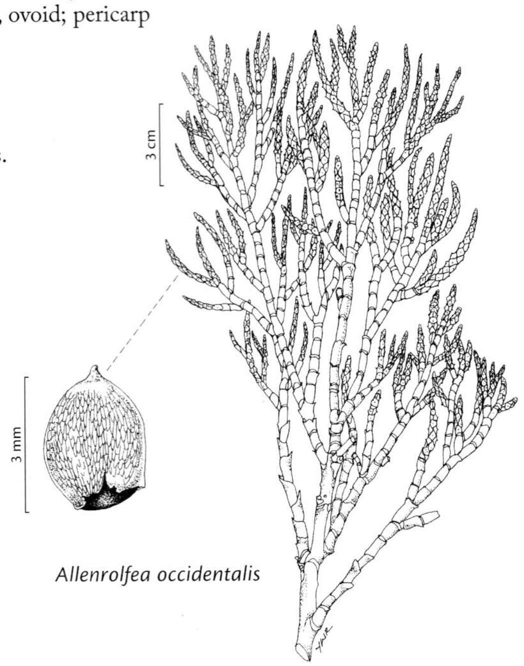

Allenrolfea occidentalis

ALLENROLFEA

AMARANTHUS Amaranth

Herbs annual, occasionally short-lived perennial, monoecious, glabrous or pubescent. **Stems** prostrate to erect, branched, occasionally simple. **Leaves** alternate; blades ovate to linear, flat, midveins extending to short spine or awn at tip, margins entire, flat or crisped, tips acute, obtuse or rounded; petioles present. **Inflorescences** terminal and axillary spikes or panicles with flowers densely clustered, or rarely axillary with solitary flowers; bracts leaf-like, often spine-tipped. **Flowers** unisexual, staminate and pistillate flowers typically distinct, perianth segments membranous or scarious. **Staminate flowers** perianth segments (2)3–5, scarious; stamens (1)2–5, filaments free. **Pistillate flowers** perianth segments (1)3–5; ovules 1, styles 0, stigmas 2–3, persistent. **Fruits** utricles, ovoid to obovoid, usually smooth, walls thin and membranous. **Seeds** lenticular to ± spherical, usually dark reddish brown to black, smooth, shiny.

Worldwide. ~70 species; 6 species treated in *Flora*.

Amaranthus species have long been cultivated as cereal grains and ornamentals.

Several cultivated and exotic species are known as short-lived waifs in Oregon: *A. deflexus*, *A. hypochondriacus*, and *A. viridus*.

Key to *Amaranthus* species:

1. Plants prostrate, much branched, forming mats.
   2. Perianth segments of pistillate flowers 1–3, only 1 well developed; seeds dark red-brown and shiny; petioles ≤½ as long as blade . . . . . ***A. californicus***
   2' Perianth segments of pistillate flowers 3–5, all well developed; seeds dark brown to black, shiny to somewhat dull; petioles ≥½ as long as blade . . . . . ***A. blitoides***

1' Plants predominantly erect, branched or not, not forming mats.
   3. Inflorescences exclusively axillary clusters of flowers, green, greenish white, or yellow . . . . . ***A. albus***
   3' Inflorescences terminal and occasionally axillary panicles, green to red.
      4. Perianth segments spoon-shaped to obovate, overlapping, becoming reflexed with age, tips rounded to notched . . . . . ***A. retroflexus***
      4' Perianth segments linear to lance-oblong, not overlapping, not reflexed, tips acute or awned.
         5. Bracts 4–7 mm; inflorescence branches few and rigid-erect . . . . . ***A. powellii***
         5' Bracts 2–4 mm; inflorescence branches many, lax and nodding . . . . . ***A. hybridus***

Amaranthus albus Linnaeus (p. 65)

White pigweed, tumbling pigweed

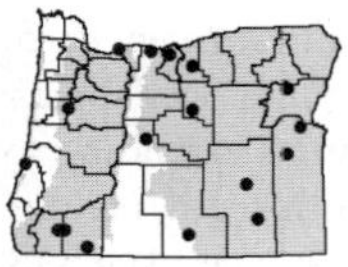

Plants glabrous to sparsely whitish-viscid pubescent, 1–10 dm; larger plants forming tumbleweeds. **Stems** usually erect, sometimes nearly prostrate, spreading and branched, green to whitish. **Leaves** blades obovate to narrowly spatulate, 1–8 cm, bases wedge-shaped, margins flat to distinctly wavy-crisped, midveins extending to a short weak spine, tips rounded; petioles 5–40 mm. **Inflorescences** axillary clusters of flowers, green, greenish white, or yellow; staminate and pistillate flowers intermixed in clusters; bracts rigid, oblong to lanceolate, 2–3(3.5) mm. **Staminate flowers** few, perianth segments 3; stamens 3. **Pistillate flowers** perianth segments 3; styles 3, erect. **Fruits** ellipsoid-ovoid, 1.5 mm, rugose at bases. **Seeds** lenticular, 0.6–1 mm, black, smooth, shiny. ***2n***=32.

Disturbed areas, fields. Flowering Jun–Oct. 0–1300 m. BR, BW, Casc, Col, Est, Lava, Owy, Sisk, WV. CA, ID, NV, WA; nearly worldwide. Exotic.

Amaranthus albus is a highly variable taxon depending on environmental conditions. It can be confused with *A. blitoides* when young and in moist areas where it is known to have a more procumbent habit.

Amaranthus blitoides S. Watson (p. 65)

Prostrate pigweed, tumbleweed amaranth

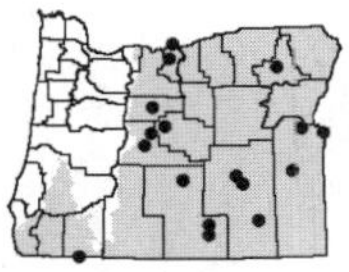

Plants glabrous. **Stems** prostrate to ascending, much branched from bases, mat-forming, 2–6(10) dm. **Leaves** blades obovate to spatulate, 1–3(4) cm, margins entire, flat to wavy or crisped, bases wedge-shaped, midveins extending to a short weak spine, tips obtuse to rounded; petioles 7–30 mm, ≥½ as long as blade. **Inflorescences** axillary clusters of flowers, green; staminate and pistillate flowers intermixed; bracts leaf-like, narrow and thin, 1–3(4) mm. **Staminate flowers** few; perianth segments 3(4), 1.5–2.5 mm, scarious; stamens 3. **Pistillate flowers** perianth segments (3)4–5, unequal, 1–3 mm; styles 3, becoming reflexed. **Fruits** ovoid to elliptic, 1.7–2.5 mm, bases smooth, rugose at tips; dehiscence circumscissile. **Seeds** lenticular, elliptic, plump, 1.1–1.7 mm, dark brown to black, shiny to somewhat dull. ***2n***=32.

Disturbed areas, riverbanks, dunes, fields. Flowering Jun–Sep. 50–1700 m. BR, BW, Col, ECas, Lava, Owy, Sisk. CA, ID, NV, WA; nearly worldwide. Native?

See also *A. albus*.

Amaranthus californicus (Moquin-Tandon) S. Watson (p. 65)

California amaranth

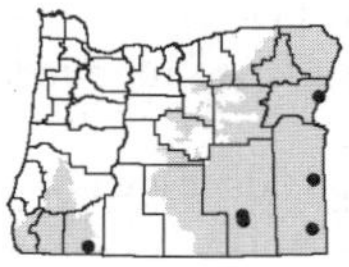

Plants glabrous. **Stems** prostrate, much branched and mat-forming, 1–5 dm, often whitish. **Leaves** blades oblong, obovate, or oblanceolate, 3–20 mm, bases wedge-shaped, margins entire, plane, flat to sometimes undulate, midveins extending to a spine, tips rounded to subacute; petioles 2–6(8)mm, ≤½ as long as blade. **Inflorescences** axillary clusters of flowers, staminate and pistillate flowers intermixed; bracts small and inconspicuous, 1–2 mm. **Staminate flowers** perianth segments (2)3, equal to subequal, 1–1.5 mm; stamens 3. **Pistillate flowers** perianth segments 1–3, unequal, only 1 well developed, 1–1.2 mm; styles 3, erect. **Fruits** subglobose, 1–1.2 mm, smooth, wrinkled near tips; dehiscence irregular to circumscissile. **Seeds** lenticular, obovate, 0.7–1 mm, dark red-brown, shiny.

Mud flats, dry ponds, ditches, gravel bars, disturbed areas. Flowering Jun–Sep. 50–1500 m. BR, BW, Owy, Sisk. CA, ID, NV, WA; scattered in western North America. Native.

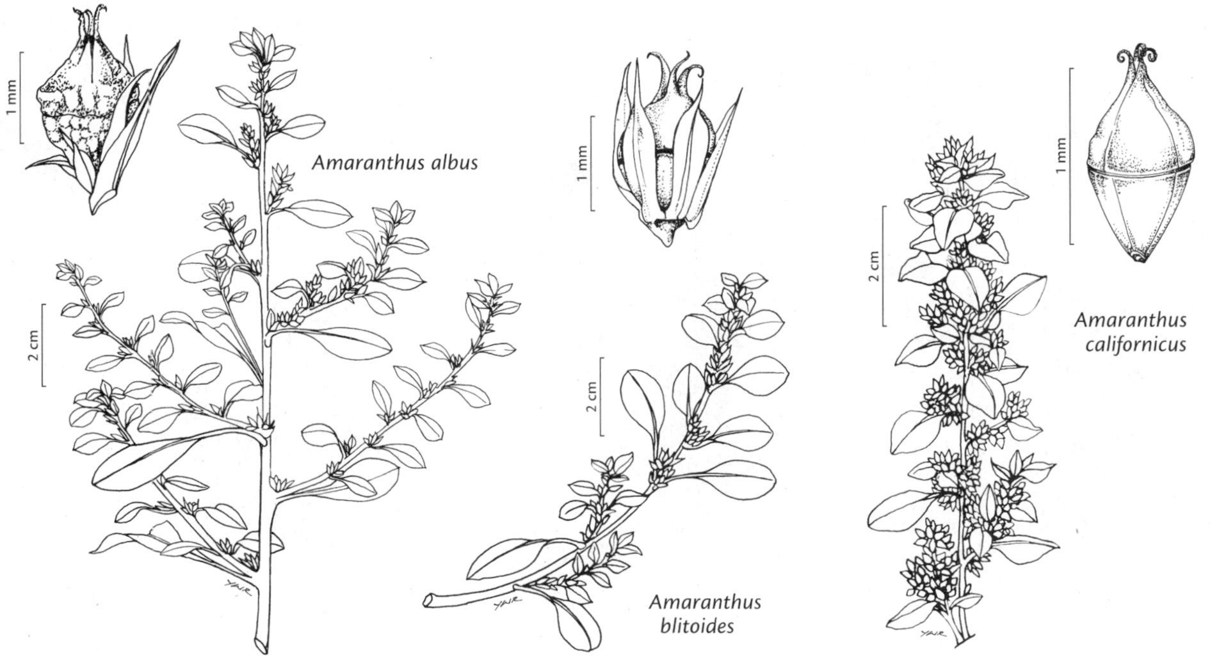

AMARANTHUS

Amaranthus hybridus Linnaeus

Slim amaranth, slender pigweed

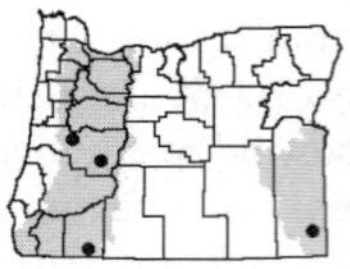

Plants mostly glabrous, villous distally when young. **Stems** erect, branched or not, 3–12(20) dm, green or red. **Leaves** blades rhombic-ovate to widely lanceolate, (2)4–12(15) cm, bases wedge-shaped, margins entire, plane, tips acute to obtuse, midveins extending to a spine; petioles 1–5(10) cm. **Inflorescences** terminal and axillary, tightly clustered, lax and nodding, dull green to red, villous; bracts 2–4 mm, midveins extending to a spine. **Flowers** perianth segments 5, mostly similar in staminate and pistillate flowers, erect, not overlapping or reflexed, linear to lance-oblong, midveins extending to spine, tips acute. **Staminate flowers** generally at inflorescence tips, perianth segments of nearly equal length, 2–3 mm; stamens 5. **Pistillate flowers** perianth segments of equal length or 1 slightly longer, 1.5–2 mm; styles 3, erect. **Fruits** ovoid, 1.5–2.2 mm, rugose at tips, smooth below; dehiscence circumscissile. **Seeds** lenticular, widely orbicular to rounded, 1–1.4 mm, red-brown to black, shiny. ***2n***=32.

Disturbed areas, streambanks, roadsides, fields. Flowering May–Oct. 0–700 m. Casc, Owy, Sisk, WV. CA, ID, NV, WA; throughout North America; scattered worldwide. Native?

Amaranthus hybridus has been known to hybridize with *A. retroflexus.*

Amaranthus powellii S. Watson (p. 66)

Powell's amaranth

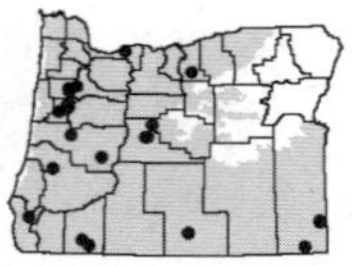

Plants mostly glabrous, occasionally villous near tips. **Stems** erect, branched or not, 2.5–15(20) dm, green or reddish purple. **Leaves** blades rhombic-ovate to broadly lanceolate, 3–6(8) cm, bases wedge-shaped, margins entire, flat to undulate, midveins extending to a short spine, tips obtuse to cuneate; petioles ≥ blade, 3–6 cm. **Inflorescences** mostly terminal, sometimes axillary, rigid-erect panicles with few branches, green to pale green; bracts numerous, lanceolate to awl-shaped, 4–7 mm. **Flowers** perianth segments 3–5, similar in staminate and pistillate flowers, erect, not overlapping or reflexed, broadly linear to lanceolate, midveins extending to a spine, tips acute. **Staminate flowers** perianth segments unequal to subequal, generally at tips of inflorescences; stamens 3–5. **Pistillate flowers** perianth segments 1.5–3.5 mm; styles 3, spreading. **Fruits** subglobose to ovoid, 2–3 mm, smooth to slightly rugose at tips; dehiscence circumscissile. **Seeds** subglobose to lenticular, 1–1.4 mm, black, shiny. ***2n***=32, 34.

Fields, disturbed areas, roadsides, streambanks. Flowering Jun–Oct. 50–1900 m. BR, Casc, Col, CR, ECas, Lava, Owy, Sisk, WV. CA, ID, NV, WA; nearly worldwide. Native.

Amaranthus retroflexus Linnaeus (p. 66)

Redroot amaranth, rough pigweed

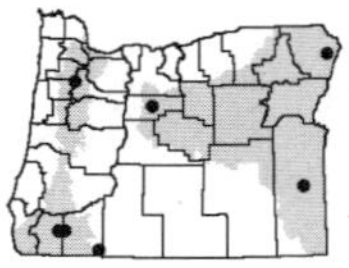

Plants densely to moderately villous, especially in distal portions. **Stems** erect, 1.5–15(20) dm, simple or branched distally, green or reddish. **Leaves** blades ovate to rhombic-ovate, 2–15 cm, bases wedge-shaped, margins entire, flat to crisped, midveins extending to a short spine, tips acute to obtuse; petioles 2.5–8 cm. **Inflorescences** terminal and axillary, erect to somewhat reflexed tightly clustered panicles, green to silvery green; bracts numerous,

Amaranthus powellii

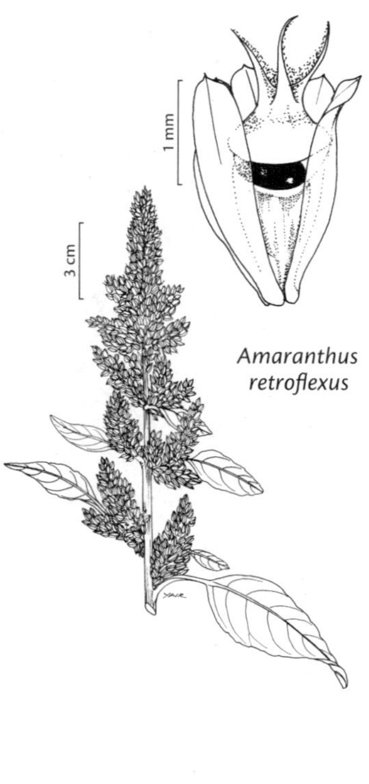

Amaranthus retroflexus

Atriplex canescens var. *canescens*

lanceolate to awl-shaped, 3–6 mm. **Flowers** perianth segments similar in staminate and pistillate flowers, spoon-shaped to obovate, overlapping, reflexed with age, equal to subequal, midveins extending to a short spine, tips rounded to notched. **Staminate flowers** perianth segments 4–5, 2–3 mm; stamens (3)4–5. **Pistillate flowers** perianth segments 5, (2)2.5–3.5 mm; stigmas 3. **Fruits** ovoid to elliptic, 1.5–2.5 mm, smooth to rugose at bases; dehiscence usually circumscissile. **Seeds** lenticular, ovoid to obovoid, 1–1.3 mm, dark brown to black, shiny. **2*n***=34.

Disturbed areas, fields, roadsides, streambanks. Flowering Jun–Sep. 0–1500 m. BW, Lava, Owy, Sisk, WV. CA, ID, NV, WA; worldwide. Exotic?

See also *A. hybridus*.

ATRIPLEX Saltbush, orache

Herbs, **subshrubs**, or **shrubs** annual or perennial, monoecious or dioecious, glabrous or farinose with short, inflated hairs. **Stems** erect to prostrate or spreading, woody or herbaceous, usually branched, branches sometimes forming spine-like appendages in shrubs. **Leaves** alternate, opposite below, sometimes hastate, margins entire or toothed or lobed, flat; sessile or petiolate. **Inflorescences** axillary or terminal clusters of flowers, sometimes forming panicles or spikes. **Flowers** unisexual, sometimes intermixed in monoecious plants, staminate and pistillate flowers distinct; perianths not scarious. **Staminate flowers** usually with 5-parted perianths, stamens (3)5; bracteoles absent. **Pistillate flowers** usually without perianth and enclosed by 2 fruiting, glabrous to short-pubescent bracteoles (but with deeply lobed, 5-parted perianths in *A. hortensis*), sometimes dimorphic. **Fruits** with membranous, free pericarp. **Seeds** usually vertical, sometimes horizontal, of 1 kind or sometimes dimorphic, brown or black, shiny.

Worldwide. ~250 species; 14 species treated in *Flora*.

The common name, saltbush, refers to the fact that many species of *Atriplex* are tolerant of salty soils and retain salt in their leaves.

Desert-holly (*A. hymenelytra*) is known only from one collection, from southern Harney County, as a short-lived waif.

Key to *Atriplex* species:

1. Plants perennial shrubs or subshrubs, predominantly dioecious.
    2. Subshrubs, much branched at bases; flowering stems with few branches, often forming mats . . . . ***A. gardneri*** var. ***falcata***
    2' Woody shrubs, branched throughout, forming spherical bushes.
        3. Fruiting bracteoles with 4 wings, plants lacking spines; leaf blades linear to oblanceolate . . ***A. canescens*** var. ***canescens***
        3' Fruiting bracteoles without wings, some branches becoming distinct rigid spines; leaf blades obovate to suborbicular or elliptic . . . . ***A. confertifolia***

1' Plants annual herbs, monoecious, rarely dioecious.
    4. Leaves green on both surfaces, sometimes red to yellowish at bases, glabrous to sparsely scaly.
        5. Margins of fruiting bracteoles entire, faces smooth or net-veined.
            6. Pistillate flowers of 2 kinds, some lacking a calyx with 2 bracteoles, others with a 5-lobed perianth without bracteoles; bracteoles net-veined; leaves ovate to deltoid . . . . ***A. hortensis***
            6' Pistillate flowers of 1 kind; bracteoles not net-veined; leaves linear to lanceolate-oblong or triangular-hastate to triangular-lanceolate.
                7. Bracteoles of 2 sizes on same plant, orbicular to ovate, tips rounded; leaves triangular-hastate to triangular-lanceolate . . . . ***A. micrantha***
                7' Bracteoles monomorphic, diamond-shaped, ovate, to lance-oblong, tips acute; leaves linear to lanceolate-oblong . . . . ***A. gmelinii*** var. ***gmelinii***
        5' Margins of fruiting bracteoles toothed or wavy, faces smooth or tuberculate.
            8. Lower leaves narrowly lanceolate or narrowly oblong, 0.2–1.5(2.5) cm wide, bases cuneate . . . . ***A. patula***
            8' Lower leaves triangular-ovate to ovate-hastate or triangular-hastate, 1–10 cm wide, bases widely cuneate to truncate.
                9. Leaf blades thick to succulent, triangular-ovate to ovate-hastate . . . . ***A. dioica***
                9' Leaf blades thin, rarely thick, triangular-hastate . . . . ***A. prostrata***
    4' Leaves grayish and usually farinose, at least on lower surface.
        10. Leaf margins wavy to coarsely dentate; seeds dimorphic, brown and shiny black. . . . . ***A. rosea***
        10' Leaf margins entire or minutely toothed; seeds of 1 kind, brown.
            11. Fruiting bracteoles ≥ 6 mm wide . . . . ***A. argentea*** var. ***hillmanii***
            11' Fruiting bracteoles < 6 mm wide.
                12. Fruiting bracteoles ≤ 1.3 mm wide . . . . ***A. pusilla***
                12' Fruiting bracteoles ≥ 1.5 mm wide.
                    13. Leaves 10–27 mm wide . . . . ***A. truncata***
                    13' Leaves 1.5–5.5 mm wide . . . . ***A. powellii***

Atriplex argentea Nuttall var. *hillmanii* M.E. Jones

Hillman's silverscale

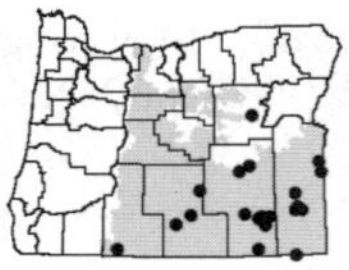

Herbs annual, monoecious, rarely dioecious, 0.5–4 dm, gray-farinose, becoming glabrate. **Stems** spreading to ascending, densely branched, forming low clumps. **Leaves** alternate above, opposite below, blades broadly ovate to triangular or rhombic to suborbicular or semi-hastate, 0.8–3 cm, bases broadly cuneate to truncate, margins entire to minutely toothed, tips rounded, obtuse or acute, surfaces gray-farinose, midveins prominent; sessile or petioles to 6 mm. **Inflorescences** axillary clusters of flowers and terminal spikes. **Staminate flowers** perianth 4- or 5-parted. **Pistillate flowers** perianths absent; enclosed by 2 fruiting bracteoles. **Fruiting bracteoles** dimorphic, 1 kind smooth-faced, turbinate, and shallowly 3–5-toothed, the other kind bur-like, with thick, longitudinal, spine-like protrusions from faces and coarse teeth, subspherical to reniform in outline; sessile or with a short stipe to 2 mm. **Seeds** 1–1.5 mm, brown. **$2n$**=18, 36, 54.

Dry open areas, deserts, sagebrush communities. Flowering May–Jul. 800–1300 m. BR, ECas, Lava, Owy. CA, NV. Native.

Atriplex canescens (Pursh) Nuttall var. *canescens* (p. 66)

Four-winged saltbush, shadscale

Atriplex canescens (Pursh) Nutt. ssp. *canescens*

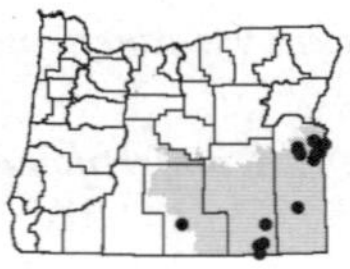

Shrubs dioecious, rarely monoecious, 8–20 dm, densely gray-farinose to glabrate. **Stems** erect, woody, branched throughout, forming spherical bush as wide or wider than tall. **Leaves** alternate; blades linear to oblanceolate, 0.8–5 × 0.2–0.8 cm, margins entire, tips obtuse to rounded or sometimes retuse, surfaces densely white-scaly; sessile or subsessile. **Staminate inflorescences** in terminal and axillary spikes, clusters of flowers 2–3 mm wide. **Pistillate inflorescences** in terminal and axillary panicles. **Staminate flowers** pale yellow. **Pistillate flowers** perianths absent; enclosed by 2 fruiting bracteoles. **Fruiting bracteoles** 8–25 mm, surfaces smooth to reticulate; with 4 prominent dentate to entire wings. **Seeds** 1.5–2.5 mm, brown.

Slopes, desert, sagebrush communities. Flowering Jun–Jul. 700–1600 m. BR, Owy. CA, ID, NV; northeast to Alberta, east to SD, southeast to TX, south to Mexico. Native.

Atriplex confertifolia (Torrey & Frémont) S. Watson

Spiny saltbush, sheepfat

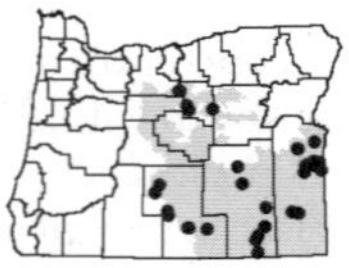

Shrubs dioecious, 3–10(20) dm, densely gray-farinose, stems becoming glabrate and yellow-brown. **Stems** erect, woody, branched throughout, forming a spherical bush, becoming distinct spines. **Leaves** alternate; blades obovate to suborbicular or elliptic, 0.8–2.5 cm, bases cuneate to truncate, margins entire, tips obtuse to rounded, sometimes mucronate; petioles 1–4 mm. **Inflorescences** axillary and terminal clusters of flowers in spikes or short panicles to 10 × 2–4 mm. **Staminate flowers** yellow to silvery. **Pistillate flowers** perianths absent; enclosed by 2 fruiting bracteoles. **Fruiting bracteoles** 5–11(13) mm, margins entire to few-toothed, faces smooth; wings absent. **Seeds** 1.5–2 mm, reddish brown. **$2n$**=18, 36, 54.

Dry slopes, desert, sagebrush communities. Flowering Apr–Jun. 500–1500 m. BR, Lava, Owy. CA, ID, NV; northeast to ND, east to WY, southeast to TX. Native.

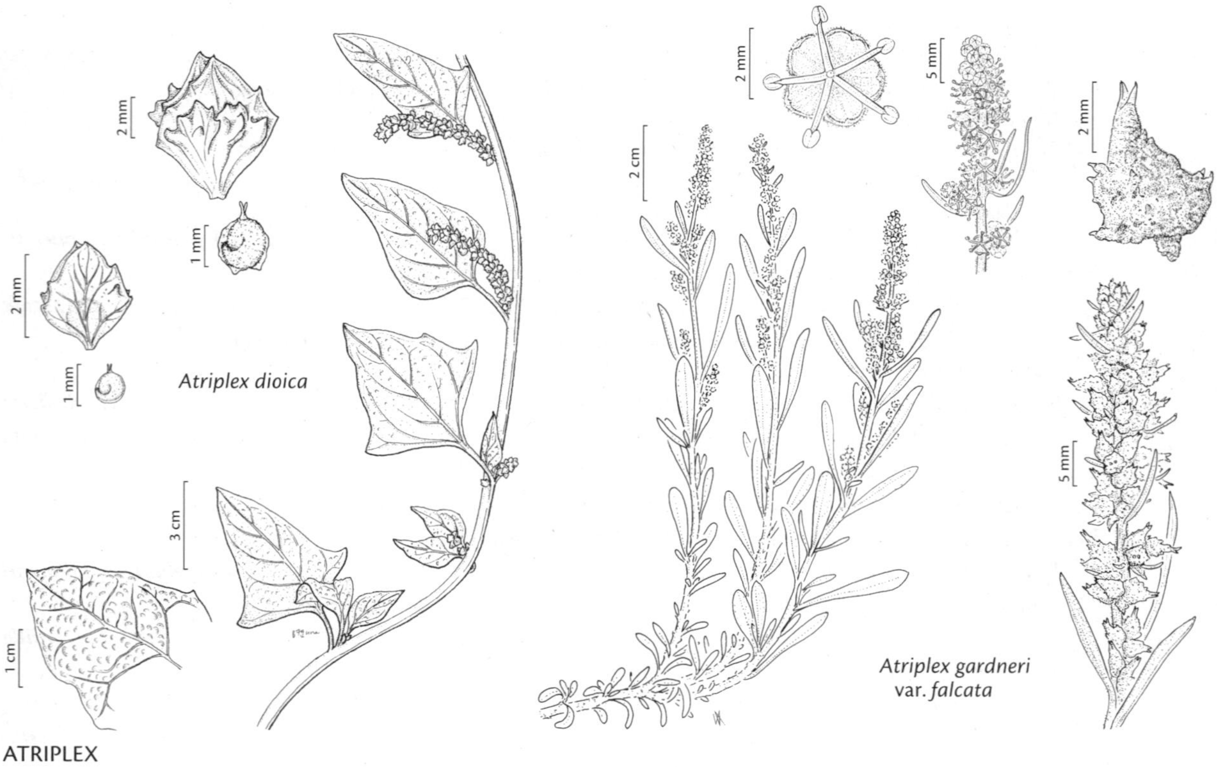

ATRIPLEX

Atriplex dioica Rafinesque (p. 68)

Rillscale, thickleaf orache

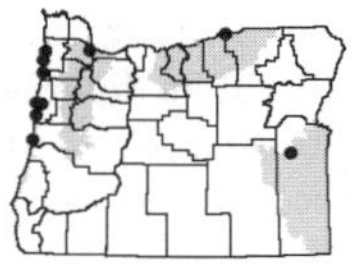

Herbs annual, monoecious, 3–15 dm, sparsely farinose, becoming green and glabrous. **Stems** erect, sometimes decumbent, branched, ribbed and striate. **Leaves** alternate above, opposite at least below; blades lanceolate to linear-lanceolate, lower leaves triangular-ovate or ovate-hastate, 2–10 × 1–5(6.5) cm, green on both surfaces, red to yellowish at base, thick to succulent, bases broadly cuneate to truncate, margins entire or minutely dentate or wavy, tips acute or obtuse, green to somewhat red, surfaces glabrous to sparsely scaly; petioles 0–2(3) cm, upper leaves often sessile. **Inflorescences** clusters of short, axillary or terminal spikes, leafless except at base, staminate and pistillate flowers usually intermixed. **Staminate flowers** perianths 5-parted. **Pistillate flowers** perianths absent; enclosed by 2 fruiting bracteoles. **Fruiting bracteoles** broadly triangular-hastate to ovate, inflated and spongy, margins few-toothed, tips acute, faces smooth or with 2 tubercles; sessile. **Seeds** dimorphic, 1–2 mm, black, shiny or 1.8–3 mm, brown. ***2n***=36, 54.

Seasonally moist ground, coastal beaches, riverbanks, roadsides, ditches. Flowering Jul–Oct. 0–100 m. Col, Est, WV. CA, ID, NV, WA; throughout Canada and most of US. Native.

Atriplex gardneri (Moquin-Tandon) D. Dietrich var. *falcata* (M.E. Jones) S.L. Welsh (p. 68)

Nuttall's saltbush

Atriplex nuttallii S. Watson var. *falcata* M.E. Jones

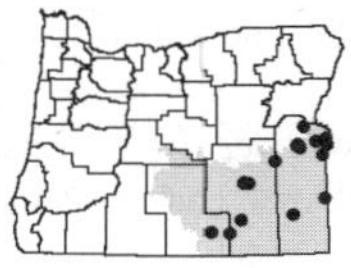

Subshrubs dioecious or occasionally monoecious, 1.5–4.5 dm, densely gray-green-farinose to glabrate. **Stems** decumbent to ascending, much branched from bases, forming a low bush or mat 1.5–7 dm wide. **Leaves** alternate; blades linear, linear-oblanceolate to spatulate, 1–5 × 0.2–0.8 cm, bases attenuate, margins entire, tips obtuse or rounded, occasionally acute; sessile. **Staminate inflorescences** in axillary clusters 2–3 mm wide or single-branched panicles 4–8 mm. **Pistillate inflorescences** in tight axillary clusters, spikes, or panicles, 3–4 mm wide. **Staminate flowers** perianths 5-parted, yellow or brown. **Pistillate flowers** perianths absent; enclosed by 2 fruiting bracteoles. **Fruiting bracteoles** lanceolate, fully fused with a single tooth at tips, faces smooth or tuberculate; sessile or with stipes to 4.5 mm. **Seeds** 1–1.5 mm, brown to tan. ***2n***=18, 36.

Dry ground, sagebrush communities, volcanic ash. Flowering May–Aug. 600–1600 m. BR, Owy. CA, ID, NV, WA; northeast to MT, east to WY, southeast to UT. Native.

Atriplex gmelinii C.A. Meyer var. *gmelinii* (p. 69)

Gmelin's saltbush

Atriplex patula L. var. *obtusa* (Cham.) C.L. Hitchc.

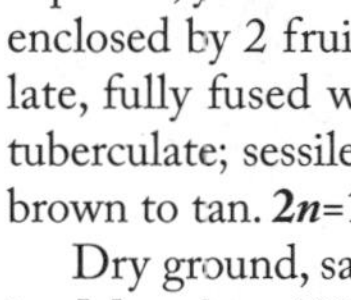

Herbs annual, monoecious, 0.5–7.5 dm, ± glabrous. **Stems** sprawling to erect, striate; branched from bases to occasionally simple, proximal branches elongated and prostrate. **Leaves** alternate, lowermost opposite; blades linear to lanceolate-oblong, 0.9–10 cm, green on both surfaces, bases rounded to tapered, rarely cuneate to hastate, margins entire, tips acute or acuminate; petioles 0.5–1.5 cm. **Inflorescences** clusters of flowers arranged in loose spikes. **Staminate flowers** perianths 5-lobed. **Pistillate flowers** perianths absent; enclosed by 2 fruiting bracteoles. **Fruiting bracteoles** monomorphic, diamond-shaped or ovate to lance-oblong, margins entire, tips acuminate to acute, faces smooth; sessile or short stipitate. **Seeds** dimorphic, convex, 1.5–2 mm, black, shiny or flat, 1.7–3 mm, brown. ***2n***=54.

Salt marshes, coastal areas. Flowering Jun–Sep. 0–50 m. Est. CA, WA; north to AK. Native.

Atriplex hortensis Linnaeus (p. 70)

Orache, garden orache

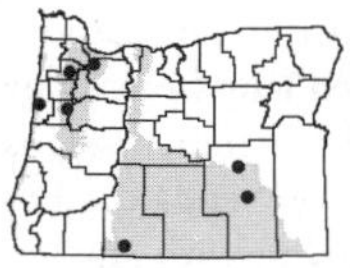

Herbs annual, monoecious, 5–20 dm, green to grayish or sometimes reddish purple, somewhat farinose when young, becoming glabrate. **Stems** erect to somewhat ascending; much branched from bases. **Leaves** alternate, lowermost opposite; blades ovate to deltoid, 5–7(20) cm, green on both surfaces, bases cuneate to cordate, margins entire to sometimes irregularly toothed, tips obtuse, surfaces glabrous to sparsely scaly; petioles 0.5–4(7) cm. **Inflorescences** panicles of clustered spikes, staminate and pistillate flowers typically intermixed. **Staminate flowers** perianths 5-parted. **Pistillate flowers** dimorphic, 1 kind with a deeply lobed, 5-parted perianth, not enclosed by bracteoles, the other lacking a perianth, enclosed by 2 bracteoles. **Fruiting bracteoles** fused only near base, suborbicular, margins entire, faces net-veined; sessile. **Seeds** dimorphic, from flowers without perianth horizontal, 1.5–2.2 mm, black, shiny; from flowers with bracteoles vertical, 3–4.5 mm, dull brown to yellowish. ***2n***=18.

Waste places, roadsides, ditches, moist, often alkaline soils. Flowering Jul–Oct. 0–1300 m. BR, ECas, Est, WV. CA, ID, NV, WA; north to AK, northeast to Quebec, southeast to NM. Asia, Europe. Exotic.

Atriplex hortensis was likely introduced to our region as a garden vegetable.

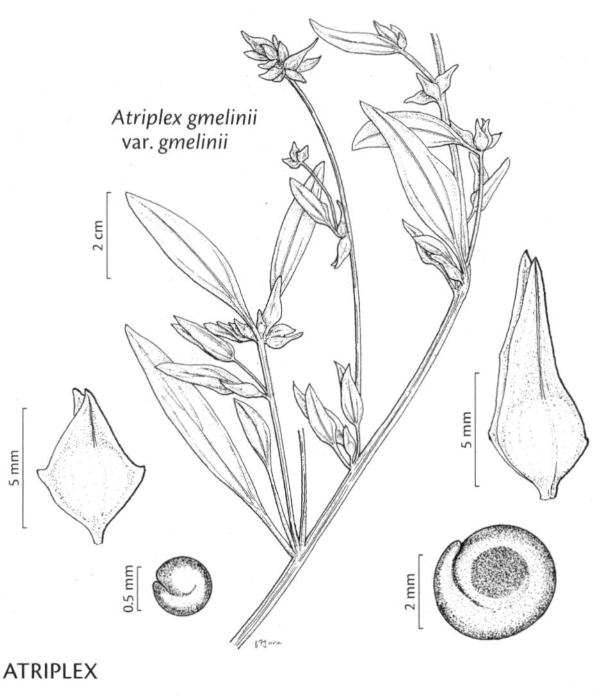

ATRIPLEX

Atriplex micrantha Ledebour

Twoscale saltbush

Atriplex heterosperma Bunge

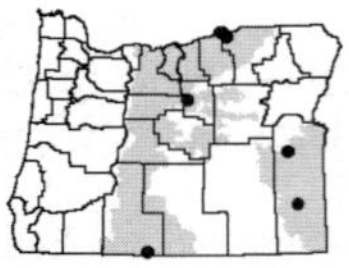

Herbs annual, monoecious, 4–17(20) dm, sparsely farinose, becoming glabrate. **Stems** erect, much branched, sometimes striate. **Leaves** mostly alternate, lowermost opposite; blades triangular-hastate to triangular-lanceolate, 1–12 cm, green on both surfaces, bases truncate to broadly cuneate, margins entire to irregularly dentate, tips acute, glabrous to sparsely scaly; petioles 20–30(40) mm. **Inflorescences** elongated spikes, sometimes arranged in panicles, staminate and pistillate flowers intermixed. **Staminate flowers** perianths 5-parted. **Pistillate flowers** dimorphic, of 2 sizes, both lacking a perianth; with 2 bracteoles. **Fruiting bracteoles** fused just at base, broadly ovate to orbicular, 1.5–6 mm, margins entire, tips rounded; faces smooth; substipitate. **Seeds** dimorphic, with larger bracteoles 2–2.5(3) mm, brown; with smaller bracteoles 1–1.5 mm, black to dark brown, shiny. **2*n***=36.

Streambanks, lakeshores, seasonally wet areas, ditches, sand dunes. Flowering Jul–Sep. 50–1300 m. Col, ECas, Lava, Owy. CA, ID, NV, WA; north to British Columbia, northeast to Quebec, southeast to NM; Asia, Europe. Exotic.

Atriplex patula Linnaeus (p. 70)

Spear orache, spearscale

Atriplex patula L. var. *patula*

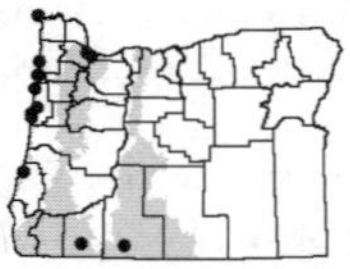

Herbs annual, monoecious, rarely dioecious, (1.5)3–15 dm, sparsely farinose to glabrous. **Stems** mostly erect, sometimes decumbent, branched, usually striate or green. **Leaves** mostly alternate, lowermost sometimes opposite; blades narrowly lanceolate to narrowly oblong, lower leaves sometimes ovate-hastate, 1–5(8) × 0.2–1.5(2.5) cm, green on both surfaces, bases cuneate, margins entire, rarely weakly dentate or serrate, glabrous to sparsely scaly; petioles 0–1.5(2.5) cm, upper leaves sometimes sessile. **Inflorescences** clusters of flowers arranged in interrupted spikes or short panicles. **Staminate flowers** perianths 5-parted. **Pistillate flowers** perianths absent; with 2 bracteoles. **Fruiting bracteoles** fused at base, rhombic-hastate or ovate, 2–7 mm, bases cuneate, margins toothed or wavy on upper ⅔, sometimes sparsely so; faces tuberculate; sessile. **Seeds** dimorphic, 1–2 mm, black, shiny or 1.8–3 mm, brown. **2*n***=36.

Coastal salt flats and marshes, roadsides and ditches. Flowering Jul–Oct. 0–1400 m. ECas, Est, Sisk, WV. CA, ID, NV, WA; throughout North America; Asia, Europe. Exotic.

Atriplex powellii S. Watson

Powell's saltweed

Atriplex powellii S. Watson var. *powellii*

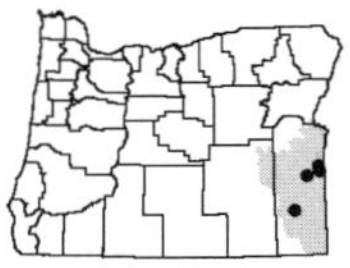

Herbs monoecious, rarely dioecious, 0.5–1.6(3) dm, gray-farinose. **Stems** erect, branched, reddish with age. **Leaves** alternate throughout; blades rhombic-ovate to ovate, 5–30 × 1.5–5.5 mm, bases cuneate to cordate, margins entire, tips obtuse to acute, gray-farinose on both or just lower surfaces; petioles becoming short to subsessile distally, 2–20(25) mm. **Inflorescences** axillary clusters, staminate and pistillate flowers typically intermixed, distal clusters occasionally staminate. **Staminate flowers** perianths 4- or 5-parted. **Pistillate flowers** perianths absent; bracteoles 2. **Fruiting bracteoles** fused to tips, oval to obovate, (1.5)2.5–5 mm, lower ½–⅔ with winged margins and faces thick-tuberculate, upper faces smooth, tips with 1 tooth and 2 rounded lobes; sessile. **Seeds** monomorphic, 1.5–2 mm, brown. **2*n***=18.

Deserts, sagebrush communities. Flowering Apr–Oct. 900–1200 m. Owy. ID, NV; northeast to Alberta, east to NE, southeast to TX. Native.

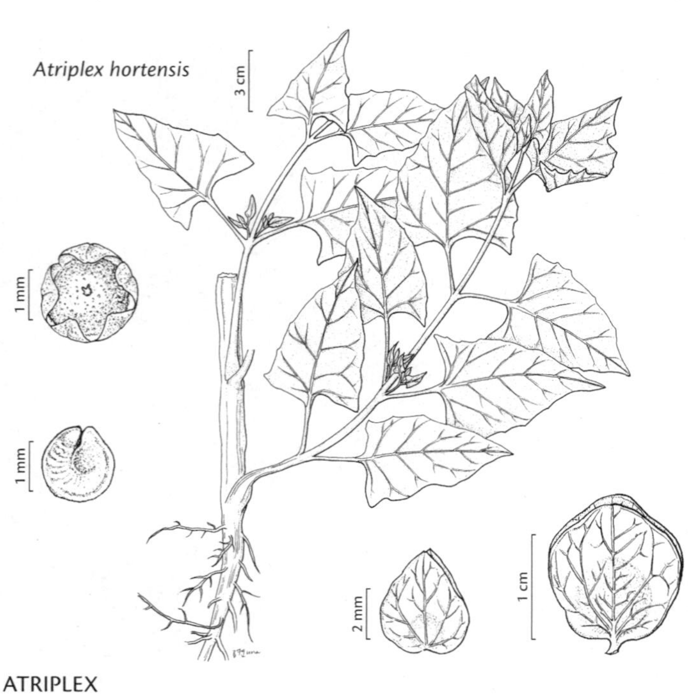

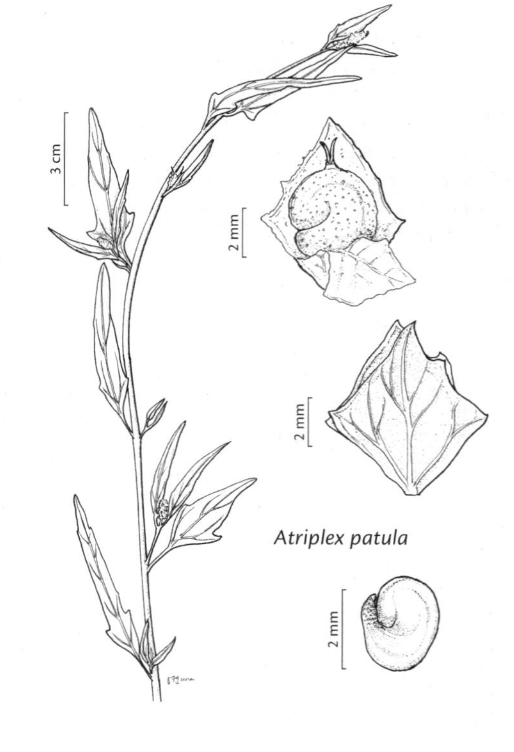

ATRIPLEX

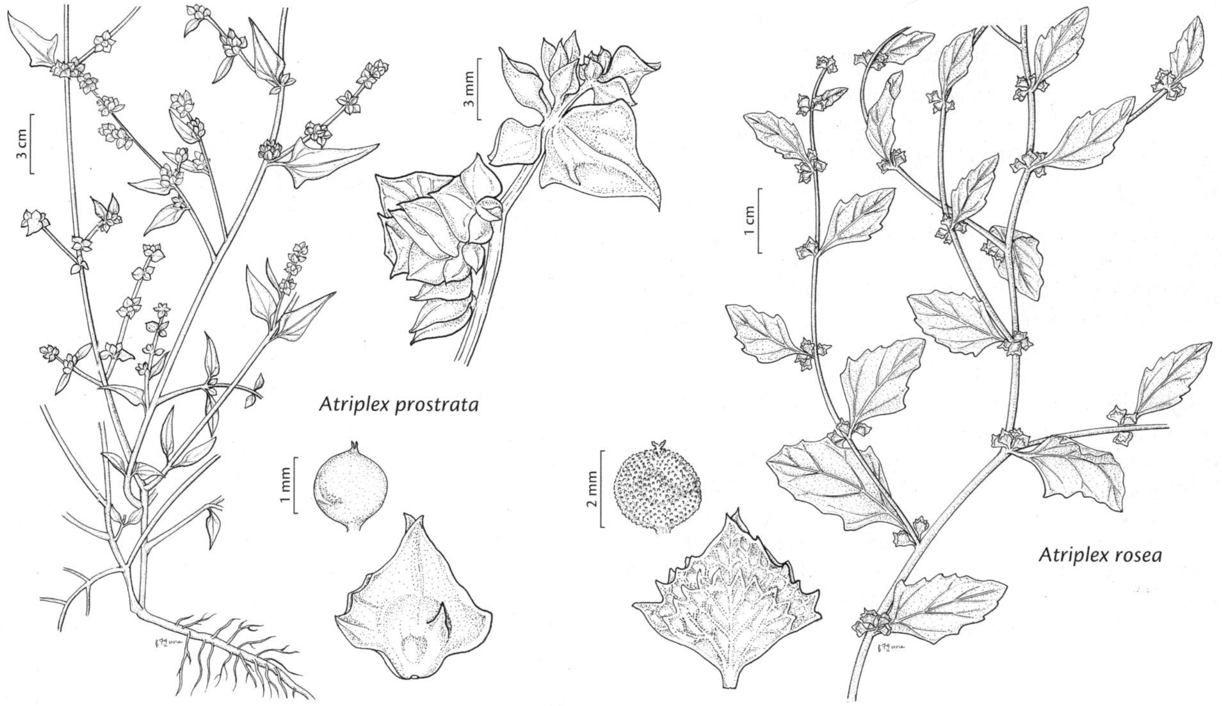

ATRIPLEX

Atriplex prostrata de Candolle (p. 71)

Fat hen, thinleaf orache

Atriplex hastata L., *Atriplex patula* L. var. *hastata* (L.) A. Gray

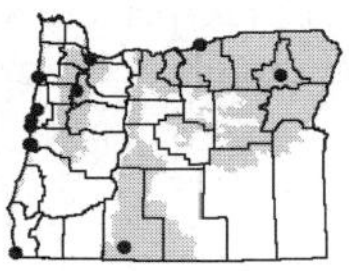

Herbs annual, monoecious, 1–10(20) dm, sparsely farinose, becoming glabrous and green. **Stems** erect to decumbent or prostrate, branching, green, sometimes striate. **Leaves** opposite at least proximally, alternate above; blades triangular-hastate, 2–12 × 3–10 cm, green on both surfaces, thin, rarely thick, bases broadly cuneate to truncate, margins irregularly dentate or serrate, tips obtuse to rounded or acute, glabrous to sparsely scaly; petioles (0)1–3 cm. **Inflorescences** clusters of flowers in spikes or occasionally terminal panicles, staminate and pistillate flowers intermixed. **Staminate flowers** perianths 4- or 5-parted. **Pistillate flowers** perianths absent; enclosed by 2 fruiting bracteoles. **Fruiting bracteoles** lower half fused, triangular-hastate to triangular-ovate, upper margins toothed, tips acute, faces smooth or with 2 tubercles; sessile or with stipe to 5 mm. **Seeds** dimorphic, 1–1.2 mm, black, shiny or 1.5–2 mm, brown. **2*n***=18.

Salt marshes, streambanks, roadsides. Flowering Jul–Sep. 0–1400 m. BW, Col, ECas, Est, WV. CA, ID, NV, WA; north to British Columbia, east to Newfoundland, southeast to SC, south to Mexico. Native.

Atriplex pusilla (Torrey) S. Watson

Dwarf orache

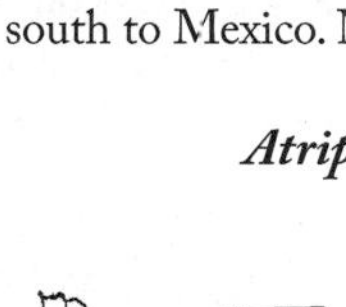

Herbs annual, monoecious, 0.5–2 dm, gray-farinose. **Stems** erect, branched throughout, somewhat spreading, often reddish and glabrate with age. **Leaves** alternate; blades ovate to oblanceolate, 0.4–1.3 × 1.5–4.5(5.5) mm, bases cuneate, margins entire, tips acute, gray-farinose at least on lower surface; sessile to subsessile distally. **Inflorescences** axillary, solitary flowers or staminate flowers in pairs. **Staminate flowers** perianths (4)5-lobed. **Pistillate flowers** perianths absent; enclosed by 2 bracteoles. **Fruiting bracteoles** similar to bracteate leaves, fused to tips, ovate, 1.2–2 mm, margins entire, tips acute to acuminate, faces smooth; sessile to subsessile. **Seeds** monomorphic, 0.8–1 mm, brown.

Deserts, hillsides, valley bottoms. Flowering Jun–Jul. 1200–1500 m. BR. CA, ID, NV. Native.

Atriplex rosea Linnaeus (p. 71)

Red orache, tumbling orache

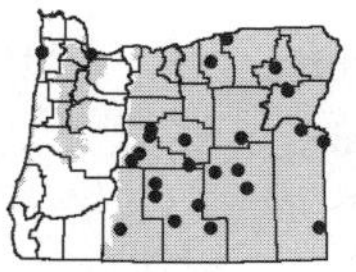

Herbs annual, monoecious, 1–10(20) dm, grayish farinose to green and glabrate. **Stems** erect to ascending, branches simple to much branched throughout, sometimes forming tumbleweeds. **Leaves** alternate; blades ovate to lanceolate, (1)1.5–6 × 1–3(3.5) cm, bases broadly cuneate to rounded, sometimes hastate on lower leaves, margins wavy to coarsely dentate, tips obtuse, farinose at least on lower surface; petioles short to subsessile, (0)2–13 mm. **Inflorescences** axillary few-flowered clusters and terminal spikes. **Staminate flowers** perianths 4–5-lobed. **Pistillate flowers** perianths absent; enclosed by 2 bracteoles. **Fruiting bracteoles** fused to ~ halfway, rhombic to triangular rhombic, bases broadly cuneate to sometimes subhastate, margins toothed in upper ⅔, tips acute to obtuse; faces smooth or short tuberculate; sessile or with stipe to 5 mm. **Seeds** dimorphic, vertical, 1–2 mm, black, shiny or 2–2.5 mm, brown, dull. **2*n***=18.

Disturbed areas, roadsides. Flowering Jul–Sep. 0–1500 m. BR, BW, Col, ECas, Est, Lava, Owy, WV. CA, ID, NV, WA; scattered throughout most of North America; Asia, Europe. Exotic.

Atriplex truncata (Torrey) A. Gray (p. 72)

Wedgescale orache, truncate saltbrush

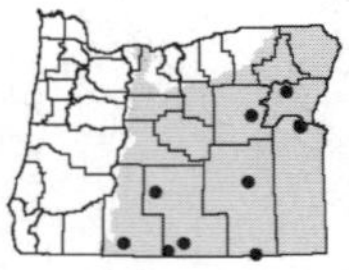

Herbs annual, monoecious, 1.5–5(9) dm, gray-farinose. **Stems** erect to ascending, much branched and spreading. **Leaves** alternate, lowermost sometimes opposite; blades triangular-ovate or rhombic to suborbicular, 1.2–3.2 × 1–2.7 cm, bases truncate, subtruncate, subhastate, or rounded, margins entire, tips rounded to truncate, gray-farinose on both surfaces; sessile or short-petiolate distally to 6 mm. **Inflorescences** axillary clusters of flowers arranged in short spikes, upper clusters staminate or sometimes mixed with pistillate flowers. **Staminate flowers** perianths 4–5-lobed. **Pistillate flowers** perianths absent, enclosed by 2 bracteoles. **Fruiting bracteoles** fused to tips, shield-shaped, 2.4–3 mm, tips truncate with 3 shallow teeth, faces smooth or with 1–2 short tubercles; sessile to substipitate. **Seeds** monomorphic, 1–1.5 mm, brown. **2*n***=18.

Meadows, near lakes or streams. Flowering Jun–Aug. 300–1500 m. BR, BW, ECas, Lava, Owy. CA, ID, NV, WA; north into British Columbia, northeast to Saskatchewan, east to WY, southeast to NM. Native.

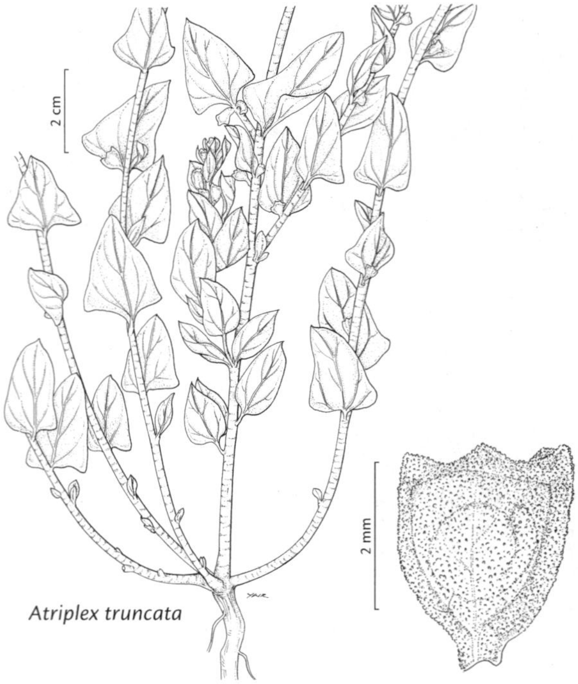

Atriplex truncata

ATRIPLEX

BASSIA Smotherweed

Herbs densely hairy to glabrous. **Stems** simple to much branched. **Leaves** alternate; blades filiform, linear to lanceolate, flat, margins entire; sessile. **Inflorescences** axillary solitary flowers or clusters of short spikes, densely hairy; typically with 1 or 2 greenish bracts. **Flowers** bisexual or functionally only pistillate or staminate, staminate and pistillate flowers not obviously distinct; perianths 5-lobed, divided to ~ half the length, completely enclosing fruit; stamens (3)5; stigmas 2(3). **Fruits** flattened dorsiventrally, pericarp membranous and loose. **Seeds** horizontal.

Worldwide. ~25 species; 2 species treated in *Flora*.

Key to *Bassia* species:

1. Perianth segments puberulent, each segment with a hooked spine adaxially . . . . ***B. hyssopifolia***

1' Perianth segments glabrous, each segment with a membranous transverse wing adaxially . . . . ***B. scoparia***

Bassia hyssopifolia (Pallas) Kuntze (p. 73)

Fivehook bassia, hyssop bassia

Echinopsilon hyssopifolium (Pall.) Moq.

Plants annual, 2–10(20) dm, densely tomentose. **Stems** erect, much branched or simple, ribbed or angular. **Leaves** cauline; blades linear to narrowly oblanceolate, 1–4 cm, bases attenuate, midveins prominent, tips acute. **Inflorescences** axillary clusters of spike-like panicles or solitary flowers in leaf axils, bisexual and pistillate flowers often intermixed. **Flowers** perianth segments puberulent, each segment with a stout, hooked adaxial spine, spine 1–2 mm. **Fruits** flattened dorsiventrally, plano-convex, 1–2.5 mm in diameter. **Seeds** ovoid, 1–1.5 mm wide. **2*n***=18.

Disturbed areas, roadsides, meadows, salt marshes, dunes. Flowering May–Nov. 50–1500 m. BR, BW, Col, Lava, Owy. CA, ID, NV, WA; western North America, scattered in eastern US; Asia, Europe. Exotic.

The small hooks on the perianth of *B. hyssopifolia* attach to clothing and animal hair, functioning as an aid to dispersal.

Bassia scoparia (Linnaeus) A.J. Scott

Mock cypress, burning bush

Kochia scoparia (L.) Schrad., *Kochia scoparia* (L.) Schrad. ssp. *scoparia*

Plants annual, 3–10(40) dm, pilose to villous, sometimes glabrate with age below. **Stems** erect, branched. **Leaves** cauline; blades oblanceolate, 0.8–4 cm, bases narrowly attenuate, tips acuminate. **Inflorescences** axillary clusters of 3(5) flowers or solitary flowers subtended by leaf-like or ciliate-reduced bracts. **Flowers** perianth segments urn-shaped to globose, glabrous, each segment with a membranous transverse adaxial wing. **Fruits** flattened dorsiventrally. **Seeds** ovoid, 1.5–2.5 mm.

Disturbed areas, roadsides, fields. Flowering Aug–Oct. 0–1500 m. BW, Col, ECas, Lava, Sisk, WV. CA, ID, NV, WA; throughout most of North America; Asia, Europe. Exotic.

A phylogenetic study by Kadereit and Freitag (2011) transferred this species from *Kochia* to *Bassia*.

BLITUM Blitum

Herbs annual, glabrous to farinose when young. **Stems** prostrate to ascending, simple or branched from base, ultimate branches not thread-like, usually fleshy. **Leaves** cauline, alternate; blades triangular-lanceolate, oblanceolate or spatulate, flat, succulent, margins entire to hastate, tips obtuse to rounded; petioles present. **Inflorescences** axillary clusters of bisexual flowers, or of bisexual and pistillate flowers; bracts leaf-like. **Flowers** perianth segments 0–1 on outer flowers, 1–3 on central flowers, greenish; stamens 0–1(2); stigmas 2, fused at base; sessile. **Fruits** mostly flattened utricles, subspherical; pericarp loose when dry, sometimes fleshy, reticulate or papillose. **Seeds** vertical, lenticular, brown to black, smooth.

Asia, Australia, Europe, North America, South America. 12 species; 2 species treated in *Flora*.

Key to *Blitum* species:

1. Leaf margins hastate-lobed near base; fruits 1.1–1.5 mm, walls whitish, minutely pitted . . . . ***B. nuttallianum***

1' Leaf margins entire near base; fruits 0.5–0.7 mm, walls pale brown, turning black, minutely papillate . . . . ***B. spathulatum***

Blitum nuttallianum Schultes (p. 73)

Nuttall's povertyweed

Monolepis nuttalliana (Schult.) Greene

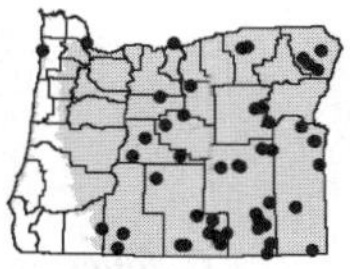

Stems 5–20(50) cm, sparsely farinose when young, becoming glabrate. **Leaves** blades broadly triangular, lanceolate to linear, 1–3(4) cm, margins hastate-lobed near base, sometimes with a few teeth above or entire, tips obtuse to rounded; petioles 0.1–3(6) cm, longest near base, distal leaves nearly sessile. **Inflorescences** axillary clusters of 5–15 flowers. **Flowers** perianths a single spatulate to obtuse bract-like segment, 1 mm; stamens 0 or 1. **Fruits** membranous, 1.1–1.5 mm; walls whitish, minutely pitted. **Seeds** broadly ovoid, laterally flattened, 0.5–1.1 mm, dark brown to black. **2*n***=18.

River and lake margins, vernal pools, disturbed areas, dry rocky soils, slopes. Flowering Apr–Sep. 0–2400 m. BR, BW, Casc, Col, ECas, Est, Lava, Owy, WV. CA, ID, NV, WA; throughout most of Canada and US, northern Mexico. Native.

Blitum spathulatum (A. Gray) S. Fuentes (p. 73)

Prostrate povertyweed, beaver povertyweed

Monolepis spathulata A. Gray

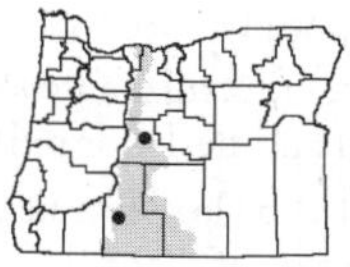

Stems 2–15(20) cm, sparsely farinose to glabrous. **Leaves** blades oblanceolate to spatulate, 3–25 mm, margins entire, tips obtuse to rounded; petioles 1–4(6) mm, longest near base, distal leaves nearly sessile. **Inflorescences** axillary clusters of 4–15 flowers, outer flowers typically pistillate. **Flowers** outer pistillate and perianth segments 0, inner bisexual and perianth segments 2–3; segments spatulate to obovate, 0.5 mm; stamens 0–2. **Fruits** membranous, 0.5–0.7 mm; walls pale brown, turning black, minutely papillate. **Seeds** 0.3–0.7 mm, brown, shiny. **2*n***=18.

Streambanks, meadows. Flowering May–Aug. 1500–2100 m. ECas. CA, NV; south to Mexico. Native.

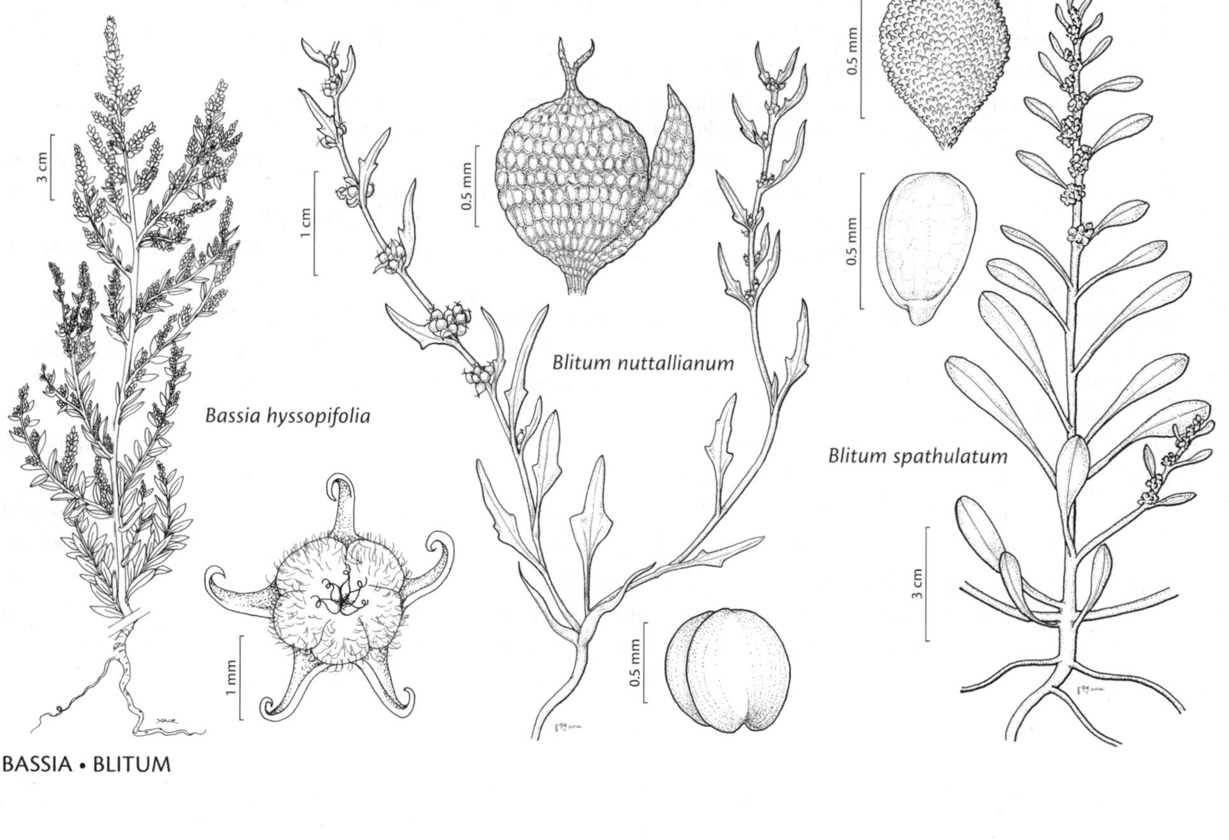

BASSIA • BLITUM

CHENOPODIUM Goosefoot

Herbs often tinged with red or purple at maturity, glabrous or farinose. **Stems** prostrate to erect, usually branched, often angled or ribbed. **Leaves** alternate, simple, flat; main stem leaves largest; distal leaves typically smaller and more narrow; petioles present. **Inflorescences** clusters of flowers or sometimes solitary, arranged in axillary or terminal spikes or panicles of spikes, glabrous or farinose; with or without subtending leaf-like bracts. **Flowers** bisexual, perianths 3–5-parted; styles short, stigmas 2(5); short-pedicellate to sessile. **Fruits** utricles or achenes, often enclosed in calyx. **Seeds** vertical or horizontal, lenticular; endosperm adherent.

Worldwide. ~100 species; 14 species treated in *Flora.*

Chenopodium is a taxonomically difficult group. Plants can be challenging to identify in their vegetative or young flowering stages. Species are most easily distinguished when mature inflorescences are available.

Several members of *Chenopodium* have been collected in our area on ballast or are known only as short-lived waifs. These include *Chenopodium incanum* var. *occidentale, C. murale, C. polyspermum* var. *acutifolium, C. strictum* var. *glaucophyllum, C. subglabrum,* and *C. vulvaria.*

Key to *Chenopodium* species:

1. Leaf margins dentate or sinuate-dentate above.
  2. Fruits vertical (or vertical and horizontal in *C. glaucum*); stamens 1–4(5).
    3. Lower leaf surface white- or gray-farinose . . . . **C. glaucum** (in part)
    3' Lower leaf surface glabrous to sparsely farinose.
      4. Flowers in unbranched spikes or axillary, flower clusters globose to subglobose, 3–10 mm wide; perianth segments red and fleshy or red to greenish and membranous.
        5. Flower clusters subtended by leaf-like bracts throughout inflorescences; flowers maturing from bases to tips of plant . . . . **C. foliosum**
        5' Flower clusters not subtended by leaf-like bracts in distal ½ of inflorescences; flowers maturing uniformly throughout plant . . . . **C. capitatum** var. ***parvicapitatum***
      4' Flowers in branched spikes, flower clusters loosely rounded to indistinct, 1–5 mm wide; perianth segments green, not fleshy.
        6. Perianth segments fused nearly to tip; stems much branched from near base. . . . . **C. chenopodioides**
        6' Perianth segments fused only at the base; stems branching from a main stem . . . . **C. rubrum**
  2' Fruits horizontal; stamens 5 (except 1–5 in *C. glaucum*).
    7. Flowers in open panicles, arranged singly or occasionally in few-flowered clusters; plants glabrous. . . . . **C. simplex**
    7' Flowers in spikes or spicate panicles of tight flower clusters; slightly farinose to farinose.
      8. Stems prostrate to erect, 0.7–2.5(4) dm; leaf blades 8–25(30) mm; stamens usually 3 . . . . . . **C. glaucum** (in part)
      8' Stems erect, 1–15 dm; leaf blades 10–60(120) mm; stamens 5.
        9. Inflorescences usually not subtended by leaf-like bracts, individual flower clusters crowded and indistinct; median keel of perianth segments indistinct or absent; pericarp of fruits smooth or minutely pitted . . . . **C. album**
        9' Inflorescences often subtended by leaf-like bracts, individual flower clusters spread out and distinct; median keel of perianth segments prominent; pericarp of fruits distinctly honeycomb pitted. . . . . **C. berlandieri** var. ***zschackei***

1' Leaf margins entire above.
  10. Leaves lanceolate to rhombic-ovate or broadly deltoid-ovate, blades (3)4–40 mm wide.
    11. Stems dichotomously branched; leaf blades lanceolate to rhombic-ovate, 4–10(20) mm . . . . **C. nevadense**
    11' Stems not dichotomously branched; leaf blades lanceolate to ovate or broadly deltoid-ovate, (10)15–50 mm.
      12. Leaf blades broadly deltoid-ovate. . . . . **C. fremontii**
      12' Leaf blades lanceolate to narrowly ovate.
        13. Blades 15–35 mm, broadly lanceolate to ovate, < 3 × as long as wide . . . . **C. atrovirens**
        13' Blades 20–50 mm, lanceolate, ~3 × as long as wide . . . . **C. pratericola**
  10' Leaves linear to narrowly lanceolate, blades 1–4(6.5) mm wide.
    14. Stems branching from near base, branches mostly of equal length; blades 3-veined . . . . **C. desiccatum**
    14' Stems branching throughout, branches shorter than main stem; blades 1-veined . . . . **C. leptophyllum**

Chenopodium album Linnaeus

Lamb's quarters, pigweed

Chenopodium album L. var. *album*

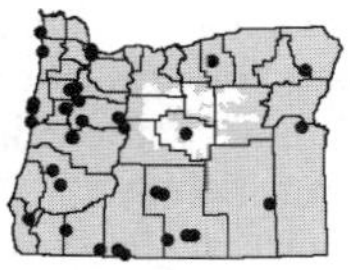

Plants annual, farinose. **Stems** erect, simple to much branched from a main axis, 2–7(15) dm, sometimes reddish; branches ascending. **Leaves** blades ovate-lanceolate, rhombic-ovate or lanceolate, 10–60(120) mm, bases cuneate, margins irregularly sinuate-dentate, rarely subentire, tips rounded or obtuse, often with a minute mucronate tip, sometimes acute, surfaces lower white-farinose, upper green and glabrate; petioles 5–40 mm. **Inflorescences** axillary and terminal panicles, erect or ascending; clusters tight and indistinct, usually not subtended by leaf-like bracts. **Flowers** perianths 5-parted, divided nearly to base, main keel indistinct or absent; stamens 5. **Fruits** pericarp smooth or minutely pitted. **Seeds** horizontal, black. ***2n***=54.

Disturbed areas, cultivated fields, roadsides, ditches. Flowering Jun–Oct. 0–1600 m. All ecoregions except Lava. CA, ID, NV, WA; worldwide. Exotic.

This species is commonly regarded as a troublesome agricultural weed in our area.

Chenopodium atrovirens Rydberg (p. 75)

Green goosefoot

Chenopodium fremontii S. Watson var. *atrovirens* (Rydb.) Fosberg

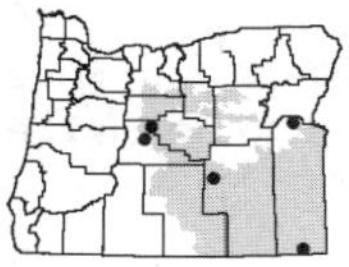

Plants annual, farinose, sometimes sparsely so. **Stems** erect, branching from a main axis along full length, 2–4.5 dm, striate and obtusely angled. **Leaves** blades broadly lanceolate to ovate, 15–25(35) × 4–17 mm, < 3 × as long as wide, bases cuneate, sometimes slightly hastate near base, margins entire, tips rounded with small mucronate tip; petioles 6–13(20) mm. **Inflorescences** clusters of flowers arranged in axillary spikes and terminal panicles of spikes. **Flowers** perianths 5-parted; lobes divided ~67% of length, strongly keeled on back; stamens 5. **Fruits** 1–1.4 mm in diameter. **Seeds** horizontal, shiny dark brown to blackish. ***2n***=18.

Deserts, open ground. Flowering Jun–Aug. 500–2000 m. BR, Lava, Owy. CA, ID, NV, WA; north to British Columbia, northeast to Saskatchewan, east to WY, southeast to NM, south to Mexico. Native.

Chenopodium berlandieri Moquin-Tandon var. ***zschackei*** (Murr) Murr (p. 75)

Berlandier's pigweed, pitseed goosefoot, lamb's quarters

Chenopodium album L.

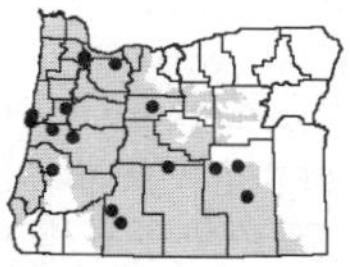

Plants annual, farinose. **Stems** erect, much branched to simple, branches ascending, 1–10(15) dm. **Leaves** blades deltoid to rhombic, 17–50 mm, bases cuneate to rounded, margins subhastate below, sinuate-dentate above, tips obtuse or rounded with a small mucronate tip, surfaces lower farinose, upper glabrate to glabrous; petioles 2–90 mm. **Inflorescences** erect to ascending terminal and axillary panicles, flower clusters tight but distinct along axils, often subtended by a lanceolate, leaf-like bract with entire margins. **Flowers** perianths 5-parted, divided nearly to base, main keel prominent; stamens 5. **Fruits** pericarp distinctly honeycomb-pitted. **Seeds** horizontal, brown to black. ***2n***=36.

Disturbed areas, fields. Flowering May–Sep. 0–1900 m. BR, Casc, CR, ECas, Est, Lava, WV. CA, ID, NV, WA; throughout North America. Native.

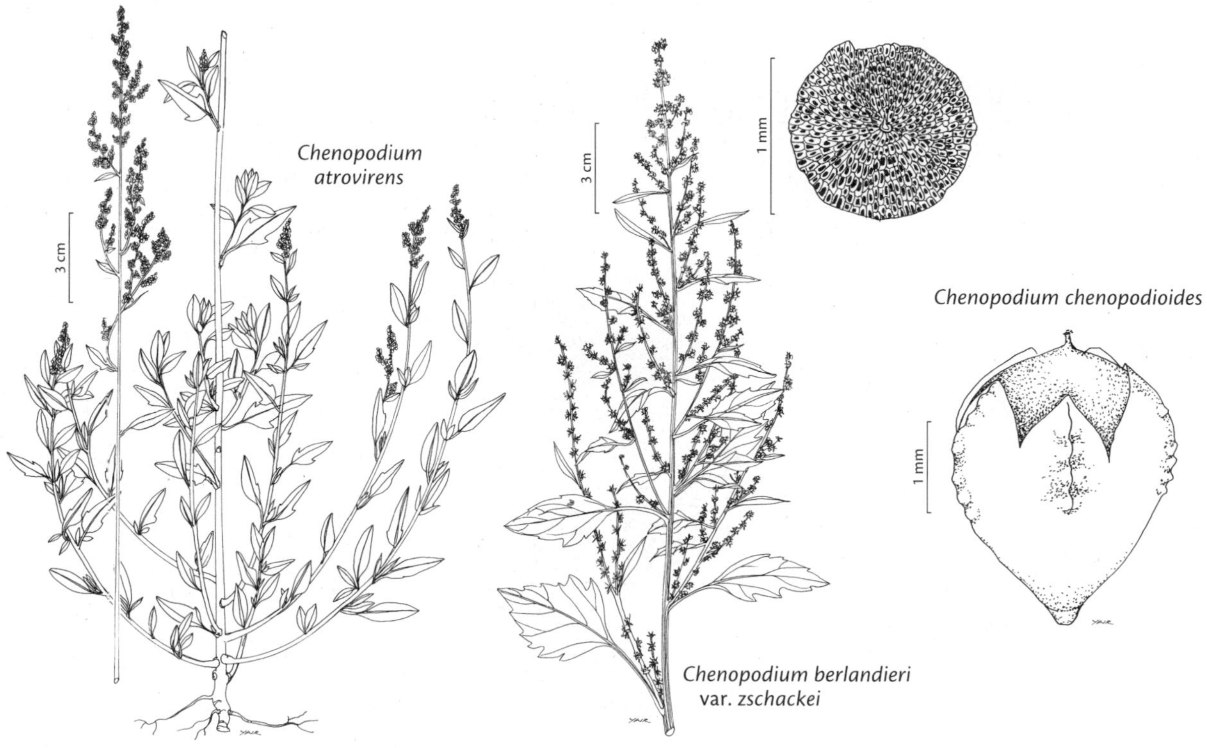

CHENOPODIUM

Chenopodium capitatum (Linnaeus) Ascherson var. ***parvicapitatum*** S.L. Welsh

Mountain strawberry blite, blite goosefoot

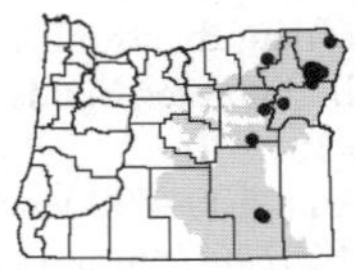

Plants annual, glabrous. **Stems** decumbent to ascending, 1–many branches from base, 1.5–6(10) dm. **Leaves** blades lanceolate, ovate or triangular, 25–100 mm, bases cuneate to truncate, hastate or not, margins shallow to minute-dentate, tips acute; petioles 15–100 mm. **Inflorescences** terminal unbranched spikes, with globose axillary flower clusters; clusters 3–10 mm wide; leaf-like bracts only in proximal half of inflorescence. **Flowers** maturing uniformly throughout plant; perianths 3–4(5)-parted, divided ~50% of length, 3–5 mm wide, red or greenish, membranous; stamens 1. **Fruits** 0.9–1 mm. **Seeds** vertical, black. ***2n***=18.

Forests, streambanks, slopes. Flowering Jun–Jul. 1100–2400 m. BR, BW. CA, ID, NV; north to AK, northeast to Quebec, east to Newfoundland, southeast to NM. Native.

Chenopodium capitatum var. *capitatum* has been recorded in Oregon only once, in 1888, at an "Oregon Experiment Station" (exact location unknown). It differs from variety *parvicapitatum* in its larger, fleshy flowers and strongly dentate-toothed leaves.

Chenopodium chenopodioides (Linnaeus) Aellen (p. 75)

Low goosefoot

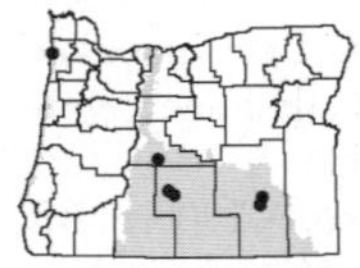

Plants annual, sometimes reddish, mostly glabrous. **Stems** prostrate to ascending, much branched from near base, 0.3–2 dm. **Leaves** blades deltoid to rhombic-ovate, 10–25(35) mm, sometimes hastate at base, margins shallowly dentate, tips rounded to obtuse, lower surface sparsely farinose; petioles 6–15(20) mm, winged, wings sometimes extending to stem. **Inflorescences** terminal and axillary spikes of loosely rounded to indistinct flower clusters; clusters 2.3–4.5 mm wide, each subtended by a leaf-like bract. **Flowers** perianths fused nearly to tip, 0.8–1.2 mm, green, not membranous or fleshy, shallowly 3–4-toothed, enclosing fruit at maturity; stamens 1 or 2. **Fruits** broadly ovoid, laterally flattened, pericarp loose and free. **Seeds** vertical, dark brown, shiny.

Shores of lakes or ponds, marshes, fields. Flowering Jul–Sep. 100–1500 m. BR, ECas, Est. CA, NV, WA; east to CO, southeast to UT; Europe. Exotic.

Chenopodium desiccatum A. Nelson

Aridland goosefoot

Chenopodium leptophyllum (Moq.) S. Watson var. *oblongifolium* S. Watson

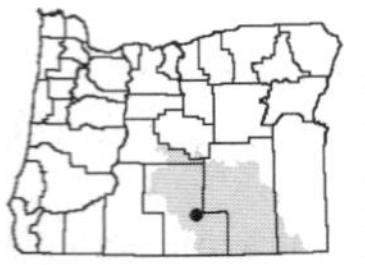

Plants annual, farinose. **Stems** erect to spreading, much branched from near base, 1–3(4) dm, farinose to glabrate, striate; branches of mostly equal length. **Leaves** blades narrowly lanceolate or narrowly oblong, 7–20(25) × 1.2–3.5(4.5) mm, bases cuneate, margins entire, 3-veined, tips acute, upper surface glabrous or glabrate and green; petioles 2–4 mm. **Inflorescences** clusters of flowers in terminal and axillary panicles. **Flowers** perianths 5-parted, divided ~50% of length, strongly keeled, concealing fruit at maturity. **Fruits** 0.7–1 mm in diameter, pericarp loose and membranous. **Seeds** horizontal, dark brown to blackish, shiny. ***2n***=18.

Disturbed areas, open ground, streambanks. Flowering May–Sep. 50–2000 m. BR. CA, ID, NV; north to British Columbia, east to NE, southeast to NM. Native.

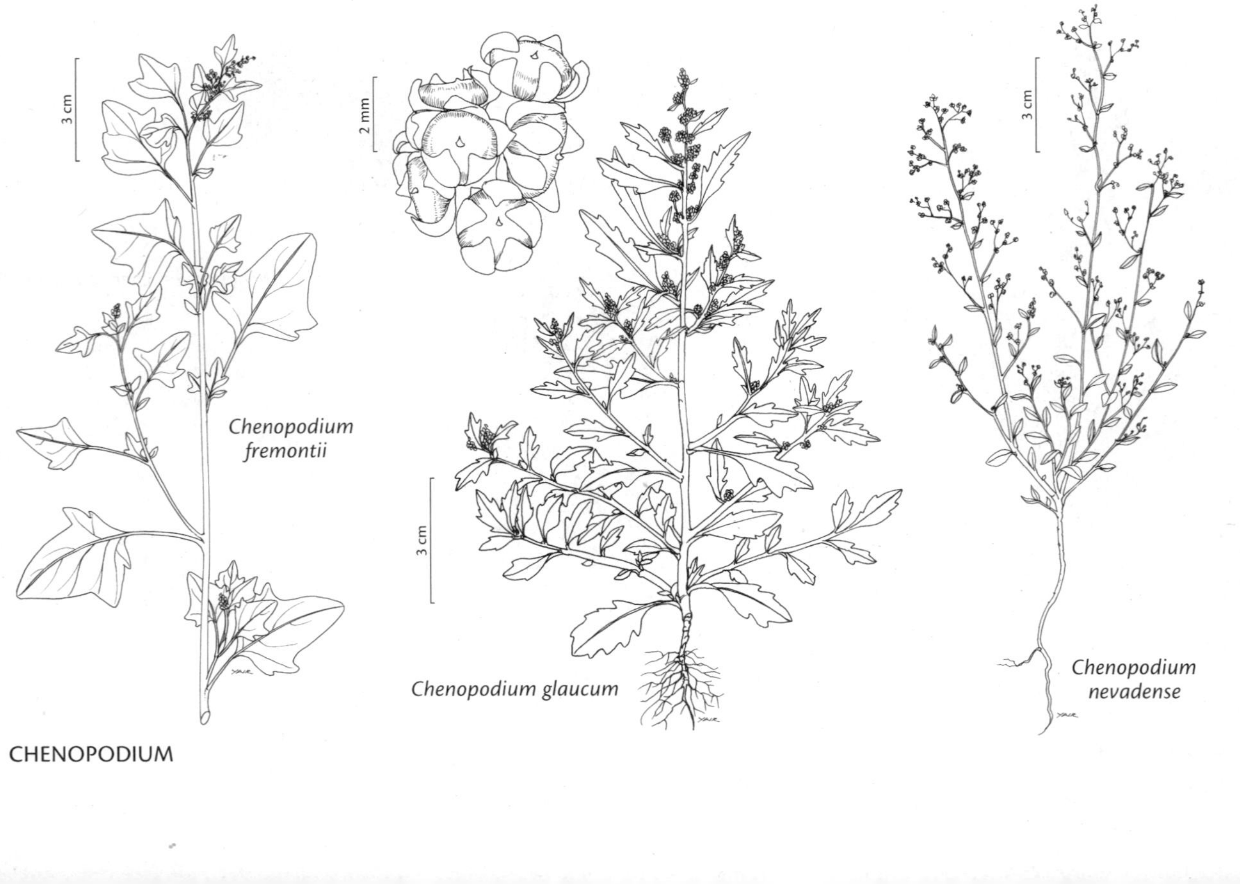

CHENOPODIUM

Chenopodium foliosum (Moench) Ascherson
Leafy goosefoot

Chenopodium virgatum (L.) Ambrosi

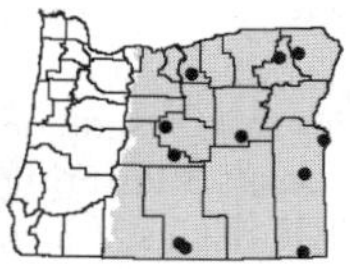

Plants annual, glabrous. **Stems** erect to ascending, branched or simple, 0.5–8 dm, sometimes tinged with red. **Leaves** blades triangular-lanceolate, 15–55(75) mm, bases sharply hastate or not, cuneate to subtruncate, margins irregularly serrate-dentate, tips acute, surfaces green and glabrous; petioles 15–80(100) mm. **Inflorescences** terminal and axillary, in large globose to subglobose spikes, 3–5(9) mm wide, each subtended by a leaf-like bract. **Flowers** maturing from bases to tips of branches; perianths 3–5-parted, divided ~50% of length, fleshy and red at maturity; stamens 1. **Fruits** 1.1–1.4 mm. **Seeds** vertical, dark red-brown.

Disturbed areas, roadsides, fields. Flowering May–Jul. 500–2000 m. BR, BW, Col, ECas, Lava, Owy. CA, ID, NV, WA; scattered in North America; Europe. Exotic.

Chenopodium fremontii S. Watson (p. 76)
Fremont's goosefoot

Chenopodium fremontii S. Watson var. *fremontii*

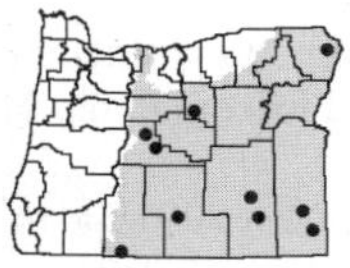

Plants annual, farinose, sometimes sparsely so. **Stems** erect, branching from a main axis along full length, not dichotomously branched, 2–8(10) dm. **Leaves** blades broadly deltoid-ovate, 15–35(45) × 13–30(40) mm, bases broadly cuneate to truncate, margins rounded, hastate-lobed near base, sometimes with 2 teeth, entire above, tips rounded with small mucronate tip; petioles 7–20(30) mm. **Inflorescences** in axillary and terminal spikes. **Flowers** perianths 5-parted, divided ~50% of length; stamens 5. **Fruits** 0.9–1.3 mm, pericarp loose and membranous. **Seeds** horizontal, black, shiny. **2*n***=18.

Slopes, talus, riparian areas, sagebrush communities, roadsides, ditches. Flowering Jul–Sep. 800–1600 m. BR, BW, ECas, Lava, Owy. CA, ID, NV, WA; north to British Columbia, northeast to Manitoba, east to SD, southeast to TX, south to Mexico. Native.

Chenopodium fremontii can occupy a wide range of habitats and is known to occur in abundance following a fire.

Chenopodium glaucum Linnaeus (p. 76)
Oakleaf goosefoot

Chenopodium glaucum L. var. *salinum* (Standl.) B. Boivin, *Chenopodium salinum* Standl.

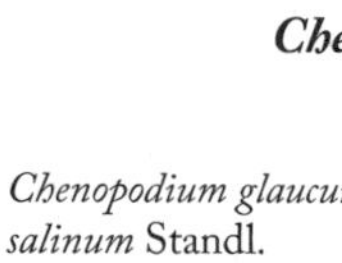

Plants annual, sparsely farinose. **Stems** prostrate to erect, branched or simple when young, 0.7–2.5(4) dm. **Leaves** blades lanceolate to oblong or ovate, 8–25(30) mm, bases cuneate, margins sinuate-dentate, tips obtuse to rounded, surfaces lower white- or gray-farinose, upper glabrous and green; petioles 3–15 mm. **Inflorescences** short axillary and terminal spikes, occasionally panicles of spikes; clusters tight, 0.8–2.5 mm wide. **Flowers** perianths 3–5-parted, divided nearly to base; stamens (1)3(5). **Fruits** 0.7–1.2 mm. **Seeds** vertical and horizontal, round, reddish brown. **2*n***=18.

Lakeshores, riverbanks, mud flats. Flowering Jun–Sep. 50–1400 m. BR, Col, Est. ID; east to Newfoundland, southeast to FL. Exotic?

Chenopodium leptophyllum (Moquin-Tandon) S. Watson
Slimleaf goosefoot

Chenopodium leptophyllum (Moq.) S. Watson var. *leptophyllum*

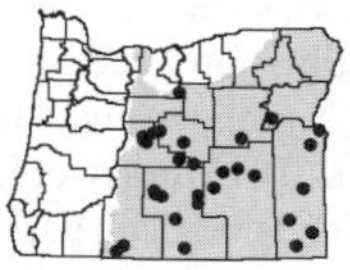

Plants annual, farinose, sometimes sparsely so. **Stems** erect, single or branching throughout, 1–5(10) dm; branches shorter than main stem, striate. **Leaves** blades linear to narrowly lanceolate, 8–22 × 1–3(3.5) mm, bases attenuate, margins entire, 1-veined, tips obtuse or rounded; petioles 1–4(7) mm. **Inflorescences** clusters of flowers arranged in terminal and axillary panicles. **Flowers** perianths 5-parted, divided ~50% of length, slightly keeled, exposing fruit at maturity; stamens 5. **Fruits** 0.8–1.1 mm, pericarp membranous and adherent to seed. **Seeds** horizontal, black. **2*n***=18.

Roadsides, cliffs, sagebrush communities. Flowering Jun–Sep. 500–2000 m. BR, BW, ECas, Lava, Owy. CA, ID, NV, WA; north to Yukon, northeast to Quebec, southeast to LA, south to Mexico. Native.

Chenopodium nevadense Standley (p. 76)
Nevada goosefoot

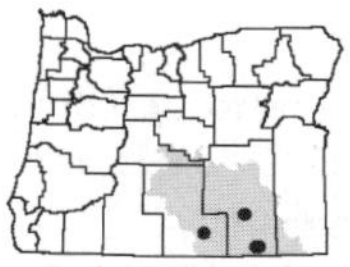

Plants annual, farinose when young, becoming glabrate. **Stems** erect, dichotomously branched, 0.5–3 dm. **Leaves** blades lanceolate to rhombic-ovate, sometimes suborbicular, 4–10(20) × 3–6(10) mm, bases cuneate, margins entire, sometimes subhastate, tips rounded; petioles 1.5–5.5 mm. **Inflorescences** cymose-panicles, flowers solitary or in few-flowered clusters. **Flowers** perianths 5-parted, fused near base, slightly keeled, enclosing fruit; stamens 5. **Fruits** 0.7–0.9 mm. **Seeds** horizontal, black.

Scrubland, flats, bottomlands. Flowering May–Jul. 1200–2000 m. BR. NV, CA. Native.

Chenopodium pratericola Rydberg (p. 78)
Narrowleaf goosefoot

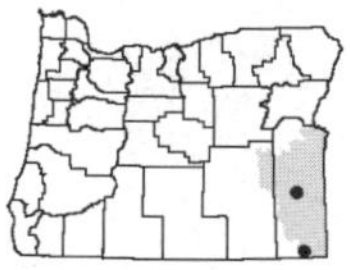

Plants annual, farinose. **Stems** erect, simple or branching from a main axis above, not dichotomously branched, 2–8 dm, striate, densely to sparsely farinose. **Leaves** blades lanceolate, 20–50 × (3)4–17 mm, ~3 × as long as wide, bases cuneate, sometimes with hastate lobes, margins entire, tips acute to obtuse with a small mucronate tip, surfaces lower farinose, upper glabrous to glabrate; petioles 4–10 mm. **Inflorescences** clusters of flowers arranged in axillary and terminal spikes. **Flowers** perianths 5-parted, divided ~50% of length, keeled on back; stamens 5. **Fruits** 0.9–1.3 mm, pericarp loose and membranous. **Seeds** horizontal, black, shiny.

Dry, open areas, sagebrush and juniper communities. Flowering May–Oct. 800–2000 m. Owy. CA, ID, NV, WA; throughout North America. Native.

Chenopodium rubrum Linnaeus

Red goosefoot, red pigweed

Plants annual. **Stems** branching from a main stem, often reddish. **Leaves** blades linear to rhombic-ovate to lanceolate, bases cuneate to attenuate, margins shallowly sinuate-dentate to deeply dentate, tips obtuse to rounded, lower surface sometimes sparsely farinose; petioles 5–40 mm. **Inflorescences** terminal and axillary spikes of loosely rounded to indistinct flower clusters; clusters 1.5–4 mm in diameter, occasionally subtended by a leaf-like bract. **Flowers** perianths with 3 or 4 segments, divided nearly to base, green, not fleshy; stamens 1 or 2(3). **Fruits** ovoid, laterally flattened, pericarp loose and free. **Seeds** vertical, dark reddish brown to nearly black. **2*n***=18.

2 varieties.

Key to *Chenopodium rubrum* varieties:

1\. Plants erect; leaves 25–75 mm, margins deeply dentate . . . . ***C. r.*** var. ***rubrum***

1' Plants prostrate and sprawling to ascending; leaves 12–25(35) mm, margins shallowly sinuate-dentate . . . . ***C. r.*** var. ***humile***

Chenopodium rubrum Linnaeus var. ***humile*** (Hooker) S. Watson

Low red goosefoot, low red pigweed

Chenopodium humile Hook.

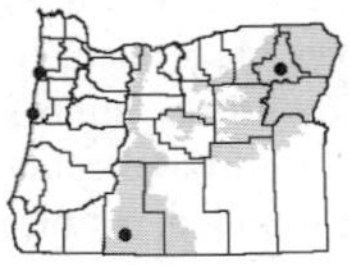

Stems prostrate and sprawling to ascending; branches 0.3–2 dm. **Leaves** 12–25(35) mm, margins shallowly sinuate-dentate.

Moist, alkaline soils, flats and bottomlands, brackish beaches and shorelines. Flowering Jun–Aug. 800–1400 m. BW, ECas, Est. ID, NV; east to UT. Native.

Chenopodium rubrum Linnaeus var. ***rubrum*** Linnaeus (p. 78)

Red goosefoot, red pigweed

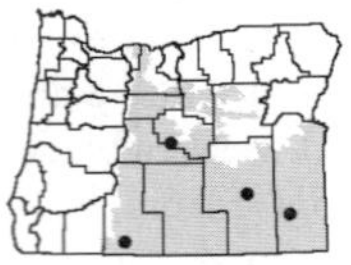

Stems erect, 1–9 dm. **Leaves** 25–75 mm, margins deeply dentate.

Moist, alkaline soils, salt marshes, muddy shorelines of ponds and lakes. Flowering Jun–Aug. 800–1400 m. BR, ECas, Lava, Owy. CA, ID, NV, WA; most of Canada, northern and western US, Mexico. Native.

Chenopodium simplex (Torrey) Rafinesque (p. 78)

Mapleleaf goosefoot, sowbane

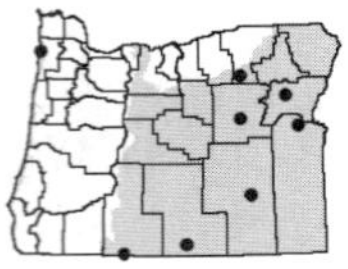

Plants annual, glabrous, sometimes sparsely farinose when very young. **Stems** branched laterally from a main axis, 3–15 dm, angled and often striate. **Leaves** blades triangular to broadly ovate, thin, 45–100(170) mm, bases rounded, truncate, or subcordate, margins prominently sinuate-dentate, veins prominent, tips acuminate to acute; petioles 1.2–4(6) cm. **Inflorescences** axillary and terminal open cymose panicles, flowers solitary or occasionally in loose clusters. **Flowers** perianths 5-parted, divided nearly to base, with prominent midrib; stamens 5. **Fruits** strongly dorsiventrally flattened with acute angled margins. **Seeds** horizontal, black, shiny. **2*n***=18.

Roadsides, fields, woods. Flowering Jul–Aug. 0–1800 m. BR, BW, ECas, Est, Lava, Owy. CA, ID, NV, WA; throughout Canada and US. Native.

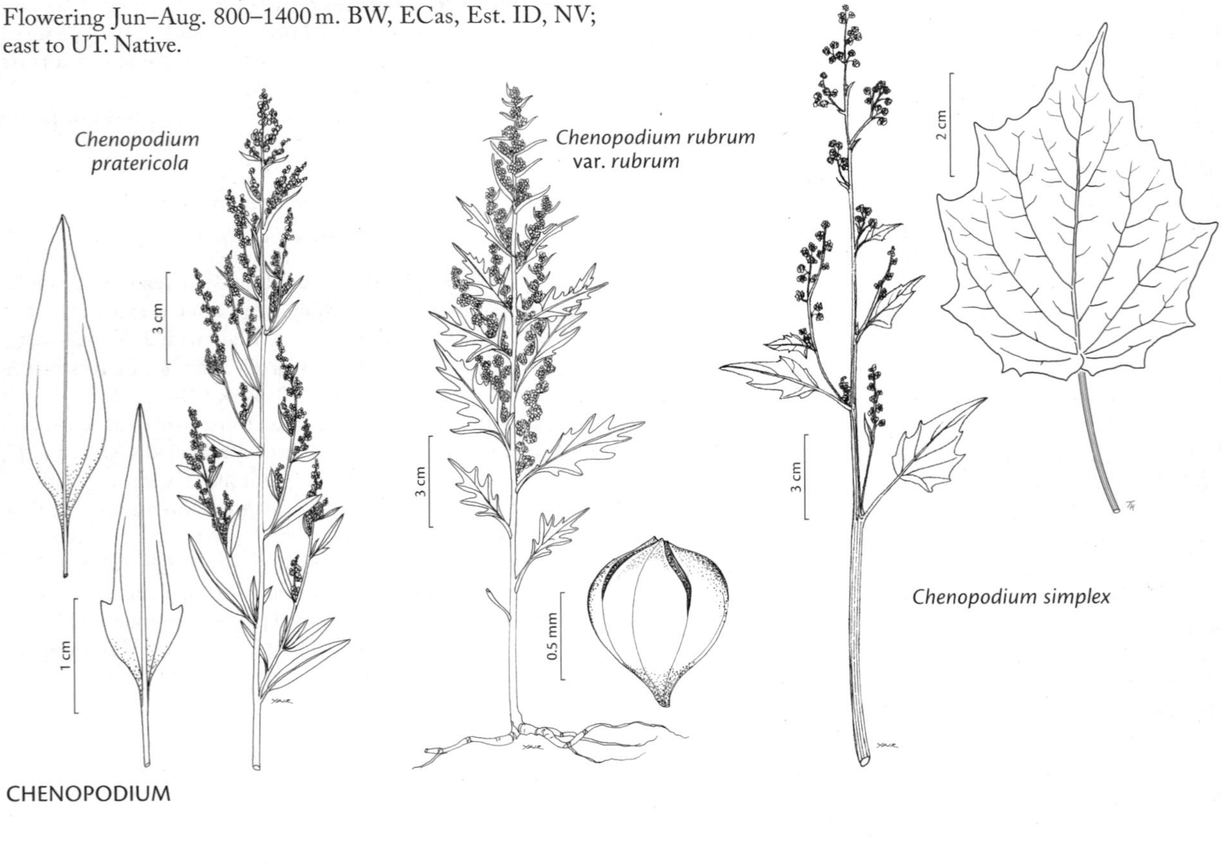

CHENOPODIUM

CORISPERMUM Bugseed

Herbs annual, with branched sometimes stellate hairs, becoming glabrous in age. **Stems** erect or ascending, branched, not jointed, not armed, not fleshy. **Leaves** alternate; blades flat to occasionally convolute, margins entire, tips acute; sessile. **Inflorescences** terminal spikes, each flower solitary in axils of leaf-like bracts. **Flowers** bisexual, 3(5); ovaries superior, stigmas 2. **Fruits** plano-convex; bases of styles forming a beak-like structure at tip of fruit; wings usually present; pericarp smooth. **Seeds** vertical.

Asia, Europe, North America. ~65 species; 3 species treated in *Flora*.

Key to *Corispermum* species:

1. Fruit wings absent or < 0.1 mm wide; plants sparsely to densely to villous, eventually glabrate .... ***C. villosum***

1' Fruit wings distinct, 0.2–0.5 mm wide; plants glabrous to sparsely villous.

  2. Fruits yellow-green to brown, wings translucent and thin........................ ***C. americanum*** var. ***americanum***

  2' Fruits black to deep olive-green, with greenish, semitransparent wings .... ***C. pacificum***

Corispermum americanum (Nuttall) Nuttall var. ***americanum*** (p. 79)

American bugseed, common bugseed

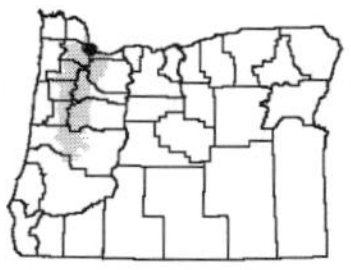

Plants 10–35 cm, glabrous or sparsely villous, with dendritic or stellate hairs. **Stems** much branched at and above base. **Leaves** linear to narrowly linear, plane or sometimes folded, 15–35 × 1–3 mm. **Inflorescences** lax, narrow, interrupted or condensed distally; bracts 5–20 mm. **Flowers** perianth segments 1. **Fruits** 2–4(4.5) × 0.2–0.3 mm, yellow-green to brown; wings 0.2–0.3 mm wide, translucent, thin.

Sandy soils, dunes, waste areas. Flowering Jul–Sep. 0–300 m. WV. CA, ID, NV, WA; scattered throughout Canada and US except eastern seaboard, Mexico. Native.

Corispermum pacificum Mosyakin (p. 79)

Pacific bugseed

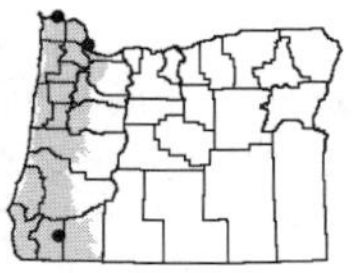

Plants (5)15–40 cm, glabrous or sparsely villous, with dendritic or stellate hairs. **Stems** branched from base. **Leaves** narrowly lanceolate to linear, ± flat, (10)15–25 × 3–7(9) mm. **Inflorescences** compact and dense, sometimes lax; bracts 15–25 mm. **Flowers** perianth segments 1. **Fruits** 3–4 × 0.3–0.6 mm, black to deep olive-brown; wings 0.2–0.5 mm wide, greenish, semitransparent. **2*n***=18.

Sandy shores, dunes. Flowering Jul–Oct. 0–400 m. CR, Sisk, WV. ID, WA. Native.

Occasional hybrids of *C. pacificum* and *C. villosum* have been documented in Oregon. These hybrids typically have aborted fruits.

Corispermum villosum Rydberg (p. 79)

Hairy bugseed

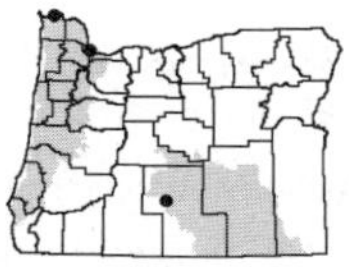

Plants (5)10–30(45) cm, sparsely to densely villous, with dendritic or stellate hairs, eventually glabrate. **Stems** much branched from base. **Leaves** linear-oblanceolate to linear, mostly plane, (10)15–35 × 2–3 mm. **Inflorescences** compact and dense toward tips; bracts 5–15(25) mm. **Flowers** perianth segments 1. **Fruits** 2.2–3 × 1.4–2 mm, yellow-brown to greenish with red to whitish mottling, tips acute; wings absent or < 0.1 mm wide.

Shores, streambanks, waste places. Flowering Jul–Sep. 0–1300 m. BR, CR, WV. ID, NV, WA; north to Northwest Territories, northeast to Quebec, east to IL, southeast to NM. Native.

See also *C. pacificum*.

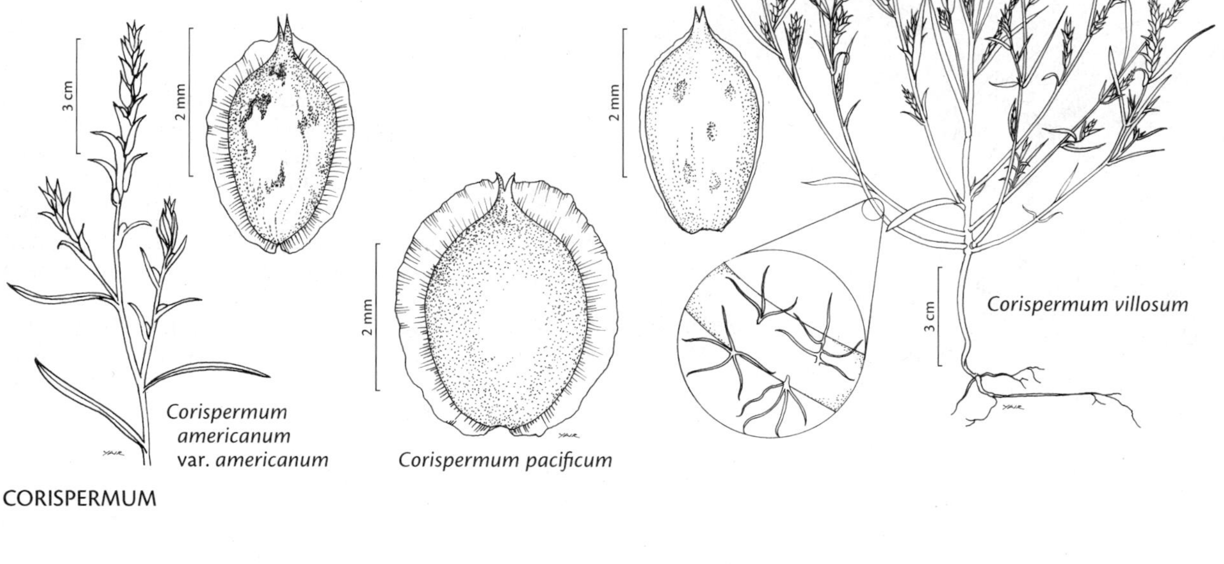

CORISPERMUM

DYSPHANIA Mexican tea, wormseed

Herbs annual or rarely short-lived perennial, usually aromatic (malodorous), with stalked yellow-resinous glandular hairs. **Stems** erect to ascending, sometimes prostrate, much branched to simple. **Leaves** alternate; blades lance-oblong to ovate, flat, margins entire or pinnately lobed, dentate or serrate; petiolate or sessile. **Inflorescences** spikes, panicles, or cymes with clusters of flowers, glabrous to hairy; bracts leaf-like or absent. **Flowers** bisexual, rarely unisexual, staminate and pistillate flowers (when present) not obviously distinct; perianth segments 4–5, usually distinct to near base; stamens 1–5; stigmas 1–3, generally sessile. **Fruits** achenes, often enclosed by perianth; pericarp free to attached. **Seeds** vertical or horizontal, < 1 mm wide, brown or red-brown to black.

Worldwide. ~32 species; 3 species treated in *Flora.*

Dysphania multifida is a South American native known from historic collections on ballast in Portland. It differs from the species treated below by having a sprawling habit and perianths fused near to the tip.

Key to *Dysphania* species:

1. Lower principal blades 1–2.5 cm, lobes rounded . . . . *D. pumilio*
1' Lower principal blades 2–11 cm; lobes acute to rounded or margins dentate to serrate.
 2. Lower principal blades 5–11 cm; inflorescences spikes, with spherical flower clusters . . . . *D. ambrosioides*
 2' Lower principal blades 2–5 cm; inflorescences branched cymes or panicles, flowers in spherical clusters or not. . . . . *D. botrys*

Dysphania ambrosioides (Linnaeus) Mosyakin & Clemants (p. 80)

Mexican tea, wormseed, epazote

Chenopodium ambrosioides L., *Chenopodium ambrosioides* L. var. *ambrosioides, Chenopodium ambrosioides* L. var. *anthelminticum* (L.) A. Gray, *Dysphania anthelmintica* (L.) Mosyakin & Clemants

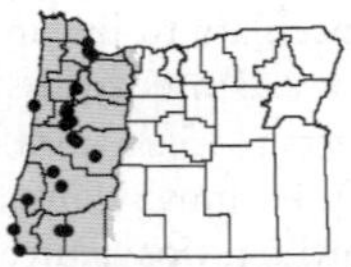

Plants 2.5–13 dm. **Stems** usually erect, often branched, gland-dotted, rarely glabrous. **Leaves** blades lanceolate to ovate, 1.5–11 cm, lower principal blades 5–11 cm, margins dentate to serrate, surfaces glandular. **Inflorescences** terminal and axillary spikes, 3–7 cm; spherical clusters 1.5–4 mm wide; bracts leaf-like. **Flowers** perianth segments 4–5, lobed halfway to base, glabrous, sparsely glandular or puberulent. **Fruits** glandular; pericarp free from seed. **Seeds** vertical or horizontal. **2*n***=32.

Disturbed areas, streambanks, sandy soils, beaches. Flowering Jul–Oct. 0–400 m. Casc, CR, Est, Sisk, WV. CA, WA; eastern Canada and US; Central America. Exotic.

Dysphania botrys (Linnaeus) Mosyakin & Clemants (p. 80)

Jerusalem oak, feather geranium

Chenopodium botrys L.

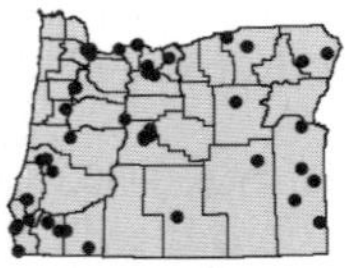

Plants 1–4(6) dm. **Stems** ascending to erect, branched, with short-stalked glandular hairs. **Leaves** blades ovate to elliptic, 1–6 cm, lower principal blades 2–5 cm, margins irregularly pinnately lobed, glandular beneath. **Inflorescences** axillary or terminal branched cymes or panicles, 10–35 cm; flowers in spherical clusters or not; bracts usually absent above. **Flowers** perianth segments 5, parted nearly to base, somewhat enclosing fruit at maturity, glandular; short-pedicelled. **Fruits** glabrous, pericarp adherent to seed. **Seeds** horizontal. **2*n***=18.

Riverbanks, disturbed areas, fields. Flowering Jun–Sep. 0–1500 m. All ecoregions. CA, ID, NV, WA; scattered throughout southern Canada and US; Asia, Europe. Exotic.

Dysphania pumilio (R. Brown) Mosyakin & Clemants

Clammy goosefoot, Australian goosefoot, small crumbweed

Chenopodium pumilio R. Br.

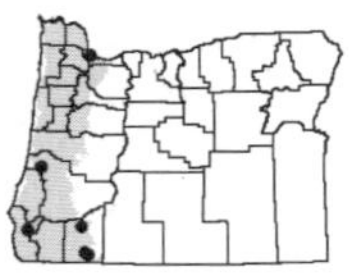

Plants 1–4.5 dm. **Stems** prostrate to ascending, with nearly sessile, glandular hairs. **Leaves** blades elliptic to lanceolate, 0.5–2.5 cm, lower principal blades 1–2.5 cm, margins shallowly pinnately lobed to wavy-dentate, lobes acute to rounded, glandular beneath, sparsely glandular to glandular above. **Inflorescences** axillary spikes of spherical clusters, clusters 2–6 mm wide; bracts leaf-like. **Flowers** perianth segments 5, parted nearly to base, enclosing fruit at maturity, glandular. **Fruits** glabrous, pericarp adherent to seed. **Seeds** vertical.

Disturbed areas, riverbanks. Flowering Jul–Oct. 0–700 m. CR, Sisk, WV. CA, NV, WA; scattered in North America; Australia. Exotic.

Dysphania ambrosioides

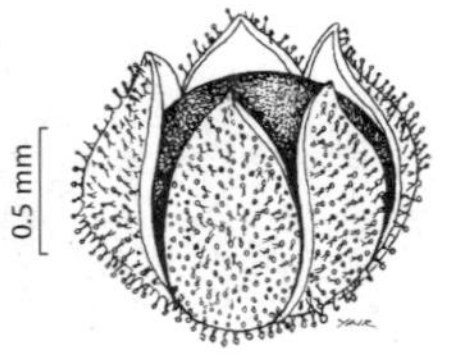

Dysphania botrys

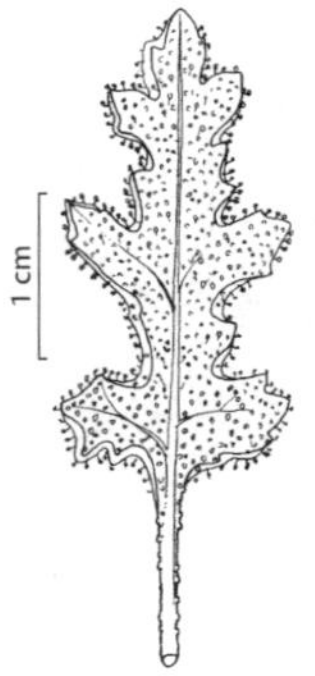

DYSPHANIA

GRAYIA Hopsage

Shrubs monoecious or dioecious, rounded, scaly-puberulent, glabrate with age; hairs branched. **Stems** erect, woody, much branched, not jointed; lateral branches spine-like with age. **Leaves** alternate, with rounded buds in axils, flat, spatulate to oblanceolate, somewhat succulent, bases tapered, margins entire, tips rounded to obtuse; short-petiolate. **Inflorescences** spike-like terminal clusters. **Flowers** unisexual. **Staminate flowers** in clusters of 2–5 in axil of bracts, perianths 4(5)-parted, enclosing stamens; stamens 4–5. **Pistillate flowers** 1–few per cluster, perianths absent; stigmas 2, exserted; subtending bracteoles 2, glabrous, when fruiting fully fused, flattened, and samara-like. **Fruits** utricles; pericarp free; enclosed by flattened bracts. **Seeds** vertical, lenticular, brown.

Western United States. 4 species; 1 species treated in *Flora*.

A systematic study by Zacharias & Baldwin (2010) separated formerly monotypic *Grayia* into four species. *Grayia spinosa* was previously the only species in the genus but remains the only species present in Oregon.

Grayia spinosa (Hooker) Moquin-Tandon (p. 81)

Hopsage

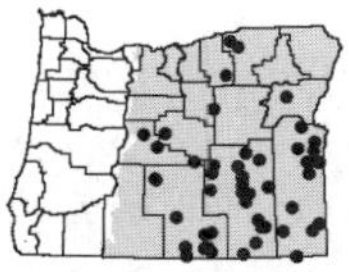

Plants dioecious, rarely monoecious. **Stems** 3–10(15) dm, red-brown, peeling into strips with age. **Leaves** 1–2.5 cm, green, tips often white. **Inflorescences** bracts of pistillate inflorescence 3–10 mm. **Staminate flowers** 1.5–2 mm. **Pistillate flowers** bracteoles 7.5–14 mm, margins entire, glabrous. **Fruits** winged, wings thickened near margin, yellow-green or red to pink-tinged. **Seeds** 1.5–2 mm. **2*n***=36.

Scrubland, juniper woods. Flowering Apr–Jun. 100–1800 m. BR, BW, Col, ECas, Lava, Owy. CA, ID, NV, WA; northeast to MT, east to WY, southeast to NM. Native.

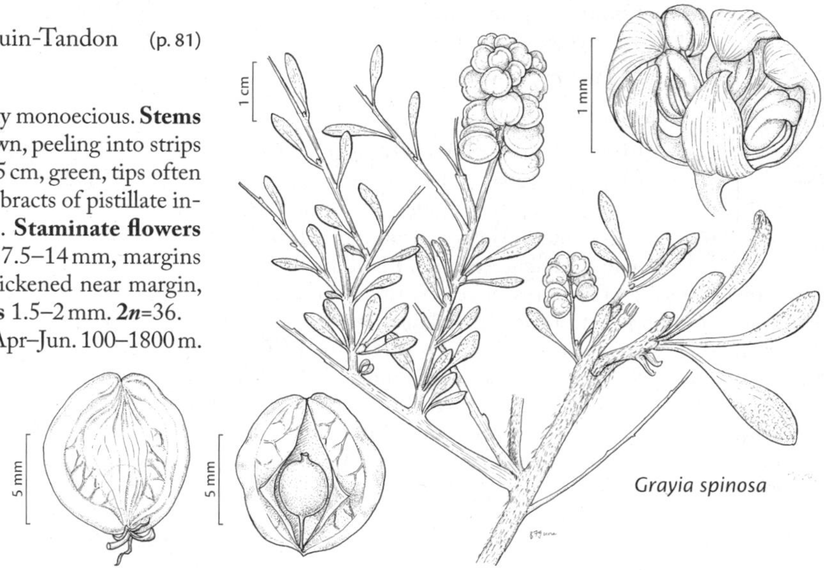

GRAYIA

HALOGETON Halogeton

Herbs annual, glabrous except for occasional axillary tufts of long hairs. **Stems** erect to spreading, branched from base, not jointed, somewhat fleshy. **Leaves** alternate, subterete to terete or club-shaped, tips obtuse with a bristle or spine; sessile. **Inflorescences** axillary in dense clusters; bracts 0–2, leaf-like. **Flowers** bisexual or pistillate; perianths deeply 5-parted, enclosing fruit, fan-like, membranous wings at tips; stamens 5 or absent, stigmas 2; not subtended by bracteoles. **Fruits** utricles, round, pericarp somewhat adherent to seed. **Seeds** vertical, round to flattened, brown to black.

Asia, Europe, North America. 5 species; 1 species treated in *Flora*.

Halogeton glomeratus (M. Bieberstein) C.A. Meyer (p. 81)

Halogeton

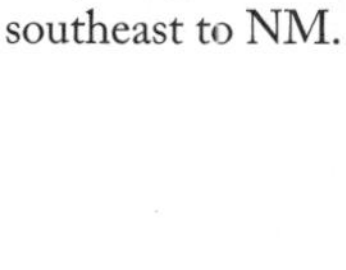

Plants 1–4 dm. **Stems** main branch erect, lateral decumbent to spreading. **Leaves** becoming withered to deciduous in fruit, 4–17 × 1–1.5 mm; bristles at tips 1–2 mm. **Inflorescences** flower clusters many; bracts 1.5–2 mm. **Flowers** perianth segments clawed, 2–3 × 2–4 mm. **Fruits** 0.5–2 mm. **2*n***=18.

Disturbed areas, roadsides, scrublands. Flowering Jun–Aug. 1100–1500 m. BR, Owy. CA, ID, NV, WA; northeast to MT, east to NB, southeast to NM. Exotic.

Concentrated oxalates make *H. glomeratus* toxic to livestock. Plants can thrive in soils with high salinity.

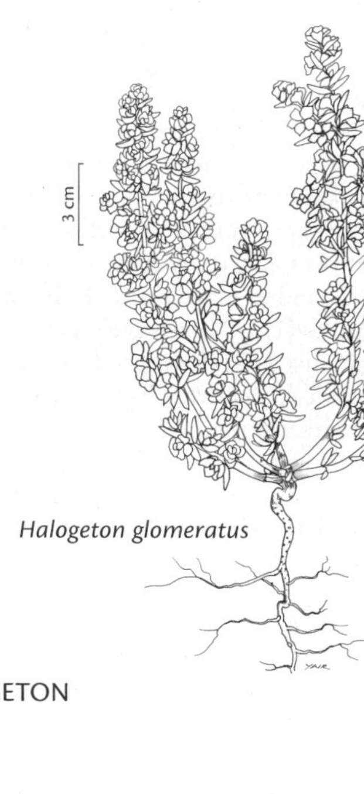

HALOGETON

KRASCHENINNIKOVIA Winter fat

Subshrubs monoecious or dioecious, densely tomentose with stellate hairs. **Stems** erect, not jointed or fleshy; branches woody toward base. **Leaves** alternate; blades linear to lanceolate, flat, not fleshy or terete, margins entire, surfaces densely stellate-tomentose; petiolate. **Inflorescences** spike-like, terminal clusters. **Flowers** unisexual. **Staminate flowers** perianths 4-parted, stamens 4, bracteoles absent. **Pistillate flowers** bracteoles partially fused with 2 lobes at tip; perianths 0; stigmas 2. **Fruits** utricles, vertical, ovate, flat, subtended by densely hirsute bracteoles. **Seeds** vertical, brown with dense white hairs.

Asia, Europe, North America. 3 species; 1 species treated in *Flora.*

This genus is named for Stephan Krasheninnikov, an 18th century Russian botanist and explorer.

Krascheninnikovia lanata (Pursh) A. Meeuse & A. Smit (p. 82)

Winter fat, winter sage

Eurotia lanata (Pursh) Moq.

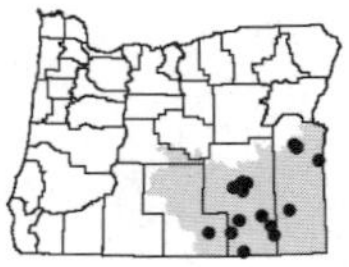

Plants 15–50 cm; hairs becoming brownish with age. **Leaves** 1–4 × 1.5–5 mm, margins typically inrolled; petioles 1.5–3.5 mm. **Inflorescences** 3–19 cm. **Staminate flowers** many, stamens exserted. **Pistillate flowers** 1–4, toward bases of inflorescences; fruiting bracteoles 4–7.5 mm, lobes horn-like. **Fruits** 2.5–3.5 mm. **2*n***=18, 36.

Dry hillsides and flats. Flowering May–Aug. 700–1700 m. BR, Owy. CA, ID, NV, WA; northeast to Manitoba, east to NB, southeast to TX, south to Mexico. Native.

Krascheninnikovia lanata is important winter forage for wildlife and livestock and is occasionally cultivated as an ornamental.

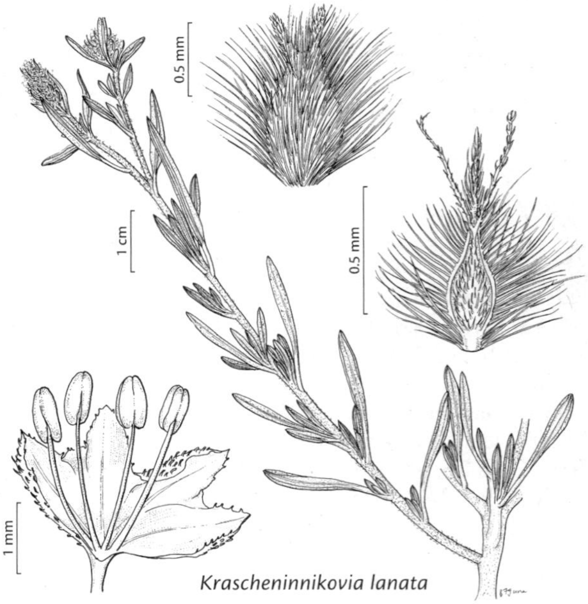

KRASCHENINNIKOVIA

MICROMONOLEPIS Povertyweed

Genus composed of one species, *M. pusilla.*

Micromonolepis pusilla (Torrey) Ulbrich (p. 82)

Tiny povertyweed, red povertyweed

Monolepis pusilla Torr. ex S. Watson

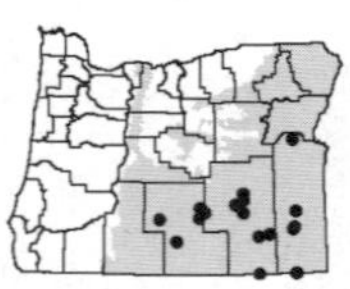

Herbs annual, monoecious, 4–14(20) cm, farinose when young, becoming glabrate. **Stems** erect, slender, dichotomously branched forming a rounded plant, reddish; ultimate branches thread-like. **Leaves** blades elliptic, oblong or somewhat spatulate, 2.5–8 mm, succulent, flat, margins entire, bases sometimes hastate, tips obtuse to rounded. **Inflorescences** solitary flowers or rarely in clusters of 2–4 flowers in leaf axils and terminal branches. **Flowers** unisexual, perianths reduced to (1)2–3 bract-like segments; sessile. **Staminate flowers** stamens 1–2. **Pistillate flowers** styles 1, stigmas 2. **Fruits** subspherical, somewhat flattened, 0.5–0.8 mm, minutely tuberculate; pericarp membranous. **Seeds** vertical, lenticular, dark brown, shiny.

Alkaline flats, lake margins. Flowering Apr–Jul. 900–1500 m. BR, BW, ECas, Owy. CA, ID, NV, WA; east to WY, southeast to CO. Native.

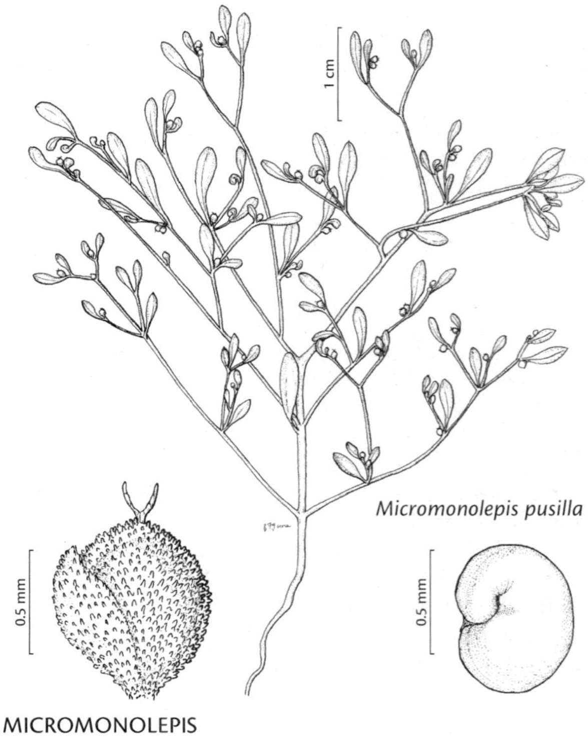

MICROMONOLEPIS

NEOKOCHIA Kochia

Herbs or **subshrubs** perennial. **Stems** erect to ascending, simple or only branched at base, not jointed or fleshy. **Leaves** alternate, nearly opposite proximally, terete to semi-terete, margins entire, tips rounded; sessile. **Inflorescences** spikes, simple or branched; bracteoles leaf-like. **Flowers** bisexual or pistillate; perianth segments 5, with prominent horizontal membranous wings on back, pubescent; stamens 5; stigmas 2–3; sessile. **Fruits** utricles, compressed-spherical; perianth segments covering fruit at maturity; pericarp free and membranous; perisperm absent. **Seeds** wedge-shaped, dull brown, somewhat ribbed.

Western North America. 2 species; 1 species treated in *Flora.*

The genus *Neokochia* is composed of what were formerly the North American representatives of *Kochia.*

Neokochia americana (S. Watson) G.L. Chu & S.C. Sanderson (p. 83)

Red sage, hairy kochia

Kochia americana S. Watson, *Kochia vestita* (S. Watson) Rydb.

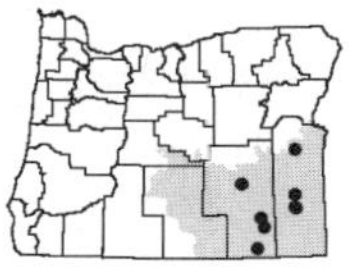

Subshrubs villous-tomentose, 8–40 cm; sprouting from roots. **Stems** white-tomentose, occasionally glabrous in age. **Leaves** blades overlapping, linear, 1–2 × 1–2 mm, with long slender hairs to nearly glabrous. **Inflorescences** 2–5 flowers per bract axil; bracts 4–15 mm. **Flowers** calyx lobes white-tomentose; wings fan-shaped.

Deserts, alkaline flats, dry lake margins. Flowering May–Aug. 800–1400 m. BR, Owy. CA, ID, NV; northeast to MT, east to WY, southeast to TX. Native.

NITROPHILA Niterwort

Herbs perennial, glabrous; with rhizomes. **Stems** decumbent to erect, branched at base, not jointed, not armed, not fleshy. **Leaves** opposite, linear to ovate, fleshy, tips acute to pointed; sessile or clasping at base. **Inflorescences** axillary, 1-flowered and sessile, or 2–3-flowered and short-pedicelled; with 2 subtending bracts. **Flowers** bisexual, perianth segments 5(7), lobes erect and overlapping, persistent in fruit; stamens 5; stigmas 2; subtending bracteoles 1 or 2, of equal length. **Fruits** utricles; pericarp free from seed and membranous. **Seeds** vertical, lenticular, black to brown, dull.

Central America, Mexico, South America, western United States. 5 species; 1 species treated in *Flora.*

Nitrophila occidentalis (Moquin-Tandon) S. Watson (p. 83)

Niterwort, borax weed

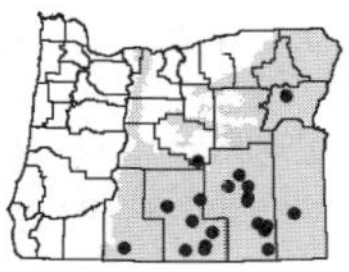

Stems decumbent to erect, 7–20(30) cm. **Leaves** 5–20(30) mm, scale-like on decumbent plants. **Flower** perianth segments 2–3.3 mm, pink to white with age. **Seeds** 1–2 mm.

Meadows, alkaline flats. Flowering Jun–Jul. 1000–1500 m. BR, BW, ECas, Owy. CA, NV, WA; southeast to UT, south to Mexico. Native.

Neokochia americana

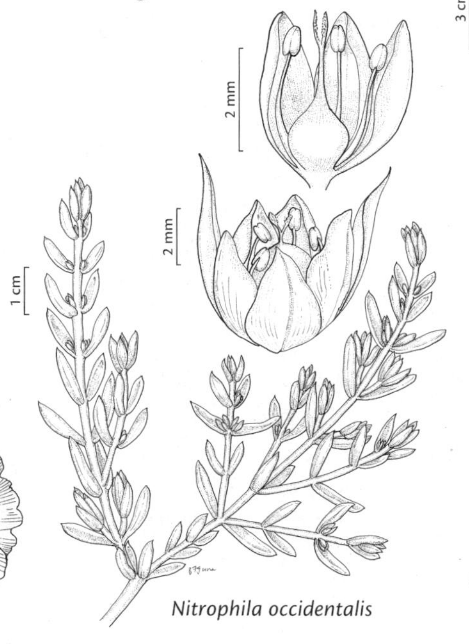

Nitrophila occidentalis

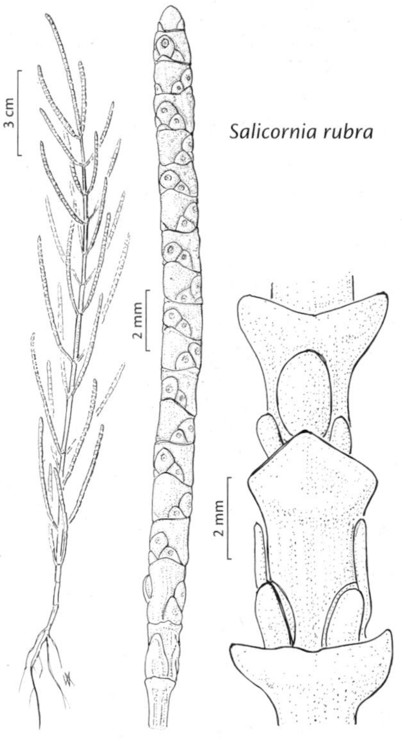

Salicornia rubra

NEOKOCHIA • NITROPHILA • SALICORNIA

SALICORNIA Glasswort

Herbs annual, glabrous. **Stems** erect, occasionally prostrate, simple to branching, green, becoming red with age, fleshy, jointed. **Leaves** opposite, reduced to fleshy scales, fused at bases and enclosing stems, margins entire and somewhat scarious; sessile. **Inflorescences** spikes, terminal at tips of stems, appearing jointed, forming fleshy segments, weakly torulose; each fertile segment consisting of opposite, 3-flowered clusters, sunken in depressions of rachis, adherent to internode, each separated by a flap of internodal tissue; bracts enclosing bases of cymes. **Flowers** bisexual, radially symmetric; perianth segments 3, persistent in fruit, mostly connate, fleshy; stamens (0)1–2; styles 2. **Fruits** utricles; pericarp membranous. **Seeds** vertical, ellipsoid, yellow-brown, hairy; perisperm absent.

Worldwide. 10 species; 1 species treated in *Flora.*

Salicornia rubra A. Nelson (p. 83)

Red glasswort

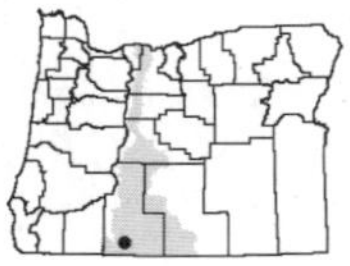

Stems (1)5–25 cm, green with red at bases and tips to completely red at fruiting; ultimate branches typically short. **Inflorescences** 0.5–3(5) cm, with 4–10(19) fertile segments; segments 2–4.5 mm, length ≈ width. **Flowers** 1.1–1.2 mm wide; anthers commonly not exserted, dehiscing with flowers. **Seeds** 1–2 mm. **2*n***=18.

Alkaline soils, salt marshes. Flowering Jul–Oct. 1200–1300 m. ECas. CA, ID, NV, WA; north to AK, east to MN, southeast to NM. Native.

This species is known in Oregon from a few collections in the southern part of state.

SALSOLA Russian thistle, tumbleweed

Plants sparsely papillose to hispid or glabrous. **Stems** erect, rarely ascending, not jointed or fleshy. **Leaves** mostly alternate, subterete to terete, tips with a short spine or bristle; sessile. **Inflorescences** 1(3)-flowered spikes, in axis of bracts or reduced leaves; bracts alternate, awl-shaped to ovate, tipped by a spine. **Flowers** bisexual; perianth segments 5, persistent, glabrous, with prominent abaxial wings; stamens 5; stigmas 2(3); subtended by 2 bracteoles. **Fruits** utricles, covered by perianth at maturity, pericarp adherent. **Seeds** horizontal, orbicular, black or brown; perisperm absent.

Worldwide. ~130 species; 1 species treated in *Flora.*

Salsola kali ssp. *pontica* has been collected once in Oregon, on ballast in 1902.

Salsola tragus Linnaeus (p. 84)

Prickly Russian thistle, tumbleweed

Salsola kali L. ssp. *kali*

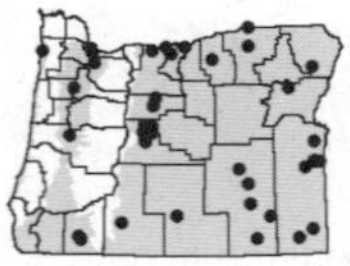

Herbs annual, (0.5)1–10 dm. **Stems** much branched from base; branches curved, lower branches occasionally ± prostrate. **Leaves** filiform or narrowly linear, 15–50 × < 1 mm; spines at tips < 1.5 mm. **Inflorescences** lower margin of bracts with a narrow translucent wing. **Flowers** wings membranous, scalloped or toothed; bracteoles distinct or united at base. **Fruits** persistent. **2*n***=36.

Disturbed areas, roadsides, lakeshores, riverbanks, deserts. Flowering Jun–Sep. 0–2000 m. BR, BW, Col, ECas, Est, Lava, Owy, Sisk, WV. CA, ID, NV, WA; worldwide. Exotic.

The mature stems of this species often break off at the base forming "tumbleweeds," travelling long distances and dispersing seeds.

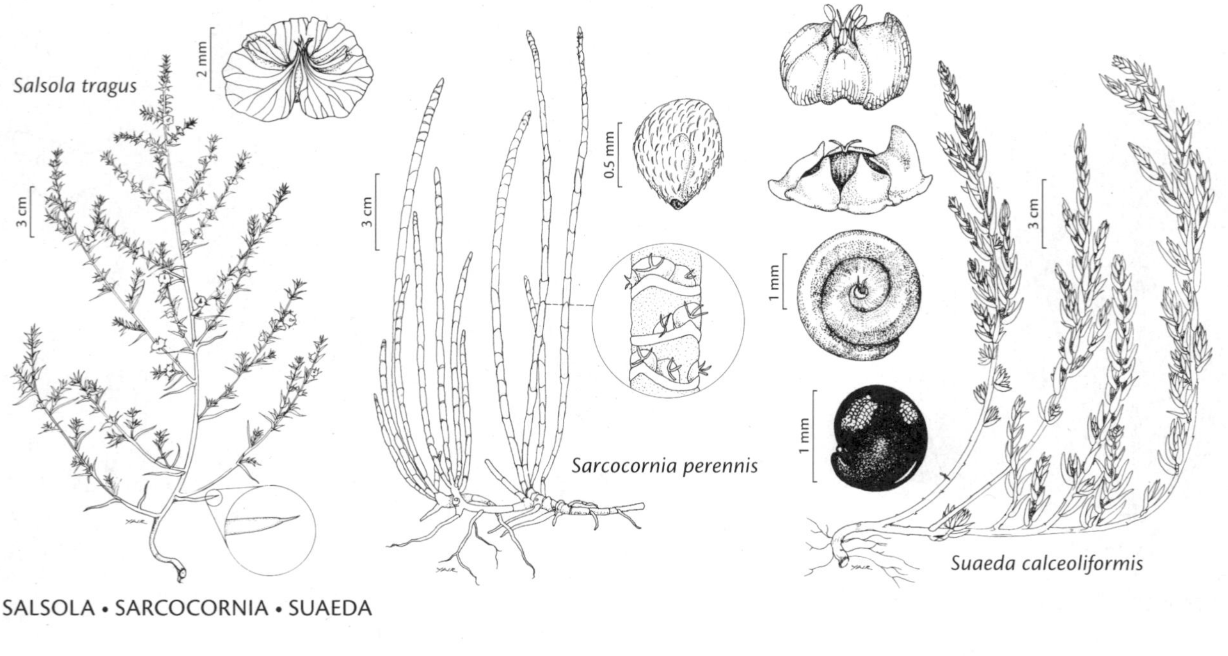

SALSOLA • SARCOCORNIA • SUAEDA

SARCOCORNIA Pickleweed

Subshrubs grayish green, glabrous. **Stems** jointed and fleshy when young, becoming woody and not jointed with age, not armed; some stems with terminal inflorescences, others vegetative. **Leaves** opposite, reduced, decurrent to stem forming fleshy joints; blades triangular fleshy projections at tips of joints, margins scarious. **Inflorescences** terminal and lateral 3–5-flowered cymes, embedded in depressions of rachis, adherent to distal branches in opposite pairs. **Flowers** bisexual or unisexual; perianth segments persistent in fruit, united except at tip, fleshy; stamens 1–2; styles 2–3. **Fruits** utricles, ellipsoid; pericarp membranous. **Seeds** vertical, ellipsoid; seed coat membranous, light brown, covered with slender curved or hooked hairs.

Worldwide. 20–24 species; 1 species treated in *Flora.*

Taxonomic revision within *Sarcocornia* in North America, particularly addressing the morphological differences and range distributions between *S. pacifica* and *S. perennis*, is warranted.

Sarcocornia perennis (Miller) A.J. Scott (p. 84)

Pickleweed, Pacific swampfire, woody glasswort

Salicornia ambigua Michx., *Salicornia pacifica* Standl., *Salicornia perennis* Mill.

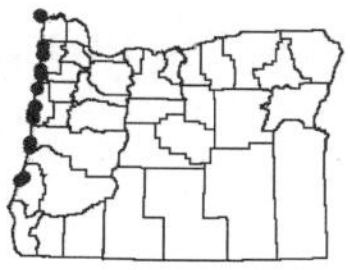

Plants prostrate and creeping, usually rooting forming mats to 1 m in diameter. **Stems** branched, 1–2(3) dm; larger fleshy segments 10–25 × 2–3 mm. **Inflorescences** terminal spikes, 1–2.5 cm; flowers larger toward inflorescence tips; fertile segments 7–14, 1.6–3.1 × 2.9–4.4 mm. **Flowers** 1.3–3.1 mm wide; anthers 0.8–1 mm. **Seeds** 1.1–1.3 mm, most hairs > 0.1 mm. **2*n*=18.**

Coastal salt marshes, tidal flats. Flowering Jul–Oct. 0–50 m. Est. CA, WA; north to Alaska. Native.

There have been reports of occurrences of *S. perennis* on coasts of the northeastern United States. These have since been identified as likely stunted or poorly developed specimens of *S. pacifica.*

SUAEDA Seablite

Shrubs, subshrubs, or **herbs** annual or perennial, glabrous to sometimes puberulent. **Stems** simple or branched, spreading to erect, glabrous, glaucous or pubescent, not jointed or fleshy. **Leaves** blades linear to narrowly lanceolate, subterete to terete, succulent to fleshy, margins entire; sessile to short-petiolate. **Inflorescences** cymes, 1–12-flowered clusters, usually in panicles of spikes; bracts leaf-like. **Flowers** bisexual, occasionally unisexual; perianth lobes 5, rounded and somewhat fleshy, with or without wings or appendages; stigmas 2–3; bracteoles subtending flowers 1–3. **Fruits** utricles, enclosed in calyx; pericarp waxy, becoming membranous. **Seeds** horizontal or vertical, lenticular.

Worldwide. ~110 species; 2 species treated in *Flora.*

Key to *Suaeda* species:

1\. Plants herbs, annual; flowers with wings or appendages at base. . . . . ***S. calceoliformis***

1' Plants perennial; flowers without wings or appendages . . . . ***S. nigra***

Suaeda calceoliformis (Hook) Moquin-Tandon (p. 84)

Pahute weed, horned seablite

Suaeda depressa (Pursh) S. Watson, *Suaeda depressa* (Pursh) S. Watson var. *depressa*, *Suaeda depressa* (Pursh) S. Watson var. *erecta* S. Watson, *Suaeda maritima* (L.) Dumort., *Suaeda occidentalis* S. Watson

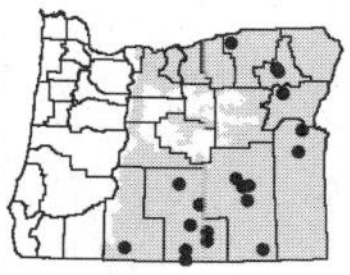

Herbs annual, 0.5–8(10) dm. **Stems** prostrate to erect, simple or branched, glabrous or glaucous. **Leaves** generally alternate; blades linear, (5)10–40 mm; sessile. **Inflorescences** generally dense with 3–7 flowers per cluster, bracts ≤ leaves. **Flowers** bisexual, bilateral, with wings or horned appendages at base; perianths 1–4 mm including appendages. **Seeds** lenticular, 0.8–1.7 mm, black, shiny or dull brown. **2*n*=36.**

Alkaline or saline wetland areas. Flowering Jun–Sep. 100–1500 m. BR, BW, Col, ECas, Owy. CA, ID, NV, WA; throughout Canada and US except southeastern seaboard. Native.

Suaeda nigra (Rafinesque) J.F. Macbride

Tall seablite, shrubby seablite

Suaeda fruticosa (L.) Forssk., *Suaeda intermedia* S. Watson, *Suaeda moquinii* (Torr.) Greene, *Suaeda torreyana* S. Watson

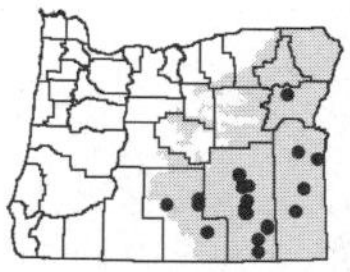

Shrubs or **subshrubs**, spreading to erect, 2–15 dm. **Stems** branches spreading from a usually woody base, glabrous, glaucous, villous or loosely hirsutulous. **Leaves** generally alternate; blades linear to narrowly lanceolate, 5–30 mm; petioles 1 mm or less. **Inflorescences** generally open with 1–12 flowers per cluster; bracts < leaves. **Flowers** mostly bisexual, occasionally pistillate or staminate, radial, without wings or appendages; perianths 0.7–2 mm. **Seeds** lenticular, 0.5–2 mm, black, shiny. **2*n*=18.**

Alkaline or saline areas, deserts. Flowering May–Aug. 700–1400 m. BR, BW, Owy. CA, ID, NV, WA; northeast to Saskatchewan, east to NE, southeast to TX, south to Mexico. Native.

ANACARDIACEAE Cashew family

Stephen C. Meyers

Shrubs, **vines**, or **trees** perennial, usually dioecious, sometimes flowers bisexual and pistillate, aromatic, with an acrid juice or milky sap. **Leaves** alternate, simple or ternately compound or pinnately compound; petioles present; stipules absent or minute. **Inflorescences** axillary or terminal, panicles or racemes. **Flowers** radial; sepals 5, fused at base; petals 5, distinct; stamens 5, distinct, absent or reduced in pistillate flowers; ovaries superior, reduced or absent in staminate flowers, styles 3, distinct or fused near base, stigmas 3, capitate. **Fruits** drupes, somewhat flattened; stones 1-seeded.

Asia, Europe, North America, South America. ~70 genera; 2 genera treated in *Flora*.

Cotinus coggygria (European smoke tree) is reported to have escaped cultivation in a few localities in Oregon. These shrubs or small trees are distinguished by the long hairs of the panicles, which are translucent and diffuse light, thus creating a "smoke-like" effect.

Key to Anacardiaceae genera:

1. Upper side of leaves dull; inflorescences terminal; fruits orange-red to red, sparsely to densely hairy; leaves simple or ternately compound or pinnately compound. . . . . **RHUS**

1' Upper side of leaves shiny; inflorescences axillary; fruits white to light yellow or light yellow-green, ± glabrous; leaves ternately compound. . . . . **TOXICODENDRON**

RHUS Sumac

Shrubs or **small trees** deciduous, usually dioecious, sometimes flowers bisexual and pistillate, often forming colonies; with long branched rhizomes. **Branches** glabrous to puberulent, interiors often pithy. **Leaves** simple, ternately or pinnately compound; upper side of leaves dull. **Inflorescences** dense terminal panicles or appearing as small clusters of flowers, usually developing before leaves expand. **Flowers** anthers ovoid; styles ≈ in length, sometimes fused near base. **Fruits** red, sparsely to densely hairy. $2n$=30.

Temperate regions in the Northern Hemisphere. ~150 species; 2 species treated in *Flora*.

Key to *Rhus* species:

1. Leaves simple or ternately compound; leaflets 0.5–3.5 cm; fruits sparsely hairy . . . . ***R. aromatica***

1' Leaves pinnately compound; leaflets 5–12 cm; fruits densely hairy . . . . ***R. glabra***

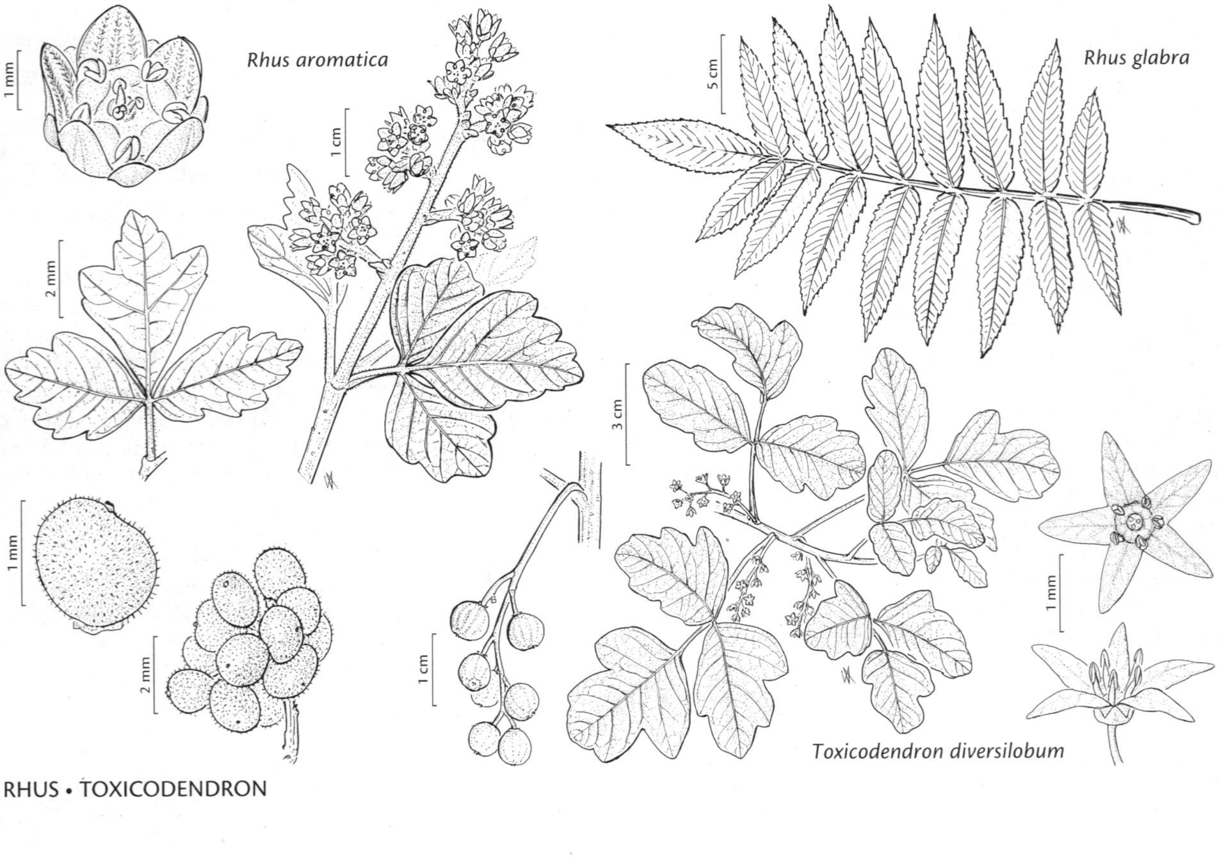

RHUS • TOXICODENDRON

Rhus aromatica Aiton (p. 86)

Fragrant sumac, skunkbush, aromatic sumac

Rhus trilobata Nutt., *Rhus trilobata* Nutt. var. *trilobata*

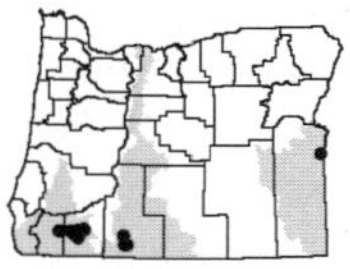

Shrubs 0.5–2.5 m, erect to ascending; branches glabrous to puberulent. **Leaves** simple to ternately compound; leaflets or lobes 3(5), obovate to quadrate, 0.5–3.5 cm, margins crenate, undersides glabrous to tomentose. **Inflorescences** terminal panicles or appearing as small clusters of flowers, 1–6 cm. **Flowers** sepals 1–2 mm, yellowish to reddish brown or reddish purple, tips rounded to acute; petals 2–3.5 mm, yellow. **Fruits** 5–8 mm, orange-red to red, sparsely hairy.

Roadsides, old fields, streamsides, scrublands, openings in dry forests. Flowering Mar–May. 300–1400 m. ECas, Owy, Sisk. CA, ID, NV, WA; throughout United States, midwestern Canada, northern Mexico. Native.

Some authorities consider plants east of the Rocky Mountains a separate species, designating those plants as *R. aromatica* while referring to ours as *R. trilobata.* The common names of this species refer to the aromatic leaves.

Rhus glabra Linnaeus (p. 86)

Smooth sumac

Rhus glabra L. var. *occidentalis* Torr.

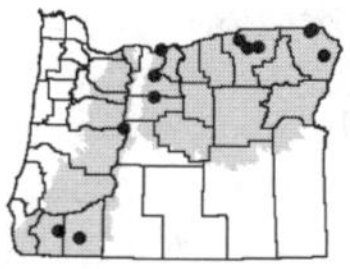

Shrubs or **small trees** 1–4 m, erect to ascending; branches glabrous to sparsely pubescent. **Leaves** pinnately compound; leaflets lance-elliptic to wide-elliptic, 5–12 cm, margins serrate, undersides glabrous, glaucous. **Inflorescences** terminal panicles, 10–25 cm. **Flowers** sepals 1.5–2.5 mm, greenish yellow, tips acute; petals 2–3 mm, greenish yellow. **Fruits** 4–6 mm, red, densely hairy.

Roadsides, railroads, fencerows, openings in forests. Flowering May–Jul. 300–1300 m. BW, Casc, Col, Lava, Sisk. CA, ID, NV, WA; throughout US, southern Canada, northern Mexico. Native.

TOXICODENDRON Poison ivy

Shrubs or **vines** deciduous, dioecious; often rhizomatous, climbing by masses of aerial roots or not. **Leaves** ternately compound, thin to leathery, margins entire, coarsely toothed, crenate, or lobed, resinous; uppersides of leaves shiny; leaflets 3, rarely up to 5. **Inflorescences** axillary panicles or racemes, open, drooping, developing after leaves expand. **Flowers** petals glabrous; anthers ovoid to oblong; styles ≈ in length, sometimes fused near base. **Fruits** globose, white to light yellow or light yellow-green, papery to leathery with age, ± glabrous.

Eastern Asia, North America, South America. 15 species; 2 species treated in *Flora*.

Species of this genus have a toxic resin on their leaves, stems, and fruits that can cause severe contact dermatitis. Extreme caution is urged when in the vicinity of these species.

Key to *Toxicodendron* species:

1. Leaflet tips rounded to obtuse, rarely some leaflets subacute; plants mainly west of and within the Cascade Mountains . . . . ***T. diversilobum***

1' Leaflet tips acute to acuminate, rarely some leaflets rounded; plants east of the Cascade Mountains . . . . ***T. rydbergii***

Toxicodendron diversilobum (Torrey & A. Gray) Greene (p. 86)

Poison oak

Rhus diversiloba Torr. & A. Gray

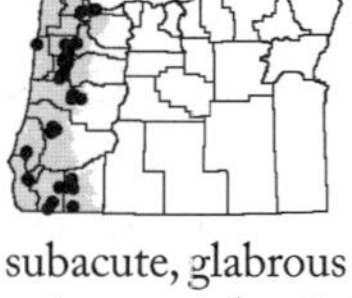

Shrubs 0.5–3 m or **vines** 0.5–20 m; branches and twigs beige, glabrous to sparsely pubescent; sometimes climbing by masses of aerial roots. **Leaves** leaflets round to oblong, 3–10(20) cm, tips rounded to obtuse, rarely subacute, glabrous or sometimes with a line of hairs along main veins; petioles 2–10 cm. **Flowers** sepals 0.5–2 mm, green to yellow-green; petals 1–3 mm, white to yellowish green, veins purple. **Fruits** 3–6 mm, white to light yellow. **2*n***=30.

Woods, canyons, slopes, fencerows, disturbed areas. Flowering May–Jun. 0–1500 m. CR, Est, Sisk, WV. CA, NV, WA; north to British Columbia, south to Mexico. Native.

The range of *T. diversilobum* overlaps slightly with that of *T. rydbergii* in the Columbia Gorge. In that area, hybrids have been reported.

Toxicodendron rydbergii (Small) Greene

Western poison ivy

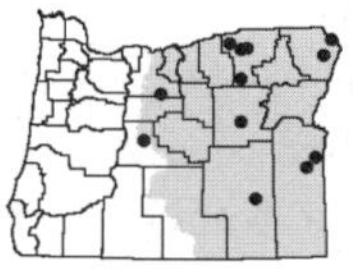

Shrubs or rarely climbing **vines** 0.5–3 m; branches and twigs glabrous to sparsely pubescent; aerial roots not present. **Leaves** leaflets suborbicular to ovate, 4–12 cm, tips acute to acuminate, rarely rounded, under surface sparsely strigose, upper surface glabrous; petioles 2–16 cm. **Flowers** sepals 0.5–2 mm; petals 1–3 mm, pale to yellowish green. **Fruits** 3–6 mm, white to light yellow or light yellow-green. **2*n***=30.

Woods, canyons, slopes, fencerows, disturbed areas. Flowering May–Jun. 50–1500 m. BR, BW, Col, Lava, Owy. ID, NV, WA; throughout northern and western US (except CA), southern Canada, Yukon. Native.

This species is sometimes considered to be a variety of, or fully merged within, *T. radicans* (eastern poison ivy). See also discussion for *T. diversilobum*.

APIACEAE Carrot or parsley family

Stephen C. Meyers (except where noted)

Herbs annual or perennial. **Roots** fibrous, tuberous, or taprooted. **Stems** often hollow. **Leaves** basal and cauline, simple or compound, generally alternate; sheaths present, not inflated to prominently inflated; petioles usually present; stipules usually not present. **Inflorescences** simple or compound umbels or heads; generally peduncled; involucral bracts present or absent; involucel bractlets present or absent. **Flowers** bisexual, rarely unisexual; sepals absent, minute, or well developed; petals 5, small; stamens 5; pistils 1, ovaries inferior, styles 2. **Fruits** schizocarps, composed of 2 mericarps; ribs 10, 4 lateral, 6 dorsal; wings present or absent.

Worldwide. ~300 genera; 31 genera treated in *Flora*.

Old references may refer to this family as Umbelliferae.

Several apiaceous species have been introduced with ballast or as escapes from cultivation. These include *Aegopodium podagraria* (goutweed), *Anethum graveolens* (dill), *Apium graveolens* (celery), *Chaerophyllum temulum* (rough chervil), *Coriandrum sativum* (cilantro), and *Pimpinella saxifraga* (burnet saxifrage). *Myrrhis odorata* (sweet cicely) has been reported historically, and again recently in Portland. Further and diligent monitoring of this potential introduction is advised.

Key to Apiaceae genera:

1. Leaves and inflorescences with spiny teeth . . . . . **ERYNGIUM**

1' Leaves and inflorescences without spiny teeth.

- 2. Leaves simple, margins entire.
  - 3. Plants of rocky areas and meadows, ascending to erect; leaves 3–15 mm wide . . . . . **BUPLEURUM**
  - 3' Plants of wet coastal soils or river or streambanks, prostrate and creeping; leaves 1–5 mm wide . . . . **LILAEOPSIS**
- 2' Leaves compound, or if simple then margins toothed.
  - 4. Flowers bisexual and staminate in same head; inflorescences with round, compact umbellets . . . . . **SANICULA**
  - 4' Flowers all bisexual (sometimes bisexual and staminate in *Lomatium*); inflorescences with round, compact umbellets or not.
    - 5. Flowers yellow or ochroleucous.
      - 6. Leaf segments thread-like; plants with a strong anise scent . . . . . **FOENICULUM**
      - 6' Leaf segments not thread-like; plants not strongly scented or if scented not anise.
        - 7. Most leaf blades cordate to oval or round, some cauline blades occasionally broadly lanceolate, shallowly toothed . . . . . **ZIZIA**
        - 7' Most leaf blades not cordate, oval, or round, shallowly toothed or not.
          - 8. Plants caulescent; stems strongly ridged . . . . . **PASTINACA**
          - 8' Plants ± acaulescent (sometimes caulescent in *Lomatium*); stems, if ridged, not strongly.
            - 9. Fruits wingless . . . . . **TAUSCHIA**
            - 9' Fruits with wings.
              - 10. Dorsal ribs of fruits winged . . . . . **CYMOPTERUS** (in part)
              - 10' Dorsal ribs of fruits not winged . . . . . **LOMATIUM** (in part)
    - 5' Flowers white, green, purple, or pink.
      - 11. Fruits with hooks, barbs, bristles, or prickles (but smooth in *Anthriscus sylvestris*).
        - 12. Fruits needle-like, beaks greatly exceeding bodies in length . . . . . **SCANDIX**
        - 12' Fruits not needle-like, beaks not exceeding bodies in length.
          - 13. Involucral bracts similar to leaves . . . . . **YABEA**
          - 13' Involucral bracts absent or not similar to leaves.
            - 14. Involucral bracts 1–2-pinnate . . . . . **DAUCUS**
            - 14' Involucral bracts absent or simple.
              - 15. Fruits 8–25 mm . . . . . **OSMORHIZA** (in part)
              - 15' Fruits 2–8(10) mm.
                - 16. Fruits with beaks (sometimes weakly developed, but present); sheathes conspicuous, somewhat to moderately inflated . . . . . **ANTHRISCUS**
                - 16' Fruits without beaks; sheathes inconspicuous, no more than slightly inflated . . . . . **TORILIS**
      - 11' Fruits glabrous or hairy, without hooks, barbs, bristles, or prickles.

17. Plants ± acaulescent.
  18. Leaflets/segments broad, thick, fleshy, tomentose below . . . . **GLEHNIA**
  18' Leaflets/segments not broad, thick, fleshy, or tomentose below.
    19. Fruits 3–4 mm; leaves ternately compound; roots tuberous; plants of vernally moist areas (near melting snowbanks, gravelly meadows, open slopes) . . . . **OROGENIA**
    19' Fruits 4–25 mm; leaves ternately compound or not; taprooted or roots tuberous; plants not of vernally moist areas.
      20. Dorsal ribs of fruits winged . . . . **CYMOPTERUS** (in part)
      20' Dorsal ribs of fruits not winged . . . . **LOMATIUM** (in part)
17' Plants caulescent.
  21. Plants wiry in appearance, habitat disturbed areas . . . . **PETROSELINUM**
  21' Plants usually not wiry in appearance, but if so not in disturbed areas.
    22. Leaflets/segments broad (length ≤ width) . . . . **HERACLEUM**
    22' Leaflets/segments not broad (length > width).
      23. Stems distinctly purple-spotted; usually pungent-scented . . . . **CONIUM**
      23' Stems, if purple, not distinctly spotted; usually not pungent-scented.
        24. Involucral bracts ≥ 5.
          25. Leaf margins toothed.
            26. Fruits 3–6 mm; sepals well developed . . . . **PERIDERIDIA** (in part)
            26' Fruits 1–3 mm; sepals absent or minute and triangular.
              27. Fruit ribs inconspicuous and thread-like, wings absent . . . . **BERULA**
              27' Fruit ribs with thick, corky wings . . . . **SIUM**
          25' Leaf margins entire . . . . **PERIDERIDIA** (in part)
        24' Involucral bracts 0–4 (but sometimes 4–8 in *Lomatium dissectum*).
          28. Umbels round and compact . . . . **SPHENOSCIADIUM**
          28' Umbels not round and compact.
            29. Sepals well developed.
              30. Fruit ribs corky . . . . **OENANTHE**
              30' Fruit ribs not corky . . . . **PERIDERIDIA** (in part)
            29' Sepals absent or minute (sepals minute and broad in *Oxypolis*).
              31. Fruit ribs corky.
                32. Stylopodia well developed; plants 3–20 dm . . . . **CICUTA**
                32' Stylopodia absent or obscure; plants typically < 5 dm . . . . **LOMATIUM** (in part)
              31' Fruit ribs not corky.
                33. Fruit wings absent.
                  34. Fruits 8–25 mm, linear or clavate; rays 2–12 . . . . **OSMORHIZA** (in part)
                  34' Fruits 3–7 mm, ovate to oblong; rays 5–25 . . . . **LIGUSTICUM** (in part)
                33' Fruit wings present.
                  35. Dorsal and lateral fruit wings of equal width.
                    36. Leaf margins serrate, rarely lobed . . . . **ANGELICA** (in part)
                    36' Leaf margins entire to finely lobed or toothed . . . . **LIGUSTICUM** (in part)
                  35' Dorsal and lateral fruit wings of unequal width or dorsal wings absent.
                    37. Leaf margins serrate or coarsely toothed, rarely lobed.
                      38. Plants taprooted; sepals absent or minute, not broad; plants on average larger, 5–20 dm . . . . **ANGELICA** (in part)
                      38' Plants tuberous; sepals minute but broad; plants on average smaller, 4–12 dm . . . . **OXYPOLIS**
                    37' Leaf margins entire, deeply dissected or lobed.
                      39. Stylopodia well developed . . . . **CONIOSELINUM**
                      39' Stylopodia absent or obscure . . . . **LOMATIUM** (in part)

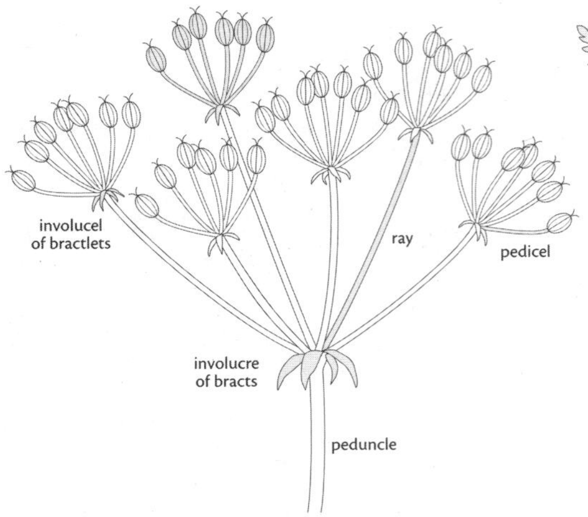

Compound Umbel

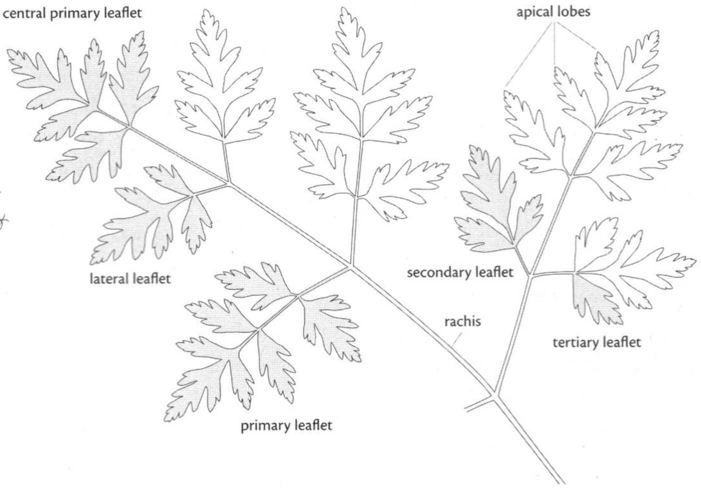

Typical Apicaeae Leaf

ANGELICA Angelica

Herbs perennial. **Roots** taprooted. **Stems** erect, hollow. **Leaves** basal and cauline, compound, 2–3-ternate-pinnate, triangular-ovate in outline, margins serrate and occasionally lobed; leaflets wide; sheaths slightly to prominently inflated; petioles short to well-developed. **Inflorescences** terminal or axillary compound umbels; peduncled; involucres absent; rays 5–65; involucel bractlets 0–15. **Flowers** sepals 0 or minute; petals obovate, white, rarely pink to purple, tips rounded. **Fruits** oblong to round, flattened or not, glabrous to hairy; ribs unequal; marginal wings wider than or equal to others.

Circumboreal. ~60 species; 6 species treated in *Flora*.

Key to *Angelica* species:

1. Leaf undersides tomentose . . . . ***A. hendersonii***

1' Leaf undersides glabrous, occasionally with scattered hairs on veins.
  2. Involucel bractlets absent or rarely 1–3 inconspicuous bractlets present.
    3. Ovaries glabrous . . . . ***A. arguta***
    3' Ovaries hairy.
      4. Leaflets 1.5–5(6) cm; plants of eastern Oregon . . . . ***A. canbyi***
      4' Leaflets (2)3–11 cm; plants of western Oregon . . . . ***A. tomentosa***
  2' Involucel bractlets conspicuous (5–15 mm).
    5. Fruits 2–4 mm, flattened; leaf rachises bent conspicuously; plants inland. . . . . ***A. genuflexa***
    5' Fruits 4–9 mm, not flattened; leaf rachises not bent; plants coastal. . . . . ***A. lucida***

Angelica arguta Nuttall (p. 91)

Shining angelica, sharptooth angelica

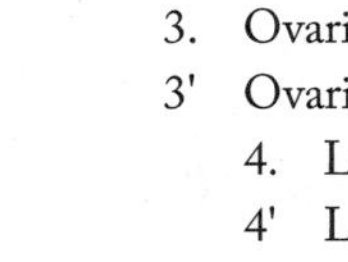

Plants 6–20 dm, glabrous, glaucous or not. **Leaves** 10–35 cm, thin, glabrous but occasionally with scattered hairs on veins on underside; leaflets lanceolate to ovate, 2–11 cm. **Inflorescences** peduncles 10–45 cm; rays 15–60, 2–11 cm; involucel bractlets 0(1–3), inconspicuous; pedicels 1–7 mm. **Flowers** ovaries glabrous. **Fruits** oblong to ovate, 4–9 mm. **2*n***=22.

Wet meadows, roadsides, wooded areas. Flowering May–Aug. 0–2800 m. BW, Casc, CR, ECas, Est, Sisk, WV. CA, ID, NV, WA; north to British Columbia, northeast to Alberta, east to WY. Native.

Angelica arguta is part of a group of described species, with nearly identical morphology, which in our area also includes *A. canbyi* and *A. tomentosa*. Here we tentatively accept these species but highly encourage further research, including sampling (molecular and morphological) from throughout each taxon's range.

Angelica canbyi J.M. Coulter & Rose

Canby's angelica

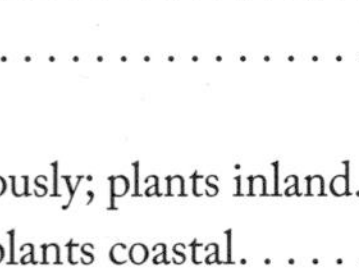

Plants 5–15 dm, glabrous, glaucous or not. **Leaves** 10–30 cm, thin, glabrous but occasionally with scattered hairs on veins on underside; leaflets lanceolate to ovate, 1.5–5(6) cm. **Inflorescences** peduncles 10–40 cm; rays 5–40, 2–9 cm; involucel bractlets 0(1–3), inconspicuous; pedicels 1–7 mm. **Flowers** ovaries hairy. **Fruits** oblong to ovate, 4–7 mm. **2*n***=22.

Wet meadows, roadsides, wooded areas. Flowering Jun–Aug. 500–1800 m. BW, Col. WA. Native.

See *A. arguta* discussion.

Angelica genuflexa Nuttall

Kneeling angelica

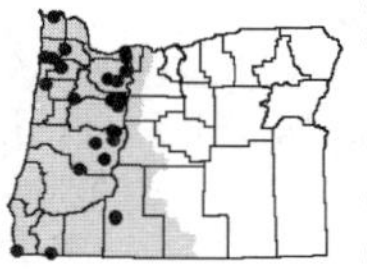

Plants 8–20 dm, glabrous, glaucous. **Leaves** 10–40 cm, glabrous but occasionally with scattered hairs on veins on underside; rachises bent abaxially at bases of some pairs of primary leaflets; leaflets lanceolate to ovate, 3–11 cm. **Inflorescences** peduncles 10–45 cm; rays 25–50, 2–8 cm; involucel bractlets conspicuous, 5–15 mm; pedicels 2–9 mm. **Flowers** ovaries glabrous. **Fruits** subglobose to ovate, flattened, 2–4 mm. ***2n***=22.

Wet meadows, bogs, streambanks. Flowering Jul–Sep. 0–1800 m. Casc, CR, ECas, Est, Sisk, WV. CA, WA; north to AK; East Asia. Native.

Both the common and species names refer to the bent leaves of this species resembling a bent knee.

Angelica hendersonii J.M. Coulter & Rose (p. 91)

Henderson's angelica

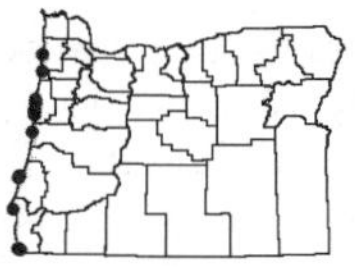

Plants 7–20 dm, nearly glabrous to tomentose. **Leaves** 10–35 cm, thick, underside tomentose; leaflets lanceolate to ovate, (2)3–11 cm. **Inflorescences** peduncles 10–45 cm; rays 20–65, 2–11 cm; involucel bractlets 0–3, 3–10 mm; pedicels 1–4 mm. **Flowers** ovaries tomentose. **Fruits** oblong to ovate, 5–9 mm. ***2n***=22.

Coastal bluffs and dunes. Flowering Jun–Aug. 0–100 m. Est. CA, WA. Native.

Angelica lucida Linnaeus (p. 91)

Seawatch

Plants 6–18 dm, glabrous, glaucous. **Leaves** 10–40 cm, glabrous; rachises not reflexed, lanceolate to ovate, 3–11 cm; sheaths often well-developed and prominent. **Inflorescences** peduncles 10–45 cm; rays 15–45, 2–8 cm; involucel bractlets conspicuous, 5–15 mm; pedicels 3–14 mm. **Flowers** ovaries glabrous. **Fruits** oblong, cylindric, not flattened, 4–9 mm. ***2n***=28.

Coastal bluffs, beaches. Flowering Apr–Aug. 0–100 m. CR, Est. CA, WA; coasts of northwestern and northeastern North America; East Asia. Native.

Angelica tomentosa S. Watson

California angelica

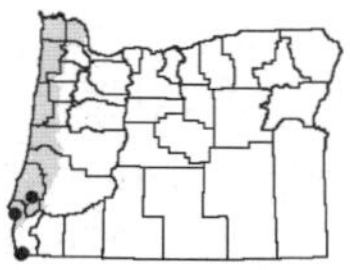

Plants 6–20 dm, glabrous, glaucous or not. **Leaves** 10–35 cm, thin, glabrous but occasionally with scattered hairs on veins on underside; leaflets lanceolate to ovate, (2)3–11 cm. **Inflorescences** peduncles 10–45 cm; rays 10–60, 2–9 cm; involucel bractlets 0(1–3), inconspicuous; pedicels 1–7 mm. **Flowers** ovaries hairy. **Fruits** oblong to ovate, 5–10 mm. ***2n***=22.

Wet meadows, roadsides, wooded areas. Flowering Jun–Aug. 0–1000 m. CR, Est. CA. Native.

Contrary to the name, no parts of this plant are tomentose. See also *A. arguta* discussion.

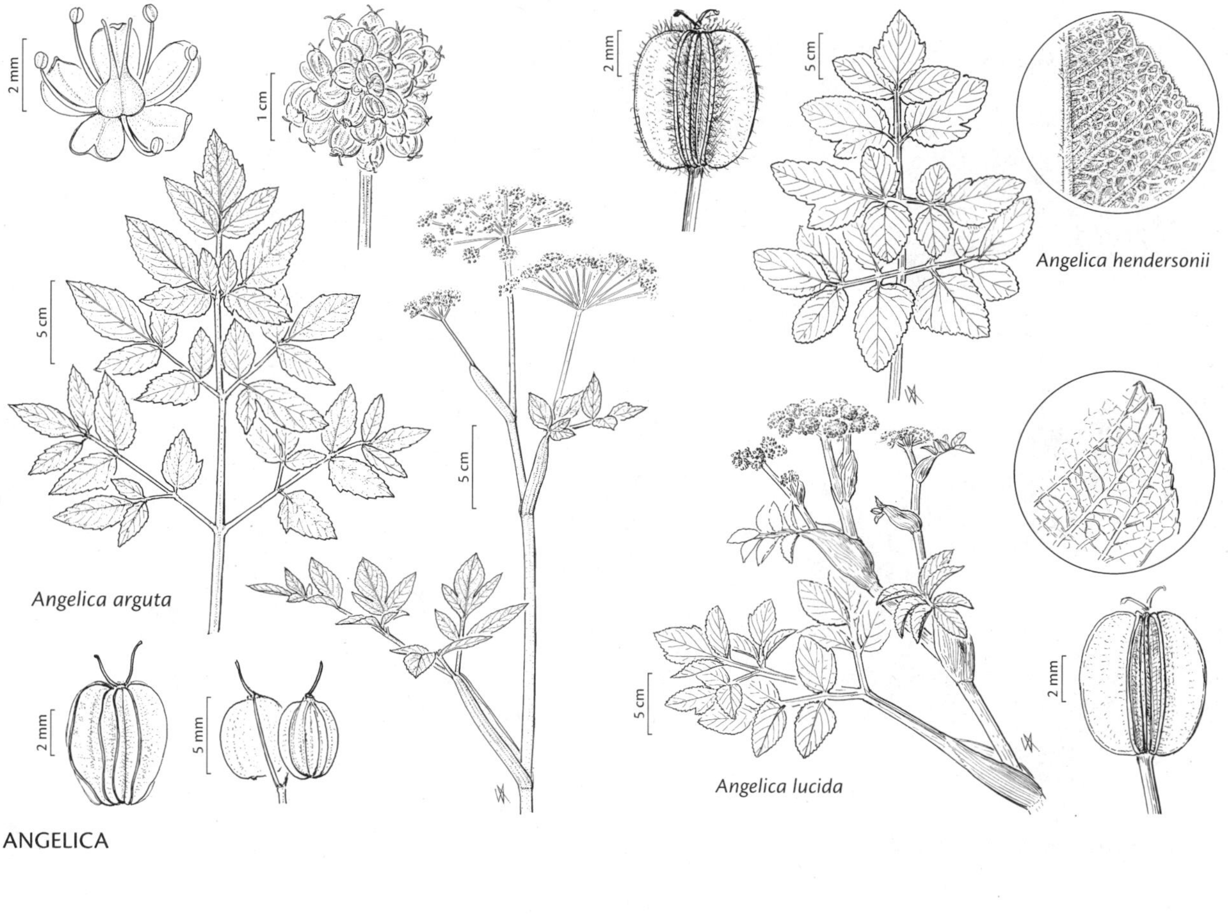

ANGELICA

ANTHRISCUS Chervil

Herbs perennial, biennial, or annual. **Roots** taprooted. **Stems** erect, branched, glabrous. **Leaves** basal and cauline, basal leaves usually absent at flowering, pinnately or ternately compound or dissected, margins dentate to incised; segments linear-oblong to ovate; sheaths somewhat to moderately inflated; petioles present. **Inflorescences** compound umbels, spreading; involucral bracts 0; rays 3–15, unequal in length; involucel bractlets 3–5, linear to lanceolate, shorter than flowering stalks; umbellets 4–14(16)-flowered, outer flowers sometimes bilateral. **Flowers** sepals absent; petals oblong-obovate, white, tips rounded. **Fruits** lanceolate to ovate, slightly flattened; beak well developed or not; ribs absent or weakly developed.

Africa, Asia, Europe, North America. 15 species; 2 species treated in *Flora*.

Key to *Anthriscus* species:

1. Flowers 4–6 per umbellet; rays 3–6(7); fruits bristly, ovate; plants annual . . . . *A. caucalis*

1' Flowers 5–14(16) per umbellet; rays 4–15; fruits smooth, lanceolate to narrowly oblong-ovate; plants perennial or biennial. . . . . *A. sylvestris*

Anthriscus caucalis Bieberstein (p. 92)

Bur chervil

Anthriscus scandicina (Weber ex F.H. Wigg.) Mansf.

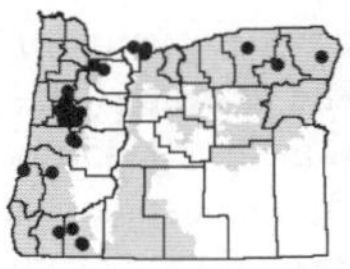

Plants annual, 4–9 dm. **Leaves** 2–10(12) cm. **Inflorescences** peduncles 0.5–2.5 cm; rays 3–6(7), 1–3 cm; umbellets 4–6-flowered; pedicels 0–8 mm. **Fruits** ovate, 3–4 mm, bristly; beaks well developed. **2*n***=14.

Roadsides, disturbed areas, streambanks, wet areas. Flowering Mar–Jun. 0–1100 m. BW, Col, CR, ECas, Est, Sisk, WV. CA, ID, WA; scattered throughout North America; Asia, Europe. Exotic.

The ovate, bristly fruits of *A. caucalis* easily distinguish this species from *A. sylvestris,* which has narrower, smooth fruits.

Anthriscus sylvestris (Linnaeus) Hoffmann

Wild chervil

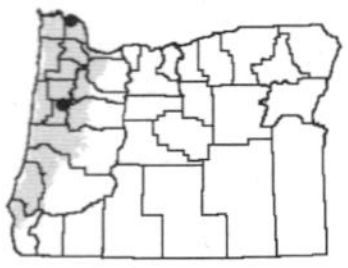

Plants perennial or biennial, 5–10 dm. **Leaves** 3–15 cm. **Inflorescences** peduncles 1–5 cm; rays 4–15, 1–3.5 cm; umbellets 5–14(16)-flowered; pedicels 0–6 mm. **Fruits** lanceolate to narrowly oblong-ovate, 4–8(10) mm, smooth; beaks sometimes weakly developed.

Roadsides wet areas. Flowering Apr–Jun. 0–300 m. CR, WV. ID, WA; scattered in northwestern and northeastern North America; Asia, Europe. Exotic.

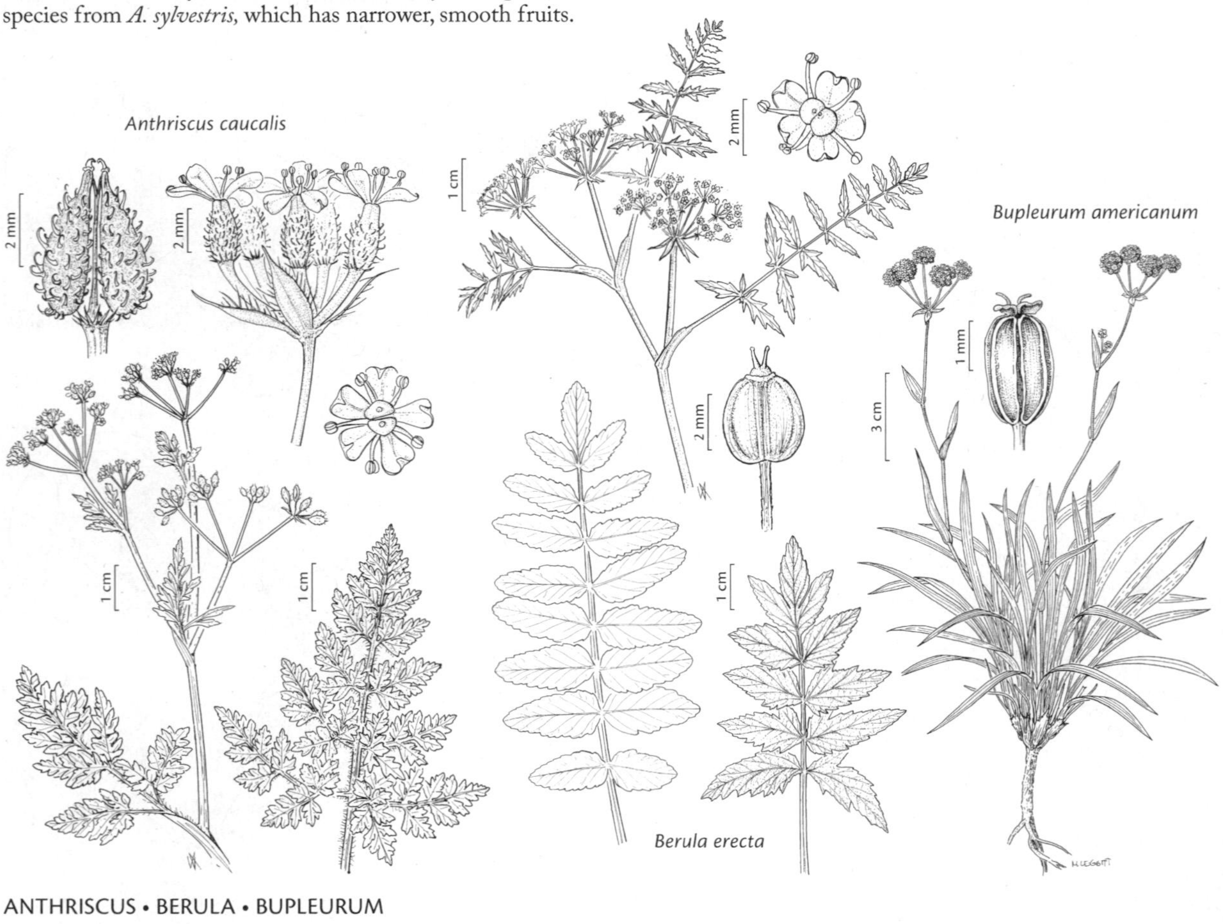

ANTHRISCUS • BERULA • BUPLEURUM

BERULA Water parsnip

Genus composed of one species, *B. erecta*.

Berula erecta (Hudson) Coville (p. 92)

Cutleaf water parsnip

Berula erecta (Huds.) Coville var. *incisa* (Torr.) Cronquist

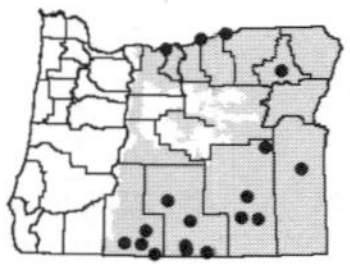

Herbs perennial, partially submerged or not, 2–9 dm, glabrous. **Roots** fibrous, stolons often present. **Stems** simple or branched, ascending or erect, hollow. **Leaves** cauline, alternate, pinnately compound, lower usually dissected more than upper, upper smaller and narrower; leaflets 7–25, oblong to ovate, 1–7 × 0.3–2.5(4) cm, nearly entire to lobed or serrate, tips rounded to acute; sheaths not inflated; petioles 4–10 cm. **Inflorescences** terminal compound umbels; involucral bracts 5–10, lanceolate, 2–12 mm; rays 6–15, 1–2(4) cm, unequal in length; involucel bractlets 4–8, lanceolate, 2–6 mm; pedicels 2–5 mm. **Flowers** sepals minute and triangular; petals white. **Fruits** orbicular, somewhat flattened, 1–2 mm, glabrous; ribs inconspicuous and thread-like; wings absent. $2n$=12, 18.

Wet areas, shallow water. Flowering Jun–Sep. 50–1500 m. BR, BW, Col, ECas, Owy. CA, ID, NV, WA; north to British Columbia, east to NY, southeast to AR and TX, south to Mexico; Asia, Europe. Native.

Some authorities separate North American plants from those in Eurasia based on minor and unreliable leaf characters. If this taxonomy is accepted, however, ours would be *B. erecta* var. *incisa*.

BUPLEURUM Thorow wax

Herbs perennial, glabrous, somewhat glaucous. **Roots** taprooted. **Stems** ascending to erect. **Leaves** basal and cauline, simple, margins entire; sheaths absent or not well developed and not inflated; basal leaves petiolate, cauline leaves sessile. **Inflorescences** terminal and axillary compound umbels; involucral bracts 2–6, leafy; rays 4–10(14), unequal in length; involucel bractlets 3–12; umbellets 10–20+-flowered. **Flowers** sepals small and inconspicuous; petals ovate to obovate, yellow, rarely purplish. **Fruits** narrowly oblong to narrowly elliptic, slightly flattened, glabrous, glaucous; ribs raised but not winged.

Asia, Europe, North America. ~200 species; 1 species treated in *Flora*.

Bupleurum americanum J.M. Coulter & Rose (p. 92)

American thorow wax

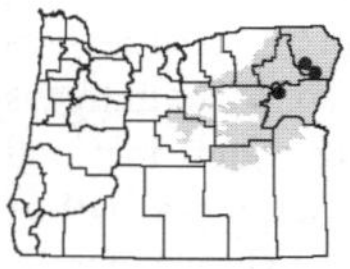

Plants perennial, 5–50 cm. **Stems** several, branching from caudex. **Leaves** simple, linear to lance-linear, 2–12(15) cm × 3–15 mm. **Inflorescences** compact, densely flowered, 0.5–1.5 cm wide; involucral bracts lanceolate, 5–15 mm; rays 0.5–5 cm; involucel bractlets ovate, 2–5 mm; pedicels minute or absent. **Fruits** oblong to narrowly elliptic, 2–4 mm; oil tubes many. $2n$=28.

Meadows, rocky areas. Flowering Jul–Aug. 2200–2800 m. BW. ID; north to AK, east to WY. Native.

Bupleurum americanum is the only species of the genus native to North America. It is one of the two apiaceous species in Oregon with simple and entire leaves (*Lilaeopsis occientalis* being the other).

CICUTA Water hemlock

Richard P. McNeill

Herbs perennial. **Roots** usually tuberous, thickened with transverse partitions creating hollow chambers. **Stems** 1–3, 3–20 dm, erect, hollow, occasionally thickened above roots, striate, glabrous, often glaucous, branched above. **Leaves** basal and cauline, pinnately compound, generally reduced above; leaflets entire to serrate, secondary veins (lateral) directed to sinus between teeth. **Scapes** round, hollow, striate. **Inflorescences** terminal and lateral compound umbels; involucral bracts 0 or few, narrow, inconspicuous; involucel bractlets several, narrow or rarely absent. **Flowers** perfect; sepals minute; corollas white or greenish; carpophores bifid to base, deciduous; styles 2, spreading, stylopodia well developed, depressed or low-conic. **Fruits** orbicular to ovate, glabrous; ribs low, corky; 1 oil tube in interval between ribs and 1–2 oil tubes on commissure.

Northern Hemisphere. 4 species; 3 species treated in *Flora*.

Cicuta is likely the most toxic genus of plants native to North America, north of Mexico. It contains the alcohol cicutoxin, which acts as a potassium channel blocker in T-lymphocytes and a gamma-aminobutyric acid antagonist in the central nervous system (Nelson et al. 2007). The chambered base and tuberous roots are the most toxic parts of the plants. Livestock deaths are more likely in the spring when the immature plant can be pulled from the ground and the entire plant is consumed.

The geographic distribution of *Cicuta* likely encompasses North America from approximately Mexico City to the Arctic Ocean, but many of these areas are poorly collected or the information is not readily available. *Cicuta virosa* is likely the ancestral taxon. It is distributed circumboreally and has 22 pairs of chromosomes that are small in comparison to all other species in the genus. In North America, it has been collected in Alaska, western Canada, and Quebec. Taxa with large chromosomes produce significantly more cicutoxin than taxa with medium or small chromosomes (Lee & Downie 2006; Mulligan 1980).

When keying members of this genus, the use of mature fruits is recommended.

Key to *Cicuta* species:

1. Flowers/fruits usually aborted; leaf axils bulbiferous; leaflets linear, terminal leaflets generally > 5 × as long as wide . . . . ***C. bulbifera***

1' Flowers usually form mature fruits; leaf axils non-bulbiferous; leaflets lanceolate-linear to lanceolate-ovate, terminal leaflets generally < 5 × as long as wide.

   2. Fruit generally orbicular, evenly and abruptly constricted along the commissure, dorsal ribs wider than oil tubes in intervals between ribs; underside of leaves rough-textured and areas enclosed by tertiary veins generally elongate; stem not purple-spotted . . . . ***C. douglasii***

   2' Fruit generally ovate, unevenly or not constricted along the commissure, dorsal ribs generally equal to or narrower than oil tubes in intervals between ribs; underside of leaves smooth and areas enclosed by tertiary veins generally rounded or square; stem often purple-spotted . . . ***C. maculata* var. *angustifolia***

Cicuta bulbifera Linnaeus

Bulb-bearing water hemlock

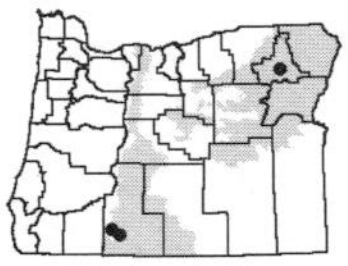

Plants 3–10 dm; generally solitary. **Leaves** 1–2-pinnate, may be reduced above; leaflets linear, margins entire or with a few teeth, 5–40 × 0.5–15 mm, terminal leaflets generally > 5 × longer than wide, when visible, lateral veins directed to sinus between teeth; 1 or more bulbils in leaf axils. **Inflorescences** often wanting; rays 1–2.5 cm. **Flowers** styles < 1 mm. **Fruits** orbicular, 1.5–2 × 1.5–2 mm, unevenly constricted along commissure, most aborting before maturity; dorsal ribs wider than oil tubes in intervals between ribs. **2*n***=22.

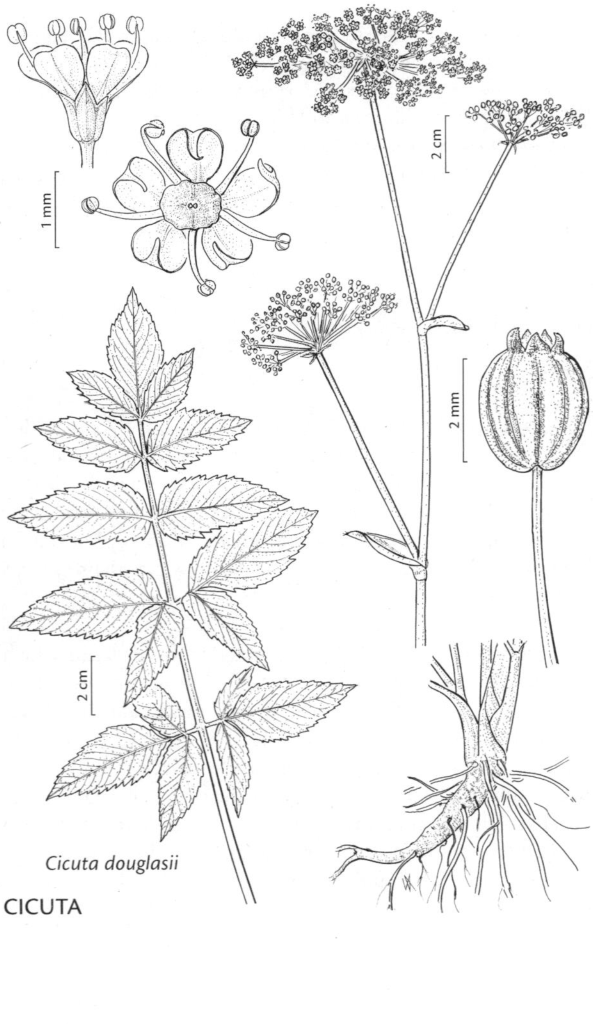

Cicuta douglasii

CICUTA

Wetlands, riparian areas, shallow standing water. Flowering Aug–Sep. 700–1400 m. BW, ECas. ID, WA; northern North America. Native.

Cicuta bulbifera reproduces primarily by bulbils in the leaf axils and produces the lowest amount of cicutoxin in the genus. These characteristics suggest some form of genetic incompatibility. The chromosomes of *Cicuta bulbifera* are intermediate in size between *C. virosa* and *C. maculata*. It is likely the result of a hybridization event and requires further study to clarify the evolutionary history. A sample contained 0.01 mg of cicutoxin per gram of root material (Mulligan 1980).

Cicuta douglasii (de Candolle) J.M. Coulter & Rose (p. 94)

Western water hemlock

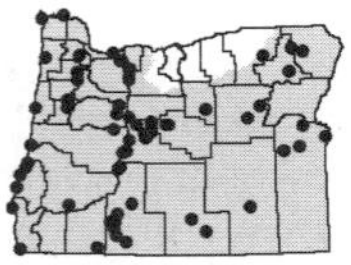

Stems 5–20 dm. **Leaves** 1–3-ternate-pinnate; leaflets lanceolate-linear to lanceolate-ovate, 30–100 × 6–59 mm, terminal leaflets generally < 5 × longer than wide, margins serrate, underside of leaves rough-textured and areas enclosed by tertiary veins generally elongate. **Inflorescences** rays 2–6 cm. **Flowers** styles > 1 mm. **Fruits** generally orbicular, 2–4 × 2–4 mm, evenly and abruptly constricted along commissure; dorsal ribs wider than oil tubes in intervals between ribs, oil tubes often reduced to thin thread-like dark strip between ribs. **2*n***=44.

Wetlands, riparian areas, shallow standing water. Flowering May–Aug. 0–2000 m. All ecoregions except Col. CA, ID, NV, WA; western North America. Native.

Cicuta douglasii is a tetraploid, likely resulting from an allopolyploid event between *C. virosa* and *C. maculata* (Lee & Downie 2006). It has 22 small chromosomes from *C. virosa* and 22 large chromosomes from *C. maculata*. *Cicuta douglasii* can cause significant livestock deaths, especially in the spring when the young plant can be pulled from the ground and the entire plant is consumed. A sample contained 0.76 mg of cicutoxin per gram of root material (Mulligan 1980).

Cicuta maculata Linnaeus var. ***angustifolia*** Hooker

Spotted water hemlock

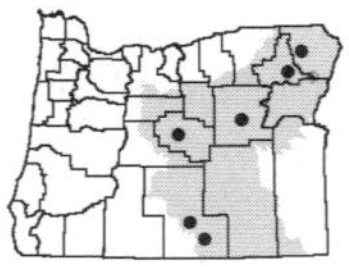

Stems often purple-spotted, especially on lower portions; 10–15 dm. **Leaves** 1–2-pinnate; leaflets lanceolate-linear to lanceolate-ovate, 20–100 × 6–35 mm, terminal leaflets generally < 5 × longer than wide, margins serrate, underside of leaves smooth and areas enclosed by tertiary veins generally rounded or square. **Inflorescences** rays 2–4.5 cm. **Flowers** styles < 1 mm. **Fruits** generally ovate, 2–5 × 2–4 mm wide, unevenly or not constricted along commissure; dorsal rib width generally equal to or narrower than oil tubes in intervals between ribs. **2*n***=22.

Wetlands, riparian areas, shallow standing water. Flowering Jul–Aug. 800–1600 m. BR, BW, Lava. CA, ID, NV, WA; northern and western North America. Native.

This species has four varieties that display significant overlap in characters. Collections from southeastern Oregon may appear to be intermediate between varieties *angustifolia*, *bolanderi*, and *maculata*. *Cicuta maculata* var. *angustifolia* can cause significant livestock deaths, especially in the spring when the young plant can be pulled from the ground and the entire plant is consumed. It has 22 large chromosomes. A sample contained 0.65 mg of cicutoxin per gram of root (Mulligan 1980).

CONIOSELINUM Hemlock parsley

Herbs perennial. **Roots** taprooted or fleshy-fibrous. **Stems** usually single, erect, branched, glabrous. **Leaves** cauline, pinnately or ternate-pinnately decompound, leaflets well defined or not and deeply cleft, margins dissected or deeply lobed; sheaths inflated; petioles present. **Inflorescences** compound umbels, usually terminal; involucral bracts 0–4, narrow; rays 9–30, subequal in length; involucel bractlets narrow and membranous but well developed; umbellets many-flowered. **Flowers** sepals absent or minute; petals white. **Fruits** oblong to ovate, dorsally flattened, glabrous; all or only lateral ribs winged.

Asia, Europe, North America. 10 species; 2 species treated in *Flora*.

Conioselinum is so named because of its resemblance to both *Conium* and *Selinum*.

Key to *Conioselinum* species:

1. All ribs of fruits winged; leaflets generally well defined; plants of Coast Range or coastal areas . . . . ***C. pacificum***

1' Lateral ribs of fruits winged, dorsal ribs raised but not winged; leaflets well defined or not; plants of the Wallowa Mountains . . . . ***C. scopulorum***

Conioselinum pacificum (A. Gray) J.M. Coulter & Rose (p. 95)

Pacific hemlock parsley

Conioselinum chinense (L.) Britton, Sterns & Poggenb., *Conioselinum gmelinii* (Cham. & Schltdl.) Steud.

Plants 2–10 dm. **Leaves** 2–3-pinnately compound, triangular-ovate in outline, 7–30(35) cm, veins sometimes glaucous; leaflets generally well defined, ovate to oblong, 1–8 cm; petioles 2–15 cm. **Inflorescences** involucral bracts (if present) 5–25 mm; rays 1.5–4.5 cm; involucel bractlets 4–11 mm; pedicels 4–8 mm. **Fruits** 5–8 mm; lateral wings well developed, dorsal wings narrower but distinct. **2*n***=44.

Coastal bluffs, rocky and sandy beaches, wet forest. Flowering May–Sep. 0–1000 m. CR, Est. CA, WA; north to AK. Native.

While the leaflets of *C. pacificum* are often more developed than those of *C. scopulorum*, the species can be difficult to distinguish. Reluctantly, geography is the best choice to use as indicator of species.

Mainly a coastal species, *C. pacificum* is found inland on rare occasions. In Oregon, the only known inland population is located on Saddle Mountain in Clatsop County.

Conioselinum scopulorum (A. Gray) J.M. Coulter & Rose

Rocky Mountain hemlock parsley

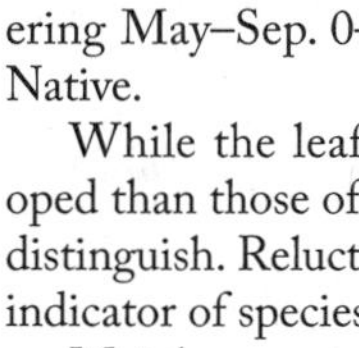

Plants 2–12 dm. **Leaves** 2–3-pinnately compound, triangular-ovate to ovate in outline, 6–35 cm; leaflets well defined or not, ovate to oblong, 1–6 cm; petioles 1–15 cm. **Inflorescences** involucral bracts (if present) 5–25 mm; rays 2–4 cm; involucel bractlets 4–10 mm; pedicels 3–6 mm. **Fruits** 3.5–6 mm; lateral ribs of fruit winged, dorsal ribs raised but not winged. **2*n***=22.

Wet forests, streamsides. Flowering Jul–Sep. 1200–2100 m. BW. Rocky Mountain region. Native.

The population of *C. scopulorum* found in the Wallowa Mountains is disjunct from the normally Rocky Mountain range of this species.

Conioselinum pacificum

CONIOSELINUM

CONIUM Poison hemlock

Herbs biennial, glabrous. **Roots** thick taproots. **Stems** branched, erect, leafy, hollow, distinctly purple-spotted. **Leaves** cauline, alternate or opposite, simple pinnate, pinnately or ternate-pinnately dissected, or compound, ovate in outline; leaf segments lanceolate to ovate, lobed to serrate; sheaths not or slightly inflated; petiolate. **Inflorescences** terminal and axillary compound umbels; involucral bracts 3–6; rays 10–20, unequal in length; involucel bractlets 2–7; umbellets 5–20-flowered. **Flowers** sepals absent; petals white. **Fruits** broadly ovate, slightly flattened, glabrous; ribs raised, often wavy, almost winged.

Southern Africa, Asia, Europe, North America. 2 species; 1 species treated in *Flora*.

Conium maculatum Linnaeus (p. 96)

Poison hemlock

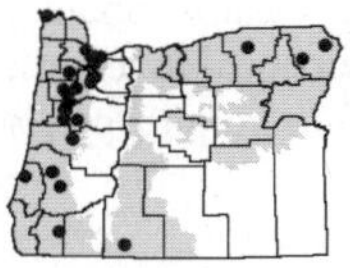

Plants usually pungent-odored, 5–30 dm. **Leaves** 10–60 cm; blades 5–30 cm; petioles 5–30 cm. **Inflorescences** peduncles 2–8 cm; involucral bracts lanceolate to ovate, 3–15 mm; rays 1–5 cm, unequal; involucel bractlets lanceolate to ovate, 1–2 mm, fused at base. **Flowers** pedicels 1–5 mm. **Fruits** 2–3 mm. **2*n***=22.

Wet disturbed areas. Flowering May–Sep. 0–1500 m. BW, Col, CR, ECas, Est, Sisk, WV. CA, ID, NV, WA; nearly worldwide. Exotic.

This species is most likely the plant ("hemlock") used as a poison to execute the Greek philosopher Socrates.

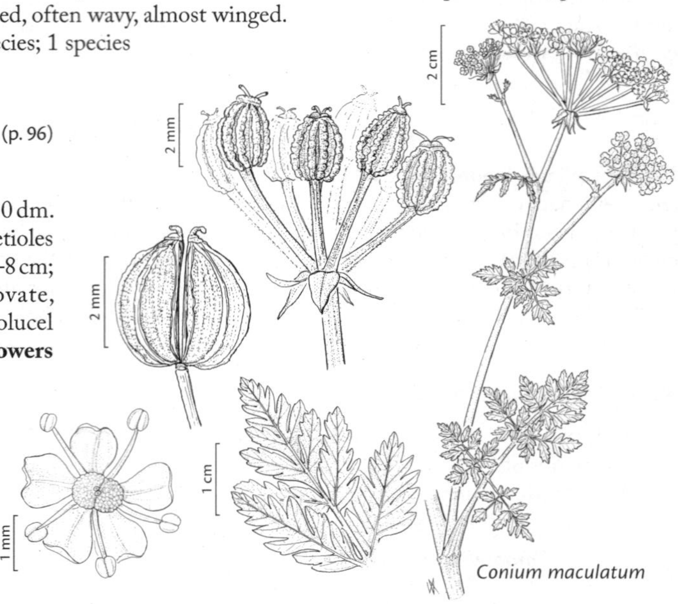

CONIUM

CYMOPTERUS Spring parsley

Herbs perennial, glaucous or not. **Roots** taprooted. **Stems** absent or appearing as short. **Leaves** basal or low cauline, pinnately or ternately compound or dissected, oblong to widely ovate in outline, margins entire or lobed; leaflets linear to obovate, usually tipped with a short spine; sheaths not inflated; petioles present. **Inflorescences** terminal compound umbels; peduncled; involucral bracts absent or poorly developed; rays 2–22; involucel bractlets 5–10. **Flowers** sepals usually absent or minute, sometimes developed; petals oblong to obovate, white, yellow, or purple, tips narrowed. **Fruits** oblong to elliptic, somewhat flattened; ribs subequal to unequal; dorsal ribs winged, lateral rib wing width ≥ others.

Western North America. ~50 species; 7 species treated in *Flora*.

The generic boundaries of *Cymopterus* have been debated for decades. Extremes of taxonomic zeal merge the group with *Lomatium* or divide it into three or more genera. Here we take a circumspect approach and recognize *Cymopterus* in its most traditional form.

Key to *Cymopterus* species:

1. Involucral bracts and involucel bractlets white, showy . . . . . *C. purpurascens*

1' Involucral bracts and involucel bractlets green, not showy.

 2. Leaf segments orbiculate to widely-oval, as long as wide or nearly so, rarely ovate and ≤ 1.5 × as long as wide; plants 3–9(10) cm . . . . . *C. corrugatus*

 2' Leaf segments linear to ovate or rectangular to triangular and skeleton-like, usually > 1.3 × as long as wide; plants 3–35 cm.

  3. Basal leaf sheaths from previous years persistent on caudices.

   4. Plants 15–35 cm; rays 0.5–6 cm, very unequal; sepals well developed, 0.5–1 mm; petals yellow . . . . . *C. petraeus*

   4' Plants 3–20 cm; rays 0.1–0.4 cm, subequal to unequal; sepals not well developed, < 0.5 mm; petals white . . . . . *C. nivalis*

  3' Basal leaf sheaths from previous years absent.

   5. Plants 10–60 cm; calyces well developed; leaves 5–25 cm . . . . . *C. terebinthinus*

   5' Plants 3–15(25) cm; calyces absent or minute; leaves 2–8 cm.

    6. Leaf segments 1–3(4) mm wide; carpophores absent . . . . . *C. glomeratus*

    6' Leaf segments 0.5–1(2) mm wide; carpophores well developed . . . . . *C. longipes* var. *ibapensis*

Cymopterus corrugatus M.E. Jones (p. 98)

Corrugated spring parsley, wrinkle-wing spring parsley

Rhysopterus plurijugus J.M. Coult. & Rose

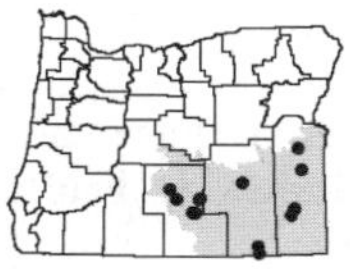

Plants 3–9(10) cm, glabrous, slightly glaucous or not; with 1–2 pseudoscapes arising from crown of a thick taproot. **Leaves** basal, 1–2-pinnately divided, oblong-ovate in outline, 2–7 cm; leaf segments orbiculate to widely-oval, ~ 1 × as long as wide, rarely ovate and ≤ 1.5 × as long as wide. **Inflorescences** peduncles 1–4, 2–8 cm; involucral bracts absent or minute, green; rays 4–11, 0–1 cm, unequal; involucel bractlets 5–10, not showy, linear, green, connate at base. **Flowers** sepals absent or minute; petals white; carpophores absent. **Fruits** 4–8 mm; wings wrinkled, 0.5–1.5 mm wide, lateral wing width ≥ dorsal.

Sandy soil, sand dunes, sagebrush areas. Flowering Apr–Jul. 700–1600 m. BR, Owy. ID, NV. Native.

Cymopterus glomeratus (Nuttall) de Candolle

Plains spring parsley

Cymopterus acaulis (Pursh) Raf. var. *acaulis*, *Cymopterus acaulis* (Pursh) Raf. var. *greeleyorum* J.W. Grimes & P.L. Packard

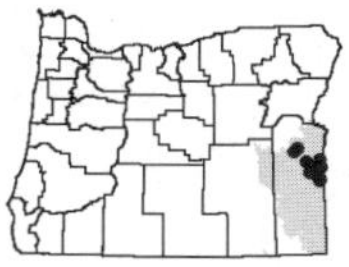

Plants 4–12 cm, glabrous, not glaucous; with 1–2 pseudoscapes arising from crown of an elongate and tapering taproot. **Leaves** basal, sometimes appearing low cauline, 2–3-pinnately or ternate-pinnately divided, ovate to oblong-ovate in outline, 2–8 cm; leaf segments oblong, 1–3(4) mm wide, usually > 1.3 × as long as wide, usually distinct. **Inflorescences** peduncles 1–5, 2–8 cm; involucral bracts absent or minute; rays 4–11, 0–0.7 cm, unequal; involucel bractlets 5–10, not showy, oblong to ovate, green, connate at base. **Flowers** sepals absent or minute; petals white; carpophores absent. **Fruits** 5–10 mm; wings wavy to wrinkled, 0.5–2 mm wide, lateral wing width > dorsal.

Dry sandy, gravelly, or clay soil. Flowering Mar–May. 700–1300 m. Owy. ID; northeast to Manitoba, east to MN, southeast to OK. Native.

Cymopterus longipes S. Watson var. *ibapensis* (M.E. Jones) Cronquist

Sprawling spring parsley

Cymopterus watsonii (J.M. Coult. & Rose) M.E. Jones

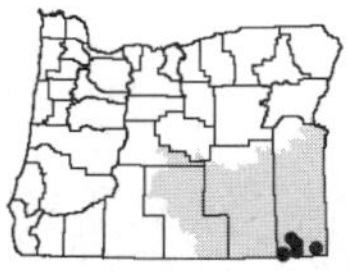

Plants 3–15(25) cm, glabrous to papillate-scabrous, usually glaucous, rarely not; with 1–2(3) pseudoscapes arising from crown of an elongate and thick taproot. **Leaves** basal or appearing low cauline, 2–3-pinnately or ternate-pinnately divided, ovate to oblong-ovate in outline, 2–8 cm; leaf segments oblong to ovate, 0.5–1(2) mm wide, usually > 1.3 × as long as wide, usually distinct. **Inflorescences** peduncles 3–12(15), 2–13(15) cm; involucral bracts absent or minute, green; rays 4–14, 0–4 cm, subequal to unequal; involucel bractlets 5–8, not showy, linear-oblong to oblong, green, connate at base. **Flowers** sepals absent or minute; petals white; carpophores well developed. **Fruits** 5–8 mm; wings wavy, 0.5–2 mm wide, lateral wing width ≥ dorsal.

Dry sandy, gravelly, rocky, or clay soil. Flowering May–Jun. 1400–1800 m. BR, Owy. ID, NV; southeast to UT. Native.

Some authorities treat this taxon as a full species.

Cymopterus nivalis S. Watson

Elko spring parsley

Cymopterus bipinnatus S. Watson

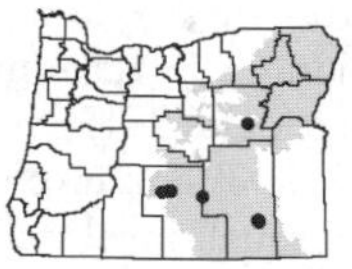

Plants 3–20 cm, glabrous to granular-scabrous; with branched caudices from a thick taproot, Basal leaf sheaths from previous years persistent on caudices. **Leaves** basal, 2–3-pinnately divided, ovate to oblong-ovate in outline; blades 1–5 cm; leaf segments linear-oblong to oblong, 0.3–1 mm wide, > 1.3 × as long as wide, distinct or indistinct; petioles 1–5 cm. **Inflorescences** peduncles 1–18, 3–20 cm; involucral bracts absent; rays 2–8, 0.1–0.4 cm, subequal to unequal; involucel bractlets 5–8, not showy, linear to linear-oblong, green, connate at base, rarely free. **Flowers** sepals not well developed, < 0.5 mm; petals white; carpophores well developed. **Fruits** (3)4–6 mm; wings slightly wavy or not, 0.5–2 mm wide, lateral wing width ≈ dorsal.

Rocky or gravelly areas, dry meadows. Flowering Jun–Aug. 1200–2900 m. BR, BW. ID, NV; northeast to MT, east to WY. Native.

Cymopterus petraeus M.E. Jones

Rock spring parsley

Cymopterus terebinthinus (Hook.) Torr. & A. Gray var. *petraeus* (M.E. Jones) S. Goodrich, *Pteryxia petraea* (M.E. Jones) J.M. Coult. & Rose

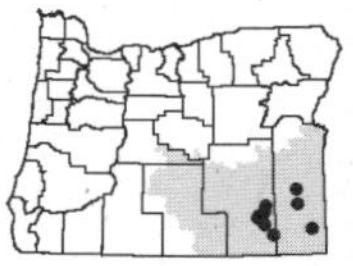

Plants 15–35 cm, glabrous; with branched caudices from a thick taproot, Basal leaf sheaths from previous years persistent on caudices. **Leaves** basal, low cauline, or cauline, 2–3-pinnately or ternate-pinnately divided, narrowly oblong to oblong in outline; blades 3–55 cm; leaf segments linear to linear-oblong, 0.2–1 mm wide, > 1.3 × as long as wide, distinct; petioles 2–12 cm. **Inflorescences** peduncles 1–15, 5–25 cm; involucral bracts absent; rays 2–8, 0.5–6 cm, very unequal; involucel bractlets 5–8, not showy, linear to linear-oblong, green, free or somewhat connate at base. **Flowers** sepals well developed, 0.5–1 mm; petals yellow; carpophores well developed. **Fruits** 4–7 mm; wings slightly wavy or not, 0.5–2 mm wide, lateral wing width ≥ dorsal.

Rocky areas and crevices. Flowering Apr–Jul. 100–2800 m. BR, Owy. CA, ID, NV; southeast to NM. Native.

Cymopterus purpurascens (A. Gray) M.E. Jones

Widewing spring parsley

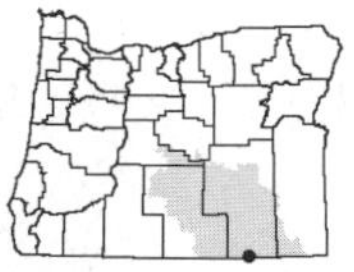

Plants 5–17 cm, glabrous; arising from a fleshy, thick taproot. **Leaves** basal, 2–3-pinnately or ternately dissected, oblong-ovate in outline; blades 1–5 cm; leaf segments, 2–5 × 2–8 mm, indistinct; petioles 1–4 cm. **Inflorescences** peduncles 1–4(5), 1.5–7 cm; involucral bracts showy, white, connate at base; rays 3–5, 0.4–1 cm, unequal to subequal; involucel bractlets showy, white, connate at base. **Flowers** sepals absent or minute; petals purple; carpophores absent. **Fruits** 7–15 mm; wings slightly wavy or not, 2–4 mm wide, lateral wing width ≥ dorsal.

Sagebrush areas. Flowering Apr–Jun. 1300–1600 m. BR. CA, ID, NV; southeast to NM. Native.

Cymopterus terebinthinus (Hooker) Torrey & A. Gray

Aromatic spring parsley, turpentine wavewing

Plants 10–60 cm, glabrous; branching caudices arising from a thick taproot. **Leaves** basal and low cauline, 2–3-pinnately or ternate-pinnately dissected, widely ovate in outline, 5–25 cm; blades 3–15 cm; leaf segments linear to oblong, 0.5–1.5 mm wide, usually > 1.3 × as long as wide, distinct; petioles 2–15 cm. **Inflorescences** 1–7 cm wide; peduncles 1–8, 8–30 cm; involucral bracts absent; rays 8–22, 0.5–7 cm, unequal; involucel bractlets 5–10, not showy, linear to linear-oblong, green, free at base. **Flowers** sepals well developed, 0.5–1.5 mm; petals yellow; carpophores well developed. **Fruits** 5–11 mm; wings wavy or not, 0.5–2 mm wide, lateral wing width ≥ dorsal.

Western North America. 4 varieties; 2 treated in *Flora*.

Key to *Cymopterus terebinthinus* varieties:

1. Inflorescences 2–7 cm wide; fruit wings wavy . . . . *C. t.* var. ***terebinthinus***

1' Inflorescences 1–3 cm wide; fruit wings not or slightly wavy . . . . *C. t.* var. ***foeniculaceus***

Cymopterus terebinthinus (Hooker) Torrey & A. Gray var. ***foeniculaceus*** (p. 98)

Fennel spring parsley, fennel wavewing

Pteryxia terebinthina (Hook.) J.M. Coult. & Rose var. *foeniculacea* (Nutt. ex Torr. & A. Gray) Mathias

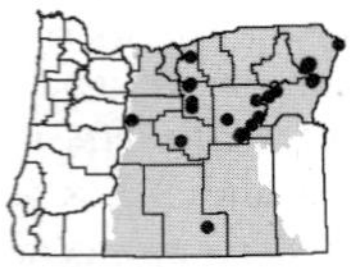

Leaves leaf segments linear to linear-oblong. **Inflorescences** 1–3 cm wide. **Fruits** wings not or slightly wavy.

Dry, rocky or sandy areas, sagebrush. Flowering Apr–Aug. 400–2900 m. BR, BW, Col, ECas, Lava. ID, WA; northeast to MT, southeast to UT. Native.

Cymopterus terebinthinus (Hooker) Torrey & A. Gray var. ***terebinthinus***

Aromatic spring parsley, turpentine wavewing

Pteryxia terebinthina (Hook.) J.M. Coult. & Rose var. *terebinthina*

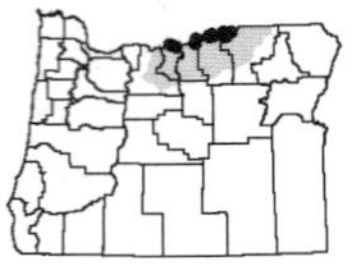

Leaves leaf segments linear-oblong to oblong. **Inflorescences** 2–7 cm wide. **Fruits** wings wavy.

Rocky or sandy areas. Flowering Apr–Jun. 50–700 m. Col. WA. Native.

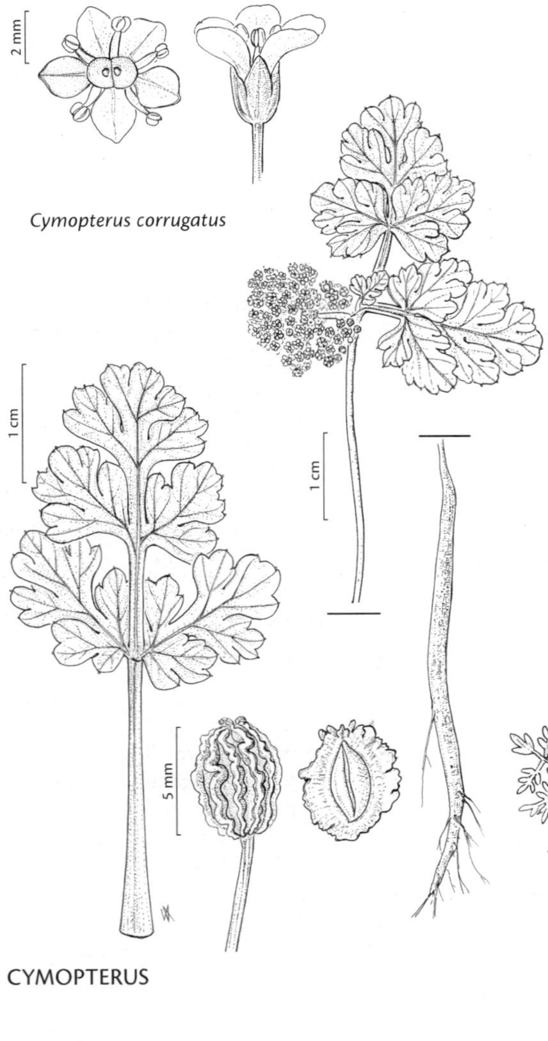

Cymopterus corrugatus

Cymopterus terebinthinus var. *foeniculaceus*

CYMOPTERUS

DAUCUS Wild carrot

David M. Spooner and Philipp W. Simon

Herbs annual or biennial, 0.5–15 dm. **Roots** tough whitish to tan taproots with a distinct odor of carrot. **Stems** and peduncles glabrous to pubescent to scabrous. **Leaves** basal and cauline, ternately dissected; blades triangular in outline, membranous, lobes linear to ovate to elliptic, margins entire to lobed. **Inflorescences** compound umbels, flat-topped to globose, peripheral flowers asymmetric and slightly larger; involucral bracts firm, 1–2-pinnate, linear to subulate; involucel bractlets mostly simple and subulate; umbels in fruit flat-topped or changing shape, with outer rays bending upward and inward, forming a "bird's nest" that temporarily holds the seeds; pedicels present. **Flowers** perfect or androgynous, with primary inflorescence often gynomonoecious; sepals absent; petals white or rarely pinkish, central flower or group of flowers white to pink to purple; carpophores well developed. **Fruits** schizocarps, ovoid, somewhat dorsally compressed, lateral ribs and 2 ridges on each face winged and bearing hooked, barbed, or simple prickles, surfaces tuberculate, sometimes with fine bristles; oil ducts under secondary ribs, 2 on commissural side.

Cosmopolitan. 26 species; 2 species treated in *Flora.*

Molecular studies suggest the need to reassess generic boundaries within the Apiaceae; molecular data place *Daucus* in a clade containing some species of the genera *Agrocharis*, *Athamanta*, *Cryptotaenia*, *Melanoselinum*, *Monizia*, *Pachyctenium*, *Pseudorlaya*, and *Tornabenea* (Spalik & Downie 2007). A major focus of future research on *Daucus* must resolve outgroup relationships, but here we treat the genus as traditionally circumscribed.

Key to *Daucus* species:

1. Plants slender annuals; 0.5–6(9) dm; involucral bract segments linear to lanceolate, frequently exceeding umbel; complete compound umbel ~1–6 cm wide, with 3–15 (primary) rays; central flowers white . . . . ***D. pusillus***

1' Plants coarse biennials, sometimes annuals; 2–15 dm; involucral bract segments subulate to linear, not exceeding umbel; complete compound umbel ~5–12 cm wide, with 7–30+ (primary) rays; central flower(s) white or rose to purple . . . . ***D. carota***

Daucus carota Linnaeus (p. 99)

Queen Anne's lace, wild carrot

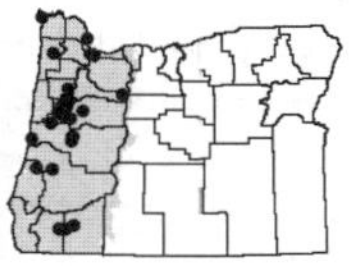

Plants biennial, sometimes annual, 2–15 dm, coarse. **Inflorescences** flat-topped to highly concave, ~5–12 cm wide, with 7–30+ (primary) rays; central flower(s) white or rose to purple; involucral bract segments subulate to linear, not exceeding umbel. **2*n***=22.

Roadsides and disturbed sites in open moist places. Flowering May–Sep. 0–1400 m. Casc, CR, Est, Sisk, WV. CA, ID, NV, WA; worldwide. Exotic.

Daucus carota and *D. pusillus* are clearly distinct species differing by chromosome number, habitat, phenology, and morphology, and are easily distinguished in the field. Our key and description are for typical plants, but small and late-flowering plants of *D. carota* or atypical morphologies of either species sometimes make it difficult to distinguish these two species on herbarium sheets.

Daucus carota is especially variable, and different taxonomists have treated it as a complex containing seven species to a single species with two to ten subspecies (Pujadas Savalà 2002; Small 1978). Our ongoing field and common garden studies document tremendous variation in *D. carota* for plant height, leaf dissection, peduncle length, flowering time, corolla color, inflorescence bract morphology, and inflorescence shape, and such characters often vary within populations. *Daucus carota* is a common and aggressive weed in many places in the eastern half of the United States but is rarer in the west in areas of lower rainfall or warmer climate.

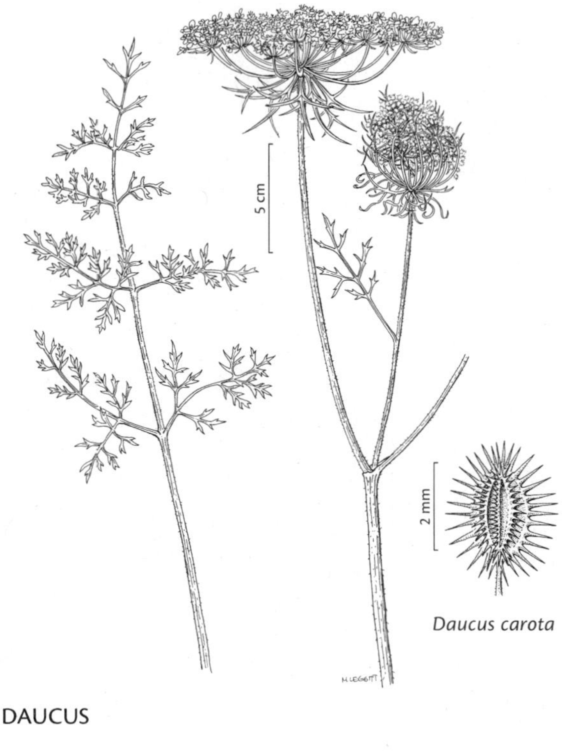

Daucus carota

DAUCUS

Daucus pusillus Michaux

Rattlesnake weed, American wild carrot

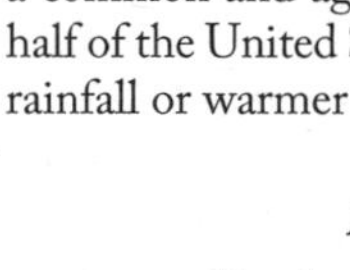

Plants annual, 0.5–6(9) dm, slender. **Inflorescences** flat-topped to slightly convex, ~1–6 cm wide, of 3–15 (primary) rays; central flowers white; involucral bract segments linear to lanceolate, frequently exceeding umbel. **2*n***=18.

Dry open prairies and grasslands. Flowering May–Jul. 0–900 m. Casc, CR, Est, Sisk, WV. CA, ID, WA; southern and western North America; Pacific Islands, South America. Native.

ERYNGIUM Eryngo, sea holly, button-celery

Herbs perennial. **Roots** fibrous. **Stems** branched above, just above base or basally, prostrate to erect, glabrous. **Leaves** basal and cauline, simple to pinnately- or palmately-lobed, margins entire to toothed, teeth usually spiny; sheaths short, not inflated; petioles present or not. **Inflorescences** terminal or axillary heads, usually in cymes or sometimes panicles; involucral bracts 1–several, entire or sharp-toothed, spine-tipped; rays 0; involucel bractlets 1–several, spine-tipped; pedicels absent. **Flowers** sepals persistent, ovate to oblong, spine-tipped; petals oblanceolate to ovate, white to blue or purple; anthers and styles green or sometimes blue. **Fruits** globose to obovate to elliptic, typically slightly laterally flattened, scaly; ribs 0; wings 0.

Temperate and subtropical regions in Asia, Australia, Europe, North America, South America. ~230 species; 3 species treated in *Flora*.

Key to *Eryngium* species:

1. Petals bright blue or purple; flowering heads blue to bluish purple; stems usually branched above . . . . ***E. articulatum***

1' Petals white; flowering heads greenish; stems branched just above base or basally.

  2. Basal leaves linear to linear-lanceolate; plants branched basally; plants mostly east of the Cascade Crest . . ***E. petiolatum***

  2' Basal leaves lanceolate to obovate; plants mostly branched just above base but sometimes branching basally; plants mostly west of the Cascade Crest . . . . ***E. alismifolium***

Eryngium alismifolium Greene

Modoc eryngo, plantain-leaved eryngo

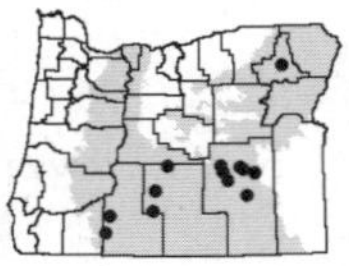

Stems 5–30 cm, decumbent to erect, branched mostly just above base but sometimes branching basally. **Leaves** mostly basal, 5–30 cm, upper leaves much reduced; blades lanceolate to narrowly obovate, 2–10 × 0.5–3.5 cm, margins spiny-toothed; petioles 3–30 cm, flattened, margins usually spiny-toothed. **Inflorescences** heads spherical, 5–12 mm wide, greenish; involucral bracts lanceolate to oblanceolate, 4–16 mm; involucel bractlets linear to oblanceolate, 2–5 mm, sometimes slightly bluish-tinted. **Flowers** sepals 3–5 mm, whitish or white-margined, spine-tipped; petals 1–4 mm, white, with brown midveins, tips notched. **Fruits** ovoid, 1.5–2.5 mm; covered with lanceolate, acuminate scales.

Vernal pools, pond and lake margins, seasonally wet meadows, ditches. Flowering Jun–Sep. 1200–1800 m. BR, BW, Casc, ECas. CA, ID, NV. Native.

While similar to *E. petiolatum*, that species is usually found only west of the Cascade Mountains, while *E. alismifolium* is found in eastern parts of the state.

Eryngium articulatum Hooker

Bee-thistle eryngo, beethistle, blue-topped eryngo

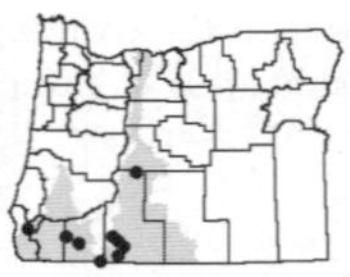

Stems 3–11 cm, erect, usually branched above, rarely just above base or basally. **Leaves** basal and caulescent, 1.5–35 cm, reduced higher on plant; blades lanceolate to narrowly obovate, 2–10 × 0.5–3.5 cm, margins sparsely toothed to spiny-toothed; petioles 0–30 cm, flattened, margins entire. **Inflorescences** heads spherical to ovoid, 10–20(25) mm wide, blue to bluish purple; involucral bracts lanceolate to oblanceolate, 5–25 mm; involucel bractlets linear to oblanceolate, 2–5 mm, bluish-tinged. **Flowers** sepals 2–4 mm, blue, acuminate; petals 1–4 mm, blue to bright blue, rarely purple. **Fruits** narrowly elliptic to ovoid, 2–3.5 mm; covered with lanceolate, acuminate scales. **2*n***=32.

Seasonally wet areas near streams, lakes, and ponds. Flowering Jun–Sep. 50–1600 m. ECas, Sisk. CA, ID, WA. Native.

The blue heads of this species easily distinguish it among our *Eryngium* taxa.

Eryngium petiolatum Hooker (p. 100)

Oregon eryngo, rush-leaf eryngo, coyote thistle

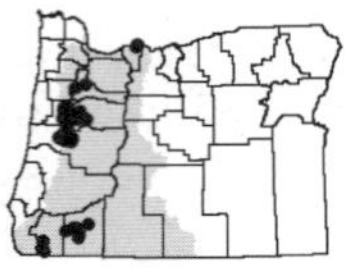

Stems 10–60(80) cm, erect, branched basally. **Leaves** basal leaves linear to narrowly elliptic, flat to round, 5–30 cm, margins entire to sparsely spiny-toothed; petioles 2–25 cm; cauline leaf blades lanceolate to elliptic, 2–7 × 0.5–1.5 cm, margins spiny-toothed; petioles 0–25 cm. **Inflorescences** heads spherical, 7–20 mm wide, greenish; involucral bracts lanceolate to oblanceolate, 5–25 mm; involucel bractlets linear to oblanceolate, 1–5 mm, sometimes slightly bluish-tinged. **Flowers** sepals 1–4 mm, acuminate; petals 1–3 mm, white. **Fruits** ovoid, 1–2 mm; covered with tapering scales. **2*n***=32.

Vernal pools, stream sides, wet meadows. Flowering Apr–Aug. 50–700 m. Casc, ECas, Sisk, WV. WA. Native.

See *E. alismifolium*.

ERYNGIUM

FOENICULUM Fennel

Herbs biennial or short-lived perennial, with a strong anise scent, glabrous, glaucous. **Roots** thick taproots. **Stems** erect, branched, leafy, solid, often ridged. **Leaves** cauline, alternate, pinnately dissected; leaf segments elongate, thread-like; sheaths somewhat to conspicuously inflated; petiolate. **Inflorescences** terminal and axillary compound umbels; involucral bracts 0; rays 15–40; involucel bractlets 0. **Flowers** sepals absent; petals yellow. **Fruits** elliptic-oblong, slightly flattened, glabrous; ribs prominent, equal in width; wings absent.

Worldwide. 1–4 species; 1 species treated in *Flora*.

Foeniculum vulgare Miller (p. 101)

Fennel

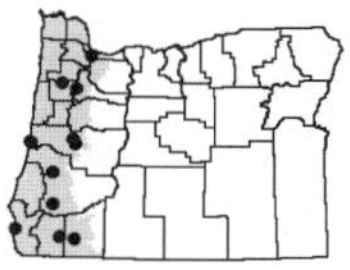

Plants 1–20 dm. **Stems** branched above. **Leaves** 1–5 × 2–4 dm; blades triangular-ovate in outline; leaf segments 4–40 mm; petioles 5–12 cm. **Inflorescences** rays 1–7 cm, unequal in length; pedicels 1–7 mm. **Fruits** 3–4 mm. **2*n***=22.

Disturbed areas, roadsides. Flowering Jun–Nov. 0–500 m. CR, Est, Sisk, WV. CA, NV, WA; scattered throughout North America; nearly worldwide. Exotic.

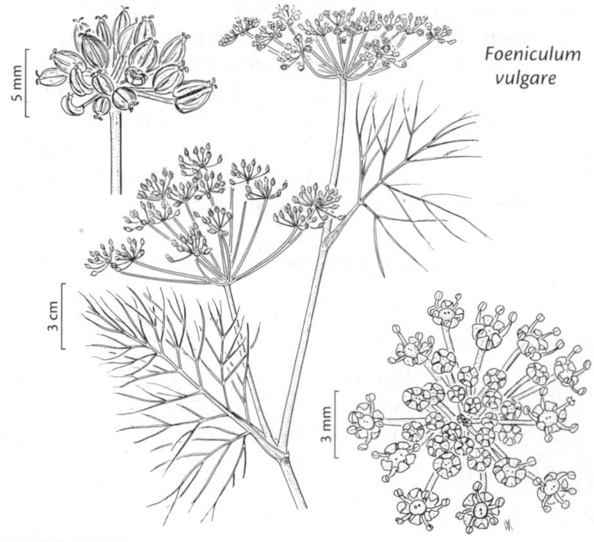

FOENICULUM

GLEHNIA Silvertop

Herbs perennial, low growing, often prostrate. **Roots** elongated, fleshy taproot. **Stems** 0 or short and buried in sand. **Leaves** basal, ternate or ternate and pinnate; leaflets broad, thick and fleshy, deeply lobed and toothed; sheaths somewhat to much inflated and usually buried in sand; petiolate. **Inflorescences** dense terminal compound umbels; involucral bracts 0–5; rays 5–15, unequal in length; involucel bractlets 3–7; umbellets 5–25-flowered. **Flowers** sepals minute; petals obovate, white. **Fruits** round to oblong, flattened; ribs equal in length, with corky wings.

Coastal north Pacific. 2 species; 1 species treated in *Flora*.

Glehnia leiocarpa Mathias (p. 101)

Beach silvertop

Glehnia littoralis (A. Gray) F. Schmidt ex Miq. ssp. *leiocarpa* (Mathias) Hultén

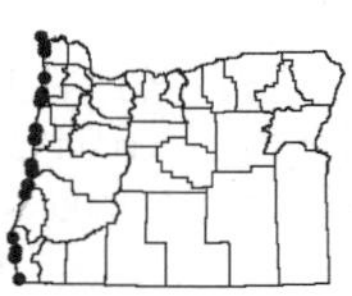

Leaves broadly elliptic to ovate in outline, 4–15 × 3–12 cm, glabrous above, tomentose below; leaflets broadly ovate in outline, 2–7 × 1–5 cm; petioles 2–13 cm. **Inflorescences** peduncles 2–9 cm, tomentose; involucral bracts (if present) linear to lanceolate, 3–10 mm; rays 1–4.5 cm, tomentose; involucel bractlets lanceolate, 3–8 mm; pedicels 1–3 mm. **Fruits** 5–13 mm, glabrous or with few hairs. **2*n***=12.

Coastal dunes and beaches. Flowering May–Aug. 0–50 m. Est. CA, WA; north to AK. Native.

Considered in some floras as a subspecies of *Glehnia littoralis*, *G. leiocarpa* differs from that species, which occurs on the Asian Pacific coast and has consistently hairy fruits.

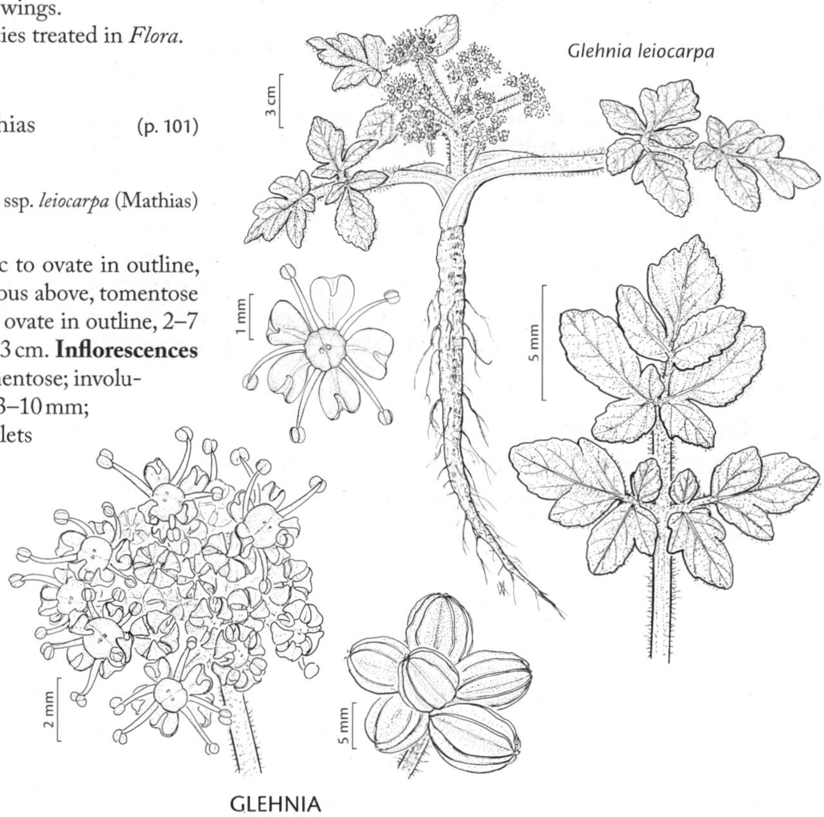

GLEHNIA

HERACLEUM Cow parsnip, hogweed

Herbs perennial, glabrous to tomentose. **Roots** fibrous or from a taproot. **Stems** erect, stout, hollow. **Leaves** mostly cauline, sometimes basal, alternate, ternately or pinnately compound; leaflets broad, toothed and deeply lobed; sheaths inflated; petiolate. **Inflorescences** terminal and axillary compound umbels; involucral bracts 5–10; rays 10–30, unequal in length; involucel bractlets 5–10; umbellets many-flowered. **Flowers** sepals absent or minute; petals obovate, white, those of outer flowers larger and usually bifid. **Fruits** elliptic to oval or obcordate, slightly flattened; ribs unequal in length; wings broad.

Circumboreal. ~80 species; 1 species treated in *Flora*.

Heracleum mantegazzianum, giant hogweed, has been reported as escaped in Portland and possibly Eugene. No persistent populations are currently known, however. If seen again, this species can be recognized by its pinnately compound leaves (versus the ternately compound leaves of *H. maximum*).

Heracleum maximum W. Bartram (p. 102)

Common cow parsnip

Heracleum lanatum Michx.

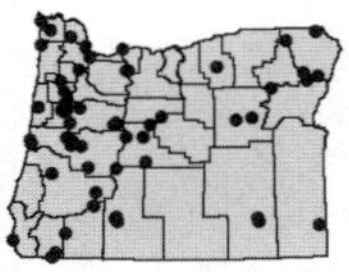

Plants 1–3 m, strongly scented; stems grooved. **Leaves** round to reniform in outline, 20–50 × 20–50 cm; leaflets broadly ovate in outline, 10–35 × 10–35 cm, lateral leaflets narrower than central; petioles 10–40 cm. **Inflorescences** peduncles 5–20 cm; involucral bracts linear, 0.5–2 cm; rays 5–10 cm; involucel bractlets linear, 0.3–1.8 cm; pedicels 1–2 cm. **Fruits** 7–12 mm, glabrous to slightly pubescent. **2*n***=12.

Moist areas. Flowering Apr–Aug. 0–2600 m. All ecoregions. CA, ID, NV, WA; throughout northern North America. Native.

The sap of *Heracleum maximum* and some other species contains photosensitive chemicals that can cause a painful rash after contact with the skin and exposure to ultraviolet light.

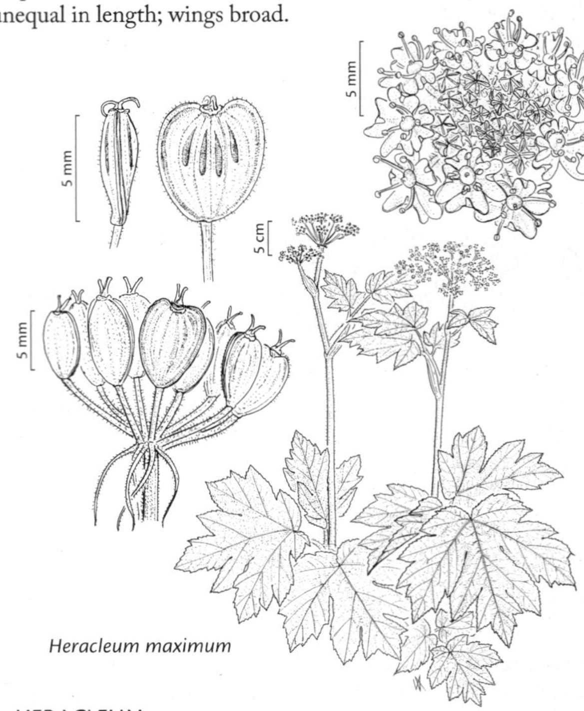

Heracleum maximum

HERACLEUM

LIGUSTICUM Licorice-root, lovage

Herbs perennial, glabrous to minutely scabrous or slightly pubescent. **Roots** moderately thick taproots. **Stems** branched, erect, leafy or not. **Leaves** basal only or basal and cauline, usually ternate-pinnate, with or without well-defined leaflets, margins entire to lobed or finely toothed; leaflets linear to oblong; sheaths somewhat inflated; petioles present or not. **Inflorescences** terminal and axillary compound umbels; involucral bracts 0–4, linear; rays 5–30, unequal; involucel bractlets 0 or few and linear; pedicels unequal. **Flowers** sepals absent or minute; petals obovate, white, rarely pinkish. **Fruits** ovate to elliptic-oblong, slightly flattened, glabrous; ribs thread-like to prominent, ≈ width; narrow wings usually present (but not in *L. apiifolium*).

Temperate regions in the Northern Hemisphere. ~20 species; 6 species treated in *Flora*.

Our *Ligusticum* species are often challenging to differentiate. The combination of both overlapping ranges and morphological characters may make identification to species questionable. While here we include all species of *Ligusticum* traditionally accepted to be present in Oregon, further taxonomic study elucidating more biologically meaningful species boundaries is highly encouraged.

Key to *Ligusticum* species:

1\. Cauline leaves 2–3, well developed.

  2\. Ribs of fruit without wings; plants generally in or west of the Cascade Mountains . . . . ***L. apiifolium***

  2' Ribs of fruit with narrow wings; plants of southwestern Oregon . . . . ***L. californicum***

1' Cauline leaves 0–3, reduced (rarely 1 nearly well developed).

  3\. Plants (5)6–12 dm; rays (9)10–30.

    4\. Most ultimate leaf segments ≥ 3 mm wide . . . . ***L. canbyi***

    4' Most ultimate leaf segments < 3 mm wide . . . . ***L. filicinum***

  3' Plants 1–6 dm; rays 5–15(18).

    5\. Most ultimate leaf segments ≥ 3 mm wide; often with 1–2 reduced cauline leaves; fruits 4–7 mm. . . . . ***L. grayi***

    5' Most ultimate leaf segments < 3 mm wide; cauline leaves 1 and reduced or absent; fruits 3–5 mm . . . ***L. tenuifolium***

Ligusticum apiifolium (Nuttall) A. Gray (p. 103)

Celery-leaved licorice-root, parsley-leaved lovage

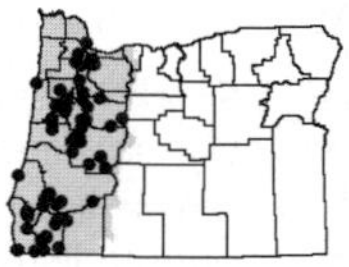

Plants 3–15 dm, glabrous to slightly pubescent. **Stems** sometimes branched. **Leaves** basal and cauline, ternate-pinnate; blades 5–25 cm; leaflets 1.5–5 cm, margins deeply pinnately lobed; cauline leaves 2–3, well developed, ≈ basal near base, but progressively smaller higher; petioles 0–30 cm. **Inflorescences** peduncles 10–35 cm; rays 12–25, 2–7 cm; pedicels 5–10 mm. **Flowers** sepals absent or minute. **Fruits** ovate to oblong, 3–5 mm; ribs thread-like; wings not present. ***2n***=44.

Meadows, woodlands. Flowering May–Sep. 0–1800 m. Casc, CR, Est, Sisk, WV. CA, WA. Native.

Among our *Ligusticum* species, the combination of well-developed cauline leaves and fruits without wings makes this species one of the most easily identifiable. *Ligusticum californicum* is very similar but has fruits with narrow wings.

Ligusticum californicum J.M. Coulter & Rose

California licorice-root, California lovage

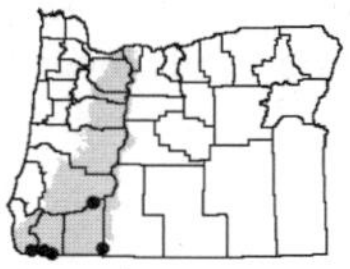

Plants 5–12 dm, glabrous. **Stems** sometimes branched. **Leaves** basal and cauline, ternate-pinnate; blades 5–20 cm; leaflets 1.5–4 cm, margins shallowly to deeply pinnately lobed; cauline leaves 2–3, well developed, ≈ basal near base, but progressively smaller higher; petioles 0–35 cm. **Inflorescences** peduncles 10–40 cm; rays 15–30, 2–7 cm; pedicels 5–10 mm. **Flowers** sepals absent or minute. **Fruits** ovate to oblong, 3–6 mm; ribs low but conspicuous; wings narrow. ***2n***=22.

Chaparral, woodlands. Flowering May–Aug. 50–1600 m. Casc, Sisk. CA. Native.

See also *L. apiifolium*.

Ligusticum canbyi J.M. Coulter & Rose

Canby's licorice-root

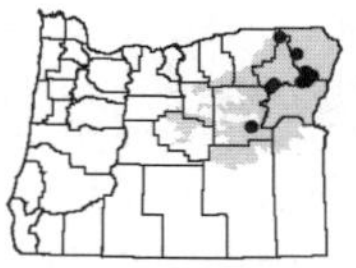

Plants (5)6–12 dm, glabrous to slightly scabrous. **Stems** usually branched above. **Leaves** basal and cauline, ternate-pinnate; blades 1–16 cm; leaflets 0.5–5 cm, most ultimate leaf segments 3 mm wide or greater, margins shallowly to deeply pinnately lobed; cauline leaves 0–3, reduced, rarely 1 nearly well developed; petioles 0–35 cm. **Inflorescences** peduncles 4–40 cm; rays (9)10–30, 2–5 cm; pedicels 2–10 mm. **Flowers** sepals minute. **Fruits** ovate to elliptic-oblong, 4–6 mm; ribs low but conspicuous; wings narrow.

Meadows, open slopes, woods. Flowering Jun–Aug. 700–2400 m. BW. ID, WA; north to British Columbia, northeast to MT, east to WY. Native.

Ligusticum filicinum S. Watson

Fern-leaved licorice-root, fern-leaf lovage

Ligusticum filicinum S. Watson var. *filicinum*

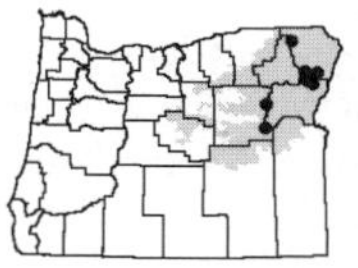

Plants (5)6–12 dm, glabrous to slightly scabrous. **Stems** usually branched above. **Leaves** basal and cauline, ternate-pinnate; blades 5–20 cm; leaflets 0.3–5 cm, most ultimate leaf segments < 3 mm wide, margins shallowly to deeply pinnately lobed; cauline leaves 0–3, reduced; petioles 0–35 cm. **Inflorescences** peduncles 5–40 cm; rays 10–25, 2–6 cm; pedicels 2–10 mm. **Flowers** sepals minute. **Fruits** ovate to elliptic-oblong, 5–7 mm; ribs low but conspicuous; wings narrow.

Meadows, open slopes, woods. Flowering Jun–Aug. 600–2500 m. BW. ID; northeast to MT, southeast to CO. Native.

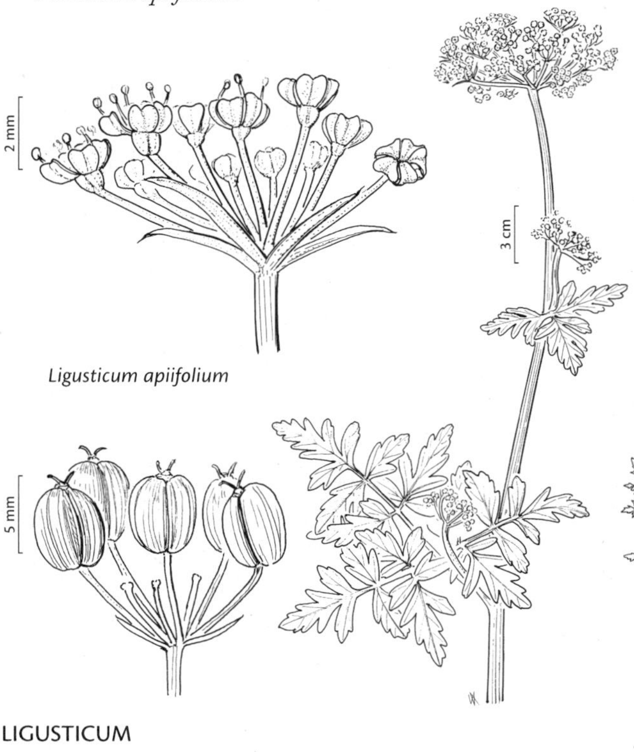

Ligusticum apiifolium

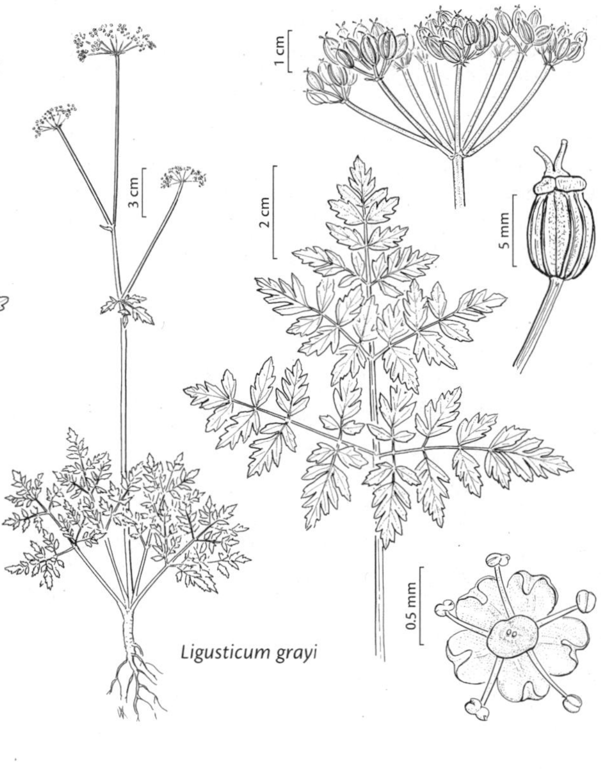

Ligusticum grayi

LIGUSTICUM

Ligusticum grayi J.M. Coulter & Rose (p. 103)

Gray's licorice-root, Gray's lovage

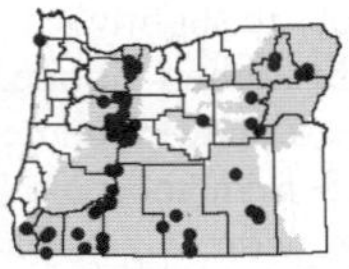

Plants 2–6 dm, glabrous. **Stems** often branched above. **Leaves** basal and cauline, ternate-pinnate; blades 5–25 cm; leaflets 1–3 cm, most ultimate leaf segments ≥ 3 mm wide, margins shallowly to deeply pinnately lobed; petioles 1–25 cm. **Inflorescences** peduncles 5–30 cm; rays 5–15(18), 1–5 cm; pedicels 2–10 mm. **Flowers** sepals minute. **Fruits** oblong to elliptic-oblong, 4–7 mm; ribs low but conspicuous; wings narrow. **2*n***=22, 44.

Meadows, open slopes, woods. Flowering May–Sep. 400–2800 m. BR, BW, Casc, ECas, Est, Sisk. CA, ID, NV, WA; northeast to MT, southeast to UT. Native.

Ligusticum tenuifolium S. Watson

Slender licorice-root, slender-leaf lovage

Ligusticum filicinum S. Watson var. *tenuifolium* (S. Watson) Mathias & Constance

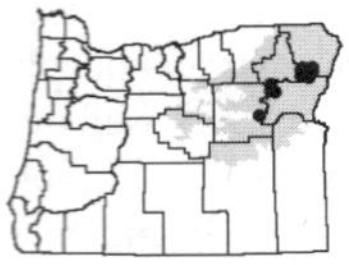

Plants 1–6 dm, glabrous to slightly scabrous. **Stems** usually not branched. **Leaves** basal only or basal and 1 reduced cauline leaf, ternate-pinnate; blades 2–16 cm; leaflets 0.5–3 cm, most ultimate leaf segments < 3 mm wide, margins deeply pinnately lobed, rarely shallowly; petioles 0.5–20 cm. **Inflorescences** peduncles 5–40 cm; rays 5–15, 1.5–4 cm; pedicels 2–10 mm. **Flowers** sepals minute. **Fruits** ovate to elliptic-oblong, 3–5 mm; ribs low but conspicuous; wings narrow.

Meadows, open slopes, woods. Flowering Jul–Sep. 1500–2800 m. BW. ID; northeast to MT, southeast to CO. Native.

LILAEOPSIS Grasswort

Herbs perennial, glabrous. **Roots** rhizomes with fibrous roots, rooting at nodes. **Stems** prostrate and creeping. **Leaves** 1–several, basal or tufted along rhizomes, cylindric or flattened, hollow, margins entire; sheaths not or slightly inflated; petioles absent. **Inflorescences** axillary simple umbels; involucral bracts several, inconspicuous. **Flowers** sepals minute; petals white, tips acute. **Fruits** ovate to obovate, slightly flattened, glabrous; ribs present; winged.

Australia, Africa, North America, South America. 13 species; 1 species treated in *Flora*.

Lilaeopsis occidentalis J.M. Coulter & Rose (p. 104)

Western grasswort, western lilaeopsis

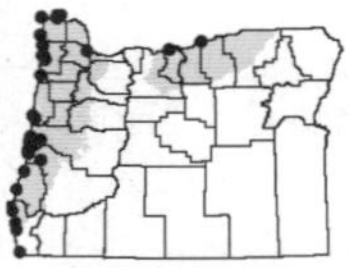

Rhizomes tan to yellow. **Leaves** linear to narrowly lanceolate, 2–18 cm × 1–5 mm, with 1–12 distinct partitions. **Inflorescences** umbels 3–12-flowered; peduncles 0.5–6 cm; pedicels 1–6 mm. **Fruits** 1.5–2.5 mm; lateral ribs wide, thickened. **2*n***=22.

Salty or brackish soils, occasionally on inland riverbanks. Flowering Jun–Aug. 0–100 m. Col, CR, Est, WV. CA, WA; north to AK. Native.

Although usually found in coastal habitats, this species is occasionally found in upstream rivers and creeks, sometimes at considerable distance away from the ocean. It is one of the two Apiaceous species, in Oregon, with simple and entire leaves (*Bupleurum americanum* being the other).

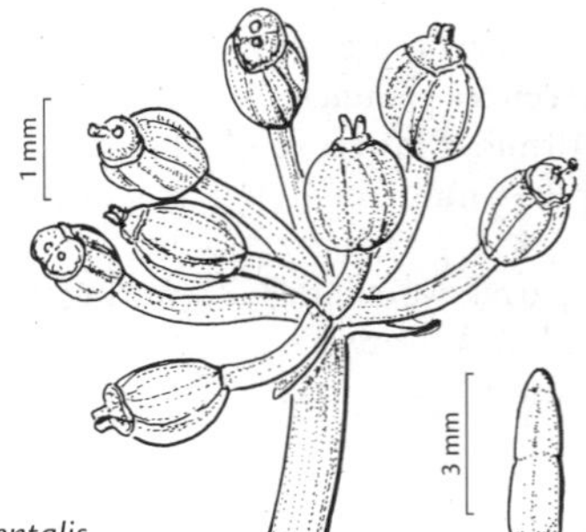

Lilaeopsis occidentalis

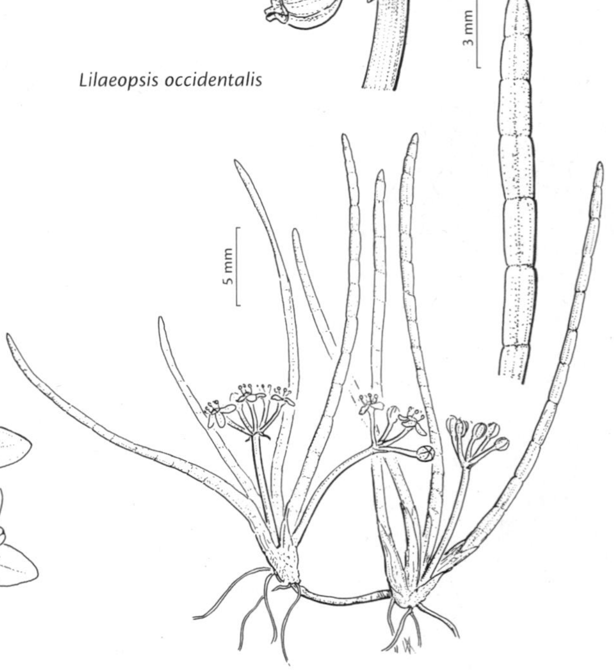

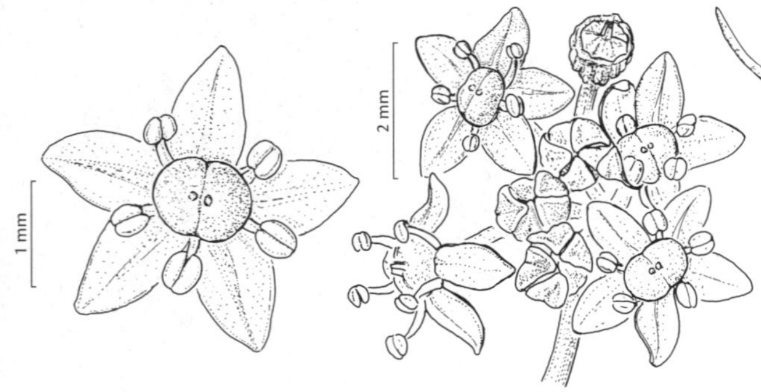

LILAEOPSIS

LOMATIUM Biscuitroot, desert parsley, lomatium

Jason A. Alexander

Herbs perennial, variously pubescent; from taproots or tubers. **Caudices** simple, few-branched, or multicipital. **Stems** 0–many, unbranched to few-branched. **Leaves** basal leaves divided at or above tip of sheath into 3–5 primary rachises, lateral primaries (in some species) equal in size and shape to secondary leaflets (and leaf appearing pinnate), each primary leaflet ternate or pinnate, with 1–9 secondary leaflet pairs along each rachis; cauline leaves 0–few. **Inflorescences** compound umbels; umbels expanded abruptly, forming attachment point of rays; flowers sometimes bisexual and staminate in same umbellet; involucral bracts present or absent, acute or acuminate; rays unequal, spreading, ascending or erect; involucel bractlets present or absent. **Flowers** calyx lobes 0 or obscure; petals 5, free, generally ovate or spatulate, yellow, ochroleucous, white or purple, tips narrowed; stamens 5; pistils 1. **Fruits** 1-seeded mericarps, dry, dehiscent from persistent, bifid carpophore; dorsal ribs thread-like, not winged, lateral ribs thin- or thick-winged.

Western North America. ~75 species; 44 species treated in *Flora.*

Leaf measurements are from fully expanded, mature leaves. Leaves from early in the season are not fully expanded and have much smaller measurements and variable shapes. Peduncle, ray, pedicel, and fruit measurements correspond to individuals in a mature fruiting stage. Fruit measurements should be made at the widest point of the mature fruit. Fruit body measurements are of the central section only, not including the lateral wings.

Key to *Lomatium* species:

1. Ultimate apical lobes of primary and secondary leaflets 1–8 cm wide.
  2. Peduncles inflated and hard at base of umbel; ultimate apical lobes 3(4–7)-toothed at tip . . . . ***L. nudicaule***
  2' Peduncles broad but not inflated and hard, texture of tip same as remainder of peduncle below, umbel expanded abruptly, forming attachment point of rays; ultimate apical lobes entire or with 4–many teeth.
    3. Leaf margins erose, dentate, or serrate, teeth pungent-tipped . . . . ***L. howellii*** (in part)
    3' Leaf margins entire or lobed, teeth, if present, cuspidate or mucronulate, not pungent-tipped.
      4. Body of mericarp 1.3–4 mm wide; pedicels 2–8 mm; ultimate apical lobes 10–120 × 0.4–12 mm . . . . ***L. triternatum*** (in part)
      4' Body of mericarp 4–6 mm wide; pedicels 7–12 mm; ultimate apical lobes 20–50 × 10–30 mm . . . ***L. californicum***

1' Ultimate apical lobes of primary and secondary leaflets < 1 cm wide.
  5. Plants cespitose, from stout taproot surmounted by matted, multicipital caudex; pedicels 0.5–2 mm.
    6. Basal leaves puberulent, hirsutulous, or villosulous, lateral primary leaflets much less than half length of central primary leaflet, ultimate apical lobes 2–4 × 0.6–0.8 mm . . . . ***L. oreganum***
    6' Basal leaves glabrous, glabrate, or sparsely scabrous, lateral primary leaflets at least half to subequal length of central primary leaflet, ultimate apical lobes 5–15 × 0.5–3 mm.
      7. Umbels 1–3-rayed; ultimate apical lobes 2–3 mm wide . . . . ***L. greenmanii***
      7' Umbels 5–12-rayed; ultimate apical lobes 0.5–2 mm wide . . . . ***L. cusickii*** (in part)
  5' Plants not cespitose, caudex simple, branched, or multicipital but not matted; pedicels 0 or up to 15 mm.
    8. Leaf margins erose, dentate, or serrate, teeth pungent-tipped . . . . ***L. howellii*** (in part)
    8' Leaf margins (not including tips) entire or lobed, teeth, if present, cuspidate or mucronulate, not pungent-tipped.
      9. Ultimate apical lobes 10–120 mm.
        10. Pedicels 0 or up to 4 mm.
          11. Ultimate apical lobes obtuse or rounded.
            12. Fruits oval or suborbicular, wings 0.5–1.5 mm wide; roots a globose tuber, elongate and slender portion above tuber, if present, < 2 cm; petals white; anthers purple . . . . ***L. piperi*** (in part)
            12' Fruits ovate, oblong, or narrowly elliptic, wings 0.15–0.6 mm wide; roots a slender and elongate taproot or a bifurcate tuber with an elongate upper section; petals yellow, ochroleucous, or white; anthers yellow or ochroleucous . . . . ***L. pastorale*** (in part)
          11' Ultimate apical lobes acuminate, acute, or subacute.
            13. Basal leaves with sparsely pubescent, scabrous, or scaberulose midnerves, blade surfaces variable, from mostly glabrous, glabrate, or sparsely scabrous above and below to sparsely pubescent throughout; fruit wings 0.7–3 mm wide . . . . . ***L. triternatum*** (in part)
            13' Basal leaves glabrous throughout; fruit wings 0.15–0.6 mm wide.
              14. Ultimate apical lobes 0.6–1.5 mm wide; fruits linear or narrowly oblong . . . . ***L. leptocarpum*** (in part)
              14' Ultimate apical lobes 1–9 mm wide; fruits ovate, oblong, or narrowly elliptic . . . . ***L. pastorale*** (in part)

10' Pedicels 4–20 mm.
 15. Ultimate apical lobes 0.1–1 mm wide.
  16. Fruits 18–24 mm, body 5.5–8 mm wide; petals purple or reddish purple; anthers purple . . . . **L. columbianum** (in part)
  16' Fruits 5–16 mm, body 1–6 mm wide; petals yellow or ochroleucous; anthers yellow or ochroleucous.
   17. Roots a globose tuber; peduncles 4–9 cm; involucel bractlets 0.3–0.8 mm wide . . . . ***L. farinosum*** var. ***hambleniae*** (in part)
   17' Roots a slender, elongate taproot, an irregularly thickened, elongate taproot, or a stout, irregularly thickened taproot; peduncles 7–60 cm; involucel bractlets 0 or 0.1–0.6 mm wide.
    18. Basal leaves biternate or 1–3-pinnate, primary leaflets ternate or pinnate, 0–6 secondary leaflet pairs along each rachis; ultimate apical lobes 10–120 × 0.4–12 mm . . . . ***L. triternatum*** (in part)
    18' Basal leaves 2–4 pinnate, primary leaflets pinnate with 5–8 secondary leaflet pairs along each rachis; ultimate apical lobes 0.7–15 × 0.1–0.7 mm.
     19. Ultimate lobes spread evenly from base to tip of rachilla, intercostal areas along rachilla more or less equal, leaflets appearing sessile, without a naked intercostal region between rachilla and basal-most pinnule lobes . . . . ***L. brunsfeldianum***
     19' Ultimate apical lobes diffuse at base, more numerous toward tip of leaflets, intercostal areas along rachilla longer toward base, leaflets appearing petiolulate, with a naked intercostal region between rachilla and basal-most pinnule lobes.
      20. Peduncles 6–8 mm wide (1 cm below tip); ultimate pinnule lobes (4.5)5–8 mm; ultimate apical lobes in cross section subterete, (6.5)7–15 mm. . . . . ***L. klickitatense***
      20' Peduncles 2–6 mm wide (1 cm below tip); ultimate pinnule lobes 1–5(6) mm; ultimate apical lobes in cross section flat or subterete, 1–7(9) mm . . . . ***L. papilioniferum*** (in part)
 15' Ultimate apical lobes 1–9 mm wide.
  21. Fruit wings 1–3 mm wide.
   22. Fruits 15–25 mm, body excluding wing 4–6 mm wide, pedicels 6–17 mm; peduncles 30–80+ cm; involucel bractlets 5–10 × 0.3–1 mm . . . . ***L. suksdorfii***
   22' Fruits 6–16 mm, body excluding wing 1.2–4 mm wide, pedicels 2–9 mm; peduncles 10–35 cm; involucel bractlets, if present, 0.7–6 × 0.1–0.4 mm.
    23. Peduncles and rays glabrate, scaberulose, or puberulent; ultimate apical lobes 10–120 mm; involucel bractlets 0–12, 1–6 mm . . . . ***L. triternatum*** (in part)
    23' Peduncles and rays glabrous; ultimate apical lobes 2–30 mm; involucel bractlets 0–2, 0.7–2 mm . . . . ***L. laevigatum*** (in part)
  21' Fruit wings 0.15–1 mm wide.
   24. Fruit bodies 4–6 mm wide; roots a stout, irregularly thickened taproot; caudices simple or multicipital; peduncles stout, 20–50 cm . . . . ***L. dissectum*** (in part)
   24' Fruit bodies 1–2.5 mm wide; roots a slender and elongate taproot or variously tuberous; caudices simple or 2–7-branched; peduncles slender, 2–20 cm (up to 33 cm in robust forms of *L. pastorale*).
    25. Basal leaves with sparsely pubescent, scabrous, or scaberulose midnerves, blade surfaces variable, from mostly glabrous, glabrate, or sparsely scabrous above and below to sparsely pubescent throughout . . . . ***L. triternatum*** (in part)
    25' Basal leaves glabrous throughout.
     26. Involucel bractlets 0.
      27. Plants caulescent; peduncles 1 terminal and 1–4 axillary; fruits narrowly oblong . . . . ***L. ambiguum***
      27' Plants acaulescent or subacaulescent; peduncles 1–10(19); fruits ovate, oblong, or narrowly elliptic. . . . . ***L. pastorale*** (in part)
     26' Involucel bractlets 1–12.
      28. Roots a slender and elongate taproot or a bifurcate tuber with an elongate upper section; umbels 3–23-rayed in fruit; ultimate apical lobes 1–9 mm wide; pedicels 0.5–7 mm. . . . . ***L. pastorale*** (in part)
      28' Roots a globose tuber with an elongate and slender portion above tuber < 4 cm; umbels 2–6-rayed in fruit; ultimate apical lobes 0.5–3 mm wide; pedicels 5–20 mm . . . . ***L. farinosum*** var. ***hambleniae*** (in part)

9' Ultimate apical lobes 0.7–15 mm (up to 15 mm in robust forms of *L. cusickii*, *L. engelmannii*, *L. rollinsii*, *L. gormanii*; immature *L. triternatum* with leaves not fully expanded and mature may also key here).

29. Involucel bractlets fused, disk-like . . . . **L. watsonii**

29' Involucel bractlets separate, not disk-like.

30. Peduncles canescent, hirsutulous, pubescent, puberulent, or villous.

31. Involucel bractlets 2–4 mm wide; fruit bodies (excluding wings) 1–2 mm wide. . . . . **L. utriculatum** (in part)

31' Involucel bractlets 0.1–2 mm wide; fruit bodies 2–4 mm wide (or as narrow as 1.3 mm in some forms of *L. triternatum*).

32. Ultimate apical lobes 5–15 mm.

33. Lateral primary leaflets less than half length of central primary leaflet, ultimate apical lobes acute, subacute or rounded at tips; fruit 6–8 mm, pedicels 5–15 mm. . . . . **L. rollinsii**

33' Lateral primary leaflets half to subequal length of central primary leaflet, ultimate apical lobes acute or acuminate; fruit 7–16 mm, pedicels 2–8 mm . . . . **L. triternatum** (in part)

32' Ultimate apical lobes 0.8–5 mm.

34. Fruit wings 0.5–1 mm wide.

35. Ultimate apical lobes glabrous or glabrate above, sparsely pubescent, puberulent or hirsutulous marginally and below; roots a globose thickened tuber abruptly tapered to an elongate and slender upper portion . . . . **L. bentonitum**

35' Ultimate apical lobes moderately to densely hirsutulous, pubescent, or puberulent throughout; roots an elongate and slender taproot or a globose thickened or irregularly thickened and elongate tuber tapered to elongate and slender upper portion < 10 cm . . . . **L. ravenii** var. ***paiutense***

34' Fruit wings 1–3.2 mm wide (as narrow as 0.8 mm in slightly immature fruits in *L. foeniculaceum* var. *macdougalii*).

36. Petals white or ochroleucous; anthers purple.

37. Fruit bodies excluding wings 2–3.5 mm wide; involucel bractlets 3–12 mm, margins toothed or 2–3-lobed. . . . . **L. macrocarpum** (in part)

37' Fruit bodies excluding wings 2.7–5 mm wide; involucel bractlets 2–5 mm, margins entire or tips with 1–2 small lobes. . . . . **L. nevadense** var. ***nevadense***

36' Petals yellow; anthers yellow, ochroleucous, or purple.

38. Peduncles 1–2, 3–6 cm; petal margins minutely ciliate . . . . **L. foeniculaceum** (in part)

38' Peduncles 1–30, 5–30 cm; petal margins glabrous.

39. Plants caulescent or short-caulescent, cauline leaves 1–4, stems 2–15 cm, pseudoscapes 1–10 cm; ultimate apical lobes 1–5 mm; involucel bractlets 3–12 mm; fruit body 7–20 mm, wings 1–2.5 mm wide, thin or thickened. . . . . **L. macrocarpum** (in part)

39' Plants acaulescent or short-caulescent, cauline leaves 0–1, stems < 1 cm, pseudoscapes 1–6 cm; ultimate apical lobes 0.8–2.5 mm; involucel bractlets 2–4 mm; fruit body 5–9 mm, wings 0.8–1.5 mm wide, thin . . . . **L. foeniculaceum** (in part)

30' Peduncles glabrous, glabrate, or sparsely scaberulose, muriculate, or papillose.

40. Involucel bractlets 1–5 mm wide, tips rounded or 2–6-lobed, apical lobes entire or 2–3-toothed.

41. Stems 10–50 cm, 2–3-branched above first node, first internode subterranean, obscured by leaf sheaths; pseudoscapes absent; plants caulescent with 1 terminal and 1–4 axillary peduncles; involucel bractlets 3–many-lobed or -toothed . . . . **L. utriculatum** (in part)

41' Stems absent or 2–20 cm, simple or 2–3-branched, first internode at or above ground level; pseudoscapes, if present, subterranean; plants acaulescent, subacaulescent, or caulescent with 1–10 peduncles (all basal *L. bradshawii*); involucel bractlets entire (in *L. cous*) or entire or 2–3-lobed.

42. Ultimate apical lobes 0.8–4 mm wide; umbels rays 5–20, 1–8 cm; fruit wings thin, 1–1.5 mm, pedicels 0.8–3 mm; plants acaulescent, subacaulescent, or caulescent . . . . **L. cous**

42' Ultimate apical lobes 0.3–0.6 mm wide; umbel rays 1–7, 5–13 cm; fruit wings corky-thickened, 1–2 mm, pedicels 2–5 mm; plants acaulescent . . . . **L. bradshawii**

40' Involucel bractlets 0 or 0.1–1 mm wide (up to 1.5 in *L. dissectum*, *L. leptocarpum*, *L. vaginatum*), tips acuminate, acute, or subacute or obtuse, entire (or 2-lobed in some formsof *L. columbianum*).

43. Fruit bodies 4–8 mm wide.
    44. Fruits 18–24 mm, body 5–8 mm wide.
        45. Involucel bractlets 7–11 mm; ultimate apical leaflet lobes linear or narrowly oblong, 4–18 mm; pedicels 4–9 mm; petals purple or reddish purple; anthers purple . . . . ***L. columbianum*** (in part)
        45' Involucel bractlets 2–5 mm; ultimate apical leaflet lobes linear or narrowly oblong, 2–8 mm; pedicels 0.5–3 mm; petals reddish purple or yellow; anthers purple or yellow . . . . ***L. dissectum*** (in part)
    44' Fruits 6–18 mm, body 4–6 mm wide.
        46. Roots stout and irregularly thickened; basal leaf sheaths from previous years weathering into a dense, fibrous thatch at or below ground level.
            47. Ultimate apical lobes 0.1–0.6 mm wide; fruit wings 0.5–2 mm, thin . . . . ***L. papilioniferum*** (in part)
            47' Ultimate apical lobes 1–4 mm wide; fruit wings 0.5–1.5 mm, corky-thickened. . . . . ***L. dissectum*** (in part)
        46' Roots tuberous or a slender or irregularly thickened and elongate taproot, not stout; basal leaf sheaths from previous years absent or weathering into a sparse, fibrous, chaffy, or chartaceous thatch.
            48. Involucel bractlets 0–4, 0.5–1.5 mm; primary leaflets with 2–4 secondary leaflet pairs below terminal one; fruit oblong or elliptic . . . . ***L. martindalei*** (in part)
            48' Involucel bractlets 4–8, 3–10 mm; primary leaflets with 3–6 secondary leaflet pairs below terminal one; fruit elliptic or suborbicular . . . . ***L. donnellii*** (in part)

43' Fruit bodies 1–4 mm wide.
    49. Peduncles and stems stout and inflated; petals purple or reddish purple; anthers purple . . . . ***L. minus***
    49' Peduncles and stems slender or stout, but not inflated; petals white, ochroleucous, or yellow; anthers ochroleucous, yellow, or purple.
        50. Fruits sparsely to moderately papillose, puberulent, or scaberulose.
            51. Rays 6–12, peduncles 5–16 cm; plants caulescent, cauline leaves 1–3; petals yellow, anthers ochroleucous or yellow. . . . . ***L. vaginatum*** (in part)
            51' Rays 1–6, peduncles 5–7 cm; plants acaulescent or subacaulescent, cauline leaves (when present) inconspicuous and bract-like; petals white, anthers purple. . . . . ***L. gormanii***
        50' Fruits glabrous or glabrate.
            52. Ultimate apical lobes obtuse or rounded (a few subacute lobes in *L. hendersonii*, *L. serpentinum*, *L. piperi*).
                53. Ultimate apical lobes 5–20 mm; lateral primary leaflets subequal in length to central primary leaflet . . . . ***L. piperi*** (in part)
                53' Ultimate apical lobes 0.7–7 mm; lateral primary leaflets half or less length of central primary leaflet (but laterals at least half central primary in length in *L. serpentinum*).
                    54. Basal leaflets 2–3 pinnate, secondary leaflets pinnate or pinnatifid (3–7-lobed) and tertiary leaflets absent or entire; fruit 4–8 mm, pedicels 1–5 mm.
                        55. Fruit bodies 1–2 mm wide; ultimate apical lobes 1.5–3 mm wide; peduncles in fruit 6–13 cm, rays 0.5–2(2.5) cm . . . . ***L. hendersonii***
                        55' Fruit bodies 2–3.5 mm wide; ultimate apical lobes 2–3.5 mm wide; peduncles in fruit 4–6 cm, rays 0.4–1 cm. . . . . ***L. ochocense***
                    54' Basal leaflets 3–4 pinnate, tertiary leaflets pinnate or pinnatifid (3–20-lobed); fruit 6–16 mm, pedicels 3–15 mm.
                        56. Plants acaulescent or subacaulescent.
                            57. Involucel bractlets 0–4, 0.1–0.3 mm wide; primary leaflets with 2–4 secondary leaflet pairs along each rachis, tertiary leaflets pinnatifid (3–20-lobed) . . ***L. martindalei*** (in part)
                            57' Involucel bractlets 3–8, 0.3–1 mm wide; primary leaflets with 4–8 secondary leaflet pairs along each rachis, tertiary leaflets entire, pinnate, or pinnatifid (2–6-lobed).
                                58. Fruit bodies 1–2 mm wide; petals yellow; anthers ochroleucous or yellow. . . ***L. serpentinum***
                                58' Fruit bodies 3–4 mm wide; petals white; anthers purple. ***L. canbyi***
                        56' Plants caulescent.
                            59. Involucel bractlets 2–10, 0.5–1.5 mm wide; primary leaflets with 4–6 secondary leaflet pairs below terminal one, tertiary leaflets entire or pinnatifid (2–6-lobed) . . . . ***L. vaginatum*** (in part)
                            59' Involucel bractlets 0–4, 0.1–0.3 mm wide; primary leaflets with 2–4 secondary leaflet pairs below terminal one, tertiary leaflets pinnatifid (3–20-lobed) . . . . ***L. martindalei*** (in part)

52' Ultimate apical lobes acuminate, acute, subacute, or mucronulate (some with rounded lobes in *L. donnellii*).
60. Umbels 1–3-rayed in fruit.
61. Fruit wings 1–2 mm.
62. Involucel bractlets 0–4, 0.5–0.8 mm wide; peduncles 6–15 cm; ultimate apical lobes 2–10 mm. . . . . **L. peckianum**
62' Involucel bractlets 0–2, 0.1–0.2 mm wide; peduncles 10–30 cm; ultimate apical lobes 2–3.4 mm . . . . **L. engelmannii** (in part)
61' Fruit wings 0.5–1 mm (but up to 1.2 mm in *L. erythrocarpum*).
63. Fruits green, red, or tan, dull; peduncles 10–20 cm in fruit; ultimate apical lobes 0.8–1.5 mm wide . . . . **L. roseanum** (in part)
63' Fruits red and glossy; peduncles 1–5.5 cm in fruit; ultimate apical lobes 0.3–0.5 mm wide. . . . **L. erythrocarpum**
60' Umbels 3–25-rayed in fruit.
64. Basal leaf sheaths from previous years weathering into a fibrous thatch at or below ground level; previous year's peduncles persistent in thatch as gray stalks; caudices branched or multicipital (only in young plants does caudex appear simple).
65. Pedicels 1–5 mm; basal leaflets biternate or pinnate with 1–2 secondary leaflet pairs along each rachis; petals purple or white; anthers ochroleucous or yellow. . . . . **L. cusickii** (in part)
65' Pedicels 5–15 mm; basal leaflets 2–4 pinnate; petals yellow; anthers ochroleucous or yellow.
66. Ultimate apical lobes 1–7(9) × 0.1–0.6 mm, tertiary leaflets pinnate, quaternary leaflets entire or pinnatifid (3–30-lobed). . . . . **L. papilioniferum**
66' Ultimate apical lobes 2–30 × 1–3 mm, tertiary leaflets entire or pinnatifid (2–3-lobed), quaternary absent . . . . **L. laevigatum** (in part)
64' Basal leaf sheaths from previous years absent or weathering into a sparse, fibrous, chartaceous, or chaffy thatch at base of stem or pseudoscape; caudices simple or 2–6-branched.
67. Fruit wings 0.4–1 mm.
68. Fruits oval or orbicular, 6–8 mm; involucel bractlets 0.3–0.8 mm wide; umbels 2–6-rayed in fruit, rays 2–4 cm . . . . **L. roseanum** (in part)
68' Fruits linear or narrowly oblong, 9–13 mm; involucel bractlets 0.5–1.5 mm wide; umbels 4–10-rayed in fruit, rays 1–8 cm . . . . **L. leptocarpum** (in part)
67' Fruit wings 1–3 mm.
69. Ultimate apical lobes 1–5 mm wide.
70. Involucel bractlets 0–4, 0.5–1.5 mm; primary leaflets with 2–4 secondary leaflet pairs along each rachis; fruit oblong or elliptic . . . . **L. martindalei** (in part)
70' Involucel bractlets 4–8, 3–10 mm; primary leaflets with 3–6 secondary leaflet pairs along each rachis; fruit elliptic or suborbicular. . . . . **L. donnellii** (in part)
69' Ultimate apical lobes 0.1–1 mm wide (up to 1.2 mm wide in robust forms of *L. hallii*).
71. Petals white or purple; anthers ochroleucous or white; peduncles 1–2, umbel rays 1–3 cm; pedicels 2–8 mm; ultimate apical lobes 2–3.4 mm . . . . **L. engelmannii** (in part)
71' Petals ochroleucous or yellow; anthers ochroleucous or yellow; peduncles 1–6, umbel rays 2–10 cm; pedicels 1–15 mm; ultimate apical lobes 2–8 mm.
72. Pedicels 1–5 mm; fruit wings corky-thickened. . . . . **L. cookii**
72' Pedicels 5–15 mm; fruit wings thin.
73. Lateral primary leaflets half to subequal length of central primary leaflet; involucel bractlets 4–8 . . . . **L. donnellii** (in part)
73' Lateral primary leaflets less than half length of central primary leaflet; involucel bractlets 0–4 . . . . **L. hallii**

Lomatium ambiguum (Nuttall) J.M. Coulter & Rose (p. 110)

Swale desert parsley, Wyeth biscuitroot

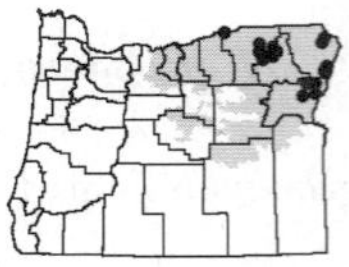

Plants caulescent. **Roots** tuberous above with slender or irregularly moniliform, elongate taproots. **Caudices** simple, with old basal leaves mostly withered post-anthesis; basal leaf sheaths from previous years not persistent as a thatch. **Stems** 6–50 cm, first internode subterranean, 2–3-branched above first node. **Basal leaves** glabrous; primary leaflets pinnate, 1–4 secondary leaflet pairs along each rachis, laterals subequal to central primary in length; secondaries entire or pinnatifid (2–3-lobed); tertiaries entire or absent; ultimate apical lobes linear or narrowly oblong, 10–50 × 1–4 mm, tips subacute, obtuse, or rounded. **Cauline leaves** similar to basal. **Inflorescences** peduncles, 1 terminal and 1–4 axillary, slender, 2–8 cm, glabrous; involucral bracts 0; rays 5–17, 1.5–8 cm, glabrous; involucels absent; umbellets 15–25-flowered; pedicels 4–8 mm. **Flowers** petals yellow; anthers ochroleucous or yellow. **Fruits** narrowly oblong, 8–10 mm, glabrous; body 1–2 mm; wings 0.4–0.7 mm; vittae 1 in intervals, 2 on commissure. **2*n***=22.

Open areas on rocky slopes and flats, mesic slopes, scablands, scrublands, meadows, forests, river drainages. Flowering May–Jul. 50–2000 m. BW, Col. ID; north to British Columbia, east to WY, southeast to UT. Native.

Lomatium bentonitum K. Carlson & D. Mansfield

Bentonite biscuitroot

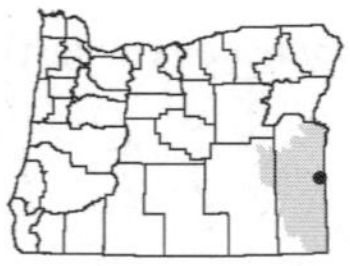

Plants acaulescent or subacaulescent. **Roots** globose thickened tubers abruptly tapered to an elongate and slender upper portion < 10 cm. **Caudices** simple or 2–7-branched; basal leaf sheaths from previous years weathering into a sparse thatch of fibrous, chaffy, or chartaceous scales at base of pseudoscapes. **Stems** 0.5–1 cm; pseudoscapes 1–4 cm, subterranean, obscured by leaf sheaths. **Basal leaves** glabrous or glabrate above, sparsely pubescent, puberulent, or hirsutulous below, or lobes glabrous or glabrate above and midnerves and margins of leaflet rachilla sparsely pubescent, puberulent, or hirsutulous; primary leaflets pinnate, 3–6 secondary leaflet pairs along each rachis, laterals less than half central primary in length, secondaries pinnate or pinnatifid; tertiaries entire or pinnatifid (2–3-lobed); ultimate apical lobes elliptic, obovate, oblong, 1–4 × 0.5–1.5 mm, tips acute, subacute, rounded, mucronulate or not. **Cauline leaves** 0–1, similar to basal. **Inflorescences** peduncles 1–6, decumbent, spreading, ascending, or erect, 3–10 cm, hirsutulous, pubescent, or puberulent; involucral bracts 0; rays 3–20, 1–5 cm, sparsely hirsutulous or puberulent; involucel bractlets 4–8, linear or lanceolate, 2–5 × 0.3–1 mm, glabrous, glabrate, or sparsely puberulent, margins scarious; umbellets 10–20-flowered; pedicels 2–10 mm, sparsely hirsutulous or puberulent. **Flowers** petals white or purple; anthers purple. **Fruits** oval or suborbicular, 4–8 mm, sparsely hirsutulous or puberulent; body 2–3 mm wide; wings 0.5–1 mm, thin; vittae obscure.

Clay badlands. Flowering Apr–May. 800–1200 m. Owy. Native.

This species might be better placed as a variety within *L. ravenii*, as there are few morphological differences between the two taxa. In recent studies, phylogenetic analyses using small segments of non-coding DNA regions have not resolved whether this species (and some of the more problematic taxa in *Lomatium*) are robust species. More extensive genetic studies combined with ecological and demographic studies will likely be necessary to determine if this species is an ecotype of *L. ravenii* or a reproductively isolated, cryptic species.

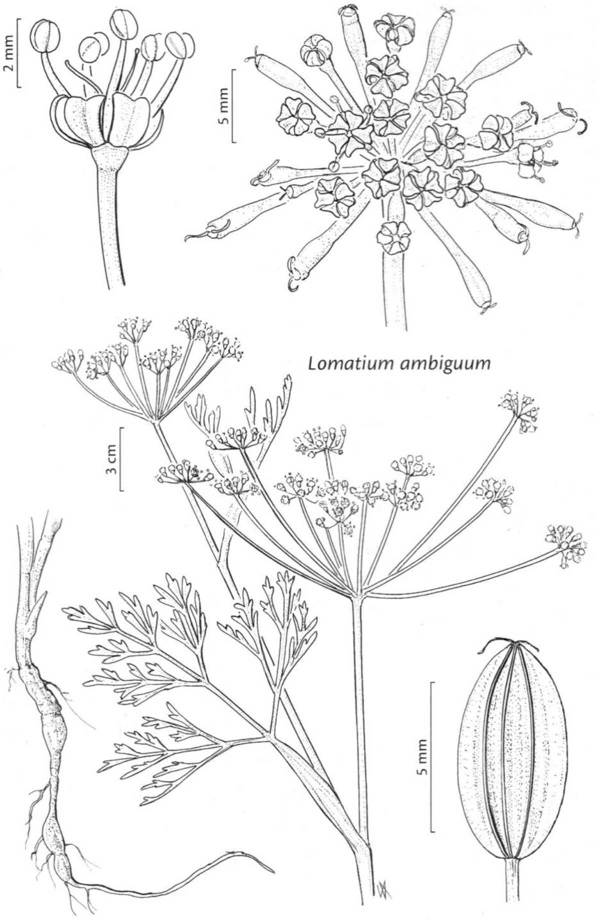

LOMATIUM

Lomatium bradshawii (Rose ex Mathias) Mathias & Constance

Bradshaw's desert parsley, Bradshaw's lomatium

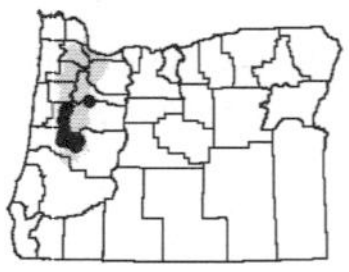

Plants acaulescent. **Roots** elongate or thickened, slender taproots, or irregularly thickened tubers with elongate upper section < 1 cm and tapering to an elongate and slender taproot at base. **Caudices** simple or 2–4-branched; basal leaf sheaths from previous years absent or weathering into a sparse thatch of a few, loose fibers and chaffy or chartaceous, elongate scales at base of pseudoscape; previous year's peduncles sometimes persistent in thatch as gray stalks. **Stems** absent; pseudoscapes 1–15 cm, subterranean, obscured by leaf sheaths. **Basal leaves** glabrous; primary leaflets pinnate, 4–6 secondary leaflet pairs along each rachis, laterals subequal to central primary in length; secondaries pinnate; tertiaries entire or pinnatifid (2–4-lobed); ultimate apical lobes linear, 2–10 × 0.3–0.6 mm, tips acute, acuminate, or subacute, mucronulate or not. **Cauline leaves** 0; **Inflorescences** peduncles 1–4, basal, 15–45 cm, glabrous below, glabrate, sparsely scaberulose, or muriculate above; involucral bracts 0; rays 1–7, 5–13 cm, glabrous, glabrate or sparsely scaberulose or muriculate on ridges; involucel bractlets 4–10, oblanceolate, oblong, or obovate, 3–6 × 0.8–3 mm, apical lobes entire or 2–3-toothed, margins scarious; umbellets 15–25-flowered; pedicels 2–5 mm. **Flowers** petals ochroleucous or yellow; anthers ochroleucous or yellow. **Fruits** oblong, elliptic, or oval, 8–16 mm, glabrous; body 2–3.5 mm wide; wings corky-thickened, 1–2 mm; vittae obscure.

Seasonally wet prairie remnants. Flowering Mar–May. 50–300 m. WV. WA. Native.

This taxon is endemic to the Willamette Valley ecoregion. The populations in Clark County, Washington, discovered in 1994, are in a northern extension of this ecoregion. This taxon and *L. cookii* are northern extensions of the *Lomatium marginatum* species complex, which is mostly restricted to vernal pool habitats in the Central Valley of California. If lined up side-by-side with California specimens, they would be nearly indistinguishable from some of the many variants of *L. marginatum* and *L. caruifolium*.

Lomatium brunsfeldianum Kemper & R.P. McNeill

Brunsfeld's desert parsley

Plants caulescent. **Roots** stout, irregularly thickened taproots. **Caudices** simple or multicipital; basal leaf sheaths from previous years weathering into a thatch of fibers and chaffy scales at base of stem. **Stems** 40–90 cm. **Basal leaves** glabrous, glabrate, scaberulose, or muriculate; primary leaflets pinnate with 5–8 secondary leaflet pairs along each rachis, laterals at least half to subequal to central primary in length; secondaries pinnate; tertiaries pinnate; quaternaries entire or pinnatifid, ultimate leaflet pinnules lobed throughout, with ± equal intercostal regions along rachilla, between rachilla and basal-most pinnule (appearing sessile), intercostal areas ± obscured by pinnule lobes; ultimate apical lobes linear, 0.7–4.2 × 0.1–0.7 mm, tips acute or acuminate, mucronulate or not. **Cauline leaves** 1–2, similar to basal. **Inflorescences** peduncles 1–10+, slender to stout, not inflated, 4–40 cm × 2–6 mm (1 cm below tip), glabrous, glabrate, or sparsely scaberulose or muriculate above; involucral bracts 0; rays 5–16, 1–13 cm, glabrate, scaberulose or muriculate; involucel bractlets 0; umbellets 5–20-flowered; pedicels 3–10 mm. **Flowers** petals yellow; anthers yellow or ochroleucous. **Fruits** narrowly oblong or narrowly elliptic, 11–16 mm, glabrous; body 3–4 mm; wings 0.3–1 mm, thin; vittae 1–2 in intervals, 2–6 on commissure.

Talus slopes, rock outcrops, cliff faces. Flowering Apr–Jun. 900–1100 m. BW. ID. Native.

In Oregon, this taxon is known only from one historical and one modern collection from the Hells Canyon vicinity in Baker County. More collections from Oregon are needed to determine the range of variability of the morphology of this taxon.

A long-overlooked name by Marcus E. Jones may take priority over *L. brunsfeldianum* at the species level if its affinity can be resolved. Jones collected the type of *Leptotaenia filicinum* in August of 1899 in Six Devils Mountains region in Adams County, Idaho (in 1899 within Washington County). No combinations for this taxon have been made within *Lomatium*, nor was it combined into *Cogswellia* by Jones in his later revision. This locality is within the range of several similar *Lomatium* species.

Other than the type publication, this type has been associated with *Lomatium grayi* (herein treated as *L. papilioniferum*) and *L. bicolor*. However, based on fruit morphology, it appears to be more similar to *L. brunsfeldianum*. Annotations by Mathias and Constance on the types of this taxon at POM and UTC propose a new combination, "*Lomatium grayi* var. *filicinum*," but this name seems to not have been published, nor was Jones' basionym placed in synonymy for any of the *Lomatium* species treated by Mathias or Constance. Some of the confusion surrounding this type is due to the time of year it was collected, which was in August when most of the leaves were already dried and withered. Jones determined it originally as *Peucedanum bicolor* (=*Lomatium bicolor*). He also stated that this species "has the leaves of *Peucedanum grayi*, the fruit nearly that of *P. bicolor* but is flat and thin" (Jones 1902). In the type publication of *L. brunsfeldianum*, Kemper & McNeil (2012) state that it is "similar in leaf structure, and in mericarp size and shape" to *Lomatium bicolor*, but it can be distinguished from *L. grayi* "by the structure and size of the mericarps." The few leaflet fragments on the type suggest the evenly spaced pinnule lobes of *L. brunsfeldianum*. The fruit on the type are narrow and curved with narrow wings, characters Jones identified in his description as diagnostic. It appears that Jones' interpretation of the morphology on his types of *L. filicinum* and McNeill's interpretation of the diagnostic characters of *L. brunsfeldianum* are congruent. A new combination in *Lomatium* for Jones' *Leptoteania filicinum* and the required nomenclatural re-arrangements await more definitive results from research in progress on populations of *Lomatium brunsfeldianum* in western Idaho and northeastern Oregon.

Lomatium californicum (Nuttall) Mathias & Constance (p. 112)

California lomatium

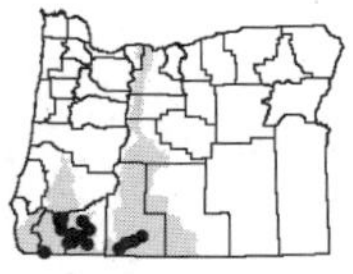

Plants caulescent. **Roots** stout and thickened, elongate taproots or stout and irregularly thickened taproots. **Caudices** 2–3-branched or multicipital; basal leaf sheaths from previous years weathering into a fibrous thatch at or below ground level; peduncles persistent in thatch as gray stalks. **Stems** 30–60 cm; pseudoscapes absent. **Basal leaves** glabrous; primary leaflets pinnate, 1–3 secondary leaflet pairs along each rachis, laterals subequal to central primary in length; secondaries pinnate or ternate; tertiaries entire or pinnatifid; ultimate apical lobes obovate, to 50 × 10–30 mm, lobes shallow and rounded, cuspidate or not, tips 4–7-lobed. **Cauline leaves** 1–3, similar to basal. **Inflorescences** peduncles 1–3 (more in robust plants), ascending or erect, stout but not inflated, 7–40 cm, glabrous; involucral bracts 0; rays 8–15, 3–8 cm, glabrous; involucel bractlets 0–6, linear, 2–3 × 0.1–0.4 mm, margins scarious or not; umbellets 10–30-flowered; pedicels 7–12 mm. **Flowers** petals yellow; anthers ochroleucous or yellow. **Fruits** oblong, oval, elliptic, 8–15 mm, glabrous; body 4–6 mm wide; wings 1–2 mm, thin or slightly thickened; vittae 3–4 in intervals, 6–10 on commissure. **2*n*=22.**

Gravelly or rocky slopes, talus, rock outcrops, scrubland, oak forests, woodlands, conifer forests. Flowering Apr–Aug. 300–1600 m. ECas, Sisk. CA. Native.

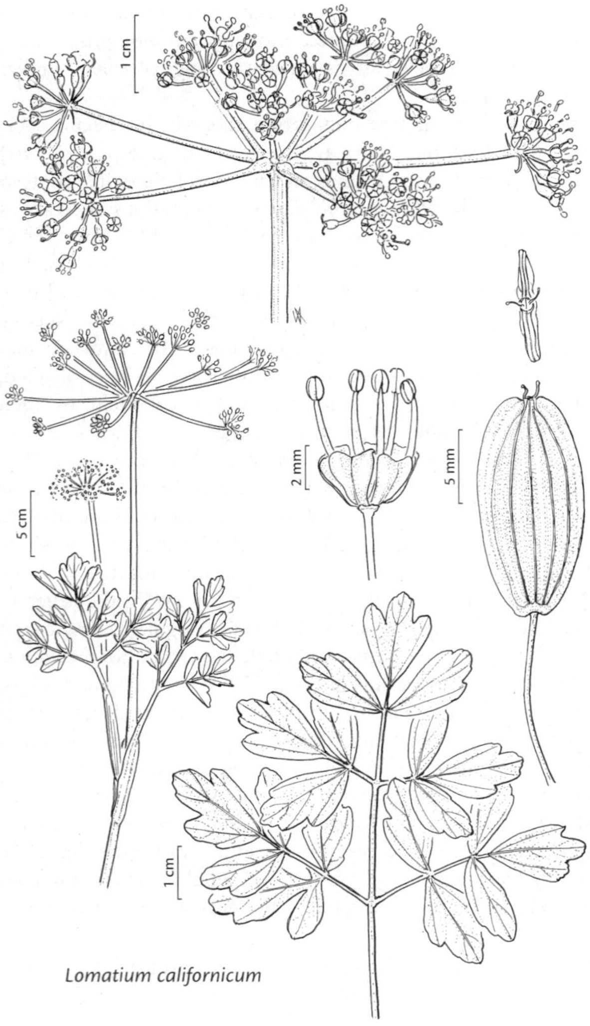

Lomatium californicum

LOMATIUM

Lomatium canbyi (J.M. Coulter & Rose) J.M. Coulter & Rose

Canby's lomatium

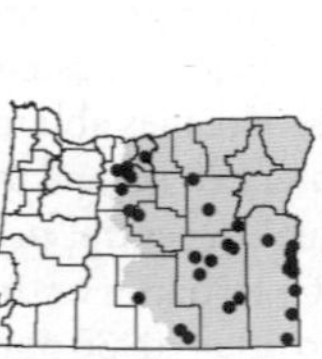

Plants acaulescent. **Roots** globose tubers, elongate and slender portion above tubers < 4 cm, slender and elongate taproots below tubers easily detached. **Caudices** simple or 2–5-branched; basal leaf sheaths from previous years slowly weathering into a thatch of chaffy or chartaceous scales at base of pseudoscape. **Stems** absent; pseudoscapes 2–5 cm, subterranean. **Basal leaves** glabrous, glaucous; primary leaflets pinnate, 4–6 secondary leaflet pairs along each rachis, laterals up to half of central primary in length; secondaries pinnate or pinnatifid; tertiaries entire or pinnatifid (2–3-lobed); ultimate apical lobes elliptic, oblong, spatulate, 3–6.3 × 0.7–1.2 mm, tips obtuse or rounded, shortly cuspidate. **Cauline leaves** 0. **Inflorescences** peduncles 1–5, 5–17 cm, glabrous; involucral bracts 0; rays 5–17, 1–6 cm, glabrous; involucel bractlets 3–7, spatulate, narrowly oblanceolate, 2.5–5 × 0.5–1 mm; umbellets 10–20-flowered; pedicels 6–16 mm. **Flowers** petals white; anthers purple. **Fruits** elliptic, obelliptic, 8–15 mm, glabrous; body 3–4 mm wide; wings 1–2.5 mm, thin, wing wider at fruit base; vittae 1–3 in intervals, 2–6 on commissure. ***2n***=22.

Gravel, clay, sandy or rocky slopes and flat, scrubland. Flowering Mar–May. 300–2200 m. BR, BW, Col, Lava, Owy. CA, ID, WA; NC. Native.

Lomatium columbianum Mathias & Constance

Columbia desert parsley, purple lomatium

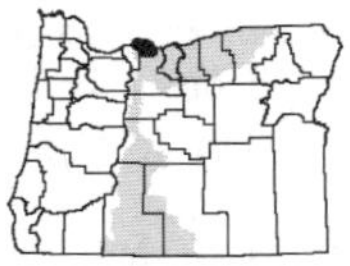

Plants caulescent. **Roots** stout and thickened, elongate taproots. **Caudices** simple, 2–3-branched, multicipital; basal leaf sheaths from previous years weathering into a fibrous thatch at or below ground level; peduncles persistent in thatch as gray stalks. **Stems** stout but not inflated, 5–20 dm; pseudoscapes absent. **Basal leaves** glabrous; primary leaflets pinnate, 4–7 secondary leaflet pairs along each rachis (mature basal leaves may have more pairs), laterals subequal to central primary in length; secondaries pinnate; tertiaries pinnate; quaternaries pinnate or pinnatifid (3–10-lobed); ultimate apical lobes linear or narrowly oblong, 4–18 × 0.5–1.5 mm, tips acute, acuminate, or subacute, mucronulate or not. **Cauline leaves** 1–3, similar to basal. **Inflorescences** peduncles 1–5 (more in robust plants), stout but not inflated, 30–80 cm; involucral bracts 0; rays 6–20, stout, 3–20 cm, glabrous; involucel bractlets 4–10, linear or lanceolate, 7–11 × 0.5–1 mm, unlobed or 2-lobed, margins narrowly scarious or not; umbellets 10–30-flowered; pedicels 4–9 mm. **Flowers** petals purple or reddish purple; anthers purple. **Fruits** oblong, oval, glabrous, 18–24 mm; body 5.5–8 mm wide; wings 1.5–2.5 mm, corky-thickened; vittae 2–3 in intervals, 3–4 on commissure.

Gravelly or rocky slopes and flats, talus, rock outcrops, scrubland, oak forests, conifer forests. Flowering Mar–Apr. 0–900 m. Col, ECas. WA. Native.

This taxon is endemic to the eastern Cascade slopes and foothills ecoregion and the edge of the Cascade Mountains ecoregion in Oregon and Washington (north of the mouth of the Klickitat River and south of the mouth of the Hood River). The large fruits are diagnostic and unmistakable. However, outside this region (in the northern part of the Columbia River drainage or on the Columbia Basin), *Lomatium dissectum* var. *dissectum* (with reddish flowers), *L. suksdorfii*, *L. tuberosum*, and *L. triternatum* (large leaf variants) are commonly misidentified as *L. columbianum* when collected in flowering or vegetative states.

Lomatium cookii Kagan

Cook's lomatium, agate desert parsley

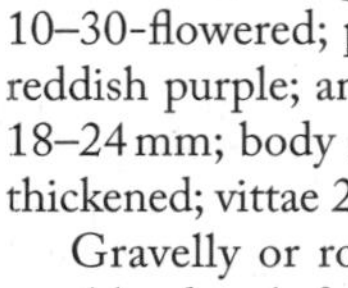

Plants acaulescent. **Roots** slender or irregularly thickened elongate taproots. **Caudices** simple or 2–8-branched; basal leaf sheaths from previous years absent or weathering into a sparse thatch of a few, loose fibers or chaffy scales at base of pseudoscape. **Stems** absent; pseudoscapes 1–4 cm, subterranean, obscured by leaf sheaths. **Basal leaves** glabrous; primary leaflets pinnate, 4–6 secondary leaflet pairs along each rachis, laterals subequal to central primary in length; secondaries pinnate; tertiaries entire or pinnatifid (2–4-lobed); ultimate apical lobes linear or narrowly oblong, 2–5 × 0.3–0.7 mm, tips acute, subacute, acuminate, mucronulate or not. **Cauline leaves** 0. **Inflorescences** peduncles 1–4,

15–40 cm, glabrous below, glabrate or sparsely scaberulose or muriculate above; involucral bracts 0; rays 1–12, 2–10 cm, glabrous, glabrate, or sparsely scaberulose or muriculate on ridges; involucel bractlets 4–6, linear or oblanceolate, 3–10 × 0.5–1 mm, margins scarious; umbellets 10–20-flowered; pedicels 1–5 mm. **Flowers** petals ochroleucous or yellow; anthers ochroleucous or yellow. **Fruits** oblong to elliptic, 8–14 mm, glabrous; body 2–3 mm wide; wings 1–2 mm, corky-thickened; vittae obscure.

Vernal pools, seasonally-wet meadows. Flowering Apr–May. 300–500 m. Sisk. Native.

See discussion for *Lomatium bradshawii*.

Lomatium cous (S. Watson) J.M. Coulter & Rose (p. 113)

Cous, cous-root desert parsley

Lomatium montanum J.M. Coult. & Rose, *Lomatium circumdatum* (S. Watson) J.M. Coult. & Rose

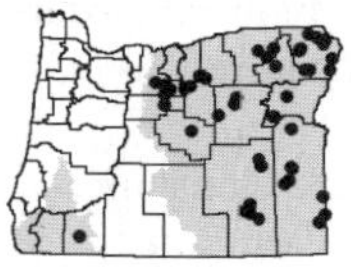

Plants acaulescent, subacaulescent, or caulescent. **Roots** slender, elongate taproots or globose tubers with an elongate and slender portion below and an elongate and slender portion above (tuber, if present, < 2 cm). **Caudices** simple; basal leaf sheaths from previous years weathering into a sparse thatch of fibers and chaffy or chartaceous scales at base of pseudoscape. **Stems** absent or 2–20 cm, simple or 2–3-branched above pseudoscape, glabrous or sparsely papillose or scaberulose; pseudoscapes 1–3 cm, subterranean. **Basal leaves** glabrous, glaucous, or sparsely papillate; primary leaflets pinnate, 3–6 secondary leaflet pairs along each rachis, laterals one half or less central primary in length; secondaries pinnate or pinnatifid; tertiaries entire or pinnatifid (2–6-lobed); ultimate apical lobes oval or oblong, 2–10 × 0.8–4 mm, tips acute, acuminate, subacute, or rounded, mucronulate or not. **Cauline leaves** 0–3, similar to basal. **Inflorescences** peduncles 1–10, 5–27 cm, glabrous or sparsely puberulent, papillose, or scaberulose; involucral bracts 0 or umbels subtended by a sheathing bract; rays 5–20, 1–8 cm, glabrous or glabrate; involucel bractlets 4–12, obovate or oblanceolate, 1.7–5 × 0.8–3.2 mm, margins scarious; umbellets 10–20-flowered; pedicels 0.8–3 mm, glabrous, glabrate, or sparsely muriculate to scaberulose. **Flowers** petals yellow; anthers ochroleucous or yellow. **Fruits** oblong, 6–12 mm; body 1.8–3 mm; wings 1–1.5 mm; vittae 1–4 in intervals, 4–7 on commissure. **2*n***=22.

Open areas on rocky slopes and flats, scablands, scrublands, meadows, forests. Flowering Mar–Jul. 200–2800 m. BR, BW, Col, Lava, Owy, Sisk. ID, NV, WA; northeast to Saskatchewan, east to WY. Native.

Lomatium cusickii (S. Watson) J.M. Coulter & Rose

Cusick's lomatium, Cusick's biscuitroot

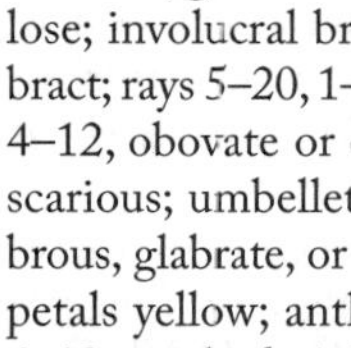

Plants caulescent, subacaulescent, or cespitose and caulescent (older plants with multicipital caudices). **Roots** slender and fusiform, slender and elongate, or thickened and elongate taproots. **Caudices** simple, 2–3-branched, or multicipital and cespitose; basal leaf sheaths from previous years weathering into a fibrous, chaffy, or chartaceous thatch at or below ground level. **Stems** 10–20 cm; pseudoscapes 0.5–2 cm, subterranean, obscured by leaf sheaths.

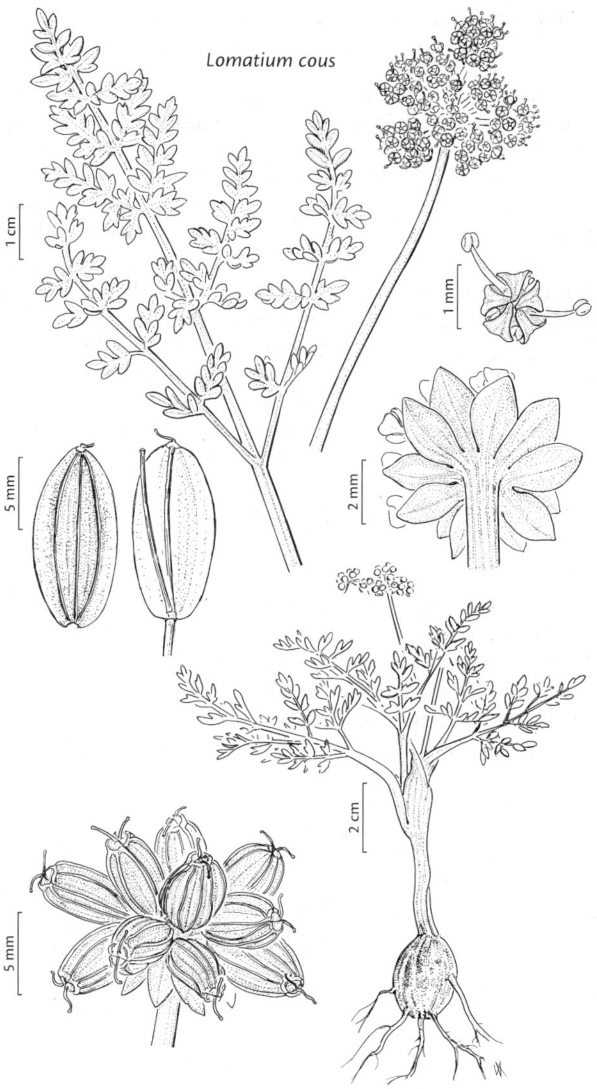

LOMATIUM

Basal leaves glabrous, glabrate, or sparsely scaberulose to muriculate; primary leaflets biternate or pinnate with 1–2 secondary leaflet pairs along each rachis, laterals subequal to central primary in length; secondaries entire, ternate, or pinnatifid (2–4-lobed); tertiaries absent; ultimate apical lobes linear or narrowly elliptic, 5–15 × 0.5–2 mm, tips acute or acuminate, mucronulate. **Cauline leaves** 1, similar to basal. **Inflorescences** peduncles 1–3, 5–15 cm, glabrate or sparsely scaberulose, papillose, or muriculate on ridges; involucral bracts 0; rays 5–12, 1–5 cm, glabrous or glabrate; involucel bractlets 2–4, linear or lanceolate, 2–4 × 0.5–1 mm, margins scarious; umbellets 10–20-flowered; pedicels 1–5 mm. **Flowers** petals purple or white; anthers ochroleucous or yellow. **Fruits** oblong, 10–14 mm, glabrous; body 1.5–3 mm wide; wings 1.5–2 mm, thin; vittae 1–3 in intervals, 5 on commissure.

Mountain ridges and slopes, rocky and gravelly areas. Flowering Jun–Aug. 1100–2800 m. BW. ID, WA; northeast to MT. Native.

Lomatium dissectum (Nuttall) Mathias & Constance

Fernleaf biscuitroot, fernleaf desert parsley

Plants caulescent. **Roots** stout, irregularly thickened taproots. **Caudices** simple, 2–3-branched, or multicipital; basal leaf sheaths from previous years weathering into a fibrous thatch at or below ground level; peduncles persistent in thatch as gray stalks. **Stems** stout but not inflated; pseudoscapes absent. **Basal leaves** primary leaflets pinnate, 4–7 secondary leaflet pairs along each rachis (mature basal leaves may have more pairs), laterals half central primary in length. **Cauline leaves** 1–3, similar to basal, tips mucronulate. **Inflorescences** peduncles 1–4 (more in robust plants), ascending or erect, stout but not inflated, glabrous; involucral bracts 0 or 4–8, linear, narrow bracts, or umbels subtended by a leaf-like sheathing bract; rays 10–30, 3–12 cm, glabrous; involucel bractlet margins scarious. **Fruits** wings corky-thickened; vittae obscure. ***2n***=22.

Key to *Lomatium dissectum* varieties:

1. Basal leaflets 2–3 pinnate, ultimate apical lobes 2–8 mm; pedicels 0.5–3 mm; petals yellow or reddish purple; anthers purple or yellow . . . . *L. d.* var. ***dissectum***

1' Basal leaflets 3–4 pinnate, ultimate apical lobes 5–30 mm; pedicels 4–20 mm; petals yellow; anthers ochroleucous or yellow . . . . *L. d.* var. ***multifidum***

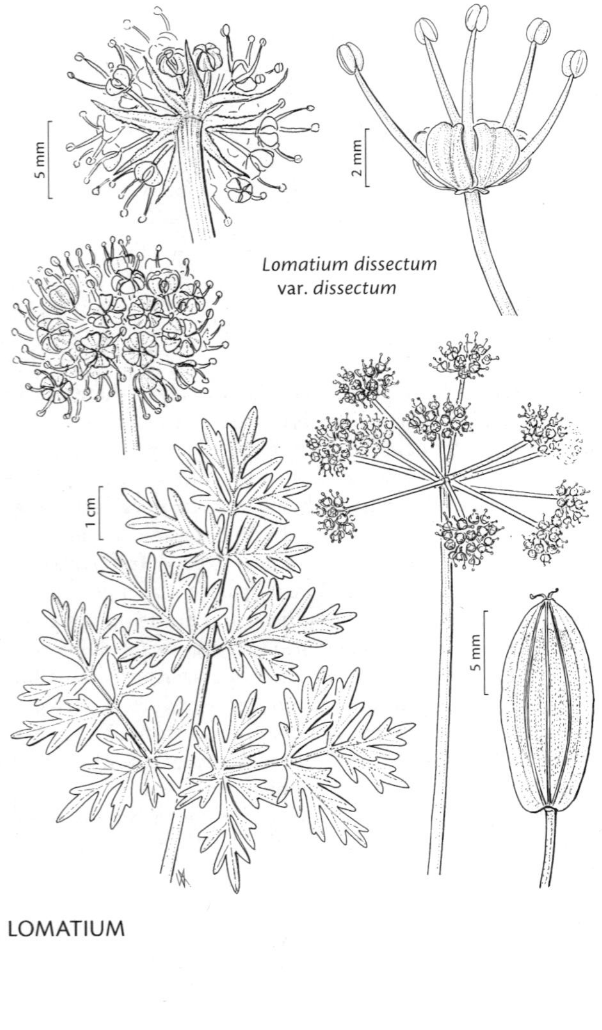

LOMATIUM

Lomatium dissectum (Nuttall) Mathias & Constance var. ***dissectum*** (p. 114)

Fernleaf biscuitroot, fernleaf desert parsley

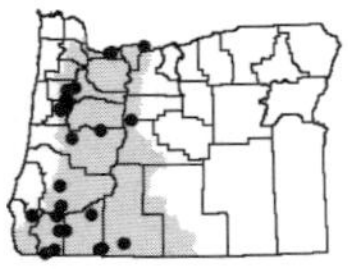

Stems 5–20 dm. **Basal leaves** leaflets 2–3 pinnate, pubescent, scaberulose or muriculate on midnerves and margins below, sparsely pubescent, scaberulose or muriculate, or glabrous or glabrate above; secondary leaflets ternate or pinnate; tertiaries pinnatifid (6–20-lobed); ultimate apical lobes linear or narrowly oblong, 2–8 × 1–4 mm, tips acute, subacute, obtuse or rounded. **Inflorescences** peduncles 20–80 cm; involucel bractlets 0–6(8), linear or lanceolate, 2–5 × 0.3–1 mm; umbellets 10–50-flowered; pedicels 0.5–3 mm. **Flowers** petals yellow or reddish purple; anthers purple or yellow. **Fruits** oblong to oval, 12–20 mm, glabrous; body 5–8 mm wide; wings 0.5–1.5 mm.

Gravelly or rocky slopes, rock outcrops, open flats in scrubland, woodland, oak forests, conifer forests. Flowering Mar–Jun. 50–1800 m. Casc, ECas, Sisk, WV. WA. Native.

Lomatium dissectum (Nuttall) Mathias & Constance var. ***multifidum*** (Nuttall) Mathias & Constance

Carrotleaf biscuitroot

Lomatium dissectum (Nutt.) Mathias & Constance var. *eatonii* (J.M. Coult. & Rose) Cronquist

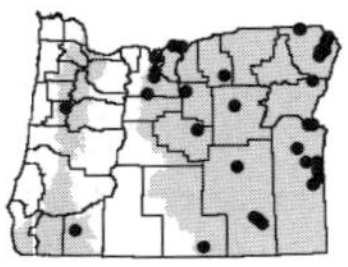

Stems 5–15 dm. **Basal leaves** leaflets 3–4 pinnate, scaberulose or muriculate only on midnerves and margins below, glabrous or glabrate above; secondary leaflets pinnate; tertiaries pinnate or pinnatifid; quaternary entire or pinnatifid (2–4-lobed); ultimate apical lobes linear, narrowly oblong, or narrowly elliptic, 5–30 × 1–3.5 mm, tips acute or subacute. **Inflorescences** peduncles 20–50 cm; involucel bractlets 2–6, linear or lanceolate, 2–7 × 0.3–1.5 mm; umbellets 20–30-flowered. **Flowers** petals yellow; anthers ochroleucous or yellow; pedicels 4–20 mm. **Fruits** oblong, elliptic, 9–17 mm, glabrous; body 4–6 mm wide; wings 0.7–1 mm.

Gravelly or rocky slopes, rock outcrops, open flats in scrubland, woodland, conifer forests. Flowering Apr–Jun. 50–2200 m. BR, BW, Col, Lava, Owy, Sisk, WV. CA, ID, NV, WA; north to British Columbia, south to Mexico. Native.

Lomatium donnellii (J.M. Coulter & Rose) J.M. Coulter & Rose

Donnell's lomatium

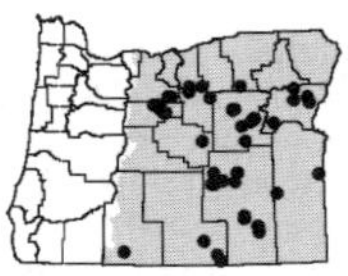

Plants acaulescent, subacaulescent, or caulescent. **Roots** elongate and slender taproots, or elongate and irregularly thickened taproots. **Caudices** simple or 2–4-branched; basal leaf sheaths from previous years absent or weathering into a few fibers or sparsely chartaceous or chaffy thatch at base of pseudoscape. **Stems** 4–10 cm, internodes short, < 2 cm, concealed by sheaths or only upper 5–10 mm visible; pseudoscapes 1–4 cm, subterranean, upper section not obscured by leaf sheaths. **Basal leaves** glabrous; primary leaflets pinnate, 3–6 secondary leaflet pairs along each rachis, laterals half to subequal to central primary in length; secondaries pinnate or pinnatifid; tertiaries pinnate or pinnatifid (3–8-lobed); ultimate apical lobes linear, oblong, narrowly oblong, elliptic, or oval, 1–8 × 0.5–3 mm, tips acute, acuminate, subacute, rounded, or mucronulate. **Cauline leaves** 0–4, similar to basal. **Inflorescences** peduncles 1–4, 7–30 cm, glabrous; involucral bracts 0; rays 4–15, 1–7 cm, glabrous; involucel bractlets 4–8, linear or lanceolate, 3–10 × 0.2–1 mm, margins narrowly scarious or not; umbellets 10–30-flowered; pedicels 5–15 mm. **Flowers** petals ochroleucous or yellow; anthers ochroleucous or yellow. **Fruits** elliptic or suborbicular, 5–12 mm, glabrous; body 3–5 mm wide; wings 1–2 mm, thin; vittae 2–6 in intervals, 4–6 on commissure.

Open rocky flats, rocky ridges, rock outcrops, dry meadows, scrublands and woodlands. Flowering Mar–Jun. 500–1900 m. BR, BW, Col, ECas, Lava, Owy. Native.

Reports of this taxon from Idaho and Washington are likely based on misidentifications. *Lomatium plummerae* has been delimited as a synonym of *L. donnellii* (Cronquist 1997). Reports from California and Nevada are based on this nomenclatural arrangement. *Lomatium plummerae* is delimited as a separate species herein, and it ranges south of the Oregon flora region.

Lomatium engelmannii Mathias

Engelmann's lomatium, Engelmann's desert parsley

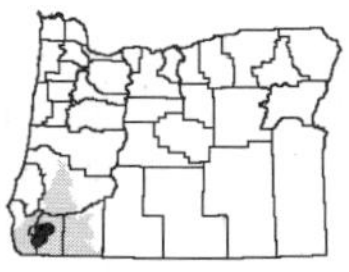

Plants acaulescent. **Roots** elongate and slender taproots. **Caudices** simple or 2–3-branched, subterranean; basal leaf sheaths from previous years weathering into a thatch of fibers or chartaceous scales at base of pseudoscape. **Stems** absent; pseudoscapes absent or 1–3 cm, subterranean, obscured by leaf sheaths. **Basal leaves** glabrous, glabrate, or sparsely pubescent on abaxial surface or along midveins; primary leaflets pinnate, 4–6 secondary leaflet pairs along each rachis, laterals half to subequal to central primary in length; secondaries pinnatifid; tertiaries entire or pinnatifid (2–5-lobed); ultimate apical lobes elliptic or oblong, 2–3.4 × 0.7–1 mm, tips acute, apiculate, or mucronulate. **Cauline leaves** 0. **Inflorescences** peduncles 1–2, 10–30 cm, glabrous or glabrate; involucral bracts 0; rays 2–12, 1–3 cm, glabrous; involucel bractlets 0–2, 2–3 × 0.1–0.2 mm, margins not scarious; umbellets 10–20-flowered; pedicels 2–8 mm. **Flowers** petals white or purple; anthers ochroleucous or white. **Fruits** oblong, obovate, or obelliptic, 7–14 mm, glabrous; body 2–4 mm wide; wings 1–2 mm, thin; vittae 1–2 in intervals, 2–6 on commissure. **$2n$**=22.

Gravelly or rocky serpentine slopes in conifer forests. Flowering May–Aug. 700–1400 m. Sisk. CA. Native.

This species is often confused with *L. tracyi*. The pedicel length in each of the two species is more variable than reported in previous floras, and *L. engelmannii* has been found to have involucel bractlets in some individuals. The flowers in *L. tracyi* are yellow while those in *L. engelmannii* are white or purple.

Lomatium erythrocarpum Meinke & Constance

Red-fruited lomatium

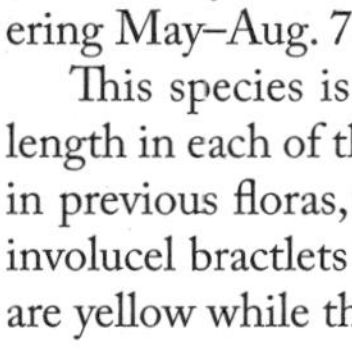

Plants acaulescent or subacaulescent. **Roots** elongate and slender taproots. **Caudices** simple or 2–3-branched; basal leaf sheaths from previous years not persistent as a thatch. **Stems** 2–9 cm; pseudoscapes 1.5–7(10) cm, subterranean, obscured by leaf sheaths. **Basal leaves** glabrous and glaucous; primary leaflets pinnate, 3–5 secondary leaflet pairs along each rachis, laterals half central primary in length; secondaries pinnatifid; tertiaries entire or pinnatifid (2–3-lobed); ultimate apical lobes lanceolate or oblong, 3–6 × 0.3–0.5 mm, tips acute or subacute. **Cauline leaves** 0–1. **Inflorescences** peduncles 1–4, spreading or ascending, 1–5.5 cm, glabrous; involucral bracts 0; rays 1–3, 0.1–0.7 cm, glabrous; involucel bractlets 0–8, linear or lanceolate, 1–4.5 × 0.1–0.4 mm, margins narrowly scarious; umbellets 5–20-flowered; pedicels 3–7 mm. **Flowers** petals white or purple; anthers purple. **Fruits** oblong or elliptic, 7–9 mm, red, glabrous, glossy; body 2–3 mm wide; wings 0.8–1.2 mm, thin; vittae 3–5 in intervals, 4–6 on commissure.

Steep, gravelly or talus slopes, granodiorite substrates, shrub-steppe and subalpine communities. Flowering Jun–Jul. 2300–2600 m. BW. Native.

Since its discovery in the 1980s, this taxon has not been found outside of the range reported in the type publication. It is found only in six small populations on Elkhorn Ridge in the Blue Mountains (Baker County).

Lomatium farinosum (Geyer ex Hooker) J.M. Coulter & Rose var. *hambleniae* (Mathias & Constance) Schlessman (p. 116)

Hamblen's lomatium

Lomatium hambleniae Mathias & Constance

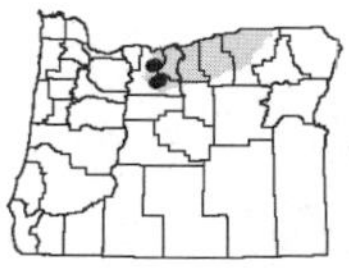

Plants acaulescent or subacaulescent. **Roots** globose tubers with an elongate and slender portion below, elongate and slender portion above tubers < 4 cm. **Caudices** simple; basal leaf sheaths from previous years absent or weathering into a sparse thatch of a few, loose fibers at base of pseudoscape. **Stems** absent or < 10 mm, 2-branched above pseudoscape; pseudoscapes 2–4 cm, subterranean, obscured by leaf sheaths. **Basal leaves** glabrous; primary leaflets ternate or pinnate, 0–3 secondary leaflet pairs along each rachis, laterals subequal to central primary in length; secondaries entire or pinnatifid (2–5-lobed); tertiaries absent; ultimate apical lobes linear, narrowly elliptic, 10–35 × 0.5–3 mm, tips acute, subacute, rounded, mucronulate or not. **Cauline leaves** 0–1, similar to basal. **Inflorescences** peduncles 1–3, slender, 4–9 cm, glabrous; involucral bracts 0; rays 2–6, 1–3 cm, glabrous; involucel bractlets 2–8, 1–4 × 0.3–0.8 mm, margins scarious; umbellets 10–15-flowered; pedicels 5–20 mm. **Flowers** petals yellow; anthers ochroleucous or yellow. **Fruits** oblong or elliptic, 5–8 mm, glabrous; body 1.5–2.5 mm wide; wings 0.4–0.6 mm, thin; vittae 1–4 in intervals, 6–8 on commissure. **$2n$**=22.

Gravelly or rocky slopes, flats in scablands and foothills. Flowering Mar–Apr. 600–1000 m. Col. WA. Native.

Crossing studies indicate that var. *hambleniae* and var. *farinosum* are not separate species. They produce fertile hybrids both in the greenhouse and in nature. Although there is a portion of the range where both the yellow-flowered variety and the typical white-flowered variety are found exclusively, both color variants and intermediates are found where the ranges overlap; a pattern of geographic distribution and intermediacy fits well within the concept of a variety.

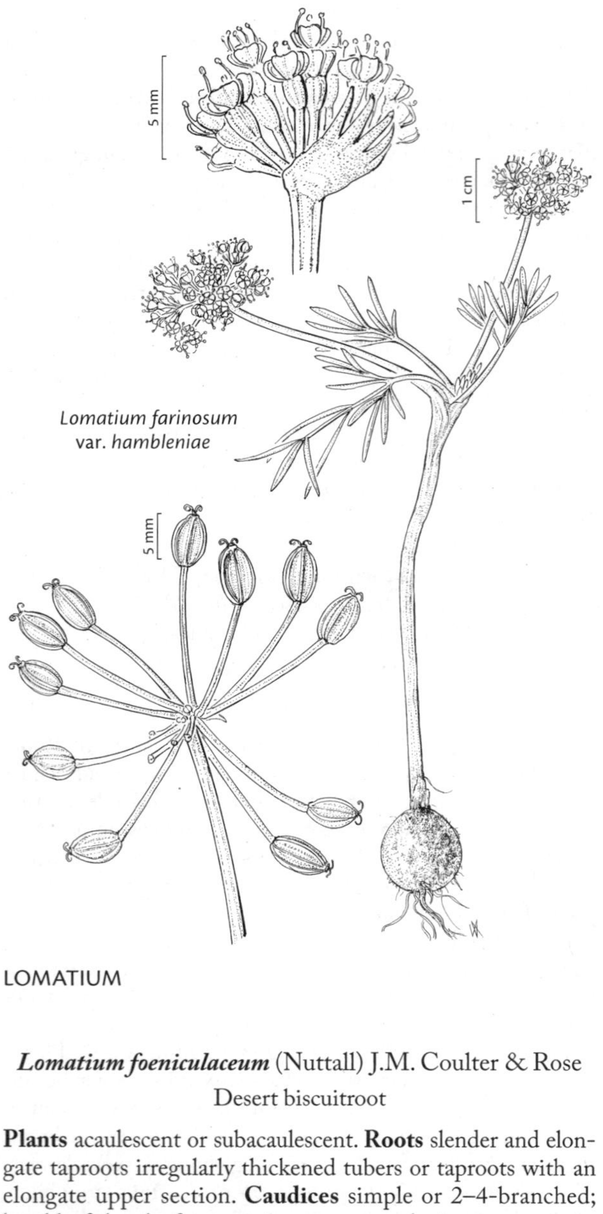

LOMATIUM

Lomatium foeniculaceum (Nuttall) J.M. Coulter & Rose

Desert biscuitroot

Plants acaulescent or subacaulescent. **Roots** slender and elongate taproots irregularly thickened tubers or taproots with an elongate upper section. **Caudices** simple or 2–4-branched; basal leaf sheaths from previous years weathering into a sparse thatch of fibrous or chartaceous scales at base of pseudoscape. **Stems** pseudoscapes subterranean, obscured by thatch and leaf sheaths. **Leaves** primary leaflets pinnate, 3–6 secondary leaflet pairs along each rachis, laterals less than half central primary in length; ultimate apical lobes elliptic, oval, oblong. **Inflorescences** decumbent, spreading, ascending, or erect; involucral bracts 0 or 1 narrow bract; involucel bractlets 4–10; umbellets 10–30-flowered; pedicels glabrate, hirsutulous or puberulent. **Flowers** petals yellow. **Fruits** sparsely to densely hirsutulous or puberulent; wings thin; vittae 1–4 in intervals, 2–4 on commissure.

Key to *Lomatium foeniculaceum* varieties:

1. Peduncles 1–2, 3–6 cm; petal margins ciliate . . . . ***L.f.*** var. ***fimbriatum***

1' Peduncles 1–6, 5–15 cm; petal margins glabrous. . . . . ***L.f.*** var. ***macdougalii***

Lomatium foeniculaceum (Nuttall) J.M. Coulter & Rose var. *fimbriatum* (W.L. Theobald) B. Boivin

Desert biscuitroot, fringed desert parsley

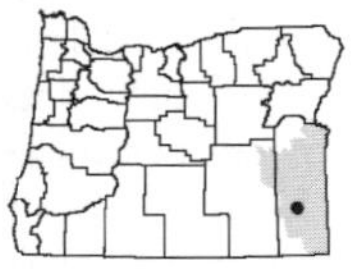

Plants acaulescent. **Roots** taproots elongate. **Stems** absent; pseudoscapes 1–2 cm. **Basal leaves** densely hirsutulous, pubescent, or puberulent; secondary leaflets pinnate; tertiaries pinnatifid (2–8-lobed); ultimate apical lobes 1–2 × 0.3–0.8 mm, tips acute, subacute, mucronulate. **Cauline leaves** 0. **Inflorescences** peduncles 1–2, 3–6 cm, hirsutulous, pubescent, or puberulent; rays 2–6, 1–4 cm, sparsely to moderately hirsutulous or puberulent; involucel bractlets, linear or lanceolate, 2–4 × 0.5–0.8 mm, glabrate, hirsutulous or puberulent, margins scarious; pedicels 3–5 mm. **Flowers** petals ciliate-margined; anthers yellow or purplish. **Fruits** elliptic or suborbicular, 6–11 mm; body 3–4 mm wide; wings 1–2 mm. **2*n***=22.

Slopes, flats, meadows. Flowering Apr–Jun. 1000–1200 m. Owy. CA, NV; southeast to UT. Native.

Lomatium foeniculaceum (Nuttall) J.M. Coulter & Rose var. *macdougalii* (J.M. Coulter & Rose) Cronquist

MacDougal's biscuitroot

Lomatium jonesii J.M. Coult. & Rose, *Lomatium foeniculaceum* (Nutt.) J.M. Coult. & Rose ssp. *macdougalii* (J.M. Coult. & Rose) W.L. Theob., *Lomatium macdougalii* J.M. Coult. & Rose

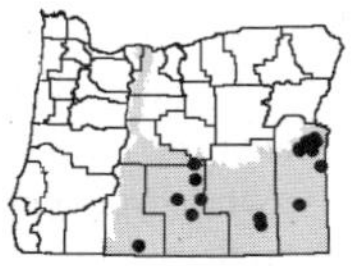

Plants acaulescent or subacaulescent. **Roots** taproots elongate, upper section (if present) 1–2 cm. **Stems** 0.5–1 cm; pseudoscapes 1–6 cm. **Basal leaves** sparsely to densely hirsutulous, pubescent, or puberulent; secondary leaflets pinnate or pinnatifid; tertiaries entire or pinnatifid (2–6-lobed); ultimate apical lobes 0.8–2.5 × 0.3–1.5 mm, tips acute, subacute or rounded, mucronulate or not. **Cauline leaves** 0–1, similar to basal. **Inflorescences** peduncles 1–6, 5–15 cm, hirsutulous, pubescent, or puberulent; rays 2–18, 1–6 cm, sparsely to moderately hirsutulous or puberulent; involucel bractlets linear or lanceolate, sometimes with 1 or 2 shallow lobes along upper margin, 2–4 × 0.5–1.2 mm, hirsutulous or puberulent, margins scarious; pedicels 3–11 mm. **Flowers** petals glabrous; anthers ochroleucous or yellow. **Fruits** elliptic, suborbicular or obovate, 5–9 mm; body 2–4 mm wide; wings 0.8–1.5 mm.

Slopes, flats, meadows. Flowering Apr–Jun. 700–1700 m. BR, ECas, Owy. CA, ID, NV, WA; northeast to MT, east to WY, southeast to NM. Native.

Lomatium gormanii (Howell) J.M. Coulter & Rose (p. 117)

Gorman's lomatium, Gorman's biscuitroot

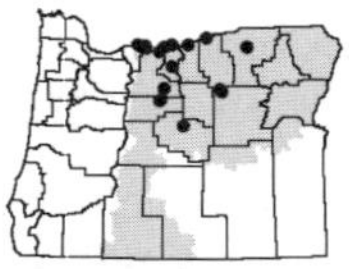

Plants acaulescent or subacaulescent. **Roots** globose tubers. **Caudices** simple; basal leaf sheaths from previous years absent or weathering into a sparse thatch of fibers and chaffy or chartaceous scales at base of pseudoscape. **Stems** absent or < 1 cm; pseudoscapes 2–6 cm, subterranean. **Basal leaves** glabrous; primary leaflets ternate or pinnate, 0–3 secondary leaflet pairs along each rachis, laterals subequal to central primary in length; secondaries entire or pinnatifid (2–3-lobed); tertiaries absent; ultimate apical lobes linear or narrowly oblong, 5–15 × 0.5–1.5 mm, tips subacute or rounded. **Cauline leaves** 0 or 1 bract-like. **Inflorescences** peduncles 1–2, decumbent, 5–7 cm, glabrous or sparsely puberulent or papillose along ridges just below distal end; involucral bracts 0; rays 1–6, 0.6–2 cm, glabrous, glabrate, sparsely papillose or puberulent; involucel bractlets 2–6, linear or lanceolate, 1–3.5 × 0.2–0.7 mm, margins narrowly scarious; umbellets 10–20-flowered; pedicels 0.5–2.5 mm. **Flowers** petals white; anthers purplish. **Fruits** elliptic, oval, 4–7 mm, sparsely to moderately papillose, puberulent, or scaberulose; body 1–1.5 mm; wings 0.5–0.7 mm; vittae 3–4 in intervals, 4–6 on commissure. ***2n***=22.

Xeric to mesic areas on rocky slopes and flats in scablands, hillsides, foothills, scrublands. Flowering Mar–May. 0–1100 m. BW, Col, ECas, Lava. ID, WA. Native.

This species grows sympatrically in Oregon with *L. piperi* and is easily confused with that species. The characteristics of mature fruit of this taxon are poorly known since few botanists collect this plant in a mature fruiting stage. The only distinguishing character between *L. gormanii* and caulescent forms of *L. piperi* is the papillate hairs on the fruit body. It is not known if the fruits are consistently smaller with shorter wings than *L. piperi* or if this is due to the prevalence of immature fruit on specimens. More collections are needed to determine the range of variability of the mature fruit of this taxon.

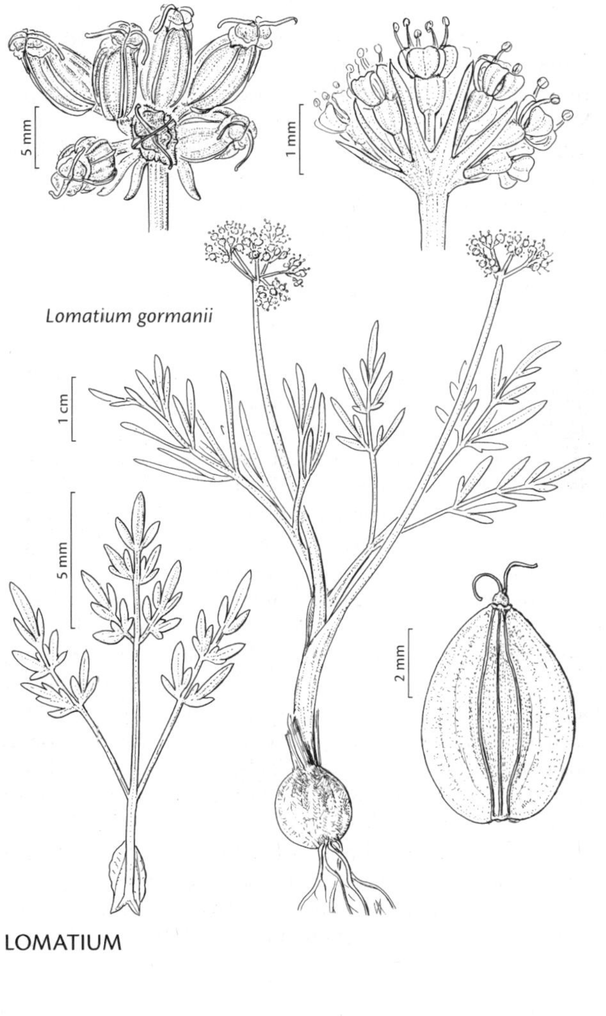

LOMATIUM

Lomatium greenmanii Mathias

Greenman's lomatium, Greenman's biscuitroot

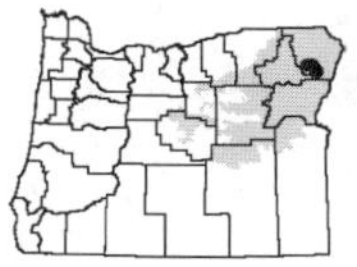

Plants cespitose and short-caulescent or caulescent. **Roots** elongate, irregularly thickened taproots. **Caudices** multicipital; basal leaf sheaths from previous years weathering into a dense, fibrous thatch at or below ground level; previous year's peduncles long-persistent in thatch as gray stalks. **Stems** 2–8 cm; pseudoscapes absent or 1–2 cm, obscured by thatch. **Basal leaves** glabrous; primary leaflets pinnate, 1–3 secondary leaflet pairs along each rachis, laterals half central primary in length; secondaries pinnate or pinnatifid; tertiaries entire or 2-lobed; ultimate apical lobes elliptic or oblong, 5–13 × 2–3 mm, tips acute, mucronulate. **Cauline leaves** 1, much reduced. **Inflorescences** peduncles 1–3 per caudex branch, 3–10 cm, glabrous; involucral bracts 0 or umbels subtended by a leaf-like bract; rays 1–3, 0.1–0.5 cm, glabrous; involucel bractlets 1–4, linear, 2–4 × 0.1–0.2 mm, margins narrowly scarious or not; umbellets 5–10-flowered; pedicels 0.5–2 mm. **Flowers** petals yellow, rarely white or purple; anthers white or purple. **Fruits** oblong, glabrous, (3)4–5 mm; body 1–2 mm wide; wings 0.4–0.7 mm, thin; vittae 1 in intervals, 2 on commissure.

Gravelly or rocky ridges, outcrops, slopes in subalpine forests and meadows, on granite-, greenstone-, or basalt-derived substrates. Flowering Jul–Aug. 2400–2800 m. BW. Native.

This taxon was thought to be extinct until a population was rediscovered on Mount Howard, in the Wallowa Mountains, in 1975. The type locality "head of Keystone Creek" has never been relocated. It is likely a name that was locally used in 1900 when Cusick collected the type of this taxon.

Lomatium hallii (S. Watson) J.M. Coulter & Rose (p. 118)

Hall's lomatium, Hall's biscuitroot

Lomatium nelsonianum J.F. Macbr.

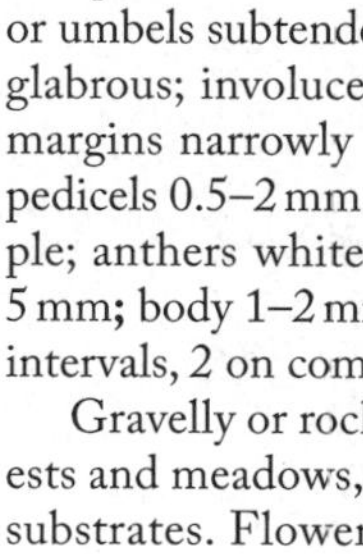

Plants acaulescent or subacaulescent. **Roots** slender and elongate or irregularly thickened taproots divided into 2–5 elongate and slender roots at base. **Caudices** simple or 2–5-branched; basal leaf sheaths from previous years weathering into a sparse thatch of a few loose fibers or chaffy to chartaceous scales at base of pseudoscape. **Stems** 0.5–5 cm, internodes concealed by leaf sheaths or only upper 5–10 mm visible; pseudoscapes absent or 0.5–4 cm, subterranean. **Basal leaves** glabrous or sparsely scaberulose along leaflet margins; primary leaflets pinnate, 4–8 secondary leaflet pairs along each rachis, laterals less than half central primary in length; secondaries pinnate; tertiaries pinnatifid (2–7-lobed); ultimate apical lobes linear, narrowly oblong, elliptic, oblong, 1.5–4 × 0.6–1.2 mm, tips acute, subacute, mucronulate. **Cauline leaves** 0–3, similar to basal. **Inflorescences** peduncles 1–6, 10–25 cm, glabrous, glabrate, or sparsely papillose, scaberulose or muriculate at distal end; involucral bracts 0; rays 4–17, 1–7 cm, glabrous, glabrate, or sparsely papillose, scaberulose or muriculate along ridges; involucel bractlets 0–4, linear or lanceolate, 1–8 × 0.4–1 mm, margins narrowly scarious; umbellets 10–25-flowered; pedicels 5–15 mm. **Flowers** petals ochroleucous or yellow; anthers ochroleucous or yellow. **Fruits** elliptic,

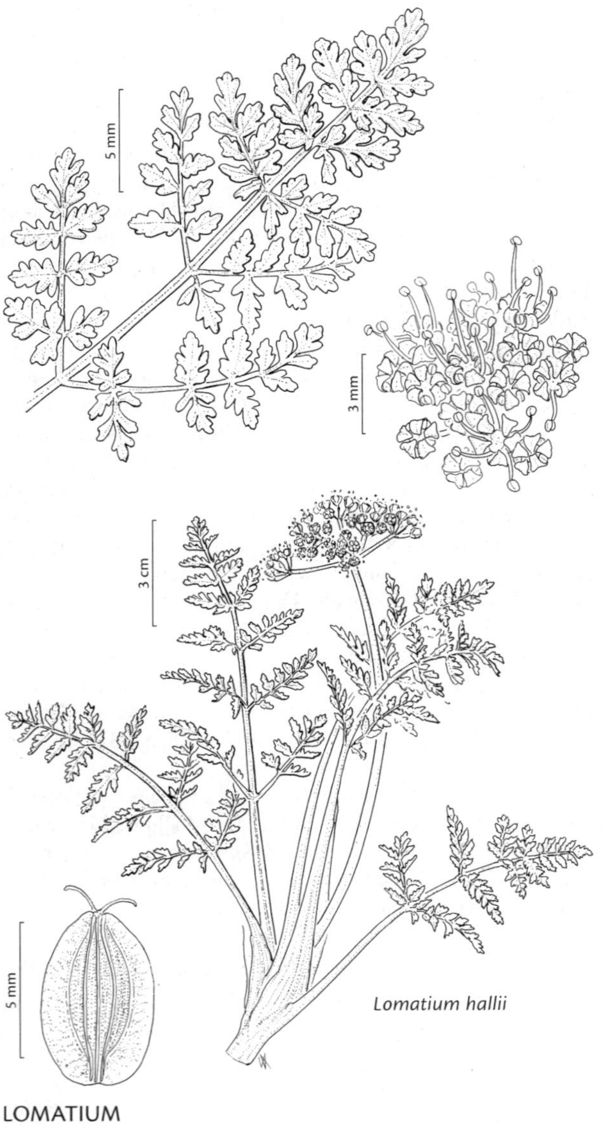

LOMATIUM

oblong, 7–12 mm, glabrous; body 2–4 mm wide; wings 1–2 mm, thin; vittae 2–3 in intervals, 4–6 on commissure. **2*n*=22.**

Rocky slopes, rock outcrops, oak forests, conifer forests. Flowering Jan–Aug. 0–1900 m. Casc, CR, Sisk, WV. CA. Native.

Reports of *Lomatium hallii* from Washington are based on misidentifications.

Lomatium hendersonii (J.M. Coulter & Rose) J.M. Coulter & Rose

Henderson's lomatium, Henderson's biscuitroot

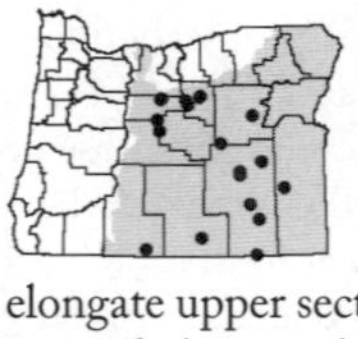

Plants acaulescent or subacaulescent. **Roots** globose or irregularly thickened tubers with elongate upper section < 1 cm and basal section abruptly tapered to a slender, elongate taproot, or irregularly thickened tubers with elongate upper section < 1 cm and bifurcate basal section that is tapered along each axis to slender taproot. **Caudices** simple or 2–3-branched; basal leaf sheaths from previous years weathering into a sparse thatch of a few, loose fibers or chaffy scales at base of pseudoscape. **Stems** absent or < 3 cm; pseudoscapes 1–3 cm, subterranean, obscured by leaf sheaths. **Basal leaves** glabrous; primary leaflets pinnate, 1–4 secondary leaflet pairs along each rachis, laterals less than half of central primary in length; secondaries pinnate or pinnatifid (3–6-lobed); tertiaries absent or entire; ultimate apical lobes elliptic, oval, obovate, 2–4.5 × 1.5–3 mm, tips obtuse, subacute, rounded, mucronulate. **Cauline leaves** 0–1, similar to basal. **Inflorescences** peduncles 1–6, 6–13 cm, glabrous; involucral bracts 0; rays 2–6, 0.5–2(2.5) cm, glabrous; involucel bractlets 3–6, linear, lanceolate, narrowly elliptic, 2–4 × 0.3–0.8 mm, margins narrowly scarious; umbellets 5–15-flowered; pedicels 1–5 mm. **Flowers** petals ochroleucous or yellow; anthers ochroleucous or yellow. **Fruits** ovate, oval, or orbicular, 4–6 mm, glabrous; body 1–2 mm wide; wings thin, 0.2–0.8 mm; vittae 1 in intervals, 2 on commissure.

Open gravelly or rocky slopes and flats in scrubland and woodland, on igneous-derived substrates. Flowering Mar–Jun. 400–2200 m. BR, BW, ECas, Lava, Owy. CA. Native.

The type publication of *Lomatium hendersonii* stated that the flowers were white. Mathias did not see the types and retained Coulter and Rose's interpretation. The types of *Leptotaenia leibergii* (basionym of *L. roseanum*) were reported to be yellow-flowered. If it can be confirmed that the types of *L. hendersonii* were white and that there are extant populations in the John Day River Valley that are white-flowered *L. hendersonii*, the status of this taxon would need to be re-evaluated. *Lomatium hendersonii* is found mostly in the mountains of central Oregon, east of the Cascades and southwest of the Wallowas with disjunct outliers in southern Oregon and northern California. The reports from Owyhee County, Idaho, and Washoe County, Nevada, are likely misidentifications of *L. roseanum*.

Lomatium hendersonii is so rarely collected that the characters of the mature fruit are poorly known. More specimens of this taxon in fruiting stage are needed to determine if the wing and body are narrower than those in *L. roseanum*.

Lomatium howellii (S. Watson) Jepson

Howell's biscuitroot, Howell's lomatium

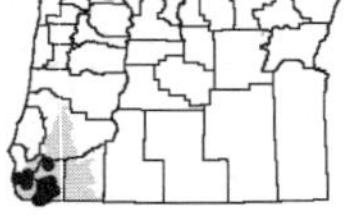

Plants acaulescent or subacaulescent. **Roots** elongate and slender taproots or irregularly thickened taproots divided into 2–5 elongate and slender taproots at base. **Caudices** simple or 2–3-branched; basal leaf sheaths from previous years weathering into a few fibers or a sparsely chaffy or chartaceous thatch at base of pseudoscape. **Stems** absent or 1–4 cm; pseudoscapes absent or < 2 cm. **Basal leaves** glabrous or glaucous; primary leaflets ternate or pinnate, 0–2 secondary leaflet pairs along each rachis, laterals less than half central primary in length; secondaries ternate or pinnatifid (4–5-lobed); tertiaries absent; ultimate apical lobes suborbicular or obovate, 5–30 × 5–30 mm, margins erose, dentate, or serrate, teeth pungent, 3–5-lobed, central lobe entire or shallowly lobed, tips rounded, obtuse, truncate. **Cauline leaves** 0–1, similar to basal. **Inflorescences** peduncles 1–2, 10–40 cm, glabrous or glaucous; involucral bracts 0; rays 10–15, 2.5–6 cm, glabrous; involucel bractlets 2–6, linear or lanceolate, 3–10 × 0.3–1 mm, margins narrowly scarious or not; umbellets 20–30-flowered; pedicels 8–12 mm. **Flowers** petals ochroleucous or yellow; anthers ochroleucous or yellow. **Fruits** elliptic, suborbicular,

7–11 mm, glabrous; body 3–4 mm wide; wings 1–2 mm, thin; vittae 2–3 in intervals, 9 on commissure. ***2n***=22.

Rocky slopes, open gravelly sites, chaparral or conifer forests, serpentine substrates. Flowering Apr–Jul. 300–1400 m. Sisk. CA. Native.

The broadest leaf forms of *L. martindalei* are often misidentified as *L. howellii*. Although their overall leaflet shapes are identical, the pungent-tipped teeth on the ultimate lobes of *L. howellii* are diagnostic and not present in any other species, which are at most mucronulate-tipped.

Lomatium klickitatense J.A. Alexander & W. Whaley

Klickitat biscuitroot, Klickitat lomatium

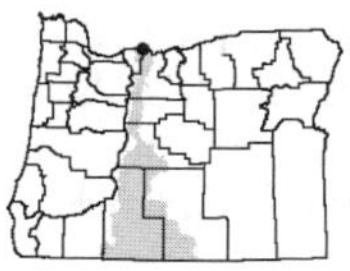

Plants subacaulescent or caulescent. **Roots** stout, irregularly thickened taproots. **Caudices** simple or multicipital; basal leaf sheaths from previous years weathering into a dense thatch of fibers and chaffy scales at base of stem; previous year's peduncles long-persistent in thatch as gray stalks. **Stems** 4–15 dm. **Basal leaves** glabrous; primary leaflets pinnate, 5–8 secondary leaflet pairs along each rachis, laterals at least half to subequal to central primary in length; secondaries pinnate; tertiaries pinnate; quaternaries entire or pinnatifid (3–20-lobed), ultimate leaflet pinnules not leafy to base, with a naked intercostal region between rachilla and basal-most pinnule (appearing like a petiolule), (4.5)5–8 mm; ultimate apical lobes linear, in cross section subterete, (6.5)7–15 × 0.1–0.4 mm, tips acuminate. **Cauline leaves** 1–3 (more in robust plants), glabrous, similar to basal. **Inflorescences** peduncles 1–10+, stout, not inflated, 30–60 cm, 6–8 mm wide (1 cm below tip), glabrous; involucral bracts 0; rays 16–30, 4–12 cm, glabrous; involucel bractlets 0–10, linear or lanceolate, 2–10 × 0.1–0.6 mm, margins narrowly scarious; umbellets 10–40-flowered; pedicels 9–15 mm. **Flowers** petals yellow; anthers yellow or ochroleucous. **Fruits** oblong or elliptic, 9–16 mm, glabrous; body 3–6 mm; wings 1–2 mm, thin; vittae 1–2 in intervals, 2–6 on commissure.

Talus slopes, rock outcrops, cliff faces, rocky hillsides, open rocky sites in the forest understory. Flowering May. 50–100 m. ECas. WA. Native.

In Oregon, this taxon has been mostly overlooked or collected when the leaflets are not fully expanded. It is found commonly on slopes and hillsides along the Columbia River near the town of Hood River and on slopes in the Klickitat River drainage in Washington. This taxon is replaced abruptly by *L. papilioniferum* eastward near the town of Mosier and westward just east of Mitchell Point. It is not known how far south this taxon extends, but *L. papilioniferum* is found just outside this subregion in Tygh Valley. The distribution of *L. klickitatense* suggests that it is a Pleistocene relict. Modern contact with *L. papilioniferum* may be a result of that taxon's post-Pleistocene range expansion westward as the cliffs and talus along the Columbia Gorge became drier, hotter, and more exposed.

Lomatium laevigatum (Nuttall) J.M. Coulter & Rose

Smooth lomatium

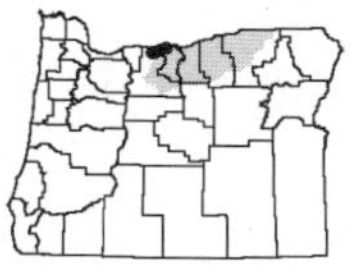

Plants caulescent or subacaulescent. **Roots** thickened, elongate taproots. **Caudices** simple, 2–3-branched or multicipital; basal leaf sheaths from previous years weathering into a fibrous thatch at or below ground level; previous year's peduncles persistent in thatch as gray stalks. **Stems** 20–40 cm; pseudoscapes absent. **Basal leaves** glabrous; primary leaflets pinnate, 1–4 secondary leaflet pairs along each rachis, laterals at least half of central primary in length; secondaries pinnate; tertiaries entire or pinnatifid (2–3-lobed); ultimate apical lobes linear, narrowly oblong, narrowly elliptic, 2–30 × 1–3 mm, tips acute or subacute, mucronulate. **Cauline leaves** 1–2, similar to basal. **Inflorescences** peduncles 1–4 (or more in robust plants), ascending or erect, stout, 10–30 cm, glabrous; involucral bracts 0; rays 8–20, 1–4.5 cm, glabrous; involucel bractlets 0–2, 0.7–2 × 0.1–0.3 mm, margins not scarious; umbellets 15–25-flowered; pedicels 5–9 mm. **Flowers** petals yellow; anthers ochroleucous or yellow. **Fruits** oblong, elliptic, 6–10 mm, glabrous; body 1.5–3 mm wide; wings 1–2 mm, thin; vittae 1 in intervals, 2 on commissure.

Cliffs, ledges, crevices, rocky slopes, outcrops in scablands and scrublands, on basalt-derived substrates. Flowering Apr–May. 50–700 m. Col. WA. Native.

Lomatium leptocarpum (Torrey & A. Gray) J.M. Coulter & Rose

Gumbo lomatium, slender-fruited lomatium

Lomatium ambiguum (Nutt.) J.M. Coult. & Rose var. *leptocarpum* (Torr. & A. Gray) Jeps., *Lomatium bicolor* (S. Watson) J.M. Coult. & Rose var. *leptocarpum* (Torr. & A. Gray) Schlessman

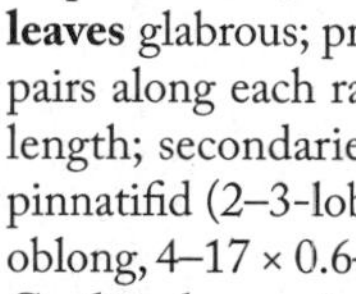

Plants acaulescent. **Roots** globose tubers with an elongate upper section < 1 cm and abruptly tapering to elongate and slender taproots at base, or irregularly thickened tubers abruptly tapered to caudex. **Caudices** simple or 2–6-branched; basal leaf sheaths from previous years absent or weathering into a sparse thatch of a few, loose fibers or chaffy scales at base of pseudoscape. **Stems** absent; pseudoscapes 1–4 cm, subterranean, obscured by leaf sheaths. **Basal leaves** glabrous; primary leaflets pinnate, 3–6 secondary leaflet pairs along each rachis, laterals subequal to central primary in length; secondaries pinnate or pinnatifid; tertiaries entire or pinnatifid (2–3-lobed); ultimate apical lobes linear or narrowly oblong, 4–17 × 0.6–1.5 mm, tips acute, acuminate, mucronulate. **Cauline leaves** 0. **Inflorescences** peduncles 1–4, decumbent, spreading, ascending, or erect, 7–30 cm, glabrous or sparsely scaberulose or papillose along ridges at distal end; involucral bracts 0; rays 4–10, ascending or erect, 1–8 cm, glabrous; involucel bractlets 4–10, 3–6 × 0.5–1.5 mm, margins scarious; umbellets 10–20-flowered; pedicels 1–4 mm. **Flowers** petals yellow; anthers ochroleucous or yellow. **Fruits** linear or narrowly oblong, 9–13 mm, glabrous, body 1–2.5 mm wide; wings 0.4–0.6 mm, thin or slightly thickened; vittae 1 in intervals, 2–4 on commissure.

Slopes, flats, meadows, streambanks. Flowering Mar–Jul. 100–2700 m. BR, BW, Col, ECas, Owy. CA, ID, NV, WA; northeast to MT, east to WY, southeast to AZ. Native.

Lomatium macrocarpum (Hooker & Arnott) J.M. Coulter & Rose

Large-fruited desert parsley, bigseed lomatium

Lomatium artemisiarum (Piper) Piper, *Lomatium elllipticum* (Torr. & A. Gray) J.M. Coult. & Rose, *Lomatium macrocarpum* (Hook. & Arn.) J.M. Coult. & Rose var. *artemisiarum* Piper, *Lomatium macrocarpum* (Hook. & Arn.) J.M. Coult. & Rose var. *ellipticum* (Torr. & A. Gray) Jeps., *Lomatium macrocarpum* (Hook. & Arn.) J.M. Coult. & Rose var. *semivittatum* Piper

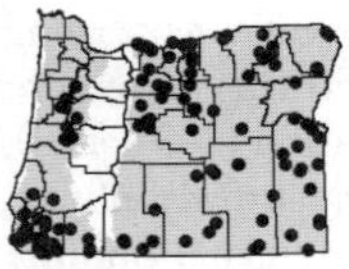

Plants caulescent or subacaulescent. **Roots** elongate and slender taproots or elongate and thickened taproots abruptly tapered to caudex or elongate and irregularly thickened tubers with an elongate and slender upper portion 2–12 cm and elongate and slender taproots at base. **Caudices** simple or 2–4-branched; basal leaf sheaths from previous years absent or weathering into a fibrous, sparsely chartaceous, or chaffy thatch at base of pseudoscape. **Stems** 2–15 cm, puberulent or villous; pseudoscapes 1–10 cm, subterranean, obscured by leaf sheaths. **Basal leaves** 3–4-pinnate, glabrous, glaucous, glabrate, or sparsely pubescent, puberulent, scabrous, or villous; primary leaflets pinnate, 3–8 secondary leaflet pairs along each rachis, laterals less than half to greater than half central primary in length; secondaries pinnate or pinnatifid; tertiaries pinnate or pinnatifid; quaternaries absent, entire, or pinnatifid (3–20-lobed); ultimate apical lobes linear, oblong, narrowly oblong, elliptic, oval, 1–5 × 0.4–1 mm, tips acute, acuminate, or subacute, mucronulate or not. **Cauline leaves** 1–4, similar to basal. **Inflorescences** peduncles 1–30, ascending or erect, 7–30 cm, sparsely to densely puberulent, pubescent, or villous; involucral bracts 0 or umbels subtended by a leaf-like bract; rays 5–15, 2–6 cm, puberulent or villous; involucel bractlets 2–10, linear to lanceolate, 3–12 × 0.3–2 mm, puberulent or villous, margins entire, toothed or 2–3-lobed, tips acute or acuminate, apical lobes acute, margins entire or 2–3-toothed, narrowly scarious or not; umbellets 10–30-flowered; pedicels, 2–10 mm, glabrous, glabrate, or sparsely puberulent. **Flowers** petals white, ochroleucous, yellow or purple; anthers white or purple. **Fruits** oblong, narrowly oblong, elliptic, or oval, 7–20 mm, glabrous, glabrate, or sparsely puberulent; body 2–3.5 mm wide; wings 1–2.5 mm, thin or thickened; vittae 1–3 in intervals, 2–6 on commissure. ***2n***=22.

Mainly open gravelly sites and rocky slopes, but widespread across a diverse range of habitat types and plant communities. Flowering Mar–Jun. 0–2300 m. BR, BW, Col, CR, ECas, Lava, Owy, Sisk, WV. CA, ID, NV, WA; north to British Columbia, northeast to Manitoba, east to SD, southeast to UT. Native.

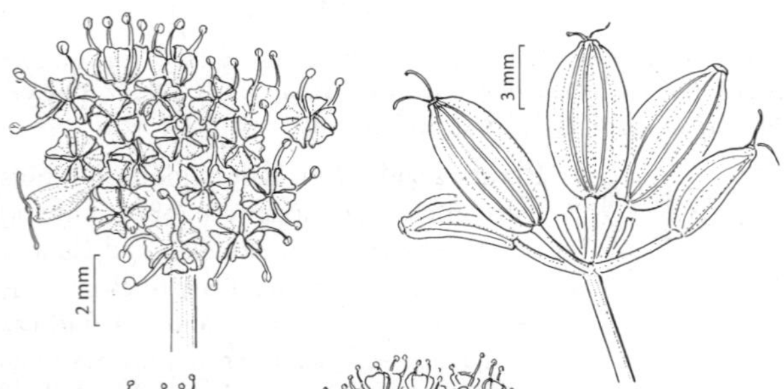

Lomatium martindalei

LOMATIUM

Lomatium martindalei (J.M. Coulter & Rose) J.M. Coulter & Rose (p. 120)

Cascade desert parsley, Coast Range lomatium, Martindale's lomatium

Lomatium angustatum (J.M. Coult. & Rose) H. St. John, *Lomatium angustatum* (J.M. Coult. & Rose) H. St. John var. *flavum* G.N. Jones, *Lomatium martindalei* (J.M. Coult. & Rose) J.M. Coult. & Rose var. *angustatum* (J.M. Coult. & Rose) J.M. Coult. & Rose, *Lomatium martindalei* (J.M. Coult. & Rose) J.M. Coult. & Rose var. *flavum* (G.N. Jones) Cronquist, *Lomatium martindalei* (J.M. Coult. & Rose) J.M. Coult. & Rose var. *martindalei*, *Peucedanum martindalei* J.M. Coult. & Rose

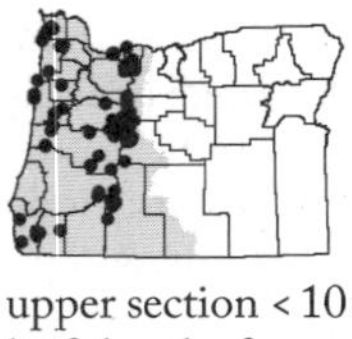

Plants acaulescent, subacaulescent, or caulescent. **Roots** elongate, slender or irregularly thickened taproots divided into 2–5 elongate and slender roots at base; irregularly thickened taproots with an elongate upper section < 10 cm. **Caudices** simple or 2–3-branched; basal leaf sheaths from previous years weathering into a sparse thatch of a few loose fibers and chaffy or chartaceous scales at base of pseudoscape. **Stems** 3–20 cm, lower nodes concealed by leaf sheaths or only upper 5–10 mm visible; pseudoscapes absent or 0.5–2 cm, subterranean. **Basal leaves** glabrous, thinly glaucous, or sparsely scaberulose or muriculate along leaflet margins and midveins, thin or thick and somewhat fleshy; primary leaflets pinnate, 2–4 secondary leaflet pairs along each rachis, laterals at least half central primary in length; secondaries pinnate or pinnatifid; tertiaries pinnatifid (3–20-lobed); ultimate apical lobes narrowly oblong, elliptic, oblong, oval, or obovate, 1–7 × 1–4 mm, tips acute, subacute, obtuse, rounded or mucronulate, entire or 2–8-toothed. **Cauline leaves** 0–3, similar to basal. **Inflorescences** peduncles 1–4, 2–20 cm, glabrous, glabrate, or sparsely scaberulose or muriculate along ridges just below distal end; involucral bracts 0; 4–16, rays 1–6 cm, glabrous, glabrate, or sparsely scaberulose or muriculate along ridges; involucel bractlets 0–4, linear, lanceolate, 0.5–1.5 × 0.1–0.3 mm, margins not scarious; umbellets 10–25-flowered; pedicels 3–8 mm. **Flowers** petals ochroleucous or yellow; anthers ochroleucous or yellow. **Fruits** elliptic or oblong, 6–16 mm, glabrous; body 1.5–4 mm wide; wings 1–2.5 mm, thin; vittae 1–2 in intervals, 2–4 on commissure. ***2n***=22.

Open areas and rocky slopes in meadows, woodland, conifer forests. Flowering Apr–Aug. 200–2500 m. Casc, CR, ECas, Est, Sisk, WV. CA, WA; north to British Columbia Native.

The types of *Peucedanum martindalei* var. *martindalei* and *P. m.* var. *angustatum* refer to a single set of Howell specimens at US that all appear to be duplicates of the same collection. As a result, there seems to be persistent confusion in application of those names to variants found on herbarium specimens. Other than fruit size, it is not possible to assign the two leaf forms described in the type publication to a specific variety. Based on this ambiguity, variety *angustatum* is recognized as a synonym of *L. martindalei*. Variety *flavum* is also recognized herein as a synonym of this taxon since both yellow and ochroleucous forms are found scattered throughout the range of this species. A more accurate definition of the leaf morphology of variety *flavum* is required before this taxon should be recognized as a separate variety. Currently, we consider it a coastal bluff, coastal-montane, and subalpine Cascadian ecotype of *L. martindalei* with slightly thicker leaves.

Lomatium minus (Rose ex Howell) Mathias & Constance

John Day Valley desert parsley

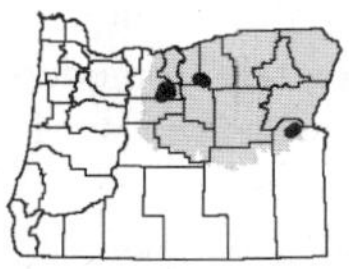

Plants caulescent or subacaulescent. **Roots** stout and irregularly thickened tubers, basal section abruptly tapered into slender, elongate taproots or branched into 2–3 elongate and slender taproots or a stout and thickened, elongate taproot. **Caudices** simple or 2–3-branched; basal leaf sheaths from previous years weathering into a fibrous thatch at or below ground level; peduncles persistent in thatch as gray stalks. **Stems** 5–30 cm, mostly subterranean, above ground portion stout and inflated; pseudoscapes absent. **Basal leaves** glabrous; primary leaflets pinnate, 4–7 secondary leaflet pairs along each rachis, laterals half to subequal to central primary in length; secondaries pinnate; tertiaries pinnate; quaternaries pinnate or pinnatifid (3–20-lobed); ultimate apical lobes linear, narrowly oblong, 1–2.5 × 0.3–0.7 mm, tips acute, acuminate, or subacute, mucronulate or not. **Cauline leaves** 1, similar to basal. **Inflorescences** peduncles 1–3, decumbent, spreading, or ascending, stout and inflated, widest at or below middle and tapering to tip, 5–15 cm, glabrous; involucral bracts 0; rays 6–15, 1–5 cm, glabrous; involucel bractlets 4–8, linear or lanceolate, 3–9 × 0.2–0.5 mm, margins narrowly scarious or not; umbellets 10–20-flowered; pedicels 3–6 mm. **Flowers** petals purple or reddish purple; anthers purple. **Fruits** oblong or elliptic, 12–16 mm, glabrous; body 2.4–4 mm wide; wings 0.8–1.5 mm, slightly corky-thickened; vittae 1 in intervals, 3–4 on commissure.

Gravelly or rocky slopes, talus and rock outcrops in scrubland. Flowering Mar–May. 700–1300 m. BW, Col, Lava. Native. ❧

This taxon is endemic to the Columbia Basin and the adjacent foothills of the Blue Mountains. Specimens from north of the Columbia River in Washington or south of the Blue Mountains in Oregon are misidentifications of *L. tuberosum* or *L. papilioniferum*, which have nearly identical leaflet structures. When this species is in late flowering to fruiting stage, the inflated peduncles are diagnostic and unmistakable.

Lomatium nevadense (S. Watson) J.M. Coulter & Rose var. ***nevadense***

Nevada desert parsley, Nevada biscuitroot

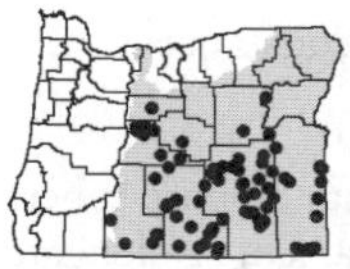

Plants acaulescent, subacaulescent, or caulescent. **Roots** elongate and slender taproots, irregularly thickened taproots with an elongate and slender upper portion 2–15 cm. **Caudices** simple or 2–6-branched; basal leaf sheaths from previous years weathering into a sparse thatch of fibrous, chaffy, or chartaceous scales at base of pseudoscape. **Stems** 1–5 cm, sparsely to moderately hirsutulous, pubescent or puberulent; pseudoscapes 1–8 cm, subterranean (in robust plants, upper portion above ground), basal portion obscured by leaf sheaths. **Basal leaves** sparsely to moderately hirsutulous, pubescent, or puberulent; primary leaflets pinnate, 3–7 secondary leaflet pairs along each rachis, laterals less than half central primary in length; secondaries pinnate or pinnatifid; tertiaries entire or pinnatifid (2–5-lobed); ultimate apical lobes narrowly oblong, linear, elliptic, 2–4 × 0.5–2 mm, tips acute, subacute, mucronulate. **Cauline leaves** 0–1, similar to basal. **Inflorescences** peduncles 1–5, 5–30 cm, hirsutulous, pubescent, or puberulent; involucral bracts 0; rays 1–15, 2–5 cm, sparsely to moderately pubescent or puberulent; involucel bractlets 4–8, linear or lanceolate, 2–5 × 0.4–1.5 mm, glabrate or sparsely puberulent, margins scarious, sometimes with 1 or 2 shallow lobes along upper margin; umbellets 10–20-flowered; pedicels 4–15 mm, glabrate or sparsely puberulent. **Flowers** petals white or ochroleucous; anthers purple. **Fruits** elliptic or suborbicular, 6–11 mm, glabrous, glabrate, hirsutulous or puberulent; body 2.7–5 mm wide; wings 1.5–3.2 mm, thin; vittae 2–4 in intervals, 4–8 on commissure. $2n=22$.

Rocky or gravelly slopes, flats. Flowering Mar–Jun. 700–2300 m. BR, BW, ECas, Lava, Owy. CA, ID, NV; southeast to NM. Native.

Lomatium nudicaule (Pursh) J.M. Coulter & Rose

Barestem lomatium, barestem biscuitroot

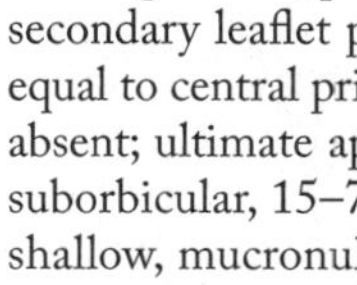

Plants acaulescent. **Roots** stout and thickened, elongate taproots or stout and irregularly thickened taproot, occasionally branched below. **Caudices** simple or 2–3-branched; basal leaf sheaths from previous years absent or weathering into a fibrous and chartaceous or chaffy thatch at base of pseudoscape. **Stems** absent; pseudoscapes 0.5–1 cm, subterranean, obscured by leaf sheaths. **Basal leaves** glabrous; primary leaflets entire, ternate, or pinnate, 0–3 secondary leaflet pairs along each rachis, laterals half to subequal to central primary in length; secondaries entire; tertiaries absent; ultimate apical lobes lanceolate, ovate, oval, elliptic, or suborbicular, 15–70 × 10–60 mm, tips 3(4–7)-toothed, teeth shallow, mucronulate. **Cauline leaves** 0. **Inflorescences** peduncles 1–2, ascending or erect, inflated and hard at base of umbel in fruit, 10–35 cm, glabrous; involucral bracts 0 or umbels subtended by a leaf-like bract; rays 10–20, 10–35 cm, glabrous; involucel bractlets 0; umbellets 10–30-flowered; pedicels 3–10 mm. **Flowers** petals yellow; anthers ochroleucous or yellow. **Fruits** oblong, 9–14 mm, glabrous; body 2–4 mm wide; wings 0.5–1 mm, thin or slightly thickened; vittae 1–4 in intervals, 4–7 on commissure. $2n=22$.

Open sites in scrublands, rocky slopes, woodlands, conifer forests. Flowering Mar–Jul. 0–2200 m. All ecoregions. CA, ID, NV, WA; north to British Columbia, southeast to UT. Native.

Lomatium ochocense Helliwell & Constance

Ochoco lomatium

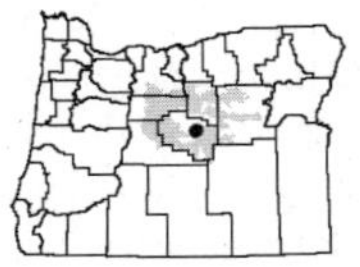

Plants acaulescent. **Roots** globose tubers with an irregularly thickened, or elongate and slender upper portion 2–4 cm. **Caudices** simple or 2–3-branched; basal leaf sheaths from previous years absent or weathering into a thatch of a few, loose fibers or chartaceous scales at base of pseudoscape. **Stems** absent; pseudoscapes 2–4 cm, subterranean. **Basal leaves** glabrous and glaucous, thick or somewhat succulent; primary leaflets pinnate, 1–3 secondary leaflet pairs along each rachis, laterals half central primary in length; secondaries pinnatifid (4–7-lobed); tertiaries absent; ultimate apical lobes obovate or oblong, 2–6 × 2–3.5 mm, tips subacute, obtuse, or rounded, mucronulate. **Cauline leaves** 0. **Inflorescences** peduncles 1–2, spreading or ascending, 4–6 cm, glabrous; involucral bracts 0; rays 2–6, 0.4–1 cm, glabrous; involucel bractlets 4–8, elliptic or lanceolate, 2–4 × 0.4–1 mm, margins scarious; umbellets 10–15-flowered; pedicels 2–4 mm. **Flowers** petals yellow; anthers ochroleucous or yellow. **Fruits** elliptic, 5–8 mm, glabrous; body 2–3.5 mm wide; wings 0.5–1 mm, thin; vittae 2–3 in intervals, 4 on commissure.

Gravelly or rocky outcrops or slopes in scablands, on basalt-derived substrates. Flowering Apr–May. 1300–1500 m. Lava. Native.

More collections from Oregon are needed to determine the range of variability of the morphology of this taxon.

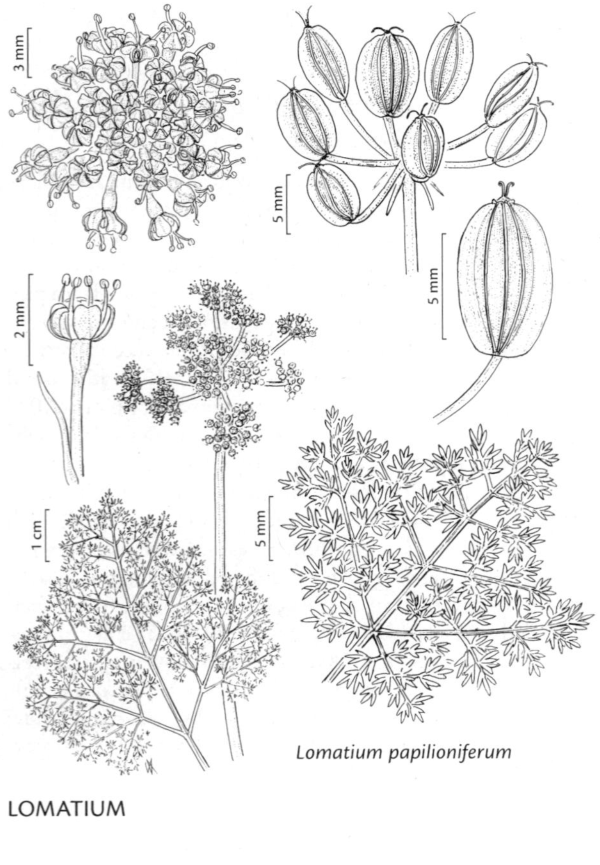

Lomatium papilioniferum

Lomatium oreganum (J.M. Coulter & Rose) J.M. Coulter & Rose

Blue Mountain biscuitroot, Oregon lomatium, Wallowa lomatium

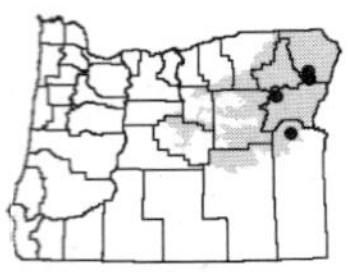

Plants densely cespitose and acaulescent. **Roots** elongate, irregularly thickened taproots. **Caudices** multicipital; basal leaf sheaths from previous years weathering into a dense, fibrous thatch at or below ground level; previous year's peduncles long-persistent in thatch as gray stalks. **Stems** absent; pseudoscapes absent or 1–2 cm, obscured by thatch. **Basal leaves** puberulent, hirsutulous, or villosulous; primary leaflets pinnate, 2–3 secondary leaflet pairs along each rachis, laterals less than half central primary in length; secondaries pinnate or pinnatifid; tertiaries entire or pinnatifid (2–3-lobed); ultimate apical lobes oblanceolate, elliptic, obelliptic, or oblong, 2–4 × 0.6–0.8 mm, tips acute, subacute, or obtuse, mucronulate or not. **Cauline leaves** 0. **Inflorescences** peduncles 1–3 per caudex branch, 1–8 cm, puberulent or villosulous; involucral bracts 0 or umbels subtended by a leaf-like bract; rays 1–2, 0.1–0.5 cm, glabrate or puberulent; involucel bractlets 1–4, linear, 3–4 × 0.5–0.7 mm, margins not scarious; umbellets 5–10-flowered; pedicels 0.5–2 mm, glabrate or sparsely puberulent. **Flowers** petals yellow; anthers ochroleucous or yellow. **Fruits** oblong, (3)4–5 mm, sparsely to moderately puberulent or villosulous; body 1–2 mm wide; wings 0.1–0.5 mm, thin; vittae 2–3 in intervals, 4 on commissure.

Sandy crevices, gravel, talus slopes, granitic-derived substrates. Flowering Apr–Jul. 2200–3000 m. BW. Native.

Lomatium papilioniferum J.A. Alexander & W. Whaley (p. 122)

Butterfly-bearing biscuitroot

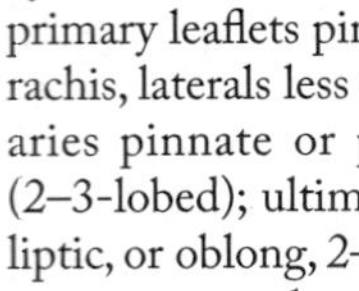

Plants acaulescent, subacaulescent, or caulescent. **Roots** stout, irregularly thickened taproots. **Caudices** simple or multicipital; basal leaf sheaths from previous years weathering into a dense thatch of fibers and chaffy scales at base of stem; previous year's peduncles long-persistent in thatch as gray stalks. **Stems** 4–8 dm. **Basal leaves** 2–4-pinnate, glabrous; primary leaflets pinnate with 5–8 secondary leaflet pairs along each rachis, laterals at least half to subequal to central primary in length; secondaries pinnate; tertiaries pinnate; quaternaries entire or pinnatifid (3–30-lobed); ultimate leaflet pinnules not leafy to base, with a naked intercostal region between rachilla and basal-most pinnule (appearing like a petiolule), 1–5(6) mm, intercostal areas not obscured by pinnule lobes; ultimate apical lobes linear, lanceolate, deltoid, in cross section dorsiventrally compressed (appearing flat) or subterete, 1–7(9) × 0.1–0.6 mm, tips acute or acuminate. **Cauline leaves** 0–2 (or more in robust plants), similar to basal. **Inflorescences** peduncles 1–10+, slender to stout, not inflated, 7–60 cm × 2–6 mm (1 cm below distal end), glabrous; involucral bracts 0; rays 6–40, 1–20 cm, glabrous; involucel bractlets 0–10, linear or lanceolate, 2–8 × 0.1–0.6 mm, margins narrowly scarious; umbellets 10–40-flowered; pedicels 6–15 mm. **Flowers** petals yellow; anthers yellow or ochroleucous. **Fruits** oblong or elliptic, 6–13 mm, glabrous; body 2–6 mm; wings 0.5–2 mm, thin; vittae 1–2 in intervals, 2–6 on commissure.

Diverse habitat types but most commonly talus slopes, rock outcrops, cliff faces, rocky hillsides. Flowering Mar–Jul. 0–2200 m. BR, BW, Casc, Col, ECas, Lava, Owy. CA, ID, NV, WA; north to British Columbia. Native.

Most populations in Oregon that were previously identified as *Lomatium grayi* are actually *L. papilioniferum*. Reports of *L. salmoniflorum* in Oregon are misidentified specimens of this taxon or *Cymopterus terebinthinus*, due to a similar leaf morphology and growth form. Neither *L. grayi* nor *L. salmoniflorum* occur in Oregon.

The leaf morphology of *Lomatium papilioniferum* is also similar to *L. brunsfeldianum*. This taxon can be distinguished from *L. papilioniferum* by the ultimate leaflet pinnules that are leafy to the base with intercostal areas 1 mm in length or less (resulting in the pinnules being crowded with lobes throughout and obscuring the rachillae) and by the leaflet lobes that are mostly flat and long.

See also the discussions for *L. brunsfeldianum* and *L. klickitatense*.

Lomatium pastorale Darrach & D.H. Wagner (as "*pastoralis*")

Umatilla biscuitroot

Lomatium tarantuloides Darrach & Hinchliff

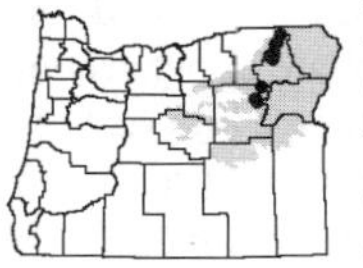

Plants acaulescent or subacaulescent. **Roots** slender and elongate taproots or bifurcate tubers with an elongate upper section. **Caudices** simple or 2–7-branched; old basal leaf sheaths from previous years absent or weathering into a sparse thatch of chaffy or chartaceous scales at base of pseudoscape. **Stems** when present, 5–10 cm, glabrous; pseudoscapes subterranean, 1.5–8 cm. **Basal leaves** glabrous, primary leaflets pinnate, 0–3 secondary leaflet pairs along each rachis, laterals subequal to central primary in length; secondaries pinnate or pinnatifid; tertiaries absent, entire or pinnatifid (2–5-lobed); ultimate apical lobes linear, oblong, elliptic, spatulate, oblanceolate, 5–60 × 1–9 mm, tips acute, subacute, obtuse, or rounded. **Cauline leaves** 0 or similar to basal. **Inflorescences** peduncles 1–10(19), prostrate, decumbent, or ascending, slender, 2–20(33) cm, glabrous, glabrate, or sparsely scaberulose or papillose along distal ridges; involucral bracts 0; rays 3–23, 1–11 cm, glabrous; involucel bractlets 0–12, linear, lanceolate, elliptic, oblanceolate or spatulate, 0.6–5 × 0.1–1.3 mm, glabrous, margins narrowly scarious or not; umbellets 5–34-flowered; pedicels 0.5–7 mm. **Flowers** petals yellow, white, or ochroleucous; anthers yellow or ochroleucous. **Fruit** ovate, oblong or narrowly elliptic, 4–12 mm, glabrous; body 1–2.2 mm wide; wings 0.15–0.6 mm, thin; vittae 1–5 in intervals, 1–3 on commissure.

Open areas, rocky or gravelly flats and slopes in scrublands, scablands, woodlands, conifer forests, basalt- or serpentine-derived substrates. Flowering Jun–Jul. 1500–2400 m. BW. Native.

The status of *L. pastorale* and *L. tarantuloides* as distinct species is still unresolved. The morphological description of *L. pastorale* herein is greatly abbreviated and more rigorously defined from that of the original type description of *L. pastorale* (Darrach & Wagner 2011) and *L. tarantuloides* (Darrach & Hinchliff 2014). The morphological averages in those descriptions were used to differentiate *L. pastorale* and *L. tarantuloides* and other similar species (*L. leptocarpum* and *L. ambiguum*). Unfortunately, the averages fail to distinguish the taxa in any consistent manner, due to the majority of values greatly overlapping. Ultimately, *L. pastorale* and *L. tarantuloides* can only be distinguished by the more prostrate angle of the stem relative to the root. Since neither of these species have undergone demographical or population genetic studies, one cannot determine if this is actually consistent difference or a character found in both taxa but at different frequencies due to environmental influences. The relationship between *L. pastorale* and *L. tarantuloides* appears to approach *L. leptocarpum*, as proposed in previous treatments, or to forms from southern Washington that have been determined recently as *L. brevifolium* (A taxon that has been delimited as a synonym or variety of *L. triternatum* in previous floras. There are many ecotypes and morphotypes of *L. triternatum*, including ones with long narrow fruits similar to *L. leptocarpum*.). Further complicating matters, the only consistent character that can be used to distinguish between *L. pastorale*, *L. tarantuloides*, and *L. triternatum* are the two former species' nearly glabrous foliage, peduncles, and rays. While *L. pastorale* is recognized here, it might be better placed as a glabrous variety of *L. triternatum*. *Lomatium tarantuloides* is herein recognized as an ecotype of *L. pastorale*.

Lomatium peckianum Mathias & Constance

Peck's desert parsley

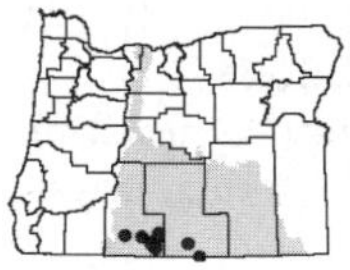

Plants acaulescent. **Roots** elongate and slender taproots or irregularly thickened taproots. **Caudices** simple or 2–3-branched; basal leaf sheaths from previous years weathering into a sparse thatch of a few, loose fibers or chaffy scales at base of pseudoscape. **Stems** absent; pseudoscapes 1–4 cm, subterranean, obscured by leaf sheaths. **Basal leaves** glabrous, glabrate, or sparsely scaberulose or muriculate; primary leaflets pinnate, 3–6 secondary leaflet pairs along each rachis, laterals subequal to central primary in length; secondaries pinnate or pinnatifid; tertiaries entire or pinnatifid (2–3-lobed); ultimate apical lobes linear, elliptic, oblong, 2–10 × 0.5–1.5 mm, tips acute or acuminate, mucronulate. **Cauline leaves** 0. **Inflorescences** peduncles 1–3, 6–15 cm, glabrous, glabrate, or sparsely scaberulose or muriculate along ridges; involucral bracts 0; rays 1–3, ascending or erect, 0.5–3 cm, glabrous; involucel bractlets 0–4, linear, lanceolate to elliptic, 2–5 × 0.5–0.8 mm, margins narrowly scarious; umbellets 10–15-flowered; pedicels 4–9 mm. **Flowers** petals ochroleucous or white; anthers ochroleucous or purple. **Fruits** elliptic, oblong or oval, 6–13 mm, glabrous, glabrate, or sparsely scaberulose; body 2–3 mm wide; wings 1–2 mm, thin; vittae 2–4 in intervals, 6–8 on commissure. **2*n***=22.

Open sandy sites, rocky flats, scrublands, woodlands, conifer forests. Flowering Apr–Jun. 1400–1800 m. BR, ECas. CA. Native.

One population of *Lomatium peckianum* in Lake County is near the Nevada-California border along Twelve-mile Creek. Although this taxon is not yet known from Nevada, the proximity of this population to that state suggests it may be present there.

Lomatium piperi J.M. Coulter & Rose

Indian biscuitroot, Piper's desert parsley

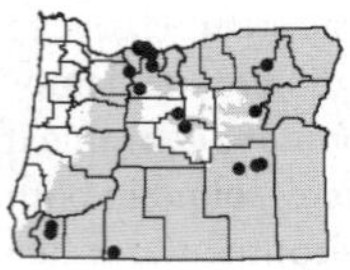

Plants acaulescent, subacaulescent, or caulescent. **Roots** globose tubers, elongate and slender portion above tuber, if present, < 2 cm. **Caudices** simple; basal leaf sheaths from previous years absent or weathering into a sparse thatch of fibers and chaffy or chartaceous scales at base of pseudoscape. **Stems** 5–20 cm; pseudoscapes 1–4 cm, subterranean, 2-branched at tip. **Basal leaves** glabrous; primary leaflets ternate or pinnate, 0–3 secondary leaflet pairs along each rachis, laterals equal to or slightly shorter than central primary in length; secondaries entire or pinnatifid (2–3-lobed); tertiaries absent; ultimate apical lobes linear or narrowly oblong, 5–20 × 0.5–2 mm, tips rounded or subacute. **Cauline leaves** 0–2, similar to basal. **Inflorescences** peduncles 1 terminal and 1–2 axillary, decumbent, spreading, ascending, or erect, glabrous, sparsely muriculate, or papillose along ridges just below tip; involucral bracts 0; rays 3–10, 5–10 cm, glabrous; involucel bractlets 2–6, linear, 0.7–2 × 0.1–0.3 mm, margins narrowly scarious; umbellets 10–20-flowered; pedicels 0.5–2 mm. **Flowers** petals white; anthers purple. **Fruits** oval or suborbicular, 4–9 mm, glabrous; body 1–2.5 mm; wings 0.5–1.5 mm; vittae 1–8 in intervals, 3–7 on commissure. ***2n***=22.

Mesic areas on slope and flats in scablands, hillsides, foothills, scrublands. Flowering Feb–May. 0–1700 m. BR, BW, Casc, Col, ECas, Owy, Sisk. CA, ID, WA. Native.

White-flowered plants with involucel bractlets 1–2.2 mm wide, determined as *Lomatium piperi* in California, may be misidentified.

Lomatium ravenii Mathias & Constance (p. 124) var. *paiutense* K. Carlson & D. Mansfield

Raven's biscuitroot, Lassen parsley

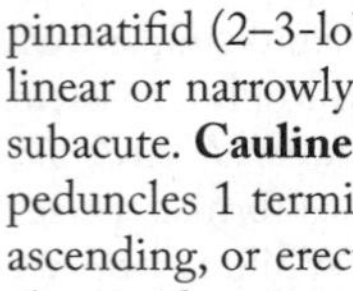

Plants acaulescent or subacaulescent. **Roots** elongate and slender taproots, globose thickened or irregularly thickened and elongate tubers tapered to elongate and slender upper portion < 10 cm. **Caudices** simple or 2–3-branched; basal leaf sheaths from previous years weathering into a sparse thatch of fibrous, chaffy, or chartaceous scales at base of pseudoscape. **Stems** 0.5–1 cm; pseudoscapes 1–4 cm, subterranean, obscured by leaf sheaths. **Basal leaves** moderately to densely hirsutulous, pubescent, or puberulent throughout; primary leaflets pinnate, 3–6 secondary leaflet pairs along each rachis, laterals less than half central primary in length; secondaries pinnate; tertiaries pinnatifid (2–6-lobed); ultimate apical lobes elliptic, obovate, oblong, 1–3 × 0.5–1 mm, tips acute, subacute, rounded, mucronulate or not. **Cauline leaves** 0–1, similar to basal. **Inflorescences** peduncles 1–6, 3–17 cm, hirsutulous, pubescent, or puberulent; involucral bracts 0; rays 1–18, 1–5 cm, sparsely to moderately hirsutulous or puberulent; involucel bractlets 4–8, linear or lanceolate, 2–5 × 0.3–1 mm, margins scarious, surfaces hirsutulous or puberulent; umbellets 10–20-flowered; pedicels sparsely hirsutulous or puberulent, 2–10 mm. **Flowers** petals white or purple; anthers purple. **Fruits** oval or suborbicular, 4–8 mm, sparsely hirsutulous or puberulent; body 2–3 mm wide; wings 0.5–1 mm, thin; vittae obscure. ***2n***=22.

Rocky or gravelly slopes or flats in scrubland. Flowering Apr–May. 800–1800 m. BR, BW, Owy. CA, ID, NV. Native.

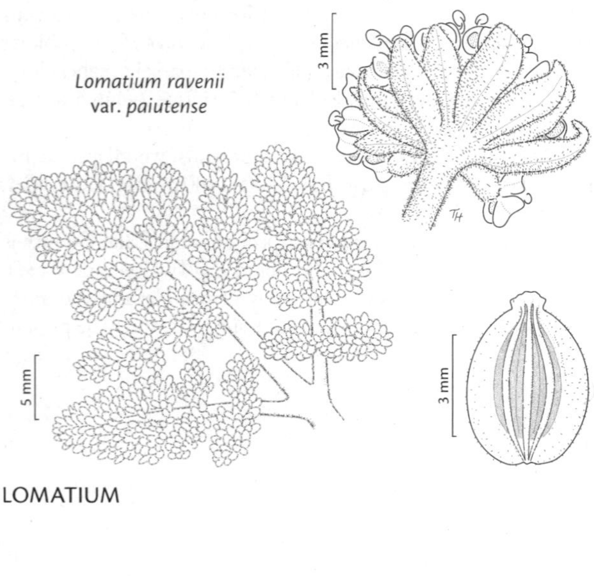

LOMATIUM

Lomatium rollinsii Mathias & Constance

Rollins' biscuitroot, Rollins' desert parsley

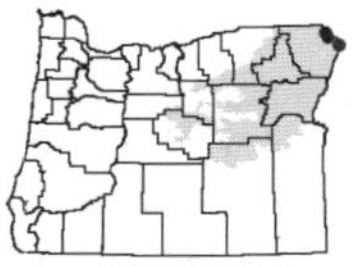

Plants caulescent. **Roots** elongate and slender taproots or globose or irregularly thickened tubers with an elongate upper section < 2 cm, tapered abruptly at base into an elongate and slender taproot. **Caudices** simple; basal leaf sheaths from previous years absent or weathering into a few fibers or a sparsely chartaceous thatch at base of stem. **Stems** 10–30 cm; pseudoscapes absent. **Basal leaves** glabrous, glabrate or sparsely puberulent or pubescent; primary leaflets pinnate, 2–6 secondary leaflet pairs along each rachis, laterals less than half central primary in length; secondaries entire, pinnate, or pinnatifid (3–5-lobed); tertiaries absent or entire; ultimate apical lobes linear, narrowly elliptic, or narrowly oblong, 5–15 × 0.5–2 mm, tips acute, subacute, or rounded. **Cauline leaves** 1–3, similar to basal. **Inflorescences** peduncles 1–4, ascending or erect, 8–20 cm, glabrate below or sparsely to moderately pubescent or puberulent above; involucral bracts 0; rays 1–6, 1–5 cm, glabrate or sparsely puberulent; involucel bractlets 0–6, linear, 2–4 × 0.2–0.5 mm, margins scarious; umbellets 7–15-flowered; pedicels 5–15 mm. **Flowers** petals ochroleucous or yellow; anthers ochroleucous or yellow. **Fruits** elliptic, oblong, 6–8 mm, glabrous; body 2–3 mm wide; wings 0.5–1 mm, thin; vittae 1–2 in intervals, 4 on commissure. ***2n***=22.

Open gravelly sites, rocky flats, grasslands, scrublands, igneous-derived substrates. Flowering Mar–May. 200–500 m. BW. ID, WA. Native.

Lomatium leptocarpum and *L. triternatum* specimens, in early flower, are often misidentified as this taxon when poorly collected (i.e., shriveled, poorly pressed, or leaf and flower clippings).

Lomatium roseanum Cronquist

Adobe parsley, adobe lomatium

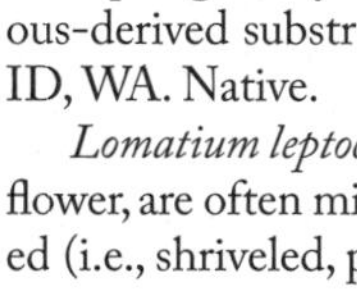

Plants acaulescent or subacaulescent. **Roots** globose or irregularly thickened tubers with elongate upper section < 1 cm and basal section abruptly tapered to slender, elongate taproot, or irregularly thickened tubers with elongate upper section < 1 cm and bifurcate basal section that is tapered along each axis to slender taproot. **Caudices** simple or 2–6-branched; basal leaf sheaths from previous years weathering into a sparse thatch of a few, loose fibers or chaffy scales at base of pseudoscape. **Stems** absent or < 3 cm; pseudoscapes 1–3 cm, subterranean, obscured by leaf sheaths. **Basal leaves** glabrous; primary leaflets pinnate, 1–4 secondary leaflet pairs along each rachis, laterals less than half of central primary in length; secondaries pinnate or pinnatifid (3–6-lobed); tertiaries absent or entire; ultimate apical lobes elliptic, oblong, oblanceolate, 1–4.5 × 0.8–1.5 mm, tips acute, subacute or mucronulate. **Cauline leaves** 0–2, similar to basal. **Inflorescences** peduncles 1–15, 10–20 cm glabrous; involucral bracts 0; rays 2–6, 2–4 cm, glabrous; involucel bractlets 3–6, linear, lanceolate to narrowly elliptic, 2–4 × 0.3–0.8 mm, margins narrowly scarious; umbellets 5–15-flowered; pedicels 3–6 mm. **Flowers** petals yellow; anthers purple or yellow. **Fruits** oval or orbicular, 6–8 mm, green, red or tan, glabrous, dull; body 2–3 mm wide; wings 0.5–1 mm, thin; vittae 1 in intervals, 2 on commissure.

Rocky slopes and talus in scrublands and sparse woodlands, igneous-derived substrates. Flowering Apr–May. 1600–1800 m. BR. CA, ID, NV. Native.

Lomatium roseanum may be a robust, desert ecotype of *L. hendersonii* and may warrant placement as a variety within that species. Growing at low elevations (with a longer growing season) may have resulted in the more robust form found in the type specimen of *Leptotaenia leibergii* (basionym of *L. roseanum*). The only consistent differentiating feature of *L. roseanum* is the ultimate apical lobes, which are acute and sometimes mucronulate. *Lomatium hendersonii* has obtuse and rounded ultimate lobes with only a few lateral lobes that are acute or mucronulate.

Lomatium serpentinum (M.E. Jones) Mathias

Snake Canyon desert parsley, sweetscented biscuitroot

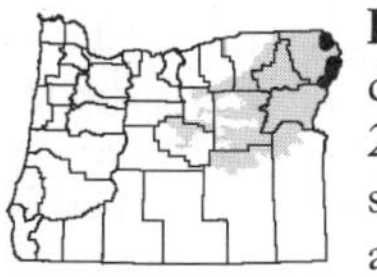

Plants acaulescent. **Roots** thickened or slender, elongate taproots. **Caudices** simple, 2-branched or multicipital; basal leaf sheaths from previous years weathering into a sparse thatch of loose fibers and chaffy or chartaceous scales at base of pseudoscape. **Stems** absent; pseudoscapes 1–3 cm, subterranean, obscured by leaf sheaths. **Basal leaves** glabrous, glabrate, or sparsely scaberulose or papillose adaxially; primary leaflets pinnate, 5–8 secondary leaflet pairs along each rachis, laterals at least half central primary in length; secondaries pinnate; tertiaries pinnate or pinnatifid (3–6-lobed); ultimate apical lobes oblong, elliptic, oval, obovate, obelliptic, 0.7–5 × 0.7–2.5 mm, tips subacute, rounded, or obtuse, mucronulate or shortly cuspidate. **Cauline leaves** 0. **Inflorescences** peduncles 1–4, ascending or erect, 5–20 cm, glabrous, glabrate, scaberulose or muriculate along ridges; involucral bracts 0 or umbels subtended by a narrow bract; rays 3–17, 2–10 cm, glabrous; involucel bractlets 4–8, linear, lanceolate, 2–5 × 0.3–1 mm, glabrous, glabrate, or sparsely puberulent, margins narrowly scarious; umbellets 10–30-flowered; pedicels 5–15 mm. **Flowers** petals yellow; anthers ochroleucous or yellow. **Fruits** oblong, elliptic, 6–10 mm, glabrous; body 1–2 mm wide; wings 0.7–1.5 mm, thin; vittae 1 in intervals, 2 on commissure.

Talus slopes, rock outcrops, cliff faces. Flowering Apr–Jul. 200–1500 m. BW. ID, WA. Native.

Lomatium suksdorfii (S. Watson) J.M. Coulter & Rose (p. 125)

Suksdorf's desert parsley, Suksdorf's lomatium

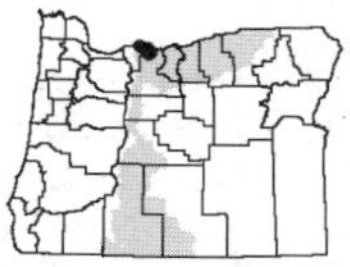

Plants caulescent. **Roots** stout and thickened, elongate taproots or stout and irregularly thickened taproots. **Caudices** 2–3-branched or multicipital; basal leaf sheaths from previous years weathering into a fibrous thatch at or below ground level; peduncles persistent in thatch as gray stalks. **Stems** 9–20 dm, stout but not inflated; pseudoscapes absent. **Basal leaves** glabrous; primary leaflets pinnate, 3–7 secondary leaflet pairs along each rachis (mature basal leaves may have more pairs, but these are not usually collected for specimens), laterals less than half of central primary in length; secondaries pinnate or pinnatifid; tertiaries entire or pinnatifid (2–4-lobed); ultimate apical lobes linear, narrowly oblong, narrowly elliptic, 10–30 × 1–3 mm, tips acute, acuminate, or subacute, mucronulate or not. **Cauline leaves** 1–3, similar to basal. **Inflorescences** peduncles 1–4 (more in robust plants), ascending or erect, stout but not inflated, 30–80+ cm; involucral bracts 0; rays 8–20, 3–11 cm, glabrous; involucel bractlets 4–6, linear or lanceolate, 5–10 × 0.3–1 mm, unlobed or 2-lobed, margins narrowly scarious or not; umbellets 10–40-flowered; pedicels 6–17 mm. **Flowers** petals yellow; anthers ochroleucous or yellow. **Fruits** oblong, 15–25 mm, glabrous; body 4–6 mm wide; wings 1–3 mm, thin; vittae 1–4 in intervals, 2 on commissure.

Gravelly or rocky slopes, talus, rock outcrops, scrubland, oak forests, conifer forests. Flowering Apr–Jun. 50–100 m. Col, ECas. WA. Native.

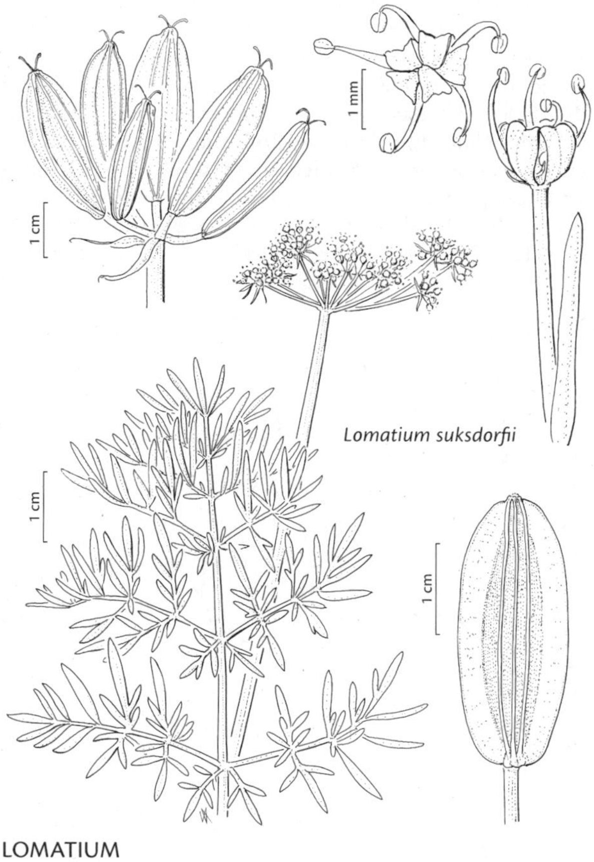

LOMATIUM

Lomatium triternatum (Pursh) J.M. Coulter & Rose (p. 126)

Nineleaf lomatium, broad nineleaf lomatium, broad-fruit lomatium, Packard's lomatium

Lomatium alatum (J.M. Coult. & Rose) J.M. Coult. & Rose, *Lomatium packardiae* Cronquist, *Lomatium platycarpum* J.M. Coult. & Rose, *Lomatium robustius* (J.M. Coult. & Rose) J.M. Coult. & Rose, *Lomatium simplex* (Nutt. ex S. Watson) J.F. Macbr., *Lomatium simplex* (Nutt. ex S. Watson) J.F. Macbr. var. *leptophyllum* (Hook.) Mathias, *Lomatium simplex* (Nutt. ex S. Watson) J.F. Macbr. var. *simplex*, *Lomatium triternatum* (Pursh) J.M. Coult. & Rose var. *alatum* (J.M. Coult. & Rose) Mathias, *Lomatium triternatum* (Pursh) J.M. Coult. & Rose var. *anomalum* (M.E. Jones ex J.M. Coult. & Rose) Mathias, *Lomatium triternatum* (Pursh) J.M. Coult. & Rose var. *macrocarpum* (J.M. Coult. & Rose) Mathias, *Lomatium triternatum* (Pursh) J.M. Coult. & Rose ssp. *platycarpum* (Torr.) Cronquist, *Lomatium triternatum* (Pursh) J.M. Coult. & Rose ssp. *triternatum*, *Lomatium triternatum* (Pursh) J.M. Coult. & Rose var. *triternatum*

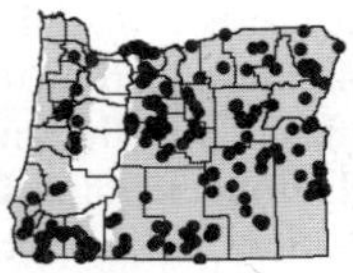

Plants acaulescent, subacaulescent, or caulescent. **Roots** slender, elongate taproots or irregularly thickened, elongate taproots with an elongate and slender upper portion 1–6 cm. **Caudices** simple, 2–3-branched or multicipital; basal leaf sheaths from previous years weathering into a sparse thatch of a few, loose fibers or chaffy scales at base of stem or pseudoscape; previous year's peduncles sometimes persistent in thatch as gray stalks. **Stems** absent or 10–20 cm; pseudoscapes absent or < 1 cm, obscured by leaf sheaths and thatch. **Basal leaves** biternate or 1–3-pinnate, glabrous, glabrate or sparsely puberulent or scaberulose along midnerves and margins, sparsely scabrous or pubescent below and glabrate or pubescent on midnerves above; primary leaflets ternate or pinnate, 0–6 secondary leaflet pairs along each rachis, laterals half to subequal to central primary in length; secondaries pinnate, pinnatifid, ternate or entire; tertiaries absent, entire, pinnatifid (2–3-lobed); ultimate apical lobes linear, narrowly elliptic, or narrowly oblong, 10–120 × 0.4–12 mm, tips acute or acuminate, mucronulate or not. **Cauline leaves** 0–2, similar to basal. **Inflorescences** peduncles 1–6, 10–35 cm, glabrate or sparsely to moderately scaberulose or puberulent; involucral bracts 0; rays 3–25, 1.5–12 cm, glabrate, sparsely scaberulose, or puberulent; involucel bractlets 0–12, linear or lanceolate, 1–6 × 0.1–0.4 mm, glabrous, entire, margins narrowly scarious or not; umbellets 10–70-flowered; pedicels 2–8 mm, glabrous, glabrate, scaberulose on ridges. **Flowers** petals ochroleucous or yellow; anthers ochroleucous or yellow. **Fruits** oblong, narrowly oblong, elliptic or obovate, 7–16 mm, glabrous, glabrate, or sparsely scaberulose or puberulent; body 1.3–4 mm wide; wings 0.7–3 mm, thin; vittae 1 in intervals, 2 on commissure.

Diverse habitat types but most commonly on open slopes, meadows, rocky hillsides. Flowering Feb–Jul. 0–2600 m. BR, BW, Col, CR, ECas, Lava, Owy, Sisk, WV. CA, ID, NV, WA. north to British Columbia, northeast to Alberta, east to CO, southeast to AZ. Native.

Lomatium triternatum is a complex with many ecotypes and morphotypes. Numerous varieties have been named, and some have been delimited at the species level. The ecotypes that have been named include: 1) acaulescent or caulescent plants with pubescent, short narrow apical leaflets lobes (1–4 mm wide) [= *L. packardiae*]; 2) caulescent plants with short narrow apical leaflets and lobes (1–2 mm wide) that are variously pubescent [= *L. tamanitchii*; *L. triternatum* var. *brevifolium*]; 3) caulescent plants with broad, short leaflets (10–50 × 2–8 mm) [= *L. anomalum*]; 4) caulescent plants from the west side of the Cascades with very long and broad leaflets (50–120 × 1–12 mm) [= *L. triternatum* var. *macrocarpum*]; 5) plants with strictly ternate lateral leaflets with lobes long and narrow (20–90 × 0.4–1.5 mm) and fruit wings (2–3 mm) equal to or wider than the body [= *L. simplex*]. All of these forms grade into each other imperceptibly, and often more than one form can be found on a herbarium sheet.

Although Lesica and Kittelson (2013) make a robust argument for the separation of *L. anomalum* from *L. triternatum*, the condition of the types of *L. anomalum* prevents the reliable assignment of this name to any of the named ecotypes of *L. triternatum*. The types are in a mature fruiting stage with dried and withered leaves. Additionally, the leaves are fragmented, and the length and width of the leaflet lobes cannot be determined. The name needs epitypification before it can be applied to forms in the remainder of the range of *L. triternatum*, regardless of how other workers have applied this name to specific morphological variants in the past.

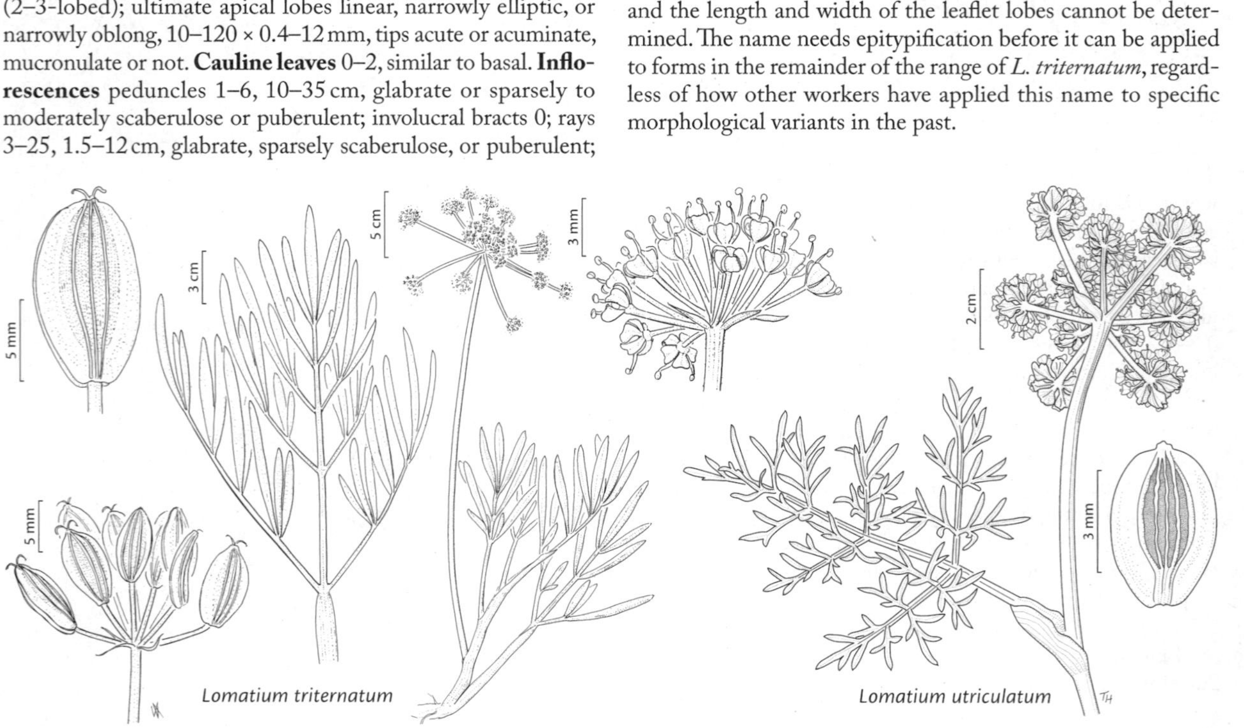

Lomatium triternatum *Lomatium utriculatum*

Lomatium utriculatum (p. 126)
(Nuttall ex Torrey & A. Gray) J.M. Coulter & Rose

Spring gold, common lomatium, fine-leaved desert parsley, pomo-celery lomatium

Lomatium utriculatum (Nutt. ex Torr. & A. Gray) J.M. Coult. & Rose var. *anthemifolium* Jeps., *Lomatium utriculatum* (Nutt. ex Torr. & A. Gray) J.M. Coult. & Rose var. *glabrum* Jeps., *Lomatium utriculatum* (Nutt. ex Torr. & A. Gray) J.M. Coult. & Rose var. *utriculatum*

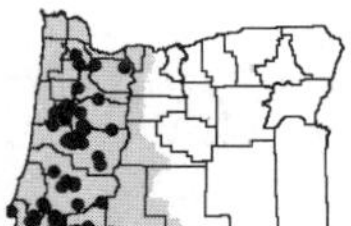

Plants caulescent. **Roots** slender, elongate taproots. **Caudices** simple; basal leaf sheaths from previous years weathered into a weak thatch of fibers or chaffy scales at base of stem. **Stems** 10–50 cm, first internode subterranean, obscured by leaf sheaths, 2–3-branched above first node; pseudoscapes absent. **Basal leaves** readily decomposed, mostly withered post-anthesis, glabrous, glabrate, or sparsely pubescent; primary leaflets pinnate, 3–5 secondary leaflet pairs along each rachis, laterals half to subequal to central primary in length; secondaries pinnate or pinnatifid; tertiaries entire or pinnatifid (2–3-lobed); ultimate apical lobes elliptic, oblong, oval, or obovate, 2–5 × 0.4–1.5 mm, tips acute or acuminate, mucronulate. **Cauline leaves** 1–3, similar to basal. **Inflorescences** peduncles 1 terminal and 1–4 axillary, ascending or erect, 5–32 cm, glabrous, glabrate, or sparsely pubescent; involucral bracts 0 or umbels subtended by a leaf-like or narrow to broad sheathing bract; rays 5–13, 1–9(12) cm, glabrous; involucel bractlets 2–10, obovate, 2–9 × 2–4 mm, 3–many-lobed or -toothed, margins narrowly scarious; umbellets 10–20-flowered; pedicels 2–9 mm. **Flowers** petals yellow; anthers ochroleucous or yellow. **Fruits** obelliptic or obovate, 5–9 mm, glabrous, glabrate, or sparsely puberulent; body 1–2 mm; wings broader at base of fruit than tip, 0.8–1.5 mm, thin; vittae 1–4 in intervals, 2–6 on commissure. **2*n***=22.

Open areas on hillsides, slopes, meadows, scrubland, forests. Flowering Jan–Jul. 0–2600 m. Casc, CR, ECas, Est, Sisk, WV. CA, WA; north to British Columbia. Native.

Lomatium vaginatum J.M. Coulter & Rose

Broadsheath lomatium, broadsheath desert parsley

Lomatium utriculatum (Nutt. ex Torr. & A. Gray) J.M. Coult. & Rose var. *papillatum* (L.F. Hend.) Mathias

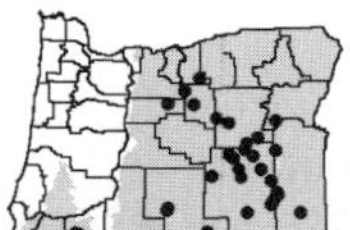

Plants caulescent. **Roots** slender, elongate taproots, tuberous thickenings if present irregular. **Caudices** simple; basal leaf sheaths from previous years weathered into a weak thatch of fibers or chaffy scales at base of stem. **Stems** 10–45 cm, first internode subterranean, 2–3-branched above first node; pseudoscapes absent. **Basal leaves** readily decomposed, mostly withered post-anthesis, glabrous, glabrate, or sparsely scaberulose; primary leaflets pinnate, 4–6 secondary leaflet pairs along each rachis, laterals half to subequal to central primary in length; secondaries pinnate or pinnatifid; tertiaries entire or pinnatifid (2–6-lobed); ultimate apical lobes elliptic, oblong, oval, or obovate, 1.5–3 × 0.8–1.5 mm, tips rounded and mucronulate, rarely only rounded or subacute and not mucronulate. **Cauline leaves** 1–3, similar to basal. **Inflorescences** peduncles 1 terminal and 1–4 axillary, ascending or erect, 5–16 cm, glabrous, glabrate, or sparsely scaberulose; involucral bracts 0 or umbels subtended by a leaf-like or narrow to broad sheathing bract; rays 6–12, 1–4.5 cm, glabrous; involucel bractlets 2–10, oblanceolate or obovate, 2–4 × 0.5–1.5 mm, tips acute, margins narrowly scarious; umbellets 15–20-flowered; pedicels 3–8 mm. **Flowers** petals yellow; anthers ochroleucous or yellow. **Fruits** elliptic, oval, ovate, 8–12 mm, glabrous, glabrate, sparsely papillose, or scaberulose; body 2–3 mm; wings 1.5–2.5 mm, thin; vittae 3–4 in intervals, 4 on commissure. **2*n***=22.

Open areas and rocky slopes in meadows, scrubland, conifer forests. Flowering Apr–Jun. 300–1700 m. BR, BW, Col, ECas, Lava, Owy, Sisk. CA, ID, NV. Native.

Lomatium watsonii (J.M. Coulter & Rose) J.M. Coulter & Rose

Watson's desert parsley, Watson's lomatium

Lomatium frenchii Math. & Const.

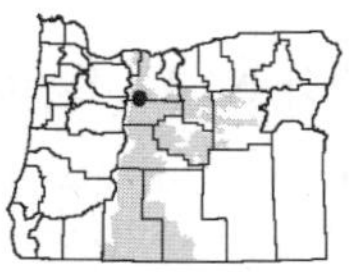

Plants acaulescent or subacaulescent. **Roots** globose tubers with slender taproots at base and elongate upper section < 1 cm, or irregularly thickened tubers abruptly tapered to caudex, or elongate and slender taproots. **Caudices** simple or 2–3-branched; basal leaf sheaths from previous years absent or weathering into a few fibers or sparsely chartaceous or chaffy thatch at base of pseudoscape. **Stems** absent or < 2 cm; pseudoscapes 0.5–2 cm, subterranean, obscured by leaf sheaths. **Basal leaves** glabrous, glabrate or sparsely villous, pubescent, or puberulent; primary leaflets pinnate, 3–6 secondary leaflet pairs along each rachis, laterals at least half central primary in length; secondaries pinnate or pinnatifid; tertiaries entire or pinnatifid (2–6-lobed); ultimate apical lobes linear, oblong, elliptic, 0.7–3 × 0.5–1.5 mm, tips subacute or rounded, mucronulate or not. **Cauline leaves** 0–1, similar to basal. **Inflorescences** peduncles 1–4, 7–15 cm, glabrous; involucral bracts 0; rays 1–6, 0.5–4 cm, glabrous; involucel bractlets 4–7, disk-like or when separate linear, elliptic, or lanceolate, bractlets fused into an incomplete, involucellar disk, disk segments formed from fusion of 2–12 bracts, tips 2–12-toothed, each segment cleft nearly to base, 2–4 segments per umbellet, sometimes with 2–5 bractlets separate, cleft to near base and 2–6 bractlets fused into 2–3-toothed segments, 1–3 × 0.5–5 mm, margins scarious; umbellets 10–20-flowered; pedicels 0.5–2 mm. **Flowers** petals ochroleucous or yellow; anthers ochroleucous or yellow. **Fruits** oval, ovate, or elliptic, 6–10 mm, glabrous, glabrate, or puberulent; body 2–3 mm wide; wings 0.5–1 mm, thin; vittae 1–4 in intervals, 6 on commissure. **2*n***=22.

Open gravelly sites and rocky flats in grasslands, scrublands, woodlands, basalt-derived substrates. Flowering Mar–May. 700–1100 m. ECas, Lava. WA. Native.

Populations from the southwestern part of the range of *Lomatium watsonii* with glabrous fruit have been called *L. frenchii*. These populations are found along the ecotone between the Eastern Cascades Slopes and Foothills ecoregion and the Blue Mountains and Columbia Basin ecoregions. There are likely glabrous populations in the far edge of the Eastern Cascades Foothills ecoregion in Washington also. These populations appear to grade imperceptibly into the pubescent *L. watsonii* populations in the Columbia Basin. Further population-level morphological investigation along this ecotone might result in evidence warranting the delimitation of *L. frenchii* as a variety of *L. watsonii*.

OENANTHE Water dropwort

Herbs perennial, glabrous. **Roots** fibrous or tuberous, occasionally rooting from nodes. **Stems** decumbent to ascending. **Leaves** cauline and basal, (1)2(3)-pinnately compound; sheaths not inflated; petiolate. **Inflorescences** terminal and axillary compound umbels; involucral bracts 0–4; rays 10–20, usually subequal, sometimes unequal; involucel bractlets 8–20. **Flowers** sepals well developed; outer flowers bisexual or staminate; petals white. **Fruits** cylindric, ovoid to round, slightly flattened, glabrous; ribs low, broad, corky; wings absent.

Africa, Asia, Europe, North America. ~30 species; 1 species treated in *Flora*.

Oenanthe sarmentosa de Candolle (p. 128)
Water parsley

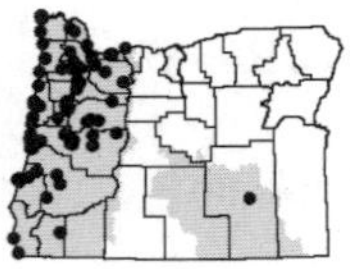

Plants 3–15 dm. **Leaves** 1–6 dm; blades 5–30 × 5–25 cm; leaflets ovate, 1–6 cm; petioles 1–3 dm. **Inflorescences** peduncles 5–12 cm; involucral bracts (if present) linear, 1–6 mm; rays 1–4 cm; involucel bractlets linear, 3–5 mm; pedicels 1–5 mm. **Flowers** sepals 0.5–1 mm, lanceolate. **Fruits** 2–4 mm. **2*n***=44.

Marshes, in or occasionally near ponds and streams. Flowering May–Sep. 0–1300 m. BR, Casc, CR, Est, Sisk, WV. CA, WA; north to AK. Native.

OROGENIA Indian potato, turkey peas

Herbs perennial; roots tuberous, fleshy. **Stems** 0 or short and mostly below ground, glabrous, glaucous. **Leaves** basal, 1–3-ternate; segments linear to linear-lanceolate, margins entire; sheaths not inflated; petioles present. **Inflorescences** compound umbels, initially compact, more open when fruiting; involucre 0; rays 2–12, very unequal in length; involucel bractlets 0 or few, minute, linear; umbellets 3–18-flowered. **Flowers** sepals absent; petals obovate, white. **Fruits** oblong to ovate, slightly flattened, glabrous; ribs corky.

Western North America. 2 species.

Key to *Orogenia* species:

1. Fruits with inconspicuous intermediate ribs; roots fusiform; umbels 3–12-rayed . . . . ***O. fusiformis***
1' Fruits with distinct intermediate ribs; roots globose; umbels 1–8-rayed . . . . ***O. linearifolia***

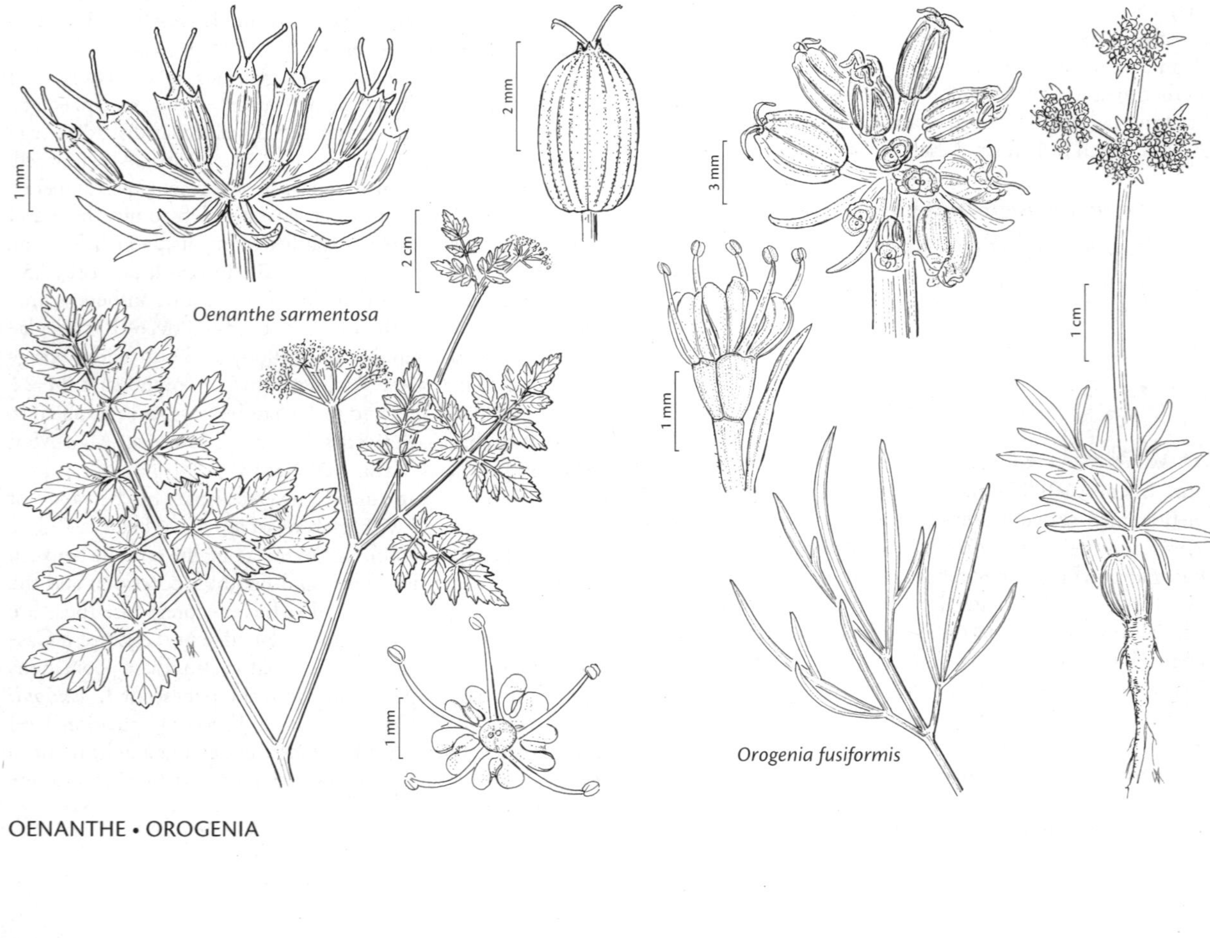

OENANTHE • OROGENIA

Orogenia fusiformis S. Watson (p. 128)

California Indian potato, mountain orogenia, turkey peas

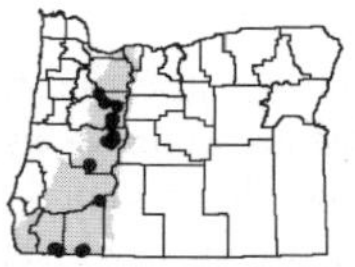

Plants 5–15 cm. **Roots** fusiform, 3–12(15) cm. **Leaves** 2–4(5), 3–10(15) cm, glabrous to somewhat scabrous; segments 1–5(8) cm; petioles 3–5 cm. **Inflorescences** rays 3–12, 5–50 mm; pedicels 0–2 mm. **Fruits** 3–4 mm; with inconspicuous intermediate ribs.

Near melting snowbanks, gravelly meadows, open slopes. Flowering May–Jul. 1200–2200 m. Casc, Sisk. CA, ID; northeast to MT. Native.

Although roots are the most reliable morphological trait to distinguish *O. fusiformis* from *O. linearifolia*, they may not always be easily available (especially on herbarium specimens). In lieu of roots, the intermediate ribs on the fruits are the next best indication of species.

Orogenia linearifolia S. Watson

Great Basin Indian potato, narrow-leaved orogenia

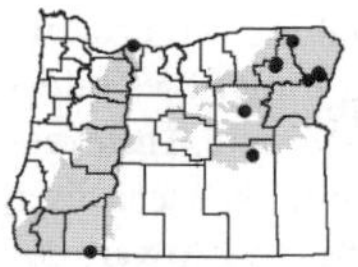

Plants 5–15 cm. **Roots** globose, 1–2 cm. **Leaves** 2–3(5), 3–10(11) cm, glabrous to somewhat scabrous; segments 10–45 mm; petioles 3–5 cm. **Inflorescences** rays 1–8, 5–50 mm; pedicels 0–2 mm. **Fruits** 3–4 mm; with distinct intermediate ribs.

Near melting snowbanks, gravelly meadows, open slopes. Flowering Feb–May. 0–2100 m. BW, Casc, Sisk. ID, WA; east to WY, southeast to CO. Native.

OSMORHIZA Sweet cicely, sweetroot

Katie E. Mitchell

Herbs perennial, glabrous, puberulent or finely hispid, with an anise-like odor. **Roots** thick taproots. **Stems** sparsely branched. **Leaves** alternate, ternately or pinnately 2–3 times compound, terminal divisions ovate to lanceolate; petiolate. **Inflorescences** compound, loose umbels; involucral bracts 0–4, narrow; rays 2–12; involucel bractlets 0 or few, narrow. **Flowers** sepals 0 or obscure; corollas white, greenish white, pink, or purple; stylopodia conic or broad and flat. **Fruits** linear or clavate; bases obtuse or attenuate, glabrous or hispid; shallowly ribbed.

Asia, North America, South America. 10 species; 4 species treated in *Flora*.

Phylogenetic studies hypothesize an Asian origin of the genus, with dispersal and radiation in the Americas (Wen et al. 2002).

Within each species, there is striking variability in pubescence, leaf shape, size, and the degree and number of leaf divisions. Unless there are purple flowers (*Osmorhiza purpurea*) or pinnately compound leaves (*O. occidentalis*), fruits will be essential for species identification. Bristly, long attenuate fruits of several species aid seed dispersal by attaching to fur and clothing.

Key to *Osmorhiza* species:

1. Fruits glabrous, fruit bases obtuse; leaves usually 2 times pinnate. . . . . ***O. occidentalis***

1' Fruits bristly, fruit bases long-attenuate; leaves 2–3 times ternate.

 2. Primary and secondary umbel branches widely spreading to reflexed; fruits convexly narrowed to rounded or abruptly pointed tips, not at all beak-like . . . . ***O. depauperata***

 2' Primary and secondary umbel branches spreading to ascending; fruits concavely narrowed to beak-like or depressed-conic, somewhat beak-like tips.

  3. Corollas pink to purple (greenish white); fruits 8–15 mm, tips depressed-conic, somewhat beak-like. . . ***O. purpurea***

  3' Corollas greenish white; fruits 12–25 mm, tips beak-like . . . . ***O. berteroi***

Osmorhiza berteroi de Candolle (p. 130)

Mountain sweet cicely

Osmorhiza chilensis Hook. & Arn.

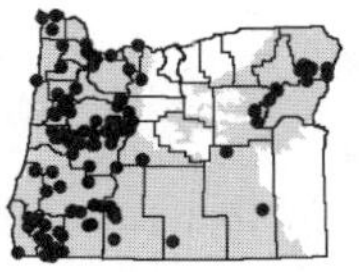

Plants 3–12 dm, glabrous or puberulent. **Leaves** biternate. **Inflorescences** rays 3–8, spreading to ascending. **Flowers** corollas greenish white; stylopodia conic. **Fruits** linear-oblong to linear-fusiform, 12–25 mm, long-attenuate at base, concavely narrowed to a beak-like tip, bristly. **2*n***=22.

Forests. Flowering Mar–Jul. 0–2300 m. BR, BW, Casc, CR, ECas, Est, Sisk, WV. CA, ID, NV, WA; north to AK, east across most Canadian provinces and northern US states, southeast to NM; South America. Native.

Osmorhiza depauperata Philippi (p. 130)

Blunt-fruited sweet cicely, bluntseed sweetroot

Osmorhiza obtusa (J.M. Coult. & Rose) Fern.

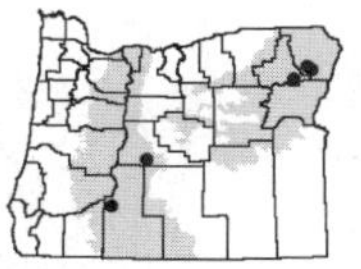

Plants 1.5–8 dm, glabrous or finely hispid. **Leaves** bi- or triternate. **Inflorescences** rays 2–5, widely spreading to reflexed. **Flowers** corollas greenish white; stylopodia short conic to depressed-conic. **Fruits** clavate, 10–18 mm, long-attenuate at base, convexly narrowed to a rounded or abruptly pointed tip, not at all beak-like, bristly. **2*n***=22.

Forests. Flowering Jun–Jul. 600–2100 m. BW, Casc, ECas. CA, ID, NV, WA; north to AK, east across most Canadian provinces and northern US states, southeast to NM; South America. Native.

Osmorhiza occidentalis (Nuttall) Torrey (p. 130)

Western sweet cicely

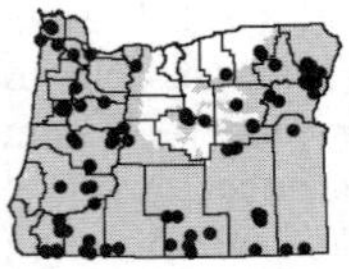

Plants 4–12 dm, glabrous or puberulent. **Leaves** usually bipinnate, sometimes biternate, pinnate, or pinnate-ternate. **Inflorescences** rays 5–16, spreading to ascending, ascending-erect with maturity. **Flowers** corollas yellow or greenish yellow; stylopodia broad and flat. **Fruits** linear-oblong to linear-fusiform, 12–22 mm, obtuse at base, abruptly and concavely narrowed to a somewhat beak-like tip, glabrous. **2*n***=22.

Thickets, open slopes, damp meadows, open areas in woods. Flowering Apr–Aug. 0–2600 m. BR, BW, Casc, CR, ECas, Est, Owy, Sisk, WV. CA, ID, NV, WA; north to British Columbia, northeast to Alberta, east to CO. Native.

Osmorhiza purpurea (J.M. Coulter & Rose) Suksdorf (p. 130)

Purple sweet cicely

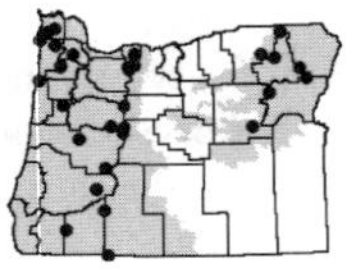

Plants 2–6 dm, glabrous or puberulent. **Leaves** bi- or triternate. **Inflorescences** rays 3–6, spreading to ascending. **Flowers** corollas pink to purple, rarely greenish white; stylopodia broad and flat. **Fruits** linear-oblong to linear-fusiform, 8–15 mm, long-attenuate at base, concavely narrowed to a depressed-conic, somewhat beak-like tip, bristly. **2*n***=22.

Forests, meadows, streambanks. Flowering May–Jul. 0–2300 m. BW, Casc, CR, ECas, Est, Sisk, WV. CA, ID, WA; north to AK, northeast to Alberta. Native.

OXYPOLIS Cowbane, oxypolis

Herbs perennial, glabrous. **Roots** tubers, clustered. **Stems** erect or ascending, branched, often glaucous. **Leaves** basal and cauline, alternate, 1-pinnate compound or ternate; sheaths not or slightly inflated; long-petiolate to ± sessile near top of plant. **Inflorescences** terminal and axillary compound umbels; involucral bracts 0–4; rays 12–24, unequal in length; involucel bractlets 0–few; umbellets 10–many-flowered. **Flowers** sepals minute and broad, not persistent in fruit; petals white. **Fruits** oblong to ovate, flattened, glabrous; ribs present; thin lateral wings present.

North America. 6 species; 1 species treated in *Flora*.

Oxypolis occidentalis J.M. Coulter & Rose (p. 130)

Western cowbane, western oxypolis

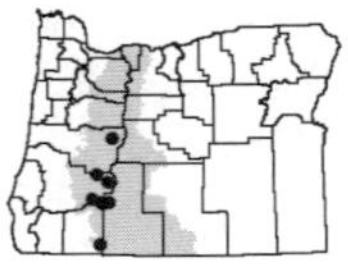

Plants 4–12 dm. **Leaves** 10–40 cm; leaflets 5–13, lanceolate to ovate, 3–9 cm, bases narrowed, tips usually acute but sometimes rounded, upper ½–⅔ serrate to coarsely toothed, lower ⅓–½ entire. **Inflorescences** peduncles 6–30 cm; involucral bracts (when present) 5–20 mm, linear-lanceolate with lighter-colored margins; rays 2–8 cm; involucel bractlets linear-lanceolate with lighter colored margins; pedicels 3–11 mm. **Fruits** 5–6 mm. **2*n***=36.

Wet meadows, wet forests, streambanks, swamps. Flowering Jul–Aug. 1000–1600 m. Casc, ECas. CA; British Columbia. Native.

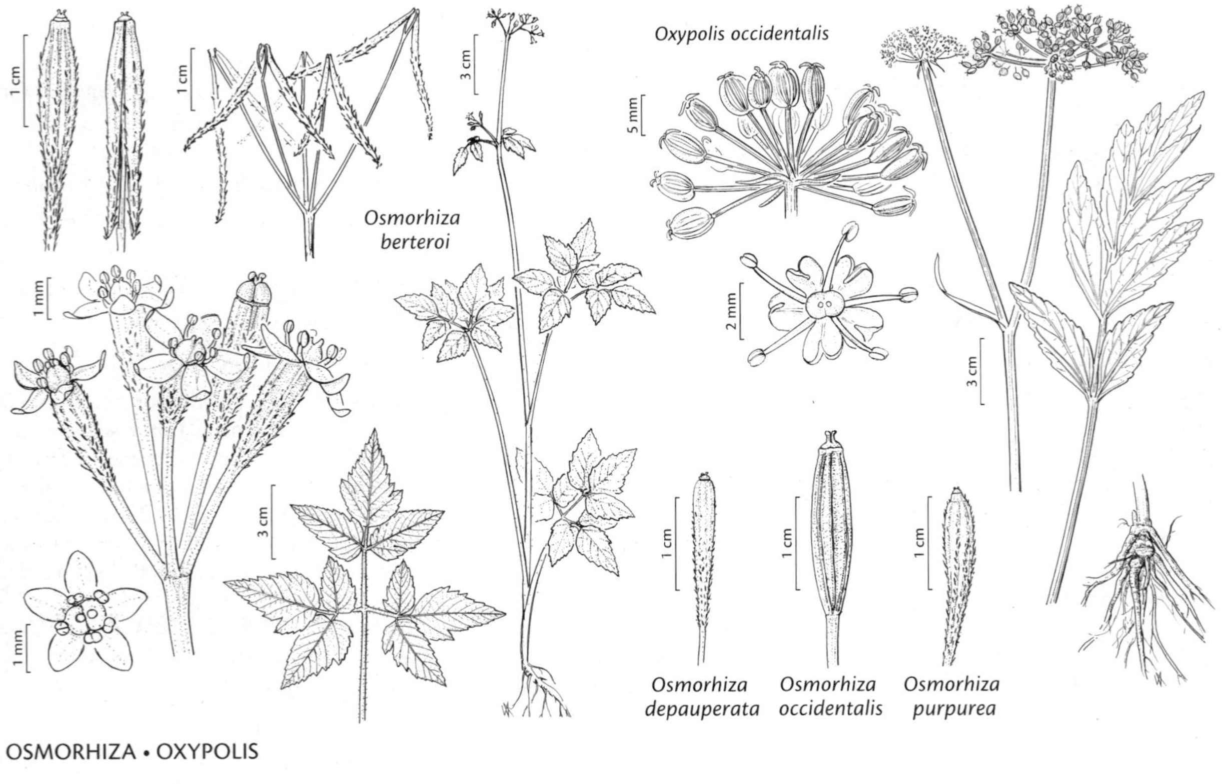

OSMORHIZA • OXYPOLIS

PASTINACA Parsnip

Herbs glabrous to moderately short-hairy. **Roots** thick taproot. **Stems** leafy, strongly ridged. **Leaves** basal and cauline, alternate, 1–2-pinnately compound; leaflets usually shallowly to deeply lobed; sheaths not or slightly inflated, upper sheaths more inflated than lower; short- to long-petiolate. **Inflorescences** terminal and axillary compound umbels in loose clusters; involucral bracts 0, rays 5–many, unequal in length; involucel bractlets 0; umbellets 5–many-flowered. **Flowers** sepals absent; petals ovate or tapered to an acute tip. **Fruits** elliptic to oblong-obovate, flattened dorsally, tan to greenish brown, glabrous; oil tubes between ribs reddish brown; lateral ribs with wings.

Asia, Europe, North America. 14 species; 1 species treated in *Flora*.

Pastinaca sativa Linnaeus (p. 131)

Common parsnip

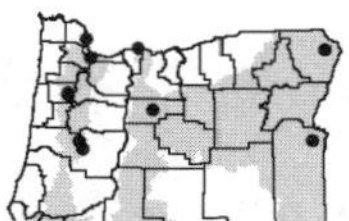

Plants biennial, 3–10(15) dm. **Stems** stout. **Leaves** elliptic to ovate in outline, 5–50 cm, length ≈ 2 × width (upper leaves highly reduced); leaflets lanceolate, to ovate to nearly circular, 1–20 × 1–10 cm, margins coarsely toothed. **Inflorescences** umbellets often in clusters of 3–5; rays 2–10 cm; pedicels 1–6 mm. **Flowers** petals yellow. **Fruits** 5–7 mm. **2*n***=22.

Roadsides, wet disturbed sites. Flowering Jun–Aug. 0–800 m. BW, ECas, Lava, Owy, Sisk, WV. CA, ID, NV, WA; throughout North America; Asia, Europe. Exotic.

Cultivated for its edible taproot, all North American *P. sativa* originated from plants brought from Eurasia.

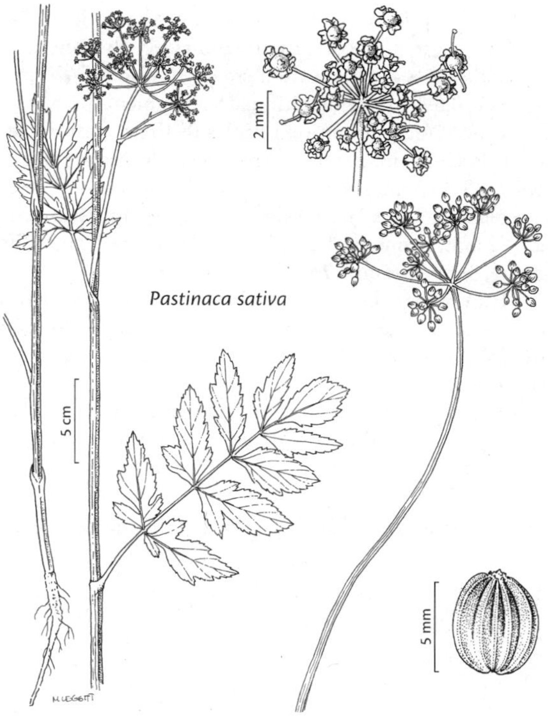

PASTINACA

PERIDERIDIA Yampah, false caraway

Herbs perennial. **Roots** tuberous or fibrous. **Stems** erect, glabrous, glaucous. **Leaves** basal and cauline, dissected or compound, ternate, pinnate or ternate-pinnate, lanceolate to triangular-ovate in outline, margins entire or toothed; ultimate leaf segments linear to lance-linear; sheaths not or only slightly inflated or appearing absent; petioles 0 to well developed. **Inflorescences** terminal or axillary compound umbels; peduncled; involucral bracts 0–12, margins narrow, membranous; rays 5–23(32); involucel bractlets 4–13, margins membranous. **Flowers** sepals well developed; petals obovate, white, tips narrow, 1–several-veined. **Fruits** elliptic to orbicular, scarcely to slightly flattened, glabrous; ribs usually shallow, generally not corky (corky in *P. howellii*); wings absent or narrow.

Eastern Asia, western United States. 15 species; 5 species treated in *Flora*.

Key to *Perideridia* species:

1. Leaf segments 10–40 mm wide, margins toothed; styles < 0.5 mm; roots fibrous . . . . . ***P. howellii***
1' Leaf segments 0.5–6 mm wide, margins entire; styles ≥ 1 mm; roots tuberous.
  2. Most leaves dissected, terminal leaf segments usually longer than others, lower leaves often crowded . . . . . ***P. bolanderi*** ssp. ***bolanderi***
  2' Most leaves compound, terminal leaf segments similar in size to others, lower leaves not crowded.
    3. Involucel bractlets linear-lanceolate, not bristly and generally well developed, ≥ 0.5 mm wide . . . . . ***P. oregana***
    3' Involucel bractlets linear, bristle-like, < 0.5 mm wide.
      4. Roots dark red to reddish brown; leaf segments 0.5–2 mm wide; tuberous roots 2–15-clustered . . . . . ***P. erythrorhiza***
      4' Roots brown to dark brown; leaf segments 0.5–8 mm wide; tuberous roots 1 or 2–3-clustered . . . . . ***P. gairdneri***

Perideridia bolanderi (A. Gray) A. Nelson & Macbride ssp. ***bolanderi*** (p. 132)

Bolander's yampah, olasi

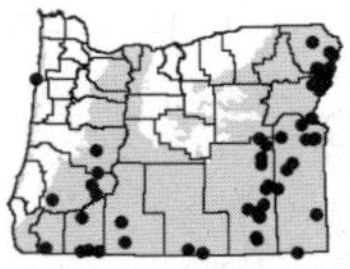

Plants 1.5–8 dm. **Roots** tuberous, 1 or 2–3-clustered, 1–7 cm. **Stems** solitary, attached to root below ground level. **Leaves** most ternate-pinnately dissected, lower leaves often crowded; blades 5–20 cm; leaf segments 0.5–4 mm wide, terminal leaf segments usually longer than others, margins entire; petioles 0–10 cm. **Inflorescences** peduncles 2–20 cm; involucral bracts 8–12, lanceolate, 3–12 mm; rays 9–23, 1–2 cm, subequal; involucel bractlets 4–7, linear to lanceolate, 3–9 × 1–3 mm wide; pedicels 1–4 mm. **Flowers** petals 1–2 mm; filaments 1.5–2.5 mm; styles 1.5–2.5 mm. **Fruits** oblong, 3–6 × 2–3 mm. ***2n***=38.

Meadows, rocky and gravelly areas, coniferous forests, streambanks, sagebrush. Flowering May–Aug. 300–2700 m. BR, BW, Casc, ECas, Est, Owy, Sisk. CA, ID, NV; east to WY. Native.

The other named taxon within *P. bolanderi* (ssp. *involucrata*) is a slight variant that occurs in the western foothills of the Sierra Nevada.

Perideridia erythrorhiza T.I. (Piper) Chuang & Constance

Red root yampah

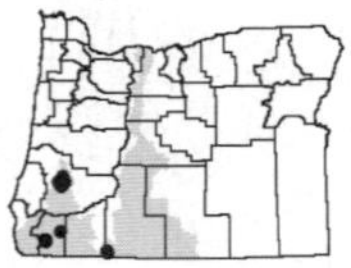

Plants 5–8 dm. **Roots** tuberous, 2–15-clustered, 3–10 cm, dark red to reddish brown. **Stems** solitary. **Leaves** compound, pinnate or ternate, lower leaves not crowded; blades 5–20 cm; leaf segments 0.5–2 mm wide, margins entire, terminal leaf segments similar in size to others; petioles 0–12 cm. **Inflorescences** peduncles 4–12 cm; involucral bracts 0–2, linear, 5–15 mm; rays 6–11, 1–5 cm, unequal; involucel bractlets 4–6, linear, bristle-like, 2–3 × 0.1–0.4 mm; pedicels 1–5 mm. **Flowers** petals 0.6–2 mm; filaments 1.5–2.5 mm; styles 1.5–2.5 mm. **Fruits** ellipsoid to ovate, 3–6 × 2–4 mm. ***2n***=38.

Meadows, edges of forests. Flowering Jun–Aug. 100–1600 m. ECas, Sisk. Native.

An Oregon endemic, the most reliable character to distinguish *P. erythrorhiza* from other *Perideridia* species is its clustered, reddish brown tuberous roots. See also *P. gairdneri*.

Perideridia gairdneri (Hooker & Arnott) Mathias

Gairdner's yampah, western false caraway

Perideridia lemmonii (J.M. Coult. & Rose) T.I. Chuang & Constance, *Perideridia gairdneri* (Hook. & Arn.) Mathias ssp. *borealis* T.I. Chuang & Constance, *Perideridia montana* (Blank.) Dorn

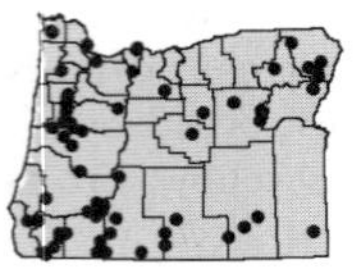

Plants 3–12 dm. **Roots** tuberous, 1 or 2–3-clustered, 1.5–8 cm, brown to dark brown. **Stems** solitary or rarely branched. **Leaves** compound, pinnate or ternate, lower leaves not crowded; blades 5–35 cm; leaf segments 0.5–8 mm wide, margins entire, terminal leaf segments similar in size to others; petioles 0–15 cm. **Inflorescences** peduncles 3–20 cm; involucral bracts 0–2, linear, 5–20 mm; rays (4)6–16(20), 1–7 cm, unequal; involucel bractlets 6–13, linear, bristle-like, 1–4 × 0.1–0.4 mm; pedicels 1–8(10) mm. **Flowers** petals 0.8–1.8 mm; filaments 1.5–2.5 mm; styles 1–2.5 mm.

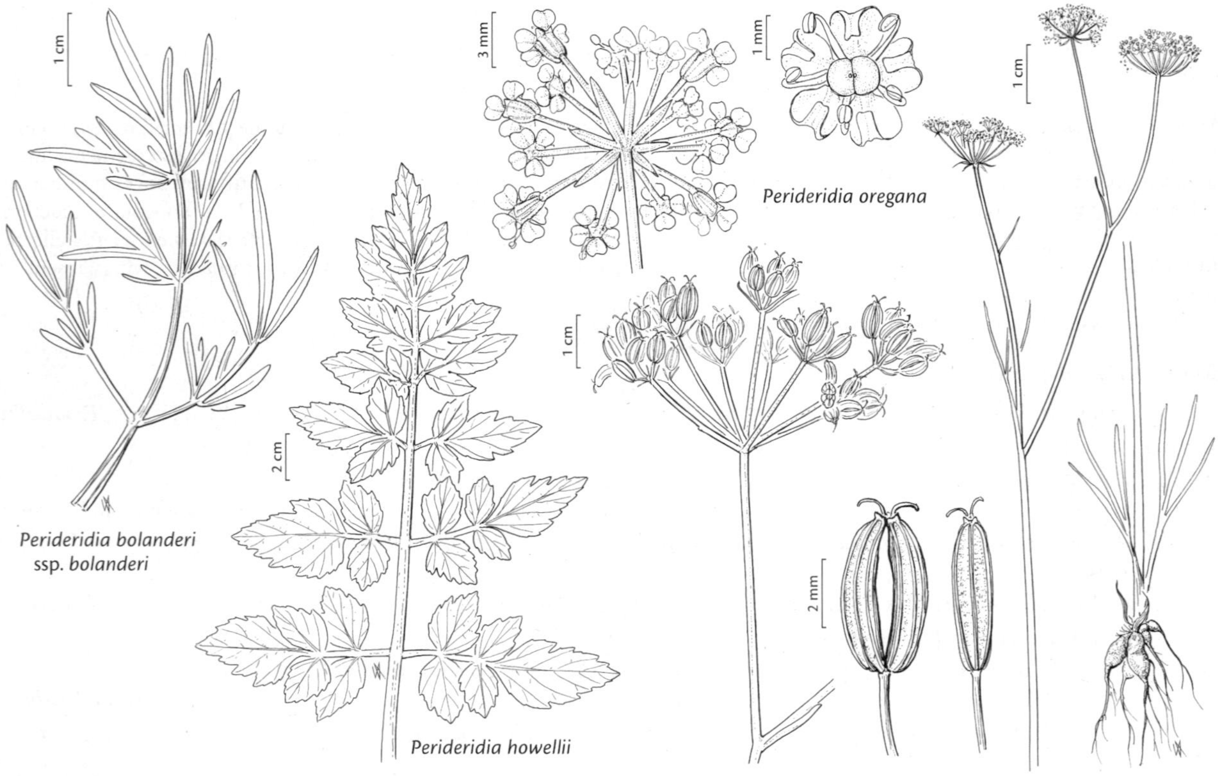

PERIDERIDIA

Fruits globose to subglobose, 2–4.5 mm. **2*n***=38, 40, 80, 120.

Moist meadows, wetlands, coniferous forests. Flowering Jun–Sep. 0–2300 m. All ecoregions except Est. CA, ID, NV, WA; north to British Columbia, northeast to Alberta, east to SD, southeast to CO. Native.

Perideridia gairdneri is the most widespread and variable species of the genus in our flora. Morphologically, the species greatly resembles both *P. erythrorhiza* and *P. oregana*. It differs chiefly from those species by only the color of its roots (*P. erythrorhiza*) and the development of its involucel bractlets and chromosome numbers (*P. oregana*). Given the similarities of the three species, study of their potential conspecificity may be warranted.

Populations of *P. gairdneri* on the California coast with branched stems have been described as *P. g.* ssp. *gairdneri*. If this taxon is recognized, our plants would be *P. g.* ssp. *borealis*.

Perideridia howellii (J.M. Coulter & Rose) Mathias (p. 132)

Howell's yampah

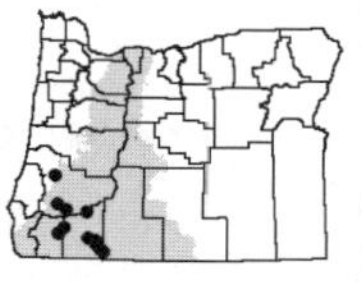

Plants 8–16 dm. **Roots** 2–4(6)-clustered, 1–4 cm, brown to dark brown, fibrous. **Stems** solitary. **Leaves** compound, pinnate or ternate, lower leaves not crowded; terminal leaf segments similar in size to others; blades 5–50 cm; leaf segments 10–40 mm wide, toothed; petioles 0–20 cm. **Inflorescences** peduncles 5–25 cm; involucral bracts 8–12, linear-spatulate, 5–20 mm; rays 15–25, 1.5–7 cm, subequal; involucel bractlets 7–13, lanceolate, 5–12 × 0.4–1.5 mm; pedicels 5–15 mm. **Flowers** petals 1–1.5 mm; filaments 1.5–2.5 mm; styles 0.1–0.4 mm. **Fruits** oblong, 3–6 mm. **2*n***=40.

Meadows, streambanks. Flowering Jun–Aug. 200–2000 m. Casc, ECas, Sisk. CA. Native.

Perideridia oregana (S. Watson) Mathias (p. 132)

Oregon yampah, eppaw

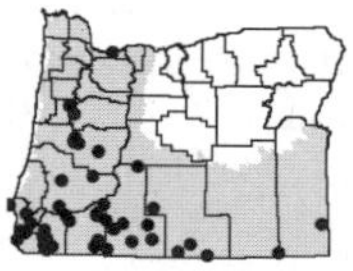

Plants 2–10 dm. **Roots** tuberous. **Stems** solitary. **Leaves** compound, pinnate or ternate, lower leaves not crowded; terminal leaf segments similar in size to others; blades 3–30 cm; leaf segments 0.5–6 mm wide, margins entire; petioles 0–12 cm. **Inflorescences** peduncles 3–20 cm; involucral bracts 0–10, linear to linear-lanceolate, 3–14 mm; rays 10–30, 1–5 cm, unequal to subequal; involucel bractlets 4–8, linear-lanceolate, 1–6 × 0.5–1.5 mm, generally well developed and not bristle-like. **Flowers** petals 1–1.5 mm; filaments 1–2 mm; styles 1–2 mm; pedicels 1–6 mm. **Fruits** oblong, 2–6 mm. **2*n***=16, 18, 20, 26.

Meadows, seeps, rocky and gravelly areas, coniferous forests. Flowering Jun–Sep. 0–2300 m. BR, Casc, CR, ECas, Owy, Sisk, WV. CA, WA. Native.

See discussion for *P. gairdneri*.

PETROSELINUM Parsley

Herbs biennial, glabrous, wiry in appearance. **Roots** slender taproots. **Stems** branched, ascending, sometimes square, solid. **Leaves** cauline, 1-pinnately compound; sheaths not inflated; petiolate. **Inflorescences** terminal and axillary compound umbels; involucral bracts 2–5; rays 2–5, unequal; involucel bractlets 2–5; umbellets 2–8-flowered. **Flowers** sepals minute. **Fruits** ovoid, slightly to moderately flattened, glabrous; ribs prominent, broad; wings absent.

Africa, Asia, Europe, North America. 2 species; 1 species treated in *Flora*.

Petroselinum segetum (Linnaeus) Koch (p. 133)

Garden parsley

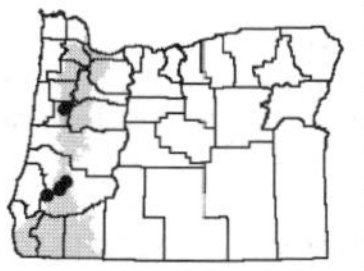

Plants 2–10 dm, glaucous. **Leaves** linear-oblong to narrowly oblong in outline, 3–20 cm; blades lanceolate to linear-oblong in outline, 2–15 cm; leaflets lanceolate to linear-oblong in outline, 0.3–2 cm, lobed to serrate; petioles 1–8 cm. **Inflorescences** involucral bracts linear, 5–30 mm; rays 3–30 mm; involucel bractlets linear to narrowly lanceolate, 1–5 mm; pedicels 1–10 mm. **Flowers** petals white. **Fruits** 2–4 mm.

Disturbed areas, roadsides. Flowering Jun–Sep. 50–500 m. Sisk, WV. Europe. Exotic.

While common parsley (*Petroselinum crispum*), a native of Eurasia, is naturalized nearly worldwide, *P. segetum* is known to have naturalized outside its native range (western Europe) only in Oregon. Despite the widespread naturalization of *P. crispum* in North America, it has never been recorded in Oregon.

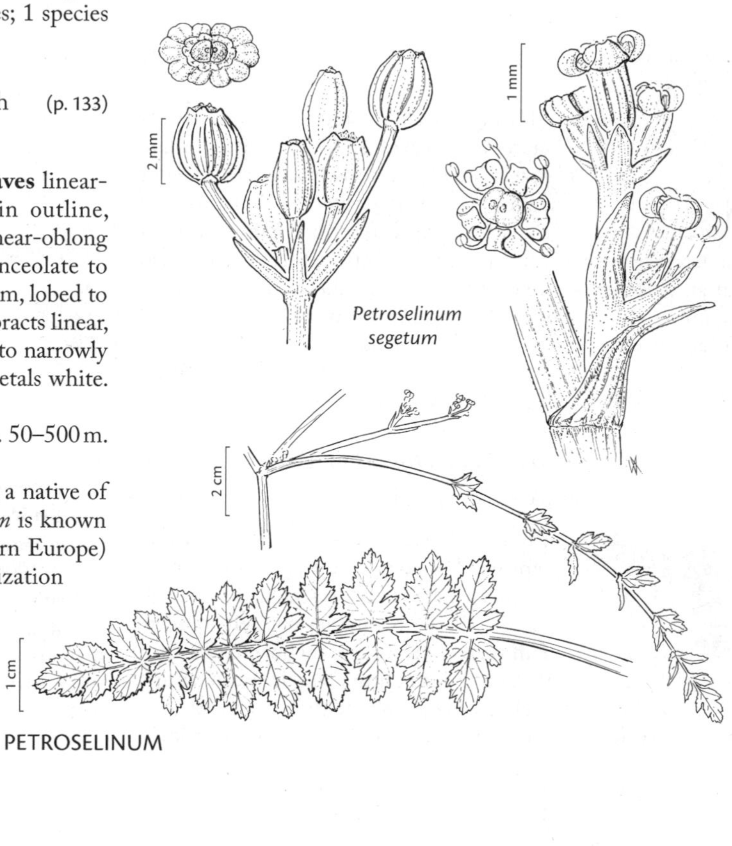

PETROSELINUM

SANICULA Snakeroot, sanicle

Rory Nichols

Herbs perennial. **Roots** taprooted or tuberous. **Stems** spreading to erect. **Leaves** simple or compound, ternately, palmately or pinnately lobed or dissected; blades orbicular to ovate in outline, margins entire or toothed; sheaths slightly inflated. **Inflorescences** compound umbels composed of round, compact clusters; flowers bisexual and staminate in same head; involucral bracts foliaceous; rays 1–6, unequal, involucel bractlets 4–10, inconspicuous or prominent; bisexual flowers pedicellate or not, staminate flowers generally long-pedicellate. **Flowers** calyx lobes generally fused; petals yellow or purple. **Fruits** oblong-ovate, somewhat compressed, prickly or not. **Seeds** flat or grooved.

Temperate areas nearly worldwide. ~40 species; 7 species treated in *Flora*.

Sanicula is commonly referred to as snakeroot as it was used historically to treat rattlesnake bites.

Sanicula taxa are occasionally confused with those in *Lomatium*. *Sanicula* species are best distinguished from *Lomatium* by their bisexual and staminate flowers in the same head.

Key to *Sanicula* species:

1. Leaves compound.
    2. Leaves ternate or palmate . . . . **S. crassicaulis** (in part)
    2' Leaves pinnate.
        3. Fruits with hooked prickles; staminate flowers inconspicuous; plants throughout Oregon . . . . **S. graveolens**
        3' Fruits without prickles; staminate flowers conspicuous; plants of southwestern Oregon . . . . **S. tuberosa**

1' Leaves simple, often deeply lobed or dissected.
    4. Bracts yellowish, conspicuous, easily distinguishable at arm's length; plants coastal . . . . **S. arctopoides**
    4' Bracts green, inconspicuous, indistinguishable from flower head at arm's length; plants generally inland.
        5. Leaves palmately or ternately lobed or dissected.
            6. Plant robust, 2.5–10 dm; blades 2–12 cm; plants of western Oregon . . . . **S. crassicaulis** (in part)
            6' Plants delicate, 1–33 dm; blades 1–4 cm; plants of southwestern Oregon . . . . **S. laciniata**
        5' Leaves pinnately, bipinnately lobed or dissected.
            7. Corollas purple (pale yellow); staminate flowers inconspicuous; plants of western Oregon. . . . . **S. bipinnatifida**
            7' Corollas yellow; staminate flowers conspicuous; plants of southwestern Oregon . . . . **S. peckiana**

Sanicula arctopoides Hooker & Arnott

Footsteps of spring, bear's foot sanicle

Sanicula crassicaulis Poepp. ex DC. var. *howellii* (J.M. Coult. & Rose) Mathias

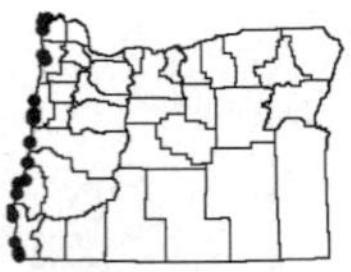

Plants prostrate to erect, 10–45 cm. **Roots** taprooted. **Leaves** simple, palmately or ternately lobed or narrowly dissected; blades deltoid to broadly ovate, 3–5.5 cm, margins coarsely toothed to serrate. **Inflorescences** bracts yellowish, conspicuous, easily distinguishable from head at arm's length, entire or lobed; bisexual flower pedicels 1 mm; staminate flower pedicels 3–5 mm. **Flowers** corollas yellow. **Fruits** prickles hooked to somewhat curled.

Dunes, beaches, coastal bluffs. Flowering Mar–Jul. 0–50 m. Est. CA, WA; north to British Columbia. Native.

Sanicula bipinnatifida Douglas (p. 135)

purple sanicle, purple snakeroot

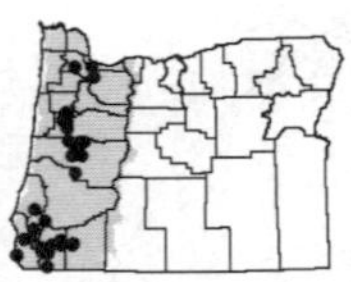

Plants robust, erect, (5)10–50 cm. **Roots** taprooted. **Leaves** simple, pinnately or bipinnately lobed or dissected; blades ovate to deltoid, 5–14 cm, margins coarsely toothed. **Inflorescences** bracts inconspicuous, not easily distinguishable from head at arm's length, green, entire; bisexual flower pedicels absent; staminate flower pedicels 1.5–2 mm. **Flowers** corollas purple, rarely pale yellow; staminate flowers inconspicuous, shorter than fruit length. **Fruits** prickles hooked.

Serpentine areas, meadows, open areas, forests, roadsides. Flowering Apr–Jun. 50–700 m. Casc, CR, Sisk, WV. CA, WA; north to British Columbia, south to Mexico. Native.

Sanicula crassicaulis Poeppig (p. 135)

Pacific snakeroot, western snakeroot

Sanicula crassicaulis Poepp. ex DC. var. *crassicaulis*, *Sanicula crassicaulis* Poepp. ex DC. var. *tripartita* (Suksd.) H. Wolff

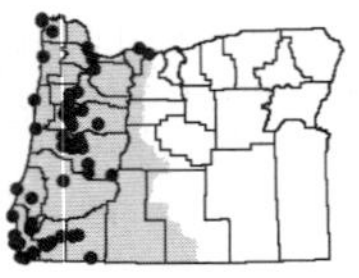

Plants robust, erect, 2.5–10 dm. **Roots** taprooted. **Leaves** simple or appearing compound, ternately or occasionally palmately lobed or dissected; blades orbicular-cordate, obovate, sometimes broadly so, 2–12 cm, margins coarsely toothed. **Inflorescences** bracts inconspicuous, not easily distinguishable from head at arm's length, green, entire. **Flowers** corollas yellow; staminate flowers few, inconspicuous. **Fruits** prickles sickle-shaped.

Streambanks, mixed forest, forest edges. Flowering Mar–Jun. 0–1000 m. Casc, CR, ECas, Est, Sisk, WV. CA, WA; north to British Columbia, south to Mexico; South America. Native.

Sanicula crassicaulis is the most common and conspicuous snakeroot in western Oregon. It generally has simple, more or less palmate leaves, but occasionally some specimens will have such dissected leaves they will appear compound.

Sanicula graveolens Poeppig (p. 135)

Sierra sanicle

Sanicula graveolens Poepp. var. *graveolens*, *Sanicula graveolens* Poepp. var. *septentrionalis* (Greene) Mathias

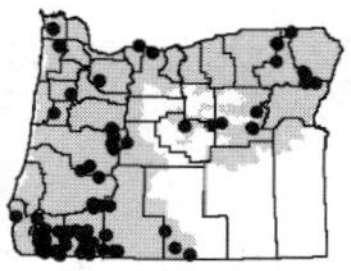

Plants low-spreading when young, erect with maturity, 4–30 cm. **Roots** taprooted or tuberous. **Leaves** compound, pinnate or bipinnate, lobed to dissected; blades broadly ovate, 1–6 cm, margins broadly toothed to lobed. **Inflorescences** bracts conspicuous, easily distinguishable at arm's length, lobed; bisexual flower pedicels absent; staminate flower pedicels 1–3 mm. **Flowers** corollas yellow; staminate flowers inconspicuous, shorter than fruit. **Fruits** prickles hooked.

Serpentine, sandy or rocky areas, open coniferous forests, meadows. Flowering Mar–Jul. 50–2800 m. BW, Casc, Col, CR, ECas, Sisk, WV. CA, ID, NV, WA; north to British Columbia, northeast to MT; South America. Native.

This species is the only *Sanicula* taxon found in eastern Oregon.

Sanicula laciniata Hooker & Arnott

Coast snakeroot

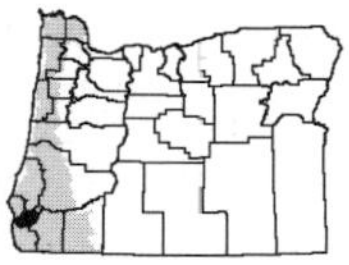

Plants delicate, 1–33 dm. **Roots** taprooted. **Leaves** simple, ternately lobed or somewhat dissected; blades broadly ovate, occasionally orbicular, 1–4 cm, margins serrate. **Inflorescences** involucral bracts inconspicuous, not easily distinguishable from head at arm's length, green, entire; bisexual flower pedicels absent; staminate flower pedicels ~1 mm. **Flowers** corollas yellow; staminate flowers inconspicuous, shorter than fruit. **Fruits** prickles curved to somewhat curly.

Dry to moist meadows, open forests. Flowering Mar–Jun. 50–200 m. CR, Sisk. CA. Native.

In Oregon, *S. laciniata* is restricted to the Rogue River watershed, although it is more common and wide-ranging in California.

Sanicula peckiana J.F. Macbride

Peck's snakeroot

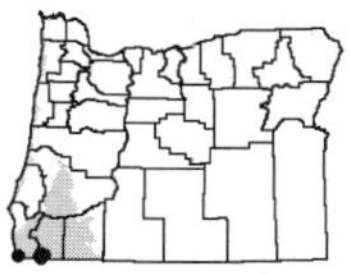

Plants spreading to erect, 10–45 cm. **Roots** taprooted. **Leaves** simple, pinnate or bipinnate, lobed to deeply dissected; blades broadly ovate, 4–10 cm, margins coarsely toothed. **Inflorescences** involucral bracts inconspicuous, not easily distinguishable from head at arm's length, green, entire; bisexual flower pedicels absent, staminate flower pedicels 2–5 mm. **Flowers** corollas yellow; staminate flowers conspicuous, extending beyond fruit. **Fruits** prickles hooked.

Dry rocky areas, serpentine, open forests. Flowering May–Jun. 400–1000 m. Est, Sisk. CA. Native.

Sanicula peckiana is our snakeroot most closely associated with serpentine soils, although *S. bipinnatifida*, *S. graveolens*, and *S. tuberosa* are sometimes found in similar habitats.

Sanicula tuberosa Torrey

turkey pea

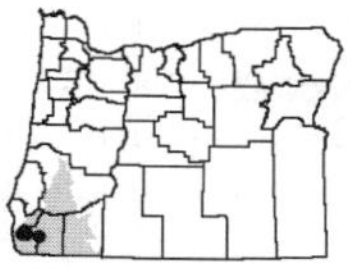

Plants delicate, erect to somewhat spreading, 10–30 cm. **Roots** tuberous. **Leaves** compound, pinnate or bipinnate, lobed to deeply dissected; blades deltoid to ovate, 1–4 cm, margins usually toothed. **Inflorescences** involucral bracts inconspicuous, not easily distinguishable from head at arm's length, entire; bisexual flower pedicels absent; staminate flower pedicels 2–4 mm. **Flowers** corollas yellow; staminate flowers conspicuous, extending beyond fruit. **Fruits** prickles absent.

Roadsides, dry meadows, serpentine. Flowering Apr–Jul. 400–1400 m. Sisk. CA; south to Mexico. Native.

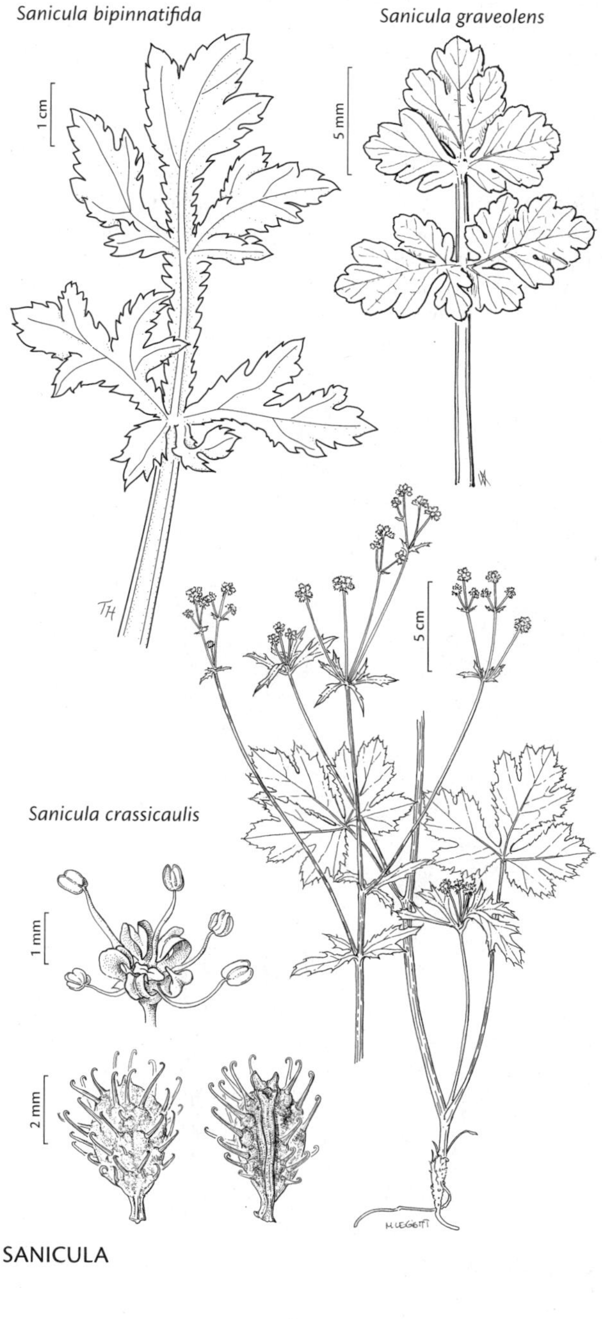

SANICULA

SCANDIX Scandix

Herbs annual, sparsely hispid but lower stems and inner sheaths sometimes densely hispid. **Roots** slender taproots. **Stems** branched, spreading to ascending or erect, leafy. **Leaves** cauline, pinnately dissected; segments thread-like to linear; sheaths not to somewhat inflated; petiolate. **Inflorescences** terminal and axillary, umbels simple or compound; involucral bracts 0(1–2); rays 1–3, subequal to unequal; involucel bractlets 2–5; umbellets 2–10-flowered. **Flowers** sepals absent; petals unequal in outer flowers, outermost petals longer than inner. **Fruits** linear and needle-like, somewhat flattened; beaks greatly exceeding bodies in length; ribs prominent, low; wings absent.

Mediterranean, North America. ~15 species; 1 species treated in *Flora.*

Scandix pecten-veneris Linnaeus (p. 136)

Venus' comb, shepherd's needle

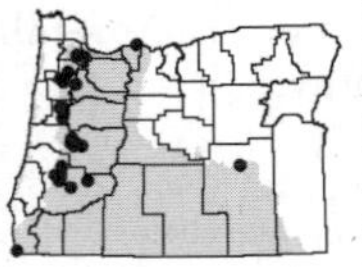

Plants 1–5 dm. **Leaves** 2–3-pinnate, lanceolate-oblong to oblong in outline, 3–20 cm; blades 2–10 cm; segments 0.5–1.5 mm wide; petioles 2–10 cm. **Inflorescences** peduncles 1–6 cm; involucral bracts (if present) lance-linear to obovate, 2–4 mm; rays 1–6 cm; involucel bractlets lance-linear to obovate, 2–5 mm, tips usually cleft; pedicels 2–3 mm. **Flowers** petals white. **Fruits** 2–9 cm; body 0.5–1.5 cm, ribbed; beaks 2–7 cm, margins barbellate.

Fields, disturbed areas, roadsides. Flowering Apr–Jun. 0–1000 m. BR, Casc, ECas, Est, Sisk, WV. CA, WA; scattered across North America; Europe. Exotic.

The long, linear beaks of the fruits of *S. pecten-veneris* easily distinguish the species from all other Apiaceous taxa within Oregon.

SIUM Water parsnip

Herbs perennial, glabrous. **Roots** fibrous. **Stems** leafy. **Leaves** cauline, alternate (basal leaves sometimes present at flowering), simple pinnate, pinnately compound or decompound, primary leaves 1-pinnate; leaflets sometimes lobed; sheaths not or slightly inflated; long-petiolate to sessile. **Inflorescences** terminal and axillary compound umbels; involucral bracts 6–10, well developed; rays 10–20, unequal in length; involucel bractlets 4–8; umbellets many-flowered. **Flowers** sepals absent or minute and triangular; petals white. **Fruits** circular to broadly elliptic, slightly flattened, glabrous; wings short, thick, corky.

Northern temperate regions, South Africa, South America. 8 species; 1 species treated in *Flora.*

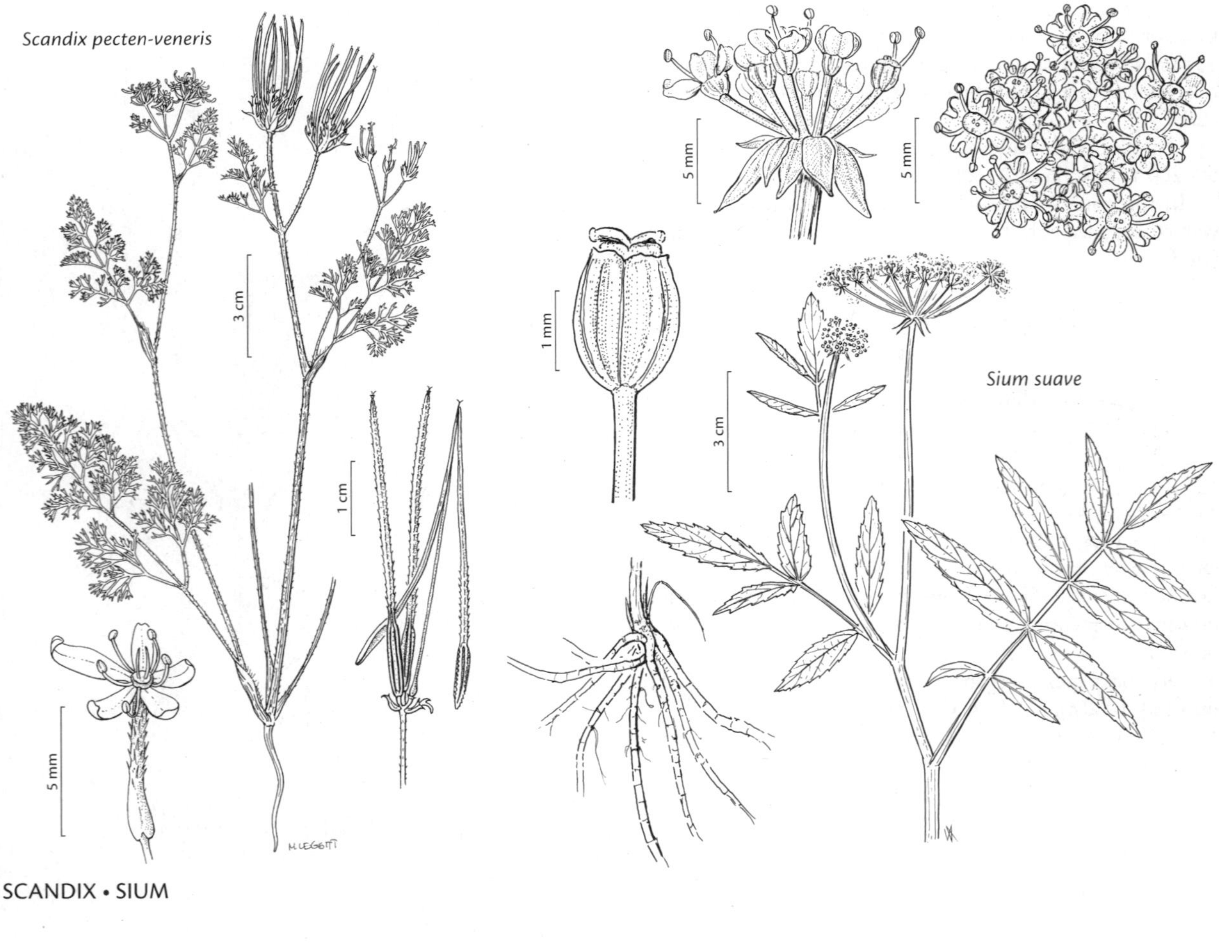

SCANDIX • SIUM

Sium suave Walter (p. 136)

Hemlock water parsnip

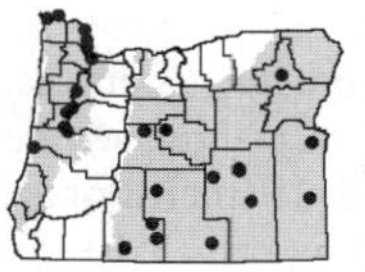

Plants 0.5–15(20) dm. **Stems** often branched, ribbed, hollow. **Leaves** 20–40 cm; leaflets (3)7–15, linear to lanceolate to broadly ovate, 1–10 cm, base narrow or rounded, margins finely to sharply toothed, tips usually narrowed to a sharp point; submerged leaves linear, deeply and finely dissected; lower leaves short- to long-petioled, upper leaves ± sessile. **Inflorescences** involucral bracts linear to lanceolate, 3–15 mm, sometimes with lighter colored, papery margins; rays 1–4 cm; involucel bractlets linear to lanceolate, 1–3 mm; pedicels 2–6(9) mm. **Fruits** 2–3 mm. **2*n***=12.

Swamps, shallow water. Flowering Jun–Sep. 0–1600 m. BR, BW, CR, ECas, Lava, Owy, WV. CA, ID, NV, WA; throughout North America. Native.

Morphologically similar to species in *Cicuta*, *S. suave* differs in having well-developed involucral bracts and once-pinnate primary leaves (versus absent or poorly developed involucral bracts and bi- to tripinnate primary leaves).

SPHENOSCIADIUM Ranger's buttons, woolly-head parsnip

Genus composed of one species, *S. capitellatum*.

Some floras include *Sphenosciadium* within *Angelica*. This is based on a large, multi-generic molecular study that included only one sample, and one gene (nrITS), per taxon (Spalik et al. 2004). The clade of the resulting phylogenetic tree, containing *Sphenosciadium*, has poor statistical support. For these reasons we are choosing to treat *Sphenosciadium* as distinct.

Sphenosciadium capitellatum A. Gray (p. 137)

Ranger's buttons, woolly-head parsnip

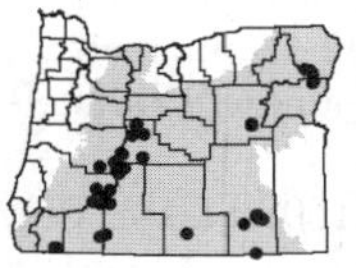

Herbs perennial, 4–15 dm; from a simple or branched caudex. **Roots** usually taprooted. **Stems** glabrous. **Leaves** basal and cauline, pinnately or bipinnately compound, 10–60 cm, glabrous to scabrous; leaflets linear to lanceolate to oblong, highly variable in shape, 2–12 cm, sometimes lobed at base, margins toothed or entire; cauline sheaths inflated; petioles 2–35 cm. **Inflorescences** terminal and axillary compound umbels with compact, head-like umbellets; involucral bracts 0; rays 4–18, 1–6 cm, usually unequal in length, densely tomentose; involucel bractlets numerous, linear; umbellets (1)2–14-flowered, spherical, 5–12 mm wide, densely tomentose, distinct; pedicels 0–2 mm. **Flowers** sepals inconspicuous; petals white or sometimes purplish. **Fruits** wedge-shaped, 5–8 mm, tomentose; winged. **2*n***=22.

Moist meadows, streambanks. Flowering Jun–Sep. 900–2800 m. BR, BW, Casc, ECas, Lava, Sisk. CA, ID, NV. Native.

The well-separated and distinct, densely tomentose, globose to subglobose umbellets are unique among our umbelliferous species.

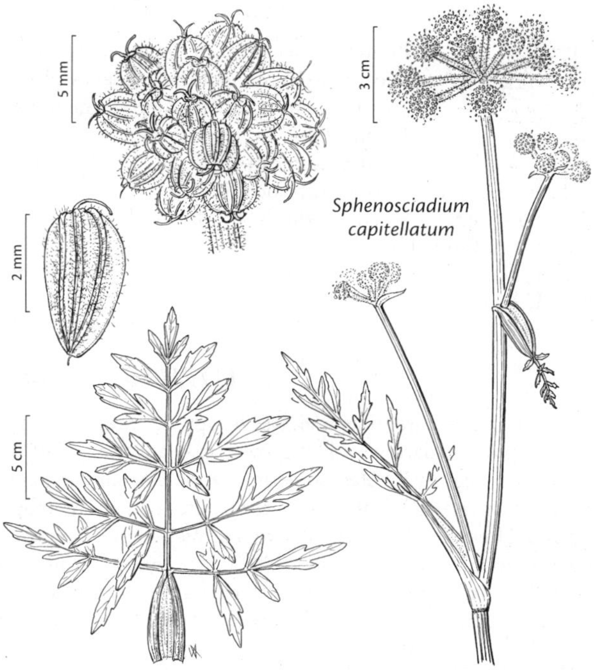

SPHENOSCIADIUM

TAUSCHIA Tauschia, umbrellawort

Herbs perennial, glabrous or minutely scabrous, glaucous or not. **Roots** tuberous or taproots. **Stems** acaulescent to short-caulescent. **Leaves** pinnate or ternate, oblong to ovate in outline, margins serrate, lobed or entire; sheaths not or slightly inflated; petioles present. **Inflorescences** terminal compound umbels; peduncled; involucral bracts 0; rays (1)2–20; involucel bractlets 1–5. **Flowers** sepals minute to well developed; petals oblanceolate to obovate, 0.5–1.5 mm, yellow, tips narrow; styles slender. **Fruits** round to oblong, slightly flattened laterally, glabrous; ribs usually thin and thread-like; wings absent.

Western North America, northern South America. 25 species; 4 species treated in *Flora*.

Key to *Tauschia* species:

1. Leaf margins entire . . . . ***T. stricklandii***

1' Leaf margins serrate or lobed.

  2. Peduncles 0.5–2 cm; petioles 1–3 cm; bractlets leaf-like, serrate . . . . ***T. howellii***

  2' Peduncles 15–50 cm; petioles 2–15 cm; bractlets not leaf-like, linear, entire.

    3. Plants minutely scabrous; rays (8)10–20; pedicels 3–15 mm . . . . ***T. kelloggii***

    3' Plants glabrous; rays 3–12; pedicels 1–3 mm . . . . ***T. glauca***

Tauschia glauca (J.M. Coulter & Rose) Mathias & Constance (p. 139)

Glaucous tauschia, glaucous umbrellawort

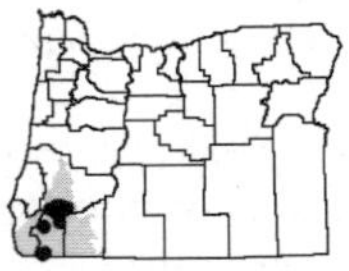

Plants acaulescent to short-caulescent, 20–40 cm, glabrous, glaucous. **Leaves** blades pinnate or ternate, 4–12 cm; leaflets 1–2 cm, margins serrate or lobed; petioles 2–12 cm. **Inflorescences** peduncles 15–40 cm; rays 3–12, 2–12 cm, unequal; involucel bractlets 1–4, not leaf-like, linear, 3–8 mm, margins entire; pedicels 1–3 mm. **Flowers** sepals minute. **Fruits** globose to subglobose, 2–3 mm. ***2n***=22.

Serpentine soils in chaparral, forests, roadsides. Flowering Apr–Jun. 300–800 m. Sisk. CA. Native.

The preference of this species for serpentine soils is unique within the genus.

Tauschia howellii (J.M. Coulter & Rose) J.F. Macbride

Howell's tauschia, Howell's umbrellawort

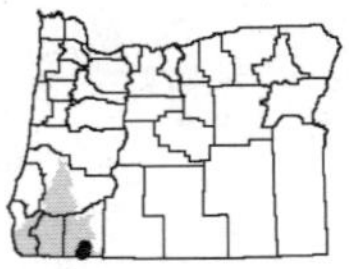

Plants short-caulescent or appearing acaulescent, 5–10 cm, glabrous. **Leaves** blades pinnate or ternate, 1.5–3 cm; leaflets 0.5–1.5 cm, margins serrate; petioles 1–3 cm. **Inflorescences** peduncles 0.5–2 cm; rays 3–5, 0.5–1.6 cm, unequal; involucel bractlets 2–5, leaf-like, 8–20 mm, margins serrate; pedicels 0.5–4 mm. **Flowers** sepals prominent. **Fruits** oblong, 2–4 mm.

Dry slopes, conifer forests. Flowering Jun–Jul. 2000–2300 m. Sisk. CA. Native.

Tauschia kelloggii (A. Gray) J.F. Macbride

Kellogg's tauschia, Kellogg's umbrellawort

Plants acaulescent, 20–70 cm, minutely scabrous. **Leaves** blades pinnate or ternate, 5–20 cm; leaflets 0.5–2.5 cm, margins serrate and often lobed; petioles 5–15 cm. **Inflorescences** peduncles 20–50 cm; rays (8)10–20, 2–12 cm, unequal; involucel bractlets 1–4, not leaf-like, linear, 3–8 mm, margins entire; pedicels 3–15 mm. **Flowers** sepals minute. **Fruits** globose to subglobose, 2–5 mm. ***2n***=22.

Meadows, scrub, forests. Flowering Apr–Jul. 300–1300 m. Sisk. CA. Native.

Tauschia stricklandii (J.M. Coulter & Rose) Mathias & Constance

Strickland's tauschia, Strickland's umbrellawort

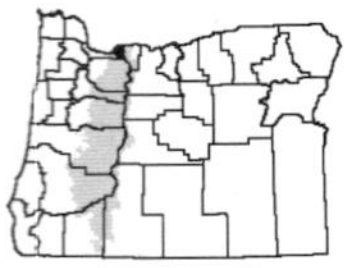

Plants acaulescent, 5–20 cm, glabrous. **Leaves** blades pinnate or ternate, 3–10 cm; leaflets 1–3 cm, margins entire; petioles 2–15 cm. **Inflorescences** peduncles 5–20 cm; rays 1–6, 0.4–1.5 cm, unequal to subequal; involucel bractlets 2–5, not leaf-like, linear, 1–2 mm, margins entire; pedicels 0.5–3 mm. **Flowers** sepals minute. **Fruits** ellipsoid to subglobose, 2–3 mm.

Meadows. Flowering May–Aug. 800–1100 m. Casc. WA. Native.

In Oregon, this species is currently known only from a relatively narrow geographic region in eastern Multnomah County.

TORILIS Hedge parsley, sock-destroyer

Herbs annual or rarely biennial. **Roots** taproots. **Stems** ascending or erect, pubescent. **Leaves** basal and cauline, 1- to 2-pinnate or ternate-pinnate, narrowly oblong-ovate to ovate-triangular in outline, usually deeply dissected, margins toothed or lobed; sheaths not or only slightly inflated, sometimes appearing absent; petioles of basal and lower leaves nearly sessile to long; petioles of upper leaves short, ± sessile. **Inflorescences** terminal or axillary compound umbels; peduncled or sessile; involucral bracts 0–10; rays 2–10; involucel bractlets 2–10, hair-like to narrowly elliptic, longer than pedicels. **Flowers** sepals absent or triangular and minute; petals obovate to fan-shaped, usually white, rarely pinkish or purple, tips rounded or notched. **Fruits** oblong-elliptic to ovate, flattened laterally; bristles present, green to bluish purple; ribs shallow; wings absent.

Asia, Africa, Europe, North America, South America. 15 species; 3 species treated in *Flora*.

Key to *Torilis* species:

1. Umbels lateral, opposite leaves; rays absent or ≤ 2 mm, inconspicuous . . . . ***T. nodosa***

1' Umbels terminal or axillary; rays present, > 2 mm, conspicuous.

 2. Involucral bracts 0–1(2); fruits 3–5 mm. . . . . ***T. arvensis***

 2' Involucral bracts 3–10; fruits 2–4 mm . . . . ***T. japonica***

Torilis arvensis (Hudson) Link (p. 139)

Field hedge parsley, tall sock-destroyer

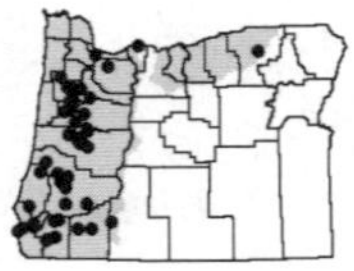

Plants annual, 2–10 dm. **Stems** erect, slender, branched. **Leaves** blades ovate to lance-ovate in outline, 3–12 cm; petioles 1–7 cm. **Inflorescences** peduncles 2–12 cm; terminal or axillary; involucral bracts 0–1(2), linear or hair-like, rarely narrowly triangular, 1–12 mm; rays (2)3–10, 0.5–2.5 cm; involucel bractlets linear or hair-like, 2–4 mm; pedicels 1–4 mm. **Flowers** petals 1–3 mm. **Fruits** oblong-ovoid, 3–5 mm; bristles straight to slightly arched, spreading, covering entire fruit or absent on inner side. ***2n***=12.

Disturbed areas. Flowering May–Jul. 0–1200 m. Casc, Col, CR, Est, Sisk, WV. CA, ID, NV, WA; north to British Columbia, throughout most of US; Europe, South America. Exotic.

Some authorities recognize two subspecies of *T. arvensis* as existing in Oregon (*arvensis* and *purpurea*). The subspecies are differentiated by the number of rays (four to ten versus two to four, respectively). The common presence of both morphologies in single populations suggests recognition of these subspecies is not warranted.

Locally, many populations contain plants in which the bristles on the fruits are absent on the inner side. Further investigation as to the geographic extent and/or taxonomic importance may be warranted.

Torilis japonica (Houttuyn) de Candolle

Japanese hedge parsley, Japanese sock-destroyer

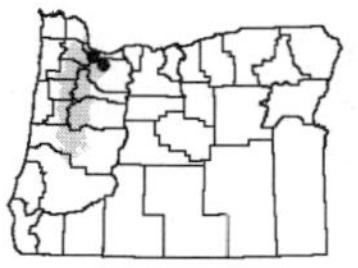

Plants annual or rarely biennial, 2–8 dm. **Stems** erect, slender, branched. **Leaves** blades ovate to lance-ovate in outline, 4–11 cm; petioles 1.5–6 cm. **Inflorescences** terminal or axillary; peduncles 1.5–15 cm; involucral bracts 3–10, linear or hair-like, rarely narrowly triangular, 1–8 mm; rays 5–10, 0.5–1.5 cm; involucel bractlets 2–4, linear or hair-like, 1–3 mm; pedicels 1–3 mm. **Flowers** petals 1–3 mm. **Fruits** oblong-ovoid, 2–4 mm; bristles straight to slightly arched, spreading, covering entire fruit. **2*n***=16.

Disturbed areas. Flowering Jun–Sep. 0–200 m. WV. WA; scattered throughout North America; Africa, Asia, Europe. Exotic.

Collected once in 1910, at the ballast grounds in Linnton (Multnomah County), this species did not appear again in Oregon until 1982 in Clackamas County.

Torilis nodosa (Linnaeus) Gaertner

Knotted hedge parsley, short sock-destroyer

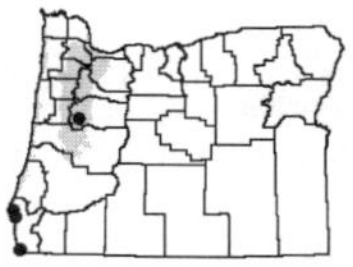

Plants annual, 1–5(7) dm. **Stems** decumbent to ascending, slender, branched or not. **Leaves** blades with deeply pinnatifid segments, oblong to elliptic in outline, 3–9 cm; petioles 0–10 mm. **Inflorescences** peduncles 0–2.5 cm; involucral bracts 0–1, linear or hair-like, 1–4 mm; rays 2–4, 0–2 mm, inconspicuous; involucel bractlets 2–4, linear, 1–4 mm; pedicels 0–3 mm. **Flowers** petals 1–3 mm. **Fruits** ovate, 2–3 mm; bristles ascending and somewhat incurved. **2*n***=24.

Disturbed areas. Flowering May–Jul. 0–100 m. Est, WV. CA; widely scattered in North America; Africa, Asia, Europe, South America. Exotic.

The sessile to nearly sessile inflorescences of *T. nodosa* are unique among our *Torilis* species.

YABEA Hedge parsley

Genus composed of one species, *Y. microcarpa*.

Yabea microcarpa (Hooker) Koso-Poljansky (p. 140)

California hedge parsley

Caucalis microcarpa Hook. & Arn.

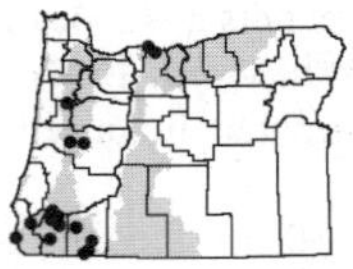

Herbs annual, 0.4–4 dm, hirsute. **Roots** slender taproots. **Stems** branched, erect, leafy. **Leaves** cauline, 1–3-pinnately or ternate-pinnately compound, narrowly ovate to ovate in outline; blades 2–6 cm; segments linear to narrowly oblong, 0.5–2 mm wide; sheaths not inflated; petioles 1–3.5 cm. **Inflorescences** terminal and axillary, simple or compound umbels; peduncles 2–10 cm; involucral bracts 2–5, similar to leaves, 1–5 cm; rays 1–9, 1–8 cm, unequal; involucel bractlets 1–5, narrowly lance-linear to lanceolate, 1–8 mm, tips lobed; umbellets 1–6(9)-flowered; pedicels 1–10 mm, unequal. **Flowers** sepals well developed; petals white. **Fruits** oblong, somewhat flattened, 3–7 mm; wide ribs with hooked prickles, lower ribs with shorter, narrow, straight prickles; wings absent. **2*n***=12.

Streamsides, meadows, woods. Flowering Apr–May. 0–1200 m. Col, ECas, Est, Sisk, WV. CA, ID, NV, WA; north to British Columbia, southeast to NM, south to Mexico. Native.

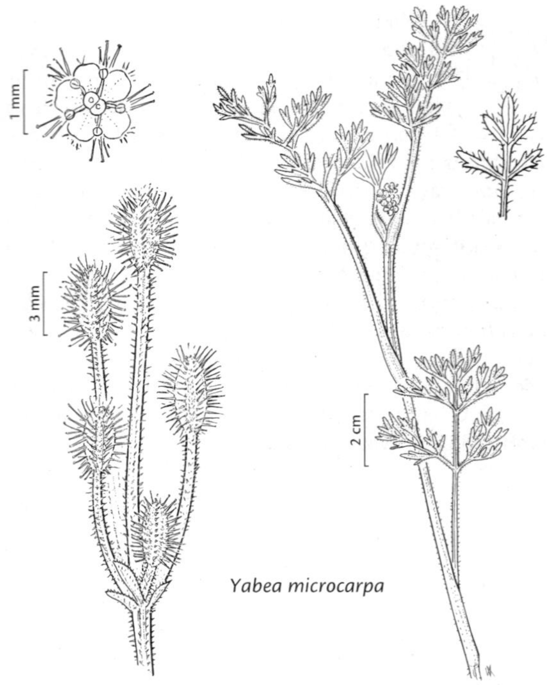

YABEA

ZIZIA Zizia

Herbs perennial, glabrous or rarely hairy at nodes. **Roots** fleshy, fibrous. **Stems** ascending to erect. **Leaves** basal and cauline, simple or ternately lobed or compound; blades mostly cordate to oval or round, some cauline blades occasionally broadly lanceolate, margins toothed, with a slender white to brown border; sheaths not or scarcely inflated; petioled to nearly sessile. **Inflorescences** terminal and axillary compound umbels; involucral bracts 0; rays 10–18, unequal in length; involucel bractlets 3–9, often small and inconspicuous; umbellets 10–20+-flowered. **Flowers** sepals small, triangular, scale-like; petals obovate, yellow. **Fruits** ovate to elliptic, flattened, yellowish brown, glabrous; ribs 5.

North America. 4 species; 1 species treated in *Flora*.

Zizia aptera (A. Gray) Fernald (p. 140)

Golden Alexanders

Zizia aptera var. *occidentalis* Fern.

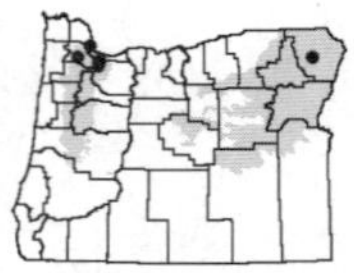

Plants 2–6 dm. **Stems** clustered on compact caudex. **Roots** fleshy-fibrous. **Basal leaves** simple to ternate, 2–13 cm, sometimes lobed at base; petioles. **Cauline leaves** few, lobed or compound, similar in length to basal leaves but upper leaves reduced; petioles short to nearly sessile. **Inflorescences** compact, densely flowered, 2–5 cm wide; rays 0.5–5 cm; pedicels 0–3 mm, central flower ± sessile. **Fruits** 2–4 mm. **2*n***=22.

Moist meadows, streambanks, roadsides. Flowering May–Jun. 50–1300 m. BW, WV. ID, NV, WA; north to British Columbia, northeast to Quebec, southeast to FL. Native.

Two varieties of *Z. aptera* have been described (*aptera* and *occidentalis*). They are loosely distinguished by the degree of margin teeth coarseness and leaf thickness. If accepted, most plants seen in our area would likely be variety *occidentalis*.

ZIZIA

APOCYNACEAE Dogbane or milkweed family

Mark Fishbein

Herbs, subshrubs, shrubs, vines, or **trees** perennial, deciduous or evergreen; cryptophytes or chamaephytes; all parts of shoots exuding white latex when damaged. **Roots** tuberous, taprooted, or fibrous. **Stems** erect, spreading, or trailing. **Leaves** cauline, simple, opposite, herbaceous or leathery, margins entire; multicellular glands (colleters) on upper base of blade sometimes present; stipules 0; interpetiolar colleters sometimes present. **Inflorescences** extra-axillary or axillary, but may appear terminal; umbels, cymes, or solitary. **Flowers** bisexual, radial; sepals 5, united; petals 5, united, lobes contorted or valvate in bud; corollas rotate with reflexed lobes, urceolate, or tubular-funnelform; may have scale-like nectaries at base; stamens 5, epipetalous, may be connate and adnate to styles, anthers with apical scale-like appendages; ovaries free, superior to partially inferior, carpels 2, ovules few–many. **Fruits** follicles, single or paired. **Seeds** few to many, may be comose. Nearly worldwide. 376 genera; 3 genera treated in *Flora*.

The milkweed subfamilies of Apocynaceae (Asclepiadoideae, Secamonoideae, Periplocoideae) were long recognized as the milkweed family Asclepiadaceae. Inclusion of these taxa in Apocynaceae is widely accepted. *Asclepias* is the only milkweed genus present in the flora of Oregon.

The Apocynaceae possess diverse secondary compounds some of which are potent toxins or pharmacologically active. The indole alkaloids vincristine and vinblastine present in *Catharanthus roseus* are currently the only effective drug treatment for Hodgkin's disease. Toxic cardiac glycosides in *Asclepias* are sequestered by several species of milkweed-feeding insects for use in their own defense, most famously by the monarch butterfly.

Key to Apocynaceae genera:

1. Plants evergreen; stems trailing; flowers solitary; corollas funnelform with somewhat narrow tube, blue or purple . . **VINCA**

1' Plants deciduous; stems erect to spreading; flowers in umbels or cymes; corollas campanulate, urceolate, cylindric, or rotate with reflexed lobes, green, white, pink, yellow, or violet.

  2. Corollas campanulate, urceolate, or cylindric, lobes shorter than fused portion; stamens and corona included in corolla; fruits linear, terete, > 15 × as long as wide wide, drooping . . . . **APOCYNUM**

  2' Corollas rotate with reflexed lobes, lobes much longer than fused portion; stamens and corona exposed, the corona highly conspicuous; fruits fusiform, < 7 × as long as wide wide, erect . . . . **ASCLEPIAS**

APOCYNUM Dogbane

Herbs perennial, deciduous; cryptophytes. **Roots** fibrous from rhizomes. **Stems** erect. **Leaves** opposite, simple, tips apiculate. **Inflorescences** terminal and axillary cymes. **Flowers** bisexual, radial; sepals 5, united; petals 5, united, lobes contorted in bud; corollas campanulate, urceolate, or cylindric, lobes shorter than fused portion, stamens and corona included; coronas of minute scales on inside of tube; nectaries scale-like, at base of ovaries; stamens 5, epipetalous, anthers coherent; carpels 2, ovules many, styles united but ovaries free. **Fruits** drooping, linear, sometimes falcate, terete, > 15 × as long as wide wide, often paired. **Seeds** many, comose.

Asia, Europe, North America. ~4 species; 2 species treated in *Flora*.

The taxonomic history of *Apocynum* is convoluted with anywhere between 2 and 50 species recognized in past treatments. A conservative approach that seems justified based on patterns of morphological and genetic variation is adopted here. Both species are highly variable in many characteristics, including degree of branching, leaf shape, hairiness, the posture and shape of the follicles, and corolla dimensions. Some of this variation is likely the result of geographic differentiation within species with subsequent gene flow. Ongoing hybridization and introgression is highly likely. Nonetheless, the species are usually readily distinguishable, especially in the field.

A putative hybrid between the two American species is found sporadically throughout the ranges of both species (including Oregon), sometimes as large populations that represent the locally dominant form of *Apocynum*. These entities evidently represent numerous independent origins of hybrid populations that probably include later generation backcrosses to one or both parental species. These entities are collectively named *Apocynum* ×*floribundum* and can be recognized by expression of intermediate or chimeric combinations of traits characteristic of the parental species. Suffice it to say, the taxonomy of this genus warrants further study.

Key to *Apocynum* species:

1. Leaves drooping; inflorescences borne above vegetative branches; corollas campanulate, 4.5–9 mm, white or pink and often with pink stripes, lobes reflexed . . . . *A. androsaemifolium*

1' Leaves spreading to ascending; inflorescences overtopped by some vegetative branches; corollas urceolate or cylindric, 2.5–5 mm, white, cream, yellowish cream, or greenish, lobes erect to ascending . . . . . . *A. cannabinum*

Apocynum androsaemifolium Linnaeus (p. 142)

Spreading dogbane

Apocynum androsaemifolium L. var. *androsaemifolium*, *Apocynum androsaemifolium* L. var. *incanum* A. DC., *Apocynum androsaemifolium* L. var. *pumilum* A. Gray, *Apocynum pumilum* (A. Gray) Greene var. *pumilum*, *Apocynum pumilum* (A. Gray) Greene var. *rhomboideum* (Greene) Bég. & Beloserky

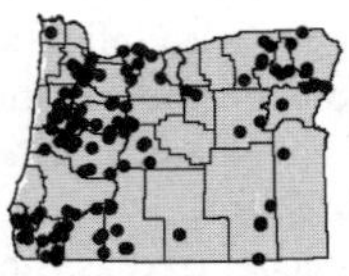

Stems erect or ascending, to 50 cm, glabrous or tomentose; much branched, a clearly dominant stem often lacking. **Leaves** drooping; blades ovate, oblong, or elliptic, glabrous or tomentose; petiolate to subsessile. **Inflorescences** cymes, overtopping vegetative branches. **Flowers** corollas campanulate, 4.5–9 mm, white or pink, often pink-striped; lobes reflexed, > 25% as long as fused portion. **Fruits** to 15 cm. **Seeds** with coma to 2 cm.

Moist and dry sites, slopes, riverbanks, roadsides, prairies, steppe, savannas, woodlands, forests, disturbed ground. Flowering May–Sep. 0–2200 m. All ecoregions except Est. CA, ID, NV, WA; throughout Canada and US, northern Mexico. Native.

Short-statured plants with especially showy flowers, some of which may also have densely hairy herbage, have been recognized as *A. androsaemifolium* var. *pumilum* or as a distinct species. These forms are particularly common in Oregon but are intermixed with the more widespread and common forms of this species and do not appear worthy of taxonomic recognition.

See description of the genus for a discussion of the common occurrence of putative hybrids between this and the following species.

Apocynum cannabinum Linnaeus

Indian hemp, common dogbane, hemp dogbane

Apocynum cannabinum L. var. *cannabinum*, *Apocynum cannabinum* L. var. *glaberrimum* A. DC., *Apocynum cannabinum* L. var. *pubescens* (Mitchell) A. DC., *Apocynum cannabinum* L. var. *suksdorfii* (Greene) Bég. & Beloserky, *Apocynum sibiricum* Jacq., *Apocynum sibiricum* Jacq. var. *salignum* (Greene) Fernald, *Apocynum suksdorfii* Greene

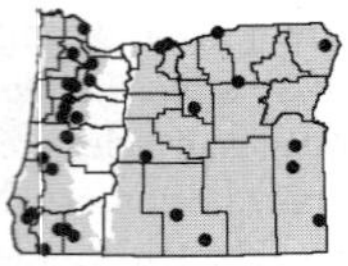

Stems erect or ascending, to 70 cm, glabrous to tomentose; much branched, especially toward apex, a clearly dominant stem usually present. **Leaves** spreading to ascending; blades lanceolate, ovate, oblong, or elliptic, sometimes with clasping base, glabrous to tomentose; petiolate to sessile. **Inflorescences** cymes, overtopped by some vegetative branches. **Flowers** corollas urceolate to cylindric, 2.5–5 mm, white, cream, yellowish cream, or greenish; lobes erect to ascending, < 25% as long as fused portion. **Fruits** drooping, to 20 cm. **Seeds** coma to 3 cm.

Wet meadows, fields, riverbanks, slopes, ditches, roadsides, disturbed ground. Flowering May–Sep. 50–2100 m. BR, BW, Col, CR, ECas, Lava, Owy, Sisk, WV. CA, ID, NV, WA; throughout Canada and US, northern Mexico. Native.

Plants with clasping leaves have been recognized as *A. sibiricum*. This form is widespread throughout the range of the species and is especially common in Oregon. With present knowledge it seems best to consider such plants morphological variants that are not accorded taxonomic recognition. See description of the genus for a discussion of the common occurrence of putative hybrids between this and the preceding species. As suggested by the common names, this species was an important fiber resource for American first peoples.

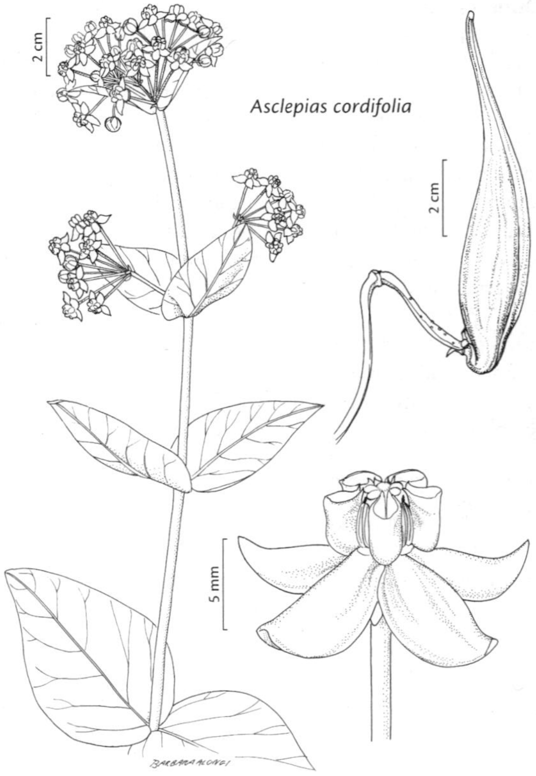

ASCLEPIAS Milkweed

Herbs perennial, deciduous; cryptophytes. **Roots** fibrous, rhizomatous, or taprooted. **Stems** erect to spreading. **Leaves** opposite or whorled, simple, colleters present or absent. **Inflorescences** extra-axillary umbels, may appear terminal. **Flowers** bisexual, rotate; sepals 5, united; petals 5, united; corolla lobes reflexed, much longer than fused portion, valvate in bud; corona conspicuous, of 5 basally fused segments (hoods), each hood with an internal cavity, sometimes also with an exserted appendage (horn); stamens and coronas exposed; stamens connate and adnate to styles forming a gynostegium, anthers with apical scale-like appendages, pollen from each anther agglutinated into pollinia, pairs of pollinia from adjacent anthers joined by translators to a corpusculum that acts as a mechanical clip for attachment to insects; ovules many. **Fruits** usually single, erect, fusiform, < 7 × as long as wide wide. **Seeds** many, comose.

Africa, North America, South America. ~400 species; 4 species treated in *Flora*.

The plants mostly commonly referred to as "milkweeds" in North America belong to this genus. Some or all of the 250 native African species are often excluded from the genus. The species are variously familiar to many Americans as rangeland and agriculture weeds as well as colorful and interesting subjects for native flower gardens. *Asclepias* species are the primary host plants of the monarch butterfly; this fact, in conjunction with the large number of butterflies and other insects drawn to the flowers, have made them highly valued by gardeners as "butterfly" plants.

Several species not native to Oregon are commonly grown in gardens, including butterflyweed (*A. tuberosa*), blood flower (*A. curassavica*), and swamp milkweed (*A. incarnata*). Much less is known about the use of Oregon native milkweeds by monarchs than that of the milkweeds of the eastern states. Several milkweeds enliven America's roadsides in the summer with large floral displays, including *A. speciosa* in some parts of Oregon.

In 2014, a small population of *Asclepias solanoana* was reported from Josephine County. Whether this represents a long-existing population, a recent migration, or a short-lived waif population may merit further investigation.

Key to *Asclepias* species:

1. Leaves whorled, linear to narrowly-elliptic, < 3 cm wide; fruits ≤ 1 cm wide. . . . . ***A. fascicularis***

1' Leaves opposite, lance-ovate to broadly ovate or oblong, ≥ 3 cm wide; fruits ≥ 1.5 cm wide.

  2. Plants tomentose throughout; leaves not glaucous, largest blades commonly > 10 cm long; corona lobe tips long and attenuate. . . . . ***A. speciosa***

  2' Plants glabrous throughout; leaves glaucous, largest blades commonly < 10 cm long; corona lobe tips blunt or truncate.

    3. Stems < 40 cm; leaf pairs < 4 per stem; flowers < 15 per umbel; corollas yellowish green; corona lobes dark to pale violet, < 4 mm; fruits < 7 cm long. . . . . ***A. cryptoceras*** ssp. ***davisii***

    3' Stems > 40 cm; leaf pairs > 4 per stem; flowers > 20 per umbel; corollas reddish purple to violet; corona lobes white to pink, > 5 mm; fruits ≥ 7 cm . . . . ***A. cordifolia***

Asclepias cordifolia (Bentham) Jepson (p. 142)

Purple milkweed, heartleaf milkweed

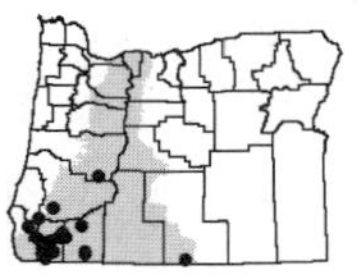

Stems usually 1–10(40) from base, often flattened below, 40–80 cm, glabrous, glaucous; usually unbranched. **Leaves** opposite, leaf pairs more than 4 per stem; blades ovate, 7–12 × 5–10 cm, largest blades commonly < 10 cm, glabrous, glaucous, bases cordate and clasping, tips obtuse to acuminate; interpetiolar colleters absent; sessile. **Inflorescences** terminal 1–3 umbels with 10+ flowers, occasional smaller umbels in upper leaf axils; peduncles 1–7 cm, rarely absent, glabrous; pedicels 2–4 cm, glabrous. **Flowers** calyx lobes lance-ovate, 3 mm, purple; corollas reflexed with spreading lobes, reddish purple to violet, lobes 6–8 mm; coronas and gynostegia borne on obconic columns 1.5 mm tall, corona lobes saccate, 3 mm, white to pink, with marginal lobes and blunt tips; horns lacking. **Fruits** erect on pedicels deflexed 180°, narrowly fusiform with attenuate tips, 7–12 × 1.5–2.0 cm, glabrous. **Seeds** broadly oval, 8 mm; with white coma to 3–3.5 cm.

Rocky, gravelly, or talus slopes, woodlands, chaparral. Flowering May–Jul. 100–1300 m. Casc, ECas, Sisk. CA, NV. Native.

This species is morphologically isolated with no apparently close relatives in the genus. The hornless corona segments are rather unusual in American *Asclepias*. The species is quite attractive in its habitat, often with bluish foliage setting off the small but colorful purple and white flowers.

Asclepias cryptoceras S. Watson ssp. *davisii* (Woodson) Woodson (p. 144)

Pallid milkweed, Davis' milkweed

Asclepias cryptoceras S. Watson var. *davisii* (Woodson) W.H. Baker

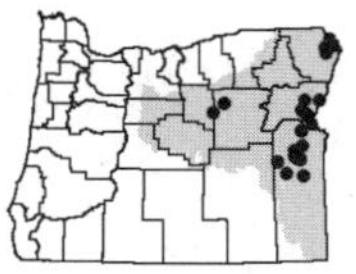

Stems usually 1–5 from base, spreading to decumbent, to 30 cm, glabrous; usually unbranched. **Leaves** opposite, leaf pairs < 4 per stem; blades broadly oval or ovate to suborbicular, 4–9 × 4–8 cm, bases rounded to subcordate, tips rounded to obtuse, glabrous, glaucous; interpetiolar colleters absent; sessile to subsessile. **Inflorescences** umbels with < 15 flowers; peduncles 0–7 cm, glabrous; pedicels 2–4 cm, glabrous. **Flowers** calyx lobes lance-ovate, 6–7 mm; corollas reflexed with spreading lobes, yellowish green, lobes 10–15 mm;

coronas and gynostegia sessile, corona lobes saccate, laterally compressed, 6–9 mm, dark to pale violet, truncate with bifid tips appressed to gynostegium; horns minute and not noticeably exserted from hood or lacking. **Fruits** pendulous from ± straight pedicels, broadly fusiform, 4–7 × 1.5–2.5 cm, with apiculate tips, glabrous. **Seeds** broadly oval, 10 mm; with pale tawny coma to 2.5 cm.

Hillsides and slopes on clay, shale, or eroded basalt. Flowering Apr–Jun. 500–1400 m. BW, Lava, Owy. CA, ID, WA. Native.

Another morphologically distinctive milkweed, *A. cryptoceras* ssp. *davisii* is unmistakable by virtue of its dwarf stature, disproportionately large yellowish and purple flowers with nearly hornless coronas, and large, round, glaucous leaves. The subspecies are almost completely allopatric with ranges that abut in eastern California and northern Nevada. They can be distinguished by the corona segment apex, which is truncate at right angles to the outer margin with relatively large bifid lobes in subspecies *davisii* and oblique with relatively small lobes in subspecies *cryptoceras*. Also, the corona of subspecies *davisii* is a relatively dull, light or dark purple, in contrast to the bright, reddish purple or maroon corona of subspecies *cryptoceras*. The typical subspecies occurs to the southwest, from eastern California to Colorado and southern Wyoming.

Asclepias fascicularis Decaisne

Narrowleaf milkweed

Asclepias mexicana auct. non Cav.

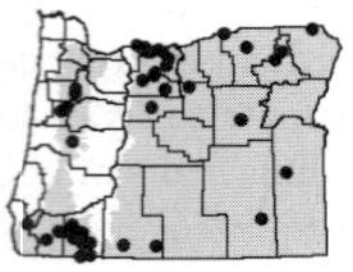

Stems usually more than 6 from base, to 1 m, inconspicuously puberulent to glabrous; little branched but commonly with short, unexpanded branches in leaf axils; rhizomatous. **Leaves** whorled (commonly 3–5 per node); blades narrowly linear to narrowly elliptic, 3–12 × 0.1–2.5 cm, bases acute to obtuse, tips obtuse to acuminate, glabrous or often inconspicuously puberulent below; interpetiolar colleters present; very shortly petiolate. **Inflorescences** umbels; peduncles 2–4 cm, inconspicuously puberulent; pedicels 0.7–1.5 cm, puberulent. **Flowers** calyx lobes narrowly deltoid, 1–1.5 mm, minutely pilose to glabrous; corollas reflexed with spreading lobes, grayish pink, rarely white, lobes 3–4 mm; coronas and gynostegia borne on cylindric columns 1 mm tall; corona lobes hood-like, 1–1.5 mm, white to pinkish, truncate at tips; horns conspicuous and arching over style apex, narrowly acicular. **Fruits** erect on erect pedicels, narrowly fusiform, 5–12 × 0.7–1 cm, glabrous. **Seeds** oval, 6–7 mm; with white coma to 2 cm.

Valleys, riparian zones, seasonally moist soils, disturbed areas. Flowering Jun–Aug. 50–1600 m. BR, BW, Col, ECas, Lava, Owy, Sisk, WV. CA, ID, NV, WA; south to Mexico. Native.

Very common in California, this species remains common through the Willamette Valley and Columbia River Gorge but otherwise becomes increasingly rare to the north. It is unmistakable in Oregon by virtue of its densely clumped habit and

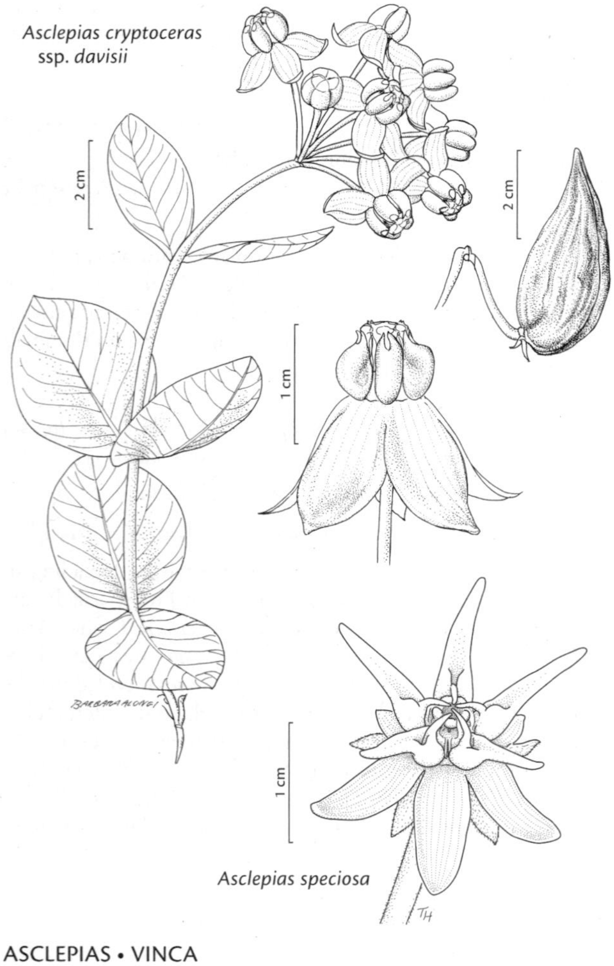

Asclepias cryptoceras ssp. davisii

Asclepias speciosa

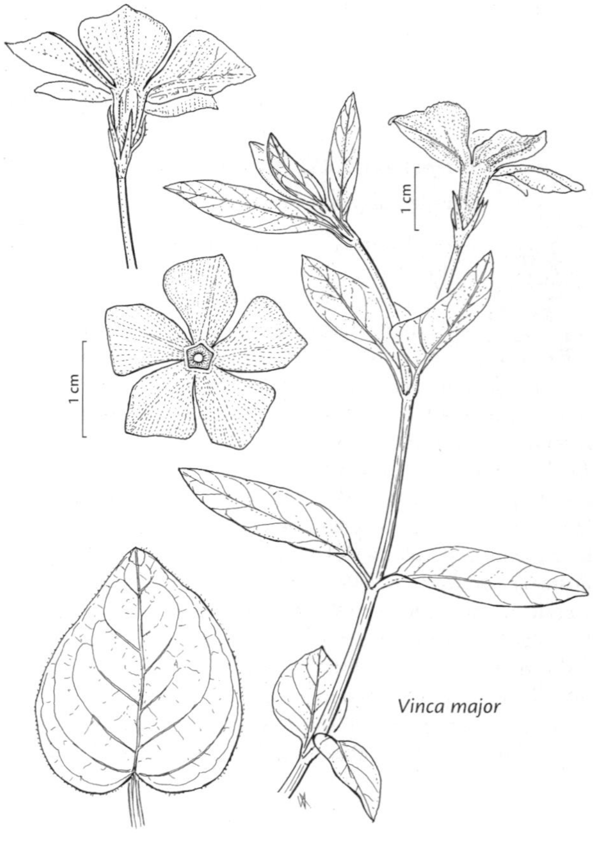

Vinca major

whorled, narrow leaves. But it is quite similar to related species found throughout the rest of North America, especially *A. subverticillata* and *A. incarnata*. The plant is rather dingy and weedy in appearance, the small, dull pinkish flowers, though numerous, not producing much visual effect. This species and *A. speciosa* are the most commonly used host plants of the monarch butterfly in western North America.

Asclepias speciosa Torrey (p. 144)

Showy milkweed

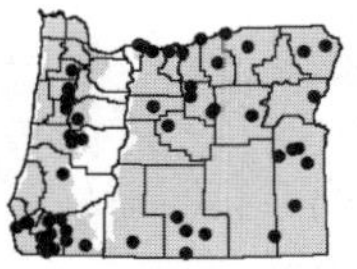

Stems usually 1–5 from base, to 1 m, tomentose; usually unbranched; often forming large colonies by rhizomatous spread. **Leaves** opposite; blades lance-ovate to oval or oblong, 6–20 × 3–14 cm, largest blades commonly > 10 cm, tomentose, bases obtuse to subcordate, tips obtuse to acute, upper surface glabrate; interpetiolar colleters present; petiolate. **Inflorescences** umbels; peduncles 1–10 cm, tomentose; pedicels 2–3 cm, tomentose. **Flowers** calyx lobes lanceolate, 5–6 mm, tomentose; corollas reflexed with spreading lobes, purplish rose to pale pink, lobes 10–15 mm; coronas and gynostegia borne on obconic columns 1 mm tall; corona lobes hood-like, laterally compressed, widely spreading, 10–14 mm, pale pink or rose, rarely white, with long-attenuate tips; horns conspicuous and arching over style apex, narrowly falciform with acicular tip. **Fruits** borne erect on pedicels deflexed 180°, fusiform, 9–12 × 2–3 cm, with attenuate tips, tomentose, commonly soft muricate. **Seeds** oval, 6–9 mm; with white coma to 3–4 cm.

Valleys, roadsides, fields, disturbed areas. Flowering Jun–Aug. 0–1600 m. BR, BW, Col, CR, ECas, Lava, Owy, Sisk, WV. CA, ID, NV, WA; north to British Columbia, northeast to Manitoba, east to MI, southeast to TX. Native.

This is the most widespread and common of the western US milkweeds, conspicuous along roadsides over much of the region. It may form dense colonies, particularly in roadside ditches. Morphologically and ecologically, it may be considered the western counterpart of the common milkweed, *A. syriaca*, a species with which it commonly hybridizes where they come in contact in the Great Plains. It is an important food source for monarch butterfly larvae. The flowers are among the largest in the genus, and the widely spreading, long-pointed, pink coronas are highly distinctive.

VINCA Periwinkle

Herbs or **subshrubs** evergreen; chamaephytes. **Stems** trailing; stoloniferous and rhizomatous. **Leaves** leathery. **Flowers** solitary in leaf axils; corollas funnelform, but approaching salverform, with rather narrow tubes and rotate limbs, usually blue or light purple (may be reddish, deep purple, or white in cultivars), lobes contorted in bud, obliquely truncate at anthesis; anthers appressed to styles, apically puberulent; carpels with scale-like nectaries at base; stigmas ring-like, apically puberulent. **Fruits** usually paired, beaked, narrowly cylindric, somewhat torulose. **Seeds** 1–4, ellipsoid, corrugated; lacking a coma.

Northwestern Africa, southwestern Asia, Europe. 6–7 species; 2 species treated in *Flora*.

Key to *Vinca* species:

1. Leaf blades ovate or broadly ovate, 2–6 cm wide, bases truncate to shallowly cordate, margins ciliate; calyx lobes 6–17 mm, ciliate; corolla tubes 12–15 mm long, limbs 30–50 mm wide . . . . ***V. major***

1' Leaf blades elliptic to rarely ovate, 0.5–2 cm wide, bases cuneate to rounded, margins glabrous; calyx lobes 2–5 mm, glabrous; corolla tubes 8–12 mm long, limbs 15–30 mm wide . . . . ***V. minor***

Vinca major Linnaeus (p. 144)

Greater periwinkle, large periwinkle

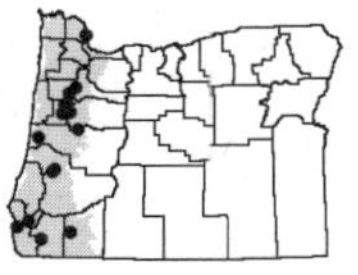

Stems arching, trailing; stoloniferous. **Leaves** blades elliptic, ovate or broadly ovate, 2–9 × 2–6 cm, bases truncate to shallowly cordate, margins ciliate, glabrous or minutely puberulent on veins of lower surface, tips obtuse to acute; shortly petiolate. **Inflorescences** axillary, flowers solitary; pedicels 2–5 cm, glabrous. **Flowers** calyx lobes narrowly triangular, 6–17 mm, ciliate; corolla tubes 12–15 mm long, limbs 30–50 mm wide. **Fruits** falcate, 2–4 cm, glabrous. **Seeds** 1–3.

Disturbed areas, streambanks, roadsides, cemeteries, forest understories. Flowering Mar–May. 0–800 m. CR, Sisk, WV. CA, ID, WA; eastern, western and southern US, British Columbia; Asia, Europe. Exotic.

Because of its clonal reproduction, *Vinca major* can be an aggressive weed requiring management, particularly in forested regions. Its common use as a ground cover in Oregon should be reconsidered, especially west of the Cascades. This species and the next are commonly confused in herbaria, and reports of *V. major* sometimes pertain to *V. minor.*

Vinca minor Linnaeus

Lesser periwinkle, common periwinkle

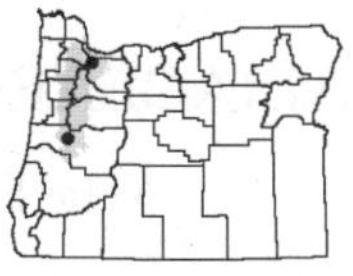

Stems trailing; stoloniferous. **Leaves** blades elliptic or rarely ovate, 2–5 × 0.5–2 cm, bases cuneate to rounded, margins glabrous, tips obtuse to acute, surfaces glabrous; shortly petiolate. **Inflorescences** axillary, flowers solitary; pedicels 1–2 cm, glabrous. **Flowers** calyx lobes narrowly elliptic, 2–5 mm, margins glabrous; corolla tubes 8–12 mm long, limbs 15–30 mm wide. **Fruits** straight, 1.5–2 cm, glabrous. **Seeds** 1–3.

Disturbed areas, streambanks, forest understories. Flowering Mar–May. 0–900 m. WV. WA; eastern North America, scattered in western North America; Asia, Europe. Exotic.

Vinca minor, though less commonly reported as an escape than *V. major*, has the potential to be a more serious invader in western Oregon. It has been observed to dominate large areas of little disturbed forest understories in the Portland area. Its true extent may be underestimated due to confusion with *V. major.* Like its congener, its use as a ground cover merits reconsideration.

AQUIFOLIACEAE Holly family

Alan T. Whittemore

Shrubs or **small trees**. **Leaves** alternate, simple, pinnately veined, short-stalked. **Flowers** unisexual, male and female on different plants, 4-merous; sepals united at base; petals free except at extreme base, spreading, white or pale green; stamens free, same number as petals and alternating with them, short and sterile on female plants; discs absent; pistils of 4 carpels (rudimentary on male plants), ovaries superior, with 4 locules, placentation axile, ovules 1(2) in each locule, styles very short, stigmas nearly sessile, shallowly 4-lobed. **Fruits** drupes with multiple stones.

Worldwide. 1 genus.

ILEX Holly

Family composed of one genus, *Ilex*.

Worldwide. ~400 species; 1 species treated in *Flora.*

Many *Ilex* species and hybrids are grown as ornamentals in Oregon, but only one is known to have escaped here.

Ilex aquifolium Linnaeus (p. 146)

English holly

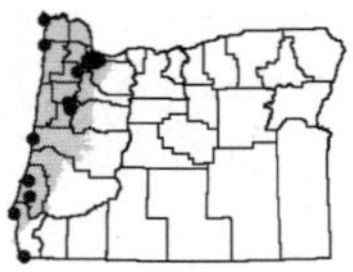

Shrubs or **small trees** evergreen, to 20 m; twigs usually finely puberulent, sometimes glabrous. **Leaves** blades elliptic, 48–108 × 30–65 mm, thick and rigid, bases obtuse or rounded, margins most often undulate, with 4–11 triangular, spine-tipped teeth on each side, teeth to 11 mm (including spine), on old trees some or all leaves with margins plane and entire, blade tips acute, acuminate, or seldom mucronate, always spine-tipped, surface glabrous, upper surface glossy, underside dull, lateral veins inconspicuous; petioles 5–15 mm, stout, with a ventral trough or groove. **Flowers** calyx lobes 0.5–1 mm; petals 3–4 mm; stamens free, spreading. **Fruits** spherical, 6–9 mm in diameter, bright red, usually with 4 stones.

Woodlands, thickets, second-growth forest, roadsides, open ground. Flowering May. 0–800 m. CR, Est, WV. CA, WA; north to British Columbia; Europe. Exotic.

English holly is widely grown as an ornamental for its glossy evergreen leaves and bright red fruits. Some cultivars produce fruit without pollination; others will have no fruit unless there is a male plant in the vicinity. The fruits are toxic; eating them will cause vomiting and abdominal pain.

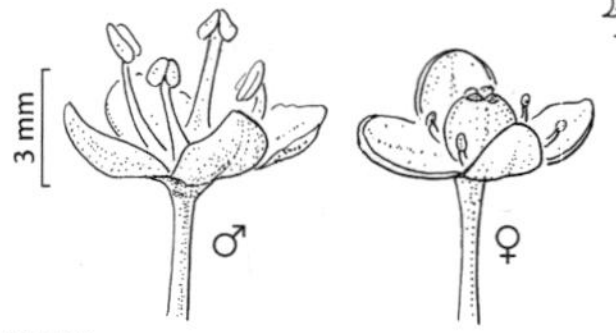

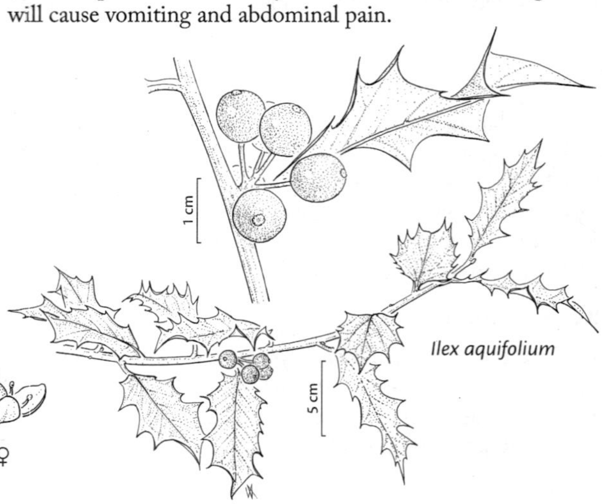

Ilex aquifolium

ILEX

ARALIACEAE Ginseng family

Katie E. Mitchell

Shrubs, vines, or **herbs** perennial. **Stems** usually branched. **Leaves** alternate, simple or compound; petioles present or not. **Inflorescences** umbels, simple or compound; bracts present. **Flowers** bisexual, radial; calyces absent or minutely 5-lobed and fused at base; petals 5, free, deciduous; stamens 5, alternate with petals; ovaries inferior, styles 1–5, persistent. **Fruits** berries.

Worldwide. ~50 genera; 4 genera treated in *Flora.*

Once considered a member of Apiaceae, *Hydrocotyle* has been transferred to Araliaceae based on molecular evidence (Chandler & Plunkett 2004).

Key to Araliaceae genera:

1. Stems erect and obviously prickly . . . . **OPLOPANAX**
1' Stems not prickly.
 2. Plants aquatic or semi-aquatic, with creeping, submerged, or floating stems; stems delicate . . . . . . **HYDROCOTYLE**
 2' Plants terrestrial, sometimes near water but not aquatic; stems not delicate.
  3. Stems erect . . . . **ARALIA**
  3' Stems climbing and trailing . . . . **HEDERA**

ARALIA Spikenard, aralia

Shrubs or **herbs** deciduous. **Stems** erect. **Leaves** alternate, pinnately compound; stipules fused to petiole base. **Inflorescences** terminal and axillary umbels in clusters of 2 or several or in panicles. **Flowers** calyces minutely 5-lobed; petals imbricate in buds; styles imbricate or connate below; pedicels articulated.

North America, Southeast Asia. ~35 species; 1 species treated in *Flora.*

Aralia californica S. Watson (p. 147)

Western aralia, elk clover, California spikenard

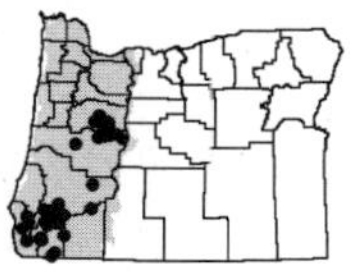

Herbs perennial; roots large; sap milky. **Stems** erect. **Leaves** 1–3 pinnate, 0.6–2 m, glabrous; leaflets 8–30 cm, ovate to oblong, serrate, bases cordate. **Inflorescences** 30–45 cm. **Flowers** ~3 × 4 mm, whitish; carpels 5; styles 5. **Fruits** round, 3–6 mm in diameter, dark purple to black. **2*n***=48.

Stream sides, ditches, moist areas. Flowering Jun–Aug. 100–1900 m. Casc, CR, Sisk, WV. CA. Native.

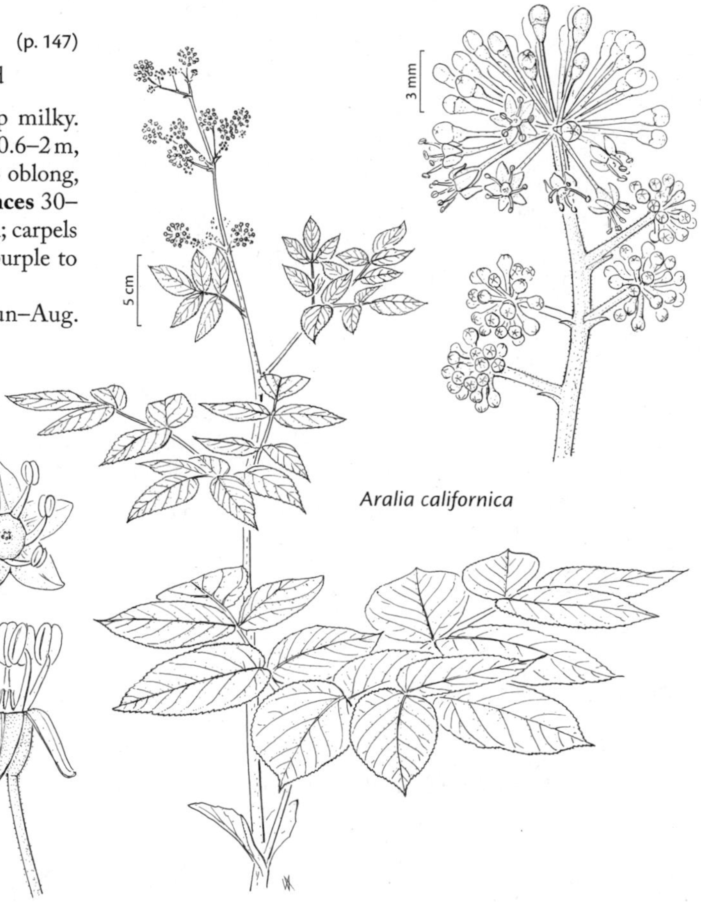

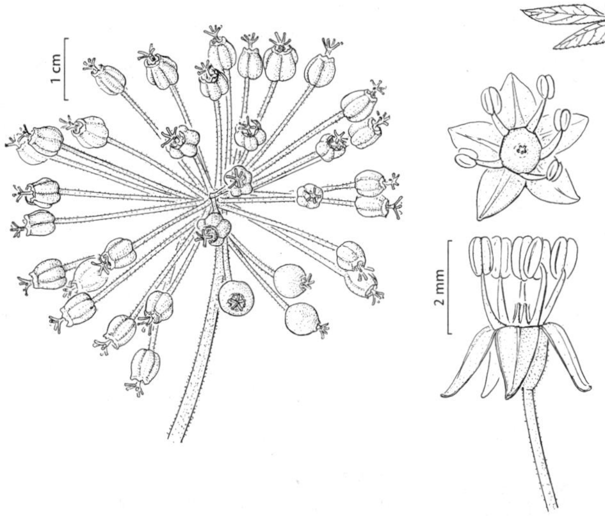

ARALIA

HEDERA Ivy

Vines evergreen, woody. **Stems** trailing or climbing by aerial roots; fertile stems branched and upright. **Leaves** alternate, simple, often palmately 3–5-lobed, sometimes entire, leathery; leaves on adult stems usually less lobed than on juvenile stems. **Inflorescences** terminal, a compact raceme of umbels. **Flowers** perfect, 5–7 mm; calyces 5-lobed, inconspicuous; petals greenish; carpels 5; styles 1. **Fruits** dark blue-black. **Seeds** 1–few, ovoid.

Worldwide. ~15 species; 2 species treated in *Flora.*

Cultivars of *H. helix* are widely variable in leaf size, shape, and color, and the leaf measurements given are valid only for wild-type, escaped plants, but trichome characters are characteristic of each species, regardless of cultivar. Trichomes, or epidermal outgrowths, are best observed on newer growth, on protected leaf undersides between main veins, under 10× or higher magnification.

Hedera colchica (Persian ivy) has been collected twice in southwestern Oregon as a garden escape. This species has unlobed or shallowly three-lobed juvenile leaves, whitish scale-like trichomes on new growth, aging to rust or coppery brown, and the center of the trichome is raised high with 12–20 very short rays depressed to the leaf surface.

Key to *Hedera* species:

1. Leaves on juvenile shoots deeply lobed, length from leaf base to first sinus 1–2 cm; trichomes with at least some rays at an angle to the leaf surface, giving a bristly appearance.. . . . . ***H. helix*** ssp. ***helix***

1' Leaves on juvenile shoots shallowly lobed or sometimes unlobed, length from leaf base to first sinus 2.5–4 cm; trichomes lying flat in one plane parallel to the leaf. . . . . ***H. hibernica***

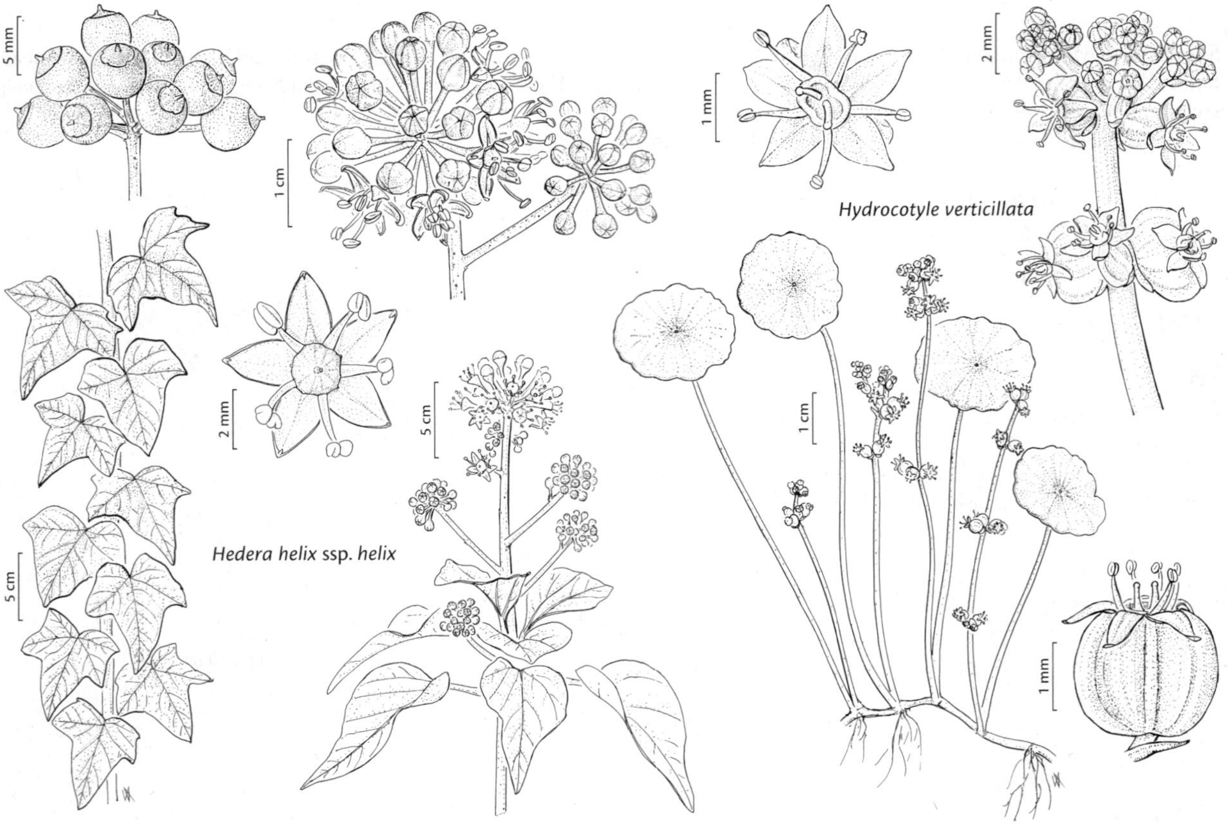

HEDERA • HYDROCOTYLE

Hedera helix Linnaeus ssp. ***helix*** (p. 148)

English ivy

Leaves on vegetative stems 6–8 nodes below apex 3–6 cm wide; length from leaf blade base to first sinus 1–2 cm; trichomes with at least some rays at an angle to leaf surface. **2*n***=48.

Forests, trees, man-made structures near inhabited areas. Flowering Aug–Nov. 0–300 m. Est, WV. CA, ID, NV, WA; worldwide. Exotic.

Of the two species naturalized in Oregon, *H. helix* is less common. In the past, most Oregon *Hedera* collections were erroneously identified as this species.

Hedera hibernica Carrière

Atlantic ivy

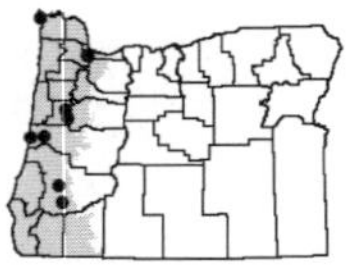

Leaves on juvenile stems 6–8 nodes below apex 6–9 cm wide; length from leaf blade base to first sinus 2.5–4 cm; trichomes with all rays parallel to leaf surface. **2*n***=96.

Forests, trees, man-made structures near inhabited areas. Flowering Aug–Nov. 0–600 m. CR, Est, Sisk, WV. WA; NC, SC; Europe. Exotic.

On herbarium specimens, the tips of a ray may curl away from the leaf surface, but none will be at an angle to the leaf surface.

Although *H. hibernica* has only been reported as escaped in Oregon, Washington, and the Carolinas, it is highly likely this is due to the misidentification of this species as *H. helix*. A more thorough examination of collections from throughout North America (and elsewhere in the world) is warranted.

HYDROCOTYLE Marsh pennywort

Herbs creeping or floating; rooting at nodes; stems delicate. **Leaves** simple; petiolate; stipules scarious. **Inflorescences** umbels, simple or proliferous. **Flowers** bisexual; calyces 5-lobed and minute or absent; petals 5, whitish; carpels 2; styles 2. **Fruits** laterally flattened, suborbicular, orbicular, or elliptic, ribbed.

Worldwide. ~100 species; 2 species treated in *Flora.*

Key to *Hydrocotyle* species:

1. Leaf blades peltate, petiole attached near the center of the abaxial side of the leaf . . . . ***H. verticillata***

1' Leaf blades not peltate, lobed to the petiole, margins sometimes overlapping . . . . ***H. ranunculoides***

Hydrocotyle ranunculoides Linnaeus f.

Floating marsh pennywort

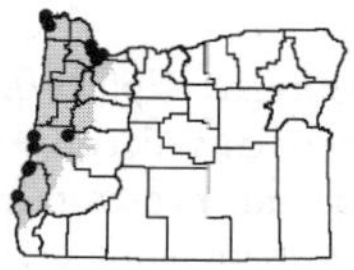

Plants glabrous, fleshy. **Stems** creeping or floating. **Leaves** reniform, 5–7-lobed, lobes crenate, 1–3.5 cm wide, not peltate; petioles 2–15 cm. **Inflorescences** simple umbels, (2)5–10-flowered, pedicels 1–2 mm, peduncles 2–4 cm, recurved in fruit. **Fruits** suborbicular to orbicular, 1.5–2 × 3 mm. ***2n***=48

Boggy ground, stagnant or slow-moving, shallow water. Flowering Mar–Aug. 0–200 m. CR, Est, WV. CA, WA; coastal and southern North America; Africa, Asia, Australia, Europe, South America. Native.

Hydrocotyle verticillata Thunberg (p. 149)

Whorled marsh pennywort

Hydrocotyle verticillata Thunb. var. *verticillata*

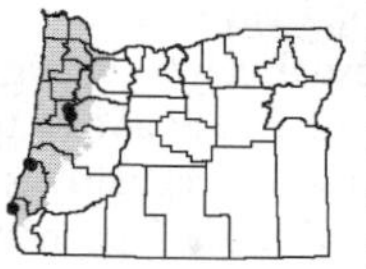

Plants glabrous. **Stems** creeping or floating. **Leaves** orbicular, crenate, rarely lobed, 1–3 cm wide, peltate; petioles 1–8 cm. **Inflorescences** proliferous umbels, each umbel up to 15-flowered, peduncles very short, pedicels 1–4 cm, flowers or fruits tightly clustered at each node. **Fruits** elliptic, 1.5–2 × 3 mm.

Boggy ground or in stagnant or slow-moving, shallow water. Flowering Apr–Sep. 0–100 m. CR, Est, WV. CA, NV; Caribbean, coastal and southern North America; Africa, Australia, Europe, HI, South America. Native.

The scientific and common names of this species refer to the whorled appearance of the inflorescence.

OPLOPANAX Devil's club

Shrubs deciduous, densely prickly. **Stems** thick and punky. **Leaves** large, palmately lobed, toothed. **Inflorescences** terminal. **Flowers** small; stamens longer than petals; ovaries 2-carpellate, styles free.

East Asia, North America. 3 species; 1 species treated in *Flora.*

The name "*Oplopanax*" comes from a Greek word meaning "armed" or "armor." These shrubs are indeed armed with long, sharp, irritating prickles.

Oplopanax horridus (Smith) Miquel (p. 149)

Devil's club

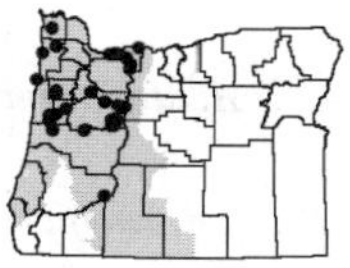

Stems densely armed with long yellow prickles, often decumbent at base, generally upright, 1–4 m. **Leaves** alternate; blades palmately 5(13)-lobed, 1–4 dm wide, margins serrate, veins prickly; stipulate. **Inflorescences** terminal, compact umbels arranged in elongate panicles or racemes. **Flowers** perfect; sepals 5; petals 5, green or greenish white; ovaries inferior, styles 2. **Fruits** flattened, 5–8 mm, red, shiny. **Seeds** 2–3.

Moist woods, talus slopes, streamsides, ditches, wet areas. Flowering May–Aug. 0–1800 m. Casc, CR, ECas, Est, WV. ID, WA; north to AK, MI, Ontario. Native.

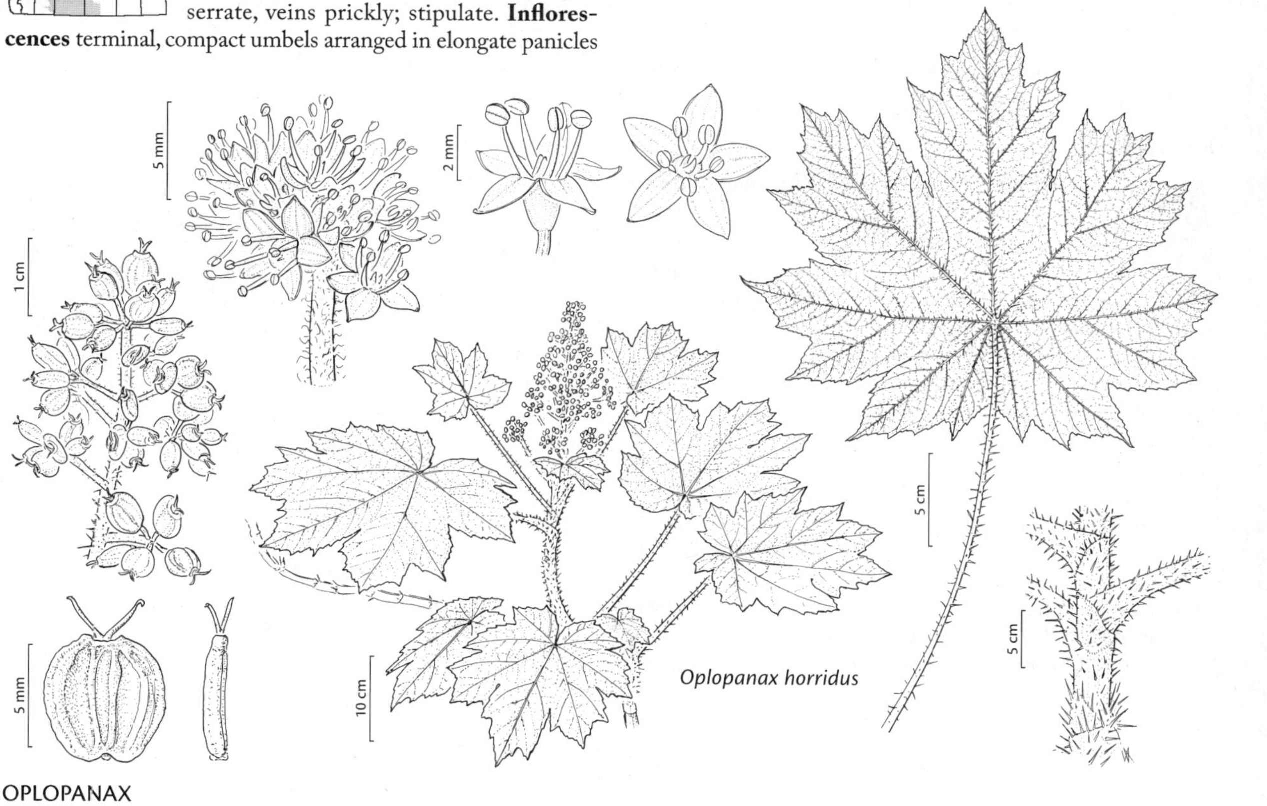

OPLOPANAX

ARISTOLOCHIACEAE Birthwort or Dutchman's pipe family

Christian Feuillet

Herbs or **vines** perennial; rhizomatous or not. **Leaves** cauline or basal; blades simple, unlobed, margins entire; petiolate; stipules 0. **Inflorescences** solitary flowers. **Flowers** bisexual; calyces enlarged, fused, petaloid, 3-lobed, tubes inflated, straight or curved; corollas reduced to scales or lacking; stamens 6 or 12, free or adnate to styles and stigmas, thereby forming gynostemia, anthers extrorse; pistils 6-carpellate, ovaries inferior. **Fruits** capsules, regularly to irregularly dehiscent. **Seeds** often flattened; endosperm copious.

Mainly tropical and temperate regions worldwide. 5 genera; 1 genus treated in *Flora.*

ASARUM Wild ginger

Herbs deciduous or semi-evergreen, usually aromatic, from underground rhizomes. **Leaves** cauline and basal, alternate, with both foliage leaves and sessile triangular scales; foliage blades reniform, surfaces pubescent at least on lower surface and on margins; petioles present. **Inflorescences** terminal, solitary flowers; bracts absent. **Flowers** sepals touching and forming a false tube; tubes straight; vestigial petals present or lacking; stamens 12, distinct. **Fruits** fleshy. **Seeds** ovoid, with fleshy appendage. **2*n*=**26.

Temperate Northern Hemisphere. ~75 species; 3 species treated in *Flora.*

The flowers of *Asarum* are mostly self-pollinated, but they are occasionally visited by mycotrophic flies. The fatty appendages (elaiosomes) on the seeds attract ants, which help disperse the seeds.

Despite its common name, *Asarum* is not closely related to the culinary species of ginger (*Zingiber officinale*). The name is derived from the ginger-like odor of the rhizomes. *Asarum* should not be consumed due it its high aristolochic acid content.

Asarum hartwegii has been collected only once, on the Little Chinquapin Mountain in Jackson County. Its variegated leaves are similar to those *A. marmoratum*, but the inner surfaces of its calyx tubes are white with red stripes while those of *A. marmoratum* are red-brown without stripes.

Key to *Asarum* species:

1. Leaf blades almost always with white or silvery variegations adaxially along veins; calyces internally dark red; rhizome internodes 2–15 mm . . . . *A. marmoratum*

1' Leaf blades never variegated; calyces internally white or light green; rhizome internodes 5–65 mm.

  2. Leaf blades cordate; calyx tubes cylindric, externally dark red to brownish purple, rarely green, lobes (11)30–75 mm . . . . *A. caudatum*

  2' Leaf blades cordate-reniform to reniform; calyx tubes subglobose to cylindric-urceolate or urceolate, externally light green, lobes 6–24 mm . . . . *A. wagneri*

Asarum caudatum Lindley (p. 150)

Western wild ginger, long-tailed wild ginger

Asarum caudatum Lindl. var. *caudatum*

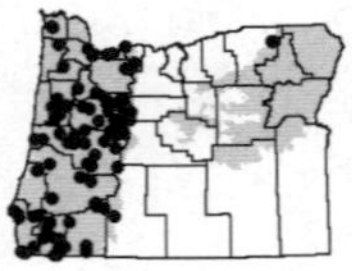

Plants semi-evergreen. **Rhizomes** horizontal, shallowly buried; internodes (5)15–65 mm. **Leaves** blades cordate, (1.5)3–8.5 × (2)4.5–12.5 cm, not variegated, marginal hairs perpendicular to margin or curved toward tip, tips usually obtuse, occasionally broadly acute, abaxially sparsely appressed-hirsute, at least proximally, adaxially glabrous or sparsely appressed-hirsute; petioles 6–19 cm. **Pedicels** 15–50 mm. **Flowers** calyx tubes cylindric, externally dark red to brownish purple, rarely green, hispid, internally white, usually with median purple stripes, with purple or rarely white hairs, lobes spreading or weakly reflexed at anthesis, (11)30–75 mm, tips filiform-attenuate, abaxially purple or greenish, sparsely hispid, adaxially purple, puberulent with crisped purple hairs.

Mesic or wet places in conifer forests or rocky areas. Flowering Mar–Aug. 0–1800 m. BW, Casc, CR, Est, Sisk, WV. CA, ID, WA; north to British Columbia, northeast to MT. Native.

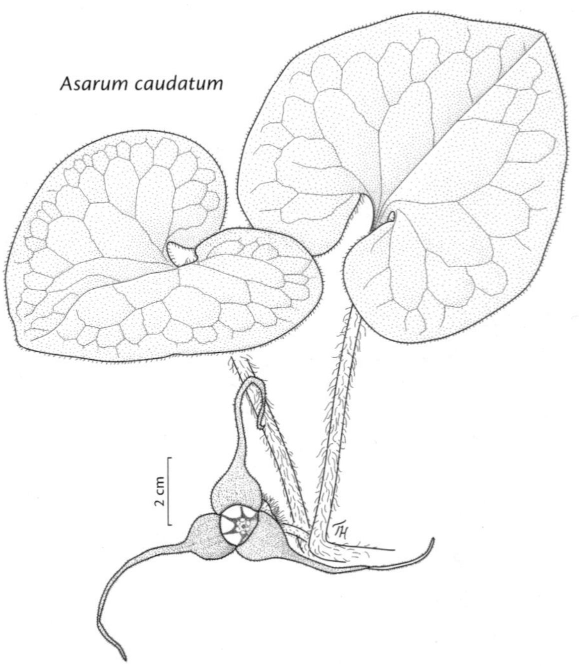

ASARUM

Asarum marmoratum Piper (p. 151)

Marbled wild ginger

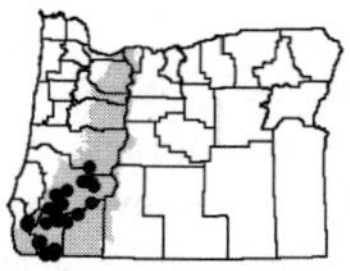

Plants semi-evergreen. **Rhizomes** erect or ascending, deeply buried; internodes 2–15 mm. **Leaves** blades cordate to cordate-reniform, (2.5)4–14 × 3–12 cm, variegated with white or silver in a broad, tapering band along midrib and in varying amounts along lateral veins, marginal hairs ± perpendicular to margin, tips acute to broadly acuminate, rarely obtuse, abaxially sparsely hispid, adaxially glabrous or sparsely hispid along veins; petioles 5–20 cm. **Pedicels** 12–19 mm. **Flowers** calyx tubes cylindric-urceolate, externally mottled red, sparsely to moderately pubescent to villous, internally dark red, with purple hairs, lobes erect or spreading at anthesis, 17–52 mm, tips filiform-attenuate, abaxially pale green, hirsute, adaxially tan or brownish green, puberulent with crisped purple hairs.

Dry or mesic forests, exposed rocky slopes, road cuts. Flowering Apr–Jul. 0–1800 m. Casc, Sisk. CA. Native.

Asarum marmoratum is found only in the Cascade and Siskiyou mountains of southwestern Oregon and extreme northwestern California.

Asarum wagneri K.L. Lu & Mesler (p. 151)

Wagner's wild ginger, green-flowered wild ginger

Asarum caudatum Lindl. var. *viridiflorum* M. Peck

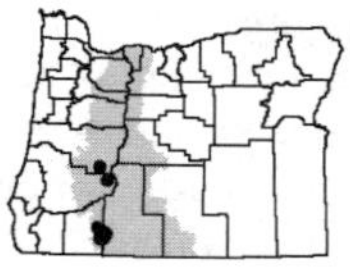

Plants deciduous. **Rhizomes** horizontal, usually deeply buried; internodes 5–21 mm. **Leaves** blades broadly reniform to cordate-reniform, (1.5)3–8 × (2.5)4–11 cm, not variegated, marginal hairs mostly curved toward tip, tips obtuse to rounded, rarely broadly acute, abaxially sparsely hirsute, adaxially sparsely hirsute only along veins; petioles 3–15 cm. **Pedicels** 8–30 mm. **Flowers** calyx tubes subglobose to cylindric-urceolate or urceolate, externally light green, sparsely to moderately hirsute, internally white or light green, bordered and occasionally striped with purple, with purple hairs, lobes spreading perpendicularly from base at anthesis, bent abruptly upward at midpoint, 6–24 mm, tips filiform-acuminate, abaxially white to pale green, sparsely villous to villous, adaxially white or light green, bordered with purple and occasionally with a purple band across base, puberulent with crisped purple-tipped hairs.

Pumice slopes, meadows, conifer forests. Flowering Jun–Aug. 900–2600 m. Casc, ECas. CA. Native.

Asarum wagneri is a rare endemic of the Cascade Mountains of southern Oregon and extreme northern California, although it may be abundant where found.

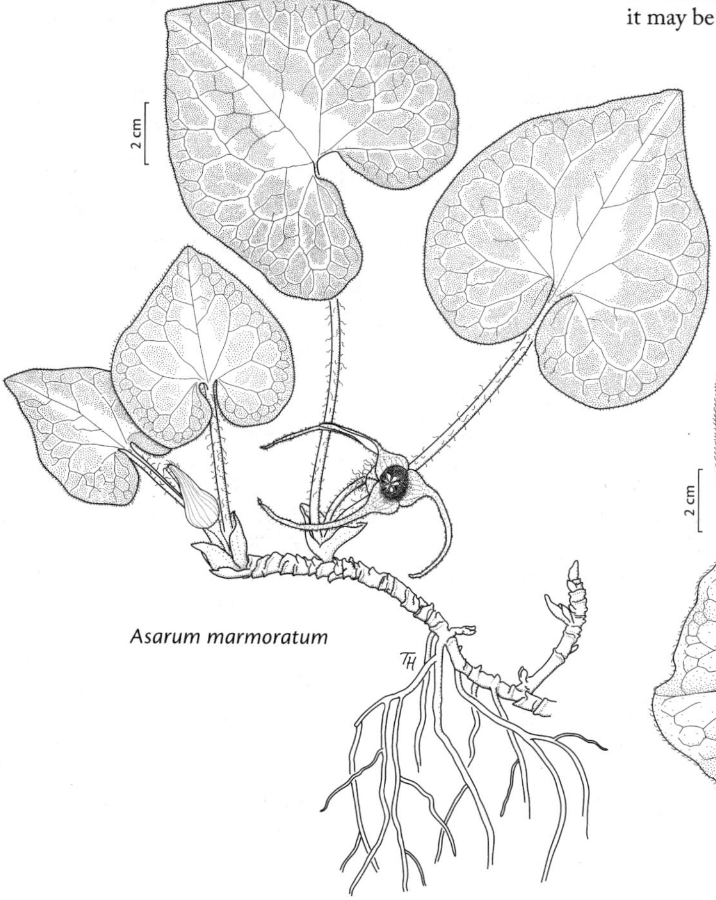

Asarum marmoratum

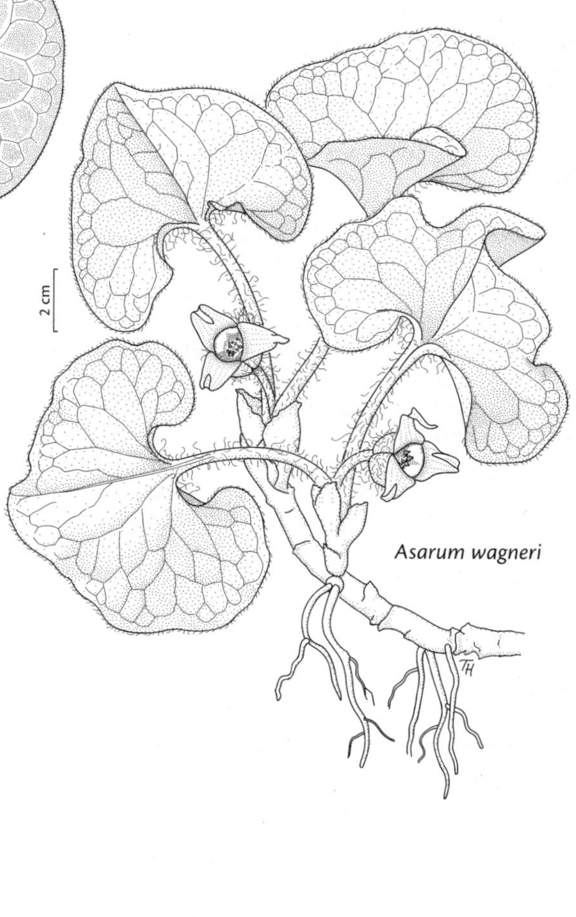

Asarum wagneri

ASARUM

ASTERACEAE Aster, sunflower, or composite family

Kenton L. Chambers (except where noted)

Herbs, subshrubs, shrubs, or rarely **trees** or **vines** annual, biennial, or perennial, terrestrial or aquatic, sometimes with resinous sap or milky latex; tap-rooted, fibrous-rooted, rhizomatous, or sometimes stoloniferous. **Stems** woody or herbaceous throughout, or only bases of stems woody and persistent. **Leaves** opposite, alternate, or whorled, basal, cauline, or both; blades simple or compound, sometimes highly dissected, margins often toothed or lobed, sometimes spiny. **Inflorescences** usually in determinate arrays, or heads solitary, terminal and/or axillary. **Heads** composed of few to many small florets. **Involucres** sometimes absent or fused into bur-like structures. **Phyllaries** in 1–many series, margins overlapping or fused to form a cup. **Receptacles** smooth or pitted, sometimes bristly. **Paleae** (when present) throughout receptacles or only at outer margins. **Florets** stamens 5, attached to tube walls by slender filaments; ovaries inferior; styles usually 1, 2-branched; florets of 3 forms: (1) **ray florets** forming circle at outer margins of receptacles, usually functionally pistillate or sterile; rays usually 3-toothed; (2) **disc florets** functionally bisexual or staminate, occupying whole receptacle or central part when ray florets present; petals fused, forming a tube with (4)5 terminal teeth; stamens 5; styles 1; (3) **ligulate florets** functionally bisexual, occupy entire receptacle, lower portions tubular, upper strap-shaped, usually spreading, 5-toothed at tips; stamens 5, styles 1. **Fruits** single-seeded achenes, indehiscent, often columnar or fusiform; occasionally beaked; pappi (when present) composed of few to many scales, awns, or bristles, persistent or deciduous.

Worldwide. ~1100 genera; 141 genera treated in *Flora*.

Asteraceae is sometimes still referred to by its former name, Compositae, and its members are often called composites referring to the characteristic inflorescence that appears to be a single flower but is in fact a composite of small flowers (*florets*) in a compact *head*.

Reproduction in Asteraceae is usually sexual by cross- or self-pollination, but some genera are facultatively or obligately asexual, propagating vegetatively by means of rhizomes, stolons, or bulblets, or making seeds with an asexually formed embryo that is genetically identical with the parent plant (apomixis).

A number of terms are used specifically in Asteraceae. Heads that have an outer ring of florets with flat, elongated corollas are called *radiate*. If all disc florets are of one kind and ray florets are absent, the heads are *discoid*. Plants with highly reduced ray corollas may be keyed as discoid. Heads with two kinds of florets (outer usually pistillate and inner bisexual) are *disciform*. Plants with heads composed of only strap-shaped ray florets are *ligulate*. The straps are called *ligules*. Ray and disc florets are called *sterile* if they don't form a fruit or *fertile* if they do. If only a pistil is present, the floret is *pistillate*, if both pistil and anthers are present and functional, it is *bisexual*, and if only the anthers are functional, it is *staminate*. *Dioecious* genera have functionally unisexual heads on separate plants.

The structure at the center of a head to which the florets are attached is the *receptacle*. Its shape varies from flat to convex, hemispheric, conic, cylindric, or globose. There are often bracts on the receptacle, situated among the florets; these are called *paleae* (sometimes referred to as *chaff*). They sometimes occur only in a ring just inside the involucres, or they may be totally absent. They may subtend the florets or may wrap around to enclose them and, at maturity, enclose the resulting fruits. Heads are usually surrounded by a group of bracts known collectively as the *involucre*. The individual bracts are termed *phyllaries*. Additionally, sometimes a group of bracts resembling an outer calyx are present. These are referred to as the *calyculus* (pl. *calyculi*).

Fruits sometimes carry a long, linear extension called a *beak*. The *pappus* (pl. *pappi*) is a modified calyx present at the summit of an achene. It is composed of *scales*, *bristles*, or *awns*, or a combination of these. Pappus scales take various forms, and sometimes small, united or separate scales form a ring at the summit called a *crown*. Awns are a stouter form of bristle and may be tapered and thickened toward the base. The scales may taper toward the tip or be blunt and lacerate, and sometimes the midrib of the scale extends as an awn or bristle. Feathery awns or bristles are termed *plumose*. Awns and bristles may also be *barbed* with sharp projections. The pappus may be *persistent* at maturity or *deciduous*.

Within Oregon, Asteraceae is the largest and most diverse plant family, with 552 described taxa. This easily surpasses the taxa counts of our next largest families (Poaceae, 333; Fabaceae, 268; Brassicaceae, 201). Due to the large number of taxa within Asteraceae, ranks other than genus are sometimes employed or referenced. The ranks include supertribes, tribes, and subfamilies, with tribes the most commonly used. Most tribes share some common morphological characteristics, but exceptions occur. As a result, we have not intentionally divided the taxa in the following key into this rank.

A number of Asteraceae species are important food, horticultural, or medicinal plants. Lettuce, artichokes, and the sweetener stevia are all members of this family. Common sunflower (*Helianthus annuus*) is the only Oregon native composite commonly grown for food. Familiar garden plants include dahlias, marigolds, asters, and chrysanthemums. Chamomile (*Matricaria chamomilla*), safflower (*Carthamus tinctorius*), and milk thistle (*Silybum marianum*) are non-natives that may occasionally escape cultivation in the state.

Many non-native species have become significant weeds in Oregon. The ubiquitous dandelion (*Taraxacum officinale*) and daisy (*Leucanthemum vulgare*) are familiar plants of lawns and disturbed areas. Some species are economically damaging enough to agricultural and range land to be listed as noxious weeds by the Oregon Department of Agriculture. These include yellow starthistle (*Centaurea solstitialis*), tansy ragwort (*Senecio jacobea*), Canada thistle (*Cirsium arvense*), diffuse knapweed (*Centaurea diffusa*), Scotch thistle (*Onopordum acanthium*), spiny cocklebur (*Xanthium spinosum*), and woolly distaff thistle (*Carduus lanatus*).

The numerous small florets in compact heads make Asteraceae inflorescences ideal for butterfly pollination, and species of *Cirsium*, *Anaphalis*, *Eucephalus*, and others are also host plants for certain species of butterflies. High nectar production in some species attracts many other pollinators as well.

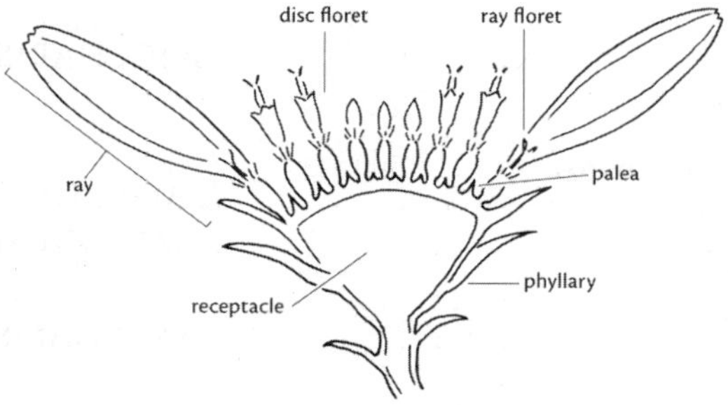

Radiate Head

Ray Floret

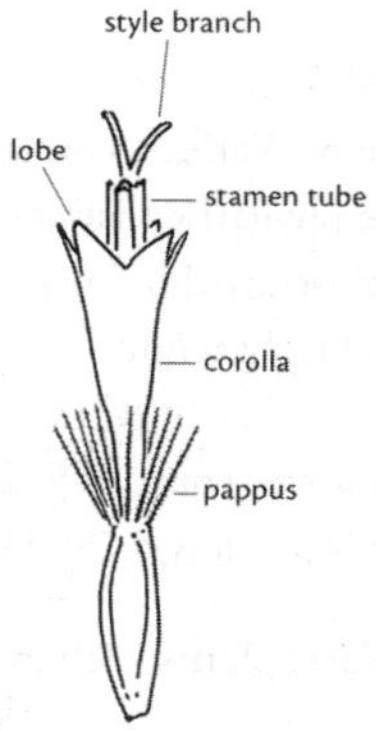

Disc Floret

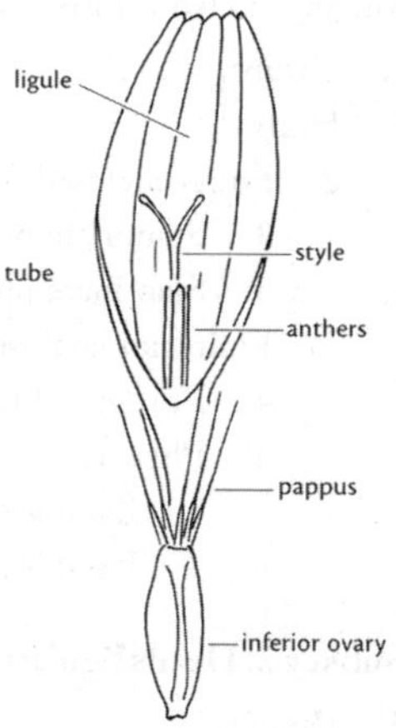

Ligulate Floret

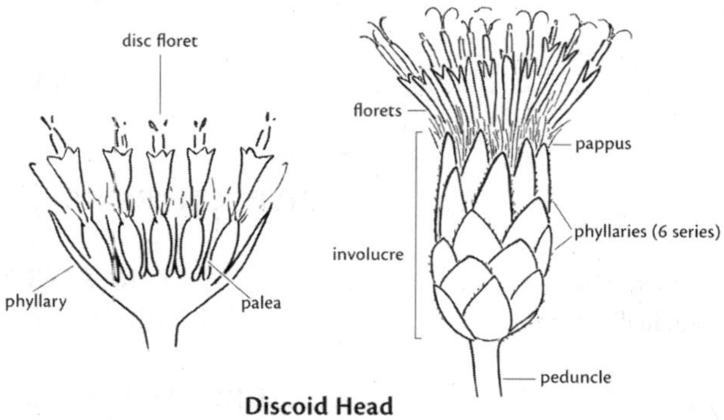

Discoid Head

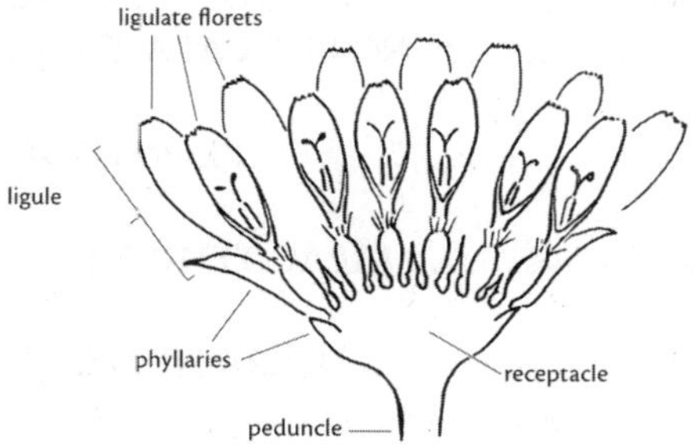

Ligulate Head

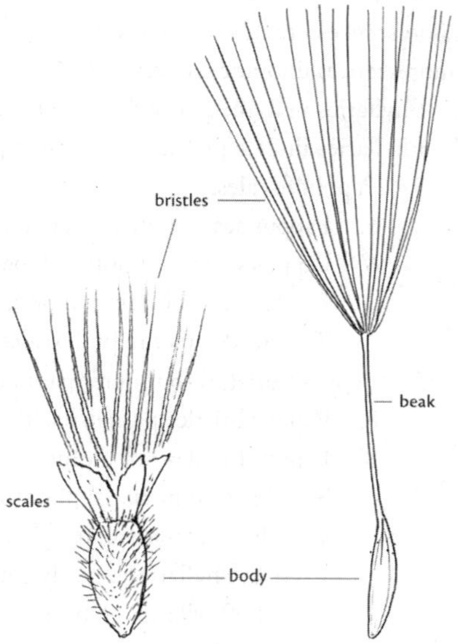

Achenes

Key to Asteraceae groups:

1. Florets all unisexual, male and female in separate heads . . . . **Subkey 1**

1' Florets all bisexual or heads a mixture of bisexual and/or unisexual florets.

  2. Heads ligulate; plants with milky latex . . . . **Subkey 2**

  2' Heads not ligulate; plants usually lacking milky latex.

    3. Plants shrubs . . . . **Subkey 3**

    3' Plants herbs or subshrubs.

      4. Stems, leaves, or involucres prickly or spiny . . . . **Subkey 4**

      4' Stems, leaves, and phyllaries not prickly or spiny.

        5. Phyllaries in 1 series, each partially, mostly or wholly enveloping subtended ray ovary . . . . **Subkey 5**

        5' Phyllaries in 1+ series, not enveloping subtended ray ovary.

          6. Heads radiate.

            7. Paleae present . . . . **Subkey 6**

            7' Paleae 0 . . . . **Subkey 7**

          6' Heads discoid or disciform.

            8. Paleae present . . . . **Subkey 8**

            8' Paleae 0 . . . . **Subkey 9**

Subkey 1: Heads unisexual.

1. Shrubs. . . . . **BACCHARIS**
1' Herbs.
  2. Fruits enclosed within burs.
    3. Staminate phyllaries ± connate . . . . **AMBROSIA**
    3' Staminate phyllaries distinct . . . . **XANTHIUM**
  2' Fruits not enclosed within burs.
    4. Phyllaries bright white . . . . **ANAPHALIS**
    4' Phyllaries not bright white.
      5. Basal leaves entire, 0.5–20 cm. . . . . **ANTENNARIA**
      5' Basal leaves lobed, 20–50 cm . . . . **PETASITES**

Subkey 2: Heads ligulate; plants with milky latex.

1. Pappi 0 . . . . **LAPSANA**
1' Pappi of awns, bristles, or scales present.
  2. Cauline leaves absent.
    3. Inflorescences branched, bracteate . . . . **HYPOCHAERIS** (in part)
    3' Inflorescences unbranched, bracts 0.
      4. Phyllaries in 2 unequal series, outer much shorter than inner . . . . **TARAXACUM**
      4' Phyllaries in 2+ equal series, or if unequal outer gradually shorter. . . . . **AGOSERIS** (in part)
        5. Pappi of scales.
          6. Pappus scales notched; peduncles inflated distally . . . . **UROPAPPUS** (in part)
          6' Pappus scales not notched; peduncles not inflated distally.
            7. Heads ± nodding in bud; scales awn-tipped, 1–11 mm . . . . **MICROSERIS** (in part)
            7' Heads erect in bud; scales bristle-tipped, 10–17 mm. . . . . **NOTHOCALAIS** (in part)
        5' Pappi of bristles (but outer pappi of scales in *Leontodon*).
          8. Pappus bristles plumose with scale-like bases . . . . **LEONTODON**
          8' Pappus bristles not plumose nor with scale-like bases.
            9. Plants annual; pappi of toothed crowns and deciduous bristles . . . . . **MALACOTHRIX** (in part)
            9' Plants perennial; pappi without crowns.
              10. Phyllaries densely glandular with black hairs. **HIERACIUM** (in part)
              10' Phyllaries glabrous or white-villous, glandular hairs (if present) yellowish.
                11. Taproots slender; pappus bristles brownish. . . . . **MICROSERIS** (in part)
                11' Taproots stout; pappus bristles white.
                  12. Outer phyllaries broadly lanceolate . . . . **NOTHOCALAIS** (in part)
                  12' Outer phyllaries narrowly lanceolate . . . . **AGOSERIS** (in part)
  2' Cauline leaves present.
    13. Pappi of scales.
      14. Pappi of crowns of short scales; florets blue, rarely purple, white, or pink . . . . **CICHORIUM**
      14' Pappi of awn- or bristle-tipped scales; florets yellow.
        15. Plants annual; pappus scales notched at tips; peduncles inflated distally. . . . . **UROPAPPUS** (in part)
        15' Plants perennial; pappus scales not notched at tips; peduncles not inflated distally.
          16. Heads ± nodding in bud . . . . **MICROSERIS** (in part)
          16' Heads erect . . . . **NOTHOCALAIS** (in part)
    13' Pappi of bristles.
      17. Pappus bristles plumose on some or all fruits.
        18. Pappus bristles interwebbed. . . . . **TRAGOPOGON**
        18' Pappus bristles not interwebbed.
          19. Fruits not beaked. . . . . **STEPHANOMERIA**
          19' Fruits beaked.
            20. Florets white. . . . . **RAFINESQUIA**
            20' Florets yellow . . . . **HELMINTHOTHECA**

17' Pappus bristles not plumose.
- 21. Stems spine-tipped .... **PLEIACANTHUS**
- 21' Stems not spine-tipped.
  - 22. Fruits flattened.
    - 23. Fruits not beaked; florets yellow........ **SONCHUS**
    - 23' Fruits beaked; florets yellow, white, or blue.
      - 24. Florets 5, yellow........ **MYCELIS**
      - 24' Florets 6–50+, yellow, white, or blue.
        - 25. Plants perennial; ligules 15–20 mm .... **MULGEDIUM**
        - 25' Plants annual or biennial; ligules 6–15 mm .... **LACTUCA**
  - 22' Fruits scarcely or not flattened.
    - 26. Inter-rib spaces with line of pits or small, rounded protuberances; beaks ≤ 1 mm........ **GLYPTOPLEURA**
    - 26' Inter-rib spaces (if present) without line of pits or small, rounded protuberances; beaks 0 or > 1 mm.
      - 27. Florets pale yellow to yellow or orange-red.
        - 28. Florets orange-red........ **HIERACIUM** (in part)
        - 28' Florets yellow, sometimes the outer ligules reddish abaxially.
          - 29. Pappus bristles connected at bases and deciduous .... **MALACOTHRIX** (in part)
          - 29' Pappus bristles not connected at bases.
            - 30. Fruits beaked.
              - 31. Cauline leaves auriculate-clasping........ **CREPIS** (in part)
              - 31' Cauline leaves not auriculate-clasping .... **CHONDRILLA**
            - 30' Fruits unbeaked.
              - 32. Plants usually cespitose, compact .... **ASKELLIA**
              - 32' Plants not cespitose or compact.
                - 33. Plants taprooted........ **CREPIS** (in part)
                - 33' Plants fibrous-rooted or rhizomatous........ **HIERACIUM** (in part)
      - 27' Florets white or pink.
        - 34. Pappi of bristles fused basally in 5 groups........ **CHAETADELPHA**
        - 34' Pappi not of bristles fused basally in 5 groups.
          - 35. Proximal leaves triangular or sagittate .... **NABALUS**
          - 35' Proximal leaves elliptic, lanceolate, oblanceolate, obovate, or oblong.
            - 36. Florets 3–4; ligules 1.5–2 mm........ **PRENANTHELLA**
            - 36' Florets 5–30+; ligules 2–4 mm .... **HIERACIUM** (in part)

Subkey 3: Shrubs.

1. Ray florets present (may be inconspicuous in *Euthamia* and *Gutierrezia*).
   - 2. Pappi of scales .... **GUTIERREZIA**
   - 2' Pappi of bristles.
     - 3. Plants rhizomatous........ **EUTHAMIA**
     - 3' Plants not rhizomatous.
       - 4. Plants mat-forming .... **NESTOTUS**
       - 4' Plants not mat-forming........ **ERICAMERIA** (in part)

1' Ray florets 0.
- 5. Plants with sagebrush aroma........ **ARTEMISIA**
- 5' Plants without sagebrush aroma.
  - 6. Fruits 10-ribbed; florets pale yellow to yellowish green .... **BRICKELLIA**
  - 6' Fruits obscurely veined, obscurely 5-ribbed or 5–12-ribbed; florets cream to yellow.
    - 7. Phyllaries in (1)2 equal series; fruits obscurely 5-ribbed .... **TETRADYMIA**

7' Phyllaries in 2–7 subequal series, or outer gradually shorter; fruits obscurely veined or 5–12-ribbed.
  8. Leaves filiform to spatulate, often stipitate-glandular or gland-dotted; fruits 5–12-ribbed . . . **ERICAMERIA** (in part)
  8' Leaves linear to lanceolate, not stipitate-glandular or gland-dotted; fruits obscurely veined . . . **CHRYSOTHAMNUS**

Subkey 4: Stems, leaves, or involucres prickly or spiny.

1. Leaf margins prickly or spiny.
  2. Stems winged.
    3. Receptacles with bristly scales . . . . **CARDUUS**
    3' Receptacles without bristly scales . . . . **ONOPORDUM**
  2' Stems not winged.
    4. Corollas white to pink, red, or purple.
      5. Leaf surfaces white-blotched . . . . **SILYBUM**
      5' Leaf surfaces not white-blotched . . . . **CIRSIUM**
    4' Corollas yellow to orange (sometimes becoming reddish in *Carthamus*).
      6. Basal leaves absent or usually withering before flowering . . . . **CARTHAMUS**
      6' Basal leaves present at flowering . . . . **CENTAUREA** (in part)
1' Leaf margins not prickly or spiny (but sometimes tips with a spine).
  7. All florets bisexual and fertile . . . . **ARCTIUM**
  7' Outer florets sterile, inner florets bisexual and fertile.
    8. Receptacles with scales; basal leaves usually withered by flowering . . . . **CRUPINA**
    8' Receptacles with bristles; basal leaves present at flowering . . . . **CENTAUREA** (in part)

Subkey 5: Herbs or subshrubs with phyllaries in 1 series, each enveloping a ray ovary.

1. Plants perennial.
  2. Ray florets absent; phyllaries 0 but paleae in 1 series resembling phyllaries . . . . **RAILLARDELLA**
  2' Ray florets present; phyllaries present.
    3. Involucres ± globose, 4–6+ mm in diameter; disc florets 5–30, functionally staminate . . . . **ANISOCARPUS**
    3' Involucres campanulate to hemispheric, 6–12+ mm in diameter; disc florets 28–65, bisexual . . . . **KYHOSIA**
1' Plants annual.
  4. Paleae subtending all or most disc florets.
    5. Disc florets bisexual, styles bristly-puberulent proximally . . . . **BLEPHARIPAPPUS**
    5' Disc florets usually functionally staminate, styles not bristly-puberulent proximally.
      6. Bracts of floral peduncles spine-tipped or apiculate; disc corollas yellow . . . . **CENTROMADIA**
      6' Bracts of floral peduncles not spine-tipped or apiculate; disc corollas white . . . . **HEMIZONIA**
  4' Paleae in 1 series between ray and disc florets.
    7. Peduncular bracts usually with 1–4+ large tack-shaped glands . . . . **CALYCADENIA**
    7' Peduncular bracts without tack-shaped glands.
      8. Ray fruits laterally compressed or somewhat 3-angled . . . . **MADIA**
      8' Ray fruits obcompressed.
        9. Plants 1–20 cm; disc florets 1–2 . . . . **HEMIZONELLA**
        9' Plants 3–150 cm; disc florets 4–100+.
          10. Disc fruits 0 . . . . **LAGOPHYLLA**
          10' Disc fruits present.
            11. Disc fruit pappi of oblong, apically obtuse scales; rays yellow . . . . **ACHYRACHAENA**
            11' Disc fruit pappi of usually linear-attenuate to subulate scales; rays white . . . . **LAYIA**

Subkey 6: Heads radiate; paleae present.

1. Paleae only between ray and disc florets . . . . **RIGIOPAPPUS**
1' Paleae throughout receptacles.
  2. Ray corollas white to pink or purple.
    3. Cauline leaves opposite . . . . **GALINSOGA**
    3' Cauline leaves alternate.
      4. Leaves simple; involucres 23–40 mm . . . . **WYETHIA** (in part)
      4' Leaves divided; involucres 3–6 mm.
        5. Leaves densely pinnately divided; inflorescences usually flat-topped clusters . . . . **ACHILLEA**
        5' Leaves 1–3-pinnatifid; inflorescences panicle-like arrays . . . . **ANTHEMIS**
  2' Ray corollas yellow or orange.
    6. Receptacles conic to columnar . . . . **RUDBECKIA**
    6' Receptacles flat to convex or hemispheric.
      7. Fruits clasped by paleae.
        8. Basal leaves absent . . . . **HELIANTHUS**
        8' Basal leaves present.
          9. Stems scapose; cauline leaves 0 or bract-like; pappi 0 . . . . **BALSAMORHIZA**
          9' Stems leafy; cauline leaves well developed; pappi of crowns, scales, and/or awns, occasionally 0.
            10. Ray florets sterile; fruits strongly flattened . . . . **HELIANTHELLA**
            10' Ray florets fertile; fruits 4-sided . . . . **WYETHIA** (in part)
      7' Fruits not clasped by paleae.
        11. Cauline leaves alternate; pappi low crowns . . . . **COTA**
        11' Cauline leaves opposite; pappi 0 or of awns or scales.
          12. Basal leaves absent; pappi of 2–4 barbed awns . . . . **BIDENS**
          12' Basal leaves present; pappi 0 or of 2 scales . . . . **COREOPSIS**

Subkey 7: Heads radiate; paleae 0.

1. Cauline leaves opposite.
  2. Phyllaries in 3–5 unequal series, outer gradually shorter; plants succulent . . . . **JAUMEA**
  2' Phyllaries in 1–3 equal series; plants not succulent.
    3. Pappi of barbellate to subplumose or plumose bristles . . . . **ARNICA**
    3' Pappi 0 or of awns or scales.
      4. Leaves woolly-tomentose or woolly pubescent abaxially . . . . **ERIOPHYLLUM** (in part)
      4' Leaves glabrous or strigillose abaxially . . . . **LASTHENIA**
1' Cauline leaves alternate.
  5. Ray corollas white to pink, violet, or purple.
    6. Plants annual.
      7. Leaves 1–4-pinnate.
        8. Fruits often with apical resin glands, inter-rib areas minutely roughened; plants not aromatic when crushed . . . . **TRIPLEUROSPERMUM**
        8' Fruits without apical resin glands, inter-rib areas not roughened; plants aromatic when crushed . . . . **MATRICARIA**
      7' Leaves simple (occasionally with linear lobes in *Centaurea*).
        9. Cauline leaves canescent . . . . **DIETERIA** (in part)
        9' Cauline leaves not canescent.
          10. Phyllaries in 6–many series . . . . **CENTAUREA**
          10' Phyllaries in 2–6 series.
            11. Pappi in 2 series (but ray fruit pappi 0 in *E. annuus* and *E. strigosus*). . . . . . **ERIGERON** (in part)
            11' Pappi in 1 series.
              12. Rays 1–2 mm; pappus bristles 6–7.5 mm . . . . **SYMPHYOTRICHUM** (in part)
              12' Rays ≤ 1 mm; pappus bristles 1.5–4.5 mm . . . . **CONYZA**

6' Plants perennial, rarely biennial.
13. Plants rhizomatous, rarely stoloniferous.
14. Plants stoloniferous . . . . **BOLTONIA**
14' Plants not stoloniferous.
15. Pappi 0.
16. Plants 2.5–20 cm; receptacles conic; rays 4–11 mm . . . . **BELLIS**
16' Plants 10–50+ cm; receptacles convex; rays 12–20+ mm . . . . **LEUCANTHEMUM**
15' Pappi of scales, bristles, or bristle-like scales.
17. Receptacles pitted.
18. Fruits 2–7-veined . . . . **ERIGERON** (in part)
18' Fruits 7–15-veined or -ribbed.
19. Ray florets 8–30+, well exceeding involucres; disc florets yellow, usually aging to purple; rhizomes not woody . . . . **EURYBIA**
19' Ray florets 1–7, slightly exceeding involucres; disc florets white to pale yellow; rhizomes woody . . . . **SERICOCARPUS**
17' Receptacles not pitted.
20. Phyllaries keeled . . . . **EUCEPHALUS** (in part)
20' Phyllaries not keeled.
21. Phyllaries glandular, often somewhat purple, bases not scarious-margined . . . . **CANADANTHUS**
21' Phyllaries rarely glandular (except *S. campestre*), rarely tinged with purple, bases usually scarious-margined . . . . **SYMPHYOTRICHUM** (in part)
13' Plants not rhizomatous.
22. Fruits flattened or compressed.
23. Phyllary midribs keeled.
24. Plants 4–25 cm; fruits silky-pubescent . . . . **IONACTIS**
24' Plants 10–130 cm; fruits glabrous to strigose or pilose . . . . **EUCEPHALUS** (in part)
23' Phyllary midribs not keeled.
25. Phyllaries often with a darkened medial or apical area; pappi outer of short bristles, crowns, or scales and inner of bristles . . . . **ERIGERON** (in part)
25' Phyllaries without darkened medial or apical area; pappi bristle-like scales . . . . **TOWNSENDIA**
22' Fruits not flattened, sometimes slightly compressed.
26. Pappi of low crowns, rarely of scales or 0 . . . . **TANACETUM** (in part)
26' Pappi usually all of bristles, or inner of bristles, outer of short bristles, crowns, or scales.
27. Pappi outer of short bristles, crowns, or scales and inner of bristles.
28. Phyllaries in 2–3 series; heads solitary or not . . . . **ERIGERON** (in part)
28' Phyllaries in 3–6 series; heads solitary . . . . **OREOSTEMMA** (in part)
27' Pappi of bristles.
29. Inflorescences solitary heads.
30. Leaves basal and cauline . . . . **OREOSTEMMA** (in part)
30' Leaves cauline . . . . **CORETHROGYNE** (in part)
29' Inflorescences pyramidal, corymb-, cyme-, panicle-, raceme-, thyrse-like, or flat-topped arrays.
31. Phyllaries in 1–2 series . . . . **SENECIO** (in part)
31' Phyllaries in 3–10 series.
32. Petioles winged; rays pink or purple; leaves cauline . . . . **CORETHROGYNE** (in part)
32' Petioles not winged; rays white to bluish purple; leaves basal and cauline (but basal often withering by flowering) . . . . **DIETERIA** (in part)

5' Ray corollas yellow or orange.
- 33. Plants annual.
  - 34. Pappi of 2 awn-tipped scales; leaves densely woolly . . . . . **EATONELLA**
  - 34' Pappi 0 or of bristles; leaves glabrous or glabrate to tomentose.
    - 35. Phyllaries in 3–4 series; pappi 0 . . . . . **GLEBIONIS**
    - 35' Phyllaries in 1(2) series; pappi 0 or of bristles.
      - 36. Leaves entire or dentate, rarely lobed; receptacles conic; pappi 0 or of minute bristles . . . . **CROCIDIUM**
      - 36' Leaves deeply incised to pinnately divided; receptacles flat to convex; pappi of bristles . . . . . **SENECIO** (in part)
- 33' Plants perennial, rarely biennial.
  - 37. Plants rhizomatous, rarely stoloniferous.
    - 38. Phyllaries in 1(2) series.
      - 39. Cauline leaves much reduced. . . . . **PACKERA** (in part)
      - 39' Cauline leaves not reduced, or if reduced then leaves triangular, or phyllary tips dark green, brown, or black, or basal leaves withering by flowering. . . . . **SENECIO** (in part)
    - 38' Phyllaries in 2–7 series.
      - 40. Plants usually aromatic.
        - 41. Rays vestigial, or ovate-elliptic; pappi 0 or low crowns or ring of scales . . . . . **TANACETUM** (in part)
        - 41' Rays linear; pappi of bristles . . . . . **INULA** (in part)
      - 40' Plants not aromatic.
        - 42. Inflorescences usually solitary heads.
          - 43. Disc florets usually purple or purple-tipped . . . . . **GAILLARDIA** (in part)
          - 43' Disc florets not purple or purple-tipped.
            - 44. Plants 30–90 cm; leaves long-petiolate . . . . . **DORONICUM**
            - 44' Plants 2–15 cm; leaves short-petiolate to sessile or clasping . . . . . **TONESTUS**
        - 42' Inflorescences flat-topped, panicle- or raceme-like arrays.
          - 45. Plants 4–20 dm; leaves narrowly elliptic-lanceolate to linear; receptacles convex . . . **EUTHAMIA**
          - 45' Plants 1–6 dm; leaves oblanceolate; receptacles flat . . . . . **COLUMBIADORIA**
  - 37' Plants not rhizomatous.
    - 46. Plants usually aromatic.
      - 47. Rays vestigial, or ovate-elliptic; pappi 0 or low crowns or ring of scales . . . . . **TANACETUM** (in part)
      - 47' Rays linear; pappi of bristles . . . . . **INULA** (in part)
    - 46' Plants not aromatic.
      - 48. Involucres usually resinous and sticky . . . . . **GRINDELIA**
      - 48' Involucres not resinous and sticky.
        - 49' Pappi of scales, rarely 0.
          - 50. Pappi of short, entire to deeply cut or lacerate scales.
            - 51. Peduncles woolly-tomentose . . . . . **ERIOPHYLLUM** (in part)
            - 51' Peduncles not woolly-tomentose . . . . . **HULSEA**
          - 50' Pappi of awn-tipped scales, rarely 0.
            - 52. Inflorescences flat-topped, raceme- or corymb-like arrays . . . . . **HYMENOXYS**
            - 52' Inflorescences solitary heads.
              - 53. Stems usually winged by decurrent leaf bases . . . . . **HELENIUM**
              - 53' Stems not winged by decurrent leaf bases. . . . . **GAILLARDIA** (in part)
        - 49' Pappi of bristles.
          - 54. Phyllaries in 1(2) series. . . . . **PACKERA** (in part)
          - 54' Phyllaries in 2–7 series.
            - 55. Pappi in 3–4 series. . . . . **HETEROTHECA**
            - 55' Pappi in 1–2 series.
              - 56. Pappi in 2 series . . . . . **ERIGERON** (in part)

56' Pappi in 1 series.
57. Phyllary midribs swollen and translucent . . . . . **SOLIDAGO** (in part)
57' Phyllary midribs not swollen and translucent.
58. Plants cespitose; receptacles convex . . . . . **STENOTUS**
89' Plants usually not cespitose; receptacles flat to convex . . . . . **PYRROCOMA**

Subkey 8: Heads discoid or disciform; paleae present.

1. Plants perennial.
2. Florets whitish to pink. . . . . **IVA**
2' Florets greenish yellow, dark maroon, or brownish purple. . . . . **RUDBECKIA**
1' Plants annual.
3. Inner florets bisexual.
4. Cauline leaves opposite, glabrous to lightly puberulent; heads 6–20 mm in diameter . . . . . **BIDENS**
4' Cauline leaves alternate, strigose- or floccose-tomentose; heads 3–5 mm in diameter. . . . . **LOGFIA**
3' Inner florets staminate.
5. Heads slightly to much exceeded by subtending bracteate leaves . . . . . **HESPEREVAX**
5' Heads not exceeded by subtending bracteate leaves.
6. Outer florets not enclosed by paleae. . . . . **CYCLACHAENA**
6' Outer florets enclosed by paleae.
7. Cauline leaves mostly opposite . . . . . **PSILOCARPHUS**
7' Cauline leaves (at least distal) alternate.
8. Inner paleae 0 . . . . . **MICROPUS**
8' Inner paleae present.
9. Outer florets 12–25+ . . . . . **STYLOCLINE**
9' Outer florets 5–11 . . . . . **ANCISTROCARPHUS**

Subkey 9: Heads discoid or disciform; paleae 0.

1. Leaves opposite.
2. Plants annual.
3. Styles becoming hardened and spike-like; fruit wings projecting distally as 2 spine-like teeth . . . **SOLIVA** (in part)
3' Styles not becoming hardened or spike-like; fruit wings not present.
4. Plants cespitose, $\leq$ 3 cm; leaves crowded; florets (1)2–3, white or purple. . . . . **DIMERESIA**
4' Plants not cespitose, 3–40 cm; florets 20–100+, yellow . . . . . **LASTHENIA**
2' Plants perennial. . . . . **ARNICA**
1' Leaves alternate.
5. Plants annual.
6. Leaves lobed.
7. Phyllaries in 2–6 unequal series.
8. Florets 10–50+, fertile . . . . . **CHAENACTIS** (in part)
8' Florets 3–7, outer sterile, inner fertile . . . . . **CRUPINA**
7' Phyllaries in 1–5 equal or subequal series.
9. Pappi 0.
10. Leaves principally cauline; styles not becoming hardened and spike-like; fruit wings (if present) not projecting distally as 2 spine-like teeth . . . . . **COTULA**
10' Leaves basal and cauline; styles becoming hardened and spike-like; fruit wings projecting distally as 2 spine-like teeth. . . . . **SOLIVA** (in part)
9' Pappi present.
11. Pappi of bristles. . . . . **ERECHTITES** (in part)
11' Pappi of scales or low crowns.
12. Pappi of scales . . . . . **CHAENACTIS** (in part)
12' Pappi of low crowns . . . . . **MATRICARIA**

6' Leaves not lobed.
13. Pappi of persistent scales or bristles.
14. Pappi of scales. . . . . **CHAENACTIS** (in part)
14' Pappi of bristles.
15. Plants with a pungent odor; inflorescences corymb-like arrays . . . . **ERECHTITES** (in part)
15' Plants without a pungent odor; inflorescences panicle- or raceme-like arrays.
16. Heads disciform, outer florets pistillate, inner bisexual; involucres 3–5 mm. . . . . **CONYZA**
16' Heads discoid, all florets bisexual; involucres 5–11 mm.
17. Phyllaries in 4–6 series, canescent to ± glandular-puberulent; leaves canescent. . . . . . **DIETERIA**
17' Phyllaries in 3–4 series, glabrous; strigose to scabrous . . . . **SYMPHYOTRICHUM**
13' Pappi of deciduous bristles.
18. Pappus bristles basally connate and deciduous in rings . . . . **GAMOCHAETA** (in part)
18' Pappus bristles not basally connate and deciduous in rings.
19. Involucres narrowly campanulate or cylindric; bracts united . . . . **EUCHITON** (in part)
19' Involucres campanulate to ovoid or hemispheric; bracts not united.
20. Plants 1–40 cm; fruits oblong. . . . . **GNAPHALIUM**
20' Plants 10–100 cm; fruits cylindric or fusiform . . . . **PSEUDOGNAPHALIUM** (in part)
5' Plants perennial, sometimes biennial.
21. Plants vines . . . . **DELAIREA**
21' Plants not vines.
22. Plants fibrous-rooted, sometimes stoloniferous.
23. Florets yellow . . . . **EUCEPHALUS** (in part)
23' Florets white, purple or purplish distally.
24. Leaves abaxially glabrous . . . . **SAUSSUREA**
24' Leaves abaxially densely white-lanate or tomentose.
25. Leaves abaxially densely white-lanate; bracts 0; fruits stipitate-glandular . . . . . . **ADENOCAULON**
25' Leaves abaxially tomentose; bracts united; fruits not glandular . . . . **EUCHITON** (in part)
22' Plants rhizomatous or taprooted.
26. Plants rhizomatous.
27. Pappi of low crowns; plants usually aromatic. . . . . **TANACETUM** (in part)
27' Pappi of bristles; plants usually not aromatic.
28. Stems finely sessile-glandular . . . . **AGERATINA**
28' Stems not glandular.
29. Leaves palmately lobed. . . . . **CACALIOPSIS**
29' Leaves not lobed.
30. Phyllaries not densely tomentose. . . . . **EUCEPHALUS** (in part)
30' Phyllaries densely tomentose.
31. One or both leaf surfaces tomentose. . . . . **LUINA** (in part)
31' Leaf surfaces glabrous or glabrate . . . . **RAINIERA**
26' Plants taprooted.
32. Pappi 0.
33. Leaves glabrous or lightly pubescent. . . . . **TANACETUM** (in part)
33' Leaves silky-tomentose or canescent . . . . **SPHAEROMERIA**
32' Pappi present.
34. Pappi of scales . . . . **TANACETUM** (in part)
34' Pappi of bristles or bristles and scales.
35. Pappi outer bristles short, inner bristles barbellate . . . . **ERIGERON**
35' Pappi not as above.
36. Phyllaries in unequal series.
37. Pappus bristles basally connate and deciduous in rings. . . . . **GAMOCHAETA** (in part)
37' Pappus bristles not basally connate and deciduous in rings.
38. Pappi of bristles and scales . . . . **HETEROTHECA**

38' Pappi of bristles.
39. Proximal leaves lobed to divided. . . . . **ACROPTILON**
39' Proximal leaves not lobed or divided.
40. Pappus bristles deciduous . . . . **PSEUDOGNAPHALIUM** (in part)
40' Pappus bristles persistent.
41. Leaves basal and cauline . . . . **PYRROCOMA**
41' Leaves cauline or principally cauline.
42. Florets green, yellow-green, or sometimes light yellow; plants often aromatic . . . . **BRICKELLIA**
42' Florets yellow or red; plants not aromatic . . . . **HAZARDIA**
36' Phyllaries in equal series.
43. Leaves pinnately compound . . . . **HYMENOPAPPUS**
43' Leaves not pinnately compound.
44. Phyllaries densely tomentose . . . . **LUINA** (in part)
44' Phyllaries not densely tomentose.
45. Phyllaries in 1–2 series . . . . **ERECHTITES** (in part)
45' Phyllaries in 3–10. . . . . **PSEUDOGNAPHALIUM** (in part)

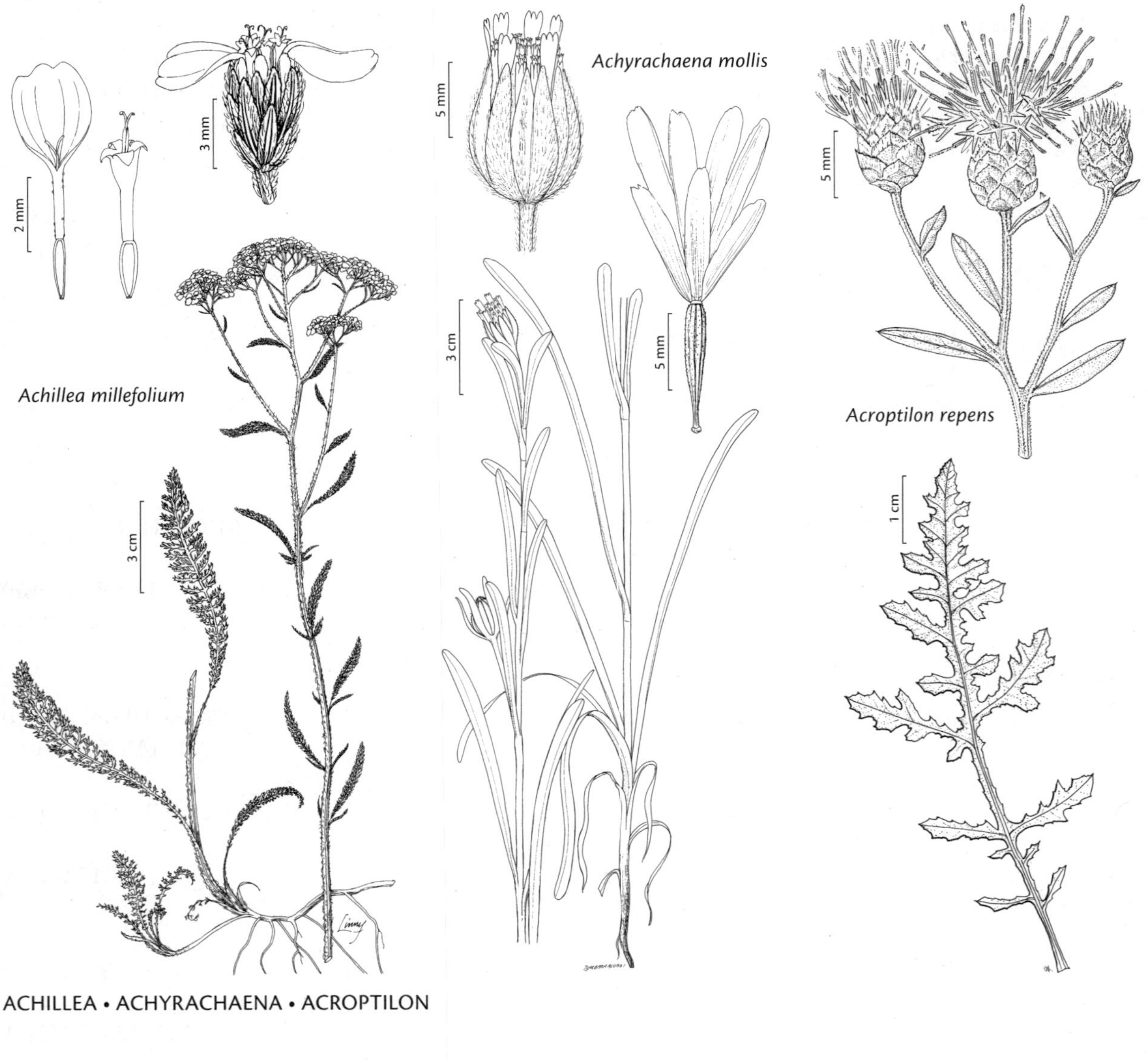

ACHILLEA • ACHYRACHAENA • ACROPTILON

ACHILLEA Yarrow

Herbs perennial; rhizomatous, rarely taprooted when young. **Stems** usually 1, usually unbranched, rarely well branched from above middle, sometimes several in a cluster from short, basal rhizomes, setose-villous, leafy throughout. **Basal leaves** densely pinnately divided into numerous lobes from a narrow rachis, lobes linear, margins entire or few-toothed or lobed, tips acute; petiolate. **Cauline leaves** gradually reduced distally, bases auriculate-clasping, surfaces sparsely to densely setose-villous; short-petiolate or sessile. **Inflorescences** terminal, usually flat-topped clusters. **Heads** radiate, numerous, small; short-pedunculate. **Involucres** cylindrical or narrowly campanulate. **Phyllaries** in 3–5 unequal series, erect in flower and fruit, broadly to narrowly lance-ovate, plane or slightly boat-shaped, colorless or with green midribs or tips, midribs distinct, sometimes keeled, surfaces sparsely to densely setose-villous; outer gradually shorter. **Receptacles** flat or rounded; paleae membranous. **Ray florets** pistillate; rays broad, well exceeding involucres. **Disc florets** bisexual; corollas cylindric, enlarged distally, lobes 5, triangular, recurved at anthesis; stamens 5, exserted from corolla; stigmas exserted from stamen tubes. **Fruits** oblong, flattened, 2 lateral veins prominent, ± winged; pappi 0.

Asia, Europe, North America. 75–130 species; 1 species treated in *Flora.*

Achillea filipendulina and *A. ptarmica* have been reported to have escaped in Oregon but are not yet established.

Achillea millefolium Linnaeus (p. 162)
Common yarrow, milfoil

Achillea borealis Bong., *Achillea lanulosa* Nutt. var. *eradiata* (Piper) M. Peck, *Achillea lanulosa* Nutt. var. *lanulosa*, *Achillea millefolium* L. var. *alpicola* (Rydb.) Garrett, *Achillea millefolium* L. var. *californica* (Pollard) Jeps., *Achillea millefolium* L. ssp. *lanulosa* (Nutt.) Piper, *Achillea millefolium* L. var. *lanulosa* (Nutt.) Piper, *Achillea millefolium* L. var. *litoralis* Ehrend. ex Nobs, *Achillea millefolium* L. var. *millefolium*, *Achillea millefolium* L. var. *pacifica* (Rydb.) G.N. Jones

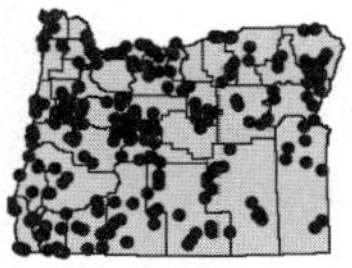

Plants 0.5–10 dm. **Leaves** 3–20 cm. **Involucres** 3.5–5.5 mm. **Ray florets** 3–5, conspicuous; rays 2–3 mm, white, sometimes pink to red. **Disc florets** 10–30; corollas 2–2.5 mm, white, slightly exserted from involucres. **Fruits** 2.5–3 mm, gray, glabrous. $2n$=36, 54.

Open woodlands, shrublands, grasslands, pastures, rocky slopes, roadsides, disturbed areas, lawns. Flowering Apr–Aug. 0–3000 m. All ecoregions. CA, ID, NV, WA; throughout North America; Asia, Europe. Native and exotic.

Achillea millefolium is unusual in the diversity of habitats that it occupies. The species is illustrative of Clausen et al.'s (1948) concept of ecotypes (ecological races), groups of local populations that are genetically adapted to a particular ecological zone or habitat. Attempts to define subspecies and varieties in the species are made nearly impossible by the continuous intergradation of populations from one habitat to another.

Diploid populations with $2n$=18, as well as higher chromosome levels up to octoploids ($2n$=72), are present in Eurasia but are not thus far known in North America.

ACHYRACHAENA Blow wives

Bruce G. Baldwin and Gerald D. Carr

Genus composed of one species, *A. mollis.*

Achyrachaena mollis Schauer (p. 162)
Blow wives

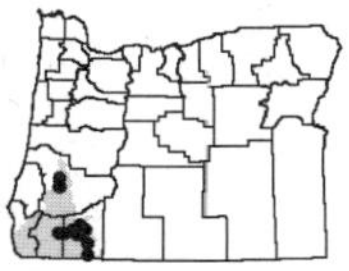

Herbs annual, 5–55 cm; self-compatible. **Stems** erect, simple or branched at base or above. **Leaves** mostly cauline; proximal opposite, distal alternate; blades linear, 2–15 cm × 1–7+ mm, margins entire or toothed, surfaces hirsute or villous and distal leaves sparsely glandular-pubescent; essentially sessile. **Inflorescences** heads borne singly or in loose, ± corymbiform arrays. **Heads** radiate. **Involucres** somewhat campanulate to cylindric or ellipsoid, 3–12+ mm in diameter. **Phyllaries** 3–8 in 1 series, basically lance-linear, herbaceous, 10–20 mm, each fully enveloping a ray ovary, falling with fruit, abaxially hirsute or villous and sparsely glandular-pubescent or eglandular. **Receptacles** flat to convex, setulose; paleate. **Paleae** falling, in 1 series between ray and disc florets, distinct. **Ray florets** 3–8, pistillate, fertile; corollas yellow, turning reddish, rays usually erect, sometimes spreading, 3–7 mm, lobes essentially parallel. **Disc florets** 4–35, bisexual and fertile; corollas 6–10 mm, yellow to reddish, tubes ≈ narrowly funnelform throats, lobes 5, lance-deltate; anthers basically dark purple; styles glabrous proximal to branches, branch tips short-bristly. **Ray fruits** obcompressed, clavate, 4.5–9(10) mm, black, 10-ribbed, straight, basal attachments centered, tips beakless, faces glabrous or somewhat scabrous; pappi 0. **Disc fruits** brown to black, similar to ray fruits, somewhat scabrous; pappi of 10 oblong, apically obtuse, white scales, outer 5 scales 3–9 mm, inner 5 scales 6–13 mm. $2n$=16.

Grassy, often seasonally wet sites, usually with clay-rich soils, in low-elevation grasslands, savannas, and open woodlands. Flowering Apr–Jun. 300–800 m. Sisk. CA; south to Mexico. Native.

During flowering, the pappus scales of each disc floret form cylinders, imparting a honeycomb-like appearance to the apices of heads. As the flowers mature, the pappus scales spread widely and eventually become very conspicuous in fruiting heads. This species is self-compatible and is not known to hybridize with other tarweeds.

ACROPTILON Russian knapweed, Turkestan thistle

Genus composed of one species, *A. repens.*

Acroptilon repens (Linnaeus) de Candolle (p. 162)

Russian knapweed, Turkestan thistle

Centaurea repens L.

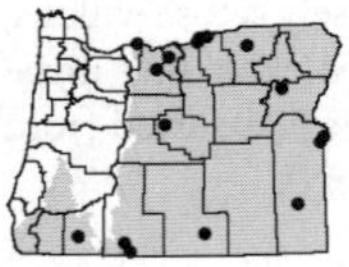

Herbs perennial, 3–10 dm; taprooted, spreading by means of buds on deep-seated lateral roots. **Stems** erect, much branched distally or from near base, floccose-tomentose; branches simple or further branched. **Leaves** cauline, alternate; proximal lanceolate, 1–15 cm, margins scabrous, irregularly lobed to deeply divided, with 1 main vein, surfaces lightly floccose to scabrous or glandular-puberulent, sometimes distinctly punctate; distal numerous, lanceolate to linear, margins dentate to entire, tips acute, surfaces glabrous to puberulent or scabrous, glandular-punctate; sessile or short-petiolate. **Inflorescences** panicle-like, of multiple solitary or loosely clustered heads terminating stem and branches. **Heads** discoid; peduncles up to 5 cm, leafless or bracteate, scabrous-puberulent. **Involucres** ovoid or globose. **Phyllaries** numerous, in 5–8 unequal series, erect in flower, spreading in fruit, pale green; outer gradually shorter, broadly lanceolate to ovate, 3–7-veined, ± truncate, tips extended, rounded or apiculate, membranous, glabrous or pilose; inner lanceolate, 9–15 mm, 3–5-veined, tips tapering. **Receptacles** flat, bristly; paleae 0. **Ray florets** 0. **Disc florets** 15–30+, bisexual, corollas exserted from involucre, tubular, expanded distally, 10–14 mm, white to pink or lavender, lobes 5, lance-linear, erect; stamens 5, well exserted, conspicuous; styles exserted from stamen tubes. **Fruits** obovoid, truncate, tapering toward base, 2–4 mm, white to grayish or brown, indistinctly veined, glabrous; pappi of numerous, sometimes deciduous bristles of differing lengths, up to 11 mm, principal slightly flattened at base, longer plumose at tip. **2*n***=26.

Cultivated fields, roadsides, streambanks, disturbed areas, clearcuts. Flowering May–Sep. 50–1400 m. BR, BW, Col, ECas, Lava, Owy, Sisk. CA, ID, NV, WA; north to British Columbia, east to Ontario, southeast to TX, south to Mexico. Exotic.

Acroptilon repens is a serious agricultural pest in the western United States. It reproduces both by seeds and by adventitious buds on deep, spreading roots (sometimes mistakenly called rhizomes).

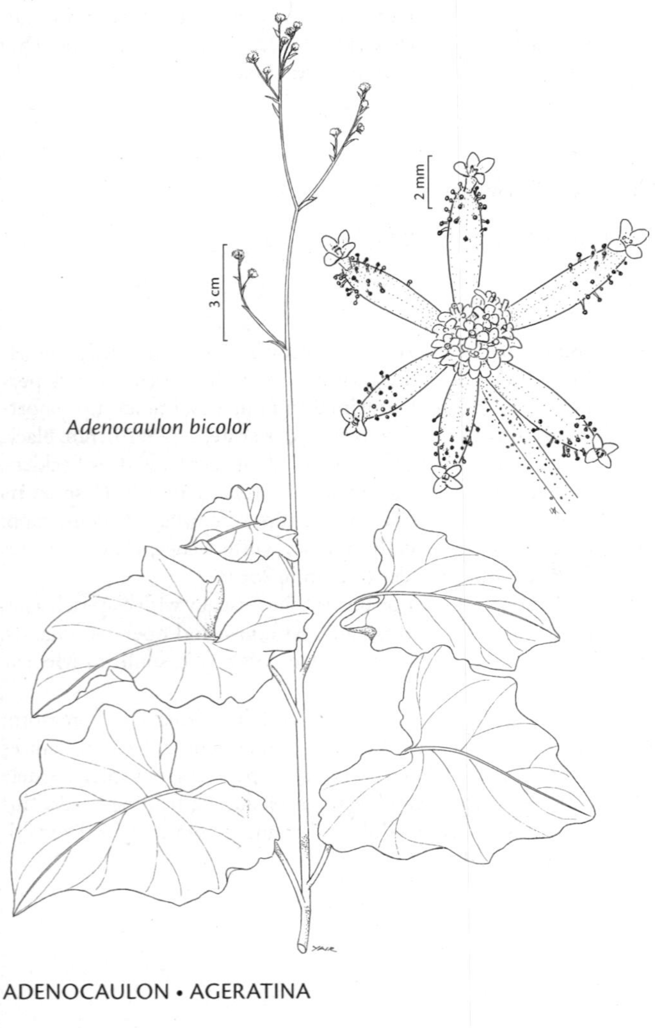

Adenocaulon bicolor

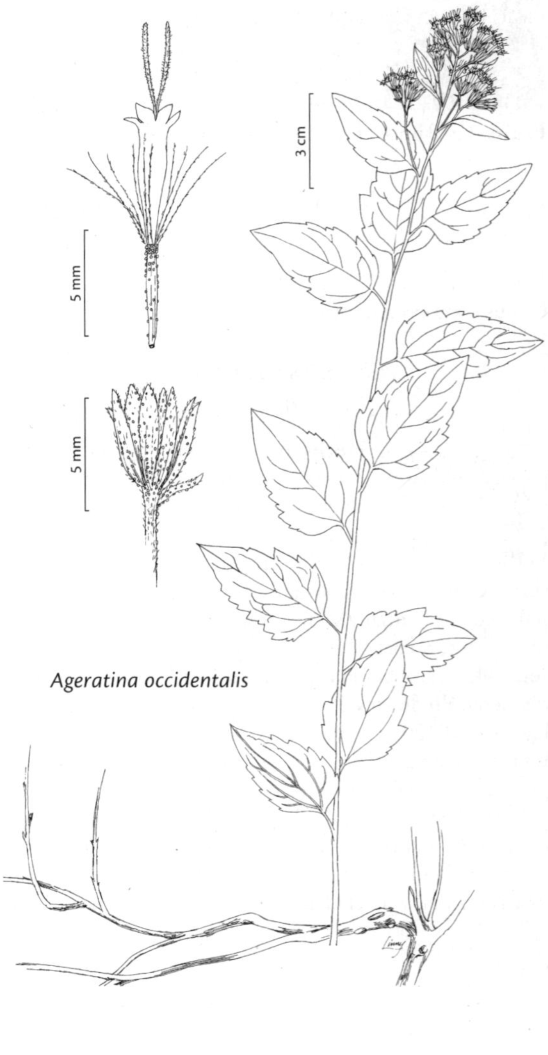

Ageratina occidentalis

ADENOCAULON • AGERATINA

ADENOCAULON Trailplant

Herbs perennial; with fibrous roots from a root-crown or short rhizome. **Stems** erect, leafy, simple or branched distally in inflorescence, lanate proximally, glandular-pubescent distally. **Leaves** basal and cauline, alternate, reduced distally; blades triangular, bases truncate or hastate, margins entire to sinuate, dentate, or irregularly lobed, tips acute to obtuse, surfaces abaxially densely white-lanate, adaxially glabrous; petioles long, narrowly wing-margined distally, lanate. **Inflorescences** panicle-like, branches few to many, spreading, glandular-pubescent. **Heads** disciform; peduncles glandular-pubescent. **Involucres** cup-shaped. **Phyllaries** in 1 series, equal, ovate, acute, free or slightly fused at base, reflexed in fruit, eventually deciduous. **Receptacles** convex; paleae 0. **Outer florets** pistillate, few, in single outer ring; corollas white, caducous, much exceeding involucre soon after anthesis, lobes 4. **Inner florets** staminate, few; corollas white, lobes 5, spreading, stamens exserted from mouth of corolla. **Fruits** club-shaped, rounded at tips, with conspicuous black, stipitate-glandular hairs distally; pappi 0.

Eastern Asia, Central America, North America, South America. 5 species; 1 species treated in *Flora.*

Adenocaulon is widely scattered in distribution, with species disjunct to Guatemala, Chile, and Argentina, as well as to eastern Asia as far south as Nepal. The sticky glands on the fruits of *Adenocaulon* are an aid to animal dispersal, and migrating birds are probably responsible for the establishment of distant populations.

Adenocaulon bicolor Hooker (p. 164)

Trailplant, pathfinder

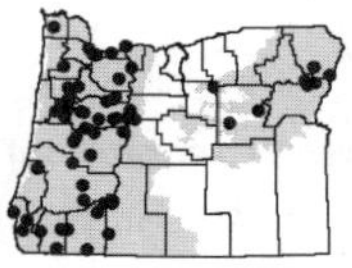

Plants 1.5–10 dm. **Leaves** 3.5–30 cm; blades 2.5–15 × 3–15 cm; petioles 1–15 cm. **Involucres** 1–2 mm. **Phyllaries** 3–5, glabrous. **Outer florets** 0.5–1.2 mm. **Inner florets** 1–2.3 mm. **Fruits** 2–6, green, 7–8 mm at maturity. **2*n***=46.

Moist, shady conifer and oak woodlands, in various soils. Flowering Jun–Aug. 0–1900 m. BW, Casc, CR, ECas, Sisk, WV. CA, ID, NV, WA; north to British Columbia, east to MI. Native.

The common names of *A. bicolor*, like the specific epithet itself, are derived from the white undersurfaces of the leaves, which become evident when plants are trampled on.

AGERATINA Snakeroot

Herbs or **subshrubs** perennial; rhizomatous. **Stems** 1–many from a woody base or rhizome, erect, simple or branched, herbaceous, sometimes woody below, finely sessile-glandular. **Leaves** cauline, alternate, not much reduced distally; blades triangular to ovate, bases truncate or rounded, margins entire to denticulate or dentate, tips acute, surfaces abaxially minutely gland-dotted, adaxially glabrous; petioles short, glabrous or lightly strigillose. **Inflorescences** panicle-like, flat-topped or not. **Heads** discoid; peduncles short, minutely sessile-glandular. **Involucres** narrowly campanulate. **Phyllaries** in 1 series or a few outer gradually shorter, lanceolate or linear, free, spreading in fruit. **Receptacles** flat to conic; paleae 0. **Disc florets** bisexual; corollas with slender tubes and widened throats, lobes 5, triangular, spreading; stamens 5, included in corolla throats or slightly exserted; stigmas long-exserted, filiform. **Fruits** columnar, truncate; pappi of numerous minutely barbed, readily deciduous bristles.

North America, South America. ~265 species; 1 species treated in *Flora.*

Ageratina is a segregate from the large, highly variable genus *Eupatorium*, with which the common name "boneset" is usually associated. The subtribal classification of tribe Eupatorieae has been thoroughly revised, based on molecular and micromorphological evidence (King & Robinson 1987), placing *Ageratina* and nine other genera in the subtribe Oxylobinae, well removed from *Eupatorium.*

Ageratina occidentalis (Hooker) R.M. King & H. Robinson (p. 164)

Western boneset, western snakeroot

Eupatorium occidentale Hook.

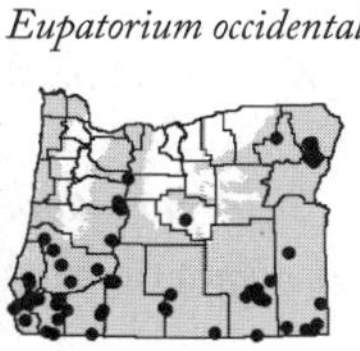

Plants 12–70 cm. **Leaves** 1.3–11 cm; blades 1–8 × 0.8–7 cm, sometimes distinctly 3-veined from base; petioles 0.2–2 cm. **Involucres** 3–5 mm. **Phyllaries** 6–14, scabridulous. **Disc florets** 9–21; corollas 5–7 mm, pink, reddish purple, or white. **Fruits** 3–4 mm, light to dark brown, sessile-glandular, strongly 4–5-veined. **2*n***=34.

Cliffs, talus, rimrock, riverbanks, gravel bars, rocky slopes, woodlands, and open sites. Flowering Jun–Sep. 0–2800 m. BR, BW, Casc, CR, ECas, Est, Owy, Sisk. CA, ID, NV, WA; northeast to MT, southeast to UT. Native.

Ageratina occidentalis is the only North American species of the genus with glandular-pubescent achenes.

AGOSERIS Agoseris, mountain dandelion

Herbs annual or perennial; sap milky; taproots stout, sometimes forming root-sprouts. **Stems** 0 or rarely 1. **Leaves** basal, in rosettes, rarely also cauline; blades linear to narrowly or broadly elliptic, oblanceolate, or spatulate, bases tapering, margins entire to denticulate or pinnately lobed, tips acuminate to rounded, surfaces glabrous to villous; petioles glabrous to villous, ± winged. **Inflorescences** solitary terminal heads; bracts 0. **Heads** ligulate. **Involucres** cylindric to broadly campanulate in flower, narrowly to broadly ovoid in fruit, glabrous or villous. **Phyllaries** in 2–several equal series or outer gradually shorter, erect, reflexing at maturity, often ciliate. **Receptacles** flat; paleae present or not. **Florets** 5–100+, bisexual; ligules usually yellow, orange, or reddish orange, longer than involucres; stamens 5. **Fruits** columnar or fusiform, usually 10-ribbed, tapering gradually or abruptly to short and stout or long and filiform beaks, or beakless; pappi of numerous white bristles.

Temperate North America and South America. 11 species; 9 species treated in *Flora*.

For the identification of specimens, it is best to have mature or nearly mature fruits, if possible. Several of the taxa are widespread, especially in and east of the Cascades, are not limited to a particular region or habitat, and may occur sympatrically. The term “fruit bodies” refers to the seed-containing part of the fruit.

Agoseris species are usually stemless, with a single flower head on each peduncle arising directly from the basal rosette. In many species, the peduncle continues to lengthen as the head matures. Presumably, this lifts the seed heads higher to where the breeze can better catch the bristly pappus to disperse the seeds.

Key to *Agoseris* species:

1. Rays orange to reddish orange, outer rays not reddish abaxially.
   2. Fruit bodies sharply tapering to beaks, rarely gradually tapered; outer phyllaries narrowly lanceolate . . . . ***A. aurantiaca*** var. ***aurantiaca***
   2' Fruit bodies gradually tapering to filiform beaks; outer phyllaries broadly lanceolate . . . . ***A. elata*** (in part)

1' Rays yellow, outer rays reddish abaxially.
   3. Plants annual . . . . ***A. heterophylla*** var. ***heterophylla***
   3' Plants perennial.
      4. Beaks of fruits > bodies.
         5. Beaks 10–25 mm.
            6. Fruit bodies tapering into bases, tips ± truncate. . . . . ***A. retrorsa***
            6' Fruit bodies tapering into bases and tips, tips not truncate . . . . ***A. grandiflora***
         5' Beaks 2–10 mm.
            7. Petioles linear or narrowly winged; leaves 2–18 cm; fruits 6–12 mm; plants coastal . . . . ***A. apargioides*** var. ***maritima*** (in part)
            7' Petioles winged; leaves 10–30 cm; fruits 12–17 mm; plants inland . . . . ***A. elata*** (in part)
      4' Beaks of fruits ≤ bodies.
         8. Plants coastal . . . . ***A. apargioides*** var. ***maritima*** (in part)
         8' Plants inland.
            9. Peduncle tips with yellow or yellow and white hairs.
               10. Peduncles (5)20–65 cm; leaves 5–30+ cm . . . . ***A. glauca*** (in part)
               10' Peduncles 1–20(30) cm; leaves 2–15(20) cm . . . . ***A. monticola***
            9' Peduncle tips glabrous or with white hairs.
               11. Peduncle tips glabrous. . . . . ***A. glauca*** (in part)
               11' Peduncle tips with hairs.
                  12. Rays equal to slightly longer than involucres; beaks 5–10 mm. . . . . ***A. elata*** (in part)
                  12' Rays longer than involucres; beaks 1–5 mm.
                     13. Leaf margins usually lobed; peduncles 4–45(50) cm . . . . ***A. parviflora***
                     13' Leaf margins usually entire; peduncles 20–50+ cm . . . . ***A. glauca*** (in part)

Agoseris apargioides (Lessing) Greene var. ***maritima*** (E. Sheldon) Cronquist (p. 167)

Seaside agoseris, Oregon agoseris

Agoseris apargioides (Less.) Cronquist ssp. *maritima* (E. Sheld.) Q. Jones ex Cronquist

Plants perennial, acaulescent or caulescent; taprooted, usually with well-developed root-crowns below ground. **Leaves** 2–18 cm; blades elliptic to oblanceolate or spatulate, margins usually denticulate to irregularly lobed, rarely entire, tips acute to broadly rounded, surfaces glabrous to villous; petioles linear or narrowly winged. **Peduncles** 1–many, erect or decumbent, 6–30 cm, lightly villous proximally, usually more densely so distally. **Involucres** campanulate in flower, broadly ovoid in fruit, 14–20 mm. **Phyllaries** inner 8–24, narrowly lanceolate, tips acute to acuminate; outer gradually shorter, margins glabrous or ciliate, tips acute to obtuse, glabrous or villous. **Florets** 25–100+, 9–18 mm, equal to slightly longer than involucres, yellow, outer reddish abaxially. **Fruits** 6–12 mm; bodies fusiform, gradually tapering to beaks, 4–6 mm, pale brown or grayish; beaks 2–7 mm, < to > bodies. $2n$=36.

Coastal dunes, sandy flats, bluffs. Flowering Jun–Oct. 0–50 m. Est. CA, WA. Native.

Two additional varieties of *Agoseris apargioides* have been described from California. These have more conspicuous rays, suggestive of an outcrossing breeding system, whereas variety *maritima* is probably self-pollinating.

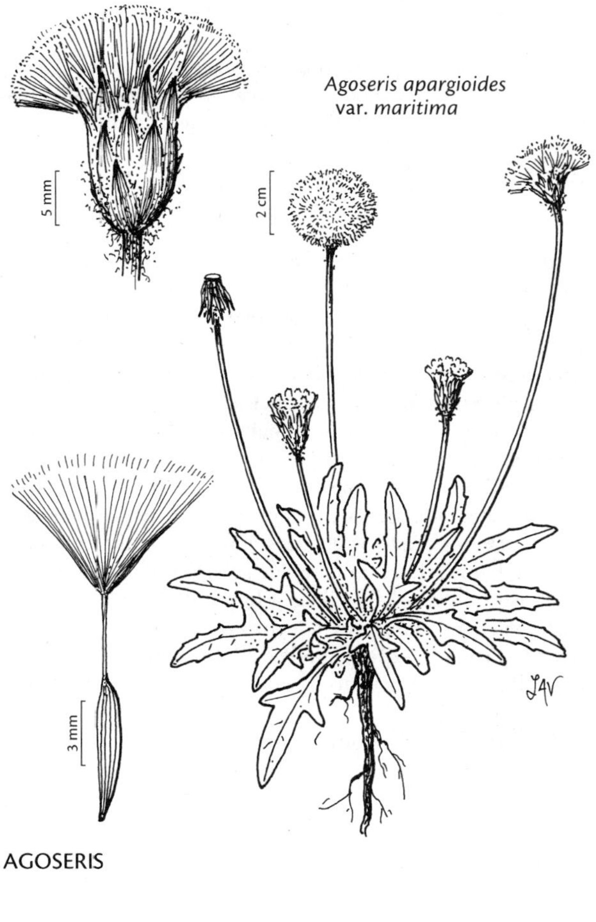

AGOSERIS

Agoseris aurantiaca (Hooker) Greene var. ***aurantiaca***

Orange agoseris

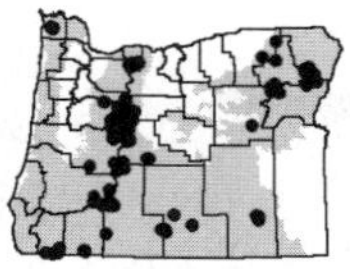

Plants perennial, acaulescent; taprooted, with single or branched root-crowns below ground. **Leaves** 4–35 cm; blades linear to narrowly or broadly oblanceolate, margins entire to denticulate or lobed, lobes linear to lanceolate, spreading or angled toward leaf base or tip, tips acute to acuminate, surfaces glabrous or lightly villous; petioles linear or winged, glabrous or lightly villous. **Peduncles** 1–many, erect, 3–70 cm, glabrous or lightly villous below, densely villous distally. **Involucres** narrowly to broadly campanulate in flower, narrowly to broadly ovoid in fruit, 10–25 mm. **Phyllaries** inner 10–24, narrowly lanceolate, tips acute to acuminate; outer narrowly lanceolate, equal or gradually shorter, margins glabrous or ciliate, tips acuminate, surfaces glabrous or villous. **Florets** 10–75+, 10–20 mm, equal to slightly longer than involucres, orange to reddish orange. **Fruits** 8–19 mm; bodies 5–8 mm, sharply tapering to beaks, rarely gradually tapering, brownish or grayish; beaks 3–11 mm, < to > bodies. $2n$=18, 36.

Forest openings, shrublands, rocky slopes, marshes, meadows, stream and lake margins. Flowering Jun–Aug. 600–2700 m. BR, BW, Casc, CR, ECas, Sisk. CA, ID, NV, WA; north to AK, northeast to Northwest Territories, east to SD, southeast to NM. Native.

Only the orange-flowered form of this species is known thus far in Oregon, but types with yellow or pinkish florets are reported from farther north and east.

Agoseris elata (Nuttall) Greene

Tall agoseris

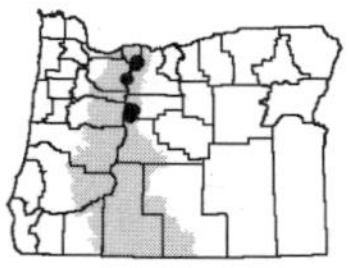

Plants perennial, acaulescent; taprooted, with single or branched root-crowns below ground. **Leaves** 10–30 cm; blades broadly to narrowly elliptic or oblanceolate, rarely linear, margins entire to sparingly denticulate to lobed, lobes linear or narrowly lanceolate, straight or arcuate, spreading or angled toward leaf base or tip, tips acute to acuminate, surfaces glabrous to lightly villous; petioles winged, glabrous to lightly villous. **Peduncles** 1–few, erect, 15–60 cm, tips with white hairs. **Involucres** broadly campanulate in flower, broadly ovoid in fruit, 15–25 mm. **Phyllaries** inner 12–35, narrowly lanceolate, tips acute to acuminate; outer equal or gradually shorter, broadly lanceolate, margins ciliate, tips acute, surfaces glabrous or villous. **Florets** 20–100+, 13–25 mm, equal to slightly longer than involucres, orange to reddish orange or yellow. **Fruits** 12–17 mm; bodies 6–8 mm, brownish or grayish, gradually tapering into beaks; beaks 5–10 mm, shorter to longer than bodies.

Montane marshes, meadows, coniferous forests, lake margins, riverbanks. Flowering Jul–Sep. 800–1500 m. Casc, ECas. CA, WA. Native.

Agoseris elata is a fertile, self-reproducing species in Oregon. It is unusual in having two flower color forms, yellow and orange-red, which occur sympatrically in the Mt. Hood region. It is probably a polyploid species of hybrid origin and may have formed from different combinations of parental species. Its long-beaked fruits are suggestive of *A. grandiflora*, its orange-red florets of *A. aurantiaca*, and its occasionally glandular scape tips of *A. glauca* var. *dasycephala*. It was first collected in 1835 in the northern Willamette Valley near Fort Vancouver, and again near Olympia, WA, in 1888, but it is presently known only from higher elevations in the Cascades of Washington and Oregon and disjunctly in the Sierra Nevada of California.

Agoseris glauca (Pursh) Rafinesque

Pale agoseris, prairie agoseris

Plants perennial, acaulescent; taprooted, with single or branched root-crowns below ground, sometimes reproducing clonally by adventitious sprouts from roots. **Leaves** 5–30+ cm; blades linear or elliptic to broadly oblanceolate, margins usually entire, occasionally toothed or lobed, lobes triangular to lanceolate, spreading or angled toward tips, tips obtuse to acute or acuminate, surfaces glabrous to puberulent or villous; petioles linear to broadly winged. **Peduncles** 1–8+, erect, (5)20–65 cm, glabrous or tips with yellow and/or white hairs. **Involucres** campanulate in flower, narrow ovoid to broadly ovoid in fruit, 1–3 cm. **Phyllaries** inner 8–30, linear or narrowly lanceolate, sometimes purple-spotted, tips acute to acuminate; outer gradually shorter, ovate to narrowly lanceolate, tips obtuse to acute, surfaces glabrous to densely villous. **Florets** 15–100+, 15–40 mm, longer than involucres, yellow, outer reddish abaxially. **Fruits** 8–12+ mm; bodies 4–8 mm, brownish or grayish, beakless or tapering into usually short beaks; beaks 1–4 mm, < bodies. $2n$=18, 36.

Coniferous forests, woodlands, shrublands, grasslands, wet meadows, rocky slopes, riverbanks. Central and western Canada and US. 6 varieties; 2 varieties treated in *Flora*.

Agoseris glauca var. *glauca*

AGOSERIS

Key to *Agoseris glauca* varieties:

1. Paleae usually present; outer phyllaries lightly to densely villous, sometimes glabrous. . . . . *A. g.* var. ***dasycephala***

1' Paleae 0; outer phyllaries glabrous . . . . . . *A. g.* var. ***glauca***

Agoseris glauca (Pursh) Rafinesque var. *dasycephala* (Torrey & A. Gray) Jepson

Arctic agoseris

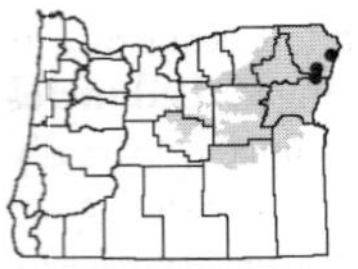

Leaves broadly oblanceolate, rarely narrowly oblanceolate, margins entire to denticulate or few-lobed, lobes triangular or lanceolate, surfaces glabrous to puberulent or villous. **Phyllaries** outer lightly to densely villous, sometimes glabrous. **Paleae** usually present.

Flowering Jun–Aug. 1300–2300 m. BW. ID, WA; north to AK, northeast to Manitoba, east to SD, southeast to AZ. Native.

Agoseris glauca (Pursh) Rafinesque var. *glauca* (p. 168)

Pale agoseris, field agoseris, prairie agoseris

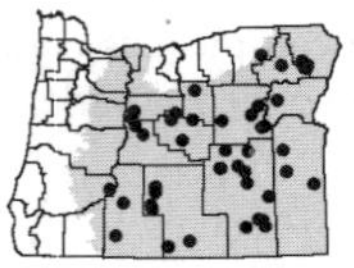

Leaves linear to narrowly or broadly oblanceolate, margins entire to denticulate or toothed, teeth triangular, surfaces glabrous. **Phyllaries** glabrous. **Paleae** 0.

Flowering Jun–Sep. 800–2400 m. BR, BW, Casc, ECas, Lava, Owy. CA, ID, NV, WA; north to British Columbia, northeast to Ontario, east to IA, southeast to NM. Native.

Hybrids have been identified between *A. glauca* var. *glauca* and both *A. monticola* and *A. parviflora*. In earlier treatments, these species were given varietal status under *A. glauca*.

Agoseris grandiflora (Nuttall) Greene

Large-flowered agoseris

Plants perennial, acaulescent; taprooted, with single or branched root-crowns below ground. **Leaves** 5–40+ cm; blades linear to narrowly or broadly oblanceolate, margins entire or variously dentate or lobed, lobes filiform to lanceolate, tips obtuse to acute or acuminate, surfaces lightly or densely villous; petioles linear or narrowly winged. **Peduncles** 1–7+, erect, 10–80+ cm, glabrous or lightly villous proximally, densely villous distally. **Involucres** campanulate in flower, ovoid in fruit, 1.5–4 cm. **Phyllaries** inner 8–30+, linear or narrowly lanceolate, tips acute to acuminate; outer gradually shorter, ovate to narrowly lanceolate, margins ciliate or not, tips obtuse to acute, sometimes broad and abruptly acuminate, surfaces glabrous to ± villous, sometimes only adaxially villous. **Florets** 40–300+, 15–30 mm, shorter to slightly longer than involucres, yellow, outer reddish abaxially. **Fruits** 10–22 mm; bodies 4–7 mm, brownish or grayish, gradually tapering into bases and tips; beaks 10–18 mm, much > bodies. $2n$=18.

Western North America. 2 varieties.

Key to *Agoseris grandiflora* varieties:

1. Leaves 10–40 mm wide, lobes linear to lanceolate; florets 100–300+ . . . . ***A. g.*** var. ***grandiflora***

1' Leaves 2–10 mm wide, lobes filiform to linear; florets 40–80+ . . . . ***A. g.*** var. ***leptophylla***

Agoseris grandiflora (Nuttall) Greene var. ***grandiflora*** (p. 169)

Large-flowered agoseris

Agoseris laciniata (Nutt.) Greene, *Agoseris plebeja* Greene

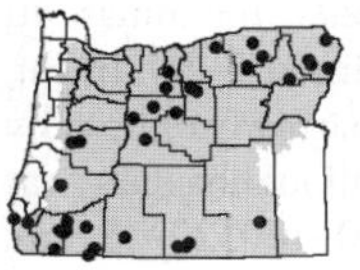

Leaves blades narrowly to broadly oblanceolate, 10–40 mm wide, lobes linear to lanceolate, tips acuminate to acute or obtuse. **Phyllaries** rarely purple-spotted. **Florets** 100–300+.

Flowering May–Jul. 0–1700 m. BR, BW, Casc, Col, ECas, Est, Lava, Sisk, WV. CA, NV, ID, WA; northeast to MT, southeast to UT. Native.

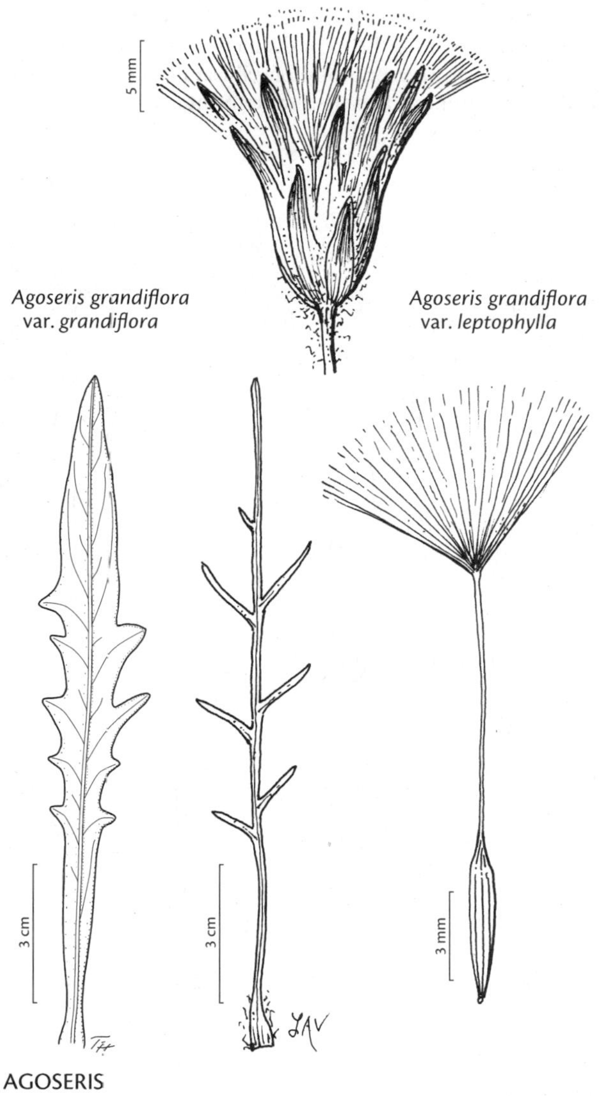

AGOSERIS

Agoseris grandiflora (Nuttall) Greene var. ***leptophylla*** G.I. Baird (p. 169)

Feather-leaf agoseris

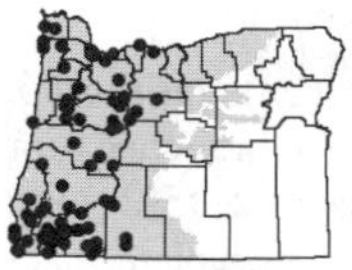

Leaves blades linear to narrowly oblanceolate, 2–10 mm wide, lobes filiform or linear, tips acuminate. **Phyllaries** sometimes purple-spotted. **Florets** 40–80+.

Flowering May–Aug. 0–1900 m. Casc, Col, CR, ECas, Est, Lava, Sisk, WV. CA, ID, WA; north to British Columbia. Native.

The varieties of *Agoseris grandiflora* represent the two extremes of leaf form in the species, and intermediates between them are frequent, especially in the southwestern Oregon region where they occur sympatrically. Otherwise, they are fairly well separated in distribution by the Cascade Range. It may be necessary to make an arbitrary judgment, based on width of the largest leaf rachis and place of occurrence, which variety a particular collection should be assigned to.

Agoseris heterophylla (Nuttall) Greene var. ***heterophylla***

Annual agoseris

Agoseris heterophylla (Nutt.) Greene var. *crenulata* Jeps., *Agoseris heterophylla* (Nutt.) Greene var. *cryptopleura* Jeps., *Agoseris heterophylla* (Nutt.) Greene ssp. *normalis* Piper, *Agoseris heterophylla* (Nutt.) Greene var. *turgida* (H.M. Hall) Jeps.

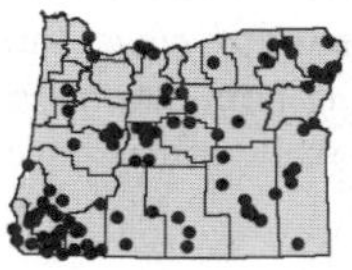

Plants annual; taprooted. **Stems** 0 or 1, to 5 cm. **Leaves** 1–15+ cm; blades linear to narrowly or broadly oblanceolate, margins usually entire, sometimes denticulate or lobed, lobes triangular to linear, spreading or angled toward tip, tips obtuse to acute or acuminate, surfaces usually scabrous or villous; sessile to broadly or narrowly petiolate. **Peduncles** 1–25+, erect or ascending, 2–30+ cm, glabrous or villous. **Involucres** cylindrical or campanulate in flower, cylindrical to ovoid in fruit, 0.5–2.5 cm. **Phyllaries** inner 5–20, linear or narrowly lanceolate, tips acute or acuminate; outer gradually shorter, lanceolate, margins ciliate or not, tips acute, surfaces glabrous to villous or ciliate with elongate glandular hairs. **Florets** 5–100+, 6–15 mm, equal to slightly longer than involucres, yellow, outer reddish abaxially. **Fruits** 7–16 mm; bodies 2–5 mm, tan, grayish, or reddish, tapering gradually or abruptly into beaks, outer sometimes with smooth or sinuate ribs or inflated bodies; beaks filiform, 5–11 mm, > bodies. **2*n***=18, 36.

Woodlands, shrublands, grasslands, open coniferous forests, ridge-tops, rocky slopes, roadsides. Flowering Apr–Jul. 0–2100 m. All ecoregions. CA, ID, NV, WA; north to British Columbia, northeast to MT, southeast to NM, south to Mexico. Native.

Like many annual, semi-weedy species, *Agoseris heterophylla* varies in vegetative form depending on growing conditions. In dry sites and thin soils, it may be extremely depauperate, with a single, short stem and one few-flowered head. In better conditions, it can be more robust, producing many heads with 100 or more fruits.

Agoseris monticola Greene

Mountain agoseris

Agoseris glauca (Pursh) Raf. var. *aspera* (Rydb.) Cronquist, *Agoseris glauca* (Pursh) Raf. var. *monticola* (Greene) Q. Jones ex Cronquist

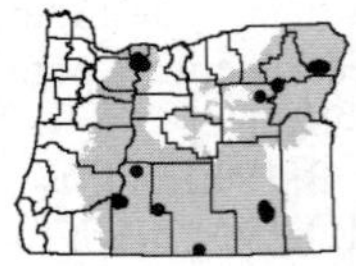

Plants perennial, acaulescent; taprooted, with well-developed single or branched root crowns below ground. **Leaves** 2–15(20) cm; blades linear to elliptic or oblanceolate, margins entire to irregularly dentate or lobed, lobes mostly triangular to linear, often angled toward base, tips obtuse to acute or acuminate, surfaces glabrous to scabrous or villous; petioles linear or narrowly winged. **Peduncles** 1–15+, erect or ascending, 1–20(30) cm, tips with yellow, glandular hairs, sometimes also white hairs. **Involucres** cylindrical to campanulate in flower, narrowly to broadly ovoid in fruit, 1–2 cm. **Phyllaries** inner 12–22, narrowly lanceolate, acute or acuminate; outer gradually shorter, lanceolate to triangular, margins ciliate or not, tips acute, surfaces glabrous to ± villous and/or minutely glandular. **Florets** 10–40, 15–35 mm, longer than involucres, yellow, outer reddish abaxially. **Fruits** 6–10 mm; bodies 6–8 mm, light brown or grayish, beakless or tapering into a short beaks; beaks 1–3 mm, < bodies. ***2n***=18, 36.

Alpine ridges, slopes, meadows, coniferous woods, brushy flats. Flowering Jul–Sep. 1300–2900 m. BR, BW, Casc, ECas. CA, NV, WA. Native.

Agoseris monticola is typically a species of higher elevations, often occurring as dwarfed plants in ecologically extreme environments. Intergrades are found between it and *A. glauca* var. *glauca* and *A. parviflora*, occasionally forming populations in which some plants have glabrous involucres while others have the minute glandular hairs of *A. monticola*.

Agoseris parviflora (Nuttall) D. Dietrich

Sagebrush agoseris, steppe agoseris

Agoseris glauca (Pursh) Raf., *Agoseris glauca* (Pursh) Raf. var. *laciniata* (D.C. Eaton) Smiley, *Agoseris glauca* (Pursh) Raf. var. *parviflora* (Nutt.) Rydb.

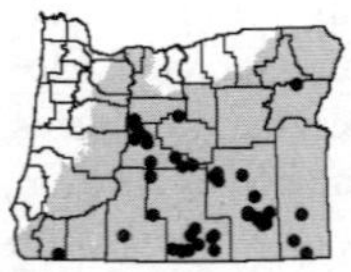

Plants perennial, acaulescent; taprooted, with well-developed single or branched root crowns below ground. **Leaves** 4–30 cm; blades linear to narrowly oblanceolate, margins usually lobed, sometimes entire or dentate, lobes triangular or more often linear, acute, spreading or angled toward base or tip, tips acute or acuminate, surfaces glabrous to scabrous or villous; petioles linear or narrowly to broadly winged. **Peduncles** 1–5+, erect, rarely decumbent, 4–45(50) cm, tips with white hairs. **Involucres** campanulate in flower, ovoid in fruit, 1–3 cm. **Phyllaries** inner 6–18, narrowly lanceolate, tips acute to acuminate; outer gradually shorter, lanceolate, margins ciliate or not, tips acute to acuminate, glabrous or variably villous especially near base and adaxially. **Florets** 25–100, 15–35 mm, longer than involucres, yellow, outer reddish abaxially. **Fruits** 8–12 mm; bodies 6–8 mm, light brown or grayish, tapering into a beak; beaks 2–5 mm. ***2n***=18.

Open woods, shrublands, grasslands, bluffs. Flowering May–Aug. 900–2600 m. BR, BW, Casc, ECas, Lava, Owy, Sisk. CA, ID, NV, WA; east to MN, southeast to NM. Native.

Some intergradation is evident between *A. parviflora* and *A. glauca* var. *glauca*, but the former is usually found in drier habitats and reaches higher elevations in the mountains than the latter.

Agoseris retrorsa (Bentham) Greene (p. 170)

Spearleaf agoseris

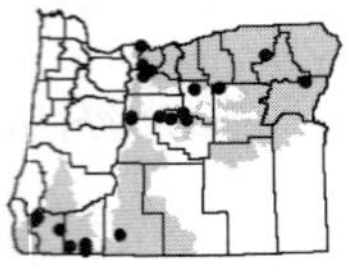

Plants perennial, acaulescent; taprooted, with single or branched root-crowns below ground. **Leaves** 10–30+ cm; blades narrowly elliptic or oblanceolate, margins regularly dentate with triangular lobes or pinnate with numerous lobes, lobes linear, acute, spreading or more often angled toward base, tips acute or acuminate, surfaces lightly or densely villous; petioles linear or narrowly winged. **Peduncles** 1–6+, erect, 15–60+ cm, villous to densely lanate proximally, lightly or densely villous distally. **Involucres** campanulate in flower, ovoid in fruit, 2–6 cm. **Phyllaries** inner 8–20, linear, acute or acuminate; outer gradually shorter, narrowly or broadly lanceolate, acute or acuminate, glabrous or villous abaxially, not glandular, margins usually ciliate. **Florets** 10–100, 20–35 mm, equal to slightly longer than involucres, yellow, outer reddish abaxially. **Fruits** 22–30 mm; bodies 5–6.5 mm, light brown or grayish, tapering toward bases, tips ± truncate; beaks 15–25 mm, > bodies. ***2n***=18.

Open coniferous forests, oak woodlands, shrublands, grasslands, roadsides. Flowering May–Jul. 100–2100 m. BW, Col, ECas, Sisk. CA, NV, WA; southeast to UT. Native.

Agoseris retrorsa is the most distinctive perennial species in the genus, having exceptionally long, linear inner phyllaries and fruits whose truncate tips taper abruptly into very long, filiform beaks.

Agoseris retrorsa

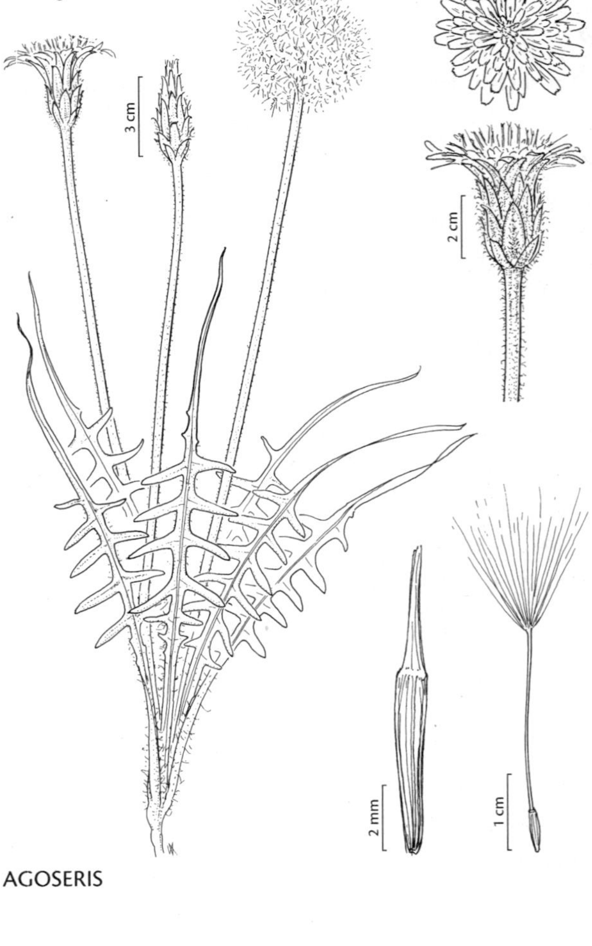

AGOSERIS

AMBROSIA Ragweed

Herbs or **shrubs** annual or perennial; taprooted or rhizomatous. **Stems** erect, prostrate, or decumbent, simple or branched, variously pubescent. **Leaves** cauline, alternate or opposite; blades simple to lobed or compound, scabrous to hispid, sericeous, or villous; usually petiolate. **Inflorescences** panicles, with heads spicate or racemose at ends of usually leafy branches. **Heads** unisexual, discoid; staminate heads usually distal to pistillate. **Involucres** of staminate heads cup- or saucer-shaped; of pistillate heads ovoid. **Phyllaries** of staminate heads 5–15+, in 1 series, ± connate, glabrous to hispid or sericeous; of pistillate heads 12–30+, outer 5–8 in 1 series, separate or connate, united into hard burs, usually with tubercles or spines, enclosing 1–5 florets. **Receptacles** conical; paleate. **Paleae** sometimes villous-pubescent or stipitate-glandular. **Florets** 5–50+; corollas funnelform, white or reddish, lobes 5, erect or recurved; stamens slightly exserted from corollas. **Fruits** enclosed within burs.

Europe, North America. 40+ species; 4 species treated in *Flora*

Waifs of *Ambrosia grayi* and *A. trifida* have been collected in Oregon, but thus far these species are not yet established.

Key to *Ambrosia* species:

1. Plants perennial; stems prostrate or decumbent; plants of coastal habitats . . . . *A. chamissonis*

1' Plants annual or perennial; stems erect; plants of inland habitats.

  2. Plants perennial . . . . *A. psilostachya*

  2' Plants annual.

    3. Burs 2–3 mm, spines or tubercles 3–5+, 0.1–0.5+ mm; staminate involucres lacking midribs . . . . *A. artemisiifolia*

    3' Burs 3–5 mm, spines 8–18+, 2–5 mm; staminate involucres with black midribs . . . . *A. acanthicarpa*

Ambrosia acanthicarpa Hooker

Bur ragweed, annual bur-sage

Franseria acanthicarpa (Hook.) Coville

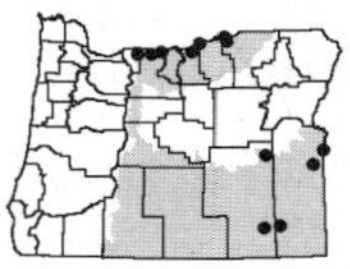

Plants annual, 1–8 dm. **Stems** erect, branched, pustulose and hispid to minutely or strongly strigose. **Leaves** mostly alternate, basal pair(s) sometimes opposite; blades deltate to lance-ovate in outline, 1.5–4 cm, usually deeply pinnatifid, segments lobed or divided, ultimate margins entire or toothed, surfaces strigose-hispid or sericeous; petiolate or distal-most sessile. **Involucres** shallowly cup-shaped. **Phyllaries** of staminate heads connate or separate, spreading, 1.5–3 mm, often with black midribs and lighter margins, surfaces minutely hispid; of pistillate heads erect, separate, surfaces lightly hispid. **Staminate florets** 6–15+. **Pistillate florets** 1. **Burs** spherical or ovoid, 3–5 mm, glabrous or minutely pubescent; spines 8–18+, 2–5(7) mm. **2*n***=36.

Desert slopes and flats, riverbanks, streambanks, sandbars, roadsides, disturbed areas. Flowering Jul–Oct. 50–1300 m. BR, Col, ECas, Owy. CA, ID, NV, WA; northeast to Manitoba, east to NE, southeast to TX, south to Mexico. Native.

Ambrosia artemisiifolia Linnaeus (p. 171)

Annual ragweed, common ragweed

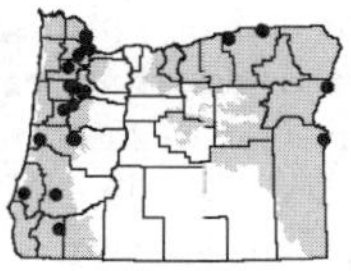

Plants annual, 1–12+ dm. **Stems** erect, simple to much branched, hispid or strigose to villous, sometimes glabrate proximally. **Leaves** mostly alternate, basal pair(s) sometimes opposite; blades lanceolate or elongate-deltate in outline, 2–15 cm, deeply pinnatifid, segments lobed or entire, ultimate margins entire or toothed, surfaces glabrous to hispid or strigose; petiolate or distal-most sessile. **Involucres** cup-shaped. **Phyllaries** of staminate heads connate, lacking midribs, tips entire or shallowly lobed, surfaces glabrous or hispid; of pistillate heads erect, separate, marginally hispid. **Staminate florets** 12–20+. **Pistillate florets** 1. **Burs** obovoid, 2–3 mm, lightly hispid or pilose; spines or tubercles 3–5+, attached near middle or subterminal, triangular, 0.1–0.5+ mm. **2*n***=34, 36.

Cultivated fields, pastures, riverbanks, roadsides, disturbed areas. Flowering Jun–Oct. 0–1200 m. BW, Col, CR, Owy, Sisk, WV. CA, ID, NV, WA; throughout North America. Exotic.

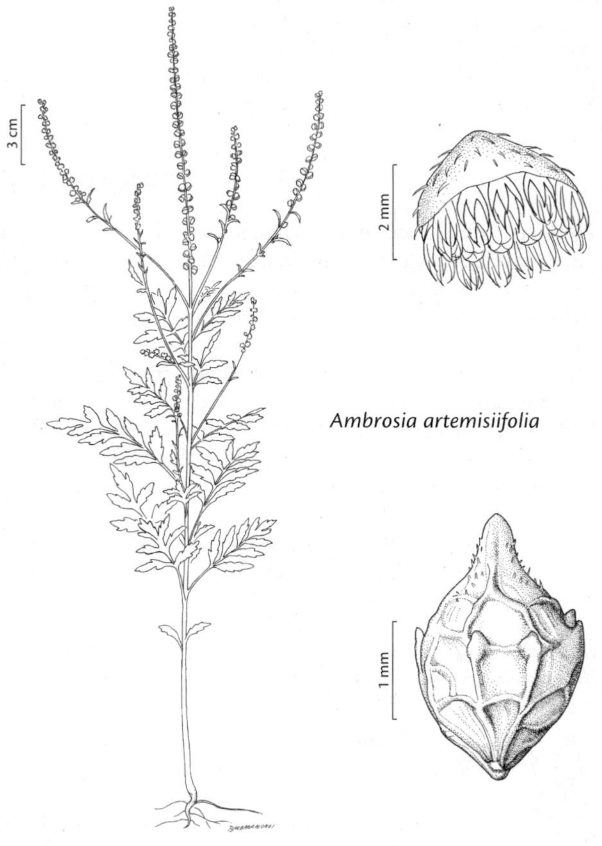

Ambrosia artemisiifolia

AMBROSIA

Ambrosia chamissonis (Lessing) Greene
Beach bur-sage

Plants perennial, 5–40 dm, mat-forming; taprooted or from root crowns. **Stems** prostrate or decumbent, simple to much branched, ± densely pilose to villous or strigose. **Leaves** alternate; blades lanceolate, broadly elliptic, obovate, or triangular, surfaces strigose-canescent; petiolate. **Involucres** cup-shaped. **Phyllaries** of staminate heads connate, sometimes with black midribs, tips shallowly lobed, surfaces glabrous or strigose; of pistillate heads erect, separate, marginally glabrous or pilose. **Staminate florets** 8–30+. **Pistillate florets** 1. **Burs** ovoid, 5–10 mm, variously pubescent; spines flattened, spreading or arched outward, 0.5–2 mm. **2*n***=36.

Western North America. 2 varieties.

Key to *Ambrosia chamissonis* varieties:

1. Leaves triangular to ovate or lanceolate, 1–2-times pinnately lobed . . . . . . *A. c.* var. ***bipinnatisecta***

1' Leaves broadly elliptic to obovate, entire to serrate or shallowly lobed . . . . *A. c.* var. ***chamissonis***

Ambrosia chamissonis var. ***bipinnatisecta*** (Lessing) J.T. Howell
Cutleaf beach bur

Franseria chamissonis Less. ssp. *bipinnatisecta* (Less.) Wiggins & Stockw. *Franseria chamissonis* Less. var. *bipinnatisecta* Less.

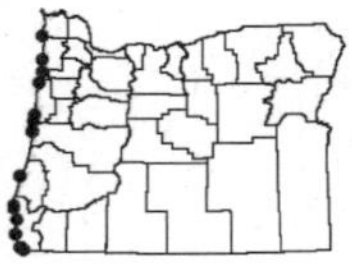

Leaves triangular to ovate or lanceolate, 1–2-times pinnately lobed.

Coastal dunes, sandy flats, bluffs. Flowering Jun–Sep. 0–50 m. Est. CA, WA; north to British Columbia. Native.

Ambrosia chamissonis var. ***chamissonis***
Silver beach-weed, silver beach bur

Franseria chamissonis Less. ssp. *chamissonis*, *Franseria chamissonis* Less. var. *chamissonis*

Leaves broadly elliptic to obovate, entire to serrate or shallowly lobed.

Coastal dunes, sandy flats. Flowering Jun–Oct. 0–50 m. Est. CA, WA; north to British Columbia. Native.

Ambrosia psilostachya de Candolle
Western ragweed, perennial ragweed

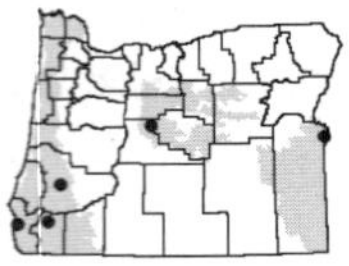

Plants perennial, 1–10+ dm; from creeping rhizome-like root-stock. **Stems** erect, simple to well branched, hispid- or strigose-pubescent. **Leaves** mostly alternate, proximal often opposite; blades lanceolate to elongate-deltate in outline, 2–10+ cm, deeply pinnatifid, segments mostly entire or toothed, sometimes few-lobed, surfaces usually densely strigose; petiolate or distal ones sessile. **Involucres** cup-shaped, often tapered below. **Phyllaries** of staminate heads connate, lacking midribs, tips entire or shallowly lobed, surfaces minutely hispid; of pistillate heads erect, separate, surfaces minutely hispid. **Staminate florets** 5–30+. **Pistillate florets** 1. **Burs** obovoid to globose, 3–6 mm, hirsute; tubercles 0–6, in a single subterminal series, stout, conic, 0.5–1 mm. **2*n***=18, 27, 36, 72, 108, 144.

Roadsides, disturbed areas, riverbanks. Flowering Jul–Oct. 0–900 m. CR, Lava, Owy, Sisk. CA, ID, NV, WA; throughout most of Canada and US, northern Mexico. Native.

ANAPHALIS Everlasting

Herbs perennial, dioecious or subdioecious; rhizomatous. **Stems** 1, erect, unbranched or occasionally much branched, floccose-pubescent. **Leaves** mostly cauline, alternate, linear to narrowly or broadly oblong-lanceolate, tapering toward base, gradually reduced in size distally, margins entire, often revolute, tips acute to acuminate, surfaces abaxially floccose-tomentose, adaxially glabrous to lightly or densely pubescent; basal and lower cauline often withering by flowering time; sessile. **Inflorescences** panicle-like, branches short to elongate, leafless or few-bracted. **Heads** discoid or disciform, ± unisexual, in compact terminal clusters; peduncles short or absent, tomentose. **Involucres** subglobose. **Phyllaries** in 8–12 series, unequal, erect in flower, spreading in fruit, bright white; inner linear to lanceolate or deltate; outer tomentose at bases. **Receptacles** naked, flat or convex. **Outer florets** pistillate; corollas elongate-filiform; styles well exserted. **Inner florets** staminate; corollas tubular; stamens 5, included. **Fruits** oblong, papillate; pappi of separate, fine, deciduous bristles.

Asia, North America. ~110 species; 1 species treated in *Flora*.

Anaphalis margaritacea (Linnaeus) Bentham & Hooker f. (p. 173)
Pearly everlasting

Anaphalis margaritacea (L.) Benth. & Hook. f. var. *margaritacea*, *Anaphalis margaritacea* (L.) Benth. & Hook. f. var. *occidentalis* Greene, *Anaphalis margaritacea* (L.) Benth. & Hook. f. var. *subalpina* A. Gray

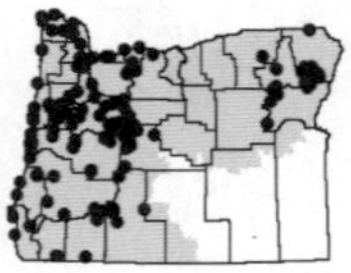

Plants dioecious, 1–10+ dm. **Leaves** 2–12 cm, 1-veined. **Involucres** 5–7 × 6–8(10) mm. **Outer florets** corollas 3–4 mm, yellow. **Inner florets** corollas ~3 mm, pale yellow. **Fruits** 0.5–1 mm, stipe-like at base; pappus bristles 10–20, 3–4 mm. **2*n***=28.

Forests, meadows, grassy hillsides, beaches, deflation plains, clearcuts, streambanks, rock outcrops, talus, roadsides, from sea level to alpine, in various soils. Flowering May–Oct. 0–2500 m. BW, Casc, Col, CR, ECas, Est, Lava, Sisk, WV. CA, ID, NV, WA; north to AK, east to SC, ME, adjacent Canada, south to northern Mexico; Asia, introduced in Europe. Native.

At first glance, the bright white phyllaries of this species may be mistaken for petals.

ANCISTROCARPHUS Groundstar

Herbs annual. **Stems** erect or prostrate, often clumped or matted, simple to much branched basally and/or distally, floccose-tomentose. **Leaves** basal and cauline, alternate; blades linear to narrowly oblanceolate, margins entire, tips acute, surfaces strigose to floccose-tomentose; sessile or tapering to slender petioles. **Inflorescences** solitary or clustered terminal heads, equaled or exceeded by bracteate leaves, depauperate plants sometimes with a single inflorescence at ground level. **Heads** disciform, ovoid to ellipsoid. **Involucres** 0 or vestigial. **Receptacles** mushroom-shaped; paleate. **Paleae** inner boat-shaped, open, with curved, spinose tips. **Outer florets** pistillate, corollas 0; style branches exserted, directed adaxially, completely enclosed in saccate paleae. **Inner florets** staminate, corollas tubular, distally reddish, 4–5-lobed, lobes minute, triangular, erect; stamens 4–5, included in corollas; ovaries vestigial, styles exserted from corollas. **Fruits** obovoid, compressed, retained within enclosing paleae; pappi 0.

Western North America. 2 species; 1 species treated in *Flora*

Ancistrocarphus filagineus A. Gray (p. 173)

Woolly fishhooks, hooked groundstar, woolly hook-fruit

Stylocline filaginea (A. Gray) A. Gray, *Stylocline filaginea* (A. Gray) A. Gray var. *depressa* Jeps.

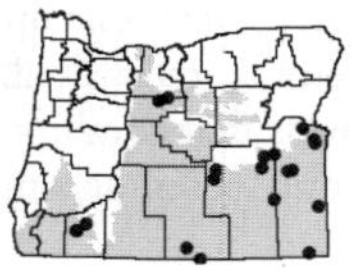

Plants 1–10+ cm. **Leaves** 3–15 mm. **Heads** 3.5–5 mm. **Outer florets** 5–11; bracts 2.5–3 mm, floccose-tomentose. **Inner florets** 5(6); corollas 1–1.5 mm; bracts 3.5–4 mm including spine, persistent, widely spreading in fruit, sometimes exposing a purple receptacle, glabrous or floccose-tomentose. **Fruits** 1.4–2 mm, brown with proximal black bands, glabrous.

Juniper woodlands, shrublands, grasslands, vernal pools, lakesides. Flowering Apr–Jun. 300–1900 m. BR, ECas, Lava, Owy, Sisk. CA, ID, NV; south to Mexico. Native.

Plants found in the Siskiyou Range are more likely to have an erect growth form, while those in eastern Oregon tend to be prostrate and clump-forming.

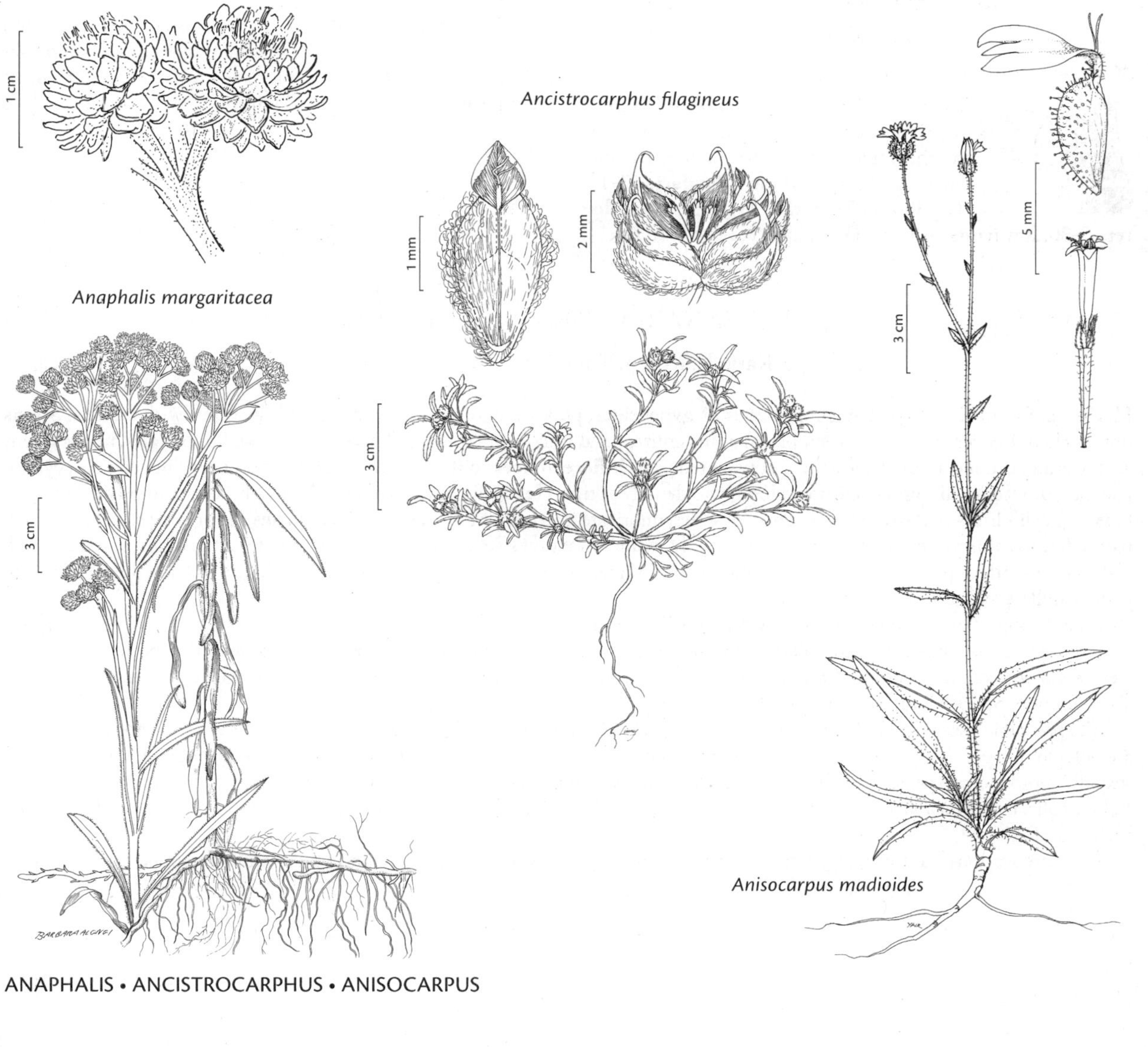

ANAPHALIS • ANCISTROCARPHUS • ANISOCARPUS

ANISOCARPUS Tarweed

Bruce G. Baldwin and Gerald D. Carr

Herbs perennial; self-compatible. **Stems** erect, generally branched from bases or throughout. **Leaves** basal and cauline; proximal opposite, sometimes in rosettes, distal alternate; blades oblong to linear, lance-linear, or oblanceolate, margins entire or toothed, surfaces hirsute to strigose or pubescent, rarely distal leaves stipitate-glandular; essentially sessile. **Inflorescences** corymbiform or racemiform or heads borne singly. **Heads** radiate. **Involucres** ± globose, 4–6+ mm in diameter. **Phyllaries** in 1 series, lanceolate to lance-attenuate, herbaceous, each fully enveloping subtended ray ovary proximally, falling with fruit, margins ciliolate, abaxial faces stipitate-glandular, sometimes also hirsutulous. **Receptacles** flat to convex, glabrous or setulose; paleate. **Paleae** falling, in 1 series, between ray and disc florets, usually connate. **Ray florets** pistillate, fertile; corollas yellow, rays deeply 3-lobed. **Disc florets** functionally staminate; corollas yellow, tubes < funnelform throats, lobes 5, deltate; anthers yellow; styles glabrous proximal to branches, branch tips densely hairy. **Ray fruits** compressed, clavate, somewhat arcuate, black, basal attachments centered, tips beaked, beaks offset adaxially, 0.2–0.3 mm, faces glabrous or hairy; pappi 0 or of crowns. **Disc fruits** 0; pappi of 5–8 linear, lanceolate, or quadrate, fimbrillate, or erose scales. **2*n***=14.

Far western Canada and United States. 2 species; 1 species treated in *Flora.*

Following Baldwin (1999), *Anisocarpus* comprises two species that previously resided in different genera and tribes of Asteraceae. The results of several multidisciplinary studies that support this union, as well as characterize the evolutionary relationship between this genus and other Pacific Coast Madiinae and the Hawaiian silversword alliance, are synthesized in Carlquist et al. (2003). *Anisocarpus madioides* and the northern Californian *A. scabridus* are the only perennial herbaceous tarweeds that combine non-scapiform capitulescences, radiate heads, ellipsoid or spherical involucres, and yellow anthers, and are the only perennials in Madiinae with **2*n***=14.

Anisocarpus madioides Nuttall (p. 173)
Woodland tarweed

Madia madioides (Nutt.) Greene

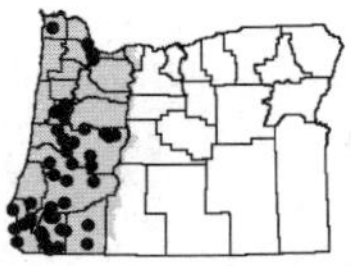

Plants 12–75 cm. **Leaf blades** 4–13 cm × 5–15 mm, dark green, margins entire or toothed, tips acute. **Heads** radiate. **Involucres** ± globose, 4–6+ mm. **Phyllaries** 7–15. **Ray florets** 7–15, rays 4–10 mm. **Disc florets** 5–30. **Ray fruits** 3–5 mm. **Disc pappi** 0.2–1.5 mm. **2*n***=14.

Forests, woodlands. Flowering May–Jul. 0–1400 m. Casc, CR, Est, Sisk, WV. CA, WA; north to British Columbia. Native.

Like some other self-compatible tarweeds, *A. madioides* ranges beyond the California Floristic Province and is the most widespread perennial in the Madiinae.

ANTENNARIA Pussytoes, everlasting

Katie E. Mitchell and Stephen C. Meyers

Herbs, rarely **subshrubs** perennial, dioecious or gynoecious, pubescent, often tomentose or pilose; often with stolons. **Stems** unbranched. **Leaves** mostly basal, simple, margins entire; basal usually in rosettes, sometimes narrowed to petioles, rarely absent at flowering; cauline alternate, usually gradually reduced. **Inflorescences** solitary heads, or up to 100+ heads in corymb-, cyme-, panicle-, or raceme-like arrays, clustered loosely to densely at tips. **Heads** discoid. **Involucres** hemispheric or campanulate to turbinate or cylindric. **Phyllaries** in 3–6+ unequal series, imbricate, scarious at least distally. **Receptacles** flat or convex, naked, pitted; paleae 0. **Disc florets** staminate or pistillate, usually white, sometimes yellow, pink or red. **Fruits** terete or slightly compressed, glabrous, smooth or papillate; staminate pappi usually clavate, sometimes capillary, serrulate to barbellate or barbellulate; pistillate pappi capillary, naked to barbellulate.

Asia, Europe, North America, South America. ~45 species; 16 species treated in *Flora.*

Antennaria is a genus with a complicated taxonomic past that reflects the complex relationships among the species. Many species of *Antennaria* can reproduce asexually via stolons and through agamospermy, and one is sometimes viviparous. Polyploidy is also common. Species limits are controversial.

The polyploid asexual complexes, or agamospecies, in Oregon are *A. howellii* and *A. microphylla*, while *A. media* and *A. pulvinata* are polyploid in part (Chmielewski 1993; Bayer & Stebbins 1987). While the agamospecies are entirely or predominantly pistillate, sexually reproducing species usually have both staminate and pistillate plants in a population in approximately equal numbers. When collecting *Antennaria,* note the relative abundance of staminate and pistillate clones or plants. Collecting individuals of each is useful.

Flags are scarious leaf tips that are characteristic of some species. They are usually found on the middle to upper cauline leaves.

Key to *Antennaria* species:

1. Heads 1 per flowering stem.
   2. Adaxial leaf surfaces green, glabrous; leaf tips emarginate to obtuse; stems 5–12 cm . . . . ***A. suffrutescens***
   2' Adaxial leaf surfaces gray-tomentose; leaf tips acute; stems 0.5–4 cm.
      3. Stolons filiform, 3–10 cm, ascending . . . . ***A. flagellaris***
      3' Stolons absent . . . . ***A. dimorpha***

1' Heads usually 2–many per flowering stem.
   4. Basal leaves not rosette-like or absent at flowering; stolons absent.
      5. Involucres mostly glabrous.
         6. Phyllaries distally light brown to olive; heads 2–8(10) per flowering stem. . . . . ***A. stenophylla***
         6' Phyllaries distally white to silvery white; heads 10–100+ per flowering stem . . . . ***A. luzuloides***
      5' Involucres pubescent at least proximally.
         7. Basal leaves withered or absent at flowering. . . . . ***A. geyeri***
         7' Basal leaves present at flowering.
            8. Stems 15–35(50) cm; basal leaves 3–5-veined; phyllaries with dark brown spot or band near base of scarious portion . . . . ***A. anaphaloides***
            8' Stems 3–15(20) cm; basal leaves 3-veined; phyllaries without dark brown spot or band near base of scarious portion . . . . ***A. lanata***
   4' Basal leaves usually forming rosettes; stolons often present.
      9. Basal leaves adaxial surface markedly less pubescent than abaxial surface, often glabrous.
         10. Inflorescences raceme- or panicle-like; basal leaves 10–40 mm wide . . . . ***A. racemosa***
         10' Inflorescences corymb-like; basal leaves 2–12 mm wide.
            11. Basal leaves 9–12 mm wide; phyllaries distally white to light brown. . . . . ***A. howellii*** (in part)
            11' Basal leaves 2–10 mm wide; phyllaries distally pink, red, green, yellow, gray, or white, rarely brown. . . . . ***A. microphylla*** (in part)
      9' Basal leaves about equally pubescent on both surfaces.
         12. Stolons slightly woody, ascending to erect; cauline leaves not flagged; inflorescences with 3–8 heads per stem . . . . ***A. umbrinella***
         12' Stolons herbaceous, horizontal to ascending; cauline leaves flagged or not; inflorescences with 2–75 heads per stem.
            13. Phyllaries distally brown, olive, gray, yellow, or blackish.
               14. Stems 2–7 cm, usually stipitate-glandular . . . . ***A. pulvinata***
               14' Stems 5–30 cm, rarely stipitate-glandular.
                  15. Heads 3–20 per stem; stems 5–30 cm; phyllaries distally yellow, gray, or rarely brown . . . . ***A. microphylla*** (in part)
                  15' Heads 2–5(9) per stem; stems 5–13 cm; phyllaries distally dark brown, olive, or blackish . . . . ***A. media***
            13' Phyllaries distally white, cream, green, red, or pink.
               16. Phyllaries with distinct dark brown or black spot or band near base of scarious portion; cauline leaves not flagged . . . . ***A. corymbosa***
               16' Phyllaries without distinct dark spot or band near base of scarious portion; cauline leaves flagged or not.
                  17. Heads 10–75 per stem in panicle-like arrays; phyllaries distally white to cream; involucres 4–5 mm; populations with ~ equal numbers of male and female plants . . . . ***A. argentea***
                  17' Heads 3–20 per stem in corymb-like arrays; phyllaries distally white, cream, red, or pink; involucres 4–11 mm; populations predominantly or entirely female.
                     18. Middle and distal cauline leaves usually not flagged; phyllaries distally white, cream, green, red, or pink . . . . ***A. microphylla*** (in part)
                     18' Middle and distal cauline leaves flagged; phyllaries distally cream to white . . . . ***A. howellii*** (in part)

Antennaria anaphaloides Rydberg

Tall pussytoes, pearly pussytoes

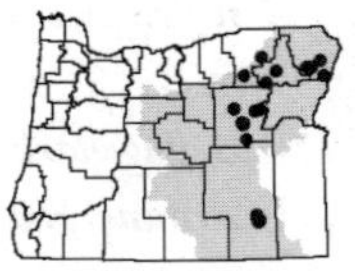

Herbs dioecious; stolons absent. **Stems** 15–35(50) cm, sometimes glandular. **Basal leaves** not rosette-like, narrowly oblanceolate to narrowly elliptic, 25–150(200) × 4–20(25) mm, 3–5-veined, tips mucronate, surfaces gray-pubescent. **Cauline leaves** oblanceolate to linear, 10–80 mm, tips acute to acuminate, usually flagged. **Heads** 8–30(50+) per stem, in corymb-like arrays. **Involucres** densely tomentose proximally; staminate heads (4)5–6.5 mm; pistillate heads 4.5–7 mm. **Phyllaries** with dark brown spot or band near base of scarious portion, distally white or cream, sometimes pink to red, tips acute to acuminate. **Fruits** 1–1.8 mm, glabrous. **2*n***=28.

Meadows, openings in shrublands or forests. Flowering May–Jul. 900–2800 m. BR, BW, Lava. NV, ID, WA; north to British Columbia, northeast to Saskatchewan, east to WY, southeast to CO. Native.

Antennaria argentea Bentham

Silvery pussytoes, silvery everlasting

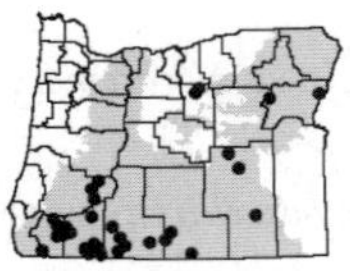

Herbs dioecious; stolons often present. **Stems** 18–40 cm, not glandular. **Basal leaves** usually forming rosettes, oblanceolate to elliptic, 20–50 × 4–20 mm, 1–3-veined, tips acute, surfaces gray-tomentose. **Cauline leaves** lanceolate, 15–45 mm, tips acute to acuminate, distal leaves flagged. **Heads** 10–75 per stem, in panicle-like arrays. **Involucres** mostly glabrous; staminate heads 4–5 mm; pistillate heads 4–5 mm. **Phyllaries** distally silvery white, tips acute. **Fruits** 1–1.5 mm, glandular. **2*n***=28.

Openings in forests, rocky slopes. Flowering Apr–Jul. 300–1800 m. BR, BW, Casc, ECas, Sisk. CA, NV. Native

There is disagreement in the literature about whether this species is stoloniferous, rhizomatous, or neither.

Antennaria corymbosa E.E. Nelson

Meadow pussytoes, flat-topped pussytoes

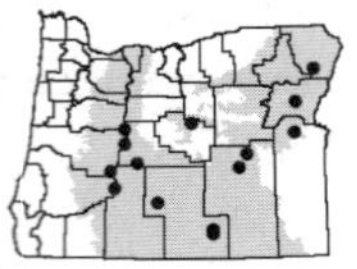

Herbs dioecious; stolons 1–10 cm. **Stems** 6–15(20) cm, not glandular. **Basal leaves** usually forming rosettes, spatulate, 18–45 × 2–4 mm, 1-veined, tips mucronate, surfaces gray-tomentose. **Cauline leaves** linear, 8–13 mm, tips acuminate, not flagged. **Heads** 3–7 per stem, in corymb-like arrays. **Involucres** mostly glabrous to pubescent; staminate heads 4–5.3 mm; pistillate heads 4–5 mm. **Phyllaries** with a distinct dark brown or black spot or band near base of scarious portion, distally white or cream, tips acute. **Fruits** 0.5–1 mm, slightly papillate. **2*n***=28.

Montane to alpine meadows. Flowering May–Aug. 1200–2600 m. BR, BW, Casc, ECas. CA, ID, NV, WA; northeast to MT, east to WY, southeast to NM. Native.

Antennaria dimorpha (Nuttall) Torrey & A. Gray

Low pussytoes

Antennaria dimorpha (Nutt.) Torr. & A. Gray var. *dimorpha*

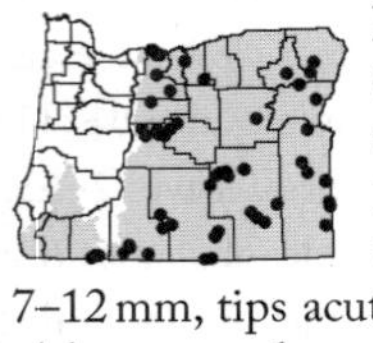

Herbs dioecious; stolons absent. **Stems** 0.5–4 cm, not glandular. **Basal leaves** linear to narrowly spatulate, 8–11 × 1–1.2 mm, 1-veined, tips acute, surfaces gray-tomentose. **Cauline leaves** linear to oblanceolate, 7–12 mm, tips acute, not flagged. **Heads** solitary. **Involucres** glabrous to pubescent; staminate heads 6–8 mm; pistillate heads 10–11 mm. **Phyllaries** distally light brown, tips acute-acuminate. **Fruits** 2–3.5 mm, pubescent. **2*n***=28, 56.

Dry ground, often in rocky or sandy soils. Flowering Mar–Jun. 50–2300 m. BR, BW, Col, ECas, Lava, Owy, Sisk. CA, ID, NV, WA; north to British Columbia, northeast to Saskatchewan, east to NE, southeast to NM. Native.

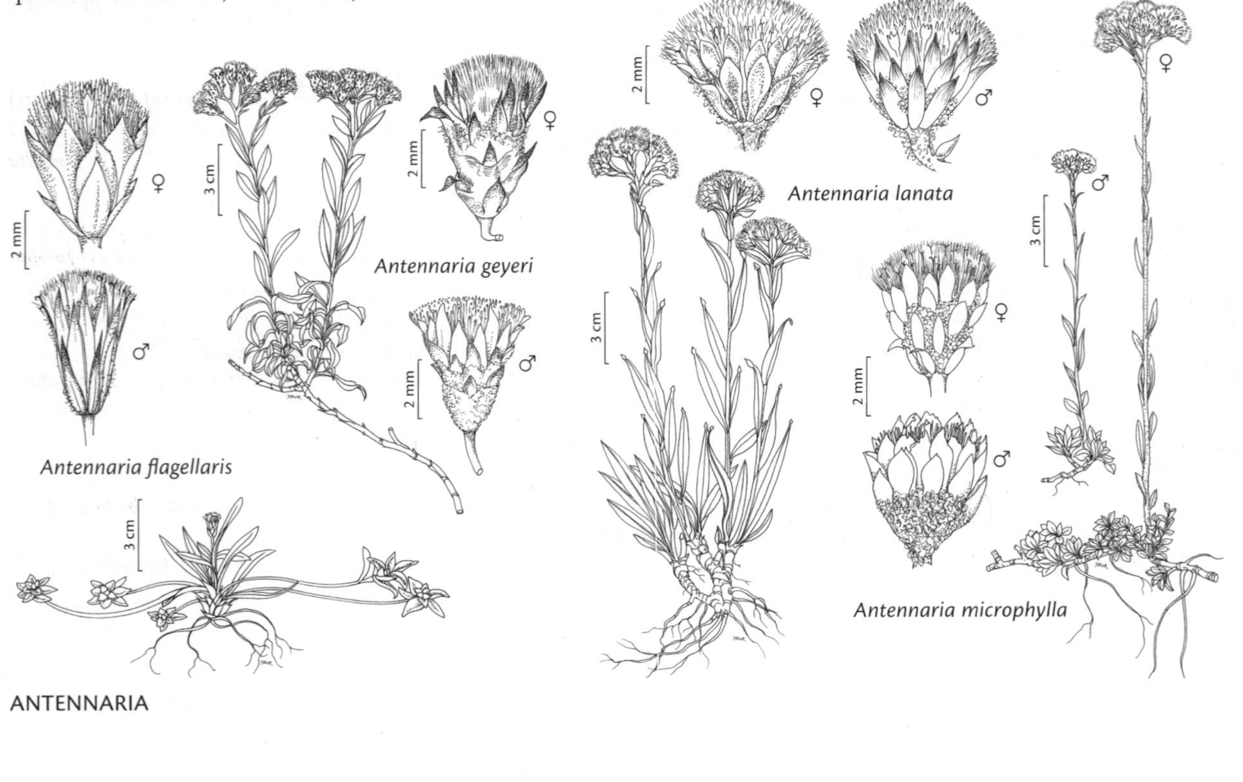

ANTENNARIA

Antennaria flagellaris (A. Gray) A. Gray (p. 176)

Flagellate pussytoes, stoloniferous everlasting

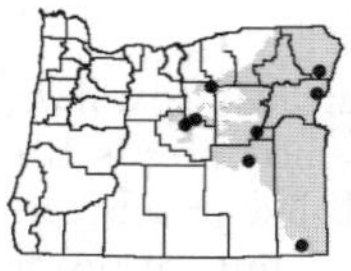

Herbs dioecious; stolons filiform, 3–10 cm, ascending, leafy only at tips. **Stems** 0.5–1.5 cm, not glandular. **Basal leaves** linear to oblanceolate, 16–18 × 1.5–2 mm, 1-veined, tips acute, surfaces gray-tomentose. **Cauline leaves** linear to oblanceolate, 7–15 mm, tips acute, not flagged. **Heads** solitary. **Involucres** loosely lanate basally; staminate heads 6–7 mm; pistillate heads 7–9 mm. **Phyllaries** distally brown or blackish to whitish, tips acute. **Fruits** 2–3 mm, papillate. **2*n***=28.

Open slopes, dry meadows, playas. Flowering Apr–Jun. 1200–2000 m. BW, Owy. CA, ID, NV, WA; north to British Columbia, east to WY. Native.

Antennaria geyeri A. Gray (p. 176)

Pinewoods pussytoes, Geyer's pussytoes

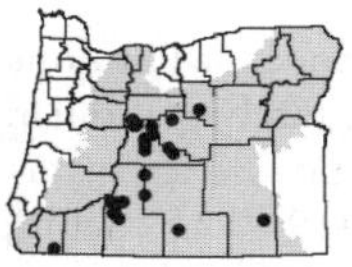

Herbs dioecious; stolons absent. **Stems** 3–14 cm, often glandular. **Basal leaves** not rosette-like, withered or absent at flowering. **Cauline leaves** linear-lanceolate to cuneate-oblanceolate, 11–35 × 2–6 mm, tips acute, not flagged. **Heads** 3–25 per stem, in corymb-like or panicle-like arrays. **Involucres** densely tomentose; staminate heads 6–8 mm; pistillate heads 6–8 mm. **Phyllaries** distally pink, red, light brown or white, tips acute. **Fruits** 2–2.5 mm, papillate, pubescent. **2*n***=28.

Dry slopes, flats, openings in forests. Flowering Jun–Aug. 900–2200 m. BR, BW, Casc, ECas, Lava, Sisk. CA, NV, WA. Native.

Antennaria howellii Greene

Howell's pussytoes

Herbs gynoecious, staminate plants uncommon; stolons 1–8(12) cm. **Stems** (6)8–35 cm, not glandular. **Basal leaves** usually forming rosettes, oblanceolate, cuneate-oblanceolate, ovate, or spatulate, (14)20–50(65) × 2.5–20 mm, 1–3-veined, tips mucronate, surfaces abaxially tomentose, adaxially glabrous to gray-pubescent. **Cauline leaves** linear, 8–40 mm, tips acute to acuminate, distal sometimes flagged. **Heads** 4–15 per stem, in corymb-like arrays. **Involucres** mostly glabrous to pubescent; staminate heads 6–6.5 mm; pistillate heads 6–11 mm. **Phyllaries** distally white to light brown, tips acute to acuminate, rarely rounded. **Fruits** 0.8–2 mm, sometimes papillate. **2*n***=56, 84, 140.

Canada and northern US. 4 subspecies; 3 subspecies treated in *Flora.*

Key to *Antennaria howellii* subspecies:

1. Basal leaves adaxially glabrous to very sparsely pubescent. . . . . ***A. h.*** ssp. ***howellii***

1' Basal leaves adaxially gray-pubescent to floccose-glabrate.

  2. Basal leaf petioles distinct; leaves along stolons ≈ those at tips. . . . . . ***A. h.*** ssp. ***neodioica***

  2' Basal leaf petioles indistinct; leaves along stolons < those at tips . . . . . ***A. h.*** ssp. ***petaloidea***

Antennaria howellii Greene ssp. *howellii*

Howell's pussytoes

Antennaria neglecta Greene var. *howellii* (Greene) Cronquist

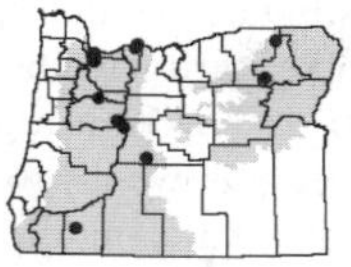

Plants with stolons 1–4 cm, leaves along stolons < those at tips. **Stems** 15–30 cm. **Basal leaves** 25–40 × 9–12 mm, 1–3-veined, tips mucronate, surfaces abaxially tomentose, adaxially glabrous to very sparsely pubescent; petioles usually distinct, sometimes indistinct. **Cauline leaves** 20–40 mm, not flagged. **Heads** 5–12 per stem. **Involucres** staminate heads 4–6 mm; pistillate heads 6–7.5 mm. **Phyllaries** distally white to light brown. **Fruits** 1.5–2 mm, papillate. **2*n***=56, 84.

Open woods, meadows, slopes. Flowering Apr–Jun. 0–2000 m. BW, Casc, ECas, Sisk, WV. CA, ID, WA; north to Yukon, northeast to Saskatchewan, east to NB. Native

Antennaria howellii Greene ssp. *neodioica* (Greene) R.J. Bayer

Field pussytoes

Antennaria neglecta Greene var. *attenuata* (Fernald) Cronquist, *Antennaria neglecta* Greene var. *neodioica* (Greene) Cronquist

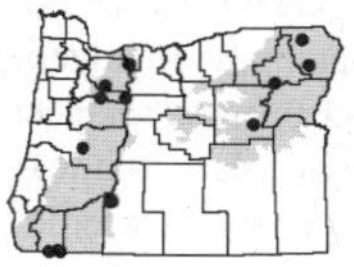

Plants with stolons 3–8(12) cm, leaves along stolons ≈ those at tips. **Stems** (6)15–35 cm. **Basal leaves** 14–48 × 2.5–20 mm, 1(3)-veined, tips mucronate, surfaces abaxially tomentose, adaxially gray-pubescent to floccose-glabrate; petioles distinct. **Cauline leaves** 8–35 mm, distal flagged. **Heads** 4–8(13) per stem. **Involucres** staminate heads 5–7 mm; pistillate heads 6–9 mm. **Phyllaries** distally cream or white. **Fruits** 0.9–1.5 mm, papillate. **2*n***=56, 84, 140.

Open woods, meadows, slopes. Flowering May–Aug. 200–1900 m. BW, Casc, Sisk. ID, WA; north to Northwest Territories, northeast to Newfoundland, southeast to NC. Native.

Antennaria howellii Greene ssp. *petaloidea* (Fernald) R.J. Bayer

Howell's pussytoes

Antennaria neglecta Greene var. *attenuata* (Fernald) Cronquist, *Antennaria pedicellata* Greene

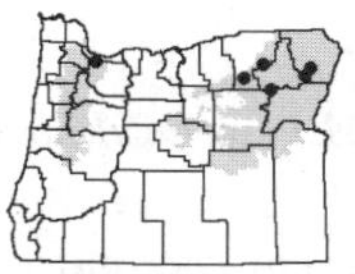

Plants with stolons 4–9 cm, with leaves along stolons < those at tips. **Stems** 8–35 cm. **Basal leaves** 14–45(65) × 5–20 mm, 1(3)-veined, tips mucronate, surfaces abaxially tomentose, adaxially gray-pubescent to floccose-glabrate; petioles indistinct. **Cauline leaves** 10–35 mm, middle and distal flagged. **Heads** 4–10(15) per stem. **Involucres** staminate heads 6–6.5 mm; pistillate heads 7–11 mm. **Phyllaries** distally cream or white. **Fruits** 0.8–1.7 mm, papillate. **2*n***=56, 84, 140.

Open woods, meadows, slopes. Flowering May–Jun. 900–1600 m. BW, WV. WA; northeastern and northwestern North America. Native.

Antennaria lanata (Hooker) Greene (p. 176)

Woolly pussytoes, woolly everlasting

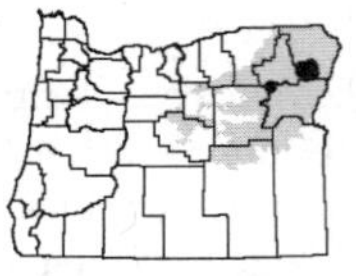

Herbs dioecious; stolons absent. **Stems** 3–15(20) cm, sometimes stipitate glandular. **Basal leaves** not rosette-like, narrowly oblanceolate, 10–60(100) × 3–12 mm, 3-veined, tips acute, surfaces gray-woolly to tomentose. **Cauline leaves** linear, 5–40 mm, tips acute, middle and distal flagged. **Heads** 3–9 per stem, in corymb-like arrays. **Involucres** pubescent proximally; staminate heads 4.5–6 mm; pistillate heads 5–8 mm. **Phyllaries** proximally dark or light brown to olive, distally light brown, cream, or white, tips acute to acuminate. **Fruits** 1–1.6 mm, glabrous. **2*n*=**28.

Open woods, meadows, slopes. Flowering Jun–Sep. 2100–3000 m. BW. CA, ID, WA; north to British Columbia, northeast to Alberta, east to WY. Native.

This species is easily distinguished by its woolly pubescence and prominent flags on the middle and distal cauline leaves.

Antennaria luzuloides Torrey & A. Gray

Woodrush pussytoes, small-flowered everlasting

Herbs dioecious, sometimes viviparous; stolons absent. **Stems** 7–35(50) cm, sometimes stipitate glandular. **Basal leaves** not rosette-like, linear to narrowly spatulate or narrowly oblanceolate, 18–55 × 1–8(10) mm, 1–3-veined, tips acuminate, surfaces gray-tomentose. **Cauline leaves** narrowly oblanceolate to linear, 5–60 mm, tips acute, flagged. **Heads** 10–100+ per stem, raceme-like, panicle-like, corymb-like, or subcapitate arrays. **Involucres** glabrous; staminate heads 3.5–5.5 mm; pistillate heads 3.5–6.5 mm. **Phyllaries** proximally green or golden brown, distally white, tips acute. **Fruits** 1–2 mm, papillate or strigose. **2*n*=**28.

Western North America. 2 subspecies.

Antennaria luzuloides is unusual among the Asteraceae in its ability to propagate asexually by plantlets in the leaf axils.

Key to *Antennaria luzuloides* subspecies:

1. Heads 10–30 per stem, in raceme- or panicle-like arrays; staminate involucres 3.5–4 mm, pistillate involucres 3.5–4.5 mm . . . . *A. l.* ssp. ***aberrans***

1' Heads 10–100+ per stem, in corymb-like or subcapitate arrays; staminate involucres 4–5.5 mm, pistillate involucres 5–6.5 mm . . . . *A. l.* ssp. ***luzuloides***

Antennaria luzuloides Torrey & A. Gray ssp. ***aberrans*** (E.E. Nelson) R.J. Bayer & Stebbins

Small-flowered everlasting

Antennaria luzuloides Torr. & A. Gray var. *microcephala* (A. Gray) Cronquist, *Antennaria microcephala* A. Gray

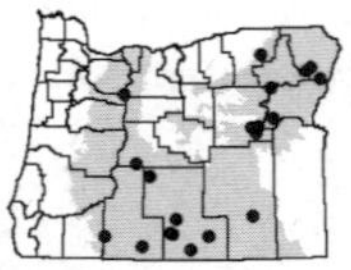

Basal leaves linear, 1–5(7) mm wide, 1(3)-veined. **Heads** 10–30 per stem, in raceme-like or panicle-like arrays. **Involucres** staminate heads 3.5–4 mm; pistillate heads 3.5–4.5 mm. **Phyllaries** proximally green, distally white. **Fruits** 1–1.5 mm, strigose.

Meadows, rocky areas. Flowering May–Aug. 1100–2400 m. BR, BW, Casc, ECas. CA, NV. Native.

Antennaria luzuloides Torrey & A. Gray ssp. ***luzuloides***

Woodrush pussytoes, silvery brown everlasting

Antennaria luzuloides Torr. & A. Gray var. *luzuloides*

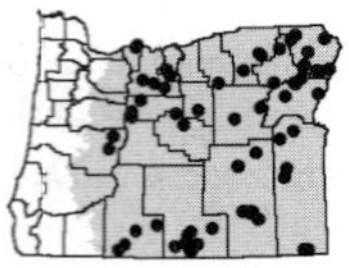

Basal leaves linear to narrowly spatulate, 2–5(10) mm wide, (1)3-veined. **Heads** 10–100+ per stem, in corymb-like or subcapitate arrays. **Involucres** staminate heads 4–5.5 mm; pistillate heads 5–6.5 mm. **Phyllaries** proximally golden brown, distally white. **Fruits** 1–2 mm, papillate.

Openings in forests, meadows, open hillsides, flats. Flowering May–Aug. 500–2300 m. BR, BW, Casc, Col, ECas, Lava, Owy. CA, ID, NV, WA; north to British Columbia, northeast to Alberta, east to WY, southeast to CO. Native.

Antennaria media Greene

Alpine pussytoes

Antennaria alpina (L.) Gaertn. var. *media* (Greene) Jeps.

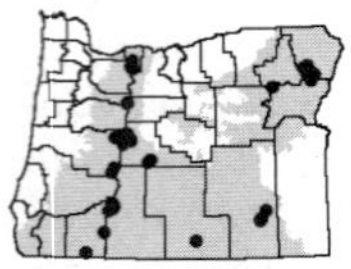

Herbs dioecious or gynoecious; stolons 1–4 cm. **Stems** 5–13 cm, rarely glandular. **Basal leaves** usually forming rosettes, spatulate to oblanceolate, 6–19 × 2.5–6 mm, 1-veined, tips mucronate, surfaces gray-pubescent. **Cauline leaves** linear, 5–20 mm, tips acute, not flagged. **Heads** 2–5(9) per stem, in corymb-like arrays. **Involucres** glabrous to pubescent proximally; staminate heads (3.5)4.5–6.5 mm; pistillate heads 4–8 mm. **Phyllaries** distally dark brown, olive, or blackish, tips acute to acuminate. **Fruits** 0.6–1.6 mm, glabrous or papillate. **2*n*=**56, 98, 112.

Alpine meadows, basins, ridges. Flowering Jun–Aug. 1800–3100 m. BR, BW, Casc, ECas, Sisk. CA, ID, NV, WA; north to AK, northeast to MT, east to WY, southeast to NM. Native.

Antennaria microphylla Rydberg (p. 176)

Littleleaf pussytoes, rosy pussytoes

Antennaria rosea Greene, *Antennaria rosea* Greene ssp. *arida* (E.E. Nelson) R.J. Bayer, *Antennaria rosea* Greene ssp. *confinis* (Greene) R.J. Bayer, *Antennaria rosea* Greene ssp. *rosea*

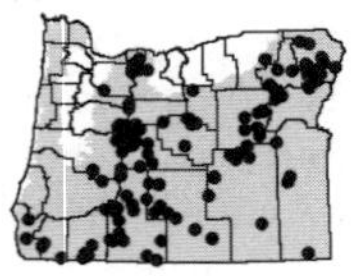

Herbs dioecious or gynoecious; stolons herbaceous, horizontal to ascending, 1–10 cm. **Stems** 5–30 cm, rarely stipitate glandular distally. **Basal leaves** usually forming rosettes, spatulate to oblanceolate, 8–40 × 2–10 mm, 1-veined, tips mucronate, surfaces usually gray-pubescent, adaxially sometimes green and glabrous. **Cauline leaves** linear, 5–36 mm, tips acute to subulate, usually not flagged. **Heads** 3–20 per stem, in corymb-like arrays. **Involucres** mostly glabrous to pubescent; staminate heads 4–8 mm; pistillate heads 4–10 mm. **Phyllaries** distally white to yellow, green, red, pink, gray, or rarely brown, tips rounded to acuminate. **Fruits** 0.7–1.5 mm, glabrous to papillate. **2*n*=**28, 42, 56, 70.

Rocky areas, meadows, forests, disturbed areas. Flowering May–Sep. 200–3000 m. BR, BW, Casc, CR, ECas, Lava, Owy, Sisk. CA, ID, NV, WA; north to AK, northeast to Greenland, east to NE, southeast to NM. Native.

Here we take a more inclusive approach and include within this species plants that some floras treat as *A. rosea.*

Antennaria pulvinata Greene

Pulvinate pussytoes, aromatic pussytoes

Antennaria aromatica Evert, *Antennaria rosea* Greene ssp. *pulvinata* (Greene) R.J. Bayer

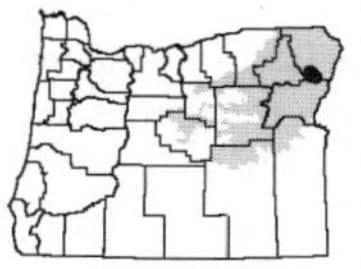

Herbs dioecious or gynoecious; stolons 1–6 cm. **Stems** 2–7 cm, usually stipitate-glandular. **Basal leaves** usually forming rosettes, cuneate-spatulate, rarely oblanceolate, 5–16 × 3–10 mm, 1-veined, tips mucronate, surfaces gray-pubescent, often glandular. **Cauline leaves** linear, 3–14 mm, tips acute, not flagged. **Heads** 2–5 per stem, in corymb-like arrays. **Involucres** mostly glabrous; staminate heads 4.5–6.5 mm; pistillate heads 3.5–4.5 mm. **Phyllaries** distally light to dark brown or olive, tips acute to acuminate. **Fruits** 0.9–2 mm, papillate. **2*n***=28, 56, 84.

Limestone talus. Flowering Aug. 2400–3000 m. BW. CA, ID, WA; north to AK, northeast to Newfoundland, east to MI, southeast to CO. Native.

Oregon specimens of *A. pulvinata* have varying degrees of glandularity. This species includes both sexual diploid and apomictic polyploid populations that some floras may refer to as *A. aromatica* and *A. rosea* ssp. *pulvinata*, respectively (Chmielewski 1993).

Antennaria racemosa Hooker

Racemose pussytoes

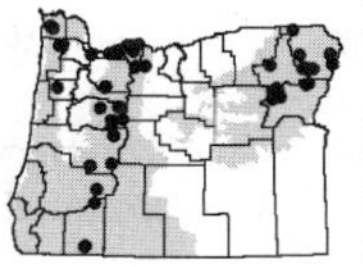

Herbs dioecious; stolons 3–8 cm. **Stems** 12–50 cm, distally stipitate-glandular. **Basal leaves** usually forming rosettes, elliptic to oblong, 30–100 × 10–40 mm, prominently 3-veined, tips mucronate, surfaces abaxially white- or gray-tomentose, adaxially glabrous. **Cauline leaves** linear, 10–30 mm, tips acute to obtuse, not flagged. **Heads** 3–12 per stem, raceme- or panicle-like arrays. **Involucres** mostly glabrous; staminate heads 4–8 mm; pistillate heads 7–9 mm. **Phyllaries** distally light brown or white, tips acute to acuminate. **Fruits** 1–1.5 mm, glabrous to papillate. **2*n***=28.

Forests, rocky or grassy slopes. Flowering May–Aug. 100–2100 m. BW, Casc, CR, ECas, Sisk. CA, ID, WA. North to British Columbia, northeast to Alberta, east to WY. Native.

Within the genus, *A. racemosa* is a very distinctive species with broad basal leaves that are adaxially glabrous and green, abaxially densely white-tomentose, and with a raceme- or panicle-like arrangement of heads.

Antennaria stenophylla (A. Gray) A. Gray

Narrow-leaved pussytoes

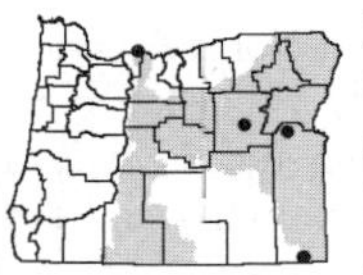

Herbs dioecious; stolons absent. **Stems** (3)10–15 cm, not glandular. **Basal leaves** not rosette-like, linear to narrowly oblanceolate, 15–50 × 1–2(4) mm, 1-veined, tips acute, surfaces gray-tomentose. **Cauline leaves** linear, 5–60 mm, tips acute to acuminate, distal flagged. **Heads** 2–8(10) per stem, in subcapitate arrays. **Involucres** glabrous, sometimes pubescent proximally; staminate heads 4–5 mm; pistillate heads 4–6.5 mm. **Phyllaries** light brown to olive, tips rounded to acuminate. **Fruits** 1–1.8 mm, glandular-puberulent. **2*n***=56.

Sagebrush, dry hillsides. Flowering May–Jun. 900–2000 m. BW, ECas, Lava, Owy. ID, NV, WA. Native.

Antennaria suffrutescens Greene

Siskiyou everlasting, shrubby pussytoes

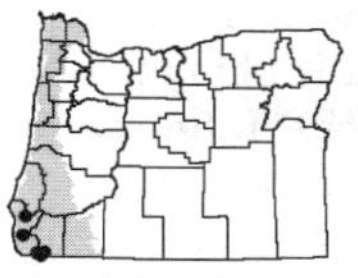

Subshrubs dioecious; stolons absent. **Stems** 5–12 cm, glandular. **Basal leaves** spatulate, 5–12 × 2–4 mm, 1-veined, tips emarginate to obtuse, surfaces abaxially tomentose, adaxially green and glabrous, sometimes glandular. **Cauline leaves** 5–12 × 0.5–2.5 mm, tips emarginate to obtuse, not flagged. **Heads** solitary. **Involucres** pubescent proximally; staminate heads 5–9 mm; pistillate heads 5–8 mm. **Phyllaries** distally white, tips acute to acuminate. **Fruits** 1–2 mm, papillate. **2*n***=28.

Coniferous forests, serpentine. Flowering May–Jul. 400–1600 m. CR, Sisk. CA. Native.

Antennaria umbrinella Rydberg (p. 179)

Umber pussytoes

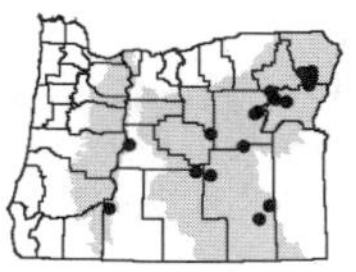

Herbs or **subshrubs** dioecious; stolons slightly woody, ascending to erect, 5–15 cm. **Stems** 7–16 cm, not glandular. **Basal leaves** usually forming rosettes, narrowly spatulate to cuneate, 10–17 × 2–5.4 mm, 1-veined, tips mucronate, surfaces gray-tomentose. **Cauline leaves** linear, 8–18 mm, tips acute, not flagged. **Heads** 3–8 per stem, in corymb-like arrays. **Involucres** mostly glabrous to pubescent; staminate heads 3–6 mm; pistillate heads 4–6.5 mm. **Phyllaries** distally white, light yellow, or light brown, often streaked with pink, tips acute to acuminate. **Fruits** 0.5–1.2 mm, glabrous. **2*n***=28, 56.

Sagebrush, coniferous forests, meadows. Flowering Jun–Sep. 1000–2800 m. BR, BW, Casc, Lava. CA, ID, NV, WA; north to British Columbia, northeast to Saskatchewan, east to WY, southeast to AZ. Native.

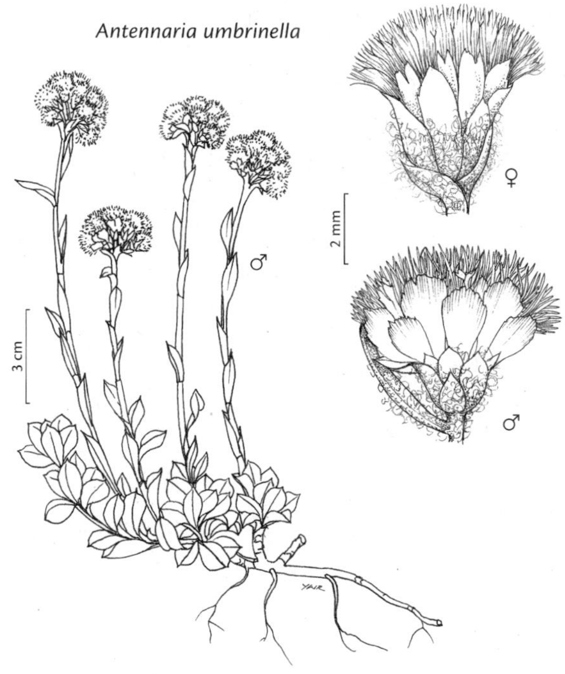

ANTENNARIA

ANTHEMIS Chamomile

Herbs annual or biennial, aromatic or not. **Stems** erect or basally decumbent, simple to much branched, principal branches sometimes all basal, glabrous to strigillose or villous-puberulent. **Leaves** basal and cauline, alternate, basal usually withered by flowering time; blades 1–3-times pinnatifid, with linear or filiform, acute rachises and segments, glabrous to strigillose or villous-puberulent, often with fringed auricles; sessile or petiolate. **Inflorescences** panicle-like arrays, terminal on stem and branches; peduncles long or short, strigillose or villous-puberulent. **Heads** radiate. **Involucres** bowl-shaped or hemispheric. **Phyllaries** in 3–5 series, erect in flower and fruit, glabrous to villous-puberulent; outer equal or slightly shorter than inner, midribs green or brownish, margins membranous, tips acute or obtuse; inner equal, lanceolate or oblong. **Receptacles** hemispheric or narrowly conic; paleate throughout or distally, exceeding involucre in fruit. **Ray florets** pistillate, fertile or sterile; rays white, ovate-elliptic or oblong, reflexed and persisting in fruit, tips acute or obtuse, sometimes 2–3-toothed. **Disc florets** bisexual; corollas tubular, gradually or abruptly widened distally, usually yellow, lobes 5, erect or spreading, triangular; stamens 5, included in corolla tubes; stigma lobes included in corolla tubes. **Fruits** obovoid or obconic, sometimes 4-angled, 9–10-veined, glabrous, smooth or roughened; pappi 0 or of low crowns. **2*n***=18.

Southwestern Asia, Australia, Europe, North Africa, North America. 175 species; 2 species treated in *Flora.*

The yellow chamomile, *Anthemis tinctoria*, has been transferred to *Cota* as *C. tinctoria*.

Key to *Anthemis* species:

1. Leaf rachises and segments linear; ray florets fertile; receptacles paleate throughout, paleae 3–4+ mm; fruit surfaces smooth . . . . **_A. arvensis_**

1' Leaf rachises and segments filiform; ray florets sterile; receptacles paleate distally, paleae 2–3+ mm; fruit surfaces roughened . . . . **_A. cotula_**

Anthemis arvensis Linnaeus (p. 180)

Corn chamomile, field chamomile

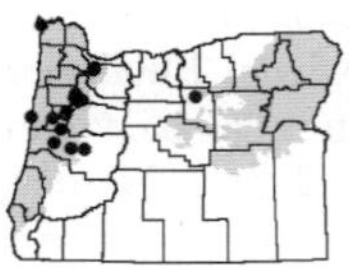

Plants annual or biennial, not strongly aromatic, 1–9 dm. **Stems** often with principal branching at or near base. **Leaves** 1–10 cm, 1-veined from base; cauline slightly reduced distally, rachises and segments linear. **Heads** 5–15 mm including receptacles; peduncles 3–20+ cm. **Involucres** 4–6 mm. **Receptacles** paleate throughout. **Paleae** oblanceolate or narrowly linear, 3–4+ mm, tips mucronate or acuminate. **Ray florets** 5–20, fertile; rays 5–15 mm, white, sometimes pink-tinged, tips obtuse, often 2–3-toothed. **Disc florets** numerous; corollas 2–4 mm. **Fruits** 1.7–2+ mm, brown, usually smooth.

Forest openings, roadsides, fields, disturbed areas. Flowering May–Jul. 0–600 m. BW, CR, Est, WV. CA, ID, WA; scattered throughout North America; Europe. Exotic.

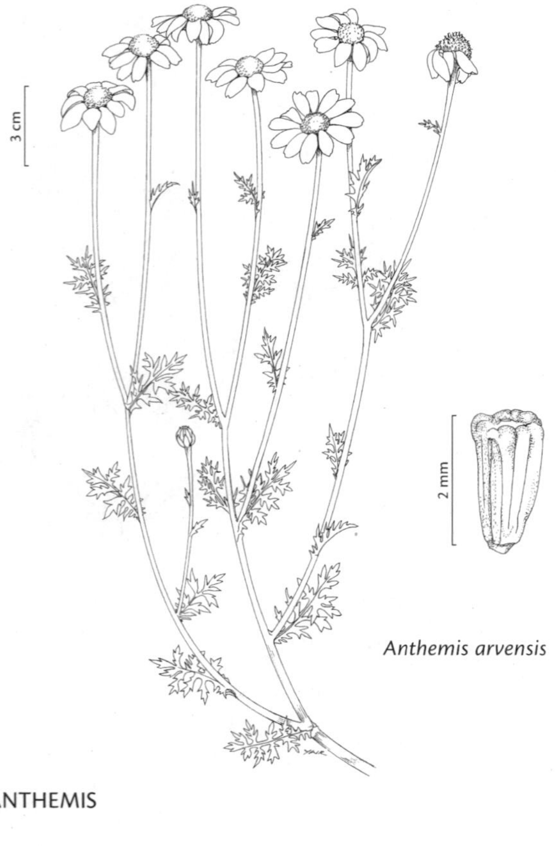

Anthemis arvensis

Anthemis cotula Linnaeus

Dog-fennel, mayweed chamomile, stinking chamomile

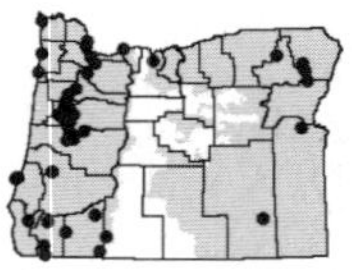

Plants annual, strongly aromatic, 0.5–4+ dm. **Stems** usually branched throughout, or only distally. **Leaves** 0.8–8 cm, 1-veined from base, cauline slightly reduced distally, rachises and segments filiform. **Heads** 4–10 mm including receptacles. **Involucres** 3–4 mm; peduncles 2–15 cm. **Receptacles** paleate distally. **Paleae** narrowly linear, 2–3+ mm, tips acuminate. **Ray florets** 10–15, sterile; rays 5–15+ mm, white, tips acute, or obtuse and 2–3-toothed. **Disc florets** numerous; corollas 2–2.5 mm. **Fruits** 1.3–2 mm, brown, surfaces roughened.

Woodlands, riverbanks, roadsides, fields, disturbed areas. Flowering May–Aug. 0–1800 m. BR, BW, Casc, Col, CR, Est, Owy, Sisk, WV. CA, ID, NV, WA; throughout North America; Asia, Europe. Exotic.

The distinctions between *Anthemis arvensis* and *A. cotula* in plant odor and width of leaf segments are subjective and can best be judged by a comparison of known specimens. The species sometimes grow together, but *A. cotula* is more common.

ARCTIUM Burdock

Herbs biennial or short-lived perennial. **Stems** erect, stout, branched distally, floccose-tomentose or minutely puberulent. **Leaves** basal and cauline, alternate; blades triangular-ovate, bases truncate or cordate, margins irregularly crenate to dentate or denticulate, tips acute or obtuse, surfaces abaxially appressed-tomentose or minutely puberulent, adaxially ± glabrous; long- to short-petiolate. **Inflorescences** leafy-bracted, raceme- or corymb-like clusters. **Heads** discoid; sessile or pedunculate. **Involucres** spherical or ovoid. **Phyllaries** numerous, in many series; outer spreading to reflexed; inner erect, chartaceous, linear, stiff, with spiny, hooked tips. **Receptacles** flat, bristly; paleae 0. **Florets** bisexual; corollas tubular, linear below, widened and cylindric distally, exserted from involucres, pink to purple (occasionally white), surfaces glabrous or puberulent, lobes 5, narrowly triangular, erect; stamens 5, partly exserted; stigmas exserted. **Fruits** obovoid or oblong-ellipsoid, compressed, truncate, glabrous, with 4 distinct veins and 4 or more minor; pappi of short, non-plumose, deciduous bristles.

Worldwide. 10 species; 3 species treated in *Flora.*

The common name, burdock, comes from the hooked tips of the spiny phyllaries. Mature dry fruiting heads fall off whole, and the hooked phyllaries catch on passing animals, effectively dispersing the seeds.

Key to *Arctium* species:

1. Phyllaries densely cobwebby throughout . . . . *A. tomentosum*

1' Phyllaries usually glabrous or lightly cobwebby distally.

  2. Involucres 2–4.5 cm in diameter; inflorescences usually long-peduncled, corymb-like clusters; petioles mostly solid . . . . *A. lappa*

  2' Involucres 1–2.5 cm in diameter; inflorescences usually short-peduncled, raceme-like clusters; petioles mostly hollow . . . . *A. minus*

Arctium lappa Linnaeus

Greater burdock

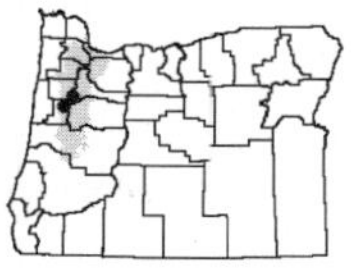

Plants 10–30 dm. **Stems** minutely puberulent, often glandular, sometimes floccose at nodes. **Leaves** basal 40–100+ cm, blades 15–80 cm, surfaces abaxially minutely glandular-puberulent; cauline reduced distally; petioles mostly solid, glandular-puberulent. **Heads** in corymb-like clusters; peduncles 2–7 cm, sometimes bracteate near heads. **Involucres** 2–4.5 cm in diameter. **Phyllaries** glabrous throughout or more cobwebby distally, lightly floccose-tomentose basally. **Disc florets** 30+; corollas 9–14 mm, purple to pink, rarely white glabrous or glandular-puberulent. **Fruits** 6–7.5 mm, light brown, often mottled; pappi 2–5 mm. **2*n***=36.

Roadsides, disturbed areas. Flowering Jun–Oct. 50–500 m. WV. CA, ID, WA; scattered in North America. Asia, Europe. Exotic.

Arctium minus (Hill) Bernhardi (p. 181)

Common burdock, lesser burdock

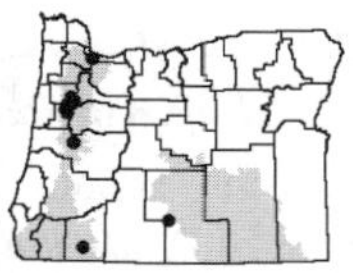

Plants 3–30 dm. **Stems** minutely glandular-puberulent, sometimes lightly floccose. **Leaves** basal 45–100 cm, blades 30–60 cm, surfaces abaxially thinly gray-tomentose; cauline reduced distally; petioles mostly hollow, glandular-puberulent, often lightly floccose. **Heads** in raceme-like clusters distally on stems and branches; sessile or short-peduncled. **Involucres** 1–2.5 cm in diameter. **Phyllaries** glabrous throughout or cobwebby distally, rarely cobwebby throughout, often floccose basally. **Disc florets** 30+; corollas 7–12 mm, pink, purple, or white, glabrous or glandular-puberulent. **Fruits** 5–8 mm, brown, often mottled; pappi 1–3.5 mm. **2*n***=36.

Roadsides, riverbanks, disturbed areas. Flowering Jul–Sep. 0–1500 m. BR, Sisk, WV. CA, ID, WA; throughout North America; Asia, Europe. Exotic.

Arctium tomentosum Miller

Woolly burdock, cotton burdock

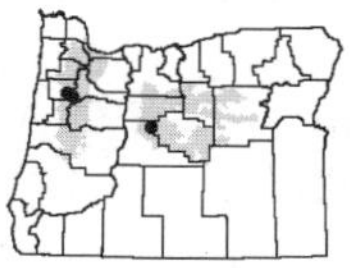

Plants 3–25 dm. **Stems** minutely glandular-puberulent, sometimes floccose distally. **Leaves** basal 40–55 cm, blades 30–40 cm, surfaces abaxially appressed white-tomentose; cauline reduced distally; petioles often floccose, glandular-puberulent. **Heads** short- or long-peduncled. **Involucres** 1.5–2.5 mm in diameter. **Phyllaries** floccose and cobwebby throughout. **Disc florets** 30+; corollas 9–13 mm, purple or rarely white, minutely glandular. **Fruits** 5–8 mm, light brown; pappi 1–3 mm. **2*n***=36.

Roadsides, disturbed areas. Flowering Jun–Oct. 0–1000 m. Lava, WV. Scattered in northern North America; Asia, Europe. Exotic.

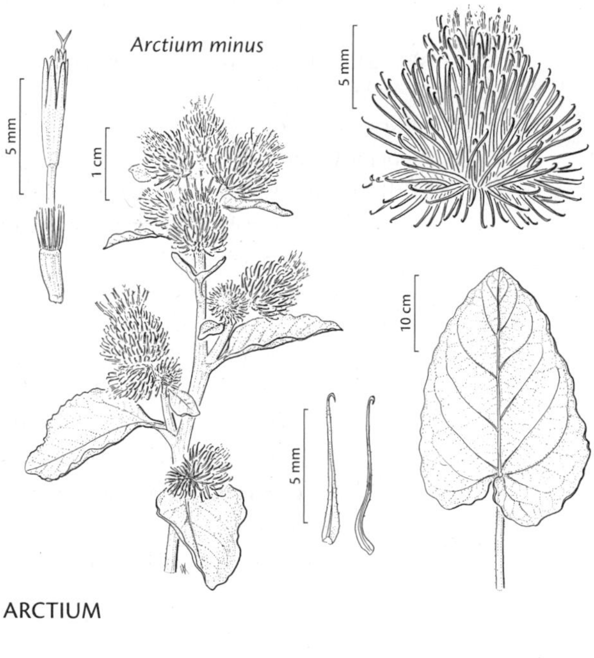

ARCTIUM

ARNICA Arnica

Herbs perennial; usually rhizomatous, sometimes strongly so. **Stems** erect, simple or branched, sometimes with basal offshoots or accessory, non-flowering basal rosettes, glabrous or pubescent, often glandular. **Leaves** basal and cauline, opposite, triangular to ovate, lanceolate, or elliptic, margins entire to dentate or serrate, tips obtuse or acute, surfaces glabrous to pubescent and/or glandular; sessile or petiolate. **Inflorescences** solitary terminal heads, sometimes few to numerous additional heads in axils of upper bracts and leaves, or on bracteate axillary branches. **Heads** radiate or discoid, usually erect. **Involucres** campanulate, hemispheric, or turbinate. **Phyllaries** few to numerous in 1–2 equal series, ovate to lanceolate or narrowly lance-elliptic, surfaces usually pubescent, often stipitate-glandular, sometimes with a tuft of adaxial hairs at tips. **Receptacles** convex; paleae 0. **Ray florets** 0 or pistillate, yellow to orange. **Disc florets** bisexual; corollas tubular, linear below, widened and cylindric distally, yellow or rarely cream, moderately or densely pilose and stipitate-glandular, 5-lobed, lobes triangular, recurved; style branches slightly exserted from stamen tubes. **Fruits** cylindric, conic, or fusiform, gray or blackish, 5–10-veined, glabrous to densely strigose or hirsute, glandular-pubescent or not; pappi of 10–50 barbellate, subplumose, or plumose bristles.

Circumboreal. 29 species; 17 treated in *Flora*.

Arnica species exhibit considerable morphologic variability as a result of frequent polyploidy and apomixis, often making identification difficult.

Key to *Arnica* species:

1. Heads discoid.
    2. Cauline leaves in 5–10 pairs, sessile . . . . . **A. viscosa**
    2' Cauline leaves in 1–5 pairs, some petiolate.
        3. Basal leaves usually withering by flowering; cauline petioles not winged; pappi straw-colored . . . **A. parryi** (in part)
        3' Basal leaves usually not withering by flowering; cauline petioles often winged; pappi white.
            4. Basal petioles not or scarcely winged; achenes puberulent . . . . . **A. discoidea**
            4' Some basal petioles distinctly winged; achenes gland-dotted, sometimes also hispid, rarely glabrous. . . . . **A. spathulata**

1' Heads radiate.
    5. Cauline leaf veins parallel from bases.
        6. Cauline leaves not reduced distally . . . . . **A. rydbergii**
        6' Cauline leaves reduced distally.
            7. Bases of basal leaves with woolly tufts; disc corollas with glandular and non-glandular trichomes . . . . **A. fulgens**
            7' Bases of basal leaves lacking woolly tufts; disc corollas with glandular trichomes only . . . . . **A. sororia**
    5' Cauline leaf veins branching laterally.
        8. Cauline leaves in 5–10 pairs.
            9. Phyllaries with conspicuous tuft of white trichomes at tips . . . . . **A. chamissonis**
            9' Phyllaries without tuft of trichomes at tip.
                10. Cauline leaves mostly 5–10 × as long as wide, margins entire to obscurely or minutely denticulate . . . . . **A. longifolia**
                10' Cauline leaves mostly 2–4 × as long as wide, margins crenulate to serrate-dentate or denticulate . . . . . **A. lanceolata** ssp. ***prima***
        8' Cauline leaves in (1)2–4(6) pairs.
            11. Pappi brownish.
                12. Basal leaves well developed; involucres broadly campanulate to hemispheric . . . . . **A. mollis**
                12' Basal leaves usually reduced; involucres turbinate . . . . . **A. diversifolia**
            11' Pappi white or straw-colored.
                13. Most cauline leaves sessile . . . . . **A. latifolia**
                13' Most cauline leaves petiolate.
                    14. Ray florets 0 . . . . . **A. parryi** (in part)
                    14' Ray florets 5–14, developed.
                        15. Cauline leaves glabrous to scabrous . . . . . **A. cernua**
                        15' Cauline leaves puberulent, villous, glandular, stipitate-glandular, or spiculate.
                            16. Stems usually much branched . . . . . **A. gracilis**
                            16' Stems usually simple.
                                17. Cauline leaf margins usually dentate or serrate proximally. . . . . **A. cordifolia**
                                17' Cauline leaf margins entire proximally . . . . . **A. nevadensis**

Arnica cernua Howell

Nodding arnica, serpentine arnica

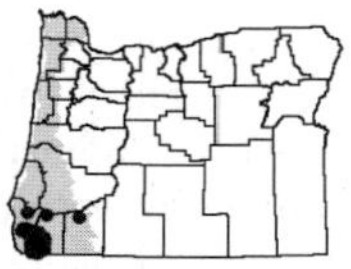

Plants (0.5)1–4 dm; strongly rhizomatous. **Stems** simple, rarely branched below, glabrous or sparsely strigose-villous, often stipitate-glandular distally. **Basal leaves** small, usually withering early. **Cauline leaves** (1)3–4 pairs; blades ovate, obovate, elliptic, or lanceolate, 1.5–6 cm, often reddish purple, ± thick or leathery, bases sometimes truncate or cordate, margins entire to crenate-dentate, veins branching laterally, serrate, or lobed, tips acute or obtuse, surfaces glabrous to scabrous; usually petiolate, rarely winged, distal-most short-petiolate or sessile. **Heads** nodding in bud, erect in flower, radiate. **Peduncles** scabrous, ± pilose, or stipitate-glandular. **Involucres** campanulate, 8–18 mm. **Phyllaries** broadly lanceolate, tips acute, surfaces pilose and stipitate-glandular. **Ray florets** 5–10; rays 6–10 mm, yellow. **Disc florets** 20–50; corollas 8–12 mm. **Fruits** columnar-fusiform, 6–8 mm, gray or blackish, strigose to hirsute; pappus bristles 5–7 mm, white, barbellate to subplumose. ***2n***=38.

Dry, open coniferous forests, shrublands, on serpentine soils. Flowering Apr–Jun. 200–1600 m. CR, Sisk. CA. Native.

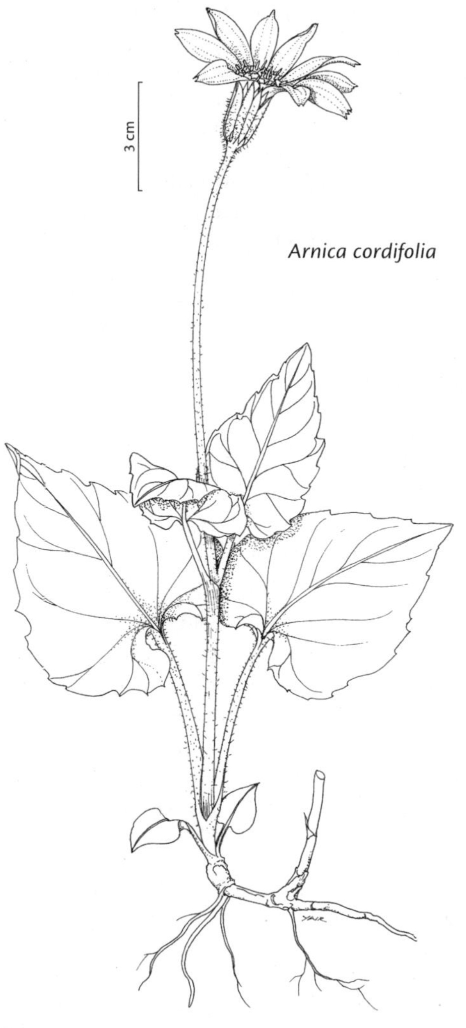

ARNICA

Arnica chamissonis Lessing

Chamisso arnica, meadow arnica

Arnica chamissonis Less. ssp. *foliosa* (Nutt.) Maguire, *Arnica chamissonis* Less. var. *foliosa* (Nutt.) Maguire, *Arnica chamissonis* Less. ssp. *incana* (A. Gray) Maguire, *Arnica chamissonis* Less. var. *incana* (A. Gray) Hultén

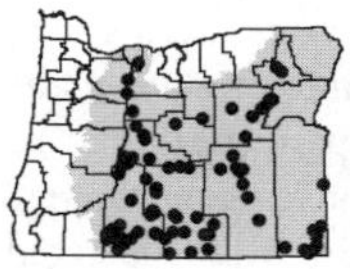

Plants (0.5)1–10 dm; strongly rhizomatous. **Stems** simple or branched distally, puberulent to villous or woolly-tomentose, sometimes glabrate proximally. **Basal leaves** small, usually withering early. **Cauline leaves** 4–10 pairs, well developed; blades narrowly to broadly lanceolate or elliptic, 2–18 cm, margins entire to denticulate or dentate, veins branching laterally, tips acute or obtuse, surfaces glabrous to puberulent, setose, villous, or woolly-tomentose, not glandular; proximally short-petiolate, distally sessile, sometimes winged. **Heads** radiate. **Peduncles** puberulent to tomentose, sometimes minutely glandular. **Involucres** campanulate, 8–13 mm. **Phyllaries** lance-linear to lance-elliptic or oblanceolate, tips acute, with conspicuous tuft of white trichomes, surfaces puberulent to tomentose, sometimes minutely glandular. **Ray florets** 8–13; rays 5–20 mm, yellow. **Disc florets** 30–50+; corollas 7–11 mm, yellow. **Fruits** columnar, 4–6 mm, blackish, glabrous to minutely puberulent; pappus bristles 4–6 mm, white or straw-colored, barbellate to subplumose. ***2n***=38, 57, 76.

Moist meadows, wet forests, streambanks. Flowering May–Sep. 800–2300 m. BR, BW, Casc, ECas, Lava, Owy. CA, NV, ID, WA; throughout Canada and western US. Native.

Arnica cordifolia Hooker (p. 183)

Heart-leaved arnica

Arnica cordifolia Hook. var. *cordifolia*, *Arnica cordifolia* Hook. var. *pumila* (Rydb.) Maguire

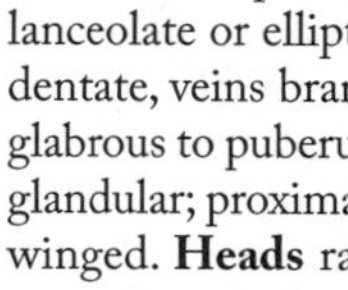

Plants (1)2–5+ dm; strongly rhizomatous. **Stems** usually simple, stipitate-glandular to lightly or densely pilose or villous. **Basal leaves** usually in separate sterile rosettes; long-petiolate. **Cauline leaves** 2–4(6) pairs, reduced distally; blades ovate or triangular, 2–15 cm, bases cordate to subcordate or truncate, margins usually dentate or serrate, veins branching laterally, tips acute, surfaces puberulent, sparsely villous, or minutely glandular; petiolate. **Heads** radiate. **Peduncles** stipitate-glandular and pilose or villous, tomentose below heads. **Involucres** campanulate, 15–20(30) mm. **Phyllaries** broadly or narrowly lance-elliptic or oblanceolate, tips acute, surfaces pilose or villous, sometimes glabrous distally or glandular-puberulent. **Ray florets** 6–13; rays 15–30 mm, yellow. **Disc florets** 20–90; corollas 9–12 mm, yellow. **Fruits** columnar-fusiform, 5–8(10) mm, dark gray or blackish, spiculate or hirsute; pappus bristles 5–9 mm, white, barbellate. ***2n***=38, 57, 76, 95, 114.

Forests, shrublands, meadows, streambanks. Flowering Apr–Aug. 50–2900 m. BR, BW, Casc, Col, CR, ECas, Lava, Sisk. CA, ID, NV, WA; north to AK, northeast to Quebec, east to SD, southeast to NM. Native.

Arnica discoidea Bentham

Rayless arnica

Arnica discoidea Benth. var. *eradiata* (A. Gray) Cronquist, *Arnica grayi* A. Heller, *Arnica parviflora* A. Gray ssp. *alata* (Rydb.) Maguire, *Arnica parviflora* A. Gray ssp. *parviflora*

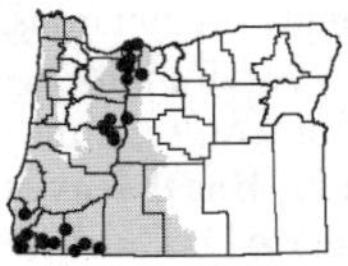

Plants 1.5–6 dm; rhizomatous. **Stems** simple or branched, villous and stipitate-glandular. **Basal leaves** in 1 pair, triangular to subcordate; petioles not or scarcely winged. **Cauline leaves** (1)2–4 pairs, often clustered near base; blades triangular to ovate or lanceolate, 2–9 cm, bases truncate to subcordate, margins denticulate to dentate or serrate, tips acute to obtuse, surfaces glabrous to pilose or villous; petioles usually broadly winged. **Heads** discoid. **Peduncles** villous, sparingly to densely stipitate-glandular. **Involucres** campanulate, 8–13 mm. **Phyllaries** broadly to narrowly elliptic-lanceolate or linear, tips acute, surfaces densely pilose or villous and stipitate-glandular. **Ray florets** 0. **Disc florets** 20–50; corollas 5–10 mm, yellow. **Fruits** columnar-fusiform, 5–8 mm, gray or blackish, puberulent; pappus bristles 6–8 mm, white, barbellate to subplumose. **2*n***=38, 57, 76.

Open mixed woodlands, coniferous forests, shrublands, rocky slopes and ridges, roadsides, serpentine. Flowering May–Aug. 100–1800 m. Casc, CR, ECas, Sisk. CA, NV, WA. Native.

Arnica diversifolia Greene

Sticky arnica

Arnica ovata Greene

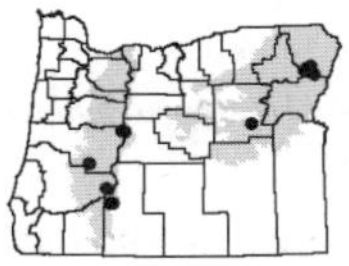

Plants 1.5–4 dm; rhizomatous. **Stems** single or branched, usually glandular-puberulent, sometimes glabrate proximally or sparsely scabrous-pilose. **Basal leaves** usually much reduced, withering early, round-ovate; petioles broadly winged. **Cauline leaves** 2–3(4) pairs, distal 1–2 pairs reduced and bract-like; blades ovate to lanceolate, 2–8 cm, bases truncate or attenuate, margins denticulate or serrate-dentate, rarely entire, veins branching laterally, tips acute or obtuse, surfaces glabrous to glandular-puberulent, often gland-dotted; short-petiolate or sessile. **Heads** radiate. **Peduncles** glandular-puberulent. **Involucres** turbinate, 10–15 mm. **Phyllaries** lanceolate to elliptic or linear, tips acute, surfaces pilose and stipitate-glandular. **Ray florets** 8–15; rays 10–20 mm, yellow. **Disc florets** 15–40; corollas 7–10 mm, yellow. **Fruits** columnar-fusiform, 5–6.5 mm, brown or blackish, hispidulous; pappus bristles 7–10 mm, brownish, subplumose. **2*n***=57, 76.

Open pine forests, grasslands, bogs, subalpine talus, cliffs, ridges. Flowering Jul–Sep. 1700–3000 m. BW, Casc. CA, ID, WA; north to AK, northeast to MT, southeast to UT. Native.

Arnica fulgens Pursh

Hillside arnica, orange arnica

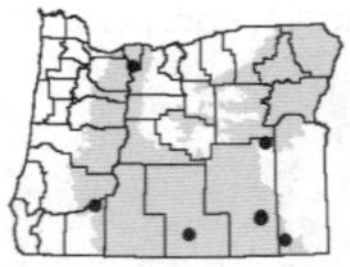

Plants 1–5 dm; rhizomatous. **Stems** simple, densely glandular-puberulent, setulose distally. **Basal leaves** well developed; petioles winged. **Cauline leaves** 1–3 pairs, much reduced distally; blades oblanceolate or lance-elliptic to linear, 2.5–8 cm, proximal bases woolly-tufted, margins entire, veins parallel from bases, tips acute, surfaces minutely stipitate-glandular, often gland-dotted; sessile. **Heads** radiate. **Peduncles** densely glandular-puberulent. **Involucres** hemispheric, 10–16 mm. **Phyllaries** elliptic-oblong to broadly lanceolate, tips acute or obtuse, surfaces densely stipitate-glandular, setulose. **Ray florets** 8–16; rays 12–30 mm, orange-yellow. **Disc florets** 15–60; corollas 7–10 mm, yellow, with glandular and non-glandular trichomes. **Fruits** columnar-fusiform, 4–7 mm, gray or blackish, densely hirsute; pappus bristles 5–7 mm, white or straw-colored, barbellate. **2*n***=38, 57.

Moist meadows, open woodlands. Flowering Jun–Aug. 1600–2800 m. BR, BW, Casc, ECas. CA, NV, ID, WA; north to British Columbia, northeast to Manitoba, east to SD, southeast to NM. Native.

The disc corolla pubescence traits differing between *Arnica fulgens* and *A. sororia* may be difficult to see except under high magnification.

Arnica gracilis Rydberg

Slender arnica

Arnica latifolia Bong. var. *gracilis* (Rydb.) Cronquist

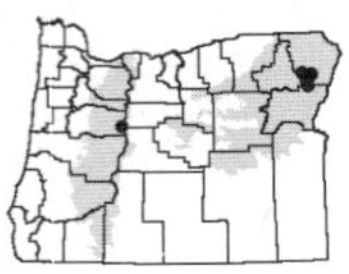

Plants 1–3 dm; rhizomatous. **Stems** usually much branched, minutely glandular-puberulent, sometimes glabrate proximally. **Basal leaves** sometimes reduced and withering early, usually in separate sterile rosettes. **Cauline leaves** 2–3 pairs; blades triangular to ovate or lanceolate, 2–6 cm, bases truncate to rounded or attenuate, margins entire to denticulate to serrate-dentate, veins branching laterally, tips acute, surfaces glandular or spiculate; usually petiolate, sometimes sessile. **Heads** radiate. **Peduncles** glabrate or minutely glandular-puberulent. **Involucres** turbinate-campanulate, 6–12 mm. **Phyllaries** elliptic-lanceolate or oblanceolate, tips acute, surfaces glabrate to strigose or glandular-puberulent. **Ray florets** 5–12; rays 10–20 mm, yellow. **Disc florets** 10–25; corollas 4–6 mm, yellow. **Fruits** columnar-fusiform, 4.5–7 mm, black, hirsute or spiculate-glandular; pappus bristles 5–7 mm, white, barbellate. **2*n***=57, 76.

Open pine forests, meadows, cliffs, rocky slopes and ridges. Flowering Jun–Aug. 1400–2800 m. BW, Casc. ID, WA; north to Northwest Territories, northeast to MT, east to WY, southeast to CO. Native.

Arnica lanceolata Nuttall ssp. *prima* (Maguire) Strother & S.J. Wolf (p. 187)

Clasping arnica, streambank arnica

Arnica amplexicaulis Nutt., *Arnica amplexicaulis* Nutt. var. *amplexicaulis*, *Arnica amplexicaulis* Nutt. var. *piperi* H. St. John & F.A. Warren

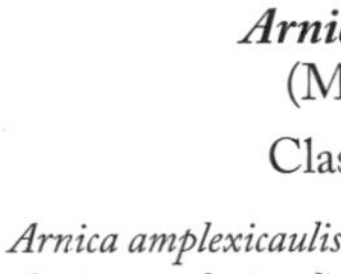

Plants 1–9 dm; rhizomatous. **Stems** simple or branched, glabrous to villous or tomentose. **Basal leaves** small, usually withering early. **Cauline leaves** 4–10 pairs; blades ovate to lanceolate, elliptic-lanceolate, or oblanceolate, 2–12 cm, mostly 2–4 × as long as wide, bases attenuate to auriculate, sometimes connate-clasping, margins denticulate, crenulate, or serrate-dentate, rarely entire, veins branching laterally, tips acute, surfaces glabrous or spiculate to villous on margins and abaxial veins; short-petiolate or sessile. **Heads** radiate. **Peduncles** pilose, glandular-puberulent. **Involucres** campanulate to hemispheric, 10–18 mm. **Phyllaries**

broadly or narrowly lanceolate to lance-elliptic or linear, surfaces stipitate-glandular, sometimes pilose, often tomentose at bases. **Ray florets** 5–17; rays 12–20 mm, yellow. **Disc florets** 30–100+; corollas 5–11 mm, yellow. **Fruits** cylindric, 4–6 mm, black, spiculate or hirsute; pappus bristles 6–10 mm, brownish, subplumose. **2*n***=38, 57, 76.

Marshes, wet rocky slopes, streambanks, alluvial shores, gravel bars, lakeshores. Flowering May–Aug. 0–2500 m. BW, Casc, CR, ECas, Est, Sisk, WV. CA, ID, NV, WA; north to AK, northeast to Alberta, east to WY, southeast to UT. Native.

The specific epithet *lanceolata* has priority over *amplexicaulis*, which has commonly been applied to this species.

Arnica latifolia Bongard

(p. 187)

Broadleaf arnica, mountain arnica

Arnica latifolia Bong. var. *latifolia*

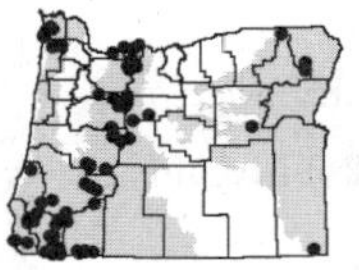

Plants 1–5 dm; strongly rhizomatous. **Stems** usually simple, sometimes branched, glabrous to puberulent or villous. **Basal leaves** withering early, usually in separate sterile rosettes; petiolate. **Cauline leaves** 2–4 pairs; blades ovate, lanceolate, or elliptic, 1.5–14 cm, bases rounded to attenuate, rarely truncate or cordate, margins entire to denticulate, dentate, or serrate, veins branching laterally, tips acute, surfaces glabrous to puberulent or strigose, adaxial surfaces sometimes densely and minutely strigillose; most sessile, lower pair sometimes subsessile. **Heads** radiate. **Peduncles** glabrate or villous, sometimes glandular-puberulent at apex. **Involucres** cylindric to campanulate or hemispheric, 8–15 mm. **Phyllaries** broadly or narrowly lanceolate to oblanceolate, elliptic, or linear, tips acute or acuminate, surfaces glabrous to pilose or villous, sometimes tomentulose or glandular-puberulent at bases. **Ray florets** 5–15; rays 10–28 mm, yellow. **Disc florets** 20–90; corollas 6–10 mm, yellow. **Fruits** columnar-fusiform, 5–9 mm, dark brown, glabrous to glandular-puberulent; pappus bristles 5–7 mm, white, barbellate. **2*n***=38, 76.

Coniferous forests, meadows, shrublands, cliffs, rocky talus, clearcuts, roadsides. Flowering May–Sep. 300–2300 m. BW, Casc, CR, ECas, Owy, Sisk. CA, ID, WA; north to AK, northeast to Alberta, east to WY, southeast to NM. Native.

Arnica longifolia D.C. Eaton

Longleaf arnica, seep-spring arnica

Arnica longifolia D.C. Eaton ssp. *myriadenia* (Piper) Maguire

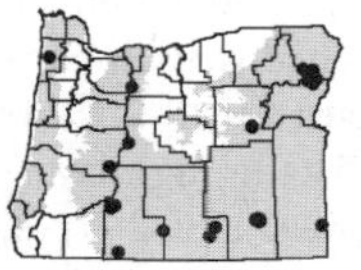

Plants 3–6+ dm; rhizomatous. **Stems** simple or branched, glandular-puberulent, sometimes glabrate proximally. **Basal leaves** small, withering early. **Cauline leaves** 5–7 pairs; blades lanceolate to narrowly lance-elliptic or linear, 3–15 cm, mostly 5–10 × as long as wide, bases truncate or attenuate, margins entire to obscurely or minutely denticulate, veins branching laterally, tips acute or acuminate, surfaces glabrous to glandular-puberulent or gland-dotted; sessile or lower short-petiolate. **Heads** radiate. **Peduncles** pilose or glandular-puberulent. **Involucres** campanulate or hemispheric, 7–10 mm. **Phyllaries** narrowly lanceolate or oblanceolate to elliptic or linear, surfaces glabrous to minutely scabrous or pilose, usually glandular-puberulent. **Ray florets** 8–13; rays 10–20 mm, yellow. **Disc florets** 15–60; corollas 7–9 mm, yellow. **Fruits** columnar, 3–7 mm, brown or black, glabrous to sparsely hispid or stipitate-glandular; pappus bristles 5–7 mm, straw-colored or brownish, barbellate or subplumose. **2*n***=57, 76.

Open coniferous forests, mixed woods, streambanks, lakesides, talus slopes, meadows. Flowering Jun–Sep. 900–3000 m. BR, BW, Casc, CR, ECas, Owy. CA, ID, NV, WA; north to British Columbia, northeast to Alberta, east to WY, southeast to CO. Native.

Arnica mollis Hooker

(p. 187)

Hairy arnica, cordilleran arnica

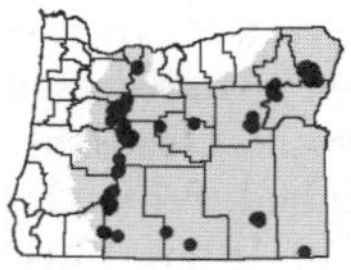

Plants 2–6 dm; rhizomatous. **Stems** simple or branched, tomentose to lightly or densely villous, usually stipitate-glandular, sometimes glabrate proximally. **Basal leaves** well developed; petiolate. **Cauline leaves** 2–4 pairs, distal pairs more developed than proximal; blades ovate to broadly or narrowly lanceolate, 3–15 cm, bases attenuate to cordate, margins entire or denticulate, veins branching laterally, tips usually acute, surfaces glabrous or villous, often glandular-puberulent; sessile or lower pair short-petiolate. **Heads** radiate. **Peduncles** glandular-puberulent, often pilose or villous. **Involucres** broadly campanulate or hemispheric, 12–14 mm. **Phyllaries** lanceolate or elliptic-lanceolate to linear, tips acute or acuminate, surfaces pilose or scabridulous, stipitate-glandular. **Ray florets** 10–22; rays 14–24 mm, yellow. **Disc florets** 20–100; corollas 7–10.5 mm, yellow. **Fruits** columnar-fusiform, 5–8 mm, brownish black, hirsute; pappus bristles 3–6 mm, brownish, plumose. **2*n***=38, 57, 76, 95, 114, 133, 152.

Moist coniferous forests, meadows, streambanks, bogs, cliffs, rocky slopes. Flowering Jun–Sep. 1000–3000 m. BR, BW, Casc, ECas, Lava, Owy. CA, ID, NV, WA; north to British Columbia, AK, northeast to Alberta, east to WY, southeast to CO. Native.

Arnica nevadensis A. Gray

Sierra arnica, Nevada arnica

Arnica tomentella Greene

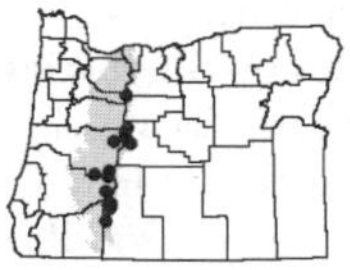

Plants 1–3 dm; rhizomatous. **Stems** usually simple, lightly or densely villous, often glandular-puberulent. **Basal leaves** usually in separate sterile rosettes; petiolate. **Cauline leaves** 1–3 pairs; blades ovate or lanceolate, 2–7 cm, bases truncate or rounded to attenuate, margins denticulate distally, entire proximally, tips acute or rounded, surfaces stipitate-glandular, scabrous to pilose on veins and margins; petiolate, sometimes distal pair sessile. **Heads** radiate. **Peduncles** pilose or villous, glandular-puberulent. **Involucres** campanulate to hemispheric, 10–17 mm. **Phyllaries** narrowly lanceolate or oblanceolate to linear, tips acute or acuminate, surfaces hispid to pilose or villous, usually glandular-puberulent. **Ray florets** 6–14; rays 15–25 mm, yellow. **Disc florets** 20–60; corollas 8–10 mm, yellow. **Fruits** ± columnar, 6–9 mm, gray or blackish, hirsute or glandular; pappus bristles 8–10 mm, white or straw-colored, barbellate or subplumose. **2*n***=76.

Open coniferous woods, rocky open slopes and ridges, cliffs. Flowering Jun–Aug. 1400–2500 m. Casc. CA, NV, WA. Native.

Arnica parryi A. Gray (p. 187)

Parry's arnica

Arnica parryi A. Gray var. *parryi, Arnica parryi* A. Gray var. *sonnei* (Greene) Cronquist

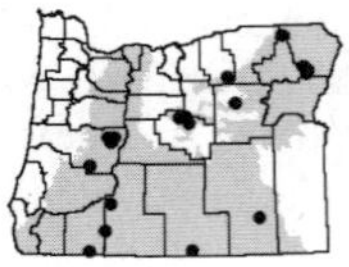

Plants 1.5–7 dm; rhizomatous. **Stems** simple to much branched, villous, stipitate-glandular distally. **Basal leaves** withered by flowering, often in separate sterile rosettes; petiolate or subsessile. **Cauline leaves** 2–4 pairs, lowermost crowded near stem bases, upper pairs much reduced; blades ovate to lanceolate or linear-elliptic, 2–10 cm, bases rounded or attenuate, margins entire or denticulate, veins branching laterally, tips acute, surfaces villous-tomentose, sparsely stipitate-glandular or gland-dotted adaxially; lowermost long-petiolate, upper sessile. **Heads** discoid or rarely obscurely radiate, nodding in bud. **Peduncles** stipitate-glandular. **Involucres** campanulate, 10–15 mm. **Phyllaries** narrowly oblanceolate to elliptic or linear, tips acute or acuminate, surfaces stipitate-glandular, sometimes also hispid. **Ray florets** 0 or obscure. **Disc florets** 20–60; corollas 8–10 mm, yellow. **Fruits** columnar-fusiform, 4–6 mm, dark brown or blackish, glabrous or hispid; pappus bristles 7–10 mm, straw-colored, barbellate. **2*n***=38, 57, 76.

Coniferous forests, meadows, ridges. Flowering Jun–Sep. 1200–2700 m. BR, BW, Casc, ECas, Sisk. CA, ID, NV, WA; north to Yukon, northeast to Alberta, east to WY, southeast to CO. Native.

Arnica rydbergii Greene

Rydberg's arnica, subalpine arnica

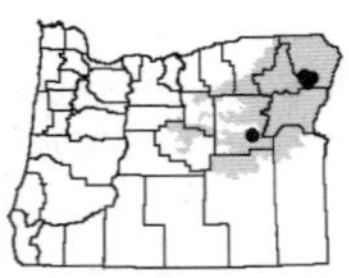

Plants 0.8–3 dm; rhizomatous. **Stems** simple, strigose or villous, usually stipitate-glandular. **Basal leaves** usually in separate sterile rosettes; petioles winged or not. **Cauline leaves** 2–4 pairs; blades lanceolate or elliptic, 2–6 cm, bases attenuate or rounded, margins entire or denticulate distally, veins parallel from bases, tips acute, surfaces sparsely to moderately pubescent; short-petiolate or sessile. **Heads** radiate. **Peduncles** stipitate-glandular, usually also villous. **Involucres** campanulate-turbinate, 7–22 mm. **Phyllaries** narrowly lanceolate to oblanceolate or elliptic, tips acute, surfaces ± pilose or tomentose at base or throughout, stipitate-glandular. **Ray florets** 6–10; rays 14–30 mm, yellow. **Disc florets** 20–60; corollas 6–9 mm, yellow. **Fruits** columnar-fusiform, 4–7 mm, brown or blackish, hirsute; pappus bristles 5–7 mm, white, barbellate. **2*n***=38, 76.

Alpine meadows, rocky slopes and ridges. Flowering Jul–Aug. 2200–2700 m. BW. ID, WA; north to British Columbia, northeast to Alberta, east to SD, southeast to CO. Native.

Arnica sororia Greene

Twin arnica, bunch arnica

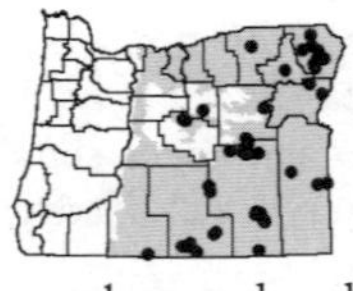

Plants 1.5–5 dm; rhizomatous. **Stems** simple or branched distally, densely glandular-puberulent throughout, sometimes lightly pilose. **Basal leaves** petioles narrow or winged. **Cauline leaves** 2–4 pairs, crowded near base, reduced distally; blades broadly or narrowly elliptic-lanceolate to linear, 3–10 cm, bases attenuate or rounded, margins entire, rarely denticulate, veins parallel from bases, tips acute, surfaces glandular-puberulent, sometimes spiculate or pilose on veins; petiolate proximally, sessile distally, petioles often winged. **Heads** radiate. **Peduncles** densely glandular-puberulent, often villous beneath heads. **Involucres** campanulate to hemispheric, 9–17 mm. **Phyllaries** lanceolate or elliptic to linear, tips acute or acuminate, surfaces stipitate-glandular or glandular-puberulent, usually pilose. **Ray florets** 9–17; rays 15–30 mm, orange-yellow. **Disc florets** 15–60; 6–10 mm, yellow, with glandular trichomes. **Fruits** tapered, 3.5–5.5 mm, black, hispid or glandular-puberulent; pappus bristles 6–9 mm, white or straw-colored, barbellate. **2*n***=38.

Coniferous forests, shrublands, grasslands, dry slopes, cliffs, roadsides. Flowering May–Jul. 300–2000 m. BR, BW, Col, ECas, Owy. CA, ID, NV, WA; north to British Columbia, northeast to Saskatchewan, east to WY, southeast to UT. Native.

Arnica spathulata Greene

Klamath arnica, spatulate arnica

Arnica spathulata Greene ssp. *eastwoodiae* (Rydb.) Maguire, *Arnica spathulata* Greene ssp. *spathulata*

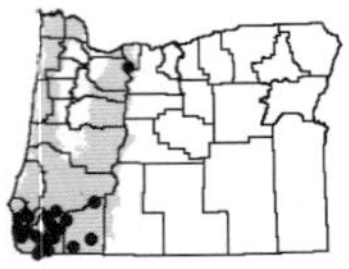

Plants 1–5 dm; rhizomatous. **Stems** simple to much branched, lightly or densely spiculate or pilose to tomentose, stipitate-glandular distally or throughout. **Basal leaves** some petioles distinctly winged. **Cauline leaves** 2–5 pairs, sometimes crowded proximally, distally reduced; blades ovate or lanceolate to linear, 1.5–13 cm, bases rounded or attenuate, margins denticulate to dentate or serrate, tips acute or obtuse, surfaces sparsely to densely villous; petiolate proximally, usually winged, sessile distally. **Heads** discoid. **Peduncles** stipitate-glandular. **Involucres** broadly to narrowly campanulate, 15–25 mm. **Phyllaries** lanceolate to oblanceolate or elliptic, tips acute, surfaces stipitate-glandular, sometimes pilose. **Ray florets** 0. **Disc florets** 15–50; corollas 8–11 mm, yellow. **Fruits** columnar or columnar-fusiform, 4–8 mm, black, gland-dotted, sometimes also hispid, rarely glabrous; pappus bristles 7–10 mm, white, barbellate. **2*n***=38, 76.

Coniferous forests, mixed woodlands, open slopes, clearcuts, on serpentine soils. Flowering May–Aug. 400–1200 m. Casc, CR, Sisk. CA. Native.

Arnica viscosa A. Gray

Shasta arnica

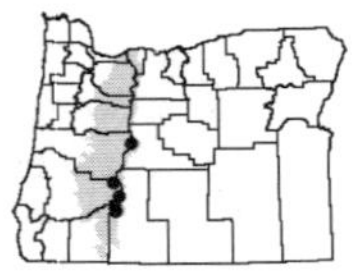

Plants 2–5 dm; rhizomatous. **Stems** simple or more often much branched, densely stipitate-glandular. **Basal leaves** usually 0. **Cauline leaves** 5–10 pairs, distally slightly reduced; blades ovate or lanceolate, 2–5 cm, bases rounded or attenuate, margins entire, rarely denticulate, tips acute, surfaces densely stipitate-glandular, sometimes pilose on abaxial veins; sessile. **Heads** discoid. **Peduncles** densely stipitate-glandular. **Involucres** campanulate, 10–20 mm. **Phyllaries** lanceolate to oblanceolate or linear, tips acute or acuminate, surfaces stipitate-glandular. **Ray florets** 0. **Disc florets** 10–30; corollas 6–10 mm, cream-colored. **Fruits** columnar-fusiform, 5–6.6 mm, gray to blackish, stipitate-glandular; pappus bristles 8–10 mm, white or straw-colored, subplumose. **2*n***=38.

Talus slopes, rock crevices. Flowering Jul–Sep. 1800–2400 m. Casc. CA. Native.

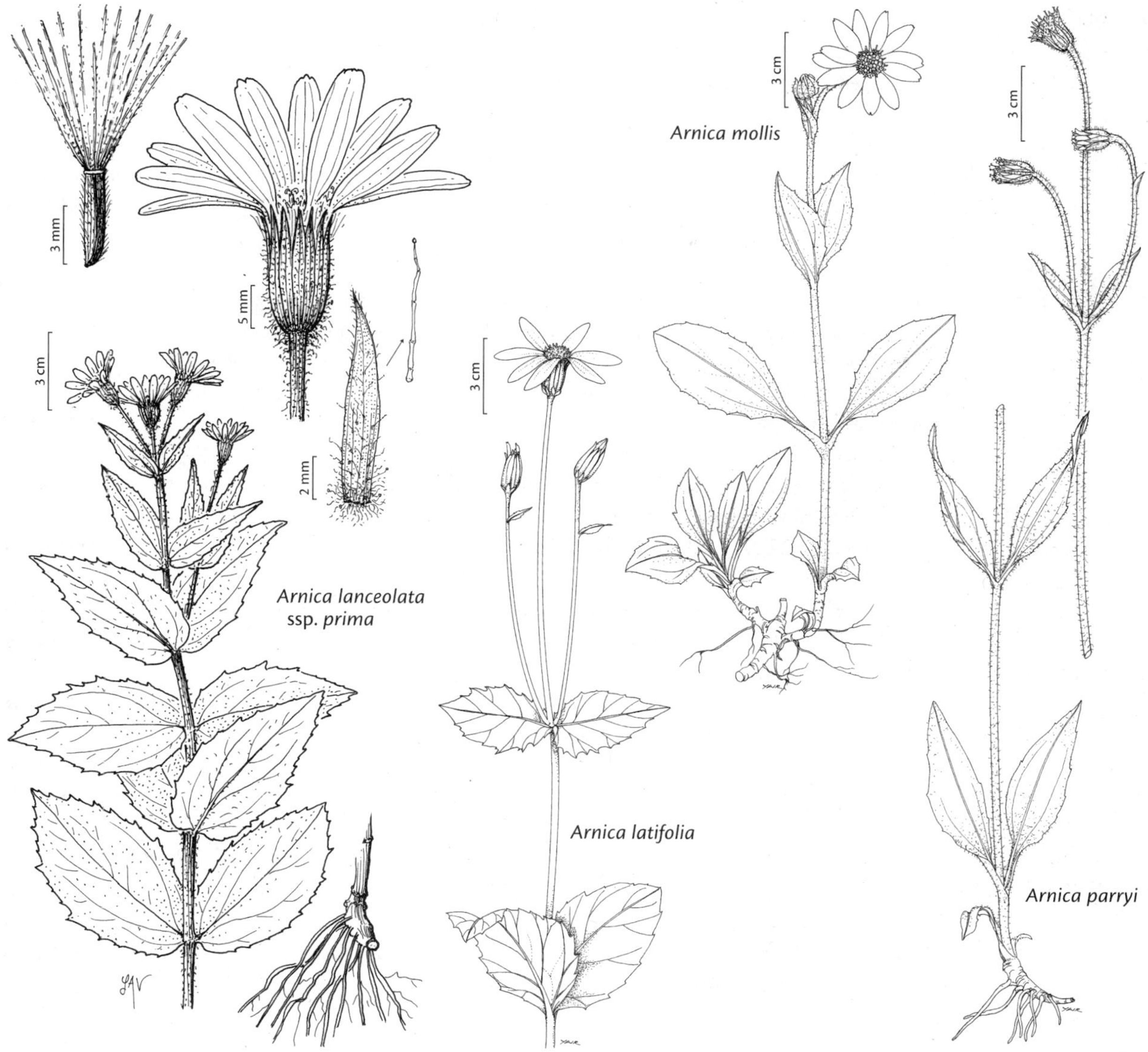

ARNICA

ARTEMISIA Sagebrush, sagewort, wormwood

Shrubs, subshrubs, or **herbs** annual, biennial, or perennial, shrubs usually evergreen, sometimes deciduous, herbs deciduous; usually aromatic. **Stems** erect or prostrate, woody or herbaceous, often much branched, glabrous to canescent or tomentose. **Leaves** alternate; blades lanceolate to ovate or wedge-shaped, margins entire, toothed, or lobed, lobes entire to serrate or with secondary lobes, surfaces glabrous to canescent or tomentose. **Inflorescences** heads usually numerous in panicle-, raceme-, or spike-like arrays or tight clusters; branches erect and close to main stem or widely spreading. **Heads** discoid or disciform. **Involucres** cylindrical to campanulate, ovoid, or nearly spherical, glabrous to canescent or tomentose, sometimes resinous-glandular. **Phyllaries** few to numerous in 2–7 unequal to equal series. **Receptacles** flat, convex, or conic, glabrous, rarely villous; paleae 0. **Disc florets** few to numerous; outer bisexual or sometimes pistillate, fertile; inner bisexual and fertile or functionally staminate; corollas tubular, cylindric proximally, widened and subglobose or funnelform distally, not much exserted from involucres, glabrous or pubescent, lobes 5, triangular; stamens 5, partly exserted from corollas; style branches slightly exserted from stamen tubes. **Fruits** fusiform, brown, 2–5-veined or veinless, glabrous or pubescent, sometimes gland-dotted; pappi 0 (minute in *A. papposa*).

Circumboreal, Africa, South America. 350–500 species; 22 species treated in *Flora*.

Artemisia tridentata and other species of sagebrush form the dominant vegetation across much of the Intermountain West, providing important shelter and food for a number of wildlife species.

Artemisia species are wind-pollinated with very small heads and florets that can be difficult to examine. The presence of terpenoids and sesquiterpene lactones gives the plants a bitter taste and strong aroma, which may protect them from herbivory.

In the past, *A. abrotanum* has been collected infrequently as a waif.

Key to *Artemisia* species:

1. Plants shrubs or subshrubs.
  2. Some lateral branches persisting and becoming spine- or thorn-like . . . . **A. spinescens**
  2' Lateral branches not becoming thorn-like.
    3. Most leaf blades entire.
      4. Leaves glabrous . . . . **A. dracunculus** (in part)
      4' Leaves canescent or tomentose. . . . . **A. cana** ssp. **bolanderi**
    3' Most leaf blades with linear or linear-oblanceolate lobes.
      5' Leaf blades deeply lobed.
        6. Inflorescences spike- or raceme-like, subtending leaves surpassing heads. . . . . **A. rigida**
        6' Inflorescences open, panicle-like, subtending leaves shorter than to equaling panicle branches. . . . . **A. tripartita** ssp. **tripartita**
      5' Leaf blades usually wedge-shaped with terminal lobes.
        7. Phyllaries widely spreading . . . . **A. papposa**
        7' Phyllaries not widely spreading.
          8. Inner phyllaries usually glabrous, resinous, shiny . . . . **A. nova**
          8' Inner phyllaries canescent, hairy or tomentose, not resinous or shiny.
            9. Florets 8–20(27). . . . . **A. spiciformis**
            9' Florets 3–8(11).
              10. Plants (0.5)1–4 dm; inflorescences spike-, raceme-, or panicle-like. . . . . **A. arbuscula**
              10' Plants 4–30 dm; inflorescences panicle-like. . . . . **A. tridentata**

1' Plants herbs.
  11. Plants annual or biennial . . . . **A. biennis**
  11' Plants perennial.
    12. Receptacles villous. . . . . **A. absinthium**
    12' Receptacles glabrous.
      13. Outer florets pistillate, inner staminate.
        14. Leaves entire, rarely some few-cleft. . . . . **A. dracunculus** (in part)
        14' Leaves usually 1–3-pinnate.
          15. Stems usually 10+ . . . . **A. pycnocephala**
          15' Stems usually 1–5 . . . . **A. campestris**
      13' Outer florets pistillate, inner bisexual.
        16. Leaves 1–5(6) cm.
          17. Leaves strongly resinous-glandular . . . . **A. packardiae**
          17' Leaves not glandular to sparsely resinous-glandular. . . . . **A. michauxiana**
        16' Leaves 2–15 cm.
          18. Plants taprooted . . . . **A. ludoviciana** (in part)
          18' Plants rhizomatous.
            19. Rhizomes short, forming emergent basal shoots that persist through winter . . . . **A. estesii**
            19' Rhizomes elongate, not forming emergent basal shoots that persist through winter.
              20. Plants 2–6(8) dm.
                21. Phyllaries densely tomentose . . . . **A. ludoviciana** (in part)
                21' Phyllaries glabrate or sparsely tomentose . . . . **A. tilesii** ssp. **unalaschcensis**
              20' Plants 5–20+ dm.
                22. Phyllaries glabrous or sparsely hairy . . . . **A. suksdorfii**
                22' Phyllaries tomentose.
                  23. Abaxial leaf surfaces tomentose . . . . **A. douglasiana**
                  23' Abaxial leaf surfaces glabrous. . . . . **A. vulgaris**

Artemisia absinthium Linnaeus

Absinthe, absinthium, wormwood

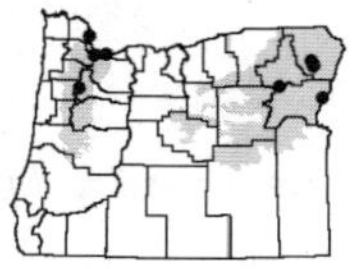

Herbs perennial, 4–12 dm; with woody caudices, fibrous-rooted. **Stems** canescent, often glabrate proximally. **Leaves** numerous, cauline, 3–8 cm, 1–3-pinnate, lobes 2–4 mm wide, toothed, surfaces finely canescent; petioles short or long. **Inflorescences** panicle-like, narrow or spreading. **Heads** disciform, arched or nodding; short-peduncled. **Involucres** broadly ovoid or hemispheric, 2–3 mm. **Phyllaries** in 3–4 ± equal series, lanceolate, tips acute or obtuse, surfaces canescent. **Receptacles** villous. **Florets** 50+, outer pistillate, inner bisexual. **Fruits** 0.4–0.6 mm, glabrous. **2*n***=18.

Coniferous woods, disturbed areas. Flowering Jun–Oct. 0–1200 m. BW, WV. ID, WA; scattered in northern North America; Europe. Exotic.

Artemisia arbuscula Nuttall

Low sagebrush, little sagebrush

Subshrubs or **shrubs** (0.5)1–4 dm. **Stems** tomentose, becoming glabrate. **Leaves** cauline, 3–10 mm; proximal usually wedge-shaped, widened distally, tips shallowly or deeply 3-toothed or -lobed, lobes ovoid to oblong, surfaces gray-canescent or tomentose; distal often linear, margins entire, tips acute or obtuse; petioles tapering, sharply cuneate or long. **Inflorescences** panicle-, raceme-, or spike-like, scarcely spreading from main axis. **Heads** discoid, erect; usually sessile, sometimes short-peduncled. **Involucres** short-cylindric to ovoid, 1.5–4(5) mm. **Phyllaries** in 4–5 unequal series, triangular to ovate or oblong, outer gradually shorter, margins membranous, surfaces canescent or finely tomentose. **Receptacles** glabrous. **Florets** 3–8(11). **Fruits** 0.7–0.8 mm, glabrous. **2*n***=18, 36.

Western United States. 3 subspecies.

Key to *Artemisia arbuscula* subspecies:

1. Lobes > ½ blade length . . . . ***A. a.*** ssp. ***thermopola***

1' Lobes ⅓–½ blade length.

  2. Leaf middle lobes often slightly > outer; involucres 1.5–2.5(3) mm in diameter . . . . ***A. a.*** ssp. ***longiloba***

  2' Leaf middle lobes ≈ outer lobes; involucres 2–4(5) mm in diameter. . ***A. a.*** ssp. ***arbuscula***

Artemisia arbuscula ssp. *arbuscula* (p. 189)

Low sagebrush, little sagebrush

Artemisia arbuscula Nutt. ssp. *longicaulis* Winward & McArthur, *Artemisia arbuscula* Nutt. var. *arbuscula*

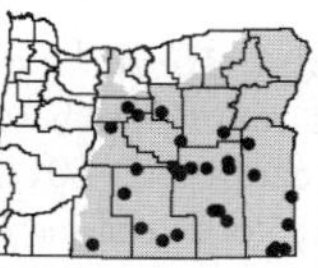

Leaves lobes ⅓–½ blade length, middle lobes ≈ outer lobes. **Involucres** 2–4(5) mm in diameter.

Shrublands, grasslands. Flowering Jun–Sep. 800–2600 m. BR, BW, ECas, Lava, Owy. CA, ID, NV, WA; northeast to MT, east to WY, southeast to NM. Native.

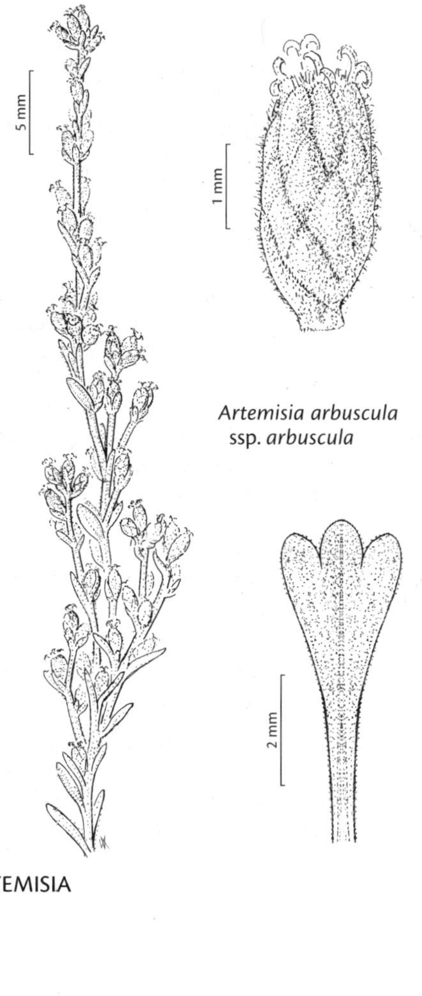

Artemisia arbuscula ssp. *arbuscula*

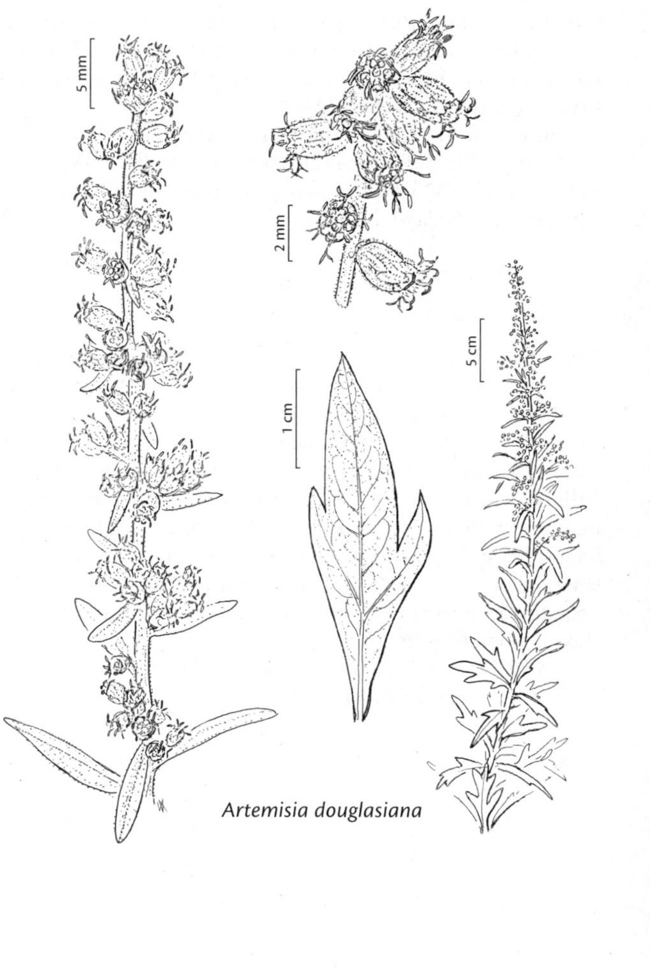

Artemisia douglasiana

Artemisia arbuscula ssp. ***longiloba*** (Osterhoout) L.M. Schultz

Early sagebrush, alkali sagebrush

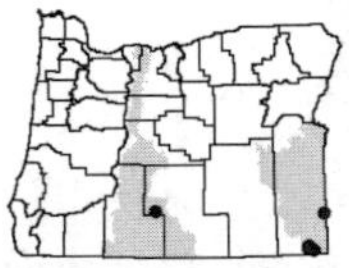

Leaves lobes ⅓–½ blade length, middle lobes often slightly > outer lobes. **Involucres** 1.5–2.5(3) mm in diameter.

Shrublands, grasslands. Flowering May–Aug. 1400–2000 m. ECas, Owy. CA, ID, NV, WA; northeast to MT, southeast to CO. Native.

Artemisia arbuscula ssp. ***thermopola*** Beetle

Cutleaf sagebrush, hot springs sagebrush

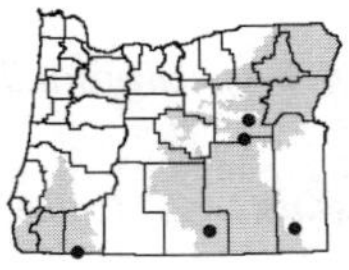

Leaves lobes > ½ blade length. **Involucres** 1.5–2.5 mm in diameter.

Openings in coniferous forests. Flowering Jun–Aug. 1400–2200 m. BR, BW, Sisk. ID; east to WY, southeast to UT. Native.

Artemisia biennis Willdenow

Biennial wormwood

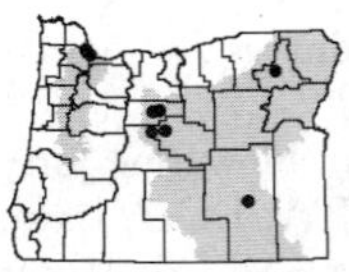

Herbs annual or biennial, 2–10+ dm; taprooted. **Stems** glabrous or minutely puberulent. **Leaves** numerous, cauline, 2–10 cm, reduced distally, 1-pinnate, lobes oblong or oblanceolate to linear, evenly or unevenly serrate, surfaces glabrous; petioles slender, sometimes with a few small leaflets. **Inflorescences** numerous tight clusters of heads. **Heads** disciform, erect; sessile or short-peduncled. **Involucres** spherical, 2–3 mm. **Phyllaries** in 2–3 equal to subequal series, ovate-elliptic, margins membranous, surfaces glabrous. **Receptacles** glabrous. **Florets** 15–50, outer pistillate, inner bisexual. **Fruits** 1–1.5 mm, glabrous. **2*n***=18.

Thickets, fields, margins of rivers and lakes, desert flats, disturbed areas. Flowering Aug–Oct. 600–1000 m. BR, BW, Lava, WV. CA, ID, NV, WA; throughout most of Canada and US; Europe, New Zealand. Native.

Artemisia campestris Linnaeus

Field sagewort

Herbs biennial or perennial, 0.5–10+ dm; taprooted. **Stems** glabrous to minutely puberulent, canescent, or tomentose. **Leaves** basal and cauline; basal forming 1 or more rosettes, ≤ 7 cm, short- or long-petiolate; cauline numerous, 2–7 cm, 2–3-pinnate, lobes linear, surfaces glabrous to canescent or tomentose. **Inflorescences** panicle-, raceme-, or spike-like. **Heads** disciform, ascending to erect; peduncled. **Involucres** spherical or hemispheric, 2.5–4 mm. **Phyllaries** in several unequal series, lanceolate to ovate, outer gradually shorter, margins membranous, tips acute or obtuse, surfaces glabrous or villous. **Receptacles** glabrous. **Florets** 15–50, outer pistillate, inner functionally staminate. **Fruits** 0.5–1 mm, glabrous. **2*n***=18, 36.

North America. 7 subtaxa; 2 subtaxa treated in *Flora.*

Key to *Artemisia campestris* subtaxa:

1. Plants 3–10+ dm; involucres 2–3 mm; inflorescences panicle-like . . . . ***A. c.*** ssp. ***pacifica***

1' Plants 0.5–3(5) dm; involucres 3–4 mm; inflorescences spike- or raceme-like . . . . ***A. c.*** var. ***wormskioldii***

Artemisia campestris ssp. ***pacifica*** (Nuttall) Hall & Clements

Pacific sagewort

Artemisia campestris L. var. *purshii* (Hook.) Cronquist, *Artemisia campestris* L. var. *scouleriana* (Besser) Cronquist

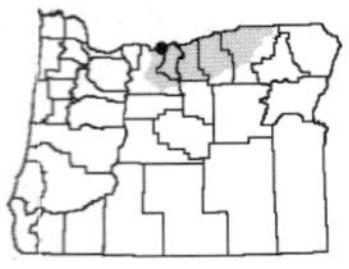

Plants 3–10+ dm. **Inflorescences** panicle-like. **Involucres** 2–3 mm.

Bluffs. Flowering Jun–Sep. 100–1700 m. Col. ID, NV, WA; north to AK, northeast to Manitoba and MT, east to NE, southeast to CO and TX. Native.

Artemisia campestris var. ***wormskioldii*** (Besser ex Hooker) Cronquist

Boreal sagewort, Columbia Islands sagewort

Artemisia borealis Pall. ssp. *borealis*, *Artemisia campestris* L. ssp. *borealis* (Pall.) H.M. Hall & Clem.

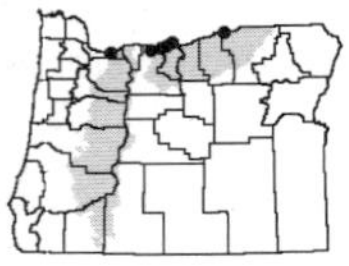

Plants 0.5–3(5) dm. **Inflorescences** spike- or raceme-like. **Involucres** 3–4 mm.

River bluffs and islands. Flowering Apr–Jun. 0–400 m. Casc, Col. WA. Native.

This taxon is found only along the Columbia River in Oregon and Washington.

Artemisia cana Pursh ssp. ***bolanderi*** (A. Gray) G.H. Ward

Silver sagebrush, Bolander's sagebrush

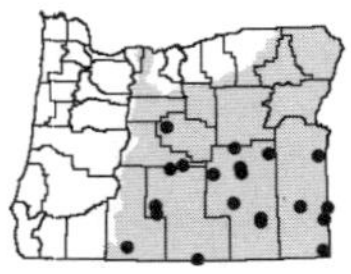

Shrubs deciduous, 2–8 dm. **Stems** canescent to tomentose, eventually glabrate. **Leaves** cauline, linear to lance-elliptic, 1–8 cm, margins entire, occasionally few leaves 3-toothed at tips, tips generally acute or acuminate, surfaces canescent or tomentose. **Inflorescences** panicle-like. **Heads** discoid, erect; short-peduncled. **Involucres** cylindric to campanulate, 3–4 mm. **Phyllaries** in several unequal series, triangular to oblong-elliptic, outer gradually shorter, margins membranous, surfaces canescent-tomentose. **Receptacles** glabrous. **Florets** 5–20. **Fruits** 1–2.5 mm, resinous-glandular. **2*n***=18, 36.

Grasslands, shrublands, mountain meadows, lakesides, riverbanks. Flowering Jun–Sep. 700–2500 m. BR, BW, ECas, Lava, Owy. CA, NV. Native.

Artemisia douglasiana Besser (p. 189)

Douglas' sagewort, Northwest mugwort

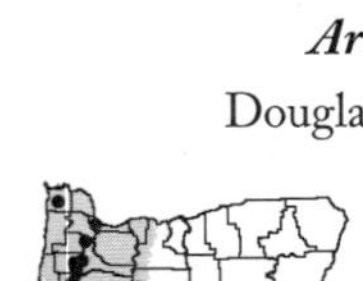

Herbs perennial, 5–20+ dm; rhizomes elongate. **Stems** canescent or tomentose. **Leaves** cauline, broadly lanceolate to elliptic or oblanceolate, 7–15 cm, gradually reduced distally, margins entire to few-toothed or lobed in distal half, surfaces abaxially densely gray-tomentose, adaxially lightly tomentose; sessile or petioles short. **Inflorescences** panicle-like. **Heads** disciform, nodding; sessile. **Involucres** campanulate or spherical, 3–4 mm. **Phyllaries** in several series, ovate or lance-elliptic, outer nearly equaling inner, margins membranous, tips obtuse, surfaces tomentose. **Receptacles** glabrous. **Florets** 15–35, outer pistillate, inner bisexual. **Fruits** 0.5–1 mm, glabrous. **2*n***=36, 54.

Woodlands, shrublands, grasslands, riverbanks, floodplains, disturbed areas. Flowering Jun–Oct. 0–2200 m. Casc, CR, ECas, Est, Sisk, WV. CA, ID, NV, WA. Native.

Some plants from eastern Oregon that formerly were assigned to *A. douglasiana* have been newly described as the diploid *A. estesii* (Chambers 2017).

Artemisia dracunculus Linnaeus (p. 191)

Tarragon, dragon sagewort

Artemisia dracunculus L. var. *dracunculus*

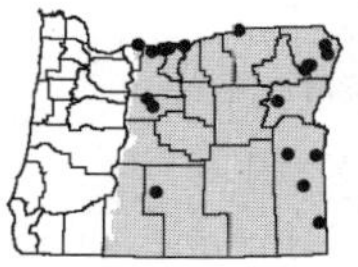

Herbs or **subshrubs** perennial, 5–15 dm; taproots woody. **Stems** glabrous or occasionally villous-puberulent. **Leaves** cauline, numerous, linear, 2–8 cm, margins entire, rarely some few-cleft, tips acute or acuminate, surfaces glabrous; sessile. **Inflorescences** panicle-like. **Heads** disciform, erect or nodding; peduncled. **Involucres** spherical or hemispheric, 2–4 mm. **Phyllaries** in several equal series, with a few short outer, lanceolate, margins membranous, tips obtuse, surfaces glabrous. **Receptacles** glabrous. **Florets** 15–45, outer pistillate, inner functionally staminate. **Fruits** 0.5–1 mm, glabrous. **2*n***=18.

Open woodlands shrublands, thickets, grasslands, riverbanks, rocky slopes and cliffs, disturbed areas. Flowering Jul–Oct. 50–1600 m. BR, BW, Col, ECas, Lava, Owy. CA, ID, NV, WA; north to AK, northeast to Quebec, east to MA, southeast to TX. Native.

Artemisia dracunculus

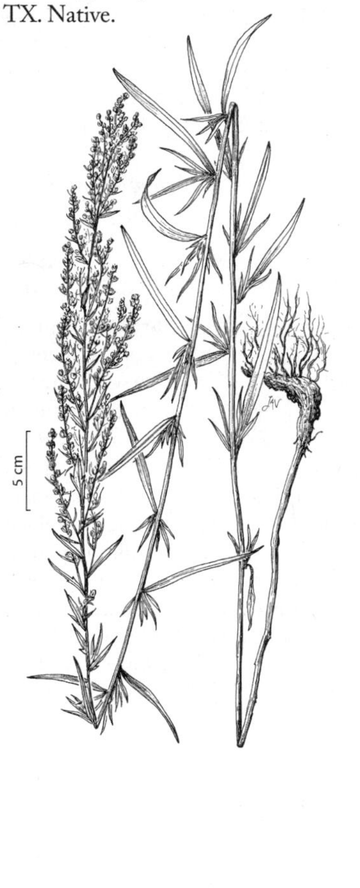

Artemisia estesii K.L. Chambers

Estes' artemisia, Estes' mugwort

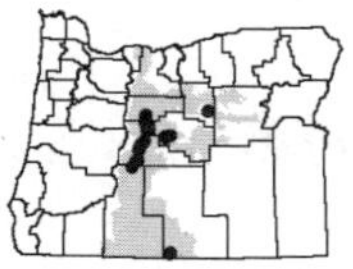

Herbs perennial, 5–15(40) dm; rhizomes short, forming emergent basal shoots that persist through winter. **Stems** densely floccose-tomentose, glabrate near base. **Leaves** cauline, 3–9 cm, regularly or irregularly 2–7-lobed or toothed, lobes entire, rarely few-toothed, surfaces abaxially densely floccose-tomentose, adaxially glabrous or lightly tomentose; sessile or petioles short. **Inflorescences** panicle-like. **Heads** disciform, erect; sessile or short-peduncled. **Involucres** campanulate, 1.5–4 mm. **Phyllaries** in 3–4 unequal series, ovate or lanceolate, outer gradually shorter, margins membranous, tips acute or obtuse, surfaces lightly tomentose. **Receptacles** glabrous. **Florets** 20–40, outer pistillate, inner bisexual. **Fruits** 0.5–1 mm, glabrous. **2*n***=18.

Marshy riverbanks. Flowering Jun–Oct. 600–1600 m. ECas, Lava. Native.

Artemisia estesii was separated from *A. douglasiana* on the basis of its diploid chromosome number, its leaf morphology, and its peculiar emergent and overwintering basal sprouts replacing lengthy underground rhizomes.

Artemisia ludoviciana Nuttall

Western wormwood, silver wormwood

Herbs perennial, 2–5(8) dm; rhizomatous or taprooted. **Stems** glabrous to canescent or tomentose. **Leaves** cauline, narrowly or broadly lanceolate to oblong or obovate, 2–11 cm, margins entire to distally toothed or pinnately lobed, lobes simple or with secondary lobes or teeth, surfaces abaxially tomentose, adaxially glabrous or tomentose; sessile or petioles short. **Inflorescences** panicle-like, rarely raceme-like. **Heads** disciform, nodding or erect; sessile or short-peduncled. **Involucres** campanulate to turbinate or hemispheric, 2–5 × 2–7 mm. **Phyllaries** in several series, lanceolate to ovate or obovate, equal or outer series gradually shorter, margins membranous, tips acute or obtuse,

Artemisia ludoviciana ssp. candicans

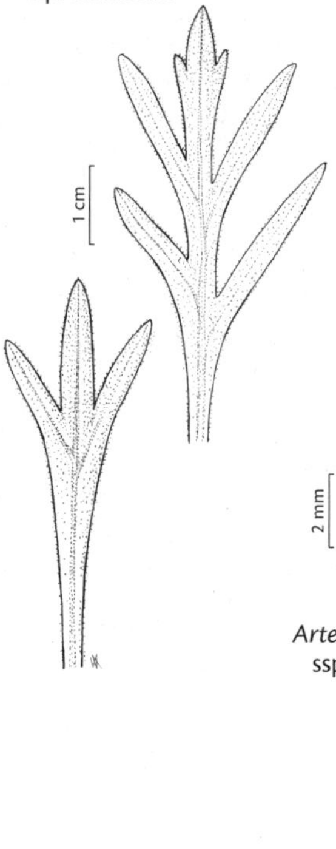

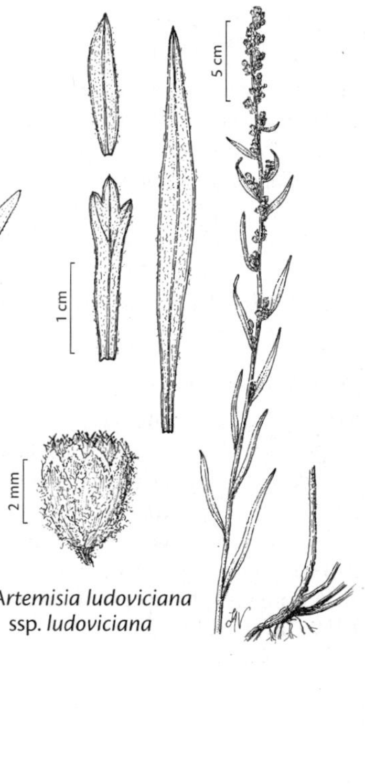

Artemisia ludoviciana ssp. ludoviciana

surfaces densely tomentose. **Receptacles** glabrous. **Florets** outer pistillate, inner bisexual. **Fruits** 0.5–1 mm, glabrous.

North America. 7 subspecies; 4 subspecies treated in *Flora*.

Key to *Artemisia ludoviciana* subspecies:

1. Plants taprooted . . . . *A. l.* ssp. ***lindleyana***

1' Plants rhizomatous.

  2. Leaves canescent to tomentose abaxially, glabrous or glabrate adaxially . . . . . *A. l.* ssp. ***incompta***

  2' Leaves canescent to tomentose.

    3. Lobes ≥ ⅓ leaf length . . . . *A. l.* ssp. ***candicans***

    3' Lobes < ⅓ leaf length . . . . . *A. l.* ssp. ***ludoviciana***

Artemisia ludoviciana ssp. ***candicans*** (Rydberg) D.D. Keck (p. 191)

Gray sagewort

Artemisia ludoviciana Nutt. var. *latiloba* Nutt.

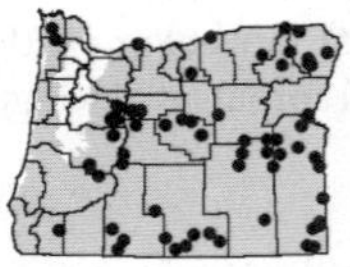

Plants with elongate rhizomes. **Stems** lightly or densely canescent to tomentose, rarely glabrate. **Leaves** lobes ≥ ⅓ leaf length, surfaces canescent to tomentose. **Inflorescences** broad panicles. **Involucres** broadly campanulate to hemispheric, 3.5–4 × 4–7 mm, surfaces tomentose. **Florets** 24–55. **2*n***=36, 54.

Coniferous woods, grasslands, streambanks, dry creek beds, rocky slopes. Flowering Jun–Sep. 50–2000 m. BR, BW, Casc, Col, CR, ECas, Lava, Owy, Sisk. CA, ID, NV, WA; north to British Columbia, northeast to Alberta, east to WY, southeast to UT. Native.

Artemisia ludoviciana ssp. ***incompta*** (Nuttall) D.D. Keck

Mountain wormwood, intermediate sagewort

Artemisia ludoviciana Nutt. var. *incompta* (Nutt.) Cronquist

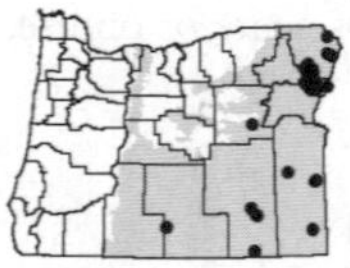

Plants with elongate rhizomes. **Stems** glabrous, less often lightly tomentose. **Leaves** lobes ≥ ⅓ leaf length, surfaces abaxially canescent to tomentose, adaxially glabrous or glabrate. **Inflorescences** narrow panicle- or raceme-like arrays. **Involucres** broadly campanulate, 3–3.5 × 2–4 mm, surfaces glabrous or tomentose. **Florets** 10–57. **2*n***=36.

Coniferous woods, shrublands, meadows, streambanks, rocky slopes, cliffs. Flowering Jun–Sep. 500–2600 m. BR, BW, ECas, Owy. CA, ID, NV, WA; north to British Columbia, northeast to Alberta, east to WY, southeast to CO. Native.

Artemisia ludoviciana ssp. ***lindleyana*** (Besser) Lesica

Columbia River mugwort

Artemisia lindleyana Besser

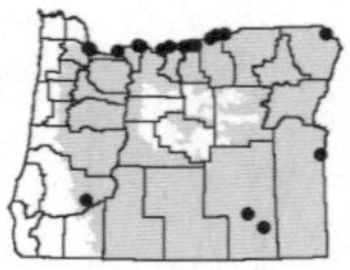

Plants taprooted. **Stems** glabrous to tomentose. **Leaves** lobes ≥ ⅓ leaf length, surfaces abaxially tomentose, adaxially glabrous or tomentose. **Inflorescences** panicle- or raceme-like. **Involucres** campanulate, 2–4 × 2–3 mm, surfaces glabrous or tomentose. **Florets** 15–40. **2*n***=36.

Riverbanks, floodplains. Flowering Jun–Oct. 0–1300 m. BR, BW, Casc, Col, ECas, Owy, WV. ID, WA, north to British Columbia. Native.

Artemisia ludoviciana ssp. *lindleyana* may occasionally hybridize with other subspecies. Leaf morphology is affected by the early leaves being lost in flood conditions, with new leaves and inflorescences being formed after the water recedes.

Artemisia ludoviciana ssp. ***ludoviciana*** (p. 191)

Western mugwort

Artemisia ludoviciana Nutt. var. *ludoviciana*

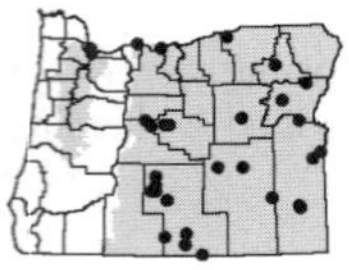

Plants with elongate rhizomes. **Stems** tomentose. **Leaves** lobes < ⅓ leaf length, surfaces tomentose. **Inflorescences** usually narrow panicle-like arrays. **Involucres** campanulate, 2–4 × 2–4 mm, surfaces tomentose. **Florets** 12–40. **2*n***=36.

Shrublands, grasslands, streambanks, disturbed areas. Flowering Jun–Sep. 0–1500 m. BR, BW, Col, ECas, Lava, Owy, WV. CA, ID, NV, WA; throughout Canada and US. Native.

Artemisia michauxiana Besser

Lemon sagewort, Michaux's mugwort

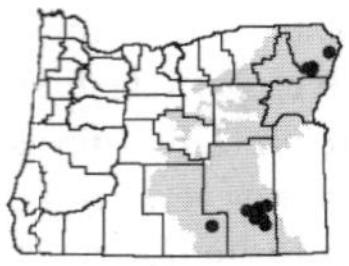

Herbs perennial, 2–7 dm; rhizomes short. **Stems** glabrous. **Leaves** cauline, often crowded, 2–4(6) cm, 1–2-pinnate, lobes linear, ultimate lobes sometimes few-toothed or lobed, surfaces abaxially glabrous or lightly tomentose, sometimes sparsely resinous-glandular; sessile or petioles short. **Inflorescences** panicle- or raceme-like. **Heads** disciform, ascending to erect; sessile or short-peduncled. **Involucres** broadly campanulate or hemispheric, 3.5–4 × 2.5–5 mm. **Phyllaries** in 2–3 equal series, deltate to ovate-elliptic, margins membranous, tips obtuse, surfaces glabrous to lightly tomentose. **Receptacles** glabrous. **Florets** 20–45, outer pistillate, inner bisexual. **Fruits** 0.5–1 mm, glabrous. **2*n***=18, 36.

Rocky slopes and benches, cliffs, subalpine meadows, streambanks. Flowering Jun–Sep. 1200–2800 m. BR, BW. CA, ID, NV, WA; north to Yukon, northeast to Saskatchewan, east to WY, southeast to CO. Native.

Artemisia nova A. Nelson

Black sagebrush

Artemisia arbuscula Nutt. ssp. *nova* (A. Nelson) G.H. Ward, *Artemisia arbuscula* Nutt. var. *nova* (A. Nelson) Cronquist

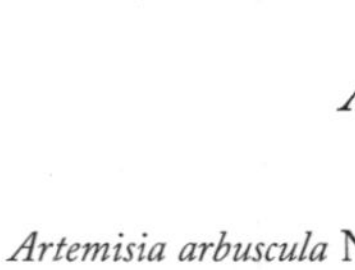

Shrubs evergreen, 1–3(5) dm. **Stems** glabrous to tomentose or glandular-puberulent. **Leaves** cauline, usually wedge-shaped, 0.2–2 cm, tips obtuse to 3-lobed, surfaces canescent to tomentose or glandular-puberulent; petioles gradually or abruptly tapered. **Inflorescences** narrow panicle-like arrays. **Heads** discoid, erect; sessile or peduncled. **Involucres** short-cylindric to ovoid, 2–3 mm. **Phyllaries** in 3–5 unequal series, ovate to elliptic, margins membranous, tips acute or obtuse, surfaces outer puberulent, inner glabrous, resinous, shiny. **Receptacles** glabrous. **Florets** 2–6. **Fruits** 1–1.5 mm, glabrous or resinous. **2*n***=18, 36.

Shrublands, flats, slopes. Flowering Jun–Sep. 1200–1500 m. BR, Owy. CA, ID, NV; northeast to MT, east to CO, southeast to NM. Native.

Artemisia packardiae J.W. Grimes & Ertter

Packard's artemisia, Packard's wormwood

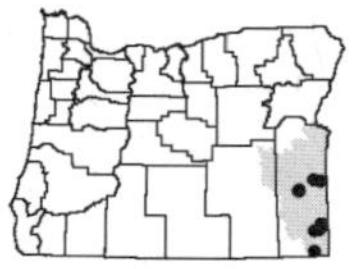

Herbs perennial, 2–6 dm; rhizomatous. **Stems** glabrous. **Leaves** cauline, 1–5 cm, 2-pinnate, lobes dentate to serrate or with secondary lobes, surfaces abaxially glabrous or rarely puberulent, strongly resinous-glandular. **Inflorescences** panicle- or raceme-like. **Heads** disciform, erect; sessile or peduncled. **Involucres** campanulate to hemispheric, 2.5–3.5 mm. **Phyllaries** in several equal series, lanceolate to ovate, tips darkly pigmented and membranous, surfaces glabrous. **Receptacles** glabrous. **Florets** 20–40+, outer pistillate, inner bisexual. **Fruits** 0.5–1.5 mm, glandular-puberulent. **2*n***=18.

Cliffs, talus, rocky riverbanks. Flowering May–Jul. 800–1800 m. Owy. ID, NV. Native.

Artemisia papposa S.F. Blake & Cronquist

Owyhee sagebrush

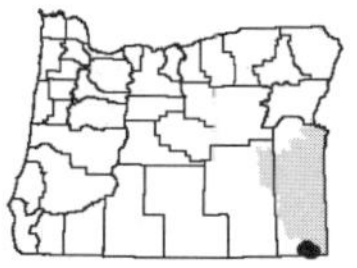

Shrubs semi-deciduous, 1.5–3.5 dm; taprooted. **Stems** canescent or tomentose. **Leaves** cauline, crowded, usually wedge-shaped, 1.5–4 cm, 3–5-lobed, lobes sometimes with secondary lobes, surfaces villous-tomentose; petiolate. **Inflorescences** raceme-like. **Heads** disciform, erect; sessile or peduncled. **Involucres** hemispheric, 2–3.5 mm. **Phyllaries** in 3–4 unequal series, widely spreading, ovate to elliptic or oblanceolate, outer shorter, margins membranous, surfaces tomentose. **Receptacles** glabrous. **Florets** 25–45, outer pistillate, inner bisexual. **Fruits** 0.3–0.5 mm, minutely glandular.

Shrublands, meadows, moist swales, edges and dry beds of streams. Flowering Jun–Jul. 1400–2000 m. Owy. ID, NV. Native.

Artemisia papposa is unique among *Artemisia* in having fruits with minute pappi.

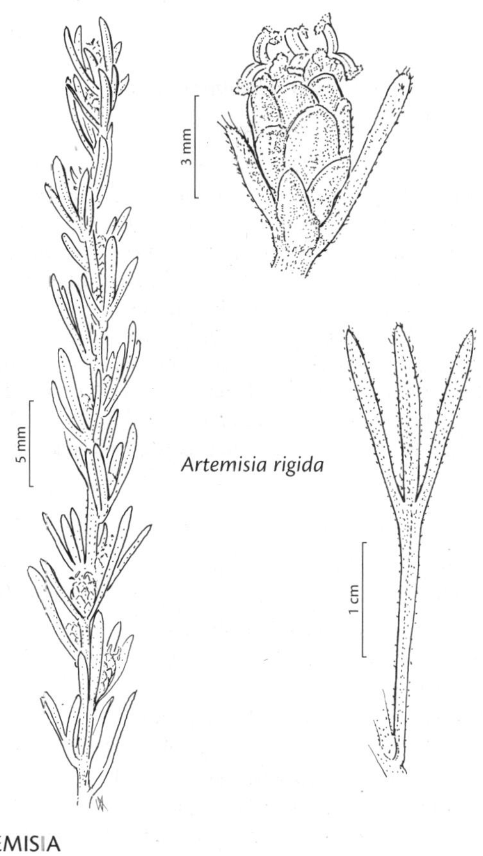

ARTEMISIA

Artemisia pycnocephala (Lessing) de Candolle

Beach wormwood, coastal sagewort

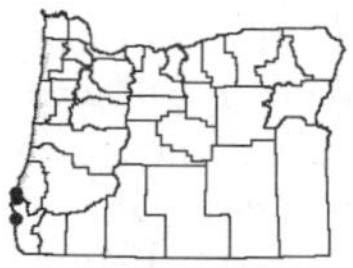

Herbs perennial, 3–7 dm; taprooted. **Stems** silky-villous. **Leaves** basal and cauline, 1–8 cm, 1–3-pinnate, surfaces silky-villous; petioles short or long. **Inflorescences** narrow panicle-like arrays. **Heads** disciform, ascending to erect; sessile or peduncled. **Involucres** broadly ovoid or hemispheric, 3–4.5 mm. **Phyllaries** in several unequal series, triangular to ovate, outer gradually shorter, surfaces densely silky-villous. **Receptacles** glabrous. **Florets** 15–45, outer pistillate, fertile, inner functionally staminate. **Fruits** 1–1.5 mm, glabrous.

Sandy beaches, dunes, coastal rocks. Flowering May–Aug. 0–200 m. Est. CA. Native.

Artemisia rigida (Nuttall) A. Gray (p. 193)

Stiff sagebrush, scabland sagebrush

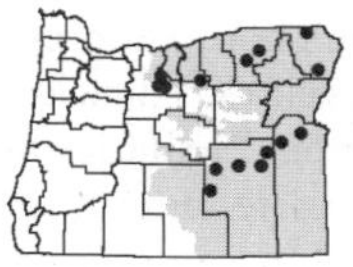

Shrubs deciduous, 2–6 dm; taprooted. **Stems** canescent or glandular-puberulent, glabrate proximally. **Leaves** cauline, 1–4 cm, deeply divided into 3–5 linear lobes, surfaces canescent or resinous-glandular; sessile or petioles present. **Inflorescences** panicle- or raceme-like. **Heads** discoid, erect; sessile. **Involucres** cylindric to narrowly campanulate, 4–6 × 2.5–3.5 mm. **Phyllaries** in 3–5 unequal series, triangular to oblanceolate, outer gradually shorter, margins membranous, surfaces canescent. **Receptacles** glabrous. **Florets** 4–8. **Fruits** 1–1.5 mm, glabrous. **2*n***=18, 36.

Shrublands, open coniferous woods. Flowering Jun–Oct. 600–1800 m. BR, BW, Col, Owy. ID, WA; northeast to MT. Native.

Artemisia spiciformis Osterhout

Snowfield sagebrush

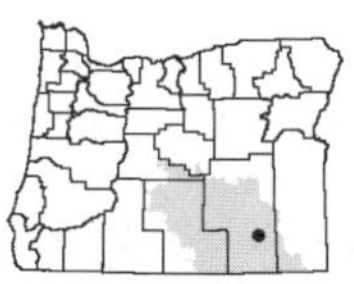

Shrubs ± deciduous, 3–8 dm; taprooted. **Stems** canescent or silky-tomentose. **Leaves** cauline, 1–5 cm; distal wedge-shaped, tips usually 3–6-toothed, surfaces canescent or tomentose; proximal linear to oblanceolate, often lobed; petioles short, tapered. **Inflorescences** panicle-, raceme-, or spike-like. **Heads** discoid, erect; sessile or peduncled. **Involucres** broadly to narrowly campanulate, 2.5–4 mm. **Phyllaries** in 2–5 unequal series, triangular to ovate or elliptic, outer gradually shorter, margins membranous, surfaces sparsely to densely hairy. **Receptacles** glabrous. **Florets** 8–20(27). **Fruits** 1–1.5 mm, glabrous or resinous. **2*n***=18, 36.

Woodlands, meadows, streambanks, open slopes. Flowering Jun–Oct. 2000–3100 m. BR. CA, ID, NV, WA; northeast to MT, east to WY, southeast to CO. Native.

Artemisia spinescens D.C. Eaton (p. 194)

Spring sagebrush, bud sagebrush

Picrothamnus desertorum Nutt.

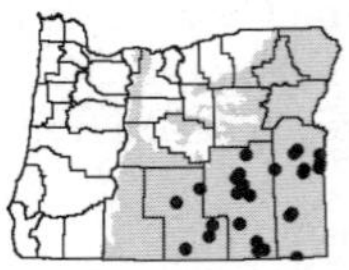

Subshrubs or **shrubs** deciduous, 0.5–5 dm; taprooted. **Stems** some becoming thorn-like, surfaces villous-tomentose. **Leaves** cauline, 0.5–2 cm, divided and cleft into narrow segments, surfaces canescent or villous-tomentose; subsessile or petioles present and expanded at base. **Inflorescences** panicle-like. **Heads** disciform, nodding to ascending; sessile or peduncled. **Involucres** broadly obconic or hemispheric, 2–5 mm. **Phyllaries** in 2–3 equal series, obovate to broadly or narrowly elliptic, surfaces lightly tomentose. **Receptacles** glabrous. **Florets** 7–20, outer pistillate, inner functionally staminate. **Fruits** 1–1.5 mm, villous-pubescent. **2*n***=18, 36.

Shrublands, open slopes, riverbanks, alkaline flats. Flowering Apr–Jul. 400–1500 m. BR, BW, ECas, Owy. CA, ID, NV; northeast to MT, east to CO, southeast to NM. Native.

Artemisia suksdorfii Piper

Suksdorf's sagewort, coastal mugwort, coastal wormwood

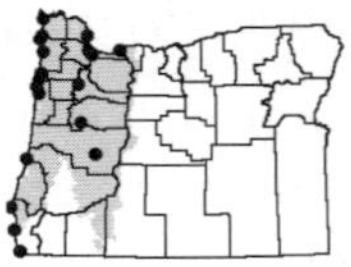

Herbs perennial, 5–15 dm; rhizomes elongate. **Stems** glabrous or gray-floccose. **Leaves** cauline, broadly or narrowly lanceolate to elliptic, 2–15 cm, margins entire or irregularly toothed or lobed above middle, tips acute, surfaces abaxially densely canescent-tomentose, adaxially green and glabrous; sessile or petioles short. **Inflorescences** panicle-like. **Heads** disciform, erect; sessile or peduncled. **Involucres** cylindric to narrowly ovoid, 3–4 mm. **Phyllaries** in 3–5 unequal series, triangular to ovate or lanceolate, outer gradually shorter, tips acute or obtuse, surfaces glabrous or sparsely hairy. **Receptacles** glabrous. **Florets** 4–12, outer pistillate, inner bisexual. **Fruits** 0.8–1.5 mm, glabrous. **2*n***=18.

Beaches, marshes, coastal headlands, riverbanks, woodlands, disturbed areas. Flowering Jun–Sep. 0–300 m. Casc, CR, Est, WV. CA, WA; north to British Columbia. Native.

Artemisia tilesii Ledebour ssp. *unalaschcensis* (Besser) Hultén

Aleutian mugwort, Cascade wormwood, Tilesius' wormwood

Artemisia tilesii Ledeb. var. *unalaschcensis* Besser

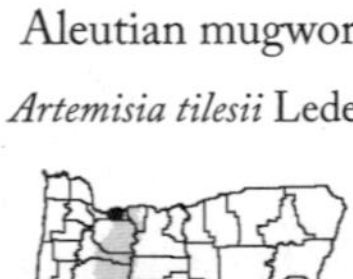

Herbs perennial, 2–6(8) dm; rhizomes elongate. **Stems** canescent to tomentose, becoming glabrate. **Leaves** cauline, linear to lanceolate, 3–10 cm, margins entire to serrate or 2–6-lobed, tips acute or acuminate, surfaces densely tomentose abaxially, green and glabrous adaxially; sessile or petioles short. **Inflorescences** panicle-like. **Heads** disciform, nodding to ascending; sessile or peduncled. **Involucres** narrowly to broadly campanulate or hemispheric, 3–5 mm. **Phyllaries** in 3–4 unequal series, triangular to ovate, outer gradually shorter, margins membranous, tips acute or obtuse, surfaces glabrate or sparsely tomentose. **Receptacles** glabrous. **Florets** 28–50+, outer pistillate, inner bisexual. **Fruits** 1–1.5 mm, glabrous.

Riverbanks, disturbed areas. Flowering Jun–Oct. 0–2000 m. Casc. ID; north to AK, northeast to MT. Native.

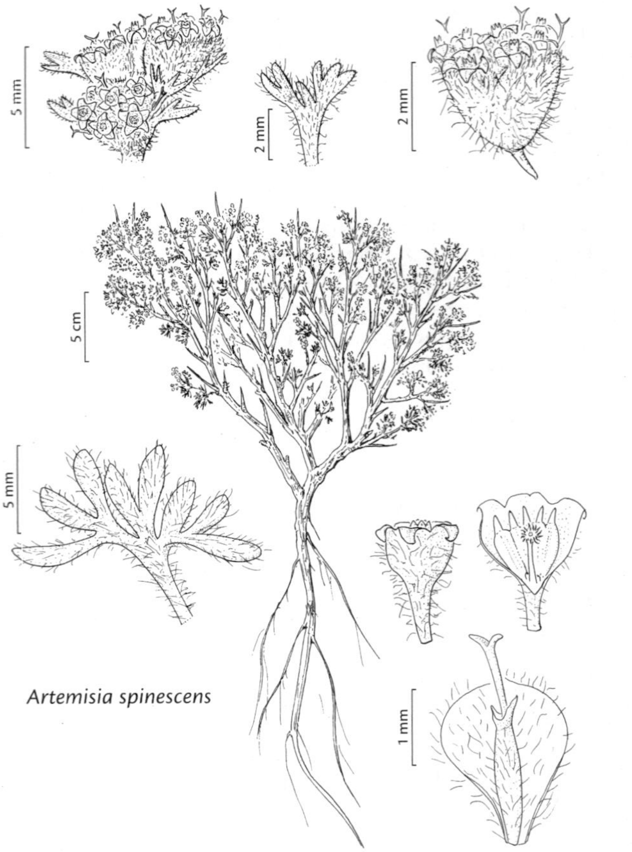

ARTEMISIA

Artemisia tridentata Nuttall

Big sagebrush

Shrubs evergreen, 2–30 dm, crowns rounded or flat-topped; taprooted. **Stems** canescent or tomentose. **Leaves** cauline, wedge-shaped, 0.5–3.5 cm, tips shallowly 3-lobed, surfaces densely gray-canescent or tomentose. **Inflorescences** panicle-like. **Heads** discoid, erect; sessile or peduncled. **Involucres** cylindric to campanulate, 1–4 mm. **Phyllaries** in several unequal series, triangular to lanceolate or elliptic, outer gradually shorter, margins membranous, tips acute or obtuse, surfaces tomentose. **Receptacles** glabrous. **Florets** 3–8(11), bisexual. **Fruits** 1–2 mm, pubescent or glabrous. **2*n***=18, 36.

Western North America. 4 subspecies; 3 subspecies treated in *Flora*.

Key to *Artemisia tridentata* subspecies:

1. Leaves > 3 × as long as wide; plants 10–30 dm; florets 3–6 . . . . *A. t.* ssp. ***tridentata***

1' Leaves < 3 × as long as wide; plants 3–8(15) dm, florets 4–8(11).

  2. Leaf blades slenderly wedge-shaped; involucres 2–3 mm; plants 4–8(15) dm, crowns flat-topped . . . *A. t.* ssp. ***vaseyana***

  2' Leaf blades broadly wedge-shaped; involucres 1–2 mm; plants 3–5(15) dm, crowns rounded . *A. t.* ssp. ***wyomingensis***

Artemisia tridentata ssp. *tridentata* (p. 195)

Big sagebrush

Artemisia tridentata Nutt. var. *tridentata*

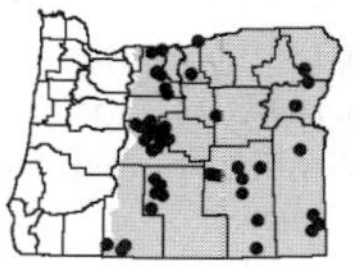

Plants 10–30 dm, crowns rounded. **Leaves** blades slenderly wedge-shaped, evenly tapered, > 3 × as long as wide. **Inflorescences** broad panicle-like arrays, (1.5)5–6 cm wide; lower branches usually long and spreading. **Involucres** 1.5–2.5 mm. **Florets** 3–6.

Valley bottoms, montane slopes. Flowering Jun–Oct. 50–2100 m. BR, BW, Col, ECas, Lava, Owy. CA, ID, NV, WA; north to British Columbia, northeast to Alberta, east to WY, southeast to NM. Native.

Artemisia tridentata ssp. *vaseyana* (Rydberg) Beetle (p. 195)

Mountain big sagebrush, Vasey sagebrush

Artemisia tridentata Nutt. var. *vaseyana* (Rydb.) B. Boivin

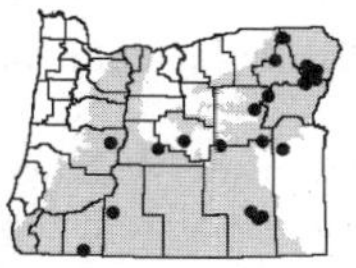

Plants 6–8(12) dm, crowns flat-topped. **Leaves** blades slenderly wedge-shaped, evenly tapered, < 3 × as long as wide. **Inflorescences** narrow panicle-like arrays, 2–4 cm wide; branches usually ascending. **Involucres** 2–3 mm. **Florets** 4–8(11).

Moist mountain meadows. Flowering Jun–Sep. 1200–2500 m. BR, BW, Casc, ECas, Sisk. CA, ID, NV, WA; north to British Columbia, northeast to Alberta, east to NE, southeast to NM. Native.

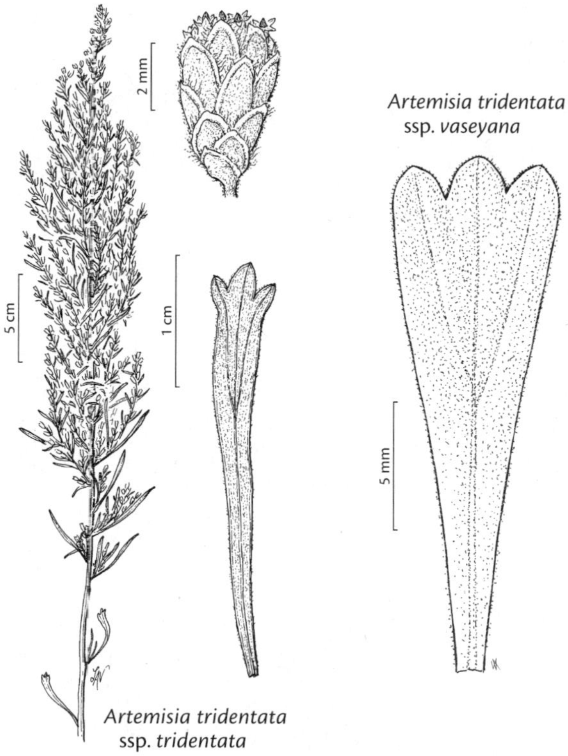

ARTEMISIA

Artemisia tridentata ssp. *wyomingensis* Beetle & A.M. Young

Wyoming sagebrush

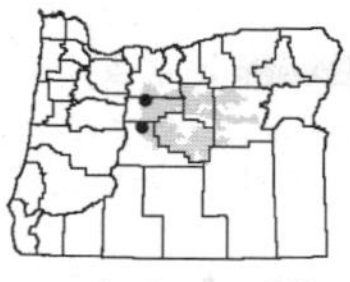

Plants 3–5(15) dm, crowns rounded. **Leaves** blades broadly wedge-shaped, tapered abruptly, < 3 × as long as wide. **Inflorescences** narrow panicle-like arrays, 1–3 cm wide; branches usually ascending. **Involucres** 1–2 mm. **Florets** 4–8.

Shrublands. Flowering Jun–Oct. 800–1900 m. Lava. CA, ID, NV; northeast to ND, east to NE, southeast to NM. Native.

Artemisia tripartita Rydberg ssp. *tripartita*

Three-tip sagebrush, cutleaf sagebrush

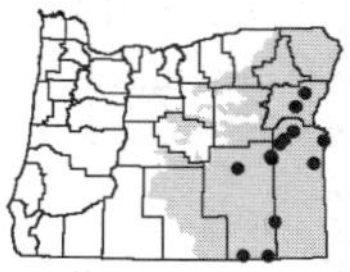

Shrubs deciduous, 2–15+ dm. **Stems** with numerous vegetative branches; flowering stems woody, simple or branched, canescent. **Leaves** cauline, wedge-shaped to linear, deeply divided into 3–5 linear lobes, surfaces canescent. **Inflorescences** spike- or panicle-like. **Heads** discoid, erect; sessile or pedunculate. **Involucres** broadly or narrowly campanulate, 2–4 mm. **Phyllaries** in 3–4 unequal series, triangular to lanceolate or oblong, outer gradually shorter, margins membranous, tips obtuse, surfaces canescent or innermost glabrate. **Receptacles** glabrous. **Florets** 3–11, bisexual. **Fruits** 1.5–2.5 mm, glabrous, resinous. **2*n***=18, 36.

Shrublands. Flowering Jun–Oct. 600–1600 m. BR, BW, Owy. ID, NV, WA; north to British Columbia, northeast to MT, east to WY, southeast to CO. Native.

Artemisia vulgaris Linnaeus

Common mugwort, lobed wormwood

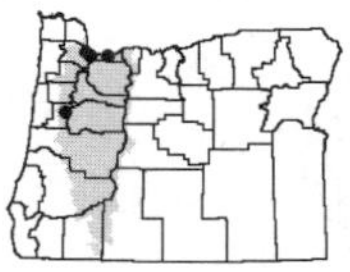

Herbs perennial, 5–15 dm; rhizomes elongate. **Stems** glabrous or tomentose. **Leaves** cauline, 2–12 cm, regularly or irregularly pinnately lobed, lobes toothed or again lobed, tips acute, surfaces abaxially glabrous, adaxially glabrous or tomentose; sessile or petiolate. **Inflorescences** panicle-like. **Heads** disciform, erect; sessile or peduncled. **Involucres** cylindric or campanulate, 3–4 mm. **Phyllaries** in several equal or unequal series, lanceolate or oblong, outer gradually shorter or sometimes exceeding inner, margins membranous, tips acute or obtuse, surfaces tomentose. **Receptacles** glabrous. **Florets** 20–30, outer pistillate, inner bisexual. **Fruits** 0.5–1.2 mm, glabrous, sometimes resinous. **2*n***=18, 36, 40, 44.

Disturbed areas. Flowering Jun–Oct. 0–2000 m. Casc, WV. ID, WA; widely scattered in North America; Asia, Europe. Exotic.

ASKELLIA Hawksbeard

Herbs perennial, usually cespitose, compact, sap milky. **Stems** 1–few, arising from caudices or terminating slender rhizomes, simple or with numerous branches from near base. **Leaves** basal and cauline; usually slender-petioled. **Inflorescences** panicle-like, often compact, sometimes well exserted from rosettes, abundantly branched within leaf clusters, forming a low, dense mass of heads. **Heads** ligulate. **Involucres** cylindrical. **Phyllaries** in 2 series; outer basal, loosely spreading at maturity; inner linear-lanceolate, tips acute. **Receptacles** flat; paleae 0. **Florets** bisexual; corollas yellow, purplish abaxially; stamens 5. **Fruits** narrowly cylindrical, tapering slightly toward base and apex, not beaked, glabrous, ribs 10(13), smooth or lightly roughened distally; pappi of numerous hair-like bristles, deciduous, free.

Asia, North America. ~7 species; 1 species treated in *Flora.*

In E. B. Babcock's monograph of *Crepis* (1947), a group of seven species was placed in section *Ixeridopsis*, one of them ranging from Central Asia to North America, as far east as Labrador and Newfoundland, one in the northern Rocky Mountains region, and five species centered in the mountains of south-central Asia. This group was segregated from *Crepis* by Weber (1984) and given the name *Askellia*. However, in *Flora of North America* (Bogler 2006), *Askellia* is not mentioned, and the name *Crepis nana* is used for Oregon's representative species. In transferring *C. nana* to *Askellia nana*, Weber overlooked an earlier legitimate epithet based on *Prenanthes pygmaea* (type from Siberia), making the new combination *Askellia pygmaea* necessary (Chambers & Meyers 2011). The separation of *Askellia* from *Crepis* is justified by DNA evidence (Enke & Gemeinholzer 2008) as well as morphology and chromosome number. Its base number of x=7 is not known in any species of *Crepis*, for example. The remaining native American species of *Crepis* are based on the polyploid chromosome number n=11.

Askellia pygmaea (Ledebour) K.L. Chambers & S.C. Meyers ssp. ***ramosa*** (Babcock) K.L. Chambers & S.C. Meyers (p. 196)

Dwarf hawksbeard

Crepis nana Richardson ssp. *ramosa* Babc., *Crepis nana* Richardson var. *ramosa* (Babc.) Cronquist

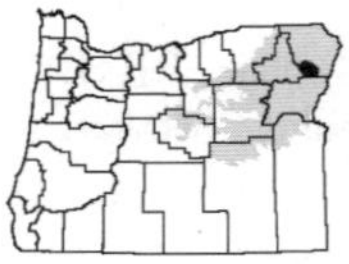

Plants 3–15 cm, glabrous; spreading by rhizomes or adventitious buds on taproot. **Leaves** 1.5–6 cm, reduced distally, blades elliptic-ovate or oblanceolate, margins entire, surfaces glabrous. **Inflorescences** heads usually clustered among basal leaves. **Involucres** (7)9–13 mm, glabrous. **Phyllaries** outer basal, deltate; inner 6–10, equal. **Florets** 8–12, ligules barely exceeding involucres. **Fruits** 4–7 mm, tan, ribs smooth; pappi white. **2*n***=14.

Rocky cliffs, talus, gravelly slopes. Flowering May–Sep. 2100–3000 m. BW. CA, ID, WA; north to British Columbia. Native.

This description is based on specimens from the Wallowa Mountains, Oregon. The type of subspecies *ramosa*, from the Olympic Mountains, Washington, is a taller plant with a more open inflorescence and often-toothed leaves. The usual habitat of this subspecies is talus slopes, where its rhizomes are an adaptation to the unstable substrate. It is rare in the Washington Cascade Range (Chelan County), absent from the Oregon Cascades, but well known in the high Sierra Nevada of California to as far south as Los Angeles County. Subspecies *pygmaea*, which is taprooted and not rhizomatous, is found in arctic tundra and on gravel riverbanks and outwash fans, not on mountainside talus.

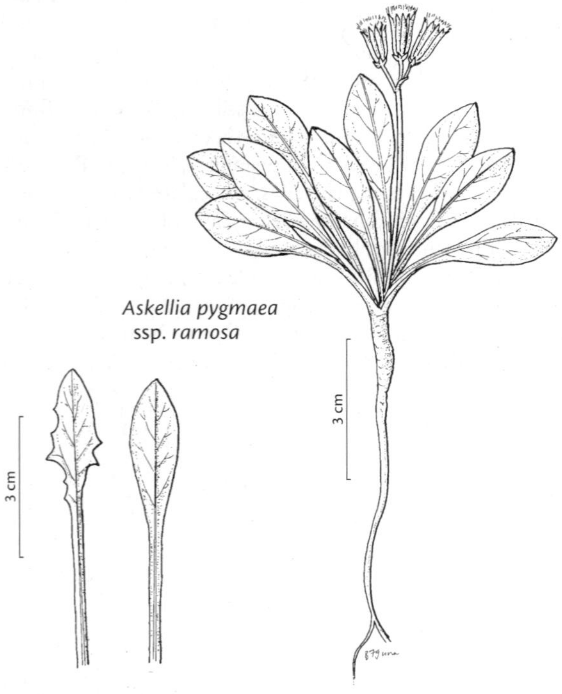

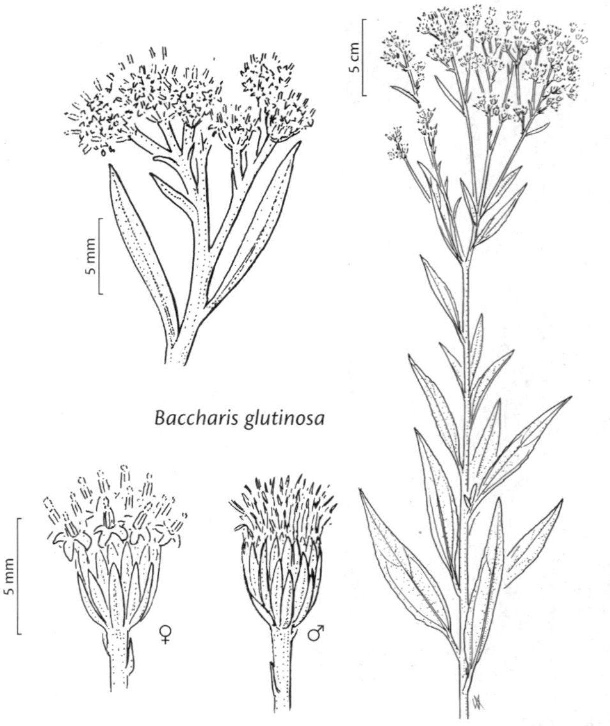

BACCHARIS Baccharis, false willow

Scott D. Sundberg

Trees, shrubs, subshrubs, herbs perennial, dioecious, bases woody; sometimes rhizomatous. **Stems** usually erect or ascending, rarely prostrate, usually striate-angled, rarely terete and smooth, glabrous, glabrate, hispidulous, or villous, often resinous. **Leaves** cauline, alternate; blades usually reduced distally, margins entire or coarsely serrate, 1- or 3-veined, surfaces usually glabrous, rarely hispidulous or villous, often gland-dotted and resinous; sessile or petiolate. **Inflorescences** usually panicle-like or corymb-like arrays. **Heads** unisexual, discoid. **Involucres** campanulate or hemispheric. **Phyllaries** 20–40 in 2–5 series, ovate to lanceolate, unequal, margins usually scarious, often erose or ciliate, 1-veined, tips obtuse to acute or acuminate, surfaces usually glabrous, rarely hispid. **Receptacles** flat, dome-shaped, or conic, pitted or smooth; paleae usually 0. **Staminate florets** 10–50; corollas white to pale yellow, lobes 5, spreading-reflexed, deltate to lance-ovate, tubes about equal to narrowly funnelform throats. **Pistillate florets** 20–150; corollas filiform-tubular, whitish, lobes 5, spreading-reflexed, ± deltate to lance-ovate; style branches glabrate, flattened, appendages lacking. **Fruits** obovoid to cylindric, ± compressed, light brown, 5–10-veined, glabrous or hispid; pappi persistent or falling, of 25–50 whitish to tawny, rarely brownish, minutely barbellate, apically attenuate bristles in 1–3 series. **2*n***=18.

Mostly New World tropics and warm-temperate regions. 350–450 species; 2 species treated in *Flora*.

Baccharis comprises mostly dioecious shrubs with gland-dotted leaves, and it is sister to the polygamo-dioecious genera *Archibaccharis* and *Heterothalamus*. Species of *Baccharis* often occur in wet sites such as streambanks, washes, ditches, and marshes.

Baccharis is of little economic importance, except for species that are invasive or toxic to livestock (Boldt 1989). *Baccharis* is not generally palatable to cattle; it could become the only forage available as a result of overgrazing.

Key to *Baccharis* species:

1. Leaves lanceolate, 50–130 mm . . . . *B. glutinosa*
1' Leaves oblanceolate to obovate, 15–40 mm . . . . *B. pilularis* ssp. *consanguinea*

Baccharis glutinosa Persoon (p. 196)

Saltmarsh baccharis

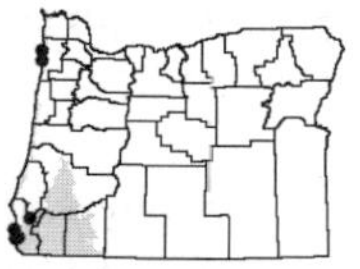

Herbs deciduous. 6–21 dm, forming colonies; rhizomatous. **Stems** erect to ascending, striate, glabrous, gland-dotted, resinous. **Leaves** blades lanceolate, 50–130 × 8–30 mm, bases tapered to petioles, margins entire or finely dentate, 1- or 3-veined, larger prominently 3-veined, tips acute to acuminate, surfaces glabrous, black gland-dotted; short-petiolate. **Inflorescences** dense, terminal, flat-topped, corymb-like arrays. **Involucres** hemispheric; of staminate heads 3.5–5 mm, of pistillate heads 3.8–4.8(6) mm. **Phyllaries** narrowly lanceolate, 2–4 mm, inner series ~ equal, margins yellowish, medians green to purplish, tips acuminate. **Staminate florets** 26–40; corollas 3.5–4 mm. **Pistillate florets** 80–150; corollas 1.7–3 mm. **Fruits** 0.6–1.5 mm, 5-veined, glandular, hispidulous distally; pappi 2.6–4(7) mm.

Moist salt marshes, coastal strands, stream edges, hillsides, railroads. Flowering Jul–Oct. 0–200 m. Est, Sisk. CA; south to Mexico. Native.

The name *B. douglasii* has previously been misapplied to this taxon. *Baccharis douglasii* occurs only in South America.

Baccharis pilularis de Candolle (p. 197) ssp. *consanguinea* (de Candolle) C.B. Wolf

Chaparral broom, coyote brush, dwarf chaparral false willow

Baccharis pilularis DC. var. *consanguinea* (DC.) Kuntze

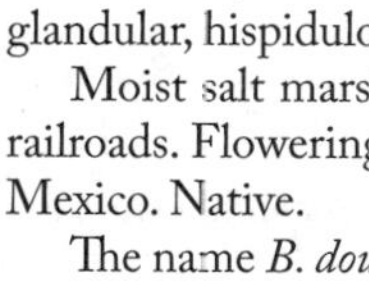

Shrubs evergreen, 1.5–45 dm, forming erect or rounded shrubs. **Stems** erect, rarely prostrate, striate-angular, dark brown, shiny, glabrous, often scurfy, usually resinous and sticky. **Leaves** blades oblanceolate to obovate, 15–40 × 2–15 mm, bases cuneate, margins entire or coarsely dentate, 1- or 3-veined, surfaces glabrous, gland-dotted, resinous; sessile or short-petiolate. **Inflorescences** panicle-like arrays. **Involucres** hemispheric to campanulate; of staminate heads 3.2–5 mm; of pistillate heads 3–6 mm. **Phyllaries** ovate to lanceolate, 1–3 mm, margins yellowish, scarious, medians yellow proximally, green distally, tips obtuse to acute or acuminate. **Staminate florets** 20–34; corollas 3–4 mm. **Pistillate florets** 19–43; corollas 2.5–3.5 mm. **Fruits** 1–2 mm, 8–10-veined, glabrous; pappi 6–9 mm.

Oak woodlands, coniferous forest openings, coastal bluffs, roadsides. Flowering Jun–Dec. 0–600 m. CR, Est, Sisk, WV. CA; south to Mexico. Native.

Baccharis pilularis can be distinguished by its small, obovate to oblanceolate leaves. Ours is the common, widespread subspecies *consanguinea*, which is typically erect. Subspecies *pilularis* is known only from exposed sandy dunes and bluffs along the central coast of California. Its growth habit is mat-like.

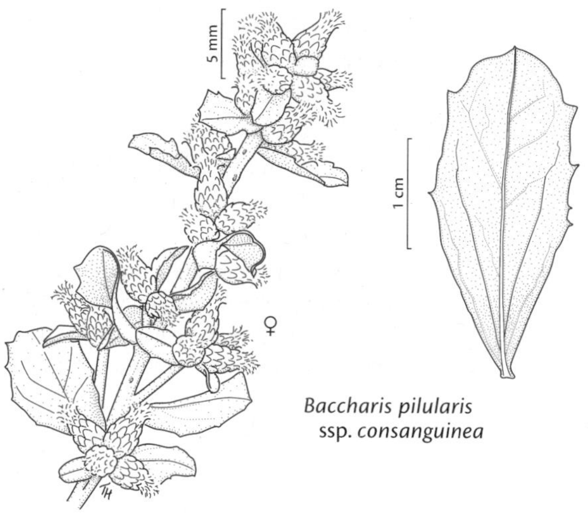

Baccharis pilularis ssp. *consanguinea*

BACCHARIS

BALSAMORHIZA Balsamroot

Abigail J. Moore

Herbs perennial; with thick, resin-producing taproots. **Stems** erect, scapose, unbranched. **Leaves** mainly basal, simple or pinnately compound, lanceolate to ovate; cauline opposite, sometimes 1 additional pair of reduced cauline leaves alternate. **Inflorescences** with 1 main head, occasionally with up to 6 axillary heads per stem. **Heads** radiate. **Involucres** broadly campanulate to hemispheric. **Phyllaries** in 2–4 series, outer equaling or surpassing inner. **Receptacles** flat to convex; paleate. **Paleae** lanceolate, acute, folded around fruits. **Ray florets** pistillate; corollas yellow or rarely pink to red. **Disc florets** bisexual; corollas yellow, 5-lobed. **Fruits** 4-sided, black, glabrous or strigillose; pappi 0.

Western Canada and United States. 12 species; 10 species treated in *Flora*.

Balsamorhiza is most closely related to *Wyethia*, from which it is distinguished by being scapose. *Balsamorhiza* is divided into two sections, both of which occur in Oregon. Section *Artorhiza* has cordate or sagittate, simple leaves; a taproot bearing multiple crowns; alternate stem leaves above the pair of opposite leaves; and often multiple heads per stem. Section *Balsamorhiza* has generally pinnately divided leaves (often simple in *B. serrata*); taproots bearing one or at most two crowns; one pair of opposite stem leaves; and a solitary head per stem.

All species of Oregon *Balsamorhiza* have a chromosome number of n=19 and hybridize when they come together except for the polyploid *B. macrophylla*. Intersectional hybridization merely produces hybrid swarms, while intrasectional hybridization tends to produce widespread clines, leading to taxonomic confusion. Hybrid swarms are easily distinguished in the field by the wide variation in leaf division and pubescence within a given population. In the herbarium, when only one plant per population is present, hybrids are best distinguished by possessing leaves that are not as deeply divided as would be expected for the most similar-looking member of section *Balsamorhiza*, or by plants with pinnately divided leaves having either multiple heads per stem or alternate, in addition to opposite, stem leaves.

Widths of heads are those of pressed specimens. Widths of pinnae are the width of full-sized pinnae from the middle of a fully developed leaf at their widest point (not including any lobes that may be present).

Key to *Balsamorhiza* species:

1. Leaves ovate, simple, bases cordate to hastate; plants forming clumps with multiple rosettes.
   2. Plants gray-tomentose; involucres densely tomentose . . . . . **B. sagittata**
   2' Plants green, villous-hirsute; involucres villous to sparsely tomentose.
      3. Ray florets persistent after drying; fruits generally strigillose, sometimes glabrous; plants usually east of Cascade Range . . . . . **B. careyana**
      3' Ray florets deciduous; fruits always glabrous; plants usually west of or in Cascade Range . . . . . **B. deltoidea**

1' Leaves lanceolate to ovate, at least some leaves pinnate or pinnatifid, bases acute to obtuse; plants with 1–few rosettes from single taproot.
   4. Margins of leaves or leaflets serrate, teeth apiculate; generally some or all leaves undivided . . . . . **B. serrata**
   4' Margins of leaves or leaflets entire, crenate, or serrate, but teeth not apiculate; leaves generally pinnate or pinnatifid.
      5. Ray florets turning from yellow to pink or red; generally with some basal leaves undivided . . . . . **B. rosea**
      5' Ray florets remaining yellow; all basal leaves pinnate or pinnatifid.
         6. Leaves once pinnate, leaflets sometimes shallowly lobed or toothed, but never pinnatifid.
            7. Leaves sericeous; plants of southwestern Oregon . . . . . **B. sericea**
            7' Leaves hispid or tomentose; plants east of Cascade Range.
               8. Leaves hispid . . . . . **B. hispidula** (in part)
               8' Leaves tomentose.
                  9. Pinnae 5–15 mm wide; rays 25–35 mm . . . . . **B. incana**
                  9' Pinnae 10–36 mm wide; rays 35–40 mm . . . . . **B. macrophylla**
         6' Leaves twice pinnate or once pinnate with deeply pinnatifid leaflets.
            10. Leaves from 1-pinnate with toothed lobes to ± 2-pinnate; hispid; phyllaries linear-lanceolate, not with long, acuminate tips . . . . . **B. hispidula** (in part)
            10' Leaves from 1-pinnate with toothed lobes to 3-pinnate; hirsute; phyllaries linear-lanceolate to ovate, sometimes with long, acuminate tips . . . . . **B. hookeri**

Balsamorhiza careyana A. Gray

Carey's balsamroot

Balsamorhiza careyana A. Gray var. *careyana*, *Balsamorhiza careyana* A. Gray var. *intermedia* Cronquist

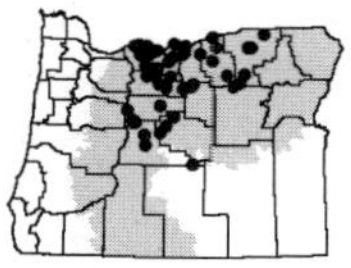

Stems forming clumps, 23–62 cm, green, sparsely villous-hirsute, more densely so below heads. **Leaves** simple, margins entire to slightly crenate, surfaces hirsutulous to hirsute; basal leaves in multiple rosettes, blades deltate, 7–30 × 4.7–16 cm, bases cordate to hastate, petioles 7.5–36 cm; cauline proximal pair generally opposite, lanceolate, distal alternate, blades tapering at both ends, 2.5–8.2 × 0.4–2.8 cm, petioles 0–7.5 cm. **Inflorescences** with 1 larger, terminal head and 1–6 smaller, axillary heads; peduncles 0–20.5 cm. **Involucres** terminal 9–15 × 15–25 mm; axillary 9–13 × 6–12 mm. **Phyllaries** ovate-lanceolate, 8–34 × 4–7 mm; outer often tapered to caudate or with acuminate tips, villous to tomentose. **Paleae** 11–13 mm. **Ray florets** 8–20 on large heads, 5–9 on small heads, yellow, persistent after drying; rays 18–26 × 5–11 mm. **Disc florets** 6.5–8.5 mm. **Fruits** 5.5–7.5 × 2–2.5 mm, generally strigillose, sometimes glabrous. **2*n***=38.

Open areas, open pine woodlands. Flowering Apr–Jul. 0–1300 m. BW, Casc, Col, ECas, Lava. WA. Native.

Balsamorhiza careyana usually grows east of the Cascades but intergrades broadly with *B. deltoidea* in the Cascades and along the Columbia River. Pure *B. careyana* is easily distinguished by a combination of strigillose fruit, ray florets that are persistent after drying, and generally more heads per stem than *B. deltoidea*. Cascade specimens can be difficult to determine and have been separated as *B. careyana* var. *intermedia* in other treatments. The two species should perhaps be considered varieties of the same species.

Balsamorhiza deltoidea Nuttall

Deltoid balsamroot

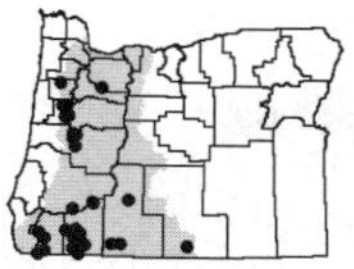

Stems forming clumps, 14–54 cm, green, villous-hirsute, with shorter, glandular hairs. **Leaves** simple, margins entire to toothed, teeth rounded and apiculate, surfaces abaxially sparsely hirsute, adaxially sparsely hirsute, villous-hirsute on veins; basal leaves in multiple rosettes, blades deltate, 7.5–27 × 6–16 cm, bases cordate or hastate, petioles 10–27 cm; cauline blades 3–7.5 × 1.1–2.7 cm, petioles 0–8.5 cm. **Inflorescences** with 1 larger, terminal head and 0–2 smaller, axillary heads; peduncles 1–12 cm. **Involucres** terminal 10–15 × 20–33 mm; axillary 8–10 × 11–17 mm. **Phyllaries** ovate, 10–32 × 4–9 mm, villous, more densely so at bases; outer phyllaries often lanceolate, longer than inner. **Paleae** ~7 mm. **Ray florets** deciduous, 11–21 on large heads, 6–9 on small heads, yellow; rays 18–35 × 4–17 mm. **Disc florets** 5–8 mm. **Fruits** 9 × 1.5 mm, glabrous. **2*n***=38.

Grasslands, pine or oak woodlands. Flowering Mar–Jul. 50–2200 m. Casc, ECas, Sisk, WV. CA, WA; north to British Columbia. Native.

Balsamorhiza deltoidea grows west of the Cascade Range and intergrades with *B. careyana* in the Cascades and along the Columbia River. (See discussion under *B. careyana*.) It also hybridizes with *B. sericea* in Josephine County. Hybrids between *Balsamorhiza deltoidea* and *B. hookeri* have been given the name *B.* ×*terebinthacea* but are not expected to be found in Oregon, given that the ranges of the parental species do not overlap here. The specific epithet refers to the deltoid shape of the leaves.

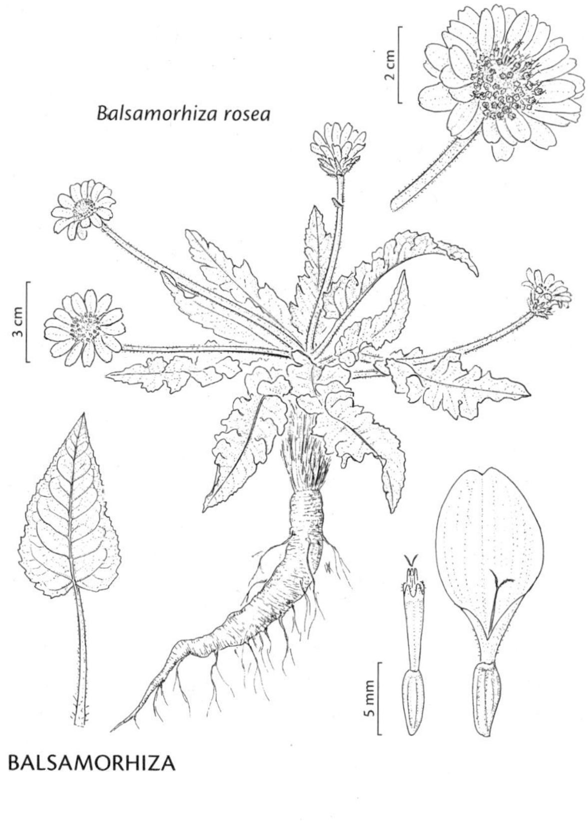

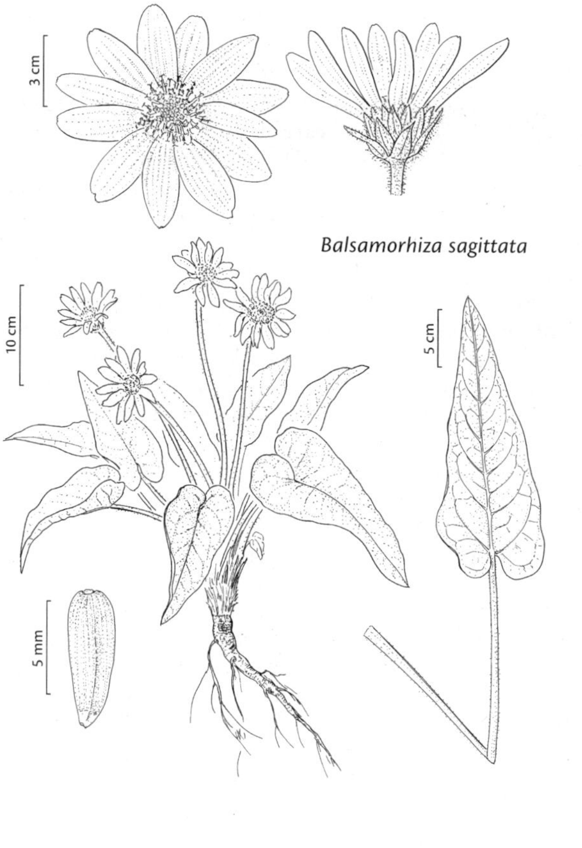

BALSAMORHIZA

Balsamorhiza hispidula W.M. Sharp

Hispid balsamroot

Balsamorhiza hookeri Nutt. var. *hispidula* (W.M. Sharp) Cronquist

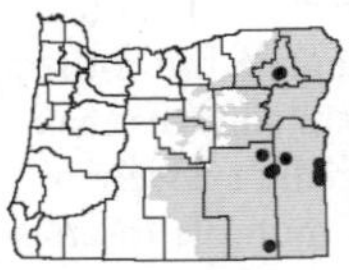

Stems 8–25 cm, villous-hirsute; from single taproot. **Leaves** ovate-lanceolate, 1-pinnately divided with pinnae again lobed, often with only a few lobes at tips or divided, margins entire, ciliate, surfaces hispid; basal leaves in 1–few rosettes, blades 5–18 × 3–7.5 cm, pinnae 2–8 mm wide, petioles 2–9 cm; cauline leaves in one opposite pair, blades 1.5–5 × 0.4–2 cm, petioles 1.5–4 cm. **Inflorescences** with 1 terminal head. **Involucres** 10–15 × 17–22 mm. **Phyllaries** linear to lanceolate, 14–17 × 2–3 mm, with acuminate or caudate tips, ciliate, villous-hirsute. **Paleae** 7–8 mm. **Ray florets** deciduous, 10–13, yellow; rays 20–25 × 5–10 mm. **Disc florets** 6–7 mm. **Fruits** 6–7 mm, glabrous.

Dry, often rocky, open areas. Flowering Apr–Jun. 800–1700 m. BR, BW, Owy. ID, NV; northeast to MT, east top WY, southeast to AZ. Native.

While *B. hispidula* is quite distinct from *B. hookeri* throughout most of its range, it intergrades completely with *B. hookeri* in Oregon. Although the two species have been kept separate in this treatment, most Oregon specimens of *B. hispidula* show some evidence of gene flow from *B. hookeri*. *Balsamorhiza hispidula* also hybridizes with *B. sagittata*. The specific epithet refers to the pubescence of the plants.

Balsamorhiza hookeri Nuttall

Hooker's balsamroot

Balsamorhiza hookeri Nutt. var. *hirsuta* (Nutt.) A. Nelson, *Balsamorhiza hookeri* Nutt. var. *neglecta* (W.M. Sharp) Cronquist, *Balsamorhiza hookeri* Nutt. var. *platylepis* (W.M. Sharp) Cronquist, *Balsamorhiza hirsuta* Nutt., *Balsamorhiza macrolepis* W.M. Sharp var. *platylepis* (W.M. Sharp) Ferris, *Balsamorhiza platylepis* W.M. Sharp

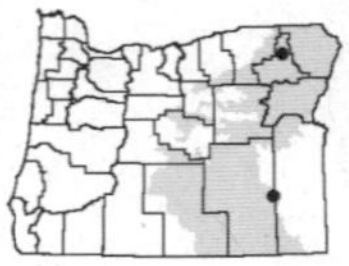

Stems 11–47 cm, hirsute; from single taproot. **Leaves** lanceolate to ovate-lanceolate, 1-pinnately divided with some pinnae toothed or lobed to 3-pinnately lobed, margins entire to crenate or serrate, surfaces hirsute to hirsute-sericeous; basal leaves in 1–few rosettes, blades 12–24 × 4–12 cm, pinnae 1–10 mm wide, petioles 4–15 cm; cauline blades 3–14.5 × 0.6–4 cm, petioles 4–11 cm. **Inflorescences** with 1 terminal head. **Involucres** 8–15 × 15–25 mm. **Phyllaries** linear-lanceolate to ovate, 13–20 × 3–6 mm, with or without long-acuminate tips, villous-hirsute and ciliate, sometimes glabrous proximally. **Paleae** 11–12 mm. **Ray florets** deciduous, 10–17, yellow; rays 15–30 × 4–7 mm. **Disc florets** ~6 mm. **Fruits** 6.5 × 1.5 mm, glabrous. **2*n***=38.

Dry, open, often rocky areas. Flowering Apr–Jun. 1000–1600 m. BR, BW. CA, ID, NV, WA; southeast to UT. Native.

As treated here, *B. hookeri* consists of the members of section *Balsamorhiza* that are left over after the entities that are distinct in at least some parts of their ranges are removed. *Balsamorhiza hookeri* hybridizes with *B. hispidula* and *B. sagittata*. The species is named after the British botanist William Jackson Hooker, who initially described it as *Heliopsis balsamorhiza* and tentatively proposed (but did not validly publish, as he did not accept it) the genus *Balsamorhiza* under the supposition that it would prove to be distinct from *Heliopsis* once better material was collected.

Balsamorhiza incana Nuttall

Hoary balsamroot, woolly balsamroot

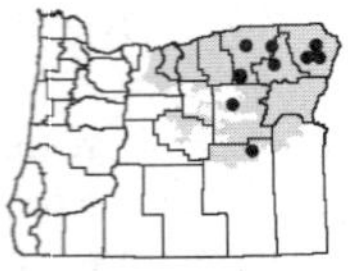

Stems 9–43 cm, tomentose; from single taproot. **Leaves** ovate-lanceolate, 1-pinnately divided, margins toothed, especially distally, sometimes also on proximally, surfaces tomentose and often lacking glands; basal leaves in 1–few rosettes, blades 11–22 × 5–8.5 cm, pinnae 5–15 mm wide, petioles 4–17 cm; cauline leaves in 1 pair, sometimes very near base of stem, blades 1.5–9.5 × 0.6–4.5 cm, petioles 4–8 cm. **Inflorescences** with 1 terminal head. **Involucres** 11–18 × 18–28 mm. **Phyllaries** lanceolate, 15–20 × 3–5 mm, tips acuminate. **Paleae** ~10 mm. **Ray florets** deciduous, 13–22, yellow; rays 25–35 × 7–10 mm. **Disc florets** 7.5–9 mm. **Fruits** 5–6 mm, glabrous. **2*n***=38.

Open, rocky areas. Flowering Apr–Jun. 300–1500 m. BW, Col. ID, WA; east to WY. Native.

Balsamorhiza incana is unusual among species of section *Balsamorhiza* in remaining fairly distinct throughout its wide range. It occurs generally north and east of the widespread *B. hispidula/B. hookeri* complex. Its hybrids with *B. sagittata* have been given the name *B.* × *tomentosa*. The specific epithet refers to the whitish appearance from the dense pubescence.

Balsamorhiza macrophylla Nuttall

Large-leaf balsamroot

Balsamorhiza hookeri Nutt. var. *idahoensis* (W.M. Sharp) Cronquist

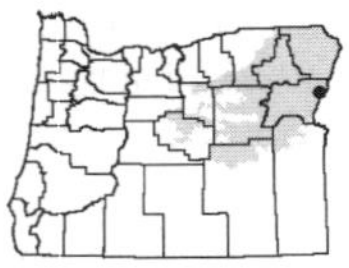

Stems up to 35 cm, tomentose; from single taproot. **Leaves** lanceolate-oblong, 1-pinnately divided, pinnae toothed, with teeth only on lower margins, surfaces tomentose; basal leaves in 1–few rosettes, blades 12–22 × 6–8 cm, pinnae 10–36 mm wide, petioles 7–20 cm; cauline blades 1.5–5 × 0.3–0.9 cm, petioles 6–9 cm. **Inflorescences** with 1 terminal head. **Involucres** 14–18 × 25–30 mm. **Phyllaries** ovate-lanceolate, 20–25 × 5–6 mm, tips acuminate, tomentose. **Paleae** 10–12 mm. **Ray florets** 14–16, yellow; rays 35–40 × 8–11 mm. **Disc florets** ~9 mm. **Fruits** 8–9 mm, glabrous. **2*n***=100 ± 2.

Dry, open areas. Flowering May–Jul. 700–1400 m. BW. ID; east to WY. Native.

True *B. macrophylla* is a decaploid centered on the Wasatch Mountains in Utah and Idaho. The plants in Oregon and western Idaho are also high polyploids, but they differ morphologically from classic *B. macrophylla* in their denser pubescence and the shape of their leaves. These eastern plants have been given the name *B. macrophylla* var. *idahoensis*. However, further study will likely show that they have a different parentage than *B. macrophylla s.s.* The specific epithet "macrophylla" means large-leaved, as the leaves are much larger than those of the other members of section *Balsamorhiza*.

Balsamorhiza rosea A. Nelson & J.F. Macbride (p. 199)

Rosy balsamroot

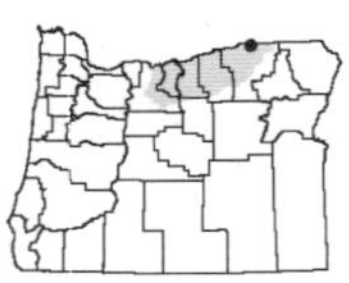

Stems 5–15 cm, villous-hirsute to villous below heads; from single taproot. **Leaves** oblong-lanceolate, pinnately lobed from 1/4 to 4/5 to midrib, occasionally some leaves undivided, margins toothed to crenate with

pointed teeth, surfaces villous-hirsute, more densely so abaxially than adaxially; basal leaves in 1–few rosettes, blades 2.5–7 × 1.5–2.8 cm, pinnae 6–12 mm wide, petioles 1.5–4.5 cm; cauline leaves in 1 opposite pair or 0, blades 0.7–2 × 0.3–1 cm, petioles 3–4 cm. **Inflorescences** with 1 terminal head. **Involucres** 10–13 × 20–25 mm. **Phyllaries** linear to linear-lanceolate, 12–15 × 1–2 cm, villous. **Paleae** 11–13 mm. **Ray florets** ~13, yellow when young, turning from yellow to pink or red; rays 10–15 × 7–8 mm. **Disc florets** ~6 mm. **Fruits** 5–6 mm, strigillose. ***2n***=38.

Open, rocky areas. Flowering Apr–May. 300–400 m. Col. WA. Native.

Balsamorhiza rosea occurs almost entirely in Washington, with one known locality in Oregon.

Balsamorhiza sagittata (Pursh) Nuttall (p. 199)

Arrowleaf balsamroot

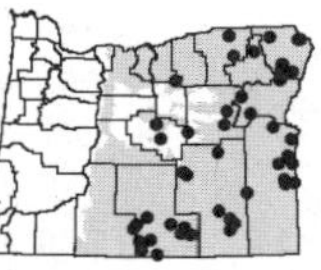

Stems forming clumps, 20–65 cm, gray-tomentose. **Leaves** simple, ovate, bases cordate or sagittate, margins entire, surfaces tomentose, more densely so abaxially than adaxially; basal leaves in multiple rosettes, blades 7–27 × 4.5–14.5 cm, petioles 8–39 cm; cauline blades 3–5.5 × 0.5–0.9 cm, petioles 0.7–5.5 cm. **Inflorescences** with 1 larger, terminal head, and rarely 1–3 smaller; axillary peduncles 0.2–13 cm. **Involucres** terminal 10–15 × (10)15–30 mm; axillary 8 × 8–12 mm. **Phyllaries** lanceolate, 10–32 × 3–7 mm, outer often longer, densely tomentose, especially basally. **Paleae** 9–11 mm. **Ray florets** 8–19 on large heads, 5–9 on small heads, yellow; rays 25–40 × 7–18 mm. **Disc florets** 6–8 mm. **Fruits** 6–7 × 1.2–2 mm, glabrous. ***2n***=38.

Open areas, edges of forests, open woodlands. Flowering Apr–Jul. 300–2100 m. BR, BW, Col, ECas, Owy. CA, ID, NV, WA; north to British Columbia, northeast to Alberta, east to SD, southeast to AZ. Native.

This is the most widespread species of *Balsamorhiza*. It hybridizes with every species of section *Balsamorhiza* with which its range overlaps but is itself morphologically quite uniform and easy to distinguish from the other members of section *Artorhiza*. Its hybrids with *B. incana* have been given the name *B.* × *tomentosa*.

Balsamorhiza sericea W.A. Weber (p. 201)

Silky balsamroot, silvery balsamroot

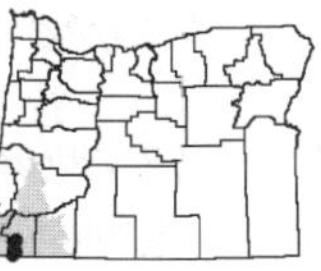

Stems 6–38 cm, sericeous; from single taproot. **Leaves** ovate-lanceolate, 1-pinnately divided almost or entirely to midribs, often more deeply divided proximally than distally, margins entire or few-toothed, occasionally some lower leaves serrate on lower margins, surfaces sericeous, more densely so abaxially than adaxially; basal leaves in 1–few rosettes, blades 3–18 × 1–7 cm, pinnae 6–16 mm wide, petioles 3–17 cm; cauline leaves in 1 pair, opposite, reduced, blades 2.5–4 × 1–1.6 cm, petioles 2.5–4 cm. **Inflorescences** with 1 terminal head. **Involucres** 12–15 × 15–22 mm. **Phyllaries** ovate, 10–17 × 5–9 mm, tips acuminate, sericeous. **Paleae** 13–14 mm. **Ray florets** 13–14, yellow; rays 12–20 × 5–8 mm. **Disc florets** ~9 mm. **Fruits** 6–7 × 1.5 mm, glabrous.

Serpentine. Flowering Apr–May. 400–600 m. Sisk. CA. Native.

Balsamorhiza sericea is endemic to serpentine-influenced soils and is thus far known only from Josephine County, Oregon, and adjacent Siskiyou and Trinity counties, California. The specific epithet "sericea" refers to the pubescence.

Balsamorhiza serrata A. Nelson & J.F. Macbride

Serrate balsamroot

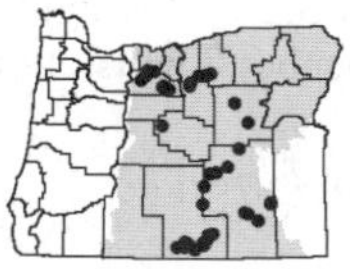

Stems 5–34 cm, sparsely villous-tomentose; from single taproot. **Leaves** ovate-lanceolate, undivided to pinnatifid, rarely with some leaves pinnately compound, margins sharply serrate, tips apiculate, surfaces sparsely hispid; basal leaves in 1–few rosettes, blades 4.5–15 × 2–5.5 cm, pinnae (when present) 4–11 mm wide, petioles 1–9(13) cm; often one or more bract-like basal leaves present; cauline leaves of 1 pair (when present), opposite, blades 1–4(5.5) × 0.6–1.5 cm, petioles 2–5.5(6.5) cm. **Inflorescences** with 1 terminal head. **Involucres** 13–15 × 10–30 mm. **Phyllaries** linear-lanceolate, 11–18 × 2–4 mm, tips acuminate, sparsely to densely villous-tomentose. **Paleae** 11–12 mm. **Ray florets** 8–21, yellow; rays 13–24 × 4–9 mm. **Disc florets** ~8 mm. **Fruits** 6–7 × 1.5–2 mm, glabrous.

Dry open areas, serpentine. Flowering Apr–Jun. 600–2100 m. BR, BW, Col, ECas, Lava. CA, NV, WA. Native.

Balsamorhiza serrata is distinct among members of the genus *Balsamorhiza* in generally having simple leaves with serrate margins and tapered (not sagittate) bases. Some plants have pinnately divided leaves. Further work may show that these plants are actually hybrids between *B. serrata* and *B. hookeri* or *B. hispidula*. *Balsamorhiza serrata* is known to hybridize with *B. sagittata*.

Balsamorhiza sericea

BALSAMORHIZA

BELLIS Daisy

Herbs perennial; rhizomatous and fibrous-rooted. **Stems** erect and simple or 0; flowering stems leafless, pilose- or strigose-pubescent. **Leaves** basal, sometimes cauline, 1–3-veined, cauline alternate; blades broadly oblanceolate or elliptic, margins entire to dentate, teeth short, angled toward tips, tips obtuse or rounded, surfaces glabrous or pilose-hispid, especially abaxially; petiolate. **Inflorescences** solitary terminal heads. **Heads** radiate, ± erect. **Involucres** hemispheric. **Phyllaries** in 2–3 equal series, erect, lanceolate, distinctly 1-veined, tips acute, surfaces pilose-hispid. **Receptacles** conic; paleae 0. **Ray florets** pistillate, rays well exceeding involucres, white to pink or pink-tipped. **Disc florets** bisexual; corollas tubular, widened distally, yellow, lobes 5, erect, triangular; stamens 5, included in corollas; stigma lobes exserted from stamen tubes. **Fruits** obovate, compressed, marginally 2-ribbed; pappi 0. $2n$=18.

Worldwide. 15 species; 1 species treated in *Flora*.

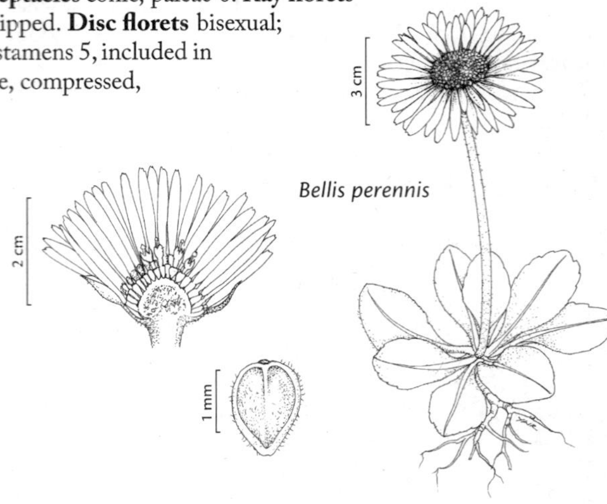

BELLIS

Bellis perennis Linnaeus (p. 202)

English daisy, lawn daisy

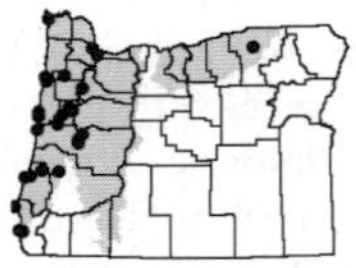

Plants 2.5–20 cm. **Leaves** 1–8 cm; petioles 0.5–5 cm, sometimes wing-margined. **Involucres** 4–6 mm. **Ray florets** 30–50+, rays 4–11 mm. **Disc florets** numerous; corollas ~1.5 mm, exserted from involucres. **Fruits** 1–2 mm, gray, minutely pilose-hispid.

Lawns, pastures, roadsides, waste areas. Flowering Mar–Jun. 0–900 m. Casc, Col, CR, Est, WV. CA, ID, WA; nearly worldwide. Exotic.

BIDENS Beggars-ticks

Herbs annual or perennial, aquatic or terrestrial. **Stems** erect, ascending, decumbent, or floating, simple or branched, glabrous to lightly puberulent or minutely pustulose. **Leaves** cauline, opposite, simple or pinnately compound, sometimes finely divided, glabrous; sessile or petiolate. **Inflorescences** usually corymb-like arrays or heads solitary and terminal. **Heads** radiate, discoid, or ± disciform. **Involucres** hemispheric or campanulate to cylindric. **Phyllaries** in ~2 series; outer bracteate, herbaceous, sometimes a few bracts enlarged and foliaceous; inner erect, firm, elliptic-lanceolate, ± equal, margins membranous, veins many, parallel. **Receptacles** flat or convex; paleate. **Ray florets** present or 0, neuter or pistillate and sterile; corollas yellow to orange. **Disc florets** bisexual; corollas tubular, linear below, gradually or abruptly widened distally, exserted from involucres, 5-lobed, lobes triangular, partly exserted from corollas; style branches exserted from stamen tubes. **Fruits** obcompressed, flat, or 3–4-angled, cuneate to oblanceolate, obovate, or fusiform, glabrous or pubescent; pappi 0 or of 2–6 stout, persistent, retrorsely barbed or ciliate awns.

Worldwide. 150–250 species; 5 species treated in *Flora*.

Key to *Bidens* species:

1. Plants aquatic; submerged leaves dissected with filiform ultimate segments . . . . ***B. beckii***

1' Plants terrestrial; leaves not dissected.

  2. Outer phyllaries spreading or reflexed; leaves mostly sessile, lowermost sometimes petiolate . . . . ***B. cernua***

  2' Outer phyllaries usually ascending or erect; leaves sessile or petiolate.

    3. Leaves 2–4-lobed, not compound . . . . ***B. tripartita***

    3' Most leaves pinnately compound.

      4. Outer whorl of phyllaries 10–16, disc florets pale yellow . . . . ***B. vulgata***

      4' Outer whorl of phyllaries mostly 5–8, disc florets orange . . . . ***B. frondosa***

Bidens beckii Torrey ex Sprengel

Water-marigold

Megalodonta beckii (Torr. ex Spreng.) Greene

Plants perennial, aquatic, 1–20 dm. **Stems** floating, simple or branched, glabrous. **Leaves** submerged dissected with filiform ultimate segments; emergent simple, ovate or lanceolate, 1–4.5 cm, margins entire to serrate or divided, tips obtuse to acuminate or attenuate, surfaces glabrous; sessile or petiolate. **Heads** solitary, terminal; peduncles (1)2–10 cm. **Involucres** hemispheric, 12–15 mm in diameter. **Phyllaries** outer 5–6, spreading or reflexed; inner 6–9 mm, glabrous. **Ray florets** 8; rays 1–2 cm, yellow. **Disc florets** 10–30; corollas 5–6 mm. **Fruits** cylindric or slightly flattened, 10–15 mm; pappi of 3–6 slightly flattened awns, 13–30+ mm, retrorsely barbed distally, smooth toward base. $2n$=26.

Lakes, ponds, slow-moving streams. Flowering Jul–Sep. 0–50 m. Est. ID, WA; north to British Columbia, northeast to Quebec, east to Nova Scotia. Native.

Bidens cernua Linnaeus (p. 203)

Nodding beggars-ticks, bur marigold

Bidens cernua L. var. *cernua*, *Bidens cernua* L. var. *elliptica* Wiegand, *Bidens cernua* L. var. *minima* (Huds.) DC.

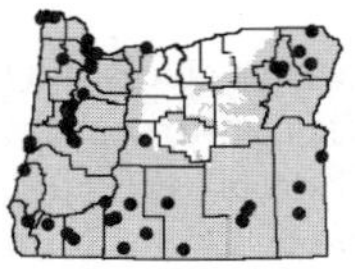

Plants annual, 0.5–10+ dm. **Stems** erect or with decumbent and ascending lower branches, glabrous or sparsely hispid. **Leaves** elliptic-lanceolate to oblanceolate or linear, 1–20+ cm, bases truncate or attenuate, margins entire to denticulate or serrate, tips acute, surfaces glabrous; sessile or lowest pairs petiolate. **Heads** solitary and terminal on stems and branches or in corymb-like arrays, erect or nodding; peduncles 1–4(10) cm. **Involucres** hemispheric or broader, 12–20 mm in diameter. **Phyllaries** outer 5–8, usually spreading or reflexed, sometimes enlarged and up to 2 cm, inner series 5–10 mm, glabrous. **Ray florets** 3–8, sometimes 0, rays 2–15+ mm, orange-yellow. **Disc florets** 10–100+, corollas orange-yellow, 3–4 mm. **Fruits** wedge-shaped, 4–8 mm, margins sometimes winged, retrorsely barbed or ciliate; pappi of (2)4 retrorsely barbed awns, 2–4 mm. **2*n***=24, 48.

Ponds, sloughs, marshes, tidal channels, gravel bars, riverbanks, irrigation ditches. Flowering Jun–Oct. 0–1600 m. BR, BW, Casc, CR, ECas, Est, Owy, Sisk, WV. CA, ID, NV, WA; north to AK, throughout US and Canada; Asia, Europe. Native.

Bidens frondosa Linnaeus (p. 204)

Leafy beggars-ticks, sticktight

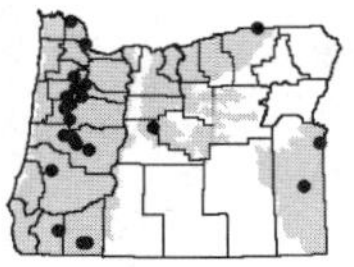

Plants annual, 1–12+ dm. **Stems** erect, simple or branched, glabrous or minutely pustulose. **Leaves** leaflets 3–5, leaflets lanceolate to lance-ovate, 2–10+ cm, margins evenly serrate or denticulate, tips acute or acuminate, surfaces glabrous; petioles 1–7 cm. **Heads** few to numerous, solitary and terminal on stems and branches or in corymb-like arrays, erect; peduncles 1–4(8) cm. **Involucres** hemispheric, 7–12 mm in diameter. **Phyllaries** outer 5–8, mostly ascending to erect, margins often sparsely setose-ciliate; inner 5–9 mm, glabrous. **Ray florets** 0 or few; rays 2–3 mm, golden-yellow. **Disc florets** 20–100+; corollas orange, 2–3 mm. **Fruits** obovate or wedge-shaped, 5–10 mm, roughly tuberculate, margins sometimes minutely barbed; pappi of 2 antrorsely or retrorsely barbed awns, 2–5 mm. **2*n***=24, 48, 72.

Mixed woods, riverbanks, edges of lake and ponds, gravel bars, irrigation ditches, cultivated fields, waste areas, roadsides. Flowering Aug–Sep. 0–1100 m. Casc, Col, CR, Lava, Owy, Sisk, WV. CA, ID, NV, WA; throughout southern Canada and US, northern Mexico; Europe. Native.

Bidens tripartita Linnaeus

Three-lobed beggars-ticks

Bidens connata Muhl. ex Willd.

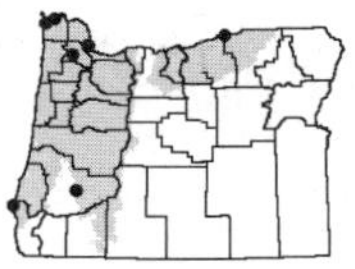

Plants annual, 1–20 dm. **Stems** erect, simple or branched, glabrous. **Leaves** simple or often ± deeply 2–4-lobed, 3–15 cm, margins entire or serrate, tips acute, surfaces glabrous; sessile or petioles 5–15(35) mm, often winged. **Heads** usually numerous, terminal on stems and branches, erect; peduncles 1–4(8) cm. **Involucres** campanulate to hemispheric, 6–12(20) mm in diameter. **Phyllaries** outer 4–5+, usually erect, sometimes spreading and foliaceous, margins entire or setose proximally; inner 5–10 mm, glabrous. **Ray florets** usually 0, rarely 5–8; rays 5–14 mm, yellow. **Disc florets** 10–100+; corollas yellow or orange, 3–5 mm. **Fruits** obovate or wedge-shaped, 3–10 mm, flat or 3–4-angled, glabrous or minutely hirsute; pappi of 2–4 retrorsely barbed awns, 1–3(6) mm. **2*n***=48.

Riverbanks, ponds, marshes, bogs, weedy sites. Flowering Aug–Oct. 0–700 m. Casc, Col, CR, Est, WV. CA, WA; north to AK, northeast to Quebec, east to Nova Scotia, southeast to GA, south to Mexico; Africa, Asia, Australia, Europe. Native.

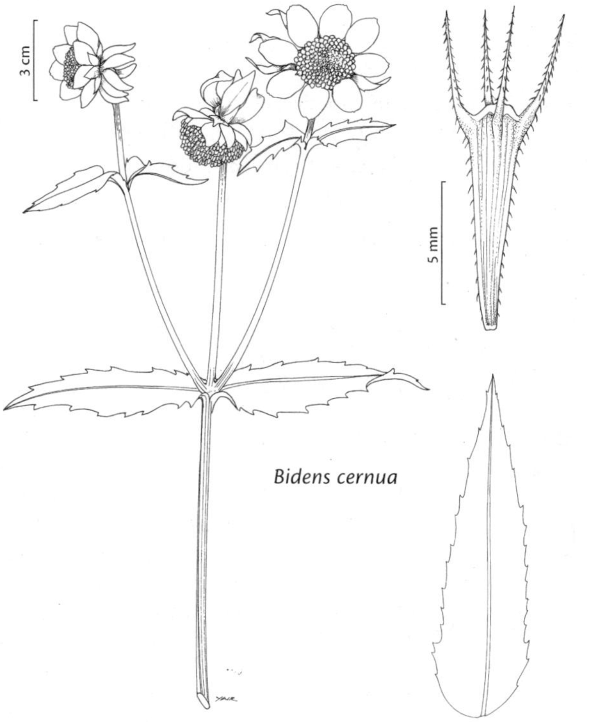

BIDENS

Bidens vulgata Greene

Tall beggars-ticks, western sticktight

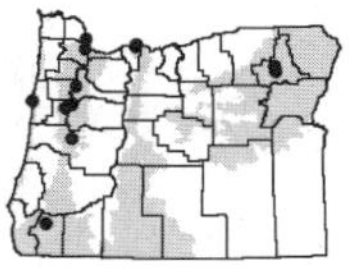

Plants annual, 1–10(15) dm. **Stems** erect, simple or branched, glabrous or minutely puberulent. **Leaves** most 3–5-pinnate, occasionally simple, 5–10 cm, margins usually denticulate or serrate, tips acute or acuminate, surfaces glabrous or minutely puberulent; petioles 1–5 cm. **Heads** few to numerous, solitary and terminal on stems and branches or in corymb-like arrays, erect. **Involucres** hemispheric, 8–10 mm in diameter. **Phyllaries** outer 10–16, ascending to erect, rarely spreading, often setose-ciliate; inner 6–9 mm, glabrous. **Ray florets** 0 or 3–5; rays 2.5–4 mm, pale yellow. **Disc florets** 40–100+; corollas yellow, 2.5–3.5 mm. **Fruits** obovate or wedge-shaped, flat, 6–12 mm, margins minutely barbed; pappi of 2 retrorsely barbed awns, 3–7 mm. **2*n***=24, 48.

Meadows, moist swales, streambanks, bogs, gravel bars, waste areas, cultivated fields. Flowering Jul–Sep. 0–900 m. BW, ECas, Est, Sisk, WV. CA, ID, NV, WA; north to British Columbia, northeast to Quebec, east to Nova Scotia, southeast to GA; Europe. Exotic?

BLEPHARIPAPPUS Rough eyelashweed

Bruce G. Baldwin and Gerald D. Carr

Genus composed of one species, *B. scaber.*

Blepharipappus scaber Hooker (p. 204)

Rough eyelashweed

Blepharipappus scaber Hook. ssp. *laevis* (A. Gray) D.D. Keck, *Blepharipappus scaber* Hook. ssp. *scaber*, *Blepharipappus scaber* Hook. var. *scaber*

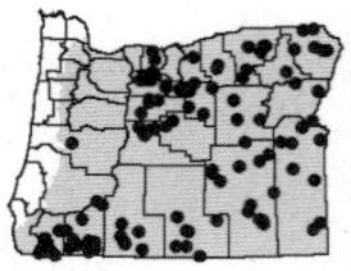

Herbs annual, 5–20(40+) cm; self-incompatible. **Stems** ± erect, simple or more commonly slender-branched, scabrous and sometimes hirsute, usually stipitate-glandular distally. **Leaves** mostly cauline; proximal opposite, most alternate; blades narrowly spatulate to linear, usually ascending, 6–25+ × 0.5–1.5 mm, margins entire, usually revolute, surfaces scabrous, hirsute, strigose, sericeous, or villous, distal leaves usually stipitate-glandular; sessile. **Inflorescences** open, corymb-like arrays or heads solitary. **Heads** radiate. **Involucres** turbinate to campanulate or hemispheric, 3–6+ mm diameter. **Phyllaries** (2)3–5(8), in 1 series, herbaceous, each partially or wholly enveloping subtended ray ovary, lanceolate or oblanceolate, 3–8 mm, abaxially somewhat hirsute and/or stipitate-glandular, falling with fruit. **Receptacles** convex, glabrous; paleae falling, subtending all or most disc florets, outer herbaceous, inner scarious proximal to herbaceous apices. **Ray florets** (2)3–5(8), pistillate, fertile; corollas whitish, rays 2–11 mm, veined with purple abaxially. **Disc florets** 6–25(60+), bisexual, fertile; corollas 2–3.5 mm, whitish, tubes ≤ funnelform throats, lobes 5, deltate; anthers basically dark purple; styles bristly-puberulent proximal to branches, branches ~0.2 mm, papillose. **Fruits** of ray and disc florets similar, nearly obconic, ± terete, 2–3.5 mm, basal attachments nearly central, faces somewhat villous, tips not beaked; pappi 0 or of 12–18(26) subulate, fimbriate to ciliate or plumose scales, 0.1–2 mm. **2*n***=16.

Openings in sagebrush scrub, pinyon-juniper woodlands, yellow-pine forests. Flowering Apr–Sep. 50–1900 m. BR, BW, Casc, Col, ECas, Lava, Owy, Sisk, WV. CA, ID, NV, WA. Native.

Blepharipappus and members of *Layia* were once treated as congeneric; *Layia* appears to be more closely related to *Lagophylla* than to *Blepharipappus* based on molecular phylogenetic data (Baldwin 2003b).

Blepharipappus scaber is unusual among self-incompatible, continental tarweeds for occurring widely in western North America while having only a relatively limited distribution in the California Floristic Province.

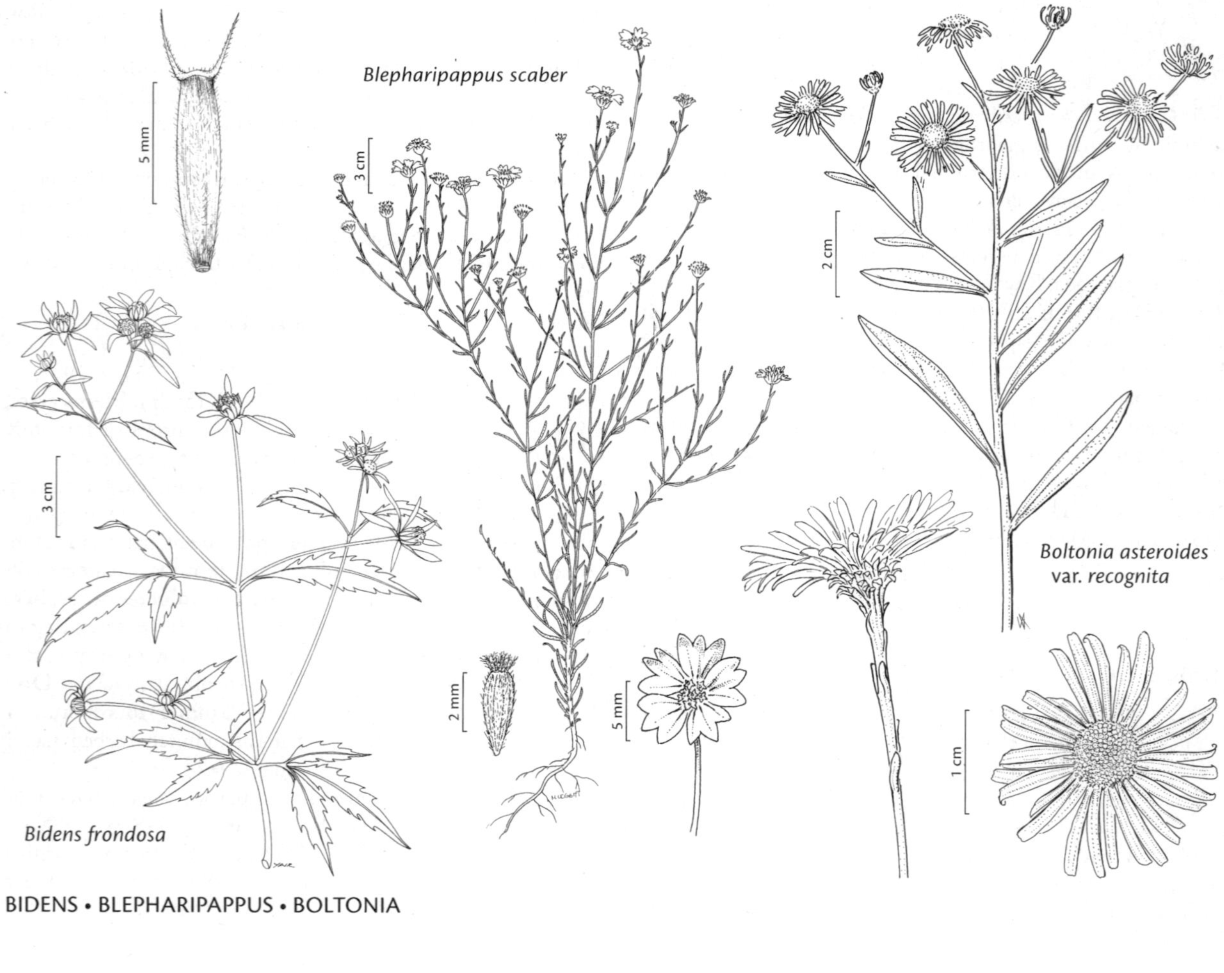

BIDENS • BLEPHARIPAPPUS • BOLTONIA

BOLTONIA Doll's-daisy

Herbs perennial; with fibrous-rooted caudices, stoloniferous. **Stems** erect, much branched, glabrous, with prominent longitudinal veins. **Leaves** cauline, alternate, lower usually withering by anthesis, reduced distally, oblong-lanceolate, margins entire, tips acute, surfaces glabrous; sessile. **Inflorescences** panicle-like, ± flat-topped; peduncles leafless or bracteate, glabrous. **Heads** radiate. **Involucres** hemispheric. **Phyllaries** in 3–6 series, erect in flower, spreading in fruit, linear-lanceolate, subequal to unequal, tips acute or acuminate. **Receptacles** conic to hemispheric, obscurely pitted; paleae 0. **Ray florets** pistillate; rays well exceeding involucres. **Disc florets** bisexual; corollas tubular, slightly widened distally, lobes 5, triangular; stamens 5, slightly exserted from corollas; style branches exserted from stamen tubes. **Fruits** flattened, obovate, truncate, narrowly winged; pappi of 2 longer and some shorter bristles.

North America. 5 species; 1 species treated in *Flora*.

Boltonia asteroides (Linnaeus) L'Héritier (p. 204) var. ***recognita*** (Fernald & Griscom) Cronquist

White doll's-daisy, false aster

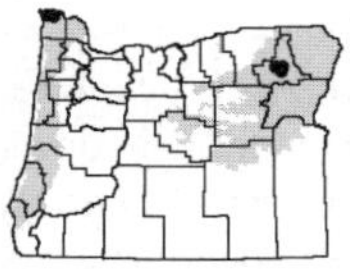

Plants 3–20 dm. **Leaves** 2–20 cm, 1–3-veined. **Involucres** 3–5 × 6–9 mm. **Heads** 15–60(80). **Ray florets** 20–60; rays linear, 5–13 mm, white to pink, purplish, or blue. **Disc florets** corollas 1.5–2 mm, slightly exceeding involucres, yellow. **Fruits** 1–3 mm, brown, minutely hispid; pappus bristles brownish.

Marshes, tide flats. Flowering Aug–Sep. 0–900 m. BW, CR. ID; midwestern Canada and US, scattered in eastern US. Exotic.

This species has long been introduced in Oregon and is currently well established in marshlands and tidal flats in Clatsop County. Horticultural cultivars of *Boltonia asteroides* are popular in nurseries.

BRICKELLIA Brickellbush, tasselflower

Shrubs, **subshrubs**, or **herbs** perennial, often aromatic; taprooted. **Stems** 1+ from basal caudices, erect or ascending, simple or branched distally or throughout, tomentulose to villous or stipitate-glandular. **Leaves** principally cauline, alternate or subopposite, ovate to lanceolate or elliptic, sometimes hastate, margins entire to denticulate or serrate, 3-veined from bases, surfaces glabrous to villous, glandular-puberulent, or gland-dotted; sessile or petiolate. **Inflorescences** raceme-like or panicle-like arrays, heads usually axillary on stems and branches or in terminal clusters, rarely solitary and terminal. **Heads** discoid, erect or nodding. **Involucres** cylindrical or campanulate. **Phyllaries** in 3–9 unequal series, erect in flower, spreading or reflexed in fruit, herbaceous or chartaceous, with 3 or more prominent, parallel veins, tips acute, erect or recurved, sometimes acuminate with flexuous tips, surfaces glabrous to puberulent, villous, or glandular-pubescent. **Receptacles** flat; paleae 0. **Florets** bisexual; corollas cylindric, filiform or gradually widened distally, slightly or well exceeding involucres, lobes 5, minute; stamens 5, included; style lobes usually well exserted. **Fruits** columnar-fusiform, truncate, 10-ribbed, glabrous to scabrous or pubescent, sometimes stipitate-glandular; pappi in 1 series of numerous persistent bristles, white or yellowish.

Central America, North America. ~100 species; 5 species treated in *Flora*.

Eupatorieae, the tribe of Asteraceae to which *Brickellia* is traditionally assigned, was long considered to be co-equal to a large and inclusive sunflower tribe, Heliantheae (e.g., Cronquist 1994). However, in a molecular phylogenetic study of helenioid Heliantheae (Baldwin et al. 2002), it was discovered that the latter tribe should itself be divided into seven separate tribes. Modified in this way, Heliantheae, in the narrow sense, stands as a well-marked DNA-based clade within this group, while Eupatorieae remains unchanged but is paired with the new tribe Madieae (tarweeds) in a separate clade. Members of Eupatorieae always lack ray florets and have disc corollas that are white, pink, purple, or pale yellowish green, not bright yellow as in many other members of the sunflower family.

Key to *Brickellia* species:

1. Heads nodding, outermost phyllary tips long-acuminate . . . . . ***B. grandiflora***

1' Heads erect or spreading, outermost phyllary tips obtuse to acute to acuminate (sometimes long-acuminate in *B. greenei*).

  2. Bracts subtending heads large and leafy . . . . . ***B. greenei***

  2' Bracts subtending heads small, not leaf-like.

    3. Leaves linear to lanceolate or elliptic, sessile or subsessile . . . . . ***B. oblongifolia***

    3' Leaves ovate to suborbiculate or triangular-ovate, usually short-petiolate.

      4. Leaf blades 1.5–7 cm; outer phyllary tips erect . . . . . ***B. californica***

      4' Leaf blades 0.5–2.4 cm; outer phyllary tips usually spreading or recurved . . . . . ***B. microphylla*** var. ***microphylla***

Brickellia californica (Torrey & A. Gray) A. Gray (p. 206)

California tasselflower, California brickellbush

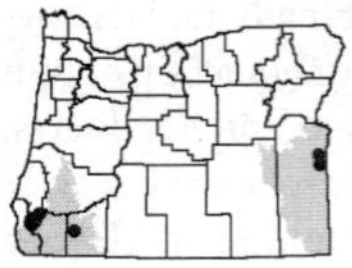

Shrubs deciduous, 5–20 dm. **Stems** puberulent, distally tomentulose, sometimes glandular-puberulent. **Leaves** blades triangular-ovate, 1.5–7 cm, bases truncate or attenuate, margins denticulate or serrate-dentate, tips acute or obtuse, surfaces glabrous or puberulent, sometimes stipitate-glandular, minutely gland-dotted; usually short-petiolate. **Heads** erect or spreading, bracteate. **Involucres** narrowly campanulate, 6–12 mm. **Phyllaries** 21–35 in 6–8 series; outer ovate to lanceolate, tips erect, obtuse to acuminate; innermost linear, tips obtuse to acuminate, glabrous or puberulent, not obviously glandular. **Disc florets** 8–18; corollas 5.5–8 mm, yellow-green. **Fruits** 2.5–3.5 mm, pale brown, scabrous; pappus bristles 6–7 mm. **2*n***=18.

Woodlands, hillsides, riverbanks. Flowering Jul–Sep. 0–1300 m. Owy, Sisk. CA, ID, NV; southeast to TX, south to Mexico. Native.

Brickellia grandiflora (Hooker) Nuttall (p. 206)

Large-flowered tasselflower, tasselflower brickellbush

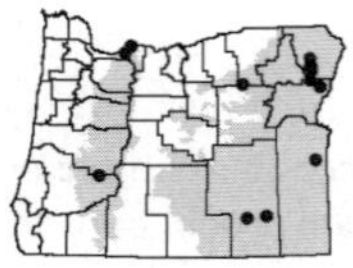

Herbs 2.5–9 dm. **Stems** puberulent, distally tomentulose. **Leaves** blades triangular-ovate to ovate or lanceolate, 1.5–10 cm, bases truncate or attenuate, margins serrate-dentate or denticulate, tips acute or acuminate, surfaces glabrous or puberulent, sometimes scabrous abaxially on veins, minutely gland-dotted; short- to long-petiolate. **Heads** nodding, bracteate. **Involucres** campanulate, 7–12 mm. **Phyllaries** 30–40 in 4–6 series; outer ovate, puberulent, tips usually flexuous, long-acuminate; inner lance-linear, tips acute to acuminate, surfaces usually glabrous, not obviously glandular. **Disc florets** 20–50; corollas 6.5–7.5 mm, yellowish green. **Fruits** 4–5 mm, brown, scabrous; pappus bristles 4–6 mm. **2*n***=18.

Coniferous woodlands, shrublands, cliffs, rocky mountainsides, riverbanks, roadsides. Flowering Jul–Sep. 50–2200 m. BR, BW, Casc, Owy. CA, ID, WA; north to British Columbia, northeast to Alberta, east to NE, southeast to TX. Native.

Brickellia greenei A. Gray

Greene's tasselflower, Greene's brickellbush

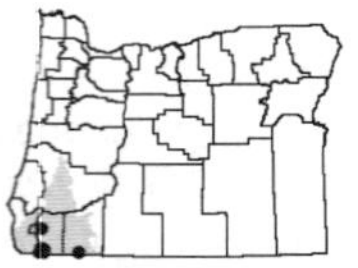

Herbs 2–5 dm. **Stems** villous, stipitate-glandular. **Leaves** blades ovate-lanceolate, 1–3 cm, bases usually attenuate, sometimes truncate, margins entire or serrate-dentate, tips acute, surfaces stipitate-glandular, minutely gland-dotted; sessile or short-petiolate. **Heads** erect, solitary, each subtended by 2–5 enlarged, leafy bracts; bracts usually ovate, stipitate-glandular. **Involucres** campanulate, 12–18 mm. **Phyllaries** (24)36–42 in 3–7 series, tips acute or sometimes long-acuminate; outer erect, lanceolate, glabrous or gland dotted; innermost ± linear, glabrous. **Disc florets** 40–60; corollas 8–10 mm, yellowish green. **Fruits** 5.5–7 mm, brown, scabrous; pappus bristles 8–10 mm. **2*n***=18.

Woodlands, shrublands, riverbanks, rocky slopes. Flowering Jul–Sep. 400–2000 m. Sisk. CA. Native.

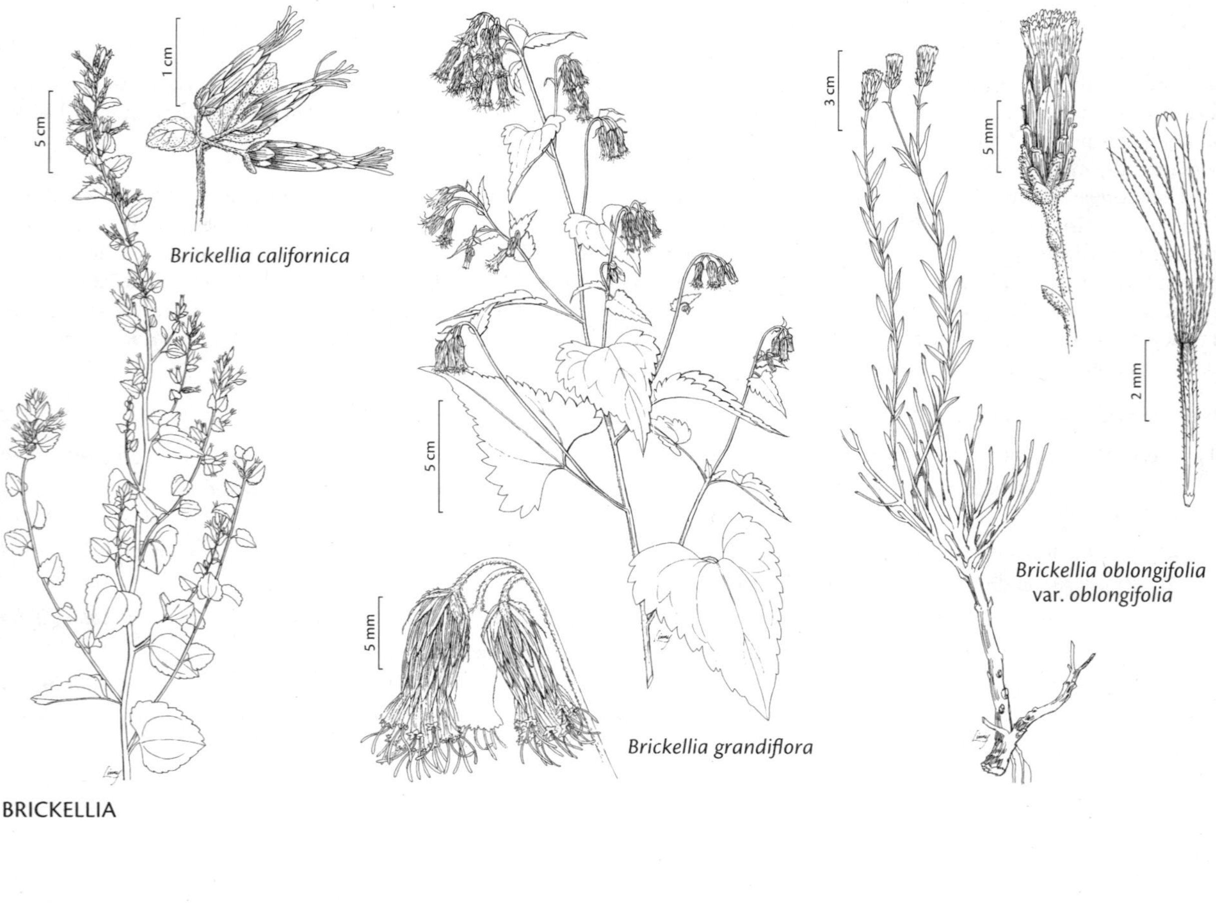

Brickellia californica

Brickellia grandiflora

Brickellia oblongifolia var. *oblongifolia*

BRICKELLIA

Brickellia microphylla (Nuttall) A. Gray var. *microphylla*

Small-leaved brickellia, littleleaf brickellbush

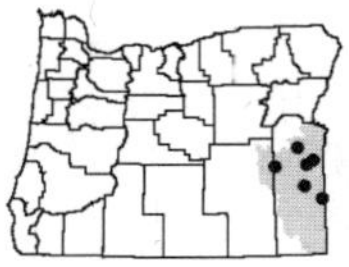

Shrubs evergreen, 3–7 dm. **Stems** glandular-puberulent. **Leaves** blades ovate or suborbiculate, bases attenuate or truncate, 0.5–2.4 cm, margins entire to denticulate or serrate-dentate, tips acute, surfaces glandular-puberulent, minutely gland-dotted; short-petiolate. **Heads** erect or spreading, bracteate. **Involucres** cylindrical or narrowly campanulate, 7–12 mm. **Phyllaries** 30–48 in 4–9 series, tips acute to acuminate; outer elliptic-lanceolate, tips usually spreading or reflexed, glandular-puberulent; inner lance-linear, glabrous. **Disc florets** 8–30; corollas 5.5–7 mm, pale yellow. **Fruits** 3.5–5 mm, brown, scabrous; pappus bristles 6–7 mm. **2*n***=18.

Shrublands, rocky slopes, cliffs, riverbanks, streambanks. Flowering Jun–Sep. 400–1500 m. Owy. CA, ID, NV, WA; southeast to UT. Native.

Brickellia oblongifolia Nuttall

Narrowleaf brickellbush, Mojave brickellbush, narrowleaf brickellia

Herbs or **subshrubs** 1–6 dm. **Stems** glandular-puberulent, sometimes glabrous near bases. **Leaves** blades linear to lanceolate or elliptic, 1–3.5 cm, margins entire, tips acute, surfaces puberulent to nearly glabrous, often minutely glandular; sessile or subsessile. **Heads** erect or spreading, bracteate. **Involucres** cylindrical or campanulate, 10–20 mm. **Phyllaries** 25–35 in 4–6 series, tips acute to acuminate; outer lanceolate, usually glandular-puberulent; inner linear, glabrous or glandular-puberulent. **Disc florets** 25–50; corollas 5–10 mm, white or pale greenish. **Fruits** 5–7 mm, brown, stipitate-glandular or setose-hispidulous; pappus bristles 8–10 mm. **2*n***=18.

Southwestern Canada and western US. 2 varieties.

Key to *Brickellia oblongifolia* varieties:

1. Fruits stipitate-glandular, sometimes also hispidulous. . . . . ***B. o.*** var. ***oblongifolia***

1' Fruits setose-hispidulous, usually lacking glands . . . . ***B. o.*** var. ***linifolia***

Brickellia oblongifolia var. *linifolia* (D.C. Eaton) B.L. Robinson

Flax-leaved brickellia

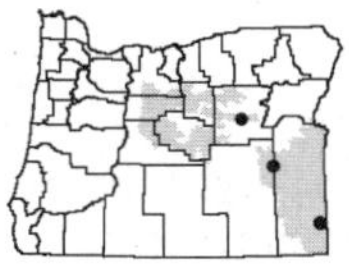

Fruits setose-hispidulous, usually lacking glands.

Rocky slopes, river floodplains, roadsides. Flowering May–Sep. 900–1300 m. Lava, Owy. CA, ID, NV; southeast to NM. Native.

Brickellia oblongifolia var. *oblongifolia* (p. 206)

Narrowleaf brickellbush, Mojave brickellbush, narrowleaf brickellia

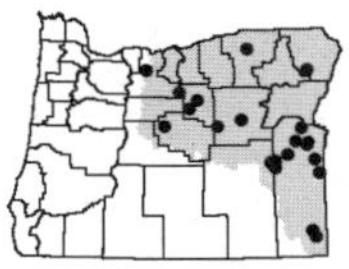

Fruits stipitate-glandular, sometimes also hispidulous.

Shrublands, rocky slopes, streambeds, gravel bars, in gravelly or sandy soils. Flowering Jun–Aug. 300–1700 m. BW, Col, Lava, Owy. ID, NV, WA; north to British Columbia, northeast to MT. Native.

CACALIOPSIS Silvercrown

Genus composed of one species, *C. nardosmia.*

Cacaliopsis nardosmia (A. Gray) A. Gray

Silvercrown

Herbs perennial, 2–10 dm; fibrous-rooted, rhizomatous. **Stems** erect, simple or branched distally, glabrous or floccose-tomentose, especially proximally. **Leaves** basal and cauline, alternate, broadly ovate or round, 10–45 cm, upper reduced, bases truncate or more often cordate, margins shallowly or more often deeply and irregularly palmately lobed, lobes dentate or lobed, secondary lobes entire or denticulate, teeth triangular, palmately veined, tips acute or obtuse, surfaces abaxially floccose-tomentose or finely tomentulose, adaxially glabrous or loosely floccose-tomentose; petiolate. **Inflorescences** heads (1)3–12 in terminal raceme-like or panicle-like arrays. **Heads** discoid; short- or long-peduncled; bracts 0 or with linear bract at base. **Involucres** campanulate in flower, ovoid in fruit, 10–20 mm. **Phyllaries** 6–22 in 1–2 series, ± equal, erect, linear-lanceolate, strongly ridged along midveins, surfaces glabrous or floccose-tomentose. **Receptacles** flat; paleae 0. **Florets** 20–50, bisexual; corollas barely exceeding involucres, yellow; stamens 5, exserted. **Fruits** cylindric, truncate, 6–8.5 mm, brown, glabrous, 12–15-veined; pappi of numerous white, minutely barbed bristles. **2*n***=60.

Western United States. 2 subspecies.

The species is quite different in its vegetative morphology from *Luina*, to which it has sometimes been assigned in the past.

The two subspecies of *Cacaliopsis nardosmia* are distinguished by minor and overlapping morphological features but have nearly separate geographic ranges. They overlap in northern Jackson County, however, where intermediate plants are found.

Key to *Cacaliopsis nardosmia* subspecies:

1. Involucres 10–12(14) mm, phyllaries 6–16; leaves glabrous adaxially . . . . ***C. n.*** ssp. ***glabrata***

1' Involucres (12)14–20 mm, phyllaries 10–22; leaves loosely floccose-tomentose adaxially, at least when young . . . . ***C. n.*** ssp. ***nardosmia***

Cacaliopsis nardosmia ssp. *glabrata* (Piper) Piper

Tall silvercrown

Luina nardosmia (A. Gray) Cronquist var. *glabrata* (Piper) Cronquist

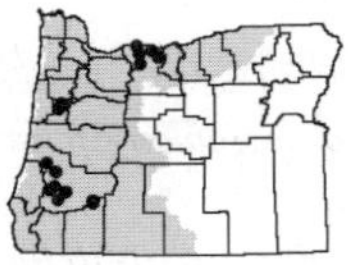

Leaves surfaces glabrous to floccose-tomentose or tomentulose abaxially, adaxially glabrous. **Involucres** 10–12(14) mm. **Phyllaries** 6–16.

Mixed conifer woodlands, pine woods, oak woods, rocky slopes, ridges, scree, roadsides. Flowering May–Jul. 50–1400 m. Casc, Col, CR, ECas, Sisk, WV. WA. Native.

Cacaliopsis nardosmia ssp. *nardosmia* (p. 208)

Silvercrown

Luina nardosmia (A. Gray) Cronquist var. *nardosmia*

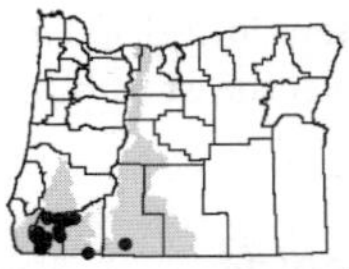

Leaves surfaces abaxially lightly or densely floccose-tomentose, sometimes glabrate, adaxially loosely floccose-tomentose, at least when young. **Involucres** (12)14–20 mm. **Phyllaries** 10–22.

Mixed conifer woodlands, grasslands, rocky hillsides, roadsides. Flowering Apr–Jun. 200–1400 m. ECas, Sisk. CA. Native.

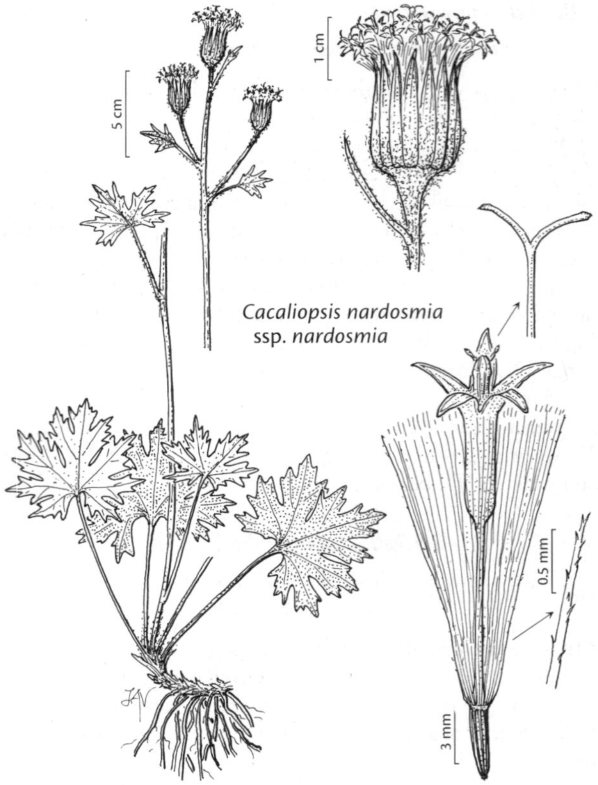

CACALIOPSIS

CALYCADENIA Western rosinweed, tackweed

Gerald D. Carr and Robert L. Carr

Herbs annual, 10–80 cm, self-incompatible. **Stems** erect, simple or more often with few to many ascending branches above the base. **Leaves** basal, mostly withering before flowering, and cauline, mostly alternate, simple; blades linear, grass-like, margins entire or sparsely toothed, revolute; sessile. **Inflorescences** heads 1–3+ per node in axillary or terminal, ± spike-like arrays. **Heads** radiate; peduncular bracts often crowded and closely subtending heads, tips with (0)1–3+ tack-shaped glands. **Phyllaries** (1)2–6(8) in 1 series, falling, each partly enveloping a ray ovary, margins often fringed with shaggy hairs, especially at tips. **Receptacles** ± flat, hairy; paleate. **Paleae** in 1 series between ray and disc flowers, ± connate, forming plicate cups, surfaces with (0)1–5+ tack-shaped glands, tips often acute to acuminate, fringed with shaggy hairs. **Ray florets** (1)2–6(8), pistillate, fertile; rays 3-lobed, lobes spreading or nearly parallel. **Disc florets** 5–25, bisexual or functionally staminate; corollas white or yellow, lobes 5, deltate to lance-ovate; anthers usually purple at anthesis; style branches well developed, short-bristly. **Ray fruits** obcompressed, somewhat 3-angled, 2.5–4 mm, abaxial sides basically 2-faced, angles between those faces ≥ 90°, adaxial sides nearly flat, glabrous or glabrate; pappi 0. **Disc fruits** narrowly clavate, 2–4 mm; pappi 0 or of 7–14 oblong or aristate scales.

California and Oregon. 10 species; 2 species treated in *Flora.*

Other than the two species extending into Oregon, *Calycadenia* is nearly confined to the Central Valley and adjacent lower elevations of bordering mountain ranges of California. At least three species, including the two in Oregon, each comprise three or more distinct chromosome races that lack clear and consistent morphologic differentiation. These chromosome races are differentiated primarily by reciprocal chromosome translocations and, in some cases, paracentric and/or pericentric inversions (Carr 2003). Collectively, the species of *Calycadenia* also represent a well-documented example of variation in haploid chromosome number (dysploidy), with molecular support for a continuous descending series from an ancestral number of n=7 to n=4 (Baldwin 2003; Carr 2003). The common name for the genus is linked to the copious quantities of aromatic resinous exudates produced by the distinctive vascularized, multicellular, tack-shaped glands found in *Calycadenia*; these have been described in detail by Carlquist (2003).

Key to *Calycadenia* species:

1. Margins of peduncular bracts conspicuously pectinate-fimbriate; central ray corolla lobes nearly equal to laterals, generally widest near or beyond middle, white to pinkish; ray fruits smooth . . . . ***C. fremontii***

1' Margins of peduncular bracts glabrous or merely minutely scabrous; central ray corolla lobes somewhat smaller than laterals, widest at base, yellow; ray fruits generally rough-wrinkled . . . . ***C. truncata***

Calycadenia fremontii A. Gray

Tufted tackweed, Fremont's western rosinweed

Calycadenia ciliosa Greene

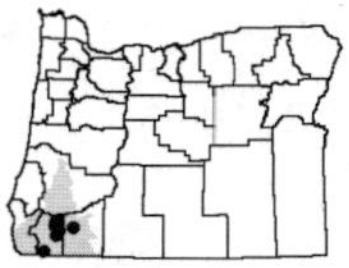

Plants 10–65 cm. **Stems** hispidulous to strigose. **Leaves** 2–8 cm, surfaces hispid to hispidulous and ± long-hairy, especially on margins and adaxial faces. **Heads** peduncular bracts narrowly elliptic to narrowly oblanceolate, 2–10 mm, hispidulous, margins strongly pectinate-fimbriate, tips ± rounded, tack-shaped glands mostly 1–3. **Phyllaries** 3–6 mm, abaxial faces somewhat hispidulous to bristly, often finely glandular, tack-shaped glands (0)1–5+. **Paleae** 4–7 mm, abaxial faces ± bristly and finely glandular, often with 1–8+ scattered tack-shaped glands of various sizes. **Ray florets** rays 4.5–7 mm, white, sometimes with pink veins and/or darker red "eye" in the throat, central lobe about equaling or narrower than laterals, mostly obovate, widest beyond middle, laterals ± symmetric; sinuses nearly equaling rays. **Disc florets** bisexual; corollas 4–5 mm, white to pinkish. **Ray fruits** smooth. **Disc fruits** appressed-hairy; pappi of 8–14 long-aristate and short-blunt, fimbriate scales alternating, 1–3 mm. ***2n***=12.

Open, dry meadows, hillsides, gravelly outwashes, often associated with serpentine substrates. Flowering Jun–Sep. 200–500 m. Sisk. CA. Native.

In California, *C. fremontii* exhibits great morphologic variation and also comprises several named chromosome races. By contrast, the Oregon populations are relatively uniform in morphology, and the only chromosome race known to occur in our flora is "Dry Creek" (Carr & Carr 1983). Although this race exhibits both white- and yellow-flowered forms in California, the latter apparently do not occur in Oregon. A single depauperate plant collected by L.F. Henderson at Sams Valley in 1930 is the only current documentation of the occurrence of *C. fremontii* in Jackson County.

Calycadenia truncata de Candolle (p. 210)

Oregon western rosinweed, smooth tackweed

Calycadenia truncata DC. var. *scabrella* (Drew) Jeps., *Calycadenia truncata* DC. ssp. *truncata*

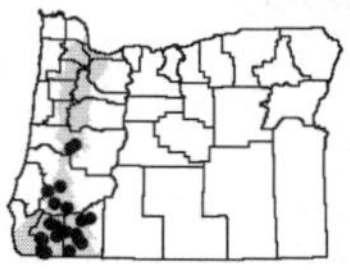

Plants 20–80 cm. **Stems** smooth, essentially glabrous. **Leaves** 2–10 cm, surfaces often minutely scabrous, otherwise nearly glabrous except for a few scattered stiff hairs up to 5 mm, distal leaves sometimes with solitary, terminal, tack-shaped glands. **Heads** peduncular bracts mostly narrowly oblong, 1–12 mm, glabrous to minutely scabrous, tips nearly truncate, tack-shaped glands (0)1, terminal. **Phyllaries** 4–6.5 mm, abaxial faces glabrous, tack-shaped glands usually 0. **Paleae** 5–8 mm, abaxial faces essentially glabrous to minutely scabrous distally, tack-shaped glands 0. **Ray florets** 2–6(8); corollas yellow, throat sometimes with red "eye"; rays (4)5–12 mm, central lobe smaller than laterals, symmetric, widest at base, laterals asymmetric; sinuses about ⅓ rays. **Disc florets** often functionally staminate; corollas 4–6 mm, yellow. **Ray fruits** generally rough-wrinkled. **Disc fruits** (when present) glabrous; pappi 0 or of 7–12, oblong, fimbriate scales, to 1 mm. ***2n***=14.

Dry, open hillsides, rocky ridges, talus, often associated with serpentine substrates. Flowering Jun–Sep. 100–800 m. Sisk, WV. CA. Native.

Calycadenia truncata is the most widespread species of the genus, ranging from Lane County, Oregon, to San Luis Obispo and Fresno counties, California. It is somewhat more widely distributed in southwestern Oregon than *C. fremontii*, but the two species co-occur in at least three Josephine County localities.

CANADANTHUS Aster

Genus composed of one species, *C. modestus*.

Canadanthus modestus (Lindley) G.L. Nesom (p. 210)

Great northern aster, few-flowered aster, giant mountain aster

Aster modestus Lindl.

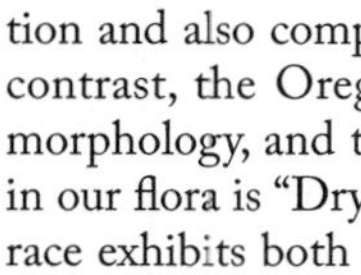

Herbs perennial, 2.5–18 dm; rhizomatous. **Stems** 1, erect, often producing thin, horizontal rhizomes near the base, often purple, glabrous proximally, villous distally, becoming glandular below the inflorescences. **Leaves** cauline, alternate, lower often withering by anthesis, oblong-lanceolate, 2–13 cm, bases usually slightly auriculate, margins entire or remotely denticulate, sometimes scabridulous, tips acute or acuminate, surfaces of proximal leaves glabrous, adaxially sometimes minutely spiculate, distal abaxially strigose to villous; sessile. **Inflorescences** few to numerous heads in open or rarely congested, raceme- or panicle-like arrays. **Heads** radiate; peduncles bracteate, glandular-puberulent. **Involucres** narrowly to broadly campanulate, 6–11 mm. **Phyllaries** 20–60+ in several series, equal, linear, green to purple, midribs evident or obscure, tips acuminate, surfaces glandular-puberulent. **Receptacles** flat; paleae 0. **Ray florets** 20–45, pistillate; rays 7–15 mm, pale to dark purple. **Disc florets** 30–65, bisexual; corollas tubular, slightly enlarged distally, slightly exceeding involucres, yellow turning purple, lobes 5, erect, triangular; stamens 5, well exserted from corollas; stigmas exserted from stamen tubes. **Fruits** obconic, 2.5–3 mm, brown, sparsely strigose, conspicuously 4–10-veined; pappi of numerous fine bristles, 5–6 mm, brownish. ***2n***=18.

Open wet meadows, marshes, seeps, riverbanks, shady woods, wet cliffs. Flowering Jul–Sep. 50–1900 m. BW, Casc, CR, ECas, WV. ID, WA; north to AK, northeast to New Brunswick. Native.

The taxonomic position of *Canadanthus* as a distinct genus from *Aster* is supported by the DNA studies of Xiang and Semple (1996) and Semple et al. (2002).

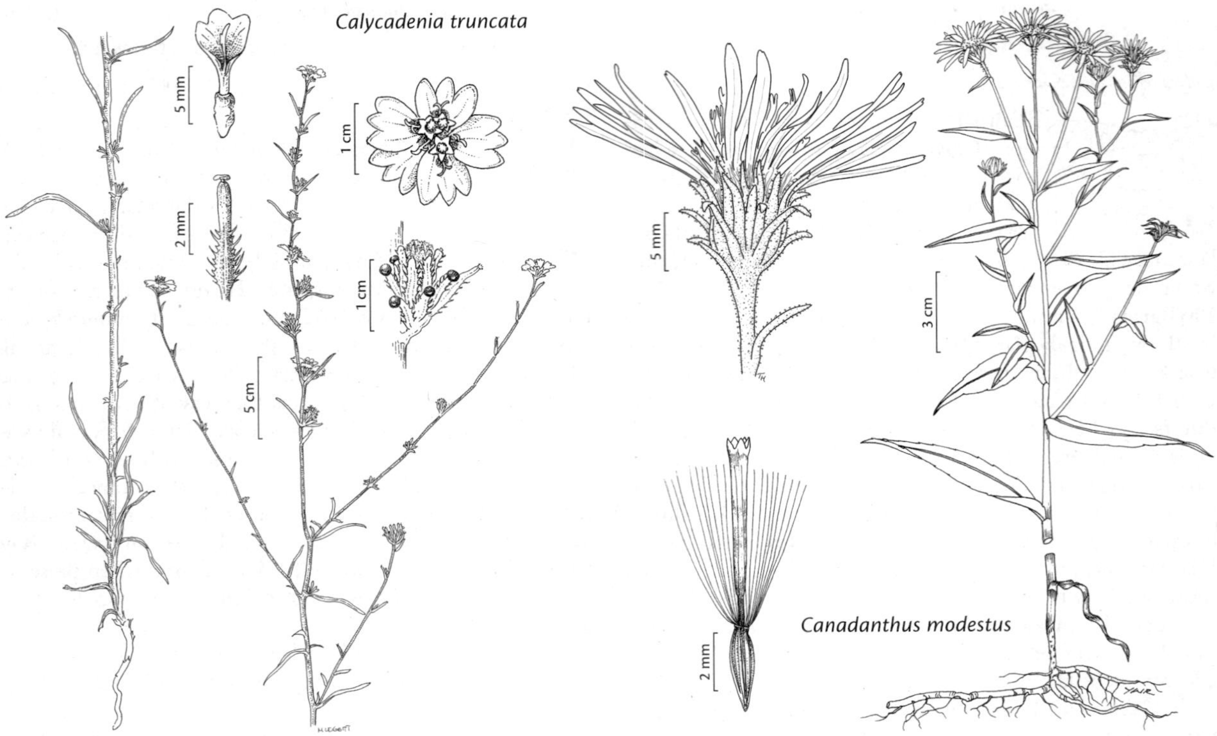

CALYCADENIA • CANADANTHUS

CARDUUS Plumeless thistle

Herbs annual or biennial. **Stems** erect, glabrate to tomentose or hispid, spiny-winged. **Basal leaves** pinnately lobed, lobe margins irregularly dentate, surfaces abaxially hispid to tomentose, adaxially punctate or hispid; petiolate. **Cauline leaves** alternate, lanceolate or oblanceolate, simple to shallowly or deeply lobed or pinnatifid, bases decurrent, with spiny wings, margins spiny, tips acute, surfaces abaxially glabrous or tomentose, adaxially glabrous to hispid-punctate; sessile. **Inflorescences** heads solitary or clustered in spike-, raceme-, or panicle-like arrays. **Heads** discoid; sessile or pedunculate. **Involucres** cylindric to spherical. **Phyllaries** many in several unequal series, proximally appressed, distally erect or reflexed, linear to ovate, tips spiny, surfaces glabrous to tomentose, rarely arachnoid-villous. **Receptacles** flat, with numerous narrow, tapering, bristly scales; paleae 0. **Florets** bisexual, fertile; corollas white to pink or purple, tubes elongate, linear, cylindrical, lobes 5, linear; styles exserted. **Fruits** attached basi-laterally on receptacles, cylindric-fusiform, bases tapered, glabrous; pappi of numerous smooth, white, persistent bristles.

Africa, Asia, Europe, North America. ~90 species; 4 species treated in *Flora*.

Carduus crispus has been collected once in Oregon (2016) in Wallowa County but does not yet appear established. It is similar to *C. acanthoides* but has densely tomentose leaves.

Key to *Carduus* species:

1. Involucres 20–60 mm in diameter; peduncles not winged . . . . **C. nutans**

1' Involucres 7–25 mm in diameter; peduncles winged or heads sessile.

  2. Involucres spherical or hemispheric . . . . ***C. acanthoides*** ssp. ***acanthoides***

  2' Involucres cylindric or ellipsoid.

    3. Phyllary margins not membranous; pappus bristles 15–20 mm . . . . ***C. pycnocephalus***

    3' Phyllary margins membranous; pappus bristles 10–15 mm . . . . ***C. tenuiflorus***

Carduus acanthoides Linnaeus ssp. *acanthoides* (p. 211)

Plumeless thistle

Plants annual or biennial, 3–20+ dm. **Basal leaves** simple or pinnate. **Cauline leaves** linear-lanceolate, 5–25 cm, deeply lobed or pinnatifid, surfaces abaxially glabrous or with curly trichomes on veins, adaxially glabrous. **Heads** solitary or in clusters of 2–5; peduncles spiny-winged. **Involucres** spherical or hemispheric, 10–25 mm in diameter. **Phyllaries** linear to narrowly-lanceolate, midribs prominent. **Disc florets** corollas 13–20 mm, purple or white. **Fruits** 2.5–3 mm, tan or brown; pappus bristles 11–13 mm. **2*n***=22.

Disturbed areas, riverbanks, edges of ponds. Flowering Jul–Oct. 100–1500 m. BW, ECas. CA, ID; scattered across southern Canada and most of US; Asia, Europe. Exotic.

Carduus nutans Linnaeus

Musk thistle, nodding thistle

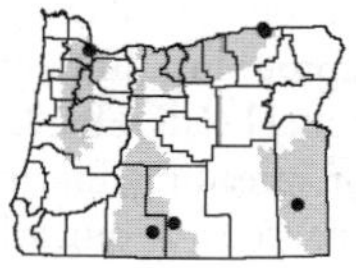

Plants annual or biennial, 4–20+ dm. **Basal leaves** 10–40 cm, 1–2-pinnate. **Cauline leaves** lanceolate to elliptic or linear, 4–25 cm, shallowly to deeply lobed or pinnate, lobe margins and tips spiny, surfaces glabrous or tomentose. **Heads** solitary or in pairs, often nodding; peduncles not winged. **Involucres** hemispheric, 20–60 mm in diameter. **Phyllaries** broadly or narrowly lanceolate, midribs prominent. **Disc florets** corollas 15–28 mm, purple. **Fruits** 4–5 mm, tan or brown; pappus bristles 13–25 mm. **2*n***=16.

Disturbed areas, clearcuts. Flowering Jun–Oct. 0–1900 m. Col, ECas, Owy, WV. CA, ID, NV, WA; throughout most of North America; Asia, Europe. Exotic.

Carduus pycnocephalus Linnaeus

Italian thistle, compact-headed thistle

Carduus pycnocephalus L. ssp. *pycnocephalus*

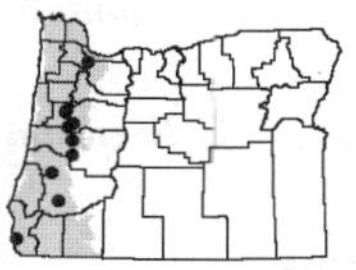

Plants annual, 2–20 dm. **Basal leaves** 10–25 cm, deeply lobed or pinnatifid. **Cauline leaves** obovate or oblanceolate to linear-elliptic, 4–15 cm, shallowly or deeply lobed or pinnatifid, lobe margins and tips spiny, surfaces abaxially tomentose, adaxially hispid-pustulose. **Heads** 2–5 per cluster; peduncles winged. **Involucres** cylindric or ellipsoid, 10–25 mm in diameter. **Phyllaries** narrowly lanceolate, margins not membranous, midribs prominent. **Disc florets** corollas 14–16 mm, purple. **Fruits** 4–6 mm, tan or brown; pappus bristles 15–20 mm. **2*n***=54, 60, 62.

Disturbed areas, waste areas, riverbanks. Flowering May–Jul. 50–700 m. CR, Sisk, WV. CA, ID; widely scattered in US; Asia, Europe. Exotic.

Carduus tenuiflorus Curtis

Slender-flowered thistle

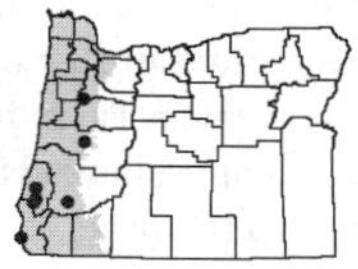

Plants annual, 2–20 dm. **Basal leaves** 15–30 cm, pinnately lobed. **Cauline leaves** obovate or oblanceolate to elliptic, 5–20 cm, shallowly to deeply lobed or pinnatifid, surfaces abaxially tomentose, adaxially hispid-pustulose. **Heads** 3–20 per cluster; sessile or short-peduncled. **Involucres** cylindric or ellipsoid, 7–25 mm in diameter. **Phyllaries** narrowly lanceolate, margins membranous, midribs inconspicuous. **Disc florets** corollas 10–14 mm, pink. **Fruits** 4–5 mm, brown; pappus bristles 10–15 mm. **2*n***=54.

Disturbed areas, riverbanks, open woods. Flowering May–Jul. 0–200 m. CR, Est, Sisk, WV. CA, WA; widely scattered in US; Asia, Europe. Exotic.

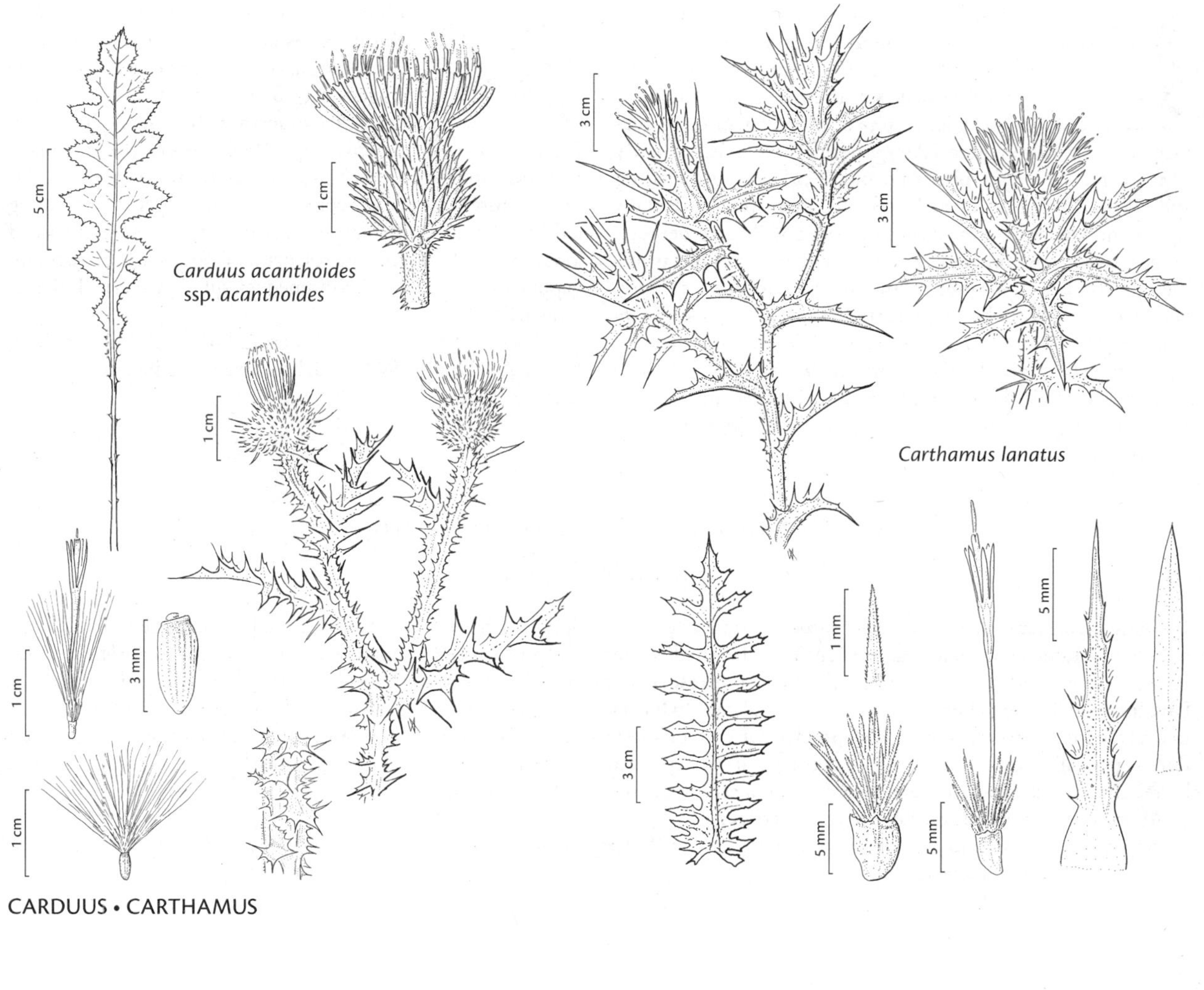

CARDUUS • CARTHAMUS

CARTHAMUS Distaff thistle

Herbs annual or perennial. **Stems** erect, branched distally or throughout, glabrous or loosely lanate-villous. **Basal leaves** absent or usually withering before flowering, margins pinnately lobed, ± spiny. **Cauline leaves** alternate; blades narrowly or broadly lanceolate to ovate, bases truncate to auriculate-clasping, margins dentate to pinnately lobed, ± spiny, surfaces both glabrous or minutely glandular-puberulent and adaxially floccose-pubescent; lower petiolate, distal sessile. **Inflorescences** solitary terminal heads. **Heads** discoid. **Involucres** ovoid to globose or hemispheric. **Phyllaries** in 5–6 series, erect to spreading; outer erect or spreading, linear to lanceolate or ovate, sometimes exceeding innermost, chartaceous proximally, 5–9-veined, distal portion herbaceous, 3-veined, spine-tipped; inner chartaceous throughout or herbaceous, acuminate, spine-tipped. **Receptacles** convex to conic, with dense covering of membranous, subulate scales; paleae 0. **Florets** bisexual; corollas tubular, slender, widened at base of throat or lobes, yellow to reddish, lobes 5, elongate, linear, erect; stamens 5, included among corolla lobes; styles well exserted from stamen tubes or not, stigma lobes minute. **Fruits** broadly oblong or cylindrical, abruptly tapered at base, truncate, with crown-like raised rims, few-angled, indistinctly veined; pappi 0 or of several series of unequal, subulate scales.

Mediterranean region. 14 species; 2 species treated in *Flora*.

A third introduced species of *Carthamus*, *C. creticus*, is credited to Oregon by Keil (2006), but the basis for this report is not known. The species differs from *C. lanatus* in its pale yellow disc corollas, the white vs. yellow color of its anthers and pollen, and the greater length of its outer phyllaries relative to the inner.

Key to *Carthamus* species:

1. Stems villous-puberulent; cauline leaf blades minutely glandular-puberulent on both sides, lightly floccose-pubescent adaxially, margins dentate or lobed, long-spined; phyllaries linear-lanceolate, tips acuminate, long-spined; disc corollas bright yellow . . . . *C. lanatus*

1' Stems glabrous; cauline leaf blades glabrous, margins dentate or denticulate, short-spined; phyllaries ovate-lanceolate, tips acute, short-spined; disc corollas yellow, becoming reddish . . . . *C. tinctorius*

Carthamus lanatus Linnaeus (p. 211)

Woolly distaff thistle

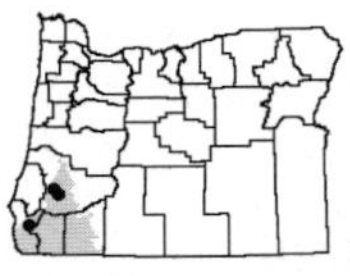

Plants annual, 4–10+ dm. **Stems** villous-puberulent. **Basal leaves** usually withering before flowering, 10–15 cm; petioles winged. **Cauline leaves** 2–15 cm, margins dentate or lobed, long-spined, with 3 main veins, surfaces both minutely glandular-puberulent, adaxially lightly floccose-pubescent. **Involucres** ovoid, 2–3.5 cm not including spine-tips. **Phyllaries** linear-lanceolate, often spinose-dentate, tips acuminate, long-spined, surfaces proximally ± tomentose, herbaceous portion minutely glandular-puberulent. **Florets** numerous; corollas 15–35 mm, yellow, tubular portions ± abruptly widened into short throats, exserted from involucres. **Fruits** 4–6 mm, gray-brown, glabrous; pappi of several series of unequal, subulate scales, 1–10 mm, brown. **2*n***=44.

Dry grasslands, fence rows, roadsides, disturbed areas. Flowering May–Sep. 50–300 m. Sisk. CA; widely scattered in North America; Europe. Exotic.

Carthamus tinctorius Linnaeus

Safflower

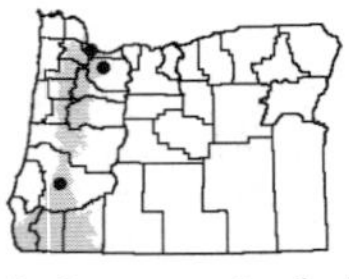

Plants annual, 3–10+ dm. **Stems** glabrous. **Basal leaves** absent. **Cauline leaves** 1.5–9 cm, margins dentate or denticulate, short-spined, with 3–5 main veins, surfaces glabrous. **Involucres** globose or hemispheric, 2–3 cm not including spine tips. **Phyllaries** ovate-lanceolate, margins entire, spinulose, tips acute, short-spined, surfaces glabrous. **Florets** numerous; corollas 20–30 mm, yellow, becoming reddish, throats 0, tubular portions abruptly widened at base of lobes, exserted from involucres. **Fruits** 7–9 mm, white, glabrous; pappi 0 or of several series of unequal, subulate scales, 1–4 mm, brown. **2*n***=24.

Fields, disturbed areas. Flowering Jun–Aug. 0–300 m. Sisk, WV. CA, ID, NV, WA; widely scattered in North America; Europe. Exotic.

CENTAUREA Knapweed, starthistle

Bridget N. Chipman

Herbs annual, biennial, or perennial. **Stems** prostrate to erect, simple to much branched, pubescent to becoming glabrous. **Leaves** basal and cauline, alternate, margins entire to toothed, sometimes spiny or 1–2-pinnate, surfaces glabrous, scabrous, villous, tomentose, or gland-dotted; petiolate or sessile. **Inflorescences** heads solitary or in corymb-like arrays. **Heads** discoid or ± radiate. **Involucres** cylindric to hemispheric or campanulate. **Phyllaries** many in 6–many series, unequal and appressed; appendages linear to ovate, with or without spines. **Receptacles** flat with bristles; paleae 0. **Florets** corollas yellow, white, blue, purple, or pink, outer florets generally sterile, slender and inconspicuous to expanded, showy, and ray-like, ± bilateral, inner bisexual, ± radial. **Fruits** ± barrel-shaped, attached ± to side, glabrous to pubescent; pappi 0 or of stiff bristles or scales.

Worldwide. ~500 species; 11 species treated in *Flora*.

Several species of *Centaurea* have been reported as locally escaped but not yet naturalized. These include *C. calcitrapa*, *C. iberica*, *C. nigrescens*, and *C.* ×*psammogena*.

Key to *Centaurea* species:

1. Corollas yellow.
  2. Heads closely subtended and obscured by leafy bracts. . . . . ***C. benedicta***
  2' Heads not subtended or obscured by leafy bracts.
    3. Central spines of outer phyllaries 5–10 mm, often purple. . . . . ***C. melitensis***
    3' Central spines of outer phyllaries 10–25 mm, straw-colored. . . . . ***C. solstitialis***

1' Corollas blue, purple, pink, or white.
  4. Outer phyllaries spine-tipped.
    5. Involucres 10–13 mm; florets 12–15 mm; corollas white, rarely pink or purple. . . . . ***C. diffusa***
    5' Involucres 7–10 mm; florets 7–12 mm; corollas pink to pale purple . . . . ***C. virgata*** ssp. ***squarrosa***
  4' Outer phyllaries not spine-tipped.
    6. Plants annual; lower leaves linear-lanceolate, ≤ 1 cm wide. . . . . ***C. cyanus***
    6' Plants perennial; lower leaves not linear-lanceolate, > 1 cm wide.
      7. Involucres 10–13 mm . . . . ***C. stoebe*** ssp. ***micranthos***
      7' Involucres 15–25 mm.
        8. Outer phyllaries not concealed by appendages; lower leaf petioles winged; upper leaves decurrent . . . . ***C. montana***
        8' Outer phyllaries obscured or somewhat obscured by appendages; lower leaf petioles not winged; upper leaves not decurrent.
          9. Outer phyllary appendages rounded, seldom triangular, margins scarious, coarsely dentate to entire . . . . ***C. jacea***
          9' Outer phyllary appendages ± triangular, margins dentate or dissected into wiry, comb-like lobes.
            10. Heads discoid; all florets fertile, not expanded or showy; phyllaries obscured by overlapping appendages . . . . ***C. nigra***
            10' Heads usually ± radiate; sterile florets occasionally present, expanded and showy; phyllaries somewhat obscured by overlapping appendages . . . . ***C. ×gerstlaueri***

Centaurea benedicta (Linnaeus) Linnaeus (p. 213)

Blessed thistle

Cnicus benedictus L.

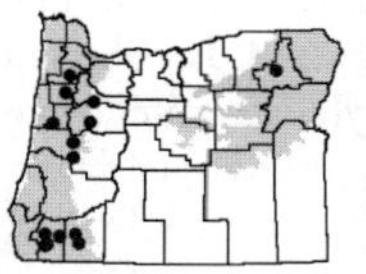

Plants annual, to 6 dm. **Stems** usually much branched, sparsely to densely tomentose. **Leaves** cauline, rarely also basal; oblanceolate to elliptic, 6–25 cm, margins shallowly to deeply pinnately lobed, dentate, armed with short weak spines, surfaces sparsely to densely tomentose, becoming glabrate, gland-dotted; petioles winged. **Inflorescences** solitary heads. **Heads** disciform; sessile; subtended and obscured by clusters of leaf-like bracts. **Involucres** ± spherical, 20–40 mm. **Phyllaries** tightly overlapping; outer ovate, green or straw-colored, with spreading spine tips, central spines 8–15 mm; inner lanceolate, purple to straw-colored, spines pinnately divided, central spines 5–15 mm; sessile. **Florets** yellow; sterile florets 3-lobed, not exceeding fertile florets; fertile florets 19–24 mm. **Fruits** 8–11 mm, straw-colored, with 20 prominent ribs; pappi in 2 series of awns, outer 9–10 mm, inner 2–5 mm. **2*n***=22.

Disturbed areas, roadsides, fields. Flowering Apr–Aug. 50–900 m. BW, CR, Sisk, WV. CA, WA; scattered in North America; nearly worldwide. Exotic.

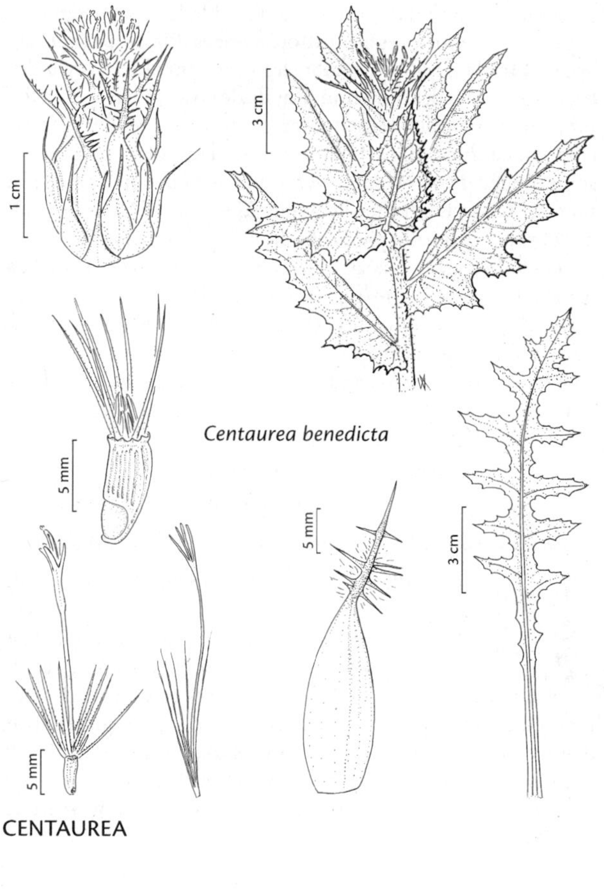

CENTAUREA

Centaurea cyanus Linnaeus

Bachelor's button, cornflower

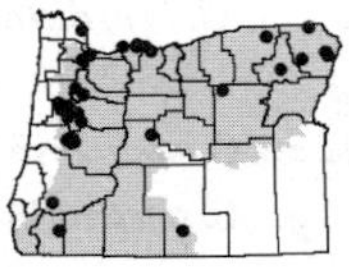

Plants annual, 2–10(12) dm. **Stems** generally 1, branched distally, loosely tomentose. **Leaves** surfaces densely tomentose to glabrate; basal blades linear-lanceolate, 3–15 cm, ≤ 1 cm wide, margins entire, shallowly dentate or with occasional linear lobes; cauline blades linear, margins entire to minutely dentate. **Inflorescences** cyme-like arrays. **Heads** ± radiate; peduncles slender. **Involucres** campanulate, 10–16 mm. **Phyllaries** green, tips and appendages dark brown to purple, margins scarious, fringed with slender teeth; outer ovate; inner oblong. **Florets** blue, rarely white, pink or purple; sterile florets enlarged and showy, 20–25 mm, margins irregular; fertile florets 10–15 mm. **Fruits** 4–5 mm, straw-colored to pale blue, finely hairy; pappi of unequal stiff bristles, 2–4 mm. **2*n***=24.

Disturbed areas, grasslands, open forests, roadsides, fields. Flowering Apr–Jul. 0–1500 m. BW, Casc, Col, ECas, Lava, Sisk, WV. CA, ID, NV, WA; throughout North America; Europe. Exotic.

Centaurea diffusa Lamarck

Diffuse knapweed, tumble knapweed

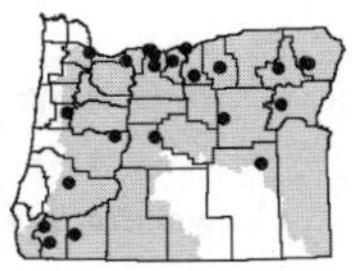

Plants annual, biennial or short-lived perennial, 2–8 dm. **Stems** branched throughout, puberulent, sparsely to densely tomentose. **Leaves** surfaces hispidulous and sparsely to densely tomentose, gland-dotted; basal and lower cauline blades 10–15 cm, 1- or 2-pinnately divided to narrowly lobed, petiolate; upper cauline blades much reduced, 1-pinnate to entire; sessile. **Inflorescences** diffuse, panicle-like arrays. **Heads** disciform; peduncles short. **Involucres** cylindric to narrowly ovoid, 10–13 mm. **Phyllaries** pale green, with a few prominent parallel veins; outer ovate to lanceolate, straw-colored, central spines 1–4 mm; inner lanceolate, lacerate or spine-tipped. **Florets** creamy white, rarely pink or purple; sterile florets ≈ fertile florets; fertile florets 12–15 mm. **Fruits** 2–3 mm, dark brown, glabrous; pappi absent or < 0.5 mm. **2*n***=18.

Disturbed areas, roadsides, gravelly or rocky sites, grasslands, woods. Flowering Jun–Aug. 0–1500 m. BW, Casc, Col, ECas, Lava, Owy, Sisk, WV. CA, ID, NV, WA; throughout North America; Europe. Exotic.

Hybrids of *C. diffusa* and *C. stoebe* ssp. *micranthos* are known to occur within Oregon.

Centaurea × *gerstlaueri* Erdner

Meadow knapweed

Centaurea × *pratensis* Thuill., *Centaurea* × *moncktonii* C.E. Britton

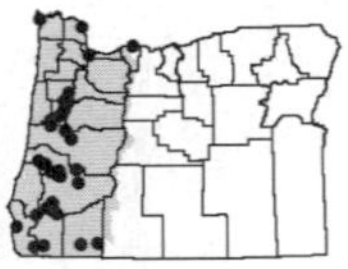

Plants perennial, 3–15 dm. **Stems** openly branched above, sparsely tomentose or villous to glabrous. **Leaves** sparsely tomentose or villous, sometimes only on veins, gland-dotted; basal and lower cauline blades oblanceolate to elliptic, 5–25 cm, > 1 cm wide, margins entire, dentate or pinnately lobed or wavy, petioles not winged; upper cauline blades oblanceolate to linear, gradually smaller, bases not decurrent, margins entire or dentate, sessile. **Inflorescences** 1–few in corymb-like arrays. **Heads** ± radiate, rarely discoid; peduncles leafy-bracted. **Involucres** ovoid to bell-shaped or hemispheric, 15–18 mm. **Phyllaries** outer lanceolate to ovate, somewhat obscured by overlapping appendages, ± triangular, margins dentate or dissected into wiry, comb-like lobes, appendages light to dark brown; inner irregularly dentate or lobed, tips truncate. **Florets** purple, rarely white; sterile florets absent or occasional, exceeding fertile florets, expanded and showy; fertile florets 15–22 mm. **Fruits** 2.5–3.5 mm, tan, finely hairy; pappi absent or bristles < 1 mm.

Disturbed areas, roadsides, streambanks, meadows, pastures, emergent wetlands, forest openings. Flowering May–Nov. 0–1500 m. Casc, CR, Est, Sisk, WV. CA, ID, WA; north to British Columbia, northeast to MT, northeastern North America; Europe. Exotic.

Centaurea × *gerstlaueri* originated as a hybrid of *C. nigra* and *C. jacea* (and possibly also *C. nigrescens*). At least some races are believed to have hybridized prior to introduction to North America. The hybrid complex includes diploids and tetraploids.

Occasional hybrids between *C.* × *gerstlaueri* and *C. solstitialis* (*C.* × *kleinii*) have been reported in Oregon.

Centaurea jacea Linnaeus

Brown knapweed

Centaurea jacea L. ssp. *jacea*, *Centaurea jacea* L. ssp. *nigra* (L.) Bonnier & Layens, *Centaurea jacea* L. nothossp. *pratensis* (W.D.J. Koch) Čelak.

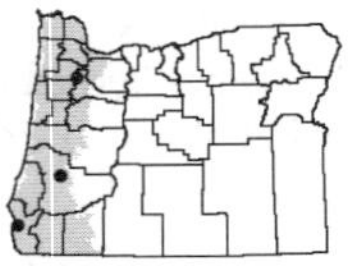

Plants perennial, 2–12(15) dm. **Stems** openly branched above, sparsely villous or tomentose to glabrate. **Leaves** scabrous, sparsely villous or tomentose to glabrate; basal and lower cauline blades oblanceolate to elliptic, 5–20 cm, > 1 cm wide, margins entire to shallowly toothed or lobed, petioles not winged; upper cauline blades linear to oblanceolate, gradually smaller, not decurrent, margins entire or minutely toothed, sessile. **Inflorescences** cyme-like arrays. **Heads** ± radiate; peduncles leafy-bracted. **Involucres** campanulate to hemispheric, 15–18 mm. **Phyllaries** outer ovate to lanceolate, green, concealed by overlapping appendages, appendages rounded, seldom triangular, light to dark brown, margins membranous, entire to coarsely dentate; inner margins irregularly dentate to lobed, tips truncate. **Florets** purple, rarely white; sterile florets expanded, exceeding fertile florets, 15–25 mm; fertile florets 15–18 mm. **Fruits** 2.5–3 mm, tan, finely hairy; pappi absent. **2*n***=22, 44.

Disturbed areas, roadsides, fields, pastures. Flowering May–Aug. 0–500 m. CR, Sisk, WV. CA, ID, WA; north to AK, northeast North America; Asia, Europe. Exotic.

Centaurea melitensis Linnaeus

Tocalote, Maltese starthistle

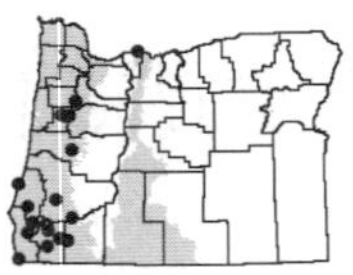

Plants annual, 1–10 dm. **Stems** branched above, sparsely tomentose, gland-dotted. **Leaves** villous, thinly to densely tomentose, gland-dotted; basal and lower cauline blades oblong to oblanceolate, 2–15 cm, margins dentate, pinnately lobed or entire, petiolate to tapering at base; upper cauline blades linear to oblong, 1–5 cm, bases decurrent, margins dentate to entire. **Inflorescences** 1–few in corymb-like arrays or heads solitary. **Heads** disciform. **Involucres** ovoid,

8–15 mm. **Phyllaries** outer ovate, bases spiny-fringed, surfaces tomentose to glabrate, central spines slender, 5–10 mm, often purple; inner entire, acute or spine-tipped. **Florets** yellow; sterile florets ≈ fertile florets; fertile florets 10–14 mm. **Fruits** 2.5–3 mm, white or light brown, finely hairy; pappi of unequal, white bristles, 1.5–3 mm. ***2n***=24.

Disturbed areas, roadsides, open woods, fields, pastures. Flowering May–Jul. 0–700 m. CR, ECas, Est, Sisk, WV. CA, ID, NV, WA; scattered in North America; Africa, Asia, Europe. Exotic.

Centaurea montana Linnaeus

Mountain cornflower, perennial cornflower, mountain bluet, montane starthistle

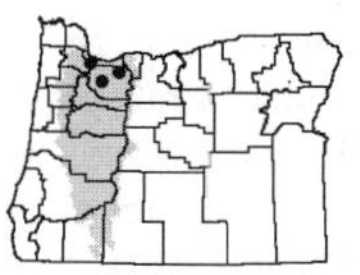

Plants perennial, 2.5–8 dm. **Stems** simple or sparsely branched, thinly tomentose to glabrate. **Leaves** thinly villous or tomentose to glabrate; basal and lower cauline blades ovate to oblong or oblanceolate, 10–30 cm, > 1 cm wide, margins entire to shallowly dentate or pinnately lobed, petioles winged; upper cauline blades ovate to oblong or lanceolate, gradually smaller, decurrent, margins entire or minutely dentate, sessile. **Inflorescences** few-headed corymb-like arrays. **Heads** ± radiate; pedunculate or sessile. **Involucres** campanulate, 20–25 mm. **Phyllaries** outer lanceolate to ovate, green, margins scarious with comb-like fringe, appendages not obscuring phyllary bodies, brown to black; inner occasionally without appendages. **Florets** sterile florets 25–45 mm, blue, rarely white, pink, or purple; fertile florets 20–30 mm, purple, anthers dark purple. **Fruits** 5–6 mm, brown to white with slightly appressed hairs; pappi of bristles, 0.5–1.5 mm.

Disturbed areas, roadsides, streambanks, open woods, sagebrush steppe. Flowering Apr–Jul. 0–500 m. Casc, WV. ID, WA; north to British Columbia, northeast to MT, northeastern North America; Europe. Exotic.

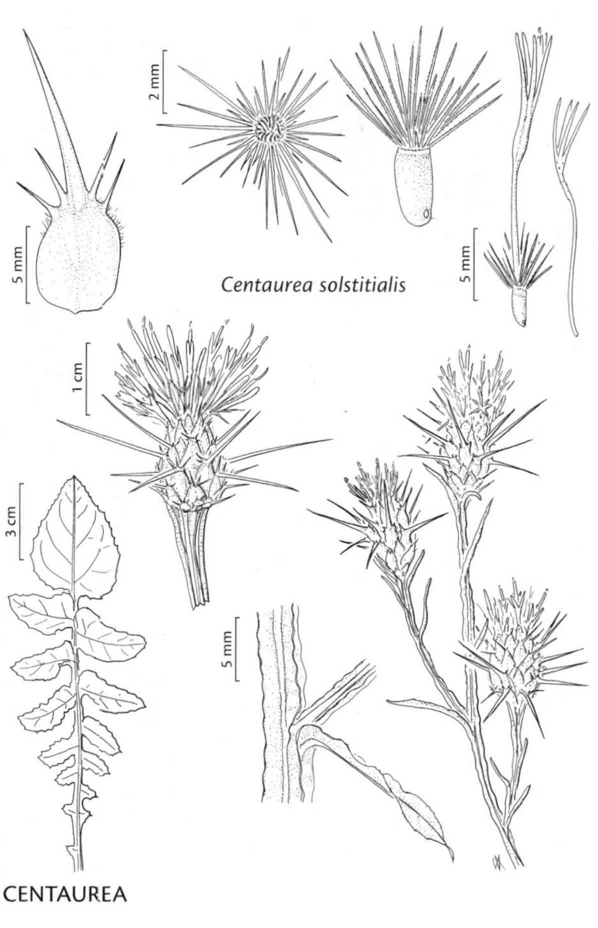

CENTAUREA

Centaurea nigra Linnaeus

Black knapweed, hardheads

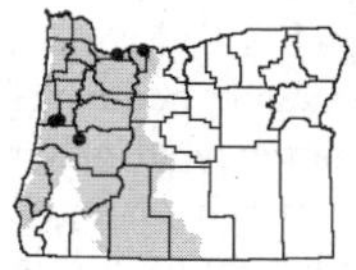

Plants perennial, 3–15 dm. **Stems** openly branched above, villous to scabrous and loosely tomentose to glabrous. **Leaves** villous, gland-dotted; basal and lower cauline blades oblanceolate to elliptic, 5–25 cm, > 1 cm wide, margins entire, shallowly dentate or pinnately lobed, petioles not winged; upper cauline blades linear to lanceolate, 3.5–8 cm, gradually smaller, bases not decurrent, margins entire or shallowly dentate, sessile. **Inflorescences** few-headed corymb-like arrays. **Heads** discoid; peduncles leafy-bracted. **Involucres** campanulate or hemispheric, 15–18 mm. **Phyllaries** outer lanceolate to ovate, obscured by overlapping appendages, margins dissected into wiry comb-like lobes, appendages erect, ± triangular, dark brown to black; inner margins irregularly toothed or lobed, tips truncate. **Florets** all fertile, 15–18 mm, purple, rarely white. **Fruits** 2.5–3 mm, tan, finely hairy; pappi of blackish, sometimes deciduous bristles, 0.5–1 mm. ***2n***=22, 44.

Disturbed areas, roadsides, fields. Flowering May–Sep. 0–500 m. Casc, CR, ECas, WV. CA, ID, WA; north to British Columbia, northeast to MT, northeastern North America; Europe. Exotic.

Centaurea solstitialis Linnaeus (p. 215)

Yellow starthistle, St. Barnaby's thistle

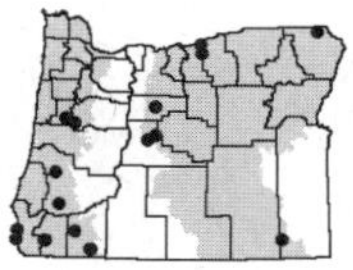

Plants annual or biennial, 1–10 dm. **Stems** 1–many, branched from bases, tomentose. **Leaves** tomentose and scabrous to bristly, bases decurrent; basal and lower cauline blades often absent at maturity, 5–15 cm, pinnately dissected or lobed, petioles winged; upper cauline blades oblong to linear, 1–10 cm, margins entire or slightly dentate, sessile. **Inflorescences** cyme-like clusters or heads solitary. **Heads** disciform; pedunculate. **Involucres** ovoid, 10–15 mm. **Phyllaries** outer ovate, bodies green, palmately spiny, central spines stout, 10–25 mm, straw-colored; inner with smaller hyaline appendages. **Florets** yellow; sterile florets ≈ fertile florets; fertile florets 13–20 mm. **Fruits** dimorphic, outer 2–3 mm, dark brown, glabrous, pappi 0; inner light brown or mottled, pappi of white bristles, 2–4 mm. ***2n***=16.

Disturbed areas, roadsides, woods, fields, pastures, streambanks. Flowering Jul–Oct. 0–2000 m. BR, BW, Col, CR, Est, Lava, Sisk, WV. CA, ID, NV, WA; throughout North America; Europe. Exotic.

Centaurea solstitialis is an aggressive weed of pastures and rangeland in western North America, often forming dense stands. This species is cumulatively toxic to horses, potentially causing a disorder known as "chewing disease."

See also *C.* ×*gerstlaueri.*

Centaurea stoebe Linnaeus ssp. ***micranthos*** (Gugler) Hayek

(p. 216)

Spotted knapweed

Centaurea maculosa Lam.

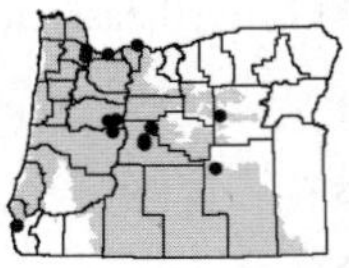

Plants perennial, 3–15 dm. **Stems** 1–many, branched from bases and/or above, thinly tomentose. **Leaves** thinly tomentose to glabrate, scabrous, gland-dotted; basal and lower cauline blades broad to narrowly lanceolate, 10–15 cm, > 1 cm wide, 1- or 2-pinnately divided, petiolate; upper cauline blades entire or pinnately lobed, sessile. **Inflorescences** panicle-like arrays. **Heads** ± radiate; pedunculate. **Involucres** ovoid, 10–13 mm. **Phyllaries** outer ovoid, green or pink-tinged with prominent parallel veins, margins fringed with slender teeth or lobes, appendages dark brown to purple on tips; inner oblong. **Florets** pink to purple, rarely white; sterile florets 15–25 mm; fertile florets 12–15 mm. **Fruits** 3–3.5 mm, white to pale brown, finely hairy; pappi of bristles 1–5 mm. **2*n***=18, 36.

Disturbed areas, roadsides, open forests, sagebrush. Flowering Jul–Sep. 0–1900 m. BR, Casc, CR, ECas, Lava, WV. CA, ID, NV, WA; throughout North America; Europe. Exotic.

Hybrids of *C. diffusa* and *C. stoebe* ssp. *micranthos* are known to occur within Oregon.

Centaurea virgata Lamarck ssp. ***squarrosa*** (Willdenow) Gugler

Squarrose knapweed

Centaurea squarrosa Willd., *Centaurea virgata* Lam. var. *squarrosa* (Willd.) Boiss.

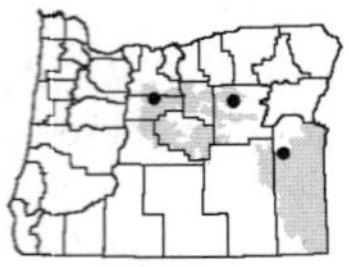

Plants perennial, 2–6 dm. **Stems** 1–many, branched from bases and/or above, loosely tomentose, scabrous, gland-dotted. **Leaves** linear to oblanceolate, finely tomentose, scabrous, gland-dotted; basal and lower cauline blades often absent at maturity, 10–15 cm, deeply 1- or 2-pinnately divided, petiolate; upper cauline blades 2–7 cm, pinnately divided to entire, sessile. **Inflorescences** panicle-like arrays. **Heads** disciform; pedunculate or sessile. **Involucres** cylindric to narrowly ovoid, 7–10 mm. **Phyllaries** outer ovate to lanceolate, green to purple-tinged, gland-dotted, central spines spreading to reflexed, 1–3 mm, straw-colored; inner lanceolate, tips acute. **Florets** pink to pale purple; sterile florets slender, not exceeding fertile florets; fertile florets 7–12 mm. **Fruits** 2.5–3.5 mm, light brown to straw-colored, glabrous; pappi of often deciduous bristles, 1–2.5 mm. **2*n***=36.

Disturbed areas, roadsides, rangeland, pastures, openings in forests. Flowering Jun–Sep. 500–1300 m. Lava, Owy. CA, NV; east to MI, southeast to UT; Asia. Exotic.

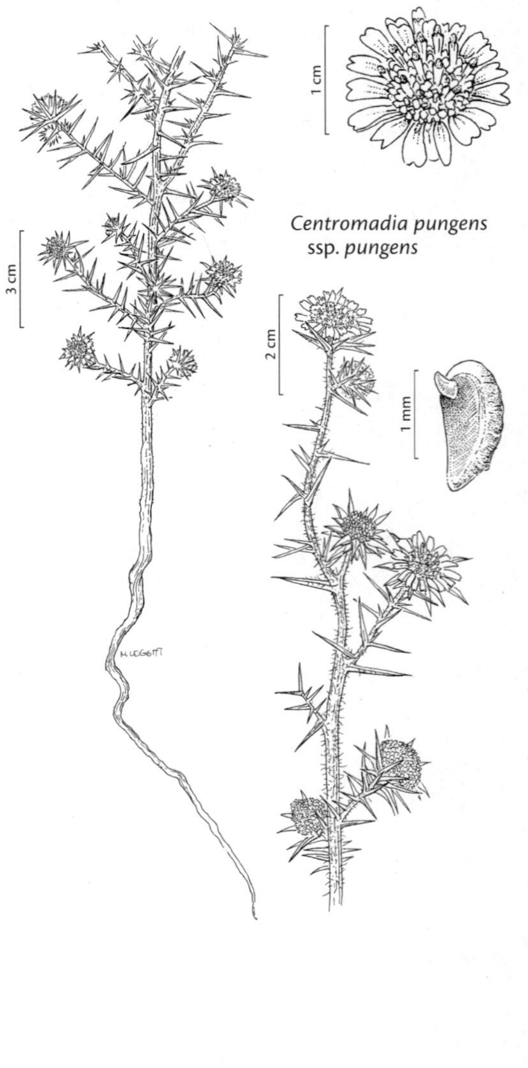

CENTAUREA • CENTROMADIA

CENTROMADIA Spikeweed

Bruce G. Baldwin and Gerald D. Carr

Herbs annual, 10–120 cm; self-incompatible. **Stems** ± erect, branches often numerous and somewhat divergent. **Leaves** mostly cauline at flowering; proximal opposite, often in winter-spring rosettes, most alternate; blades oblanceolate to linear or lance-linear, proximal usually 1–2-pinnatifid, ultimate margins toothed or entire, sometimes bristly-ciliate, tips of distal leaves usually spine-tipped, surfaces glabrous, scabrous-hirsutulous, somewhat hirsute, or villous, often glandular as well; essentially sessile. **Inflorescences** ± spike-, panicle-, or umbel-like, or heads in glomerules. **Heads** radiate. **Peduncular bracts** usually spine-tipped, sometimes apiculate. **Involucres** obconic or urceolate, 3–8+ mm in diameter, subtended by 5–12+ calyculi, usually spine-tipped bractlets. **Phyllaries** falling, 5–75+ in 1 series, lanceolate to lance-attenuate, herbaceous, each usually ½ enveloping subtended ray ovary proximally, often spine-tipped, abaxial faces scabrous-hirsutulous, hirsute, or villous and/or glandular. **Receptacles** flat to convex, setulose. **Paleae** persistent, subtending all or most disc florets, distinct, ± phyllary-like. **Ray florets** 5–75+, pistillate, fertile; corollas yellow. **Disc florets** 6–200+, usually functionally staminate, rarely bisexual and fertile; corollas 2–5 mm, yellow, tubes < funnelform throats, lobes 5, deltate; anthers reddish to dark purple or yellow to brownish; styles glabrous proximal to branches, branch tips densely hairy. **Ray fruits** 2–3 mm, somewhat compressed, abaxially gibbous, attachments basal or oblique, tips beaked or elevated adaxially, faces glabrous; pappi 0. **Disc fruits** usually 0; pappi 0, or of (6)8–12 narrowly oblanceolate to linear scales.

Northwestern Mexico, western United States. 4 species; 2 species treated in *Flora*.

Following Baldwin (1999), *Centromadia* is treated here as distinct from *Hemizonia*; *Centromadia* is more closely related to *Calycadenia*, *Deinandra*, *Holocarpha*, and *Osmadenia* than to *Hemizonia* in the strict sense. Taxa of *Centromadia* are self-incompatible and of low to high interfertility (Venkatesh 1958).

Key to *Centromadia* species:

1. At least distal leaves stalked-glandular; disc pappi of (6)8–12 narrowly oblanceolate to linear scales . . . . ***C. fitchii***

1' Leaves eglandular or only minutely glandular; disc pappi 0 . . . . ***C. pungens*** ssp. ***pungens***

Centromadia fitchii (A. Gray) Greene

Fitch's spikeweed

Hemizonia fitchii A. Gray

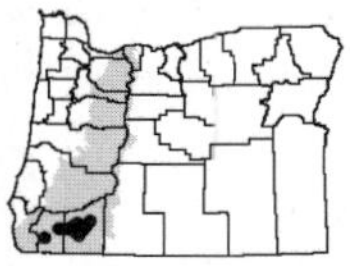

Plants 5–50 cm. **Leaves** densely villous or hirsute, all or at least distal leaves conspicuously stalked-glandular; glands yellow, brown, or black. **Involucres** 5–10 mm. **Paleae** not strongly exserted, tips blunt, long-villous. **Anthers** usually reddish to dark purple, rarely yellow to brownish. **Disc fruit** pappi of (6)8–12 narrowly oblanceolate to linear scales. $2n=26$.

Grasslands, alkaline plains, vernal pool edges and beds, disturbed areas, open woodlands, sometimes on serpentine. Flowering May–Sep. 300–800 m. Casc, Sisk. CA. Native.

Outside southwestern Oregon, *C. fitchii* occurs widely in northern and central California (except deserts). Unlike most other members of *Centromadia*, *C. fitchii* often occurs in upland, presumably well-drained sites.

Centromadia pungens (Hooker & Arnott) Greene ssp. *pungens* (p. 216)

Common spikeweed

Centromadia pungens (Hook. & Arn.) Greene ssp. *septentrionalis* (D.D. Keck) B.G. Baldwin, *Hemizonia pungens* (Hook. & Arn.) Torr. & A. Gray ssp. *pungens*, *Hemizonia pungens* (Hook. & Arn.) Torr. & A. Gray var. *pungens*, *Hemizonia pungens* (Hook. & Arn.) Torr. & A. Gray ssp. *septentrionalis* D.D. Keck, *Hemizonia pungens* (Hook. & Arn.) Torr. & A. Gray var. *septentrionalis* (D.D. Keck) Cronquist

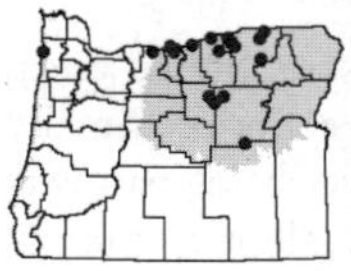

Plants 10–120 cm. **Leaves** glabrous, scabrous, or hirsute, eglandular or minutely glandular, margins sometimes ciliate, especially proximally, midribs sometimes bristly. **Involucres** 3–6 mm. **Paleae** often strongly exserted, tips somewhat attenuate, spine-tipped. **Anthers** yellow to brownish. **Disc fruit** pappi 0. $2n=18, 20$.

Open, alkaline flats, depressions, waterway banks and beds, in grasslands, saltbush scrub, disturbed areas, fallow fields. Flowering May–Oct. 0–900 m. BW, Col, Est, Lava. CA, ID, NV, WA; southeast to AZ, south to Mexico, NY. Exotic.

Subspecies *pungens* is circumscribed broadly to include ssp. *septentrionalis* (and ssp. *maritima*, of California) based on morphologic and molecular data (B.G. Baldwin, unpubl.). As treated here, *C. pungens* ssp. *pungens* occurs widely in central and northern California and is putatively introduced in southwestern California and outside the state. In Oregon, *C. pungens* ssp. *pungens* is found primarily in disturbed sites east of the Cascade summit.

This native of California has become established only in mostly disturbed sites in the northeastern quadrant of Oregon and at one site in Tillamook County. The only known vouchers for reports of *C. pungens* ssp. *pungens* from southern Oregon (Jackson County) have proved to be *C. fitchii.*

CHAENACTIS Pincushion, dusty maidens

Herbs to **subshrubs** annual, biennial, or perennial, 2–100+ cm; taprooted. **Stems** decumbent to erect, simple to much branched, glabrous to densely tomentose, sometimes glandular. **Leaves** basal, cauline, or both, alternate, linear to lanceolate or elliptic, simple to 1–2-pinnately lobed, tips acute to obtuse, surfaces glabrous or densely tomentose; sessile or petiolate. **Inflorescences** heads terminal and solitary or in cyme-, panicle-, or raceme-like arrays. **Heads** discoid, terminal on stem and branches, erect or drooping; sessile or peduncled. **Involucres** cylindric to obconic, campanulate, or hemispheric, (3)5–15(25) mm in diameter. **Phyllaries** in few to many series, lanceolate, oblanceolate, or linear, equal or unequal, surfaces glabrous to pubescent. **Receptacles** convex or flat; paleae 0. **Florets** 10–50+, bisexual, fertile, outer sometimes radially symmetric and enlarged, resembling ray florets; corollas tubes < throats, throats cylindric or funnelform, lobes 5, erect or spreading; stamens slightly to well exserted. **Fruits** club-shaped to cylindric or compressed, scabrous to sericeous or strigose, glandular or not; pappi of 2–15+ persistent, erose scales, in 1–4 equal or unequal series, sometimes 0 or of crowns.

Western North America. 18 species; 8 species treated in *Flora*.

Key to *Chaenactis* species:

1. Plants biennial or perennial.
   2. Leaves basal .... ***C. alpina***
   2' Leaves cauline or basal and cauline.
      3. Phyllaries cobwebby to woolly, stipitate-glandular .... ***C. douglasii***
      3' Phyllaries glabrous, sometimes stipitate-glandular .... ***C. suffrutescens***

1' Plants annual.
   4. Corollas bright yellow; pappi 0 or coroniform .... ***C. nevii***
   4' Corollas white, cream, or pink; pappi of scales.
      5. Pappi in 1 equal series of scales; outer corollas bilaterally symmetric .... ***C. stevioides***
      5' Pappi in 2–4 unequal series of scales; outer corollas radially symmetric.
         6. Corollas night-blooming, 9–14 mm .... ***C. macrantha***
         6' Corollas not night-blooming, 3–10 mm.
            7. Plants 1–15 cm; phyllaries subequal; pappi in 3–4 series .... ***C. cusickii***
            7' Plants 10–40 cm; phyllaries ± equal; pappi in 2 series .... ***C. xantiana***

Chaenactis alpina (A. Gray) M.E. Jones

Alpine chaenactis, alpine dusty maidens

Chaenactis douglasii (Hook.) Hook. & Arn. var. *alpina* A. Gray

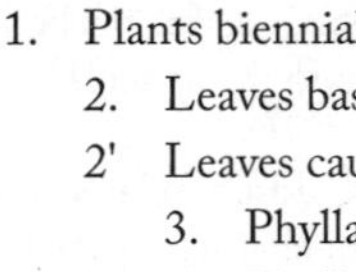

Plants perennial, 2–10+ cm. **Stems** erect, glandular-puberulent, usually proximally floccose or glabrate. **Leaves** basal, narrowly lanceolate to oblanceolate, elliptic, or linear, 1–6 cm, 2-pinnately lobed, secondary lobes becoming inrolled and disc-like with age, surfaces floccose-puberulent; petiolate. **Heads** erect; peduncled. **Involucres** campanulate, 9–14 mm. **Phyllaries** in 3–4 subequal series, linear or narrowly oblanceolate, surfaces glandular-puberulent, sometimes hispid-ciliate. **Disc florets** corollas 5–8 mm, white, cream, or pinkish, surfaces glabrous; outer radially symmetric. **Fruits** 5–8 mm, strigose-puberulent; pappi in 3–4 equal series of scales. **2*n***=12.

Alpine slopes and ridges. Flowering Jul–Sep. 2100–3100 m. BW. CA, ID; northeast to MT, east to WY, southeast to CO. Native.

Some authorities treat *C. alpina* as a variety of *C. douglasii*.

Chaenactis cusickii A. Gray

Cusick's pincushion, morning brides

Plants annual, 1–15 cm. **Stems** erect or decumbent, glabrous to floccose-pubescent. **Leaves** cauline, sometimes cauline and basal, linear or narrowly oblanceolate, 0.5–4 cm, surfaces glabrous; sessile or short-petiolate. **Heads** erect; sessile or peduncled. **Involucres** campanulate, 6–11 mm. **Phyllaries** in 3–4 subequal series, linear to narrowly elliptic or oblanceolate, surfaces glabrous to lightly floccose-puberulent. **Disc florets** corollas 3–8 mm, white or pink, surfaces glabrous; outer radially symmetric. **Fruits** 4–6 mm, strigose; pappi in 3–4 unequal series of scales. **2*n***=12.

Shrublands, barren slopes, ridges, scree. Flowering Apr–Jun. 700–1300 m. Owy. ID. Native.

Chaenactis douglasii (Hooker) Hooker & Arnott (p. 219)

Douglas' dusty maidens, hoary false yarrow

Chaenactis douglasii (Hook.) Hook. & Arn. var. *douglasii*, *Chaenactis douglasii* (Hook.) Hook. & Arn. var. *glandulosa* Cronquist

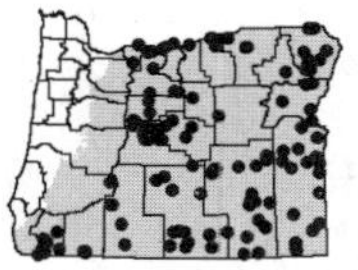

Plants biennial or perennial, 5–60 cm. **Stems** erect, floccose-tomentose, often glandular-puberulent distally. **Leaves** basal and cauline, narrowly lanceolate to oblanceolate, elliptic, or linear, 1–12 cm, 2-pinnately lobed, secondary lobes becoming inrolled and disc-like with age, surfaces tomentose; petiolate. **Heads** erect; peduncled. **Involucres** obconic or campanulate, 8–15 mm. **Phyllaries** in 2–5 unequal series, linear to narrowly elliptic or oblanceolate, surfaces cobwebby to woolly, stipitate-glandular. **Disc florets** corollas 5–8 mm, white, pink, or pale yellow, glabrous or glandular-puberulent; outer radially symmetric. **Fruits** 5–8 mm, surfaces strigose or glandular-puberulent; pappi in 3–4 series of unequal scales. **2*n***=12, 24, 36.

Woodlands, shrublands, grasslands, dry hillsides, ridge tops, riverbanks, rock outcrops, disturbed areas. Flowering Apr–Sep. 0–2300 m. BR, BW, Casc, Col, ECas, Lava, Owy, Sisk. CA, ID, NV, WA; north to British Columbia, northeast to Alberta, east to SD, southeast to NM. Native.

Chromosome studies by Mooring (1980) show that this species varies from diploid to hexaploid, with intermediate numbers resulting from hybridization. As a result, morphological differences at the diploid level become mixed into a morphological continuum among the polyploids.

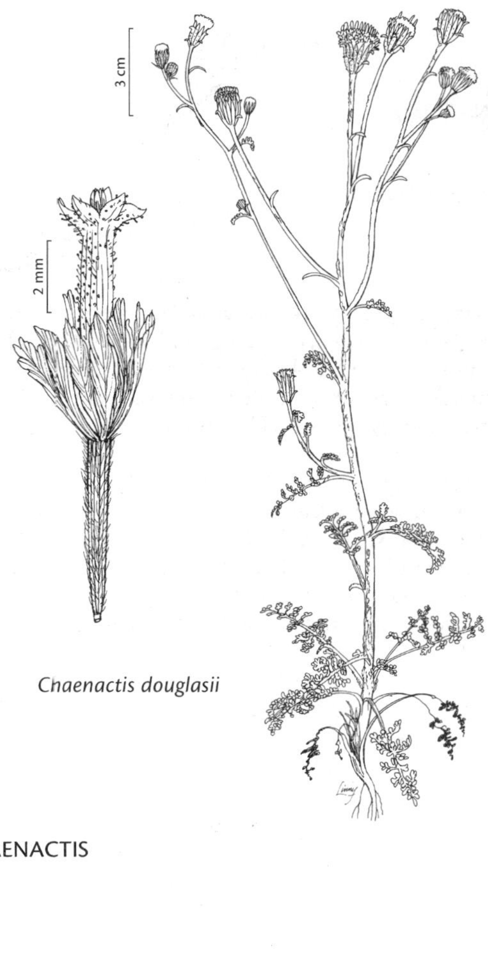

Chaenactis douglasii

CHAENACTIS

Chaenactis macrantha D.C. Eaton

Large-flowered chaenactis, Mojave pincushion, bighead dusty maidens

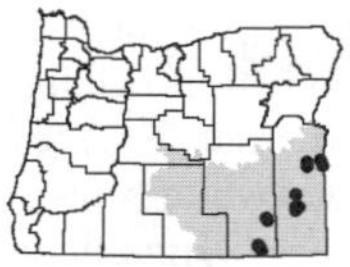

Plants annual, 2–30+ cm. **Stems** erect or decumbent, lightly to densely floccose-tomentose. **Leaves** basal and cauline, lanceolate or ovate, 1–5 cm, 1–2-pinnately lobed, surfaces glabrous or tomentose; petiolate. **Heads** usually erect, sometimes drooping in bud; peduncled. **Involucres** obconic or campanulate, 10–17 mm. **Phyllaries** in 3–4 unequal series, linear to narrowly elliptic or oblanceolate, surfaces lightly or densely tomentose. **Disc florets** corollas night-blooming, 9–14 mm, cream or pink, surfaces glandular-puberulent; outer radially symmetric. **Fruits** 5–7 mm, surfaces strigose; pappi in 2 series of unequal scales. **2*n***=12.

Shrublands, grasslands, open slopes and flats. Flowering May–Jun. 700–1400 m. BR, Owy. CA, ID, NV, WA; southeast to AZ. Native.

Chaenactis nevii A. Gray

John Day pincushion, Nevius' chaenactis

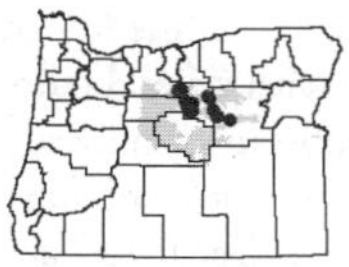

Plants annual, 10–30 cm. **Stems** erect, gland-dotted or glandular-puberulent. **Leaves** basal withering early; cauline lanceolate or elliptic, 1–5 cm, 1–2-pinnately lobed, surfaces glabrous or sometimes margins glandular-puberulent; sessile or petiolate. **Heads** erect; peduncled. **Involucres** obconic or campanulate, 6–8 mm. **Phyllaries** in 2–4 ± equal series, linear to narrowly oblanceolate, glabrous or stipitate-glandular. **Disc florets** corollas 3–7 mm, bright yellow, surfaces glabrous or puberulent; outer radially symmetric. **Fruits** 3.5–6 mm, surfaces glandular-puberulent; pappi 0 or coroniform. **2*n***=12.

Shrublands, grasslands, dry slopes and flats. Flowering May–Jun. 300–1200 m. Lava. Native.

This species is found only in the John Day Basin.

Chaenactis stevioides Hooker & Arnott

Broad-flowered chaenactis, desert pincushion, Esteve's pincushion

Chaenactis stevioides Hook. & Arn. var. *stevioides*

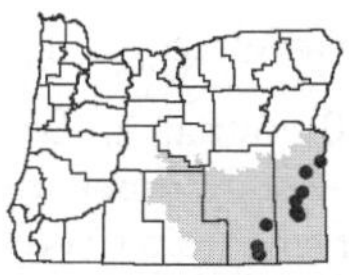

Plants annual, 5–40 cm. **Stems** erect or decumbent, glabrous or puberulent. **Leaves** basal withering early; cauline elliptic, 1–10 cm, 1–2-pinnately lobed, surfaces glabrous, sometimes puberulent near bases; petiolate. **Heads** erect; peduncled. **Involucres** obconic or campanulate, 7–12 mm. **Phyllaries** in 4–5 ± equal series, linear to narrowly elliptic or oblanceolate, surfaces glabrous or glandular-puberulent, sometimes tomentose on midribs and margins. **Disc florets** corollas 4–7 mm, white or pink, surfaces glabrous; outer bilaterally symmetric. **Fruits** 3–6.5 mm, surfaces glandular-puberulent; pappi in 1 equal series of scales. **2*n***=10.

Shrublands, grasslands, dry slopes and flats, talus. Flowering May–Jun. 600–1500 m. BR, Owy. CA, ID, NV; east to WY, southeast to NM, south to Mexico. Native.

Chaenactis suffrutescens A. Gray

Shasta chaenactis

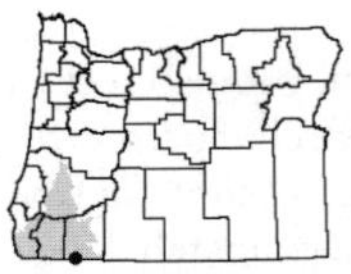

Plants perennial, (10)20–60 cm. **Stems** erect or spreading, felty white-tomentose on leafy portions. **Leaves** cauline, sometimes cauline and basal, lanceolate or deltate, 2–10 cm, 1–2-pinnately lobed, surfaces white-tomentose, lobes linear; petiolate. **Heads** erect; peduncled. **Involucres** usually obconic, sometimes broadly campanulate, 12–18 mm. **Phyllaries** in 3–5 ± equal series, linear or narrowly lance-elliptic, surfaces glabrous or stipitate-glandular. **Disc florets** corollas 8–10 mm, white or pinkish, surfaces puberulent; outer radially symmetric. **Fruits** 7–9 mm, surfaces hispid-pubescent; pappi in 3–4 equal series of scales. **2*n***=12.

Rocky or gravelly slopes, serpentine. Flowering Jun–Sep. 1600–1900 m. Sisk. CA. Native.

Chaenactis xantiana A. Gray (p. 220)

Fleshy pincushion, Xantus' pincushion

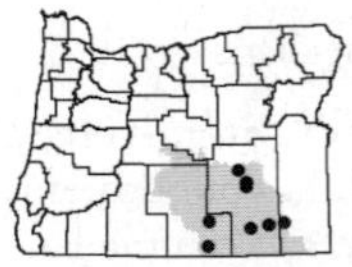

Plants annual, 10–40 cm. **Stems** erect, glabrous, sometimes puberulent below heads. **Leaves** basal withering early; cauline linear, 1–6 cm, margins entire or with several pairs of linear lobes, surfaces glabrous or puberulent. **Heads** erect; peduncled. **Involucres** obconic or campanulate, 9–16 mm. **Phyllaries** in 2–4 ± equal series, linear to narrowly lanceolate or oblanceolate, surfaces glabrous or sometimes puberulent at bases, margins or tips. **Disc florets** corollas 6–10 mm, white or pink, surfaces glabrous; outer radially symmetric. **Fruits** 5–9 mm, surfaces minutely strigose; pappi in 2 unequal series of scales. **2*n***=14.

Shrublands, grasslands, dry slopes and flats, bluffs, dunes. Flowering May–Jul. 300–1400 m. BR. CA, NV; southeast to AZ. Native.

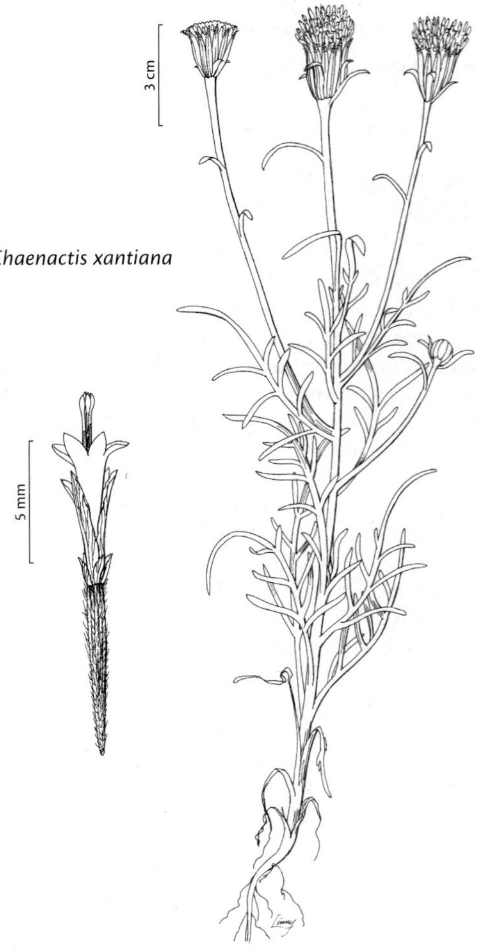

CHAENACTIS

CHAETADELPHA Skeletonweed

Genus composed of one species, *C. wheeleri.*

Chaetadelpha wheeleri A. Gray ex S. Watson (p. 221)

Wheeler's skeletonweed, Wheeler's wirelettuce

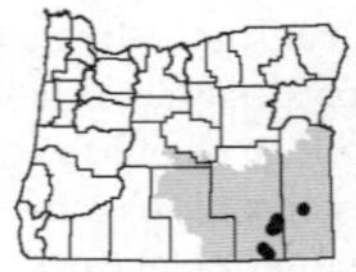

Herbs perennial, 10–40 cm; sap milky. **Stems** erect, slender, sub-woody, rhizome-like, subterranean stems arising from woody caudices, much branched, branches strongly ridged, green, appearing leafless late in the season, glabrous. **Leaves** basal withering at flowering; cauline alternate, deciduous, linear, 0.3–5 cm, reduced and bract-like distally, surfaces glabrous; sessile. **Inflorescences** solitary heads, terminal on branches. **Heads** ligulate. **Involucres** cylindrical, 11–15 × 3–4 mm. **Phyllaries** in 2 series, spreading at maturity; outer deltate, very short, tips acute; inner 5, linear, equal, tips acute, surfaces glabrous. **Receptacles** flat; paleae 0. **Florets** 5, bisexual; ligules 5–9 mm, exceeding involucres, white to lavender or pink; stamens 5. **Fruits** columnar, scarcely tapered, 8–12 mm, light tan, ribs 5, distinctly grooved between the ribs, surfaces smooth, glabrous, beakless; pappi of 5 stiff, dingy white, minutely roughened bristles and numerous shorter bristles basally fused in 5 groups. **2*n***=18.

Sagebrush scrub, alkali scrub, dunes. Flowering May–Jul. 1000–1500 m. BR, Owy. CA, NV. Native.

The growth habit of this desert herb is similar to that of *Lygodesmia juncea* and *Pleiacanthus spinosus*, with rigid, green, photosynthetic stems and narrow, early-deciduous leaves.

CHONDRILLA Skeletonweed

Herbs perennial; sap milky. **Stems** 1+ from the base, erect, usually much branched, leafy throughout, tomentulose and with sharp, retrorse hairs proximally, glabrous distally. **Basal leaves** withering by flowering; blades elliptic to oblong, margins denticulate to deeply lobed, lobes abundant, triangular, linear, tips acute, surfaces setose; petioles short, winged. **Cauline leaves** alternate, lower similar to basal; upper ± persistent, reduced, margins entire, surfaces glabrous; short-petiolate. **Inflorescences** heads solitary or in clusters on the stems and branches; sessile. **Heads** ligulate. **Involucres** cylindrical, 2.5–5 mm in diameter. **Phyllaries** in 2 unequal series; inner loosely spreading in fruit. **Receptacles** flat; paleae 0. **Florets** bisexual; ligules slightly exceeding involucres, yellow. **Fruits** tapering downward from sharply tapered tips, tips roughened and minutely spiculate, ribs 5, long-beaked; pappi of 40+ smooth bristles.

Asia, Australia, Europe, North America, South America. 25 species; 1 species treated in *Flora*.

Chondrilla juncea Linnaeus (p. 221)

Rush skeletonweed

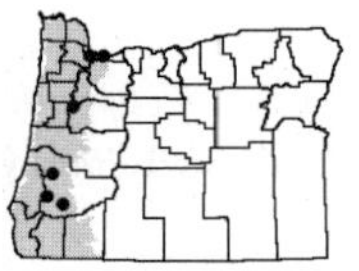

Plants 5–15 dm; with stout, deeply penetrating taproots, spreading laterally by buds on rhizome-like side roots. **Basal leaves** 5–15 cm. **Cauline leaves** upper 1–7 cm. **Involucres** 9–12 mm. **Phyllaries** surfaces tomentose; outer deltoid, all basal; inner 5–10, equal, linear-lanceolate, tips acuminate. **Florets** 6–15, ligules 10–12 mm. **Fruits** bodies 3–3.5 mm, gray or pale brown, beaks 5–6 mm. **2*n***=15, 30.

Roadsides, riverbanks, disturbed areas. Flowering Jul–Sep. 0–700 m. CR, Sisk, WV. CA, ID, WA; scattered in North America; Australia, Europe. Exotic.

Although morphologically variable, *Chondrilla juncea* is reported to reproduce only by clonal root-sprouts and asexually produced seeds. The species is listed as a noxious weed in several western states, including Oregon. Several biological control organisms have been introduced, including a gall midge, gall mite, and rust fungus.

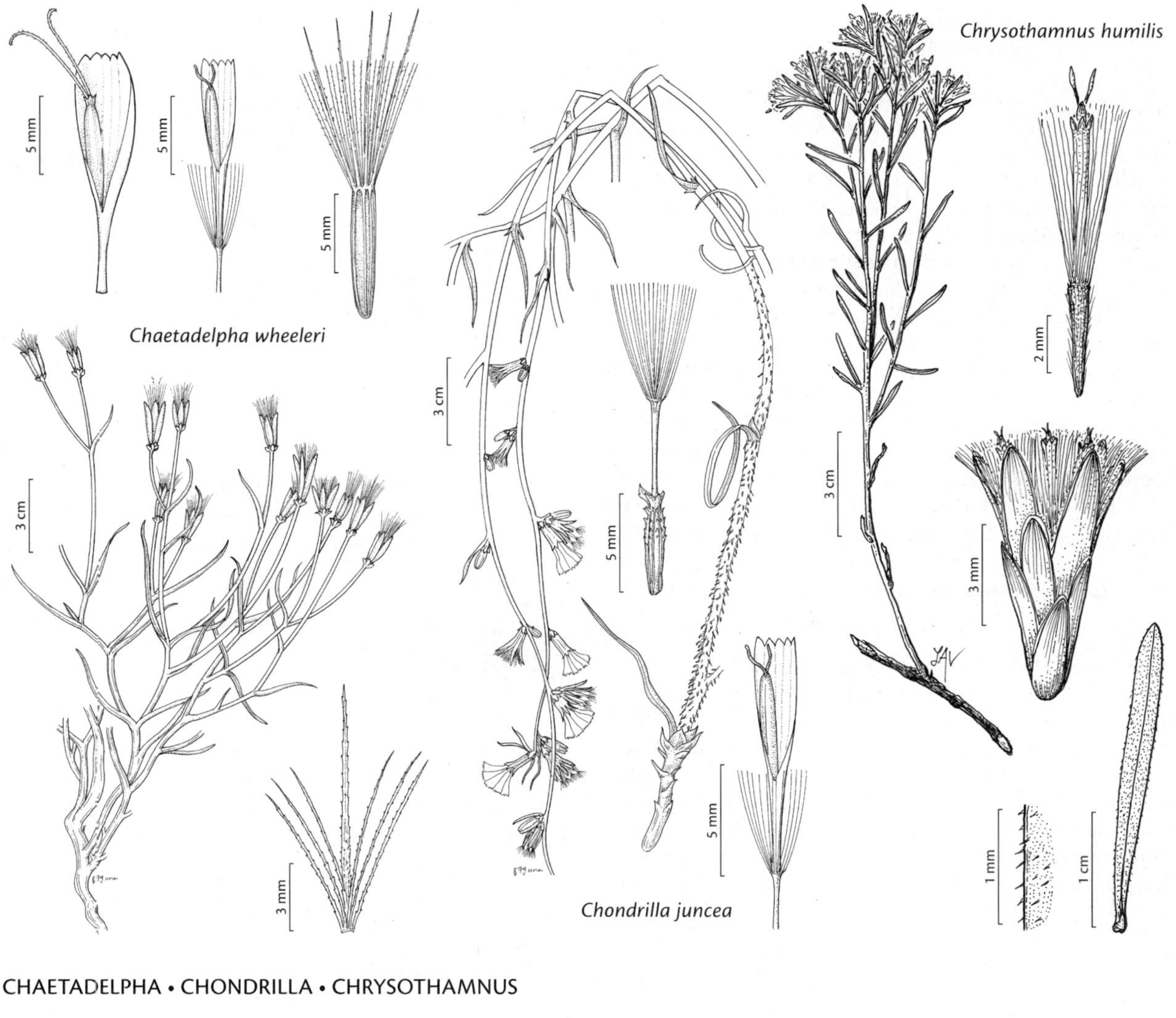

CHAETADELPHA • CHONDRILLA • CHRYSOTHAMNUS

CHRYSOTHAMNUS Rabbitbrush

Shrubs or **subshrubs** usually evergreen, sometimes deciduous (if very cold). **Stems** main stem often branched, woody; herbaceous stems usually erect, unbranched below inflorescences, sometimes persistent and branched second year, glabrous or puberulent, sometimes minutely glandular distally. **Leaves** cauline, alternate, broadly or narrowly linear to lanceolate, margins entire, tips acute, surfaces glabrous to puberulent, adaxially sometimes minutely glandular or viscid; sessile. **Inflorescences** panicle-like arrays, glabrous to puberulent or minutely glandular; peduncles present or not. **Heads** discoid. **Involucres** cylindric. **Phyllaries** in 3–6 unequal series, erect, spreading in fruit, chartaceous, midribs obscure, surfaces glabrous or puberulent. **Receptacles** convex; paleae 0. **Florets** bisexual; corollas tubular, gradually widened distally, yellow, lobes 5, triangular, erect or spreading, included to well exserted from involucres; stamens well exserted; stigma lobes included or exserted. **Fruits** cylindric, truncate, tapering toward base, obscurely veined; pappi of numerous equal bristles, often a few outer bristles are shorter.

Western North America. 9 species; 2 species treated in *Flora.*

Following Roberts and Urbatsch (2004), three groups of species of *Chrysothamnus* have been segregated into *Ericameria, Loranandersonia*, and *Cuniculotinus*, of which only *Ericameria* occurs in Oregon.

Key to *Chrysothamnus* species:

1. Subshrubs, 0.5–2(3) dm; florets 2–3(4); style branches included or slightly exserted . . . . ***C. humilis***

1' Shrubs or subshrubs, 2–10(12) dm; florets (3)4–14; style branches usually well exserted . . . . ***C. viscidiflorus***

Chrysothamnus humilis Greene (p. 221)

Truckee rabbitbrush

Chrysothamnus viscidiflorus (Hook.) Nutt. ssp. *humilis* (Greene) C.L. Hitchc. & Cronquist, *Chrysothamnus viscidiflorus* (Hook.) Nutt. var. *humilis* (Greene) Jeps.

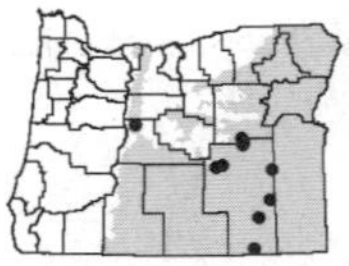

Subshrubs 0.5–2(3) dm. **Herbaceous stems** ascending or erect, minutely spiculate-puberulent. **Leaves** linear to oblanceolate, 0.5–3 cm, usually 1-veined from bases, surfaces minutely spiculate-puberulent, adaxially sometimes glandular; sessile. **Inflorescences** some clusters equaled or exceeded by 1 or more of the terminal leaves; peduncles (when present) minutely puberulent. **Involucres** 6–8 × 3–4 mm. **Phyllaries** oblong to linear-lanceolate, tips obtuse or acute, surfaces glabrous or outer slightly puberulent. **Florets** 2–3(4); corollas 6–8 mm; style branches included or slightly exserted. **Fruits** 3–6 mm, reddish brown, densely strigose; pappus bristles 2.5–4 mm, white or tan. **2*n***=18.

Brushlands, grasslands. Flowering Jul–Sep. 900–3000 m. BR, BW, ECas, Owy. CA, ID, NV, WA. Native.

In this treatment, *Chrysothamnus humilis* is limited to plants with at least some of the inflorescence clusters equaled or overtopped by one or more of the terminal stem leaves. Plants resembling *C. humilis* but with the heads standing well above the upper leaves are better placed in *C. viscidiflorus* ssp. *puberulus*.

Chrysothamnus viscidiflorus (Hooker) Nuttall

Sticky-flowered rabbitbrush

Shrubs or **subshrubs** 2–10(12) dm. **Herbaceous stems** ascending or erect, glabrous or minutely spiculate-puberulent. **Leaves** broadly or narrowly linear to lanceolate, 1–7 cm, 1- or 3–5-veined from bases, surfaces glabrous or minutely spiculate-puberulent, adaxially sometimes glandular; sessile. **Inflorescences** clusters not equaled or exceeded by terminal leaves; peduncles (when present) minutely spiculate, sometimes glandular-puberulent. **Involucres** 4–8 × 1.5–2.5 mm. **Phyllaries** oblong to linear-lanceolate, tips obtuse or acute, surfaces glabrous or with puberulent tips. **Florets** (3)4–14; corollas 4.5–6.5 mm; style branches usually well exserted. **Fruits** 2.5–4 mm, reddish brown, ± strigose; pappus bristles 3–4 mm, tan.

Western Canada and US. 5 subspecies; 3 subspecies treated in *Flora.*

As stated by Cronquist (1994), the subspecies of *Chrysothamnus viscidiflorus* are "morphologically obvious but wholly confluent…with only minimal ecogeographic significance." The pubescence features of stems and leaves hold up well in practice, however.

Key to *Chrysothamnus viscidiflorus* subspecies:

1. Plants 5–10(12) dm; stems glabrous; leaves glabrous abaxially; disc florets 4–14 . . . . ***C. v.*** ssp. ***viscidiflorus***

1' Plants 2–5 dm; stems minutely spiculate-puberulent; leaves spiculate-puberulent abaxially; disc florets usually 5.

  2. Leaves narrowly to broadly linear, 2–10 mm wide, 3–5-veined from base . . . . ***C. v.*** ssp. ***lanceolatus***

  2' Leaves narrowly linear, 1–2 mm wide, usually 1-veined from base . . . . ***C. v.*** ssp. ***puberulus***

Chrysothamnus viscidiflorus ssp. ***lanceolatus*** (Nuttall) H.M. Hall & Clements

Lanceleaf green rabbitbrush

Chrysothamnus viscidiflorus (Hook.) Nutt. var. *lanceolatus* (Nutt.) Greene

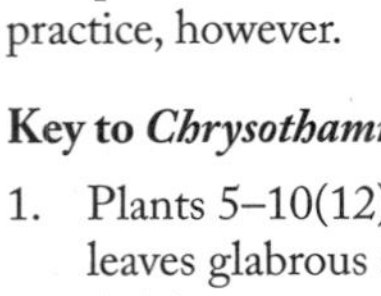

Plants 2–5 dm. **Stems** minutely spiculate-puberulent. **Leaves** narrowly to broadly linear, 2–6 mm wide, 3–5-veined from bases, margins smooth or scabridulous, surfaces abaxially spiculate-puberulent, adaxially usually glabrous or glandular-puberulent. **Florets** 5. **2*n***=18, 36.

Open slopes and flats, brushlands, grasslands, riverbanks. Flowering Jul–Sep. 50–2300 m. BR, BW, Col, Lava, Owy. CA, ID, NV, WA; north to British Columbia, northeast to MT, east to WY, southeast NM. Native.

Chrysothamnus viscidiflorus ssp. *puberulus* (D.C. Eaton) H.M. Hall & Clements

Downy green rabbitbrush

Chrysothamnus viscidiflorus (Hook.) Nutt. var. *puberulus* (D.C. Eaton) Jeps.

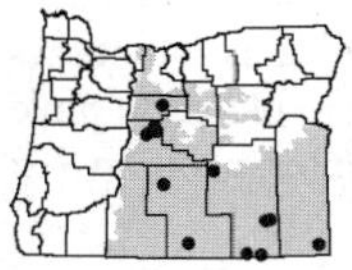

Plants 2–5 dm. **Stems** minutely spiculate-puberulent. **Leaves** narrowly linear, 1–2 mm wide, usually 1-veined from bases, margins smooth or scabridulous, surfaces adaxially spiculate-puberulent or glabrous, not glandular. **Florets** (4)5(7). **2*n***=18, 36.

Open slopes and flats, brushlands, grasslands. Flowering Jun–Sep. 800–2000 m. BR, ECas, Lava, Owy. CA, ID, NV; southeast to AZ. Native.

Chrysothamnus viscidiflorus ssp. *viscidiflorus*

Yellow rabbitbrush, sticky-flowered rabbitbrush

Chrysothamnus viscidiflorus (Hook.) Nutt. ssp. *pumilus* (Nutt.) H.M. Hall & Clem., *Chrysothamnus viscidiflorus* (Hook.) Nutt. var. *pumilus* (Nutt.) Jeps., *Chrysothamnus viscidiflorus* (Hook.) Nutt. ssp. *stenophyllus* (A. Gray) H.M. Hall & Clem., *Chrysothamnus viscidiflorus* (Hook.) Nutt. var. *stenophyllus* (A. Gray) H.M. Hall, *Chrysothamnus viscidiflorus* (Hook.) Nutt. var. *viscidiflorus*

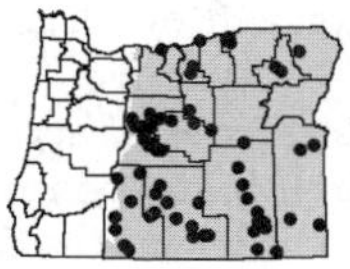

Plants 5–10(12) dm. **Stems** glabrous. **Leaves** linear to lanceolate, (0.5)1–7(10) mm wide, 1–5-veined from the base, margins often scabridulous, sometimes ciliolate, surfaces abaxially glabrous, adaxially usually glandular-puberulent or viscid. **Florets** 4–14. **2*n***=18, 36, 54.

Pine forests, woodlands, brushlands, grasslands, open slopes and flats, riverbanks, roadsides. Flowering May–Oct. 50–2800 m. BR, BW, Col, ECas, Lava, Owy. CA, ID, NV, WA; northeast to MT, east to NE, southeast to NM. Native.

CICHORIUM Chicory

Herbs perennial; sap milky; with stout taproots and sometimes branched caudices. **Stems** erect, leafy, branched distally, glabrous or pilose to hispid-pubescent. **Leaves** basal and cauline; blades oblanceolate, margins entire to shallowly to deeply lobed with broad to narrowly-winged rachises, lobes acute, usually angled toward leaf base, margins finely to coarsely toothed, tips broadly to narrowly acute, surfaces abaxially glabrous or pilose on the midribs; basal usually petiolate, winged, cauline smaller, sessile, sometimes clasping. **Inflorescences** heads clustered at nodes, sometimes 1 per node, distal on the stem and branches; bracts much reduced. **Heads** ligulate; sessile or pedunculate. **Involucres** campanulate in flower, ovoid-cylindrical in fruit. **Phyllaries** in several series; outer ovate-lanceolate, shorter, bases thickened and whitish, tips reflexed-spreading; inner linear-lanceolate, equal, tips erect. **Receptacles** flat, minutely scaly; paleae 0. **Florets** 8–25, bisexual; ligules well exceeding involucres, blue, rarely purple, white, or pink. **Fruits** oblong, 3–5-angled, glabrous, smooth, not beaked; pappi of crowns of short, whitish scales.

Asia, Africa, Europe, North America. 6 species; 1 species treated in *Flora.*

Cichorium intybus Linnaeus (p. 223)

Chicory

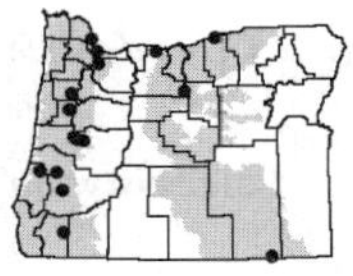

Plants 10–120 cm. **Leaves** 5–35 cm. **Involucres** 6–12 mm. **Phyllaries** 10–16, glabrous or sparsely stipitate-glandular. **Florets** 10–30; ligules 15–22 mm. **Fruits** 2–3 mm, brownish. **2*n***=18.

Roadsides, streambanks, meadows, fencerows, cultivated fields, disturbed areas. Flowering Jun–Oct. 0–1700 m. BR, Col, CR, Lava, Sisk, WV. CA, ID, NV, WA; throughout North America; Africa, Asia, Europe, South America. Exotic.

Although *Cichorium intybus* is a roadside weed, its conspicuous, blue flower heads are quite showy. They are especially visible in the morning but usually close up in the early afternoon.

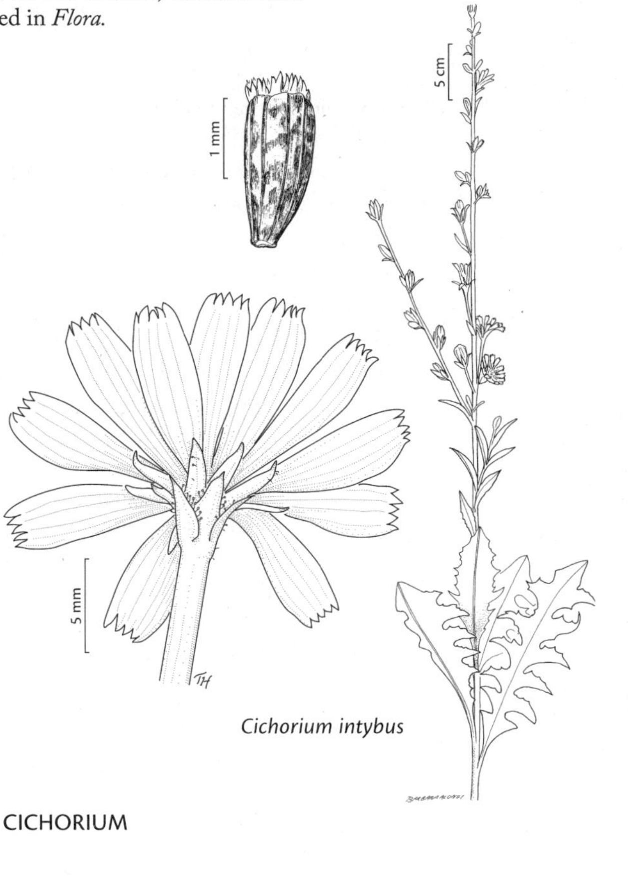

Cichorium intybus

CICHORIUM

CIRSIUM Thistle

Bridget N. Chipman

Herbs perennial or biennial; taprooted or from creeping roots. **Stems** erect, branched or simple, sometimes spiny-winged. **Leaves** basal and cauline, alternate, margins entire to deeply toothed or pinnately lobed, lobes and teeth usually spiny. **Inflorescences** heads solitary at branch tips or distal axils, or in raceme- or panicle-like arrays. **Heads** discoid; pedunculate or sessile. **Phyllaries** many in 5–20 series, imbricate or not, margins usually entire or occasionally scarious, ciliate-spiny, or erose, midveins sometimes with glutinous ridges. **Receptacles** flat to convex, very bristly; paleae 0. **Florets** corollas bilateral, white or cream-colored to pink, red, or purple; style tips elongate. **Fruits** compressed, ovoid, tan to dark brown, glabrous; attachment scar off-center; pappi in 3–5 series, persistent or deciduous in rings.

Africa, Asia, Europe, North America. ~200 species; 15 species treated in *Flora*.

The taxonomy of *Cirsium* is complicated by hybridization and a high level of morphological variation within species. Style tips are measured including the somewhat swollen nodes and fused portions of style branches.

Cirsium ochrocentrum var. *ochrocentrum* has been collected once in Oregon (2013) but does not appear to be naturalized.

Key to *Cirsium* species:

1. Plants dioecious . . . . . **C. arvense**
1' Plants with bisexual florets.
  2. Adaxial leaf surfaces with bristle-like spines . . . . . ***C. vulgare***
  2' Adaxial leaf surfaces without bristle-like spines.
    3. Phyllaries with glutinous ridges (ridges sometimes weakly developed in *C. inamoenum*).
      4. Abaxial leaf surfaces with prominent veins. . . . . ***C. cymosum***
      4' Abaxial leaf surfaces without prominent veins.
        5. Heads 10–many; corollas rose to purple, rarely white-tinged . . . . . ***C. douglasii*** var. ***breweri***
        5' Heads 1–10; corollas white or cream-colored to pale lavender or pink.
          6. Plants from creeping roots; plants of southwestern Oregon. . . . . ***C. ciliolatum***
          6' Plants taprooted, sometimes with lateral roots; plants usually of eastern Oregon.
            7. Corollas 24–50 mm, lobes 7–13 mm. . . . . ***C. undulatum***
            7' Corollas 18–32 mm, lobes 3.5–7 mm.
              8. Phyllary tips reflexed; pappi 28–34 mm. . . . . ***C. brevifolium***
              8' Phyllary tips tapering to ascending or spreading; pappi 12–25 mm . . . . . ***C. inamoenum*** var. ***inamoenum***
    3' Phyllaries without glutinous ridges (sometimes with inconspicuous glutinous ridges in *C. remotifolium*).
      9. Stems conspicuously white-tomentose (rarely glabrate in *C. occidentale* var. *venustum*) . . . . . ***C. occidentale***
      9' Stems not conspicuously white-tomentose.
        10. Styles included or exserted 1–2 mm beyond corolla. . . . . ***C. brevistylum***
        10' Styles conspicuously exserted.
          11. Corollas red, occasionally red-purple . . . . . ***C. andersonii***
          11' Corollas white to pink or purple, not red.
            12. Plants on or near Steens Mountain.
              13. Phyllary and leaf spines stout; plants caulescent, stems not stout and bushy-branched . . . . . ***C. eatonii*** var. ***peckii***
              13' Phyllary and leaf spines thin to moderately thick; plants often acaulescent or short-caulescent, sometimes caulescent with single, stout, bushy-branched stem . . . . . ***C. scariosum*** (in part)
            12' Plants not on or near Steens Mountain.
              14. Plants often acaulescent or short-caulescent, sometimes caulescent with single, stout, bushy-branched stem . . . . . ***C. scariosum*** (in part)
              14' Plants caulescent, stems not stout and bushy-branched.
                15. Outer phyllary bases < 2 mm wide. . . . . ***C. edule***
                15' Outer phyllary bases ≥ 2 mm wide. . . . . ***C. remotifolium***

Cirsium andersonii (A. Gray) Petrak

Anderson's thistle

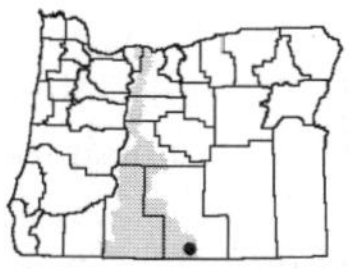

Plants perennial, (1.5)4–7(10) dm; taprooted, sometimes with creeping roots. **Stems** usually 1, sparsely arachnoid-tomentose to glabrate. **Leaves** elliptic to linear-oblong, 8–20(35) × 2–8 cm, gradually reduced above, margins coarsely dentate or 1–2-pinnate, spines 1–5 mm, surfaces abaxially ± thinly gray-tomentose, adaxially glabrous to sparsely pilose; basal usually present at flowering; clasping to petiolate. **Heads** 1–6 per branch, lateral heads widely spaced along distal portions of branches. **Involucres** broadly cylindric to narrowly campanulate, 3–5 × 2–4 cm, glabrous to thinly tomentose. **Phyllaries** without glutinous ridges, spines weak, 1–3 mm, outer short, linear-lanceolate, appressed, margins entire or spiny-ciliate, tips long-acuminate, inner linear, long, entire, tips red to purple, flat. **Florets** corollas 30–45 mm, red, occasionally red-purple, tubes 10–20 mm, throats abruptly narrowed to tubes, 10–16 mm, lobes 9–11 mm; styles conspicuously exserted, tips 3.5–5 mm, sometimes geniculate. **Fruits** 6–7 mm, brown; pappi 25–40 mm.

Open grasslands, montane woodlands. Flowering Jul–Sep. 1500–2500 m. ECas. CA, NV. Native.

Cirsium andersonii, while currently rare in Oregon, is well documented in northern California and Nevada.

Cirsium arvense (Linnaeus) Scopoli

Canada thistle, creeping thistle

Cirsium arvense (L.) Scop. var. *arvense*, *Cirsium arvense* (L.) Scop. var. *horridum* Wimm. & Grab.

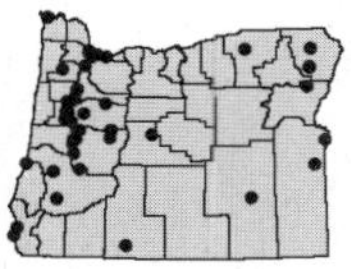

Plants perennial, dioecious, 3–15(20) dm; from deep-seated creeping roots. **Stems** 1–many, glabrous to appressed-tomentose. **Leaves** elliptic to oblong, 3–30 × 1–6 cm, margins entire to dentate or pinnately lobed, spines 1–7 mm, surfaces abaxially ± tomentose, adaxially glabrous to thinly tomentose; basal usually absent at flowering; sessile, sometimes decurrent. **Heads** (1)few–many. **Involucres** ovoid, becoming campanulate in fruit, 1–2 × 1–2(2.5) cm. **Phyllaries** strongly imbricate, with narrow glutinous ridges; outer with weak spines to 1 mm. **Florets** corollas purple, rarely white or pink; staminate florets 12–18 mm, tubes 8–11 mm, throats 1–2 mm, lobes 3–5 mm; pistillate florets 14–20 mm, tubes 10–15 mm, throats 0.5–1.5 mm, lobes 2–3 mm; style tips 1–2 mm. **Fruits** 2–4 mm, brown; pappi 13–32 mm.

Disturbed areas, cultivated fields, pastures, forest openings. Flowering Jun–Sep. 0–2400 m. All ecoregions. CA, ID, NV, WA; throughout most of North America; worldwide. Exotic.

Listed as a noxious weed throughout North America, *C. arvense* is a prolific agricultural weed.

Cirsium brevifolium Nuttall

Palouse thistle

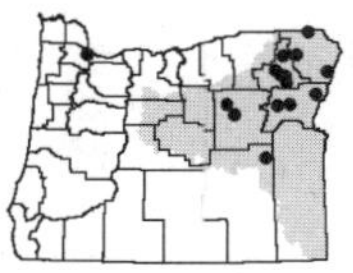

Plants perennial, 2.5–12 dm; taprooted with lateral roots. **Stems** 1–several, gray-tomentose. **Leaves** elliptic to oblanceolate, 12–45 × 2–10 cm, margins entire to deeply dentate or pinnately lobed, spines 2–6 mm, surfaces abaxially densely white-tomentose, adaxially thinly tomentose to glabrate; basal usually present at flowering; sessile, clasping, or petiolate. **Heads** 1–few. **Involucres** hemispheric to campanulate, (1.5)2–3.5 × 2–4 cm. **Phyllaries** strongly imbricate, with prominent glutinous ridges, tips reflexed, spines fine, 2–5 mm. **Florets** corollas 20–28 mm, white to cream-colored, occasionally lavender-tinged, tubes 8–13 mm, throats 7–11 mm, lobes 4–6 mm; style tips 5–6 mm. **Fruits** 5–6 mm, straw-colored to light brown; pappi 28–34 mm.

Meadows, forest edges, sagebrush areas, roadsides. Flowering Jun–Aug. 200–1500 m. BW, Lava, Owy. ID, WA. Native.

Cirsium brevistylum Cronquist

Short-styled thistle, clustered thistle

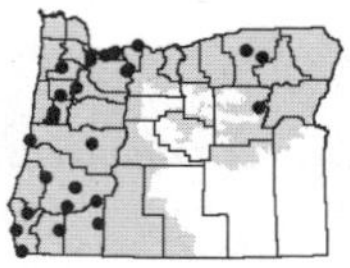

Plants annual or biennial, 2–35 dm; taprooted. **Stems** usually 1, thinly to densely villous, often with arachnoid hairs below heads. **Leaves** elliptic to oblanceolate, 15–35 × 2–10 cm, margins coarsely dentate to pinnate, lobes broadly triangular, spines slender, 3–7 mm, surfaces abaxially thinly tomentose, adaxially thinly villous, villous along veins; basal usually absent at flowering; sessile, clasping, or petiolate. **Heads** 1–many. **Involucres** hemispheric to campanulate, 2–4 × 2.5–4 cm, thinly arachnoid-tomentose. **Phyllaries** without glutinous ridges, tips ascending to spreading, spines 3–5 mm. **Florets** corollas 20–28 mm, white to pink or purple, tubes 10–17 mm, throats 4–5 mm, lobes 3–5 mm; style tips 2–4 mm, included or exserted 1–2 mm beyond corolla lobes. **Fruits** 3–4.5 mm, brown; pappi 10–22 mm.

Woodlands, meadows, coastal scrub, roadsides. Flowering Jun–Aug. 0–1300 m. BW, Casc, Col, CR, ECas, Est, Sisk, WV. CA, ID, WA; north to British Columbia, northeast to MT. Native.

Cirsium ciliolatum (L.F. Henderson) J.T. Howell

Ashland thistle

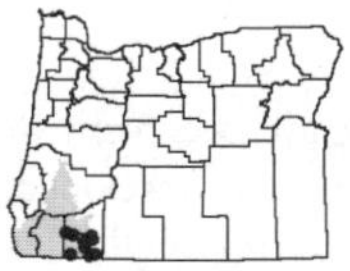

Plants perennial, 6–20 dm; from creeping roots. **Stems** 1–several, thinly arachnoid to densely white-tomentose. **Leaves** oblong-elliptic, 10–30 × 3–12 cm, margins entire to deeply dentate, lobed or pinnate, spines 1–6 mm, surfaces abaxially white-tomentose, adaxially thinly arachnoid-tomentose; basal present at flowering; sessile, clasping, or petiolate. **Heads** 1–few. **Involucres** ovoid to hemispheric, 1.5–2.5 × 1.5–3 cm, glabrate to thinly arachnoid. **Phyllaries** strongly imbricate, with prominent glutinous ridges, tips ascending to spreading, spines fine, 1–3 mm. **Florets** corollas 15–27 mm, white to lavender, tubes 7–11 mm, throats 5–11 mm, lobes 5–7 mm; style tips 5–7 mm. **Fruits** 3.5–7 mm, brown; pappi 15–20 mm.

Meadows, open woodlands, hillsides, dry or rocky ground. Flowering May–Aug. 400–1400 m. Sisk. CA. Native.

Cirsium ciliolatum is listed as endangered by the state of California.

Cirsium cymosum (Greene) J.T. Howell

Peregrine thistle

Plants biennial or perennial, 2.5–12 dm; taprooted, sometimes with creeping roots. **Stems** usually 1, sparsely to densely arachnoid-villous, often also tomentose to glabrate. **Leaves**

oblanceolate to elliptic, 10–40(50) × 3–12 cm, margins 1–2-pinnate, spines 2–7 mm, surfaces glabrate to arachnoid-tomentose, abaxial veins prominent, straw-colored; basal usually present at flowering; sessile, clasping, decurrent, or petiolate. **Heads** 1–few. **Involucres** ovoid to hemispheric or campanulate, 1.5–3.5 × 1.5–5 cm, glabrate to thinly tomentose. **Phyllaries** with glutinous ridges, tips ascending to spreading, spines fine, 2–4 mm. **Florets** corollas 20–38 mm, cream-colored to purple, tubes 8–15 mm, throats 5.5–12 mm, lobes 5.5–7 mm; style tips 4–6(7) mm. **Fruits** 5–8 mm, light to dark brown; pappi 15–25 mm.

Western United States. 2 varieties.

Key to *Cirsium cymosum* varieties:

1. Larger heads 1.5–2.5 cm in diameter; outer phyllaries shorter than inner phyllaries, glutinous ridges well developed . . . . *C. c.* var. ***canovirens***

1' Larger heads 2–3.5 cm in diameter; outer phyllaries subequal to inner phyllaries, glutinous ridges weakly developed . . . . *C. c.* var. ***cymosum***

Cirsium cymosum (Greene) J.T. Howell var. ***canovirens*** (Rydberg) D.J. Keil

Gray green thistle

Cirsium canovirens (Rydb.) Petr.

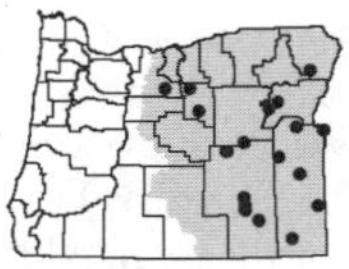

Inflorescences larger heads 1.5–2.5 cm in diameter. **Phyllaries** with well-developed glutinous ridges, appearing dark brown when dried; outer phyllaries linear-lanceolate, shorter than inner phyllaries.

Sagebrush areas, meadows, open woods. Flowering Jun–Aug. 900–2400 m. BR, BW, Col, Lava, Owy. CA, ID, NV; northeast to MT, east to WY. Native.

Cirsium cymosum (Greene) J.T. Howell var. ***cymosum***

Peregrine thistle

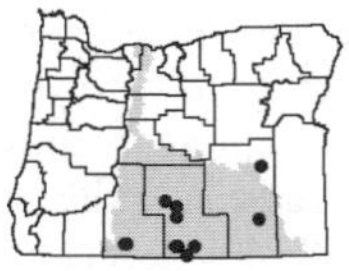

Inflorescences larger heads 2–3.5 cm in diameter. **Phyllaries** with weakly developed, narrow glutinous ridges, outer phyllaries linear-elongate, subequal to inner phyllaries.

Sagebrush areas, meadows, open woods. Flowering Jun–Jul. 1000–1600 m. BR, ECas. CA, NV. Native.

Cirsium cymosum var. *cymosum* is reported to occasionally occur on serpentine soils in California. However, there are no collections of specimens on serpentine in Oregon.

Cirsium douglasii de Candolle var. ***breweri*** (A. Gray) D.J. Keil & C.E. Turner (p. 226)

Brewer's thistle, swamp thistle

Cirsium breweri (A. Gray) Jeps., *Cirsium douglasii* DC. var. *canescens* (Petr.) J.T. Howell

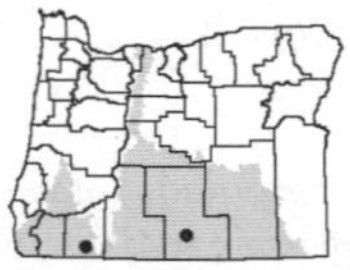

Plants biennial or short-lived perennial, 6–25 dm; taprooted. **Stems** usually 1, closely white-tomentose. **Leaves** elliptic to oblong, 10–60(110) × 2–15 cm, margins deeply pinnate or dentate, lobed or not, spines 2–7 mm, surfaces gray-tomentose; basal present at flowering; sessile, clasping, decurrent, or petiolate. **Heads** 10–many. **Involucres** ovoid to hemispheric, 1.5–2.5 × 1–3 cm, glabrate to thinly arachnoid on margins. **Phyllaries** with narrow glutinous ridges, spines 1–6(9) mm. **Florets** corollas 16–21 mm, rose to purple, rarely white-tinged, tubes 6–9 mm, throats 5–6 mm, lobes 5–6 mm; style tips 3–5 mm. **Fruits** 4–6 mm, brown, sometimes darkly speckled or striped; pappi 15–20 mm.

Wet ground, meadows, montane forests, streambanks, near springs. Flowering Jun–Sep. 1200–1900 m. BR, ECas, Sisk. CA, NV. Native.

Cirsium eatonii (A. Gray) B.L. Robinson var. ***peckii*** (L.F. Henderson) D.J. Keil

Steens Mountain thistle, Peck's thistle

Cirsium peckii L.F. Hend.

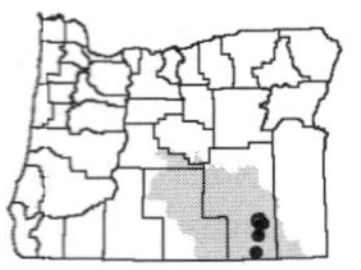

Plants perennial, 1–15 dm; taprooted, often with creeping roots. **Stems** 1–several, glabrous or with few long, woolly hairs. **Leaves** oblong, 10–30 × 2–8 cm, margins pinnately divided, dentate or spiny-toothed, spines stout, 2–12 mm, surfaces glabrous or sparsely woolly-villous; basal usually present at flowering; sessile, decurrent, or petiolate. **Heads** 1–several. **Involucres** broadly ovoid to campanulate, 2–4 × 1.5–5.5 cm, thinly arachnoid-villous. **Phyllaries** without glutinous ridges, spines stout, 7–35 mm. **Florets** corollas 20–25 mm, pale lavender to pink, tubes 6.5–8 mm, throats 8–11.5 mm, lobes 5.5–7 mm; styles conspicuously exserted, tips 3–6 mm. **Fruits** 5–7 mm, dark brown; pappi 17–22 mm.

Rocky or gravelly slopes, sagebrush areas, streambanks. Flowering Jun–Sep. 1300–3000 m. BR. NV. Native.

Cirsium eatonii var. *peckii* is known to hybridize with *Cirsium inamoenum* var. *inamoenum*, creating hybrid swarms.

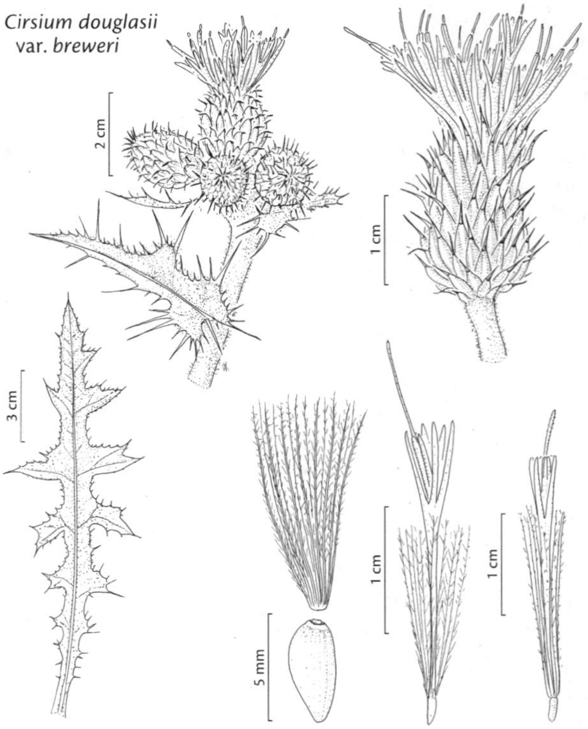

CIRSIUM

Cirsium edule Nuttall

Edible thistle

Plants biennial or monocarpic perennial, 2–35 dm; taprooted. **Stems** usually 1, sparsely to densely arachnoid-villous. **Leaves** oblong to oblanceolate, 5–50 × 1–10 cm, margins entire, sometimes undulate, coarsely dentate to deeply pinnate, spines 2–10 mm, surfaces abaxially sparsely glabrate to arachnoid-tomentose, adaxially glabrate to sparsely villous, often villous along veins; basal usually absent at flowering; sessile, clasping, or petiolate. **Heads** 1–many. **Involucres** narrowly ovoid to hemispheric or campanulate, 1.5–4 × 1.5–4 cm, thinly to densely arachnoid-tomentose. **Phyllaries** without glutinous ridges, spines slender, 1–15 mm; outer bases < 2 mm wide. **Florets** corollas 14–22 mm, purple to pink, rarely white, tubes 7–11 mm, throats 4–8 mm, lobes (2)4–7 mm; styles conspicuously exserted, tips 3–4 mm. **Fruits** 3.5–6.5 mm, light to dark brown; pappi 10–25 mm.

Northwestern North America. 3 varieties; 2 varieties treated in *Flora.*

Cirsium edule is known to form fertile hybrids with *Cirsium remotifolium* where their ranges overlap.

Key to *Cirsium edule* varieties:

1. Phyllary spines 1–5 mm; involucres 1.5–2.5 × 1.5–3 cm . . . . ***C. e.*** var. ***edule***

1' Phyllary spines 5–15 mm; involucres 2–4 × 2.5–4 cm . . . . ***C. e.*** var. ***macounii***

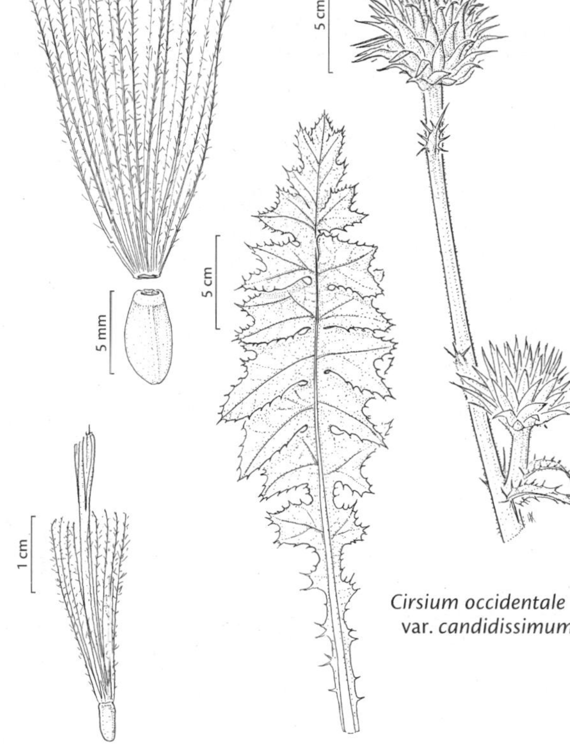

Cirsium occidentale var. *candidissimum*

CIRSIUM

Cirsium edule Nuttall var. *edule*

Indian thistle, edible thistle

Cirsium hallii (A. Gray) M.E. Jones

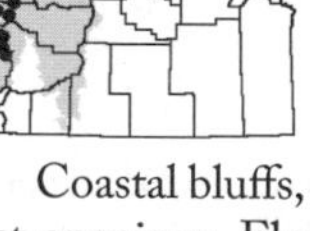

Inflorescences peduncles 0–1(4) cm. **Involucres** 1.5–2.5 × 1.5–3 cm. **Phyllaries** tips ascending to spreading, flat to needle-like, spines 1–5 mm. **Florets** corollas 14–24 mm.

Coastal bluffs, dunes, roadsides, rocky or gravelly slopes, forest openings. Flowering Jun–Aug. 0–1000 m. Casc, CR, Est, WV. WA. Native.

Cirsium edule Nuttall var. *macounii* (Greene) D.J. Keil

Macoun's thistle

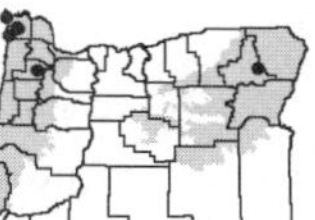

Inflorescences peduncles 0–10 cm. **Involucres** 2–4 × 2.5–4 cm. **Phyllaries** tips widely spreading, needle-like, spines 5–15 mm. **Florets** corollas (15)18–22 mm.

Damp areas, meadows, forests, rocky slopes. Flowering Jun–Oct. 0–1000 m. BR, CR, Est, WV. WA; north to AK. Native.

Cirsium inamoenum (E.L. Greene) D.J. Keil var. *inamoenum*

Jackson Hole thistle, Intermountain thistle

Cirsium subniveum Rydb.

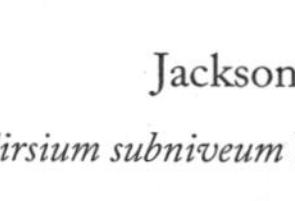

Plants biennial or perennial, (0.5)2–10 dm; taprooted. **Stems** 1–several, thinly to densely gray-tomentose. **Leaves** elliptic to oblanceolate, 10–35 × 1–7 cm, margins unlobed, dentate, or deeply pinnate, spines 2–7 mm, surfaces abaxially gray-tomentose, adaxially glabrate to thinly tomentose; basal often absent at flowering; sessile, decurrent, or petiolate. **Heads** 1–several. **Involucres** ovoid to spherical or campanulate, 2–3 × 1.5–3 cm, glabrous or arachnoid. **Phyllaries** strongly imbricate, with weakly developed, narrow glutinous ridges, tips ascending to spreading, spines 2–6 mm. **Florets** corollas 18–32 mm, white or pale lavender, tubes 7–15 mm, throats 6–10 mm, lobes 3.5–7 mm; style tips 3.5–7 mm. **Fruits** 5–8 mm, brown; pappi 12–25 mm.

Dry slopes, sagebrush areas, meadows, montane forests, woods. Flowering May–Aug. 700–1900 m. BR, BW, ECas, Lava, Owy. CA. ID, NV, WA; east to WY, southeast to UT. Native.

Cirsium occidentale (Nuttall) Jepson

Western thistle, cobwebby thistle

Plants biennial, 4–30 dm; taprooted. **Stems** usually 1, usually conspicuously white-tomentose, rarely glabrate. **Leaves** elliptic to oblanceolate, 6–40 × 1.5–10 cm, margins pinnate, lobes often with shallow teeth or again lobed, spines weak to stout, 1–15 mm, surfaces white-tomentose, adaxially sometimes thin to glabrate; basal usually present at flowering; sessile, clasping, decurrent, or petiolate. **Heads** 1–many. **Involucres** ovoid to spherical, (2)3–7 × 2–8 cm, thinly to densely arachnoid-tomentose. **Phyllaries** subequal to strongly imbricate, without glutinous ridges, linear to narrowly lanceolate, reflexed or ascending, spines weak to moderately thick, 1–10 mm. **Florets**

corollas 25–40 mm, red, rarely white or pink, tubes 8–18 mm, throats 5–10 mm, gradually narrowed to tubes, lobes 5–12 mm; style tips 4–5 mm. **Fruits** 5–7 mm, light to dark brown; pappi 20–30 mm.

Western Mexico and United States. 7 varieties; 2 varieties treated in *Flora.*

Key to *Cirsium occidentale* varieties:

1. Phyllary tips usually with connecting trichomes . . . . *C. o.* var. ***candidissimum***

1' Phyllary tips usually without connecting trichomes . . . . *C. o.* var. ***venustum***

Cirsium occidentale (Nuttall) Jepson var. ***candidissimum*** (Greene) J.F. Macbride (p. 227)

Snowy thistle, snowwool thistle

Cirsium pastoris J.T. Howell

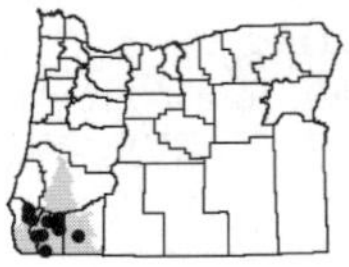

Plants 4–20 dm, densely white-tomentose. **Phyllaries** imbricate to subequal, tips usually with connecting trichomes. **Florets** corollas bright red to crimson, rarely white or pink.

Disturbed areas, roadsides, hillsides, coastal scrub, open woods, dry to seasonally wet areas. Flowering May–Aug. 50–1500 m. Sisk. CA, NV. Native.

Cirsium occidentale (Nuttall) Jepson var. ***venustum*** (Greene) Jepson

Venus thistle

Cirsium proteanum J.T. Howell

Plants 5–30 dm, usually white-tomentose, rarely glabrate. **Phyllaries** imbricate, tips usually without connecting trichomes. **Florets** corollas red, rarely white, pink or purple.

Disturbed areas, roadsides, woodlands, grasslands, streambanks. Flowering Apr–Jul. 200–1300 m. Sisk. CA, NV. Native.

Cirsium remotifolium (Hooker) de Candolle

Few-leaved thistle, weak thistle

Plants biennial or monocarpic perennial, 2–15 dm; taprooted or from creeping roots. **Stems** usually 1, slender, arachnoid-villous, sometimes thinly so. **Leaves** linear-oblong to elliptic, 7–35 × 1–15 cm, margins entire to dentate or pinnately lobed, lobes triangular-ovate to linear, spines fine, 1–6 mm, surfaces abaxially thinly to densely tomentose, adaxially glabrate to thinly arachnoid-villous; basal occasionally present at flowering; sessile, clasping, or petiolate. **Heads** few–many. **Involucres** hemispheric to campanulate, 1–2.5 × 1.5–3.5 cm, glabrous to thinly arachnoid-villous. **Phyllaries** subequal to strongly imbricate, sometimes with inconspicuous glutinous ridges, spines 1–6 mm; outer bases ≥ 2 mm wide. **Florets** corollas 16–25 mm, cream-colored to purple, tubes 6–12 mm, throats 5–10.5 mm, lobes 3.5–8 mm; styles conspicuously exserted, tips 4–6 mm. **Fruits** 4.5–5.5 mm, tan to dark brown; pappi 12–25 mm.

Western United States. 3 varieties.

Cirsium remotifolium is known to form fertile hybrids with *C. edule* where their ranges overlap.

Key to *Cirsium remotifolium* varieties:

1. Phyllaries strongly imbricate, unequal in length, outer phyllaries oblong to obovate, margins scarious . . . . *C. r.* var. ***odontolepis***

1' Phyllaries not imbricate, equal or subequal, outer phyllaries linear to narrowly oblong, margins not scarious.

 2. Phyllary margins entire; corollas cream-colored . . . . *C. r.* var. ***remotifolium***

 2' Phyllary margins ciliate with minute spreading to reflexed spines; corollas purple, rarely white. . . . . *C. r.* var. ***rivulare***

Cirsium remotifolium (Hooker) de Candolle var. ***odontolepis*** Petrak (p. 228)

Mountain thistle, fringe bract thistle

Cirsium callilepis (Greene) Jeps. var. *callilepis*, *Cirsium callilepis* (Greene) Jeps. var. *oregonense* (Petr.) J.T. Howell, *Cirsium callilepis* (Greene) Jeps. var. *pseudocarlinoides* (Petr.) J.T. Howell, *Cirsium centaureae* (Rydb.) K. Schum.

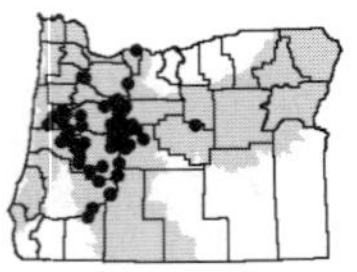

Phyllaries strongly imbricate, unequal; outer phyllaries oblong to obovate, margins scarious, erose. **Florets** corollas 16–25 mm, cream-colored or purple, tubes 7–9 mm, throats 5–10.5 mm, lobes 3.5–6.5 mm.

Meadows, grasslands, openings in woods, forests, streambanks, hillsides, brushy slopes. Flowering May–Sep. 50–1900 m. BW, Casc, CR, ECas, WV. CA. Native.

Populations of this taxon generally have either exclusively cream-colored or purple corollas, rather than mixed.

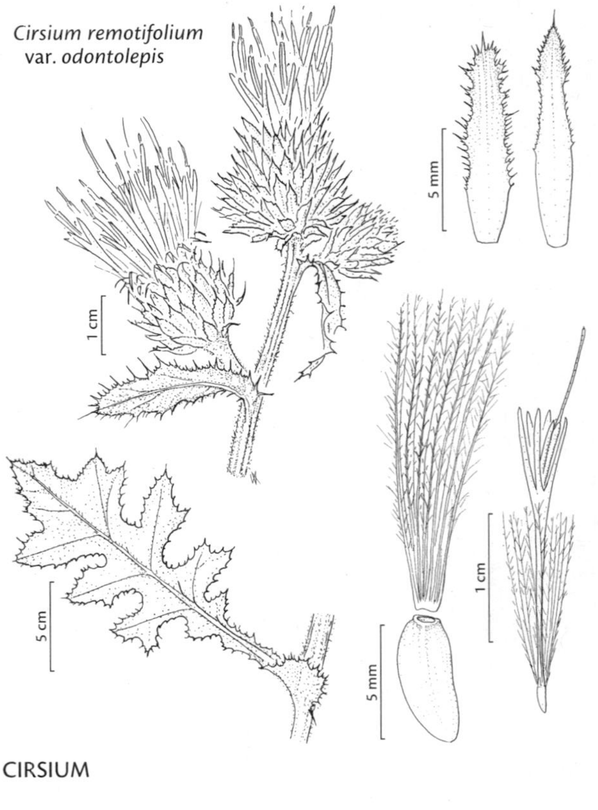

CIRSIUM

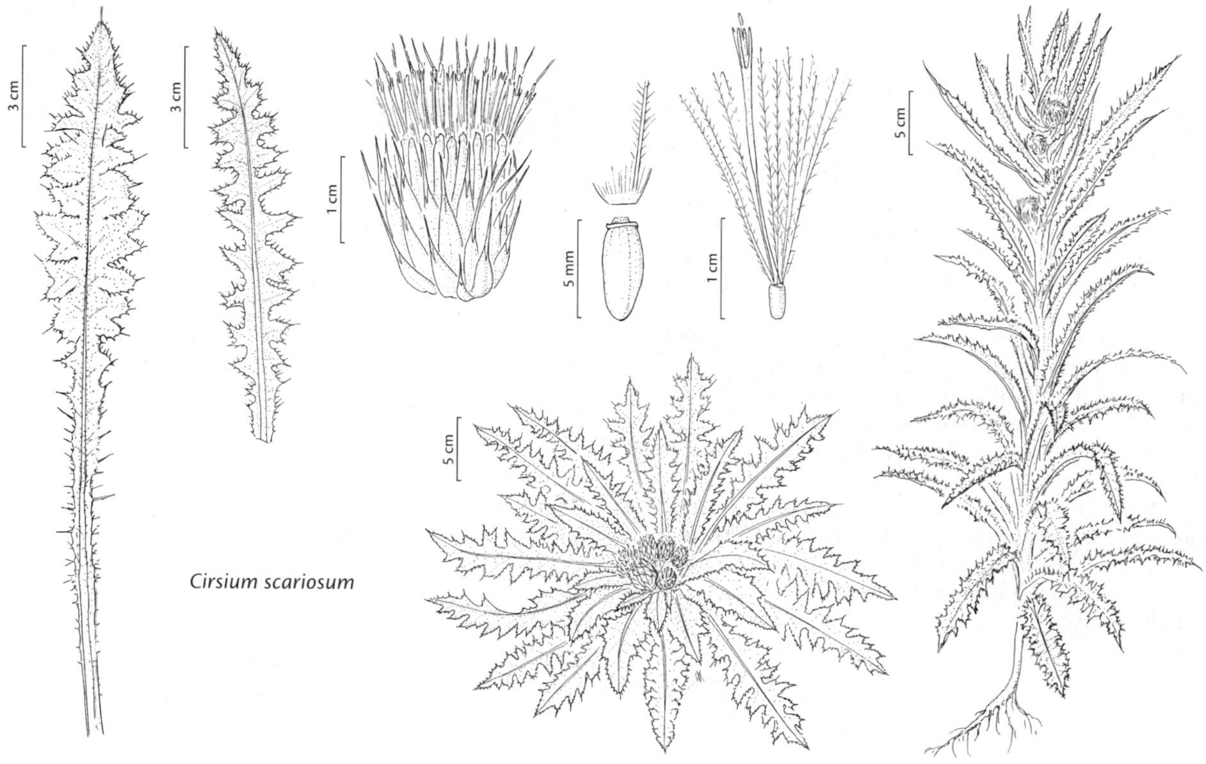

CIRSIUM

Cirsium remotifolium (Hooker) de Candolle var. *remotifolium*

Few-leaved thistle, weak thistle

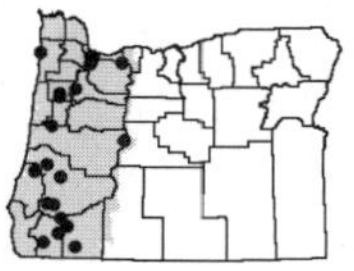

Phyllaries equal to subequal, not imbricate, linear to narrowly oblong, margins entire. **Florets** corollas 20–28 mm, cream-colored, tubes 6–12 mm, throats 8–10 mm, lobes 4.5–6.5 mm.

Meadows, grasslands, forest openings, woodlands, hillsides. Flowering May–Aug. 50–2000 m. Casc, CR, Est, Sisk, WV. CA, WA. Native.

Cirsium remotifolium (Hooker) de Candolle var. *rivulare* Jepson

Klamath thistle

Cirsium acanthodontum S.F. Blake

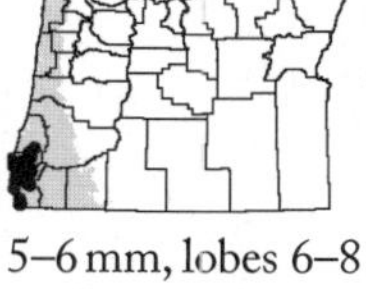

Phyllaries equal to subequal, not imbricate, linear to narrowly oblong, margins ciliate with minute spreading to reflexed spines. **Florets** corollas 20–25 mm, lavender to cream-colored, tubes 8–12 mm, throats 5–6 mm, lobes 6–8 mm.

Seaside bluffs, meadows, streambanks, old dunes, forest openings. Flowering May–Aug. 0–1300 m. CR, Est, Sisk. CA. Native.

Plants of this taxon may assume a compact growth form in maritime habitats.

Cirsium scariosum Nuttall (p. 229)

Elk thistle, meadow thistle, dinnerplate thistle

Cirsium scariosum Nutt. var. *americanum* (A. Gray) D.J. Keil, *Cirsium scariosum* Nutt. var. *robustum* D.J. Keil, *Cirsium scariosum* Nutt. var. *scariosum*, *Cirsium scariosum* Nutt. var. *toiyabense* D.J. Keil

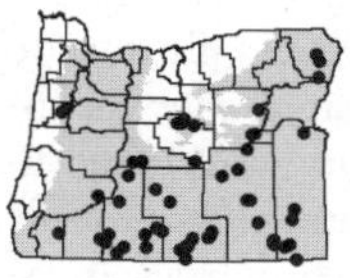

Plants biennial or monocarpic perennial, 0.5–20 dm; taprooted. **Stems** often acaulescent or short-caulescent, sometimes caulescent with a single bushy-branched stem, stout and fleshy, glabrous or sparsely arachnoid-tomentose. **Leaves** linear to elliptic, 10–40 × 3–8 cm, margins entire to deeply pinnately divided, plane or undulate, spines usually thin to moderately thick, 2–15 mm, surfaces abaxially glabrous to densely arachnoid-tomentose, adaxially glabrous to thinly tomentose or villous; basal usually present at flowering; sessile or petiolate. **Heads** 1–many. **Involucres** ovoid to hemispheric, 1–3(4) × 1.5–6 cm, glabrous to thinly arachnoid-villous. **Phyllaries** imbricate, without glutinous ridges, spines thin, 0–12 mm. **Florets** corollas 20–40 mm, white to purple, sometimes rose or pink-tinged, tubes 9–22 mm, throats 4–12 mm, lobes 4–10 mm; styles conspicuously exserted, tips 3–8 mm. **Fruits** 4–6.5 mm, light to dark brown, occasionally striate; pappi 15–35 mm.

Wet meadows, forest openings, sagebrush areas, pastures. Flowering Jun–Aug. 100–2000 m. BR, BW, Casc, ECas, Owy, Sisk, WV. CA, ID, NV, WA; north to British Columbia, northeast to Alberta, east to WY, southeast to NM, Quebec. Native.

Cirsium scariosum is known to hybridize with other native species where their ranges overlap.

Cirsium undulatum (Nuttall) Sprengel

Wavyleaf thistle

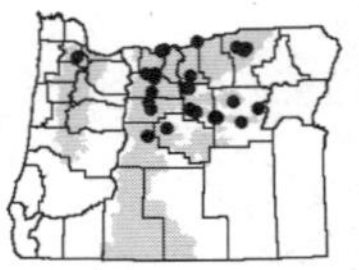

Plants perennial, 2–23 dm; taprooted, often with well-developed lateral roots. **Stems** 1–several, white-tomentose. **Leaves** elliptic to ovate, 10–40 × 1–10(14) cm, margins shallowly dentate to deeply pinnate, undulate, spines 2–12 mm, surfaces abaxially densely white-tomentose, adaxially glabrate to sparsely tomentose; basal occasionally present at flowering; sessile, clasping, or petiolate. **Heads** 1–10. **Involucres** ovoid to campanulate, 2.5–4.5 × 1.5–4.5 cm, glabrate to thinly tomentose. **Phyllaries** imbricate, with prominent glutinous ridges, spines spreading to reflexed, 1.5–5 mm. **Florets** corollas 24–50 mm, white to pale lavender or pink, tubes 12–28 mm, throats 6–14 mm, lobes 7–13 mm; style tips 5–7.5 mm. **Fruits** 6–7 mm, light to dark brown; pappi 20–40 mm.

Sagebrush areas, dry hillsides, prairies, disturbed areas, roadsides. Flowering May–Aug. 50–1400 m. Col, ECas, Lava, WV. CA, ID, NV, WA; north to British Columbia, northeast to Manitoba, east to MI, southeast to TX, south to Mexico, GA, PA. Native.

Cirsium vulgare (Savi) Tenore

Bull thistle, common thistle

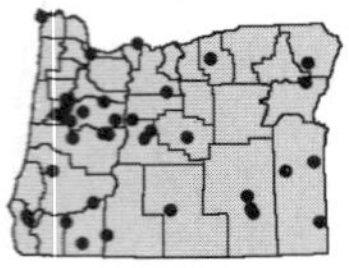

Plants biennial, 3–20 dm; taprooted. **Stems** usually single, spreading-villous, sometimes arachnoid. **Leaves** oblong-lanceolate to obovate, 15–40 × 6–15 cm, margins 1–2-pinnate, lobes lanceolate to triangular, spines 2–10(15) mm, surfaces abaxially tomentose, ± villous, adaxially glabrate to sparsely tomentose, with appressed bristle-like spines; basal absent or present at flowering; sessile, decurrent, or petiolate. **Heads** few to many. **Involucres** hemispheric to campanulate, 2–4 × 2–4 cm, thinly tomentose. **Phyllaries** without glutinous ridges, spines 1–5 mm. **Florets** corollas 25–35 mm, purple, rarely white, tubes 18–25 mm, throats 3–6 mm, lobes 5–9 mm; style tips 3.5–6 mm. **Fruits** 3–5 mm, light brown with darker streaks; pappi 20–30 mm.

Disturbed areas, pastures. Flowering Jun–Sep. 0–2000 m. All ecoregions. CA, ID, NV, WA; throughout North America; worldwide. Exotic.

Cirsium vulgare is our only species with bristle-like spines on the upper leaf surface. This species is considered a noxious weed and is unpalatable to most livestock.

COLUMBIADORIA Goldenweed

Genus composed of one species, *C. hallii.*

Columbiadoria hallii (A. Gray) G.L. Nesom (p. 230)

Hall's goldenweed

Haplopappus hallii A. Gray

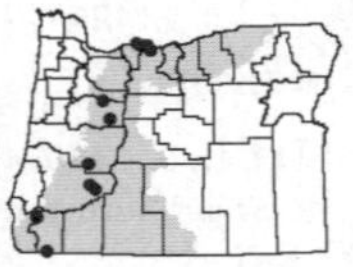

Herbs or **subshrubs** perennial, 1–6 dm; rhizomatous. **Stems** annual, 1–many, arising from rhizomes, erect, leafy throughout, unbranched below the terminal inflorescences, glabrous or puberulent. **Leaves** basal withering early; cauline alternate, oblanceolate, ≤ 6 cm, reduced distally, margins entire, papillate, tips acute, surfaces glabrous to hispidulous or scabrous, sometimes gland-dotted; short-petiolate or sessile. **Inflorescences** few to numerous heads in terminal, panicle- or raceme-like, often congested arrays, occasionally solitary. **Heads** radiate; on naked or bracteate peduncles. **Involucres** cylindrical or narrowly campanulate, 7–11 mm. **Phyllaries** 35–40 in 5–6 unequal series, erect in flower, spreading in fruit, lanceolate, outer boat-shaped, gradually shorter, pale, green- or brown-tipped, shiny, midribs visible or not, surfaces glabrous, resinous. **Receptacles** flat; paleae 0. **Ray florets** 5–8, pistillate, conspicuous; corollas 6–10 mm, blades yellow, exceeding involucres. **Disc florets** 12–25, bisexual; corollas narrowly tubular, yellow, slightly exceeding involucres, lobes 5, erect; stamens < corollas; stigmas slightly exserted from corolla tubes. **Fruits** oblong, truncate, tapering toward bases, 4–5 mm, brown, inconspicuously 8-ribbed, strigose; pappi of numerous fine bristles, 4–6 mm, yellowish or tan.

Open prairies, slopes, hillsides, ridge tops, rock outcrops. Flowering Aug–Oct. 50–1900 m. Casc, Col, ECas, Sisk. WA. Native.

Columbiadoria is one of several North American genera (including *Ericameria, Hazardia, Nestotus, Pyrrocoma, Stenotus,* and *Tonestus*) that have been segregated from the South American genus *Haplopappus.*

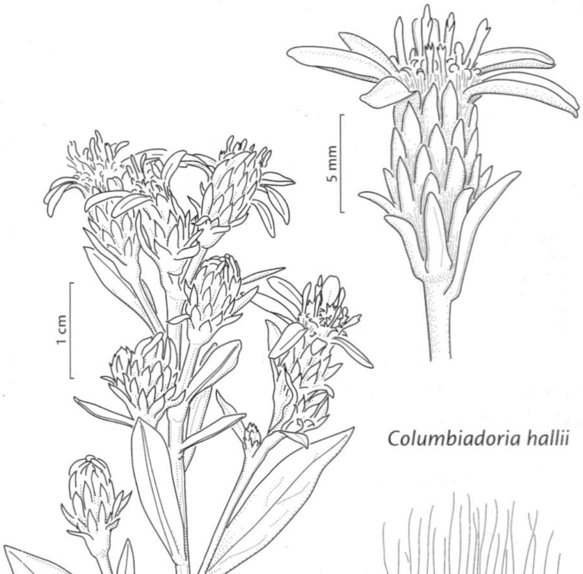

COLUMBIADORIA

CONYZA Conyza, horseweed

Herbs annual, rarely perennial; taprooted. **Stems** 1–several from base, simple to well branched, erect, glabrous to scabrous, hispidulous, or strigose. **Leaves** basal and cauline, alternate, not much reduced distally, lower often withering by anthesis; blades simple, linear or lanceolate to oblanceolate or elliptic, margins entire or serrulate, tips most often acute, surfaces glabrous to scabrous, hispidulous, or strigose; short-petiolate or sessile. **Inflorescences** heads numerous in panicle-like, rarely raceme-like, often much-branched arrays. **Heads** ± radiate to disciform; peduncles usually short, glabrous to hispidulous or strigose. **Involucres** turbinate, cylindric. **Phyllaries** in several series, erect, reflexed in fruit, outer gradually shorter, inner equal, linear or linear-lanceolate, mostly 1-veined, with brownish or greenish midline and membranous margins, tips acute or acuminate, surfaces glabrous to strigillose. **Receptacles** flat; paleae 0. **Ray florets** 0, although outer disc florets may have short rays. **Outer florets** in 1–3 series, pistillate, fertile; corollas tubular, filiform, often with short rays < 1.5 mm, styles slightly or well exserted from corolla. **Inner florets** bisexual, fertile; corollas tubular, widened distally, white, lobes 5, triangular, erect or spreading; stamens 4–5, slightly exserted from corolla. **Fruits** fusiform or oblong, compressed, glabrous or strigillose, with 2 marginal veins; pappi persistent, of numerous ± equal bristles in 1 series.

Worldwide. 25–100+ species; 2 species treated in *Flora.*

Conyza is kept separate from *Erigeron* in some treatments (Nesom & Robinson 2007; Strother 2006). Strict adherence to cladistic taxonomy, however, would support its inclusion in *Erigeron.* This hypothesis is based on the work of Noyes (2000), which shows it to be polyphyletic and nested within a large and complex *Erigeron*. Further molecular studies are needed to clarify evolutionary relationships.

Key to *Conyza* species:

1. Involucres and peduncles densely strigillose; outer florets 60–150+, white to purple, rays 0–0.3 mm; pappus bristles 3–4.5 mm . . . . *C. bonariensis*

1' Involucres usually glabrous, peduncles glabrous to strigillose; outer florets 20–45+, tan to gray-brown, rays 0.3–1 mm; pappus bristles 1.5–3 mm . . . . *C. canadensis*

Conyza bonariensis (Linnaeus) Cronquist

South American conyza, flax-leaved horseweed

Conyza floribunda Kunth

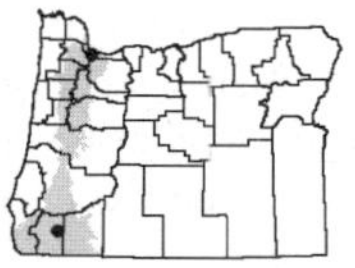

Plants 1–15 dm. **Leaves** 2–8+ cm, margins entire or serrulate, surfaces densely strigillose or scabrous. **Peduncles** 2–30 mm, densely strigillose. **Involucres** 3.5–5 mm. **Phyllaries** tips acute or acuminate, surfaces, densely strigillose. **Outer florets** 60–150+; corollas 3–3.5 mm, rays 0–0.3 mm, white to purple. **Inner florets** 8–20+. **Fruits** 1–1.5 mm, pale tan, lightly strigillose; pappus bristles 3–4.5 mm. **2*n***=54.

Roadsides, disturbed areas. Flowering Jul–Sep. 200–1100 m. Sisk, WV. CA, WA; southeastern and western US; South America. Exotic.

Conyza bonariensis has sometimes been identified as *C. floribunda*, which is a different species not present in the Pacific Northwest. According to Christy et al. (2009), *C. bonariensis* was introduced into Multnomah County around 1875–1899, but there are no recent reports from that area.

Conyza canadensis (Linnaeus) Cronquist (p. 231)

Horseweed, Canadian fleabane

Conyza canadensis (L.) Cronquist var. *canadensis*, *Erigeron canadensis* L.

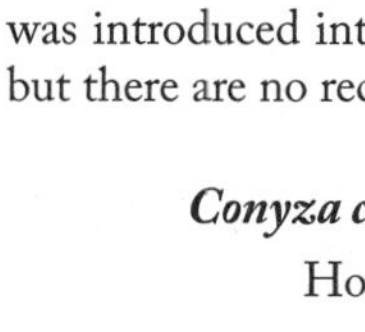

Plants 0.2–30+ dm. **Leaves** 2–10 cm, margins entire or serrulate, surfaces hispid to strigillose or scabrous, often glabrous with age. **Peduncles** 1–25 mm, glabrous to strigillose. **Involucres** 3–4 mm. **Phyllaries** tips acute, surfaces glabrous or sparsely strigillose. **Outer florets** 20–45+; corollas 2–3 mm, rays 0.3–1 mm, white. **Inner florets** 8–30+. **Fruits** 1–1.5 mm, tan to gray-brown, lightly strigillose; pappus bristles 1.5–3 mm. **2*n***=18.

Riverbanks, lake shores, woods, shrublands, clearcuts, roadsides, waste ground, disturbed areas. Flowering Jun–Sep. 0–2000 m. All ecoregions. CA, ID, NV, WA; nearly worldwide. Native.

Like many weedy species, *C. canadensis* is variable in size and habit depending on environmental conditions. It ranges from plants over 3 m tall to small, depauperate individuals only 2–3 cm. It is well separated from *C. bonariensis* by its smaller, more or less glabrous heads and shorter pappus bristles. Both species are able to reproduce through self-pollination (Noyes 2000).

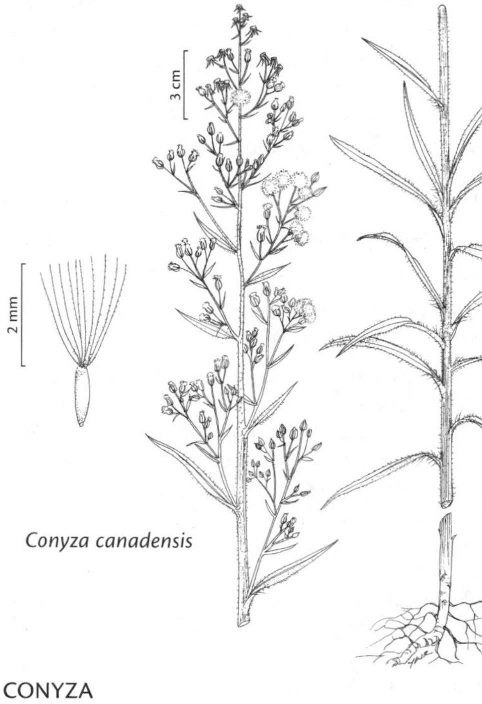

CONYZA

COREOPSIS Coreopsis, tickseed

Herbs annual, biennial, or perennial; taprooted, sometimes with rhizomes or well-developed root-crowns. **Stems** 1–several, erect, simple or branched, glabrous or proximally villous. **Leaves** basal and cauline, basal usually withering by anthesis, opposite, simple or pinnately compound, divisions of pinnate leaves linear or terminal segments narrowly lance-oblong, margins and surfaces glabrous or villous-puberulent; petiolate or sessile. **Inflorescences** heads solitary or in panicle-like arrays. **Heads** radiate; peduncles glabrous. **Phyllaries** in 2 series, erect, herbaceous, spreading in fruit; outer series somewhat shorter than inner, narrowly lanceolate to lance-linear, tips acute, margins hispid or much reduced and clustered at base of involucres, larger phyllaries with numerous parallel veins; inner series broadly lance-ovate, tips acute, surfaces glabrous; paleate. **Receptacles** flat to convex; paleate. **Paleae** narrow, membranous, deciduous. **Ray florets** pistillate or sterile; rays well exceeding involucres, dark pink to light rose or white, or yellow and sometimes marked with brown. **Disc florets** bisexual; corollas tubular, gradually widened distally, yellow, lobes 4–5, triangular, erect or spreading; stamens 5, slightly to well exserted from corolla tubes, yellow or purple; style branches exserted from stamen tubes. **Fruits** ± compressed, oblanceolate or elliptic, narrowed toward bases and tips, sometimes winged; pappi 0 or of 2 short scales.

North America. ~20 species; 2 species treated in *Flora.*

Coreopsis rosea has been reported from Sauvie Island (Multnomah County) but does not appear to have become established in Oregon.

Key to *Coreopsis* species:

1. Plants perennial; inflorescences solitary heads; outer series of phyllaries 50–~100% length of inner series; leaves villous- or glandular-puberulent; rays yellow, tips usually 4+-lobed . . . . *C. lanceolata*

1' Plants annual or biennial; inflorescences numerous heads in well-branched, panicle-like arrays; outer series of phyllaries much reduced and basal to inner series; leaves glabrous; rays yellow marked with brown, tips often 3-lobed . . . . *C. tinctoria*

Coreopsis lanceolata Linnaeus

Garden coreopsis, lance-leaved coreopsis, lanceleaf tickseed

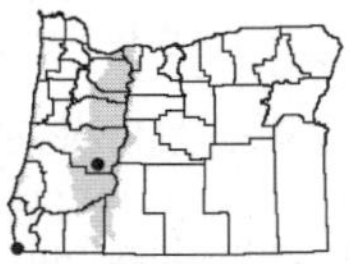

Plants perennial, 2–9 dm. **Leaves** simple, linear to elliptic-lanceolate, or pinnately compound with 2–6 broadly linear leaflets and a terminal elliptic-lanceolate segment, 4–20 cm, margins hispid-puberulent, tips acute, surfaces villous- or glandular-puberulent. **Inflorescences** solitary terminal and axillary heads. **Peduncles** 12–30 cm. **Involucres** hemispheric, 8–14 mm. **Phyllaries** outer series 50–~100% length of inner series. **Ray florets** 8–15; rays 13–30 mm, yellow, tips usually 4+-lobed. **Disc florets** corollas 4–7 mm, well exceeding involucres; stamens yellow. **Fruits** 2.5–4 mm, black, glabrous, broadly or narrowly winged; pappi of 2 short scales. **2*n***=26.

Roadsides, disturbed areas. Flowering May–Jul. 0–500 m. Casc, Est. CA, WA; British Columbia, central and eastern Canada and US. Exotic.

Coreopsis tinctoria Nuttall (p. 232)

Columbia coreopsis, calliopsis, golden tickseed

Coreopsis atkinsoniana Douglas ex Lindl., *Coreopsis tinctoria* Nutt. var. *atkinsoniana* (Douglas ex Lindl.) H.M. Parker

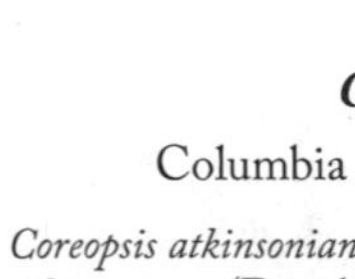

Plants annual or biennial, 2.5–12 dm. **Leaves** pinnately compound, rarely simple, 4–15 cm, with linear rachises and lobes, terminal segments linear or narrowly elliptic, margins smooth, tips acute, surfaces glabrous. **Inflorescences** panicle-like arrays. **Peduncles** 1.5–15 cm. **Involucres** hemispheric, 5–10 mm. **Phyllaries** outer series much reduced and basal to inner series. **Ray florets** 6–9; rays 7–20 mm, yellow marked with brown, tips often 3-lobed. **Disc florets** corollas 4–7 mm, well exceeding involucres; stamens yellow. **Fruits** 1.5–3 mm, black, glabrous or minutely roughened, narrowly winged or wingless; pappi 0. **2*n***=24, 26.

Riverbanks, floodplains. Flowering Jun–Oct. 0–700 m. Casc, Col, ECas, WV. CA, ID, WA; throughout most of Canada and US, northern Mexico. Native.

According to Cronquist (1994), this species is native to the Great Plains and the immediately more eastern states but has become naturalized throughout most of the rest of the country. It is considered native in Oregon and Washington, having been collected along the Columbia River by David Douglas early in the 19th century. The species may be expected elsewhere in Oregon as an escape from cultivation.

This treatment follows that of Strother (2006) by not recognizing any varieties or subspecies in *C. tinctoria*.

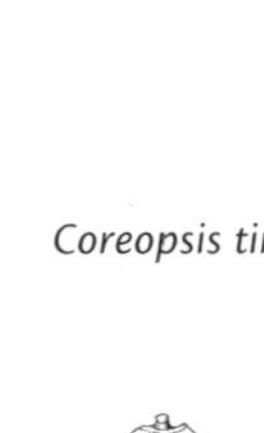

Coreopsis tinctoria

COREOPSIS

CORETHROGYNE Sandaster

Genus composed of one species, *C. filaginifolia.*

Corethrogyne filaginifolia (Hooker & Arnott) Nuttall (p. 234)

California sandaster, prostrate beach aster

Corethrogyne californica DC., *Corethrogyne filaginifolia* (Hook. & Arn.) Nutt. var. *californica* (DC.) Saroyan, *Corethrogyne obovata* Benth., *Lessingia filaginifolia* (Hook. & Arn.) M.A. Lane var. *californica* (DC.) M.A. Lane

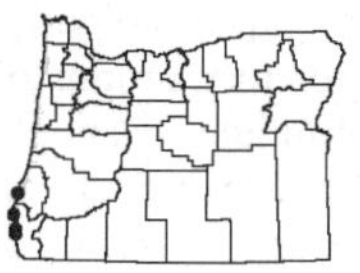

Herbs, subshrubs perennial, ± prostrate; fibrous-rooted from ± woody root-crowns. **Stems** 1–many, decumbent or erect, 5–20 cm, simple or branched, generally floccose-tomentose. **Leaves** cauline, alternate, 1.5–8 cm, gradually reduced distally, often becoming small and bracteate on inflorescence peduncles; blades oblanceolate to broadly obovate, bases tapering, with 1 main vein, distal margins and tips dentate or denticulate, tips rounded, surfaces floccose-tomentose; petioles winged. **Inflorescences** solitary heads or in loose corymb-like arrays. **Heads** radiate; peduncles elongated, ≤ 15 cm, floccose-tomentose to densely glandular-puberulent. **Involucres** hemispheric, 8–12 mm. **Phyllaries** in 3–9 series, erect, often reflexed, linear-lanceolate, green or purplish, minutely glandular-puberulent, midribs distinct, tips acute or acuminate; outer gradually shorter; inner equal. **Receptacles** flat or rounded; paleae 0. **Ray florets** 15–40, pistillate; rays ≤ 15 mm, well exceeding involucres, pink or purple. **Disc florets** 20–100, bisexual; corollas tubular, gradually widened distally, 4–8 mm, slightly exceeding involucres, yellow, lobes 5, erect, lanceolate; stamens 5, included; styles well exserted, stigmatic lobes hispid. **Fruits** cylindric or obconic, tapering toward base, 2–5 mm, puberulent, 5–7-ribbed; pappi of numerous, somewhat unequal bristles, 3–8 mm, brownish. **2*n***=10.

Beaches, dunes, coastal grasslands, pine woods, roadsides. Flowering Jun–Oct. 0–50 m. Est. CA; south to Mexico. Native.

Corethrogyne is unusually diverse in vegetative and floral characteristics, and at least 18 species names have been recognized within the genus in the past (Saroyan et al. 2000). The most recent approach (Markos & Strother 2006, 2012) has been to combine these into a single species with no subspecies or varieties. Oregon populations would belong to *C. filaginifolia* ssp. *californica*, if infraspecific taxa were recognized.

COTA Chamomile

Herbs perennial; rhizomatous and/or stoloniferous. **Stems** unbranched or with simple, distal branches, villous- or floccose-puberulent. **Leaves** basal and cauline, basal usually withered at flowering, alternate, pinnately lobed, dentate or finely serrulate, lobes linear, tips acute, surfaces glabrous or villous-puberulent; sessile or short-petiolate. **Inflorescences** panicle- or corymb-like arrays or heads solitary; peduncles leafless or proximally bracteate, villous-puberulent. **Heads** radiate. **Involucres** bowl-shaped or hemispheric. **Phyllaries** in 3–5 series, erect, spreading in fruit, silky-villous; inner equal, lanceolate, margins membranous, midribs green or brown, tips acute; outer gradually shorter, lanceolate to triangular, tips acute. **Receptacles** ± hemispheric; paleate. **Paleae** ± flat, often cartilaginous. **Ray florets** pistillate; corollas ovate-elliptic or oblanceolate, tips obtuse to emarginate or 3-toothed. **Disc florets** bisexual; corollas tubular, slightly widened distally, 5-lobed, lobes triangular, spreading; stamens 5, included; stigmas included in corolla tube. **Fruits** obconic or 4-angled, somewhat compressed, indistinctly 5–10-veined, glabrous; pappi of low crowns.

Africa, Asia, Europe, North America. 40 species; 1 species treated in *Flora.*

Molecular evidence for the separation of *Cota* from *Anthemis* was presented by Oberprieler (2002). In Oregon, is it distinguished from *Anthemis* by its perennial rather than annual habit and yellow versus white or pinkish ray corollas.

Cota tinctoria (Linnaeus) J. Gay (p. 234)

Golden marguerite, golden chamomile, yellow chamomile

Anthemis tinctoria L.

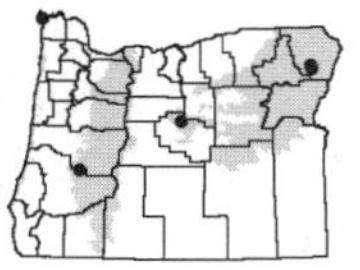

Plants 2–7 dm. **Leaves** 2–5 cm, 1-veined from bases, often with fringed auricles; cauline gradually reduced distally. **Heads** 10–15 mm; peduncles 6–18 mm. **Involucres** 5–8 mm. **Paleae** 4–5 mm. **Ray florets** 10–21(35); corollas reflexed and persisting in fruit, 5–12+ mm, yellow. **Disc florets** numerous; corollas 3–4.5 mm, yellow. **Fruits** 1.8–2.2 mm, brown. **2*n***=18.

Woodlands, roadsides, disturbed areas. Flowering Jun–Sep. 0–1300 m. BW, Casc, Est. CA, ID, WA; north to AK, east to Newfoundland, southeast to VA; Asia, Europe. Exotic.

COTULA Cotula, waterbuttons

Herbs annual or perennial; fibrous-rooted or stoloniferous. **Stems** 1–several, decumbent to erect or spreading, simple or branched, glabrous to villous or minutely glandular. **Leaves** principally cauline, alternate, simple to pinnately toothed or lobed, or pinnately compound, surfaces glabrous to villous or minutely gland-dotted. **Inflorescences** solitary terminal heads. **Heads** disciform, rarely radiate or discoid. **Involucres** spreading in fruit. **Phyllaries** in 2–3 equal series, lance-elliptic, margins membranous, 1–5-veined, glabrous or villous; inner obtuse; outer obtuse or acute. **Receptacles** flat to convex; paleae 0. **Ray florets** 0, or rarely 1–3 series of 8–80+ pistillate florets with corollas 0. **Disc florets** corollas tubular, slightly widened distally, lobes 4–5, triangular, erect; stamens 4–5, not exserted; styles ≤ 1 mm. **Fruits** obovoid or oblong, compressed, with 2 marginal veins, sometimes wing-margined; pappi 0.

Africa, Australia, Europe, North America, Pacific Islands, South America. 55 species; 2 species treated in *Flora.*

A single collection of *Cotula mexicana* has been made in Oregon, in 2009, from closely mowed greens of a golf course in Morrow County, where it was probably introduced in turf from California. It has a matted growth form, pinnately lobed leaves, and floral peduncles about as long as the leaves. Thus far it appears to be limited to this specialized type of cultivated habitat.

Key to *Cotula* species:

1. Plants annual; stems villous; leaves petiolate; phyllaries 1-veined . . . . ***C. australis***

1' Plants perennial; stems glabrous or minutely glandular; leaves sessile; phyllaries 3–5-veined . . . . ***C. coronopifolia***

Cotula australis (Sieber) Hooker f.

Australian cotula, Australian waterbuttons

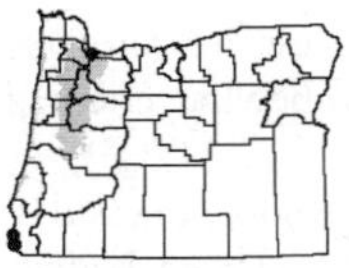

Plants annual, (2)5–20 cm. **Stems** erect or spreading, usually well branched, villous. **Leaves** obovate to spatulate, (0.5)1–4 cm, pinnately lobed or compound, margins entire or 3–5-lobed, tips acute, surfaces glabrous or villous; petiolate. **Peduncles** 0.8–7 cm. **Involucres** ovoid or hemispheric, 1.5–3 mm. **Phyllaries** obtuse, 1-veined, surfaces glabrous or villous. **Disc florets** corollas 0.75–1 mm, equaling or exceeding involucres, white. **Fruits** 1–1.2 mm, minutely glandular, brown; winged. **2*n***=36, 40.

Roadsides, disturbed areas. Flowering Jan–Jul. 0–300 m. Est, WV. CA; widely scattered in North America; Africa, Australia, South America. Exotic.

Cotula australis was found on ballast dumps in Portland as early as 1899, but it has not been reported recently from that area.

Cotula coronopifolia Linnaeus (p. 234)

Brass buttons

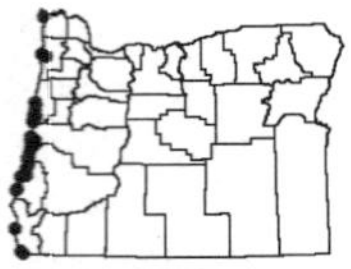

Plants perennial, 5–45 cm; usually stoloniferous. **Stems** decumbent to erect, simple to well branched, glabrous or minutely glandular. **Leaves** linear to linear-lanceolate, 1–7 cm, bases auriculate and forming a membranous sheath around stem, margins entire or few-lobed, lobes linear, tips acute, surfaces glabrous, sometimes gland-dotted; sessile. **Peduncles** 1–9 cm. **Involucres** hemispheric, 3–4 mm. **Phyllaries** 3–5-veined, surfaces glabrous; inner obtuse; outer obtuse or acute. **Disc florets** corollas 1–1.5 mm, equaling or exceeding involucres, yellow. **Fruits** 0.7–1.8 mm, glabrous or minutely glandular, brown; winged or not. **2*n***=20.

Beaches, tide marshes, moist fields, disturbed areas. Flowering Mar–Dec. 0–200 m. Est. CA, NV, WA; north to British Columbia, southeast to AZ, south to Mexico; Africa, Australia, Europe, South America. Exotic.

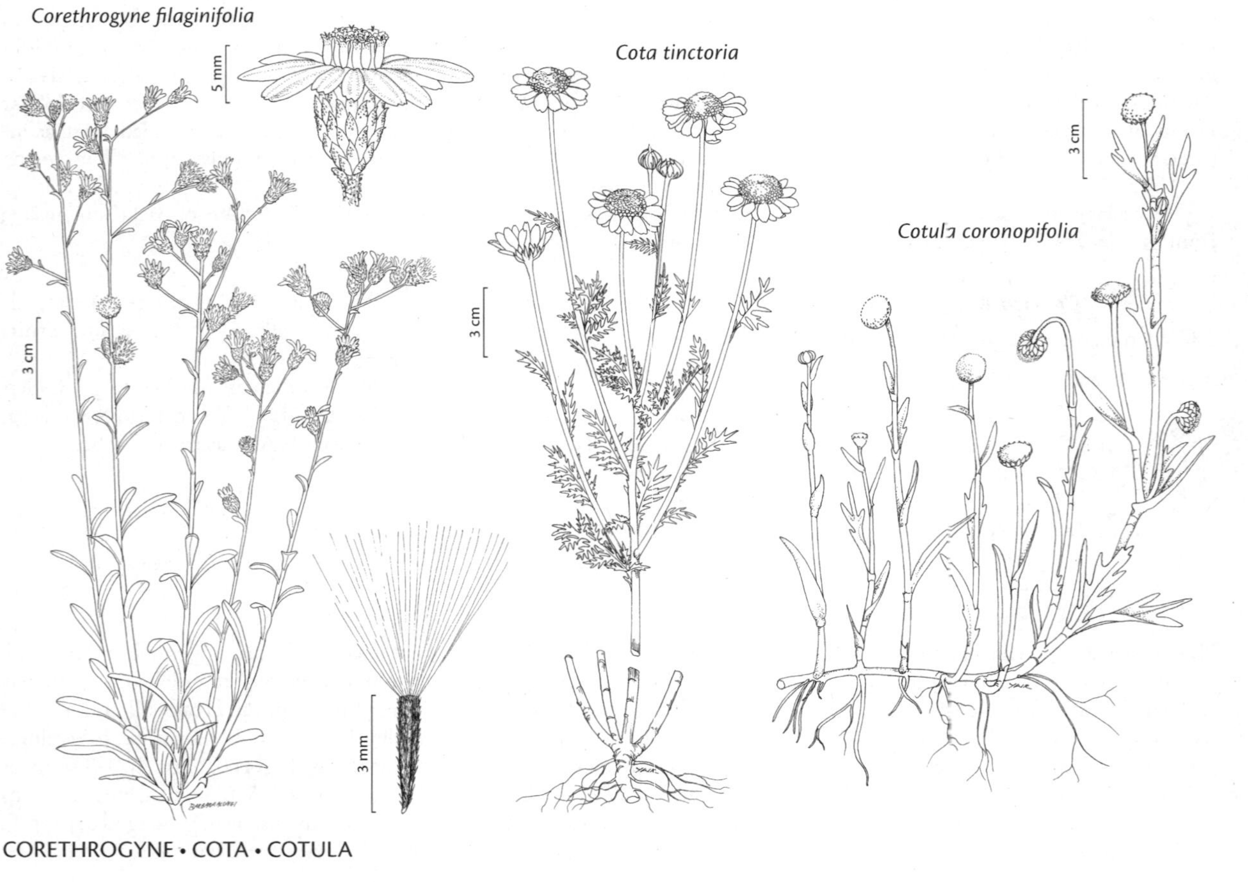

CORETHROGYNE • COTA • COTULA

CREPIS Hawksbeard

Herbs annual, biennial, or perennial, sap milky; taprooted. **Stems** erect or decumbent, simple or branched, glabrous or pubescent, or plants acaulescent with inflorescence axis arising from root crowns. **Basal leaves** sometimes absent in annual species; blades linear to lanceolate or ovate, surfaces glabrous or pubescent; usually petiolate, ± winged. **Cauline leaves** alternate, usually reduced in size and degree of lobing, sometimes auriculate-clasping. **Inflorescences** corymb-, cyme-, or panicle-like arrays, flat-topped or not. **Heads** ligulate, erect. **Involucres** cylindric to campanulate. **Phyllaries** in 1–2 series, erect, reflexing at maturity, surfaces glabrous to pubescent; outer gradually reduced in size or all short and basal; inner equal. **Receptacles** flat to convex; paleae 0. **Florets** 5–100+, bisexual; ligules usually well exceeding involucres, yellow, sometimes reddish on abaxial surfaces; stamens 5. **Fruits** cylindrical, straight or curved, sometimes tapering, ribs 10–20, smooth or roughened; beaked or not; pappi of numerous fine, white bristles.

Nearly worldwide. ~200 species; 15 species treated in *Flora*.

The rare alpine species *Crepis nana* has been transferred to the genus *Askellia* as *A. pygmaea*. There is a sharp distinction between the native, strongly taprooted perennial species, mostly of dry habitats east of the Cascades, and the short-lived, weedy, introduced species found in lawns, gardens, and other disturbed habitats, especially west of the Cascades. The perennials are unique in forming a polyploid complex with a base chromosome number of x=11 (Babcock & Stebbins 1938).

In Oregon, *C. tectorum* has been collected infrequently as a waif. It can be identified by its combination of tapered fruits and pubescent phyllaries.

Key to *Crepis* species:

1. Plants annual or biennial (rarely perennial in *C. vesicaria* ssp. *taraxacifolia*); fruits 1.5–9 mm.
    - 2. Fruits not beaked, 1.5–6 mm.
        - 3. Fruits 4–6 mm; inner phyllaries glabrous . . . . . **C. pulchra**
        - 3' Fruits 1.5–4 mm; inner phyllaries glandular.
            - 4. Fruits 1.5–2.5 mm; involucres 5–8 mm . . . . . **C. capillaris**
            - 4' Fruits 2.5–4 mm; involucres 8–10 mm . . . . . **C. nicaeensis**
    - 2' Inner fruits beaked, 3–9 mm.
        - 5. Fruits 3–5 mm, outer beaked . . . . . **C. setosa**
        - 5' Fruits 5–9 mm, outer not beaked . . . . . ***C. vesicaria*** ssp. ***taraxacifolia***

1' Plants perennial; fruits 4–11 mm.
- 6. Cauline leaves 0 or stems few-leaved only near base.
    - 7. Florets 5–8(12) . . . . . **C. pleurocarpa**
    - 7' Florets 20–50 . . . . . **C. runcinata**
- 6' Cauline leaves well developed.
    - 8. Stems usually densely setose . . . . . **C. monticola**
    - 8' Stems glabrous, tomentulose, tomentose, or glandular-hirsute (sometimes sparsely setose in *C. barbigera*).
        - 9. Inner phyllaries usually glabrous . . . . . **C. acuminata**
        - 9' Inner phyllaries finely glandular, setose, long-setose, tomentose, canescent-tomentulose, or tomentulose.
            - 10. Inner phyllaries densely long-setose with yellowish or greenish bristles . . . . . **C. barbigera**
            - 10' Inner phyllaries not long-setose, bristles (if present) usually not yellowish or greenish.
                - 11. Inner phyllaries usually finely glandular . . . . . **C. bakeri**
                - 11' Inner phyllaries setose, tomentose, canescent-tomentulose, or tomentulose.
                    - 12. Inner phyllaries usually black-setose, hairs not glandular . . . . . **C. modocensis**
                    - 12' Inner phyllaries usually not black-setose, or if setose then hairs glandular.
                        - 13. Leaves linear to narrowly lanceolate . . . . . **C. atribarba**
                        - 13' Leaves lanceolate to elliptic.
                            - 14. Heads (10)20–60, in flat-topped arrays . . . . . **C. intermedia**
                            - 14' Heads 2–30, in open arrays . . . . . **C. occidentalis**

Crepis acuminata Nuttall (p. 238)

Long-leaved hawksbeard, tapertip hawksbeard

Crepis acuminata Nutt. ssp. *acuminata*

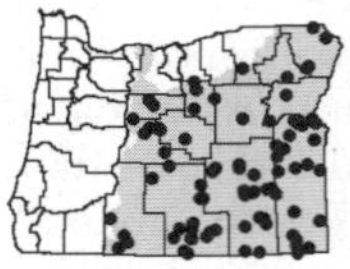

Plants perennial, 20–100 cm; with stout taproots and enlarged, woody crowns. **Stems** erect, unbranched or with distal inflorescence branches, glabrous or tomentulose. **Basal leaves** lanceolate-elliptic, rarely linear, 12–40 cm, irregularly pinnatifid, margins mostly entire, lobes elongate-deltoid to lance-linear, tips acuminate, surfaces glabrous or puberulent; petiolate. **Cauline leaves** similar to basal and reduced or linear to lanceolate; sessile. **Inflorescences** flat-topped. **Heads** 20–70. **Involucres** cylindric, 8–15 mm. **Phyllaries** outer deltoid; inner (5)7–8, linear-lanceolate, usually glabrous. **Florets** (5)7–10(12); ligules 13–18 mm. **Fruits** cylindric or tapering, 5.5–9 mm, pale brown, 10-ribbed. **2*n***=22, 33, 44, 55, 88.

Dry, rocky slopes and ridges, pine woods, riverbanks. Flowering May–Jul. 200–2400 m. BR, BW, ECas, Lava, Owy. CA, ID, NV, WA; northeast to MT, east to WY, southwest to NM. Native.

Crepis acuminata is best distinguished from the somewhat similar *C. intermedia* by its glabrous phyllaries. It commonly reproduces sexually and is a likely parent of the hybrid and completely apomictic species *C. intermedia*.

Crepis atribarba A. Heller

Slender hawksbeard

Crepis atribarba A. Heller ssp. *atribarba*, *Crepis atribarba* A. Heller ssp. *originalis* Babc. & Stebbins

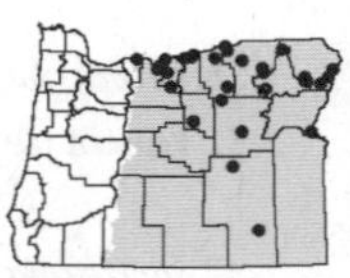

Plants perennial, 15–60 cm; with slender to stout taproots and enlarged, woody crowns. **Stems** erect, unbranched or with distal inflorescence branches, glabrous or tomentulose, sometimes minutely spiculate. **Basal leaves** linear to narrowly lanceolate, 10–35 cm, margins entire or irregularly pinnatifid, lobes slender, attenuate, tips acuminate, surfaces glabrous or puberulent; petiolate. **Cauline leaves** similar to basal; petiolate. **Inflorescences** flat-topped or not. **Heads** 3–30. **Involucres** cylindric to campanulate, 8–14 mm. **Phyllaries** outer linear-deltoid; inner 6–10, lanceolate, usually grayish-tomentose, rarely glabrous or black-setose with glandular hairs. **Florets** 6–25(35); ligules 12–20 mm. **Fruits** tapering, 3–10 mm, greenish or blackish, 12–15-ribbed. **2*n***=22, 33, 44, 55, 88.

Dry, open ground, mixed conifer woods, riverbanks. Flowering May–Jul. 50–1800 m. BR, BW, Col, ECas, Lava, Owy. ID, NV, WA; north to British Columbia, northeast to Alberta, east to NB, southeast to CO. Native.

Crepis bakeri Greene

Baker's hawksbeard

Crepis bakeri Greene ssp. *bakeri*, *Crepis bakeri* Greene ssp. *cusickii* (Eastw.) Babc. & Stebbins, *Crepis bakeri* Greene ssp. *idahoensis* Babc. & Stebbins

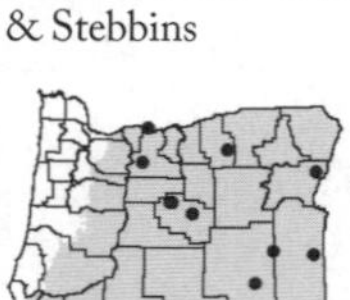

Plants perennial, 8–30 cm; with stout taproots and enlarged, woody crowns. **Stems** erect, unbranched or with distal inflorescence branches, tomentose and with short, fine, glandular hairs. **Basal leaves** narrowly to broadly oblanceolate or elliptic, 8–20 cm, margins irregularly denticulate to pinnately lobed, lobes deltate to lanceolate or linear, tips acute to obtuse, surfaces lightly tomentose and with fine glandular hairs; petiolate. **Cauline leaves** similar to basal, reduced distally; sessile. **Inflorescences** ± flat-topped. **Heads** 2–22. **Involucres** campanulate, 11–20 mm. **Phyllaries** outer lanceolate; inner 10–14, lanceolate, usually finely glandular, rarely black-setose. **Florets** 11–40+; ligules 13–28 mm. **Fruits** tapering, 5–10 mm, yellowish to dark brown, 10–12-ribbed. **2*n***=22, 33, 44, 55.

Grasslands, shrublands, open coniferous forests, riverbanks. Flowering Apr–Jul. 500–1900 m. BR, BW, Casc, Col, ECas, Lava, Owy, Sisk. CA, ID, NV, WA; southeast to UT. Native.

We are not recognizing the three described subspecies of *C. bakeri*, which are said to vary in chromosome number but differ only slightly in head size and pappus length.

Crepis barbigera Coville

Bearded hawksbeard

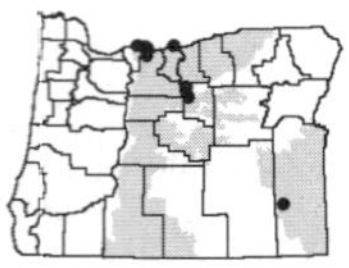

Plants perennial, 20–80 cm; with stout taproots and enlarged, woody crowns. **Stems** erect, unbranched or with distal inflorescence branches, glabrous to tomentulose, sometimes sparsely setose. **Basal leaves** narrowly or broadly oblanceolate, 10–40 cm, margins entire or toothed, shallowly to deeply lobed, lobes deltate to lanceolate or linear, tips attenuate, surfaces glabrous or sparsely tomentulose, sometimes hispid-setose; petiolate. **Cauline leaves** similar to basal, reduced distally; upper sessile. **Inflorescences** flat-topped or not. **Heads** 6–70. **Involucres** campanulate, 9–17 mm. **Phyllaries** outer deltoid to lanceolate; inner 6–10, lanceolate, canescent-tomentose and densely long-setose with yellowish or greenish bristles. **Florets** 8–25; ligules 10–12 mm. **Fruits** tapering, 7–10 mm, dark brown, 10–12-ribbed. **2*n***=44, 55, 88.

Dry hillsides, grasslands, shrublands, open coniferous woods. Flowering May–Jul. 0–900 m. Col, ECas, Lava, Owy. ID, WA. Native.

Crepis barbigera is an entirely apomictic, polyploid species of hybrid origin, the proposed parental species being *C. acuminata*, *C. atribarba*, and *C. modocensis* (Babcock & Stebbins 1938).

Crepis capillaris (Linnaeus) Wallroth (p. 238)

Smooth hawksbeard

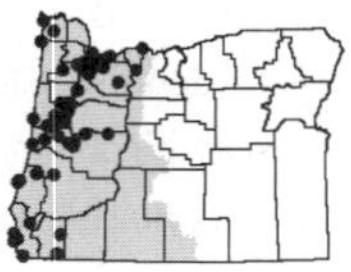

Plants annual or biennial, 10–120 cm; taproots usually slender, sometimes stout. **Stems** erect or ascending, simple or branched, glabrous or puberulent near base or throughout. **Basal leaves** rarely absent, narrowly to broadly lanceolate or oblanceolate, 5–30 cm, margins entire to toothed or lobed, lobes deltate to linear, tips obtuse or acute, surfaces glabrous or hispidulous; petiolate. **Cauline leaves** similar to basal, reduced distally; upper sessile, auriculate-clasping. **Inflorescences** open, not flat-topped. **Heads** 10–15(35). **Involucres** cylindric or campanulate, 5–8 mm. **Phyllaries** outer linear; inner 8–16, lanceolate, minutely tomentose and usually with short setose, glandular hairs. **Florets** 20–60; ligules 6–12 mm. **Fruits** narrowed at tips, 1.5–2.5 mm, yellowish or light brownish, ~10-ribbed. **2*n***=6.

Disturbed areas, lawns, gardens, meadows, roadsides, shrublands, forests, clearcuts. Flowering May–Sep. 0–1200 m. Casc, CR, ECas, Est, Sisk, WV. CA, ID, NV, WA; scattered nearly worldwide. Exotic.

The growth form and vegetative morphology of this weedy species are much affected by conditions of the habitat.

Crepis intermedia A. Gray

Intermediate hawksbeard

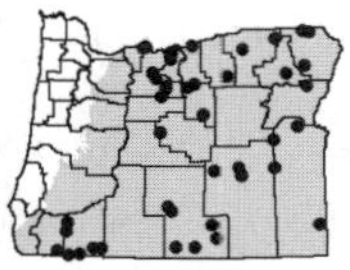

Plants perennial, 15–70 cm; with stout taproots and enlarged, woody crowns. **Stems** erect, simple or with distal inflorescence branches, lightly to densely tomentulose, rarely glabrous near bases. **Basal leaves** lanceolate or oblanceolate, rarely linear, 5–40 cm, margins toothed or lobed, lobes linear to lanceolate, tips acute or acuminate, surfaces sparsely to densely tomentulose; petiolate. **Cauline leaves** similar to basal, reduced distally; upper sessile. **Inflorescences** flat-topped or not. **Heads** (10)20–60. **Involucres** cylindric or campanulate, 10–14 mm. **Phyllaries** outer deltoid; inner 7–8(12), lanceolate, canescent-tomentulose, rarely yellow-setose, not glandular. **Florets** 7–12(16); ligules 15–30 mm. **Fruits** tapering, 5.5–9 mm, yellowish to brownish, 10–12-ribbed. **2*n***=33, 44, 55, 88.

Dry ridges and slopes, grasslands, shrublands, riverbanks, forests. Flowering May–Jul. 0–2200 m. BR, BW, Casc, Col, ECas, Lava, Owy, Sisk. CA, ID, NV, WA; north to British Columbia, northeast to Alberta, east to WY, southeast to NM. Native.

This species is entirely apomictic, polyploid, and of hybrid origin. Proposed parental species contributing to the origin of *C. intermedia* in different parts of its range are *C. acuminata, C. occidentalis, C. modocensis,* and *C. atribarba.*

Crepis modocensis Greene

Low hawksbeard, Modoc hawksbeard

Crepis modocensis Greene ssp. *modocensis, Crepis modocensis* Greene ssp. *subacaulis* (Kellogg) Babc. & Stebbins

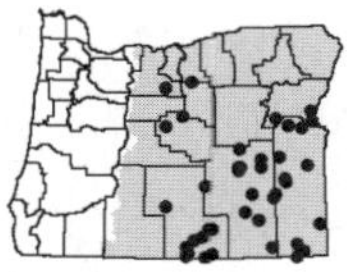

Plants perennial, (5)10–40 cm; with stout taproots and enlarged, woody crowns. **Stems** erect, unbranched or with distal inflorescence branches, lightly to densely tomentulose, often yellow-setose basally. **Basal leaves** oblanceolate or elliptic, 7–25 cm, margins toothed, shallowly or deeply lobed, lobes linear to lanceolate, tips acuminate. surfaces sparsely to densely tomentulose, often setose on midribs below; petiolate. **Cauline leaves** similar to basal; upper sessile. **Inflorescences** flat-topped or not. **Heads** 1–9. **Involucres** broadly campanulate, 10–21 mm. **Phyllaries** outer deltoid or lanceolate; inner 8–18, lanceolate, canescent-tomentulose, usually black-setose, not glandular. **Florets** 10–60; ligules 13–22 mm. **Fruits** tapering, 7–12 mm, brownish to reddish or blackish, 10-ribbed. **2*n***=22, 33, 44, 55, 66, 88.

Dry slopes and ridges, moist swales, subalpine scree, grasslands, shrublands. Flowering May–Jul. 500–2600 m. BR, BW, Col, ECas, Lava, Owy. CA, ID, NV, WA; north to British Columbia, northeast to MT, east top WY, south east to CO. Native.

As indicated by the range of chromosome numbers, *Crepis modocensis* consists of numerous hybrid, apomictic races, at least four of which have been given subspecific status. Diploid sexual forms occur only in northwestern California, southeastern Oregon, and in a disjunct area of central Washington, while apomicts occupy the rest of the wide range of the species. The two subspecific entities in Oregon differ only by pubescence features and are not being recognized in this treatment.

Crepis monticola Coville

Mountain hawksbeard

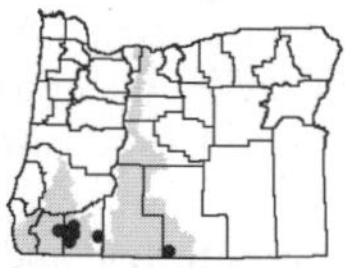

Plants perennial, 10–40 cm; with stout to slender taproots and woody crowns. **Stems** erect, simple or with distal inflorescence branches, usually densely setose, hairs gland-tipped. **Basal leaves** oblanceolate to elliptic, 10–27 cm, margins entire or toothed, shallowly to deeply and irregularly lobed, lobes lanceolate, tips acute, surfaces densely hirsute; petiolate. **Cauline leaves** similar to basal; sessile. **Inflorescences** flat-topped or not. **Heads** 1–10(20). **Involucres** broadly campanulate, 14–24 mm. **Phyllaries** outer lanceolate; inner 7–12, lanceolate, canescent and densely long-hirsute with gland-tipped hairs. **Florets** 16–20; ligules 16–21 mm. **Fruits** tapering, 5.5–9 mm, reddish brown to blackish, 10-ribbed. **2*n***=22, 33, 44, 55, 77, 88.

Open woods and thickets. Flowering May–Jul. 300–1000 m. ECas, Sisk. CA. Native.

Crepis monticola is easily recognized by the dense pubescence on its leaves, stems, and involucres.

Crepis nicaeensis Persoon

French hawksbeard, Turkish hawksbeard

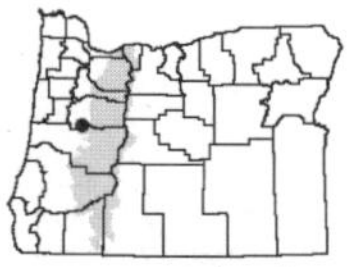

Plants annual or biennial, 25–110 cm; with slender taproots. **Stems** erect, simple or branched, hispid-pubescent near bases and below heads. **Basal leaves** narrowly oblanceolate or elliptic, 16–20 cm, margins usually toothed, deeply and irregularly lobed, tips obtuse, surfaces pubescent with straight or curled hairs; short-petiolate. **Cauline leaves** similar to basal, reduced upwards; short-petiolate or sessile, sometimes clasping. **Inflorescences** not flat-topped. **Heads** 2–15. **Involucres** campanulate, 8–10 mm. **Phyllaries** outer linear; inner 10–15, lanceolate, glandular-hispid. **Florets** 45–60; ligules 11 mm. **Fruits** narrowed at tips, 2.5–4 mm, golden brown, 10-ribbed. **2*n***=8.

Disturbed, gravelly openings in coniferous forests. Flowering May–Jul. 200–1000 m. Casc. WA; sparsely scattered in North America; Europe. Exotic.

Crepis occidentalis Nuttall (p. 238)

Western hawksbeard

Crepis occidentalis Nutt. ssp. *conjuncta* Babc. & Stebbins, *Crepis occidentalis* Nutt. ssp. *costata* (A. Gray) Babc. & Stebbins, *Crepis occidentalis* Nutt. var. *costata* A. Gray, *Crepis occidentalis* Nutt. ssp. *occidentalis, Crepis occidentalis* Nutt. ssp. *pumila* (Rydb.) Babc. & Stebbins

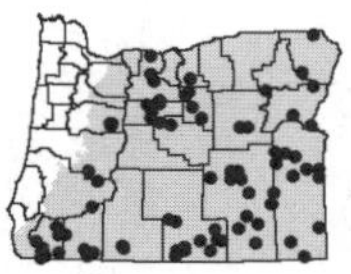

Plants perennial, (5)8–40 cm; with stout taproots and woody crowns. **Stems** erect, simple or branched, tomentulose, sometimes glandular-hirsute distally. **Basal leaves** narrowly to broadly elliptic, (5)8–40 cm, margins denticulate to shallowly or deeply lobed, lobes lanceolate, rarely triangular, tips acute or acuminate, surfaces lightly or densely tomentulose; petiolate. **Cauline leaves** similar to basal, reduced; petiolate or sessile. **Inflorescences** open, not flat-topped. **Heads** 2–30. **Involucres** cylindric or campanulate, 11–19 mm. **Phyllaries** outer deltoid; inner 7–18, lanceolate, tomentulose, rarely with blackish, setose, gland-tipped hairs. **Florets** 9–40; ligules 18–22 mm. **Fruits** narrowed at tips, 6–10 mm, light brown to blackish, 10–18-ribbed. **2*n***=22, 33, 44, 55, 66, 77, 88.

Dry hillsides, ridges, subalpine slopes, open woods, grasslands, sagebrush flats. Flowering Apr–Jul. 50–2500 m. BR, BW, Casc, Col, ECas, Lava, Owy, Sisk. CA, ID, NV, WA; north to British Columbia, northeast to Alberta, east to SD, southeast to NM. Native.

Over most of its wide range, *C. occidentalis* is polyploid and apomictic, the diploid, sexual forms being limited to small areas in northwestern California and the central Sierra Nevada (Babcock & Stebbins 1938).

Crepis pleurocarpa A. Gray

Naked-stemmed hawksbeard

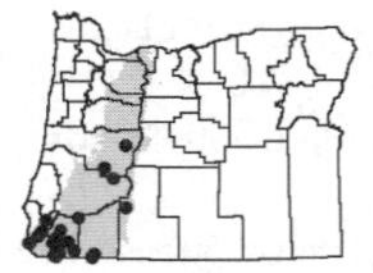

Plants perennial, 11–40(60) cm; with stout taproots and woody crowns. **Stems** simple or branched, sparsely tomentulose. **Basal leaves** elliptic or oblanceolate, 7–28 cm, margins entire or toothed, deeply and irregularly lobed, lobes triangular or lanceolate, tips attenuate, surfaces glabrous or lightly tomentulose; petiolate. **Cauline leaves** 0–1, rarely few, similar to basal; sessile or short-petiolate. **Inflorescences** flat-topped or not. **Heads** 7–10(30). **Involucres** cylindric in flower and developing fruit, 8–17 mm. **Phyllaries** outer deltoid; inner 5(8), lanceolate, tomentulose, usually glabrous along midribs. **Florets** 5–8(12); ligules 15–20 mm. **Fruits** slightly tapering, 5–8 mm, reddish brown, conspicuously 10-ribbed. **2*n***=22, 33, 44, 55, 77, 88.

Mountain slopes, ridges, rock outcrops, hillsides, meadows, shrublands, coniferous forests, serpentine. Flowering May–Aug. 400–2200 m. Casc, Sisk. CA, NV, WA. Native.

In Oregon, *C. pleurocarpa* is principally a species of the Western Cascades and Siskiyou Mountains, and in the southwestern counties it is often associated with serpentine soils.

Crepis pulchra Linnaeus (p. 240)

Small-flowered hawksbeard

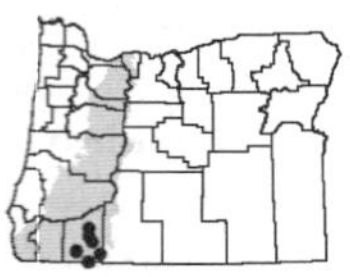

Plants annual, 10–100 cm; with slender taproots. **Stems** erect, simple or much branched proximally or distally, puberulent proximally, glabrous distally, sometimes finely glandular. **Basal leaves** narrowly or broadly oblanceolate, 2–25 cm, margins entire or lower ½ sharply lobed, tips obtuse to acute, surfaces glabrous or hispidulous; short-petiolate or sessile. **Cauline leaves** similar to basal. **Inflorescences** not flat-topped. **Heads** 10–40. **Involucres** cylindric, 8–12 mm. **Phyllaries** outer deltoid; inner 10–14, lanceolate, glabrous. **Florets** 15–30; ligules 5–12 mm. **Fruits** tapering, 4–6 mm, light brown, 10–12-ribbed. **2*n***=8.

Roadsides, creek-sides, disturbed grasslands, shrublands, woodlands. Flowering May–Aug. 700–1800 m. Casc, Sisk. Eastern and southeastern US, Quebec; Asia, Europe. Exotic.

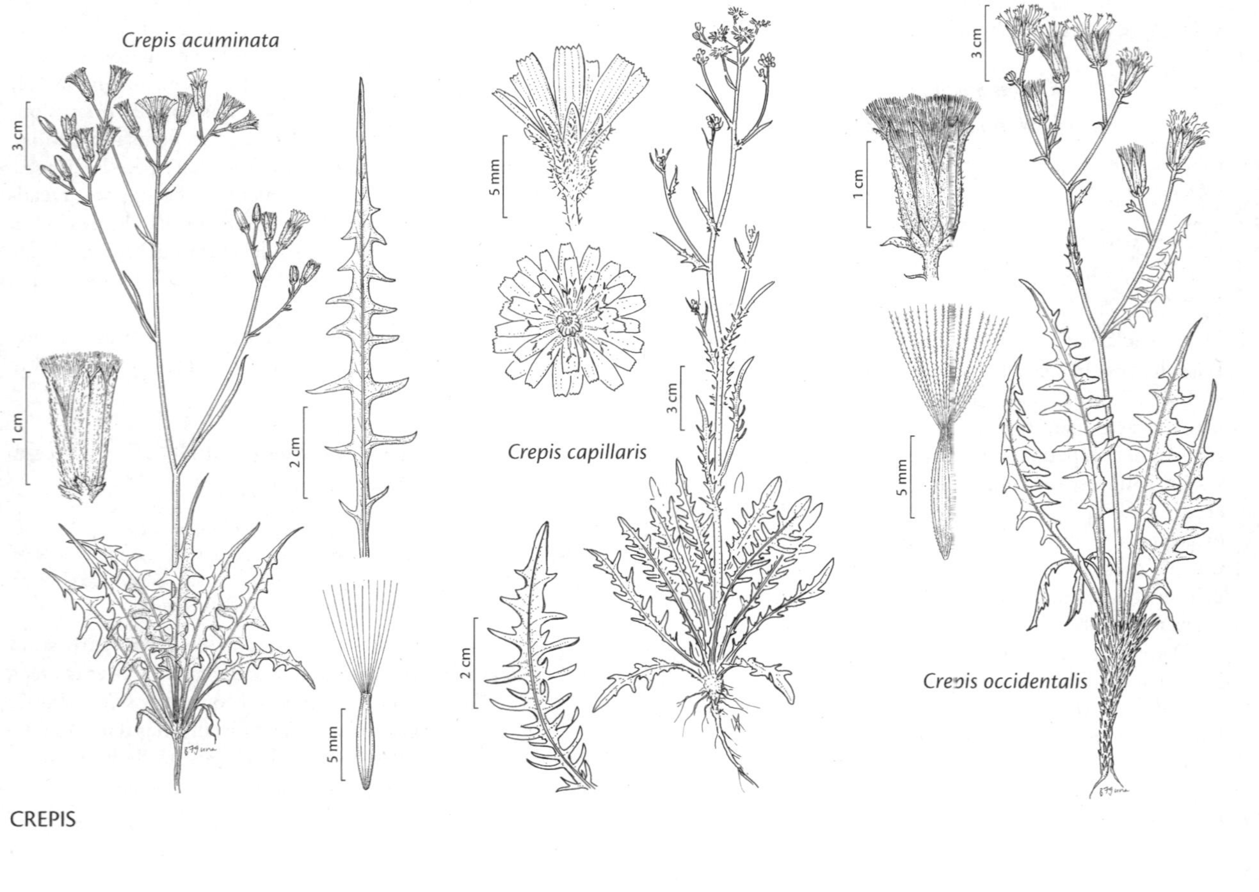

CREPIS

Crepis runcinata (E. James) Torrey & A. Gray

Meadow hawksbeard, fiddleleaf hawksbeard

Plants perennial, 15–60(80) cm; with stout taproots and woody crowns. **Stems** erect or ascending, usually simple, rarely branched from near base, spiculate in ours. **Basal leaves** narrowly to broadly oblanceolate or obovate, 5–15(30) cm, margins entire or toothed or shallowly lobed, lobes triangular, tips rounded, surfaces glabrous or hispid-puberulent; petioles short, winged. **Cauline leaves** 0–few, similar to basal, reduced. **Inflorescences** flat-topped or not. **Heads** 3–30. **Involucres** cylindric or campanulate, 8–21 mm. **Phyllaries** outer deltoid; inner 10–15, lanceolate, usually hispid-glandular. **Florets** 20–50; ligules 9–18 mm. **Fruits** narrowed at tips or tapering, 3.5–8 mm, light to dark brown, 10–13-ribbed. **2*n***=22.

Central and western North America. 7 subspecies; 2 subspecies treated in *Flora.*

Key to *Crepis runcinata* subspecies:

1. Leaves usually obovate, teeth not predominately white-tipped . . . . *C. r.* ssp. ***hispidulosa***

1' Leaves narrowly to broadly oblanceolate, teeth often predominately white-tipped . . . . *C. r.* ssp. ***imbricata***

Crepis runcinata (E. James) Torrey & A. Gray ssp. *hispidulosa* (Rydberg) Babcock & Stebbins

Meadow hawksbeard

Crepis runcinata (E. James) Torr. & A. Gray var. *hispidulosa* Howell ex Rydb.

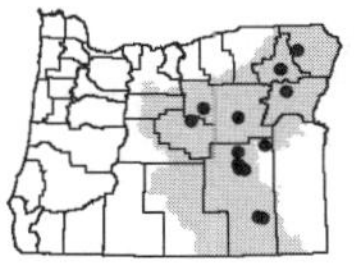

Plants 25–50 cm. **Stems** glabrous to hispid or minutely spiculate, often glandular distally. **Leaves** usually obovate, 6–25 × 2.5–8 cm, lower ½ entire to triangular-dentate, upper ½ sparingly denticulate, teeth not predominately white-tipped. **Heads** 10–30. **Phyllaries** lanceolate, tomentulose or glandular-hispid. **Fruits** narrowed at apices, not tapering, brown.

Subalpine marshes, dry or moist alkaline meadows, irrigated fields and pastures. Flowering Jun–Jul. 800–1600 m. BR, BW, Lava. ID, WA; northeast to Alberta, east to WY, southeast to CO. Native.

Crepis runcinata (E. James) Torrey & A. Gray ssp. *imbricata* Babcock & Stebbins (p. 240)

Imbricate hawksbeard, Alvord hawksbeard

Crepis runcinata (E. James) Torr. & A. Gray var. *imbricata* (Babc. & Stebbins) M. Peck

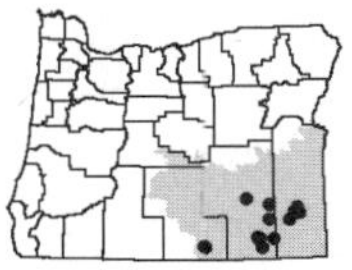

Plants 10–40 cm. **Stems** glabrous. **Leaves** narrowly to broadly oblanceolate, 3.5–20 × 0.5–3 cm, margins denticulate to serrate or short-lobed throughout, or lower ⅓ entire, lobes linear or triangular, teeth often predominately white-tipped. **Heads** 3–7. **Phyllaries** oblong, glabrous or minutely glandular. **Fruits** tapering to apices, reddish brown.

Alkaline meadows, lake beds, springs. Flowering May–Jun. 1000–1400 m. BR, Owy. CA, NV. Native.

Crepis setosa Haller f.

Rough hawksbeard, bristly hawksbeard

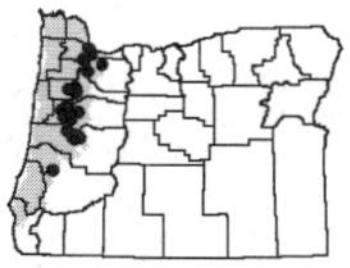

Plants annual, 15–90 cm; with slender or sometimes stout taproots. **Stems** erect, simple or branched, densely to lightly long-hispid. **Basal leaves** lanceolate to oblanceolate, 5–30 cm, margins entire or toothed or lobed, lobes deltate or lanceolate, tips acute to obtuse, surfaces lightly hispidulous; petiolate. **Cauline leaves** lanceolate, reduced upward; distal sessile, often auriculate-clasping. **Inflorescences** usually not flat-topped. **Heads** 1–20. **Involucres** cylindric or campanulate, 6–10 mm. **Phyllaries** outer lanceolate; inner 12–16, lanceolate, hispid-setose. **Florets** 20–30(37); ligules 8–10 mm. **Fruits** tapering, 3–5 mm, pale or reddish brown, 10-ribbed, beaked. **2*n***=8.

Disturbed areas, lawns, gardens, meadows, roadsides, gravel pits, shrublands. Flowering May–Aug. 0–600 m. CR, WV. CA; widely scattered in US; Europe. Exotic.

Crepis vesicaria Linnaeus ssp. *taraxacifolia* (Thuillung) Schinz & R. Keller

Weedy hawksbeard, beaked hawksbeard

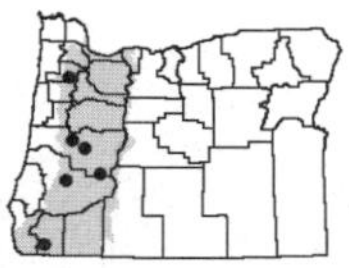

Plants annual, biennial, rarely perennial, 5–120 cm; with slender taproots. **Stems** erect, simple or branched, glabrous to spiculate, hispid, or villous. **Basal leaves** oblanceolate, 10–35 cm, margins entire or dentate, irregularly deeply lobed, lobes deltate to linear, tips obtuse or acute, surfaces abaxially usually hispid or villous, sometimes glabrous; petiolate. **Cauline leaves** similar to basal, reduced distally, sometimes lobed only below middle; sessile, auriculate-clasping. **Inflorescences** flat-topped or not. **Heads** 10–20. **Involucres** campanulate, 8–14 mm. **Phyllaries** outer deltate to lanceolate or ovate; inner 7–15, broadly to narrowly lanceolate, densely puberulent, sometimes glandular. **Florets** 40–70; ligules 6–15 mm. **Fruits** 5–9 mm, pale to medium brown, 10-ribbed; outer not beaked; inner attenuate into fine beaks, beaks 2–5 mm. **2*n***=8, 16.

Weed in disturbed areas, roadsides, fields, open woodlands, thickets, rock outcrops. Flowering May–Jul. 0–800 m. Casc, Sisk, WV. CA; British Columbia, widely scattered in eastern US; Australia, Europe, South America. Exotic.

Babcock (1947) described this as the most variable of the eight subspecies that he recognized in *Crepis vesicaria.* We are accepting the subspecific name used in this reference and in Babcock and Stebbins (1938), rather than ssp. *haenseleri* used in *Flora Europaea* (Sell 1976). Sell's rejection of the epithet ssp. *taraxacifolia* was based on an earlier, now obsolete article in the International Code of Botanical Nomenclature.

CROCIDIUM Gold star, spring gold

Genus composed of one species, *C. multicaule*.

Crocidium multicaule Hooker (p. 240)

Gold star, spring gold

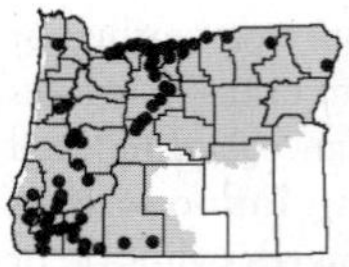

Herbs annual, 2–15(30) cm; taproots short, becoming fibrous-rooted terminally. **Stems** 1–many from basal leaf rosettes, erect, simple or branched from bases, often woolly-pubescent at nodes, otherwise glabrous. **Basal leaves** obovate to linear-elliptic, 5–40 cm, margins entire or dentate to denticulate, rarely lobed, teeth few, triangular, tips usually obtuse. **Cauline leaves** few to numerous, alternate, linear, 2–30 mm, much reduced distally, margins entire or lightly denticulate, tips acute, surfaces woolly-pubescent basally; sessile. **Inflorescences** solitary terminal heads. **Heads** radiate, nodding in bud, on cloudy days, and at night, erect in fruit; peduncles 2–12 cm, glabrous. **Involucres** ± hemispheric, 3–6+ mm wide. **Phyllaries** 4–15, in 1 equal series, erect in flower, reflexed in fruit, 3–7 mm, lobes lance-ovate, parallel-veined, tips acute, surfaces glabrous. **Receptacles** conic; paleae 0. **Ray florets** (5)8–13, pistillate; rays 3–10 mm, exceeding involucres, spreading, yellow. **Disc florets** 10–40+, bisexual; corollas narrowly tubular below, widened distally, 3–4 mm, yellow, lobes 5, spreading; stamens 5, partly exserted from corolla throats; stigmas slightly exserted from stamen tubes. **Fruits** columnar-fusiform, truncate, 1.5–2.5 mm, brown, puberulent with minute, globular trichomes that are mucilaginous when wetted, 4-ribbed; pappi 0 or of deciduous, minutely barbed bristles, 1–4 mm, white. **2*n***=18.

Open woodlands, shrublands, grasslands, moist cliffs, rock outcrops, talus, riverbanks, roadsides. Flowering Feb–Jun. 0–1500 m. BW, Casc, Col, CR, ECas, Lava, Sisk, WV. CA, ID, WA; north to British Columbia. Native.

An early blooming species, *Crocidium multicaule* often forms colorful displays of yellow blooms in sagebrush scrub and sandy grasslands and on mossy, rocky banks and cliff ledges. Its flower heads are erect and open in the sun but nod at the end of the day and on overcast days. Its heads usually have eight ray florets, but in dry conditions, depauperate individuals are often much smaller than usual and may have only five ray florets.

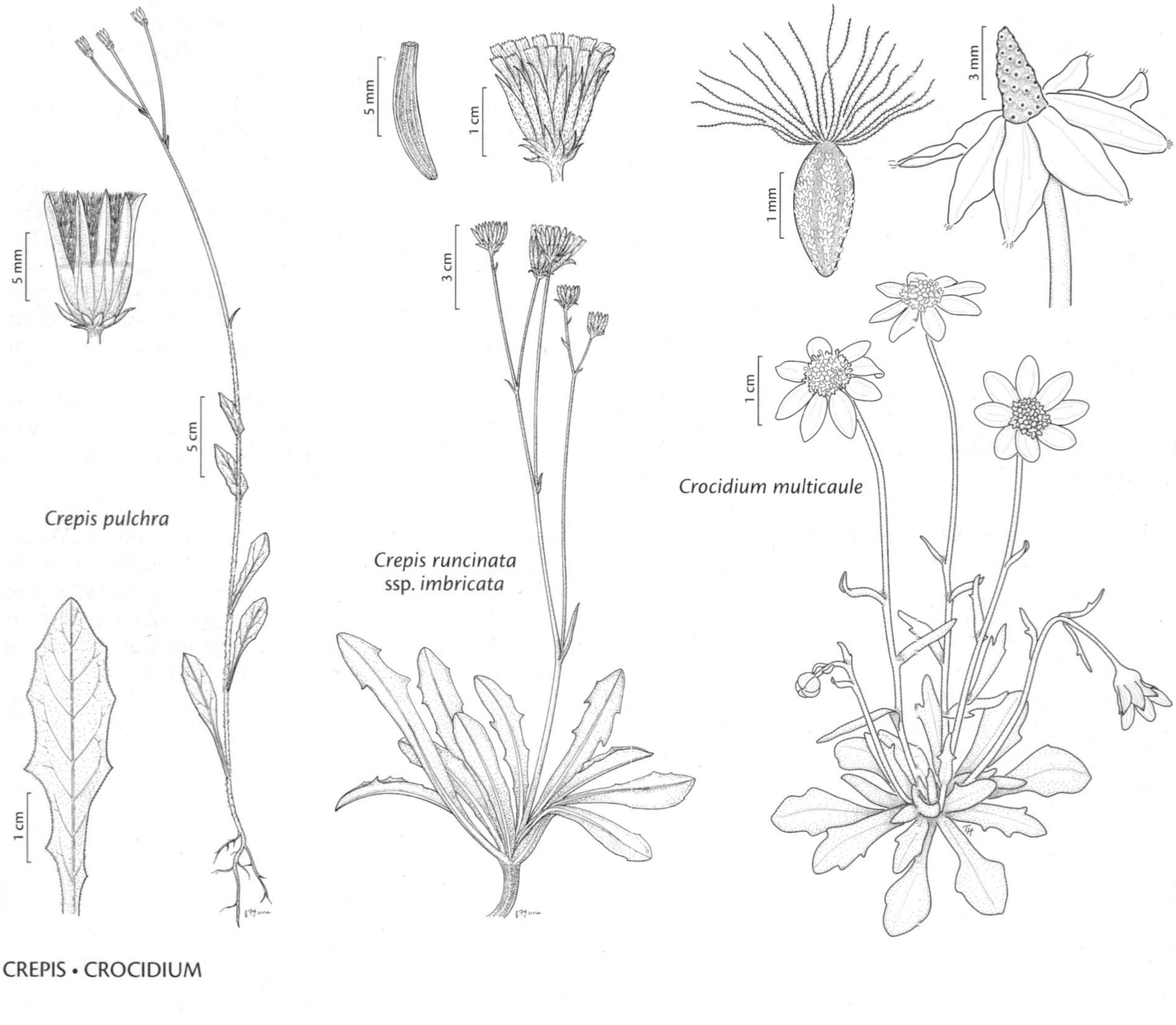

CREPIS • CROCIDIUM

CRUPINA Crupina

Cindy T. Roché

Herbs annual, slender; taprooted. **Stems** erect, openly branching above. **Leaves** basal in tight rosettes, often withering by flowering ti margins me, margins entire or toothed, surfaces scabrous, sessile; cauline alternate, pinnatisect to pinnate, margins with multicellular glochidiate hairs, petiolate. **Inflorescences** heads solitary or clustered at branch tips. **Heads** discoid. **Involucres** cylindric to ovoid. **Phyllaries** in 4–6 unequal series, graduated, lanceolate to oblong, tips acute, unarmed. **Receptacles** flat, bearing subulate scales; paleae 0. **Florets** corollas ± bilateral, pink-purple, tubes slender, gradually expanded into narrowly funnelform throats, hairy, lobes linear; anther bases short-tailed, apical appendages narrowly triangular. **Fruits** cylindrical or compressed, smooth, velutinous, with basal or lateral hilum; pappi of several outer rows of light brown to black barbellate bristles and inner shorter row of 5–10 black triangular scales.

Africa, Asia, Europe, North America. 2 species; 1 species treated in *Flora*.

Crupina vulgaris Persoon ex Cassini (p. 241)

Common crupina, bearded creeper

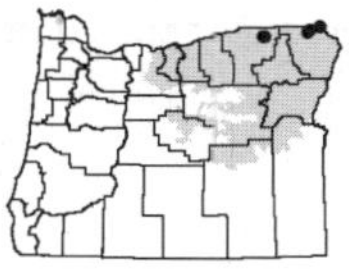

Plants 2–10 dm, branched trichomes present on both leaves and stem. **Leaves** basal leaves up to 5 × 2 cm; cauline leaves reduced distally, 1–2-pinnately divided into linear lobes, margins ciliate with bristle-tipped hairs. **Involucres** 8–20 mm. **Phyllaries** elliptic, striped green or purplish, tips acute. **Florets** outer sterile, inner fertile; corollas pink or light purple, exceeding involucres, tubes bearing hairs with walls having lateral finger-like projections. **Fruits** barrel-shaped, 3–5 × 1.5–3 mm, attachment scars basal, usually 1(2) per head; inner pappus scales erect, 0.5–1 mm; outer pappi 4–7 mm, hygroscopic. **2*n***=28.

Grasslands, meadows. Flowering May–Jun. 500–1100 m. BW, Col. CA, ID, WA; MA; Asia, Europe. Exotic.

This species was likely introduced from Spain by Basque sheepherders and moved among remote locations in the Western United States by bands of sheep in their wool. It was first reported in Idaho in 1968. Federally listed as a noxious weed in the 1980s, it is currently a Class B Noxious weed in Oregon.

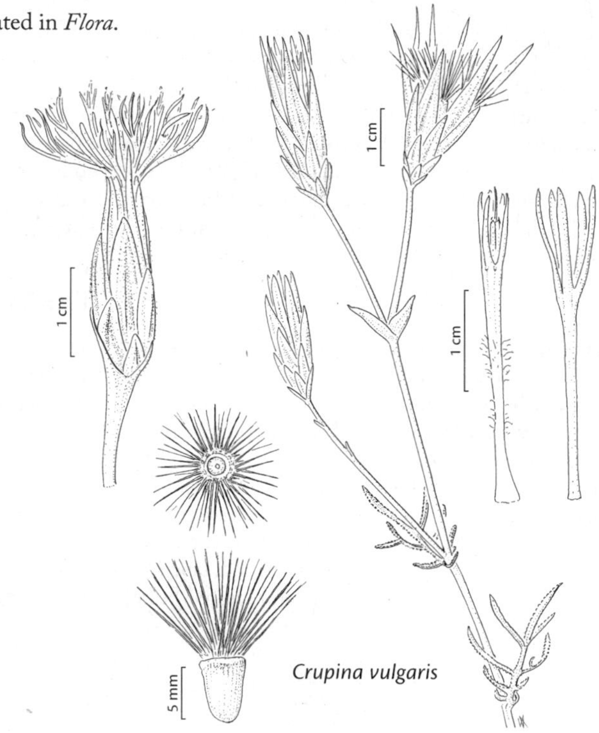

CRUPINA

CYCLACHAENA Marsh elder

Genus composed of one species, *C. xanthiifolia.*

Cyclachaena xanthiifolia (Nuttall) Fresenius (p. 242)

Tall marsh elder, bur-weed marsh elder, carelessweed, giant sumpweed

Iva xanthiifolia Nutt.

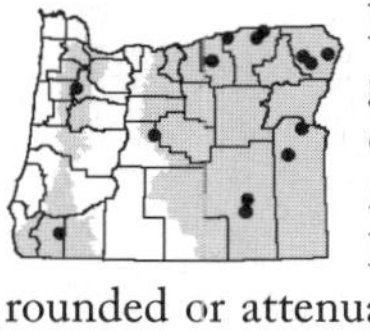

Herbs annual, 3–12(20) dm. **Stems** erect, glabrous proximally, villous distally. **Leaves** cauline, opposite or distal-most alternate, 5–20+ cm, reduced distally, broadly ovate to lanceolate, bases cordate or truncate to rounded or attenuate, margins irregularly serrate-dentate or few-lobed, veins 3-branched distally, tips acute to acuminate, abaxially minutely scabrous or canescent, adaxially glabrous or minutely scabrous, lobes and teeth triangular; petioles glabrous or villous. **Inflorescences** panicle-like, branches bracteate or leafless, short-villous, ultimate branchlets spicate-racemose, with numerous sessile or nearly sessile heads. **Heads** ± disciform. **Involucres** hemispheric, 1–3 mm. **Phyllaries** in 2–3 ± equal series, erect, lanceolate or oblanceolate, tips rounded to acuminate, surfaces scabrous or hispid. **Receptacles** convex to conic; paleae membranous, spatulate to linear. **Outer florets** 4–6, pistillate, corollas vestigial or 0. **Inner florets** 8–20, functionally staminate, corollas tubular, widened and cup-shaped distally, 2–2.5 mm, pale yellow, lobes 5, erect, triangular; stamens 5, included; styles exserted from involucres. **Fruits** obovoid, truncate, tapering toward bases, slightly flattened and 2-ribbed in lower half, 2–3 mm, black, glabrous or lightly villous apically; pappi 0. **2*n***=36.

Shrublands, pastures, riverbanks, roadsides, gardens, disturbed areas. Flowering Jul–Sep. 50–1400 m. BR, BW, Col, Lava, Owy, Sisk, WV. CA, ID, NV, WA; throughout most of Canada and US; Europe. Native.

DELAIREA Cape ivy, German ivy

Thea Jaster

Genus composed of one species, *D. odorata.*

Delairea odorata Lemaire (p. 242)

Cape ivy, German ivy

Senecio mikanioides Walp.

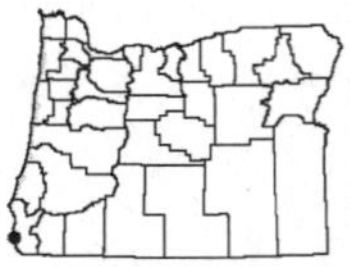

Vines perennial, 1–6+ m. **Stems** usually 1, branching, scrambling, climbing. **Leaves** cauline, alternate, widely deltate to orbiculate, 3–8(12) × 3–8(12) cm, palmately lobed or lobes less pronounced and reduced to teeth, surfaces glabrous, often shiny; petioles ± equal in length to blades. **Inflorescences** (10)20–40 heads in terminal or axillary corymb-like arrays. **Heads** discoid. **Involucres** cylindric to weakly turbinate, 2–4 mm in diameter. **Phyllaries** (7)8+ in 1–2 series, lance-linear, 3–4 mm, margins scarious. **Receptacles** convex; paleae 0. **Florets** 8–45, bisexual; corollas bright yellow, lobes 5, recurved; style branches stigmatic in 2 lines, tips ± truncate. **Fruits** 5-angled, glabrous; pappi of minutely barbed bristles, white. **2*n***=20.

Riparian woodlands, shaded scrub. Flowering Nov–Mar. 0–50 m. Est. CA; Africa, Asia, Australia, Hawaii, New Zealand. Exotic.

Delairea odorata is native to South Africa and has been introduced in New Zealand, Australia, India, Hawaii, and California, where it is considered a noxious weed, overtaking and strangling trees and obscuring the ground. The species is distributed as a house plant. It was first discovered near the coast of southern Oregon in 1997.

DIETERIA Tansyaster

Herbs annual, biennial, or short-lived perennial; taprooted. **Stems** 1–many from root-crowns, erect or ascending, canescent or glandular-puberulent or both, rarely glabrous proximally. **Basal leaves** often withering by flowering, linear to lanceolate, oblanceolate, or ovate, margins entire to serrate or dentate, surfaces glabrous to canescent. **Cauline leaves** alternate, distally ± reduced, surfaces canescent, sometimes also glandular-puberulent. **Inflorescences** well-branched panicle- or raceme-like arrays. **Heads** radiate or sometimes discoid. **Involucres** cylindrical to campanulate or hemispheric. **Phyllaries** erect in flower, spreading or reflexed in fruit, midribs evident or obscure, tips acute, surfaces usually hairy or stipitate-glandular. **Receptacles** convex; paleae 0. **Ray florets** 0–25+, pistillate, fertile or sterile, rays well exceeding involucres, bluish purple to white. **Disc florets** bisexual; corollas tubular, gradually enlarged distally, equaling or exceeding involucres, yellow, lobes 5, erect, triangular; stamens slightly exserted; stigmas exserted. **Fruits** oblanceolate or obovate, flattened, tapering toward bases, 8–12-ribbed; pappi of 40–50 bristles in 1–3 series, persistent.

Central and western Canada and United States. 3 species; 1 species treated in *Flora.*

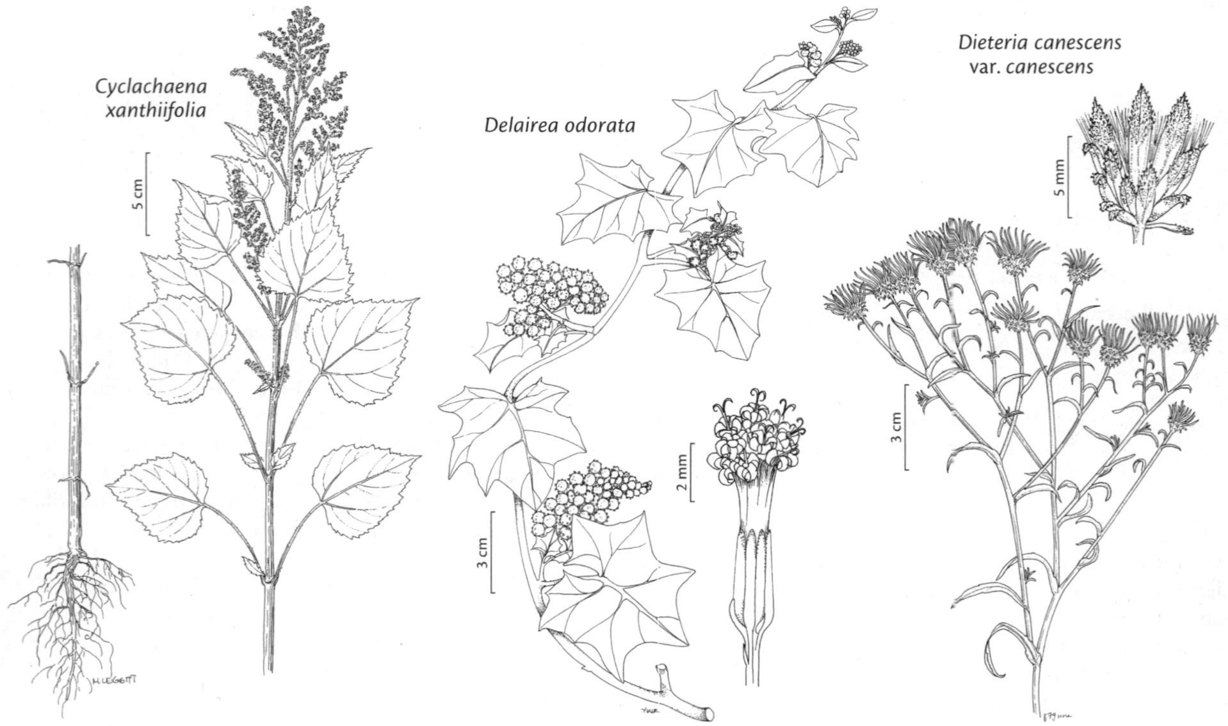

CYCLACHAENA • DELAIREA • DIETERIA

Dieteria canescens (Pursh) Nuttall

Hoary-aster, hoary tansyaster

Plants 3–50(100) cm. **Basal leaves** linear to oblanceolate or rounded, 0.5–10 cm, margins entire or dentate; usually sessile. **Cauline leaves** usually linear, sometimes obovate or oblanceolate, margins entire or denticulate; petiolate or sessile. **Involucres** 5–11 mm. **Phyllaries** in 4–10 series, canescent to ± glandular-puberulent. **Ray florets** 0 or 5–25+; corollas 5–12 mm, bluish purple to white. **Fruits** 3–4 mm, tan or brown, glabrous or sparsely strigose-pubescent; pappus bristles 5–6 mm, white or tan.

Western North America. 10 varieties; 4 varieties treated in *Flora.*

Key to *Dieteria canescens* varieties:

1. Stems stipitate-glandular . . . . . ***D. c.*** var. ***leucanthemifolia***

1' Stems not or sparsely stipitate-glandular.

  2. Stems usually 1, erect. . . . . ***D. c.*** var. ***incana***

  2' Stems 1+, ascending to erect.

    3. Ray florets 0 or if present sterile; phyllaries usually in 4–6 series . . . . ***D. c.*** var. ***shastensis***

    3' Ray florets usually present, fertile; phyllaries in 6+ series . . ***D. c.*** var. ***canescens***

Dieteria canescens var. *canescens* (p. 242)

Hoary-aster

Aster canescens Pursh, *Aster shastensis* (A. Gray) A. Gray var. *glossophyllus* Cronquist, *Aster shastensis* (A. Gray) A. Gray var. *latifolius* Cronquist, *Machaeranthera canescens* (Pursh) A. Gray var. *canescens*, *Machaeranthera shastensis* A. Gray var. *glossophylla* (Piper) Cronquist & D.D. Keck, *Machaeranthera shastensis* A. Gray var. *latifolia* (Cronquist) Cronquist & D.D. Keck

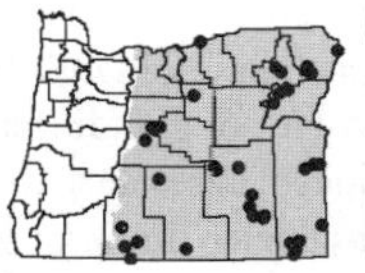

Stems 1+, ascending, canescent, sometimes sparsely stipitate-glandular distally, rarely throughout. **Inflorescences** panicle- or raceme-like arrays; bracts ± linear or narrowly oblanceolate. **Phyllaries** in 6–10 series, middle and lower series with recurved tips. **Ray florets** usually present, fertile. **2*n***=8.

Sagebrush, pine forests, meadows. Flowering Jun–Sep. 100–2800 m. BR, BW, Col, ECas, Lava, Owy. CA, ID, NV, WA; north to British Columbia, northeast to Saskatchewan east to MN, southeast to AZ. Native.

Dieteria canescens var. *canescens* is most similar to variety *shastensis* but differs in the fertility of its ray florets and usually in the more numerous series of reflexed phyllaries, especially those developed at the base of the involucres. Intermediate plants with involucres like variety *shastensis* and fertile ray florets sometimes occur. Fertility of the ray florets may be difficult to judge, especially on plants just coming into flower.

Dieteria canescens var. *incana* (Lindley) D.R. Morgan & R.L. Hartman

Tall hoary-aster

Machaeranthera canescens (Pursh) A. Gray var. *incana* (Lindl.) A. Gray

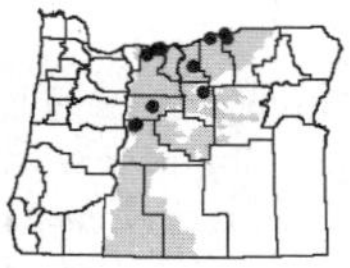

Stems usually 1, erect, canescent, sometimes stipitate-glandular distally. **Inflorescences** panicle- or raceme-like arrays; bracts linear. **Phyllaries** in 6–10 series, middle and lower series with erect or recurved tips. **Ray florets** fertile.

Sagebrush, streamsides, meadows. Flowering May–Sep. 0–1300 m. Col, ECas, Lava. CA, ID, WA; north to British Columbia. Native.

Other than stem number, the difference between variety *incana* and variety *canescens* is sometimes one of habit. It is not known whether plants typical of the two varieties may occur together in the same population.

Dieteria canescens var. *leucanthemifolia* (Greene) D.R. Morgan & R.L. Hartman

Small-leaf hoary-aster

Machaeranthera canescens (Pursh) A. Gray var. *leucanthemifolia* (Greene) S.L. Welsh

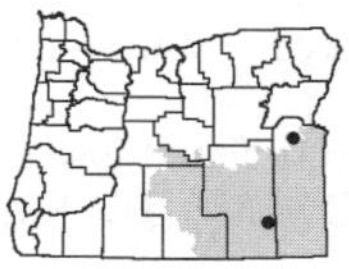

Stems 1+, erect, canescent, stipitate-glandular. **Inflorescences** panicle-like arrays; bracts oblanceolate. **Phyllaries** in 4–6 series, middle and lower series usually with recurved tips. **Ray florets** fertile. **2*n***=8.

Sagebrush, pine or juniper forests. Flowering Jul–Sep. 700–1300 m. BR, Owy. CA, NV; southeast to UT. Native.

Dieteria canescens var. *leucanthemifolia* has only rarely been collected in Oregon, its range being mostly to the south.

Dieteria canescens var. *shastensis* (A. Gray) D.R. Morgan & R.L. Hartman

Rayless Shasta aster

Aster shastensis (A. Gray) A. Gray var. *eradiatus* A. Gray, *Machaeranthera canescens* (Pursh) A. Gray var. *shastensis* (A. Gray) B.L. Turner, *Machaeranthera shastensis* A. Gray, *Machaeranthera shastensis* A. Gray var. *eradiata* (A. Gray) Cronquist & D.D. Keck, *Machaeranthera shastensis* A. Gray var. *shastensis*

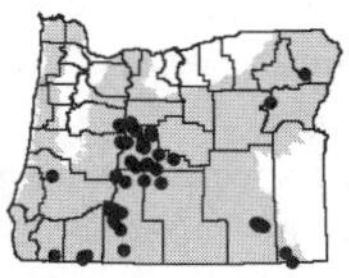

Stems 1+, ascending or erect, rarely decumbent, canescent, often sparsely stipitate glandular distally, rarely throughout. **Inflorescences** panicle- or raceme-like arrays; bracts obovate, oblanceolate, or linear. **Phyllaries** in 4–6 series, all erect or principally lower series with recurved tips. **Ray florets** 0 or if present sterile. **2*n***=8.

Sagebrush, pine forests, meadows. Flowering Jun–Sep. 400–2800 m. BR, BW, Casc, CR, ECas, Lava, Sisk. CA, NV. Native.

DIMERESIA Doublet, dimeresia

Genus composed of one species, *D. howellii*.

Dimeresia howellii A. Gray (p. 244)

Doublet, dimeresia

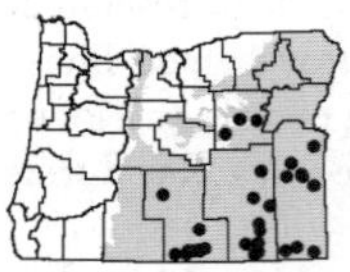

Herbs annual, cespitose, often acaulescent; taprooted. **Stems** 0 or ≤3 cm, simple or with whorls of several branches, glabrous. **Leaves** basal and cauline, crowded, usually whorled below, opposite above, spatulate to lanceolate, ovate, or broadly elliptic, 4–35 mm, margins entire, tips acute to obtuse, surfaces abaxially glabrous or lightly villous, sometimes obscurely punctate-glandular; sessile or petioles 1–12 mm. **Inflorescences** 1–many heads in 1–10 dense clusters. **Heads** discoid, erect; sessile or peduncles 0–1 mm, surfaces glabrous to woolly-villous. **Involucres** cylindric to obconic, 4–6 × 2(4) mm. **Phyllaries** 2–3 in 1 equal series, erect, deciduous in fruit, broadly elliptic or ovate, obscurely parallel-veined, tips obtuse, surfaces glabrous. **Receptacles** minute; paleae 0. **Florets** (1)2–3, corolla tubes 1.5–2 mm, white or purple, lobes 5, spreading, 0.5–2 mm, white or purple at base; stamens partly or wholly exserted; stigmas included. **Fruits** cylindric or oblanceolate, tapering to base, flattened, truncate, 2.5–3 mm, black, glabrous, with 14–20 inconspicuous ribs; pappi of basally connate, plumose scales or flattened bristles, deciduous, 3–4 mm, white. $2n$=14.

Rocky ridges, hillsides, talus slopes, sagebrush flats. Flowering May–Jul. 700–2800 m. BR, BW, ECas, Owy. CA, ID, NV. Native.

Plants of *D. howellii* may range from highly diminutive, with one to two heads, to robust, with up to ten-headed clusters, depending on the availability of moisture during the growing season. A strongly self-pollinating breeding system is suggested by the failure of the stigmatic style branches to extend beyond stamen tubes. The common name "doublet" is based on the genus name, which is taken from the Greek meaning "two parts," presumably referring to the two flowers in each head.

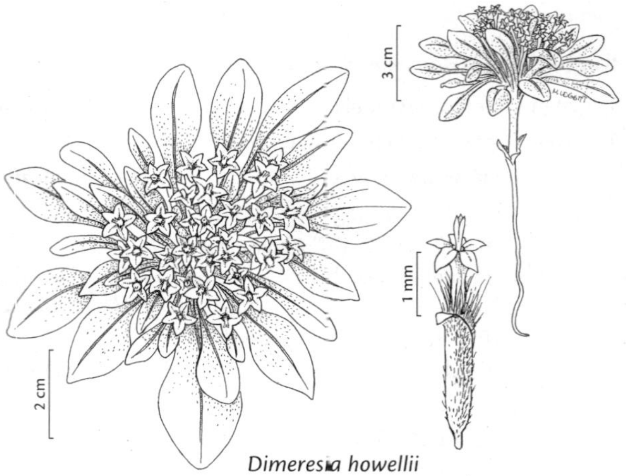

Dimeresia howellii

DIMERESIA

DORONICUM Leopard's-bane

Herbs perennial; rhizomatous. **Stems** erect, simple or few-branched distally, glabrous or pilose-tomentose, sometimes stipitate-glandular distally. **Basal leaves** ovate-orbicular with cordate bases or ovate-elliptic to lanceolate with wedge-shaped bases, margins entire to crenulate, denticulate, or dentate, surfaces pilose-hispid, sometimes glabrous to merely spiculate-puberulent or glandular; long-petiolate. **Cauline leaves** alternate; lower similar to basal, petiolate; upper broadly ovate or lanceolate, reduced, bases auriculate-clasping or not; sessile. **Inflorescences** solitary terminal heads. **Heads** radiate. **Involucres** campanulate. **Phyllaries** numerous in 2–3 equal series, erect to spreading, narrowly lanceolate, herbaceous, margins ciliate, 1–5-veined, tips long-acuminate, surfaces tomentose or spiculate-puberulent, sometimes stipitate-glandular. **Receptacles** convex to hemispheric; paleae 0. **Ray florets** pistillate, fertile; rays narrowly elliptic, spreading, well exceeding involucres, yellow, tips acute or with 3 short teeth. **Disc florets** bisexual, fertile; corollas tubular, not much enlarged distally, yellow turning brownish, lobes 5, erect or spreading, triangular; stamens partially exserted; stigmas slightly exserted. **Fruits** cylindric or obovoid, truncate, glabrous or strigillose, 5–10-ribbed; pappi of numerous bristles in 1 series, ray floret fruits lacking pappi.

Asia, Europe, North America. 35 species; 1 species treated in *Flora.*

Doronicum plantagineum has been infrequently reported in Oregon but is not established. It can be identified by its leaves, which have wedge-shaped bases and glabrous petioles.

Doronicum pardalianches Linnaeus (p. 245)

Great leopard's-bane

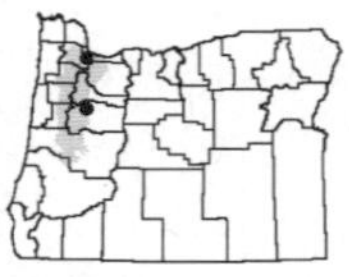

Plants 3–9 dm. **Stem** pilose-tomentose, sometimes stipitate-glandular distally. **Basal leaves** 25–35 cm; blades 10–20 cm, ovate-orbicular, bases cordate, margins crenulate or dentate, tips obtuse, surfaces pilose-hispid; petioles pilose-tomentose. **Cauline leaves** lower ovate-orbicular, bases cordate or truncate; upper ovate, bases auriculate-clasping, margins dentate or denticulate, tips acute. **Involucres** 12–15 mm. **Phyllaries** tomentose or glandular-puberulent. **Ray florets** 20–30+; rays 15–25 mm. **Disc florets** numerous; corollas 4.5–5 mm. **Fruits** 3–4 mm, dark brown; pappi 4–6 mm, white or tan. $2n$=60.

Coniferous forests, mixed woodlands, thickets, riverbanks. Flowering Apr–Aug. 100–300 m. WV. WA; north to British Columbia; Europe. Exotic.

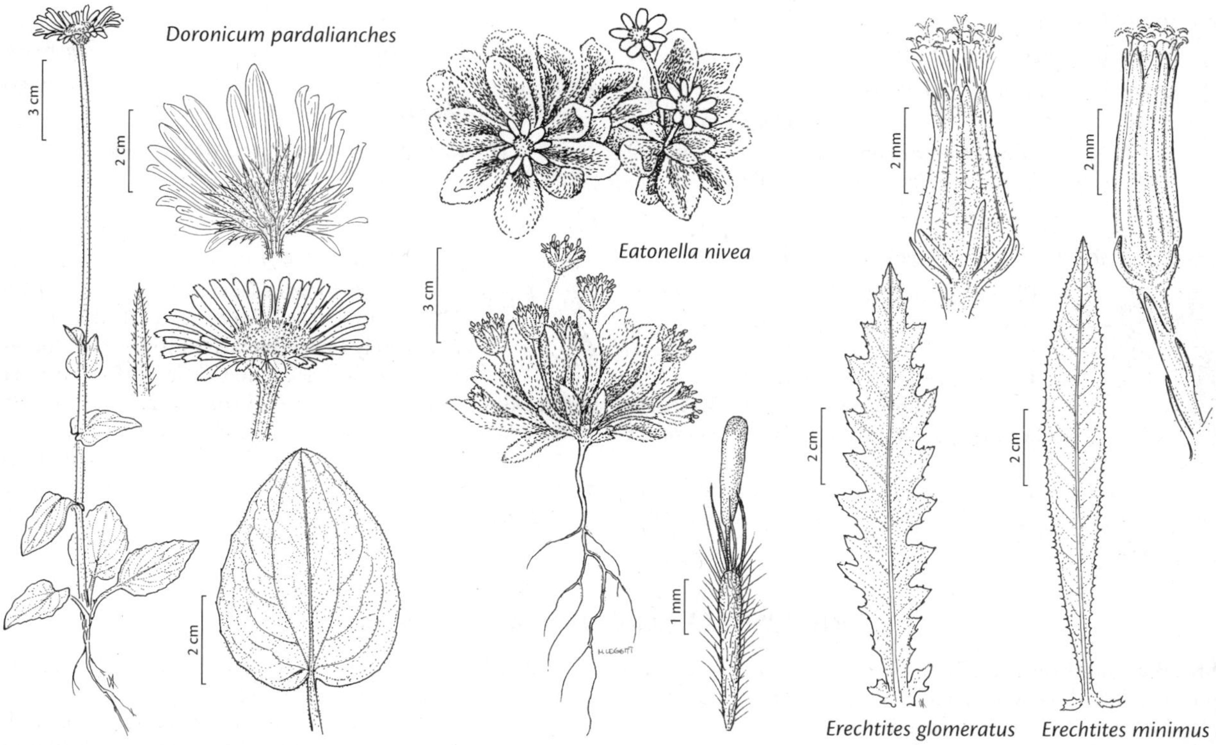

DORONICUM • EATONELLA • ERECHTITES

EATONELLA False tickhead, eatonella

Genus composed of one species, *E. nivea.*

Eatonella nivea (D.C. Eaton) A. Gray (p. 245)
White false tickhead, white eatonella

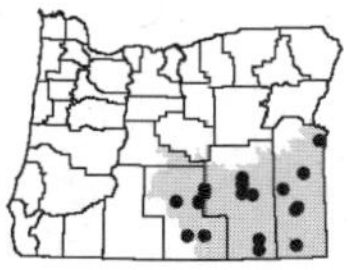

Herbs annual, cespitose, 1–5 cm; taprooted. **Stems** simple or branched, branches short, prostrate or decumbent, densely woolly. **Leaves** basal and cauline, whorled or alternate, crowded, linear to oblanceolate, 0.3–4 cm, margins entire, tips obtuse, surfaces densely woolly; sessile or petiolate. **Inflorescences** solitary terminal heads. **Heads** radiate, erect; sessile or peduncled in the leaf axils, peduncles 3.5–15 mm. **Involucres** campanulate. **Phyllaries** 5–13, in 1–2 series, erect in flower, reflexed in fruit, linear-elliptic, ± equal, 5–6 mm, tips acute, surfaces woolly-pubescent. **Receptacles** flat; paleae 0. **Ray florets** 8–13, pistillate; corollas yellow or pinkish, tubes 2.5–3 mm, rays 2–2.5 mm, slightly exserted from involucres. **Disc florets** 5–13, bisexual; corollas narrowly tubular, 2 mm, yellow, lobes 5, erect or spreading; stamens slightly exserted; styles included. **Fruits** compressed, narrowly oblong, 2.5–3.5 mm, black, tips obtuse, densely ciliate on the 2 marginal ribs; pappi of 2 awn-tipped scales set over marginal ribs, 1.5–2.5 mm, white. **2*n***=38.

Desert scrub, grasslands, ridges. Flowering Apr–Jul. 1000–1800 m. BR, Owy. CA, ID, NV, WA. Native.

Eatonella can be recognized by its low, mounded growth form and woolly pubescence. In the most recent classification of Asteraceae, the genus is assigned to tribe Madieae (tarweed tribe), which is one of 15 tribes now making up the large subfamily Helianthoideae, the sunflower group (Baldwin 2009).

ERECHTITES Burnweed

Stephen C. Meyers

Herbs annual or perennial, often with pungent odor; taprooted. **Stems** usually 1, erect. **Leaves** basal and cauline, alternate, lance-linear to lanceolate, bases often weakly clasping, margins entire to lobed or toothed, surfaces glabrous to villous; basal and proximal cauline often withered by flowering; sessile or petiolate. **Inflorescences** corymb-like arrays. **Heads** disciform. **Calyculi** 1–6(12). **Involucres** campanulate to cylindric. **Phyllaries** in 1–2 series, linear to lanceolate. **Receptacles** flat to convex; paleae 0. **Outer florets** 10–35 in 1–3 series, pistillate; corollas tubular-filiform, white to pale yellow, minutely lobed. **Inner florets** bisexual, sometimes staminate; corollas white to pale yellow, tubes > throats, lobes 4. **Fruits** obovoid to fusiform, ribbed, glabrous to puberulent; pappi of white, barbellulate bristles.

Australia, North America, Pacific Islands, South America. 12 species; 2 species treated in *Flora.*

Some floras include members of *Erechtites* within *Senecio*. *Erechtites* differs from *Senecio* by its lack of ray florets.

Key to *Erechtites* species:

1. Leaves pinnately lobed; phyllaries usually 13 . . . . . *E. glomeratus*
1' Leaves not pinnately lobed; phyllaries 7–10 . . . . . *E. minimus*

Erechtites glomeratus (Desfontaines ex Poiret) de Candolle (p. 245)

Australian burnweed

Erechtites arguta DC., *Senecio glomeratus* Desf. ex Poir.

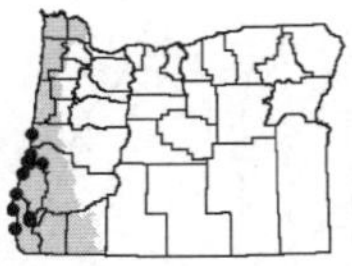

Plants (2)10–20 dm. **Leaves** lance-linear to lanceolate, 2–15 × 1–4 cm, reduced distally, margins pinnately lobed, lobes dentate. **Heads** 20–100+. **Involucres** cylindric, 4–7 mm. **Phyllaries** usually 13. **Outer florets** 25–35. **Inner florets** 8–10. **Fruits** 1–2 mm, puberulent between ribs.

Disturbed areas, mainly coastal, sometimes inland. Flowering Jun–Oct. 0–500 m. CR, Est, Sisk. CA, WA; Australia, Pacific Islands. Exotic.

Erechtites minimus (Poiret) de Candolle (p. 245)

Coastal burnweed

Senecio minimus Poir.

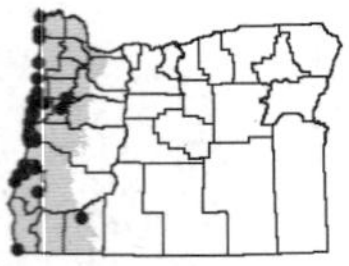

Plants (5)10–20 dm. **Leaves** narrowly lanceolate to lanceolate, 3–20 × 1–4 cm, often reduced distally, margins toothed, not pinnately lobed. **Heads** 40–100+. **Involucres** cylindric, 4–7 mm. **Phyllaries** 7–10. **Outer florets** 10–30. **Inner florets** 3–5. **Fruits** 1–2.5 mm, usually puberulent between ribs, sometimes glabrous.

Disturbed areas, mainly coastal, sometimes inland. Flowering Jun–Sep. 0–600 m. CR, Est, Sisk, WV. CA, WA; Australia, Pacific Islands. Exotic.

ERICAMERIA Goldenbush, rabbitbrush

Shrubs or **subshrubs** usually evergreen, sometimes deciduous (if very cold); taprooted. **Stems** usually erect, sometimes decumbent, much branched. **Leaves** cauline, alternate, often crowded, filiform to spatulate, glabrous to tomentose, often stipitate-glandular or gland-dotted; glands pustulose or in circular pits; petiolate or sessile. **Inflorescences** cyme-, panicle-, or raceme-like arrays, or heads solitary, usually well branched. **Heads** radiate or discoid; pedunculate or sessile. **Involucres** campanulate or obovoid. **Phyllaries** in 2–7 series, erect or with spreading or recurved tips, subequal or the outer gradually shorter, glabrous to tomentose or stipitate-glandular, sometimes resinous. **Receptacles** flat to convex, pitted; paleae 0. **Ray florets** pistillate; corollas white or yellow. **Disc florets** bisexual; corollas cylindric, well exceeding involucres, tubes $<$ throats, throats funnelform or campanulate, lobes triangular; styles usually well exserted. **Fruits** angled or cylindric, ellipsoid, or obconic, 5–12-ribbed, glabrous or pubescent, sometimes gland-dotted; pappi of numerous, usually persistent bristles in 1–3 series, white to reddish. $\mathbf{2n=18}$.

Western North America. 36 species; 9 species treated in *Flora*.

Based on molecular phylogenetic studies (Roberts & Urbatsch 2003), *Ericameria* combines species that in earlier floras were placed in *Chrysothamnus* and *Haplopappus*. *Haplopappus* is now recognized only from South America, while in Oregon, *Chrysothamnus* is reduced to two species.

Key to *Ericameria* species:

1. Ray florets 1–8; heads radiate.
   2. Stems stipitate-glandular; outer phyllaries $\geq$ inner.
      3. Heads 2–3 in leafy raceme-like arrays or heads solitary . . . . . *E. suffruticosa*
      3' Heads in congested, not leafy, cyme- to raceme-like arrays, rarely solitary . . . . . *E. greenei* (in part)
   2' Stems not stipitate-glandular (but may be gland-dotted or resinous); outer phyllaries $\approx$ or gradually $<$ inner.
      4. Ray and disc florets white . . . . . *E. resinosa*
      4' Ray and disc florets yellow.
         5. Rays 3–4 mm; leaves 0.3–1.5 cm . . . . . *E. nana*
         5' Rays 6–12 mm; leaves 1–7 cm . . . . . *E. bloomeri*
1' Ray florets 0; heads discoid.
   6. Stems with felt-like pubescence.
      7. Leaves obovate to oblanceolate, oblong or elliptic, minutely stipitate-glandular . . . . . *E. discoidea*
      7' Leaves linear or filiform to linear-oblanceolate, not stipitate-glandular (but may be gland-dotted).
         8. Phyllaries subequal . . . . . *E. parryi* var. *nevadensis*
         8' Phyllaries unequal . . . . . *E. nauseosa*
   6' Stems glabrous to lightly or finely tomentose, not felt-like.
      9. Leaves oblanceolate to spatulate . . . . . *E. greenei* (in part)
      9' Leaves filiform . . . . . *E. arborescens*

Ericameria arborescens (A. Gray) Greene

Golden-fleece

Haplopappus arborescens (A. Gray) H.M. Hall

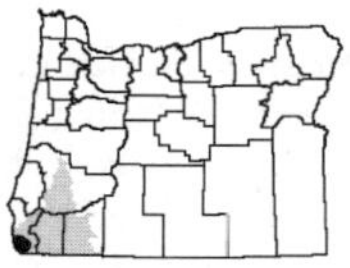

Plants 1–5 dm. **Stems** glabrous to finely tomentose, gland-dotted. **Leaves** filiform, 2–9 cm, glabrous, gland-dotted. **Inflorescences** cyme-like arrays. **Heads** discoid. **Involucres** ovoid or campanulate, 4–6 mm. **Phyllaries** in 3–4 series, lanceolate to elliptic or linear, glabrous; outer gradually < inner. **Ray florets** 0. **Disc florets** 10–25; corollas 5–7 mm, yellow. **Fruits** 2–3 mm, pubescent.

Open woods, chaparral, clearcuts, roadsides. Flowering Aug–Nov. 300–800 m. Sisk. CA. Native.

Ericameria bloomeri (A. Gray) J.F. Macbride

Rabbitbrush goldenweed, Bloomer's goldenbush

Haplopappus bloomeri A. Gray, *Haplopappus bloomeri* A. Gray var. *angustatus* A. Gray, *Haplopappus bloomeri* A. Gray var. *bloomeri*, *Haplopappus bloomeri* A. Gray var. *sonnei* (Greene) H.M. Hall

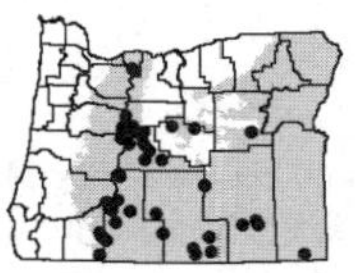

Plants 1–6 dm. **Stems** glabrous or lightly tomentose, sometimes gland-dotted. **Leaves** filiform to narrowly oblong, 1–7 cm, glabrous or lightly tomentose, often dotted with pustulose glands. **Inflorescences** panicle- or raceme-like arrays. **Heads** radiate. **Involucres** narrowly campanulate, 7–11 mm. **Phyllaries** in 3–6 series, glabrous or lightly tomentose; outer ≈ or gradually < inner. **Ray florets** 1–5; rays 6–12 mm, yellow. **Disc florets** 4–14; corollas 7–11 mm, yellow. **Fruits** 6–10 mm, glabrous or slightly villous.

Mixed forests, pine woods, shrublands, open slopes, ridges, roadsides. Flowering Jun–Sep. 900–2800 m. BR, BW, Casc, ECas, Owy. CA, ID, WA; north to British Columbia. Native.

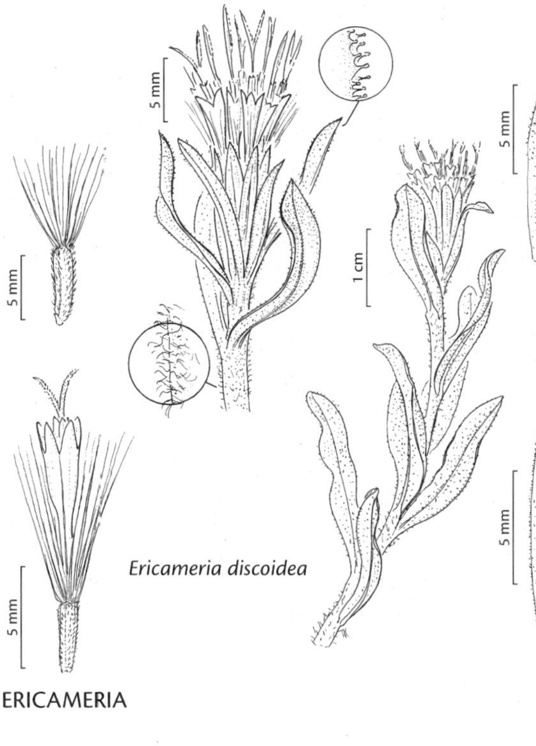

Ericameria discoidea

ERICAMERIA

Ericameria discoidea (Nuttall) G.L. Nesom (p. 247)

Discoid goldenweed, white-stem goldenbush

Haplopappus macronema A. Gray, *Haplopappus macronema* A. Gray var. *macronema*

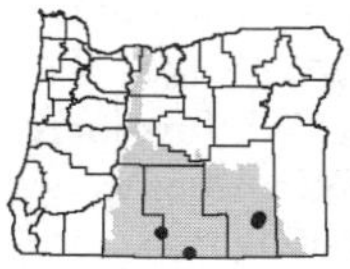

Plants 1–4 dm. **Stems** densely white, felt-like tomentose, sometimes stipitate-glandular. **Leaves** obovate to oblanceolate, oblong, or elliptic, 1–3.5 cm, minutely stipitate-glandular. **Inflorescences** cyme- or raceme-like arrays, or heads solitary. **Heads** discoid. **Involucres** obconic, 8–13 mm. **Phyllaries** in 3–4 subequal series, stipitate-glandular; inner often glabrous. **Ray florets** 0. **Disc florets** 10–25; corollas 9–11 mm, yellow. **Fruits** 8.5–11 mm, strigose or villous.

Brushlands, grasslands, dry slopes, talus, rock crevices. Flowering Jul–Sep. 2400–3000 m. BR, ECas. CA, ID, NV; northeast to MT, southeast CO. Native.

Ericameria greenei (A. Gray) G.L. Nesom

Greene's goldenbush

Haplopappus bloomeri A. Gray, *Haplopappus bloomeri* A. Gray var. *greenei* (A. Gray) Cronquist, *Haplopappus greenei* A. Gray, *Haplopappus greenei* A. Gray var. *greenei*, *Haplopappus greenei* A. Gray var. *mollis* A. Gray

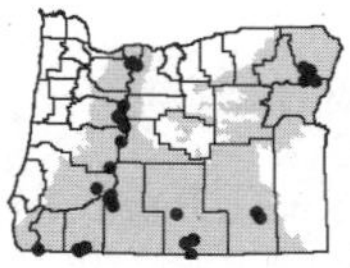

Plants 1–3 dm. **Stems** glabrous proximally, rarely lightly tomentose, minutely stipitate-glandular distally. **Leaves** oblanceolate to spatulate, 1.5–3 cm, glabrous to stipitate-glandular, rarely lightly tomentose. **Inflorescences** not leafy, in congested, cyme- or raceme-like arrays, rarely heads solitary. **Heads** radiate or discoid. **Involucres** broadly campanulate, 8–12 mm. **Phyllaries** in 3–5 series, broadly or narrowly linear; outer ≥ inner. **Ray florets** 0–7; rays 10–14 mm, yellow. **Disc florets** 7–20; corollas 10–14 mm, yellow. **Fruits** 5–7 mm, glabrous or lightly pubescent.

Open coniferous forests, shrublands, grasslands, rocky slopes, scree, pumice desert, rock outcrops, alpine ridges. Flowering Jul–Sep. 1000–2800 m. BR, BW, Casc, ECas, Sisk. CA, ID, WA. Native.

Ericameria nana Nuttall

Dwarf goldenbush, rubber weed

Haplopappus nanus (Nutt.) D.C. Eaton

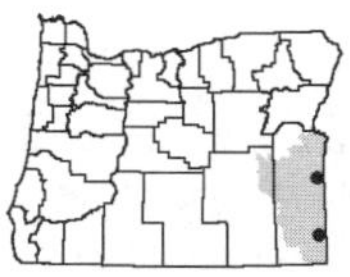

Plants 0.5–5 dm. **Stems** glabrous, resinous, shiny. **Leaves** narrowly oblanceolate to elliptic or linear, 0.3–1.5 cm, glabrous, glandular-resinous. **Inflorescences** congested cyme-like arrays, rarely heads solitary. **Heads** radiate. **Involucres** narrowly obconic, 5.5–7.5 mm. **Phyllaries** in 5–9 unequal series, linear-oblong; outer gradually < inner. **Ray florets** 1–7; rays 3–4 mm, yellow. **Disc florets** 4–8; corollas 4.5–6.5 mm, yellow. **Fruits** 4–5.5 mm, glabrous or sericeous.

Shrublands, open slopes, rock outcrops, cliffs. Flowering Jun–Oct. 1000–1400 m. Owy. CA, ID, NV; northeast to MT, southeast to UT. Native.

Ericameria nauseosa (Pallas ex Pursh) G.L. Nesom & G.I. Baird

Rubber rabbitbrush

Plants 1–25 dm. **Stems** densely white, felt-like tomentose, sometimes glabrate distally. **Leaves** filiform to linear-oblanceolate, 1–7 cm, glabrous or tomentose, often gland-dotted. **Inflorescences** cyme-like arrays. **Heads** discoid. **Involucres** narrowly obconic, 6–16 mm. **Phyllaries** in 5–7 unequal series, linear-oblong, chartaceous, surfaces glabrous or lightly tomentose; outer gradually < inner. **Ray florets** 0. **Disc florets** 4–6; corollas 6–12 mm, yellow. **Fruits** 3–8 mm.

Western North America. 21 varieties; 4 varieties treated in *Flora*.

Key to *Ericameria nauseosa* varieties:

1. Plants 1–3 dm . . . . ***E. n.*** var. ***nana***

1' Plants 3–25 dm.

  2. Phyllaries glabrous . . . . ***E. n.*** var. ***oreophila***

  2' Phyllaries (at least outermost) minutely tomentose.

    3. Disc corolla lobes 0.5–1 mm . . ***E. n.*** var. ***hololeuca***

    3' Disc corolla lobes 1.1–2.1 mm . . ***E. n.*** var. ***speciosa***

Ericameria nauseosa (Pallas ex Pursh) G.L. Nesom & G.I. Baird var. *hololeuca* (A. Gray) G.L. Nesom & G.I. Baird

White rabbitbrush

Chrysothamnus nauseosus (Pall. ex Pursh) Britton ssp. *hololeucus* (A. Gray) H.M. Hall & Clem.

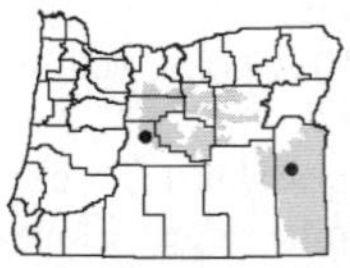

Plants 3–12 dm. **Leaves** 1–7 cm, tomentose or glabrate. **Phyllaries** (at least the outermost) minutely tomentose. **Disc florets** corolla lobes 0.5–1 mm; non-stigmatic portion of style branches < stigmatic portion. **Fruits** densely pubescent.

Shrublands, dry creek beds. Flowering Jun–Sep. 1000–2000 m. Lava, Owy. CA, ID, NV; southeast to NM. Native.

Ericameria nauseosa (Pallas ex Pursh) G.L. Nesom & G.I. Baird var. *nana* (Cronquist) G.L. Nesom & G.I. Baird

Little rabbitbrush

Chrysothamnus nauseosus (Pall. ex Pursh) Britton ssp. *nanus* (Cronquist) D.D. Keck, *Chrysothamnus nauseosus* (Pall. ex Pursh) Britton var. *nanus* Cronquist

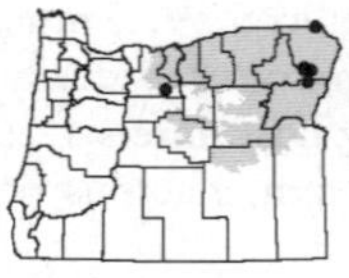

Plants 1–3 dm. **Leaves** 1–6 cm, tomentose. **Phyllaries** (at least outermost) sparsely tomentose. **Disc florets** corolla lobes 1.6–2.5 mm; non-stigmatic portion of style branches < stigmatic portion. **Fruits** densely pubescent.

Conifer woods, shrublands, grassy meadows. Flowering Jul–Sep. 600–2200 m. BW, Col. ID, WA. Native.

Ericameria nauseosa (Pallas ex Pursh) G.L. Nesom & G.I. Baird var. *oreophila* (A. Nelson) G.L. Nesom & G.I. Baird

Great Basin rabbitbrush

Chrysothamnus nauseosus (Pall. ex Pursh) Britton var. *artus* (A. Nelson) Cronquist, *Chrysothamnus nauseosus* (Pall. ex Pursh) Britton ssp. *consimilis* (Greene) H.M. Hall & Clem., *Chrysothamnus nauseosus* (Pall. ex Pursh) Britton var. *oreophilus* (A. Nelson) H.M. Hall & Clem.

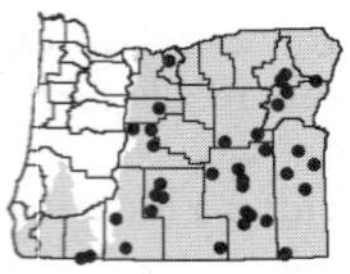

Plants 5–25 dm. **Leaves** 2–6 cm, glabrate. **Phyllaries** glabrous. **Disc florets** corolla lobes 1.5–2.5 mm; non-stigmatic portion of style branches ≤ stigmatic portion. **Fruits** pubescent.

Pine forests, shrublands, grasslands, alkali flats, streambanks, roadsides. Flowering Jul–Sep. 500–2000 m. BR, BW, Col, ECas, Lava, Owy, Sisk. CA, ID; northeast to MT, east to WY, southeast to NM. Native.

Ericameria nauseosa (Pallas ex Pursh) G.L. Nesom & G.I. Baird var. *speciosa* (Nuttall) G.L. Nesom & G.I. Baird

Showy rabbitbrush

Chrysothamnus nauseosus (Pall. ex Pursh) Britton ssp. *albicaulis* (Nutt.) H.M. Hall & Clem., *Chrysothamnus nauseosus* (Pall. ex Pursh) Britton var. *albicaulis* (Nutt.) Rydb., *Chrysothamnus nauseosus* (Pall. ex Pursh) Britton var. *speciosus* (Nutt.) H.M. Hall

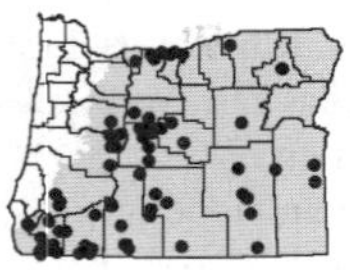

Plants 2–20 dm. **Leaves** 3–7 cm, usually tomentose. **Phyllaries** (at least the outermost) minutely tomentose. **Disc florets** corolla lobes 1.1–2.1 mm; non-stigmatic portion of style branches > stigmatic portion. **Fruits** densely pubescent.

Conifer forests, shrublands, grasslands, dry hillsides, ridgetops, streambanks, irrigated fields, roadsides. Flowering Jul–Nov. 50–2500 m. BR, BW, Casc, Col, ECas, Lava, Owy, Sisk. CA, ID, NV, WA; north to British Columbia, northeast to Alberta, east to WY, southeast to CO. Native.

Ericameria parryi (A. Gray) G.L. Nesom & G.I. Baird var. *nevadensis* (A. Gray) G.L. Nesom & G.I. Baird

Nevada rabbitbrush

Chrysothamnus parryi (A. Gray) Greene, *Chrysothamnus parryi* (A. Gray) Greene ssp. *nevadensis* (A. Gray) H.M. Hall & Clem., *Chrysothamnus parryi* (A. Gray) Greene var. *nevadensis* (A. Gray) Kittell

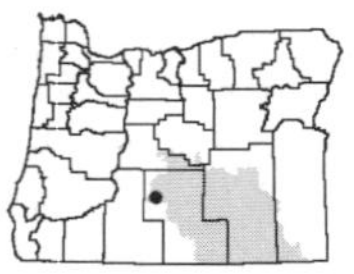

Plants 2–6 dm. **Stems** densely white, felt-like tomentose. **Leaves** linear, 1.5–4 cm, minutely tomentose, sometimes gland-dotted. **Inflorescences** congested cyme- or raceme-like arrays. **Heads** discoid. **Involucres** cylindrical or narrowly obovoid, 11–15 mm. **Phyllaries** in 3–5 series, subequal, tips slender, tapering, adaxial surfaces and margins tomentose; inner chartaceous; outer chartaceous below. **Ray florets** 0. **Disc florets** 4–6; corollas 9–10.5 mm, yellow. **Fruits** 3–8 mm, sericeous.

Open conifer forest, shrublands. Flowering Jul–Sep. 1200–1500 m. BR. CA, NV; southeast to AZ. Native.

Ericameria resinosa Nuttall (p. 249)

Columbia goldenweed

Haplopappus resinosus (Nutt.) A. Gray

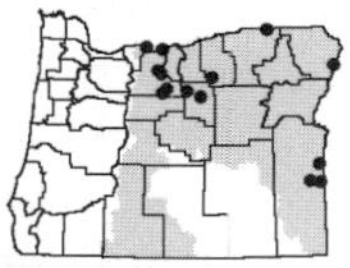

Plants 0.8–4 dm. **Stems** glabrous, resinous. **Leaves** filiform, 0.5–3 cm, glabrous, resinous. **Inflorescences** loose cyme-like arrays. **Heads** radiate. **Involucres** campanulate to obovoid, 5.5–8 mm. **Phyllaries** in 5–7 unequal series, deltoid to lance-linear, chartaceous, glabrous, resinous; outer gradually < inner. **Ray florets** 3–7; rays 5.5–7 mm, white. **Disc florets** 10–15; corollas 6–8 mm, white. **Fruits** 4.5–5 mm, sericeous.

Rock crevices, basalt cliffs. Flowering Jun–Oct. 100–1300 m. BW, Col, ECas, Lava, Owy. ID, WA. Native.

Ericameria suffruticosa (Nuttall) G.L. Nesom

Shrubby goldenweed, single-headed goldenweed

Haplopappus suffruticosus (Nutt.) A. Gray

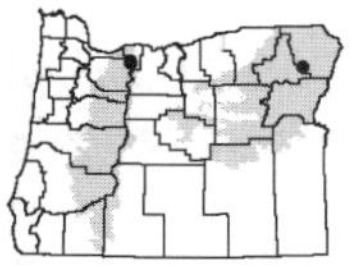

Plants 1–4 dm. **Stems** becoming stipitate-glandular. **Leaves** oblanceolate or elliptic, 0.5–4 cm, glabrous or stipitate-glandular, often gland-dotted. **Inflorescences** leafy, 2–3-headed, raceme-like arrays or heads solitary. **Heads** radiate. **Involucres** campanulate, 9–15 mm. **Phyllaries** in 3–5 subequal series, broadly to narrowly lanceolate, chartaceous below, glabrous or minutely glandular; outer ≥ inner. **Ray florets** 1–8; rays 8–14 mm, yellow. **Disc florets** 15–40; corollas 8.5–10.5 mm, yellow. **Fruits** 5.5–8 mm, pubescent.

Open slopes, moraines, scree, ridges. Flowering Jul–Sep. 2200–3000 m. BW, Casc. CA, ID, NV; northeast to MT, east to WY. Native.

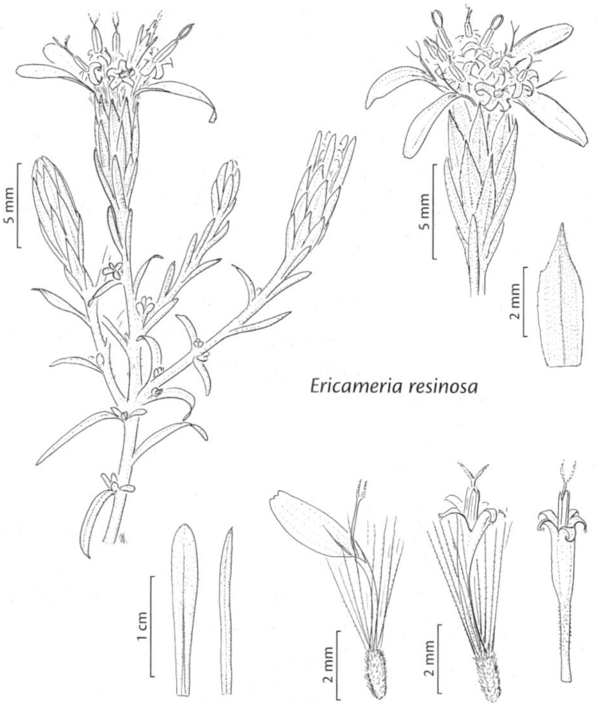

ERICAMERIA

ERIGERON Fleabane, daisy

James P. Riser II and Stephen C. Meyers

Herbs, rarely **subshrubs** or **small shrubs** annual, biennial, or perennial. **Stems** usually erect to ascending, occasionally decumbent, glabrous to densely hirsute, glandular or eglandular. **Leaves** basal and/or cauline, basal sometimes withering by flowering, alternate, broadly obovate to oblanceolate to narrowly spatulate to linear, margins entire, lobed, or toothed, surfaces glabrous to densely hirsute, glandular or eglandular; cauline well developed or ± reduced distally; sessile or petiolate. **Inflorescences** heads solitary on scapes or scape-like stems or few to many in corymb-, raceme-, or panicle-like arrays. **Heads** radiate, disciform, or discoid. **Involucres** turbinate or campanulate to hemispheric. **Phyllaries** usually in 2–3 series, numerous, often with a darkened medial or apical area, surfaces glabrous to densely hirsute, glandular or eglandular. **Receptacles** flat to conic, pitted; paleae 0. **Ray florets** 0 or in 1(2) series, pistillate; rays usually strap-shaped, sometimes filiform, sometimes very short or lacking, white, pink, lavender, purple, blue, yellow, or cream-colored. **Disc florets** usually bisexual; corollas short, yellow. **Fruits** ± oblong, flattened to ± cylindric, 2–7-veined; pappi of 2 series, outer of crowns, scales, or short bristles, inner of usually antrorsely barbellate, often accrescent, bristles, sometimes lacking.

Temperate areas nearly worldwide. ~390 species; 40 species treated in *Flora*.

Erigeron needs a modern generic revision. There is little consistency in the amount of variation warranting specific status. Often, there is greater morphological variation within species (particularly among varieties) than between species. Many species are defined by morphological minutia and/or a single character (e.g., direction of stem hairs).

Heads with both peripheral pistillate (ray) florets (in one or two series) and central disc florets (radiate heads) are most common. Heads lacking obvious ray florets may have inconspicuous or missing pistillate rays (disciform heads: *E. aphanactis*, *E. chrysopsidis* var. *austiniae*) or the pistillate flowers may actually be absent (discoid heads: *E. bloomeri*, *E. inornatus*, *E. petrophilus*). Pistillate flowers lacking conspicuous rays may be obscured by the involucres or adjacent disc floret pappi. Close inspection under magnification is needed before determining that the pistillate flowers are truly missing.

Some individuals, especially if sterile, depauperate, or of poor quality, are not identifiable without direct comparison with reference material, if at all. The rays of pistillate flowers may change color upon drying; it is important to record flower color at time of collection. Due to shrinkage and curling of the pistillate flower rays on dried specimens, care must be taken when assessing whether rays are strap-shaped (usually greater than 0.5 mm wide) or filiform (less than 0.5 mm wide).

Key to *Erigeron* species:

1. Basal leaves spatulate or obovate, margins lobed or dissected.
   2. Basal leaves 5–60 mm, (1)2–3-ternately lobed . . . . **_E. compositus_**
   2' Basal leaves 6–25 mm, 3–5-lobed . . . . **_E. vagus_**

1' Leaves various, margins not dissected or lobed.
   3. Ray florets lacking.
      4. Heads solitary; cauline leaves abruptly reduced.
         5. Fruits 3–5 mm, surfaces glabrous with few sericeous hairs at distal ends . . . . **_E. bloomeri_**
         5' Fruits 2–3 mm, moderately strigose . . . . **_E. chrysopsidis_** (in part)
      4' Heads > 1, in corymb-like arrays (rarely heads solitary in *E. petrophilus* var. *petrophilus*); cauline leaves not abruptly reduced.
         6. Phyllaries eglandular; plants 20–65 cm . . . . **_E. inornatus_** var. **_inornatus_**
         6' Phyllaries minutely glandular; plants 4–25 cm.
            7. Phyllaries moderately to densely hispid-villous; basal leaves persistent . . . . **_E. aphanactis_** var. **_aphanactis_**
            7' Phyllaries nearly glabrous; basal leaves withering by flowering . . . . **_E. petrophilus_** var. **_petrophilus_**
   3' Ray florets present.
      8. Ray fruit pappi lacking.
         9. Phyllaries usually with dark medial area, with or without orangish stripe, surfaces glabrous to sparsely villous . . . . **_E. annuus_**
         9' Phyllaries pale yellow with greenish to purplish brown stripe, surfaces hirsute . . . . **_E. strigosus_** var. **_strigosus_**
      8' Ray fruit pappi present.
         10. Rays ≤ 0.5 mm wide (rarely up to 0.8 mm in *E. acris* var. *kamtschaticus*).
            11. Ray florets in 2 series.
               12. Plants 20–60 cm; outer ray floret series 75–150 . . . . **_E. acris_** var. **_kamtschaticus_**
               12' Plants 4–20 cm; outer ray floret series 30–70 . . . . **_E. nivalis_**
            11' Ray florets in 1 series.
               13. Basal leaves persistent . . . . **_E. lonchophyllus_**
               13' Basal leaves withering by flowering.
                  14. Ray corollas purple to lavender . . . . **_E. divergens_**
                  14' Ray corollas white to pink . . . . **_E. philadelphicus_**
         10' Rays > 0.5 mm wide.
            15. Phyllaries with yellow-orange to orange medial stripe.
               16. Basal leaves withering by flowering . . . . **_E. speciosus_**
               16' Basal leaves persistent.
                  17. Leaves basal and cauline . . . . **_E. pumilis_** var. **_intermedius_**
                  17' Leaves mostly basal, cauline few.
                     18. Ray florets white to cream, occasionally fading to pink . . . . **_E. disparipilus_** (in part)
                     18' Ray florets lavender to dark purple.
                        19. Fruits glabrate to sparsely strigose . . . . **_E. latus_** (in part)
                        19' Fruits densely sericeous . . . . **_E. poliospermus_** var. **_poliospermus_** (in part)
            15' Phyllaries without a medial stripe or stripe not yellow-orange to orange.
               20. Basal leaves withering by flowering.
                  21. Plants of the Willamette Valley . . . . **_E. decumbens_**
                  21' Plants not of the Willamette Valley.
                     22. Heads usually solitary; ray florets white to cream; plants of the Wallowa Mountains.
                        23. Ray florets 60–110 . . . . **_E. coulteri_** (in part)
                        23' Ray florets 25–50 . . . . **_E. disparipilus_** (in part)
                     22' Heads 1–7 in corymb-like arrays; ray florets lavender to purple; plants not of the Wallowa Mountains.
                        24. Stems glabrate to strigose, hairs usually appressed . . . . **_E. foliosus_** var. **_confinis_**
                        24' Stems hirsute, hairs usually spreading and slightly reflexed . . . . **_E. klamathensis_**

20' Basal leaves persistent.
25. Plants fibrous-rooted.
26. Plants not rhizomatous; ray florets yellow. . . . . ***E. aureus***
26' Plants usually rhizomatous; ray florets white to purple.
27. Stems usually minutely glandular. . . . . ***E. glaucus***
27' Stems usually eglandular.
28. Plants 3–15 cm. . . . . ***E. simplex***
28' Plants 15–70 cm.
29. Fruits 1.5–2 mm.
30. Phyllaries glabrous . . . . ***E. cervinus***
30' Phyllaries hirsute or villous. . . . . ***E. coulteri*** (in part)
29' Fruits 2–4 mm.
31. Phyllaries glabrous.
32. Ray florets lavender to purple . . . . ***E. glacialis*** var. ***glacialis***
32' Ray florets white . . . . ***E. howellii***
31' Phyllaries hirsute to villous.
33. Involucres 5–7 × 13–17 mm. . . . . ***E. aliceae***
33' Involucres 7–9 × 17–24 mm. . . . . ***E. peregrinus*** var. ***peregrinus***
25' Plants taprooted.
34. Ray florets yellow.
35. Leaf bases sheathing. . . . . ***E. linearis***
35' Leaf bases not sheathing . . . . ***E. chrysopsidis*** (in part)
34' Ray florets white to purple.
36. Phyllaries eglandular. . . . . ***E. elegantulus*** (in part)
36' Phyllaries glandular.
37. Some basal leaf margins serrate, dentate, or lobulate distally.
38. Cauline leaves reduced distally; ray florets white . . . . ***E. cascadensis*** (in part)
38' Cauline leaves little to not reduced distally; ray florets pink to purple . . . . ***E. oreganus***
37' All basal leaf margins entire.
39. All basal leaf tips rounded. . . . . ***E. cascadensis*** (in part)
39' Basal leaf tips usually acute to acuminate, some may be rounded.
40. Plants of Curry County . . . . ***E. stanselliae***
40' Plants not of Curry County.
41. Basal leaves narrowly-lanceolate to spatulate.
42. Involucres 3–5 × 7–11 mm; ray florets 15–24; disc corollas 2–4 mm; fruits 1–3 mm . . . . ***E. tener***
42' Involucres 5–9 × 9–21 mm; ray florets 20–35; disc corollas 4–5 mm; fruits 2–4 mm.
43. Fruits glabrate to sparsely strigose . . . . ***E. latus*** (in part)
43' Fruits densely sericeous . . . . ***E. poliospermus*** var. ***poliospermus*** (in part)
41' Basal leaves filiform to linear or narrowly-lanceolate.
44. Basal leaves 0.5–2 mm wide.
45. Heads solitary; phyllaries eglandular . . . . ***E. elegantulus*** (in part)
45' Heads 1–5 in corymb-like arrays; phyllaries glandular. . . . . ***E. filifolius***
44' Basal leaves 2–10 mm wide.
46. Phyllaries usually with dark medial area; plants 15–40 cm. . . . . ***E. corymbosus***
46' Phyllaries without dark medial area; plants 4–28 cm . . . . ***E. eatonii***

Erigeron acris Linnaeus var. ***kamtschaticus*** (de Candolle) Herder

Bitter fleabane

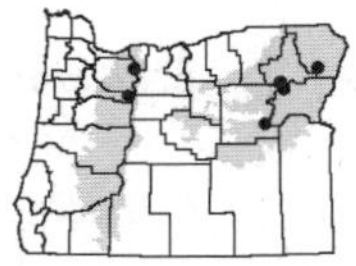

Plants annual, biennial, or short-lived perennial, 20–60 cm; usually fibrous-rooted, rarely taprooted, with usually simple caudices. **Stems** erect, nearly glabrous to sparsely villous, occasionally minutely glandular. **Basal leaves** usually persistent, narrowly spatulate to oblanceolate, 40–100 × 5–10 mm, margins entire, surfaces glabrous to remotely hirsute, mostly eglandular. **Cauline leaves** narrowly oblanceolate to elliptic, 10–60 × 2–(5)8 mm, reduced distally. **Heads** 1–35+ in corymb-like arrays, radiate. **Involucres** 5–6 × 7–13 mm. **Phyllaries** usually in 2–3 series, usually with a purplish medial area, surfaces moderately to densely villous especially basally, minutely glandular. **Ray florets** in 2 series; outer 75–150, erect, filiform, 0.3–0.5 × 3–5 mm, white, pink, or purplish; inner many fewer than outer, tubular. **Disc florets** corollas 4–6 mm. **Fruits** 2–3 mm, sparsely strigose; dense inner pappi of numerous barbellate bristles.

Roadsides, trail sides, streambanks. Flowering Jul–Aug. 1000–1700 m. BW, Casc. ID, WA; north to AK, northeast to Newfoundland, east to MI, southeast to CO; Asia. Native.

Erigeron aliceae Howell

Alice Eastwood's fleabane

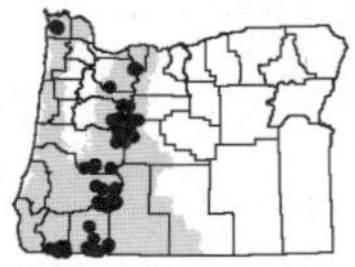

Plants perennial, 30–70 cm; fibrous-rooted, rhizomatous, caudices sometimes branching. **Stems** erect, nearly glabrous or sparsely strigose often becoming more so distally, eglandular. **Basal leaves** usually persistent, lanceolate to oblanceolate, 50–120 × 10–30 mm, margins entire, sometimes coarsely toothed with 1–5 pairs of teeth, midveins prominent, rarely with 2 prominent veins, surfaces remotely to moderately strigose. **Cauline leaves** well developed, lanceolate to narrowly elliptic, 10–80 × 3–14 mm, distally reduced, bases clasping. **Heads** 1–few, radiate. **Involucres** 5–7 × 13–17 mm. **Phyllaries** in 2–3 series, with dark green medial area, surfaces white woolly-villous, glandular. **Ray florets** 30–80, white to lavender to purple; rays 4–12 × 1–1.5 mm. **Disc florets** corollas 2–4 mm. **Fruits** 2–3 mm, moderately to densely strigose; inner pappi of numerous barbellate bristles.

Meadows, ridges, prairies. Flowering Jun–Sep. 400–2200 m. Casc, CR, ECas, Sisk. CA, WA. Native.

This is a variable species that can be confused with *Erigeron glacialis*. The woolly-villous phyllaries best distinguish *E. aliceae* from *E. glacialis*.

Erigeron annuus (Linnaeus) Persoon

Eastern daisy fleabane, annual fleabane

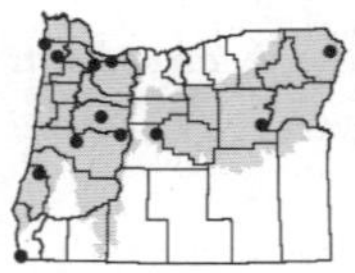

Plants annual, 40–110 cm; fibrous-rooted or weakly taprooted. **Stems** erect, sparsely to moderately hirsute, usually becoming strigose distally, eglandular. **Basal leaves** usually withering by flowering, broadly lanceolate to oblanceolate, 50–120 × 5–25 mm, margins coarsely serrate or nearly entire, surfaces sparsely strigose. **Cauline leaves** lanceolate, oblanceolate, or oblong, 15–80 × 2–15+ mm, often reduced distally, margins entire to coarsely serrate, surfaces sparsely strigose. **Heads** 5–many in corymb-like or panicle-like arrays, radiate. **Involucres** 3–5 × 5–13 mm. **Phyllaries** in 2–3 series, usually with dark medial area, with or without an orangish stripe, surfaces glabrous to sparsely villous, often minutely glandular. **Ray florets** 60–110, white to pale lavender; rays 2–8 × 0.5–1 mm. **Disc florets** corollas 2–3 mm. **Fruits** 1–2 mm, sparsely strigose; disc fruit pappi of numerous barbellate bristles; ray fruit pappi lacking. **2*n***=27.

Roadsides, fields, disturbed areas. Flowering Jul–Sep. 0–1000 m. BW, Casc, CR, Est, Lava, WV. CA, ID, WA; scattered in western North America, throughout eastern and midwestern North America; Asia, Europe. Exotic.

Erigeron aphanactis (A. Gray) Greene var. ***aphanactis***

Rayless shaggy fleabane

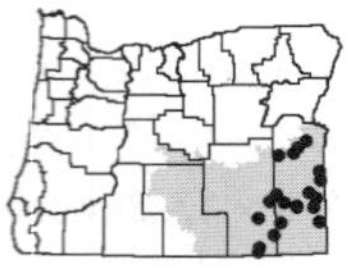

Plants perennial, 4–17 cm; taprooted with branching caudices. **Stems** erect to slightly decumbent, moderately to densely hirsute, glandular. **Basal leaves** persistent, narrowly oblanceolate to nearly linear, 10–60 × 1–4 mm, margins entire, surfaces densely hirsute to hispid, frequently glandular. **Cauline leaves** narrowly oblanceolate becoming linear, 5–30 × 1–3 mm, ± reduced distally. **Heads** 1–4 in scape- or corymb-like arrays, disciform. **Involucres** 3–5 × 6–11 mm. **Phyllaries** in 2–3 series, usually with dark green or yellow medial area, surfaces moderately to densely hispid-villous, usually densely minutely glandular. **Outer florets** 30–40, pale yellow; corollas tubular, rays lacking or minute, often shorter than involucres. **Inner florets** corollas 3–5 mm. **Fruits** 1.5–2 mm, sparsely strigose; inner pappi of numerous barbellate bristles.

Dry rocky slopes, basalt. Flowering May–Jul. 800–1600 m. BR, Owy. CA, NV; southwest to NM. Native.

Erigeron aureus Greene

Golden fleabane

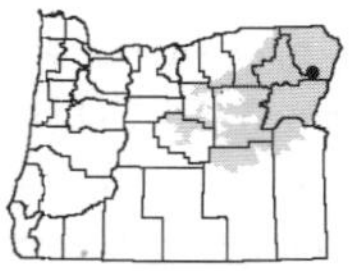

Plants perennial, 5–10 cm; fibrous-rooted, with short, usually unbranched caudices. **Stems** erect, hirsute to villous, minutely glandular. **Basal leaves** persistent, spatulate to oblanceolate, 10–50 × 5–10 mm, margins entire, surfaces moderately to densely strigose, eglandular to minutely glandular. **Cauline leaves** 5–10 × 1–5 mm abruptly reduced, often a single linear bract. **Heads** 1, radiate. **Involucres** 4–6 × 10–15 mm. **Phyllaries** in 2–3 series, often purplish colored, surfaces moderately to densely woolly-villous, sparsely glandular. **Ray florets** 40–60, bright yellow; rays 4–6 × 0.5–1.5 mm. **Disc florets** corollas 3–4 mm. **Fruits** 2–3 mm, moderately strigose; inner pappi of numerous barbellate bristles. **2*n***=18.

Ridges, talus, rocky slopes. Flowering Jul–Aug. 1600–2600 m. BW. WA; north to British Columbia, northeast to Alberta. Native.

In Oregon, this native species is currently known only from a single collection in the Eagle Cap Wilderness, Wallowa Mountains.

Erigeron bloomeri A. Gray

Bloomer's fleabane

Plants perennial, 3–22 cm; taprooted with branching caudices. **Stems** erect, glabrous to densely strigose, eglandular. **Basal leaves** persistent, linear to very narrowly oblanceolate, 10–80 × 0.7–3 mm, bases white-indurate, often tinged with purple, margins entire, surfaces glabrous to densely strigose, eglandular. **Cauline leaves** linear, 1–5 × 0.5–2 mm, abruptly reduced. **Heads** 1, discoid. **Involucres** 5–8 × 6–16 mm. **Phyllaries** in 2–3 series, with dark medial area, surfaces glabrous to densely woolly-villous, minutely glandular to eglandular. **Ray florets** 0. **Disc florets** corollas 4–6 mm, pale to bright yellow, lobes often purple-tipped. **Fruits** 3–5 mm, glabrous with few sericeous hairs at distal ends below inner pappi; inner pappi of numerous barbellate bristles.

Western United States. 2 varieties.

Key to *Erigeron bloomeri* varieties:

1. Leaves and stems strigose; phyllaries strigose to villous . . . . ***E. b.*** var. ***bloomeri***

1' Leaves and stems glabrous; phyllaries glabrous to sparsely hirsute . . . . ***E. b.*** var. ***nudatus***

Erigeron bloomeri A. Gray var. *bloomeri*

Bloomer's fleabane, scabland fleabane

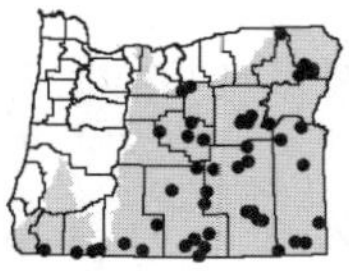

Stems strigose. **Leaves** strigose. **Phyllaries** strigose to villous. **2*n*=18.**

Talus, shale, meadows, sagebrush, serpentine. Flowering May–Aug. 700–3100 m. BR, BW, ECas, Lava, Owy, Sisk. CA, ID, NV. Native.

Erigeron bloomeri A. Gray var. *nudatus* (A. Gray) Cronquist

Bloomer's fleabane, Waldo daisy

Stems glabrous. **Leaves** glabrous. **Phyllaries** glabrous to sparsely hirsute.

Ridges, slopes, chaparral, serpentine. Flowering May–Jul. 400–2300 m. Sisk. CA. Native.

This variety is restricted to the Siskiyou Mountains of southern Oregon and northern California.

Erigeron cascadensis A. Heller (p. 253)

Cascade fleabane

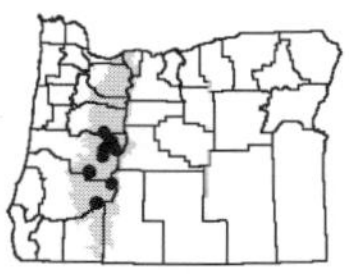

Plants perennial, 5–20 cm; taprooted with with woody, branching caudices. **Stems** ascending to erect, mostly glabrous to sparsely hirsute, sparsely minutely glandular. **Basal leaves** persistent in distinct rosettes, obovate to spatulate, 10–90 × 5–15 mm, margins entire or occasionally weakly dentate on distal portions, surfaces glabrous to remotely strigose, occasionally remotely glandular. **Cauline leaves** elliptic to oblanceolate, 7–25 × 2–6 mm, reduced distally. **Inflorescences** 1–3 heads, terminal and axillary. **Heads** radiate. **Involucres** 3–8 × 9–15 mm. **Phyllaries** in 2–3 series, often with purple medial or apical area, surfaces nearly glabrous or sparsely to moderately villous, eglandular or glandular. **Ray florets** 30–50, white to occasionally purplish; rays 4–7 × 1–2 mm. **Disc florets** corollas 3–4 mm. **Fruits** 2–3 mm, sparsely strigose; inner pappi of numerous barbellate bristles.

Shaded rocky slopes, cliffs, crevices, outcrops. Flowering May–Aug. 1200–2200 m. Casc. Native.

This species is endemic to rocky areas in the Cascades from Douglas and Klamath counties only as far north as Marion County.

Erigeron cervinus Greene

Siskiyou fleabane, Siskiyou daisy

Erigeron delicatus Cronquist

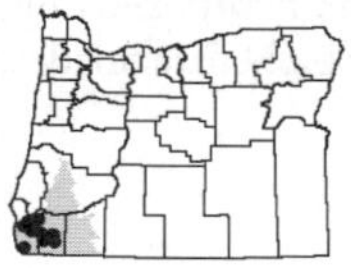

Plants perennial, 10–45 cm; fibrous-rooted, rhizomatous, occasionally with branching caudices. **Stems** erect or ascending, glabrous and mostly eglandular. **Basal leaves** persistent, oblanceolate to obovate, 10–100 × 5–15 mm, margins entire, surfaces glabrous, eglandular. **Cauline leaves** oblanceolate to elliptic, 10–50 × 2–10 mm, gradually reduced distally. **Heads** 1–4 in loose corymb-like arrays, radiate. **Involucres** 4–6 × 10–13 mm. **Phyllaries** in 2–3 series, usually with dark medial area, surfaces glabrous, densely glandular. **Ray florets** 20–40; corollas white to purple; rays 3–7 × 0.5–1.5 mm. **Disc florets** corollas 3–5 mm. **Fruits** 1.5–2 mm, moderately strigose; inner pappi of numerous barbellate bristles.

Streamsides, river bars, wet rocky banks, moist seeps. Flowering May–Aug. 0–1700 m. Sisk. CA. Native.

Erigeron cascadensis

ERIGERON

Erigeron chrysopsidis A. Gray

Dwarf yellow fleabane

Plants perennial, 3–20 cm; taprooted, usually with many woody, branching caudices. **Stems** erect, hirsute, eglandular to minutely glandular. **Basal leaves** persistent in dense clusters, linear to narrowly oblanceolate, 10–80 × 1–3 mm, margins entire with hispid hairs, surfaces hirsute to hispid, sometimes nearing strigose. **Cauline leaves** abruptly reduced to (0)1–3 linear bracts. **Heads** 1, radiate or disciform. **Involucres** 4–7 × 10–16 mm. **Phyllaries** in 2–3 series, sparsely to moderately hispid-villous, often minutely glandular. **Ray florets** 20–60; corollas with well-developed yellow rays, 4–8 × 0.5–2 mm, or tubular and short, or rays lacking. **Disc florets** corollas 4–5 mm. **Fruits** 2–3 mm, moderately strigose; inner pappi of numerous barbellate bristles.

Western United States. 3 varieties.

Key to *Erigeron chrysopsidis* varieties:

1. Rays 0 or minute. . . . . *E. c.* var. ***austiniae***

1' Rays 4–11 mm.

  2. Rays 4–7 mm; involucres 3.5–5.5 mm; stems usually strigose . . . *E. c.* var. ***brevifolius***

  2' Rays 8–11 mm; involucres 5–7 mm; stems hirsute-villous . . . . *E. c.* var. ***chrysopsidis***

Erigeron chrysopsidis A. Gray var. *austiniae* (Greene) G.L. Nesom

Austin's fleabane

Erigeron austiniae Greene, *Erigeron chrysopsidis* A. Gray ssp. *austiniae* (Greene) Cronquist

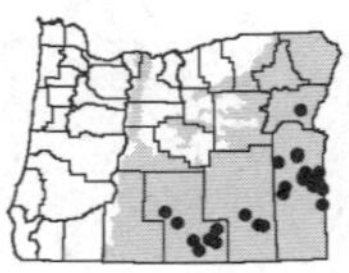

Stems hirsute-villous. **Involucres** 4.5–6 mm. **Ray florets** rays 0 or minute. **2*n***=18.

Ridges, crevices, sagebrush. Flowering May–Jul. 700–1900 m. BR, BW, ECas, Owy. CA, ID, NV. Native.

Erigeron chrysopsidis A. Gray var. *brevifolius* Piper

Alpine yellow daisy, Wallowa daisy

Erigeron chrysopsidis A. Gray ssp. *brevifolius* Piper

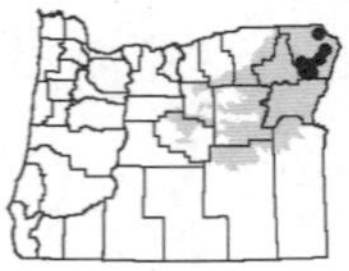

Stems usually strigose. **Involucres** 3.5–5.5 mm. **Ray florets** rays 4–7 mm.

Cliffs, rocky areas, alpine slopes. Flowering Jun–Aug. 1500–3100 m. BW. Native.

This variety is endemic to the Wallowa Mountains.

Erigeron chrysopsidis A. Gray var. *chrysopsidis*

Golden daisy, golden fleabane

Erigeron chrysopsidis A. Gray ssp. *chrysopsidis*

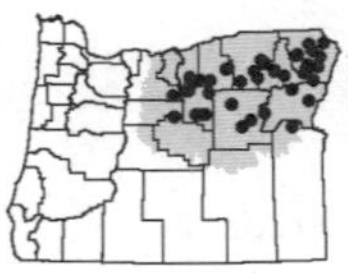

Stems hirsute-villous. **Involucres** 5–7 mm. **Ray florets** rays 8–11 mm.

Rocky areas, sagebrush. Flowering May–Aug. 700–2500 m. BW, Col, Lava. WA. Native.

Erigeron compositus Pursh (p. 254)

Fernleaf fleabane, dwarf mountain fleabane, cutleaf daisy

Erigeron compositus Pursh var. *compositus*, *Erigeron compositus* Pursh var. *glabratus* Macoun

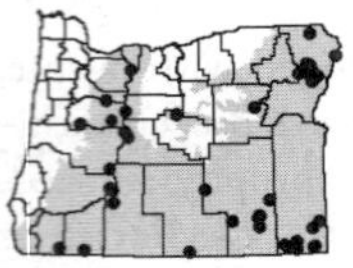

Plants perennial, 5–20 cm; taprooted with branching caudices. **Stems** erect, glabrous to sparsely hirsute and becoming more so distally, usually densely glandular. **Basal leaves** persistent, spatulate to obovate, 5–60 × 5–20 mm, margins (1)2–3-ternately lobed or dissected, surfaces moderately to densely strigose or hispid, usually minutely glandular. **Cauline leaves** 5–20 × 1–2 mm, abruptly reduced to linear bracts. **Heads** 1, radiate, sometimes disciform. **Involucres** 2–8 × 6–18 mm. **Phyllaries** in 2–3 series, usually with dark purple or green medial area, surfaces sparsely to densely hirsute, usually minutely glandular. **Ray florets** 20–60+, sometimes missing or reduced to tubes, white to pink to purple; rays 0–8 × 0–1.5 mm. **Disc florets** corollas 3–5 mm. **Fruits** 2–3 mm, densely strigose; inner pappi of numerous barbellate bristles. **2*n***=18, 36, 45, 54.

Rocky slopes, ridges. Flowering May–Sep. 600–3000 m. BR, BW, Casc, ECas, Owy, Sisk. CA, ID, NV, WA; north to AK, northeast to Greenland, east to SD, southeast to NM; Asia. Native.

Individuals with once-ternately lobed leaves have been separated as variety *glabratus*, but there does not seem to be any geographic pattern to this classification. Polyploids are known within this species and reduction of rays is associated with increases in ploidy.

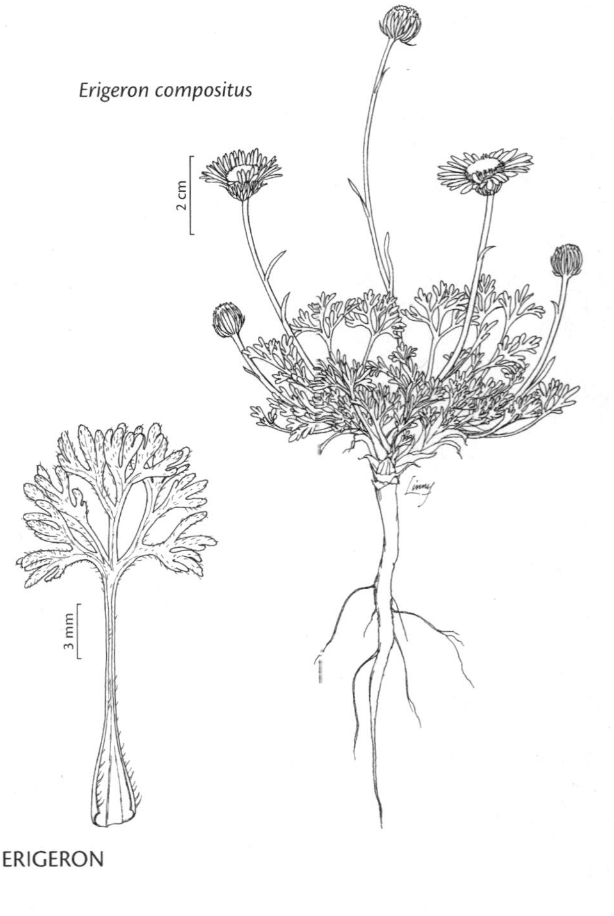

ERIGERON

Erigeron corymbosus Nuttall

Longleaf fleabane, foothill daisy

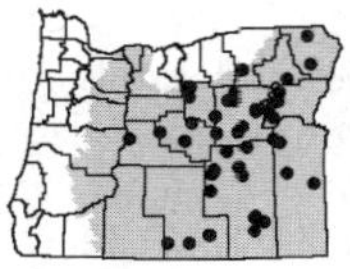

Plants perennial, 15–40 cm; taprooted, often with slender, branching caudices. **Stems** erect or ascending, moderately hirsute to villous, eglandular. **Basal leaves** usually persistent, linear to narrowly oblanceolate, 50–200 × 2–10 mm, margins entire, 1–3 prominent veins, tips usually acute, surfaces moderately to densely strigose, eglandular. **Cauline leaves** linear, 15–60 × 1–5 mm, gradually reduced distally. **Heads** 1–5(10) in corymb-like arrays, radiate. **Involucres** 4–7 × 8–14 mm. **Phyllaries** in 2–3 series, usually with dark medial area, surfaces moderately to densely hirsute to villous, occasionally glandular. **Ray florets** 30–60, white to purple; rays 2.5–14 × 1–1.5 mm. **Disc florets** corollas 4–5 mm. **Fruits** 2–3 mm, sparsely strigose; inner pappi of numerous barbellate bristles. **2*n***=18.

Dry open hillsides, sagebrush, pine forests. Flowering Jun–Aug. 200–2500 m. BR, BW, Casc, ECas, Lava, Owy. ID, WA; north to British Columbia, northeast to MT, east to WY, southeast to UT. Native.

Erigeron coulteri Porter

Coulter's fleabane

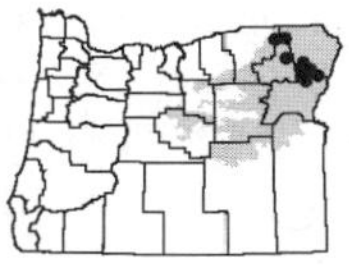

Plants perennial, 15–55 cm; roots fibrous, rhizomatous, occasionally with branching caudices. **Stems** erect, sparsely villous, eglandular. **Basal leaves** occasionally withering by flowering, broadly oblanceolate-elliptic, 40–150 × 10–25 mm, margins entire or occasionally with poorly developed teeth, surfaces sparsely strigose or hirsute, eglandular. **Cauline leaves** lanceolate to elliptic, 10–80 × 2.5–25(30) mm, gradually reduced distally, bases clasping, margins entire or with minute teeth. **Heads** 1(3), radiate. **Involucres** 5–8 × 11–17 mm. **Phyllaries** in 2–3 series, surfaces moderately to densely hirsute or villous, eglandular to minutely glandular. **Ray florets** 60–110, white; rays 8–15 × 0.5–1 mm. **Disc florets** corollas 3–5 mm. **Fruits** 1.5–2 mm, sparsely strigose; inner pappi of numerous barbellate bristles. **2*n***=18.

Moist coniferous forests, wet meadows. Flowering Jun–Aug. 1100–2000 m. BW. CA, ID, NV; northeast to MT, east to WY, southeast to NM. Native.

Erigeron decumbens Nuttall

Willamette fleabane, Willamette Valley daisy

Erigeron decumbens Nutt. ssp. *decumbens*, *Erigeron decumbens* Nutt. var. *decumbens*

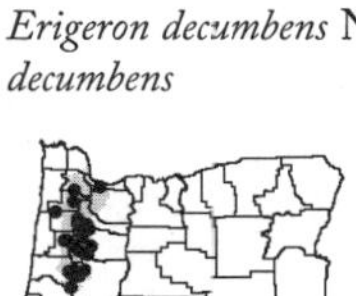

Plants perennial, 15–55 cm; taprooted, with upright, branching caudices. **Stems** few to many, erect or decumbent, remotely to moderately strigillose, eglandular. **Basal leaves** usually withering by flowering, linear to narrowly oblanceolate, 30–120 × 3–5 mm, margins entire, surfaces sparsely strigillose to minutely strigose, eglandular. **Cauline leaves** linear to narrowly oblanceolate, 10–100 × 1–3 mm, gradually reduced distally, margins entire, usually 3-veined, surfaces lightly to moderately strigose. **Heads** 1–5 in corymb-like arrays, radiate. **Involucres** 3–5 × 7–15 mm. **Phyllaries** in 2–3 series, usually with dark medial area, moderately to densely hirsute or strigose, eglandular. **Ray florets** 25–50, white to lavender; rays 5–8(12) × 1–2 mm. **Disc florets** corollas 3–4 mm. **Fruits** 1–2 mm, sparsely to moderately strigose; inner pappi of numerous barbellate bristles.

Prairies. Flowering May–Jul. 50–300 m. WV. Native.

Erigeron decumbens is federally listed as an endangered species. It is an Oregon endemic known only from scattered prairie remnants in the Willamette Valley where it inhabits seasonally flooded bottomland prairies as well as well-drained upland prairies. It has likely been extirpated from parts of its former range in Clackamas, Washington, and Yamhill counties.

Erigeron disparipilus Cronquist

White cushion fleabane, Snake River daisy

Erigeron davisii (Cronquist) G.L. Nesom, *Erigeron engelmannii* A. Nelson, *Erigeron engelmannii* A. Nelson ssp. *davisii* Cronquist, *Erigeron engelmannii* A. Nelson var. *davisii* (Cronquist) Cronquist

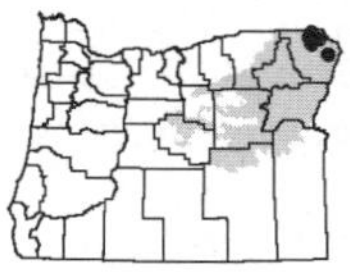

Plants perennial, 8–22 cm; taprooted, with branching caudices. **Stems** erect, sparsely to moderately hirsute to strigose, eglandular. **Basal leaves** persistent, narrowly spatulate to narrowly oblanceolate-elliptic, 30–80 × 3–5 mm, margins entire, with 1 prominent vein, surfaces moderately to densely strigose, eglandular. **Cauline leaves** few, linear, 10–40 × 1–3 mm, reduced distally. **Heads** usually 1, radiate. **Involucres** 3.5–8 × 12–18 mm. **Phyllaries** in 2–3 series, usually with dark green medial area with or without orange stripe, surfaces moderately to densely hirsute or strigose, moderately to densely glandular. **Ray florets** 25–50, white or cream, occasionally fading to pink; rays 4–8 × 1–3 mm. **Disc florets** corollas 3–4 mm. **Fruits** 1.5–3 mm, sparsely to moderately strigose; inner pappi of numerous barbellate bristles.

Open rocky areas, ridges. Flowering Jun–Jul. 1100–1800 m. BW. ID, WA. Native.

Some floras recognize plants with stem hairs in the same orientation (versus mixed) as *E. davisii*. The ranges and habitat of these taxa are, however, sympatric. Until further evidence shows otherwise, there is no compelling reason to recognize *E. davisii* as distinct from *E. disparipilus*.

Erigeron divergens Torrey & A. Gray

Spreading fleabane, diffuse daisy

Erigeron divergens Torr. & A. Gray var. *divergens*

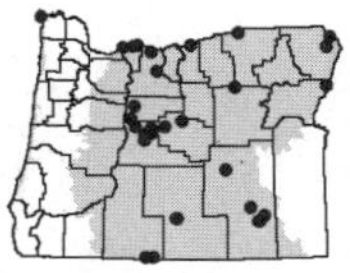

Plants annual or short-lived perennial, 12–50+ cm; taprooted, with short, branching caudices. **Stems** erect, ascending to decumbent, densely pilose to sparsely hirsute, minutely glandular at least distally. **Basal leaves** persistent or withering by flowering, spatulate, 20–80 × 5–10 mm, margins entire or with 1–2 coarse teeth, surfaces moderately to densely strigose, usually eglandular. **Cauline leaves** numerous and well developed, elliptic, obovate, or linear, 5–30 × 1–4 mm, usually slightly reduced distally, surfaces densely strigose to villous. **Heads** 1–20+ in corymb-like arrays, radiate. **Involucres** 2–4 × 6–10 mm. **Phyllaries** in 2–3 series, usually with dark green or brown medial area and lighter colored stripe, surfaces moderately to densely hirsute or strigose, moderately to densely glandular. **Ray florets** 75–150, light purple to lavender; rays 2–5 × 0.25–0.6 mm. **Disc florets** corollas 2–3 mm.

Fruits 0.75–1 mm, glabrous to very sparsely strigose; inner pappi of numerous barbellate bristles. **2*n***=18, 27, 36.

Dry open areas, roadsides, streambanks. Flowering May–Sep. 50–2200 m. BR, BW, Casc, Col, ECas, Est, Lava. CA, ID, WA; north to British Columbia, northeast to Alberta, east to SD, southeast to TX, south to Mexico. Native.

Erigeron eatonii A. Gray

Eaton's fleabane

Plants perennial, 4–28 cm; taprooted, caudices branching or not. **Stems** erect to ascending or decumbent, usually strigose to occasionally villous, becoming woolly distally, sometimes minutely glandular. **Basal leaves** persistent, linear to narrowly oblanceolate, 20–150 × 2–10 mm, margins entire, usually with 3 distinct basal veins, tips usually rounded to acute or acuminate, surfaces strigose to hirsute, eglandular. **Cauline leaves** gradually to abruptly reduced distally, sometimes becoming linear and bract-like. **Heads** 1–3 in scape- or corymb-like arrays, radiate. **Involucres** 4–6 × 8–15 mm. **Phyllaries** in 2–3 series, surfaces sparsely sericeous to densely villous, sometimes minutely glandular. **Ray florets** 20–45, white to purple; rays 3–8 × 0.8–2.5 mm. **Disc florets** corollas 3–4 mm. **Fruits** 2–3 mm, glabrate to remotely strigose; inner pappi of numerous barbellate bristles.

Western United States. 6 varieties; 3 varieties treated in *Flora*.

Key to *Erigeron eatonii* varieties:

1. Phyllaries moderately to densely glandular . . . . ***E. e.*** var. ***lavandulus***

1' Phyllaries eglandular to sparsely glandular.

  2. Heads 1–4; phyllary hairs 0.3–0.8 mm . . . . ***E. e.*** var. ***plantagineus***

  2' Heads 1(2); phyllary hairs 1–2 mm . . ***E. e.*** var. ***villosus***

Erigeron eatonii A. Gray var. ***lavandulus*** Strother & Ferlatte

Eaton's lavender fleabane

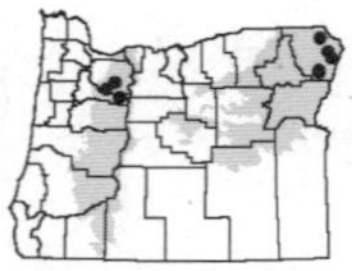

Heads (1)2–4(6). **Phyllaries** sparsely to moderately villous, moderately to densely glandular, hairs 0.5–1.5 mm. **2*n***=18.

Meadows, sagebrush. Flowering May–Aug. 1000–2000 m. BW, Casc. ID. Native.

Erigeron eatonii A. Gray var. ***plantagineus*** (Greene) Cronquist

Eaton's plantain-leaved fleabane, plantain-leaved daisy

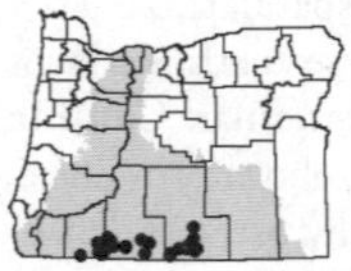

Heads 1–4. **Phyllaries** sparsely to moderately hirsutulous, eglandular to sparsely glandular, hairs 0.3–0.8 mm. **2*n***=18.

Meadows, sagebrush, serpentine. Flowering May–Jul. 1000–2200 m. BR, Casc, ECas, Sisk. CA. Native.

Erigeron eatonii A. Gray var. ***villosus*** (Cronquist) Cronquist

Eaton's shaggy fleabane

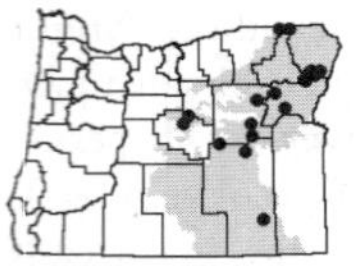

Heads 1(2). **Phyllaries** moderately to densely villous, eglandular to sparsely glandular, hairs 1–2 mm.

Meadows, sagebrush, pine forests. Flowering Jun–Aug. 1200–2600 m. BR, BW. ID, WA. Native.

Erigeron elegantulus Greene

Volcanic fleabane, dwarf blue fleabane

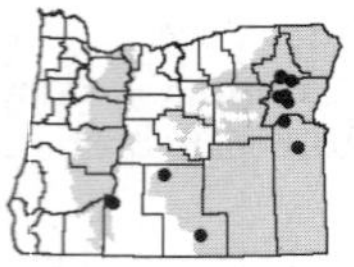

Plants perennial, 4–10 cm; taprooted, with branching caudices. **Stems** erect to basally ascending, sparsely to moderately strigose, eglandular. **Basal leaves** persistent, linear to filiform, 20–50 × 0.5–1 mm, white, bases sheathing, margins entire, tips round to acute or acuminate, surfaces sparsely to moderately strigose. **Cauline leaves** abruptly reduced and usually restricted to basal ⅓ of stem, bases sheathing. **Heads** 1, radiate. **Involucres** 3–5 × 8–11 mm. **Phyllaries** in 3–4 unequal series, surfaces sparsely strigose, eglandular. **Ray florets** 15–30, lavender to purple; rays 3–8 × 0.7–1.3 mm. **Disc florets** corollas 2.5–3.5 mm. **Fruits** 1–2 mm, nearly glabrous to sparsely strigose; inner pappi of numerous barbellate bristles. **2*n***=27.

Rocky areas, sagebrush, coniferous forests. Flowering Jun–Aug. 700–2500 m. BR, BW, Casc, Owy. CA. Native.

Erigeron filifolius (Hooker) Nuttall (p. 257)

Threadleaf fleabane

Erigeron filifolius (Hook.) Nutt. var. *filifolius*, *Erigeron filifolius* (Hook.) Nutt. var. *robustior* M. Peck

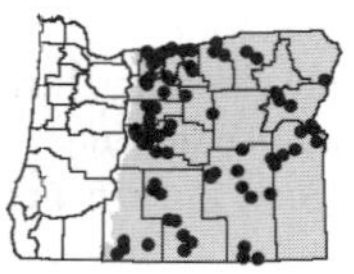

Plants perennial, 9–40 cm; taprooted, with branching caudices. **Stems** erect, moderately to densely strigose, eglandular. **Basal leaves** persistent, linear to filiform, 20–80 × 0.5–2 mm, margins entire, tips rounded to acute or acuminate, surfaces moderately to densely strigose. **Cauline leaves** gradually reduced distally or not. **Heads** 1–5 in corymb-like arrays, radiate. **Involucres** 4–6 × 7–11 mm. **Phyllaries** in 2–4 series, surfaces moderately to densely strigose or villous, sometimes nearly glabrous distally, densely minutely glandular. **Ray florets** 20–40, white to purple; rays 3–9 × 0.7–1.5 mm. **Disc florets** corollas 2–4 mm. **Fruits** 1–3 mm, sparsely strigose; inner pappi of numerous barbellate bristles. **2*n***=18.

Rocky areas, sagebrush, juniper and pine forests. Flowering Apr–Aug. 50–2200 m. BR, BW, Col, ECas, Lava, Owy. CA, ID, NV, WA; north to British Columbia, northeast to MT, southeast to UT. Native.

If subspecific taxa were recognized, Oregon specimens would correspond to the typical variety.

Erigeron foliosus Nuttall var. ***confinis*** (Howell) Jepson
Leafy fleabane

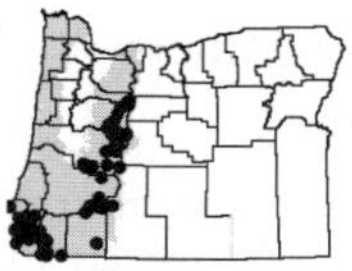

Plants perennial, 10–30 cm; taprooted, with woody, branching caudices. **Stems** erect or ascending, often widely spreading, glabrate to moderately strigose, hairs usually appressed. **Basal leaves** usually withering early. **Cauline leaves** narrowly oblanceolate, occasionally linear, 10–40 × 1–3 mm, surfaces glabrate to moderately strigose. **Heads** 1–7 in loose, corymb-like arrays, radiate. **Involucres** 4–8 × 8–16 mm. **Phyllaries** in 3–4 series, surfaces glabrous to weakly hirsute, usually densely glandular. **Ray florets** 20–30, lavender to purple; rays 3–8 × 0.5–2 mm. **Disc florets** corollas 3–6 mm. **Fruits** 1.5–2.5 mm, sparsely strigose; inner pappi of numerous barbellate bristles.

Rocky areas. Flowering May–Aug. 0–2000 m. Casc, CR, Est, Sisk. CA. Native.

Erigeron glacialis (Nuttall) A. Nelson var. ***glacialis*** (p. 257)
Subalpine fleabane, peregrine fleabane

Erigeron peregrinus (Pursh) Greene, *Erigeron peregrinus* (Pursh) Greene var. *angustifolius* (A. Gray) Cronquist, *Erigeron peregrinus* (Pursh) Greene ssp. *callianthemus* (Greene) Cronquist, *Erigeron peregrinus* (Pursh) Greene var. *callianthemus* (Greene) Cronquist, *Erigeron peregrinus* (Pursh) Greene var. *eucallianthemus* Cronquist, *Erigeron peregrinus* (Pursh) Greene var. *scaposus* (Torr. & A. Gray) Cronquist

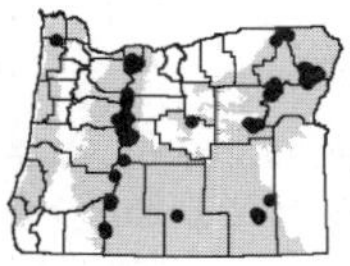

Plants perennial, 20–50 cm; fibrous-rooted, often rhizomatous, caudices usually simple. **Stems** erect, nearly glabrous, or sparsely hirsute proximally to densely strigillose or pilose distally, eglandular. **Basal leaves** persistent, linear-oblanceolate to broadly oblanceolate, 30–150 × 8–30 mm, margins entire, surfaces glabrous, eglandular. **Cauline leaves** lanceolate to oblanceolate, 10–80 × 3–30 mm, reduced distally, bases occasionally clasping, surfaces glabrous. **Heads** 1(4), radiate. **Involucres** 5–8 × 12–15 mm. **Phyllaries** in 2–3 series, with dark medial area, glabrous, densely glandular. **Ray florets** 30–60, lavender to purple; rays 5–9 × 1.5–4 mm. **Disc florets** corollas 4–6 mm. **Fruits** 2–4 mm, sparsely to moderately strigose; inner pappi of numerous barbellate bristles.

Meadows, streamsides, lakeshores, bogs. Flowering Jun–Sep. 800–3200 m. BR, BW, Casc, CR, ECas. CA, ID, NV, WA; north to AK, northeast to Alberta, east to WY, southeast to NM. Native.

Erigeron glaucus Ker Gawler
Seaside fleabane, seaside daisy

Plants perennial, 3–30 cm; fibrous-rooted, rhizomatous, caudic,es usually simple, occasionally branching. **Stems** decumbent, procumbent or ascending, sparsely to moderately hirsute, usually minutely glandular. **Basal leaves** persistent, obovate to broadly spatulate, 15–40 × 5–25 mm, margins entire, surfaces usually glabrous and eglandular. **Cauline leaves** similar to basal, 7–30 × 3–15 mm, gradually reduced distally. **Heads** 1(3), radiate. **Involucres** 5–10 × 10–21 mm. **Phyllaries** in 3–4 series, surfaces moderately to densely villous, sparsely to moderately glandular. **Ray florets** 70–100, lavender to purple; rays 6–9 × 1–2 mm. **Disc florets** corollas 4–6 mm. **Fruits** 2–3 mm, sparsely to moderately strigose; inner pappi of numerous barbellate bristles. **2*n***=18.

Coastal dunes, bluffs. Flowering May–Aug. 0–500 m. Est. CA. Native.

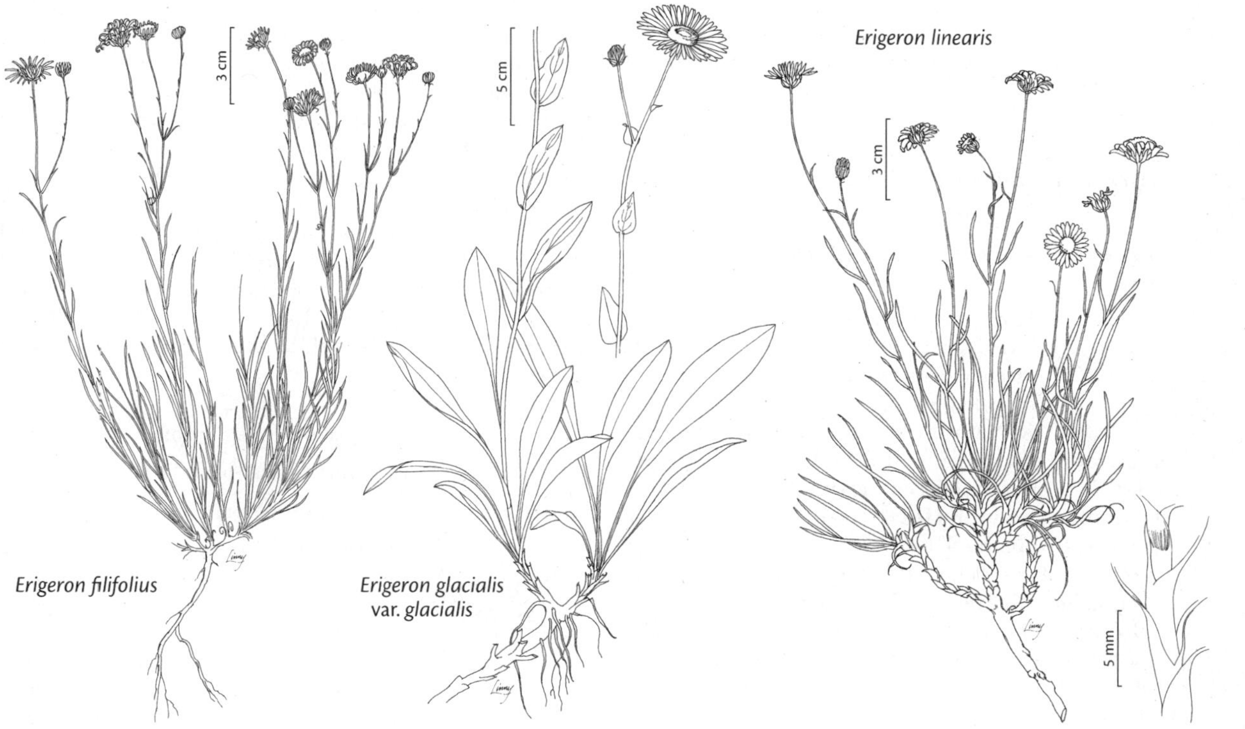

ERIGERON

Erigeron howellii (A. Gray) A. Gray

Howell's fleabane

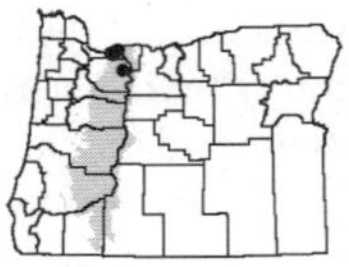

Plants perennial, 20–45 cm; fibrous-rooted, rhizomatous, caudices branching or not. **Stems** erect, glabrous to sparsely and minutely strigillose, eglandular. **Basal leaves** persistent, broadly spatulate, 40–150 × 10–35 mm, margins entire or with few large teeth, surfaces glabrous, eglandular. **Cauline leaves** lanceolate to elliptic, 15–80 × 10–35 mm, gradually reduced distally, margins entire or with few coarse teeth, surfaces often ciliate or glandular. **Heads** 1, radiate. **Involucres** 5–9 × 15–23 mm. **Phyllaries** in 2–3 series, with prominent central veins, surfaces glabrous, densely glandular. **Ray florets** 25–50, white; rays 8–20 × 2.5–4 mm. **Disc florets** corollas 4–5 mm. **Fruits** 2.5–3 mm, sparsely to moderately strigose; inner pappi of numerous barbellate bristles.

Rocky slopes, open woods. Flowering May–Jul. 100–1500 m. Casc. WA. Native.

Erigeron inornatus (A. Gray) A. Gray var. *inornatus*

Lava rayless fleabane, California rayless daisy, unadorned fleabane

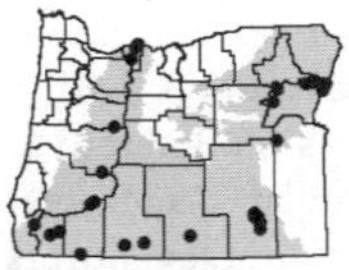

Plants perennial, 20–65 cm; taprooted, caudices branching. **Stems** decumbent to erect, glabrous or becoming strigose basally, eglandular. **Basal leaves** withering by flowering. **Cauline leaves** linear to narrowly oblong or elliptic, 10–60 × 2–8 mm, surfaces glabrous to moderately strigose. **Heads** 2–15+ in corymb-like arrays, discoid. **Involucres** 4–7 × 8–13 mm. **Phyllaries** in 3–5 unequal series, usually with a dark medial area and orange-colored stripe, margins usually membranous, surfaces glabrous, eglandular. **Ray florets** 0. **Disc florets** corollas 4–5 mm. **Fruits** 2–3 mm, nearly glabrous to sparsely strigose; inner pappi of numerous barbellate bristles. $2n$=18.

Roadsides, rocky hillsides, dry meadows, prairies. Flowering Jun–Aug. 100–2300 m. BR, BW, Casc, ECas, Sisk. CA, NV, WA. Native.

Erigeron klamathensis (G.L. Nesom) G.L. Nesom

Klamath fleabane

Erigeron breweri A. Gray, *Erigeron breweri* A. Gray var. *klamathensis* G.L. Nesom

Plants perennial, 5–25 cm; taprooted, caudices branching. **Stems** erect to decumbent-ascending, hirsute, eglandular, hairs usually spreading and slightly reflexed. **Basal leaves** withering by flowering. **Cauline leaves** oblong to narrowly oblanceolate, 10–30 × 1–4 mm, slightly reduced distally or not, margins entire, surfaces moderately to densely hirsute, eglandular. **Heads** 1–4 in loose corymb-like arrays, radiate. **Involucres** 4–6 × 8–14 mm. **Phyllaries** in 2–3 unequal series, surfaces glabrous to very sparsely hirsute, densely minutely glandular. **Ray florets** 20–40, lavender to purple; rays 4–6 × 0.7–2 mm. **Disc florets** corollas 3.5–5 mm. **Fruits** 2–3 mm, sparsely to densely strigose; inner pappi of numerous barbellate bristles.

Rocky areas, ridges, crevices. Flowering Jun–Sep. 500–1500 m. Sisk. CA. Native.

This species is currently known only from the Siskiyou Mountains of southern Oregon and northern California.

Erigeron latus (A. Nelson & J.F. Macbride) Cronquist

Broad fleabane, Owyhee fleabane

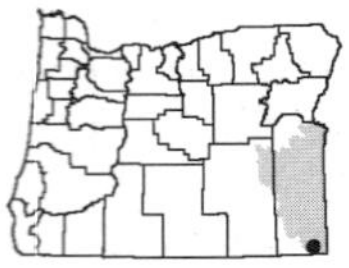

Plants perennial, 5–10 cm; taprooted, caudices branching. **Stems** erect, sparsely to moderately hirsute, densely minutely glandular. **Basal leaves** persistent, narrowly oblanceolate to spatulate, 15–50 × 1.3–4.5 mm, margins entire, tips rounded to acute or acuminate, surfaces moderately to densely hirsute to nearly hispid, minutely glandular. **Cauline leaves** few, similar to basal, usually abruptly reduced distally. **Heads** 1, radiate. **Involucres** 6–9 × 13–21 mm. **Phyllaries** in 2–3 series, with dark medial area, sometimes with orange stripe, surfaces sparsely to moderately hispid-hirsute, densely minutely glandular. **Ray florets** 20–35, lavender to purple; rays 6–11 × 1–2 mm. **Disc florets** 4–5 mm. **Fruits** 2–4 mm, glabrate to sparsely strigose; inner pappi of numerous barbellate bristles.

Open areas, sagebrush, streambanks. Flowering May–Jul. 1600–2000 m. Owy. ID, NV. Native.

Erigeron latus and *E. poliospermus* are very similar with the exception of the vestiture of the fruits. Further research is needed to determine if continued recognition of two distinct species is warranted.

Erigeron linearis (Hooker) Piper (p. 257)

Desert yellow fleabane, lineleaf fleabane

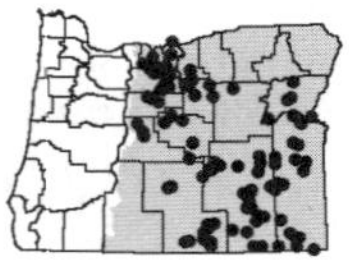

Plants perennial, 7–20 cm; taprooted, caudices branching. **Stems** erect or basally ascending, moderately to densely strigose, eglandular. **Basal leaves** persistent, linear to very narrowly oblanceolate, 15–80 × 0.5–2 mm, bases sheathing and light- or white-colored, margins entire, surfaces moderately to densely strigose, eglandular. **Cauline leaves** linear, 5–20 × 0.5–1 mm, usually reduced distally, bases sheathing. **Heads** 1(2), radiate. **Involucres** 5–8 × 9–13 mm. **Phyllaries** in 2–3 series, surfaces moderately to densely strigose, hirsute, or manicate, minutely glandular. **Ray florets** 20–40, bright yellow; rays 5–7 × 1–2 mm. **Disc florets** corollas 3–5 mm. **Fruits** 2–3 mm, sparsely strigose; inner pappi of numerous barbellate bristles. $2n$=18, 27, 36, 45.

Rocky areas, sagebrush, juniper forests. Flowering May–Jul. 200–2500 m. BR, BW, Col, ECas, Lava, Owy. CA, ID, NV, WA; north to British Columbia, northeast to MT, east to WY, southeast to UT. Native.

Erigeron lonchophyllus Hooker (p. 259)

Short-rayed fleabane, spear-leaved fleabane

Trimorpha lonchophylla (Hook.) G.L. Nesom

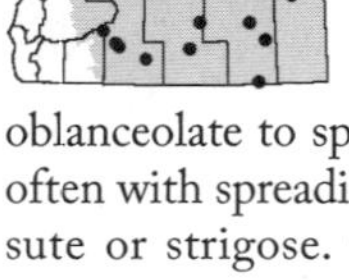

Plants biennial or short-lived perennial, 4–33 cm; usually fibrous-rooted, caudices simple. **Stems** erect, sparsely to densely hirsute, eglandular or occasionally with densely minute glands. **Basal leaves** persistent, oblanceolate to spatulate, 10–100 × 4–10 mm, margins entire, often with spreading cilia, surfaces glabrous to moderately hirsute or strigose. **Cauline leaves** narrowly elliptic, oblong,

ERIGERON

oblanceolate to linear, 10–100 × 1–5 mm, reduced distally, bases often clasping, surfaces glabrous to moderately hirsute or strigose. **Heads** 1–10 in raceme-like arrays, radiate. **Involucres** 4–7 × 5–11 mm. **Phyllaries** in 2–3 series, surfaces sparsely to densely hirsute, glandular or eglandular. **Ray florets** 40–100, white to pink; rays 2–5 × 0.2–0.4 mm. **Disc florets** corollas 3–5 mm. **Fruits** 1.5–2.5 mm, sparsely to moderately strigose; inner pappi of numerous barbellate bristles. ***2n***=18.

Floodplains, lakesides, gravel bars, roadsides, marshes, seeps. Flowering Jun–Aug. 800–2300 m. BR, BW, Casc, ECas, Lava, Owy. CA, ID, NV, WA; north to AK, northeast to Nunavut, east to Quebec, southeast to NM. Native.

Erigeron nivalis Nuttall (p. 259)

Northern daisy

Erigeron acris L., *Erigeron acris* L. var. *debilis* A. Gray, *Trimorpha acris* (L.) Gray var. *debilis* (A. Gray) G.L. Nesom

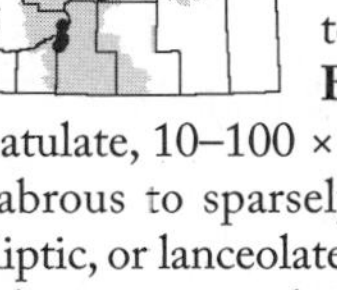

Plants biennial or short-lived perennial, 4–20 cm; fibrous-rooted to weakly taprooted, caudices branched. **Stems** erect, sparsely to moderately hirsute, minutely glandular. **Basal leaves** persistent, oblanceolate to spatulate, 10–100 × 3–15 mm, margins entire, ciliate, surfaces glabrous to sparsely strigose. **Cauline leaves** oblanceolate, elliptic, or lanceolate, 5–50 × 1–5 mm, reduced distally, surfaces glabrous or sparsely strigose. **Heads** 1–10 in corymb-like arrays, radiate. **Involucres** 4–7 × 7–13 mm. **Phyllaries** in 2–3 series, surfaces nearly glabrous to sparsely hirsute, occasionally villous basally, densely glandular. **Ray florets** in 2 series; outer of 30–70 erect, filiform rays, 0.3–0.5 × 3–6 mm, white to light pink; inner of tubular florets, fewer than outer series. **Disc florets** corollas 4–6 mm. **Fruits** 1.5–2.5 mm, sparsely to moderately strigose; inner pappi of numerous barbellate bristles. ***2n***=18.

Gravelly areas, mossy rocks, lava beds, talus slopes, cliffs. Flowering May–Aug. 1500–2500 m. BW, Casc, ECas. CA, ID, WA; north to AK, northeast to Northwest Territories, east to WY, southeast to NM. Native.

Erigeron oreganus A. Gray

Gorge fleabane, Gorge daisy, Oregon fleabane

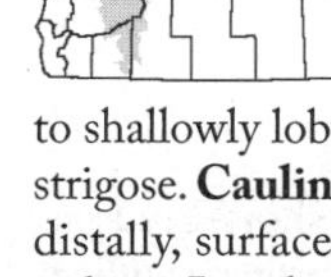

Plants perennial, 5–15 cm; taprooted, caudices unbranched. **Stems** decumbent to erect, hirsute with unequal hairs, sparsely glandular. **Basal leaves** persistent, obovate, 15–90 × 5–20 mm, margins coarsely serrate to shallowly lobulate, surfaces sparsely to moderately hirsute to strigose. **Cauline leaves** obovate to elliptic, little or not reduced distally, surfaces sparsely to moderately hirsute. **Heads** 1–4, radiate. **Involucres** 4–7 × 9–15 mm. **Phyllaries** in 2–3 series, pale green with dark green medial stripe, surfaces remotely hirsute, strongly stipitate-glandular. **Ray florets** 30–60, pink to purple; rays 4–5 × 0.5–1 mm. **Disc florets** corollas 3–5 mm. **Fruits** 2–3 mm, moderately strigose; inner pappi of several unbarbed to weakly barbellate twisted bristles.

Shady cliffs, rocky slopes, crevices. Flowering May–Aug. 100–400 m. Casc. WA. Native.

This uncommon species is found only in the Columbia River Gorge.

Erigeron peregrinus (Pursh) Greene var. *peregrinus*

Wandering fleabane

Erigeron peregrinus (Pursh) Greene ssp. *peregrinus*

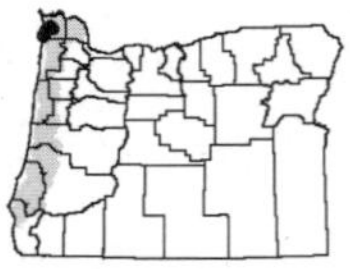

Plants perennial, 15–50 cm; fibrous-rooted, rhizomatous, caudices usually unbranched. **Stems** erect or basally ascending, basally glabrate or remotely villous, becoming densely villous to manicate distally, eglandular. **Basal leaves** persistent, broadly elliptic, oblanceolate, or spatulate, 20–150 × 8–30 mm, margins usually with a few coarse teeth and densely ciliate, surfaces glabrous to remotely hirsute, eglandular. **Cauline leaves** lanceolate, 10–50 × 5–15 mm, gradually reduced distally, bases clasping, margins entire or with a few coarse teeth and densely ciliate, surfaces usually glabrous. **Heads** 1, radiate. **Involucres** 7–9 × 17–24 mm. **Phyllaries** in 2–3 series, margins usually ciliate, surfaces sparsely to moderately hirsute to villous, usually glandular distally with stalked glands. **Ray florets** 30–60, white to purple; rays 8–15 × 1.5–4 mm. **Disc florets** corollas 3.5–4.5 mm. **Fruits** 2.5–3 mm, strigose; inner pappi of numerous barbellate bristles.

Meadows, clearings. Flowering Jun–Aug. 600–1000 m. CR. WA; north to AK, northeast to Alberta. Native.

In Oregon, this species is currently known only from isolated peaks in Clatsop and Tillamook counties.

Erigeron petrophilus Greene var. *petrophilus*

Rock-loving fleabane, cliff fleabane

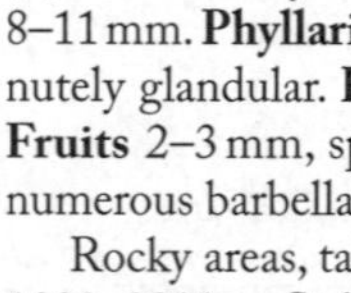

Plants perennial, 5–25 cm; taprooted, caudices branched. **Stems** decumbent-ascending or erect, sparsely hirsute to densely lanate, densely minutely glandular. **Basal leaves** withering by flowering. **Cauline leaves** narrowly oblanceolate-elliptic to oblong, 10–30 × 1–4 mm, surfaces sparsely to moderately hirsute to densely lanate or manicate, usually densely minutely glandular. **Heads** (1)2–5 in corymb-like arrays, discoid. **Involucres** 6–8 × 8–11 mm. **Phyllaries** in 3–5 series, surfaces nearly glabrous, minutely glandular. **Ray florets** 0. **Disc florets** corollas 3–5 mm. **Fruits** 2–3 mm, sparsely strigose to glabrous; inner pappi of numerous barbellate bristles. **2*n***=18.

Rocky areas, talus, crevices, serpentine. Flowering Jun–Sep. 1800–2300 m. Sisk. CA. Native.

Erigeron philadelphicus Linnaeus (p. 259)

Philadelphia fleabane

Erigeron philadelphicus L. var. *philadelphicus*

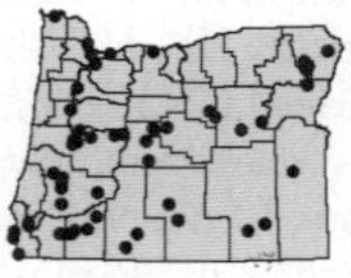

Plants annual, biennial, or short-lived perennial, 10–80 cm; roots fibrous, caudices usually simple. **Stems** erect, sparsely to moderately hirsute to lanate, minutely glandular. **Basal leaves** withering by flowering, oblanceolate to nearly spatulate, 20–170 × 5–35 mm, margins nearly entire, coarsely serrate, lobulate, or shallowly dentate, surfaces slightly to moderately strigose or hirsute, eglandular. **Cauline leaves** oblanceolate to lanceolate to elliptic or oblong, 5–130 × 3–25 mm, distally reduced, bases clasping, surfaces eglandular. **Heads** (1)3–many in corymb-like arrays, radiate. **Involucres** 3–6 × 6–15 mm. **Phyllaries** in 2–3 series, surfaces sparsely to moderately hirsute, minutely glandular. **Ray florets** 150–250, white to pink; rays 2–7 × 0.2–0.5 mm. **Disc florets** corollas 2–3 mm. **Fruits** 1–1.5 mm, sparsely to moderately strigose; inner pappi of numerous barbellate bristles. **2*n***=18.

Disturbed areas, streambanks, fields, meadows, forest openings. Flowering Apr–Sep. 0–1900 m. All ecoregions. CA, ID, NV, WA; nearly worldwide. Native.

Erigeron poliospermus A. Gray var. *poliospermus* (p. 260)

Purple cushion fleabane, hairy-seeded daisy

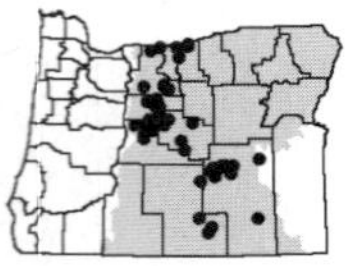

Plants perennial, 2–20 cm; taprooted, caudices usually branched, sometimes simple. **Stems** erect, moderately to densely hirsute, minutely glandular. **Basal leaves** persistent, narrowly oblanceolate to spatulate, 10–60 × 1–4 mm, margins entire, tips rounded to acute or acuminate, surfaces densely hispid-hirsute to hispid-strigose. **Cauline leaves** few, similar to basal, abruptly reduced. **Heads** 1, radiate. **Involucres** 5–8 × 9–18 mm. **Phyllaries** in ± 2–3 series, with dark green medial area, occasionally with orange stripe, surfaces sparsely to densely hirsute, minutely glandular. **Ray florets** 25–50, lavender to dark purple; rays 3–10 × 0.5–2 mm. **Disc florets** corolla 4–5 mm. **Fruits** 2–3 mm, densely sericeous; inner pappi of numerous barbellate bristles.

Rocky, gravelly or sandy areas, sagebrush, juniper or pine woods. Flowering Apr–Jul. 50–2000 m. BR, BW, Col, ECas, Lava. ID, WA; north to British Columbia. Native.

See also *E. latus*.

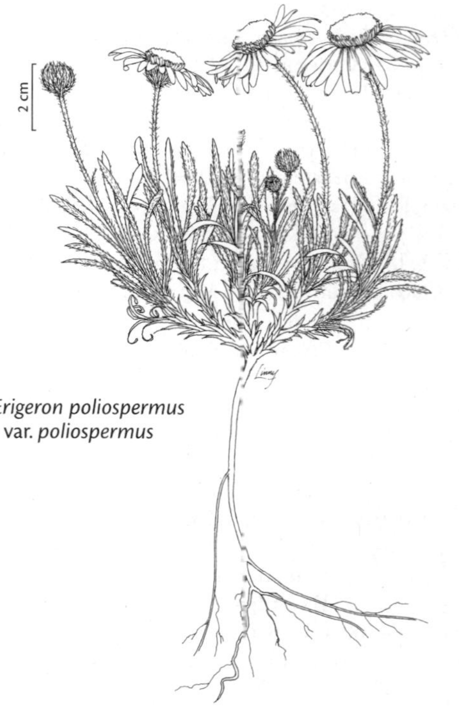

Erigeron poliospermus var. *poliospermus*

ERIGERON

Erigeron pumilus Nuttall var. ***intermedius*** Cronquist

Shaggy fleabane

Erigeron pumilus Nutt. var. *euintermedius* Cronquist, *Erigeron pumilus* Nutt. var. *gracilior* Cronquist, *Erigeron pumilus* Nutt. ssp. *intermedius* Cronquist

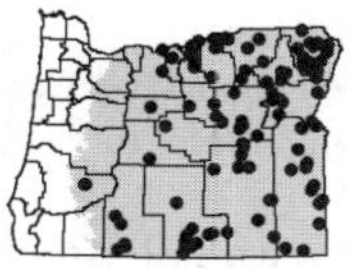

Plants perennial, 7–40 cm; taprooted, caudices branching. **Stems** erect, moderately to densely hispid, often glandular. **Basal leaves** persistent, narrowly oblanceolate, 20–90+ × 1–5 mm, margins entire and hispid ciliate, surfaces moderately to densely hispid. **Cauline leaves** oblanceolate, becoming narrowly elliptic or linear, gradually reduced distally. **Heads** 1–5+ in scape- to corymb-like arrays, radiate. **Involucres** 4–6 × 7–13 mm. **Phyllaries** in 2–4 series, yellow-green with orange to purplish medial stripe, surfaces with scattered to dense hispid hairs. **Ray florets** 40–80, white to purple or bluish; rays 4–9 × 0.5–1 mm. **Disc florets** 2–5 mm. **Fruits** 1–2 mm, strigose; inner pappi of 10–27 barbellate bristles. **2*n***=18, 36.

Dry plains, rocky slopes, sagebrush, grassy slopes, openings in pine forests. Flowering May–Aug. 50–2400 m. BR, BW, Casc, Col, ECas, Lava, Owy. CA, ID, NV, WA; north to British Columbia, northeast to MT, east to WY, southeast to UT. Native.

Erigeron simplex Greene

One-stem fleabane

Erigeron grandiflorus Hook.

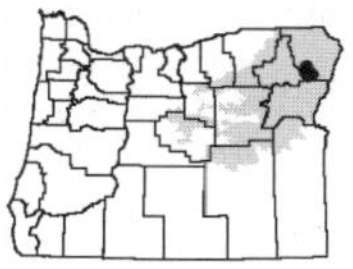

Plants perennial, 3–15 cm; fibrous-rooted, rhizomatous, caudices branched or not. **Stems** erect, sparsely to moderately villous, eglandular. **Basal leaves** persistent, oblanceolate to spatulate, 50–350 × 30–90 mm, margins entire, sometimes moderately ciliate, surfaces glabrous to glabrate. **Cauline leaves** obovate to linear, abruptly reduced distally, surfaces usually hirsute. **Heads** 1, radiate. **Involucres** 3–7 × 8–16 mm. **Phyllaries** in 2 series, margins usually ragged, surfaces moderately to densely villous. **Ray florets** 30–50+, lavender to purple; rays 5–10 × 0.6–2.2 mm. **Disc florets** 2–3 mm. **Fruits** 1–2 mm, sparsely strigose; inner pappi of numerous barbellate bristles.

Rocky areas, meadows. Flowering Jul–Aug. 2400–3000 m. BW. ID, NV; northeast to MT, east to WY, southeast to CO, NM. Native.

Erigeron speciosus (Lindley) de Candolle

Showy fleabane, showy daisy

Erigeron speciosus (Lindl.) DC. var. *speciosus*

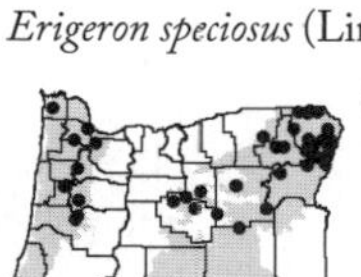

Plants perennial, 15–75 cm; rhizomatous, fibrous-rooted, caudices branching. **Stems** erect, surfaces glabrous to rarely remotely hirsute, sometimes minutely glandular distally. **Basal leaves** usually withering by flowering, oblanceolate to spatulate. **Cauline leaves** lanceolate, 20–80 × 5–15 mm, gradually reduced distally, bases ± clasping, margins entire, usually long-ciliate, surfaces glabrous, rarely remotely strigose. **Heads** 1–5+ in corymb-like arrays, radiate. **Involucres** 3–7 × 11–16 mm. **Phyllaries** in 2–3+ series, usually with an orange stripe, surfaces basally hirsute to villous, occasionally minutely glandular. **Ray florets** 40–115+, lavender to purple; rays 7–11 × 0.7–1.5 mm. **Disc florets** 2–3 mm. **Fruits** 2–3 mm, glabrous to sparsely strigose; inner pappi of numerous barbellate bristles.

Conifer forests, moist slopes, riverbanks, wet meadows. Flowering Jun–Sep. 50–2500 m. BR, BW, CR, WV. ID, NV, WA; north to British Columbia, northeast to Alberta, east to NE, southeast to NM. Native.

Erigeron stanselliae K.L. Chambers (p. 261)

Stansell's fleabane

Plants perennial, 7–25 cm; tap-rooted, caudices few-branched. **Stems** erect to decumbent, glabrous to sparsely strigose, sometimes minutely glandular. **Basal leaves** linear to narrowly oblanceolate, 40–120 × 2–5 mm, margins entire, tips rounded to acute or acuminate, surfaces glabrous to sparsely strigose. **Cauline leaves** linear to narrowly elliptic, 5–40 × 1–3 mm, gradually or abruptly reduced distally, surfaces glabrous to loosely strigose. **Heads** 1–2(4), radiate. **Involucres** 5–7 × 5–11 mm. **Phyllaries** in 2–3 series, surfaces sparsely to densely pilose or strigose, minutely glandular. **Ray florets** 14–36, white to pinkish; rays 6–10 × 1–1.5 mm. **Disc florets** 3–4 mm. **Fruits** 3–4 mm, sparsely to densely strigose; inner pappi of numerous barbellate bristles.

Serpentine areas. Flowering Jun–Jul. 700–900 m. Sisk. Native.

First described in 2011, this serpentine endemic has thus far only been collected a few times in Curry County. The species is named in honor of noted Oregon botanist, Veva Stansell.

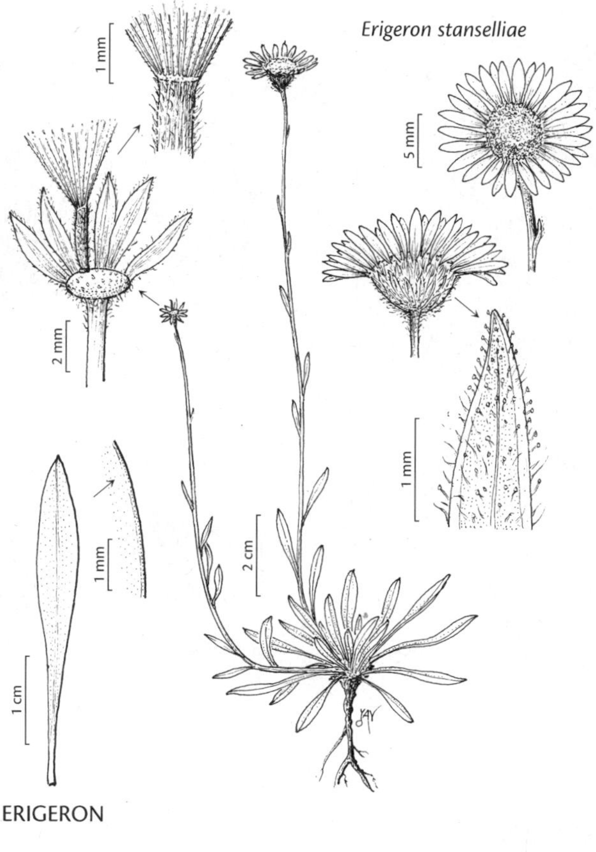

ERIGERON

Erigeron strigosus Muhlenberg ex Willdenow var. ***strigosus*** (p. 262)

Daisy fleabane, branching daisy, prairie fleabane

Erigeron strigosus Muhl. ex Willd. var. *septentrionalis* (Fernald & Wiegand) Fernald

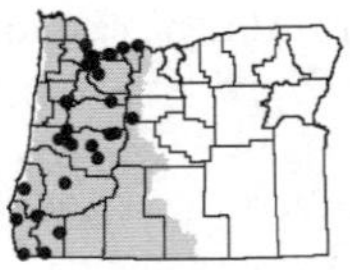

Plants annual or biennial, 50–127 cm; fibrous-rooted, occasionally with simple caudices. **Stems** erect, sparsely to moderately hirsute or strigose, hairs terete or flattened, eglandular. **Basal leaves** usually withering by flowering, oblanceolate to oblong, 40–100 × 8–18 mm. **Cauline leaves** linear to oblanceolate, 10–50 × 2–10 mm, reduced distally, margins entire or with a few small coarse teeth, surfaces sparsely to moderately strigose. **Heads** 10–50+ in corymb- to panicle-like arrays, radiate. **Involucres** 2–4 × 6–9 mm. **Phyllaries** in 2–3+ series, pale yellow with greenish to purplish brown stripe, surfaces hirsute. **Ray florets** 50–110+, white to pink; rays 2–5 × 0.5–1 mm. **Disc florets** 1–3 mm. **Fruits** 0.75–1.5 mm, strigose; disc fruit pappi of numerous barbellate bristles; ray fruit pappi lacking.

Disturbed areas, fields, forest edges. Flowering May–Sep. 0–1000 m. Casc, CR, ECas, Sisk, WV. CA, ID, WA; throughout most of Canada and US; Europe, Hawaii. Native.

Erigeron tener (A. Gray) A. Gray (p. 262)

Slender fleabane

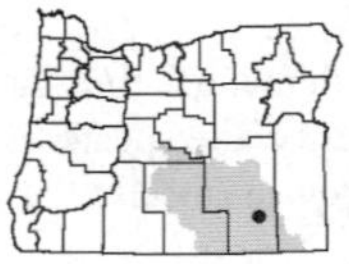

Plants perennial, 3–15 cm; taprooted, caudices branching. **Stems** erect, sparsely strigose, sometimes minutely glandular. **Basal leaves** persistent, narrowly oblanceolate-elliptic to spatulate, 5–50 × 1–4 mm, margins entire, tips acute, surfaces densely strigose. **Cauline leaves** few, linear, 5–20 × ~1 mm, abruptly reduced distally. **Heads** 1(3), radiate. **Involucres** 3–5 × 7–11 mm. **Phyllaries** medial areas usually purple, surfaces glabrous to moderately hirsute, densely glandular. **Ray florets** 15–24, blue to purple; rays 3–6 × 0.5–1 mm. **Disc florets** corollas 2–4 mm. **Fruits** 1–3 mm, sparsely strigose; inner pappi of numerous barbellate bristles.

Rocky areas, talus, slops, crevices. Flowering May–Aug. 1700–2500 m. BR. CA, ID, NV; northeast to MT, east to WY, southeast to AZ. Native.

In Oregon, *Erigeron tener* is thus far known only from Steens Mountain.

Erigeron vagus Payson

Rambling fleabane

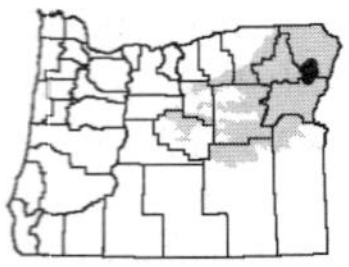

Plants perennial, 3–10 cm; taprooted, caudices branching. **Stems** erect, nearly glabrous to remotely hirsute, minutely glandular. **Basal leaves** persistent, spatulate with 3–5 distal lobes, 6–25 × 3–10 mm, margins entire, surfaces moderately to densely hirsute. **Cauline leaves** absent or reduced to linear bracts. **Heads** 1, radiate. **Involucres** 5–7 × 10–13 mm. **Phyllaries** in 2 series, purplish to green, with yellow-orange medial stripe, surfaces remotely hirsute to basally villous. **Ray florets** 25–40, white to lavender to purple; rays 6–9 × 0.5–1.5 mm. **Disc florets** 2–3 mm. **Fruits** 2–3 mm, sparsely strigose; inner pappi of numerous barbellate bristles. **2*n***=18.

Rocky areas, slopes. Flowering Jun–Aug. 1500–3100 m. BW. CA, NV; southeast to CO. Native.

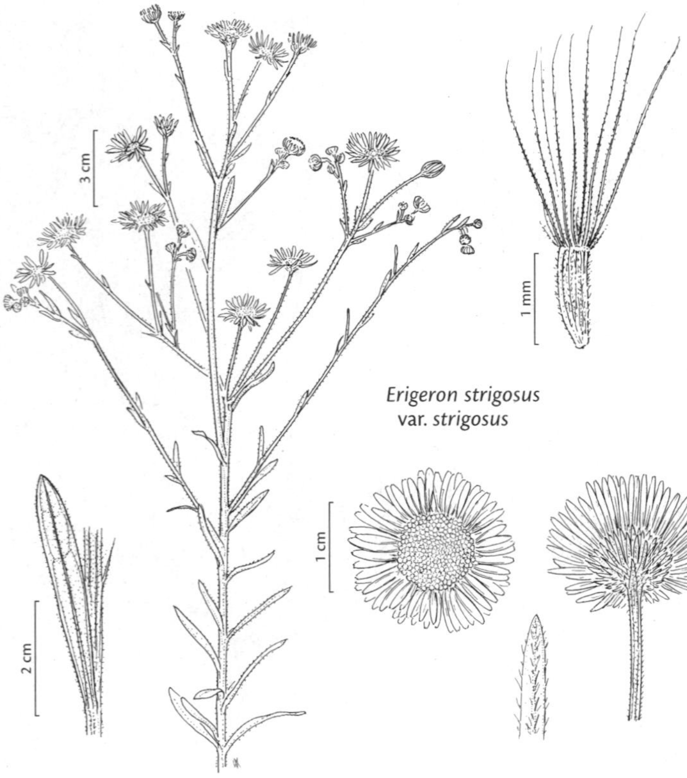

Erigeron strigosus var. *strigosus*

Erigeron tener

ERIGERON

ERIOPHYLLUM Woolly sunflower

Herbs, subshrubs, shrubs annual or perennial; taprooted. **Stems** erect or decumbent, branched proximally, distally, or throughout, woolly-pubescent. **Leaves** cauline, usually opposite, upper sometimes alternate, lanceolate or oblanceolate to linear, bases auriculate-clasping, margins usually pinnately toothed or lobed, rarely entire, tips acute, surfaces usually somewhat woolly; sessile or petiolate. **Inflorescences** raceme- or panicle-like arrays, or heads solitary. **Heads** radiate, rarely discoid, erect; peduncles bractless or few-bracted. **Involucres** campanulate or hemispheric, 3–12+ mm in diameter. **Phyllaries** in few equal series, erect, linear-lanceolate, midribs evident, tips acute or obtuse, surfaces woolly-pubescent. **Receptacles** flat to columnar; paleae 0. **Ray florets** pistillate, fertile; rays slightly to well exceeding involucres, bright yellow, rarely white or pale yellow, lobes distinct to minute or absent. **Disc florets** bisexual; corollas tubular, narrowed in lower half, exceeding involucres, yellow, puberulent, lobes 5, triangular, erect or spreading; stamens slightly exserted; stigmas slightly exserted. **Fruits** linear-clavate or flattened, ribbed or angled, truncate, glabrous or pubescent; pappi 0 or of 6–12 persistent, lacerate scales in 1–2 series.

Western North America. 13 species; 2 species treated in *Flora*.

Key to *Eriophyllum* species:

1. Plants herbs; stems not woody; heads solitary or few in open panicles; range widespread. ***E. lanatum***

1' Plants subshrubs, stems woody near bases; heads numerous in tight clusters; range southwestern coast. . . ***E. staechadifolium***

Eriophyllum lanatum (Pursh) J. Forbes

Woolly sunflower, Oregon sunshine

Herbs perennial, 1–6 dm. **Stems** not woody. **Leaves** opposite, proximal often alternate, 1–8 cm, surfaces woolly-pubescent, or adaxially glabrate and green; sessile or petiolate. **Heads** solitary or few in open panicle-like arrays; peduncles 3–30 cm, woolly-tomentose. **Involucres** 3–11 mm. **Phyllaries** in 2–3 equal series, lance-linear, woolly tomentose. **Ray florets** (0)5–15; corollas 5–20 mm, yellow. **Disc florets** 20–200. **Fruits** 2–5 mm, brown; pappi of 6–12 erose or lacerate scales, 0.3–2 mm.

10 varieties; 6 varieties treated in *Flora*.

Key to *Eriophyllum lanatum* varieties:

1. Upper leaf margins plane, rarely slightly revolute.
   2. Ray florets (5)8(10) . . . . *E. l.* var. ***integrifolium***

   2' Ray florets 10–15 . . . . *E. l.* var. ***lanceolatum***

1' Upper leaf margins revolute.
   3. Involucres usually 7–10 mm in diameter.
      4. Upper leaves mostly alternate; rays 6–9(16) mm . . . . *E. l.* var. ***achilleoides***

      4' Upper leaves mostly opposite; rays 8–20 mm . . . . *E. l.* var. ***leucophyllum***

   3' Involucres usually 10–15 mm in diameter.
      5. Fruits 2.5–3.5(4) mm; peduncles mostly 10–30 cm . . . . *E. l.* var. ***grandiflorum***

      5' Fruits 3.5–5 mm; peduncles mostly 5–15(20) cm. . . . . *E. l.* var. ***lanatum***

Eriophyllum lanatum var. *achilleoides* (de Candolle) Jepson

Yarrow-leaf woolly sunflower

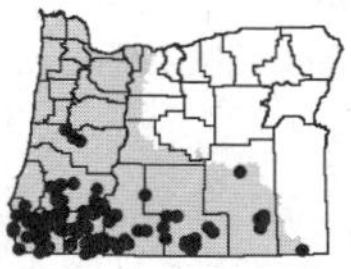

Leaves upper mostly alternate, margins usually revolute. **Involucres** 8–10 mm in diameter. **Ray florets** 9–13; rays 6–9(16) mm. **Fruits** 2–3.5 mm. **2*n***=16, 32.

Chaparral, forest openings. Flowering Apr–Aug. 0–2300 m. BR, Casc, CR, ECas, Est, Sisk, WV. CA, NV. Native.

This variety intergrades northward with variety *leucophyllum*.

Eriophyllum lanatum var. *grandiflorum* (A. Gray) Jepson (p. 264)

Large-flowered woolly sunflower

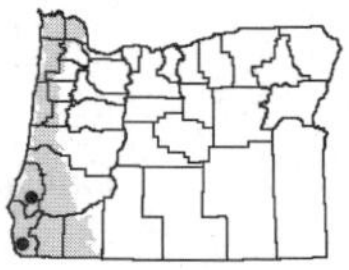

Leaves upper mostly alternate, margins revolute. **Involucres** 10–15 mm in diameter. **Ray florets** 10–13(15); rays 10–20 mm. **Fruits** 2–3.5(4) mm. **2*n***=16, 32, 48, 64.

Grasslands, forest openings. Flowering May–Jul. 0–1200 m. CR, Sisk. CA. Native.

Eriophyllum lanatum var. *integrifolium* (Hooker) Smiley

Oregon sunshine

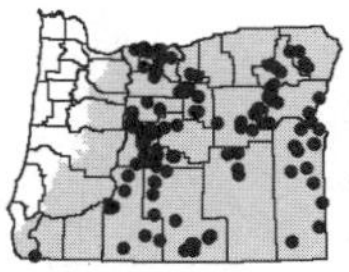

Leaves upper mostly alternate, margins plane, rarely slightly revolute. **Involucres** 6–10 mm in diameter. **Ray florets** (5)8(10); rays 6–12 mm. **Fruits** (5)2.5–5 mm. **2*n***=16, 32, 48, 64.

Coniferous forests, juniper woodlands, sagebrush scrub, grasslands, subalpine slopes, roadsides. Flowering May–Aug. 50–2800 m. BR, BW, Casc, Col, ECas, Lava, Owy, Sisk. CA, ID, NV, WA; northeast to MT, east to WY, southeast to UT. Native.

This variety intergrades with variety *lanatum*.

Eriophyllum lanatum var. *lanatum*

Eastern woolly sunflower

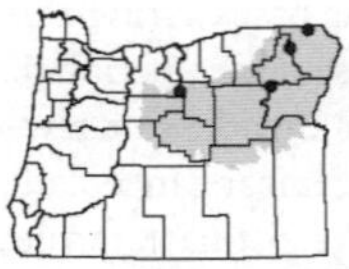

Leaves upper alternate, margins revolute. **Involucres** 10–15 mm in diameter. **Ray florets** (8)11–13; rays 10–20 mm. **Fruits** 3.5–5 mm. **2*n***=16, 32.

Coniferous forests, sagebrush scrub, grasslands, riverbanks, subalpine slopes. Flowering May–Aug. 1000–2500 m. BW, Lava. ID, WA; northeast to MT. Native.

This variety intergrades with variety *integrifolium*. Reports of variety *lanatum* from west of the Cascades are now referred to variety *leucophyllum*.

Eriophyllum lanatum var. *lanceolatum* (Howell) Jepson

Lance-leaf woolly sunflower

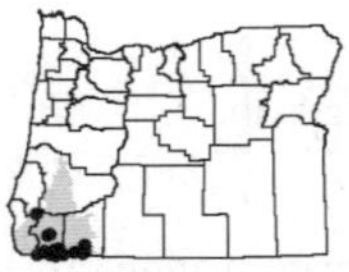

Leaves upper alternate, margins plane. **Involucres** 12–15 mm in diameter. **Ray florets** 10–15; rays 7–10(15) mm. **Fruits** 2–3.5 mm. **2*n***=16, 32.

Coniferous forests, mixed woodlands, subalpine slopes, serpentine. Flowering Jun–Sep. 100–2300 m. Sisk. CA. Native.

Eriophyllum lanatum var. *leucophyllum* (de Candolle) W.R. Carter

Western woolly sunflower

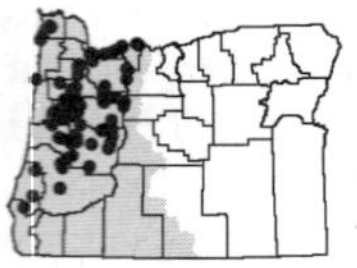

Leaves upper mostly opposite, margins revolute. **Involucres** 7–11 mm in diameter. **Ray florets** 8–13; rays 8–20 mm. **Fruits** 2–3(4) mm. **2*n***=16, 32.

Coniferous forest, shrublands, grasslands, rocky cliffs, talus, roadsides, edges of cultivated fields. Flowering May–Aug. 0–1600 m. Casc, CR, ECas, Est, Sisk, WV. WA; north to British Columbia. Native.

Plants of this variety were earlier referred to variety *lanatum*.

Eriophyllum staechadifolium Lagasca (p. 264)

Seaside woolly sunflower

Eriophyllum staechadifolium Lag. var. *artemisiifolium* (Less.) J.F. McBride

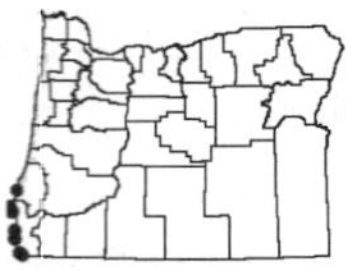

Subshrubs 3–15 dm. **Stems** woody near bases. **Leaves** cauline, 2–5 cm, surfaces abaxially densely white-tomentose, adaxially glabrous or lightly tomentose; sessile or short-petiolate. **Heads** numerous in tight clusters; peduncles 0.3–5 cm, woolly-tomentose. **Involucres** 5–8 mm. **Phyllaries** in 2–3 series, lance-linear, woolly-tomentose. **Ray florets** (0)5–9; corollas 3–5 mm, yellow. **Disc florets** 25–40. **Fruits** 3–4 mm, brown; pappi of 8–12 scales, ~1 mm. **2*n***=30.

Coastal bluffs and dunes, shrublands, roadsides. Flowering Jun–Jul. 0–50 m. Est. CA. Native.

The species name is sometimes mistakenly spelled *stoechadifolium*. A form with entire leaves occurs in California.

Eriophyllum lanatum var. *grandiflorum*

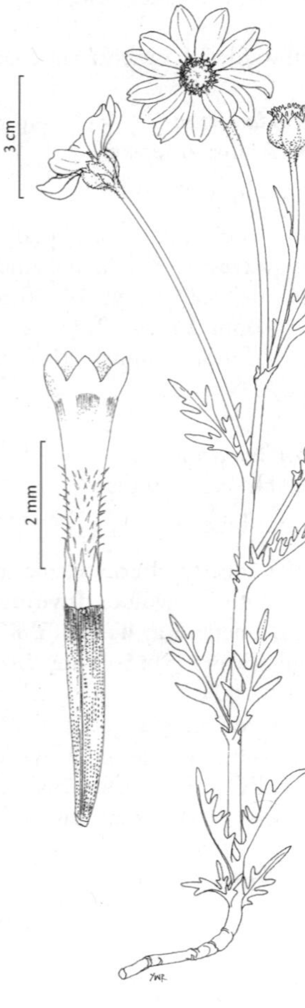

Eriophyllum staechadifolium

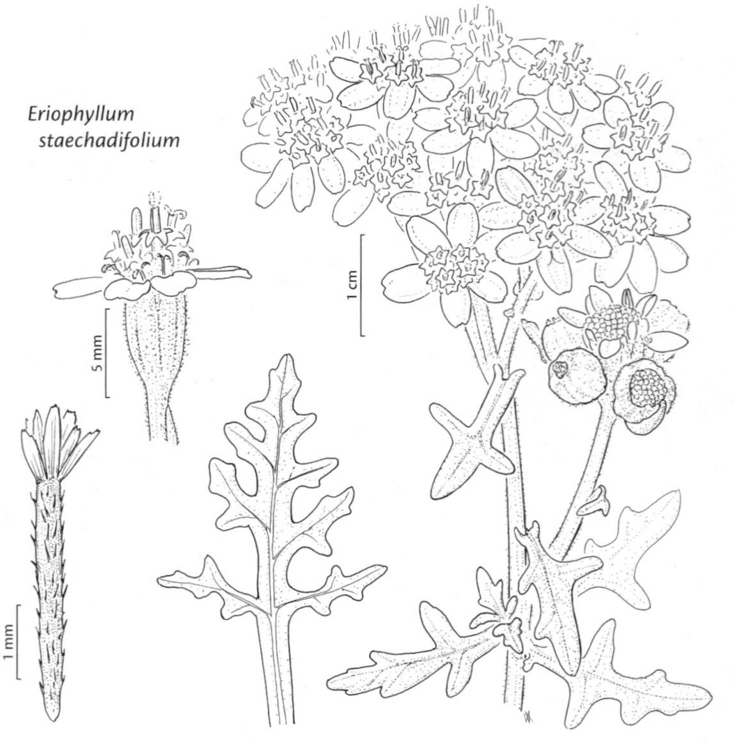

ERIOPHYLLUM

EUCEPHALUS Aster

Geraldine A. Allen

Herbs perennial; fibrous-rooted from caudices or short rhizomes. **Stems** ascending to erect, glabrate to glandular-pubescent or woolly or cottony-pubescent. **Leaves** cauline, alternate, linear-lanceolate to ovate or elliptic, ± entire, lowermost often reduced and deciduous by flowering; sessile. **Inflorescences** heads solitary or in corymb-, raceme- or panicle-like arrays; branches often with reduced leaves or bracts. **Heads** radiate or discoid. **Involucres** cylindrical to campanulate or turbinate. **Phyllaries** in 3–6 series, erect to spreading, usually imbricate, bases ± chartaceous, 1-veined, midribs raised, keeled, tips green, often reddish at margins. **Receptacles** ± flat; paleae 0. **Ray florets** 0–21, pistillate; corollas white to pink or violet-purple. **Disc florets** 10–25, bisexual; corollas tubular, gradually enlarged distally, yellow, lobes 5, triangular, erect or reflexed. **Fruits** ± obconic, flattened, 1–2-ribbed, sometimes with 1–2 additional veins on each face; pappi of 30–50 barbellate bristles in 2–3 series.

Western Canada and United States. 10 species; 7 species treated in *Flora*.

Several species of *Eucephalus* intergrade extensively in southwestern Oregon, including *E. glabratus, E. ledophyllus, E. tomentellus, E. vialis*, and possibly *E. breweri*. Of these, three (*E. breweri, E. ledophyllus,* and *E. vialis*) are well marked elsewhere in their range. The remaining two (*E. glabratus* and *E. tomentellus*) are poorly marked and appear to represent indistinct intermediates along a morphological gradient between *E. ledophyllus* and *E. vialis.*

Key to *Eucephalus* species:

1. Ray florets ≥ 5.
    2. Rays white or sometimes pink-tinged; heads 1(2–5); leaves 1.5–3 cm. . . . . **E. gormanii**
    2' Rays violet-purple; heads 3–35; leaves 2–7 cm.
        3. Leaves linear-oblong to linear-lanceolate, surfaces ± glandular-pubescent. . . . . **E. elegans**
        3' Leaves lanceolate-elliptic to oblong, abaxially ± woolly-pubescent or cottony-pubescent, adaxially glabrous to sparsely pubescent. . . . . **E. ledophyllus**

1' Ray florets 0–4.
    4. Outer phyllaries ≥ ⅔ as long as involucres . . . . **E. breweri**
    4' Outer phyllaries < ⅔ as long as involucres.
        5. Leaves 4–10 cm, glabrous above, glandular-pubescent below; plants 5–12(13) dm. . . . . **E. vialis**
        5' Leaves 2.5–7 cm, glabrous to sparsely pubescent above, glabrous or woolly-pubescent or cottony-pubescent below; plants 3–8(10) dm.
            6. Leaves glabrous to sparsely glandular-pubescent on both faces . . . . **E. glabratus**
            6' Leaves glabrous to glabrate above, woolly-pubescent or cottony-pubescent below . . . . **E. tomentellus**

Eucephalus breweri (A. Gray) G.L. Nesom

Brewer's aster

Aster breweri (A. Gray) Semple, *Chrysopsis breweri* A. Gray

Plants 2–7 dm. **Stems** ascending to erect, glabrous to sparsely pubescent or glandular. **Leaves** linear-lanceolate to ovate, 2–6 cm × 5–12 mm, surfaces glabrate to ± glandular-pubescent or woolly-pubescent. **Inflorescences** corymb- to panicle- or raceme-like arrays. **Heads** 3–15, discoid. **Involucres** turbinate-cylindric, 5–10 mm. **Phyllaries** in 3–4 series, linear-lanceolate to lanceolate-oblong, subequal, tips acuminate, surfaces glabrate to ± glandular or woolly-pubescent; outer ≥ ⅔ as long as involucres. **Ray florets** 0. **Fruits** ± strigose; pappus bristles in 2 series, barbellate. **2*n***=18.

Open coniferous woods, meadows. Flowering Jul–Sep. 1200–1800 m. Sisk. CA, NV. Native.

Eucephalus elegans Nuttall (p. 266)

Elegant aster

Aster perelegans A. Nelson & J.F. Macbr.

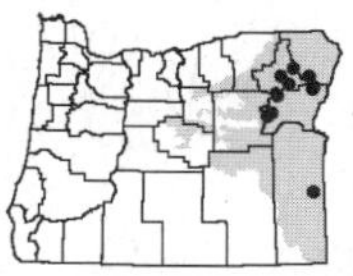

Plants 3–7 dm. **Stems** erect, sparsely pubescent to glandular-pubescent. **Leaves** linear-oblong to linear-lanceolate, 2–6 cm × 3–10 mm, surfaces ± glandular-pubescent. **Inflorescences** corymb- to raceme-like arrays. **Heads** 3–8(20), radiate. **Involucres** turbinate-cylindric, 6–9 mm. **Phyllaries** in 3–5 series, imbricate, ovate, tips acute, surfaces ± glandular-pubescent. **Ray florets** 5–8; corollas violet-purple. **Fruits** ± strigose; pappus bristles in 2 series, barbellate. **2*n***=18.

Open forests, meadows. Flowering Jul–Aug. 800–2500 m. BW, Owy. ID, NV; northeast to MT, east to WY, southeast to CO. Native.

Eucephalus glabratus (Greene) Greene

Siskiyou aster, smooth aster

Aster siskiyouensis A. Nelson & J.F. Macbr.

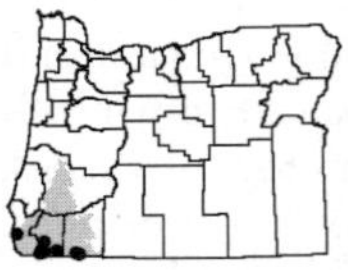

Plants 3–7 dm. **Stems** erect, glabrous to sparsely glandular-pubescent. **Leaves** lanceolate to lanceolate-ovate, 3–6 cm × 5–20 mm, surfaces ± glabrous. **Inflorescences** corymb- to raceme-like arrays. **Heads** 3–20, discoid or radiate. **Involucres** campanulate, 7–10 mm. **Phyllaries** in 3–5 series, strongly imbricate, lanceolate to narrowly ovate, tips acute, surfaces ± glabrous; outer < ⅔ as long as involucres. **Ray florets** 0–2(4); corollas violet-purple. **Fruits** ± pilose; pappus bristles in 2 series, ± smooth. ***2n***=18.

Dry open forests, chaparral. Flowering Jul–Sep. 400–2000 m. Sisk. CA. Native.

This species hybridizes with *E. tomentellus* and *E. vialis*.

Eucephalus gormanii Piper (p. 266)

Gorman's aster

Aster gormanii (Piper) S.F. Blake

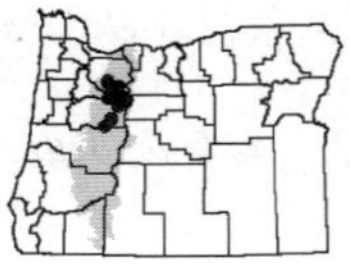

Plants 1–4 dm. **Stems** ascending to erect, sparsely glandular-pubescent. **Leaves** lanceolate-ovate to elliptic, 1.5–3 cm × 4–10 mm, surfaces ± glandular-pubescent. **Inflorescences** solitary terminal heads or corymb- to raceme-like arrays. **Heads** 1(2–5), radiate. **Involucres** turbinate, 6–8 mm. **Phyllaries** in 2–4 series, imbricate, lanceolate to ovate, tips acute, surfaces glabrous to glandular-pubescent. **Ray florets** 5–13; corollas white or sometimes pink-tinged. **Fruits** pilose; pappus bristles in 2 series, smooth to barbellate. ***2n***=18.

Open to somewhat shaded rocky slopes. Flowering Jul–Aug. 1100–1900 m. Casc. Native.

This species is endemic to a small area of the Cascade Mountains in Clackamas, Marion, and Linn counties. It occasionally hybridizes with *E. ledophyllus*.

Eucephalus ledophyllus (A. Gray) Greene

Cascade aster

Plants 2–8 dm. **Stems** erect, sparsely glandular-pubescent or pilose. **Leaves** lanceolate-elliptic to oblong, 2.5–7 cm × 5–20 mm, surfaces abaxially ± woolly-pubescent or cottony-pubescent, adaxially glabrous to sparsely pubescent. **Inflorescences** corymb- to raceme-like arrays. **Heads** 3–20(35), radiate. **Involucres** turbinate, 7–11 mm. **Phyllaries** in 3–5 series, imbricate, lanceolate to ovate, tips acute, surfaces puberulent to glandular-pubescent. **Ray florets** 5–21; corollas violet-purple. **Fruits** pilose; pappus bristles in 2 series, smooth to ± barbellate. ***2n***=18.

Western United States. 2 varieties.

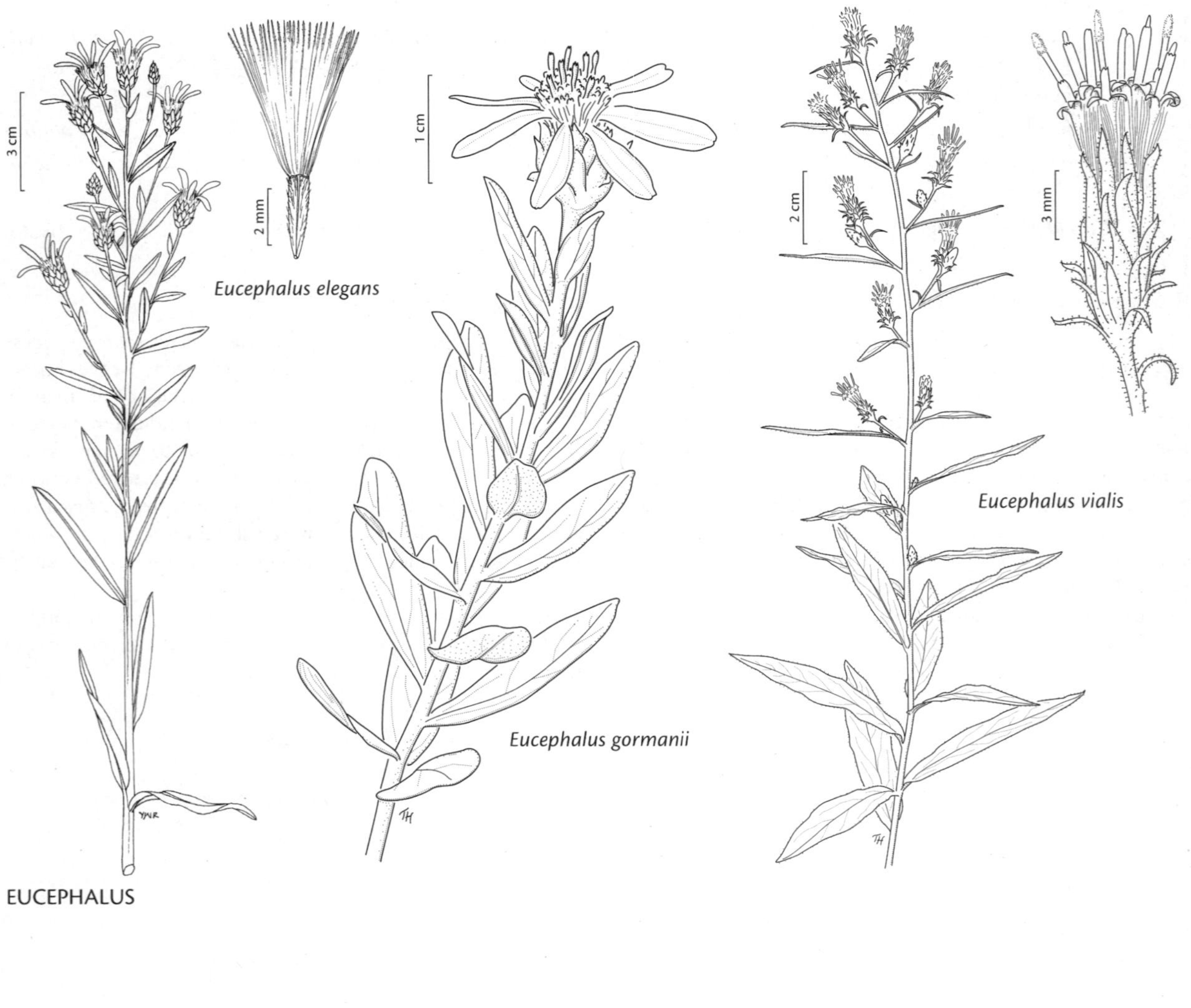

EUCEPHALUS

Key to *Eucephalus ledophyllus* varieties:

1. Peduncles glabrous to glandular or with short straight hairs; ray florets usually 5–8 . . . . *E. l.* var. ***covillei***
1' Peduncles ± woolly- or cottony-pubescent; ray florets usually 13–21 . . *E. l.* var. ***ledophyllus***

Eucephalus ledophyllus (A. Gray) Greene var. ***covillei*** (Greene) G.L. Nesom

Coville's Cascade aster

Aster ledophyllus (A. Gray) A. Gray var. *covillei* (Greene) Cronquist

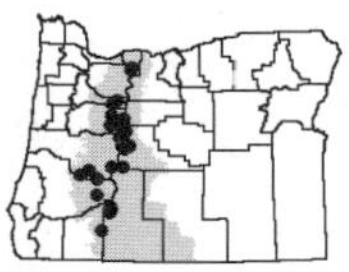

Stems ± glandular-pubescent. **Peduncles** glabrous to glandular or with short straight hairs. **Ray florets** usually 5–8.

Meadows, openings in montane coniferous forests. Flowering Jul–Sep. 1100–2200 m. Casc, ECas. WA. Native.

Intermediates between *E. ledophyllus* var. *covillei* and rayless species (*E. glabratus, E. tomentellus, E. vialis*) are common in southwestern Oregon.

Eucephalus ledophyllus (A. Gray) Greene var. ***ledophyllus***

Northern Cascade aster

Aster ledophyllus (A. Gray) A. Gray var. *ledophyllus*

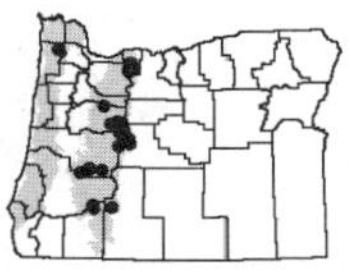

Stems glandular-pubescent to pilose. **Peduncles** ± woolly- or cottony-pubescent. **Ray florets** usually 13–21.

Meadows, open coniferous forests. Flowering Jul–Sep. 900–2500 m. Casc, CR. CA, WA. Native.

Eucephalus tomentellus (Greene) Greene

Hairy rayless aster, brickellbush aster

Aster brickellioides Greene

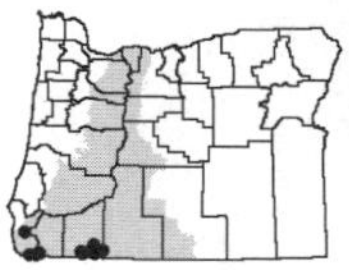

Plants 4–8(10) dm. **Stems** erect, ± cottony-pubescent. **Leaves** lanceolate to elliptic, 2.5–7 cm × 7–20 mm, surfaces abaxially woolly-pubescent or cottony-pubescent, adaxially glabrous to sparsely pubescent. **Inflorescences** panicle- to raceme-like arrays. **Heads** 3–20(40), discoid or radiate. **Involucres** turbinate, 7–10 mm. **Phyllaries** in 3–5 series, strongly imbricate, narrowly oblong to ovate, tips acute, surfaces glandular-pubescent to tomentose; outer < ⅔ as long as involucres. **Ray florets** 0, occasionally 1–4; corollas violet-purple. **Fruits** glabrous to pilose; pappus bristles in 2 series, ± barbellate. $2n$=18.

Open woods, rocky cliffs. Flowering Jul–Sep. 500–2200 m. Casc, ECas, Sisk. CA. Native.

This species hybridizes with *E. glabratus* and *E. ledophyllus.*

Eucephalus vialis Bradshaw (p. 266)

Wayside aster

Aster vialis (Bradshaw) S.F. Blake

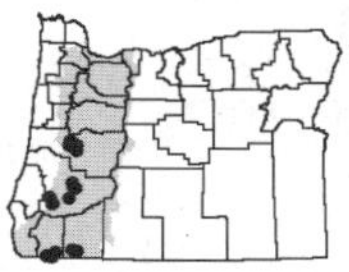

Plants 5–12(13) dm. **Stems** erect, pilose to glandular-pubescent. **Leaves** lanceolate to elliptic, 4–10 cm × 15–30 mm, surfaces abaxially glandular-pubescent, adaxially glabrous. **Inflorescences** panicle- to raceme-like arrays. **Heads** 5–50(120), discoid. **Involucres** turbinate, 7–10 mm. **Phyllaries** in 3–6 series, strongly imbricate, linear to linear-oblong, subequal, tips acute to acuminate, surfaces glabrate to glandular-pubescent; outer nearly as long as inner, < ⅔ as long as involucres. **Ray florets** 0. **Fruits** pilose; pappus bristles in 2 series, smooth to ± barbellate. $2n$=18.

Dry open woodland. Flowering Jun–Sep. 100–1700 m. Casc, Sisk, WV. CA. Native.

This species hybridizes with *E. glabratus* in the southern part of its range.

EUCHITON Cudweed

Herbs annual or perennial; taprooted or fibrous-rooted, sometimes stoloniferous. **Stems** decumbent to erect, simple or branched, tomentose. **Leaves** alternate, basal and proximal cauline sometimes withering before flowering, linear to oblanceolate, ± reduced distally, margins entire, tips acute or obtuse, surfaces abaxially tomentose, adaxially glabrous or lightly villous; petiolate or sessile. **Inflorescences** numerous heads in tight terminal clusters; bracts ± united, abaxially tomentose. **Heads** disciform; sessile. **Involucres** narrowly campanulate or cylindric, spreading in fruit. **Phyllaries** in several series, erect, lanceolate to ovate, equal or outer gradually shorter, sometimes red-spotted, surfaces glabrous, membranous or partly herbaceous. **Receptacles** flat, smooth; paleae 0. **Outer florets** 16–150, pistillate; corollas distally purplish, elongate-filiform. **Inner florets** 1–7, bisexual; corollas purple or distally purplish. **Fruits** ovoid-fusiform, minutely strigillose; pappi of unequal bristles, deciduous.

Asia, Australia, North America, Pacific Islands. 17 species; 2 species treated in *Flora.*

Key to *Euchiton* species:

1. Plants perennial; bases of cauline leaves partially clasping; inflorescence bracts shorter than or slightly surpassing clusters; inner florets 3–7 per head . . . . ***E. gymnocephalus***
1' Plants annual; bases of cauline leaves not clasping; most inflorescence bracts well surpassing clusters; inner florets 1 per head . . . . ***E. sphaericus***

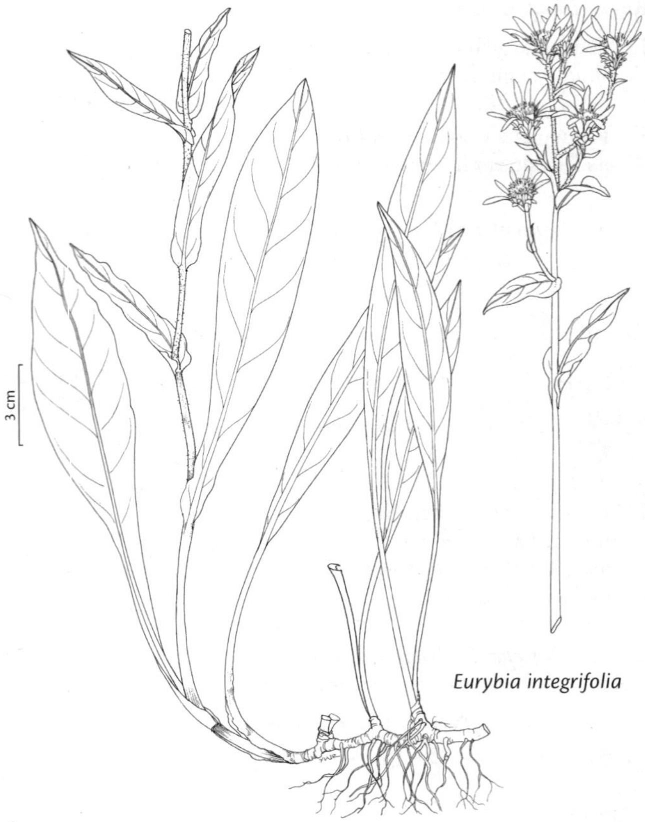

EUCHITON • EURYBIA

Euchiton gymnocephalus (de Candolle) Holub

Creeping cudweed

Gnaphalium collinum Labill.

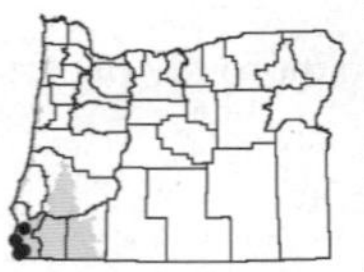

Plants perennial, 0.5–4 dm; fibrous-rooted, usually stoloniferous. **Leaves** 1–22 cm; cauline bases partially clasping. **Inflorescences** bracts 2–3, shorter than or slightly surpassing clusters. **Involucres** 4–4.5 mm. **Phyllaries** ± herbaceous, membranous-margined, often red-spotted dorsally, tips acute or obtuse. **Outer florets** 40–60; corollas 2.5–3 mm. **Inner florets** 3–7; corollas 2.5–3 mm. **Fruits** 0.8 mm, brown; pappi 1–2.5 mm, white. **2*n***=14.

Road banks, disturbed areas. Flowering May–Sep. 0–300 m. Est, Sisk. CA; Australia, Pacific Islands. Exotic.

Euchiton sphaericus (Willdenow) Anderberg (p. 268)

Star cudweed

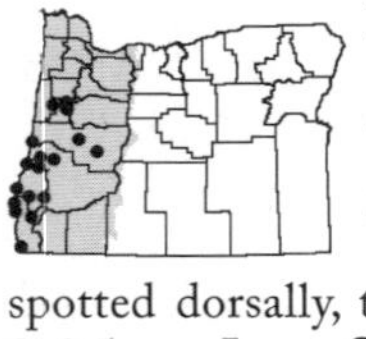

Plants annual, 1.5–6.5 dm; taprooted. **Leaves** 2–6 cm, cauline bases not clasping. **Inflorescences** bracts 4–8, most well surpassing clusters. **Involucres** 2.5–3 mm. **Phyllaries** membranous, sometimes red-spotted dorsally, tips acute. **Outer florets** 16–26; corollas 2–2.5 mm. **Inner florets** 1; corollas 2–2.5 mm. **Fruits** 0.8 mm, brown; pappi 2–2.5 mm, white. **2*n***=14.

Roadsides, riverbanks, disturbed areas. Flowering Jun–Oct. 0–400 m. Casc, CR, Est, Sisk, WV. CA; Asia, Australia, Pacific Islands. Exotic.

EURYBIA Aster

Herbs perennial; often rhizomatous. **Stems** erect or ascending, simple or branched distally, less often branched proximally, hispidulous or villous-puberulent, sometimes densely glandular. **Leaves** basal and cauline or all cauline, alternate, simple, obovate or oblanceolate to ovate, lanceolate, or linear, bases sometimes auriculate-clasping, margins entire to denticulate or serrate-dentate, tips acute, surfaces smooth or scabrous; petiolate or sessile. **Inflorescences** corymb-like arrays, rarely heads solitary; bracts present or not. **Heads** radiate, erect. **Involucres** campanulate or hemispheric. **Phyllaries** in few to many series, erect, equal or outer progressively shorter, herbaceous or chartaceous with herbaceous tips, midribs evident or obscure, tips and margins sometimes lined with purple. **Receptacles** flat to slightly convex, pitted; paleae 0. **Ray florets** pistillate, fertile; rays well-exceeding involucres, white to light or dark violet. **Disc florets** bisexual; corollas tubular, gradually enlarged distally, equaling or exceeding involucres, yellow, usually aging to purple, lobes 5, triangular, erect or spreading; stamens exserted; stigmas exserted. **Fruits** cylindro-obconic to fusiform, truncate, glabrous or strigillose, ribs 7–15+; pappi of numerous bristles in 2–4 series, reddish, brownish, or yellowish.

Asia, Europe, North America. 23 species; 4 species treated in *Flora*.

Eurybia is one of several North American Asteraceae genera that were formerly placed in *Aster* but now have been segregated as separate genera. Molecular evidence (Noyes & Rieseberg 1999) supports a common New World origin for these genera, independent of *Aster* and other Eurasian taxa.

Key to *Eurybia* species:

1. Stems glandular-puberulent distally.
    2. Leaves basal and cauline, margins entire . . . . . ***E. integrifolia***
    2' Leaves cauline, margins denticulate or serrate-dentate . . . . . ***E. conspicua***

1' Stems hispid-puberulent or villous distally, not glandular.
    3. Leaves abaxially glabrous to sparsely villous; ray corollas light to dark violet . . . . . ***E. merita***
    3' Leaves abaxially scabrous; ray corollas white or pale violet . . . . . ***E. radulina***

Eurybia conspicua (Lindley) G.L. Nesom

Showy aster

Aster conspicuus Lindl.

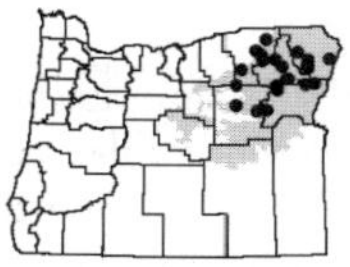

Plants 3–10 dm. **Stems** villous proximally, glandular-puberulent distally. **Leaves** cauline, lanceolate to oblanceolate, ovate, or broadly elliptic, 4–15 cm, bases attenuate or truncate, often auriculate, margins denticulate or serrate-dentate distally, surfaces glabrous or scabridulous; sessile. **Heads** peduncles 5–50 mm, glandular-puberulent. **Involucres** 9–12 mm. **Phyllaries** in 4–5 series, chartaceous below, becoming herbaceous above or only at tips, lanceolate, margins ciliate, surfaces puberulent. **Ray florets** 12–35; corollas 8–15 mm, blue or violet. **Disc florets** 45–55. **Fruits** 3–4 mm, tan or brown, glabrous; pappus bristles 9–10 mm.

Coniferous forests, mixed woodlands, shrublands, riverbanks, roadsides. Flowering Jul–Sep. 400–2000 m. BW. WA; north to British Columbia; northeast to Manitoba, east to SD. Native.

Eurybia integrifolia (Nuttall) G.L. Nesom (p. 268)

Thick-stemmed aster

Aster integrifolius Nutt.

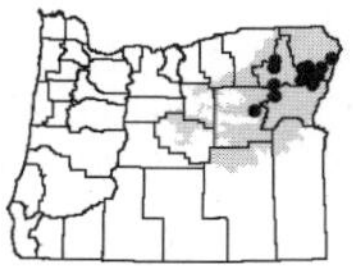

Plants 1.2–6 dm. **Stems** glabrous proximally, glandular-puberulent distally. **Leaves** basal and cauline, lanceolate to oblanceolate or elliptic, 3–20 cm, bases often truncate and auriculate-clasping, margins entire, surfaces glabrous; basal petiolate, cauline usually sessile. **Heads** peduncles 5–30 mm, glandular-puberulent. **Involucres** 8–14 mm. **Phyllaries** in 3–4 series, chartaceous below becoming herbaceous above or only at tips, narrowly lanceolate to linear, often purplish, margins rarely ciliate, surfaces glandular-puberulent. **Ray florets** 8–25; corollas 10–15 mm, purple-violet. **Disc florets** 20–50. **Fruits** 4–4.5 mm, brown, puberulent; pappus bristles 7–8 mm. **2*n***=18.

Coniferous woods, grasslands, open slopes, riverbanks, roadsides. Flowering Jul–Sep. 1200–2400 m. BW. CA, ID, NV, WA; northeast to MT, east to WY, southeast to UT. Native.

Eurybia merita (A. Nelson) G.L. Nesom

Subalpine aster

Aster sibiricus L., *Aster sibiricus* L. var. *meritus* (A. Nelson) Raup

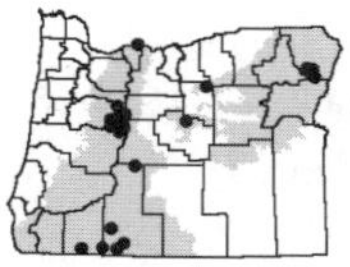

Plants 0.3–4 dm. **Stems** often purple, glabrous or villous-puberulent, not glandular. **Leaves** basal and cauline, obovate or oblanceolate to lanceolate or linear, 2–10 cm, bases acute to truncate, sometimes auriculate, margins entire to denticulate or serrate-dentate, surfaces abaxially glabrous to sparsely villous, adaxially glabrous to densely strigose; sessile or short-petiolate. **Heads** peduncles 3–30 mm, scabrous to villous-puberulent. **Involucres** 6–9 mm. **Phyllaries** in 4–7 series, chartaceous below becoming herbaceous above or only at tips, lanceolate, rarely oblanceolate, margins lined with purple (except some alpine plants in Wallowa Mountains), ciliate, surfaces glabrous or puberulent. **Ray florets** 10–30+; corollas 7–15 mm, light to dark violet. **Disc florets** 25–50+. **Fruits** 3.5 mm, brownish, strigillose; pappus bristles 5–6 mm. **2*n***=36.

Coniferous forests, woodlands, shrublands, rock outcrops, talus slopes, alpine meadows. Flowering Jul–Aug. 900–3000 m. BW, Casc, ECas, Sisk. CA, ID, WA; north to British Columbia, northeast to MT, east to SD, southeast to UT. Native.

Eurybia radulina (A. Gray) G.L. Nesom

Rough-leaved aster

Aster radulinus A. Gray

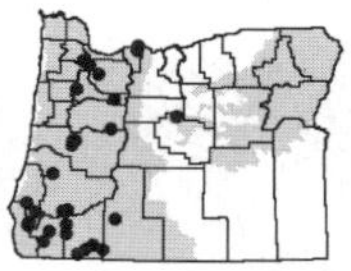

Plants 1–7 dm. **Stems** often purple, villous-puberulent distally or throughout, not glandular. **Leaves** basal and cauline, obovate or oblanceolate to lanceolate or elliptic, 1–12 cm, bases sometimes truncate-auriculate, margins denticulate to dentate-serrate, rarely entire, surfaces abaxially scabrous, adaxially scabrous, strigose; lower petiolate, upper sessile. **Heads** peduncles 3–45 mm, villous-puberulent. **Involucres** 6–9 mm. **Phyllaries** in 4–6 series, chartaceous below, herbaceous above or only at tips, lanceolate or elliptic-linear, margins sometimes lined with purple, ciliate or not, surfaces glabrous or puberulent. **Ray florets** 10–15; corollas 7–15 mm, white to pale violet. **Disc florets** 30–70. **Fruits** 3–3.5 mm, brownish, strigillose; pappus bristles 2.7–3 mm. **2*n***=18.

Woodlands, grasslands, hillsides, ridge tops, rocky slopes, riverbanks, roadsides. Flowering Jun–Oct. 0–1800 m. BW, Casc, CR, ECas, Sisk, WV. CA, WA; north to British Columbia. Native.

EUTHAMIA Goldentop

Herbs or **subshrubs** perennial; rhizomatous. **Stems** usually 1, erect, branches few to many, sometimes woody proximally, leafy throughout, glabrous. **Leaves** cauline, alternate, slightly reduced distally, sometimes with tufts of linear leaves on short axillary branches; blades narrowly elliptic-lanceolate to linear, margins entire, veins 3–5, parallel from bases, tips acute or acuminate, surfaces glabrous, ± glandular-punctate; sessile. **Inflorescences** numerous small heads clustered on branch tips in panicle-like arrays, sometimes ± flat-topped. **Heads** radiate. **Involucres** cylindrical or narrowly campanulate. **Phyllaries** in 4–6 unequal series, erect, spreading in fruit, narrowly lanceolate or linear; outer gradually shorter, colorless or green-tipped, outermost often green throughout, midribs indistinct, surfaces glabrous, resinous. **Receptacles** convex; paleae 0. **Ray florets** pistillate; rays exceeding involucres, yellow. **Disc florets** bisexual; corollas cylindric, widened in distal half, yellow, lobes 5, narrowly triangular, spreading to recurved at anthesis; stamens 5, well exserted from corollas; stigmas exserted from stamen tubes. **Fruits** cylindrical, truncate, tapering to bases, inconspicuously ribbed; pappi of numerous fine bristles.

Asia, Europe, North America. 5 species; 1 species treated in *Flora*.

Euthamia graminifolia has been recorded in Oregon as an adventive species in cranberry bogs in Curry County. Its native range is to the north and east of Oregon, extending from Washington through Canada, to the central and eastern United States. The specimens collected in Oregon resemble *E. graminifolia* in the shape of their inflorescences and involucral phyllaries. However, the specimens do not have the kind of robust growth needed in order to rule out environmental variation in *E. occidentalis* as the source of these traits.

Euthamia occidentalis Nuttall (p. 270)

Western goldentop, western goldenrod

Solidago occidentalis (Nutt.) Torr. & A. Gray

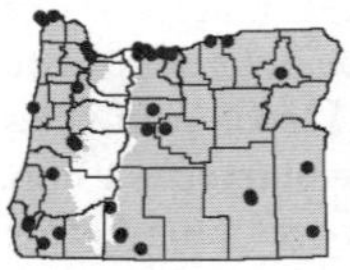

Plants 4–20 dm. **Leaves** 4–12 cm, gland-dotted. **Involucres** 3.5–5 mm. **Phyllaries** outer tips obtuse to acute; inner tips acute to acuminate. **Ray florets** 15–25, inconspicuous; rays 2–2.5 mm. **Disc florets** 8–15; corollas 3–4 mm, equaling involucres. **Fruits** 0.75–1 mm, pale brown, strigose; pappus bristles 3–4 mm, white. **2*n***=18.

Moist flats, slopes, open woods, riverbanks, fresh or saline marshes, roadside ditches, deflation plains. Flowering Jul–Oct. 0–1600 m. All ecoregions except Casc. CA, ID, NV, WA; north to British Columbia, northeast to MT, east to NE, southeast NM, south to Mexico. Native.

Euthamia occidentalis is best distinguished from the related *E. graminifolia* by its overall more elongate inflorescence, the proximal branches being shorter than the distal ones, and not forming a mostly flat-topped unit as in the latter species. However, on less robust individuals having only a few clusters of heads, this difference may not be so clear.

Euthamia occidentalis

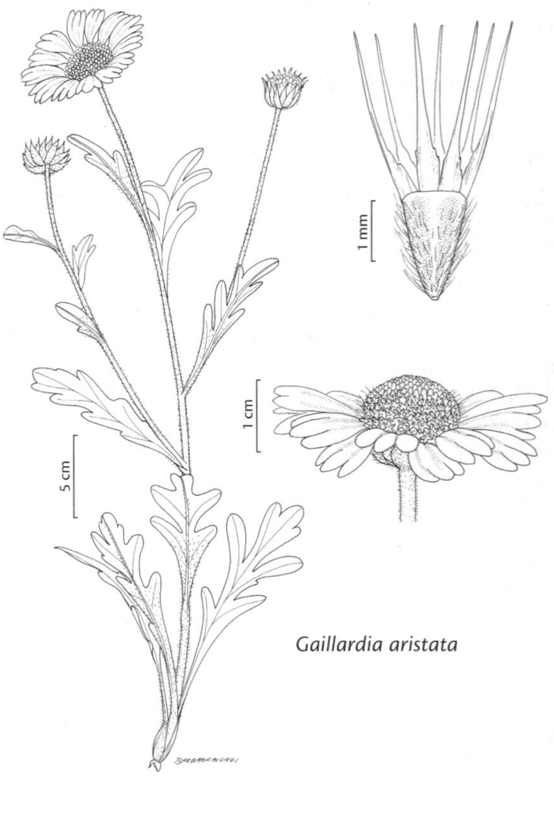

Gaillardia aristata

EUTHAMIA • GAILLARDIA

GAILLARDIA Blanket flower

Herbs annual or perennial; taprooted, sometimes rhizomatous. **Stems** 1–many, erect, slender, unbranched to much branched, glabrous to strigillose, puberulent, or villous-pubescent. **Leaves** basal and cauline, alternate, lanceolate to oblanceolate or elliptic, simple or pinnately few-lobed, bases attenuate or truncate, sometimes auriculate, margins entire to sparingly denticulate or dentate, 1–3-veined, tips usually acute, surfaces tomentose to hispid or strigillose-pubescent, densely gland-dotted, scabridulous; basal petiolate, cauline petiolate or sessile, petioles sometimes narrowly winged. **Inflorescences** solitary, terminal heads. **Heads** usually radiate. **Involucres** convex or hemispheric. **Phyllaries** in 2–4 ± equal series, spreading to reflexed, narrowly lanceolate to linear, margins often ciliate, sometimes purple-lined, 1–3-veined, tips acute or acuminate, surfaces tomentose or densely strigillose. **Receptacles** convex to hemispheric, smooth or pitted; paleae 0. **Ray florets** usually sterile; corollas inversely triangular, deeply 3-lobed, spreading, well exceeding involucres. **Disc florets** bisexual; corollas tubular, gradually enlarged distally, puberulent, lobes 5, erect; stamens partially exserted; stigmas slightly or well exserted. **Fruits** obconic, truncate, 4-angled; pappi of awn-tipped scales in 1–2 series, rarely 0.

North America, South America. 15–17 species; 1 species treated in *Flora.*

Gaillardia pulchella has been collected once in Oregon (Linn County, 1992) but has not become naturalized.

Gaillardia aristata Pursh (p. 270)

Blanket flower, great-flowered gaillardia

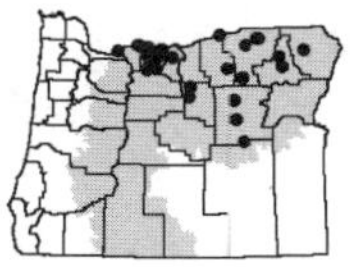

Plants perennial, 2–8 dm. **Leaves** lanceolate to oblanceolate or elliptic, 2.5–15 cm, margins entire to pinnately lobed. **Heads** 1.5–2.5 mm in diameter. **Phyllaries** 5–15 mm. **Ray florets** 6–18+; corollas 15–35 mm, yellow, sometimes purple-tinged. **Disc florets** 60–100+; corollas 5–7 mm, usually purple or purple-tipped, sometimes yellow. **Fruits** 2.5–6 mm, brownish, hispid-pubescent; pappi 5–6 mm. **2*n***=34, 68.

Conifer forests, mixed woodlands, grasslands, hillsides, riverbanks, roadsides. Flowering May–Sep. 50–1600 m. BW, Casc, Col, ECas, Lava. CA, ID, WA; north to Yukon, northeast to Quebec, east to MA, southeast to NM. Native.

GALINSOGA Gallant soldiers

Dominic Maze

Herbs annual. **Stems** erect, rarely decumbent, striate. **Leaves** cauline, opposite; blades lanceolate to elliptic to broadly ovate, margins entire or serrulate to serrate, 3-veined, surfaces nearly glabrous to pilose or glandular-pilose; petiolate. **Inflorescences** leafy-bracted, cyme-like clusters, generally small. **Heads** radiate, rarely discoid. **Involucres** hemispheric to campanulate. **Phyllaries** 6–9(16), in 2(3) series, herbaceous or scarious. **Receptacles** conical; paleate. **Paleae** persistent or deciduous, united in groups of 2 or 3 to adjacent proximal phyllary, each complex often enclosing a ray achene, convex to folded, lanceolate to obovate, scarious, margins entire or 2- or 3-lobed. **Ray florets** 0–8(15), pistillate, fertile; corollas white or dull white to pinkish, lobes 0–3. **Disc florets** 5–50(150), bisexual, fertile; corollas yellow, pilose, tubes < cylindric throats, lobes 5, deltate; anthers yellow, rarely brown; style branch tips acute. **Fruits** glabrous or strigose; pappi of (0)14–20 scales, persistent, white or gray, fimbriate, sometimes awned.

Worldwide. 14 species; 1 species treated in *Flora.*

Galinsoga parviflora var. *parviflora* was collected on ballast at Linnton in 1911 but is not currently known in the state.

Galinsoga quadriradiata Ruiz & Pavón (p. 272)

Hairy gallant soldiers, hairy galinsoga, shaggy soldier, quickweed

Galinsoga ciliata (Raf.) S.F. Blake

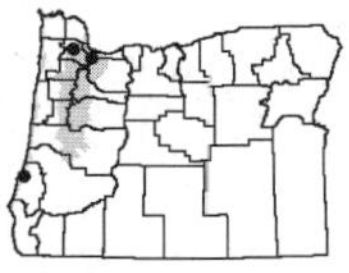

Plants 8–62 cm. **Leaves** 2–6 cm. **Phyllaries** deciduous, usually with sparse stipitate glands. **Paleae** inner narrower, entire or 2- or 3-lobed, lobes to ⅓ of total lengths, blunt. **Ray florets** corollas (0)5–8, rays 0.9–2.5 mm, white. **Disc florets** corollas 8–50. **Fruits** ray achenes usually shed with subtending phyllary plus 2 or 3 adjacent paleae. **2*n***=32, (48, 64).

Riverbanks, urban areas. Flowering Jun–Nov. 0–100 m. Est, WV. CA, WA; throughout most of North America; nearly worldwide. Exotic.

Both *Galinsoga parviflora* and *G. quadriradiata* are late-flowering, worldwide weeds and easily confused with one another (and reportedly hybridize). *Galinsoga parviflora* is distinguished by persistent and usually glabrous phyllaries with inner paleae lobes greater than one-third of their total length.

“Gallant soldiers” is probably derived from Cockney rhyming slang of the generic epithet, which honors Ignacio Mariano Martinez de Galinsoga, physician to Spanish queen consort Maria Luisa of Parma, director of the Real Jardín Botánico de Madrid, and founder of the Real Academia Nacional de Medicina.

GAMOCHAETA Cudweed

Herbs annual, biennial, or perennial; fibrous-rooted, sometimes taprooted or short-rhizomatous. **Stems** erect or decumbent, unbranched, floccose-tomentose. **Leaves** basal and cauline, alternate, simple, obovate to oblanceolate or linear, reduced distally, bases cuneate to ± cordate, margins entire, 1-veined, surfaces abaxially floccose-tomentose, adaxially lightly pubescent to nearly glabrous; sessile or with winged petioles. **Inflorescences** dense, rounded or cylindrical, spike-like arrays, sometimes more open and extending well down stems. **Heads** disciform. **Involucres** campanulate. **Phyllaries** erect in flower, spreading in fruit; inner equal, lanceolate or elliptic; outer gradually shorter, ovate to triangular. **Receptacles** glabrous; paleae 0. **Outer florets** 50–130, pistillate, yellowish, sometimes purplish distally. **Inner florets** 2–7, bisexual, tubular, scarcely lobed, yellowish, sometimes purplish distally. **Fruits** oblong, slightly flattened; pappi of numerous, basally connate bristles, deciduous in rings.

Nearly worldwide. ~50 species; 1 species treated in *Flora*.

Gamochaeta is a segregate of the formerly large and diverse genus *Gnaphalium*.

Gamochaeta ustulata (Nuttall) Holub (p. 272)

Purple cudweed

Gamochaeta purpurea (L.) Cabrera, *Gnaphalium purpureum* L., *Gnaphalium purpureum* L. var. *purpureum*

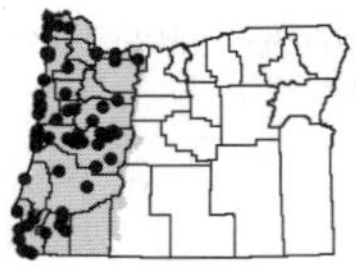

Plants annual, biennial, or perennial, 0.4–45 cm. **Leaves** basal usually withering before flowering; cauline blades spatulate to oblanceolate, 1–10 cm. **Inflorescences** bracts shorter than to slightly surpassing flower clusters. **Involucres** narrowly campanulate, 4.5–5 mm. **Phyllaries** in 4–6 series, chartaceous, becoming brownish in fruit, sometimes green along midlines, tips sharply acute, surfaces tomentose. **Florets** corollas 1–2 mm. **Fruits** 0.7–0.8 mm, tan or brownish; pappi white.

Forests, grasslands, ocean bluffs, dunes, riverbanks, roadsides, disturbed areas. Flowering May–Aug. 0–1100 m. Casc, CR, Est, Sisk, WV. CA, WA; north to British Columbia. Native.

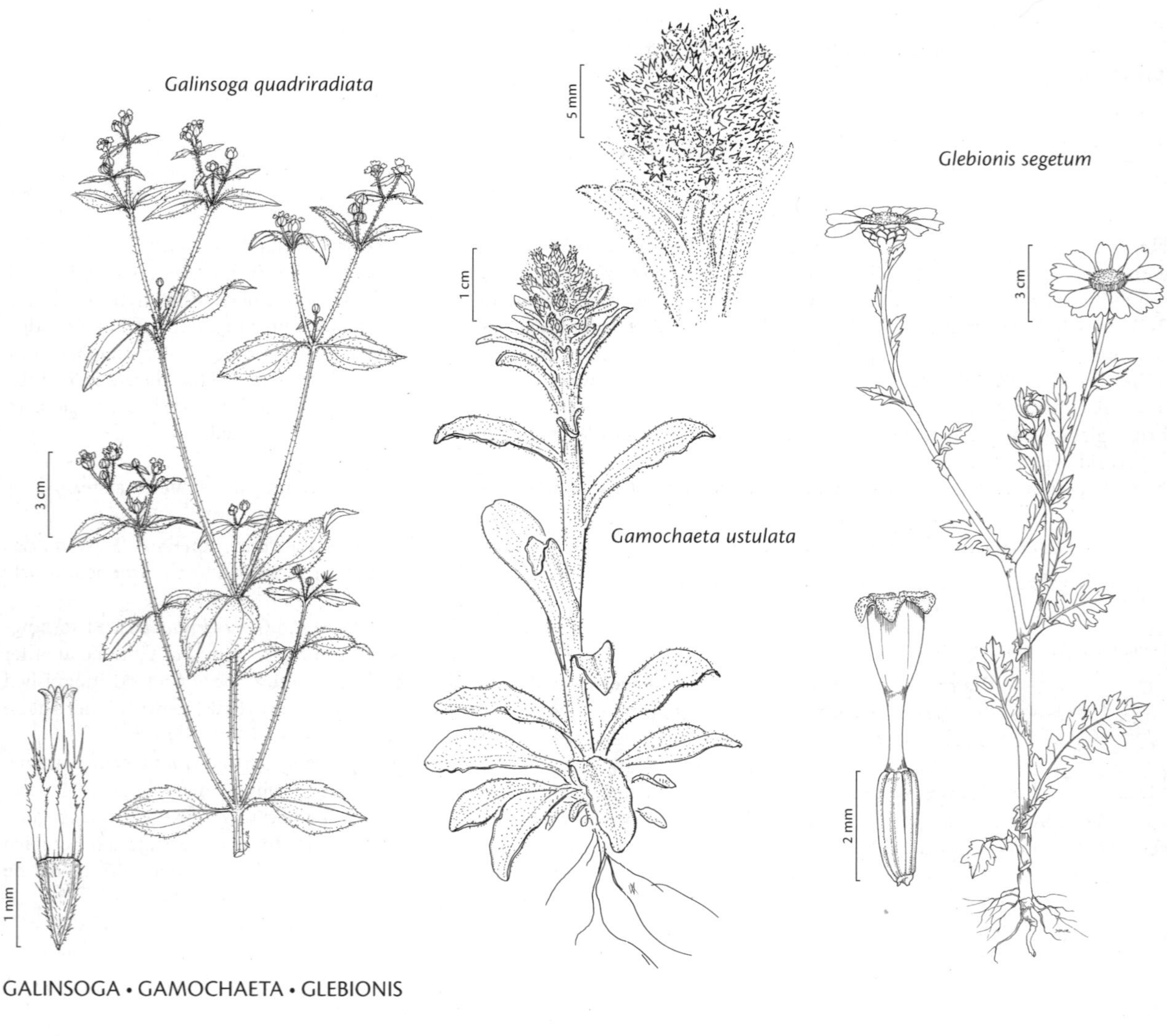

GALINSOGA • GAMOCHAETA • GLEBIONIS

GLEBIONIS Glebionis

Herbs annual; taprooted. **Stems** erect, branched, glabrous. **Leaves** mostly cauline, alternate, lanceolate to oblanceolate, tips acute or obtuse, surfaces glabrous; sessile or with broadly winged petioles. **Inflorescences** solitary, terminal heads. **Heads** radiate. **Involucres** hemispheric. **Phyllaries** in 3–4 series, erect in flower, spreading in fruit, inner obovate, 3–7-veined, tips enlarged membranous; outer broadly elliptic to triangular, often with brownish margins and membranous tips. **Receptacles** convex to hemispheric; paleae 0. **Ray florets** pistillate; corollas broadly elliptic, tips truncate, often 3-toothed. **Disc florets** bisexual; corollas tubular, widened distally, 5-lobed, lobes triangular, reflexed; stamens slightly exserted; stigma lobes equaling or slightly exceeding anther tubes. **Fruits** dimorphic; ray fruits elliptic-ovoid; disc fruits distinctly or obscurely 10-ribbed; pappi 0.

Africa, Asia, Europe, North America. 2 species; 1 species treated in *Flora*.

A second species of *Glebionis*, *G. coronaria* (crown daisy), occurs in California and might be looked for in Oregon. It differs from *G. segetum* in its usually two- to three-times-pinnately lobed leaves.

Glebionis is one of several genera that previously were included in *Chrysanthemum* but that mostly are well separated from each other phylogenetically (Watson et al. 2000). The name *Chrysanthemum* is now limited to the cultivated florists' chrysanthemum, *C. indicum*.

Glebionis segetum (Linnaeus) Fourreau (p. 272)

Corn marigold

Chrysanthemum segetum L.

Plants 1–6+ dm. **Leaves** lanceolate or oblanceolate, 2.5–6+ cm, sometimes 1-pinnate, margins entire to dentate, 1–5-veined. **Heads** 2–6 cm in diameter. **Ray florets** 10–20+; corollas 8–20 mm, yellow. **Disc florets** 50–100+; corollas 3–4 mm, yellow. **Ray fruits** weakly 3-angled, 2–3 mm, scarcely winged. **Disc fruits** compressed-columnar, 2–3 mm, brown to blackish. **2*n***=18.

Beaches, roadsides, disturbed areas. Flowering May–Sep. 0–50 m. Est. CA; AK, eastern US; Africa, Asia, Europe. Exotic.

GLYPTOPLEURA Glyptopleura

Herbs annual, sap milky; taprooted. **Stems** prostrate, leafy, forming compact, cushion-like growth, simple or branched, glabrous, warty. **Leaves** basal and cauline, oblanceolate-elliptic, pinnate with short, closely spaced lobes, denticulate with white border, tips obtuse, surfaces glabrous; petioles winged. **Inflorescences** solitary, terminal heads or 2–3 in cyme-like clusters. **Heads** ligulate; peduncles short. **Involucres** campanulate in flower, ovoid in fruit. **Phyllaries** in 1–2 series, equal, erect, lanceolate to linear, often purple-tinged, margins coarsely ciliate, tips acute. **Receptacles** ± flat, smooth, glabrous; paleae 0. **Florets** bisexual; ligules equaling or slightly exceeding involucres. **Fruits** columnar, slightly curved, bases tapering, tips flared and cup-shaped, ribs 4–5, smooth, inter-rib spaces with line of pits or small, rounded protuberances; short-beaked; pappi of several series of deciduous or persistent, fine, white, non-plumose bristles. **2*n***=18.

Western United States. 2 species; 1 species treated in *Flora*.

Glyptopleura marginata D.C. Eaton (p. 273)

Carveseed

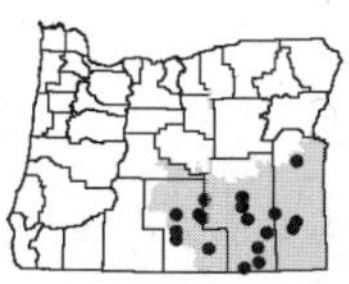

Plants 1–6 cm. **Leaves** 1–6 cm. **Involucres** 10–14 mm. **Phyllaries** 5–14, glabrous. **Florets** 8–18; ligules 4–10 mm, white to pale yellow, aging pink-purple. **Fruits** 3–5 mm, straw-colored, glabrous; beak 0.5–1 mm.

Desert scrub, dunes, meadows, washes, lake margins. Flowering May–Jun. 1000–1700 m. BR, Owy. CA, ID, NV; southeast to UT. Native.

Glyptopleura marginata and *G. setulosa* (not in Oregon) were earlier considered varieties of a single species. However, they differ consistently in length of florets, those of *G. setulosa* being much showier and long-exserted, and perhaps therefore in breeding system as well. Recent references place them as distinct species, *G. marginata* having the wider range of the two.

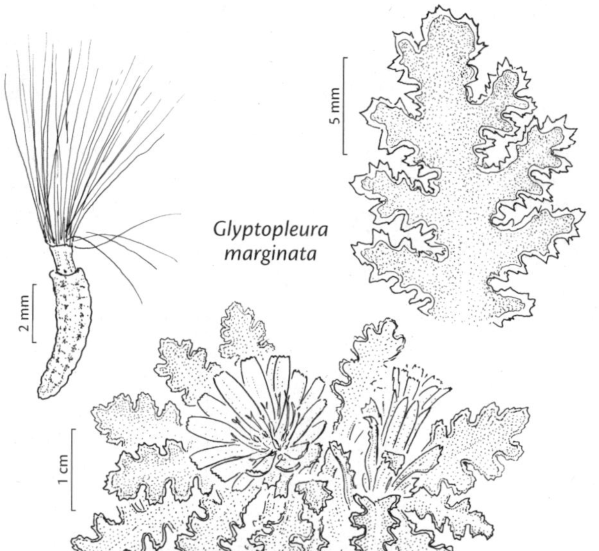

GLYPTOPLEURA

GNAPHALIUM Cudweed

Herbs annual or biennial; taprooted or fibrous-rooted. **Stems** prostrate, decumbent, or erect, sometimes short and clump-forming, simple to well branched, lightly to densely floccose- or pilose-tomentose. **Leaves** basal usually withered by flowering; cauline alternate, narrowly oblanceolate to elliptic or linear, margins entire, 1-veined, tips obtuse or acute, surfaces strigillose to floccose- or pilose-tomentose; sessile or winged short-petiolate. **Inflorescences** axillary or terminal, tight clusters; bracts linear to elliptic. **Heads** disciform. **Involucres** campanulate or hemispheric. **Phyllaries** in 3–5 series, erect, spreading in fruit; inner equal, oblong-lanceolate, tips obtuse or acute, white; outer ovate to triangular, gradually shorter. **Receptacles** flat, smooth; paleae 0. **Outer florets** 40–80, pistillate, white to purplish. **Inner florets** 4–7, bisexual, minutely 5-lobed, white to purplish. **Fruits** oblong, glabrous or papillate; pappi of 8–12 equal, deciduous bristles.

Nearly worldwide. ~38 species; 2 species treated in *Flora*.

Key to *Gnaphalium* species:

1. Bracts oblanceolate to elliptic, shorter to slightly longer than clusters; some inner phyllary tips obtuse . . . . ***G. palustre***

1' Bracts linear to narrowly oblong, longer than clusters; inner phyllary tips acute . . . . ***G. uliginosum***

Gnaphalium palustre Nuttall

Lowland cudweed, western marsh cudweed

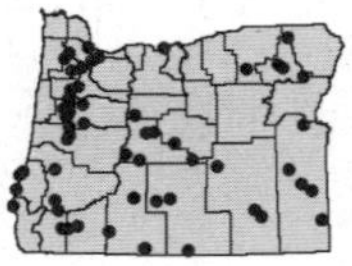

Plants annual, 2–40 cm. **Leaves** broadly linear to elliptic or oblanceolate, 0.5–3.5 cm. **Inflorescences** bracts oblanceolate to elliptic, shorter than to slightly longer than clusters. **Involucres** 2.5–4 mm. **Phyllaries** becoming pale brown in fruit, basally tomentose, distally glabrous; some inner tips obtuse. **Fruits** 0.5 mm, light brown; pappi white. **2*n***=14.

Mixed woodlands, shrublands, grasslands, lake margins, riverbanks, pastures, dunes, marshes, dry beds of vernal pools, roadsides, disturbed areas. Flowering May–Oct. 0–2800 m. All ecoregions. CA, ID, NV, WA; north to British Columbia, northeast to Saskatchewan, east to NE, southeast to NM, south to Mexico. Native.

Gnaphalium uliginosum Linnaeus (p. 274)

Marsh cudweed

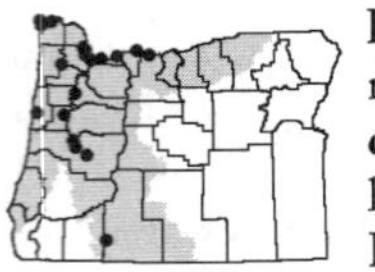

Plants annual, 1–20+ cm. **Leaves** linear or narrowly oblanceolate, 1–5 cm. **Inflorescences** bracts linear to narrowly oblong, longer than clusters. **Involucres** 2–4 mm. **Phyllaries** becoming brown in fruit, basally tomentose, distally glabrous; inner tips acute. **Fruits** 0.5–0.6 mm, light brown; pappi white. **2*n***=14.

Grasslands, riverbanks, dry lakebeds, roadsides, disturbed areas. Flowering Jun–Oct. 0–1100 m. Casc, Col, CR, ECas, Est, WV. ID, WA; north to AK, northeast to Greenland, southeast to VA; Europe. Exotic.

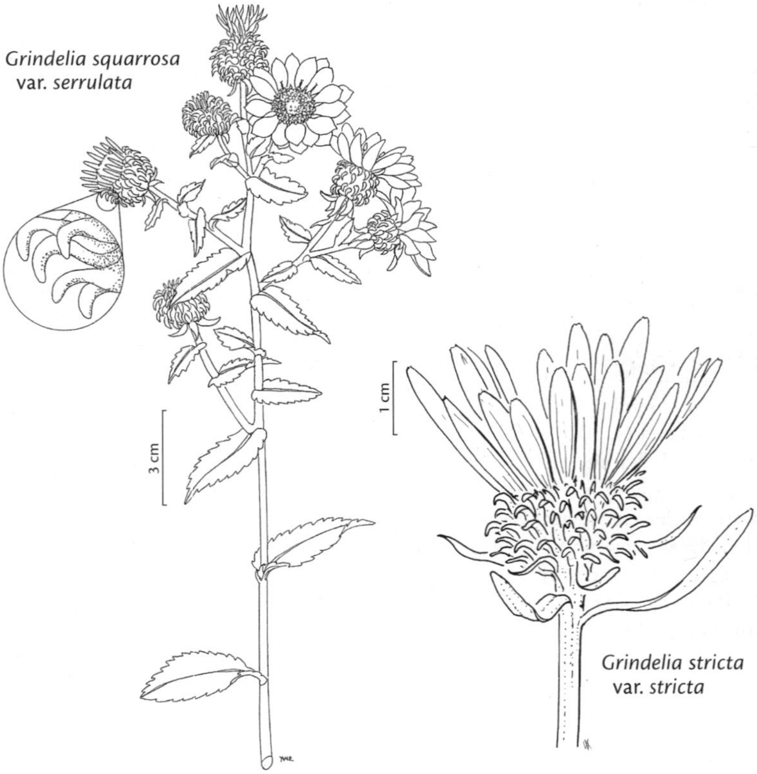

GNAPHALIUM • GRINDELIA

GRINDELIA Gumplant, gumweed

Abigail J. Moore

Herbs biennial or perennial; taprooted. **Stems** prostrate to erect, few- to many-branched, usually glabrous. **Leaves** basal and cauline, alternate, simple, reduced distally, surfaces usually glabrous and glandular-punctate; sessile. **Inflorescences** terminal, corymb-like arrays with branches diverging at a 20–60° angle from main stem; bracts similar in shape to distal leaves but smaller. **Heads** radiate or discoid. **Involucres** campanulate to hemispheric, usually resinous and sticky, usually widening with age. **Phyllaries** in multiple series, linear to lanceolate, chartaceous in basal ¼–¾, green at tips, tips usually loose or recurved, surfaces glandular-punctate and usually glabrous. **Receptacles** flat or convex, ± pitted; paleae 0. **Ray florets** pistillate; corollas yellow, obscurely 3-lobed, much exceeding involucres. **Disc florets** bisexual; corollas yellow, 5-lobed. **Fruits** faintly to strongly longitudinally grooved, with crown of 3–4 bumps, white to brown, glabrous; central fruits straight; marginal fruits curved; pappi of 3–4 twisted, deciduous awns, tapering from base to tip, margins entire to serrulate.

North America, South America. ~60 species; 5 species and 1 hybrid treated in *Flora*.

North American *Grindelia* has had several quite divergent taxonomic treatments in recent years. *Grindelia* exhibits substantial ecological and morphological diversity, with the morphological diversity often being correlated with ecology. However, much of this variation is at the level of individual populations, so that it is not always clear how the populations should be grouped into species, and not all individuals can be keyed out unambiguously, especially when distribution and habitat are not considered.

Key to *Grindelia* species:

1. Stems with multicellular, non-glandular, crinkled, matted hairs; cauline leaves with upright, glandular hairs, sometimes also with non-glandular hairs, lacking punctate glands . . . . ***G. integrifolia***

1' Stems usually glabrous to sparsely strigose, or with fine multicellular hairs, or with hairs below heads; cauline leaves glabrous, subglabrous, with scattered hairs, and/or glandular-punctate.

  2. At least some outer phyllaries coiled > 270° (with tips pointing back toward head instead of down or away from head).

    3. Cauline leaves serrate, with rounded teeth, each tipped by a rounded, yellow bump; heads radiate . . . . ***G. squarrosa*** var. ***serrulata***

    3' Cauline leaves entire or serrate with pointed teeth; heads radiate or discoid . . . . ***G. nana***

  2' Outer phyllaries spreading, erect, or recurved, but none coiled > 270°.

    4. Upper cauline leaves widest at bases.. . . . . ***G. integrifolia*** × ***G. nana*** var. ***nana***

    4' Upper cauline leaves widest at tips or ± same width throughout; heads borne singly.

      5. Stems prostrate, ascending, or erect; upper cauline leaves usually widest at tips. . . . . ***G. stricta*** var. ***stricta***

      5' Stems erect; upper cauline leaves ± same width throughout. . . . . ***G. hirsutula***

Grindelia hirsutula Hooker & Arnott

Hairy gumweed

Grindelia hirsutula Hook. & Arn. var. *davyi* (Jeps.) M.A. Lane

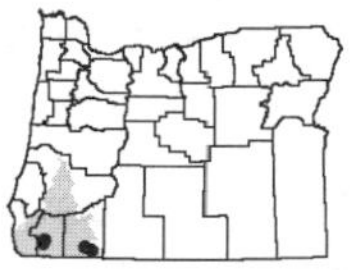

Plants perennial. **Stems** erect, 30–75 cm, glabrous to sparsely strigose. **Basal leaves** similar to cauline but larger and more strongly oblanceolate, up to 13 × 1.7 cm, usually present at flowering. **Cauline leaves** oblong to oblanceolate, 1–9 × 0.3–0.8 cm, bases narrowly cordate to cuneate, margins entire to sparsely toothed, tips rounded or obtuse, sometimes apiculate, surfaces glabrous and glandular-punctate; distal reduced in length, usually ± same width throughout. **Heads** 3–15 per main stem, borne singly, radiate; peduncles 0.2–14 cm. **Involucres** 6–12 × 9–21 mm. **Phyllaries** in 4–5 series, glabrous; tips of outer erect to recurved 90°; tips of inner usually erect. **Ray florets** 10–24; rays 7–9 × 1.8–2.4 mm. **Disc florets** 5.5–6.5 mm. **Fruits** 4.5–5 × 1.5–2 mm, white or light tan; crowns 0–0.2 mm; pappus awns ~5 mm. **2*n***=12, 24.

Open serpentine areas. Flowering Jul–Sep. 0–1600 m. Sisk. CA; British Columbia. Native.

Oregon plants of this species are distinct from other *Grindelia* in the state due to their rounded, mostly untoothed leaves, phyllaries that are usually erect, and distribution in serpentine areas in the southwestern part of the state. They are also distinct from the *G. hirsutula* from California, which have larger, usually tomentose heads and toothed leaves.

Grindelia integrifolia de Candolle

Puget Sound gumplant, Willamette Valley gumweed

Grindelia integrifolia DC. var. *integrifolia*

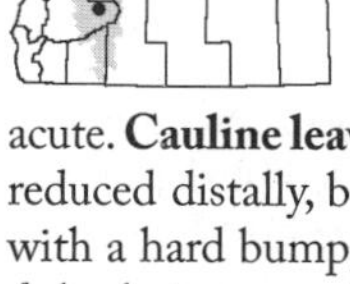

Plants perennial. **Stems** erect, 40–63 cm, with multicellular, crinkled, matted hairs, and shorter gland-tipped hairs. **Basal leaves** narrowly oblanceolate, narrowed to base, 4–17 × 0.6–3 cm, tips obtuse to abruptly acute. **Cauline leaves** lanceolate, 4–12.5 × 0.8–3.5 cm, gradually reduced distally, bases cordate, margins entire to serrate, teeth with a hard bump, tips acuminate, surfaces with upright, glandular hairs, sometimes also with non-glandular hairs, lacking punctate glands. **Heads** 3–32 per main stem, borne singly or in tight clusters of 2–3, radiate; peduncles 0–15 cm. **Involucres** 5–10 × 7–22 mm. **Phyllaries** in 3–4 series, with glandular hairs; tips of outer spreading; tips of inner erect to spreading. **Ray florets** 18–28; rays 8–12 × 2.1–2.8 mm. **Disc florets** 4.5–6.2 mm. **Fruits** 3.5–4 × 1.8–2 mm, white to brown; crowns 0.05–0.3 mm; pappus awns 2–3 mm. **2*n***=12.

Freshwater marshes. Flowering Jun–Sep. 0–300 m. Casc, WV. Native.

This species is easily identified by long, multicellular hairs on the stems that are visible with the naked eye and stalked glands on the leaves. It hybridizes extensively with *G. nana* almost throughout its range (see separate discussion of hybrids).

Grindelia integrifolia de Candolle × *Grindelia nana* var. *nana* Nuttall

Willamette Valley hybrid gumplant

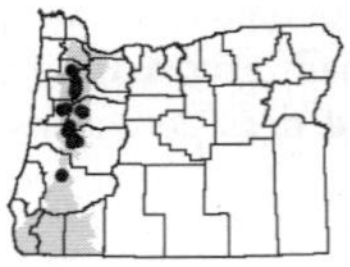

Plants perennial. **Stems** erect, 40–72 cm, usually glabrous or with a few sparse hairs below heads. **Basal leaves** often withered at flowering, narrowed to base, 7–27 × 0.5–3 cm. **Cauline leaves** oblong to lanceolate, 2–11 × 0.3–2 cm, reduced distally, bases cordate, margins entire to serrate, tips acute to acuminate, surfaces subglabrous and glandular; upper widest at bases. **Heads** 2–40 per main stem, often clustered, radiate; peduncles 0.2–11 cm. **Involucres** 5–12 × 10–22 mm. **Phyllaries** in 4 series, glabrous; tips of outer erect to recurved < 270°; tips of inner erect to recurved 90°. **Ray florets** 18–24; rays 8–12 × 1.5–3 mm. **Disc florets** ~6 mm. **Fruits** 3.5–4 × 1.8–2 mm, tan, faintly longitudinally grooved; crowns < 0.1 mm; pappus awns 3–3.5 mm.

Seasonally moist open habitats. Flowering Jun–Oct. 50–200 m. Sisk, WV. Native.

These plants are likely a combination of first-generation hybrids and later-generation hybrids that may have displaced one or both parents in the local area. They can be distinguished from *G. integrifolia* by having leaves that are almost glabrous instead of glandular hairy and from *G. nana* by having leaves that are widest at their cordate bases instead of having leaves that are widest distally or that have a more or less uniform width.

Grindelia nana Nuttall

Low gumweed

Plants perennial. **Stems** erect, 12–80 cm, glabrous. **Basal leaves** sometimes withered at flowering, similar to cauline but larger and more narrowed to bases, to 16 cm. **Cauline leaves** oblong to oblanceolate, reduced distally, margins entire to deeply serrate, often only lower leaves toothed, often short-ciliate with hairs 0.1–0.2 mm, tips rounded, acute, or acuminate, surfaces glabrous and glandular-punctate. **Heads** 1–23 per main stem, discoid or radiate; peduncles 0.6–10.5 cm. **Phyllaries** in 4–5 series, tightly coiled, glabrous. **Ray florets** 0 or 11–28. **Disc florets** 5.5–6.5 mm. **Fruits** 2–5 × 1–1.5 mm, white to brown, faintly longitudinally grooved; crowns to 0.5 mm; pappus awns 4.5–5 mm. **2*n***=12.

2 varieties.

Grindelia nana can be distinguished from all other Oregon species (except *G. squarrosa*) by its tightly coiled phyllaries and from *G. squarrosa* by having leaves with pointed teeth or entire margins, instead of leaves with rounded teeth.

Key to *Grindelia nana* varieties:

1. Heads radiate . . . . **G. n.** var. ***nana***
1' Heads discoid . . . . **G. n.** var. ***discoidea***

Grindelia nana Nuttall var. *discoidea* (Nuttall) A. Gray

Columbia gumplant

Grindelia columbiana (Piper) Rydb., *Grindelia nana* Nutt. ssp. *columbiana* Piper

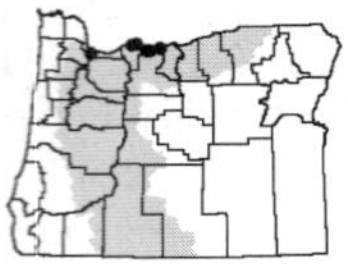

Cauline leaves 1.5–15 1.0.2–1.6 cm, bases usually slightly cordate, sometimes cuneate. **Heads** discoid. **Involucres** 6–10 × 10–19 mm. **Phyllaries** sometimes continuing down stem up to 10 mm below head and gradually transitioning to stem leaves, outer tips recurved 270–450°, inner tips recurved 90–180°. **Ray florets** 0.

Dry, open, rocky areas. Flowering Jul–Sep. 0–400 m. Casc, Col, ECas, WV. WA. Native.

This is the only discoid *Grindelia* in Oregon. It is endemic to the areas near the Columbia River in Oregon and Washington.

Grindelia nana Nuttall var. *nana*

Idaho gumweed

Grindelia nana Nutt. var. *integrifolia* Nutt., *Grindelia nana* Nutt. ssp. *nana*

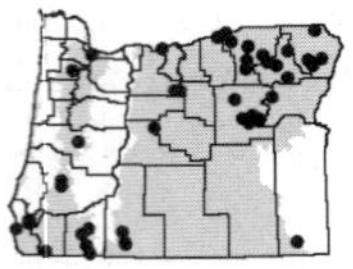

Cauline leaves 1–10 × 0.2–1.9 cm, narrowed to base. **Heads** radiate. **Involucres** 7–15 × 10–21 mm; **Phyllaries** tips of outer recurved (90)180–360°, some at least to 270°, of inner erect to recurved 180°. **Ray florets** 11–28; rays 7–14 × 1.5–3 mm.

Open areas, roadsides, streambanks, clearings in forests, serpentine. Flowering May–Oct. 100–2000 m. BR, BW, Col, ECas, Est, Lava, Sisk, WV. CA, ID, NV, WA; northeast to MT, southeast to UT. Native.

Like many species of *Grindelia*, *G. nana* is somewhat variable and can be difficult to distinguish from both *G. hirsutula* and *G. stricta*. In particular, plants growing along streams tend to be larger in all respects and to have less tightly coiled phyllaries. Differences in habitat can be helpful when morphology is ambiguous, with *G. hirsutula* largely confined to serpentine-influenced substrates in the southwestern corner of the state and *G. stricta* in salt marshes and other coastal habitats. *Grindelia nana* is the most widespread species in Oregon; however it appears to be getting displaced by *G. squarrosa*, at least in more disturbed habitats.

Grindelia squarrosa (Pursh) Dunal var. *serrulata* (Rydberg) Steyermark (p. 274)

Curlycup gumweed, resinweed

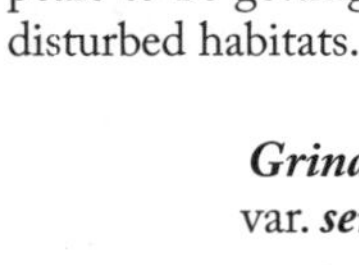

Plants biennial. **Stems** erect or ascending, 11–52 cm, glabrous. **Basal leaves** oblanceolate to spatulate, to 16 × 3 cm. **Cauline leaves** oblong to oblanceolate (lower sometimes oblanceolate to spatulate), sometimes narrowest in middle, 1.4–6 × 0.4–1.5 cm, reduced distally, bases cordate, margins usually serrate, teeth rounded, tipped with minute, rounded, yellow bumps, tips acute or rounded, surfaces glabrous and glandular-punctate. **Heads** 4–39 per main stem, radiate; peduncles (0.3)1–10 cm. **Involucres** (4)6–9 × (5)9–17 mm. **Phyllaries** in 3–6 series, glabrous; tips of outer recurved 180–360°, some at least to 270°; tips of inner erect to recurved 180°. **Ray florets** 17–48; rays 6–9 × 1.4–3 mm. **Disc florets**

5–6 mm. **Fruits** 1.9–3.3 × 0.9–1.2 mm, white to tan; crowns ~0.1 mm; pappus awns 3–6 mm. **2*n***=12.

Dry, usually disturbed, open areas. Flowering Jul–Sep. 50–1400 m. BW, Lava, Owy. CA, ID, NV, WA; north to British Columbia, east to SD, southeast to NM. Native.

Although *G. squarrosa* is similar to *G. nana* in the form of its inflorescence and can sometimes be difficult to distinguish, the two species are rather distantly related and are unlikely to hybridize. *Grindelia squarrosa* appears to be increasing throughout its range and taking over areas that were occupied until recently by *G. nana* (and other species in other states).

Grindelia stricta de Candolle var. ***stricta*** (p. 274)

Coastal gumweed, Oregon gumplant, Oregon gumweed

Grindelia arenicola Steyerm., *Grindelia integrifolia* DC. var. *macrophylla* (Greene) Cronquist, *Grindelia stricta* DC. ssp. *stricta*

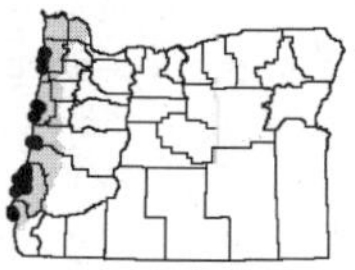

Plants perennial. **Stems** prostrate, ascending, or erect, 13–63 cm, mostly glabrous but with fine multicellular hairs below heads. **Basal leaves** (if present) similar to lower cauline leaves. **Cauline leaves** oblong to oblanceolate to spatulate, 1–10.5 × 0.5–2.8 cm, usually reduced in size and sometimes more oblong on side branches, margins entire or remotely serrate, ciliate, tips rounded, surfaces glabrous or with scattered hairs, glandular-punctate; upper usually widest distally, rarely ± same width throughout. **Heads** 1–14 per main stem, borne singly, radiate; peduncles 1.5–20 cm. **Involucres** 6–10 × 9–21 mm. **Phyllaries** in 2–5 series, glabrous; tips of outer spreading to recurved 180°; tips of inner erect to somewhat loose. **Ray florets** 18–35; rays 7–18 × 1.8–3.4 mm. **Disc florets** 5.2–6.5 mm. **Fruits** 4–7 × 1.2–2.4 mm, dark brown; crowns to 0.5 mm; pappus awns 3.5–5 mm. **2*n***=24.

Salt marshes, coastal bluffs, coastal sand dunes, tidal flats. Flowering Jun–Oct. 0–50 m. CR, Est. CA, WA; north to AK. Native.

There are two somewhat distinct forms of *G. stricta* var. *stricta* in Oregon that may possibly deserve recognition as separate varieties. One form (corresponding to Steyermark's *G. stricta*) is upright and has cauline leaves that are much narrower at the base than at the tip and are more or less the same throughout the plant, only becoming reduced in size on branches bearing heads. This form grows in salt marshes. The other form (corresponding to Steyermark's *G. arenicola*) is prostrate with ascending side branches and has leaves that are at least half as wide at the base as they are at the tips. These plants usually have two distinct categories of leaves: wider leaves with more teeth on the prostrate main branches, and narrower leaves with fewer teeth on the side branches. This form is found in habitats with more open vegetation, such as sand dunes.

GUTIERREZIA Snakeweed, matchweed

Herbs, shrubs, or **subshrubs** annual or perennial; taprooted with woody caudices. **Stems** several to many, erect, simple or branched, minutely hispidulous or glabrate proximally, hispidulous and glandular-viscid distally. **Leaves** alternate, linear, tips acute, surfaces minutely hispidulous, often punctate-glandular; proximal nodes sometimes producing axillary fascicles of secondary leaves, especially on second-year stems. **Inflorescences** heads solitary and terminal, or in clusters in tight or open arrays. **Heads** radiate; peduncles glandular-viscid. **Phyllaries** in several series, erect in flower, slightly spreading in fruit, lanceolate to deltate, obscurely veined, surfaces glandular-viscid; inner equal; outer gradually shorter. **Receptacles** flat to conic, pitted; paleae 0. **Ray florets** 1–30, pistillate, yellow or white. **Disc florets** 1–150, bisexual; corollas tubular below, widened and funnel-shaped distally, yellow or white, lobes 5, triangular, recurved; stamens partly exserted; style branches slightly exserted. **Fruits** oblong-cylindrical, truncate, tapering toward bases, obscurely veined; pappi of linear-oblong or ovate-lanceolate, ± deciduous scales.

North America, South America. 28 species; 1 species treated in *Flora*.

Gutierrezia sarothrae (Pursh) Britton & Rusby (p. 277)

Broom snakeweed, matchweed, kindlingweed

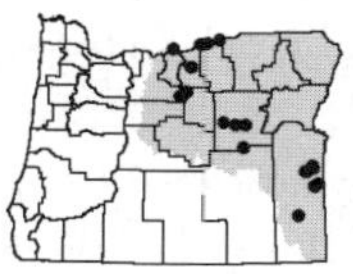

Subshrubs 5–60+ cm. **Leaves** basal and proximal cauline withering by flowering; cauline 1–5 cm, sometimes reduced distally, with 1–3 main veins. **Inflorescences** flat-topped, corymb-like arrays. **Heads** 3–6; peduncles ≤5 mm. **Involucres** 3–5 mm. **Ray florets** 3–8; corollas 2–5 mm, yellow. **Disc florets** 3–9; corollas 2.5–3 mm, yellow, slightly exserted from involucres. **Fruits** 0.8–2.2 mm, densely white-sericeous; pappi scales 0.5–1.5 mm. **2*n***=8, 16, 32.

Grasslands, shrublands, roadsides. Flowering Jun–Oct. 50–1500 m. BW, Col, Lava, Owy. CA, ID, NV, WA; throughout most of western North America. Native.

Gutierrezia sarothrae

GUTIERREZIA

HAZARDIA Bristleweed, goldenweed

Herbs or **subshrubs** perennial; taprooted. **Stems** branched or unbranched, glabrous to tomentose or villous. **Leaves** cauline, alternate, ovate to oblong or oblanceolate, bases truncate or auriculate, margins entire to denticulate or evenly dentate, tips obtuse to acute, surfaces glabrous, sparsely puberulent, or glandular-scabrous; sessile. **Inflorescences** heads solitary or in spike-, raceme-, or cyme-like arrays. **Heads** radiate or discoid, erect. **Involucres** campanulate. **Phyllaries** in 5–9 unequal series, erect, spreading in fruit, narrowly lanceolate to linear, chartaceous, tips acute to acuminate, surfaces glandular-puberulent. **Receptacles** flat; paleae 0. **Ray florets** 0 or 3–25. **Disc florets** bisexual; corollas yellow, narrowly funnelform, lobes 5, triangular, erect; stamens ± exserted; stigma lobes exserted. **Fruits** 5-angled, truncate, tapering toward bases, 8–14-veined; pappi of smooth bristles.

Western North America. 13 species; 1 species treated in *Flora.*

Hazardia whitneyi (A. Gray) Greene var. *discoidea* (J.T. Howell) W.D. Clark (p. 279)

Whitney's bristleweed, Whitney's goldenweed

Haplopappus whitneyi A. Gray, *Haplopappus whitneyi* A. Gray ssp. *discoideus* (J.T. Howell) D.D. Keck

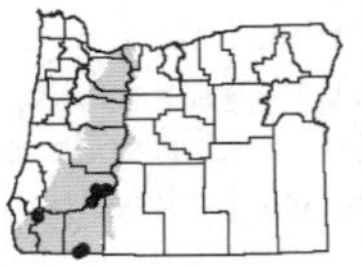

Plants 2–5 dm. **Leaves** 2–5 × 0.5–1.5 cm. **Involucres** 9–13 mm. **Heads** discoid. **Disc florets** 15–30; corollas 8–10 mm. **Fruits** 5–10 mm, surfaces glabrous; pappus bristles 7–9 mm, reddish brown.

Open coniferous forests, rocky slopes and ridges. Flowering Jul–Sep. 1300–2400 m. Casc, Sisk. CA. Native.

Prior to the study by Clark (1979), *Hazardia* was usually included in *Haplopappus*, a South American genus no longer recognized in North America. According to Clark (2006), additional molecular studies of generic relationships in Astereae may yet prove that *Hazardia* should be reassigned to *Haplopappus*, giving this taxon a bipolar New World distribution once again.

HELENIUM Sneezeweed

Herbs annual, perennial; taprooted or fibrous-rooted from stout caudices. **Stems** erect or decumbent, branched or unbranched, glabrous to puberulent or tomentose, sometimes minutely glandular, usually winged by decurrent leaf bases. **Leaves** basal and cauline, alternate, ovate to broadly or narrowly lanceolate, oblanceolate, oblong, or elliptic, bases attenuate or truncate, usually decurrent, margins entire to denticulate or serrate-dentate, tips obtuse to acute or acuminate, surfaces glabrous or puberulent, densely gland-dotted; lower petiolate, often winged or auriculate-clasping, upper sessile. **Inflorescences** solitary heads or in panicle- or corymb-like arrays. **Heads** radiate, occasionally discoid. **Involucres** ± globose or hemispheric to ovoid. **Receptacles** conic, ± globose, hemispheric, or ovoid, pitted; paleae 0. **Phyllaries** in 1–3 ± equal series, linear to lanceolate, sometimes basally connate, tips acuminate, surfaces strigillose or puberulent to tomentulose. **Ray florets** pistillate; corollas well exceeding involucres, adaxially glabrous, abaxially puberulent and gland-dotted, shallowly to deeply 3-lobed. **Disc florets** bisexual; corollas tubular, gradually enlarged distally, lobes 5; stamens exserted; stigmas exserted. **Fruits** columnar or obconic, truncate, 4–10-veined; pappi of aristate scales in 1 series.

North America, South America. 32 species; 3 species treated in *Flora.*

Some older reports (approximately 100 years ago or more) of *H. puberulum* in Oregon have been documented. The species does not, however, appear to have become established.

Key to *Helenium* species:

1. Some leaves with denticulate or serrate-dentate margins; stems often with 4+ branches . . . . ***H. autumnale***
1' All leaf margins entire; stems unbranched or with 1–3 branches.
    2. Pappi 1.3–2.5 mm; heads 1–20; peduncles sparsely to moderately hairy . . . . ***H. bigelovii***
    2' Pappi 2.5–4.5 mm; heads 1(3); peduncles usually densely hairy . . . . ***H. bolanderi***

Helenium autumnale Linnaeus (p. 279)

Common sneezeweed

Helenium autumnale L. var. *autumnale*, *Helenium autumnale* L. var. *grandiflorum* (Nutt.) Torr. & A. Gray, *Helenium autumnale* L. var. *montanum* (Nutt.) Fernald

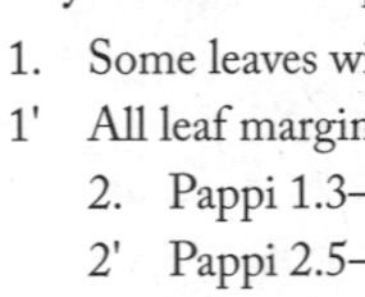

Plants perennial, 2–13 dm. **Stems** often with 4+ branches, glabrous or puberulent. **Leaves** basal usually withered by flowering; cauline lanceolate-elliptic or oblanceolate, 3–13 cm, bases attenuate or truncate, decurrent, margins generally entire but with some denticulate or serrate-dentate, tips acute, surfaces glabrous or minutely glandular. **Heads** 5–70+, radiate; peduncles moderately to densely hairy. **Involucres** 8–23 mm in diameter. **Phyllaries** 5–12 mm, linear, surfaces puberulent. **Ray florets** 8–21; corollas 10–23 mm, yellow. **Disc florets** 200–400+; corollas 2.5–5 mm, yellow, sometimes brownish distally. **Fruits** 1–2 mm, brown, sparsely hairy; pappi of 5–7 aristate scales, 0.5–1.5 mm. **2*n***=32, 34, 36.

Riverbanks, lake shores, wet meadows, swamps, marshes. Flowering Jul–Oct. 0–1500 m. Casc, Col, CR, ECas, Sisk, WV. CA, ID, NV, WA; throughout most of North America. Native.

This species was formerly divided into variety *grandiflorum* and variety *montanum*, but the morphological differences between varieties are too minor and intergrading to merit taxonomic recognition.

Helenium bigelovii A. Gray

Tall sneezeweed, Bigelow's sneezeweed

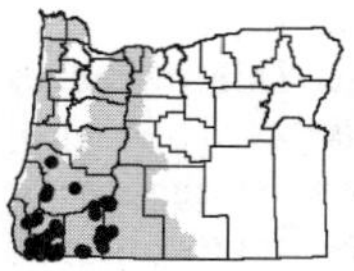

Plants perennial, 3–13 dm. **Stems** unbranched or with 1–3 branches, usually glabrous or puberulent distally to sparsely tomentose below heads, rarely densely long-tomentose distally. **Leaves** oblanceolate to lanceolate or elliptic, 7–30 cm, bases attenuate or truncate, decurrent, margins entire, tips acute or acuminate, surfaces glabrous. **Heads** 1–20, radiate; peduncles sparsely to moderately hairy. **Involucres** (14)17–22(25) mm in diameter. **Phyllaries** linear, 4–7 mm, surfaces puberulent. **Ray florets** 14–20; corollas 12–25 mm, yellow. **Disc florets** 250–500+; corollas 3–4.5 mm, yellow, sometimes brownish to purplish distally. **Fruits** 1.8–2.4 mm, brown, glabrous or hispid-pubescent; pappi of 6–8 aristate scales, 1.3–2.5 mm. **2*n***=32.

Meadows, bogs, swamps, thickets, seepage slopes, streambanks, serpentine. Flowering May–Aug. 0–2300 m. Casc, CR, ECas, Sisk. CA. Native.

Helenium bolanderi A. Gray

Coast sneezeweed

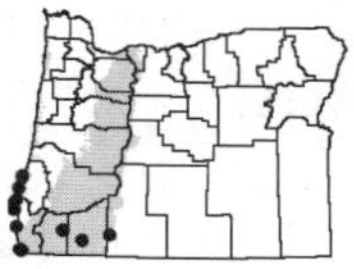

Plants perennial, 2–10+ dm. **Stems** unbranched or with 1–3 branches, glabrous or puberulent with densely long-tomentose below heads. **Leaves** obovate, oblanceolate, lanceolate, or elliptic, 8–30 cm, bases attenuate or truncate, decurrent, margins entire, tips acute or obtuse, surfaces glabrous or minutely glandular-puberulent. **Heads** 1(3), radiate; peduncles usually densely hairy. **Involucres** 18–34 mm in diameter. **Phyllaries** 4–11 mm, linear to narrowly lanceolate, surfaces puberulent or tomentose. **Ray florets** 15–30; corollas 10–25 mm, yellow, sometimes purple-streaked. **Disc florets** 300–750+; corollas 4–5.5 mm, yellow, sometimes brown to purple distally. **Fruits** 1.5–2.5 mm, brown, hispid-pubescent; pappi of 6–8 aristate scales, 2.5–4.5 mm. **2*n***=32.

Coastal bluffs, meadows, thickets, bogs. Flowering Jun–Sep. 0–2000 m. Casc, Est, Sisk. CA. Native.

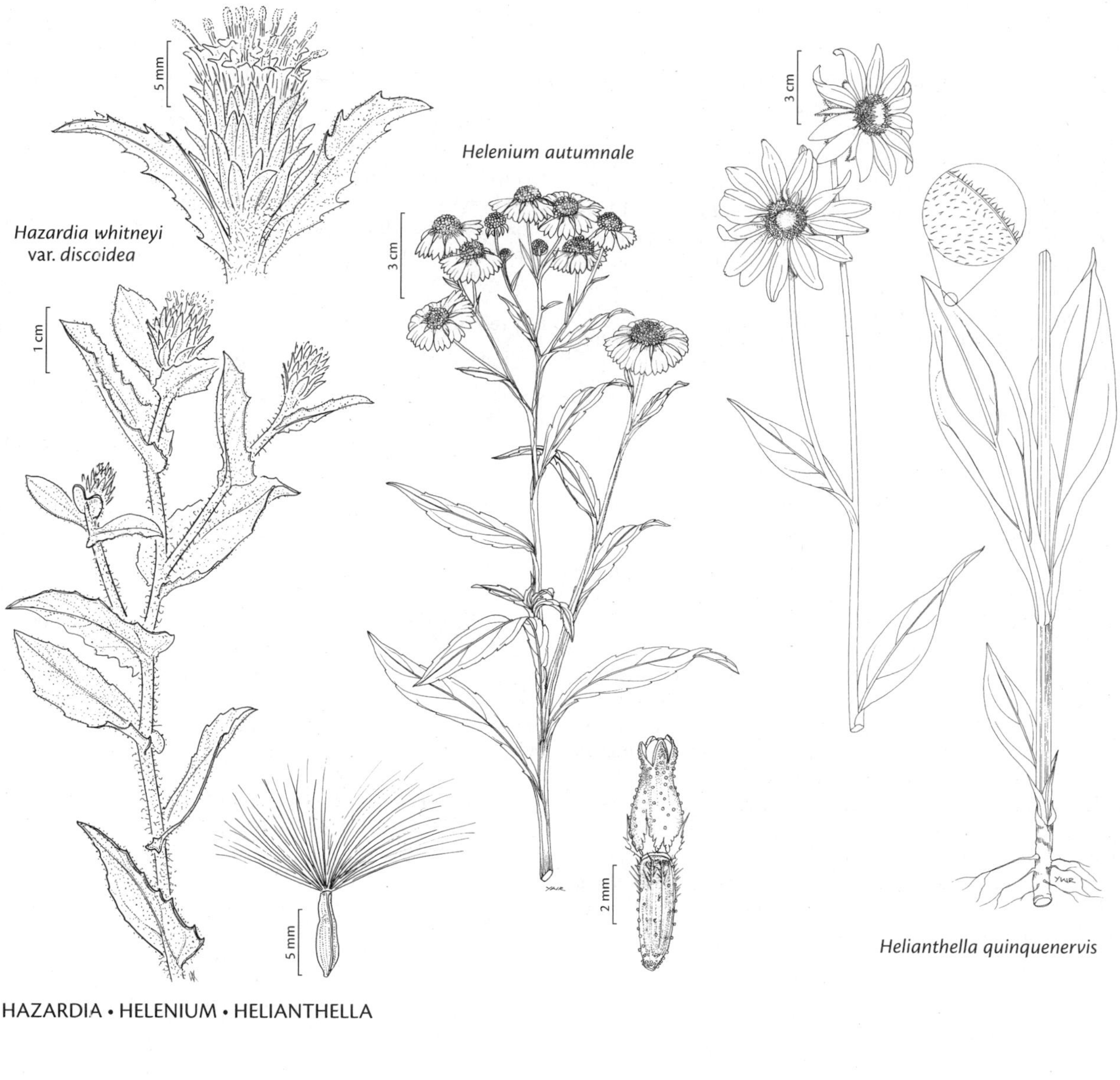

HELIANTHELLA Little sunflower, false sunflower, helianthella

Herbs perennial; from stout caudices. **Stems** erect, branched or unbranched, glabrous to puberulent, strigillose, or pilose-villous. **Leaves** basal or basal and cauline, opposite proximally, becoming alternate distally, narrowly or broadly elliptic to lanceolate or oblanceolate, bases attenuate, margins entire, tips acute, surfaces glabrous to strigillose or hispid; sessile or petiolate. **Inflorescences** heads solitary, terminal or in corymb-like arrays. **Heads** radiate. **Involucres** campanulate to hemispheric. **Phyllaries** in 3–6 equal series, narrowly lanceolate, margins ciliate, obscurely 1–3-veined, tips attenuate to acuminate, surfaces glabrous to strigillose or puberulent abaxially. **Receptacles** convex, bracteate; paleate. **Paleae** folded around fruits. **Ray florets** sterile. **Disc florets** bisexual; corollas cylindric, scarcely enlarged distally, lobes 5, triangular; stamens partly or wholly exserted; stigmas exserted. **Fruits** usually clasped by and falling with paleae, strongly flattened, obovate or obcordate, often notched at tips, glabrous or strigillose, sometimes narrowly winged; pappi 0 or of 2 awns and a minute, lacerate crown, persistent.

Western North America. 9 species; 3 species treated in *Flora*.

Key to *Helianthella* species:

1. Involucres 25–40+ mm in diameter. . . . . ***H. quinquenervis***
1' Involucres 15–25(30) mm in diameter.
  2. Proximal cauline leaves > central; fruit surfaces glabrous. . . . . ***H. californica*** var. ***nevadensis***
  2' Proximal cauline leaves < central; fruit surfaces strigillose. . . . . ***H. uniflora***

Helianthella californica A. Gray var. *nevadensis* (Greene) Jepson

Nevada helianthella, California helianthella

Helianthella californica A. Gray ssp. *nevadensis* (Greene) W.A. Weber

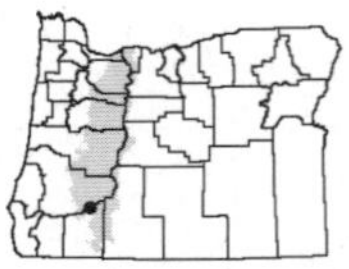

Plants 2–5+ dm. **Stems** ± glabrous proximally, puberulent distally. **Leaves** linear-elliptic or narrowly oblanceolate, 5–30 cm, margins smooth or scabrous, surfaces glabrous or strigillose; proximal cauline leaves > central; petiolate. **Involucres** 15–25 mm in diameter. **Phyllaries** 10–25 mm, margins ciliate, surfaces glabrous or strigillose. **Paleae** chartaceous, relatively firm. **Ray florets** 10–16; corollas 8–20 mm, yellow. **Disc florets** corollas 5–6 mm, yellow. **Fruits** strongly compressed, obovate, 7–8 mm, brown, surfaces glabrous; narrowly winged at tips; pappi 0 or of 2 short awns and a crown of lacerate scales.

Montane grasslands. Flowering May–Jul. 600–2000 m. Casc. CA. Native.

Helianthella quinquenervis (Hooker) A. Gray (p. 279)

Nodding helianthella, five-nerved helianthella

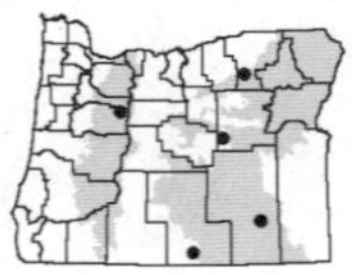

Plants 4–13 dm. **Stems** ± glabrous proximally, strigose to pilose or villous distally. **Leaves** broadly or narrowly oblanceolate, 5–35 cm, margins smooth or scabrous, surfaces glabrous strigillose or hispid; proximal cauline leaves > central; petiolate or sessile. **Involucres** 25–40+ mm in diameter. **Phyllaries** 12–20 mm, margins ciliate, surfaces glabrous. **Paleae** scarious, relatively soft. **Ray florets** 13–21; corollas 25–30+ mm, yellow. **Disc florets** corollas 4.5–5.5 mm, yellow. **Fruits** narrowly obovate, strongly compressed, 8–9 mm, brown, strigillose; not winged; pappi of 2 short awns and a minute, lacerate crown. **2*n***=15.

Woodlands, grasslands, riverbanks. Flowering Jun–Aug. 1300–2300 m. BR, BW, Casc. ID, NV; northeast to MT, east to SD, southeast to NM. Native.

Helianthella uniflora (Nuttall) Torrey & A. Gray

Rocky Mountain helianthella, oneflower helianthella

Plants 4–10 dm. **Stems** strigillose to sparsely puberulent or pilose-hirsute. **Leaves** lanceolate-elliptic, 5–30 cm, margins smooth or scabrous; surfaces glabrous to strigillose or hispid; proximal cauline leaves < central; petiolate or sessile. **Involucres** 15–25(30) mm in diameter. **Phyllaries** 5–20 mm, margins ciliate or not, surfaces glabrous or puberulent. **Paleae** chartaceous, relatively firm. **Ray florets** 11–15+; corollas 15–40 mm, yellow. **Disc florets** corollas 5–6 mm, yellow. **Fruits** narrowly obovate, compressed, 6–7 mm, blackish, strigillose; not winged; pappi of 2 short awns and a minute, lacerate crown. **2*n***=30.

2 varieties.

Key to *Helianthella uniflora* varieties:

1. Rays 30–40 mm; phyllaries glabrous abaxially . . . . ***H. u.*** var. ***douglasii***
1' Rays 20–30 mm; phyllaries puberulent abaxially . . . . ***H. u.*** var. ***uniflora***

Helianthella uniflora (Nuttall) Torrey & A. Gray var. *douglasii* (Torrey & A. Gray) W.A. Weber

Douglas' helianthella

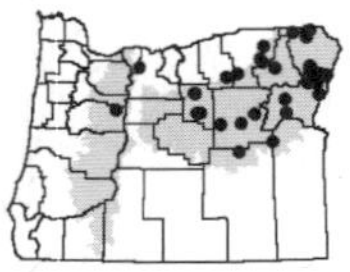

Involucres (15)20–25(30) mm in diameter. **Phyllaries** abaxially glabrous. **Ray florets** rays 30–40 mm.

Mixed conifer forests, pine woodlands, grasslands, alpine meadows, ridgetops, talus slopes. Flowering Jun–Aug. 800–2500 m. BW, Casc, Lava. ID, WA; north to British Columbia. Native.

Helianthella uniflora (Nuttall) Torrey & A. Gray var. *uniflora*

Rocky Mountain helianthella, common little sunflower

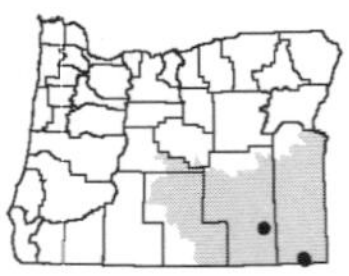

Involucres 15–20(25) mm in diameter. **Phyllaries** abaxially puberulent. **Ray florets** rays 20–30 mm.

Brushlands. Flowering Jun–Aug. 1700–2300 m. BR, Owy. ID; northeast to MT, east to WY, southeast to AZ. Native.

HELIANTHUS Sunflower

Herbs annual or perennial. **Stems** erect or decumbent, simple to well branched, glabrous to pustulose-hispid, pilose, or villous. **Leaves** cauline, opposite or alternate, surfaces pustulose-hispid to strigose or pilose; petiolate, often becoming sessile distally. **Inflorescences** heads solitary or in cyme- or panicle-like arrays. **Heads** radiate, rarely discoid, usually erect. **Involucres** hemispheric, campanulate, or cylindric. **Phyllaries** in 1–several series, ± equal, herbaceous. **Receptacles** flat to rounded; paleate. **Paleae** rectangular-oblanceolate, folded around fruits, tips acute or blunt. **Ray florets** (0)5–30+, sterile; rays usually yellow. **Disc florets** bisexual, fertile; corollas cylindric, tubular, equaling or exceeding involucres, lobes 5, yellow or red to purple; stamens usually exserted; style lobes exserted. **Fruits** usually clasped by and falling with paleae, compressed-quadrangular, obovoid, glabrous or pubescent; pappi of 2 scales plus 0–8 shorter scales, deciduous.

North America. 52 species; 4 species treated in *Flora*.

Helianthus species get their common name, sunflower, from their apparent ability to track the sun. Although they do follow light while in bud, mature flowers eventually all develop to face east, toward the morning sun.

Helianthus petiolaris and *H. tuberosus* have been sporadically collected or observed in Oregon as waifs.

Key to *Helianthus* species:

1. Plants annual; disc floret lobes usually reddish.
   2. Phyllaries 5–8 mm wide, ovate to lance-ovate; plants 10–30 dm . . . . ***H. annuus***
   2' Phyllaries 3–5 mm wide, lanceolate to lance-ovate; plants 6–15 dm . . . . ***H. bolanderi***

1' Plants perennial; disc floret lobes yellow.
   3. Plants not rhizomatous . . . . ***H. cusickii***
   3' Plants rhizomatous . . . . ***H. nuttallii*** ssp. ***nuttallii***

Helianthus annuus Linnaeus (p. 281)

Common sunflower

Helianthus annuus L. ssp. *annuus*, *Helianthus annuus* L. ssp. *lenticularis* (Douglas ex Lindl.) Cockerell

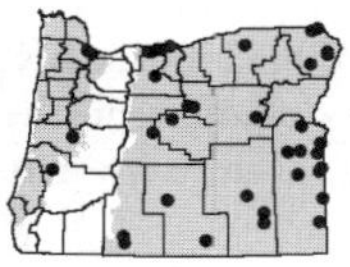

Plants annual, 10–30 dm. **Stems** erect, pustulose-hispid. **Leaves** mostly alternate, proximal pairs usually opposite, triangular to ovate or lanceolate, 3–30+ cm, bases truncate or rounded to attenuate, margins entire to serrate-dentate or denticulate, tips acute, surfaces spiculate or pustulose-hispid, sometimes setose on veins; petioles 1–7 cm. **Heads** solitary to numerous. **Involucres** hemispheric, 15–40+ mm in diameter. **Phyllaries** 20–30(10) in 3+ series, ovate to lance-ovate, 10–25 × 5–8 mm, margins hispid-ciliate, tips acuminate, surfaces scabrous or hispid, rarely glabrous; outer gradually shorter. **Paleae** glabrous or hispid, inner tips 3-toothed. **Ray florets** 12–30+; rays 25–50 mm. **Disc florets** 50+; corollas 5–8 mm, throats ± bulbous at base, gradually widened distally, lobes usually reddish; anthers dark; style appendages yellow to dark. **Fruits** 3–10+ mm, glabrate; pappi of 2 lanceolate scales, 2–3.5 mm, with 0–4 additional smaller scales. $2n=34$.

Shrublands, pastures, banks of rivers and lakes, marshes, roadsides, disturbed areas. Flowering Jun–Sep. 0–1600 m. BR, BW, Col, CR, ECas, Lava, Owy, WV. CA, ID, NV, WA; throughout North America. Native.

Helianthus annuus is one of the few North American natives to become a major cultivated crop. Sunflower seeds were exported from North America as early as the 16th century and are grown in much of the world both for their seeds and oil.

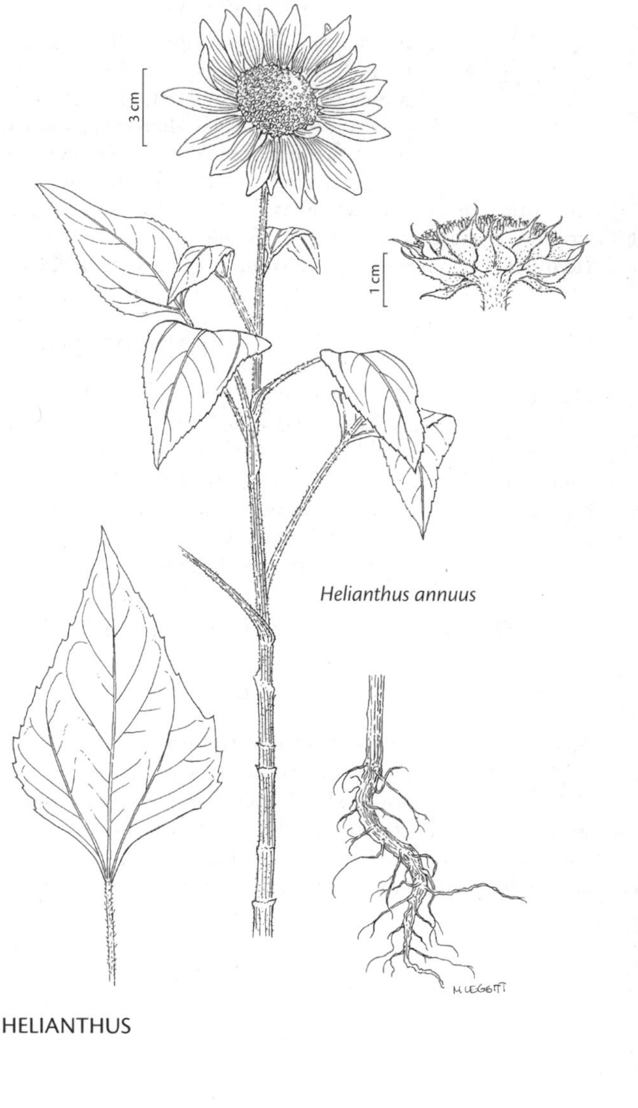

HELIANTHUS

Helianthus bolanderi A. Gray

Bolander's sunflower

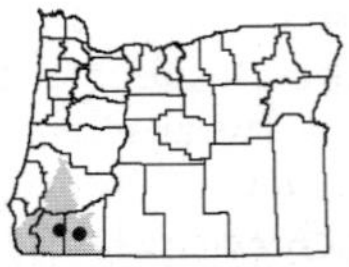

Plants annual, 6–15 dm. **Stems** erect, pustulose-hispid to setose or villous. **Leaves** mostly alternate, ovate to lanceolate or linear-lanceolate, 3–15 cm, bases wedge-shaped or attenuate, margins entire or denticulate to serrulate, tips acute, surfaces hispid or strigose; petiolate, becoming sessile distally, petioles 1–4 cm. **Heads** solitary to numerous. **Involucres** hemispheric, 17–25 mm diameter. **Phyllaries** 10–18 in 2–3 series, lanceolate or lance-ovate, 9–25 × 3–5 mm, margins strigose-ciliate, tips acuminate, surfaces pustulose-hispid to strigose or silky-villous; outer nearly equal to inner or sometimes much enlarged and bract-like. **Paleae** glabrous, inner tips 3-toothed. **Ray florets** 12–17; rays 14–20 mm. **Disc florets** 75+; corollas 5–7 mm, throats slightly widened distally, lobes usually reddish. **Fruits** 3.5–4.5 mm, glabrate; pappi of 2 lanceolate scales, 1.5–3 mm. **2*n***=34.

Open fields, roadsides. Flowering Jul–Oct. 300–1400 m. Sisk. CA. Native.

Helianthus cusickii A. Gray

Cusick's sunflower, turnip-root sunflower

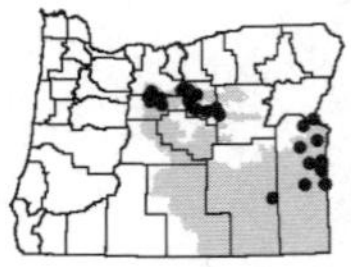

Plants perennial, 6–12 dm; with thick taproots or root-crowns, not rhizomatous. **Stems** erect or decumbent, hispid to strigose or setose. **Leaves** alternate, proximal pairs often opposite, lanceolate to elliptic-lanceolate or linear, 3–15 cm, bases rounded or attenuate, margins entire, tips acute or acuminate, surfaces glabrous to hispid, strigose, or setose; petioles 0–1 cm. **Heads** solitary to numerous. **Involucres** hemispheric, 12–28 mm in diameter. **Phyllaries** 12–18 in 3–5 series, lance-linear, 11–18 × 1–3 mm, margins hispid-ciliate, tips attenuate, surfaces pilose or strigose. **Paleae** sparsely pubescent, inner tips entire or 3-toothed. **Ray florets** 8–21; rays 10–40 mm. **Disc florets** 35+; corollas 6.5–7.5 mm, throats cylindric, lobes yellow. **Fruits** 4–5 mm, glabrous; pappi of 2 slender scales, 3–4 mm. **2*n***=34.

Shrublands, hillsides, roadsides. Flowering May–Aug. 100–1900 m. BR, Lava, Owy. CA, ID, NV, WA. Native.

Helianthus nuttallii Torrey & A. Gray ssp. *nuttallii*

Nuttall's sunflower, cordilleran sunflower

Helianthus nuttallii Torr. & A. Gray var. *nuttallii*

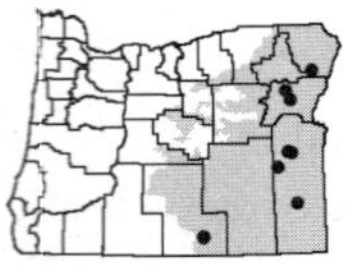

Plants perennial, 10–40 dm; rhizomatous. **Stems** erect, usually glabrous, occasionally pilose. **Leaves** opposite or distally alternate, broadly to narrowly lanceolate or elliptic, 3–20 × 0.5–4 cm, bases attenuate, margins entire or crenate-denticulate, tips acute or acuminate; petioles 0–1.5 cm. **Heads** usually few, sometimes numerous. **Involucres** hemispheric, 10–20 mm in diameter. **Phyllaries** 30–38 in 3–5 series, lance-linear, 8–15 × 1.5–3 mm, margins hispid-ciliate, tips conspicuously attenuate, surfaces minutely strigose; outer sometimes longer than inner. **Paleae** glabrous, inner tips blunt or acute. **Ray florets** 10–20; rays 5–7 mm. **Disc florets** 50+; corollas 5–7 mm, throats cylindric, lobes yellow. **Fruits** 3–5 mm, glabrous; pappi of 2 lanceolate scales, 2–4.5 mm, and 0–2 shorter scales. **2*n***=34.

Subalpine meadows, riverbanks, swamps, open slopes, roadsides, disturbed areas. Flowering Jun–Aug. 800–2600 m. BR, BW, Owy. CA, ID, NV, WA; north to British Columbia, northeast to Manitoba, east to NE, southeast OK, south to Mexico. Native.

HELMINTHOTHECA Ox-tongue

Herbs annual or biennial, sap milky; taprooted. **Stems** usually 1, erect, leafy throughout, branched or unbranched, pubescent with stiff, sharp, erect hairs. **Leaves** basal and cauline, broadly elliptic or oblanceolate, bases often clasping, margins usually dentate, tips broadly acute, surfaces hispid; petiolate or sessile. **Inflorescences** 5–50+ heads in terminal panicle- or raceme-like arrays. **Heads** ligulate. **Involucres** campanulate in flower, broadly ovoid in fruit. **Phyllaries** in 1 series, linear to lance-linear, tips acuminate. **Receptacles** flat, pitted, glabrous; paleae 0. **Florets** bisexual. **Fruits** ovoid-elliptic, wrinkled, glabrous to finely hairy; beaks ≥ bodies; pappi of white barbellate or plumose scales.

Europe, North America. 4 species; 1 species treated in *Flora*.

Helminthotheca echioides (Linnaeus) Holub (p. 283)

Bristly ox-tongue

Picris echioides L.

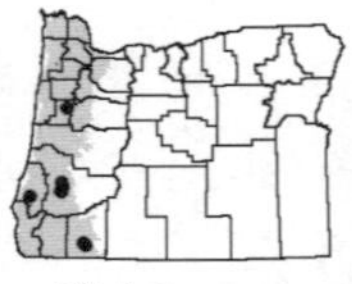

Plants 10–100+ cm. **Leaves** 5–25 cm. **Calyculi** 5, ovoid to deltate. **Involucres** 12–15 mm. **Phyllaries** abaxially bristly. **Florets** 20–40+; ligules 15–20 mm, yellow, often reddish abaxially. **Fruits** bodies 2.5–3 mm, reddish brown or outermost pale tan; beaks 2.5–4 mm. **2*n***=10.

Roadsides, disturbed areas. Flowering Jun–Sep. 0–600 m. CR, Sisk, WV. CA; widely scattered in North America; Europe. Exotic.

This species' roadside habit may encourage further spreading to other parts of the state.

HEMIZONELLA Miniature tarweed

Bruce G. Baldwin and Gerald D. Carr

Genus composed of one species, *H. minima.*

Hemizonella minima (A. Gray) A. Gray (p. 283)

Miniature tarweed, small-headed tarweed, least tarweed, opposite-leaved tarweed

Madia minima (A. Gray) D.D. Keck

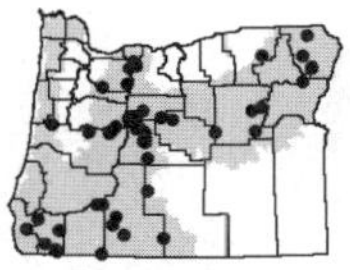

Herbs annuals, 1–20 cm; self-compatible. **Stems** ± erect, branches often divaricate, hispidulous, stipitate-glandular. **Leaves** basal and/or cauline, opposite proximally, alternate distally, often in clusters of 2–3+ immediately proximal to branches; blades linear, 5–25 × (0.5)1–3 mm, margins entire or toothed, surfaces hirsute, distal glandular-puberulent; sessile. **Inflorescences** heads in subumbel-like arrays, glomerules or solitary. **Heads** radiate; peduncles filiform, short. **Involucres** ± obovoid, (1)2–4 × (1)2–4 mm. **Phyllaries** 3–5 in 1 series, each mostly or wholly enveloping a ray ovary, mostly oblanceolate, herbaceous, abaxially hirsute and glandular-hirsutulous. **Receptacles** flat to convex, glabrous or sparsely setulose; paleae in 1 series between ray and disc florets, falling, connate. **Ray florets** 3–5, pistillate, fertile; corollas pale yellow, rays 0.5–1 mm. **Disc florets** 1–2, bisexual, fertile; corollas 1–2.5 mm, pale yellow, pubescent, tubes about equaling funnelform throats, lobes 5, deltate; anthers yellow; styles glabrous proximal to branches, branch tips densely hairy. **Ray fruits** obcompressed, arcuate, 1.8–2.8 mm, black, sparsely hispidulous or glabrate, basal attachments centered, tips minutely beaked, beaks straight, oriented adaxially, 0.1–0.15 mm; pappi 0. **Disc fruits** nearly terete, clavate, 1.8–2.8 mm, black, somewhat hispidulous, tips minutely beaked, beaks straight, oriented vertically, 0.1–0.15 mm; pappi 0. **2*n***=32.

Gravelly or rocky, usually open sites, scrublands, meadows, forests. Flowering May–Aug. 500–2000 m. BW, Casc, CR, ECas, Lava, Sisk. CA, ID, NV, WA; north to British Columbia, northeast to MT. Native.

Hemizonella is treated as distinct from *Madia* based on morphological and molecular evidence (Baldwin 2003a, b), including having somewhat hairy fruits versus those of *Madia* species, which are glabrous. It is most similar to *Madia exigua* but is usually a more well-branched plant with shorter peduncles. *Hemizonella minima* is self-compatible, like most other tarweeds that are distributed widely in western North America.

Helminthotheca echioides

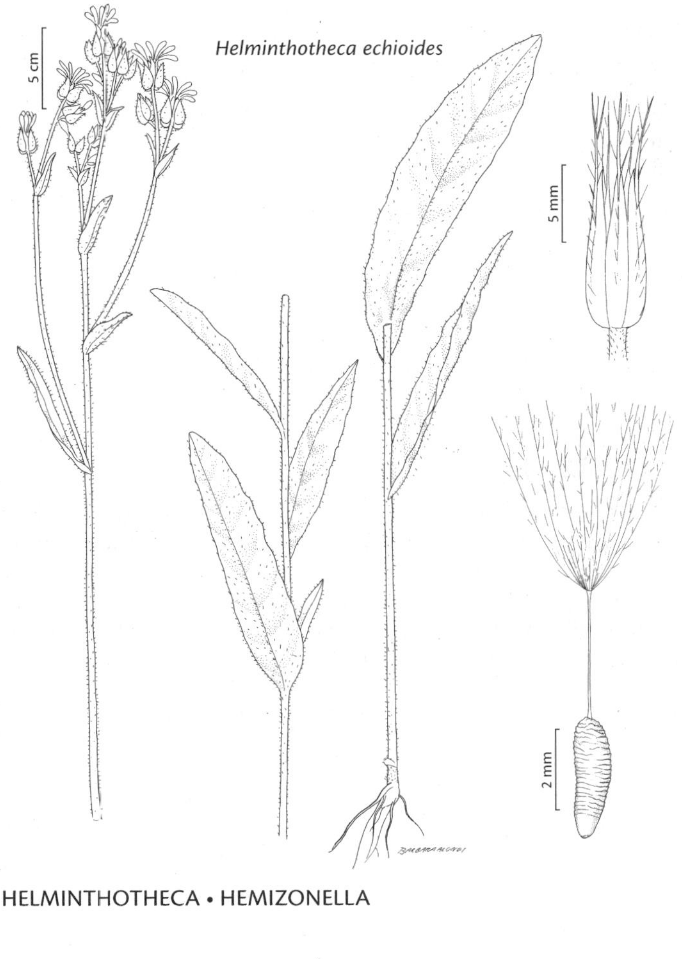

Hemizonella minima

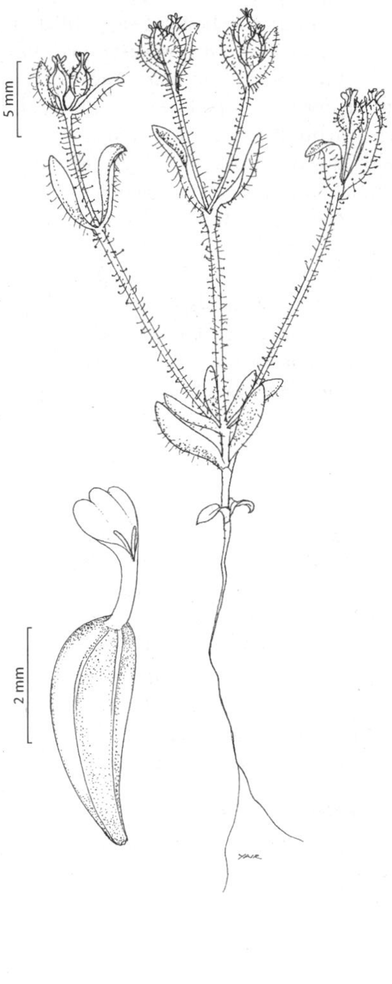

HELMINTHOTHECA • HEMIZONELLA

HEMIZONIA Hayfield tarweed

Bruce G. Baldwin and Gerald D. Carr

Genus composed of one species, *H. congesta*. 6 subspecies; 1 subspecies treated in *Flora.*

Hemizonia congesta de Candolle ssp. ***clevelandii*** (Greene) Babcock & H.M. Hall (p. 284)

Cleveland's hayfield tarweed

Hemizonia clevelandii Greene

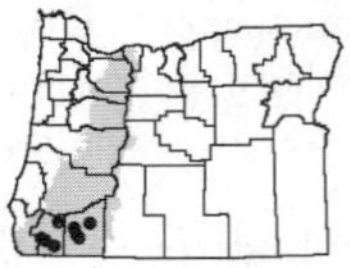

Herbs annuals, 5–80 cm; self-incompatible. **Stems** erect, simple or more commonly branched distally. **Leaves** basal and cauline; opposite proximally or forming winter-spring rosettes sometimes present at flowering, alternate distally; blades narrowly elliptic to linear or linear-lanceolate, 50–175+ × 2–8(12+) mm, margins serrulate or entire, surfaces hirsute to villous and often hispidulous or strigose, usually all or only distal leaves stipitate-glandular; sessile. **Inflorescences** heads solitary or in spike-raceme- to spike-panicle-like arrays. **Heads** radiate; peduncles 0 or 1–4 mm; bracts not or barely surpassing phyllaries. **Involucres** hemispheric to somewhat urceolate or globose, 3.5–12 × 3–8+ mm. **Phyllaries** 5–8 in 1 series, herbaceous, linear to lanceolate or oblanceolate, 3–7 mm, usually ½ enveloping a subtended ray ovary proximally, tips usually < bodies, abaxially hirsute and hispidulous to strigillose, stipitate-glandular. **Receptacles** flat to conic, glabrous; paleae connate, forming cells around all or most individual disc florets, scarious. **Ray florets** 5–8, pistillate, fertile; corollas white, rays 5–12 mm, often purple-veined abaxially. **Disc florets** 5–60+, functionally staminate; corollas 2.5–3.5 mm, white, tubes < funnelform throats, lobes 5, deltate; anthers ± dark purple; styles glabrous proximal to branches, branches free or proximally connate, tips densely hairy. **Ray fruits** nearly round to ± 3-angled in cross section, usually widely 2-faced abaxially, flattened to slightly bulging adaxially, 2–3.5 mm, black, basal attachments oblique, tips sometimes beaked, beaks inconspicuous, glabrous; pappi 0. **Disc fruits** 0. **2*n***=28.

Dry sites, grasslands, openings in chaparral and woodlands. Flowering Jun–Aug. 50–700 m. Casc, Sisk. CA. Native.

Hemizonia, as often circumscribed to include *Centromadia* and *Deinandra* (California, Mexico), is polyphyletic. Molecular phylogenetic studies have indicated that *Centromadia* and *Deinandra* are more closely related to *Calycadenia, Holocarpha* (California), and *Osmadenia* (California, Mexico) than to *Hemizonia* in the strict sense. *Hemizonia* is most closely related to *Blepharizonia* (California), also with 2*n*=28 (Baldwin 2003b).

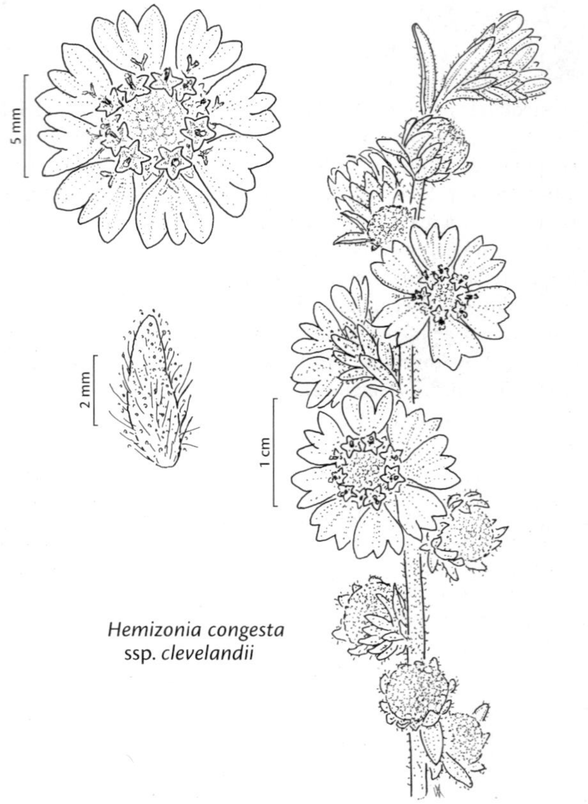

Hemizonia congesta ssp. *clevelandii*

HEMIZONIA

HESPEREVAX Evax

Herbs annual. **Stems** simple or branched, sometimes acaulescent, woolly-tomentose. **Leaves** basal and cauline, alternate, ± whorled below heads, linear to narrowly oblanceolate or ± round, often becoming spatulate distally, margins entire, tips acute to obtuse, surfaces woolly-tomentose; sessile or petiolate. **Inflorescences** heads solitary or in cyme- or spike-like arrays. **Heads** disciform, campanulate or cylindric, slightly to much exceeded by subtending bracteate leaves. **Involucres** 0. **Phyllaries** 0 or vestigial. **Receptacles** conic to cylindric; paleae persistent. **Outer florets** 3–25, pistillate; corollas obscure. **Inner florets** 2–12, staminate; corolla lobes (3)4(5), unequal. **Fruits** obovoid, compressed, brown with proximal black band, glabrous; pappi 0.

California and Oregon. 3 species; 2 species treated in *Flora.*

Plants of this genus are subject to environmental variation, with some individuals being vigorous and much branched, while others are depauperate, single-headed, and short.

Key to *Hesperevax* species:

1. Heads solitary or in ± tight clusters, subtended by 5–12 bracteate leaves . . . . ***H. acaulis***

1' Heads usually in ± loose clusters, subtended by 1–4 bracteate leaves . . . . ***H. sparsiflora*** var. ***brevifolia***

Hesperevax acaulis (Kellogg) Greene

Stemless dwarf evax, stemless dwarf-cudweed

Plants 1–12 cm, erect, prostrate, or acaulescent. **Leaves** oblanceolate to obovate, largest 7–22(32) × 2–4(5) mm. **Heads** solitary or 2–8 in ± tight clusters, cylindric to campanulate, 3–8 mm; subtended by 5–12 bracteate leaves. **Receptacles** 0.8–1.9 × 0.8–1.7 mm. **Outer paleae** 1–3.5 mm. **Inner paleae** 1–4 mm. **Fruits** 0.6–1.6 mm.

California and Oregon. 3 varieties; 2 varieties treated in *Flora*.

Key to *Hesperevax acaulis* varieties:

1. Plants prostrate or acaulescent; largest leaves 7–12 × 0.5–2 mm; fruits 0.6–0.8 mm . . . . ***H. a.*** var. ***acaulis***

1' Plants erect; largest leaves (9)12–22(32) × 2–4(5) mm; fruits 1–1.6 mm . . . . ***H. a.*** var. ***robustior***

Hesperevax acaulis (Kellogg) Greene var. ***acaulis***

Stemless dwarf evax

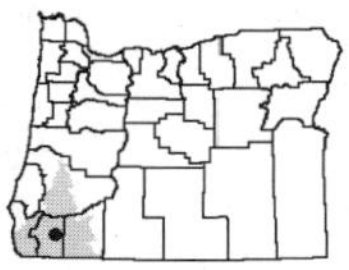

Plants 1–2 cm, prostrate or acaulescent. **Leaves** oblanceolate, largest 7–12 × 0.5–2 mm. **Fruits** 0.6–0.8 mm.

Chaparral shrubland edges, barren soil. Flowering Mar–May. 300–400 m. Sisk. CA. Native?

This variety of *Hesperevax acaulis* is known in Oregon from a single locality near Grants Pass, which is disjunct by over 400 km from the nearest California populations in the Sierra Nevada foothills.

Hesperevax acaulis (Kellogg) Greene var. ***robustior*** Morefield

Robust dwarf evax

Evax acaulis (Kellogg) Greene

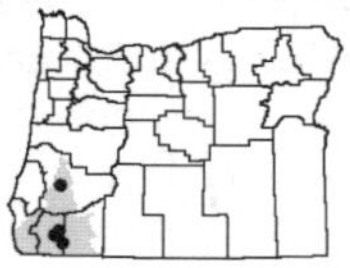

Plants 2–12 cm, erect. **Leaves** oblanceolate to obovate, largest (9)12–22(32) × 2–4(5) mm. **Fruits** 1–1.6 mm.

Shrublands, grasslands, roadsides. Flowering Apr–Jun. 100–500 m. Sisk. CA. Native.

Hesperevax sparsiflora (A. Gray) Greene var. ***brevifolia*** (A. Gray) Morefield (p. 285)

Seaside evax, short-leaved evax, shortleaf dwarf-cudweed

Evax sparsiflora (A. Gray) Jeps.

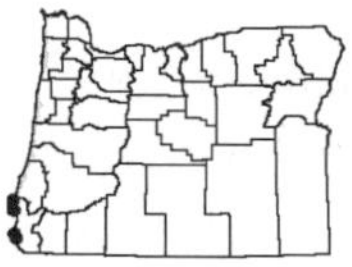

Plants 1–9 cm, erect. **Leaves** ± round, largest 5–12+ × 3–5(6) mm. **Heads** usually 3–5 in ± loose clusters, cylindric, 2–3.5 mm; subtended by 1–4 bracteate leaves. **Receptacles** 1.4–1.8 × 0.2–0.4 mm. **Outer paleae** 2–4.5 mm. **Inner paleae** 1–1.8 mm. **Fruits** 1–1.7 mm.

Coastal bluffs and dunes, grasslands, woodlands. Flowering Apr–Jul. 0–100 m. Est. CA. Native.

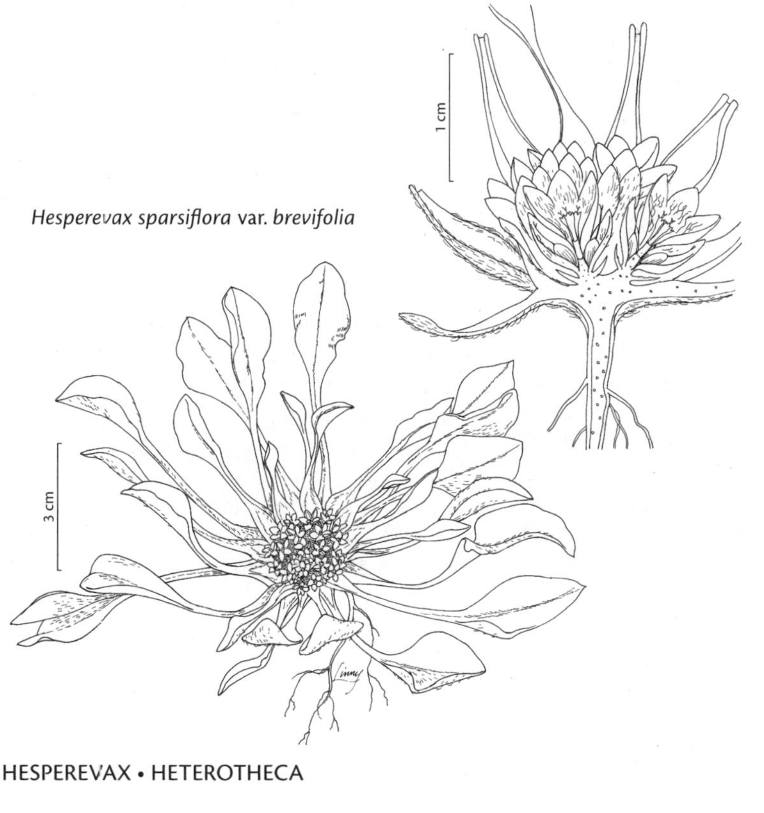

Hesperevax sparsiflora var. *brevifolia*

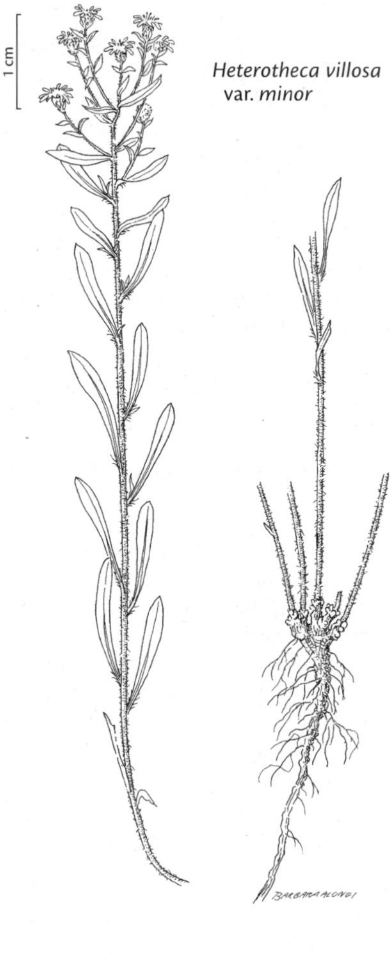

Heterotheca villosa var. *minor*

HETEROTHECA Goldenaster

Herbs annual or perennial. **Stems** 1+, erect or ascending, branched or unbranched, pilose to strigose-hispid or tomentose, glandular or not. **Leaves** mostly cauline, alternate, broadly to narrowly lanceolate, oblanceolate, oblong, or elliptic, margins entire, tips acute, surfaces pilose to strigose-hispid or tomentose, abaxially sometimes glabrous, sometimes minutely glandular; sessile or petiolate. **Inflorescences** heads solitary or in raceme- or panicle-like arrays. **Heads** radiate or discoid; sometimes subtended by leafy bracts. **Involucres** campanulate or hemispheric. **Phyllaries** in 3–7 unequal series, erect in flower, spreading or reflexed in fruit, margins membranous, tips acute, surfaces glabrous or pubescent, minutely stipitate glandular or not; outer gradually shorter. **Receptacles** flat to slightly convex, pitted; paleae 0. **Ray florets** 0–20(30), pistillate; corollas well exceeding involucres, yellow. **Disc florets** bisexual; corollas tubular, gradually enlarged distally, equaling or exceeding involucres, yellow, lobes 5, elongate-triangular, erect; stamens included or slightly exserted; stigmas well exserted. **Fruits** fusiform or obconic, tapering toward bases, obscurely ribbed; pappi in 3–4 series, outer of scales, inner of bristles.

North America. 28 species; 2 species treated in *Flora*.

Key to *Heterotheca* species:

1. Ray florets 0 . . . . ***H. oregona***
1' Ray florets (7)10–30(35). . . . . ***H. villosa***

Heterotheca oregona (Nuttall) Shinners

Oregon goldenaster, rayless goldenaster

Chrysopsis oregona (Nutt.) A. Gray, *Chrysopsis oregona* (Nutt.) A. Gray var. *oregona*, *Chrysopsis oregona* (Nutt.) A. Gray var. *rudis* (Greene) Jeps., *Heterotheca oregona* (Nutt.) Shinners var. *oregona*, *Heterotheca oregona* (Nutt.) Shinners var. *rudis* (Greene) Semple

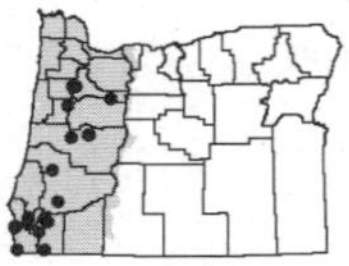

Plants perennial, 1.5–10 dm; from branched caudices. **Stems** sometimes woody at bases, surfaces lightly strigose-hispid, often finely puberulent, glandular or not. **Leaves** 1–5 cm, linear, narrowly or broadly lanceolate, oblong, or oblanceolate, surfaces usually lightly strigose-hispid, abaxially sometimes glabrous, rarely minutely glandular; usually sessile, sometimes with short, narrowly winged petioles. **Heads** 1–15, discoid; peduncles 1–7 cm. **Involucres** 7.5–14 mm. **Phyllaries** in 4–6 series; outer long-triangular or lanceolate, surfaces glabrous or lightly setose, sometimes minutely glandular; inner lance-linear, surfaces glabrous or minutely glandular. **Ray florets** 0. **Disc florets** 15–50; corollas 6–12 mm. **Fruits** 2–5 mm, densely strigose; pappus scales linear, 0.1–0.5 mm, bristles 5–7.5 mm, yellowish or brownish. $2n=18$.

Dry slopes, riverbanks, gravel bars, roadsides, in sandy or gravelly soils. Flowering Jun–Oct. 0–500 m. Casc, CR, Est, Sisk, WV. CA, WA. Native.

Heterotheca villosa (Pursh) Shinners

Hairy goldenaster

Plants perennial, 1–5+ dm; taprooted. **Stems** not woody at base, surfaces lightly or densely pilose to strigose-hispid or tomentose, minutely glandular or not. **Leaves** 1–6 cm, linear, broadly or narrowly lanceolate, oblanceolate, or oblong, surfaces strigillose or silky-puberulent, sometimes hispid, minutely glandular or not; proximally petiolate to distally sessile. **Heads** 1–16(40), radiate; peduncles 1–5 cm. **Involucres** 5–10 mm. **Phyllaries** in 4–7 series; outer linear or elongate-triangular to lanceolate, surfaces usually strigillose; inner linear, surfaces glabrous to strigillose or hispid, sometimes minutely glandular. **Ray florets** (7)10–30(35); corollas 4–11 mm. **Disc florets** 10–50+; corollas 5–6 mm. **Fruits** 2–4 mm, strigillose-silky; pappus scales linear, 0.3–1 mm, bristles 6–7 mm, yellowish or brownish. $2n=18, 36$.

North America. 8–9 varieties; 2 varieties treated in *Flora*.

Key to *Heterotheca villosa* varieties:

1. Stems and distal leaves minutely glandular-puberulent . . . . ***H. v.*** var. ***minor***
1' Stems and distal leaves eglandular . . . . . . ***H. v.*** var. ***villosa***

Heterotheca villosa (Pursh) Shinners var. ***minor*** (Hooker) Semple (p. 285)

Hispid goldenaster

Chrysopsis villosa (Pursh) Nutt. ex DC. var. *hispida* (Hook.) A. Gray, *Chrysopsis villosa* (Pursh) Nutt. ex DC. var. *viscida* A. Gray, *Heterotheca villosa* (Pursh) Shinners var. *hispida* (Hook.) V.L. Harms

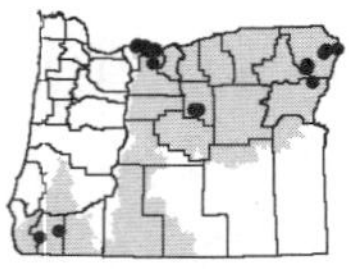

Stems minutely glandular-puberulent. **Leaves** distal minutely glandular puberulent.

Pine woodlands, brushlands, grasslands, riverbanks, dry slopes, roadsides. Flowering May–Sep. 50–1700 m. BW, Col, ECas, Lava, Sisk. CA, ID, NV, WA; north to British Columbia, northeast to Manitoba, east to WI, southeast to NM. Native.

The minute glandular trichomes of this variety are difficult to distinguish without high magnification. However, because of them, the plants may feel viscid to the touch.

Heterotheca villosa (Pursh) Shinners var. ***villosa***

Hairy goldenaster

Chrysopsis villosa (Pursh) Nutt. ex DC. var. *villosa*, *Heterotheca villosa* (Pursh) Shinners var. *foliosa* (Nutt.) V.L. Harms

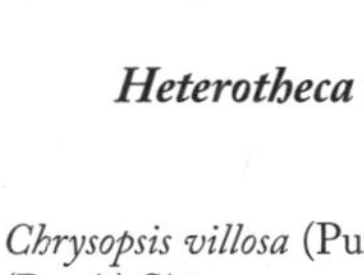

Stems eglandular. **Leaves** eglandular.

Woodlands, grasslands, riverbanks, dry slopes, sandbars, roadsides. Flowering Jun–Aug. 0–1400 m. Col, ECas, Lava, Sisk, WV. ID, WA; north to British Columbia, east to IL, MI, NE, adjacent Canada. Native.

HIERACIUM Hawkweed

Herbs perennial, sap milky; tap- or fibrous-rooted, sometimes stoloniferous or rhizomatous. **Stems** erect, simple or branched, glabrous or pubescent. **Leaves** basal and cauline or cauline, elliptic, lanceolate, oblanceolate, obovate, or oblong, margins entire to denticulate or dentate, rarely pinnatifid, surfaces glabrous or pubescent; short-petiolate or sessile, sometimes auriculate-clasping. **Inflorescences** panicle- or raceme-like arrays. **Heads** ligulate, erect. **Involucres** cylindric to campanulate in flower, ovoid in developing fruit. **Phyllaries** erect, reflexed in fruit, surfaces glabrous to densely glandular with black hairs; inner linear or lanceolate; outer gradually reduced or all basal. **Receptacles** flat, pitted, surfaces glabrous; paleae 0. **Florets** bisexual, 5–100+; ligules exceeding involucres, white, yellow, or reddish orange, sometimes drying purplish. **Fruits** columnar or sometimes enlarged toward bases, usually 10-ribbed; pappi of numerous white, yellowish, or brownish bristles.

Cosmopolitan. 250–1000+ species; 14 species treated in *Flora.*

This genus is notorious for the extent to which some of its members may hybridize and reproduce by asexual seeds (apomixis). Large numbers of such self-perpetuating variants have been named as distinct species, especially in Europe and Asia, leading to a complex taxonomy that is reflected in the various names applied to the adventive taxa in North America (Strother 2006).

In describing the types of pubescence, the term tomentulose is used for a covering of short, dense, usually white hairs that may or may not be branched or stellate. This branching is on such a small scale that it cannot be readily seen with a normal hand lens. The term pilose-setose is used for long, straight, usually somewhat stiff hairs, which may grade into softer, more wavy hairs. Stipitate-glandular pubescence describes long or short, straight hairs with a distinct gland at the tip.

Descriptions of basal and/or cauline leaf presence or absence refer to plants at flowering.

Hieracium piloselloides has been reported once in Oregon (Umatilla County, 2003) but is not established. It can be identified by a combination of having rhizomes and glabrous leaves.

Key to *Hieracium* species:

1. Flowers white, cream, or reddish orange.
    2. Flowers white or cream; stolons absent . . . . . ***H. albiflorum***
    2' Flowers reddish orange; stolons present . . . . . ***H. aurantiacum***

1' Flowers yellow, ligules often reddish on outer surface.
    3. Plants stoloniferous.
        4. Heads 1(2–3). . . . . ***H. pilosella***
        4' Heads 5–30 . . . . . ***H. caespitosum*** (in part)
    3' Plants not stoloniferous.
        5. Leaves cauline, rarely a few basal.
            6. Phyllaries densely long-setose; leaf surfaces usually glabrous . . . . . ***H. longiberbe***
            6' Phyllaries glabrous; leaf surfaces usually minutely hispid, often long-hispid . . . . . ***H. umbellatum***
        5' Leaves basal or basal and cauline.
            7. Leaves denticulate to dentate.
                8. Leaf bases tapered . . . . . ***H. lachenalii***
                8' Leaf bases truncate or cordate . . . . . ***H. murorum***
            7' Leaves usually entire or some leaves bluntly toothed.
                9. Plants rhizomatous . . . . . ***H. caespitosum*** (in part)
                9' Plants not rhizomatous.
                    10. Phyllaries glabrous or sparsely pilose-setose . . . . . ***H. bolanderi***
                    10' Phyllaries not glabrous and with obvious pubescence.
                        11. Leaf surfaces glabrous. . . . . ***H. triste***
                        11' Leaf surfaces usually pubescent (rarely glabrous abaxially in *H. parryi*).
                            12. Stems, leaves, and phyllaries densely pilose-villous . . . . . ***H. horridum***
                            12' Stems, leaves, and phyllaries not densely pilose-villous.
                                13. Phyllaries lightly to moderately tomentulose, without black hairs . . . . . ***H. greenei***
                                13' Phyllaries stipitate-glandular or pilose-setose, with black hairs.
                                    14. Phyllaries stipitate-glandular . . . . . ***H. parryi***
                                    14' Phyllaries pilose-setose. . . . . ***H. scouleri***

Hieracium albiflorum Hooker (p. 288)

White-flowered hawkweed

Hieracium siskiyouense M. Peck

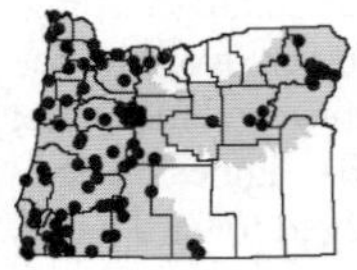

Plants 10–100(120) cm; taprooted. **Stems** erect, simple or branched, usually long-pilose proximally, becoming glabrous distally. **Leaves** basal and cauline, elliptic or oblanceolate, 3–15 cm, margins entire to denticulate or sinuate-dentate, tips acute or obtuse, surfaces pilose-setose marginally, sometimes nearly glabrous; basal petiolate, cauline reduced upwards, sessile, sometimes auriculate-clasping. **Inflorescences** raceme- or panicle-like arrays, often much branched; bracteate. **Involucres** cylindric or campanulate in flower, ovoid in fruit, 6–11 mm. **Phyllaries** linear-lanceolate, surfaces glabrous to sparsely pilose, sometimes sparingly glandular or tomentulose; inner 8–15; outer > inner. **Florets** 5–30+; ligules 2–4 mm, white or cream. **Fruits** columnar, 2.5–4 mm, dark brown. **2*n***=18.

Open forests, brushlands, meadows, rocky hillsides and cliffs, dunes, disturbed areas. Flowering Jun–Sep. 0–2200 m. BW, Casc, CR, ECas, Est, Lava, Sisk, WV. CA, ID, NV, WA; north to AK, northeast to Manitoba, east to SD, southeast to CO, south to Mexico. Native.

Hieracium albiflorum has been identified as having a self-pollinating breeding system (Guppy 1978). In spite of its morphological distinctness from the other native taxa of the Pacific Northwest, natural hybrids have been observed or suspected between it and species such as *H. scouleri* and *H. greenei.* Hybrids with yellow-flowered taxa such as these are difficult to detect due to the fact that they have a yellow flower color. Back-cross hybrids may also be found, with pale yellow flowers suggestive of *H. albiflorum* parentage.

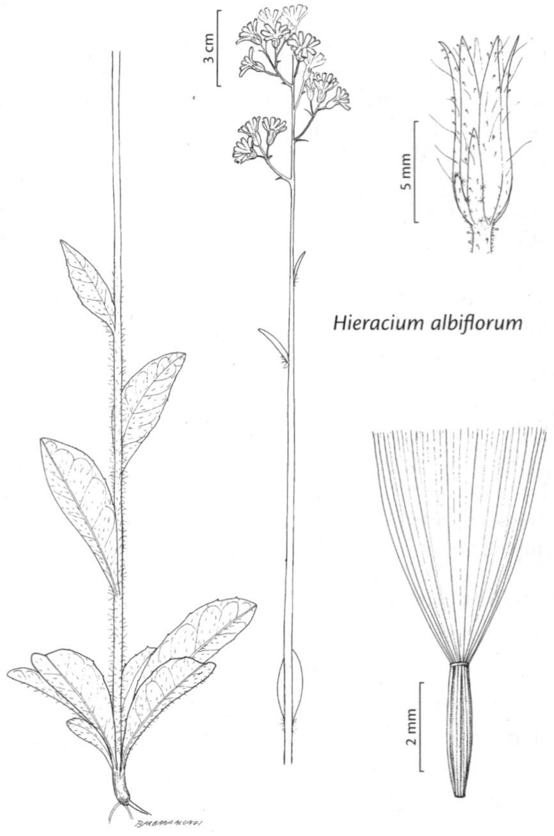

HIERACIUM

Hieracium aurantiacum Linnaeus

Orange hawkweed, king-devil, devil's paintbrush

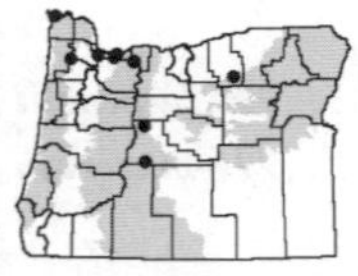

Plants 15–60 cm; taprooted, stoloniferous. **Stems** erect, simple, pilose-hispid, usually also stipitate-glandular, tomentulose distally. **Leaves** basal, sometimes 1 near stem base, oblanceolate, 5–20 cm, margins entire, tips acute or obtuse, surfaces pilose-setose; petioles short, winged. **Inflorescences** raceme- or panicle-like arrays; bracteate or not. **Involucres** campanulate in flower, ovoid in fruit, 6–8 mm. **Phyllaries** linear-lanceolate, surfaces blackish stipitate-glandular, usually setose-hispid; inner 13–30; outer > inner. **Florets** 25–100+; ligules 10–15 mm, reddish orange, drying purplish. **Fruits** columnar, 1.2–2 mm, dark brown. **2*n***=18, 27, 36, 45, 54, 63, 72.

Disturbed areas, roadsides, clearcuts, riverbanks, pastures. Flowering Jun–Aug. 100–1400 m. BW, Casc, CR, ECas. CA, ID, NV, WA; scattered in North America; Europe. Exotic.

Hieracium aurantiacum is a conspicuous and well-known weed over much of the United States and southern Canada. Its scattered distribution in Oregon indicates an ability to disperse by seed, and once established it may persist and spread by both seeds and stolons.

Hieracium bolanderi A. Gray

Bolander's hawkweed

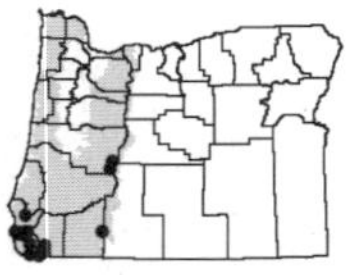

Plants 5–50 cm; taprooted. **Stems** simple or branched, glabrous or tomentulose, sometimes pilose-hispid near bases. **Leaves** basal and sometimes a few reduced cauline, obovate or oblanceolate, 2–10+ cm, margins entire, rarely remotely denticulate, tips acute or obtuse, surfaces pilose-hispid, abaxially sometimes glabrous; basal short-petiolate, cauline sessile, often winged. **Inflorescences** raceme- or panicle-like arrays; bracteate. **Involucres** campanulate in flower, ovoid in fruit, 7–10 mm. **Phyllaries** linear-lanceolate, surfaces glabrous or sparsely pilose-setose; inner 8–14; outer > inner. **Florets** 6–12; ligules 7–15 mm, yellow. **Fruits** columnar or somewhat enlarged toward base, 5–5.5 mm, brown or straw-colored.

Ridge-tops, wooded slopes, cliffs, talus, road-banks, serpentine. Flowering Jun–Aug. 400–2400 m. Casc, CR, Sisk. CA. Native.

Collections of *Hieracium bolanderi* from the southern Cascades may show tomentulose involucres suggestive of hybridization with *H. scouleri* or *H. greenei.*

Hieracium caespitosum Dumortier

Yellow king-devil

Hieracium pratense Tausch

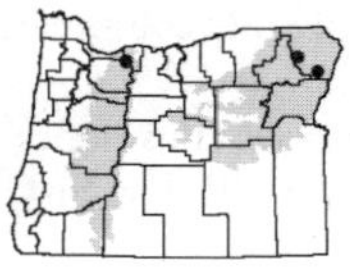

Plants 20–90 cm; rhizomatous, sometimes stoloniferous. **Stems** erect, simple, pilose-hispid, stipitate-glandular, sometimes tomentulose below heads. **Leaves** basal and sometimes a few cauline near base, oblanceolate, 5–15+ cm, margins entire, tips acute or obtuse, surfaces pilose-hispid or villous, sometimes nearly glabrous; petioles short, usually winged. **Inflorescences** raceme- or panicle-like arrays; bracteate or not. **Heads** 5–30. **Involucres** campanulate in flower, ovoid in fruit, 5–9 mm. **Phyllaries** inner 12–18+, linear, surfaces black stipitate-glandular and usually setose-hispid, sometimes tomentulose; outer > inner. **Florets** 25–50+; ligules 8–12+ mm, yellow. **Fruits** columnar, 1.5–2 mm, dark brown. **2*n***=18, 27, 36, 45.

Disturbed areas, pastures, open woods, roadsides, riverbanks. Flowering Jun–Jul. 800–1900 m. BW, Casc. ID, WA; widely scattered in Canada and US; Europe. Exotic.

Hieracium greenei A. Gray

Greene's hawkweed

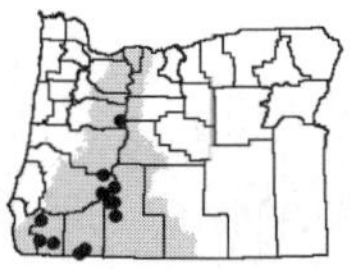

Plants 10–60 cm; taprooted. **Stems** erect, simple or branched, tomentulose, sometimes pilose near base. **Leaves** basal and sometimes 1 cauline, oblanceolate, 3–15 cm, margins mostly entire, some usually blunt-toothed, tips usually acute, surfaces densely pilose-hispid; petioles sometimes winged. **Inflorescences** panicle-like arrays; bracteate. **Involucres** cylindric in flower and fruit, 9–12 mm. **Phyllaries** linear-lanceolate, surfaces lightly to moderately tomentulose; inner 6–10; outer gradually shorter. **Florets** 5–10+; ligules 8–10 mm, yellow. **Fruits** columnar, 3–3.5 mm, dark brown. **2*n***=18.

Rocky slopes and outcrops, roadcuts, forest openings. Flowering Jun–Aug. 600–2500 m. Casc, ECas, Sisk. CA. Native.

Occasional specimens are suggestive of hybridization with *Hieracium bolanderi*. The molecular phylogenetic studies by Gaskin and Wilson (2007) placed *H. greenei* as the sister taxon to *H. parryi*.

Hieracium horridum Fries

Shaggy hawkweed

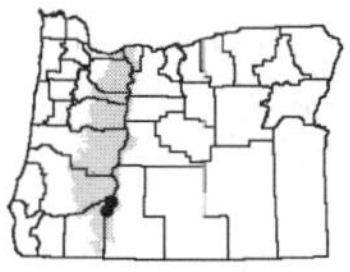

Plants 10–30 cm; taprooted. **Stems** simple or branched, densely pilose-villous. **Leaves** basal and cauline, broadly elliptic to oblanceolate, 2–15 cm, margins entire, tips acute or obtuse, surfaces densely pilose-villous; petioles winged. **Inflorescences** panicle-like arrays; bracteate. **Involucres** ± cylindric in flower and fruit, 6–9 mm. **Phyllaries** lanceolate, surfaces densely pilose-villous; inner 12–15; outer > inner. **Florets** 6–15; ligules 8–9 mm, yellow. **Fruits** columnar, 3–3.5 mm, reddish brown. **2*n***=18.

Alpine rock outcrops, gravelly slopes, meadows. Flowering Jun–Oct. 1500–2300 m. Casc. CA, NV. Native.

Hieracium lachenalii Suter

Lachenal's hawkweed

Hieracium vulgatum Fr.

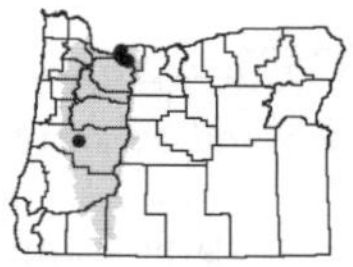

Plants 30–70 cm; taprooted. **Stems** simple or branched, usually tomentulose or spiculate-glandular. **Leaves** basal and cauline, or sometimes cauline, broadly elliptic, 5–10 cm, bases tapered, margins dentate, lower strongly toothed, surfaces glabrous or lightly pilose; basal petiolate, cauline petiolate or sessile. **Inflorescences** panicle- or raceme-like arrays; bracteate or not. **Involucres** campanulate in flower, ovoid in fruit, 7–10 mm. **Phyllaries** linear-lanceolate, surfaces black stipitate-glandular, sometimes tomentulose; inner 14–20+; outer gradually shorter. **Florets** 40–80+; ligules 13–18 mm, yellow. **Fruits** columnar, 2.5–3 mm, black. **2*n***=27.

Disturbed areas, roadsides. Flowering Jun–Aug. 0–1400 m. Casc, WV. WA; sparsely scattered in North America; Europe. Exotic.

Plants of *H. lachenalii* produce abundant fruits, probably by apomixis since its chromosome number is triploid.

Hieracium longiberbe Howell

Long-bearded hawkweed

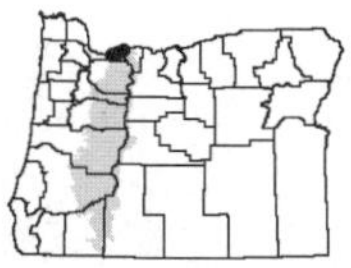

Plants 10–60 cm; taprooted. **Stems** simple or branched, glabrous or pilose. **Leaves** cauline, rarely a few basal early, spatulate to oblanceolate, 5–15 cm, bases attenuate, margins entire or rarely denticulate, surfaces glabrous to lightly or densely pilose-setose; usually sessile. **Inflorescences** panicle- or raceme-like arrays; bracteate or not. **Involucres** campanulate in flower, ovoid in fruit, 7–10 mm. **Phyllaries** linear-lanceolate, surfaces densely long-setose with brown or black hairs; inner 10–16; outer gradually shorter. **Florets** 12–30; ligules 10–15 mm, yellow. **Fruits** columnar, 3–4 mm, brown.

Rocky slopes and cliffs, open woods. Flowering May–Aug. 0–600 m. Casc. WA. Native.

This species is limited to the Columbia River Gorge and its immediate vicinity. Its distinctive, densely pilose-setose involucres, lacking either tomentulose or stipitate-glandular pubescence, distinguish it from the closely allied species *Hieracium scouleri*, a widespread taxon whose numerous pubescence variants are not correlated with any particular geographical areas. *Hieracium longiberbe* and *H. scouleri* were shown to be sister taxa in a molecular phylogenetic study by Gaskin and Wilson (2007).

Hieracium murorum Linnaeus

Wall hawkweed

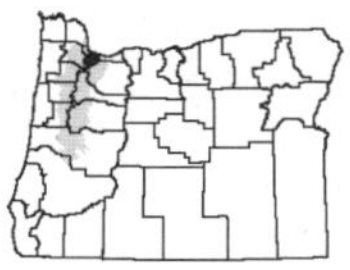

Plants 25–90 cm; taprooted. **Stems** simple or branched, setose-pilose proximally, becoming tomentulose and hispid-glandular distally. **Leaves** basal or basal and cauline, ovate or broadly elliptic, 5–50 cm, bases truncate or cordate, margins denticulate to dentate, lower usually strongly toothed, tips acute or obtuse. surfaces abaxially pilose, adaxially glabrous or lightly pilose; basal petiolate, cauline sessile. **Inflorescences** panicle-like arrays; bracteate.

Involucres campanulate in flower, ovoid in fruit, 7–9 mm. **Phyllaries** linear-lanceolate, surfaces densely stipitate-glandular, sometimes tomentulose; inner 8–16; outer gradually shorter. **Florets** 30–50+; ligules 15–20 mm, yellow. **Fruits** columnar, 2.5–3 mm, dark brown. ***2n***=27.

Disturbed areas, roadsides. Flowering Jun–Aug. 50–200 m. WV. AK, British Columbia, northeastern US and adjacent Canada; Europe. Exotic.

The common name refers to the translation of the Latin *murorum*, meaning "of walls."

Hieracium parryi Zahn
Parry's hawkweed

Hieracium scouleri Hook. var. *nudicaule* (A. Gray) Cronquist

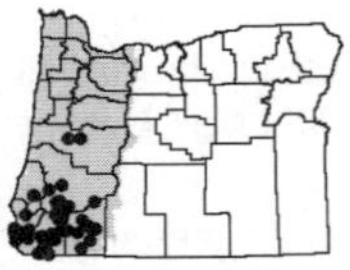

Plants 10–55 cm; taprooted, caudices sometimes branched below ground. **Stems** simple or branched, setose-pilose proximally, occasionally glandular-hairy, becoming hispid-glandular distally. **Leaves** basal or basal and cauline, oblanceolate or narrowly elliptic, 3–18 cm, bases attenuate, margins usually entire, rarely denticulate, surfaces hispid-setose, abaxially rarely glabrous; basal petiolate, cauline sessile. **Inflorescences** panicle- or raceme-like arrays; bracteate or not. **Involucres** campanulate in flower, ovoid in fruit, 10–12 mm. **Phyllaries** linear-lanceolate, surfaces stipitate-glandular with black hairs; inner 10–20+; outer gradually shorter. **Florets** 30–60+; ligules 10–17 mm, yellow. **Fruits** columnar, 2.5–3 mm, brown.

Open woodlands, shrublands, grasslands, ridges, talus slopes, serpentine. Flowering Jun–Aug. 0–1600 m. Casc, CR, Est, Sisk, WV. CA. Native.

The vast majority of collections of *Hieracium parryi* are from southwestern Oregon counties, and reports from counties east of the Cascades are suspected to be misidentified populations of *H. scouleri*.

Hieracium pilosella Linnaeus
Mouse-ear hawkweed

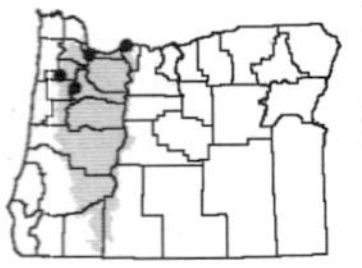

Plants 4–35 cm; taprooted, stoloniferous. **Stems** simple, tomentulose and lightly stipitate-glandular throughout or only distally, sometimes pilose proximally. **Leaves** basal or occasionally 1 or 2 cauline near bases, oblanceolate, 2–10+ cm, bases attenuate, margins entire, surfaces glabrous or pilose-setose, sometimes tomentulose below;

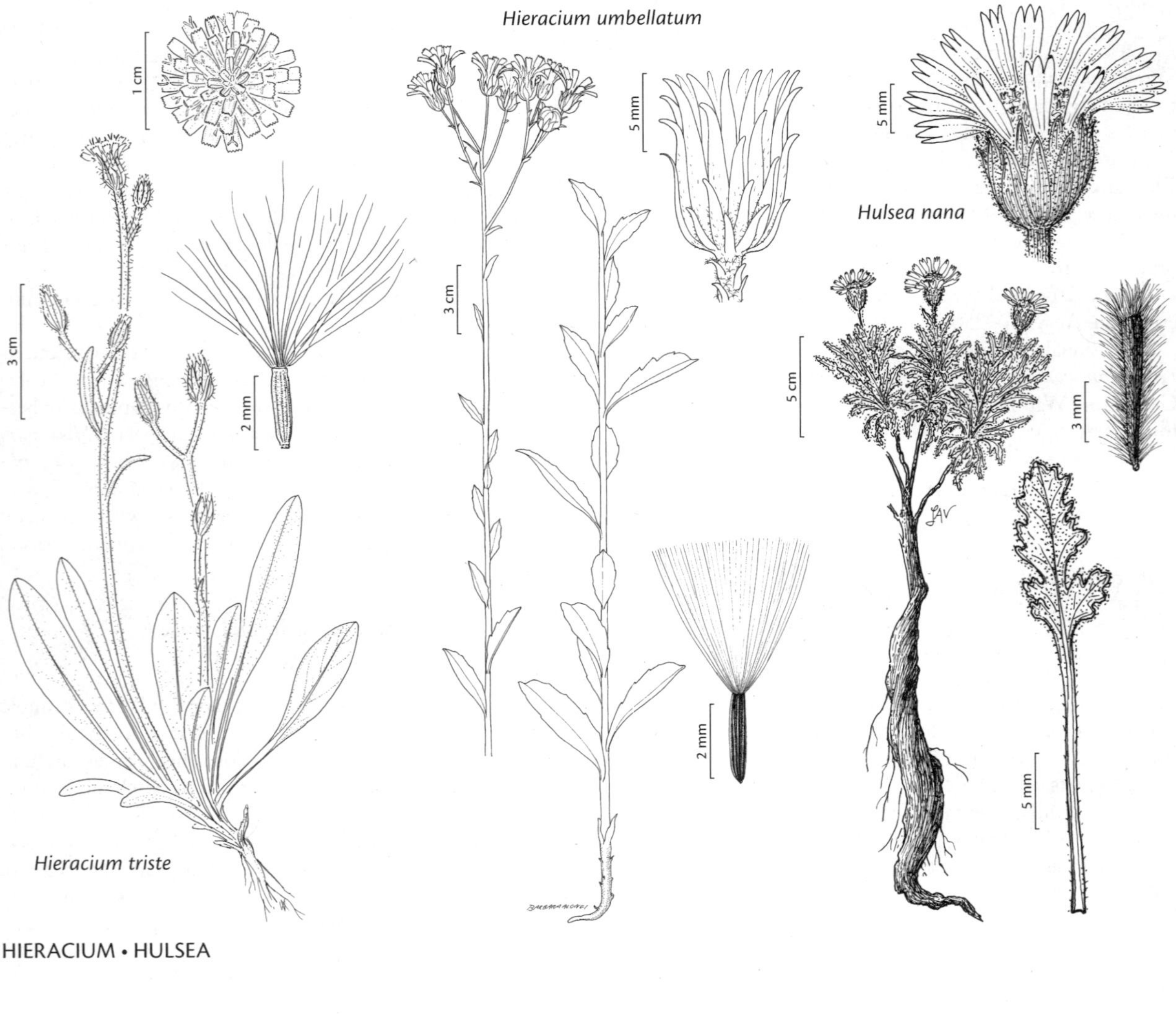

petioles short, winged. **Inflorescences** in raceme- or panicle-like arrays. **Heads** 1(2–3). **Involucres** campanulate in flower, ovoid in fruit, 7–9 mm. **Phyllaries** linear-lanceolate, surfaces densely stipitate-glandular with black hairs, tomentulose; inner 18–30+; outer gradually shorter. **Florets** 60–120+; ligules 10–15 mm, yellow. **Fruits** columnar, 1.5–2 mm, dark brown. **2*n***=18, 36, 45, 54, 63.

Grasslands, roadsides, lawns, disturbed areas. Flowering May–Jun. 0–300 m. Casc, WV. WA; widely scattered in North America; Europe. Exotic.

Hieracium scouleri Hooker

Scouler's hawkweed, woolly weed

Hieracium albertinum Farr, *Hieracium chapacanum* Zahn, *Hieracium cusickii* Gand., *Hieracium cynoglossoides* Arv.-Touv., *Hieracium nudicaule* (A. Gray) A. Heller, *Hieracium scouleri* Hook. var. *albertinum* (Farr) G.W. Douglas & G.A. Allen, *Hieracium scouleri* Hook. var *scouleri*

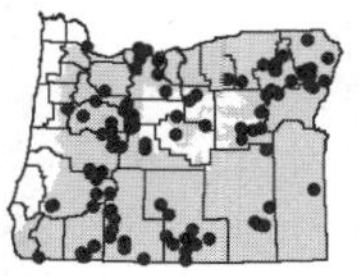

Plants 30–100 cm; taprooted. **Stems** simple or branched, glabrous to tomentulose, puberulent, or pilose-setose, sometimes glandular-setose distally. **Leaves** basal or basal and cauline, narrowly or broadly oblanceolate or elliptic, 2.5–30 cm, bases attenuate, margins entire, rarely denticulate, surfaces pilose-setose and stellate-pubescent, rarely glabrous; basal petioles often winged, cauline usually sessile. **Inflorescences** panicle- or raceme-like arrays; bracteate or not. **Involucres** campanulate in flower, ovoid in fruit, 5–12 mm. **Phyllaries** linear-lanceolate, surfaces pilose-setose with black hairs; inner 12–20+; outer gradually shorter. **Florets** 15–45+; ligules 8–20 mm, yellow. **Fruits** columnar, 2.5–3 mm, dark brown. **2*n***=18.

Forests, shrublands, grasslands, rocky ridges and slopes, roadsides. Flowering Jun–Sep. 0–2600 m. BR, BW, Casc, Col, ECas, Owy, Sisk, WV. CA, ID, NV, WA; north to British Columbia, northeast to Alberta, east to WY. Native.

Hieracium scouleri is here given a broad circumscription to include such segregate taxa as *Hieracium albertinum* and *H. cynoglossoides*, which are sometimes also recognized at varietal rank.

Hieracium triste Willdenow ex Sprengel (p. 290)

Alpine hawkweed, slender hawkweed

Hieracium gracile Hook., *Hieracium gracile* Hook. var. *densifloccosum* (Zahn) Cronquist, *Hieracium gracile* Hook. var. *gracile*, *Hieracium triste* Willd. ex Spreng. var. *gracile* (Hook.) A. Gray

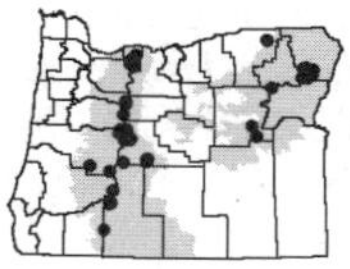

Plants (2.5)4–30 cm; taprooted. **Stems** erect, simple, tomentulose, sometimes glabrous proximally. **Leaves** basal, broadly or narrowly elliptic to oblanceolate, 3–12+ cm, margins entire, rarely obscurely denticulate, tips acute or obtuse, surfaces usually glabrous, rarely stipitate-glandular; petiolate. **Inflorescences** raceme-like arrays or heads solitary; bracteate or not. **Involucres** cylindric or campanulate, ovoid in fruit, 6–10 mm. **Phyllaries** linear-lanceolate, surfaces stellate-tomentulose, black-villous, stipitate-glandular; inner 12–20+; outer gradually shorter. **Florets** 20–60+; ligules 7–11 mm, yellow. **Fruits** columnar, 1.5–3.5 mm, dark brown. **2*n***=18.

Subalpine meadows, rocky slopes, streamsides, forests. Flowering Jul–Sep. 1300–3000 m. BW, Casc, ECas. CA, ID, WA; north to AK, northeast to Northwest Territories, east to WY, southeast to NM. Native.

Hieracium umbellatum Linnaeus (p. 290)

Narrowleaf hawkweed, umbellate hawkweed

Plants 30–100+ cm; fibrous-rooted. **Stems** usually simple or few-branched, glabrous or with villous, hispid, or puberulent hairs, sometimes puberulent distally. **Leaves** cauline, elliptic-lanceolate, 2–10 cm, reduced upwards, margins entire or remotely denticulate, surfaces hispid; sessile. **Inflorescences** panicle- or raceme-like arrays; bracteate or not. **Involucres** campanulate in flower, broadly ovoid in fruit, 8–12 mm. **Phyllaries** linear-lanceolate, surfaces glabrous; inner 12–22+; outer gradually shorter. **Florets** 30–80+; ligules 15–20 mm, yellow. **Fruits** columnar, 2–3 mm, dark brown. **2*n***=18, 27.

Disturbed areas, roadsides, forest edges, thickets, sandy areas. Flowering Jun–Sep. 200–1600 m. WV. WA; north to AK, northeast to Greenland, east to MA, southeast to WV; Asia, Europe. Native.

HULSEA Hulsea

Herbs perennial; taprooted, often cespitose, underground caudices simple or branched. **Stems** erect or decumbent, simple or proximally branched, glabrous to villous, often glandular-puberulent. **Leaves** cauline, alternate, numerous and often clustered at bases, becoming bracteate upward, ovate to oblong or oblanceolate, margins dentate or lobed, lobes short, triangular to lanceolate, tips obtuse or toothed, surfaces puberulent to villous, often minutely glandular; sessile or petiolate, petioles slender to broadly winged. **Inflorescences** solitary, terminal heads. **Heads** radiate, erect. **Involucres** obconic to hemispheric. **Phyllaries** of 2+ series, ± equal or with a few outermost reduced, reflexed in fruit, linear to lanceolate, tips acute to acuminate, surfaces puberulent to villous or woolly, often minutely glandular. **Receptacles** flat to convex; paleae 0. **Ray florets** pistillate; rays slightly to well exceeding involucres, yellow. **Disc florets** bisexual; corollas tubular, expanded distally, yellow, lobes 5, erect, triangular; stamens 5, partly or wholly exserted; styles exserted. **Fruits** oblong, flattened, obscurely ribbed, not beaked; pappi of several short, entire to deeply cut scales, white.

Western North America. 7 species; 2 species treated in *Flora*.

Hulsea has a contractile root that draws the caudex downward each year to maintain the terminal growing point at the surface of the ground.

Key to *Hulsea* species:

1. Phyllaries 30–50+, outer phyllaries usually narrowly linear; leaves and involucres rarely glandular-puberulent. . . . . ***H. algida***

1' Phyllaries 14–30, outer phyllaries usually lanceolate; leaves and involucres usually glandular-puberulent. . . . . ***H. nana***

Hulsea algida A. Gray

Alpine hulsea, fleshy hulsea

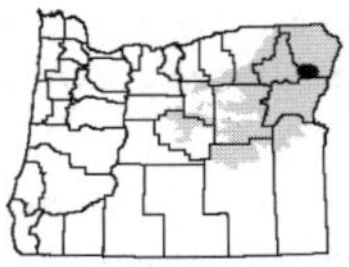

Plants 4–15 cm. **Leaves** oblanceolate, 1–8 cm, margins dentate, teeth triangular, tips obtuse or denticulate, surfaces usually lightly to densely villous, rarely glandular-puberulent; sessile or petiolate. **Involucres** obconic to hemispheric, 10–20 × 12–25 mm. **Phyllaries** 30–50+, outer usually narrowly linear, inner lance-linear, tips sharply acute to acuminate, margins and often abaxial surfaces ± densely villous to woolly-pubescent, rarely glandular-puberulent. **Ray florets** 25–55; rays 6–13 mm. **Disc florets** 60+; corollas 4–10 mm. **Fruits** 6–10 mm, black, densely pilose; pappi 1–1.5 mm.

Granitic alpine and subalpine slopes. Flowering Jul–Sep. 2400–3000 m. BW. CA, ID, NV; northeast to MT, east to WY. Native.

Some plants of *H. algida* in the Wallowa Mountains vary in the direction of *H. nana*. Further study of Wallowa County *Hulsea* populations, perhaps involving molecular analysis of selected plants, may help clarify the taxonomic status of these species.

Hulsea nana A. Gray (p. 290)

Dwarf hulsea

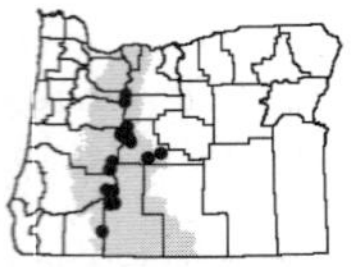

Plants 3–20 cm. **Leaves** oblanceolate to elliptic-oblong, 1–12 cm, margins dentate or lobed, teeth triangular, lobes triangular or lanceolate, tips obtuse or denticulate, surfaces usually glandular-puberulent, sometimes also pilose to lightly or densely villous; sessile or petiolate. **Involucres** obconic, 7–13 × 8–12 mm. **Phyllaries** 14–30, outer usually lanceolate, rarely linear, inner lanceolate, tips acute to acuminate, surfaces pilose to lightly or densely villous, usually glandular-puberulent. **Ray florets** 12–30; rays 7–14 mm. **Disc florets** 50–100+; corollas 9–12 mm. **Fruits** 6–9 mm, black, pilose; pappi 0.5–2 mm, white.

Ridges, slopes, flats, volcanic pumice. Flowering Jul–Sep. 1400–3200 m. Casc, ECas. CA, WA. Native.

HYMENOPAPPUS Hymenopappus, woollywhites

Herbs biennial or perennial; from stout underground caudices. **Stems** erect, branched or unbranched, tomentose. **Leaves** basal and cauline, pinnately compound, with simple or pinnately forked linear segments, surfaces tomentose or glabrate; basal clustered, cauline alternate, becoming bracteate distally; petiolate. **Inflorescences** panicle- or raceme-like arrays. **Heads** radiate or discoid, erect; peduncles tomentulose. **Involucres** globose or hemispheric. **Phyllaries** in 2–3 ± equal series, erect in flower, reflexed in fruit, obovate, margins membranous, 1-veined, tips blunt, surfaces tomentulose. **Receptacles** flat or convex; paleae 0. **Ray florets** 0 or 8, pistillate, fertile. **Disc florets** bisexual; corollas tubular, expanded distally, yellow or white, glandular-puberulent below, ± glabrous distally, lobes 5, triangular, strongly recurved; stamens included to well exserted, sometimes minutely glandular distally; stigma lobes exserted. **Fruits** cylindrical, tapering toward bases, tips truncate, prominently 8–10-veined; pappi of 12–20 short, triangular scales.

North America. 11–14 species; 1 species treated in *Flora*.

Hymenopappus filifolius Hooker var. *filifolius* (p. 292)

Columbia cutleaf, fineleaf hymenopappus

Hymenopappus gloriosus A. Heller

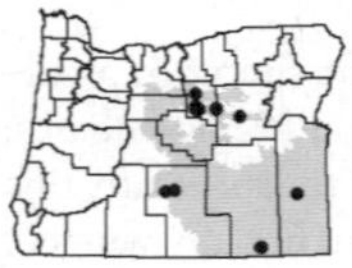

Plants 1–6(10) dm. **Leaves** 2-pinnate, 4–18 cm. **Heads** (2)6–40, discoid. **Involucres** 5–12 mm. **Ray florets** 0. **Disc florets** 15–50+; corollas 3–4 mm, yellow. **Fruits** 4–6 mm, brown, surfaces white-pilose. **2*n***=34.

Open sites in brushlands, grasslands, barrens. Flowering May–Aug. 300–1700 m. BR, Lava, Owy. ID, WA. Native.

Hymenopappus filifolius is widespread in the central and western United States, with up to 13 varieties recognized by some authors.

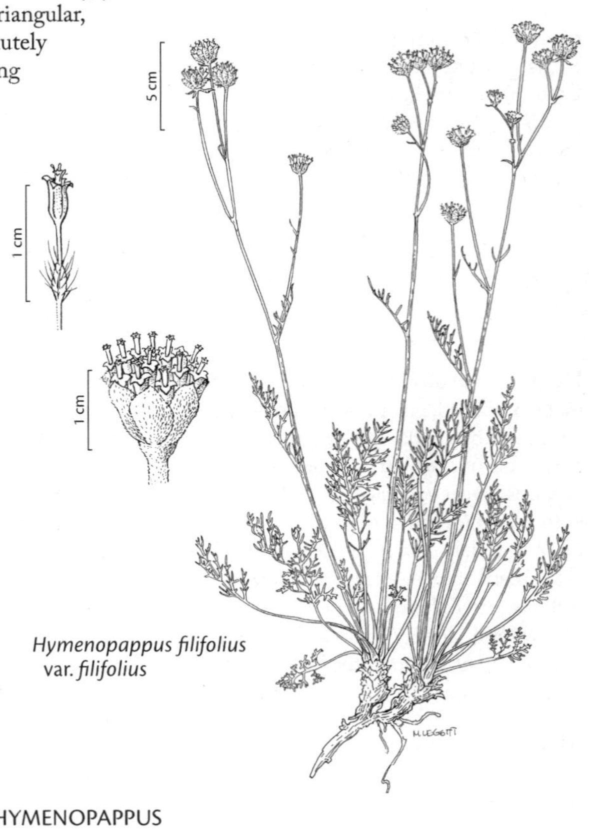

Hymenopappus filifolius var. *filifolius*

HYMENOPAPPUS

HYMENOXYS Rubberweed, bitterweed

Herbs annual, biennial, or perennial, aromatic; taprooted, often with woody caudices. **Stems** erect, branched or unbranched, glabrous to canescent-puberulent or villous. **Leaves** basal and cauline, alternate, broad to narrow, margins entire or few-lobed, lobes linear, gradually reduced distally, tips acute to obtuse, surfaces glabrous to puberulent or villous, usually gland-dotted; petiolate. **Inflorescences** flat-topped, raceme- or corymb-like arrays. **Heads** radiate or discoid, erect; peduncles glabrous to puberulent or villous, sometimes minutely glandular. **Involucres** globose or hemispheric. **Phyllaries** in 2–4 equal series, erect in flower, reflexed in fruit, linear or lanceolate; inner membranous; outer herbaceous throughout or distally. **Receptacles** usually hemispheric, spheroid, ovoid, or conic, smooth or pitted; paleae 0. **Ray florets** pistillate; corollas well exceeding involucres. **Disc florets** bisexual; corollas tubular, gradually widened distally, minutely gland-dotted, lobes 5, erect or cupped inward; stamens included or partly exserted; stigma lobes included or slightly exserted. **Fruits** obpyramidal or quadrangular, truncate, tapered toward base, distinctly or obscurely veined; pappi of tapering, awn-tipped scales.

North America, South America. 25 species; 2 species treated in *Flora*.

Key to *Hymenoxys* species:

1. Leaves usually 2–6-lobed; basal and lower cauline petioles not winged; ray florets 8–13. . . . . *H. cooperi*

1' Leaves entire, rarely slightly dentate; basal and lower cauline petioles winged; ray florets 13–26 . . . . *H. hoopesii*

Hymenoxys cooperi (A. Gray) Cockerell

Cooper's golden-flower, Cooper's rubberweed, ragged rustlers

Hymenoxys cooperi (A. Gray) Cockerell var. *canescens* (D.C. Eaton) K.L. Parker

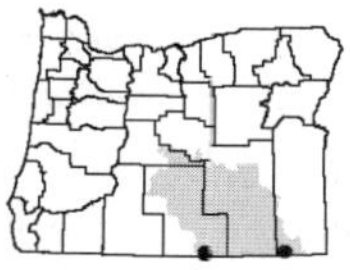

Plants biennial or perennial, 1–4 dm. **Stems** canescent-puberulent. **Leaves** 1.5–10 cm, usually 2–6-lobed, surfaces canescent-puberulent or villous, lobes linear; petioles not winged. **Involucres** 4–9 mm, surfaces glabrous or villous, gland-dotted. **Ray florets** 8–13; corollas 1–2 cm, yellow. **Disc florets** 30–100+; corollas 2.5–4.5 mm, yellow, well exserted. **Fruits** obpyramidal, 1.5–3.5 mm, black, veins indistinct, strigose; pappi 1.5–3.5 mm, silvery. **2*n***=30.

Brushlands, grasslands, ridge tops, slopes, flats. Flowering May–Aug. 1400–1800 m. BR. CA, ID, NV; southeast to AZ. Native.

If subspecific taxa of this species are recognized, ours would refer to variety *canescens*.

HYMENOXYS • HYPOCHAERIS • INULA

Hymenoxys hoopesii (A. Gray) Bierner (p. 293)

Owl's claw, orange sneezeweed

Dugaldia hoopesii (A. Gray) Rydb., *Helenium hoopesii* A. Gray

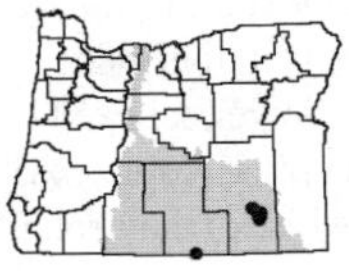

Plants perennial, 1.5–10 dm. **Stems** glabrous or villous. **Leaves** 5–30 cm, margins entire, rarely slightly dentate, surfaces glabrous or villous; basal and lower cauline petioles winged. **Involucres** 6–10 mm, surfaces villous, gland-dotted. **Ray florets** 13–26; corollas 2–4 cm, yellow-orange to orange. **Disc florets** 100–300+; corollas 4–5.5 mm, yellow-orange, well exserted. **Fruits** quadrangular, 3.5–4.5 mm, veins distinct, dark brown, strigose; pappi 2.5–4 mm, silvery. **2*n***=30.

Grasslands, brushlands, meadows, hillsides, streambanks, lakesides. Flowering Jun–Sep. 1600–3000 m. BR, ECas. CA, ID, NV; southeast to NM. Native.

HYPOCHAERIS Cat's-ear

Herbs annual or perennial, sap milky; taprooted. **Stems** erect, leafless except for scaly bracteoles in inflorescences, branched, glabrous or proximally rough-hairy. **Leaves** basal, oblanceolate, bases tapering and petiole-like, margins irregularly dentate or lobed, tips acute or obtuse, surfaces glabrous or rough-hairy. **Inflorescences** terminal, branched, with solitary heads; bracteate. **Heads** ligulate. **Involucres** narrowly to broadly campanulate in flower, cylindrical to ovoid in fruit. **Phyllaries** 10–30 in 4–5 graduated series, erect, reflexed in fruit, surfaces glabrous or midribs with stiff, sharp hairs. **Receptacles** flat to convex; paleae membranous. **Florets** bisexual; ligules slightly to much exceeding involucres, white or yellow. **Fruits** tapering toward bases and tips or only toward bases, brown, finely roughened, glabrous, ribs 10+; pappi of stiff, plumose bristles, tawny or dull white, shorter outer bristles barbed.

Africa, Asia, Europe, North America, South America. 50–100 species; 2 species treated in *Flora.*

Key to *Hypochaeris* species:

1. Plants annual; leaves glabrous or with scattered stiff hairs on midribs below; ligules slightly exceeding involucres; outer fruits beakless, inner beaked . . . . ***H. glabra***

1' Plants perennial; leaves hispid; ligules much exceeding involucres; fruits all beaked . . . . ***H. radicata***

Hypochaeris glabra Linnaeus

Smooth cat's-ear

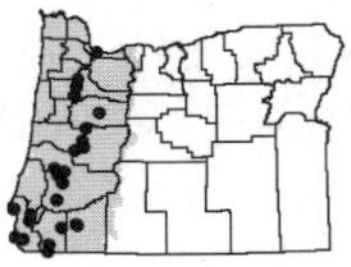

Plants annual, 10–50 cm. **Leaves** 1–10 cm, margins entire to irregularly dentate or lobed, sparsely ciliate, tips usually broadly acute, surfaces glabrous or with scattered stiff hairs on midribs below. **Phyllaries** 10–20, 2–16 mm, glabrous. **Florets** 5–40; ligules 5–10 mm, slightly exceeding involucres, white or yellowish, outer red abaxially. **Fruit** bodies 3–5 mm; outer cylindrical, truncate, beakless; inner fusiform with slender beaks; pappi tawny. **2*n***=10, 12.

Open forests, shrublands, grasslands, roadsides, riverbanks. Flowering Apr–Jun. 0–1300 m. Casc, CR, Est, Sisk, WV. CA, WA; widely scattered in North America; nearly worldwide. Exotic.

Hypochaeris radicata Linnaeus (p. 293)

Hairy cat's-ear

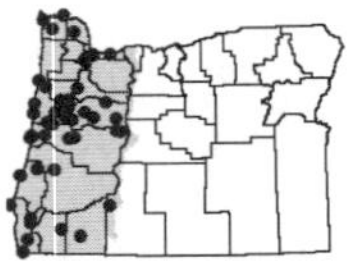

Plants perennial, 10–60 cm. **Leaves** 5–30 cm, margins hispid, often with large, triangular teeth or down-pointing lobes, tips usually obtuse, surfaces white-hispid. **Phyllaries** 20–30, 3–20 mm, glabrous or hispid on midveins. **Florets** 13–100+; ligules 10–25 mm, much exceeding involucres, yellow, outer red abaxially. **Fruit** bodies 3–5 mm, fusiform, beaks slender; pappi dull white. **2*n***=8.

Open forests, grasslands, dunes, roadsides, disturbed areas, lawns. Flowering Apr–Oct. 0–1900 m. Casc, CR, Est, Sisk, WV. CA, ID, NV, WA; scattered in North America; Asia, Europe. Exotic.

Because of its yellow ligules and globose heads of beaked fruits with plumose pappi, this species is often mistaken for the common dandelion (*Taraxacum officinale*), whose flowering stems, however, are hollow and always unbranched.

INULA Inula, yellowhead

Herbs perennial, usually aromatic; taproots thick, sometimes rhizomatous. **Stems** erect, simple or few-branched, pilose- or hispid-puberulent. **Leaves** basal and cauline, alternate, ovate-lanceolate to broadly elliptic-lanceolate, bases truncate or cordate-clasping, margins densely denticulate, tips acute, surfaces abaxially puberulent, adaxially glabrous to minutely hispid or pilose; proximal petiolate, distal sessile. **Inflorescences** heads solitary or in corymb-like arrays. **Heads** radiate; peduncles hispid-pilose. **Involucres** broadly hemispheric. **Phyllaries** in 4–7 equal series, tips acute; inner narrow; outer broadly ovate. **Receptacles** flat or convex, smooth or alveolate; paleae 0. **Ray florets** pistillate; rays linear, much exceeding involucres. **Disc florets** bisexual; corollas tubular, gradually widened distally, lobes 5, triangular; stamens included; stigma lobes slightly exserted. **Fruits** linear-fusiform, 10–15-veined; pappi of numerous, basally connate bristles or narrow scales.

Asia, Europe, North America. ~100 species; 1 species treated in *Flora.*

Inula helenium Linnaeus (p. 293)

Elecampane, elecampane inula

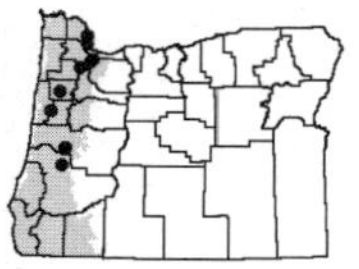

Plants 5–20 dm. **Leaves** lanceolate to elliptic or ovate, 10–40 × 5–20+ cm, bases cordate, clasping, margins denticulate to serrate, tips acute, surfaces abaxially velvety-woolly, adaxially sparsely pubescent. **Involucres** (20)30–40 mm in diameter. **Ray florets** (15)50–100+; rays linear, (10)20–30 mm, yellow. **Disc florets** 50–200+; corollas 10–12 mm, yellow. **Fruits** 3–4 mm, brown, surfaces glabrous; pappus bristles 6–10 mm, brownish. $2n$=10.

Swamps, roadside ditches, fields, streamsides, disturbed areas. Flowering Jun–Aug. 0–300 m. CR, Sisk, WV. WA; north to British Columbia; UT, eastern Canada and US; Asia, Europe. Exotic.

This tall, conspicuous species is a frequent garden escape.

IONACTIS Aster, ankle-aster

Herbs perennial; taproots stout, usually with much-branched, underground caudices. **Stems** erect, few to many, leafy throughout, unbranched, tomentulose. **Leaves** cauline, alternate, linear to oblong-elliptic, margins entire, spiculate, 1-veined, tips acute, surfaces minutely to densely spiculate; sessile. **Inflorescences** solitary, terminal heads or 2–3 in corymb-like arrays. **Heads** radiate, erect; peduncles tomentulose. **Involucres** campanulate. **Phyllaries** in several series, erect in flower, spreading in fruit, linear-lanceolate, midribs keeled, margins smooth or ciliate, tips acute or acuminate; outer gradually shorter, green or purplish, surfaces lightly to densely spiculate or strigose; inner ± equal, often chartaceous-margined, surfaces glabrous or spiculate. **Receptacles** flat, pitted; paleae 0. **Ray florets** pistillate, rays well exceeding involucres, blue or violet. **Disc florets** bisexual; corollas tubular, gradually enlarged distally, lobes 5, erect, triangular; stamens partly or fully exserted; stigmas exserted. **Fruits** flattened, oblanceolate, tapering toward bases, inconspicuously veined; pappi of numerous elongate bristles, with whorl of short outer bristles at bases.

North America. 5 species; 1 species treated in *Flora*.

Ionactis, now limited to *I. alpina* and four other species (Nesom & Leary 1992), is one of more than seven North American genera recently segregated from *Aster*, which now has only two recognized species in the New World (Brouillet 2006).

IONACTIS • IVA • JAUMEA

Ionactis alpina (Nuttall) Greene (p. 295)

Crag aster, lava aster, lava ankle-aster

Aster scopulorum A. Gray

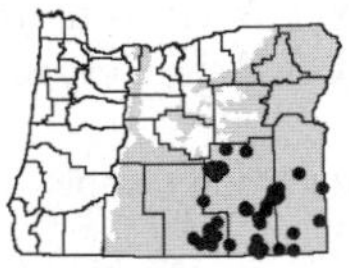

Plants 4–25 cm. **Leaves** 4–18 mm. **Inflorescences** solitary, terminal heads. **Involucres** 7–11 mm. **Ray florets** 8–13; corollas 6–15 mm. **Disc florets** corollas 5–6 mm, slightly exserted from involucres, yellow. **Fruits** 4–6 mm, brownish, surfaces silky-pubescent; inner pappus bristles 4–6 mm, white or tan, outer ≤ 1 mm. ***2n***=18, 36.

Dry, open slopes, rimrocks, flats. Flowering May–Jul. 1000–2500 m. BR, BW, ECas, Owy. CA, ID; northeast to MT, east to WY, southeast to UT. Native.

IVA Marsh elder

Herbs or **shrubs** taprooted, often rhizomatous. **Stems** 1–many, simple or branched, lightly strigose-hispid with minute, sessile, brownish glands. **Leaves** cauline, opposite, becoming alternate distally, reduced upwards, bases rounded or tapering, tips acute, surfaces dotted with sessile, brownish glands; sessile or petiolate. **Heads** discoid, nodding; peduncles glabrous or puberulent. **Involucres** campanulate or hemispheric. **Phyllaries** in 1–3+ series, surfaces glabrous or lightly scabrous, lobes erect, triangular to rounded. **Receptacles** flat to hemispheric; paleate. **Paleae** linear to wedge-shaped, ± membranous, sometimes 0. **Staminate florets** corollas funnelform, whitish to pink, lobes 5. **Pistillate florets** corollas inconspicuous, tubular, whitish. **Fruits** obovoid, ± 3-angled, tapering toward bases, 2 lateral angles becoming more prominent toward the base; pappi 0.

Asia, Europe, North America. 9 species; 1 species treated in *Flora*.

Iva axillaris Pursh (p. 295)

Poverty weed, deep-root

Iva axillaris Pursh ssp. *robustior* (Hook.) Bassett

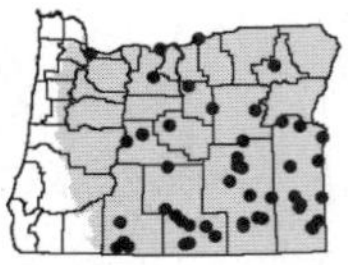

Herbs perennial, 1–6 dm. **Leaves** narrowly ovate to elliptic or spatulate, 10–50 × 3–15 mm, margins entire, rarely toothed, surfaces sericeous to scabrid. **Heads** unisexual, solitary in leaf axils; peduncles 0.5–3(6) mm. **Involucres** 2–3 mm. **Staminate flowers** 4–10(20); corollas 2–2.5 mm, equaling or slightly exceeding involucres, yellow. **Pistillate flowers** 3–8; corollas 1–1.5 mm, exserted. **Fruits** 2.5–3 mm, blackish, minutely gland-dotted. ***2n***=36, 54.

Pine woods, shrublands, riverbanks, lake shores, ponds, roadsides, disturbed areas. Flowering May–Sep. 0–1900 m. BR, BW, Casc, Col, ECas, Lava, Owy, WV. CA, ID, NV, WA; north to British Columbia, northeast to Manitoba, east to WI, southeast to TX. Native.

This species is wind-pollinated, a somewhat unusual characteristic within Asteraceae.

JAUMEA Jaumea

Herbs perennial, succulent; rhizomatous or with leafy stolons. **Stems** erect or lax and spreading, simple or branched distally, glabrous. **Leaves** cauline, opposite, linear to narrowly oblanceolate, margins entire, tips acute or obtuse, surfaces glabrous; petioles indistinct. **Inflorescences** solitary, terminal heads on stems and distal branches. **Heads** radiate, rarely discoid; peduncles glabrous. **Involucres** campanulate. **Phyllaries** 12–15 in 3–5 unequal series; outer ovate to round, gradually shorter, tips obtuse; inner erect, ± broadly lanceolate, tips obtuse, with numerous parallel veins. **Receptacles** conic; paleae 0. **Ray florets** pistillate, fertile; corollas exceeding involucres, rays yellow. **Disc florets** bisexual, fertile; corollas narrowly tubular below, widened distally, yellow, lobes 5, erect, triangular; stamens partly exserted; stigmas slightly exserted. **Fruits** columnar-fusiform, truncate; pappi 0 or rarely of 1–5 narrow, persistent scales.

North America, South America. 2 species; 1 species treated in *Flora*.

This genus has a disjunct distribution between North and South America, ranging from British Columbia to Baja California and reoccurring as a different species in Atlantic Coastal Argentina and Uruguay (Panero 2007).

Jaumea carnosa (Lessing) A. Gray (p. 295)

Fleshy jaumea

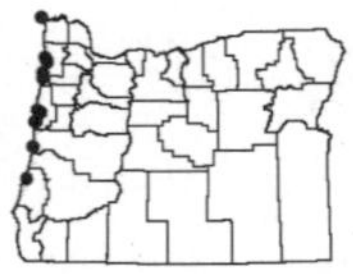

Plants 2–30 cm. **Leaves** 15–60 × 2–6 mm. **Involucres** 7–12 mm, glabrous. **Phyllaries** inner 4–8; outer 5–7. **Heads** peduncles 0.5–4 cm. **Ray florets** (0)3–10; rays 6–9 mm. **Disc florets** 15–50; corollas 5–6 mm. **Fruits** 4–5 mm, brown, glabrous, ± 10-ribbed. ***2n***=38.

Coastal salt marshes, sloughs, beaches, mudflats. Flowering Jun–Oct. 0–50 m. Est. CA, WA; north to British Columbia, south to Mexico. Native.

Jaumea carnosa is sometimes a host for dodder (*Cuscuta*). The paleopolyploid chromosome number of *n*=19 is unusual, also being found in such taxonomically isolated Helianthoid genera as *Arnica* and *Venegasia* (Baldwin 2009).

KYHOSIA Tarweed

Bruce G. Baldwin and Gerald D. Carr

Genus composed of one species, *K. bolanderi.*

Kyhosia bolanderi (A. Gray) B.G. Baldwin (p. 298)

Bolander's tarweed

Madia bolanderi (A. Gray) A. Gray

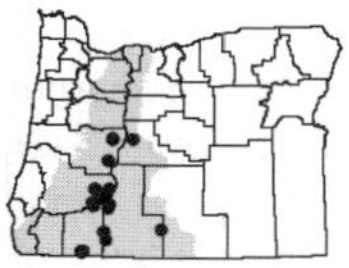

Herbs perennial, 5–12 dm; rhizomatous; self-incompatible. **Stems** erect, usually branched distally. **Leaves** basal and cauline; basal in rosettes; cauline opposite proximally, alternate distally; blades linear to lance-linear; proximal 5–35 cm × 4–15 mm, reduced distally, margins entire, surfaces hirsute; distal glandular-hirsutulous; sessile. **Inflorescences** heads solitary and terminal or in loose, ± corymb-like arrays. **Heads** radiate. **Involucres** ± campanulate to hemispheric, 6–12+ mm in diameter. **Phyllaries** 8–12 in 1 series, each mostly or wholly enveloping a ray floret, lanceolate to lance-linear, 7–14 mm, herbaceous, abaxially hirsute and glandular-hirsutulous. **Receptacles** flat to convex, glabrous; paleae in 1 series between ray and disc florets, falling, weakly connate or distinct. **Ray florets** 8–12, pistillate, fertile; corollas bright yellow, rays 7–13 mm. **Disc florets** 28–65, bisexual and fertile; corollas 5–8 mm, bright yellow, pubescent, tubes < funnelform throats, lobes 5, deltate; anthers dark purple; styles glabrous proximal to branches, branch tips densely hairy. **Ray fruits** compressed, clavate, arcuate, 5–7 mm, black, basal attachments centered, tips beakless, glabrous or hispidulous; pappi 0 or of crowns, to 0.7 mm. **Disc fruits** ± terete, straight or arcuate, 5–9 mm, brown to black, hispidulous; pappi of 5–10 lanceolate to subulate scales, stramineous to purplish, ciliate to plumose, 1–5 mm. **2*n***=12.

Wet meadows, streambanks. Flowering Jul–Sep. 1200–2200 m. Casc, ECas, Sisk. CA, NV. Native.

Kyhosia has been treated as congeneric with *Madia*, which is evidently more closely related to the Hawaiian silversword alliance (*Argyroxiphium*, *Dubautia*, and *Wilkesia*) than to *Kyhosia*. The similarity of head architecture between *Kyhosia* and *Argyroxiphium* noted by Carlquist (1959) and the production of hybrids between *Kyhosia* and *Dubautia* suggest that a fairly close relationship also exists between *Kyhosia* and the Hawaiian Madiinae (Carlquist et al. 2003). *Kyhosia* is the only perennial tarweed with **2*n***=12.

Kyhosia bolanderi is unusual among continental tarweeds for occurring in wet, montane habitats. Heads often close during the day, presumably in response to heat or drought stress.

LACTUCA Lettuce

Herbs annual or biennial, sap milky; taprooted. **Stems** 1, erect, usually not branched below inflorescences, glabrous or hairy. **Leaves** basal and cauline early, mostly cauline at flowering, alternate, surfaces glabrous or hairy; sessile or petiolate. **Inflorescences** in spike-, raceme-, or panicle-like arrays. **Heads** ligulate, erect. **Involucres** cylindrical or campanulate in flower, ovoid in fruit. **Phyllaries** in 2–several unequal series, erect, reflexed at maturity, deltate to linear or lanceolate, tips acute or acuminate. **Receptacles** flat or convex; paleae 0. **Florets** bisexual; ligules exceeding involucres, blue, yellow, or white. **Fruits** flattened, tapering toward bases and tips, glabrous, strongly veined, with 1–several weaker veins on each face; beaked; pappi of numerous deciduous or persistent bristles, white or brown.

Northern temperate regions. ~75 species; 3 species treated in *Flora.*

Lactuca canadensis and *L. ludoviciana* have been infrequently collected in Oregon as waifs.

The numbers in the vein descriptions below refer collectively to both strong and lesser veins.

Key to *Lactuca* species:

1. Florets blue to white (rarely yellow); leaf bases don't clasp stem; beaks of fruits (0)0.1–0.5 mm, stout; pappi brownish . . . . ***L. biennis***

1' Florets yellow; leaf bases clasp stem; beaks of fruits 2–6 mm, very slender; pappi white.

  2. Leaves linear or narrowly lanceolate, margins not prickly. . . . . ***L. saligna***

  2' Leaves oblanceolate to elliptic or obovate, margins usually prickly. . . . . ***L. serriola***

Lactuca biennis (Moench) Fernald (p. 298)

Tall blue lettuce

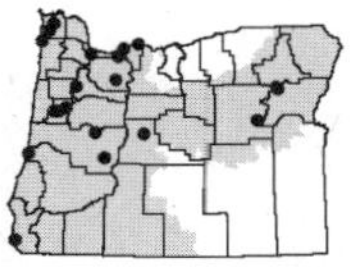

Plants annual or biennial, (15)75–200+ cm. **Stems** glabrous. **Leaves** lanceolate to ovate, 5–40 × 4–15 cm, usually much reduced distally, margins entire or denticulate, usually deeply pinnately lobed, surfaces glabrous or main veins hairy abaxially, lobes broad to narrow; proximally short-petiolate, winged, distally sessile. **Inflorescences** 10–40+ heads in panicle-like arrays; peduncles short, sometimes minutely bracteolate. **Involucres** 7–12 mm. **Phyllaries** inner lanceolate, surfaces glabrous; outer deltate to lanceolate, gradually shorter. **Florets** (15)20–30+; ligules 10–15 mm, blue to white, rarely yellow. **Fruits** bodies narrowly elliptic, 4–5 mm brown, often mottled, each face 5–7-veined, smooth; beaks stout, (0)0.1–0.5 mm; pappus bristles deciduous, brownish. **2*n***=34.

Woodlands, shrublands, riverbanks, cliffs, roadsides, waste areas. Flowering Jul–Sep. 0–1300 m. BW, Casc, CR, ECas, Est, Lava, Sisk, WV. CA, ID, WA; north to AK, northeast to Newfoundland, east to MA, southeast to VA. Native.

Lactuca saligna Linnaeus

Willow lettuce

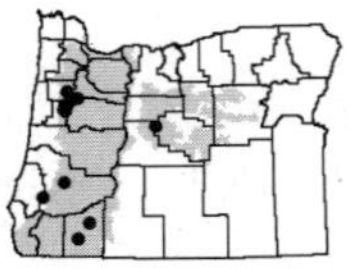

Plants annual, 15–100+ cm. **Stems** usually glabrous. **Leaves** linear to narrowly lanceolate, 5–15+ × 1–7 cm, reduced below inflorescences, bases with 2 downward-pointing basal lobes clasping stems, margins entire or denticulate or with a few sharp, linear lobes, surfaces glabrous; sessile. **Inflorescences** 10–30+ heads in panicle- or raceme-like arrays; peduncles short. **Involucres** 6–13 mm. **Phyllaries** inner, lance-linear, surfaces glabrous; outer deltate to lance-linear, gradually shorter. **Florets** 6–12(20+); ligules 8–12 mm, yellow, sometimes bluish-tinged. **Fruits** bodies broadly elliptic, 2.5–3.5 mm, brown, each face 5–7-veined, minutely roughened; beaks slender, (2)5–6 mm; pappus bristles persistent or deciduous, white. **2*n***=18.

Openings in woods, fields, roadsides, disturbed areas. Flowering Jul–Sep. 0–800 m. Casc, Lava, Sisk, WV. CA, NV, WA; throughout eastern and southern North America; Europe. Exotic.

Lactuca serriola Linnaeus (p. 298)

Prickly lettuce

Lactuca scariola L. var. *integrata* Gren. & Godr., *Lactuca scariola* L. var. *scariola*

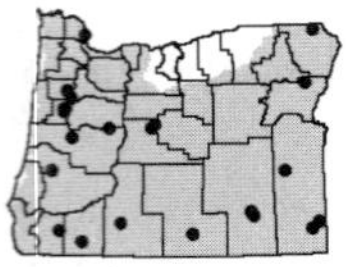

Plants annual, (15)30–100+ cm. **Stems** glabrous to prickly-bristly. **Leaves** oblanceolate to elliptic-oblong or obovate, 5–30 × 1–10 cm, much reduced distally, margins entire to denticulate or pinnately few-lobed, bases basally broadened and clasping stems, margins denticulate, usually prickly, surfaces glabrous or midribs with stout prickles below, lobes broad, often somewhat curved toward base; sessile. **Inflorescences** 10–30+ heads in panicle- or rarely raceme-like arrays; peduncles short to long, sometimes bracteolate. **Involucres** 9–12 mm. **Phyllaries** inner lance-linear, surfaces glabrous; outer deltate to lance-linear, gradually or abruptly shorter. **Florets** 12–20+; ligules 6–12 mm, yellow, drying bluish. **Fruits** bodies broadly elliptic, 2.5–4 mm, grayish to tan, each face 5–9-veined, sometimes slightly winged distally, smooth except minutely bristled near bases; beaks slender, 2.5–4 mm; pappus bristles deciduous, white. **2*n***=18.

Grasslands, open hillsides, roadsides, disturbed areas. Flowering Jul–Sep. 0–1800 m. BR, BW, Casc, CR, ECas, Lava, Owy, Sisk, WV. CA, ID, NV, WA; worldwide. Exotic.

Lactuca serriola is the most common exotic wild lettuce in Oregon. In some older references it is referred to as *L. scariola*, an invalid spelling variation.

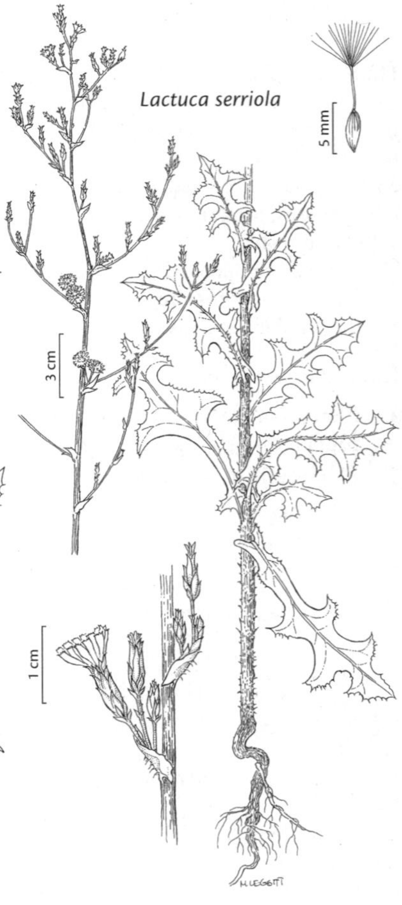

LAGOPHYLLA Hareleaf

Bruce G. Baldwin and Gerald D. Carr

Herbs annual. **Stems** ± erect. **Leaves** mostly cauline, most alternate, opposite proximally; blades narrowly elliptic to linear, or oblanceolate to spatulate proximally, some with petioloid bases, margins usually entire, sometimes toothed proximally; sessile. **Inflorescences** panicle-like arrays or heads in glomerules. **Heads** radiate. **Involucres** obovoid to obconic, 3–5 mm in diameter, subtended by calyculi of 2–5 bractlets. **Phyllaries** 5 in 1 series, herbaceous, linear to oblanceolate, each wholly enveloping a subtended ray ovary proximally. **Receptacles** flat to convex, densely hirsutulous. **Paleae** in ring between ray and disc flowers, distinct or proximally connate, scarious. **Ray florets** 5, pistillate, fertile; corollas yellow, rays often veined red to purple abaxially. **Disc florets** 6, functionally staminate; corollas yellow, tubes < funnelform throats, lobes 5, deltate; anthers dark purple; styles glabrous proximal to branches, branches connate, tips sparsely hairy. **Ray fruits** ± obcompressed, attachments basal, glabrous; beakless; pappi 0. **Disc fruits** 0; pappi 0.

Western North America. 5 species; 1 species treated in *Flora.*

Molecular phylogenetic data have indicated that *Lagophylla* is sister to *Layia* (Baldwin 2003).

Lagophylla ramosissima Nuttall (p. 299)

Slender hareleaf, common rabbitleaf

Lagophylla ramosissima Nutt. ssp. *ramosissima*

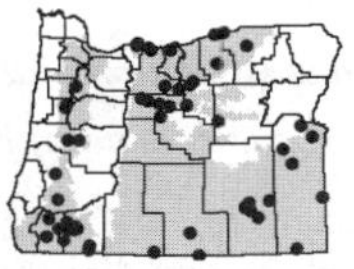

Plants 1–10(15) dm; branching excurrent or pseudo-dichotomous. **Leaves** grayish, surfaces hirsute to strigose, sericeous, or villous, distal abaxially stipitate-glandular, glands whitish or yellowish. **Calyculi** ≈ or slightly exceeding phyllaries, often stipitate-glandular abaxially, marginal hairs pilose-hirsute. **Phyllaries** 4–7 mm, rarely glandular abaxially, pilose-hirsute on angles, hairs somewhat spreading to antrorsely curved, 0.5–1+ mm. **Ray florets** rays 3–6 mm. **Fruits** dull to somewhat glossy, weakly striate. **2*n***=14.

Grasslands, openings in chaparral, scrub, woodlands, forests. Flowering Apr–Sep. 0–1800 m. BR, Col, ECas, Lava, Owy, Sisk, WV. CA, ID, NV, WA; northeast to MT. Native.

The self-compatible *Lagophylla ramosissima* is easily overlooked in the field; leaves often wither before or during flowering, and heads often close at midday in response to heat or drought stress (Thompson 1983). Rays often shrivel greatly in pressed specimens, making lengths difficult to assess.

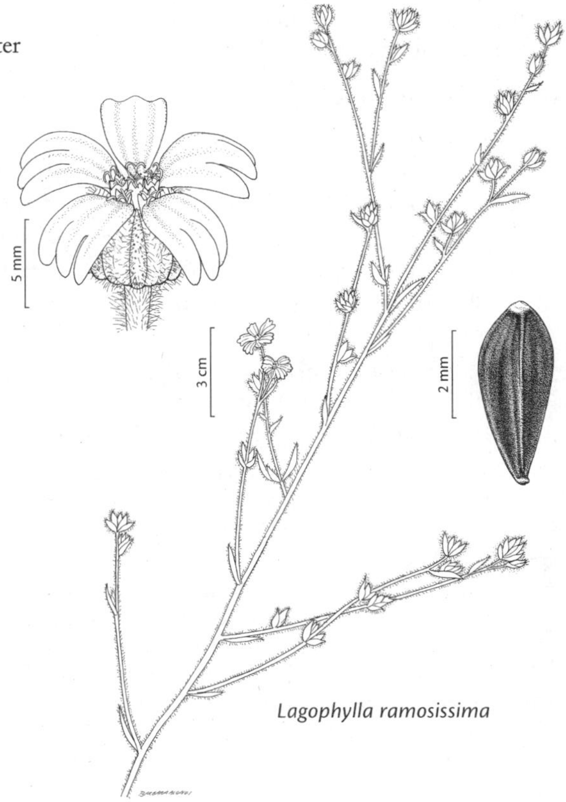

LAGOPHYLLA

LAPSANA Nipplewort

Genus composed of one species, *L. communis.*

Lapsana communis Linnaeus (p. 301)

Nipplewort

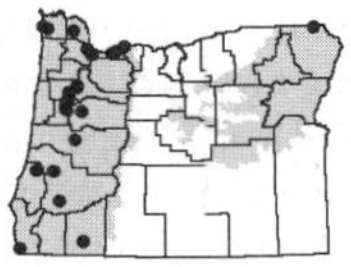

Herbs annual, 15–150 cm, sap milky; fibrous-rooted. **Stems** erect, leafy throughout, branched distally, glabrous or with scattered soft, white hairs. **Leaves** ovate, 1–15+ × 1–10 cm, margins usually irregularly dentate, surfaces glabrous or lightly hairy; petioles often with 1 or more pairs of leaflets. **Inflorescences** terminal panicle-like arrays. **Heads** 5–50+, ligulate; peduncles sometimes minutely bracteolate. **Involucres** narrowly ovoid. **Phyllaries** 8–10, in 1 series with a few small outer, erect, lance-oblong, 3–9 mm, midribs thickened toward bases, tips acute, surfaces glabrous. **Receptacles** flat, smooth, glabrous; paleae 0. **Florets** 6–15, bisexual; ligules 4–12 mm, well exceeding involucres, yellow; stamens 5. **Fruits** tapering toward bases and tips, curved, 3–5 mm, outer sometimes much longer than inner, pale brown, glabrous, ribs ~20; pappi 0. **2*n***=12, 14, 16.

Forests, roadsides, disturbed areas. Flowering Jun–Oct. 0–1300 m. BW, Casc, CR, Est, Sisk, WV. CA, ID, WA; north to AK, northeast to Newfoundland, east to MA, southeast to VA; Asia, Europe. Exotic.

This species is a common introduced weed of gardens and disturbed areas.

LASTHENIA Goldfields

Herbs annual or perennial; taprooted, occasionally rhizomatous or fibrous-rooted. **Stems** erect or decumbent, simple or branched, glabrous, strigillose, or villous. **Leaves** mostly cauline, opposite, linear to elliptic or oblanceolate, margins entire to regularly or irregularly dentate or lobed, sometimes strigose-ciliate, tips acute to obtuse, surfaces glabrous or strigillose; proximal pairs sometimes basally united and sheathing the stem; sessile or indistinctly petiolate. **Inflorescences** heads solitary or in corymb-like arrays. **Heads** radiate, rarely discoid; peduncles glabrous, strigillose, or villous. **Involucres** hemispheric to obconic or broadly cylindric. **Phyllaries** in 1–2 equal series, tips acute to obtuse. **Receptacles** conic to hemispheric; paleae 0. **Ray florets** usually present, sometimes inconspicuous or absent, pistillate; rays slightly to well exceeding involucres, yellow. **Disc florets** bisexual; corollas narrowly tubular below, widened distally, yellow, lobes 4–5, erect or spreading; stamens included or slightly exserted; stigmas slightly exserted. **Fruits** linear to columnar-fusiform, truncate, indistinctly ribbed; pappi 0 or of few to many persistent awns or scales.

Western North America and South America. 17 species; 4 species treated in *Flora*.

For a relatively small genus, *Lasthenia* is unusually diverse and is divided into seven sections (Chan & Ornduff 2006; Chan et al. 2002). In years of abundant rainfall, some form expanses of yellow in California pasturelands, hence the common name "goldfields." Most species are endemic to California, while a single species is found only in Chile.

Key to *Lasthenia* species:

1. Phyllaries ⅔ united; ray florets inconspicuous . . . . ***L. glaberrima***

1' Phyllaries not united; ray florets conspicuous or inconspicuous.

  2. Leaves regularly or irregularly dentate or lobed; ray florets inconspicuous; plants of coastal habitats . . . . ***L. maritima***

  2' Leaves entire; ray florets conspicuous; plants of coastal habitats or not.

    3. Plants annual, taprooted; plants not of coastal habitats. . . . . ***L. californica*** ssp. ***californica***

    3' Plants perennial, often rhizomatous; plants of coastal habitats. . . . . ***L. ornduffii***

Lasthenia californica ssp. ***californica*** de Candolle ex Lindley (p. 301)

Slender goldfields, California goldfields

Baeria chrysostoma Fisch. & C.A. Mey. ssp. *chrysostoma*, *Baeria chrysostoma* Fisch. & C.A. Mey. ssp. *gracilis* (DC.) Ferris, *Baeria chrysostoma* Fisch. & C.A. Mey. var. *gracilis* (DC.) H.M. Hall

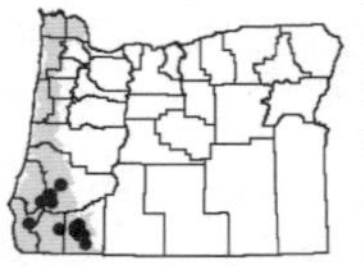

Plants annual, 2–36 cm; taproots slender. **Stems** usually strigillose. **Leaves** linear to narrowly oblanceolate, 3–70 × 0.2–2 mm, margins entire, often strigose-ciliate, tips acute, surfaces glabrous or strigillose. **Involucres** fusiform to campanulate, 3–6 mm, surfaces strigose. **Phyllaries** 3–12, not united, lanceolate, tips acute. **Heads** radiate; peduncles 1–18 cm, surfaces strigillose. **Ray florets** 4–15, conspicuous; corollas 2–10 mm, yellow. **Disc florets** 2–30+; corollas 2–4 mm, yellow; stamens exserted; styles exserted. **Fruits** 2–3 mm, black or gray, glabrous or strigillose; pappi of 0–7 brown or white aristate awns, 2–5 mm. **2*n***=16, 32, 48.

Open woodlands, shrublands, rocky slopes, grasslands, near vernal pools. Flowering Apr–May. 0–900 m. CR, Sisk. CA. Native.

Plants of *L. californica* are subject to extreme reduction in size when growing in thin, dry soil.

Lasthenia glaberrima de Candolle (p. 301)

Smooth lasthenia, smooth goldfields

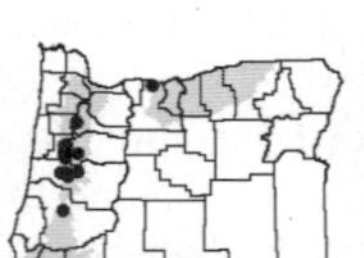

Plants annual, 5–30 cm; fibrous-rooted. **Stems** glabrous. **Leaves** linear, 8–60 × 0.5–4 mm, margins entire, tips blunt, surfaces glabrous. **Involucres** hemispheric or campanulate, 4–7 mm. **Phyllaries** 6–20, ⅔ united, triangular, lobes free, tips acute, surfaces glabrous or strigillose distally. **Heads** radiate; peduncles 0.5–15 cm, surfaces glabrous or minutely puberulent. **Ray florets** 5–12, inconspicuous; corollas 1–1.5 mm, yellow, included or slightly exceeding involucres. **Disc florets** 3–30; corollas 1–2 mm, yellow, lobes 4; stamens included; styles included. **Fruits** 3–4 mm, black or grayish, strigillose; pappi of 5–10 fimbriate scales, 1–2 mm. **2*n***=10.

Wet meadows, marshy areas, ditches, vernal pools. Flowering May–Jul. 50–400 m. Col, Sisk, WV. CA, WA. Native.

The short stamens and styles of *Lasthenia glaberrima* are indicative of a self-pollinating breeding system (Ornduff 1963, 1966).

The leaf tips of *L. glaberrima* are often described as "blunt" (Chan & Ornduff 2006), but more than this, each blunt tip appears to consist of a small, specialized gland, perhaps functioning for water secretion. This structure is less obvious or absent in the leaves of other Oregon *Lasthenia* taxa. During growth early in the season, when submerged in water, the stem of *L. glaberrima* develops a brownish, wrinkled epidermis, in contrast to the smooth, green stem later produced aerially.

Lasthenia maritima (A. Gray) M.C. Vasey

Seaside goldfields, maritime goldfields

Baeria maritima (A. Gray) A. Gray, *Baeria minor* (DC.) Ferris ssp. *maritima* (A. Gray) Ferris, *Lasthenia minor* (DC.) Ornduff, *Lasthenia minor* (DC.) Ornduff var. *maritima*

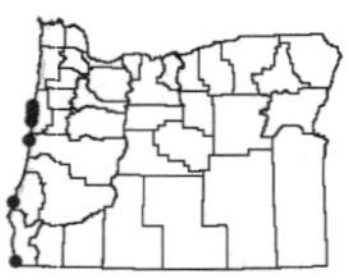

Plants annual, 3–40 cm; taproots slender. **Stems** lightly villous proximally, densely villous distally. **Leaves** linear to narrowly or broadly elliptic, 15–55 × 2–18 mm, margins regularly or irregularly dentate or lobed, teeth triangular, lobes linear, tips obtuse, surfaces glabrous to lightly villous. **Involucres** hemispheric, 4–6 mm, surfaces lightly villous. **Phyllaries** 5–8, not united, ovate, tips acute to obtuse. **Heads** radiate; peduncles 0.4–6 cm, surfaces villous. **Ray florets** few, inconspicuous; corollas 3–5 mm, yellow, slightly exceeding involucres. **Disc florets** 20–100+; corollas 1.5–2 mm, yellow;

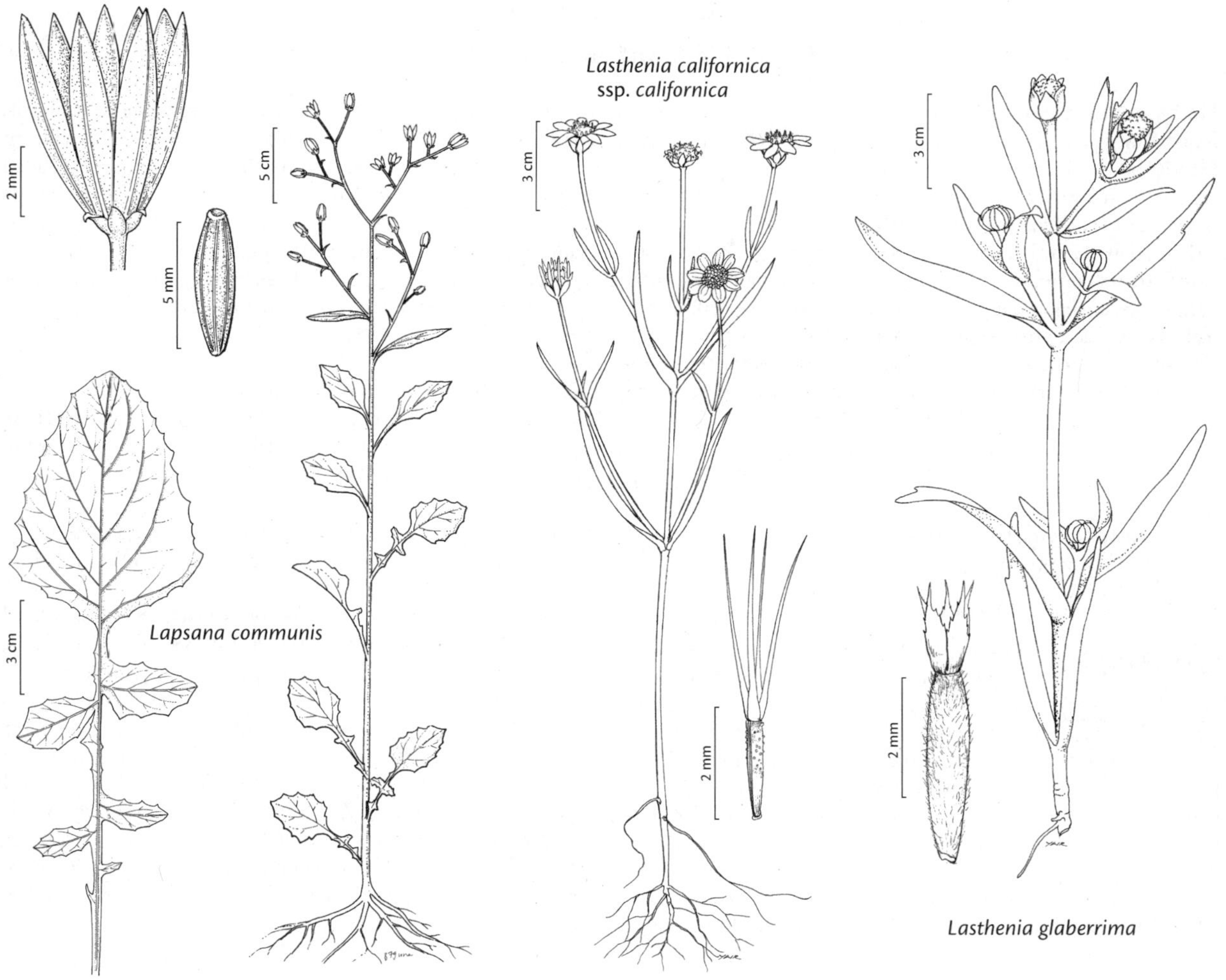

LAPSANA • LASTHENIA

stamens included; styles included. **Fruits** 2.5–3 mm, black or grayish, minutely puberulent; pappi of 0–10+ scales and awns, up to 2 mm. ***2n***=8.

Rocky headlands, offshore islands. Flowering May–Jul. 0–50 m. Est. CA, WA, north to BC. Native.

The peculiar habitat of this species was first noted by Ornduff (1966), who considered it an adaptation to sites rich in bird guano. Like *L. glaberrima*, *L. maritima* is a self-pollinating species whose fruits are probably dispersed by migrating birds.

Lasthenia ornduffii R. Chan

Large-flowered goldfields, Ornduff's lasthenia

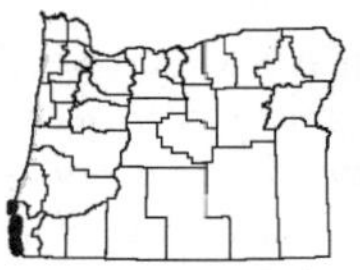

Plants perennial, often clump-forming, 2.5–30 cm; often rhizomatous. **Stems** glabrous or strigillose. **Leaves** linear to oblong, 5–40 × 0.8–5 mm, margins entire, often coarsely strigose-ciliate proximally, tips acute or obtuse, surfaces glabrous or strigillose. **Involucres** hemispheric, 5–10 mm, surfaces strigillose. **Phyllaries** 6–15, not united, elliptic to ovate, tips acute. **Heads** radiate; peduncles 0.5–8 cm, surfaces strigillose. **Ray florets** 8–16, conspicuous; corollas 5–9 mm, yellow. **Disc florets** 10–70+; corollas 1–3 mm, yellow; stamens slightly exserted; styles slightly exserted. **Fruits** 3–4 mm, gray, glabrous; pappi of 0–4 aristate awns, 1–3 mm. ***2n***=32.

Beaches, dunes, bluffs, rocky headlands. Flowering Jun–Aug. 0–200 m. Est. Native.

Lasthenia ornduffii is endemic to Curry County. It differs from its nearest relative, *L. californica* ssp. *macrantha* of northwestern California, in its tetraploid chromosome number, the latter taxon being consistently hexaploid. Plants often occur in windswept sites on bluffs and headlands and are dwarfed and compact in their growth. Often many stems arise together from the rhizome, leading to a clumped habit. Taller plants may occur in more protected sites.

LAYIA Tidytips

Bruce G. Baldwin and Gerald D. Carr

Herbs annual. **Stems** usually ± erect, some with numerous spreading-ascending branches, laterals often surpassing central axis. **Leaves** mostly cauline; most alternate, opposite proximally; sessile. **Inflorescences** loose, ± corymbiform arrays or heads solitary. **Heads** radiate. **Involucres** ± hemispheric to campanulate. **Phyllaries** in 1 series, herbaceous, each wholly enveloping a subtended ray ovary proximally, lanceolate to lance-attenuate or oblanceolate, abaxially hirsute or papillate-scabrous, often glandular. **Receptacles** flat to convex, setulose. **Ray florets** pistillate, fertile. **Disc florets** bisexual, fertile; corollas yellow, tubes < funnelform throats, lobes 5, deltate; anthers dark purple or yellow to brownish; styles glabrous proximal to branches, branch tips hairy and papillose. **Ray fruits** obcompressed, clavate, somewhat arcuate, basal attachments central; beakless; pappi 0. **Disc fruits** ± clavate, somewhat strigose to sericeous; pappi of 9–15 lance-attenuate to subulate scales, nearly equal, rarely unequal, white to tawny, surfaces glabrous, scabrous, or each proximally plumose and often adaxially woolly. **2*n***=16.

Western North America. 14 species; 1 species treated in *Flora*.

Layia is evidently most closely related to *Lagophylla*, based on molecular phylogenetic data (Baldwin 2003b).

Additional species of *Layia* have been ascribed to Oregon, each with one or two historical or contemporaneous records. These species include *L. carnosa*, *L. chrysanthemoides*, *L. fremontii*, *L. platyglossa*, and *L. septentrionalis*. Insufficient documentation for the nature of their occurrences and/or their current status in Oregon is available to warrant full treatments for them here.

Layia glandulosa (Hooker) Hooker & Arnott (p. 303)

White layia, whitedaisy tidytips

Layia glandulosa (Hook.) Hook. & Arn. ssp. *glandulosa*

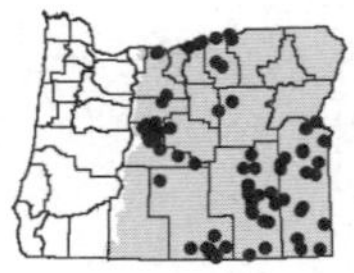

Plants 3–60 cm, glandular, sometimes spice-scented; self-incompatible. **Stems** often uniformly dark purple, hispid, distally glandular. **Leaves** linear to obovate to linear, 4–80 mm, surfaces hirsute to strigose; basal usually somewhat irregularly toothed to lobed; distal stipitate-glandular. **Involucres** 4–11 × 3–11+ mm. **Phyllaries** 3–14, tips often shorter than folded bases, surfaces hirsute, glandular. **Paleae** in 1 series between ray and disc florets. **Ray florets** 3–14; rays 3–22 mm, white. **Disc florets** 17–100+; corollas 3.5–6.5 mm; anthers yellow to brownish. **Ray fruits** glabrous. **Disc fruits** pappi of 10–15 usually linear-attenuate to subulate scales, 2–5 mm, white, rarely tawny, each proximally plumose and often adaxially woolly. **2*n***=16.

Openings in scrub, woodlands, forests, grasslands, and meadows. Flowering Apr–Jul. 0–2200 m. BR, BW, Col, ECas, Lava, Owy. CA, ID, NV, WA; southeast to NM, south to Mexico. Native.

As treated here (provisionally) and previously, *L. glandulosa* corresponds to a paraphyletic group; molecular phylogenetic data indicate that Californian *L. discoidea* is most closely related to a subset of lineages in *L. glandulosa*, including yellow-rayed populations in California previously recognized as ssp. *lutea* or var. *lutea* (Baldwin 2005). Oregon populations of *L. glandulosa* consistently exhibit white rays and historically were treated as ssp. *glandulosa*.

LEONTODON Hawkbit

Herbs annual, biennial, or perennial, sap milky. **Stems** erect or ascending from a decumbent base, leafless, unbranched, mostly glabrous, hispid proximally and near terminal heads. **Leaves** basal, oblanceolate, tapered to a petiole-like base, margins dentate or with pinnate, triangular lobes, rarely entire, tips broadly acute or obtuse, surfaces hispid. **Inflorescences** heads usually solitary. **Heads** ligulate. **Involucres** campanulate in flower, ovoid in fruit. **Phyllaries** in several series, erect, reflexed in fruit; outer much shorter than inner, linear to narrowly lanceolate, tips acuminate; inner equal, lanceolate. **Receptacles** flat; paleae 0. **Florets** bisexual; ligules well exceeding involucres, yellow. **Fruits** tapering toward bases and tips, smooth on outermost fruits, roughened on inner fruits, ribs 10+, beakless or short-beaked; outer pappi of scales; inner pappi of plumose bristles with scale-like bases.

Africa, Asia, Europe, North America. ~45 species; 1 species treated in *Flora*.

Leontodon autumnalis has been reported in Oregon; however, it is not yet established and no vouchered specimens exist. It has branched, rather than scapose, stems and is most easily confused with *Hypochaeris radicata*, differing in having fusiform, non-beaked fruits and outer phyllaries that are linear-acuminate rather than ovate or lanceolate.

Leontodon saxatilis Lamarck (p. 303)

Hairy hawkbit

Leontodon leysseri (Wallr.) Beck, *Leontodon nudicaulis* (L.) Mérat ssp. *taraxacoides* (Vill.) Schinz & Thell, *Leontodon saxatilis* Lam. ssp. *saxatilis*

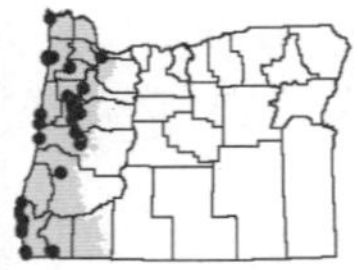

Plants perennial, 10–40 cm; taprooted, often appearing fibrous-rooted from a thickened caudex. **Leaves** 2–15 cm. **Phyllaries** 16–20, 2–10 mm, surfaces glabrous or hispid. **Florets** 15–50, ligules 8–20 mm, outer red externally. **Fruits** 4–5.5 mm, brown, beaks 0 or 0.5–1 mm. **2*n***=8.

Grasslands, dunes, ocean bluffs, marshes, river floodplains, roadsides, disturbed areas. Flowering May–Oct. 0–900 m. CR, Est, Sisk, WV. CA, NV, WA; widely scattered in North America; Europe. Exotic.

If subspecific taxa are recognized, our plants would refer to ssp. *saxatilis*, although ssp. *longirostris* is known in adjacent California and may be expected here. The latter differs in being annual or biennial and having fruits with beaks 2–3 mm.

LEUCANTHEMUM Daisy

Herbs perennial; usually rhizomatous. **Stems** 1–several, erect, simple or branched, glabrous to lightly puberulent or villous. **Leaves** basal and cauline, alternate, lanceolate to oblanceolate, obovate, or round, bases often with fringed, clasping auricles, margins serrate to dentate or pinnately lobed, tips truncate or rounded to acute or acuminate, surfaces glabrous or abaxially lightly puberulent; sessile or petiolate. **Inflorescences** heads solitary or in 2s or 3s. **Heads** usually radiate, rarely discoid; peduncles glabrous or puberulent. **Involucres** broadly hemispheric. **Phyllaries** in 3–5 series, erect in flower, spreading in fruit, herbaceous, surfaces glabrous; inner tips membranous, obtuse or acute; outer lanceolate to elongate-triangular, gradually reduced, tips acute. **Receptacles** convex; paleae 0. **Ray florets** pistillate; rays elliptic to oblanceolate, tips obtuse or truncate, sometimes 2–3-toothed. **Disc florets** bisexual; corollas tubular, widened distally, 5-lobed, lobes triangular, erect or spreading; stamens slightly exserted; stigma lobes slightly exserted. **Fruits** ellipsoid-columnar, glabrous, ribs 8–10; pappi 0.

Europe, North America. 20–40+ species; 1 species treated in *Flora*.

Small, short-lived, escaped populations of *L. maximum* and *L.* × *superbum* (cultivated Shasta daisies) have been reported in Oregon. Neither species appears established, however. *Leucanthemum maximum*, if found again, can be identified by its unlobed basal leaves. *Leucanthemum* × *superbum* is similar to *L. vulgare* but is larger and has unlobed leaves.

Leucanthemum vulgare Lamarck (p. 303)

Ox-eye daisy

Chrysanthemum leucanthemum L.

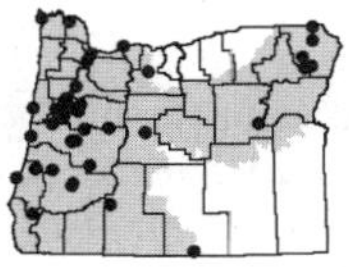

Plants 1–5+ dm. **Leaves** 2–12 cm × 2–35 mm, margins serrate to dentate, usually pinnately lobed, 1–5-veined, lobes short, sometimes with secondary lobes; petioles 1.5–3 cm. **Heads** peduncles 3–20 cm. **Involucres** 12–20+ mm in diameter. **Ray florets** 13–30+; rays 12–20+ mm, white. **Disc florets** 120–150+; corollas yellow. **Fruits** 1.2–2.5 mm, dark brown, ribs white. **2*n***=18, 36, 54, 72, 90.

Forest openings, grasslands, wetlands, riverbanks, clearcuts, roadsides, disturbed areas. Flowering May–Oct. 0–2000 m. BW, Casc, CR, ECas, Est, Lava, Sisk, WV. CA, ID, NV, WA; throughout North America; Europe. Exotic.

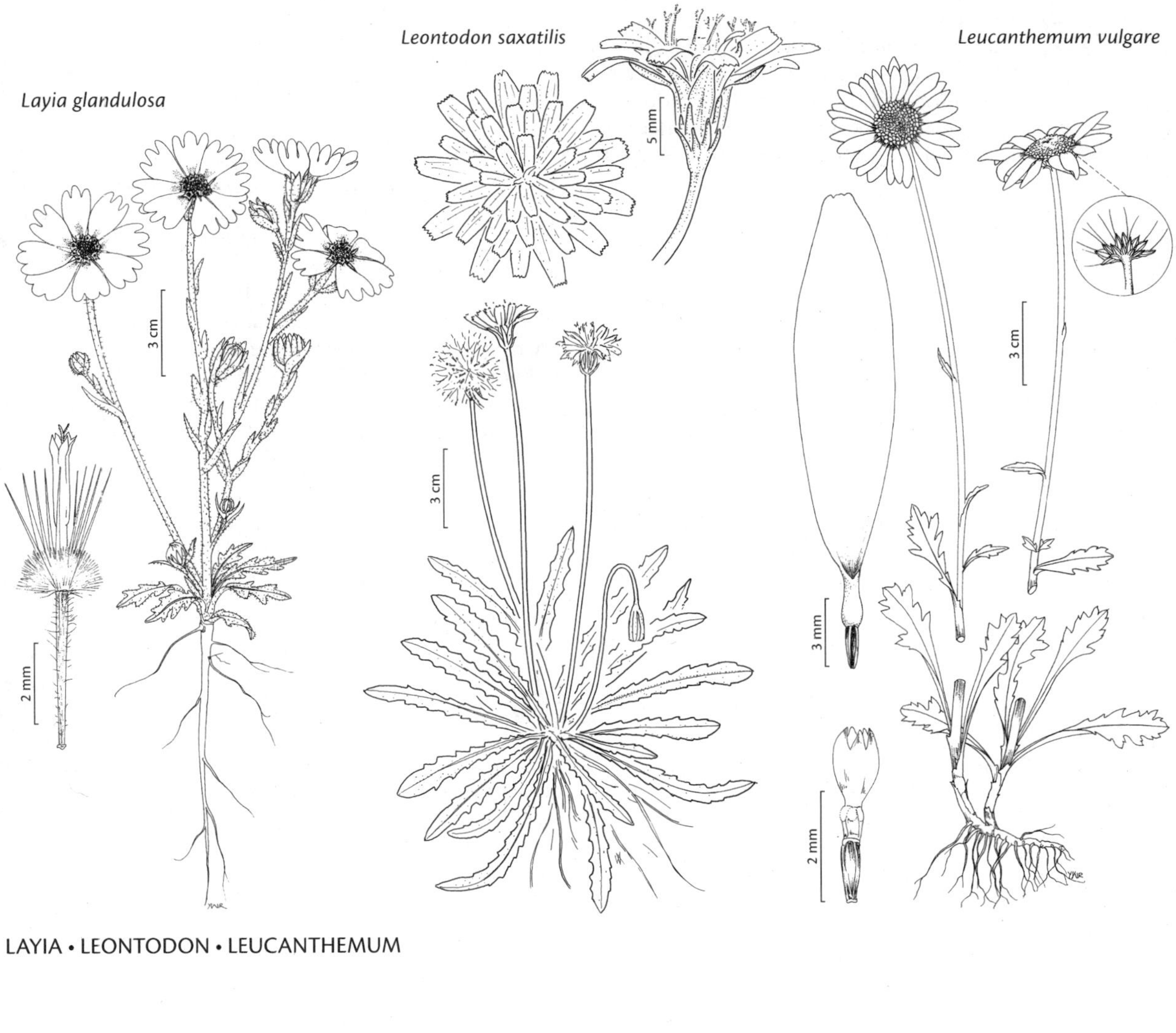

LAYIA • LEONTODON • LEUCANTHEMUM

LOGFIA Cottonrose

Stephen C. Meyers

Herbs annual. **Stems** erect or ascending, branched, strigose- or floccose-tomentose. **Leaves** cauline, alternate, linear to narrowly lanceolate or elliptic, margins entire, tips acute or acuminate, surfaces strigose- or floccose-tomentose, sometimes strigillose; sessile. **Inflorescences** raceme- or panicle-like arrays. **Heads** disciform, pyriform to cylindric. **Involucres** present or not. **Phyllaries** 0–5, sometimes palea-like. **Receptacles** fungiform to obovoid, glabrous. **Outer paleae** falling. **Inner paleae** persistent. **Outer florets** pistillate, 14–45+. **Inner florets** bisexual, 2–10. **Fruits** dimorphic; outer obovoid, compressed, surfaces smooth; inner fusiform, not compressed, usually minutely roughened; outer pappi 0; inner pappi of deciduous bristles.

Worldwide. 12 species; 3 species treated in *Flora.*

Logfia species were formerly placed in *Filago.* The name is an anagram of that genus.

Key to *Logfia* species:

1. Leaves narrowly linear; some subtending bracteate leaves of inflorescences usually well exceeding clusters of heads . . ***L. gallica***

1' Leaves narrowly lanceolate, oblanceolate, or elliptic; subtending bracteate leaves of inflorescences shorter than clusters of heads.

2. Pistillate paleae vertically ranked, helmet-shaped . . . . ***L. minima***

2' Pistillate paleae spirally ranked, not helmet-shaped . . . . ***L. arvensis***

Logfia arvensis (Linnaeus) Holub (p. 304)

Field cottonrose, field filago

Filago arvensis L.

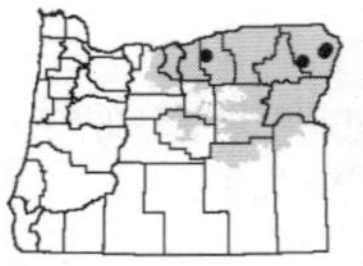

Plants 3–50+ cm. **Leaves** 4–20 × 1–4 mm, narrowly lanceolate, oblanceolate, or elliptic, tips acute. **Inflorescences** subtending bracteate leaves shorter than clusters of heads. **Heads** 2–10(13), pyriform to cylindric, 4–6 × 3–5 mm. **Phyllaries** 0 or 1–4, unequal, palea-like. **Outer paleae** spirally ranked, not helmet-shaped. **Inner paleae** 7–9, in 2 series. **Inner florets** 3–4; corollas 2–3 mm, red to purple. **Fruits** 0.6–1 mm, brown; pappus bristles falling, 2–4 mm, white. **2*n*=28.**

Disturbed areas, creek banks, roadsides. Flowering Jun–Aug. 300–1400 m. BW, Col. ID, WA; north to British Columbia, east to WI; Africa, Asia, Europe. Exotic.

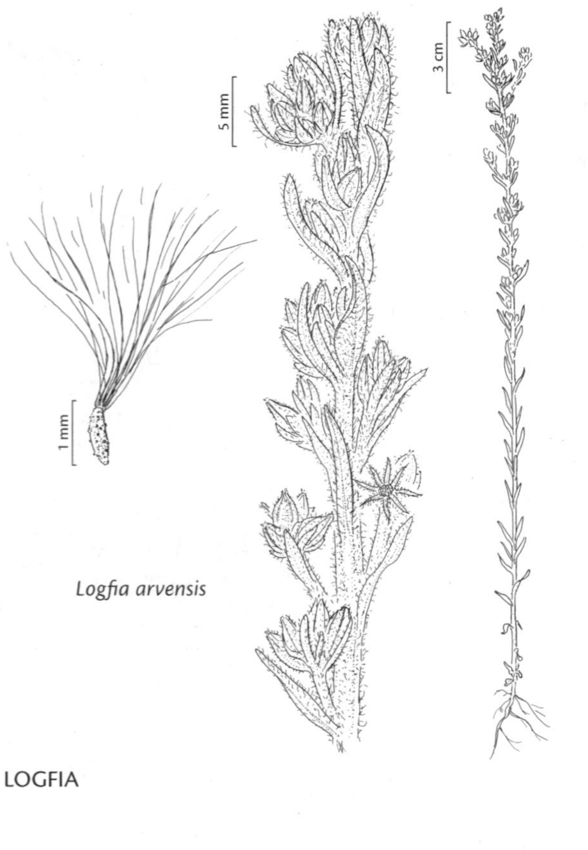

Logfia arvensis

LOGFIA

Logfia gallica (Linnaeus) Cosson & Germain de Saint-Pierre

Narrow-leaved cudweed

Filago gallica L.

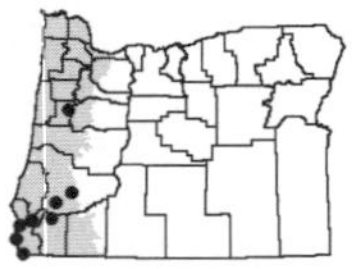

Plants 3–40+ cm. **Leaves** 5–45 × 0.5–2 mm, narrowly linear, tips acuminate. **Inflorescences** some subtending bracteate leaves usually well exceeding clusters of heads. **Heads** 2–10(14), cylindric, (3)3.5–4.5 × 2–3 mm. **Phyllaries** 5, equal, not palea-like. **Outer paleae** vertically ranked, helmet-shaped. **Inner paleae** 4–6, in 1 series. **Inner florets** 2–5; corollas 2–3 mm, brown to yellow. **Fruits** 0.8–1 mm, brown; pappus bristles falling, 2–3 mm, white. **2*n*=28.**

Meadows, creek banks, roadsides, disturbed areas. Flowering May–Jun. 0–700 m. CR, Est, Sisk, WV. CA, ID; widely scattered worldwide. Exotic.

Logfia minima (Smith) Dumortier

Small cudweed, little cottonrose

Filago minima (Sm.) Pers.

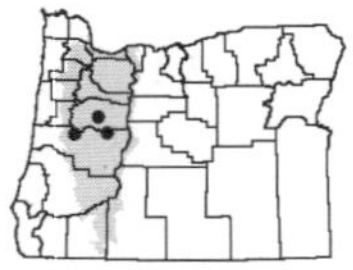

Plants 2–25 cm. **Leaves** 4–12 × 0.5–1.5 mm, narrowly lanceolate or elliptic, tips acute. **Inflorescences** subtending bracteate leaves shorter than clusters of heads. **Heads** 3–7, pyriform, 3–3.5 × ~2 mm. **Phyllaries** 5, equal, palea-like. **Outer paleae** vertically ranked, helmet-shaped. **Inner paleae** 7–9, in 2 series. **Inner florets** 3–5; corollas 1.5–2 mm, brown to yellow. **Fruits** 0.6–1 mm, brown; pappus bristles falling, 2.2–3 mm, white. **2*n*=28.**

Woodlands, riverbanks, roadsides, disturbed areas. Flowering Jun–Aug. 0–400 m. Casc, WV. MA, NY, PA; Asia, Europe. Exotic.

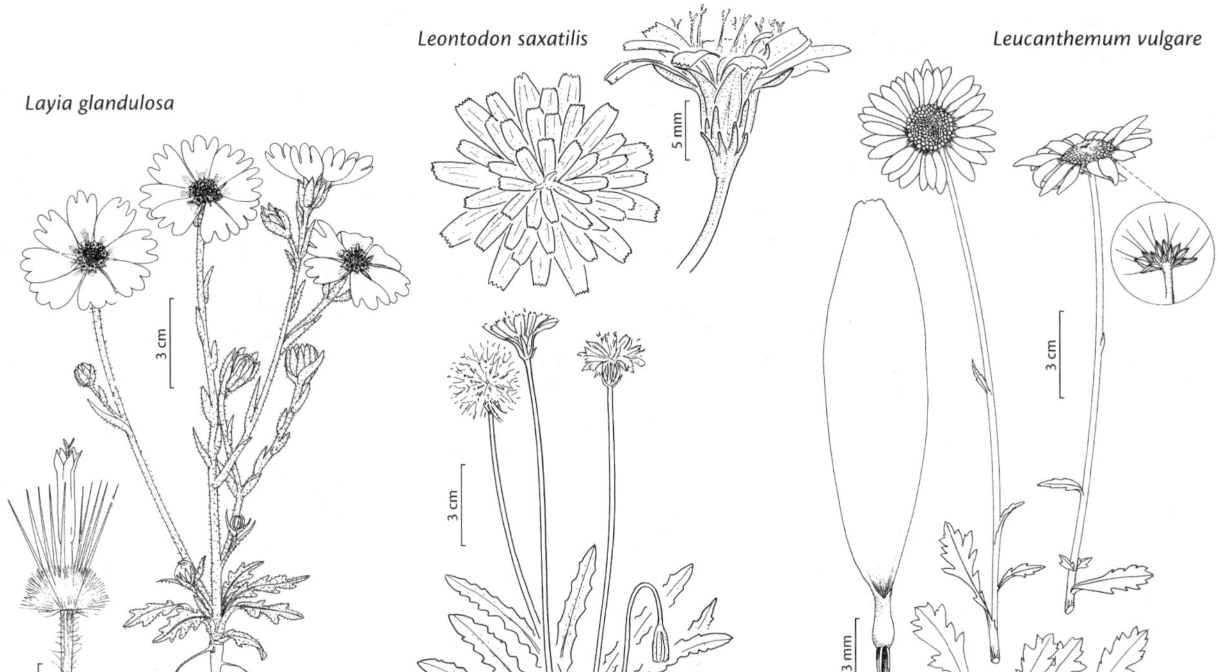

LUINA Luina, silverback

Herbs perennial; from branched caudices, sometimes taprooted or rhizomatous. **Stems** 1–many, simple or branched, tomentose, sometimes glabrate. **Leaves** cauline, alternate, ovate to broadly or narrowly elliptic, tips acute, surfaces abaxially densely white-tomentose, adaxially lightly tomentose to ± glabrous; sessile or petiolate. **Inflorescences** heads in terminal corymb- or umbel-like arrays. **Heads** discoid; peduncles bracted or not. **Involucres** cylindric. **Phyllaries** in 1 equal series, erect, linear-lanceolate, strongly ridged along midveins, surfaces densely tomentose. **Receptacles** flat to convex; paleae 0. **Florets** bisexual; corollas exceeding involucres; stamens yellow, well exserted. **Fruits** cylindric-fusiform, glabrous; pappi of numerous white, minutely barbed bristles.

Western North America. 2 species.

Key to *Luina* species:

1. Leaves ovate to broadly elliptic, adaxially glabrate to glabrous, sessile; involucres 5–9 mm; corollas white . . . . . . ***L. hypoleuca***

1' Leaves lanceolate to narrowly elliptic, adaxially white-tomentose, mostly petiolate; involucres 9–10 mm; corollas yellow. . . . . ***L. serpentina***

Luina hypoleuca Bentham (p. 305)

Silverback luina, littleleaf luina, littleleaf silverback

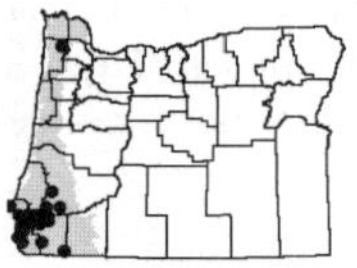

Plants 8–60 cm. **Stems** spreading to erect. **Leaves** ovate to broadly elliptic, 18–65 × 7–40 mm, margins entire or remotely denticulate, surfaces adaxially glabrate to glabrous, often shiny; sessile. **Involucres** 5–9 mm. **Phyllaries** 6–10. **Florets** 11–23; corollas 8–11 mm, white or cream. **Fruits** 3–4 mm, brown, 9–10-veined. ***2n***=60.

Rocky cliffs, ridge tops, talus, streambanks, roadsides, serpentine. Flowering May–Aug. 0–1700 m. CR, Sisk. CA, WA; north to British Columbia. Native.

The frequent occurrence of *L. hypoleuca* in crevices of rock cliffs is facilitated by its deep-seated taproots or taproot-like rhizomes.

Luina serpentina Cronquist (p. 305)

Colonial luina, creeping silverback

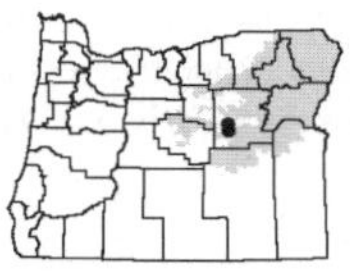

Plant 15–30 cm. **Stems** decumbent to erect. **Leaves** lanceolate to narrowly elliptic, 50–120 × 6–20 mm, margins entire, surfaces adaxially white-tomentose; proximal petiolate, distal sessile. **Involucres** 9–10 mm. **Phyllaries** 10–16. **Florets** 13–26; corollas 10–12 mm, yellow. **Fruits** 5–6 mm, brown, 12–15-veined.

Open rocky outcrops and slopes, serpentine. Flowering Jun–Jul. 1000–2200 m. BW. Native.

Luina serpentina is a narrowly endemic species limited to serpentine substrates and well separated geographically from *L. hypoleuca*.

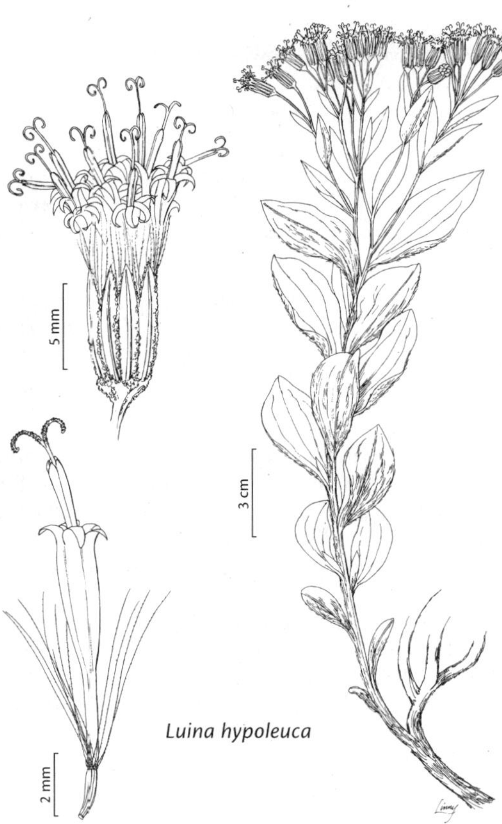

Luina hypoleuca

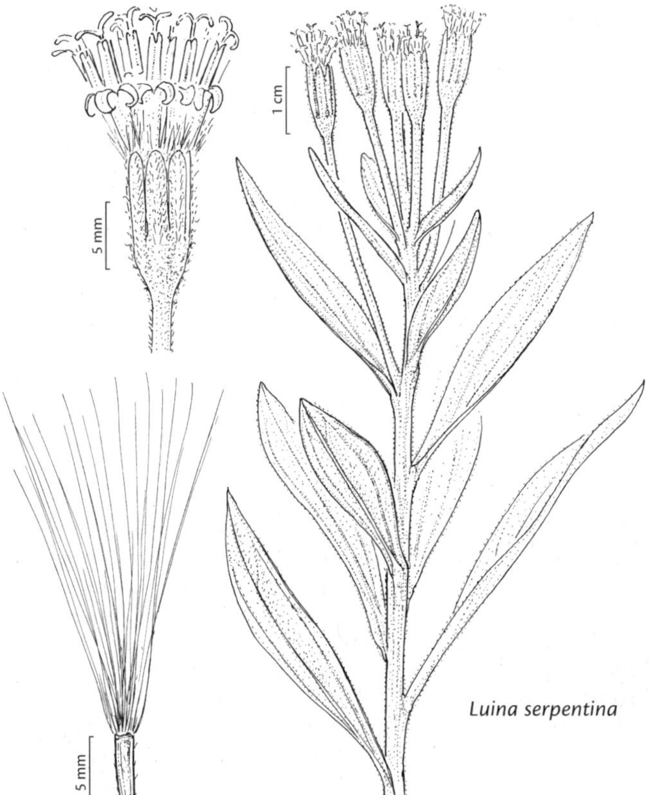

Luina serpentina

MADIA Tarweed

Bruce G. Baldwin and Gerald D. Carr

Herbs annual, 5–250 cm. **Stems** erect. **Leaves** mostly cauline at flowering, opposite proximally, alternate distally; blades linear to lanceolate or oblong-linear, margins usually entire, rarely toothed, surfaces hirsute to strigose, usually glandular-pubescent; sessile. **Inflorescences** corymb-, panicle-, raceme-, or spike-like arrays or heads in glomerules. **Heads** usually radiate (sometimes discoid in *M. glomerata*). **Involucres** ellipsoid, depressed-globose, globose, obconic, ovoid, or urceolate, 1–10+ mm in diameter. **Phyllaries** 1–22 (sometimes 0 in *M. glomerata*), in 1 series, lanceolate to lance-attenuate, herbaceous, each mostly or wholly enveloping a subtended ray ovary proximally, abaxially hirsute, usually glandular. **Receptacles** flat to convex, glabrous or setulose. **Paleae** in 1 series between ray and disc florets, persistent, ± phyllary-like, ± connate or distinct. **Ray florets** 1–22 (sometimes 0 in *M. glomerata*), pistillate, fertile; corollas yellowish (sometimes purplish red in *M. sativa*, sometimes with maroon bases in *M. elegans*). **Disc florets** 1–80+, bisexual and fertile or functionally staminate; corollas usually yellow, sometimes purplish, tubes ≤ funnelform throats, lobes 5, deltate; anthers ± dark purple or yellow to brownish; styles glabrous proximal to branches, branch tips densely hairy to sparsely papillose and hairy. **Ray fruits** compressed or somewhat 3-angled, clavate, often arcuate, basal attachments central or offset, glabrous; sometimes beaked; pappi 0. **Disc fruits** similar to ray fruits, sometimes obovoid, often ± straight, basal attachments central; not beaked; pappi 0.

Hawaii, North America, South America. 10 species; 6 species treated in *Flora.*

Madia is more narrowly circumscribed here than in treatments by D.D. Keck (1959) and others. Molecular phylogenetic data have indicated that *Madia* in those earlier senses is not monophyletic (Baldwin 2003b). As treated here, *Madia* comprises all members of Keck's informal "section Madia" except *M. minima* (= *Hemizonella*). Most species are reportedly either cross-incompatible or intersterile (Clausen 1951).

Key to *Madia* species:

1. Heads showy (plants self-incompatible) or not showy (plants self-compatible); disc florets functionally staminate, ovaries aborting, walls remaining pallid, membranous; odor resinous.
   2. Rays greenish yellow to yellow, 4–11 mm; ray fruits somewhat 3-angled, abaxial sides broadly rounded, angles between adaxial faces about 45–70° . . . . . ***M. citriodora***
   2' Rays bright yellow, sometimes with maroon bases, 4–20 mm; ray fruits compressed, abaxially slightly rounded, angles between adaxial faces about 15–23° . . . . . ***M. elegans***

1' Heads not showy (plants self-compatible); disc florets bisexual, ovaries developing, walls becoming dark, rigid; odor acrid or resinous.
   3. Heads usually in glomerules, sometimes in corymb- or panicle-like arrays; involucres narrowly ovoid or ellipsoid; ray florets 0–3; odor acrid . . . . . ***M. glomerata***
   3' Heads in crowded or open, corymb-, panicle-, raceme-, or spike-like arrays; involucres depressed-globose, ovoid, or urceolate; ray florets 1–16+; odor acrid or resinous.
      4. Plants 1–30(50) cm; involucres 2.5–5 mm; phyllary glands golden yellow, tips 0.5–1.5 mm, erect to inflexed, often sulcate; rays 0.7–1 mm; anthers yellow to brownish; odor resinous . . . . . ***M. exigua***
      4' Plants (0.3)6–100(240) cm; involucres 5–16 mm; phyllary glands black, purple, or yellow, tips 1–4+ mm, ± reflexed, flat; rays 1.5–8 mm; anthers ± dark purple; odor acrid or resinous.
         5. Stems distally glandular-pubescent; heads in open raceme- or panicle-like arrays; involucres 5–10 mm; odor resinous . . . . . ***M. gracilis***
         5' Stems glandular-pubescent throughout; heads in usually crowded panicle-, raceme-, or spike-like arrays; involucres 6–16 mm; odor acrid . . . . . ***M. sativa***

Madia citriodora Greene

Lemon-scented tarweed

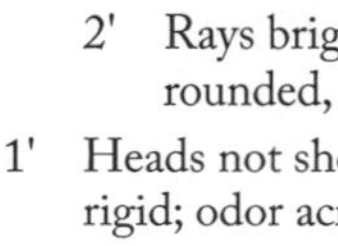

Plants 10–70 cm, odor resinous; self-compatible. **Stems** proximally villous to hirsute, distally glandular-pubescent, glands purple; lateral branches often surpassing main stems. **Leaves** linear, 2–9 cm × 1–10 mm. **Inflorescences** open, ± corymb-like arrays. **Heads** not showy. **Involucres** somewhat ovoid to hemispheric, 6–8 mm. **Phyllaries** ± villous or hirsute, often sparsely glandular-pubescent, tips usually somewhat erect, flat, glands purple. **Paleae** mostly connate ≥ ½ their lengths. **Ray florets** 5–12; corollas greenish yellow to yellow, rays 4–11 mm. **Disc florets** 8–50+, functionally staminate; corollas 2–3 mm, pubescent; anthers ± dark purple; ovaries aborting, walls remaining pallid, membranous. **Ray fruits** somewhat 3-angled, black or brown, sometimes mottled, glossy, abaxial sides broadly rounded, adaxial sides 2-faced, angles between faces about 45–70°; ± not beaked. **Disc fruits** 0. **2*n***=16.

Openings in woodlands, forests, shrublands, disturbed areas, streambanks. Flowering Apr–Jul. 100–1600 m. BW, Col, ECas, Lava, Sisk. CA, NV, WA. Native.

Madia citriodora is sometimes closely associated and often confused with *M. gracilis*.

Madia elegans D. Don (p. 308)

Common madia, showy tarweed

Madia elegans D. Don ssp. *densifolia* (Greene) D.D. Keck, *Madia elegans* D. Don var. *densifolia* (Greene) Jeps., *Madia elegans* D. Don ssp. *elegans*, *Madia elegans* D. Don var. *elegans*, *Madia elegans* D. Don ssp. *vernalis* D.D. Keck

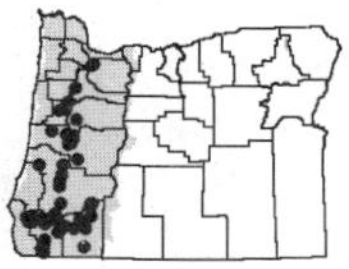

Plants 6–250 cm, odor resinous; self-incompatible. **Stems** proximally villous to hirsute, distally glandular-pubescent, glands yellowish, purple, or black; lateral branches sometimes surpassing main stems. **Leaves** lanceolate to linear, 3–20 cm × 2–20 mm. **Inflorescences** open, ± corymb-like arrays. **Heads** showy. **Involucres** ± globose to campanulate, 4.5–12 mm. **Phyllaries** somewhat hirsute or villous, usually glandular-pubescent, tips erect or reflexed, flat, glands yellowish, purple, or black. **Paleae** mostly connate ≥ ½ their lengths. **Ray florets** (2)5–22; corollas bright yellow, sometimes with maroon bases, rays 4–20 mm. **Disc florets** 25–80+, functionally staminate; corollas 2.5–5 mm, pubescent; anthers ± dark purple; ovaries aborting, walls remaining pallid, membranous. **Ray fruits** compressed, black or brown, sometimes mottled, dull or glossy, abaxial sides slightly rounded, adaxial sides 2-faced, angles between faces about 15–23°; ± not beaked. **Disc fruits** 0. **2*n***=16.

Grasslands, meadows, open sites in shrublands, forests, disturbed areas. Flowering Apr–Sep. 0–2700 m. Casc, CR, Sisk, WV. CA, NV, WA; south to Mexico. Native.

Madia elegans is unusually variable in morphology, ecology, and phenology. Molecular data have indicated that D.D. Keck's (1959) infraspecific taxonomy for *M. elegans* needs revision (B.G. Baldwin, unpubl.). Putative natural (sterile) hybrids with *M. sativa* have been collected from northern California.

Madia exigua (Smith) A. Gray (p. 308)

Little tarweed, threadstem madia

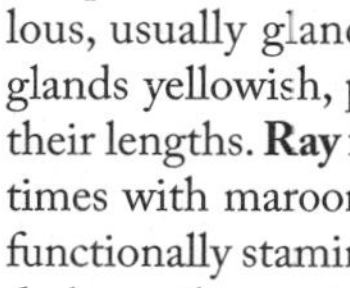

Plants 1–30(50) cm, odor resinous; self-compatible. **Stems** hirsute and glandular-pubescent, glands yellowish or purple; lateral branches seldom surpassing main stems. **Leaves** linear, 0.2–4 cm × 0.5–2 mm. **Inflorescences** open, corymb-, panicle-, or raceme-like arrays; peduncles ± filiform. **Heads** not showy; peduncles elongating. **Involucres** depressed-globose, 2.5–5 mm. **Phyllaries** ± hirsute and glandular-pubescent, glands golden yellow, tips ± erect, sulcate. **Paleae** connate ≥ ½ their lengths. **Ray florets** 1–8; corollas pale yellow, rays 0.7–1 mm. **Disc florets** 1(2), bisexual, fertile; corollas 1–1.8 mm, glabrous; anthers yellow to brownish; ovaries developing, walls becoming dark, rigid. **Ray fruits** compressed, strongly arcuate, black or brown, dull; beaks adaxially offset, curved. **Disc fruits** obovoid, weakly compressed. **2*n***=32.

Openings in grasslands, meadows, shrublands, forests, disturbed areas. Flowering May–Aug. 0–2000 m. All ecoregions. CA, ID, NV, WA; north to British Columbia, northeast to MT, south to Mexico. Native.

Morphologically, *M. exigua* is somewhat similar to *Hemizonella minima*, which, unlike *M. exigua*, has subumbelliform arrays of heads and obcompressed, sparsely hairy fruits. Also, the peduncles of *M. exigua* grow quite long in relation to the flower head as it matures, while those of *H. minima* remain short.

Madia glomerata Hooker

Cluster tarweed, mountain tarweed

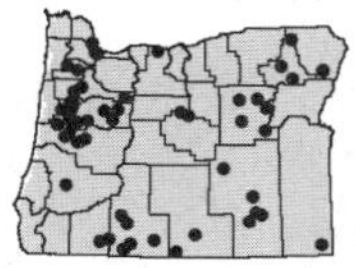

Plants 5–120 cm, odor acrid; self-compatible. **Stems** proximally villous to hispid, distally glandular-pubescent, glands yellowish or black; lateral branches sometimes surpassing main stems. **Leaves** linear to linear-lanceolate, 2–10 cm × 2–7 mm. **Inflorescences** usually crowded glomerules, sometimes corymb- or panicle-like arrays. **Heads** not showy. **Involucres** narrowly ovoid or ellipsoid, 5.5–9 mm. **Phyllaries** ± pilose and glandular-pubescent, glands yellowish or black, tips erect or reflexed, flat. **Paleae** distinct. **Ray florets** 0–3; corollas greenish yellow to purplish, rays 1–3 mm. **Disc florets** 1–5(12), bisexual, fertile; corollas 3–4.5 mm, pubescent; anthers ± dark purple; ovaries developing, walls becoming dark, rigid. **Ray fruits** compressed, black, dull; not beaked. **Disc fruits** similar to ray fruits. **2*n***=28.

Openings in grasslands, meadows, swales, shrublands, forests, edges of marshes, lakes, or watercourses, disturbed areas. Flowering Jun–Sep. 50–2500 m. All ecoregions except Est. CA, ID, NV, WA; north to AK, northeast to Quebec, east to ME, southeast to NM. Native.

Madia glomerata has the most extensive North American distribution of any species in Madiinae.

Madia gracilis (J.E. Smith) D.D. Keck & J.C. Clausen ex Applegate

Slender tarweed, grassy tarweed, common tarweed, gumweed

Madia gracilis (Sm.) D.D. Keck & J.C. Clausen ex Applegate ssp. *gracilis*

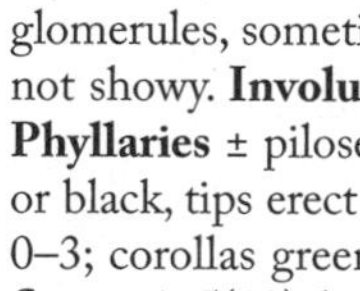

Plants 10–200+ cm, odor resinous; self-compatible. **Stems** proximally pilose to hirsute, distally glandular-pubescent, glands yellowish, purple, or black; lateral branches seldom surpassing main stems. **Leaves** linear to oblong, 1–10(15) cm × 1–8(10) mm. **Inflorescences** open, panicle- or raceme-like arrays, rarely heads congested. **Heads** not showy. **Involucres** depressed-globose to urceolate, 5–10 mm. **Phyllaries** sometimes hirsute, finely or coarsely glandular-pubescent, glands yellowish, purple, or black, tips erect or somewhat reflexed, flat. **Paleae** connate ≥ ½ their lengths. **Ray florets** 3–10; corollas lemon-yellow or greenish yellow, rays 1.5–8 mm. **Disc florets** 2–16+, bisexual, fertile; corollas 2.5–5 mm, pubescent; anthers ± dark purple; ovaries developing, walls becoming dark, rigid. **Ray fruits** compressed, black, purple, or mottled, dull; ± not beaked. **Disc fruits** similar to ray fruits. **2*n***=32, 48.

Open or partially shaded slopes or flats in grasslands, meadows, shrublands, forests, disturbed areas, streambanks. Flowering May–Aug. 0–2100 m. All ecoregions. CA, ID, NV, WA; north to British Columbia, northeast to MT, southeast to UT, south to Mexico. Native.

Madia gracilis often co-occurs with *M. sativa.* The two species are easily confused because of morphological similarities and inherent variation within both, especially when presented with only ill-prepared or incomplete herbarium material. Despite the tendency of *M. gracilis* to flower earlier than *M. sativa*, the two species are partially interfertile (Clausen 1951); hybridization may explain the difficulty of identifying some individuals.

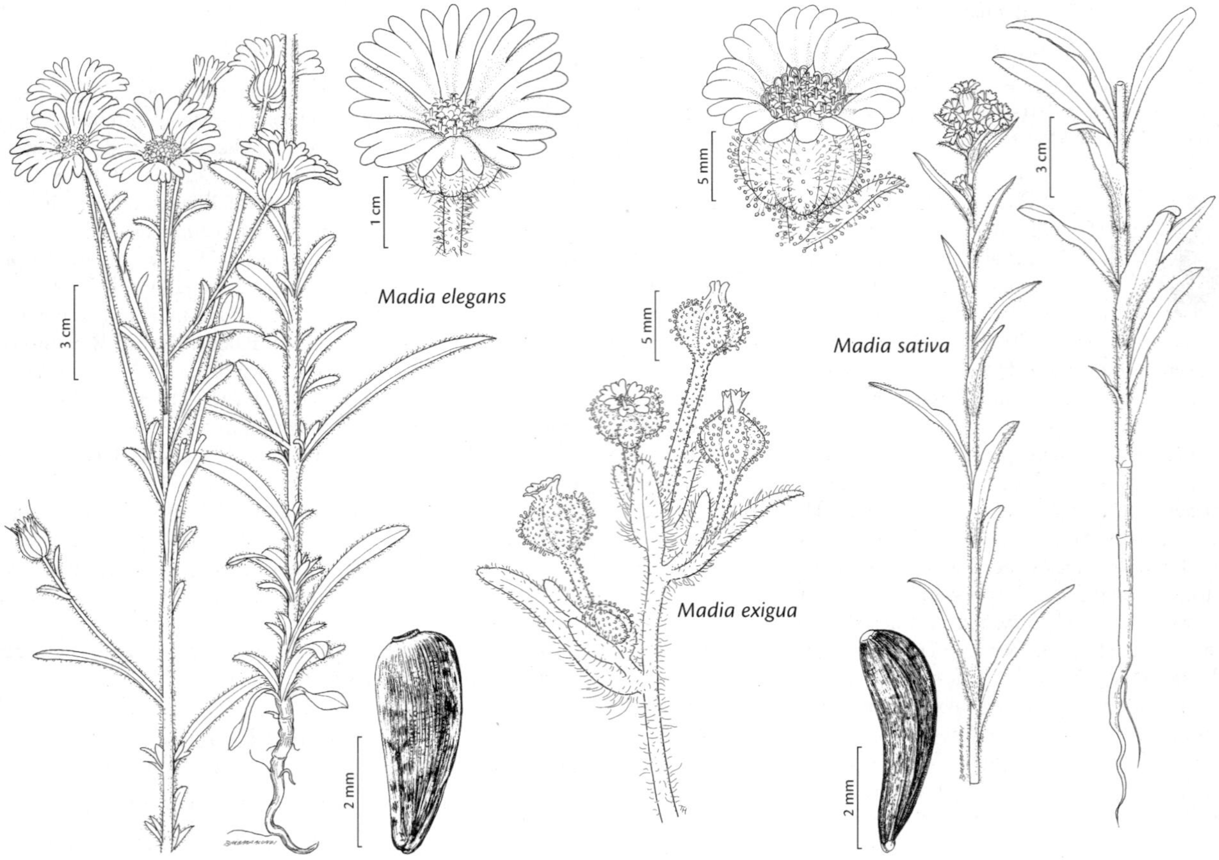

MADIA

Madia sativa Molina (p. 308)

Coast tarweed

Madia capitata Nutt., *Madia sativa* Molina ssp. *capitata* (Nutt.) Piper, *Madia sativa* Molina var. *congesta* Torr. & A. Gray, *Madia sativa* Molina ssp. *sativa*, *Madia sativa* Molina var. *sativa*

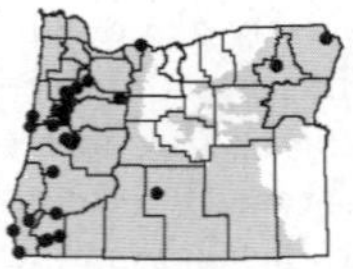

Plants 10–200+ cm, odor acrid; self-compatible. **Stems** hirsute and glandular-pubescent, glands yellowish, purple, or black; lateral branches rarely surpassing main stems. **Leaves** broadly linear to lanceolate to linear-oblong, 2–18 cm × 3–18(29) mm. **Inflorescences** usually crowded, panicle-, raceme-, or spike-like arrays, heads often distally congested. **Heads** not showy. **Involucres** ovoid to urceolate, 6–16 mm. **Phyllaries** hirsute and glandular-pubescent, glands yellowish, purple, or black, tips erect or somewhat reflexed, flat. **Paleae** connate ≥½ their lengths. **Ray florets** (5)8–13; corollas greenish yellow or sometimes purplish red abaxially or throughout, rays 1.5–4 mm. **Disc florets** 11–45+, bisexual, fertile; corollas 2–5 mm, pubescent; anthers ± dark purple; ovaries developing, walls becoming dark, rigid. **Ray fruits** compressed, black or brown, sometimes mottled, dull; not beaked. **Disc fruits** similar to ray fruits. **2*n***=32.

Grasslands, openings in shrublands and woods, disturbed sites, streambanks, roadsides. Flowering Jun–Aug. 0–1500 m. BR, BW, Casc, CR, ECas, Est, Sisk, WV. CA, WA; north to British Columbia, scattered in eastern North America, HI; South America. Native.

Sampled populations of *M. sativa* from California are somewhat divergent in DNA sequences from sampled Chilean populations, in apparent conflict with earlier suggestions that *M. sativa* was recently introduced to North America from South America by Europeans (Baldwin, unpubl.). *Madia sativa* has been cultivated for seed oil in Africa, Asia Minor, Europe, and South America (Zardini 1992).

MALACOTHRIX Desertdandelion, malacothrix

Herbs annual or perennial, sap milky; taprooted, rarely rhizomatous. **Stems** 1–numerous, erect, rarely prostrate, branched or unbranched, glabrous or pubescent. **Leaves** basal and cauline, reduced distally, often pinnately lobed, surfaces glabrous or pubescent; sessile or petiolate. **Inflorescences** heads solitary or in panicle-like arrays. **Heads** ligulate, nodding or erect. **Involucres** campanulate. **Phyllaries** in 2–several unequal series, erect in flower, reflexed in fruit, deltate to linear or lanceolate, margins membranous, tips acute or acuminate. **Receptacles** flat to convex, pitted or smooth, glabrous or bristly; paleae 0. **Florets** bisexual; ligules slightly to well exceeding involucres. **Fruits** columnar, sometimes tapering toward base and tips, glabrous, distinctly 5–15-ribbed; pappi of persistent toothed crowns and/or deciduous bristles connected at bases.

Western North America, South America. 20 species; 4 species treated in *Flora*.
This genus is best developed in California, where 18 of the 20 species occur.

Key to *Malacothrix* species:

1. Ligules exceeding involucres by 1–2 mm . . . . *M. stebbinsii*
1' Ligules exceeding involucres by 6–13 mm.
    2. Basal leaves not fleshy; fruits 5-ribbed . . . . *M. glabrata*
    2' Basal leaves fleshy; fruits 15-ribbed.
        3. Cauline leaf bases sagittate or auriculate . . . . *M. sonchoides*
        3' Cauline leaf bases strap-shaped . . . . *M. torreyi*

Malacothrix glabrata (D.C. Eaton) A. Gray

Smooth desertdandelion

Malacothrix californica DC., *Malacothrix californica* DC. var. *glabrata* A. Gray ex D.C. Eaton

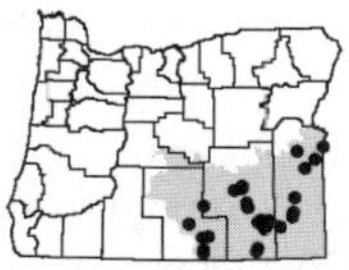

Plants annual, 5–40 cm. **Stems** unbranched or few-branched distally, glabrous or lightly woolly-pubescent proximally. **Leaves** 5–15 cm, pinnatifid with linear lobes, surfaces glabrous; basal not fleshy. **Heads** solitary. **Involucres** 9–17 mm, surfaces glabrous. **Phyllaries** 15–25+; outer deltate to linear-lanceolate, gradually shorter or much reduced. **Florets** 30–100+; ligules 15–20+ mm, white to pale yellow; outermost exceeding involucres by 9–13 mm. **Fruits** 2–3 mm, straw-colored to brown, smooth, ribs 5; pappi of toothed crowns and deciduous bristles. **2*n***=14.

Sagebrush scrub, dunes, open flats and slopes. Flowering May–Jul. 600–1400 m. BR, Owy. CA, ID, NV; southeast to NM, south to Mexico. Native.

Malacothrix sonchoides (Nuttall) Torrey & A. Gray (p. 310)

Sow-thistle desertdandelion, sow-thistle malacothrix

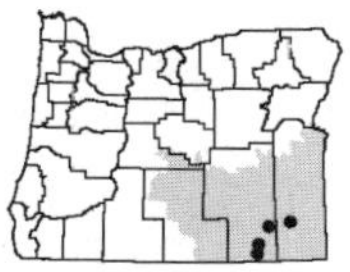

Plants annual, 10–40 cm. **Stems** sometimes branched near bases, glabrous. **Leaves** 2–12 cm, surfaces glabrous, margins dentate or with numerous linear to deltate or ovate lobes, marginal teeth acute, with minute spicules; basal fleshy; cauline bases sagittate or auriculate. **Heads** few to numerous. **Involucres** 7–11 mm, surfaces glabrous. **Phyllaries** 8–24; outer deltate to lanceolate, gradually shorter or much reduced. **Florets** 50–115; ligules 10–15 mm, lemon-yellow; outermost exceeding involucres by 6–10 mm. **Fruits** 1.8–3 mm, brown, smooth, ribs 15; pappi of toothed crowns and deciduous bristles. **2*n***=14.

Dunes and sandy flats, open areas, alkali scrub. Flowering May–Jul. 1000–1600 m. BR, Owy. CA, NV; east to WY, southeast to NM. Native.

Malacothrix sonchoides resembles the more common *M. torreyi* and for a time was confused with that species in Oregon. It appears to be limited to sandy soils, while *M. torreyi* is found on a variety of other substrates as well.

Malacothrix stebbinsii W.S. Davis & P.H. Raven

Stebbins' desertdandelion, Stebbins' malacothrix

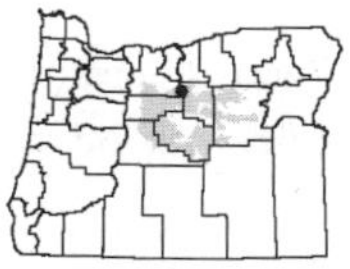

Plants annual, 5–60 cm. **Stems** branched near base and distally, glabrous or short-pubescent proximally. **Leaves** 1.5–8 cm, margins dentate or with short, narrow lobes, surfaces glabrous, marginal teeth not spiculate. **Heads** few, sometimes solitary. **Involucres** 6–10 mm, surfaces glabrous. **Phyllaries** 8–16; outer deltate to lanceolate, much shorter. **Florets** 20–70; ligules 6–7 mm, yellow, rarely white; outermost exceeding involucres by 1–2 mm. **Fruits** 1.7–2.3 mm, brown, smooth, ribs 15; pappi of toothed crowns and 1 persistent bristle. **2*n***=28.

Shrublands and open slopes and flats, streambanks. Flowering Apr–Jun. 700–800 m. Lava. CA, NV; southeast to AZ, south to Mexico. Native?

The only known Oregon location for *Malacothrix stebbinsii*, in Wheeler County, is widely disjunct from the main range of the species in the deserts of California, Nevada, Arizona, and New Mexico.

Malacothrix torreyi A. Gray

Torrey's desertdandelion, Torrey's malacothrix

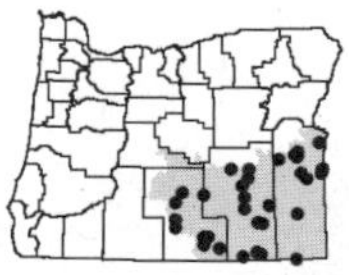

Plants annual, (4)5–30 cm. **Stems** branched near base and distally, glabrous or sparsely glandular-hairy. **Leaves** 2–12 cm, margins dentate or with pinnate, usually narrow, often dentate lobes, sometimes glandular-hairy, surfaces glabrous, marginal teeth often with minute spicules; basal fleshy; cauline bases strap-shaped. **Heads** few, sometimes solitary. **Involucres** 8–14 mm, surfaces glabrous or sparsely glandular-hairy. **Phyllaries** 8–24, linear to deltate or lanceolate, outer gradually reduced. **Florets** 50–100; ligules 14–20 mm, yellow; outermost exceeding involucres by 7–10 mm. **Fruits** 2.5–4 mm, pale brown, smooth, ribs 15; pappi of toothed crowns and deciduous bristles. **2*n***=14.

Sagebrush scrub, juniper woodlands, grasslands, on flats, hillsides, bluffs. Flowering May–Jul. 700–1600 m. BR, Owy. CA, ID, NV, WA; northeast to MT, east to WY, southeast to AZ. Native.

MATRICARIA Mayweed, chamomile

Herbs annual, often aromatic when crushed; taprooted. **Stems** erect or decumbent, simple to much branched, glabrous or sparingly puberulent. **Leaves** mostly cauline, alternate, often aromatic when crushed, bases sometimes with fringed auricles, 1–3-pinnate, segments linear to filiform; sessile or short-petiolate. **Inflorescences** heads solitary or in panicle-like arrays. **Heads** radiate or discoid; peduncles glabrous. **Involucres** campanulate or hemispheric. **Phyllaries** in 2–5 equal series, erect in flower, erect or spreading in fruit, oblong-lanceolate, margins membranous, midribs brown, tips obtuse; outer shorter. **Receptacles** hemispheric to conic, well exceeding involucres; paleae 0. **Ray florets** 0–25, pistillate; corollas ovate-elliptic or oblanceolate, white, tips acute or obtuse and few-toothed. **Disc florets** bisexual; corollas 4–5-lobed, yellow or greenish yellow, lobes triangular, erect or spreading, tubes ± dilated distally or throughout; stamens included; stigmas included or exserted. **Fruits** cylindric or obconic, 3–5-ribbed, glabrous, often with 2 round or elongate, distal resin glands on ribs; pappi 0 or of low crowns.

Worldwide. 7 species; 3 species treated in *Flora.*

Key to *Matricaria* species:

1. Ray florets 10–25; fruits 0.7–0.9 mm, resin glands 0 . . . . *M. chamomilla*

1' Ray florets 0; fruits 1–1.5 mm, resin glands present.

  2. Leaves aromatic when crushed; fruits with 2 linear resin glands . . . . *M. discoidea*

  2' Leaves not aromatic when crushed; fruits with 2 circular or ovate resin glands . . . . *M. occidentalis*

Matricaria chamomilla Linnaeus

Wild chamomile, sweet false chamomile

Matricaria recutita L.

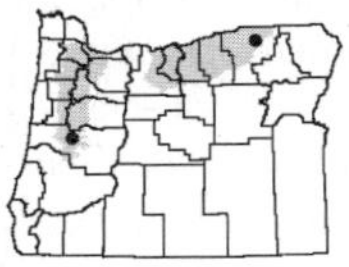

Plants 10–60+ cm. **Leaves** aromatic when crushed, 1–6+ cm, gradually reduced distally, bases often with fringed auricles. **Heads** radiate, 5–10 mm; peduncles 1–12 cm, bractless or few-bracted. **Involucres** 2–4 mm. **Ray florets** 10–25; corollas reflexed and persisting in fruit, 5–10 mm. **Disc florets** 250–500+; corollas 1.5–1.8 mm. **Fruits** 0.7–0.9 mm, brownish, resin glands 0; pappi 0 or of low crowns. **2*n***=18.

Roadsides, ditch-banks, disturbed areas. Flowering May–Jul. 0–1200 m. Col, WV. CA, WA; widely scattered in North America; Asia, Europe. Exotic.

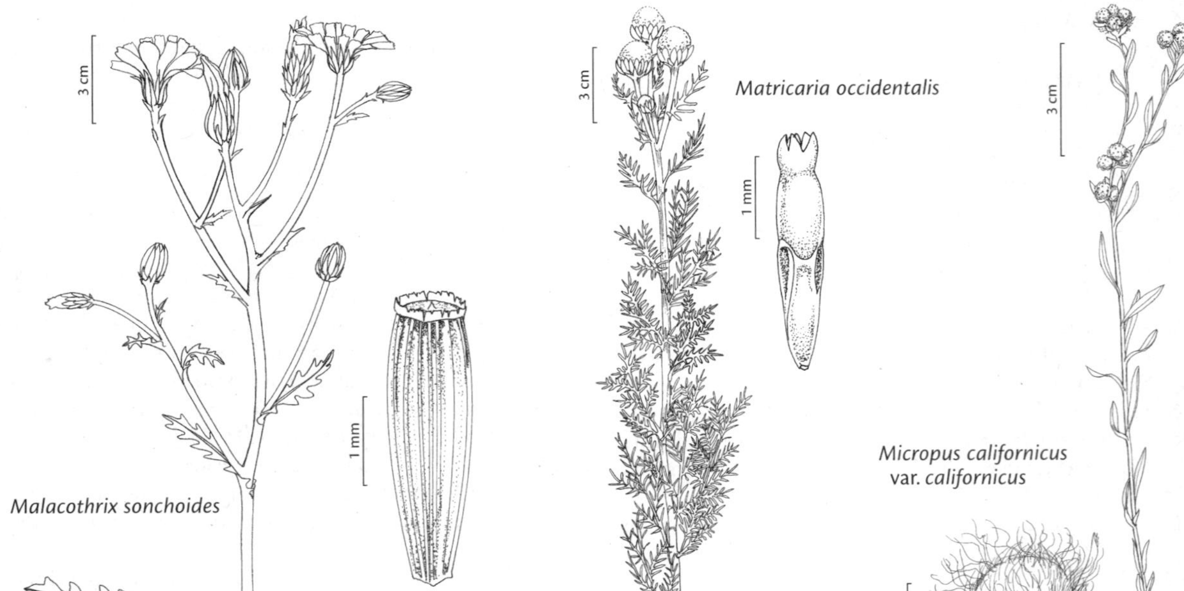

MALACOTHRIX • MATRICARIA • MICROPUS

Matricaria discoidea de Candolle

Pineapple weed

Chamomilla suaveolens (Pursh) Rydb., *Matricaria matricarioides* (Less.) Porter

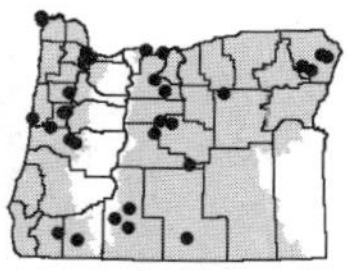

Plants 2–45 cm. **Leaves** aromatic when crushed, 0.5–5 cm, not much reduced distally, bases often with fringed auricles. **Heads** discoid, 3–10 mm; peduncles 1–5 cm, ± bracteate. **Involucres** 2–4.5 mm. **Ray florets** 0. **Disc florets** 100–500+; corollas 1–1.3 mm, greenish yellow. **Fruits** 1–1.5 mm, pale brown, resin glands 2, linear, distal or running entire length of ribs; pappi of low crowns. **2*n***=18.

Roadsides, river floodplains, meadows, disturbed areas, waste places. Flowering Apr–Aug. 0–2000 m. BR, BW, Col, CR, ECas, Est, Lava, Sisk, WV. CA, ID, NV, WA; throughout North America; Asia, Australia, Europe. Native.

Although a native species, *Matricaria discoidea* occurs most often in weedy, disturbed areas. It was previously known as *M. matricarioides*, a name that is now considered to be illegitimate.

Matricaria occidentalis Greene (p. 310)

Valley mayweed

Chamomilla occidentalis (Greene) Rydb.

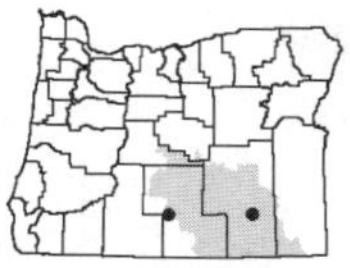

Plants 5–50+ cm. **Leaves** not aromatic when crushed, 1–4+ cm, gradually reduced distally, bases with fringed auricles or not. **Heads** discoid, 5–15 mm; peduncles 1–8 cm, ± bracteate. **Involucres** 3–4.5 mm. **Ray florets** 0. **Disc florets** 400–750+; corollas 1–2 mm, yellow or greenish yellow. **Fruits** 1–1.5 mm, tan or brownish, resin glands circular or ovate, distal on ribs; pappi of 2-lobed crowns. **2*n***=18.

Alkali flats, meadows, vernal pools. Flowering Apr–Jul. 1100–1400 m. BR. CA. Native.

Matricaria occidentalis is widespread and sometimes weedy in California. Oregon occurrences are disjunct and rare, although it may be overlooked due to its similarity in habit to *M. discoidea*.

MICROPUS Cottonseed

Herbs annual. **Stems** simple to much branched, floccose. **Leaves** basal and cauline; basal withered by flowering time; cauline alternate, linear, not much reduced distally, tips acute, surfaces strigillose to floccose-pubescent. **Inflorescences** panicle-like arrays. **Heads** 1–5, disciform. **Involucres** in 1 series. **Phyllaries** 4–6, inconspicuous. **Receptacles** depressed-spheric or obovoid, surfaces glabrous. **Outer paleae** usually winged. **Inner paleae** 0. **Outer florets** 4–12, pistillate, enclosed by paleae; styles exserted. **Inner florets** 2–5, staminate; corolla lobes 4–5; stamens included. **Fruits** falling at maturity, obovoid, curved, veins obscure, glabrous, smooth, shiny; pappi 0 or of 1–5 bristles.

Africa, Asia, Europe, North America. 5 species; 1 species treated in *Flora*.

Micropus californicus Fischer & C.A. Meyer (p. 310) var. ***californicus***

Slender cottonseed, q-tips

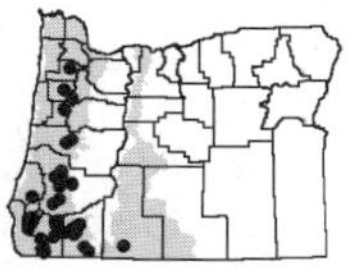

Plants 4–40 cm. **Leaves** 3–20 mm, 1-veined. **Heads** depressed-spheric, 3–4 × 4.5–6 mm; sessile. **Outer paleae** 2–4 mm, woolly. **Inner florets** corollas 1–2 mm, distally reddish. **Fruits** 1.5–2.5 mm, brown; pappi 0 or of 1 bristle, 1–1.5 mm.

Openings in mixed woodlands, shrublands, grasslands, hillsides, pastures, roadsides, serpentine. Flowering Apr–Jun. 0–1400 m. CR, ECas, Sisk, WV. CA; south to Mexico. Native.

A second variety of *Micropus californicus*, var. *subvestitus*, is known principally from the central California coast and central Sierra Nevada foothills. It has silky-tomentose paleae and smaller florets and fruits than var. *californicus*.

MICROSERIS Microseris, silver puffs

Herbs annual or perennial; sap milky; with elongated, fleshy taproots. **Stems** erect, scapose or branched, pubescence of mealy hairs drying as minute white scales. **Leaves** basal or basal and cauline, alternate, linear to lanceolate or oblanceolate, margins entire to toothed or pinnately lobed, surfaces glabrous or mealy-pubescent. **Inflorescences** solitary, terminal. **Heads** ligulate, ± nodding in bud; peduncles long, leafy or not. **Involucres** narrowly to broadly campanulate in flower, slenderly ovoid to globose in fruit. **Phyllaries** 5–40 in 3–5 series, surfaces glabrous or with mealy pubescence. **Receptacles** flat to low-convex, pitted, glabrous; paleae 0. **Florets** bisexual; ligules slightly shorter to much exceeding involucres, yellow or white, outer often reddish externally. **Fruits** columnar, tapering downward, sometimes narrowed distally, light to dark brown or purplish, ribs ~10; outer sometimes villous; pappi of triangular to round or lanceolate, white to brownish, awn-tipped scales, or of brownish bristles.

Australia, New Zealand, western North America, South America. 14 species; 7 species treated in *Flora*.

The annual species of *Microseris* form a tightly knit complex of diploids and tetraploids centered in California. Two alloploid species, derived from hybridization between two of these diploids and *Uropappus lindleyi*, have been separated as the genus *Stebbinsoseris* (Jansen et al. 1991). The perennial taxa form a closely allied group of geographically separate but potentially interfertile species. These were earlier given the genus name *Scorzonella*, distinguished by their longer ligules and mostly self-incompatible breeding system, whereas the annuals consistently have short ligules and are self-pollinating.

Key to *Microseris* species:

1. Plants perennial; ligules well exceeding involucres.
    2. Plants scapose; caudices often horizontal; pappi of 24–48 barbed bristles; habitat sphagnum bogs . . . . ***M. borealis***
    2' Plants not scapose; caudices not horizontal; pappi of 5–30 scales with plumose to minutely barbed awns; habitat not sphagnum bogs.
        3. Pappi of 9–30 scales.
            4. Pappus scales 3.5–10 mm . . . . ***M. laciniata*** (in part)
            4' Pappus scales 0.5–3(5) mm.
                5. Pappus scales 0.5–2 mm, white, awns barbed to subplumose. . . . . ***M. laciniata*** (in part)
                5' Pappus scales 1–3(5) mm, silvery white, awns plumose . . . . ***M. nutans***
        3' Pappi of 5–10 scales.
            6. Pappus scales 3–6 mm. . . . . ***M. howellii***
            6' Pappus scales 0.5–3(4) mm. ***M. laciniata*** (in part)

1' Plants annual; ligules shorter than or slightly exceeding involucres.
    7. Pappus scales linear-lanceolate, 4–11 mm . . . . ***M. acuminata***
    7' Pappus scales lanceolate to ovate, 1–6.5 mm.
        8. Fruit ribs scabrous . . . . ***M. douglasii*** ssp. ***douglasii***
        8' Fruit ribs smooth. . . . . ***M. bigelovii***

Microseris acuminata Greene

Sierra foothills microseris, Sierra foothills silver puffs

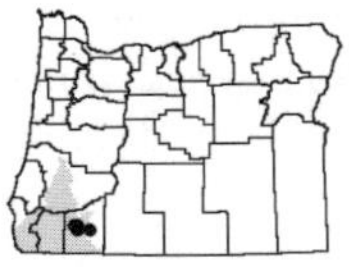

Plants annual, 5–35 cm; taproots slender. **Leaves** basal, 3–20 cm, linear to narrowly lanceolate, bases tapering, margins often pinnate with linear lobes, tips acuminate, surfaces glabrous or mealy-pubescent. **Peduncles** leafless, surfaces glabrous or mealy-pubescent. **Involucres** 10–22 mm, glabrous. **Phyllaries** in 2 unequal series, often purple-lined on midveins; inner equal, lanceolate, tips acute or acuminate; outer deltoid-lanceolate, < inner. **Florets** 5–50; ligules shorter than or slightly exceeding involucres, yellow. **Fruits** constricted below tips, 4.5–7 mm, brown, ribs scabrous. **Pappi** scales 5, linear-lanceolate, 4–11 mm, light brown, awns finely barbed. **2*n*=18.**

Open grassy, stony flats, slopes, near vernal pools. Flowering Apr–Jun. 300–600 m. Sisk. CA. Native.

Microseris acuminata, one of two tetraploid annual species, is common in the Sacramento Valley and adjacent foothills in California. It is disjunct to similar habitats in a small area of Jackson County north of Medford.

Microseris bigelovii (A. Gray) Schultz-Bipontinus

Coastal microseris, coastal silver puffs

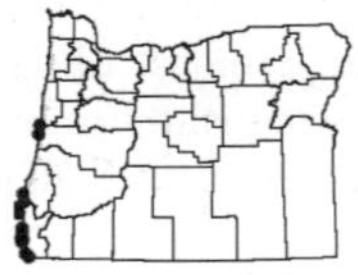

Plants annual, 5–30 cm; taproots slender. **Leaves** basal, 3–25 cm, linear to narrowly lanceolate or oblanceolate, bases tapering, margins often pinnate with broad to linear lobes, tips acuminate to bluntly rounded, surfaces glabrous or mealy-pubescent. **Peduncles** leafless, mealy-pubescent. **Involucres** 5–14 mm, glabrous. **Phyllaries** in 2 unequal series, often purple-lined on midveins; inner equal, lanceolate, tips acute or acuminate; outer deltoid-lanceolate, < inner. **Florets** 5–75; ligules shorter than or slightly exceeding involucres, yellow. **Fruits** 2.5–5.5 mm, brown or purplish, ribs smooth; outer white-hairy. **Pappi** scales 5, lanceolate to ovate, 1–4 mm, brown to blackish, awns finely barbed. **2*n*=18.**

Sandy or rocky sites, dunes, coastal bluffs, bird-nesting islands. Flowering Apr–Jul. 0–100 m. Est. CA, WA; north to British Columbia. Native.

This is the only strictly coastal annual species of the genus. It is disjunct from Lincoln County to the southern tip of Vancouver Island, British Columbia, and adjacent San Juan Islands, Washington, probably due to long-range dispersal by birds. In parts of its range, it resembles *Agoseris maritima* in favoring coastal headlands and islands near seabird roosting sites.

Microseris borealis (Bongard) Schultz-Bipontinus (p. 314)

Northern microseris, apargidium, bog microseris

Apargidium boreale (Bong.) Torr. & A. Gray

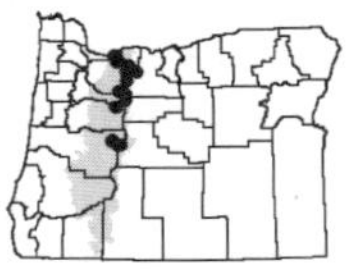

Plants perennial, 15–70 cm; taproots slender, fleshy, from often horizontal caudices. **Leaves** basal, 5–30 cm, linear to narrowly oblanceolate, bases tapering, margins entire or sparsely toothed, tips acute or acuminate, surfaces glabrous. **Peduncles** leafless, glabrous. **Involucres** 10–18 mm, glabrous. **Phyllaries** in several series; inner equal, lanceolate, tips acuminate, some pubescent with black, appressed hairs; outer lanceolate. **Florets** 18–50; ligules well exceeding involucres, yellow. **Fruits** 4–8 mm, brown, ribs smooth. **Pappi** bristles 24–48, brownish, barbed. **2*n*=18.**

Sphagnum bogs. Flowering Jun–Sep. 100–1700 m. Casc. CA, WA; north to AK. Native.

Microseris borealis was once placed in its own genus (*Apargidium*), based on the scapose, perennial habit and non-scaly pappi. It can be hybridized with other *Microseris* species, both annual and perennial, and by DNA analysis it is placed in a clade with the other perennial taxa (Jansen et al. 1991). From British Columbia north, it is found in bog habitats on the coast as well as in the mountains.

Microseris douglasii (de Candolle) Schultz-Bipontinus ssp. *douglasii* (p. 314)

Douglas' microseris

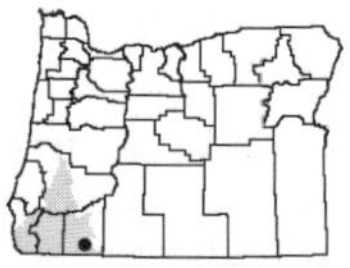

Plants annual, 5–15 cm; taproots slender. **Leaves** basal, 2.5–15 cm, linear to narrowly lanceolate, bases tapering, margins often pinnate with linear lobes, tips acute to acuminate, surfaces glabrous or mealy-pubescent. **Peduncles** leafless. **Involucres** 7–15 mm, glabrous to mealy-pubescent. **Phyllaries** in 2 unequal series, often purple-lined on midveins; inner equal, lanceolate, tips acute or acuminate; outer deltoid, < inner. **Florets** 5–75; ligules shorter than or slightly exceeding involucres, yellow or white. **Fruits** slightly constricted below tips, 4–8 mm, light or dark brown, ribs scabrous; outer sometimes white-hairy. **Pappi** scales ≤ 5, lanceolate to ovate, 1–6.5 mm, tan to blackish, awns finely barbed. **2*n*=18**.

Grasslands, near vernal pools. Flowering Mar–May. 400–1000 m. Sisk. CA. Native.

This subspecies of *M. douglasii* is widespread from the Sacramento Valley southward in California. In the late 19th century, it was collected once in Jackson County by Thomas Howell but has not been seen since. It is an early spring annual that favors clay soil and might be sought on Grizzly Peak or vicinity.

Microseris howellii A. Gray

Howell's microseris

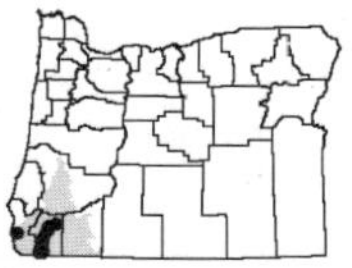

Plants perennial, 10–50 cm; taproots fleshy, from short caudices. **Leaves** basal and cauline, 10–30 cm, linear to narrowly oblanceolate, bases tapering, margins entire to toothed or pinnate with linear, often curved lobes, tips acuminate, surfaces glabrous or lightly mealy-pubescent. **Peduncles** few-bracted near bases, glabrous or mealy-pubescent below heads. **Involucres** 8–17 mm. **Phyllaries** in several series, purple-spotted; inner equal, lanceolate, tips acuminate, surfaces usually pubescent with black, appressed hairs; outer sometimes unequal, deltoid to lanceolate, tips acuminate. **Florets** 8–30; ligules well exceeding involucres, yellow. **Fruits** 4–7 mm, brown, ribs smooth. **Pappi** scales 5–10, 3–6 mm, silvery white, awns smooth or finely barbed. **2*n*=18**.

Open, usually rocky, serpentine grasslands and woodlands. Flowering May–Jun. 300–1100 m. Sisk. Native.

Microseris howellii is endemic to Oregon, occurring principally on moist serpentine outcrops. Its long pappi scales differentiate it from other nearby perennial taxa, from which it is ecologically separated by its adaptation to exposed, rocky habitats. *Microseris laciniata* ssp. *laciniata* sometimes shows similarly elongated pappi scales where it comes into contact with ssp. *detlingii*.

Microseris laciniata (Hooker) Schultz-Bipontinus

Cut-leaved microseris

Plants perennial, 15–120 cm, 1–several fleshy taproots arising from a short caudex. **Leaves** basal and cauline, 10–50 cm, linear to broadly lanceolate or oblanceolate, margins entire or toothed or pinnate with sharp-tipped, curved lobes, bases tapering, tips acute to acuminate, surfaces glabrous or mealy-pubescent. **Peduncles** often bracted proximally, glabrous or mealy-pubescent. **Involucres** 10–30 mm. **Phyllaries** in several series, sometimes purple-spotted; inner equal, lanceolate, tips acute or acuminate, external surfaces with blackish appressed hairs, internal surfaces with short, white, appressed hairs; outer unequal, linear to deltoid, ovate, or lanceolate. **Florets** 13–300; ligules well exceeding involucres, yellow, some outer reddish externally. **Fruits** 3.5–8 mm, brownish, ribs smooth; outer often scabrous. **Pappi** scales 5–24, deltoid to lanceolate, usually silvery to dull white.

Microseris laciniata is the most variable perennial species of the genus.

4 subspecies.

Key to *Microseris laciniata* subspecies:

1. Stems simple or rarely single-branched at the bases; pappus scales 9–15, 3.5–10 mm, awns barbed to subplumose . . . *M. l.* ssp. ***detlingii***

1' Stems branched; pappus scales 5–24, 0.5–3(4) mm, awns smooth to barbed.

  2. Phyllaries often purple-spotted, ovate-lanceolate to broadly ovate, 2.5–9 mm wide, generally glabrous, pappus scales 5–10. . . . . *M. l.* ssp. ***laciniata***

  2' Phyllaries rarely purple-spotted, linear to deltoid, ovate or lanceolate, 0.5–2.5 mm wide, often mealy-pubescent, sometimes black-pubescent, pappus scales 5–24.

    3. Pappus scales 9–24, awns distinctly barbed . . . . *M. l.* ssp. ***siskiyouensis***

    3' Pappus scales 5–10, awns smooth to distinctly barbed. . *M. l.* ssp. ***leptosepala***

Microseris laciniata ssp. *detlingii* K.L. Chambers

Detling's microseris

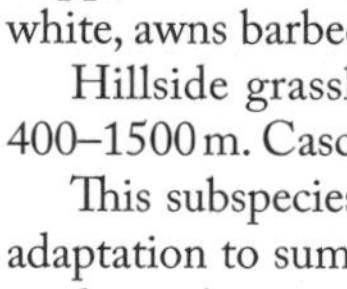

Taproots usually single, much elongated. **Stems** unbranched or rarely single-branched. **Leaves** margins entire or toothed, sometimes undulate, sometimes pinnate. **Peduncles** glabrous or sparingly mealy-pubescent. **Phyllaries** sometimes purple-spotted; inner lanceolate, tips acuminate, surfaces glabrous or black-pubescent; outer ovate to lanceolate, mostly ≥ 4 mm wide, surfaces glabrous. **Pappi** scales 9–15, linear-lanceolate, 3.5–10 mm, silvery to dull white, awns barbed to subplumose. **2*n*=18**.

Hillside grasslands and shrublands. Flowering May–Jul. 400–1500 m. Casc, Sisk. CA. Native.

This subspecies' unusually elongated taproot is probably an adaptation to summer drought in its clay soil habitat. It intergrades with ssp. *laciniata* north of Medford (Chambers 2004).

Microseris laciniata (Hooker) Schultz Bipontinus ssp. ***laciniata*** (p. 314)

Cut-leaved microseris

Scorzonella laciniata (Hook.) Nutt. var. *laciniata*, *Scorzonella laciniata* (Hook.) Nutt. var. *pratensis* (Greene) Jeps., *Scorzonella procera* Greene

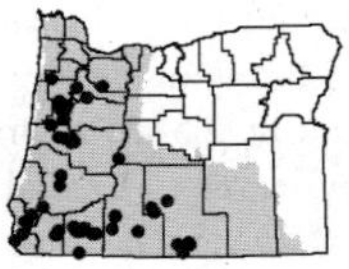

Taproots 1+, short. **Stems** branched. **Leaves** margins entire to toothed or more often pinnate with tapering, sharp-tipped lobes. **Peduncles** usually glabrous. **Phyllaries** often purple-spotted; inner lanceolate, tips acuminate, surfaces glabrous or black-pubescent, rarely tips mealy-pubescent; outer ovate-lanceolate to broadly ovate, 2.5–9 mm wide, surfaces usually glabrous, rarely mealy-pubescent. **Pappi** scales 5–10, deltoid, rarely lanceolate, 0.5–3(4) mm, white, awns smooth. **2*n***=18.

Valley prairies to mid-elevation grasslands and open woodlands. Flowering May–Jul. 50–2000 m. BR, Casc, CR, ECas, Sisk, WV. CA, WA. Native.

Subspecies *laciniata* is characteristic of remnant prairies of the mid-Willamette Valley. It is arbitrarily separated from ssp. *leptosepala* by a minimum 2.5 mm width of the outermost involucral bracts, a character showing complete intergradation from one subspecies to the other. Additionally, ssp. *leptosepala* often has more involucral pubescence than ssp. *laciniata*, especially in plants of the southwestern counties. Subspecies *laciniata* develops its longest pappus scales in Jackson County, where it is in genetic contact with ssp. *detlingii*, but retains its smooth pappus awns.

Microseris laciniata ssp. ***leptosepala*** (Nuttall) K.L. Chambers

Slender-bracted microseris

Scorzonella laciniata (Hook.) Nutt. var. *bolanderi* (A. Gray) Jeps., *Scorzonella leachiana* M. Peck, *Scorzonella leptosepala* Nutt.

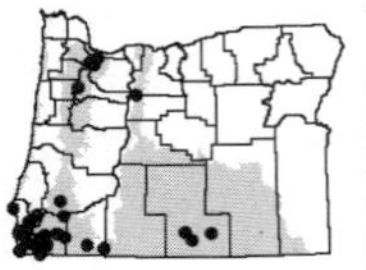

Taproots 1+, short. **Stems** branched. **Leaves** margins entire or pinnate with linear lobes. **Peduncles** glabrous or mealy-pubescent below heads. **Phyllaries** rarely purple-spotted; inner lanceolate, tips acuminate, surfaces usually black-pubescent; outer linear to deltoid or lanceolate, 0.5–2.5 mm wide, surfaces glabrous or mealy-pubescent. **Pappi** scales 5–10, deltoid, 0.5–2 mm, white, awns smooth to barbed. **2*n***=18.

Grasslands and open woodlands, serpentine. Flowering May–Aug. 50–2000 m. BR, ECas, Est, Sisk, WV. CA. Native.

Subspecies *leptosepala* is most common in the Siskiyous, with historical records from Tillamook Prairie and the vicinity of Portland, and with disjunct records from Jefferson and Lake counties. In the upper Illinois River region, it may occur together with ssp. *siskiyouensis*, in populations with plants whose pappi parts range in number from five to more than ten and whose pappi awns have the distinctly barbed character of the latter subspecies (Chambers 2004). Its key distinction from ssp. *laciniata* is in the shape and size of the outermost involucral bracts.

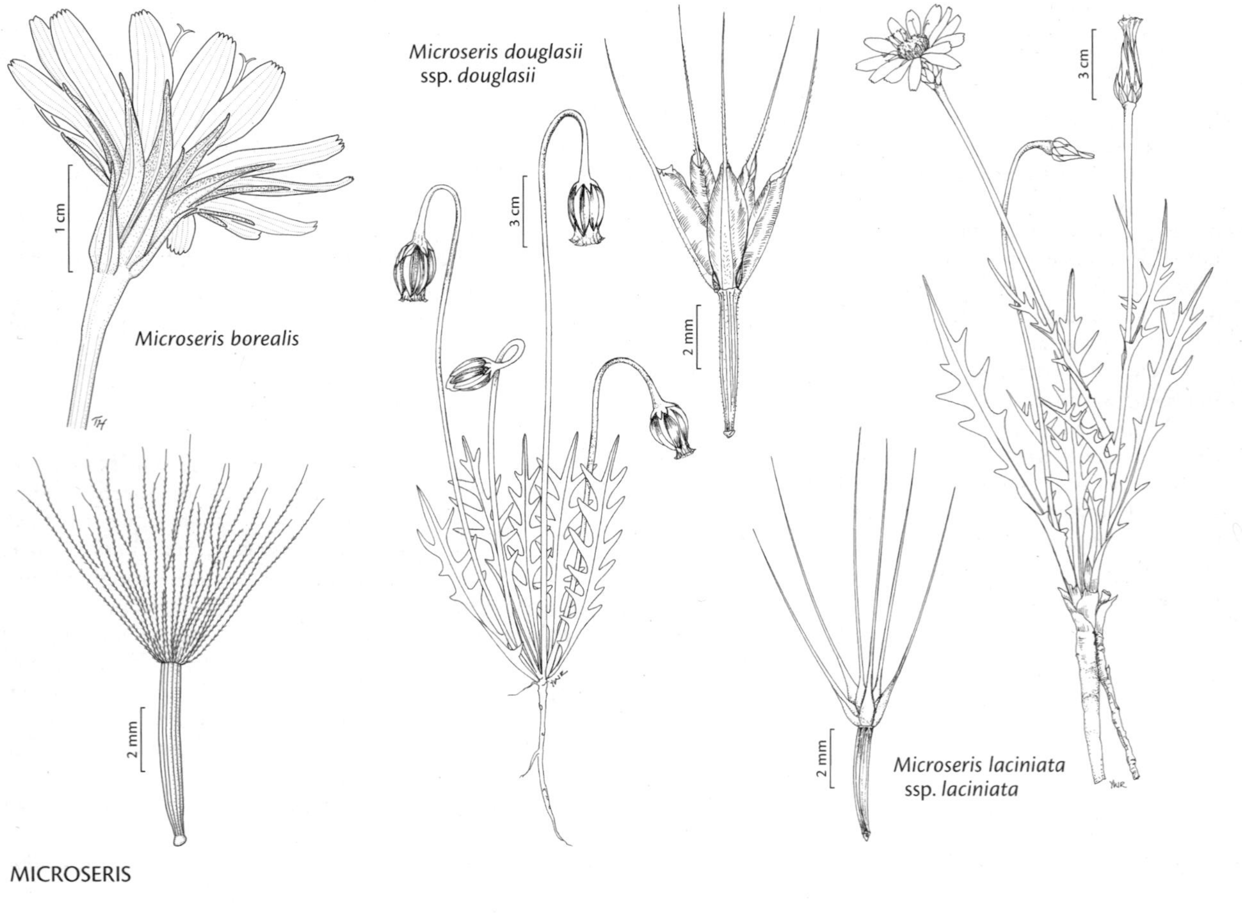

MICROSERIS

Microseris laciniata ssp. *siskiyouensis* K.L. Chambers
Siskiyou microseris

Taproots 1+, short. **Stems** branched. **Leaves** linear or narrowly lanceolate, margins entire or pinnate with linear lobes. **Peduncles** glabrous or mealy-pubescent below heads. **Phyllaries** not purple-spotted; inner lanceolate, tips acuminate, surfaces glabrous or mealy-pubescent, sometimes black-pubescent; outer linear to narrowly deltoid, 0.5–2.5 mm wide, surfaces glabrous or mealy-pubescent. **Pappi** scales 9–24, deltoid, 0.5–2 mm, white, awns barbed. **2*n***=18.

Rocky slopes and openings in woods, serpentine. Flowering May–Jul. 100–800 m. Sisk. CA. Native.

Except for the distinctive ssp. *detlingii*, the usual number of pappus parts in *M. laciniata* is five to ten. However, plants of ssp. *siskiyouensis* have numbers ranging from 9–24 and awns more barbed than is typical of sspp. *laciniata* and *leptosepala*. These traits are perhaps attributable to past genetic contact with *M. nutans*, a species that does not at present co-occur with *M. laciniata*. Subspecies *siskiyouensis* sometimes occurs in mixed populations with ssp. *leptosepala*.

Microseris nutans (Hooker) Schultz-Bipontinus
Nodding microseris, nodding silver puffs

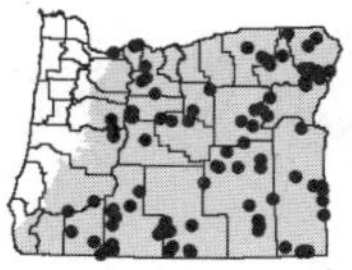

Plants perennial, 10–70 cm; taproots fleshy, from short caudices. **Leaves** basal and cauline, 5–30 cm, linear to oblanceolate or broadly elliptic, bases tapering, rarely obtuse, margins entire to toothed or pinnate with linear, curved lobes, tips acuminate, surfaces glabrous or lightly mealy-pubescent. **Peduncles** naked or few-bracted, surfaces glabrous or mealy-pubescent below head. **Involucres** 8–22 mm. **Phyllaries** in a few series, inner series equal, lanceolate, tips acuminate, usually sparsely black-pubescent; outer linear to deltoid-acuminate, < inner, rarely purple-spotted. **Florets** 10–75; ligules well exceeding involucres, yellow. **Fruits** 3.5–8 mm, brownish, ribs smooth; outer often scabrous. **Pappi** scales 15–30, linear to lanceolate or oblong, 1–3(5) mm, silvery white, awns plumose. **2*n***=18.

Grasslands, oak woodlands, montane meadows, woods, and rocky slopes, pine forests, sagebrush shrublands. Flowering May–Jul. 100–2900 m. BR, BW, Casc, Col, ECas, Lava, Owy, Sisk. CA, ID, NV, WA; north to British Columbia, east to SD and CO. Native.

This is the most widespread perennial species of *Microseris*. It is distinctive for its silvery white pappus scales.

MULGEDIUM Lettuce

Herbs perennial, sap milky; rhizomatous. **Stems** usually 1, erect, branched or unbranched, glabrous. **Leaves** basal and cauline or mostly cauline, linear to lanceolate or lance-elliptic, margins entire, dentate, or with spreading or down-pointed entire lobes, tips acute, surfaces glabrous; sessile or short-petiolate. **Inflorescences** few to numerous heads in terminal panicle- or rarely raceme-like arrays. **Heads** ligulate. **Involucres** campanulate, 2–5 mm in diameter. **Phyllaries** in 1–2 series, erect, surfaces glabrous; inner equal; outer gradually or abruptly shorter. **Receptacles** flat, pitted; paleae 0. **Florets** bisexual; ligules well exceeding involucres. **Fruits** flattened, lanceoloid, tapering toward bases and tips, minutely roughened, ribs 8–10; usually short-beaked; pappi of numerous fine, white bristles.

Asia, Europe, North America. 15 species; 1 species treated in *Flora*.

Mulgedium oblongifolium (Nuttall) Reveal (p. 315)
Blue lettuce

Lactuca pulchella (Pursh) DC., *Lactuca tatarica* (L.) C.A. Mey. ssp. *pulchella* (Pursh) Stebbins, *Lactuca tatarica* (L.) C.A. Mey. var. *pulchella* (Pursh) Breitung

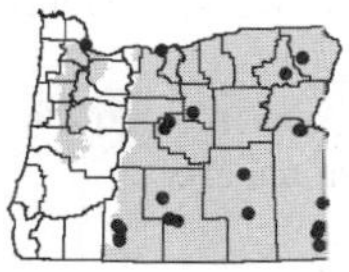

Plants 15–100 cm. **Leaves** 3–18 cm. **Heads** (1)3–100+. **Phyllaries** 8–13+, 12–20 mm, linear-lanceolate, tips acute. **Florets** 10–50+; ligules 15–20 mm, blue. **Fruits** 4–7 mm, dark brown; beaks 0.5–2 mm, whitish. **2*n***=18.

Open forests, shrublands, grasslands, riverbanks, moist roadsides. Flowering Jun–Sep. 0–1900 m. BR, BW, Col, ECas, Lava, Owy, WV. CA, ID, NV, WA; throughout most of Canada and US. Native.

Mulgedium oblongifolium, formerly placed in *Lactuca*, is the only representative of the genus in North America. It is distinguished from the native, biennial species of *Lactuca* by its stout, long-lived rhizomes, and from most *Lactuca* species by its short-beaked fruits and blue flowers.

The name used here replaces the better known *Mulgedium pulchellum*, an illegitimate name.

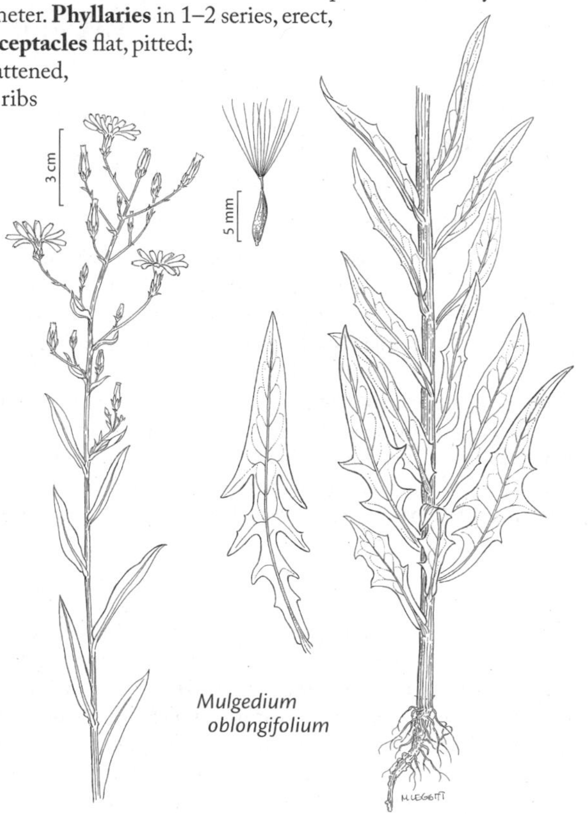

MULGEDIUM

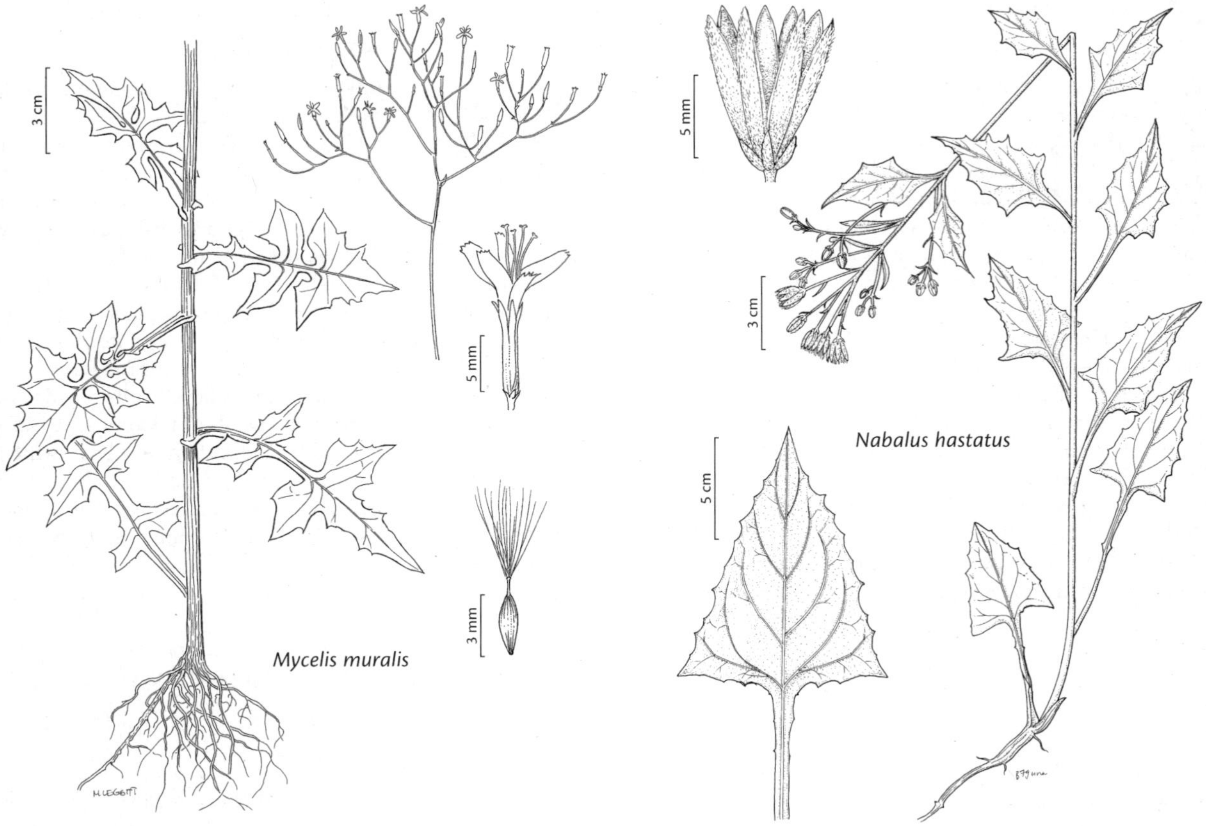

MYCELIS • NABALUS

MYCELIS Wall lettuce

Genus composed of one species, *M. muralis.*

Mycelis muralis (Linnaeus) Dumortier (p. 316)

Wall lettuce

Lactuca muralis (L.) Fresen.

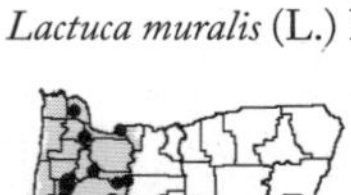

Herbs annual or perennial, 1–10+ dm, sap milky; fibrous-rooted from thickened caudices. **Stems** erect, branched distally, glabrous, hollow. **Leaves** basal and cauline, mostly cauline at flowering, 3–30 cm, oblanceolate to spatulate, with large, 3-lobed, terminal segments and 2–6 paired or alternate smaller lobes below, margins irregularly dentate, surfaces glabrous; proximal petiolate with expanded, clasping bases, distal sessile. **Inflorescences** 10–300+ heads in much branched, panicle-like arrays. **Heads** ligulate. **Involucres** narrowly cylindrical, 1–2 mm in diameter. **Phyllaries** 4–5, in 1(2) series, erect, 7–11 mm, linear, tips acute or obtuse, surfaces glabrous. **Receptacles** flat to convex; paleae 0. **Florets** 5, bisexual; ligules 10–12 mm, exceeding involucres, yellow. **Fruits** tapering gradually toward bases and abruptly toward short-beaked tips, flattened, bodies 2.5–3 mm, black, surfaces smooth to scaberulose, ribs ~20; beaks 0.5–1 mm; pappi of numerous fine, deciduous, white bristles. $2n=18$.

Shaded sites, forests, shrublands, roadsides, disturbed areas. Flowering Jun–Oct. 0–1700 m. Casc, CR, WV. WA; widely scattered in North America; Europe. Exotic.

This species has become increasingly common in recent decades and will undoubtedly spread into additional areas.

NABALUS Rattlesnakeroot

Herbs perennial, sap milky; taprooted, often with slender rhizomes. **Stems** usually simple, erect, glabrous proximally, woolly-pubescent distally. **Leaves** basal and cauline, alternate, surfaces usually glabrous; petiolate or sessile. **Inflorescences** panicle-like arrays. **Heads** ligulate, nodding or erect. **Involucres** cylindrical to campanulate in flower, cylindrical to ovoid in fruit. **Phyllaries** in 2 unequal series, reflexed at maturity; inner linear-lanceolate, tips acute to acuminate; outer linear or deltate, shorter than inner, tips acute. **Receptacles** convex, smooth, glabrous; paleae 0. **Florets** bisexual; ligules exceeding involucres. **Fruits** columnar, often tapering toward bases and tips, glabrous, obscurely 4–5-ribbed; pappi of numerous smooth or minutely roughened bristles.

Asia, North America. ~17 species; 1 species treated in *Flora.*

The 14 North American species formerly treated in *Prenanthes* are now segregated as the genus *Nabalus*, based on molecular phylogenetic research (Liu et al. 2013; Zhang et al. 2011).

Nabalus hastatus (Lessing) A. Heller (p. 316)

Western rattlesnakeroot

Prenanthes alata (Hook.) D. Dietr., *Prenanthes lessingii* Hultén

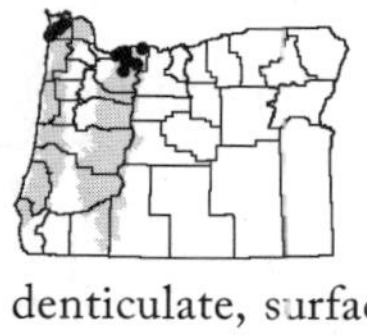

Plants 15–80 cm. **Leaves** 3–30 cm, reduced upwards; distal elliptic-lanceolate, margins entire to dentate or denticulate, surfaces glabrous or puberulent on midveins; proximal triangular or sagittate, margins dentate or denticulate, surfaces glabrous. **Inflorescences** bracteolate. **Involucres** narrowly campanulate, 10–13 mm. **Phyllaries** 6–9, glabrous or minutely woolly-pubescent. **Florets** 7–16; ligules 9–16 mm, white, rarely purplish. **Fruits** 4–7 mm, brown to light tan, smooth; pappi brownish. **2*n*=16.**

Moist, shady cliffs, streambanks, woodlands, edges of forests and shrublands. Flowering Jul–Sep. 200–1000 m. Casc, CR. ID, WA; north to AK, northeast to Alberta. Native.

Nabalus hastatus is one of three western American representatives of the genus, which, in North America, is best developed in the central and eastern United States and adjacent Canada.

NESTOTUS Goldenweed

Subshrubs mat-forming; taprooted. **Stems** prostrate to erect, branched, herbaceous to woody, gray-puberulent. **Leaves** basal and cauline, sometimes clustered basally or terminally on sterile shoots, alternate, linear, bases tapering, margins entire, veins 1, tips acute, surfaces gray-puberulent; sessile. **Inflorescences** solitary, terminal heads. **Heads** radiate; peduncles gray-puberulent, sometimes distally glandular. **Involucres** campanulate. **Phyllaries** in 2–3 series, erect in flower, reflexed in fruit, ± equal, linear-lanceolate, tips acute, surfaces minutely gray-puberulent or scabridulous-glandular; inner membranous-margined. **Receptacles** convex, pitted; paleae 0. **Ray florets** pistillate; rays well exceeding involucres. **Disc florets** bisexual; corollas tubular, widened distally, lobes 5, triangular, spreading; stamens exserted; stigma lobes slightly exserted. **Fruits** turbinate, slightly flattened, obscurely 8–10-veined; pappi of numerous, equal bristles.

Northwestern North America. 2 species; 1 species treated in *Flora.*

Nestotus stenophyllus (A. Gray) Urbatsch, R.P. Roberts, & Neubig (p. 317)

Narrow-leaved goldenweed

Haplopappus stenophyllus A. Gray, *Stenotus stenophyllus* (A. Gray) Greene

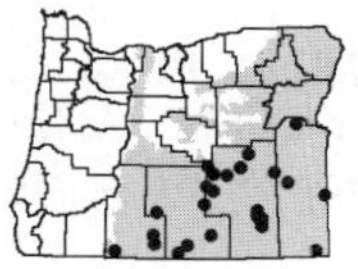

Plants 2–15 cm. **Leaves** 0.5–2.5 cm. **Involucres** 5–9 mm. **Ray florets** 8–12; corollas 7–11 mm, yellow. **Disc florets** corollas 5–7 mm, slightly exserted from involucres, yellow. **Fruits** 4–7 mm, tan, silky-puberulent; pappus bristles 6–7 mm, whitish. **2*n*=18.**

Open slopes, flats, ridge-tops, dry, rocky soil, often with sagebrush. Flowering May–Sep. 900–2800 m. BR, BW, ECas, Owy. CA, ID, NV, WA. Native.

Nestotus stenophyllus was earlier classified in the genera *Haplopappus* or *Stenotus* until phylogenetic studies (Roberts et al. 2005) led to its segregation as a distinct genus.

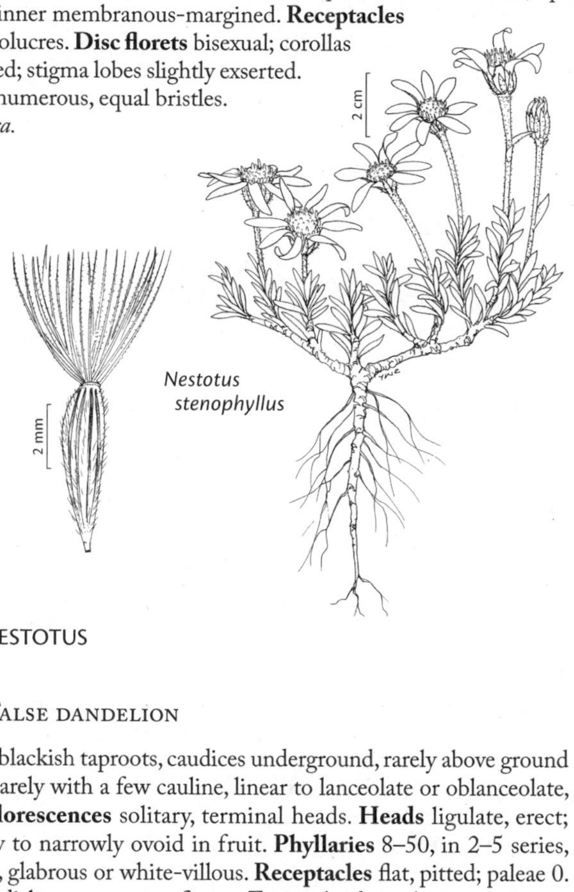

NESTOTUS

NOTHOCALAIS False dandelion

Herbs perennial, 3–45 cm, sap milky; with stout, often branched and blackish taproots, caudices underground, rarely above ground and leafy. **Stems** 1–5+, scape-like, glabrous to villous. **Leaves** basal, rarely with a few cauline, linear to lanceolate or oblanceolate, margins entire to pinnately lobed, surfaces glabrous or ciliate. **Inflorescences** solitary, terminal heads. **Heads** ligulate, erect; peduncles not bracteate. **Involucres** campanulate in flower, broadly to narrowly ovoid in fruit. **Phyllaries** 8–50, in 2–5 series, equal or some outer slightly shorter, erect in flower, reflexing in fruit, glabrous or white-villous. **Receptacles** flat, pitted; paleae 0. **Florets** bisexual; ligules much exceeding involucres, yellow, often reddish on outermost florets. **Fruits** slender, columnar, truncate or distally narrowed, brown or gray, smooth or distally scabrous, ribs 10; pappi of barbellate bristles or narrow, acuminate, white, bristle-tipped scales.

Central and western North America. 4 species; 2 species treated in *Flora.*

The genus was earlier confused with both *Microseris* and *Agoseris*, due to variability in its pappus. It forms a natural unit when excluded from both those genera (Chambers 1955). DNA studies by Lee et al. (2003) placed *Nothocalais* and *Agoseris* in a clade that is sister to one containing *Uropappus, Microseris,* and *Stebbinsoseris*. The two species of *Nothocalais* occurring in Oregon are the most divergent in the genus in regards to both morphology and habitat.

Key to *Nothocalais* species:

1. Pappi of 30–50 minutely barbed bristles; leaves entire, toothed or lobed; phyllaries and peduncles glabrous . . . . . ***N. alpestris***

1' Pappi of 10–30 narrow, tapering, bristle-tipped scales; leaves entire or wavy-margined; phyllaries and peduncles glabrous or white-villous . . . . ***N. troximoides***

Nothocalais alpestris (A. Gray) K.L. Chambers

Alpine lake false dandelion

Agoseris alpestris (A. Gray) Greene, *Agoseris barbellulata* Greene, *Microseris alpestris* (A. Gray) Q. Jones ex Cronquist

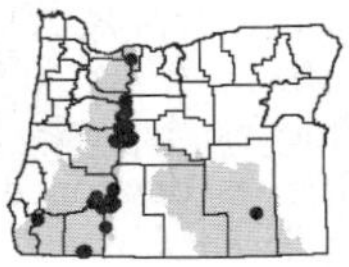

Leaves linear to broadly oblanceolate, 3–20 cm, bases tapering, margins entire, toothed, or pinnate with spreading or backward-pointing lobes, tips acute to rounded, surfaces usually glabrous. **Heads** peduncles 3–45 cm, glabrous. **Involucres** 10–20 mm. **Phyllaries** 8–35, equal, ovate-lanceolate, tips acuminate, surfaces glabrous, minutely purple-spotted. **Florets** 13–75; ligules 10–20 mm. **Fruits** 5–10 mm, somewhat narrowed in distal 1–2 mm; pappi of 30–50 minutely barbed bristles, 6–10 mm. **2*n***=18.

Subalpine meadows, rocky slopes, pumice flats, woodlands. Flowering May–Sep. 1300–2800 m. BR, Casc, Sisk. CA, WA. Native.

This species is easily confused with various species of *Agoseris*, which can lead to false reports of range extensions. Its frequently pinnately lobed leaves are not found in any species of *Agoseris*, however.

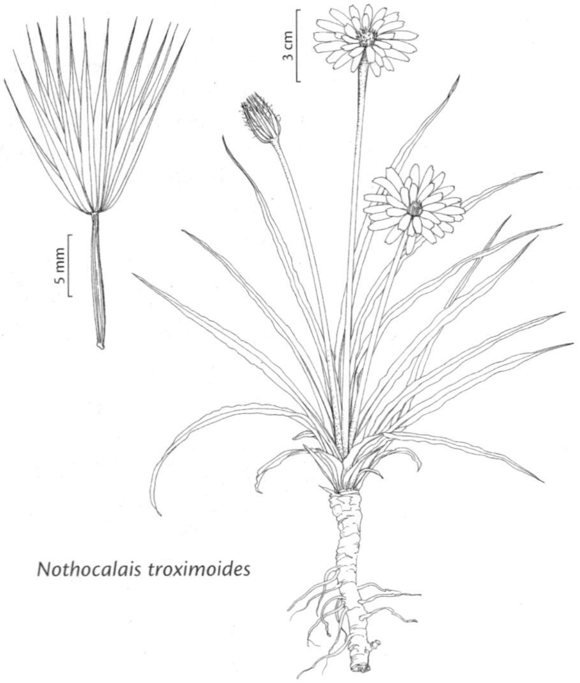

Nothocalais troximoides

NOTHOCALAIS

Nothocalais troximoides (A. Gray) Greene (p. 318)

Sagebrush false dandelion

Microseris troximoides A. Gray, *Scorzonella troximoides* (A. Gray) Jeps.

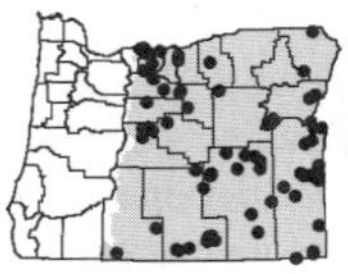

Leaves 7–30 cm, linear to linear-lanceolate, bases straight or tapering, margins entire or wavy, tips acuminate, surfaces glabrous or villous, often ciliate. **Heads** peduncles 5–45 cm, glabrous or white-villous. **Involucres** 14–30 mm. **Phyllaries** 8–25, equal or some outer slightly shorter, lanceolate, tips acuminate, surfaces glabrous or white-villous, usually purple-striped along midveins, sometimes finely purple-spotted. **Florets** 13–90; ligules 10–25 mm. **Fruits** 7–13 mm, sometimes narrowed distally; pappi of 10–30 narrow, tapering, bristle-tipped scales, 10–17 mm. **2*n***=18.

Flats and hillsides in grasslands, sagebrush and juniper/sagebrush steppe, open ponderosa pine woodlands. Flowering Apr–Jun. 0–2400 m. BR, BW, Col, ECas, Lava, Owy. CA, ID, NV, WA, north to British Columbia, northeast to MT, east to WY, southeast to UT. Native.

Populations in Malheur County and southwestern Idaho are unusual in showing finely purple-dotted, rather than striped, phyllaries. This may be due to previous hybridization with *N. nigrescens*.

ONOPORDUM Cotton thistle

Herbs biennial; taprooted. **Stems** 1, erect, stout, branched distally, spiny-winged. **Leaves** basal first year, cauline second year, alternate, blades lanceolate or ovate-lanceolate, reduced distally, with 1 main vein, bases decurrent with spiny wings, margins regularly or irregularly lobed and dentate, lobes triangular, tips acute, spine-tipped, surfaces floccose-tomentose or glandular-puberulent; sessile. **Inflorescences** panicle- or rarely raceme-like arrays. **Heads** discoid; peduncles floccose-tomentose or glandular-puberulent, often spiny-winged. **Involucres** hemispheric to ovoid. **Phyllaries** in 8–10+ series; inner equal; outer gradually shorter, lanceolate to linear, chartaceous, midribs obscure, tips acute or acuminate, spine-tipped. **Receptacles** flat to convex, pitted; paleae 0. **Florets** bisexual; corolla tubes long, slender, widened at tips, lobes 5, linear, erect; stamens included; style branches fused, exserted. **Fruits** obovoid, glabrous, cross-wrinkled; pappi of numerous, unequal bristles, united at bases, deciduous.

Nearly worldwide. 40+ species; 1 species treated in Flora.

Onopordum tauricum has been reported once as occurring in Oregon, in Klamath County. It does not appear naturalized, however. If seen again, it can be recognized by its lanceolate phyllaries with relatively short spines (up to 3 mm).

Onopordum acanthium Linnaeus ssp. ***acanthium*** (p. 320)

Scotch thistle, cotton thistle

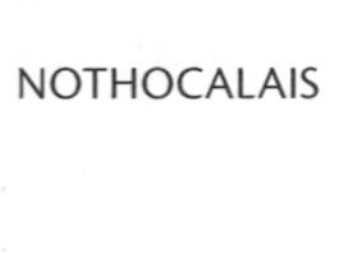

Plants 5–40 dm. **Leaves** 5–60 cm, surfaces floccose-tomentose. **Heads** 2–3; peduncles 2–25 cm. **Involucres** 2–3 cm. **Phyllaries** linear to linear-lanceolate, spines ≤ 9 mm, surfaces floccose-tomentose, inner distally glandular-puberulent. **Florets** corollas 20–25 mm, exceeding involucres, purple or white. **Fruits** 4–5 mm, black; pappus bristles 7–9 mm, reddish. **2*n***=34.

Shrublands, grasslands, roadsides, disturbed areas, cultivated fields. Flowering May–Aug. 0–1500 m. BW, Col, CR, Lava, Owy. CA, ID, NV, WA; scattered throughout southern Canada and US; Asia, Australia, Europe. Exotic.

Scotch thistle can form impenetrable patches in rangeland and is listed as a Class B Noxious weed in Oregon.

OREOSTEMMA Aster

Herbs perennial, cespitose; taprooted. **Stems** decumbent to ascending or erect, rarely branched, glabrous, occasionally strigose proximally and lanate-villous distally. **Basal leaves** persistent, oblanceolate to linear, tapering to narrow petiole-like bases, 3–9-veined from bases, surfaces glabrous to sometimes strigose, villous, or stipitate-glandular. **Cauline leaves** alternate, much reduced and bracteate distally or throughout, linear or narrowly lanceolate, margins entire, tips acute or obtuse; sessile. **Inflorescences** solitary, terminal heads. **Heads** radiate, erect; peduncles lanate-villous. **Involucres** campanulate. **Phyllaries** 25–40, in 3–6 series, erect, spreading in fruit; outer equaling inner or gradually shorter; outer linear to narrowly oblong or lanceolate, 1-veined, tips acute, surfaces strigose or lanate-villous. **Receptacles** flat; paleae 0. **Ray florets** pistillate; rays well exceeding involucres. **Disc florets** bisexual; corollas gradually widened distally, yellow, lobes 5, triangular, erect; stamens included or exserted; stigma lobes well exserted. **Fruits** narrowly cylindric or compressed, truncate, tapering toward bases, 7–10-veined; pappi of 25–40 ± equal bristles, sometimes with a shorter outer series.

Western United States. 3 species; 1 species treated in *Flora.*

Oreostemma alpigenum (Torrey & A. Gray) Greene

Alpine aster, tundra aster, mountain crown

Plants 2–60+ cm. **Leaves** 1–25+ cm, surfaces glabrous to villous. **Involucres** 5–12 mm. **Ray florets** 8–30+, conspicuous; corollas 7–15 mm, reddish violet or purple, occasionally white. **Disc florets** 10–40+; corollas 5–7 mm, equaling or exserted from involucres, yellow. **Fruits** brown, glabrous or ± strigillose; pappus bristles 4.5–6.5 mm, tan.

Western United States. 3 varieties.

In Oregon, the three varieties of *Oreostemma alpigenum* are principally allopatric, although plants resembling var. *andersonii* occasionally can be found in moist meadow habitats within the range of var. *alpigenum* to as far north as Mt. Hood. Additionally, plants with linear leaves like those of var. *haydenii* are sometimes found in mixed populations with broad-leaved forms of var. *alpigenum* at drier sites in the Cascades.

Key to *Oreostemma alpigenum* varieties:

1. Fruits strigillose . . . . ***O. a.*** var. ***andersonii***

1' Fruits glabrous throughout or glabrous proximally and strigillose distally.

 2. Basal leaves usually oblanceolate, sometimes linear, tips obtuse to round . . . . ***O. a.*** var. ***alpigenum***

 2' Basal leaves usually linear, sometimes narrowly oblanceolate, tips acute . . . ***O. a.*** var. ***haydenii***

Oreostemma alpigenum var. ***alpigenum***

Alpine aster, tundra aster

Aster alpigenus (Torr. & A. Gray) A. Gray ssp. *alpigenus*, *Aster alpigenus* (Torr. & A. Gray) A. Gray var. *alpigenus*

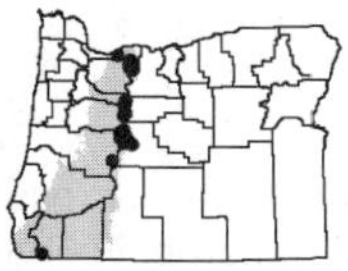

Stems decumbent to erect, 5–25 cm. **Basal leaves** usually oblanceolate, sometimes linear, to 15 mm wide, tips obtuse to round. **Fruits** glabrous proximally, usually strigillose distally.

Open slopes, meadows, stream and lake margins. Flowering Jun–Sep. 900–2800 m. Casc, Sisk. WA. Native.

Oreostemma alpigenum var. ***andersonii*** (A. Gray) G.L. Nesom

Anderson's aster

Aster alpigenus (Torr. & A. Gray) A. Gray ssp. *andersonii* (A. Gray) Onno, *Aster alpigenus* (Torr. & A. Gray) A. Gray var. *andersonii* (A. Gray) M. Peck

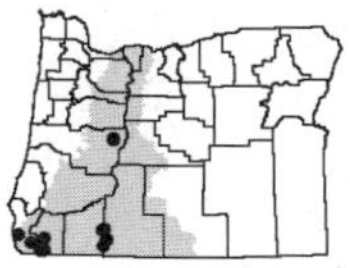

Stems erect, 5–40+ cm. **Basal leaves** usually linear, sometimes narrowly oblanceolate, to 12 mm wide, tips acute. **Fruits** strigillose.

Swamps, bogs, fens, wet meadows. Flowering Jun–Sep. 400–2100 m. Casc, ECas, Sisk. CA, NV. Native.

Oreostemma alpigenum var. ***haydenii*** (Porter) G.L. Nesom (p. 320)

Hayden's aster

Aster alpigenus (Torr. & A. Gray) A. Gray ssp. *haydenii* (Porter) Cronquist, *Aster alpigenus* (Torr. & A. Gray) A. Gray var. *haydenii* (Porter) Cronquist

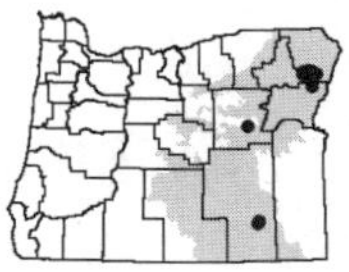

Stems erect to decumbent or prostrate with upturned heads, 2–12 cm. **Basal leaves** usually linear, sometimes narrowly oblanceolate, to 5 mm wide, tips acute. **Fruits** glabrous throughout or glabrous proximally and strigillose distally.

Alpine and subalpine ridgetops, dry slopes, talus, meadows, lake shores. Flowering Jul–Sep. 1600–3000 m. BR, BW. ID, NV; northeast to MT, east to WY. Native.

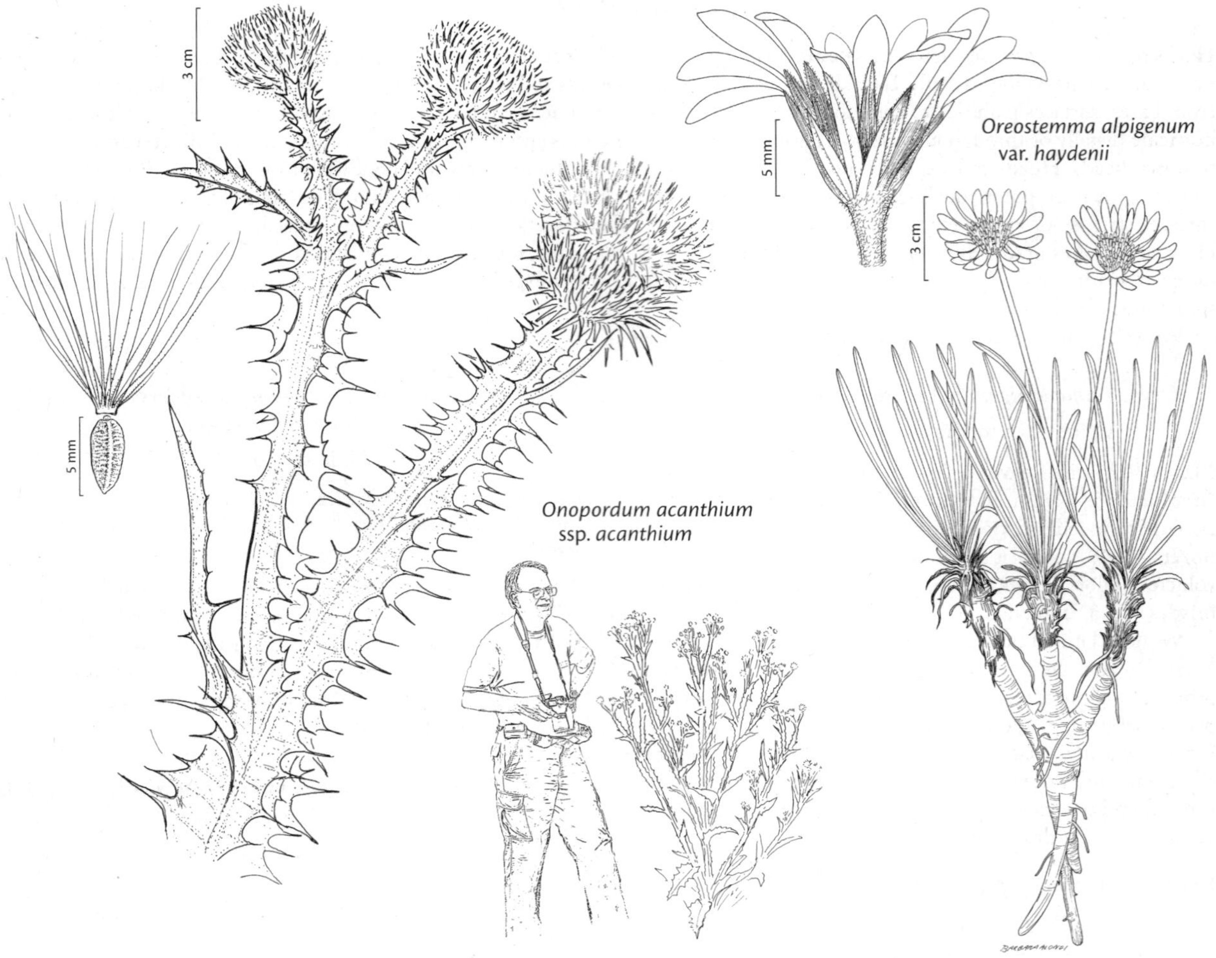

ONOPORDUM • OREOSTEMMA

PACKERA Ragwort

Debra K. Trock

Herbs annual, biennial, or perennial, 3–100+ cm; caudices rhizomatous or taprooted, often with thin, branched, fibrous roots. **Stems** single or clustered, erect, glabrous or pubescent. **Leaves** alternate, basal and cauline; blades elliptic, lanceolate, linear, lyrate, orbiculate, ovate, reniform, or spatulate, often pinnately lobed to pinnatifid or pinnatisect, margins entire or crenate, dentate, or serrate to denticulate, pinnately veined; cauline much reduced; upper leaves bract-like; petiolate or sessile. **Inflorescences** terminal heads, in compact or open corymb-, cyme-, or subumbel-like arrays, rarely heads solitary. **Heads** radiate or discoid. **Calyculi** 0 or 1–5+ bractlets. **Involucres** campanulate to cylindric. **Phyllaries** persistent, 8, 13, or 21 in 1 series. **Receptacles** flat; paleae 0. **Ray florets** 0 or 5–13, pistillate, fertile, pale yellow to orange-red. **Disc florets** 20–80+, bisexual, fertile; corollas pale yellow to orange-red, tubes ≤ campanulate, lobes 5, erect to recurved, usually ± deltate; style branch tips usually truncate. **Fruits** cylindric, tan to dark brown, glabrous or pubescent, ribs usually 5–10; pappi of white, barbellulate bristles.

North America. 66 species; 10 species treated in *Flora*.

Packera is a segregate of *Senecio*. It contains species that were traditionally referred to as "aureoid senecios," an informal group that was first recognized by Asa Gray. Base chromosome numbers of x=22 or 23 (as opposed to x=10) and a suite of morphological characters have traditionally been used in the circumscription of the genus. More recent molecular and palynological studies lend good support for the recognition of the genus.

Species boundaries in the genus are notoriously imprecise, and polyploidy is common. Some species hybridize to some extent with others whenever they come into contact, while others are never known to hybridize. Authors of some studies on the more widespread species propose dividing the species into many varieties based on subtle differences in vegetative morphology, which are often overlapping and inconsistent, and on populational differences due to polyploidy. While recognizing the value of such work in understanding evolutionary relationships, I have taken the view that splitting these species into many varieties based on these difficult characters does little to help with their identification in the field, and these proposed varieties are not supported by molecular data.

Key to *Packera* species:

1. Plants < 15 cm.
    2. Inflorescences of 1–4+ heads.
        3. Basal leaf blades reniform to subreniform, margins crenate, sometimes wavy. . . . . ***P. porteri***
        3' Basal leaf blades lanceolate, ovate, obovate, oblanceolate, elliptic, or spatulate, margins entire to dentate or crenate-dentate.
            4. Peduncles without bracts or bracts inconspicuous; involucres campanulate; phyllaries glabrous; plants of alpine and subalpine meadows in damp areas . . . . ***P. subnuda*** var. ***subnuda*** (in part)
            4' Peduncles with red-tinged, tomentose bracts; involucres cylindric; phyllaries densely tomentose; plants on serpentine in open woodlands . . . . ***P. hesperia***
    2' Inflorescences of 4–15+ heads.
        5. Stems densely tomentose or canescent; leaf margins entire, irregularly undulate or weakly dentate; inflorescences of 8–15+ heads in corymb-like arrays. . . . . ***P. cana*** (in part)
        5' Stems glabrous or sparsely pubescent; leaf margins deeply dissected or pinnatifid; inflorescences of 4–10 heads in subumbellate cyme-like arrays. . . . . ***P. flettii*** (in part)

1' Plants usually > 15 cm.
    6. Leaves distinctly lobed.
        7. Caudices fibrous-rooted and branching; basal and lower cauline leaves with 1–3 pairs of lateral lobes, ultimate margins shallowly to deeply incised . . . . ***P. bolanderi***
        7' Caudices taprooted, suberect to erect; basal and lower cauline leaves with 2–6+ pairs of lateral lobes, ultimate margins irregularly and coarsely serrate-dentate . . . . ***P. eurycephala*** var. ***eurycephala***
    6' Leaves various but not distinctly lobed.
        8. Stems sparsely to densely tomentose to canescent or glabrate.
            9. Plants usually < 30 cm; leaf blades ovate or elliptic to lanceolate, margins entire, irregularly undulate, or weakly dentate; habitat open plains, sagebrush, dry rocky slopes, crevices . . . . ***P. cana*** (in part)
            9' Plants > 30 cm; leaf blades narrowly lanceolate to oblanceolate, margins entire to shallowly toothed, slightly revolute; habitat streams, roadsides, clearings, disturbed areas, coniferous woodlands. . . . . ***P. macounii***
        8' Stems glabrous to sparsely tomentose or pubescent near base or in axils of leaves.
            10. Plants > 50 cm; leaves broadly lanceolate to ovate, sometimes sub-hastate, bases truncate to subcordate or obtuse; upper cauline leaves gradually reduced, becoming sessile and clasping . . . . ***P. pseudaurea*** var. ***pseudaurea***
            10' Plants generally < 50 cm; leaves oblanceolate, elliptic, ovate, obovate, orbiculate, or sub-lyrate, bases tapering or contracted; upper cauline leaves absent or abruptly or gradually reduced, petiolate or sessile, not or weakly clasping.
                11. Leaves thick, turgid, upper cauline leaves gradually reduced . . . . ***P. streptanthifolia***
                11' Leaves not obviously thick and turgid, cauline leaves absent or abruptly reduced.
                    12. Ultimate leaf margins dissected or pinnatifid; heads 4–10+; ray florets 5 or 8 . . . . ***P. flettii*** (in part)
                    12' Leaf margins entire to crenate-dentate; heads 1–3; ray florets 13. . . . . ***P. subnuda*** var. ***subnuda*** (in part)

Packera bolanderi (A. Gray) W.A. Weber & Á. Löve
Bolander's ragwort

Plants perennial, 20–50+ cm; caudices fibrous-rooted, branched. **Stems** 1–3, loosely clustered, glabrous. **Basal leaves** blades orbiculate to lyrate, bases cordate to sub-cordate, ultimate margins shallowly to deeply incised, with distinct lateral lobes in 1–3 pairs, midribs usually winged; petiolate. **Cauline leaves** lower orbiculate to lyrate, with distinct lateral lobes in 1–3 pairs; middle and upper cauline reduced, lanceolate, incised; lower petiolate, upper sessile or clasping. **Heads** 8–15+, in corymb- or umbel-like arrays, radiate; peduncles glabrous; bracts absent or inconspicuous. **Calyculi** conspicuous. **Involucres** campanulate. **Phyllaries** 8, 13 or 21, 5–8 mm, dark green, surfaces glabrous or sparsely pubescent. **Ray florets** 8 or 13; rays 10–15+ mm. **Disc florets** 30–45+; corolla tubes 3–4 mm, limbs 3.5–4.5 mm. **Fruits** 2.5–3.5 mm, glabrous; pappi 5–7 mm. ***2n***=46.

The two varieties of *Packera bolanderi* have been treated as species by some authors. However, the morphological characteristics used to distinguish them vary widely within some populations. Although the morphological distinction is weak, habitat and elevation records lend support for recognizing two varieties.

Key to *Packera bolanderi* varieties:

1. Basal leaves relatively thick and turgid; phyllaries 13 or 21, 6–8+ mm, hairy basally. . . . . ***P. b.*** var. ***bolanderi***

1' Basal leaves relatively thin; phyllaries 8 or 13(21), 5–6 mm, glabrous . . . . ***P. b.*** var. ***harfordii***

Packera bolanderi var. ***bolanderi*** (A. Gray) W.A. Weber & Á. Löve (p. 324)

Bolander's ragwort, Bolander's groundsel

Senecio bolanderi A. Gray var. *bolanderi*

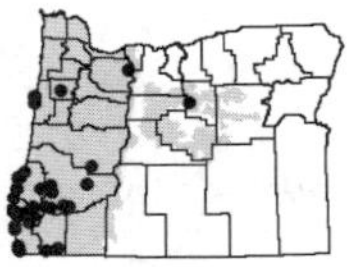

Basal leaves relatively thick and turgid. **Phyllaries** 13 or 21, 6–8+ mm, basally hairy, hairs tan or brownish.

Open areas in coastal forests, wet rocky cliffs. Flowering May–Aug. 0–1700 m. Casc, CR, Est, Lava, Sisk, WV. CA. Native.

Packera bolanderi var. ***harfordii*** (Greenman) Trock & T.M. Barkley

Harford's ragwort

Senecio bolanderi A. Gray var. *harfordii* (Greenm.) T.M. Barkley, *Senecio harfordii* Greenm.

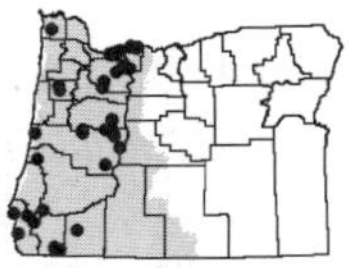

Basal leaves relatively thin. **Phyllaries** 8 or 13(21), 5–6 mm, glabrous.

Rocky streambanks, shaded woodlands. Flowering Jun–Aug. 0–1500 m. Casc, CR, ECas, Sisk, WV. WA. Native.

Packera cana (Hooker) W.A. Weber & Á. Löve (p. 324)

Woolly groundsel

Senecio canus Hook.

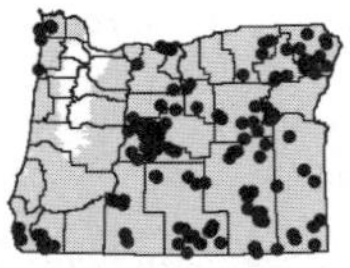

Plants perennial, 10–30+ cm; caudices suberect, stout, branched. **Stems** 1, densely tomentose or canescent. **Basal leaves** blades ovate or elliptic to lanceolate, bases tapering, margins entire, irregularly undulate, or weakly dentate, surfaces abaxially persistently lanate-pubescent, adaxially lanate to glabrate; petiolate. **Cauline leaves** similar to basal; upper gradually reduced; petiolate or sessile and weakly clasping. **Heads** 8–15+, in corymb-like arrays, radiate; peduncles usually densely lanate, sometimes glabrate; bracteate. **Calyculi** inconspicuous, surfaces tomentose. **Involucres** campanulate. **Phyllaries** 13 or 21, 5–8+ mm, green, surfaces densely tomentose. **Ray florets** 8–10(13); rays 8–10+ mm. **Disc florets** 35–50+; corolla tubes 2.5–3.5 mm, limbs 3.5–4.5 mm. **Fruits** 2.5–3.5+ mm, glabrous; pappi 4.5–7 mm. **2*n***=46, 92.

Open plains, sagebrush, dry rocky slopes, crevices in granitic and limestone outcrops. Flowering May–Aug. 0–3000 m. All ecoregions except WV. CA, ID, NV, WA; north to British Columbia, northeast to Manitoba, east to MN, southeast to NM. Native.

Packera cana grows in a wide variety of habitats. Vegetative morphology is relatively uniform throughout the range, except that alpine specimens are notably dwarfed and often mistaken for *P. werneriifolia*, which differs by its scapiform habit. In Oregon, *P. cana* can also be mistaken for *P. macounii*. However, the latter is not consistently canescent and has a taprooted caudex and leaves that are narrower and frequently revolute. Several specimens in Curry and Josephine counties that have been reported as *Packera cana* are more likely *P. macounii* as they are cited as growing on serpentine.

Packera eurycephala var. ***eurycephala*** (Torrey & A. Gray) W.A. Weber & Á. Löve

Widehead groundsel, cutfleaf butterweed

Senecio eurycephalus Torr. & A. Gray ex A. Gray var. *eurycephalus*

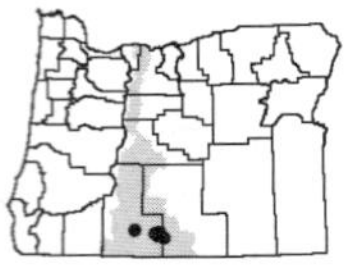

Plants perennial, 20–50+ cm; caudices taprooted, suberect to erect, woody. **Stems** (1)2–6, clustered, densely tomentose to glabrate. **Basal leaves** blades lyrate, ovate, or obovate, bases tapering, with distinct lateral lobes in 2–6+ pairs, ultimate margins irregularly and coarsely serrate-dentate, soon withering; petiolate. **Cauline leaves** lower similar to basal but with terminal lobes smaller than laterals, gradually reduced; upper margins incised; sessile. **Heads** 5–20+, in open, cyme-like arrays, rarely compact, radiate; peduncles sparsely to densely tomentose; bracteate. **Calyculi** inconspicuous. **Involucres** campanulate. **Phyllaries** 13 or 21, 8–10+ mm, green, tips sometimes yellow, surfaces densely tomentose basally, glabrate distally. **Ray florets** 8 or 13; rays 10–15 mm. **Disc florets** 35–50+; corolla tubes 3.5–4.5 mm, limbs 2.5–3.5 mm. **Fruits** 4.5–5 mm; pappi 6–7 mm. **2*n***=46.

Dry disturbed areas, roadsides, wooded areas, dry or drying streambeds. Flowering May–Jul. 1300–1600 m. ECas. CA, NV. Native.

Packera eurycephala var. *eurycephala* is relatively common in northern and central California, but only a few occurrences have thus far been recorded in southern Oregon.

Packera flettii (Wiegand) W.A. Weber & Á. Löve

Flett's groundsel

Senecio flettii Wiegand

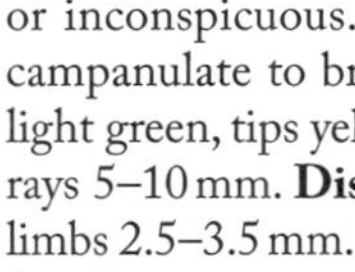

Plants perennial, 10–40+ cm; rhizomes horizontal to erect. **Stems** 1 or 2–5, loosely clustered, glabrous or leaf axils sparsely pubescent. **Basal leaves** blades obovate to ovate or sub-lyrate, bases tapering, margins deeply dissected or pinnatifid, crenate to crenate-dentate; petiolate. **Cauline leaves** lower similar to basal; middle and upper absent or abruptly reduced distally; sessile. **Heads** 4–10+, in subumbellate, cyme-like arrays, radiate; peduncles glabrous; bracts absent or inconspicuous. **Calyculi** 0 or inconspicuous. **Involucres** campanulate to broadly cylindric. **Phyllaries** (8)13, 4–7 mm, light green, tips yellowish, surfaces glabrous. **Ray florets** 5 or 8; rays 5–10 mm. **Disc florets** 20–30+; corolla tubes 2.5–3.5 mm, limbs 2.5–3.5 mm. **Fruits** 1–1.5 mm, glabrous; pappi 4–4.5 mm. **2*n***=40.

Exposed slopes, rocky or gravelly soils. Flowering Jun–Jul. 300–1100 m. CR. WA. Native.

In Oregon, *P. flettii* is known only from the Coast Range and is rarely collected there. It is also rare in Washington but has a much wider distribution. This species has a unique chromosome number for the genus (2*n*=40) and is not known to hybridize with other species.

Packera hesperia (Greene) W.A. Weber & Á. Löve

Western ragwort, Siskiyou butterweed, serpentine groundsel

Senecio hesperius Greene

Plants perennial, 7–15 cm; caudices fibrous-rooted, relatively slender. **Stems** 1, glabrous or sparsely tomentose. **Basal leaves** blades ovate, oblanceolate, or spatulate, bases tapering, sometimes obtuse, margins entire, subentire, or dentate; petiolate. **Cauline leaves** lower similar to basal; middle and upper abruptly reduced and bract-like; sessile and clasping. **Heads** 1–4+, in corymb-like arrays; peduncles densely lanate-tomentose, bracts red-tinged. **Calyculi** conspicuous. **Involucres** cylindric, bractlets narrow, red-tinged. **Phyllaries** (13)21, green, tips red or dark red, surfaces densely tomentose basally. **Ray florets** (8)13; rays 6–10+ mm, deep yellow. **Disc florets** 35–50+; corolla tubes 2.5–3.5 mm, limbs 3–4 mm. **Fruits** 1.5–2 mm, glabrous; pappi 5–6 mm. **2*n*=**46.

Serpentine, open woodlands. Flowering Apr–Jun. 300–1400 m. Sisk. CA. Native.

Until collections of this species were made in Del Norte County in California, *Packera hesperia* was known only from two counties in southern Oregon. It has since been listed as a species of conservation concern in California.

Packera macounii (Greene) W.A. Weber & Á. Löve

Siskiyou Mountains ragwort, Puget butterweed, long-rayed groundsel

Senecio fastigiatus Schwein. ex Elliott, *Senecio macounii* Greene

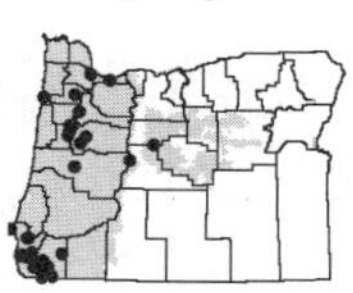

Plants perennial, 30–40+ cm; taprooted, caudices ± branching, becoming rhizomatous. **Stems** 1, sparsely to densely tomentose or glabrate. **Basal leaves** blades narrowly lanceolate to oblanceolate, bases tapering, margins entire or shallowly toothed, slightly revolute; petiolate. **Cauline leaves** gradually reduced; upper bract-like; lower petiolate, upper sessile. **Heads** 6–15+, in corymb-like arrays; peduncles sparsely tomentose to glabrate, bracts absent or inconspicuous. **Calyculi** 0 or inconspicuous. **Involucres** cylindric to narrowly campanulate. **Phyllaries** 13 or 21, 5–7+ mm, green, surfaces glabrous. **Ray florets** 8(13); rays 8–10+ mm. **Disc florets** 30–40+; corolla tubes 2.5–3.5 mm, limbs 2–3 mm. **Fruits** 3–4.5 mm, glabrous; pappi 4–5 mm. **2*n*=**46, 92.

Streams, roadsides, clearings, disturbed areas, coniferous woodlands in rocky soils, serpentine. Flowering Apr–Jul. 50–1400 m. Casc, CR, Est, Lava, Sisk, WV. CA, WA; north to British Columbia. Native.

Packera macounii is similar in overall morphology to *P. cana*. However, its leaves are narrower and commonly revolute. Although it is often cited as being collected on serpentine soils, it is not restricted to them.

Packera porteri (Greene) C. Jeffrey

Porter's butterweed, Porter's groundsel

Senecio porteri Greene

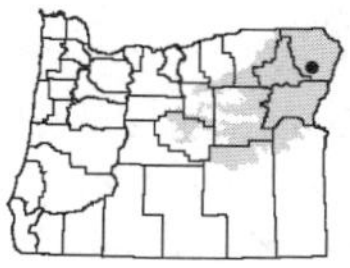

Plants perennial, 3–10+ cm; caudices with fleshy, branching rhizomes. **Stems** 1, glabrous. **Basal leaves** blades reniform to subreniform, relatively turgid, bases abruptly contracted to cordate, margins crenate, sometimes wavy, abaxially cyanic; petiolate. **Cauline leaves** bract-like, margins entire. **Heads** 1; peduncles glabrous, bracts absent. **Calyculi** conspicuous. **Involucres** campanulate, bractlets often cyanic. **Phyllaries** 13 or 21, 8–10+ mm, deep red, surfaces glabrous. **Ray florets** 8 or 13; rays 8–10+ mm. **Disc florets** 40–50+; corolla tubes 2.5–3.5 mm, limbs 2.5–3.5 mm. **Fruit** 1.5–2.5 mm, glabrous; pappi 4–5.5 mm.

Steep, alpine talus slopes. Flowering Jul–Aug. 2100–2500 m. BW. WA; CO. Native.

Only a single collection locality for *Packera porteri* is known from Oregon. In August of 1899, W.C. Cusick collected several duplicates of this species from "alpine ridges of the Wallowa Mountains, near the lake." He noted at the time that "few plants were seen." It has not been collected there since, although there is one additional observation from 1978. This species is more frequently collected (although still rare) on high peaks in the Rocky Mountains of central Colorado, nearly always being collected at over 3,600 m. *Packera porteri* is likely infrequently collected because it occurs only on steep talus slopes, using extensive fleshy rhizomes to find purchase in a precarious habitat. In this author's opinion, it is one of the most interesting and elusive species in the genus and certainly worthy of concerted efforts to determine its current conservation status.

Packera pseudaurea var. ***pseudaurea*** (Rydberg) W.A. Weber & Á. Löve

Falsegold groundsel, streambank butterweed

Senecio pseudaureus Rydb. var. *pseudaureus*

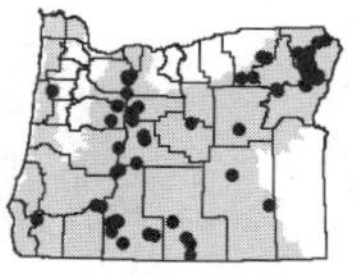

Plants perennial, 50–70+ cm; fibrous-rooted, caudices simple or branched, horizontal or erect. **Stems** usually 1, sometimes 2–4 and clustered, glabrous or sparsely tomentose near bases. **Basal leaves** blades usually broadly lanceolate to ovate, sometimes sub-hastate, bases truncate to subcordate or obtuse, margins crenate, denticulate, bluntly dentate, sharply dentate, subentire or pinnately lobed to laciniate; petiolate. **Cauline leaves** similar to basal; upper gradually reduced; lower petiolate, upper becoming sessile and clasping. **Heads** 5–20+, in open or congested subumbel-like arrays; peduncles glabrous, bracteate. **Calyculi** inconspicuous. **Involucres** campanulate. **Phyllaries** (13)21(30), 3–8 mm, light green, surfaces glabrous. **Ray florets** 0, 8 or 13; rays 6–10+ mm. **Disc florets** 70–80+; corolla tubes 2.5–3.5 mm, limbs 2–3 mm. **Fruits** 1–3 mm, glabrous; pappi 4.5–5.5 mm. **2*n*=**40, 44, 80.

Damp streambanks, wet meadows, open wet woodlands. Flowering Jun–Aug. 700–2300 m. BR, BW, Casc, CR, ECas, Lava, Sisk. CA, ID, NV, WA; north to British Columbia, northeast to Manitoba, east to WY, southeast to CO. Native.

Packera pseudaurea var. *pseudaurea* is a robust and distinctive plant and is relatively common throughout its range. It is one of the few species of *Packera* that flowers late in the summer.

Packera streptanthifolia (Greene) W.A. Weber & Á. Löve

Rocky Mountain groundsel, Rocky Mountain butterweed, cleftleaf groundsel

Packera streptanthifolia (Greene) W.A. Weber & Á. Löve var. *streptanthifolia*, *Senecio leonardii* Rydb., *Senecio streptanthifolius* Greene

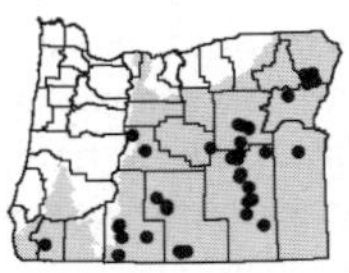

Plants perennial, 20–50+ cm; caudices weak or stout, horizontal or suberect and fibrous-rooted. **Stems** 1 or 2–5, clustered, glabrous or sometimes sparsely pubescent near base and in leaf axils. **Basal leaves** blades oblanceolate to orbiculate, thick, turgid, bases tapering to contracted, margins dentate, dissected, entire, subentire or weakly lobulate, surfaces usually glabrous, sometimes pubescent; petiolate. **Cauline leaves** similar to basal, gradually reduced; petiolate or sessile. **Heads** 2–20+, in corymb- or subumbel-like arrays; peduncles glabrous or sparsely tomentose, bracteate. **Calyculi** conspicuous. **Involucres** campanulate. **Phyllaries** (8)13 or 21, 4–7+ mm, green, tips sometimes anthocyanic, surfaces glabrous. **Ray florets** 8 or 13; rays 5–10 mm. **Disc florets** 35–60; corolla tubes 2–4 mm, limbs 2.5–4 mm. **Fruits** 1–2.5 mm, glabrous; pappi 3–6 mm. **2*n***=46, 92.

Forests, open meadows, and valleys in dry to damp loamy soils. Flowering May–Aug. 400–2900 m. BR, BW, ECas, Lava, Owy, Sisk. ID, NV, WA; north to Yukon, northeast to Saskatchewan, east to WY, southeast to NM. Native.

Packera streptanthifolia includes weakly defined "phases" that have been treated as distinct species or as varieties. Characteristics used to define those taxa often overlap and are difficult to distinguish; some phases grade into each other.

Packera subnuda var. ***subnuda*** (de Candolle) Trock & T.M. Barkley (p. 324)

Cleftleaf groundsel, alpine meadow groundsel, pale groundsel

Senecio cymbalarioides H. Buek, *Senecio subnudus* DC.

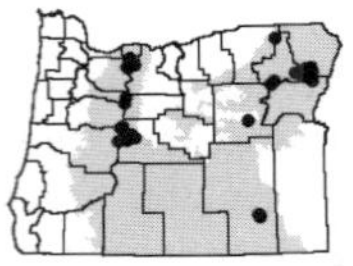

Plants perennial, 10–30+ cm; fibrous-rooted or rhizomatous, caudices erect to horizontal. **Stems** 1(2–3), loosely clustered, glabrous. **Basal leaves** blades lanceolate, ovate, obovate, or elliptic, bases tapering, margins entire, subentire, or crenate-dentate. **Cauline leaves** similar to basal, abruptly reduced; lower petiolate, upper sessile and often weakly clasping. **Heads** 1–3, sometimes subtended by 2 large bractlets; peduncles without bracts or bracts inconspicuous. **Calyculi** conspicuous, bractlets frequently anthocyanic. **Involucres** campanulate. **Phyllaries** 13 or 21, 5–8+ mm, green, rarely with cyanic tips, surfaces glabrous. **Ray florets** 13, rays 7–12+ mm. **Disc florets** 40–55+, corolla tubes 2.5–3.5 mm, limbs 2.5–3.5 mm. **Fruits** 1.5–2.5 mm, glabrous; pappi 4–6 mm. **2*n***=46.

Alpine and subalpine meadows, lake shores, streambanks, damp or marshy areas. Flowering Jun–Sep. 1000–2800 m. BR, BW, Casc, ECas. CA, ID, NV, WA; north to British Columbia, northeast to Alberta. Native.

Packera subnuda var. *subnuda* is usually scapiform and frequently has only a single large head. It is not likely to be mistaken for any other *Packera* species in Oregon.

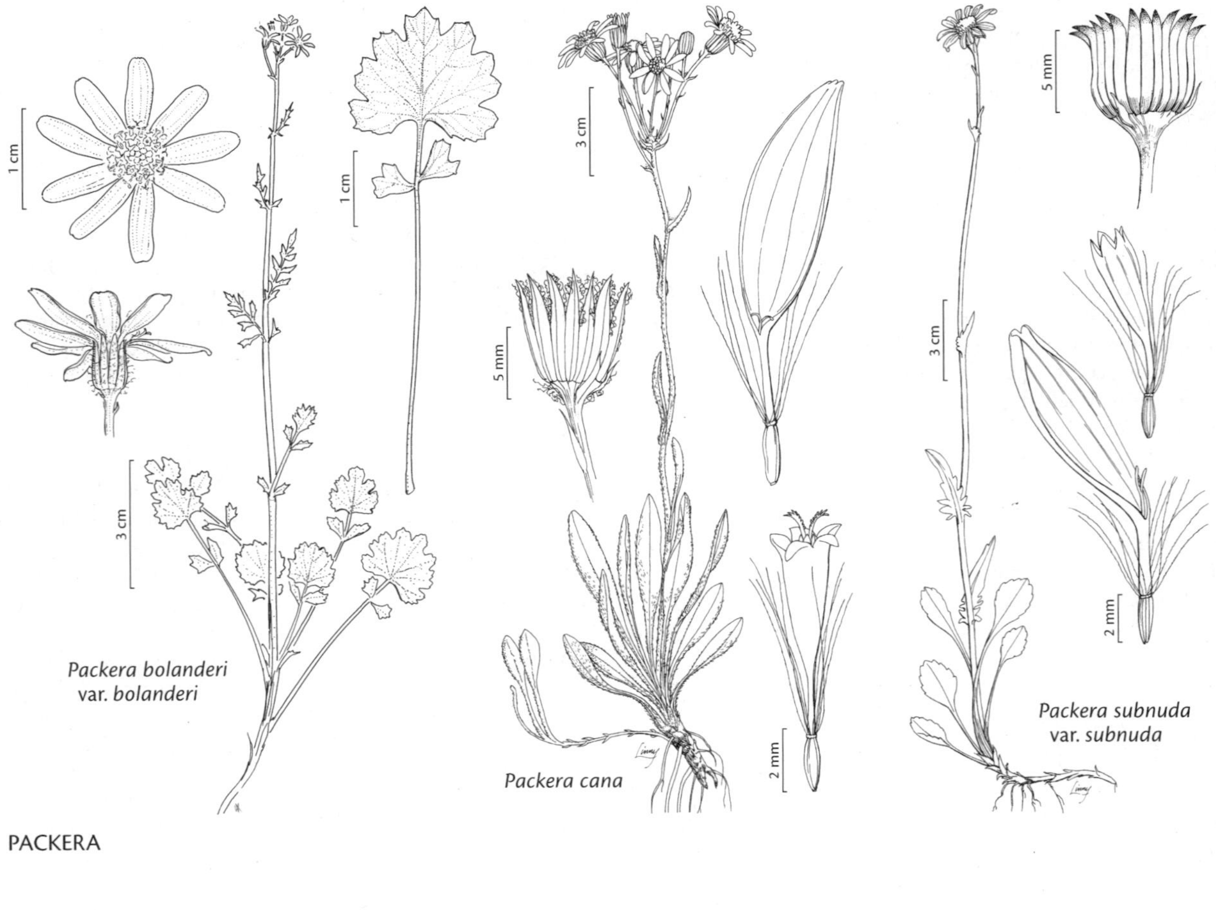

PACKERA

PETASITES Coltsfoot

Stephen C. Meyers

Herbs perennial, polygamodioecious; rhizomatous. **Stems** 1–several, erect, unbranched below inflorescences, tomentose to pilose, puberulent, or nearly glabrous, often stipitate-glandular distally. **Basal leaves** deltate to orbiculate, margins entire to toothed or lobed, surfaces abaxially tomentose, adaxially glabrous to tomentulose and glabrate; petiolate. **Cauline leaves** alternate, reduced and bract-like; sessile. **Inflorescences** corymb-, panicle-, or raceme-like arrays, rarely heads solitary; peduncles present. **Staminate heads** radiate or disciform, with sterile, styliferous ray florets and staminate disc florets. **Pistillate heads** radiate or disciform, with pistillate ray florets and staminate disc florets. **Involucres** obconic to turbinate. **Phyllaries** 1–3 equal series, herbaceous, margins membranous, often purplish, 1–7-veined, tips acute or acuminate, surfaces strigillose or puberulent, sometimes glandular. **Receptacles** flat to convex; paleae 0. **Fruits** cylindric or fusiform, glabrous, 5–10-ribbed; pappi of 60–100 white bristles.

Asia, Europe, North America. ~18 species; 1 species treated in *Flora*.

Some ornamental species of *Petasites* are known to have locally escaped in Oregon but are not established. *Petasites fragrans* has been reported once in Benton County. It can be identified by the combination of its unlobed, irregularly toothed leaves and purplish-tinged to purple corollas. Similarly, *P. japonicus* has been reported once from Portland. It can be distinguished by its large leaves (up to 150 cm wide) and white corollas.

Petasites frigidus (Linnaeus) Fries

Coltsfoot

Plants 1–6+ dm. **Basal leaves** orbicular, reniform, deltate, hastate, sagittate, or cordate, 20–50 × 20–50 cm, bases cordate, margins subentire to dentate, lobed or not; petioles to 50 cm. **Staminate heads** 2–50; ray florets 1–70, rays linear to narrowly oblong, 1–12 mm, white to pink or purple; disc florets 10–80, corollas filiform, 3–5 mm, white to pink or purple. **Pistillate heads** 5–60; ray florets 30–130, linear to narrowly oblong, 0.5–6 mm, white to pink or purple; disc florets 1–12, corollas filiform, 0.5–5 mm, white to pink or purple. **Phyllaries** linear. **Fruits** 1.5–3.5 mm, brown; pappi 5–17 mm.

Asia, Europe, North America. 4 varieties; 2 varieties treated in *Flora*.

Key to *Petasites frigidus* varieties:

1. Basal leaves deltate, hastate, sagittate, or cordate, shallowly lobed . . . . *P. f.* var. ***frigidus***

1' Basal leaves orbicular to reniform, deeply lobed . . . . *P. f.* var. ***palmatus***

Petasites frigidus var. *frigidus* (p. 326)

Alpine coltsfoot

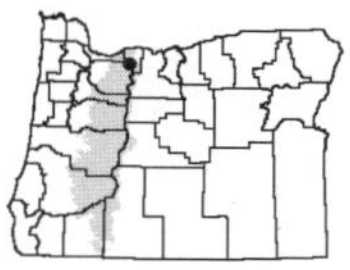

Basal leaves deltate, hastate, sagittate, or cordate, shallowly lobed. **2*n***= 60, 90.

Moist alpine and subalpine areas. Flowering Feb–Apr. 100–1900 m. Casc. WA; north to AK, northeast to Nunavut. Native.

Petasites frigidus var. *palmatus* (Aiton) Cronquist (p. 326)

Sweet coltsfoot, western coltsfoot

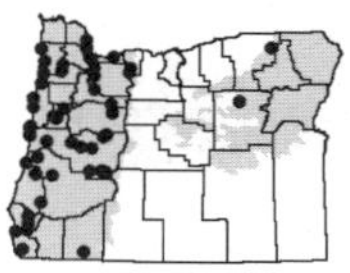

Basal leaves orbicular to reniform, deeply lobed. **2*n***=60, 61, 62.

Moist forests, streambanks, roadsides, marshes. Flowering Mar–Jun. 0–1400 m. BW, Casc, CR, Est, Sisk, WV. CA, ID, WA; north to Yukon, northeast to Nunavut, east to Nova Scotia. Native.

PLEIACANTHUS Skeletonweed

Genus composed of one species, *P. spinosus*.

Pleiacanthus spinosus (Nuttall) Rydberg (p. 326)

Spiny skeletonweed

Lygodesmia spinosa Nutt., *Stephanomeria spinosa* (Nutt.) Tomb

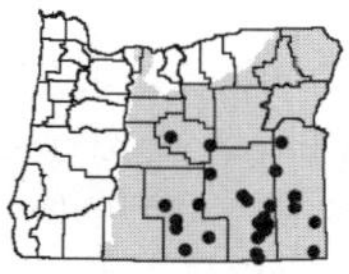

Subshrubs perennial, deciduous, 1–8 dm, sap milky; taprooted. **Stems** photosynthetic, erect, slender, woody at bases, much branched, glabrous but with enlarged, wooly basal bud-scales. **Leaves** linear, 1–5 cm, surfaces glabrous. **Inflorescences** panicle-like arrays or heads solitary. **Heads** ligulate; peduncles 1–4 mm. **Involucres** cylindrical, 7–12 mm. **Phyllaries** in 1–2 series, surfaces glabrous; inner equal, ovate-lanceolate, tips acute; outer unequal, ovate to deltate, < inner, margins smooth or sparsely short-ciliate, tips acute. **Receptacles** flat to convex, smooth, glabrous; paleae 0. **Florets** 3–5, bisexual; ligules 12–16 mm, well exceeding involucres, pink to reddish purple, rarely white. **Fruits** columnar, tapering to base, 6–8 mm, light tan, glabrous, distinctly grooved between ribs, ribs 5; pappi of numerous, dingy white, minutely roughened bristles. **2*n***=16.

Juniper woodlands, sagebrush scrub, sand dunes, open fields, hillsides. Flowering Jun–Sep. 400–2200 m. BR, BW, ECas, Lava, Owy. CA, ID, NV; northeast to MT, southeast to AZ. Native.

This species was previously assigned to either *Lygodesmia* or *Stephanomeria*, but a phylogenetic study (Lee et al. 2002) separates it as a monotypic genus.

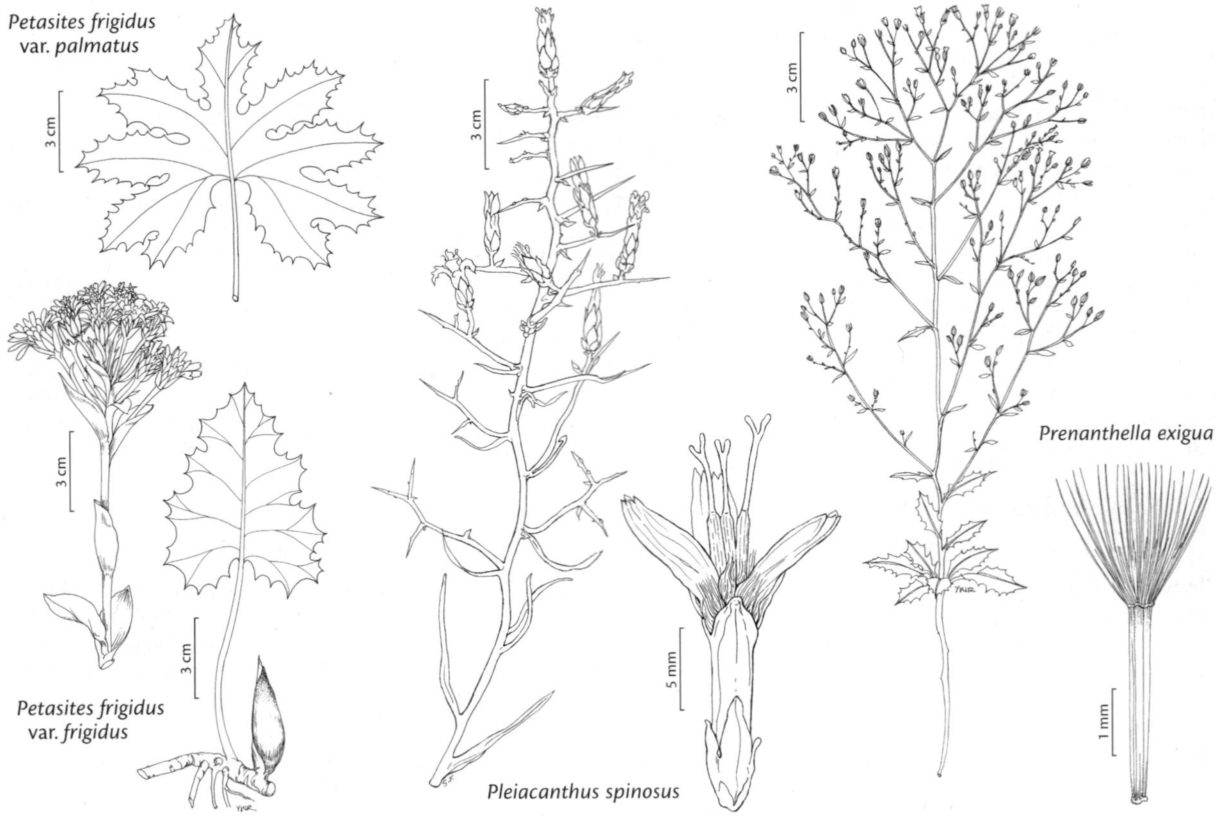

PETASITES • PLEIACANTHUS • PRENANTHELLA

PRENANTHELLA Brightwhite

Genus composed of one species, *P. exigua*.

Prenanthella exigua (A. Gray) Rydberg (p. 326)
Brightwhite

Lygodesmia exigua (A. Gray) A. Gray

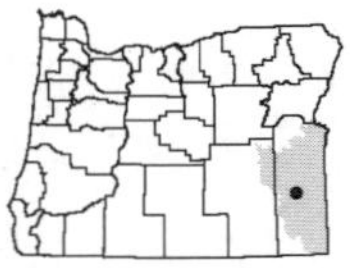

Herbs annual, 5–30 cm, sap milky; taproots slender. **Stems** erect, proximally leafy, glabrous or minutely glandular-pubescent, branched near bases or distally. **Leaves** basal and cauline, blades broadly to narrowly oblanceolate, 0.5–3 cm, reduced distally to bracts, margins irregularly dentate to lobed, surfaces glabrous, teeth spine-tipped. **Inflorescences** panicle-like arrays, much branched; peduncles bracteolate. **Heads** 10–150+, ligulate. **Involucres** ovoid. **Phyllaries** 3–5, in 1 series, lanceolate, tips acute, surfaces glabrous or minutely glandular. **Receptacles** flat to ± convex, smooth, glabrous; paleae 0. **Florets** 3–4, bisexual; ligules 1.5–2 mm, slightly exceeding involucres, pink or white. **Fruits** columnar, tapering downward, 2.5–3.5 mm, pale brown, glabrous, 5-ribbed and grooved, ribs minutely roughened; pappi of basally joined, smooth, white bristles in 2 unequal series, longest bristles 2–3 mm. **2*n***=14.

Sandy, gravelly, or clay soils, desert washes and slopes, sagebrush-juniper steppe. Flowering Mar–Jun. 800–1100 m. Owy. CA, NV; southeast to TX, south to Mexico. Native.

PSEUDOGNAPHALIUM Cudweed

Herbs annual, biennial, or perennial; taprooted, often aromatic. **Stems** 1–several from base, erect to decumbent or prostrate, simple or branched, ± floccose-tomentose, glandular or not. **Leaves** basal and cauline, or mostly cauline, alternate, linear to linear-lanceolate, lanceolate, oblong, or oblanceolate, gradually reduced distally, bases decurrent or not, sometimes clasping, margins entire, 1-veined, tips acute or obtuse, surfaces one or both tomentose, sometimes adaxially stalked-glandular; sessile. **Inflorescences** glomerules in corymb- or panicle-like arrays; bracts present or not. **Heads** disciform. **Involucres** ovoid or hemispheric. **Phyllaries** in 3–10 series, erect in flower, spreading in fruit, chartaceous; inner equal, linear to lanceolate or ovate, tips acute or obtuse; outer ± equal or gradually shorter, ovate to triangular, surfaces glabrous or tomentose. **Receptacles** flat, smooth; paleae 0. **Outer florets** pistillate; corollas red- or yellow-tipped. **Inner florets** bisexual; corollas tubular, linear, yellow to red, minutely 5-lobed. **Fruits** cylindric or fusiform, surfaces smooth or minutely papillate; pappi of 10–12 equal, deciduous bristles.

Worldwide. ~100 species; 5 species treated in *Flora*.

Key to *Pseudognaphalium* species:

1. Leaf surfaces contrasting in color, abaxial gray-tomentose, adaxial green. . . . . *P. macounii*
1' Leaf surfaces similar in color, both greenish to gray.
  2. Stems stalked-glandular. . . . . *P. californicum*
  2' Stems not stalked-glandular.
    3. Leaf bases not clasping . . . . *P. thermale*
    3' Leaf bases clasping.
      4. Involucres 3–4 mm; pistillate corollas red-tipped; bisexual flowers 5–10 . . . . *P. luteoalbum*
      4' Involucres 4–6 mm; pistillate corollas yellow-tipped; bisexual flowers 10–30. . . . . *P. stramineum*

Pseudognaphalium californicum (de Candolle) Anderberg
California cudweed, California everlasting

Gnaphalium californicum DC.

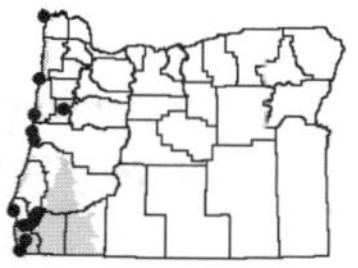

Plants annual, biennial, or perennial, 2–7 dm. **Stems** floccose-tomentose, stalked-glandular. **Leaves** 1–10 cm, linear to narrowly lanceolate or oblanceolate, bases clasping stem or not, decurrent or not, surfaces villous-tomentose and stalked-glandular. **Involucres** 5–7 mm. **Phyllaries** in 7–10 series, white, surfaces glabrous. **Outer florets** 100–140; corolla tips yellow. **Inner florets** 5–12. **Fruits** 0.4–0.6 mm, brown; pappi white. **2*n*=28.**

Marshes, meadows, cliffs, coastal bluffs, riverbanks, roadsides, clearcuts, disturbed areas. Flowering Jun–Sep. 0–400 m. Est, Sisk. CA; south to Mexico. Native.

Pseudognaphalium luteoalbum (Linnaeus) Hilliard & B.L. Burtt (p. 328)
Weedy cudweed

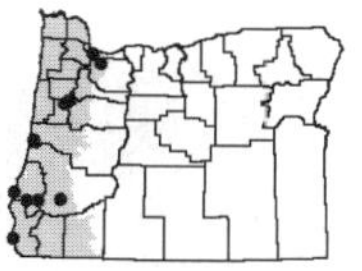

Plants annual, 1.5–4 dm. **Stems** floccose-tomentose, not glandular. **Leaves** 1.5–5 cm, broadly or narrowly linear to oblanceolate, bases clasping, not decurrent, surfaces floccose-tomentose, not glandular. **Involucres** 3–4 mm. **Phyllaries** in 3–4 series, green or brownish, surfaces glabrous. **Outer florets** 130–160; corolla tips red. **Inner florets** 5–10. **Fruits** 0.3–0.5 mm, brown; pappi white. **2*n*=14, 16, 28.**

Riverbanks, roadsides, disturbed areas. Flowering May–Aug. 0–300 m. CR, Est, Sisk, WV. CA, NV, WA; southeast to FL, south to Mexico; Africa, Asia, Australia, Europe, Pacific Islands. Exotic.

Pseudognaphalium macounii (Greene) Kartesz
Sticky cudweed, winged cudweed

Gnaphalium macounii Greene

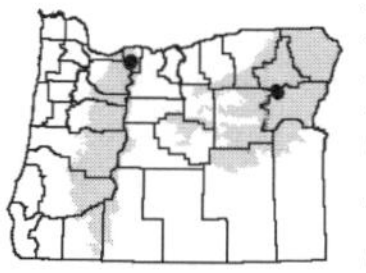

Plants annual or biennial, 4–10 dm. **Stems** pilose- or floccose-pubescent, stalked-glandular. **Leaves** 2–10 cm, linear to lanceolate or oblanceolate, bases decurrent, surfaces contrasting in color, abaxially gray-tomentose, sometimes stalked-glandular on midveins, adaxially green, glabrous or scarcely tomentose, usually stalked-glandular. **Involucres** 4.5–6 mm. **Phyllaries** in 4–5 series, white or brownish, surfaces glabrous. **Outer florets** 50–100; corolla tips yellow. **Inner florets** 5–12. **Fruits** 0.4–0.5 mm, brown; pappi white.

Open woodlands, grasslands, riverbanks, roadsides. Flowering Jul–Sep. 1000–2000 m. BW, Casc. CA, ID, WA; north to British Columbia, northeast to Nova Scotia, east to MA, southeast to NM, south to Mexico. Native.

Pseudognaphalium stramineum (Kunth) Anderberg
Cotton batting cudweed, cotton batting plant

Gnaphalium chilense Spreng., *Gnaphalium stramineum* Kunth

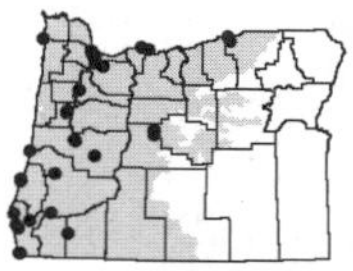

Plants annual or biennial, 1–8 dm. **Stems** floccose-tomentose, not glandular. **Leaves** 1.5–12 cm, linear to lanceolate or oblanceolate, bases clasping, usually not decurrent, surfaces floccose-tomentose, not glandular. **Involucres** 4–6 mm. **Phyllaries** in 4–5 series, white or brownish, surfaces glabrous. **Outer florets** 150–200; corolla tips yellow. **Inner florets** 10–30. **Fruits** 0.4–0.6 mm, brown; pappi brownish. **2*n*=28.**

Shrublands, grasslands, ocean bluffs, beaches, dunes, marshes, riverbanks, roadsides, disturbed areas. Flowering Feb–Oct. 0–1000 m. Casc, Col, CR, ECas, Est, Lava, Sisk, WV. CA, ID, NV, WA; north to British Columbia, northeast to MT, east to NE, southeast to TX, south to Mexico, NC, NY, SC, VA; South America. Native.

Pseudognaphalium thermale (E.E. Nelson) G.L. Nesom (p. 328)
Northwestern rabbit-tobacco, Wright's cudweed

Gnaphalium canescens DC. ssp. *thermale* (E.E. Nelson) Stebbins & D.J. Keil, *Gnaphalium microcephalum* Nutt., *Gnaphalium microcephalum* Nutt. var. *thermale* (E.E. Nelson) Cronquist, *Gnaphalium thermale* E.E. Nelson

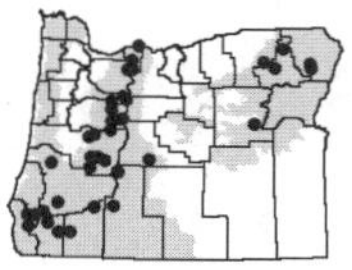

Plants perennial, 2–9 dm. **Stems** floccose-tomentose, not glandular. **Leaves** 1.5–9 cm, linear to oblanceolate to narrowly oblong, bases decurrent on stem, not clasping, surfaces floccose-tomentose, sometimes minutely glandular beneath longer trichomes. **Involucres** 4–6 mm. **Phyllaries** in 3–4 series, white, surfaces glabrous. **Outer florets** 35–55; corolla tips yellow. **Inner florets** 4–7. **Fruits** 0.5–0.6 mm, brown; pappi white.

Conifer forests, shrublands, grasslands, riverbanks, lakesides, dry slopes, roadsides, disturbed areas. Flowering Jun–Oct. 0–2000 m. BW, Casc, CR, ECas, Sisk. CA, ID, NV, WA; north to British Columbia, northeast to Alberta, east to WY, southeast to UT. Native.

Pseudognaphalium thermale is here recognized at the species rank rather than its earlier placement as a subspecies of *P. canescens* or a variety of *P. microcephalum*.

PSILOCARPHUS Woollyheads

Herbs annual. **Stems** erect, spreading, or prostrate, often clump-forming, simple to well branched, branches often in pairs beneath terminal heads, floccose-tomentose. **Leaves** basal and cauline, mostly opposite, sometimes alternate distally, becoming bracteate and whorled below heads, linear to ovate, lanceolate, elliptic, oblanceolate, or obovate, margins entire, tips acute or obtuse, surfaces floccose-tomentose; sessile. **Inflorescences** heads solitary or in tightly clustered glomerules of 2–4 heads. **Heads** disciform, ± spherical. **Phyllaries** 0. **Involucres** 0. **Receptacles** obovoid, surfaces glabrous. **Outer paleae** wings inflexed. **Inner paleae** 0. **Outer florets** 20–100+, pistillate, enclosed by paleae; corolla tubes filiform; styles exserted. **Inner florets** 2–10, staminate; corollas 4–5-lobed; stamens included. **Fruits** obovoid or cylindric, glabrous; pappi 0.

North America, South America. 5 species; 4 species treated in *Flora*.

Key to *Psilocarphus* species:

1. Outer paleae 2.5–4 mm.
   2. Distal leaves subtending heads mostly lanceolate to ovate, widest in lower ⅔, usually 8–15 mm . . . . *P. brevissimus* var. *brevissimus*
   2' Distal leaves subtending heads mostly oblanceolate to linear, widest in distal ⅓, usually 10–35 mm . . . . *P. elatior*

1' Outer paleae 1.2–2.7 mm.
   3. Leaves linear or linear-oblanceolate, mostly 6–12 × as long as wide . . . . *P. oregonus*
   3' Leaves oblong or oblanceolate to ovate or elliptic, mostly 1.5–6 × as long as wide . . . . *P. tenellus*

Psilocarphus brevissimus Nuttall var. *brevissimus* (p. 328)

Dwarf woollyheads

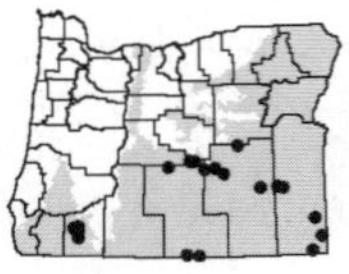

Plants 1–8 cm. **Stems** usually much branched from base and clump-forming, sometimes simple. **Cauline leaves** 2–20 mm; distal, mostly lanceolate to ovate, widest in lower ⅔, usually 8–15 mm. **Heads** largest 6–9 mm in diameter. **Receptacles** lobed. **Outer paleae** 8–100, 2.5–4 mm. **Inner corollas** 0.8–1.6 mm, lobes usually 5. **Fruits** obovoid, 1–2 mm, brown. **2*n***=28.

Woodlands, grasslands, lakesides, alkaline flats, dry beds of vernal pools. Flowering May–Jul. 300–2000 m. BR, BW, ECas, Owy, Sisk. CA, ID, NV, WA; north to British Columbia, northeast to Saskatchewan, east to WY, southeast to UT, south to Mexico; South America. Native.

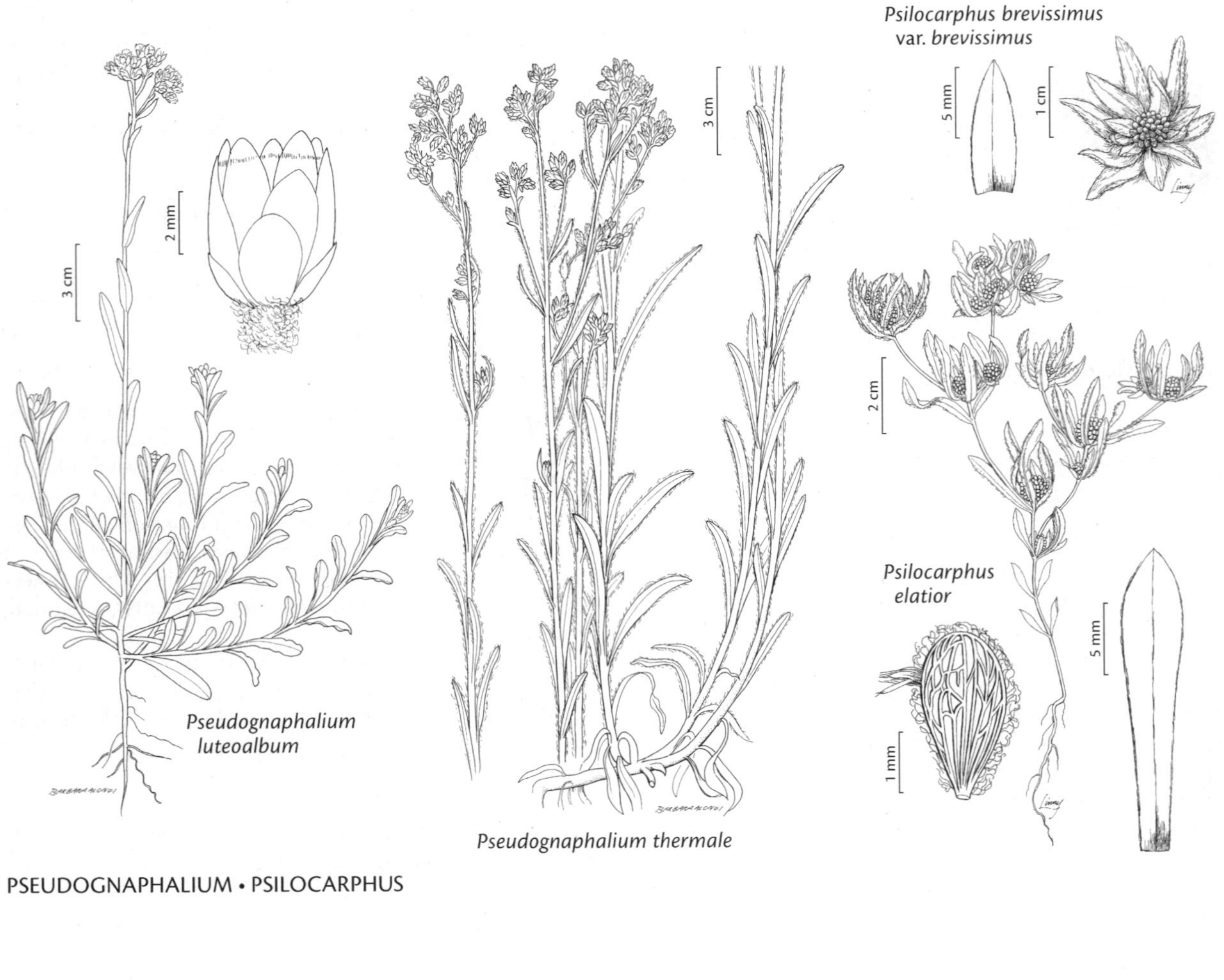

PSEUDOGNAPHALIUM • PSILOCARPHUS

Psilocarphus elatior (A. Gray) A. Gray (p. 328)

Tall woollyheads

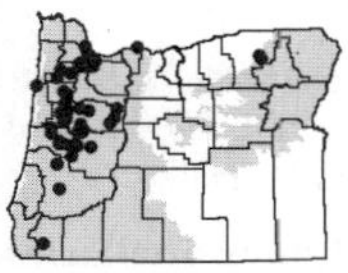

Plants 2–15 cm. **Stems** simple or more often much branched from base or throughout, spreading or clump-forming. **Cauline leaves** 3–35 mm; distal, mostly oblanceolate to linear, widest in distal ⅓, usually 10–35 mm. **Heads** largest 6–8 mm in diameter. **Receptacles** unlobed. **Outer paleae** (20)50–80, 2.5–3.8 mm. **Inner corollas** 1.3–1.9 mm, lobes 5. **Fruits** cylindric, 1–1.7 mm, brown.

Forests, woodlands, grasslands, riverbanks, dry vernal pools, roadsides, disturbed areas. Flowering May–Aug. 0–1600 m. BW, Casc, CR, ECas, Sisk, WV. CA, ID, WA; north to British Columbia. Native.

Psilocarphus oregonus Nuttall

Oregon woollyheads

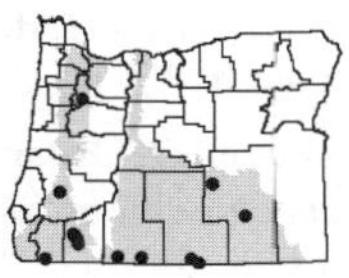

Plants 1–10 cm. **Stems** often much branched from base and clump-forming. **Cauline leaves** 3–18 mm, linear or narrowly oblanceolate, mostly 6–12 × as long as wide. **Heads** largest 4–6 mm in diameter. **Receptacles** unlobed. **Outer paleae** 20–80, 1.5–2.7 mm. **Inner corollas** 0.7–1.4 mm, lobes usually 4. **Fruits** cylindric, 0.6–1.2 mm, brown.

Openings in forests, woodlands, shrublands, grasslands, dry beds of vernal pools. Flowering May–Jul. 100–1900 m. BR, ECas, Sisk, WV. CA, ID, NV, WA. Native.

Psilocarphus tenellus Nuttall

Slender woollyheads

Psilocarphus tenellus Nutt. var. *tenellus*

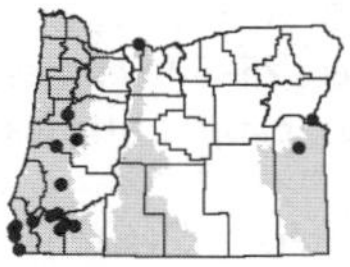

Plants 2–8 cm. **Stems** usually much branched throughout, prostrate, clump-forming or spreading, up to 15 cm. **Cauline leaves** 3–12 mm, oblong or oblanceolate to obovate or elliptic, mostly 1.5–6 × as long as wide. **Heads** 3–5.5 mm in diameter. **Receptacles** unlobed. **Outer paleae** 25–50, 1.2–2.5 mm. **Inner corollas** 0.8–1.5 mm, lobes usually 5. **Fruits** obovoid, 0.6–1.2 mm, brown.

Grasslands, beaches, riverbanks, sandbars, roadsides, disturbed areas. Flowering May–Aug. 0–1300 m. CR, ECas, Est, Owy, Sisk, WV. CA, ID, WA; north to British Columbia, south to Mexico. Native.

PYRROCOMA Goldenweed

Herbs perennial; taproots stout. **Stems** 1–many, decumbent to ascending or erect, simple to well branched, glabrous to villous-tomentose, hispid, or glandular-pubescent. **Basal leaves** lanceolate to elliptic or linear, margins entire to mucronate-dentate or serrate, tips acute or acuminate, usually mucronate, surfaces glabrous to villous-tomentose, hispid, and/or stipitate-glandular; petiolate. **Cauline leaves** usually similar to basal, alternate, sometimes reduced upwards; sessile or petiolate, sometimes auriculate-clasping. **Inflorescences** corymb-, panicle-, raceme-, or spike-like arrays, or heads solitary; bracts present or not. **Heads** disciform or radiate. **Involucres** campanulate, hemispheric, or globose. **Phyllaries** in 2–6 equal or unequal series, erect, herbaceous or scarious basally and marginally, ovate to lanceolate or linear, tips acute or acuminate, surfaces glabrous to villous-tomentose, hispidulate, or stipitate-glandular, midribs evident. **Receptacles** flat to convex; paleae 0. **Ray florets** pistillate; rays equaling to well exceeding involucres, yellow, lobes minute or absent. **Disc florets** bisexual; corollas tubular, narrowed in lower half, exceeding involucres, yellow, surfaces glabrous, lobes 5, erect; stamens slightly exserted; stigmas exserted. **Fruits** linear to oblong, tapering to bases, glabrous or sericeous, ribbed or angled; pappi of numerous brownish, persistent bristles in 1 series.

Western Canada and United States. 14 species; 8 species treated in *Flora*.

Like several other genera in the flora, *Pyrrocoma* was previously included in a large and variable genus *Haplopappus*. However, for nomenclatural reasons, *Haplopappus* is not recognized in North America by most authors.

Key to *Pyrrocoma* species:

1. Stems and leaves glandular . . . . *P. hirta*
1' Stems and leaves not glandular.
  2. Ray florets 0 or inconspicuous . . . . *P. carthamoides*
  2' Ray florets conspicuous.
    3. Heads subtended by leafy bracts . . . . *P. radiata*
    3' Heads not subtended by leafy bracts.
      4. Heads 3–20(30).
        5. Inflorescences in corymb- or panicle-like arrays . . . . *P. lanceolata*
        5' Inflorescences in raceme-like arrays.
          6. Disc corollas 5–8 mm; fruits 2–5 mm . . . . *P. racemosa*
          6' Disc corollas 7–11 mm; fruits 4.5–6.5 mm . . . . *P. scaberula*
      4' Heads 1(2).
        7. Basal leaves 2–5 mm wide, margins entire . . . . *P. howellii*
        7' Basal leaves 3–15 mm wide, margins denticulate to sharply dentate, rarely entire . . . . *P. uniflora*

Pyrrocoma carthamoides Hooker

Large-flowered goldenweed

Plants 3–80 cm. **Stems** erect or decumbent-ascending, hispidulous or puberulent, rarely glabrous. **Basal leaves** oblanceolate to broadly or narrowly elliptic, 5–20 cm × 5–35 mm, margins entire or denticulate, surfaces glabrous or puberulent. **Cauline leaves** sessile or petiolate. **Heads** solitary or 1–3 in raceme-like arrays, discoid or radiate; subtended by leafy bracts; peduncles 1–50 mm. **Involucres** 13–25 mm. **Phyllaries** in 3–5 series, ovate to narrowly lanceolate, margins sometimes sharply muricate, surfaces glabrous or puberulent. **Ray florets** 0–30, inconspicuous, rarely exceeding the involucres. **Disc florets** 25–50, corollas 9–14 mm. **Fruits** 3–5.5 mm, brown, glabrous; pappus bristles 6–9 mm.

Western Canada and United States. 3 varieties; 2 varieties treated in *Flora.*

Key to *Pyrrocoma carthamoides* varieties:

1. Plants 30–80 cm; outer phyllaries broadly oblong or ovate, acute, margins often sparsely muricate. . . . . *P. c.* var. ***carthamoides***

1' Plants 3–40 cm; outer phyllaries narrowly lanceolate or elliptic to linear, margins not muricate . . . . *P. c.* var. ***cusickii***

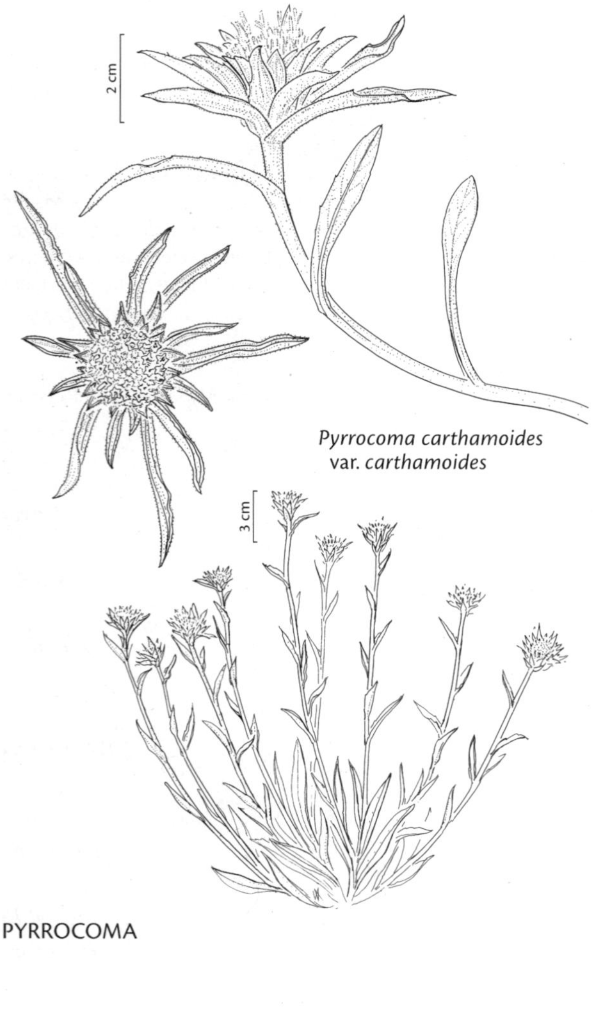

Pyrrocoma carthamoides var. *carthamoides*

PYRROCOMA

Pyrrocoma carthamoides var. *carthamoides* (p. 330)

Large-flowered goldenweed, Columbia goldenweed

Haplopappus carthamoides (Hook.) A. Gray ssp. *carthamoides*, *Haplopappus carthamoides* (Hook.) A. Gray var. *carthamoides*, *Haplopappus carthamoides* (Hook.) A. Gray ssp. *rigidus* (Rydb.) H.M. Hall

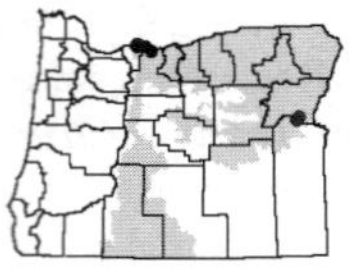

Plants 30–80 cm, erect. **Outer phyllaries** broadly oblong or ovate, acute, margins often sparsely muricate. **2*n*=**12.

Dry prairies, hillsides, roadsides, in sandy or gravelly soils. Flowering Jun–Aug. 100–1600 m. BW, Col, ECas. ID, WA; north to British Columbia, northeast to MT, east to WY. Native.

Pyrrocoma carthamoides var. *cusickii* (A. Gray) Kartesz & Gandhi

Cusick's goldenweed

Haplopappus carthamoides (Hook.) A. Gray ssp. *cusickii* (A. Gray) H.M. Hall, *Haplopappus carthamoides* (Hook.) A. Gray var. *cusickii* A. Gray

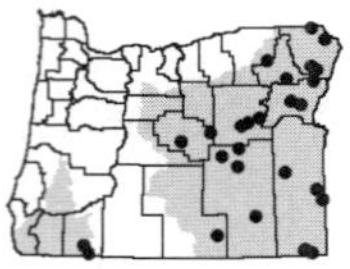

Plants 3–40 cm, decumbent to erect. **Outer phyllaries** narrowly lanceolate or elliptic to linear, margins not muricate. **2*n*=**12.

Coniferous and mixed hardwood forests, brushlands, grassy hillsides, lava beds. Flowering Jun–Aug. 900–2800 m. BR, BW, Lava, Owy, Sisk. CA, ID, WA, NV. Native.

Pyrrocoma hirta (A. Gray) Greene

Tacky goldenweed

Haplopappus hirtus A. Gray, *Haplopappus hirtus* A. Gray ssp. *hirtus*, *Haplopappus hirtus* A. Gray var. *hirtus*, *Haplopappus hirtus* A. Gray ssp. *lanulosus* (Greene) H.M. Hall, *Haplopappus hirtus* A. Gray var. *lanulosus* (Greene) M. Peck, *Haplopappus hirtus* A. Gray ssp. *sonchifolius* (Greene) H.M. Hall, *Haplopappus hirtus* A. Gray var. *sonchifolius* (Greene) M. Peck, *Haplopappus subviscosus* (Greene) S.F. Blake, *Pyrrocoma hirta* (A. Gray) Greene var. *hirta*, *Pyrrocoma hirta* (A. Gray) Greene var. *lanulosa* (Greene) Mayes ex G.K. Br. & D.J. Keil, *Pyrrocoma hirta* (A. Gray) Greene var. *sonchifolia* (Greene) Kartesz & Gandhi, *Pyrrocoma lanceolata* (Hook.) Greene var. *subviscosa* (Greene) Mayes ex G.K. Br. & D.J. Keil

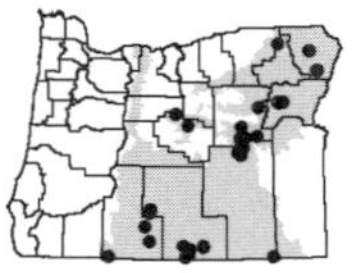

Plants 10–60 cm. **Stems** erect or ascending, usually tomentose, lightly to densely stipitate glandular. **Basal leaves** lanceolate to oblong or elliptic, 3–20 cm × 5–35 mm, margins denticulate to sharply serrate-dentate, surfaces villous to tomentose, stipitate-glandular, sometimes ciliate. **Cauline leaves** bases broadly attenuate to truncate; sessile. **Heads** 3–7 in raceme-like arrays, radiate; peduncles 5–50 mm. **Involucres** 7–13 mm. **Phyllaries** in 3–5 series, narrowly lanceolate to linear, margins entire, surfaces tomentose to scabrous and gland-dotted. **Ray florets** 10–30, conspicuous; corollas 6–12 mm. **Disc florets** 30–60, corollas 5–7 mm. **Fruits** 3–4.5 mm, brown, puberulent; pappus bristles 5–7 mm. **2*n*=**12, 24.

Conifer forests, shrublands, grasslands, roadsides. Flowering Jun–Aug. 400–1900 m. BR, BW, ECas. CA, ID, WA; southeast to UT. Native.

Pyrrocoma howellii (A. Gray) Greene

Marsh goldenweed, thinleaf goldenweed, linear pyrrocoma

Haplopappus uniflorus (Hook.) Torr. & A. Gray ssp. *howellii* (A. Gray) H.M. Hall, *Haplopappus uniflorus* (Hook.) Torr. & A. Gray var. *howellii* (A. Gray) M. Peck, *Haplopappus uniflorus* (Hook.) Torr. & A. Gray ssp. *linearis* D.D. Keck, *Pyrrocoma linearis* (D.D. Keck) Kartesz & Gandhi

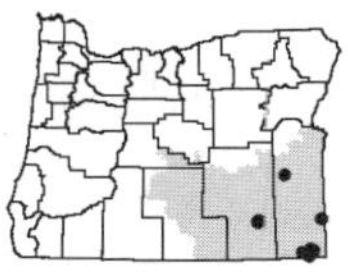

Plants 5–20 cm. **Stems** erect or decumbent-ascending, villous-tomentose. **Basal leaves** linear, 3–12 cm × 2–5 mm, margins entire, surfaces glabrous or villous-tomentose. **Cauline leaves** bases widened; sessile. **Heads** solitary, radiate; peduncles 10–80 mm. **Involucres** 7–11 mm. **Phyllaries** in 3–5 series, narrowly lanceolate to elliptic or linear, margins entire, surfaces tomentose. **Ray florets** 20–50, conspicuous; corollas 5–10 mm. **Disc florets** 30–60; corollas 5–8 mm. **Fruits** 2–4 mm, brown, puberulent; pappus bristles 3–4 mm.

Shrublands, grasslands, marshes, dry river beds. Flowering May–Jul. 1400–2800 m. BR, Owy. ID, NV. Native.

Pyrrocoma lanceolata (Hooker) Greene (p. 332)

Lanceleaf goldenweed

Haplopappus lanceolatus (Hook.) Torr. & A. Gray, *Haplopappus lanceolatus* (Hook.) Torr. & A. Gray ssp. *lanceolatus*, *Haplopappus lanceolatus* (Hook.) Torr. & A. Gray ssp. *tenuicaulis* (D.C. Eaton) H.M. Hall, *Pyrrocoma lanceolata* (Hook.) Greene var. *lanceolata*

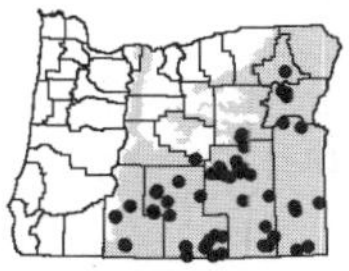

Plants 5–60 cm. **Stems** erect or decumbent-ascending, glabrous to lightly tomentose. **Basal leaves** broadly or narrowly lanceolate to oblong or elliptic, rarely linear, 3–30 cm × 3–25 mm, surfaces glabrous or villous-tomentose, margins entire to denticulate or sharply serrate-dentate. **Cauline leaves** reduced distally; usually sessile. **Heads** 5–20(30) in corymb- or panicle-like arrays, radiate; peduncles 5–70 mm. **Involucres** 4–11 mm. **Phyllaries** usually in 3–5 series, deltate to linear, margins entire, surfaces glabrous or villous. **Ray florets** 10–45, conspicuous; corollas 5–10 mm. **Disc florets** 20–100; corollas 5–10 mm. **Fruits** 3.5–4.5 mm, brown, puberulent; pappus bristles 5–7 mm. **2*n***=12, 24, 36.

Conifer forests, shrublands, alkaline meadows, dry lake beds, sand dunes, disturbed areas. Flowering Jun–Aug. 700–2000 m. BR, BW, ECas, Owy. CA, ID, NV; northeast to Saskatchewan, east to WY, southeast to CO. Native.

Pyrrocoma racemosa (Nuttall) Torrey & A. Gray

Clustered goldenweed

Plants 20–80 cm. **Stems** erect or ascending, glabrous or lightly tomentose. **Basal leaves** broadly to narrowly lanceolate, oblong, or elliptic, 7–25 cm × 5–35 mm, margins entire or denticulate, surfaces glabrous. **Cauline leaves** reduced upwards, sessile or petiolate. **Heads** 3–15 in raceme-like arrays, radiate; peduncles 0–25 mm. **Involucres** 5–15 mm. **Phyllaries** in 3–6 series, margins entire, deltate to linear, surfaces glabrous or lightly tomentose. **Ray florets** 7–35, conspicuous; corollas 5–12 mm. **Disc florets** 20–60; corollas 5–8 mm. **Fruits** 2–5 mm, brown, puberulent; pappus bristles 6–8 mm.

The three varieties of *Pyrrocoma racemosa* in Oregon are geographically separate but can be difficult to tell apart morphologically due to intergradation of various key morphological characteristics.

Key to *Pyrrocoma racemosa* varieties:

1. Involucres 10–15 mm . . . . ***P. r.*** var. ***racemosa***

1' Involucres 6–9 mm.

  2. Phyllaries sparsely tomentose. . . . . . ***P. r.*** var. ***congesta***

  2' Phyllaries glabrous . . . . ***P. r.*** var. ***paniculata***

Pyrrocoma racemosa var. ***congesta*** (Greene) Mayes ex G.K. Brown & D.J. Keil

Serpentine goldenweed

Haplopappus racemosus (Nutt.) Torr. ssp. *congestus* (Greene) H.M. Hall

Plants 20–80 cm. **Inflorescences** heads crowded, in congested, spike-like arrays. **Involucres** 7–9 mm. **Phyllaries** sparsely tomentose.

Open conifer forests, grasslands, dry slopes, creek beds, serpentine. Flowering Jul–Sep. 300–1000 m. Sisk. CA. Native.

Pyrrocoma racemosa var. ***paniculata*** (Nuttall) Kartesz & Gandhi

Panicled goldenweed

Haplopappus racemosus (Nutt.) Torr. ssp. *brachycephalus* (A. Nelson) H.M. Hall, *Haplopappus racemosus* (Nutt.) Torr. ssp. *duriusculus* (Greene) H.M. Hall, *Haplopappus racemosus* (Nutt.) Torr. ssp. *glomeratus* (Nutt.) H.M. Hall, *Haplopappus racemosus* (Nutt.) Torr. var. *glomerellus* A. Gray, *Haplopappus racemosus* (Nutt.) Torr. ssp. *halophilus* (Greene) H.M. Hall, *Haplopappus racemosus* (Nutt.) Torr. var. *paniculatus* (Nutt.) Cronquist

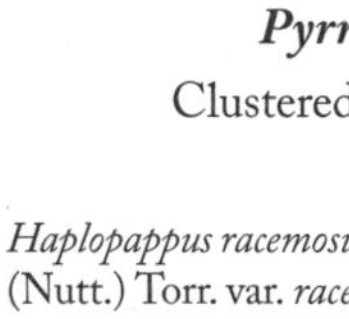

Plants 20–60 cm. **Inflorescences** heads not crowded or crowded distally, in narrowly panicle-, raceme-, or spike-like arrays. **Involucres** 6–8 mm. **Phyllaries** glabrous. **2*n***=12, 24.

Grasslands, dry to moist meadows, roadsides. Flowering Jul–Aug. 800–1600 m. BR, BW, ECas. CA, ID, NV; southeast to UT. Native.

Pyrrocoma racemosa var. ***racemosa***

Clustered goldenweed, racemed goldenweed, racemose pyrrocoma

Haplopappus racemosus (Nutt.) Torr. ssp. *racemosus*, *Haplopappus racemosus* (Nutt.) Torr. var. *racemosus*

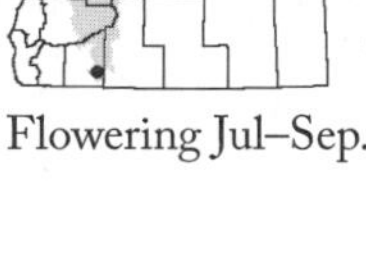

Plants 30–70 cm **Inflorescences** heads not crowded or crowded distally, in raceme-like arrays. **Involucres** 10–15 mm. **Phyllaries** glabrous.

Marshes, grasslands, open conifer forests. Flowering Jul–Sep. 50–300 m. Casc, WV. CA. Native.

Pyrrocoma radiata Nuttall

Snake River goldenweed

Haplopappus carthamoides (Hook.) A. Gray var. *maximus* A. Gray, *Haplopappus radiatus* (Nutt.) Cronquist

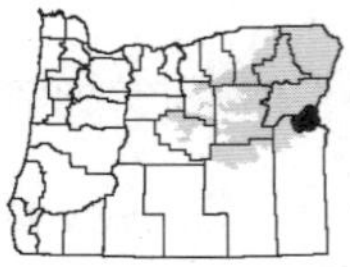

Plants 30–90 cm. **Stems** erect, glabrous. **Basal leaves** broadly to narrowly elliptic, 20–50 cm × 40–80 mm, margins entire, sometimes slightly undulate, surfaces glabrous. **Cauline leaves** margins sometimes serrate-dentate; sessile or petiolate. **Heads** 1–12 in corymb-like arrays, radiate; subtended by leafy bracts; peduncles 5–80 mm. **Involucres** 20–32 mm. **Phyllaries** in 5–8 series, ovate to elliptic or oblong, margins entire, surfaces glabrous. **Ray florets** 15–50, conspicuous; corollas 7–15 mm. **Disc florets** 35–60; corollas 5–12 mm. **Fruits** 6–12 mm, brown, glabrous; pappus bristles 9–12 mm. **2*n***=36.

Grasslands, ridge-tops, dry slopes. Flowering Jun–Aug. 600–1700 m. BW. ID. Native.

This species is listed as Endangered under the Oregon Endangered Species Act.

Pyrrocoma scaberula Greene

Scabrous goldenweed, Palouse goldenweed

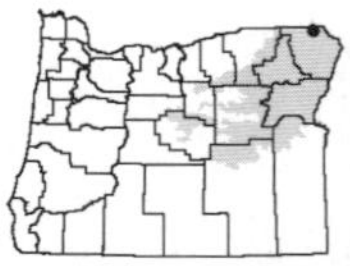

Plants 30–70 cm. **Stems** erect, usually glabrous, sometimes minutely scabrid distally. **Basal leaves** oblanceolate or narrowly elliptic, 7–20 cm × 6–19 mm, margins entire, surfaces glabrous to sparsely villous or tomentose. **Cauline leaves** reduced distally; sessile or petiolate. **Heads** 4–12 in raceme-like arrays, radiate; peduncles 1–20 mm. **Involucres** 6–15 mm. **Phyllaries** in 4–6 series, oblanceolate or oblong, margins entire, sometimes hispid-scabrous, surfaces glabrous, sometimes resinous-dotted. **Ray florets** 12–25, conspicuous; corollas 6–10 mm. **Disc florets** 35–50; corollas 7–11 mm. **Fruits** 4.5–6.5 mm, brown, surfaces pubescent; pappus bristles 6–8 mm.

Canyon grasslands, sandy or gravelly soils. Flowering Jun–Sep. 700–1400 m. BW. ID, WA. Native.

Pyrrocoma scaberula was formerly merged with *P. liatriformis*.

Pyrrocoma uniflora (Hooker) Greene

Plantain goldenweed, one-flowered goldenweed

Haplopappus uniflorus (Hook.) Torr. & A. Gray ssp. *uniflorus*, *Haplopappus uniflorus* (Hook.) Torr. & A. Gray var. *uniflorus*, *Pyrrocoma uniflora* (Hook.) Greene var. *uniflora*

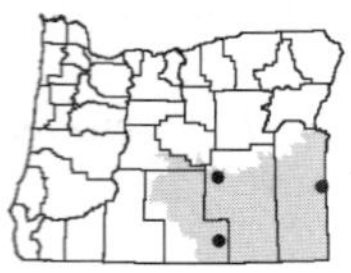

Plants 7–35 cm. **Stems** erect or ascending, lightly to densely villous-tomentose. **Basal leaves** narrowly elliptic to broadly linear, 3–12 cm × 3–15 mm, margins denticulate to sharply dentate, rarely entire, surfaces glabrous or villous-tomentose. **Cauline leaves** linear to narrowly lanceolate, margins entire; sessile. **Heads** 1(2), radiate; peduncles 10–100 mm. **Involucres** 7–12 mm. **Phyllaries** in 3–6 series, linear or narrowly oblanceolate, margins ciliate, surfaces glabrous to tomentose. **Ray florets** 20–50, conspicuous; corollas 7–11 mm. **Disc florets** 30–60; corollas 5–8 mm. **Fruits** 2–4 mm, brown, pubescent; pappus bristles 5–7 mm. **2*n***=12.

Forests, shrublands, grasslands, meadows, ridges. Flowering May–Jul. 1000–1900 m. BR, Owy. CA, ID; north to Northwest Territories, east to WY, southeast to CO. Native.

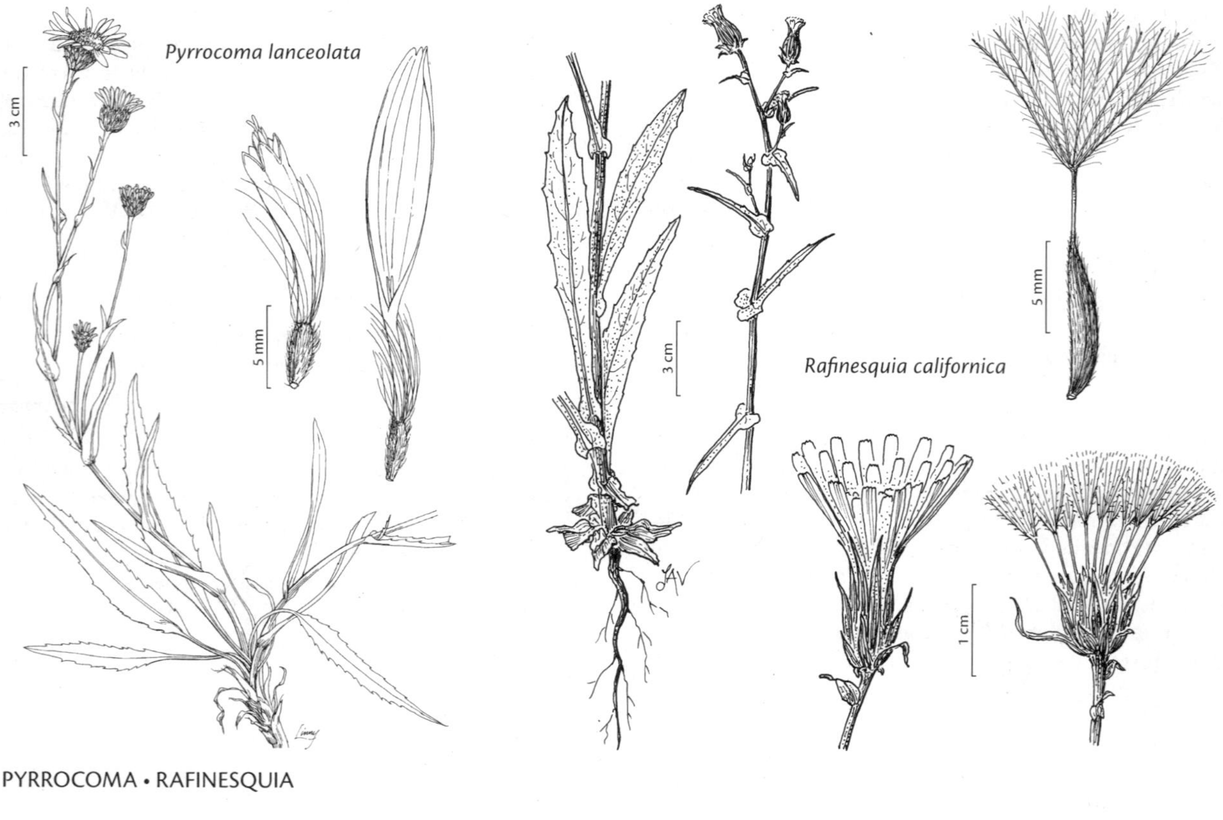

PYRROCOMA • RAFINESQUIA

RAFINESQUIA Rafinesque's chicory, plumeseed

Herbs annual, sap milky; taprooted. **Stems** erect, leafy, ± glaucous, unbranched or branched distally. **Leaves** basal and cauline, elliptic-lanceolate, bases usually enlarged and clasping, margins irregularly toothed or lobed, sometimes a few entire, tips obtuse to acute or acuminate, surfaces glabrous; sessile or short-petiolate. **Inflorescences** in open, panicle-like arrays, simple or branched; bracts often numerous. **Heads** ligulate. **Involucres** campanulate in flower, ovoid in fruit. **Phyllaries** in 1 series, erect in flower, reflexed in fruit, tips acuminate; inner equal, lanceolate; outer lanceolate to linear, < inner. **Receptacles** flat, smooth, glabrous; paleae 0. **Florets** bisexual; ligules exceeding involucres. **Fruits** slender, tapering to bases and tips, glabrous, ribs 5; beaks slender; pappi of 5–20 ± persistent, dingy white, plumose bristles. **2*n***=16.

Western Mexico and United States. 2 species; 1 species treated in *Flora*.

Rafinesquia californica Nuttall (p. 332)

California chicory, California plumeseed, desert plumeseed

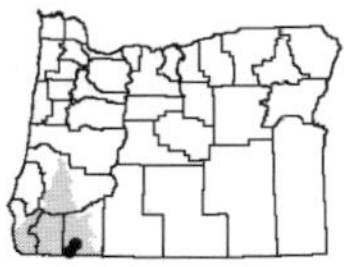

Plants 2–15 dm. **Leaves** 2–20 cm. **Involucres** 10–22 mm. **Phyllaries** glabrous. **Florets** 5–35; ligules 15–22 mm, white, outer sometimes purplish. **Fruits** bodies 4–7 mm, tan to brown, smooth to roughened; beaks 5–7 mm; pappus bristle barbs straight.

Oak woodlands, chaparral, desert scrub. Flowering May–Jul. 500–1400 m. Sisk. CA, NV; southeast to AZ, south to Mexico. Native.

RAILLARDELLA Raillardella

Bruce G. Baldwin and Gerald D. Carr

Herbs perennial, 3–35 cm; caudices branching, rhizomes present or not. **Stems** erect. **Leaves** mostly basal rosettes, some opposite; blades lanceolate or oblanceolate to linear, bases somewhat petioloid, margins usually entire, sometimes denticulate, surfaces sericeous or sparsely scaberulose and/or stipitate-glandular; sessile. **Inflorescences** solitary heads, scapiform; bracts 0, or occasionally 1–2. **Heads** discoid. **Involucres** cylindric to somewhat campanulate, 3–20 mm in diameter. **Phyllaries** 0. **Receptacles** flat or convex, surfaces glabrous or setulose. **Paleae** usually in 1 series, persistent, peripheral and constituting "involucres," connate or distinct, herbaceous, linear-lanceolate, sometimes overlapping, densely glandular-pubescent. **Florets** 7–44, bisexual; corollas yellow to yellow-orange, tubes < funnelform throats, lobes 5, deltate; anthers yellowish; styles glabrous proximal to branches, branch tips densely hairy. **Ray fruits** 0. **Disc fruits** terete, somewhat clavate, 5–10 mm, basal attachments central, tips not beaked, strigose; pappi of 8–30, ± subulate, ciliate to plumose scales.

Western United States. 3 species; 2 species treated in *Flora*.

Raillardella represents an early diverging lineage of tarweeds; putatively ancestral morphological, ecological, and chromosomal characteristics are shared with *Arnica*, the sister group of Madiinae (Baldwin & Wessa 2000; Baldwin 2003b).

Key to *Raillardella* species:

1. Leaves silvery, silky-hairy, sometimes sparsely and obscurely glandular near bases, mostly 10–70 × 2–7 mm; plants 3–15 cm . . . . ***R. argentea***

1' Leaves green, conspicuously stipitate-glandular, sometimes sparsely scaberulose, mostly 20–110 × 4–14 mm; plants 6–35 cm . . . . ***R. scaposa***

Raillardella argentea (A. Gray) A. Gray (p. 334)

Silky raillardella

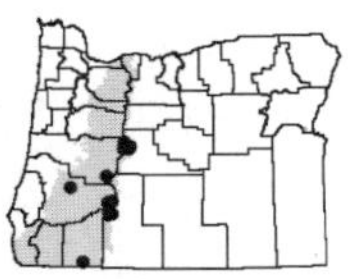

Plants 3–15 cm. **Leaves** mostly 10–70 × 2–7 mm, silvery, surfaces silky-hairy, sometimes sparsely and obscurely glandular near bases. **Florets** 7–26; corollas 6–11 mm, yellow. **2*n***=34, 36.

Dry, open, gravelly sites in conifer forests, semi-barren subalpine and alpine slopes, flats. Flowering Jul–Sep. 1500–2800 m. Casc, Sisk. CA, NV. Native.

Raillardella scaposa (A. Gray) A. Gray

Stem raillardella, greenleaf raillardella

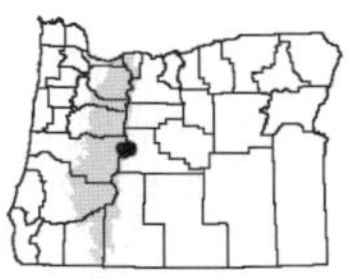

Plants 6–35 cm. **Leaves** mostly 20–110 × 4–14 mm, green, surfaces conspicuously stipitate-glandular, sometimes sparsely scaberulose. **Florets** 7–44; corollas 7.5–12 mm, yellow to yellow-orange. **2*n***=68, 70.

Dry to wet, often sandy sites, meadows, open conifer forest, semi-barren subalpine and alpine slopes and flats. Flowering Jul–Sep. 1500–2200 m. Casc. CA, NV. Native.

In Oregon, *R. scaposa* is very rare, known from a few closely spaced sites within the general range of *R. argentea*. In spite of overlapping ranges, the two species are not known to occur together, and each retains a relatively uniform set of distinguishing features. However, putative hybrids between *R. argentea* and *R. scaposa* have been noted in California, where populations of the two species are sometimes closely juxtaposed.

RAINIERA Rainiera

Genus composed of one species, *R. stricta.*

Rainiera stricta (Greene) Greene (p. 334)

Rainiera, tongue-leaved luina

Luina stricta (Greene) B.L. Rob.

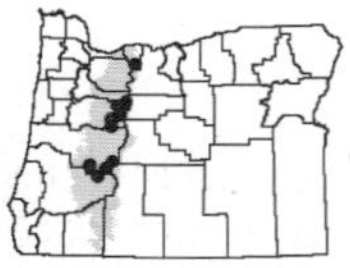

Herbs perennial, 3–10 dm; with thick rhizomes or root-crowns. **Stems** 1, erect, stout, leafy, unbranched, glabrous. **Leaves** basal and cauline, alternate, oblanceolate to elliptic, 6–20 × 2–9 cm, reduced distally, margins entire or toothed, tips acute or acuminate, surfaces glabrous or glabrate; petioles 1–12 cm, sessile distally. **Inflorescences** terminal, raceme-like arrays; bractless or with 1 bract at bases. **Heads** 20–80+, discoid. **Involucres** cylindric, 5–9 mm. **Phyllaries** 3–8, in 1 equal series, erect, linear-lanceolate, strongly ridged along midveins, surfaces densely tomentose. **Receptacles** flat; paleae 0. **Florets** 4–8, bisexual; corollas yellow, sometimes purple-tinged, exceeding the involucres; stamens exserted. **Fruits** cylindric-fusiform, 4–6 mm, pale brown, 10–20-veined with 3–4 prominent veins; pappi of numerous, minutely barbed bristles in 3–4 series.

Subalpine conifer forests, open slopes, meadows. Flowering Jun–Sep. 700–2200 m. Casc. WA. Native.

RIGIOPAPPUS Bristlehead

Genus composed of one species, *R. leptocladus.*

Rigiopappus leptocladus A. Gray (p. 334)

Bristlehead

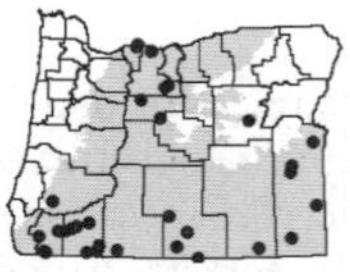

Herbs annual, 2–45 cm; taprooted. **Stems** erect, simple to much branched, glabrous to puberulent. **Leaves** principally cauline, ± linear, 5–30+ × 1–2(5) mm, margins entire, tips acute, surfaces puberulent. **Inflorescences** solitary, terminal heads. **Heads** radiate, erect; sessile. **Involucres** 4–7 mm. **Phyllaries** 5–20+, in 2–3 series, subequal, linear to lanceolate. **Receptacles** ± flat, obscurely pitted; paleate. **Paleae** only between ray and disc florets. **Ray florets** 3–13+, pistillate; corollas 1–3 mm, yellow or purplish. **Disc florets** 5–70+, bisexual; corollas narrowly tubular, 1.5–2.5 mm, yellow, lobes 3–4, erect; stamens included; styles slightly exserted. **Fruits** cylindric, narrowed at bases and tips, 4–5.5 mm, brown, transversely wrinkled, strigillose, obscurely few-ribbed; pappi of 3–5 narrow, tapering, awn-like scales, sometimes reduced or 0 on ray florets, 1–6 mm, brownish.

Open woodlands, shrublands, grasslands, dry flats, rocky slopes, ridges, serpentine. Flowering May–Jul. 100–1800 m. BR, Casc, Col, ECas, Lava, Owy, Sisk. CA, ID, NV, WA. Native.

The bare, wiry inflorescence branches are a recognizable trait of *Rigiopappus*. As in many other dryland annuals, plant size can be much reduced under poor growing conditions. Depauperate individuals often have some branches with small, nonfunctional flowering heads.

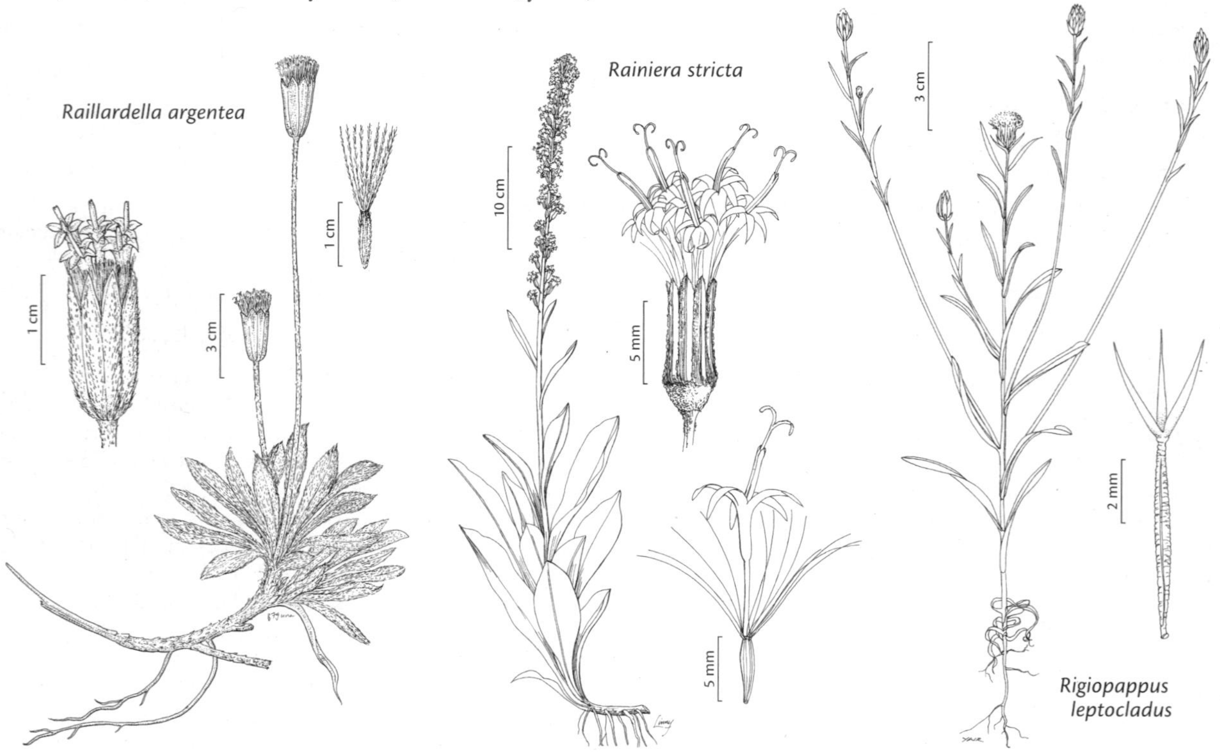

RAILLARDELLA • RAINIERA • RIGIOPAPPUS

RUDBECKIA Coneflower

Herbs perennial; fibrous-rooted from root-crowns or stout rhizomes. **Stems** usually single, erect, unbranched or few-branched distally, usually glabrous, sometimes pubescent distally. **Leaves** basal and cauline, alternate, ovate, lanceolate, or oblong-elliptic, bases attenuate to truncate or cordate, margins entire to coarsely dentate or regularly serrate-dentate, simple to deeply lobed or compound, tips acute or acuminate, surfaces glabrous to puberulent, strigose, villous, or scabridulous; proximal petiolate, distal sessile. **Inflorescences** heads solitary or in corymb- or panicle-like arrays. **Heads** radiate or discoid. **Involucres** hemispheric to rotate. **Phyllaries** in 2(3) series, herbaceous, oblong-lanceolate to broadly or narrowly linear, margins smooth or scabridulous, tips acute or obtuse, surfaces glabrous to tomentose or silky-puberulent. **Receptacles** conic to columnar; paleate. **Paleae** folded around fruits, margins ciliate. **Ray florets** 0–15, neuter; rays elliptic to oblong, yellow. **Disc florets** 300–500, bisexual; corollas tubular, lobes 5, erect, triangular; stamens included or slightly exserted; stigmas slightly exserted. **Fruits** ± columnar, 4-angled, truncate; pappi 0 or short, scaly crowns.

Europe, North America. 23 species; 3 species treated in *Flora*.

Rudbeckia hirta var. *pulcherrima* (black-eyed Susan), a popular cultivar, occasionally escapes but has not yet become established in Oregon.

Key to *Rudbeckia* species:

1. Ray florets present; leaves glaucous. . . . . ***R. glaucescens***

1' Ray florets 0; leaves not glaucous.

  2. Some or many leaves pinnatifid to pinnate . . . . ***R. alpicola***

  2' Leaves not pinnatifid to pinnate, rarely some shallowly lobed. . . . . ***R. occidentalis***

Rudbeckia alpicola Piper

Wenatchee Mountains coneflower

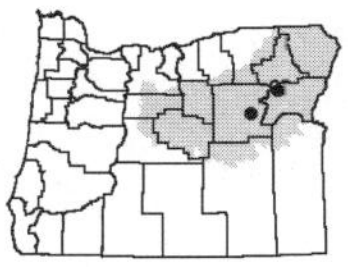

Plants 5–15 dm. **Stems** glabrous nearly throughout, becoming tomentulose beneath heads. **Basal leaves** ovate to elliptic or deltate, 25–70 × 6–40 cm, bases rounded to acute, margins entire to serrate, some or many leaves pinnatifid to pinnate, tips acute. **Cauline leaves** similar to basal, 10–50 × 5–35 cm. **Heads** discoid, solitary or 2–10 in corymb-like arrays. **Phyllaries** 5–60 mm, surfaces puberulent to scabrous or nearly glabrous. **Receptacles** columnar 30–60 × 15–25 mm. **Ray florets** 0. **Disc florets** 300–500; corollas 4–6 mm, brownish purple. **Fruits** 3.5–5 mm, black; pappi low, 2–4-toothed crowns.

Montane forests, shrublands, meadows, riverbanks. Flowering Jul–Aug. 1000–2100 m. BW, Lava. CA, WA. Native.

This species was cited by Cronquist (1994) as *Rudbeckia occidentalis* var. *montana* from the Elkhorn Range, Baker County, widely disjunct from the known range of that taxon in Colorado and Utah. These Oregon collections, and others from Steens Mountain, are here referred to *R. alpicola*, which according to Urbatsch and Cox (2006) is limited to the Wenatchee Mountains, Washington. However, *R. alpicola* had previously been reported from the Cascade Range in California as well (Ferris 1960). Additional collections are needed to clarify its distribution in Oregon.

Rudbeckia glaucescens Eastwood

Waxy coneflower

Rudbeckia californica A. Gray, *Rudbeckia californica* A. Gray var. *glauca* S.F. Blake

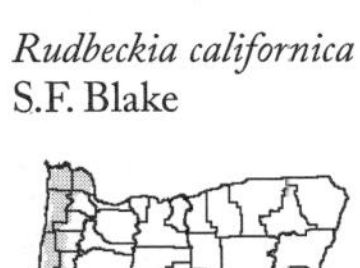

Plants 5–15 dm. **Stems** glabrous nearly throughout, sometimes pilose or scabrous beneath heads. **Basal leaves** lanceolate to elliptic, 20–50 × 4–10 cm, bases attenuate, margins entire to sparsely serrate, tips acute, surfaces glaucous. **Cauline leaves** similar to basal, 10–25 × 2–8 cm. **Heads** radiate, solitary or 2–10 in corymb-like arrays. **Phyllaries** 5–20 mm, surfaces glabrous. **Receptacles** conic or columnar, 25–40 × 15–20 mm. **Ray florets** 7–15; rays 25–40 mm. **Disc florets** 250–400; corollas 3–4 mm, yellowish green. **Fruits** 3.5–5 mm, black; pappi partly fused, scaly crowns. **2*n***=36.

Seeps, bogs, marshes, serpentine. Flowering Jun–Sep. 100–1400 m. CR, Sisk. CA. Native.

Rudbeckia occidentalis Nuttall (p. 336)

Western coneflower

Rudbeckia occidentalis Nutt. var. *occidentalis*

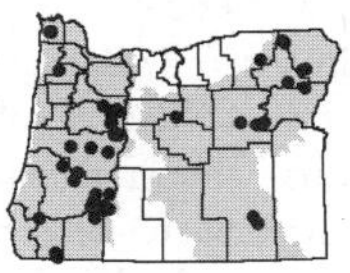

Plants 5–20 dm. **Stems** glabrous or puberulent nearly throughout, becoming tomentulose beneath heads. **Basal leaves** lanceolate to broadly ovate, 10–30 × 3–10 cm, bases attenuate to rounded, margins entire to serrate, rarely some shallowly lobed, tips acute. **Cauline leaves** similar to basal, 5–20 × 2–9 cm. **Heads** discoid, solitary or 2–5 in corymb-like arrays. **Phyllaries** 5–35 mm, surfaces glabrous to strigose or tomentulose. **Receptacles** columnar, 25–60 × 12–25 mm. **Ray florets** 0. **Disc florets** 200–500; corollas 4–6 mm, greenish yellow proximally, dark maroon distally. **Fruits** 3.5–5 mm, black; pappi fused, scaly crowns. **2*n***=36.

Forest openings, shrublands, meadows, marshes, bogs, streambanks, clearcuts, roadsides. Flowering Jun–Sep. 300–2800 m. BR, BW, Casc, CR, Lava, Sisk, WV. CA, ID, NV, WA; northeast to MT, east to WY, southeast to UT. Native.

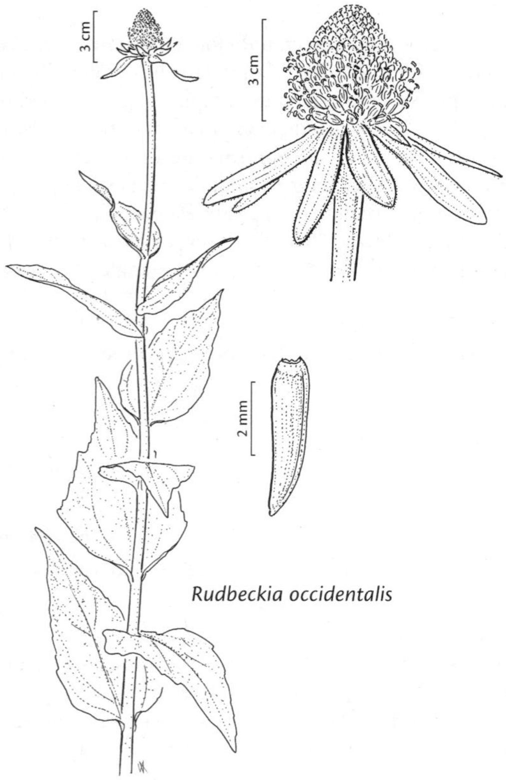

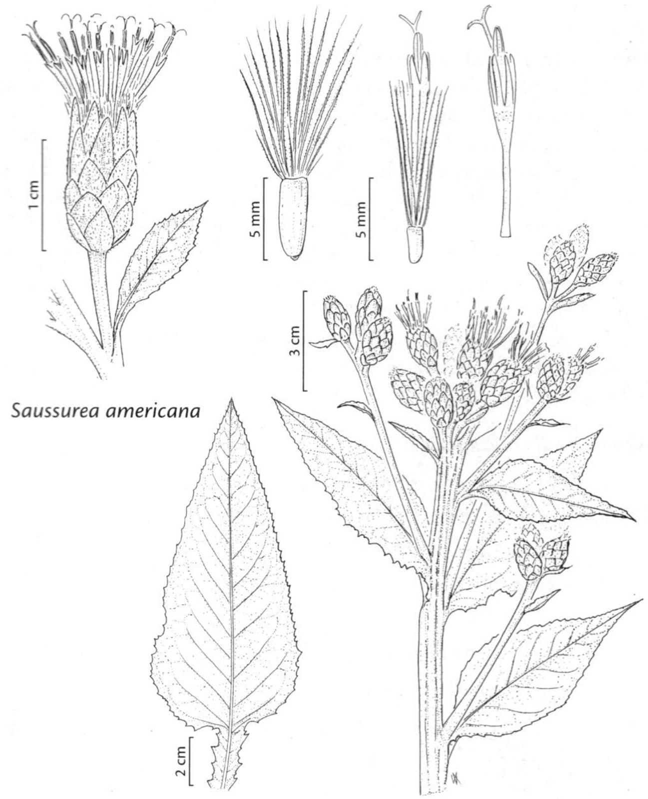

RUDBECKIA • SAUSSUREA

SAUSSUREA Saw-wort

Herbs perennial; fibrous-rooted from caudices. **Stems** 1–several, erect, leafy, glabrous or sparingly tomentose, unbranched below inflorescences. **Leaves** basal and cauline; basal withering before flowering; cauline alternate, broadly to narrowly lanceolate, gradually reduced distally, bases cordate to truncate or attenuate, margins denticulate to sharply dentate, tips acute to acuminate, surfaces glabrous; sessile or petiolate. **Inflorescences** in corymb-like arrays. **Heads** discoid; peduncles short, glabrous, sometimes minutely glandular. **Involucres** ovoid to campanulate or turbinate. **Phyllaries** erect in flower, spreading in fruit; inner linear-lanceolate, tips tomentose; outer lanceolate to ovate, gradually shorter, usually with reddish margins and tips, surfaces glabrous to tomentose. **Receptacles** flat or convex; paleae 0. **Florets** bisexual; corollas tubular, expanded distally, lobes 5, erect to recurved, linear; stamens well exserted; stigma lobes exserted. **Fruits** oblong-cylindrical, truncate, abruptly tapering toward bases, obscurely veined; pappi of usually plumose bristles in 2 series; outer short, deciduous; inner longer, persistent, connate basally.

Australia, Asia, Europe, North America. ~400 species; 1 species treated in *Flora*.

Saussurea americana D.C. Eaton (p. 336)

American saw-wort

Plants 3–12+ dm. **Leaves** 4–20 cm; proximal and mid cauline petioles to 6 cm, winged, sometimes decurrent. **Heads** 5–30+; peduncles 0–4 cm. **Involucres** 10–15 mm. **Phyllaries** in 5 unequal series. **Florets** 8–20+; corollas 11–13 mm, purple, rarely white, exserted from involucres. **Fruits** 4–6 mm, surfaces glabrous; outer pappus bristles sometimes not plumose.

Montane forests, meadows, streambanks, springs. Flowering Jul–Aug. 1300–2400 m. BW, Sisk. CA, ID, WA; north to AK, northeast to Alberta. Native.

SENECIO Groundsel, ragwort, butterweed

Debra K. Trock

Herbs, **subshrubs**, **shrubs**, **trees**, and **vines** annual, biennial, or perennial, 3–100(250+) cm; caudices taprooted, fibrous-rooted, often fleshy, or with branched or unbranched, suberect to creeping rhizomes. **Stems** 1+, erect to lax, glabrous or pubescent, simple or branched. **Leaves** basal and/or cauline, alternate, ovate, deltate, oblanceolate, lanceolate, linear, or filiform, rarely orbiculate, sometimes palmately or pinnately lobed, ultimate margins entire to serrate or toothed, some with callous denticles on the teeth, palmately to pinnately veined, surfaces glabrous or hairy, often glabrate with age; petiolate or sessile, often clasping. **Inflorescences** corymb-, cyme-, panicle-, raceme-like, or flat-topped arrays, rarely heads solitary, usually terminal, sometimes from axils of distal leaves. **Heads** radiate or discoid, sometimes nodding. **Calyculi** 0 or 1–8+ bractlets, mostly ¼–½ the length of phyllaries. **Involucres** mostly cylindric or turbinate, sometimes campanulate. **Phyllaries** 5, 8, 13, or 21 in 1(2) equal or subequal series, persistent. **Receptacles** flat to convex; paleae 0. **Ray florets** (0)5, 8, 13, or 21, pistillate, fertile; rays yellow, rarely yellowish white. **Disc florets** (5)13–100+, bisexual, fertile; corollas yellow, rarely yellowish white, reddish, or purplish, tubes ≤ campanulate throats, lobes 5, erect to recurved, usually ± deltate; style branches with tips usually truncate-penicillate. **Fruits** cylindric or prismatic, tan to brown, glabrous or pubescent, usually 5-ribbed; pappi of white to straw-colored, barbellate to smooth bristles.

Nearly worldwide. 1000+ species; 13 species treated in *Flora*.

Senecio is one of the largest genera of flowering plants and includes nearly all growth forms from herbs to shrubs, trees, succulents, aquatics, and vines. Many species produce alkaloids to deter predators, but there are several species of lepidopterans whose larval forms use *Senecio* species as host plants. There are many species that are common annual or perennial weeds, but there are also several succulents that are frost intolerant.

Researchers working in this large genus have, for many years, recognized informal "species groups" but were resistant to splitting up the genus. More recently with more thorough chromosome and palynological work, along with increasingly refined molecular work, segregation of many of these informal groups into new genera is easily supported.

Senecio viscosus has been collected on ballast in Portland and putatively in Lake County in the early part of the 20th century.

Key to *Senecio* species:

1. Plants annuals, with taproots.
    2. Stems densely pubescent . . . . **S. sylvaticus**
    2' Stems glabrous to glabrate to sparsely and/or unevenly tomentose.
        3. Phyllaries usually 13, tips green; ray florets 8 or rarely 13 . . . . ***S. ertterae***
        3' Phyllaries 21, tips usually black; ray florets 0. . . . . ***S. vulgaris***

1' Plants perennial or biennial, with simple or branched rhizomatous, sub-rhizomatous, or woody caudices or with button-like caudices with thick fibrous roots.
    4. Caudices button-like with fleshy, fibrous, unbranched roots.
        5. Stems arachnoid-tomentose or villous, becoming glabrate.
            6. Basal leaves ovate to oblanceolate, margins denticulate or sharply toothed to subentire; ray florets 0, rarely 1–2 . . . . ***S. aronicoides***
            6' Basal leaves elliptic, lanceolate, linear, oblanceolate, rounded-deltate, or suborbiculate, rarely suborbicular, margins entire or dentate; ray florets usually 5, rarely 0. . . . . ***S. integerrimus***
        5' Stems glaucous, glabrous, or glabrate, young plants may be sparsely hairy.
            7. Leaves glaucous, thick, fleshy-turgid; margins entire to denticulate; habitat swamps and alkaline sites . . . . ***S. hydrophilus***
            7' Leaves glabrous, not glaucous, firm; margins dentate to denticulate; habitat wet hillsides and meadows, not alkaline sites . . . . ***S. hydrophiloides***
    4' Caudices woody or ± woody, branching, rhizomatous or sub-rhizomatous.
        8. Leaves evenly distributed along stem with basal leaves usually withering before flowering; heads 10–90+, in compound corymb-like or flat-topped to panicle-like arrays.
            9. Stems sparsely and unevenly tomentose; leaves ovate to broadly ovate, usually 1–3-pinnate; phyllaries 13, tips black or dark green; ray florets ~13 . . . . ***S. jacobea***
            9' Stems glabrous or slightly tomentose; leaves narrowly lanceolate, lanceolate, or sublinear, not pinnate; phyllaries ~8, tips green, rarely black; ray florets ~5 . . . . ***S. serra*** var. ***serra***
        8' Leaves mostly basal or evenly distributed along stems, basal leaves present at flowering; heads (1)3–30(60), in simple corymb- or ± raceme-like arrays.
            10. Plants 1–5(7) dm; leaves not reduced distally, thick and fleshy, glabrous; heads usually ≤12.
                11. Phyllaries with black tips; leaf margins dentate with callous denticles to subentire . . . . ***S. crassulus***
                11' Phyllaries with green tips; leaf margins laciniate, dentate, or subentire . . . . ***S. fremontii*** var. ***fremontii***
            10' Plants (2)3–12(20) dm; leaves reduced distally, not fleshy, tomentose to nearly glabrous; heads usually > 10.
                12. Leaves narrowly to widely triangular, margins dentate, rarely subentire; phyllaries (8)13(21), tips green, rarely black; ray florets ~8 . . . . ***S. triangularis***
                12' Leaves oblanceolate to elliptic, margins weakly dentate or denticulate with callous denticles; phyllaries (13)21, tips black or dark brown; ray florets ~13 . . . . ***S. sphaerocephalus***

Senecio aronicoides de Candolle (p. 338)

Rayless groundsel, rayless ragwort

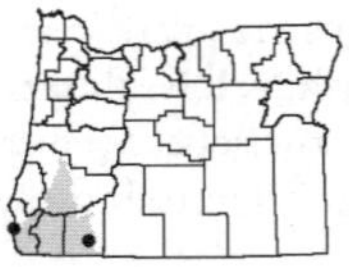

Herbs perennial, 3–9 dm; caudices button-like, with fleshy, fibrous, unbranched roots. **Stems** 1, sparsely and loosely to finely arachnoid-tomentose, becoming unevenly glabrate. **Leaves** basal and proximal cauline ovate to oblanceolate, bases tapering, margins denticulate or sharply toothed to subentire; middle and distal cauline smaller, sessile and weakly clasping. **Inflorescences** corymb-like arrays; bracts 2–3 mm, conspicuous. **Heads** (6)15–30+. **Calyculi** 0 or 1–3+ oblong to linear bractlets. **Involucres** campanulate. **Phyllaries** 8 or 13, 4–8 mm, green, often with black tips, surfaces glabrous to sparsely and irregularly tomentose. **Ray florets** 0(1–2); rays 4–7 mm. **Disc florets** 40–75+. **Fruits** 1.5–2.5 mm, glabrous. **2*n*=40.**

Open woodlands, foothills, montane forests. Flowering May–Jul. 900–1800 m. Sisk. CA. Native.

Senecio crassulus A. Gray

Mountain meadow butterweed, thick-leaved groundsel

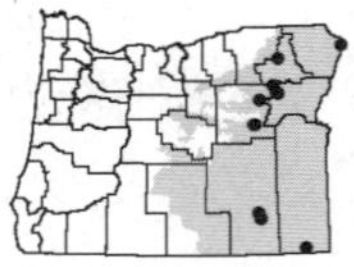

Herbs perennial, (1.5)2–5(7) dm; caudices ± woody, branching, rhizomatous. **Stems** 1–4, glabrous. **Leaves** basal present at flowering; cauline evenly distributed along stems, thick and fleshy, lanceolate to subelliptic, bases tapering, margins dentate with callous denticles to subentire, surfaces glabrous, sometimes with broadly winged petioles; proximal often early deciduous; mid-cauline sometimes larger, sessile and weakly clasping; distal smaller, sessile and clasping. **Inflorescences** corymb-like arrays; bracts 2–4 mm, conspicuous. **Heads** (1)4–12. **Calyculi** (1)3–6 linear to filiform bractlets, (1)2–6 mm, with or without black tips. **Involucres** broadly campanulate. **Phyllaries** (8)13 or 21, 5–9 mm, with black tips, surfaces densely tomentose. **Ray florets** 8 or 13; rays 5–12 mm. **Disc florets** 55–65; deep golden to orange-yellow. **Fruits** 2–2.5 mm, glabrous. **2*n*=40.**

Wet to drying hillsides and meadows, open forests. Flowering May–Aug. 800–3000 m. BR, BW, Owy. ID, NV; northeast to MT, east to SD, southeast to NM. Native.

Senecio ertterae T.M. Barkley

Ertter's groundsel

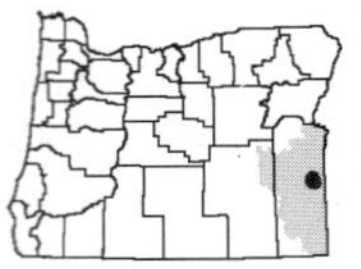

Herbs annual, 2–5(6+) dm; taproots twisting, thin. **Stems** usually 1, rarely clustered, subsucculent, unevenly tomentose to glabrate. **Leaves** basal withering early; cauline oblanceolate to spatulate, bases tapering, margins deeply incised; petioles winged; distal sessile and bract-like, progressively reduced. **Inflorescences** cyme-like arrays; bracts 0 or few and inconspicuous. **Heads** 6–10(20). **Calyculi** 1–5+ linear bractlets, to 2 mm. **Involucres** widely cylindric. **Phyllaries** usually 13, 6–7 mm, tips green, surfaces densely floccose-tomentose. **Ray florets** 8, rarely 13; rays 5–6 mm. **Disc florets** 45–55. **Fruits** 2–3 mm, usually hairy along the ribs, sometimes glabrous. **2*n*=40.**

Talus slopes, volcanic tuff. Flowering May–Sep. 800–1400 m. Owy. Native. ✿

This species was not recognized until the late 1970s. It is a very rare endemic found only in Malheur County. It has been added to the Center for Plant Conservation's National Collection of Endangered Plants.

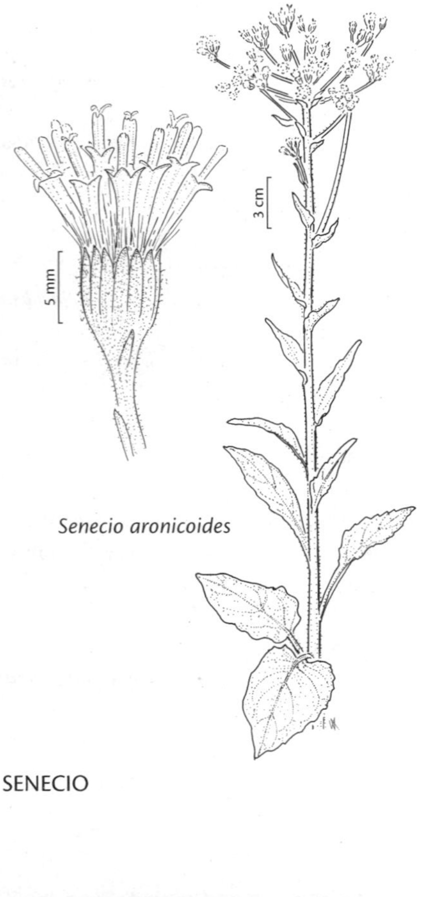

Senecio aronicoides

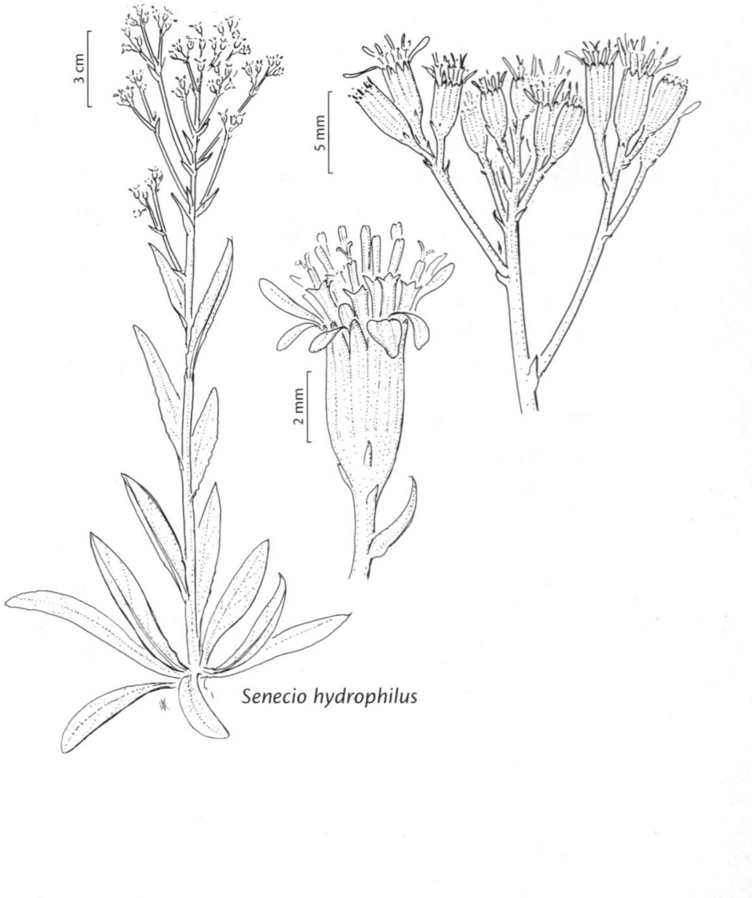

Senecio hydrophilus

Senecio fremontii Torrey & A. Gray var. *fremontii*

Dwarf mountain ragwort

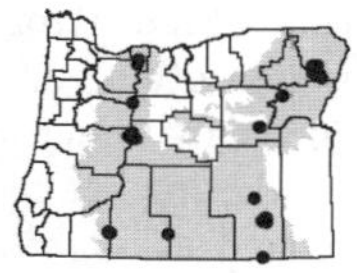

Herbs perennial, 1–2 dm; caudices spreading, sometimes woody, branching, sub-rhizomatous. **Stems** 1, or 2–4+ and clustered, spreading at bases, arching upward, often purple-tinged, glabrous. **Leaves** basal present at flowering, cauline evenly distributed along stems, ovate or obovate to oblanceolate, thick and fleshy, bases truncate or tapering, margins laciniate, dentate, or subentire, surfaces glabrous; petiolate. **Inflorescences** corymb-like arrays; bracts conspicuous. **Heads** (1)3–5. **Calyculi** 0 or 1–5 usually linear bractlets, sometimes foliaceous. **Involucres** campanulate. **Phyllaries** (8)13, 7–10 mm, tips green, surfaces glabrous. **Ray florets** 8; rays 8–12 mm. **Disc florets** 40–50+. **Fruits** 2.5–3 mm, glabrous or with short stiff hairs on ribs. **2*n***=40, 40+, 80.

Rocky subalpine and alpine areas. Flowering Jul–Sep. 1600–2900 m. BR, BW, Casc, ECas. CA, ID, WA; north to British Columbia, northeast to Alberta, east to WY, southeast to UT. Native.

Senecio hydrophiloides Rydberg

Stout meadow groundsel, sweet marsh butterweed

Senecio foetidus Howell var. *foetidus*, *Senecio foetidus* Howell var. *hydrophiloides* (Rydb.) T.M. Barkley, *Senecio oreganus* Howell

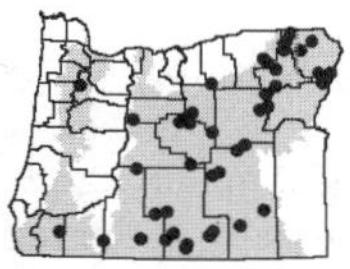

Herbs perennial or biennial, 3–10(14) dm; caudices erect, button-like, with fleshy, fibrous, unbranched roots. **Stems** 1, sometimes 2–4 and clustered, sometimes reddish tinged, glabrous or glabrate, young plants sometimes sparsely hairy. **Leaves** mostly basal, cauline reduced distally, elliptic to broad lanceolate, firm, bases tapering, margins dentate to denticulate, surfaces glabrous; petioles often distinctly winged, distal sessile and bract-like. **Inflorescences** loose or congested corymb-like arrays; bracts 0 or inconspicuous. **Heads** (6)15–30+. **Calyculi** 2–5 bractlets, inconspicuous, < 2 mm. **Involucres** widely cylindric to campanulate. **Phyllaries** 8, 13 or 21, 4–9 mm, tips black. **Ray florets** 0 or 3–8; rays 5–10 mm. **Disc florets** 30–45. **Fruits** 2–3 mm, glabrous. **2*n***=40.

Damp hillsides, meadows, seeps. Flowering Jun–Jul. 1000–1900 m. BR, BW, ECas, Lava, Sisk, WV. CA, ID, NV, WA; north to British Columbia, northeast to Alberta, east to WY, southeast to UT. Native.

Plants in the western part of this species' range tend to have single stems, more congested inflorescences, and more ray florets and have been treated by some authors as a distinct species (*S. foetidus*). Those further east have more stems, a looser inflorescence, and fewer ray florets. T.M. Barkley (1978) noted that these two forms intergrade so completely that they should be treated as a single species.

Senecio hydrophilus Nuttall (p. 338)

Alkali marsh butterweed, water groundsel

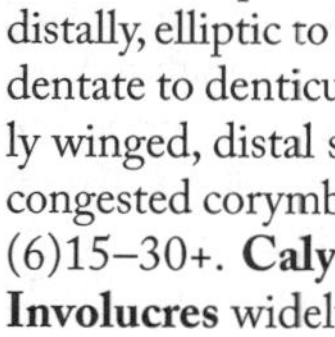

Herbs perennial or biennial, 4–10(20) dm; caudices short and button-like, with fleshy, fibrous, unbranched roots. **Stems** 1, or 2–4 and loosely clustered, often bluish green, usually glaucous and glabrous. **Leaves** progressively reduced distally, thick, fleshy-turgid; basal and lower cauline elliptic to oblanceolate, bases tapered, margins entire to denticulate, ± petiolate; middle and upper cauline bract-like, surfaces glaucous, sessile. **Inflorescences** compound corymb-like arrays; bracts 0. **Heads** 20–40(80). **Calyculi** 2–4+ bractlets, inconspicuous, usually < 2 mm. **Phyllaries** 8 or 13, 5–8 mm, tips usually black. **Involucres** cylindric or narrowly urn-shaped. **Ray florets** 5, sometimes 0; rays 3–5 mm. **Disc florets** 20–35. **Fruits** 2.5–3 mm, glabrous. **2*n***=40.

Standing water in marshes, swampy places, alkaline sites. Flowering Jun–Aug. 0–2100 m. BR, BW, Casc, CR, ECas, Lava, WV. CA, ID, NV, WA; north to British Columbia, northeast to MT, east to SD, southeast to CO. Native.

Senecio hydrophilus and *S. hydrophiloides* can be very difficult to distinguish on herbarium specimens. There are, however, significant ecological differences, with the former species restricted to alkaline soils and the latter found in damp but not alkaline places.

Senecio integerrimus Nuttall

Western groundsel, lambstongue ragwort

Herbs perennial or biennial, (1)2–7 dm; caudices button-like, with fleshy, fibrous, unbranched roots. **Stems** 1, loosely arachnoid-tomentose, or villous, glabrate with age. **Leaves** mostly basal, cauline progressively reduced distally, ± petiolate; basal and lower cauline elliptic, lanceolate, linear, oblanceolate, rounded-deltate, rarely suborbiculate, bases tapered or truncate to cordate, margins entire or dentate; distal becoming sessile and bract-like. **Inflorescences** corymb-like arrays; bracts 0 or inconspicuous. **Heads** 6–15(30+). **Calyculi** 1–5+, inconspicuous, usually < 2 mm. **Involucres** urn-shaped. **Phyllaries** (8)13 or 21, tips usually black, sometimes green. **Ray florets** 5, rarely 0; rays 6–15(20) mm, yellow or white to pale yellow. **Disc florets** 35–45. **Fruits** 2.5–3 mm, usually glabrous, sometimes hairy along ribs. **2*n***=40, 80.

Western North America. 5 varieties; 3 varieties treated in *Flora*.

Senecio integerrimus var. *exaltatus*

SENECIO

Key to *Senecio integerrimus* varieties:

1. Leaf blades ovate to rounded-deltate, sometimes obovate to oblanceolate; ray florets white or pale yellow . . . . . . ***S. i.*** var. ***ochroleucus***

1' Leaf blades elliptic to lanceolate or oblanceolate, rarely rounded-deltate or suborbiculate; ray florets yellow.

   2. Phyllaries linear to awl-shaped, 8–12 mm, tips green, rarely minutely black; rays 5–8 mm. . . . . ***S. i.*** var. ***major***

   2' Phyllaries lanceolate, (4)5–10 mm, tips black; rays 6–15 mm . . . . ***S. i.*** var. ***exaltatus***

Senecio integerrimus var. ***exaltatus*** (Nuttall) Cronquist (p. 339)

Tall western groundsel, lambstongue groundsel, Columbia ragwort

Senecio integerrimus Nutt. var. *vaseyi* (Greenm.) Cronquist, *Senecio vaseyi* Greenm.

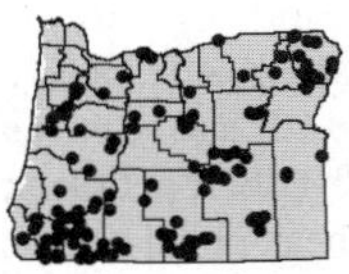

Leaves elliptic to lanceolate or oblanceolate, rarely rounded-deltate or suborbiculate; basal and proximal indistinctly petiolate. **Heads** 6–15(30+). **Phyllaries** lanceolate, (4)5–10 mm, tips black. **Ray florets** 5, rarely 0; rays 6–15 mm, yellow.

Open woodlands, sagebrush, meadows, grasslands. Flowering Apr–Jul. 50–2700 m. All ecoregions except Est. CA, ID, NV, WA; north to British Columbia, northeast to Saskatchewan, east to WY, southeast to CO. Native.

This is the most variable and widespread variety of the species. Eradiate plants have been recognized by some authors as variety *vaseyi*. However, both radiate and eradiate plants can be found within the same populations.

Senecio integerrimus var. ***major*** (A. Gray) Cronquist

Bigheaded groundsel

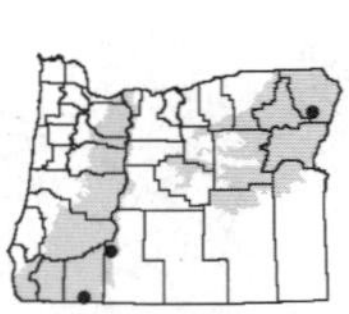

Leaves usually lanceolate to oblanceolate; basal and proximal cauline distinctly petiolate. **Heads** 6–12(30+). **Phyllaries** linear to awl-shaped, 8–12 mm, tips green, rarely minutely black. **Ray florets** usually 5; rays 5–8 mm, yellow.

Open areas in coniferous forests, sagebrush. Flowering Apr–Jul. 2000–2600 m. BW, Casc, Sisk. CA, NV. Native.

Senecio integerrimus var. ***ochroleucus*** (A. Gray) Cronquist

White western groundsel

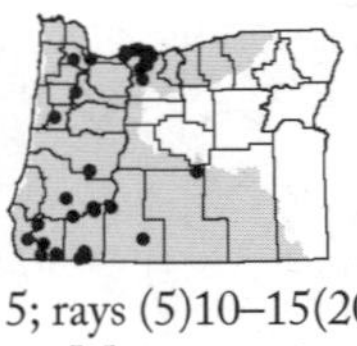

Leaves ovate to rounded deltate, sometimes obovate to oblanceolate; basal and proximal cauline distinctly petiolate. **Heads** (3)8–12(40+). **Phyllaries** linear, 7–10 mm, tips green or minutely black. **Ray florets** usually 5; rays (5)10–15(20) mm, white to pale yellow.

Moist openings in coniferous forests. Flowering Apr–Jul. 0–2000 m. BR, Casc, Col, CR, ECas, Sisk, WV. CA, ID, WA; north to British Columbia, northeast to Alberta. Native.

Senecio jacobea Linnaeus (p. 341)

Tansy ragwort, stinking willie

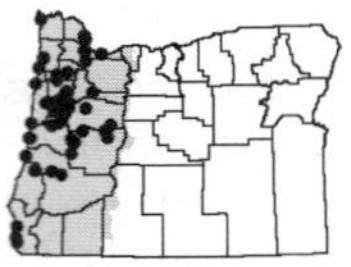

Herbs perennial, 2–8(10) dm; caudices ± woody, branching, rhizomatous. **Stems** 1, or rarely 2–4 and loosely clustered, often purple-tinged, sparsely and unevenly tomentose. **Leaves** basal usually withering before flowering; cauline ± evenly distributed along stems, distal slightly smaller, ovate to broadly ovate, bases tapered, usually 1–3-pinnate, ultimate margins dentate, lobes obovate to spatulate; petiolate. **Inflorescences** compound corymb-like arrays; bracts conspicuous but small. **Heads** (10)20–60+. **Calyculi** 2–6 bractlets, inconspicuous, usually < 2 mm. **Involucres** widely cylindric or urn-shaped. **Phyllaries** 13, 3–4(5), tips black or dark green. **Ray florets** ~13; rays 8–12 mm. **Disc florets** 60–70+. **Fruits** 0.75–1.3 mm, sparsely hairy or glabrous. $2n$=40.

Disturbed areas, pastures, roadsides. Flowering May–Sep. 0–1600 m. Casc, CR, Est, Sisk, WV. CA, ID, WA; north to British Columbia, northeast to MT, northeastern North America; Europe. Exotic.

Senecio jacobea is an introduced weed, originally from Europe. It establishes in places with cool, wet summers and is particularly toxic to livestock. In most states where it occurs, this species has been declared a noxious weed.

Senecio serra Hooker var. ***serra***

Tall ragwort, tall butterweed, butterweed groundsel

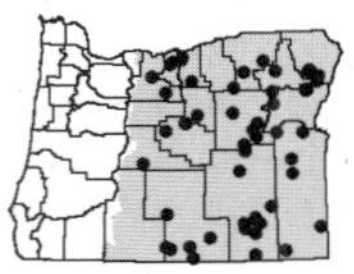

Herbs perennial, 4–10(25) dm; caudices woody, branching, rhizomatous. **Stems** 1 or 2–4 and loosely clustered, glabrous or lightly tomentose. **Leaves** basal usually withering before flowering; cauline evenly distributed along stems, distal smaller and bract-like, narrowly lanceolate, lanceolate or sublinear, bases tapered, margins dentate to subentire; petiolate or subsessile. **Inflorescences** flat-topped to panicle-like arrays; bracts conspicuous. **Heads** 40–90+. **Calyculi** 2–6 linear to filiform bractlets, 0.5–5 mm. **Involucres** narrowly campanulate. **Phyllaries** ~8, 4–6 mm, tips green, rarely black. **Ray florets** ~5; rays 5–7 mm. **Disc florets** 10–20+. **Fruits** 1.5–2.5 mm, glabrous. $2n$=40.

Streambanks, meadows, open places in coniferous forests, sagebrush. Flowering Jun–Aug. 400–2500 m. BR, BW, Col, ECas, Lava, Owy. CA, ID, NV, WA; northeast to MT, east to WY, southeast to CO. Native.

Senecio sphaerocephalus Greene

Mountain marsh butterweed, ballhead ragwort

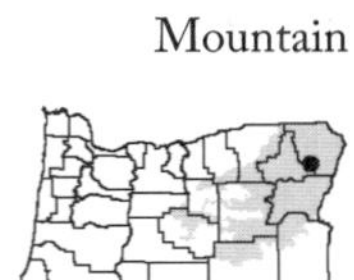

Herbs perennial, (3)4–8 dm; caudices ± woody, branching, with stout rhizomes. **Stems** 1, tomentose or unevenly glabrate. **Leaves** mostly basal, basal present at flowering; cauline reduced distally, distal much reduced and bract-like, oblanceolate to elliptic, bases tapered, margins weakly dentate or denticulate with callous denticles, surfaces sparsely tomentose to nearly glabrous; petiolate. **Inflorescences** corymb-like arrays; bracts conspicuous. **Heads** (3)5–24, usually > 10. **Calyculi** 1–5 bractlets, 2–3 mm. **Involucres** campanulate. **Phyllaries** (13)21, 3–7 mm, tips black or dark brown. **Ray florets** ~13; rays 6–10 mm. **Disc florets** 25–35+. **Fruits** 1.5–2.5 mm, usually hairy, sometimes glabrous. **2*n***=40.

Open meadows, near conifer forests. Flowering Jun–Jul. 1100–1700 m. BW. ID, NV; northeast to MT, east to WY, southeast to CO. Native.

Although *Senecio sphaerocephalus* has been collected in Oregon, it is relatively rare within the state and more abundant in the more easterly parts of its range.

Senecio sylvaticus Linnaeus

Woodland ragwort, wood groundsel

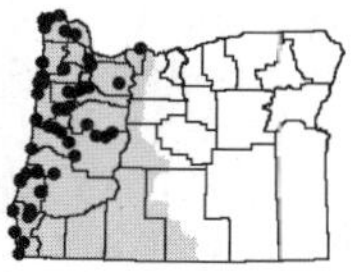

Herbs annual, (1.5)3–8+ dm; from fibrous taproots. **Stems** 1, densely pubescent with abundant curly hairs. **Leaves** evenly distributed along stems, obovate to oblong in outline, bases tapering, usually 1–2-pinnately lobed or divided, ultimate margins dentate; proximal petiolate, distal clasping and bract-like. **Inflorescences** corymb-like arrays; bracts 0 or inconspicuous. **Heads** 12–24. **Calyculi** 0 or 1–5+ linear to filiform bractlets, 2–3 mm. **Involucres** cylindric. **Phyllaries** 13+, rarely 21, 4–7+ mm, tips green or minutely black. **Ray florets** 0, rarely 1–8; rays 1–2+ mm. **Disc florets** 40–50. **Fruits** 1.5–2.5 mm, sparsely pubescent on ribs. **2*n***=40.

Disturbed open wooded areas. Flowering May–Oct. 0–1200 m. Casc, CR, ECas, Est, Sisk, WV. CA, WA; north to British Columbia, scattered in eastern North America; Asia, Europe. Exotic.

Senecio sylvaticus is a native of Eurasia and favors cool, damp climates. In overall appearance, it resembles the more common *S. vulgaris*. However, the latter has distinctive black-tipped phyllaries, while *S. sylvaticus* has phyllaries that are generally green-tipped. This species is well established in the Pacific Northwest and in parts of eastern Canada. Elsewhere it is sporadic.

Senecio triangularis Hooker (p. 341)

Arrowleaf groundsel

Senecio gibbonsii Greene, *Senecio triangularis* Hook. var. *angustifolius* G.N. Jones, *Senecio triangularis* Hook. var. *triangularis*

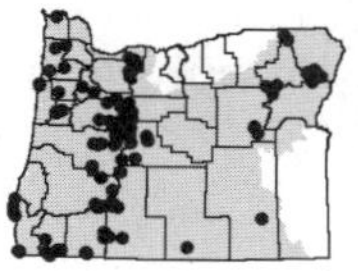

Herbs perennial, (2)5–12(20) dm; caudices branching, woody, rhizomatous. **Stems** 1 or 2–4 and loosely clustered, glabrous or sparsely pubescent. **Leaves** basal present at flowering; cauline ± evenly distributed along stems, gradually reduced distally, narrowly to widely triangular, bases tapered to truncate or cordate, margins dentate, rarely

SENECIO

subentire, surfaces sparsely tomentose to nearly glabrous; proximal petiolate, distal sessile. **Inflorescences** corymb- or ± raceme-like arrays; bracts conspicuous. **Heads** 10–30(60). **Calyculi** 2–6 bractlets, inconspicuous, rarely > 2 mm. **Involucres** cylindric to narrowly campanulate. **Phyllaries** (8)13(21), 6–10 mm, tips green, rarely black. **Ray florets** ~8; rays 9–15 mm. **Disc florets** 35–45. **Fruits** 2.5–3.5 mm, glabrous. **2*n***=40, 80.

Open areas, rocky streambanks, wet meadows, bogs, damp places in coniferous forests. Flowering May–Sep. 0–2800 m. BR, BW, Casc, CR, ECas, Est, Lava, Sisk, WV. CA, ID, NV, WA; north to AK, northeast to MT, east to WY, southeast to NM. Native.

Some plants with narrow, subentire leaves that taper to the petioles have been found in acidic bogs in Washington and Oregon. A few authors recognize these as variety *angustifolia*. However, this author has observed several specimens (specifically indicated as being collected in bogs) that had both narrower leaves and the wider leaves more typical of variety *triangularis* on the same plant.

Senecio vulgaris Linnaeus (p. 341)

Common groundsel, old-man-in-the-spring

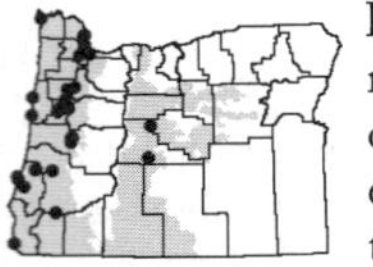

Herbs annual, (1)2–5(6+) dm; caudices taprooted. **Stems** usually 1, rarely 2–3, glabrous or sparsely and unevenly tomentose. **Leaves** evenly distributed along stems, oblanceolate to narrowly obovate or ovate, bases tapering, margins lobulate to dentate, ultimate margins secondarily denticulate; proximal petiolate, distal sessile. **Inflorescences** loose corymb-like arrays; bracts few, inconspicuous. **Heads** 8–20. **Calyculi** 2–4(6+) bractlets, 2–3 mm. **Involucres** urceolate. **Phyllaries** 21, 4–6 mm, tips usually black. **Ray florets** 0. **Disc florets** (30)55–65. **Fruits** 2–2.5 mm, sparsely to densely pubescent, rarely glabrous. **2*n***=40.

Disturbed areas. Flowering Feb–Oct. 0–1000 m. CR, ECas, Est, Lava, Sisk, WV. CA, ID, NV, WA; throughout North America; Asia, Europe. Exotic.

SERICOCARPUS White-topped aster

Herbs perennial; with woody rhizomes. **Stems** ascending to erect, leafy, usually glabrous proximally, minutely spiculate distally. **Leaves** cauline, alternate, oblanceolate or oblong, rarely linear, gradually reduced distally, margins entire, 1–5-veined, tips acute, surfaces minutely spiculate; sessile or tapering to petiole-like bases. **Inflorescences** corymb-like arrays, branched or not. **Heads** radiate. **Involucres** cylindric or narrowly campanulate. **Phyllaries** in 3–5 series, erect in flower, reflexed in fruit; inner unequal, linear; outer gradually shorter. **Receptacles** slightly convex, pitted; paleae 0. **Ray florets** 1–7, pistillate; rays slightly exceeding involucres, white. **Disc florets** bisexual; corollas tubular, gradually widened distally, white to pale yellow, lobes 5, triangular, spreading; stamens included or slightly exserted; stigma lobes exserted. **Fruits** cylindrical or slightly compressed, truncate, tapering toward bases, 7–10-veined; pappi of 25–50 ± equal bristles.

North America. 5 species; 2 species treated in *Flora*.

Key to *Sericocarpus* species:

1. Ray florets 3–7; rays 3–5 mm; plants 25–120 cm; inflorescences usually branched. . . . . ***S. oregonensis*** ssp. ***oregonensis***
1' Ray florets 1–2(3); rays 1–3 mm; plants 10–45 cm; inflorescences usually not branched . . . . ***S. rigidus***

Sericocarpus oregonensis Nuttall ssp. *oregonensis* (p. 343)

Oregon white-topped aster

Aster oregonensis (Nutt.) Cronquist, *Aster oregonensis* (Nutt.) Cronquist ssp. *oregonensis*

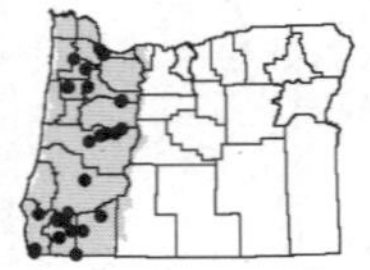

Plants 25–120 cm. **Leaves** 2–10 cm, usually 1-veined. **Inflorescences** usually branched; bracts lanceolate to ovate, hairy. **Involucres** 7–10 mm. **Ray florets** 3–7; rays 3–5 mm. **Disc florets** 9–15(20); corollas 3–7 mm, exserted from involucres, lobes 1–2 mm; anthers purple. **Fruits** 2–4 mm, surfaces densely white-strigose; pappus bristles 5–7 mm, white. **2*n***=18.

Open places in mixed woodlands, grasslands, ridgetops, riverbanks, floodplains, roadsides, disturbed areas. Flowering Jul–Sep. 50–1400 m. Casc, CR, Sisk, WV. CA, WA. Native.

A second subspecies, ssp. *californica*, is found in the Sierra Nevada Mountains of California. It is distinguished from ssp. *oregonensis* by its densely hirsute stem, with spreading, many-celled trichomes.

Sericocarpus rigidus Lindley

Rigid white-topped aster

Aster curtus Cronquist

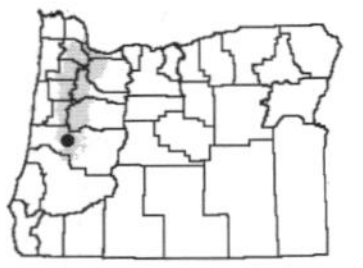

Plants 10–45 cm. **Leaves** 2–5 cm, usually 3–5-veined. **Inflorescences** usually not branched; bracts ovate, puberulent. **Involucres** 5–9 mm. **Ray florets** 1–2(3); rays 1–3 mm. **Disc florets** 9–20; corollas 4–6 mm, exserted from involucres, lobes 0.5–1 mm; anthers purple. **Fruits** 2–4 mm, surfaces densely white-strigose; pappus bristles 6–7 mm, white.

Grasslands. Flowering Jul–Sep. 50–400 m. WV. WA; north to British Columbia. Native.

Sericocarpus rigidus is a rare species of remnant Willamette Prairie grasslands. It is classified as a threatened species by the Oregon Department of Agriculture.

SILYBUM Milk thistle

Herbs annual or biennial; taprooted. **Stems** erect, simple or branched distally, glabrous or lightly floccose-tomentose. **Leaves** basal and cauline, alternate, margins irregularly dentate to deeply incised or lobed, lobes ± entire to dentate, teeth and lobes spine-tipped, tips acute or obtuse, surfaces glabrous or minutely puberulent, white-blotched adaxially; proximal petiolate, distal sessile and auriculate-clasping. **Inflorescences** solitary heads, terminal or axillary. **Heads** discoid. **Involucres** ovoid to globose or hemispheric. **Phyllaries** in 4–6 series, erect in flower, spreading in fruit, chartaceous, distally ± herbaceous, surfaces glabrous; inner equal, lanceolate, tips with or without terminal spines; outer lanceolate to ovate or round, gradually shorter, margins spine-fringed, tips acute to rounded. **Receptacles** flat, covered with whitish bristles; paleae 0. **Florets** bisexual; corollas tubular, abruptly widened distally, lobes 5, erect, elongate, linear; stamens included or slightly exserted; styles exserted, stigma lobes minute. **Fruits** obovoid, tapering toward bases, truncate; pappi of numerous unequal, basally united bristles, deciduous.

Mediterranean, North America. 2 species; 1 species treated in *Flora.*

Silybum marianum (Linnaeus) Gaertner (p. 343)
Milk thistle

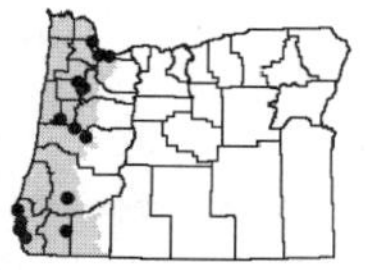

Plants 1.5–30 dm. **Leaves** 6–50+ cm, with 1 main vein. **Involucres** 3–6 cm including spine-tips. **Florets** 25–100+; corollas 25–35 mm, exserted from involucres, pink to purple. **Fruits** 6–8 mm, brown to purple, mottled, surfaces glabrous; pappi 15–20 mm. **2*n***=18.

Roadsides, riverbanks, fields, disturbed areas. Flowering Apr–Sep. 0–300 m. CR, Est, Sisk, WV. CA, NV, WA; widely scattered in North America; Mediterranean. Exotic.

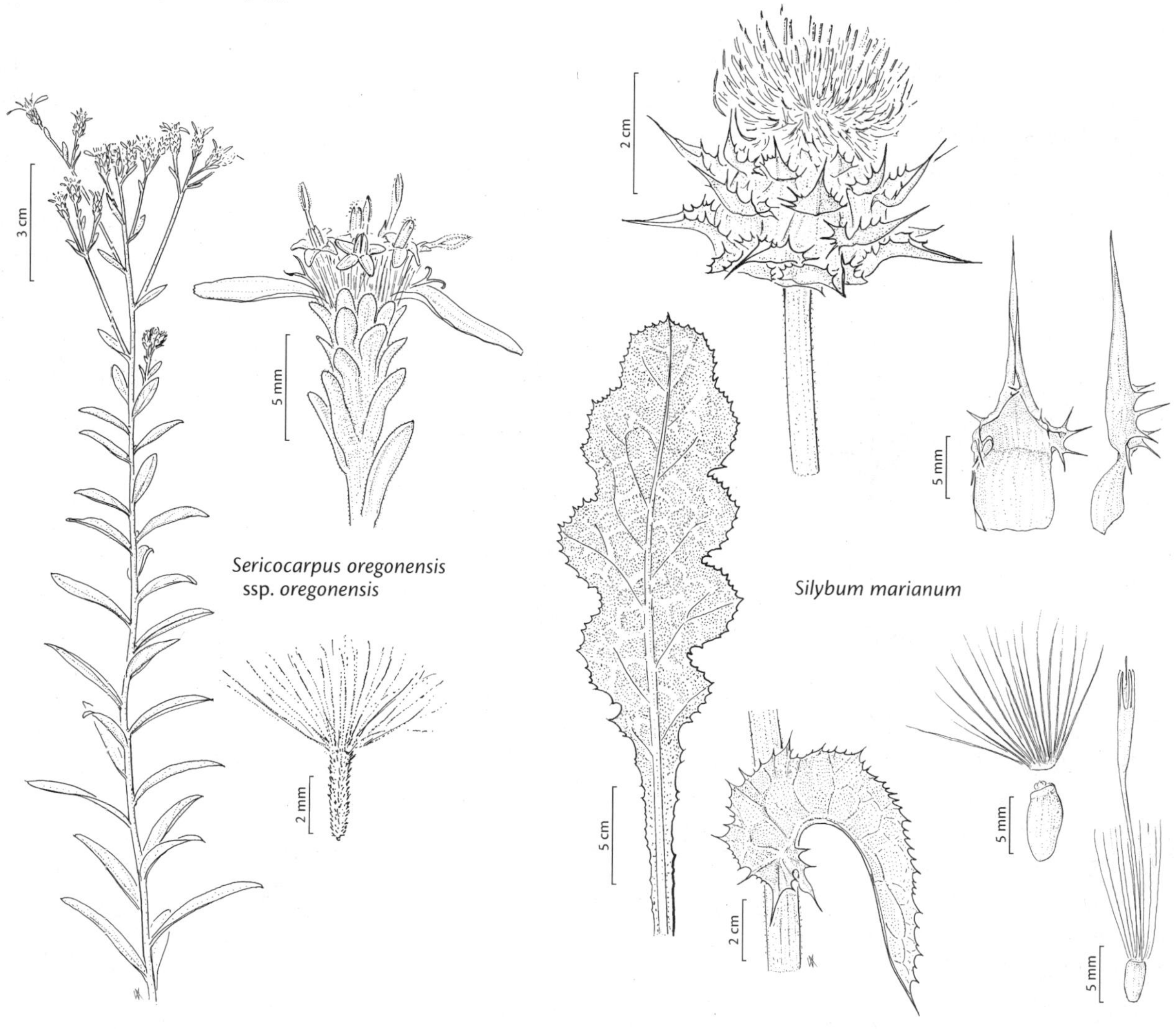

SERICOCARPUS • SILYBUM

SOLIDAGO Goldenrod

Herbs perennial; fibrous-rooted from woody root-crowns or rhizomes. **Stems** 1–few from bases, decumbent to ascending or erect, branched principally in inflorescences. **Leaves** alternate, simple, margins entire or dentate, surfaces glabrous to pubescent, sometimes resinous; sessile or petiolate. **Inflorescences** pyramidal, flat-topped, corymb-, panicle-, or thyrse-like arrays. **Heads** usually radiate, sometimes discoid, small. **Involucres** cylindric or campanulate. **Phyllaries** usually in 3–5 unequal series, midribs swollen and translucent, sometimes with 2–5 additional nerves, surfaces glabrous or puberulent, sometimes minutely glandular or resinous. **Receptacles** plane or convex; paleae 0. **Ray florets** 0 or 2–20+, pistillate; rays narrow, yellow, rarely white. **Disc florets** bisexual; corollas cylindric, widened distally, exceeding involucres, yellow, lobes 5, triangular or lance-linear; styles exserted. **Fruits** obconic or cylindric, slightly compressed, glabrous or strigose, 8–10-ribbed; pappi of persistent bristles, rarely with minute outer scales.

Asia, Europe, North America, South America. ~100 species; 9 species treated in *Flora*.

Key to *Solidago* species:

1\. Leaves and phyllaries copiously resinous .... ***S. spathulata***
1' Leaves and phyllaries not or slightly resinous.
  2\. Leaves and phyllaries slightly resinous.......... ***S. simplex***
  2' Leaves and phyllaries not resinous.
    3\. Stems strigillose to puberulent.
      4\. Inflorescences corymb-like arrays.......... ***S. multiradiata***
      4' Inflorescences pyramidal or thyrse-like arrays.
        5\. Leaves basal and cauline .... ***S. velutina***
        5' Leaves cauline.
          6\. Inflorescence bracts 0 or bracteate only at lower 1–3 nodes .... ***S. elongata***
          6' Inflorescence bracts generally present above lower 1–3 nodes .... ***S. lepida***
    3' Stems glabrous.
      7\. Leaves cauline .... ***S. gigantea***
      7' Leaves basal and cauline.
        8\. Rhizomes slender, often elongate.......... ***S. missouriensis***
        8' Rhizomes short, stout .... ***S. spectabilis***

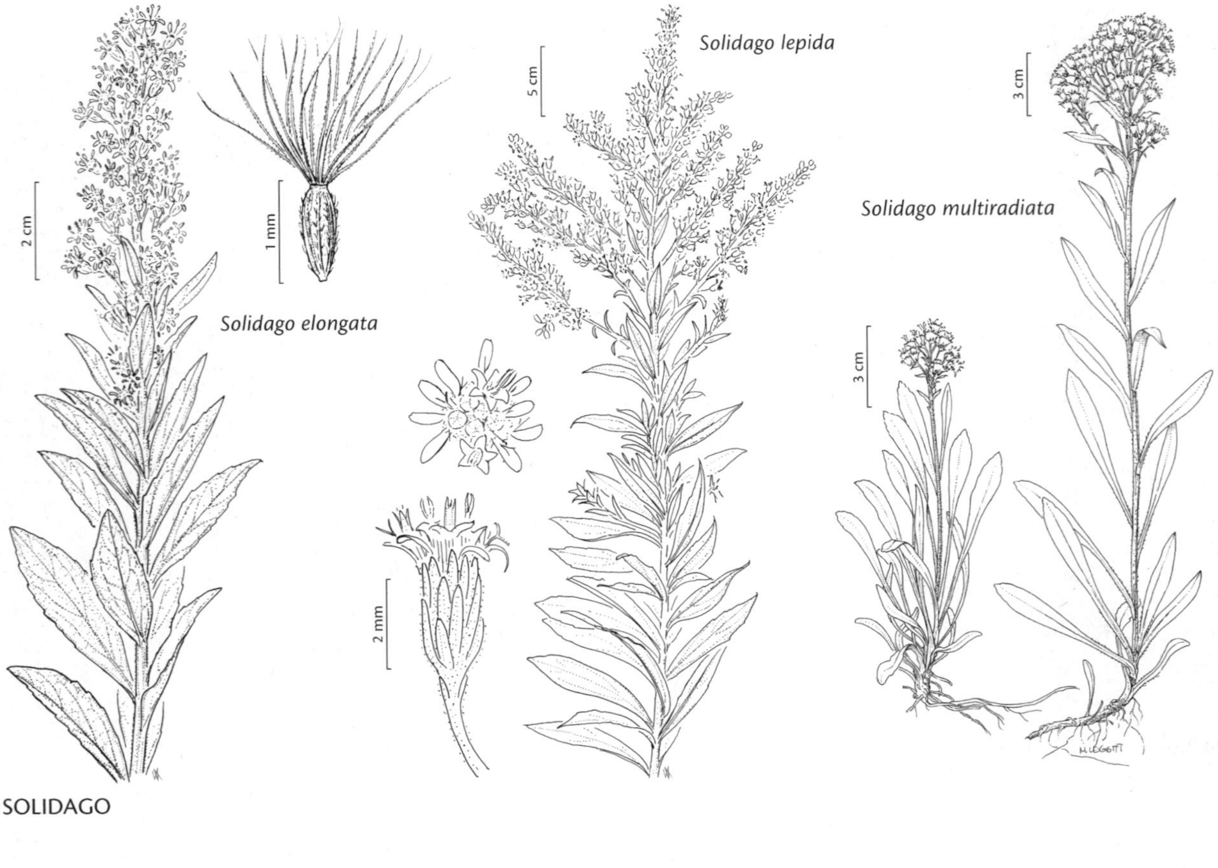

SOLIDAGO

Solidago elongata Nuttall (p. 344)

West Coast goldenrod, narrow goldenrod

Solidago canadensis L., *Solidago canadensis* L. ssp. *elongata* (Nutt.) D.D. Keck, *Solidago canadensis* L. var. *salebrosa* (Piper) M.E. Jones

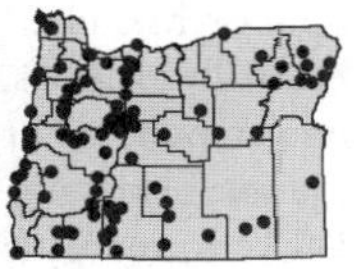

Plants 2–15 dm; rhizomatous. **Stems** simple to well branched, puberulent, sometimes glabrate proximally. **Leaves** cauline; blades broadly to narrowly oblanceolate or elliptic, 2–10 cm, reduced distally, margins entire to denticulate or serrulate, tips acute, surfaces glabrous to sparsely strigose; sessile or short-petiolate. **Inflorescences** thyrse-like or narrow pyramidal arrays; bracts 0 or bracteate at only 1–3 lowest nodes. **Heads** sessile or pedunculate. **Involucres** narrowly campanulate, 2.5–4 mm. **Phyllaries** in 3–5 series, linear-lanceolate, surfaces glabrous; outer gradually shorter. **Ray florets** 7–15; rays yellow. **Disc florets** 3–15; corollas 3–4 mm. **Fruits** 1–1.5 mm, strigose. $2n$=18, 36.

Open woods, meadows, dunes, riverbanks, roadsides. Flowering Jun–Oct. 0–2200 m. All ecoregions. CA, WA. Native.

Solidago gigantea Aiton

Giant goldenrod, smooth goldenrod

Solidago gigantea Aiton var. *serotina* (Kuntze) Cronquist

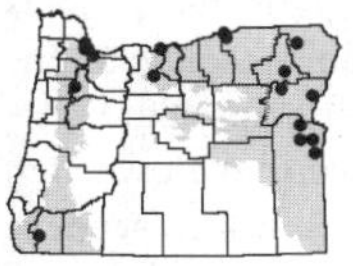

Plants 5–20 dm; rhizomatous. **Stems** simple or branched, glabrous, glaucous. **Leaves** cauline; blades broadly to narrowly oblanceolate or elliptic, 4–15 cm, margins entire to denticulate or serrulate, tips acute, surfaces glabrous, nerves pilose; sessile or short-petioled. **Inflorescences** thyrse-like arrays, rarely club-shaped; bracts reduced distally. **Heads** pedunculate. **Involucres** campanulate, 2–5 mm. **Phyllaries** in 3–5 series, linear-lanceolate, surfaces glabrous; outer gradually shorter. **Ray florets** 7–20. **Disc florets** 5–15; corollas 2.5–4.5 mm. **Fruits** 1.3–1.5 mm, slightly strigose. $2n$=18, 36, 54.

Meadows, riverbanks, wetlands, roadsides, disturbed areas. Flowering Jul–Sep. 0–1000 m. BW, Col, Owy, Sisk, WV. CA, ID, NV, WA; throughout North America. Native.

Solidago lepida de Candolle (p. 344)

Western goldenrod, Rocky Mountain goldenrod

Solidago canadensis L., *Solidago canadensis* L. ssp. *salebrosa* (Piper) D.D. Keck, *Solidago canadensis* L. var. *salebrosa* (Piper) M.E. Jones, *Solidago lepida* DC. var. *lepida*, *Solidago lepida* DC. var. *salebrosa* (Piper) Semple

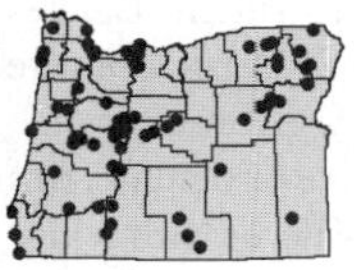

Plants 2–15 dm; rhizomatous. **Stems** simple or branched, densely puberulent, often glabrate proximally. **Leaves** cauline; blades broadly to narrowly oblanceolate or elliptic, 4–15 cm, margins entire to denticulate or serrulate, tips acute, surfaces glabrous to sparsely villous, distal leaves sometimes minutely glandular; sessile or short-petioled. **Inflorescences** thyrse-like or pyramidal arrays; bracts generally present above lower 1–3 nodes, reduced distally. **Heads** sessile to short- or long-pedunculate. **Involucres** campanulate, 2–5 mm. **Phyllaries** in 3–5 series, linear-lanceolate, surfaces glabrous or minutely glandular; outer gradually shorter. **Ray florets** 7–20. **Disc florets** 2–13; corollas 2–5 mm. **Fruits** 0.5–1.2 mm, strigillose. $2n$=18, 36, 54.

Forests, open woods, meadows, pastures, bogs, marshes, riverbanks. Flowering Jun–Oct. 0–2000 m. All ecoregions. CA, ID, NV, WA; throughout Canada and western US. Native.

In Oregon, varieties *lepida* and *salebrosa* are not geographically or ecologically separate and show a blending of morphology. Here they are merged into the species *S. lepida*.

Solidago missouriensis Nuttall

Missouri goldenrod

Solidago missouriensis Nutt. var. *extraria* A. Gray, *Solidago missouriensis* Nutt. var. *missouriensis*, *Solidago missouriensis* Nutt. var. *tolmieana* (A. Gray) Cronquist

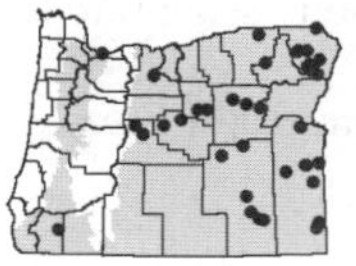

Plants 1.5–8 dm; rhizomes slender, often elongate. **Stems** simple, glabrous throughout or minutely puberulent in inflorescences. **Leaves** basal and cauline; blades oblanceolate or elliptic to linear, 1.5–25 cm, reduced distally, margins entire, rarely denticulate, sometimes ciliate, tips acute, surfaces glabrous; basal usually present at flowering; sessile or with wing-margined petioles. **Inflorescences** panicle-like or club-shaped arrays, sometimes flat-topped; bracts reduced distally. **Heads** pedunculate. **Involucres** campanulate, 2.5–4.5 mm. **Phyllaries** in 3–5 series, linear-lanceolate or lance-elliptic, surfaces glabrous; outer gradually shorter. **Ray florets** 5–15. **Disc florets** 6–20; corollas 2–4 mm. **Fruits** 1–2 mm, glabrous or strigose. $2n$=18, 36.

Coniferous forest, mixed woods, shrublands, grasslands, talus slopes, riverbanks, roadsides. Flowering Jun–Oct. 300–1700 m. BR, BW, Col, ECas, Lava, Owy, Sisk, WV. ID, NV, WA; north to British Columbia, northeast to Quebec, east to MI, southeast to MS. Native.

Solidago multiradiata Aiton (p. 344)

Northern goldenrod

Solidago multiradiata Aiton var. *scopulorum* A. Gray

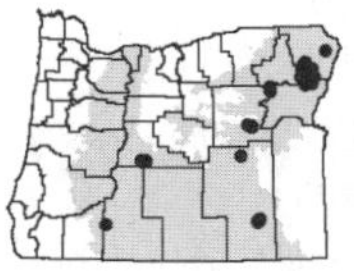

Plants 0.3–5+ dm; with simple or branched root-crowns. **Stems** simple, sometimes basally clustered, puberulent, often glabrate proximally. **Leaves** basal and cauline; blades narrowly to broadly lanceolate or oblanceolate, 1–17 cm, reduced distally, margins entire, rarely denticulate, usually ciliate, tips acute or obtuse, surfaces glabrous; basal usually withering by flowering; petiolate proximally, sessile distally. **Inflorescences** round-topped corymb-like arrays; bracts present. **Heads** pedunculate. **Involucres** campanulate, 4–8 mm. **Phyllaries** in 3–5 series, lanceolate to linear-lanceolate, surfaces glabrous; some outer shorter. **Ray florets** 12–18. **Disc florets** 10–30+; corollas 3–5 mm. **Fruits** 1.5–4 mm, surfaces strigose. $2n$=18, 36.

Open woodlands, meadows, creeksides, lake shores, swamps, cliffs, rock outcrops, talus. Flowering Jul–Sep. 0–3000 m. BR, BW, Casc, ECas. CA, ID, NV, WA; throughout northern and western North America. Native.

Solidago simplex Kunth

Sticky goldenrod, Mt. Albert goldenrod

Plants 0.5–8 dm; with simple or branched root-crowns. **Stems** simple, sometimes basally clustered, glabrous or puberulent, sometimes stipitate-glandular distally. **Leaves** basal and cauline; blades round, broadly elliptic, spatulate, orbiculate, or linear, 1–10 cm, margins entire to denticulate or dentate, tips rounded to obtuse or acute, surfaces glabrous, slightly resinous; basal sometimes present at flowering. **Inflorescences** panicle-like arrays; bracts present. **Heads** sessile to pedunculate. **Involucres** campanulate, 3–7 mm. **Phyllaries** in 3–5 series, lanceolate or linear-lanceolate, surfaces glabrous, slightly resinous; outer gradually shorter. **Ray florets** 7–15. **Disc florets** 6–30; corollas 4–5.5 mm. **Fruits** 2–3 mm, strigose. **2*n*=18.**

North America. 11 varieties; 2 varieties treated in *Flora*.

Key to *Solidago simplex* varieties:

1. Basal leaves usually spatulate, rarely broadly elliptic; plants 0.5–1.5 dm . . . . ***S. s.*** var. ***nana***

1' Basal leaves orbiculate to broadly elliptic; plants 1–5+ dm . . . . ***S. s.*** var. ***simplex***

Solidago simplex Kunth var. ***nana*** (A. Gray) G.S. Ringius

Alpine goldenrod, dwarf goldenrod

Solidago decumbens Greene, *Solidago spathulata* DC. var. *nana* (A. Gray) Cronquist

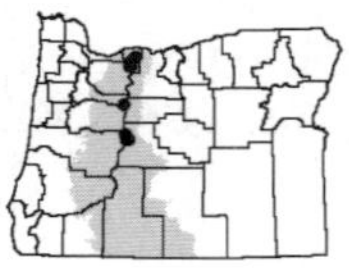

Plants 0.5–1.5 dm. **Leaves** basal usually spatulate, rarely broadly elliptic, margins usually dentate; petioles narrowly winged, sessile distally.

Montane tundra above tree-line, moraines, dry slopes. Flowering Jul–Aug. 1800–2500 m. Casc, ECas. WA; north to British Columbia. Native.

Solidago simplex Kunth var. ***simplex***

Sticky goldenrod

Solidago spathulata DC. ssp. *glutinosa* (Nutt.) D.D. Keck, *Solidago spathulata* DC. var. *neomexicana* (A. Gray) Cronquist

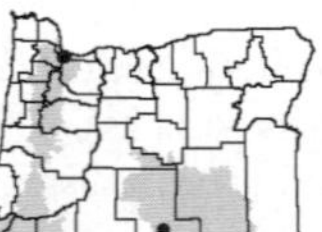

Plants 1–5+ dm. **Leaves** basal orbiculate to broadly elliptic, margins entire to denticulate; petioles linear or narrowly winged, sessile distally.

Shrublands, grasslands, sunny riverbanks, roadsides. Flowering Aug–Sep. 0–1300 m. BR, Sisk, WV. ID, NV, WA; scattered across northern and western North America. Native.

Solidago spathulata de Candolle

Dune goldenrod, coast goldenrod

Solidago spathulata DC. ssp. *spathulata*, *Solidago spathulata* DC. var. *spathulata*

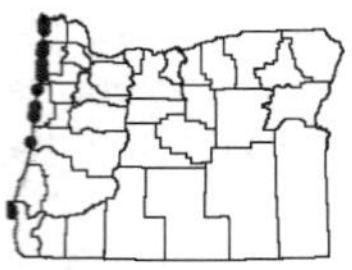

Plants 1–3 dm; with simple or branched root-crowns. **Stems** simple, usually glabrous, sometimes strigose or glandular-puberulent distally. **Leaves** basal and cauline; blades obovate or oblanceolate, 2–12 cm, margins denticulate or dentate, tips rounded to acute, surfaces glabrous, copiously resinous; basal sometimes present at flowering; sessile, or with wing-margined or linear petioles. **Inflorescences** dense thyrse- or panicle-like arrays; bracts present. **Heads** pedunculate. **Involucres** campanulate, 4–7 mm. **Phyllaries** in 3–5 series, lanceolate to elliptic or linear, surfaces glabrous, copiously resinous; outer gradually shorter. **Ray florets** 4–10. **Disc florets** 10–18; corollas 4–6 mm. **Fruits** 2–3 mm, strigose. **2*n*=18.**

Beaches, dunes, coastal forests and meadows, bluffs, headlands. Flowering Jul–Sep. 0–100 m. Est. CA. Native.

Solidago spectabilis (D.C. Eaton) A. Gray

Showy goldenrod, basin goldenrod

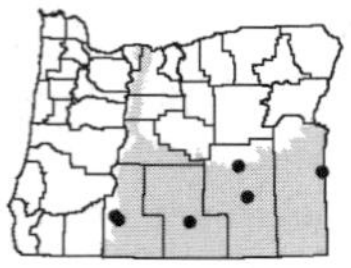

Plants 3–20 dm; rhizomes short, stout. **Stems** usually simple, glabrous. **Leaves** basal and cauline; blades oblanceolate to linear, 1–30 cm, margins entire, tips acute, surfaces glabrous; basal present at flowering; sessile, or with wing-margined or linear petioles. **Inflorescences** thyrse- or panicle-like arrays, sometimes flat-topped; bracts present, sometimes minute. **Heads** mostly short-pedunculate. **Involucres** broadly to narrowly campanulate, 3–5 mm. **Phyllaries** in 3–5 series, elliptic-lanceolate or linear, surfaces glabrous; outer gradually shorter. **Ray florets** 6–15. **Disc florets** 8–20; corollas 1.5–4.5 mm. **Fruits** 1.5–2.5 mm, strigose. **2*n*=18.**

Meadows, bogs, hot springs, streambanks, lake shores. Flowering Jun–Aug. 1200–2200 m. BR, ECas, Owy. CA, ID, NV; southeast to AZ. Native.

Solidago velutina de Candolle

Velvety goldenrod, California goldenrod

Solidago californica Nutt., *Solidago velutina* DC. ssp. *californica* (Nutt.) Semple, *Solidago velutina* DC. ssp. *sparsiflora* (A. Gray) Semple

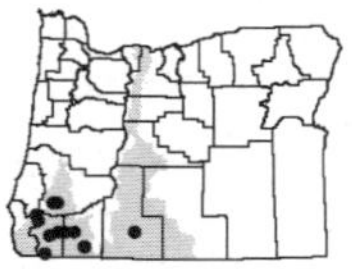

Plants 1.5–8+ dm; rhizomes slender, creeping. **Stems** simple or well branched distally, strigillose to minutely puberulent. **Leaves** basal and cauline; blades 2–10 cm, obovate to oblanceolate or elliptic, glabrous or minutely puberulent, margins entire to denticulate or dentate, tips acute or obtuse; basal present at flowering; sessile, or with wing-margined or linear petioles. **Inflorescences** thyrse-like or pyramidal arrays; bracts minute. **Heads** mostly sessile or short-pedunculate. **Involucres** cylindric or campanulate, 3.5–6 mm. **Phyllaries** in 4–6 series, narrowly lanceolate to elliptic or linear, glabrous or outermost puberulent; outer gradually shorter. **Ray florets** 6–12. **Disc florets** 5–17; corollas 3.5–6 mm. **Fruits** 1–3 mm, strigose. **2*n*=18, 36.**

Mixed forests, grasslands, riverbanks, disturbed areas. Flowering Jun–Oct. 100–1400 m. ECas, Sisk. CA, ID, NV; east to NE, southeast to TX, south to Mexico. Native.

SOLIVA Burrweed

Herbs annual. **Stems** erect or prostrate, sometimes clustered and clump-forming, usually much branched, villous or strigillose-puberulent. **Leaves** basal and cauline, alternate, obovate to spatulate, margins entire or dentate, 2-pinnati-palmately compound, bases sometimes expanded and subclasping, surfaces glabrous to strigillose-puberulent, lobes acute; petiolate. **Inflorescences** solitary, axillary heads. **Heads** disciform; sessile. **Involucres** hemispheric. **Phyllaries** in 1–2 series, equal or outer slightly shorter, erect in flower, spreading in fruit, herbaceous, ovate-lanceolate, 1–3-nerved, tips acute. **Receptacles** flat to hemispheric or conic; paleae 0. **Outer florets** pistillate; corollas 0. **Inner florets** staminate; corollas slenderly tubular below, widened and cup-shaped distally, 3–4-lobed, lobes triangular, erect; styles persistent, becoming hardened and spine-like, unbranched. **Fruits** compressed, wings projecting distally as spine-like teeth; pappi 0.

Worldwide. 55 species; 1 species treated in *Flora*.

Soliva sessilis Ruiz & Pavón (p. 347)
Field burrweed, prickly soliva

Soliva pterosperma (Juss.) Less.

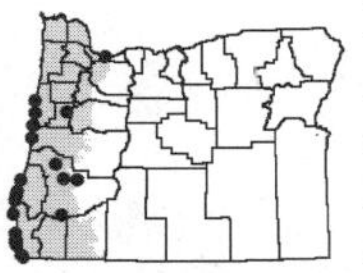

Plants 2–20+ cm. **Stems** purplish. **Leaves** 5–30 mm. **Involucres** 2–5 mm in diameter, surfaces villous or strigillose. **Outer florets** 5–15+. **Inner florets** 2–8; corollas 1.5–2.5 mm, white or yellow. **Fruits** obovate to lanceolate, 2.5–3+ mm, glabrous to scabrous or hispid, brown.

Roadsides, lawns, disturbed areas. Flowering Apr–Jul. 0–400 m. CR, Est, Sisk, WV. CA, WA; southern US; Europe, South America. Exotic.

Soliva sessilis often occurs as a lawn weed in Oregon. Like some other annual weeds, it varies greatly in size and habit depending on growth conditions. Its flat, winged, spined fruits are a distinctive trait.

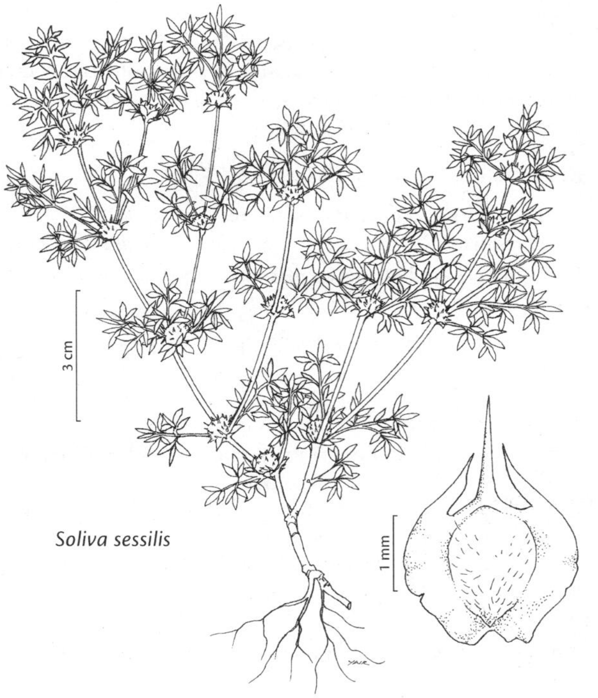

SOLIVA

SONCHUS Sow-thistle

Herbs annual, biennial, or perennial, sap milky; taprooted, sometimes rhizomatous and/or stoloniferous. **Stems** erect, unbranched or distally branched, leafy, glabrous, sometimes stipitate-glandular distally. **Leaves** basal and cauline, or cauline, lanceolate to oblong, bases often with auricles, margins dentate, usually pinnately lobed, teeth usually prickle-tipped; basal usually petiolate and winged, cauline sessile or petiolate. **Inflorescences** corymb- or umbel-like arrays. **Heads** ligulate; peduncles glabrous to glandular. **Involucres** campanulate in flower, ovoid in fruit. **Phyllaries** in 3–5+ series; inner equal, lanceolate; outer progressively shorter, lance-linear, midribs often thickened toward bases. **Receptacles** flat to convex, ± pitted, glabrous; paleae 0. **Florets** bisexual; ligules exceeding involucres. **Fruits** narrowly obovate, flattened, tapering to bases and tips, glabrous, ribs 6–10; pappi of numerous delicate, deciduous, plumose, white bristles.

Nearly worldwide. 50–60+ species; 3 species treated in *Flora*.

Key to *Sonchus* species:

1. Plants perennial, rhizomatous, sometimes stoloniferous; stems solid . . . . ***S. arvensis***

1' Plants annual or biennial, taprooted; stems hollow.

  2. Ligules usually < floret tubes; auricles with rounded lobes, coiled against stems; fruits smooth . . . . ***S. asper***

  2' Ligules ≈ floret tubes; auricles with acute lobes, not coiled against stems; fruits rugulose or tuberculate . . . . . ***S. oleraceus***

Sonchus arvensis Linnaeus

Perennial sow-thistle

Plants perennial, 10–150+ cm; rhizomatous, sometimes stoloniferous. **Stems** solid. **Leaves** 3–35 × 2–15 cm, pinnately lobed, lobes deltate; auricles straight or coiled against stems, lobes rounded. **Involucres** 10–18 mm. **Phyllaries** 25–50, surfaces glabrous to hispid or stipitate-glandular. **Florets** 30–200+; ligules 20–30 mm, ≈ tubes, yellow. **Fruits** 2.5–3 mm, dark brown, roughened. **2*n***=36, 54.

2 subspecies.

Both subspecies of *S. arvensis* have been the target of extermination efforts in some states, due to their potential as crop pests. The pubescence difference between the subspecies seems minor but is clear and consistent.

Key to *Sonchus arvensis* subspecies:

1. Peduncles and phyllaries stipitate-glandular . . . . ***S. a.*** ssp. ***arvensis***

1' Peduncles and phyllaries glabrous . . . ***S. a.*** ssp. ***uliginosus***

Sonchus arvensis ssp. *arvensis*

Corn sow-thistle

Sonchus arvensis L. var. *arvensis*

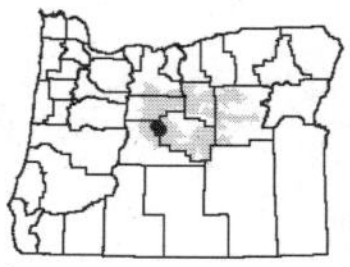

Peduncles stipitate-glandular. **Phyllaries** stipitate-glandular.

Wet, disturbed sites, fields, ditches, streambanks. Flowering Jul–Sep. 100–1600 m. Lava. CA, ID, NV, WA; throughout North America; Europe. Exotic.

Sonchus arvensis ssp. *uliginosus* (M. Bieberstein) Nyman

Marsh sow-thistle

Sonchus arvensis L. var. *glabrescens* Günther, Grab. & Wimm., *Sonchus uliginosus* M. Bieb.

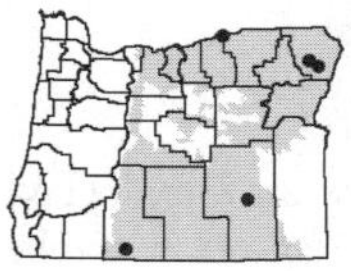

Peduncles glabrous. **Phyllaries** glabrous.

Wet, disturbed sites, fields, ditches, streambanks. Flowering Jul–Sep. 0–1300 m. BR, BW, Col, ECas. CA, NV, ID, WA; throughout North America; Europe. Exotic.

Sonchus asper (Linnaeus) Hill (p. 348)

Prickly sow-thistle

Sonchus asper (L.) Hill ssp. *asper*

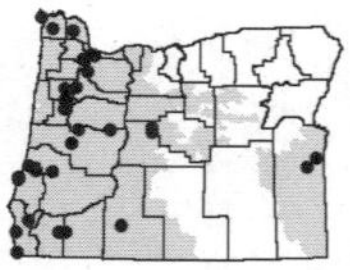

Plants annual or biennial, 10–120+ cm; taprooted. **Stems** hollow. **Leaves** 5–30 × 1–15 cm, usually pinnately lobed, lobes deltate; auricles coiled against stems, lobes rounded. **Involucres** 9–15 mm. **Phyllaries** 20–50, surfaces usually glabrous, sometimes hispid or stipitate-glandular. **Florets** 20–150+; ligules 10–15+ mm, usually < tubes, yellow. **Fruits** 2–3 mm, straw-colored to reddish brown, smooth. **2*n***=18.

Open woods, meadows, hillsides, dunes, disturbed sites, streambanks. Flowering May–Sep. 0–1500 m. Casc, CR, ECas, Est, Lava, Owy, Sisk, WV. CA, ID, NV, WA; throughout North America; Europe. Exotic.

Sonchus oleraceus Linnaeus

Common sow-thistle

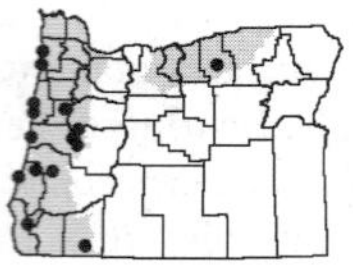

Plants annual or biennial, 10–150+ cm; taprooted. **Stems** hollow. **Leaves** 5–35 × 1–15 cm, usually pinnately lobed, lobes deltate to lanceolate; auricles straight, not coiled against stem, lobes acute. **Involucres** 9–13+ mm. **Phyllaries** 12–50+, surfaces glabrous or stipitate-glandular. **Florets** 15–150+; ligules 10–30 mm, ≈ tubes, yellow. **Fruits** 2.5–3.8 mm, dark brown, rugulose or tuberculate. **2*n***=32, 36.

Hillsides, roadsides, streambanks, fields, disturbed areas. Flowering Apr–Oct. 0–1400 m. Col, CR, Est, Sisk, WV. CA, ID, NV, WA; throughout North America; Europe. Exotic.

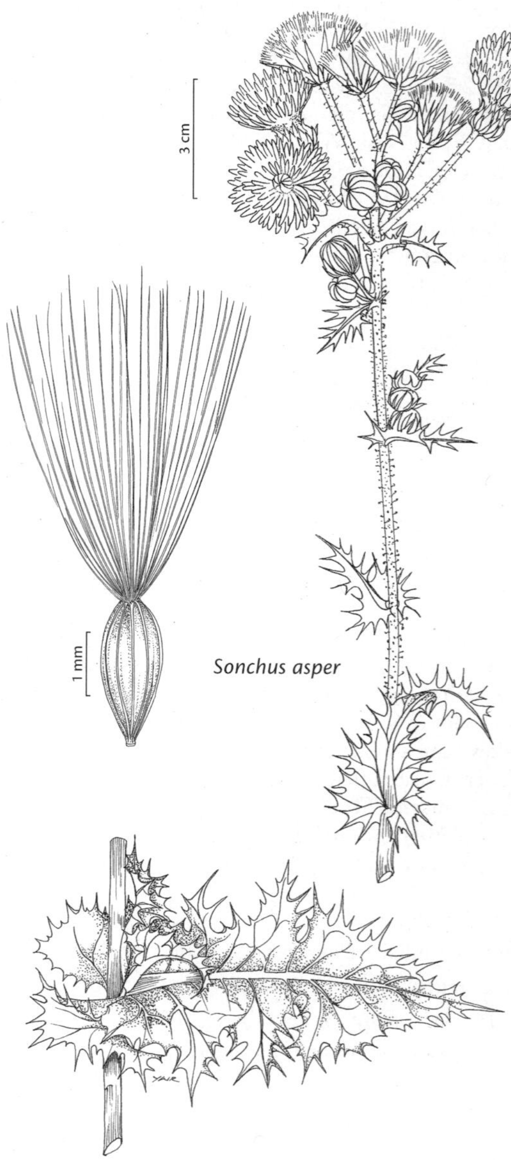

SONCHUS

SPHAEROMERIA Chickensage, false sagebrush

Herbs or **subshrubs** perennial; taprooted. **Stems** decumbent to ascending or erect, branched or unbranched, glabrous or puberulent. **Leaves** basal and cauline or cauline, alternate, ± reduced distally, margins entire to deeply lobed or compound, 1–3-veined, tips acute or obtuse, surfaces silky-tomentose or canescent; proximal petiolate, distal sessile. **Inflorescences** corymb- or panicle-like arrays, rarely heads solitary. **Heads** disciform; peduncles silky-tomentose or lanate-puberulent. **Involucres** globose. **Phyllaries** in 2–3 subequal or unequal series, chartaceous, sometimes a few lanceolate, margins membranous, surfaces silky-tomentose or canescent; inner obovate; outer obovate, some usually lanceolate. **Receptacles** rounded to nearly flat, glabrous or hispid; paleae 0. **Outer florets** pistillate, 4–15+, filiform. **Inner florets** 30–50+, bisexual; corollas tubular, abruptly widened from bases, exserted from involucres, lobes 5, erect or spreading; stamens included or slightly exserted; stigma lobes included or barely exserted. **Fruits** oblong or obovoid, truncate, tapering toward bases, obscurely or distinctly 2–10-ribbed, sometimes becoming mucilaginous when wet; pappi 0.

Western Mexico and United States. 9 species; 2 species treated in *Flora*.

Key to *Sphaeromeria* species:

1. Plants subshrubs; receptacles glabrous; fruits not becoming mucilaginous when wet . . . . **S. cana**

1' Plants herbs; receptacles hispid; fruits becoming mucilaginous when wet. . . . . ***S. potentilloides*** var. ***potentilloides***

Sphaeromeria cana (D.C. Eaton) A. Heller (p. 349)

Gray chickensage

Tanacetum canum D.C. Eaton

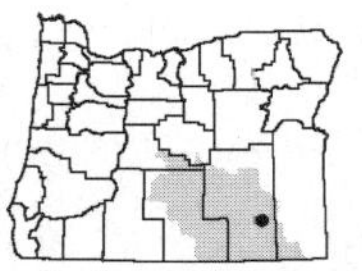

Subshrubs 10–60 cm. **Stems** erect, canescent. **Leaves** cauline, linear to linear-oblanceolate, 1–3 cm, surfaces lanate-puberulent, margins entire to deeply lobed, lobes linear, tips obtuse or acute. **Heads** 3–12 in corymb-like arrays, rarely solitary. **Involucres** 3–4 mm. **Phyllaries** 10–12, obovate, sometimes a few outer lanceolate, surfaces canescent. **Receptacles** glabrous. **Florets** corollas 1.5–2 mm, ochroleucous, lobes puberulent. **Fruits** 1.8–2 mm, glabrous or inconspicuously resinous-glandular, not becoming mucilaginous when wet. **2*n***=18.

Cliffs, rocky areas. Flowering Jun–Sep. 1800–2200 m. BR. CA, NV. Native.

The only current known Oregon localities for *S. cana* are at higher elevations on Steens Mountain.

Sphaeromeria potentilloides (A. Gray) A. Heller var. ***potentilloides***

Five-finger chickensage, cinquefoil false sagebrush

Tanacetum potentilloides A. Gray, *Tanacetum potentilloides* A. Gray var. *potentilloides*

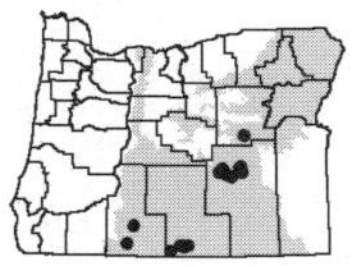

Herbs 5–30 cm. **Stems** arcuate-ascending, lower portion ± decumbent, silky-tomentose. **Leaves** basal 3–10 cm, usually 2-pinnately lobed, surfaces silky-tomentose; cauline leaves similar to basal, usually 1-pinnately lobed. **Heads** 3–20 in corymb- or panicle-like arrays. **Involucres** 2.5–4 mm. **Phyllaries** 8–12, obovate, surfaces silky-tomentose. **Receptacles** hispid. **Florets** corollas 2–2.5 mm, yellow, lobes glabrous. **Fruits** 0.8–1.5 mm, glabrous, becoming mucilaginous when wet. **2*n***=18.

Grasslands, fields, moist meadows. Flowering May–Jul. 1200–1500 m. BR, BW, ECas. CA, NV. Native.

A second variety of *S. potentilloides*, var. *nitrophila*, occurs in Idaho and Nevada, differing from var. *potentilloides* in its fewer, larger heads and basal leaves that are only one-pinnately lobed.

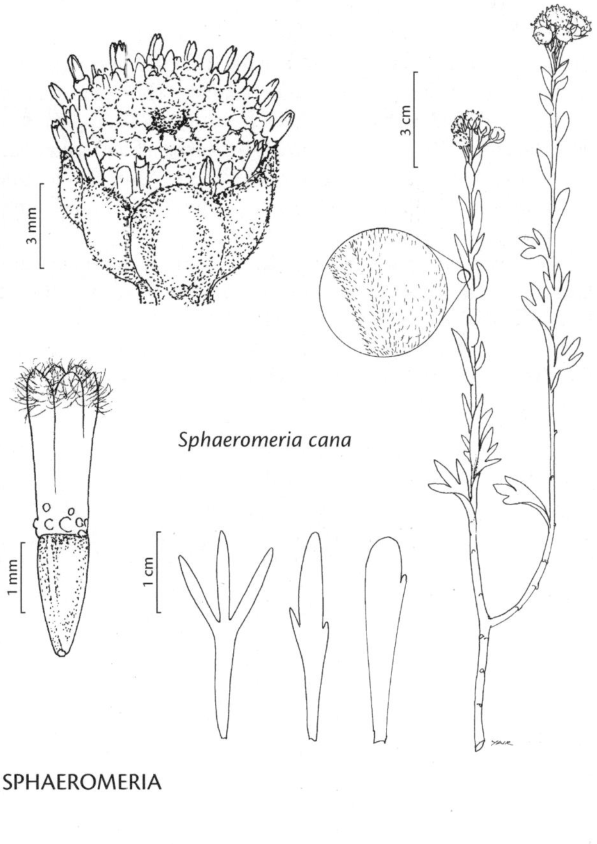

SPHAEROMERIA

STENOTUS Goldenweed

Herbs perennial, cespitose; with stout taproots and woody caudices. **Stems** erect from caudex branches, simple or few-branched, minutely scabrous-puberulent or lanate-villous. **Leaves** basal or basal and cauline, alternate, broadly to narrowly oblanceolate or linear, bases tapering, margins entire, tips acute, surfaces glabrous to scabrous-puberulent or lanate-villous; petiolate. **Inflorescences** usually solitary heads. **Heads** radiate; peduncles spiculate or lanate-villous, sometimes stipitate-glandular. **Involucres** campanulate to hemispheric. **Phyllaries** in 3–5 series, linear to narrowly or broadly lanceolate, tips acute, surfaces glabrous to minutely scabrous or lanate-villous; inner equal, sometimes membranous; outer equaling inner or gradually shorter. **Receptacles** convex; paleae 0. **Ray florets** pistillate; rays well exceeding involucres, yellow. **Disc florets** bisexual; corollas tubular, gradually widened distally, yellow, lobes 5, triangular, erect; stamens ± exserted; stigma lobes well exserted. **Fruits** cylindrical-fusiform or flattened, truncate, tapering toward bases, 6–12-veined; pappi of numerous, ± equal bristles in 1 series.

Western North America. 4 species; 2 species treated in *Flora.*

Key to *Stenotus* species:

1. Leaves rigid, surfaces glabrous or minutely scabrous; stems scabrous-puberulent . . . . ***S. acaulis***
1' Leaves lax, flexible, surfaces ± lanate-villous; stems lanate-villous . . . . ***S. lanuginosus*** var. ***lanuginosus***

Stenotus acaulis (Nuttall) Nuttall (p. 350)

Stemless goldenweed, cushion stenotus

Haplopappus acaulis (Nutt.) A. Gray

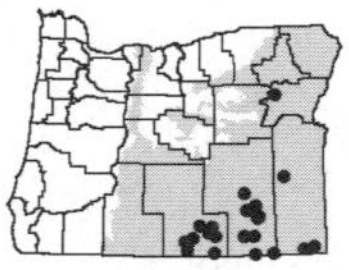

Plants 2–20 cm, tightly cespitose. **Stems** leafless or leafy proximally, naked or bracteate distally, scabrous-puberulent. **Leaves** narrowly to broadly oblanceolate, rarely linear, 1–5(10) cm, rigid, surfaces glabrous or minutely scabrous; petioles short, sometimes indistinct. **Involucres** 6–10 mm. **Phyllaries** outer broadly lanceolate, < inner, distinctly 1–5-nerved, surfaces glabrous or minutely scabrous. **Ray florets** 5–15, rays 6–17 mm. **Disc florets** 15–30+; corollas 5–10 mm, exceeding involucres. **Fruits** 3–5 mm, brown, strigillose, rarely glabrous; pappus bristles 5–7 mm, white. **2*n***=18, 36.

Open slopes and flats, basalt ledges, rocky ridgetops. Flowering May–Jul. 1200–2900 m. BR, BW, ECas, Owy. CA, ID, NV, WA; northeast to MT, east to WY, southeast to AZ. Native.

Stenotus lanuginosus (A. Gray) Greene var. ***lanuginosus*** (p. 350)

Woolly goldenweed, woolly stenotus

Haplopappus lanuginosus A. Gray, *Haplopappus lanuginosus* A. Gray var. *lanuginosus*

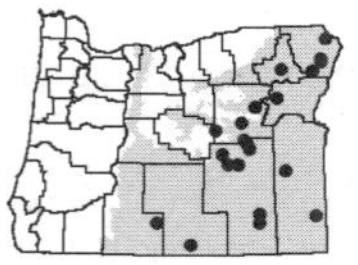

Plants 5–20 cm, tightly or loosely cespitose. **Stems** leafy proximally, leafy or bracteate distally, lanate-villous. **Leaves** linear to narrowly oblanceolate, 1–10 cm, lax, flexible, surfaces ± lanate-villous; petioles usually distinct. **Involucres** 6–12 mm. **Phyllaries** outer linear to narrowly or broadly lanceolate, ≈ inner or a few ± shorter, obscurely or distinctly 1–3-nerved, surfaces glabrous or lanate-villous. **Ray florets** 7–18+; rays 8–14 mm. **Disc florets** 20–40+; corollas 6–9 mm, equaling to slightly exceeding involucres. **Fruits** 3–6 mm, tan, strigillose or setose; pappus bristles 5–9 mm, white.

Pine woods, shrublands, grasslands, open, rocky slopes and flats. Flowering May–Aug. 900–2800 m. BR, BW, ECas, Owy. CA, ID, NV, WA. Native.

A second variety, var. *andersonii*, occurs in Idaho and Montana. It differs in its narrower phyllaries.

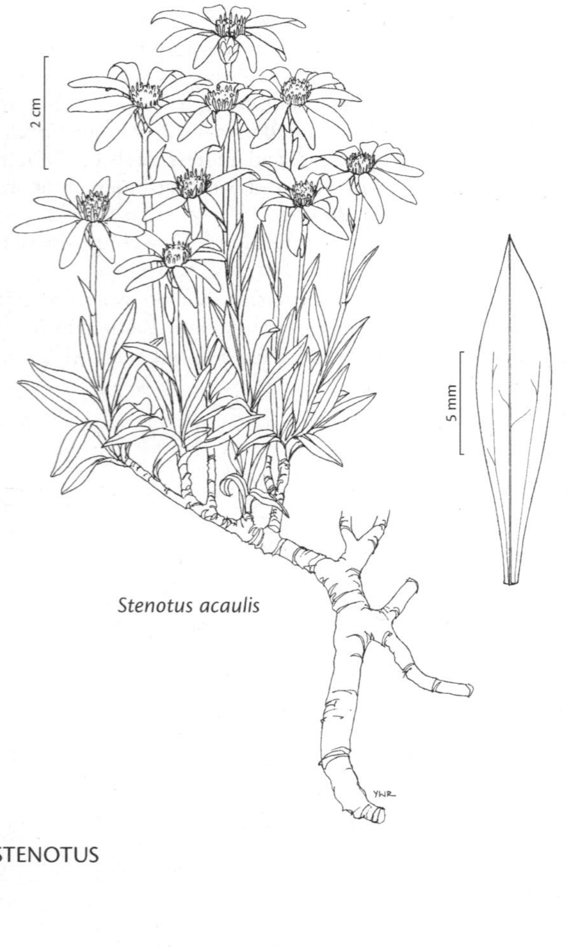

Stenotus acaulis

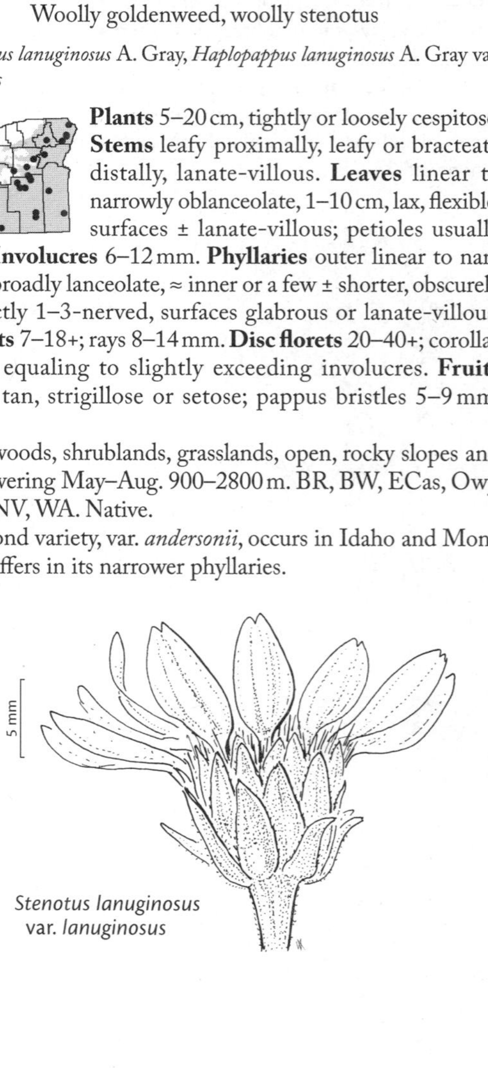

Stenotus lanuginosus var. lanuginosus

STEPHANOMERIA Wirelettuce

Herbs annual or perennial, sap milky; taprooted, sometimes with below-ground caudices and slender, creeping rhizomes. **Stems** photosynthetic, erect, simple or more often abundantly branched; branches spreading or ascending, leafy or bracteate. **Leaves** basal and cauline or cauline, often withering early, oblanceolate to linear, margins lobed, dentate, or entire; usually sessile, sometimes petiolate. **Inflorescences** panicle- or raceme-like arrays, sometimes heads solitary and terminal. **Heads** ligulate. **Calyculi** 0 or 4–8 appressed or reflexed bractlets. **Involucres** narrowly cylindrical to campanulate. **Phyllaries** usually 5–12 in 1 series, deltate to lanceolate or linear; inner equal; outer gradually reduced or all basal, reflexed at maturity or not. **Receptacles** flat, usually smooth, glabrous; paleae 0. **Florets** bisexual, ligules equaling or exceeding involucres. **Fruits** cylindrical or oblong, tapering slightly toward bases and tips, smooth or roughened, glabrous, ribs 5; not beaked; pappi of 5–40 barbed or plumose bristles, free or basally united in groups, persistent or deciduous, sometimes only distal portions breaking off. **2*n***=16.

Western North America. 16 species; 7 species treated in *Flora*.

The annual species of *Stephanomeria* are well understood taxonomically. The perennial species have not been equally well studied, but their relationships are better understood since the removal of two species formerly placed in *Stephanomeria* to the separate genera *Pleiacanthus* and *Munzothamnus* (Lee et al. 2002).

Key to *Stephanomeria* species:

1. Plants perennial.
    2. Florets (7)8–13(14); leaves persistent at flowering; rhizomatous. . . . . **S. lactucina**
    2' Florets 3–6; leaves usually withered by flowering; usually taprooted. . . . . **S. tenuifolia**

1' Plants annual.
    3. Fruits roughened but not grooved between ribs . . . . . ***S. virgata*** ssp. ***pleurocarpa***
    3' Fruits grooved between ribs.
        4. Pappus bristles ± plumose to base.
            5. Florets 9–15; outer phyllaries loosely reflexed . . . . . ***S. elata***
            5' Florets 5; outer phyllaries strongly reflexed . . . . . ***S. paniculata***
        4' Pappus bristles plumose in distal 50–80%, not to base.
            6. Florets 5–11; ligules 10–15 mm; fruits 2.5–3 mm . . . . . ***S. exigua*** ssp. ***coronaria***
            6' Florets 5–6; ligules 8–9.5 mm; fruits 3.3–3.8 mm . . . . . ***S. malheurensis***

Stephanomeria elata Nuttall
Nuttall's wirelettuce

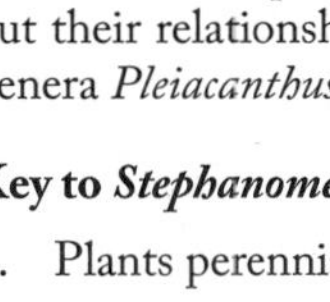

Plants annual, 50–150 cm. **Stems** usually glabrous. **Leaves** basal and cauline; basal withering by flowering, linear or narrowly oblanceolate, 2–10 cm, margins sparingly toothed or lobed, surfaces glabrous or puberulent; cauline reduced to bractlets in inflorescences. **Inflorescences** panicle- or raceme-like arrays, sometimes heads solitary. **Peduncles** 3–7 mm. **Involucres** cylindrical or narrowly campanulate, 5–7 mm, surfaces glabrous or minutely puberulent. **Phyllaries** inner 6–12, lanceolate; outer basal, deltate to lanceolate, gradually shorter, loosely reflexed. **Florets** 9–15; ligules 8–10 mm, pink to peach-colored. **Fruits** 3.5–4.5 mm, tan to dark brown, grooved and roughened between ribs; pappus bristles 17–20+, white or tan, plumose to base, deciduous. **2*n***=32.

Forest openings, chaparral, dry hillsides, rock outcrops, streambanks, roadsides. Flowering Jul–Sep. 100–1100 m. Sisk. CA; south to Mexico. Native.

Stephanomeria elata is a tetraploid species originating from hybrids between *S. exigua* and *S. virgata*. In earlier Oregon floristic references, it was confused taxonomically with the latter species. *Stephanomeria elata* is distinguished from that species by the presence of a groove between the ribs of the fruits that is absent in *S. virgata*.

Stephanomeria exigua Nuttall ssp. *coronaria* (Greene) Gottlieb
Small wirelettuce

Stephanomeria exigua Nutt. var. *pentachaeta* (D.C. Eaton) H.M. Hall

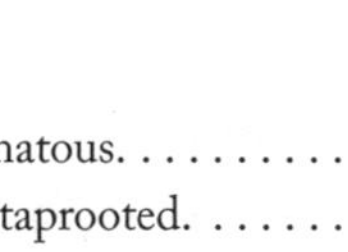

Plants annual, 10–50+ cm. **Stems** glabrous or sparsely pubescent. **Leaves** basal and cauline; basal withering by flowering, linear to oblanceolate, 3–10 cm, margins toothed or sparingly lobed, surfaces glabrous; cauline reduced to bractlets in inflorescences, mostly linear. **Inflorescences** panicle- or raceme-like arrays. **Peduncles** 1–40+ mm. **Involucres** cylindrical to narrowly campanulate, 5–10 mm, surfaces glabrous or puberulent. **Phyllaries** inner 4–8, linear-lanceolate; outer mostly basal, deltate to lanceolate, loosely reflexed. **Florets** 5–11; ligules 10–15 mm, pink to peach-colored. **Fruits** 2.5–3 mm, light tan to dark brown, grooved and roughened between ribs; pappus bristles 5–24, white or tan, plumose on distal 60–80%, distal portion deciduous. **2*n***=16.

Juniper woodlands, sagebrush scrub, open slopes, flats, dunes, cultivated fields. Flowering May–Sep. 100–1500 m. BR, ECas, Lava, Owy. CA, ID, NV; south to Mexico. Native.

There is considerable intergradation between the subspecies of *Stephanomeria exigua*. In eastern Oregon, some plants may show intermediate characteristics between subspecies *coronaria* and subspecies *exigua*. Typical subspecies *coronaria* from the California Coast Range has heads on short (10 mm or less) peduncles, forming a raceme on the inflorescence branches; however, the intermediate plants have a more panicle-like

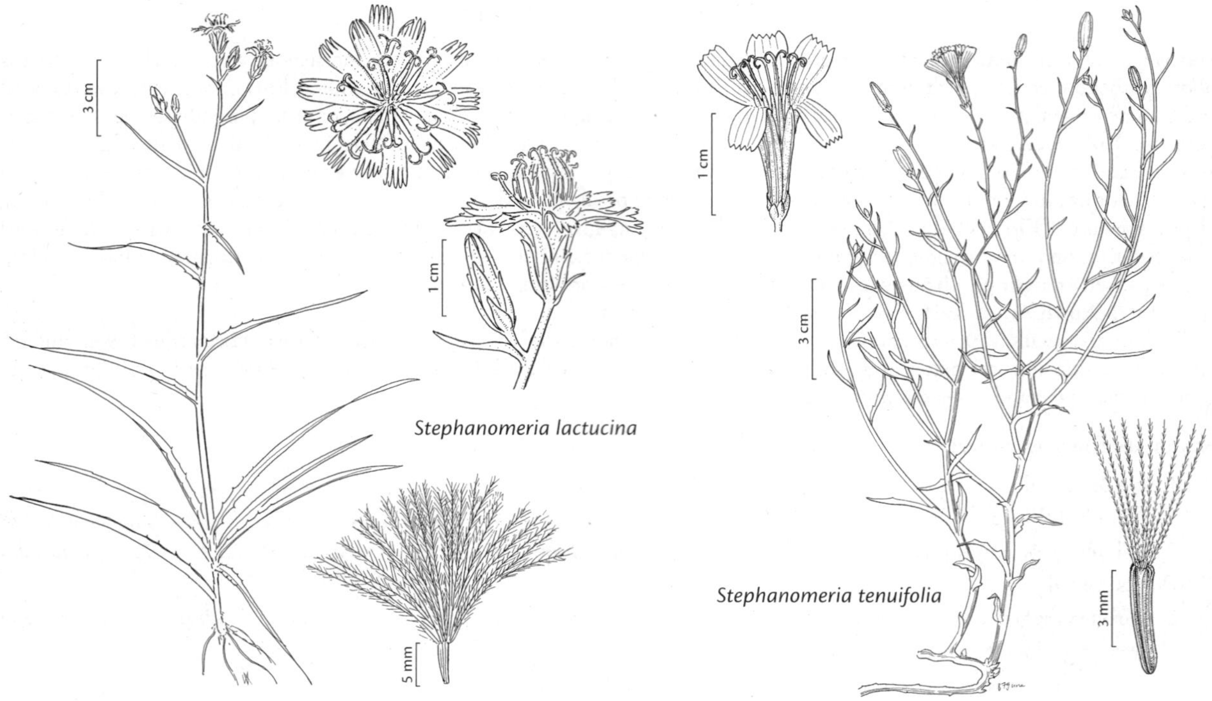

STEPHANOMERIA

arrangement with some heads on peduncles more than 10 mm long mixed with those on shorter peduncles. To simplify this treatment, we are considering such plants to be referable to subspecies *coronaria*, since no typical subspecies *exigua* (with all peduncles more than 10 mm) has been identified in the state.

Stephanomeria lactucina A. Gray (p. 352)

Large-flowered wirelettuce, woodland wirelettuce

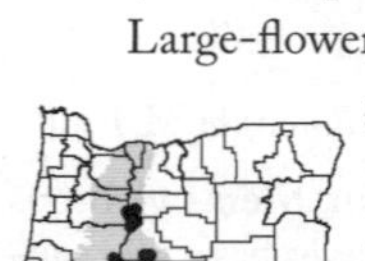

Plants perennial, 5–70 cm; rhizomatous. **Stems** glabrous or sparsely pubescent. **Leaves** cauline, persistent at flowering, scarcely reduced distally, linear or narrowly lanceolate, 3–8 cm, margins entire or denticulate, surfaces glabrous or sparsely pubescent; sessile or short-petiolate. **Inflorescences** panicle- or raceme-like arrays, sometimes heads solitary. **Peduncles** 3–10 mm. **Involucres** campanulate, 9–15 mm, surfaces glabrous or minutely puberulent. **Phyllaries** inner 6–12, narrowly lanceolate; outer basal, lanceolate, gradually shorter, loosely reflexed. **Florets** (7)8–13(14); ligules 12–20 mm, pink or light purple. **Fruits** 5–6 mm, light tan, grooved and smooth between ribs; pappus bristles 25–30, white or light tan, plumose to base, persistent. ***2n***=16.

Open flats and hillsides in coniferous forests. Flowering Jul–Aug. 900–2300 m. Casc, ECas. CA. Native.

Stephanomeria malheurensis Gottlieb

Malheur wirelettuce

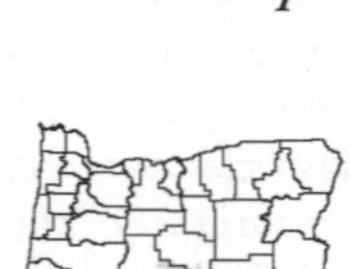

Plants annual, 10–60 cm. **Stems** glabrous. **Leaves** basal and cauline; basal usually withering by flowering, 5–7 cm, oblanceolate, margins entire to irregularly lobed, surfaces glabrous; cauline much reduced, bract-like. **Inflorescences** panicle- or raceme-like arrays. **Peduncles** 3–15 mm. **Involucres** cylindrical, 8–9.5 mm, surfaces glabrous. **Phyllaries** inner 5–6, linear-lanceolate; outer mostly basal, deltate, loosely reflexed. **Florets** 5–6; ligules 8–9.5 mm, light to dark pink. **Fruits** 3.3–3.8 mm, tan or light brown, grooved and roughened between ribs; pappus bristles 9–12(15), light tan, plumose on distal 50–60%, persistent or distal portion deciduous. ***2n***=16.

Sagebrush scrub. Flowering Jul–Aug. 1300–1400 m. BR. Native.

Stephanomeria malheurensis is a recently evolved species derived from *S. exigua*, with which it coexists at its only known location in Harney County. It is morphologically distinct from the parent species, differing in fruit size, floret number, and ligule length. Additionally, *S. malheurensis* is self-compatible and sets viable fruits without insect pollination, whereas *S. exigua* requires cross-pollination for successful fruit production.

Stephanomeria paniculata Nuttall

Stiff-branched wirelettuce, tufted wirelettuce

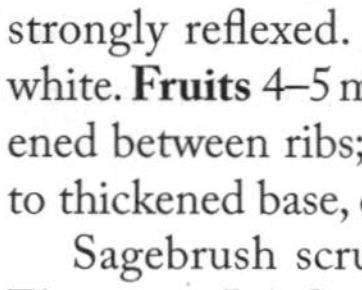

Plants annual, 20–100 cm. **Stems** usually glabrous. **Leaves** basal and cauline, usually withering before flowering; basal oblanceolate, 6–10 cm, margins dentate or sometimes with basal lobes, surfaces glabrous; cauline reduced upwards, linear, margins toothed or entire. **Inflorescences** panicle-like arrays. **Peduncles** 0–5(10) mm. **Involucres** cylindrical or narrowly ovate, 6–9 mm, surfaces glabrous. **Phyllaries** inner 5–8, linear-lanceolate; outer basal, deltate to ovate, strongly reflexed. **Florets** 5; ligules 6–10 mm, dark pink to white. **Fruits** 4–5 mm, tan to brown, lightly grooved and roughened between ribs; pappus bristles 15–18, tan, plumose nearly to thickened base, deciduous above bases. ***2n***=16.

Sagebrush scrub, open flats and slopes, disturbed areas. Flowering Jul–Sep. 50–1400 m. BW, Col, ECas, Lava, Owy. CA, ID, WA. Native.

Stephanomeria tenuifolia (Rafinesque) H.M. Hall (p. 352)

Narrow-leaved wirelettuce

Stephanomeria tenuifolia (Torr.) H.M. Hall var. *myrioclada* (D.C. Eaton) Cronquist, *Stephanomeria tenuifolia* (Torr.) H.M. Hall var. *tenuifolia*

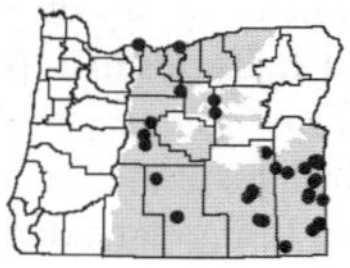

Plants perennial, 20–70 cm; usually taprooted. **Stems** glabrous. **Leaves** basal and cauline, usually withered by flowering; basal narrowly linear, 5–8 cm, margins entire or dentate, surfaces glabrous; cauline reduced upwards, narrowly linear or thread-like. **Inflorescences** raceme-like arrays. **Peduncles** 0–30 mm. **Involucres** cylindrical, 5–14 mm, surfaces glabrous. **Phyllaries** inner (3)4–6, linear or linear-lanceolate; outer basal, deltate to lanceolate, loosely reflexed. **Florets** 3–6; ligules 6–15 mm, dark pink to white. **Fruits** 3–6 mm, light tan or gray, grooved and smooth between ribs; pappus bristles 15–25, white, ± plumose to base, persistent. **2*n***=16.

Rocky cliffs, stony hillsides, lava flows, ridge tops, barren slopes, road cuts. Flowering Jun–Sep. 100–2400 m. BR, Col, ECas, Lava, Owy. CA, ID, NV, WA; north to British Columbia, northeast to Saskatchewan, east to WY, southeast to TX. Native.

Stephanomeria tenuifolia is diverse in growth form, sometimes appearing intricately branched and shrub-like and other times few-stemmed, slender, and divergently branched. It is frequently associated with crevices in rock cliffs and lava flows, as well as dry, open, stony ground.

Stephanomeria virgata Bentham ssp. ***pleurocarpa*** (Greene) Gottlieb

Wand wirelettuce, tall wirelettuce

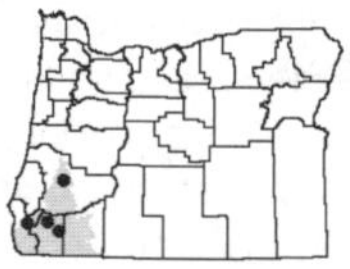

Plants annual, 10–120+ cm. **Stems** usually glabrous. **Leaves** basal and cauline; basal withering by flowering, oblanceolate, 3–10 cm, margins dentate or lobed, surfaces usually glabrous; cauline linear, much reduced. **Inflorescences** raceme-like arrays. **Peduncles** 0–5(10) mm. **Involucres** cylindrical, 6–8 mm, surfaces glabrous. **Phyllaries** inner (4)5–8, linear-lanceolate; outer basal, deltate to lanceolate, erect with spreading tips. **Florets** 5–9; ligules 8–12 mm, dark pink to white. **Fruits** 2.5–3.5 mm, tan to dark brown, roughened but not grooved between obscure ribs; pappus bristles 23–28, white, plumose to thickened base, deciduous. **2*n***=16.

Open forests, shrublands, hillsides, riverbanks, roadsides. Flowering Jul–Sep. 100–1100 m. Sisk. CA, NV. Native.

STYLOCLINE Neststraw, stylocline

Herbs annual. **Stems** mostly erect, lateral often decumbent, simple to much branched, floccose-tomentose. **Leaves** cauline, alternate, narrowly lanceolate to oblanceolate, becoming bracteate below heads, margins entire, surfaces silky-tomentose; sessile or petiolate and winged. **Inflorescences** tightly clustered cyme- or panicle-like arrays. **Heads** disciform, ovoid or ellipsoid. **Involucres** 0 or inconspicuous. **Phyllaries** 0 or 1–4 and vestigial, equal or subequal. **Receptacles** usually cylindric; paleate. **Outer paleae** falling, ovate to boat-shaped, > inner paleae, winged. **Inner paleae** falling, linear-lanceolate to oblanceolate. **Outer florets** pistillate, 12–25+, enclosed by paleae. **Inner florets** staminate, 2–5; corolla lobes 5, equal. **Fruits** obovoid, compressed, retained within paleae; pappi 0 or of few bristles.

Western Mexico and United States. 7 species; 1 species treated in *Flora*.

Stylocline psilocarphoides M. Peck (p. 353)

Baretwig neststraw, Malheur stylocline

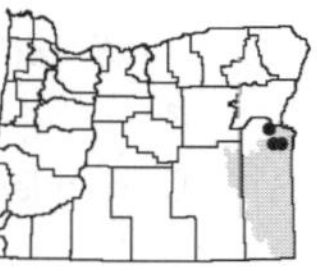

Plants 0.5–10 cm. **Leaves** 2–20 mm, tips acute, mucronate. **Heads** 3–10 × 1–2 mm. **Phyllaries** 0 or 0.1–0.5 mm. **Receptacles** 2–3 mm. **Outer paleae** 2–4 mm, surfaces densely floccose-pubescent, wings terminal or subterminal, sometimes brownish. **Inner florets** corollas 1.3–1.5 mm. **Fruits** 1.1–1.6 mm, dark brown, glabrous; pappi 0 or of 1–3 bristles, 1.1–1.5 mm.

Brushlands, open slopes and flats. Flowering Mar–Jun. 100–2000 m. Owy. CA, ID, NV; southeast to UT. Native.

Although discovered in Oregon, *Stylocline psilocarphoides* is more abundant to the south, in the Great Basin region and Mojave Desert. The breeding system of *Stylocline* is principally by self-pollination, with the styles of the pistillate flowers being positioned close to the open corollas of the staminate flowers. The dense pubescence of the pistillate paleae aids in fruit dispersal.

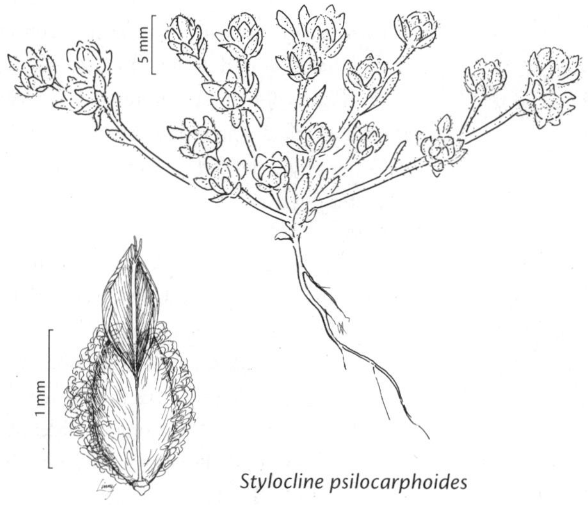

Stylocline psilocarphoides

STYLOCLINE

SYMPHYOTRICHUM Aster

Geraldine A. Allen

Herbs annual or perennial; taprooted or from caudices or short to elongate rhizomes. **Stems** generally erect, glabrous to densely pubescent, sometimes glandular-pubescent. **Leaves** basal and cauline, alternate, linear to lanceolate or elliptic or obovate, generally entire, basal leaves senescent or persistent at flowering; petiolate to sessile. **Inflorescences** compact to open panicle- or raceme-like arrays; branches usually bracteolate. **Heads** radiate or rarely discoid. **Involucres** cylindrical to campanulate. **Phyllaries** in 3–6 series, appressed to spreading, imbricate to subequal or outer bracts longer, bases with pale scarious margins and tips green, or the outer foliaceous and green throughout. **Receptacles** flat to slightly convex; paleae 0. **Ray florets** (0)10–50(100), pistillate; rays white to pink or violet-purple. **Disc florets** (6)10–50(100), bisexual; corollas yellow, tubular, tubes < throats, lobes triangular, erect or reflexed. **Fruits** obconic, somewhat flattened, 3–5-veined, brown to purplish; pappi of (20)25–40 barbellate bristles in 1 series.

Asia, North America, South America. ~90 species; 15 species treated in *Flora.*

Symphyotrichum is well known for its taxonomic difficulty, a reflection of multiple factors including the prevalence of polyploid complexes (in Oregon, mainly involving species with n=8) and the influence of phenology and environment on many morphological characters.

Symphyotrichum novae-angliae, an occasional garden escape, is separable from *S. campestre* by its larger heads with usually 50–75 rays (versus 15–30 rays for *S. campestre*).

Key to *Symphyotrichum* species:

1. Plants annual, taprooted; rays 0 or 1–2 mm.
 2. Rays 0 . . . . **S. ciliatum**
 2' Rays 1–2 mm . . . . **S. frondosum**

1' Plants perennial, from rhizomes or caudices; rays ≥ 4 mm.
 3. Involucres and peduncles glandular . . . . **S. campestre**
 3' Involucres and peduncles not glandular.
  4. Involucres 2.5–5 mm; rays 10–20; outer phyllaries minutely mucronate; leaves 1–6 cm, tips mucronate to spinose . . . . ***S. ericoides* var. *pansum***
  4' Involucres 4–20 mm; rays 13–60; outer phyllaries obtuse to acute, not mucronate; leaves 1–20 cm, tips not mucronate to spinulose.
   5. Basal leaves withering by flowering.
    6. Rays usually violet or purple, sometimes white; inflorescences open, with at least some branches > 10 cm.
     7. Outer phyllaries obtuse or acute, < inner; plants generally coastal . . . . **S. chilense**
     7' Outer phyllaries acute, ≤ inner; plants widespread.
      8. Stems glabrous or with pubescence in lines from leaf bases; inflorescence bracts often > peduncles; phyllaries usually appressed; plants of wet meadows, streambanks, and open coniferous forests . . . . ***S. lanceolatum* var. *hesperium***
      8' Stems glabrous to pubescent, hairs not in lines; inflorescence bracts < peduncles; phyllaries often spreading; plants of disturbed habitats . . . . **S. subspicatum**
    6' Rays usually pink to white; inflorescences narrow, with branches generally < 10 cm.
     9. Plants 4–10 dm; ray florets 20–40; leaves 6–20 cm; outer phyllaries ≈ inner. . . . . **S. bracteolatum**
     9' Plants 3–6 dm; ray florets 14–35; leaves 1–6 cm; outer phyllaries < inner . . . . **S. hallii**
   5' Basal leaves usually present at flowering, or if withering cauline leaf bases clasping.
    10. Outer phyllaries obtuse to rounded; stems strigose at least above . . . . **S. ascendens**
    10' Outer phyllaries acute (sometimes outer in *S. spathulatum* obtuse); stems glabrous to pubescent.
     11. Phyllaries appressed; leaves glabrous, glaucous . . . . ***S. laeve* var. *geyeri***
     11' Phyllaries not appressed; leaves glabrous to pubescent, not glaucous.
      12. Leaves ≥ 7 × as long as wide; outer phyllaries scarious-margined.
       13. Inflorescences much branched; cauline leaves elliptic to oblanceolate. . . . . **S. hendersonii**
       13' Inflorescences usually few-branched; cauline leaves linear to narrowly elliptic . . **S. spathulatum**
      12' Leaves < 7 × as long as wide; phyllaries not scarious-margined.
       14. Basal leaves withering by flowering; cauline leaves sessile, clasping. . . . . **S. cusickii**
       14' Basal leaves present at flowering; cauline leaves subpetiolate to sessile, not clasping . . . . **S. foliaceum**

Symphyotrichum ascendens (Lindley) G.L. Nesom (p. 355)

Long-leaved aster, Intermountain aster, western aster

Aster ascendens Lindl., *Aster chilensis* Nees, *Aster chilensis* Nees ssp. *adscendens* (Lindl.) Cronquist

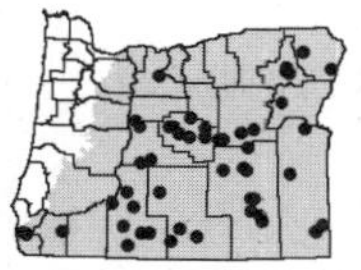

Plants perennial, 2–6 dm; from elongate rhizomes. **Stems** ascending to erect, strigose at least above. **Leaves** margins entire, tips acute, surfaces glabrous to strigose; basal usually present at flowering, oblanceolate, 5–15 cm, petiolate; cauline oblong to narrowly oblanceolate, 3–12 cm, reduced upward, subpetiolate to sessile. **Inflorescences** open panicle-like arrays. **Heads** 3–25(40). **Involucres** 4–7 mm. **Phyllaries** in 3–5 unequal series, bases scarious-margined, tips green, surfaces strigose; inner oblong, tips acute; outer narrowly obovate, < inner, tips obtuse to rounded. **Ray florets** 15–40; rays 8–14 mm, violet. **Fruits** ± strigose; pappus bristles 4–7 mm, whitish. ***2n***=26, 52.

Montane meadows, open ponderosa pine forests, damp places in sagebrush steppe. Flowering Jul–Sep. 700–2300 m. BR, BW, Casc, Col, ECas, Lava, Owy, Sisk. CA, ID, NV, WA; north to British Columbia, northeast to Saskatchewan, east to NB, southeast to NM, south to Mexico. Native.

This species is the only *Symphyotrichum* with a base chromosome number of *x*=13. It is an allopolyploid derived from hybridization between *S. spathulatum* (*x*=8) and the Great Plains species *S. falcatum* (*x*=5).

Symphyotrichum bracteolatum (Nuttall) G.L. Nesom (p. 355)

Eaton's aster, Oregon aster

Aster eatonii (A. Gray) Howell, *Symphyotrichum eatonii* (A. Gray) G.L. Nesom

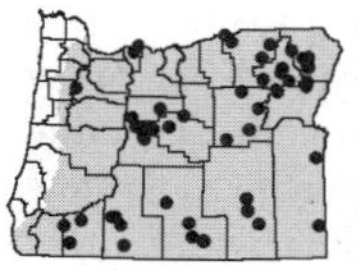

Plants perennial, 4–10 dm, from short rhizomes. **Stems** ascending to erect, often reddish at bases, glabrous or sometimes sparsely pubescent above. **Leaves** narrowly lanceolate, margins generally entire, tips acute, glabrous to sparsely pubescent, petiolate; basal withering by flowering, 6–20 cm; cauline 5–15 cm, sessile. **Inflorescences** narrow panicle-like arrays, branches generally < 10 cm. **Heads** numerous. **Involucres** 5–8 mm. **Phyllaries** in 3–5 subequal series, often spreading, bases scarious-margined, tips green, or green throughout, surfaces glabrous or sparsely pubescent; inner oblong, tips acute; outer oblanceolate, tips obtuse. **Ray florets** 20–40; rays 7–15 mm, pink to white, sometimes violet. **Fruits** pubescent; pappus bristles 5–6 mm, whitish. ***2n***=16, 32, 48, 64.

Damp meadows, open riparian areas. Flowering Jul–Oct. 50–2200 m. BR, BW, Casc, Col, ECas, Lava, Owy, Sisk, WV. CA, ID, NV, WA; north to British Columbia, northeast to Saskatchewan, east to NB, southeast to NM, south to Mexico. Native.

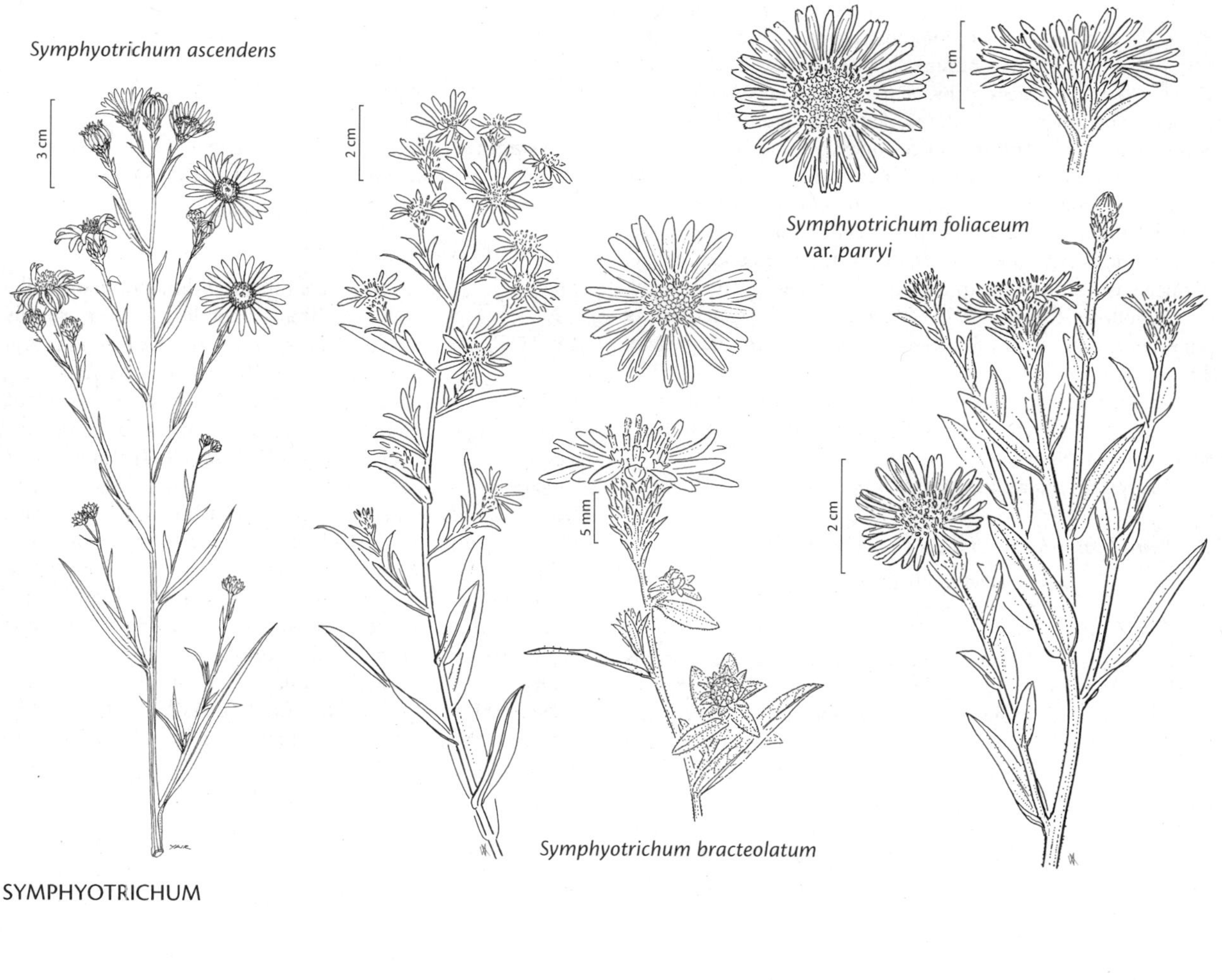

SYMPHYOTRICHUM

Symphyotrichum campestre (Nuttall) G.L. Nesom
Western meadow aster

Aster campestris Nutt., *Aster campestris* Nutt. var. *bloomeri* A. Gray, *Aster campestris* Nutt. var. *campestris*

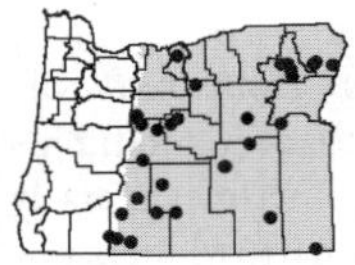

Plants perennial, 1–4 dm; from elongate rhizomes. **Stems** ascending to erect, generally strigose, glandular. **Leaves** linear to narrowly oblanceolate, 2–6 cm, margins entire, tips obtuse to acute, mucronate to minutely spinulose, surfaces glabrous to stipitate-glandular; basal withering by flowering; sessile. **Inflorescences** narrow to spreading panicle-like arrays. **Heads** 1–10(30); peduncles glandular. **Involucres** 5–8 mm, glandular. **Phyllaries** in 3–4 subequal to unequal series, spreading to reflexed, linear to lanceolate, bases scarious-margined, tips acute, green, surfaces stipitate-glandular. **Ray florets** 15–30; rays 6–15 mm, violet. **Fruits** pubescent; pappus bristles 4–6 mm, brownish. **2*n***=10.

Meadows, open dry coniferous forests, disturbed areas. Flowering Jul–Sep. 200–1900 m. BR, BW, Col, ECas, Lava, Owy. CA, ID, NV, WA; north to British Columbia, northeast to Alberta, east to WY, southeast to NM, south to Mexico. Native.

Symphyotrichum chilense (Nees) G.L. Nesom
Common California aster, Pacific aster

Aster chilensis Nees, *Aster chilensis* Nees ssp. *chilensis*, *Aster chilensis* Nees var. *chilensis*

Plants perennial, 4–10 dm; from elongate rhizomes. **Stems** ascending to erect, glabrous to hirsute. **Leaves** oblanceolate, margins entire to finely toothed, tips acute, surfaces glabrous to sparsely puberulent; basal withering by flowering, 5–20 cm, petiolate; cauline 4–15, subpetiolate to sessile. **Inflorescences** open, panicle-like arrays, at least some branches 10–20 cm. **Heads** numerous. **Involucres** 5–8 mm. **Phyllaries** in 3–5 unequal to subequal series, often spreading, surfaces glabrous to puberulent; inner linear-oblong, bases scarious-margined, tips acute, green; outer oblanceolate, < inner, bases scarious-margined or sometimes green throughout, tips obtuse. **Ray florets** 15–40; rays 9–15 mm, violet. **Fruits** pubescent; pappus bristles 4–8 mm, whitish to pale brown. **2*n***=48, 64.

Grasslands, salt marshes, coastal dunes, disturbed areas. Flowering Jun–Oct. 0–400 m. CR, Est. CA, WA; north to British Columbia. Native.

This species intergrades with *S. subspicatum*.

Symphyotrichum ciliatum (Ledebour) G.L. Nesom
Rayless alkali aster

Aster brachyactis S.F. Blake

Plants annual, 1–7 dm; taprooted. **Stems** ascending to erect, often red-tinged, glabrous. **Leaves** margins entire to serrulate, tips acute, surfaces strigose to scabrous; basal withering by flowering, spatulate, 2–20 cm, petiolate; cauline linear to oblanceolate, 3–10 cm, sessile. **Inflorescences** narrow, crowded panicle-like arrays. **Heads** numerous, discoid. **Involucres** 5–7 mm. **Phyllaries** in 3–4 subequal or equal series, linear to oblanceolate, bases usually scarious-margined, tips acute, green, surfaces glabrous. **Ray florets** 0. **Fruits** hirsute to strigose, purplish to grayish; pappus bristles 4–6 mm, white to pinkish. **2*n***=14.

Moist saline or alkaline soils, disturbed places. Flowering Aug–Oct. 50–2000 m. Col. ID, WA; north to AK, northeast to Quebec, east to NY, southeast to NM; Asia, Europe. Exotic.

Symphyotrichum cusickii (A. Gray) G.L. Nesom
Cusick's aster

Aster cusickii A. Gray, *Aster foliaceus* Lindl. ex DC., *Aster foliaceus* Lindl. ex DC. var. *cusickii* (A. Gray) Cronquist, *Aster foliaceus* Lindl. ex DC. var. *lyallii* (A. Gray) Cronquist

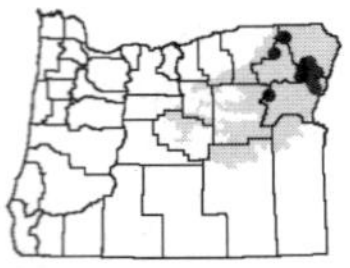

Plants perennial, 4–10 dm; from elongate rhizomes. **Stems** ascending to erect, glabrous to puberulent. **Leaves** < 7 × as long as wide, margins entire, tips acute, surfaces glabrous to puberulent; basal withering by flowering, broadly elliptic to obovate, 1–20 cm, petiolate; cauline lower withering by flowering, oblanceolate to obovate, 7–12 cm, bases strongly clasping, sessile. **Inflorescences** congested to open, panicle-like arrays, or sometimes heads solitary. **Heads** 1–numerous. **Involucres** 10–20 mm. **Phyllaries** in 4–5 equal or subequal series, surfaces glabrous to finely puberulent; inner linear-lanceolate, bases scarious-margined, tips acute, green; outer foliaceous, broadly lanceolate to oblanceolate, > inner, bases not scarious-margined, tips acute. **Ray florets** 25–55; rays 10–18 mm, violet. **Fruits** pubescent; pappus bristles 5–9 mm, white. **2*n***=16, 32.

Damp meadows, open dry forests, near springs or streams. Flowering Jul–Sep. 1200–2500 m. BW. ID, WA; northeast to MT, east to WY. Native.

This species intergrades with *S. foliaceum* and *S. spathulatum*.

Symphyotrichum ericoides (Linnaeus) G.L. Nesom var. ***pansum*** (S.F. Blake) G.L. Nesom
White heath aster

Aster pansus (S.F. Blake) Cronquist

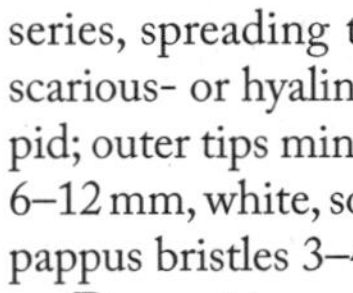

Plants perennial, 2–8 dm; from corm-like caudices. **Stems** ascending to erect, sparsely to densely hispid-strigose. **Leaves** all except uppermost usually withering by flowering, linear to oblong, oblanceolate or spatulate, 1–6 cm, margins entire, tips acute, mucronate to spinulose, surfaces glabrous to sparsely pubescent; sessile. **Inflorescences** elongate to pyramidal, sometimes secund, panicle-like arrays. **Heads** 1–200. **Involucres** 2.5–5 mm. **Phyllaries** in 3–4 unequal series, spreading to reflexed, oblanceolate to spatulate, bases scarious- or hyaline-margined, surfaces sparsely to densely hispid; outer tips minutely mucronate. **Ray florets** 10–18(20); rays 6–12 mm, white, sometimes pinkish or bluish. **Fruits** strigillose; pappus bristles 3–4 mm, whitish. **2*n***=10.

Dry prairies, grasslands, salt flats, disturbed areas. Flowering Jul–Sep. 100–1000 m. BW, Col, Lava. ID, WA; north to Northwest Territories, northeast to Ontario, east to WY, southeast to AZ. Native.

Symphyotrichum foliaceum (Lindley ex de Candolle) G.L. Nesom

Leafybract aster

Plants perennial, 0.5–4(6) dm; from elongate rhizomes. **Stems** ascending to erect, glabrous to sparsely pubescent. **Leaves** < 7 × as long as wide, margins generally entire, surfaces usually glabrous; basal present at flowering, elliptic to obovate, 3–20 cm, tips acute or sometimes rounded, petiolate; cauline elliptic to obovate, 3–12 cm, bases cuneate to sometimes rounded, subpetiolate to sessile. **Inflorescences** open, panicle-like arrays. **Heads** 1–12. **Involucres** 6–16 mm. **Phyllaries** in 4–6 subequal series, oblanceolate to oblong or linear, tips acute, surfaces glabrous or puberulent; inner bases scarious-margined, tips green; outer often leaf-like, bases not scarious-margined. **Ray florets** 15–50(60); rays 8–18 mm, violet-purple. **Fruits** pubescent; pappus bristles 5–8 mm, white to pale brown. **2*n***=16, 32, 48, 64.

Varieties of *S. foliaceum* intergrade with *S. spathulatum* and *S. cusickii* where their ranges overlap.

Western North America. 4 varieties; 2 varieties treated in *Flora*.

Key to *Symphyotrichum foliaceum* varieties:

1. Plants 0.5–2 dm; heads 1(4); phyllaries usually purple-tinged . . . . ***S. f.*** var. ***apricum***

1' Plants 1–4(6) dm; heads (1)2–12; phyllaries green. . . . . ***S. f.*** var. ***parryi***

Symphyotrichum foliaceum (Lindley ex de Candolle) G.L. Nesom var. ***apricum*** (A. Gray) G.L. Nesom

Alpine leafybract aster

Aster foliaceus Lindl. ex DC. var. *apricus* A. Gray

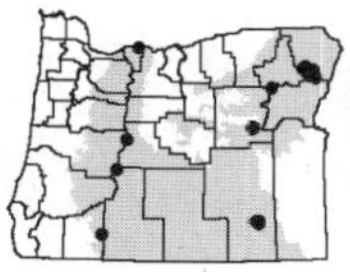

Plants 0.5–2 dm. **Leaves** generally 3–10 cm. **Heads** 1(4). **Phyllaries** usually purple-tinged, subequal or the outer slightly longer.

Alpine meadows. Flowering Jul–Sep. 1800–2900 m. BR, BW, Casc, ECas. CA, ID, WA; north to British Columbia, northeast to MT, east to WY. Native.

Symphyotrichum foliaceum (Lindley ex de Candolle) G.L. Nesom var. ***parryi*** (D.C. Eaton) G.L. Nesom (p. 355)

Parry's leafybract aster

Aster foliaceus Lindl. ex DC. var. *frondeus* A. Gray, *Aster foliaceus* Lindl. ex DC. var. *parryi* (D.C. Eaton) A. Gray

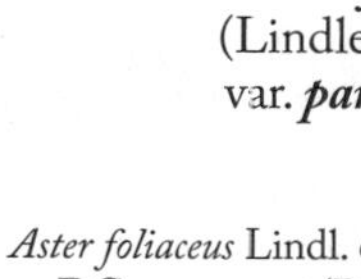

Plants 1–4(6) dm. **Leaves** generally 5–12 cm. **Heads** (1)2–12. **Phyllaries** green, subequal.

Montane and subalpine meadows, forest openings. Flowering Jun–Oct. 700–2700 m. BR, BW, Casc, ECas, Lava, Owy. CA, ID, NV, WA; north to British Columbia, northeast to MT, east to WY. Native.

Symphyotrichum frondosum (Nuttall) G.L. Nesom

Alkali aster, short-rayed aster

Aster frondosus (Nutt.) Torr. & A. Gray

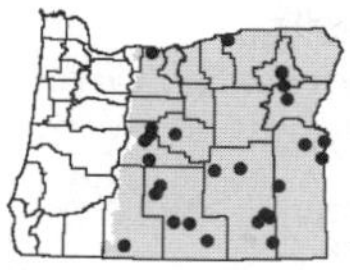

Plants annual, 1–6 dm; taprooted. **Stems** decumbent to erect, surfaces glabrous. **Leaves** margins entire to serrulate, tips acute to obtuse, surfaces glabrous; basal withering by flowering, oblanceolate to spatulate, 2–12 cm, petiolate; cauline lower withering by flowering, linear to oblanceolate, 1–8 cm, sessile. **Inflorescences** narrow, ± crowded, panicle-like arrays. **Heads** numerous. **Involucres** 5–9 mm. **Phyllaries** in 3–4 subequal to unequal series, ± spreading, oblong to oblanceolate or obovate, bases scarious-margined, tips green, surfaces glabrous. **Ray florets** 90–110; rays 1–2 mm, barely exceeding the disc florets, pinkish white to purplish white. **Fruits** strigillose; pappus bristles 6–7.5 mm, white to yellowish. **2*n***=14.

Marshes, lake edges. Flowering Jun–Oct. 50–1600 m. BR, BW, Col, ECas, Lava, Owy. CA, ID, NV, WA; north to British Columbia, northeast to Alberta, east to WY, southeast to NM. Native.

Symphyotrichum hallii (A. Gray) G.L. Nesom

Hall's aster

Aster chilensis Nees, *Aster chilensis* Nees ssp. *hallii* (A. Gray) Cronquist, *Aster hallii* A. Gray

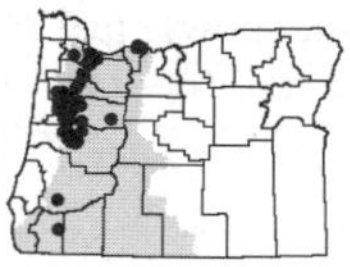

Plants perennial, 3–6 dm; from elongate rhizomes. **Stems** ascending to erect, glabrous or sometimes puberulent above. **Leaves** narrowly oblanceolate to linear, 1–6 cm, margins entire, tips acute, surfaces glabrous to sparsely scabrous; basal withering by flowering, petiolate; cauline lower withering by flowering, sessile. **Inflorescences** narrow, panicle-like arrays, branches < 10 cm. **Heads** numerous. **Involucres** 4–7 mm. **Phyllaries** in 5–6 unequal series, narrowly oblanceolate to linear; outer appressed to spreading, < inner, bases scarious-margined, tips obtuse, green, or green throughout, surfaces glabrous. **Ray florets** 14–35; rays 7–12 mm, white or occasionally pink. **Fruits** pubescent; pappus bristles 5–7 mm, white. **2*n***=16, 32, 48.

Moist or dry grasslands, meadows. Flowering May–Oct. 0–500 m. Casc, ECas, Sisk, WV. WA. Native.

Symphyotrichum hendersonii (Fernald) G.L. Nesom

Kootenai aster, Lyall's aster

Aster foliaceus Lindl. ex DC., *Aster foliaceus* Lindl. ex DC. var. *lyallii* (A. Gray) Cronquist

Plants perennial, 7–15 dm; from elongate rhizomes. **Stems** ascending to erect, glabrous or sometimes pubescent above. **Leaves** ≥ 7 × as long as wide, 6–15(20) cm, margins entire, tips acute, surfaces glabrous to sparsely pubescent; basal usually present at flowering, oblanceolate, petiolate; cauline elliptic to oblanceolate, reduced upward, sessile. **Inflorescences** open, panicle-like arrays, usually much branched. **Heads** numerous. **Involucres** 5–10 mm. **Phyllaries** in 4–5 unequal series, narrowly oblong to linear, bases scarious-margined, tips acute, green, surfaces glabrous. **Ray**

florets 25–45, rays 9–15 mm, violet. **Fruits** pubescent; pappus bristles 4–5 mm, white. ***2n***=16, 32.

Meadows, forest openings, stream margins. Flowering Jul–Sep. 1000–1900 m. Sisk. CA, ID, WA. Native.

Symphyotrichum laeve (Linnaeus) Á. Löve & D. Löve var. ***geyeri*** (A. Gray) G.L. Nesom

Smooth aster, smooth blue aster

Aster laevis L.

Plants perennial, 2–8 dm; from elongate rhizomes. **Stems** erect, sometimes reddish at base, glabrous. **Leaves** lanceolate to ovate, margins entire or sometimes serrulate, tips acute, surfaces glabrous, ± glaucous; basal usually present at flowering, sometimes withering, (5)8–20 cm, petioles ± winged; cauline 8–15 cm, auriculate-clasping at base, subsessile to sessile. **Inflorescences** open, often flat-topped, panicle-like arrays; peduncles glaucous. **Heads** numerous. **Involucres** 5–7 mm. **Phyllaries** in 4–6 unequal series, appressed, oblong to lanceolate or linear-lanceolate, bases scarious-margined, tips acute, green, surfaces glabrous, green zones diamond-shaped. **Ray florets** 13–25, 7–12 mm, violet-purple. **Fruits** ± glabrous; pappus bristles 5–7 mm, brownish or sometimes pink-tinged. ***2n***=48.

Grasslands, open dry forests, roadsides. Flowering Aug–Sep. 0–1600 m. Est. CA, ID, NV, WA; north to Yukon, northeast to Ontario, southeast to TX. Native.

Symphyotrichum lanceolatum (Willdenow) (p. 358) G.L. Nesom var. ***hesperium*** (A. Gray) G.L. Nesom

Western willow aster, marsh aster, white panicle aster

Aster hesperius A. Gray, *Aster lanceolatus* Willd. ssp. *hesperius* (A. Gray) J. Semple & J. Chmiel.

Plants perennial, 6–16 dm; from elongate rhizomes. **Stems** erect, glabrous to more commonly pubescent in decurrent lines from leaf bases. **Leaves** linear-lanceolate to narrowly oblanceolate, 5–16 cm, margins entire to shallowly serrate, tips acute to mucronate, surfaces glabrous to sparsely scabrous; basal withering by flowering, petiolate; cauline sessile. **Inflorescences** open, panicle-like arrays, at least some branches > 10 cm; bracts often > peduncles. **Heads** numerous. **Involucres** 4–8 mm. **Phyllaries** in 4–6 unequal series, usually appressed, linear-lanceolate to linear, bases scarious-margined, tips acute to acuminate, green, ± glabrous; outer phyllaries ≤ inner. **Ray florets** 18–45; rays 4–10 mm, usually purple, sometimes white. **Fruits** sparsely pubescent; pappus bristles 4–6 mm, whitish to pale brown. ***2n***=64.

Wet meadows, streambanks, open conifer forests. Flowering Jul–Sep. 600–1200 m. Lava, Owy. CA, ID, NV, WA; north to Northwest Territories, northeast to Quebec, east to MN, southeast to TX. Native.

Symphyotrichum spathulatum (Lindley) G.L. Nesom

Western mountain aster

Aster occidentalis (Nutt.) Torr. & A. Gray

Plants perennial, 2–8 dm; from elongate rhizomes. **Stems** ascending to erect, glabrous to sparsely pubescent. **Leaves** ≥ 7 × as long as wide, margins entire, tips acute, surfaces glabrous to sparsely puberulent; basal leaves present at flowering, narrowly elliptic to obovate, 5–15 cm, petiolate; cauline linear to narrowly elliptic, 3–6(12) cm, subpetiolate to sessile. **Inflorescences** open, panicle-like arrays, usually few-branched. **Heads** 3–50. **Involucres** 5–10 mm. **Phyllaries** in 3–5 unequal series, linear to narrowly oblanceolate, bases scarious-margined, tips acute, sometimes ± obtuse, green, surfaces glabrous to sparsely puberulent. **Ray florets** 15–40; rays 9–15 mm, violet. **Fruits** pubescent; pappus bristles 5–7 mm, whitish.

3 varieties.

Key to *Symphyotrichum spathulatum* varieties:

1. Leaves ± linear . . . . ***S. s.*** var. ***yosemitanum***

1' Leaves elliptic to lanceolate, sometimes obovate.

  2. Plants 2–5(6) dm; heads 3–10 . . ***S. s.*** var. ***spathulatum***

  2' Plants 4–8 dm; heads 10–50. . . ***S. s.*** var. ***intermedium***

Symphyotrichum spathulatum (Lindley) G.L. Nesom var. ***intermedium*** (A. Gray) G.L. Nesom

Larger western mountain aster

Aster occidentalis (Nutt.) Torr. & A. Gray var. *intermedius* A. Gray, *Aster spathulatus* Lindl. var. *intermedius* (A. Gray) Cronquist

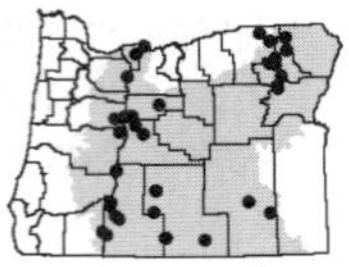

Plants 4–8 dm. **Leaves** elliptic to lanceolate, sometimes obovate. **Heads** 10–50. ***2n***=32, 48, 64.

Grasslands, montane meadows, openings in conifer forests. Flowering Jul–Oct. 200–2000 m. BR, BW, Casc, ECas, Lava. CA, ID, NV, WA; north to British Columbia, northeast to MT, east to WY. Native.

Symphyotrichum lanceolatum var. *hesperium*

SYMPHYOTRICHUM

Symphyotrichum spathulatum (Lindley) G.L. Nesom var. ***spathulatum***

Western aster, western mountain aster

Aster occidentalis (Nutt.) Torr. & A. Gray var. *occidentalis*, *Aster spathulatus* Lindl. var. *spathulatus*

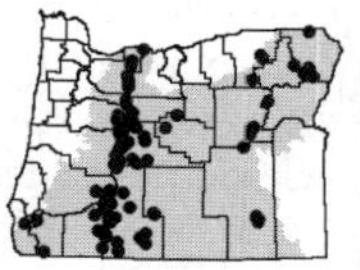

Plants 2–5(6) dm. **Leaves** elliptic to lanceolate. **Heads** 3–10. ***2n***=16, 32, 48, 64.

Montane meadows, open aspen and conifer forests. Flowering Jul–Sep. 400–2700 m. BR, BW, Casc, ECas, Lava, Sisk. CA, ID, NV, WA; north to Northwest Territories, east to WY, southeast to NM. Native.

Symphyotrichum spathulatum (Lindley) G.L. Nesom var. ***yosemitanum*** (A. Gray) G.L. Nesom

Western bog aster

Aster occidentalis (Nutt.) Torr. & A. Gray var. *yosemitanus* (A. Gray) Cronquist, *Aster paludicola* Piper

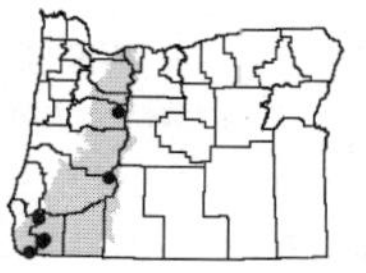

Plants 3–8 dm. **Leaves** ± linear. **Heads** 3–25. ***2n***=16, 32.

Damp meadows in oak woodlands and conifer forests. Flowering Jul–Sep. 400–1700 m. Casc, Sisk. CA. Native.

Symphyotrichum subspicatum (Nees) G.L. Nesom (p. 359)

Douglas' aster

Aster subspicatus Nees, *Aster subspicatus* Nees var. *grayi* (Suksd.) Cronquist, *Aster subspicatus* Nees var. *subspicatus*

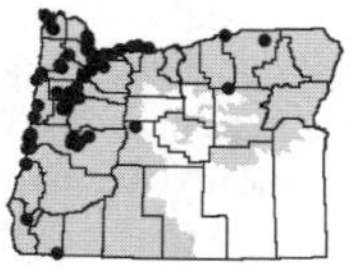

Plants perennial, 4–12 dm; from elongate rhizomes. **Stems** ascending to erect, ± glabrous below, sparsely to densely pubescent above. **Leaves** obovate to lanceolate or elliptic, (3)5–15 cm, margins entire to shallowly toothed, tips acute or sometimes obtuse, surfaces ± pubescent; basal withering by flowering, petiolate; cauline sessile. **Inflorescences** narrow to broad panicles, at least some inflorescence branches > 10 cm; bracts < peduncles. **Heads** numerous. **Involucres** 6–10 mm. **Phyllaries** in 4–6 unequal series, often spreading, oblanceolate to linear, bases scarious-margined, tips acute or sometimes obtuse, green, surfaces sparsely to densely pubescent; outer phyllaries ≤ inner. **Ray florets** 18–40; rays 10–20 mm, violet. **Fruits** sparsely pubescent; pappus bristles 7–9 mm, whitish to pale brown or reddish. ***2n***=48, 64, 80, 96.

Disturbed areas. Flowering Jun–Oct. 0–1000 m. BW, Casc, Col, CR, ECas, Est, Sisk, WV. CA, ID, WA; north to AK, northeast to Alberta. Native.

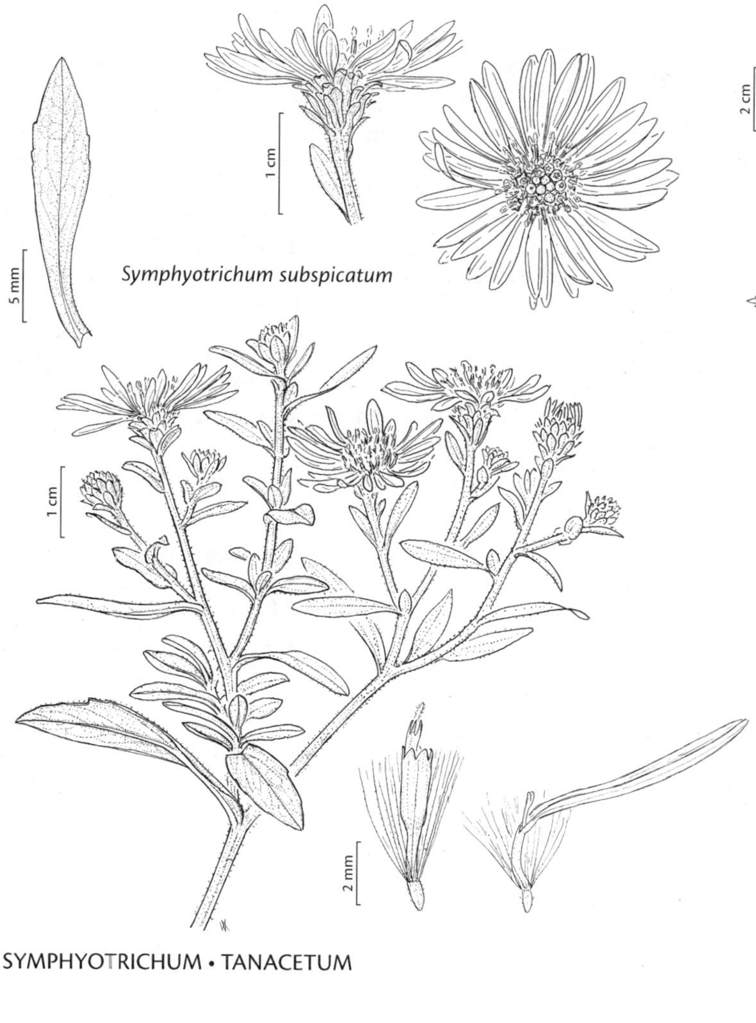

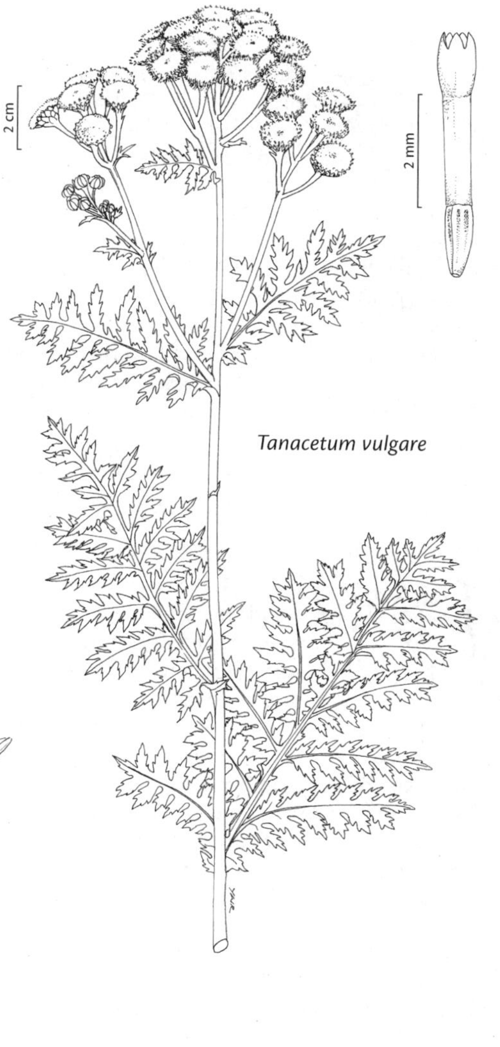

SYMPHYOTRICHUM • TANACETUM

TANACETUM Tansy

Herbs perennial, usually aromatic; taprooted, usually rhizomatous. **Stems** erect, simple to distally much branched, glabrous to villous or puberulent. **Leaves** basal and cauline, alternate, bases often with lobed, fringed, or clasping auricles, margins entire to dentate, often 1–3-times pinnately lobed, surfaces glabrous, strigillose, floccose, or lightly to densely villous, abundantly punctate-glandular; basal often withered by flowering; sessile or petiolate. **Inflorescences** clusters in panicle-like arrays. **Heads** usually radiate, sometimes disciform; peduncles glabrous to strigillose or villous. **Involucres** campanulate or hemispheric. **Phyllaries** in 3–6 series, surfaces glabrous, villous, or minutely glandular; inner equal, linear-lanceolate to lanceolate or ovate; outer equal or gradually shorter, lanceolate, ovate or triangular, margins membranous or chartaceous, tips acute or obtuse. **Receptacles** flat to rounded or conic, sometimes exceeding involucres in fruit; paleae 0. **Ray florets** pistillate, sometimes vestigial or 0; rays ovate-elliptic, tips acute or truncate, sometimes 2–4-lobed. **Disc florets** numerous, bisexual; corollas tubular, linear or gradually widened distally, 4–5-lobed, lobes triangular; stamens included or slightly exserted; stigma lobes at tips of stamen tubes. **Fruits** obconic or columnar, glabrous or gland-dotted, veins 5–10+; pappi usually of low crowns, rarely of scales or 0.

Asia, Europe, North Africa, North America. 160 species; 4 species treated in *Flora*.

Key to *Tanacetum* species:

1. Leaves not lobed or sometimes few-lobed near bases . . . . *T. balsamita*

1' Leaves 1–3–pinnately lobed.

  2. Rays white; leaves 1–2–pinnately lobed, primary lobes in 2–3(4) pairs . . . . *T. parthenium*

  2' Rays 0 or if present yellow; leaves 2–3–pinnately lobed, primary lobes 4–24 pairs.

    3. Leaves glabrous or lightly pubescent; heads 20–200; involucres 5–10 mm in diameter . . . . *T. vulgare*

    3' Leaves villous to woolly, sometimes glabrous; heads 2–20+; involucres 8–22+ mm in diameter . . . . . . *T. bipinnatum*

Tanacetum balsamita Linnaeus
Costmary

Chrysanthemum balsamita (L.) Baill.

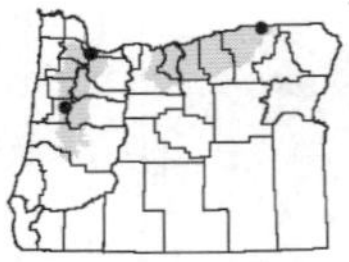

Plants 3–8+ dm. **Leaves** ovate-lanceolate or elliptic, 2–35 × 2–10 cm, margins entire or few-lobed near bases, surfaces glabrous or strigillose. **Heads** 3–60; peduncles 0–8 mm, surfaces strigillose. **Involucres** 3–4 × 3–8 mm. **Phyllaries** chartaceous-margined; inner often membrane-tipped. **Ray florets** usually 0, rarely 12–15, 4–7 mm, white. **Disc florets** corollas 2–2.5 mm, yellow. **Fruits** 1.5–2 mm, brown, gland-dotted, often puberulent, 5–8-ribbed; pappi of low crowns. **2*n***=18, 54.

Roadsides, disturbed areas. Flowering Jul–Sep. 0–500 m. Col, WV. CA, ID, NV, WA; widely scattered in Canada and US; Asia, Europe. Exotic.

Tanacetum bipinnatum (Linnaeus) Schultz-Bipontinus
Dune tansy

Tanacetum camphoratum Less., *Tanacetum douglasii* DC.

Plants 1.5–5+ dm. **Leaves** oblanceolate or elliptic in outline, 3–35 × 2–10 cm, margins often curled, 2–3-times pinnately lobed, surfaces villous to woolly, sometimes glabrous, primary lobes in 6–24 pairs, secondary lobes linear to ovate or obovate. **Heads** 2–20+, peduncles 0.3–12 cm, villous. **Involucres** 4–5 × 8–22+ mm. **Phyllaries** with membranous margins; inner not membrane-tipped. **Ray florets** 0–25; rays 1–6 mm, yellow. **Disc florets** corollas 2–2.5 mm, yellow. **Fruits** 2–3 mm, tan, gland-dotted, 3–5-ribbed; pappi of low crowns or a ring of scales. **2*n***=54.

Dunes, sandy flats, bluffs, grassy slopes. Flowering May–Aug. 0–100 m. Est. CA, WA; north to AK, northeast to Newfoundland; Asia, Europe. Native.

Formerly treated as *T. camphoratum*, this taxon, as well as the related *T. huronense*, has been combined into *T. bipinnatum*.

Tanacetum parthenium (Linnaeus) Schultz-Bipontinus
Feverfew, featherfew

Chrysanthemum parthenium (L.) Bernh.

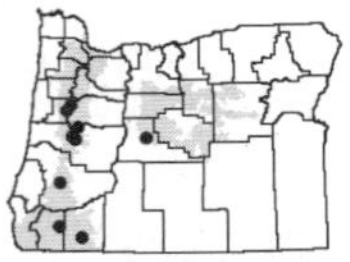

Plants 2–6+ dm. **Leaves** ovate to lanceolate in outline, 2–12 × 1–4 cm, margins often dentate, 1–2-pinnately lobed, surfaces glabrous or minutely pilose, primary lobes in 2–3(4) pairs, secondary lobes ovate to oblanceolate. **Heads** 5–25; peduncles 1–10 cm, glabrous or puberulent. **Involucres** 3–3.5 × 5–12 mm. **Phyllaries** with chartaceous margins; inner not membrane-tipped. **Ray florets** 10–40; rays 3–6 mm, white. **Disc florets** corollas 1–3 mm, yellow. **Fruits** 1–2 mm, brown, glabrous, 5–8-ribbed; pappi 0 or of low crowns. **2*n***=18.

Woodlands, riverbanks, roadsides, disturbed areas. Flowering Jun–Oct. 0–1200 m. Lava, Sisk, WV. ID, NV, WA; widely scattered in North America; worldwide. Exotic.

Tanacetum vulgare Linnaeus (p. 359)
Common tansy

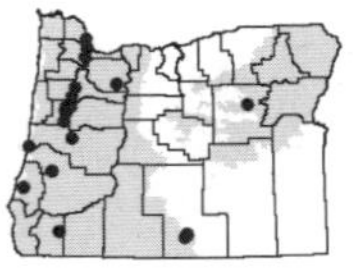

Plants 4–15 dm. **Leaves** elliptic in outline, 2–25 × 2–10 cm, margins usually denticulate, 2-pinnately lobed, surfaces glabrous or lightly pubescent, primary lobes in 4–10 pairs, secondary lobes linear to oblanceolate or triangular. **Heads** 20–200; peduncles 2–15 mm, glabrous or villous. **Involucres** 3–5 × 5–10 mm. **Phyllaries** with chartaceous margins; inner often membrane-tipped. **Ray florets** 15–25, vestigial, yellow. **Disc florets** corollas 1.5–2 mm, yellow. **Fruits** 1–2 mm, brown, 4–5-ribbed, glabrous or gland-dotted; pappi 0 or of low crowns. **2*n***=18.

Grasslands, riverbanks, roadsides, disturbed areas. Flowering Jul–Sep. 0–1400 m. BW, Casc, CR, ECas, Sisk, WV. CA, ID, NV, WA; worldwide. Exotic.

TARAXACUM Dandelion

Herbs perennial, sap milky; taprooted. **Stems** erect or spreading, unbranched. **Leaves** basal, oblong to oblanceolate, margins dentate to pinnate with sharp, linear to lanceolate or triangular lobes, lobes spreading or directed toward bases; usually petiolate, petioles ± narrowly winged. **Inflorescences** solitary heads. **Heads** ligulate, erect. **Involucres** campanulate in flower, broadly to narrowly ovoid in fruit. **Phyllaries** in 2 unequal series; inner linear or narrowly lanceolate; outer deltate-acuminate to lanceolate, much < inner. **Receptacles** flat; paleae 0. **Florets** bisexual; ligules exceeding involucres. **Fruits** obovoid or oblanceoloid, tapering toward bases, glabrous, sharply scabrous or roughened, ribs 10–15; slenderly long-beaked; pappi of numerous fine, persistent bristles.

Worldwide. 60–2000 species; 3 species treated in *Flora*.

Taraxacum is a well-known "problem" genus taxonomically, due to the excessive number of species that have been described. In North America, the proliferation of names has been greatest in Alaska and Canada. Polyploidy and apomixis, leading to clonal-style reproduction, are responsible for the plethora of names.

Key to *Taraxacum* species:

1. Phyllaries usually with small, knob-like or horn-like projections at tips; outer erect . . . . ***T. ceratophorum***
1' Phyllaries without projections at tips; outer reflexed.
  2. Petioles slightly winged distally; fruits red to reddish brown. . . . . ***T. erythrospermum***
  2' Petioles narrowly winged; fruits tan to brown, olive, or gray . . . . ***T. officinale***

Taraxacum ceratophorum (Ledebour) de Candolle (p. 361)

Horned dandelion

Taraxacum paucisquamosum M. Peck

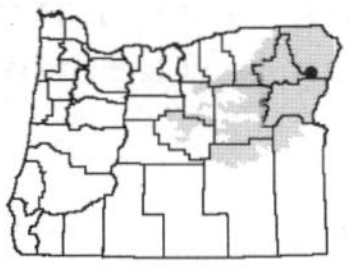

Plants 3–30+ cm. **Stems** glabrous or sparsely villous. **Leaves** narrowly lanceolate to linear-oblong, 2–20+ × 0.5–4 cm, margins dentate to pinnately or irregularly lobed, surfaces usually glabrous, sometimes villous proximally; sessile to petiolate and winged. **Involucres** 8–18 mm, surfaces glabrous. **Phyllaries** 10–15, usually with small, knob-like or horn-like projections at tips; outer erect, < inner. **Florets** 35–80+; ligules 12–25 mm, yellow. **Fruits** bodies 2.5–4 mm, tan to brown or olive, ribs ~15; beaks 4.5–14 mm; pappus bristles white. **2*n***=16, 32, 40, 48.

Moist rocky slopes, wet meadows, early snow-melt areas. Flowering Mar–Oct. 0–1800 m. BW, Col, CR, ECas, Est, Sisk, WV. CA, ID, NV, WA; north to AK, northeast to Greenland, southeast to NM. Native.

Taraxacum erythrospermum Besser

Red-seeded dandelion

Taraxacum laevigatum (Willd.) DC.

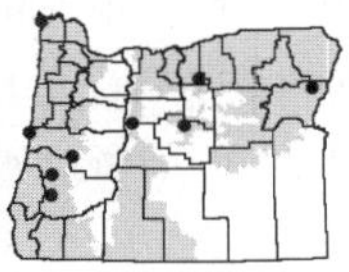

Plants 5–30+ cm. **Stems** glabrous or sparsely villous. **Leaves** obovate to oblanceolate, 3–25 × 1–4 cm, margins irregularly toothed, deeply pinnatifid, surfaces glabrous or lightly villous; petioles slightly winged distally. **Involucres** 3.5–10 mm, surfaces glabrous. **Phyllaries** 18–20; outer reflexed, < inner, without projections at tips. **Florets** 15–60+; ligules 12–20+ mm, yellow; outer with gray or purple stripe. **Fruits** bodies 2.5–3 mm, red to reddish brown, ribs ~15; beaks 5–8.5 mm; pappus bristles white. **2*n*** =12, 16, 24, 32.

Open, rocky sites, disturbed areas, roadsides, lawns. Flowering Mar–Oct. 0–1800 m. BW, Col, CR, ECas, Est, Sisk, WV. CA, ID, WA; throughout most of North America; Europe. Exotic.

Taraxacum officinale F.H. Wiggers

Common dandelion

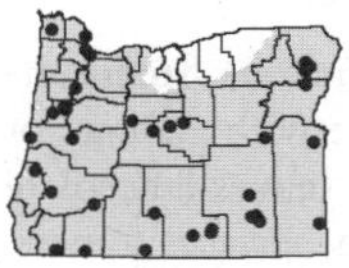

Plants 5–40+ cm. **Stems** glabrous or sparsely villous, hollow. **Leaves** oblanceolate to obovate, 4–45 × 1–10 cm, margins toothed to lobed, surfaces glabrous or lightly villous; petioles narrowly winged. **Involucres** 12–25 mm, surfaces glabrous. **Phyllaries** 13–18; outer reflexed, < inner, without projections at tips. **Florets** 20–80+; ligules 15–30 mm, yellow. **Fruits** bodies 2.5–4 mm, tan to brown, olive, or gray, ribs 4–12; beaks 7–9 mm; pappus bristles white. **2*n***=16, 24, 32, 40.

Disturbed areas, riverbanks, pastures, forests, roadsides, lawns. Flowering Mar–Sep. 0–2800 m. All ecoregions except Col. CA, ID, NV, WA; worldwide. Exotic.

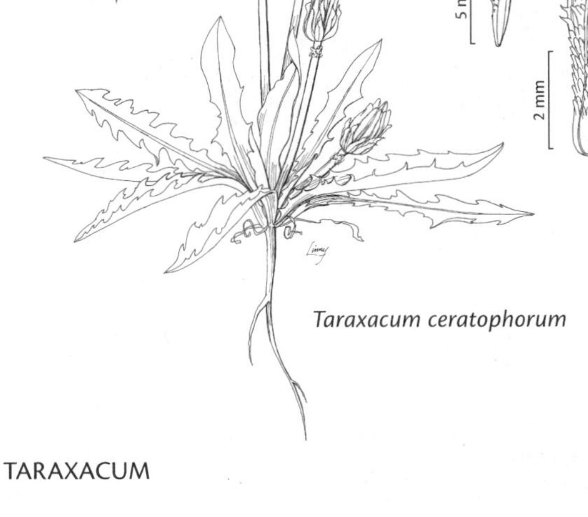

Taraxacum ceratophorum

TARAXACUM

TETRADYMIA Horsebrush

Stephen C. Meyers

Shrubs deciduous; taprooted, often rhizomatous. **Stems** usually erect, usually densely tomentose, sometimes with glabrate or floccose stripes, sometimes spiny. **Leaves** cauline, alternate; primary leaves sometimes becoming spines; secondary leaves clustered in axils of primary, filiform, linear, narrowly awl-shaped, lanceolate or oblanceolate, margins entire, surfaces glabrous to tomentose, sometimes glabrate. **Inflorescences** corymb-like arrays. **Heads** discoid, erect; peduncles tomentose. **Involucres** turbinate or hemispheric to cylindric. **Phyllaries** 4–6, in (1)2 equal series, lanceolate to oblong or ovate. **Receptacles** flat, pitted; paleae 0. **Florets** 4–8, bisexual; corollas narrowly tubular below, widened distally, yellow, rarely cream, lobes 5, linear, spreading, with 1–3 longitudinal oil ducts; stamens exserted; stigma lobes exserted. **Fruits** obconic to fusiform, brown, glabrous to hirsute or densely pubescent; pappi of bristles or slender, tapering scales.

Western North America. 10 species; 3 species treated in *Flora.*

Key to *Tetradymia* species:

1. Stems spiny, without glabrate or floccose stripes; primary leaves becoming spines . . . . ***T. spinosa***

1' Stems unarmed, with glabrate or floccose stripes; primary leaves not becoming spines.

  2. Primary leaves lanceolate to oblanceolate, tomentose to sericeous; stems with glabrate or floccose stripes. . . . ***T. canescens***

  2' Primary leaves narrowly awl-shaped, glabrous to sparsely floccose; stems with glabrate stripes . . . . ***T. glabrata***

Tetradymia canescens de Candolle (p. 362)

Gray horsebrush, spineless horsebrush

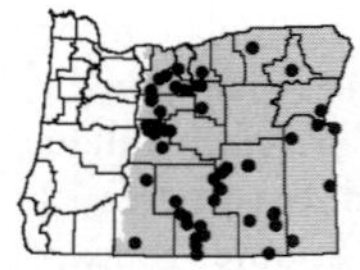

Plants 1–8 dm; not rhizomatous. **Stems** unarmed, densely tomentose with glabrate or floccose stripes. **Primary leaves** lanceolate to oblanceolate, 7–40 mm, surfaces tomentose to sericeous. **Secondary leaves** similar to primary, 2–20 mm. **Heads** 3–8. **Involucres** turbinate to cylindric, 6–10 mm. **Phyllaries** 4(5), lanceolate to oblong, surfaces tomentose. **Florets** 4(5); corollas 5–15 mm, yellow, rarely cream. **Fruits** 3–5 mm, glabrous to hirsute; pappi of 100–150 bristles, 6–10 mm. **2*n***=60, 62, 90, 120.

Sagebrush, juniper forests, pine forests. Flowering May–Aug. 300–2100 m. BR, BW, Col, ECas, Lava, Owy. CA, ID, NV, WA; north to British Columbia, northeast to MT, east to WY, southeast to NM. Native.

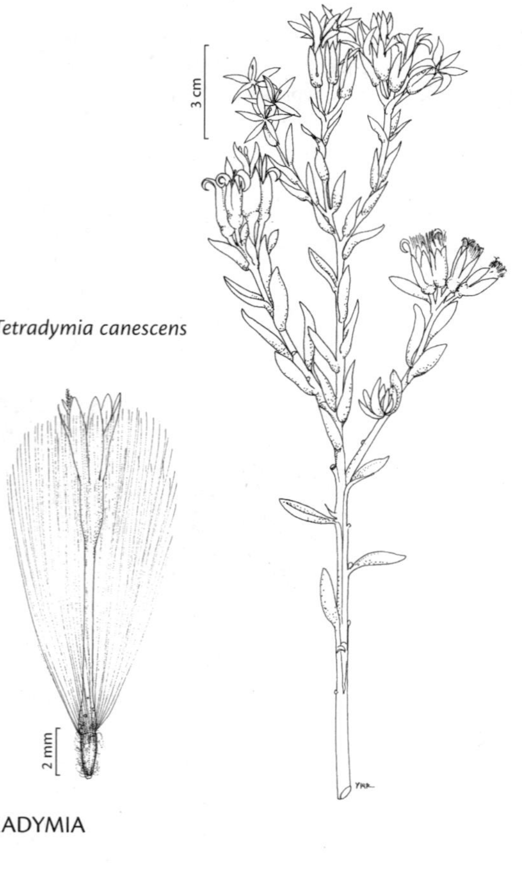

Tetradymia canescens

TETRADYMIA

Tetradymia glabrata Torrey & A. Gray

Littleleaf horsebrush

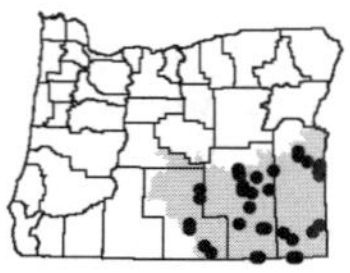

Plants 2–12 dm, not rhizomatous. **Stems** unarmed, densely tomentose with glabrate stripes. **Primary leaves** narrowly awl-shaped, 4–18 mm, surfaces glabrous to sparsely floccose. **Secondary leaves** filiform to linear, 2–13 mm, surfaces glabrous. **Heads** 3–7. **Involucres** turbinate, 7–10 mm. **Phyllaries** 4(5), surfaces glabrous to tomentose. **Florets** 4; corollas 6–10 mm, yellow, rarely cream. **Fruits** 2.5–3.5 mm, hirsute; pappi of 100–140 bristles, 6–10 mm. **2*n***=60, 62, 120, 180.

Sagebrush, juniper forests. Flowering May–Jul. 700–1700 m. BR, Owy. CA, ID, NV; southeast to UT. Native.

Tetradymia spinosa Hooker & Arnott (p. 363)

Cotton-thorn horsebrush, catclaw horsebrush

Tetradymia spinosa Hook. & Arn. var. *longispina* M.E. Jones, *Tetradymia spinosa* Hook. & Arn. var. *spinosa*

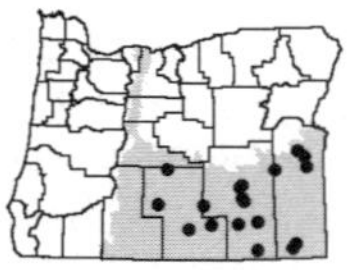

Plants 1–12 dm; rhizomatous. **Stems** spiny, densely tomentose. **Primary leaves** becoming recurved spines, 5–30 mm. **Secondary leaves** linear to oblanceolate, 3–25 mm, surfaces glabrous to glabrate. **Heads** 1–2. **Involucres** hemispheric, 6–12 mm. **Phyllaries** 4–6, surfaces glabrous to tomentose. **Florets** 5–8; corollas 6–10 mm, yellow. **Fruits** 6–8 mm, densely pubescent; pappi of 20–30 scales, 6–9 mm. **2*n***=60.

Sagebrush, scrub, juniper forests. Flowering May–Jul. 400–1800 m. BR, ECas, Owy. CA, ID, NV; northeast to MT, east to WY, southeast to NM. Native.

TONESTUS Goldenweed

Herbs perennial, cespitose; taprooted or rhizomatous. **Stems** 1, erect, usually unbranched, glandular-puberulent. **Leaves** basal and cauline, alternate, linear to oblong or spatulate, bases sometimes clasping, tips obtuse or acute, surfaces stipitate-glandular; sessile or short-petiolate. **Inflorescences** heads usually solitary or in raceme- or corymb-like arrays. **Heads** radiate; peduncles glandular-puberulent. **Involucres** campanulate. **Phyllaries** in 3–4 series, broadly to narrowly oblong or lanceolate, tips acute, surfaces glandular-puberulent; inner sometimes membranous; outer nearly equaling inner or gradually shorter. **Receptacles** convex; paleae 0. **Ray florets** pistillate; rays well exceeding involucres, yellow. **Disc florets** bisexual; corollas tubular, gradually widened distally, slightly exserted, lobes 5, triangular, erect or spreading; stamens slightly exserted; stigma lobes exserted. **Fruits** cylindrical, truncate, tapering toward bases, unequally 8–12-veined; pappi of numerous equal bristles.

Western North America. 4–6 species; 1 species treated in *Flora*.

Tonestus lyallii (A. Gray) A. Nelson (p. 363)
Lyall's goldenweed

Haplopappus lyallii A. Gray

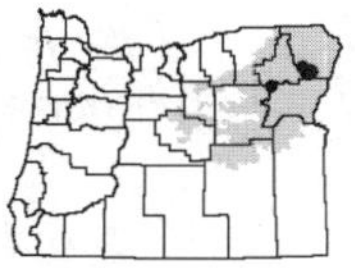

Plants 2–9(15) cm. **Leaves** 3–7 × 0.3–1 cm, margins entire; cauline gradually reduced distally. **Inflorescences** usually solitary heads. **Involucres** 6–11 mm. **Ray florets** 10–30; rays elliptic, 5.5–10 mm. **Disc florets** 25–60+; corollas 6.5–8 mm, yellow. **Fruits** 3–5 mm, brown, glabrous or distally puberulent; pappus bristles 5.5–7.5 mm, white. **2*n***=18.

Alpine and subalpine meadows, talus, rocky areas. Flowering Jul–Sep. 2100–3000 m. BW. CA, ID, NV, WA; north to British Columbia, northeast to Alberta, east to WY, southeast to CO. Native.

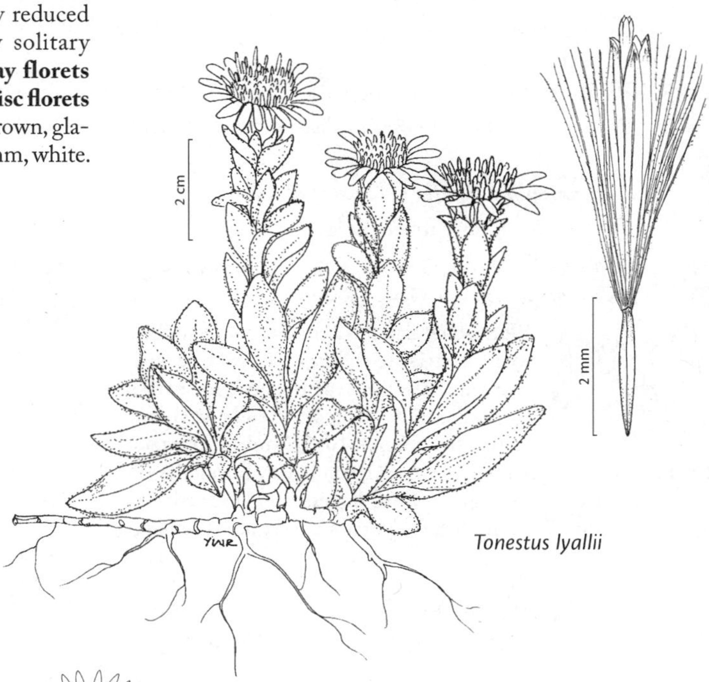

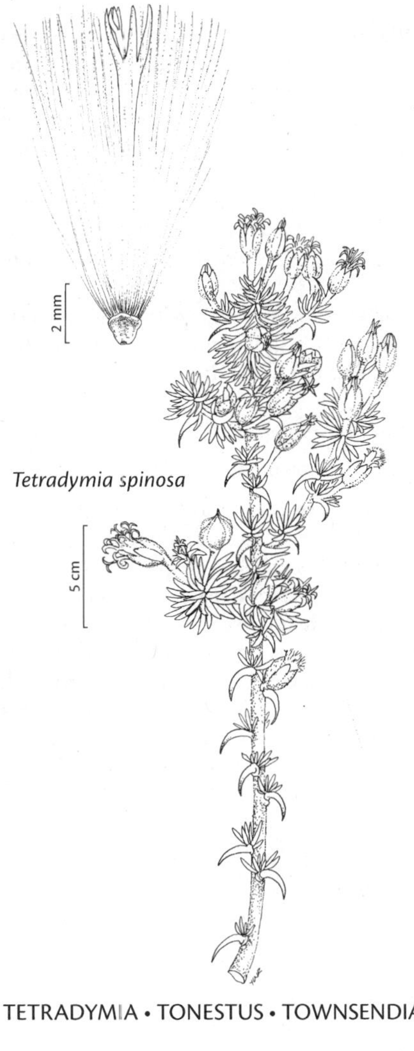

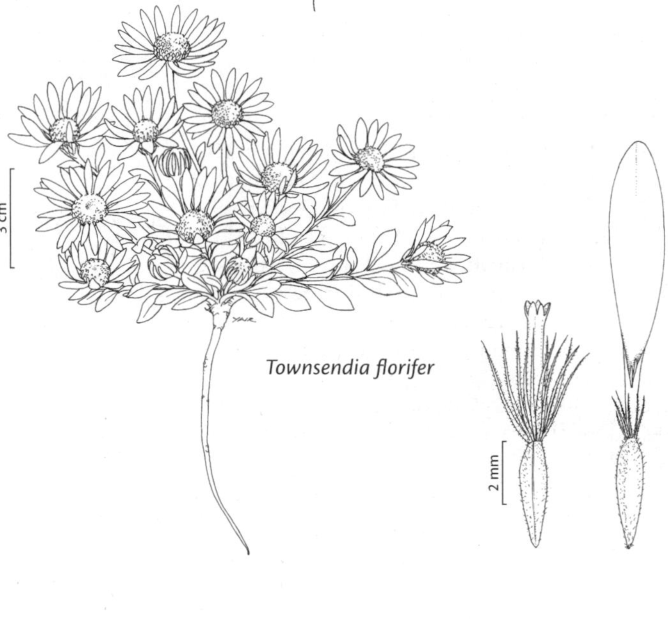

TETRADYMIA • TONESTUS • TOWNSENDIA

TOWNSENDIA Townsendia

Stephen C. Meyers

Herbs biennial or perennial, sometimes cushion-forming; taprooted. **Stems** 1–numerous, decumbent to erect, strigose to hispid. **Leaves** basal and cauline, linear to spatulate, surfaces glabrous to glabrate or strigose; usually petiolate. **Inflorescences** heads solitary, terminal or on scapiform peduncles. **Heads** radiate. **Involucres** hemispheric or campanulate. **Phyllaries** in 3–7 series, erect to spreading, narrowly lanceolate to lance-ovate, margins narrowly membranous, surfaces glabrous to strigose; inner equal; outer gradually shorter. **Receptacles** flat or convex; paleae 0. **Ray florets** pistillate; rays narrowly oblanceolate or oblong, white to pink or purple. **Disc florets** bisexual; corollas cylindric, usually slightly enlarged distally, yellow, lobes 5, triangular, erect; stamens included or partially exserted. **Fruits** obovate or oblanceolate, compressed, glabrous to pubescent; pappi of lanceolate or narrow, tapering, bristle-like scales.

Western North America. 26–27 species; 4 species treated in *Flora*.

Key to *Townsendia* species:

1. Phyllaries 6–9(10) mm.
   2. Plants 2–12 cm; disc florets 100–150; plants not of the Wallowa Mountains . . . . *T. florifer*
   2' Plants 1–3 cm; disc florets 30–70; plants of the Wallowa Mountains . . . . *T. montana*

1' Phyllaries 9–16 mm.
   3. Phyllaries 40–65 in 4–7 series; disc florets 150–200; plants of the Wallowa Mountains . . . . *T. parryi*
   3' Phyllaries 15–35 in 3–4 series; disc florets 40–110; plants of southeastern Oregon . . . . *T. scapigera*

Townsendia florifer (Hooker) A. Gray (p. 363)

Showy townsendia

Townsendia florifer (Hook.) A. Gray var. *florifer*, *Townsendia florifer* (Hook.) A. Gray var. *watsonii* (A. Gray) Cronquist

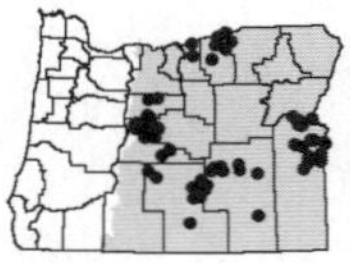

Plants biennial, 2–12 cm. **Stems** decumbent to erect, strigose to hispid. **Leaves** linear to spatulate, 10–50 × 2–6 mm, surfaces sparsely to moderately strigose. **Heads** at tips of stems. **Involucres** 10–25 mm in diameter. **Phyllaries** 25–35, in 3–4 series, 6–9(10) mm. **Ray florets** 12–35; rays 8–20 mm, white to pink. **Disc florets** 100–150; corollas 3–6 mm. **Fruits** 3–5 mm, hairy; ray pappi 2–6 mm; disc pappi 4–7 mm.

Sagebrush, pine forests, juniper forests. Flowering Apr–Aug. 50–1500 m. BR, BW, Col, ECas, Lava, Owy. ID, NV, WA; northeast to MT, east to WY, southeast to UT. Native.

Townsendia montana M.E. Jones

Mountain townsendia

Townsendia montana M.E. Jones var. *montana*

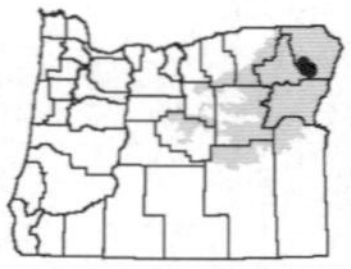

Plants perennial, 1–3 cm. **Stems** erect, strigose. **Leaves** oblanceolate to spatulate, 10–35 × 2–5 mm, surfaces strigose, rarely glabrous. **Heads** sessile or on scapiform peduncles. **Involucres** 8–15 mm in diameter. **Phyllaries** 15–35, in 3–4 series, 6–9 mm. **Ray florets** 12–25; rays 6–15 mm, white to lavender. **Disc florets** 30–70; corollas 3–6 mm. **Fruits** 3–5 mm, glabrous to hairy; ray pappi 5–6 mm; disc pappi 5–7 mm. $2n$=18, 36.

Alpine meadows, cliffs, rocky ridges and slopes. Flowering Jul–Aug. 2100–3000 m. BW. ID; northeast to MT, east to WY, southeast to CO. Native.

Townsendia parryi D.C. Eaton

Parry's townsendia

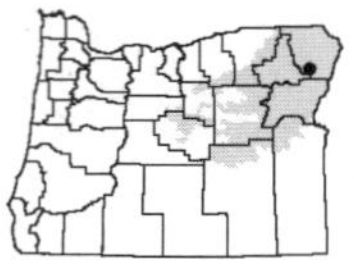

Plants biennial or perennial, 2–8 cm. **Stems** decumbent to erect, strigose. **Leaves** oblanceolate to spatulate, 15–35 × 2–5 mm, surfaces glabrate to strigose. **Heads** at tips of stems or on scapiform peduncles. **Involucres** 10–30 mm in diameter. **Phyllaries** 40–65, in 4–7 series, 10–16 mm. **Ray florets** 20–60; rays 7–17 mm, purple or bluish purple, rarely white or pink. **Disc florets** 150–200; corollas 4–6 mm. **Fruits** 3–5 mm, hairy; ray pappi 3–6 mm; disc pappi 3–6 mm. $2n$=18.

Meadows, rocky areas, talus. Flowering May–Aug. 1500–2700 m. BW. ID; north to British Columba, northeast to Alberta, east to WY, southeast to UT. Native.

Townsendia scapigera D.C. Eaton

Tufted townsendia

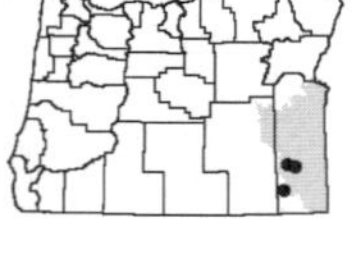

Plants perennial, sometimes biennial, 2–8 cm. **Stems** ascending to erect, strigose. **Leaves** oblanceolate to spatulate, 15–40 × 2–7 mm, surfaces sparsely to moderately strigose. **Heads** sessile or on scapiform peduncles. **Involucres** 10–25 mm in diameter. **Phyllaries** 15–35, in 3–4 series, 9–16 mm. **Ray florets** 15–30; rays 7–16 mm, pink to purple. **Disc florets** 40–110; corollas 3–6 mm, yellow. **Fruits** 4–6 mm, hairy; ray pappi 3–6 mm; disc pappi 5–7 mm. $2n$=18.

Sagebrush, meadows. Flowering May–Sep. 1000–1700 m. Owy. CA, ID, NV; southeast to UT. Native.

TRAGOPOGON Goatsbeard, salsify

Stephen C. Meyers

Herbs biennial, sap milky; taprooted. **Stems** usually 1, erect, often branched proximally. **Leaves** basal and cauline, alternate, linear and somewhat grass-like, margins entire, tips straight or recurved, surfaces glabrous to tomentose or floccose, sheathing; sessile. **Inflorescences** solitary heads, terminal at ends of branches. **Heads** ligulate, closing by midday; peduncles inflated distally or not, without bracts. **Involucres** conic to campanulate. **Phyllaries** usually 8–13 in 1 series, linear to triangular-lanceolate, tips acute. **Receptacles** convex, glabrous; paleae 0. **Florets** ligules yellow or purple. **Fruits** linear, terete or angled, 5–10-ribbed; beaks slender, ≥ bodies; pappi in 1 series of plumose bristles; bristles united at bases, interwebbed, subequal to unequal.

Africa, Asia, Europe, North America. 50–150 species; 4 species treated in *Flora.*

The heads of all *Tragopogon* species open early in the morning and generally close by midday. As a result of the closed heads and linear leaves, plants at first glance are often mistaken as grasses.

Key to *Tragopogon* species:

1. Corollas purple distally, yellow proximally . . . . ***T. mirus***
1' Corollas wholly purple or yellow.
   2. Corollas purple. . . . . ***T. porrifolius***
   2' Corollas yellow.
      3. Corollas bright yellow; peduncles distally inflated; fruits 2.5–3.5(4) cm; outer ligules not surpassing phyllaries . . . . ***T. dubius***
      3' Corollas pale yellow; peduncles slightly or not inflated; fruits 1.5–2.5 cm; outer ligules equaling or surpassing phyllaries . . . . ***T. pratensis***

Tragopogon dubius Scopoli

Yellow salsify

Tragopogon major Jacq.

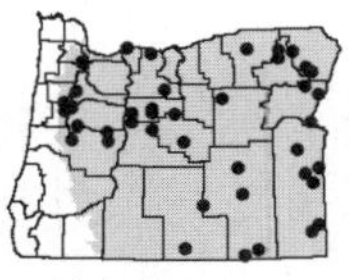

Plants 3–10 dm. **Leaves** floccose when young, glabrous with maturity, tips straight. **Peduncles** inflated distally. **Involucres** conic. **Florets** corollas bright yellow; outer ligules not surpassing phyllaries. **Fruits** 2.5–3.5(4) cm. **2*n***=12.

Roadsides, fields, disturbed areas. Flowering May–Jul. 0–2300 m. BR, BW, Casc, Col, ECas, Lava, Owy, WV. CA, ID, NV, WA; throughout North America; Europe. Exotic.

See also *T. pratensis* and *T. mirus* discussions.

Tragopogon mirus Ownbey (p. 366)

Remarkable goatsbeard

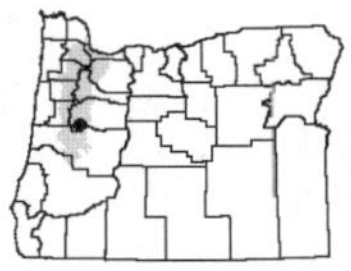

Plants 4–15 dm. **Leaves** tomentose or floccose when young, glabrous with maturity, tips straight. **Peduncles** inflated distally. **Involucres** conic. **Florets** corollas purple distally, yellow proximally; outer ligules ± surpassing phyllaries. **Fruits** 1.2–3 cm. **2*n***=24.

Fields, disturbed areas. Flowering May–Jul. 100–1000 m. WV. ID, WA; AZ. Native.

Tragopogon mirus is an allotetraploid species. It was formed as a result of a hybridization event between *T. dubius* and *T. porrifolius*, with a subsequent chromosome doubling among the offspring of said hybrids. As a consequence of the chromosome doubling, the newly formed allotetraploid plants cannot back cross with the parental species, thus leading to the creation of a new and novel species. Allotetraploid speciation, involving *T. dubius* and *T. porrifolius*, and resulting in the formation of *T. mirus*, has occurred on several separate occasions in the Palouse region of Idaho and Washington, once in Arizona near Flagstaff, and once in Oregon (Soltis et al. 2012). Although populations of *T. dubius* and *T. porrifolius* are sympatric in Europe, where these species are native, no *T. mirus* has been reported from that continent.

Tragopogon porrifolius Linnaeus

Salsify, oyster plant

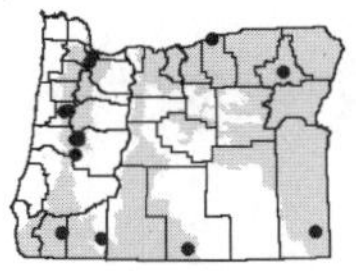

Plants 4–12 dm. **Leaves** glabrous, tips straight. **Peduncles** inflated distally. **Involucres** conic. **Florets** corollas purple; outer ligules not surpassing phyllaries. **Fruits** 1–2 cm. **2*n***=12.

Roadsides, fields, disturbed areas. Flowering Apr–Jun. 50–1500 m. BW, Col, ECas, Owy, Sisk, WV. CA, ID, NV, WA; throughout most of North America; Europe. Exotic.

See also *T. mirus* discussion.

Tragopogon pratensis Linnaeus

Meadow salsify, Jack-go-to-bed-at-noon

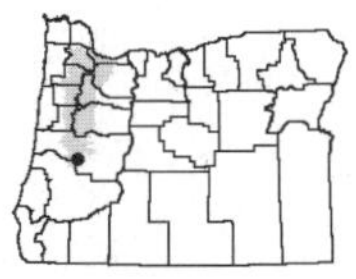

Plants 1.5–9 dm. **Leaves** tomentose or floccose when young, glabrous with maturity, tips recurved. **Peduncles** slightly or not inflated. **Involucres** usually campanulate. **Florets** corollas pale yellow; outer ligules equaling or surpassing phyllaries. **Fruits** 1.5–2.5 cm. **2*n***=12.

Roadsides, fields, disturbed areas. Flowering May–Jul. 0–1500 m. WV. CA, ID, NV, WA; throughout most of North America; Europe. Exotic.

Similar to *Tragopogon mirus*, allotetraploid hybrid speciation has occurred between *T. pratensis* and *T. dubius*. The result (*T. miscellus*) has been documented only near Pullman, WA. Because the parental species and *T. miscellus* are similar morphologically (all with yellow flowers), populations of the latter may be overlooked. *Tragopogon miscellus* can be fairly easily identified, however, by the combination of leaves with recurved tips and outer ligules not surpassing the phyllaries.

TRIPLEUROSPERMUM Mayweed

Herbs annual, biennial, or perennial; taprooted. **Stems** usually much branched, glabrous or minutely puberulent. **Leaves** basal and cauline, alternate, 1–3-pinnate, ultimate segments filiform to linear, margins crenate or serrate, surfaces glabrous; proximal petiolate, distal sessile. **Inflorescences** corymb-like arrays or heads solitary; peduncles glabrous. **Heads** radiate. **Involucres** campanulate to hemispheric. **Phyllaries** in 3–5 subequal series, broadly lanceolate to ovate, margins membranous, midribs dark green or brownish, tips acute. **Receptacles** rounded to hemispheric, well exceeding involucres; paleae 0. **Ray florets** pistillate; rays lanceolate to ovate, often 2–3-toothed, tips acute or obtuse. **Disc florets** bisexual; corollas tubular, widened near tips, 5-lobed, lobes triangular, spreading; stamens exserted; stigmas exserted. **Fruits** obovoid or obconic, prominently 3–5-ribbed, inter-rib areas minutely roughened, often with 1–5 circular to elongate, apical resin glands; pappi 0 or of low crowns.

Africa, Asia, Europe, North America. 40 species; 1 species treated in *Flora*.

Tripleurospermum inodorum (Linnaeus) Schultz-Bipontinus (p. 366)

Scentless mayweed, scentless chamomile

Matricaria maritima L.

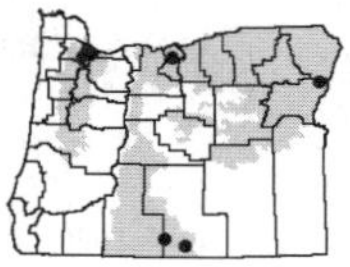

Plants annual, rarely biennial or perennial, 5–50+ cm. **Stems** ascending to erect. **Leaves** 1.5–10 cm; basal withering by flowering; cauline gradually reduced. **Heads** 10–200; peduncles 3–15 cm. **Involucres** 3–4.5 mm. **Ray florets** 10–25; rays reflexed and persisting in fruit, 7–15+ mm, white. **Disc florets** numerous; corollas 1–2.5 mm, yellow. **Fruits** 1.5–2 mm, pale to dark brown, resin glands ± circular. **2*n***=18, 36.

Roadsides, clearcuts, disturbed areas, cultivated fields. Flowering Jun–Oct. 0–1700 m. BW, Col, ECas, WV. CA, ID, NV, WA; scattered throughout most of Canada and US. Exotic.

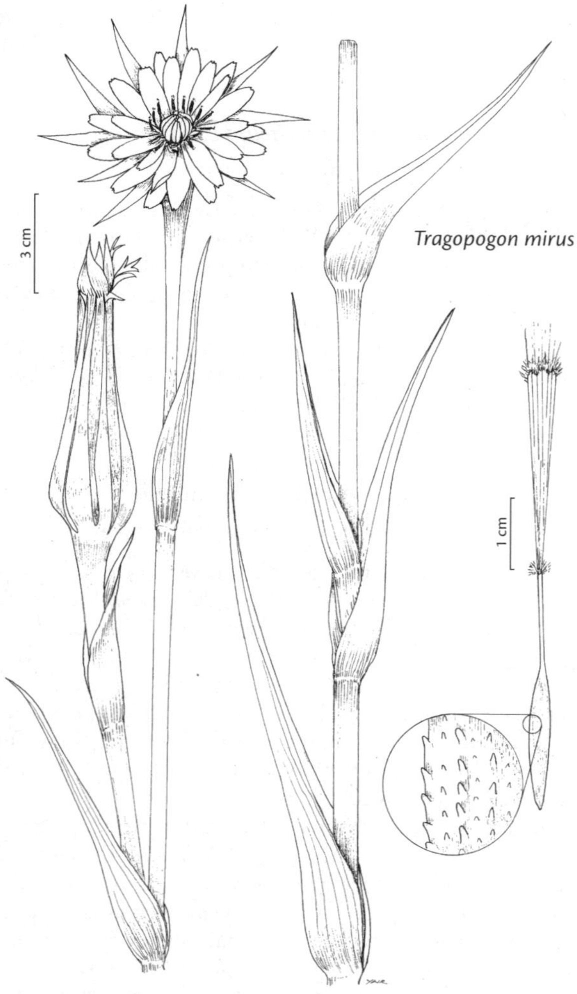

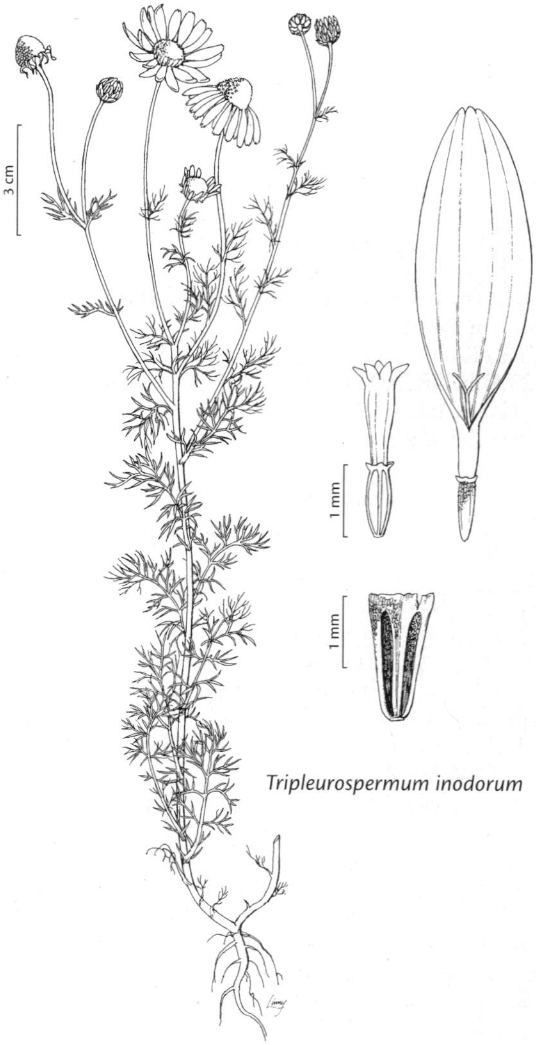

TRAGOPOGON • TRIPLEUROSPERMUM

UROPAPPUS Silver puffs

Genus composed of one species, *U. lindleyi.*

Uropappus lindleyi (de Candolle) Nuttall (p. 367)

Lindley's silver puffs, linearleaf microseris

Microseris lindleyi (DC.) A. Gray, *Microseris linearifolia* (Nutt.) Sch. Bip., *Uropappus linearifolius* Nutt.

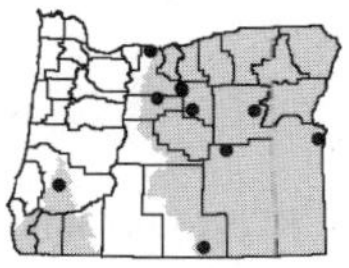

Herbs annual, 5–70 cm; sap milky; taproots slender. **Stems** erect, initially short, often elongating, leafy with age, proximally lightly white-villous, distally usually reddish and glabrous. **Leaves** basal and cauline, alternate, 5–30 cm, usually pinnatifid with slender, acuminate rachises and linear lobes, white-villous proximally on margins and upper surfaces, usually glabrous distally, often reddish. **Inflorescences** heads solitary, terminal and axillary. **Heads** ligulate, erect; peduncles swollen and hollow distally, 5–40 cm, usually reddish, surfaces glabrous but often white-puberulent when young, sometimes minutely glandular proximally. **Involucres** 10–40 mm, often reddish, surfaces glabrous. **Phyllaries** 5–26, in 3–4 series, erect, lanceolate, reflexing at maturity; outer ~½ as long as inner. **Receptacles** flat or convex; paleae 0. **Flowers** 5–150, bisexual; ligules equaling or barely exceeding involucres, yellow, outer often reddish abaxially. **Fruits** narrowly fusiform, slender, tapering to a short slender beak, 7–17 mm, blackish, 10-ribbed, ribs scabrous to hispidulous; pappi of 5 lanceolate scales, 5–15 mm, silvery, notched at tip, notches 1–2 mm, terminating in short, delicate bristles. **2*n***=18.

Interior grasslands, shrublands, open woodlands. Flowering Mar–Jun. 50–1400 m. BR, BW, Col, Lava, Owy, Sisk. CA, ID, NV, WA; north to British Columbia, south to TX, south to Mexico. Native.

The northernmost occurrence of *Uropappus lindleyi* is on the islands east of southern Vancouver Island, probably a result of long-range achene dispersal by birds. In its northern range, *Uropappus lindleyi* is a very distinct species, but in California, it is linked morphologically to *Microseris* through the two intergeneric alloploid hybrid species of *Stebbinsoseris.*

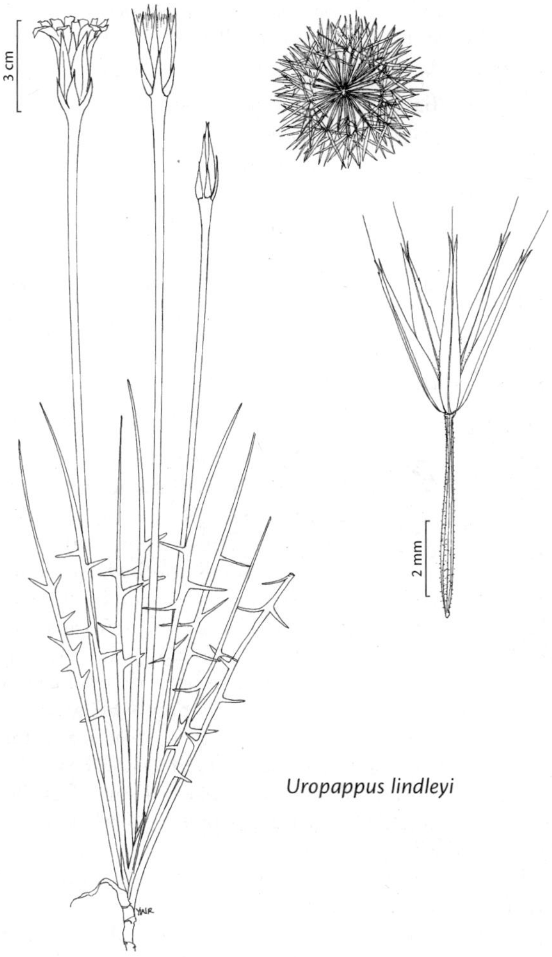

Uropappus lindleyi

UROPAPPUS

WYETHIA Mule's ears

Abigail J. Moore

Herbs perennial; taproots thick, resin-producing. **Stems** erect or ascending, leafy, simple or branched. **Leaves** basal and cauline, alternate, simple, linear-lanceolate to ovate; basal > cauline. **Inflorescences** heads solitary or in corymb- or raceme-like arrays, usually with 1 terminal main head, sometimes with 1–5 axillary heads per stem. **Involucres** broadly campanulate to hemispheric. **Receptacles** flat to convex; paleate. **Paleae** lanceolate, acute, folded around fruits. **Ray florets** pistillate; rays yellow or white, 3-lobed, much exceeding involucres. **Disc florets** bisexual; corollas yellow, 5-lobed. **Fruits** 4-sided, black, glabrous to strigillose; pappi jagged crowns and 1+ longer scales. **2*n***=38.

Western Mexico and western United States. 14 species; 5 species treated in *Flora.*

All species of *Wyethia* are diploid with a chromosome base number of 19 and are capable of hybridizing when they come into contact; however, hybrids are generally rare. *Wyethia* is most closely related to *Balsamorhiza,* from which it is distinguished by having cauline leaves.

Here, head width measurements are of pressed specimens.

The genus was named by Thomas Nuttall after Nathaniel Wyeth, who led the expedition on which the type species of the genus, *W. helianthoides,* was collected.

Key to *Wyethia* species:

1. Plants entirely glabrous; phyllaries glabrous, not ciliate . . . . *W. amplexicaulis*

1' Plants with hairs on the leaves or stems; phyllaries tomentose or ciliate.

  2. Plants tomentose, leaves sometimes becoming glabrate, but phyllaries and area below the heads remaining tomentose; phyllaries not ciliate . . . . *W. mollis*

  2' Plants with long or short hairs but not tomentose; phyllaries ciliate.

    3. Basal leaves elliptic to oblanceolate, generally > 4 × as long as wide; ray flowers yellow or orange-yellow . . . . *W. angustifolia*

    3' Basal leaves lanceolate to elliptic to ovate, generally < 3 × as long as wide; ray flowers white or light yellow.

      4. Rays white; basal leaf blades generally < 20 cm; phyllaries 2–4 mm wide, linear; leaves sparsely villous-hirsute and ciliate . . . . *W. helianthoides*

      4' Rays light yellow; basal leaf blades > 20 cm; phyllaries 3–6 mm wide, linear to lanceolate; leaves often glabrous, but generally sparsely hirsute on midrib and margins . . . . *W.* × *cusickii*

Wyethia amplexicaulis (Nuttall) Nuttall (p. 368)

Northern mule's ears

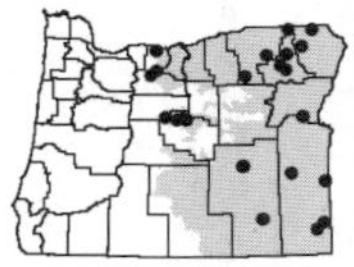

Plants glabrous. **Stems** erect, (15)30–62 cm, with sessile glands. **Basal leaves** elliptic to oblanceolate, rarely lanceolate, (18)22–46 × 3.6–12 cm, tapering to both ends, margins entire to remotely toothed, surfaces glabrous, often shiny with resin; petioles 0–16 cm. **Cauline leaves** blades lanceolate to elliptic, 7–21 × 1.8–6 cm, margins entire; sessile. **Inflorescences** terminal heads 1, larger, axillary heads 2–5, smaller; peduncles 0.2–7.5 cm. **Involucres** terminal 20–35 × 20–25 mm, axillary 12–21 × 11–20 mm. **Phyllaries** ovate-lanceolate to lanceolate, 16–30 × 4–12 mm, surfaces glabrous, not ciliate. **Paleae** 15–18 mm. **Ray florets** larger heads 11–21, smaller heads 7–15, yellow; rays 22–44 × 5–13 mm. **Disc florets** 9–11 mm. **Fruits** 7–8 × 1.3 mm, glabrous; pappus scales ovate, 1–2 mm; crowns 0.5–1 mm.

Dry, open areas. Flowering May–Jul. 100–2300 m. BR, BW, Col, Owy. ID, NV, WA; southeast to CO. Native.

Wyethia amplexicaulis generally grows in drier areas than does *W. helianthoides*, but the two species form the hybrid *W.* × *cusickii* where they meet. "Amplexicaulis" refers to the cauline leaves, which are generally sessile and more or less clasp the stem.

Wyethia angustifolia (de Candolle) Nuttall

Narrow-leaved mule's ears

Wyethia angustifolia (DC.) Nutt. var. *angustifolia*

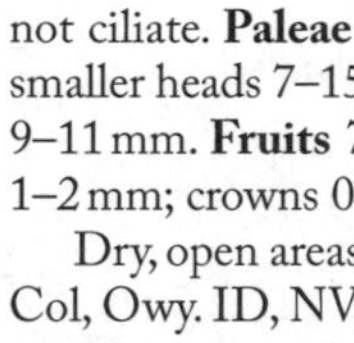

Plants not tomentose. **Stems** decumbent to ascending, (9)20–96 cm, villous-hirsute, more densely so at bases of stems. **Basal leaves** elliptic to oblanceolate, 8–47 × 1.3–8.3(10.2) cm, generally > 4 × as long as wide, tapered to both ends, but more strongly tapered to bases, margins entire or toothed, surfaces hirsutulous, villous-hirsute on midribs, sometimes resinous and shiny; petioles 0–19.4 cm. **Cauline leaves** lanceolate to oblanceolate, 3–38 × 0.7–6 cm; petioles 0–6 cm. **Inflorescences** terminal heads 1(3), often with smaller axillary heads; peduncles 2–20 cm. **Involucres** (15)20–45 × (10)15–22 mm. **Phyllaries** linear to lanceolate, 10–27(35) × 2–7 mm, hirsutulous, ciliate; cilia ~1 mm. **Paleae** 12–17 mm. **Ray florets** 10–17, yellow to orange-yellow; rays 21–38 × 5–14 mm. **Disc florets** 7–11 mm. **Fruits** 7–8 × 1.2–2 mm, strigillose; pappus scales lanceolate, 2–8 mm; crowns 0.7–1.1 mm.

Open, grassy areas, coastal bluffs, open woodlands, serpentine. Flowering May–Jul. 0–1600 m. Casc, ECas, Est, Sisk, WV. CA, WA. Native.

"Angustifolia" means narrow-leaved.

Wyethia amplexicaulis

WYETHIA

Wyethia* × *cusickii Piper

Hybrid mule's ears

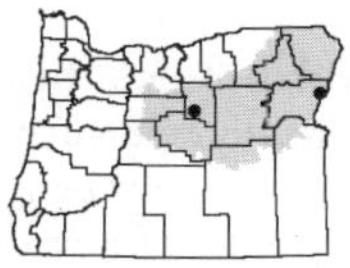

Plants not tomentose. **Stems** erect, 32–45 cm, sparsely to very sparsely villous-hirsute. **Basal leaves** lanceolate to elliptic to ovate, 21–26 × 6.5–11.5 cm, generally < 3 × as long as wide, bases sometimes decurrent, margins entire or with small teeth, surfaces glabrous but generally sparsely hirsute on midribs and margins; petioles 2–13 cm. Cauline leaves lanceolate to oblong, 6–18 × 2–6 cm; petioles 0–4 cm. **Inflorescences** terminal heads 1, occasionally with 1–4 reduced, axillary heads; peduncles up to 1.5 cm. **Involucres** terminal 30–45 × 13–15 mm, axillary ~12 × 10 mm. **Phyllaries** linear to lanceolate, 17–28 × 3–6 mm, surfaces ciliate. **Ray florets** ~15, light yellow; rays 35–45 × 8–10 mm. **Fruits** 7–11 × 1–2 mm, strigillose apically; pappus scales lanceolate; crowns 0.3–1 mm.

Open, grassy areas. Flowering May–Jun. 800–1700 m. BW, Lava. ID, WA. Native.

Wyethia × *cusickii* is the hybrid between *W. amplexicaulis* and *W. helianthoides* and can be expected wherever their ranges overlap. It ranges from something that looks very much like *W. amplexicaulis*, but with ciliate phyllaries, to something that looks much like *W. helianthoides* but is mostly glabrous.

Wyethia helianthoides Nuttall

White-headed wyethia

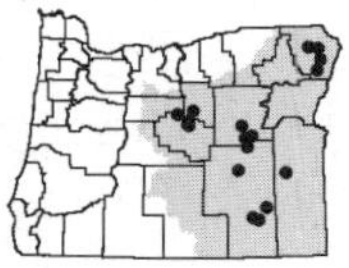

Plants not tomentose. **Stems** erect, 10–38(42) cm, moderately densely villous-hirsute. **Basal leaves** ovate to oblong, 9–20(21) × 2.6–11 cm, generally < 3 × as long as wide, decurrent or not, margins entire, surfaces sparsely villous-hirsute and ciliate; petioles 3.5–7 cm. **Cauline leaves** lanceolate to oblanceolate, 3–18 × 0.9–5 cm; petioles 1–4.5 cm. **Inflorescences** terminal heads 1. **Involucres** 23–40 × 12–16 mm. **Phyllaries** linear, 16–24 × 2–4 mm, surfaces hirsutulous, ciliate. **Paleae** 15–19 mm. **Ray florets** 11–20, white, rays 25–42 × 10–12 mm. **Disc florets** 9–11 mm. **Fruits** 8–11 × 1.2–1.5 mm, strigillose; pappus scales lanceolate, 2–5 mm; crowns 0.3–0.5 mm.

Open, grassy, somewhat moist areas. Flowering May–Jul. 800–2300 m. BR, BW, Lava, Owy. ID, NV; east to WY. Native.

Wyethia helianthoides generally grows in moister areas than does *W. amplexicaulis*. The two species form the hybrid *W.* × *cusickii* when they meet. "Helianthoides" means sunflower (*Helianthus*)-like.

Wyethia mollis A. Gray (p. 369)

Woolly mule's ears

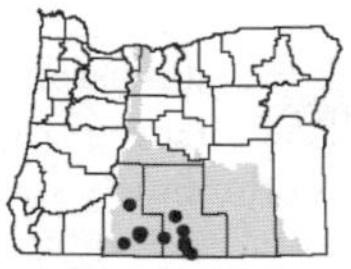

Plants tomentose, denser below heads and at bases. **Stems** erect, 30–72 cm. **Basal leaves** ovate-lanceolate to obovate-oblanceolate, 18–32.5 × 5.5–15.3 cm, sometimes ± decurrent, margins entire, surfaces tomentose, becoming glabrate; petioles 10.5–19 cm. **Cauline leaves** ovate-lanceolate to obovate-oblanceolate, 5–21 × 2–7.5 cm; petioles 0.5–8.6 cm. **Inflorescences** terminal heads 1, larger, axillary heads 0–3, sometimes reduced; peduncles 1.9–5 cm. **Involucres** terminal 17–30 × 10–20 mm, axillary 15–25 × 10–15 mm. **Phyllaries** ovate-lanceolate to lanceolate, 17–34 × 5–10 mm, surfaces tomentose, not ciliate. **Paleae** 15–18 mm. **Ray florets** larger heads 8–13, smaller heads 7–10, yellow; rays 20–35 × 6–9 mm. **Disc florets** ~9 mm. **Fruits** 9–12 × 1.5–2 mm, loosely villous; pappus scales subulate, 1–3, 2–3 mm, ciliate; crowns 0.3–1 mm.

Grassy areas. Flowering Jun–Aug. 1300–2000 m. BR, ECas. CA, NV. Native.

"Mollis" means soft, in reference to the hairs.

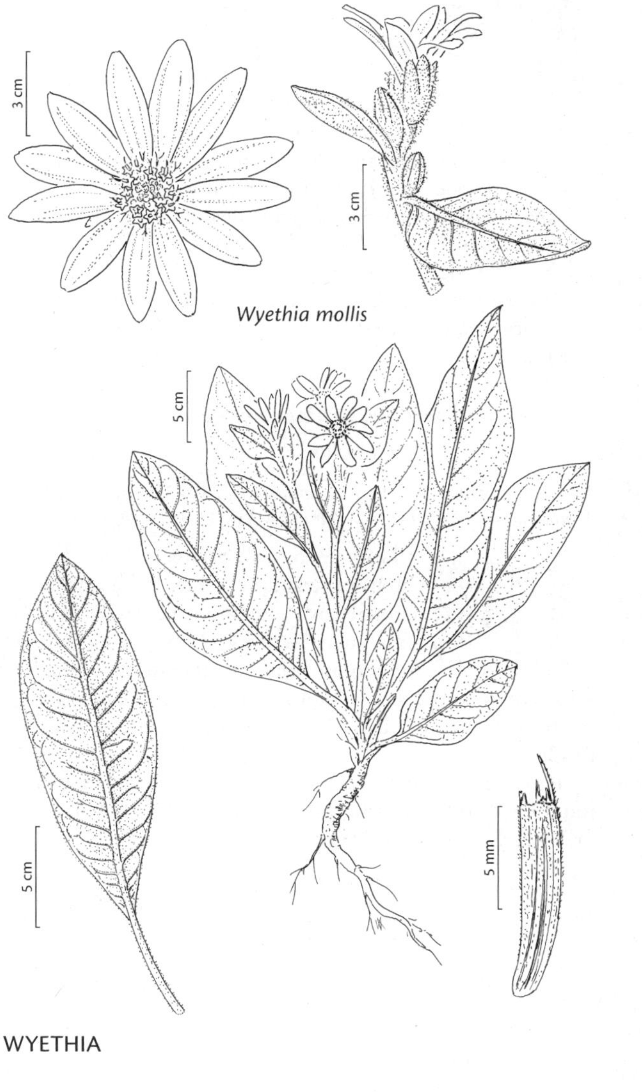

WYETHIA

XANTHIUM Cocklebur

Stephen C. Meyers

Herbs annual. **Stems** erect, branched, glabrous to scabrous, puberulent, hispidulous, or tomentose, spineless or with 2–3-branched spines at nodes. **Leaves** mostly cauline, usually alternate, proximal sometimes opposite, narrowly lanceolate to broadly triangular, bases attenuate to truncate or cordate, glabrous or minutely scabrous, margins entire to toothed or lobed, tips acute; proximal petiolate, distal petiolate or sessile. **Inflorescences** raceme- or spike-like arrays, or heads solitary. **Heads** unisexual, arrays of mixed sexuality or not, discoid; staminate 3–5 mm in diameter; pistillate 4–22 mm in diameter. **Staminate involucres** spherical, becoming disc-shaped. **Pistillate involucres** ellipsoid. **Staminate phyllaries** 6–20 in 1–3 series, often poorly formed. **Pistillate phyllaries** 30–80, in 6–15 series, proximal with erect, straight or hooked, horn-like spines. **Staminate florets** 20–60; corollas funnel-shaped, white, stamens exserted, lobes 5. **Pistillate florets** 2; corollas 0. **Fruits** fusiform, black, enclosed by pistillate involucres developing into 2-chambered burs with hard prickles; pappi 0.

Worldwide. 3 species; 2 species treated in *Flora*.

Key to *Xanthium* species:

1. Stems with 2- or 3-branched spines at the nodes; leaves narrowly lanceolate to elliptic; burs 1–1.5 cm, prickles slender . . . . *X. spinosum*

1' Stems not spiny; leaves broadly ovate to suborbicular or broadly triangular; burs 1.5–3 cm, prickles stout . . . . *X. strumarium*

Xanthium spinosum Linnaeus

Spiny cocklebur

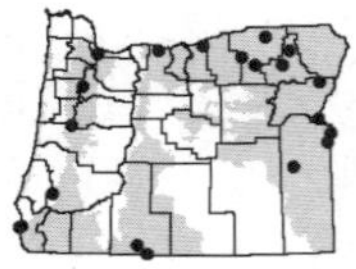

Plants 1–8+ dm. **Stems** puberulent to hispidulous or tomentose, with 2–3-branched spines at the nodes. **Leaves** blades narrowly lanceolate to elliptic, 2–15 × 1–4 cm, bases cuneate or attenuate, margins entire to toothed or lobed, tips acute or acuminate, surfaces abaxially densely strigose, adaxially glabrate to sparsely strigose. **Burs** 1–1.5 cm, prickles slender; beaks 0–1. **2*n***=36.

Roadsides, disturbed areas, streambanks, fields. Flowering Jun–Oct. 0–1600 m. BW, Col, ECas, Est, Owy, Sisk, WV. CA, ID, NV, WA; nearly worldwide. Exotic.

The seeds and seedlings of this species are poisonous and may cause death of livestock. It is listed as a Class B Noxious weed in Oregon.

Xanthium strumarium Linnaeus (p. 370)

Common cocklebur

Xanthium strumarium L. var. *canadense* (Mill.) Torr. & A. Gray, *Xanthium strumarium* L. var. *glabratum* (DC.) Cronquist

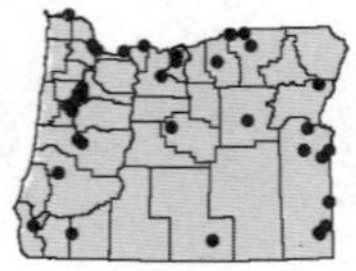

Plants 1–15+ dm. **Stems** glabrous to scabrous or strigillose, spines 0. **Leaves** blades broadly ovate to suborbicular or broadly triangular, 3–20 × 2–15 cm, bases truncate or cordate, margins dentate to lobed, tips acute, surfaces abaxially scabrous to strigose, adaxially glabrous to sparsely strigose. **Burs** 1.5–3 cm, prickles stout; beaks 2. **2*n***=36.

Roadsides, disturbed areas, streambanks, fields. Flowering Jun–Oct. 0–1600 m. All ecoregions except Est. CA, ID, NV, WA; nearly worldwide. Native.

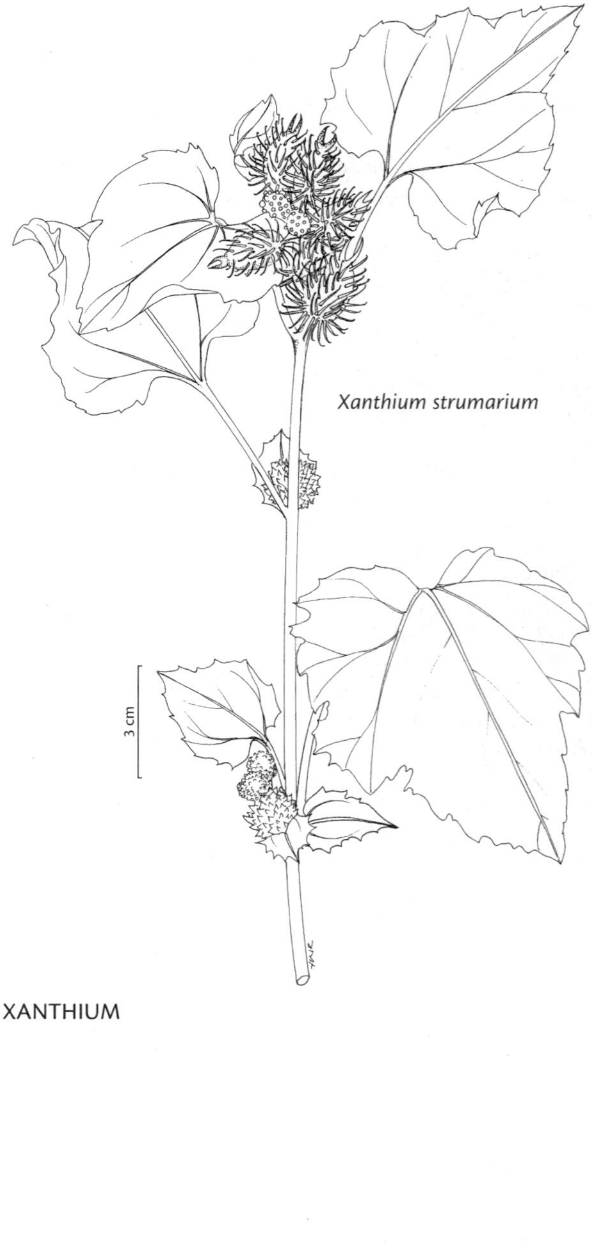

XANTHIUM

BALSAMINACEAE Balsam family

Peter F. Zika

Plants annual; roots fibrous. **Stems** herbaceous, succulent, often hollow, erect or slightly creeping at base, simple or branching. **Leaves** cauline, alternate, opposite or whorled, few–many, simple; sessile or petiolate; stipules absent or reduced to glands. **Inflorescences** racemes and irregular panicles. **Flowers** bisexual, greatly reduced if cleistogamous, bilaterally symmetric if outcrossing; sepals 3, with 2 reduced and greenish, 1 enlarged, petaloid, ± saccate and tube-forming, often spurred; petals 5 but lateral pairs fused, bilobed, so appearing as 3, upper petal notched; stamens 5, anthers often fused around style; ovaries superior, 5-carpellate, styles 1, stigmas 5-lobed, short. **Fruits** capsules, locules 5, exploding when jostled. **Seeds** smooth to ridged, rugose, papillose, or warty.

Cosmopolitan. 2 genera; 1 genus treated in *Flora.*

IMPATIENS Jewelweed, touch-me-not, balsam

Herbs 0.2–3.3 m. **Stems** glabrous. **Leaves** thin, margins entire to crenate or serrate, teeth sometimes gland-tipped, tips blunt or acute to acuminate or cuspidate, surfaces glabrous. **Inflorescences** cleistogamous flowers solitary or in small panicles in lower leaf axils, often produced late in season; outcrossing flowers axillary in small to large racemes or panicles, often on upper half of stem; leafy-bracteate or not. **Flowers** cleistogamous flowers inconspicuous and similar, greenish white; outcrossing flowers often pendulous and showy, white, yellow, orange, pink, or purple, with contrasting markings within floral tube and often at mouth serving as nectar guides for pollinators. **Fruits** narrow, similar but often smaller if derived from cleistogamous flowers, green, glabrous, smooth to ridged. **Seeds** narrowly ovate to oblong, green to blackish.

Cosmopolitan. 1000+ species; 7 taxa treated in *Flora.*

Identification of *Impatiens* relies on the color and shape of the outcrossing flowers; taxa native to North America cannot be distinguished by their cleistogamous flowers or their foliage. It is suggested to carefully note flower colors and separately press extra flowers and some seeds for a packet to accompany herbarium specimens. Only the mature outcrossing flowers are discussed in the species descriptions and keys. Spur lengths are given as if the spur was straightened, although this can be difficult to measure when the spur is curved or hook-shaped. The first and last outcrossing flowers of the season may be atypical in size or shape; be cautious keying if there are very few flowers present in the colony or on an herbarium sheet.

Key to *Impatiens* species:

1. Flowers not spurred, yellow-orange.
   2. Flowers irregularly and coarsely spotted on petals and within tubes, spots and blotches usually dark orange, all flowers on a plant spotted and unspurred . . . . . ***I. ×pacifica*** (in part)
   2' Flowers not spotted or splotched, tubes and petals uniformly yellow-orange . . . . . ***I. ecornuta***

1' Flowers spurred, yellow-orange, pink, purple, or white.
   3. Flowers gradually tapered to a straight or gently curved nectar spur; with dark yellow markings within floral tube and often at mouth.
      4. Distance between pedicel and base of spur 10–15 mm; spurs 10–20 mm; flowers pink to purple and white, lacking dark reddish lines within the dark yellow markings . . . . . ***I. balfourii***
      4' Distance between pedicel and base of spur 1–5 mm; spurs 3–9 mm; flowers pale yellow and white, often with dark reddish lines within the dark yellow markings . . . . . ***I. parviflora***
   3' Flowers abruptly tapered to a strongly curved or bent nectar spur; with orange, red, purple, pink or dark yellow markings within floral tube and often at mouth, or the flowers without dark markings.
      5. Petioles glandular at base; leaf tips acuminate to cuspidate; flowers white to purple or purple and orange; seeds not strongly ridged, papillose striate to rugose; exotic.
         6. Upper stem leaves opposite or whorled; floral tubes white, pink or purple, never orange, with dark yellow, dark pink or dark purple splotches or spots within; petals white, pink or purple; nectar spurs relatively small, 2–7 mm . . . . . ***I. glandulifera***
         6' Upper stem leaves alternate; floral tubes orange, with dark orange lines within; petals pink to purple; nectar spurs prominent, 8–14 mm . . . . . ***I. bicolor***
      5' Petioles eglandular at base; leaf tips blunt to acute; flowers primarily yellow-orange, never purple and orange; seeds few-ridged, otherwise smooth to faintly papillose; exotic and native.
         7. Floral tube usually yellow-orange, with coarse, primarily ventral spots and splotches of dark red to dark orange, mouth of tube often spotted or splotched; rare floral forms can be white or pale yellow with dark yellow markings, or pink with dark pink markings . . . . . ***I. capensis*** (in part)
         7' Floral tube and mouth unspotted, uniformly colored.
            8. Flowers in a population uniformly spurred and unspotted; pollen fertile . . . . . ***I. capensis*** (in part)
            8' Flowers in a population a mix of spurred and unspurred, spotted and unspotted; pollen fertility reduced; all flowers on a plant unspotted and spurred . . . . . ***I. ×pacifica*** (in part)

Impatiens balfourii Hooker f.

Kashmir balsam, Balfour's touch-me-not

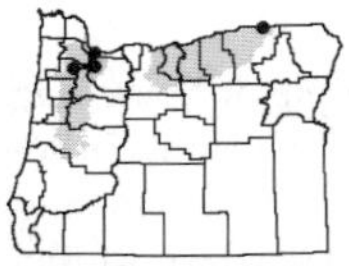

Plants to 1 m. **Leaves** alternate; blades lanceolate to elliptic or ovate, serrate, tips acuminate, glabrous; petiole bases with low domed glands. **Inflorescences** 1–14-flowered; peduncles 5–12 cm. **Flowers** gradually tapered to straight or gently curved spur, primarily bicolored, white and pink to purple, with dark yellow markings within white, pink or purple tube and often at mouth; distance between pedicel and base of spur 10–15 mm; spurs 10–20 mm. **Fruits** linear-oblong to oblanceolate, 20–40 mm. **Seeds** ovate to oblong-ovate, longitudinally striate-papillose. ***2n***=14.

Disturbed ground, forested or rocky slopes, scree, ditches. Flowering Jul–Sep. 50–200 m. Col, WV. CA, WA; WI; Asia. Exotic.

Impatiens balfourii is invasive at several locations in the Portland area, and control efforts are underway.

Impatiens bicolor Royle

Himalayan balsam, bicolored jewelweed

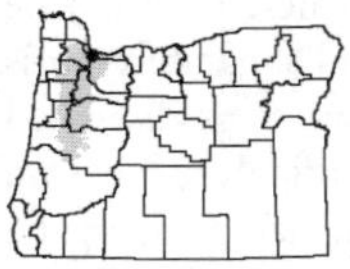

Plants to 1.5 m. **Leaves** alternate; blades lanceolate to elliptic, oblong, or ovate, crenate, tips acuminate; petiole bases with low domed glands. **Inflorescences** 1–120-flowered; peduncles 1–25 cm. **Flowers** abruptly tapered to strongly curved spur, primarily bicolored, orange and pink to purple, with dark orange lines within orange tube and at mouth; distance between pedicel and base of spur 11–17 mm; spurs 8–14 mm. **Fruits** narrowly oblanceolate, 25–35 mm. **Seeds** ovate to oblong-ovate or pyriform, smooth to rugose.

Forested streambanks, damp ravine slopes. Flowering Jul–Aug. 50–100 m. WV. Asia. Exotic.

Impatiens bicolor flowers are partly purple and partly orange, which is a unique combination in the *Flora*.

Impatiens capensis Meerburgh (p. 372)

Spotted jewelweed, spotted touch-me-not

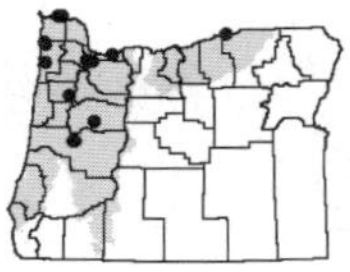

Plants to 2 m. **Leaves** alternate; blades ovate to elliptic, sparsely crenate to dentate, tips blunt to acute; petiole bases eglandular. **Inflorescences** 1–6-flowered; peduncles 1–4 cm. **Flowers** abruptly tapered to strongly curved spur, primarily colored yellow-orange, unspotted or with coarse dark orange to dark red ventral markings within yellow-orange tube and at mouth, rarely white with yellow markings or pale yellow to pink with dark pink markings; distance between pedicel and base of spur 15–23 mm; spurs 5–25 mm; pollen fertile. **Fruits** oblanceolate to narrowly obovate, 15–25 mm. **Seeds** ± oblong, few-ridged, otherwise smooth or faintly papillose. ***2n***=20.

Shores, damp slopes, swales, freshwater intertidal, wet forest, floating logs, ditches. Flowering Jun–Oct. 0–300 m. Casc, Col, CR, Est, WV. CA, WA; most of southern Canada, eastern US. Exotic.

Impatiens capensis is often confused with *I. noli-tangere*, a boreal species found no closer than northern Washington that differs in the shape of the floral spur and has finer dorsal spotting on the floral tube. *Impatiens capensis* is similar to the smaller-flowered *I. aurella*, just north and east of Oregon, which has a more variable taper to its spur and is more frequently unspotted. Our populations of *I. capensis* are considered exotic, with the first collection in 1957. The species is native in central and eastern North America. Wherever *I. capensis* comes into contact with native *I. ecornuta*, hybrids are formed, mediated by pollinating bees.

Impatiens ecornuta Gerry Moore, Zika & Rushworth

Spurless jewelweed, spurless touch-me-not

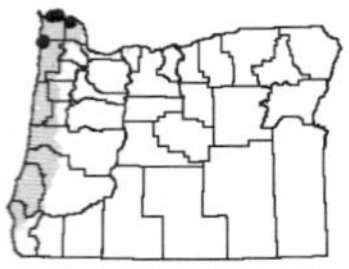

Plants to 1.2 m. **Leaves** alternate; blades ovate to elliptic, crenate to dentate, tips acute to blunt; petiole bases eglandular. **Inflorescences** 1–6-flowered; peduncles 1–4 cm. **Flowers** sac-like, uniformly yellow-orange, without dark markings within tube or at mouth; without a spur. **Fruits** linear-oblong to oblanceolate, 12–20 mm. **Seeds** oblong to elliptic-oblong or ovate-oblong, longitudinally ridged on corners, faces ± smooth.

Shores, marshes, freshwater intertidal, streams, damp ground, floodplains. Flowering Jul–Sep. 0–50 m. CR, Est. ID, WA; north to British Columbia, northeast to MT. Native.

This species hybridizes with introduced *Impatiens capensis* to produce stable populations of *I.* × *pacifica*. The name *I. ecalcarata* is illegitimate and is replaced by *I. ecornuta*.

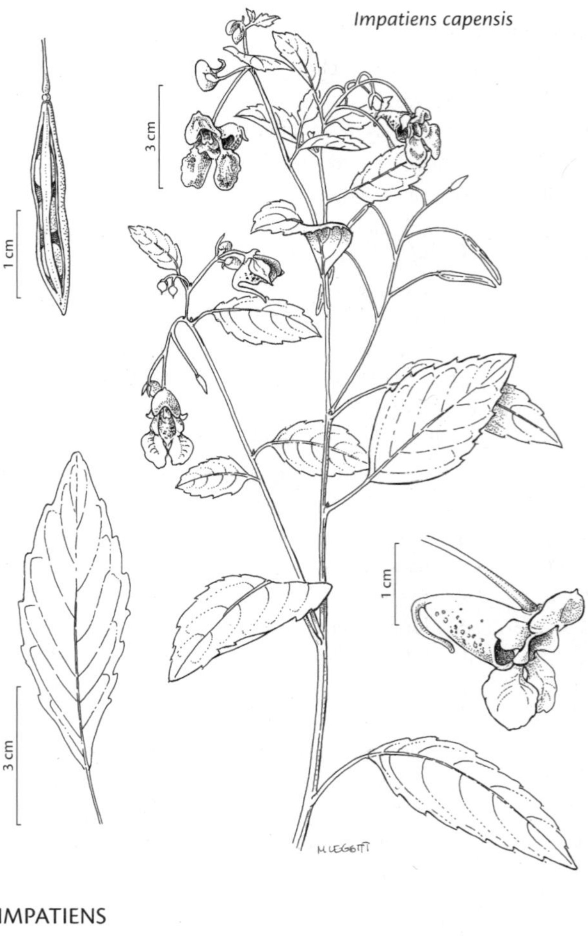

IMPATIENS

Impatiens glandulifera Royle

Policeman's helmet, Himalayan balsam

Impatiens roylei Walp.

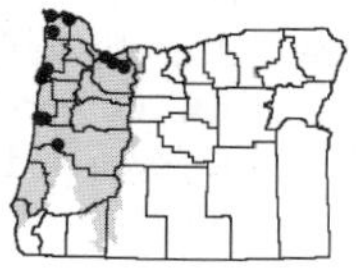

Plants to 3.3 m. **Leaves** opposite or whorled, sometimes a few alternate near inflorescence; blades lanceolate to elliptic, finely serrate, tips acuminate; petiole bases with columnar or oblong glands. **Inflorescences** 3–12-flowered; peduncles 5–15+ cm. **Flowers** abruptly tapered to a spur strongly bent at base, then gently curved, primarily purple, pink, or white, with dark pink to dark purple or yellow ventral splotches or spots within tube or at mouth; spurs 2–7 mm. **Fruits** oblanceolate to obovate, 14–30 mm. **Seeds** ovate to oblong-ovate or pear-shaped, densely striate-papillose. ***2n***=18.

Wet forest, shores, floodplains, marshes, damp ground. Flowering Jul–Oct. 0–500 m. Casc, CR, Est, WV. CA, ID, WA; north to British Columbia, eastern North America; Asia. Exotic.

Impatiens glandulifera is an increasingly invasive ornamental.

Impatiens ×pacifica Zika

Pacific jewelweed, Pacific touch-me-not

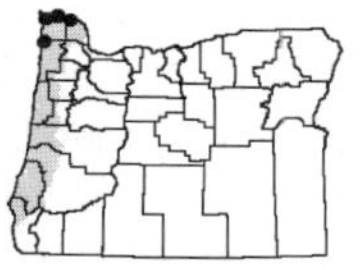

Plants to 2.2 m. **Leaves** alternate; blades ovate to elliptic, crenate to dentate, tips acute to blunt; petiole bases eglandular. **Inflorescences** 1–4-flowered; peduncles 1–3 cm. **Flowers** primarily yellow-orange, uniform on an individual but expressed as 2 separate forms in a population; spotted spurless forms, sac-like, with irregular and coarse, usually dark orange markings within tube or at mouth; spotless spurred forms abruptly tapered to strongly curved spur, without dark markings within tube or at mouth; distance between pedicel and base of spur 14–18 mm; pollen with reduced fertility. **Fruits** narrowly oblong to narrowly oblanceolate, 12–20 mm. **Seeds** oblong to narrowly ovate or narrowly elliptic, longitudinally ridged on corners, faces ± smooth.

Shores, marshes, freshwater intertidal, damp ground, floodplains. Flowering Aug–Oct. 0–50 m. CR, Est. WA. Native.

Impatiens ×pacifica is a hybrid always found mixed among its parents, exotic *I. capensis* and native *I. ecornuta*, and combining their floral characters in one of two ways. Hybrid plants bear only one sort of flower: either the spotted, spurless floral form or the spotless, spurred floral form. Both forms are usually common in a population. The spotless, spurred floral form closely resembles the spotless color form of *I. capensis*, which is invariably spurred, rare in Oregon, and, so far, does not occur in mixed populations that include spurless plants.

Impatiens parviflora de Candolle

Small balsam

Plants to 1 m. **Leaves** alternate; blades ovate to elliptic, finely to coarsely serrate, tips acuminate; petiole bases with columnar glands. **Inflorescences** 1–12-flowered; peduncles 2–10 cm. **Flowers** gradually tapered to a straight or gently curved spur, primarily pale yellow and white, with dark yellow throat and dark yellow markings within tube and often at mouth, often with dark reddish lines within dark yellow markings; 1–5 mm between pedicel and base of spur; spurs 3–9 mm. **Fruits** linear-oblong to oblanceolate, 14–30 mm. **Seeds** oblong to elliptic-oblong or ovate-oblong, densely striate-papillose. ***2n***=26.

Mesic slopes, open forest. Flowering Jun–Sep. 50–200 m. WV. WA; north to British Columbia, eastern Canada, WI; Asia. Exotic.

BERBERIDACEAE Barberry family

Alan T. Whittemore

Shrubs or **herbs** perennial, rhizomatous. **Stems** sometimes spiny. **Leaves** basal or alternate, simple or pinnately or ternately compound. **Inflorescences** racemes, panicles, umbels, or spikes. **Flowers** radially symmetric, bisexual, hypogynous; hypanthium absent, sometimes with 3–9 bractlets forming an epicalyx appressed to calyx; sepals 6, free, sometimes falling as flowers open, or absent; petals 6, free, showy, producing nectar, or absent; stamens 6–18, free and distinct, often same number as petals and opposite them, anthers opening by apical flaps or longitudinal slits; pistils 1, superior, placentation basal or lateral, styles short or obsolete. **Fruits** berries or capsules. **Seeds** 1–10 per fruit, sometimes bearing an aril.

East Africa, Asia, Europe, North America, South America. 15 genera; 3 genera treated in *Flora.*

The structure of the pistil in Berberidaceae can be confusing, and its homologies are uncertain. Berberidaceae taxa have a single pistil with a single unlobed stigma and an ovary having a single locule. Most taxa have one or a group of basal ovules. A few taxa (including *Vancouveria*) have more complex pistils, but their structure is difficult to interpret. It is uncertain whether the pistil evolved from a single carpel, like the pistils of Ranunculaceae, or from several fused carpels, as in Papaveraceae (Cronquist 1981).

Key to Berberidaceae genera:

1. Shrubs (aerial stems sometimes very short); leaves all cauline, simple or pinnately compound (occasionally a few leaves 3-foliolate with the terminal leaflet stalked and lateral leaflets sessile) . . . . **BERBERIS**

1' Acaulescent herbs; leaves all basal, 1–3 times ternately compound (rarely 5-foliolate in *Vancouveria chrysantha*).

  2. Leaves 3-foliolate, leaflets sessile or on 1–3 mm petiolules; flowers in dense spikes, perianths absent . . . . . . **ACHLYS**

  2' Leaves usually 2–3 times ternately compound (occasionally 3-foliolate in *V. chrysantha*), leaflets on 12–35 mm petiolules; flowers in lax racemes or panicles, perianths present . . . . **VANCOUVERIA**

ACHLYS Vanilla-leaf

Herbs perennial, deciduous, acaulescent, 2.5–5 dm; rhizomes slender, long-creeping, dark brown, woody, 1–5 mm thick, older parts naked except for lateral roots, younger parts clothed in persistent leaf bases, each tip with 1–2 leaves and 1 scape. **Leaves** basal, 3-foliolate, soft, thin, 16–31 cm, yellow-green; leaflets with shallow rounded (seldom acute) lobes, central leaflets obtriangular, with 1(3) main veins, lateral leaflets asymmetrically fan-shaped, with 3(4) main veins; petioles slender. **Inflorescences** dense spikes terminating a slender, erect, naked peduncle. **Flowers** a whorl of stamens surrounding base of ovary, bracteoles, sepals, and petals absent; stamens 8–10(13), dehiscing by 2 apically hinged flaps, white; ovaries ellipsoidal, ovules marginal, placenta developed only near base of ovary, stigma asymmetric, sessile on ovary. **Fruits** follicle-like, hemispheric, purplish red or brown, puberulent, dehiscing transversely. **Seeds** 1, brown, aril absent.

Japan, western North America. 2 species.

The common name refers to the pleasant fragrance of the dried leaves. In the past, it was popular to hang clusters of dried *Achlys* leaves in rooms. They not only scented the air pleasantly, but they were said to repel flies and mosquitoes. Prior to the studies of Fukuda (1967) and Fukuda and Baker (1970), North American *Achlys* were always treated as a single species, and older specimens are always identified as *A. triphylla*. For this reason, older reports of *A. triphylla* are unreliable—they could represent either species. Fukuda's work made it clear that there are actually two species in Oregon that are genetically quite distinct from one another, and well-developed specimens are morphologically distinguishable. The key and descriptions give typical ranges of variation for the species, but the characters are variable and identifications are often difficult, especially with specimens that are fragmentary.

Key to *Achlys* species:

1. Central leaflet blade 5–9 × 4–8 cm, with 3–5 shallow lobes; stamens 2.8–4.3 mm; fruits red-purple. . ***A. triphylla*** ssp. ***triphylla***

1' Central leaflet blade 8.4–13 × 7–13 cm, with (4)6–9 shallow lobes; stamens 4–5 mm; fruits brown . . . . ***A. californica***

Achlys californica Fukuda & H.G. Baker (p. 374)

Deer's-foot

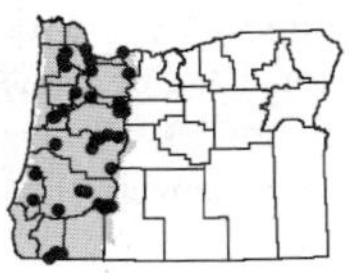

Leaves petiolules to 2 mm; central leaflets 8.4–13 × 7–13 cm, bases obtuse, tips truncate with (4)6–9 shallow, rounded lobes; lateral leaflets with 7–12 shallow, rounded lobes. **Inflorescences** 3–6 cm; peduncles 28–41 cm. **Flowers** stamens 4–5 mm. **Fruits** 4–5 mm, brown. **2*n*=24.**

Forests. Flowering Apr–Jul. 0–1700 m. Casc, CR, Sisk, WV. CA, WA; north to British Columbia. Native.

Achlys triphylla (Smith) de Candolle ssp. *triphylla* (p. 374)

Northern vanilla-leaf, sweet-after-death

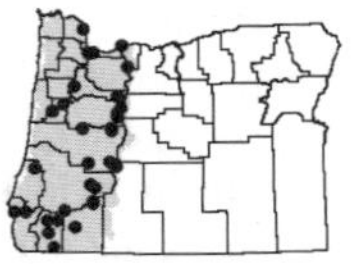

Leaves petiolules 1–3 mm; central leaflets 5–9 × 4–8 cm, bases obtuse or broadly acute, tips truncate with 3–5 shallow, rounded lobes; lateral leaflets with 6–14 shallow, rounded (seldom acute) lobes. **Inflorescences** 2.3–4.4 cm; peduncles 23–31 cm. **Flowers** stamens 2.8–4.3 mm. **Fruits** 3–4 mm, red-purple. **2*n*=12.**

Forests. Flowering Apr–Aug. 50–1900 m. Casc, CR, Sisk, WV. CA, WA; north to British Columbia. Native.

The other variety of this species, *Achlys triphylla* ssp. *japonica*, is endemic to Japan.

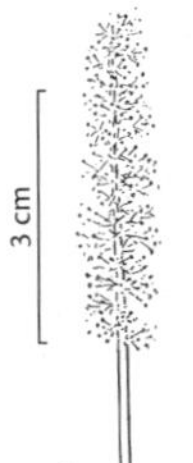

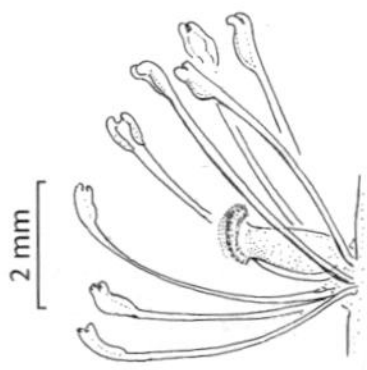

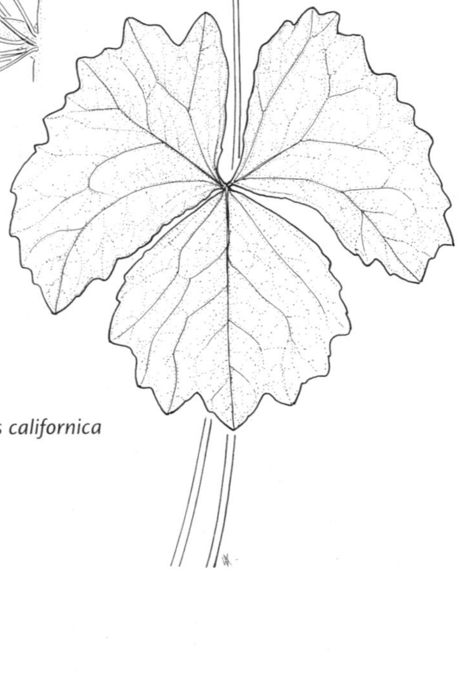

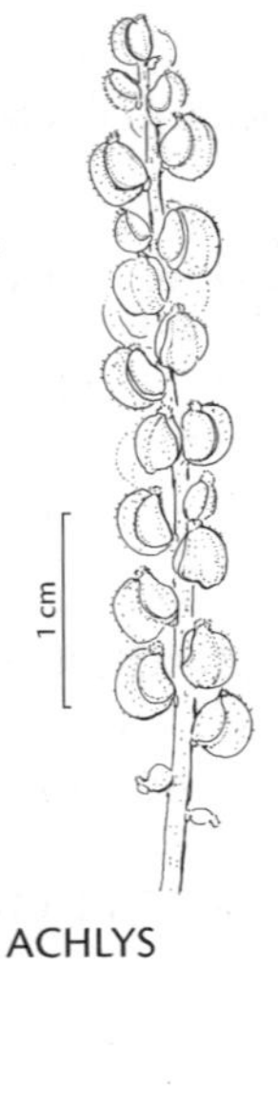

Achlys californica

Achlys triphylla ssp. triphylla

ACHLYS

BERBERIS Barberry, Oregon-grape

Shrubs or **subshrubs** evergreen (waifs may be deciduous), 1–45(80) dm; rhizomes present or absent, short or long. **Leaves** alternate, pinnately compound (simple in some waifs); leaflets ovate to oblong-ovate, lance-ovate, or elliptic, toothed or shallowly lobed, pinnately or palmately veined; petioles stout, 0.1–9 cm. **Inflorescences** usually dense racemes (lax racemes or umbels in some waifs). **Flowers** with 3 caducous scale-like bracteoles; sepals 6, falling immediately after anthesis, yellow; petals 6, flat, yellow; ovaries symmetrically club-shaped, ovules sub-basal, style central. **Fruits** berries, spherical to cylindric-ovoid or ellipsoid, red, blue, or purple, juicy; seeds 1–10, tan to red-brown or black, aril absent.

Almost worldwide. ~500 species; 7 species treated in *Flora*.

Many species of *Berberis* are grown as ornamental shrubs. However, many ornamental barberries and Oregon-grapes are of hybrid origin. The berries of many species are edible and frequently used for jam and jelly. In many exotic barberry species, the leaves on the main stems are modified as spines. The foliage leaves, which are simple, are clustered on short shoots in the axils of these spines. Several simple-leaved species are grown as garden plants in Oregon. In addition to *Berberis darwinii* (fully described below), two other species with simple leaves and spiny stems have been known to escape in Oregon. Both differ from *B. darwinii* in having thin, flexible leaves with 15–30 slender flexible spines about 1 mm on their margins. *Berberis julianae* has dark green, glossy, oblanceolate to narrowly elliptic evergreen leaves, and flowers in sessile umbels; its fruit is narrowly cylindrical, purple and very glaucous. *Berberis julianae* is often grown as an ornamental; it has escaped and become locally established in a wet prairie in Lane County. In addition, the common barberry of Europe, *B. vulgaris*, was recently found growing as a waif at one site in Harney County. Common barberry is deciduous, with obovate to oblanceolate leaves, flowers in lax racemes, and red, oblong-ellipsoidal fruit. It was widely cultivated at one time, but cultivation of *B. vulgaris* has been discontinued because it is an alternate host for the black stem-rust of wheat (*Puccinia graminis*). Our native species all have spineless stems and compound leaves, and horticultural sources sometimes separate the compound-leaved species as a distinct genus, *Mahonia*. The distinction is clear in Oregon, but intermediate species are found in other parts of the world, and *Mahonia*, though often recognized in horticultural works, is seldom accepted by botanists.

Key to *Berberis* species:

1. Leaves simple; stems tomentose, spiny; inflorescences 10–20-flowered . . . . ***B. darwinii***

1' Leaves pinnately compound; stems glabrous, unarmed; inflorescences 25–70-flowered.

- 2. Leaflets 4–6-veined from base; bud scales (13)20–44 mm, persistent; filaments lacking appendages. . . . . ***B. nervosa***
- 2' Leaflets 1(3)-veined from base; bud scales 2–5 mm, deciduous; distal end of each filament with a pair of recurved teeth.
  - 3. Cells of abaxial epidermis of leaf plane, thus leaflets smooth and somewhat shiny abaxially.
    - 4. Terminal leaflets 1.7–2.5 × as long as wide; lateral leaflets lance-ovate or lance-elliptic; petioles 1–6 cm; widespread in Oregon . . . . ***B. aquifolium***
    - 4' Terminal leaflets 1.3–1.9 × as long as wide; lateral leaflets ovate or lance-ovate; petioles 0.5–4.5(7.5) cm; extreme southwestern Oregon only . . . . ***B. pinnata*** ssp. ***pinnata***
  - 3' Cells of abaxial epidermis of leaf strongly bulging, thus leaflets microscopically papillate and very dull abaxially.
    - 5. Leaves thin and flexible, teeth 6–24 per margin, 0.1–0.25 mm thick; plants of Cascades and eastern Oregon . . . . ***B. repens***
    - 5' Leaves thick and rigid, teeth 2–12 per margin, 0.2–0.5 mm thick; plants of southwestern Oregon.
      - 6. Plant 3–8 dm; adaxial leaf surface glossy, each leaf margin with 6–12 teeth; inflorescences 3–10 cm; fruits 7–10 mm . . . . ***B. piperiana***
      - 6' Plant 1–4 dm; adaxial leaf surface dull, more or less glaucous, each leaf margin with 2–10 teeth; inflorescences 2–4 cm; fruits 5–8 mm. . . . . ***B. pumila***

Berberis aquifolium Pursh (p. 376)

Tall Oregon-grape, shining Oregon-grape

Berberis aquifolium Pursh var. *aquifolium*, *Mahonia aquifolium* (Pursh) Nutt.

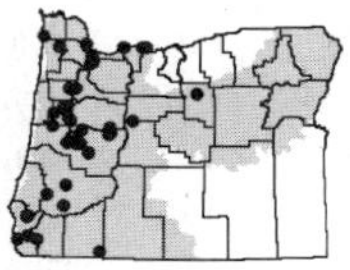

Shrubs 3–30 dm. **Stems** usually monomorphic, seldom with short axillary shoots, twigs glabrous; bud scales 4–8(14) mm, deciduous; spines absent. **Leaves** compound; leaflets 5–9, thin and flexible or rather rigid, glossy and smooth abaxially, green and often glossy adaxially; terminal leaflets 5.1–8.7(14.5) × 2.4–4.5(5.5) cm, 1.7–2.5 × as long as wide; lateral leaflets lance-ovate to lance-elliptic, 1(3)-veined from base, bases obtuse or truncate (rarely weakly cordate), margins plane or undulate, each margin with 5–21 0–2 mm teeth, tipped with spines to 0.8–2.2 × 0.2–0.3 mm, tips acute or sometimes obtuse or rounded; petioles 1–6 cm. **Inflorescences** 30–60-flowered, 3–11 cm; bracts rounded or obtuse, sometimes apiculate. **Stamens** filaments with a distal pair of recurved teeth. **Fruits** oblong-ovoid, 6–10 mm, blue, glaucous.

Open woods, shrublands. Flowering Feb–May. 0–1500 m. BW, Casc, CR, ECas, Lava, Sisk, WV. CA, ID, WA; north to British Columbia, northeast to MT. Native.

Berberis aquifolium is the state flower of Oregon. It is widely grown as an ornamental, and it may be expected to escape from cultivation in areas where it has been planted. *Berberis aquifolium* from nursery sources are often introgressed with genes from other species. In most areas, *B. aquifolium* is an erect shrub 1–3 m tall, but low forms occur in exposed places in the Cascades and the Columbia River gorge. These forms have been confused with *B. repens*, but they are clearly *B. aquifolium* from their leaf characters. *Berberis aquifolium* does hybridize with *B. repens* where they come together on the east side of the Cascades, and some populations in this area appear intermediate between these two species.

Berberis darwinii Hooker

Darwin's barberry

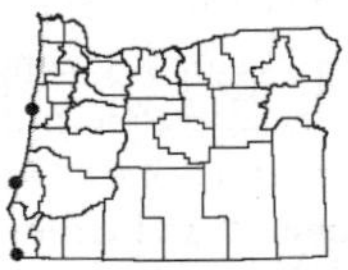

Shrubs 10–30 dm. **Stems** dimorphic, with short axillary shoots, twigs densely tomentose; bud scales 2–4 mm, deciduous; spines present, pedately 5–9-fid. **Leaves** simple, blades obovate, 1-veined from base, 1.7–3.0 × 0.9–1.2 cm, thick and rigid, bases acute or acuminate, margins reflexed, undulate, each margin with 2–4 1–3 mm teeth or shallow lobes, tipped with 1.2–1.6 × 0.2–0.3 mm spines, glossy and smooth abaxially, glossy and green adaxially, tips obtuse or rounded in outline; petioles 0.1–0.3 cm. **Inflorescences** 10–20-flowered, 3–4 cm; bracts acuminate. **Stamens** filaments without lateral teeth. **Fruits** globose, 6–7 mm, dark purple.

Shrublands, open coastal woodlands. Flowering Mar–Jun. 0–100 m. Est. CA, WA; South America. Exotic.

Berberis darwinii is a popular hedge plant in coastal gardens, and it has escaped in several places along the U. S. Pacific coast. Once established, it can spread by underground runners to form thickets.

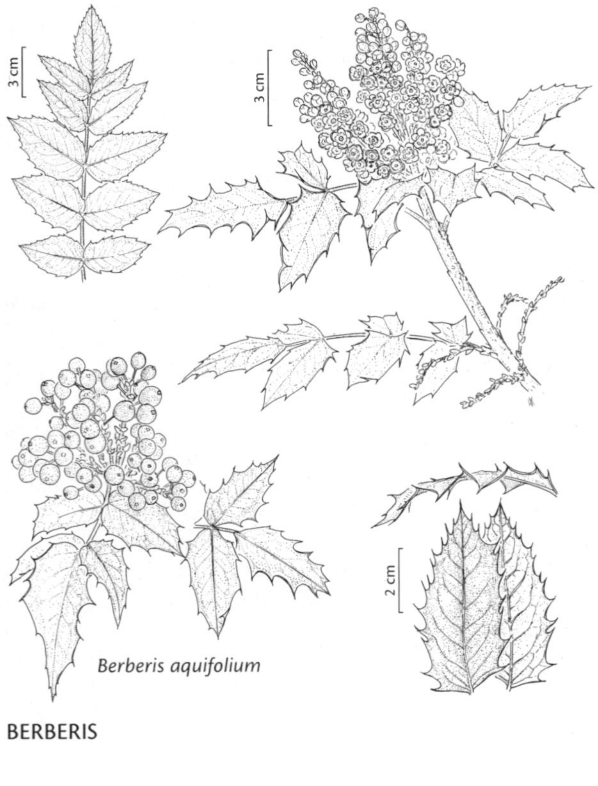

BERBERIS

Berberis nervosa Pursh (p. 377)

Cascade Oregon-grape

Mahonia nervosa (Pursh) Nutt.

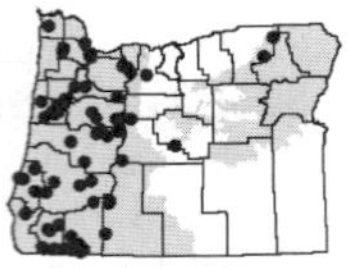

Shrubs 1–6 dm. **Stems** monomorphic, without short axillary shoots, twigs glabrous; bud scales (13)20–44 mm, persistent; spines absent. **Leaves** compound; leaflets 9–21, thin and ± flexible, smooth but rather dull abaxially, dull and somewhat glaucous adaxially; terminal leaflet 2.9–8.4 × 1.2–4.8 cm, 1.8–3.2 × as long as wide; lateral leaflets lance-ovate to ovate, 4–6-veined from base, bases rounded to cordate, margins plane, each margin with 6–13 teeth 1–2(3) mm, tipped with spines to 1.0–2.4 × 0.1–0.2 mm, tips acute or broadly acuminate; petioles 2–11 cm. **Inflorescences** 30–70-flowered, 6–17 cm; bracts acute, obtuse or rounded. **Stamens** filaments without teeth. **Fruits** oblong-ovoid or globose, 8–11 mm, blue, glaucous.

Open or shaded woods, rocky areas. Flowering Mar–Jun. 0–2200 m. BW, Casc, CR, ECas, Est, Sisk, WV. CA, ID, WA; north to British Columbia. Native.

Berberis pinnata Lagasca ssp. *pinnata*

California Oregon-grape

Mahonia pinnata (Lag.) Fedde

Shrubs 3–16 dm. **Stems** usually monomorphic, seldom with short axillary shoots, twigs glabrous; bud scales 3–7 mm, deciduous; spines absent. **Leaves** compound; leaflets (3)5–13, thin but ± rigid, glossy and smooth abaxially, glossy and green adaxially; terminal leaflet 2.6–6.2 × 2.0–4.5 cm, 1.3–1.9 × as long as wide; lateral leaflets ovate to broadly lanceolate, 1(3)-veined from base, bases broadly obtuse, truncate or weakly cordate, margins undulate or crisped, each margin with 5–22 0–2 mm teeth, tipped with spines, 1.2–3.0 × 0.2–0.3 mm, tips acute to rounded-obtuse; petioles 0.5–4.5(7.5) cm. **Inflorescences** 25–50-flowered, 2–5 cm; bracts rounded to broadly acute (sometimes apiculate). **Stamens** filaments with a distal pair of recurved teeth. **Fruits** oblong-ovoid to subglobose, 6–7 mm, blue, glaucous.

Stabilized sand dunes. Flowering Mar–May. 0–100 m. Est. CA; south to Mexico. Native.

In Oregon, *Berberis pinnata* is known only along the immediate coast from Gold Beach southward.

Berberis piperiana (Abrams) McMinn (p. 377)

Piper's Oregon-grape

Mahonia piperiana Abrams

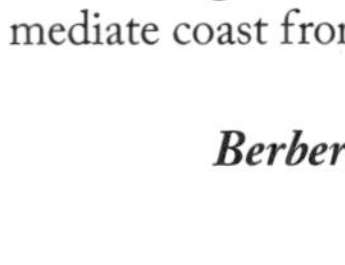

Shrubs 3–8 dm. **Stems** monomorphic, without short axillary shoots, twigs glabrous; bud scales 3–8 mm, deciduous; spines absent. **Leaves** compound; leaflets 5–9, rather thick and ± rigid, dull and papillose abaxially, glossy and green adaxially; terminal leaflet 3.6–9.5 × 2.0–5.2 cm, 1.3–2.1 × as long as wide; lateral leaflets lance-elliptic to elliptic-ovate or narrowly oblong, 1(3)-veined from base, bases rounded to obtuse or truncate, margins undulate, each margin with 6–12 1–4 mm teeth, tipped with 1.6–2.8 × 0.2–0.3 mm spines, tips acute to rounded-obtuse; petioles 1–6 cm.

Inflorescences 25–60-flowered, 3–10 cm; bracts obtuse or rounded. **Stamens** filaments with a distal pair of recurved teeth. **Fruits** oblong-ovoid, 7–10 mm, dark blue, glaucous.

Open woods, shrubby areas. Flowering Mar–Jun. 50–1800 m. Casc, ECas, Sisk, WV. CA. Native.

Berberis pumila Greene

Pygmy Oregon-grape, dwarf western barberry

Mahonia pumila (Greene) Fedde

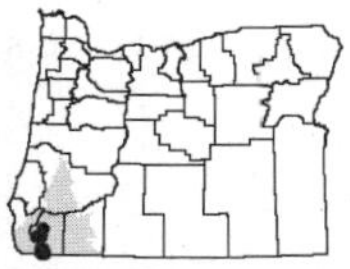

Shrubs 1–4 dm. **Stems** monomorphic, without short axillary shoots, twigs glabrous; bud scales 3–6 mm, deciduous; spines absent. **Leaves** compound; leaflets 3–9, thick and rigid, dull and papillose abaxially, dull and glaucous adaxially; terminal leaflet 4.0–8.0 × 2.0–5.0 cm, 1.3–1.9 × as long as wide; lateral leaflets ovate to oblong-ovate or elliptic, 1(3)-veined from base, bases obtuse (rarely truncate), margins plane or undulate, each margin with 2–10 1–3 mm teeth, tipped with 1.6–3.0 × 0.3–0.4(0.5) mm spines, tips obtuse or rounded (rarely broadly acuminate) in outline; petioles 0.5–4 cm. **Inflorescences** 30–45-flowered, 2–4 cm; bracts rounded or obtuse, sometimes apiculate. **Stamens** filaments with a distal pair of recurved teeth. **Fruits** oblong-ovoid to globose, 5–8 mm, dark blue, glaucous.

Open woods, shrublands, rocky or gravelly areas. Flowering Mar–Jun. 400–1600 m. Sisk. CA. Native.

Berberis repens Lindley (p. 377)

Creeping Oregon-grape

Berberis aquifolium Pursh var. *repens* (Lindl.) Scoggan, *Mahonia repens* (Lindl.) G. Don

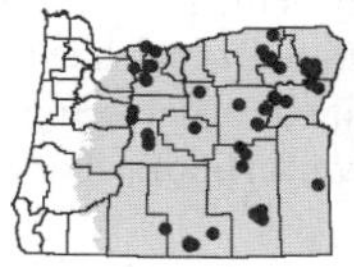

Shrubs 0.2–2(6) dm. **Stems** monomorphic, usually without short axillary shoots, twigs glabrous; bud scales 3–8 mm, deciduous; spines absent. **Leaves** compound; leaflets (3)5–7, thin and flexible, dull and papillose abaxially, dull (rarely glossy) and somewhat glaucous adaxially; terminal leaflet 3.2–9.5 × 2.3–6.0 cm, 1.2–2.2(2.5) × as long as wide; lateral leaflets ovate or elliptic, 1(3)-veined from base, bases rounded to obtuse or truncate, margins plane, each margin with 6–24 0.5–3 mm teeth, tipped with 0.6–2.8 × 0.1–0.25 mm spines, tips rounded (rarely obtuse or even broadly acute); petioles (1)3–9 cm. **Inflorescences** 25–50-flowered, 3–10 cm; bracts rounded to obtuse or broadly acute. **Stamens** filaments with a distal pair of recurved teeth. **Fruits** oblong-ovoid, 6–10 mm, blue, glaucous.

Open woods, shrublands, grasslands. Flowering Apr–Aug. 0–2300 m. BR, BW, Casc, Col, ECas, Lava, Owy. CA, ID, NV, WA; north to British Columbia, east to SD, southeast to TX. Native.

See also *B. aquifolium*.

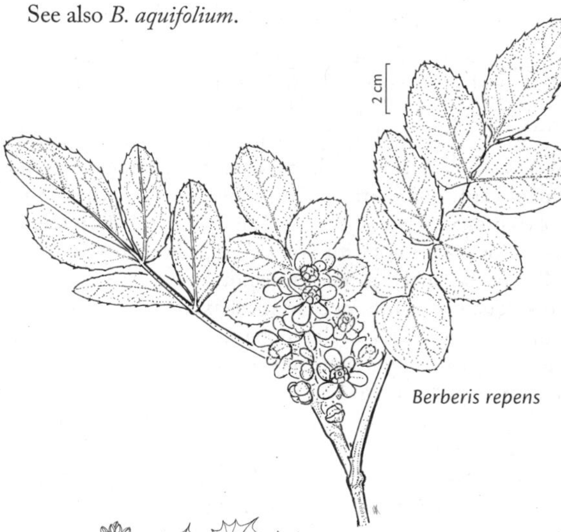

Berberis repens

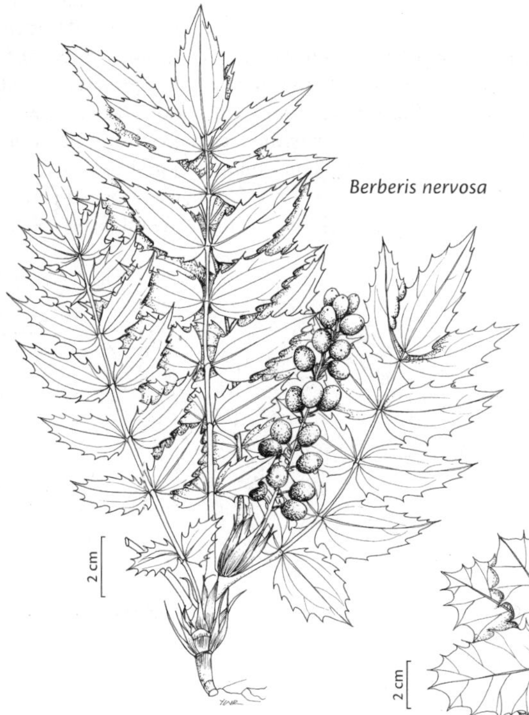

Berberis nervosa

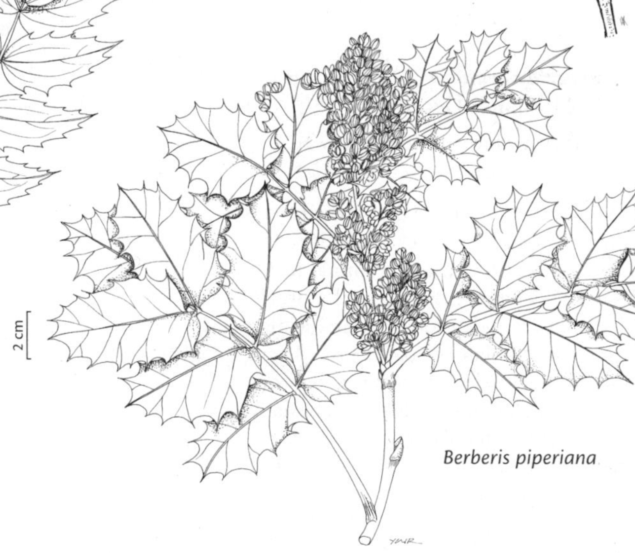

Berberis piperiana

VANCOUVERIA Inside-out flower

Herbs perennial, acaulescent, evergreen or deciduous, 1–5 dm; rhizomes slender, long-creeping, light or dark brown, woody, 1–4 mm thick, older parts naked except for lateral roots, tips clothed in brown scales; each rhizome tip with 1–8 leaves and 1 inflorescence. **Leaves** basal, 2–3-ternately compound (sometimes 3–5-foliolate in *V. chrysantha*); leaflets rhomboidal or rounded-pentagonal to ovate or oblong, shallowly 3-lobed, palmately veined, each leaflet on a 12–35 mm petiolule; petioles slender, 5–31 cm. **Inflorescences** open racemes or panicles. **Flowers** with 6–9 sepal-like bracteoles; sepals 6, white to yellow; petals 6, white to yellow, hooded with tip reflexed or flat, producing nectar; stamens 6, dehiscing by 2 apically hinged flaps; ovaries asymmetrically ellipsoidal, ovules marginal, style lateral. **Fruits** follicle-like, brown, cylindrical or ellipsoidal, dehiscing by 2 valves. **Seeds** 4–7, black to reddish brown, aril whitish, covering lower part of seed.

Western North America. 3 species.

Vancouveria is closely related to the Eurasian genus *Epimedium*. Both of these genera are cultivated as flowering ground covers in Oregon gardens.

Key to *Vancouveria* species:

1. Deciduous, leaves falling as fruit ripens; margins of leaflets thin; pedicels not glandular. ***V. hexandra***

1' Evergreen; margins of leaflets thickened; pedicels glandular-hairy.

 2. Petals yellow, apical part reflexed; fruits asymmetrically cylindrical, glandular-pubescent, 8–10 × 3–4 mm . . . . ***V. chrysantha***

 2' Petals white or tinged with purple, plane; fruits ellipsoidal, glabrous, 3–4 × 1–2 mm . . . . ***V. planipetala***

Vancouveria chrysantha Greene

Golden inside-out flower, yellow vancouveria, Siskiyou inside-out flower

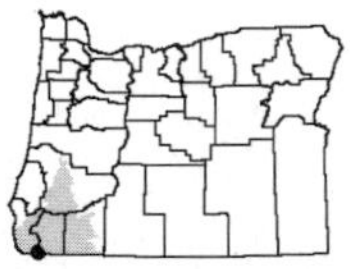

Plants evergreen. **Leaves** twice ternately or once pinnately compound; leaflets 5–9, tough, terminal leaflet 29–34 × 25–41 mm, unlobed or with three shallow rounded lobes, otherwise entire or crenate, margin narrowly cartilaginous-thickened; petioles 7–17 cm, developing purplish brown pigmentation, at least in places. **Inflorescences** racemes (or sometimes with a branched pedicel, therefore technically few-branched panicles), glandular-pubescent. **Flowers** yellow, sepals 6–8 mm, pubescent abaxially; petals 4–5 mm, hooded; filaments glandular-pubescent; ovaries glandular-pubescent. **Fruits** asymmetrically cylindrical, 8–10 × 3–4 mm, glandular-pubescent. $2n$=12.

Forests, chaparral. Flowering May–Jun. 300–1000 m. Sisk. CA. Native.

This species is endemic to the Klamath region in southwestern Oregon and northwestern California.

Vancouveria hexandra (p. 378)
(Hooker) C. Morren & Decaisne

Northern inside-out flower, white inside-out flower, northern vancouveria

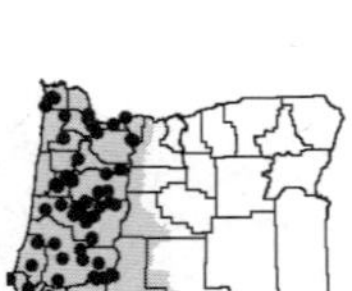

Plants deciduous, leaves falling when fruits mature. **Leaves** 2–3 times ternately compound; leaflets 9–27, soft, terminal leaflet 32–60 × 31–78 mm, entire except for usually 3–5 shallow rounded lobes (lateral leaflets seldom unlobed), margin not thickened; petioles 7–31 cm, straw-colored. **Inflorescences** panicles or racemes, glabrous. **Flowers** white, sepals 5–8 mm, glandular-pubescent abaxially; petals 4–6 mm, hooded; filaments glandular-pubescent; ovaries glandular-pubescent. **Fruits** asymmetrically cylindrical, 8–10 × 3–4 mm, glandular-pubescent. $2n$=12.

Forests, chaparral. Flowering Apr–Aug. 0–1900 m. Casc, CR, ECas, Est, Sisk, WV. CA, WA. Native.

Vancouveria planipetala Calloni

Redwood-ivy, redwood inside-out flower, small inside-out flower, small-flowered vancouveria

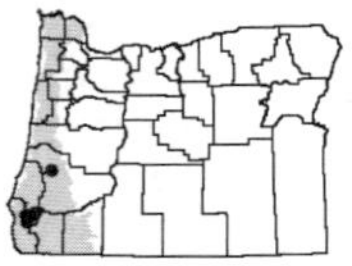

Plants evergreen. **Leaves** twice ternately or pinnately compound (3–5 segments at each level), leaflets 9–19, tough, terminal leaflet 22–41 × 27–42 mm, entire or crenate and with 3 shallow rounded lobes (lateral leaflets sometimes unlobed), margin narrowly cartilaginous-thickened; petioles 5–11(21) cm, developing red-brown pigmentation, at least in places. **Inflorescences** panicles, seldom racemes, with gland-tipped hairs. **Flowers** white, sepals 3–5 mm, glabrous abaxially; petals 2–3 mm, plane; filaments glabrous; ovaries glabrous. **Fruits** ellipsoidal, 3–4 × 1–2 mm, glabrous. $2n$=24.

Forests, open woodlands. Flowering May–Jul. 50–1200 m. CR, Sisk. CA. Native.

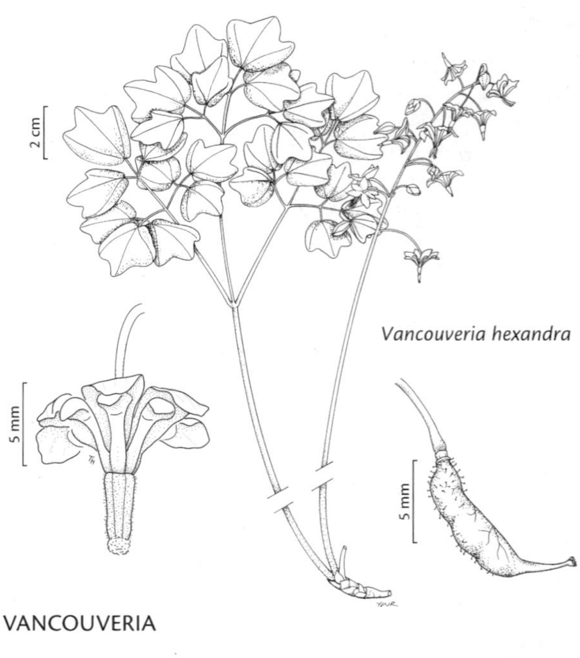

VANCOUVERIA

BETULACEAE Birch family

Alan T. Whittemore

Shrubs or **trees** deciduous. **Leaves** alternate, simple, pinnately veined, margins toothed; stipules present, often deciduous. **Inflorescences** unisexual, emerging before leaves, male and female on same plant; staminate flowers in pendent catkins, 1–3 near tips of previous years' twigs, immature catkins formed the fall before flowering in most taxa; pistillate flowers in catkins or solitary in leaf axils. **Flowers** radially symmetric. **Staminate flowers** calyces absent or of about 4 tiny, scale-like sepals; corollas absent; stamens mostly 4–6, free; pistillodes present or absent. **Pistillate flowers** calyces rudimentary or absent; corollas and staminodes absent; ovaries inferior, with 2 locules that are usually united above, placentation axile, styles 2 (sometimes united basally). **Fruits** nuts or tiny 2-winged samaras, surrounded by bracts, either grouped into cone-like infructescences (*Alnus* and *Betula*) or 1–3 together (*Corylus*).

Northern Hemisphere, South America. 6 genera; 3 genera treated in *Flora*.

Betulaceae form an important part of the woody vegetation in many parts of Oregon. Many species are tolerant of flooding, waterlogging, and cold, and they are a common component of the woody cover on streambanks and wet bottomlands, especially in the mountains. Some are very tolerant of disturbance, and they are often found on recently logged sites. The seeds of all our species are eaten by birds and other wildlife.

Key to Betulaceae genera:

1. Fruits spherical or depressed-spherical nuts 10 mm, in clusters of 2–3, each enclosed in an involucre; leaves elliptic to almost circular. . . . . **CORYLUS**

1' Fruits samaras 2–3 mm, almost hidden between overlapping bracts to form a smooth, cone-like infructescence; leaves ovate, elliptic, or triangular.

  2. Bracts subtending fruit papery and deciduous, so infructescences disintegrate at maturity; bark of some species peeling in thin, papery sheets. . . . . **BETULA**

  2' Bracts subtending fruit woody and persistent, so mature infructescences resemble small woody cones; bark persistent, never peeling. . . . . **ALNUS**

ALNUS Alder

Shrubs or **trees** to 25 m; bark pale gray, persistent, smooth or tardily furrowed; buds resinous. **Leaves** blades ovate or elliptic, 57–130 × 39–93 mm, surfaces abaxially glabrous or pubescent; petioles glabrous or pubescent, sometimes with spreading hairs. **Pistillate catkins** dense, on twigs with 2–5 catkins in open terminal racemes or panicles (usually on a short side branch). **Fruits** lenticular 2-winged samaras, 2–4 mm wide, aggregated in dense catkins resembling small woody cones, each subtended by 1 bract, bracts free, undivided, thick and woody, persistent on axis.

Asia, Europe, North America, South America. ~25 species; 4 species treated in *Flora*.

Species of *Alnus* fix nitrogen in root nodules that contain the filamentous bacterium *Frankia*. This allows these trees to invade habitats where the soil is deficient in nitrogen compounds. Alder bark produces a red dye that has been used in Oregon and many other parts of the world. *Alnus rhombifolia* and *A. rubra* are often planted as fast-growing street trees.

In addition to our native species, the European black alder, *Alnus glutinosa*, has been reported to have escaped in Oregon, but is not yet established. It is a small tree; its leaves are very broadly obovate or almost circular, truncate or broadly emarginate, sharply toothed, with very prominent tufts of hair in the axils of the secondary veins. Black alder has been known to escape in the eastern United States and eastern Canada, but this is the first report of this species from western North America.

Key to *Alnus* species:

1. Shrubs; fine teeth of leaf margins sharp.

  2. Winter buds stalked, covered by 2–3 valvate scales; buds of catkins of both sexes develop from buds in summer, the year before flowering and visible all winter; leaves neither glossy nor sticky . . . . ***A. incana*** ssp. ***tenuifolia***

  2' Winter buds sessile, covered by 4–6 overlapping scales; male catkins develop from buds in summer, the year before flowering and visible all winter, female catkins develop in spring, shortly before flowering; young leaves abaxially glossy and sticky . . . . ***A. alnobetula*** ssp. ***sinuata***

1' Trees; fine teeth of leaf margins blunt.

  3. Coarse teeth scarcely larger than fine teeth; leaf margins plane; fruit wings grading into body of fruit . . . ***A. rhombifolia***

  3' Coarse teeth much larger than fine teeth; immediate leaf margins folded under; fruit wings sharply set off from body of fruit . . . . ***A. rubra***

Alnus alnobetula (Ehrhart) K. Koch ssp. ***sinuata***

Sitka alder

Alnus sinuata (Regel) Rydb., *Alnus viridis* (Chaix) DC. ssp. *fruticosa* (Rupr.) Nyman, *Alnus viridis* (Chaix) DC. ssp. *sinuata* (Regel) A. Löve & D. Löve

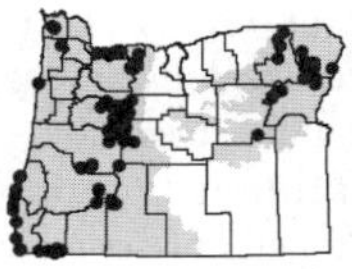

Shrubs or **small trees** to 5(10) m; twigs glabrous or very finely, inconspicuously puberulent (rarely sessile-glandular). **Winter buds** narrowly ovoid or ellipsoid, acute, resinous, scales several, overlapping, < buds; sessile. **Leaves** blades ovate, 57–85 × 43–69 mm, thin and flexible, bases rounded, shallowly cordate, or broadly obtuse, margins plane, teeth coarsely triangular or rarely obsolete, 2–3 mm, finely and sharply serrulate, teeth fine, narrowly and sharply acuminate, 1 mm, secondary veins 6–9 on each side, tips obtuse or broadly acute, surfaces abaxially resinous and glossy, glabrous or with spreading hairs along sides of midrib in proximal leaf; petioles 8–18 mm, glabrous or very finely, inconspicuously puberulent (rarely sessile-glandular). **Catkins** male catkins develop in summer, open the following spring as leaves unfurl, 4–10 cm × 6–9 mm; female catkins develop in spring shortly before flowering. **Fruiting "cones"** short-cylindrical, 13–18 × 8–10 mm; peduncles of terminal cones 11–22 mm long, 0.6–0.8 mm thick; fruit wings 0.7–1.4 mm wide, membranous, translucent, sharply set off from body of fruit.

Streambanks, bogs, coastal bluffs, wooded slopes and thickets, rocky ridges, alpine slopes. Flowering Mar–Jun. 0–2300 m. BW, Casc, CR, ECas, Est, Sisk, WV. CA, ID, WA; north to AK, east to WY. Native.

Furlow (1997) separates some Oregon specimens as *Alnus alnobetula* ssp. *fruticosa*, which he says differs from ssp. *sinuata* by having thicker, darker green leaves that are acute to short-acuminate with the marginal teeth uniform in size. He says ssp. *sinuata* has thin, light or yellowish green leaves that are always acuminate and have large and small teeth along the margins. I have been unable to distinguish two taxa among the Oregon specimens.

Alnus incana (Linnaeus) Moench ssp. ***tenuifolia*** (Nuttall) Breitung

Mountain alder

Alnus incana (L.) Moench var. *occidentalis* (Dippel) C.L. Hitchc., *Alnus tenuifolia* Nutt.

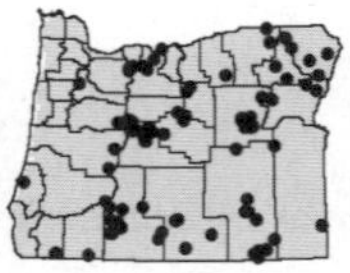

Shrubs or **small trees** to 12 m; twigs pubescent. **Winter buds** cylindrical, rounded, very resinous (scales hard to see before bud opens), scales 2–3, valvate, each scale as long as bud; stalked. **Leaves** blades ovate or elliptic, 70–87 × 57–63 mm, thin and flexible, bases truncate or very broadly obtuse, margins sharply serrate, teeth coarse, triangular, 3–7 mm, secondary veins 9–11 on each side of midrib, tips obtuse, surfaces dull and dry, not glossy, pubescent when expanding, later in the season hairs mostly confined to major veins abaxially; petioles 15–19 mm, pubescent. **Catkins** of both sexes develop in summer, open the following spring before leaves unfurl; male catkins 2–4 from buds at tips of previous year's stems, 6–7 cm × 6–8 mm. **Fruiting "cones"** ellipsoidal, 13–18 × 9–10 mm; peduncles of terminal cones 3–11 mm long, 1.0–1.4 mm thick; fruit wings 0.2–0.5(0.7) mm wide, opaque, grading into body of fruit.

Streambanks, lakeshores, edges of bogs. Flowering Mar–Jun. 200–2000 m. All ecoregions except Est. CA, ID, NV, WA; north to AK, northeast to Saskatchewan, southeast to NM. Native.

Alnus rhombifolia Nuttall

White alder

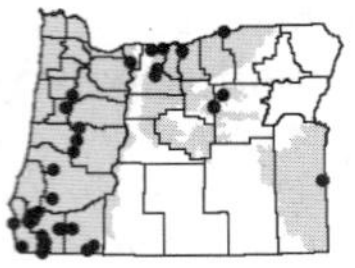

Trees to 20 m; twigs glabrous or pubescent. **Winter buds** cylindrical, rounded, resinous, scales 2, valvate, each scale as long as bud; stalked. **Leaves** blades elliptic or ovate, 57–102 × 39–63 mm, firm, bases obtuse, margins plane, coarse teeth absent or triangular, to 2 mm, finely serrulate except at leaf base, fine teeth bluntly triangular, secondary veins 8–10 on each side, tips rounded or obtuse, surfaces abaxially pubescent, at least on veins, adaxially dry or somewhat sticky, inconspicuously appressed-pubescent; petioles 9–21 mm, puberulent or with spreading hairs. **Catkins** of both sexes develop in summer, open the following spring before leaves unfurl; male catkins, 3–5 in open terminal racemes, 7–9 cm × 4–5 mm. **Fruiting "cones"** cylindrical or ovoid-cylindrical, 11–17 × 7–9 mm; peduncles of terminal cones 5–13 mm long, 1.0–1.8 mm thick; fruit wings 0.2–0.6 mm wide, opaque, grading into body of fruit.

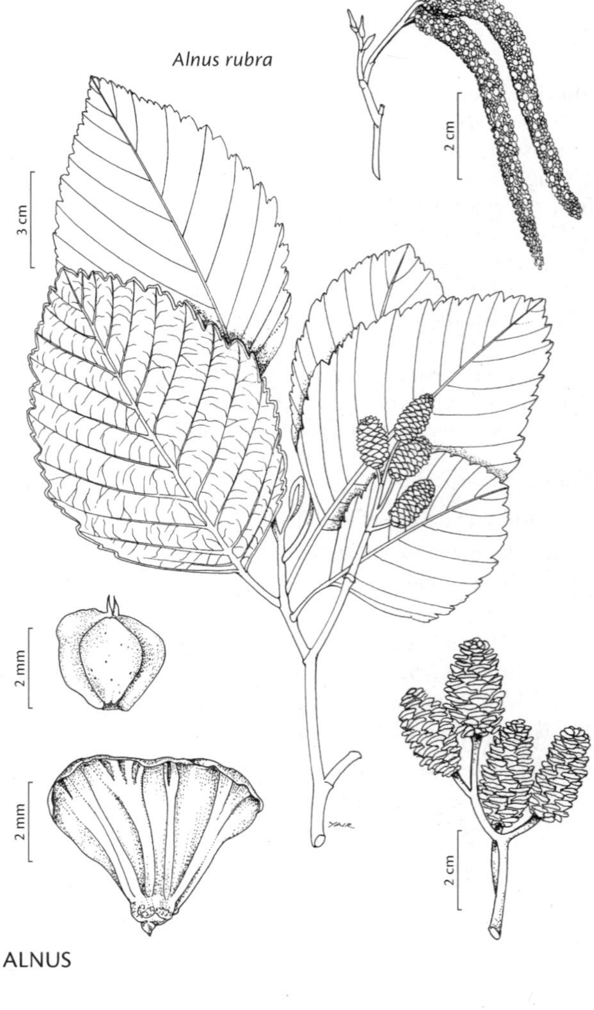

ALNUS

Riverbanks, streambanks. Flowering Feb–May. 50–1200 m. Casc, Col, CR, Est, Lava, Owy, Sisk, WV. CA, ID, WA; northeast to MT. Native.

White alder is often planted as a street tree.

Alnus rubra Bongard (p. 380)

Red alder

Alnus oregona Nutt., *Alnus oregona* Nutt. var. *oregona*, *Alnus oregona* Nutt. var. *pinnatisecta* Starker

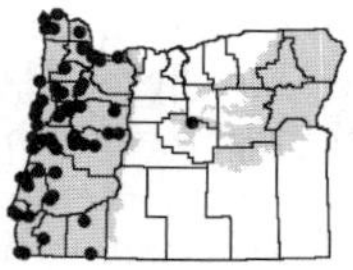

Trees to 25 m; twigs glabrous or pubescent. **Winter buds** cylindrical, rounded, resinous, scales 2–3, valvate, each scale as long as bud; stalked. **Leaves** blades ovate, 75–130 × 51–93 mm, firm, bases obtuse or truncate, margins folded under, coarse teeth rounded or triangular, 3–7 mm (sometimes deeper and forming shallow lobes), finely serrulate with low, blunt teeth, secondary veins 9–13 on each side, tips obtuse or weakly acuminate, surfaces abaxially dry or seldom resinous, pubescent (sometimes late in season only on and around veins, blades becoming glabrate), adaxially glabrous; petioles 12–22(30) mm, glabrous. **Catkins** of both sexes develop in summer, open the following spring before leaves unfurl; male catkins to 12 cm × 7 mm. **Fruiting "cones"** ovoid, 15–24 × 9–14 mm; peduncles of terminal cones 5–13 mm long, 1.0–1.8 mm thick; fruit wings 0.2–0.6 mm wide, opaque, grading into body of fruit.

Riverbanks, streambanks, bogs, moist forests. Flowering Feb–Apr. 0–1600 m. BW, Casc, CR, Est, Sisk, WV. CA, ID, WA; north to AK. Native.

Red alder is the most important commercial hardwood of the Pacific Northwest. The wood is used for furniture and cabinetry. Red alder is often the first species to colonize following logging of moist conifer forests. When red alder colonizes such sites, it often prevents the establishment of Douglas-fir.

Trees with lobed leaves are met with occasionally.

BETULA Birch

Shrubs or **trees** to 25 m; bark white, coppery or reddish to dark brown (seldom dull gray), persistent or peeling in thin, papery sheets; buds resinous or not. **Leaves** blades ovate, obovate, or triangular to almost circular, 15–93 × 12–66 mm, surfaces glabrous or with sparse spreading long hairs; petioles glabrous or pubescent, sometimes with spreading hairs. **Pistillate flowers** in dense catkins, solitary on short axillary spur shoots. **Fruits** lenticular, 2-winged samaras, 1.5–6 mm wide, aggregated in dense catkins, each subtended by 1 bract; bracts free, deeply 3-lobed, thin, papery, falling with fruit.

Asia, Europe, and North America. ~35 species; 5 species treated in *Flora.*

Birches are notoriously difficult, taxonomically. The species are very variable, making it difficult to characterize them and difficult to recognize hybrids. Polyploidy is widespread, the species are often able to hybridize, and the biology of the species is complex: multiple chromosome numbers have been reported from *B. papyrifera* (Li et al. 1996; Dugle 1966), and Williams and Arnold (2001) provide evidence for gene flow between the diploid *B. occidentalis* and the polyploid *B. papyrifera*. Full understanding of Oregon birches will require research to correlate ploidy level with morphology and perhaps other markers. Until such studies have been carried out, the classification must remain tentative.

Key to *Betula* species:

1. Twigs pendent; leaves triangular to triangular-ovate, strongly acuminate; marginal teeth very unequal in size, with teeth 2–3 mm terminating the secondary veins and much smaller serrations between them . . . . ***B. pendula***

1' Twigs ascending; leaves ovate, acute or short-acuminate; marginal teeth almost uniform in size (irregular in size in *B. papyrifera*), 1–2(3) mm.

  2. Leaf blades firm, 15–22 mm wide (to 31 mm on vigorous leaders), margins crenate or with obtuse teeth.

    3. Twigs very finely puberulent with hairs to 0.1 mm; low shrubs to 3 m; crenations of leaf margins ~1 mm deep . . . . ***B. glandulosa***

    3' Twigs finely puberulent but also with spreading hairs to 0.3 mm; shrubs to 6 m; crenations of leaf margins 1–2 mm deep . . . . ***B. pumila***

  2' Leaf blades thin, 20–66 mm wide, margins with sharp teeth.

    4. Shrubs or small trees to 10(15) m; bark coppery or reddish to dull gray, not exfoliating; leaves 26–52 mm, obtuse or weakly acuminate, marginal teeth almost uniform in size. . . . . ***B. occidentalis***

    4' Trees to 20 m; bark white, peeling off in papery sheets; leaves 47–93 mm, short-acuminate, marginal teeth irregular in size . . . . ***B. papyrifera***

Betula glandulosa Michaux (p. 382)

Resin birch

Betula glandulosa Michx. var. *glandulosa*

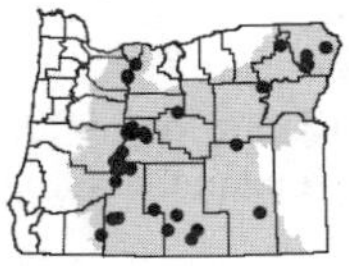

Shrubs to 3 m; bark dark brown with pale lenticels, dull, not exfoliating. **Twigs** ascending, glabrous or puberulent with spreading hairs to 0.1 mm, dry or rarely glutinous near apex, with few to many resin-blisters. **Leaves** blades elliptic to almost circular, 15–28 × 12–21 mm (on fertile branches; up to 37 × 31 mm on sterile branches), green above, light green beneath, firm but flexible, bases obtuse or rounded (seldom broadly acute), margins crenate, crenations obtuse or rounded, 8–15 on a side, usually all approximately same size, ~1 mm, secondary veins 3–4(5) on each side, tips broadly rounded, surfaces abaxially glabrous or both sides vernicose; petioles 3–7 mm, glabrous. **Pistillate catkins** 15–24 mm; bracts with a long, narrow central lobe and short, broad lateral lobes. **Fruits** narrowly winged.

Wet meadows, bottomlands, swamps, streambanks. Flowering May–Jul. 300–2200 m. BR, BW, Casc, ECas, Lava. CA, ID, WA; north to AK, east to Greenland. Native.

Betula glandulosa and *B. pumila* are morphologically similar and difficult to distinguish. Elsewhere in North America, *B. glandulosa* is diploid and *B. pumila* is tetraploid, but no chromosome counts have been made from the Pacific Slope populations. Dugle (1966) has reported *B. glandulosa* × *B. pumila* (*B.* × *sargentii*) and *B. glandulosa* × *B. occidentalis* (*B.* × *eastwoodiae*) from other parts of western North America, but neither hybrid has been confirmed from Oregon.

Betula occidentalis Hooker (p. 382)

Water birch

Betula fontinalis Sarg., *Betula occidentalis* Hook. var. *inopina* (Jeps.) C.L. Hitchc., *Betula occidentalis* Hook. var. *occidentalis*

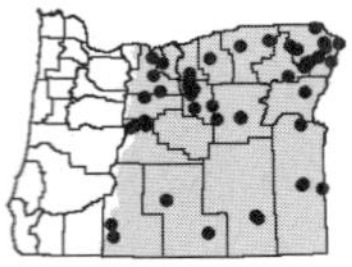

Shrubs or **small trees** to 10(15) m; bark coppery or reddish to dull gray with pale lenticels, shiny, not exfoliating. **Twigs** ascending, glabrous or pubescent, when young dry or glutinous, with prominent resin-blisters. **Leaves** blades ovate to almost circular, 26–52 × 20–52 mm, green above, very pale green beneath, thin, bases broadly rounded to truncate, margins sharply toothed throughout, teeth mostly acuminate, almost uniform in size, 1–2(3) mm, secondary veins 4–6 on each side, tips obtuse or weakly acuminate, surfaces glabrous or with sparse spreading long hairs; petioles 5–23 mm, glabrous or with a few spreading hairs when young. **Pistillate catkins** 20–35 mm; bracts with lateral lobes shorter and broader than middle lobe. **Fruits** broadly winged, wing often as broad as seed.

Riverbanks, streambanks. Flowering Apr–Jul. 300–2200 m. BR, BW, Col, ECas, Lava, Owy. CA, ID, NV, WA; north to AK, east to Ontario. Native.

Betula occidentalis hybridizes with *B. papyrifera*, and the hybrid, *Betula* × *utahensis*, is known from logged-off land in Union County. Williams and Arnold (2001) provide evidence of gene flow between the two species, although water birch is diploid and paper birch is polyploid.

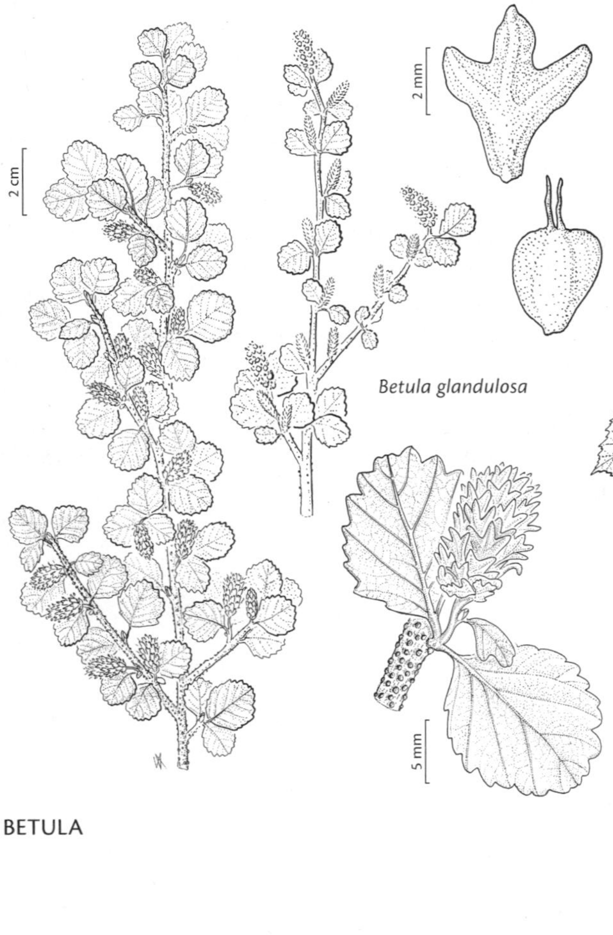

Betula glandulosa

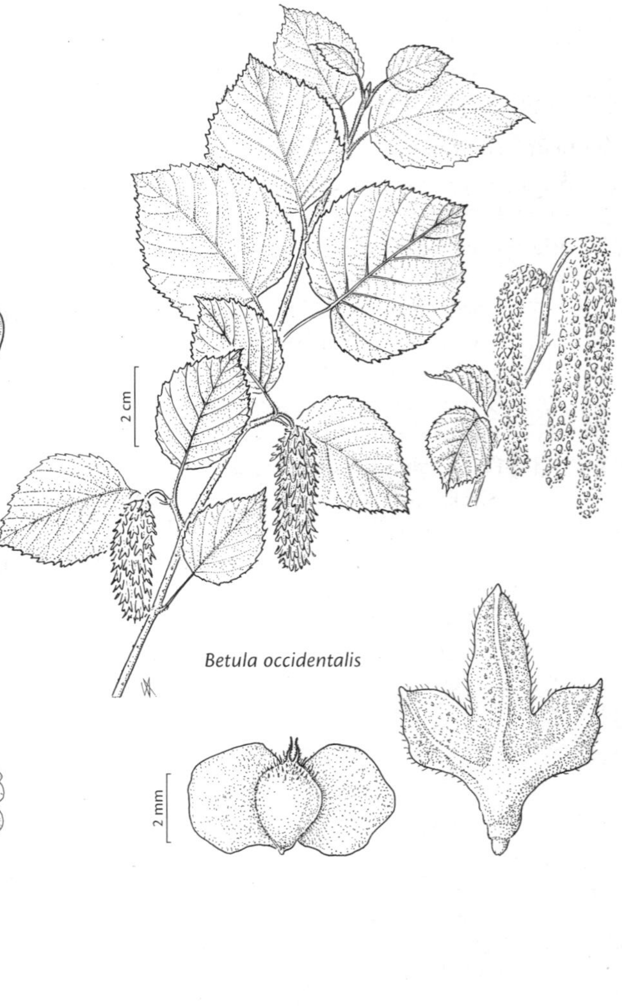

Betula occidentalis

BETULA

Betula papyrifera Marshall (p. 383)

Paper birch

Betula papyrifera Marshall var. *commutata* (Regel) Fernald, *Betula papyrifera* Marshall var. *subcordata* (Rydb.) Sarg.

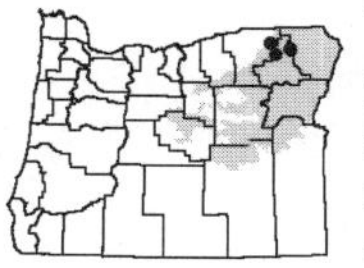

Trees to 20 m; bark white with dark lenticels, peeling off in papery sheets. **Twigs** ascending, glabrous or with scattered coarse hairs, resin-blisters few or none. **Leaves** blades usually narrowly ovate, sometimes ovate, 47–93 × 34–66 mm, green above, very pale green beneath, thin, bases broadly obtuse or rounded, seldom truncate, margins sharply toothed throughout, teeth mostly acuminate, irregular in size, largest teeth 1–2 mm, secondary veins 5–8 on each side, tips short-acuminate, surfaces glabrous except for spreading long hairs on midrib (at least in axils of secondary veins abaxially); petioles 17–30 mm, glabrous or with a few coarse hairs. **Pistillate catkins** 25–28 mm; bracts with a long, narrow central lobe and short, broad lateral lobes. **Fruits** broadly winged, wing at least as broad as seed.

Riverbanks. Flowering Mar–May. 300–1100 m. BW. ID, WA, north to Yukon, east to Newfoundland. Native.

In Oregon, paper birch is known only from a few sites along the Minam River in the Wallowa Mountains and one site on the Umatilla River in the northern Blue Mountains. Several ploidy levels have been reported from *B. papyrifera* (Li et al. 1996; Dugle 1966), but no chromosome counts are available from Oregon populations.

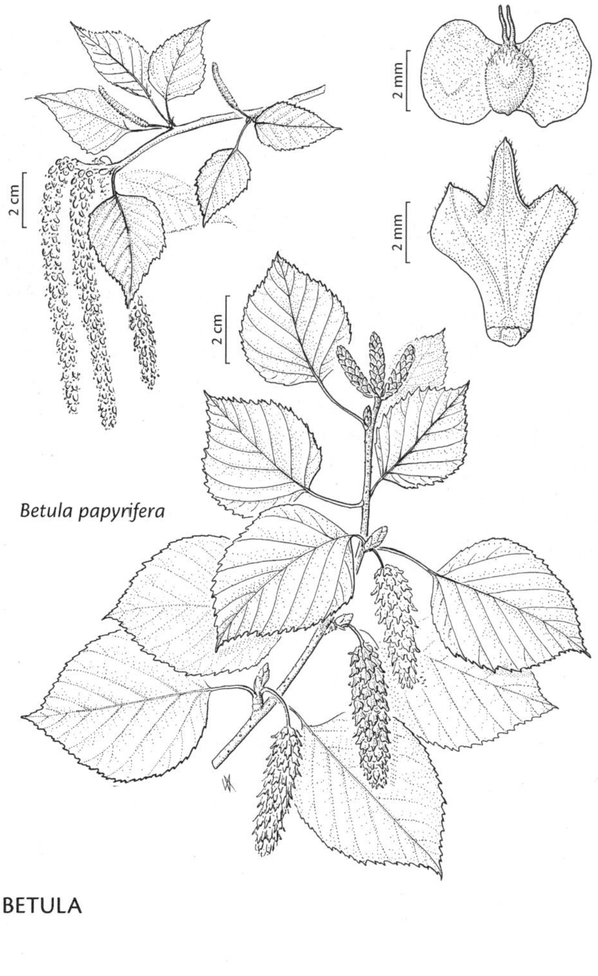

Betula papyrifera

BETULA

Betula pendula Roth

European white birch

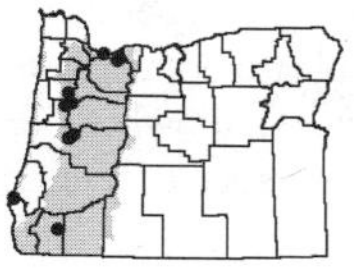

Trees to 25 m; bark white with dark lenticels, peeling off in papery sheets. **Twigs** pendent, glabrous or very finely puberulent, sometimes glutinous when young but without evident resin-blisters. **Leaves** blades triangular to triangular-ovate, thin, green above, pale green beneath, 41–57 × 31–50 mm, bases obtuse or truncate, margins irregularly serrate, teeth (12)15–27 on each side, acute, each side with 5–7 large teeth, 2–3 mm, terminating main veins, other teeth very much smaller, secondary veins 5–7 on each side, tips strongly acuminate, surfaces glabrous; petioles 15–25 mm, glabrous or very finely puberulent. **Pistillate catkins** 25–35 mm; bracts with a broad, triangular or ovate central lobe and very broad lateral lobes. **Fruits** broadly winged, wing broader than seed.

River bottoms, forests, disturbed areas. Flowering Apr–May. 0–900 m. Casc, Est, Sisk, WV. WA; north to British Columbia, east to MA; Asia, Europe. Exotic.

Weeping birch is a popular shade tree, widely planted in western Oregon, and occasionally escaping. It differs from all our native species in its pendent twigs and its triangular to triangular-ovate, strongly acuminate leaves. Plants with deeply laciniate leaves are sometimes cultivated, but this morphology is not known from escaped plants.

Betula pumila Linnaeus

Bog birch

Betula glandulosa Michx. var. *hallii* (Howell) C.L. Hitchc., *Betula pumila* L. var. *glandulifera* Regel

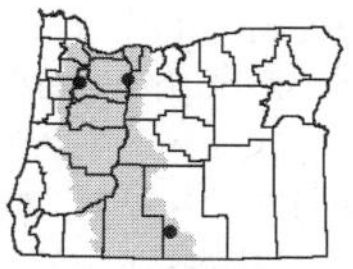

Shrubs to 6 m; bark dark brown with pale lenticels, dull, not exfoliating. **Twigs** ascending, finely puberulent with hairs to 0.1 mm and spreading hairs ~0.3–0.4 mm, when young dry or glutinous, with scattered resin-blisters. **Leaves** blades obovate, elliptic, or almost circular, 22–40 × 15–31 mm, green above, light green beneath, firm but flexible, bases obtuse or rounded, margins crenate from base to tip with 14–19 strong rounded crenations on each margin, usually all approximately same size, 1–2 mm, secondary veins obscure, 4–6 on each side, tips broadly obtuse or rounded, surfaces glabrous; petioles 4–7 mm, glabrous or puberulent. **Pistillate catkins** 15–23 mm; bracts with lateral lobes often shorter and broader than middle lobe. **Fruits** very narrowly winged.

Boggy margins of lakes. Flowering May–Jun. 0–1300 m. Casc, ECas, WV. CA, ID, WA; north to Yukon, east to Newfoundland. Native.

Many specimens of *B. pumila* from eastern North America have longer, denser pubescence on the twigs and few or no glandular blisters. Our western plants, with sparser pubescence and more numerous glands, are sometimes separated as a distinct taxon (*B. hallii* or *B. pumila* var. *glandulifera*; Dugle 1966; Hitchcock & Cronquist 1964), but the two forms intergrade. Furlow (1997) argues against recognizing segregate taxa.

CORYLUS Hazel

Shrubs or **small trees** 1–5 m; bark pale gray, persistent, becoming furrowed; buds not resinous. **Leaves** blades elliptic to almost circular, 42–105 × 35–99 mm, surfaces soft-pubescent; petioles pubescent with spreading simple, glandular and non-glandular hairs. **Pistillate flowers** not in catkins, usually 2–3 clustered near apices of twigs. **Fruits** globose nuts, 10–15 mm wide, each surrounded by 2 bracts; bracts ± fused to form an involucre, shallowly to very deeply lacerate, thin.

Asia, Europe, North America. ~15 species; 2 species treated in *Flora*.

Key to *Corylus* species:

1. Involucre of 2 bracts, fused only at base, nut exposed; glandular hairs of twigs and petiole much > non-glandular hairs . . . . **C. avellana**

1' Involucre tubular, bracts fused for their whole length and completely enclosing nut; glandular hairs of twigs and petiole much < non-glandular hairs. . . . . **C. cornuta** var. ***californica***

Corylus avellana Linnaeus

European hazel

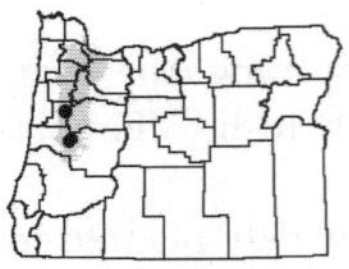

Plants 3–5 m. **Leaves** blades 70–105 × 60–99 mm, dark green above, green beneath, bases narrowly cordate, margins with 4–6 coarse triangular or acuminate teeth mostly in distal half and 2 orders of fine sharp serration around whole margin, tips abruptly short-acuminate, abaxially soft-pubescent, felty to the touch, surfaces adaxially sparsely pubescent, smooth to the touch; petioles 8–19 mm, pubescent with spreading simple, glandular and non-glandular hairs, glandular hairs much > non-glandular hairs. **Staminate catkins** sometimes solitary, usually 2–4 on a short leafless twig from bud of previous year's growth, 25–55 × 4–5 mm. **Fruits** usually 2–3 clustered at tips of twigs, ovoid or subspherical, 10–15 × 10–15 mm, brown; bracts separate, deeply laciniate, 12–17 mm, loosely surrounding nut, ≤ nut and not concealing it, finely canescent.

Riparian woodlands, pond shores. Flowering Jan–Mar. 0–200 m. WV. WA; north to British Columbia; southwestern Asia, Europe. Exotic.

Corylus avellana is an important commercial crop in Oregon. It is grown for its nuts, known as hazelnuts or filberts. Over 99% of the hazelnuts produced in the United States are grown in the Willamette Valley. Most of the world supply, however, comes from Turkey. Two other Eurasian species are important in the hazelnut industry. The Turkish hazel, *C. colurna*, has a longer petiole, and the involucre, which is covered with bristly glandular hairs, is lacerate almost to its base into lanceolate lacinae that have many sharp, lanceolate lateral teeth. It is a tree up to 10 m tall, and it does not sucker from the roots, so it has a single trunk (sometimes 2 or 3 trunks arising from the same point). Because it doesn't sucker, it is widely used as a rootstock for grafting commercial hazels.

Corylus maxima, a suckering shrub, has a long tubular involucre, somewhat like the native *C. cornuta* but scarcely contracted above the nut. Some commercial hazelnut cultivars are hybrids between *C. avellana* and *C. maxima*. All three of these species (*C. avellana*, *C. colurna*, and *C. maxima*) are susceptible to the fungus *Anisogramma anomala*, known as eastern filbert blight. This fungus, native to the eastern United States, was introduced to Oregon in the late twentieth century.

Corylus cornuta Marshall var. ***californica*** (A. de Candolle) W.M. Sharp (p. 385)

California hazel

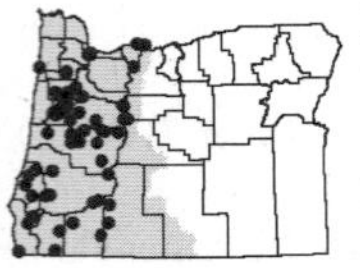

Plants 1–4 m. **Leaves** blades 42–95 × 35–90 mm, light yellow-green above, paler beneath, bases narrowly truncate to cordate, margins with 4–6 coarse triangular teeth in distal half and 2 orders of fine sharp serration around whole margin, tips obtuse or abruptly short-acuminate, surfaces very soft-pubescent, felty to the touch; petioles 4–14 mm, soft-pubescent with long spreading, glandular and non-glandular hairs, glandular hairs much < non-glandular hairs or rarely absent. **Staminate catkins** solitary from buds on previous year's growth, 25–45 × 5–6 mm. **Fruits** usually 2–3 clustered at tips of twigs, each tightly enclosed in an involucre; bracts fused to form an envelope with a spherical base, ~10–15 × 10–15 mm, abruptly contracted to a tubular beak 12–26 × 5–10 mm.

Slopes, streambanks, lakeshores. Flowering Jan–May. 0–1800 m. Casc, CR, ECas, Est, Sisk, WV. CA, WA; north to British Columbia. Native.

Hazelnuts are collected by native jays and rodents (mostly squirrels and chipmunks). The California hazel is resistant to the eastern filbert blight. *Corylus cornuta* var. *cornuta* is found from eastern Washington and British Columbia east to the Atlantic coast. It differs in having a very long tubular beak (more than twice the length of the fruit) and the leaves always having acuminate tips and lacking glandular hairs on the petioles. Specimens of *C. cornuta* from Oregon occasionally lack glandular hairs on the petioles, but they never have long-beaked involucres and cannot be referred to as variety *cornuta*.

BIGNONIACEAE Catalpa or trumpetvine family

Raul Gutierrez

Trees, shrubs, or **vines** deciduous. **Leaves** simple, opposite or whorled; petioles present; stipules 0. **Inflorescences** terminal panicles. **Flowers** perfect; sepals 5, fused; corollas 5-lobed; stamens 2, adnate to corolla; ovaries superior, unilocular, bicarpellate, placentation parietal, styles 1, terminal, stigmas 1. **Fruits** capsules, dehiscent, bivalved. **Seeds** many; winged.

Mostly pantropical. ~109 genera; 1 genus treated in *Flora*.

Bignoniaceae is composed mostly of trees or lianas that are popular as ornamentals, especially in the tropics. A few representatives of the family make it into temperate areas. Bignoniaceae is the only family in Oregon with trees that have opposite, simple leaves and large trumpet-shaped corollas.

CATALPA Catalpa, catawba

Trees to 18 m. **Leaves** ovate to obcordate, bases truncate or cordate, margins entire, tips acuminate. **Flowers** conspicuous; calyces bilabiate, sometimes splitting irregularly; corollas bilabiate; stamens included; styles slightly > stamens. **Fruits** cylindric, very long and narrow. **Seeds** oblong; wing tissue fraying along edge.

East Asia, eastern North America, West Indies. 11 species; 1 species treated in *Flora*.

Species of *Catalpa* are popular as ornamentals and have some limited use as timber. *Catalpa* shares an East Asian-eastern North American disjunction with other genera such as *Campsis* (Bignoniaceae), *Carya* (Juglandaceae), and *Sassafras* (Lauraceae).

Catalpa bignonioides Walter (p. 385)

Southern catalpa, common catalpa

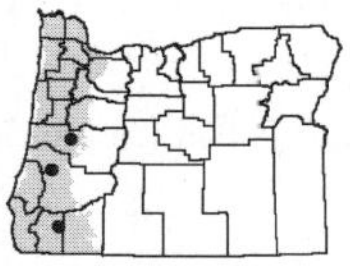

Plants to 15 m. **Bark** flaking into thin scales, light brown. **Leaves** often whorled; blades 1–2 dm; petioles ≤ 1.5 dm. **Inflorescences** to 20 cm. **Flowers** to 4 cm; corollas tubular or funnelform, white, internal surface of tube with purple or magenta specks and yellow nectar glands. **Fruits** to 4 dm, valves flattening after dehiscence. **Seeds** to 4.5 mm wide.

Roadsides, wet forests, streamsides. Flowering Jun–Jul. 50–500 m. CR, Sisk, WV. CA; eastern North America, scattered in western North America. Exotic.

The specific epithet refers to the plant's likeness to *Bignonia*, the type genus of Bignoniaceae. *Catalpa speciosa* is also commonly cultivated and can be distinguished from *C. bignonioides* by its smaller flowers.

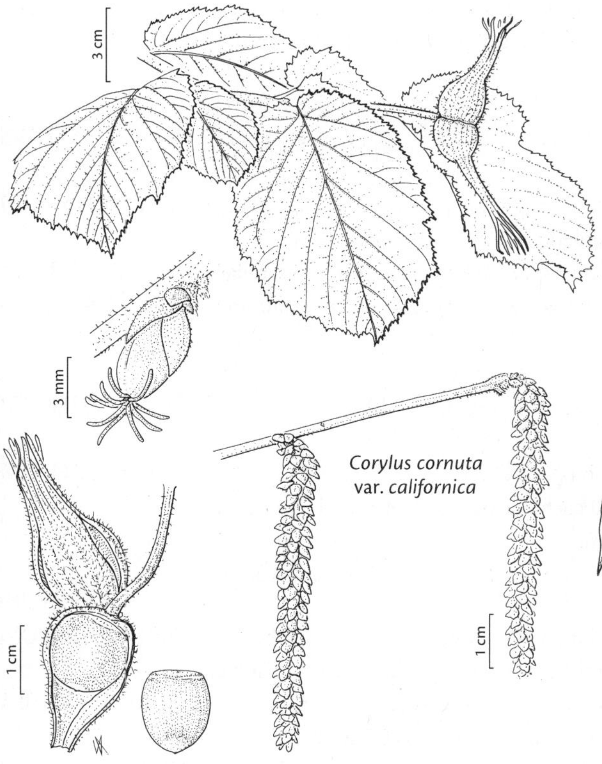

CORYLUS • CATALPA

BORAGINACEAE Forget-me-not or borage family

(see also Ehretiaceae, and Heliotropiaceae and Namaceae, Volume 3)

Herbs, **shrubs**, or **small trees** annual, biennial, or perennial, often with bristly hairs; hairs often with calcified, silicified, or pustulose bases; roots staining red when dried or not. **Stems** prostrate to erect. **Leaves** cauline and often basal, usually alternate, usually simple; margins usually entire; sessile or petiolate; stipules absent. **Inflorescences** generally scorpioid cymes, coiled in flower, usually elongating in fruit, or flowers 1–2 and axillary; bracts present or not. **Flowers** bisexual, radially or occasionally bilaterally symmetric; sepals 5, usually connate at base; corollas 5-lobed; fornices present or not; stamens 5, adnate to petals; ovaries superior, styles 1, not persistent. **Fruits** nutlets, 1–4, smooth to rough, prickly, barbed, bristly, or not.

Tropical and temperate regions worldwide. ~120 genera; 27 genera treated in *Flora*.

Some authorities combine Boraginaceae with Hydrophyllaceae. Here, however, the families are treated separately. Furthermore, although their members have morphological similarities to those within Boraginaceae, families Ehretiaceae, Heliotropiaceae, and Namaceae are treated here as distinct.

Key to Boraginaceae genera:

1\. Anthers dark brown; fruits erect. . . . . **BORAGO**

1' Anthers not dark brown; fruits erect or spreading.

2\. Nutlets spreading.

3\. Corollas blue to red-purple; plants biennial or perennial.

4\. Cauline leaves petiolate . . . . **ADELINIA**

4' Cauline leaves sessile.

5\. Cauline leaves lanceolate to elliptic . . . . **CYNOGLOSSUM**

5' Cauline leaves oblanceolate to spatulate . . . . **ANDERSONGLOSSUM**

3' Corollas white; plants annual.

6\. Nutlets smooth or rough, without bristles . . . . **SONNEA** (in part)

6' Nutlets with bristles.

7\. Cauline leaves usually alternate, opposite proximally; calyces usually bilaterally symmetric, lobes < nutlets . . . . **PECTOCARYA**

7' Cauline leaves opposite; calyces radially symmetric, lobes > nutlets . . . . **GRUVELIA**

2' Nutlets erect.

8\. Inflorescences axillary with 1–2 flowers; calyx lobes with 1–2 teeth basally . . . . **ASPERUGO**

8' Inflorescences axillary and/or terminal cymes or cymules (but solitary axillary flowers often present below in *Mertensia*); calyx lobes without teeth basally.

9\. Nutlets with thick basal rim.

10\. Flowers rotate . . . . **PENTAGLOTTIS**

10' Flowers funnelform to campanulate.

11\. Styles exserted; corolla throats expanded . . . . **SYMPHYTUM**

11' Styles included; corolla throats not expanded.

12\. Flowers radially symmetric; nutlets rugose and tuberculate; plants perennial . . . . **ANCHUSA**

12' Flowers bilaterally symmetric; nutlets rugose but not tuberculate; plants annual . . . . . **LYCOPSIS**

9' Nutlets without basal rim.

13\. Attachment scars basal.

14\. Flowers bilaterally symmetric . . . . **ECHIUM**

14' Flowers radially symmetric.

15\. Plants dwarfed, forming small cushions . . . . **ERITRICHIUM**

15' Plants not dwarfed or forming small cushions (some plants may be prostrate).

16\. Nutlets smooth or slightly pitted.

17\. Corollas blue, sometimes white, pink, purple, or yellow, dark yellow beneath white spots at bases of lobes . . . . **MYOSOTIS**

17' Corollas green-yellow to yellow, not dark yellow at bases of lobes . . **LITHOSPERMUM**

16' Nutlets usually rough and tubercled, or rugose, verrucose, and/or papillate.

18\. Corollas usually blue, rarely white, without a yellow center . . . . **MERTENSIA**

18' Corollas white or white with a yellow center, rarely bluish.

19\. Corollas usually rotate with a yellow center . . . . **PLAGIOBOTHRYS** (in part)

19' Corollas funnelform to salverform, center not yellow . . . . **BUGLOSSOIDES**

13' Attachment scars medial, submedial, or lateral.
- 20. Nutlets without a ventral groove (except in *Amsinckia carinata*).
  - 21. Corollas bright yellow to orange . . . . **AMSINCKIA**
  - 21' Corollas white and/or blue, with or without yellow center.
    - 22. Inflorescences glomerate, not uncoiling in fruit distally . . . . **SONNEA** (in part)
    - 22' Inflorescences not glomerate, uncoiling normally in fruit distally.
      - 23. Nutlets with barbed prickles.
        - 24. Pedicels recurved or deflexed in fruit; plants perennial . . . . **HACKELIA**
        - 24' Pedicels erect or ascending in fruit; plants annual . . . . **LAPPULA**
      - 23' Nutlets without barbed prickles.
        - 25. Nutlet scars elongate along ventral keels . . . . **AMSINCKIOPSIS**
        - 25' Nutlet scars not elongate along ventral keels . . . . **PLAGIOBOTHRYS** (in part)
- 20' Nutlets with a ventral groove.
  - 26. Plants annual, often cushion-like; roots red or purple when dry; flower bracts present.
    - 27. Calyces with sepals fused basally; gynobases < mature nutlets . . . . **GREENEOCHARIS**
    - 27' Calyces with sepals distinct; gynobases > mature nutlets . . . . **EREMOCARYA**
  - 26' Plants annual, biennial, or perennial, not cushion-like; roots not red or purple when dry; flower bracts usually absent.
    - 28. Plants biennial or perennial; nutlet tips broadly rounded to obtuse, ventral groove tips not extending to nutlet tips . . . . **OREOCARYA**
    - 28' Plants annual or perennial, rarely biennial tips extending nearly to nutlet tips . . . . **CRYPTANTHA**

ADELINIA Hound's tongue

James I. Cohen

Genus composed of one species, *A. grandis*.

Adelinia grandis (Douglas ex Lehmann) J.I. Cohen (p. 387)

Pacific hound's tongue, grand hound's tongue

Cynoglossum grande Douglas ex Lehm.

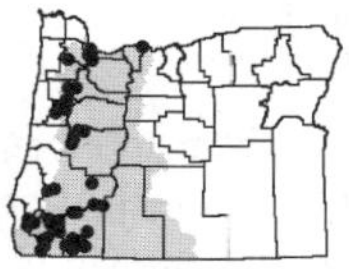

Herbs perennial, erect, to 1 m; taprooted. **Stems** glabrous. **Leaves** basal and cauline, alternate, simple; blades deltate or rounded at base (sometimes uppermost one sessile and narrow-elliptic), 5–21 × 2–10 cm, margins entire, tips acute, surfaces glabrous to strigose or sericeous; basal leaves long-petiolate, cauline leaves petiolate. **Inflorescences** terminal scorpioid cymes, 15–25-flowered; bracts 0; pedicels not recurved even in fruit. **Flowers** radially symmetric; calyces 5-lobed to base with linear to lanceolate lobes; corollas salverform, 8–12 mm, blue to purple, glabrous, 5-lobed, lobes elliptic, flared; fornices present, white; stamens 5, included, anthers ellipsoid; ovaries superior, 4-parted, styles included, stigmas 2, terminal. **Fruits** 4 or fewer by abortion, spreading, ovoid, brown to yellow, prickles barbed at tips; attachment scars broad-triangular. **2*n***=24.

Forests, meadow edges. Flowering Mar–Jun. 0–1700 m. Casc, ECas, Sisk, WV. CA, WA. Native.

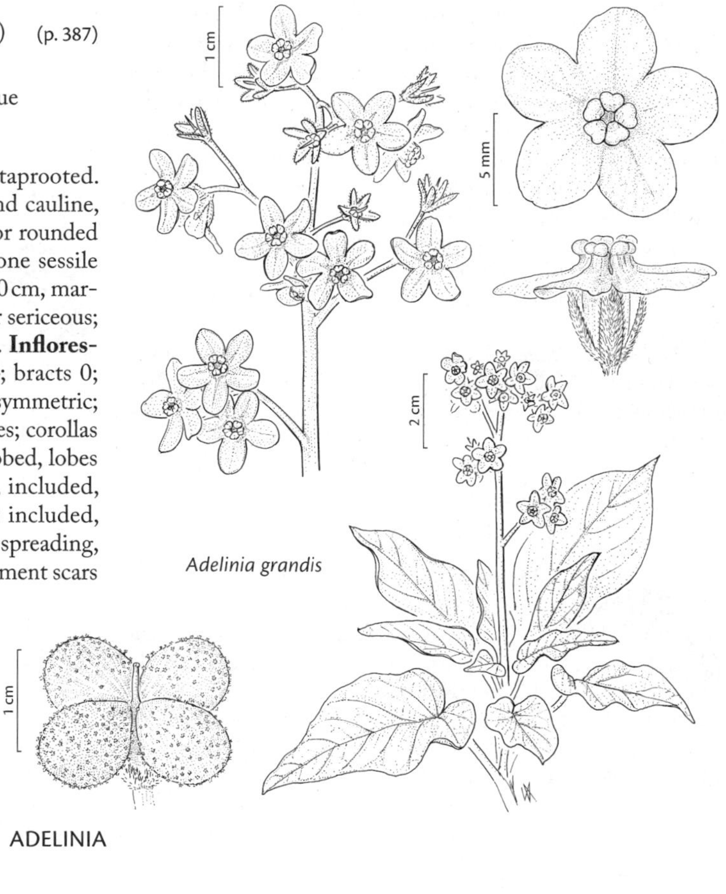

ADELINIA

AMSINCKIA Fiddleneck

Christian Feuillet

Herbs annual. **Stems** usually covered with stiff, erect hairs, often with raised bases. **Leaves** basal and cauline, surfaces usually covered with stiff, erect hairs, often with raised bases; sessile or short-petiolate below. **Inflorescences** terminal, elongate, spike-like cymes, coiled at tips. **Flowers** receptacles conic or elongate; calyx lobes without basal teeth in fruit; corollas bright yellow to orange, tubes cylindric, generally not constricted, appendages in throat generally lacking, limbs spreading. **Fruits** erect, adaxially keeled distal to scar, attachment scars lateral and raised, lacking a thick rim, tubercles present or not, sharp or not.

North America, South America. 15 species; 5 species treated in *Flora*.

Key to *Amsinckia* species:

1. Leaves erose-denticulate, somewhat succulent; fruits 1–2 mm; coastal habitats. . . . . ***A. spectabilis*** var. ***spectabilis***

1' Leaves generally entire, not succulent; fruits 2–7 mm; various habitats.

  2. Calyx lobes 2–4 from partial fusion of some lobes, unequal in width.

    3. Fruits smooth, without tubercles, 4–7 mm, with an adaxial slit-like groove . . . . ***A. carinata***

    3' Fruits covered with tubercles, 2.5–4(5) mm, without an adaxial slit-like groove. . . . . ***A. tessellata*** var. ***tessellata***

  2' Calyx lobes 5, subequal to equal in width.

    4. Corolla throats somewhat closed by 5 hairy appendages, tubes constricted in basal half; stamens attached below middle of corolla tube. . . . . ***A. lycopsoides***

    4' Corolla throats open, glabrous, without appendages, not constricted; stamens attached above middle of corolla tube . . . . ***A. menziesii***

Amsinckia carinata A. Nelson & J.F. Macbride (p. 389)

Malheur Valley fiddleneck

Amsinckia vernicosa Hook. & Arn. var. *vernicosa*

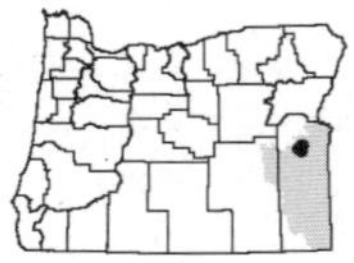

Plants 1–2 dm. **Stems** glaucous below, with sparse bristles above; branched at base or above. **Leaves** linear-oblong to lance-ovate, tips acute, surfaces hispid-ciliate, with sparse hairs from light-colored pustules. **Flowers** calyx lobes 2–4, unequal in width, some fused below middle; corollas ~8 mm, deep yellow to dark orange with deeper orange sinuses, tubes basally 10-veined, limbs 3–6 mm in diameter; stamens attached near apex of corolla tube; styles reaching throat. **Fruits** trigonous, 4–7 mm, smooth, without tubercles, shiny, abaxially somewhat concave and dull carinate, adaxially sharply keeled with a slit-like groove, scars ~1 mm. **2*n***=24.

Rocky soil. Flowering Apr–Jun. 700–1100 m. Owy. Native.

Amsinckia carinata is known only from fewer than ten small populations in Malheur County. The most closely related *Amsinckia* species is *A. vernicosa*, a central and southern California endemic with tubes that are 20-veined.

Amsinckia lycopsoides Lehmann (p. 389)

Tarweed fiddleneck, bugloss fiddleneck

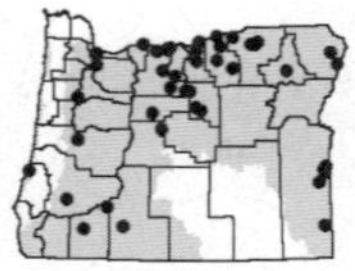

Plants 1–6 dm. **Stems** erect to procumbent, hirsute with short, retrorse hairs; branched at base or above. **Leaves** narrow-lanceolate to ovate, margins entire, surfaces hispid-hirsute. **Flowers** calyx lobes 5, ± equal in width, not fused above base; corollas 7–11 mm, deep yellow, often with 5 darker spots, tubes exserted, basally 10-veined, throats constricted in basal half, somewhat closed by 5 hairy appendages, limbs 4–10 mm in diameter; stamens attached below middle of corolla tube; styles short, included. **Fruits** 2.5–3 mm, usually not ridged, tubercles sharp, adaxial slits sinuous, wider near base, scars elliptic to round. **2*n***=30.

Dry grasslands, woods, disturbed areas. Flowering Apr–Jul. 0–1500 m. BW, Casc, Col, ECas, Est, Lava, Owy, Sisk, WV. CA, ID, WA; north to British Columbia, scattered across US. Native.

Amsinckia menziesii (Lehmann) A. Nelson & J.F. Macbride (p. 389)

Menzies' fiddleneck, rancher's fiddleneck

Amsinckia intermedia Fisch. & C.A. Mey., *Amsinckia menziesii* (Lehm.) A. Nelson & J.F. Macbr. var. *intermedia* (Fisch. & C.A. Mey.) Ganders, *Amsinckia menziesii* (Lehm.) A. Nelson & J.F. Macbr. var. *menziesii*, *Amsinckia micrantha* Suksd., *Amsinckia retrorsa* Suksd.

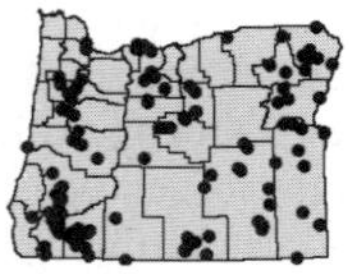

Plants 1–7 dm. **Stems** decumbent to erect, with short, soft, retrorse hairs; branched or not. **Leaves** linear to oblong, margins entire, surfaces sparsely hispid to hispid-hirsute. **Flowers** calyx lobes 5, subequal to equal in width, not fused above base; corollas 4–11 mm, pale yellow to orange-yellow, usually with darker spots, tubes included to exserted, basally 10-veined, throats open, not constricted, glabrous, without appendages, limbs 1–10 mm in diameter; stamens attached above middle of tube, anthers usually appressed to stigma; styles included or reaching throat. **Fruits** 2–3.5 mm, occasionally ridged, tubercles sharp, scars elevated, oval to round. **2*n***=16, 26, 30, 34, 38.

Prairies, disturbed areas, forest edges. Flowering Mar–Jul. 0–2300 m. All ecoregions. CA, ID, NV, WA; western Canada and US, scattered in eastern US; Australia. Native.

Some floras separate two additional species (or varieties) from *A. menziesii*. If accepted, *A. intermedia* would be identified by having larger corollas (7–11 mm long) and ascending to erect habit, whereas *A. menziesii* has smaller corollas (4–7 mm long) and a decumbent to ascending habit. *Amsinckia retrorsa* would be identified by its longer bristles and shorter retrorse hairs, whereas

A. menziesii is lacking retrorse hairs. The three taxa, however, overlap greatly in morphology. Furthermore, the ranges and habitats are nearly identical. Treatment of *A. intermedia* and *A. retrorsa* as varieties may be warranted with further research.

Amsinckia spectabilis Fischer & C.A. Meyer var. *spectabilis* (p. 389)

Seaside fiddleneck, woolly breeches

Amsinckia scouleri I.M. Johnst.

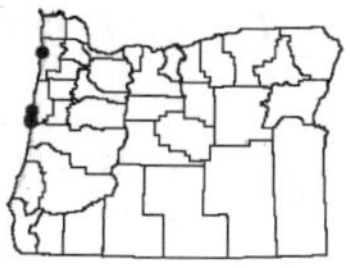

Plants 1–6 dm. **Stems** usually decumbent, sparsely to moderately hispid. **Leaves** linear-lanceolate to broadly lanceolate, somewhat succulent, margins erose-denticulate, surfaces sparsely to moderately hispid, abaxial hairs from pustules. **Flowers** calyx lobes usually 2–3, unequal in width, rarely 5 and equal in width, approximately half-fused, dark-ciliate; corollas 7–15 mm, yellow, tubes basally 10-veined, limbs 5–12 mm in diameter; stamens unevenly attached near throat, anthers not appressed to not fused above base of stigma; styles included. **Fruits** 1–2 mm, brown, often ridged, tubercles acute or obtuse, scars nearly median, ovate, without a slit-like groove. **2*n***=10.

Beach dunes, sandy bluffs. Flowering Apr–Aug. 0–100 m. Est. CA, ID, NV, WA; north to British Columbia, south to Mexico. Native.

This species was once found in a few sites along the northern coast, but despite being common in both California and Washington, it hasn't been collected or reported in Oregon since 1939.

Amsinckia tessellata A. Gray var. *tessellata* (p. 389)

Bristly fiddleneck, tessellate fiddleneck, checker fiddleneck, devil's lettuce

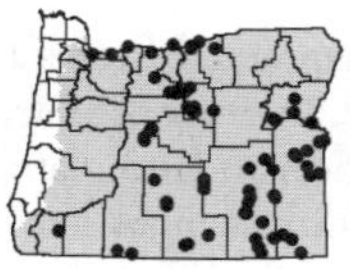

Plants 1.5–6 dm. **Stems** usually erect to ascending, usually branched at base and above, hispid with bristles. **Leaves** linear- to ovate-lanceolate, entire, surfaces hispid, hairs thin, pustulose at base. **Flower** calyx lobes 3–4, unequal in width, some fused below middle; corollas 8–12 mm, tubes 5–10 mm, 20-veined near base, limbs 2–6 mm in diameter, yellow, dark orange in center; stamens attached near throat of corolla tube, anthers appressed to stigma; styles reaching anthers. **Fruits** 2.5–4(5) mm, mottled, ridged or not, tubercles rounded, cobblestone-like, scars nearly median, ovate, without a slit-like groove. **2*n***=24.

Rocky or sandy soils. Flowering Apr–Jul. 0–1800 m. BR, BW, Casc, Col, ECas, Lava, Owy, Sisk, WV. CA, ID, NV, WA; north to British Columbia, southeast to NM, south to Baja California; MO; South America. Native.

The specific name refers to the tessellated (mosaic-like) covering of tubercles on the fruits.

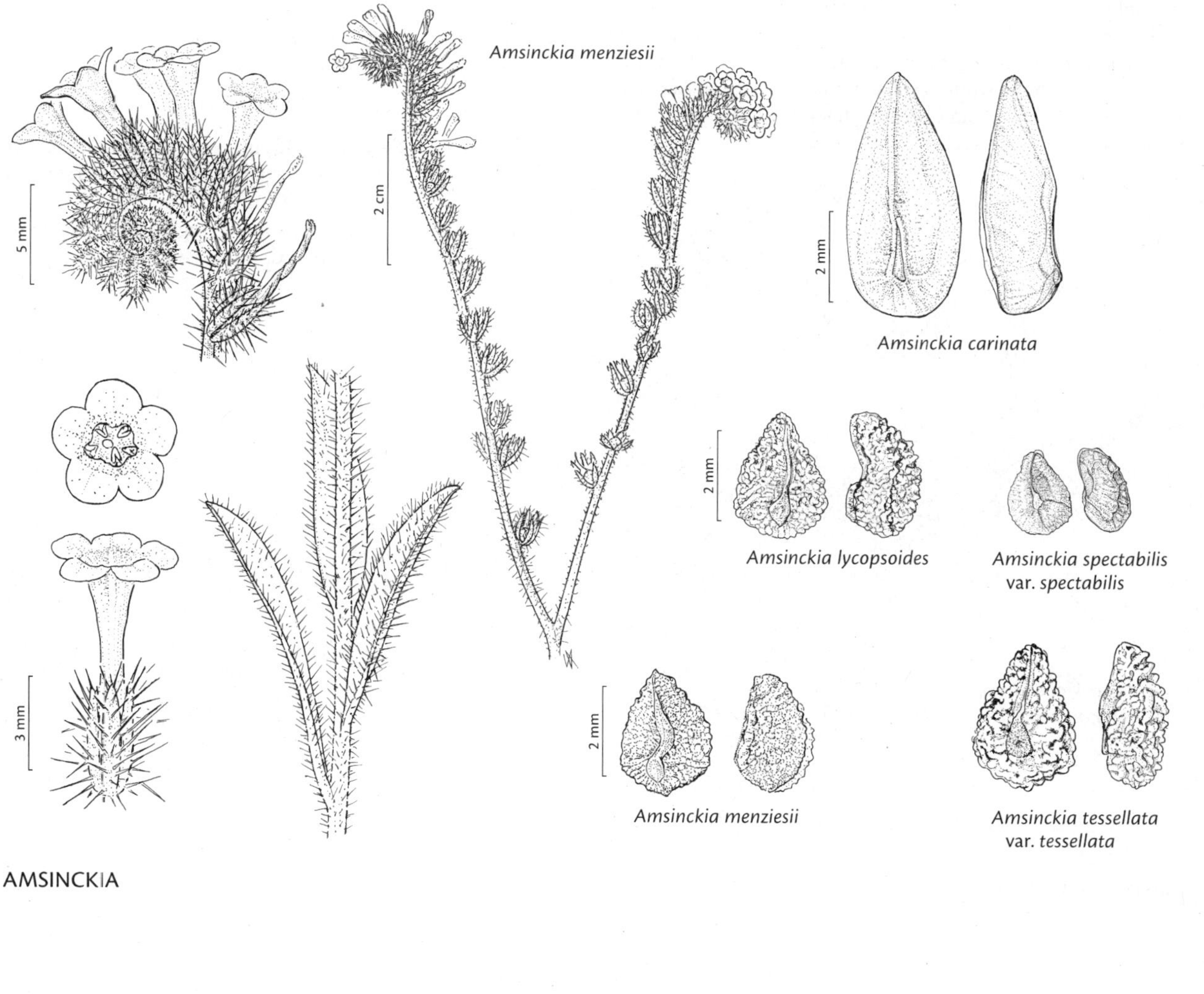

AMSINCKIA

AMSINCKIOPSIS Popcornflower

C. Matt Guilliams, Kristen E. Hasenstab-Lehman, and Ronald B. Kelley

Genus composed of one species, *A. kingii.*

Amsinckiopsis kingii (S. Watson) Guilliams, Hasenstab-Lehman, & B.G. Baldwin

Great Basin popcornflower

Herbs annual, 10–40 cm, hirsute and villous; taprooted. **Stems** 1 to several, ascending to erect. **Leaves** cauline, alternate, simple; basal blades oblanceolate; cauline blades linear to lanceolate, 2–6 cm, surfaces hirsute to hispid; sessile. **Inflorescences** scorpioid cymules, circinate in bud, glomerate or somewhat elongating in fruit; bracteate at base, proximal bracts sometimes exceeding flowers. **Flowers** calyces 5–6 mm, lobes lanceolate; corolla limbs 4–7 mm in diameter, white; fornices yellow. **Fruits** usually 4, ± ovate, 2.5–3 mm, abaxial surfaces sometimes with medial and lateral keels, irregularly rugose and papillate; attachment scars on adaxial surfaces, positioned along raised adaxial keel, elongate, narrowly triangular to narrowly lanceolate, length ~½ nutlet, scar tips acute.

Western North America. 2 varieties.

Amsinckiopsis, formerly included in *Plagiobothrys* section *Amsinckiopsis*, is more closely related to *Amsinckia* than to the majority of taxa of *Plagiobothrys* (Simpson et al. 2017; Guilliams 2015; Hasenstab-Lehman & Simpson 2012). Therefore, former *P. kingii* was transferred to a new genus so that only monophyletic groups are recognized taxonomically (Guilliams et al. in prep). Taxa of *Amsinckiopsis* are distinctive in subtribe Amsinckiinae in having the combination of hirsute to hispid vestiture throughout, white corollas, nutlets with rugose and papillate sculpturing, and an elongate attachment scar positioned along a raised, adaxial keel.

Amsinckiopsis kingii var. harknessii

Key to *Amsinckiopsis kingii* varieties:

1. Inflorescences glomerate in fruit; stems 5–15 cm, ± equal in height . . . . *A. k.* var. ***harknessii***

1' Inflorescences somewhat elongating in fruit; stems 10–40 cm, central ± taller . . . . *A. k.* var. ***kingii***

Amsinckiopsis kingii (S. Watson) Guilliams, Hasenstab-Lehman, & B.G. Baldwin var. *harknessii* (Greene) Guilliams, Hasenstab-Lehman, & B.G. Baldwin

(p. 390)

Northern Great Basin popcornflower

Plagiobothrys harknessii (Greene) A. Nelson & J.F. Macbr., *Plagiobothrys kingii* (S. Watson) A. Gray var. *harknessii* (Greene) Jeps.

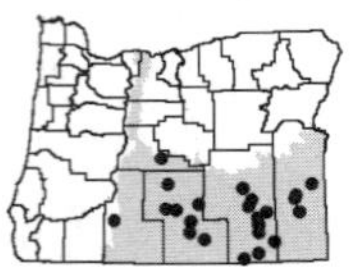

Stems 5–15 cm, ± equal in height. **Leaves** < 1 cm wide. **Inflorescences** glomerate in fruit.

Dry, open slopes, juniper woodland, sagebrush scrub. Flowering Apr–Jun. 800–2300 m. BR, ECas, Owy. CA, NV; southeast to UT. Native.

Amsinckiopsis kingii (S. Watson) Guilliams, Hasenstab-Lehman, & B.G. Baldwin var. *kingii*

(p. 390)

Southern Great Basin popcornflower

Plagiobothrys kingii (S. Watson) A. Gray var. *kingii*

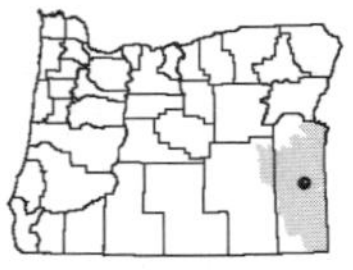

Stems 10–40 cm, central ± taller. **Leaves** < 2 cm wide. **Inflorescences** somewhat elongating in fruit.

Dry, open slopes, sagebrush scrub, juniper woodland. Flowering May–Jun. 900–2300 m. Owy. CA, NV. Native.

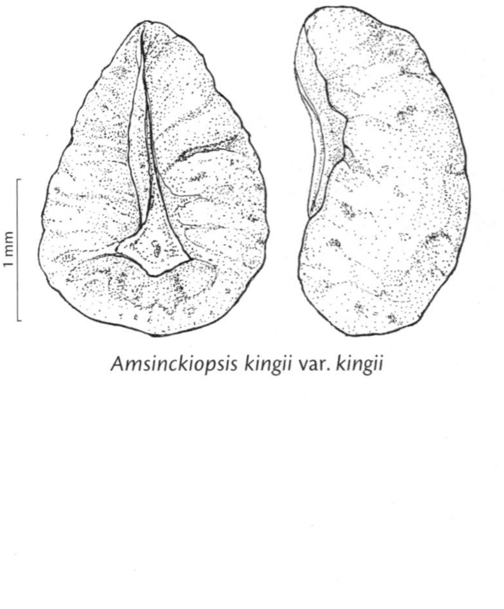

Amsinckiopsis kingii var. kingii

ANCHUSA Alkanet, bugloss

Richard R. Halse

Herbs hispid; taproots slender to stout. **Stems** simple to branched. **Leaves** alternate, simple; petiolate below, becoming sessile to ± clasping distally. **Inflorescences** axillary and terminal cymes, elongating and straightening in age; pedicels erect or ascending. **Flowers** bisexual, radially symmetric, funnelform to salverform; calyces 5-lobed, shallowly to deeply divided; corollas 5-lobed, purple to blue to white, throat closed by 5 scales; stamens 5, included or slightly exserted; ovaries deeply 4-lobed, styles 1, included, stigmas capitate. **Fruits** 1–4, erect, tan or brown, rugose and tuberculate, with stipe-like basal attachment surrounded by a prominent thickened rim; receptacles flattened.

Africa, Asia, Europe, North America. ~35 species; 2 species treated in *Flora*.

Both species of *Anchusa* in Oregon are cultivated as ornamentals and are probably escapees from gardens. A third species, *Anchusa capensis*, a native of South Africa, was collected once in Oregon in 1925. It has not been collected since and is best considered a waif. Compared to *A. officinalis*, it has narrower leaves and bracts and calyces much shorter than the corolla tube.

Key to *Anchusa* species:

1. Calyx lobes linear, > 2 × as long as calyx tube; corolla limbs 10–20 mm in diameter; fruits ± oblong. . . . . ***A. azurea***

1' Calyx lobes lanceolate, < 2 × as long as calyx tube; corolla limbs 6–15 mm in diameter; fruits obliquely ovoid. . . ***A. officinalis***

Anchusa azurea Miller

Showy bugloss, Italian bugloss, garden bugloss

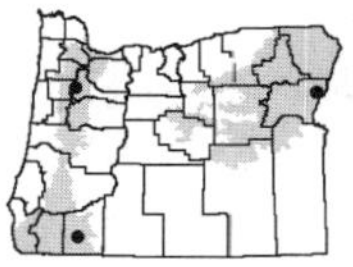

Plants perennial. **Stems** erect, mostly simple, 15–150 cm. **Leaves** 5–30 × 1.5–5 cm, margins entire; basal oblanceolate; cauline lanceolate to oblong. **Flowers** calyces divided almost to base into linear acute lobes, 6–10 mm in flower, to 18 mm in fruit, lobes > 2 × as long as calyx tube; corollas violet to deep blue, tubes straight, 6–11 mm, lobes equal, 8–10 mm, limbs 10–20 mm in diameter; stamens inserted at top of tube. **Fruits** erect, ± oblong. **2*n***=32.

Waste ground, disturbed areas. Flowering May–Aug. 0–900 m. BW, Sisk, WV. CA, ID, WA; scattered in North America; Europe. Exotic.

Anchusa azurea belongs to subgenus *Buglossum* and may be a separate genus. *Anchusa azurea* is a garden plant originally from Europe, western Asia, and eastern Maghreb. In temperate regions, it has escaped and become a widely distributed noxious weed.

Anchusa officinalis Linnaeus (p. 391)

Common bugloss

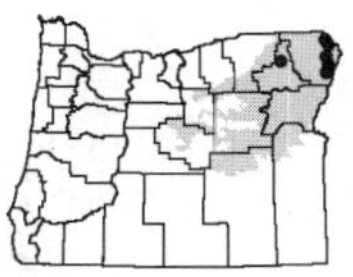

Plants perennial, rarely biennial. **Stems** erect, simple or branched, 20–170 cm. **Leaves** 50–160 × 7–40 mm, margins entire to obscurely undulate; basal oblanceolate, cauline linear to oblanceolate; petioles winged, slightly auriculate at base. **Flowers** calyces divided to mid-length or almost to base into lanceolate acute lobes, 4–7 mm in flower, 9–12 mm in fruit, lobes < 2 × as long as calyx tube; corollas violet or reddish to dark blue, rarely white, tubes straight, 5–11 mm, lobes equal, 2–6 mm, limbs 6–15 mm in diameter; stamens inserted in upper half of tube. **Fruits** obliquely ovoid. **2*n***=16.

Hillsides, disturbed areas, roadsides. Flowering May–Jul. 300–1100 m. BW. CA, ID, WA; scattered in North America; Asia, Europe. Exotic.

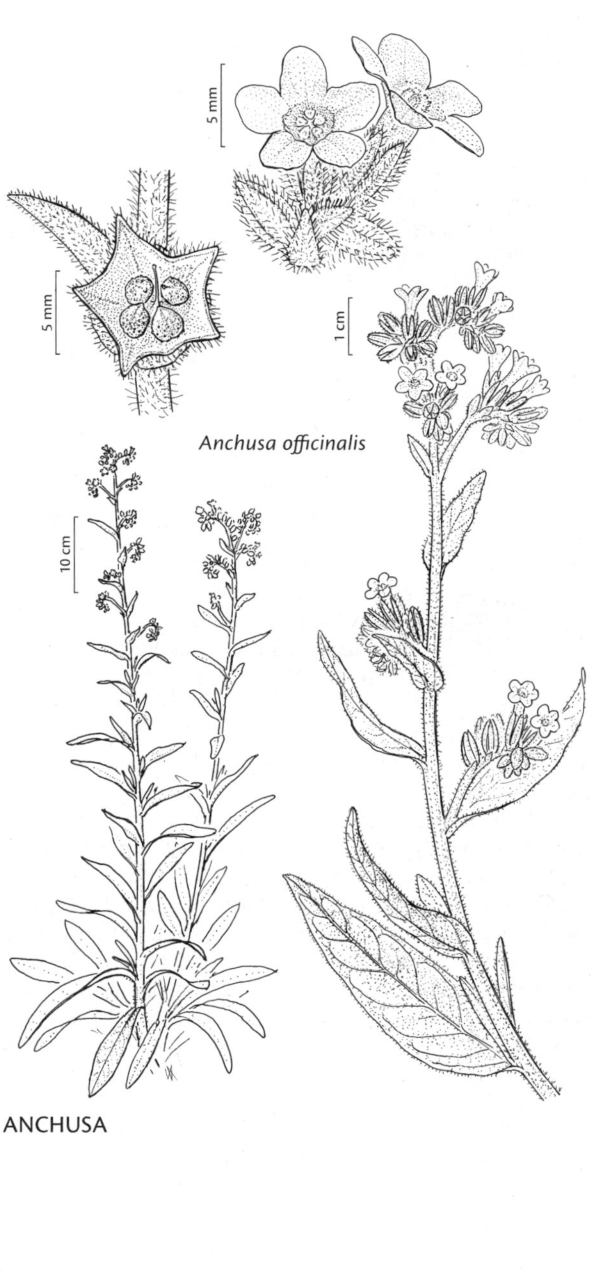

ANCHUSA

ANDERSONGLOSSUM Hound's tongue

James I. Cohen

Herbs perennial, erect. **Stems** hispid, strigose, pilose, or sericeous. **Leaves** basal and cauline, simple; sessile. **Inflorescences** terminal scorpioid cymes. **Flowers** bisexual; sepals persistent, linear; corollas radially symmetric; stamens 5, included; styles 1, included or exserted, persistent, stigmas 2, terminal. **Fruits** 4 or fewer by abortion, spreading.

North America. 2 species; 1 species treated in *Flora*.

Andersonglossum is a small genus of species that were formerly members of *Cynoglossum*.

Andersonglossum occidentale (A. Gray) J.I. Cohen (p. 392)

Western hound's tongue

Cynoglossum occidentale A. Gray

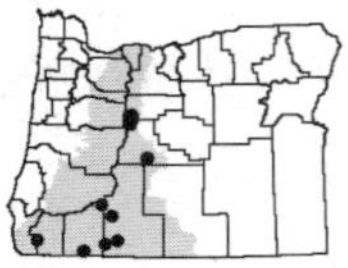

Plants to 50 cm; taprooted. **Leaves** alternate, 3–38 × 0.5–11 cm, margins entire, surfaces strigose to sericeous; cauline oblanceolate to spatulate. **Inflorescences** 15–25-flowered; bracts 0; pedicels recurved in fruit. **Flowers** sepals 5-lobed to base, lobes linear; corollas salverform, 5-lobed, purple to red, glabrous, lobes erect to flared, elliptic; fornices white to red to purple; anthers ellipsoid; ovaries 4-parted, styles > 4 mm. **Fruits** ovoid, brown to yellow, prickles barbed at tips; attachment scars broad-triangular.

Conifer forests. Flowering May–Jul. 900–1900 m. Casc, ECas, Sisk. CA. Native.

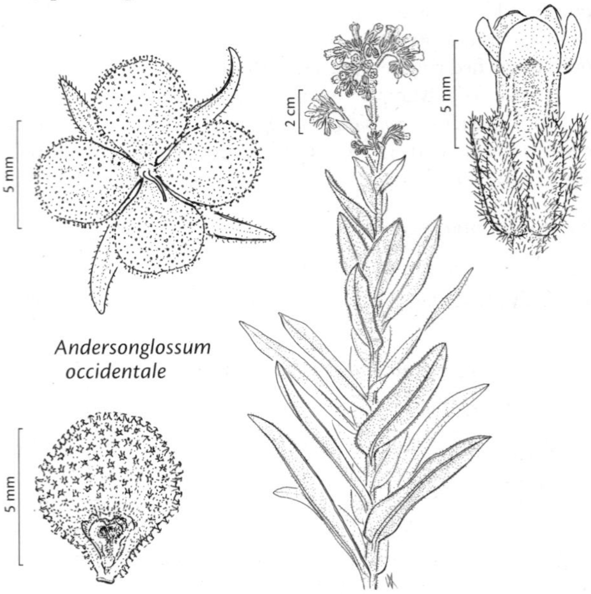

ANDERSONGLOSSUM

ASPERUGO Madwort, catchweed

Richard R. Halse

Genus composed of one species, *A. procumbens.*

Asperugo procumbens Linnaeus (p. 392)

Madwort

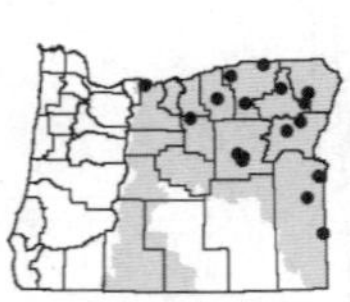

Herbs annual, prostrate to decumbent to weakly climbing-scrambling; taproots slender. **Stems** simple to many-branched, angled or ribbed, 1–12 dm, retrorsely prickly-hispid mainly on angles. **Leaves** forming basal rosette when young, soon withering; cauline leaves opposite (subopposite) to partly alternate or whorled; blades ovate to oblanceolate to elliptic, 1–10 × 0.4–4.5 cm, margins entire to remotely denticulate, tips subacute to obtuse, surfaces scabrous-hispid; petiolate proximally to sessile (subsessile) distally. **Inflorescences** axillary, in or near axils of leaves or bracts or in forks of branches, 1–2 flowers per axil; pedicels short, stout, 1–2 mm in flower, 2–6 mm in fruit, elongate and recurved in fruit. **Flowers** tubular to campanulate; calyces 5-lobed, divided about halfway, becoming strongly accrescent, compressed, 2–3 mm, papery, reticulate, veiny and to 15 × 10–20 mm in fruit, prickly hispid, hairs curved or hooked, lobes with 1 or 2 small teeth on each side near base or in sinus; corollas 5-lobed, 2–3 mm, blue to purplish; fornices prominent; stamens 5, included, inserted near middle of tube; ovaries deeply 4-lobed, styles 1, included, stigmas capitate. **Fruits** 2–4, erect, laterally attached above middle, elongate-ovoid, obliquely compressed, 2.5–4 mm, tan to brown; scars small, finely tuberculate; receptacles elongate-conic. $2n$=48.

Fields, riparian areas including alkaline pond margins, roadsides, disturbed areas. Flowering May–Jul. 100–1700 m. BW, Col, ECas, Lava, Owy. CA, ID, NV, WA; north to AK, northeast to Greenland, east to PA; South America, Asia, Europe. Exotic.

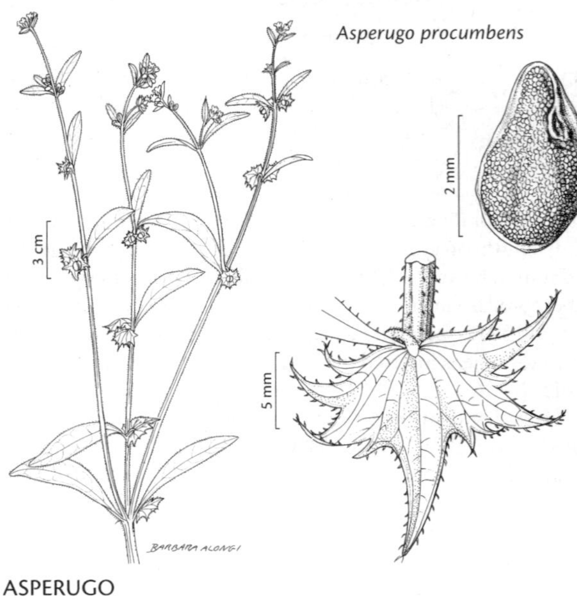

ASPERUGO

BORAGO Borage

Stephen C. Meyers

Herbs annual, hispid. **Stems** erect, curving downward with age. **Leaves** cauline, alternate, margins entire; petiolate to sessile. **Inflorescences** terminal, helicoid cymes; leafy bracts present; pedicels elongate, curving downward in fruit. **Flowers** radially symmetric; calyces 5-lobed; corollas 5-lobed, blue, glabrous; fornices well developed; stamens 5, exserted, anthers surrounding, but not fused to style, dark brown. **Fruits** 4, erect, obovoid, tubercled; with a swollen basal attachment.

Europe, North Africa, North America. 4 species; 1 species treated in *Flora.*

Borago officinalis Linnaeus (p. 393)

Common borage

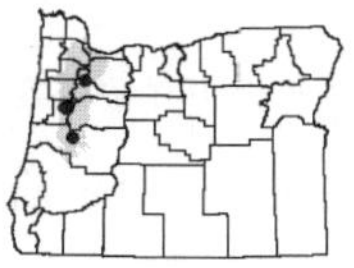

Plants 2–6 dm; taprooted. **Stems** branched. **Leaves** elliptic to ovate, 3–20 × 2–8 cm, surfaces hairy, bristly at base; proximal leaves petiolate, distal leaves sessile. **Flowers** calyces 5–15 mm; corollas 18–20 mm in diameter; anthers 5–10 mm, dark brown. **Fruits** 4–6 mm. **2*n***=16.

Disturbed areas, roadsides, fields. Flowering Apr–Aug. 0–200 m. WV. CA, WA; scattered throughout southern Canada and US; Europe. Exotic.

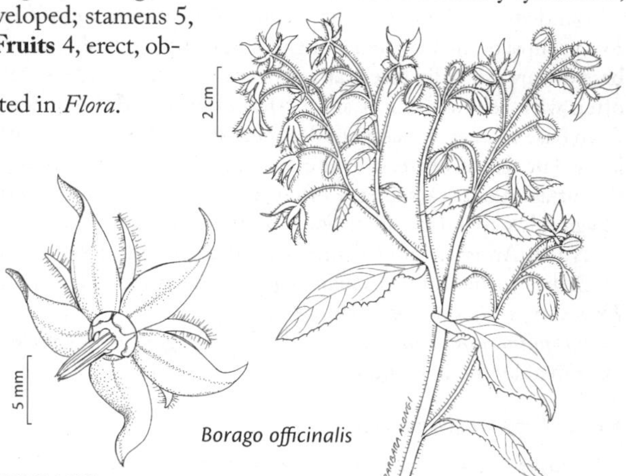

Borago officinalis

BORAGO

BUGLOSSOIDES Gromwell

James I. Cohen

Herbs perennial or annual. **Stems** simple or much branched, strigose. **Leaves** alternate, simple, margins entire, midveins conspicuous. **Inflorescences** terminal cymes; bracts present; pedicels erect to appressed. **Flowers** radially symmetric; calyces 5-lobed, divided almost to base, becoming larger in fruit; corollas funnelform to salverform, throats with 5 longitudinal bands of hairs; fornices absent; stamens 5, inserted below middle of tube, included; ovaries deeply 4-lobed, styles 1, included, stigmas 2-lobed; receptacles flattened. **Fruits** erect, 1–4; attachment scars basal.

Worldwide. ~15 species; 1 species treated in *Flora.*

Buglossoides is native to Eurasia and Africa, but some species are common weeds across the globe.

Buglossoides arvensis (Linnaeus) I.M. Johnston (p. 393)

Corn gromwell, field gromwell

Lithospermum arvense L.

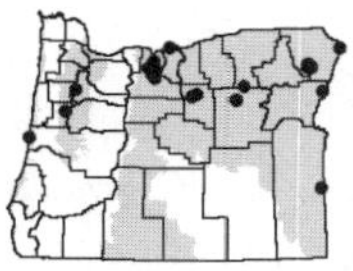

Plants annual, strigose. **Stems** erect, 0.5–9 dm, central stem usually largest. **Leaves** basal-most sometimes deciduous; blades oblong-spatulate or obovate to oblanceolate to linear, 0.5–6.5 × 0.2–1.5 cm, margins plane to revolute, tips acute; basal distinctly petiolate and spatulate; cauline sessile or petiole short and ill defined. **Inflorescences** elongate with age, crowded in flower, open and remote in fruit; pedicels 1–5 mm. **Flowers** calyx lobes linear to linear-lanceolate, 2–5 mm in flower, 5–13 mm in fruit, unequal, usually much > nutlets; corollas white, rarely bluish, fading to yellowish when dried, tubes 4–9 mm, strigose externally, limbs 2–4 mm in diameter. **Fruits** erect, ovoid, 2.5–4 × 1.5–2.5 mm, white to tan to brown, rarely gray, rugose and verrucose with prominent ventral keel. **2*n***=28.

Fields, pastures, roadsides, disturbed areas, open slopes, hillsides, rocky ridges, cliffs. Flowering Apr–Jul. 0–1500 m. BW, Col, ECas, Est, Lava, Owy, WV. CA, ID, NV, WA; throughout most of North America; Asia, Europe. Exotic.

This is one of the most widespread and morphologically variable species of *Buglossoides* and grows in many different habitats and soil types.

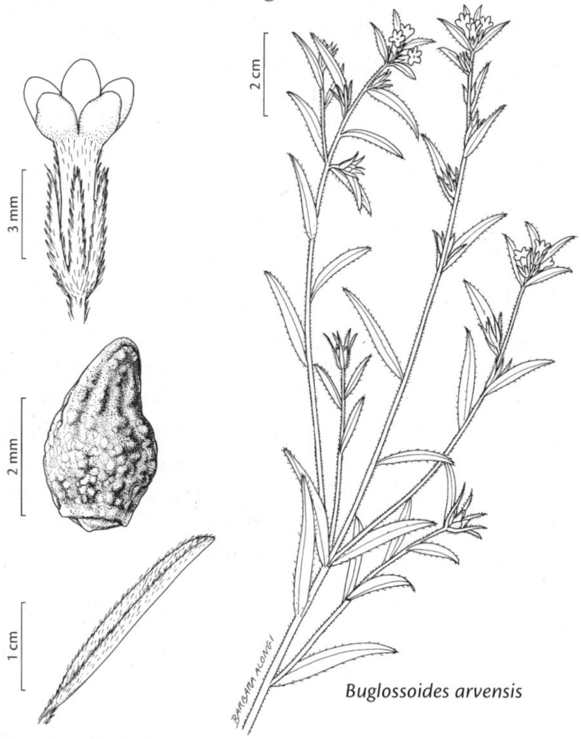

Buglossoides arvensis

BUGLOSSOIDES

CRYPTANTHA Cryptantha, cat's eye

Michael G. Simpson and Ronald B. Kelley

Herbs annual, erect or ascending to rarely prostrate or decumbent, usually strigose to spreading-hairy, pustulose or not. **Leaves** basal and cauline, simple, elliptic, linear, lanceolate, oblanceolate to rarely spatulate, usually reduced in size distally, bases cuneate to rounded, margins entire, often ciliate, tips acute to rounded; sessile. **Inflorescences** terminal, scorpioid cymules, solitary or in up to 5 groups; branches often elongated in fruit; bracts absent or present. **Flowers** bisexual, scented or not; sepals 5, lobes separate to base or rarely basally fused, lanceolate to linear, often persistent; corollas radially symmetric, rotate to funnelform, lobes 5, white; often with yellow fornices at junction of limb and throat; stamens 5, attached to upper corolla throat, included; styles 1, arising from base of ovary lobes, included, persistent, stigmas 1; flower bracts usually absent. **Fruits** 1–4, homomorphic or heteromorphic, lanceolate, lance-ovate, or ovate to triangular, surfaces smooth or papillate, tuberculate, and/or muricate, spinal ridges present or not; attachment scars lateral, often referred to as "ventral grooves," extending nearly to nutlet tips, margins abutted entire length to variably gapped for all or portion of scar, often with areole and bifid-forked at base margin.

North America, South America. ~105 species; 18 species treated in *Flora.*

A diagnostic feature of many *Cryptantha* species is the ventral groove of the nutlet, but this is also found in the segregate genera *Eremocarya, Greeneocharis,* and *Oreocarya.*

Cryptantha muricata, reported from Oregon, is not present in the state. This report was based on misidentified specimens, examined by the first author.

Key to *Cryptantha* species:

1. Nutlets all smooth or only obscurely roughened.
 2. Nutlets asymmetric, attachment scars strongly off-center, appearing marginal . . . . ***C. affinis***
 2' Nutlets symmetric, attachment scars centered.
  3. Some calyx hairs curved or hooked; nutlets 1.
   4. Corolla limbs (1)2–4(6) mm in diameter; nutlet attachment scar edges abutted to overlapping entire length, rarely forming tiny areole at base; styles extending to < 50% length of mature nutlets. . . . ***C. flaccida***
   4' Corolla limbs 0.5–2 mm in diameter; nutlet attachment scar edges abutted or overlapping near apex, widening into conspicuous areole at base; styles extending to 50–70% length of mature nutlets. . . . ***C. rostellata***
  3' Calyx hairs not curved or hooked; nutlets 1–4.
   5. Flower bracts usually present; branches generally prostrate to decumbent . . . . ***C. leiocarpa***
   5' Flower bracts usually absent, rarely 1–2; branches generally erect to ascending (but decumbent to erect in *C. gracilis*).
    6. Nutlet margins sharp-angled to flattened into narrow rim, especially toward apex; nutlets (3)4, narrowly lance-ovate; corolla limbs 0.5–1.5 in diameter . . . . ***C. watsonii***
    6' Nutlet margins usually rounded, sometimes angled basally or obscurely; nutlets 1–4, narrowly lance-ovate to ovate; corolla limbs 0.5–6 mm in diameter.
     7. Calyx abaxial surfaces generally with 1 hair type, usually tufted long soft-hirsute, lobes basally connate, midvein/midrib hairs soft, with longer, coarser but not hispid hairs; nutlets 1(2+), 3-sided, adaxially 2-planed-convex.
      8. Corolla limbs 1–2.5 mm in diameter; all cauline leaves alternate; calyces 2–3 mm in fruit; stem hairs mostly spreading-hirsute . . . . ***C. gracilis***
      8' Corolla limbs 2–6 mm in diameter; proximal-most cauline leaves usually opposite; calyces 3–4 mm in fruit; stem hairs strigose and ascending-hirsute . . . . ***C. milobakeri***
     7' Calyx abaxial surfaces with 2 hair types, fine, strigose to finely hirsute hairs and coarser, hispid hairs, lobes separate to base, midvein/midrib hairs stiff and hispid; nutlets 1–4, not 3-sided, adaxially flat to convex.
      9. Corolla limbs 0.5–1.5 mm in diameter; nutlets (2)3–4.
       10. Nutlets usually ovate, tips not elongate, acute, abaxial surfaces flattened, especially toward base, low, rounded spinal ridges present at extreme apex, attachment scar edges bifid-forked but not gapped at base, areoles absent . . . . ***C. torreyana*** var. ***torreyana***
       10' Nutlets usually lance-ovate, tips elongate, narrowly acute to acuminate, abaxial surfaces convex, spinal ridges absent, attachment scar edges widening into triangular areole at base, bifid-forked at extreme base . . . . ***C. fendleri***
      9' Corolla limbs (2.5)3–8 mm in diameter; nutlets 1–2(3).
       11. Nutlets widely lance-ovate to narrowly ovate, not basally swollen; calyx longest hairs 2 mm . . . . ***C. grandiflora***
       11' Nutlets narrowly lance-ovate, basally swollen; calyx longest hairs 1 mm . . . . ***C. hispidula***

1' Nutlets, or at least 1, variously roughened, surfaces bearing papillae and/or tubercles.
12. Nutlets usually heteromorphic, 3 marginally winged and 1 unwinged, rarely homomorphic with all nutlets winged . . . . **C. *pterocarya*** var. ***pterocarya***
12' Nutlets usually homomorphic with margins rounded or angled, not winged.
13. Nutlets 1(2–3).
14. Calyces recurved arched, 2.5–3.5 mm in fruit; nutlets incurved bent . . . . ***C. recurvata***
14' Calyces straight not arched, (3.5)4.5–6 mm in fruit; nutlets straight . . . . ***C. intermedia*** (in part)
13' Nutlets 3–4.
15. Corolla limbs (3)4–11 mm in diameter; inflorescence cymules usually in 3s or paired, occasionally in 4s or 5s . . . . ***C. intermedia*** (in part)
15' Corolla limbs 1–2 mm in diameter; inflorescence cymules usually solitary to in groups of as many as 3.
16. Stems generally strigose only, rarely sparsely ascending to spreading-hirsute.
17. Calyx margins distally with short recurved hispid hairs; nutlets ovate, tips short-acuminate, surfaces papillate and sparsely low-rounded tuberculate . . . . ***C. simulans***
17' Calyx margins with dense, ascending, fine-hirsute hairs, not recurved; nutlets lanceolate to lance-ovate, tips narrowly acute to acuminate, surfaces minutely tuberculate-spinulose . . . . ***C. scoparia***
16' Stems both strigose and spreading-hirsute to hispid.
18. Nutlets minutely tuberculate-spinulose, spinules present . . . . ***C. echinella***
18' Nutlets low-rounded tuberculate, spinules absent . . . . ***C. ambigua***

Cryptantha affinis (A. Gray) Greene (p. 400)

Side-grooved cryptantha, slender cryptantha, quill cryptantha

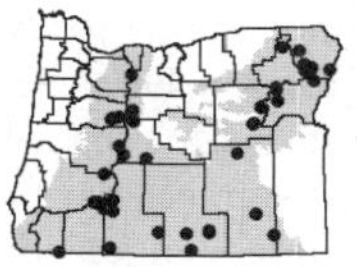

Plants 5–30(40) cm. **Stems** erect, branches loosely ascending from base and beyond, strigose and spreading-hirsute. **Leaves** basal rosette sometimes present, withering at anthesis; cauline scattered, proximal-most pair opposite, oblanceolate to oblong or narrowly elliptic, 10–40(50) × 3–6(8) mm, slightly reduced in size distally, tips obtuse-rounded, surfaces inclined- to ascending-hirsute, minutely pustulose. **Inflorescences** cymules solitary or paired, 4–10-flowered, 2–8(15) cm, usually much elongated in fruit, proximal flowers not touching; bracts present at base, similar to distal leaves; pedicels elongating in fruit, 0.5–1 mm. **Flowers** unscented, ascending; calyces tardily deciduous at maturity, asymmetric, 1.5–2 mm at anthesis, 2.5–3.5(4) mm in fruit, lobes separate to base, lanceolate, margins appressed to ascending, hirsute, midribs slightly thickened, spreading-hispid, tips erect to slightly spreading, abaxial surfaces appressed- to ascending-hirsute, lobes opposite rachises with enlarged hispid hairs; corollas rotate to funnelform, tubes 1–2 mm, limbs 1–2 mm in diameter; fornices white, minute; gynobases extending to 50% and styles to 50–70% length of mature nutlets; flower bracts subtending only proximal flowers. **Fruits** 4, homomorphic, ovate, diagonally compressed, strongly asymmetric, 1 side appearing narrower, 1.8–2.5 mm, brownish, usually mottled, bases obtuse-pointed, margins rounded, tips acute, surfaces essentially smooth, shiny, abaxial surfaces low-convex; spinal ridges absent; attachment scars (ventral groove) strongly off-center, appearing marginal, edges not raised, abutted entire length, sometimes short bifid-forked. **2*n***=24.

Open areas, coniferous forests, scrublands. Flowering Jun–Aug. 800–2000 m. BR, BW, Casc, ECas, Sisk. CA, ID, NV, WA; north to British Columbia, northeast to Alberta, east to SD, southeast to CO. Native.

The nutlets of *C. affinis* are distinctive in being highly asymmetric, the attachment scar (ventral groove) appearing marginal.

Cryptantha ambigua (A. Gray) Greene (p. 400)

Wilkes' cryptantha, basin cryptantha, obscure cryptantha

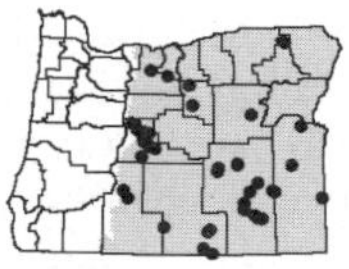

Plants 10–30(45) cm. **Stems** erect, solitary or several-branched from base, branches usually many, loosely ascending to occasionally decumbent, strigose and spreading-hirsute. **Leaves** basal leaves withering at maturity, mostly spatulate; cauline linear, narrowly oblong, or spatulate-oblanceolate, 20–40(50) × 1–3 mm, little reduced in size distally, midveins thickened abaxially, tips acute or rounded, surfaces appressed- to ascending-hirsute pustulose. **Inflorescences** cymules solitary, occasionally paired, 5–15 cm in fruit, proximal-most flowers usually not touching; bracts present near base of cymules, similar to distal leaves; pedicels 0.5(1) mm in fruit. **Flowers** ascending to spreading, tardily deciduous at maturity; calyces slightly asymmetric, lance-ovoid to urceolate, bases truncate to rounded, constricted distally, 1.5–2.5 mm at anthesis, 4–6(7) mm in fruit, lobes separate to base, linear to lanceolate, often yellowish, margins densely ascending-hirsute, midribs slightly thickened, widely spreading hispid, hairs pustule-based, tips erect to occasionally spreading, abaxial surfaces densely appressed- to inclined-hirsute, adaxial surfaces appressed short-hirsute near tip; corollas funnelform, tubes 1.5–2 mm, limbs 1–2 mm in diameter; fornices minute; gynobases extending to 65% and styles to 75% length of mature nutlets; flower bracts absent or occasionally subtending proximal-most flowers. **Fruits** 3–4, homomorphic, ovate, somewhat compressed, (1.4)1.6–2 mm, grayish to brownish, bases truncate, margins rounded, tips acute to short acuminate, surfaces somewhat glossy, scattered low-rounded tuberculate, papillate between tubercles, occasionally becoming smoothish toward base, abaxial surfaces low convex, spinal ridges usually absent but occasionally obscurely present, adaxial surfaces 2-planed convex, attachment scars centered, edges not raised, usually abutted entire length, occasionally narrowly gapped, widely bifid-forked, but usually not gapped at base.

Gravelly or sandy soil of open slopes, open areas of shrublands, grasslands, coniferous forests, pinyon-juniper woodlands. Flowering Jun–Aug. 100–1900 m. BR, BW, Col, ECas, Lava, Owy. CA, ID, NV, WA; north to British Columbia, east to SD, southeast to CO. Native.

Cryptantha ambigua is distinguished by its small corolla size, strigose and spreading-hirsute stems, and 3–4 ovate nutlets with low, rounded tubercles. This species can introgress with populations of *Cryptantha torreyana* var. *torreyana*, their close relationship recently confirmed by molecular phylogenetic studies (Mabry & Simpson 2018; Simpson et al. 2017).

Cryptantha echinella Greene (p. 400)

Prickly cryptantha

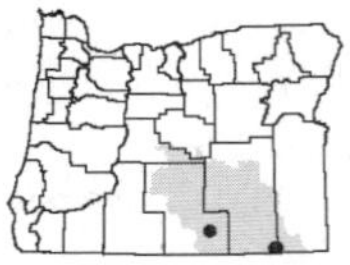

Plants 5–35 cm, usually pale green. **Stems** erect, occasionally without clear central stem, unbranched or branched throughout, branches usually few, ascending, strigose and spreading-hirsute. **Leaves** usually few, oblanceolate, linear, or narrowly oblong, 10–50 × 2–5 mm, tips acute to rounded, surfaces appressed- to ascending-hirsute, pustulose, pustules larger on margins and abaxial surfaces. **Inflorescences** cymules solitary, rarely paired, 1–5(10) cm in fruit, proximal-most flowers usually not touching; bracts occasionally 1–2 at cymule base, resembling vegetative leaves; pedicels 0–0.5 mm in fruit. **Flowers** ascending to spreading, promptly deciduous at maturity; calyces symmetric, cylindric to ovoid, bases widely rounded, slightly constricted distally, 1.5–2.5 mm at anthesis, 4–6 mm in fruit, lobes separate to base, linear to lanceolate, margins often densely whitish ascending-hirsute, midribs slightly thickened and ascending to inclined coarse-hispid, hairs pustule-based, tips erect to slightly recurved, abaxial surfaces finely appressed-hirsute, adaxial surfaces glabrous proximally, appressed short-hirsute near tips; corollas funnelform, tubes 1–1.5 mm, limbs 1–2 mm in diameter; fornices minute; gynobases extending to 75% and styles to slightly < length of mature nutlets; flower bracts absent. **Fruits** (3)4, homomorphic, lance-ovate, usually compressed, 1.8–2.2 mm, grayish to brownish, occasionally mottled, bases truncate, margins rounded, tips narrowly acute rounded, slightly acuminate, surfaces dull, uniformly minutely tuberculate-spinulose, spinule projections short, cylindric, transparent, and apically rounded to sharp-tipped, abaxial surfaces low convex, usually flat, spinal ridges absent to obscurely present, adaxial attachment scars centered, edges not raised, abutted entire length or variably narrowly gapped, widely bifid-forked at base, occasionally gapped to form minute areole.

Openings in shrubland, coniferous forest, juniper woodlands. Flowering May–Aug. 1600–2200 m. BR. CA, ID, NV. Native.

Cryptantha echinella is distinguished by its dense, finely tuberculate-spinulose nutlets, which have short, narrowly cylindric, and transparent spinules.

Cryptantha fendleri (A. Gray) Greene (p. 400)

Sand dune cryptantha, Fendler's cryptantha

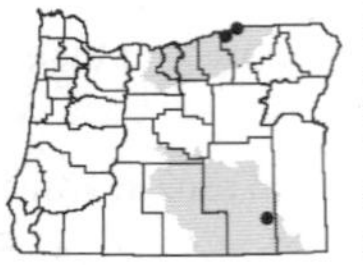

Plants 10–50 cm. **Stems** rigidly erect with prominent central axis, usually many-branched distally, branches rigid, usually ascending, densely strigose and inclined to spreading-hirsute. **Leaves** many, linear to narrowly oblanceolate, 10–50 × 1–4 mm, tips obtuse to narrowly acute, margins and surfaces appressed- to ascending-hirsute, abaxial hairs pustulose. **Inflorescences** cymules solitary to paired (rarely in 3s), 2–12 cm, proximal-most flowers not touching; bracts absent or sparsely reduced at base; pedicels usually not elongating in fruit, 0–0.5 mm. **Flowers** unscented, ascending; calyces deciduous at maturity, slightly asymmetric, cylindric to narrowly urceolate, usually constricted distally, 1.5–2 mm at anthesis, 3.5–6(7.5) mm in fruit, lobes separate to base, linear to lance-linear, separate to base, midribs thickened, with stiff and hispid hairs, tips slightly spreading, abaxial surfaces with 2 hair types, fine, strigose to finely hirsute hairs and coarser hispid hairs, adaxial surfaces glabrous at extreme base, appressed and densely white-hirsute in distal region; corollas funnelform, tubes 1 mm, limbs 0.5–1(1.5) mm in diameter; fornices absent or minute, white; gynobases extending to 70% and styles to about length of mature nutlets; flower bracts absent or rarely 1–2. **Fruits** usually 4, those toward inflorescence axis always developing, homomorphic, usually lance-ovate, rarely ovate, 1.4–2 mm, brown, gray, or mottled, bases rounded to narrowly truncate, margins rounded to obscurely angled, tips elongate, narrowly acute to acuminate, surfaces smooth, glossy, abaxial surfaces convex, spinal ridges absent, adaxial surfaces flat to slightly plano-convex, attachment scars centered, edges not raised, abutted entire length, widening into triangular areole at base, bifid at extreme base.

Sandy soils, sagebrush, pinyon pine-sagebrush, shrub communities. Flowering May–Jul. 100–2000 m. BR, Col. CA, ID, NV, WA; northeast to Saskatchewan, east to SD, southeast to NM. Native.

Cryptantha fendleri can be distinguished by its small corollas; four, smooth, usually lance-ovate nutlets; and rigid, ascending branches, with a prominent central axis.

Cryptantha flaccida (Douglas ex Lehmann) Greene (p. 400)

Weak-stemmed cryptantha, pale cryptantha

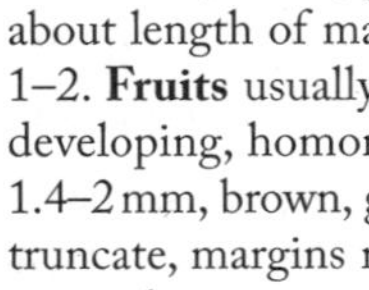

Plants 15–50 cm. **Stems** erect, slender, wiry, branched distally to throughout, branches ascending, pallid-strigose. **Leaves** only proximal-most pair occasionally opposite, linear, 20–60 × 1–2(3) mm, tips acute-rounded, surfaces closely strigose to ascending-hirsutulous, inconspicuously pustulose. **Inflorescences** cymules solitary or up to 5 in a group, (3)4–8(16) cm, much elongated in fruit, proximal-most flowers not touching; bracts absent; pedicels not elongated in fruit, < 0.5 mm. **Flowers** usually unscented, strictly appressed; calyces persistent at maturity, asymmetric, narrowly ovoid to urceolate, constricted distally, 2–3 mm at anthesis, 3–4.5(5) mm in fruit, lobes separate to base, linear, separate to base, margins sparsely finely hirsute-ciliate, midribs thickened, tips usually spreading, abaxial surfaces densely covered with recurved, hook-tipped, pustulose hairs, especially along midrib, hairs longer away from inflorescence axis, adaxial surfaces glabrous basally, appressed short-hirsute apically; corollas rotate, tubes 1–2 mm, limbs (1)2–4(6) mm in diameter; fornices yellow; styles < 50% length of mature nutlets; flower bracts absent. **Fruits** 1, positioned away from inflorescence axis, lance-ovate, not compressed, symmetric, 1.8–2.8 mm, brown to mottled, proximal half globose-swollen, bases narrowly truncate, somewhat pointed, margins rounded, tips long-acuminate, surfaces smooth, shiny, abaxial and adaxial surfaces strongly convex, spinal ridges absent; attachment scars centered, edges usually raised, abutted to overlapped entire length, rarely forming tiny areole at base.

Sandy, rocky, occasionally serpentine derived soils; slopes or flats of shrub, grasslands, or open woodland communities. Flowering Apr–Jul. 0–1200 m. BW, Col, CR, ECas, Lava, Sisk, WV. CA, ID, NV, WA. Native.

Cryptantha flaccida is distinguished by having recurved calyx hairs and a single, smooth nutlet with a swollen base and long-acuminate tips. It is similar to *C. rostellata* (see discussion).

Cryptantha gracilis Osterhout (p. 400)

Narrowstem cryptantha, slender cryptantha

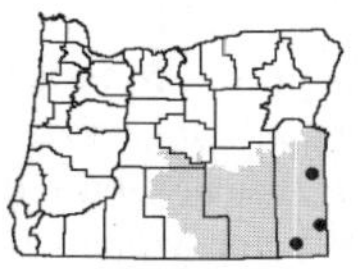

Plants 10–35 cm. **Stems** erect, ultimate stems slender, unbranched or branched throughout, branches decumbent to erect, mostly spreading-hirsute. **Leaves** cauline alternate, linear to oblanceolate, (5)10–50 × 1–3 mm, tips rounded to obtuse, surfaces spreading-hirsute, pustulose. **Inflorescences** cymules solitary or paired, occasionally in 3s, 0.5–4.5(10) cm, initially somewhat glomerate but elongating somewhat in fruit, proximal-most flowers usually touching; reduced bracts occasionally at base; pedicels slightly elongating in fruit, 0.5 mm. **Flowers** horizontal to inclined; calyces promptly deciduous at maturity, slightly asymmetric, ellipsoid to ovoid with acutely obconic base, 1.5–2 mm at anthesis, 2–3 mm in fruit, lobes basally connate, lanceolate to narrowly elliptic, basally connate, sometimes connate proximally, margins densely long ascending white hirsute, midribs slightly thickened, hairs soft, with longer, coarser but not hispid hairs, tips usually erect, abaxial surfaces with 1 hair type, usually tufted long soft-hirsute, hairs soft, adaxial surfaces glabrous basally, appressed short-hirsute apically; corollas funnelform to rotate, tubes 1 mm, limbs 1–2.5 mm in diameter; fornices yellow; gynobases extending to 50% and styles to 75% length of mature nutlets; flower bracts absent. **Fruits** usually 1, positioned toward inflorescence axis, (seldom 2 or 3, then additional ones tending to be smaller and imperfectly developed), homomorphic in more than 1, lance-ovate, 3-sided, symmetric, 1.7–2.1 mm, brown, bases truncate, margins rounded, tips narrowly acute to slightly acuminate, surfaces smooth, shiny, abaxial surfaces low convex, often slightly concave basally, spinal ridges absent, adaxial surfaces 2-planed-convex, occasionally swollen proximally, narrowed distally and appearing somewhat vertically ridged, mid-line regions straight or angled outwardly above nutlet middle; attachment scars centered, edges not raised, abutted to slightly gapped especially at base, occasionally overlapping, margins short bifid-forked at extreme base, areoles absent or minutely triangular.

Sandy, gravelly, or clay soils, juniper or pinyon pine-juniper communities. Flowering Mar–Jun. 800–1600 m. BR, Owy. CA, ID, NV, WA; east to WY, southeast to NM. Native.

Cryptantha gracilis is morphologically distinctive in having long, soft-hirsute calyx hairs, a relatively small corolla, and a usually single, smooth, strongly three-sided nutlet. It appears to be a close relative of *C. milobakeri* (see discussion).

Cryptantha grandiflora Rydberg (p. 400)

Clearwater cryptantha, large-flowered cryptantha

Cryptantha intermedia var. *grandiflora* (Rydb.) Cronquist

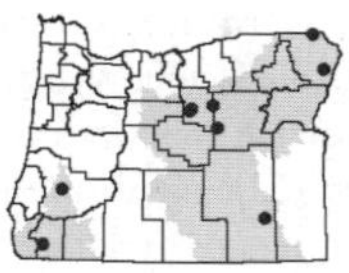

Plants 5–35 cm. **Stems** erect to ascending, branching mostly distally, occasionally from base, strigose and spreading-hirsute. **Leaves** basal rosette usually present, well developed, withering at anthesis, smaller than proximal cauline leaves, surfaces more prominently pustulose, linear to oblong, 15–50(70) × 1–5 mm, midveins impressed, tips obtuse to rounded, surfaces abaxially inclined coarse-hirsute on thickened midveins, appressed- to ascending-hirsute on adjacent regions, inclined-hirsute on margins, pustulose, adaxially ascending-hirsute, pustulose or not. **Inflorescences** cymules usually in 3s, occasionally solitary or paired, 2–15(20) cm, not evidently 2-rowed, proximal-most flowers not touching; bracts rarely present, 1 at cymule base; pedicels 0–0.5(1) mm. **Flowers** scented, inclined to ascending; calyces tardily deciduous at maturity, asymmetric, narrowly ovoid to urceolate, bases truncate or narrowly rounded, slightly constricted or not distally, 2–3 mm at anthesis, 3.5–5.5 mm in fruit, lobes separate to base, lanceolate to linear, loosely connivent, midribs thickened with coarse, stiff, spreading hispid hairs especially proximally, longest hairs 2 mm, tips erect to slightly spreading, abaxial surfaces with 2 hair types, fine, strigose to finely hirsute hairs and coarser hispid hairs, adaxial surfaces appressed sparsely to densely hirsute near tip; corollas rotate, tubes 2–3 mm, limbs 4–8 mm in diameter; fornices yellow; gynobases extending to 50% and styles to 75% length of mature nutlets; flower bracts absent. **Fruits** 1–2(3), homomorphic, lance-ovate to narrowly ovate, symmetric, 1.8–2.8 mm, gray to brown-mottled, bases truncate, not swollen, margins rounded to sharp-angled basally, tips acute, surfaces smooth, shiny, spinal ridges absent, abaxial surfaces usually flat to occasionally low 2-planed convex; attachment scars centered, edges not raised, usually abutted entire length, widely bifid-forked, but not gapped, at base.

Rocky, clay, or volcanic-derived soils, flats or slopes of grasslands, shrub communities. Flowering May–Jul. 400–1600 m. BR, BW, Lava, Sisk. ID, NV, WA. Native.

Cryptantha grandiflora is often treated as a variety of *C. intermedia* and has the three-merous cymules typical of that taxon. It is treated here as a separate species because of its consistently smooth nutlets, only one or two per fruit, and relatively limited distribution in the Pacific Northwest.

Cryptantha hispidula Greene ex Brand (p. 400)

Napa cryptantha

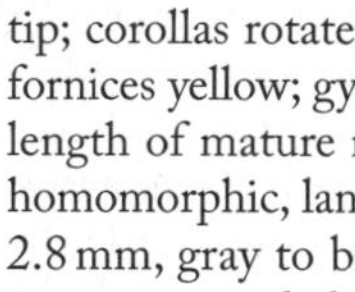

Plants 10–40(50) cm. **Stems** usually erect, slender, usually many-branched throughout, branches ascending, sparsely strigose and short, spreading-hirsute. **Leaves** linear to narrowly oblanceolate, 5–20 × 1–2 mm, midribs carinate abaxially, tips rounded to obtuse, surfaces sparsely appressed to spreading-hirsute, conspicuously pustulose. **Inflorescences** cymules paired or in 3s, 5–7(10) cm, much elongated in fruit, proximal-most flowers not touching; bracts usually absent; pedicels not elongating in fruit, < 0.5 mm. **Flowers** ascending to appressed; calyces asymmetric, narrowly urceolate, 1.5–2 mm at anthesis, (2.5)3–4.5 mm in fruit, lobes separate to base, linear to narrowly lanceolate, connivent distally, midribs of outer 3 lobes thickened, with many proximal spreading, stiff, hispid hairs, longest hairs 1 mm, tips reclined distally, abaxial surfaces with 2 hair types, fine, strigose to finely hirsute hairs and coarser hispid hairs, adaxial surfaces glabrous basally, densely appressed hirsute distally; corollas rotate, tubes 1–1.5 mm, limbs (2.5)3–6 mm in diameter; fornices yellow; gynobases extending to 50% and styles to almost 100% length of mature nutlets; flower bracts absent. **Fruits** 1(2), positioned away from inflorescence axis, homomorphic, narrowly lance-ovate, symmetric, 1.7–2.2 mm, usually mottled gray-brown, bases truncate, swollen, margins rounded, tips abruptly acuminate, surfaces smooth, shiny, abaxial surfaces convex, spinal ridges absent, adaxial surfaces low convex to convex 2-planed; attachment scars centered, edges not raised, abutted or occasionally slightly gapped, short bifid-forked at base.

Gravelly, serpentine-derived soils. Flowering Mar–Jun. 600–800 m. Sisk. CA. Native.

Cryptantha hispidula is a serpentine endemic. It differs morphologically from other species in having short calyx hairs, relatively large corollas, and generally one smooth, lance-ovate, basally swollen nutlet.

Cryptantha intermedia (A. Gray) Greene

Common cryptantha, Clearwater cat's eye, nievitas

Plants 10–60+ cm. **Stems** erect to ascending, branching mostly distally, occasionally from base, spreading-hirsute or strigose and hirsute. **Leaves** well-developed basal rosette usually present, withering at anthesis, smaller than proximal cauline leaves, surfaces more prominently pustulose, linear to lanceolate to oblong or oblanceolate, 15–50(70) × 1–5 mm, midveins impressed, tips acute to obtuse to rounded, surfaces abaxially inclined coarse-hirsute on thickened midveins, appressed- to ascending-hirsute on adjacent regions, inclined hirsute on margins, pustulose, adaxially ascending-hirsute, pustulose or not. **Inflorescences** cymules usually in 3s or paired, occasionally in 4s or 5s, 2–15(20) cm, not evidently 2-rowed, proximal-most flowers not touching; bracts absent or rarely present at cymule base; pedicels 0–1.5 mm in fruit. **Flowers** scented, appressed to spreading, deciduous to persistent at maturity; calyces somewhat asymmetric, usually cylindric to ovoid, straight, bases truncate or narrowly rounded, 2–4 mm at anthesis, 3.5–6 mm in fruit, lobes separate to base, lanceolate to linear, margins appressed-hirsute proximally, midribs thickened with coarse spreading hispid hairs, especially proximally, hairs occasionally spine-tipped, tips erect to slightly spreading, abaxial surfaces sparsely to densely ascending-hirsute, adaxial surfaces sparsely to densely appressed hirsute near tip; corollas rotate, tubes 2–3 mm, limbs (3)4–11 mm in diameter; fornices pale yellow to yellow; gynobases extending to 50–75% and styles to 75% to slightly > 100% length of mature nutlets; flower bracts absent. **Fruits** 1–4, homomorphic, lance-ovate to ovate, straight, 1.4–2.8 mm, grayish to brownish, bases rounded to truncate, margins rounded to basally sharp-angled, tips acute to acuminate, surfaces papillate (rarely lacking) and tuberculate, abaxial surfaces convex, spinal ridges absent, adaxial surfaces usually flat to occasionally low, 2-planed convex, attachment scars centered, edges partially raised or not, abutted entire length or variously gapped, bifid-forked but not gapped, or flare-gapped at base forming areole.

Western US. 3 varieties; 2 varieties treated in *Flora*.

Key to *Cryptantha intermedia* varieties:

1. Nutlets usually 1–2, surfaces variably papillate and sparsely low-tuberculate . . . . **C. i.** var. ***hendersonii***

1' Nutlets usually 3–4, surfaces consistently papillate and densely coarse-tuberculate to muricate . . . . **C. i.** var. ***intermedia***

Cryptantha intermedia (A. Gray) Greene var. *hendersonii* (A. Nelson) Jepson & Hoover

Palouse cryptantha

Cryptantha hendersonii (A. Nelson) Piper ex J.C. Nelson

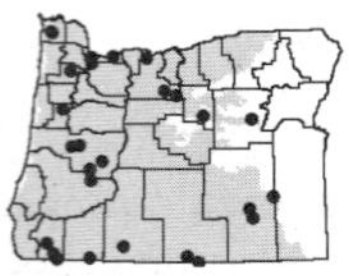

Fruits 1–2(4), solitary nutlet, if present, positioned toward inflorescence axis, ovate to lance-ovate, 2.2–2.5 mm, brown to dark grayish, bases truncate to rounded, tips acute, surfaces dull or shiny, low-rounded tuberculate, usually densely papillate, papillae rarely absent, adaxial surfaces attachment scar edges not raised, usually abutted entire length, widely bifid-forked, but usually not gapped at base.

Gravelly, basalt, clay, or serpentine soil, chaparral, grasslands, slopes, open woodlands. Flowering May–Jul. 0–2500 m. BR, Casc, Col, CR, ECas, Lava, Sisk, WV. CA, ID, WA. Native.

Variety *hendersonii* is interpreted as the northern phase of *Cryptantha intermedia*, with generally fewer nutlets per flower (typically one or two), surfaces usually densely papillate and scattered low-tuberculate, and the attachment scar usually abutted the entire length. The geographic boundary between *C. i.* var. *hendersonii* and var. *intermedia* is ill-defined. The former might warrant species recognition.

Cryptantha intermedia (A. Gray) Greene var. *intermedia* (p. 399, 400)

Common cryptantha, Clearwater cat's eye, nievitas

Cryptantha fragilis M. Peck

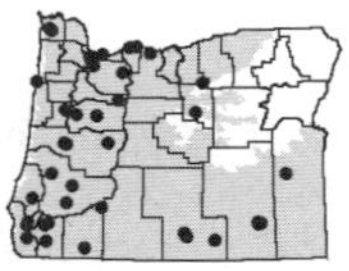

Fruits usually 3–4, lance-ovate to ovate, 1.4–2.1(2.2) mm, grayish or brownish, often mottled, bases rounded to narrowly truncate, tips acute to attenuate, surfaces usually densely papillate and coarse tuberculate or muricate, adaxial surface attachment scar edges raised near tip, usually narrowly gapped near tip or abutted entire length, flare-gapped at base forming triangular areole.

Rocky or clay soil, chaparral, coastal sage scrub, desert scrub, grasslands, slopes, woodland openings. Flowering May–Jul. 0–1600 m. BR, Casc, Col, CR, ECas, Lava, Owy, Sisk, WV. CA, ID, NV, WA; southeast to AZ. Native.

Cryptantha intermedia var. *intermedia* is one of the most common and most collected *Cryptantha* taxa, intergrading with other species in parts of its range. This complex needs more study.

Cryptantha leiocarpa (Fischer & C.A. Meyer) Greene (p. 400)

Beach cryptantha, seaside cryptantha

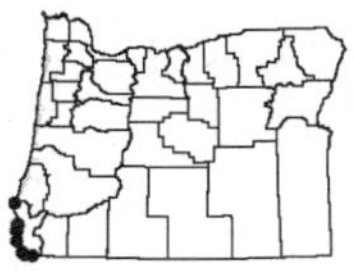

Plants 5–20(40) cm. **Stems** laxly branched, often mat-forming and without clear central axis, many-branched from base, branches generally decumbent or prostrate, strigose, hairs appressed to ascending. **Leaves** crowded at base, linear to spatulate-oblanceolate, 10–35 × 2–5 mm, tips rounded, surfaces densely appressed- to ascending-hirsute, hairs usually pustulose. **Inflorescences** cymules solitary or paired, rarely in 3s, terminal and axillary; terminal densely flowered, 3–6(8) cm, short-elongated in fruit; axillary 1–3-flowered, not elongated in fruit with proximal-most flowers usually touching; bracts present at base of cymules; pedicels not elongated in fruit, < 0.5 mm. **Flowers** unscented, ascending to inclined; calyces tardily deciduous at maturity, usually symmetric, ovoid to narrowly urceolate, 1.5–2 mm at anthesis, 2.5–4 mm in fruit, lobes separate to base, linear-lanceolate, basally connate, margins sparsely finely hirsute, midribs thickened, pustulose-hispid, tips erect to slightly spreading, abaxial surfaces appressed- to ascending-hirsute, adaxial surfaces glabrous basally, appressed short-hirsute at extreme tip; corollas funnelform, tubes 1 mm long, limbs 1–2.5(3.5) mm in diameter; fornices pale yellow; gynobases extending to 70% and styles to 100% or more length of mature nutlets; flower bracts usually present. **Fruits** (3)4, usually homomorphic, ovate, symmetric, 1.6–2 mm, mottled gray to brown, bases usually rounded, margins rounded, tips acute, surfaces smooth, shiny, abaxial surfaces convex, spinal ridges absent, adaxial surfaces flat to low convex; attachment scars centered, edges not raised, abutted entire length, short bifid-forked gapped at base, areoles absent.

Sandy soil, coastal dunes, beaches. Flowering May–Jul. 0–50 m. Est. CA. Native.

Cryptantha leiocarpa is distinctive in its prostrate to decumbent stem habit, bracteate flowers, and fruits with four smooth, ovate-acute nutlets.

Cryptantha intermedia var. *intermedia*

CRYPTANTHA

Cryptantha milobakeri I.M. Johnston (p. 400)

Milo Baker's cryptantha

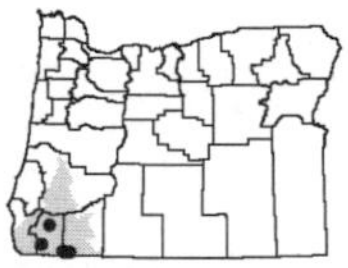

Plants 10–50 cm. **Stems** erect to ascending, usually branching throughout, strigose and ascending-hirsute. **Leaves** proximal-most usually opposite, narrowly oblanceolate to oblong, 5–25(30) × (1)2–4(5) mm, midribs thickened abaxially, sunken adaxially, tips rounded to obtuse, surfaces appressed- to ascending-hirsute, pustulose. **Inflorescences** cymules paired or occasionally in 3s, 5–10(15) cm, usually short elongated in fruit, proximal-most flowers overlapping or not; bracts absent or occasionally 1 near base; pedicels not elongating in fruit, < 0.5 mm. **Flowers** spreading; calyces promptly deciduous at maturity, asymmetric, urceolate to ovoid with narrowly acute-obconic base, 1.5–2 mm at anthesis, 3–4 mm in fruit, lobes basally connate, linear to narrowly lanceolate, basally connate, midribs not appreciably thickened, hairs soft, with longer coarse but not hispid hairs, tips erect to slightly spreading, margins and abaxial surfaces with 1 hair type, usually tufted long soft-hirsute, adaxial surfaces glabrous basally, appressed short-hirsute distal end; corollas rotate, tubes 1–1.5 mm, limbs 2–6 mm in diameter; fornices white or yellow; gynobases extending to 50% and styles to 75% length of mature nutlet; bracts absent. **Fruits** 1(2+), positioned away from inflorescence axis, homomorphic, lance-ovate, generally 3-sided, narrowed distally, symmetric, 1.5–2 mm, dark brown, bases widely truncate, margins rounded, tips narrowly acute to acuminate (beaked), surfaces smooth, shiny, abaxial surfaces low convex to basally transversely flat, spinal ridges absent, adaxial surfaces obtusely 2-planed-convex and somewhat swollen proximally, narrowed distally and appearing somewhat vertically ridged, mid-line regions angled outwardly beyond nutlet middle; attachment scars centered, edges raised, abutted entire length or slightly gapped proximally, occasionally overlapping, bifid-forked at base, areoles absent.

Gravelly or serpentine-derived soil, grasslands, chaparral, forest openings. Flowering May–Jul. 400–1400 m. Sisk. CA. Native.

Cryptantha milobakeri is similar to *C. gracilis* in calyx fusion, shape, vestiture, and in general nutlet morphology, both with nutlets strongly three-sided and adaxially two-planed. The two species differ in nutlet position, calyx size, corolla limb diameter, stem vestiture, and distribution.

Cryptantha pterocarya (Torrey) Greene var. ***pterocarya*** (p. 400)

Wingnut cryptantha, winged cryptantha

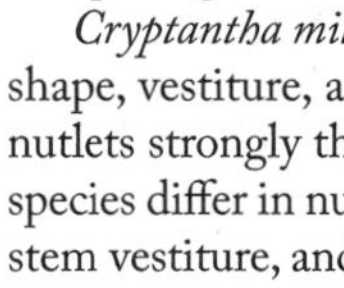

Plants 5–50 cm, green to grayish green. **Stems** erect, slender to stout, unbranched to branched throughout, branches few to occasionally many, strigose and ascending to spreading short hirsute. **Leaves** basal rosette usually not well developed, linear, oblong, or lanceolate, 5–40 ×

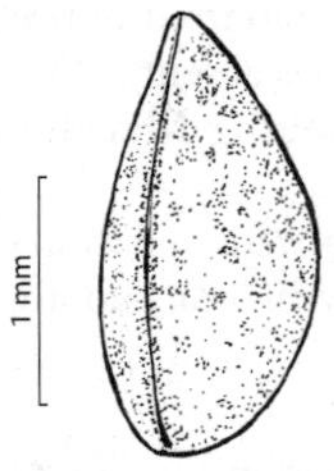

Cryptantha affinis

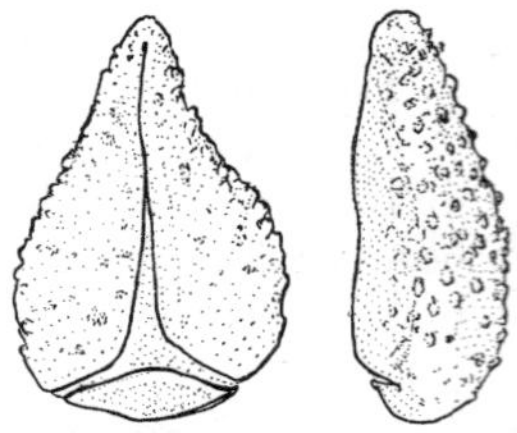

Cryptantha ambigua

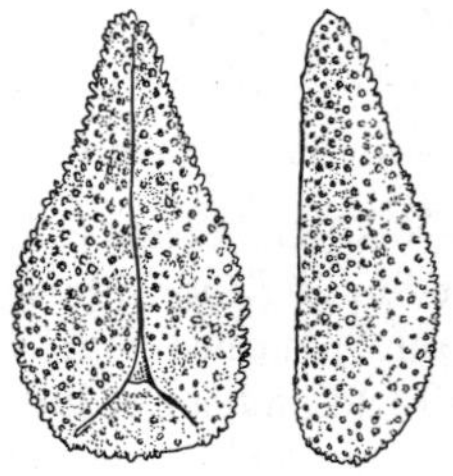

Cryptantha echinella

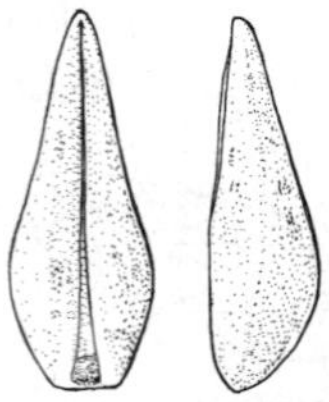

Cryptantha fendleri

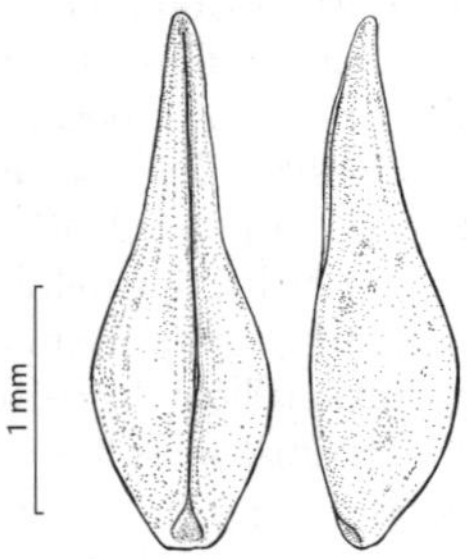

Cryptantha flaccida

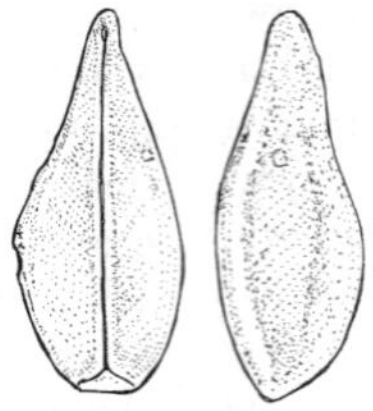

Cryptantha gracilis

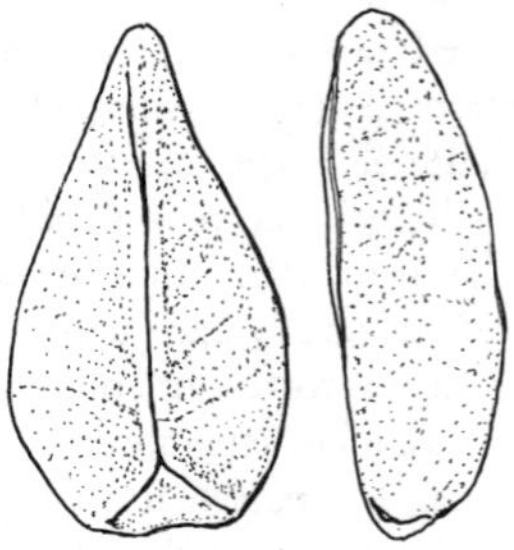

Cryptantha grandiflora

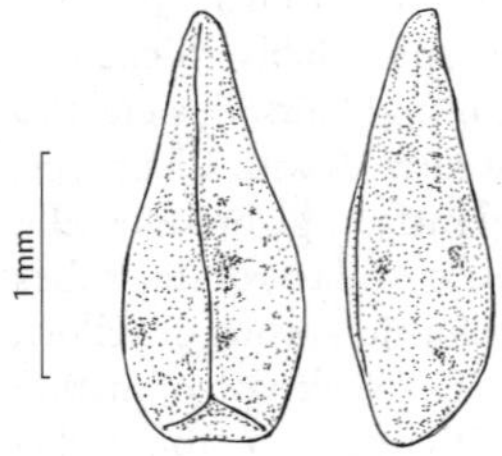

Cryptantha hispidula

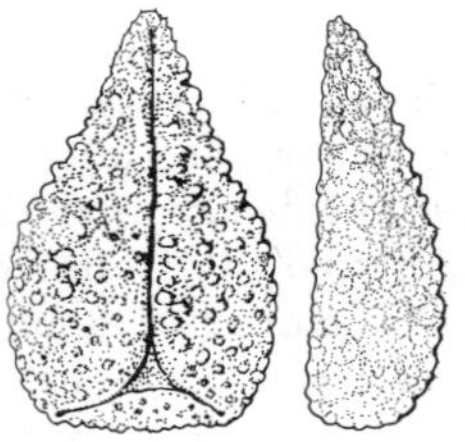

Cryptantha intermedia var. *intermedia*

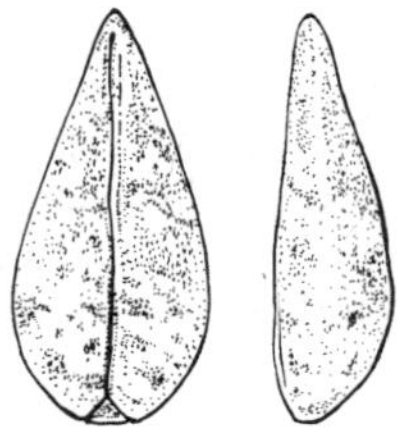

Cryptantha leiocarpa

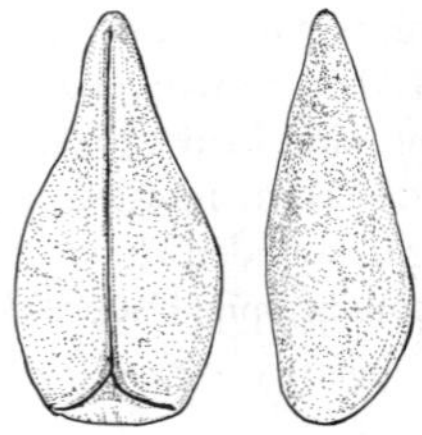

Cryptantha milobakeri

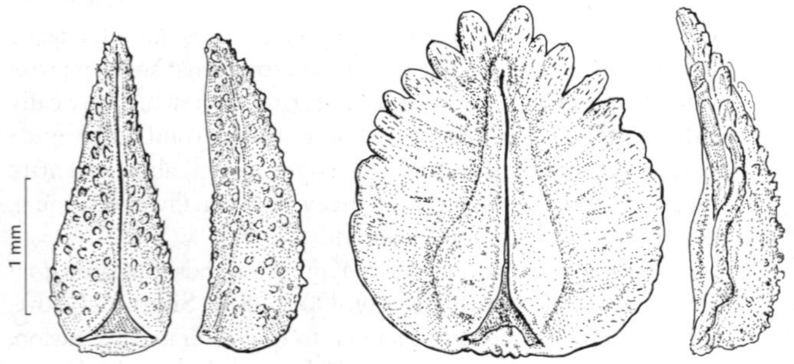

Cryptantha pterocarya var. *pterocarya*

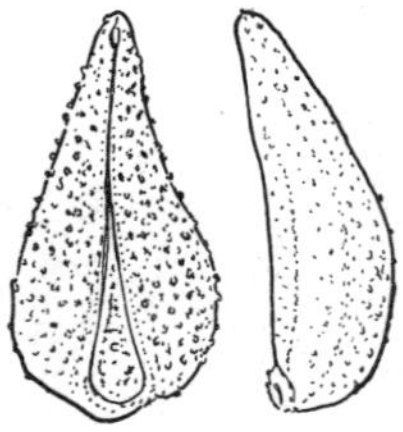

Cryptantha recurvata

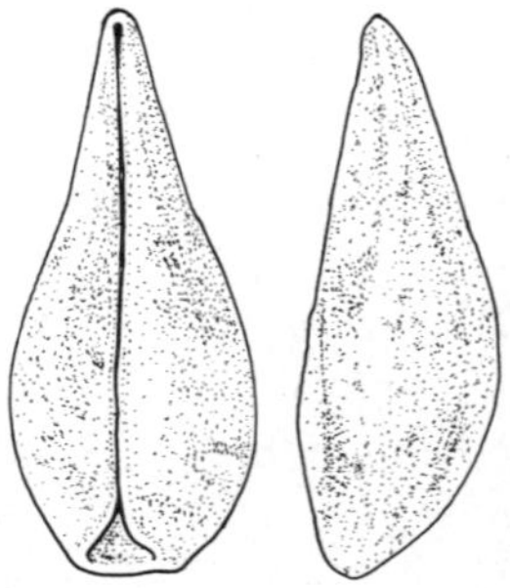

Cryptantha rostellata

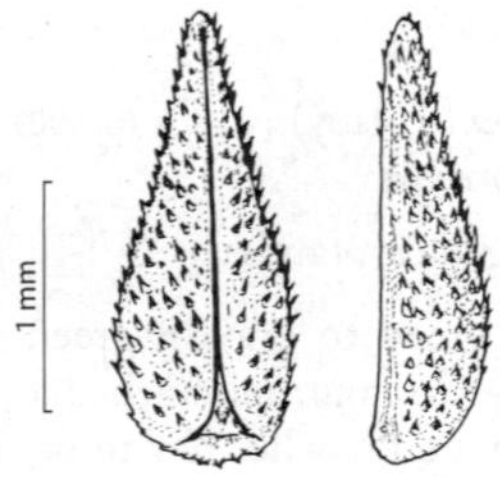

Cryptantha scoparia

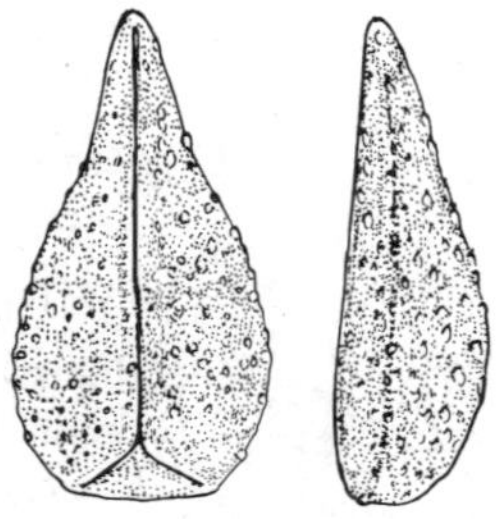

Cryptantha simulans

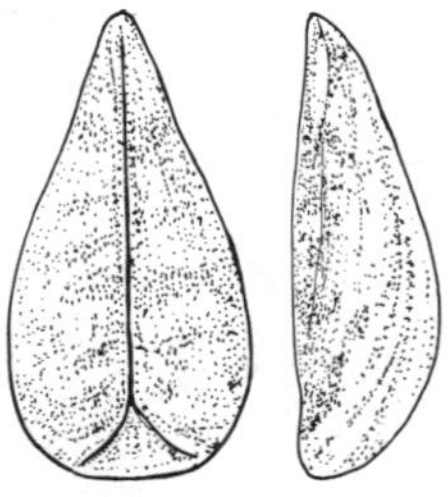

Cryptantha torreyana var. *torreyana*

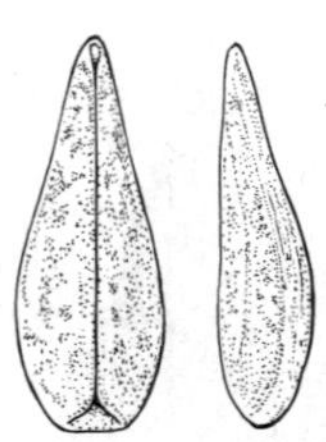

Cryptantha watsonii

CRYPTANTHA

1–3 mm, margins hirsute, tips rounded to obtuse, surfaces appressed- to ascending-hirsute, pustulose, more coarsely so abaxially. **Inflorescences** cymules paired or in 3s, occasionally solitary, glomerate to loosely flowered in fruit, 2–6(12) cm, proximal-most flowers touching or not; bracts absent; pedicels usually not elongating in fruit, 0–0.5(0.8) mm. **Flowers** spreading to occasionally ascending, usually deciduous at maturity; calyces symmetric, usually widely ovoid, 2–3 mm at anthesis, 3–4.5 mm in fruit, lobes separate to base, ovate, margins sparsely to densely ascending-hirsute, midribs thickened, inclined-hispid, pustulose, tips connivent, abaxial surfaces densely appressed-hirsute, adaxial surfaces glabrous; corollas rotate to rarely funnelform, tubes 1–2 mm, limbs 1–2.5(3) mm in diameter; fornices minute, pale yellow or white, or absent; gynobases extending to 70% length of nutlets, sessile, not basally elongated into stipe; styles to 100% length of mature nutlets; flower bracts absent. **Fruits** (3)4, heteromorphic, 1.8–2.8 mm, gray, brown, or mottled gray and brown, with odd nutlet adaxial and more firmly attached to gynobase, margins sharp-angled and lacking wing, (2)3 consimilar nutlet margins, wide-winged, > 0.6 mm wide on 1 side, or uncommonly homomorphic with all nutlet margins wide-winged, surfaces usually densely papillate and tuberculate, adaxial surfaces convexly 2-planed, sometimes glabrous or near glabrous, bases truncate, spinal ridges absent, attachment scars centered, edges not raised, abutted entire length or slightly triangular-gapped at base; wings when present, entire or variously scalloped or dissected, not extending along base of nutlet body. $2n$=24.

Sandy, gravelly, volcanic, or clay soil, desert scrub, washes, sagebrush, woodland openings. Flowering Apr–Jun. 200–1700 m. BR, Col, Lava, Owy. CA, ID, NV, WA; southeast to NM. Native.

Cryptantha pterocarya is readily recognized by its widely winged nutlets. In a recent treatment, the species has been classified into three varieties. Variety *pterocarya*, the only one found in Oregon, is the most widespread and has both heteromorphic (forma *pterocarya*) and homomorphic (forma *pseudocycloptera* M.E. Mabry & M.G. Simpson) forms, but the latter apparently does not occur in Oregon (Mabry & Simpson 2016).

Cryptantha recurvata Coville (p. 400)

Bent-nut cryptantha, curvenut cryptantha

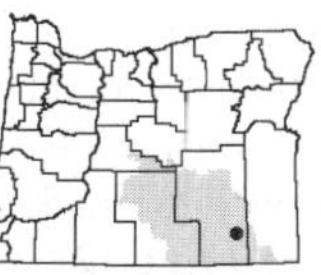

Plants 5–40 cm. **Stems** erect to ascending, slender, unbranched to branched from base or throughout, branches ascending, strigose throughout, proximal stem sometimes spreading-hirsute. **Leaves** linear to narrowly oblanceolate, 10–30 × 1–5 mm, tips rounded to obtuse, surfaces abaxially appressed- to ascending-hirsute, pustule-based, midveins thickened, adaxially sparsely appressed to ascending-hirsute. **Inflorescences** cymules usually paired or solitary, (0.5)2–10 cm, flowers not touching; bracts absent or occasionally 1–2 at base of cymule, leafy; pedicels 0–0.5 mm in fruit. **Flowers** spreading to reclined, tardily deciduous at maturity; calyces asymmetric, cylindric, recurved-arched, 1.5–2 mm at anthesis, 2.5–3.5 mm in fruit, lobes separate to base, narrowly linear, connivent distally, margins appressed to ascending, finely hirsute, tips erect to slightly recurved, abaxial surfaces appressed- to ascending-hirsute, midribs thickened, spreading-hispid, lobes near inflorescence axis with enlarged, yellowish, hispid hairs, adaxial surfaces glabrous basally, appressed short-hirsute apically; corollas rotate, tubes 1 mm, limbs 0.5–1(1.5) mm in diameter; fornices absent; ovules 2, gynobases extending to 50% and styles to 75% length of mature nutlets; flower bracts absent. **Fruits** 1(2), single nutlet positioned toward inflorescence axis, lanceolate, 1.6–2 mm, rarely with smaller and imperfectly developed second nutlet, brown, dull, strongly incurved-bent, bases rounded, margins rounded, tips acuminate, surfaces densely low tuberculate, abaxial surfaces convex, spinal ridges absent or obscure, adaxially flat to low convexly 2-planed; attachment scars centered, edges not raised, usually abutted or overlapping near tip, flare gapped at base into small, narrowly triangular areole, slightly bifid-forked at base.

Rocky, gravelly, or sandy soils, desert scrub. Flowering Mar–Jun. 700–1600 m. BR. CA, ID, NV; southeast to NM. Native.

Cryptantha recurvata is easily recognized by a usually solitary, strongly bent nutlet, and strongly recurved bent fruiting calyx. Other distinctive features are the spreading to often reclined flower angle relative to the rachis and the presence of only two ovules (four being the almost universal number in the genus), one fertile, one typically abortive.

Cryptantha rostellata (Greene) Greene (p. 400)

Red-stemmed cryptantha, beaked cryptantha

Cryptantha rostellata (Greene) Greene var. *rostellata*

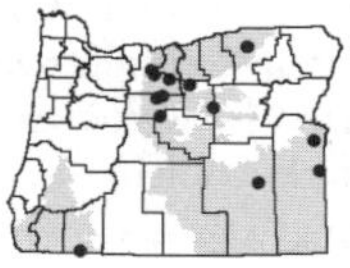

Plants 5–20+ cm. **Stems** erect to ascending, stout, stiff, usually reddish, few-branched distally to occasionally throughout, strigose, canescent. **Leaves** usually several opposite pairs below inflorescence, notably persistent, often conduplicate, linear to narrowly oblanceolate, somewhat thick, 10–40 × 1–2 mm, tips rounded to obtuse, surfaces appressed-strigose, pustulose throughout. **Inflorescences** cymules solitary or paired, stiff, 2–4 cm, somewhat elongated in fruit, proximal-most flowers not touching; pedicels not elongated in fruit, < 0.5 mm. **Flowers** unscented, appressed to ascending, flower bracts absent; calyces somewhat persistent, asymmetric, ovoid to urceolate, 2–3 mm at anthesis, 3–3.5(4) mm in fruit, lobes separate to base, lanceolate, connivent at middle beyond nutlet, margins finely ascending-hirsute, tips erect to recurved or slightly spreading, midribs with stout, hispid, recurved hairs occasionally basally encrusted with exudate, proximal midrib hairs longer, less curved, abaxial surfaces sparsely appressed-hirsute, adaxial surfaces basally glabrous, apically appressed short-hirsute; corollas rotate to funnelform, tubes 1.5 mm, limbs 0.5–2 mm in diameter; fornices minute, pale yellow; styles 50–70% length of mature nutlets; bracts absent. **Fruits** 1, positioned away from inflorescence axis, lance-ovate, symmetric, 2.3–2.8 mm, bases truncate, margins rounded, tips acuminate, surfaces smooth, shiny, abaxial surfaces strongly convex, spinal ridges absent, adaxial surfaces flat; attachment scars centered, edges raised or not, variably gapped proximally, widening at base into conspicuous areole, abutted or overlapping near tip.

Rocky or basalt soil, slopes, shrublands, open meadows, woodlands. Flowering Apr–Jun. 300–1200 m. BR, Col, Lava, Owy, Sisk. CA, ID, WA. Native.

Cryptantha rostellata is very similar to *C. flaccida*, the former differing in having a generally smaller corolla, more open basal attachment scar, and a longer style extension.

Cryptantha scoparia A. Nelson (p. 400)

desert cryptantha, gray cryptantha

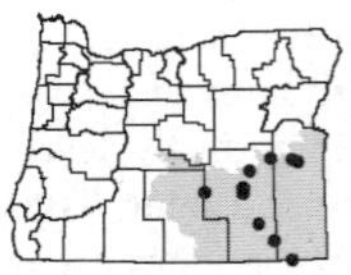

Plants 5–30 cm. **Stems** primary axis often ill-defined, branches erect to ascending, slender, densely appressed strigose and sometimes sparsely spreading-hirsute. **Leaves** linear to lanceolate, 10–40(50) × 1–3 mm, tips rounded to obtuse, surfaces appressed- to ascending-hirsute, coarser hairs pustulose, pustules larger on margins and abaxial surfaces. **Inflorescences** cymules solitary or paired, occasionally crowded, elongating to 5–9 cm, proximal-most flowers usually not touching; bracts absent; pedicels 0.5–1 mm in fruit. **Flowers** ascending to usually appressed, usually persistent at maturity; calyces symmetric, cylindric, bases rounded, constricted below tip, 1.5–2.5 mm at anthesis, 4–6 mm in fruit, lobes separate to base, linear to narrowly lanceolate, one lobe usually longer, margins densely ascending fine-hirsute, midribs thickened, spreading coarse-hirsute proximally, hairs pustule-based, tips erect to slightly recurved, abaxial surfaces ascending white-hirsute, adaxial surfaces glabrous proximally, appressed short-hirsute near tip; corollas rotate to funnelform, tubes 1.5–2 mm, limbs 1–1.5 mm in diameter; fornices absent; gynobases extending to 75% and styles to 100% length of mature nutlets; flower bracts absent. **Fruits** 4, homomorphic, lanceolate to lance-ovate, 1.6–2.1 mm, gray with brown mottling, bases truncate, margins somewhat sharp angled at base, tips narrowly acute to acuminate, surfaces minutely tuberculate-spinulose, spinules ascending, sharp-tipped, abaxial surfaces low convex, spinal ridges absent, adaxial surfaces flattened, attachment scars centered, edges not raised, abutted entire length or narrowly gapped proximally, gradually bifid-forked and flare-gapped at extreme base.

Sandy, clay, or talus substrate, slopes or flats, shrublands, woodlands. Flowering Jun–Jul. 1200–1600 m. BR, Owy. CA, ID, NV, WA; east to WY. Native.

Cryptantha scoparia can be readily identified by its nutlet morphology, the four nutlets per fruit having a distinctive tuberculate-spinulose sculpturing, the spinules ascending and sharp-tipped.

Cryptantha simulans Greene (p. 400)

Pinewoods cryptantha, pine cryptantha

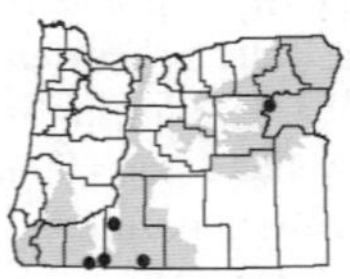

Plants 5–45 cm, pale green. **Stems** slender, branches few from primary stem, usually strigose only, occasionally ascending-hirsute. **Leaves** few, scattered, proximal-most persistent during flowering, linear to narrowly oblanceolate, 10–50(70) × 2–6 mm, margins hirsute, tips acute to rounded, surfaces finely appressed- to ascending-hirsute, more coarsely so abaxially, hairs minutely pustulose. **Inflorescences** cymules usually paired or in 3s, rarely solitary, few-flowered, 2–15(22) cm, proximal-most flowers not touching; bracts absent; pedicels 0.5–1 mm in fruit. **Flowers** appressed to ascending, promptly deciduous at maturity; calyces slightly asymmetric, cylindric to urceolate, bases rounded, constricted distally, 2–3 mm at anthesis, 4–7(8) mm in fruit, lobes separate to base, narrowly lanceolate, apical half linear, margins distally with many short, recurved to reclined hispid hairs, midribs thickened, distally with many short recurved to reclined hispid hairs, tips spreading, abaxial surfaces basally densely appressed hirsute, adaxial surfaces proximally glabrous, distally finely appressed hirsute; corollas funnelform, tubes 1–1.5 mm, limbs 1.5–2 mm in diameter; fornices minute; gynobases extending to 50% and styles to slightly < 100% length of mature nutlets; flower bracts absent. **Fruits** 3–4, homomorphic, ovate, compressed, 1.9–2.2(2.5) mm, grayish or brownish, bases truncate, margins rounded, tips short-acuminate, surfaces papillate and sparsely low-rounded tuberculate, abaxial surfaces low-rounded convex, spinal ridges absent, adaxial surfaces flat, attachment scars centered, edges not raised, abutted, widely bifid-forked but not gapped at base, areoles absent.

Basalt-derived or loess substrates, slopes, banks, open shrublands, open coniferous woodlands. Flowering Jun–Jul. 1300–1800 m. BW, ECas, Sisk. CA, ID, NV, WA. Native.

Cryptantha simulans is often found in forest understories. It is readily recognized by the stem hairs being appressed only, the narrow distal calyx lobes with recurved, hispid hairs, and the flattened, ovate nutlets sparsely covered with low-rounded tubercules.

Cryptantha torreyana (A. Gray) Greene (p. 400) var. ***torreyana***

Torrey's cryptantha

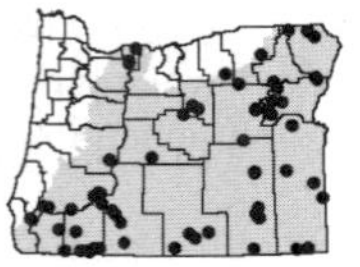

Plants 5–40 cm. **Stems** erect, occasionally lax, unbranched or branched throughout, branches erect to ascending, sparsely to densely strigose and ascending to spreading-hirsute. **Leaves** linear, narrowly oblong, or oblanceolate, 20–50(70) × 3–6 mm, tips obtuse to acute, surfaces appressed- to inclined-hirsute, abaxial midribs carinate, some hairs inconspicuously pustulose. **Inflorescences** cymules usually paired, 4–8(15) cm, usually much elongated in fruit, or often also terminally congested and conspicuously glomerate, proximal-most flowers touching or not; bracts usually absent; pedicels not elongating in fruit, 0.5 mm. **Flowers** unscented, ascending to spreading; calyces tardily deciduous at maturity, slightly asymmetric, ovoid to cylindric or narrowly urceolate, bases widened, usually constricted distally, calyces 2–2.5 mm at anthesis, 4–6.5(8) mm in fruit, lobes separate to base, usually lanceolate to linear-lanceolate, margins strigose, tips spreading, abaxial surfaces with 2 hair types, fine, strigose to finely hirsute hairs and coarser hispid hairs, midribs slightly thickened, hairs inclined, stiff, hispid, adaxial surfaces glabrous basally, densely appressed-hirsute apically; corollas rotate to funnelform, tubes 1 mm, limbs 0.5–1.5 mm in diameter; fornices minute, white, or absent; gynobases extending to 50% and styles to 75% length of mature nutlets, rarely equal; flower bracts absent. **Fruits** (2)3–4, homomorphic, ovate, symmetric, (1.6)1.7–2.2(2.5) mm, grayish or usually mottled gray and brown, bases broadly rounded to truncate, margins rounded, tips acute, surfaces smooth, shiny, abaxial surfaces flattened, especially toward base, low rounded spinal ridge present at extreme tip, adaxial surfaces 2-planed convex; attachment scars centered, edges occasionally raised near tip, abutted entire length, bifid-forked but not gapped at base, areoles absent.

Rocky, gravelly, or sandy/sandy-loam substrate, slopes or flats, meadows, grasslands, shrublands, coniferous woodlands. Flowering May–Aug. 200–2500 m. BR, BW, Casc, ECas, Lava, Owy, Sisk. CA, ID, NV, WA; north to AK, northeast to ND; southeast to CO. Native.

This taxon is one of the most common *Cryptantha* in Oregon. Introgression has been documented between *C. torreyana* var. *torreyana* and *C. ambigua* in the Pacific Northwest, resulting in plants with a mix of habit and nutlet characteristics. *Cryptantha torreyana* var. *pumila*, with relatively small calyces and nutlets, is restricted to the San Francisco Bay area of California.

Cryptantha watsonii (A. Gray) Greene (p. 400)

Watson's cryptantha

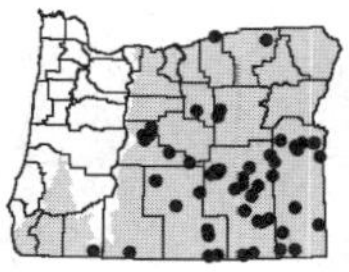

Plants 10–35 cm. **Stems** slender, unbranched to few-branched and erect to ascending, strigose and spreading-hirsute. **Leaves** linear to oblanceolate, 5–40(50) × 1–4(5) mm, midribs ridged abaxially, sunken adaxially, tips rounded to obtuse, surfaces ascending to inclined white-hirsute, pustulose throughout. **Inflorescences** cymules solitary or paired, 1–4(6) cm, flowers dense at anthesis, open in fruit, proximal-most flowers usually not touching; bracts occasionally 1–few at base, linear, lanceolate, or oblanceolate; pedicels not elongating in fruit, < 0.5 mm. **Flowers** ascending; calyces deciduous at maturity, symmetric, narrowly ovoid, 1.5–2 mm at anthesis, 2.5–3.5(4) mm in fruit, lobes separate to base, lanceolate, margins densely ascending finely white-hirsute, midribs slightly thickened, pustulose-hispid, tips usually connivent, abaxial surfaces appressed- to ascending-hirsute, adaxial surfaces basally glabrous, apically appressed short-hirsute; corollas usually funnelform, tubes 1 mm, limbs 0.5–1.5 mm in diameter; fornices minute, pale yellow to white; gynobases extending to 70% and styles to 100% length of nutlets; flower bracts usually absent. **Fruits** (3)4, homomorphic, narrowly lance-ovate, compressed, symmetric, 1.4–2 mm, gray-tan or mottled, bases short-truncate, margins sharp-angled to flattened into narrow rim, especially toward tip, tips narrowly acute, surfaces smooth, shiny, abaxial surfaces flattened, spinal ridges absent, adaxial surfaces convexly 2-planed; attachment scars centered, edges abutted or partially overlapping, usually not raised, short bifid-forked at base, areoles absent to very small triangular.

Sandy, gravelly, or rocky substrates, banks, shrublands, slopes, coniferous woodlands, meadows. Flowering May–Jul. 600–2500 m. BR, BW, Col, ECas, Lava, Owy, Sisk. CA, ID, NV, WA; north to British Columbia, northeast to MT; southeast to CO. Native.

Cryptantha watsonii is distinctive in having a small corolla and four (occasionally three) smooth nutlets with sharp-angled to flattened margins, especially toward the tips.

CYNOGLOSSUM Hound's tongue

Barbara Montoya and James I. Cohen

Herbs erect. **Stems** hispid. **Leaves** basal and cauline, simple. **Inflorescences** scorpioid cymes; bracts 0; pedicellate. **Flowers** bisexual, homostylous; sepals 5, persistent, linear; corollas radially symmetric salverform, sympetalous; petals 5, lobes flared to erect, margins entire; stamens 5, included, attached to petals; styles 1, included, arising from base of ovary lobes, persistent. **Fruits** 4 or fewer by abortion, spreading, lanceolate to ovoid to orbicular, yellow to dark brown, barbed, attached below middle.

Worldwide. ~75 species; 1 species treated in *Flora*.

Two other species native to the United States and Canada previously in *Cynoglossum* have been transferred to other genera (see *Adelinia* and *Andersonglossum*). The cauline leaves of *Adelinia* are long-petiolate rather than sessile, and those of *Andersonglossum* are oblanceolate to spatulate versus elliptic to lanceolate. Consequently, *C. officinale* is the only species of *Cynoglossum* present in the flora. It is a widespread weed, having a wider geographic distribution than any of the native species previously included in *Cynoglossum*.

Cynoglossum officinale Linnaeus (p. 403)

Common hound's tongue, gypsy flower, beggar's lice

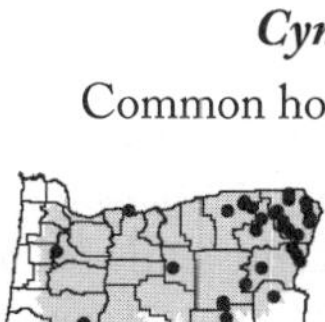

Plants biennial or short-lived perennial, 3–12 dm. **Basal leaves** spatulate, 8–22 × 1.6–4 cm, bases acute, margins entire, tips acute, surfaces abaxially sericeous, adaxially sericeous to strigose. **Cauline leaves** lanceolate to elliptic, 3.4–24 × 0.6–4.5 cm, bases truncate to acute, margins entire, tips acute, surfaces abaxially sericeous, adaxially sericeous to sparse sericeous. **Inflorescences** pedicels 0–20 mm. **Flowers** corollas red to purple, glabrous, limbs 5–9 mm in diameter; styles 1.5–2 mm. **Fruits** 4–7 × 4–5 mm. **2*n*=24**.

Disturbed areas. Flowering May–Aug. 50–1900 m. BW, Casc, Col, ECas, Lava, WV. CA, ID, NV, WA; cosmopolitan. Exotic.

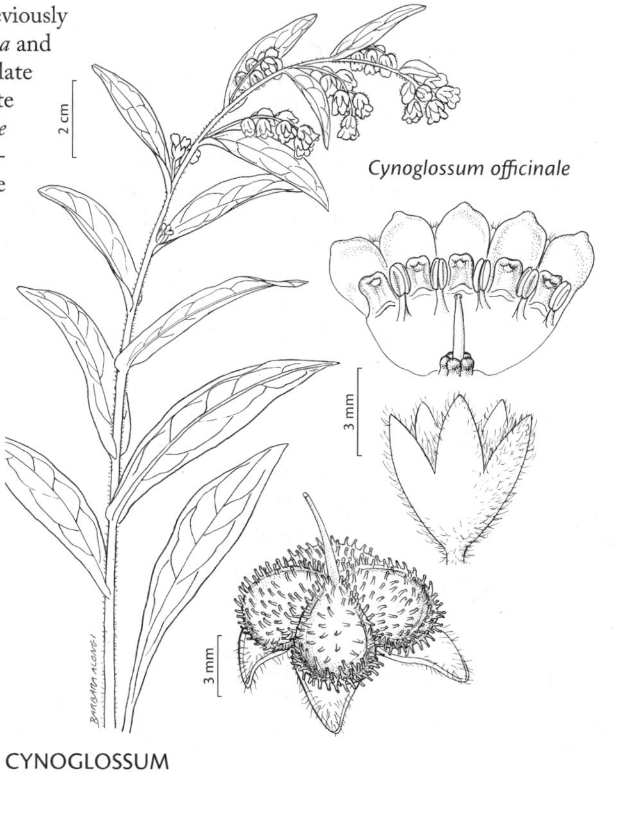

CYNOGLOSSUM

ECHIUM Echium

Christian Feuillet

Herbs annual or biennial; vegetative parts hairy or with bristly trichomes. **Leaves** basal and cauline, margins entire. **Inflorescences** helicoid cymes, branched, ascending, large; bracts present; pedicels short. **Flowers** corollas ± bilabiate, usually pink, purple, or blue, rarely white at anthesis; 2 stamens exserted, 3 included; styles > stamens, stigmas bifid. **Fruits** 4, ± erect; attachment scars basal.

Europe, Macaronesia, North Africa, North America. ~60 species; 1 species treated in *Flora*.

Echium candicans has been reported from Curry County (2015) but is not yet naturalized. Likewise, un-naturalized *E. vulgare* has been reported from Linn (1954) and Umatilla (2002) counties.

Echium plantagineum Linnaeus (p. 404)

Paterson's curse, purple viper's bugloss, salvation Jane

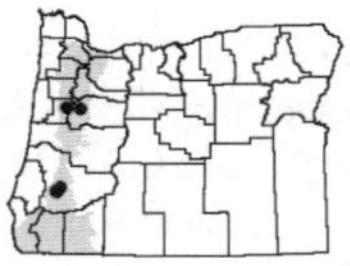

Plants to 8 dm. **Leaves** elliptic to oblong, 2–15 cm, surfaces hairy, bristly at base; sessile. **Inflorescences** with well-developed branches mostly > 10 cm. **Flowers** sepals half length of corolla tube, green; corollas pinkish in bud, becoming blue or purple-blue or rarely white, sub-glabrous outside except on veins and margins, tubes 15–25 mm long, limbs up to 30 mm in diameter; anthers small, pollen blue. **Fruits** ovate, 3-angled, black, rough. **2*n***=16.

Disturbed areas. Flowering May–Jul. 50–400 m. Sisk, WV. CA; scattered in northeastern US and Canada; Africa, Asia, Australia, Europe. Exotic.

Echium plantagineum is a noxious weed. A high concentration of pyrrolizidine alkaloids makes this species poisonous to livestock, especially non-ruminants.

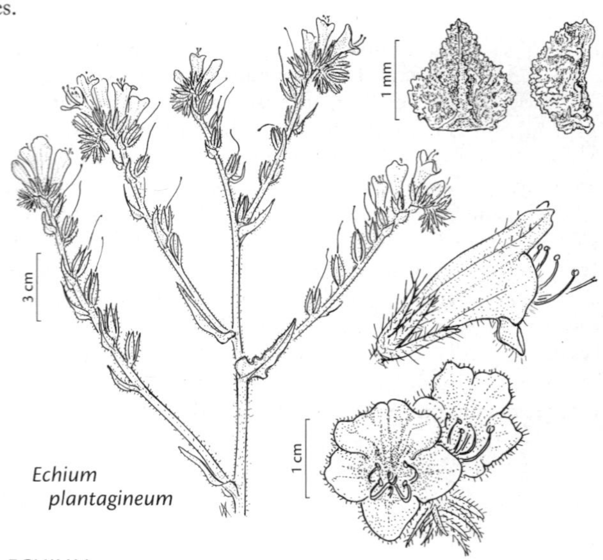

ECHIUM

EREMOCARYA Redroot

Michael G. Simpson and Ronald B. Kelley

Herbs annual, often cushion-like, bases usually red-purple; taprooted, roots red to purple when dry. **Stems** erect to ascending, branched from base and throughout; branches slender, antrorse-strigillose. **Leaves** alternate, linear, oblong, or oblanceolate, margins entire, hairs pustulose; sessile. **Inflorescences** terminal and axillary scorpioid cymules, axes and flowers bracteate; bracts leaf-like; pedicels up to 1.5 mm in fruit. **Flowers** unscented, radially symmetric, homostylous, inclined to appressed; calyces symmetric, sepals 5, distinct, linear-oblong, abaxially appressed to inclined-hirsute; corollas deciduous, radially symmetric, white, lobes 5, rotate to salverform; fornices absent or present and prominent, yellow; ellipsoid bodies in upper tube adjacent to fornices present or absent; stamens included; gynobases narrow, terete, > mature nutlet length and resembling style, styles not evident; flower bracts usually exceeding calyx. **Fruits** (3)4, homomorphic or heteromorphic with 1 papillate and others smooth, erect, lance-ovate, bases and margins rounded, tips beaked or not, both surfaces smooth or finely tuberculate, spinal ridge absent; attachment scars a lateral ventral groove, edges abutted or slightly gapped, minutely forked and slightly gapped at base.

Northwestern Mexico, western and southwestern United States. 2 species; 1 species treated in *Flora*.

Molecular phylogenetic studies (Simpson et al. 2017; Hasenstab-Lehman & Simpson 2012) have split *Cryptantha* in the broad sense into five genera: *Cryptantha* (*s.s.*, as treated here), *Eremocarya*, *Greeneocharis*, *Johnstonella*, and *Oreocarya*. *Eremocarya* is distinguished by its elongate gynobase that protrudes beyond the nutlets and is terminated by a sessile stigma, consistently red to red-purple staining roots and stem bases, generally short and often cushion-like stature, flowers consistently bracteate, and calyces with sepals distinct. *Eremocarya* nutlets are exceptionally variable in surface sculpturing within a given taxon or even specimen.

Eremocarya micrantha (Torrey) Greene var. ***micrantha*** (p. 405)

Desert redroot

Cryptantha micrantha (Torr.) I.M. Johnst.

Plants 2.5–10 cm, often wider than tall. **Flowers** corolla limbs 0.5–1.5 mm in diameter, center white or inconspicuously yellow; fornices, fornix bodies absent. **Fruits** erect, 1–1.1 mm, tips acute. **2*n***=24.

Flats, washes, sandy, silty, or fine gravelly soil, desert shrub associations. Flowering May–Jul. 1100–1600 m. BR, Owy. CA, NV; southeast to TX, south to Mexico. Native.

Eremocarya micrantha was shown to be distinct in a number of features from the other species in the genus, *E. lepida* (Simpson et al. 2014). A second, larger-flowered variety of *E. micrantha*, variety *pseudolepida*, occurs only in north-central Baja California, Mexico, and possibly Arizona (Simpson et al. 2016).

ERITRICHIUM Alpine forget-me-not

Christian Feuillet

Herbs perennial, dwarfed, cespitose. **Leaves** hairy. **Inflorescences** terminal, condensed cymes, often > stems. **Flowers** radially symmetric; calyx lobes free or barely fused at base; corollas blue; fornices developed; filaments attached well down tube, anthers included. **Fruits** 1–4, erect, somewhat rough; attachment scars basal.

Cold temperate and arctic areas in Northern Hemisphere. 50–76 species; 1 species treated in *Flora.*

A collection in the herbarium of Whitman College, identified as *E. howardii*, is most likely misidentified; *Eritrichium howardii* occurs east of the Continental Divide and has densely strigose leaves that lack tufts or fringes at the tips.

Eritrichium nanum (Villars) Schrader ex Gaudin var. ***elongatum*** (Rydberg) Cronquist (p. 405)

Arctic alpine forget-me-not

Eritrichium elongatum (Rydb.) W. Wight

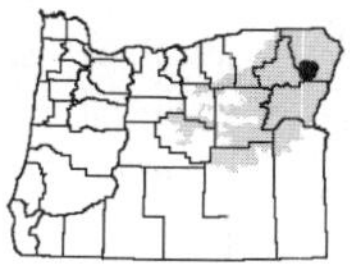

Plants forming small cushions, up to 10 cm. **Stems** short, villous or loosely strigose. **Leaves** ovate to narrowly oblong, 5–10 × 1–2 mm, long-hairy with a tuft or a fringe at tips. **Flowers** corollas blue and usually yellow or whitish in middle, limbs 4–8 mm in diameter, lobes (4)5. **Fruits** glabrous; flanges entire. $2n$=44, 46.

Rocky areas. Flowering Jul–Aug. 2300–3000 m. BW. ID, WA; northeast to MT, east to WY, southeast to NM. Native.

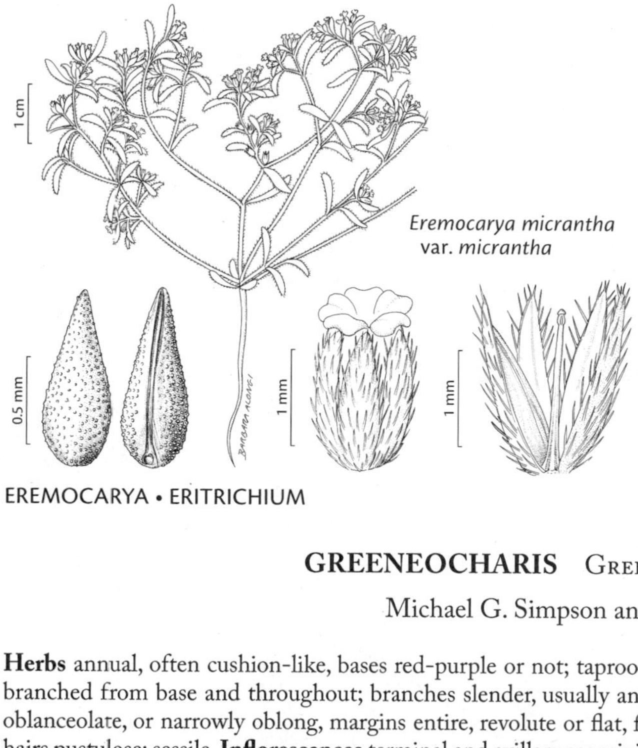

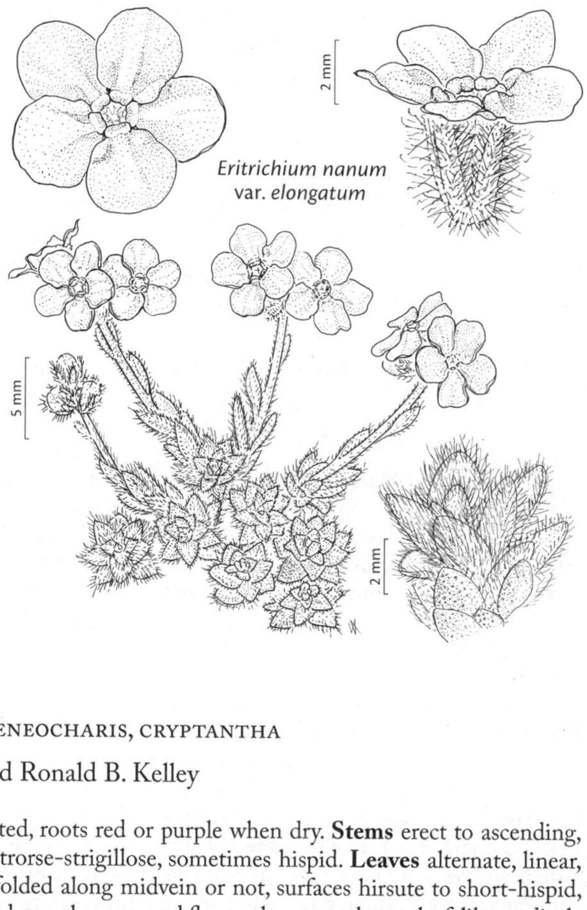

EREMOCARYA • ERITRICHIUM

GREENEOCHARIS Greeneocharis, cryptantha

Michael G. Simpson and Ronald B. Kelley

Herbs annual, often cushion-like, bases red-purple or not; taprooted, roots red or purple when dry. **Stems** erect to ascending, branched from base and throughout; branches slender, usually antrorse-strigillose, sometimes hispid. **Leaves** alternate, linear, oblanceolate, or narrowly oblong, margins entire, revolute or flat, folded along midvein or not, surfaces hirsute to short-hispid, hairs pustulose; sessile. **Inflorescences** terminal and axillary scorpioid cymules, axes and flowers bracteate; bracts leaf-like; pedicels 0.5 mm or less in fruit. **Flowers** unscented, radially symmetric, homostylous, inclined to appressed; sepals fused basally, symmetric, circumscissile proximally in fruit, lobes and distal tube falling as one unit, basal part persistent, scarious, and cup-like in fruit, abaxial surfaces strigose and spreading-hirsute; corollas deciduous, radially symmetric, white, tubes exserted from calyces, salverform, 2–3 mm, lobes flat; fornices present; stamens included; gynobases extending to 70% and styles to 90–100% of mature nutlets; flower bracts equal to or exceeding calyx. **Fruits** (2)3–4, usually homomorphic, rarely heteromorphic (then odd-sized nutlet away from inflorescence axis), erect, triangular-ovate to lance-ovate, bases truncate, margins sharp-angled or lineate-rimmed, tips beaked or not, both surfaces smooth or coarsely papillate, abaxially convex below, compressed apically, spinal ridge absent, adaxially convex or convexly 2-planed.

Northwestern Mexico, South America, western United States. 2 species; 1 species treated in *Flora.*

A recently resurrected segregate from *Cryptantha* in the broad sense, *Greeneocharis* is distinguished by its basally synsepalous, circumscissile calyx. It is similar to *Eremocarya* in its often cushion-like stature, bracteate flowers and inflorescences, and often red roots or stem bases. However, molecular phylogenetic studies place these two genera well apart (Simpson et al. 2017; Hasenstab-Lehman & Simpson 2012).

Greeneocharis circumscissa (Hooker & Arnott) Rydberg var. ***circumscissa*** (p. 406)

Cushion greeneocharis, cushion cryptantha, matted cryptantha

Cryptantha circumscissa (Hook. & Arn.) I.M. Johnst. var. *circumscissa*

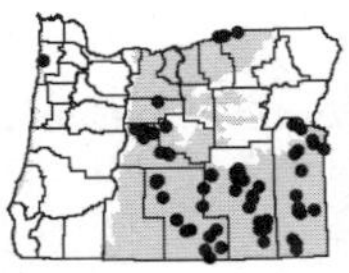

Plants 2.5–8 × 4–15 cm. **Leaves** 3–15 × 0.5–1.5 mm. **Flowers** calyces 3–4 mm in fruit, styles extending to nutlet tips; corolla limbs 1–2(4) mm in diameter, lobes oblong to obovate, margins not overlapping; fornices inconspicuous, pale yellow to white. **Fruits** 1.2–1.8 mm.

Barren areas, open shrub, sandy, gravelly, or ashy soil. Flowering May–Aug. 0–1800 m. BR, Col, ECas, Est, Lava, Owy. CA, ID, NV, WA; east to WY, southeast to AZ, south to Mexico; South America. Native.

A second variety, *G. c.* var. *rosulata*, consists of diminutive plants restricted to barren, granitic-derived gravel at high elevation in a limited locality of the Sierra Nevada, California.

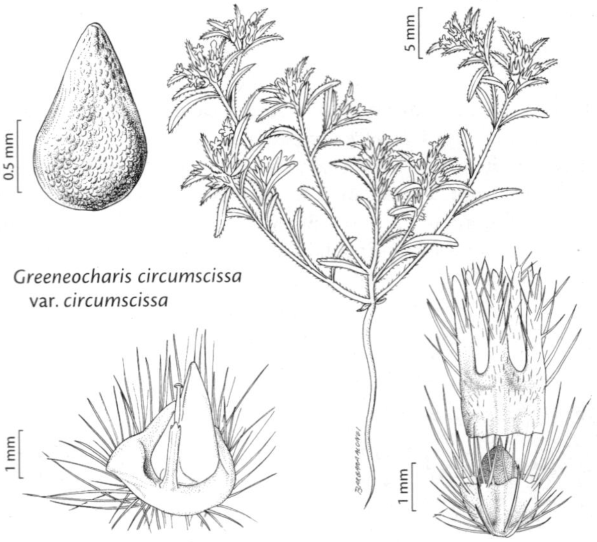

Greeneocharis circumscissa var. *circumscissa*

GREENEOCHARIS

GRUVELIA Gruvelia

C. Matt Guilliams

Genus composed of one species *G. pusilla*.

Gruvelia pusilla A. de Candolle (p. 406)

Little gruvelia

Pectocarya pusilla (A. DC.) A. Gray

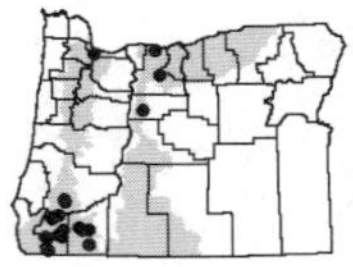

Herbs annual, 3–38 cm; taprooted. **Stems** usually erect, cinereous-strigose, branched at base, sometimes simple or branched distally. **Leaves** basal rosettes usually absent; basal (when present) and proximal cauline leaves opposite, decussate, connate at base, distal cauline leaves alternate; blades linear to linear spatulate, 3–20 × 0.5–2 mm, surfaces setulose to hispidulous; petioles indistinct. **Inflorescences** scorpioid cymes; pedicels reflexed to ascending and 0.3–1 mm in flower, recurved and 1–2.3 mm in fruit. **Flowers** calyces radially symmetric, 4.3–7.5 mm in diameter in fruit, lobes exceeding nutlets, appressed short-hairy, with long, stiff bristles distally; corollas funnel-shaped, white; fornices present; ovaries usually 4-lobed, equal, styles attached to receptacles, unbranched, persistent, stigmas 1, globose; gynobases conical-pyramidal. **Fruits** spreading, radially symmetric, monomorphic, equally divergent, widely trowel-shaped, 1.5–3 mm, margins with narrow rim bearing hooked bristles.

Gravelly to rocky soils, open areas, grasslands, woodlands, roadsides. Flowering Mar–May. 50–1000 m. Col, ECas, Sisk, WV. CA, WA; South America. Native.

Recent treatments of Boraginaceae usually include *Gruvelia pusilla* in *Pectocarya s.l.* Molecular phylogenetic studies show a broadly circumscribed *Pectocarya s.l.* is not monophyletic.

Gruvelia pusilla is amphitropically distributed between western North America and central Chile. Although this pattern is not uncommon in Boraginaceae, *G. pusilla* is considered to be non-native in Chile where it has been only seldom collected and then only in areas near major cities.

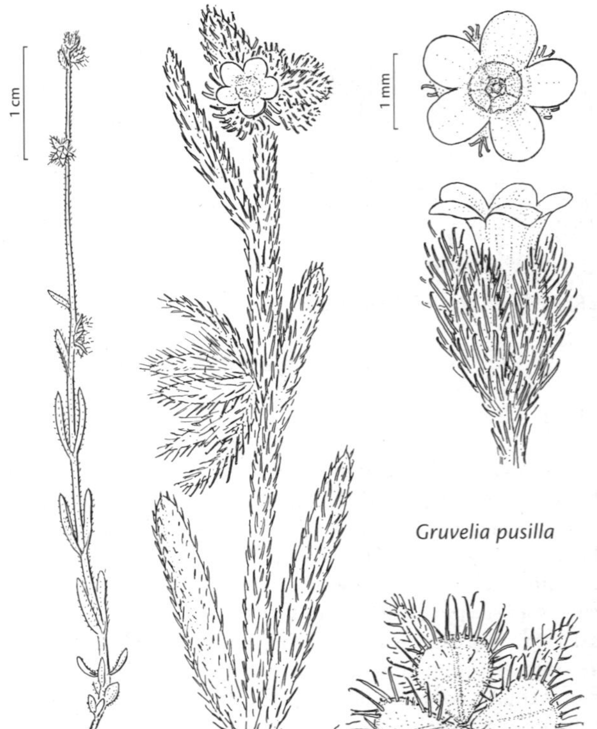

Gruvelia pusilla

GRUVELIA

HACKELIA Stickseed

Erika Saenz, Luciano Mancera, and James I. Cohen

Herbs perennial, erect, to 1 m. **Leaves** basal and cauline, simple, of various types, bases acute to truncate, margins entire, midveins evident, other orders of veins sometimes evident, tips acuminate to acute to rounded; sessile or petiolate. **Inflorescences** terminal and axillary, scorpioid cymes with a single main axis; sessile; bracts linear to lanceolate to elliptic, bases truncate to acute, margins entire, planar or revolute, midveins evident, tips acute to acuminate; pedicels recurved or deflexed in fruit. **Flowers** bisexual; sepals 5, lanceolate to linear, persistent, deeply lobed, hairy; corollas radially symmetric, salverform, rarely campanulate; petals 5, blue to white, sometimes pink, tubes blue, sometimes yellow to yellow and white, lobes flared, elliptic, margins entire; fornices present; stamens 5, attached to petals, included; styles 1, arising from base of ovary lobes, included, persistent. **Fruits** 4 or fewer by abortion, erect, ovoid, sometimes ellipsoid, lanceolate, or narrowly deltoid, brown to yellow; attachment scars submedial; barbed prickles (glochids) present, brown to yellow.

Asia, Europe, North America, South America. ~45 species; 10 species treated in *Flora*.

Glochids are a diagnostic feature on the nutlets of many *Hackelia* species. These are short, hair-like prickles with barbed tips. *Hackelia mundula* has recently been reported in Oregon, highly disjunct from its normal range in central California. If found again it can be identified from other *Hackelia* species by its pink flowers.

Key to *Hackelia* species:

1. Basal leaf blades ovate to oblong . . . . **_H. ophiobia_**

1' Basal leaf blades lanceolate to elliptic.

2. Dark blue spots evident at base of corolla lobes.

3. Stems 3–5 mm wide, glabrous toward the base, extremely sparsely strigillose toward the apex . . . . **_H. cronquistii_**

3' Stems 2–3 mm wide, strigose throughout. . . . . **_H. patens_**

2' Dark blue spots not present at base of corolla lobes.

4. Stems 1–3 (rarely more).

5. Corollas campanulate; plants of northeastern Oregon . . . . **_H. hispida_**

5' Corollas salverform; range various.

6. Nutlets with (16)30–80 glochids . . . . **_H. californica_**

6' Nutlets with 7–34 glochids.

7. Stems 4–7 mm wide; corolla lobes < 3 mm; anthers < 0.6 mm; plants of eastern Oregon. . **_H. floribunda_**

7' Stems 2–5 mm wide; corolla lobes > 2 mm; anthers > 0.5 mm; plants of central and north-central Oregon or Josephine County.

8. Plant pubescence soft; plants of low-elevation grassy and rocky slopes, central and north-central Oregon . . . . **_H. diffusa_**

8' Plant pubescence stiff; plants of mid-elevation open slopes, Josephine County. . . . . **_H. setosa_**

4' Stems 3–14 (rarely more).

9. Corolla tubes blue to white; glochids 0.7–2.3 mm . . . . **_H. cusickii_**

9' Corolla tubes yellow; glochids (1)2–3 mm . . . . **_H. micrantha_**

Hackelia californica I.M. Johnston

California stickseed

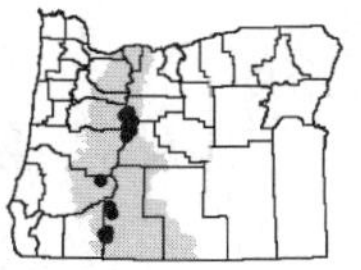

Stems 1–3 or rarely more, strigillose to strigose. **Basal leaves** blades lanceolate to elliptic, 3.9–13.1 × 0.4–2.5 cm, bases truncate, tips acute to rounded, surfaces abaxially strigillose to strigose to sericeous, adaxially strigillose to sericeous. **Cauline leaves** blades spatulate to lanceolate to elliptic, (5.5)6.5–14.6 × 0.3–3.6 cm, bases acute to truncate, tips acute to obtuse, rarely rounded, surfaces strigillose to scabrous. **Inflorescences** bracts lanceolate, rarely elliptic or ovate, 0.8–11 × 0.3–3(4) mm, bases truncate, rarely acute, tips acute, surfaces scabrous to strigillose; pedicels 4–15 mm. **Flowers** corollas salverform, white, rarely bluish white; styles 0.8–3.5 mm. **Fruits** ovoid, 2.5–5.5 × 2–6 mm; glochids (16)30–80, 1–4 mm, brown, occasionally yellow. ***2n***=24.

Meadows, forest openings. Flowering Jun–Aug. 900–2300 m. Casc, ECas. CA. Native.

Hackelia californica exhibits great variation in the shape of the cauline leaves as well as the type of hairs on the stems and leaves. This species is easily identified because each nutlet has many glochids, frequently more than twice as many as any other species in Oregon.

Hackelia cronquistii J.L. Gentry (p. 408)

Cronquist's stickseed

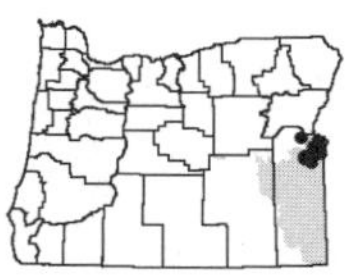

Stems 3–5 mm wide, glabrous toward base, sparsely strigillose toward apex. **Basal leaves** blades lanceolate to elliptic, 4.6–18.4 × 0.4–2.4 cm, bases acuminate, tips acute, surfaces abaxially hispidulous, adaxially hispidulous to strigillose. **Cauline leaves** blades linear, (1.1)1.7–9.3 × 0.2–0.6(0.9) cm, bases truncate, tips acute, surfaces abaxially strigillose to hispidulous, adaxially strigillose. **Inflorescences** bracts linear, 1–4 × 0.2–1 mm, bases truncate, tips acuminate, surfaces strigillose; pedicels 1–9 mm. **Flowers** corollas salverform, white, blue spots near base of lobes; styles (0.4)1.5–2.8 mm. **Fruits** lanceolate, 2–4 × 2–4(5) mm; glochids 8–20(30), 0.4–3 mm, beige.

Sandy soils, sagebrush. Flowering Apr–Jun. 600–1100 m. Owy. ID. Native.

Hackelia cronquistii is distributed across a small geographic range in southeastern Oregon and southwestern Idaho. This species is putatively closely related to *H. patens*. *Hackelia cronquistii* bears a stem that is glabrous toward the base and sparsely strigose toward the apex. In contrast, the stem of *H. patens* is strigose throughout.

Hackelia cusickii (Piper) Brand (p. 408)

Cusick's stickseed

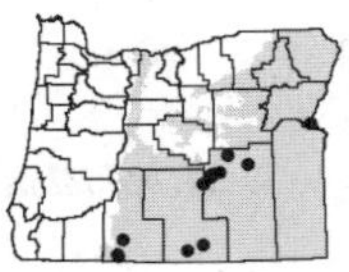

Stems 3–14 or rarely more, strigillose to strigose. **Basal leaves** blades lanceolate to elliptic, (2.9)4.4–13.4 × 0.25–1.5 cm, bases truncate, tips acute, surfaces strigillose to strigose. **Cauline leaves** blades elliptic to lanceolate, 2.4–9.3 × 0.1–1.5 cm, bases acute to truncate, tips acute, surfaces strigillose to strigose. **Inflorescences** bracts lanceolate, 1–19.5 × 0.3–3.5 mm, bases truncate, rarely acute, tips acute, surfaces strigillose to strigose; pedicels 2–12 mm. **Flowers** corollas salverform, blue, occasionally white with blue tint, tubes blue to white; styles 1–3.5 mm. **Fruits** narrowly deltoid, 2–4 × 1.2–4 mm; glochids 8–35, 0.7–2.3 mm, brown to light yellow. **2*n***=48.

Dry areas with junipers. Flowering May–Jun. 600–2000 m. BR, BW, ECas, Owy. CA, NV. Native.

Hackelia cusickii develops very narrow cauline leaves, several basal leaves, and a woody caudex wrapped in old stems and leaf bases.

Hackelia diffusa (Douglas ex Lehmann) I.M. Johnston

Spreading stickseed

Plants pubescence soft. **Stems** 1–3 or rarely more, 2–5 mm wide, pilose and sericeous. **Basal leaves** blades lanceolate to elliptic, 5.5–18 × 0.5–2.5 cm, bases attenuate, rarely secondary veins evident, tips acute, surfaces sericeous. **Cauline leaves** blades lanceolate, 1.4–10.5(15.5) × 0.3–1.8(2.1) cm, bases truncate, rarely rounded, rarely secondary veins evident, tips acute to acuminate, rarely rounded, surfaces sericeous to strigose. **Inflorescences** bracts lanceolate, 1–11(25) × 0.5–3(5) mm, bases truncate, tips acute, surfaces sericeous; pedicels 1–9 mm. **Flowers** corollas salverform, blue to white, lobes > 2 mm, tubes yellow; anthers > 0.5 mm; styles 2–3 mm. **Fruits** ovoid, 3–5 × 2–5 mm; glochids 7–34, 0.5–3 mm, brown to beige to pale yellow.

Northwestern North America. 4 varieties; 2 varieties treated in *Flora*.

Hackelia diffusa is a leafy species known for its long, soft and spreading pubescence present throughout the plant. The flowers of *H. diffusa* have relatively wide corollas, especially in comparison to other species of the genus in Oregon.

Key to *Hackelia diffusa* varieties:

1. Basal leaves 0.5–1.5 cm wide; corollas white to cream, 2–4 mm . . . . ***H. d.*** var. ***cottonii***

1' Basal leaves 1.6–2.1 cm wide; corollas blue, 5–6 mm . . . . ***H. d.*** var. ***diffusa***

HACKELIA

Hackelia diffusa var. ***cottonii*** (Piper) R.L. Carr

Spreading stickseed

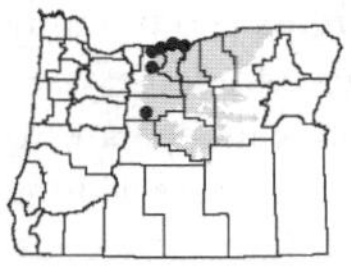

Basal leaves 0.5–1.5 cm wide. **Corollas** white to cream, 2–4 mm.

Grassy and rocky slopes. Flowering Apr–Jun. 100–600 m. Col, Lava. WA. Native.

Hackelia diffusa var. ***diffusa***

Spreading stickseed

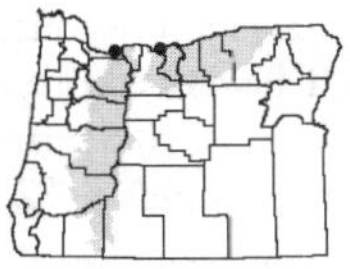

Basal leaves 1.6–2.1 cm wide. **Corollas** blue, 5–6 mm.

Grassy and rocky slopes. Flowering May–Jul. 0–700 m. Casc, Col. WA; north to British Columbia. Native.

Hackelia floribunda (Lehmann) I.M. Johnston (p. 410)

Manyflower stickseed

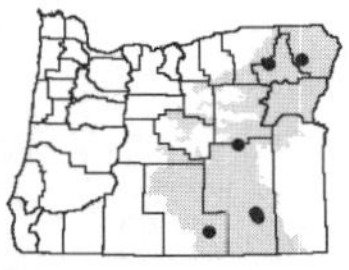

Stems 1–3 or rarely more, 4–7 mm wide, pilose to hispid toward base, densely strigose to densely strigillose toward apex. **Basal leaves** blades lanceolate to elliptic, (5)9–17.5(23.1) × (0.6)1–2.2 cm, bases truncate, rarely secondary veins evident, tips acute sometimes rounded, rarely obtuse, surfaces strigose to densely strigose. **Cauline leaves** blades spatulate to elliptic to oblanceolate, 1–13(15) × 0.4–1.8 cm, bases truncate to acute, rarely secondary veins evident, tips acute, rarely rounded, surfaces abaxially densely strigose, adaxially strigillose to densely strigose. **Inflorescences** bracts lanceolate to linear, 0.1–7.5(10) × (0.1)0.5–2.5(4) mm, bases truncate, rarely acute, tips acute, surfaces abaxially strigose to densely strigose, scabrous, sericeous, or pilosulose; pedicels 1.5–9 mm. **Flowers** corollas salverform, blue, lobes < 3 mm, tubes yellow; anthers < 0.6 mm; styles 1–3 mm. **Fruits** ovoid to ovoid-lanceolate; 1.8–4 × 1–2.8(3) mm; glochids 10–17, 1.2–3(3.5) mm, brown to yellow.

Open, moist areas. Flowering Jun–Aug. 900–2500 m. BR, BW. CA, ID, NV, WA; north to British Columbia, northeast to Quebec and southeast to TX. Native.

Hackelia floribunda is one of the most common and widespread species in the genus across its range. Stems of this species are relatively thick and arise from a short taproot. Other distinguishing features include ephemeral basal leaves and small anthers, never measuring more than 0.6 mm long. These are some of the shortest anthers of any species of *Hackelia* in Oregon.

Hackelia hispida (A. Gray) I.M. Johnston

Showy stickseed

Hackelia hispida (A. Gray) I.M. Johnst. var. *hispida*

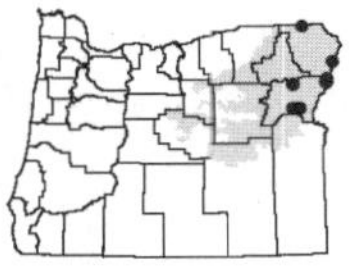

Stems 1–3 or rarely more, sparsely strigose to glabrous at base, densely strigose toward apex. **Basal leaves** blades lanceolate to elliptic, (2.8)4–18.8(21.2) × 0.4–2.9 cm, bases truncate, tips acute, rarely rounded, surfaces strigose to densely strigose. **Cauline leaves** blades lanceolate to elliptic, 0.9–7.8 × 0.4–1.6 cm, bases acute to truncate, tips acute, surfaces abaxially strigose to hispid, adaxially strigose to densely strigose. **Inflorescences** bracts lanceolate, rarely elliptic, 1.5–30 × 0.75–8 mm, bases acute to truncate, tips acute to acuminate, surfaces abaxially strigose, adaxially strigose to glabrous; pedicels 4–10 mm. **Flowers** corollas campanulate, cream-colored, sometimes with brown lines; styles 2–5 mm. **Fruits** ovoid, (3)4–5 × 2–6 mm; glochids 12–38, 0.75–1.8 mm, yellow to brown.

Rocky or talus slopes. Flowering Apr–Jun. 400–1000 m. BW. ID, WA. Native.

Hackelia hispida has several unique characteristics among the species of *Hackelia* in Oregon including campanulate, cream-colored corollas and leaves with a uniformly strigose to densely strigose indumentum. The nutlets of this species are rather large, most measuring between 4–5 mm in length.

Hackelia micrantha (Eastwood) J.L. Gentry (p. 410)

Jessica sticktight, blue stickseed, meadow stickseed

Hackelia jessicae (E.A. McGregor) Brand

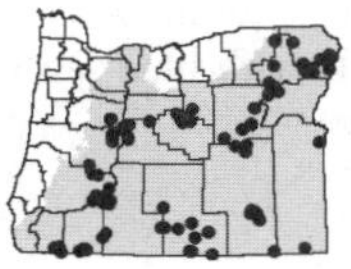

Stems 3–14 or rarely more, pilose and sericeous. **Basal leaves** blades lanceolate to elliptic, (8.1)9.3–20.5(25.5) × 0.8–3 cm, bases truncate, secondary veins evident, tips acute, surfaces sericeous. **Cauline leaves** blades lanceolate, rarely lanceolate-linear or elliptic, 1.5–12.5(15.2) × 0.3–1.5(2.2) cm, bases truncate, secondary veins evident, tips acute to acuminate, surfaces sericeous. **Inflorescences** bracts lanceolate, 3–23(31) × 1–5(7) mm, bases truncate, tips acute, surfaces sericeous; pedicels 4–15 mm. **Flowers** corollas salverform, blue, tubes yellow; styles 2–3 mm. **Fruits** ovoid, 3–5 × (2)3–5 mm; glochids 8–43(54), (1)2–3 mm, brown to yellow or purple, rarely beige. **2*n***=24.

Dry meadows to moist forests. Flowering May–Aug. 600–2800 m. BR, BW, Casc, ECas, Lava, Owy, Sisk. CA, ID, NV, WA; north to AK, south east to CO. Native.

Hackelia micrantha is an abundant and widespread species of *Hackelia* and can inhabit a variety of ecosystems.

Hackelia ophiobia R.L. Carr

Three Forks stickseed, Owyhee River stickseed

Stems 1–2 mm wide, sparsely strigose. **Basal leaves** blades ovate to oblong, 2.6–7.5 × 1.1–4.1 cm, bases truncate to cordate, secondary veins evident, tips acute to rounded, surfaces abaxially sericeous, adaxially strigose. **Cauline leaves** blades elliptic to ovate, 1.6–5.7 × 0.9–2.2 cm, surfaces strigillose to strigose, bases truncate to rounded, secondary veins evident, tips acute to acuminate. **Inflorescences** bracts elliptic to lanceolate, 1–11 × 0.3–5 mm, bases truncate, rarely rounded, tips acute to acuminate, surfaces abaxially sparsely strigillose to strigose, rarely glabrous, adaxially strigillose to strigose; pedicels 6–22 mm. **Flowers** corollas salverform, white to blue, tubes yellow; styles 0.9–1.1 mm. **Fruits** ovoid, 2–3.2 × 2–4 mm; glochids 9–23, 0.5–1.2 mm, brown to light yellow-green.

Shaded rock outcrops. Flowering May–Jul. 1100–1500 m. Owy. ID, NV. Native.

Hackelia patens (Nuttall) I.M. Johnston

Spotted stickseed

Hackelia diffusa (Lehm.) I.M. Johnst. var. *coerulescens* (Rydb.) I.M. Johnst., *Hackelia patens* (Nutt.) I.M. Johnst. var. *patens*

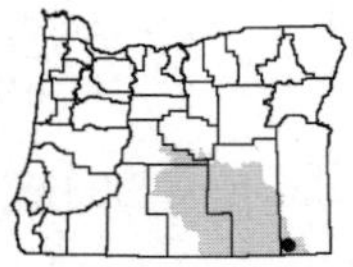

Stems 2–3 mm wide, strigose. **Basal leaves** blades lanceolate to elliptic, 5.7–9.5 × 1.1–2.2 cm, bases attenuate, tips acute to rounded, surfaces abaxially strigose, rarely sericeous, adaxially strigose. **Cauline leaves** blades lanceolate to elliptic, 2.2–5.9 × 0.4–0.6 cm, bases truncate, tips acute, surfaces abaxially strigillose to strigose, adaxially strigose. **Inflorescences** bracts lanceolate, 5–13 × 1.5–3 mm, bases truncate to rounded, tips acute to acuminate, surfaces strigose to strigillose; pedicels 5 mm. **Flowers** corollas salverform, white with blue spots near base of lobes; styles 1.3–3 mm. **Fruits** 2–3 × 3–4 mm, ovoid; glochids 8–20, 0.3–1.5 mm, brown to yellow.

Rocky slopes. Flowering May–Jun. 1900–2500 m. BR. ID, NV; southeast to CO. Native.

Hackelia patens is one of two species of the genus in Oregon (including *H. cronquistii*) that bears a corolla with blue spots near the base of the lobes.

Hackelia setosa I.M. Johnston

Bristly stickseed

Plants pubescence stiff. **Stems** 1–3 or rarely more, 2–5 mm wide, strigose to densely sericeous. **Basal leaves** blades lanceolate to elliptic, 1.5–8.2 × 0.3–1.1 cm, bases acuminate, tips sharply acute, surfaces sericeous. **Cauline leaves** blades lanceolate to linear-lanceolate, 1.1–7.9 × 0.1–1 cm, bases truncate, tips acute to acuminate, surfaces abaxially densely long-strigose, adaxially strigose to densely strigose. **Inflorescences** bracts linear, 1–12 × 0.2–4 mm, truncate, tips acuminate, surfaces abaxially densely to sparsely sericeous, adaxially densely sericeous; pedicels 1–9 mm. **Flowers** corollas salverform, blue, lobes > 2 mm, tubes white; anthers > 0.5 mm; styles 1.8–2.5 mm. **Fruits** ovoid, 4–5 × 4–5 mm; glochids 8–32, (0.4)1–2 mm, brown. **2*n***=48.

Open slopes. Flowering Jun–Jul. 1200–1700 m. Sisk. CA. Native.

Hackelia setosa is most closely aligned with *H. diffusa*. The two taxa can be distinguished by the cauline leaves of *H. setosa*, which tend to be smaller and have denser pubescence compared to those of *H. diffusa*.

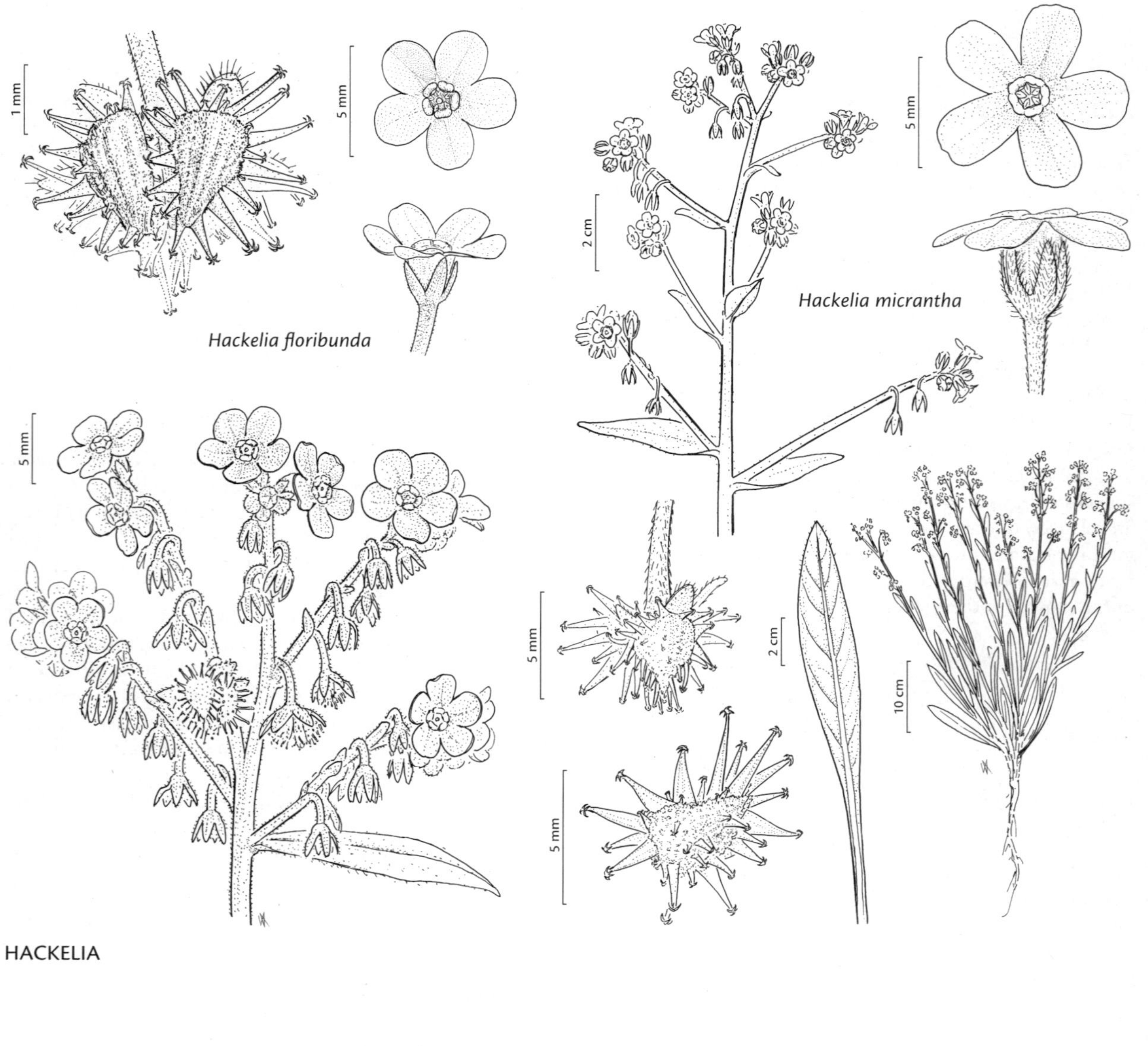

HACKELIA

LAPPULA Stickseed, tickweed

Stephen C. Meyers

Herbs annual; taprooted. **Stems** erect, branched or not. **Leaves** basal and cauline, margins entire; sessile to short-petiolate below, sessile above. **Inflorescences** terminal, raceme-like cymes, elongating with maturity; pedicels erect or ascending in fruit. **Flowers** calyces 5-lobed nearly to base; corollas small, funnelform, 5-lobed, purple or white; fornices present; stamens included; styles usually surpassing nutlets, entire. **Fruits** 4, erect, ovate, with 1 or more rows of marginal barbed prickles (glochids); attachment scars lateral, ovate.

Northern Hemisphere. 12–17 species; 2 species treated in *Flora*.

Older floras may include the species of this genus within *Hackelia*.

Key to *Lappula* species:

1. Marginal prickles of nutlets in 1 row . . . . *L. redowskii*

1' Marginal prickles of nutlets in 2(3) rows . . . . *L. squarrosa*

Lappula redowskii (Hornemann) Greene (p. 411)

Western stickseed, western tickweed

Lappula redowskii (Hornem.) Greene var. *cupulata* (A. Gray) M.E. Jones, *Lappula redowskii* (Hornem.) Greene var. *desertorum* (Greene) I.M. Johnst., *Lappula redowskii* (Hornem.) Greene var. *redowskii*

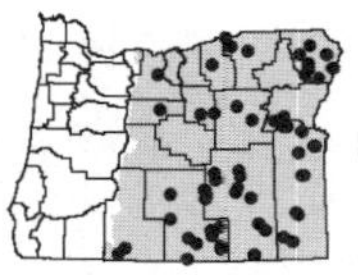

Plants 0.5–4 dm. **Leaves** oblanceolate to linear, 1–5 cm. **Flowers** calyces 1–2 mm; corollas 2–3 × 1.5–2.5 mm. **Fruits** 2–3 mm; marginal prickles in 1 row, distinct at base or united to form a cup-like rim.

Disturbed areas, fields, grasslands, rocky areas. Flowering Mar–Jul. 0–2300 m. BR, BW, Col, ECas, Lava, Owy. CA, ID, NV, WA; throughout western North America; Asia, Europe. Native and exotic?

Based mainly on geography, some authorities treat North American plants as a separate species (*L. occidentalis*). Furthermore, a more extreme fragmentation of *L. redowskii* has been adopted by some in which the taxon is divided into as many as eight separate species in North America alone, based mostly on subtle nutlet characteristics (Rolfsmeier 2013).

In some floras, two varieties of *L. redowskii* are recognized, based on marginal prickles that are distinct at the base (variety *redowskii*) or united to form a cup-like rim (variety *cupulata*). The varieties have no geographic or habitat differences. Additionally, plants with intermediate morphology or both types of morphology are not uncommon.

Some authors have theorized that both native and introduced populations of *L. redowskii* are present in North America. Evidence for this is speculative, however. More study is needed.

Lappula squarrosa (Retzius) Dumortier

European stickseed, bristly-fruited tickweed

Lappula echinata Gilib.

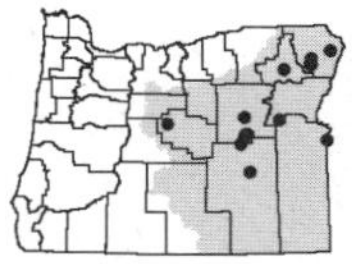

Plants 0.5–4 dm. **Leaves** oblanceolate to linear, 1–6 cm. **Flowers** calyces 2–3 mm; corollas 2–3 × 1.5–3 mm. **Fruits** 2–3 mm; marginal prickles in 2(3) rows, outer row usually < inner, distinct at base.

Disturbed areas, fields, grasslands, rocky areas. Flowering Jun–Aug. 600–1600 m. BR, BW, Lava, Owy. CA, ID, NV, WA; scattered throughout most of North America. Exotic?

Somewhat similar to *L. redowskii*, the nativity of North American populations of *L. squarrosa* are questionable. Current speculation suggests Rocky Mountain populations may be native, but others within North America are introduced.

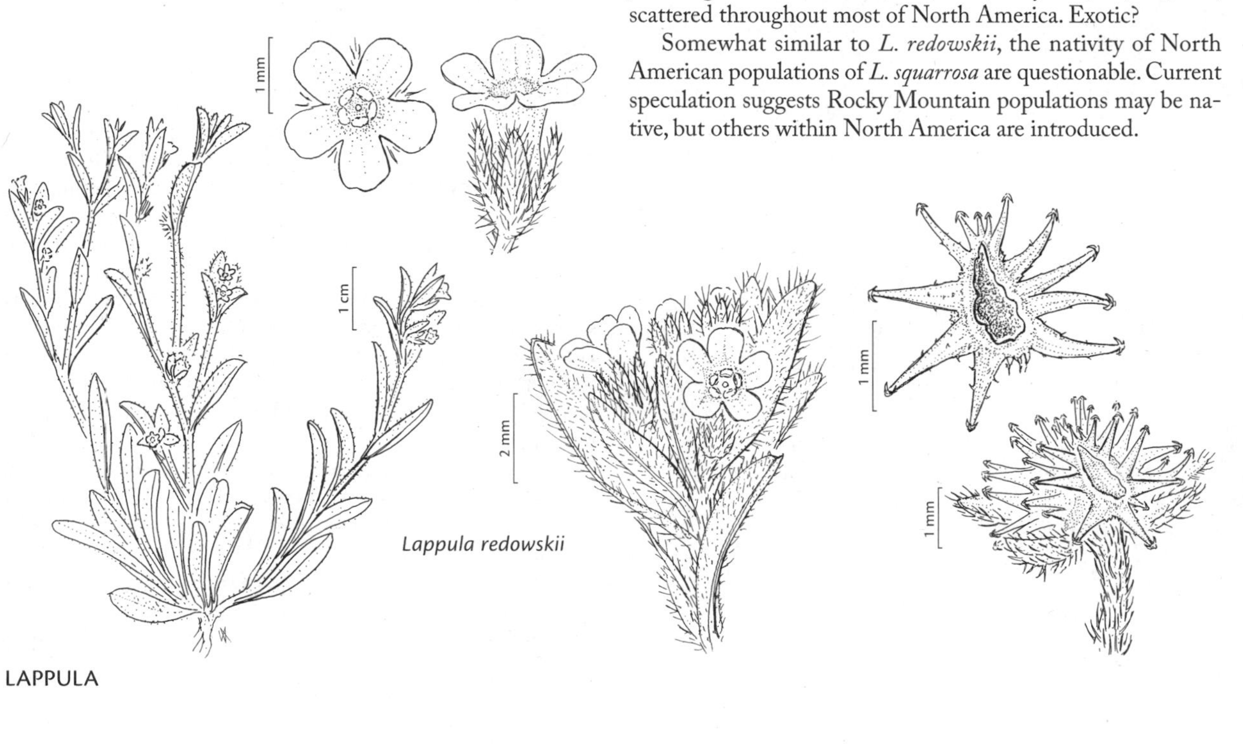

LAPPULA

LITHOSPERMUM Gromwell, puccoon, stoneseed

James I. Cohen

Herbs perennial, erect, up to 6 dm. **Leaves** simple, linear to elliptic to lanceolate, bases acute to rounded to truncate, margins entire, flat or revolute, midveins evident, tips acuminate to acute to rounded; sessile. **Inflorescences** bracts linear, lanceolate or elliptic, bases rounded to truncate to acute, margins entire, flat or revolute, tips acute to acuminate. **Flowers** radially symmetric, homostylous or heterostylous; sepals 5, lanceolate to linear, persistent; corollas sympetalous, funnelform to salverform, green-yellow to yellow, lobes 5, erect to ascending or reflexed, orbicular to elliptic, margins entire; stamens 5, attached to petals, included; styles 1, arising from base of ovary lobes, included to exserted up to 1 mm from apex of corolla tube, persistent. **Fruits** 4 or fewer by abortion, erect, ovoid, white to tan, smooth to slighty pitted; attachment scars basal.

Southern Africa, Asia, Europe, North America, northern South America. ~60 species; 2 species treated in *Flora*.

Key to the *Lithospermum* species:

1. Plants heterostylous; corollas funnelform, yellow, tubes 9–15 mm, lobes reflexed . . . . ***L. californicum***
1' Plants homostylous; corollas salverform to salverform-funnelform, green-yellow to yellow, tubes 1–7 mm, lobes erect to ascending . . . . ***L. ruderale***

Lithospermum californicum A. Gray (p. 412)

California stoneseed, California gromwell

Stems hispid to sparse hispid. **Leaves** cauline; blades linear to elliptic to lanceolate, rarely ovate, 1.4–7.5 × 0.4–1.7(2.1) cm, bases acute to rounded to truncate, margins entire, flat, tips acute, surfaces hispid to strigose, rarely densely hispid, abaxial midveins hispid. **Inflorescences** with a prominent terminal cyme; bracts ovate to lanceolate, sometimes elliptic, 1–5(6) × 0.3–2.5 cm, bases acute to truncate to rounded, rarely cordate, margins entire, tips acute, surfaces strigose to hispid, abaxial midveins hispid; pedicels to 10 mm. **Flowers** heterostylous; corollas funnelform, yellow, villous, tubes 9–15 mm, lobes reflexed; styles included. **Fruits** 4–6 × 3–4 mm, white, smooth to slightly pitted.

Pine and pine-oak forests, chaparral. Flowering Apr–Jun. 300–1300 m. Sisk. CA. Native.

In Oregon, this is the only heterostylous species of *Lithospermum*. This species appears to be a facultative serpentinophile. Individuals of *L. californicum* that grow on serpentine soil appear to be much shorter than those that do not.

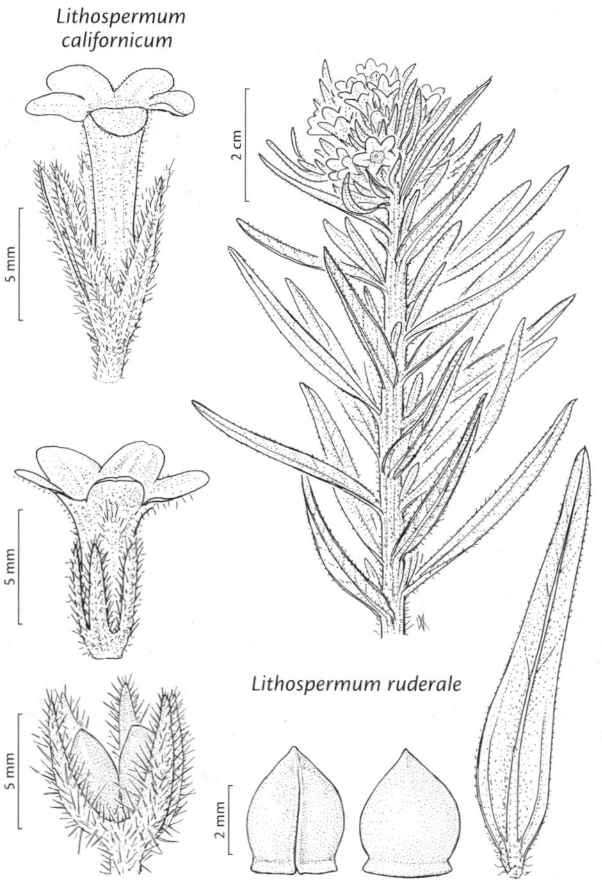

LITHOSPERMUM

Lithospermum ruderale Douglas ex Lehmann (p. 412)

Western stoneseed, western gromwell, Columbia puccoon

Lithospermum lanceolatum Rydb., *Lithospermum pilosum* Nutt.

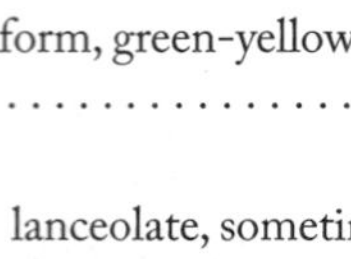

Stems pilose to sericeous to strigose, rarely hispid. **Leaves** cauline; blades lanceolate to elliptic, 2.3–7.5(9) × 0.1–1(1.4) cm, bases acute to rounded to truncate, margins entire, revolute, midveins evident, rarely a pair of secondary veins present toward base of leaf, tips acute to acuminate, surfaces strigose to sericeous. **Inflorescences** with numerous, axillary cymes; bracts lanceolate to elliptic, sometimes linear, 0.7–5(6) × 0.1–0.8 cm, bases truncate to rounded to acute, margins entire, revolute, sometimes flat, surfaces strigose to sericeous, rarely hispid, midveins evident, tips acute to acuminate; pedicels to 4 mm. **Flowers** homostylous; corollas salverform to salverform-funnelform, yellow to yellow-green, villous toward apex, tubes 1–7 mm; lobes erect to ascending; styles included, rarely exserted 1 mm from corolla tube. **Nutlets** 4–5(6) × 3–5 mm, white to light tan, smooth to slightly pitted.

Chaparral and open forests. Flowering Apr–Jul. 50–2200 m. BR, BW, Col, ECas, Lava, Owy. CA, ID, NV, WA; north to British Columbia, northeast to Saskatchewan, east to WY, southeast to CO. Native.

This is the only species of *Lithospermum* that produces numerous, axillary cymes rather than a prominent terminal cyme. These inflorescences give the plants an unusual shape for a species of *Lithospermum*.

LYCOPSIS Small bugloss

Richard R. Halse

Herbs annual; taprooted. **Stems** hispid-strigose. **Leaves** alternate; blades simple, lanceolate to oblong, surfaces sparsely hispid; petioles present or absent. **Inflorescences** apical, bracteate cymes on central axes, branches slightly elongating and straightening with age; pedicels ascending. **Flowers** bilaterally symmetric, funnelform to campanulate; calyces 5-lobed, lobes slightly unequal; corollas usually blue, rarely whitish, tubes curved to bent at middle, throats closed by 5 white-papillose fornices, lobes 5, unequal; stamens 5, included, 2 inserted near base of tube, 3 just below middle; gynobases flattened, much < nutlets; ovaries deeply 4-lobed, styles 1, gynobasic, included, stigmas bilobed, conical-capitate. **Fruits** 1–4, erect, homomorphic, ovoid, oblique, rugose; basal attachment stipe-like, surrounded by a basal, rounded, thickened rim. **2*n***=16, 48.

Worldwide. 2 species; 1 species treated in *Flora*.

Lycopsis arvensis Linnaeus (p. 413)

Small bugloss, annual bugloss

Anchusa arvensis (L.) M. Bieb.

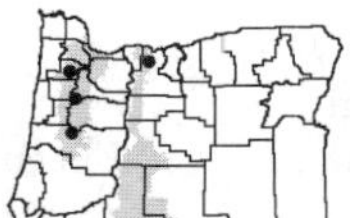

Stems ascending to erect, simple or branched, 10–80 cm. **Leaves** blades linear-lanceolate to oblanceolate to oblong, (1.5)3–14 × 0.5–2(3) cm, margins undulate, entire to ± dentate; petiolate proximally, sessile and/or ± clasping distally. **Inflorescences** bracts > or < calyces; pedicels 0–2 mm in flower, to 10 mm in fruit. **Flowers** calyces divided almost to base, lobes linear-lanceolate, acute, 4–5 mm in flower, to 10(15) mm in fruit. **Fruits** ovoid, 1.5–2 × 3–4 mm, grayish brown to gray, rugose, tuberculate. **2*n***=48.

Disturbed areas, fields, roadsides. Flowering Apr–Jul. 50–600 m. ECas, WV. CA, ID, WA; southern Canada, northwestern, northern, and northeastern US; Africa, east Asia, Australia, Europe, South America. Exotic.

Lycopsis arvensis is apparently scattered widely and sporadically across North America. In some states and provinces, it may have been collected once or twice and may no longer be present. This species is listed as a noxious weed in Washington.

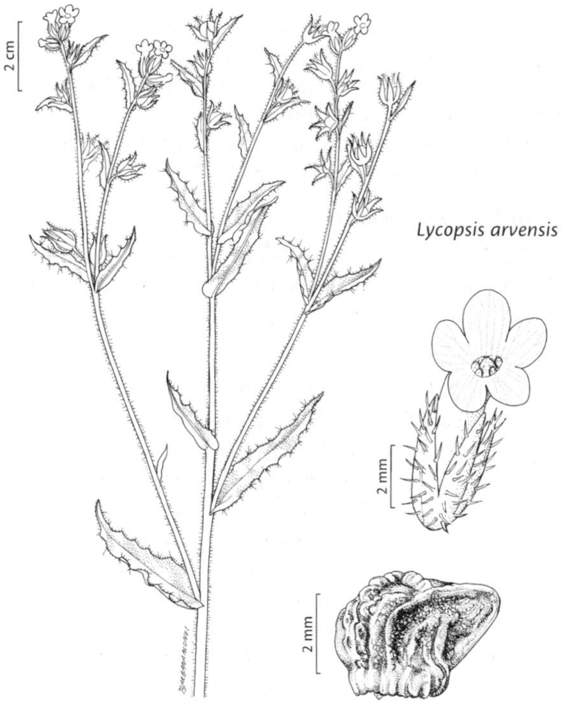

LYCOPSIS

MERTENSIA Bluebells, lungwort

Mare Nazaire

Herbs perennial, glabrous, strigose, hispid, or hirsute, hairs often with pustulose bases; roots usually thick, stout taproots; caudices usually branched, occasionally corm-like or tuber-like. **Stems** erect, ascending, or decumbent, 5–150 cm, frequently purplish at base. **Leaves** basal (frequently absent) and cauline; basal arising from caudices independent of flowering stems; cauline alternate, similar to basal but smaller, lowermost reduced to a scarious sheath; blades ovate, elliptic, linear, or lanceolate, thin or subcoriaceous and somewhat fleshy, margins entire, lateral veins conspicuous or not, surfaces ciliate or not; petiolate to sessile. **Inflorescences** compound cymes, panicle-like, few- to many-flowered, lowest flowers often solitary in leaf axils or lowest branches subtended by leaves; pedicels erect, arching, nodding, arcuate, or slightly reflexed, often elongating in fruit. **Flowers** radially symmetric; calyces 5-merous, cleft to near base or sometimes with a well-developed campanulate tube exceeding length of lobes, strigose, glabrous, or pustulose, often accrescent in fruit, lobes usually keeled, often ciliate along margin; corollas 5-merous, tubular, funnelform, or campanulate, throat and limb often abruptly expanded, sometimes subcylindrical, pink or pink-purple in bud, pale blue, blue, or dark blue, rarely white; fornices alternating with stamens, below anthers, occasionally above; stamens included, filaments inserted at or below fornices, filiform or conspicuously expanded and ligulate; usually homostylous, occasionally heterostylous, styles gynobasic, included or exserted, bases often dilated slightly, stigmas entire, capitate. **Fruits** (3)4, 0.8–5 mm, attached laterally to gynobase, ovate to tetrahedral, off-white, pale brown, or dark brown, usually rugulose and papillate, occasionally smooth and shiny; attachment scars medial or submedial, oblique and elevated, sometimes twisted to one side; keel above attachment scar winged or evidently raised.

Asia, Europe, North America. 58 species; 9 species treated in *Flora*.

Mertensia has long been considered taxonomically difficult. The taxonomic confusion stems largely from continuous morphological variation among species, possibly due in part to incomplete lineage sorting, hybridization, and because selected characters, especially inflorescence architecture and flowers, frequently vary with the stage of development. Recent molecular phylogenetic investigations (Nazaire & Hufford 2014) have provided important insights on the taxonomic issues that have long surrounded broadly circumscribed taxa, resulting in narrow circumscriptions of several species, the resurrection of names formerly treated as synonyms, and the elevation of selected varieties to specific status.

In previous treatments of *Mertensia*, selected characters were emphasized to delimit species and subgeneric taxonomy. However, many characters used to hypothesize relationships are labile and do not necessarily delineate close relationships among species as proposed by earlier workers. Therefore, a cautionary approach is advised in the use of some characters. North American *Mertensia* species are best distinguished with a suite of characters rather than with a solitary attribute. While the key below may conveniently group selected taxa based on shared characters, be aware that this does not necessarily suggest close evolutionary relationships among selected species.

Previous treatments of *Mertensia* have referenced the corolla as consisting of a "tube" (i.e., the portion of the corolla below the fornices) and a "limb" (i.e., the portion of the corolla above the fornices). This terminology may be appropriate for some members of Boraginaceae where the corolla lobes (i.e., limb) begin at the point of the fornices; however, for *Mertensia*, the "tube" technically continues above the level of the fornices and terminates with the lobes. The key and species descriptions below will adhere to the following terms: "tube" to refer to the portion of the corolla below the fornices, "throat" to refer to the dilated portion of corolla tube above the level of the fornices, and "limb" used in reference to the lobe(s) of the corolla.

Descriptions in the key reference typical attributes of the species at the height of anthesis. Stamen filament width measurements should be taken at the proximal end.

Key to *Mertensia* species:

1. Caudices corm-like, ellipsoid to globose; cauline leaves opposite distally; corollas campanulate, flaring from near the base, not sharply distinguished into tube and throat .... ***M. bella***

1' Caudices branched or absent; cauline leaves alternate throughout; corollas not campanulate, tube and throat distinct.

  2. Plants 40–150 cm when well developed; leaves thin, lateral venation conspicuous.

    3. Cauline leaf blades broadly ovate to cordate, adaxial surfaces scabrous; anthers 3.5–5 mm .... ***M. subcordata***

    3' Cauline leaf blades lanceolate, elliptic, ovate, or obovate, adaxial surfaces glabrous, pustulose, or strigose; anthers 1.2–3.5 mm.

      4. Corollas tubular with funnelform throat, 15.5–21 mm, corolla tubes 8.3–13.7 mm, 1.2–1.9 × length of combined throat and limb .... ***M. umbratilis***

      4' Corollas tubular with dilated to funnelform throat and limb, 8–17 mm, corolla tubes 3.7–8 mm, 0.6–1.1 × length of combined throat and limb.

        5. Stems, leaves, pedicels usually glaucous; pedicels sparsely pustulose, pustules sometimes supporting short strigose hairs, sometimes glabrous; calyx lobes oblong to ovate, 1.3–2 mm, tips of calyx lobes obtuse to subacute .... ***M. ciliata*** var. ***ciliata***

        5' Stems, leaves, pedicels not glaucous; pedicels strigose to spreading hispid; calyx lobes lanceolate, linear-lanceolate, or linear-oblong, 1.3–6 mm, tips of calyx lobes acute to subacute or obtuse.

          6. Calyces 2.2–2.6(2.9) mm, cleft 67–75% the length to base .... ***M. brachycalyx***

          6' Calyces 3–7 mm, cleft 75–80% the length to base .... ***M. paniculata***

  2' Plants ≤ 40 cm; leaves usually thick, lateral venation inconspicuous.

    7. Roots shallow-seated, short, tuberous, easily detached; caudices absent; corolla tubes 2–3.1 × length of combined throat and limb .... ***M. longiflora***

    7' Taproots woody, slender to stout; caudices present, branched; corolla tubes 0.8–2 × length of combined throat and limb.

      8. Stems pubescent, hairs spreading to retrorse; leaf surfaces canescent, abaxially densely strigose to spreading-pubescent, adaxially densely strigillose to spreading-pubescent .... ***M. cusickii***

      8' Stems glabrous; leaf surfaces frequently glaucous, abaxially glabrous, adaxially pustulose .... ***M. foliosa***

Mertensia bella Piper

(p. 415)

Oregon bluebells, Oregon lungwort

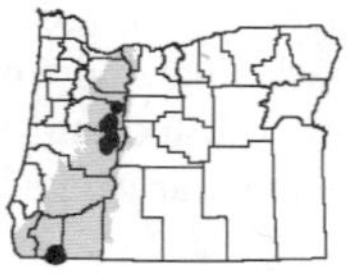

Plants 1–4 dm; roots fibrous; caudices corm-like, ellipsoid to globose. **Stems** 1 to few, erect, slender, mostly glabrous. **Leaves** pale green, thin, margins finely ciliate, lateral venation conspicuous, surfaces abaxially glabrous or sparsely strigose, adaxially strigillose; basal blades frequently absent; cauline opposite distally, blades ovate to broadly ovate, 3–7.5 × 1.5–3.5 cm, length 1.7–2.9 × width, bases rounded to cuneate, tips acute, petioles 0.1–2.5 cm. **Inflorescences** mostly open, branched; cymes 3–11-flowered; pedicels 4–9 mm, densely strigose. **Flowers** calyces 2–3(3.4) mm, cleft 67–75% the length to base, abaxial surface sparsely to densely strigose, lobes lanceolate to deltoid-lanceolate, 1.3–2 mm, margins densely strigose-ciliate, tips acute or subacute; corollas campanulate, not clearly distinguished into tube and throat, flaring from near base, 5–10 mm, blue, rarely white, glabrous within, tubes 0.6–1.4 mm, 0.1–0.2 × length of combined throat and limb, not exceeding calyx lobes; fornices inconspicuous, reduced to a slight thickening; filaments inserted 0–0.2 mm below fornices, filiform, 0.5–1.3 × 0.1–0.2 mm, anthers 1–1.5 × 0.2–0.6 mm, 0.9–1.8(2.6) × length of filaments; styles 2.5–5(6) mm, stigmas included, surpassing anthers. **Fruits** 1.5–3 mm, rugulose, papillate.

Montane to subalpine meadows, moist open slopes, streambanks, lakeshores, rocky, springy soils. Flowering Jun–Jul. 600–2000 m. Casc, Sisk. CA, ID; northeast to MT. Native.

Endemic to the Pacific Northwest, *Mertensia bella* is disjunct between eastern (western Montana/central Idaho) and western (southwestern Oregon/northern California) populations, a common distribution pattern of mesic coniferous forests of the Pacific Northwest. *Mertensia bella* possesses characters distinct from other species in the genus, including its corm-like caudices (corm-like roots often incorrectly applied), slender stems, campanulate corollas, and inconspicuous fornices.

Mertensia brachycalyx Piper (p. 415)

Short-sepal bluebells

Mertensia paniculata (Aiton) G. Don, *Mertensia paniculata* (Aiton) G. Don var. *borealis* (J.F. Macbr.) L.O. Williams

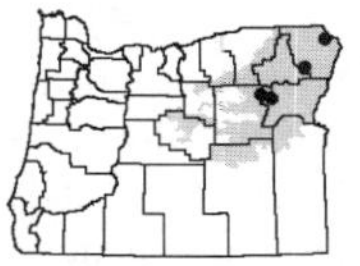

Plants 4–15 dm; taproots woody, stout; caudices branched. **Stems** 1 to several, erect to ascending, glabrous. **Leaves** green, thin, margins ciliate, lateral venation conspicuous, surfaces abaxially glabrous or sparsely spreading-pubescent, hairs spreading; basal blades frequently absent; cauline alternate, blades elliptic to ovate, 4–12.5 × 1.4–5.4 cm, length 1.8–3.4 × width, bases rounded to cuneate, tips acute, petioles 0.5–5.5 cm, sessile distally. **Inflorescences** open, branched; cymes 3–12-flowered; pedicels 3–16 mm, strigose to spreading-hispid. **Flowers** calyces 2.2–2.6(2.9) mm, cleft 67–75% the length to base, abaxial surface sparsely strigose to glabrous, lobes lanceolate to deltoid, or linear-oblong, 1.3–2(2.2) mm, margins short strigose-ciliate, tips acute to subacute or obtuse; corollas tubular with dilated to funnelform throat, tube and throat distinct, 9–12.8 mm, blue to pale blue, glabrous within, tubes 3.7–5.9 mm, 0.6–1 × length of combined throat and limb, exceeding calyx lobes; fornices conspicuous; filaments inserted 0.2–0.5 mm below fornices, ligular, 1.1–2.8 × 0.4–1 mm, anthers 2.3–3.4 × 0.5–0.9 mm, 1.1–1.8(2.2) × length of filaments; styles 8–13.2 mm, stigmas included, surpassing anthers. **Fruits** 3.2–4.3 mm, rugulose, papillate.

Wet meadows, talus slopes, streambanks, riverbanks. Flowering Jun–Aug. 1300–2000 m. BW. ID, WA; northeast to MT. Native.

Mertensia brachycalyx was formerly treated as a synonym of *M. paniculata* var. *borealis*, but it is distinguished from that taxon by its smaller calyx with acute to subacute or obtuse calyx lobe tips (vs. acute to acuminate in *M. paniculata* var. *borealis*).

Mertensia ciliata (James ex Torrey) G. Don var. *ciliata*

Tall fringed bluebells, mountain bluebells, streamside bluebells

Mertensia ciliata (James ex Torr.) G. Don var. *stomatechoides* (Kellogg) Jeps.

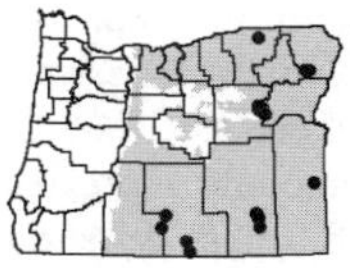

Plants 4–10(15) dm; taproots woody, stout; caudices branched. **Stems** 1 to several, erect, glabrous, frequently glaucous. **Leaves** green, thin, margins strigose-ciliate, lateral venation conspicuous, surfaces often glaucous, abaxially glabrous, adaxially pustulose, otherwise glabrous; basal blades frequently absent, ovate, lanceolate, or elliptic, 3–11 ×

MERTENSIA

2–6 cm, bases cordate to rounded, tips acute, petioles 4–17 cm; cauline alternate, blades lanceolate, elliptic, or ovate, 1.5–9 × 1–5 cm, length 2–3.5 × width, bases cuneate to rounded, tips acute, petioles 0.3–4.2 cm, sessile distally. **Inflorescences** compact to somewhat open, branched or not; cymes 4–18-flowered; pedicels 2–12.7 mm, sparsely pustulose, pustules sometimes supporting short strigose hairs, sometimes glabrous, usually glaucous. **Flowers** calyces 1.5–3(4) mm, cleft 67–75% the length to base, abaxial surface glabrous, occasionally sparsely strigose at base, lobes oblong to ovate, 1.3–2 mm, margins ciliate, short-strigose or hispidulous, tips subacute to obtuse; corollas tubular with dilated to funnelform throat, tube and throat distinct, 9.6–17 mm, blue, rarely white, glabrous or sparsely pubescent within, tubes 4.6–8 mm, 0.8–1.1 × length of combined throat and limb, exceeding calyx lobes; fornices conspicuous; filaments inserted 0.2–0.7 mm below fornices, ligular, 1.3–2.7 × (0.4)0.6–0.9 mm, anthers 1.2–2.5 × 0.5–0.9 mm, 0.8–1.5 × length of filaments; styles 8.6–12.4 mm, stigmas included to barely exserted, surpassing anthers. **Fruits** 2–4 mm, rugulose, papillate, occasionally bearing minute spinose appendages along margin and dorsal surface. **2*n***=24, 48.

Streambanks, riverbanks, subalpine meadows, aspen woods, coniferous forests, seeps, marshy areas. Flowering May–Jul. 500–2800 m. BR, BW, Col, ECas, Owy. ID; northeast to MT, southeast to NM. Native.

Mertensia ciliata is among the most geographically widespread and ecologically variable species of North American *Mertensia*. Molecular evidence (Nazaire & Hufford 2014), as well as an examination of herbarium specimens, indicates putative hybridization between *M. ciliata* and other *Mertensia* species, including *M. bakeri*, *M. foliosa*, and *M. tweedyi*.

Mertensia cusickii Piper

Cusick's bluebells

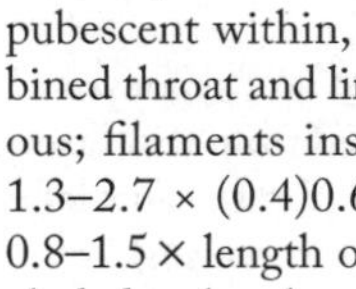

Plants 2–4 dm; taproots woody, slender to stout; caudices branched. **Stems** 1 to several, erect, densely pubescent, hairs spreading to retrorse. **Leaves** thick, margins ciliate, lateral venation inconspicuous, strigose to spreading-pubescent, surfaces abaxially densely strigose to spreading-pubescent, adaxially densely strigillose to spreading pubescence; basal blades frequently absent, ovate-elliptic, 6–10 × 1.5–4 cm, bases attenuate, tips subacute, petioles 4–8 cm; cauline alternate, blades lanceolate, elliptic, or obelliptic, 2.5–9 × 1–1.9 cm, length (3)4–5.4 × width, bases attenuate to cuneate, tips acute, petioles 0.8–4 cm, sessile distally. **Inflorescences** compact to somewhat open, branched; cymes 3–13-flowered; pedicels 3.7–12 mm, densely hispid. **Flowers** calyces 3.5–4.8 mm, cleft 67–75% the length to base, abaxial surface densely strigose-sericeous, lobes lanceolate to deltoid-lanceolate, 2.3–4.1 mm, margins ciliate, with long strigose-sericeous hairs, tips acute; corollas tubular with dilated to funnelform throat, tube and throat distinct, (7)9–14.4 mm, pale blue, glabrous, seldom pubescent within, tubes 3.9–6.1 mm, 0.8–1.2 × length of combined throat and limb, exceeding calyx lobes; fornices conspicuous; filaments inserted 0.2–0.5 mm below fornices, ligular, 1.7–2.7 × 0.6–0.9 mm, anthers 1.8–2.3 × 0.6–0.8 mm, 0.8–1.2 × length of filaments; styles 8–13 mm, stigmas included, surpassing anthers. **Fruits** 2–4 mm, rugulose, papillate.

Open slopes, meadows, streambanks, woodlands. Flowering May–Jul. 1200–2400 m. BR, BW, ECas. CA, NV. Native.

Putative hybrids between *M. cusickii* and *M. umbratilis* have been reported where their distributions overlap (Nazaire & Hufford 2014).

Mertensia foliosa A. Nelson

Oblongleaf bluebells, sagebrush bluebells

Mertensia oblongifolia (Nutt.) G. Don

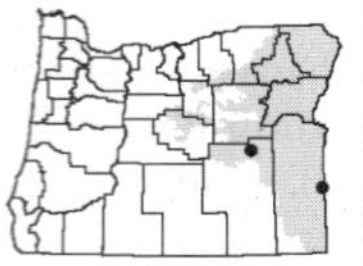

Plants 1–4 dm; taproots woody, frequently stout and deep seated; caudices branched. **Stems** (1) several–many, erect to ascending, glabrous. **Leaves** green to gray-green, thick, margins short strigose-ciliate or more often pustulose, lateral venation inconspicuous, surfaces frequently glaucous, abaxially glabrous, adaxially pustulose; basal blades elliptic, oblanceolate, or lance-ovate, 2.5–11 × 0.7–6 cm, bases attenuate, tips subacute or obtuse, petioles 2–10 cm; cauline alternate, blades elliptic, oblanceolate or broadly lanceolate, 2–7.5 × 0.5–2 cm, length 3–7 × width, bases attenuate to cuneate, tips subacute or obtuse, petioles 0.5–3 cm, sessile distally. **Inflorescences** compact, branched or not; cymes 5–13-flowered; pedicels 1.5–9 mm, glabrous, pustulose, pustules occasionally bearing short, strigose hairs. **Flowers** calyces 2.5–6 mm, cleft 75–80% the length to base, abaxial surface strigose, or pustulose at base, otherwise glabrous, lobes lanceolate, 2.5–4 mm, margins strigose-ciliate, hairs short, tips acute to acuminate; corollas tubular with dilated to funnelform throat, tube and throat distinct, 10–16.5 mm, blue, glabrous within, tubes 5.5–9.5 mm, 1.4–2 × length of combined throat and limb, exceeding calyx lobes; fornices conspicuous; filaments inserted 0.2–0.4 mm below fornices, ligular, 2–3.3 × 0.5–1 mm, anthers 1.4–2.5 × 0.4–0.7 mm, 0.6–1.3 × length of filaments; styles 9–15 mm, stigmas included, ± surpassing anthers. **Fruits** 2–4 mm, rugulose, papillate. **2*n***=24, 48.

Open slopes, dry meadows, vernally moist habitats, sagebrush communities. Flowering May–Jul. 1300–2000 m. BW, Owy. CA, ID, NV, WA; northeast to ND, east to SD, southeast to NM. Native.

Mertensia foliosa has long been known by the taxon name *M. oblongifolia* var. *nevadensis*; however, this combination is an illegitimate name since *M. nevadensis* was first published by A. Nelson (1904) and later treated as a variety of *M. foliosa* by J.F. Macbride (1916). *Mertensia foliosa* is among the most common, widespread, and morphologically variable species of western North American *Mertensia* and has presented considerable taxonomic confusion. This confusion has been exacerbated by the numerous names that have been applied as well as the inclusive treatment of the taxon by L.O. Williams (1937). Anecdotal and molecular evidence (Nazaire & Hufford 2014) indicate putative hybrids between *M. foliosa* and several other species, including *M. ciliata* and *M. paniculata*.

Mertensia longiflora Greene (p. 418)

Long-flowered bluebells, small bluebells, trumpet bluebells, long-flowered lungwort

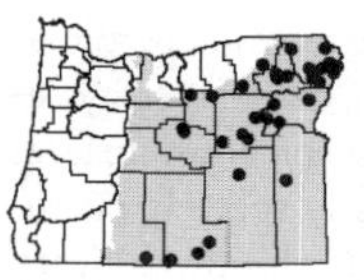

Plants 0.6–2.5(3.5) dm; roots shallow-seated, short, tuberous, easily detached; caudices absent. **Stems** 1 to few, erect, glabrous to strigose. **Leaves** green, thick, margins ciliate, lateral venation inconspicuous, surfaces abaxially glabrous, rarely pubescent, adaxially strigose to glabrous; basal blades rarely present, oval to spatulate, 2–5 × 1.2–2.5 cm, bases cuneate, tips subacute, obtuse, or rounded, petioles 2–5 cm; cauline alternate, blades elliptic to oblong-lanceolate, 2–6 × 0.5–3 cm, length 1.6–3.5 × width, bases cuneate, tips subacute, subsessile to sessile. **Inflorescences** compact, branched or not; cymes 5–9-flowered; pedicels 0.5–6 mm, glabrous. **Flowers** calyces 4–7 mm, cleft 67–75% the length to base, abaxial surface glabrous, lobes lanceolate to deltoid-lanceolate, 3–4 mm, margins short-ciliate, tips acute; corollas tubular with dilated throat, tube and throat distinct, (10)15–20(25) mm, blue, glabrous within, tubes 7–14 mm, 2–3.1 × length of combined throat and limb, exceeding calyx lobes; fornices conspicuous; filaments inserted 0.1–0.4 mm below fornices, ligular, 1.5–3 × 0.4–1 mm, anthers 1–2.2 × 0.5–0.8 mm, (0.5)0.8–1.1 × length of filaments; styles 14–19 mm, stigmas included to scarcely exserted, surpassing anthers. **Fruits** 2–4 mm, rugulose, papillate. **2*n***=24.

Sagebrush foothills, prairie grasslands, woodlands, rocky slopes. Flowering Apr–Jun. 500–2400 m. BR, BW, ECas, Lava, Owy. CA, ID, WA; north to British Columbia, northeast to Alberta. Native.

Leaf pubescence appears to be polymorphic in *M. longiflora*, as the degree of pubescence can be highly variable among and within populations. The corolla tube to throat/limb ratio in *M. longiflora* is among the largest in the genus.

Mertensia paniculata (Aiton) G. Don

Northern bluebells, tall bluebells, tall lungwort

Plants (3)4–10 dm; woody taprooted or rhizomatous; caudices branched. **Stems** 1 to numerous, erect to ascending, glabrous to sparsely spreading-pubescent. **Leaves** green, very thin, margins ciliate with spreading hairs, lateral venation conspicuous; basal blades elliptic-lanceolate to ovate-subcordate, 3–20 × 1–14 cm, bases rounded to cuneate, tips acute to acuminate, petioles 3–10 cm; cauline alternate, blades ovate, elliptic or lanceolate, 1.5–14 × 0.5–7 cm, length 1.6–4.2 × width, bases cuneate, tips acute to acuminate, petioles 1–2 cm, sessile distally. **Inflorescences** open, branched; cymes 3–12-flowered; pedicels 3–17 mm, strigose to spreading-hispid. **Flowers** calyces 3–7 mm, cleft 75–80% length to base, abaxial surface strigose to glabrous, lobes lanceolate, 3–6 mm, margins ciliate-strigose, tips acute; corollas tubular with dilated to funnelform throat, tube and throat distinct, 8–16 mm, blue, rarely white, pubescent or glabrous within, tubes 4.5–7 mm, 0.6–1 × length of combined throat and limb, generally exceeding calyx lobes; fornices conspicuous; filaments inserted 0.3–1 mm below fornices, ligular, 1.5–3 × 0.3–1 mm, anthers 2–3.5 × 0.5–1 mm, 1–1.9(2.3) × length of filaments; styles 9–17 mm, stigmas included to scarcely exserted, surpassing anthers. **Fruits** 3–4.5 mm, rugulose, papillate. **2*n***=24, 72.

Canada, northwestern United States. 3 varieties; 2 varieties treated in *Flora*.

Key to *Mertensia paniculata* varieties:

1. Calyces 3.2–4.9 mm; corollas usually glabrous within; stems usually glabrous; adaxial surfaces of cauline leaves glabrous or pustulose, seldom strigose; branches of inflorescence glabrous . . . . ***M. p.*** var. ***borealis***

1' Calyces (4)4.9–6.5 mm; corollas pubescent within; stems sparsely pubescent with spreading to crisped hairs, occasionally glabrous; adaxial surfaces of cauline leaves short strigose; branches of inflorescence strigose to spreading-pubescent . . . . ***M. p.*** var. ***paniculata***

Mertensia paniculata (Aiton) G. Don var. *borealis* (J.F. Macbride) L.O. Williams

Northern bluebells

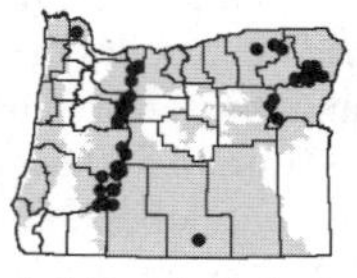

Plants 5–10 dm. **Stems** usually glabrous. **Leaves** cauline blades ovate to broadly ovate, surfaces abaxially strigose to spreading pubescent, adaxially glabrous or pustulose, seldom strigose; middle cauline blades 4–10 × 1.5–3 cm, length 1.8–2.8 × width. **Inflorescences** branches glabrous. **Flowers** calyces 3.2–4.9 mm, strigose, tips of lobes acute to acuminate; corollas glabrous within, occasionally pubescent.

Montane meadows, coniferous forests, stream and riverbanks, rocky slopes. Flowering May–Aug. 100–2400 m. BR, BW, ECas, Casc, Col, CR. ID, WA; north to British Columbia, northeast to MT. Native.

In C.V. Piper's (1906) treatment of Pacific Northwest *Mertensia*, he distinguished several species that L.O. Williams (1937) later treated as synonyms of *M. paniculata* var. *borealis*. Williams considered the great variability in pubescence (ranging from glabrous to strigose) as a reason to lump Piper's taxa. Recent studies based on molecular evidence (Nazaire & Hufford 2014) recovered variety *borealis* entities in two separate clades (provisionally treated as *M. brachycalyx* and *M. laevigata*). Further examination of herbarium specimens indicates that these two lineages may conform better to *M. brachycalyx* and *M. paniculata* var. *borealis*. A conservative approach is warranted until additional research is undertaken.

Mertensia paniculata (Aiton) G. Don var. *paniculata*

Northern bluebells, tall bluebells, tall lungwort

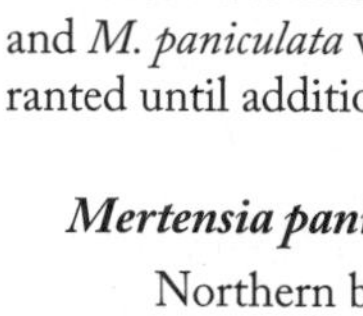

Plants (3)4–10 dm. **Stems** sparsely pubescent with spreading to crisped hairs, occasionally glabrous. **Leaves** cauline blades elliptic to lanceolate, surfaces abaxially strigose to spreading-pubescent, adaxially short-strigose; middle cauline blades 4–10 × 1.5–3 cm, length 1.9–3.8 × width. **Inflorescence** branches strigose to spreading-pubescent. **Flowers** calyces (4)4.9–6.5 mm, strigose, tips of lobes acute; corollas pubescent within.

Montane meadows, aspen woods, moist forests, stream and riverbanks. Flowering May–Jul. 600–2000 m. BW, Casc, ECas. ID, WA; north to AK, east to Quebec. Native.

Variety *paniculata* is among the most widespread and common variant in the species.

Mertensia subcordata Greene (p. 418)

Broadleaf bluebells

Mertensia platyphylla A. Heller, *Mertensia platyphylla* A. Heller var. *subcordata* (Greene) L.O. Williams

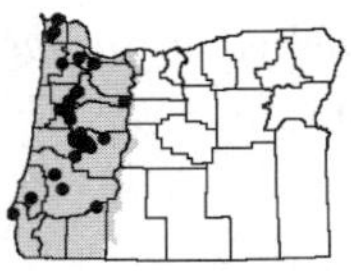

Plants 4–9 dm; woody taprooted or rhizomatous; caudices branched. **Stems** 1 to few, erect, glabrous. **Leaves** green, very thin, margins strigose-ciliate, lateral venation conspicuous, surfaces abaxially glabrous to sparsely hirsute, adaxially scabrous; basal blades ovate-subcordate, 12–15 × 8–10 cm, bases cordate, tips acute, petioles 12–25(27) cm; cauline alternate, blades broadly ovate to cordate, 7.5–13.5 × 3–8 cm, length 1.5–2.3 × width, bases rounded to cordate, tips acute; petioles 1–7 mm. **Inflorescences** rather open, branched; cymes 3–8-flowered; pedicels 2–40 mm, sparsely strigose. **Flowers** calyces 2.5–4.5 mm, cleft 67–80% the length to base, abaxial surface sparsely strigose, lobes deltoid, 1.8–3.5 mm, margins ciliate-strigose, tips subacute; corollas tubular with dilated to funnelform throat, tube and throat distinct, 10–16.5 mm, blue, glabrous within, tubes 4–7 mm, 0.5–0.8 × length of combined throat and limb, exceeding calyx lobes; fornices conspicuous; filaments inserted 0.5–0.8 mm below fornices, ligular, 1.8–3.5 × 1–2 mm, anthers 3.5–5 × 0.9–1.4 mm, (1.1)1.8–2.3 × length of filaments; styles 9–16 mm, stigmas exserted, surpassing anthers. **Fruits** 4–5.5 mm, rugulose, papillate.

Streambanks, moist to wet woods, damp thickets. Flowering Apr–Jun. 0–1000 m. Casc, CR, Est, Sisk, WV. Native.

Formerly treated as a variety of *M. platyphylla*, *M. subcordata* has a smaller calyx with deltoid lobes (vs. lanceolate to linear-lanceolate) than observed in *M. platyphylla*.

Mertensia umbratilis Greenman

Shade bluebells, shade lungwort

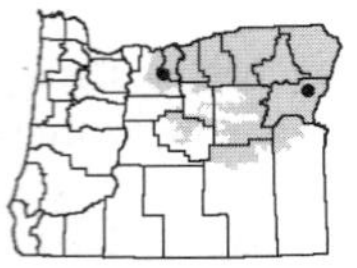

Plants (3.5)4–6 dm; woody taprooted; caudices branched. **Stems** 1 to few, erect, glabrous. **Leaves** green, thin, margins strigose-ciliate, lateral venation conspicuous, surfaces abaxially glabrous, adaxially sparsely pustulose to short-strigose; basal blades lanceolate, elliptic or ovate, 9–12.5 × 2.8–5 cm, bases attenuate to cuneate, tips acute, petioles 9–11.5 cm; cauline alternate, blades elliptic, ovate, to obovate, 2.8–11 × 1.5–3.8 cm, length 2–5.5 × width, bases cuneate, tips acute, petioles 1.5–6 cm, sessile distally. **Inflorescences** compact to somewhat open, branched or not; cymes 3–12-flowered; pedicels 2.4–11.3 mm, glabrous to sparsely pustulose. **Flowers** calyces 3.6–5.5 mm, cleft 67–75% to the base, abaxial surface pustulose at base, otherwise glabrous, lobes lanceolate to ovate, 3.1–4.6 mm, margins strigose-ciliate, tips acute; corollas tubular with funnelform throat, tube and throat distinct, 15.5–21 mm, pale blue, glabrous within, tubes 8.3–13.7 mm, 1.2–1.9 × length of combined throat and limb, exceeding calyx lobes; fornices conspicuous; filaments inserted 0.2–0.5 mm below fornices, ligular, 3.3–4 × 0.8–1.2 mm, anthers 1.8–2.4 × 0.5–0.9 mm, 0.4–0.7 × length of filaments; styles 14.4–17.2 mm, stigmas included to scarcely exserted, surpassing anthers. **Fruits** 3.6–4.6 mm, rugulose, papillate.

Rocky slopes, sagebrush communities. Flowering Apr–Jun. 1100–1400 m. BW, Col. WA. Native.

This species has a relatively narrow distribution and is presumed to occur from northeastern Oregon to central Washington. Relatively few specimens have been identified correctly as *M. umbratilis*. According to C.L. Hitchcock et al. (1955–1969), *M. umbratilis* was the result of hybridization and subsequent introgression between *M. oblongifolia* and *M. paniculata*. Continued recognition of *M. umbratilis* at the species level is questionable. More study is required.

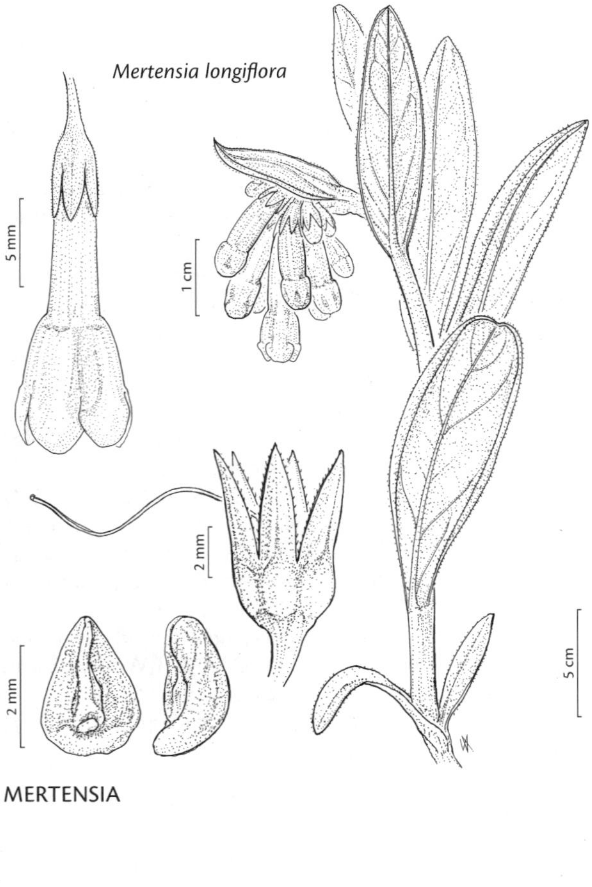

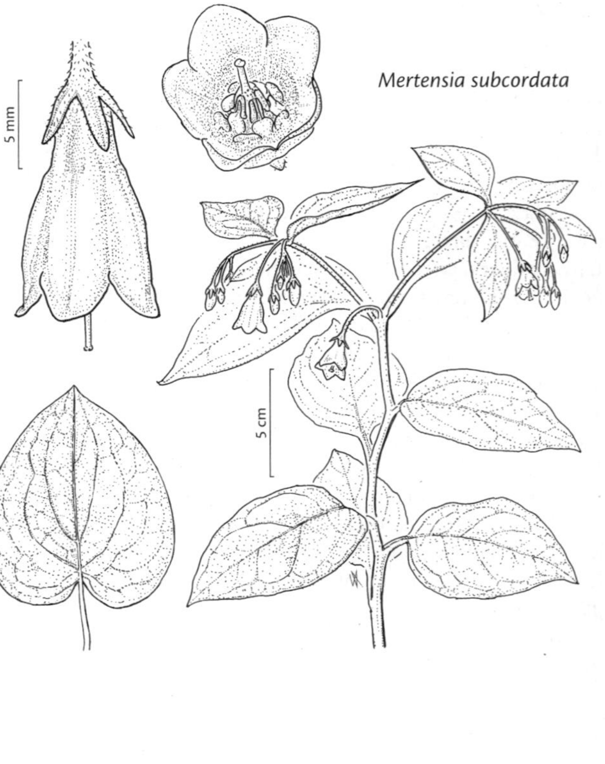

MERTENSIA

MYOSOTIS Forget-me-not

Christian Feuillet

Herbs annual, biennial, or perennial; roots generally fibrous. **Leaves** basal and cauline, simple, margins entire, unlobed. **Inflorescences** terminal and often axillary, scorpioid, 1-sided cymes. **Flowers** radially symmetric; calyces 5-lobed, lobes erect; corollas tubular with 5 spreading lobes, usually blue or grayish blue, sometimes white, pink, purple, or yellow, usually with white spots at base of lobes, dark yellow beneath white spots, margins of lobes curving upward and exposing narrow white triangle, 10-branched star surrounding fornices and throat, limbs blue or white, rarely purple or pink, white beneath. **Fruits** 4, erect, usually smooth, with raised margins; attachment scars basal.

Asia, Europe, New Zealand, North America, South America. ~50 species; 7 species treated in *Flora*.

Escaped *Myosotis sylvatica* has been recorded once in Oregon (1986), near Corbett, Multnomah County.

Key to *Myosotis* species:

1. Stems, leaves, and calyx tubes strigose; trichomes of calyx tubes not hooked.
    2. Calyces divided for < 40% of length, tubes narrowly campanulate or cylindric at base; corolla limbs 2–5 mm wide, grayish blue . . . . *M. laxa*
    2' Calyces divided for > 40% of length, tubes broadly campanulate or funnel-shaped; corolla limbs 5–10 mm wide, pink, maturing to intense blue . . . . *M. scorpioides*

1' Stems, leaves, and calyx tubes hirsute; some or all trichomes of calyx tubes hooked.
    3. Most cauline leaves petiolate, but some subsessile; fruiting pedicels ≈ calyces . . . . *M. latifolia*
    3' Cauline leaves sessile to subsessile; fruiting pedicels < or > calyces, but not equal.
        4. Pedicels > calyces, rarely < . . . . *M. arvensis*
        4' Pedicels > calyces.
            5. Calyx lobes unequal; corolla fornices small, greenish white . . . . *M. verna*
            5' Calyx lobes equal; corolla fornices well developed, yellow.
                6. Inflorescences bearing flowers entire length; corolla limbs pale blue . . . . *M. micrantha*
                6' Inflorescences bearing flowers only on upper ½; corolla limbs yellow when young, maturing to white or blue . . . . *M. discolor*

Myosotis arvensis (Linnaeus) Hill

Field forget-me-not

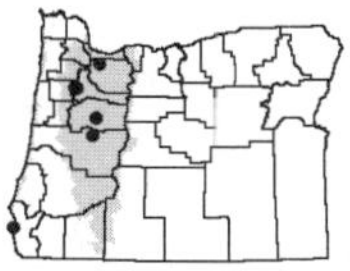

Plants annual. **Stems** erect, 1–4 dm, softly hirsute. **Leaves** blades narrowly lanceolate, elliptic or oblanceolate, tips acute to rounded, surfaces softly hirsute; basal petiolate, cauline sessile. **Inflorescences** pedicels > calyces, rarely shorter. **Flowers** calyx tubes hirsute with erect hooked trichomes, lobes equal; corolla limbs 3–5 mm wide, grayish blue with white star in center; fornices yellow when young, becoming white. **Fruits** surpassing style, brown to black.

Open, well-drained areas. Flowering May–Jul. 0–300 m. Casc, Est, WV. ID, WA; southern Canada, northern US; Europe. Exotic.

Myosotis discolor Persoon (p. 419)

Changing forget-me-not, yellow and blue forget-me-not

Myosotis versicolor (Pers.) J.E. Sm.

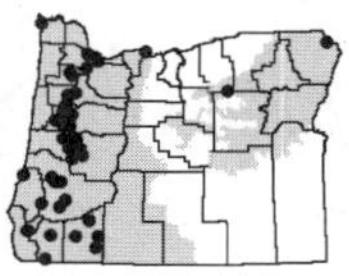

Plants annual or perennial. **Stems** erect, 1–5 dm, softly hirsute. **Leaves** blades oblanceolate, surfaces softly hirsute; basal petiolate, cauline sessile. **Inflorescences** pedicels < calyces. **Flowers** calyx tubes hirsute with erect hooked trichomes, lobes equal; corolla limbs 1–4 mm wide, yellow when young, maturing white or blue, with small white marking at base of lobes; fornices well developed, yellow. **Fruits** not surpassing style, dark brown to black. $2n$=64.

Disturbed areas, freshwater wetlands. Flowering Apr–Jun. 0–1500 m. BW, Casc, CR, ECas, Est, Sisk, WV. CA, ID, NV, WA; throughout Canada and US except Midwest; Europe. Exotic.

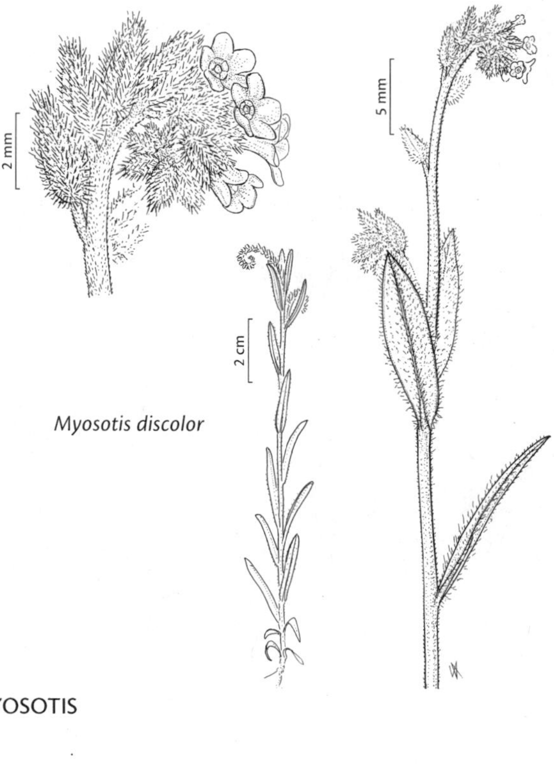

Myosotis discolor

MYOSOTIS

Myosotis latifolia Poiret

African forget-me-not, broad-leaf forget-me-not

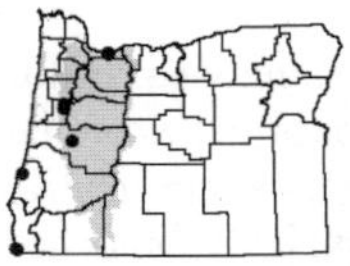

Plants perennial. **Stems** erect, up to 6 dm, woody at base, softly hirsute. **Leaves** blades oval, tips rounded to obtuse, surfaces hirsute; most leaves petiolate, but some cauline subsessile. **Inflorescences** fruiting pedicels ≈ calyces. **Flowers** calyx tubes hirsute with erect hooked trichomes; corolla limbs blue or pink, up to 10 mm wide, with white star in center; fornices orange, becoming white. **Fruits** surpassing style, dark brown. **2*n***=18.

Disturbed areas. Flowering Apr–Jul. 0–600 m. Casc, Est, WV. CA, WA; Africa. Exotic.

Myosotis laxa Lehmann (p. 420)

Small-flowered forget-me-not, tufted forget-me-not

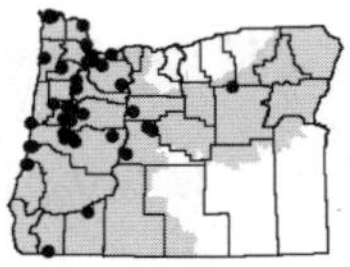

Plants annual, biennial, or perennial. **Stems** erect, 1–4 dm, strigose. **Leaves** blades narrowly elliptic, tips usually acute, sometimes rounded, surfaces strigose; basal petiolate or not, cauline sessile. **Inflorescences** pedicels > calyces. **Flowers** calyces divided for < 40% of length, tubes narrowly campanulate or cylindric at base, surfaces strigose, trichomes not hooked; corolla limbs 2–5 mm wide, grayish blue, with white star in center; fornices yellow. **Fruits** surpassing style, brown to black. **2*n***=80.

Freshwater wetlands. Flowering May–Oct. 0–2000 m. BW, Casc, CR, ECas, Est, Lava, Sisk, WV. CA, ID, NV, WA; throughout Canada and US except Midwest. Native.

Myosotis micrantha Pallas (p. 420)

Blue forget-me-not, mouse ear

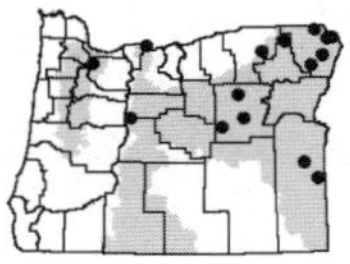

Plants annual. **Stems** erect, up to 2 dm, hirsute. **Leaves** oblanceolate to elliptic to oblong, tips usually rounded, surfaces hirsute; sessile or subsessile. **Inflorescences** pedicels < calyces. **Flowers** calyx tubes hirsute with erect hooked trichomes, lobes equal; corolla limbs pale blue, 1–2 mm wide, lobes ascending; fornices well developed, yellow. **Fruits** surpassing style, dark brown. **2*n***=36–40.

Disturbed areas. Flowering Apr–Jun. 0–1700 m. BW, ECas, Lava, Owy, WV. CA, ID, NV, WA; throughout Canada and US except Midwest; Asia, Europe. Exotic.

Myosotis scorpioides Linnaeus

Common forget-me-not, water scorpion grass

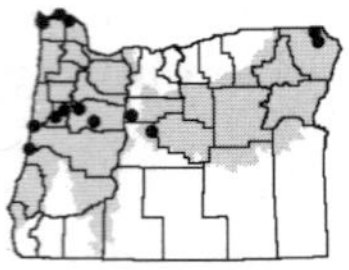

Plants perennial. **Stems** erect, 2–6 dm, strigose. **Leaves** blades elliptic to narrowly elliptic, surfaces strigose; sessile. **Inflorescences** pedicels > calyces. **Flowers** calyces divided for > 40% of length, tubes broadly campanulate or funnel-shaped, strigose; corolla limbs pink, maturing to intense blue, 5–10 mm wide, with white star in center; fornices yellow. **Fruits** ≤ style, rarely surpassing, black. **2*n***=64.

Freshwater wetlands. Flowering May–Sep. 0–2100 m. BW, Casc, CR, Est, Lava, WV. CA, ID, NV, WA; southern Canada, northern and eastern US; Europe. Exotic.

Myosotis verna Nuttall

Spring forget-me-not

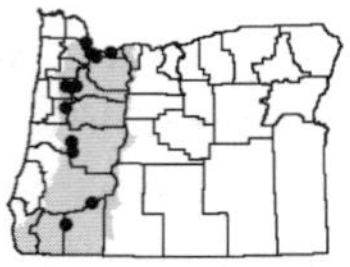

Plants annual or biennial. **Stems** 1.5–3 dm, hirsute. **Leaves** blades oblanceolate, tips usually rounded, surfaces hirsute; basal shortly petiolate, cauline sessile. **Inflorescences** pedicels < calyces. **Flowers** calyx tubes hirsute with erect hooked trichomes, lobes unequal; corollas white or light purple, ~3 mm wide; fornices small, greenish white. **Fruits** surpassing style, brown.

Dry foothills and lowlands. Flowering May–Jun. 0–1400 m. Casc, Sisk, WV. CA, ID, WA; southern Canada, throughout US. Native.

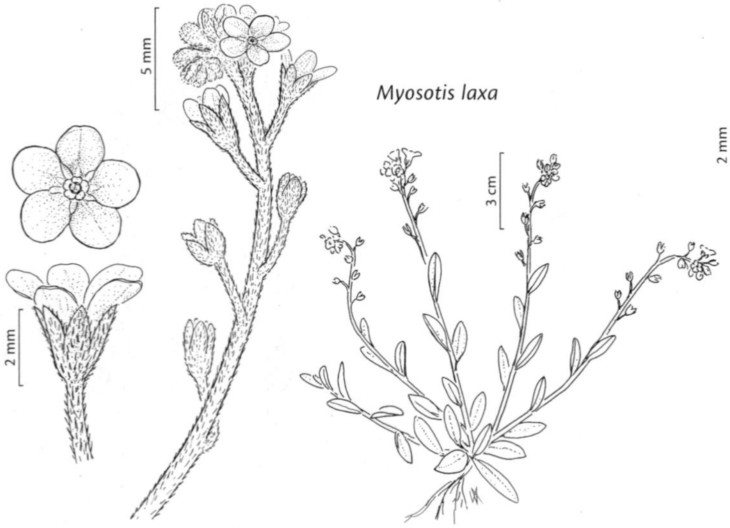

Myosotis laxa

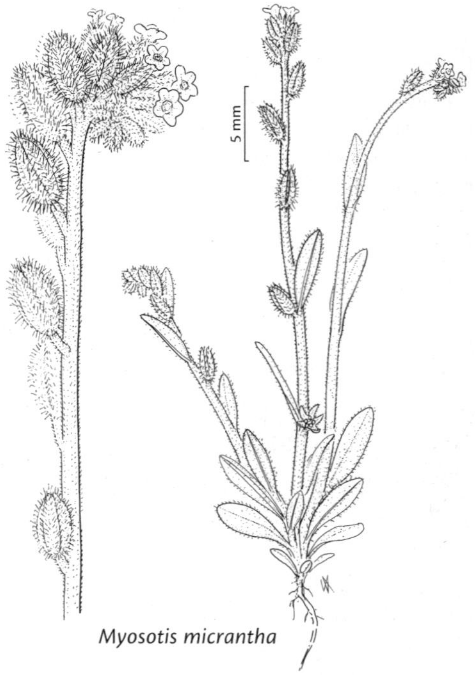

Myosotis micrantha

MYOSOTIS

OREOCARYA Oreocarya

Christian Feuillet

Herbs perennial, rarely biennial; ± branched; branched caudex topped by rosettes of leaves with obvious indumentum, taprooted. **Stems** ascending to erect, hairs simple or divaricate, unicellular. **Leaves** basal or tufted on short stems, usually spatulate or oblanceolate, margins entire, hairs simple or divaricate, unicellular. **Inflorescences** terminal, spike-like, scorpioid cymes; branches often elongated in fruit; bracts absent or present. **Flowers** calyces persistent, lobes free or fused at base; corollas salverform, white or yellow, tubes <, =, or > calyces; stamens included, anthers subsessile to sessile; styles nearly reaching to exceeding mature nutlets, stigmas entire; flower bracts absent. **Fruits** 1–4, smooth, rugose, or scabrous, margins not elevated, tips broadly rounded to obtuse, ventral grooves not extending to nutlet tips; attachment scars lateral.

North America, South America. 63 species; 8 species treated in *Flora*.

Members of *Oreocarya* have most often been treated as a section or a subgenus of *Cryptantha*. However, recent molecular studies suggest that other genera, including *Amsinckia* and *Plagiobothrys*, are more closely related to *Cryptantha* than *Oreocarya*.

Key to *Oreocarya* species:

1. Stems of 2 kinds: sterile stems with terminal clusters of leaves and fertile stems with terminal inflorescences . . . ***O. sobolifera***
1' Stems of 1 kind: fertile stems with terminal inflorescences.
  2. Adaxial surface of nutlets ± smooth.
    3. Stems with some bristly hairs; nutlet scars closed . . . ***O. schoolcraftii***
    3' Stems strigose or villosulous and strigose; nutlet scars open . . . ***O. subretusa***
  2' Adaxial surface of nutlets rugose, tuberculate, or ridged.
    4. Scars of nutlets open some distance from the base, narrowly triangular to triangular near base.
      5. Leaves strigose-pustulose and tomentose . . . ***O. humilis***
      5' Leaves finely strigose or strigose and appressed-setulose, pustulose hairs lacking or inconspicuous . . . ***O. propria***
    4' Scars of nutlets ± closed, linear to linear-lanceolate near base.
      6. Styles ≤ mature nutlets . . . ***O. interrupta***
      6' Styles > mature nutlets.
        7. Corolla limbs 8–12 mm wide; plants biennials or short-lived perennials . . . ***O. glomerata***
        7' Corolla limbs 4–8 mm wide; plants perennials . . . ***O. spiculifera***

Oreocarya glomerata (Fraser f. ex Pursh) Greene (p. 422)

Buttecandle, cockscomb oreocarya

Cryptantha celosioides (Eastw.) Payson, *Cryptantha sheldonii* (Brand) Payson, *Oreocarya celosioides* Eastw.

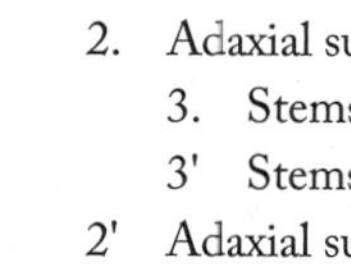

Plants biennial or short-lived perennial, cespitose to loosely cespitose, 10–50 cm. **Stems** of one kind, abundantly strigose, hairs divaricate. **Leaves** basal and cauline similar, spatulate to oblanceolate, 2–8 cm, smaller distally, margins slightly strigose-ciliate, surfaces abundantly strigose, usually with bristles. **Inflorescences** bracts inconspicuous. **Flowers** calyces 3–5 mm, longer in fruit, lobes linear-lanceolate, tips acute, surfaces densely strigose; corollas white, tubes ≤ calyces, limbs 8–12 mm wide; fornices white; styles exceeding mature nutlets. **Fruits** broadly lanceolate, 3–5 mm, tips subacute to acute, adaxial surfaces tuberculate, rarely also rugose; attachment scars closed, linear near base.

Rocky areas. Flowering Apr–Aug. 300–2500 m. BR, BW, Col, Lava, Owy. CA, ID, NV, WA; north to British Columbia, northeast to Saskatchewan, east to SD, southeast to CO. Native.

Oreocarya glomerata is nearly indistinguishable from *O. spiculifera*. The only reliable morphological feature is the size of the corollas (*O. glomerata* corollas are usually larger).

Oreocarya humilis (A. Gray) Greene (p. 422)

Low oreocarya

Cryptantha humilis (A. Gray) Payson

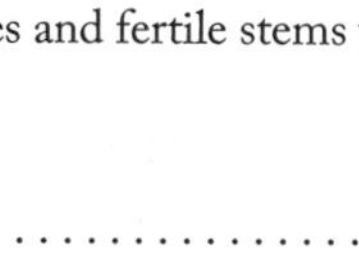

Plants perennial, cespitose, 5–30 cm; caudices woody, branched. **Stems** of one kind, strigose, with weak bristles. **Leaves** basal and cauline similar, obovate to oblanceolate, 1–6 cm, smaller distally, tips usually obtuse, surfaces strigose-pustulose and tomentose, with appressed or slightly spreading bristles. **Inflorescences** bracts inconspicuous. **Flowers** calyces 3–5 mm, 5–11 mm in fruit, > nutlets, lobes linear, tips acute, surfaces densely strigose with slender bristles; corollas white, tubes ≈ calyces, limbs 8–10 mm wide; fornices yellow; styles exceeding mature nutlets. **Fruits** ovate to lanceolate, 3–4.5 mm, tips acute or obtuse, surfaces somewhat glossy, adaxial surfaces tuberculate to muricate-tuberculate with a few irregular ridges; attachment scars narrowly open to open, narrowly triangular near base.

Alpine ridges. Flowering May–Jun. 1000–1600 m. BR, Owy. CA, ID, NV; northeast to MT, southeast to AZ. Native.

Oreocarya interrupta Greene

Bristly oreocarya

Cryptantha interrupta (Greene) Payson

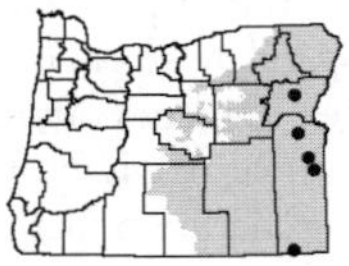

Plants perennial, 5–40 cm; caudices branched. **Stems** of one kind, strigose with white hairs. **Leaves** basal and cauline similar, oblanceolate, 2–5 cm, smaller distally, tips obtuse or acute, surfaces densely strigose or tomentose, pustulose, with appressed, inconspicuous bristles, less marked adaxially. **Inflorescences** bracts conspicuous at bases. **Flowers** calyces 2–4 mm, 5–8 mm in fruit, > nutlets, strigose, lobes linear to linear-lanceolate; corollas white, tubes < calyces, limbs 5–7 mm wide; fornices pale yellow; styles ≤ mature nutlets. **Fruits** narrowly lanceolate, 3 mm, tips obtuse to subacute, surfaces dull to slightly glossy, adaxially tuberculate and muricate; attachment scars ± closed, linear to linear-lanceolate near base.

Open woods, sagebrush. Flowering May–Jun. 1000–1200 m. BR, BW, Owy. ID, NV; southeast to UT. Native.

Oreocarya propria A. Nelson & J.F. Macbride (p. 422)

Malheur oreocarya, Idaho oreocarya

Cryptantha propria (A. Nelson & J.F. Macbr.) Payson

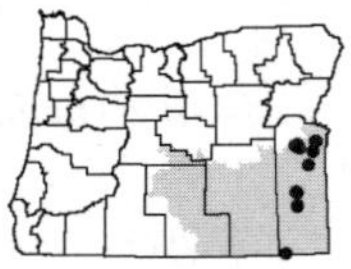

Plants perennial, cespitose, 10–30 cm; taproots strongly lignified. **Stems** of one kind, strigose. **Leaves** oblanceolate to spatulate, surfaces abaxially finely and uniformly strigose, adaxially finely strigose and scattered appressed-strigose or uniformly strigose, pustulose hairs lacking or inconspicuous. **Inflorescences** bracts inconspicuous. **Flowers** calyces 4–6 mm, 7–13 mm in fruit, surfaces densely strigose-hirsute, lobes linear-lanceolate; corollas white, tubes < calyces, limbs 6–8 mm wide; fornices yellow; styles exceeding mature nutlets. **Fruits** ovate-lanceolate, 3–4 mm, adaxial surfaces tuberculate or short-muricate; attachment scars narrowly open, usually narrowly triangular near base.

Clay soil in deserts. Flowering May–Jun. 700–1100 m. BR, Owy. ID. Native.

Oreocarya schoolcraftii (Tiehm) R.B. Kelley

Schoolcraft's oreocarya

Cryptantha schoolcraftii Tiehm

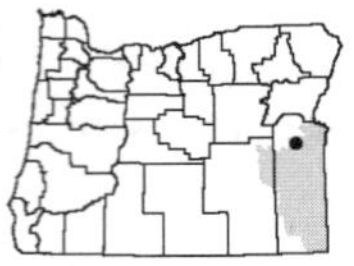

Plants short-lived perennial, sometimes biennial, 6–25 cm; caudices unbranched. **Stems** of one kind, with spreading, bristly hairs and soft, shorter hairs. **Leaves** surfaces densely gray-pubescent with bristly and soft shorter hairs; basal leaves oblanceolate to spatulate. **Inflorescences** bracts conspicuous at bases. **Flowers** calyces 2–4 mm, 5–7 mm in fruit; corollas white, tubes ≈ calyces, limbs spreading flat, 3–6 mm wide; fornices yellow; styles exceeding mature nutlets. **Fruits** lance-ovate, 2.2–3.2 mm, adaxial surfaces shiny, ± smooth; attachment scars elongate, closed.

Desert, sagebrush scrub. Flowering May–Jul. 800–1000 m. Owy. CA, NV. Native.

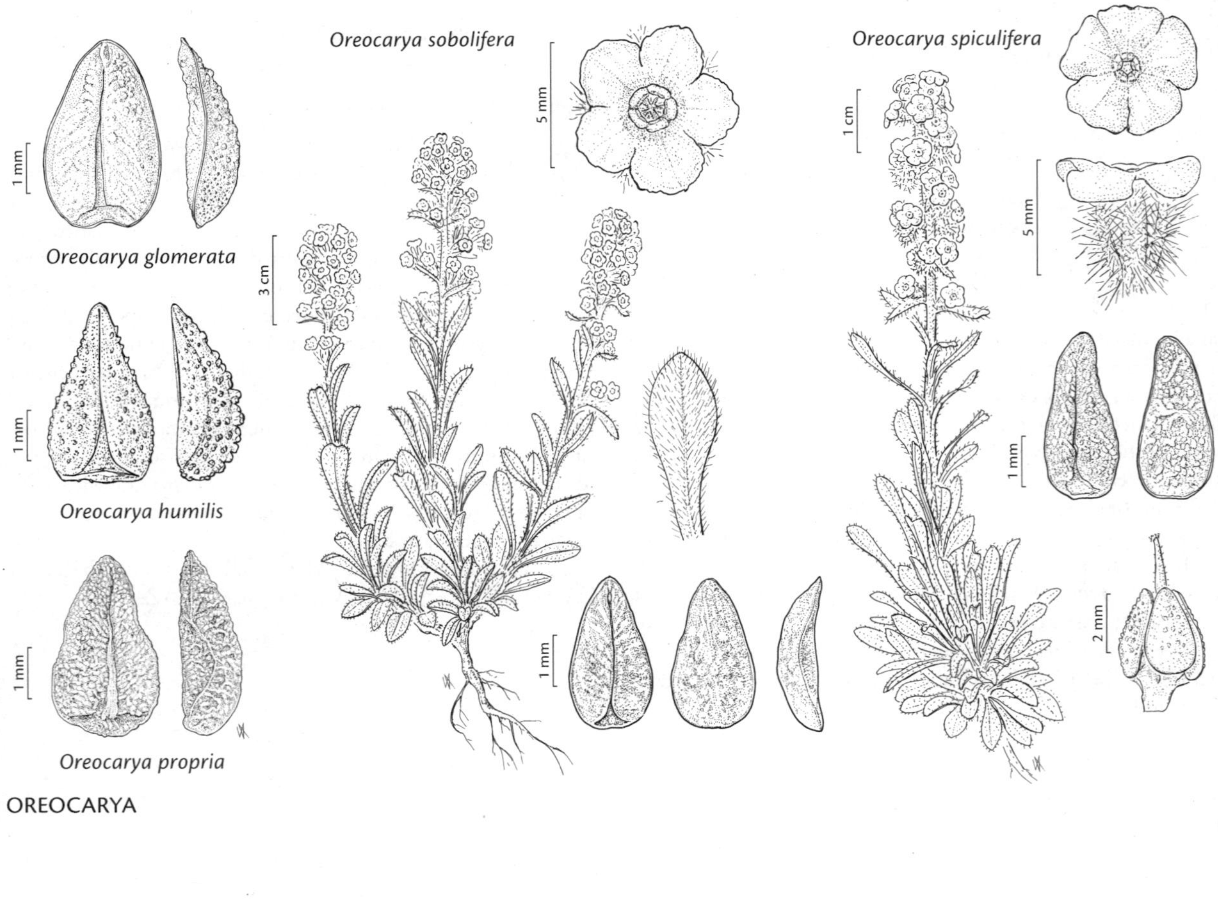

OREOCARYA

Oreocarya sobolifera (Payson) R.B. Kelley (p. 422)

Waterton Lakes oreocarya

Cryptantha sobolifera Payson

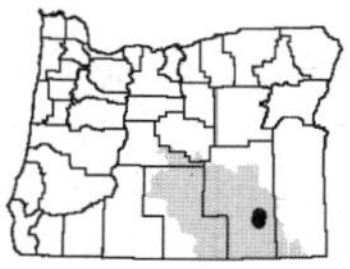

Plants perennial, 5–15 cm; caudices branched. **Stems** of two kinds, hirsute, strigose; sterile stems short, with terminal clusters of leaves; fertile stems taller, with terminal inflorescences. **Leaves** spatulate to oblanceolate, surfaces strigose and hirsute, somewhat tomentose, conspicuously pustulose. **Inflorescences** bracts inconspicuous to conspicuous. **Flowers** calyces 3–5 mm, 6–7 mm in fruit, strigose, lobes linear-lanceolate; corollas white, tubes ≤ calyces, limbs 6–8 mm wide; fornices yellow; styles exceeding mature nutlets. **Fruits** ovate-lanceolate, 2.5–3.5 mm, adaxial surfaces ± smooth; attachment scars closed, linear near base.

Cliffs, talus. Flowering Jul–Aug. 2500–3100 m. BR. ID, NV; northeast to MT. Native.

Oreocarya sobolifera is the only species of *Oreocarya* in Oregon with sterile stems.

Oreocarya spiculifera Piper (p. 422)

Snake River oreocarya, bristly oreocarya

Cryptantha spiculifera (Piper) Payson

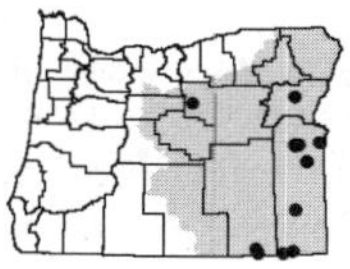

Plants perennial, cespitose, 20–30 cm. **Stems** of one kind, strigose. **Leaves** narrowly oblanceolate, densely strigose and spreading strigose, pustulose, strongly strigose-ciliate at margins. **Inflorescences** bracts inconspicuous. **Flowers** calyces 4–5 mm, 6–10 mm in fruit, lobes linear-lanceolate, surfaces strigose and spreading strigose; corollas white, tubes < calyces, limbs 4–8 mm wide; fornices yellow; styles exceeding mature nutlets. **Fruits** lanceolate to ovate-lanceolate, 2.8–4 mm, adaxial surfaces rugose and tuberculate; attachment scars ± straight, subulate, ± closed, linear to linear-lanceolate near base.

Dry areas, sandy or clay soils. Flowering May–Jun. 300–1700 m. BR, BW, Lava, Owy. ID, NV, WA; northeast to MT, southeast to UT, AK. Native.

See *O. glomerata* discussion.

Oreocarya subretusa (I.M. Johnston) Abrams

Crater Lake oreocarya

Cryptantha subretusa I.M. Johnst.

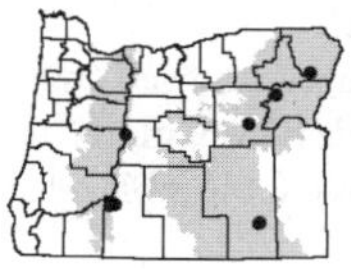

Plants perennial, cespitose, 5–20(25) cm. **Stems** of one kind, strigose. **Leaves** spatulate, surfaces tomentose and weakly strigose, conspicuously pustulose abaxially, less so adaxially. **Inflorescences** bracts inconspicuous. **Flowers** calyces 3–4 mm, 6–9 mm in fruit, lobes lanceolate, surfaces strigose and subtomentose; corollas white, tubes ≈ calyces, limbs 4–6 mm wide; fornices yellow; styles exceeding mature nutlets. **Fruits** lanceolate, 2.5–4 mm, adaxial surfaces ± smooth, inconspicuously tuberculate; attachment scars open, subulate or narrowly linear.

Talus slopes of volcanic origin, pumice. Flowering Jul–Aug. 2000–3000 m. BR, BW, Casc. CA, NV. Native.

PECTOCARYA Pectocarya, combseed, comb-bur

C. Matt Guilliams and Ronald B. Kelley

Herbs annual; taprooted. **Stems** cinereous-strigose. **Leaves** basal rosettes usually absent, cauline usually alternate, opposite proximally, surfaces usually strigose and appressed-setulose; ± sessile. **Inflorescences** 1–20(25) cm; cymules usually terminal and solitary, occasionally paired or branched, scorpioid and circinate, few- to usually many-flowered; flowers usually with persistent subtending bracts. **Pedicels** 0.2–0.6 mm at anthesis, erect to ascending or recurved, usually free from mature nutlets, rarely adnate. **Flowers** homostylous; calyces deeply 5-parted, usually bilaterally symmetric, 4 lobes aligned with nutlets, lobes erect to ascending, divergent, usually equal in size and shape, usually linear to lanceolate, (0.5)0.7–1.2(1.7) mm at anthesis, usually < mature nutlets; corollas usually funnelform to tubular, 1–2(2.5) mm, limbs white, tubes pale yellow; fornices usually present, ± yellow initially, usually minute; ovaries 4-lobed, styles short. **Fruits** (3)4, spreading, bilaterally symmetric, arranged in 2 pairs, usually in same plane, yellowish brown; attachment scar proximally ± round, irregular; gynobases low-pyramidal.

Argentina, Chile, western North America. 13 species; 2 species treated in *Flora*.

Fruit and nutlet features are variable and taxonomically useful in *Pectocarya*. Nutlet characters such as size, shape, wing development, and margin ornamentation vary between taxa. In addition, these nutlet characters can vary discontinuously between basal and cauline fruits on an individual, termed fruit dimorphism. When fruits are dimorphic, nutlets in basal fruits are usually smaller and/or with reduced wing development and reduced margin ornamentation when compared to nutlets in cauline fruits. These basal fruits are from cleistogamous flowers that may be partly to completely buried in the soil at anthesis. Fruit monomorphism is the common condition, where all fruits are ± identical, cauline, and arise from chasmogamous flowers.

Nutlet features can also vary within a single fruit on an individual, termed nutlet heteromorphism. When nutlets in a fruit are heteromorphic, it is usually the nutlets positioned nearer to the inflorescence axis that are smaller and/or less ornamented when compared to the nutlets positioned away from the inflorescence axis. Nutlets are homomorphic when they are ± identical within a single fruit, similar in size, shape, wing development, and/or margin ornamentation.

Key to *Pectocarya* species:

1. Nutlet bodies ± oblong; calyces ± bilaterally symmetric, lobes < nutlets . . . . ***P. penicillata***

1' Nutlet bodies round to ± obovate; calyces radially symmetric, lobes > nutlets . . . . ***P. setosa***

Pectocarya penicillata (Hooker & Arnott) A. de Candolle (p. 424)

Northern pectocarya, sleeping combseed, winged pectocarya

Pectocarya linearis (Ruíz & Pav.) DC. var. *penicillata* (Hook. & Arn.) M.E. Jones

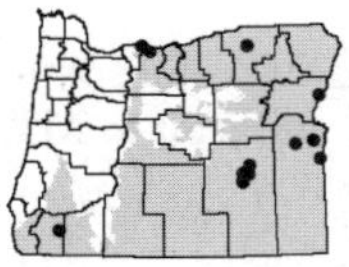

Stems prostrate to laxly ascending, 2–25 cm. **Leaves** linear, 10–40 × 0.5–2 mm. **Pedicels** ascending, 1.2–2.5 mm in fruit. **Flowers** chasmogamous; calyces ± bilaterally symmetric, lobes 1.5–2.2 mm in fruit, < nutlets; corolla limbs 0.5–1.5 mm in diameter. **Fruits** monomorphic, flat; nutlets ± homomorphic, straight, ± oblong, 1.6–2.5 mm; wings thick and partially revolute, ± hourglass-shaped, (0.3)0.5–0.8 mm wide, lobed along each long edge, margins ± pectinate, distally tufted with uncinate hairs and bristles, central portion usually glabrous, proximal portion occasionally bristly. ***2n***=24.

Sandy to fine gravelly soils, sagebrush scrub, disturbed areas, grassland. Flowering Apr–Jun. 0–1500 m. BR, BW, Col, ECas, Owy, Sisk. CA, ID, NV, WA; north to British Columbia, east to WY, south to Mexico. Native.

The most broadly distributed of the *Pectocarya* taxa, *P. penicillata* is easily identified by its flat, bilaterally symmetric fruits of homomorphic nutlets, these bearing more or less hourglass-shaped wings with a distal tuft of uncinate bristles on both the wing and body.

Pectocarya setosa A. Gray (p. 424)

Round-nut pectocarya, stiff-stemmed comb-bur, moth combseed

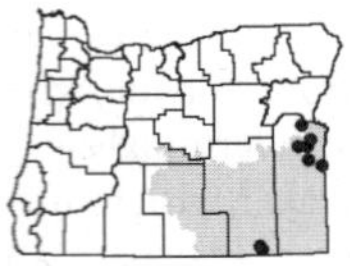

Stems ascending to erect, simple or branched, (2)4–10(23) cm. **Leaves** linear to linear-spatulate, 5–30 × 0.2–2.3 mm. **Pedicels** ascending to recurved, 0.7–2.8 mm in fruit. **Flowers** chasmogamous; calyces radially symmetric, lobes 3–4.5 mm in fruit, > nutlets, strigillose and sparsely setose; corolla limbs 0.5–1.5(2) mm in diameter. **Fruits** monomorphic, flat; nutlets usually heteromorphic, rarely homomorphic, straight, round to ± obovate, (1.5)2–4 mm, abaxial surface with scattered uncinate bristles; wings of 3 consimilar nutlets thin and scarious, spreading, 0.4–0.8 mm wide, surface sparsely to densely hairy, wing margins ± entire, with ± scattered uncinate bristles; wings of dissimilar lower nutlets nearer to inflorescence axis narrower, < 0.2 mm wide, or absent. ***2n***=24.

Sandy to coarse gravelly soils, often from granitic substrates, in washes, bajadas, slopes, roadsides, desert and sagebrush scrub, pinyon-juniper woodlands, dry grasslands. Flowering May–Jul. 800–1900 m. BR, Owy. CA, ID, NV, WA; southeast to NM, south to Mexico. Native.

I.M. Johnston (1924) recognized three varieties within *Pectocarya setosa*: variety *aptera*, variety *genuina*, and variety *holoptera*. Johnston stated that variety *genuina* is the typical form, with usually three of the four nutlets having a wide, scarious wing. Variety *aptera* is a homomorphic, unwinged form from mid-elevation desert transition habitats in southern California, while variety *holoptera,* from the western Mojave Desert, has fruits with all four nutlets equally winged.

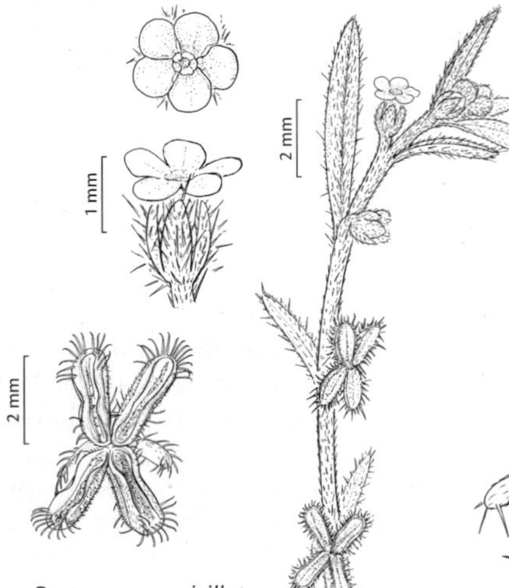

Pectocarya penicillata

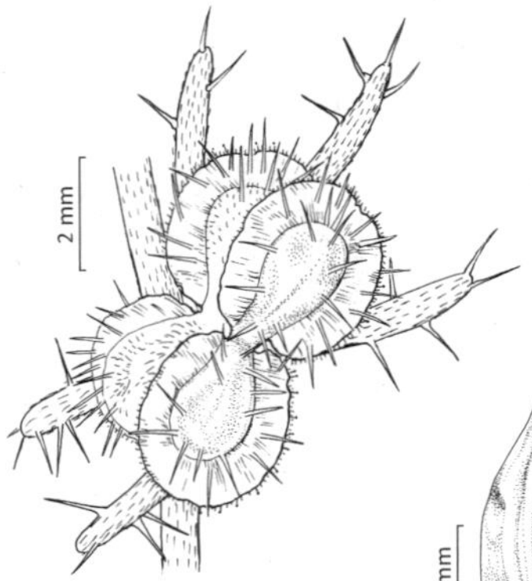

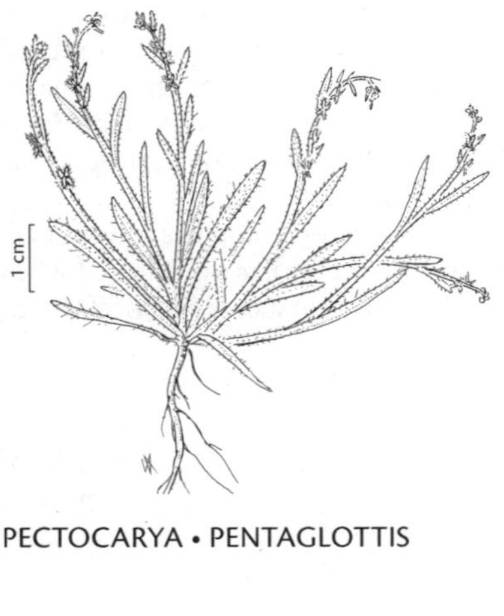

Pectocarya setosa

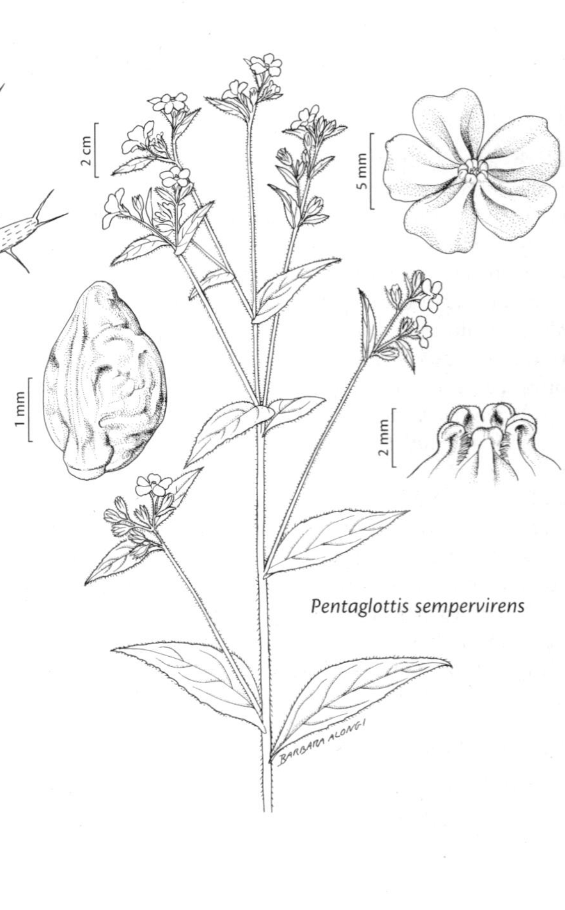

Pentaglottis sempervirens

PENTAGLOTTIS Alkanet

Richard R. Halse

Genus composed of one species, *P. sempervirens.*

Pentaglottis sempervirens (L.) Tausch ex L.H. Bailey (p. 424)

Green alkanet, evergreen bugloss

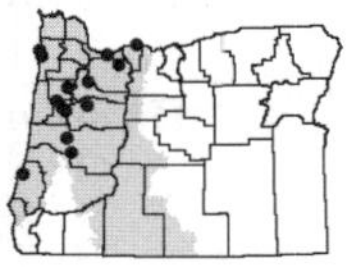

Herbs perennial, ascending to erect, hispid; taproots stout, deep. **Stems** simple to branched, 0.3–1.5 m, hispid. **Leaves** basal and proximal leaves 10–40 cm, ovate-oblong or ovate, with conspicuous venation, hispid; cauline alternate, 6–10 cm, margins entire to denticulate, tips acuminate, surfaces hispid; long-petiolate proximally, sessile distally. **Inflorescences** axillary and terminal bracteate cymes, 2(3) subcapitate cymes at end of peduncle; peduncles 2–11 cm; each cyme subtended by small leaf-like, lanceolate to ovate bract; pedicels ascending, 0–4 mm. **Flowers** 5–10(15) per cyme, rotate, radially symmetric; calyces 2.5–5 mm in flower, 5–10 mm in fruit, lobes 5, divided almost to base, linear-lanceolate, entire; corollas bright blue, 5-lobed, tubes funnelform to cylindric, 4–6 mm, throats closed by 5 ovate to oblong, hairy, white scales inserted at top of tube; stamens 5, inserted in distal tube, included; ovaries deeply 4-lobed; styles 1, gynobasic, included, stigmas capitate. **Fruits** 1–4, erect, ovoid, concave at base with thickened collar-like ring, 1.5–4 mm, tan to whitish, shortly stalked; attachment scars triangular; receptacles flattened, much < nutlets. **2*n*=22.**

Roadsides, riverbanks, ditches, waste areas, shady forested slopes. Flowering Apr–Jun. 0–400 m. Casc, CR, ECas, Est, WV. WA; north to British Columbia; Europe. Exotic.

PLAGIOBOTHRYS Popcornflower

C. Matt Guilliams, Kristen Hasenstab-Lehman, and Ronald B. Kelley

Herbs annual, rarely perennial, usually glabrate to strigose, villous, hirsute, and/or hispid, red-staining or not. **Roots** fibrous or taproots. **Stems** branched at bases or above, usually < 5(7) dm, rooting at nodes in some aquatic species. **Leaves** monomorphic or dimorphic in some aquatic species; monomorphic leaves or non-aquatic phase dimorphic leaves cauline or basal and cauline, flat, usually not fistulose, linear to oblanceolate, 0.5–10 cm, usually reduced distally; aquatic phase dimorphic leaves basal, terete, fistulose; generally sessile. **Inflorescence**s scorpioid cymules, appearing raceme- or spike-like, circinate in bud, usually elongate in fruit; bracts 0–many. **Flowers** calyx lobes fused at or usually below middle, sometimes accrescent, 1.5–10 mm in fruit; corollas rotate to funnel-shaped or cylindric, white or white with yellow throat, tubes usually ± yellow inside, limbs 1–12 mm in diameter; fornices white to yellow, prominent to minute; stamens included; gynobases low-pyramidal. **Fruits** usually 4, ± flat to strongly arched, narrowly lanceolate to widely ovate, rarely cross-shaped, 0.7–3.5 mm, rarely on narrow adaxial stalks or short basal stipes, abaxial surfaces with central longitudinal ridge, transverse ridges or ribs, transverse grooves, tubercles, prickles, bristles, or rarely smooth, adaxial surfaces usually with keel above attachment scar. **Attachment scars** near base to middle, oblique, or rarely on basal nutlet surfaces, usually raised, rarely on short adaxial stalk or basal stipe, linear to ± round, sometimes deep-concave to ± pit-like.

Northeast Asia, Australia, Europe, North America, South America. ~65 species; 20 species treated in *Flora.*

As in other Boraginaceae, features of the individual mericarps, or nutlets, have been used extensively in *Plagiobothrys* for taxonomic circumscription. Due, in part, to recent diversification in the genus, many taxa are separated by relatively small morphological differences. Fully mature nutlets are critical for identification; use of immature nutlets may result in an incorrect determination. In some species, the nutlet closest to inflorescence axis is often more firmly attached, larger, uniquely textured, and may have a different attachment scar morphology than the other three nutlets. The keys and descriptions here focus on the three more loosely attached nutlets. Finally, the position of the attachment scar varies among taxa from the middle of the adaxial or inner nutlet surface to the basal surface of the nutlet. We refer to an attachment scar position exactly between the adaxial and basal surfaces as oblique, in keeping with earlier treatments.

Corolla throat and appendage color may change from yellow to white after pollination; corolla size may decrease through the flowering season.

Morphological intergradation is common in some species and species groups, and many lower taxa and names in synonymy require further study, especially in *Plagiobothrys* section *Allocarya.* This is true of plants in the "*Plagiobothrys scouleri* complex," with members treated here including *P. bracteatus*, *P. cognatus*, *P. cusickii*, *P. figuratus*, *P. hirtus*, *P. hispidulus*, *P. reticulatus*, and *P. scouleri.* Taxonomic treatments for this difficult group have varied; for example, Hitchcock et al. (1959) recognized a broadly circumscribed *P. scouleri s.l.* that included the names *P. cognatus*, *P. cusickii*, *P. hispidulus*, *P. reticulatus,* and many others not recognized here as synonyms. We elect to recognize a greater number of lower taxa in this group, understanding that further study may result in increased or reduced support for these taxa.

Key to *Plagiobothrys* species:

1. Lower cauline leaves alternate, occasionally opposite, occasionally crowded; basal rosettes present; plants usually taprooted.
    2. Nutlet attachment scars deep-concave, pit-like . . . . **P. fulvus** var. **campestris**
    2' Nutlet attachment scars convex, rarely shallow-concave.
        3. Calyces circumscissile; nutlets round-ovate, abaxial surfaces transversely ribbed, spaces between ribs wide and flat, transverse grooves absent . . . . **P. nothofulvus**
        3' Calyces not circumscissile; nutlets cross-shaped, abaxial surfaces transversely grooved, grooves narrow, papillate.
            4. Inflorescences not coiled toward tip, not elongate in fruit; calyces 5–7 mm; nutlets 2–3 mm; flowers generally < 10 per branch . . . . **P. shastensis**
            4' Inflorescences strongly coiled toward tip, elongate in fruit; calyces usually 3–4 mm, rarely to 5 mm; nutlets 1–2 mm; flowers generally > 10 per branch . . . . **P. tenellus**

1' Lower cauline leaves opposite; basal rosettes 0; plants usually fibrous-rooted, becoming taprooted.
    5. Plants perennial; stems stout.
        6. Plants greenish to whitish, densely villous; prostrate to decumbent; inflorescences usually solitary. . **P. mollis** var. **mollis**
        6' Plants bright green, proximally glabrous, distally spreading-hirsute; usually erect proximally, sometimes reclining; inflorescences usually paired . . . . **P. hirtus** (in part)
    5' Plants annual; stems usually slender (but may be stout in *P. cognatus*).
        7. Nutlet attachment scars 25–60% fruit length, ± deep-concave.
            8. Nutlet abaxial surfaces usually with stout prickles, rarely absent, barb-tipped, hair-like bristles absent.
                9. Nutlets shiny, prickles few, on abaxial ridges, rarely absent . . . . **P. austiniae**
                9' Nutlets ± dull, prickles many, throughout abaxial surface . . . . **P. greenei**
            8' Nutlet abaxial surfaces usually without stout prickles, occasionally bristled. . . . . **P. glyptocarpus** var. **glyptocarpus**
        7' Nutlet attachment scars < 20% fruit length, usually ± flat or ± concave.
            10. Nutlet attachment scars on basal surfaces, generally on short stipes; calyx lobe midveins ± stout.
                11. Stems prostrate to decumbent; calyx lobes generally bent in same direction, generally strongly bent upward; nutlets narrowly lanceolate to lanceolate . . . . **P. leptocladus**
                11' Stems ascending to erect; calyx lobes erect to spreading, not bent in same direction; nutlets lanceolate to lance-ovate . . . . **P. stipitatus**
            10' Nutlet attachment scars near bases on adaxial surfaces or oblique, not on short stipes; calyx lobe midveins generally slender.
                12. Stem hairs spreading.
                    13. Inflorescences usually paired; bracts absent; corolla limbs 8–14 mm in diameter; nutlets lance-ovate to widely ovate . . . . **P. hirtus** (in part)
                    13' Inflorescences solitary; bracts many, dense throughout; corolla limbs 2–5 mm in diameter; nutlets compressed-lanceolate. . . . . **P. salsus**
                12' Stem hairs appressed.
                    14. Nutlet adaxial ridges in troughs.
                        15. Nutlets usually 4, (0.7)1–1.9 mm, dull . . . . **P. reticulatus** var. **reticulatus**
                        15' Nutlets 1–2, ± 1.5 mm, shiny . . . . **P. lamprocarpus**
                    14' Nutlet adaxial ridges not in troughs.
                        16. Nutlet attachment scars usually oblique.
                            17. Plants green; stems not stout; nutlets lance-ovate, usually ± symmetric; transverse ridges many, high, crowded, especially toward tips . . . . **P. bracteatus**
                            17' Plants pale green; stems stout; nutlets ± oblong-ovate, asymmetric; transverse ridges few, low, scattered . . . . **P. cognatus**
                        16' Nutlet attachment scars usually on adaxial surfaces, near or above bases.
                            18. Nutlet attachment scars narrow-elliptic to ± oblong; corolla limbs 1–2 mm in diameter.
                                19. Nutlets plump, ± symmetric, sharp-roughened, often also minute-bristled or scabrous, dull, adaxial ridges at middle, generally not folded over to 1 side . . . **P. hispidulus**
                                19' Nutlets ± flat, asymmetric, smooth to rounded-roughened, rarely bristled or scabrous, shiny, adaxial ridges beyond middle, ± folded over to 1 side below . . **P. cusickii**
                            18' Nutlet attachment scars widely lanceolate to ± ovate; corolla limbs 2–8 mm in diameter.
                                20. Inflorescences usually paired; corolla limbs 4–8 mm in diameter; nutlet attachment scars on adaxial surface, above bases. . . . . **P. figuratus**
                                20' Inflorescences usually solitary; corolla limbs 2–5 mm in diameter; nutlet attachment scars on adaxial surface, near bases. . . . . **P. scouleri**

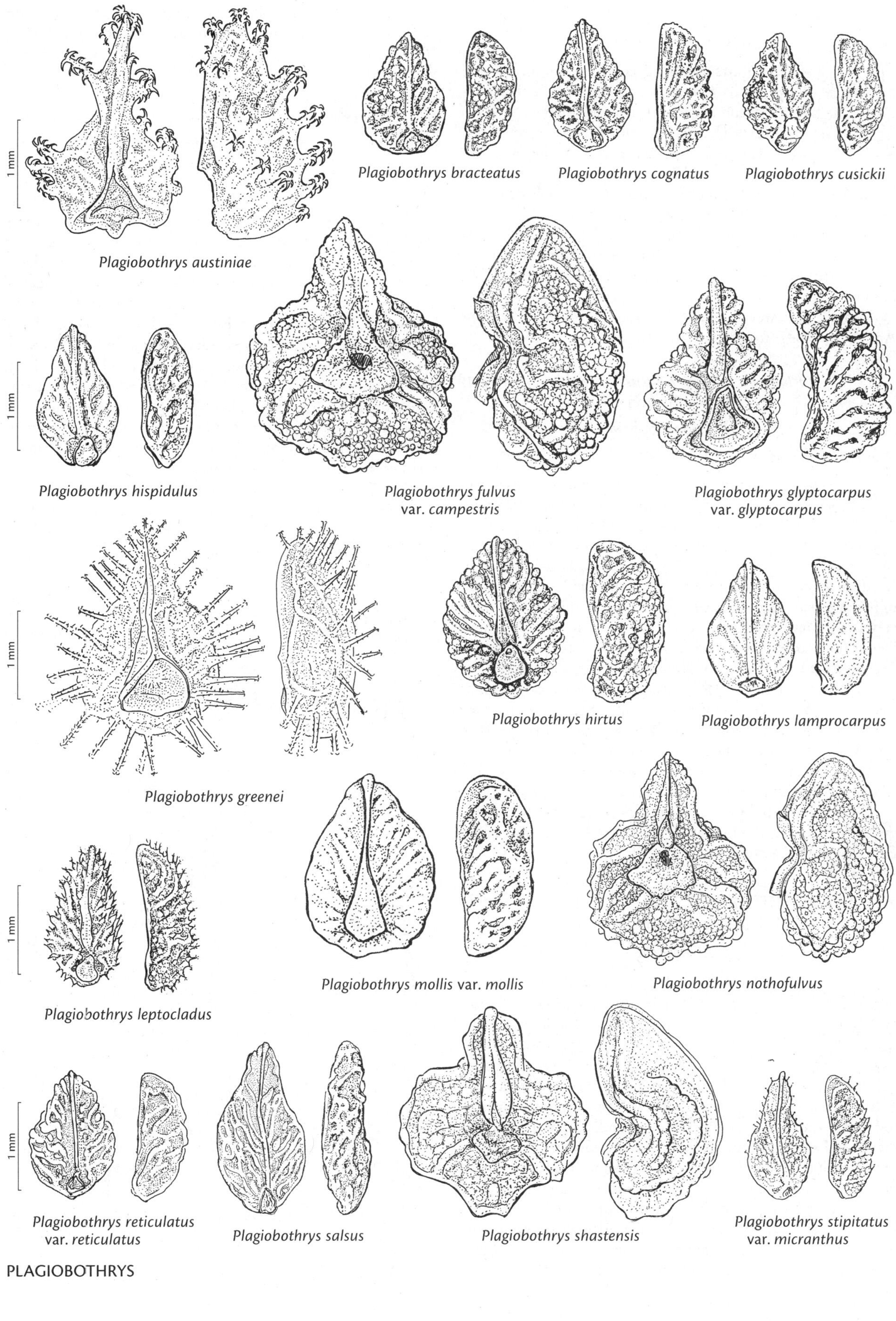

PLAGIOBOTHRYS

Plagiobothrys austiniae (Greene) I.M. Johnston (p. 427)

Austin's popcornflower

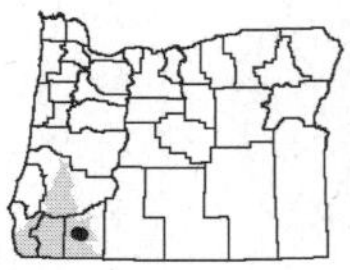

Plants annual, green, strigose; usually fibrous-rooted, becoming taprooted. **Stems** decumbent to erect, usually slender, 1–4 dm. **Basal leaves** 0. **Cauline leaves** lower opposite, 1–5 cm; upper generally alternate. **Inflorescences** bracts near base. **Flowers** calyces 5–8 mm; corolla limbs 1–2.5 mm in diameter; fornices yellow. **Fruits** ovate, 2–3 mm, ± granular, shiny, abaxial prickles few, rarely absent, on ridges, stout, barb-tipped, ± 0 adaxially, abaxial and lateral ridges prominent, transverse ridges 0 to obscure. **Attachment scars** on adaxial surfaces, near bases, ± triangular, deep-concave, 25–60% fruit length.

Vernal pools, wet areas in thin rocky, clay soils. Flowering Apr–May. 300–700 m. Sisk. CA. Native.

Plagiobothrys bracteatus (Howell) I.M. Johnston (p. 427)

Bracted popcornflower

Allocarya bracteata Howell

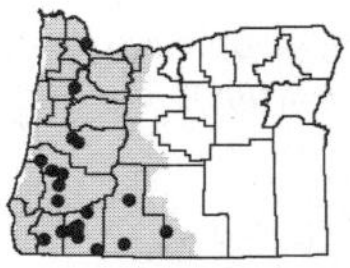

Plants annual, green, sparse- to dense-strigose; usually fibrous-rooted, becoming taprooted. **Stems** generally ascending, occasionally decumbent, usually slender, usually 1–4 dm, hairs appressed. **Basal leaves** 0. **Cauline leaves** lower opposite, 3–10 cm, hairs bulbous-based; upper alternate. **Inflorescences** bracts below middle. **Flowers** calyces 2–4 mm; lobes strigose; corolla limbs 1–3 mm in diameter; fornices pale yellow. **Fruits** ± symmetric, ± flat, lance-ovate, 1–1.8 mm, ± dull, margins with narrow borders, abaxial ridges short, near tips, lateral ridges ± arched, transverse ridges high, many ± crowded, especially toward tips, interspaces narrow, tubercled, generally glabrous, rarely bristly, adaxial ridges near middle, not folded to 1 side, not in troughs. **Attachment scars** ± sunken by surrounding ridges, oblique, ± ovate to wide-triangular, usually ± flat or ± concave, < 20% fruit length.

Common; vernal pools, wet places in grasslands, coastal sage scrub, chaparral. Flowering Apr–Jul. 0–1800 m. Casc, CR, ECas, Sisk, WV. CA; south to Mexico. Native.

Plagiobothrys cognatus (Greene) I.M. Johnston (p. 427)

Cognate popcornflower

Allocarya cognata Greene, *Plagiobothrys hispidulus* (Greene) I.M. Johnst.

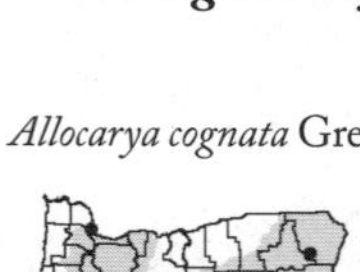

Plants annual, pale green, strigose; usually fibrous-rooted, becoming taprooted. **Stems** prostrate to occasionally ascending, stout or slender, 0.5–2+ dm, hairs appressed. **Basal leaves** 0. **Cauline leaves** lower opposite, 2–7 cm; upper alternate. **Inflorescences** bracts below middle to ± throughout. **Flowers** calyces 2–4 mm, lobes strigose to spreading-hairy, midveins generally slender; corolla limbs 1–2 mm in diameter. **Fruits** asymmetric, ± flat, oblong-ovate, 1.2–2.0 mm, brown, dull or shiny, abaxial ridges low, short, near tips, not in troughs, lateral ridges obscure, transverse ridges few, low, scattered, interspaces wide, tubercled or papillate-dentate and scabrous-bristled, adaxial ridges beyond middle, generally folded to 1 side below. **Attachment scars** not sunken, oblique, triangular, generally ± flat, < 20% fruit length.

Wet, muddy areas, moist shallow vernal depressions in sagebrush scrub, sparse coniferous forests. Flowering Apr–Aug. 200–2300 m. BR, BW, Casc, ECas, Owy, WV. CA, ID, NV, WA; northeast to MT, southeast to AZ. Native.

Plagiobothrys cusickii (Greene) I.M. Johnston (p. 427)

Cusick's popcornflower

Allocarya cusickii Greene, *Plagiobothrys hispidulus* (Greene) I.M. Johnst.

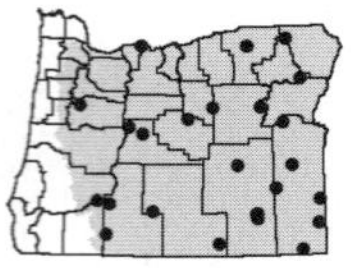

Plants annual, green, strigose; usually fibrous-rooted, becoming taprooted. **Stems** erect to occasionally prostrate, ± branched above base, usually slender, 0.5–3+ dm, hairs appressed. **Basal leaves** 0. **Cauline leaves** lower opposite, 3–10 cm; upper alternate. **Inflorescences** bracts below middle to throughout. **Flowers** calyces 1.5–4 mm, lobes strigose, midveins generally slender; corolla limbs 1–2 mm in diameter. **Fruits** asymmetric, ± flat, ovate, 1–1.9 mm, shiny, smooth to rounded-roughened, rarely bristled or scabrous, abaxial ridges short, near tips, lateral ridges obscure, transverse ridges ± evenly spaced, interspaces ± tubercled, adaxial ridges beyond middle, not in troughs, ± folded to 1 side below. **Attachment scars** on adaxial surfaces, near base, sunken, occasionally ± oblique, ± oblong, generally hollow, margins thin, ± incurved.

Moist places in meadows, sagebrush flats. Flowering Apr–Aug. 0–2500 m. BR, BW, Casc, Col, ECas, Lava, Owy, WV. CA, ID, NV, WA; southeast to UT. Native.

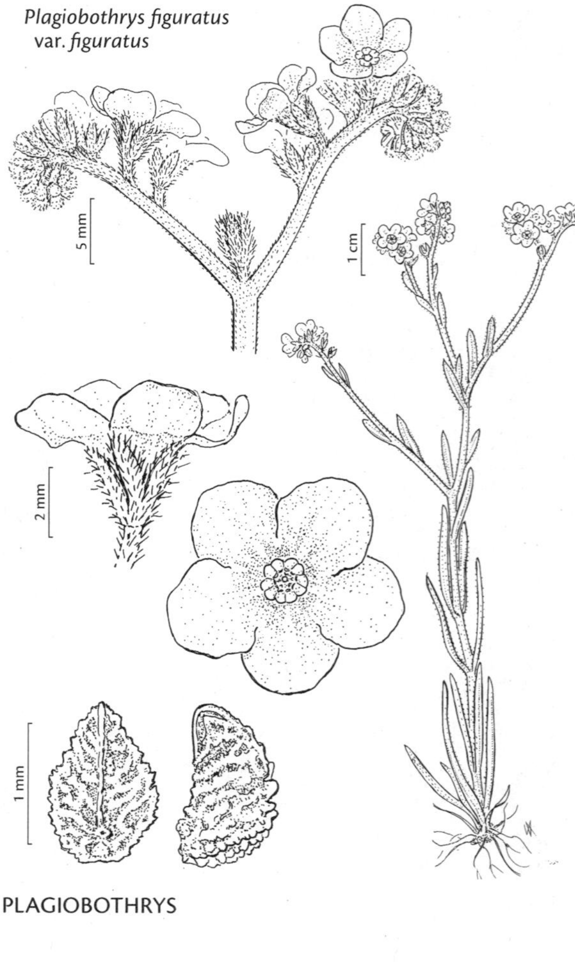

PLAGIOBOTHRYS

Plagiobothrys figuratus (Piper) I.M. Johnston

Fragrant popcornflower

Plants annual, green, sparsely to usually densely strigose; usually fibrous-rooted, becoming taprooted. **Stems** erect, usually slender, 1.5–4.5 dm, hairs appressed. **Basal leaves** 0. **Cauline leaves** lower opposite, 6–15 mm; upper alternate. **Inflorescences** usually paired, sometimes in 3s, usually with a flower in the fork; bracts absent or 1–few at base; pedicels 0.5–2 mm. **Flowers** calyces 3–5 mm; lobes linear-lanceolate, densely spreading-villous, strongly to slightly orange-hairy, midveins generally slender; corolla limbs 4–8 mm in diameter; fornices yellow. **Fruits** ovate, 1–1.5 mm, abaxial ridges prominent, not in troughs, transverse ridges irregular, coarsely papillate. **Attachment scars** on adaxial surfaces, near base, ovate.

Northwestern North America. 2 varieties.

Key to *Plagiobothrys figuratus* varieties:

1. Nutlet abaxial surfaces conspicuously alveolate, ridges high, prominent, regular . . . . *P. f.* var. ***corallicarpus***

1' Nutlet abaxial surfaces not conspicuously alveolate, ridges low, not prominent, irregular . . . . *P. f.* var. ***figuratus***

Plagiobothrys figuratus (Piper) I.M. Johnston var. ***corallicarpus*** (Piper) K.L. Chambers

Fragrant popcornflower

Allocarya hirta Greene ssp. *corallicarpa* (Piper) Abrams

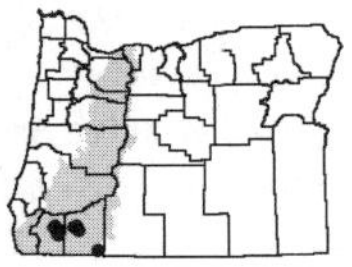

Fruits abaxial surfaces conspicuously alveolate, ridges high, prominent, regular. Moist open spaces. Flowering May–Jun. 300–1200 m. Casc, Sisk. Native.

Plagiobothrys figuratus (Piper) I.M. Johnston var. ***figuratus*** (p. 428)

Fragrant popcornflower

Allocarya figurata Piper

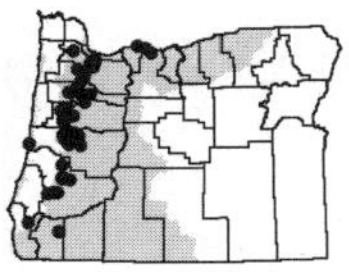

Fruits abaxial surfaces not conspicuously alveolate, ridges low, not prominent, irregular. Wet meadows, seeps. Flowering Apr–Jul. 0–900 m. Casc, Col, ECas, Est, Sisk, WV. WA; north to British Columbia. Native.

Plagiobothrys fulvus (Hooker & Arnott) I.M. Johnston var. ***campestris*** (Greene) I.M. Johnston (p. 427)

Fulvous popcornflower

Plagiobothrys campestris Greene

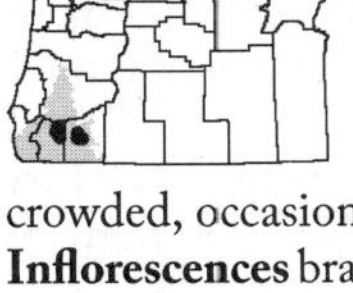

Plants annual, green, tomentose, some hairs spreading, long, rough; usually taprooted. **Stems** erect, 3–6 dm. **Basal leaves** in persistent rosettes, oblong to wide-oblanceolate, 2–10 cm. **Cauline leaves** lower occasionally crowded, occasionally opposite; upper alternate, few, elliptic. **Inflorescences** bracts 0–few, near base; pedicels 1–3 mm. **Flowers** calyces 5–6 mm; lobes densely brown-hairy, midveins green, obscured by hair; corolla limbs 3–4 mm in diameter; fornices yellow. **Fruits** 2–4, ± flat, triangular-ovate, ± narrowed below tips, 2.5–3.5 mm, brown, abaxial papillae 0–few, transverse ribs 0–few, narrow, straight, interspaces wide. **Attachment scars** on adaxial surfaces, near middle, ± round, deep-concave, pit-like.

Grasslands, open woodland, loamy, sandy, or gravelly soil. Flowering Apr–Jun. 200–400 m. Sisk. CA. Native.

Plagiobothrys glyptocarpus (Piper) I.M. Johnston var. ***glyptocarpus*** (p. 427)

Sculptured popcornflower

Allocarya glyptocarpa Piper ssp. *glyptocarpa*

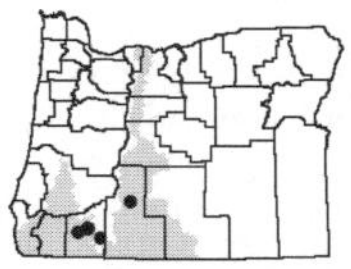

Plants annual, green, strigose; usually fibrous-rooted, becoming taprooted. **Stems** ascending to erect, usually slender, 1–5 dm. **Basal leaves** 0. **Cauline leaves** lower opposite, 4–8 cm; upper alternate. **Inflorescences** bracts near base. **Flowers** calyces 3–5 mm, lengthening in fruit; corolla limbs 5–9 mm in diameter; fornices yellow. **Fruits** ± flat, ± ovate, 1.5–2.8 mm, ± dull, abaxial and transverse ridges narrow, high, ± toothed, occasionally bristled, interspaces wide, granular. **Attachment scars** on adaxial surfaces, near bases, ± triangular, deep-concave, ± 40% fruit length, flanked by crowded wrinkles.

Common in moist places including vernal pools, streambanks, seeps, openings in grasslands and woodlands. Flowering Mar–May. 500–900 m. ECas, Sisk. CA. Native.

Plagiobothrys greenei (A. Gray) I.M. Johnston (p. 427)

Greene's popcornflower

Allocarya greenei (A. Gray) Greene

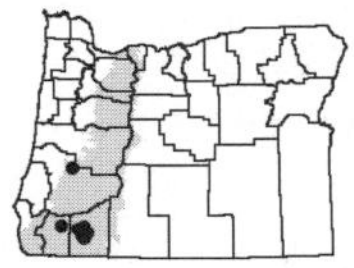

Plants annual, green, strigose; usually fibrous-rooted, becoming taprooted. **Stems** decumbent to erect, usually slender, 1–4 dm. **Basal leaves** 0. **Cauline leaves** lower opposite, 1–5 cm; upper alternate. **Inflorescences** bracts near base. **Flowers** calyces 5–8 mm; corolla limbs 1–2.5 mm in diameter; fornices pale yellow. **Fruits** ovate, ± narrowed near tips, 1.5–3.1 mm, ± dull, granular, abaxial prickles many, throughout surface, stout, ± barb-tipped, 0 adaxially, abaxial transverse ridges 0. **Attachment scars** on adaxial surfaces, near bases, triangular, deep-concave, 25–60% fruit length.

Wet sites, grasslands, woodlands. Flowering Apr–Jun. 300–1000 m. Casc, Sisk. CA. Native.

Plagiobothrys hirtus (Greene) I.M. Johnston (p. 427)

Rough popcornflower, hairy popcornflower

Allocarya hirta Greene ssp. *hirta*

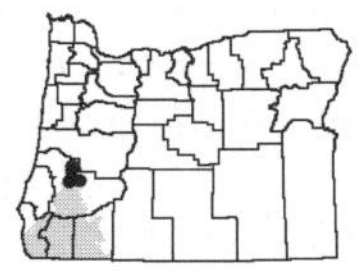

Plants annual or perennial, green; usually fibrous-rooted, becoming taprooted. **Stems** usually erect proximally, sometimes reclining, stout or slender, 3–7 dm, spreading-hirsute above, mostly glabrous below, rooting at nodes in some populations. **Basal leaves** 0. **Cauline leaves** seasonally dimorphic; lower opposite, 1–2.5 cm; upper alternate. **Inflorescences** usually paired; bracts absent; pedicels 1–2 mm. **Flowers** calyces 5–6 mm, lobes erect to somewhat spreading, densely setose-hirsute, midveins generally slender;

corolla limbs 8–14 mm in diameter; fornices yellow, fading white. **Fruits** lance-ovate to widely ovate, usually 1.6–2.0 mm, brown to dark brown, abaxial surfaces coarsely papillate to tuberculate, transverse ridges irregular, sometimes reticulate, keels usually well defined. **Attachment scars** on adaxial surfaces, near bases, ovate to deltate, usually ± flat or ± concave, < 20% fruit length.

Ditches, swamps, swales, seasonal wet meadows. Flowering May–Jun. 0–400 m. Sisk. Native.

Plagiobothrys hispidulus (Greene) I.M. Johnston (p. 427)

Eastside popcornflower, harsh popcornflower

Allocarya hispidula Greene, *Plagiobothrys scouleri* (Hook. & Arn.) I.M. Johnst., *Plagiobothrys scouleri* (Hook. & Arn.) I.M. Johnst. var. *penicillatus* (Greene) Cronquist

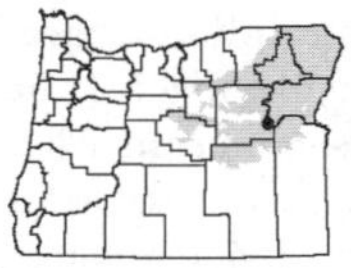

Plants annual, green, strigose; usually fibrous-rooted, becoming taprooted. **Stems** prostrate to erect, usually slender, 0.5–4 dm, hairs appressed. **Basal leaves** 0. **Cauline leaves** lower opposite, 1–5 cm; upper alternate. **Inflorescences** bracts generally below middle. **Flowers** calyces 1.5–3 mm; lobes ± spreading, midveins generally slender; corolla limbs 1–2 mm in diameter; fornices yellow. **Fruits** ± symmetric, ovate, plump, 1.2–2.0 mm, brown to black, dull, sharp-roughened, often also minutely bristled or scabrous, abaxial ridges short, near tips, papillate-dentate, lateral ridges 0, transverse ridges narrow, scattered, papillate-dentate, interspace margins wide, granular to papillate-dentate, adaxial ridges at middle, generally not folded, not in troughs. **Attachment scars** on adaxial surfaces, near bases, ± sunken, ± narrow-elliptic, usually ± flat or ± concave, < 20% fruit length.

Moist to drying meadows, flats, conifer forest openings. Flowering May–Aug. 1200–2000 m. BW. CA, ID, NV, WA; north to AK, northeast to Quebec, east to IL, southeast to NM, south to Mexico. Native.

Plagiobothrys lamprocarpus (Piper) I.M. Johnston (p. 427)

Shinyfruit popcornflower

Allocarya lamprocarpa Piper

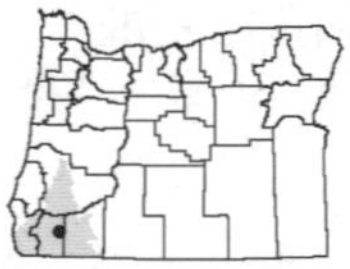

Plants annual, green, strigose; usually fibrous-rooted, becoming taprooted. **Stems** erect, usually slender, 1–3 dm, hairs appressed. **Basal leaves** 0. **Cauline leaves** lower opposite, 1–2 cm; upper alternate. **Inflorescences** bracts near base; pedicels very short. **Flowers** calyces 2 mm, lobes narrowly-lanceolate, midveins generally slender; corolla limbs 1–1.5 mm in diameter. **Fruits** 1–2, incurved, broadly ovate, ~1.5 mm, shiny, abaxial ridges prominent, transverse ridges low and broad, irregularly papillate to tuberculate, adaxial ridges in troughs. **Attachment scars** near bases, usually ± flat or ± concave, < 20% fruit length.

Wet areas. Flowering Jun. 200–400 m. Sisk. Native.

This species is known only from the type locality in Josephine County. It has not been collected in almost 100 years and is presumed extinct.

Plagiobothrys leptocladus (Greene) I.M. Johnston (p. 427)

Slender-branched popcornflower, alkali popcornflower

Allocarya leptoclada Greene

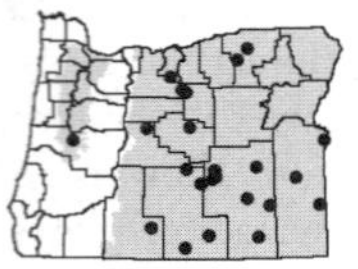

Plants annual, green, sparsely to densely strigose; usually fibrous-rooted, becoming taprooted. **Stems** prostrate to decumbent, usually slender, 1–3+ dm. **Basal leaves** 0. **Cauline leaves** lower opposite, 3–10 cm; upper alternate. **Inflorescences** bracts sparse to common throughout; pedicels 0–1+ mm. **Flowers** calyces 2 mm, 5–8 mm in fruit, lobes generally strongly bent upward, narrow, midveins stout; corolla limbs 1–2 mm in diameter; fornices pale yellow. **Fruits** attached ± loosely, lanceolate, 1–2.1 mm, gray to brown, dull, margins rounded, entire, abaxial ridges only at tips, transverse ridges 0–few, above middle, ± flat below, tubercled, often bristly, adaxial ridges shallowly V-shaped, granular. **Attachment scars** basal, usually ± flat or ± concave, < 20% fruit length, generally on short stipes.

Vernal pools, wet areas. Flowering May–Jul. 100–1800 m. BR, BW, Col, ECas, Lava, Owy, WV. CA, ID, NV, WA; east to ND, southeast to UT. Native.

Plagiobothrys mollis (A. Gray) I.M. Johnston var. *mollis* (p. 427)

Soft popcornflower

Allocarya mollis (A. Gray) Greene var. *mollis*

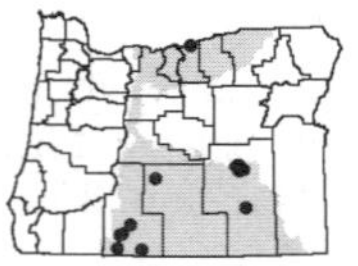

Plants perennial, greenish to whitish, densely villous; usually fibrous-rooted, becoming taprooted. **Stems** prostrate to decumbent, stout, 1–3 dm, rooting at nodes. **Basal leaves** 0. **Cauline leaves** lower opposite, 4–8 cm; upper opposite or alternate. **Inflorescences** bracts absent. **Flowers** calyces 4–5 mm; corolla limbs 5–10 mm in diameter; fornices yellow. **Fruits** ovate, 1.5–2.4 mm, gray, abaxial transverse ridges irregular. **Attachment scars** near middle, ovate or triangular.

Moist, alkaline places in sagebrush scrub. Flowering May–Aug. 800–1700 m. BR, Col, ECas. CA, NV. Native.

Plagiobothrys nothofulvus (A. Gray) A. Gray (p. 427)

Rusty popcornflower

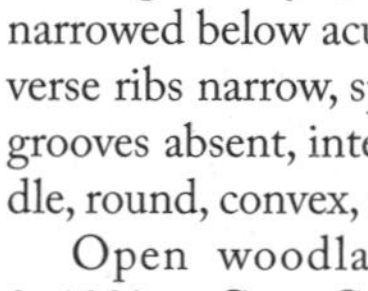

Plants annual, spreading-hirsute, red-staining; usually taprooted. **Stems** ± erect, 2–7 dm, ± red. **Basal leaves** in rosettes, 3–10 cm. **Cauline leaves** few, alternate, lower occasionally crowded, occasionally opposite. **Inflorescences** paired or in 3s; bracts 0 or near base. **Flowers** calyces circumscissile, 2–3 mm; lobes generally curved over fruit; corolla limbs (3)4–9 mm in diameter; fornices yellow. **Fruits** generally 3, ± arched, round-ovate, 2.0–2.7 mm, abruptly narrowed below acute tips, abaxial and lateral ridges and transverse ribs narrow, spaces between ribs wide and flat, transverse grooves absent, interspaces wide. **Attachment scars** near middle, round, convex, rarely shallow-concave.

Open woodlands, grasslands. Flowering Mar–Jun. 0–1000 m. Casc, Col, CR, ECas, Sisk, WV. CA, WA; north to British Columbia, south to Mexico. Native.

Plagiobothrys reticulatus (Piper) I.M. Johnston var. *reticulatus* (p. 427)

Netted popcornflower

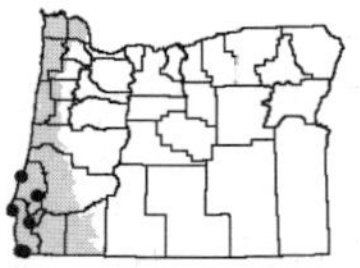

Plants annual, green, strigose; usually fibrous-rooted, becoming taprooted. **Stems** ascending to erect, occasionally decumbent, usually slender, 1–4 dm, hairs appressed. **Basal leaves** 0. **Cauline leaves** lower opposite, 3–8 cm; upper alternate. **Inflorescences** bracts below middle. **Flowers** calyces 2–4 mm; lobes appressed- to spreading-hairy, midveins generally slender; corolla limbs 1.5–3.5 mm in diameter; fornices yellow. **Fruits** usually 4, ± ovate, (0.7)1–1.9 mm, brown to gray, dull, abaxial transverse ridges generally irregularly net-like, interspace tubercles generally sparse or small, lateral ridges ± 0, adaxial ridges in troughs. **Attachment scars** sunken, oblique, narrow-elliptic to triangular, concave, < 20% fruit length, bordered by generally wide troughs.

Moist places, meadows in forest, grasslands, beaches. Flowering May–Jun. 0–100 m. CR, Est, Sisk. CA. Native.

Plagiobothrys salsus (Brandegee) I.M. Johnston (p. 427)

Desert popcornflower

Allocarya salsa Brandegee

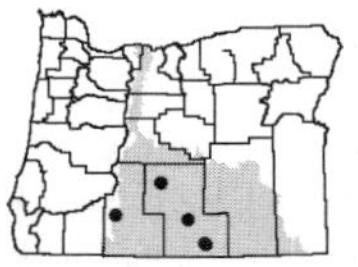

Plants annual, green, sparsely spreading-hirsute; usually fibrous-rooted, becoming taprooted. **Stems** decumbent to erect, 0.6–2.5 dm. **Basal leaves** 0. **Cauline leaves** lower opposite, 3–6 cm, hairs ± densely bulbous-based; upper alternate. **Inflorescences** solitary; bracts many, ± dense throughout; pedicels 0–1 mm. **Flowers** calyces 2–5 mm, lobe midveins generally slender; corolla limbs 2–5 mm in diameter; fornices yellow. **Fruits** asymmetric, compressed-lanceolate, 1–2.4 mm, brown, ± shiny, abaxial ridges obscure, lateral ridges 0, transverse ridges few, low, above middle, curved or irregular, granular to tubercled, adaxial ridges low, entire fruit length. **Attachment scars** generally oblique, ± ovate, usually ± flat or ± concave, < 20% fruit length.

Moist, saline, alkaline mudflats, marshes. Flowering Mar–Jul. 600–1500 m. BR, ECas. CA, NV. Native.

Plagiobothrys scouleri (Hooker & Arnott) I.M. Johnston (p. 431)

Scouler's popcornflower

Allocarya granulata Piper, *Plagiobothrys granulatus* (Piper) I.M. Johnst. *Plagiobothrys scouleri* (Hook. & Arn.) I.M. Johnst. var. *scouleri*

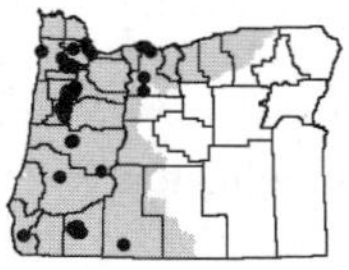

Plants annual, green, sparsely strigose; usually fibrous-rooted, becoming taprooted. **Stems** ascending to erect, usually slender, 1–5 dm, hairs stifly erect. **Basal leaves** 0. **Cauline leaves** lower opposite, 4–12 cm; upper alternate. **Inflorescences** usually solitary; bracts 0 or near base; pedicels usually 1 mm, rarely to 2 mm. **Flowers** calyces 2.5–4 mm, lobe midveins generally slender; corolla limbs 2–5 mm in diameter; fornices yellow. **Fruits** ovate, 1–1.5 mm, abaxial ridges from tip to at least middle, not in troughs, transverse ridges present, sometimes ± reticulate, usually more prominent above middle, granular to tuberculate. **Attachment scars** on adaxial surfaces, near base, ovate.

Vernal pools, wet areas. Flowering May–Jul. 0–1400 m. Casc, Col, CR, ECas, Est, Sisk, WV. CA, WA; north to British Columbia. Native.

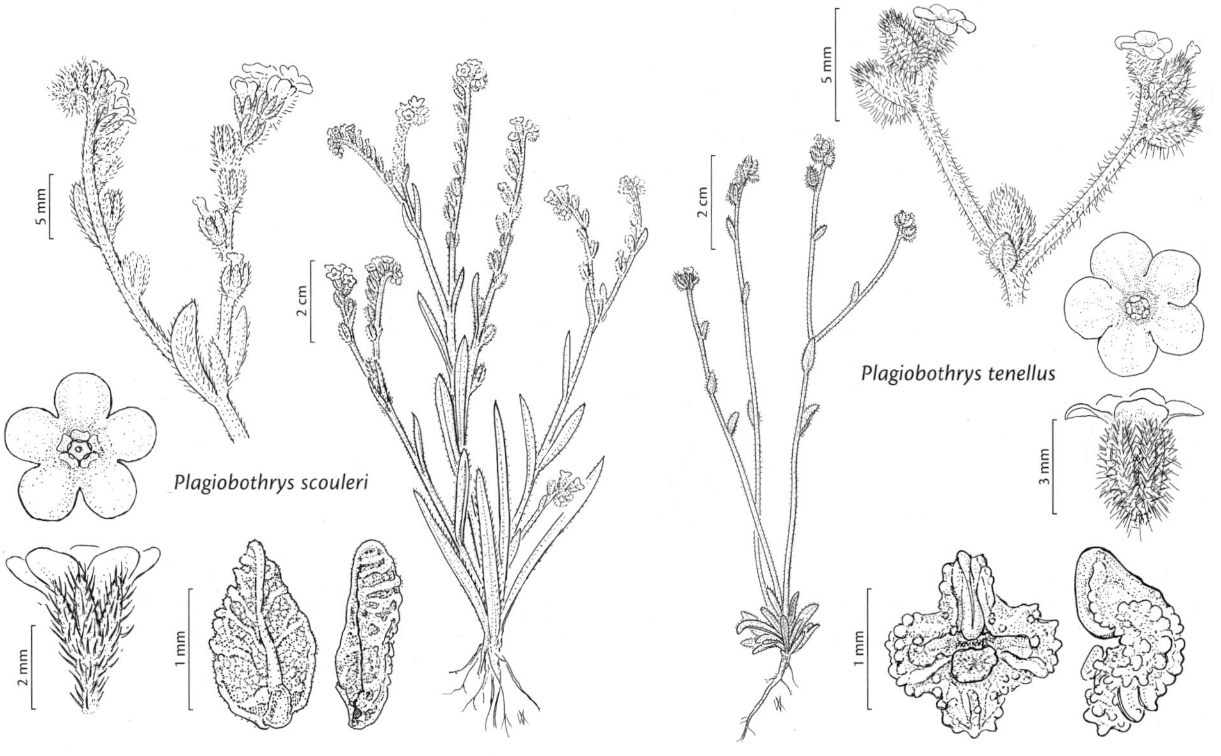

PLAGIOBOTHRYS

Plagiobothrys shastensis Greene ex A. Gray (p. 427)

Shasta popcornflower

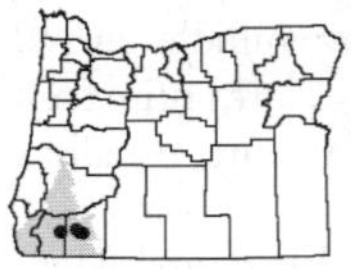

Plants annual, spreading-pubescent, occasionally red-staining on stems and lower leaves; usually taprooted. **Stems** 1–few, erect, 0.5–3 dm. **Basal leaves** in rosettes, 1–3 cm. **Cauline leaves** few, alternate, lower occasionally crowded, occasionally opposite. **Inflorescences** usually in 2s, not coiled toward tip, not elongate in fruit; flowers generally < 10 per branch; bracts ± throughout. **Flowers** calyces not circumscissile, 5–7 mm; corolla limbs 1–3 mm in diameter; fornices yellow. **Fruits** strongly arched, cross-shaped, 2–3 mm, gray to tan, shiny, abaxial transverse ridges wide, generally papillate, grooves narrow. **Attachment scars** on adaxial surfaces, near middle, convex, rarely shallow-concave.

Dry slopes, flats, grasslands, woodlands. Flowering May. 300–1000 m. Sisk. CA. Native.

Plagiobothrys stipitatus (Greene) I.M. Johnston

Stalked popcornflower

Plants annual, green, short-strigose; usually fibrous-rooted, becoming taprooted. **Stems** ascending to erect, usually slender, often fleshy, ± fistulose, 1–5 dm. **Basal leaves** 0. **Cauline leaves** lower opposite, 2–11 cm, hairs ± bulbous-based; upper alternate. **Inflorescences** bracts below middle; pedicels 0–3 mm. **Flowers** calyces 3 mm, 5–8 mm in fruit; lobes spreading to erect, not bent in same direction, narrow, linear to lanceolate, midveins stout; corolla limbs 2–12 mm in diameter; fornices yellow. **Fruits** attached ± loosely, ± flat, lanceolate to lance-ovate, 1–2.2 mm, tan, ± dull, margins rounded, obscure, abaxial ridges narrow, obscure below middle, transverse ridges ± above middle, ± flat below, rounded, wrinkled, tubercled, adaxial ridges deeply V-shaped, tubercled. **Attachment scars** basal, rarely ± oblique, usually ± flat or ± concave, < 20% fruit length, on short stipes.

Western United States. 2 varieties.

Key to *Plagiobothrys stipitatus* varieties:

1. Corolla limbs 2–4 mm in diameter . . . *P. s.* var. ***micranthus***

1' Corolla limbs 5–12 mm in diameter . . . . *P. s.* var. ***stipitatus***

Plagiobothrys stipitatus (Greene) I.M. Johnston var. ***micranthus*** (Piper) I.M. Johnston (p. 427)

Stalked popcornflower

Allocarya stipitata Greene ssp. *micrantha* Piper

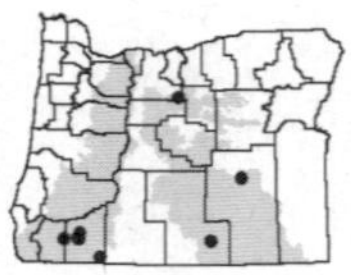

Flowers corolla limbs 2–4 mm in diameter. **Fruits** 1.1–1.9 mm.

Vernal pools, wet sites in grasslands, conifer forest, sagebrush scrub. Flowering Apr–Jun. 300–1800 m. BR, Casc, Lava, Sisk. CA, NV. Native.

Plants intermediate between *P. stipitatus* var. *micranthus* and *P. bracteatus* have been collected in California.

Plagiobothrys stipitatus (Greene) I.M. Johnston var. ***stipitatus***

Stalked popcornflower

Allocarya stipitata Greene ssp. *stipitata*

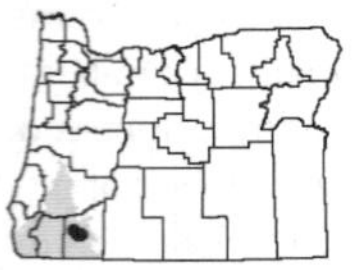

Flowers corolla limbs 5–12 mm in diameter. **Fruits** 1.6–2.2 mm.

Vernal pools, wet sites in grasslands. Flowering Apr–May. 300–500 m. Sisk. CA. Native.

Plagiobothrys tenellus (Nuttall) A. Gray (p. 431)

Pacific popcornflower, slender popcornflower

Plagiobothrys asper Greene

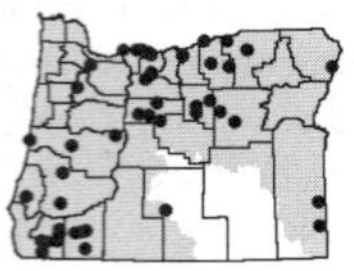

Plants annual, spreading-pubescent, rarely red-staining; taprooted. **Stems** 1–many, erect, 0.5–3 dm. **Basal leaves** in rosettes, 1–5 cm. **Cauline leaves** few, alternate, lower occasionally crowded, occasionally opposite. **Inflorescences** strongly coiled toward tip, elongate in fruit; flowers generally > 10 per branch; bracts near base. **Flowers** calyces 3–4(5) mm, corolla limbs 2–4 mm in diameter; fornices ± pale yellow. **Fruits** strongly arched, cross-shaped, 1–2 mm, shiny, abaxial and lateral ridges smooth to tubercled, transverse ridges wide, grooved, generally papillate, grooves narrow, often obscure. **Attachment scars** on adaxial surfaces, near middle, ± round, convex, rarely shallow-concave.

Dry, open slopes, grasslands, chaparral, woodlands. Flowering Mar–Jun. 0–1400 m. All ecoregions except BR. CA, ID, NV, WA; north to British Columbia, southeast to AZ, south to Mexico. Native.

This a variable species. See also *P. shastensis*.

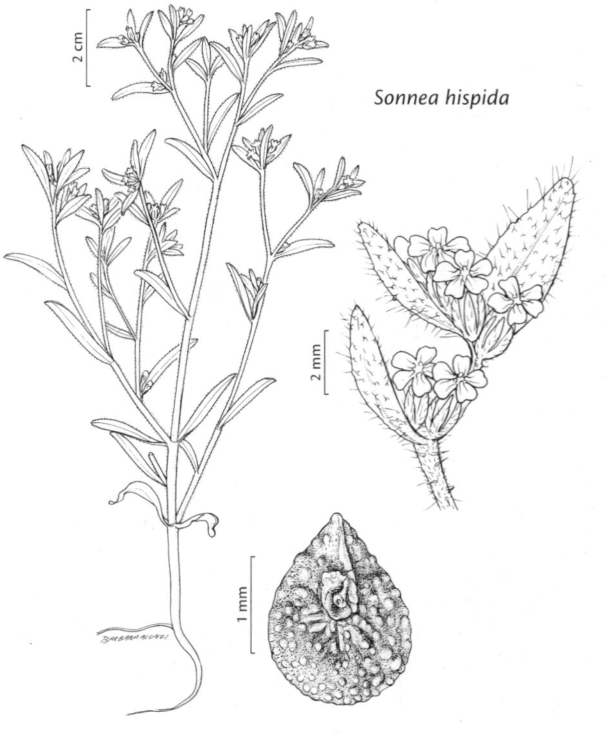

SONNEA

SONNEA Sonnea

C. Matt Guilliams

Herbs annual; taproots sometimes staining purple. **Stems** usually ascending, sometimes erect, 5–22 cm, hispid and tomentulose; densely branched throughout. **Leaves** basal early withering; cauline alternate, sometimes clustered distally in age, ovate to linear-oblong, 1–4(5) cm, surfaces hispid, sparsely short-tomentose, hispid hairs with pustulose bases; sessile. **Inflorescences** axillary and terminal, dense, scorpioid cymes, glomerate and not uncoiling in fruit distally, usually surpassed by subtending leaves; bracteate proximally; pedicels spreading to erect, filiform, usually < 1 mm. **Flowers** sepals lanceolate, 2–3 mm; corollas funnelform to rotate, white; fornices light yellow; stamens included; ovaries 4-lobed, stigmas included, globose; gynobases pyramidal. **Fruits** 1–4, ovoid, 1–3 mm, if 1 or 2 usually spreading, if 3 or 4 usually erect, monomorphic; attachment scars medial, elevated on caruncle of soft tissue, at or beyond middle of inner/adaxial surface (when nutlets erect), or abaxial surfaces (when nutlets spreading); caruncles globose when fresh, white, mucilaginous when wet.

Western United States. 2 species; 1 species treated in *Flora*.

Molecular phylogenetic studies support segregation of *S. glomerata* and *S. hispida* from *Plagiobothrys* and other members of the Amsinckiinae.

Sonnea hispida (A. Gray) Greene (p. 432)

Cascade popcornflower, bristly sonnea

Plagiobothrys hispidus A. Gray

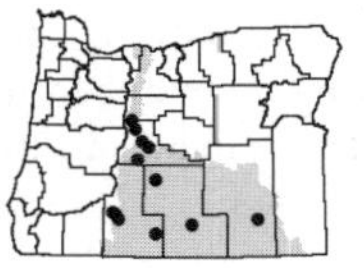

Flowers corollas 1.5–3 mm, limbs 0.7–1.8(2.5) mm. **Fruits** dark-colored, not smooth, dull; if 1 or 2, usually horizontally oriented, distal abaxial surfaces not rugose, keels on outer/abaxial surfaces obscure, 1–2.4 mm, less roughened; if 3 or 4, usually vertically oriented, distal abaxial surface rugose, keels on outer/abaxial surfaces not obscure, 1–2 mm, more roughened; attachment scars at or above middle of nutlet on inner/adaxial surface.

Sandy to gravelly soils, open areas, sagebrush scrub, yellow pine forests. Flowering Jun–Aug. 1100–1700 m. BR, ECas. CA, NV. Native.

Sonnea hispida is a summer-blooming annual of dry, high elevation areas of northeastern California, western Nevada, and south-central Oregon.

SYMPHYTUM Comfrey

Christian Feuillet

Herbs perennial; from a fleshy taproot. **Stems** ascending to erect. **Leaves** cauline, lanceolate to ovate, surfaces bristly; petiolate below to sessile above. **Inflorescences** terminal or axillary cymes, curved downward. **Flowers** calyces deeply lobed, bristly; corollas campanulate, throats expanded, with 5 small, scale-like appendages; stamens included; styles 1, exserted, twisted below tip; stigmas capitate, unlobed. **Fruits** erect, ovoid, with flat scar surrounded by thick, ring-like rim.

Asia, Europe, North America. 35 species; 1 species treated in *Flora*.

Symphytum officinale Linnaeus (p. 433)

Common comfrey

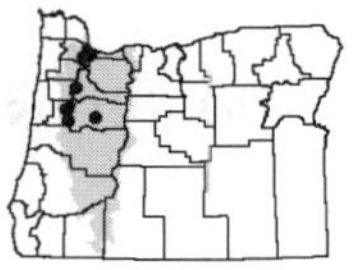

Plants 5–12 dm, generally bristly. **Stems** branched, internodes winged. **Leaves** ovate-lanceolate to lanceolate, 5–30 cm, bases decurrent, tips acuminate; petioles winged. **Inflorescences** many-flowered cymes. **Flowers** drooping; calyces 5–8 mm, lobes lanceolate in flower, becoming triangular in fruit; corollas 12–19 mm, purple-violet to white or cream, appendages linear-lanceolate to lanceolate; filaments as wide as anthers. **Fruits** 4–6 mm, black, smooth. $2n$=24, 26, 36, 40, 48, 54.

Fields, roadsides, disturbed areas. Flowering May–Jun. 0–400 m. Casc, WV. CA, ID, WA; scattered across North America; Asia, Europe. Exotic.

Symphytum officinale

SYMPHYTUM

BRASSICACEAE Mustard family

Ihsan A. Al-Shehbaz

Herbs, rarely **woody subshrubs** or **shrubs** annual, biennial, or perennial; taprooted or with caudices, sometimes rhizomatous; trichomes unicellular, simple, 2–many-rayed, stellate, dendritic, or malpighiaceous; stalked or sessile; multicellular glands with uniseriate or multiseriate stalks present or absent. **Stems** leafy or leafless, sometimes absent. **Leaves** basal and/or cauline, simple or compound, margins entire or variously divided; petiolate or sessile; without stipules. **Inflorescences** racemes, corymbs, or panicles, sometimes flowers solitary from a basal rosette. **Flowers** bisexual, hypogynous, radially or rarely bilaterally symmetric; sepals 4, in 2 decussate pairs, usually free; petals 4, forming a cross, rarely absent; stamens (2 or 4)6, in 2 whorls, tetradynamous, rarely equal in length or in 3 pairs of unequal length, median filaments free or rarely united, anthers dithecal, dehiscing by longitudinal slits, pollen 3(10)-colpate, trinucleate; pistils 1, 2-carpelled, ovaries superior, placentation parietal or rarely apical, septa complete, perforated, or lacking, styles 1, stigmas entire or 2-lobed; nectar glands receptacular. **Fruits** typically 2-valved capsules (siliques or silicles), dehiscent or indehiscent, terete, 4-angled, flattened parallel to or perpendicular to septum; replums rounded or rarely winged. **Seeds** uniseriately or biseriately arranged in each locule, rarely aseriate, mucilaginous or not when wetted; winged or wingless; cotyledons incumbent or accumbent, conduplicate.

Cosmopolitan. 340 genera; 59 genera treated in *Flora*.

This family includes many economically important crop plants that are grown as vegetables, sources of vegetable oils (canola), or condiments (table mustard, horseradish, wasabi), or ornamentals. It also includes some 130 species of weeds, including *Arabidopsis thaliana*, a model organism in experimental and molecular biology.

The limits of Brassicaceae (Cruciferae) have not changed in the past two centuries because of characteristic floral and fruit morphology. Generic delimitation within the family is often difficult because of the reliance on fruit characters to separate genera. The ovule number per ovary is a taxonomically important character, which can easily be determined by counting the number of seeds and aborted ovules in the mature fruit. The use of seed number alone may be taxonomically unreliable because many ovules fail to develop into seeds, especially late in the season due to the lack of sufficient pollen.

Key to Brassicaceae groups:

1. Cauline leaves auriculate, sagittate, or amplexicaul . . . . **Subkey 1**

1' Cauline leaves not auriculate, sagittate, or amplexicaul, sometimes absent.

  2. At least some trichomes branched . . . . **Subkey 2**

  2' Trichomes absent or simple . . . . **Subkey 3**

Subkey 1: Cauline leaves auriculate, sagittate, or amplexicaul.

1. Stamens 2 . . . . **LEPIDIUM** (in part)

1' Stamens 4 or 6.

  2. Plants aquatic, rooting from lower nodes; leaves compound . . . . **NASTURTIUM**

  2' Plants terrestrial, rarely aquatic, not rooting from lower nodes; all leaves simple.

    3. Plants with at least some branched trichomes.

      4. Fruits pyriform, obovoid, obdeltoid, or obcordate silicles.

        5. Fruits pyriform or obovoid, terete, valves with caudate tips extended onto styles . . . . **CAMELINA**

        5' Fruits obcordate to obdeltoid, angustiseptate, valve tips not extended onto styles . . . . **CAPSELLA**

      4' Fruits linear or lanceolate siliques.

        6. Fruits terete or nearly quadrangular; flowers yellow or creamy white . . . . **TURRITIS**

        6' Fruits latiseptate; flowers white, pink, or purple.

          7. Fruits straight, erect to ascending, usually appressed or subappressed to rachis, 0.8–1.8(2) mm wide; stems branched; trichomes 2-forked . . . . **ARABIS**

          7' Fruits arcuate-falcate, curved, or rarely straight, reflexed, pendent, horizontal, or divaricate, if subappressed then seeds subbiseriate, usually > 2 mm wide; stems branched; trichomes usually more than 2-forked, dendritic, or submalpighiaceous.

            8. Trichomes simple, forked, or dendritic; ovules and seeds 20+ per fruit . . . . **BOECHERA**

            8' Trichomes exclusively dendritic; ovules and seeds (6)8–16(18) per fruit . . . . . **PHOENICAULIS**

    3' Plants glabrous or with simple trichomes only.

      9. Fruits 1-seeded, indehiscent.

        10. Fruits oblong-oblanceolate to oblong siliques, wings entire, rayless; flowers yellow . . . . **ISATIS**

        10' Fruits orbicular or suborbicular silicles, wings perforate, incised, or rayed; flowers white or purplish . . . . **THYSANOCARPUS**

      9' Fruits 2- to many-seeded, dehiscent or rarely indehiscent.

11. Fruits silicles.
  12. Seeds 2(4) per fruit; fruits indehiscent . . . . **LEPIDIUM** (in part)
  12' Seeds ≥ 4 per fruit; fruits dehiscent.
    13. Fruits terete, wingless, obtuse or acute at apex . . . . **RORIPPA** (in part)
    13' Fruits angustiseptate, winged, notched at apex.
      14. Seeds smooth; plants perennial. . . . . **NOCCAEA**
      14' Seeds concentrically striate; plants annual . . . . **THLASPI**
11' Fruits siliques.
  15. Fruits segmented; seeds globose; cotyledons conduplicate . . . . **BRASSICA**
  15' Fruits unsegmented; seeds ovoid to oblong; cotyledons incumbent or accumbent.
    16. Gynophores (0.8)1–2(2.5) cm . . . . **STANLEYA**
    16' Gynophores obsolete or very rarely to 0.7 cm.
      17. Stamens in 3 pairs of unequal length; calyces urceolate or campanulate . . . . **STREPTANTHUS**
      17' Stamens tetradynamous or subequal; calyces tubular.
        18. Petals white, lavender, or purple; gynophores (0.5)1–6(7) mm . . . . **THELYPODIUM**
        18' Petals yellow or rarely absent; gynophores obsolete.
          19. Cauline leaves entire, cordate-amplexicaul at base; fruits 4-angled or terete; cotyledons incumbent . . . . **CONRINGIA**
          19' Cauline leaves lobed or dentate, auriculate at base; fruits terete or latiseptate, rarely slightly 4-angled; cotyledons accumbent.
            20. Seeds uniseriate, reticulate; stems angled . . . . **BARBAREA**
            20' Seeds biseriate, with small bumps; stems terete . . . . **RORIPPA** (in part)

Subkey 2: Cauline leaves not auriculate, sagittate, or amplexicaul; some trichomes branched.

1. Cauline leaves absent.
  2. Seeds aseriate; ovules 2–4 per ovary; cotyledons incumbent . . . . **CUSICKIELLA**
  2' Seeds biseriate; ovules 4–88 per ovary; cotyledons accumbent.
    3. All flowers chasmogamous, petaliferous; plants annual or perennial. . . . . **DRABA** (in part)
    3' Mid- to late-season flowers cleistogamous and apetalous; plants annual . . . . **TOMOSTIMA** (in part)
1' Cauline leaves present.
  4. Upper cauline leaves 1–3-pinnatifid or -pinnatisect, or pectinate.
    5. Plants annual or biennial; petals yellow. . . . . **DESCURAINIA**
    5' Plants perennial; petals white or pink.
      6. Leaves rigid, pectinate, becoming spiny; ovules and seeds 30–46 per fruit . . . . **POLYCTENIUM**
      6' Leaves soft, 1- or 2-pinnatifid or -pinnatisect, not spiny; ovules and seeds 4–18 per fruit . . . **SMELOWSKIA**
  4' Upper cauline leaves entire or dentate.
    7. Fruits siliques.
      8. All trichomes sessile, primarily malpighiaceous . . . . **ERYSIMUM**
      8' At least some trichomes stalked, forked, stellate, or dendritic, never malpighiaceous.
        9. Stigmas prominently 2-lobed, lobes decurrent, connivent.
          10. Petals 0.6–1(1.2) cm; fruits 4-angled, 2.5–5.5(7) cm. . . . . **STRIGOSELLA**
          10' Petals (1.3)1.5–2(2.2) cm; fruits terete, (4.5)6–10(14) cm . . . . **HESPERIS**
        9' Stigmas entire or rarely slightly lobed, lobes neither decurrent nor connivent.
          11. Fruits terete; cotyledons incumbent . . . . **ARABIDOPSIS**
          11' Fruits latiseptate; cotyledons accumbent.
            12. Fruits (1.7)2–8.5 cm; seeds winged, rarely wingless, uniseriate, rarely subbiseriate.
              13. Seed wings distal or absent; fruits 1–2.2 mm wide, straight, ascending to divaricate, not reflexed, falcate, or secund . . . . **ARABIS**
              13' Seed wings continuous; fruits 1.5–7 mm wide, reflexed, falcate, or secund, or rarely erect . . . . **BOECHERA**
            12' Fruits 0.5–1.6(2) cm; seeds wingless, biseriate.
              14. All flowers chasmogamous, petaliferous; plants annual or perennial. . . . . . . **DRABA** (in part)
              14' Mid- to late-season flowers cleistogamous, apetalous; plants annual . . **TOMOSTIMA** (in part)

7' Fruits silicles.
 15. Inflorescences secund . . . . . **ATHYSANUS**
 15' Inflorescences not secund.
  16. Trichomes malpighiaceous . . . . . **LOBULARIA**
  16' Trichomes simple, stellate, or forked.
   17. Fruits indehiscent, nutlet-like, appressed to rachis, with beak-like curved styles . . . . . **EUCLIDIUM**
   17' Fruits dehiscent, if not then bladdery, not appressed to rachis, without beak-like styles.
    18. Petals obcordate, deeply bifid . . . . . **BERTEROA**
    18' Petals variously shaped but not obcordate, obtuse, rarely emarginate.
     19. Leaf trichomes stellate.
      20. Filaments unappendaged; ovules 4–28(40) per ovary; seeds biseriate; septa usually with apical veins . . . . . **PHYSARIA**
      20' Filaments usually winged or appendaged (except *Alyssum alyssoides*); ovules 2 or 4 per ovary; seeds aseriate; septa veinless.
       21. Plants annual; ovules 4 per ovary . . . . . **ALYSSUM**
       21' Plants perennial; ovules 2 per ovary . . . . . **ODONTARRHENA**
     19' Leaf trichomes simple, forked, or cruciform.
      22. Fruits angustiseptate; cauline leaves petiolate . . . . . **HORNUNGIA**
      22' Fruits latiseptate or terete; cauline leaves sessile.
       23. All flowers petaliferous, chasmogamous; plants perennial or annual . . . . . **DRABA** (in part)
       23' At least mid- to late-season flowers apetalous, cleistogamous; plants annual.
        24. Fruits elliptic or narrowly so, glabrous; leaf trichomes cruciform; ovules 8–16(20) per ovary . . . . . **ABDRA**
        24' Fruits obovate to obovate-oblong, pubescent; leaf trichomes simple and 2–4-rayed; ovules (28)40–70(82) per ovary . . . . . **TOMOSTIMA** (in part)

Subkey 3: Cauline leaves not auriculate, sagittate, or amplexicaul; trichomes absent or simple.

1. Cauline leaves absent.
 2. Flowers solitary; leaves flat and broad; plants terrestrial . . . . . **IDAHOA**
 2' Flowers in racemes; leaves subulate; plants ± aquatic . . . . . **SUBULARIA**
1' Cauline leaves present.
 3. Fruits 20–30(35) mm wide . . . . . **LUNARIA**
 3' Fruits 1–3.5 mm wide.
  4. Plants with multicellular glands; styles 1–2.2 cm . . . . . **CHORISPORA**
  4' Plants without glands; styles < 1 cm.
   5. Fruits silicles.
    6. Fruits 2-segmented . . . . . **CRAMBE**
    6' Fruits unsegmented.
     7. Fruits terete or slightly angustiseptate, apically wingless; seeds and ovules 8–24 per fruit.
      8. Fruits terete, 2–4(5) mm wide; petals 2–4(5) mm; seeds papillate . . . . . **COCHLEARIA**
      8' Fruits slightly angustiseptate, (1)1.4–2.2 mm wide; petals 0.6–1.2 mm; seeds smooth . . . . . **HORNUNGIA**
     7' Fruits strongly angustiseptate, often apically winged; seeds and ovules 2 or 4 per fruit.
      9. Seeds 2 per fruit; flowers radially symmetric; filaments unappendaged . . . . . **LEPIDIUM**
      9' Seeds 4 per fruit; flowers bilaterally symmetric; filaments basally appendaged . . . . . **TEESDALIA**
   5' Fruits siliques.
    10. Petal veins brown or purple, much darker than rest of blade.
     11. Fruits corky, valvular segment seedless, terminal segment seeded . . . . . **RAPHANUS**
     11' Fruits not corky, valvular segments seeded, terminal segment seedless . . . . . **ERUCA**
    10' Petal veins same color as rest of blade.

12. Fruits corky, indehiscent; plants coastal . . . . **CAKILE**
12' Fruits not corky, dehiscent; plants coastal or inland.
13. Replums strongly flattened; fruit valves dehiscing elastically and becoming spirally coiled . . . . **CARDAMINE**
13' Replums terete, rounded, or slightly flattened; fruit valves neither dehiscing elastically nor coiled.
14. Fruits slightly to strongly latiseptate, not terete or 4-angled; seeds winged at least distally.
15. Stamens equal in length, subequal, or tetradynamous; styles obsolete or 0.1–0.5(0.8) mm.
16. Fruits sessile; all leaves entire or very rarely dentate; stamens tetradynamous; petal claws entire . . . . **BOECHERA**
16' Fruits on gynophores (1)2–8(10) mm; at least lower leaves hastate, lyrate, or sinuate; stamens subequal; petal claws dentate to laciniate . . . . **CHLOROCRAMBE**
15' Stamens in 3 pairs of unequal length; styles (0.5)1–8 mm.
17. Annual; fruits strongly reflexed, 1.2–2(2.2) mm wide; valves united apically with style and replum; cotyledons incumbent . . . . **STREPTANTHELLA**
17' Perennial; fruits straight or curved inwards, 2.5–3.5 mm wide; valves readily separated from style and replum; cotyledons accumbent. . . . . **STREPTANTHUS**
14' Fruits terete or 4-angled, not latiseptate; seeds wingless.
18. Petal claws and filament bases pubescent; plants with woody bases, sometimes shrubs or subshrubs . . . . **STANLEYA**
18' Petal claws and filament bases glabrous; annual, biennial, or perennial herbs without woody bases.
19. Petals white, lavender, purple, pink, greenish, or brown.
20. Leaves compound; plants aquatic, rooting from lower nodes . . . . **NASTURTIUM**
20' Leaves simple; plants terrestrial, not rooting from lower nodes.
21. Cauline leaves cordate or deltoid; seeds concentrically striate . . . . **ALLIARIA**
21' Cauline leaves not cordate or deltoid; seeds reticulate or nearly smooth.
22. Stamens subequal; anthers linear to narrowly oblong, circinately coiled when dehisced; gynophores slender, narrower than fruit, (0.5)1–8.5 mm . . **THELYPODIUM**
22' Stamens tetradynamous or in 3 pairs of unequal length; anthers oblong to ovate, not coiled when dehisced; gynophores as stout as fruit base, obsolete or < 1 mm.
23. Stigmas 2-lobed; petal claws well differentiated from blades; plants perennial or biennial . . . . **CAULANTHUS**
23' Stigmas entire; petal claws undifferentiated; plants annual. . . . . **GUILLENIA**
19' Petals yellow.
24. Fruits appressed to rachis.
25. Fruit tips acute or obtuse, terminal segment 1- or 2-seeded . . . . **HIRSCHFELDIA**
25' Fruit tips acuminate to subulate, terminal segment seedless.
26. Seeds globose; petals (6)7.5–11(13) mm . . . . **BRASSICA** (in part)
26' Seeds oblong; petals 2.5–4 mm . . . . **SISYMBRIUM** (in part)
24' Fruits not appressed to rachis.
27. Fruits unsegmented; cotyledons incumbent or accumbent.
28. Fruit valves 1-veined or veinless; stigmas entire; cotyledons accumbent . . . . **RORIPPA**
28' Fruit valves 3-veined; stigmas 2-lobed; cotyledons incumbent . . . . . **SISYMBRIUM** (in part)
27' Fruits 2-segmented; cotyledons conduplicate.
29. Fruit valves prominently 3–7-veined . . . . **SINAPIS**
29' Fruit valves distinctly 1-veined.
30. Seeds biseriate, ovoid to ellipsoid; fruits latiseptate . . . . **DIPLOTAXIS**
30' Seeds uniseriate, globose; fruits terete or 4-angled . . . . **BRASSICA** (in part)

ABDRA Whitlow-grass, abdra

Herbs annual. **Trichomes** short-stalked to sessile, cruciform. **Basal leaves** not rosulate, simple, margins entire; petiolate. **Cauline leaves** bases not auriculate, margins entire; sessile or short-petiolate. **Inflorescences** bracts 0. **Flowers** chasmogamous and petaliferous early in season, cleistogamous and apetalous in mid or late season; sepals caducous, free, erect, oblong, equal, lateral pair not saccate; petals spatulate, > sepals, white, claws undifferentiated from blades, petal tips obtuse; nectar glands confluent, subtending filament bases; stamens 6, tetradynamous, anthers ovate, not apiculate; septa complete; styles to 0.1 mm, stigmas capitate, entire. **Fruits** dehiscent silicles or short siliques, latiseptate, unsegmented; valves veinless; replums rounded. **Seeds** biseriate, ovoid.

United States. 2 species; 1 species treated in *Flora*.

For the delimitation of *Abdra* and *Tomostima*, see the discussion under *Draba*.

Abdra brachycarpa (Nuttall) Greene (p. 438)

Short pod whitlow-grass

Draba brachycarpa Nutt. ex Torr. & A. Gray

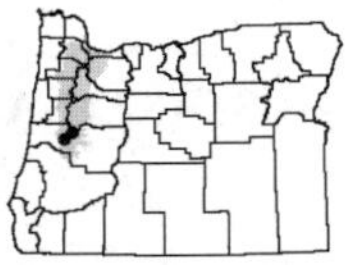

Stems (3)4–19(22) cm. **Basal leaves** oblanceolate to obovate, 5–20 × 2–8 mm. **Cauline leaves** (4)6–11, lanceolate to oblong or linear; petioles to 5 mm, not ciliate. **Inflorescences** (20)25–65(74)-flowered, rachises not flexuous; fruiting pedicels horizontal to divaricate-ascending, straight, (1)1.5–4(5) mm, pubescent. **Flowers** sepals 0.7–1.2(1.5) mm, green or pink, pubescent; petals 2–3 × 0.8–1.1 mm; ovules 8–16(20) per ovary, styles 0.05–0.1 mm. **Fruits** elliptic or narrowly so, flattened, not twisted, (2)2.5–5(6) × 0.9–1.4(1.9) mm, glabrous. **Seeds** ovoid, flattened, 0.5–0.7 × 0.4–0.5 mm; winged. ***2n***=16.

Open woods, cedar glades, pastures, lawns, roadsides, disturbed sites. Flowering Mar–May. 100–1000 m. WV. Central and southern US, scattered in western US. Native.

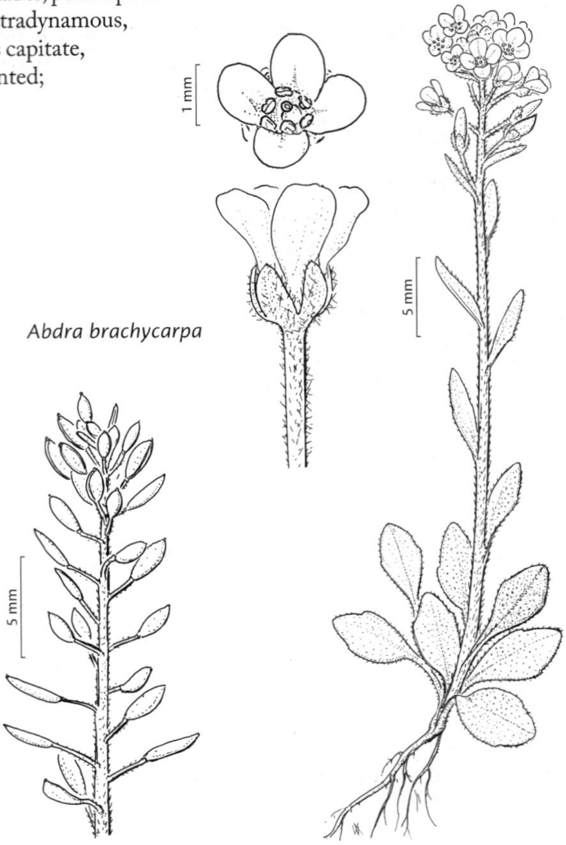

ABDRA

ALLIARIA Garlic mustard

Genus composed of one species, *A. petiolata*.

Alliaria petiolata (Marschall von Bieberstein) Cavara & Grande (p. 440)

Garlic mustard

Alliaria officinalis Andrz. ex M. Bieb.

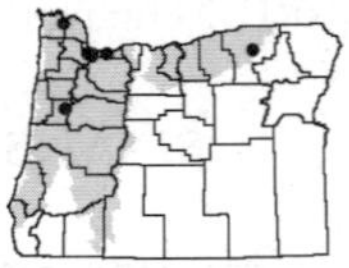

Herbs annual or biennial, with garlicky smell when crushed, glabrous or pilose. **Trichomes** absent or simple. **Stems** erect, (1.5)3–9(13) dm, simple or branched above. **Basal leaves** reniform or cordate, (0.6)1.5–8(11) cm wide, bases cordate, margins crenate, dentate, or sinuate. **Cauline leaves** cordate or deltoid, to 15 × 15 cm, bases cordate or truncate, not auriculate, margins dentate; petioles 3–16(22) cm. **Inflorescences** bracts 0 or only lowermost few flowers with bracts; fruiting pedicels divaricate or ascending, (2)3–10(15) mm, nearly as thick as fruit. **Flowers** sepals erect, caducous, oblong, (2)2.5–3.5(4.5) mm, glabrous, bases of lateral pair not saccate; petals oblanceolate, (2.5)4–8(9) × (1.5)2–3(3.5) mm, > sepals, white, claws obscurely differentiated from blades; nectar glands confluent, subtending bases of all stamens; stamens 6, slightly tetradynamous, filaments free, anthers tips obtuse; ovules 14–22 per ovary, septa complete, styles obsolete or (0.2)1–2(3) mm, stigmas capitate, entire. **Fruits** dehiscent siliques, divaricate-ascending, linear, 4-angled or subterete, (2)3–7(8) cm × 1.2–2.5 mm, unsegmented; valves with prominent midveins, glabrous, subtorulose; replums rounded; sessile. **Seeds** narrowly oblong, 2–4.5 × 0.7–2 mm, dark brown or black, concentrically striate, not mucilaginous when wetted; wingless; cotyledons incumbent. ***2n***=42.

Roadsides, trails, railroad tracks, waste places, fields, riverbanks, woodlands, bluffs, thickets, disturbed areas, floodplains. Flowering Apr–May. 0–800 m. Casc, Col, CR, WV. ID, WA; north to AK, northeast to Quebec, southeast to GA; Africa, Asia, Europe. Exotic.

This species is listed as a noxious weed in 46 states.

ALYSSUM Madwort, alyssum

Herbs annual. **Trichomes** stellate, stalked or sessile, with simple or branched rays, rarely mixed with simple and forked. **Basal leaves** rosulate or not, simple, margins entire; petiolate or sessile. **Cauline leaves** bases not auriculate, margins entire; petiolate or sessile. **Inflorescences** bracts 0. **Flowers** sepals ovate or oblong, erect or ascending, caducous or persistent, lateral pair not saccate; petals suborbicular, obovate or spatulate, yellow or white, tips obtuse or emarginate; nectar glands 4, lateral; stamens 6, tetradynamous, filaments wingless or winged, appendaged or not, toothed or not, anthers ovate or oblong, apiculate or not; ovules 4 per ovary, placentation apical, septa complete, styles distinct, stigmas capitate, entire. **Fruits** dehiscent silicles, strongly latiseptate, unsegmented; valves veinless, pubescent or glabrous, smooth; replums rounded; sessile. **Seeds** aseriate, flattened, mucilaginous or not when wetted; winged or wingless; cotyledons accumbent.

Northern Africa, Asia, Europe. 114 species; 3 species treated in *Flora*.

Alyssum is a taxonomically difficult genus. Both flowers and mature fruit are needed for the determination of species.

Alyssum murale is now assigned to the genus *Odontarrhena*. Reports of *Alyssum corsicum* (*Odontarrhena corsica*) are apparently based on immature plants of *O. muralis*.

Key to *Alyssum* species:

1. Fruits glabrous or glabrate . . . . *A. desertorum*

1' Fruits with stellate trichomes.

  2. Sepals persistent; filaments not toothed, winged, or appendaged . . . . *A. alyssoides*

  2' Sepals caducous; filaments toothed, winged, or appendaged . . . . *A. simplex*

Alyssum alyssoides (Linnaeus) Linnaeus

Pale alyssum, pale madwort, small alyssum

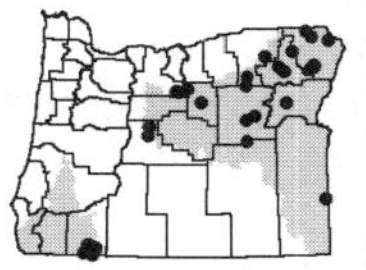

Plants canescent with appressed, sessile, 6–10-rayed stellate trichomes, mixed with simple and forked on pedicels and sepals. **Stems** 1 to several from base, erect, ascending, or decumbent, 5–35(50) cm, simple or branched above. **Cauline leaves** narrowly oblanceolate or linear, sometimes spatulate or obovate, 3–4(4.5) cm × (0.5)1–3.5(5) mm, bases attenuate or cuneate; subsessile. **Inflorescences** fruiting pedicels divaricate or ascending, slender, straight, 2–5(6) mm. **Flowers** sepals oblong, (1.5)2–3 mm, persistent; petals linear to linear-oblanceolate, rarely obovate, 2–3(4) × 0.3–0.7(1) mm, white or pale yellow, often persistent, tips emarginate; filaments 1–1.5 mm, slender, not toothed, winged, or appendaged; styles 0.3–0.6(1) mm. **Fruits** orbicular, (2)3–4(5) mm in diameter, tips emarginate or truncate, surfaces with sparse stellate trichomes; valves strongly flattened at margins, uniformly inflated at center. **Seeds** oblong or ovate, compressed, 1.1–2 × 0.7–1.1 mm, narrowly margined all around. **2*n***=32.

Roadsides, railroad tracks, sagebrush flats, grassy areas, fields, disturbed sites, limestone ledges or bluffs. Flowering Apr–Jul. 300–1600 m. BW, Lava, Owy, Sisk. CA, ID, NV, WA; north to British Columbia, east to Newfoundland, southeast to LA; Africa, Asia, Europe. Exotic.

Alyssum desertorum Stapf

Desert alyssum, desert madwort

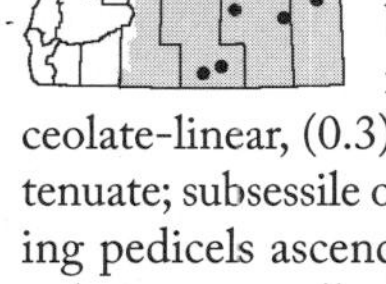

Plants canescent except fruit with appressed, sessile, 8–20-rayed stellate trichomes. **Stems** erect, ascending, or decumbent, (2)5–18(28) cm, simple or branched from base. **Cauline leaves** linear to oblanceolate-linear, (0.3)0.5–2.5(3) cm × (0.5)1–3(4) mm, bases attenuate; subsessile or lowermost petiolate. **Inflorescences** fruiting pedicels ascending or subdivaricate, (1)1.5–3.5(4.5) mm, pubescence stellate. **Flowers** sepals caducous, oblong, 1.4–1.8(2) mm, stellate pubescent; petals caducous, oblanceolate, 2–2.5 × 0.5–0.6 mm, pale yellow, tips obtuse or retuse; filaments 1–1.8(2) mm, median pairs narrowly winged at base, not toothed, lateral pair with a broadly winged apically 2-toothed appendage; styles 0.3–0.7 mm. **Fruits** orbicular, 2.5–4(4.5) mm in diameter, tips shallowly emarginate, surfaces glabrous or glabrate; valves not veined, broadly flattened at margin, uniformly inflated at middle. **Seeds** ovate, slightly compressed, 1.2–1.5 × 0.9–1.1 mm, margined or not. **2*n***=32.

Roadsides, sagebrush flats, deserts, rocky areas, fields, disturbed sites. Flowering Apr–Jun. 700–1700 m. BR, ECas, Lava, Owy. CA, ID, NV, WA; north to British Columbia, northeast to Manitoba, east to MI, southeast to NM; Asia, Europe. Exotic.

Alyssum simplex Rudolphi (p. 440)

Small alyssum

Alyssum minus (L.) Rothm. ssp. *micranthum* (C.A. Mey.) Breistr.

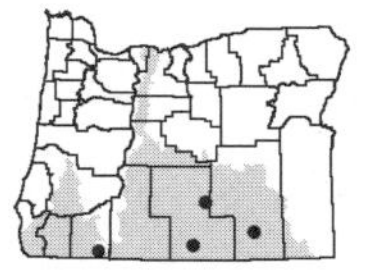

Plants canescent with appressed, sessile, 5–10-rayed stellate trichomes, sometimes with unequal rays. **Stems** often several from base, erect or ascending, (3)7–30(40) cm, simple or branched above. **Cauline leaves** oblanceolate, obovate-spatulate, or elliptic-lanceolate, (0.5)0.7–2.2(3) cm × (1)2–6(8) mm, bases attenuate; subsessile or lowermost petiolate. **Inflorescences** fruiting pedicels divaricate, stout, straight, (2)3–5(6) mm. **Flowers** sepals oblong, 1.7–2.3 mm, stellate-pubescent, caducous; petals oblanceolate, (1.8)2–2.8(3.2) × 0.6–0.8 mm, pale yellow, caducous, tips often emarginate; filaments 1.4–2 mm, median pairs winged, apically 1- or 2-toothed, lateral pair with a broadly winged, apically 2-cleft appendage; styles 0.7–1.6 mm. **Fruits** orbicular, (3.5)4–6.5(7) mm in diameter, tips truncate or emarginate; valves inflated at middle, broadly flattened at margin, surfaces with coarse, stellate, equal- or unequal-rayed trichomes. **Seeds** ovate, slightly compressed, 1.6–2 × 1.2–1.5 mm, margined or not. **2*n***=16.

Sandy slopes, sagebrush areas, fields, roadsides. Flowering May–Jun. 900–1600 m. BR, ECas, Sisk. CA, NV; scattered in western US; Africa, Asia, Europe. Exotic.

ARABIDOPSIS Mouse-ear cress, rockcress

Herbs with stolons or woody caudices. **Trichomes** simple, mixed with stalked, 1–4-forked. **Basal leaves** rosulate, simple, margins entire, toothed, or pinnately lobed. **Cauline leaves** bases not auriculate, margins entire, dentate, or rarely lyrate; petiolate or subsessile. **Inflorescences** bracts 0. **Flowers** sepals erect or ascending, caducous, oblong, lateral pair subsaccate or not; petals obovate, spatulate, or oblanceolate, white, pink, or purple, tips obtuse or emarginate, clawed or not; nectar glands confluent, subtending bases of filaments; stamens 6, slightly tetradynamous, filaments slender, anthers oblong, obtuse; ovules 15–80 per ovary, septa complete, styles obsolete or to 1 mm, stigmas capitate, entire. **Fruits** dehiscent siliques, terete or latiseptate, linear, unsegmented; valves glabrous, veinless or midveins distinct; replums rounded. **Seeds** uniseriate, plump or flattened, mucilaginous or not when wetted; wingless or margined; cotyledons accumbent or rarely incumbent.

Asia, Europe, North America. 10 species; 1 species treated in *Flora.*

Arabidopsis thaliana (Linnaeus) Heynhold (p. 443)

Thale cress, mouse-ear cress

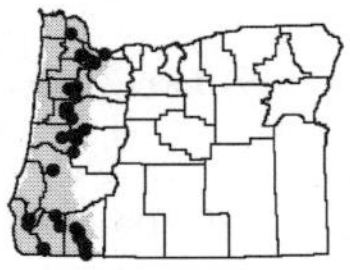

Plants annual. **Stems** 1 or few from base, erect, (2)5–30(50) cm, simple or branched above, with simple and forked trichomes basally, glabrous apically. **Basal leaves** obovate, spatulate, ovate, or elliptic, 8–35(45) × (1)2–10(15) mm, margins entire or dentate, surfaces adaxially with predominantly simple and stalked forked trichomes; short-petiolate. **Cauline leaves** lanceolate, linear, oblong, or elliptic, (4)6–18(25) × 1–6(10) mm, margins entire or rarely few-toothed; subsessile. **Inflorescences** fruiting pedicels divaricate, slender, 3–10(15) mm. **Flowers** sepals 1–2(2.5) mm; petals spatulate, 2–3.5(4) × 0.5–1.5 mm, white, attenuate to a short claw; filaments white, 1.5–2 mm; ovules 40–70 per ovary, styles to 0.5 mm. **Fruits** terete, (8)10–15(18) × 0.5–0.8 mm, smooth; valves with distinct midveins. **Seeds** ellipsoid, plump, 0.3–0.5 mm, light brown; cotyledons incumbent. $2n=10$.

Shale and serpentine barrens, gravel, sandy areas, sandstone, railroad tracks and embankments, lawns, fields, sandy areas, limestone ledges, roadsides, grassy flats, riverbanks, open pastures, prairies, floodplains, woods. Flowering Mar–Jun. 0–1600 m. CR, Sisk, WV. CA, ID, WA; scattered throughout North America; Asia, Europe. Exotic.

Arabidopsis thaliana is the most widely used model organism in plant biology. It is an ideal organism in experimental biology because of its small size (tens can be grown in a pot), autogamous breeding system (induced mutations are expressed in two generations), ability to grow on various synthetic media, fast growth cycle (from seed germination to fruit maturity in four to six weeks), and its small chromosome number (second lowest within flowering plants).

Alliaria petiolata

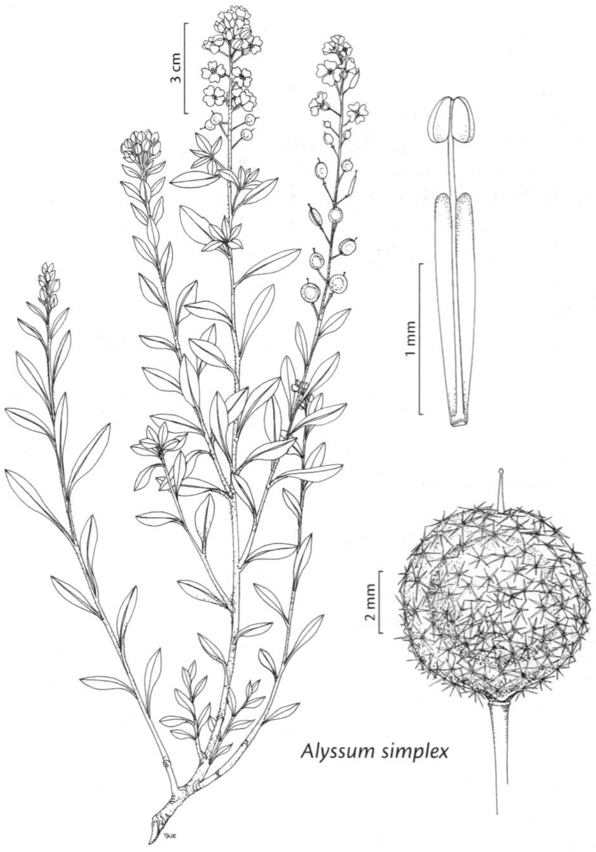

Alyssum simplex

ARABIS Rockcress

Herbs annual, biennial, or perennial. **Trichomes** stalked, stellate or forked, sometimes simple. **Basal leaves** rosulate, simple, margins entire or dentate, rarely lyrate-pinnatifid; petiolate. **Cauline leaves** bases auriculate to amplexicaul or not, margins entire or dentate; sessile or rarely petiolate. **Inflorescences** bracts 0. **Flowers** sepals erect or ascending, caducous, ovate or oblong, lateral pair saccate or not; petals spatulate, oblong, or oblanceolate, white, pink, or purple, claws < sepals, petal tips obtuse or emarginate; nectar glands confluent, rarely median glands absent; stamens 6, tetradynamous, filaments slender, anthers ovate or oblong, obtuse; ovules 12–110 per ovary, septa complete, styles obsolete or distinct, stigmas capitate, entire or 2-lobed. **Fruits** dehiscent siliques, latiseptate, linear, unsegmented; valves glabrous, smooth or torulose, midveins obscure or prominent; replums rounded. **Seeds** uniseriate or biseriate, oblong or orbicular, flattened, not mucilaginous when wetted; winged or margined or neither; cotyledons accumbent.

Africa, Asia, Europe, North America. 60 species; 8 species treated in *Flora*.

North American accounts of *Arabis* (tribe Arabideae), especially those of Mulligan (1996), Rollins (1993), and earlier workers, broadly circumscribed the genus to include species currently assigned to *Arabidopsis* (tribe Camelineae), *Boechera* (tribe Boechereae), and *Turritis* (tribe Turritideae). These accounts also recognized the strictly European, diploid ($2n$=16) *A. hirsuta* as native to North America, despite the fact that our plants are tetraploids or octoploids and quite distinct morphologically (Al-Shehbaz et al. 2010).

Key to *Arabis* species:

1. Cauline leaf bases auriculate or subcordate.
  2. Petals 3.5–5(5.5) mm; lateral sepals not saccate; fruits 0.8–1(1.2) mm wide; cauline leaves (7)10–45(61), usually overlapping . . . . ***A. pycnocarpa***
  2' Petals (5.5)6.5–9(10) mm; lateral sepals saccate; fruits 1.2–1.8(2) mm wide; cauline leaves (2)4–12(18), often well spaced, rarely overlapping. . . . . ***A. eschscholtziana***

1' Cauline leaf bases cuneate.
  3. Petals white (rarely pale pink), 7–11 mm; sepals 2.5–4 mm; plants of northern Oregon.
    4. Fruits divaricate, 1–1.5 mm wide; trichomes of basal and lowermost cauline leaves cruciform or 3-rayed stellate . . . . ***A. crucisetosa***
    4' Fruits suberect to ascending, 1.7–2.2 mm wide; trichomes of basal and lowermost cauline leaves simple or 2-rayed, rarely glabrous . . . . ***A. furcata***
  3' Petals pink to magenta or purple, 8–18(20) mm; sepals 4–8 mm; plants of southwestern Oregon.
    5. Trichomes of basal and lowermost cauline leaves cruciform or 3-rayed stellate . . . . ***A. modesta***
    5' Trichomes of basal and lowermost cauline leaves simple, 2-rayed, or absent, if 3-rayed stellate trichomes present then few in number.
      6. Plants glabrous or with a few simple trichomes along margins of basal leaves; simple trichomes not bulbous-based; fruits 2–4 cm. . . . . ***A. mcdonaldiana***
      6' Plants pubescent with simple and forked trichomes; simple trichomes often bulbous-based; fruits (30)35–65 mm.
        7. Cauline leaves 10–30(60) mm; fruits slightly torulose; few 3-rayed stellate trichomes present. . . ***A. oregana***
        7' Cauline leaves 4–10(15) mm; fruits not torulose; no 3-rayed stellate trichomes present. . . . . . ***A. aculeolata***

Arabis aculeolata Greene

Waldo rockcress

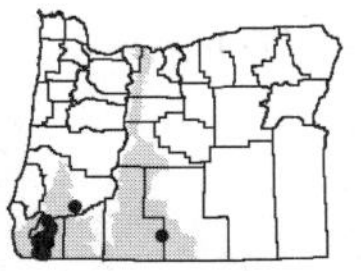

Plants perennial; pubescent with simple and stalked, forked trichomes, trichomes often bulbous-based; caudices present. **Stems** (6)15–35(45) cm, simple. **Basal leaves** rosulate, blades oblanceolate to obovate, (5)10–30(40) × (2)3–6(10) mm, margins entire, repand, or obtusely dentate, ciliate, surfaces pubescent or rarely subglabrous; petioles 3–15 mm. **Cauline leaves** 3–6(7), oblong, 4–10(15) × (1)2–4 mm, bases cuneate, margins entire or repand; sessile. **Inflorescences** fruiting pedicels ascending to suberect, 8–12(15) mm. **Flowers** sepals oblong, 4–8 mm, purple, lateral pair saccate; petals spatulate, (9)10–18(20) × (2.5)3.5–6(8) mm, purple, tips obtuse; ovules 24–36 per ovary, styles (0.7)1–2 mm. **Fruits** ascending to suberect, 35–65 × 1.5–2 mm; valves glabrous, not torulose, with prominent midvein. **Seeds** uniseriate, oblong, 1.5–2.3 × 1–1.3 mm; wings 0.1–0.3 mm wide. **$2n$**=32.

Rocky hillsides, serpentine slopes. Flowering Mar–Jun. 200–1800 m. ECas, Sisk. CA. Native.

A few old collections made in the 1940s appear to be intermediate between *Arabis aculeolata* and *A. modesta*. However, without experimental hybridization and molecular studies, this theory remains untested.

Arabis crucisetosa Constance & Rollins

Cross-haired rockcress, wetsoil rockcress

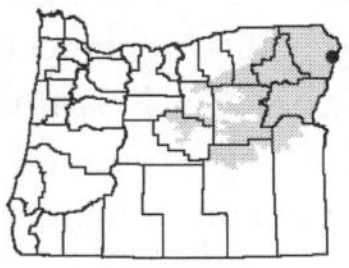

Plants perennial; caudices present. **Stems** erect, (10)15–40(50) cm, simple, glabrous or sparsely pubescent below middle. **Basal leaves** rosulate, blades oblanceolate to obovate, (6)10–35(45) × (3)5–15(20) mm, margins entire or dentate, not ciliate, surfaces glabrous or sparsely pubescent, trichomes short-stalked, forked and simple; petioles 5–25(30) mm. **Cauline leaves** 3–4(6), linear, oblong, or oblanceolate, 10–30(40) × (1.5)3–7(10) mm, bases cuneate, margins entire or rarely dentate, surfaces glabrous or sparsely pubescent at least along margin with short-stalked, forked and simple trichomes; sessile. **Inflorescences** fruiting pedicels ascending to divaricate, (5)8–20 mm. **Flowers** sepals oblong, 2.5–4 mm, greenish, glabrous, lateral pair saccate; petals oblanceolate, 7–9 × 2.5–3.5 mm, white, tips rounded; ovules 14–28 per ovary, styles 0.5–1.3 mm. **Fruits** divaricate, 17–35(40) × 1–1.5 mm; valves glabrous, slightly torulose, with obscure midvein. **Seeds** uniseriate, oblong, 1–1.2 × 0.5–0.6 mm; wingless. ***2n***=32.

Steep rocky bluffs, open woods, canyon margins. Flowering Apr–Jun. 400–1800 m. BW. ID, WA. Native.

Arabis eschscholtziana Andrzejowski

Eschscholtz's hairy rockcress

Arabis hirsuta (L.) Scop. var. *eschscholtziana* (Andrz.) Rollins, *Arabis hirsuta* (L.) Scop. var. *glabrata* Torr. & A. Gray

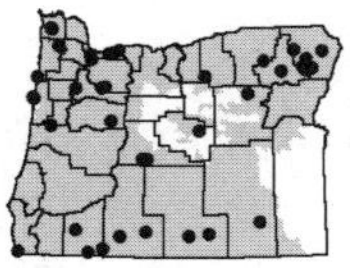

Plants biennial or perennial, hirsute with simple or stalked, forked trichomes; caudices present. **Stems** erect, 20–70(100) cm, rarely subglabrous. **Basal leaves** rosulate, blades spatulate to oblanceolate, 12–100 × 7–30 mm, margins entire or dentate, pubescent or glabrous; petioles 7–30 mm, ciliate or not. **Cauline leaves** (2)4–12(18), well spaced or rarely overlapping, ovate to oblong or lanceolate, 15–65(100) × 5–30 mm, bases auriculate; sessile. **Inflorescences** fruiting pedicels erect to ascending, 3.5–10(15) mm. **Flowers** sepals oblong, 3.5–5 mm, lateral pair saccate; petals linear-oblanceolate to spatulate, (5.5)6.5–9(10) × 1.5–3 mm, white, rarely pale pink, obtuse; ovules 54–80 per ovary, styles (0.1)0.3–1 mm. **Fruits** erect to ascending, often appressed to rachis, 35–65 × 1.2–1.8(2) mm; valves glabrous, torulose, midvein prominent. **Seeds** uniseriate, oblong or suborbicular, 1–1.8 × 0.9–1.3 mm; wings distal, to 0.2 mm wide, or absent. ***2n***=32, 64.

Wooded slopes, meadows, rocky areas, crevices, ledges, moist banks and grounds. Flowering Apr–Jul. 0–2200 m. BR, BW, Casc, Col, CR, ECas, Est, Sisk, WV. ID, WA; northeast to MT, AK. Native.

Arabis furcata S. Watson

Fork-haired rockcress

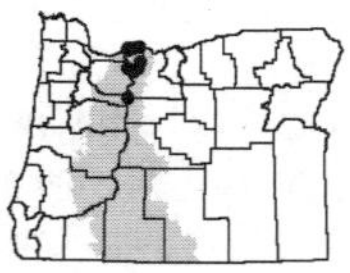

Plants perennial; caudices present. **Stems** erect, (7)10–38(45) cm. **Basal leaves** rosulate, blades oblanceolate, spatulate, or obovate, (7)12–30 × 5–17(22) mm, margins entire or dentate, glabrous or ciliate, surfaces sparsely pubescent or rarely glabrous, trichomes short-stalked, forked and simple; petioles (4)10–30(45) mm. **Cauline leaves** (2)3–5(6), linear, oblong, obovate, or oblanceolate, 7–30(40) × 3–8 mm, bases cuneate, margins entire or rarely dentate, surfaces lowermost leaves with simple or short-stalked forked trichomes, rarely glabrous; sessile. **Inflorescences** fruiting pedicels suberect to divaricate, (2)4–17(22) mm. **Flowers** sepals oblong, 3–4 mm, green; petals spatulate, 7–11 × 2.5–4 mm, white, tips rounded; ovules 14–26 per ovary, styles 0.5–1.2(1.6) mm. **Fruits** suberect to ascending, not appressed to rachis, (20)25–40(46) × 1.7–2.2 mm; valves glabrous, slightly torulose, midvein prominent. **Seeds** uniseriate, oblong-ovate, (1.5)1.8–2.5(3) × 1–1.3 mm; wings distal, (0.2)0.4–0.8(1) mm wide.

Alpine meadows, cliffs, open slopes, ridge crests. Flowering May–Aug. 0–2400 m. Casc, ECas. WA. Native.

Arabis mcdonaldiana Eastwood (p. 443)

McDonald's rockcress

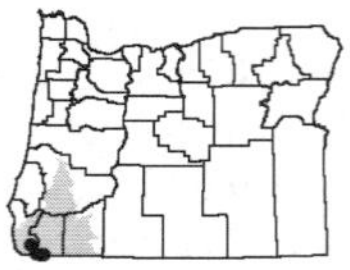

Plants perennial; caudices present. **Stems** erect, (6)15–30(40) cm, glabrous. **Basal leaves** rosulate, blades oblanceolate to obovate, (5)10–30(40) × (2)3–6(10) mm, margins entire or obtusely dentate, surfaces glabrous or with a few simple trichomes, trichomes not bulbous-based; petioles 3–15mm, glabrous. **Cauline leaves** (2)3–5(6), oblong, 3–10(12) × 1–3 mm, bases cuneate, margins entire or repand, glabrous; sessile. **Inflorescences** fruiting pedicels ascending to suberect, 3–10(13) mm. **Flowers** sepals oblong, 4–8 mm, purple, lateral pair saccate; petals spatulate, 8–16 × 2–5 mm, purple, tips obtuse; ovules 20–34 per ovary, styles 0.3–1.5 mm. **Fruits** ascending to suberect, 20–40 × 1.5–2 mm; valves glabrous, not torulose, midvein prominent. **Seeds** uniseriate, oblong, 1.5–2.2 × 1–1.3 mm; wings distal, 0.1–0.2 mm wide, or absent.

Serpentine soil. Flowering Apr–Jun. 400–1500 m. Sisk. CA. Native.

Arabis modesta Rollins

Rogue Canyon rockcress, modest rockcress

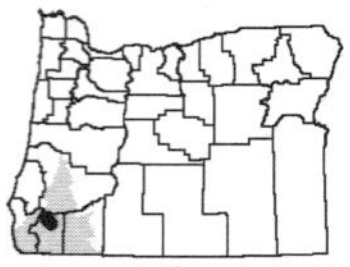

Plants perennial; caudices present. **Stems** erect, 22–55(67) cm, pubescent or rarely subglabrous above. **Basal leaves** rosulate, blades oblanceolate to obovate, (8)20–45(60) × (4)10–20(25) mm, margins entire or dentate, not ciliate, surfaces pubescent with short-stalked, cruciform or 3-rayed stellate trichomes; petioles 10–40(50) mm, rarely ciliate basally. **Cauline leaves** (2)4–6(9), oblong or ovate, 10–35(45) × (3)6–10(20) mm, bases cuneate, margins entire or dentate, surfaces pubescent, uppermost leaves with short-stalked, cruciform or 3-rayed stellate trichomes; sessile. **Inflorescences** fruiting pedicels ascending to divaricate, 7–18(25) mm. **Flowers** sepals oblong, (4)5–6.5(8) mm, purple, lateral pair saccate; petals spatulate, (10)12–16(20) × 4–6(7) mm, purple, tips rounded; ovules 20–34 per ovary, styles 0.5–1(1.5) mm. **Fruits** suberect to divaricate, (28)35–60 × 1.5–2 mm; valves glabrous, slightly torulose, midvein prominent. **Seeds** uniseriate, oblong, 1.7–2.2 × 0.9–1.2 mm; wings distal, 0.2–0.5 mm wide. ***2n***=32.

Rocky canyon walls, moist shaded banks, talus, basaltic bluffs, slopes. Flowering Apr–May. 200–900 m. Sisk. CA. Native.

Arabis oregana Rollins

Oregon rockcress

Arabis purpurascens Howell

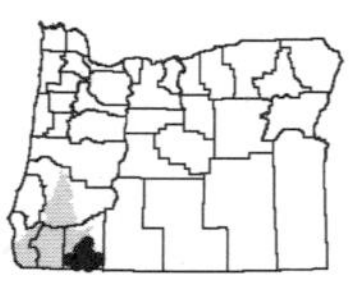

Plants perennial, pubescent with simple and long-stalked, forked trichomes and 3-rayed, stalked trichomes, trichomes often bulbous-based; caudices present. **Stems** erect, (6)17–45(50) cm, hirsute or rarely subglabrous. **Basal leaves** rosulate, blades oblanceolate to obovate, (10)20–50(90) × (5)8–18(22) mm, margins entire or dentate, ciliate, surfaces pubescent or glabrous; petioles 5–20(30) mm, ciliate. **Cauline leaves** 3–6(7), oblong, 10–30(60) × 2–7(15) mm, bases cuneate, margins entire or dentate; sessile. **Inflorescences**

fruiting pedicels ascending to erect, 5–10 mm. **Flowers** sepals oblong, (5)6–8 mm, purple, lateral pair saccate; petals spatulate, (10)12–15(16) × (3)4–5 mm, purple; ovules 24–30 per ovary, styles 0.5–1 mm. **Fruits** erect to suberect, (30)45–60 × 1.5–2 mm; valves glabrous, slightly torulose, midveins prominent. **Seeds** uniseriate, oblong, 1.8–2.2 × ~1 mm; wings distal, ~0.1 mm wide, or absent. **2*n***=32.

Chaparral, steep banks, moist granitic soil, rocky hillsides. Flowering Mar–Jun. 400–1700 m. Sisk. CA. Native.

Arabis pycnocarpa M. Hopkins (p. 443)

Creamflower rockcress

Arabis hirsuta (L.) Scop. var. *pycnocarpa* (M. Hopkins) Rollins, *Arabis pycnocarpa* M. Hopkins var. *pycnocarpa*

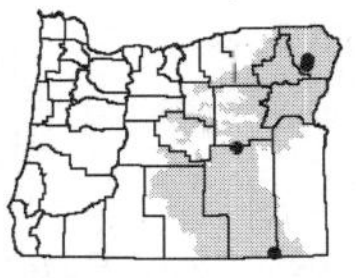

Plants biennial or perennial, hirsute basally with simple trichomes; caudices present. **Stems** erect, 10–80 cm. **Basal leaves** rosulate, blades spatulate, oblanceolate, or oblong, (8)15–80 × (5)10–25 mm, margins entire or dentate, surfaces pubescent with simple or forked trichomes; petioles 5–20 mm, ciliate or not. **Cauline leaves** (7)10–45(61), usually overlapping, ovate to oblong, (10)15–60(80) × (1)3–20(25) mm, bases subcordate or auriculate, surfaces hirsute or adaxially glabrate; sessile. **Inflorescences** fruiting pedicels erect to ascending, (2)3–8(12) mm. **Flowers** sepals oblong, 2.5–4 mm, lateral pair not saccate; petals linear-oblanceolate to spatulate, 3.5–5(5.5) × 1–2(2.5) mm, white, tips obtuse; ovules (54)60–86 per ovary, styles (0.2)0.5–1(1.3) mm. **Fruits** erect to ascending, often appressed to rachis, (35)40–60(65) × 0.8–1(1.2) mm; valves glabrous, torulose, midveins obscure or distinct. **Seeds** uniseriate, oblong or suborbicular, (0.8)1–1.5(1.7) × 0.8–1.3 mm; wings continuous, to 0.2 mm wide. **2*n***=32.

Ledges, bluffs, cliffs, open woods, streamsides, meadows, upland prairies, grassy swales, hillsides, bottom lands, gravel bars, river bottoms. Flowering Mar–Jul. 1100–2300 m. BR, BW. CA, ID, NV, WA; throughout most of Canada and US. Native.

Hopkins (1937) divided the species into two varieties, *pycnocarpa* and *adpressipilis*. The latter is less widespread (IL, IN, IA, KS, MO, NY, OH, PA, TN, VA, and Ontario) and differs by having pilose (vs. hirsute) lower stems with appressed branched (vs. spreading simple) trichomes.

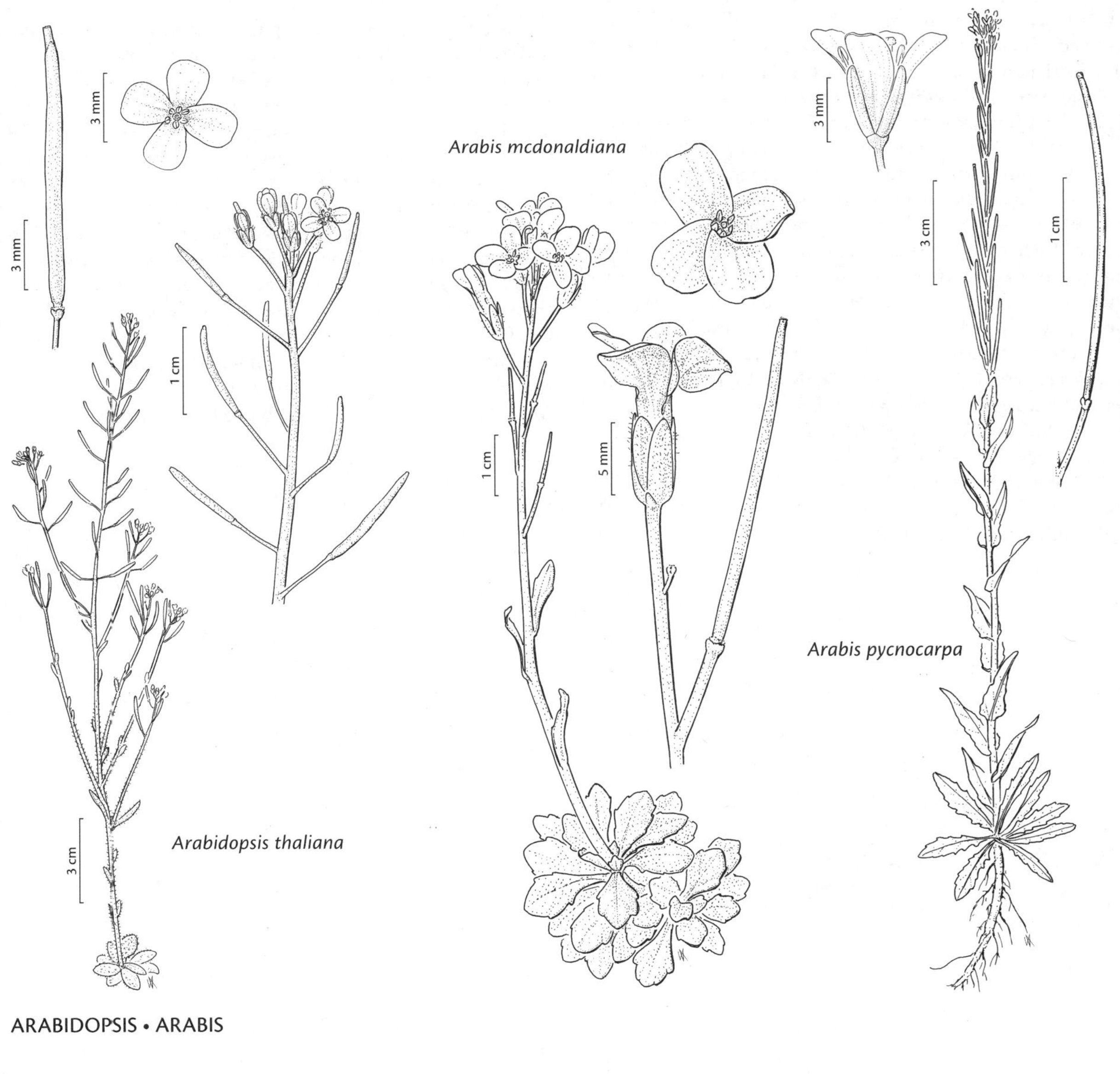

ARABIDOPSIS • ARABIS

ATHYSANUS Athysanus

Herbs annual. **Trichomes** simple, mixed with short-stalked, forked, 3-rayed, or cruciform. **Basal leaves** not rosulate, simple, margins entire or dentate; petiolate. **Cauline leaves** bases not auriculate; short-petiolate to sessile. **Inflorescences** secund, lax; bracts 0. **Flowers** sepals oblong, suberect, caducous, lateral pair not saccate; petals rudimentary, spatulate to oblong, much < sepals, rarely absent or > sepals, white, indistinctly clawed; nectar glands lateral, 1 on each side of lateral stamens; stamens 6, subequal, filaments free, anthers obtuse; ovules 2–12 per ovary, septa present or absent, styles obsolete, stigmas capitate, entire. **Fruits** indehiscent or very tardily dehiscent silicles, latiseptate, pendulous, unsegmented, orbicular or obovate to elliptic, wingless; valves not veined; replums rounded. **Seeds** uniseriate or aseriate, oblong, flattened, not mucilaginous when wetted; wingless; cotyledons accumbent.

Western North America. 2 species.

Key to *Athysanus* species:

1. Fruits 2–2.5 mm, with hooked trichomes, septa absent, 1-seeded, indehiscent . . . . ***A. pusillus***

1' Fruits 3–5 mm, without hooked trichomes, with septa, 2+-seeded, tardily dehiscent . . . . ***A. unilateralis***

Athysanus pusillus (Hooker) Greene (p. 444)

Dwarf athysanus, sandweed

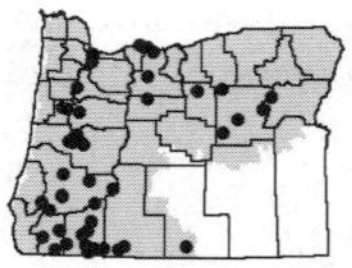

Plants pubescent, sometimes glabrous above, (2)5–30(50) cm. **Basal leaves** oblanceolate, obovate, or oblong, 3–15(20) × (1)2–6(8) mm, margins entire or dentate, uniformly pubescent with cruciform and 3-rayed trichomes, sometimes mixed adaxially with coarse, simple trichomes; petioles 1–5 mm. **Cauline leaves** 1–6, similar to basal leaves. **Inflorescences** fruiting pedicels slender, (1.5)2–4(6) mm. **Flowers** sepals oblong, 0.5–1 mm; petals oblong, 1.5–3 × 0.5–1 mm; ovules 2(4) per ovary, septa absent, styles 0.05–0.2 mm. **Fruits** indehiscent, orbicular to obovate, 2–2.5 mm, unilocular, not twisted, bases cuneate or obtuse, tips rounded or less frequently emarginate, pubescent with simple, hooked trichomes to 0.5 mm, these sometimes restricted to fruit margins or frequently mixed with much smaller, sessile cruciform or 3-rayed trichomes, very rarely glabrous; valves thin. **Seeds** 1, (0.8)0.9–1.1(1.2) × (0.5)0.7–0.9(1) mm. **2*n***=26.

Sandy and gravelly flats, rock outcrops, cliffs, ledges, grassy slopes, chaparral, glades in woodlands, floodplains, limestone, serpentine, sandstone. Flowering Feb–Jun. 0–1900 m. BW, Casc, Col, CR, ECas, Lava, Sisk, WV. CA, ID, NV, WA; north to British Columbia, northeast to MT, southeast to AZ, south to Mexico. Native.

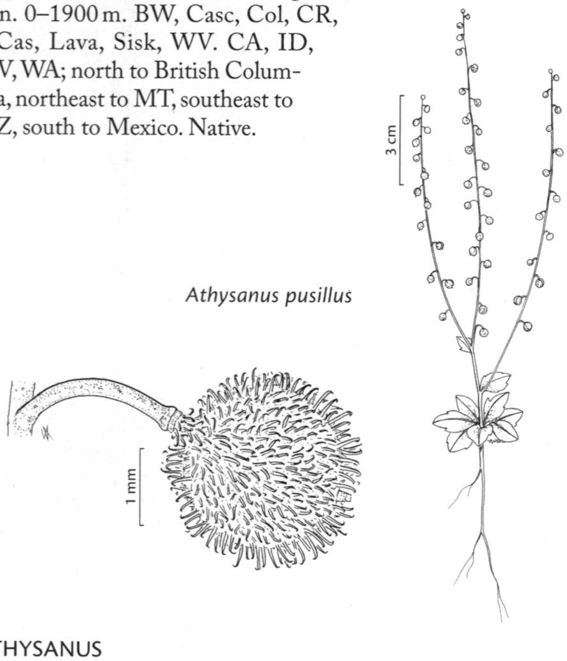

Athysanus pusillus

Athysanus unilateralis (M.E. Jones) Jepson

Spreading athysanus, heterodraba

Heterodraba unilateralis (M.E. Jones) Greene

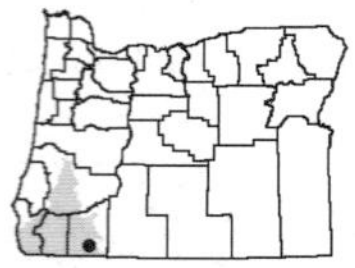

Plants pubescent, (3)7–25(35) cm. **Basal leaves** oblanceolate, obovate, or oblong, (3)7–20(22) × (2)4–7 mm, margins entire or 1-toothed, pubescent with cruciform and fewer 3-rayed trichomes; subsessile or with petioles to 3 mm. **Cauline leaves** 2–5, similar but smaller than basal, cuneate. **Inflorescences** fruiting pedicels stout, 1–2(3) mm. **Flowers** sepals oblong, 0.6–1 mm; petals 1.3–1.7 × 0.5–0.8 mm; ovules 6–12 per ovary, septa complete, membranous, styles 0.1–0.2 mm. **Fruits** tardily dehiscent, latiseptate, elliptic, obovate, or orbicular, flat or slightly twisted, 3–5 mm, glabrous or uniformly pubescent with subsessile or short-stalked cruciform and 3-rayed trichomes, these sometimes mixed with stouter simple subsetose trichomes to 0.6 mm, wingless; valves thickened, often with prominent veins. **Seeds** 2+, 0.9–1.2 × 0.6–0.8 mm.

Open grassy slopes and flats, floodplains, friable clay, gypsum clay slopes, adobe and heavy gumbo clay. Flowering Feb–May. 100–900 m. Sisk. CA; south to Mexico. Native.

BARBAREA Winter cress, yellowrocket, upland cress

Herbs biennial or perennial, rarely annual; rhizomatous or with caudices. **Trichomes** simple or absent. **Stems** angled. **Basal leaves** rosulate or not, simple, margins pinnatisect, lyrate, or dentate; petiolate. **Cauline leaves** bases auriculate, margins lobed or dentate; sessile. **Inflorescences** bracts present or not. **Flowers** sepals erect or spreading, caducous or persistent, ovate or oblong, lateral pair often subsaccate; petals > sepals, blades spatulate or oblanceolate, yellow, rarely creamy white, claws obscurely differentiated, petal tips obtuse or rounded; nectar glands 4, median tooth-like, lateral annular; stamens 6, tetradynamous, filaments slender, anthers oblong, obtuse; ovules 10–52 per ovary, septa complete, styles obsolete or to 5 mm, stigmas entire or 2-lobed. **Fruits** dehiscent siliques, terete or latiseptate, rarely 4-angled, linear, unsegmented; valves smooth or torulose, midveins prominent; replums rounded. **Seeds** 10+ per fruit, uniseriate, ovoid to oblong, seed coat reticulate, not mucilaginous when wetted; winged or wingless; cotyledons accumbent.

Worldwide. 22 species; 3 species treated in *Flora*.

Key to *Barbarea* species:

1\. Fruits (45)53–70(80) mm; seeds (34)38–48(52) per fruit; fruiting pedicels nearly as thick as fruit. . . . . ***B. verna***

1' Fruits (7)15–40(45) mm; seeds 18–36 per fruit; fruiting pedicels often narrower than fruit.

  2\. Styles (1)1.5–3(3.5) mm, slender; tips of sepals and auricles of upper cauline leaves glabrous . . . . ***B. vulgaris***

  2' Styles 0.2–1(2) mm, stout; tips of sepals and/or auricles of upper leaves with hairs, rarely glabrous . . . . ***B. orthoceras***

Barbarea orthoceras Ledebour (p. 445)

American winter cress

Barbarea orthoceras Ledeb. var. *orthoceras*, *Barbarea orthoceras* Ledeb. var. *dolichocarpa* Fernald

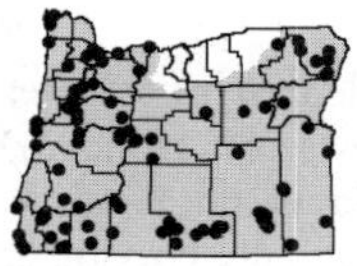

Plants biennial or perennial. **Stems** (1)2–6(10) dm, glabrous. **Basal leaves** glabrous or ciliate basally, blades (1)1.5–6(11) cm, lyrate-pinnatifid, with (1)2–4(5) lateral lobes,, oblong or ovate, 2–10 × 1–5 mm, terminal lobes > laterals, 15–50 × 10–25 mm; petioles (0.5)1–7(15) cm. **Cauline leaves** lyrate-pinnatifid, with 1–4 lateral lobes, uppermost incised or pinnatifid, auricles with hairs or rarely glabrous; sessile. **Inflorescences** fruiting pedicels erect or ascending, (2)3–6(7) mm, narrower than fruits. **Flowers** sepals erect, oblong, 2.5–3.5 mm, yellow, tips rounded and with hairs or rarely glabrous, lateral pair slightly saccate; petals oblanceolate, 5–7(8) × (1.5)2–3 mm, yellow; styles stout, 0.2–1(2) mm in fruit. **Fruits** (25)31–40(45) × 1.5–2 mm, terete to subquadrangular, torulose, erect to ascending, rarely appressed to rachis. **Seeds** (24)26–36 per fruit, ovate or oblong, plump, 1.2–1.5 × 0.9–1 mm; wingless. **2*n***=16.

Alpine meadows, boggy ground, grasslands, sandbars, scree, rocky cliffs, forests, riverbanks, streambanks, disturbed areas. Flowering Mar–Aug. 0–2700 m. All ecoregions except Col. CA, ID, NV, WA; throughout Canada, northern and western US; Asia. Native.

Barbarea verna (Miller) Ascherson

Early winter cress, Belle Isle cress, land cress, upland cress, scurvy cress

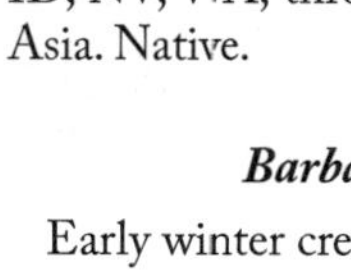

Plants biennial or perennial, glabrous except for leaves. **Stems** erect, (1)2.5–8 dm. **Basal leaves** 1.5–11 cm, pinnatifid to pinnatisect, lateral lobes (3)6–10, oblong or ovate, 4–30 × 1–10 mm, terminal lobes 10–50 mm; petioles (0.5)1–6(8) cm. **Cauline leaves** oblong to lanceolate, bases auriculate, margins pinnatisect, rarely toothed, lateral lobes 1–4; mostly sessile but lowermost petiolate. **Inflorescences** fruiting pedicels divaricate to ascending, (2)3–6(7) mm, nearly as thick as fruits. **Flowers** sepals oblong, 3–5 mm, yellow, glabrous, lateral pair slightly saccate; petals oblanceolate to spatulate,

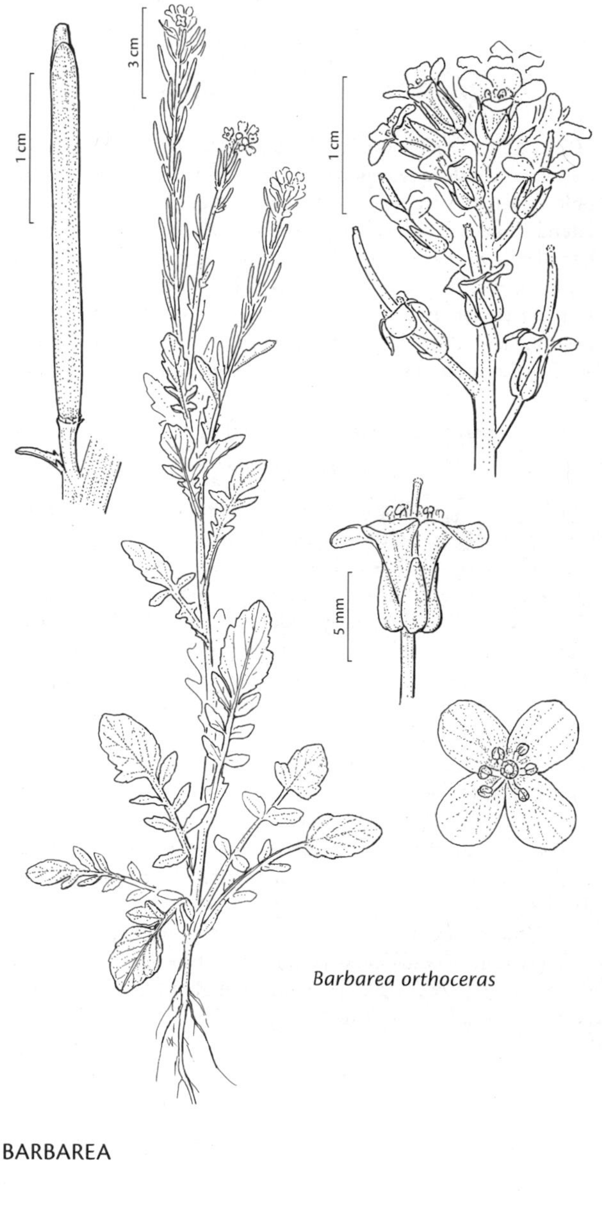

Barbarea orthoceras

BARBAREA

(5)6–7(8.5) × 1.5–3 mm, yellow, truncate or emarginate; styles stout, 0.2–1(2) mm in fruit. **Fruits** erect to ascending, subterete to latiseptate, torulose, (45)53–70(80) × 1.5–2 mm. **Seeds** (34)38–48(52) per fruit, uniseriate, oblong or quadrate, plump, 1.8–2.5 × 1.4–1.6 mm; wingless. ***2n***=16.

Fields, limestone glades, rocky outcrops, roadsides, waste grounds, railroad embankments, disturbed sites. Flowering Mar–Jul. 0–1600 m. WV. CA, WA; scattered in eastern and far western North America; Asia, Europe. Exotic.

Barbarea vulgaris W.T. Aiton

Garden yellowrocket, bitter winter cress, creasy greens

Barbarea vulgaris W.T. Aiton var. *brachycarpa* Rouy & Foucaud

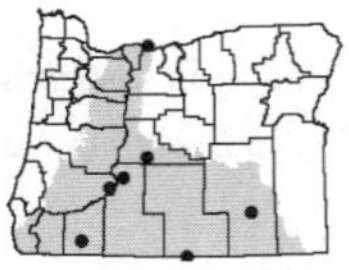

Plants biennial or rarely perennial, glabrous or pubescent. **Stems** erect, (1.5)2–9(12) dm. **Basal leaves** (1)2–8(10) cm, lyrate-pinnatifid, margins entire or dentate, glabrous or ciliate, lateral lobes 1–3(5), oblong or ovate, 3–20(40) × 1–8(15) mm, terminal lobes ovate or suborbicular, (7)15–45(70) × (4)10–30(50) mm; petioles (0.5)2–10(17) cm. **Cauline leaves** lowermost similar to basal, upper with glabrous auricles, undivided, dentate, or rarely subentire; sessile. **Inflorescences** fruiting pedicels divaricate to erect, 3–7 mm, narrower than fruits. **Flowers** sepals erect, oblong, 3–4.5(5) mm, yellow; petals spatulate or oblanceolate, (5)6–9(10) × 1.5–2.5(3.5) mm, yellow, tips rounded and glabrous; styles slender, (1)1.5–3(3.5) mm in fruit. **Fruits** erect to ascending, terete, compressed, or 4-angled, torulose, (7)15–30 × 1.2–2 mm. **Seeds** 18–24(28) per fruit, uniseriate, ovate to oblong, plump, 1.2–1.5 × 1–1.2 mm; wingless. ***2n***=16.

Riverbanks, damp grasslands, fields, disturbed sites, waste places, roadsides. Flowering Apr–Jun. 50–2100 m. BR, Casc, ECas, Sisk. CA, ID, WA; throughout Canada and US; Africa, Asia, Europe. Exotic.

BERTEROA Hoary alyssum, false madwort

Herbs annual or biennial. **Trichomes** stellate and simple. **Basal leaves** not rosulate, simple, margins entire, dentate, or sinuate; petiolate. **Cauline leaves** bases not auriculate, margins usually entire; sessile. **Inflorescences** bracts 0. **Flowers** sepals caducous, oblong, lateral pair not saccate; petals obcordate, > sepals, white or rarely yellow, short-clawed, tips deeply bifid; nectar glands 4, lateral, 1 on each side of lateral stamen; stamens 6, tetradynamous, filaments of median pairs unappendaged or laterally 1-toothed, lateral pair with a basal tooth-like appendage, anthers oblong, obtuse; ovules 4–16 per ovary, septa complete, styles to 4 mm, stigmas capitate, obscurely 2-lobed. **Fruits** dehiscent silicles, latiseptate, oblong, elliptic, ovate, obovate, or orbicular, unsegmented; valves veinless, stellate or glabrous; replums rounded. **Seeds** biseriate, suborbicular or lenticular, plump or flattened, minutely reticulate, not mucilaginous when wetted; winged or wingless; cotyledons accumbent.

Northern Hemisphere. 5 species; 1 species treated in *Flora*.

Berteroa incana (Linnaeus) de Candolle (p. 446)

Hoary alyssum

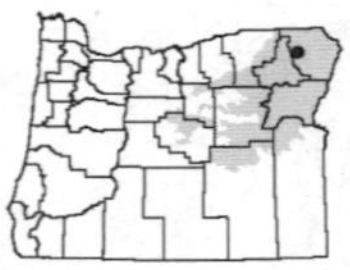

Plants trichomes appressed stellate mixed with simple. **Stems** erect, (2)3–8(11) dm. **Basal leaves** oblanceolate, (2.5)3.5–8(10) cm. **Cauline leaves** similar to basal, gradually reduced above. **Inflorescences** fruiting pedicels erect to ascending, appressed to rachis, (4)5–9(12) mm. **Flowers** sepals erect-ascending, 2–2.5 mm; petals obcordate, (4)5–6.5(8) mm, white, lobes oblong, (1)1.5–3 × 0.5–1.5 mm; median filament pairs appendaged or not, lateral pair appendage 0.3–1 mm; styles 1–4 mm, sparsely pubescent basally. **Fruits** oblong or elliptic, slightly inflated, (4)5–8.5(10) × (2)2.5–4 mm; valves pubescent with stellate trichomes sometimes with unequal rays. **Seeds** lenticular or ovate-orbicular, slightly flattened, 1–2.3 mm in diameter, narrowly margined. ***2n***=16.

Disturbed areas, pastures, hillsides, floodplains, woodlands, meadows, grasslands, fields, riverbanks. Flowering May–Sep. 0–2500 m. BW. CA, ID, NV, WA; throughout most of Canada and US; Asia, Europe. Exotic.

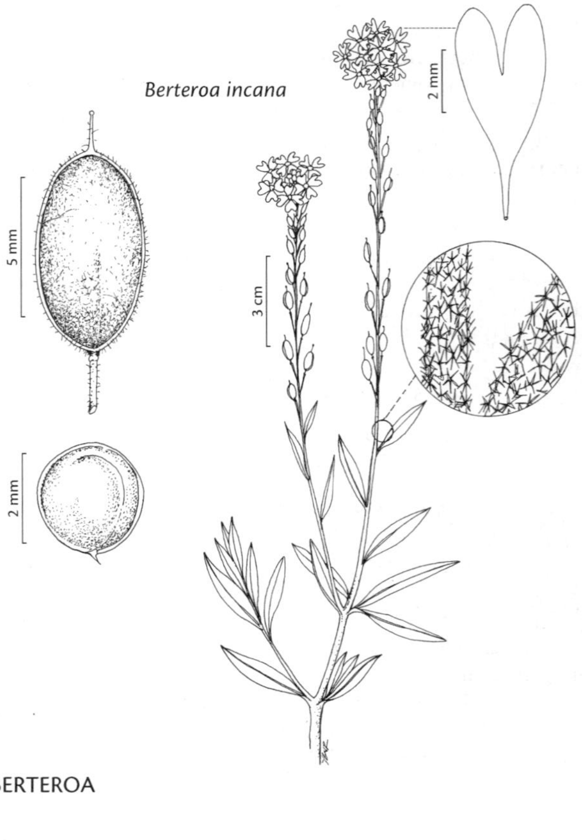

BERTEROA

BOECHERA Rockcress

Herbs biennial or perennial; caudices present. **Trichomes** simple and/or branched, stalked or sessile, 2–14-rayed, forked or dendritic, rarely malpighiaceous. **Basal leaves** rosulate or not, simple, margins entire or dentate; petiolate. **Cauline leaves** bases usually auriculate, margins entire or dentate; sessile or short-petiolate. **Inflorescences** racemes; bracts 0. **Flowers** sepals ovate or oblong, lateral pair slightly saccate or not; petals spatulate or oblanceolate, white, pink, or purple, claws < sepals or undifferentiated, petal tips obtuse; nectar glands confluent, subtending bases of all stamens; stamens 6, tetradynamous, anthers ovate or oblong, obtuse, pollen ellipsoid (sexual plants) or spheroid (apomictic); ovules 8–216 per ovary, septa complete, styles obsolete or distinct, stigmas capitate, entire. **Fruits** dehiscent siliques, linear, rarely oblong or lanceolate, latiseptate, unsegmented; valves smooth or torulose; replums rounded. **Seeds** 20+ per fruit, uniseriate or biseriate, oblong or orbicular, flattened, not mucilaginous when wetted; winged or rarely wingless; cotyledons accumbent.

Eastern Asia, North America. ~110 species; 33 treated in *Flora*.

Boechera is by far the most taxonomically complex genus within the family. Over 60% of the species are sexual diploids (n=7) with small ellipsoid pollen, while the rest are apomictic triploids (n=21) with larger globose pollen (though some species are both sexual and apomictic). Wherever the ranges of diploids overlap they hybridize, and the production of unreduced pollen eventually leads to the formation of apomictic triploid hybrids. The hybrids are also stable, often widespread, and produce viable apomictic seeds. As such, they should be treated as distinct species.

In order to correctly identify species, careful study of the trichome types, as well as pollen morphology (at least under a dissecting microscope with high magnification), is often required.

Key to *Boechera* species:

1. Fruits erect, appressed to subappressed to rachis.
    2. Basal leaf trichomes exclusively malpighiaceous, sessile. . . . . ***B. stricta***
    2' Basal leaf trichomes simple and/or 2–4-rayed, stalked or subsessile.
        3. Lower stems pubescent.
            4. Lower stems with at least some simple trichomes; ovules 24–40 per ovary . . . . ***B. paupercula***
            4' Lower stems pubescent with exclusively branched trichomes; ovules 60–140 per ovary . . . . ***B. pratincola***
        3' Lower stems glabrous.
            5. Petals purple; margins of basal leaves with sessile trichomes; ovules 64–134 per ovary . . . . ***B. calderi***
            5' Petals white to lavender; margins of basal leaves with stalked trichomes; ovules (34)42–64 per ovary . . ***B. lyallii***

1' Fruits ascending, divaricate, horizontal, or reflexed, not appressed to rachis, rarely appressed when sharply reflexed.
    6. Fruiting pedicels and fruits distinctly reflexed.
        7. Fruiting pedicels recurved at bases.
            8. Lower stems glabrous; ovules 46–60 per ovary. . . . . ***B. hastatula***
            8' Lower stems densely pubescent; ovules (60)70–116 per ovary. . . . . ***B. retrofracta***
        7' Fruiting pedicels usually not recurved at bases.
            9. All cauline leaves not auriculate at bases . . . . ***B. pendulocarpa***
            9' At least some cauline leaves auriculate at bases.
                10. Fruits pubescent.
                    11. Petals white to lavender, 5–9 × 0.8–1.8 mm; plants short-lived perennials, with non-woody caudices; sepals pubescent . . . . ***B. puberula***
                    11' Petals purple or rarely lavender, 9–14 × 1.5–3 mm; plants long-lived perennials, with woody caudices; sepals glabrous . . . . ***B. subpinnatifida***
                10' Fruits glabrous.
                    12. Fruits 2.5–6 mm wide; ovules 14–30 per ovary.
                        13. Petals 8–11 mm; fruit 2.5–3.5 mm wide; ovules 14–22 per ovary. . . . . ***B. rollei***
                        13' Petals 4.5–6 mm; fruit 3–6 mm wide; ovules 20–30 per ovary . . . . ***B. suffrutescens***
                    12' Fruits 1.5–2.5(3) mm wide; ovules (24)34–162 per ovary.
                        14. Fruits appressed to rachis, lower stems glabrous or rarely with few simple trichomes . . ***B. rectissima***
                        14' Fruits not appressed to rachis, lowers stems pubescent with at least some branched trichomes.
                            15. Lower stems pubescent with exclusively 4–10-rayed trichomes . . . . ***B. cobrensis***
                            15' Lower stems pubescent with simple and 2–4-rayed trichomes.
                                16. Fruiting pedicels abruptly curved at base; petals 5–6 mm; cauline leaves 7–33 . . . . ***B. pinetorum***
                                16' Fruiting pedicels slightly curved at base; petals 5–8 mm; cauline leaves 14–60 . . . . ***B. pauciflora*** (in part)

6' Fruiting pedicels and fruits ascending, divaricate, horizontal, descending, or widely pendent.
- 17. Fruits secund.
  - 18. Fruits 2.5–3.5 mm wide; ovules 44–76(104) per ovary; seeds subbiseriate; petals 6–8 mm . . . . ***B. drepanoloba***
  - 18' Fruits 1.5–2.3(3) mm wide; ovules 28–54 per ovary; seeds uniseriate; petals 3.5–6 mm.
    - 19. Basal leaves not ciliate at base; fruits 3–5.5 cm . . . . ***B. paddoensis***
    - 19' Basal leaves ciliate at base; fruits 1.6–4 cm.
      - 20. Fruiting pedicels 4–11 mm; fruits 2–3 mm wide, with undulate, non-parallel margins; seeds 2–2.5 × 1.7–2 mm . . . . ***B. horizontalis***
      - 20' Fruiting pedicels 2–6 mm; fruits 1.6–2.3 mm wide, with straight, parallel margins; seeds 1.3–2 × 1–1.5 mm . . . . ***B. lemmonii***
- 17' Fruits not secund.
  - 21. Lower stems pubescent with exclusively simple trichomes . . . . ***B. cusickii***
  - 21' Lower stems glabrous or pubescent with mixture of simple and branched trichomes.
    - 22. Fruits (3)3.5–7 mm wide; seeds 3–8 mm, wings 0.8–2.5 mm wide.
      - 23. Lower stems and sepals pubescent . . . . ***B. platysperma***
      - 23' Lower stems and sepals glabrous.
        - 24. Margins of basal leaves ciliate with 2- or 3(4)-rayed and simple trichomes; ovules 18–30 per ovary; plants apomictic, pollen spheroid . . . . ***B. covillei***
        - 24' Margins of basal leaves ciliate with only 2-rayed and simple trichomes; ovules 10–20 per ovary; plants sexual, pollen ellipsoid . . . . ***B. howellii***
    - 22' Fruits 1–2.5 mm wide; seeds 1–2.5 mm, wings 0.05–0.5 mm wide.
      - 25. Cauline leaves not auriculate; basal and cauline leaves glabrous . . . . ***B. davidsonii***
      - 25' Cauline leaves auriculate; basal and lowermost cauline leaves with simple and stalked 2–8-rayed trichomes.
        - 26. Lower stems pubescent with exclusively branched trichomes . . . . ***B. acutina***
        - 26' Lower stems pubescent with mixture of simple and branched trichomes, rarely glabrous.
          - 27. Plants biennial or short-lived, non-cespitose, perennial without woody caudices.
            - 28. Branched trichomes on basal leaves sessile or subsessile; basal leaves ciliate; flowers apomictic, pollen spheroid . . . . ***B. divaricarpa***
            - 28' Branched trichomes on basal leaves stalked; basal leaves ciliate or not; flowers sexual, pollen ellipsoid.
              - 29. Petals dark reddish purple drying to indigo, 1.5–2 mm wide; lower stems subglabrous or sparsely pubescent with trichomes to 0.15 mm; ovules 80–100 per ovary . . . . ***B. atrorubens***
              - 29' Petals purple, lavender, or white, 2–5 mm wide; lower stems densely pubescent with trichomes to 1.5 mm; ovules 90–170 per ovary . . . . ***B. sparsiflora*** (in part)
          - 27' Plants long-lived, perennial, with woody caudices or distinctly cespitose.
            - 30. Basal leaves with at least some 7–8-rayed trichomes . . . . ***B. microphylla***
            - 30' Basal leaves with 2–6-rayed trichomes.
              - 31. Styles 0.8–1.5 mm . . . . ***B. cascadensis***
              - 31' Styles obsolete or 0.1–0.3(0.5) mm.
                - 32. Petals 5–8 × 0.5–2.5 mm; flowers apomictic, pollen spheroid; fruiting pedicels horizontal to descending, rarely ascending . . . . ***B. pauciflora*** (in part)
                - 32' Petals 7–13 × 2–5 mm; flowers sexual, pollen ellipsoid; fruiting pedicels ascending to divaricate.
                  - 33. Lower stems with 2–4-rayed and simple trichomes; basal leaves 1–3 mm wide; petals deep purple to magenta . . . . ***B. koehleri***
                  - 33' Lower stems with 2-rayed and simple trichomes; basal leaves 3–12 mm wide; petals white, lavender, or purple.
                    - 34. Fruiting pedicels with simple and 2-rayed trichomes, rarely glabrous; ovules 48–96 per ovary; stems 0.6–3.5(4.5) dm; cauline leaves 5–15(28) . . . . ***B. breweri***
                    - 34' Fruiting pedicels glabrous or pubescent with simple trichomes; ovules 90–170 per ovary; stems 3–8 dm; cauline leaves (9)15–35 . . . . ***B. sparsiflora*** (in part)

Boechera acutina (Greene) Windham & Al-Shehbaz
Sharp-podded rockcress

Arabis acutina Greene, *Arabis* × *divaricarpa* A. Nelson, *Arabis* × *divaricarpa* A. Nelson var. *interposita* (Greene) Rollins

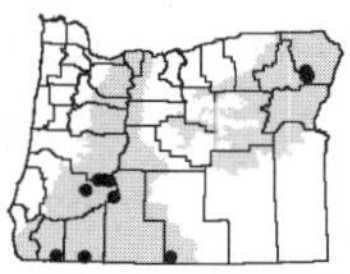

Plants biennial or short-lived perennial; with or without caudices. **Stems** 1.5–6 dm, lower parts with many sessile, 2–4-rayed trichomes, upper parts sparsely pubescent or rarely glabrous. **Basal leaves** oblanceolate, 1.5–6 mm wide, margins entire or rarely denticulate, ciliate proximally with simple trichomes, surfaces pubescent with sessile, 3–5(6)-rayed trichomes. **Cauline leaves** 2–20(38), often concealing stem proximally, bases auriculate. **Inflorescences** 5–20(36)-flowered, mostly unbranched; fruiting pedicels erect to ascending, mostly straight, 3–10 mm, with appressed, branched trichomes or glabrous. **Flowers** apomictic, ascending at anthesis; sepals pubescent; petals 5–7 × 0.9–1.3 mm, white or lavender, glabrous; pollen spheroid; ovules 46–100 per ovary, styles 0.3–1 mm. **Fruits** ascending, not appressed to rachis, not secund, straight, 2.5–7.5 cm × 1.2–1.8 mm, margins parallel, glabrous. **Seeds** uniseriate, 1.8–2.2 × 1–1.4 mm; wings continuous, 0.1–0.4 mm wide.

Meadows, gravelly slopes, open forests. Flowering Jun–Jul. 1200–2200 m. BW, Casc, ECas, Sisk. CA, ID, NV, WA. Native.

Boechera atrorubens (Suksdorf ex Greene) Windham & Al-Shehbaz (p. 450)
Purple sicklepod rockcress

Arabis sparsiflora Nutt., *Arabis sparsiflora* Nutt. var. *atrorubens* (Greene) Rollins

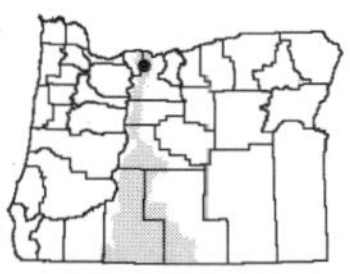

Plants short-lived perennial; with non-woody caudices. **Stems** 0.8–6 dm, lower parts subglabrous to sparsely pubescent with simple and short-stalked, 2-rayed trichomes to 0.15 mm, upper parts glabrous. **Basal leaves** oblanceolate, 4–10 mm wide, margins usually dentate, not ciliate, surfaces densely pubescent with short-stalked, 3- or 4-rayed trichomes. **Cauline leaves** 7–25, concealing stem proximally, bases auriculate, uppermost glabrous. **Inflorescences** 6–20-flowered, mostly unbranched; fruiting pedicels ascending, usually straight, 5–10 mm, glabrous or with spreading, simple trichomes. **Flowers** sexual, ascending at anthesis; sepals pubescent; petals 7–9 × 1.5–2 mm, dark reddish purple drying to indigo, glabrous; pollen ellipsoid; ovules 80–100 per ovary, styles 0.2–0.5 mm. **Fruits** ascending, not appressed to rachis, not secund, curved or straight, (4)6–12 cm × 1.7–2.2 mm, margins parallel, glabrous. **Seeds** uniseriate, 1.2–1.7 × 0.9–1.2 mm; wings distal and lateral, 0.1–0.2 mm wide.

Rocky summits and sandy loam on sagebrush slopes. Flowering Apr–May. 600–1500 m. ECas. WA. Native.

Boechera breweri (S. Watson) Al-Shehbaz
Brewer's rockcress

Plants long-lived perennial; with woody caudices. **Stems** 0.6–3.5(4.5) dm, lower parts with simple and sometimes long-stalked, 2-rayed trichomes, upper parts sparsely pubescent or glabrous. **Basal leaves** oblanceolate, 3–7(11) mm wide, margins entire or dentate, ciliate proximally, surfaces densely pubescent with simple and long-stalked, 2–4-rayed trichomes. **Cauline leaves** 5–15(28), often concealing stem proximally, bases auriculate, uppermost pubescent. **Inflorescences** 7–20(30)-flowered; fruiting pedicels ascending to divaricate-ascending, straight, 3–25 mm, with spreading, simple and 2-rayed trichomes or rarely glabrous. **Flowers** sexual, ascending at anthesis; sepals pubescent; petals 7–12 × 2–4 mm, purple or rarely lavender, glabrous; pollen ellipsoid; ovules 48–96 per ovary, styles 0.05–0.3 mm. **Fruits** ascending to divaricate-ascending, not appressed to rachis, not secund, 3.5–10 cm × 1.5–2.2 mm, margins parallel, glabrous or rarely sparsely pubescent. **Seeds** uniseriate, 1.2–1.7 × 1–1.2 mm; wings continuous, 0.1–0.2 mm wide.

California and Oregon. 2 subspecies.

Key to *Boechera breweri* subspecies:

1. Stems 0.6–2 dm, moderately to sparsely pubescent distally; fruiting pedicels 3–8(12) mm, pubescent; fruits 3.5–7.5 cm . . . . ***B. b.*** ssp. ***breweri***
1. Stems 1.8–3.5(4.5) dm, usually glabrous distally; fruiting pedicels (7)10–25 mm, glabrous or with few scattered hairs; fruits (4.5)7–10 cm . . . . ***B. b.*** ssp. ***shastaensis***

Boechera breweri (S. Watson) Al-Shehbaz ssp. ***breweri*** (p. 450)
Brewer's rockcress

Arabis breweri S. Watson var. *breweri*

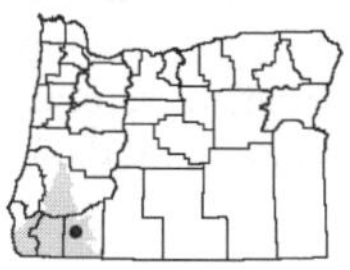

Stems 0.6–2 dm, often densely pubescent proximally, moderately to sparsely pubescent distally. **Inflorescences** fruiting pedicels 3–8(12) mm, pubescent. **Fruits** 3.5–7.5 cm.

Rocky outcrops, ledges, talus. Flowering Mar–Jul. 300–1800 m. Sisk. CA. Native.

Boechera breweri (S. Watson) Al-Shehbaz ssp. ***shastaensis*** Windham & Al-Shehbaz
Brewer's rockcress

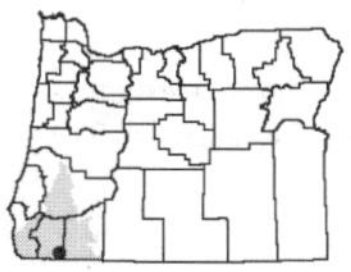

Stems 1.8–3.5(4.5) dm, moderately to sparsely pubescent proximally, usually glabrous distally. **Inflorescences** fruiting pedicels (7)10–25 mm, glabrous or with few scattered hairs. **Fruits** (4.5)7–10 cm. $2n$=14.

Rocky areas in woodlands. Flowering Mar–Jun. 300–1200 m. Sisk. CA. Native.

Boechera calderi (G.A. Mulligan) Windham & Al-Shehbaz
Calder's rockcress

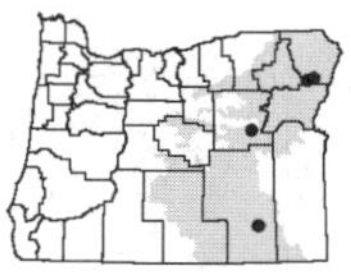

Plants mostly short-lived perennial, glabrous except for leaves; with non-woody caudices. **Stems** 1–4.5 dm, glabrous. **Basal leaves** usually oblanceolate, 1.5–6 mm wide, margins entire, surfaces sparsely pubescent with sessile, malpighiaceous and 3- or 4-rayed trichomes mostly restricted to margin, ciliate at petiole bases. **Cauline leaves** 5–17, often concealing stem proximally, bases auriculate, margins entire, uppermost glabrous. **Inflorescences** 10–25-flowered, unbranched; fruiting pedicels erect, straight, 5–10 mm, glabrous. **Flowers** apomictic, erect at anthesis; sepals glabrous; petals 6–9 × 1.5–2.5 mm, purple; pollen spheroid; ovules 64–134 per ovary, styles obsolete or rarely to 0.2 mm. **Fruits** erect, appressed to rachis, not secund, straight, 3.5–6.5 cm × 1.8–2.5 mm, margins parallel, glabrous. **Seeds** biseriate or subbiseriate, 1.8–2.2 × 1–1.8 mm; wings continuous, 0.2–0.5 mm wide.

Meadows, exposed rocky ridges, open forests near timberline. Flowering Jun–Aug. 2000–2900 m. BR, BW. CA, ID, WA; north to Yukon, northeast to Northwest Territories, east to MT, WY. Native.

Boechera cascadensis Windham & Al-Shehbaz

Cascade rockcress

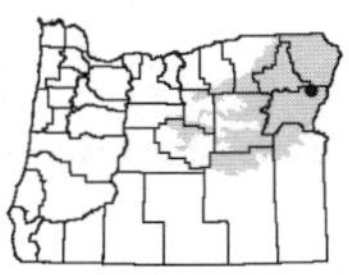

Plants long-lived perennial, cespitose; with somewhat woody caudices. **Stems** 0.5–2.2 dm, lower parts with few simple and short-stalked, 2- or 3-rayed trichomes, upper parts glabrous. **Basal leaves** linear-oblanceolate, 0.7–2 mm wide, margins entire, ciliate proximally, surfaces densely pubescent with simple and short-stalked, (2)3–6-rayed trichomes. **Cauline leaves** 4–6, not concealing stem, bases auriculate, uppermost glabrous. **Inflorescences** 3–11-flowered, unbranched; fruiting pedicels ascending to divaricate-ascending, straight, 3–8 mm, glabrous. **Flowers** apomictic, ascending at anthesis; sepals glabrous or sparsely pubescent; petals 5–6 × 1–1.7 mm, lavender, glabrous; pollen spheroid; ovules 58–80 per ovary, styles 0.8–1.5 mm. **Fruits** ascending to divaricate-ascending, not appressed to rachis, not secund, straight, 3.5–6.2 cm × 1.2–1.5 mm, margins parallel, glabrous. **Seeds** uniseriate, 1.1–1.3 × 0.9–1 mm; wings continuous, 0.05–0.1 mm wide.

Cliffs and rocky slopes in subalpine areas. Flowering Jun–Jul. 1300–2000 m. BW. WA. Native.

A rare Pacific Northwest endemic, this species has not been collected since 1936.

Boechera cobrensis (M.E. Jones) Dorn (p. 450)

Cobre rockcress, masonic rockcress

Arabis cobrensis M.E. Jones, *Arabis cobrensis* M.E. Jones var. *cobrensis*

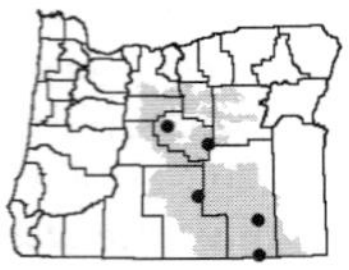

Plants long-lived perennial, often cespitose; with somewhat woody caudices. **Stems** (1.2)2.5–6 dm, lower parts with short-stalked, 4–10-rayed trichomes, upper parts sparsely pubescent to glabrous. **Basal leaves** linear to narrowly oblanceolate, 1–4 mm wide, margins entire, surfaces densely pubescent with short-stalked, 4–10-rayed trichomes, ciliate near petiole bases with simple trichomes. **Cauline leaves** 5–10, not concealing stem, bases auriculate, uppermost usually sparsely pubescent. **Inflorescences** 10–25-flowered; fruiting pedicels divaricate-ascending to horizontal, often straight proximally, usually strongly recurved or reflexed distally, 4–17 mm, with appressed, branched trichomes or rarely glabrous. **Flowers** sexual, divaricate-ascending at anthesis; sepals pubescent; petals 3.5–6 × 0.7–1 mm, white to lavender; pollen ellipsoid; ovules 34–60(64) per ovary, styles 0.05–0.2 mm. **Fruits** pendent, not appressed to rachis, not secund, straight or slightly curved, 2.5–5.5 cm × 1.7–2.5 mm, margins parallel, glabrous. **Seeds** uniseriate, 1.4–1.8 × 1–1.2 mm; wings continuous, 0.25–0.5 mm wide. $2n$=14.

Sandy soil, under shelter of shrubs in semi-desert communities. Flowering May–Jun. 1300–2600 m. BR, Lava. CA, ID, NV; east to WY. Native.

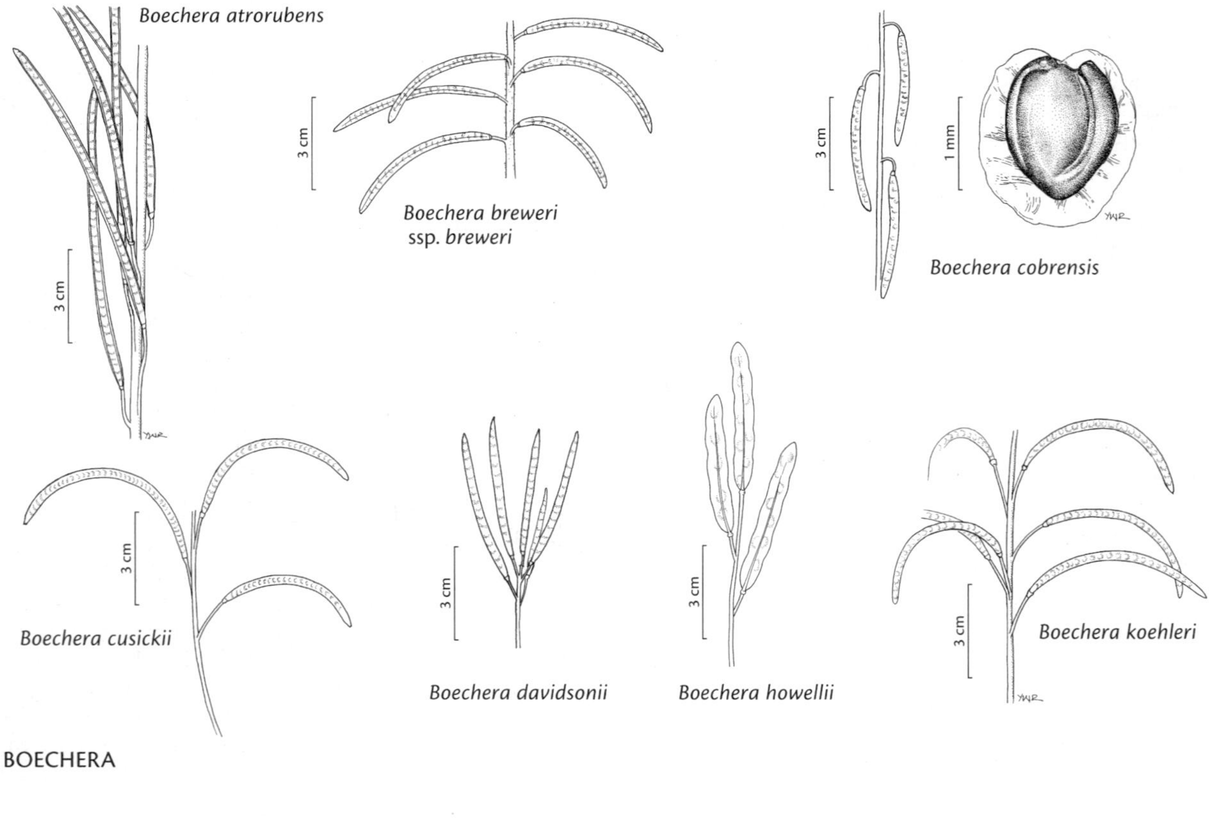

BOECHERA

Boechera covillei (Greene) Windham & Al-Shehbaz

Coville's rockcress

Arabis covillei Greene, *Arabis platysperma* A. Gray, *Arabis platysperma* A. Gray var. *howellii* (S. Watson) Jeps.

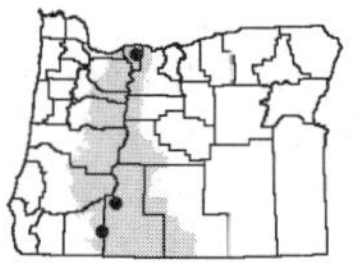

Plants long-lived perennial; with woody caudices. **Stems** 0.5–2.5 dm, glabrous. **Basal leaves** oblanceolate, 1.5–5 mm wide, margins entire, ciliate proximally and marginally with simple trichomes and distally with short-stalked 2- or 3(4)-rayed trichomes. **Cauline leaves** 2–7, not concealing stem, bases auriculate or not, uppermost glabrous. **Inflorescences** 2–9-flowered; fruiting pedicels ascending to suberect, straight, 6–20 mm, glabrous. **Flowers** apomictic, ascending at anthesis; sepals glabrous; petals 5–6 × 1–2 mm, lavender to purplish; pollen spheroid; ovules 18–30 per ovary, styles 0.2–1 mm. **Fruits** ascending to suberect, rarely appressed to rachis, not secund, straight, 3.5–5 cm × 3–5 mm, margins slightly undulate or rarely parallel, glabrous. **Seeds** uniseriate, 4–5 × 3–4 mm; wings continuous, 1–1.5 mm wide.

Rocky slopes in alpine meadows, open coniferous forests. Flowering Jul–Aug. 2200–2600 m. Casc, ECas. CA, ID, NV. Native.

Boechera cusickii (S. Watson) Al-Shehbaz (p. 450)

Cusick's rockcress

Arabis cusickii S. Watson

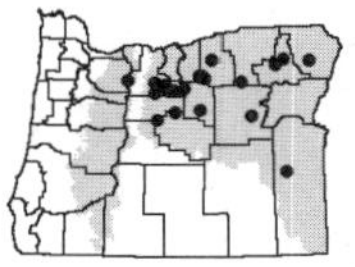

Plants long-lived perennial, often cespitose; with mostly woody caudices. **Stems** 1–2.5(3.8) dm, lower parts with simple trichomes only, upper parts sparsely pubescent or glabrous. **Basal leaves** linear to narrowly oblanceolate, 0.5–2 mm wide, margins entire, ciliate with simple trichomes, surfaces moderately pubescent with simple and long-stalked, 2-rayed trichomes. **Cauline leaves** 10–30, often concealing stem proximally, bases auriculate or not, uppermost sparsely pubescent. **Inflorescences** 5–14-flowered, unbranched; fruiting pedicels ascending, straight, 4–15(20) mm, glabrous or with few simple trichomes. **Flowers** sexual, ascending at anthesis; sepals pubescent; petals 6–10 × 2–3 mm, white to lavender, glabrous; pollen ellipsoid; ovules 56–106 per ovary, styles 0.1–0.5 mm. **Fruits** ascending to divaricate-ascending, not appressed to rachis, not secund, straight or curved, 4–8(9.5) cm × 2.3–3 mm, margins parallel, glabrous. **Seeds** uniseriate, 1.5–2.5 × 1.5–2 mm; wings continuous, 0.1–0.2 mm wide.

Rocky slopes, sagebrush hills, basaltic bluffs, gravelly hillsides, crevices, outcrops of volcanic rock. Flowering Apr–May. 600–1700 m. BW, Casc, Col, Lava, Owy. ID, NV, WA. Native.

Boechera davidsonii (Greene) N.H. Holmgren (p. 450)

Davidson's rockcress

Arabis davidsonii Greene

Plants long-lived perennial, cespitose; caudices woody; branches thickly covered by crowded, persistent leaf bases. **Stems** usually 0.6–2.3 dm, glabrous. **Basal leaves** oblanceolate, 3.5–14 mm wide, margins entire or rarely slightly dentate, not ciliate, surfaces glabrous. **Cauline leaves** 3–10, not concealing stem, not auriculate, glabrous. **Inflorescences** 4–24-flowered; fruiting pedicels ascending, straight, 3–18 mm, glabrous. **Flowers** sexual, ascending at anthesis; sepals glabrous; petals 6–10 × 2.5–4 mm, white to lavender, claws entire; stamens tetradynamous, pollen ellipsoid; ovules 28–50 per ovary, styles 0.1–0.5(0.8) mm. **Fruits** ascending to divaricate-ascending, not appressed to rachis, not secund, straight to slightly curved, 2.5–7 cm × 1.5–2.5 mm, margins parallel, glabrous. **Seeds** uniseriate, 1.8–2.2 × 0.8–1.7 mm; wings distal or continuous, 0.1–0.5 mm wide.

Ledges, crevices of rock outcrops. Flowering Apr–Jul. 1200–2500 m. BW. CA, NV. Native.

Boechera divaricarpa (A. Nelson) Á. Löve & D. Löve

Spreading-pod rockcress

Arabis × *divaricarpa* A. Nelson, *Arabis* × *divaricarpa* A. Nelson var. *divaricarpa*

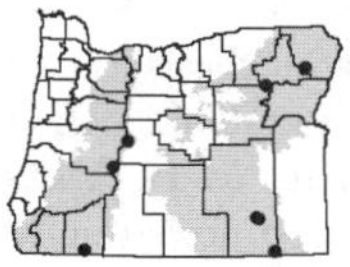

Plants biennial or short-lived perennial; without woody caudices. **Stems** (1.5)3–9 dm, lower parts glabrous or with sessile, 2–4-rayed and few simple trichomes, upper parts glabrous. **Basal leaves** oblanceolate, 2–10 mm wide, margins entire or rarely denticulate, ciliate with simple trichomes, surfaces pubescent with sessile or subsessile, 2–6-rayed trichomes. **Cauline leaves** (10)15–56, concealing stem proximally, bases auriculate, uppermost glabrous. **Inflorescences** 12–40(65)-flowered; fruiting pedicels divaricate-ascending to horizontal, straight, 5–12 mm, glabrous. **Flowers** apomictic, divaricate-ascending at anthesis; sepals glabrous or with scattered trichomes; petals 6–9 × 1.5–3 mm, purple or rarely lavender; pollen spheroid; ovules 114–142 per ovary, styles 0.05–0.2 mm. **Fruits** divaricate-ascending to horizontal, not appressed to rachis, not secund, straight, (4.5)5.5–11 cm × 1.7–2.5 mm, margins parallel, glabrous. **Seeds** uniseriate to subbiseriate, 1.4–2 × 1–1.5 mm; wings continuous, 0.1–0.2 mm wide. **2*n***=21.

Talus slopes, rock outcrops, gravelly hillsides in sagebrush, shrublands, open conifer forests. Flowering Jun–Jul. 1500–2800 m. BR, BW, Casc, Sisk. CA, ID, NV, WA; northeast to MT, east to WY. Native.

Boechera drepanoloba (Greene) Windham & Al-Shehbaz

Soldier rockcress

Arabis lemmonii S. Watson var. *drepanoloba* (Greene) Rollins

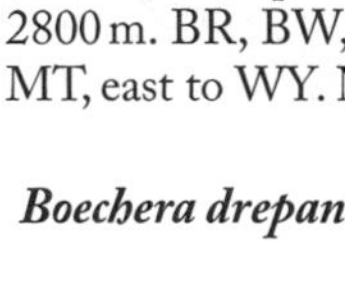

Plants short- to long-lived perennial; with mostly non-woody caudices. **Stems** 1–4 dm, lower parts glabrous or with few sessile or subsessile, 2- or 3-rayed trichomes, upper parts glabrous. **Basal leaves** oblanceolate, 2–6 mm wide, margins entire, surfaces sparsely to densely pubescent with subsessile, 2–6-rayed trichomes, ciliate on petioles. **Cauline leaves** 3–15, often concealing stem proximally, bases auriculate, uppermost usually glabrous. **Inflorescences** 5–25-flowered; fruiting pedicels horizontal to ascending, straight or curved, 4–8 mm, glabrous. **Flowers** apomictic, ascending at anthesis; sepals glabrous or with scattered trichomes; petals 6–8 × 1.5–2.5 mm, purple to lavender; pollen spheroid; ovules 44–76(104) per ovary, styles 0.05–0.5 mm. **Fruits** horizontal to ascending, secund, straight to slightly curved, 3–5.8 cm

× 2.5–3.5 mm, margins parallel, glabrous. **Seeds** subbiseriate, 1.5–2.2 × 1.2–1.6 mm; wings continuous, 0.3–0.7 mm wide.

Rocky slopes, talus. Flowering Jul–Aug. 1700–2800 m. Casc. NV; British Columbia, north to AK, northeast to Alberta, east to WY, southeast to UT. Native.

Boechera hastatula (Greene) Al-Shehbaz

Hells Canyon rockcress

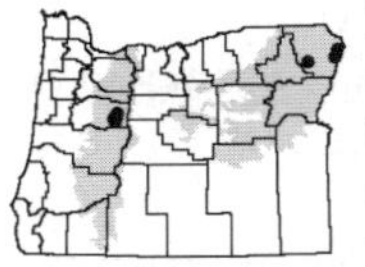

Plants short- to long-lived perennial; often with woody caudices. **Stems** 1–3 dm, glabrous. **Basal leaves** linear-oblanceolate to oblanceolate, 1–3(5) mm wide, margins entire or minutely toothed, surfaces pubescent with short-stalked, 3–6-rayed trichomes or glabrous, rarely ciliate near petiole bases with few simple trichomes. **Cauline leaves** 4–15, not concealing stem, bases auriculate, uppermost glabrous. **Inflorescences** 12–40-flowered; fruiting pedicels reflexed, 4–8 mm, abruptly recurved at bases, glabrous. **Flowers** apomictic or sexual, divaricate to pendent at anthesis; sepals pubescent; petals 5–6 × ~1 mm, lavender; apomictic pollen spheroid, sexual pollen ellipsoid; ovules 46–60 per ovary. **Fruits** reflexed, usually appressed to rachis, often secund, straight, 3–4.2 cm × 1–1.5 mm, margins parallel, glabrous. **Seeds** uniseriate, 1–1.3 × 0.8–1 mm; wings continuous, 0.05–0.1 mm.

Igneous rock outcrops. Flowering Jun–Aug. 1500–2700 m. BW, Casc. Native.

Boechera horizontalis (Greene) Windham & Al-Shehbaz

Crater Lake rockcress

Arabis suffrutescens S. Watson, *Arabis suffrutescens* S. Watson var. *horizontalis* (Greene) Rollins

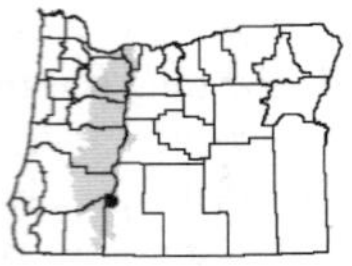

Plants long-lived perennial, cespitose; with woody caudices. **Stems** 1–3.5 dm, lower parts with few short-stalked, 3–6-rayed trichomes, upper parts glabrous. **Basal leaves** narrowly oblanceolate, 1–5 mm wide, margins entire, ciliate near bases with simple trichomes, surfaces densely pubescent with short-stalked, 3–6-rayed trichomes. **Cauline leaves** 3–13, often concealing stem proximally, bases auriculate, uppermost glabrous. **Inflorescences** 5–32-flowered; fruiting pedicels horizontal to descending, straight or slightly curved downward, 4–11 mm, glabrous. **Flowers** apomictic, divaricate-ascending at anthesis; sepals pubescent; petals 5–6 × 1.5–2 mm, lavender to purple; pollen spheroid; ovules 40–54 per ovary, styles 0.2–0.5 mm. **Fruits** horizontal or descending, not appressed to rachis, secund, straight, 2–4 cm × 2–3 mm, margins undulate, non-parallel, glabrous. **Seeds** uniseriate, 2–2.5 × 1.7–2 mm; wings continuous, 0.5–1 mm wide.

Dry pumice slopes. Flowering Jul–Aug. 2100–2800 m. Casc. Native.

Molecular evidence suggests that *Boechera horizontalis* is an apomictic species that arose through hybridization between *B. lemmonii* and *B. suffrutescens* (Windham & Al-Shehbaz 2007).

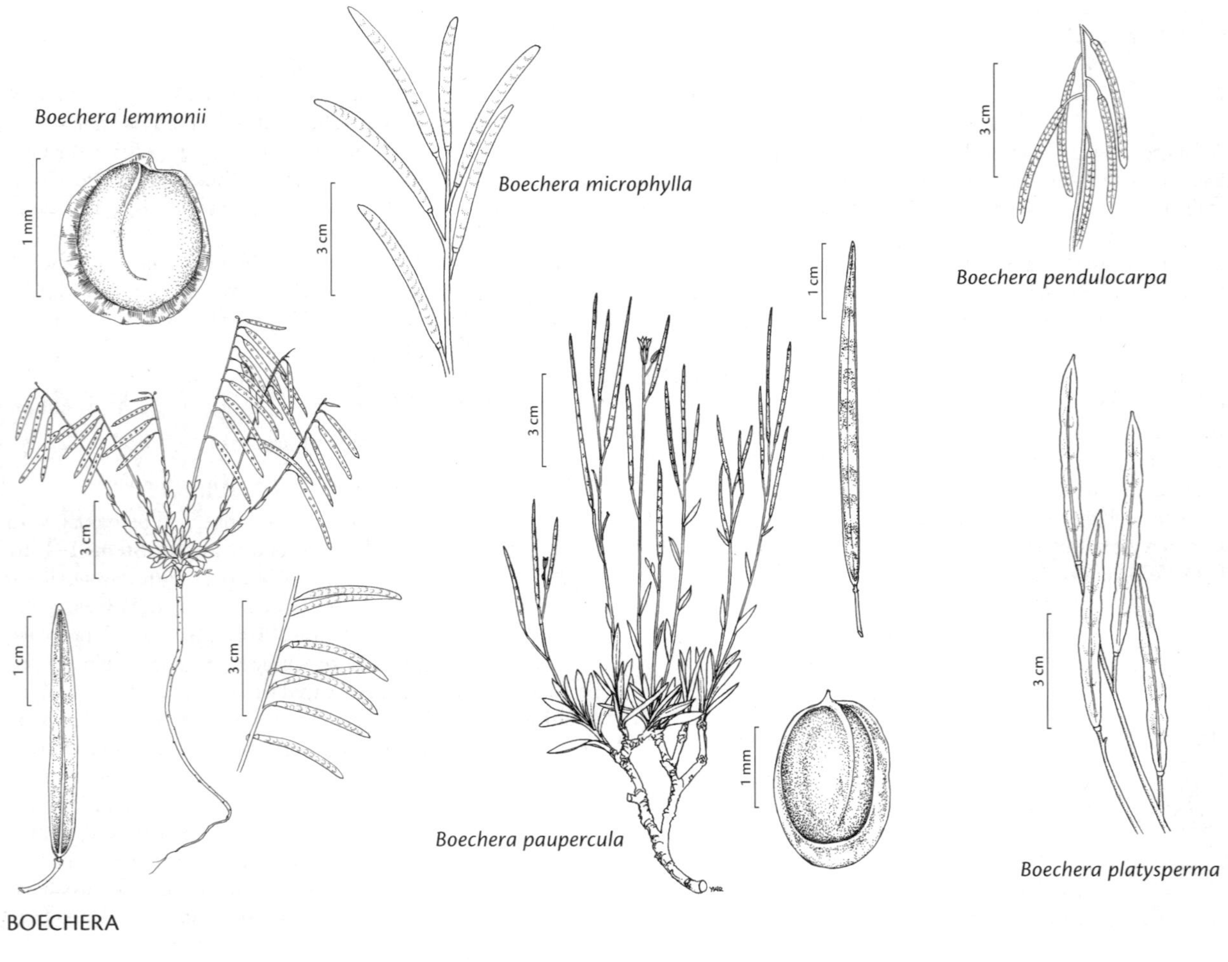

BOECHERA

Boechera howellii (S. Watson) Windham & Al-Shehbaz (p. 450)

Howell's flatseed rockcress

Arabis howellii S. Watson, *Arabis platysperma* A. Gray, *Arabis platysperma* A. Gray var. *howellii* (S. Watson) Jeps.

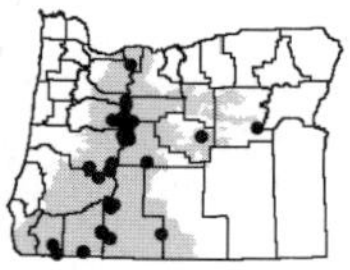

Plants long-lived perennial; with mostly woody caudices. **Stems** 0.6–2(3) dm, glabrous. **Basal leaves** narrowly oblanceolate, 1–7 mm wide, margins entire, ciliate proximally and marginally with simple and a few short-stalked, 2-rayed trichomes, glabrous. **Cauline leaves** 2–4, not concealing stem, bases auriculate or rarely not, uppermost glabrous or rarely ciliate proximally. **Inflorescences** 2–5-flowered, unbranched; fruiting pedicels ascending, straight, 4–10 mm, glabrous. **Flowers** sexual, erect to ascending at anthesis; sepals glabrous; petals 4–8 × 1–2.5 mm, white, aging to purple; pollen ellipsoid; ovules 10–20 per ovary, styles 0.05–0.3 mm. **Fruits** suberect to ascending, usually not appressed to rachis, not secund, straight to curved, 2.5–6.5 cm × (3)3.5–7 mm, margins often somewhat undulate and non-parallel, glabrous. **Seeds** uniseriate, 3–6 × 2–4 mm; wings continuous, 1.3–2.5 mm wide.

Rock outcrops, talus slopes, gravelly soil in alpine and subalpine habitats. Flowering Jul–Aug. 1500–2600 m. Casc, ECas, Lava, Sisk. CA, ID, NV. Native.

Boechera koehleri (T.J. Howell) Al-Shehbaz (p. 450)

Shrubby rockcress

Arabis koehleri Howell var. *koehleri*, *Arabis koehleri* Howell var. *stipitata* Rollins

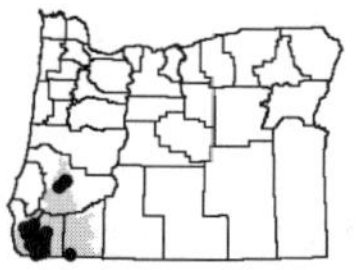

Plants long-lived perennial, sometimes subshrubs; with woody caudices. **Stems** 0.8–4.5 dm, lower parts with stalked, simple and 2–4-rayed trichomes, upper parts glabrous. **Basal leaves** narrowly oblanceolate, 1–3 mm wide, margins entire, ciliate proximally with simple trichomes to 1 mm, surfaces moderately to sparsely pubescent with simple and stalked, 2–4-rayed trichomes. **Cauline leaves** 3–17(30), rarely concealing stem, bases auriculate, uppermost glabrous. **Inflorescences** 6–20(35)-flowered; fruiting pedicels divaricate-ascending to horizontal, straight, 10–18 mm, glabrous or with few appressed, branched trichomes. **Flowers** sexual, ascending at anthesis; sepals pubescent; petals 8–12 × 2.5–4 mm, purple to magenta; pollen ellipsoid; ovules 58–94 per ovary, styles 0.05–0.2 mm. **Fruits** divaricate-ascending to widely pendent, not appressed to rachis, not secund, curved or rarely straight, 5–7.5 cm × 1.8–2.5 mm, margins parallel, glabrous. **Seeds** uniseriate, 1.3–1.8 × 1.2–1.5 mm; wings continuous, 0.1–0.2 mm wide.

Serpentine and limestone outcrops. Flowering Mar–Jun. 100–1300 m. Sisk. CA. Native.

Boechera lemmonii (S. Watson) W.A. Weber (p. 452)

Lemmon's rockcress

Arabis lemmonii S. Watson var. *lemmonii*, *Arabis lemmonii* S. Watson

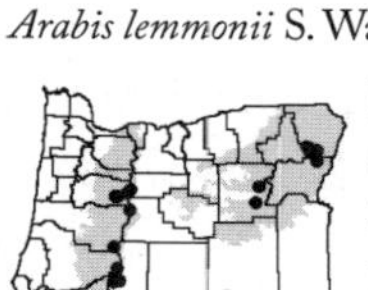

Plants long-lived perennial, somewhat cespitose; with woody caudices. **Stems** 0.5–2(2.5) dm, lower parts glabrous or with scattered short-stalked, 2–6-rayed trichomes. **Basal leaves** oblanceolate to obovate, margins usually entire, surfaces densely to sparsely pubescent with short-stalked, 3–9-rayed trichomes, ciliate on petioles. **Cauline leaves** 2–8(10), not concealing stem, bases auriculate or not, uppermost glabrous or sparsely pubescent. **Inflorescences** 3–12(17)-flowered; fruiting pedicels divaricate-ascending to slightly descending, slightly recurved or rarely straight, 2–6 mm, glabrous or with few appressed, branched trichomes. **Flowers** sexual or apomictic, divaricate-ascending at anthesis; sepals glabrous or sparsely pubescent; petals 3.5–6 × 1–1.5 mm, purple to lavender; sexual pollen ellipsoid, apomictic pollen spheroid; ovules 28–40(44) per ovary, styles 0.1–0.2 mm. **Fruits** divaricate-ascending to slightly descending, not appressed to rachis, secund, (1.7)2–4 cm × 1.6–2.3 mm, margins parallel, glabrous. **Seeds** uniseriate, 1.3–2 × 1–1.5 mm; wings continuous, 0.1–0.5 mm wide. $2n$=14.

Cliffs, talus slopes, gravelly soil in alpine and subalpine habitats. Flowering Jun–Aug. 1300–3000 m. BW, Casc. CA, ID, NV, WA; north to AK, east to WY, southeast to UT. Native.

Boechera lyallii (S. Watson) Dorn

Lyall's slender rockcress

Arabis lyallii S. Watson, *Arabis lyallii* S. Watson var. *lyallii*

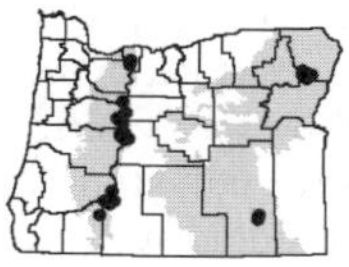

Plants long-lived perennial, cespitose; with woody caudices. **Stems** 0.3–1.5(2) dm, glabrous. **Basal leaves** linear-oblanceolate, 1–5(8) mm wide, margins entire, usually ciliate at least proximally with simple and/or short-stalked, 2- or 3-rayed trichomes, or rarely youngest leaves of sterile shoots pubescent with 4–6-rayed trichomes, margins with stalked trichomes. **Cauline leaves** 1–5, mostly not concealing stem, bases auriculate or rarely not, glabrous. **Inflorescences** 2–10(15)-flowered; fruiting pedicels erect, straight, 3–8(15) mm, glabrous. **Flowers** sexual or apomictic, erect at anthesis; sepals glabrous; petals 6–8.5 × 1.5–3 mm, white to lavender; sexual pollen ellipsoid, apomictic pollen spheroid; ovules 34–64 per ovary, styles 0.1–0.7 mm. **Fruits** erect, appressed to rachis, not secund, straight, 3–5.6 cm × 1.5–2.5 mm, margins parallel, glabrous. **Seeds** subbiseriate, 1.5–2 × 1–1.5 mm; wings continuous, 0.3–0.5 mm wide.

Talus slopes, cliffs, gravelly soil in alpine and subalpine habitats. Flowering Jun–Aug. 1700–3000 m. BR, BW, Casc. CA, ID, NV, WA; north to Yukon, northeast to MT, southeast to UT. Native.

Boechera microphylla (Nuttall) Dorn (p. 452)

Small-leaved rockcress

Arabis microphylla Nutt. ex Torr. & A. Gray, *Arabis microphylla* Nutt. ex Torr. & A. Gray var. *microphylla*, *Boechera microphylla* (Nutt.) Dorn var. *microphylla*

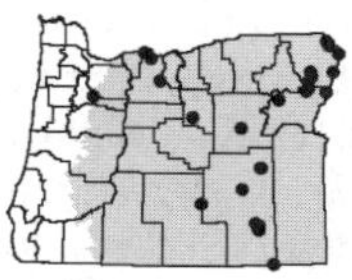

Plants long-lived perennial, cespitose; with woody caudices. **Stems** 0.8–2.8(3.5) dm, lower parts with few simple and subsessile, 2-rayed trichomes, upper parts glabrous. **Basal leaves** oblanceolate to linear-oblanceolate, 1–4(6) mm wide, margins entire or subapically few-toothed, ciliate proximally with simple trichomes, surfaces densely pubescent with short-stalked, 4–8-rayed trichomes. **Cauline leaves** 2–6, not concealing stem, bases auriculate, uppermost glabrous. **Inflorescences** 5–17-flowered; fruiting pedicels ascending to divaricate-ascending, straight, 4–14 mm, glabrous. **Flowers** sexual or apomictic, ascending at anthesis; sepals glabrous or pubescent; petals 3.5–5.5 × 1–1.8 mm, lavender to purple; sexual pollen ellipsoid, apomictic pollen spheroid; ovules 52–72 per ovary, styles 0.05–0.3 mm. **Fruits** ascending to divaricate-ascending, not appressed to rachis, not secund, straight to slightly curved, 3–7 cm × 1–1.5 mm, margins parallel, glabrous. **Seeds** uniseriate, 1–1.5 × 0.7–1 mm; wings often distal, ~0.1 mm wide. ***2n***=14.

Rocky slopes, cliffs, sagebrush, mountain shrub, open conifer forests. Flowering Mar–Jul. 0–3000 m. BR, BW, Casc, Col, ECas, Lava, Owy. CA, ID, NV, WA; northeast to MT, southeast to UT. Native.

Boechera paddoensis (Rollins) Windham & Al-Shehbaz

Mt. Adams rockcress

Arabis lemmonii S. Watson, *Arabis lemmonii* S. Watson var. *paddoensis* Rollins

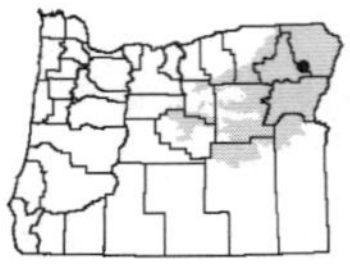

Plants long-lived perennial, cespitose; with mostly woody caudices. **Stems** 1–2.5 dm, glabrous. **Basal leaves** narrowly oblanceolate, 1–3(6) mm wide, margins entire or rarely few-toothed near tips, not ciliate, surfaces pubescent with short-stalked, 4–8-rayed trichomes. **Cauline leaves** 4–8, not concealing stem, bases auriculate, uppermost glabrous. **Inflorescences** 5–12(18)-flowered; fruiting pedicels divaricate-ascending or rarely horizontal, straight, 2–5(10) mm, glabrous. **Flowers** apomictic, ascending at anthesis; sepals glabrous; petals 4–5.5 × 1–1.5 mm, lavender to purple; pollen spheroid; ovules 42–54 per ovary, styles 0.05–0.1 mm. **Fruits** divaricate-ascending or rarely horizontal, not appressed to rachis, secund (often weakly so), straight, 3–5.5 cm × 1.5–2 mm, margins parallel, glabrous. **Seeds** uniseriate, 1.1–1.5 × 0.9–1.3 mm; wings continuous, 0.1–0.3 mm wide.

Rocky ridges. Flowering Jul–Aug. 1500–2100 m. BW. WA. Native.

Boechera pauciflora (Nuttall) Windham & Al-Shehbaz

Hairystem rockcress

Arabis campyloloba Greene, *Arabis sparsiflora* Nutt., *Arabis sparsiflora* Nutt. var. *subvillosa* (S. Watson) Rollins, *Boechera sparsiflora* (Nutt.) Dorn var. *subvillosa* (S. Watson) Dorn

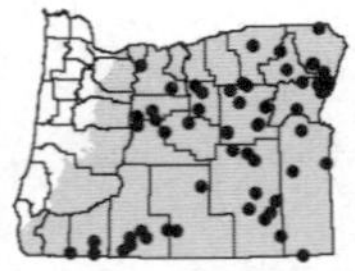

Plants short- to long-lived perennial; occasionally with woody caudices. **Stems** (1.4)3–11.2 dm, lower parts with abundant simple trichomes mixed with stalked, 2(3)-rayed trichomes, upper parts sparsely pubescent. **Basal leaves** oblanceolate, 3–10 mm wide, margins dentate or rarely entire, occasionally ciliate with simple or 2–5-rayed, stalked trichomes. **Cauline leaves** 14–60, often concealing stem proximally, bases auriculate, uppermost glabrous or sparsely pubescent. **Inflorescences** 17–60-flowered; fruiting pedicels horizontal to descending, rarely ascending, slightly curved at base, 4–13 mm, with spreading 2- or 3-rayed trichomes. **Flowers** apomictic, divaricate at anthesis; sepals pubescent; petals 5–8 × 0.5–2.5 mm, lavender to whitish; pollen spheroid; ovules 80–162 per ovary, styles 0.05–0.5 mm. **Fruits** horizontal, widely pendent, not appressed to rachis, not secund, curved, 5.5–10.5 cm × 1.5–2.2 mm, margins parallel, glabrous. **Seeds** uniseriate, 1.4–1.8 × 1–1.4 mm; wings continuous, 0.1–0.3 mm wide. ***2n***=21.

Rocky soil in sagebrush areas, mountain shrub communities, edges of conifer forests. Flowering Apr–Jul. 300–2800 m. BR, BW, Casc, Col, ECas, Lava, Owy, Sisk. CA, WA; north to British Columbia, northeast to MT, southeast to UT. Native.

Boechera paupercula (Greene) Windham & Al-Shehbaz (p. 452)

Small-flowered rockcress

Arabis lyallii S. Watson, *Arabis lyallii* S. Watson var. *nubigena* (J.F. Macbr. & Payson) Rollins

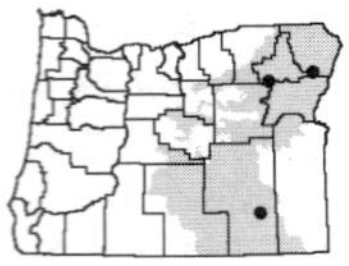

Plants long-lived perennial, cespitose; often with woody caudices. **Stems** 0.3–1.5 dm, lower parts with abundant simple and short-stalked, 2–4-rayed trichomes, upper parts glabrous or sparsely pubescent. **Basal leaves** linear-oblanceolate, 1–3(5) mm wide, margins entire, ciliate proximally, densely pubescent with simple and short-stalked, 2–6-rayed trichomes. **Cauline leaves** 2–6, rarely concealing stem, not or rarely auriculate, uppermost glabrous. **Inflorescences** 3–8-flowered; fruiting pedicels erect, straight, 3–9 mm, glabrous or sparsely pubescent. **Flowers** sexual, erect at anthesis; sepals glabrous or sparsely pubescent; petals 4–6(7) × 1–2 mm, lavender to purplish; pollen ellipsoid; ovules 24–40 per ovary, styles 0.2–1 mm. **Fruits** erect, appressed to rachis, not secund, straight, 2.5–5.5 cm × 1.5–1.7(2) mm, margins parallel, glabrous. **Seeds** uniseriate or rarely subbiseriate, 1.5–2(2.5) × 1–1.4 mm; wings continuous, 0.3–1 mm wide.

Talus slopes, gravelly soil, rocky outcrops. Flowering May–Sep. 2300–2800 m. BR, BW. CA, ID, WA; east to WY. Native.

Boechera pendulocarpa (A. Nelson) Windham & Al-Shehbaz (p. 452)

Dropseed rockcress

Arabis holboellii Hornem. var. *pendulocarpa* (A. Nelson) Rollins

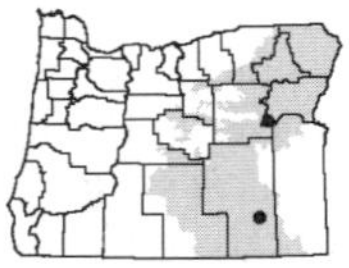

Plants short- to long-lived perennial; with mostly non-woody caudices. **Stems** 0.6–3 dm, lower parts with abundant simple and stalked, 2–4-rayed trichomes, upper parts glabrous or sparsely pubescent. **Basal leaves** narrowly oblanceolate, 1.5–4(5) mm wide, margins entire, surfaces densely pubescent with short-stalked, 4–8-rayed trichomes, often ciliate on petioles with simple trichomes. **Cauline leaves** 6–17, occasionally concealing stem proximally, bases not auriculate, uppermost glabrous or sparsely pubescent. **Inflorescences** 4–11-flowered; fruiting pedicels arched, 3–8(12) mm, recurved at bases, glabrous or with few appressed, branched trichomes. **Flowers** sexual, ascending at anthesis; sepals pubescent; petals 4–6 × 1–1.5 mm, white or lavender; pollen ellipsoid; ovules 66–92 per ovary, styles 0.3–0.5 mm. **Fruits** pendent, not appressed to rachis, not secund, usually straight, (2)2.5–3.8 cm × 1.5–2.2 mm, margins parallel, glabrous. **Seeds** subbiseriate, 1–1.2 × 0.7–0.9 mm; wings distal, 0.05–0.1 mm wide. ***2n***=14.

Rock outcrops, gravelly slopes in sagebrush, open conifer forests, subalpine meadows. Flowering Apr–Aug. 1300–2400 m. BR, BW. CA, ID, NV, WA; north to Yukon, northeast to MT, east to WY, southeast to UT. Native.

Boechera pinetorum (Tidestrøm) Windham & Al-Shehbaz

Holboell's rockcress, woodland rockcress

Arabis holboellii Hornem., *Arabis holboellii* Hornem. var. *pinetorum* (Tidestr.) Rollins, *Boechera holboellii* (Hornem.) Á. Löve & D. Löve

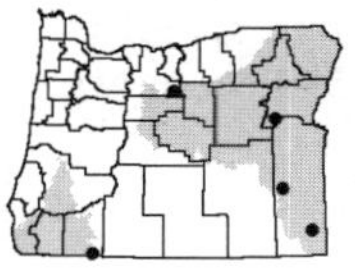

Plants short-lived perennial; with non-woody caudices. **Stems** 2–8(9.6) dm, lower parts with many simple and short-stalked, 2–4-rayed trichomes, upper parts sparsely pubescent or glabrous. **Basal leaves** oblanceolate, 2–8(11) mm wide, margins entire or denticulate, ciliate proximally with simple trichomes, surfaces sparsely to densely pubescent with short-stalked, 2–5-rayed trichomes mixed with fewer simple trichomes. **Cauline leaves** 7–33, often concealing stem proximally, bases auriculate, uppermost sparsely pubescent or glabrous. **Inflorescences** 15–63-flowered; fruiting pedicels reflexed, abruptly recurved near bases, 4–12 mm, glabrous or with few spreading, simple and 2-rayed trichomes. **Flowers** apomictic, divaricate to pendent at anthesis; sepals pubescent; petals 5–6 × 1–1.5 mm, white to lavender; pollen spheroid; ovules 70–110 per ovary, styles 0.05–0.4 mm. **Fruits** reflexed to closely pendent, not appressed to rachis, not secund, straight or curved, (4.5)5.5–8.5 cm × 1.5–2 mm, margins parallel, glabrous. **Seeds** uniseriate, 1.5–1.8 × 1.2–1.5 mm; wings continuous, 0.15–0.3 mm wide.

Rock outcrops and gravelly soil in meadows and open conifer forests. Flowering May–Sep. 400–1700 m. BW, Lava, Owy, Sisk. CA, ID, NV, WA. Native.

Boechera platysperma (A. Gray) Al-Shehbaz (p. 452)

Pioneer rockcress, broad-seeded rockcress

Arabis platysperma A. Gray, *Arabis platysperma* A. Gray var. *platysperma*

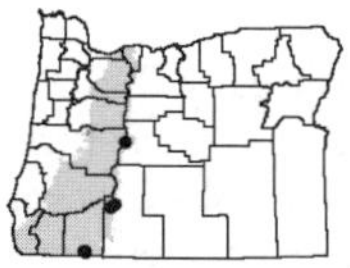

Plants long-lived perennial; with woody caudices. **Stems** 0.6–2.8(3.5) dm, lower parts pubescent with short-stalked, 2–5-rayed trichomes, upper parts glabrous or sparsely pubescent. **Basal leaves** narrowly oblanceolate, 3–7(10) mm wide, margins entire, surfaces densely pubescent with short-stalked, 2–5-rayed trichomes, ciliate near petiole bases with simple or spurred trichomes. **Cauline leaves** 3–7(12), not concealing stem, bases not auriculate, uppermost glabrous to sparsely pubescent. **Inflorescences** 2–7-flowered, unbranched; fruiting pedicels ascending, straight, 3–13 mm, glabrous or with few appressed, branched trichomes. **Flowers** sexual, ascending at anthesis; sepals pubescent; petals 4–6 × 1–2 mm, white, aging to purple; pollen ellipsoid; ovules 16–44 per ovary, styles 0.05–1 mm. **Fruits** ascending, not appressed to rachis, not secund, straight to slightly curved, (2.5)4–8.5 cm × (3)3.5–5.5 mm, margins undulate, non-parallel, glabrous. **Seeds** uniseriate, 3–8 × 2–3.5 mm; wings continuous, 0.8–2.5 mm wide. **2*n***=14.

Gravelly soil in dry pine forests, rock outcrops, lodgepole-chaparral woodlands. Flowering Jun–Aug. 1600–2600 m. Casc, Sisk. CA, ID, NV. Native.

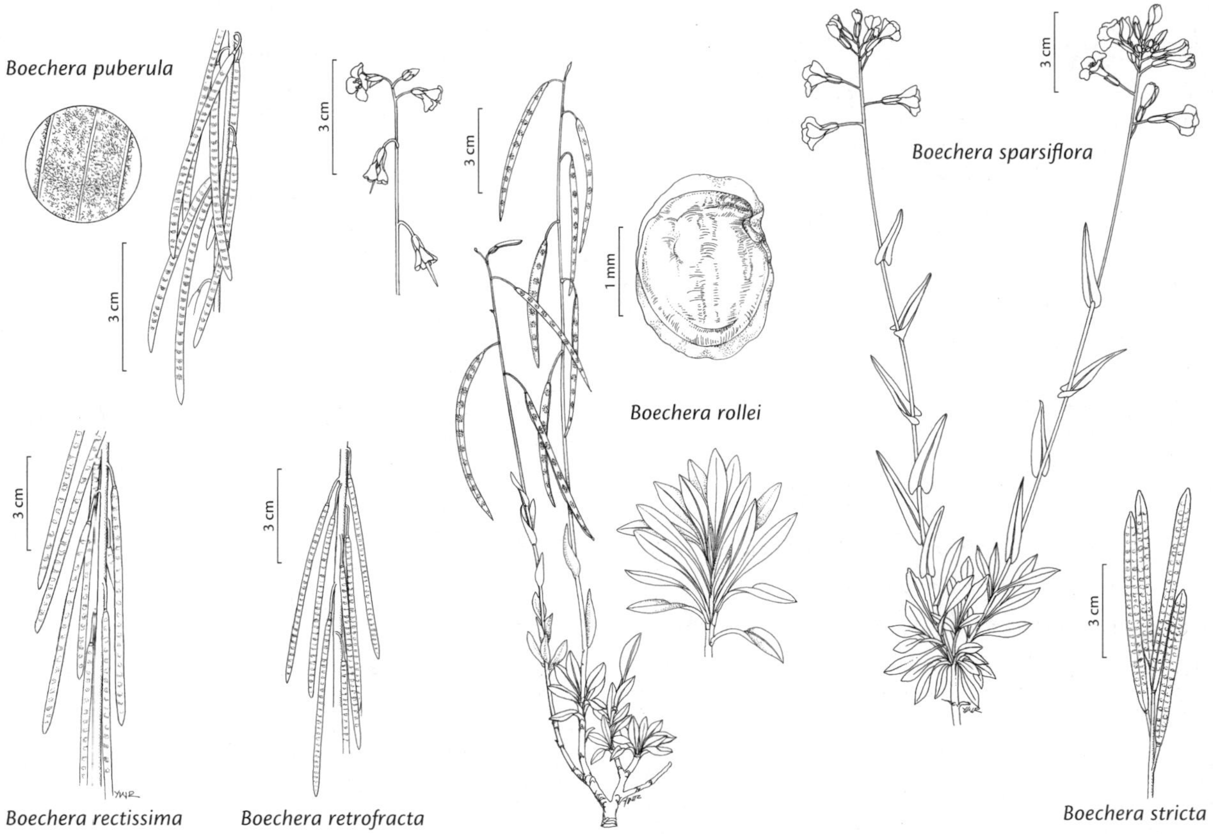

BOECHERA

Boechera pratincola (Greene) Windham & Al-Shehbaz
Meadow rockcress

Arabis × *divaricarpa* A. Nelson

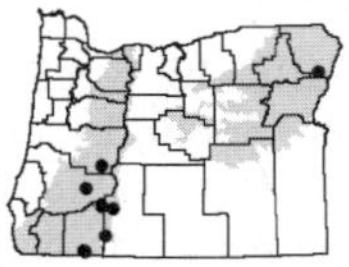

Plants mostly short-lived perennial; with non-woody caudices lacking crowded, persistent leaf bases. **Stems** usually 2–6 dm, lower parts with sessile, 2- or 3-rayed trichomes, upper parts glabrous. **Basal leaves** oblanceolate, 1.5–5(7) mm wide, margins entire or rarely denticulate, surfaces pubescent with sessile, 2–4(6)-rayed trichomes, ciliate at petiole bases. **Cauline leaves** (7)13–42, concealing stem proximally, bases auriculate, uppermost sparsely pubescent. **Inflorescences** (7)14–45-flowered; pedicels 5–12 mm, erect, straight, glabrous. **Flowers** apomictic, ascending at anthesis; sepals glabrous; petals 6–10 × 1.5–2 mm, lavender, glabrous; pollen spheroid; ovules 60–140 per ovary, styles 0.1–0.3 mm. **Fruits** erect, appressed to rachis, not secund, straight, 4–6.5 cm × 1.5–2 mm, margins parallel, glabrous. **Seeds** subbiseriate, 1.4–2 × 1–1.4 mm; wings continuous, 0.1–0.35 mm wide.

Rocky slopes, soil patches in open areas, forest edges. Flowering Jun–Aug. 1500–2200 m. BW, Casc, Sisk. CA, ID, NV. Native.

Boechera puberula (Nuttall) Dorn (p. 455)
Hoary rockcress

Arabis puberula Nutt.

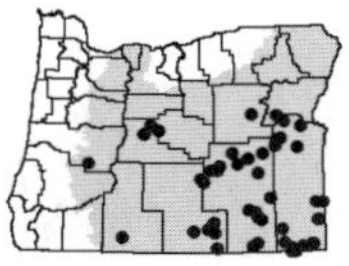

Plants short-lived perennial; with non-woody caudices. **Stems** (1)2–6.3 dm, pubescent with many short-stalked, 3–8-rayed trichomes. **Basal leaves** oblanceolate to linear-oblanceolate, 1.5–5 mm wide, margins usually dentate, not ciliate, surfaces densely pubescent with short-stalked, 5–12-rayed trichomes. **Cauline leaves** 7–45(65), concealing stem proximally, bases of some leaves auriculate, uppermost pubescent. **Inflorescences** 10–40(64)-flowered; fruiting pedicels arched, recurved above horizontal to ascending bases, 4–10 mm, with appressed, branched trichomes. **Flowers** sexual, ascending to descending at anthesis; sepals pubescent; petals 5–9 × 0.8–1.8 mm, white to lavender; pollen ellipsoid; ovules 38–64 per ovary, styles 0.01–0.1 mm. **Fruits** closely pendent, rarely appressed to rachis, not secund, mostly straight, 3–6.5 cm × 1.9–2.2 mm, margins parallel, pubescent. **Seeds** uniseriate, 1.4–1.8 × 1–1.4 mm; wings continuous, 0.1–0.3 mm wide. **2*n***=14.

Rocky slopes, ledges, gravelly hillsides in sagebrush, pinyon-juniper forests. Flowering Apr–Jul. 800–3000 m. BR, BW, Casc, ECas, Lava, Owy. CA, ID, NV, WA; southeast to UT. Native.

Boechera rectissima (Greene) Al-Shehbaz (p. 455)
Bristly-leaved rockcress

Arabis rectissima Greene, *Arabis rectissima* Greene var. *rectissima*

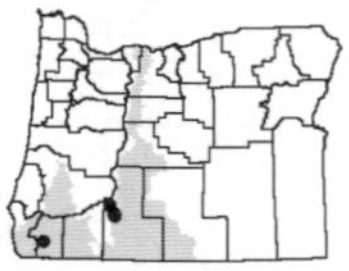

Plants biennial or short-lived perennial; with or without caudices. **Stems** 2–8(10) dm, glabrous or rarely lower parts with few simple trichomes. **Basal leaves** oblanceolate, 3–8(13) mm wide, margins entire or denticulate, ciliate and pubescent with simple trichomes, these often mixed with short-stalked, 2- or 3-rayed trichomes. **Cauline leaves** 6–30(45), often concealing stem proximally, bases auriculate, uppermost ciliate or sparsely pubescent. **Inflorescences** 10–30(50)-flowered; fruiting pedicels reflexed, abruptly recurved at bases but otherwise straight, 4–10 mm, glabrous. **Flowers** sexual, pendent at anthesis; sepals pubescent; petals 3–4 × 0.7–1.2 mm, white; pollen ellipsoid; ovules 46–80 per ovary, styles 0.2–0.8 mm. **Fruits** strongly reflexed, appressed to rachis, not secund, straight, 5–9 cm × 1.8–2.5 mm, margins parallel, glabrous. **Seeds** uniseriate, 1.8–2.1 × 1–1.5 mm; wings continuous, 0.2–0.3 mm wide.

Rocky slopes in open conifer forests. Flowering May–Jul. 1300–1700 m. ECas, Sisk. CA, NV. Native.

Boechera retrofracta (Graham) Á. Löve & D. Löve
Secund rockcress, reflexed rockcress

Arabis holboellii Hornem. var. *retrofracta* (Graham) Rydb., *Arabis secunda* Howell, *Boechera holboellii* (Hornem.) Á. Löve & D. Löve, *Boechera holboellii* (Hornem.) Á. Löve & D. Löve var. *secunda* (Howell) Dorn

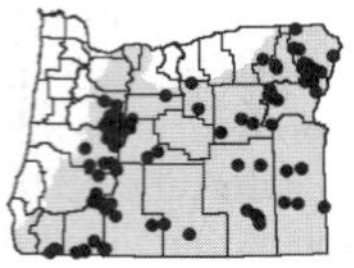

Plants biennial or short-lived perennial; with or without caudices. **Stems** 1.5–7(10.5) dm, lower parts densely pubescent with short-stalked, 2–8-rayed trichomes, upper parts sparsely pubescent. **Basal leaves** oblanceolate, 2–7 mm wide, margins entire or shallowly dentate, surfaces densely pubescent with short-stalked, 5–10-rayed trichomes, occasionally ciliate at petiole bases. **Cauline leaves** 15–40, concealing stem proximally, bases auriculate or very rarely not, uppermost pubescent. **Inflorescences** 15–80(140)-flowered; fruiting pedicels recurved at bases. **Flowers** sexual, pendent at anthesis; sepals pubescent; petals 4–8 × 0.8–2.2 mm, white to lavender; pollen ellipsoid; ovules (60)70–116 per ovary. **Fruits** strongly reflexed, usually appressed to rachis, occasionally somewhat secund, straight, 3.5–9 cm × 0.9–1.8 mm, margins parallel, glabrous or pubescent. **Seeds** uniseriate, 1–1.4 × 0.9–1.2 mm; wings continuous, to 0.1 mm wide. **2*n***=14.

Open hillsides, sagebrush, sandy slopes, rock outcrops, oak woodlands, gravel bars, sandy banks in grasslands, open conifer forests. Flowering Apr–Aug. 700–2800 m. BR, BW, Casc, ECas, Lava, Owy, Sisk. CA, ID, NV, WA; north to AK, northeast to Quebec, east to WY, southeast to UT. Native.

Boechera rollei (Rollins) Al-Shehbaz (p. 455)
Rolle's rockcress

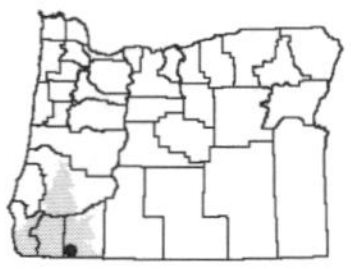

Plants long-lived perennial; with woody caudices. **Stems** 1.5–2.5 dm, glabrous. **Basal leaves** oblanceolate, 3–8 mm wide, margins entire, ciliate proximally with simple trichomes 0.2–0.7 mm, surfaces glabrous or sparsely pubescent with short-stalked, 2–4-rayed trichomes. **Cauline leaves** 6–12, occasionally concealing stem proximally, bases auriculate, uppermost glabrous. **Inflorescences** 3–7-flowered; fruiting pedicels reflexed, not geniculate at base, 4–6 mm, glabrous. **Flowers** sexual, ascending to descending at anthesis; sepals glabrous; petals 8–11 × 2–2.5 mm, white to lavender; pollen ellipsoid; ovules 14–22 per ovary. **Fruits** pendent, not appressed to rachis, not secund, straight to slightly curved, 3.5–6 cm × 2.5–3.5 mm, margins undulate, glabrous. **Seeds** uniseriate, 3–4 × 1.5–2 mm; wings distal and proximal, 0.3–0.6 mm wide.

Peridotite rocks on sparsely forested slopes. Flowering Aug. 1600–1800 m. Sisk. CA. Native.

Boechera sparsiflora (Nuttall) Dorn (p. 455)

Sicklepod rockcress

Arabis sparsiflora Nutt., *Arabis sparsiflora* Nutt. var. *sparsiflora*, *Boechera sparsiflora* (Nutt.) Dorn var. *sparsiflora*

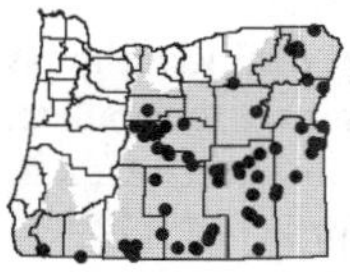

Plants biennial or short-lived perennial; without woody caudices. **Stems** 3–8 dm, lower parts densely pubescent with simple and short-stalked, 2-rayed trichomes to 1.5 mm, upper parts glabrous or sparsely pubescent. **Basal leaves** oblanceolate or linear-oblanceolate, 3–12 mm wide, margins entire or rarely dentate, ciliate proximally, surfaces densely pubescent with simple and short-stalked, 2–5-rayed trichomes. **Cauline leaves** (9)15–35, often concealing stem proximally, bases auriculate, uppermost glabrous or ciliate. **Inflorescences** 12–50-flowered; fruiting pedicels ascending or horizontal, straight or slightly recurved, 3–10(18) mm, with spreading, mostly simple trichomes, rarely glabrous. **Flowers** sexual, ascending at anthesis; sepals pubescent; petals 7–13 × 2–5 mm, lavender to purple, rarely white; pollen ellipsoid; ovules 90–170 per ovary, styles 0.1–0.3 mm. **Fruits** ascending or rarely horizontal, not appressed to rachis, not secund, usually curved, 5–13 cm × 1.7–2 mm, margins parallel, glabrous. **Seeds** uniseriate, 1.5–2 × 1–1.5 mm; wings continuous, 0.1–0.2 mm wide. **$2n$**=14.

Clay hills, meadows, rocky slopes, sandy soil in sagebrush and mountain shrub communities, open conifer forests. Flowering Apr–Jun. 400–2500 m. BR, BW, ECas, Lava, Owy, Sisk. CA, ID, NV, WA; north to British Columbia, southeast to UT. Native.

Boechera stricta (Graham) Al-Shehbaz (p. 455)

Drummond's rockcress

Arabis drummondii A. Gray

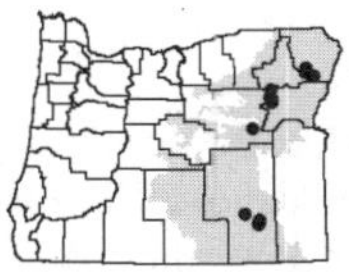

Plants mostly short-lived perennial, rarely biennial; with non-woody caudices. **Stems** 1.5–8(10.2) dm, lower parts glabrous or with few sessile, malpighiaceous trichomes, upper parts glabrous. **Basal leaves** oblanceolate, 1.5–8(14) mm wide, margins entire or rarely dentate, pubescent with malpighiaceous trichomes, very rarely ciliate with few simple trichomes at petiole bases, occasionally glabrous. **Cauline leaves** 6–52, concealing stem proximally, bases auriculate, uppermost glabrous. **Inflorescences** 8–35(80)-flowered; fruiting pedicels erect, straight, 5–18(25) mm, glabrous. **Flowers** sexual, erect at anthesis; sepals glabrous; petals (5)7–11 × 1.5–2.7 mm, white to lavender; pollen ellipsoid; ovules 110–216 per ovary, styles 0.1–0.3 mm. **Fruits** erect, appressed to rachis, not secund, straight, 4–9(10.2) cm × 1.5–3.5 mm, margins parallel, glabrous. **Seeds** biseriate, 1.3–2.2 × 1–1.6 mm; wings distal and lateral, 0.3–0.8 mm wide. **$2n$**=14.

Rocky slopes, gravelly soil in sagebrush and mountain shrub communities, open conifer and hardwood forests, alpine meadows. Flowering May–Aug. 700–2800 m. BR, BW. CA, WA; throughout Canada and western US. Native.

Boechera subpinnatifida (S. Watson) Al-Shehbaz

Ashy rockcress

Arabis subpinnatifida S. Watson

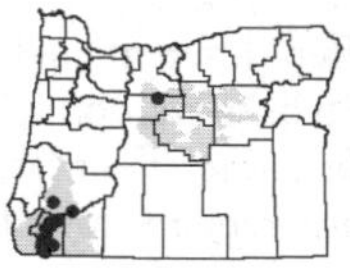

Plants long-lived perennial; with woody caudices. **Stems** 1–4(5) dm, lower parts with short-stalked, 2–6-rayed trichomes, upper parts sparsely pubescent. **Basal leaves** narrowly oblanceolate, 1–4(5) mm wide, margins prominently dentate to subpinnatifid, entire on those of sterile shoots, surfaces densely pubescent with short-stalked, (2)4–9-rayed trichomes, ciliate near petiole bases with simple or 2-rayed trichomes. **Cauline leaves** (10)20–60, often concealing stem throughout, bases uppermost auriculate and sparsely to moderately pubescent. **Inflorescences** 8–30-flowered; fruiting pedicels reflexed, strongly recurved, 5–15 mm, with appressed, branched trichomes. **Flowers** sexual, divaricate-ascending to pendent at anthesis; sepals glabrous; petals 9–14 × 1.5–3 mm, purple or rarely lavender; pollen ellipsoid; ovules 24–42 per ovary, styles 0.5–1 mm. **Fruits** pendent, not appressed to rachis, not secund, straight to slightly curved, (3.5)5–8 cm × 2–3 mm, margins parallel, pubescent. **Seeds** uniseriate, 2.5–3.5 × 1.5–2.2 mm; wings continuous or at both ends, 0.4–0.8 mm wide. **$2n$**=14.

Talus, rock outcrops, gravelly soil, sagebrush-grassland communities. Flowering Mar–Jun. 100–1800 m. Lava, Sisk. CA, ID, NV; southeast to UT. Native.

Boechera suffrutescens (S. Watson) Dorn

Woody rockcress

Arabis suffrutescens S. Watson, *Arabis suffrutescens* S. Watson var. *suffrutescens*

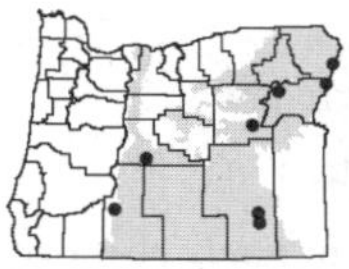

Plants long-lived perennial; with woody caudices. **Stems** 1–4(5) dm, lower parts glabrous or rarely with short-stalked, 2-rayed trichomes, upper parts glabrous. **Basal leaves** narrowly oblanceolate, 1.5–6 mm wide, margins entire, not ciliate or rarely with simple trichomes, surfaces glabrous or sparsely to densely pubescent with short-stalked, 2–6-rayed trichomes. **Cauline leaves** (4)6–12, occasionally concealing stem proximally, bases auriculate, uppermost glabrous. **Inflorescences** 6–12-flowered; fruiting pedicels reflexed, usually abruptly recurved at bases but otherwise straight, 4–18 mm, glabrous. **Flowers** sexual or apomictic, pendent at anthesis; sepals glabrous; petals 4.5–6 × 2–2.5 mm, purple or whitish with rose tips; sexual pollen ellipsoid, apomictic pollen spheroid; ovules 20–30 per ovary, styles 0.4–1.2 mm. **Fruits** reflexed to closely pendent, occasionally appressed to rachis, often secund, straight to somewhat curved, 3–7(8) cm × 3–6 mm, margins undulate and non-parallel, glabrous. **Seeds** uniseriate, 2.5–5.5 × 1.8–3.5 mm; wings continuous, 0.8–1.5 mm wide.

Rocky slopes, gravelly soil, sagebrush communities. Flowering Jun–Aug. 1600–2400 m. BR, BW, ECas. CA, ID, NV, WA. Native.

BRASSICA Cabbage, mustard

Herbs annual, biennial, or rarely perennial. **Trichomes** absent or simple. **Leaves** basal leaves usually rosulate, simple, margins entire, dentate, or lyrate-pinnatifid; cauline leaves bases auriculate to amplexicaul or neither; petiolate or sessile. **Inflorescences** bracts 0. **Flowers** sepals oblong or ovate, erect, ascending, or rarely spreading, caducous, lateral pair usually saccate; petals > sepals, yellow, rarely white, blades obovate, claws often differentiated from blades; nectar glands confluent or not, median glands present; stamens 6, tetradynamous; anthers oblong or ovate, obtuse; ovules (4)10–50 per ovary, septa complete, styles distinct; stigmas entire or 2-lobed. **Fruits** dehiscent siliques, linear, terete or latiseptate, rarely 4-angled, 2-segmented, tips acuminate to subulate; valves prominently 1-veined, glabrous, torulose; replums rounded; terminal segment filiform, conic, or rarely cylindrical, seedless. **Seeds** 4+ per fruit, uniseriate, globose, seed coat reticulate, mucilaginous or not when wetted; wingless; cotyledons conduplicate.

Worldwide. 38 species; 5 species treated in *Flora*.

Brassica species are the most significant oilseed plants and vegetables in the Brassicaceae. Human selection has produced numerous cultivars, generally recognized as varieties or subspecies. Many cultivars are grown in Oregon, but only a few have become naturalized.

Extensive molecular studies on the tribe Brassiceae (Arias & Pires 2012; Hall et al. 2011; Warwick & Sauder 2005) have clearly demonstrated that the genera *Brassica, Diplotaxis*, and *Erucastrum* are polyphyletic, and the majority of their species should be transferred to other genera.

Although *Brassica elongata* is well established in Nevada, it is known in Oregon only from a 1911 ballast collection in Linnton. It differs from the five *Brassica* species that are naturalized in the state by having fruit with a gynophore 1.5–5 mm long, whereas the others have sessile fruit.

Key to *Brassica* species:

1. Petals (15)18–25(30) mm; plants glabrous; sepals erect; all filaments erect at base. . . . . *B. oleracea*

1' Petals (5)7–16(18) mm; plants glabrous or sparsely pubescent basally; sepals spreading or ascending; filaments of lateral stamens curved at base.

  2. Upper cauline leaves sessile, bases amplexicaul or auriculate.

    3. Flowers overtopping buds; petals (5)7–10(13) mm; plants rarely glaucous. . . . . *B. rapa*

    3' Flowers not overtopping buds; petals (9)10–16(18) mm; plants often glaucous. . . . . *B. napus*

  2' Upper cauline leaves petiolate or subsessile, bases not auriculate.

    4. Fruits (0.5)1–2.5(2.7) cm, 4-angled, subappressed to rachis; fruiting pedicels (2)3–5(6) mm. . . . . *B. nigra*

    4' Fruits (2)3–5(6) cm, terete or slightly flattened, not appressed to rachis; fruiting pedicels (5)8–15(20) mm . . . . *B. juncea*

Brassica juncea (Linnaeus) Czernajew

Chinese mustard, Indian mustard

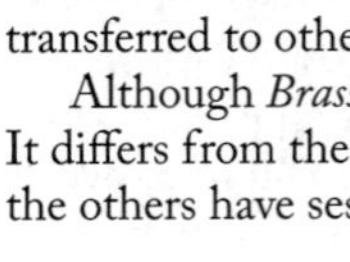

Plants annual, pubescent or rarely glabrous, glaucous or not. **Stems** erect, (2)3–10(18) dm, branched above. **Basal leaves** and lowermost cauline (4)6–30(80) × 1.5–15(28) cm, margins lyrate-pinnatifid to pinnatisect, incised, dentate, or entire, terminal lobes larger than 1–3 lateral lobes on each side; long-petiolate. **Upper cauline leaves** oblanceolate, oblong, lanceolate, or linear, to 10 cm, bases not auriculate; petiolate or subsessile. **Inflorescences** fruiting pedicels straight, divaricate, (5)8–15(20) mm. **Flowers** sepals spreading, oblong, (3.5)4–6(7) mm; petals ovate to obovate, (6.5)8–11(13) × 5–7.5 mm, yellow, claws 3–6 mm; filaments 4–7 mm, lateral curved at base. **Fruits** divaricate to ascending, not appressed to rachis, terete or slightly flattened, (2)3–5(6) cm × 3–4(5) mm; valvular segments (1.5)2–4.5 cm, 6–15(20)-seeded per locule; valves with prominent midveins, slightly torulose; terminal segments conical, (4)5–10(15) mm, seedless; sessile. **Seeds** 1–1.7 mm in diameter. $\mathbf{2n}$=36.

Roadsides, disturbed areas, cultivated and abandoned fields. Flowering Apr–Sep. 0–300 m. BR, WV. CA, ID, NV, WA; worldwide. Exotic.

Brassica juncea is an allotetraploid derived from *B. nigra* (2*n*=16) and *B. rapa* (2*n*=20).

Brassica napus

Canola, rutabaga, rapeseed

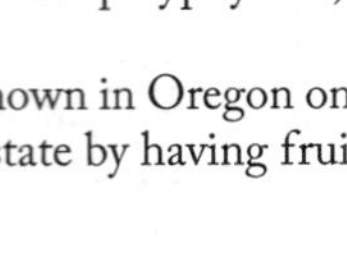

Plants annual or biennial, glabrous or basally sparsely hirsute, often glaucous. **Stems** 3–15 dm. **Basal leaves** and lowermost cauline 5–25(40) × 2–7(10) cm, margins pinnately lobed, lyrate, or undivided, entire or dentate, terminal lobes larger than 1–6 lateral lobes on each side; petiolate. **Upper cauline leaves** to 8 × 3.5 cm, bases amplexicaul or auriculate; sessile. **Inflorescences** fruiting pedicels divaricate, (10)12–23(30) mm. **Flowers** not overtopping buds; sepals ascending, oblong, (5)6–10 mm; petals obovate, (9)10–16(18) × (5)6–9(10) mm, bright or pale yellow, claws 5–9 mm; filaments (5)7–10 mm, lateral curved at base; styles obsolete. **Fruits** divaricate to ascending, terete or slightly 4-angled, (3.5)5–9.5(11) cm × (2.5)3.5–5 mm; valvular segments (3)4–8.5(9.5) cm, 12–20(30)-seeded per locule; valves with prominent midvein, slightly torulose or smooth; terminal segments conical, (0.5)0.9–1.6 cm, seedless or 1-seeded. **Seeds** (1.2)1.5–2.5(3) mm in diameter. $\mathbf{2n}$=38.

Roadsides, disturbed areas, waste places, cultivated and abandoned fields, garden escape from cultivation. Flowering May–Sep. 0–800 m. Sisk. CA, ID, NV, WA; worldwide. Exotic.

Brassica napus is an allotetraploid derived from hybridization between the *B. oleracea* complex (*n*=9) and *B. rapa* (*n*=10). It is widely cultivated worldwide for canola seed oil.

Brassica nigra (Linnaeus) W.D.J. Koch (p. 459)

Black mustard

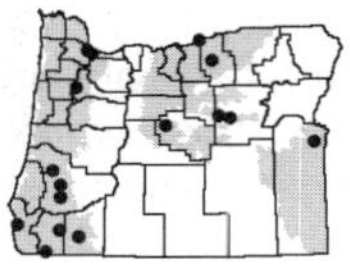

Plants annual, hirsute at least basally. **Stems** 3–20(31) dm. **Basal leaves** and lowermost cauline 6–30 × 1–10 cm, margins lyrate-pinnatifid to pinnatisect, terminal lobes larger than 1–3 lateral lobes on each side; petiolate. **Upper cauline leaves** lanceolate to linear-oblong, to 5 × 1.5 cm, bases not auriculate; petiolate. **Inflorescences** fruiting pedicels straight, slender, erect to ascending, subappressed to rachis, (2)3–5(6) mm. **Flowers** sepals oblong, 4–6(7) mm, spreading to ascending; petals ovate, (6)7.5–11(13) × (2.5)3–4.5(5.5) mm, yellow, claws 3–6 mm; filaments 3.5–5 mm, lateral curved at base. **Fruits** 4-angled, subappressed to rachis, (0.5)1–2.5(2.7) cm × (1.5)2–3(4) mm; valvular segments (0.4)0.8–2(2.5) cm, 2–5(8)-seeded per locule; valves with prominent midveins, slightly torulose; terminal segments style-like, (1)2–5(6) mm, seedless; sessile. **Seeds** globose, 1.2–2 mm in diameter. **2*n***=16.

Roadsides, disturbed areas, fields, orchards. Flowering Apr–Oct. 0–1000 m. Col, CR, Lava, Owy, Sisk, WV. CA, ID, NV, WA; worldwide. Exotic.

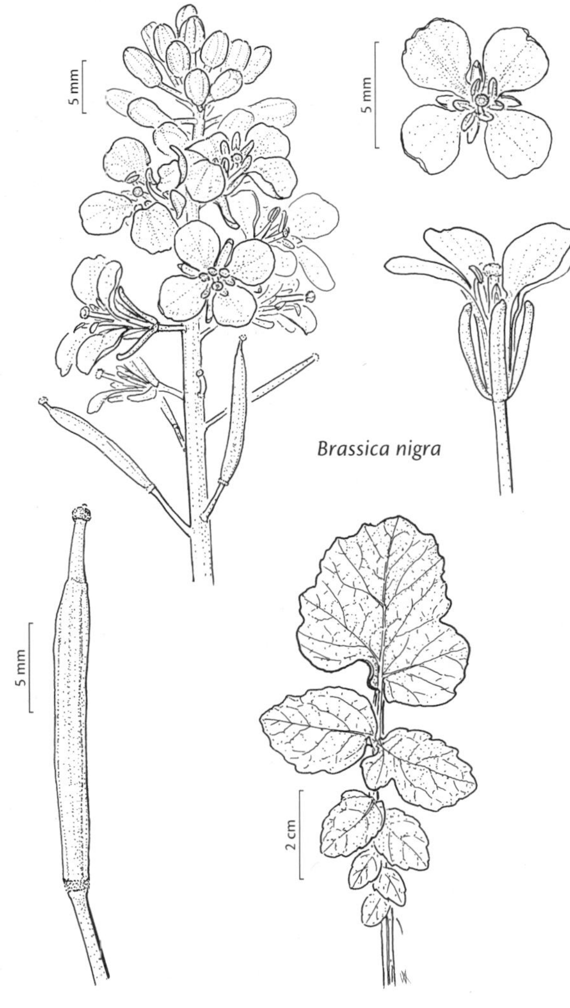

BRASSICA

Brassica oleracea Linnaeus

Cabbage

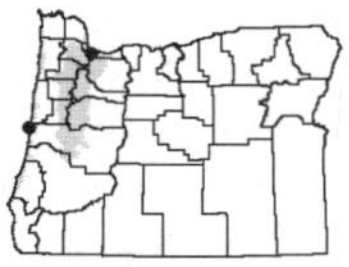

Plants biennial or perennial, glabrous, glaucous. **Stems** (3)6–15(3) dm. **Basal leaves** and lowermost cauline to 40 × 15 cm, margins entire, dentate, or pinnately lobed, terminal lobes larger than 1–13 lateral lobes on each side; petiolate. **Upper cauline leaves** oblanceolate, ovate, to oblong, to 10 × 4 cm, bases amplexicaul, auriculate, or not; sessile. **Inflorescences** fruiting pedicels ascending to divaricate, (8)14–25(40) mm. **Flowers** sepals oblong, 8–15 mm, erect; petals ovate to elliptic, (15)18–25(30) × (6)8–12 mm, yellow or white, claws 7–15 mm; filaments 8–12 mm, erect at base; styles obsolete. **Fruits** divaricate to ascending, terete, (2.5)4–8(10) cm × (2.5)3–4(5) mm; valvular segments (2)3–7.5(9) cm, 10–20-seeded per locule; valves with prominent midveins; terminal segment conical, (3)4–10 mm, seedless or 1(2)-seeded. **Seeds** 1.5–2.5 mm in diameter. **2*n***=18.

Gardens, abandoned fields, waste places, coastal bluffs. Flowering May–Aug. 0–300 m. Est, WV. CA, WA; eastern North America; Africa, Asia, Australia, Europe. Exotic.

Brassica oleracea has been cultivated since ancient times, and many common vegetables have been cultivated from this species, including broccoli (var. *italica*), Brussels sprouts (var. *gemmifera*), cabbage (var. *capitata*), cauliflower (var. *botrytis*), collard greens (var. *viridis*), kale (var. *sabellica*), and kohlrabi (var. *gongylodes*).

Brassica rapa Linnaeus

Field mustard, turnip

Brassica campestris L.

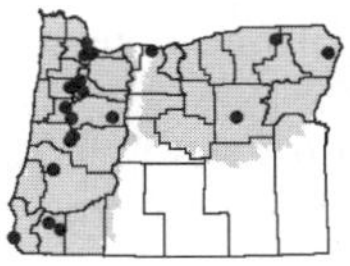

Plants annual or biennial, glabrous or sparsely pubescent basally, rarely glaucous. **Stems** 3–12(19) dm. **Basal leaves** and lowermost cauline (5)10–40(60) × 3–10 (20) cm, margins entire, dentate, or pinnatisect, with terminal lobes larger than 1–6 lateral lobes on each side; petiolate. **Upper cauline leaves** ovate to lanceolate, 2–8(12) × 0.8–3 cm, bases amplexicaul or auriculate; sessile. **Inflorescences** fruiting pedicels ascending to divaricate, (5)10–25(30) mm. **Flowers** sepals oblong, (3)4–6.5(8) mm, ascending; petals obovate, (5)7–10(13) × (2.5)3–6(7) mm, bright yellow, rarely pale or whitish yellow, claws 3–7 mm; anthers oblong, 1.5–2 mm, filaments 4–6(7) mm, lateral curved at base; styles obsolete. **Fruits** divaricate to ascending, terete, (2)3–8(11) cm × 2–4(5) mm; valvular segments (1.3)2–5(7.5) cm, 8–15-seeded per locule; valves with prominent midveins; terminal segments conical, (0.3)1–2.5(3.5) cm, seedless or rarely 1-seeded; sessile. **Seeds** 1–1.8 mm in diameter. **2*n***=20.

Abandoned fields, roadsides, disturbed areas, gardens. Flowering Apr–Oct. 0–1500 m. BW, Casc, Col, CR, Est, Lava, Sisk, WV. CA, ID, NV, WA; worldwide. Exotic.

Brassica rapa includes many cultivars. These are recognized as varieties or subspecies, including turnip (ssp. *rapa*), Chinese mustard or bok choy (ssp. *chinensis*), and Chinese cabbage or petsai (ssp. *pekinensis*).

CAKILE Searocket

Herbs annual or rarely perennial, succulent. **Trichomes** absent. **Leaves** cauline, simple, fleshy, bases not auriculate, margins pinnatifid; petiolate. **Inflorescences** bracts 0. **Flowers** sepals erect, caducous, ovate or oblong, lateral pair not saccate; petals rarely aborted, obovate to spatulate, > sepals, white or lavender, clawed; nectar glands 4, median and lateral; stamens 6, tetradynamous, anthers ovate to oblong, not apiculate; ovules (1)2(4) per ovary, septa membranous, appressed to 1 side, styles absent, stigmas entire or slightly 2-lobed. **Fruits** siliques or silicles, obovoid, oblong, or lanceolate, 2-segmented, becoming corky and dry; segments indehiscent, both falsely 1-loculed, usually 1-seeded; lower segments persistently attached to fruiting pedicel, terete, sometimes laterally horned; terminal segments beaked, terete, 4-angled, deciduous by a transverse articulation; valves and replums not distinguishable. **Seeds** aseriate; wingless; cotyledons accumbent or rarely incumbent.

Asia, Australia, Europe, North America. 7 species; 2 species treated in *Flora*.

Key to *Cakile* species:

1. Lower fruit segments without lateral horns; petals 1.5–3 mm wide; lower leaves not pinnatifid . . . . ***C. edentula*** ssp. ***edentula***

1' Lower fruit segments with 2 opposite lateral horns; petals 3–6 mm wide; lower leaves often pinnatifid . . . . ***C. maritima*** ssp. ***maritima***

Cakile edentula (Bigelow) Hooker ssp. ***edentula*** (p. 460)

American searocket

Cakile edentula (Bigelow) Hook. var. *californica* (A. Heller) Fernald, *Cakile edentula* (Bigelow) Hook. var. *edentula*

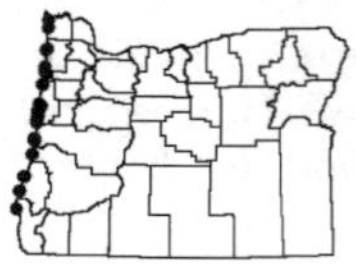

Stems erect, 2–8 dm, much branched. **Leaves** spatulate to ovate, fleshy, margins dentate, crenate, or sinuate, lower leaves not pinnatifid; upper smaller, oblanceolate, margins dentate. **Inflorescences** 1–2 dm in fruit; fruiting pedicels 1.8–8 mm. **Flowers** sepals ovate, 3.5–5 mm; petals well developed or reduced to bristles, 4.6–9.7 × 1.5–3 mm, white to pale lavender, not distinctly clawed. **Fruits** 4-angled, 12–24 × 5–9 mm; beaks flattened, acute, blunt, or rarely retuse at apex; terminal segments turbinate, plump, 7–15 × 5–9 mm; lower segments without lateral horns, terete, 5–10 mm. **2*n***=18.

Sandy beaches. Flowering Jun–Oct. 0–50 m. Est. CA, WA; north to AK, eastern North America; Asia, Australia. Exotic.

Cakile maritima Scopoli ssp. ***maritima***

European searocket

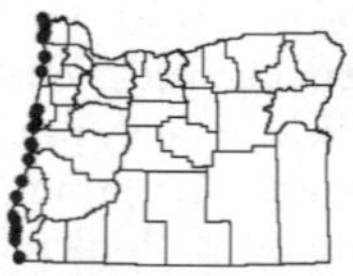

Stems erect to prostrate, 1–8 dm. **Leaves** lower leaves broadly ovate to lanceolate, to 10 cm, margins often deeply pinnatifid, sometimes sinuately lobed; lateral lobes 4–8, 5–15 mm. **Inflorescences** to 4 dm in fruit; rachises straight; fruiting pedicels 1.5–7 mm. **Flowers** sepals ovate, 4–5.5 mm; petals 8–14 × 3–6 mm, lavender or rarely white, clawed. **Fruits** hastate, with 2 opposite, often reflexed lateral horns on upper end of lower segment, 12–27 × 4–8 mm; terminal segments conical, usually 4-angled, tips acute or blunt and usually compressed; lower segments with 2 opposite lateral horns, conical, becoming pedicel-like when aborted. **2*n***=18.

Sandy beaches. Flowering May–Oct. 0–50 m. Est. CA, WA; north to British Columbia, south to Mexico; Asia, Europe. Exotic.

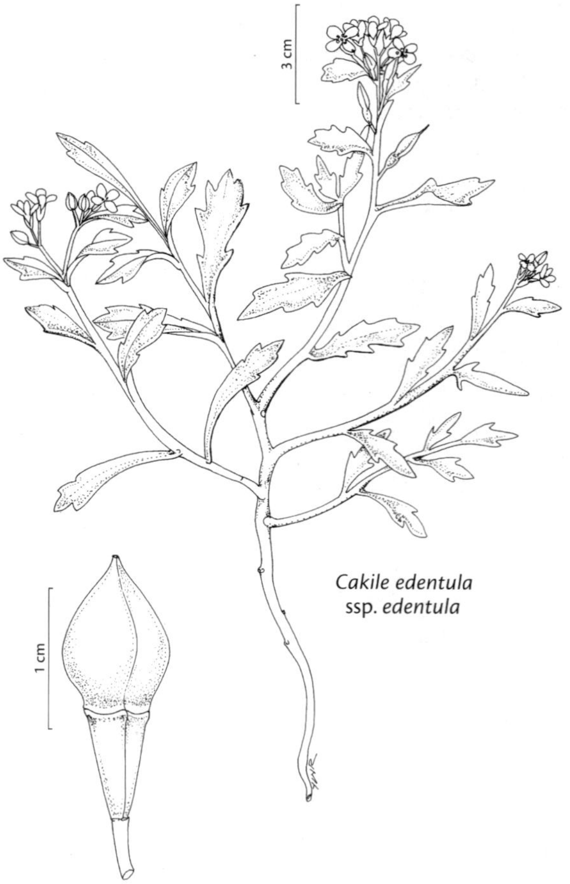

CAKILE

CAMELINA Flaxweed, false flax, gold-of-pleasure

Herbs annual or biennial. **Trichomes** simple and stalked to sessile, forked to subdendritic. **Basal leaves** rosulate or not, simple, margins entire or toothed, rarely lobed; petiolate. **Cauline leaves** bases auriculate or sagittate, margins entire or dentate; sessile. **Inflorescences** racemes; bracts 0. **Flowers** sepals erect to ascending, caducous, oblong or ovate, lateral pair not saccate; petals spatulate, > sepals, yellow to white, claws somewhat differentiated from blades; nectar glands 4, lateral, 1 on each side of lateral stamen; stamens 6, in 3 pairs of unequal length, anthers ovate or oblong, tips obtuse; ovules 8–25 per ovary, septa complete, styles 1.5–4 mm, stigmas capitate, entire. **Fruits** dehiscent silicles, pyriform or obovoid, terete, slightly latiseptate, unsegmented; valves leathery, flattened, tips abruptly caudate and extending 1–2.5 mm onto style and appearing as part of it, connate at margins; replums rounded, concealed by connate valve margins. **Seeds** biseriate, copiously mucilaginous when wetted; wingless; cotyledons incumbent.

Worldwide. 8 species; 3 species treated in *Flora*.

Key to *Camelina* species:

1. Fruits 7–9(13) mm, valves prominently veined; seeds (1.5)1.8–2.5 mm; stems basally glabrous or almost exclusively with minute branched trichomes . . . . *C. sativa*

1' Fruits 3.5–7 mm, valves obscurely veined; seeds 0.8–1.5 mm; stems basally with small branched and simple trichomes.

  2. Petals (2.5)3–4(6) mm; basal leaves withered at anthesis . . . . *C. microcarpa*

  2' Petals (5)6–8(9) mm; basal leaves persisting after anthesis . . . . *C. rumelica*

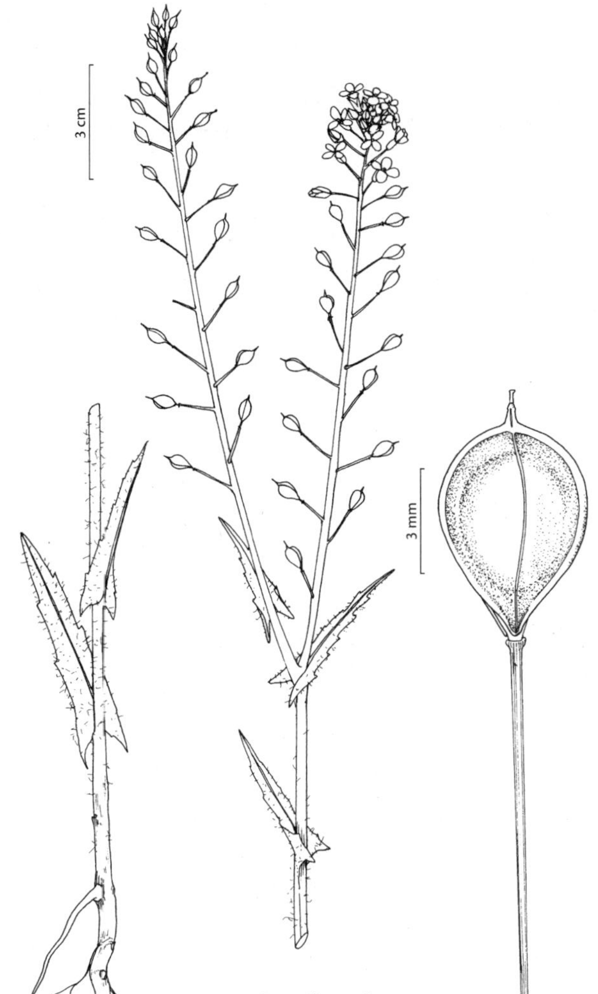

Camelina microcarpa

CAMELINA

Camelina microcarpa Andrzejowski (p. 461)

Littlepod false flax, hairy false flax

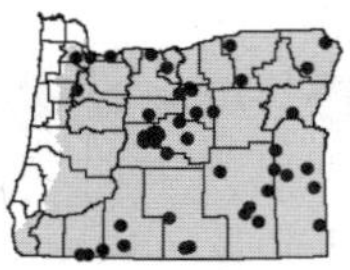

Plants annual. **Stems** (0.8)2–8(10) dm, basally with small branched and simple trichomes to 3.5 mm, glabrate above. **Basal leaves** withered at anthesis. **Cauline leaves** lanceolate, narrowly oblong, or linear-lanceolate, middle (0.8)1.5–5.5(7) × 0.1–1(2) cm, bases sagittate or minutely auriculate, margins entire or rarely remotely denticulate, surfaces pubescent primarily with simple trichomes, often subciliate. **Inflorescences** fruiting pedicels ascending, 4–14(17) mm. **Flowers** sepals 2–3.5 mm; petals oblanceolate, (2.5)3–4(6) × 1–2 mm, yellow to white; styles 1–3.5 mm. **Fruits** pyriform to narrowly so, 3.5–5(7) × 2–4(5) mm, tips acute; valves obscurely veined, narrowly winged at margin. **Seeds** oblong, 0.8–1.4(1.5) × 0.5–0.6 mm, reddish brown or brown. **2*n***=40.

Fields, meadows, prairies, forest margins, roadsides, open woods. Flowering Apr–Jul. 0–1800 m. BR, BW, Casc, Col, ECas, Lava, Owy, Sisk, WV. CA, ID, NV, WA; throughout Canada and US; Africa, Asia, Europe. Exotic.

Camelina rumelica Velenovský

Graceful false flax

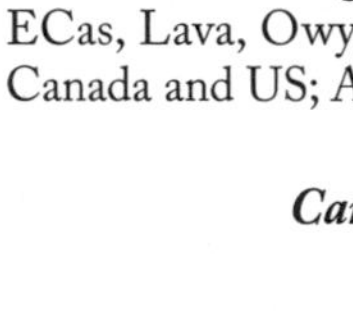

Plants annual. **Stems** 1.5–4(6) dm, basally with small branched and simple trichomes, glabrate above. **Basal leaves** persisting after anthesis. **Cauline leaves** lanceolate to oblong, middle (1)2–6(9) × 0.2–1(2) cm, bases sagittate or minutely auriculate, margins entire or denticulate, surfaces pubescent primarily with simple trichomes, often subciliate. **Inflorescences** fruiting pedicels ascending to divaricate, 7–10(14) mm. **Flowers** sepals (2.7)3–4(4.5) mm; petals oblanceolate, (5)6–8(9) × 1.5–2 mm, white to yellow; styles 2–3 mm. **Fruits** pyriform to obovoid, 5–7 × 3.5–5 mm, acute; valves obscurely veined, narrowly winged at margin. **Seeds** oblong, 1.2–1.5 × 0.5–0.6 mm, brown. **2*n***=12, 26.

Fields, roadsides, waste places. Flowering May–Jun. 100–1500 m. BR, Lava, Owy. NV; scattered in western US; Asia, Europe. Exotic.

Camelina sativa (Linnaeus) Crantz
Flaxweed, false flax, gold-of-pleasure

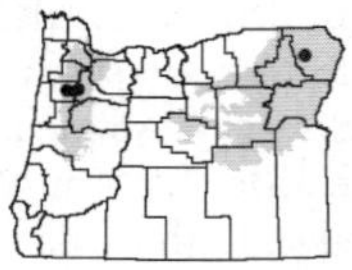

Plants annual or biennial. **Stems** (1.2)3–10(13) dm, basally glabrous or with minute branched trichomes. **Basal leaves** often withered at anthesis. **Cauline leaves** lanceolate, narrowly oblong, or linear-lanceolate, middle (1)2–7(9) cm × 2–10(15) mm, bases sagittate or strongly auriculate, margins entire or rarely remotely denticulate, surfaces glabrate or sparsely pubescent with primarily forked trichomes. **Inflorescences** fruiting pedicels ascending, (8)12–20(27) mm. **Flowers** sepals 2–3(4) mm; petals oblanceolate, (3.5)4–6 × 1–1.5 mm, yellow; styles 1.2–2.5 mm. **Fruits** pyriform or broadly obovoid, 7–9(13) × 4–5(6) mm, often subtruncate at apex; valves prominently veined, narrowly winged at margin. **Seeds** oblong, (1.5)1.8–2.5 × 0.7–1 mm, dark brown. ***2n***=40.

Grassy areas, fields, waste places. Flowering May–Jun. 100–1000 m. BW, WV. ID, WA; scattered worldwide. Exotic.

CAPSELLA Shepherd's purse

Herbs annual or biennial. **Trichomes** sessile and 3–5-rayed stellate, sometimes mixed with simple or forked. **Basal leaves** rosulate, simple, margins usually pinnately lobed, lyrate, or runcinate, rarely entire or dentate; petiolate. **Cauline leaves** bases auriculate or amplexicaul, margins entire, dentate, or sinuate; sessile. **Inflorescences** bracts 0. **Flowers** sepals erect or ascending, caducous, ovate or oblong, lateral pair not saccate; petals obovate or spatulate, > or < sepals, white, pink, or reddish, claws differentiated from blades, petal tips obtuse; nectar glands 4, lateral, 1 on each side of lateral stamen; stamens 6, tetradynamous, anthers ovate or oblong, obtuse; ovules (12)20–40 per ovary, septa complete, styles < 1 mm, stigmas capitate, entire. **Fruits** dehiscent silicles, strongly angustiseptate, obdeltoid to obdeltoid-obcordate, unsegmented; valves prominently veined, strongly keeled; replums rounded. **Seeds** uniseriate, oblong, mucilaginous when wetted; wingless; cotyledons incumbent.

Worldwide. 5 species; 1 species treated in *Flora.*

Capsella bursa-pastoris (Linnaeus) Medikus (p. 462)
Shepherd's purse

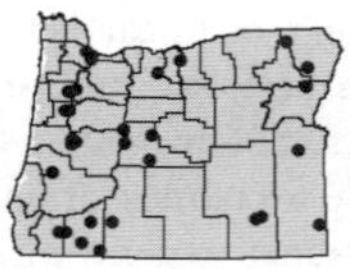

Plants mostly sparsely to densely pubescent. **Stems** (2)10–50(70) cm, simple or branched. **Basal leaves** blades oblong or oblanceolate, (0.5)1.5–10(15) × 0.2–2.5(5) cm, margins pinnatisect, pinnatifid, runcinate, lyrate, dentate, or entire; petioles 0.5–4(6) cm. **Cauline leaves** narrowly oblong, lanceolate, or linear, 1–5.5(8) × 0.1–1.5(2) cm. **Inflorescences** fruiting pedicels (3)5–15(20) mm, divaricate, usually straight. **Flowers** sepals green or reddish, 1.5–2 mm; petals obovate, (1.5)2–4(5) × 1–1.5 mm, white, rarely pinkish; styles 0.2–0.7 mm. **Fruits** (3)4–9(10) × (2)3–7(9) mm, tips emarginate or truncate; valves with subparallel lateral veins, glabrous. **Seeds** 0.9–1.1 × 0.4–0.6 mm, brown. ***2n***=32.

Roadsides, vineyards, pastures, waste areas, fields, barren gravel, orchards, lawns, cultivated ground, mountain slopes. Flowering Jan–Oct. 0–2100 m. All ecoregions except Est. CA, ID, NV, WA; worldwide. Exotic.

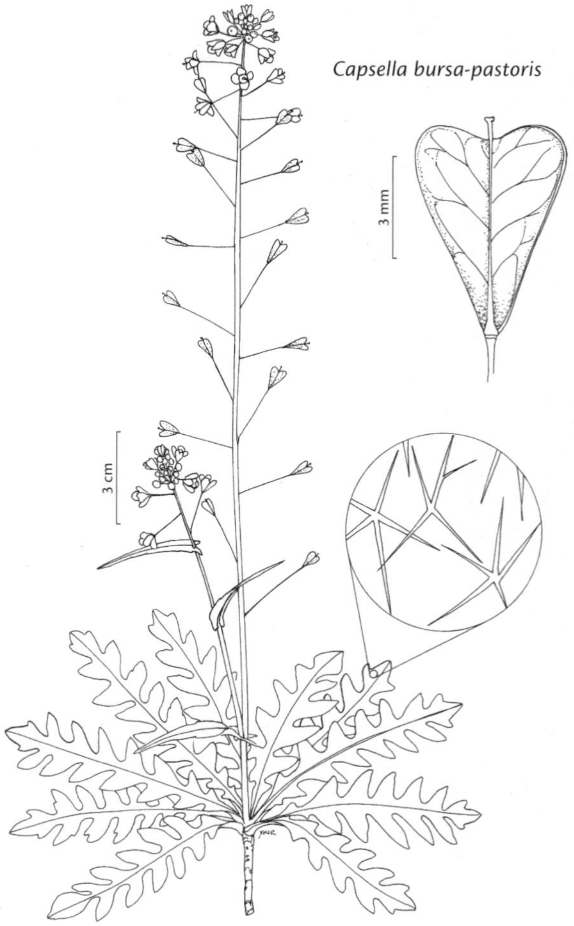

CAPSELLA

CARDAMINE Bittercress

Herbs annual or perennial. **Trichomes** absent or simple. **Basal leaves** rosulate or not, simple, trifoliolate, or pinnately or palmately compound, margins entire, toothed, or 1–3-pinnatisect, or palmately lobed; petiolate. **Rhizomal leaves** similar to basal but appearing along rhizomes (separate from flowering stems). **Cauline leaves** alternate, rarely opposite or whorled, simple or compound, bases cuneate or auriculate to sagittate, margins entire, dentate, or lobed; petiolate or sessile. **Inflorescences** racemes; bracts usually 0. **Flowers** sepals erect to spreading, caducous, ovate or oblong, lateral pair saccate or not; petals obovate, spatulate, oblong, or oblanceolate, white, pink, purple, or violet, claws absent or differentiated from blades, petal tips obtuse or emarginate; nectar glands confluent, subtending bases of all stamens; stamens 6, tetradynamous, rarely 4 and equal, anthers ovate to linear, obtuse; ovules 4–80 per ovary, septa complete, styles distinct, stigmas capitate, entire. **Fruits** linear siliques, latiseptate, unsegmented; valves not veined, dehiscent elastically and becoming spirally coiled; replums strongly flattened. **Seeds** uniseriate, flattened, mucilaginous or not when wetted; cotyledons accumbent.

Worldwide. ~200 Species; 14 species treated in *Flora.*

Key to *Cardamine* species:

1. Inflorescences with bracts throughout . . . . *C. pattersonii*

1' Inflorescences without bracts.

- 2. At least some cauline leaves simple, rarely absent.
  - 3. Plants with distinct caudices; rhizomes absent; cauline leaves absent or 1(2) . . . . *C. bellidifolia*
  - 3' Plants without caudices; rhizomes present; cauline leaves 2–23.
    - 4. Rhizomes fleshy, (3)4–18 mm in diameter; rhizomal leaves present; cauline leaves 2–5; fruits 2–3 mm wide . . . . *C. californica* (in part)
    - 4' Rhizomes not fleshy, 1–3 mm in diameter; rhizomal leaves often absent; cauline leaves (3)5–17(23); fruits 1.2–2 mm wide . . . . *C. cordifolia*
- 2' Cauline leaves pinnately or palmately compound, sometimes pinnatisect and appearing compound.
  - 5. Plants annual or biennial, not rhizomatous.
    - 6. Basal leaves rosulate, persistent at anthesis.
      - 7. Stamens 4(6); fruiting pedicels erect to ascending; fruits often appressed to rachis, glabrous; seeds narrowly margined . . . . *C. hirsuta*
      - 7' Stamens 6; fruiting pedicels divaricate-ascending; fruits usually not appressed to rachis, sparsely pubescent or glabrous; seeds usually not margined . . . . *C. oligosperma*
    - 6' Basal leaves not rosulate, often withered at anthesis.
      - 8. Lateral leaflets of cauline leaves narrowly oblong, linear, or filiform, 1–3 mm wide; fruits 0.6–0.9 mm wide . . . . *C. parviflora*
      - 8' Lateral leaflets of cauline leaves orbicular, ovate, elliptic, or oblong, 4–25 mm wide; fruits 0.8–1.5 mm wide . . . . *C. pensylvanica*
  - 5' Plants perennial, rhizomatous.
    - 9. Rhizomes cylindrical, not producing fragile segments; stem base above ground as wide as at its attachment to rhizome.
      - 10. Sepals 1.2–1.5 mm; petals 2–2.5 × 0.8–1 mm . . . . *C. holmgrenii*
      - 10' Sepals 2–5 mm; petals 3.5–15 × 1.5–8 mm.
        - 11. Rhizomal leaves absent; ovules 14–28 per ovary; petals 3.5–6(7) × 1.5–2.5(3) mm . . . . *C. breweri*
        - 11' Rhizomal leaves present; ovules 10–16 per ovary; petals 8–15 × 4–8 mm . . . . *C. angulata*
    - 9' Rhizomes tuberous, segmented, producing fragile segments; stem base above ground much wider than at its attachment to rhizome.
      - 12. Cauline leaves 5–11-foliolate; rhizomal leaves 5–13-foliolate . . . . *C. penduliflora*
      - 12' At least some cauline leaves 3-foliolate; rhizomal leaves simple or 3(7)-foliolate.
        - 13. Cauline leaves 3–7-foliolate; sepals 1.7–2 mm; petals 4–6 mm; styles 0.5–1.5 mm . . . . *C. occidentalis*
        - 13' Cauline leaves 3(5)-foliolate; sepals 3.5–5.5 mm; petals 8–15 mm; styles 2–8 mm.
          - 14. Cauline leaves 2–5; stems (20)27–60(70) cm; ovules 12–22 per ovary; petals white or pink . . . . *C. californica* (in part)
          - 14' Cauline leaves 1–3; stems 5–20(30) cm; ovules 8–16 per ovary; petals purple to pale pink, rarely white . . . . *C. nuttallii*

Cardamine angulata Hooker

Angled bittercress, seaside bittercress, angle-leaved bittercress

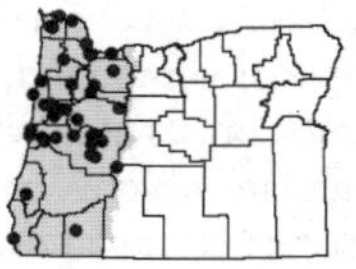

Plants perennial, glabrous or pubescent; rhizomes slender, cylindrical, to 2 mm in diameter. **Stems** (15)25–85(100) cm, bases above ground as wide as its attachment to rhizome, hirsute basally. **Rhizomal leaves** 3(5)-foliolate, (4)7–20(22) cm, bases cuneate, rarely subreniform or obtuse, margins 3–5(7)-lobed or -toothed, puberulent, terminal leaflets ovate to lanceolate, 1.5–7(9) cm. **Cauline leaves** (3)4–8, pinnately or palmately compound, 3(5)-foliolate, bases not auriculate, pubescent along margins, terminal leaflets ovate to lanceolate, 2–7 × 0.6–4 cm. **Inflorescences** bracts 0; fruiting pedicels ascending to divaricate, (9)12–25 mm. **Flowers** sepals oblong, 2.5–4 mm, lateral pair saccate; petals white or rarely pinkish, obovate, 8–15 × 4–8 mm, clawed, tips rounded or emarginate; ovules 10–16 per ovary, styles (0.5)1–4 mm. **Fruits** 15–32 × 1.4–2 mm, glabrous. **Seeds** oblong, 1.8–2.3 × 1–1.2 mm. ***2n***=40.

Wet meadows, moist grounds, thickets, streambanks, swampy or damp woods. Flowering Apr–Jun. 0–1400 m. Casc, CR, Est, Sisk, WV. CA, WA; north to AK. Native.

Cardamine bellidifolia Linnaeus (p. 464)

Alpine bittercress

Cardamine bellidifolia L. var. *pachyphylla* Coville & Leiberg

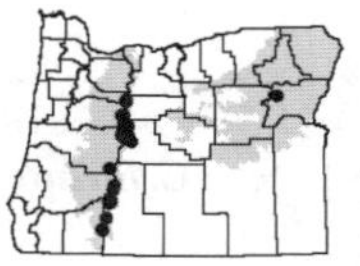

Plants perennial, glabrous, somewhat cespitose; rhizomes absent, caudex branches slender. **Stems** 1–8(14) cm, subscapose. **Basal leaves** simple, rosulate, (0.6)1.2–5(7) cm, blades ovate to oblong, rarely oblanceolate to obovate, (0.4)0.8–1.7(2.5) × (0.2)5–1(1.6) cm, bases cuneate to obtuse, margins entire or obtusely small-toothed. **Cauline leaves** absent or 1(2), simple, bases not auriculate. **Inflorescences** bracts 0; fruiting pedicels ascending to erect, 3–6(8) mm. **Flowers** sepals oblong, 2–3(4) mm, erect, lateral pair not saccate; petals white, oblanceolate, 4–5.5(7) × 1.3–2(2.5) mm, not clawed, tips rounded or emarginate; ovules 8–18 per ovary, styles 0.5–3 mm. **Fruits** (8)13–28(37) × 1.3–2 mm, glabrous. **Seeds** oblong, 1.5–2 × 0.9–1.2 mm. ***2n***=16.

Marshy grounds, cliffs, talus slopes, moist rock crevices, alpine brooks. Flowering Jun–Sep. 1600–3200 m. BW, Casc. CA, WA; northern and western North America, Greenland; Asia, Europe. Native.

Cardamine breweri S. Watson

Brewer's bittercress

Cardamine breweri S. Watson var. *breweri*, *Cardamine breweri* S. Watson var. *orbicularis* (Greene) Detling

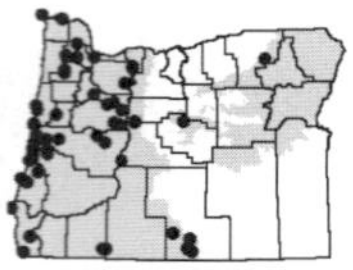

Plants perennial, glabrous or pubescent basally; rhizomes slender, cylindrical, rarely slightly thickened at stem base, 1–3(4) mm in diameter. **Stems** (6)15–60(70) cm, bases above ground as wide as its attachment to rhizome. **Rhizomal leaves** absent. **Cauline leaves** 3–8(11), pinnately or palmately compound, 3–5-foliolate or rarely only terminal leaflets present, 2.5–12(13.5) cm, bases truncate, rounded, or cordate, not auriculate, margins dentate, sinuate, or ≤ 11-lobed, sometimes puberulent, terminal leaflets ovate to orbicular or subcordate, 1.5–4(5) × 1.5–3.5(5) cm, lateral leaflets (when present) smaller than terminal. **Inflorescences** bracts 0; fruiting pedicels ascending to divaricate, (7)10–20 mm. **Flowers** sepals oblong, 2–3(3.8) mm, erect, lateral pair not saccate; petals white, oblanceolate, 3.5–6(7) × 1.5–2.5(3) mm, not clawed, tips rounded or subemarginate; anthers oblong, 0.7–1 mm; ovules 14–28 per ovary, styles 0.2–1.5(2.5) mm. **Fruits** 15–35 × 1–1.5 mm, glabrous. **Seeds** oblong, 1–1.6 × 0.9–1.1 mm.

Wet meadows, streambanks, within streams, lake shores, swamps, ponds. Flowering May–Jul. 0–2000 m. BW, Casc, CR, ECas, Est, Sisk, WV. CA, ID, NV, WA; north to British Columbia, northeast to MT, east to WY, southeast to CO. Native.

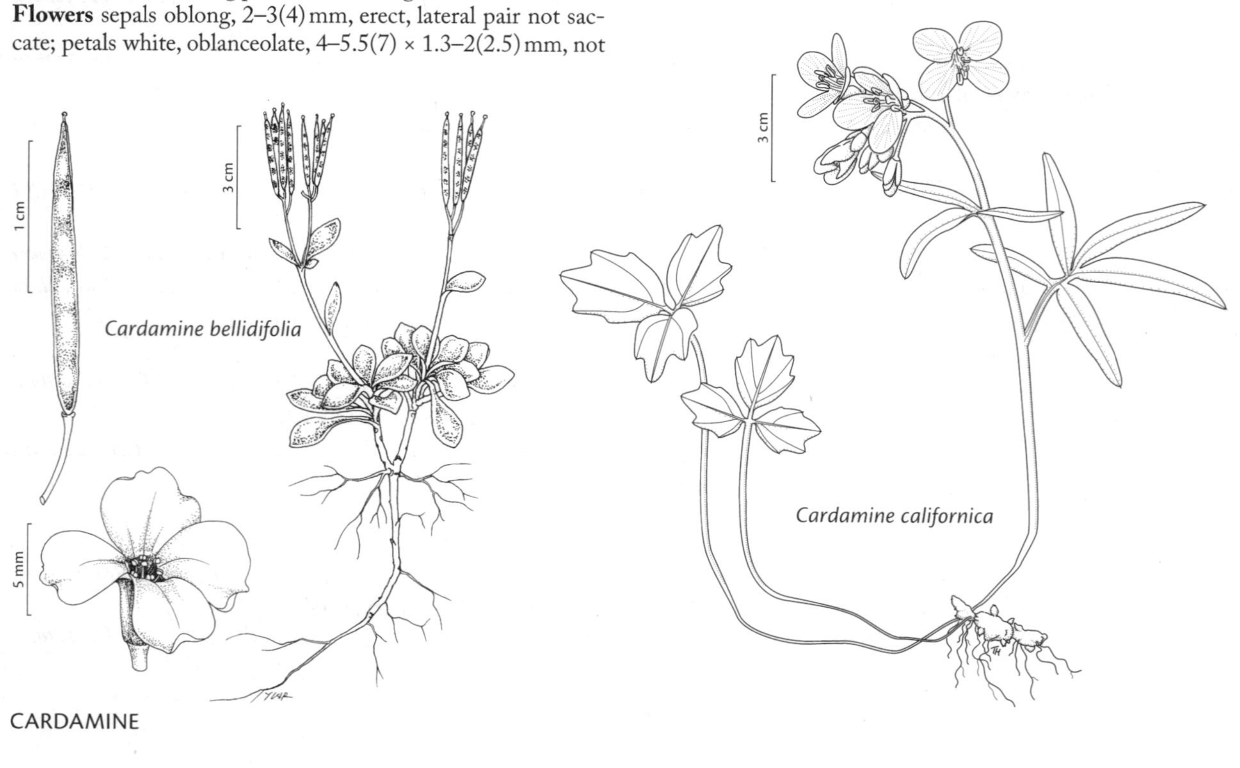

CARDAMINE

Cardamine californica (Nuttall) Greene (p. 464)

Milkmaids, California toothwort

Cardamine californica (Nutt.) Greene var. *californica*, *Cardamine californica* (Nutt.) Greene var. *integrifolia* (Nutt. ex Torr. & A. Gray) Rollins, *Cardamine californica* (Nutt.) Greene var. *sinuata* (Greene) O.E. Schulz, *Cardamine integrifolia* (Nutt.) Greene var. *integrifolia*, *Cardamine integrifolia* (Nutt.) Greene var. *sinuata* (Greene) C.L. Hitchc., *Dentaria californica* Nutt. var. *sinuata* (Greene) Detling

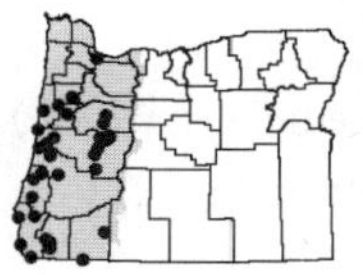

Plants perennial, glabrous or rarely pubescent; rhizomes tuberous, segmented, fleshy, globose, ovoid, or suboblong, (3)4–18 mm in diameter. **Stems** producing fragile segments, (20)27–60(70) cm, bases above ground much wider than attachment to rhizome. **Rhizomal leaves** 3(5–7)-foliolate, sometimes simple, 8–25(38) cm, blades or terminal leaflets ovate, orbicular, cordate, or reniform, (1.5)2.5–7.5(10) × (1.2)2–9(13) cm, margins entire, dentate, to shallowly sinuate, vein ends extend beyond margins, lateral leaflets similar to terminal. **Cauline leaves** 2–5, pinnately or palmately compound, rarely simple, 3–5-foliolate, terminal leaflets 1–7 × (0.5)1–4.7(6.5) cm. **Inflorescences** bracts 0; fruiting pedicels ascending to divaricate, 10–33(41) mm. **Flowers** sepals oblong, 3.5–4.5(5.5) mm, lateral pair saccate; petals obovate, 8–13(15) × 4–8 mm, white or pink, clawed, tips rounded; ovules 12–22 per ovary, styles 2–5(6) mm. **Fruits** 22–54(60) × 2–3 mm, glabrous. **Seeds** oblong to ovate, 1.7–2.8 × 1.2–1.8 mm. ***2n***=32.

Forests, moist hillsides, shady rock crevices and slopes, wooded ravines, open woods, streambanks and bottoms, canyons, cliffs. Flowering Mar–Jun. 0–2100 m. Casc, CR, Est, Sisk, WV. CA, WA; south to Mexico. Native.

Cardamine cordifolia A. Gray (p. 467)

Large mountain bittercress, heart-leaved bittercress, Lyall's bittercress

Cardamine cordifolia A. Gray var. *lyallii* (S. Watson) A. Nelson & J.F. Macbr., *Cardamine lyallii* S. Watson

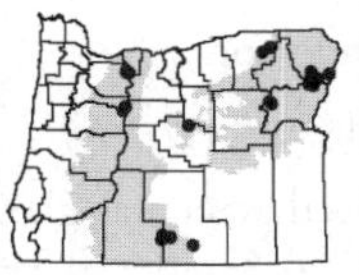

Plants perennial, glabrous or puberulent; rhizomes cylindrical, slender or stout, 1–3 mm in diameter, not fleshy. **Stems** 20–70(102) cm, glabrous or puberulent throughout. **Rhizomal leaves** rarely present, simple, 5–15 cm, blades reniform to cordate, 1.5–4.5(6) cm. **Cauline leaves** (3)5–17(23), simple or rarely lowermost 3-foliolate, blades reniform, deltoid-cordate, or ovate-cordate, (1)2–7.2(9.7) × 1–5.5(8.5) cm, somewhat fleshy, bases cordate or truncate, not auriculate, margins crenate to sinuate, vein ends extend beyond margins, surfaces glabrous or pubescent. **Inflorescences** bracts 0; fruiting pedicels divaricate to ascending, (7)10–20 mm. **Flowers** sepals erect, oblong, 2.5–4.5 mm, lateral pair subsaccate; petals broadly obovate, 7–12 × 4–6 mm, white, claws to 6 mm, petal tips rounded to emarginate; ovules 14–24 per ovary, styles 0.5–3(6) mm. **Fruits** (20)25–37(40) × 1.2–2 mm, glabrous or puberulent. **Seeds** oblong, 1.6–2 × 1–1.3 mm. ***2n***=24.

Streambanks, springs, creek bottoms, lake shores, ponds, meadows, moist hillsides, mixed coniferous forests. Flowering May–Aug. 600–2200 m. BW, Casc, ECas. CA, ID, NV, WA; north to British Columbia, east to WY, southeast to NM. Native.

Cardamine hirsuta Linnaeus

Hairy bittercress

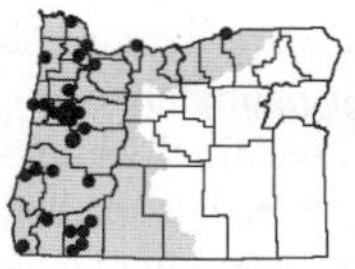

Plants annual, sparsely hirsute at least along petioles of basal leaves, often glabrous above. **Stems** erect, ascending, not flexuous, (3)10–35(45) cm. **Basal leaves** persistent at anthesis, rosulate, (5)8–15(22)-foliolate, (2)3.5–15(17) cm, terminal leaflets reniform or orbicular, 0.4–2 × 0.6–3 cm, entire, dentate, or 3–5-lobed, lateral leaflets smaller than terminal, entire to 3-lobed. **Cauline leaves** pinnately or palmately compound, bases not auriculate. **Inflorescences** bracts 0; fruiting pedicels erect to ascending, (2)3–10(14) mm. **Flowers** sepals oblong, 1.5–2.5 mm, lateral pair not saccate; petals sometimes absent, spatulate, 2.5–4.5(5) × 0.5–1.1 mm, white; stamens 4(6), lateral pair often absent; ovules 14–40 per ovary, styles 0.1–0.6(1) mm. **Fruits** (9)15–25(28) × (0.8)1–1.4 mm, often appressed to rachis, glabrous. **Seeds** oblong or subquadrate, 0.9–1.3(1.5) × 0.6–0.9(1.1) mm, narrowly margined. ***2n***=16.

Disturbed sites, roadsides, clearings, mixed woods, meadows, fields, grassy areas. Flowering Mar–Jul. 0–1000 m. Casc, Col, CR, ECas, Est, Sisk, WV. CA, WA; scattered worldwide. Exotic.

Cardamine holmgrenii Al-Shehbaz

Holmgren's bittercress

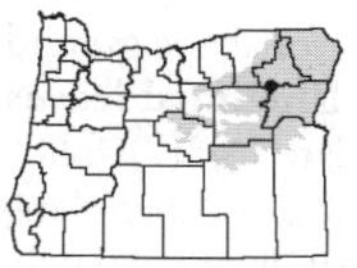

Plants perennial, glabrous; rhizomes slender, 0.5–1 mm in diameter. **Stems** erect to ascending, 15–20 cm, bases above ground as wide as attachment to rhizomes. **Rhizomal leaves** absent. **Cauline leaves** 3 or 4, pinnately or palmately compound, 3- or 5-foliolate, bases not auriculate, terminal leaflets obovate to lanceolate or linear in uppermost leaves, 1.3–2.3 × 0.2–0.8 cm, lateral leaflets linear to linear-lanceolate, 0.9–1.5 × 0.1–0.3 cm, margins entire. **Inflorescences** bracts 0; fruiting pedicels ascending to suberect, 5–14 mm. **Flowers** sepals erect, oblong, 1.2–1.5 mm, lateral pair not saccate; petals oblanceolate, 2–2.5 × 0.8–1 mm, white, tips rounded, not clawed; ovules 16–24 per ovary, styles 0.5–0.7 mm. **Fruits** 10–20 × ~1 mm, glabrous. **Seeds** oblong, 1.2–1.5 × 0.8–1 mm, brown.

Boggy slopes. Flowering Jul–Aug. 2000–2200 m. BW. Native.

This species is known only from the type locality in the Blue Mountains of Baker County.

Cardamine nuttallii Greene

Nuttall's toothwort, slender toothwort

Cardamine nuttallii Greene var. *covilleana* (O.E. Schulz) Rollins, *Cardamine nuttallii* Greene var. *dissecta* (O.E. Schulz) Rollins, *Cardamine nuttallii* Greene var. *gemmata* (Greene) Rollins, *Cardamine nuttallii* Greene var. *nuttallii*, *Cardamine pulcherrima* Greene var. *pulcherrima*, *Cardamine pulcherrima* Greene var. *tenella* (Pursh) C.L. Hitchc., *Dentaria gemmata* (Greene) Howell, *Dentaria tenella* Pursh var. *pulcherrima* (Greene) Detling, *Dentaria tenella* Pursh var. *quercetorum* (Howell) Detling, *Dentaria tenella* Pursh var. *tenella*

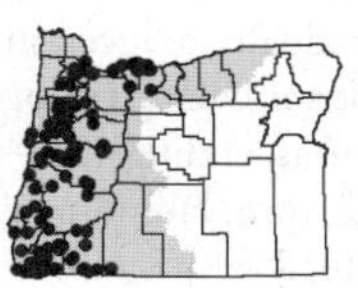

Plants perennial, glabrous or pubescent; rhizomes slender, tuberous, segmented, producing fragile segments, ovoid, oblong, or cylindrical, 2–5 mm in diameter, nodes fleshy. **Stems** erect, 5–20(30) cm, simple, bases above ground much wider than attachment to rhizome.

Rhizomal leaves simple or 3(5)-foliolate, (3)4–20(25) cm, blades or terminal leaflets reniform, suborbicular, ovate, or oblong, (0.9)1.3–4(5.2) × (0.8)1.2–5(7) cm, often with purple spots, bases cordate to obtuse, margins apiculate-dentate or 5–7-lobed. **Cauline leaves** 1–3, pinnately or palmately compound, 3(5)-foliolate, bases not auriculate, terminal leaflets ovate, oblong, or linear, (0.5)1–3.5(6) cm, lateral leaflets smaller, entire, dentate, or lobed. **Inflorescences** bracts 0; fruiting pedicels ascending to divaricate, 10–35 mm. **Flowers** sepals erect, oblong, 3.5–5 mm, lateral pair saccate; petals obovate, 10–15 × 4–7.5 mm, purple to pale pink, rarely white, not clawed, tips rounded; ovules 8–16 per ovary, styles 4–8 mm. **Fruits** 25–56 × 2–2.3 mm, glabrous. **Seeds** oblong, 2–2.5 × 1.4–1.6 mm.

Damp woods, shaded bottomlands and hillsides, mossy slopes, streamsides. Flowering Mar–Jul. 0–2000 m. Casc, Col, CR, ECas, Est, Sisk, WV. CA, WA; north to British Columbia. Native.

Cardamine occidentalis (S. Watson ex B.L. Robinson) T.J. Howell

Western bittercress, big western bittercress

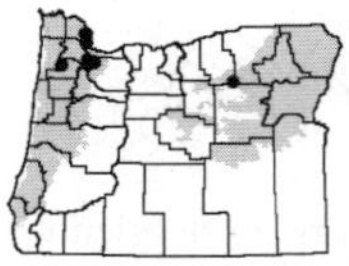

Plants perennial, glabrous or hirsute; rhizomes tuberous, segmented, producing fragile segments, ovoid or globose at stem base, 3–10 mm in diameter, fleshy. **Stems** not flexuous, 10–50 cm, bases above ground much wider than attachment to rhizome. **Rhizomal leaves** simple or 3(5)-foliolate. **Basal leaves** not rosulate, (3)5(7)-foliolate, 2–10 cm, terminal leaflets orbicular, ovate, or subcordate, 0.5–2 × 0.7–2.5 cm. **Cauline leaves** 3–7, pinnately or palmately compound, 5–7-foliolate, becoming 3-foliolate upwards, bases not auriculate, terminal leaflets obovate to oblanceolate, 0.5–2.6 × 0.3–1.3 cm, margins entire or dentate, lateral leaflets smaller. **Inflorescences** bracts 0; fruiting pedicels divaricate to ascending, 7–18 mm. **Flowers** sepals oblong, 1.7–2 mm, lateral pair not saccate; petals oblanceolate, 4–6 × 1.5–2 mm, white, not clawed; ovules 18–40 per ovary, styles 0.5–1.5 mm. **Fruits** 15–33 × 1.7–2.2 mm, glabrous or sparsely pubescent. **Seeds** ovate, 1–1.6 × 1–1.2 mm. ***2n*** =64.

Muddy ground, lake margins, shallow streams, meadows. Flowering Mar–May. 0–1500 m. BW, CR, WV. CA, WA; north to AK. Native.

Cardamine oligosperma Nuttall

Little western bittercress, few-seed bittercress

Cardamine oligosperma Nutt. var. *oligosperma*

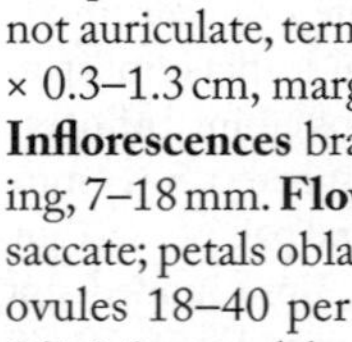

Plants annual or biennial, hirsute at least proximally. **Stems** not flexuous, (5)8–32(41) cm. **Basal leaves** persistent at anthesis, rosulate, pinnately compound, 5–9(13)-foliolate, 2–8.5(11) cm, terminal leaflets orbicular, ovate, or oblong, 0.4–1.5(2.3) × 0.3–1(1.3) cm, bases cordate to rounded, margins entire, dentate, or 3–5-lobed. **Cauline leaves** 3–8, pinnately compound, gradually reduced in size and number of leaflets upwards, bases not auriculate. **Inflorescences** bracts 0; fruiting pedicels divaricate-ascending, (2)3–9(12) mm. **Flowers** sepals oblong, 1.3–1.8(2) mm, lateral pair not saccate; petals spatulate to oblanceolate, 2.5–3.5 × 0.9–1.5 mm, white, not clawed; stamens 6; ovules 16–36(42) per ovary, styles 0.4–1(1.5) mm. **Fruits** usually not appressed to rachis, (13)16–28 × 1–1.7 mm, sparsely pubescent or glabrous. **Seeds** oblong, 1–1.6 × 0.8–1.2 mm, brown, usually not margined. ***2n***=16.

Meadows, streambanks, creek bottoms, lake shores, wooded slopes. Flowering Mar–Jul. 0–2200 m. Col, CR, ECas, Est, Sisk, WV. CA, ID, NV, WA; north to AK, northeast to Northwest Territories, east to WY, southeast to CO, south to Mexico; NY. Native.

Cardamine parviflora Linnaeus (p. 467)

Sand bittercress, small-flowered bittercress, narrow-leaved bittercress

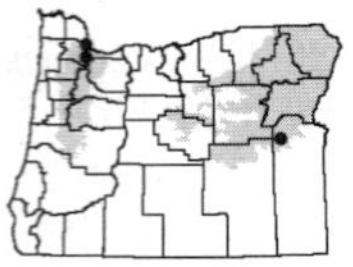

Plants annual, slender, glabrous or pubescent. **Stems** somewhat flexuous, (5)10–30(40) cm. **Basal leaves** often withered by anthesis, not rosulate, pinnately (5)7–13(17)-foliolate, (2)4–10 cm, terminal leaflets linear, oblong, oblanceolate, obovate, or suborbicular, (1)3–10 × 1–7 mm, margins entire, or 3(5)-toothed or -lobed, lateral leaflets similar to terminal. **Cauline leaves** pinnately or palmately compound, (5)9–15(17)-foliolate, bases not auriculate, lateral leaflets narrowly oblong, linear, or filiform, 1–3 mm wide, terminal leaflets filiform, linear, or narrowly oblong, 3–10(16) × 0.3–3 mm, margins entire or rarely 1–3-toothed. **Inflorescences** bracts 0; fruiting pedicels divaricate or ascending, 4–10 mm. **Flowers** sepals oblong, 1–1.5(2) mm; petals oblanceolate, (1.5)1.8–2.5(3) × 0.4–0.8(1) mm, white; stamens 6, anthers ovate, 0.2–0.4 mm; ovules 20–50 per ovary, styles 0.3–0.7(1) mm. **Fruits** (5)10–20(25) × 0.6–0.9 mm, glabrous. **Seeds** oblong-ovate, 0.6–0.9 × 0.4–0.6 mm, margined or not. ***2n***=16.

Roadsides, riverbanks, rocky crests and outcrops, dry woods, fallow fields, marsh and swamp margins, floodplains, waste grounds, ledges and cliffs. Flowering Apr–Jun. 0–1300 m. BW, WV. WA; throughout southern Canada, central and eastern US. Native.

Cardamine pattersonii L.F. Henderson

Saddle Mountain bittercress

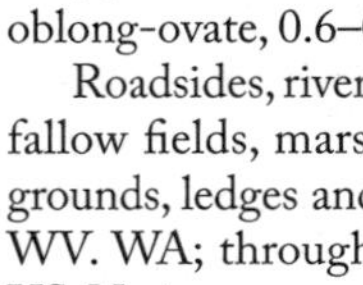

Plants annual or short-lived perennial, glabrous; rhizomes cylindrical, slender, 0.5–1.5 mm in diameter. **Stems** not flexuous, 6–30 cm. **Basal leaves** rosulate or not, 3–5-foliolate, 1–6 cm, terminal leaflets obovate to orbicular or subcordate, 0.3–1.5(2) × 0.25–1.6(1.8) cm, bases obtuse or cordate, margins entire, dentate, or sinuate, lateral leaflets considerably smaller. **Cauline leaves** 2–4 below raceme, 3–5-foliolate, bases not auriculate. **Inflorescences** with bracts throughout; fruiting pedicels divaricate-ascending, (10)15–30(45) mm. **Flowers** sepals oblong, 2–3 mm, lateral pair not saccate; petals obovate, 6–9 × 3–4 mm, purple or pink, not clawed; ovules 14–20 per ovary, styles 2–4 mm. **Fruits** 20–30 × 1–1.5 mm, glabrous. **Seeds** oblong to ovate, 1.7–2.2 × 1–1.5 mm, distally winged.

Moist mossy cliffs, rocky slopes, banks. Flowering Apr–Jun. 200–1000 m. CR. Native.

The species is known only from a limited number of collections, all made in northwestern Oregon.

Cardamine penduliflora O.E. Schulz (p. 467)

Willamette Valley bittercress, wetlands bittercress

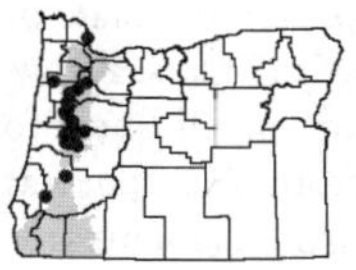

Plants perennial, glabrous; rhizomes tuberous, segmented, producing fragile segments, 4–9(11) mm in diameter, fleshy. **Stems** erect or basally decumbent, 20–60(75) cm, bases above ground much wider than attachment to rhizome. **Rhizomal leaves** 5–13-foliolate, (4)10–18(25) cm, terminal leaflets oblong, elliptic, or ovate, (0.4)0.7–1.7(2) cm, bases cuneate or obtuse, margins entire or obscurely 3-lobed, lateral leaflets subsessile. **Cauline leaves** 2–6, pinnately compound, 5–11-foliolate, 3–10 cm, bases not auriculate, terminal leaflets narrowly ovate, oblong, or oblanceolate, 1.5–3.5 × 0.2–1.5 cm, margins entire, toothed, or 3-lobed. **Inflorescences** bracts 0; fruiting pedicels ascending to divaricate, (10)20–40(60) mm. **Flowers** sepals oblong to ovate, 3.5–5 mm, erect, lateral pair saccate; petals obovate, 12–16 × 6–8 mm, white, not clawed, tips rounded or subemarginate; ovules 12–24 per ovary, styles 4–6 mm. **Fruits** 25–45 × 1.4–2 mm, glabrous. **Seeds** oblong, 1.8–2 × 1–1.5 mm.

Shallow pools, wet grounds, marshes, meadows, creeks, channels, swampy woods. Flowering Apr–May. 0–400 m. Sisk, WV. Native. ✾

Cardamine pensylvanica Muhlenberg ex Willdenow

Pennsylvania bittercress, Quaker bittercress

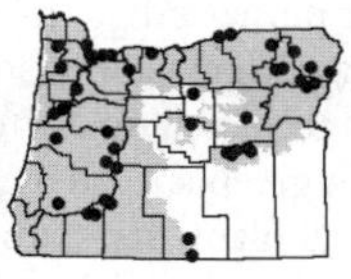

Plants annual or biennial, glabrous or hirsute. **Stems** (5)15–55(70) cm, glabrous above. **Basal leaves** withered by anthesis, not rosulate, 4–15 cm. **Cauline leaves** (3)5–20(35), pinnately (5)7–13(19)-foliolate, sometimes pinnatisect; middle leaves 2–11 cm, bases not auriculate, often sparsely hirsute, lateral leaflets orbicular, ovate, elliptic, or oblong, 4–25 mm wide, terminal leaflets suborbicular, obovate, oblanceolate, or elliptic, 1.3–3(4) × 0.6–2.5 cm, margins entire to repand or obscurely 3–5-lobed, subsessile or with petiolule to 10 mm; uppermost leaves narrower and with fewer lobes or leaflets. **Inflorescences** bracts 0; fruiting pedicels divaricate to ascending, (3)4–10(13) mm. **Flowers** sepals oblong, (1)1.3–2.3 mm; petals spatulate to oblanceolate, 2–3.5(4) × 0.8–1.5 mm, white, not clawed; stamens 6; ovules 40–80 per ovary, styles 0.5–1 mm. **Fruits** (14)17–27(32) × 0.8–1.1 mm, glabrous. **Seeds** oblong to ovate, 0.7–1.1 × 0.5–0.8 mm. ***2n***=32, 64.

Marshes, streams, swamps, seepage, lake margins, ditches, waste ground. Flowering Apr–Jun. 0–2100 m. BW, Casc, Col, CR, ECas, Sisk, WV. CA, ID, NV, WA; throughout most of Canada and US. Native.

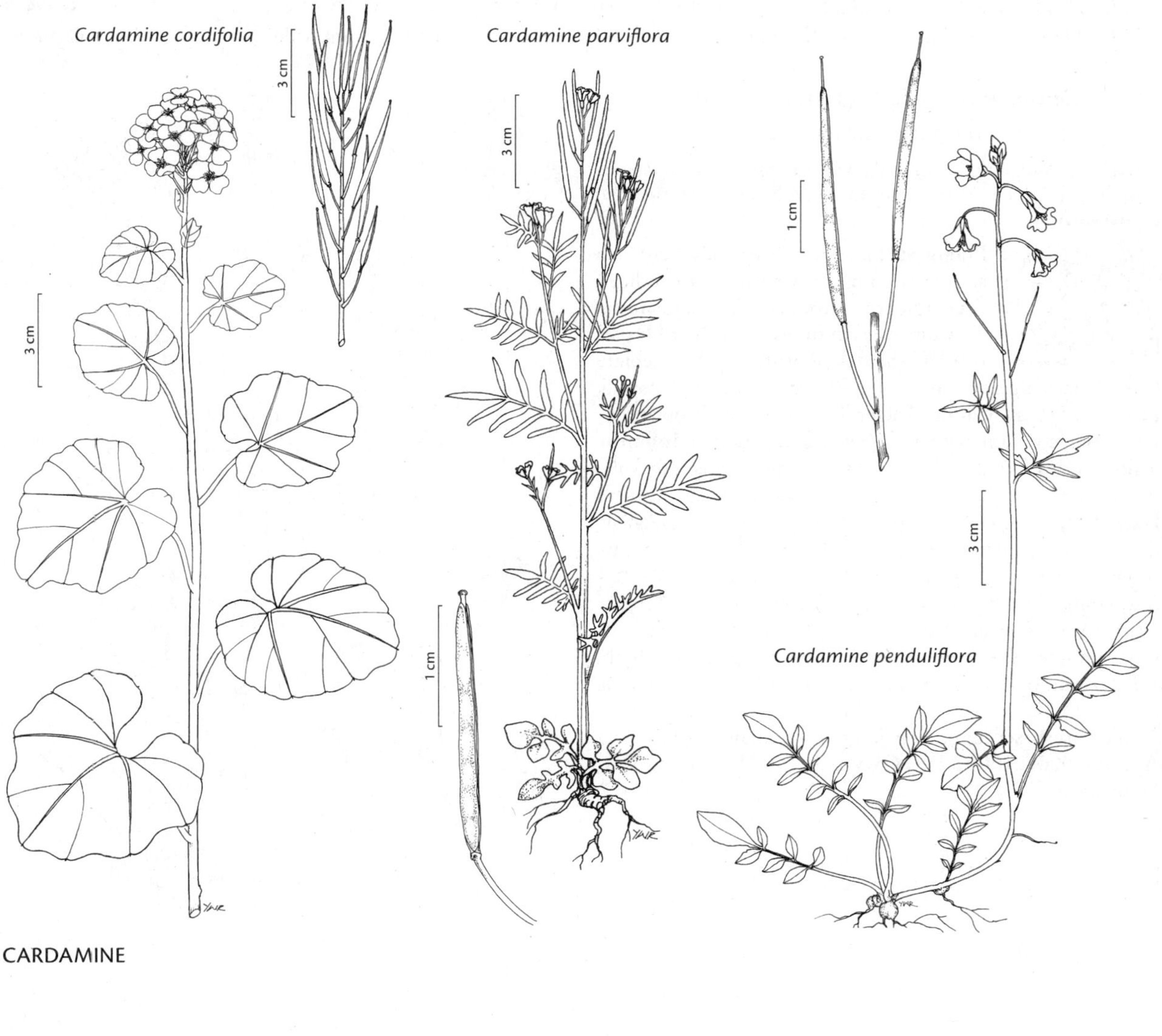

CARDAMINE

CAULANTHUS Wild cabbage, jewelflower

Herbs biennial or perennial; caudices present. **Trichomes** absent or simple, rarely also stalked, 2-rayed. **Basal leaves** rosulate or not, simple, margins entire, dentate, lobed, or pinnatifid; petiolate. **Cauline leaves** bases auriculate to sagittate or not; petiolate or sessile. **Inflorescences** sometimes terminated with cluster of sterile flowers; bracts 0, rarely lowermost flowers with bracts. **Flowers** sepals erect, caducous, oblong, ovate, or lanceolate, usually forming urceolate calyces, lateral pair saccate or not; petals linear to oblanceolate, > sepals, brown, purple, white, or greenish, often channeled and crisped, claws well differentiated from blades; nectar glands mostly confluent, subtending bases of all filaments; stamens 6, tetradynamous or in 3 pairs of unequal length, filaments free or median pairs united, anthers oblong, not coiled when dehisced; ovules 24–210 per ovary, septa complete, styles to 4 mm, stigmas 2-lobed and lobes opposite valves or replum, or entire; gynophore obsolete or < 1 mm. **Fruits** dehiscent siliques, rarely latiseptate or angustiseptate, unsegmented, linear, terete; replums rounded. **Seeds** uniseriate, smooth or reticulate, not mucilaginous when wetted; wingless; cotyledons incumbent.

Northwestern Mexico, western United States. 14 species; 3 species treated in *Flora*.

The limits of *Caulanthus* have fluctuated substantially over the past century, and I have treated the genus as distinct (Al-Shehbaz et al. 2010) or merged it with a more inclusive *Streptanthus* (Al-Shehbaz 2012). There is a tremendous overlap in floral morphology between *Caulanthus* and *Streptanthus*, and the former has been separated from the latter mainly by having terete (vs. latiseptate) fruit and incumbent (vs. accumbent) cotyledons. Recent molecular phylogenetic studies (Cacho et al. 2014) amply demonstrated that limits of the above genera ought to be re-drawn, although *Streptanthus* remains to be split into a number of segregates based on further molecular and detailed morphological studies in progress.

Key to *Caulanthus* species:

1. Stems fusiform, strongly inflated; inflorescences with a terminal cluster of sterile flowers; stigmas strongly 2-lobed, lobes connivent . . . . ***C. crassicaulis***

1' Stems cylindrical, not or rarely slightly inflated; inflorescences without a terminal cluster of sterile flowers; stigmas slightly 2-lobed.

  2. Plants perennial, with woody caudices; ovules 46–58 per ovary; fruits 2.2–2.8 mm wide . . . . ***C. major***

  2' Plants biennial, without woody caudices; ovules 152–198 per ovary; fruits 1–1.5 mm wide . . . . ***C. pilosus***

Caulanthus crassicaulis (Torrey) S. Watson
Thickstem wild cabbage

Caulanthus crassicaulis (Torr.) S. Watson var. *crassicaulis*, *Caulanthus crassicaulis* (Torr.) S. Watson var. *glaber* M.E. Jones, *Streptanthus crassicaulis* Torr.

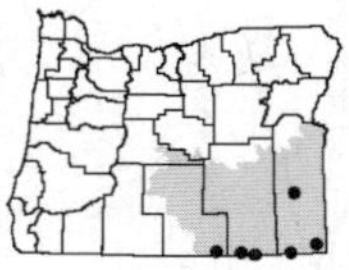

Plants perennial, glabrous or pubescent with simple trichomes; with woody caudices. **Stems** hollow, strongly inflated, fusiform, 2–10 dm, to 3 cm in diameter. **Basal leaves** rosulate, blades obovate to oblanceolate 0.3–4.5 cm, margins entire, dentate-sinuate, lyrate, or pinnatifid; petioles 0.5–9 cm. **Cauline leaves** linear to narrowly oblanceolate, bases not auriculate, margins entire; petiolate. **Inflorescences** with a terminal cluster of sterile flowers; bracts 0; fruiting pedicels ascending, 1–5 mm. **Flowers** sepals ovate to lanceolate, 7.5–14 mm, creamy white, purplish, or greenish, densely pilose; petals 10–15 mm, blades 3–5 × 1.2–2.5 mm, brown or purple, not crisped, claws oblanceolate, 7–10 × 2–2.5 mm; filaments tetradynamous, free; ovules 98–126 per ovary, styles to 0.6 mm, stigmas strongly 2-lobed, lobes connivent, opposite valves. **Fruits** erect to ascending, terete, 4.5–14 cm × 2–2.5 mm; valve midveins obscure. **Seeds** 1.5–4 × 1.5–2 mm. **2*n***=28.

Sagebrush scrub, pinyon-juniper woodlands. Flowering Apr–Jul. 900–2100 m. BR, Owy. CA, ID, NV; east to WY, southeast to CO. Native.

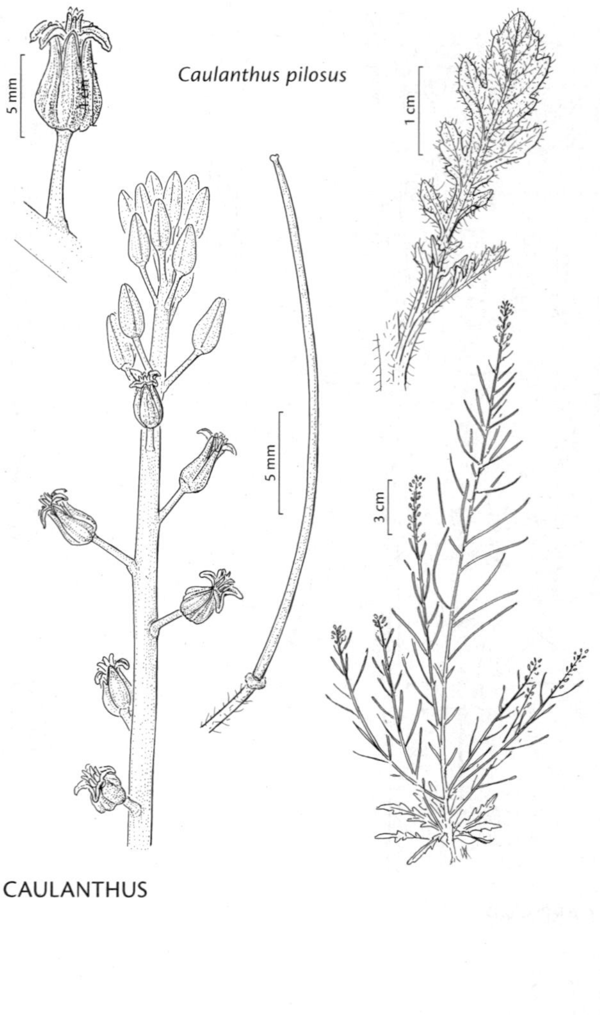

CAULANTHUS

Caulanthus major (M.E. Jones) Payson

Slender wild cabbage, Nevada wild cabbage

Caulanthus major (M.E. Jones) Payson var. *nevadensis* Rollins

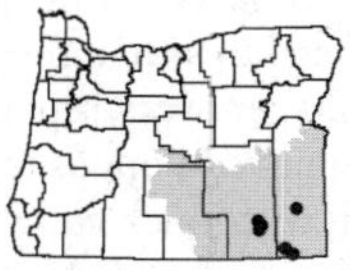

Plants perennial, glabrous or pubescent with simple trichomes; with woody caudices. **Stems** cylindrical, hollow, rarely slightly inflated, 2–10 dm. **Basal leaves** rosulate, blades 1–14 × 0.5–2.5 cm, margins entire, dentate-sinuate, or pinnatifid-runcinate; petioles 0.5–9 cm. **Cauline leaves** linear to narrowly oblanceolate, bases not auriculate, margins entire; short-petiolate. **Inflorescences** without a terminal cluster of sterile flowers; bracts 0; fruiting pedicels ascending, 1–6 mm. **Flowers** sepals ovate to lanceolate, 6.5–9.5 mm, creamy white or purple; petals 11–17 mm, purple, blades 4–7 × 1–1.7 mm, not crisped, claws oblanceolate, 7–13 × 2.5–3.5 mm; filaments tetradynamous, free, anthers oblong, 4–6 mm; ovules 46–58 per ovary, styles to 0.4 mm, stigmas slightly 2-lobed, lobes opposite valves. **Fruits** erect to ascending, terete, 4.5–12 cm × 2.2–2.8 mm. **Seeds** 2–3.5 × 1.3–1.8 mm. **2*n*=28.**

Margins of montane forests, sagebrush, pinyon-juniper woodlands. Flowering May–Aug. 1600–2200 m. BR, Owy. CA, NV; southeast to UT. Native.

Caulanthus pilosus S. Watson (p. 468)

Hairy wild cabbage, chocolate drops

Streptanthus pilosus (S. Watson) Jeps.

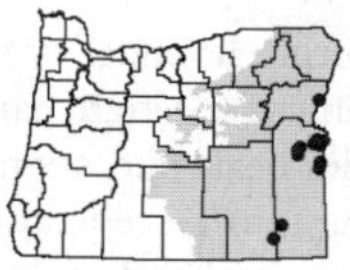

Plants biennial, pilose with simple trichomes; without woody caudices. **Stems** cylindrical, not inflated, 2–12 dm. **Basal leaves** rosulate, blades 2–24 × 0.5–9 cm, margins pinnatifid, pinnatisect, or rarely dentate-sinuate; petioles 1–8 cm. **Cauline leaves** linear to oblanceolate, bases not auriculate, margins entire or dentate; short-petiolate. **Inflorescences** without a terminal cluster of sterile flowers; bracts 0 or lowermost flowers with bracts; fruiting pedicels ascending, 4–18 mm. **Flowers** sepals erect, not keeled, narrowly ovate to lanceolate, 4.5–9.5 mm, dark purple in bud becoming paler or greenish, pilose; petals 7–12 mm, blades 3–4 × 1–1.5 mm, purple, crisped, claws oblanceolate to spatulate, 4–9 × 1–2 mm; filaments tetradynamous, free; ovules 152–198 per ovary, styles obsolete or rarely to 1 mm, stigmas slightly 2-lobed. **Fruits** ascending to divaricate, terete, 2–18 cm × 1–1.5 mm. **Seeds** 1–2 × 0.7–1 mm.

Flats, rocky slopes, scrub and sagebrush communities, pinyon-juniper woodlands. Flowering Apr–Jun. 600–1100 m. BR, BW, Owy. CA, ID, NV; southeast to UT. Native.

CHLOROCRAMBE Spearhead

Genus composed of one species, *C. hastata.*

Chlorocrambe hastata (S. Watson) Rydberg (p. 469)

Spearhead

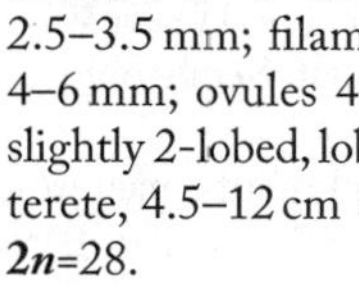

Herbs perennial; with woody caudices. **Trichomes** absent. **Stems** erect, 5–15 dm. **Basal leaves** not rosulate, soon withered. **Cauline leaves** usually hastate, deltoid to lanceolate, 5–20 × 3–7 cm, not auriculate, margins entire or subapically denticulate, petiolate; lowermost leaves lyrate or sinuate, petioles 6–15 cm; uppermost leaves narrowly lanceolate, entire. **Inflorescences** flowers reflexed at anthesis; fruiting pedicels divaricate or slightly reflexed, often curved upward, 5–12 mm, wider than gynophores; bracts 0 or lowermost few flowers with bracts. **Flowers** sepals caducous, narrowly lanceolate, 4–6 mm, yellowish green, lateral pair not saccate; petals 5–7 mm, slightly > sepals, white, blades linear, 2.5–4 × 0.5–1 mm, slightly crisped, claws 2–3.5 × 1–1.7 mm, distinctly wider than blades, petal margins dentate to laciniate; stamens 6, subequal, well exserted, anthers linear, apiculate, coiled after dehiscence; ovules 40–60 per ovary, styles obsolete or rarely to 0.5 mm, stigmas capitate, entire; gynophore (1)2–8(10) mm. **Fruits** dehiscent siliques, linear, slightly flattened, divaricate, unsegmented, 4–13 cm × 1.5–2.5 mm, long-stipitate; valves with prominent midveins, subtorulose; replums rounded. **Seeds** uniseriate, oblong, 3–5 × 1.7–2.2 mm, not mucilaginous when wetted; distally winged, wings to 0.7 mm wide; cotyledons accumbent.

Mountain slopes and canyons, thickets, shady damp areas, stony and brushy hillsides. Flowering Jun–Aug. 1500–2000 m. BW. UT. Native.

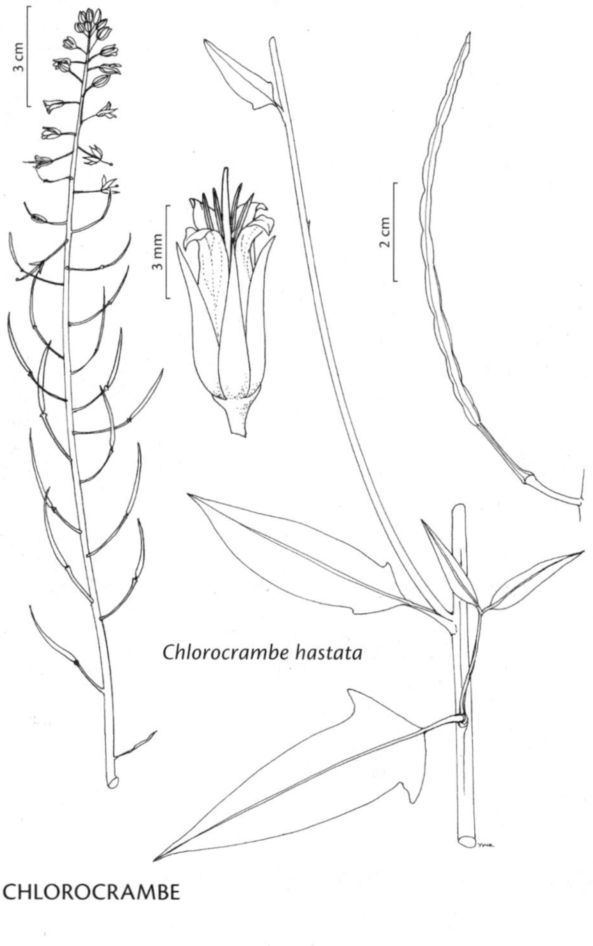

CHLOROCRAMBE

CHORISPORA Chorispora, purple mustard

Herbs with stipitate, multicellular glands. **Basal leaves** rosulate or not, margins sinuate-dentate, pinnatisect, rarely entire; petiolate. **Cauline leaves** absent or present, bases not auriculate. **Inflorescences** several-flowered, rarely flowers on solitary pedicels from axils of rosette leaves; bracts 0. **Flowers** sepals erect, caducous, ovate, oblong, or linear, lateral pair saccate; petals broadly obovate, obcordate, or oblanceolate, > sepals, yellow, purple, lavender, or white, claws strongly differentiated from blades, petal tips emarginate or obtuse; nectar glands 2 or 4, lateral; stamens 6, tetradynamous, anthers oblong or linear, obtuse; ovules 5–30 per ovary, septa becoming corky and splitting at middle, styles beak-like, stigmas conical, 2-lobed, lobes decurrent, strongly connivent, opposite replum. **Fruits** siliques, linear, terete, breaking into 1-seeded units, lomentaceous or torulose; units indehiscent, with thick corky or woody wall; replums flattened, persistent; sessile. **Seeds** uniseriate, oblong, not mucilaginous when wetted; wingless; cotyledons accumbent.

Africa, Asia, Europe, North America, South America. 11 species; 1 species treated in *Flora*.

Chorispora tenella (Pallas) de Candolle (p. 470)

Chorispora, purple mustard, crossflower

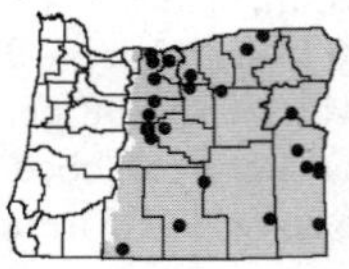

Plants annual, with multicellular glands. **Trichomes** absent or simple. **Stems** (5)10–40(56) cm. **Basal leaves** not rosulate, oblanceolate or oblong, (1.5)2.5–8(13) × (0.4)0.8–2(3) cm, margins dentate to sinuate, surfaces glandular; petioles (0.5)1–2(4) cm. **Cauline leaves** margins entire or dentate; subsessile. **Inflorescences** fruiting pedicels divaricate, stout, (2)3–5 mm, glandular. **Flowers** sepals linear, (3)4–5(6) mm, purplish; petals oblanceolate, 8–10(12) × 1–2 mm, purple, lavender, or rarely white, claws 6–7 mm, petal tips obtuse; styles 10–22 mm. **Fruits** cylindrical, slightly curved upward, with 8–12 constrictions on each side, (14)18–25(30) × 1.5–2 mm, corky, glandular or rarely eglandular. **Seeds** oblong, 1–1.4 × 0.8–1 mm. $2n$=14.

Waste places, roadsides, railroad embankments, pastures, fields, grassy slopes. Flowering Apr–Jun. 50–1800 m. BR, BW, Col, ECas, Lava, Owy. CA, ID, NV, WA; north to British Columbia, northeast to Saskatchewan, east to MA, southeast to MS; Africa, Asia, Europe, South America. Exotic.

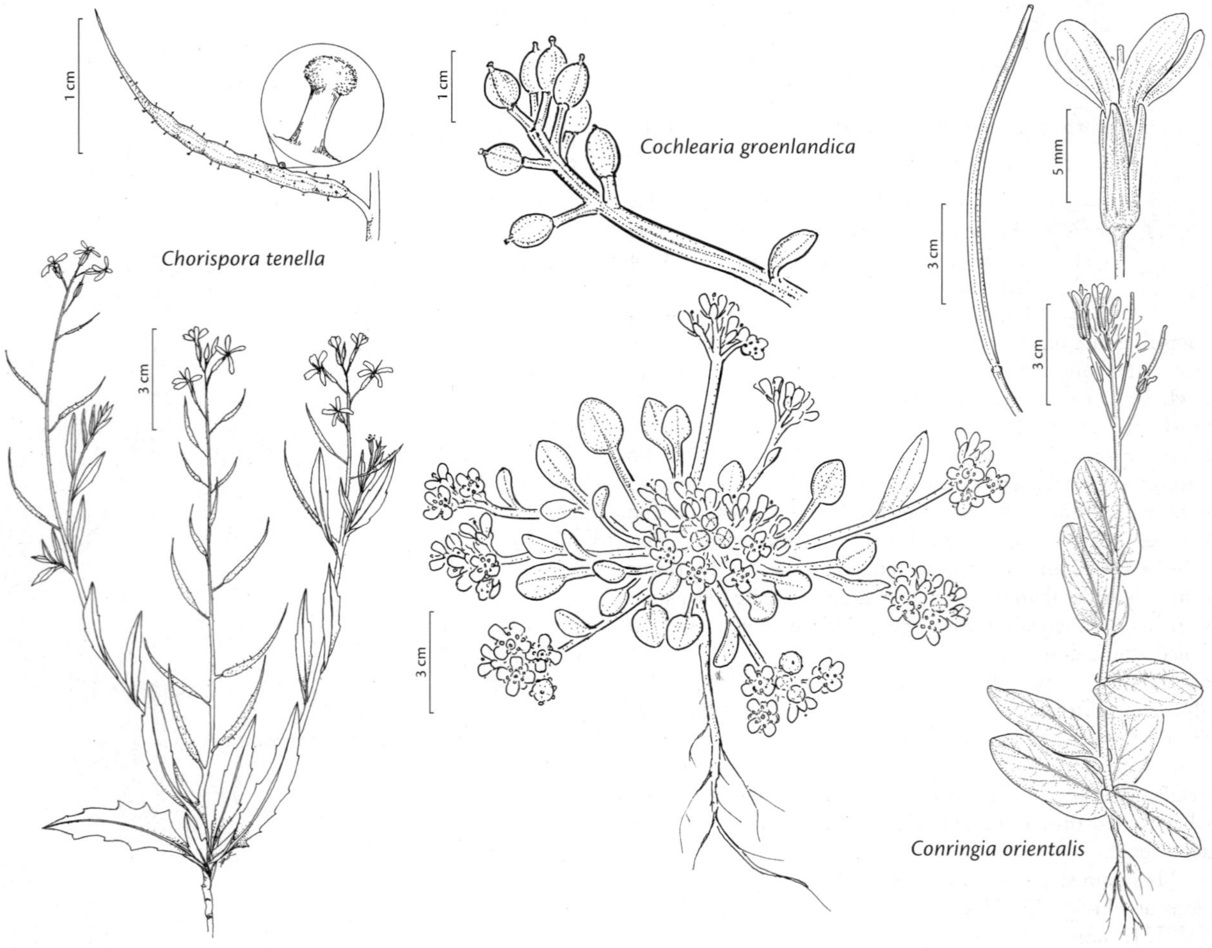

CHORISPORA • COCHLEARIA • CONRINGIA

COCHLEARIA Scurvygrass

Herbs annual, biennial, rarely perennial, glabrous; roots slender. **Trichomes** absent. **Basal leaves** often rosulate, simple, margins entire or dentate; petiolate. **Cauline leaves** simple, bases sometimes auriculate or amplexicaul, margins entire or dentate; petiolate or sessile. **Inflorescences** bracts 0. **Flowers** sepals ascending to spreading, caducous, lateral pair not saccate; petals oblanceolate, oblong, ligulate, or elliptic, white, claws undifferentiated from blades; nectar glands lateral, 1 on each side of lateral stamens; stamens 6, subequal or slightly tetradynamous; anthers ovate; septa complete, fenestrate, or absent; styles short or obsolete, stigmas entire. **Fruits** dehiscent silicles, terete, unsegmented; valves with distinct midveins, sometimes inflated; replums rounded; sessile. **Seeds** biseriate, ellipsoid or ovoid, papillose, not mucilaginous when wetted; wingless; cotyledons accumbent or rarely incumbent.

Asia, Europe, North America. 20 species, 1 species treated in *Flora.*

Cochlearia groenlandica Linnaeus (p. 470)

Greenland scurvygrass, Danish scurveygrass

Cochlearia officinalis L., *Cochlearia officinalis* L. var. *arctica* (Schltdl. ex DC.) Gelert

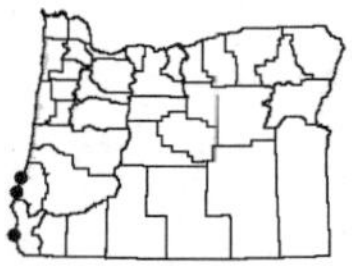

Stems erect to decumbent, (1)5–30(40) cm. **Basal leaves** deltoid to ovate, (3)7–20(25) × (2)5–20 mm, bases cuneate or occasionally truncate, margins entire, repand, or obscurely dentate; petioles (0.2)1–7(10) cm; middle and upper cauline leaves 4–20 × 1–15 mm, bases cuneate, not auriculate; short-petiolate to sessile. **Inflorescences** fruiting pedicels divaricate to ascending, rarely erect, (2)5–15(20) mm. **Flowers** sepals ovate or oblong, 1–2(3) mm; petals oblanceolate to spatulate, 2–4(5) × (0.5)0.8–2(3) mm; ovules 8–24 per ovary, styles 0.1–0.4 mm. **Fruits** ovoid to ellipsoid or obovoid, 3–5.5(7) × 2–4(5) mm. **Seeds** ovoid to subglobose, papillate, 1–1.5 × 0.8–1.3 mm. **2*n***=14.

Tidal flats, beaches, dunes, streambanks, peat hummocks, rocky beaches, mudflats. Flowering May–Aug. 0–50 m. Est. CA, WA; throughout most of Canada, AK, Greenland. Native.

In Oregon, this species appears to be restricted to Coos and Curry counties.

CONRINGIA Hare's-ear mustard

Herbs annual or biennial, usually glaucous. **Trichomes** absent. **Basal leaves** not rosulate, simple, margins usually entire; petiolate. **Cauline leaves** bases cordate to amplexicaul, rarely auriculate, margins entire; sessile. **Inflorescences** bracts 0. **Flowers** sepals erect to slightly ascending, caducous, oblong, forming tubular calyces, lateral pair slightly saccate; petals obovate to oblanceolate, > sepals, yellow or white, claws differentiated from blades or not; stamens 6, slightly tetradynamous, anthers oblong, not apiculate; ovules 10–50 per ovary, septa complete, styles distinct, stigmas capitate, entire or lobes opposite replum. **Fruits** dehiscent siliques, linear, 4-angled or terete, unsegmented; valves with prominent midveins, glabrous, torulose; replums rounded; sessile. **Seeds** uniseriate, oblong or ovoid, papillose, copiously mucilaginous when wetted; wingless; cotyledons incumbent.

Worldwide. 6 species; 1 species treated in *Flora.*

Conringia orientalis (Linnaeus) Dumortier (p. 470)

Hare's-ear mustard, hedge cabbage

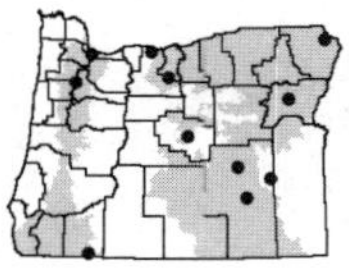

Stems simple, (1)3–7 dm. **Basal leaves** obovate or oblanceolate, 5–10 cm. **Cauline leaves** lanceolate, oblong, or rarely elliptic, (1.5)3.5–11(15) × (0.7)2–2.5(5) cm. **Inflorescences** fruiting pedicels, straight or curved upward, somewhat stout, (8)10–15(22) mm. **Flowers** sepals 6–8 mm; petals 7–12 × 2–3 mm, pale yellow, claws 6–8 mm, petal tips obtuse; styles, cylindric, 0.5–4 mm. **Fruits** strongly 4-angled to subterete, (5)8–14 cm × 2–2.5 mm, with prominent midvein, subtorulose. **Seeds** oblong, 2–2.9 × 1.2–1.5 mm. **2*n***=14.

Disturbed areas, roadsides, cultivated areas. Flowering May–Aug. 50–2000 m. BR, BW, Col, Sisk, WV. CA, ID, NV, WA; worldwide. Exotic.

CRAMBE Crambe, sea kale

Herbs rarely **subshrubs** perennial. **Trichomes** simple or absent. **Basal leaves** rosulate, margins pinnatisect, lyrate, or dentate; long-petiolate. **Cauline leaves** bases not auriculate; petiolate. **Inflorescences** often in panicles; bracts 0. **Flowers** sepals spreading to ascending, caducous, oblong or ovate, lateral pair not saccate; petals obovate or oblong, > sepals, white or rarely yellow, claws distinct; nectar glands 4, lateral; stamens 6, tetradynamous, lateral filaments wingless, unappendaged, median filaments winged, toothed, or rarely toothless, anthers ovate or oblong, obtuse; ovules 2 per ovary, septa obsolete, styles absent or rarely to 1 mm, stigmas capitate, entire. **Fruits** indehiscent, nutlet-like silicles, 2-segmented, basal segments slender, seedless, terminal segments caducous, 1(2)-seeded, globose or ovoid, terete or 4-angled, corky or woody, smooth, rugose, or reticulate. **Seeds** aseriate, pendulous on a basal funicle, globose, ovoid, or oblong, plump, not mucilaginous when wetted; wingless; cotyledons conduplicate.

Africa, Asia, Europe, North America. 35 species; 1 species treated in *Flora.*

Crambe maritima Linnaeus (p. 472)

Sea kale

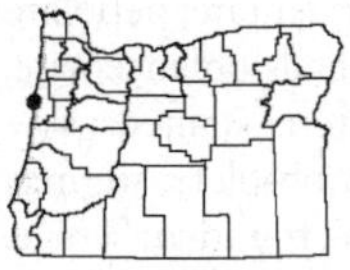

Plants succulent, glabrous, glaucous. **Stems** 20–50(75) cm, many-branched. **Leaves** basal blades ovate to oblong or elliptic, 10–42 × 7–30 cm, margins coarsely dentate or sinuate to subpinnatifid; petioles 4–16 cm. **Cauline leaves** margins sinuate-dentate to pinnately lobed. **Inflorescences** fruiting pedicels suberect to ascending or subdivaricate, (10)15–30(40) mm, stout. **Flowers** sepals broadly ovate, 3–4 mm; petals (6)8–12(15) × (4)5–7 mm, white to pale yellow; filaments 3–4 mm, median pairs with toothed wing; styles obsolete. **Fruits** lower segments 1–4 mm, slightly thicker than pedicels, terminal segments 1-seeded, subglobose to ovoid, thick, 7–14 × 6–8.2 mm, corky. **Seeds** globose or ovoid, 4–5(6) mm in diameter. ***2n***=30, 60.

Coastal beaches. Flowering May–Sep. 0–50 m. Est. CA; Africa, Europe. Exotic.

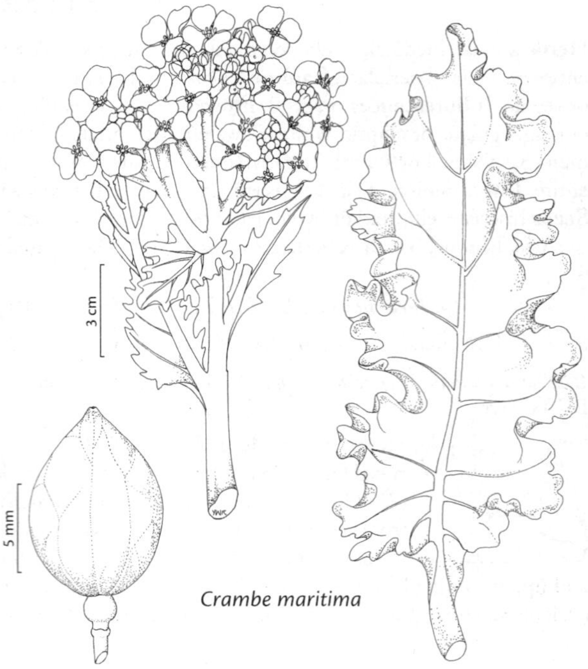

CRAMBE

CUSICKIELLA Cusickiella, Cusick's mustard

Herbs perennial, pulvinate; with woody caudices. **Trichomes** simple and stalked, forked or subdendritic. **Stems** scapose, leafless or few-leaved. **Basal leaves** rosulate, simple, margins entire, persistent; sessile. **Cauline leaves** absent or bract-like. **Inflorescences** bracts 0 or lowermost flowers with bracts. **Flowers** sepals erect to ascending, caducous, oblong, lateral pair not saccate; petals spatulate to oblanceolate, slightly > sepals, white or yellowish, short clawed; nectar glands confluent, lateral annular; stamens 6, slightly tetradynamous, anthers ovate; ovules 2–4 per ovary, styles distinct, stigmas capitate, entire. **Fruits** dehiscent silicles, ovoid to ellipsoid, terete or 4-angled, unsegmented; valves thick, leathery, with prominent or obscure midveins, sometimes keeled, not torulose, puberulent or glabrous. **Seeds** aseriate, ovoid to oblong, plump, not mucilaginous when wetted; wingless; cotyledons incumbent.

Western United States. 2 species; 1 species treated in *Flora.*

Cusickiella douglasii (A. Gray) Rollins (p. 472)

Douglas' draba, alkali cusickiella

Draba douglasii A. Gray

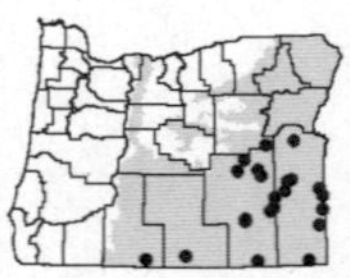

Plants puberulent with crisped trichomes. **Stems** 1–5 cm. **Basal leaves** oblanceolate to oblong, (3)5–12(14) × (0.5)1–2(2.5) mm, leathery, ciliate with simple and forked trichomes, midveins prominent, surfaces glabrous or puberulent. **Cauline leaves** absent. **Inflorescences** 2–20-flowered; bracts 0; fruiting pedicels divaricate or ascending, 2–6(8) mm. **Flowers** sepals 2.5–4 mm; petals spatulate, 3.5–6 × 1.5–2 mm, attenuate to claw-like base; styles (0.5)1–2(2.5) mm. **Fruits** terete, (2)3–5.5(7) × 2.5–3 mm; valves glabrous or puberulent, with obscure midvein. **Seeds** 2–3 × 1.5–2 mm.

Scree, serpentine ridges, rocky ridges, loose volcanic hillsides. Flowering Apr–Jun. 400–2200 m. BR, BW, ECas, Owy. CA, ID, NV, WA; southeast to UT. Native.

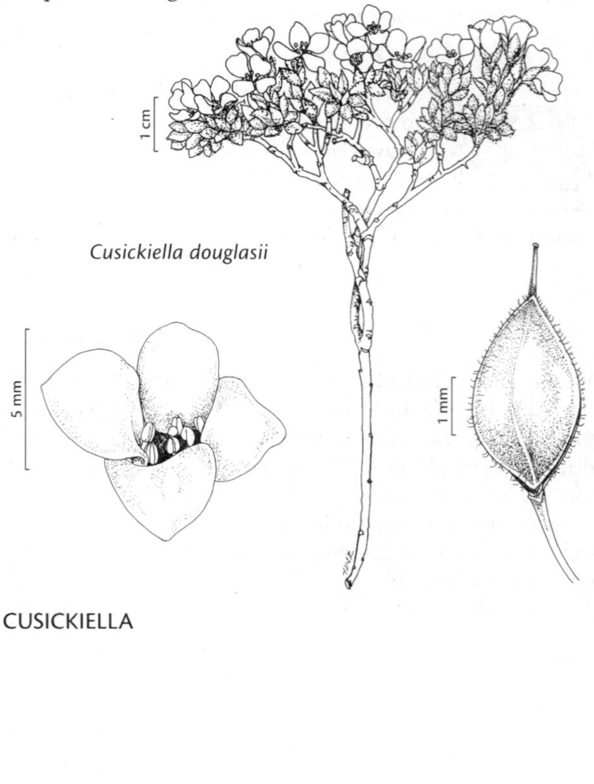

CUSICKIELLA

DESCURAINIA Tansymustard

Herbs annual or biennial, rarely **subshrubs** or **shrubs**. **Trichomes** short-stalked, dendritic, very rarely also simple, sometimes mixed with glandular, clavate papillae. **Basal leaves** rosulate, margins 1–3-pinnatisect; petiolate. **Cauline leaves** similar to basal, bases not auriculate. **Inflorescences** bracts present or not. **Flowers** sepals erect to spreading, ovate to oblong, lateral pair not saccate; petals obovate or oblanceolate, rarely oblong, < to > sepals, yellow, rarely creamy white, tips obtuse; nectar glands confluent, subtending bases of all stamens; stamens 6, tetradynamous, anthers oblong to ovate, tips obtuse; ovules 5–100 per ovary, septa complete or perforated, styles absent or distinct, stigmas capitate, entire. **Fruits** dehiscent siliques or silicles, linear, oblong, clavate, fusiform, or rarely ellipsoid or obovoid, terete, unsegmented; replums rounded. **Seeds** uniseriate or biseriate, oblong or ellipsoid, plump, mucilaginous when wetted; wingless; cotyledons incumbent.

Africa, Asia, Europe, North America, South America. 37 species; 8 species treated in *Flora*.

Descurainia is taxonomically confusing, and species readily hybridize. Many species have weedy tendencies and produce numerous seeds, which are readily dispersed by animals and machinery due to copious mucilage that exudes when wetted.

Key to *Descurainia* species:

1. Fruits fusiform, broadly ellipsoid, obovoid, clavate, or linear-clavate.
   - 2. Fruits clavate, rarely linear-clavate; ovules 16–40 per ovary; valve midveins prominent . . . . . ***D. pinnata*** ssp. ***brachycarpa***
   - 2' Fruits fusiform, broadly ellipsoid, clavate or obovoid; ovules 4–12 per ovary; valve midveins obscure.
     - 3. Fruits long-acute at both ends; styles (0.2)0.3–0.6(0.8) mm; stems (1.3)2–10.5(13.5) dm, simple at base, branched above . . . . . ***D. californica***
     - 3' Fruits obtuse apically, acute basally; styles 0.05–0.3 mm; stems (0.8)1.5–3.2(4.1) dm, branched at base and above . . . . . ***D. paradisa***

1' Fruits linear or narrowly linear.
   - 4. Fruits and fruiting pedicels appressed to rachis . . . . . ***D. incana***
   - 4' Fruits and fruiting pedicels not appressed to rachis.
     - 5. Fruit septum with a central band appearing as 2 or 3 veins; basal and lowermost cauline leaves 2- or 3-pinnatisect . . . . . ***D. sophia***
     - 5' Fruit septum not veined; basal and lowermost cauline leaves 1-pinnatisect.
       - 6. Fruits 3–8(10) mm; petals 0.7–1.2 mm; seeds 0.5–0.8 mm; plants eglandular . . . . . ***D. nelsonii***
       - 6' Fruits 8–20 mm; petals 1.7–2.8 mm; seeds 0.9–1.3 mm; plants glandular or eglandular.
         - 7. Ultimate segments of cauline leaves linear, entire; ovules 18–32 per ovary; fruits not torulose, with obscure midvein; fruiting pedicels (8)10–20(25) mm . . . . . ***D. longepedicellata***
         - 7' Ultimate segments of cauline leaves oblong to lanceolate, dentate, incised, or pinnatifid; ovules 12–24 per ovary; fruits torulose, with distinct midvein; fruiting pedicels (3)5–10(12) mm . . . . . ***D. incisa*** ssp. ***incisa***

Descurainia californica (A. Gray) O.E. Schulz (p. 475)

Sierra tansymustard, California tansymustard

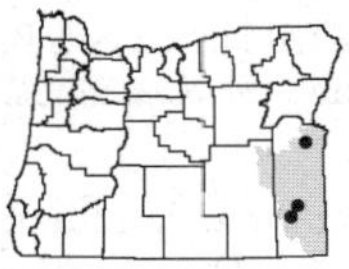

Plants annual or biennial, eglandular, sometimes glabrous distally. **Stems** erect, (1.3)2–10.5(13.5) dm, simple at base, branched above. **Basal leaves** 1.5–6 cm, pinnatisect; lateral lobes in 2–4(5) pairs, lanceolate, 5–22 × 1–5 mm, margins entire or crenate to incised or rarely lobed; petioles 0.4–4.2 cm. **Cauline leaves** upper smaller and with narrower ultimate leaf segments; sessile or short-petiolate. **Inflorescences** fruiting pedicels divaricate to ascending or suberect, slender, often straight, 3–9(11) mm. **Flowers** sepals spreading, oblong, 0.9–1.5 mm, yellowish; petals oblanceolate, 1.1–1.8 × 0.4–0.6 mm, yellow; ovules 4–12 per ovary, septa not veined, styles (0.2)0.3–0.6(0.8) mm. **Fruits** divaricate to erect, not appressed to rachis, terete, fusiform, (2)3–5(6) × (0.8)1–1.3 mm, long-acute at both ends; valves not torulose, midveins obscure. **Seeds** uniseriate, ellipsoid, 1–1.4 × 0.6–0.8 mm. **2*n*=14.**

Dry hillsides, disturbed areas, decomposed granite slopes, open woods, sagebrush, moist roadsides, gravel and talus slopes. Flowering May–Aug. 1100–1400 m. Owy. CA, ID, NV; east to WY, southeast to NM. Native.

Descurainia incana (p. 475)

(Bernhardi ex Fisch. & C.A. Meyer) Dorn

Mountain tansymustard

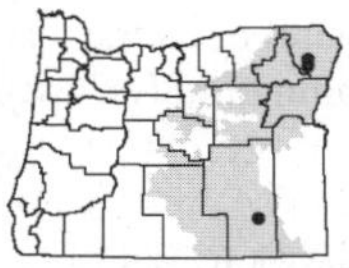

Plants biennial, eglandular or rarely glandular, finely pubescent with dendritic trichomes. **Stems** erect, (1.5)2.5–12 dm, simple below, often many-branched above. **Basal leaves** pinnate, 1.5–10(13) cm, lateral lobes pinnatifid, ultimate segments linear to oblong or narrowly lanceolate, 3–10(15) × 1–3(5) mm, margins entire; petioles 1–3.5(5.5) cm. **Cauline leaves** upper smaller and often with narrower ultimate leaf segments; sessile or short-petiolate. **Inflorescences** elongated in fruit; fruiting pedicels erect to erect-ascending, slender, straight, appressed to rachis, 2–8(11) mm. **Flowers** sepals erect, oblong, 1–1.8 mm, yellowish, sparsely pubescent; petals oblanceolate, 1.2–2 × 0.3–0.6 mm, yellow; ovules 14–22 per ovary, styles 0.1–0.4 mm. **Fruits** erect, appressed to rachis, terete, linear, (4)5–10(15) × 0.7–1.2(1.5) mm; valves glabrous, slightly torulose, midveins prominent. **Seeds** uniseriate, ellipsoid to narrowly oblong, 0.8–1.2 × 0.4–0.5 mm. **2*n*=14, 28.**

Alpine and subalpine areas, gravel and sandbars, scree, grassy slopes, prairies, steep rocky slopes, roadsides, disturbed sites, spruce-fir, pine, aspen, or sagebrush communities. Flowering May–Sep. 100–2800 m. BR, BW. CA, ID, NV, WA; throughout Canada and western US, CT, ME, MA. Native.

This species is easily distinguished from the other *Descurainia* in Oregon by having linear fruits appressed to the rachis.

Descurainia incisa (Engelmann) Britton ssp. *incisa*

Cutleaf tansymustard

Descurainia incana (Bernh. ex Fisch. & C.A. Mey.) Dorn ssp. *incisa* (Engelm. ex A. Gray) Kartesz & Gandhi, *Descurainia incisa* (Engelm.) Britton ssp. *viscosa* (Rydb.) Rollins, *Descurainia richardsonii* (Sweet) O.E. Schulz var. *incisa* (Engelm.) Detling, *Descurainia richardsonii* (Sweet) O.E. Schulz var. *sonnei* (B.L. Rob.) C.L. Hitchc., *Descurainia richardsonii* (Sweet) O.E. Schulz ssp. *viscosa* (Rydb.) Detling, *Descurainia richardsonii* (Sweet) O.E. Schulz var. *viscosa* (Rydb.) M. Peck

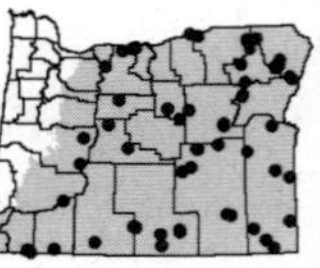

Plants annual, glandular or eglandular. **Stems** erect, (1.3)2–8.2(10.7) dm, often simple, branched above or occasionally throughout. **Basal leaves** 1-pinnatisect, lateral lobes (3)5–9 pairs, ultimate leaf segments oblong to lanceolate, margins dentate, incised, or pinnatifid; petioles 0.5–4.7 cm; **Cauline leaves** upper pubescent or glabrous; sessile or short-petiolate. **Inflorescences** fruiting pedicels ascending to divaricate, slender, straight, (3)5–10(12) mm. **Flowers** sepals erect to ascending, oblong to ovate, 1.6–2.4 mm, yellowish; petals narrowly oblanceolate, 1.7–2.8 × 0.5–0.9 mm, yellow; ovules 12–24 per ovary, septa not veined, styles 0.1–0.3 mm. **Fruits** erect to ascending, straight or slightly incurved inwards, terete, linear, 8–20 × 0.9–1.3 mm; valves glabrous, torulose, midveins distinct. **Seeds** uniseriate, oblong, 0.9–1.3 × 0.5–0.6 mm.

Disturbed sites, roadsides, streambanks, meadows, sagebrush and juniper communities, open woods, rocky cliffs, sandy areas, talus slopes. Flowering Apr–Jul. 0–2800 m. BR, BW, Casc, Col, ECas, Lava, Owy, Sisk. CA, ID, NV, WA; north to Yukon, northeast to MT, east to WY, southeast to TX, south to Mexico. Native.

Subspecies *incisa* differs from ssp. *paysonii* (southwestern United States) by having ascending to divaricate (vs. divaricate to horizontal) fruiting pedicels (3)5–10(12) mm (vs. (10)13–25(30) mm) and (3)5–9 pairs (vs. 2 or 3(4) pairs) of lateral leaf lobes that are coarsely dentate to incised or pinnatifid (vs. usually entire).

Descurainia longepedicellata (E. Fournier) O.E. Schulz

Western tansymustard, thread-stalk cutleaf tansymustard

Descurainia incisa (Engelm.) Britton ssp. *filipes* (A. Gray) Rollins, *Descurainia incisa* (Engelm.) Britton var. *filipes* (A. Gray) N.H. Holmgren, *Descurainia pinnata* (Walter) Britton ssp. *filipes* (A. Gray) M. Peck, *Descurainia pinnata* (Walter) Britton var. *filipes* (A. Gray) Detling

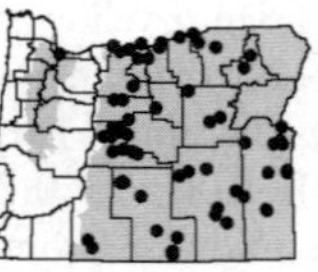

Plants annual, eglandular or rarely glandular, not canescent, often glabrous distally. **Stems** (1.5)3–6.2(8.5) dm. **Basal leaves** 1.5–7 cm, 1-pinnatisect, ultimate leaf segments entire; petioles 0.4–3.5 cm. **Cauline leaves** upper glabrous or rarely pubescent. **Inflorescences** fruiting pedicels horizontal to divaricate, slender, straight, (8)10–20(25) mm. **Flowers** sepals ascending, oblong, 1.5–2 mm, yellow; petals narrowly oblanceolate, 1.7–2.6 × 0.5–1 mm, yellow; ovules 18–32 per ovary, septa not veined, styles 0.1–0.2 mm. **Fruits** erect, straight or slightly curved inwards, terete, linear, (9)12–17 × 0.8–1.1 mm; valves glabrous, not torulose, midveins obscure. **Seeds** uniseriate, oblong, 1–1.3 × 0.6–0.7 mm. **2*n***=14.

Sandy plains and banks, sagebrush and juniper or pine communities, dry washes, open hillsides, grasslands. Flowering Apr–Jul. 0–1600 m. BR, BW, Col, ECas, Lava, Owy, WV. ID, NV, WA; north to British Columbia, northeast to MT, east to WY, southeast to AZ. Native.

Descurainia nelsonii (Rydberg) Al-Shehbaz & Goodson

Nelson's tansymustard

Descurainia pinnata (Walter) Britton ssp. *nelsonii* (Rydb.) Detling, *Descurainia pinnata* (Walter) Britton var. *nelsonii* (Rydb.) M. Peck

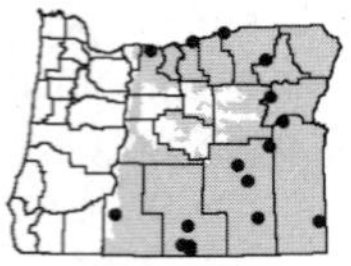

Plants annual, eglandular, not canescent, sometimes glabrous distally. **Stems** (0.7)0.9–3.2(4.5) dm. **Basal leaves** 0.8–2.5 cm, 1-pinnatisect, lateral leaflets 2–5 pairs, pinnatifid or dentate, ultimate leaf segments dentate or entire; petioles 0.5–1.5 cm. **Cauline leaves** upper smaller and often with narrower ultimate leaf segments; sessile or short-petiolate. **Inflorescences** fruiting pedicels divaricate-ascending, often at a 20–45° angle, straight, (1.5)2.5–7(10) mm. **Flowers** sepals ascending, oblong, 0.7–1.2 mm, yellowish; petals narrowly oblanceolate, 0.7–1.2 × 0.2–0.4 mm, yellow; ovules 6–12 per ovary, septa not veined, styles 0.1–0.2 mm. **Fruits** erect or ascending, terete, linear, 3–8(10) × 0.7–1 mm; valves slightly torulose or not, midveins distinct. **Seeds** uniseriate, oblong, 0.5–0.8 × 0.4–0.5 mm. **2*n***=14.

Gravelly grounds, sagebrush, wash bottoms, roadsides, silty flats. Flowering May–Jul. 50–2200 m. BR, BW, Col, ECas, Owy. CA, ID, NV, WA; north to British Columbia, northeast to MT, east to WY. Native.

Descurainia paradisa (A. Nelson & P.B. Kennedy) O.E. Schulz

Paradise tansymustard, Nevada tansymustard

Descurainia paradisa (A. Nelson & P.B. Kenn.) O.E. Schulz ssp. *paradisa*, *Descurainia paradisa* (A. Nelson & P.B. Kenn.) O.E. Schulz var. *paradisa*, *Descurainia pinnata* (Walter) Britton ssp. *paradisa* (A. Nelson & P.B. Kenn.) Detling, *Descurainia pinnata* (Walter) Britton var. *paradisa* (A. Nelson & P.B. Kenn.) M. Peck

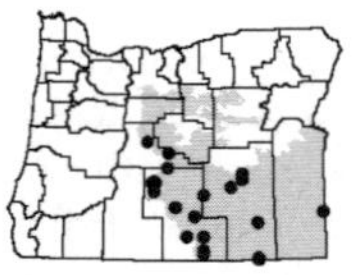

Plants annual, glandular or eglandular. **Stems** erect, (0.8)1.5–3.2(4.1) dm, often purplish, branched basally and above. **Basal leaves** and lowermost cauline soon withered, 1.5–3 cm, pinnately lobed, ultimate leaf segments oblong to linear or lanceolate, 1–5 × 0.3–1 mm, margins entire or dentate; petioles 0.3–1.3 cm. **Cauline leaves** upper smaller and often with narrower ultimate leaf segments; sessile. **Inflorescences** fruiting pedicels divaricate to ascending, straight, 2.5–7(9) mm. **Flowers** sepals spreading to ascending, oblong, 0.8–1.2 mm, pale yellow; petals oblanceolate, 0.9–1.3 × 0.2–0.5 mm, yellow; ovules 4–10 per ovary, septa not veined, styles 0.05–0.3 mm. **Fruits** divaricate to erect, not appressed to rachis, terete, obovoid to clavate, rarely broadly ellipsoid, 2–5 × 1–2 mm, bases acute, tips obtuse; valves glabrous, not torulose, midveins obscure. **Seeds** uniseriate or biseriate, oblong, 0.8–1.2 × 0.5–0.6 mm.

Shrub communities, sandy washes and dunes, roadsides. Flowering May–Jul. 1100–1500 m. BR, Lava, Owy. CA, NV. Native.

Descurainia pinnata (Walter) Britton ssp. ***brachycarpa*** (Richardson) Detling

Short-fruit western tansymustard, salty tansymustard, intermediate tansymustard

Descurainia pinnata (Walter) Britton var. *brachycarpa* (Richardson) Fern., *Descurainia pinnata* (Walter) Britton var. *halictorum* (Cockerell) M. Peck, *Descurainia pinnata* (Walter) Britton ssp. *intermedia* (Rydb.) Detling, *Descurainia pinnata* (Walter) Britton var. *intermedia* (Rydb.) C.L. Hitchc.

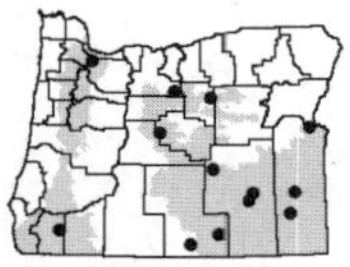

Plants annual, usually glandular, pubescent throughout, not canescent. **Stems** (0.8)1.3–5.7(9.2) dm, simple basally. **Basal leaves** and lowermost cauline 1–15 cm, 1(2)-pinnatisect, lateral leaflets 4–9 pairs, dentate to pinnatifid, ultimate leaf division linear or oblanceolate to ovate, margins entire or dentate to lobed; petioles 0.5–3.6 cm; **Cauline leaves** upper smaller, with narrower lobes. **Inflorescences** fruiting pedicels ascending, divaricate, horizontal, or descending, at 20–60(80)° angle, (7)10–18(23) mm. **Flowers** sepals spreading to ascending, oblong, 1.5–2.6 mm, yellow; petals narrowly oblanceolate, (1.7)2–3 × 0.3–1 mm, yellow; ovules 16–40 per ovary, septa not veined, styles obsolete or to 0.2 mm. **Fruits** erect to ascending, terete, clavate or rarely linear-clavate, 4–13(17) × 1.2–2.2 mm, wider distally; valves glabrous, not torulose, midveins prominent. **Seeds** biseriate, oblong, 0.6–0.9 × 0.4–0.5 mm. $2n$=14, 28.

Roadsides, gravel washes, dry slopes, sagebrush, pinyon-juniper communities, streamsides, railroad tracks and embankments, limestone ledges, dry washes, foothills, cliffs, prairies, sandy ground. Flowering Apr–Jun. 0–2000 m. BR, Lava, Owy, Sisk, WV. WA; throughout Canada and US except southeastern and southwestern US. Native.

Descurainia pinnata ssp. *brachycarpa* is easily distinguished from ssp. *glabra*, which is distributed in AZ, CA, NV, NM, UT, and Mexico, by having a pubescent and glandular (vs. glabrous and eglandular) rachis of the raceme. It is distinguished from ssp. *ochroleuca* (Wooton) Detling (southwestern US and Mexico) by being glandular and not canescent (vs. eglandular and canescent) with yellow (vs. purple) sepals, and basally simple (vs. basally branched) stems, and from ssp. *pinnata* (southeastern US) by having yellow (vs. rose) sepals, petals (1.7)2–3 mm (vs. 1–1.8 mm), and fruiting pedicels (7)10–18(23) mm held at a 20–60(80)° angle (vs. 4–14(17) mm held at a (60)70–90(110)° angle).

Descurainia sophia (Linnaeus) Webb ex Prantl

Flixweed, herb Sophia, flixweed tansymustard

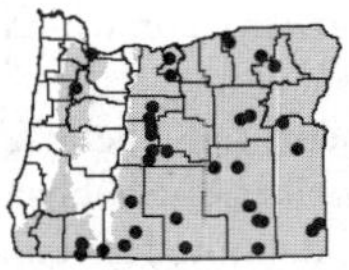

Plants annual, eglandular, pubescent or sometimes glabrous distally. **Stems** (1)2–7(10) dm. **Basal leaves** to 15 cm, 2- or 3-pinnatisect, ultimate leaf segments linear or oblong, to 10 × 2 mm, margins entire; petioles 0.1–2(3) cm. **Cauline leaves** upper smaller; sessile or short-petiolate. **Inflorescences** fruiting pedicels divaricate to ascending, (5)8–15(20) mm. **Flowers** sepals erect to ascending, oblong, 1.8–2.8 mm, yellowish; petals narrowly oblanceolate, 2–3 × 0.4–0.6 mm, yellow; ovules 20–48 per ovary, septa with a broad central longitudinal band appearing as 2 or 3 veins, styles 0.05–0.2 mm. **Fruits** divaricate-ascending to erect, not appressed to rachis, terete, straight or curved upward, narrowly linear, (12)15–27(30) × 0.5–0.8(1) mm; valves glabrous, torulose, midveins prominent. **Seeds** uniseriate, oblong, 0.7–1.3 × 0.3–0.6 mm. $2n$=28.

Roadsides, waste places, disturbed sites, railroad embankments, hillsides, mountain slopes, canyon bottoms, riverbanks, fields, lawns, deserts, sagebrush and pinyon-juniper communities. Flowering Apr–Jul. 0–2300 m. BR, BW, Col, ECas, Lava, Owy, Sisk, WV. CA, ID, NV, WA; throughout Canada and US; Africa, Asia, Europe. Exotic.

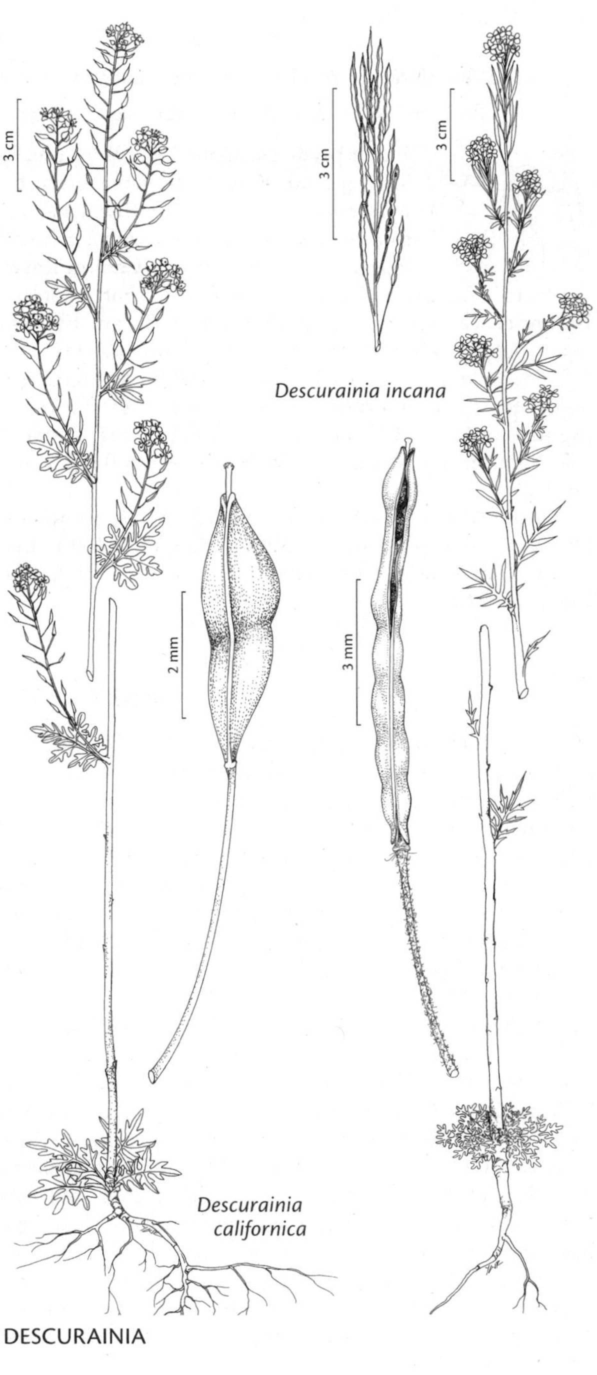

DESCURAINIA

DIPLOTAXIS Wallrocket

Herbs annual, biennial, or perennial. **Trichomes** simple. **Stems** leafy or leafless, erect or ascending. **Basal leaves** rosulate or not, margins dentate to pinnatisect; petiolate. **Cauline leaves** bases not auriculate; petiolate or sessile. **Inflorescences** racemes; bracts 0, rarely with bracts at base. **Flowers** sepals ascending, or spreading, caducous, oblong, lateral pair not saccate; petals obovate, yellow, white, or purple, tips rounded or truncate; nectar glands 2 lateral and 2 median; stamens 6, tetradynamous, anthers oblong to ovate, obtuse; ovules 12–276 per ovary, septa complete, styles obsolete or distinct, stigmas capitate or somewhat decurrently 2-lobed. **Fruits** dehiscent siliques, linear, latiseptate, 2-segmented; valvular segments usually with numerous seeds, torulose, distinctly 1-veined, distal segments seedless or 1- or 2-seeded; replums rounded. **Seeds** biseriate, ovoid to ellipsoid, mucilaginous or not when wetted; wingless; cotyledons conduplicate.

Africa, Asia, Europe, North America. 25 species; 1 species treated in *Flora*.

Diplotaxis muralis has been documented in Linnton (Multnomah County), on ballast.

Diplotaxis tenuifolia (Linnaeus) de Candolle (p. 476)

Perennial wallrocket, slimleaf wallrocket

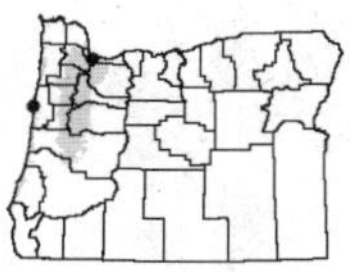

Plants perennial. **Stems** 2–7(10) dm, usually leafy, glabrate. **Basal leaves** not rosulate, elliptic to obovate, 2–15 × 1–6(8) cm, margins sinuate to deeply pinnatifid, with 2–5 lateral lobes on each side; petiolate. **Cauline leaves** similar to basal, upper with narrower segments; short-petiolate. **Inflorescences** fruiting pedicels divaricate, 8–35 mm. **Flowers** sepals oblong, 4–6 mm; petals broadly obovate, 7–11(13) × 5–8 mm, yellow, tips rounded; ovules 20–32(46) per ovary; gynophores 0.5–3 mm. **Fruits** 2–5 cm × 1.5–2.5 mm, ascending, slightly compressed, somewhat torulose, terminal segment 1.5–3 mm, seedless, stout, beak-like. **Seeds** 1–1.3 × 0.6–0.9 mm. **2*n***=22.

Disturbed areas, sandy beaches, muddy shores, wet woods, roadsides, mountain slopes. Flowering Mar–Jul. 0–100 m. Est, WV. CA, ID; widely scattered in North America; Africa, Asia, Europe. Exotic.

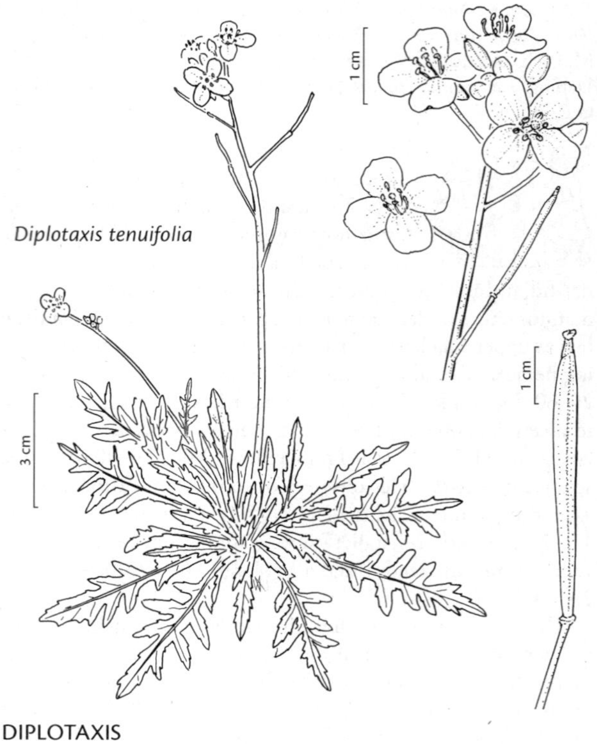

DIPLOTAXIS

DRABA Whitlow-grass, draba

Herbs annual, biennial, or perennial, rarely subshrubs; with caudices. **Trichomes** simple, forked, cruciform, stellate, malpighiaceous, or dendritic, stalked or sessile. **Basal leaves** usually rosulate, simple, margins entire or toothed; usually petiolate. **Cauline leaves** sometimes absent, bases rarely auriculate, margins entire or dentate; petiolate or sessile. **Inflorescences** racemes; bracts present or not. **Flowers** chasmogamous; sepals caducous or persistent, ovate to oblong, not saccate; petals > or < sepals, yellow, white, pink, purple, orange, or rarely red, claws differentiated or not from blades, petal tips obtuse, emarginate, or rarely bifid; nectar glands distinct or confluent and subtending bases of stamens; stamens 6, tetradynamous, anthers ovate or oblong, obtuse; ovules 4 to numerous per ovary, septa complete, styles distinct or obsolete, stigmas capitate, usually entire. **Fruits** dehiscent, silicles or siliques, ovate, lanceolate, elliptic, oblong, linear, orbicular, ovoid, or globose, latiseptate or terete, unsegmented; replums rounded. **Seeds** 4+ per fruit, biseriate, oblong, ovate, or orbicular, flattened, not mucilaginous when wetted; wingless or rarely winged; cotyledons accumbent.

Northwestern Africa, Asia, Europe, North America, South America. ~400 species; 16 species treated in *Flora*.

Draba is the largest and most complex genus in Brassicaceae. Complexity can be attributed to one or more factors (e.g., aneuploidy, apomixis, hybridization, polyploidy) and the voluminous literature covering these aspects (see Al-Shehbaz et al. 2010). As indicated in the family introduction, seed/ovule number is extremely useful taxonomically in the genus, as is flower color, which should be recorded in the field rather than from dried specimens.

Recent molecular phylogenetic studies (Jordon-Thaden et al. 2010) and subsequent nomenclatural adjustments (Al-Shehbaz 2012) have clearly shown that the earlier delimitations of *Draba*, including the account for the *Flora of North America* (Al-Shehbaz et al. 2010), recognize a polyphyletic concept that included species from somewhat remotely related genera. As delimited herein, however, *Draba* includes species that produce uniform flowers throughout the season, whereas both *Abdra* and *Tomostima* produce chasmogamous and petaliferous flowers early in the season and cleistogamous and apetalous flowers in mid or late season.

Trichome morphology is very important in the determination of *Draba* species. Species that are morphologically similar can have very different trichome morphology and chromosome numbers.

Key to *Draba* species:

1. Plants densely cespitose or pulvinate, long-lived perennial; at least some caudex branches produce sterile rosettes.
    2. Basal leaves with prominent abaxial midvein.
        3. Leaves glabrous or pubescent with simple and 2–4-rayed trichomes; fruits flattened . . . . ***D. densifolia***
        3' Leaves pubescent with pectinate trichomes, sometimes glabrous abaxially; fruits inflated at least basally . . . . ***D. oligosperma***
    2' Basal leaves with obscure abaxial midvein.
        4. Styles (1)1.6–3 mm; petals 5.5–8 mm; cauline leaves 1–4, rarely absent . . . . ***D. howellii***
        4' Styles 0.2–1 mm; petals 2–6 mm; cauline leaves absent, rarely 1.
            5. Leaves abaxially glabrous or with simple and 2-rayed trichomes . . . . ***D. cyclomorpha***
            5' Leaves abaxially with 2–6-rayed trichomes.
                6. Petals 4–6 mm; fruits flattened, 5–9(11) mm, glabrous or puberulent with simple and 2-rayed trichomes . . . . ***D. incerta***
                6' Petals 2–4 mm; fruits basally inflated, (2.5)3–4(5) mm, densely pubescent with 2–6-rayed trichomes . . . . ***D. novolympica***

1' Plants neither pulvinate nor densely cespitose, annual or short-lived perennial; caudex branches, when present, do not produce sterile rosettes.
    7. Petals deeply bifid; rachises flexuous . . . . ***D. verna***
    7' Petals obtuse or slightly emarginate (but deeply notched in *D. nemorosa*); rachises straight or rarely slightly flexuous.
        8. Petals white.
            9. Inflorescences 3–9-flowered; ovules 16–24(28) per ovary . . . . ***D. lonchocarpa***
            9' Inflorescences (5)8–47(63)-flowered; ovules (24)28–52 per ovary.
                10. Inflorescences with basal bracts; basal leaves with 4–12-rayed trichomes . . . . ***D. cana***
                10' Inflorescences without bracts; basal leaves with 3–6-rayed and fewer simple trichomes . . . . ***D. praealta***
        8' Petals yellow (fading to white in *D. crassifolia*).
            11. Petals 3.7–6 mm; ovules 10–20 per ovary; styles 0.4–2 mm; rachises of inflorescences pubescent with 2–4-rayed trichomes or hirsute.
                12. Cauline leaves 5–33; styles (0.6)1–2 mm . . . . ***D. aureola***
                12' Cauline leaves absent; styles 0.4–0.8(1) mm . . . . ***D. cusickii***
            11' Petals (1.2)1.7–3.5 mm; ovules 16–72 per ovary; styles 0–0.2 mm; rachises of inflorescences glabrous (or rarely with 2–4-rayed trichomes in *D. stenoloba*).
                13. Cauline leaves 4–12(15); fruiting pedicels 7–28(35) mm, (1.5)2–7 × as long as fruit . . . . ***D. nemorosa***
                13' Cauline leaves absent or 1–3(5); fruiting pedicels (1)3–14(20) mm, ≤ 1 × as long as fruit.
                    14. Leaves abaxially glabrous or with some simple and 2-rayed trichomes; lower stems glabrous or rarely sparsely pubescent with simple trichomes . . . . ***D. crassifolia***
                    14' Leaves abaxially pubescent with 2–4-rayed and sometimes simple trichomes; lower stems with simple and 2-rayed trichomes.
                        15. Lower stems and adaxial leaf surfaces with simple and 2–4-rayed trichomes; fruits (4)6–12(15) mm . . . . ***D. albertina***
                        15' Lower stems and adaxial leaf surfaces with 2–4-rayed trichomes; fruits (8)10–17(20) mm . . . . ***D. stenoloba***

Draba albertina Greene

Slender draba, Alaska whitlow-grass

Draba stenoloba Ledeb. var. *nana* (O.E. Schulz) C.L. Hitchc.

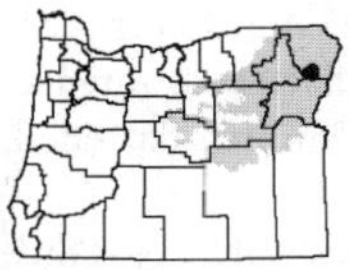

Plants annual or short-lived perennial, rarely scapose; caudices slender, without sterile rosettes. **Stems** (2.5)5–30(42) cm, lower with simple and 2-rayed trichomes, upper glabrous. **Basal leaves** rosulate, obovate to oblanceolate or linear-lanceolate, (3)10–28(35) × (1)2–6(9) mm, margins entire or denticulate, ciliate with simple trichomes, surfaces abaxially with stalked, 2–4-rayed trichomes, rarely with simple trichomes along midvein, adaxially with simple and 2-rayed trichomes, rarely glabrous; sessile or short-petiolate. **Cauline leaves** 1–3(5), rarely absent, lanceolate to elliptic or ovate, margins entire or denticulate; sessile. **Inflorescences** (2)6–30(50)-flowered, rachises straight or rarely flexuous, glabrous; fruiting pedicels divaricate-ascending or horizontal, straight or curved upward, (3)5–14(20) mm, ≈ or < fruit. **Flowers** sepals ovate, 1.4–2.1 mm; petals spatulate to oblanceolate, 2–3.2 × 0.7–1.2 mm, yellow, tips obtuse or slightly emarginate; ovules (20)24–38(44) per ovary, styles 0.01–0.12 mm. **Fruits** lanceolate to elliptic or linear, flattened, untwisted, (4)6–12(15) × (1)1.4–2.1 mm, glabrous. **Seeds** oblong, flattened, 0.7–1 × 0.4–0.5 mm; wingless. **2*n***=24.

Rocky knolls, meadows, open woodlands, pine forests, alpine slopes, disturbed areas, streambanks. Flowering Jun–Aug. 1100–2800 m. BW. CA, ID, NV, WA; north to AK, northeast to Northwest Territories, east to WY, southeast to CO. Native.

Draba aureola S. Watson (p. 478)

Great alpine whitlow-grass, Mt. Lassen draba, golden alpine draba

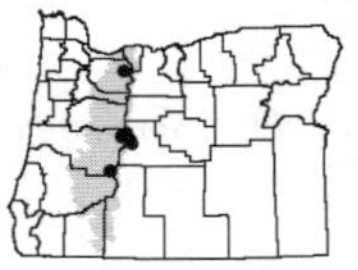

Plants short-lived perennial; caudices simple or few-branched. **Stems** 2.5–15 cm, hirsute throughout with simple and long-stalked, 2–4-rayed trichomes. **Basal leaves** rosulate, numerous, forming dense clusters, oblanceolate to linear, (7)10–25(32) × 2–3.5(5) mm, margins entire, ciliate with simple and 2-rayed trichomes, surfaces abaxially hirsute with stalked, 3–5-rayed trichomes, adaxially with simple and long-stalked, 2-rayed and fewer, smaller, 3–5-rayed trichomes; sessile. **Cauline leaves** 5–33, oblong to linear, margins entire; sessile. **Inflorescences** 12–83-flowered, rachises straight, hirsute; fruiting pedicels horizontal to divaricate, straight or rarely curved upward, (3)5–12(19) mm; bracts 0 or lowermost 1–9 flowers with bracts. **Flowers** sepals ovate, 2.5–4 mm, glabrous; petals linear to linear-oblanceolate, 4–6 × 0.5–1(1.2) mm, bright yellow, tips obtuse or slightly emarginate; ovules 10–20 per ovary, styles (0.6)1–2 mm. **Fruits** oblong or oblong-ovate, flattened, untwisted, (6)9–14(16) × 3–5 mm, pubescent with stalked, 2–4-rayed trichomes. **Seeds** oblong, flattened, 1.4–1.9 × 0.8–1.1 mm; wingless. ***2n***=20.

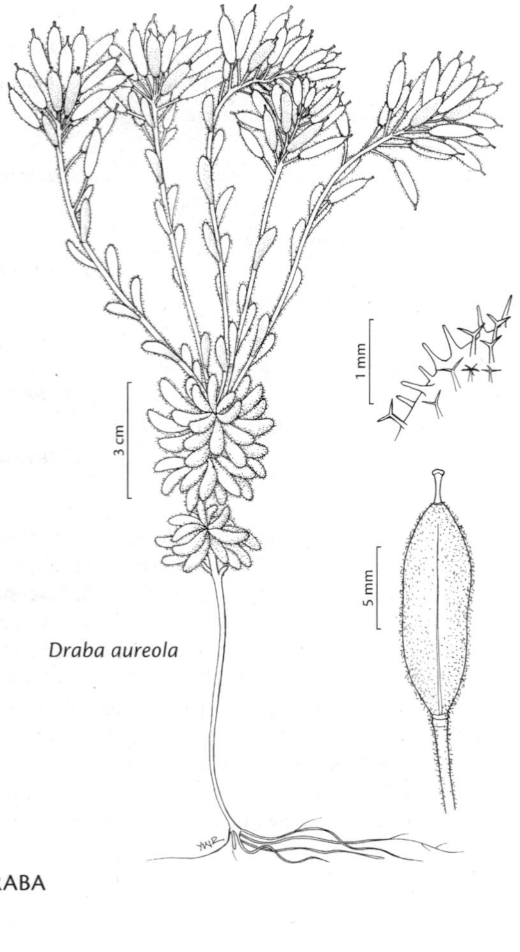

Draba aureola

DRABA

Open conifer forests , talus slopes, alpine meadows and moraines. Flowering Jul–Aug. 2300–3100 m. Casc. CA, WA. Native.

Draba cana Rydberg

Lance-leaved draba, canescent draba, Brewer's draba, cushion draba, hoary whitlow-grass

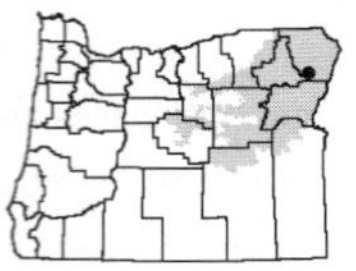

Plants perennial; with simple or short-branched caudices. **Stems** (6)10–30(38) cm, pubescent proximally with simple and 4–10-rayed trichomes. **Basal leaves** rosulate, linear to oblanceolate or oblong, (5)8–20(35) × 1.5–4(11) mm, margins entire or dentate, pubescent with short-stalked, 4–12-rayed trichomes; petiolate. **Cauline leaves** 3–10(17), not overlapping, lanceolate to ovate or oblong, margins entire or dentate, surfaces pubescent as basal leaves; sessile. **Inflorescences** (10)15–47(63)-flowered, rachises straight; fruiting pedicels suberect or ascending. **Flowers** sepals oblong to ovate, 1.5–2 mm; petals oblanceolate to spatulate, 2.3–3.7(4.5) × 0.7–1.7 mm, white, obtuse or slightly emarginate; ovules 28–48 per ovary, styles 0.1–0.6 mm. **Fruits** subappressed to rachis, linear-lanceolate to linear or very rarely ovate-oblong, flattened, slightly twisted or plane, (5)6–11 × 1.5–2(2.5) mm, pubescent with short-stalked, 3–7-rayed trichomes. **Seeds** ovoid, flattened, 0.5–0.7(0.9) × 0.3–0.5 mm; wingless. ***2n***=32.

Rock outcrops, talus, open prairie benchlands, roadsides, meadows. Flowering Jun–Aug. 2300–2500 m. BW. CA, ID, NV, WA; north to AK, throughout Canada, northern and western US, Greenland. Native.

This species is easily distinguished from the other *Draba* species in Oregon by a combination of white flowers, basally bracteate inflorescences, and an indumentum of 4–10-rayed trichomes.

Draba crassifolia Graham

Snowbed draba, Rocky Mountain draba, thick-leaved draba

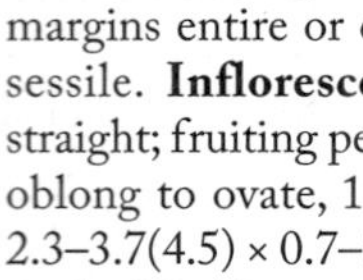

Plants annual or short-lived perennial; rarely with branched caudices. **Stems** (1)3–11(15) cm, lower glabrous or rarely sparsely pubescent with simple trichomes. **Basal leaves** rosulate, oblanceolate to obovate, (2)5–25(30) × (1)2–4(6) mm, margins entire or denticulate, sometimes ciliate, surfaces glabrous to sparsely pubescent with simple and 2-rayed trichomes; petioles ciliate. **Cauline leaves** absent or very rarely 1, oblong to ovate, margins entire. **Inflorescences** (2)4–15(25)-flowered, rachises slightly flexuous or straight, glabrous; fruiting pedicels horizontal to divaricate-ascending, straight or rarely curved upward, 3–8(10) mm, glabrous, ≈ or < fruit; bracts 0. **Flowers** sepals ovate, 1–2 mm, green or purplish, glabrous; petals oblanceolate, 1.5–2.5(3) × 0.5–1 mm, yellow, fading to white, tips obtuse or slightly emarginate; ovules 16–24(30) per ovary, styles to 0.1 mm. **Fruits** narrowly elliptic, lanceolate, or linear-lanceolate, flattened, untwisted, (3)5–10 × 1.5–2.5 mm, glabrous. **Seeds** elliptic, flattened, 0.7–0.8 × 0.4–0.5 mm; wingless. ***2n***=40.

Subalpine meadows, alpine summits, rock outcrops, talus, bare snow-melt areas. Flowering Jun–Sep. 1200–2800 m. BR. CA, ID, NV, WA; northern Canada, western Canada and US except CA; Europe. Native.

Draba cusickii B.L. Robinson ex O.E. Schulz

Steens Mountain whitlow-grass, Cusick's draba

Draba cusickii B.L. Rob. ex O.E. Schulz var. *cusickii*, *Draba sphaeroides* Payson, *Draba sphaeroides* Payson var. *cusickii* (B.L. Rob. ex O.E. Schulz) C.L. Hitchc.

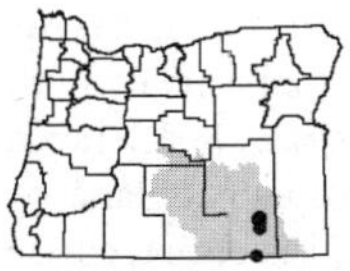

Plants perennial, loosely cespitose, scapose; caudices few- to several-branched. **Stems** few to many from caudices, 3–12 cm, unbranched, pubescent throughout with (2)–4-rayed trichomes. **Basal leaves** rosulate, oblanceolate to obovate, 4–12 × 2–5 mm, margins entire or rarely denticulate, surfaces pubescent with stalked, (2)4-rayed trichomes; short-petiolate, ciliate with 2-rayed trichomes only at petiole bases. **Cauline leaves** absent. **Inflorescences** 10–20(25)-flowered, rachises straight, pubescent as lower stems; fruiting pedicels divaricate-ascending, straight, 4–9 mm, pubescent with (2)4-rayed trichomes, not expanded at base; bracts 0. **Flowers** sepals broadly ovate, 2–2.5 mm, with short-stalked, 3- or 4-rayed trichomes; petals obovate to oblanceolate, 3.7–6 × 1.5–2.5 mm, yellow, tips obtuse or slightly emarginate; ovules 10–18 per ovary, styles 0.4–0.8(1) mm. **Fruits** lanceolate to broadly ovate, flattened, untwisted, 6–9 × 2–4.5 mm, pubescent with short-stalked, (2)4-rayed trichomes. **Seeds** oblong, flattened, 1.2–1.5 × 0.7–0.8 mm; wingless. **2*n***=26.

Basaltic outcrops, talus. Flowering Jun–Sep. 1200–3000 m. BR. NV. Native.

Draba cyclomorpha Payson

Lemmon's draba, granite draba

Draba lemmonii S. Watson, *Draba lemmonii* S. Watson var. *cyclomorpha* (Payson) O.E. Schulz

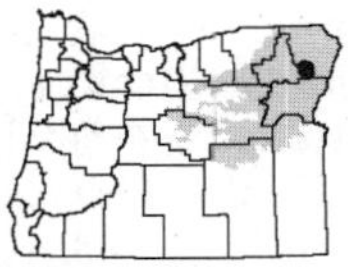

Plants perennial, cespitose, not pulvinate, scapose; caudex branches many, some terminated in sterile rosettes. **Stems** 1–6(10) cm, unbranched, pilose with simple and stalked, 2(3)-rayed, often-crisped trichomes. **Basal leaves** rosulate, oblanceolate to obovate, 4–10 × 2–5 mm, margins entire, midveins obscure, surfaces abaxially glabrous or with stalked, 2-rayed and fewer simple trichomes, adaxially glabrous or hirsute with simple trichomes; petioles short, margins ciliate with simple trichomes. **Cauline leaves** absent. **Inflorescences** 5–14(20)-flowered, rachises straight, pubescent; fruiting pedicels horizontal to divaricate-ascending, curved upward or straight, 3–8(10) mm, sparsely pilose; bracts 0. **Flowers** sepals caducous, ovate, 1.5–2.2 mm, pubescent with simple and short-stalked 2-rayed trichomes; petals caducous, spatulate, 4–5.5 × 1.5–2.5 mm, yellow; ovules 8–12 per ovary, styles 0.3–1 mm. **Fruits** oblong to oblong-ovate, flattened, untwisted, 3.5–9 × 2.5–4 mm, glabrous. **Seeds** ovoid, flattened, 1.4–1.9 × 1–1.2 mm; wingless.

Basaltic talus, stony ridges. Flowering Jul–Sep. 2300–3000 m. BW. Native. ✿

This species is endemic to the Wallowa Mountains.

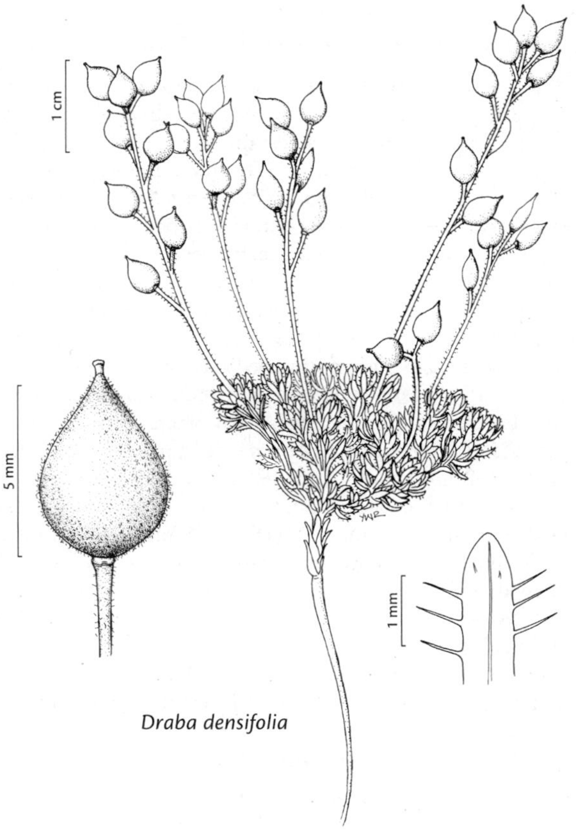

Draba densifolia

DRABA

Draba densifolia Nuttall (p. 479)

Nuttall's draba, denseleaf draba

Draba caeruleomontana Payson & H. St. John, *Draba nelsonii* J.F. Macbr. & Payson, *Draba sphaerula* J.F. Macbr. & Payson

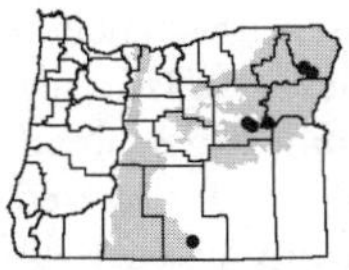

Plants perennial, cespitose, pulvinate, scapose; caudices many-branched; some branches terminated in sterile rosettes. **Stems** (0.5)1.5–10(17) cm, simple, glabrous or with simple and rarely 2–4(5)-rayed trichomes. **Basal leaves** rosulate, linear, oblong, or oblanceolate, 2.5–9(14) × 0.5–1.5(2) mm, margins entire, ciliate, midveins prominent, surfaces abaxially glabrous or with short-stalked, 2–4-rayed trichomes, adaxially glabrous or with subapical simple trichomes; sessile. **Cauline leaves** absent. **Inflorescences** 2–10(22)-flowered, rachises as glabrous or pubescent as stem; fruiting pedicels divaricate to ascending, straight, (0.7)1.5–10(25) mm, glabrous or with simple or 2–4(5)-rayed trichomes; bracts 0. **Flowers** sepals caducous, ovate, 2–3 mm; petals oblanceolate to obovate, 2–5 × 1–1.7(2) mm, pale yellow sometimes fading to white; ovules 4–12 per ovary, styles 0.3–0.6(1) mm. **Fruits** ovate to lanceolate, flattened, untwisted, straight, (2.5)3–6(8) × 2–3 mm, puberulent with simple and short-stalked, 2–5-rayed trichomes. **Seeds** oblong to ovoid, flattened, 1.2–2(2.6) × 0.9–1.2(1.4) mm; wingless. **2*n***=36.

Rocky outcrops, talus, rocky knolls, alpine ridges. Flowering Jun–Aug. 2000–2900 m. BW, ECas. CA, ID, NV, north to AK, northeast to MT, east to WY, southeast to UT. Native.

Draba howellii S. Watson

Howell's whitlow-grass, Howell's draba, rosette draba

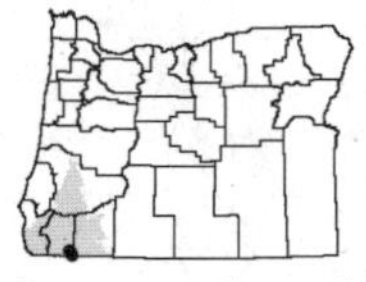

Plants perennial, cespitose; caudices somewhat surculose, few-branched, some branches terminated in sterile rosettes. **Stems** (2)4–11(15) cm, simple, glabrous or with simple and 2–4-rayed trichomes. **Basal leaves** rosulate, oblanceolate, spatulate, or obovate, 4–16(25) × (1.5)3–6(10) mm, margins entire, midveins obscure, surfaces glabrous or with stalked, 2–4-rayed trichomes, not ciliate. **Cauline leaves** 1–4, rarely absent, ovate to oblong, margins entire. **Inflorescences** (5)7–18(25)-flowered, rachises straight, pubescent or rarely glabrous; fruiting pedicels divaricate-ascending, straight, (4)7–10 mm, pubescent or glabrous; bracts 0 or sometimes lowermost 1 or 2 flowers with bracts. **Flowers** sepals ovate, 2.5–3.2 mm, glabrous or pubescent; petals oblanceolate, 5.5–8 × 1–2 mm, yellow; ovules 8–22 per ovary, styles (1)1.6–3 mm. **Fruits** lanceolate, elliptic, or broadly ovate, strongly flattened, untwisted, straight, 6–11(15) × 3–5 mm, pubescent with simple and 2–4-rayed trichomes, rarely glabrous. **Seeds** oblong, flattened, 1–1.6 × 0.8–1 mm; wingless or distally appendaged.

Rocky summits, rock crevices. Flowering Jun–Aug. 1400–1900 m. Sisk. CA. Native.

Draba howellii is known only from Josephine County in Oregon and from Siskiyou and Trinity counties in California.

Draba incerta Payson

Yellowstone draba

Draba incerta Payson var. *incerta*

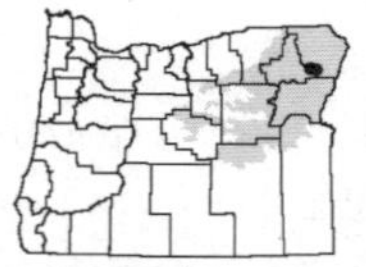

Plants perennial, cespitose, often pulvinate, scapose; caudices many-branched, some terminated in sterile rosettes. **Stems** (1.5)4–14(21) cm, simple, pubescent with simple and 2–5-rayed or mostly subpectinate trichomes, sometimes glabrous distally. **Basal leaves** rosulate, narrowly oblanceolate to linear, (4)6–17(25) × (1)1.5–3.5(5) mm, margins entire, ciliate, midveins usually obscure, surfaces pubescent with short-stalked, pectinate trichomes, at least a few 4–6-rayed, adaxially sometimes glabrous. **Cauline leaves** absent or rarely 1, linear to oblong. **Inflorescences** 3–14(30)-flowered, rachises straight, glabrous or pubescent; fruiting pedicels ascending, straight, (2.5)4–11(27) mm. **Flowers** sepals ovate, 2.5–3.5(4) mm; petals oblanceolate to obovate, 4–6 × 1.5–2.5 mm, yellow fading to white; ovules 8–16(20) per ovary, styles 0.2–0.9 mm. **Fruits** broadly ovate to lanceolate, flattened, untwisted, 5–9(11) × 2–4 mm, glabrous or puberulent with simple and 2-rayed trichomes. **Seeds** oblong, flattened, 1.1–1.5 × 0.7–1 mm; wingless. ***2n***=112.

Rock outcrops, talus, gravelly areas. Flowering Jun–Aug. 2300–2500 m. BW. ID, NV, WA; north to AK, northeast to Northwest Territories, east to WY, southeast to CO, Quebec. Native.

Draba lonchocarpa Rydberg

Lancepod whitlow-grass, lancefruit draba

Draba lonchocarpa Rydb. var. *lonchocarpa*, *Draba nivalis* Lilj., *Draba nivalis* Lilj. var. *elongata* S. Watson

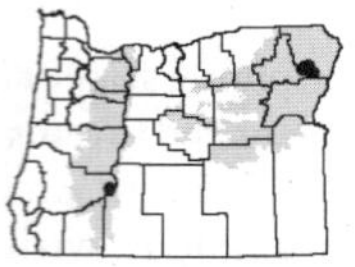

Plants perennial, cespitose, sometimes pulvinate; caudices many-branched. **Stems** (1)3–11 cm, unbranched, glabrous or with non-crisped, minutely stalked, 8–12-rayed trichomes. **Basal leaves** rosulate, oblanceolate to obovate, (1.5)3–15 × 1–3(5) mm, margins entire, ciliate with simple and 2-rayed trichomes, midveins obscure, surfaces abaxially with short-stalked, 8–12-rayed stellate trichomes, adaxially glabrous or with simple and long-stalked branched trichomes. **Cauline leaves** absent or 1(4), ovate or oblong, margins entire. **Inflorescences** 3–9-flowered, rachises often slightly flexuous; fruiting pedicels ascending, straight or curved upward, 2–9(15) mm, glabrous or pubescent; bracts 0 or rarely lowermost 1(4) flowers with bracts. **Flowers** sepals caducous, ovate, 1.5–2 mm; petals caducous, oblanceolate, 2–3.5 × 1–1.5 mm, white, tips obtuse or slightly emarginate; ovules 16–24(28) per ovary, styles 0.1–0.3(0.4) mm. **Fruits** linear, lanceolate, or narrowly oblong, flattened, plane, 6–15(18) × 1–2(3) mm, glabrous or puberulent with simple and stalked, 2-rayed trichomes. **Seeds** ovoid, flattened, 0.7–1 × 0.5–0.6 mm; wingless. ***2n***=16.

Ridges, rocky outcrops, tundra, loose talus. Flowering Jun–Jul. 1500–3000 m. BW, Casc. CA, ID, NV, WA; north to AK, northeast to MT, east to WY, southeast to CO; Europe. Native.

Draba nemorosa Linnaeus

Woodland draba, woodland whitlow-grass

Draba nemorosa L. var. *leiocarpa* Lindblom, *Draba nemorosa* L. var. *nemorosa*

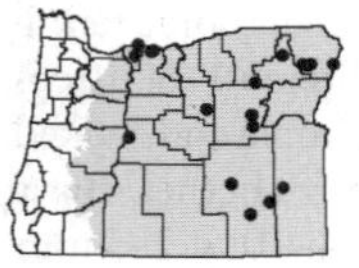

Plants annual. **Stems** (3)6–45(60) cm, pubescent with simple and stalked 2–4-rayed trichomes, glabrous distally. **Basal leaves** rosulate or not, oblong-obovate or oblanceolate, (4)10–35(50) × (2)5–15(20) mm, margins dentate or rarely subentire, surfaces pubescent with stalked, 2–4-rayed trichomes. **Cauline leaves** 4–12(15), ovate to oblong, margins dentate, surfaces pubescent; sessile. **Inflorescences** (10)18–60(92)-flowered, rachises straight, glabrous; fruiting pedicels divaricate, straight, 7–28(35) mm, (1.5)2–7 × as long as fruit, glabrous; bracts 0. **Flowers** sepals ovate, (0.7)0.9–1.6 mm; petals spatulate to oblanceolate, (1.2)1.7–2.2(2.5) × (0.4)0.6–1 mm, yellow, tips deeply notched; ovules (30)36–60(72) per ovary, styles 0.01–0.1 mm. **Fruits** oblong, elliptic, or oblong-oblanceolate, slightly latiseptate, untwisted, (3)5–8(10) × 1.5–2.5 mm; valves glabrous or puberulent with simple, antrorse trichomes. **Seeds** ovoid, 0.5–0.7(0.8) × 0.3–0.4(0.5) mm; wingless. ***2n***=16.

Open wooded slopes, rock outcrops, prairies, riverbanks, meadows, roadsides, gravelly grounds, waste places. Flowering Mar–Jul. 0–2200 m. BR, BW, Casc, Col, ECas, Lava, Owy. CA, ID, NV, WA; north to AK, northeast to Quebec, east to MI, southeast to NM; Asia, Europe. Native.

Draba novolympica Payson & H. St. John

Payson's draba, Trelease's whitlow-grass

Draba paysonii J.F. Macbr., *Draba paysonii* J.F. Macbr. var. *treleasei* (O.E. Schulz) C.L. Hitchc.

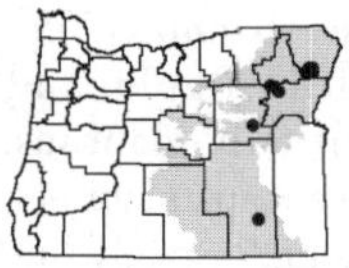

Plants perennial, cespitose, pulvinate, scapose; caudex branches many, some terminated in sterile rosettes. **Stems** 0.5–3.5 cm, unbranched, pubescent with simple and stalked, 2–5-rayed stellate trichomes. **Basal leaves** rosulate, densely imbricate, oblong to linear-oblanceolate, 2–8 × 0.5–1.5 mm, margins entire, ciliate, surfaces abaxially pubescent with stalked, 2–12-rayed stellate trichomes, adaxially with simple and fewer 2-rayed trichomes; sessile. **Cauline leaves** absent. **Inflorescences** 2–12-flowered, rachises straight, pubescent; fruiting pedicels divaricate-ascending, straight, 1–5 mm, pubescent with simple and stalked 2–5-rayed trichomes; bracts 0. **Flowers** sepals oblong, 1.5–2.5 mm, pubescent; petals oblanceolate to spatulate, 2–4 × 1.5–2 mm, bright yellow; ovules 4–8(12) per ovary, styles 0.2–0.6(0.8) mm. **Fruits** ovoid, basally inflated, symmetric, untwisted, (2.5)3–4(5) × 1.5–3.5 mm, densely pubescent with 2–6-rayed trichomes, occasionally mixed with fewer, simple trichomes. **Seeds** oblong, flattened, 1.2–1.8 × 0.8–1.1 mm; wingless. ***2n***=42.

Open knolls, talus, alpine crests, fell-fields, weathered shale, calcareous scree, rocky ground, cliffs, subalpine conifer forests. Flowering Jun–Aug. 1700–3000 m. BR, BW. CA, ID, NV, WA; north to British Columbia, northeast to Alberta, east to WY, southeast to UT. Native.

Draba oligosperma Hooker (p. 481)

Few-seeded whitlow-grass

Draba oligosperma Hook. var. *oligosperma*

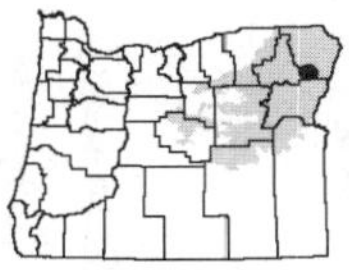

Plants perennial, cespitose, densely pulvinate, scapose; caudices many-branched, some branches terminated in sterile rosettes. **Stems** (1)2–6(10) cm, unbranched, glabrous or with sessile, pectinate trichomes. **Basal leaves** rosulate, linear to linear-oblanceolate, (2)4–11(15) × 0.4–1.5(1.8) mm, margins entire, midveins prominent abaxially, surfaces pubescent with sessile, pectinate trichomes, adaxially sometimes glabrous, rarely ciliate; sessile. **Cauline leaves** absent. **Inflorescences** 4–12(17)-flowered, rachises straight; fruiting pedicels divaricate-ascending, straight, (2)3–10(13) mm, glabrous or pubescent; bracts 0. **Flowers** sepals ovate, 1.5–3 mm, glabrous or pubescent; petals obovate, 2.5–4 × 1.5–3 mm, yellow or rarely creamy white; ovules 6–12 per ovary, styles 0.1–0.8(1.1) mm. **Fruits** ovoid to lanceolate, inflated at least basally, untwisted, 3–6(7) × 2–3.5(4) mm, puberulent with simple and sessile, often unequally 2-rayed trichomes, rarely glabrous. **Seeds** ovoid, flattened, 1.1–1.5 × 0.7–1 mm; wingless. ***2n***=32, 64.

Talus, rock outcrops, gravel benches. Flowering May–Jul. 2400–3000 m. BW. CA, ID, NV, WA; north to AK, northeast to MT, east to WY, southeast to CO. Native.

This apomictic, pulvinate species is easily distinguished from other North American species by its pectinate trichomes, prominent midveins on basal leaves, and basally inflated fruits.

Draba praealta Greene

Tall draba, tall whitlow-grass

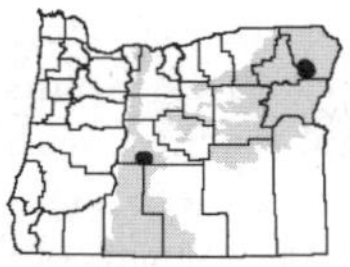

Plants annual or short-lived perennial; caudices simple or few-branched. **Stems** (5)8–32(38) cm, unbranched or branched, pubescent with 2–5-rayed and simple trichomes. **Basal leaves** rosulate, oblanceolate, (7)10–35(45) × 1.5–7(9) mm, margins dentate or entire, surfaces abaxially with stalked, 3–6-rayed trichomes, adaxially with stalked, 3–5-rayed and fewer simple trichomes, ciliate at base; subsessile. **Cauline leaves** (1)2–5(9), ovate to lanceolate, margins dentate or entire; sessile. **Inflorescences** (5)8–30(37)-flowered, rachises straight, pubescent; fruiting pedicels divaricate to ascending or suberect, not appressed to rachis, straight, (3)4–10(12) mm; bracts 0. **Flowers** sepals ovate, 1.7–2.5 mm; petals oblanceolate to spatulate, 2.8–3.5(4) × 0.8–1.2 mm, white, tips obtuse or slightly emarginate; ovules (24)30–52 per ovary, styles 0.03–0.15 mm. **Fruits** lanceolate to linear-lanceolate, flattened, erect, untwisted, (5)7–12(15) × 1.5–2.5(3) mm, puberulent with simple and short-stalked, 2–4-rayed trichomes. **Seeds** oblong, flattened, 0.8–1.1 × 0.5–0.6 mm; wingless. ***2n***=56.

Woodlands, rocky slopes, subalpine or alpine meadows, shale cliffs, grassy volcanic ridges, limestone talus. Flowering Jun–Aug. 1600–2400 m. BW, ECas. CA, ID, NV, WA; north to AK, northeast to MT, east to WY; Europe. Native.

Draba oligosperma

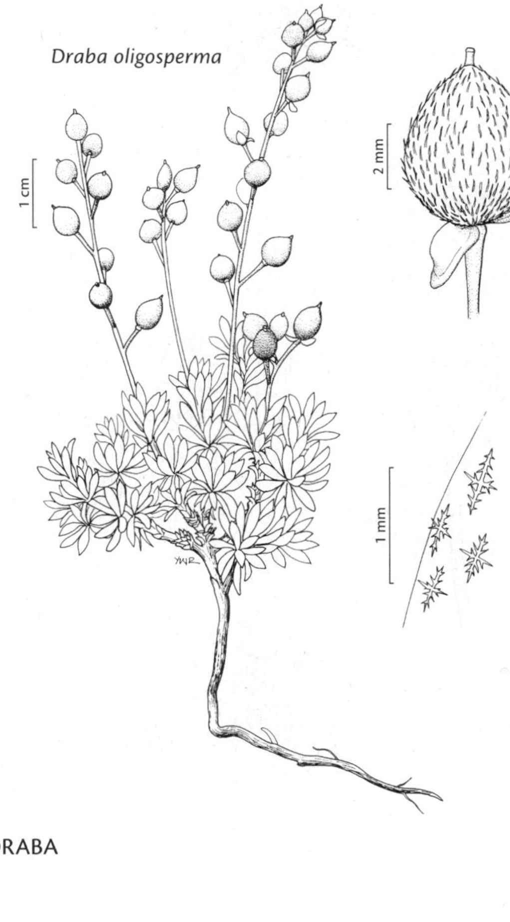

DRABA

Draba stenoloba Ledebour

Alaska draba, Alaska whitlow-grass

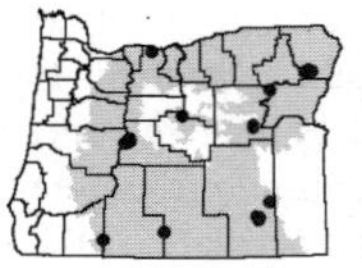

Plants biennial or short-lived, non-cespitose perennial; with simple caudices. **Stems** (2)4–30(34) cm, often unbranched, with short-stalked, 2–4-rayed trichomes, usually glabrous distally. **Basal leaves** rosulate, oblanceolate to obovate, 4–26(31) × 1.5–7(10) mm, margins entire or denticulate, ciliate, surfaces abaxially with short-stalked, 3- or 4-rayed trichomes, adaxially with simple and stalked, 2–4-rayed trichomes. **Cauline leaves** absent or 1 or 2(4) and elliptic, lanceolate, or ovate; sessile. **Inflorescences** (2)4–10(15)-flowered, rachises straight, glabrous or sparsely pubescent, terminal and axillary to cauline leaves; fruiting pedicels horizontal to divaricate-ascending, straight, 3–14(19) mm, ≈ or < fruits, glabrous or rarely pubescent with 2–4-rayed trichomes; bracts 0. **Flowers** sepals oblong, 1.5–2.5 mm, green or purplish; petals oblanceolate, 2.5–3.5 × 0.6–1 mm, yellow, tips obtuse or slightly emarginate; ovules 24–36 per ovary, styles 0–0.2 mm. **Fruits** linear or rarely linear-elliptic, flattened, untwisted, (8)10–17(20) × 1.5–2 mm, glabrous or rarely puberulent with simple trichomes. **Seeds** ovoid, flattened, 0.9–1 × 0.5–0.6 mm; wingless. **2*n***=40.

Meadows, grassy knolls, creek banks, alpine thickets. Flowering Jun–Aug. 50–2800 m. BR, BW, Casc, Col, ECas. CA, NV, WA; north to AK. Native.

Draba verna Linnaeus (p. 482)

Spring whitlow-grass, spring draba

Draba verna L. var. *aestivalis* Lej., *Draba verna* L. var. *boerhaavii* H.C. Hall, *Draba verna* L. var. *verna*

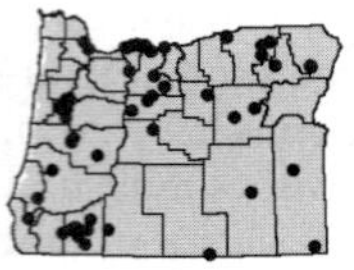

Plants annual, scapose. **Stems** (2)5–20(30) cm, basally with simple and 2(4)-rayed trichomes, distally glabrous. **Basal leaves** rosulate, obovate, spatulate, oblanceolate, lanceolate, oblong, or rarely linear, 2–18(30) × (0.5)1–5(10) mm, margins entire or dentate, surfaces pubescent with simple or stalked, 2–4-rayed trichomes. **Cauline leaves** absent. **Inflorescences** 4–20(30)-flowered, rachises flexuous, glabrous; fruiting pedicels divaricate to ascending, straight or slightly curved upward, (2)5–20(35) mm, glabrous; bracts 0. **Flowers** sepals oblong, 1–2.5 mm, green or purplish, glabrate or pubescent with simple or 2-rayed trichomes; petals (1.5)2–4.5(6) × 1–2 mm, white, tips deeply bifid; ovules (20)32–70(84) per ovary, styles 0–0.2 mm. **Fruits** obovate, oblanceolate, lanceolate, elliptic, oblong, or linear, flattened, untwisted, (2.5)4–9(12) × 1.5–2.5(3.5) mm, glabrous; **Seeds** ovoid, slightly compressed, 0.3–0.6(0.8) × 0.2–0.4 mm; wingless.

Disturbed areas, roadsides, pastures, lawns, fields, grassy hillsides. Flowering Feb–May. 0–2300 m. All ecoregions except Est. CA, ID, NV, WA; throughout North America except Midwest; Africa, Asia, Europe. Exotic.

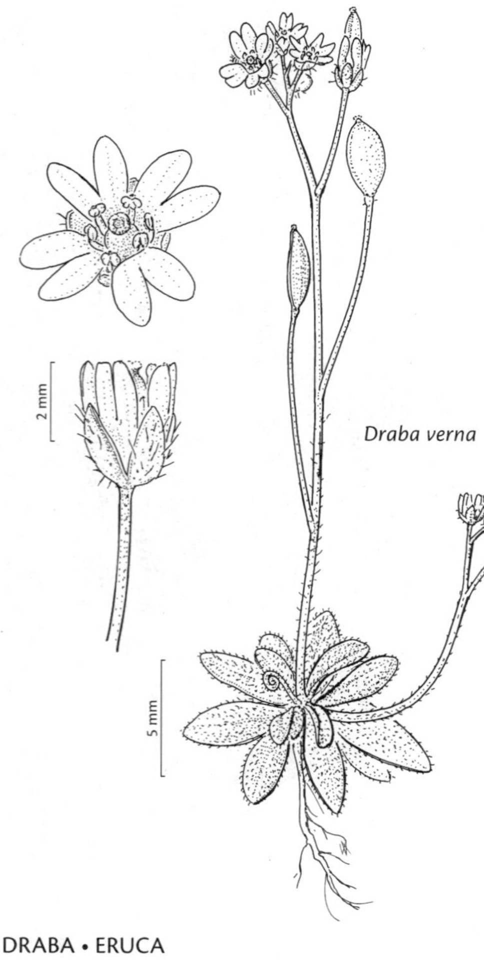

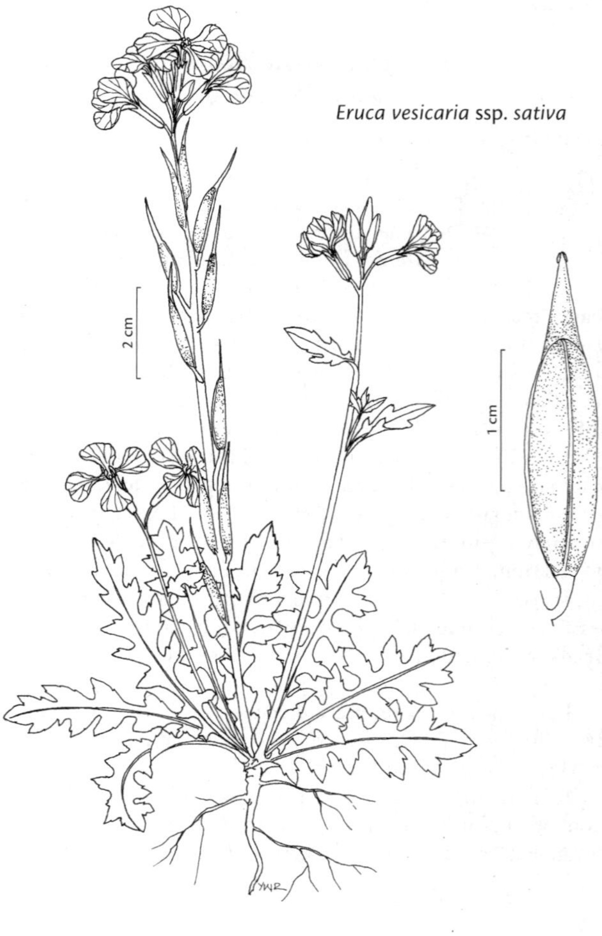

ERUCA Garden rocket, rocketsalad

Herbs annual. **Trichomes** absent or simple. **Basal leaves** lyrate or pinnatifid, rarely 2-pinnatisect or undivided; petiolate. **Cauline leaves** not auriculate, margins entire, dentate, or pinnatifid; short-petiolate or sessile. **Inflorescences** corymbose racemes; bracts 0. **Flowers** sepals caducous or persistent, erect, linear or oblong, saccate; petals cream or yellow, with brown or purple veins darker than rest of petal, claws ≈ or > sepals, petal tips obovate; nectar glands 2 median and 2 lateral; stamens 6, tetradynamous, anthers oblong or linear, obtuse; ovules 10–50 per ovary, septa complete, styles obsolete, stigmas conical, 2-lobed, lobes connivent, decurrent. **Fruits** dehiscent siliques, linear or oblong, terete, segmented; valvular segments seeded, midveins prominent, terminal segments indehiscent, seedless; replums rounded. **Seeds** biseriate, globose or ovoid, plump, mucilaginous when wetted; wingless; cotyledons conduplicate.

Africa, Asia, Europe, North America; 1–2 species; 1 species treated in *Flora.*

Eruca vesicaria (Linnaeus) Cavanilles ssp. ***sativa*** (Miller) Thellung (p. 482)

Garden rocket, rocketsalad, rocket, arugula

Eruca sativa Mill.

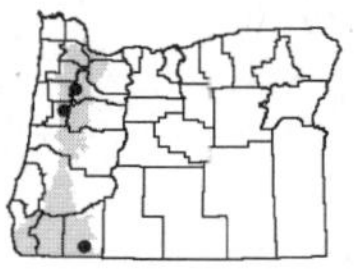

Stems (1)2–8(10) dm, retrorsely hirsute. **Basal leaves** oblanceolate, (2)4–15(20) × (1)2–4(6) cm, lateral lobes 3–9 on each side, margins entire or dentate; petioles (1)2–5(7) cm. **Cauline leaves** upper lobed or not; subsessile. **Inflorescences** fruiting pedicels erect to ascending, subappressed to rachis, 2–8(10) mm. **Flowers** sepals (6)7–10(12) mm; petals (12)15–20(26) × (4)5–7(9) mm. **Fruits** (11)15–35(40) × (2.5)3–5 mm, terminal segments (4)5–10(11) mm; valves (0.7)1–2.5(3.2) mm, glabrous or retrorsely hirsute. **Seeds** 1.6–2.5 mm in diameter. **2*n***=22.

Fields, waste places, roadsides, disturbed areas, gravelly slopes, sandy plains, open rangelands. Flowering Mar–Sep. 0–1200 m. Sisk, WV. CA, NV, WA; scattered throughout North America; Africa, Asia, Europe. Exotic.

This taxon, which is widely cultivated as a salad plant and for its seeds as a source of industrial oils, is naturalized worldwide.

ERYSIMUM Wallflower

Herbs annual, biennial, or perennial, rarely subshrubs or shrubs. **Trichomes** sessile, primarily malpighiaceous on fruits, stems and pedicels, or 3–5(8)-rayed stellate. **Basal leaves** rosulate, simple, margins entire or dentate, rarely pinnatifid; petiolate. **Cauline leaves** bases rarely auriculate, margins entire or dentate, infrequently pinnatifid; petiolate or sessile. **Inflorescences** bracts present or not. **Flowers** sepals erect, oblong or linear, lateral pair saccate or not; petals > sepals, yellow or orange, rarely white, purple, brown, or red, blades suborbicular, obovate, spatulate, or oblong, claws differentiated from blades; nectar glands 1, 2, or 4, distinct or confluent and subtending bases of all stamens; stamens 6, tetradynamous; ovules 15–100 per ovary, septa complete, styles obsolete or distinct, stigmas capitate, entire or 2-lobed. **Fruits** dehiscent siliques, linear, terete, 4-angled, latiseptate, or angustiseptate, unsegmented, smooth or torulose; valves pubescent; replums rounded. **Seeds** uniseriate or rarely biseriate, oblong, plump or flattened, mucilaginous when wetted; winged or wingless; cotyledons incumbent or rarely accumbent.

Northern Hemisphere. 150–200 species; 8 species treated in *Flora.*

Key to *Erysimum* species:

1. Petals 3–9(11) × 1–2 mm; median filaments 2–7 mm.
    2. Inside fruit valves densely pubescent; sepals 1.8–3.2 mm . . . . ***E. cheiranthoides***
    2' Inside fruit valves glabrous or rarely sparsely pubescent; sepals 4–6 mm.
        3. Fruiting pedicels narrower than fruit, 5–9(15) mm; plants biennial or short-lived perennial . . . . ***E. inconspicuum***
        3' Fruiting pedicels approximately as thick as fruit, 2–4(6) mm; plants annual. . . . . ***E. repandum***

1' Petals 12–32 × 3–16 mm; median filaments 7–18 mm.
    4. Leaf trichomes 2–3-rayed.
        5. Ovules (40)54–82 per ovary; fruits 4-angled to latiseptate. . . . . ***E. capitatum*** (in part)
        5' Ovules 24–46 per ovary; fruits latiseptate to strongly latiseptate.
            6. Plants perennial; seeds wingless or distally winged; fruits 1.5–2.7 mm wide . . . . ***E. arenicola***
            6' Plants biennial; seeds winged all around; fruits (2)2.4–3.7 mm wide. . . . . ***E. occidentale***
    4' Leaf trichomes 2–5(7)-rayed.
        7. Fruits torulose; ovules 26–44 per ovary; petals 3.5–6 mm wide . . . . ***E. perenne***
        7' Fruits not torulose; ovules (40)42–82 per ovary; petals (5)6–16 mm wide.
            8. Fruiting pedicels 2–4(6) mm; seeds broadly ovate to suborbicular, 1.5–3 mm wide . . . . ***E. concinnum***
            8' Fruiting pedicels 4–17(25) mm; seeds oblong, (0.7)1–2 mm wide . . . . ***E. capitatum*** (in part)

Erysimum arenicola S. Watson

Cascade wallflower, sand-dwelling wallflower

Erysimum arenicola S. Watson var. *torulosum* (Piper) C.L. Hitchc.

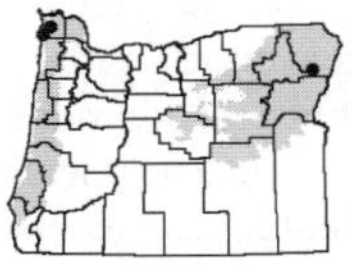

Plants perennial; caudices present. **Stems** erect, 0.4–3 dm, 1 to several from base. **Trichomes** 2(3)-rayed. **Basal leaves** oblanceolate or rarely spatulate, 2–8 cm × 3–10 mm, margins entire or dentate to pinnatifid, surfaces pubescent with 2- or 3-rayed trichomes. **Cauline leaves** margins upper entire or dentate-sinuate; sessile. **Inflorescences** fruiting pedicels divaricate-ascending, 5–8 mm, narrower than fruits. **Flowers** sepals oblong to linear-oblong, 7–10 mm, lateral pair saccate; petals 14–27 × 3.5–6 mm, blades obovate, yellow, claws 9–13 mm, petal tips rounded; median filaments 9–12 mm; ovules 24–42 per ovary, styles cylindrical, slender, 1.5–5 mm, sparsely pubescent, stigmas slightly 2-lobed. **Fruits** latiseptate, erect to ascending, straight or sometimes twisted, strongly torulose, 3–10(12) cm × 1.5–2.7 mm; valves pubescent outside with 2–4-rayed trichomes, glabrous inside. **Seeds** oblong, 2–3.5 × 0.8–1.5 mm; wingless or distally winged. **2*n***=36.

Gravel, rock crevices, talus slopes, open ridges. Flowering Apr–Aug. 800–2600 m. BW, CR. WA; north to British Columbia. Native.

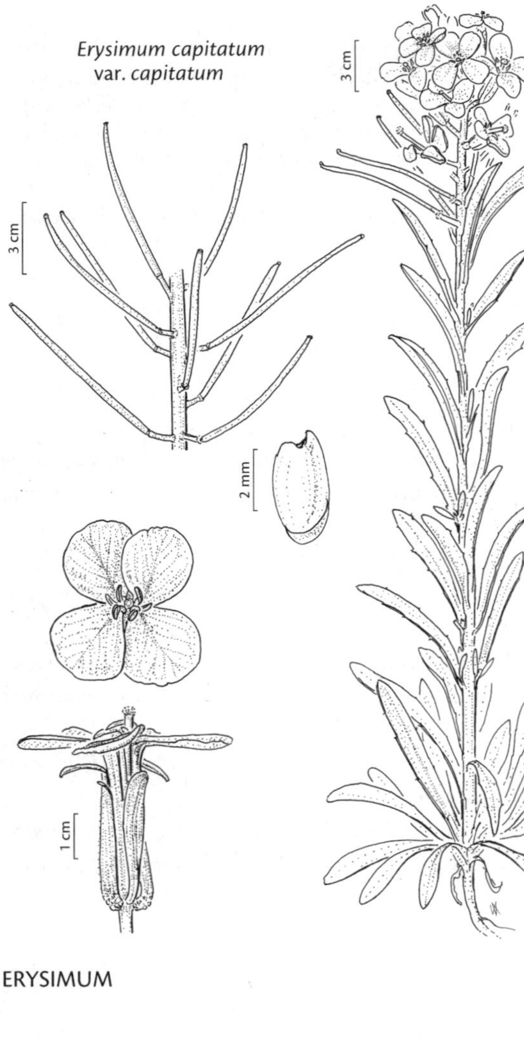

ERYSIMUM

Erysimum capitatum (Douglas ex Hooker) Greene

Western wallflower

Plants biennial or short-lived perennial. **Stems** (0.5)1.2–10(12) dm. **Trichomes** 2–4(7)-rayed. **Basal leaves** spatulate to narrowly oblanceolate or linear, flat, 2–18(27) × 0.3–1.5(3) cm, margins entire or dentate. **Cauline leaves** margins upper entire or denticulate; sessile. **Inflorescences** fruiting pedicels divaricate to ascending, 4–17(25) mm, narrower than fruits. **Flowers** sepals narrowly oblong, 7–14 mm, lateral pair saccate at base; petals suborbicular to obovate, 12–25(30) × (5)6–10(13) mm, orange to yellow, rarely lavender or purplish, claws 8–16 mm, petal tips rounded; median filaments 9–18 mm; ovules (40)54–82 per ovary, styles 0.2–2.5(3) mm, sparsely pubescent, stigmas 2-lobed. **Fruits** 4-angled to latiseptate, not torulose, 3.5–11(15) cm × 1.3–3.3 mm; valves with distinct midveins, pubescent outside with 2–5-rayed trichomes, glabrous inside. **Seeds** oblong, 1.5–4 × (0.7)1–2 mm; apically winged or wingless. **2*n***=36.

2 varieties.

Key to *Erysimum capitatum* varieties:

1. Seeds winged at least distally, rarely wings rudimentary, 2–4 × (0.8)1–2 mm; fruits latiseptate or rarely 4-angled; petals orange, sometimes orange-yellow or yellow; leaves adaxially with (2)3(7)-rayed trichomes . . . . ***E. c.* var. *capitatum***

1' Seeds wingless, 1.5–2(2.4) × 0.7–1.2 mm; fruits 4-angled or rarely latiseptate; petals yellow, rarely lavender or purplish; leaves adaxially with 2(3)-rayed trichomes . . . . ***E. c.* var. *purshii***

Erysimum capitatum var. *capitatum* (p. 484)

Western wallflower, sanddune wallflower, prairie rocket

Erysimum asperum (Nutt.) DC., *Erysimum capitatum* (Douglas ex Hook.) Greene ssp. *capitatum*

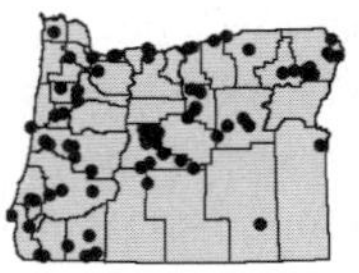

Leaves adaxially mostly with (2)3(7)-rayed trichomes. **Flowers** petals orange, sometimes orange-yellow or yellow. **Fruits** latiseptate or rarely 4-angled. **Seeds** 2–4 × (0.8)1–2 mm; winged at least distally, rarely wings rudimentary. **2*n***=36.

Woodlands, sandy mesas, chaparral clearings, hillsides, open slopes, valley bottoms, alpine areas, deserts. Flowering Apr–Aug. 0–2500 m. All ecoregions. CA, ID, NV, WA; north to AK, northeast to MT, east to MD, southeast to TX, south to Mexico. Native.

Erysimum capitatum var. *purshii* (Durand) Rollins

Pursh's wallflower

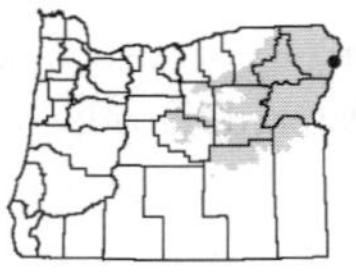

Leaves adaxially with 2(3)-rayed trichomes. **Flowers** petals yellow, rarely lavender or purplish. **Fruits** 4-angled, rarely latiseptate. **Seeds** 1.5–2(2.4) × 0.7–1.2 mm; wingless. **2*n***=36.

Dry slopes, meadows, hillsides. Flowering Apr–Sep. 1000–2700 m. BW. CA, ID, NV; northeast MT, east to WY, southeast to TX, south to Mexico. Native.

Erysimum cheiranthoides Linnaeus

Wormseed mustard, treacle mustard

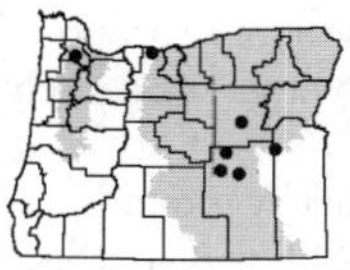

Plants annual. **Stems** (0.7)1.5–10(15) dm, ribbed. **Trichomes** 3–4(5)-rayed. **Basal leaves** withered by fruiting. **Cauline leaves** middle and upper lanceolate, linear, or elliptic-oblong, (1)2–7(11) × (0.2)0.5–1.2(2.3) cm, margins subentire, denticulate, or rarely sinuate-dentate; short-petiolate or sessile. **Inflorescences** fruiting pedicels divaricate or ascending, 5–13(16) mm, much narrower than fruits. **Flowers** sepals oblong, 1.8–3.2 mm, lateral pair not saccate; petals spatulate, 3–5.5 × 1.5–2 mm, yellow, claws 1.5–3.5 mm, petal tips rounded; median filaments 2–3.5 mm; ovules (20)30–55 per ovary, styles slender, 0.5–1.5 mm, stigmas entire or slightly 2-lobed. **Fruits** 4-angled, somewhat torulose, (1)1.5–2.5(4) cm × 1–1.3 mm; valves with distinct midveins, pubescent outside with 3–5-rayed trichomes, densely pubescent inside. **Seeds** oblong, 1–1.5 × 0.4–0.6 mm; wings absent. **2*n***=16.

Pastures, fields, waste grounds, disturbed areas, dry creek beds. Flowering Apr–Aug. 50–1400 m. BR, BW, Col, Lava, WV. CA, ID, NV, WA; throughout Canada and US; Asia, Europe. Exotic.

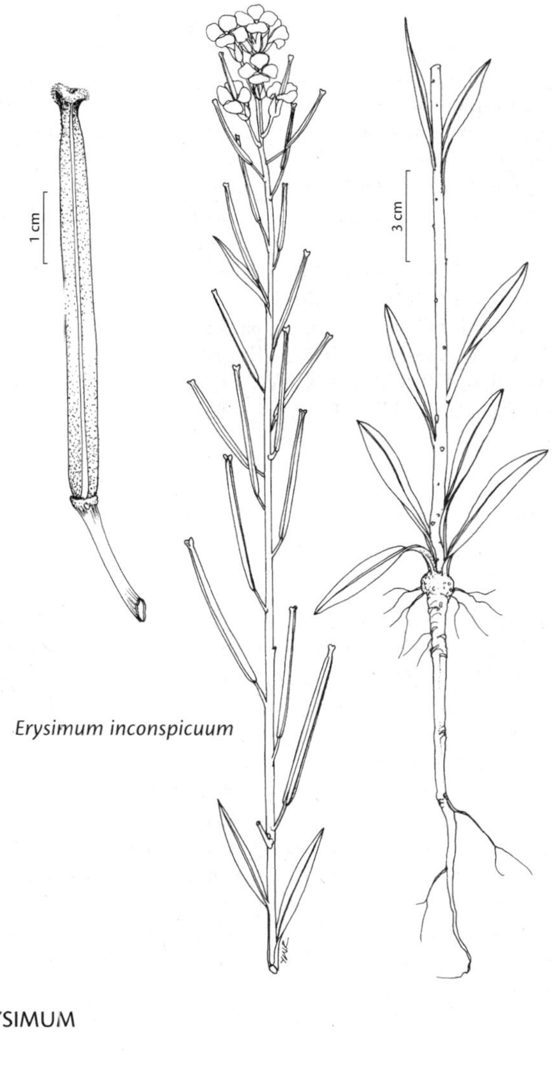

Erysimum inconspicuum

ERYSIMUM

Erysimum concinnum Eastwood

Coast wallflower, Pacific wallflower, curly wallflower

Erysimum menziesii (Hook.) Wettst. ssp. *concinnum* (Eastw.) R.A. Price

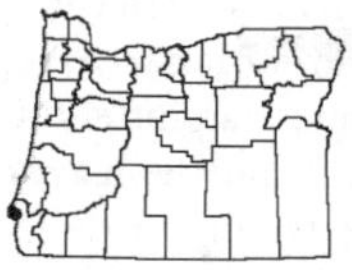

Plants biennial or short-lived perennial. **Stems** 0.4–5(7) dm. **Trichomes** 2–3(7)-rayed. **Basal leaves** spatulate to oblanceolate, 2–11 × 0.4–2 cm, margins sinuate-dentate to coarsely dentate. **Cauline leaves** margins upper entire or denticulate; sessile. **Inflorescences** fruiting pedicels ascending, 2–4(6) mm, narrower than fruits. **Flowers** sepals oblong, 8–19 mm, lateral pair saccate; petals suborbicular to broadly obovate, 15–32 × 6–16 mm, yellow to cream, claws 8–12 mm, petal tips rounded; median filaments 8–11 mm; ovules 42–68 per ovary, styles stout, 0.5–2.5 mm, sparsely pubescent, stigmas 2-lobed. **Fruits** terete when young, becoming strongly latiseptate, not torulose, (3)5–13 cm × 2.2–5 mm; valves with obscure midveins, pubescent outside with 2–5-rayed trichomes, glabrous inside. **Seeds** broadly ovate to suborbicular, (1.5)2–4 × 1.5–3 mm; wings continuous. **2*n***=36.

Dunes, coastal bluffs, prairies. Flowering Mar–Jun. 0–400 m. Est. CA. Native.

The species is known only in Oregon from Curry County and in California from Del Norte, Humboldt, Mendocino, Sonoma, and Marin counties.

Erysimum inconspicuum (S. Watson) MacMillan

(p. 485)

Small-flowered rocket, small wallflower

Erysimum inconspicuum (S. Watson) MacMill. var. *inconspicuum*

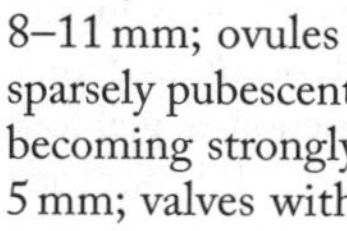

Plants biennial or short-lived perennial; caudices present. **Stems** 1.5–7 dm. **Trichomes** 2- or 3-rayed. **Basal leaves** linear to linear-oblanceolate, 1.5–6(8) cm × 2–8 mm, margins entire or dentate. **Cauline leaves** margins upper entire or denticulate; sessile. **Inflorescences** fruiting pedicels divaricate-ascending to ascending, 5–9(15) mm, narrower than fruits. **Flowers** sepals oblong to linear-oblong, 4–6 mm, lateral pair slightly saccate; petals 6–9(11) × 5–7 mm, blades oblanceolate, yellow, claws 5–7 mm, petal tips rounded; median filaments 5–7 mm; ovules 36–66 per ovary, styles stout, 0.7–3 mm, pubescent, stigmas strongly 2-lobed. **Fruits** slightly 4-angled or terete, not torulose, 3–6(7) cm × 1–1.5(1.8) mm; valves with somewhat distinct midvein, pubescent outside with 2–4-rayed trichomes, glabrous inside. **Seeds** ovoid, 1.2–1.7 × 0.8–1 mm, sometimes distally apiculate; wingless. **2*n***=54.

Abandoned fields, railroad embankments, hillsides, alkaline ground, calcareous talus and cliffs, roadsides, bluffs, crevices and ledges, gravel, prairies, rocky pastures, waste sites. Flowering May–Jul. 1200–2500 m. BR, BW. ID, NV, WA; north to AK, northeast to Quebec, east to Nova Scotia, southeast to NM. Native.

Erysimum occidentale (S. Watson) B.L. Robinson

Pale wallflower

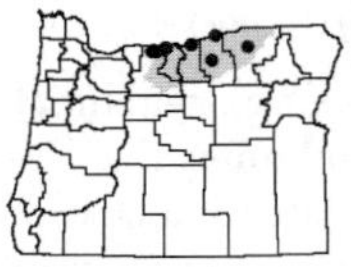

Plants biennial. **Stems** 0.5–3(6.5) dm. **Trichomes** 2–3-rayed. **Basal leaves** often withered by fruiting, linear-oblanceolate, 2–11(15) cm × (1.5)2–6 mm, margins entire or very rarely denticulate. **Cauline leaves** margins upper entire; sessile. **Inflorescences** fruiting pedicels divaricate-ascending, 4–11(15) mm, narrower than fruits. **Flowers** sepals linear-oblong, 8–13 mm, lateral pair saccate; petals obovate to broadly so, 14–22(25) × 3–7.5 mm, bright or pale yellow, claws 11–16 mm, tips rounded; median filaments 10–14 mm; ovules 34–46 per ovary, styles slender, (2)2.5–5 mm, sparsely pubescent, stigmas entire. **Fruits** strongly latiseptate, not torulose, 3–12 cm × (2)2.4–3.7 mm; valves with prominent midveins, pubescent outside with 2- and 3-rayed trichomes, glabrous inside. **Seeds** ovoid, (2)2.5–3.5 × 1.3–2.5 mm, 0.3–0.8 mm wide at distal end; winged. ***2n***=36.

Sand deposits. Flowering Mar–Jul. 0–900 m. Col. ID, WA. Native.

The species is confined to sand deposits along or near the Columbia River and its tributaries.

Erysimum perenne (S. Watson ex Coville) Abrams

Sierra wallflower

Erysimum capitatum (Douglas ex Hook.) Greene ssp. *perenne* (S. Watson ex Coville) R.A. Price, *Erysimum capitatum* (Douglas ex Hook.) Greene var. *perenne* (S. Watson ex Coville) R.J. Davis

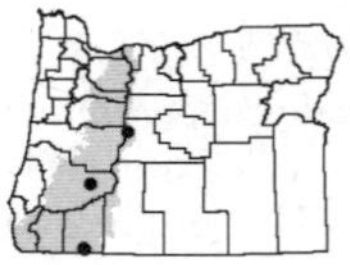

Plants perennial, rarely biennial; with slender caudices. **Stems** 0.4–6.5 dm. **Trichomes** 2–5-rayed. **Basal leaves** spatulate to broadly oblanceolate, 2.5–7 cm × 3–10 mm, margins dentate or subentire, surfaces adaxially pubescent with 2(5)-rayed trichomes. **Cauline leaves** margins upper often entire; sessile. **Inflorescences** fruiting pedicels divaricate-ascending, slender, 4–12 mm, narrower than fruits. **Flowers** sepals linear-oblong to oblong, 8–12 mm, lateral pair saccate; petals 15–22 × 3.5–6 mm, blades broadly obovate to suborbicular, yellow, claws 8–14 mm, petal tips rounded; median filaments 7–14 mm; ovules 26–44 per ovary, styles (1.5)2–5.5 mm, stigmas subentire to slightly 2-lobed. **Fruits** latiseptate, torulose, 3.8–14 cm × 1.2–3 mm; valves with distinct midveins, pubescent outside with 2- or 3(4)-rayed trichomes, glabrous to sparsely pubescent inside. **Seeds** ovoid, 2–3.4 × 1–2 mm; wingless or rarely distally winged. ***2n***=36.

Knolls, rocky slopes, alpine fell-fields, gravelly grounds, talus, granitic sand. Flowering Jun–Sep. 1700–2400 m. Casc, Sisk. CA, ID, NV. Native.

Erysimum perenne was treated as a variety or invalidly published subspecies of *E. capitatum* by Rollins (1993) and Price (1993), respectively. It is easily separated from the latter by its torulose (vs. non-torulose) and flattened (vs. 4-angled or rarely flattened) fruit, styles (1.5)2–5.5 mm (vs. 0.2–2.5(3) mm), and yellow (vs. orange to rarely yellow) petals.

Erysimum repandum Linnaeus

Spreading wallflower, repand wallflower

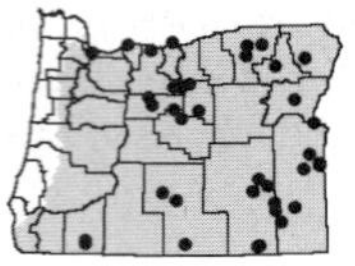

Plants annual. **Stems** (0.4)1.5–4.5(7) dm. **Trichomes** 2(3)-rayed. **Basal leaves** often withered by fruiting; middle cauline leaves linear, narrowly oblanceolate, elliptic, or oblong, (1)2–8(11) cm × (2)5–12(17) mm, margins sinuate, dentate, or repand; petiolate. **Cauline leaves** upper entire or denticulate; sessile. **Inflorescences** fruiting pedicels divaricate, stout, 2–4(6) mm, width ≈ fruit. **Flowers** sepals oblong, 4–6 mm, lateral pair not saccate; petals oblanceolate to spatulate, 6–8 × 1.5–2 mm, yellow, claws 3–6 mm, petal tips rounded; median filaments 4–6 mm; ovules (40)50–80(90) per ovary, styles stout, 1–4 mm, sparsely pubescent, stigmas slightly 2-lobed. **Fruits** 4-angled, somewhat torulose, (2)3–8(10) cm × 1.5–2 mm; valves with distinct midveins, pubescent outside with 2(3)-rayed trichomes, glabrous or rarely sparsely pubescent inside. **Seeds** oblong, 1.1–1.5 × 0.6–0.7 mm; wingless or rarely distally winged. ***2n***=16.

Roadsides, disturbed areas, pastures, fields, barren hillsides. Flowering Apr–Jun. 0–2600 m. BR, BW, Casc, Col, ECas, Lava, Owy, Sisk, WV. CA, ID, NV, WA; scattered throughout southern Canada and US; Africa, Asia, Europe. Exotic.

EUCLIDIUM Syrian mustard

Genus composed of one species, *E. syriacum.*

Euclidium syriacum (Linnaeus) W.T. Aiton (p. 487)

Syrian mustard

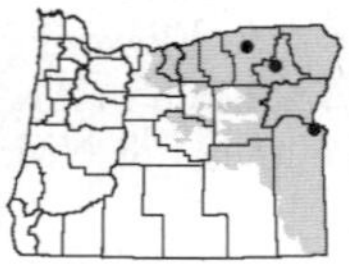

Herbs annual, scabrous pubescent throughout. **Trichomes** minutely stalked, 2-forked, mixed with fewer simple and rarely 3-forked. **Stems** (4)10–40(45) cm, rigidly branched. **Basal leaves** not rosulate, margins entire, dentate, or pinnatifid; petiolate. **Cauline leaves** oblong, lanceolate, or elliptic, (1)1.5–7(9) × (0.3)0.7–2(3) cm, bases not auriculate, margins entire, dentate; petioles (0.2)0.5–2(2.5) cm, to sessile or subsessile distally. **Inflorescences** bracts 0; fruiting pedicels appressed to rachis, stout, 0.5–1(1.2) mm, nearly half as wide as fruit. **Flowers** sepals erect, caducous, ovate to oblong, 0.6–0.9 mm, lateral pair not saccate; petals narrowly spatulate, 0.9–1.3 × 0.1–0.2 mm, slightly > sepals, white, claws slightly differentiated from blades, 0.4–0.6 mm; nectar glands 4, 1 on each side of lateral stamens; stamens 6, slightly tetradynamous, anthers ovate, tips pointed; ovules 2 per ovary, septa complete, thickened, styles persistent, subconical, beak-like, curved, stout, 1–1.8 mm, sparsely pubescent, stigmas capitate, 2-lobed, lobes not decurrent. **Fruits** nutlet-like silicles, 2-seeded, indehiscent, erect, appressed to rachis, ovoid, subterete to slightly 4-angled, 2–2.5 × 1.5–2 mm, scabrous, unsegmented; valves woody, veinless; replums strongly expanded laterally. **Seeds** aseriate, oblong, 1.3–1.7 × 0.8–1.2 mm, brown, not mucilaginous when wetted; wingless; cotyledons accumbent. ***2n***=14.

Waste places, roadsides. Flowering May–Jun. 300–1000 m. BW, Col, Owy. CA, ID, NV, WA; northeast to MT, east to NE, southeast to CO, MA; Asia, Europe. Exotic.

GUILLENIA California mustard

Herbs annual. **Trichomes** absent or simple. **Basal leaves** rosulate or not, simple, margins entire, dentate, lobed, or pinnatifid; petiolate. **Cauline leaves** bases not auriculate, margins entire or dentate, rarely lobed; petiolate or uppermost subsessile. **Inflorescences** bracts 0. **Flowers** sepals erect or spreading, caducous, oblong or ovate, not forming urceolate calyces, lateral pair not saccate; petals erect or spreading, linear, oblanceolate or spatulate, > sepals, white or lavender, neither channeled nor crisped, claws differentiated or not, narrower than blade; nectar glands mostly confluent, subtending bases of all filaments; stamens 6, tetradynamous, filaments erect to spreading, anthers oblong to ovate, not coiled when dehisced, tips obtuse; ovules 14–60 per ovary, septa complete, styles to 2 mm, stigmas entire or subentire; gynophore as stout as fruit base, obsolete or < 1 mm. **Fruits** dehiscent siliques, linear, terete, unsegmented; valves obscurely veined; replums rounded. **Seeds** uniseriate, oblong, smooth, not mucilaginous when wetted; wingless; cotyledons incumbent.

Northwestern Mexico and western United States. 3 species; 1 species treated in *Flora*.

This species complex is tremendously variable in indumentum density, thickness of the fruiting pedicels, orientation and size of fruit, number of ovules/seeds per ovary/fruit, and leaf morphology. Although no infraspecific taxa are recognized herein, detailed, population-based morphological and molecular studies may well lead to the recognition of independent species or the division of the species into several infraspecific taxa.

Guillenia lasiophylla (Hooker & Arnott) Greene (p. 487)

California mustard, Coast Range western cabbage, hairyleaf caulanthus

Caulanthus lasiophyllus (Hook. & Arn.) Payson, *Thelypodium lasiophyllum* (Hook. & Arn.) Greene

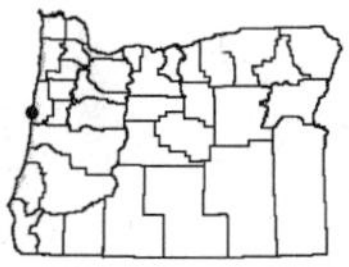

Plants hirsute, rarely subglabrous. **Stems** (0.8)2–10(16) dm. **Basal leaves** soon withered. **Cauline leaves** lanceolate, oblong, or oblanceolate, 2–12 × 0.5–5 cm, pinnatifid, lateral lobes dentate or entire, upper smaller, less divided; petioles 0.5–3 cm. **Inflorescences** fruiting pedicels strongly reflexed or spreading, slender or thick, (0.7)1–2.2(3) mm. **Flowers** sepals erect, 2–4 mm; petals oblanceolate, 2.5–5(6.5) × 0.9–1.1 mm, white to creamy white or rarely pinkish, claws undifferentiated from blades; ovules 14–60 per ovary, styles 0.5–2 mm, stigmas subentire. **Fruits** terete, straight to slightly curved, 20–48(57) × 0.7–1.2 mm; valves glabrous or pubescent. **Seeds** 0.9–1.5 × 0.6–1 mm. **2*n*=28.**

Desert flats, gravelly areas, limestone rocks, talus slopes, shrub communities, hillsides, sandy banks, disturbed sites, grassy fields, ravines. Flowering Mar–May. 0–700 m. Est. CA, NV; southeast to NM, south to Mexico. Native.

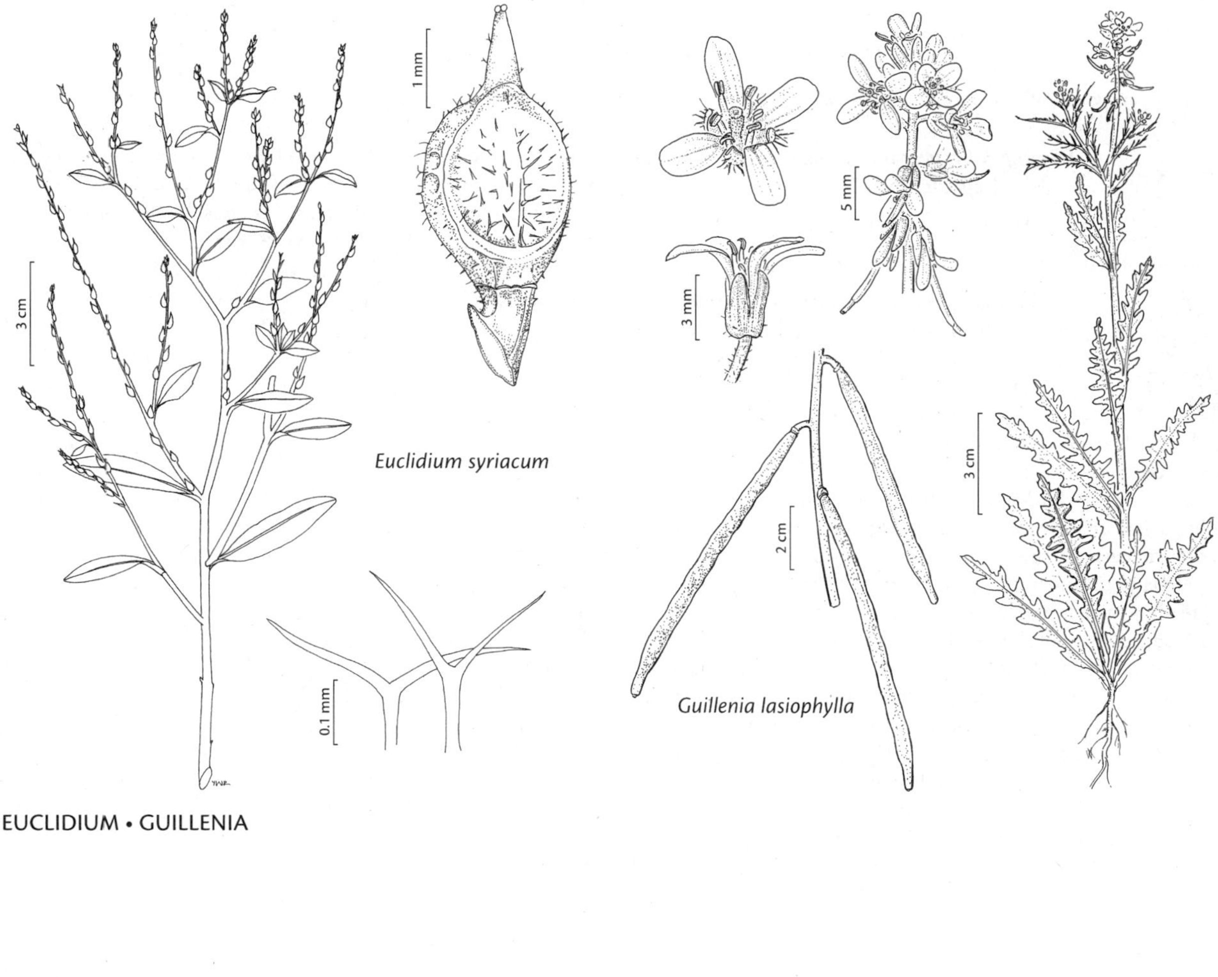

EUCLIDIUM • GUILLENIA

HESPERIS Dame's-rocket, rocket, dame's-violet

Herbs biennial or perennial; caudices present. **Trichomes** simple and/or forked, these often mixed with unicellular glands on uniseriate multicellular stalks. **Basal leaves** usually rosulate, margins entire, dentate, or pinnately divided; petiolate. **Cauline leaves** bases sometimes auriculate or amplexicaul, margins entire or dentate; petiolate or sessile. **Inflorescences** elongated corymbose racemes; bracts present or not. **Flowers** sepals erect, caducous, sometimes connivent, oblong or linear, lateral pair strongly saccate; petals obovate or oblong, much > sepals, white, lavender, purple, yellow, orange, brownish, or green, rounded, claws strongly differentiated from blades; nectar glands 2, lateral, ring-like or lunate; stamens 6, tetradynamous, anthers oblong or linear, obtuse; ovules 4–40 per ovary, septa complete, styles obsolete or short, stigmas conical, 2-lobed, lobes decurrent, connivent, or free. **Fruits** tardily dehiscent siliques, linear, terete, 4-angled, or slightly latiseptate, unsegmented; valves with prominent midveins, torulose; replums rounded. **Seeds** uniseriate, oblong, plump, not mucilaginous when wetted; wingless; cotyledons incumbent.

Africa, Asia, Europe, North America. 25 species; 1 species treated in *Flora*.

Hesperis matronalis Linnaeus (p. 488)

Dame's-rocket, dame's-violet

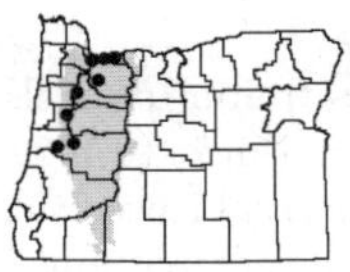

Plants pubescent. **Stems** 4–8(11) dm. **Basal leaves** withered by flowering. **Cauline leaves** middle and upper narrowly oblong, lanceolate, or broadly ovate, (2)4–15(20) × (0.4)0.8–4(6) cm, bases not auriculate, margins denticulate or entire; short-petiolate. **Inflorescences** fruiting pedicels divaricate or ascending, (5)7–17(25) mm, eglandular. **Flowers** sepals 5–8 mm; petals obovate, (13)15–20(22) × 3.5–7.5(9) mm, deep purple, lavender, pink, or white, claws 6–12 mm. **Fruits** terete, (4.5)6–10(14) cm × 2–2.5 mm; valves glabrous. **Seeds** (2.5)3–4 × 1–1.5 mm. **2*n***=24.

Abandoned fields, gardens, floodplains, waste areas, roadsides, oak glades, bluffs, railroad embankments, thickets, woodlands. Flowering May–Aug. 0–1200 m. Casc, WV. CA, ID, NV, WA; throughout Canada and US; Asia, Europe. Exotic.

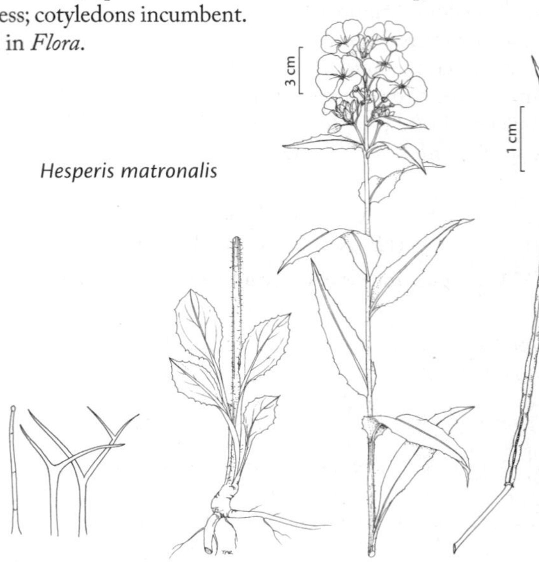

HESPERIS

HIRSCHFELDIA Mediterranean mustard

Genus composed of one species, *H. incana.*

Hirschfeldia incana (Linnaeus) Lagrèze-Fossat (p. 488)

Mediterranean mustard

Sinapis incana L.

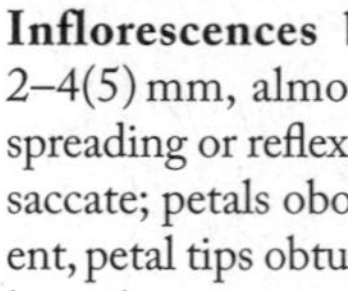

Herbs annual or biennial. **Trichomes** simple. **Stems** (2)4–15(20) dm, densely retrorse-hairy below. **Basal leaves** rosulate, lyrate to pinnatifid, (3)4–22(35) × 1.5–6(8) cm, surfaces densely pubescent, terminal lobes broadly ovate, lateral lobes 1–6(9) on each side, ovate or lanceolate, smaller than terminal lobe, margins crenate-dentate; petioles 1–4(10) cm. **Cauline leaves** similar to basal, bases not auriculate; subsessile or petiolate, uppermost leaves sessile. **Inflorescences** bracts 0; fruiting pedicels erect, stout, 2–4(5) mm, almost as thick as fruit. **Flowers** sepals widely spreading or reflexed, caducous, oblong, 3–5 mm, lateral pair not saccate; petals obovate, 5–10 × 2.5–4.5 mm, yellow, claws present, petal tips obtuse; nectar glands 4, not confluent, median and lateral; stamens 6, tetradynamous, anthers oblong, obtuse; ovules 10–22 per ovary, septa complete, styles present, stigmas capitate, entire. **Fruits** dehiscent siliques, strongly appressed to rachis, linear, terete or slightly 4-angled, 7–15(17) × 1–1.7 mm, tips acute or obtuse; valvular segments 8–20-seeded, not torulose, 6–10 mm, terminal segments indehiscent, slightly swollen, 3–6 mm, seedless; valves 3-veined; replums rounded. **Seeds** 1–2, uniseriate, ovoid or globose, 0.9–1.5 mm in diameter, mucilaginous when wetted; wingless; cotyledons conduplicate. **2*n***=14.

Roadsides, disturbed areas, canyons, creek bottoms, dry fields, open desert. Flowering Mar–Nov. 0–2000 m. CR, Est, WV. CA, NV; Africa, Asia, Europe. Exotic.

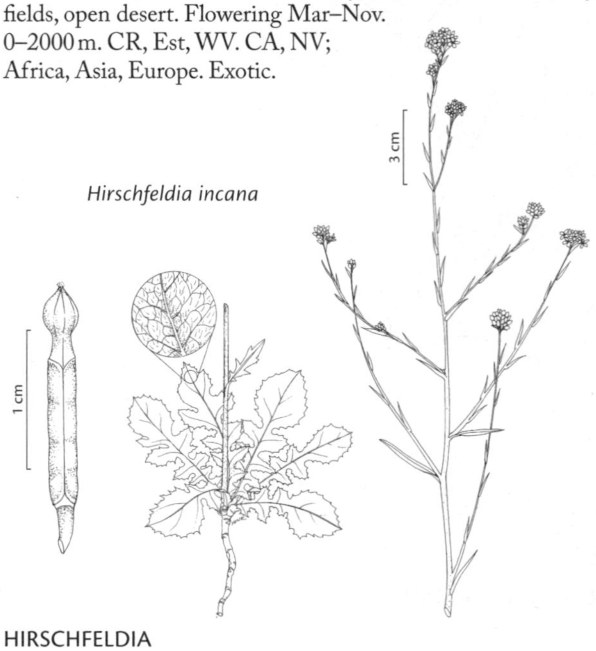

HIRSCHFELDIA

HORNUNGIA Hornungia

Herbs with branched caudices; roots slender. **Trichomes** absent, simple, or minutely forked. **Basal leaves** rosulate or not, pinnatisect, margins sometimes dentate or entire; petiolate. **Cauline leaves** sometimes absent, bases not auriculate, margins pinnatisect, pinnatifid, dentate, or entire; petiolate. **Inflorescences** bracts 0. **Flowers** sepals spreading or reflexed, caducous, ovate or oblong, lateral pair not saccate; petals obovate, oblong, or oblanceolate, > or < sepals, white, claws absent, tips obtuse or rounded; nectar glands 4, 1 on each side of lateral stamen; stamens 6, subtetradynamous, rarely 4, equal, anthers ovate, obtuse; ovules 4–20 per ovary, septa complete, styles obsolete or rarely to 0.5 mm, stigmas capitate, entire. **Fruits** dehiscent silicles, slightly angustiseptate, unsegmented; valves with prominent midveins, keeled; replums rounded. **Seeds** biseriate or aseriate when few, oblong, plump, mucilaginous or not when wetted; wingless; cotyledons incumbent or rarely accumbent.

Nearly worldwide. 3 species; 1 species treated in *Flora.*

Hornungia procumbens (Linnaeus) Hayek (p. 489)

Oval purse, prostrate hutchinsia

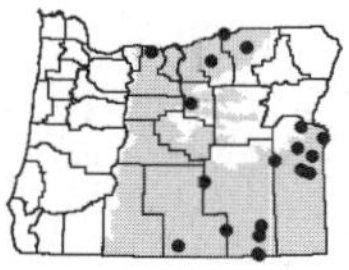

Plants annual, glabrous or puberulent with minutely forked and some simple trichomes. **Stems** erect to decumbent, (2)5–22(30) cm. **Basal leaves** rosulate or not, blades obovate, oblanceolate, or oblong, (2)10–25(40) × (1)5–11(16) mm, margins entire, dentate, or pinnatifid; petioles (2)5–12(20) mm. **Cauline leaves** similar to basal. **Inflorescences** fruiting pedicels divaricate, 3–8(12) mm. **Flowers** sepals 0.6–1.1 mm; petals 0.6–1.2 × 0.3–0.6 mm; ovules 10–24 per ovary, styles obsolete or to 0.1 mm. **Fruits** oblong, elliptic, or obovate, (2)3–4(4.5) × (1)1.4–2.2 mm, tips subtruncate to slightly emarginate; valves glabrous. **Seeds** 0.5–0.6 × 0.3–0.4 mm, smooth. **2*n***=12, 24.

Sagebrush plains, deserts, meadows, shade of bushes, disturbed habitats, saline banks, salt marshes, alkaline flats. Flowering Apr–Jul. 50–1600 m. BR, Col, ECas, Lava, Owy. CA, ID, NV, WA; north to British Columbia, northeast to Manitoba, east to WY, southeast to CO; nearly worldwide. Exotic.

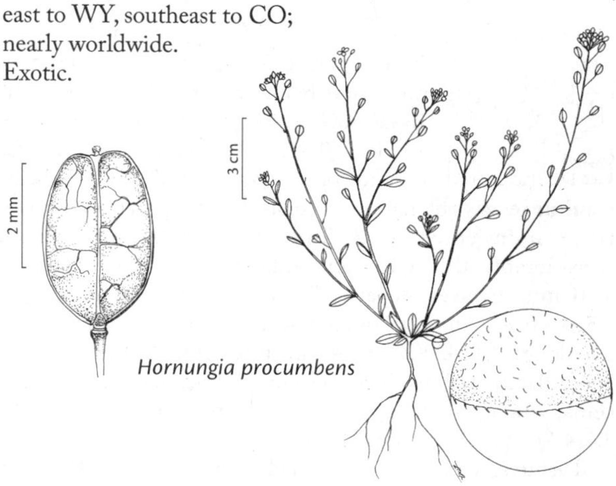

HORNUNGIA

IDAHOA Flatpod

Genus composed of one species, *I. scapigera.*

Idahoa scapigera (Hooker) A. Nelson & J.F. Macbride (p. 489)

Flatpod, scalepod, oldstem idahoa

Platyspermum scapigerum Hook.

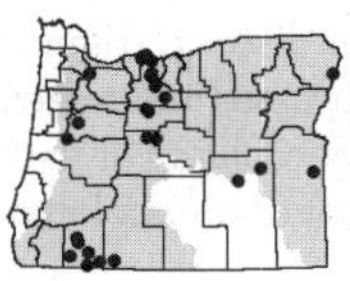

Herbs annual, scapose. **Trichomes** absent. **Stems** absent. **Leaves** all basal, simple, flat, broad, ovate, obovate, or spatulate, (3)7–20(25) × 2–8(12) mm, margins entire or lyrately lobed; petioles (2)5–20(30) mm. **Inflorescences** flowers solitary; fruiting pedicels ascending or erect, straight, (1.5)2.5–11(13) cm. **Flowers** sepals ascending to spreading, caducous, ovate to oblong, 1.5–1.8 mm, purplish, lateral pair not saccate; petals oblanceolate, 2–2.5 × 0.5–0.8 mm, slightly > sepals, white, claws undifferentiated from blades, petal tips obtuse; lateral nectar glands semiannular; stamens 6, slightly tetradynamous, anthers ovate or broadly so, obtuse; ovules 6–10(12) per ovary, septa complete, styles subconical, 0.3–0.8 mm, stigmas capitate, entire, sessile or rarely on gynophores to 1 mm, unsegmented. **Fruits** dehiscent silicles, strongly latiseptate, orbicular to orbicular-ovate, (5)6–9(10) mm, slightly longer than wide; valves rounded on both ends, veinless or obscurely veined; replums slightly flattened. **Seeds** biseriate, suborbicular, flattened, 3.5–5 mm in diameter, brown, coarsely reticulate except for wing, not mucilaginous when wetted; wings 0.8–1.2 mm wide; cotyledons accumbent. **2*n***=16.

Sagebrush slopes, gravelly hills, dry ground, scabland, rocky volcanic flats and outcrops, open moist slopes and banks, meadows, rocky hillsides. Flowering Feb–May. 0–1800 m. BW, Casc, Col, ECas, Lava, Owy, Sisk, WV. CA, ID, NV, WA; north to British Columbia, northeast to MT. Native.

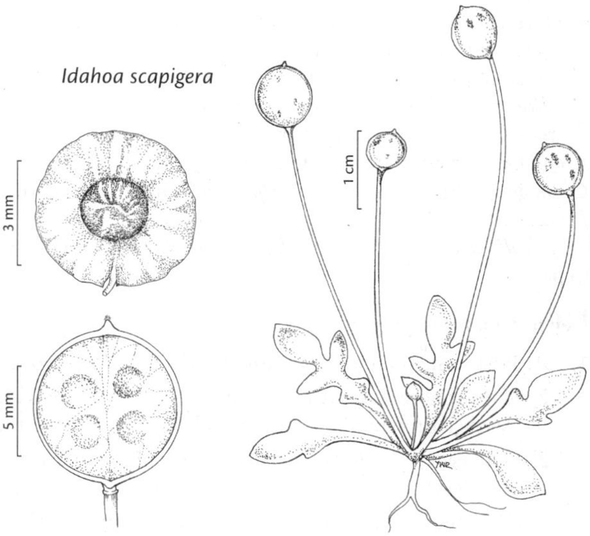

IDAHOA

ISATIS Woad

Herbs biennial, rarely annual or perennial. **Trichomes** absent or simple. **Leaves** basal leaves rosulate or not, margins entire, dentate, or rarely pinnately lobed; petiolate or rarely sessile. **Cauline leaves** bases auriculate to amplexicaul, margins entire or dentate; sessile or rarely petiolate. **Inflorescences** racemes; bracts 0. **Flowers** sepals erect or ascending, caducous, ovate or oblong, lateral pair not saccate; petals obovate, spatulate, oblong, or oblanceolate, obtuse or subemarginate, ≥ sepals, yellow, claws absent; nectar glands confluent, or 4, 2 median and 2 lateral; stamens 6, tetradynamous, anthers ovate or oblong, tips pointed or obtuse; ovules 1(2) per ovary, septa absent, styles absent, stigmas capitate, entire. **Fruits** indehiscent siliques, strongly angustiseptate, unsegmented; valves and replum united. **Seeds** 1, oblong, plump, not mucilaginous when wetted; wingless; cotyledons incumbent or accumbent.

Africa, Asia, Europe. 86 species; 1 species treated in *Flora*.

Isatis tinctoria Linnaeus (p. 490)

Woad, dyers woad

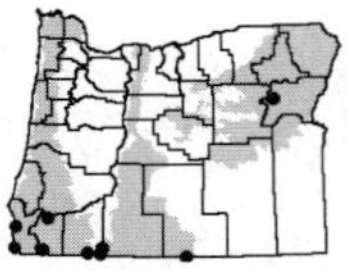

Plants glaucous, glabrous or sometimes pubescent below. **Stems** (3)4–10(15) dm. **Basal leaves** oblong or oblanceolate, (2.5)5–15(20) × (0.5)1.5–3.5(5) cm, margins entire, repand, or dentate; petioles 0.5–5.5 cm. **Cauline leaves** oblong or lanceolate, bases sagittate or auriculate, margins entire; sessile. **Inflorescences** fruiting pedicels considerably thickened and subclavate at tips, reflexed, 5–10 mm. **Flowers** sepals 1.5–2.8 mm; petals oblanceolate, 2.5–4 × 0.9–1.5 mm, yellow, obtuse. **Fruits** oblong-oblanceolate to oblong, (9)11–20(27) × 3–6 mm, black or dark brown; locules 3–6(10) mm, with distinct midveins and inconspicuous lateral veins; wings entire, rayless, 3.5–5(7) mm wide. **Seeds** 2.3–3.5(4.5) × 0.8–1 mm. **2*n***=14, 28.

Pastures, roadsides, prairies, fields, waste places, sagebrush hillsides, railroad embankments. Flowering Apr–Jun. 0–1400 m. BW, CR, ECas, Sisk. CA, ID, NV, WA; scattered in western and northeastern North America; Asia, Europe. Exotic.

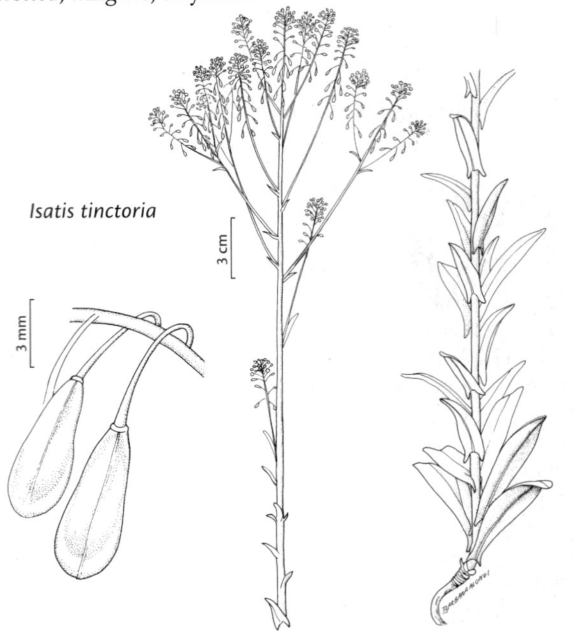

ISATIS

LEPIDIUM Peppergrass, pepperwort, cress, peppercress

Herbs annual, biennial, or perennial, sometimes subshrubs or shrubs; roots slender. **Trichomes** absent or simple. **Basal leaves** rosulate or not, margins entire, dentate, or 1–3-pinnatisect. **Cauline leaves** bases sometimes auriculate or amplexicaul; petiolate or sessile. **Inflorescences** bracts 0. **Flowers** radially symmetric; sepals ovate or oblong, lateral pair not saccate; petals sometimes rudimentary or absent, blades white, yellow, pink, or purple, obtuse or emarginate, claws absent or distinct; nectar glands 4 or 6, distinct, median glands often present; stamens 2 or 4 and equal in length, lateral or median, or 6 and tetradynamous, anthers ovate or oblong, filaments unappendaged; ovules 2 per ovary, septa complete or perforated, styles absent, obsolete, or distinct, stigmas capitate, entire or rarely 2-lobed. **Fruits** dehiscent, schizocarpic, or indehiscent silicles, angustiseptate or rarely inflated and terete, unsegmented; apically winged or wingless; replums rounded. **Seeds** 2(4) per fruit, 1(2) per locule, oblong or obovate, plump or flattened, usually mucilaginous when wetted; winged, margined, or wingless; cotyledons incumbent, rarely accumbent or diplecolobal.

Worldwide. 252 species; 18 species treated in *Flora*.

Based on extensive molecular and developmental studies (Bowman et al. 1999; Mummenhoff et al. 2001, 2009), the limits of *Lepidium* were expanded by Al-Shehbaz et al. (2002) and Al-Shehbaz & Mummenhoff (2010) to include the genera *Cardaria*, *Coronopus*, *Stroganowia*, *Stubendorffia*, and *Winklera*, all of which were found to be polyphyletic and nested within a more inclusive *Lepidium*.

Lepidium graminifolium and *L. ruderale* have been collected on ballast, in Linnton (Multnomah County), but apparently failed to persist.

Key to *Lepidium* species:

1. At least some cauline leaves sessile and auriculate, sagittate, or amplexicaul.
   2. Uppermost leaves deeply cordate-amplexicaul; basal leaves 2- or 3-pinnatifid; petals yellow; seeds narrowly winged all around; plants glabrous at least above. . . . . **L. perfoliatum**
   2' Uppermost leaves auriculate or sagittate; basal leaves entire, dentate or 1- or 2-pinnatifid; petals white; seeds wingless; plants hirsute to subglabrous.
      3. Stamens 2; petals absent or rudimentary; basal leaves 1–2-pinnatifid . . . . **L. oblongum** (in part)
      3' Stamens 6; petals distinct; basal leaves dentate to entire.

4. Plants not rhizomatous; fruits dehiscent, apically broadly winged, notched; inflorescences elongated.
5. Plants annual; fruit valves papillate except for wing; styles 0.2–0.5(0.7) mm, slightly exserted beyond apical notch of fruit or included in it . . . . ***L. campestre***
5' Plants perennial; fruit valves often not papillate; styles (0.6)1–1.5 mm, well exserted beyond apical notch of fruit. . . . . ***L. heterophyllum***
4' Plants rhizomatous; fruits indehiscent, apically wingless, not notched; inflorescences corymbose panicles.
6. Fruits flattened, cordate to subreniform; valves veined . . . . ***L. draba***
6' Fruits inflated, globose, subglobose, obovoid or obcompressed globose; valves usually not veined.
7. Fruits pubescent, globose or rarely subglobose; styles 0.5–1.5 mm . . . . ***L. appelianum***
7' Fruits glabrous, obovoid to obcompressed globose; styles (0.8)1.2–2(2.3) mm. . . . . ***L. chalepense***
1' Cauline leaves petiolate or if sessile, then not auriculate, sagittate, or amplexicaul.
8. Fruits indehiscent, didymous, 1.3–1.7 mm, valves rugose . . . . ***L. didymum***
8' Fruits dehiscent, not didymous, > (1.6)1.8 mm, valves smooth.
9. Plants perennial with woody caudices and often petiolar remains of previous years.
10. Basal leaves 1- or 2-pinnatifid or -pinnatisect . . . . ***L. montanum*** (in part)
10' Basal leaves entire, serrate, 3–5-lobed or -toothed.
11. Plants pulvinate or densely cespitose; basal leaves 3–5-toothed or -lobed; fruiting pedicels puberulent all around or rarely subglabrous . . . . ***L. davisii***
11' Plants neither pulvinate nor cespitose; basal leaves entire or serrate; fruiting pedicels glabrous or puberulent adaxially . . . . ***L. latifolium***
9' Plants annual or biennial without caudices.
12. Stamens 2.
13. Fruits reticulate-veined; sepals persistent. . . . . ***L. strictum***
13' Fruits not veined or rarely weakly veined; sepals caducous.
14. Plants hirsute . . . . ***L. oblongum*** (in part)
14' Plants puberulent or glabrous.
15. Fruits obovate to obovate-suborbicular, widest above middle; rachises of inflorescences puberulent with straight or rarely curved, usually subclavate trichomes; petals absent or rudimentary and 0.3–0.9 mm; cotyledons incumbent. . . ***L. densiflorum***
15' Fruits orbicular, widest at middle; rachises of inflorescences puberulent with curved trichomes, rarely glabrous; petals present, 1–2.5 mm, rarely rudimentary; cotyledons accumbent or rarely incumbent. . . . . . ***L. virginicum*** ssp. ***menziesii***
12' Stamens 4 or 6.
16. Stamens 6; styles obsolete or 0.2–0.7(1) mm; petals 2–4(4.3) mm; fruiting pedicels terete or rarely slightly flattened.
17. Fruiting pedicels divaricate to horizontal, not appressed to rachis; fruits 2–4.3(5) × (1.5)1.8–3.6(4) mm; cotyledons entire. . . . . ***L. montanum*** (in part)
17' Fruiting pedicels erect to ascending, appressed to rachis; fruits (4)5–6.4(7) × 3–4.5(5.6) mm; cotyledons 3-lobed. . . . . ***L. sativum***
16' Stamens 4 (or rarely 6 in *L. nitidum*); styles obsolete or rarely to 0.1 mm; petals absent or 1–2.5 mm; fruiting pedicels strongly flattened.
18. Petals present, rarely rudimentary or absent; fruit valves not or obscurely veined; fruiting pedicels not appressed to rachises . . . . ***L. nitidum***
18' Petals absent; fruit valves strongly reticulate-veined; fruiting pedicels appressed to rachises or distally slightly recurved.
19. Fruits (3)4–6 × 2.5–3.5 mm, apical notch (0.8)1–2 mm deep, open, V-shaped; fruiting pedicels (2)3–4.2 mm . . . . ***L. acutidens***
19' Fruits 2.5–3.5 × 2–2.8 mm, apical notch 0.5–0.7(0.8) mm deep, closed, often U-shaped; fruiting pedicels (1.6)1.9–2.5(3) mm . . . . ***L. dictyotum***

Lepidium acutidens (A. Gray) Howell (p. 493)

Alkali veiny peppergrass

Lepidium dictyotum A. Gray var. *acutidens* A. Gray

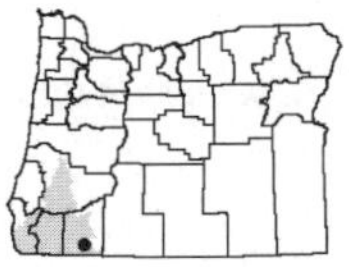

Plants annual, hirsute or puberulent. **Stems** (0.5)0.8–3 dm. **Basal leaves** not rosulate, linear and undivided or pinnatisect, (2)2.5–6.1(7.2) cm, lobes linear to narrowly oblong, (0.5)1–2(3) mm wide; petioles 1–4 cm. **Cauline leaves** linear, 1.2–5.8 cm × 0.5–2(3) mm, bases not auriculate; sessile. **Inflorescences** rachises puberulent or hirsute with straight cylindrical trichomes; fruiting pedicels erect to slightly ascending, appressed to rachis or distally slightly recurved, straight, strongly flattened, (2)3–4.2 mm, 0.4–0.6 mm wide, puberulent all around. **Flowers** sepals caducous, 0.7–1.1 mm; petals absent; stamens 4, median; styles absent or obsolete, included in apical notch. **Fruits** dehiscent, ovate to ovate-oblong, (3)4–6 × 2.5–3.5 mm, apical notch (0.8)1–2 mm deep, V-shaped, open; valves thin, smooth, glabrous or pubescent, strongly reticulate-veined; apically winged. **Seeds** ovate, 1.3–2 × 0.8–1.2 mm; cotyledons incumbent.

Alkaline flats, gullies, fields, saline vernal flats, grassy fields. Flowering Feb–May. 0–1500 m. Sisk. CA, WA; south to Mexico. Native.

Lepidium appelianum Al-Shehbaz (p. 493)

Globe-podded hoary cress, hairy whitetop

Cardaria pubescens (C.A. Mey.) Jarm., *Cardaria pubescens* (C.A. Mey.) Jarm. var. *elongata* Rollins, *Hymenophysa pubescens* C.A. Mey.

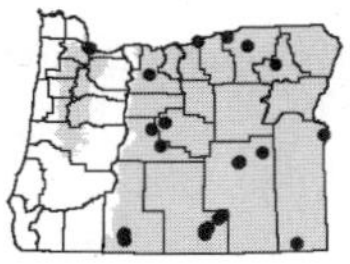

Plants perennial, densely hirsute; rhizomatous. **Stems** (1)1.5–3.5(5) dm. **Basal leaves** not rosulate, obovate to oblanceolate, (1)2–6(7) cm × 3–20 mm, margins dentate; petioles 0.5–1.5 cm. **Cauline leaves** oblong or lanceolate, 1–5(8) cm × (3)5–15(30) mm, bases sagittate, margins dentate or subentire; sessile. **Inflorescences** in corymbose panicles, rachises pubescent with often curved trichomes; fruiting pedicels divaricate to ascending, terete, straight or slightly curved, 3–9(12) mm, pubescent all around. **Flowers** sepals caducous, 1.4–2 mm; petals broadly obovate, (2.2)2.8–4 × 1–3 mm, white, claws 1–1.4 mm; stamens 6; styles 0.5–1.5 mm. **Fruits** indehiscent, globose to rarely subglobose, inflated, (2)3–4.4(5) mm in diameter, apical notch absent; valves smooth, densely puberulent, not veined; wingless. **Seeds** ovoid, 1.5–2 × 1–1.5 mm; wingless; cotyledons incumbent. **2*n***=16.

Sagebrush communities, alkaline meadows, fields, pastures, roadsides, streamsides. Flowering May–Sep. 100–1700 m. BR, BW, Col, ECas, Lava, Owy, WV. CA, ID, NV, WA; southern Canada, northern and western US; Asia, South America. Exotic.

Lepidium campestre (Linnaeus) W.T. Aiton (p. 493)

Field cress, field pepperweed

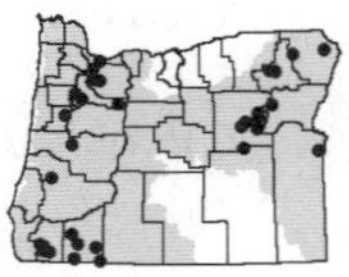

Plants annual, densely hirsute. **Stems** (0.8)1.2–5(6.3) dm. **Basal leaves** rosulate, oblanceolate or oblong, (1)2–6(8) cm × 5–15 mm, margins entire, lyrate or pinnatifid; petioles (0.5)1.5–6 cm. **Cauline leaves** oblong, lanceolate, or deltoid-lanceolate, (0.7)1–4(6.5) cm × (2)5–10(15) mm, bases sagittate or auriculate, margins dentate or subentire. **Inflorescences** much elongated, rachises hirsute with straight, spreading trichomes; fruiting pedicels horizontal, terete, straight or slightly recurved, (3)4–8(10) mm, hirsute all around. **Flowers** sepals caducous, (1)1.3–1.8 mm; petals spatulate, (1.5)1.8–2.5(3) × (0.2)0.5–0.7 mm, white; stamens 6; styles 0.2–0.5(0.7) mm, slightly exserted beyond apical notch or included in it. **Fruits** dehiscent, broadly oblong to ovate, (4)5–6(6.5) × (3)4–5 mm, incurved adaxially, apical notch (0.2)0.4–0.6 mm deep; valves papillate except for wing, not veined; apically broadly winged. **Seeds** ovoid, 2–2.3(2.8) × 1–1.4 mm; wingless; cotyledons incumbent. **2*n***=16.

Disturbed areas, roadsides, pastures, gardens, rocky slopes, forests. Flowering May–Jun. 50–1900 m. BW, Casc, CR, ECas, Lava, Owy, Sisk, WV. CA, ID, NV, WA; scattered throughout most of North America; Asia, Europe. Exotic.

Lepidium chalepense Linnaeus (p. 493)

Lens-podded hoary cress, chalapa hoary cress, lenspod whitetop

Cardaria chalepensis (L.) Hand.-Mazz., *Cardaria draba* (L.) Desv. var. *repens* (Schrenk) O.E. Schulz

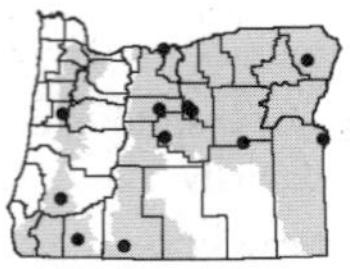

Plants perennial, densely hirsute to subglabrous; rhizomatous. **Stems** (0.8)2.1–6.6(9.2) dm. **Basal leaves** not rosulate, obovate, spatulate, or ovate, (1.8)2.5–8.6(14) × 1–3.7 cm, margins subentire or dentate; petioles 0.9–4.4 cm. **Cauline leaves** obovate, oblong, lanceolate, or oblanceolate, (1.5)2.6–9.3(13.2) cm × (7)12–31(45) mm, bases sagittate-amplexicaul or auriculate, margins dentate or entire; sessile. **Inflorescences** in corymbose panicles, rachises glabrous or puberulent with straight or curved cylindrical trichomes; fruiting pedicels ascending to horizontal, terete, straight, 5–16(19) mm, glabrous or sparsely puberulent adaxially. **Flowers** sepals caducous, 1.7–3 mm; petals obovate, 3–5 × 1.2–2.4 mm, white; stamens 6; styles (0.8)1.2–2(2.3) mm. **Fruits** indehiscent, obovoid to obcompressed globose, 3.5–5.8(7) × (3.5)4–6.2(7) mm, apical notch absent; valves smooth, glabrous, usually not veined; wingless. **Seeds** ovate, 1.5–2.3 × 1–1.3 mm; wingless; cotyledons incumbent. **2*n***=48, 80, 128.

Slopes, roadsides, fields, riverbanks, pastures, disturbed areas. Flowering Apr–Jun. 50–1400 m. BW, Col, ECas, Lava, Owy, Sisk, WV. CA, ID, NV, WA; Mountain and Pacific states, east into IA and south into TX; throughout midwestern and western Canada and US; Asia, Europe, South America. Exotic.

Lepidium davisii Rollins (p. 493)

Davis' peppergrass, Davis' pepperwort

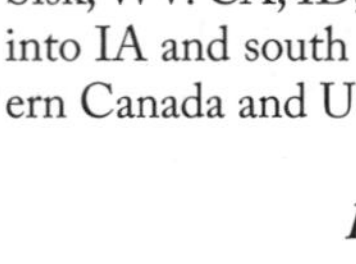

Plants cespitose perennial, puberulent; with woody, many-branched caudices covered with petiole remains of previous years. **Stems** (0.2)0.5–1(1.4) dm. **Basal leaves** often deciduous, not rosulate, spatulate to oblanceolate, 1–2.5(3.2) cm × 2–6(9) mm, margins entire or apically 3(5)-toothed or -lobed. **Cauline leaves** oblanceolate, oblong, or rarely obovate, (0.8)1.3–2.5 cm × (2)4–7 mm, bases not auriculate, margins entire or apically 3(5)-toothed; sessile. **Inflorescences** rachises puberulent with straight or curved trichomes; fruiting pedicels divaricate-ascending, terete, straight, (2.5)3–4.2(5) mm, puberulent all around or rarely subglabrous. **Flowers** sepals caducous, 1.2–2 mm; petals obovate, 2–3.2(4) ×

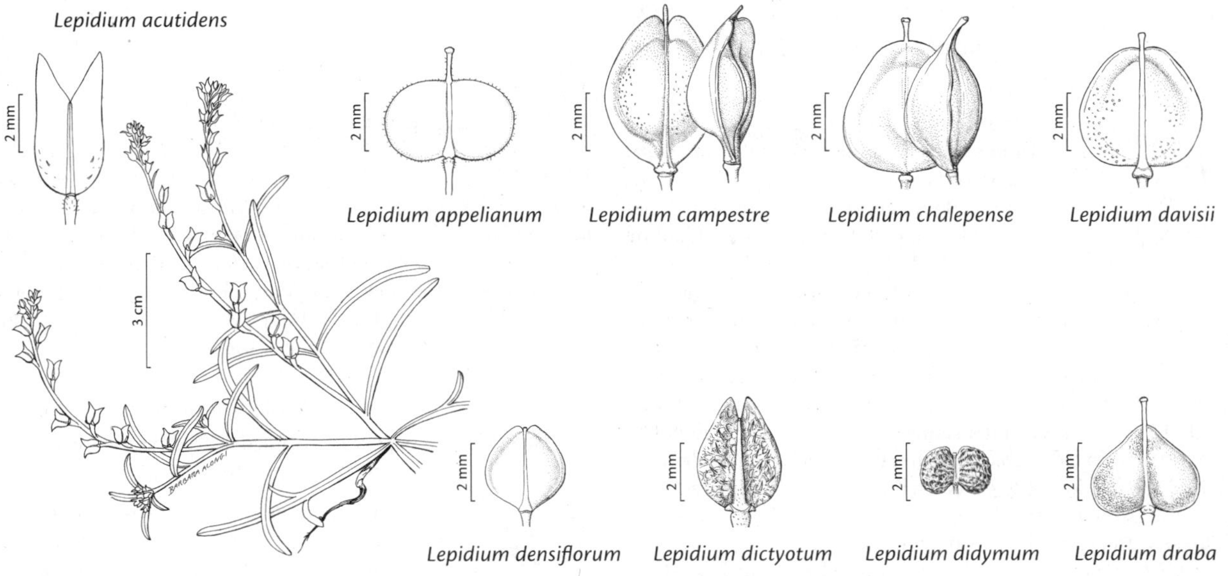

LEPIDIUM

1.5–2 mm, white, claws 0.5–1 mm; stamens 6; styles 0.5–1 mm, exserted beyond apical notch. **Fruits** dehiscent, suborbicular to broadly ovate, (2.5)3–4.3(5) × (2)2.3–4 mm, apical notch 0.1–0.4 mm deep; valves smooth, glabrous or puberulent, not veined; apically winged. **Seeds** oblong-ovate, (1.8)2–2.3 × 1–1.2 mm; cotyledons incumbent. ***2n***=32.

Playas of sagebrush plains, vernal ponds. Flowering May–Aug. 1100–1700 m. BR, Owy. ID, NV. Native.

The range of this rare species is restricted to southeastern Oregon (Malheur County), southwestern Idaho (Ada, Elmore, Owyhee, and Twin Falls counties), and northern Nevada (Elko County).

Lepidium densiflorum Schrader (p. 493)

Common pepperweed, prairie peppergrass, babyseed pepperweed, miner's pepperweed

Lepidium densiflorum Schrad. var. *bourgeaunum* (Thell.) C.L. Hitchc., *Lepidium densiflorum* Schrad. var. *densiflorum*, *Lepidium densiflorum* Schrad. var. *elongatum* (Rydb.) Thell., *Lepidium densiflorum* Schrad. var. *macrocarpum* G.A. Mulligan, *Lepidium densiflorum* Schrad. var. *pubicarpum* (A. Nelson) Thell., *Lepidium densiflorum* Schrad. var. *ramosum* (A. Nelson) Thell., *Lepidium texanum* Buckley

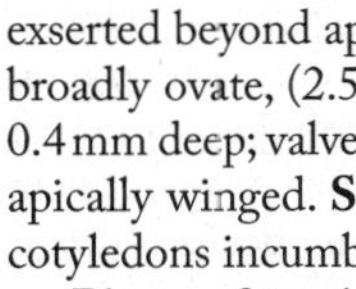

Plants annual or biennial, puberulent or glabrous. **Stems** (1)2.5–5(6.5) dm. **Basal leaves** rosulate, oblanceolate, spatulate, or oblong, (1.5)2.5–8(11) cm × 5–10(20) mm, margins serrate or pinnatifid; petioles 0.5–1.5(2) cm. **Cauline leaves** oblanceolate or linear, (0.7)1.3–6.2(8) cm × (0.5)1.5–10(18) mm, bases not auriculate, margins entire, serrate, dentate, or rarely pinnatifid; short-petiolate. **Inflorescences** rachises puberulent with straight or rarely curved, usually subclavate trichomes; fruiting pedicels divaricate-ascending to horizontal, terete, straight or slightly recurved, (1.5)2–3.5(4) mm, puberulent adaxially. **Flowers** sepals caducous, 0.5–0.8(1) mm; petals absent or rudimentary, filiform, 0.3–0.9 mm, white; stamens 2, median; styles 0.1–0.2 mm, included in apical notch. **Fruits** dehiscent, obovate to obovate-suborbicular, (2)2.5–3(3.5) × 1.5–2.5(3) mm, widest above middle, apical notch 0.2–0.4 mm deep; valves smooth, glabrous or sometimes sparsely puberulent along margin, not veined; apically winged. **Seeds** ovate, 1.1–1.3(1.5) × 0.8–0.9 mm; cotyledons incumbent. ***2n***=32.

Disturbed areas, pastures, grasslands, chaparral, sagebrush flats, gravelly hillsides, rock crevices. Flowering Apr–Aug. 0–1600 m. All ecoregions. CA, ID, WA; scattered throughout the US, Mexico; Asia, Europe. Native.

This is a highly variable species in which variation does not generally correlate with geography. Thus, recognition of varieties is neither practical nor useful.

Lepidium dictyotum A. Gray (p. 493)

Alkali pepperweed, veiny peppergrass

Lepidium dictyotum A. Gray var. *dictyotum*

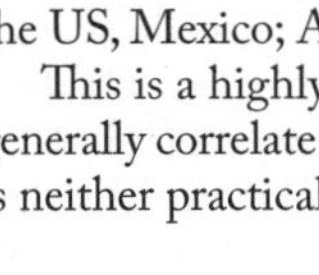

Plants annual, hirsute. **Stems** (0.15)0.3–1.3(2.1) dm. **Basal leaves** not rosulate, (1.5)2.2–5.7(7) cm, pinnatifid to pinnatisect, lobes linear to narrowly oblong, (0.5)1–2(3) mm wide; petioles 0.5–2 cm. **Cauline leaves** linear or rarely with linear lobes, 1–5 cm × 0.5–2 mm, bases attenuate, not auriculate, margins entire; sessile. **Inflorescences** rachises hirsute with straight cylindrical trichomes; fruiting pedicels erect to slightly ascending, straight or slightly recurved distally, appressed to rachis, strongly flattened, (1.6)1.9–2.5(3) × 0.4–0.8 mm, hirsute or rarely glabrous abaxially. **Flowers** sepals caducous, 0.7–1.1 mm; petals absent; stamens 4, median; styles absent or obsolete, included in apical notch. **Fruits** dehiscent, ovate, 2.5–3.5 × 2–2.8 mm, apical notch 0.5–0.7(0.8) mm deep, closed, often U-shaped; valves smooth, hirsute or glabrous, strongly reticulate-veined; apically winged. **Seeds** ovate, 1.2–1.8 × 0.8–1.2 mm; cotyledons incumbent.

Alkaline, clay or sandy flats, meadows, near hot springs, roadsides. Flowering Apr–Jun. 50–1300 m. BR, Col, ECas, Owy. CA, ID, NV, WA; southeast to UT; Mexico. Native.

Lepidium didymum Linnaeus (p. 493)

Lesser swinecress, lesser wartcress

Coronopus didymus (L.) Sm.

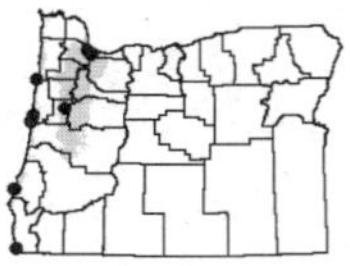

Plants annual, fetid, glabrous or pilose. **Stems** often decumbent, 1–4.5(7) dm. **Basal leaves** not rosulate, 1–6(8) cm, 1- or 2-pinnatisect, lobes entire or dentate, sometimes deeply lobed; petioles 0.5–4(6) cm. **Cauline leaves** similar to basal, 1.5–3.5(4.5) × 0.5–1.2 cm, bases not auriculate; petiolate to subsessile. **Inflorescences** rachises glabrous or pubescent with straight cylindrical trichomes; fruiting pedicels divaricate to horizontal, terete, straight or slightly recurved, 1.4–2.5(4) mm, glabrous or sparsely pubescent adaxially. **Flowers** sepals caducous, 0.5–0.7(0.9) mm; petals elliptic to linear, 0.4–0.5 × ~0.1 mm, white; stamens 2, median; styles absent or obsolete, included in apical notch. **Fruits** indehiscent, didymous, 1.3–1.7 × 2–2.5 mm, apical notch 0.2–0.4 mm deep; valves thick, rugose, glabrous, strongly veined; wingless. **Seeds** ovate, 1–1.2 × 0.7–0.8 mm; cotyledons incumbent. **2*n***=32.

Disturbed areas, fields, pastures. Flowering May–Oct. 0–300 m. Est, WV. CA, WA; north to British Columbia, eastern Canada and US, southern US; nearly worldwide. Exotic.

Lepidium draba Linnaeus (p. 493)

Heart-podded hoary cress, hoary pepperwort

Cardaria draba (L.) Desv., *Cardaria draba* (L.) Desv. var. *draba*, *Lepidium draba* L. ssp. *draba*

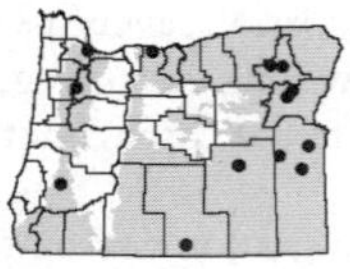

Plants perennial, hirsute to subglabrous; rhizomatous. **Stems** (0.8)2–6.5(9) dm. **Basal leaves** not rosulate, obovate, spatulate, or ovate, (1.5)3–10(15) × 1–4 cm, margins dentate or entire; petioles 1–4 cm. **Cauline leaves** ovate, elliptic, oblong, lanceolate, or obovate, (1)3–9(15) × (0.5)1–2(5) cm, bases sagittate-amplexicaul or auriculate, margins dentate or entire; sessile. **Inflorescences** corymbose panicles, rachises glabrous or puberulent with straight or curved cylindrical trichomes; fruiting pedicels ascending to horizontal, terete, straight, 5–10(15) mm, glabrous or sparsely puberulent adaxially. **Flowers** sepals caducous, 1.5–2.5 mm; petals obovate, (2.5)3–4(4.5) × (1)1.3–2(2.2) mm, white; stamens 6; styles (0.6)1–1.8(2) mm. **Fruits** indehiscent, flattened, cordate to subreniform, (2)2.5–3.7(4.3) × (3.2)3.7–5(5.6) mm, tips obtuse to subacute, apical notch absent; valves smooth, glabrous, reticulate-veined; wingless. **Seeds** ovate, 1.5–2.3 × 1–1.3 mm; wingless; cotyledons incumbent. **2*n***=32, 64.

Disturbed areas, fields, riverbanks. Flowering Apr–Aug. 0–1700 m. BR, BW, Col, ECas, Owy, Sisk, WV. CA, ID, NV, WA; throughout North America except northern Canada and southeast US; nearly worldwide. Exotic.

Lepidium heterophyllum Bentham (p. 495)

Purple-anther field pepperweed, Smith's pepperwort

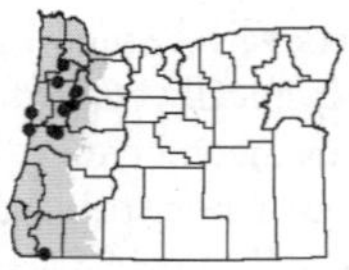

Plants perennial, hirsute; with branched caudices. **Stems** 1–5 dm. **Basal leaves** rosulate, oblanceolate or oblong-elliptic, 1–4.5 cm × 4–14 mm, margins dentate or entire; petioles 1–6.2 cm. **Cauline leaves** oblong to deltoid-lanceolate, 1–3.5 cm × 3–8 mm, bases sagittate or auriculate, margins dentate to denticulate. **Inflorescences** elongated, rachises hirsute with straight, spreading trichomes; fruiting pedicels horizontal, terete, straight or slightly recurved, 2.8–5 mm, hirsute all around. **Flowers** sepals caducous, 1.6–2.2 mm; petals spatulate, 1.8–2.8 × 0.8–1.4 mm, white; stamens 6; styles (0.6)1–1.5 mm, well exserted beyond apical notch. **Fruits** dehiscent, broadly oblong to ovate, 4–5.5 × 3.5–4 mm, incurved adaxially, apical notch 0.2–0.3 mm deep; valves often not papillate, not veined; apically broadly winged. **Seeds** ovoid, 1.8–2.2 × 1–1.2 mm; wingless; cotyledons incumbent. **2*n***=48.

Disturbed areas, fields, gardens, hillsides. Flowering Apr–Jun. 0–800 m. CR, Est, Sisk, WV. CA, WA; north to British Columbia, NJ, NY, MA; Europe. Exotic.

Lepidium latifolium Linnaeus (p. 495)

Broad-leaved peppergrass, broad-leaved pepperweed, perennial pepperweed

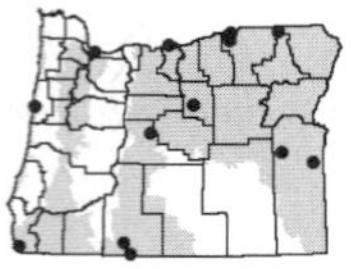

Plants perennial, glabrous or pubescent; with thick rhizomes and woody caudices. **Stems** (2)3.5–12(15) dm. **Basal leaves** not rosulate, elliptic-ovate to oblong, (2)3.5–15(25) cm × (5)15–50(80) mm, leathery, margins entire or serrate; petioles 1–9(14) cm. **Cauline leaves** oblong to elliptic-ovate or lanceolate, (1)2–9(12) cm × 3–45 mm, bases not auriculate, margins serrate or entire; sessile or petiolate. **Inflorescences** rachises glabrous or puberulent with straight cylindrical trichomes; fruiting pedicels ascending to divaricate, terete, straight or curved, 2–5(6) mm, glabrous or puberulent adaxially. **Flowers** sepals caducous, 1–1.4 mm; petals obovate, 1.8–2.5 × (0.8)1–1.3 mm, white; stamens 6; styles 0.05–0.15 mm, slightly exserted beyond apical notch. **Fruits** dehiscent, oblong-elliptic, broadly ovate or suborbicular, (1.6)1.8–2.4(2.7) × 1.3–1.8 mm, apical notch absent or rarely to 0.1 mm; valves smooth, glabrous or sparsely pilose, not veined; wingless. **Seeds** oblong, (0.8)1–1.2 × 0.6–0.9 mm; cotyledons incumbent. **2*n***=24.

Saline meadows, streambanks, fields, grasslands, disturbed places, slopes, disturbed areas, sagebrush and pinyon-juniper communities, barren hillsides, edge of marshes. Flowering May–Sep. 0–1400 m. BW, Col, ECas, Est, Lava, Owy, Sisk, WV. CA, ID, NV, WA; western North America, widely scattered eastern North America; Africa, Asia, Europe. Exotic.

Lepidium montanum Nuttall (p. 495)

Mountain pepperweed, mountain pepperwort

Lepidium montanum Nutt. var. *canescens* (Thell.) C.L. Hitchc., *Lepidium montanum* Nutt. var. *montanum*, *Lepidium montanum* Nutt. var. *nevadense* Rollins

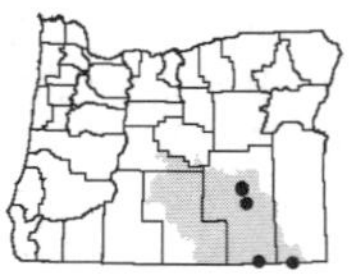

Plants annual, biennial, or perennial, cespitose or not; with woody caudices or not. **Stems** 0.4–5(7) dm. **Basal leaves** rosulate or not, (0.8)1.5–4(6) cm, 1- or 2-pinnatifid or -pinnatisect, lobes entire or dentate; petioles (0.5)1.2–5.3(7.6) cm. **Cauline leaves** bases not auriculate; short-petiolate. **Inflorescences** rachises puberulent with straight or curved, cylindrical trichomes, rarely glabrous; fruiting pedicels divaricate to horizontal, not appressed to rachis, terete, slightly recurved or somewhat sigmoid, (2.7)3.3–8.5(10) mm, puberulent adaxially. **Flowers** sepals caducous, 1.2–1.8(2.1) mm; petals spatulate to oblanceolate, 2.2–3.7(4.3) × 1.3–1.8 mm, white;

stamens 6; styles obsolete or 0.2–0.7(1) mm, exserted beyond or rarely subequaling apical notch. **Fruits** dehiscent, ovate, suborbicular, or rarely oblong, 2–4.3(5) × (1.5)1.8–3.6(4) mm, apical notch 0.1–0.3 mm deep; valves smooth, glabrous or rarely puberulent, not veined; apically winged. **Seeds** oblong, 1.2–1.8 × 0.7–1 mm; cotyledons incumbent, entire. ***2n***=32.

Sagebrush communities, cliffs, pinyon-juniper woodlands, rocky hillsides, washes, alkaline flats and lowlands, roadsides. Flowering May–Aug. 1000–2500 m. BR. CA, ID, NV; northeast to MT, east to WY, southeast to NM. Native.

Lepidium nitidum Nuttall (p. 495)

Shining pepperweed

Lepidium nitidum Nutt. var. *nitidum*, *Lepidium nitidum* Nutt. var. *oreganum* (Howell ex Greene) C.L. Hitchc.

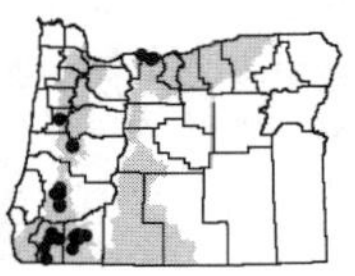

Plants annual, puberulent. **Stems** (0.4)1–3.5(4.2) dm. **Basal leaves** not rosulate, pinnatisect, (0.8)1.5–7.3(8) cm, lobes linear to filiform, rarely oblong or lanceolate, margins entire or rarely dentate; petioles (0.4)1–4.2(5) cm. **Cauline leaves** similar to basal, linear, bases not auriculate; petiolate or subsessile. **Inflorescences** rachises puberulent with straight cylindrical trichomes; fruiting pedicels 2.5–5(6.5) mm, strongly flattened, 0.5–1.1 mm wide, slightly or weakly recurved, rarely straight, divaricate, rarely ascending or suberect, not appressed to rachis, puberulent adaxially and sometimes abaxially. **Flowers** sepals caducous, (0.7)0.9–1.3(1.5) mm; petals oblanceolate, 1–2.5 × 0.2–1(1.6) mm or rarely rudimentary or absent, white; stamens 4(6), median; styles obsolete or rarely to 0.1 mm, included in apical notch. **Fruits** dehiscent, orbicular to broadly ovate, (2.5)3–5.5(6.5) × (2)2.6–5(5.4) mm, usually red at maturity, apical notch 0.3–0.7(1) mm deep; valves smooth, usually glabrous, not or obscurely veined; apically winged. **Seeds** ovate-oblong, 1.6–2.6 × 0.9–1.2 mm; cotyledons incumbent.

Meadows, pastures, dry vernal pools, gravelly slopes. Flowering Feb–May. 0–700 m. Col, ECas, Sisk, WV. CA, NV, WA; south to Mexico. Native.

The species is highly variable in fruit morphology and has been divided into three varieties by some authorities.

Lepidium oblongum Small (p. 495)

Veiny pepperweed

Lepidium oblongum Small var. *oblongum*

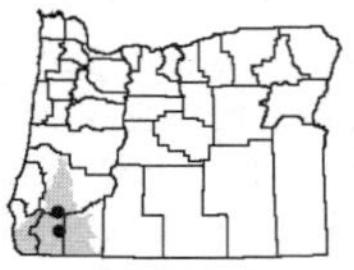

Plants annual, hirsute. **Stems** (5)1–2.4(3.2) dm. **Basal leaves** not rosulate, 0.7–3.5 cm, 1- or 2-pinnatifid, lobes entire or dentate; petioles (0.5)1–3. **Cauline leaves** 0.8–2 cm × 2–9 mm, bases auriculate or not, margins laciniate or pinnatifid; sessile, rarely short-petiolate. **Inflorescences** rachises hirsute with straight cylindrical trichomes; fruiting pedicels divaricate to horizontal, terete, slightly recurved or rarely straight, 2–3.5(5) mm, puberulent adaxially or rarely all around. **Flowers** sepals caducous to somewhat persistent, 0.7–1 mm; petals absent or rudimentary, linear-oblanceolate, 0.1–0.7 × 0.05–0.15 mm, white; stamens 2, median; styles to 0.1 mm, included in apical notch. **Fruits** dehiscent, orbicular to obovate or elliptic, 2.2–3.5 × 2–3 mm, apical notch

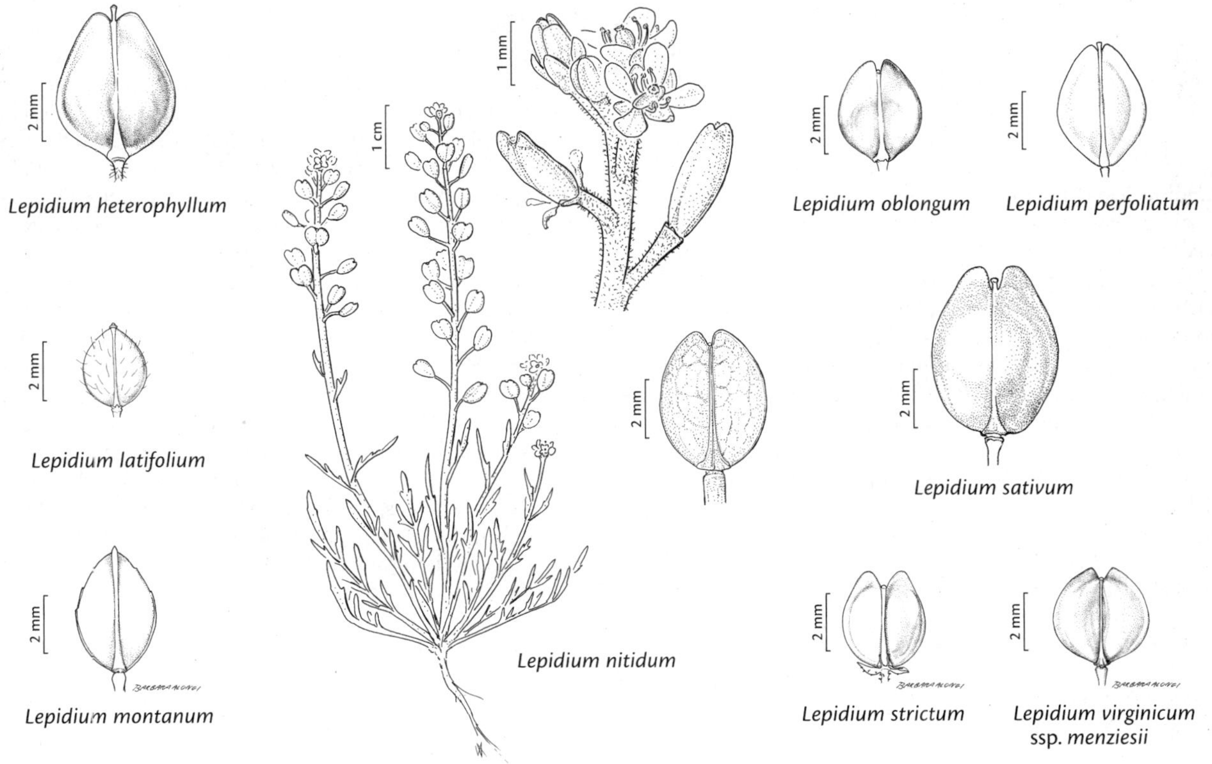

LEPIDIUM

0.2–0.3 mm deep; valves smooth, glabrous or sparsely puberulent along margin, not veined; apically winged. **Seeds** ovate, 1.2–1.6 × 0.7–1 mm; wingless; cotyledons incumbent.

Prairies, pastures, disturbed areas. Flowering Apr–Aug. 200–400 m. Sisk. CA; midwestern and southern US, Mexico. Native.

Lepidium perfoliatum Linnaeus (p. 495)

Clasping pepperweed

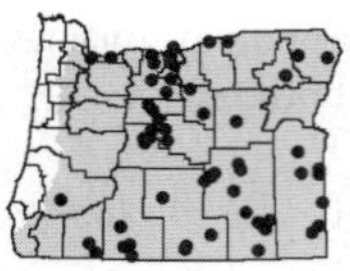

Plants annual or occasionally biennial, glaucous, glabrous above, glabrous or sparsely pubescent below. **Stems** (0.7)1.5–4.3(5.6) dm. **Basal leaves** rosulate, (1)3–8(15) cm, 2- or 3-pinnatifid, lobes linear to oblong, margins entire; petioles (0.5)1–2(4) cm. **Cauline leaves** middle and upper ovate to cordate or suborbicular, (0.5)1–3(4) × (0.5)1–2.5(3.5) cm, bases deeply cordate-amplexicaul, margins entire; sessile. **Inflorescences** rachises glabrous; fruiting pedicels divaricate-ascending to horizontal, terete, straight, 3–6(7) mm, glabrous. **Flowers** sepals caducous, 0.8–1(1.3) mm; petals narrowly spatulate, 1–1.5(1.9) × 0.2–0.5 mm, pale yellow; stamens 6; styles 0.1–0.4 mm, subequaling or slightly exserted beyond apical notch. **Fruits** dehiscent, orbicular, rhombic, or broadly obovate, 3–4.5(5) × 3–4.1 mm, apical notch 0.1–0.3 mm deep; valves smooth, glabrous, not veined; apically winged. **Seeds** ovate, 1.6–2(2.3) × 1.2–1.4 mm; narrowly winged; cotyledons incumbent. ***2n***=16.

Disturbed areas, pinyon-juniper woodlands, sagebrush flats, open deserts, fields. Flowering Apr–Jun. 0–1900 m. BR, BW, Casc, Col, ECas, Lava, Owy, Sisk, WV. CA, ID, NV, WA; throughout North America; Africa, Asia, Europe. Exotic.

Lepidium sativum Linnaeus (p. 495)

Garden cress, garden pepperwort, pepper cress

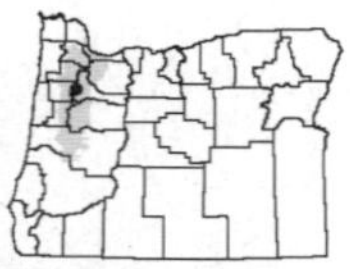

Plants annual, glabrous or sparsely pilose, often glaucous. **Stems** (1)2–8(10) dm. **Basal leaves** not rosulate, 2–8(10) cm, 1- or 2-pinnatifid or -pinnatisect, lobes ovate to oblong, margins entire or dentate; petioles 1–4 cm. **Cauline leaves** less divided, uppermost leaves often linear, bases not auriculate, margins entire; petiolate. **Inflorescences** rachises glabrous; fruiting pedicels erect to ascending, appressed to rachis, terete or rarely slightly flattened, straight, 1.5–4(6) mm, glabrous. **Flowers** sepals caducous, 1–1.8 mm; petals spatulate to obovate, 2–4 × 0.7–1.4 mm, white or lavender; stamens 6; styles 0.2–0.5(0.8) mm, included in or rarely subequaling apical notch. **Fruits** dehiscent, broadly ovate or ovate-oblong, (4)5–6.4(7) × 3–4.5(5.6) mm, apical notch 0.2–0.8 mm deep; valves smooth, glabrous, not veined; apically broadly winged. **Seeds** ovate-oblong, 2–2.7(3) × 1–1.5 mm; cotyledons incumbent, 3-lobed. ***2n***=16, 32.

Disturbed areas, cultivated areas. Flowering Apr–Aug. 0–400 m. WV. CA, WA; scattered throughout northern North America; Africa, Asia, Europe. Exotic.

Lepidium strictum (S. Watson) Rattan (p. 495)

Upright pepperweed

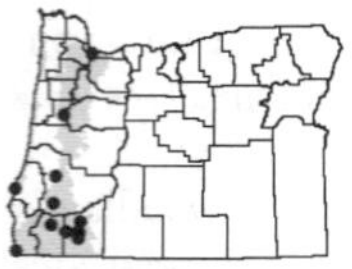

Plants annual, hirsute. **Stems** (0.4)0.7–1.7(2) dm. **Basal leaves** not rosulate, 1.5–5.6 cm, 2-pinnatifid, lobes lanceolate to oblong, margins entire; petioles 0.5–3 cm. **Cauline leaves** 0.8–3 cm × 0.3–8 mm, pinnatifid, bases not auriculate; short-petiolate. **Inflorescences** rachises puberulent with straight cylindrical trichomes; fruiting pedicels strongly recurved, suberect and subappressed at base, becoming divaricate distally, often flattened, (1)1.4–2.5(3) mm, puberulent adaxially. **Flowers** sepals persistent, 0.7–1(1.2) mm; petals rudimentary, linear, 0.2–0.5 × ~0.1 mm, white; stamens 2, median; styles obsolete or to 0.1 mm, included in apical notch. **Fruits** dehiscent, ovate-orbicular to ovate, 2.5–3.3 × 2–3 mm, apical notch 0.3–0.6 mm deep; valves smooth, glabrous or puberulent at margin, reticulate-veined; apically winged. **Seeds** oblong, 1.2–1.6 × 0.7–0.8 mm; cotyledons incumbent. ***2n***=32.

Woods, disturbed areas. Flowering Apr–Jun. 0–500 m. Est, Sisk, WV. CA; UT, CO; Chile. Exotic.

Lepidium virginicum Linnaeus ssp. ***menziesii*** (de Candolle) Thellung (p. 495)

Medium pepperweed, hairy pepperweed, Menzies' pepperweed

Lepidium menziesii DC., *Lepidium virginicum* L. var. *medium* (Greene) C.L. Hitchc., *Lepidium virginicum* L. var. *menziesii* (DC.) C.L. Hitchc., *Lepidium virginicum* L. var. *pubescens* (Greene) Schmitz

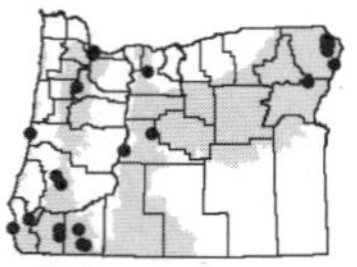

Plants annual, puberulent. **Stems** (0.6)1.5–5.5(7) dm. **Basal leaves** not rosulate, obovate to oblanceolate, (1)2.5–10(15) × 0.5–3(5) cm, margins pinnatifid, lyrate, or dentate; petioles 0.5–3.5 cm. **Cauline leaves** oblanceolate or linear, 1–6 cm × (1)3–10 mm, bases not auriculate, margins serrate or entire; petiolate. **Inflorescences** rachises puberulent with curved cylindrical trichomes, rarely glabrous; fruiting pedicels divaricate-ascending to horizontal, straight or recurved, flattened at least below apex, 2.5–4(6) × (0.2)0.3–0.4 mm, puberulent adaxially or all around, or glabrous. **Flowers** sepals caducous, (0.5)0.7–1(1.1) mm; petals rarely rudimentary, spatulate to oblanceolate, 1–2(2.5) × 0.3–0.8(1) mm, white; stamens 2, median; styles 0.1–0.2 mm, included in apical notch. **Fruits** dehiscent, orbicular, 2.5–3.5(4) mm in diameter, widest at middle, apical notch 0.2–0.5 mm deep; valves smooth, glabrous or puberulent, not veined; apically winged. **Seeds** ovate, 1.3–1.9(2.1) × 0.7–1(1.2) mm; cotyledons accumbent or rarely incumbent. ***2n***=32.

Desert shrub communities, disturbed areas, pastures, riverbanks, woods, dry mountain slopes. Flowering Mar–Jul. 0–1600 m. BW, ECas, Est, Lava, Sisk, WV. CA, ID, NV, WA; north to British Columbia, northeast to MT, east to WY, southeast to TX, south to Mexico. Native.

LOBULARIA Sweet alyssum

Herbs or sometimes **subshrubs** annual or perennial. **Trichomes** malpighiaceous. **Leaves** cauline, bases not auriculate, margins entire; sessile. **Inflorescences** bracts present or absent. **Flowers** sepals erect or spreading, caducous, ovate or oblong, lateral pair not saccate; petals spatulate, obovate, or orbicular, > sepals, white or purple, tips rounded, clawed; nectar glands 8, 4 lateral and 4 median; stamens 6, tetradynamous, anthers ovate, obtuse; ovules 2–13 per ovary, septa complete, styles 0.6 mm; stigmas capitate, entire. **Fruits** dehiscent silicles, orbicular, obovate, or elliptic, latiseptate, unsegmented; replums rounded. **Seeds** uniseriate or biseriate, ovate or orbicular, compressed, mucilaginous when wetted; cotyledons accumbent.

Worldwide. 2 species; 1 species treated in *Flora*.

Lobularia maritima (Linnaeus) Desvaux (p. 497)

Sweet alyssum

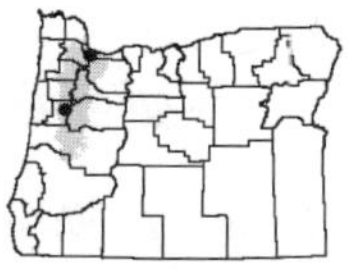

Herbs or sometimes **subshrubs** perennial. **Stems** 5–25(40) cm. **Leaves** linear, lanceolate, or oblanceolate, 10–25(42) × 1–3(7) mm, margins entire, surfaces sericeous. **Inflorescences** fruiting pedicels ascending, (3)4.5–6(9.5) mm. **Flowers** sepals 1.5–1.7(2.4) mm; petals broadly obovate, (2)2.3–2.8(3) × (1.2)1.6–2(2.5) mm; styles 0.4–0.6 mm. **Fruits** elliptic to suborbicular, (2)2.3–2.7(4.2) × (1.2)1.6–2(2.9) mm; valves 1-veined, pubescent, smooth; seeds 1 in each locule, lenticular, ovate, (1)1.2–1.4(2) × (0.7)1–1.3(1.6) mm; wingless. **2*n***=24.

Fields, roadsides, waste places. Flowering Apr–Aug. 0–100 m. WV. CA, ID, NV, WA; scattered throughout most of North America; nearly worldwide. Exotic.

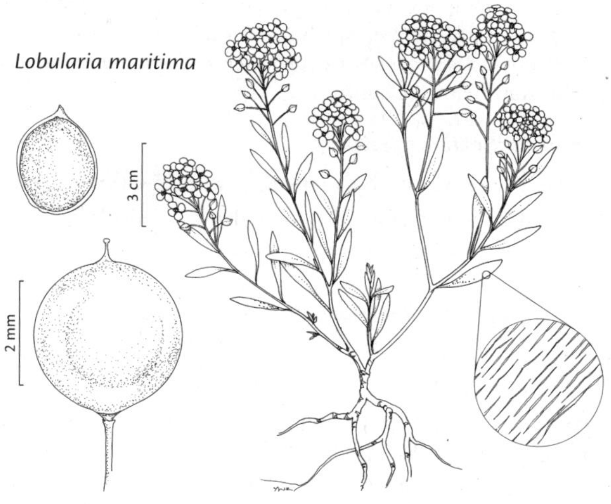

LOBULARIA

LUNARIA Honesty, money-plant, satin-flower

Herbs biennials or perennial, rarely annual. **Trichomes** simple. **Basal leaves** rosulate, margins coarsely dentate; long-petiolate. **Cauline leaves** bases not auriculate, margins dentate; petiolate. **Inflorescences** bracts 0 or rarely lowermost flowers with bracts. **Flowers** sepals erect, caducous, linear, lateral pair strongly saccate; petals obovate, obtuse, much > sepals, violet or purple, rarely lavender or white, claws strongly differentiated from blades; nectar glands lateral, annular or semi-annular; stamens 6, tetradynamous, anthers oblong or linear, obtuse; ovules 4–8 per ovary, septa complete, styles to 1 cm, stigmas 2-lobed, lobes opposite replum, connivent or not; gynophore 1–5 cm. **Fruits** dehiscent silicles, strongly latiseptate, orbicular, oblong, or lanceolate-elliptic, unsegmented; valves not veined, not torulose; replums rounded. **Seeds** biseriate, reniform to orbicular, strongly flattened, not mucilaginous when wetted; funicles united with septum; broadly winged or wingless; cotyledons accumbent.

Europe, North America. 3 species; 1 species treated in *Flora*.

Lunaria annua Linnaeus (p. 497)

Honesty, money-plant, moonwort

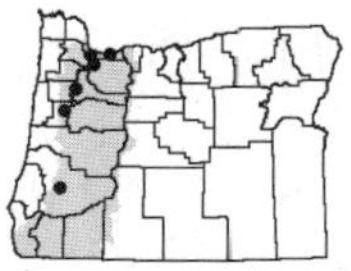

Plants annual to biennial, hispid. **Stems** (3)4–10(12) dm. **Basal leaves** broadly cordate to cordate-ovate, (1.5)3–12(18) × (1)2–8(12) cm; petioles (1.5)3–10(17) cm. **Cauline leaves** similar to basal, alternate, lowermost opposite; uppermost sessile. **Inflorescences** fruiting pedicels divaricate or ascending, (7)10–15 mm. **Flowers** sepals 6–9(10) mm; petals (15)17–25(30) × 5–10 mm, claws 6–10 mm; styles 4–10 mm; gynophore 7–18 mm. **Fruits** broadly oblong to suborbicular, 3–4.5(5) × 2–3(3.5) cm; valves rounded at tip and base. **Seeds** reniform, (6)7–10(12) × 5–9 mm. **2*n***=30.

Woods, disturbed areas, pasture margins. Flowering Mar–May. 0–1000 m. Casc, Sisk, WV. CA, ID, WA; widely scattered in North America; Europe. Exotic.

Lunaria annua is a widely cultivated ornamental known for its attractive flowers.

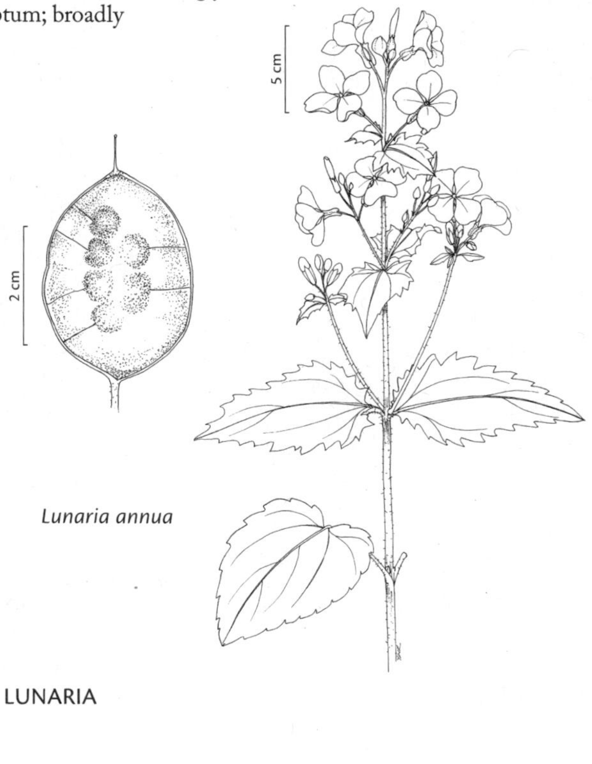

LUNARIA

NASTURTIUM Watercress

Herbs perennial; rhizomatous. **Trichomes** absent or simple. **Stems** rooting from lower nodes. **Leaves** cauline, pinnately compound, or often simple in deeply submersed plants; lateral leaflets 1–6(12) pairs, petiolulate or sessile; petioles sometimes auriculate at base. **Inflorescences** bracts 0. **Flowers** sepals erect or ascending, caducous, ovate or oblong, lateral pair not saccate; petals obovate or narrowly spatulate, > sepals, white or rarely pink, claws undifferentiated, tips obtuse; nectar glands 2, lateral; stamens 6, tetradynamous, anthers oblong, tips obtuse; ovules 25–50 per ovary, septa complete, styles obsolete or to 2 mm, stigmas capitate, entire. **Fruits** dehiscent siliques, linear, terete, unsegmented; valves obscurely veined, smooth; replums rounded. **Seeds** uniseriate or biseriate, oblong or ovoid, plump, minutely to coarsely reticulate, not mucilaginous when wetted; wingless; cotyledons accumbent.

Africa, Asia, Europe, North America. 5 species; 2 species treated in *Flora*.

Nasturtium × *sterile* (*N. microphyllum* × *N. officinale*), a sterile interspecific hybrid, is known only from two old collections (both 1916) from Benton County.

Key to *Nasturtium* species:

1. Seeds biseriate, coarsely reticulate, with 25–50(60) areoles on each side; fruits (1.8)2–2.5(3) mm wide . . . . ***N. officinale***

1' Seeds uniseriate, minutely to moderately reticulate, with (75)100–150(175) areoles on each side; fruits 1–1.5(1.8) mm wide . . . . ***N. microphyllum***

Nasturtium microphyllum Boenninghausen ex Reichenbach

One-row yellowcress, watercress

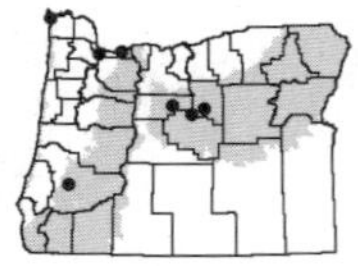

Plants glabrous or sparsely pubescent. **Stems** (1)2–10(18) dm. **Leaves** (3)5–9(11)-foliolate, (1.5)2.5–10(15) cm, bases auriculate or not; terminal leaflets suborbicular or oblong, (6)15–35(45) × (4)08–20(25) mm, bases obtuse, cuneate, or subcordate, margins entire or repand, tips obtuse; lateral leaflets smaller; sessile or petiolulate. **Inflorescences** fruiting pedicels divaricate or descending, straight or recurved, (7)9–21(25) mm. **Flowers** sepals oblong, 2.5–4 mm; petals spatulate or obovate, 4.5–6 × 1.5–3 mm, white or pink, tips rounded, attenuate to claw-like; ovules 30–38(40) per ovary, styles 0.5–1.5(2) mm. **Fruits** linear, 14–22(27) × 1–1.5(1.8) mm. **Seeds** uniseriate, ovoid, (0.8)1–1.2 × (0.6)0.7–0.8(0.9) mm, minutely to moderately reticulate, with (75)100–150(175) areoles on each side. **2*n***=64.

Lake margins, riverbanks, springs, seeps, wet meadows. Flowering May–Jul. 0–1400 m. BW, Casc, Est, Lava, Sisk. ID; scattered throughout Canada and US; Australia, Europe. Exotic.

Nasturtium officinale W.T. Aiton (p. 498)

Watercress

Rorippa nasturtium-aquaticum (L.) Hayek

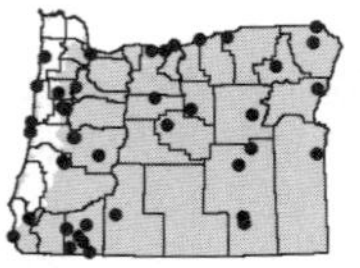

Plants glabrous or sparsely pubescent. **Stems** 1–11(20) dm. **Leaves** 3–9(13)-foliolate, (1)2–15(22) cm, bases usually auriculate; terminal leaflets suborbicular, ovate, oblong, or lanceolate, (4)10–40(50) × (3)7–25(40) mm, margins entire or repand; lateral leaflets smaller; sessile or petiolulate. **Inflorescences** fruiting pedicels divaricate or descending, straight or recurved, 5–17(24) mm. **Flowers** sepals 2–3.5 mm; petals 2.8–4.5(6) × 1.5–2.5 mm, white; ovules (28)36–50(60) per ovary, styles 0.5–1(1.5) mm. **Fruits** (6)10–18(25) × (1.8)2–2.5(3) mm. **Seeds** biseriate, ovoid, (0.8)0.9–1.1(1.3) × (0.6)0.7–0.9(1) mm, coarsely reticulate, with 25–50(60) areoles on each side. **2*n***=32.

Marshes, flowing streams, lake margins, ditches, swamps, seeps. Flowering Feb–Sep. 0–2000 m. All ecoregions except CR. CA, ID, NV, WA; throughout North America; Africa, Asia, Europe. Exotic.

NASTURTIUM

NOCCAEA Pennycress

Herbs often glaucous. **Trichomes** absent. **Basal leaves** rosulate, margins entire or dentate; petiolate. **Cauline leaves** bases auriculate, sagittate, or amplexicaul, margins entire or dentate; sessile. **Inflorescences** bracts 0. **Flowers** sepals erect, caducous, oblong or ovate, lateral pair not saccate; petals > sepals, blades obovate, spatulate, oblanceolate, oblong, or rarely linear, white, pink, or purple, claws obscurely differentiated, petal tips obtuse; nectar glands lateral; stamens 6, tetradynamous, anthers ovate or oblong, obtuse; septa complete, styles distinct, stigmas capitate, entire. **Fruits** dehiscent silicles, or rarely siliques, obcordate, obovate, obdeltoid, elliptic, or oblong, angustiseptate or rarely inflated, unsegmented; valves often keeled, smooth; replums rounded; winged or wingless. **Seeds** 2+ per fruit, uniseriate, oblong to ovate, smooth, not mucilaginous when wetted; wingless; cotyledons accumbent.

Asia, Europe, North America, South America. 128 species; 1 species treated in *Flora.*

Noccaea fendleri (A. Gray) Holub

Pennycress

Plants perennial; caudices simple or branched. **Stems** 1–45 cm. **Basal leaves** linear, oblong, oblanceolate, ovate, to spatulate, 4–30 × 2–20 mm; petioles 4–73 mm. **Cauline leaves** 2–21. **Inflorescences** fruiting pedicels horizontal to ascending, 2.5–15 mm. **Flowers** sepals 1.6–5.3 mm; petals spatulate, (3.4)4.2–13 × 1–4.9 mm, white to pinkish purple, tips obtuse; ovules 4–10 per ovary, styles (0.4)1.1–4.2 mm. **Fruits** obovate, obcordate, obdeltoid, or elliptic, 2.5–16 × 1.5–9 mm, tips obtuse, truncate, or emarginate; winged or very rarely wingless apically. **Seeds** ovate, 1.1–2.1 mm. **2*n***=14.

Northern Mexico and western United States. 5 subspecies; 2 subspecies treated in *Flora.*

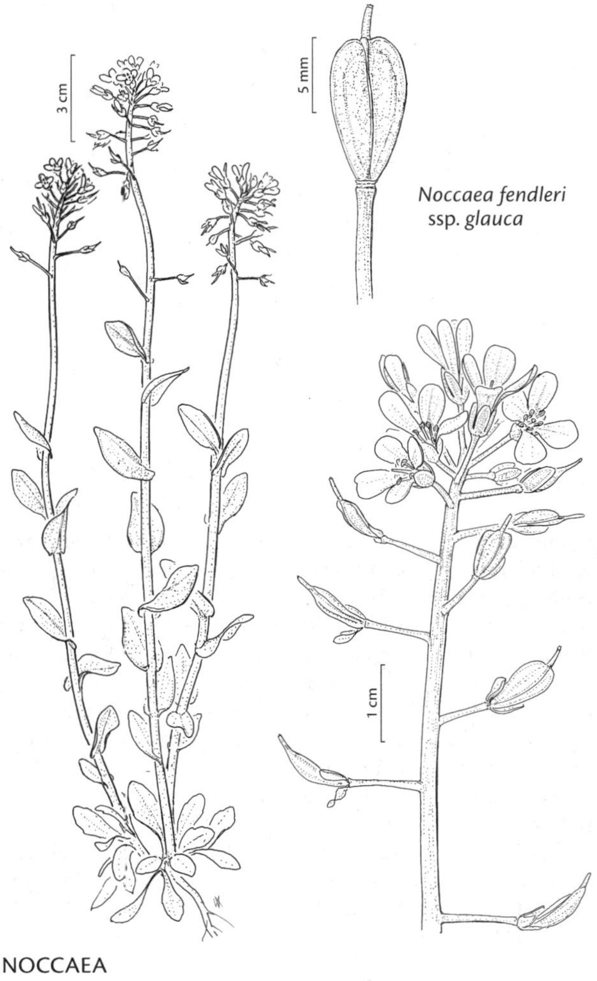

NOCCAEA

Key to *Noccaea fendleri* subspecies:

1. Seeds 4–6 per fruit; petioles of basal leaves 0.8–1(2) × as long as blades; cauline leaves 4–16(21). . . . . *N.f.* ssp. ***glauca***

1' Seeds 2–4 per fruit; petioles of basal leaves (1.7)2–3.6(4) × as long as blades; cauline leaves 2–6 . . . . *N.f.* ssp. ***siskiyouensis***

Noccaea fendleri (A. Gray) Holub ssp. *glauca* (A. Nelson) Al-Shehbaz & M. Koch (p. 499)

Alpine pennycress, rock pennycress

Thlaspi alpestre L., *Thlaspi fendleri* A. Gray var. *glaucum* (A. Nelson) C.L. Hitchc., *Thlaspi montanum* L. var. *montanum*

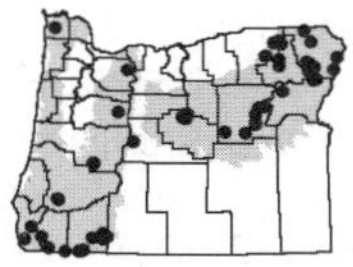

Stems (1.5)5–32(45) cm. **Basal leaves** ovate to oblong, 4–9(15) mm wide; petioles 0.8–1(2) × as long as blades. **Cauline leaves** (4)7–16(21). **Flowers** petals (3.4)4–7(8.5) × (1)1.5–2.7(4.2) mm, white or occasionally pinkish purple; styles (0.4)1–2.2(3) mm in fruit. **Fruits** obovate to obcordate, (2.5)5–8(12) × (1.5)2.6–4.5(6.6) mm, tips obtuse, truncate, or emarginate; winged or occasionally wingless. **Seeds** 4–6 per fruit. **2*n***=14, 28.

Rocky or talus slopes, streamsides, alpine or subalpine meadows, near snow banks, forest clearings. Flowering Apr–Jul. 200–2800 m. BW, Casc, CR, Lava, Sisk. CA, ID, NV, WA; northeast to MT, southeast to TX, south to Mexico. Native.

Noccaea fendleri (A. Gray) Holub ssp. *siskiyouensis* (P.K. Holmgren) Al-Shehbaz & M. Koch

Siskiyou pennycress

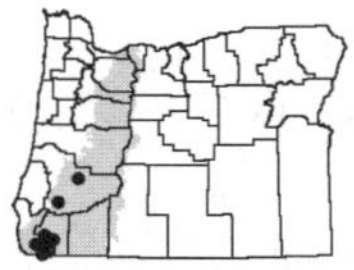

Stems (5.5)7–16(24) cm. **Basal leaves** ovate to oblong, 4–8(9) mm wide; petioles (1.7)2–3.6(4) × as long as blades. **Cauline leaves** 2–6. **Flowers** petals 3.8–6 × 1.2–2.2(2.9) mm, white; styles (0.8)1–2 mm in fruit. **Fruits** obovate to obdeltoid, (4)4.8–7(8) × (3)3.5–5.5(6) mm, tips truncate to strongly emarginate; styles (0.8)1–2 mm in fruit; often strongly winged. **Seeds** 2–4 per fruit. **2*n***=14.

Moist, open rocky serpentine slopes. Flowering Mar–Jun. 200–1900 m. Casc, Sisk. Native.

This subspecies is currently known to be endemic to Curry, Douglas, and Josephine counties.

ODONTARRHENA Odontarrhena

Herbs perennial; with woody caudices. **Stems** some branches terminated with sterile shoots and rosettes. **Trichomes** (6)8–20(24)-rayed, stellate-appressed, sessile. **Basal leaves** rosulate or not, simple, margins entire; petiolate. **Cauline leaves** bases not auriculate, margins entire; petiolate or subsessile. **Inflorescences** bracts 0. **Flowers** sepals erect to ascending, caducous, ovate to oblong, bases of lateral pair not saccate; petals obovate to oblanceolate, > sepals, yellow, claws slightly or strongly differentiated from blade, < sepals, petal tips obtuse or retuse; nectar glands 4, lateral; stamens 6, slightly tetradynamous, filaments appendage-winged, entire or toothed, anthers ovate or rarely oblong, tips obtuse; ovules 2 per ovary, septa complete, styles 1–3 mm, stigmas capitate, entire. **Fruits** dehiscent silicles, rarely indehiscent and samaroid, strongly latiseptate, unsegmented; valves veinless, glabrous or pubescent; replums rounded. **Seeds** aseriate, ovate, flattened; not or rarely mucilaginous when wetted; cotyledons accumbent.

Northern Hemisphere. 82 species; 1 species treated in *Flora*. See also *Alyssum* discussion.

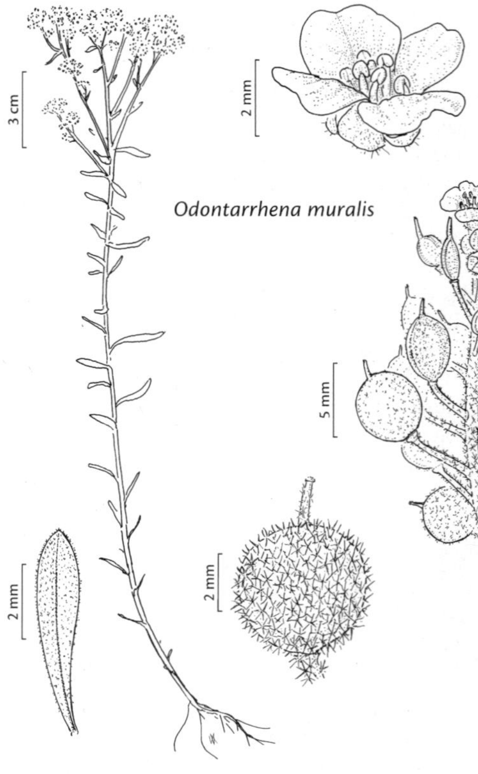

ODONTARRHENA

Odontarrhena muralis (Waldstein & Kitaibel) Endlicher (p. 500)

Yellowtuft

Alyssum murale Waldst. & Kit.

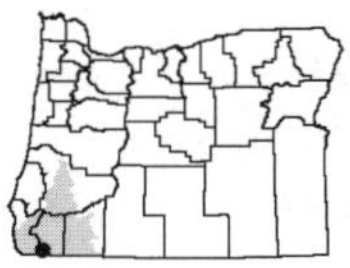

Plants cespitose, canescent or not. **Trichomes** 6–20-rayed stellate, sessile. **Stems** (2.5)3–6(7) dm, often several from caudices. **Cauline leaves** narrowly oblanceolate to linear, (5)7–25(30) × 1–5(8) mm, surfaces abaxially often more grayish than adaxially; subsessile. **Inflorescences** fruiting pedicels divaricate-ascending, slender, (2)2.5–6(8) mm. **Flowers** sepals caducous, oblong, 1.2–2 mm, stellate pubescent; petals spatulate, 2.5–3.5 × 0.5–1 mm, tips obtuse or rounded; filaments 1.5–2.5 mm, median pairs broadly winged, apically 1-toothed, lateral pair with tips subacute, basal appendage lanceolate or narrowly oblong; styles 1–2 mm. **Fruits** broadly elliptic to orbicular, 3.5–5 × 2.5–5 mm, tips obtuse or rounded; valves flattened, slightly undulate, pubescent with stellate trichomes. **Seeds** suborbicular to broadly ovate, compressed, 3–3.8 × 2–3.2 mm; wings broad, 0.5–0.9 mm wide. **2*n*=**16, 32.

Disturbed areas. Flowering May–Jul. 300–500 m. Sisk. CO, UT, widely scattered in Canada; Asia, Europe. Exotic.

PHOENICAULIS Daggerpod

Genus composed of one species, *P. cheiranthoides*.

Phoenicaulis cheiranthoides Nuttall (p. 501)

Daggerpod

Phoenicaulis cheiranthoides Nutt. ssp. *cheiranthoides*, *Phoenicaulis cheiranthoides* Nutt. ssp. *lanuginosa* (S. Watson) Abrams

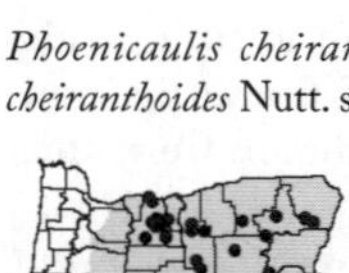

Herbs perennial; woody caudices simple or few-branched, covered with petiolar remains of previous years. **Trichomes** finely dendritic. **Stems** (8)12–25(30) cm, glabrous. **Basal leaves** persistent, linear-oblanceolate to obovate, (2)3–7(10) × (0.5)0.8–2(2.6) cm, margins entire, surfaces tomentose; petioles (1)2–5.5(8) cm. **Cauline leaves** 5–12, ovate, oblong, or linear, 7–20 × 2–6 mm, bases auriculate, margins entire; sessile. **Inflorescences** bracts 0; fruiting pedicels divaricate, (6)10–30(35) mm. **Flowers** sepals erect, caducous, oblong, 4–6 mm, glabrous, lateral pair saccate; petals spatulate to oblanceolate, 9–13(15) × 2.5–4 mm, > sepals, purple, claws undifferentiated from blades, petal tips obtuse; nectar glands confluent; stamens 6, tetradynamous, anthers oblong, 1–1.5 mm, tips obtuse; ovules (6)8–16(18) per ovary, septa complete, styles 0.5–2 mm, stigmas capitate, entire; sessile or gynophores to 2 mm. **Fruits** dehiscent siliques, latiseptate, lanceolate or linear, (20)30–60(90) × (2)3–5(6) mm, glabrous, unsegmented; valves with prominent midveins, not torulose; replums rounded. **Seeds** (6)8–16(18) per fruit, uniseriate, broadly ovate to oblong, flattened, (2)2.5–3.2 × (1.5)1.8–2.5 mm, not mucilaginous when wetted; wingless; cotyledons accumbent. **2*n*=**28.

Rocky crevices and slopes, gravelly hillsides, sagebrush scrub, sandy banks, grassy or alpine slopes. Flowering Mar–Jul. 50–2800 m. BR, BW, Casc, Col, ECas, Lava, Owy, Sisk. CA, ID, NV, WA. Native.

PHYSARIA Bladderpod, twinpod

Herbs perennial, usually pubescent; caudices present. **Trichomes** stellate or scurfy, short-stalked or sessile. **Basal leaves** rosulate, margins entire, dentate, or pinnatifid; usually petiolate. **Cauline leaves** bases not auriculate. **Inflorescences** bracts 0 or rarely basally with bracts. **Flowers** sepals erect or spreading, oblong to ovate, saccate; petals broadly ovate or spatulate, > sepals, usually yellow, clawed or not; nectar glands confluent or lateral; stamens 6, tetradynamous, anthers obtuse; ovules 4–28(40) per ovary, septa complete or perforate, usually with apical midvein extending to center, styles distinct, stigmas entire. **Fruits** dehiscent silicles, latiseptate, or angustiseptate, inflated or not, globose, ellipsoid, oblong, ovoid, or didymous, terete, unsegmented; valves not torulose; replums rounded. **Seeds** biseriate, often suborbicular, flattened, lenticular or plump, mucilaginous or not when wetted; wingless or rarely winged or margined; cotyledons accumbent.

Asia, North America, South America. 106 species; 6 species treated in *Flora*.

Recent work has shown that *Lesquerella* is indistinct from the genus *Physaria*, and both genera have been united under *Physaria*.

Key to *Physaria* species:

1. Fruit tips notched; valves inflated, bladdery or not, retaining seeds after dehiscence; replum narrower than fruit.
   2. Valves slightly inflated, not bladdery; fruits obcordate . . . . . ***P. geyeri*** ssp. ***geyeri***
   2' Valves strongly inflated, bladdery; fruits terete.
      3. Styles (4)6–8 mm; fruit valves keeled, 3-angled on back; fruiting pedicel slightly sigmoid . . . . . ***P. chambersii***
      3' Styles 1–2 mm; fruit valves rounded, not angled on back; fruiting pedicel curved upward . . . . . ***P. oregona***

1' Fruit tips not notched; valves neither inflated nor bladdery, usually shedding seeds during dehiscence; replum as wide as fruit.
   4. Fruits latiseptate distally and at margin, elliptic to obovate . . . . . ***P. occidentalis*** ssp. ***occidentalis***
   4' Fruits terete, subglobose to obovoid.
      5. Sepals ovate to elliptic; cauline leaves narrowly linear; stems erect to ascending . . . . . ***P. douglasii*** ssp. ***douglasii***
      5' Sepals lanceolate; cauline leaves obovate to elliptic; stems prostrate or decumbent, rarely ascending or erect . . . . . ***P. kingii***

Physaria chambersii Rollins

Chambers' bladderpod

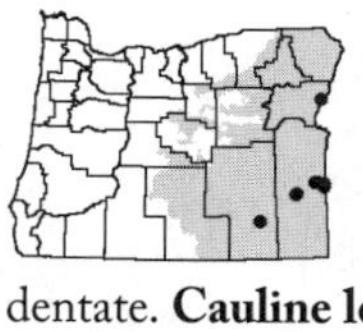

Plants cespitose, silvery; caudices present. **Trichomes** few-rayed, with umbos, rays forked, slightly tuberculate. **Stems** arising laterally, 5–15 cm. **Basal leaves** obovate to orbicular, 3–6 × 1–2 cm, margins entire or dentate. **Cauline leaves** spatulate, 1–2 cm × 3–6 mm, margins entire. **Inflorescences** fruiting pedicels divaricate, slightly sigmoid, 8–15 mm. **Flowers** sepals narrowly lanceolate, 5–8(9) mm; petals narrowly oblanceolate, 9–12 mm, yellow, claws undifferentiated; ovules 4–12 per ovary, styles (4)6–8 mm. **Fruits** didymous, terete, bases obtuse to subcordate, tips notched, pubescent, retaining seeds after dehiscence; valves keeled and 3-angled on back, flat-sided, bladdery, strongly inflated, 1–1.5 × ~1 cm; replums oblong, obtuse at tip, 4–6 × ~1 mm, narrower than fruits. **Seeds** not margined. **2*n***=8, 10, 16, 24.

Loose gravel, clay hillsides, sagebrush, pinyon-juniper communities, steep road banks. Flowering Apr–Jun. 500–1400 m. BR, BW, Owy. CA, NV; southeast to AZ. Native.

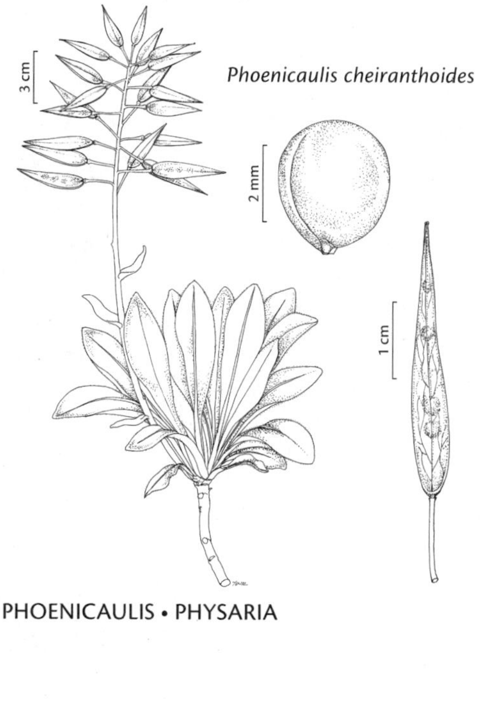

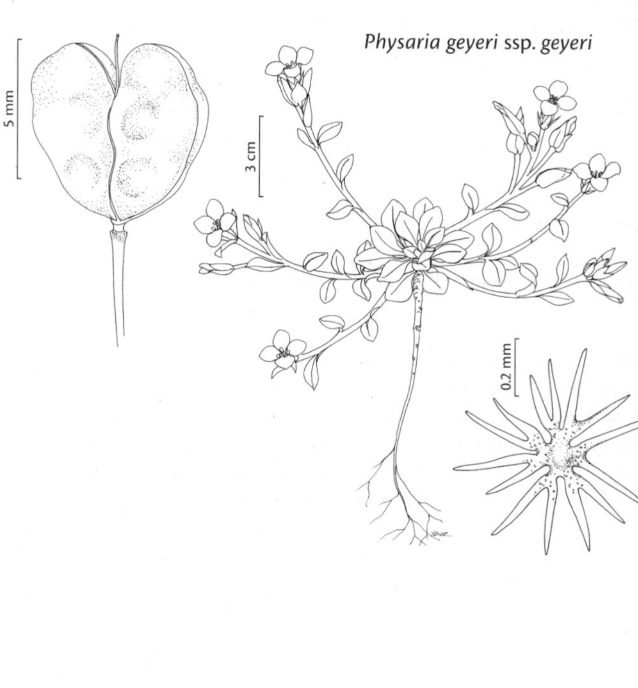

PHOENICAULIS • PHYSARIA

Physaria douglasii (S. Watson) O'Kane & Al-Shehbaz ssp. *douglasii*

Columbia bladderpod, Douglas' bladderpod

Lesquerella douglasii S. Watson

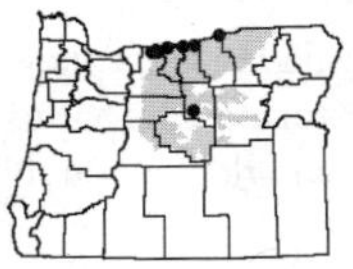

Plants densely pubescent; with simple caudices. **Trichomes** 4–6(10)-rayed, rays forked about base, tuberculate, sessile. **Stems** erect to ascending, to 45 cm. **Basal leaves** suborbicular to elliptic, 2–9.5(11.5) cm, margins entire, dentate, sinuate, or lyrate-pinnatifid. **Cauline leaves** similar to basal but narrowly linear and loosely arranged. **Inflorescences** fruiting pedicels loosely sigmoid, straight, or slightly recurved, 6–20 mm. **Flowers** sepals ovate to elliptic, (2)3.5–7.5 mm; petals 6–11 mm, yellow; ovules 4(8) per ovary, styles (1.6)3–6 mm. **Fruits** subglobose to obovoid, terete, 3–6 mm, tips not notched, usually shedding seeds during dehiscence; valves sparsely pubescent outside with sessile trichomes; replums as wide as fruits. **Seeds** not margined. ***2n***=10, 30.

Bases of cliffs, sagebrush slopes, pine woods, sandy and gravelly soil, riverbanks. Flowering Mar–May. 50–900 m. Col, Lava. ID, WA; north to British Columbia, northeast to MT. Native.

Physaria douglasii ssp. *douglasii* is distinguished from subspecies *tuplashensis* (southeastern Washington) by having sessile (vs. stalked) fruit trichomes and narrowly linear (vs. rounded) and loosely arranged (vs. imbricate) cauline leaves.

Physaria geyeri (Hooker) A. Gray ssp. *geyeri* (p. 501)

Geyer's twinpod

Physaria geyeri (Hook.) A. Gray var. *geyeri*

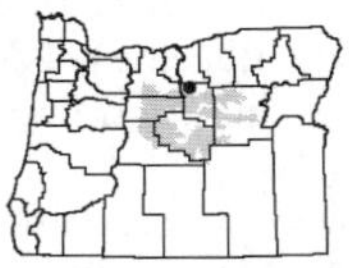

Plants cespitose, silvery pubescent; with simple caudices. **Trichomes** 6–8-rayed, rays usually forked, tuberculate. **Stems** decumbent, arising laterally, 10–30 cm. **Basal leaves** obovate, 3–7 cm, margins entire or rarely with few-toothed petioles. **Cauline leaves** oblanceolate, 1.5–3 cm, margins entire. **Inflorescences** fruiting pedicels spreading, slightly curved upward or sigmoid, 10–20 mm. **Flowers** sepals 5–7 mm, oblong; petals spatulate, 8–12 mm, yellow; ovules 4 per ovary, styles 5–7 mm. **Fruits** strongly angustiseptate, obcordate, 5–7 × 6–9 mm, sparsely pubescent with spreading trichomes, tips notched, basal sinus absent, retaining seeds after dehiscence; valves slightly inflated, not bladdery; replums ovate, 4–7 × 2–3 mm, narrower than fruits, tips acute. **Seeds** not margined. ***2n***=8.

Sandy bluffs, gravel slides, riverbanks, washes, lake shores. Flowering Apr–Jun. 300–2000 m. Lava. WA; MT. Native.

Physaria kingii (S. Watson) O'Kane & Al-Shehbaz

King's bladderpod

Plants caudices present. **Trichomes** 3–7-rayed, rays distinct or basally united, 1–3-forked at base, sessile or short-stalked. **Stems** prostrate or decumbent to ascending or erect, 5–20(40) cm. **Basal leaves** suborbicular, oblanceolate, elliptic, or rhombic, (1.2)2–6(8) cm, margins entire, sinuate, or lobed. **Cauline leaves** obovate to elliptic, 0.5–2 cm. **Inflorescences** fruiting pedicels straight, recurved, or sigmoid, 4.5–10(15) mm. **Flowers** sepals lanceolate, 4–6(7) mm; petals obovate to oblanceolate, 6–13 mm, yellow, creamy white, or white; styles 2–9 mm. **Fruits** subglobose to obovoid, terete, 3.2–9 mm, tips not notched, usually shedding seeds during dehiscence; valves densely pubescent outside, glabrous or sparsely pubescent inside; replums as wide as fruits.

Western US. 7 subspecies; 2 subspecies treated in *Flora*.

Key to *Physaria kingii* subspecies:

1. Valves glabrous inside; fruiting pedicels recurved; inflorescences secund. . . . . *P. k.* ssp. ***cobrensis***

1' Valves pubescent inside; fruiting pedicels typically sigmoid; inflorescences not secund. . . . . *P. k.* ssp. ***diversifolia***

Physaria kingii ssp. *cobrensis* (Rollins & E.A. Shaw) O'Kane & Al-Shehbaz

Cobre bladderpod

Physaria cobrensis (Rollins & E.A. Shaw) N.H. Holmgren

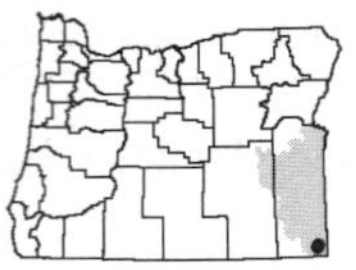

Stems ascending or sometimes erect. **Trichomes** tuberculate throughout. **Inflorescences** secund; fruiting pedicels recurved. **Flowers** ovules 4(8) per ovary. **Fruits** valves glabrous inside.

Limestone gravel, silt, rocky areas with low sagebrush. Flowering May–Jun. 1000–1800 m. Owy. ID, NV. Native.

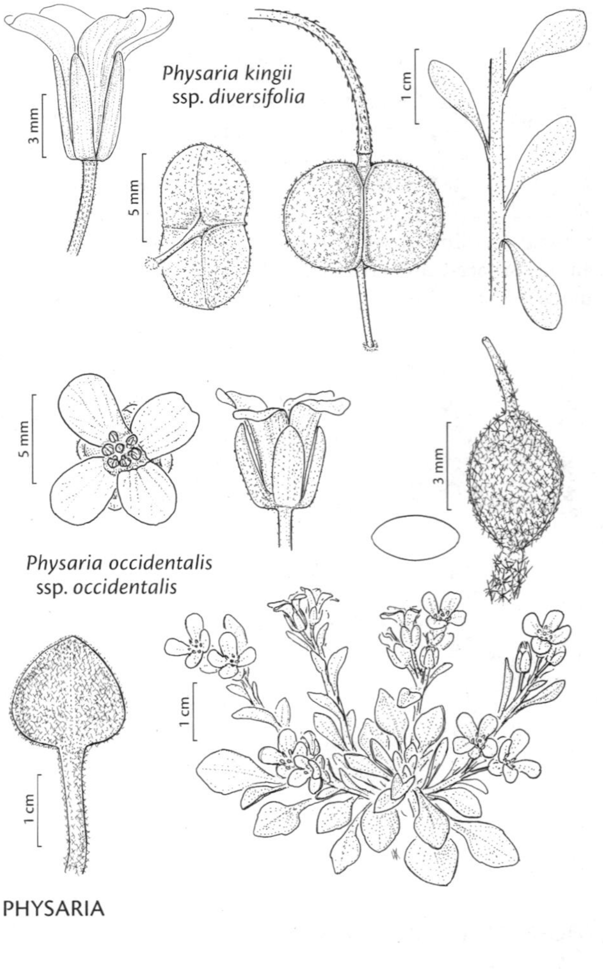

PHYSARIA

Physaria kingii ssp. *diversifolia* (Greene) O'Kane & Al-Shehbaz (p. 502)

Wallowa bladderpod

Lesquerella diversifolia Greene, *Lesquerella kingii* (S. Watson) S. Watson var. *sherwoodii* (M. Peck) C.L. Hitchc., *Lesquerella occidentalis* (S. Watson) S. Watson, *Lesquerella occidentalis* (S. Watson) S. Watson ssp. *diversifolia* (Greene) Maguire & A.H. Holmgren, *Lesquerella occidentalis* (S. Watson) S. Watson var. *diversifolia* (Greene) C.L. Hitchc., *Lesquerella sherwoodii* M. Peck

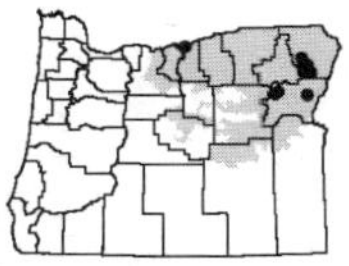

Stems prostrate or decumbent. **Trichomes** of lower layer smooth, upper layer tuberculate. **Basal leaves** margins obtusely dentate, sinuate, or lyrately lobed. **Inflorescences** not secund; fruiting pedicels typically sigmoid. **Flowers** ovules 4–6 per ovary. **Fruits** valves pubescent inside. **2*n***=10.

Sandy and gravelly soils, gravelly flood banks, rock crevices, talus slopes, cliffs. Flowering May–Aug. 300–2600 m. BW, Col. Native.

The subspecies is endemic to the Wallowa Mountains.

Physaria occidentalis (S. Watson) O'Kane & Al-Shehbaz ssp. *occidentalis* (p. 502)

Western bladderpod

Lesquerella cusickii M.E. Jones, *Lesquerella occidentalis* (S. Watson) S. Watson, *Lesquerella occidentalis* (S. Watson) S. Watson var. *cusickii* (M.E. Jones) C.L. Hitchc., *Lesquerella occidentalis* (S. Watson) S. Watson ssp. *occidentalis*, *Lesquerella occidentalis* (S. Watson) S. Watson var. *occidentalis*, *Physaria occidentalis* (S. Watson) O'Kane & Al-Shehbaz var. *occidentalis*

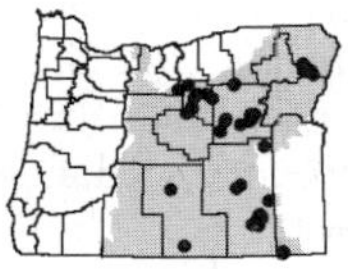

Plants perennial, densely pubescent. **Trichomes** 5–7-rayed, rays distinct, forked, tuberculate, sessile or short-stalked. **Stems** prostrate, decumbent, or erect, 3–15(30) cm. **Basal leaves** suborbicular, obovate, or elliptic, 1–8 cm, margins sinuate-dentate to entire. **Cauline leaves** oblanceolate, 0.5–1.5(2.5) cm, margins entire or remotely dentate. **Inflorescences** fruiting pedicels sigmoid or curved, 5–10(15) mm. **Flowers** sepals 4.5–7 mm; petals spatulate, 7–9(14) mm, yellow, not clawed; ovules 4–8 per ovary, styles (2)3–6.5 mm. **Fruits** latiseptate distally and at margin, elliptic to obovate, (5)6–9 mm, tips not notched, usually shedding seeds during dehiscence; valves densely pubescent outside, sparsely so on inside; replums as wide as fruits. **Seeds** not margined. **2*n***=10.

Volcanic soil, rocky ridges, loose gravelly soils, talus slopes. Flowering Apr–Aug. 400–3000 m. BR, BW, ECas, Lava. CA, ID, NV, WA. Native.

Physaria occidentalis ssp. *occidentalis* differs from ssp. *cinerascens* (western UT, NV) by having fruit strongly flattened (vs. poorly compressed) at tips and margins, 4–8 (vs. 8–12) ovules per ovary, fruit valves pubescent (vs. often glabrous) on inside, and often curved (vs. straight) styles.

Physaria oregona S. Watson

Oregon twinpod

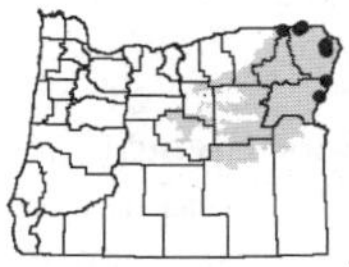

Plants cespitose, silvery pubescent; with simple caudices. **Trichomes** stipitate, with few forked rays, tuberculate. **Stems** erect to decumbent, 10–35 cm. **Basal leaves** obovate, 4–6 cm, incised or toothed along petioles. **Cauline leaves** oblanceolate, 1.5–2.5 cm, margins entire or dentate. **Inflorescences** fruiting pedicels spreading, curved upward, 10–20 mm. **Flowers** sepals oblong, 5–7 mm; petals spatulate, 9–12 mm, lemon-yellow; ovules 8 per ovary, styles 1–2 mm. **Fruits** slightly angustiseptate, terete, 10–12 mm, tips notched, basal sinus lacking, retaining seeds after dehiscence; valves strongly inflated, bladdery, rounded, not angled on back; replums lanceolate, 6–8 mm, narrower than fruits, acute at tips. **Seeds** not margined. **2*n***=8.

Serpentine areas, rocky slopes, gravelly banks, dry hillsides. Flowering Apr–Jun. 500–1800 m. BW. ID. Native.

POLYCTENIUM Combleaf

Genus composed of one species, *P. fremontii*.

Polyctenium fremontii (S. Watson) Greene (p. 504)

Desert combleaf, Washoe combleaf, two-grooved combleaf

Polyctenium fremontii (S. Watson) Greene var. *confertum* Rollins, *Polyctenium fremontii* (S. Watson) Greene var. *fremontii*, *Polyctenium williamsiae* Rollins, *Smelowskia fremontii* S. Watson, *Smelowskia fremontii* S. Watson var. *bisulcata* (Greene) O.E. Schulz, *Smelowskia fremontii* S. Watson var. *fremontii*

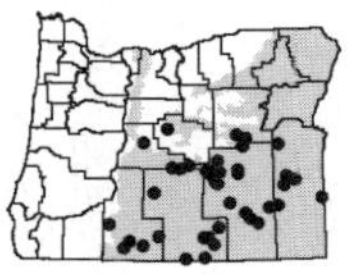

Herbs perennial, cespitose, woody at base. **Trichomes** dendritic mixed with fewer 2-rayed and larger simple trichomes. **Stems** (2)5–16(20) cm, pubescent or sometimes glabrous above. **Leaves** cauline, (0.5)1–2.2(2.8) cm, persistent as pectinate spines; lobes 2–5 per side, linear, with prominent midveins and a narrow band of blade, terminal lobes (2)4–7(10) × (0.3)0.5–1(1.5) mm; petioles 0–3(5) mm, winged. **Inflorescences** racemes; bracts 0; fruiting pedicels divaricate-ascending, (2)4–7(10) mm. **Flowers** sepals erect or ascending, caducous, oblong-ovate, 1.5–2 mm, lateral pair not saccate; petals obovate to oblanceolate, 3–4 × 1.5–2 mm, > sepals, white, aging to pink, not clawed, tips obtuse; nectar glands confluent, subtending bases of all filaments; stamens 6, tetradynamous, anther tips obtuse; ovules 30–46 per ovary, septa complete, styles (0.2)0.4–0.7(1) mm, stigmas capitate, entire. **Fruits** dehiscent siliques or silicles, angustiseptate, linear or oblong, (2)4–13(18) × 0.9–2 mm, unsegmented; valves not veined, not torulose; replums rounded. **Seeds** 30–46, uniseriate, oblong, 0.7–0.9 × 0.3–0.5 mm, not mucilaginous when wetted; wingless; cotyledons incumbent.

Sagebrush areas, dry meadows, vernal pool margins, mudflats, rocky washes, dry stream beds, gravel bars. Flowering Apr–Jul. 1100–2000 m. BR, BW, ECas, Owy. CA, ID, NV. Native.

RAPHANUS Radish

Herbs annual. **Trichomes** simple or absent. **Basal leaves** not rosulate, lyrately lobed or pinnatifid to pinnatisect; petiolate. **Cauline leaves** bases not auriculate, margins dentate; short-petiolate or subsessile. **Inflorescences** racemes; bracts 0. **Flowers** sepals erect, caducous, oblong or linear, lateral pair subsaccate; petals obovate or suborbicular, rounded or emarginate, white, yellow, pink, or purple, with brown or purple veins much darker than rest of petal, claws ≈ or > sepals; nectar glands 4, median pair oblong; stamens 6, strongly tetradynamous, anthers obtuse; ovules 2–22 per ovary, septa undifferentiated; styles slender, stigmas capitate, slightly 2-lobed. **Fruits** indehiscent siliques or silicles, linear, lanceolate, oblong, ellipsoid, or ovoid, terete or polygonal, lomentaceous, breaking into 1-seeded units or not, segmented; valvular segments seedless, aborted, about as wide as fruiting pedicel; terminal segments seeded, torulose, or moniliform, corky, smooth; replums undifferentiated. **Seeds** uniseriate, subglobose to oblong, not mucilaginous when wetted; wingless; cotyledons conduplicate.

Asia, Europe, North America. 3 species; 2 species treated in *Flora.*

Key to *Raphanus* species:

1. Petals yellow or creamy white; fruits longitudinally ribbed, constricted and breaking between seeds; roots not fleshy . . . . *R. raphanistrum*

1' Petals pink, purple or white; fruits not ribbed, not or only slightly constricted between and not breaking between seeds; roots fleshy or not. . . . . *R. sativus*

Raphanus raphanistrum Linnaeus (p. 504)

Jointed charlock

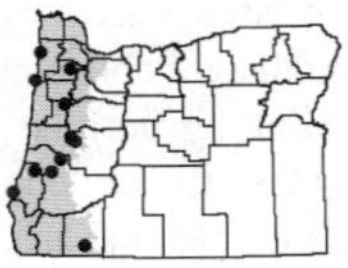

Plants without fleshy roots. **Stems** (2)3–8 dm, retrorsely hispid. **Basal leaves** 3–15(22) × 1–5 cm, margins lyrate, pinnatifid, or dentate, lateral lobes 1–4 on each side of midvein, < terminal lobes; petioles 1–6 cm. **Cauline leaves** margins uppermost dentate; subsessile. **Inflorescences** fruiting pedicels divaricate or ascending, 7–25 mm. **Flowers** sepals 7–11 mm; petals 15–25 × 4–7 mm, yellow or creamy white, claws to 1.5 cm; styles 1–5 cm. **Fruits** linear to narrowly lanceolate, valvular segments 1–1.5 mm; distal segments (1.5)2–11(14) cm × (2.5)3–8(11) mm, longitudinally ribbed, constricted and breaking between seeds. **Seeds** 2.5–3.5 × 1.8–2.5 mm. **2*n***=18.

Roadsides, disturbed areas, orchards, fields. Flowering Mar–Aug. 0–200 m. CR, Est, Sisk, WV. CA, ID, NV, WA; scattered throughout North America; Asia, Europe. Exotic.

Raphanus sativus Linnaeus

Wild radish, garden radish

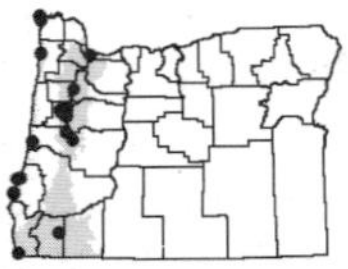

Plants with or without fleshy roots. **Stems** (1)4–13 dm. **Basal leaves** 2–60 × 1–20 cm, margins lyrate, pinnatisect, or dentate, lateral lobes 1–12 on each side of midvein, to 10 × 5 cm; petioles 1–30 cm. **Cauline leaves** margins uppermost undivided, dentate; subsessile. **Inflorescences** fruiting pedicels divaricate to ascending, 5–40 mm. **Flowers** sepals 5.5–10 mm; petals 15–25 × 3–8 mm, pink, purple or white, claws to 1.4 cm; styles 1–4 cm. **Fruits** fusiform, lanceolate, ovoid, or cylindric; valvular segments 1–3.5 mm; seed-bearing distal segments (1)3–15(25) cm × (5)7–13(15) mm, corky, not ribbed, smooth or rarely slightly constricted, not breaking between seeds. **Seeds** 2.5–4 mm. **2*n***=18.

Waste places, fields, gardens, roadsides, disturbed areas, orchards. Flowering Apr–Oct. 0–900 m. Est, Sisk, WV. CA, ID, NV, WA; north to AK, scattered throughout eastern Canada and US; Africa, Asia, Europe. Exotic.

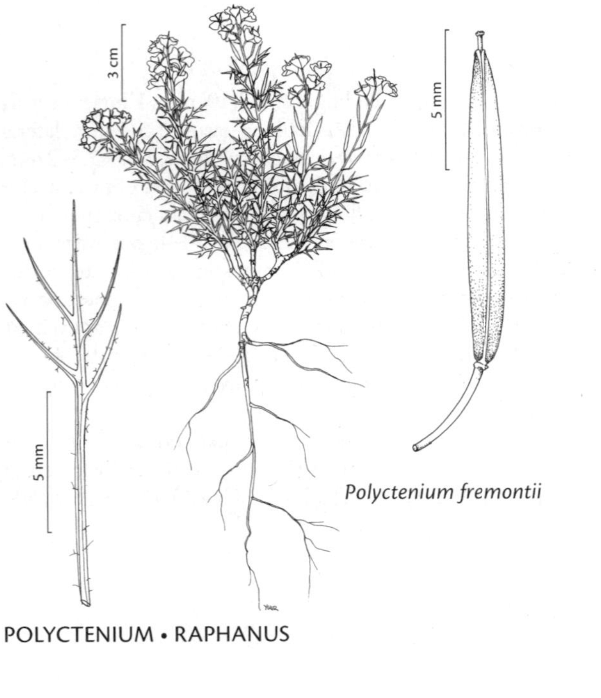

Polyctenium fremontii

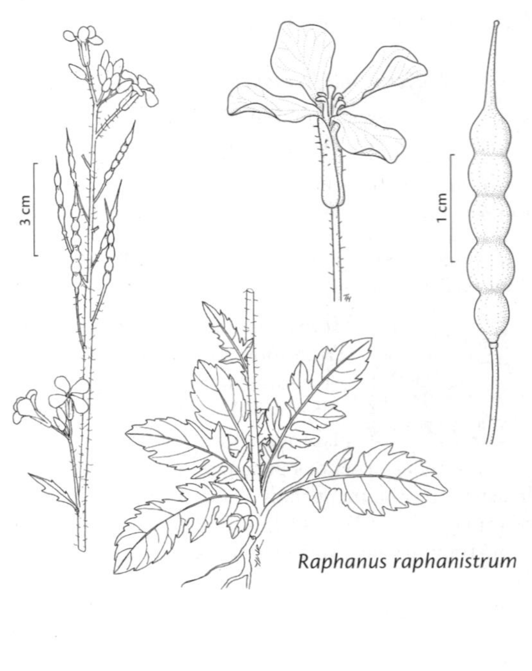

Raphanus raphanistrum

POLYCTENIUM • RAPHANUS

RORIPPA Yellowcress

Herbs annual or perennial; with caudices or rhizomes. **Trichomes** absent or simple. **Stems** terete. **Basal leaves** withered early, margins entire, dentate, or variously divided; petiolate. **Cauline leaves** bases auriculate or not, margins entire, dentate, or pinnatisect; petiolate or sessile. **Inflorescences** racemes; bracts 0 or rarely present. **Flowers** sepals erect or spreading, caducous or persistent, ovate or oblong, lateral pair saccate or not; petals rarely absent, obovate, oblanceolate, or oblong, yellow, rarely white or pink, sometimes clawed; nectar glands confluent, subtending bases of all stamens; stamens 6, tetradynamous, anthers obtuse or rarely apiculate; ovules 10–300 per ovary, septa complete or perforate, stigmas capitate, entire, or 2-lobed. **Fruits** dehiscent siliques or silicles, terete or latiseptate, linear, oblong, ovoid, to globose, tips obtuse or acute, unsegmented; valves smooth or torulose, 1-veined or veinless; replums rounded. **Seeds** 10–300, biseriate or uniseriate, oblong to ovoid, reticulate, with small bumps, mucilaginous or not when wetted; wingless or rarely winged; cotyledons accumbent.

Worldwide. ~85 species; 8 species treated in *Flora.*

A collection of *Rorippa indica* has been documented from Linnton (Multnomah County), on ballast, in 1902. This species has biseriate seeds and almost apetalous flowers.

Key to *Rorippa* species:

1. Fruit valves papillate, pilose or with vesicular trichomes, if glabrous then plants with hemispheric or clavate vesicular trichomes.
    2. Plants annual, glabrous; leaves not auriculate; sepals 0.7–1.3 mm; petals 0.5–0.8 mm; styles 0.2–0.7(1) mm . . . . ***R. tenerrima***
    2' Plants perennial, hirsute or pilose or with hemispheric or clavate vesicular trichomes; some leaves auriculate; sepals 2–3.7(4.5) mm; petals 2.7–5.3(6) mm; styles (0.7)1–3.2(3.5) mm.
        3. Sepals persistent; ovules 24–40 per ovary; plants and fruits hirsute or pilose; fruiting pedicels ascending, subappressed to rachis . . . . ***R. columbiae***
        3' Sepals caducous; ovules (30)50–82(98) per ovary; plants and sometimes fruits with hemispheric or clavate vesicular trichomes; fruiting pedicels divaricate to horizontal, sigmoid or recurved . . . . ***R. sinuata***

1' Fruit valves glabrous.
    4. Fruits globose or subglobose, 2.5–3.2 mm . . . . ***R. austriaca***
    4' Fruits linear, narrowly oblong, ovoid, ellipsoid, or pyriform, 1–13(18) mm.
        5. Plants perennial; seeds rarely produced; uppermost leaves pinnatisect; petals (2.5)3–5.5(6) × 1.5–2.5 mm. . . . . ***R. sylvestris***
        5' Plants annual or rarely short-lived and perennial; seeds always produced; uppermost leaves not pinnatisect; petals 0.6–2.5(3) × 0.2–1.5 mm.
            6. Stems erect, simple at base; petals (1.5)1.8–2.5(3) mm . . . . ***R. palustris***
            6' Stems ascending to prostrate, several from base; petals 0.6–1.8(2) mm.
                7. Fruits ovoid to pyriform; petals erect; ovules (20)30–80 per ovary . . . . ***R. curvipes***
                7' Fruits linear to narrowly oblong; petals spreading; ovules (40)50–106 per ovary . . . . ***R. curvisiliqua***

Rorippa austriaca (Crantz) Besser

Austrian yellowcress

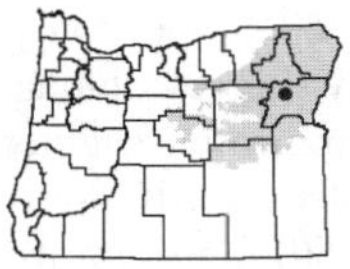

Plants perennial, glabrous, rarely pubescent on lower portions of stem; with thickened short rhizomes. **Stems** erect, 4–11(18) dm, simple, much branched above. **Basal leaves** not rosulate, pinnatifid. **Cauline leaves** lanceolate, (2.5)4–12(15) × 0.5–2(2.5) cm, bases auriculate to amplexicaul, margins entire or serrate; sessile. **Inflorescences** fruiting pedicels divaricate-ascending to horizontal, straight, 4–15 mm. **Flowers** sepals ascending, caducous, 2–3 mm; petals obovate, 3–5 × 1.7–2.5 mm; ovules 18–40 per ovary, septa complete, styles 1–1.5(2) mm. **Fruits** silicles, rarely produced, globose to subglobose, 2.5–3.2 × 1.5–2.7 mm, glabrous; valves glabrous. **Seeds** biseriate, ovoid, 0.7–0.9 mm, reddish brown. **2*n*=**16.

Lake shores, ditches, riverbanks, mudflats, marshes, floodplains, wet grasslands, fields, roadsides, disturbed areas. Flowering May–Jul. 1000–1500 m. BW. CA, ID, WA; northeast to Manitoba; scattered in southern Canada and northern and western US; Europe. Exotic.

Rorippa columbiae (B.L. Robinson) Howell

Columbia yellowcress

Rorippa calycina (Engelm.) Rydb., *Rorippa calycina* (Engelm.) Rydb. var. *columbiae* (B.L. Rob.) Rollins

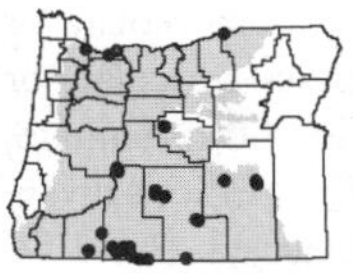

Plants perennial, pilose; with creeping roots and adventitious stems. **Stems** 1–3.2(4.1) dm. **Basal leaves** absent. **Cauline leaves** oblanceolate to oblong, 2.4–5.2 × 0.5–1.2 cm, bases often auriculate, margins sinuate to pinnatifid; short-petiolate or sessile. **Inflorescences** fruiting pedicels ascending, subappressed to rachis, (3)4–10(12) mm. **Flowers** sepals ascending, persistent, 2–3.5 mm, pilose; petals oblanceolate to spatulate, 2.7–4.2 × 0.7–1.7 mm; ovules 24–40 per ovary, septa complete, styles (0.7)1–3.2 mm, pilose. **Fruits** silicles, subglobose, oblong, or ellipsoid, straight, (1.5)2.5–5.5(7) × (1)1.7–2.8(3.5) mm, densely pilose; valves pilose. **Seeds** biseriate, ovoid-globose, 0.7–0.9 mm, yellowish brown.

Meadows, streambanks, ditches, lake and pond margins, wet fields, roadsides, gravel bars. Flowering Jun–Nov. 0–1700 m. BR, Casc, Col, ECas, Lava, Sisk, WV. CA, WA. Native.

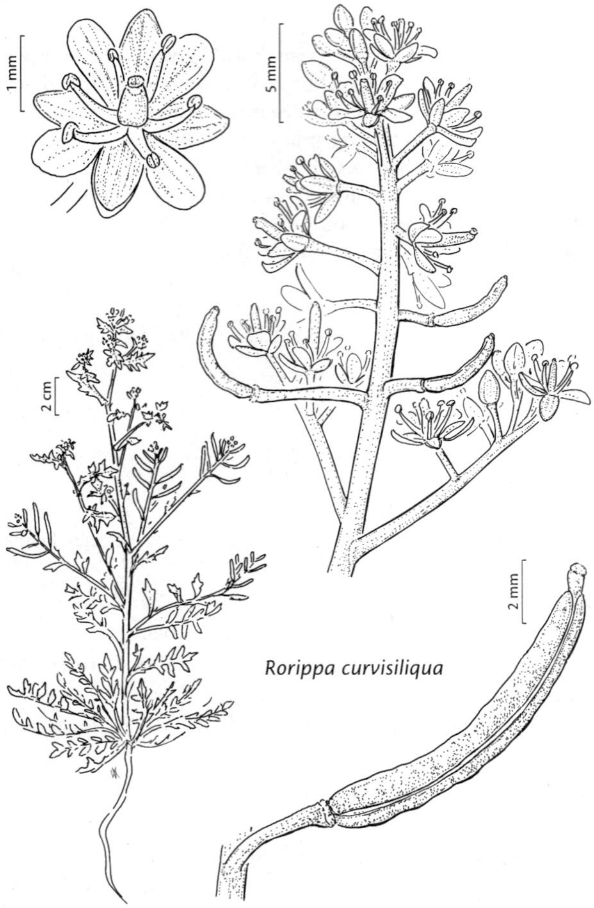

RORIPPA

Rorippa curvipes Greene

Bluntleaf yellowcress, common yellowcress

Rorippa curvipes Greene var. *curvipes*, *Rorippa curvipes* Greene var. *truncata* (Jeps.) Rollins

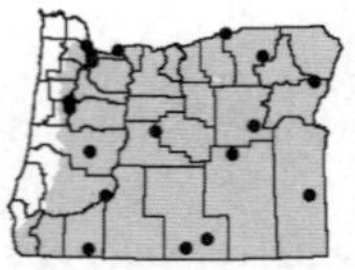

Plants annual or rarely short-lived perennial, glabrous or hirsute. **Stems** several from base, ascending to prostrate, 1–4.2(5) dm, lower parts hirsute. **Basal leaves** not rosulate. **Cauline leaves** oblong, oblanceolate, or obovate, (2)3.5–10(12) × (0.5)1–3(3.7) cm, bases auriculate or amplexicaul, margins entire to dentate, pinnatifid or sinuate; short-petiolate or sessile. **Inflorescences** fruiting pedicels divaricate-ascending to horizontal, straight or recurved, (1.2)1.7–5(8) mm. **Flowers** sepals erect, caducous, 0.8–1.8 mm; petals erect, oblanceolate to spatulate, 0.6–1.8 × 0.2–1 mm; ovules (20)30–80 per ovary, septa complete, styles 0.3–1 mm. **Fruits** silicles or siliques, ovoid to pyriform, curved, 3–8(8.8) × (0.5)1–2.5 mm, glabrous; valves glabrous. **Seeds** biseriate, cordiform, 0.5–0.7 mm, brown. ***2n***=16.

Ditches, muddy lake shores and ponds, meadows, streambanks, edges of cultivated fields, wet roadsides, seepage areas, gravel bars. Flowering May–Nov. 0–2400 m. BR, BW, Casc, Col, ECas, Lava, Owy, Sisk, WV. CA, ID, NV, WA; north to AK, northeast to Saskatchewan, east to MI, southeast to AZ, south to Mexico. Native.

Rorippa curvisiliqua (Hooker) Bessey ex Britton (p. 506)

Curvepod yellowcress, western yellowcress

Rorippa curvisiliqua (Hook.) Bessey ex Britton var. *curvisiliqua*, *Rorippa curvisiliqua* (Hook.) Bessey ex Britton var. *lyrata* (Nutt. ex Torr. & A. Gray) C.L. Hitchc., *Rorippa curvisiliqua* (Hook.) Bessey ex Britton var. *occidentalis* (Greene) Stuckey, *Rorippa curvisiliqua* (Hook.) Bessey ex Britton var. *orientalis* Stuckey, *Rorippa curvisiliqua* (Hook.) Bessey ex Britton var. *spatulata* Stuckey

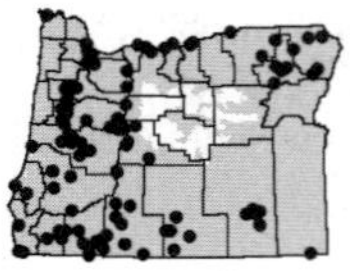

Plants annual, glabrous or sparsely hirsute. **Stems** several from base, ascending to prostrate, (0.5)1–4(6) dm. **Basal leaves** not rosulate. **Cauline leaves** (2)3–9(13) × 0.8–2(3.5) cm, bases auriculate, margins pinnatifid to pinnatisect or rarely pectinate; petiolate or sessile. **Inflorescences** fruiting pedicels divaricate-ascending to horizontal, straight, 1–4.5(9) mm. **Flowers** sepals ascending, caducous or rarely persistent, 0.8–2(2.5) mm; petals spreading, oblong to oblanceolate, 0.7–1.8(2) × 0.3–1.3 mm; ovules (40)50–106 per ovary, septa complete, styles 0.1–0.8 mm. **Fruits** siliques, linear to narrowly oblong, curved upward, 4–13(18) × 1–2 mm, glabrous; valves glabrous. **Seeds** biseriate, cordiform, 0.5–0.7 mm, brown. ***2n***=16.

Seepage areas, shores of lakes, ponds and rivers, marshy grounds, mudflats, streamsides, wet roadsides, meadows. Flowering Mar–Oct. 0–2600 m. All ecoregions except Lava. CA, ID, NV, WA; north to AK, northeast to MT, east to WY, southeast to AZ. Native.

This highly variable species was artificially split by Stuckey (1972) into seven poorly defined varieties that cannot be reliably identified because of continuous, unreliable characters.

Rorippa palustris (Linnaeus) Besser

Marsh yellowcress

Plants annual, rarely short-lived perennial, glabrous or hirsute. **Stems** erect, (0.5)1–10(14) dm, simple at base. **Basal leaves** rosulate, (4)6–20(30) × 1–5(8) cm, lyrate-pinnatisect. **Cauline leaves** (1.5)2.5–10(18) × (0.5)0.8–2.5(3) cm, bases auriculate or amplexicaul, margins entire, dentate, sinuate, or lyrate-pinnatifid; petiolate or subsessile. **Inflorescences** fruiting pedicels divaricate or reflexed, straight or curved, (2.5)3–10(14) mm. **Flowers** sepals erect, caducous, 1.5–2.4(2.6) mm; petals spatulate, (1.5)1.8–2.5(3) × 0.5–1.5 mm; ovules 20–90 per ovary, styles 0.2–1(1.2) mm. **Fruits** silicles or rarely siliques, oblong, ellipsoid, or ovoid, (3)4–10 × (1.5)1.7–3(3.5) mm, often slightly curved, glabrous; valves glabrous; septa complete. **Seeds** biseriate, ovoid or subglobose, 0.5–0.7(0.9) × 0.4–0.6(0.7) mm, brown to yellowish brown. ***2n***=32.

2 subspecies.

Rorippa palustris is by far the most variable species in the genus. Stuckey (1972) divided the species into four subspecies and eleven varieties, but all except two were artificially delimited, and the characters used to separate them show tremendous overlap.

Key to *Rorippa palustris* subspecies:

1. Stems and abaxial leaf surfaces often densely hirsute . . . . ***R. p.*** ssp. ***hispida***

1' Stems and abaxial leaf surfaces glabrous, rarely lowermost parts sparsely pubescent . . . . ***R. p.*** ssp. ***palustris***

Rorippa palustris ssp. ***hispida*** (Desvaux) Jonsell

Hairy marsh yellowcress

Rorippa islandica (Oeder) Borbás var. *hispida* (Desv.) Butters & Abbe, *Rorippa palustris* (L.) Besser var. *hispida* (Desv.) Rydb.

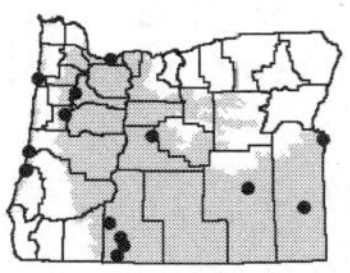

Stems and abaxial leaf surfaces often densely hirsute. **2*n***=32.

Pond and lake shores, riverbanks, wet meadows, swamps. Flowering Jun–Oct. 0–1300 m. BR, Casc, ECas, Est, Lava, Owy, WV. CA, ID, NV, WA; north to AK, northeast to Newfoundland, east to ME, southeast to NM. Native.

Rorippa palustris ssp. ***palustris***

Western bog yellowcress, smooth marsh yellowcress

Rorippa islandica (Oeder) Borbás var. *fernaldiana* Butters & Abbe, *Rorippa islandica* (Oeder) Borbás var. *glabrata* (Lunell) Butters & Abbe, *Rorippa palustris* (L.) Besser var. *glabra* (O.E. Schulz) R.L. Taylor & MacBryde, *Rorippa palustris* (L.) Besser var. *occidentalis* (S. Watson) Rollins, *Rorippa palustris* (L.) Besser ssp. *occidentalis* (S. Watson) Abrams

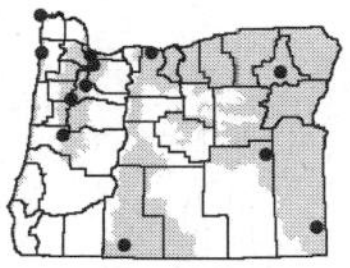

Stems and abaxial leaf surfaces glabrous, rarely lowermost parts sparsely pubescent. **2*n***=32.

Marshlands, meadows, sandbars, riverbanks, ditches, disturbed areas, shores of lakes and ponds, bogs, streamsides. Flowering Apr–Oct. 0–2500 m. BW, Col, ECas, Est, Owy, WV. CA, ID, NV, WA; worldwide. Native.

Rorippa sinuata (Nuttall) Hitchcock

Spreading yellowcress, western yellowcress

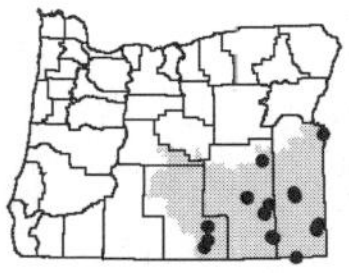

Plants perennial, pubescent with hemispheric or clavate vesicular trichomes; with rhizomes or creeping roots. **Stems** decumbent or prostrate, 1–4.2(5) dm. **Basal leaves** not rosulate, pinnatifid. **Cauline leaves** oblong, oblanceolate, or lanceolate, (1.5)2.5–6.5(9) × 0.5–1.8(2.5) cm, bases uppermost auriculate, margins pinnatifid or sinuate; sessile. **Inflorescences** fruiting pedicels divaricate to horizontal, sigmoid or recurved, 4–12(15) mm. **Flowers** sepals ascending, caducous, oblong, 2.2–3.7(4.5) mm; petals oblanceolate to spatulate, (2.7)3.2–5.3(6) × 1.5–2.5 mm; ovules (30)50–82(98) per ovary, styles 1–2.5(3.5) mm. **Fruits** siliques, oblong to lanceolate or linear, curved, (4)4.7–11.5(16) × (1)1.5–2.5 mm; valves glabrous or with vesicular trichomes; septa complete. **Seeds** biseriate, angular cordiform, 0.7–1 × 0.5–0.6 mm, yellow-brown. **2*n***=16.

Shores of ponds and lakes, streambanks, fields, ditches. Flowering May–Jul. 900–2000 m. BR, Owy. CA, ID, NV, WA; southern Canada, throughout US except eastern seaboard. Native.

Rorippa sylvestris (Linnaeus) Besser (p. 507)

Creeping yellowcress, shore yellowcress

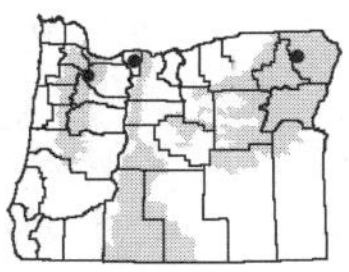

Herbs perennial, glabrous or sparsely pubescent. **Stems** (0.5)1.5–8(10) dm. **Basal leaves** (2)3.5–15(20) × (0.7)1–4.5(6) cm, bases not or rarely minutely auriculate, margins deeply pinnatisect, lateral lobes 3–6 on each side of midvein, linear, lanceolate, oblong, or ovate, dentate or subentire; petiolate. **Cauline leaves** similar to basal; bases uppermost pinnatisect, with 1–3 lateral lobes; usually subsessile. **Inflorescences** fruiting pedicels divaricate, straight, (3)4–10(12) mm. **Flowers** sepals ascending or spreading, caducous, 1.8–3(3.5) mm; petals spatulate or obovate, (2.5)3–5.5(6) × 1.5–2.5 mm; ovules 24–80 per ovary, septa complete, styles 0.5–1(1.5) mm. **Fruits** siliques, linear, 10–20(25) × (0.7)1–1.3(1.6) mm, glabrous; valves glabrous. **Seeds** rarely produced, uniseriate or subbiseriate, ovoid, 0.5–0.9 × 0.4–0.5 mm, reddish brown. **2*n***=32, 40, 48.

Washes, meadows, ditches, sandy beaches, shores of ponds and lakes, disturbed areas, fields. Flowering May–Aug. 50–900 m. BW, ECas, WV. CA, ID, WA; scattered throughout North America; Europe. Exotic.

Rorippa tenerrima Greene

Modoc yellowcress, low yellowcress

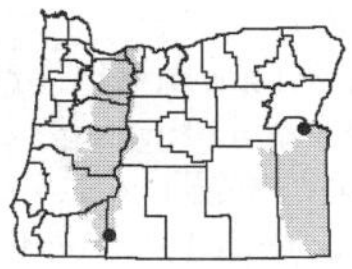

Plants annual, glabrous. **Stems** several, prostrate to decumbent, 0.7–3.5(4) dm, branched from base. **Basal leaves** not rosulate. **Cauline leaves** oblong, oblanceolate, or lanceolate, 2–9(11) × 0.7–2(3) cm, bases not auriculate, margins pinnatifid, lateral lobes entire, dentate, or sinuate; short-petiolate to sessile. **Inflorescences** fruiting pedicels ascending to divaricate, straight, (1)1.5–3.2(4.2) mm. **Flowers** sepals ascending, caducous, 0.7–1.3 mm; petals oblong to spatulate, 0.5–0.8 × 0.1–0.3 mm; ovules 20–80 per ovary, septa complete, styles 0.2–0.7(1) mm. **Fruits** siliques, curved upward, lanceolate to oblong-lanceolate, often slightly constricted at middle, 3–7(9) × (0.8)1–1.7(2) mm; valves papillate. **Seeds** biseriate, cordiform, 0.5–0.7 mm, reddish brown.

Marshes, mudflats, shores of lakes and ponds, sandbars, streamsides, riverbanks. Flowering Jun–Oct. 700–1600 m. Casc, Owy. CA, ID, NV, WA; north to British Columbia, northeast to Saskatchewan, east to MO, southeast to TX, south to Mexico. Native.

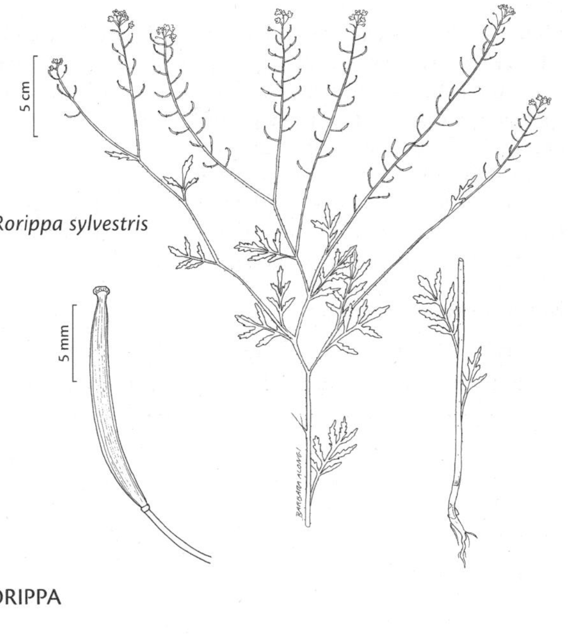

RORIPPA

SINAPIS Charlock, mustard

Herbs annual. **Trichomes** simple or absent. **Basal leaves** not rosulate, lyrate or pinnatifid to 1- or 2-pinnatisect, rarely undivided, coarsely dentate; petiolate. **Cauline leaves** bases not auriculate, margins entire or dentate; petiolate or sessile. **Inflorescences** corymbose racemes; bracts 0. **Flowers** sepals spreading or reflexed, caducous, oblong or linear, lateral pair not saccate; petals spreading, obovate, yellow, claws subequaling sepals, petal tips obtuse or emarginate; nectar glands 4, 2 median and 2 lateral; stamens 6, tetradynamous, anthers obtuse; ovules 4–20 per ovary, septa complete, styles distinct, stigmas capitate, 2-lobed. **Fruits** dehiscent siliques, linear, lanceolate or oblong, terete, 4-angled, or latiseptate, 2-segmented; valvular segments dehiscent, few-to many-seeded, torulose, valves prominently 3–7-veined; terminal segments indehiscent, seedless or 1- or 2-seeded, flattened or terete; replums rounded. **Seeds** uniseriate, globose, plump, mucilaginous or not when wetted; wingless; cotyledons conduplicate.

Worldwide. 5 species; 1 species treated in *Flora*.

Sinapis arvensis Linnaeus (p. 508)

Charlock, wild mustard

Brassica kaber (DC.) L.C. Wheeler, *Brassica kaber* (DC.) L.C. Wheeler var. *pinnatifida* (Stokes) L.C. Wheeler

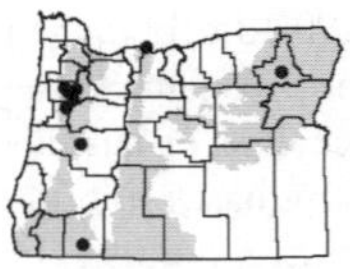

Plants (0.5)2–10(21) dm, retrorsely or spreading-hirsute or hispid, sometimes glabrous. **Basal leaves** (3)4–18(25) × 1.5–5(7) cm, lyrate or pinnatifid, with 1–4 lateral lobes on each side, sometimes undivided; petioles 1–4(7) cm. **Cauline leaves** often undivided, coarsely toothed; petiolate to subsessile. **Inflorescences** fruiting pedicels ascending or suberect, (2)3–7 mm. **Flowers** sepals (4.5)5–6(7) mm; petals (8)9–12(17) × (3)4–6(7.5) mm. **Fruits** linear, (15)20–45(57) × (1.5)2.5–4 mm; valvular segments (2)4–8(12)-seeded in each locule, terete, torulose, (6)12–35(43) mm, 3–5(7)-veined, glabrous or rarely pubescent; terminal segments conical, terete, (7)10–16 mm, much < valves, 1-seeded. **Seeds** (1)1.5–2 mm in diameter. **2*n***=18.

Roadsides, disturbed areas, fields, orchards. Flowering Feb–Jul. 0–900 m. BW, ECas, Sisk, WV. CA, ID, NV, WA; worldwide. Exotic.

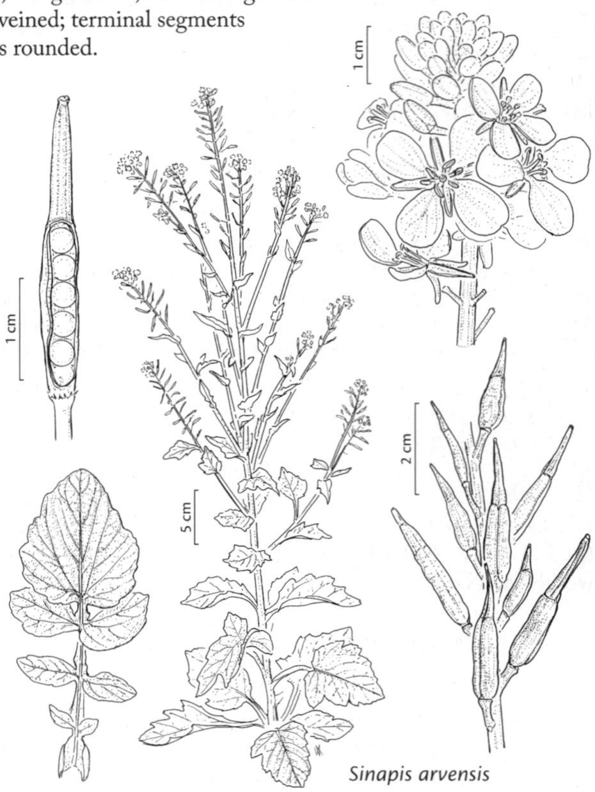

SINAPIS

SISYMBRIUM Hedge mustard, rocket

Herbs, rarely **subshrubs** annual or perennial. **Trichomes** absent or simple, very rarely branched. **Basal leaves** rosulate or not, margins entire, lyrate, or pinnately lobed; petiolate. **Cauline leaves** bases not auriculate; petiolate or sessile. **Inflorescences** racemes; bracts present or not. **Flowers** sepals erect or spreading, caducous, ovate or oblong, lateral pair sometimes subsaccate; petals obovate, spatulate, oblong or suborbicular, > sepals, yellow, rarely white or pinkish, claws often ≈ or > sepals, petal tips obtuse or emarginate; nectar glands confluent, subtending bases of all stamens; stamens 6, tetradynamous, anthers obtuse; ovules 6–160 per ovary, septa complete, styles distinct, stigmas capitate, 2-lobed. **Fruits** dehiscent siliques, linear or subulate, terete, unsegmented, tips acuminate to subulate; valves 3-veined, smooth or torulose; replums rounded; terminal segment seedless. **Seeds** uniseriate, oblong or ovoid, plump, not mucilaginous when wetted; wingless; cotyledons incumbent.

Worldwide. 47 species; 5 species treated in *Flora*.

Key to *Sisymbrium* species:

1. Fruits subulate-linear, (0.7)1–1.4(1.8) cm, appressed to rachis . . . . *S. officinale*

1' Fruits narrowly linear, 2–10(13) cm, not appressed to rachis.

 2. Fruiting pedicels nearly as thick as mature fruit . . . . *S. altissimum*

 2' Fruiting pedicels narrower than fruit.

  3. Plants perennial; uppermost leaves filiform to linear, 1–3.5(5) mm wide . . . . *S. linifolium*

  3' Plants annual; uppermost leaves variously shaped, sometimes lobed, never filiform or linear.

   4. Young fruits not overtopping flowers; petals 6–8 mm; stems densely hispid below . . . . *S. loeselii*

   4' Young fruits overtopping flowers; petals 2.5–3.5(4) mm; stems glabrous or sparsely pubescent below . . . . *S. irio*

Sisymbrium altissimum Linnaeus

Tumble mustard, Jim Hill mustard

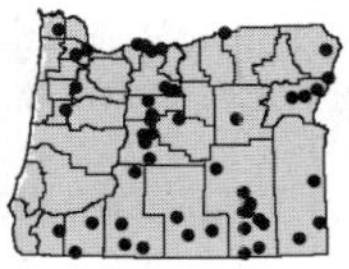

Plants annual, hirsute below to glabrous above. **Stems** (2)4–12(16) dm. **Basal leaves** rosulate, blades (2)5–20(35) × (1)2–8(10) cm, pinnatisect, pinnatifid, or runcinate, lateral lobes (3)4–6(8) per side, < terminal, margins entire, dentate, or lobed; petioles 1–10(15) cm. **Cauline leaves** lower similar to basal, uppermost smaller, pinnatisect with linear to filiform lobes. **Inflorescences** bracts 0; fruiting pedicels divaricate or rarely ascending, stout, (4)6–10(13) mm, nearly as thick as fruits. **Flowers** sepals ascending or spreading, 4–6 mm; petals spatulate, (5)6–8(10) × 2.5–4 mm, claws 3.5–6 mm; ovules 90–120 per ovary, styles subclavate, 0.5–2 mm, stigmas prominently 2-lobed. **Fruits** not appressed to rachis, narrowly linear, not torulose, terete, usually straight, stout, (4.5)6–9(12) cm × 1–2 mm; valves glabrous. **Seeds** 0.8–1 × 0.5–0.6 mm. ***2n***=14.

Fields, pastures, disturbed areas. Flowering Apr–Sep. 0–2600 m. All ecoregions except Est. CA, ID, NV, WA; throughout North America; Asia, Europe. Exotic.

Sisymbrium irio Linnaeus

London rocket

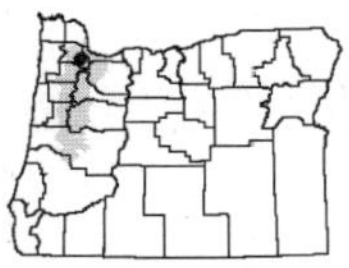

Plants annual, glabrous or sparsely pubescent at least below. **Stems** (1)2–6(7.5) dm. **Basal leaves** rosulate, blades oblanceolate or oblong in outline, (1.5)3–12(15) × (0.5)1–6(9) cm, runcinate to pinnatisect, lateral lobes (1)2–6(8) per side, < terminal lobe, margins entire, dentate, or lobed; petioles (0.5)1–4.5(6) cm. **Cauline leaves** lower similar to basal, uppermost smaller. **Inflorescences** bracts 0; fruiting pedicels divaricate or ascending, (5)7–12(20) mm, narrower than fruits. **Flowers** sepals erect, 2–2.5 mm; petals oblong-oblanceolate, 2.5–3.5(4) × 1–1.5 mm, claws 1–1.5 mm; ovules 40–90 per ovary, styles 0.2–0.5 mm, stigmas prominently 2-lobed. **Fruits** not appressed to rachis, narrowly linear, slightly torulose, terete, (2.5)3–4(5) cm × 0.9–1.1 mm, young fruits overtopping flowers; valves glabrous. **Seeds** 0.8–1 × 0.5–0.6 mm. ***2n***=14.

Pastures, disturbed areas, rocky slopes, fields, orchards, prairies. Flowering Jan–May. 0–50 m. WV. CA, NV; scattered in North America; scattered nearly worldwide. Exotic.

Sisymbrium linifolium (Nuttall) Nuttall (p. 509)

Flax-leaved plains mustard, lava cress

Schoenocrambe linifolia (Nutt.) Greene

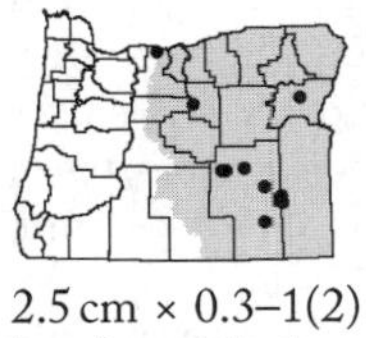

Plants perennial, glabrous or rarely sparsely pubescent basally; rhizomatous. **Stems** (1.5)3–7(11) dm. **Basal leaves** not rosulate. 1.5–6 cm, margins entire, pinnatifid or pinnatisect, lateral lobes oblong to linear, 0.5–2.5 cm × 0.3–1(2) mm. **Cauline leaves** lowermost similar to basal, middle leaves linear, filiform, or oblanceolate, margins entire, dentate, or pinnately lobed, (1)2.5–9(12) cm × 1–3.5(5) mm, uppermost leaves filiform to linear, margins entire; sessile or short-petiolate. **Inflorescences** bracts 0; fruiting pedicels divaricate to ascending, (3)5–9(11) mm, narrower than fruits. **Flowers** sepals (3)4–7 mm, ascending; petals spatulate, (6)8–12 × (1.5)2–4 mm, claws 2–5 mm; ovules 60–94 per ovary, styles 0.5–1 mm, stigmas prominently 2-lobed. **Fruits** not appressed to rachis, narrowly linear, not torulose, terete, (2.5)3.5–6.5 cm × 0.9–1.2 mm. **Seeds** 1–2 × 0.4–0.6 mm. ***2n***=14.

Rocky or gravelly hillsides, shady rock cliffs, pinyon-juniper areas, sagebrush communities, abandoned fields, sandy prairies. Flowering Apr–Jun. 500–1600 m. BR, BW, Col, Lava, Owy. ID, NV, WA; north to British Columbia, northeast to MT, east to WY, southeast to NM. Native.

Sisymbrium loeselii Linnaeus

Loesel's tumblemustard, small tumbleweed mustard, tall hedge mustard

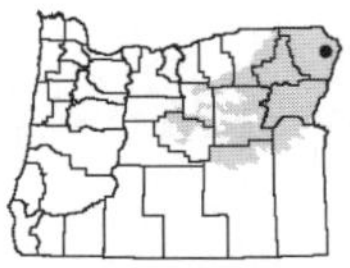

Plants annual, densely hispid below with retrorse trichomes, usually glabrous above. **Stems** (2)3.5–12(17.5) dm. **Basal leaves** rosulate, blades broadly oblanceolate, (1.5)2.5–8(12) × (1)2–5(7) cm, runcinate to lyrate-pinnatifid, lateral lobes 2–4 per side, margins entire or dentate; petioles 1–4(5) cm. **Cauline leaves** similar to basal, uppermost leaves much smaller, margins entire or toothed. **Inflorescences** bracts 0; fruiting pedicels divaricate or ascending, 5–12(15) mm, narrower than fruits. **Flowers** sepals ascending, 3–4 mm; petals spatulate, 6–8 × 2–3 mm, claws 2.5–3.5 mm; ovules 40–60 per ovary, styles stout, 0.3–0.7 mm, stigmas prominently 2-lobed. **Fruits** not appressed to rachis, narrowly linear, subtorulose, terete, curved or straight, 2–3.5(5) cm × 0.9–1.1 mm, young fruits not overtopping flowers; valves often glabrous. **Seeds** 0.7–1 × 0.5–0.6 mm. ***2n***=14.

Roadsides, disturbed areas, prairies, riverbanks, pastures. Flowering May–Sep. 600–1200 m. BW. CA, ID, NV, WA; scattered in southern Canada and northern US; Asia, Europe. Exotic.

Sisymbrium linifolium

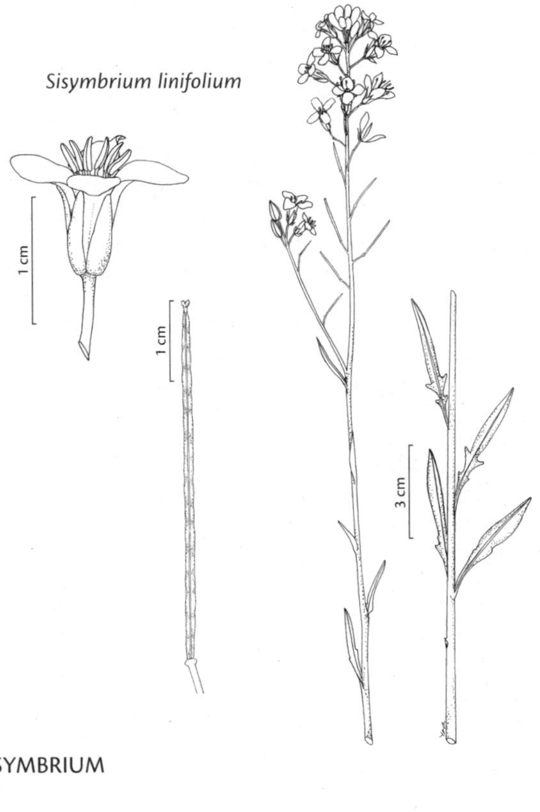

SISYMBRIUM

Sisymbrium officinale (Linnaeus) Scopoli

Hedge mustard

Sisymbrium officinale (L.) Scop. var. *leiocarpum* DC., *Sisymbrium officinale* (L.) Scop. var. *officinale*

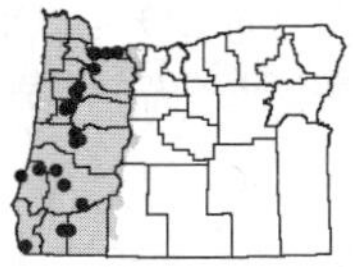

Plants annual, hirsute with retrorse trichomes, rarely subglabrous above. **Stems** 2.5–7.5(11) dm. **Basal leaves** usually rosulate, blades (2)3–10(15) × (1)2–5(8) cm, lyrate-pinnatifid, pinnatisect, or runcinate, lateral lobes 2–4(5) per side, margins entire, dentate, or lobed; petioles (1)2–7(10) cm. **Cauline leaves** similar to basal, uppermost leaves smaller, lobed, dentate, or subentire. **Inflorescences** bracts 0; fruiting pedicels erect, appressed to rachis, stout, 1.5–3(4) mm, narrower than fruits. **Flowers** sepals erect, 2–2.5 mm; petals spatulate, 2.5–4 × 1–2 mm; claws 1–2 mm; ovules 10–20 per ovary, styles (0.8)1–1.5(2) mm, stigmas slightly 2-lobed. **Fruits** appressed to rachis, slightly torulose or not, terete, stout, straight subulate-linear, (0.7)1–1.4(1.8) cm × 1–1.5 mm; valves glabrous or pubescent. **Seeds** oblong, 1–1.3 × 0.5–0.6 mm. **2*n***=14.

Fields, roadsides, pastures, disturbed areas, deserts. Flowering Mar–Aug. 0–1900 m. Casc, CR, Est, Sisk, WV. CA, ID, NV, WA; scattered throughout North America; Africa, Asia, Europe. Exotic.

SMELOWSKIA Smelowskia, false candytuft

Herbs perennial, rarely annual; with well-developed caudices. **Trichomes** dendritic, sometimes also simple and stalked forked. **Basal leaves** rosulate, 1- or 2-pinnatisect, rarely entire; petiolate. **Cauline leaves** rarely absent, bases not auriculate, margins entire or pinnatisect; petiolate or sessile. **Inflorescences** racemes; bracts present or not. **Flowers** sepals suberect to spreading, caducous or persistent, oblong to ovate, lateral pair not saccate; petals spatulate, obovate, or suborbicular, > sepals, white, pink, purple, or rarely yellow, clawed, tips rounded; nectar glands usually confluent, subtending bases of all stamens; stamens 6, tetradynamous, anthers obtuse; ovules 4–18 per ovary, septa complete or perforated; styles obsolete or distinct, stigmas capitate, entire or rarely slightly 2-lobed. **Fruits** dehiscent siliques or silicles, latiseptate or angustiseptate, linear, oblong, ovoid, obovoid, ellipsoid, fusiform, or suborbicular, terete, 4-angled, unsegmented; valves smooth; replums rounded. **Seeds** 4–18 per fruit, uniseriate or biseriate, oblong, plump, not mucilaginous when wetted; wingless; cotyledons incumbent or accumbent.

Asia, North America. 25 species; 2 species treated in *Flora*.

Key to *Smelowskia* species:

1. Sepals caducous; fruits ellipsoid, oblong or linear, 5–13 × 1.5–2 mm, cuneate on both ends, valve midveins prominent . . . . *S. americana*

1' Sepals persistent; fruits ovoid, 2–6 × 2–3 mm, obtuse on both ends, valve midveins obscure . . . . *S. ovalis*

Smelowskia americana Rydberg (p. 511)

Alpine smelowskia, Siberian smelowskia

Smelowskia calycina (Stephan ex Willd.) C.A. Mey., *Smelowskia calycina* (Stephan ex Willd.) C.A. Mey. var. *americana* (Regel & Herder) W.H. Drury & Rollins

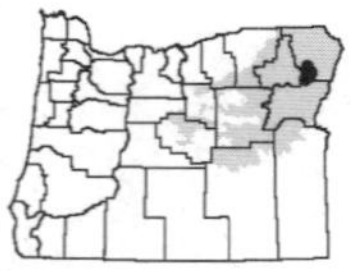

Plants cespitose; caudices branched, covered with petioles of previous seasons. **Stems** (4)6–20(27) cm, gray-pubescent with simple and dendritic trichomes. **Basal leaves** blades (0.5)1–3.5(5.2) × 0.4–1.7 cm, 1- or 2-pinnatifid or pinnatisect, ciliate; ultimate segments linear, oblong, or ovate, 2–14 × 0.5–4 mm; petioles 1–5(8) cm. **Cauline leaves** few, alternate, similar to basal but smaller; short-petiolate to sessile. **Inflorescences** fruiting pedicels suberect to ascending, subappressed to rachis, often forming < 40° angle, 4–10(14) mm, pubescent with simple and smaller dendritic trichomes; lowermost with bracts. **Flowers** sepals caducous, ovate, 2–3.5 mm; petals suborbicular to obovate, 3.5–6.5 × 1.5–3.5 mm, white, rarely pinkish or lavender, claws 1.5–3 mm; ovules 8–12(14) per ovary, styles 0.2–0.8 mm. **Fruits** subappressed to rachis or rarely ascending, ellipsoid, oblong or linear, 4-angled, cuneate at both ends, 5–13 × 1.5–2 mm; valve midveins prominent. **Seeds** 1.1–1.9 × 0.6–0.9 mm. **2*n***=12, 22.

Alpine meadows, talus and scree slopes. Flowering Jun–Aug. 2200–3000 m. BW. ID, NV, WA; north to British Columbia, northeast to Alberta, east to WY, southeast to CO. Native.

Smelowskia ovalis M.E. Jones (p. 511)

Short-fruited smelowskia, alpine false candytuft

Smelowskia ovalis M.E. Jones var. *congesta* Rollins, *Smelowskia ovalis* M.E. Jones var. *ovalis*

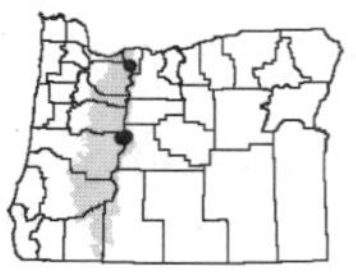

Plants cespitose, gray-pubescent with simple and smaller dendritic trichomes; caudices branched, covered with petioles of previous seasons. **Stems** 3–18 cm. **Basal leaves** blades 0.5–2.5 × 0.5–1.5 cm, pinnatisect or pinnatifid, ciliate with simple trichomes; ultimate segments obovate or oblong, 2–10 × 1.5–5 mm; petioles 1–6 cm. **Cauline leaves** few, alternate, similar to basal but smaller; short-petiolate to sessile. **Inflorescences** fruiting pedicels suberect to ascending, often forming < 40° angle, 3–10 mm; lowermost sometimes with bracts. **Flowers** sepals persistent, ovate, 2–2.5 mm; petals spatulate to obovate, 3.5–4.5 × 1.5–2.5 mm, white or rarely pinkish, claws 0.5–1.5 mm; ovules 4–8 per ovary, styles 0.2–1 mm. **Fruits** subappressed to rachis to ascending, ovoid, terete or slightly compressed, obtuse at both ends, 2–6 × 2–3 mm; valve midveins obscure. **Seeds** 1–1.5 × 0.6–0.7 mm.

Loose talus, rocky moraines, rock crevices. Flowering Jul–Aug. 2000–3000 m. Casc. CA, WA; north to British Columbia. Native.

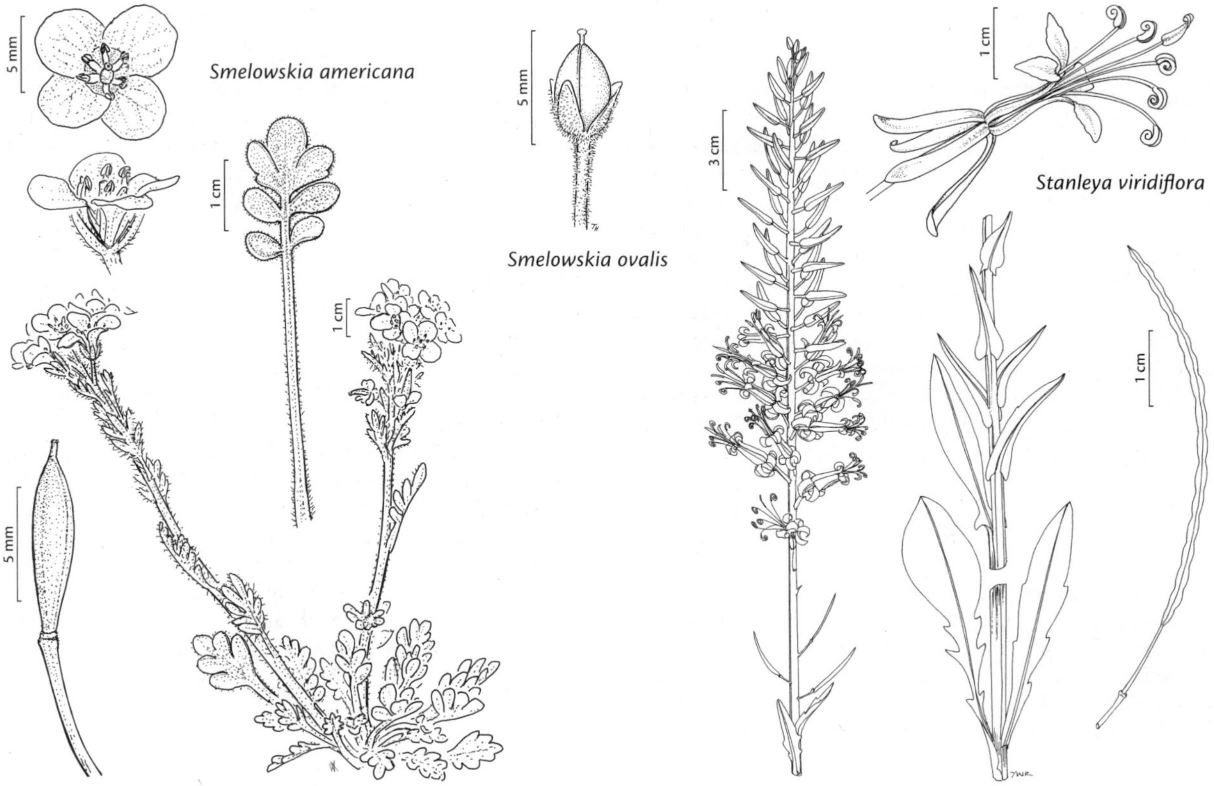

SMELOWSKIA • STANLEYA

STANLEYA Prince's plume, stanleya

Herbs or **shrubs**, annual or perennial. **Trichomes** absent or simple. **Basal leaves** rosulate, margins entire, lyrate, or 1- or 2-pinnatifid; petiolate. **Cauline leaves** bases sometimes auriculate, margins entire, dentate, or pinnatifid; petiolate or sessile. **Inflorescences** bracts 0. **Flowers** sepals spreading to reflexed, caducous, oblong-linear, lateral pair not saccate; petals ovate, elliptic, oblong or linear, yellow, rarely white, claws well differentiated from blades; nectar glands confluent, sometimes only lateral, annular; stamens 6, exserted, equal, filaments often basally papillate or pubescent, anthers linear, coiled spirally after dehiscence; septa complete, styles obsolete or distinct, to 1.7 mm, stigmas entire; gynophore (8)10–20(25) mm. **Fruits** linear siliques, dehiscent, terete or latiseptate, long-stipitate, unsegmented; valves with conspicuous midveins, glabrous, often torulose; replums rounded. **Seeds** many, uniseriate, oblong, plump, slightly mucilaginous when wetted; wingless; cotyledons accumbent to incumbent.

Western United States. 7 species; 3 species treated in *Flora*.

Key to *Stanleya* species:

1. Middle and upper cauline leaves petiolate, not auriculate or sagittate; filaments and bases of petal claws pubescent . . . . ***S. pinnata*** var. ***pinnata***

1' Middle and upper cauline leaves sessile, auriculate to sagittate; filaments and bases of petal claws glabrous.
   2. Annual or biennial; sepals 6–12 mm; fruiting pedicels 10–20(26) mm; petals linear to filiform, 0.5–1.5 mm wide, margins crisped . . . . ***S. confertiflora***

   2' Perennial; sepals 12–18 mm; fruiting pedicels 4–9(12) mm; petals narrowly oblanceolate, 1–3 mm wide, margins erose or rarely subentire . . . . ***S. viridiflora***

Stanleya confertiflora (B.L. Robinson) Howell

Biennial stanleya

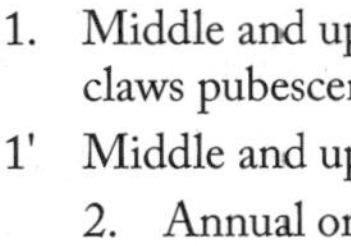

Plants annual or biennial, glabrous, glaucous. **Stems** 2–6(8) dm. **Basal leaves** obovate to ovate, margins entire; sessile. **Cauline leaves** middle (2.2)4–13(16) × (0.5)1–3(4) cm, middle and upper lanceolate, bases auriculate to sagittate; sessile. **Inflorescences** fruiting pedicels horizontal to divaricate-ascending, 10–20(26) mm. **Flowers** sepals spreading to reflexed, 6–12 mm; petals linear to filiform, (12)14–25 × 0.5–1.5 mm, yellow turning whitish, claws nearly linear, 5–11 mm, slightly wider at base, glabrous, petal margins crisped; filaments 12–17 mm, glabrous; ovules 40–62 per ovary, styles 0.6–1.7 mm; gynophore (8)10–18 mm. **Fruits** suberect to divaricate-ascending, subterete, slightly curved inward, torulose, 2–5.2(6) cm × 1.5–2.2 mm. **Seeds** 1.5–2.5 × 0.8–1 mm.

Alkaline meadows, heavy clay flats, barren clay slopes in sagebrush communities, dry sandy grounds. Flowering Apr–Jul. 600–1400 m. BR, BW, Owy. ID. Native.

Stanleya pinnata (Pursh) Britton var. ***pinnata***

Prince's plume, bush stanleya

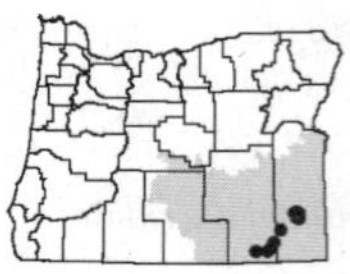

Plants sometimes subshrubs or shrubs, perennial with a woody base. **Stems** (1.2)3–12(15.3) dm, glabrous. **Basal leaves** withered by flowering. **Cauline leaves** lower and middle leaves oblanceolate to broadly lanceolate or ovate, 3–15 cm, bases not auriculate or sagittate, lyrate-pinnatifid or runcinate, petioles 0.7–6.2 cm; upper not auriculate nor sagittate, margins pinnately lobed or undivided, petiolate. **Inflorescences** fruiting pedicels horizontal to divaricate, 3–11 mm. **Flowers** sepals 8–16 mm; petals oblanceolate to oblong, 10–20 × 2–3 mm, yellow, claws 5–10 mm, wider at base, densely pubescent inside and at base; median filaments 11–28 mm, pubescent at base; ovules 22–38 per ovary, styles 0.2–0.6 mm; gynophore 10–28 mm, basally pubescent. **Fruits** spreading to divaricate or rarely ascending, terete, often strongly curved downwards, not torulose, 3–9 cm × 1.5–3 mm, smooth. **Seeds** 2.5–4 × 1.2–2 mm. ***2n***=28, 56.

Sandstone, basaltic, and volcanic rocks, chalk bluffs, talus, sagebrush and mixed shrub communities, selenium-rich areas. Flowering May–Jul. 800–1400 m. BR, Owy. CA, ID, NV; northeast to ND, east to SD, southeast to TX. Native.

Stanleya pinnata var. *pinnata* is distinguished from variety *integrifolia* (CO, KS, NV, UT, and WY) and variety *texana* (TX) by its lowermost cauline leaves that are lyrate-pinnatifid (vs. entire or rarely dentate).

Stanleya viridiflora Nuttall (p. 511)

Perennial stanleya

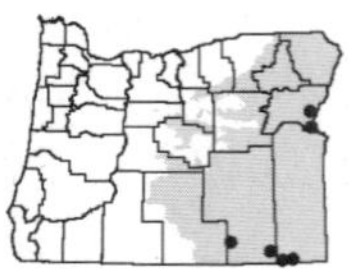

Plants perennial, glabrous, glaucous; caudices simple, with petiolar remains of previous seasons. **Stems** (2.5)4–12(14) dm. **Basal leaves** lanceolate, oblanceolate, or ovate, entire or sometimes dentate, rarely lyrate-pinnatifid, (2.2)5–18(22) × 1–4(6) cm; petioles 2–10(16) cm. **Cauline leaves** middle leaves lanceolate, (2)3.5–8.5(11) × (0.2)0.5–1.9(2.8) cm, margins entire; middle and upper cauline leaves bases auriculate to sagittate, sessile. **Inflorescences** fruiting pedicels horizontal to divaricate-ascending, 4–9(12) mm. **Flowers** sepals spreading to reflexed, 12–18 mm; petals narrowly oblanceolate, 13–20 × 1–3 mm, whitish to lemon-yellow, claws nearly linear-lanceolate, 7–11 mm, slightly wider at base, glabrous, margins erose or rarely subentire; filaments 11–20 mm, glabrous; ovules 28–50 per ovary, styles 0.1–0.3 mm; gynophore (8)11–22(25) mm. **Fruits** divaricate or descending, curved inward, subterete, torulose, 3–6(7) cm × 1.2–2 mm. **Seeds** 2–3 × 1–1.2 mm. ***2n***=28.

Limestone shale and rocks, clay knolls, red sandstone slopes, volcanic rocky slopes, sagebrush and pinyon-juniper communities. Flowering May–Jul. 600–1700 m. BR, BW, Owy. CA, ID, NV; northeast MT, east to WY, southeast to NM. Native.

STREPTANTHELLA Streptanthella

Genus composed of one species, *S. longirostris.*

Streptanthella longirostris (S. Watson) Rydberg (p. 512)

Beaked streptanthella, fiddle mustard

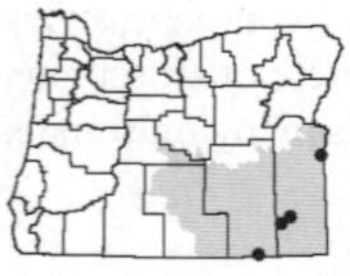

Herbs annual, often glaucous. **Trichomes** absent. **Stems** (12)20–60(75) cm. **Leaves** cauline, lanceolate to oblanceolate, 2–5.5(6.5) × 0.3–1(1.5) cm, bases not auriculate, margins entire, dentate, sinuate, or pinnatifid; sessile or short-petiolate; uppermost leaves linear. **Inflorescences** bracts 0; fruiting pedicels reflexed or rarely divaricate, recurved or rarely straight, (1)2–5(7) mm. **Flowers** sepals erect, caducous, oblong, (2)2.5–4(5) mm, lateral pair subsaccate; petals spatulate, (3.5)4–6(7) × 0.7–1.1 mm, slightly exceeding sepals, white or yellow, often with purplish veins, claws oblanceolate, 3–5 mm, > blades, petal margins crisped, tips obtuse; nectar glands lateral, semi-annular; stamens 6, in 3 pairs of unequal length, longest filaments exserted, 3–6 mm, anthers apiculate; ovules (12)16–28(34), septa complete, styles (2)3.5–6(8) mm, united with valves and replum apices, stigmas subentire. **Fruits** dehiscent siliques, latiseptate, strongly reflexed, linear, (25)35–60(70) × 1.2–2(2.2) mm, sometimes secund, unsegmented; valves with prominent midveins, not torulose; replums rounded to slightly flattened. **Seeds** uniseriate, oblong, flattened, 2–3 × 1–1.7(2) mm, not mucilaginous when wetted; wings 0.3–0.7 mm wide; cotyledons incumbent. ***2n***=28.

Roadsides, dry slopes and washes, rocky areas, sandy ridges, chaparral, deserts. Flowering Jan–Jun. 400–1400 m. BR, Owy. CA, ID, NV, WA; northeast to MT, east to WY, southeast to NM, south to Mexico. Native.

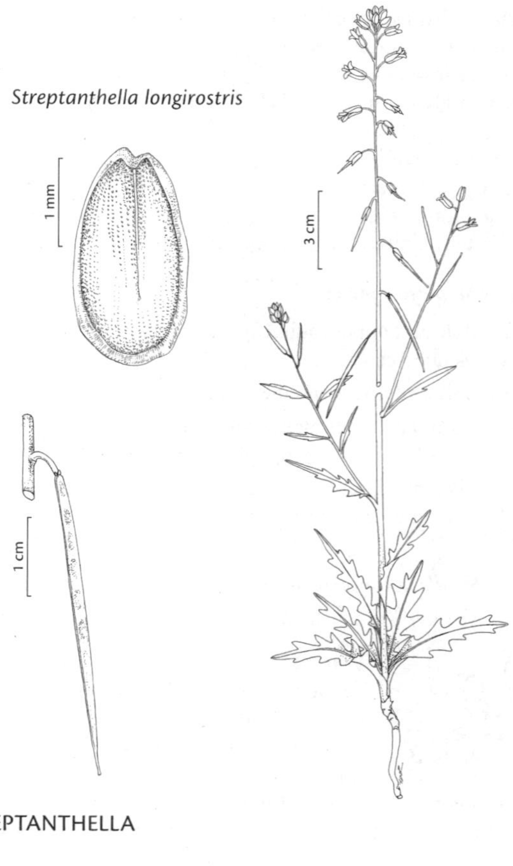

STREPTANTHELLA

STREPTANTHUS Jewelflower, twistflower

Herbs annual, biennial, or perennial; caudices present. **Trichomes** absent or simple. **Basal leaves** rosulate or not, margins entire, dentate, or lyrate-pinnatifid. **Cauline leaves** middle amplexicaul to auriculate and sessile, or not auriculate and petiolate. **Inflorescences** bracts present or not. **Flowers** radially or bilaterally symmetric; sepals erect, caducous, forming urceolate or campanulate calyces, often saccate; petals white, yellow, pink, rose, purple, violet, brownish or greenish, narrow and crisped or channeled, or broad and not crisped or channeled, claws differentiated or not; nectar glands confluent; stamens 6, in 3 pairs of unequal length, adaxial pair longest, frequently with united or free filaments and sterile anthers, sometimes all filaments free and all anthers fertile; ovules 10–120 per ovary, septa complete, styles distinct or obsolete, stigmas entire or 2-lobed. **Fruits** dehiscent siliques, latiseptate, linear, unsegmented; valves torulose or smooth; replums rounded. **Seeds** 10+ per fruit, uniseriate, ovoid to oblong, not mucilaginous when wetted; cotyledons accumbent.

Northern Mexico, western and south central United States. ~35 species; 4 species treated in *Flora*.

As delimited herein or treated by Al-Shehbaz et al. (2010), *Streptanthus s.l.* is not a monophyletic genus when studied along with the rest of the North American members of the tribe Thelypodieae (Cacho et al. 2014). Future molecular phylogenetic studies on the tribe would likely lead to the splitting of this genus into smaller segregates.

Key to *Streptanthus* species:

1. Cauline leaves petiolate, bases neither auriculate nor amplexicaul . . . . ***S. howellii***

1' Cauline leaves sessile, bases auriculate to amplexicaul.

  2. Inflorescences with bracts below or between lowermost 1 or 2 flowers . . . . ***S. tortuosus***

  2' Inflorescences with bracts 0.

    3. Fruits 3–7 mm wide; plants perennial, with woody caudices; all stamens fertile; calyces campanulate, radially symmetric; adaxial filaments free . . . . ***S. cordatus* ssp. *cordatus***

    3' Fruits 1.5–2.5 mm wide; plants annual or biennial, without caudices; adaxial stamens sterile; calyces urceolate, bilaterally symmetric; adaxial filaments united basally. . . . . ***S. glandulosus* ssp. *josephinensis***

Streptanthus cordatus Nuttall ssp. *cordatus*

Heartleaf twistflower

Streptanthus cordatus Nutt. var. *cordatus*

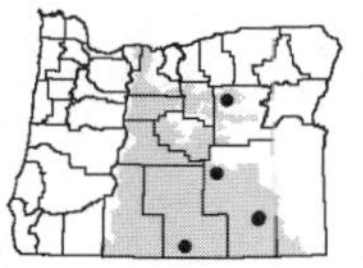

Plants perennial, mostly glabrous, glaucous; with woody caudices. **Stems** (1)3–9(11) dm. **Basal leaves** rosulate, spatulate to obovate, 1.5–8(11) cm, margins dentate, tips obtuse to rounded; petioles winged, setose-ciliate. **Cauline leaves** oblong, ovate, or suborbicular, 2–9 × 0.7–4.5(6) cm, bases auriculate to amplexicaul, tips rounded or acute; sessile. **Inflorescences** fruiting pedicels divaricate-ascending, 3–11(14) mm; bracts 0. **Flowers** sepals forming radially symmetric, campanulate calyces, not keeled, oblong, 5–12 mm, greenish brown to purple, glabrous or subapically bristly; petals 9–15 mm, purple to brownish, claws 7–10 mm, purplish, petal margins not crisped; filaments in 3 pairs of unequal length, abaxial pair 5–7 mm, adaxial pair free, 7.5–10 mm, lateral pair 4–5 mm, anthers fertile; ovules 20–38(46) per ovary, styles 0.2–3 mm, stigmas 2-lobed. **Fruits** ascending to divaricate, latiseptate, not torulose, 5–10.5(14.5) cm × 3–6(7) mm. **Seeds** broadly oblong to suborbicular, 2.5–5 × 2.2–5 mm; winged. **2*n***=28, 56.

Pinyon-juniper woodlands, sagebrush, open pine forests, dry stony slopes, sandy draws. Flowering May–Jul. 300–2800 m. BR, ECas, Lava. CA, ID, NV; east to WY, southeast to NM. Native.

Streptanthus glandulosus Hooker ssp. *josephinensis* Al-Shehbaz & M.S. Mayer

Josephine jewelflower

Streptanthus glandulosus Hook. ssp. *glandulosus*, *Streptanthus glandulosus* Hook. ssp. *secundus* (Greene) Kruckeb.

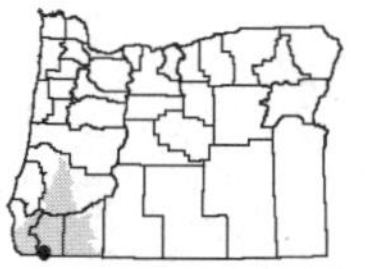

Plants annual, pubescent below, glabrous above. **Stems** 1.7–3.8 dm. **Basal leaves** not rosulate, soon withered; petiolate. **Cauline leaves** linear-lanceolate to lanceolate or oblanceolate, flat distally, 1–12 × 0.1–1.1 cm (smaller distally), bases auriculate to amplexicaul, margins dentate, sparsely pubescent; sessile. **Inflorescences** lax, secund; fruiting pedicels ascending to divaricate, straight or curved, 5–10 mm, glabrous or sparsely pubescent; bracts 0. **Flowers** sepals forming bilaterally symmetric, urceolate calyces, lanceolate, 3–5 mm, white or cream, glabrous or sparsely hirsute; petals subequal, 7–8 mm, with purple veins, margins crisped; filaments in 3 pairs, adaxial pair united basally, 5–6 mm, abaxial pair free, 3–5 mm, lateral pair 2–3 mm, abaxial and anthers fertile, adaxial anthers sterile; ovules 22–70 per ovary, styles 0.1–2.5 mm, stigmas subentire. **Fruits** divaricate to ascending, curved inwards, flattened, not torulose, 3–9 cm × 1.5–2.5 mm; valves with distinct midveins. **Seeds** ovate to oblong, 1.5–2.1 mm; winged. **2*n***=28.

Serpentine areas. Flowering May–Jul. 400–700 m. Sisk. Native.

Streptanthus glandulosus ssp. *josephinensis* is endemic to Josephine County and disjunct from the other seven California endemic subspecies.

Streptanthus howellii S. Watson (p. 514)

Howell's jewelflower, Howell's streptanthus

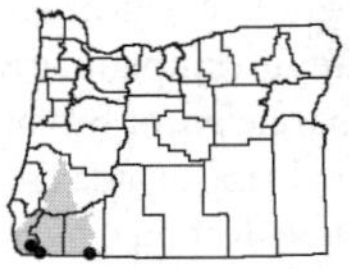

Plants perennial, glabrous; caudices present. **Stems** 3–8 dm. **Basal leaves** not rosulate, soon withered, fleshy; petiolate. **Cauline leaves** bases not auriculate or amplexicaul; petiolate; middle cauline leaves, spatulate-obovate to suborbicular, 1.5–10 × 0.5–4.5 cm, margins entire, repand, or dentate; upper narrowly oblong-oblanceolate to linear. **Inflorescences** fruiting pedicels divaricate to ascending, 0.7–1.7 cm; bracts 0. **Flowers** sepals forming subcampanulate calyces, not keeled, oblong, 5–8 mm, purplish; petals 8–12 mm, maroon to purple, claws 5–7 mm, petal margins not crisped; filaments in 3 pairs, free, abaxial pair 5–6 mm, adaxial pair 6–7 mm, lateral pair 4–5 mm, anthers fertile; ovules 24–44 per ovary, styles (0.5)1–3 mm, stigmas subentire; gynophore 0.5–5 mm. **Fruits** straight or curved inwards, not torulose, 5.5–12 cm × 2.5–3.5 mm; valves readily separated from styles and replums. **Seeds** broadly oblong to suborbicular, 3–4 × 2–3 mm; winged. ***2n***=28.

Dry serpentine slopes in open conifer-hardwood forests. Flowering Jun–Aug. 400–1700 m. Sisk. CA. Native.

The species is narrowly endemic to the Klamath Mountains in Oregon (Curry and Josephine counties) and California (Del Norte and Siskiyou counties).

Streptanthus tortuosus Kellogg (p. 514)

Mountain jewelflower, shield plant

Streptanthus tortuosus Kellogg var. *orbiculatus* (Greene) H.M. Hall, *Streptanthus tortuosus* Kellogg var. *tortuosus*

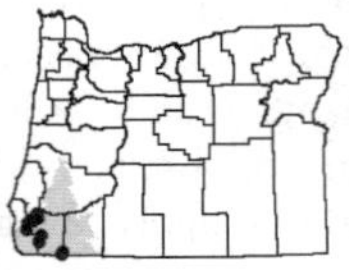

Plants biennial or short-lived perennial, glabrous, glaucous. **Stems** (0.5)1.5–12(15) dm. **Basal leaves** soon withered. **Cauline leaves** oblong, obovate, or suborbicular, (0.7)1.5–6(9) × (0.4)1–4.5 cm, margins entire or repand, bases auriculate to amplexicaul, tips acute to obtuse or rounded; sessile. **Inflorescences** fruiting pedicels divaricate to ascending, (2)3–12(17) mm; with bracts below or between lowermost 1 or 2 flowers. **Flowers** sepals forming urceolate calyces, 6–10(12) mm, obtuse to subacute, purplish, gray-green or yellowish; petals 8–14 mm, purplish or yellowish white, usually with purple veins, claws 4–10 mm, petal margins not crisped; filaments in 3 pairs, free, abaxial pair (2.5)4–7 mm, adaxial pair (5)7–11 mm, lateral pair (1.5)3–5 mm, anthers all fertile or adaxial pair subfertile and shorter; ovules 26–76(110) per ovary, styles 0.4–0.7 mm, stigmas entire. **Fruits** arcuate-spreading to pendent, flattened, torulose or smooth, (3)4–13(16) cm × 1.5–2.5(3) mm. **Seeds** oblong to orbicular, 1.5–2.5 × 0.8–1.7 mm; winged. ***2n***=28.

Rocky open slopes, open woodlands, montane forests, alpine areas. Flowering May–Aug. 200–1900 m. Sisk. CA. Native.

STREPTANTHUS • STRIGOSELLA

STRIGOSELLA Strigosella

Herbs annual. **Trichomes** simple and stalked, forked or dendritic. **Basal leaves** not rosulate, margins entire or variously dentate or pinnatifid; petiolate. **Cauline leaves** bases not auriculate; petiolate or subsessile. **Inflorescences** bracts 0. **Flowers** sepals erect, ovate to oblong, lateral pair not saccate; petals spatulate, oblong, or oblanceolate, > sepals, white, pink or purple, tips obtuse or rounded; nectar glands 4, lateral; stamens 6, tetradynamous, filaments connate in 2 pairs, anthers apiculate or not; ovules 40–80 per ovary, septa complete, styles obsolete, stigmas conical, 2-lobed, lobes decurrent, connivent, often connate, opposite replum. **Fruits** dehiscent siliques, linear, terete or 4-angled, unsegmented; valves smooth or torulose; replums rounded. **Seeds** uniseriate, oblong or ovate, not mucilaginous when wetted; wingless; cotyledons incumbent.

Africa, Asia, Europe, North America, South America. 20 species; 1 species treated in *Flora*.

Strigosella africana (Linnaeus) Botschantzev (p. 514)
African mustard

Malcolmia africana R. Br.

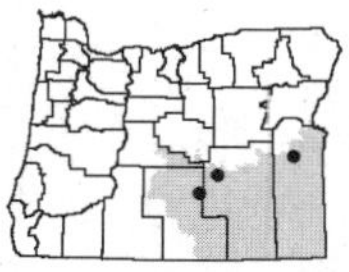

Plants pubescent, rarely glabrate. **Stems** (4)15–30(50) cm. **Cauline leaves** elliptic, oblong, or oblanceolate, (0.5)1.5–6(10) × (0.3)1–2.5(3.5) cm, margins entire, dentate, or rarely sinuate, petioles (0.1)0.6–2(3) cm; uppermost smaller; subsessile. **Inflorescences** fruiting pedicels as wide as fruit. **Flowers** sepals caducous or persistent, (3.5)4–5 mm; petals narrowly oblanceolate, 6–10(12) × 1–2 mm, pink, purple or rarely white; filaments free, styles obsolete, stigmas to 1 mm. **Fruits** divaricate-ascending, not torulose, 4-angled, linear, 2.5–5.5(7) cm × 1–1.3 mm; valves pubescent with coarse, setose and forked trichomes and much smaller, subdendritic trichomes, rarely glabrous. **Seeds** oblong, 1–1.2 × 0.5–0.6 mm. **2*n***=14, 28.

Disturbed areas, deserts, fields, sandy flats, sagebrush areas, grasslands, juniper woodlands. Flowering May–Aug. 700–900 m. BR, Owy. CA, ID, WA; northeast to MT, east to WY, southeast to AZ; Africa, Asia, Europe, South America. Exotic.

SUBULARIA Awlwort

Herbs annual, scapose. **Trichomes** absent. **Basal leaves** persistent, rosulate, margins entire; sessile. **Cauline leaves** absent. **Inflorescences** racemes; bracts 0. **Flowers** sepals ascending to erect, caducous or persistent, ovate-oblong, not saccate; petals tongue-shaped to narrowly oblanceolate, slightly exceeding sepals, white, not clawed, tips obtuse; nectar glands confluent, subtending bases of all filaments; stamens 6, subequal; ovules 4–18 per ovary, septa complete, styles absent, stigmas capitate, entire. **Fruits** dehiscent silicles, obovoid, ellipsoid or oblong, terete, unsegmented; valves not torulose; replums rounded. **Seeds** biseriate, oblong, not mucilaginous when wetted; wingless; cotyledons incumbent.

Africa, northern Europe, northern North America. 2 species; 1 species treated in *Flora*.

Subularia aquatica Linnaeus (p. 515)
ssp. ***americana*** G.A. Mulligan & Calder
American water-awlwort, awlwort

Subularia aquatica L. var. *americana* (G.A. Mulligan & Calder) B. Boivin

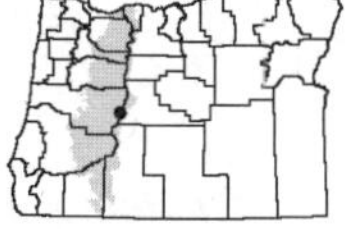

Stems (0.5)1.5–15(23) cm. **Basal leaves** subulate, (0.5)1–7(10) cm. **Inflorescences** (1)3–12(18)-flowered; fruiting pedicels ascending, usually attached to rachis at 30–40° angle, slender or slightly stout, 1–7(10) mm, sometimes terminal flower on longer pedicels to 1.8 cm. **Flowers** chasmogamous or cleistogamous; sepals persistent, (0.5)0.7–1(1.3) mm; petals 1.2–1.5 × 0.2–0.5 mm. **Fruits** broadly ellipsoid or obovoid, (1.5)2–3.5(5.5) × 1.2–2(2.5) mm. **Seeds** (4)6–12(18) per fruit, slightly compressed, 0.8–1 × 0.5–0.8 mm. **2*n***=30.

Shallow lake margins, streambanks, wet sedge meadows, muddy flats, salt marshes. Flowering Jul–Oct. 1500–1800 m. Casc. CA, ID, NV, WA; northern and western North America, Greenland. Native.

Subularia aquatica ssp. *americana* differs from ssp. *aquatica* (northern Europe and Russia) by persistent (versus caducous) sepals, fruiting pedicels ascending at a 30–40° (versus divaricate at a 50–90°) angle, and broadly ellipsoid to broadly obovoid (versus ellipsoid) fruits.

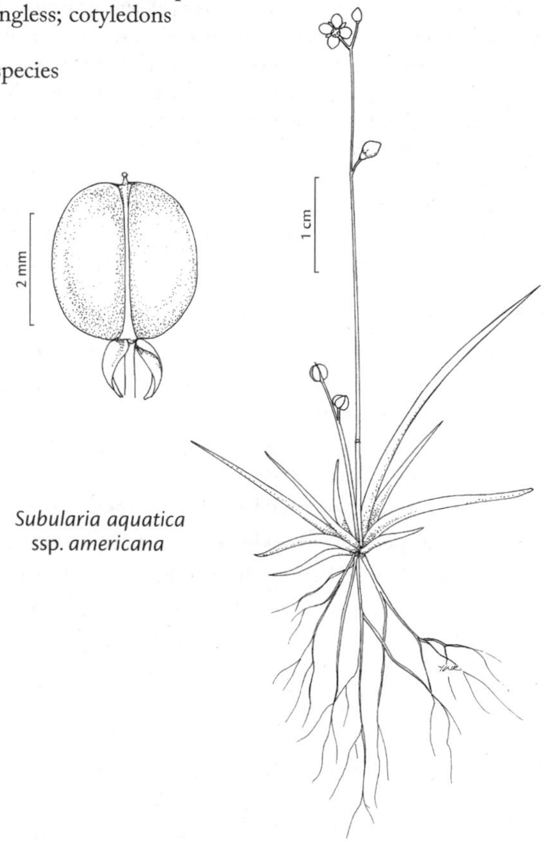

SUBULARIA

TEESDALIA Shepherd's cress

Herbs annual or perennial, scapose; roots slender. **Trichomes** simple or absent. **Basal leaves** rosulate, margins entire, dentate, lyrate, or pinnatifid; petiolate. **Cauline leaves** absent or few, margins entire or dentate; sessile. **Inflorescences** terminal racemes; bracts 0. **Flowers** sepals ascending to spreading, ovate, lateral pair not saccate; petals oblong to obovate, equal or outer pair larger, white, not clawed, tips obtuse; nectar glands 4, lateral; stamens 6 and tetradynamous or 4 and equal, filaments basally appendaged; ovules 4 per ovary, septa complete, styles obsolete, stigmas capitate, entire. **Fruits** dehiscent silicles, angustiseptate, broadly obcordate to suborbicular, apically notched, unsegmented; valves slightly keeled, apically winged; replums rounded. **Seeds** uniseriate, broadly ovate, slightly compressed, copiously mucilaginous when wetted; wingless; cotyledons accumbent.

Europe, North America. 3 species; 1 species treated in *Flora*.

Teesdalia nudicaulis (Linnaeus) W.T. Aiton (p. 516)

Shepherd's cress, teesdalia

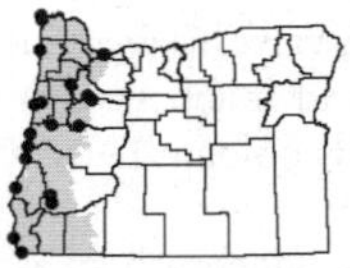

Plants glabrous or pubescent. **Stems** 5–15(20) cm. **Basal leaves** 3–15(20) × 2–4(7) mm; petioles 5–25(35) mm. **Cauline leaves** 0–4, linear to oblong, margins entire or dentate. **Inflorescences** fruiting pedicels divaricate, 3–7 mm. **Flowers** bilaterally symmetric; sepals 0.5–0.8(1) mm; outer petal pair 1.5–2.5 × 0.6–1 mm, inner pair 0.5–1 × 0.3–0.4 mm; stamens 6, tetradynamous, filaments basally appendaged, 0.4–0.5 mm. **Fruits** 3–4.5 × 3–4.5 mm. **Seeds** 1–1.2 × 0.9–1 mm, pale brown. **2*n***=36.

Sandy areas along trails, near beaches, pine clearings, disturbed areas, gravelly slopes. Flowering Mar–Jul. 0–1100 m. CR, Est, Sisk, WV. ID, WA; north to British Columbia, scattered in eastern US; Europe. Exotic.

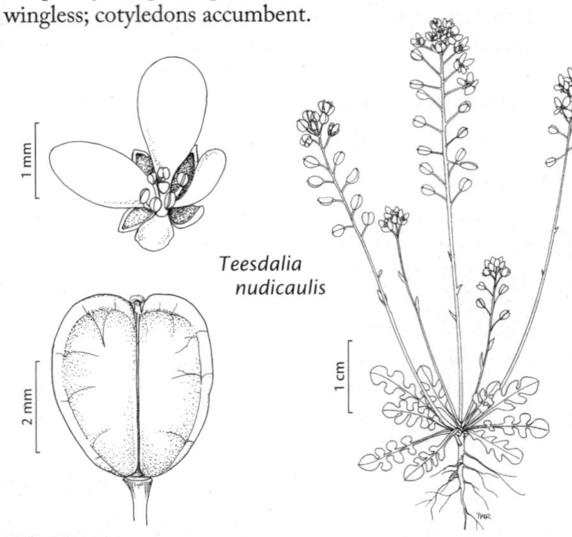

TEESDALIA

THELYPODIUM Thelypody

Herbs biennial or perennial. **Trichomes** absent or simple. **Basal leaves** rosulate or not, margins entire, dentate, laciniate, lyrate, or pinnately lobed; petiolate. **Cauline leaves** bases auriculate to sagittate or not; petiolate or sessile. **Inflorescences** elongate to corymbose racemes; bracts 0. **Flowers** sepals erect, spreading, or reflexed, ovate, oblong, or linear, forming tubular calyces, subsaccate or not; petals erect or spreading, spatulate, obovate, oblanceolate or linear, white, lavender or purple, crisped or not, claws differentiated or not; nectar glands confluent and subtending filament bases, or 2 or 4 and lateral; stamens 6, subequal or tetradynamous, anthers circinately coiled or remaining straight after dehiscence; ovules 12–128 per ovary, septa complete, styles distinct, stigmas capitate, entire; gynophore slender and narrower than fruit base, (0.5)1–8.5 mm. **Fruits** dehiscent siliques, linear, terete or slightly latiseptate, stipitate, unsegmented; valves torulose or not; replums rounded. **Seeds** 12+ per fruit, uniseriate, oblong or ovoid, plump or flattened, obscurely reticulate, not mucilaginous when wetted; wingless; cotyledons obliquely incumbent or accumbent.

Western North America. 16 species; 7 species treated in *Flora*.

Key to *Thelypodium* species:

1. Cauline leaves petiolate; petal claws widest at base.
   2. Fruiting pedicels straight, horizontal; stems solid; petals linear to linear-lanceolate; petioles of basal and lower cauline leaves glabrous . . . . *T. laciniatum*
   2' Fruiting pedicels curved upwards; stems hollow; petals oblanceolate to spatulate; petioles of basal and lower leaves ciliate . . . . *T. milleflorum*

1' Cauline leaves sessile; petal claws narrowest at base.
   3. Plants perennial; woody caudices covered with petiolar remains of previous season; stems flexuous . . . . *T. flexuosum*
   3' Plants biennial, rarely short-lived perennial; caudices naked; stems straight.
      4. Cauline leaves cuneate to attenuate at base . . . . *T. integrifolium*
      4' Cauline leaves sagittate, amplexicaul, or auriculate at base.
         5. Inflorescences densely flowered; petal margins crisped throughout; fruiting pedicels 1–2(2.5) mm . . . . *T. brachycarpum*
         5' Inflorescences lax; petal margins not or partially crisped; fruiting pedicels (2.5)3–8(14.5) mm.
            6. Flower buds oblong-linear; anthers exserted, circinately coiled after dehiscence; gynophores (1)2.5–6(7) mm . . . . *T. eucosmum*
            6' Flower buds ovate to lanceolate; anthers included, not circinately coiled; gynophores (0.2)0.5–1(3.5) mm . . . . *T. howellii*

Thelypodium brachycarpum Torrey

Short-podded thelypodium, shortpod thelypody

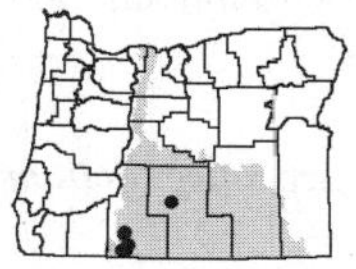

Plants biennial, glabrous or pubescent, often glaucous; caudices naked. **Stems** straight, (1.3)3.4–8.3(12) dm. **Basal leaves** oblanceolate to obovate, 3.3–14(20) × (0.5)1–3.2(4.7) cm, margins pinnately lobed to lyrate, sometimes dentate or entire; petioles (1.5)2–5.2(9) cm. **Cauline leaves** bases sagittate, margins entire or rarely dentate; sessile. **Inflorescences** spike-like, densely flowered; fruiting pedicels straight or slightly curved, horizontal to divaricate, stout, 1–2(2.5) mm. **Flowers** sepals erect, (3)3.5–5(5.5) mm; petals white, linear to narrowly oblanceolate, 8–12.5(16) × 0.3–0.5(1) mm, claws 2.5–4 mm, narrowest at base, petal margins crisped throughout; anthers apiculate; nectar glands surrounding bases of lateral stamens; ovules 12–26 per ovary, styles cylindrical, (0.2)0.5–1(2.5) mm, stigmas entire; gynophore (0.5)1–2(5) mm. **Fruits** terete, torulose, (0.8)1.2–2.7(3) cm × 1–1.5(2) mm. **Seeds** (1.3)1.5–2 × 0.8–1(1.3) mm.

Strongly alkaline meadows and desert flats. Flowering Jun–Aug. 1000–1700 m. BR, ECas. CA. Native.

The species is currently known only in Oregon from Klamath and Lake counties and in California from Napa, Shasta, and Siskiyou counties.

Thelypodium eucosmum B.L. Robinson (p. 518)

Arrowleaf thelypody

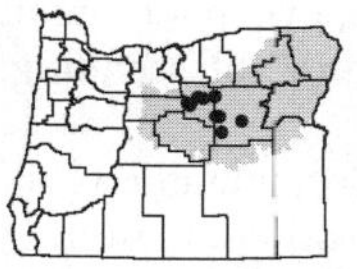

Plants biennial or rarely short-lived perennial, glabrous except petioles, glaucous; caudices naked. **Stems** straight, 2–10 dm. **Basal leaves** oblanceolate, oblong, lanceolate, or rarely ovate, (2.8)3.5–8.8(11) × (0.7)1–2.5(3.5) cm, margins entire or sinuate; petioles ciliate. **Cauline leaves** lanceolate to oblong, 1.8–4.5(6) × 0.5–1.6(2.4) cm, bases amplexicaul to auriculate; sessile. **Inflorescences** lax; fruiting pedicels horizontal to divaricate, (2.5)3–5.5(6.5) mm. **Flowers** buds oblong-linear; sepals linear-oblong, erect, 5–7(8) mm; petals spatulate to oblanceolate, (6.6)7.5–10(11.5) × 1–1.8(2) mm, purple to magenta, claws (3)3.5–5(5.5) mm, narrowest at base, petal margins not crisped; filaments subequal, (4.5)6–9(10) mm, anthers exserted, linear to narrowly oblong, circinately coiled after dehiscence; nectar glands confluent; ovules 44–58 per ovary, styles cylindrical or rarely subclavate, 0.5–1.5(2) mm, stigmas entire; gynophore (1)2.5–6(7) mm. **Fruits** terete, torulose, (2)2.4–5(6.5) cm × 0.7–1(1.3) mm. **Seeds** 0.7–1.5 × 0.5–0.8 mm.

Shady slopes and canyons, pinyon-juniper and oak-woodland communities, streamsides, riverbanks. Flowering May–Jul. 300–1600 m. BW, Lava. Native.

This Oregon endemic is currently known only from Baker, Grant, and Wheeler counties.

Thelypodium flexuosum B.L. Robinson

Spreading thelypodium, zigzag thelypody, nodding thelypody

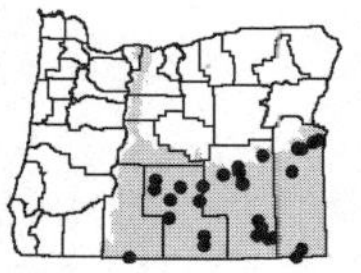

Plants perennial, glabrous; woody caudices covered with petiolar remains of previous seasons. **Stems** often subdecumbent, flexuous, 1.5–5.6(8.5) dm. **Basal leaves** lanceolate to oblanceolate, (2)3.5–16.5(20.5) × (0.5)1–2.5(4.5) cm, margins entire; petioles 1–6.5(13) cm, not ciliate. **Cauline leaves** lanceolate to linear, 1–7(11) × 0.2–0.7(1.4) cm, bases sagittate to somewhat amplexicaul; sessile. **Inflorescences** few-flowered; fruiting pedicels horizontal to divaricate-ascending, (2.5)4–9(16) mm. **Flowers** sepals oblong, erect, 3–4(4.5) mm; petals spatulate, oblanceolate, or obovate, 6–9(10) × (1.5)2–3(3.5) mm, lavender or white, claws 2–3.5(4) mm, narrowest at base, petal margins not crisped; anthers oblong, not circinately coiled; nectar glands lateral; ovules 12–30 per ovary, styles cylindrical, (0.3)1–2(3) mm, stigmas entire; gynophore 0.5(1) mm. **Fruits** erect to ascending, terete, torulose, 1–2.5(4.2) cm × 0.8–1(1.5) mm. **Seeds** (1)1.3–1.5 × 0.5–8(1) mm. **2*n***=26.

Alkaline sandy loam or clay soils, open deserts. Flowering Apr–Jun. 700–1900 m. BR, ECas, Owy. CA, ID, NV; southeast to UT. Native.

Thelypodium howellii S. Watson

Howell's thelypody

Plants biennial or rarely short-lived perennial, glabrous except petioles; caudices naked. **Stems** straight, 1–8(9.3) dm. **Basal leaves** oblanceolate, spatulate, oblong, or lanceolate, 2–10(13.5) × 1–2.3(4) cm, margins lyrate, dentate, or entire; petioles 0.5–2 cm, ciliate. **Cauline leaves** lanceolate, linear, or oblong, (0.8)1–9.7(14.5) × 0.2–1(3) cm, bases amplexicaul to sagittate; sessile. **Inflorescences** lax; fruiting pedicels straight, ascending to divaricate, (2.5)3.5–8(14.5) mm. **Flowers** buds ovate to lanceolate; sepals linear or lanceolate, erect, (3.5)5–8.5(11.5) mm; petals linear, oblanceolate to spatulate, (6)9–16.5(22) × (0.5)0.8–2.7(3) mm, lavender to purple, claws (2.5)4–7(11) mm, narrowest at base, petal margins partially crisped; median filaments free or united, anthers included, linear, (1.5)2–4(5.5) mm, not circinately coiled; nectar glands confluent, subtending bases of all stamens; ovules 22–40 per ovary, styles (0.5)1–2.7(4.3) mm, stigmas entire; gynophore (0.2)0.5–1(3.5) mm. **Fruits** terete, straight and forming a straight line with pedicel, slightly torulose, (1.5)2–6.2(8.2) cm × 1–1.3(2) mm. **Seeds** (1)1.2–1.7(2) × 0.5–0.7 mm. **2*n***=26.

2 subspecies.

Key to *Thelypodium howellii* subspecies:

1. Median filaments united for most of their length; petals linear to narrowly oblanceolate, rarely spatulate, 0.5–1.2(3) mm wide. ***T. h.*** ssp. ***howellii***

1' Median filaments free; petals spatulate, rarely oblanceolate, (1.2)1.5–2.7(3) mm wide . . . . ***T. h.*** ssp. ***spectabilis***

Thelypodium howellii ssp. ***howellii***

Howell's thelypody

Thelypodium howellii S. Watson var. *howellii*

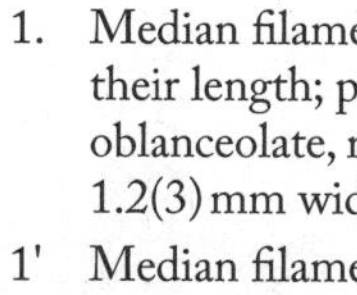

Basal leaves oblanceolate, margins lyrate, entire or rarely dentate. **Petals** linear to narrowly oblanceolate, rarely spatulate, 0.5–1.2(3) mm wide; median filaments united for most of their length. **Fruits** 1.5–4.5(7) cm. **2*n***=26.

Desert shrub communities, alkaline meadows and flats. Flowering Jun–Aug. 1200–1600 m. BR, BW, ECas. CA, WA. Native.

Thelypodium howellii ssp. *spectabilis* (M. Peck) Al-Shehbaz

Spectacular thelypody

Thelypodium howellii S. Watson var. *spectabilis* M. Peck

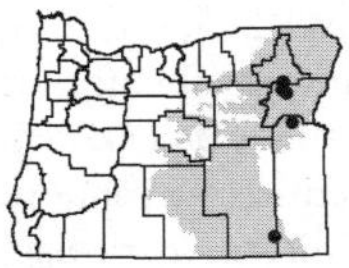

Basal leaves oblong-lanceolate to oblong, entire or repand. **Petals** spatulate or rarely oblanceolate, (1.2)1.5–2.7(3) mm wide; median filaments free. **Fruits** (1.8)3–6.5(8.2) cm. **2*n***=26.

Desert shrub communities, alkaline ground. Flowering Jun–Jul. 900–1100 m. BR, BW. Native.

Thelypodium integrifolium (Nuttall) Endlicher

Entire-leaved thelypody

Plants biennial, glabrous; caudices naked. **Stems** straight, (2)4.5–17(28) dm. **Basal leaves** oblong, lanceolate, oblanceolate to obovate, (3.7)5–31(54) × (1.2)1.6–7.8(14) cm, margins entire or dentate; petioles (0.5)1.2–10(15) cm, not ciliate. **Cauline leaves** linear to linear-lanceolate, (1)2.2–8.3(19.5) × (0.2)0.3–1.1(2.5) cm, bases cuneate to attenuate, margins entire; sessile. **Inflorescences** densely flowered; fruiting pedicels ascending, divaricate, or reflexed, (2)3–9(13) mm. **Flowers** sepals linear, erect, (3)3.5–5.5(7.5) mm; petals spatulate to oblanceolate, (4.5)6–9(13) × (0.5)0.8–1.5(2. 3) mm, white, lavender, or purple, claws (2)3–4.5(7) mm, petal margins not crisped; filaments subequal, anthers linear to narrowly oblong, not circinately coiled; nectar glands lateral; ovules 14–40 per ovary, styles cylindrical or subclavate, 0.5–1.3(3.2) mm, stigmas entire; gynophore 1–2(5.5) mm. **Fruits** terete, strongly incurved or rarely straight, (0.7)1.3–3.5(8) cm × 1–1.3(2) mm. **Seeds** (1)1.2–2(2.3) × 0.7–1(1.3) mm.

Western United States. 5 subspecies; 2 subspecies treated in *Flora.*

Key to *Thelypodium integrifolium* subspecies:

1. Fruiting pedicels stout, strongly flattened at base, 3–5(6) mm. . . . . . *T. i.* ssp. ***complanatum***

1' Fruiting pedicels slender or rarely stout, not or hardly flattened at base, (4)6–10(13) mm . . . . *T. i.* ssp. ***integrifolium***

Thelypodium integrifolium ssp. *complanatum* Al-Shehbaz

Entire-leaved thelypody

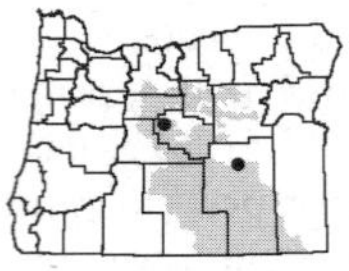

Inflorescences fruiting pedicels horizontal, straight, stout, strongly flattened at base, 3–5(6) mm. **Flowers** petals (5)6–8.5(9.5) mm, lavender or rarely white; gynophore stout, 0.5–1(3) mm. **Fruits** horizontal to ascending, straight or incurved, (0.9)1.4–2.7(3.4) cm.

Alkaline areas, desert shrub communities, canyons. Flowering Jun–Aug. 800–1300 m. BR, Lava. CA, NV; southeast to UT. Native.

Thelypodium integrifolium ssp. *integrifolium*

Entire-leaved thelypody

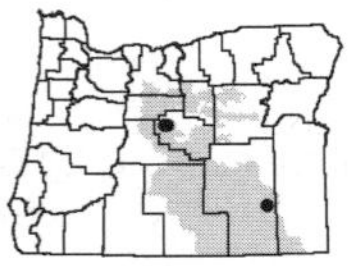

Inflorescences fruiting pedicels divaricate to divaricate-ascending, rarely horizontal, straight or slightly incurved, slender or rarely stout, not or hardly flattened at base, (4)6–10(13) mm. **Flowers** petals (4.5)5.5–8(10.5) mm, lavender to purple, rarely white; gynophore stout, 0.5–1.2(2.5) mm. **Fruits** divaricate-ascending to ascending, straight or incurved, (1)1.5–3(4) cm. **2*n***=26.

Alkaline grounds and flats, desert shrub communities, barren hillsides, mineralized grounds near hot springs. Flowering Jun–Sep. 800–1300 m. BR, Lava. ID, WA; northeast to ND, east to NE, southeast to CO. Native.

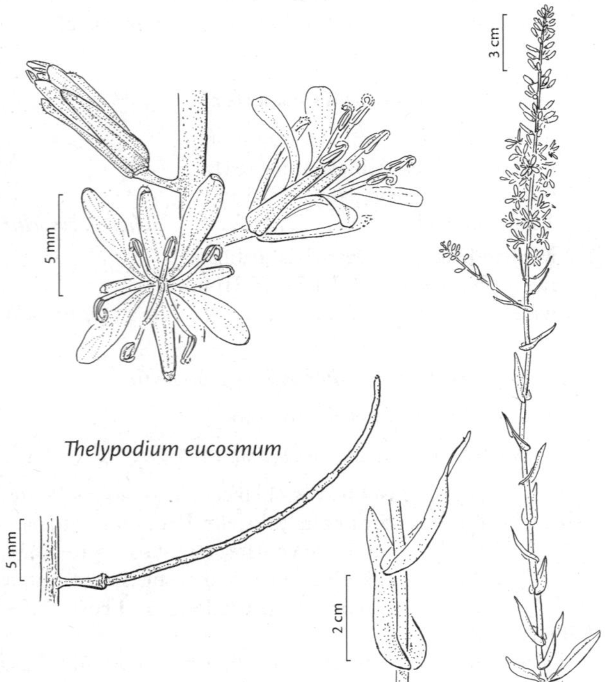

Thelypodium eucosmum

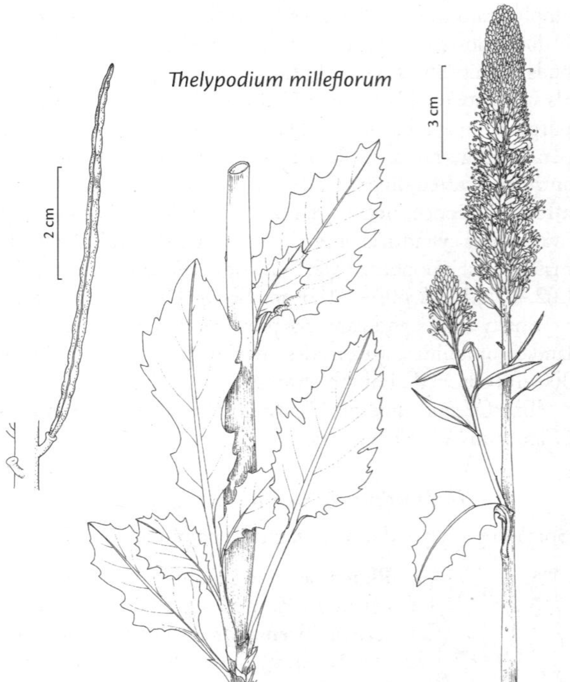

Thelypodium milleflorum

THELYPODIUM

Thelypodium laciniatum (Hooker) Endlicher

Cutleaf thelypody, thick-leaved thelypody, western thelypodium

Thelypodium laciniatum (Hook.) Endl. var. *laciniatum*, *Thelypodium laciniatum* (Hook.) Endl. var. *streptanthoides* (Leiberg) Payson

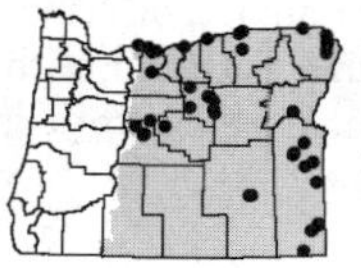

Plants biennial, glabrous. **Stems** solid, (1.3)2.6–10(14.5) dm. **Basal leaves** (4.3)6.8–24(45) × (1)1.8–8.5(14) cm, margins pinnately lobed or laciniate, glabrous; petioles (1)1.5–10(15) cm. **Cauline leaves** lower similar to basal, middle and upper not auriculate, margins entire, sinuate, or laciniate; petiolate. **Inflorescences** densely flowered; fruiting pedicels horizontal, straight, often stout, (2)3–6.5(15) mm. **Flowers** sepals narrowly oblong, erect, (3.5)4–7.5(9.5) mm, white, green or purple; petals linear to linear-lanceolate, (6)8.5–18(20) × 0.3–0.8(1.5) mm, white, aging to purple, slightly crisped, claws (2.5)3–6(7) mm, widest at base; filaments subequal; anthers linear, 2–4(5.5) mm, circinately coiled, apiculate; nectar glands confluent; ovules 56–108 per ovary, styles (0.3)0.7–2.5(4) mm, stigmas entire; gynophore 1–5(8.5) mm. **Fruits** divaricate-ascending to slightly reflexed, terete, somewhat tortuous, torulose, (2.5)3.5–10(12) cm × (0.7)1–1.5(2) mm. **Seeds** (0.7)1–1.5(1.8) × 0.5–0.7(1) mm. **2*n***=26.

Rock crevices, cliffs, rocky outcrops, among boulders, serpentine rock, and talus. Flowering Apr–Sep. 50–2000 m. BR, BW, Col, ECas, Lava, Owy. CA, ID, NV, WA; north to British Columbia. Native.

Thelypodium milleflorum A. Nelson (p. 518)

Many-flowered thelypody

Thelypodium laciniatum (Hook.) Endl. var. *milleflorum* (A. Nelson) Payson

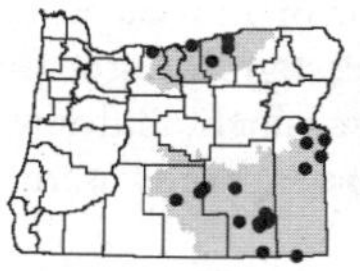

Plants biennial, glabrous except for petioles. **Stems** hollow, (1.8)4.5–13(21) dm. **Basal leaves** oblong or occasionally ovate or lanceolate, (3.8)6–23(28) × (0.9)2–6.5(10) cm, margins dentate or sinuate; petioles (1)1.8–9.5(13) cm, ciliate. **Cauline leaves** lower similar to basal, middle and upper cauline leaves lanceolate, bases not auriculate, margins dentate or entire; petiolate. **Inflorescences** densely flowered; fruiting pedicels erect distally, curved upwards, stout, (1.5)2.5–5(7) mm. **Flowers** sepals narrowly oblong, erect, (4)4.5–8(9) × 1–2 mm, white, sometimes tipped with green; petals oblanceolate to spatulate, (7)9–15(16) × 1–2 mm, white, claws (3)3.5–5.5(6) mm, widest at base; filaments subequal, anthers linear to narrowly oblong, 2.5–4.5(6) mm, circinately coiled, apiculate; nectar glands confluent; ovules 50–78 per ovary, styles (0.5)0.7–1.5(3) mm, stigmas entire; gynophore 1–4(6) mm. **Fruits** erect, subappressed to rachis, terete, somewhat tortuous, torulose, (1.8)3.3–8(10) cm × (0.7)1–1.5(1.8) mm. **Seeds** (0.7)1–1.5(2) × 0.5–0.7(1) mm. **2*n***=26.

Sand dunes, sagebrush, desert shrub communities. Flowering May–Jul. 100–1700 m. BR, Col, Owy. CA, ID, NV, WA; southeast to UT. Native.

THLASPI Pennycress

Herbs annual, often fetid when crushed. **Trichomes** absent or simple. **Basal leaves** rosulate or not, margins entire or dentate; petiolate. **Cauline leaves** bases auriculate or sagittate, margins entire or dentate; sessile. **Inflorescences** racemes; bracts 0. **Flowers** sepals erect or ascending, caducous, ovate or oblong, lateral pair not saccate; petals oblong or spatulate, white, claws undifferentiated from blades, petal tips obtuse or emarginate; nectar glands 2 or 4, lateral; stamens 6, slightly tetradynamous, anthers obtuse; ovules 6–16 per ovary, septa complete, styles obsolete or often < 0.5 mm, included in apical notch, stigmas capitate, entire. **Fruits** dehiscent silicles, strongly angustiseptate, oblong, obovate, obcordate, or suborbicular, apically notched, unsegmented; valves keeled, winged all around or only apically; replums rounded. **Seeds** 6–16 per fruit, uniseriate, ovoid, plump, not mucilaginous when wetted; wingless; cotyledons accumbent.

Africa, Asia, Europe. 6 species; 1 species treated in Flora.

Thlaspi arvense Linnaeus (p. 519)

Fanweed, field pennycress, pennycress, stinkweed

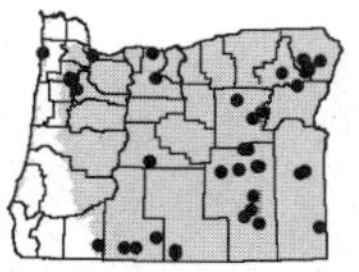

Plants glabrous. **Stems** (9)15–55(80) cm. **Basal leaves** oblanceolate, spatulate, or obovate, 10–50 × 4–23 mm; petioles 5–30 mm. **Cauline leaves** middle oblong, (5)15–40(80) × (2)5–15(25) mm. **Inflorescences** fruiting pedicels divaricate, straight or curved upward, slender, (5)9–13(15) mm. **Flowers** sepals (1.5)2–3(3.3) mm; petals (2.4)3–4.5(5) × (0.8)1.1–1.7 mm; styles 0.1–0.3 mm. **Fruits** obovate or suborbicular, (6)9–20 × (5)7–20 mm, bases obtuse or rounded, tips deeply emarginate, apical notch ~5 mm deep; wings 1–1.5 mm wide basally, 3.5–5 mm wide distally. **Seeds** (1.2)1.6–2(2.3) × 1.1–1.3 mm, blackish brown, concentrically striate. **2*n***=14.

Disturbed areas, woods. Flowering Mar–Jul. 0–2000 m. BR, BW, Casc, Col, ECas, Est, Lava, Owy, WV. CA, ID, NV, WA; throughout most of Canada and US; Asia, Europe. Exotic.

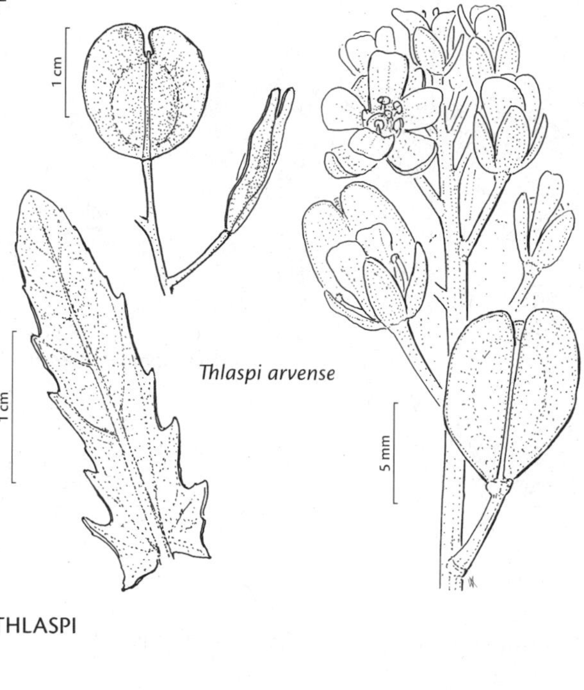

THLASPI

THYSANOCARPUS Fringepod

Herbs annual, glabrous or pubescent. **Trichomes** absent or simple. **Basal leaves** margins entire, dentate, pinnatifid, or pinnatisect; short-petiolate. **Cauline leaves** bases auriculate or not, margins as in basal leaves; sessile. **Inflorescences** racemes; bracts 0. **Flowers** sepals ascending, caducous, oblong to ovate, lateral pair not saccate; petals spatulate to oblong, ≈ or > sepals, white to purplish, not clawed; nectar glands lateral, semiannular, or 1 on each side of lateral stamen; stamens 6, slightly tetradynamous; ovules 1 per ovary, septa obsolete, styles short or obsolete, stigmas entire. **Fruits** indehiscent silicles, strongly latiseptate, pendulous, orbicular or suborbicular, unsegmented; valves with prominent midveins; replums entire or crenate, often perforated; wings flattened, with radiating rays, entire or incised, often perforate. **Seeds** aseriate, elliptic to orbicular, flattened, not mucilaginous when wetted; wingless; cotyledons accumbent.

North America. 7 species; 2 species treated in *Flora*.

Key to *Thysanocarpus* species:

1. Lower fruiting pedicels gradually recurved, 3–7(12) mm; fruits 3–6(9) mm wide, wings rayless or with rays 0.2–0.5 mm wide . . . . *T. curvipes*

1' Lower fruiting pedicels straight, 7–18 mm; fruits 7–10 mm wide, wings with rays 0.1–0.2 mm wide . . . . *T. radians*

Thysanocarpus curvipes Hooker
Sand fringepod, lacepod

Thysanocarpus curvipes Hook. var. *curvipes*, *Thysanocarpus curvipes* Hook. var. *elegans* (Fisch. & C.A. Mey.) B.L. Rob.

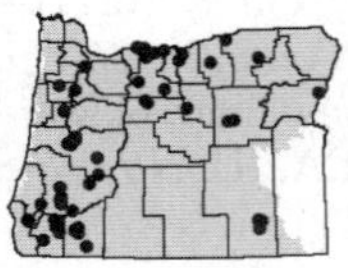

Stems 1–6(8) dm. **Basal leaves** oblanceolate to obovate, 1–6(13) cm, margins subentire to sinuate-dentate, usually hirsute, sometimes glabrous. **Cauline leaves** middle lanceolate, widest at base, bases auriculate-clasping. **Inflorescences** lower fruiting pedicels gradually recurved, 3–7(12) mm. **Fruits** flat or convex, suborbicular, 3–6(9) mm wide, with clavate or pointed trichomes, 0.2–0.4 mm, sometimes glabrous; wings entire, perforated, or incised, rayless or with rays 0.2–0.5 mm wide.

Rocky slopes, oak woodlands, streamsides, meadows, occasionally on serpentine. Flowering Mar–Jun. 0–1700 m. BR, BW, Casc, Col, CR, ECas, Lava, Sisk, WV. CA, ID, NV, WA; north to British Columbia, southeast to NM, south to Mexico. Native.

Thysanocarpus radians Bentham (p. 520)
Ribbed fringepod, large fringepod

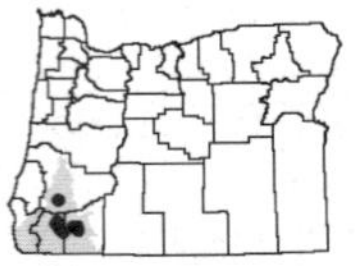

Stems 1.5–6 dm. **Basal leaves** oblanceolate, 1.5–4 cm, runcinate-pinnatifid or sinuate-dentate, glabrous or rarely sparsely hirsute with whitish hairs. **Cauline leaves** middle ovate to lanceolate, widest at base, bases auriculate to amplexicaul. **Inflorescences** lower fruiting pedicels straight, geniculately reflexed at tips, 7–18 mm. **Fruits** flat, orbicular, 7–10 mm wide, glabrous or with pointed hairs; wings entire or undulate at margins, rays prominent, 0.1–0.2 mm wide.

Meadows. Flowering Mar–May. 200–700 m. Sisk. CA. Native.

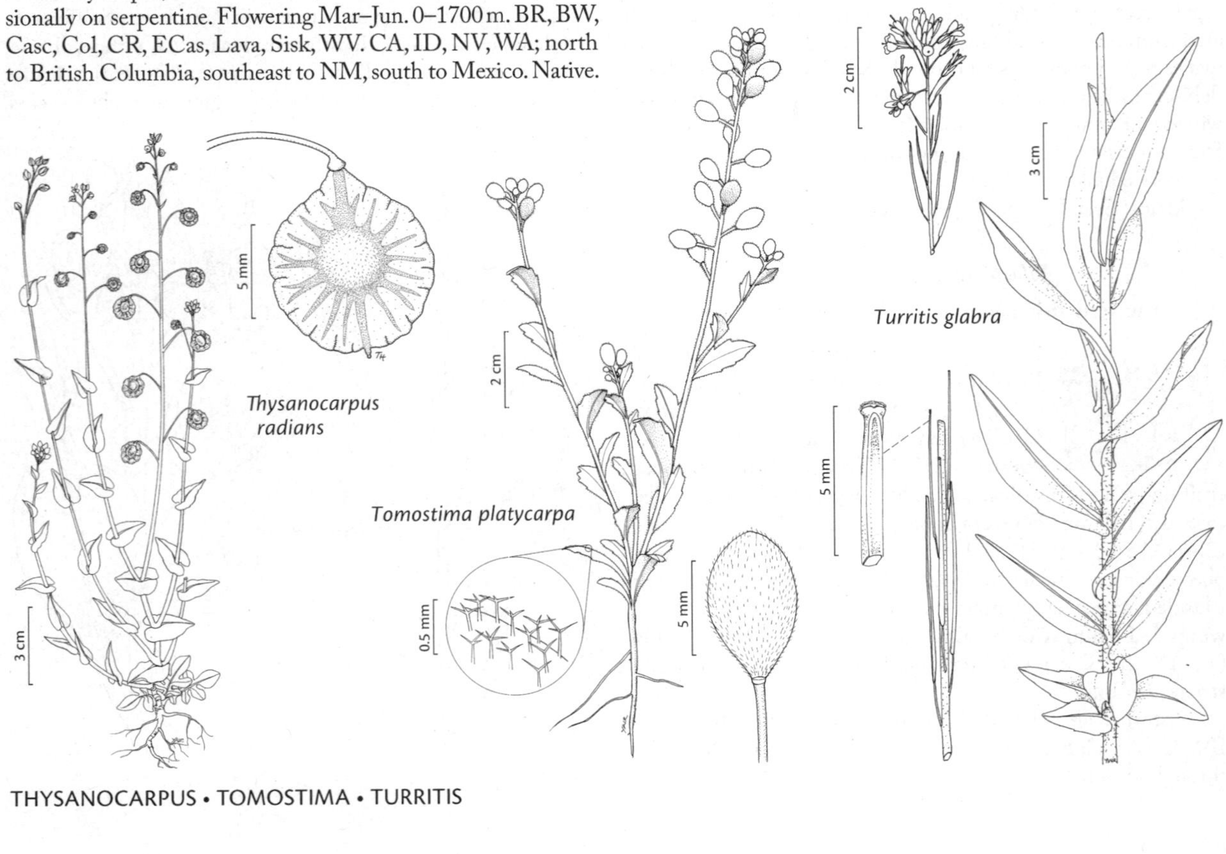

THYSANOCARPUS • TOMOSTIMA • TURRITIS

TOMOSTIMA Whitlow-grass

Herbs annual, scapose or not. **Trichomes** simple and 2–7-forked. **Basal leaves** rosulate or not, margins entire or rarely dentate; subsessile. **Cauline leaves** absent or few, bases not auriculate, margins entire or dentate; sessile. **Inflorescences** racemes; bracts 0. **Flowers** chasmogamous and petaliferous in early season, cleistogamous and apetalous in mid to late season; sepals erect, caducous, oblong, not saccate; petals obovate, spatulate, or oblanceolate, > sepals, white, not clawed, tips obtuse; nectar glands confluent, subtending bases of all stamens; stamens 6, slightly tetradynamous, anthers obtuse; ovules 12–88, septa complete, styles obsolete, stigmas capitate, entire. **Fruits** dehiscent silicles or siliques, linear, oblong, ovate, to obovate, latiseptate, flattened, unsegmented; valves obscurely veined; replums rounded. **Seeds** biseriate, ovate, not mucilaginous when wetted; wingless; cotyledons accumbent.

North America, South America. 6 species; 2 species treated in *Flora*.

Key to *Tomostima* species:

1. Fruits silicles, obovate to obovate-oblong, widest above middle; inflorescences lax in fruit . . . . ***T. platycarpa***

1' Fruits siliques, linear to linear-oblong, widest at or below middle; inflorescences subumbellate in fruit . . . . ***T. reptans***

Tomostima platycarpa (Torrey & A. Gray) Al-Shehbaz, M. Koch & Jordon-Thaden (p. 520)

Flat-podded whitlow-grass

Draba cuneifolia Nutt., *Draba cuneifolia* Nutt. var. *platycarpa* (Torr. & A. Gray) S. Watson, *Draba platycarpa* Torr. & A. Gray

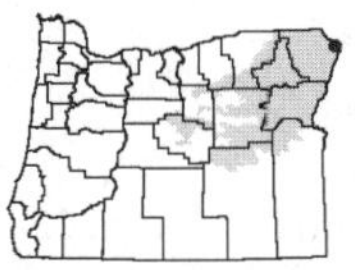

Plants hirsute with simple and stalked, 2–4-rayed trichomes. **Stems** (3)5–32(40) cm, simple or branched above base. **Basal leaves** obovate to spatulate or oblanceolate, 1–4.5 × 0.5–2.5(3) cm, margins dentate; petioles obscure. **Cauline leaves** 4–8(14), similar to basal. **Inflorescences** lax in fruit, 12–45(60)-flowered, pubescent; fruiting pedicels horizontal to divaricate, 4–9(11) mm. **Flowers** sepals 1.4–2.2 mm, pubescent; petals oblanceolate, (1.7)2.5–3.5(4) × 1–2 mm; ovules (28)40–70(82) per ovary, styles 0.01–0.2 mm. **Fruits** silicles, obovate to obovate-oblong, 5–8.2(9.5) × (2.4)2.8–3.7(4) mm, widest above middle, pubescent with antrorse, simple trichomes. **Seeds** 0.5–0.7 × 0.3–0.5 mm. **2*n***=16, 32.

Rocky hillsides, gravelly and sandy areas, dry plains, roadsides. Flowering Feb–Mar. 300–1000 m. BW. ID, WA; scattered in western and south central US, Mexico. Native.

Tomostima reptans (Lamarck) Al-Shehbaz, M. Koch & Jordon-Thaden

Carolina whitlow-grass

Draba reptans (Lam.) Fernald, *Draba reptans* (Lam.) Fernald var. *micrantha* (Nutt.) Fernald, *Draba reptans* (Lam.) Fernald var. *reptans*, *Draba reptans* (Lam.) Fernald var. *stellifera* (O.E. Schulz) C.L. Hitchc.

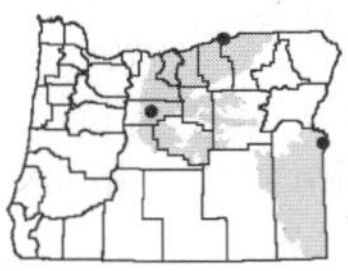

Plants pubescent with 2–4-rayed trichomes sometimes mixed with simple or spurred, often glabrous above. **Stems** (1)3–12(16) cm, simple. **Basal leaves** elliptic, spatulate, or suborbicular, 5–23(30) × 1.5–8(13) mm, margins entire. **Cauline leaves** absent or rarely 1–3 just above stem base. **Inflorescences** subumbellate in fruit, (3)5–12(16)-flowered, glabrous or rarely pubescent; fruiting pedicels horizontal to ascending, (1)2–7(9) mm. **Flowers** sepals 1.5–2.3 mm; petals spatulate, 2–4.5 × 1–1.5 mm; ovules 32–88 per ovary, styles 0.02–0.1 mm. **Fruits** siliques, linear to linear-oblong, (5)7–16(20) × 1.2–2.3 mm, widest at or below middle, glabrous or with antrorse, simple trichomes rarely mixed with fewer, 2-rayed trichomes. **Seeds** 0.5–0.8 × 0.3–0.5 mm. **2*n***=16, 32.

Rock outcrops, glades, hillsides, prairies, roadsides, disturbed sites. Flowering Feb–Jun. 100–700 m. Col, Lava, Owy. CA, ID, NV, WA; throughout southern Canada and US. Native.

TURRITIS Tower mustard

Herbs biennial, rarely short-lived perennial, glaucous above. **Trichomes** simple and stalked, forked or substellate. **Basal leaves** rosulate, margins repand, dentate, or lobed; petiolate. **Cauline leaves** bases auriculate, sagittate, or amplexicaul, margins entire; sessile. **Inflorescences** racemes; bracts 0. **Flowers** sepals erect, caducous, oblong or linear, lateral pair not saccate; petals spatulate, oblanceolate, or rarely linear, yellowish or creamy white, rarely pink or purple, not clawed, tips obtuse; nectar glands confluent, subtending bases of all stamens; stamens 6, tetradynamous, anthers obtuse; ovules 130–200 per ovary, septa complete, styles stout, stigmas capitate, subentire. **Fruits** dehiscent siliques, linear, terete or nearly quadrangular, unsegmented; valves with prominent midveins, smooth; replums rounded. **Seeds** biseriate, elliptic or orbicular, flattened, not mucilaginous when wetted; wingless or rarely narrowly winged; cotyledons accumbent.

Asia, Europe, North Africa, North America. 2 species; 1 species treated in *Flora*.

Turritis glabra Linnaeus (p. 520)

Tower mustard, tower rockcress

Arabis glabra (L.) Bernh., *Arabis glabra* (L.) Bernh. var. *glabra*

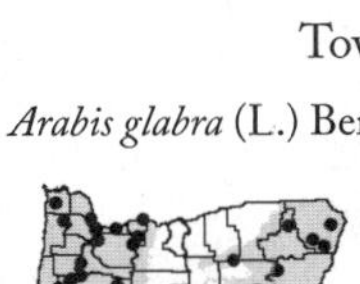

Plants pilose basally, glabrous and glaucous above. **Stems** (3)4–12(15) dm. **Basal leaves** spatulate, oblanceolate, or oblong, (4)5–12(15) × 1–3 cm. **Cauline leaves** lanceolate, oblong-elliptic, or ovate, 2–9(12) × (0.5)1–2.5(4) cm. **Inflorescences** fruiting pedicels erect, appressed to rachis, (6)7–16(20) mm, glabrous. **Flowers** sepals (2.5)3–5 mm; petals 5–8.5 × 1.3–1.7 mm; styles 0.5–0.8(1) mm. **Fruits** erect, appressed to rachis, (3)4–10(12.5) cm × 0.7–1.5 mm. **Seeds** 0.6–1.2 × 0.5–0.9 mm. **2*n***=12, 16, 32.

Forest margins, fields, roadsides, riverbanks, disturbed sites, mountain slopes, woods, meadows. Flowering May–Aug. 0–2400 m. BR, BW, Casc, CR, ECas, Est, Sisk, WV. CA, ID, NV, WA; throughout most of Canada and US; Africa, Asia, Europe. Native.

CABOMBACEAE Water-shield family

Stephen C. Meyers

Herbs perennial, aquatic; rhizomes branched; roots adventitious. **Stems** with obvious air chambers. **Leaves** alternate or opposite, submersed or floating, shield-like or finely dissected, margins entire; petiolate. **Inflorescences** solitary, axillary flowers. **Flowers** bisexual; sepals 3(4), distinct; petals 3(4); stamens 3–36(50), anthers dehiscing longitudinally; pistils 2–18, ovaries superior, stigmas head-like or linear. **Fruits** achene-like, leathery, indehiscent. **Seeds** 1–3.

Tropical and temperate areas worldwide. 2 genera.

The flowers of this family only remain open for two days. On the first day, during daylight hours, the stigmas are receptive, while the stamens remain unelongated and do not shed their pollen. The flowers close at nightfall at which time the pedicels bend, submerging the flowers. The flowers then re-emerge at daylight on the second day; however, the stigmas are not receptive, and the stamens elongate and shed pollen. This strategy prevents self-pollination.

Key to Cabombaceae genera:

1. Leaves shield-like; submerged leaves absent (but small submerged leaves may be present in young plants); young plants covered with mucilage; leaves 3–12 cm; stamens 18–36(50) .... **BRASENIA**

1' Leaves finely dissected; submerged leaves present, floating leaves only present at flowering; young plants with little or no covering of mucilage; leaves 0.5–5 cm; stamens 3–6 .... **CABOMBA**

BRASENIA Water-shield

Genus composed of one species, *B. schreberi.*

Brasenia schreberi J.F. Gmelin (p. 522)

Water-shield

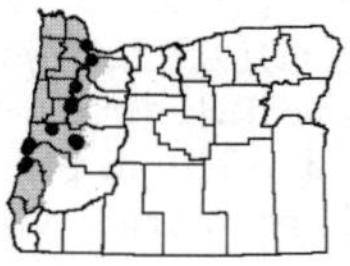

Plants covered with mucilage when young, mature plants glabrous. **Leaves** floating (but small submerged leaves may be present in young plants), alternate, shield-like, elliptic, 3–12 × 2–8 cm. **Flowers** sepals linear to narrowly ovate, 10–14 mm, outer surface green, inner surface reddish purple to dark red; petals linear to broadly linear, 12–16 mm, reddish purple to dark red; stamens 18–36(50); pistils 4–18, styles persistent, stigmas linear. **Fruits** club- to peanut-shaped, 6–10 mm. **Seeds** 1–2, ovoid, 2–4 × 2–3 mm. $2n=80$.

Ponds, lakes, and slow moving water. Flowering May–Oct. 0–1600 m. CR, Est, WV. CA, ID, WA; throughout most of North America except the intermountain and Rocky Mountain regions; worldwide, except Europe. Native.

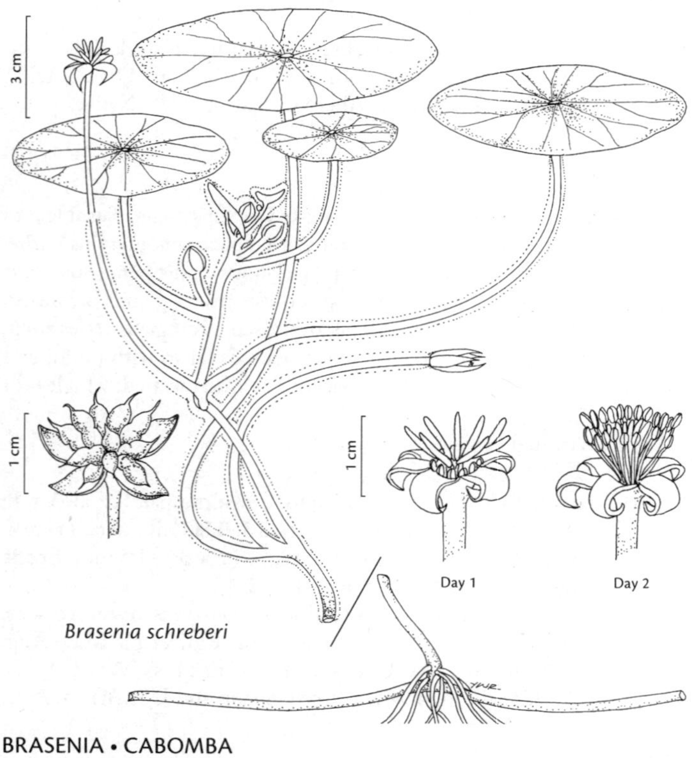

Brasenia schreberi

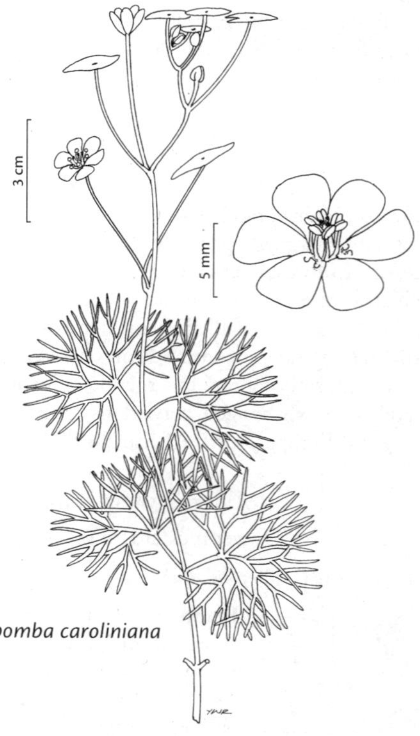

Cabomba caroliniana

CABOMBA Fanwort

Plants with little or no covering of mucilage when young, mature plants with scattered hairs. **Leaves** submerged, finely dissected, floating leaves present only at flowering, margins entire or notched basally, inconspicuous. **Flowers** sepals and petals similar. **Fruits** narrowly ovoid to peanut-shaped. **Seeds** 2–3, pendulous.

Nearly worldwide. 5 species; 1 species treated in *Flora.*

Cabomba caroliniana A. Gray (p. 522)

Carolina fanwort, green cabomba

Leaves submerged leaves opposite, 1–5 × 1–5 cm, palmately divided 3–4 times into finely dissected segments; floating leaves alternate, narrowly oblong to linear-elliptic, 5–25 × 1–2 mm. **Flowers** sepals and petals elliptic to obovate, 5–15(18) mm, petals slightly > sepals, white, yellow at base, occasionally tapered at base; stamens 3–6; pistils 2–4, stigmas head-like. **Fruits** 4–7 mm. **Seeds** oblong to ovate, 1–3 mm. **2*n***=34, 96.

Ponds, lakes, and ditches. Flowering May–Sep. 0–50 m. Est. CA, WA; eastern North America, South America; Australia. Exotic.

This species is listed as a Weed of National Significance in Australia where it is regarded as one of the continent's worst weeds.

CACTACEAE Cactus family

Stephen C. Meyers

Shrubs or **subshrubs** succulent, fleshy; roots fibrous or taprooted. **Stems** branched or not, cylindric, spherical or compressed, jointed or not; spines radial, originating from areoles. **Leaves** 0 or early deciduous. **Inflorescences** solitary flowers, 1 per areole. **Flowers** bisexual, radial, sessile; perianth parts numerous, grading from bract- or sepal-like outer tepals to petal-like inner tepals; stamens many; ovaries inferior, styles 1, stigma lobes 3. **Fruits** spiny or not. **Seeds** many.

Dry regions in Africa, Central America, North America, South America, and Sri Lanka. ~125 genera; 2 genera treated in *Flora.*

Key to Cactaceae genera:

1. Stems branched, jointed, compressed or subcylindrical; fruits spiny . . . . **OPUNTIA**

1' Stems unbranched, not jointed, spherical; fruits not spiny. . . . . **PEDIOCACTUS**

OPUNTIA Prickly-pear

Roots fibrous. **Stems** branched, jointed, compressed or subcylindrical; areoles spiny and usually surrounded with barbed hairs (glochids); wool white. **Fruits** pear-shaped, spiny, dry at maturity. **Seeds** disk-shaped, flat, yellow to gray, glabrous; girdle present, protruding 1–1.5 mm.

North America and South America. ~150 species; 2 species treated in *Flora.*

Key to *Opuntia* species:

1. Stem segments slightly compressed or subcylindrical; terminal segment easily detached when young . . . . ***O. fragilis***

1' Stem segments conspicuously compressed; terminal segment not easily detached . . . . ***O.* × *columbiana***

Opuntia* × *columbiana Griffiths (pro sp.)

Columbia prickly-pear

Opuntia polyacantha Haw.

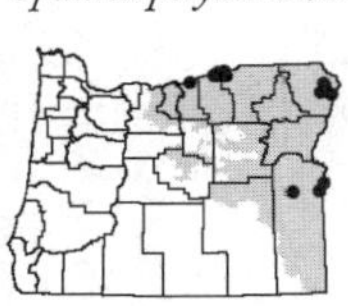

Plants prostrate, mat-forming, 10–30 cm. **Stems** segments conspicuously compressed, obovate to orbicular, 3–10 × 2–5 cm; terminal segments not easily detached; spines 2–8 per areole, yellow-gray, gray, or brown, longer spines 10–40 mm, shorter spines 3–6 mm; glochids 1–4 mm, yellow. **Flowers** 3–5 × 4–6 cm, yellow, sometimes reddish-tinged; filaments white to reddish to purple; styles white, stigmas green. **Fruits** 2–5 cm, tan to gray at maturity; areoles 10–15, with 1–12 spines. **Seeds** 5–7 mm, tan. **2*n***=66.

Sandy or gravelly soil on cliffs, canyons and sagebrush areas. Flowering May–Jul. 50–1100 m. BW, Col, Owy. ID, WA; north to British Columbia. Native.

Opuntia × *columbiana* is a hexaploid hybrid between *O. fragilis* and *O. polyacantha.* Previous Oregon collections and observations attributed to *O. polyacantha* are highly likely to be this species. No known genetically pure populations of *O. polyacantha* are known to exist in Oregon.

This hybrid was originally described as a non-hybrid species; hence the indication of the authority as "(pro sp.)."

Opuntia fragilis (Nuttall) Haworth (p. 524)

Brittle prickly-pear

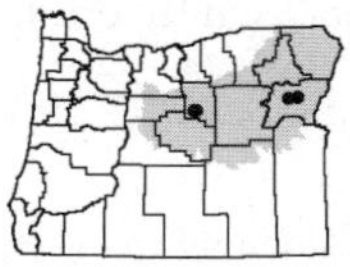

Plants prostrate, mat-forming, 5–20 cm. **Stems** segments slightly compressed or subcylindrical, elliptic to obovate, 2–5.5 × 1.5–3 cm; terminal segments easily detached when young; spines 3–8 per areole, gray with brown tips, longer spines 8–25 mm, shorter spines 1–3 mm; glochids 1–3 mm, tan to brown. **Flowers** 3–4 × 4–6 cm, yellow; filaments reddish, rarely white; styles white, stigmas green. **Fruits** 1–2.5 cm, tan at maturity; areoles 10–20, with 1–6 spines. **Seeds** 4–6 mm. **2*n*=22.**

Sandy or gravelly soil on dry hillsides and open ground. Flowering Jun–Jul. 600–1900 m. BW, Lava. CA, ID, NV, WA; north to British Columbia, east to Ontario, MI, IL, southeast to TX. Native.

PEDIOCACTUS Hedgehog cactus

Roots taproots at surface, branching, becoming fibrous. **Stems** unbranched, not jointed; areoles spiny, without barbed hairs. **Flowers** diurnal. **Fruits** globose, green to greenish yellow, turning reddish, not spiny, dry at maturity. **Seeds** ovoid, tuberculate. **2*n*=22.**

Western United States. 9 species; 1 species treated in *Flora.*

Pediocactus nigrispinus (Hochstätter) Hochstätter (p. 524)

Snowball cactus

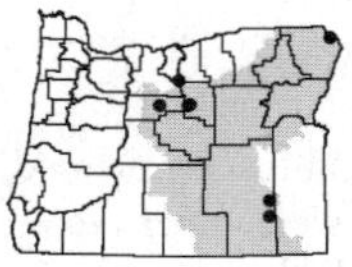

Plants often deeply seated in substrate. **Stems** subglobose to depressed globose, 5–30 × 5–15 cm; tubercles in 8–13 spiral rows; radial spines 10–30 per areole, 8–20(25) mm, light yellow to reddish brown; central spines 8–12 per areole, 15–30(35) mm, reddish brown to black, lighter colored at base. **Flowers** 1–3.5 × 2–5 cm, pale yellow to yellowish green, turning pink to purplish with age, clustered at or near top of stem; outer tepals with greenish brown stripe. **Fruits** 5–10 mm. **Seeds** 1.5–3 mm, gray to black. **2*n*=22.**

Scrub and sagebrush. Flowering Apr–Jul. 500–2300 m. BR, BW, Lava. ID, WA. Native.

Older Oregon collections of this species were attributed to *P. simpsonii.* The two species are nearly identical and are separated only by overlapping spine characteristics (shape and length) and location. Combining the species and/or relegating *P. nigrispinus* to subspecific status may be in order.

CALLITRICHIACEAE

(see Plantaginaceae, Volume 3)

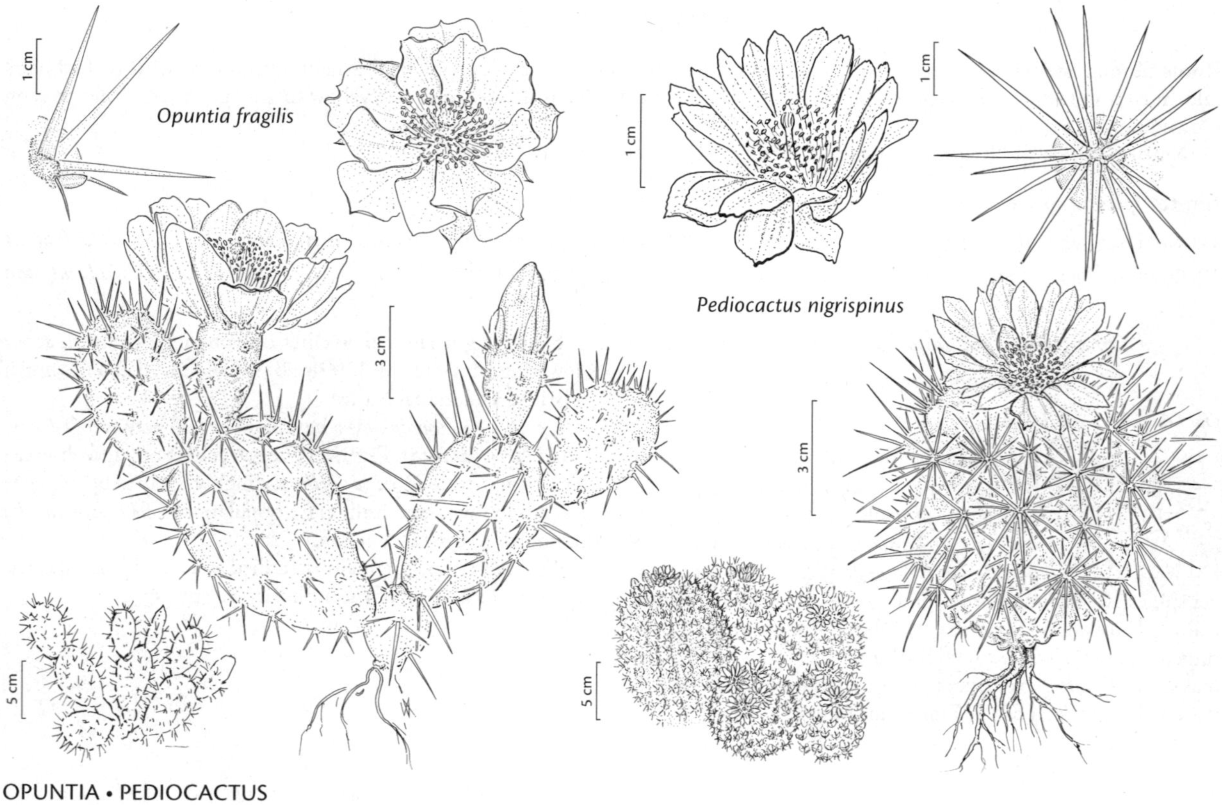

OPUNTIA • PEDIOCACTUS

CAMPANULACEAE Bellflower family

Nancy R. Morin (except where noted)

Herbs annual, biennial, or perennial. **Stems** erect to ascending. **Leaves** sometimes in basal rosette, usually also cauline, simple, alternate; petiolate or sessile; stipules 0. **Inflorescences** terminal or in axils of leaf-like or reduced bracts; cymes, panicles, racemes, spikes or flowers 1. **Flowers** bisexual, cleistogamous or open, inverted (pedicel twisted 180°) or not, radially or bilaterally symmetrical; hypanthia present, generally adnate to ovary; calyx lobes 5; corollas radially symmetric to 2-lipped or all lobes on 1 side; petals connate, or sometimes separate nearly to base, tubes deeply divided on 1 side or not, lobes 5; stamens 5, distinct or connate (anthers, filaments connate into tube or filaments connate above middle); ovaries inferior or half inferior, chambers 1–3, ovules many, styles 1, stigmas 2–3-lobed. **Fruits** capsules, open on sides or apically by pores or short valves. **Seeds** many.

Nearly worldwide. ~90 genera; 9 genera treated in *Flora.*

Subfamilies of Campanulaceae are sometimes treated as families. Recent phylogenetic work suggests that many genera, especially in Campanuloideae, are polyphyletic (Crowl et al. 2014; Wendling et al. 2011; Cellinese unpublished). Most North American taxa fall within a *Rapunculus* clade (Mansion et al. 2012), which includes the Asian genus *Asyneuma.* Within this clade, *Triodanis* is sister to the European *Legousia*; *Githopsis, Heterocodon*, and *Campanula prenanthoides* share a clade with annual North America *Campanula* species. *Campanula scouleri* is allied with *Campanula aurita, C. scabrella, C. parryi, C. piperi, C. divaricata*, and *C. lasiocarpa* in a clade that also contains Asian genera and European campanulas. Ultimately the taxa in this large clade must either be collapsed into an enormous *Campanula* genus, or species now in *Campanula* must be divided into individual genera.

Sporadically escaped *Jasione montana* has recently been recorded in Oregon. If seen again, it can be recognized by flowers in dense heads, corolla lobes cut nearly to base, and long-exserted styles.

In descriptions, the positions of flower parts are given after floral inversion, if any.

Key to Campanulaceae genera:

1. Corollas radially symmetric; filaments entirely distinct.
   2. Plants perennial .... **CAMPANULA**
   2' Plants annual.
      3. Flowers terminal, subtended (but generally not immediately so) by 1–several, generally non-leaf-like bracts, short-pedicelled .... **GITHOPSIS**
      3' Flowers opposite or in axils of leaf-like bracts, sessile.
         4. Corollas 3–5 mm, cylindric; calyx lobes widely triangular, leaf-like .... **HETEROCODON**
         4' Corollas 5–10 mm, rotate; calyx lobes narrowly triangular, not leaf-like .... **TRIODANIS**

1' Corollas bilaterally symmetric; filaments and anthers connate into tube or filaments connate into tube above middle and anthers distinct.
   5. Filaments distinct at base, connate above middle, anthers distinct; plants of dry areas .... **NEMACLADUS**
   5' Filaments and anthers connate; plants of wet areas.
      6. Capsules open on sides by slits or irregular tears.
         7. Pedicels 0; ovaries long, narrow, pedicel-like; leaves lanceolate to awl-like.......... **DOWNINGIA**
         7' Pedicels 1–4(8) mm; ovaries ellipsoid, not pedicel-like; leaves linear .... **HOWELLIA**
      6' Capsules open at top (within calyx lobes) by valves.
         8. Capsules spherical; plants biennial or perennial.......... **LOBELIA**
         8' Capsules narrowly obconic or cylindric; plants annual.......... **PORTERELLA**

CAMPANULA Harebell, bellflower

Plants perennials, glabrous or hirsutulous or pubescent; from taproot, fibrous roots or rhizomes. **Stems** reclining or erect, branched or simple. **Leaves** in basal rosette or not; cauline leaves linear, lanceolate to ovate, cordate, thin, fleshy, or leathery, margins entire or toothed or crenulate, sessile or petiolate, gradually reduced distally to bracts in inflorescence. **Inflorescences** terminal or axillary, racemes or flowers 1 or several in bract axils (head, spike, or panicle). **Flowers** not inverted; corollas radially symmetric, cylindric to funnel- or bell-shaped, cut 25–66% or nearly to base, white to deep blue, lobes narrow- to wide-triangular; filaments entirely distinct; ovaries inferior, hemispheric, spherical, or oblong to obconic. **Fruits** opening by 3 lateral pores. **Seeds** oblong or fusiform.

Northern Hemisphere. ~500–600 species; 5 species treated in *Flora.*

Campanula taxonomy is in a state of flux, and generic placement of North American Campanuloideae is uncertain, although molecular phylogenies place most of them in a *Rapunculus* clade (Mansion et al. 2012).

Key to *Campanula* species:

1. Styles much > corolla, well exserted.
   2. Corolla lobes cut ~66% length of corolla, deltate; leaves petiolate.......... ***C. scouleri***
   2' Corolla lobes cut nearly to base, linear to linear-lanceolate; leaves sessile to subsessile.......... ***C. prenanthoides***

1' Styles ≤ corolla.
  3. Capsules opening by pores near tips; corollas broadly bell-shaped, ≤ 40 mm . . . . *C. persicifolia*
  3' Capsules opening by pores near base; corollas funnel-shaped to bell-shaped, 12–30 mm.
    4. Distal cauline leaves ovate to lanceolate, 3.5–4.5 cm wide; pedicels down-curved; calyx lobes spreading to reflexed . . . . *C. rapunculoides*
    4' Distal cauline leaves lanceolate to linear, 0.4–1.7 mm wide; pedicels straight or arched; calyx lobes erect or spreading . . . . *C. rotundifolia*

Campanula persicifolia Linnaeus

Peachleaf bellflower

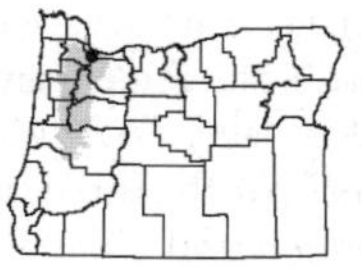

Plants glabrous; roots fibrous. **Stems** erect, simple, to 9 dm, ribbed. **Leaves** basal lanceolate to oblong-obovate, bases attenuate, margins crenate and toothed; cauline lanceolate to oblong-ovate, 1–6 cm, thin to leathery, margins crenate-denticulate, tips attenuate; petioles absent. **Inflorescences** loose racemes; pedicels short. **Flowers** calyx lobes widely spreading, acuminate, broad at base; corollas broadly bell-shaped, ≤ 40 mm, lilac blue, rarely white, lobes spreading, broadly deltate, cut ~33% of length; stamens ≤ 2 cm, filament bases ciliate; ovaries broadly obconic, 5–7 mm, styles ≤ 22 mm, stigmas straight, ≤ 10 mm, blue or whitish, ≤ corolla, separating from middle, distal 55% papillate. **Fruits** obconic, erect, furrowed, bases acute, opening by pores at tips. **Seeds** oblong, to 4 mm.

Disturbed areas. Flowering Apr–Jun. 200–800 m. WV. WA; UT and scattered locations from MI to VT and south to VA. Exotic.

Campanula persicifolia is a very showy garden escape that has sparsely naturalized.

Campanula prenanthoides Durand (p. 526)

California harebell, California bellflower

Asyneuma prenanthoides (Durand) McVaugh

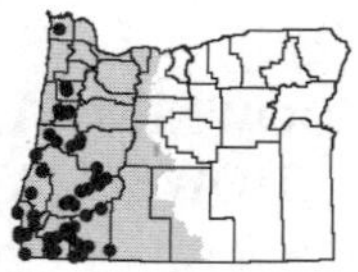

Plants glabrous or hairs sparse; taprooted. **Stems** reclining or erect, branched, 2–15 dm, 4-angled, sometimes deeply sulcate. **Leaves** cauline, sometimes also basal, lanceolate to ovate, 1–6 cm, thin to leathery, margins serrate; petioles absent or < 5 mm. **Inflorescences** flowers generally 2–4 per node, sometimes more; pedicels 2–6(20) mm. **Flowers** calyx lobes spreading to reflexed; corollas widely funnel-shaped, pale to deep blue, lobes lance-linear, 7–14 mm, cut nearly to base, erect in proximal half, reflexed distally; stamens ≤ 6 mm, filament bases ciliate; ovaries hemispheric, 2.5–5 mm, styles 15–18 mm, much > corolla, blue, well exserted, distal 55% papillate. **Fruits** hemispheric, bases cordate, strongly ribbed pores at or below middle. **Seeds** oblong, ≤ 2 mm.

Moist or dry banks and open areas, coniferous forests. Flowering Jun–Sep. 0–2600 m. Casc, CR, ECas, Est, Sisk, WV. CA. Native.

Morin (2012) treated *Campanula prenanthoides* as *Asyneuma prenanthoides* based on its sulcate stem, multiple pedicellate flowers from a node, deeply divided corollas with linear lobes, conspicuously basket-like, ribbed capsules, and long-exserted styles.

Campanula rapunculoides Linnaeus

Rover bluebell, creeping bellflower

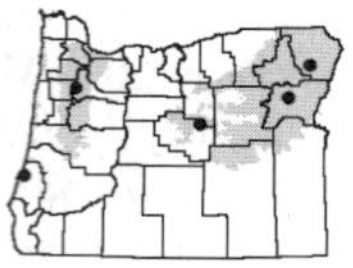

Plants glabrous, downy, or bristly; with stout, spreading rootstocks and stolons. **Stems** erect, to 10 dm. **Leaves** basal and cauline, 5–10 × 3–5 cm; proximal leaves ovate to cordate, bases cordate, margins crenate to serrate, veins conspicuous abaxially, tips acuminate; distal cauline leaves ovate to lanceolate, 3.5–4.5 cm wide, bases cuneate, tips acuminate, petiolate. **Inflorescences** racemes, flowers usually secund; bracts reduced distally; pedicels down-curved (rarely straight), 2–12 mm (to 6 cm in fruit). **Flowers**

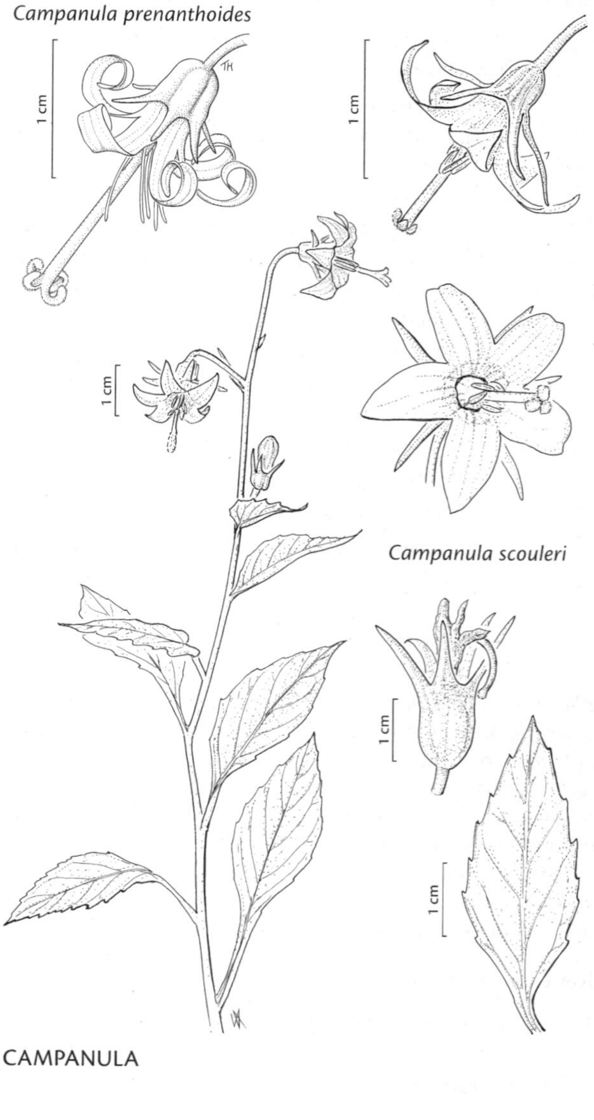

CAMPANULA

calyx lobes oblong to triangular-lanceolate, spreading to reflexed; corollas narrowly funnel-shaped to bell-shaped, ≤ 30 mm, purple-blue, lobes cut ~33% of length of corolla, erect to spreading; ovaries hemispheric, 3–4 mm, styles ≤ 30 mm, ≤ corolla, blue, distal 30% papillate. **Fruits** hemispheric, pendent, strongly ribbed, opening by pores near base.

Disturbed areas, woodlands, forests, riparian areas. Flowering May–Sep. 0–1600 m. BW, Est, WV. ID, NV, WA; throughout US and Canada except scattered in southern states; Asia, Europe. Exotic.

Campanula rapunculoides has become a serious weed, especially in the Rocky Mountains, and should be expected to become more broadly distributed in Oregon.

Campanula rotundifolia Linnaeus

American harebell, Scot's bellflower

Campanula sacajaweana M. Peck

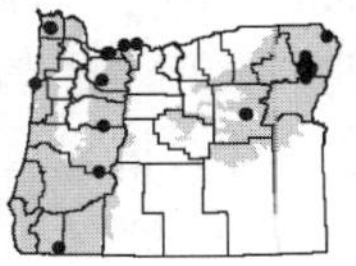

Plants glabrous, hirsutulous, or pubescent in lines; from taproots and slender rhizomes. **Stems** erect, 1–6 dm. **Leaves** basal leaves oblanceolate to nearly round, 1–2 cm, margins dentate, long-petiolate; distal cauline leaves linear to lanceolate, 30–90 × 0.4–1.7 mm, thin, margins entire or remotely serrulate, sessile. **Inflorescences** terminal racemes or panicles, flowers 1–several, buds erect, flowers nodding; pedicels straight or arched, 10–20 mm. **Flowers** calyx lobes linear to deltate, 2–11 mm, spreading to ascending; corollas narrowly bell- to funnel-shaped, 12–20 mm, pale to deep blue, rarely white, erect to spreading, lobes cut ≥ 50% of length of corolla, erect, spreading, or reflexed; stamens 7–12 mm, bases ciliate; ovaries turbinate or hemispheric, 4–6 mm, bases rounded, styles 9–17 mm, ≤ corolla (rarely longer), white to blue, distal 50% papillate. **Fruits** hemispheric, nodding, weakly ribbed, opening by pores near base.

Rocky outcrops, cliffs, talus slopes, moist slopes, openings in coniferous forests. Flowering May–Sep. 0–3000 m. BW, Casc, CR, Est, Sisk. CA, ID, WA; throughout North America except KS, KY, NV, OK, and southeastern states; circumboreal. Native.

Campanula rotundifolia is a variable and widespread species. In North America, it is most common in the northern tier of states and the Rocky Mountains. Populations in central North America and western North America south of Puget Sound were identified as belonging to a cordilleran race by Shetler (1982). He included *C. sacajaweana*, of the Wallowa Mountains, in this race and noted that it was an interesting small-flowered alpine ecotype.

Campanula scouleri Hooker (p. 526)

Scouler's harebell, pale bellflower

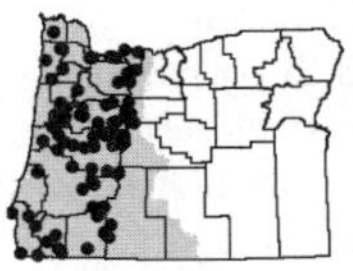

Plants glabrous or short-pubescent; from slender rootstocks and fibrous roots. **Stems** reclining to erect, 1–3 dm. **Leaves** basal and cauline; proximal leaves widely lanceolate to round, 1–6 cm, thin to leathery, serrate, petioles narrow, winged, 1–2 cm; cauline leaves lanceolate, tips attenuate, reduced distally to lance-linear bracts, sessile. **Inflorescences** racemose or paniculate, flowers few, nodding; pedicels 5–20 mm. **Flowers** calyx lobes linear to subulate, spreading to ascending; corollas widely bell-shaped, 8–15 mm, pale blue, lobes broadly deltate to lanceolate, reflexed, cut ~66% of length; stamens 4–6 mm, bases ciliate; ovaries hemispheric to obconic, 3 mm, styles straight, 12–25 mm, blue, much > corolla, well exserted, distal 20–40% papillate. **Fruits** obconic to broadly ovoid, weakly ribbed, pores near middle.

Shaded woodland, openings in coniferous forests, bluffs, streamsides, roadsides. Flowering Jun–Sep. 0–1900 m. Casc, CR, ECas, Est, Sisk, WV. CA, WA; north to Alaska. Native.

DOWNINGIA Downingia

Stephen C. Meyers

Herbs annual, glabrous; roots fibrous. **Leaves** alternate, lanceolate to awl-like, margins entire; sessile. **Inflorescences** spikes; bracts present; pedicels 0 (but ovaries long and pedicel-like). **Flowers** sepals 5, unequal; corollas bilaterally symmetric, 2-lipped, lower lip 3-lobed, upper lip 2-lobed; filaments and anthers connate, lower 2 anthers with tufts of hairs; ovaries inferior, long, narrow, pedicel-like, styles 1, stigmas 2-lobed. **Fruits** elongate, sub-cylindrical, opening on sides by slits or irregular tears. **Seeds** many, small, brown.

Western North America and South America. 14–16 species; 6 species treated in *Flora*.

Key to *Downingia* species:

1. Corollas somewhat inconspicuous, 3–7 mm; plants of southeastern Oregon . . . . *D. laeta*
1' Corollas showy, 7–20 mm; plants throughout Oregon.
  2. Anther tubes strongly bent (> 45°) and greatly exserted from corolla tube.
    3. Lower corolla lips with prominent purple band or spots . . . . *D. insignis*
    3' Lower corolla lips without prominent purple band or spots (but sometimes approaching prominent in *D. bacigalupii*).
      4. Corollas 3-colored (blue, white, and yellow) . . . . *D. bacigalupii*
      4' Corollas 2-colored (blue and white) . . . . *D. elegans*
  2' Anther tubes not or weakly bent (< 45°) and included or only partially exserted from corolla tube.
    5. Anther bristles twisted together; bases of lower lobe with knob-like projections . . . . *D. bicornuta*
    5' Anther bristles not twisted together; bases of lower lobe without knob-like projections . . . . *D. yina*

Downingia bacigalupii Weiler

Bach's calicoflower, Bach's downingia

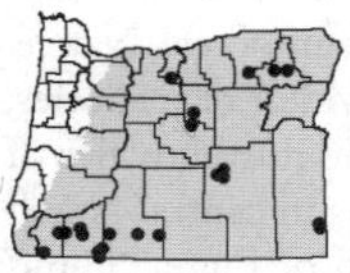

Plants 5–25(35) cm; stems erect. **Leaves** lanceolate, 10–20 mm, tips acute. **Flowers** sepals 4–8 mm; corollas 5–20 mm, blue, tubes 2.5–5 mm, lateral depressions between lower lobes much > upper; lower lobes 6–10 mm, with white and yellow spots, occasionally with diffuse purple spots near throat; upper lobes 4–10 mm; anther tubes strongly bent (45–90°) and greatly exserted from corolla tube; ovaries unilocular. **Fruits** 2–6 cm. **2*n***=22.

Margins of lakes and streams, wet meadows, ditches. Flowering May–Aug. 400–1900 m. BR, BW, Casc, Col, ECas, Lava, Owy, Sisk. CA, ID, NV. Native.

Although differing in chromosome number from *D. insignis* (2*n*=24 vs. 20), *D. bacigalupii* is nearly identical morphologically. Specimens of *D. bacigalupii* occasionally exhibit a pattern of purple spotting near the corolla throat that very closely (if not completely) matches that of *D. insignis*. This, unsatisfactorily, leaves geography as the best character to determine identification. More study of the exact relationship between these taxa is warranted.

Downingia bicornuta A. Gray

Doublehorn downingia

Downingia bicornuta A. Gray var. *bicornuta*

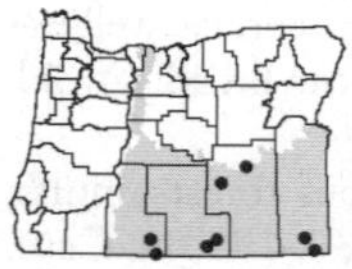

Plants 5–25(35) cm; stems erect, occasionally decumbent. **Leaves** narrowly lanceolate to lanceolate, 10–20 mm, tips acute. **Flowers** sepals 4–9 mm; corollas 8–20 mm, blue to purple-blue, tubes 2.5–4.5 mm, lateral depressions between lower lobes > upper; lower lobes 5–10 mm, with white and yellow or green spots, prominent purple band near throat with knob-like projections; upper lobes 4–8 mm; anther tubes not or weakly bent (< 45°) and included or only partially exserted from corolla tube; anther bristles twisted together; ovaries bilocular. **Fruits** 2–5 cm. **2*n***=24.

Vernal pools, margins of lakes and streams, wet meadows, ditches. Flowering May–Jul. 1200–2100 m. BR, ECas, Owy. CA, ID, NV. Native.

Two poorly differentiated varieties of *D. bicornuta* have been described. If accepted, ours is the nominate variety.

Downingia elegans (Douglas ex Lindley) Torrey (p. 529)

Showy downingia, elegant downingia

Downingia elegans (Douglas ex Lindl.) Torr. var. *brachypetala* (Gand.) McVaugh, *Downingia elegans* (Douglas ex Lindl.) Torr. var. *elegans*

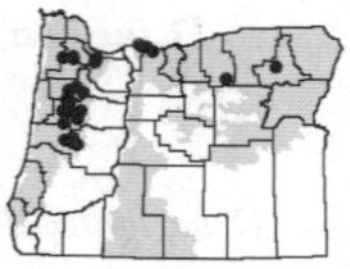

Plants 8–35(40) cm; stems erect, occasionally decumbent. **Leaves** linear to lanceolate, 10–20 mm, tips obtuse to acute. **Flowers** sepals 4–8 mm; corollas 7–18 mm, blue or rarely white, tubes 2–5 mm, lateral depressions between lower lobes much > upper; lower lobes 5–11 mm, with white spots, rarely white spots barely evident and lobes appearing nearly completely blue, without purple band or spots near throat; upper lobes 4–8 mm; anther tubes strongly bent (45–90°) and greatly exserted from corolla tube; ovaries unilocular. **Fruits** 2–5 cm. **2*n***=20.

Vernal pools, margins of lakes and ponds, wet meadows, ditches. Flowering May–Oct. 50–1600 m. BW, Col, CR, ECas, WV. CA, ID, NV, WA; north to British Columbia. Native.

Some specimens of *D. yina* without yellow spots can be confused with *D. elegans*. The exserted and bent anther tube of *D. elegans* (versus the included or partially exserted tube of *D. yina*) is a reliable character to distinguish the species, however.

Downingia insignis Greene

Cupped downingia, harlequin calicoflower

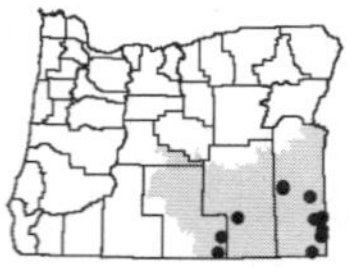

Plants 5–25(30) cm; stems erect, ascending or decumbent. **Leaves** lanceolate to oblanceolate, 7–15(20) mm, tips obtuse to acute. **Flowers** sepals 3–8 mm; corollas 8–15 mm, blue to purple-blue, tubes 3–5 mm, lateral depressions between lower lobes > upper; lower lobes 5–9 mm, with white and yellow spots, prominent purple band or spots near throat; upper lobes 6–10 mm; anther tubes strongly bent (45–90°) and greatly exserted from corolla tube; ovaries bilocular. **Fruits** 3–8 cm. **2*n***=22.

Vernal pools, margins of lakes and streams, ditches. Flowering Jun–Jul. 1000–1900 m. BR, Owy. CA, ID, NV. Native.

See comments under *D. bacigalupii*.

Downingia laeta (Greene) Greene

Great Basin downingia

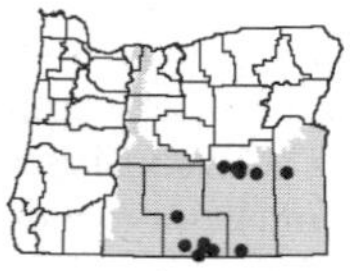

Plants 5–20(25) cm; stems erect or occasionally decumbent. **Leaves** lanceolate to oblanceolate, 5–20 mm, tips obtuse to acute. **Flowers** sepals 2–7 mm; corollas 3–7 mm, white, pink, blue or purple, tubes 1–2 mm, lateral depressions between lower lobes ≈ to upper; lower lobes 2–5 mm, with white and yellow spots, prominent purple band or spots near throat; upper lobes 2.5–4 mm; anther tubes nearly straight or weakly bent (< 45°) and included or partly exserted from corolla tube; ovaries bilocular. **Fruits** 2–4.5 cm. **2*n***=22.

Vernal pools, margins of lakes and streams, ditches. Flowering Jun–Aug. 1100–1800 m. BR, ECas, Owy. CA, ID, NV; northeast to Saskatchewan, southeast to UT. Native.

Downingia yina Applegate (p. 529)

Cascade downingia, Willamette downingia

Downingia yina Applegate var. *major* McVaugh, *Downingia pulcherrima* Peck, *Downingia willamettensis* M. Peck, *Downingia yina* Applegate var. *yina*

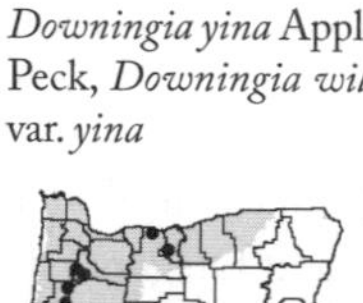

Plants 5–25(30) cm; stems erect or occasionally decumbent. **Leaves** lanceolate to oblanceolate, 5–20(25) mm, tips obtuse to acute. **Flowers** sepals 3–8 mm; corollas 7–10(15) mm, blue to purple-blue, tubes 3.5–6 mm, lateral depressions between lower lobes ≥ upper; lower lobes 5–8 mm, with white and yellow spots, rarely with only white spots, prominent, diffuse or no purple band or spots near throat, without knob-like projections; upper lobes 4–7 mm; anther tubes nearly straight or weakly bent (< 45°) and included or partly exserted from corolla tube, anther bristles not twisted together; ovaries unilocular. **Fruits** 1.5–3 cm. **2*n***=12, 16, 20.

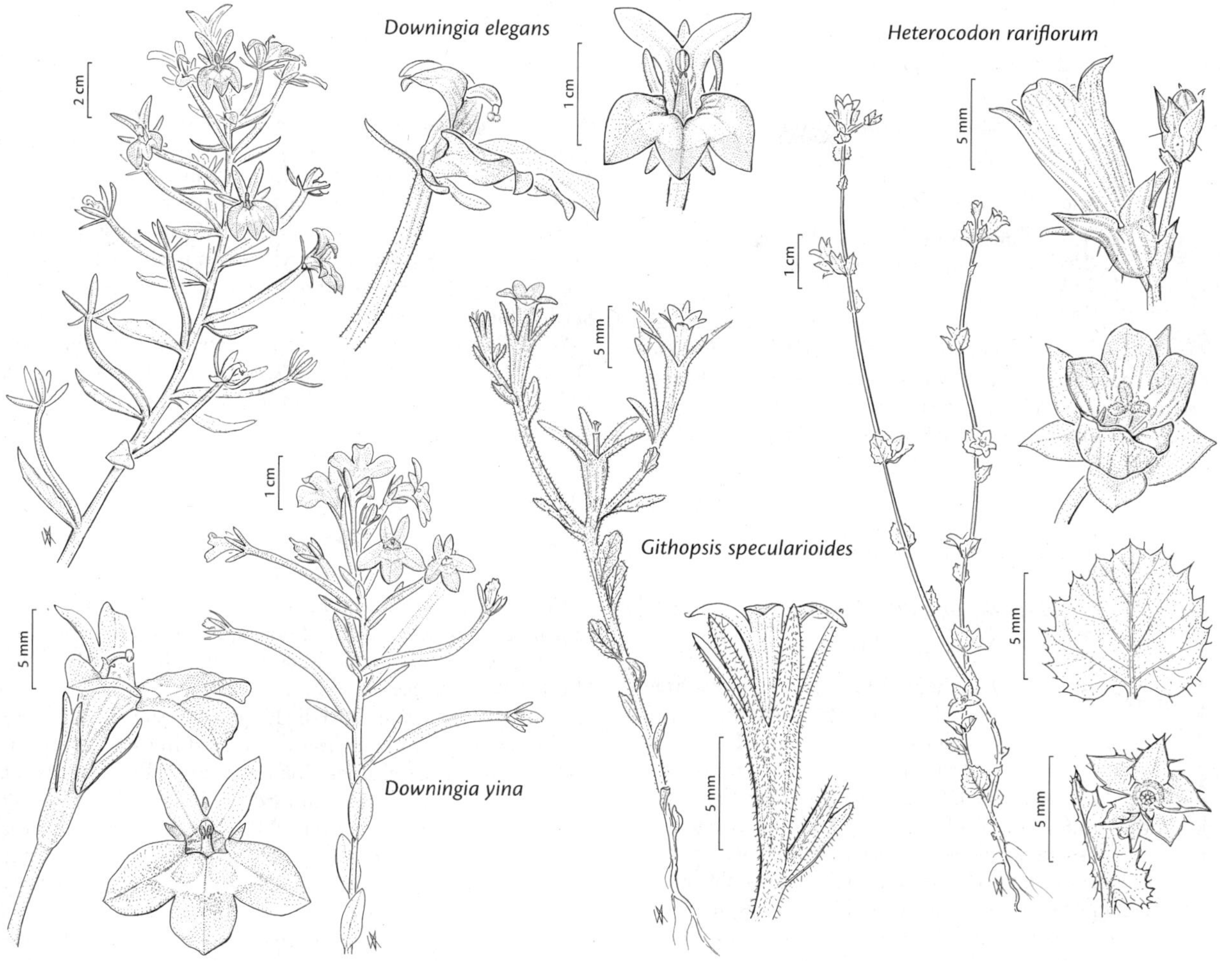

DOWNINGIA • GITHOPSIS • HETEROCODON

Vernal pools, margins of lakes and streams, wet meadows, ditches. Flowering Apr–Aug. 50–2300 m. BR, Casc, Col, CR, ECas, Owy, Sisk, WV. CA, WA. Native.

Based on molecular and cytological data, Schultheis (2010) divided *D. yina* into three species (*D. pulcherrima*, *D. willamettensis* and *D. yina*). The three species are morphologically identical, leaving only geography for reliable identification. Rather than presenting several cryptic species here, we are pragmatically accepting only *D. yina*.

See also comments under *D. elegans*.

GITHOPSIS Bluecup

Plants annuals, glabrous to hairy; roots fibrous. **Stems** erect, branched or simple. **Leaves** basal leaves ovate; cauline leaves widely linear to ovate, serrate; sessile. **Inflorescences** flowers terminal, subtended (but generally not immediately so) by 1–several, generally non-leaf-like bracts; short-pedicelled. **Flowers** not inverted; corollas radially symmetric, funnel-shaped, blue, throat white, lobes deltate, cut 25(66)% to base; ovaries inferior, obconic, stigmas 3-lobed. **Fruits** open at top irregularly by tears where style falls off, within persistent calyx lobes. **Seeds** angular-fusiform, ~1 mm.

Western North America. 4 species; 1 species treated in *Flora*.

Githopsis specularioides Nuttall (p. 529)

Common bluecup

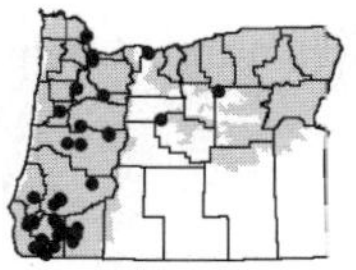

Plants glabrous or hairy. **Stems** erect, 2–40 cm. **Leaves** 4–20 mm. **Inflorescences** bracts lanceolate to oblong, 3–12 mm, > 5 mm apart; pedicels 1–3 mm, rarely 0. **Flowers** corollas funnel-shaped, 4.5–14 mm, 2 × longer than wide, lobes ≤ tube, deep blue; filament bases narrow, sparsely ciliate; ovaries obconic, 4–14 mm, ~3 × longer than wide, bases long-tapered, top narrowed, ovary ribs 10, equal, styles 2.5–6.5 mm, distal 50% papillate.

Chaparral, oak woodland, grassy areas, slopes, yellow pine forests, sometimes on serpentine soil. Flowering Apr–Jul. 0–1500 m. BW, Casc, Col, CR, Sisk, WV. CA, WA; north to British Columbia, northeast to MT. Native.

HETEROCODON Heterocodon

Genus composed of one species, *H. rariflorum.*

Heterocodon rariflorum Nuttall (p. 529)

Rareflower heterocodon

Specularia rariflorum (Nutt.) McVaugh

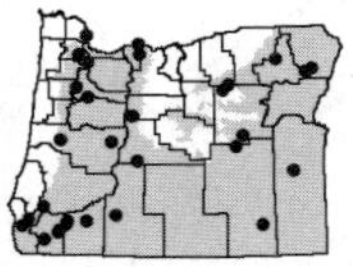

Plants annual; roots fibrous. **Stems** erect, simple or branched from base, 5–30 cm, 4-angled, thin, sparsely hairy. **Leaves** cauline, round-cordate, 2–10 mm, thin, margins serrate; sessile. **Inflorescences** terminal and at nodes, opposite or in leaf axils; flowers 1; pedicels 0. **Flowers** proximal flowers cleistogamous, distal flowers chasmogamous; flowers not inverted; calyx lobes widely triangular, 2–4 mm, leaf-like, toothed; corollas of chasmogamous flowers radially symmetric, cylindric, cut 25% or less to base, 3–5 mm; tubes white to pale blue; lobes ascending to suberect, spreading to erect at tips, triangular, ~1.5 mm, deep blue; stamens ~1.5 mm, filaments linear; ovaries short-oblong, 2–3 mm, papery, generally with many long, stiff hairs, styles to 3 mm, distal 25% papillate, stigmas 3-lobed. **Fruits** opening by lateral pores near base. **Seeds** elliptic, 0.5 mm.

Vernally wet places, meadows, streambanks. Flowering Jun–Oct. 0–2100 m. BR, BW, Casc, ECas, Owy, Sisk, WV. CA, ID, NV, WA; north to British Columbia, northeast to MT, southeast to CO. Native.

HOWELLIA Howellia

Genus composed of one species, *H. aquatilis.*

Howellia aquatilis A. Gray (p. 531)

Water howellia

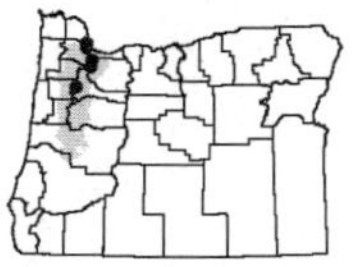

Plants annual, glabrous; roots fibrous. **Stems** submersed or floating, ascending to erect, branched or simple, 10–60 cm. **Leaves** cauline, alternate, sometimes opposite or whorled, linear, 10–45 × 0.5–1.5 mm, margins entire or minutely toothed; sessile. **Inflorescences** in leaf or bract axils, scattered, flowers 1; bracts 1 per flower; pedicels stout, 1–4(8) mm. **Flowers** some cleistogamous; flowers inverted; calyx lobes narrowly triangular, 15–40% length of ovary; corollas 0 in proximal flowers or not, bilaterally symmetric, white, tubes narrowly cylindric, 2–2.7 mm, split adaxially to base, ± 2-lipped, abaxial lobes 3, 1–1.3 mm, adaxial lobes 2, narrowly strap-shaped, 1–1.3 mm, spreading; stamens connate into tube, anthers 0.2–0.9 mm, 2 shorter anthers minutely appendaged; ovaries ellipsoid, 5–13 mm, stigmas capitate, 2-lobed. **Fruits** opening on sides by slits or irregular tears, ellipsoid, 5–13 mm long, 1–2 mm in diameter, 1-chambered, depressed-conic at tips. **Seeds** cylindric, 2–4 mm, shiny, chestnut-brown.

Shallow, stagnant ponds. Flowering Apr–Jul. 0–100 m. WV. CA, ID, WA; northeast to MT. Native.

Howellia aquatilis is federally listed as threatened. This species was thought to have been extirpated in Oregon until it was relocated in Clackamas County in 2013. The most recent collection in Oregon had been made in 1935, and virtually all other collections were from the 1800s. The plant grows in areas that were once associated with glacial potholes and former river oxbows that flood in the spring, but usually dry at least partially by late summer. It is often found in shallow water (one to two meters) and on the edges of deep ponds that are partially surrounded by deciduous trees such as black cottonwood and aspen (Lesica 1988). Recent genetic studies on populations in California have shown that there is little genetic diversity within or between populations, although more genetic diversity was found within individual ponds (Schierenbeck & Phipps 2010).

LOBELIA Lobelia

Stephen C. Meyers

Plants biennial or perennial. **Leaves** sessile. **Inflorescences** bracteate racemes. **Flowers** calyces 5-lobed; corollas bilaterally symmetric, strongly 2-lipped with spreading lobes, lower lip 3-lobed, upper lip 2-lobed; stamens exserted, filaments and anthers connate; ovaries inferior or half inferior, styles 1, exserted, stigmas 2-lobed. **Fruits** spherical, opening at top (within calyx lobes) by valves.

Worldwide. ~350 species; 1 species treated in *Flora.*

Lobelia dortmanna Linnaeus (p. 531)

Water lobelia, Dortmann's cardinalflower

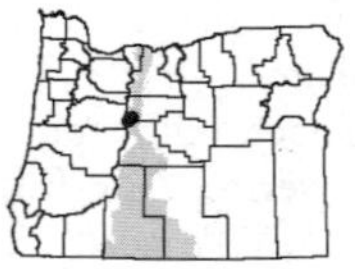

Herbs aquatic, glabrous; stems 0. **Leaves** in a basal rosette, linear, cylindric, 2–6 cm, fleshy, entire, hollow. **Scapes** 1–6(10) dm; bracts reduced and often filiform. **Flowers** 1–11; calyces becoming cylindric in fruit, lobes triangular to linear, 1–4 mm; corollas 8–20 mm, white or light blue, lobes linear, lower lip pubescent; anthers pubescent. **Fruits** 5–10 mm. **Seeds** 0.5–1 mm.

Edges of lakes and ponds. Flowering Jul–Aug. 100–1500 m. ECas. WA; north to Alaska, east across Canada to Great Lakes and northeastern states. Native.

In Oregon, this species is known only from near Camp Sherman, Jefferson County.

NEMACLADUS Threadplant

Plants annual, glabrous or puberulent; taprooted. **Stems** decumbent or erect; branched or not. **Leaves** basal; sessile or petioles short. **Inflorescences** raceme-like; bracts 1 per flower; pedicels 3.5–12 mm. **Flowers** usually not inverted; calyx lobes linear to triangular; corollas bilaterally symmetric, 2-lipped, lobes 5; filaments distinct at base, connate into tube around style distally, appendages attached to a stalk or directly on 2 adjacent filaments, each with 2–6 cells, anthers distinct, all alike; ovaries hemispheric to obconic, nectary glands 3, on free part of ovary, stigmas 2-lobed, papillate. **Fruits** hemispheric to fusiform, open at top by 2 valves (circumscissile). **Seeds** elliptic to oblong.

Western North America. 18 species; 2 species treated in *Flora*.

Key to *Nemacladus* species:

1. Plants fleshy; capsule bases rounded; calyx lobes spreading or reflexed; corolla lobes 1–1.5 mm . . . . *N. rigidus*

1' Plants not fleshy; capsule bases acute; calyx lobes erect; corolla lobes ~0.7 mm . . . . *N. capillaris*

Nemacladus capillaris Greene
Common threadplant

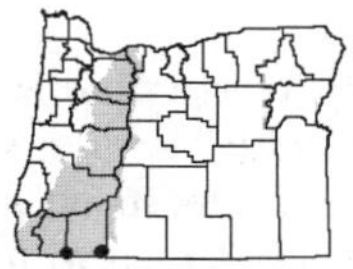

Plants not fleshy. **Stems** stiffly ascending, slender, 7–18 cm, often dark reddish; 1st branches generally 1.5 cm from base. **Leaves** ovate, 3–15 mm, bases narrowed abruptly to short petiole, margins entire or obscurely crenate. **Inflorescences** axis zigzag, especially in fruit; bracts spreading to erect, lanceolate to elliptic or ovate, 1–3 mm, appressed to pedicel; pedicels spreading, straight or slightly curved, 8–12 mm long, 0.1–0.15 mm in diameter. **Flowers** hypanthia turbinate, ~1 mm; calyx lobes elliptic to lanceolate, erect, ~0.5 mm; corollas white tinged with pink, lobes oblanceolate, glabrous; adaxial lobes 3, erect, ~0.7 mm, with pink spot on tip; filaments declined, ~1 mm, tips curved, glabrous, filament appendage cells 2, directly attached to bases of filaments, 0.1 mm, blunt, anthers 0.1–0.2 mm; ovaries half inferior. **Fruits** 1.5–2.5 mm, bases acute, tips rounded. **Seeds** widely elliptic, 0.5–0.7 mm, narrowly zigzag-ridged.

Dry slopes, burned areas, volcanic outcrops. Flowering Jun–Jul. 500–1700 m. Casc, Sisk. CA. Native.

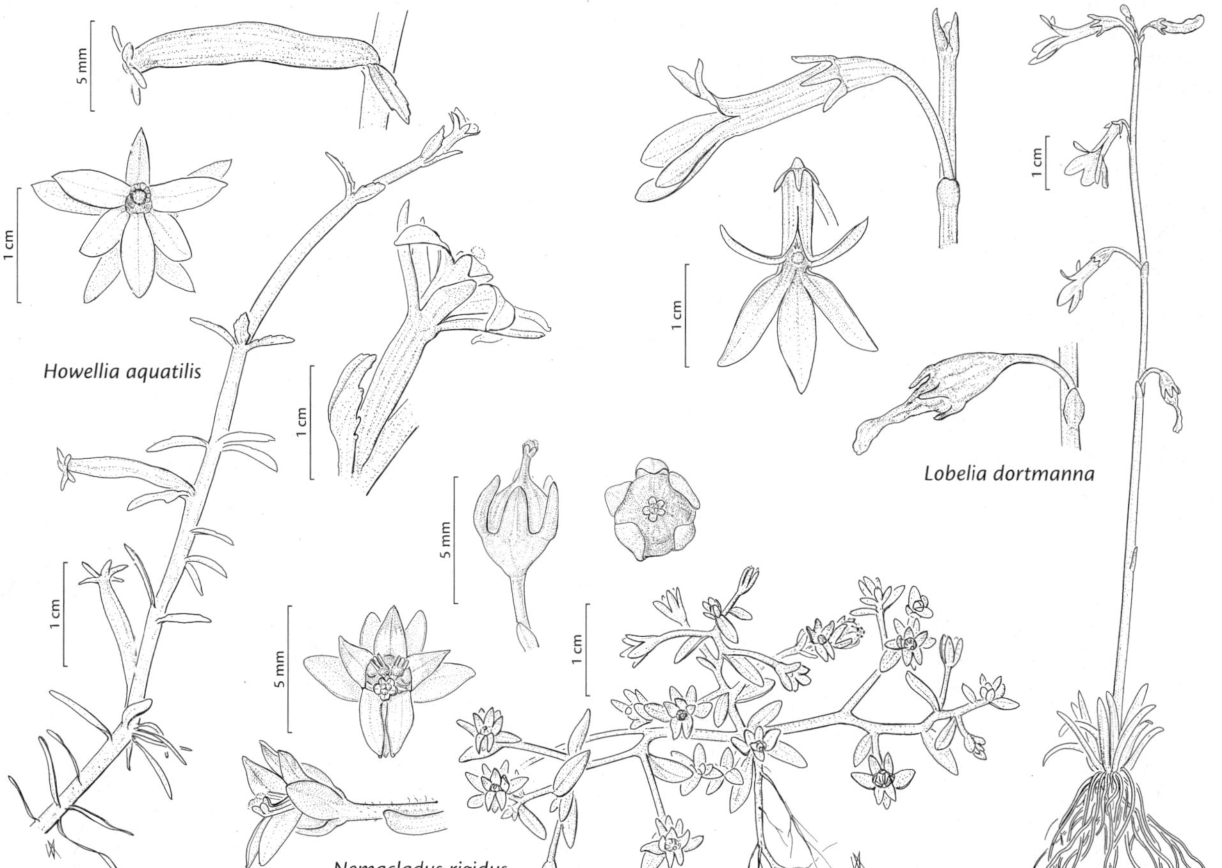

HOWELLIA • LOBELIA • NEMACLADUS

Nemacladus rigidus Curran (p. 531)

Stoutstem threadplant

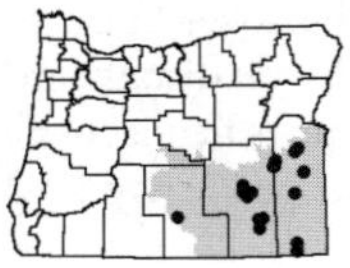

Plants fleshy. **Stems** simple or branched from base, spreading to erect, stout, 2–4(9) cm, maroon or dark green marked with maroon, bases shiny. **Leaves** elliptic to oblanceolate, 5–10 mm, fleshy, surfaces hairy. **Inflorescences** axis strongly zigzagged; bracts spreading or reflexed, widely elliptic, 2–3 mm; pedicels spreading, straight (curved in age), 3.5–8.5 mm long, 0.2 mm in diameter. **Flowers** hypanthia ~1 mm; calyx lobes spreading or reflexed; middle abaxial lobes linear-elliptic, ~0.7 mm; flanking abaxial lobes 2, widely triangular, 1–4 mm; adaxial lobes 2; corollas divided to base, 2-lipped, white; abaxial lobes 2, oblong, acute, 0.4–0.5 mm; adaxial lobes 3, erect, ovate, 1–1.5 mm, glabrous, veins and tips maroon; filaments declined, 1.2–1.6 mm, tips curved, glabrous, cells wide, blunt, anthers 0.2–0.3 mm; ovaries ± inferior in fruit. **Fruits** 3–4 mm, tips pointed, bases rounded. **Seeds** elliptic, 0.6–0.7 mm, widely zigzag-ridged.

Dry areas, bare soil, sand, volcanic outcrops. Flowering May–Jul. 800–1700 m. BR, Owy. CA, ID, NV. Native.

Nemacladus rigidus is widespread in the Intermountain region. The ovary seems to develop into mature capsules even after being pressed.

PORTERELLA Porterella

Genus composed of one species, *P. carnosula.*

Porterella carnosula (Hooker & Arnott) Torrey (p. 532)

Fleshy porterella

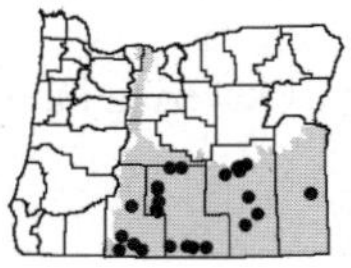

Plants annual, emergent or terrestrial, glabrous; roots fibrous. **Stems** erect, simple or branched from base, 2–30 cm. **Leaves** cauline, narrowly ovate (aerial) to narrowly triangular (submersed), (5)12–15 × (1)3–5 mm, margins entire or few-toothed; sessile. **Inflorescences** racemes; bracts 1 per flower, ascending, leaf-like; pedicels in fruit 1–3 cm. **Flowers** fragrant; calyx tubes 1–2.5 mm, lobes elongate in fruit, narrowly triangular; corollas bilaterally symmetric, > calyx lobes, blue, with symmetric yellow spot on lower lip; tubes cylindric, 4–5 mm, abaxial lip round, spreading, 3–8 mm, with 2 low ridges, adaxial lobes 2, erect, narrowly triangular, 1–2 mm; stamens connate, anthers all ~2 mm, tufted, short anthers 2, each with horn-like appendage ~0.5 mm; ovaries obconic to cylindric, stigmas cup- or plate-like, papillate, 2-lobed. **Fruits** narrowly obconic to cylindric, 5–10 mm, tips acute or rounded, valvate. **Seeds** ~1 mm, smooth and finely striate.

Moist, grassy roadsides, lake and pond edges, seasonally wet ponds. Flowering Jun–Aug. 100–2500 m. BR, ECas, Owy. CA, ID, NV, WA; southeast to AZ. Native.

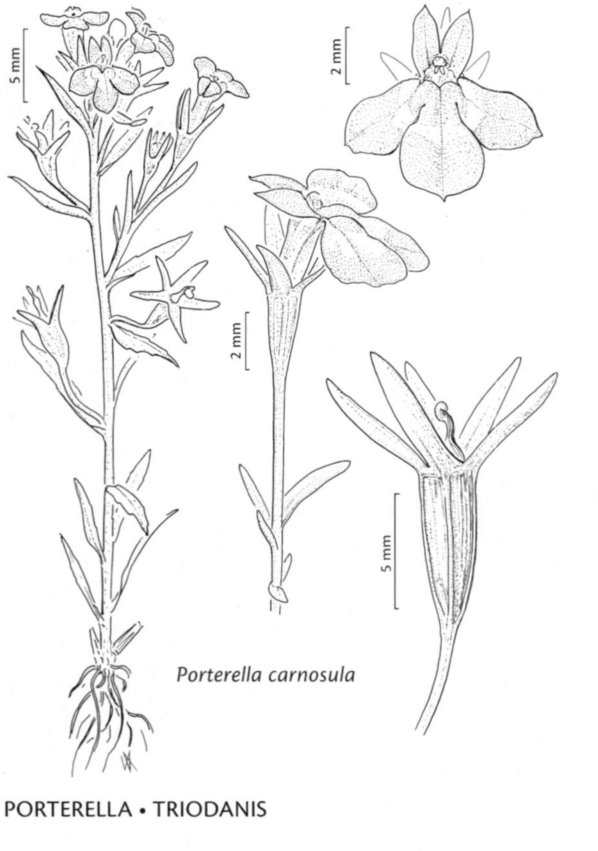

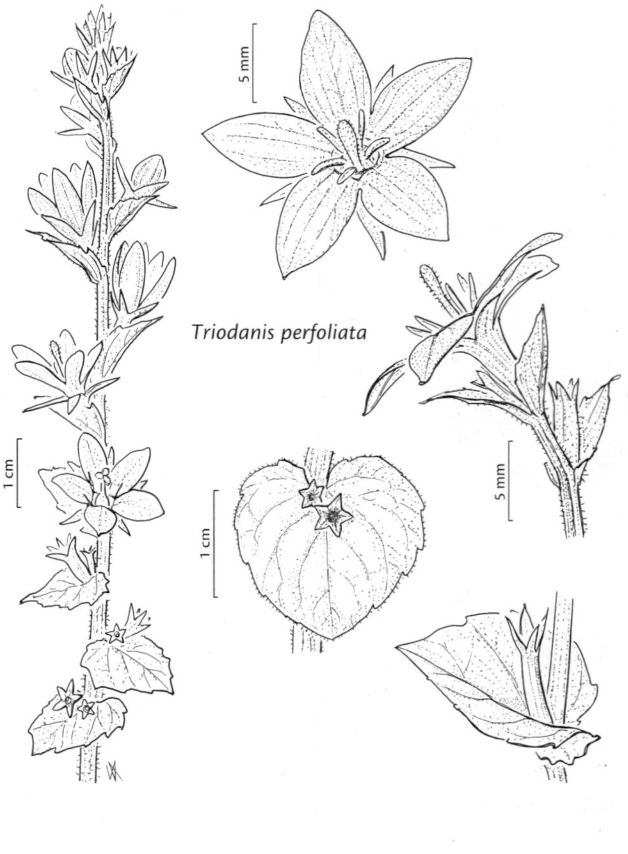

PORTERELLA • TRIODANIS

TRIODANIS Venus' looking-glass

Plants annual, sparsely hairy; hairs stiff, reflexed; roots fibrous. **Stems** simple or branched from base, erect, 4-angled, 5–40 cm. **Leaves** cauline, thin, serrate or crenate; sessile but lowest short-petiolate. **Inflorescences** flowers 1–several in axils of leaf-like bracts; sessile. **Flowers** proximal flowers cleistogamous; calyx lobes spreading, narrowly triangular, not leaf-like; corollas radially symmetric, rotate, lobes deep blue to blue-violet; stamen bases wide, ciliate; ovaries inferior, styles with distal half papillate, 2–3-lobed. **Fruits** opening by lateral pores. **Seeds** widely elliptic, ~0.5 mm.

Mediterranean and North America. 7–8 species; 2 species treated in *Flora*.

Key to *Triodanis* species:

1. Ovaries elliptic to ovoid, narrowed at top; leaves widely lanceolate to ovate, bases not clasping stem . . . . ***T. biflora***

1' Ovaries oblong to obovoid, not narrowed at top; leaves round-cordate, bases clasping stem . . . . ***T. perfoliata***

Triodanis biflora (Ruiz & Pavon) Greene

Small Venus' looking-glass

Specularia biflora (Ruíz & Pav.) Fisch. & C.A. Mey.

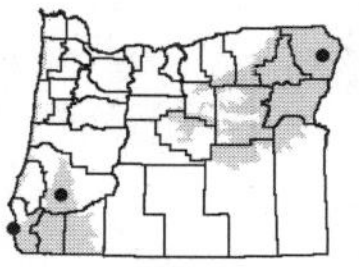

Plants 5–40 cm, retrorsely hispid distally. **Leaves** widely lanceolate to ovate, 10–30 mm, narrowed to base, bases not clasping stem, margins crenate, ciliolate, veins inconspicuous abaxially, tips acute. **Inflorescences** flowers 1–4 per node. **Flowers** chasmogamous only in upper 1–3 bract axils, otherwise cleistogamous; calyx lobes deltoid to deltoid-ovate; corollas of chasmogamous flowers 5–9 mm, blue to violet, lobes 4–7 mm; corollas of cleistogamous flowers 0.5 mm; stamens ~2.5 mm; ovaries elliptic to ovoid, 4.5–7 mm, narrowed at top, styles 3–3.5 mm. **Fruits** oval or roundish, pores near tips.

Disturbed areas. Flowering May–Jul. 0–700 m. BW, Est, Sisk. CA; southern, central, and northeastern US, Mexico; South America. Native.

Triodanis perfoliata (Linnaeus) Nieuwland (p. 532)

Clasping Venus' looking-glass

Specularia perfoliata (L.) A. DC.

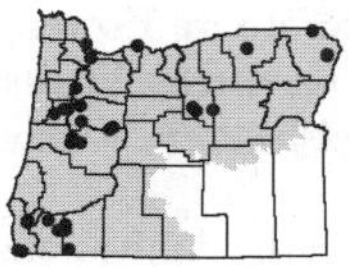

Plants 1–4 dm, spreading-hirsute. **Leaves** round-cordate, 5–20 × 5–25 mm, bases clasping stem, margins crenate-dentate, revolute, tips blunt. **Inflorescences** clusters of 3–5 flowers (rarely 1). **Flowers** cleistogamous in proximal leaves, chasmogamous in most distal bract axils; calyx lobes lanceolate, acuminate; corollas of chasmogamous flowers 8–10 mm, tubes 2–4 mm, white, lobes 6–8 mm, deep blue; stamens ~3.5 mm; ovaries oblong to obovoid, not narrowed at top, 5–10 mm, styles ~4.5 mm. **Fruits** broadly elliptic, pores near or below middle.

Disturbed areas. Flowering May–Jul. 0–1300 m. BW, Casc, Col, CR, ECas, Est, Lava, Sisk, WV; throughout most of US, British Columbia, northeastern Canada; South America. Native.

CANNABACEAE Hemp family

Alan T. Whittemore (except where noted)

Trees, shrubs, herbs, or **vines** perennial or annual; with a surface covering of unbranched hairs, these often scabrous or with swollen bases, glandular hairs sometimes present. **Leaves** all alternate or lower opposite, simple, unlobed or palmately lobed or compound. **Inflorescences** terminal or axillary. **Flowers** unisexual or bisexual, if unisexual then plants dioecious, radially symmetrical; sepals 4–5, free or united; petals absent; stamens 4–5; pistils 2-carpellate, ovaries superior, locules 1, with a solitary apical ovule, styles deeply bifid. **Fruits** drupes or achenes.

Worldwide. 11 genera; 3 genera treated in *Flora*.

Key to Cannabaceae genera:

1. Plants trees or shrubs; leaves alternate, simple and unlobed; flowers staminate and bisexual on same twig; fruits drupes, perianths not persistent . . . . **CELTIS**

1' Plants herbs or vines; lower leaves opposite, palmately lobed or compound; flowers all unisexual; fruits achenes, generally enclosed in persistent perianths.

   2. Plants herbs; lower leaves palmately compound; pistillate inflorescences spike-like clusters, erect to spreading . . . . **CANNABIS**

   2' Plants vines; lower leaves simple, palmately lobed; pistillate inflorescences head- or cone-like clusters, generally pendent . . . . **HUMULUS**

CANNABIS Hemp

Stephen C. Meyers

Genus composed of one species, *C. sativa.*

Cannabis sativa Linnaeus (p. 534)

Hemp, marijuana

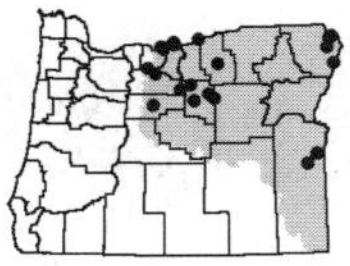

Herbs annual, dioecious, aromatic, erect, branched, 5–40 dm, pubescent with resinous glands, fibrous. **Leaves** abaxially light green, adaxially dark green; leaflets 3–7(9), linear to narrowly elliptic, 4–15 × 0.5–2 cm, coarsely toothed; lower leaves opposite, palmately compound; upper leaves alternate, palmately compound or simple. **Staminate inflorescences** in short panicle- or spike-like clusters, on leafless branches. **Pistillate inflorescences** in spike-like clusters, erect to spreading, on leafy branches. **Staminate flowers** 2–5 mm. **Pistillate flowers** perianth fused, persistent. **Fruits** achenes, ovoid, 3–5 mm, pale green to brownish green, usually purple-mottled. **2*n***=20.

Disturbed areas, usually escaped from nearby cultivation. Flowering Jul–Nov. 0–700 m. CR, Est, Owy, WV. CA, ID, NV, WA; throughout North America; nearly worldwide. Exotic.

Cannabis sativa has long been cultivated in Oregon as an illegal cash crop. Recent legalization, and the inevitable increase in cultivation, may result in the documentation of more escaped populations. This species is divided into three subtaxa by some authorities (*indica, ruderalis,* and *sativa*).

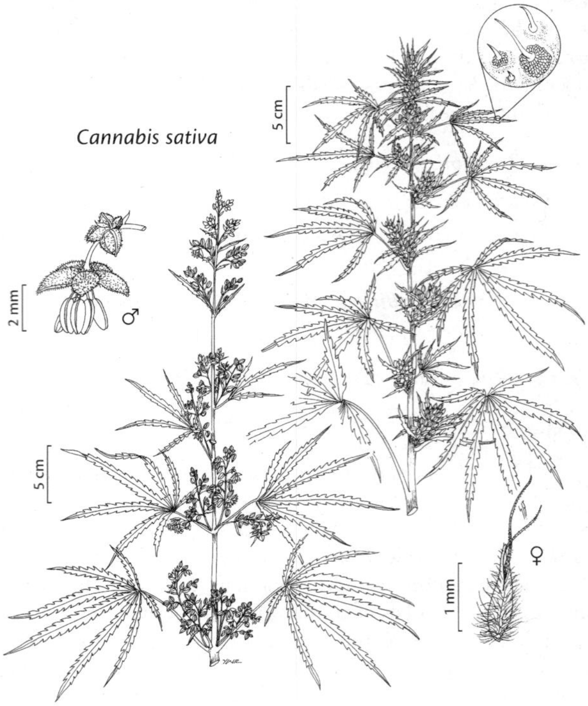

CANNABIS

CELTIS Hackberry

Trees or **shrubs** bark smooth or with irregular corky warts and/or ridges, pale ash gray. **Leaves** alternate, simple, palmately 3-veined from base. **Flowers** male and bisexual, appearing before leaves in spring, perianths not persistent; male flowers clustered at base of current year's growth; bisexual flowers solitary in axils. **Fruits** smooth drupes, with thin layer of sweet-tasting flesh and large, stony endocarp.

Worldwide. ~70 species; 1 species treated in *Flora.*

Celtis reticulata Torrey (p. 534)

Netleaf hackberry

Celtis douglasii Planch.

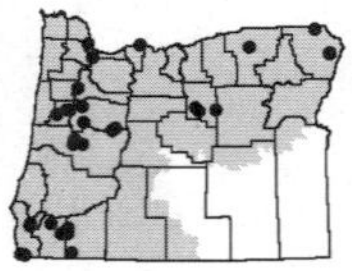

Shrubs or **small trees** deciduous, 3–5(10) m; leaders spreading or arching, central leader not formed; twigs crisp-hispid, sometimes glandular, older twigs glabrate. **Leaves** triangular-ovate, 30–74 × 18–44 mm, abaxially light green, adaxially dull dark green, thick, rigid, bases truncate or weakly cordate, rarely broadly rounded, margins serrate, teeth (1)7–32 on each side, veins and veinlets deeply impressed above, strongly projecting abaxially, tips acute or weakly acuminate, surfaces with loosely appressed stiff hairs on veins, often with axillary tufts near leaf base; juvenile leaves ovate or elliptic, bright green, thin, with short, blunt tips; leaves of vigorous leaders to 120 mm, toothed, strongly acuminate; petioles 3–9 mm. **Inflorescences** pedicels 6–10(18) mm, crisp-hispid especially near base. **Fruits** subglobose, 7–8 mm in diameter, purplish brown.

Rocky slopes and banks, sandy or rocky terraces, often near streams and rivers, or in arid areas. Flowering

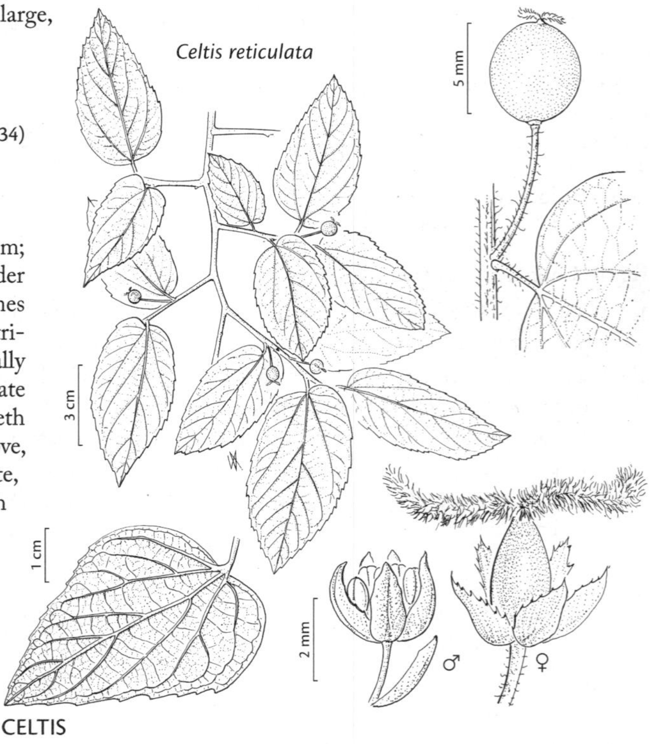

CELTIS

Mar–Jun. 50–900 m. BW, Col, Lava, Owy. CA, ID, NV, WA; east to WY, southeast to KS and LA. Native.

Celtis reticulata is triploid and apomictic; very little viable pollen is produced, and seed is generally produced without sexual recombination. The form of the shrub is variable in different situations. Shrubs that are growing vigorously often produce arching leaders with very large leaves, while shrubs that are maintaining their size usually remain compact, with only smaller leaves. Leaves on different parts of the plant may be of very different size, shape, and serration. The main description above applies to leaves that are subtending a pedicel, which are consistent in size and form. Hackberries, however, often do not flower every year, and leaves from non-flowering stem segments vary greatly in size and shape.

HUMULUS Hop, hops

Stephen C. Meyers

Vines perennial, dioecious; rhizomatous. **Leaves** simple, margins toothed; upper leaves alternate, unlobed; lower leaves opposite, palmately lobed. **Flowers** perianths persistent. **Fruits** achenes, enclosed by membranous perianth, subtended by papery bracts.

Nearly worldwide. 2 species; 1 species treated in *Flora*.

Humulus lupulus Linnaeus (p. 535)

Hop, common hop, hops

Humulus lupulus L. var. *lupulus*

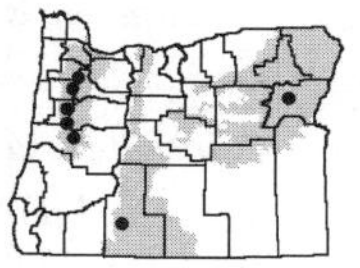

Plants 1–8+ m, sparsely to moderately pubescent, rough and prickly. **Leaves** usually palmately lobed, cordate to ovate in outline, 3–15 × 2–14 cm. **Staminate inflorescences** panicles, 3–15 cm. **Pistillate inflorescences** head- or cone-like clusters, 1–5 cm, generally pendent. **Flowers** sepals 1–2.5 mm, with yellow to orange glands. **Fruits** ovoid to round, 2–3 × 2–2.5 cm, yellowish brown, smooth. ***2n***=20, 40.

Disturbed areas, usually escaped from nearby cultivation. Flowering Apr–Aug. 50–1300 m. BW, ECas, WV. CA, ID, NV, WA; throughout North America; nearly worldwide. Exotic.

Several varieties of *Humulus lupulus* have been described. Given the large number of cultivars currently grown in Oregon, we have chosen to describe only one "generic" taxon. With nearly 7000 acres devoted to hop cultivation, mainly as an ingredient in beer, Oregon ranks second only to Washington in commercial hop production.

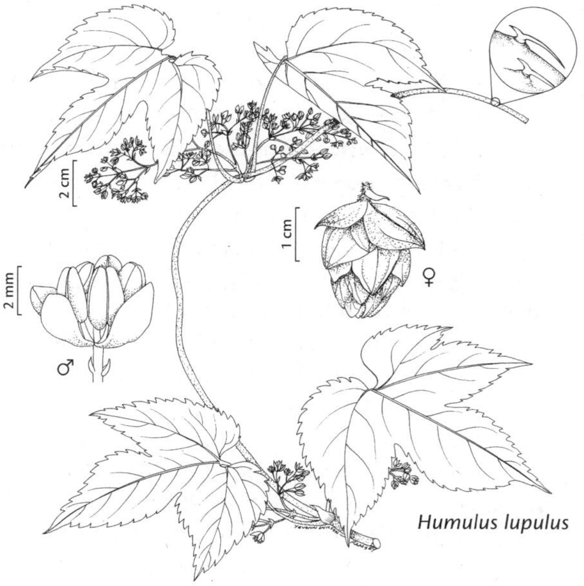

HUMULUS

CAPPARACEAE

(see Cleomaceae)

CAPRIFOLIACEAE Honeysuckle family

(see also Linnaeaceae, Viburnaceae, Volume 3)

Christian Feuillet

Shrubs, **herbs**, or **vines** erect, climbing, or prostrate. **Leaves** opposite; simple, margins entire to somewhat lobed. **Inflorescences** cymes, sometimes reduced to 2 flowers, usually short. **Flowers** calyces small, generally 5-lobed; corollas generally 5-lobed. **Fruits** berries or drupes, fleshy.

Mostly North Temperate Zone, also tropical Asia, subtropical Central America and North America. 5 genera; 2 genera treated in *Flora*.

Caprifoliaceae is treated here in its narrow sense. For *Linnaea*, see Linnaeaceae; for *Sambucus* and *Viburnum*, see Viburnaceae (both families in Volume 3).

Key to Caprifoliaceae genera:

1. Corollas spurred at bases and/or bilaterally symmetric, ± bilabiate, tubes cylindric; fruits red, blue, or black berries, seeds generally > 2 . . . . **LONICERA**

1' Corollas sometimes slightly swollen at bases, radially symmetric, not bilabiate, tubes bell-shaped to salverform; fruits white to pink drupes, with 2 1-seeded nutlets . . . . **SYMPHORICARPOS**

LONICERA Honeysuckle

Shrubs or **vines** deciduous or evergreen, erect or twining. **Leaves** 1 or 2 pairs beneath inflorescence often fused, margins entire; petioles usually short; stipules generally lacking. **Inflorescences** terminal or on axillary peduncles, spicate, often reduced to a few flowers, or flowers in a pair subtended by 0–6 bracts. **Flowers** corollas ± bilabiate, bilaterally symmetric and/or spurred at bases, tubes cylindric; ovaries with 2–3 chambers. **Fruits** berries, usually roundish, red, blue, or black. **Seeds** usually > 2.

Mostly northern temperate, especially eastern Asia and eastern North America. ~180 species; 9 species treated in *Flora*.

Currently known only as an occasional escaped waif from cultivation, plants originally identified as *Lonicera tatarica* are actually *Lonicera × bella*, a hybrid between *L. tatarica* and *L. morrowii*. More than ten other species and several cultivars have been or are cultivated in Oregon as ornamentals or fruit crops but have not naturalized.

While ornamentals and exotics are often fragrant, our natives are not.

Key to *Lonicera* species:

1. Flowers paired on axillary peduncles; plants erect.
   2. Ovaries of a flower pair at least partly fused or appearing so.
      3. Bracts 1–3 pairs, narrowly lanceolate; fruits tightly enclosed by inner, fused bracts, thereby appearing fully fused . . . . **L. cauriana**
      3' Bracts lacking or minute; fruits not enclosed by bracts.
         4. Fruits fused generally more than halfway to tips, axis of 2 fruits ≤ 90° angle . . . . **L. conjugialis**
         4' Fruits fused at bases, axis of 2 fruits at ~180° angle . . . . **L. utahensis**
   2' Ovaries of a flower pair free.
      5. Bracts leaf-like, forming an involucre . . . . **L. involucrata**
      5' Bracts not leaf-like, not forming an involucre . . . . ***L.* × *bella***

1' Flowers in ± interrupted spikes on axillary peduncles or at end of branches; plants twining or trailing or vining (except erect in *Lonicera interrupta*).
   6. Uppermost pairs of leaves not fused around stem . . . . **L. periclymenum**
   6' Uppermost pairs of leaves fused around stem.
      7. Leaves 5–10 cm; corollas 15–50 mm; inflorescences dense, short spikes; corollas orange to red, weakly bilabiate . . . . **L. ciliosa**
      7' Leaves 2–8 cm; corollas 8–16 mm; inflorescences ± long, interrupted spikes; corollas pink or white to cream-yellow, strongly bilabiate.
         8. Stipules ± conspicuous, at least near inflorescence; corollas hairy to glandular-hairy . . . . **L. hispidula**
         8' Stipules lacking; corollas glabrous . . . . **L. interrupta**

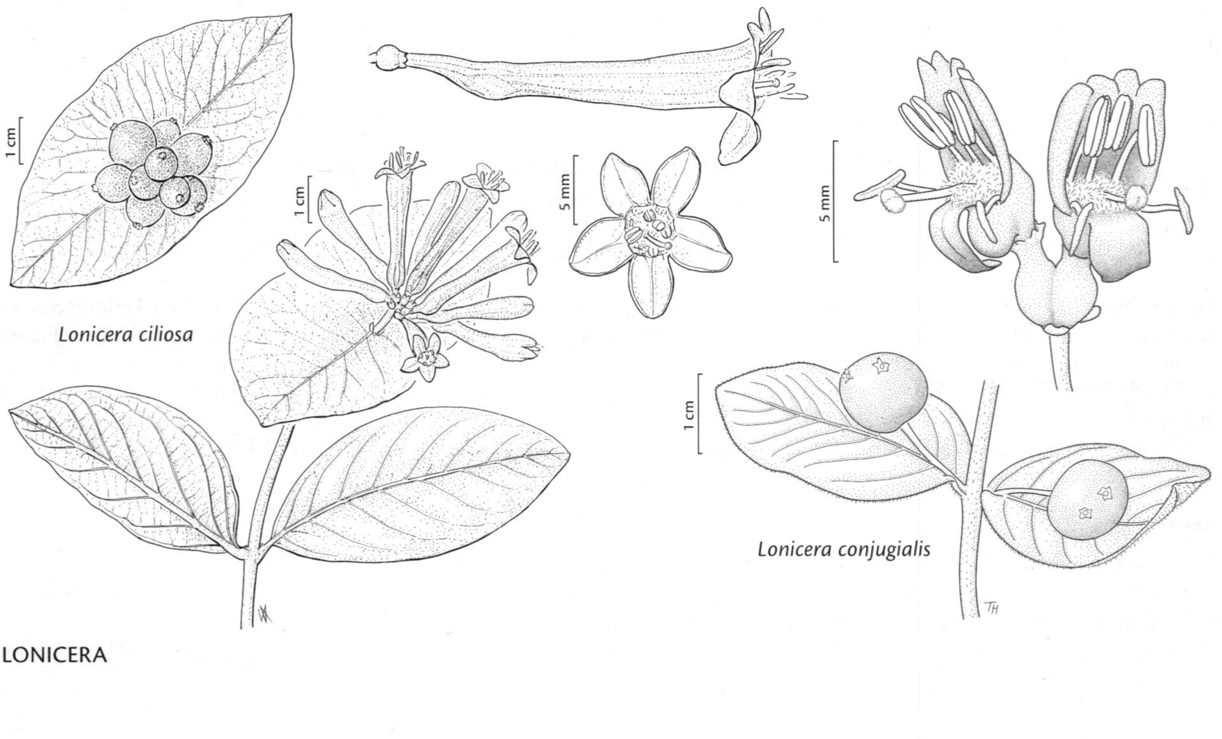

Lonicera ciliosa

Lonicera conjugialis

LONICERA

Lonicera* × *bella Zabel

Bella honeysuckle

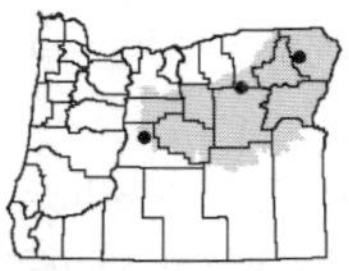

Shrubs ± evergreen, erect, up to 6 m; twigs hollow. **Leaves** ovate to ovate-elliptic, 2.5–7.5 cm, bases rounded to subcordate, margins entire, tips obtuse to rounded, mucronate; petioles very short. **Inflorescences** flowers paired on axillary peduncles; bracts not leaf-like, not forming an involucre. **Flowers** corollas pink, often fading to dull yellow, tubes nearly cylindric, slightly bent; ovaries of a flower pair free. **Fruits** red or orange, fused at bases in pairs, tips divergent in a pair.

Floodplains, rocky pastures. Flowering May–Jun. 600–1200 m. BW, Lava. CA, WA; scattered mainly in eastern North America; Europe. Exotic.

Plants within this species are known to have corollas from white to red. In Oregon, they are mainly pink, often fading to dull yellow.

Lonicera cauriana Fernald

Western sweetberry honeysuckle

Lonicera caerulea L. var. *cauriana* (Fern.) B. Boivin

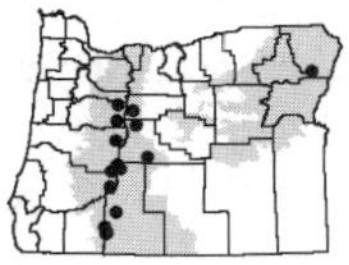

Shrubs deciduous, erect, up to 1 m; young stems puberulent. **Leaves** oblong-ovate, 2–5 cm, bases obtuse to rounded, margins ciliate, tips round to obtuse. **Inflorescences** axillary pairs of flowers; peduncles ~2 mm; bracts in 1–3 pairs, narrowly lanceolate, inner fused, tightly surrounding ovaries. **Flowers** corollas obscurely or weakly bilabiate, 6–9 mm, pale yellow; anthers exserted; ovaries of a flower pair partly fused or appearing so, tightly enclosed by inner fused bracts, styles nearly exserted, glabrous. **Fruits** in pairs, fully fused, appearing as single berry, orbicular to elliptic, ~8 mm, reddish purple, with a slight bloom.

Bogs, wet meadows. Flowering May–Jul. 300–2200 m. BW, Casc, ECas. CA, ID, NV, WA; north to British Columbia, east to WY. Native.

Lonicera cauriana can be differentiated from the European cultivated *L. caerulea* by its small, orbicular to elliptic berries that are approximately 8 mm long (versus berries elongated, more or less parallel-sided, approximately 10 mm long) and are reddish purple with only a slight bloom (versus purplish blue with a conspicuous bloom).

Lonicera ciliosa (Pursh) Poiret ex de Candolle (p. 536)

Orange honeysuckle

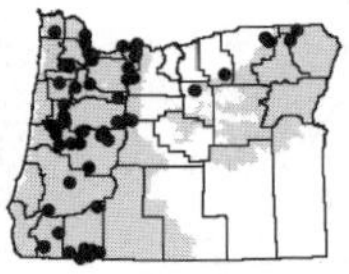

Vines evergreen, trailing to vining up to 10 m; twigs hollow. **Leaves** elliptic to ovate, 5–10 cm, bluish green below, green above, bases obtuse, uppermost pairs fused around stem, margins ciliate, tips rounded to acute, mucronate; petioles ~5 mm. **Inflorescences** on axillary peduncles or at end of branches, spicate, short, dense, ± interrupted, with ~20 flowers in 2–4 whorls. **Flowers** corollas weakly bilabiate, narrow-trumpet shaped, 15–50 mm, orange to red, lobes short; stamens reaching tips of corolla lobes or exserted; styles pubescent, stigmas exserted. **Fruits** ~10 mm, orange red, translucent, glaucous. **Seeds** many.

Forests, thickets. Flowering Apr–Jul. 0–1900 m. BW, Casc, CR, ECas, Sisk, WV. CA, ID, NV, WA; north to British Columbia, northeast to MT, southeast to AZ. Native.

The long, red, tubular flowers of *L. ciliosa* are pollinated by hummingbirds.

Lonicera conjugialis Kellogg (p. 536)

Purple-flower honeysuckle

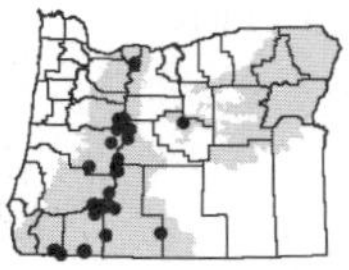

Shrubs deciduous, erect, 0.6–1.8 m; young stems puberulent. **Leaves** elliptic to rounded, 2–8 cm, bases acute, tips round to acute. **Inflorescences** axillary pairs of flowers; peduncles 1.4–3 cm; bracts lacking or minute. **Flowers** calyces fused in basal ½ in pair, lobes minute; corollas strongly bilabiate, short-tubular, widening and opening distally, 4–7 mm, dark red; anthers reaching corollas sinuses, filaments hairy basally; ovaries of pair usually half fused, styles hairy in basal ⅔; stigmas exserted. **Fruits** generally fused more than halfway to tips, 6–8 mm, bright red, translucent, tips divergent, not enclosed by bracts, axis of fruits ≤ 90° angle.

Rocky areas, streambanks, wet meadows. Flowering May–Aug. 100–2300 m. BW, Casc, ECas, Sisk. CA, ID, NV, WA. Native.

Lonicera hispidula (Lindley) Douglas ex Torrey & A. Gray

Hairy honeysuckle, pink honeysuckle

Lonicera hispidula (Lindl.) Douglas ex Torr. & A. Gray var. *hispidula*

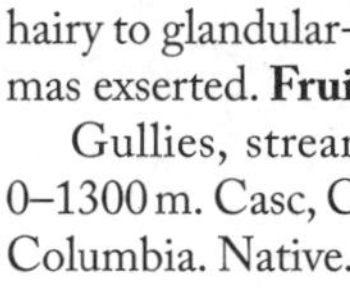

Vines ± evergreen, sprawling to twining, 1.6–6 m; young stems puberulent. **Leaves** oblong to ovate, 4–8 cm, bases obtuse to subcordate, upper 2 pairs fused around stem, others usually auriculate at base, margins ciliate, tips obtuse; stipules ± conspicuous, at least near inflorescence. **Inflorescences** axillary or at end of branches, spicate, with flowers in a dense whorl; peduncles densely glandular. **Flowers** corollas strongly bilabiate, 12–16 mm, usually pink, hairy to glandular-hairy; anthers exserted, filaments hairy; stigmas exserted. **Fruits** orbicular, ~8 mm, red.

Gullies, streamsides, woodlands. Flowering May–Aug. 0–1300 m. Casc, CR, Est, Sisk, WV. CA, WA; north to British Columbia. Native.

Lonicera interrupta Bentham

Chaparral honeysuckle

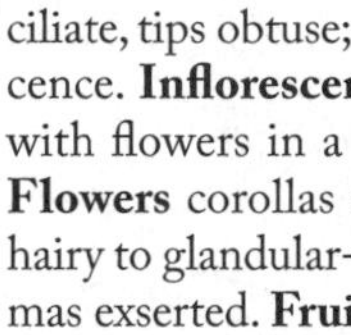

Shrubs deciduous, erect; stems rigid, woody, ~0.3 m; young stems glabrous or puberulent; branches climbing or sprawling. **Leaves** elliptic to round, 2–2.5 cm, bases acute, uppermost pairs of leaves fused around stem, tips usually rounded; stipules lacking. **Inflorescences** spicate, long, interrupted. **Flowers** corollas strongly bilabiate, 8–10 mm, white to cream-yellow, glabrous; anthers exserted; stigmas exserted. **Fruits** ~10 mm, red.

Dry ridges, floodplains. Flowering Jun–Aug. 400–1100 m. Sisk. CA; AZ. Native.

Although currently uncommon in Oregon, this species may become more widespread as a result of climate change.

Lonicera involucrata (Richardson) Banks ex Sprengel

Twinberry, twinberry honeysuckle, bearberry honeysuckle

Shrubs deciduous, erect, 0.6–3.5 m; young stems sparsely hairy, with stalked glands. **Leaves** elliptic to ovate, 3–12 cm, bases rounded or acute, tips acute to acuminate. **Inflorescences** axillary pairs of flowers; bracts in 2 pairs, leaf-like, forming an involucre. **Flowers** corollas bilabiate, 12–18 mm, yellowish, hairy; stamens usually included; ovaries of a flower pair free. **Fruits** 6–10 mm, black in an enlarged red or purple involucre.

Key to *Lonicera involucrata* varieties:

1. Plants < 1 m; corollas yellow, tubes somewhat wider above middle; stigmas exserted; mountains above 1000 m or escaped near habitations . . . . ***L. i.*** var. ***involucrata***

1' Plants 1.5–3.5 m; corollas yellow tinged with orange or red, tubes cylindric; stigmas included or somewhat exserted; coastal under 100 m . . . . ***L. i.*** var. ***ledebourii***

Lonicera involucrata var. ***involucrata*** (p. 538)

Black twinberry

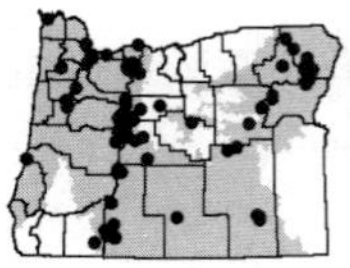

Plants 0.6–0.9 m. **Leaves** not leathery. **Flowers** corollas yellow, tubes somewhat wider above middle; stigmas exserted. **2*n***=18.

Streamsides, moist places. Flowering Apr–Jun. 0–2400 m. BR, BW, Casc, CR, ECas, Est, WV. CA, ID, NV, WA; north to Alaska, Canada, Greenland, east to MT, WI, MI, southeast to NM and northwestern Mexico. Native.

Lonicera involucrata var. ***ledebourii*** (Eschscholtz) Jepson

Coastal twinberry

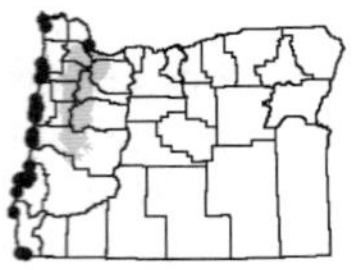

Plants 1.5–3.5 m. **Leaves** blades leathery. **Flowers** corollas yellow tinged with orange or red, tubes cylindric; stigmas included or somewhat exserted.

Moist places. Flowering Mar–Aug. 0–100 m. Est, WV. CA, WA. Native.

Lonicera periclymenum Linnaeus

European honeysuckle, woodbine

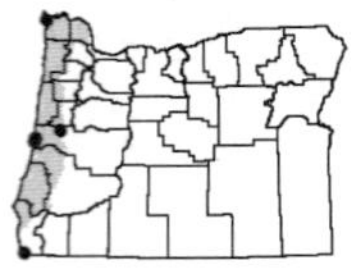

Vines evergreen, twining, up to 7 m; young stems reddish. **Leaves** elliptic, bases obtuse to rounded, margins wavy, tips acute, uppermost leaves not fused around stem. **Inflorescences** axillary clusters of 3–30 flowers. **Flowers** corollas yellowish to white, pink when young, tubes sub-cylindric; stamens exserted; styles exserted. **Fruits** pyriform, translucent red. **2*n***=18, 36.

Thickets, streamsides. Flowering Jun–Sep. 0–300 m. CR, Est. WA; scattered in northeastern North America; Africa, Europe. Exotic.

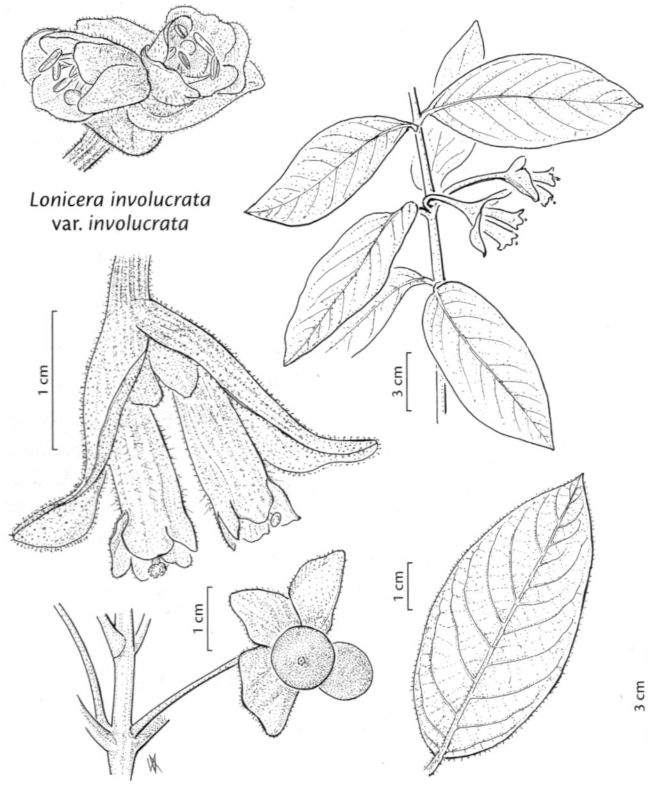

Lonicera involucrata var. *involucrata*

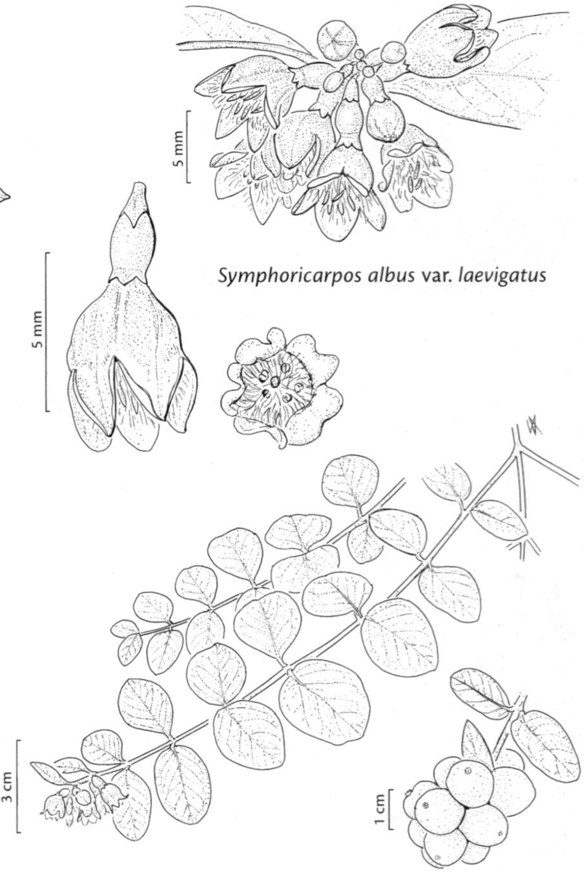

Symphoricarpos albus var. *laevigatus*

LONICERA • SYMPHORICARPOS

Lonicera utahensis S. Watson

Utah honeysuckle

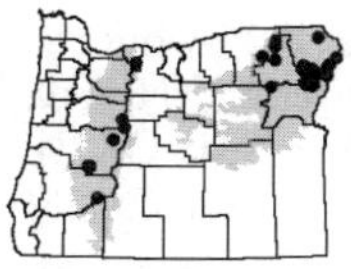

Shrubs deciduous, erect, 0.5–2 m. **Leaves** blades elliptic to ovate or oblong, 2–8 cm, bases sometimes subcordate, tips obtuse to rounded, glabrous above. **Inflorescences** axillary pairs of flowers; peduncles 0.5–1.5 cm, longer in fruit, bearing a pair of tiny bracts at tips. **Flowers** corollas obscurely bilabiate, 10–20 mm, light yellow to nearly white, tubes hairy within, short-spurred at bases, lobes much shorter; stamens glabrous, anthers at throat level; ovaries of a pair divergent, partly fused when young; styles glabrous, slightly exserted. **Fruits** fused at bases, not enclosed by bracts, ~1 cm, bright red to salmon pink, axes of fruits at ~180° angle.

Moist wooded or sometimes open slopes. Flowering Apr–Jul. 800–2300 m. BW, Casc. ID, WA; north to British Columbia and Alberta, southeast to AZ. Native.

SYMPHORICARPOS Snowberry, waxberry, ghostberry

Shrubs deciduous, erect to trailing. **Leaves** opposite, simple, margins entire to somewhat lobed; petioles short; stipules lacking. **Inflorescences** axillary or terminal, clusters or flowers solitary. **Flowers** corollas radially symmetric, sometimes slightly swollen at bases or on one side, tubes bell-shaped to salverform; stamens inserted on corolla; ovaries inferior, 4-chambered, 2-seeded, ovules pendulous. **Fruits** berry-like drupes, white to pink, with 2 1-seeded nutlets.

Central America, North America, western China. ~15 species; 4 species treated in *Flora*.

The name snowberry comes from the white, berry-like fruits borne by most species. These fruits are valuable food for wildlife, and the plants are browsed by mammals and provide valuable cover. The leaves are eaten by caterpillars of the snowberry checkerspot (*Euphadryas colon*) and other related butterflies.

Key to *Symphoricarpos* species:

1. Corolla tubes elongate-campanulate to tubular-funnel-shaped, symmetric, not ventricose, lobes much < tube; fruits ovoid to ellipsoid.
   2. Corollas tubes salverform, with 1 nectary at base . . . . **S. longiflorus**
   2' Corolla tubes elongate-campanulate or funnel-shaped, with 5 nectaries at base . . . . **S. rotundifolius**

1' Corolla tubes short-campanulate, often slightly asymmetrically ventricose, lobes ≈ or slightly > tube; fruits subglobose.
   3. Plants erect, 2–25 dm; flowers in clusters of 8–16 . . . . **S. albus** var. **laevigatus**
   3' Plants trailing, 3–10 dm; flowers in clusters of 2–8 . . . . **S. mollis**

Symphoricarpos albus (Linnaeus) S.F. Blake (p. 538) var. *laevigatus* (Fernald) S.F. Blake

Common snowberry

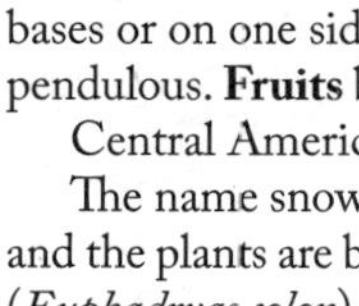

Plants erect, 2–25 dm. **Stems** young branches ± glabrous; bark of old stems gray, smooth. **Leaves** oval, 15–30 mm, bases acute, margins entire or few-toothed or lobed, tips acute to slightly obtuse, surfaces glabrous or sparsely ciliate; petioles 2–4 mm. **Inflorescences** terminal clusters of 8–16 flowers; bracts glabrous. **Flowers** corollas 5–7 mm, white to pink, densely white villous within, tubes short-campanulate, often slightly asymmetrically ventricose, pubescent within, lobes ≈ tube; stamens included; styles included, 2–3 mm, glabrous. **Fruits** subglobose, 5–15 mm, white, in terminal clusters; nutlets oval, 4–6 mm. **2*n***=54.

Shaded slopes, forests, streambanks. Flowering May–Aug. 0–2500 m. BW, Casc, Col, CR, ECas, Est, Lava, Sisk, WV. CA, ID, WA; north to AK, northeast to MT, southeast to CO, eastern US. Native.

The abundant attractive fruit and the value to wildlife of this species make it a popular addition to native plant gardens. It is also used in restoration and is one of the first species to recolonize a site after a fire.

Symphoricarpos longiflorus A. Gray

Desert snowberry, fragrant snowberry

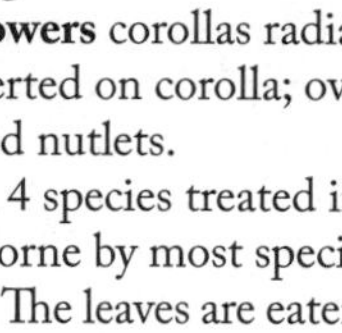

Plants erect to arching, 5–15(20) dm. **Stems** young stems glaucous; small twigs often persisting and somewhat thorn-like; old branches bark whitish, fibrous. **Leaves** narrowly elliptic to oblanceolate, 4–12(20) mm, stiff, bases narrowed, margins with 0–2 teeth, tips acute to obtuse, surfaces glaucous, glabrous to short-hairy; petioles 1–3 mm. **Inflorescences** in leaf axils with 1–2 flowers or terminal few-flowered racemes. **Flowers** very fragrant; corollas symmetric, not ventricose, tubes salverform, 7–15 mm, pale to deep pink, glabrous inside and outside, lobes 2–4 mm, rounded, spreading; stamens inserted at throat, filaments 0–0.5 mm, anthers tips usually shortly exserted; styles included, ~ ½ length of corolla tube, hairy above middle; corolla bases with 1 small nectary. **Fruits** ellipsoid, 8–10 mm, white; nutlets ~0.5 cm.

Dry, rocky areas, stream beds. Flowering May–Jul. 1000–1900 m. BR, Owy. CA, ID, NV; southeast to CO, TX, and Mexico. Native.

Symphoricarpos mollis Nuttall

Creeping snowberry, sharpleaf snowberry, trailing snowberry, trip vine

Symphoricarpos mollis Nutt. var. *acutus* A. Gray, *Symphoricarpos mollis* Nutt. ssp. *hesperius* G.N. Jones, *Symphoricarpos mollis* Nutt. var. *hesperius* (G.N. Jones) Cronquist, *Symphoricarpos mollis* Nutt. var. *mollis*

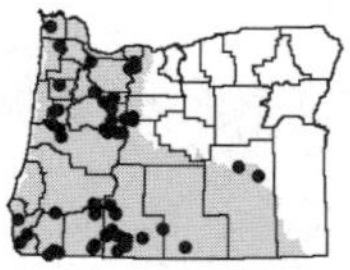

Plants trailing, 3–10 dm; often rooting from trailing stems. **Stems** young stems finely puberulent; older branches gray, shredding; petioles 1–4 mm. **Leaves** oval to orbicular, 10–40 mm, bases obtuse, margins ciliate, entire but often lobed on trailing stems, tips obtuse, surfaces abaxially short pilose, adaxially denser pilose; petioles 1–4 mm. **Inflorescences** terminal clusters of 2–8 flowers. **Flowers** corollas campanulate, pinkish to pink, tubes short-campanulate, often slightly asymmetrically ventricose, somewhat swollen or not, 3–5 mm, pubescent within, lobes 2–4 mm, sparsely hairy inside, ≈ or slightly > tube; stamens reaching tips of lobes; styles glabrous, reaching throat. **Fruits** subglobose, 4–10 mm, white.

Slopes, openings in forests. Flowering May–Aug. 100–2100 m. BR, Casc, CR, ECas, Sisk, WV. CA, ID, NV, WA; north to British Columbia, southeast to NM. Native.

Symphoricarpos rotundifolius A. Gray (p. 540)

Roundleaf snowberry

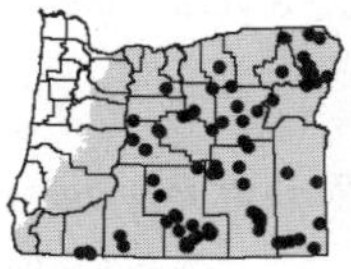

Plants erect or straggling, 6–12 dm. **Stems** young stems soft pubescent with straight hairs, older branches brownish. **Leaves** suborbicular to broadly ovate, 10–40 mm, margins entire to lobed, tips obtuse or rounded, often mucronate, surfaces pubescent, sometimes becoming glabrous; petioles 1–3 mm. **Inflorescences** terminal 1-sided short spikes, drooping, or small clusters in apical axils. **Flowers** corollas tubular-funnel-shaped, 6–12 mm, light pink, tubes funnel-shaped, 2–4 × length of lobes, pilose below, with 5 small basal glandular nectaries; stamens reaching throats; styles included, glabrous. **Fruits** ovoid to ellipsoid, 8–12 cm, white.

Slopes, openings in forests, meadows. Flowering May–Aug. 500–2800 m. BR, BW, Cas, ECas, Lava, Owy, Sisk. CA, ID, NV, WA; east to WY, southeast to TX, south to Mexico. Native.

Some texts refer to plants with elongate-campanulate tubes as *S. oreophilus* var. *utahensis*. Even if recognized, however, this name is illegitimate. Furthermore, other segregates of this species are difficult, if not impossible, to distinguish and are not recognized here.

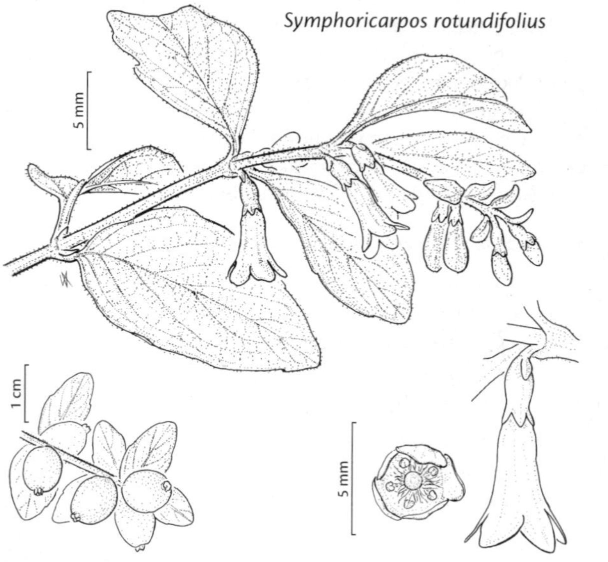

SYMPHORICARPOS

CARYOPHYLLACEAE Pink family

Richard K. Rabeler and Ronald L. Hartman

Herbs winter annual, annual, biennial, or perennial. **Stems** erect to prostrate. **Leaves** opposite, pseudoverticillate, whorled, or rarely alternate, simple; petioles often present; stipules present or 0. **Inflorescences** terminal or axillary cymes, thyrses, or capitula, or flowers 1; bracts foliaceous, scarious, or absent; involucel bracts (epicalyx) immediately subtending calyx occasionally present. **Flowers** bisexual or occasionally unisexual, radial; hypanthium, when present, urceolate, cup-, disk-, or dish-shaped, sometimes abruptly expanded distally; sepals 4 or 5, distinct or connate basally or into a cup or tube; petals (0)4 or 5, distinct; stamens 1–10, in 1 or 2 whorls, arising from base of ovary, nectariferous disk or hypanthium rim, staminodes usually 0 or 1–10; pistils 1, ovaries superior, 1-locular or rarely 2-locular proximally or 3–5 locular, placentas free-central, basal, or axile in proximal half; styles 1–5(6), distinct or connate proximally; stigmas 2–5(6), often linear along adaxial surface of styles (or branches). **Fruits** capsules or utricles; carpophore sometimes present. **Seeds** 1–150(500), often laterally compressed, brown or black.

Worldwide, especially in the Mediterranean region and North Temperate Zone. ~100 genera; 30 genera treated in *Flora*.

Caryophyllaceae has traditionally been divided into three subfamilies: Paronychioideae (stipules present, sepals distinct), Alsinoideae (stipules absent, sepals mostly distinct), and Caryophylloideae (stipules absent, sepals connate into a cup or tube). Recent molecular studies have shown this scheme does not reflect phylogeny within the family; recognition of at least ten tribes is more realistic (see Harbaugh et al. 2010).

Besides the taxa described below, species in two additional genera have been collected on ballast. *Myosoton aquaticum*, a native of Europe and Asia, was collected on ballast near Portland in 1902. It is similar to *Stellaria media* but has minute glandular hairs on both the stems and calyces as well as 5–6 styles and a capsule opening by 5–6 bifid valves. It is known from a few localities in Washington, Idaho, and British Columbia. *Paronychia franciscana*, a native of Chile known from around San Francisco Bay, has been collected on four occasions in Oregon, twice on ballast near Portland, twice in Curry County in 1924 and 1926. This plant is most similar to *Cardionema ramosissimum* but has sepals with a shorter awn (less than 1 mm) that are not densely woolly.

Key to Caryophyllaceae genera:

1. Stipules present, lanceolate to ovate, triangular, or bristle-like, mostly scarious, connate (except *Corrigiola*).
   2. Fruits utricles; seeds 1; petals 0 or < 0.6 mm.
      3. Leaves alternate to rarely subopposite; staminodes petaloid, ovate to oblong. . . . . **CORRIGIOLA**
      3' Leaves opposite (or uppermost often alternate in *Herniaria*); staminodes not petaloid, subulate to narrowly oblong.
         4. Sepal awns 0; stipules 0.4–1.5 mm . . . . **HERNIARIA**
         4' Sepal awns 1.5–4 mm; stipules 1–8 mm . . . . **CARDIONEMA**
   2' Fruits capsules; seeds 5–150+; petals rarely 0.
      5. Leaves oblanceolate to obovate, petioles usually present; styles 1, 3-branched . . . . **POLYCARPON**
      5' Leaves awl-like to linear, petioles 0; styles 3–5, free.
         6. Leaves opposite but appearing whorled due to axillary clusters of 16–30 per node; styles and valves 5 . . . . **SPERGULA**
         6' Leaves opposite, often with axillary clusters of 2–4+ per node but not appearing whorled; styles and capsule valves 3.
            7. Stipules partly fused to leaf bases, tips bristle-like; flowers 1(2), axillary; sepals with lateral spurs; petals 0 or vestigial . . . . **LOEFLINGIA**
            7' Stipules lanceolate to broadly triangular, not partially fused to leaf bases; flowers few–many, in terminal cymes or sometimes solitary and axillary; sepals lack spurs; petals 5. . .**SPERGULARIA**

1' Stipules absent.
   8. Sepals usually fused into a prominent tube or sometimes a cup, lobes or teeth < tube (except *Agrostemma*).
      9. Styles 3–5; capsule valves or teeth 3–6, 8, or 10.
         10. Plants densely covered with relatively long usually silky hairs; styles (4)5; flowers bisexual.
            11. Calyx lobes > tube, broadly linear, often ≥ petals . . . .**AGROSTEMMA**
            11' Calyx lobes < tube, linear to lanceolate, < petals . . . . **LYCHNIS**
         10' Plants glabrous (sometimes glaucous) or if pubescent then not with silky hairs; styles 3(4) or if 5 then flowers imperfect and plants dioecious.
            12. Flowers in terminal capitate cymes or tight paniculate clusters; calyx tubes glabrous, narrowly funnelform or club-shaped, dilated above midpoint; capsule stalks 7–8 mm . . . . **ATOCION**
            12' Flowers in terminal or axillary, open cymes, or flowers rarely solitary; calyx tubes glabrous or pubescent, tubular to campanulate; capsule stalks 0.5–4(6) mm, rarely absent . . . . **SILENE**
      9' Styles 2; capsule valves or teeth generally 4.
         13. Flower(s) subtended or enclosed by 2–6 bracts.
            14. Involucral bracts green, not scarious, linear to lanceolate, often < 1 mm wide, subtending flowers; calyx veins 20–25 or 40, tubes green or reddish, commissures between adjacent sepals absent. . . . . **DIANTHUS**
            14' Involucral bracts red- to brown-scarious, widely ovate, usually 5–12 mm wide, enclosing flowers; calyx veins 15, tubes green or reddish, commissures between adjacent sepals veinless, scarious. . . . . **PETRORHAGIA**
         13' Flower or flowers not individually subtended by bracts.
            15. Calyx tubes 1–3 mm, cup-shaped, commissures between adjacent sepals veinless, scarious . . . .**GYPSOPHILA**
            15' Calyx tubes 6–17 mm, oblong-cylindric to ovoid, commissures between adjacent sepals absent.
               16. Plants perennial; inflorescences congested; pedicels 1–6 mm; calyces terete, not angled; petal appendages 2 . . . . **SAPONARIA**
               16' Plants annual; inflorescences open; pedicels 5–40+ mm; calyces 5-angled; petal appendages 0. . . . .**VACCARIA**
   8' Sepals free or at most basally connate.
      17. Fruits utricles enclosed by a prominent, persistent, hard hypanthium; flowers inconspicuous, sessile or subsessile; petals 0 . . . .**SCLERANTHUS**
      17' Fruits capsules; hypanthia if present inconspicuous, not enclosing fruit; flowers mostly conspicuous, usually pedicellate; petals present or 0.

18. Petals absent or rudimentary.
  19. Fruits cylindric, opening by 8 or 10 teeth.
    20. Teeth 8, capsules ≤ sepals . . . . **MOENCHIA** (in part)
    20' Teeth 10, capsules > sepals . . . . **CERASTIUM** (in part)
  19' Fruits ovoid to globose, opening by valves or teeth.
    21. Fruit valves 2 × number of styles. . . . . **STELLARIA** (in part)
    21' Fruit valves = number of styles.
      22. Sepals 4 or 5, styles and fruit valves 4 or 5 . . . . **SAGINA** (in part)
      22' Sepals 5, styles and fruit valves 3 . . . . **SABULINA** (in part)
18' Petals present, 4 or 5(6).
  23. Petals 2-lobed or divided nearly to base.
    24. Fruits cylindric, opening by (6 or 8)10 teeth.
      25. Styles 3(4), capsules straight. . . . . **DICHODON**
      25' Styles 5, capsules generally curved in upper half . . . . **CERASTIUM** (in part)
    24' Fruits spherical to ovoid, not curved, opening by 6(8 or 10) valves.
      26. Petals 2-lobed < 20% of length; fruits spherical, valves 2–3 times coiled; seeds 1–3; rhizomes with tuber-like thickenings, 3–12 mm in diameter . . . . **PSEUDOSTELLARIA**
      26' Petals 2-lobed > 20% of length, or often divided nearly to base; fruits ovoid or spherical to cylindric-oblong, valves ascending to recurved, not coiled; seeds 3–20+; rhizomes, when present, without tuber-like thickenings. . . . . **STELLARIA** (in part)
  23' Petals entire, emarginate, irregularly toothed, or notched.
    27. Leaves conspicuously fleshy; petal tips emarginate; plants maritime . . . . **HONCKENYA**
    27' Leaves not conspicuously fleshy; petal tips entire, irregularly toothed, or notched; plants of non-maritime habitats.
      28. Petal tips irregularly toothed; inflorescences umbel-like; fruits opening by 6 teeth. . . . . **HOLOSTEUM**
      28' Petal tips entire or notched; inflorescences not umbel-like (except *Eremogone congesta*); fruits opening by 8 or 10 teeth or 3–6 valves.
        29. Capsules cylindric, opening by 8 or 10 teeth.
          30. Petals 4, tips entire; fruit teeth 8, capsules ≤ sepals . . . . **MOENCHIA** (in part)
          30' Petals 5, tips notched; fruit teeth 10, capsules > sepals . . . . **CERASTIUM** (in part)
        29. Capsules ovoid to ovoid-conic, ellipsoid, urceolate, to globose, opening by 3–6 valves or teeth.
          31. Styles 4 or 5; fruit valves 4 or 5. . . . . **SAGINA** (in part)
          31' Styles 3; fruit valves 3 or 6 or teeth 6.
            32. Fruit valves 3; ovary sutures 3.
              33. Sepals hooded, tips obtuse and margins incurved; seeds obscurely sculptured. . . . . **CHERLERIA**
              33' Sepals not hooded, tips acute to acuminate, sometimes obtuse, margins not incurved; seeds tuberculate or papillate . . . . **SABULINA** (in part)
            32' Fruit teeth or valves 6; ovary sutures 6.
              34. Leaves needle-like to narrowly linear, often congested at or near bases of flowering stems, tips generally sharp-pointed . . . . **EREMOGONE**
              34' Leaves ovate or oblanceolate to elliptic, not congested at or near bases of flowering stems, tips obtuse or rounded to acute or acuminate.
                35. Plants perennial, with a slender, branched rhizome; leaves oblanceolate to elliptic; seed appendages ~ 0.7 mm, elliptic, spongy. . . . **MOEHRINGIA**
                35' Plants annual, with a slender taproot; leaves ovate; seed appendages absent. . . . . **ARENARIA**

AGROSTEMMA Corn-cockle

Plants annual. **Stems** erect. **Leaves** opposite; linear to narrowly lanceolate; petioles 0; stipules 0. **Inflorescences** terminal cymes or flowers 1; bracts present. **Flowers** bisexual; sepals 5, fused into prominent tube, ovoid to widely cylindric, green, hairy, strongly 10-veined, commissures between adjacent sepals 1-veined, green, lobes > tube; petals 5, purple-red, limbs entire or notched; appendages 0; stamens 5, borne at ovary base, staminodes 0; styles and stigmas 5. **Fruits** capsules, opening by 5 ascending teeth. **Seeds** 30–60, widely ovate, 3–3.5 mm, black, tuberculate.

Worldwide. 2 species; 1 species treated in *Flora*.

Agrostemma githago Linnaeus var. ***githago*** (p. 544)

Common corn-cockle, corn campion

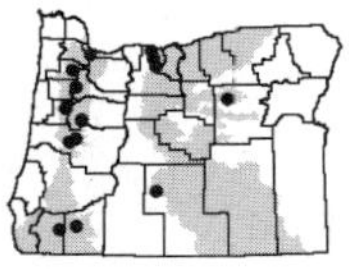

Stems 30–90+ cm, simple or sparingly branched above, covered with dense, long, silky, appressed hairs. **Leaves** 5–15 cm. **Inflorescences** axillary, solitary flowers. **Flowers** calyx tubes 12–17 mm, lobes broadly linear, 12–50 mm; petals 24–40 mm, limbs exserted, obovate, 10–20 mm, rounded to truncate.

Disturbed areas; roadsides, cultivated and fallow fields. Flowering May–Aug. 0–1500 m. BR, Col, Lava, Sisk, WV. CA, ID, WA; across Canada and US, except northern and western Canada and southwestern US; worldwide. Exotic.

Agrostemma githago probably occupies only a fraction of the distribution stated above. As in much of North America, this species is now less commonly found in Oregon since improvements in seed cleaning have reduced its presence as a common weed in wheat fields. The only collections made since 1960 that we have seen came from a roadside in Wasco County in 1984 and a yard waste burn site in Josephine County in 1998.

ARENARIA Sandwort

Plants annual; taproots slender. **Stems** erect to ascending. **Leaves** opposite, ovate, not congested at or near bases of flowering stems, tips acute to acuminate; petioles present (basal leaves) or 0; stipules 0. **Inflorescences** terminal cymes; bracts present. **Flowers** bisexual; sepals 5, free, stipitate-glandular; petals 5, white, tips entire; stamens 10, borne at ovary base, staminodes 0; ovary sutures 6, styles 3. **Fruits** capsules, ovoid to ovoid-conic, opening by 6 ascending to recurved teeth. **Seeds** 10–15, reniform, 0.5–0.6 mm, grayish black, tuberculate, appendages absent.

Temperate regions, especially mountains of Asia and Europe, Asia Minor, Mediterranean region, and Andean South America. ~135 species; 1 species treated in *Flora.*

We follow Harbaugh et al. (2010) and Sadeghian et al. (2015) in recognizing the distant relationship between *Arenaria* and *Eremogone*. In this narrow sense, *Arenaria* is represented by one introduced taxon while our six native, needle-leaved species are now placed in *Eremogone.*

Arenaria montana (rock sandwort) was collected in a deserted lot in Washington County in 1977. This species would be easily distinguished from *A. serpyllifolia* by the large petals (up to 11 × 5 mm). This taxon is often cultivated, especially in rock gardens, and is not otherwise known to escape in North America.

Arenaria serpyllifolia Linnaeus var. ***serpyllifolia*** (p. 544)

Thyme-leaf sandwort

Arenaria serpyllifolia L. ssp. *serpyllifolia*

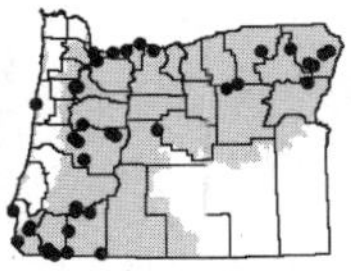

Stems 3–25 cm, branched near base, hairy; hairs minute, down-curved. **Leaves** ovate, 2–7 mm. **Flowers** sepals lanceolate to ovate-lanceolate, 2.5–3 mm, to over 4 mm in fruit; petals oblong, 0.6–2.7 mm, tips obtuse to rounded.

Roadsides, hillsides, riverbanks, disturbed areas, sandy or rocky ground. Flowering Mar–Jul. 0–1800 m. BW, Casc, Col, ECas, Est, Lava, Sisk, WV. CA, ID, NV, WA; throughout most of North America; northern Africa, Asia, Europe. Exotic.

First collected in the Umpqua Valley by Thomas Howell in 1887, *A. serpyllifolia* has become a widespread plant in disturbed areas of Oregon.

ATOCION Catchfly

Plants annual or biennial. **Stems** erect. **Leaves** opposite, oblong-lanceolate to elliptic or ovate; petioles 0; stipules 0. **Inflorescences** terminal, capitate cymes or flowers in tight paniculate clusters; bracts present. **Flowers** bisexual; sepals 5, fused, green or reddish, glabrous, tubes dilated above midpoint, weakly 10-veined, commissures between adjacent sepals 1-veined, lobes < tube; petals 5, pink or purple, limbs entire or emarginate, appendages 2; stamens 5, borne at ovary base, staminodes 0; styles and stigmas 3(4). **Fruits** capsules, opening by 6(8) ascending to slightly recurved teeth; capsule stalks 7–8 mm. **Seeds** kidney-shaped, 0.5–0.7 mm, brown, papillate.

Asia, Europe, North America. 5 species; 1 species treated in *Flora.*

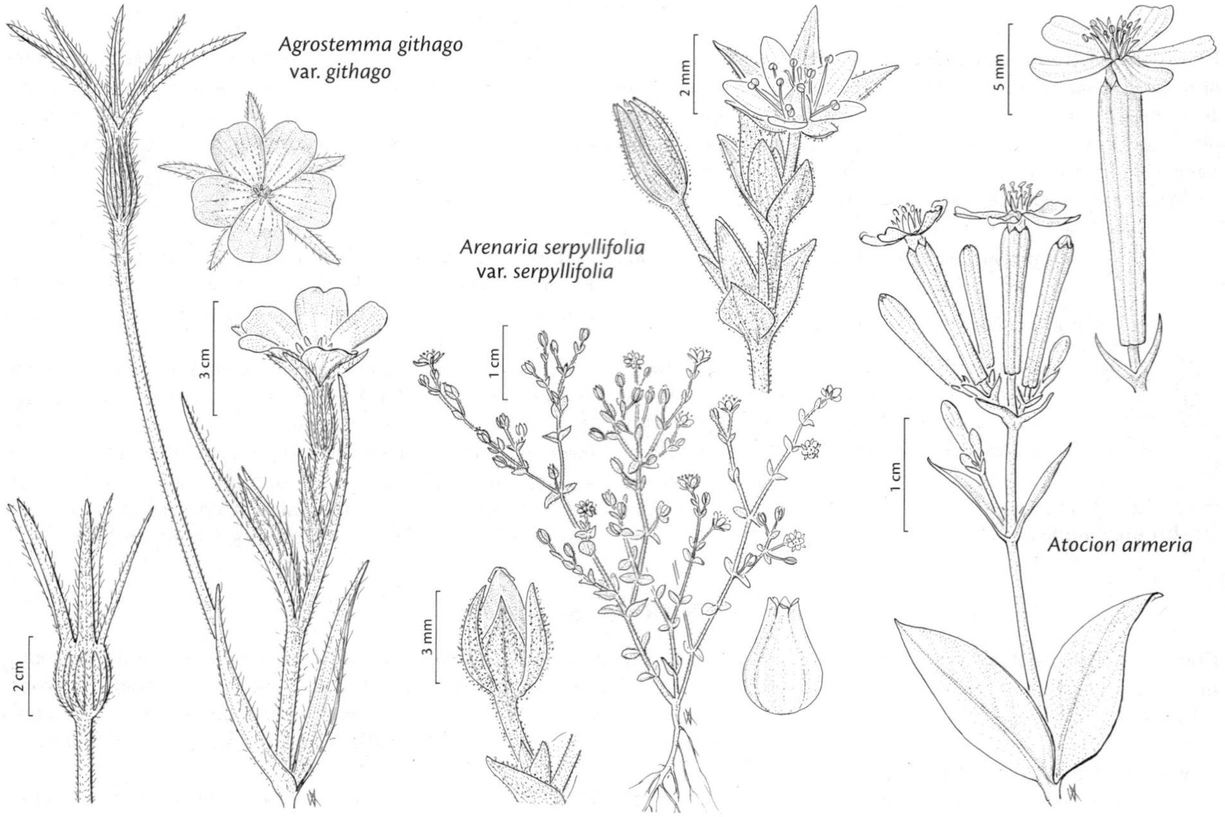

AGROSTEMMA • ARENARIA • ATOCION

Atocion armeria (Linnaeus) Rafinesque (p. 544)
Sweet William catchfly

Silene armeria L.

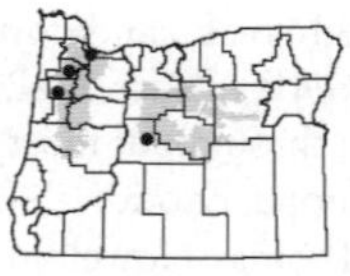

Stems (10)20–40 cm, usually simple, glabrous, rarely stipitate-glandular apically. **Leaves** 1–6 cm. **Flowers** sepals fused into funnelform or club-shaped tube, 13–17 mm, lobes ovate-triangular, ~1 mm; petals 10–15 mm, limbs exserted, oblanceolate to spatulate, ~5 mm. ***2n***=24.

Disturbed areas, flower bed edges. Flowering Jun–Sep. 50–1200 m. Lava, WV. CA, WA; north to British Columbia, UT, scattered in eastern North America; Europe. Exotic.

We follow Oxelman et al. (2001) in separating *Atocion* from *Silene,* recognizing that several small genera (*Atocion* and *Lychnis* in the Oregon flora) form a monophyletic group separate from *Silene. Atocion armeria* appears to be a somewhat recent introduction to Oregon, the first specimen we have seen was collected in 1954 in Polk County.

CARDIONEMA Sandcarpet

Plants perennial. **Stems** horizontal. **Leaves** opposite, needle-like; petioles 0; stipules 2 per node, lanceolate to ovate, scarious. **Inflorescences** axillary cymes or flowers 1; bracts present. **Flowers** bisexual, sepals 5, free, densely woolly, each with spine-tipped awn; petals 5, white, scale-like; stamens 3–5, arising on hypanthium rim, staminodes not petaloid, subulate to narrowly oblong; styles and stigmas 2. **Fruits** utricles, indehiscent. **Seeds** 1, narrowly ovate, 1.4–1.6 mm, tan with red spot at hilum, smooth.

Western North America and South America. 6 species; 1 species treated in *Flora.*

Cardionema ramosissimum (Weinmann) A. Nelson & J. F. Macbride (p. 546)
Sandmat, sandcarpet

Stems 5–30+ cm, much branched from base, often obscured by stipules. **Leaves** 5–13 mm; stipules 1–8 mm. **Flowers** sepals (excluding awn) 1.2–2.8 mm, awns 1.5–4 mm; petals 0.3–0.5 mm.

Sandy beaches and dunes, grassy meadows. Flowering Apr–Jul. 0–200 m. Est. CA, WA; Mexico; coastal South America. Native.

CERASTIUM Mouse-ear chickweed

Plants annual or perennial. **Stems** erect to prostrate and mat-forming. **Leaves** opposite, linear to ovate; petioles present (basal leaves) or 0; stipules 0. **Inflorescences** terminal cymes, open or dense; bracts present; usually pedicellate. **Flowers** bisexual; sepals 5, free, hairy to glandular-hairy; petals (0)5, white, generally divided 50% of their length or notched, tips usually 2-lobed; stamens 5 or 10, borne at ovary base, staminodes 0; styles 5. **Fruits** cylindric, upper half generally curved, > sepals, opening by 10 spreading to recurved teeth. **Seeds** 15–150+, angular to reniform, 0.4–1.5 mm, pale to red-brown, papillate-tuberculate.

Worldwide. ~180 species; 11 species treated in *Flora.*

Examination of both bract margins and the hairs on the sepals is critical to correct identification of *Cerastium* species. As an example, the only Oregon specimens of *C. nutans* that we examined while preparing this treatment were gathered before 1914. Most that we have seen labeled with that name are misidentified, instead being *C. fontanum* ssp. *vulgare*, a common, introduced plant that can readily be distinguished from *C. nutans* by scarious bract margins (versus green) and long, often non-glandular (versus sparse short, peg-like, and glandular) sepal hairs.

Cerastium tomentosum (snow-in-summer), a plant similar to *C. arvense* but with leaves and stems covered with dense, white, non-glandular hairs, is commonly cultivated; all of the specimens we have seen except two are clearly from intentional plantings.

Key to *Cerastium* species:

1. All or uppermost (distal) bracts with scarious, white to translucent, margins.
    2. Petals showy, 2–3 × length of sepals; axillary leaves usually present in basal and mid-stem leaf axils.
        3. Stems 5–25 cm; all bract margins scarious; sepals 4.2–6(7) mm; petals 7.5–9 mm . . . . ***C. arvense*** ssp. ***strictum***
        3' Stems 10–45 cm; lower bracts green, not scarious; sepals 6–9 mm; petals 10–15 mm . . . . ***C. viride***
    2' Petals about 1(2) × length of sepals; axillary leaves usually absent.
        4. Petals deeply 2-lobed; stamens 10; fruits curved, 5.5–13 mm.
            5. Petals 1(2) × length of sepals; sepals densely glandular hairy; plants of rocky, alpine areas . . . . ***C. beeringianum***
            5' Petals (0.6)1 × length of sepals; sepals hairy, rarely glandular; plants of grassy or disturbed areas at lower elevations . . . . ***C. fontanum*** ssp. ***vulgare***
        4' Petals notched or shallowly lobed; stamens 5; fruits somewhat curved, 4–8 mm.
            6. All bract margins broadly scarious (sometimes including entire apical half); petal veins not branched . . . . ***C. semidecandrum***
            6' Uppermost bract margins narrowly scarious; petal veins branched . . . . ***C. pumilum***

1' All bracts with green, non-scarious, margins.
    7. Sepal hairs mostly non-glandular sometimes also glandular, hairs extending beyond sepal tip.
        8. Plants covered with long, silvery hairs; flowers in open cymes, pedicels 2–3+ × length of sepals . . . ***C. brachypetalum***
        8' Plants hairy, hairs not long and silvery; flowers in a dense, or slightly open, cyme, pedicels ≤length of sepals . . . . ***C. glomeratum***
    7' Sepal hairs glandular, hairs not extending beyond sepal tip.
        9. Sepals densely covered with glandular hairs; stamens 5; capsules straight . . . . ***C. dichotomum***
        9' Sepals with sparse short, peg-like glandular hairs; stamens 10; capsules curved.
            10. Pedicels in fruit 2–5 × length of sepals, apically hooked, not deflexed at base . . . . ***C. nutans*** var. ***nutans***
            10' Pedicels in fruit 1.0–1.5 × length of sepals, not apically hooked, often deflexed at base . . . . ***C. brachypodum***

Cerastium arvense Linnaeus ssp. *strictum* Gaudin

(p. 546)

Field chickweed, field mouse-ear chickweed

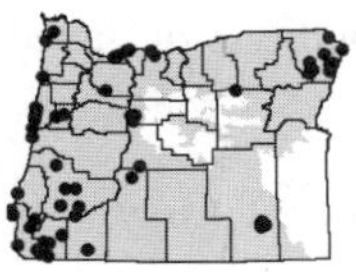

Plants perennial, usually not flowering in 1st year. **Stems** 5–25 cm; non-flowering stems mat-forming; flowering stems erect, hairs glandular above, not glandular and somewhat longer below. **Leaves** on flowering stems linear or lanceolate, 5–25 mm, not succulent; axillary leaves present in basal and mid-stem leaf axils. **Inflorescences** open cymes; all bract margins white, scarious; pedicels in fruit 1–4 × length of sepals. **Flowers** sepals 4.2–6(7) mm, hairs glandular, rarely extending beyond sepal tip; petals shallowly 2-lobed, 7.5–9 mm, 2 × length of sepals; stamens 10. **Fruits** curved, 7.5–11 mm. **Seeds** 1–1.2 mm. **2*n***=36.

Wet meadows and bogs, rocky slopes, gravel bars, sandy beaches and dunes, coniferous forests. Flowering Mar–Aug. 0–3000 m. BR, BW, Casc, Col, CR, ECas, Est, Sisk, WV. CA, ID, NV, WA; throughout most of North America; Europe, South America. Native.

Cerastium arvense is closely related to *C. viride*, and the two species can be very difficult to separate reliably. Both species occur west of the Cascades; *C. arvense* is less common than *C. viride* in coastal habitats and is often somewhat smaller in many dimensions with narrower, non-succulent leaves.

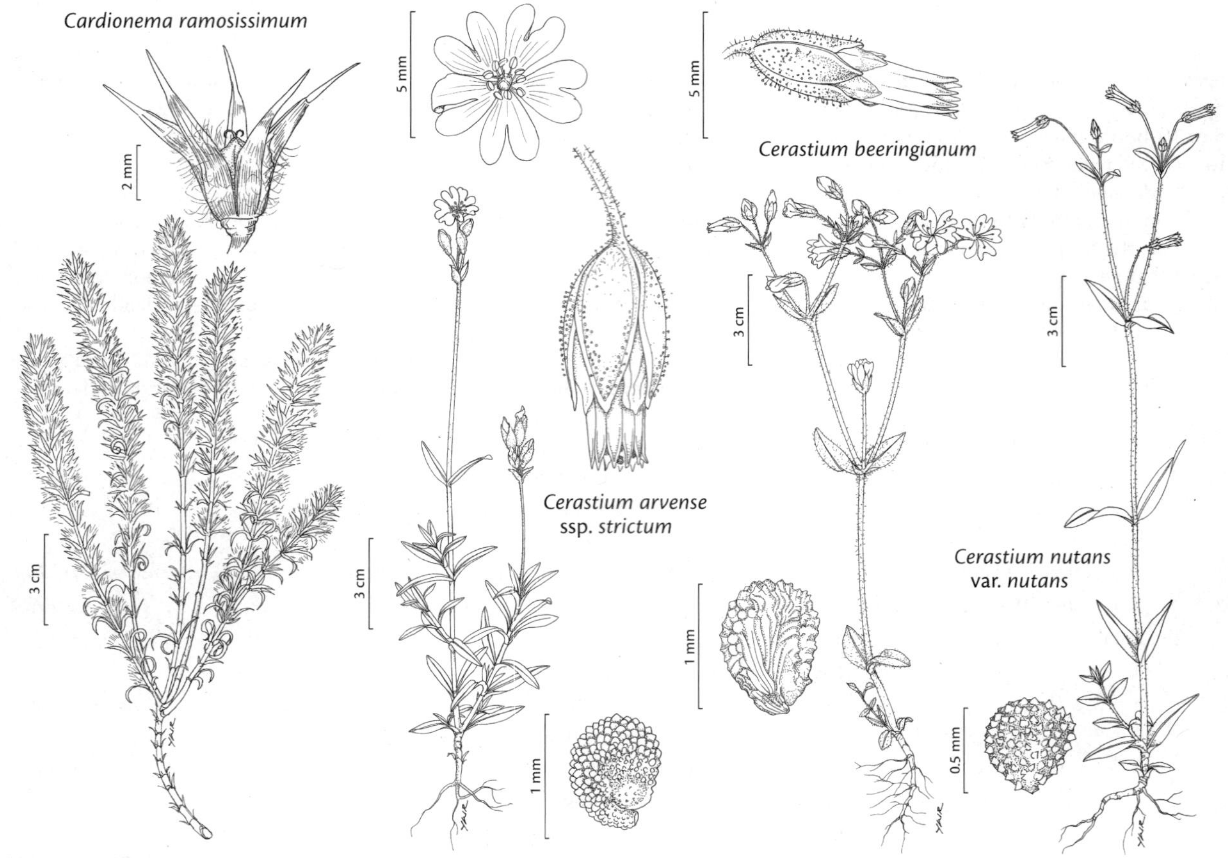

CARDIONEMA • CERASTIUM

Cerastium beeringianum Chamisso & Schlechtendal (p. 546)

Bering mouse-ear chickweed

Cerastium beeringianum Cham. & Schltdl. var. *capillare* Fernald & Wiegand

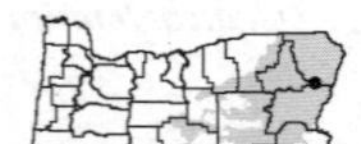

Plants perennial, usually not flowering in 1st year. **Stems** 10–20 cm; non-flowering stems mat-forming; flowering stems erect, hairs glandular or not. **Leaves** on flowering stems lanceolate to elliptic, 5–15 mm; axillary leaves absent. **Inflorescences** open cymes; uppermost bract margins white, scarious; pedicels in fruit 0.3–2.2 × length of sepals. **Flowers** sepals 4.5–7 mm, hairs densely glandular, not extending beyond sepal tip; petals deeply 2-lobed, 6–12 mm, 1(2) × length of sepals; stamens 10. **Fruits** curved, 5.5–12 mm. **Seeds** 0.7–1.1 mm. ***2n***=72.

Rocky, alpine areas. Flowering Jul–Aug. 2500–3000 m. BR, BW. CA, ID, WA; western and northern North America. Native.

Cerastium beeringianum and *C. arvense* both occur at high elevation sites but can be distinguished by the presence (*C. arvense*) or absence (*C. beeringianum*) of axillary leaf clusters.

Cerastium brachypetalum Persoon

Gray mouse-ear chickweed

Plants annual. **Stems** erect, 5.5–30 cm, hairs long, silvery white, not glandular. **Leaves** elliptic to ovate, 4–20 mm; axillary leaves absent. **Inflorescences** open cymes; all bract margins green, not scarious; pedicels in fruit 2–3+ × length of sepals. **Flowers** sepals 3–5 mm, hairs dense, non-glandular or with some glandular hairs, extending beyond sepal tip; petals shallowly 2-lobed, 2–3 mm, 0.6–0.8 × length of sepals; stamens 10. **Fruits** somewhat curved, 5–8 mm. **Seeds** 0.5 mm. ***2n***=72.

Road banks. Flowering Apr–Jun. 100–600 m. WV. WA; southeastern US; Asia, Europe. Exotic.

Cerastium brachypetalum has recently appeared in the Pacific Northwest but has been known in the southeastern United States since at least 1954.

Cerastium brachypodum (Engelmann ex A. Gray) B.L. Robinson

Short-stalked mouse-ear chickweed

Plants annual. **Stems** erect, 5–20 cm, hairs glandular. **Leaves** lanceolate to narrowly obovate 5–30 mm; axillary leaves absent. **Inflorescences** dense to open cymes; all bract margins green, not scarious; pedicels in fruit

1.0–1.5 × length of sepals, not apically hooked, often deflexed at base. **Flowers** sepals 3–4.5 mm, hairs sparse, short, glandular, peg-like, not extending beyond sepal tip; petals shallowly 2-lobed, 3–5 mm, 1–1.5 × length of sepals; stamens 10. **Fruits** curved, 6–12 mm. **Seeds** 0.4–0.7 mm. ***2n***=34.

Streambanks. Flowering Mar–Jun. 900–1100 m. Owy. ID, WA; throughout most of North America. Native.

Cerastium brachypodum and *C. nutans* are closely related species, both near the western edges of their native range. Neither has been collected recently in Oregon.

Cerastium dichotomum Linnaeus

Forked mouse-ear chickweed

Cerastium siculum Guss.

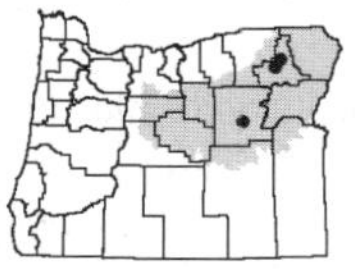

Plants annual. **Stems** erect, 7–30 cm, hairs glandular. **Leaves** lanceolate to oblong, 10–35 mm; axillary leaves usually absent. **Inflorescences** dense cymes; all bract margins green, not scarious; pedicels in fruit 0.3–0.9 × length of sepals. **Flowers** sepals (6)8.5–12 mm, hairs dense, short, glandular, not extending beyond sepal tip; petals shallowly 2-lobed, 8–10 mm, ~1 × length of sepals; stamens 5. **Fruits** straight, 12–22 mm. **Seeds** 0.9–1.3 mm. ***2n***=38.

Pastures, fields, slopes. Flowering May–Jun. 800–900 m. BW, Lava. CA, WA; Europe. Exotic.

First collected in Oregon in 1969, *C. dichotomum* has been known in the Pacific Northwest since its discovery near Pullman, Washington in the 1950s.

Cerastium fontanum Baumgarten ssp. ***vulgare*** (Hartman) Greuter & Burdet

Common mouse-ear chickweed

Cerastium holosteoides Fr.

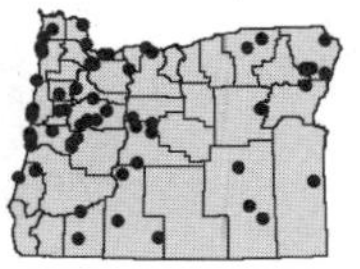

Plants perennial, often flowering in first year (appearing as an annual). **Stems** 6–45 cm; non-flowering stems mat-forming; flowering stems erect, hairs usually not glandular, rarely glandular. **Leaves** lanceolate to oblong, 8–25 mm; axillary leaves absent. **Inflorescences** open cymes; uppermost bract margins white, scarious; pedicels in fruit 1–4 × length of sepals. **Flowers** sepals 4.5–7 mm, hairs non-glandular or sometimes glandular, not extending beyond sepal tip; petals deeply 2-lobed, 4–5 mm, (0.6)1 × length of sepals; stamens 10. **Fruits** curved, (6.5)8–13 mm. **Seeds** 0.6–0.9 mm. ***2n***=126, 136, 144.

Disturbed areas, yards, beaches, wet meadows, open woods. Flowering Mar–Aug. 0–2600 m. All ecoregions. CA, ID, NV, WA; nearly worldwide. Exotic.

Cerastium fontanum and *C. glomeratum*, long known as *C. vulgatum* and C. *viscosum* respectively, are the most common and widely distributed of the introduced *Cerastium* species. Both have at some time been confused with each other and other members of the genus.

A few specimens have been seen from Oregon that exhibit a brownish hirsute pubescence on the stems and pedicels and bracts with very narrow scarious margins; the flowers match *C. fontanum.*

Cerastium glomeratum Thuillier

Sticky mouse-ear chickweed

Cerastium glomeratum Thuill. var. *apetalum* (Dumort.) Fenzl., *Cerastium glomeratum* Thuill. var. *glomeratum*, *Cerastium viscosum* L. var. *viscosum*

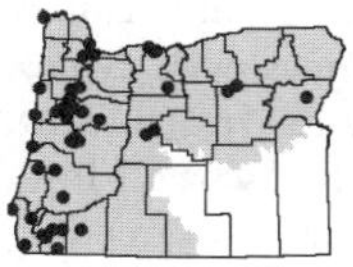

Plants annual. **Stems** ascending to erect, 3–45 cm, hairs both non-glandular and glandular (especially distally). **Leaves** lanceolate or oblanceolate to ovate, 5–30(35) mm; axillary leaves absent. **Inflorescences** dense or slightly open cymes; all bract margins green, not scarious; pedicels in fruit 0.5–0.9 × length of sepals (lowest may be longer). **Flowers** sepals 3–5 mm, hairs non-glandular and glandular, extending beyond sepal tip; petals shallowly 2-lobed, 3–5 mm, 0.7–1(1.2) × length of sepals, or rarely 0; stamens 10. **Fruits** curved, (3.5)5–10 mm. **Seeds** 0.4–0.6 mm. ***2n***=72.

Disturbed areas, fields, roadsides, oak woodlands, shores. Flowering Mar–Aug. 0–1600 m. BW, Casc, Col, CR, ECas, Est, Lava, Sisk, WV. CA, ID, NV, WA; north to AK, eastern North America; Europe. Exotic.

Cerastium nutans Rafinesque var. ***nutans*** (p. 546)

Nodding mouse-ear chickweed

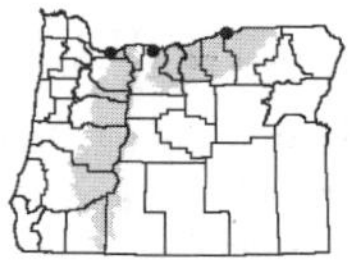

Plants annual. **Stems** erect, 10–50 cm, hairs not glandular below, glandular above. **Leaves** oblong-lanceolate to linear-lanceolate, 10–60 mm; axillary leaves usually absent. **Inflorescences** open cymes; all bract margins green, not scarious; pedicels in fruit 2–5+ × length of sepals, apically hooked, not deflexed at base. **Flowers** sepals 4–6 mm, hairs sparse, short, glandular, peg-like, not extending beyond sepal tip; petals shallowly 2-lobed, (3)5–7 mm, (0.7)1–1.5 × length of sepals; stamens 10. **Fruits** curved, (5)8–13 mm. **Seeds** 0.5–0.8 mm. ***2n***=36.

Meadows, riverbanks. Flowering Apr–Jun. 0–1800 m. Casc, Col. ID, WA; throughout North America except New England states. Native.

As noted under *C. brachypodum*, *C. nutans* is a native species that has only been collected infrequently in Oregon.

Cerastium pumilum Curtis

European chickweed, dwarf mouse-ear chickweed

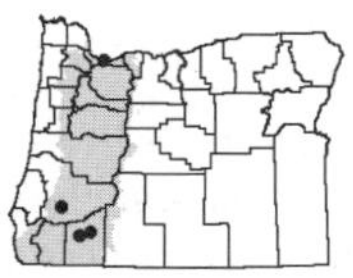

Plants annual. **Stems** erect, 2–15 cm, hairs both not glandular and glandular. **Leaves** oblong to ovate, 3–20 mm; axillary leaves usually absent. **Inflorescences** open cymes; uppermost bract margins white, narrowly scarious; pedicels in fruit 1.0–1.5 × length of sepals. **Flowers** sepals 3–5 mm, hairs densely short-glandular, not extending beyond sepal tip; petals notched or shallowly 2-lobed, 3–5 mm, ~1 × length of sepals, veins branched; stamens 5. **Fruits** somewhat curved, 5–8 mm. **Seeds** 0.6–0.7 mm. ***2n***=72.

Pastures, silty shores. Flowering Apr–Jun. 0–800 m. Casc, Sisk, WV. WA; north to British Columbia, eastern US; Asia, Europe. Exotic.

This species has been gathered only infrequently since it was first noted in Oregon in 1951. It is slightly more common than *C. semidecandrum*, an exotic species with which it is both easily confused and sometimes joins in disturbed areas.

Cerastium semidecandrum Linnaeus

Five-stamen mouse-ear chickweed

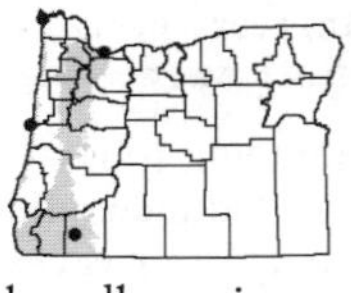

Plants annual. **Stems** erect, 2–20 cm, hairs both not glandular and glandular. **Leaves** elliptic-oblong to ovate, 3–20 mm; axillary leaves absent. **Inflorescences** open cymes; uppermost bract margins translucent, broadly scarious; pedicels in fruit 1–2 × length of sepals. **Flowers** sepals 3–5 mm, hairs densely short-glandular, not extending beyond sepal tip; petals notched or shallowly 2-lobed, 3–5 mm, 1–1.5 × length of sepals, veins unbranched; stamens 5. **Fruits** somewhat curved, 4–7 mm. **Seeds** 0.4–0.6 mm. ***2n***=36.

Silty shores, roadsides. Flowering Apr–May. 0–50 m. Est, Sisk, WV. ID, WA; north to British Columbia, eastern US; Asia, Europe. Exotic.

Except for a collection on ballast near Portland in 1912, the relatively few collections of *C. semidecandrum* have been made since 1941.

Cerastium viride A. Heller

Western field mouse-ear chickweed

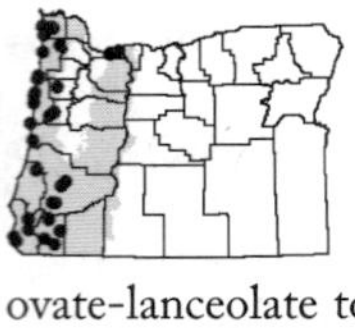

Plants perennial, usually not flowering in 1st year. **Stems** 10–45 cm; non-flowering stems mat-forming; flowering stems erect, hairs glandular above, hairs not glandular, longer below. **Leaves** on flowering stems ovate-lanceolate to narrowly oblong, 10–45 mm, often somewhat succulent; axillary leaves usually present in basal and mid-stem leaf axils. **Inflorescences** open cymes; lower bract margins green, not scarious, uppermost bract margins white, scarious; pedicels in fruit 2–4+ × length of sepals. **Flowers** sepals 6–9 mm, hairs densely glandular, rarely extending beyond sepal tip; petals shallowly 2-lobed, 10–15 mm, 2–3 × length of sepals; stamens 10. **Fruits** nearly straight, 9–16 mm. **Seeds** 1–1.5 mm. ***2n***=72.

Coastal banks and dunes, rocky slopes, open woods. Flowering Mar–Jul. 0–1300 m. Casc, CR, Est, Sisk. CA. Native.

Cerastium viride is closely related to *C. arvense,* and the two species can be very difficult to reliably separate; *C. viride* plants are often somewhat larger than *C. arvense*, with broader, somewhat succulent leaves. In Oregon, populations are chiefly coastal where *C. arvense* occurs less frequently.

CHERLERIA Sandwort

Plants perennial, cespitose or mat-forming. **Stems** trailing; flowering stems ± erect. **Leaves** opposite, needle-like to subulate; petioles 0; stipules 0. **Inflorescences** terminal solitary flowers or occasionally in 2–3-flowered cymes; bracts present. **Flowers** bisexual; sepals 5, free, hooded via incurved margins, stipitate-glandular, tips obtuse; petals 5, white, tips entire; stamens 10, arising from hypanthium, staminodes 0; ovary sutures 3, styles 3. **Fruits** capsules, narrowly ellipsoid, opening by 3 incurved to recurved valves. **Seeds** subspherical to reniform, 0.6–0.7 mm, red-tan, obscurely sculptured.

North temperate and arctic regions, especially central Asia and western North America. 20 species; 2 species treated in *Flora.*

We follow Dillenberger and Kadereit (2014) in their treatment of *Minuartia s.s.* as being a European genus that includes only two of the sections of subgenus *Minuartia* as defined by McNeill (1962) and Moore and Dillenberger (2017) in treating *Cherleria.*

Key to *Cherleria* species:

1. Sepals (or just tips) glabrous to sparsely hairy, distinctly less hairy than pedicels, recurved in fruit; leaves straight, ± flat in cross section; petals 1.2–1.5 × length of sepals; moist areas . . . . ***C. biflora***

1' Sepals and pedicels equally hairy, with same indumentum as pedicels, erect in fruit; leaves ± outwardly curved, strongly keeled, triangular in cross section; petals ≥ 2 × length of sepals; drier areas. . . . . ***C. obtusiloba***

Cherleria biflora (Linnaeus) A.J. Moore & Dillenberger

Mountain sandwort

Minuartia biflora (L.) Schinz & Thell.

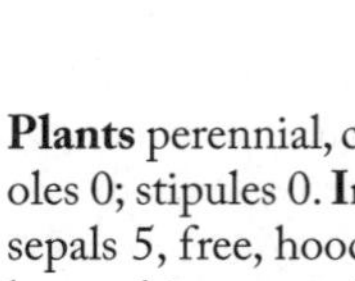

Stems flowering stems erect, 2–10 cm, glabrous or rarely retrorsely-hairy, slightly woody proximally, rarely not. **Leaves** straight, ± flat in cross section, 5–10 × 0.7–2 mm, flexible, ± evenly spaced; axillary leaves mostly absent. **Flowers** sepals 3.5–4.5 mm, veins 3, prominent proximally, sepals (or just tips) glabrous to sparsely hairy, distinctly less hairy than pedicels, recurved in fruit; petals 1.2–1.5 × length of sepals.

Moist rocky ridges and slopes, granitic gravel, mossy bogs. Flowering Jul–Sep. 2000–3000 m. BW. CA, ID, NV, WA; western North America; Asia, Europe. Native.

Cherleria obtusiloba (Rydberg) A.J. Moore & Dillenberger (p. 549)

Alpine sandwort, alpine stitchwort, twinflower sandwort

Arenaria obtusiloba (Rydb.) Fernald, *Minuartia obtusiloba* (Rydb.) House

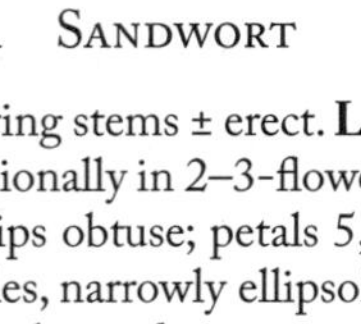

Stems flowering stems erect, 1–12 cm, stipitate-glandular, woody proximally. **Leaves** ± outwardly curved, strongly keeled, triangular in cross section, 1.5–8 × 0.4–1 mm, flexible, ± evenly spaced; axillary leaves present. **Flowers** sepals 2.9–4 mm, veins 3, prominent proximally, sepals and pedicels equally hairy, with same indumentum as pedicels, erect in fruit; petals ≥ 2 × length of sepals.

Rocky summits, talus and limestone cliffs. Flowering Jul–Aug. 1900–3000 m. BW, Casc, Lava. ID, NV, WA; western North America. Native.

CORRIGIOLA Strapwort

Plants annual or biennial. **Stems** decumbent to ascending. **Leaves** alternate or rarely subopposite, sometimes with a basal rosette, linear-oblanceolate to spatulate; petioles minute (subopposite leaves) or 0; stipules 2 per node, ovate, scarious. **Inflorescences** terminal or axillary cymes; bracts present. **Flowers** bisexual; sepals 5, free, glabrous or minutely hairy; petals 0; stamens 5, scale-like, arising on hypanthium rim, staminodes 5, petaloid, ovate to oblong; styles and stigmas 3. **Fruits** utricles, partly enclosed by persistent calyx/hypanthium, indehiscent. **Seeds** 1, reniform or subglobose, 0.6–0.7 mm, white with tan band, finely papillose.

Nearly worldwide. 10 species; 1 species treated in *Flora*.

Corrigiola litoralis Linnaeus ssp. ***litoralis*** (p. 549)
Shore whitlow-wort

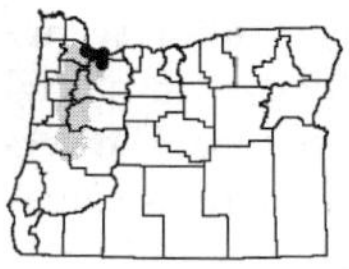

Stems 5–50 cm, simple or branched near base, glabrous, glaucous. **Leaves** 5–30 mm. **Flowers** sepals 1–1.5 mm; staminodes white to pink, sometimes red-tipped. **2*n***=16, 18, 32, 34.

Gravel shores. Flowering Jun–Sep. 0–50 m. WV. British Columbia, MD; Asia, Europe. Exotic.

While most Oregon collections of *C. litoralis* were made between 1911 and 1927 on ballast lands near Lower Albina (Portland), the species was rediscovered in 1992 near Portland.

DIANTHUS Pink, carnation

Plants annual, biennial or perennial. **Stems** ascending to erect. **Leaves** opposite, linear to lanceolate or oblanceolate, petioles present (basal leaves) or 0 (cauline leaves); stipules 0. **Inflorescences** terminal, open or dense cymes, sometimes flowers 1; bracts present; involucral bracts 2–6, linear to lanceolate, long-tapered, usually < 1 mm wide, green, not scarious, subtending flowers. **Flowers** bisexual; sepals 5, fused, glabrous to hairy, tubes prominent, cylindric, greenish or reddish, 20–40-veined, commissures between adjacent sepals absent, lobes < tube; petals 5, pink or reddish with white dots or white to pink or violet, sometimes with a darker center, limbs toothed; appendages 0; stamens 10, fused with petals to stalk below ovary, staminodes 0; styles and stigmas 2. **Fruits** capsules, opening by 4 ascending teeth. **Seeds** 40–100+, shield-shaped, 1.1–2.5 mm, black, papillate.

Africa, Asia, Europe, North America. ~325 species; 2 species treated in *Flora*.

Dianthus deltoides, a perennial species with involucral bracts ⅓–½ as long as the calyx tubes and pedicellate flowers, might be expected in Oregon; it has been collected occasionally in California, Montana, and Washington.

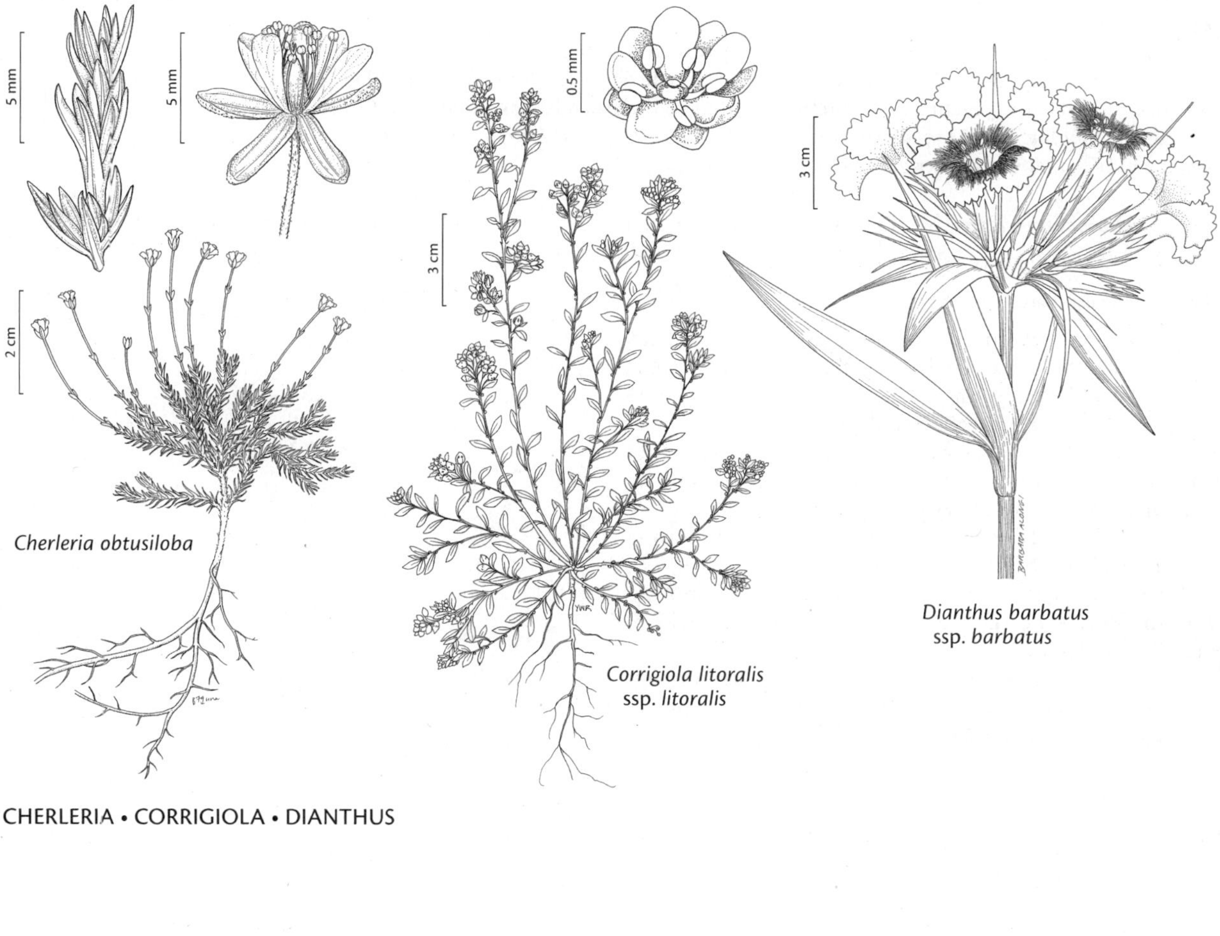

CHERLERIA • CORRIGIOLA • DIANTHUS

Key to *Dianthus* species:

1. Cauline leaves chiefly linear; inflorescences 3–6-flowered, open cymes, sometimes flowers 1; calyces moderately hairy . . . . ***D. armeria*** ssp. ***armeria***

1' Cauline leaves lanceolate to ovate; inflorescences 5–20+-flowered, dense heads, rarely flowers 1; calyces glabrous (lobe margins sometimes ciliate) . . . . ***D. barbatus*** ssp. ***barbatus***

Dianthus armeria Linnaeus ssp. ***armeria***

Deptford pink, grass pink

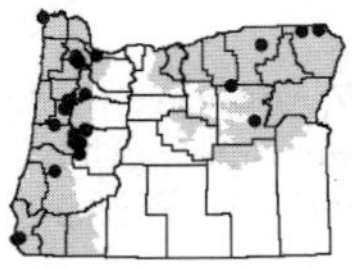

Plants annual or biennial. **Stems** 15–70 cm. **Leaves** basal leaves lanceolate to oblanceolate; cauline leaves chiefly linear. **Inflorescences** open cymes, flowers (1)3–6; involucral bracts linear to lanceolate, mostly ≥ calyx tube. **Flowers** calyces 1.5–2 cm, moderately hairy, hairs long, ± appressed, 20–25-veined, lobes long-tapered; petal limbs 4–5 mm, pink or rose with white dots. **Seeds** 1.1–1.4 mm. **2*n***=30.

Disturbed areas, roadsides, fields, woods. Flowering Apr–Sep. 0–1900 m. BW, Col, CR, Est, Sisk, WV. CA, ID, WA; north to British Columbia, most of eastern North America; central and southwestern Asia, Chile, Europe, Hawaii. Exotic.

First collected in Oregon in 1902 from Mt. Tabor in Multnomah County, *D. armeria* was noted in widely scattered sites by the 1920s but not commonly collected until the 1960s.

Dianthus barbatus Linnaeus ssp. ***barbatus*** (p. 549)

Sweet William

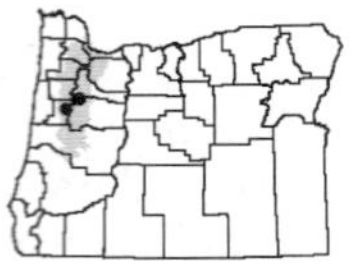

Plants perennial. **Stems** 30–60 cm. **Leaves** basal leaves lanceolate to oblanceolate; cauline leaves lanceolate to ovate. **Inflorescences** dense cymes, flowers (1)5–20+; involucral bracts linear, mostly ≥ calyx tube. **Flowers** calyces 1.5–1.8 cm, glabrous, 40-veined, lobe margins sometimes ciliate, tips acute to short-tapered; petal limbs 6–10 mm, white to pink, purple, violet, or 2-colored. **Seeds** 2–2.5(2.7) mm. **2*n***=30.

Wooded areas, roadside, fields. Flowering May–Jul. 0–500 m. WV. CA, ID, WA; north to British Columbia, scattered in eastern North America; southern Europe. Exotic.

The occurrence of *D. barbatus* in Oregon is defined by three collections, only one of which has been made since 1990. This species is sometimes cultivated in North America and may persist and/or escape from such plantings.

DICHODON Chickweed

Plants annual. **Stems** erect. **Leaves** opposite, linear to linear-oblong; petioles 0; stipules 0. **Inflorescences** terminal cymes, bracts present. **Flowers** bisexual, sepals 5, free, hairs glandular; petals 5, white, tips shallowly 2-lobed; stamens 10, borne at ovary base, staminodes 0; styles 3(4). **Fruits** capsules, straight, cylindric, > sepals, opening by 6(8) erect to spreading teeth. **Seeds** ovate, 0.6 mm, pale brown, tuberculate.

Asia, Europe, North America. 5 species; 1 species treated in *Flora*.

We are following molecular evidence illustrated in Scheen et al. (2004) and Greenberg and Donoghue (2011) in segregating *Dichodon* from *Cerastium*. The expanded sampling in the Greenberg and Donoghue (2011) study showed *Dichodon* clustering with *Holosteum*, together forming a clade that was a sister group to *Cerastium* and *Moenchia*.

Dichodon viscidus (M. Bieberstein) Holub (p. 550)

Anomalous mouse-ear chickweed

Cerastium dubium (Bast.) Guépin

Plants annual. **Stems** erect, 10–40 cm, hairs glandular. **Leaves** linear to linear-oblong, 10–30 mm, not covered with dense, white hairs. **Inflorescences** open cymes; all bract margins green, not scarious; pedicels in fruit 0.5–3 × length of sepals. **Flowers** sepals 5–6 mm, hairs short-glandular, not extending beyond sepal tip; petals shallowly 2-lobed, 5–8 mm, 1.5 × length of sepals; stamens 10. **Fruits** straight, 8–11 mm. **Seeds** 0.6 mm. **2*n***=36, 38.

River and lake shores, roadsides. Flowering Apr–May. 0–100 m. WV. Mult; ID, WA; scattered in south-central US; Asia, Europe. Exotic.

First collected in North America in 1966 in Clark County, Washington, *D. viscidum* is currently known from a few sites in Multnomah County, first gathered in 1987.

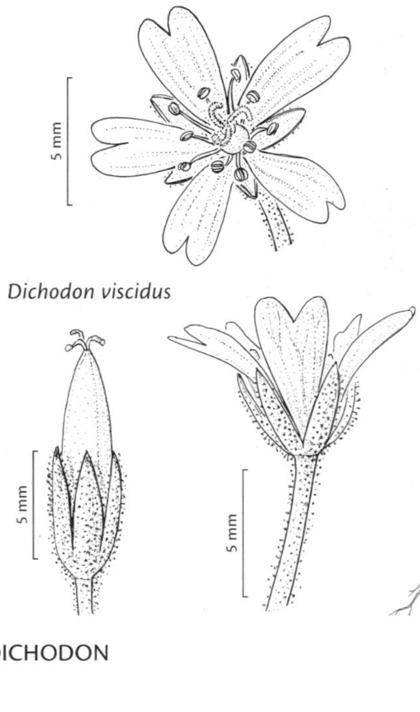

DICHODON

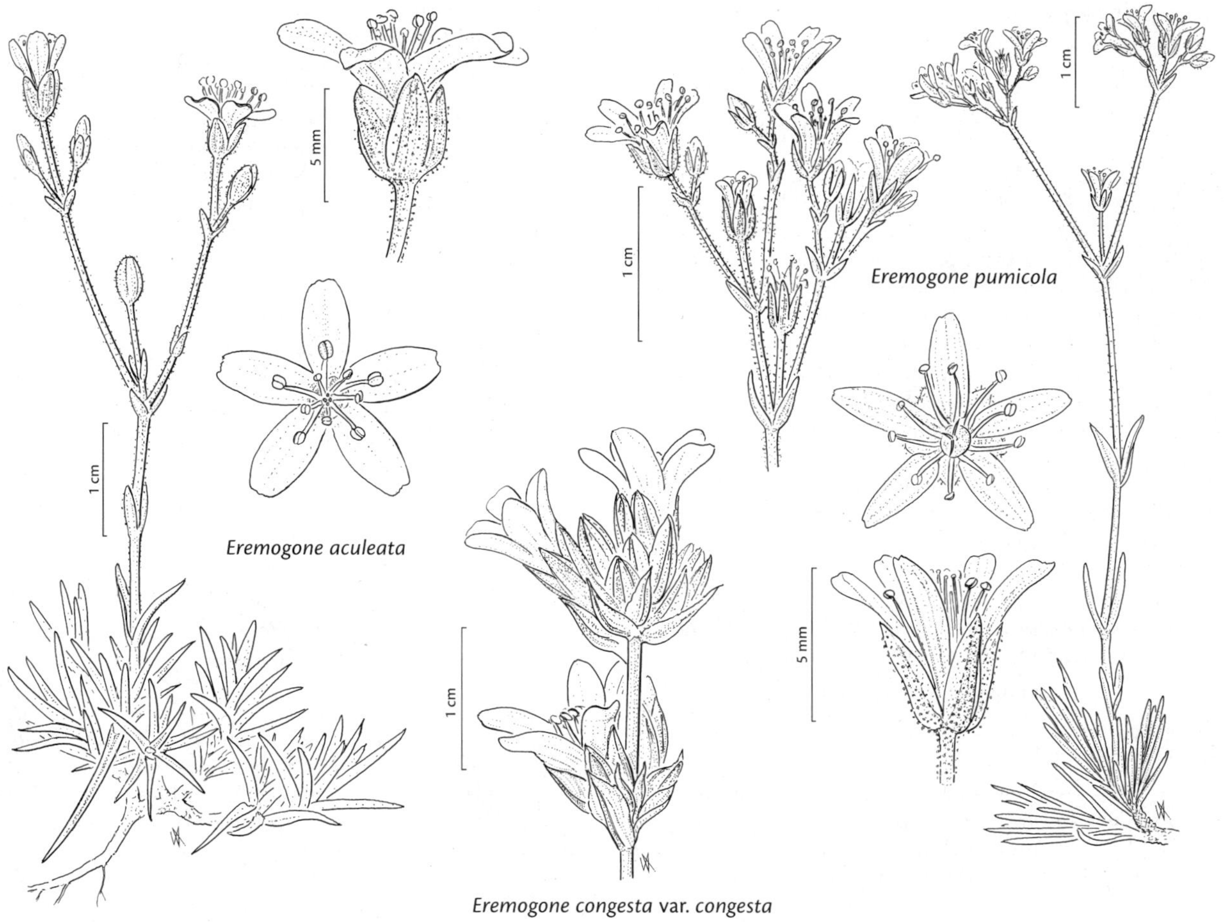

EREMOGONE

EREMOGONE Sandwort

Plants perennial. **Stems** prostrate (non-flowering) or ascending to erect (flowering), sometimes mat-forming. **Leaves** opposite, needle-like to narrowly linear, often congested at or near bases of flowering stems, tips sharp pointed; petioles 0; stipules 0. **Inflorescences** terminal, open or congested cymes; bracts present. **Flowers** bisexual; sepals 5, ± free, glabrous to glandular-hairy; petals 5, white, tips entire or slightly notched; stamens 10, borne on hypanthium, staminodes 0; ovary sutures 6; styles 3. **Fruits** capsules, ovoid to urceolate, opening by 6 ascending to recurved teeth. **Seeds** 1–10, elliptic-ovoid, oblong, or sometimes spherical, 1.2–3(3.2) mm, grayish, brown, yellowish brown, or blackish, smooth to tuberculate.

North temperate regions, especially in Asia, Europe, and western North America. ~95 species; 6 species treated in *Flora.*

Key to *Eremogone* species:

1. Inflorescences usually dense or somewhat open capitate or subcapitate cymes.
    2. Plants green; cauline leaves 3–5 pairs, not closely overlapping; sepals 3–6.5 mm . . . . **_E. congesta_**
    2' Plants bluish green; cauline leaves 6–10 pairs, closely overlapping; sepals 5–12 mm . . . . **_E. franklinii_**

1' Inflorescences usually open cymes (flowers may be as few as 1–3).
    3. Basal leaves (2)4–8 cm, straight or often curving in one direction; capsule valves glandular-hairy . . . . **_E. capillaris_ var. _americana_**
    3' Basal leaves (0.5)1–3.5 cm, recurving in multiple directions; capsule valves glabrous.
        4. Sepal tips acute to acuminate . . . . **_E. kingii_ var. _glabrescens_**
        4' Sepal tips obtuse to rounded or barely acute.
            5. Basal leaves rigid, not fleshy; cauline leaves abruptly reduced above basal pair . . . . **_E. aculeata_**
            5' Basal leaves generally flexuous, fleshy; cauline leaves all of similar size . . . . **_E. pumicola_**

Eremogone aculeata (S. Watson) Ikonnikov (p. 551)

Prickly sandwort

Arenaria aculeata S. Watson, *Arenaria aculeata* S. Watson var. *aculeata*, *Arenaria aculeata* S. Watson var. *uintahensis* (A. Nelson) M. Peck, *Arenaria salmonensis* L.F. Hend., *Arenaria uintahensis* A. Nelson

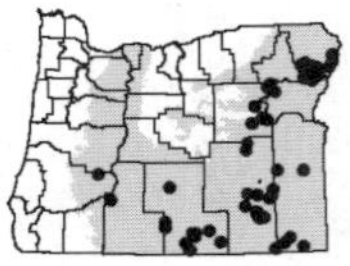

Plants mat-forming, glaucous. **Stems** 7–20 cm, glandular-hairy. **Leaves** basal ascending or arching, 10–35 mm, rigid, not fleshy, tips spinose; cauline 1–3 pairs, not closely overlapping, abruptly reduced above basal pair. **Inflorescences** open; flowers 5–25+; pedicels 3–25 mm, glandular-hairy. **Flowers** sepals 3–4.5 mm, to 6 mm in fruit, tips obtuse to rounded, rarely abruptly acute; petals 4.5–10 mm. **Fruits** 5–9 mm, valves glabrous. **Seeds** elliptic-oblong, compressed, 2–2.5 mm, yellow-tan to gray; tubercles low, rounded, elongate. **2*n*=22.**

Rocky, gravelly, or lava slopes, coniferous forests, high meadows. Flowering May–Sep. 1200–3000 m. BR, BW, Casc, ECas, Owy. CA, ID, NV, WA; northeast to MT, southeast to UT. Native.

Eremogone capillaris (Poiret) Fenzl var. ***americana*** (Maguire) R.L. Hartman & Rabeler

Slender mountain sandwort

Arenaria capillaris Poir. var. *americana* (Maguire) R.J. Davis, *Arenaria formosa* (Fisch.) Regel

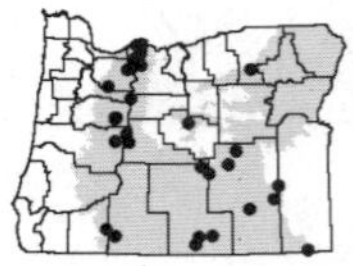

Plants tufted or sometimes mat-forming, green. **Stems** 5–20(30) cm, glabrous or glabrous below and glandular-hairy above. **Leaves** basal straight or often curved in one direction, 20–80 mm, flexuous, not fleshy, tips spinose; cauline 2–5 pairs, gradually reduced above. **Inflorescences** open; flowers 1–12(18); pedicels (2)5–20 mm, glandular-hairy. **Flowers** sepals 3–4.7 mm, to 5 mm in fruit, tips broadly rounded to barely acute; petals 6–10 mm. **Fruits** 5–8 mm, valves glandular-hairy. **Seeds** elliptic to ovoid, (1.2)1.5–2(2.5) mm, brown to black, tuberculate. **2*n*=22.**

Coniferous forests, rocky and gravelly slopes. Flowering May–Sep. 100–2800 m. BR, BW, Casc, ECas. ID, NV, WA; north to British Columbia, northeast to Alberta and MT. Native.

Eremogone capillaris var. *americana* is known from forested regions of the northern Cascades. Specimens identified as this species in eastern Oregon are likely from populations that Baad (1969) noted as intermediate between *E. capillaris* var. *americana* and *E. kingii* var. *glabrescens*. While they may bear some resemblance to this taxon, most lack glandular hairs on the capsules and pedicels and are better considered as *E. kingii* var. *glabrescens* or occasionally *E. congesta* var. *prolifera*.

Eremogone congesta (Nuttall) Ikonnikov

Ballhead sandwort

Plants tufted, green. **Stems** 5–40(50) cm, generally glabrous. **Leaves** basal erect to somewhat spreading, 20–80(140) mm, flexuous or rigid, fleshy or generally not, tips sharply acute to spinose; cauline 3–5 pairs, not closely overlapping, gradually reduced above. **Inflorescences** generally dense, capitate or subcapitate; flowers 3–50+; pedicels 0 or 0.1–6(15) mm, generally glabrous. **Flowers** sepals 3.5–6 mm, not enlarging in fruit, tips rounded to obtuse or acute to acuminate or spinose; petals 5–8 mm. **Fruits** 3.5–6 mm, valves glabrous. **Seeds** widely elliptic to ovoid, 1.4–3 mm, red-brown, tuberculate.

Eremogone congesta var. *cephaloidea*, known chiefly from eastern Washington and adjacent Idaho, may exist in Oregon. It would resemble var. *congesta* but with a pyramidal inflorescence of flowers with acute- to acuminate-tipped sepals.

3 varieties.

Key to *Eremogone congesta* varieties:

1. Pedicels 1–6(15) mm; inflorescences subcapitate to open . . . . ***E. c.*** var. ***prolifera***
1' Pedicels 0–0.2(2) mm; inflorescences dense, capitate, rounded or pyramidal.
   2. Leaves 2–3 cm, needle-like, fleshy . . ***E. c.*** var. ***crassula***
   2' Leaves 3–8(14) cm, filiform, not fleshy . . . . ***E. c.*** var. ***congesta***

Eremogone congesta (Nuttall) Ikonnikov var. ***congesta*** (p. 551)

Ballhead sandwort

Arenaria congesta Nutt. var. *congesta*

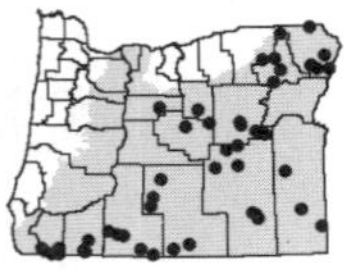

Leaves basal 30–80 mm, not fleshy. **Inflorescences** tightly dense, capitate, rounded; pedicels 0–0.2(2) mm. **Flowers** sepals 3.5–4.2 mm, tips rounded to obtuse. **2*n*=22.**

Gravelly or sandy soil, open grassy or forested slopes. Flowering May–Aug. 900–2800 m. BR, BW, Casc, ECas, Lava, Owy, Sisk. CA, ID, NV, WA; northeast to MT, east to CO. Native.

Eremogone congesta (Nuttall) Ikonnikov var. ***crassula*** (Maguire) R.L. Hartman & Rabeler

Ballhead sandwort

Arenaria congesta Nutt. var. *crassula* Maguire

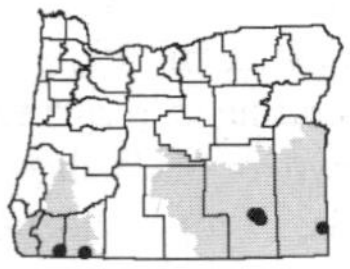

Leaves basal 20–30 mm, fleshy. **Inflorescences** tightly dense, capitate, rounded; pedicels 0–0.1 mm. **Flowers** sepals 3.5–4 mm, tips obtuse.

Dry ridges, rocky areas. Flowering Jun–Jul. 1300–3000 m. BR, Owy, Sisk. CA. Native.

Eremogone congesta (Nuttall) Ikonnikov var. ***prolifera*** (Maguire) R.L. Hartman & Rabeler

Ballhead sandwort

Arenaria burkei Howell, *Arenaria congesta* Nutt. var. *prolifera* Maguire

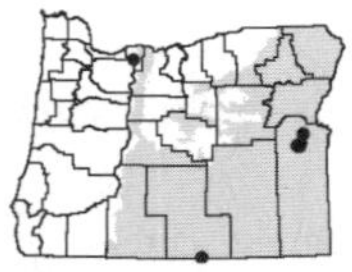

Leaves basal (20)30–80 mm, not fleshy. **Inflorescences** subcapitate to ± open; pedicels 1–6(15) mm. **Flowers** sepals 4.5–6 mm, tips acute to acuminate or spinose.

Diatomite or ashy hills, sandy areas. Flowering May–Aug. 800–2400 m. BR, BW, ECas, Owy. ID, NV, WA; Alberta. Native.

Eremogone franklinii (Douglas ex. Hooker) R.L. Hartman & Rabeler

Franklin's sandwort

Plants mostly tufted, bluish green. **Stems** 3–10(15) cm, glabrous. **Leaves** basal arching, (6)10–20 mm, rigid, not fleshy, tips spinose; cauline 6–10 pairs, closely overlapping, not reduced above. **Inflorescences** subcapitate, dense or somewhat open; flowers 3–45+; pedicels 0.1–4 mm, glabrous. **Flowers** sepals 5–12 mm, not enlarging in fruit, tips acuminate; petals 7–9 mm. **Fruits** 2.3–3.3 mm, valves glabrous. **Seeds** pyriform, 1.2–1.7 mm, tuberculate.

2 varieties.

Key to *Eremogone franklinii* varieties:

1. Inflorescences tightly congested, pedicels rarely exceeding 1 mm; sepals (7)8.5–12 mm; petals < sepals. . . . . ***E. f.*** var. ***franklinii***

1' Inflorescences somewhat congested, pedicels 1–4 mm; sepals 5–8 mm; petals slightly ≥ sepals . . . . ***E. f.*** var. ***thompsonii***

Eremogone franklinii (Douglas ex. Hooker) R.L. Hartman & Rabeler var. *franklinii*

Franklin's sandwort

Arenaria franklinii Douglas ex Hook. var. *franklinii*

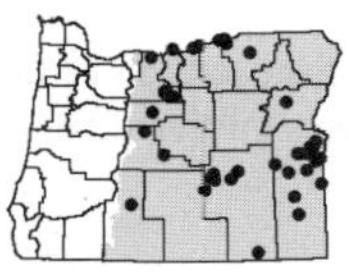

Inflorescences tightly congested; pedicels rarely exceeding 1 mm. **Flowers** sepals (7)8.5–12 mm; petals 7 mm, < sepals.

Sand hills, rocky slopes, sagebrush steppe. Flowering Apr–Jun. 50–1600 m. BR, BW, Col, ECas, Lava, Owy. ID, NV, WA. Native.

Eremogone franklinii (Douglas ex. Hooker) R.L. Hartman & Rabeler var. *thompsonii* (M. Peck) R.L. Hartman & Rabeler

Thompson's sandwort

Arenaria franklinii Douglas ex Hook. var. *thompsonii* M. Peck

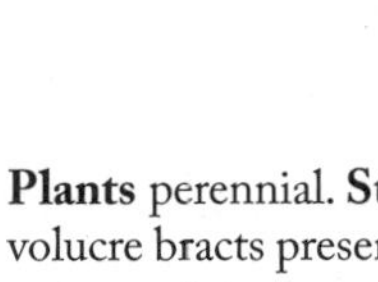

Inflorescences somewhat congested; pedicels 1–4 mm. **Flowers** sepals 5–8 mm; petals 7–9 mm, slightly ≥ sepals.

Dry sandy soil. Flowering May–Jun. 800–1100 m. Col. WA. Native.

Variety *thompsonii* is known from only a few sites in Gilliam County as well as two in Washington (Benton and Grant counties).

Eremogone kingii (S. Watson) Ikonnikov var. *glabrescens* (S. Watson) Dorn

King's sandwort

Arenaria kingii (S. Watson) M.E. Jones var. *glabrescens* (S. Watson) Maguire

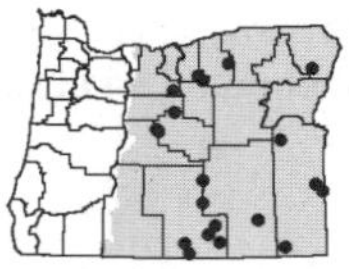

Plants tufted, green. **Stems** 1–20(25) cm, glandular-hairy or glabrous. **Leaves** basal erect, ascending, or somewhat spreading, recurving in multiple directions, 3–20 mm, not fleshy, tips generally spinose; cauline (1–2)4+ pairs, not closely overlapping, sometimes reduced above. **Inflorescences** open; flowers (1–2)3–13; pedicels 2–15 mm, glabrous or densely glandular-hairy. **Flowers** sepals 2.8–5 mm, barely enlarging in fruit, tips acute to acuminate; petals 4–7 mm. **Fruits** 4.5–7 mm, valves glabrous. **Seeds** spherical or elliptic-oblong to ovoid, 1.2–2.1 mm, red-brown to dark purple or black, tuberculate, sometimes papillate on dorsal ridge.

Rocky slopes, summits, dry foothills, basalt flats. Flowering Apr–Jul. 300–2600 m. BR, BW, Col, ECas, Lava, Owy. CA, ID, NV; southeast to UT. Native.

Some plants in eastern Washington may appear to be intermediate with *E. capillaris* var. *americana* but usually lack the glandular hairs found on the capsules and pedicels of that species.

Eremogone pumicola (Coville & Leiberg) Ikonnikov (p. 551)

Crater Lake sandwort, pumice sandwort

Arenaria pumicola Coville & Leiberg

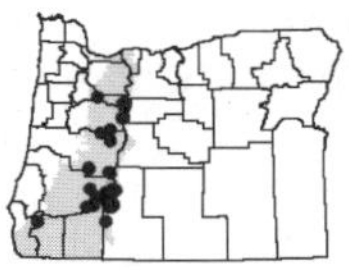

Plants tufted, green. **Stems** 10–20 cm, glandular-hairy. **Leaves** basal ascending to somewhat spreading, flexuous, recurving in multiple directions, 15–35 mm, generally fleshy, tips blunt to apiculate; cauline (2)3–5 pairs, not closely overlapping, not reduced above. **Inflorescences** open; flowers 7–40+; pedicels 3–20 mm, glandular-hairy. **Flowers** sepals 3–3.8 mm, 4–4.5 mm in fruit, tips obtuse or rounded; petals 2–4.5 mm. **Fruits** 4.5–5.5 mm, valves glabrous. **Seeds** oblong to pyriform, 1.8–2.4 mm, smooth to tuberculate.

Rocky flats, talus slopes, pumice soils, coniferous forests. Flowering May–Sep. 1300–2400 m. Casc, Sisk. Native.

Eremogone pumicola is endemic to Oregon.

GYPSOPHILA Baby's-breath

Plants perennial. **Stems** erect. **Leaves** opposite, lanceolate; petioles 0; stipules 0. **Inflorescences** terminal panicle-like cymes; involucre bracts present. **Flowers** bisexual; sepals 5, fused below, glabrous, calyx tubes cup-shaped, 1–3 mm, green and white, veins 5, not prominent, commissures veinless, scarious, calyx lobes < tube; petals 5, white, limbs entire or notched; appendages 0; stamens 10, borne on nectariferous disk, staminodes 0; styles and stigmas 2. **Fruits** capsules, opening by 4 ascending to recurved teeth. **Seeds** 2–5, reniform, 1.3–1.6 mm, black, coarsely tuberculate.

Asia, Europe, North America. ~150 species; 1 species treated in *Flora*.

Gypsophila paniculata Linnaeus (p. 554)
Baby's-breath

Gypsophila paniculata L. var. *paniculata*

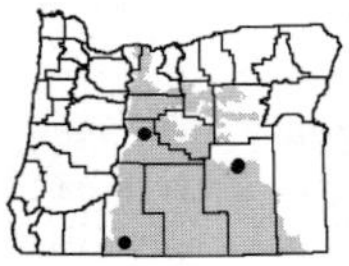

Plants perennial; from stout rhizomes. **Stems** 50–90 cm. **Leaves** lanceolate, 2–9 mm wide. **Inflorescences** highly branched, flowers many. **Flowers** petals 1–4 mm, white. **2*n***=34, 68.

Disturbed areas. Flowering Jun–Sep. 900–1400 m. BR, ECas, Lava. CA, ID, NV, WA; northern North America; Asia, Europe. Exotic.

Although known in all adjacent states and designated as a noxious weed in California and Washington, *G. paniculata* has apparently not become a major pest in Oregon. First collected in Klamath County in 1956, it is still only known from three sites.

HERNIARIA Rupturewort, herniary

Plants annual. **Stems** horizontal to ascending. **Leaves** opposite below, alternate above, oblanceolate to elliptic or suborbicular; petioles 0; stipules 2 per node, ovate to deltate, scarious. **Inflorescences** dense axillary cymes; bracts present. **Flowers** bisexual; sepals 5, free, glabrous or hairy, not awned, margins green, not scarious; petals 0; stamens 2–3, arising on hypanthium rim, staminodes 4–5, subulate to narrowly oblong, not petaloid, hair-like; styles 2, proximally united 33% of their length, stigmas 2. **Fruits** utricles, indehiscent. **Seeds** 1, ovoid to lenticular, 0.5–0.6 mm, dark red-brown, smooth, margin with prominent rim.

Africa, Asia, Mediterranean region, North America, Andean South America. 45 species; 2 species treated in *Flora*.

Key to *Herniaria* species:

1. Plants light to yellowish green; stems glabrous or puberulent; calyces not bur-like, sepals glabrous . . . . ***H. glabra***

1' Plants grayish green; stems densely covered with spreading hairs; calyces bur-like, sepals densely hairy . . . . ***H. hirsuta*** var. ***cinerea***

Herniaria glabra Linnaeus
Smooth rupturewort, green carpet

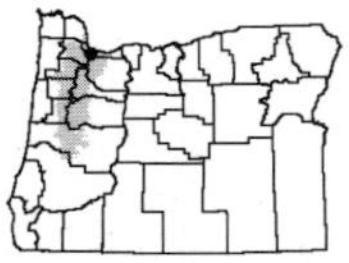

Plants light to yellowish green. **Stems** 5–35 cm, much branched from base, glabrous or puberulent. **Leaves** 3–7(10) mm; stipules 0.5–1.5 mm. **Flowers** calyces not bur-like, sepals 0.5–0.6 mm, glabrous. **2*n***=18, 36, 54, 72.

Disturbed gravelly areas. Flowering May–Sep. 0–50 m. WV. UT; northeastern US and adjacent Canada; Asia, Europe. Exotic.

Herniaria glabra was collected for the first time in the Pacific Northwest in Portland in 2014. It is sometimes cultivated as a lawn substitute.

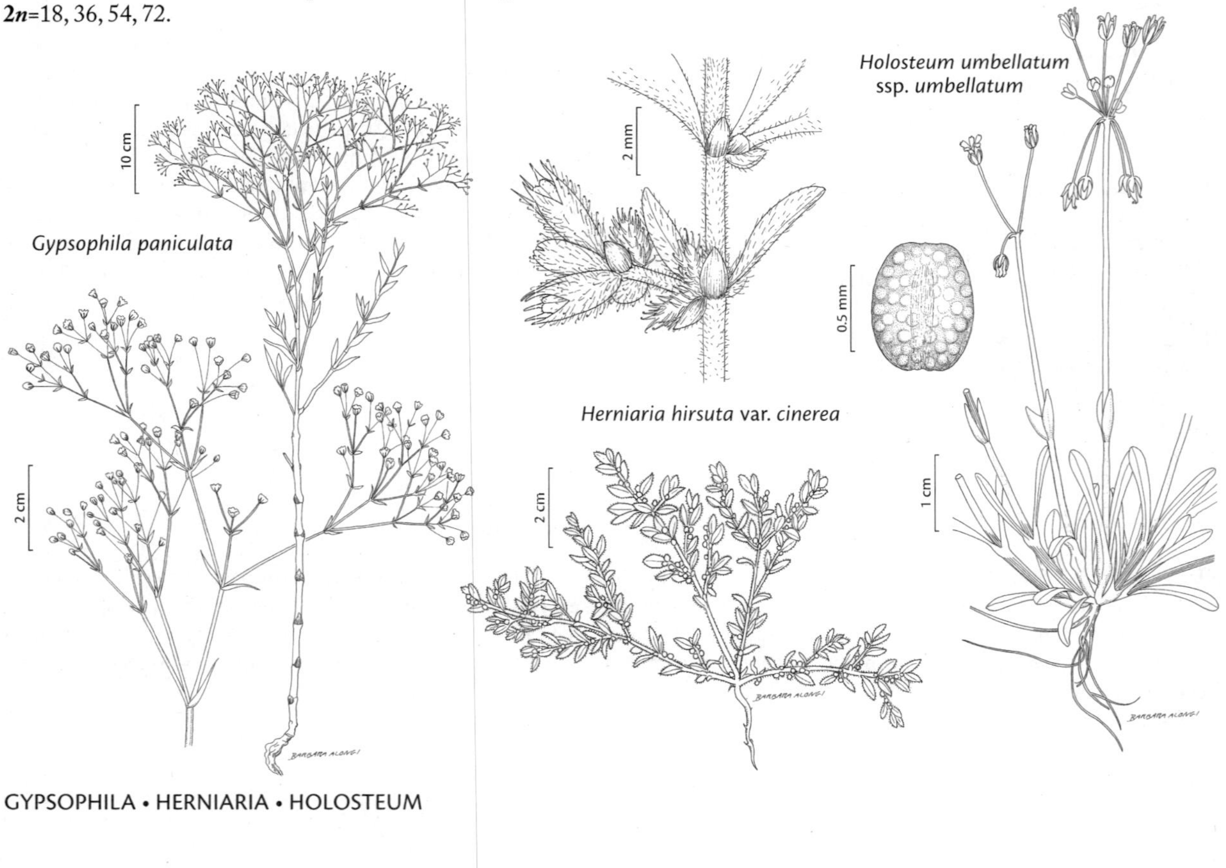

GYPSOPHILA • HERNIARIA • HOLOSTEUM

Herniaria hirsuta Linnaeus (p. 554)
var. ***cinerea*** (de Candolle) Loret & Barrandon

Hairy rupturewort

Herniaria hirsuta L. ssp. *hirsuta*

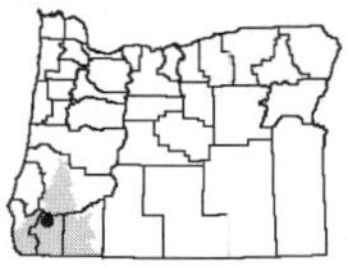

Plants grayish green. **Stems** 4–20 cm, much branched from base, hairs dense, spreading. **Leaves** 3–12 mm; stipules 0.4–1 mm. **Flowers** calyces bur-like, sepals 0.6–1.2 mm, covered with 2 kinds of hairs, long hairs 50–67% as long as sepals, tips of some or all hooked or tightly coiled, short hairs 25–33% as long as sepals, tips recurved, margins green and not scarious. **2*n***=36.

River sandbars. Flowering May–Sep. 200–500 m. Sisk. CA; AZ, MD; Africa, Asia, Mediterranean Europe. Exotic.

Herniaria hirsuta var. *cinerea* has been collected along the Rogue River on three occasions. It was collected once in Washington in 1979, but the population did not persist.

HOLOSTEUM Jagged chickweed

Plants annual. **Stems** ascending to erect. **Leaves** opposite and basal; basal oblanceolate to spatulate; cauline narrowly oblanceolate to oblong-ovate; petioles present (basal leaves) or 0; stipules 0. **Inflorescences** terminal, umbel-like cymes; bracts present. **Flowers** bisexual; sepals 5, free, glabrous; petals 5, tips irregularly toothed; stamens 3–5, borne at ovary base, staminodes 0; styles and stigmas 3. **Fruits** capsules, opening by 6 recurved teeth. **Seeds** 35–60, shield-shaped, 0.7–0.8 mm, brown, papillate.

Africa, Asia, Mediterranean Europe, North America. 3–4 species; 1 species treated in *Flora.*

Holosteum umbellatum Linnaeus (p. 554)
ssp. ***umbellatum***

Jagged chickweed

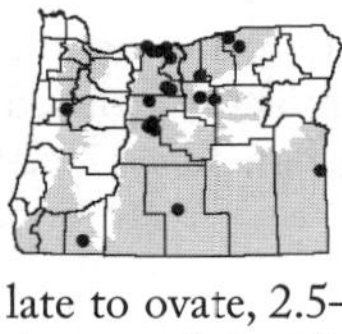

Stems 5–25 cm, unbranched, glabrous or glandular-pubescent more often proximally, often glaucous. **Leaves** 2–30 mm. **Inflorescences** bracts whorled; pedicels reflexed in flower, erect in fruit. **Flowers** sepals lanceolate to ovate, 2.5–4.5 mm; petals narrowly elliptic, 3.3–5 mm, tips rounded with 1–several irregular teeth. **2*n***=20, 40.

Disturbed areas, roadsides, railroad gravel, riverbanks, rocky or sandy areas. Flowering Mar–Jun. 0–1500 m. BR, Col, ECas, Lava, Owy, Sisk, WV. CA, ID, NV, WA; north to British Columbia, scattered in eastern North America; Africa, Asia, Mediterranean Europe. Exotic.

Known in North America since 1856, the first collections of *H. umbellatum* in the Pacific Northwest were made in Umatilla County in 1923. At least two populations (e.g., *Joyal 463*, OSC) are infected with an ovary smut (likely to be *Microbotryum holostei*). Infected plants are sterile and the "normal" contents of the ovary are replaced with reddish fungal spores.

HONCKENYA Seaside sandwort

Plants perennial. **Stems** prostrate to decumbent, flowering stems ascending or somewhat erect. **Leaves** opposite, conspicuously fleshy, elliptic-ovate to elliptic or oblanceolate; petioles 0; stipules 0. **Inflorescences** terminal cymes or flowers axillary and solitary; bracts present. **Flowers** unisexual; sepals 5(6), free; petals 5(6), white, tips emarginate; stamens 10, borne from nectariferous disk, staminodes 0 (male flowers) or 1–several (pistillate flowers); styles 3–5. **Fruits** capsules, globose, opening by 3 spreading valves. **Seeds** 3–15, compressed obovoid, 0.5–0.6 mm, reddish or yellowish brown, rugulose.

Temperate and arctic regions in Asia, Europe and North America. 1 species.

Honckenya peploides (Linnaeus) Ehrhart (p. 555)
ssp. ***major*** (Hooker) Hultén

Seaside sandwort, seaside sandplant

Arenaria peploides L. var. *major* Hook.

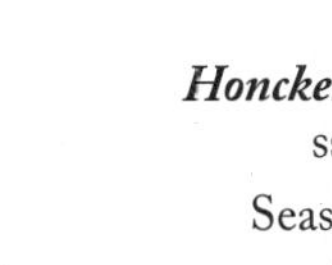

Stems 10–30(50) cm, simple or few-branched, glabrous. **Leaves** (10)20–46 mm. **Flowers** sepals narrowly ovate to elliptic, 4–6 mm; petals spatulate to oblanceolate, 0.8–2 mm (pistillate flowers) or 2.5–6 mm (staminate flowers). **2*n***=48, 64, 66, 68.

Sand dunes and coastal beaches. Flowering May–Sep. 0–50 m. Est. WA; north to AK; northeastern Asia. Native.

Honckenya peploides ssp. *major* occupies coastal areas along the northern Pacific Ocean, while the three other subspecies adjoin either the Arctic (one subspecies) or Atlantic oceans (two subspecies).

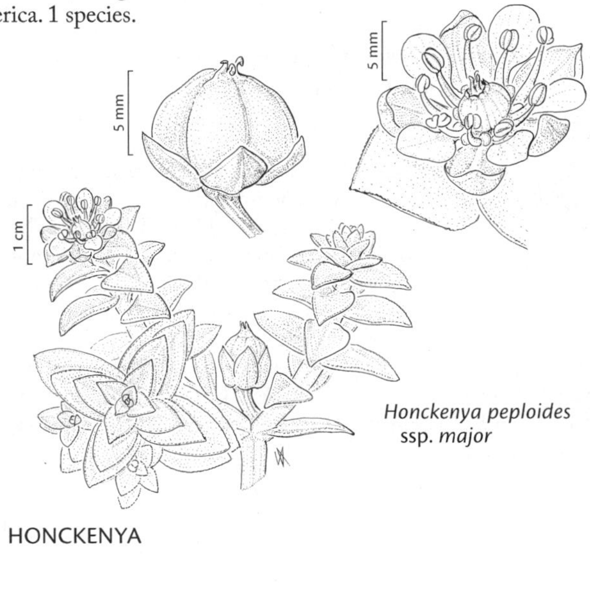

Honckenya peploides
ssp. *major*

HONCKENYA

LOEFLINGIA Loeflingia

Plants annual. **Stems** erect to horizontal. **Leaves** opposite, with axillary clusters of 2–4+ per node, awl-like to oblong; petioles 0; stipules 2 per node, scarious, partly fused to leaf bases, tips bristle-like. **Inflorescences** axillary, flowers 1(2); bracts 0. **Flowers** bisexual, sepals 5, usually free, glandular-hairy, with 2 lateral spurs; petals 0 or vestigial; stamens 3–5, arising from base of ovary, staminodes 0; styles 3, free, stigmas 3. **Fruits** 3-angled capsules, opening by 3 slightly recurved valves. **Seeds** 25–35, tear-shaped, 0.4–0.7 mm, tan with red-brown band on curved edge, minutely papillate on flat edge.

Northern Africa, southwestern Asia, southwestern Europe, North America. 7 species; 1 species treated in *Flora*.

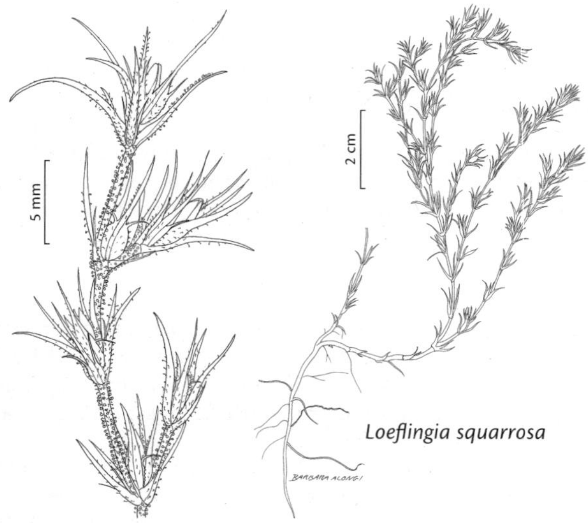

LOEFLINGIA

Loeflingia squarrosa Nuttall (p. 556)

Spreading-leaved loeflingia, California loeflingia, spreading pygmyleaf

Loeflingia squarrosa Nutt. var. *artemisiarum* (Barneby & Twisselm.) Dorn

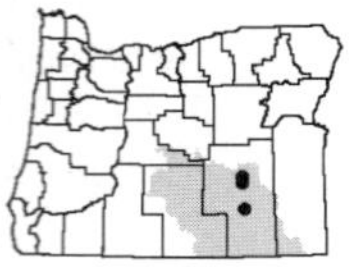

Stems 1–12 cm, much branched at base, glandular-hairy. **Leaves** 2–7 mm. **Flowers** cleistogamous; sepals linear to lanceolate, 2.7–6 mm.

Pasture, roadsides, sandy areas. Flowering May–Jul. 1200–1300 m. BR. CA, NV, WA; east to NE, southeast to AR and TX, south to northern Mexico. Native.

Plants in Oregon were formerly known as *Loeflingia squarrosa* var. *artemesiarum*. In our study of *Loeflingia* for *Flora of North America* (Hartman & Rabeler 2005), we found that the four varieties that Barneby and Twisselmann had recognized were more appropriately considered to be geographical races within a single species.

LYCHNIS Campion

Plants biennial or usually perennial. **Stems** erect. **Leaves** opposite; basal leaves oblanceolate with petioles present; cauline leaves lanceolate to narrowly elliptic, sessile; stipules 0. **Inflorescences** terminal cymes; bracts present. **Flowers** bisexual; sepals 5, fused, densely silky-hairy to tomentose, tubes prominent, grayish white, strongly 10-veined (obscured by hairs), commissures between adjacent sepals 1-veined, grayish green, lobes < tube; petals 5, red-purple, limbs widely notched; appendages 2; stamens 5, borne at ovary base, staminodes 0; styles and stigmas (4)5. **Fruits** capsules, opening by (4)5 ascending teeth. **Seeds** globose, 1–1.5 mm, gray to black-purple, tuberculate.

Northern Africa, Asia, Europe, North America. 30 species; 1 species treated in *Flora*.

Lychnis coronaria (Linnaeus) Desrousseaux (p. 556)

Mullein pink, rose campion

Silene coronaria (L.) Clairv.

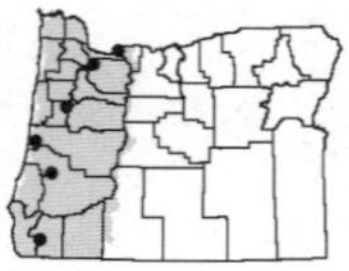

Stems 40–100 cm, sparingly branched above, covered with dense, silky hairs. **Leaves** basal 9–15 cm; cauline 5–15 cm. **Flowers** calyx lobes twisted, linear to lanceolate, 4–7 mm, tubes elliptic to ovoid, 12–15 mm; petals 22–30 mm, limbs exserted, rounded, 10–15 mm; appendages awl-like, 2–4 mm, thickened. $2n$=24.

Roadsides, railroad rights-of-way, woods. Flowering Jun–Aug. 0–1300 m. Casc, CR, Sisk, WV. CA, ID, WA; north to British Columbia, northeast to MT, eastern North America; Europe. Exotic.

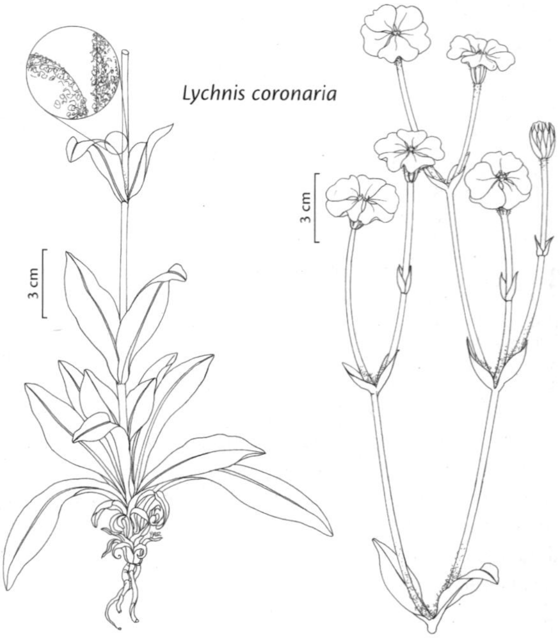

LYCHNIS

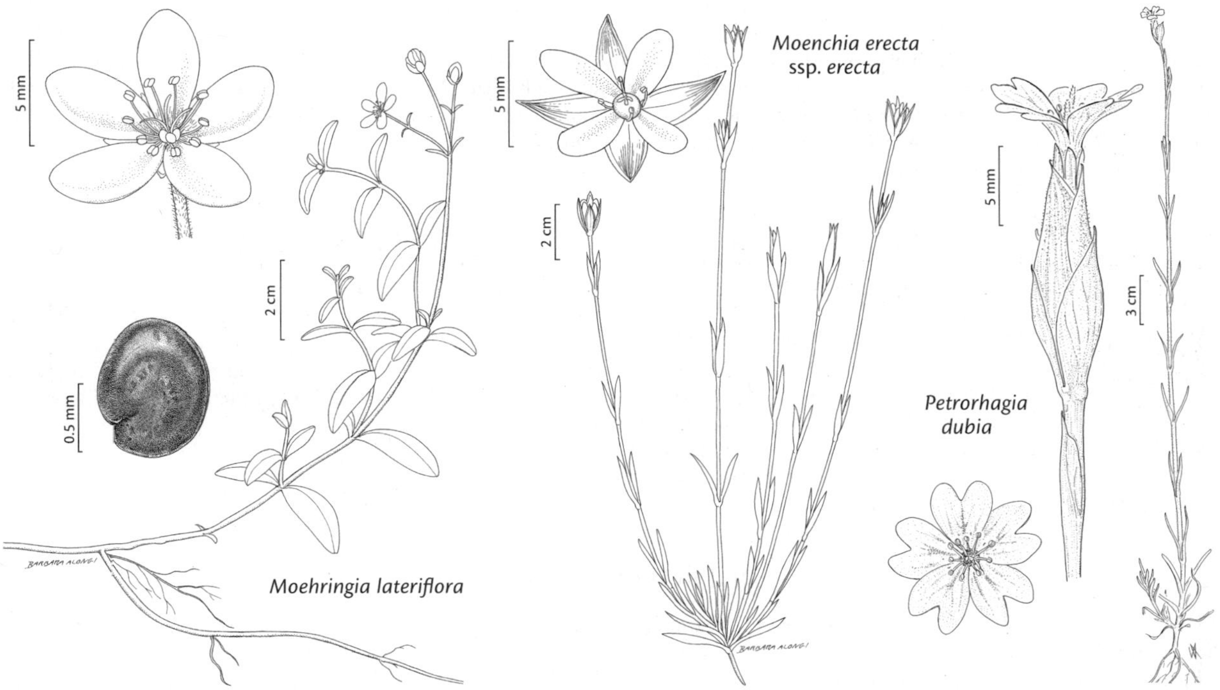

MOEHRINGIA • MOENCHIA • PETRORHAGIA

MOEHRINGIA Sandwort

Plants perennial; rhizomes branched, slender. **Stems** ascending or decumbent to erect. **Leaves** opposite, not congested at or near bases of flowering stems; oblanceolate to elliptic, tips obtuse or rounded or acute; petioles present or 0; stipules 0. **Inflorescences** terminal or axillary cymes; bracts present. **Flowers** bisexual; sepals 5, free, glabrous to ciliate; petals 5, white, tips entire; stamens 10, borne on nectariferous disk, staminodes 0; ovary sutures 6; styles 3. **Fruits** capsules, ovoid or subglobose, opening by 6 revolute teeth. **Seeds** 2–6, widely elliptic or reniform, 1–2.2 mm, red-brown, smooth or minutely tuberculate; appendage elliptic, spongy.

North temperate Asia, Europe, North America. 25 species; 2 species treated in *Flora.*

Key to *Moehringia* species:

1. Stems uniformly retrorsely-hairy; sepals 1.7–2.8(3) mm, tips mostly rounded or obtuse, herbaceous portion of sepals oblong to elliptic; petals ~2 × as long as sepals. . . . . ***M. lateriflora***

1' Stem hairs minute, peg-like, spreading; sepals (2.8)3–6 mm, tips acute to acuminate, herbaceous portion of sepals lanceolate; petals 0.75–1.5 × as long as sepals . . . . ***M. macrophylla***

Moehringia lateriflora (Linnaeus) Fenzl (p. 557)

Blunt-leaved sandwort

Arenaria lateriflora L.

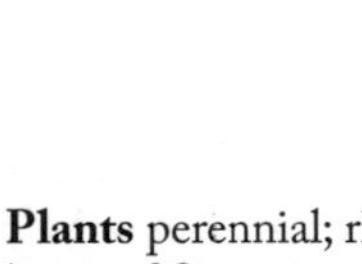

Stems ascending or decumbent, terete, 5–30 cm, often branched, uniformly retrorsely-hairy. **Leaves** broadly elliptic to oblong-elliptic or oblanceolate, 6–30 mm, tips obtuse or rounded. **Inflorescences** bracts 1–3 mm, margins scarious. **Flowers** sepals 1.7–2.8(3) mm, midrib not keeled, herbaceous portion of sepals oblong to elliptic, margins narrowly scarious, tips mostly obtuse or rounded; petals 3–6 mm, ~2 × as long as sepals. **Seeds** reniform, 1 mm, smooth. **2*n***=48.

Damp woods, moist slopes. Flowering May–Aug. 200–1700 m. BR, BW, ECas, Lava, WV. ID, NV, WA; northern North America; Asia, Europe. Native.

Moehringia macrophylla (Hooker) Fenzl

Big-leaved sandwort

Arenaria macrophylla Hook.

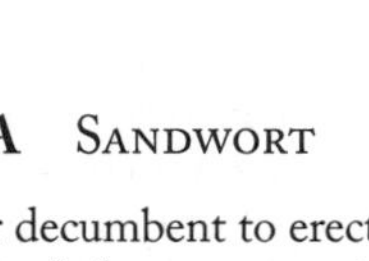

Stems ascending to erect, angled or grooved, 2–18 cm; hairs minute, peg-like, spreading. **Leaves** lanceolate to elliptic, 15–50 mm, tips acute. **Inflorescences** bracts 1–4 mm, margins widely scarious, minutely ciliate. **Flowers** sepals (2.8)3–6 mm, midrib keeled, herbaceous portion of sepals lanceolate, margins scarious, tips acute to acuminate; petals 2–6 mm, 0.75–1.5 × as long as sepals. **Seeds** widely elliptic, 1.5–2.2 mm, minutely tuberculate. **2*n***=48.

Coniferous woods, hillsides and dry slopes, roadside banks. Flowering Feb–Aug. 0–2500 m. BW, Casc, CR, ECas, Lava, Sisk, WV. CA, ID, WA; scattered across northern and western US and most of Canada; Asia. Native.

MOENCHIA Upright chickweed

Plants annual. **Stems** erect. **Leaves** opposite and basal, linear to oblanceolate; petioles present (basal leaves) or 0; stipules 0. **Inflorescences** terminal cymes or flowers 1–few; bracts present; usually pedicellate. **Flowers** bisexual; sepals 4, free, glabrous; petals (0)4, white, tips entire; stamens 4 or 8, borne at ovary base, staminodes 0; styles and stigmas 4. **Fruits** capsules, cylindric, ≤ sepals, opening by 8 recurved teeth. **Seeds** 35–55, reniform, 0.7–0.9 mm, red-brown, papillate.

Mediterranean Europe, North America. 3 species; 1 species treated in *Flora*.

Moenchia erecta (Linnaeus) P. Gaertner, B. Meyer & Scherbius ssp. ***erecta*** (p. 557)

Munchkin chickweed

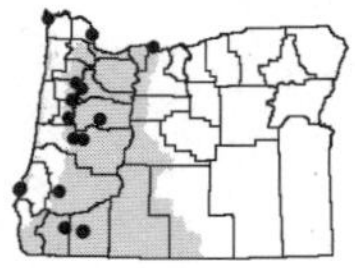

Stems 2.5–15 cm, simple or sparingly branched above, glabrous, glaucous. **Leaves** basal leaves oblanceolate, 5–18 mm; cauline leaves linear to linear-lanceolate, 3–12 mm. **Flowers** sepals lanceolate, 3.8–7 mm; petals lanceolate, 2.5–6 mm, tips blunt to rounded. **Fruits** straight, 5–7 mm. **2*n***=38.

Roadsides, moist meadows, rocky flats. Flowering Apr–May. 0–400 m. Casc, ECas, Est, Sisk, WV. CA, WA; north to British Columbia, IL, MT, SC; Europe. Exotic.

First collected in the 1830s in Baltimore and Philadelphia, most *Moenchia erecta* collections in North America have been made since the late 1940s (including the first in Oregon in 1948) from widely scattered sites in the Pacific Northwest.

PETRORHAGIA Pink

Plants annual. **Stems** erect. **Leaves** opposite, bases sheathing, linear to linear-lanceolate to oblanceolate; petioles 0; stipules 0. **Inflorescences** terminal, head-like cymes; bracts 2–6, widely ovate, usually 5–12 mm wide, red- to brown-scarious, enclosing flowers. **Flowers** bisexual; sepals 5, fused, glabrous to sparsely hairy, tubes prominent, cylindric, greenish or reddish, 15-veined, commissures between adjacent sepals veinless, scarious, lobes < tube; petals 5, pink to purplish, often with one or several darker veins, limbs entire or 2-lobed; styles and stigmas 2. **Fruits** capsules, opening by 4 ascending to recurved valves. **Seeds** 8–15, shield- or helmet-shaped, 1–1.8 mm, black-brown to black, reticulate or conic-papillate.

Central Asia, Mediterranean region, North America. 33 species; 1 species treated in *Flora*.

Two additional species, closely resembling and related to *Petrorhagia dubia*, are to be expected in Oregon. Thus far known in Oregon from a single collection from Jackson County, *Petrorhagia prolifera* (childing pink) is distinguished from *P. dubia* by a shorter (1–2 mm) leaf sheath, petals that lack darkened veins, and reticulate, shield-shaped seeds generally 1.1–1.8 mm. *Petrorhagia nanteuilii*, a stable hybrid between *P. dubia* and *P. prolifera*, is known in North America from two sites in northern California and one in southern British Columbia and might be expected in Oregon. The taxon is best noted by having a 3–4 mm leaf sheath, petals with 1–3 darkened veins, and tuberculate, shield-shaped seeds generally 1.5–1.8 mm—features that are intermediate between the two parent taxa.

Petrorhagia dubia (Rafinesque) G. López & Romo (p. 557)

Hairy pink

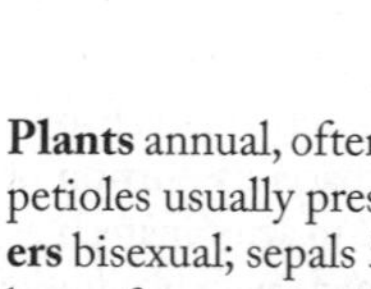

Stems 9.5–60 cm, middle internodes often densely glandular-tomentose. **Leaves** sheaths (3)4–9 mm, about 2–3 × longer than wide; blades linear to oblong, lowermost oblanceolate, 10–60 mm. **Inflorescences** involucral bracts mucronate. **Flowers** petals obcordate to 2-lobed, pink or pinkish, veins 5–6, darker pink. **Seeds** helmet-shaped, 1–1.4 mm, conic-papillate. **2*n***=30.

Sandy or gravelly river shores. Flowering Apr–May. 100–500 m. CR, Sisk. CA; TX to MS; northern Africa, southern Europe. Exotic.

Petrorhagia dubia is likely a recent introduction to southern Oregon; the first collection we have seen was made in 1991. It may have spread northward from California, where it was first collected in the late 1920s.

POLYCARPON Manyseed

Plants annual, often matted. **Stems** horizontal to erect. **Leaves** opposite, appearing in whorls of 4 or not, oblanceolate to obovate; petioles usually present; stipules 2 per node, lanceolate to triangular, scarious. **Inflorescences** axillary cymes; bracts absent. **Flowers** bisexual; sepals 5, free, glabrous, keeled and awned; petals 5, white, entire or notched; stamens 3–5, fused at base, arising from base of ovary, staminodes 0; styles 1, 3-branched, stigmas 3. **Fruits** capsules, opening by 3 incurving valves. **Seeds** 8–15, obliquely triangular, 0.4–0.5 mm, brown, granular.

Africa, Asia, Europe, western North America, South America. 15 species; 1 species treated in *Flora*.

Polycarpon tetraphyllum (Linnaeus) Linnaeus ssp. ***tetraphyllum*** (p. 559)

Four-leaved allseed, fourleaf manyseed

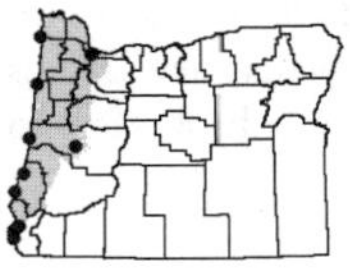

Stems 3–17 cm, often much branched, especially above, glabrous. **Leaves** blades 4–12 mm. **Flowers** sepals 1.5–2.5 mm, awns widely triangular; petals linear to elliptic, 0.4–0.8 mm.

Gravel river bars, roadsides, sidewalk cracks. Flowering May–Aug. 0–200 m. CR, Est, WV. CA; British Columbia, widely scattered in southeastern US; scattered worldwide. Exotic.

Similar to the earliest reports of *P. tetraphyllum* in eastern North America, this taxon was first collected in Oregon in 1880 along the Columbia River and on ballast near Portland, most recently in 1916. After a nearly 100-year gap, a number of populations have recently been found.

PSEUDOSTELLARIA Sticky starwort

Plants perennial; rhizomes 3–12 mm in diameter, with tuber-like thickenings. **Stems** ascending to erect. **Leaves** opposite, linear to lanceolate or elliptic; petioles 0; stipules 0. **Inflorescences** terminal or axillary cymes; bracts present. **Flowers** bisexual; sepals 5, free, glandular-hairy; petals 5, 2-lobed with v-shaped notch, white; stamens 10, borne at ovary base, staminodes 0; styles and stigmas 3. **Fruits** capsules, spherical, opening by 6 valves that are coiled 2–3 times. **Seeds** 1–3, broadly elliptic, 2–3.4 mm, red-brown, tuberculate with stipitate glands.

Eastern Asia, southeastern Europe, western United States. 21 species; 1 species treated in *Flora.*

Pseudostellaria jamesiana (Torrey) W.A. Weber & R.L. Hartman (p. 559)

Tuber starwort, sticky starwort

Stellaria jamesiana Torr.

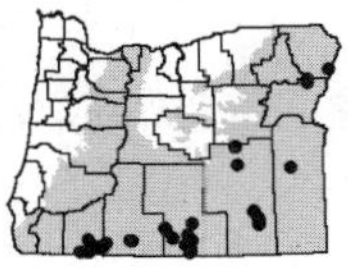

Stems 12–45 cm, simple to much branched, glandular-hairy, at least distally. **Leaves** 15–150 mm. **Flowers** sepals lanceolate to ovate, 3–7 mm; petals obdeltoid, 5–9.5 mm, tips blunt to rounded, 2-lobed. **2*n***=96.

Meadows, dry understory of aspen and pine forests, moist banks, rocky slopes. Flowering May–Aug. 1000–2300 m. BR, BW, Casc, ECas, Owy, Sisk. CA, ID, NV, WA; north to British Columbia, northeast to MT, east to CO, southeast to TX; Native.

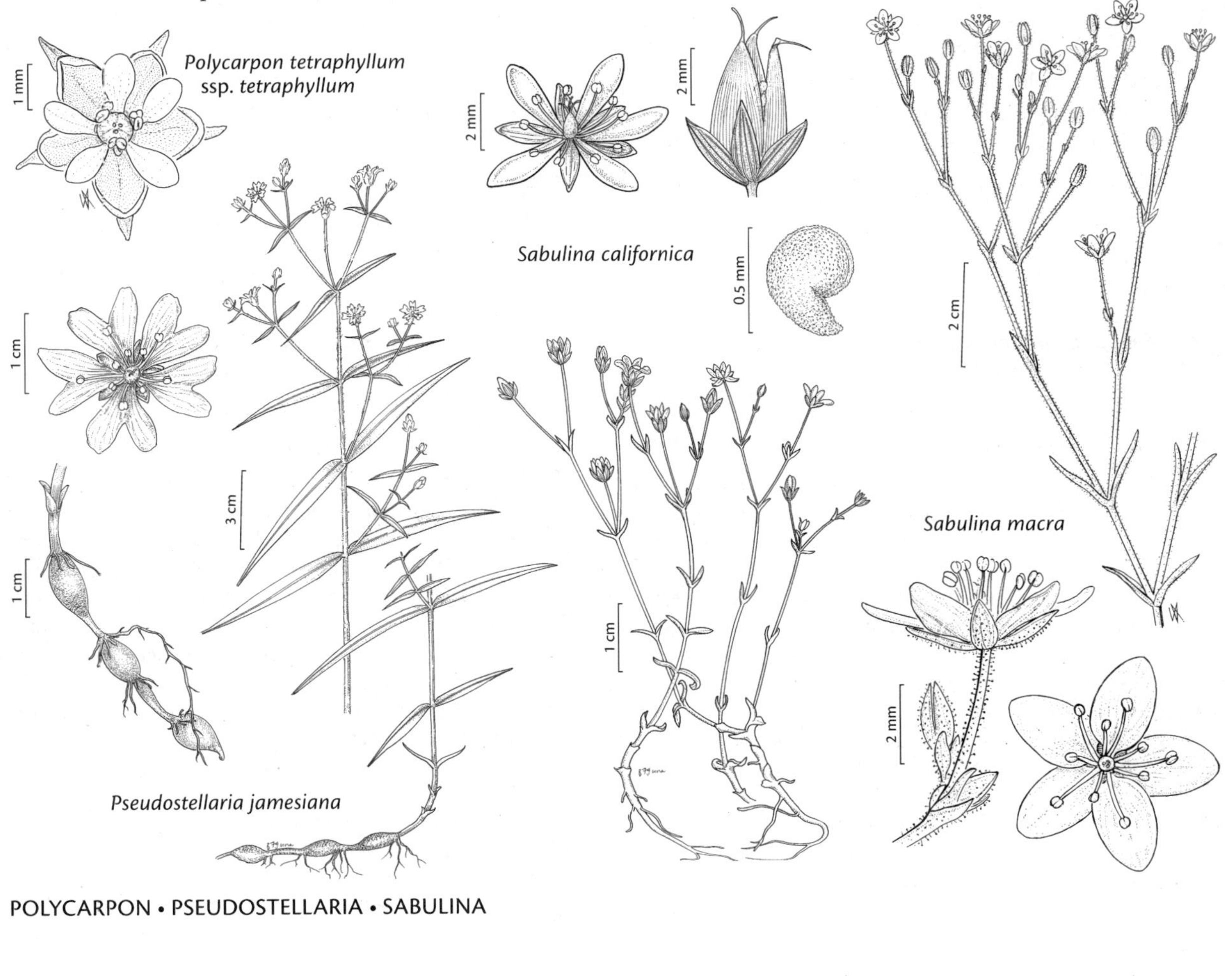

POLYCARPON • PSEUDOSTELLARIA • SABULINA

SABULINA Sandwort

Plants annual or perennial. **Stems** spreading and often mat-forming or ascending to erect. **Leaves** opposite, linear-filiform or needle-like to subulate or lanceolate-oblong; petioles 0; stipules 0. **Inflorescences** terminal cymes or sometimes flowers solitary, bracts present. **Flowers** bisexual; sepals 5, free, not hooded, glabrous to stipitate-glandular, margins not incurved, tips acute to acuminate or sometimes obtuse; petals (0)5, white, tips entire; stamens 10, arising from hypanthium, staminodes 0; ovary sutures 3; styles 3. **Fruits** capsules, ovoid or ellipsoid, opening by 3, incurved to recurved valves. **Seeds** 1–25, spherical or subspherical to reniform, 0.4–2.2 mm, brown to red-brown, tuberculate or papillate.

Asia Minor, Mediterranean region, North America, western South America. ~65 species; 9 species treated in *Flora.*

We follow Dillenberger & Kadereit (2014) in their treatment of *Minuartia s.s.* as being a European genus. Nine of our ten *Minuartia* species are now included in *Sabulina*, but, given the morphological diversity of those species, further work is needed to see if this concept should be retained or actually be split further. The tenth species, *Minuartia obtusiloba*, is treated in *Cherleria*.

Key to *Sabulina* species:

1. Plants annual, not cespitose or mat-forming.
    2. Pedicels stipitate-glandular; leaves (5)10–30 mm.
        3. Cauline leaves with evident axillary fascicles; sepals densely stipitate-glandular, veins 3, prominent; seeds 0.4–0.6 mm . . . . ***S. macra***
        3' Cauline leaves with few (proximal) or 0 axillary fascicles; sepals stipitate-glandular at least proximally, veins obscure or 1(3); seeds 1.3–2 mm.
            4. Leaves evenly spaced, thread-like, 0.2–0.4 mm wide, flexible, becoming curled; seeds red-brown, winged . . . . ***S. douglasii***
            4' Leaves mostly near bases of stems, linear-lanceolate, 1–1.5(2.5) mm wide, rigid, recurved; seeds brown to black-brown, not winged . . . . ***S. howellii***
    2' Pedicels glabrous; leaves 1.5–7(10) mm.
        5. Plants 1–5 cm; petals 0.5–1 × length of sepals or 0; sepal veins 1, or 3, faint, only near base . . . . ***S. pusilla***
        5' Plants usually 5–25 cm; petals 1–1.8 × length of sepals; sepal veins 3, faint, or 3(5), very prominent.
            6. Petals 1.5–1.8 × length of sepals; sepal veins 3, faint, lateral pair rarely extending to tip . . . . ***S. californica***
            6' Petals 1–1.3 × length of sepals; sepals veins 3(5), very prominent, lateral pair converging with midrib at tip . . . . ***S. cismontana***

1' Plants perennial, cespitose to mat-forming.
    7. Pedicels glabrous; flowers solitary . . . . ***S. austromontana***
    7' Pedicels stipitate-glandular; inflorescences 3+-flowered.
        8. Plants cespitose or mat-forming, stems ascending to erect; taproots not woody; leaves 3-veined below . . . . ***S. rubella***
        8' Plants mat-forming, rhizomes elongate and stems or stolons trailing; taproots woody, thickened; leaves 1-veined below . . . . ***S. nuttallii***

Sabulina austromontana (S.J. Wolf & Packer) Dillenberger & Kadereit

Rocky Mountain sandwort

Minuartia austromontana S.J. Wolf & Packer

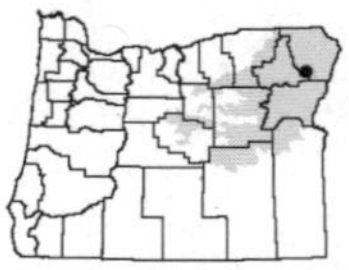

Plants perennial, mat-forming; taproots not woody. **Stems** spreading to erect, 3–15 cm, green, glabrous. **Leaves** linear to subulate, 3–10 × 0.1–1 mm, straight to outwardly curved, flexible, evenly spaced; axillary leaves present. **Inflorescences** flowers solitary, terminal; pedicels glabrous. **Flowers** sepals 2–3 mm, glabrous, veins 3; petals usually 0 or if present < sepals. **Seeds** subspherical, 0.6–1 mm, brown, tuberculate, not winged. **2*n***=30.

Boggy banks, rocks. Flowering Jun–Sep. 1200–2800 m. BW. ID; British Columbia, northeast to Alberta, MT, east to WY. Native.

Oregon populations of *S. austromontana* represent a disjunction for this Rocky Mountain taxon.

Sabulina californica (A. Gray) Dillenberger & Kadereit (p. 559)

California sandwort

Arenaria californica (A. Gray) W.H. Brewer, *Minuartia californica* (A. Gray) Mattf.

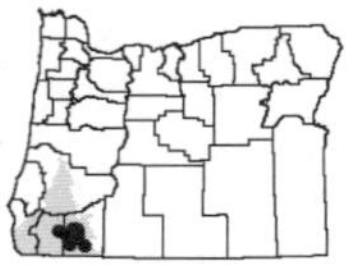

Plants annual; taproots thread-like. **Stems** widely spreading to erect, (1)2–12 cm, simple or much branched, green, glabrous. **Leaves** linear to subulate or narrowly oblong, 2–5 × 0.2–1.5 mm, straight, flexible, evenly spaced; axillary leaves absent. **Inflorescences** 5–7(9)-flowered, rarely solitary; pedicels glabrous. **Flowers** sepals 2.5–2.8 mm, stipitate-glandular proximally, veins 3, faint, lateral pair rarely extending to tip; petals 1.5–1.8 × length of sepals. **Seeds** subspherical, 0.4–0.5 mm, red-brown, papillate, not winged. **2*n***=26.

Gravelly, sandy slopes and flats, oak-madrone woodlands, grasslands, chaparral. Flowering Mar–May. 0–700 m. Sisk. CA. Native.

Sabulina cismontana (Meinke & Zika) Dillenberger & Kadereit

Cismontane sandwort

Minuartia cismontana Meinke & Zika

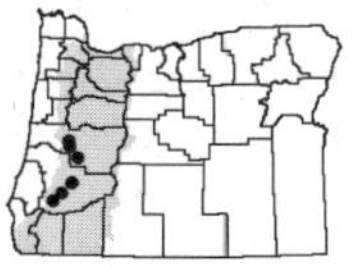

Plants annual; taproots thread-like. **Stems** widely spreading to erect, (5)8–20(25) cm, much branched, green or red-purple, glabrous. **Leaves** lance-attenuate to linear, 2–7(9) × 0.5–1.2(1.8) mm, straight to outwardly curved, flexible, evenly spaced; axillary leaves occasionally present. **Inflorescences** 5–20-flowered; pedicels glabrous. **Flowers** sepals 3.2–5.5 mm, glabrous, veins 3(5), very prominent, lateral pair converging with midrib at tip; petals 1–1.3 × length of sepals. **Seeds** reniform, 0.7–1 mm, brown to reddish, papillate, not winged.

Serpentine slopes. Flowering Apr–Jun. 300–1000 m. Casc, Sisk, WV. CA. Native.

Similar to *S. californica*, *S. cismontana* is found only on serpentine substrates in Oregon.

Sabulina douglasii (Torrey & A. Gray) Dillenberger & Kadereit

Douglas' stitchwort

Arenaria douglasii Fenzl ex Torr. & A. Gray, *Minuartia douglasii* (Torr. & A. Gray) Mattf.

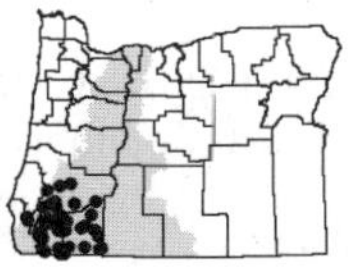

Plants annual; taproots thread-like. **Stems** erect or spreading, 4–30 cm, simple or often branched from base, green or purple, finely stipitate-glandular at least distally. **Leaves** thread-like, 5–30 × 0.2–0.4 mm, becoming curled, flexible, evenly spaced; cauline leaves with few (proximal) or 0 axillary fascicles; axillary leaves present proximally. **Inflorescences** 7–15+-flowered; pedicels stipitate-glandular. **Flowers** sepals 2.5–3.7 mm, stipitate-glandular proximally, veins 1(3); petals 1.7–2.1 × length of sepals. **Seeds** spherical, 1.3–2 mm, red-brown, tuberculate, winged.

Dry grassy areas, roadsides, forested slopes, flats in yellow pine-chaparral, often on serpentine. Flowering Apr–Aug. 100–1500 m. Casc, ECas, Sisk. CA; AZ. Native.

Sabulina howellii (S. Watson) Dillenberger & Kadereit

Howell's sandwort

Arenaria howellii S. Watson, *Minuartia howellii* (S. Watson) Mattf.

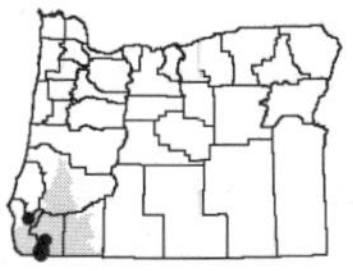

Plants annual; taproots not woody. **Stems** erect to spreading, 12–30 cm, simple or often branched from base, green or purple in fruit, finely stipitate-glandular. **Leaves** linear-lanceolate, (5)10–15 × 1–1.5(2.5) mm, recurved, rigid, mostly near stem bases; cauline leaves with few (proximal) or 0 axillary fascicles; axillary leaves absent. **Inflorescences** 5–25+-flowered; pedicels stipitate-glandular. **Flowers** sepals 1.9–3 mm, stipitate-glandular proximally, veins 1(3), obscure proximally; petals 1.8–2.3 × length of sepals. **Seed** subspherical, 1.4–1.7 mm, brown to black-brown, tuberculate, not winged.

Rocky chaparral, Jeffrey pine woodlands, serpentine slopes. Flowering Apr–Jun. 300–600 m. Sisk. CA. Native.

Most similar to *S. douglasii*, *S. howellii* has a more limited distribution and is only known from southern Oregon and the Klamath Range of California (chiefly Del Norte County).

Sabulina macra (A. Nelson & J.F. McBride) (p. 559) Dillenberger & Kadereit

Slender sandwort, slender stitchwort

Arenaria stricta Michx., *Arenaria stricta* Michx. var. *puberulenta* (M. Peck) C.L. Hitchc., *Arenaria tenella* Nutt., *Minuartia tenella* (Nutt.) Mattf.

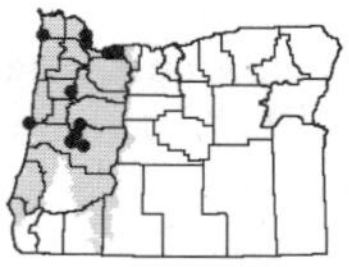

Plants annual; taproots thread-like. **Stems** erect, 5–25 cm, slender, simple or branched, green, stipitate-glandular at least distally. **Leaves** subulate to narrowly lanceolate, 5–17 × 0.5–1.5 mm, straight to outwardly curved, flexible, evenly spaced; cauline leaves with evident axillary fascicles; axillary leaves present. **Inflorescences** 7–25+-flowered; pedicels stipitate-glandular. **Flowers** sepals 2.5–3 mm, densely stipitate-glandular, veins 3, prominent; petals 1.5–2 × length of sepals. **Seeds** subspherical, 0.4–0.6 mm, brown, tuberculate, not winged. **2*n***=24.

Dry gravelly areas, rocky slopes, mossy balds. Flowering May–Jul. 0–1000 m. Casc, CR, Est, WV. WA; north to British Columbia. Native.

Sabulina nuttallii (Pax) Dillenberger & Kadereit

Nuttall's sandwort

Plants perennial; rhizomes elongate and stems or stolons trailing; taproots woody. **Stems** 2–20 cm, mat-forming; flowering stems ascending to erect; trailing stems to 60 cm, green, densely stipitate-glandular. **Leaves** needle-like to subulate, 5–20 × 0.3–1.5 mm, straight-appressed to arcuate-spreading, rigid, evenly spaced, 1-veined below; axillary leaves present. **Inflorescences** (3)6–30-flowered; pedicels stipitate-glandular. **Flowers** sepals 3.5–7 mm, stipitate-glandular, veins 1 or 3; petals 0.7–1.6 × length of sepals. **Seeds** oblong, 1.5–2.2 mm, red-brown to dark brown, tuberculate, not winged.

4 varieties.

Key to *Sabulina nuttallii* varieties:

1. Leaf blade tips acute to acuminate; sepal tips acuminate.
   2. Petals 0.5–1.2 × length of sepals; sepals lanceolate or narrowly lanceolate . . . . ***S. n.*** var. ***nuttallii***
   2' Petals 1.1–1.6 × length of sepals; sepals ovate to lanceolate . . . . ***S. n.*** var. ***gregaria***

1' Leaf blade and sepal tips spinescent.
   3. Leavs arcuate-spreading, 10–20 mm; sepal veins (1)3, equally prominent . . ***S. n.*** var. ***fragilis***
   3' Leaves appressed or sometimes arcuate-spreading, 5–7 mm; sepal veins 1(3), lateral pair 0 or less prominent . . . . ***S. n.*** var. ***gracilis***

Sabulina nuttallii (Pax) Dillenberger & Kadereit var. ***fragilis*** (Maguire & A.H. Holmgren) Dillenberger & Kadereit

Brittle sandwort

Arenaria nuttallii Pax ssp. *fragilis* Maguire & A.H. Holmgren, *Arenaria nuttallii* Pax var. *fragilis* (Maguire & A.H. Holmgren) C.L. Hitchc., *Minuartia nuttallii* (Pax) Briq. ssp. *fragilis* (Maguire & A.H. Holmgren) McNeill, *Minuartia nuttallii* (Pax) Briq. var. *fragilis* (Maguire & A.H. Holmgren) Rabeler & R.L. Hartman

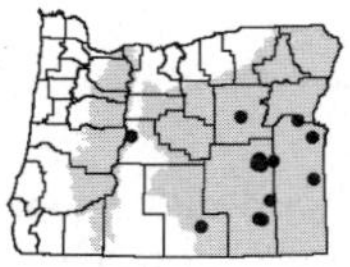

Leaves arcuate-spreading, 10–20 mm, tips spinescent. **Flowers** sepal veins (1)3, equally prominent, tips spinescent; petals 0.8–1.1 × length of sepals.

Stony, serpentine and talus slopes, sage prairies. Flowering May–Jul. 900–2800 m. BR, BW, Casc, Lava, Owy. CA, NV, WA. Native.

Sabulina nuttallii (Pax) Dillenberger & Kadereit var. ***gracilis*** (B.L. Robinson) Dillenberger & Kadereit

Brittle sandwort

Arenaria nuttallii Pax var. *gracilis* B.L. Rob. *Minuartia nuttallii* (Pax) Briq. ssp. *gracilis* (B.L. Rob.) McNeill, *Minuartia nuttallii* (Pax) Briq. var. *gracilis* (B.L. Rob.) Rabeler & R.L. Hartman

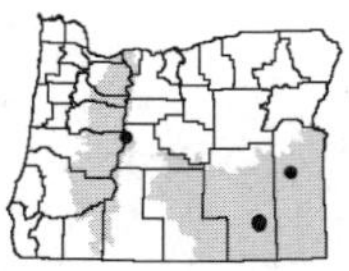

Leaves straight-appressed, sometimes arcuate-spreading, 5–7 mm, tips spinescent. **Flowers** sepal veins 1(3), lateral pair 0 or less prominent, tips spinescent; petals 0.7–0.9 × length of sepals.

Barren rock summits, grassy, open hillsides. Flowering May–Aug. 1200–2900 m. BR, Casc, Owy. CA, NV. Native.

Sabulina nuttallii (Pax) Dillenberger & Kadereit var. ***gregaria*** (A. Heller) Dillenberger & Kadereit

Brittle sandwort

Arenaria nuttallii Pax var. *gregaria* (A. Heller) Jeps., *Minuartia nuttallii* (Pax) Briq. ssp. *gregaria* (A. Heller) McNeill, *Minuartia nuttallii* (Pax) Briq. var. *gregaria* (A. Heller) Rabeler & R.L. Hartman

Leaves straight-appressed or arcuate-spreading, 5–10 mm, tips acute to acuminate. **Flowers** sepals ovate to lanceolate, veins 1 or 3, lateral pair 0 or less prominent, tips acuminate; petals 1.1–1.6 × length of sepals.

Gravelly, rocky, serpentine or talus slopes. Flowering Jun–Aug. 1000–2200 m. CR, Sisk. CA. Native.

Sabulina nuttallii var. *gregaria* is the only one of the four varieties found in southwestern Oregon, the others being chiefly eastern.

Sabulina nuttallii (Pax) Dillenberger & Kadereit var. ***nuttallii***

Nuttall's sandwort

Arenaria nuttallii Pax ssp. *nuttallii*, *Arenaria nuttallii* Pax var. *nuttallii*, *Minuartia nuttallii* (Pax) Briq. var. *nuttallii*

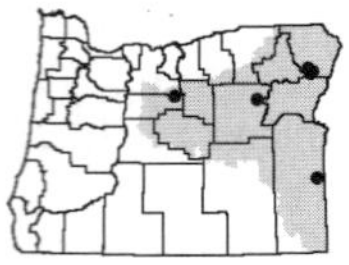

Leaves straight or slightly recurved, 5–7 mm, tips acute to acuminate. **Flowers** sepals lanceolate or narrowly lanceolate, veins (1)3, equally prominent, tips acuminate; petals 0.5–1.2 × length of sepals.

Bare talus slopes. Flowering May–Aug. 800–3000 m. BW, Lava, Owy. ID, WA; north to British Columbia, east to CO. Native.

Sabulina pusilla (S. Watson) Dillenberger & Kadereit

Annual sandwort

Arenaria pusilla S. Watson, *Minuartia pusilla* (S. Watson) Mattf.

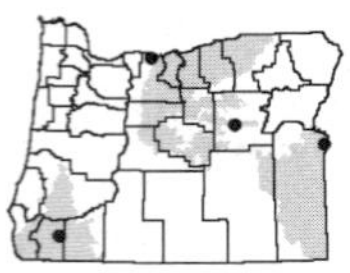

Plants annual; taproots thread-like. **Stems** spreading to erect, 1–5 cm, slender, simple or branched, green, glabrous. **Leaves** subulate to lanceolate, 1.5–5 × 0.2–1.5 mm, straight, flexible; veins generally 1, often faint. **Inflorescences** 2–9-flowered; pedicels glabrous. **Flowers** sepals 1.5–3.5 mm, glabrous, veins 1, or 3, faint, only near base; petals 0.5–1 × length of sepals or 0. **Seeds** reniform, 0.4–0.7 mm, purple-brown, papillate, not winged.

Sandy meadows. Flowering Apr–May. 50–1700 m. Col, Lava, Owy, Sisk. CA, ID, NV, WA; north to British Columbia, CO. Native.

A minute plant, *S. pusilla* is often overlooked.

Sabulina rubella (Wahlenberg) Dillenberger & Kadereit (p. 563)

Reddish sandwort, beautiful sandwort, boreal stitchwort

Arenaria propinqua Richardson, *Arenaria rubella* (Wahlenb.) Sm., *Minuartia rubella* (Wahlenb.) Hiern

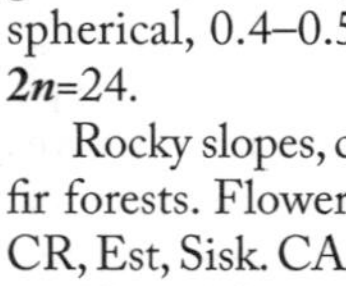

Plants perennial, cespitose; taproots not woody. **Stems** ascending to erect, 2–8(10) cm, green, densely stipitate-glandular. **Leaves** needle-like, 1.5–10 × 0.3–0.8 mm, straight, flexible, mostly near stem base, 3-veined below; axillary leaves present. **Inflorescences** 3–7+-flowered cyme or rarely solitary, terminal; pedicels densely stipitate-glandular. **Flowers** sepals 2.5–3.2 mm, stipitate-glandular, veins 3; petals 0.8–1 × length of sepals. **Seeds** subspherical, 0.4–0.5 mm, red-brown, tuberculate, not winged. **2*n*=24**.

Rocky slopes, cliffs, ridges, gravel creek bars, andesite ledges, fir forests. Flowering May–Sep. 100–3000 m. BR, BW, Casc, CR, Est, Sisk. CA, ID, NV, WA; throughout Canada and western United States; circumboreal. Native.

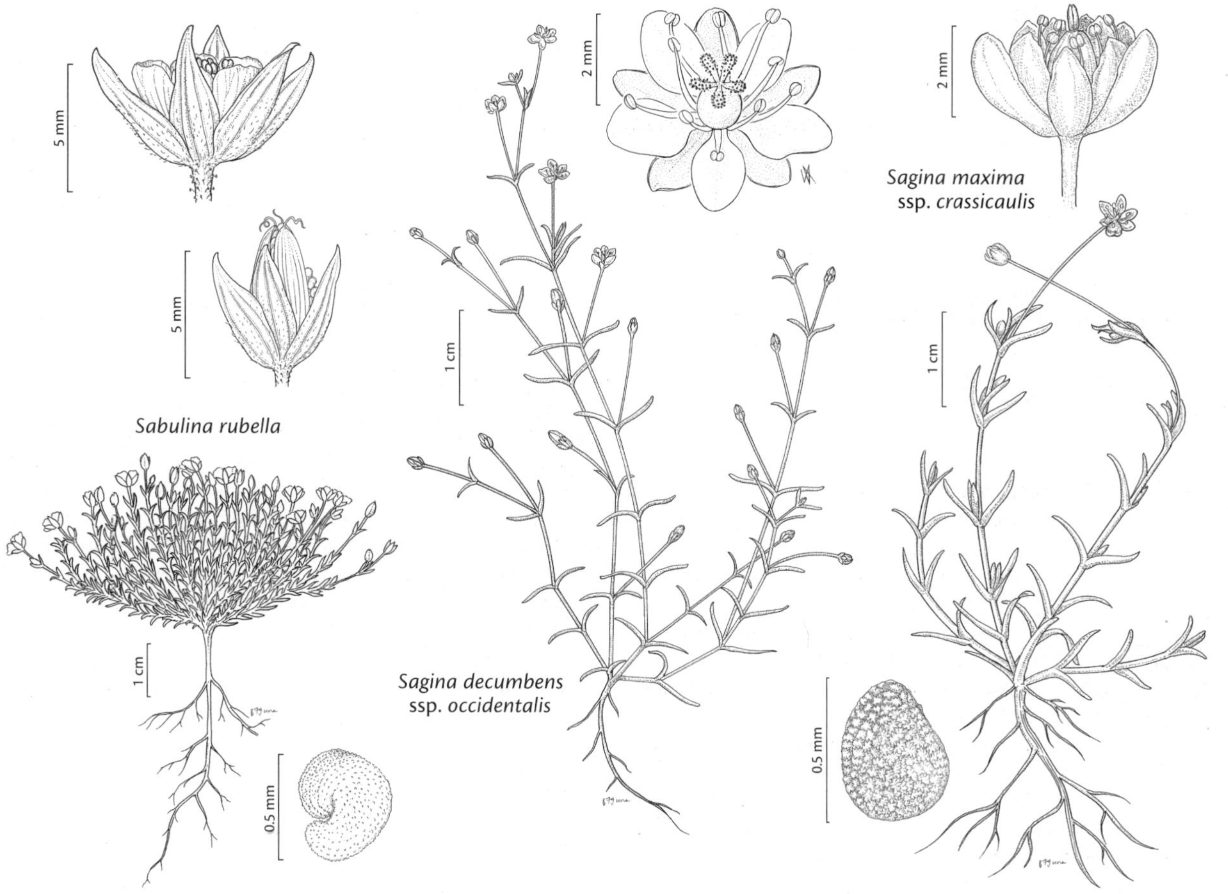

SABULINA • SAGINA

SAGINA Pearlwort

Plants annual or perennial. **Stems** erect or ascending to decumbent or spreading. **Leaves** opposite, linear, fleshy or not; petioles 0; stipules 0. **Inflorescences** terminal or axillary cymes; bracts present. **Flowers** bisexual; sepals 4 or 5, free, glabrous to glandular-hairy; petals 0, 4, or 5, white, tips entire; stamens 4, 5, or 10, borne at ovary base, staminodes 0; styles 4–5. **Fruits** capsules, globose to ovoid, opening by 4 or 5 ascending to recurved valves. **Seeds** ~125, obliquely triangular and compressed with a dorsal groove or reniform and plump without a dorsal groove, 0.2–0.5 mm, brown or red-brown, smooth to tuberculate.

North temperate regions and scattered tropical mountains. ~25 species; 5 species treated in *Flora*.

Sagina japonica might be expected in Oregon. Known from several recent collections in British Columbia, the only Oregon collections are from ballast near Portland in 1899–1902. It would key near *S. apetala* but can be distinguished by the fleshy leaves, five-merous flowers that have petals, and plump, reniform seeds that lack a dorsal groove.

Sagina alexandrae (formerly *subulata*) has been collected twice on Steens Mountain in Harney County. It is similar to *Sagina saginoides*, but its leaf margins are minutely glandular, and its leaf bases form a scarious cup (versus leaf margins glabrous and leaf bases not forming a cup). It is sold in the horticultural trade as Scotch moss or Irish moss.

Key to *Sagina* species:

1. Plants annual; pedicels and/or sepals often glandular-hairy; sterile basal rosettes absent.
    2. Upper leaves minutely ciliate near base; sepals 4(5); petals 0 or sometimes 4, minute. . . . . **S. apetala**
    2' Upper leaves glabrous; sepals usually 5; petals usually 5, equaling sepals . . . . ***S. decumbens*** ssp. ***occidentalis***

1' Plants perennial, glabrous; sterile basal rosettes often present.
    3. Leaves fleshy; sepals 2.5–3.5 mm; seeds reniform, plump . . . . ***S. maxima*** ssp. ***crassicaulis***
    3' Leaves not fleshy; sepals 1.3–2.5 mm; seeds obliquely triangular, laterally compressed.
        4. Sepals 4(5), petals 0 or 25–50% as long as sepals; stamens usually 4. . . . . ***S. procumbens***
        4' Sepals 5, petals 75–100% as long as sepals; stamens usually 10. . . . . ***S. saginoides***

Sagina apetala Arduino

Common pearlwort

Sagina apetala Ard. var. *barbata* Fenzl

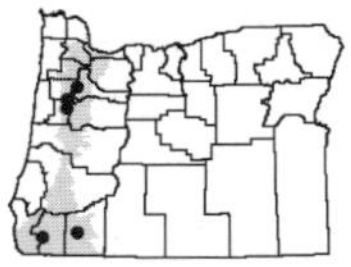

Plants annual. **Stems** erect to decumbent, (1.5)3–8 cm, glandular-hairy above; sterile basal rosettes absent. **Leaves** 3–9(12) mm, usually straight, not fleshy, upper minutely ciliate near base. **Inflorescences** pedicels 2–8(12) mm, glandular-hairy. **Flowers** sepals 4(5), 1.5–2 mm, glandular-hairy, usually appressed in fruit; petals 0 or sometimes 4, minute; stamens usually 4. **Seeds** obliquely triangular, laterally compressed with a dorsal groove, 0.2–0.3 mm, brown, roughened or papillate. **2*n***=12.

Sandy and gravelly waste areas. Flowering Apr–Jun. 0–500 m. Sisk, WV. CA; north to British Columbia, KS; Europe. Exotic.

Sagina decumbens (Elliott) Torrey & A. Gray ssp. ***occidentalis*** (S. Watson) G.E. Crow (p. 563)

Western pearlwort

Sagina occidentalis S. Watson

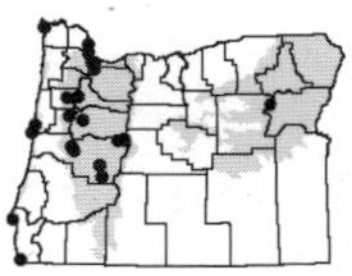

Plants annual. **Stems** erect or ascending, (2)4–16 cm, glabrous or glandular-hairy above; sterile basal rosettes absent. **Leaves** 4–20 mm, not fleshy, glabrous. **Inflorescences** pedicels 2–14(20) mm, usually straight, glabrous or glandular-hairy. **Flowers** sepals usually 5, 1.7–2.1(2.5) mm, glandular-hairy, usually appressed in fruit; petals usually 5, equaling sepals; stamens 5 or 10. **Seeds** obliquely triangular, laterally compressed with dorsal groove, ~0.4 mm, brown, smooth to slightly roughened.

Moist creek banks and rocks, sand and gravel fill. Flowering Apr–Aug. 0–1900 m. BW, Casc, Est, WV. CA, ID, WA; north to AK, northeast to MT, southeast to AZ; Europe. Native.

Sagina maxima A. Gray ssp. ***crassicaulis*** (S. Watson) G.E. Crow (p. 563)

Fleshy pearlwort, thick-stemmed pearlwort

Sagina crassicaulis S. Watson

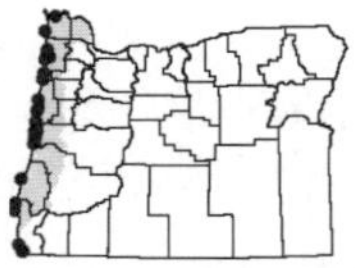

Plants perennial. **Stems** spreading to decumbent, 3–18 cm, glabrous; sterile basal rosettes present. **Leaves** 7–22 mm, fleshy, glabrous. **Inflorescences** pedicels 5–25 mm, usually straight, glabrous. **Flowers** sepals 5, 2.5–3.5 mm, usually appressed in fruit; petals 5, 75–100% as long as sepals; stamens 10. **Seeds** reniform, plump without a dorsal groove, 0.4–0.5 mm, red-brown, smooth or slightly roughened. **2*n***=46, 66.

Moist coastal bluffs and dunes. Flowering May–Oct. 0–100 m. CR, Est. CA, WA; north to AK; eastern Asia. Native.

Sagina procumbens Linnaeus

Procumbent pearlwort

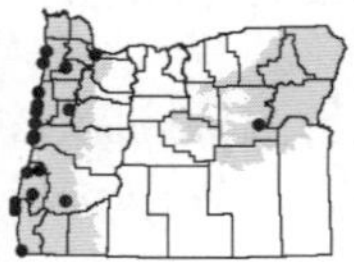

Plants perennial. **Stems** decumbent to ascending, 2–18 cm, glabrous; sterile basal rosettes often present. **Leaves** 3–10(20) mm, not fleshy, glabrous. **Inflorescences** pedicels 5–20(30) mm, usually recurved in flower, straight in fruit, glabrous. **Flowers** sepals 4(5), 1.3–2 mm, spreading to ascending in fruit; petals 0 or 4(5), 25–50% as long as sepals; stamens usually 4. **Seeds** obliquely triangular, laterally compressed with a dorsal groove, 0.3–0.4 mm, brown, smooth or slightly roughened.

Moist coastal bluffs and dunes, lawns and flower beds, roadsides. Flowering May–Oct. 0–600 m. BW, CR, Est, Sisk, WV. CA, WA, ID; scattered across North America; Europe. Exotic.

Sagina saginoides (Linnaeus) H. Karsten

Alpine pearlwort, arctic pearlwort

Sagina saginoides (L.) H. Karst. var. *hesperia* Fernald

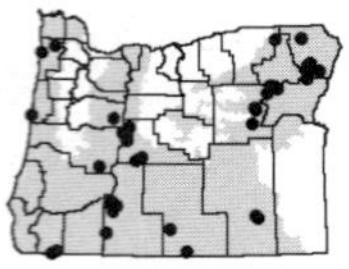

Plants perennial. **Stems** ascending or decumbent, (1)2–12 cm, glabrous; sterile basal rosettes often present. **Leaves** (3)5–15 mm, not fleshy, glabrous. **Inflorescences** pedicels 10–30 mm, recurved in flower, straight in fruit, glabrous. **Flowers** sepals 5, 1.5–2.5 mm, usually appressed in fruit; petals 5, 75–100% as long as sepals; stamens usually 10. **Seeds** obliquely triangular, laterally compressed with a dorsal groove, 0.2–0.4 mm, brown, smooth or slightly roughened. **2*n***=22.

Moist meadows and creek banks. Flowering Jun–Aug. 0–2800 m. BR, BW, Casc, CR, ECas, Est, Sisk. CA, ID, NV, WA; western North America; circumboreal. Native.

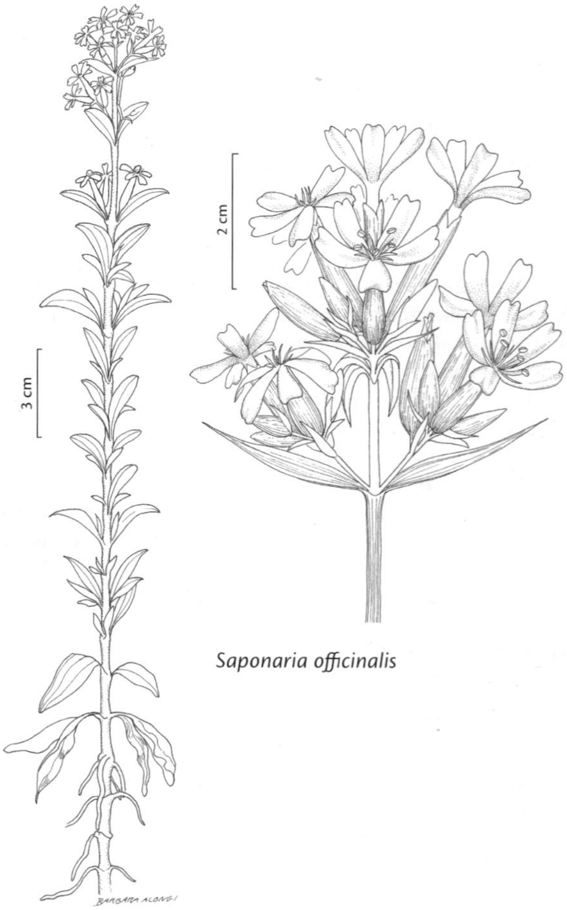

Saponaria officinalis

SAPONARIA

SAPONARIA Soapwort

Plants perennial. **Stems** erect or trailing, procumbent, or ascending, much branched. **Leaves** opposite, spatulate to ovate; petioles present (basal and some cauline leaves) or absent (cauline); stipules absent. **Inflorescences** terminal congested cymes; involucral bracts present; pedicels 1–6 mm. **Flowers** bisexual; sepals 5, fused, glabrous or glandular-pubescent, tubes prominent, terete, oblong-cylindric, 6–17 mm, greenish, reddish or purple, 20-veined, commissures between adjacent sepals absent, lobes <tube; petals 5(10 in some cultivars), red or pink to white, limbs entire to obcordate, appendages borne at bases of petal limbs 2; stamens 10, fused with petals to stalk below ovary, staminodes 0(present in some cultivars); styles and stigmas 2. **Fruits** capsules, opening by 4 ascending to recurved valves. **Seeds** 15–75, reniform, 1.6–2 mm, purplish to black, papillate.

Northern Africa, central and western Asia, Europe, North America. ~40 species; 2 species treated in *Flora.*

Key to *Saponaria* species:

1. Stems erect, 30–90 cm; calyces 15–25 mm, glabrous or virtually so; fruits ~15–20 mm . . . . ***S. officinalis***

1' Stems trailing, procumbent, or ascending, 5–25 cm; calyces 7–12 mm, glandular-pubescent; fruits 6–8 mm. . . . . ***S. ocymoides***

Saponaria ocymoides Linnaeus

Rock soapwort, saponaria

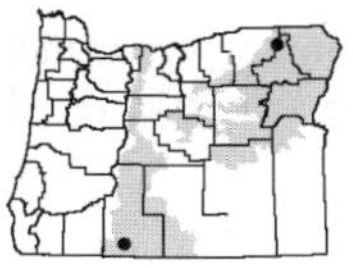

Stems trailing, procumbent, or ascending, 5–25 cm, much branched. **Leaves** spatulate to ovate-lanceolate, 0.6–2.5 cm. **Inflorescences** cymes spreading, lax. **Flowers** calyces not cleft, 7–12 mm, usually purple, glandular-pubescent; petals red or pink to white, limbs 8–15 mm. **Fruits** 6–8 mm. **Seeds** 1.6–2 mm. **2*n***=28.

Roadsides. Flowering May–Jul. 1300–1400 m. BW, ECas. CA; widely scattered in North America; Europe. Exotic.

Sometimes cultivated, *Saponaria ocymoides* has been collected twice in Oregon, first in 1956.

Saponaria officinalis Linnaeus (p. 564)

Bouncing-bet

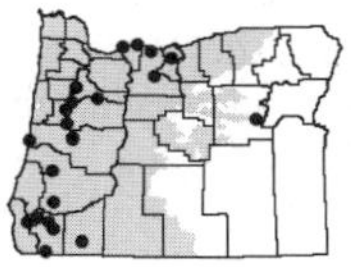

Stems erect, 30–90+ cm, simple or branched above. **Leaves** oblanceolate to ovate, 3–10 cm. **Inflorescences** cymes often dense. **Flowers** calyces often cleft, 15–25 mm, green or sometimes reddish, glabrous or virtually so; petals pink, limbs 8–15 mm. **Fruits** ~15–20 mm. **Seeds** 1.5–2 mm. **2*n***=28.

Roadsides, meadows, slopes, pastures, riverbanks. Flowering Jun–Aug. 0–1500 m. Casc, Col, CR, ECas, Est, Lava, Sisk, WV. CA, ID, NV, WA; most of North America; southern Europe. Exotic.

Sometimes cultivated, this species was first collected in Oregon in 1911.

SCLERANTHUS Knawel

Plants annual or biennial. **Stems** horizontal to erect. **Leaves** opposite, linear to needle-like; petioles 0; stipules 0. **Inflorescences** terminal or axillary cymes; bracts present; flowers sessile or subsessile. **Flowers** bisexual, inconspicuous; sepals 5, free, glabrous; petals 0; stamens 2–10, arising on hypanthium rim, staminodes 0 or 5–8; styles and stigmas 2. **Fruits** utricles, indehiscent, enclosed in a prominent, persistent, hard, 10-ribbed hypanthium. **Seeds** 1, widely ovate, 1.4–1.6 mm, tan except for a red crescent near acute tip, smooth.

Worldwide. 10 species; 1 species treated in *Flora.*

Scleranthus annuus Linnaeus ssp. ***annuus*** (p. 565)

Annual knawel, German knotgrass

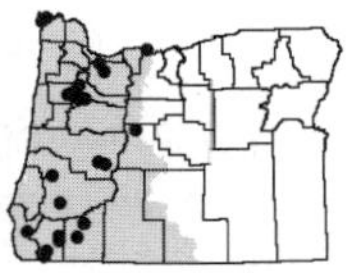

Stems 4–20 cm, often much branched, covered with fine, recurved hairs usually in lines. **Leaves** 4–25 mm. **Flowers** sepals narrowly triangular to awl-shaped, 1.5–4 mm. **2*n***=44.

Roadsides, dry areas, lawns and gardens, sandbars, fields, dried vernal pools. Flowering Jan–Dec. 0–1000 m. Casc, CR, ECas, Est, Sisk, WV. CA, ID, WA; southern Canada, eastern US; worldwide. Exotic.

While some collections of this taxon were made as early as 1918, most collections have been gathered since 1948.

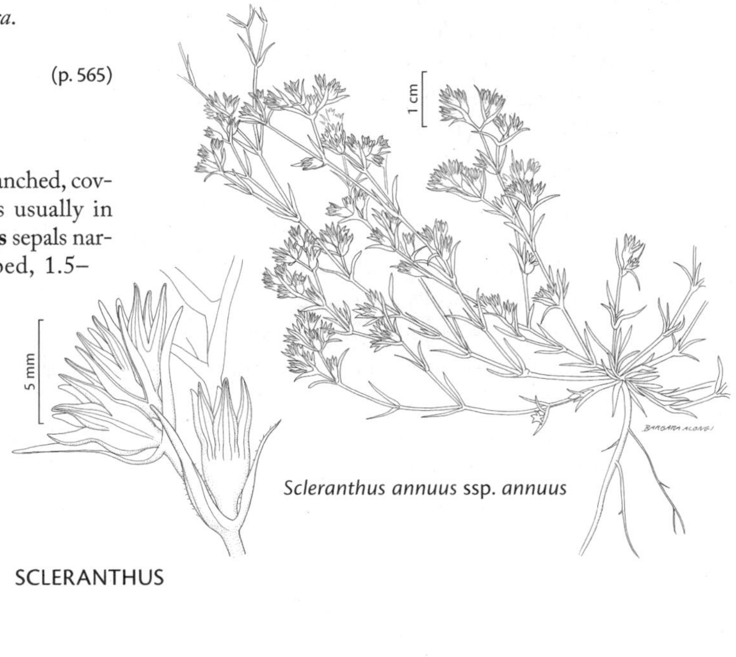

Scleranthus annuus ssp. *annuus*

SCLERANTHUS

SILENE Catchfly, campion

Plants annual, biennial or perennial. **Stems** decumbent or ascending to erect. **Leaves** opposite, usually linear to oblanceolate, sometimes to elliptic or broadly ovate; petioles present (some basal leaves) or 0 (cauline leaves); stipules 0. **Inflorescences** terminal or axillary, open cymes or rarely flowers solitary; bracts present. **Flowers** bisexual, rarely unisexual (plants dioecious); sepals 5, fused, tubular to campanulate, hairy or glabrous, tubes prominent, green, whitish, and/or purplish, glabrous to pubescent, (8)10- or 20–30-veined, commissures between adjacent sepals 1-veined, green or often pale, <tube; petals (0)5, white, yellowish white, greenish white, pink, bright pink, lavender, or purplish, rarely bright red; appendages (0)2(6); limbs entire, emarginate, notched or usually 2, 4(6+)-lobed; stamens 10, borne at ovary base, staminodes 0; styles and stigmas 3(4–5). **Fruits** capsules, opening by (3–5)6(8–10) ascending or recurved teeth; capsule stalks 0.5–4(6) mm, rarely absent. **Seeds** 15–100+, reniform to globose, 0.5–3 mm, gray, yellowish brown, brown, or black, tuberculate or papillate.

Africa, Asia, Europe, North America, South America. ~700 species; 25 species treated in *Flora*.

In addition to the species described below, *Silene dichotoma* ssp. *dichotoma* (forked catchfly) was collected in 1906 near Portland, likely on ballast. It is most similar to *Silene gallica*, but has a larger (5–12 mm) bilobed petal limb. Although listed as occurring in Oregon in *Flora of North America* (Morton 2005), we have not located a specimen of *S. pendula* or *S. verecunda* to confirm these reports. *Silene pendula* is an annual plant with a ten-veined calyx and showy bright pink flowers and may rarely escape from cultivation. *Silene verecunda* is a perennial with a ten-veined calyx that is distinctly narrowed at the base in fruit and greenish white to pink two-lobed petals.

The petal limbs in *Silene* species are often divided into a varying number (and size) of lobes. In species where the lobes are often of two different sizes (e.g., *S. hookeri, S. parryi*), they can be either described as having four lobes, with the outer being smaller than the inner, or as having two lobes, each with a lateral tooth; the first definition is being followed in this treatment, except in *S. douglasii* var. *oraria* where the lateral "tooth" is far smaller than the lobes and should not mistaken for an additional lobe.

Key to *Silene* species:

1. Plants annual or biennial; taproots generally slender.
  2. Calyx tubes clearly 20–30-veined.
    3. Styles 5; petal limbs ± entire or 2-lobed; capsule teeth 5, splitting to 10; flowers pistillate . . . . . . ***S. latifolia*** (in part)
    3' Styles 3; petal limbs entire or notched; capsule teeth 6; flowers bisexual . . . . . . ***S. conoidea***
  2' Calyx tubes clearly (8)10-veined.
    4. Calyx tubes 4–10 mm, lobes 1–2.5 mm.
      5. Inflorescences 1-sided; pedicels and calyces hairy . . . . . . ***S. gallica***
      5' Inflorescences not 1-sided; pedicels and calyces glabrous . . . . . . ***S. antirrhina***
    4' Calyx tubes 12–22 mm, lobes 3–13 mm.
      6. Calyx tube veins not clearly netted above middle of sepal, lobes 3–6 mm, tips obtuse to acuminate; flowers staminate . . . . . . ***S. latifolia*** (in part)
      6' Calyx tube veins netted above middle of sepal, lobes 6–13 mm, tips acuminate; flowers bisexual . . . . . . ***S. noctiflora***

1' Plants perennial, arising from a rhizome or branched, generally woody, caudex.
  7. Flowers unisexual; styles 5 in pistillate flowers.
    8. Petals white; lower stem hirsute . . . . . . ***S. latifolia*** (in part)
    8' Petals bright pink; hairs on lower stem long, spreading . . . . . . ***S. dioica***
  7' Flowers bisexual; styles 3(5).
    9. Pale commissures absent between obscure calyx veins.
      10. Calyces glabrous.
        11. Calyces not inflated in fruit; petals bright pink, rarely white . . . . . . ***S. acaulis***
        11' Calyces clearly inflated in fruit, papery; petals white . . . . . . ***S. vulgaris***
      10' Calyces hairy, glandular or not.
        12. Calyces glandular-viscid; seeds winged . . . . . . ***S. spaldingii***
        12' Calyces puberulent, glandular or not; seeds not winged.
          13. Flowers ascending to erect; petal limbs 2-lobed . . . . . . ***S. menziesii***
          13' Flowers nodding to spreading; petal limbs 4–6(8)-lobed . . . . . . ***S. campanulata*** ssp. ***glandulosa***
    9' Pale commissures present between usually prominent calyx veins.

14. Petal limbs generally 4–6-lobed, lobes of similar size or inner lobes larger than outer.
  15. Upper cauline leaves overlapping with, or exceeding, pedicels. . . . . **S. hookeri**
  15' Upper cauline leaves not overlapping pedicels.
    16. Basal leaves densely tufted.
      17. Calyces tubular, not enlarging in fruit . . . . **S. bernardina** (in part)
      17' Calyces campanulate or broadly cylindric and becoming campanulate, enlarged or inflated in fruit.
        18. Leaves fleshy . . . . **S. grayi**
        18' Leaves not fleshy.
          19. Veins in sepal lobes thickened and broader apically than commissural veins; commissural veins not forking/fusing with lobe veins; petal limbs 2.5–5.5 mm . . . . **S. nuda**
          19' Veins in sepal lobes not thickened or broader apically than commissural veins; commissural veins fork above and fuse with veins in lobes; petal limbs 5–7 mm . . . . . **S. parryi** (in part)
    16' Basal leaves not densely tufted.
      20. Flowers nodding; stamens exserted; styles long-exserted, both exceeding petal claws; petal claws short-woolly at base . . . . **S. lemmonii**
      20' Flowers spreading to erect, rarely nodding; stamens and styles equaling petal claws; petal claws glabrous or ciliate at base.
        21. Petal lobes 4–6, 2 inner ≤ outer.
          22. Petal appendages 2, deeply divided; petal claws ciliate at base; corolla lobes 4 . . . **S. bernardina** (in part)
          22' Petal appendages 4–6, entire; petal claws glabrous at base; corolla lobes 4–6 (some splitting further) . . . . **S. oregana**
        21' Petal lobes 4(6), 2 inner > outer.
          23. Cauline leaves in 1–12 pairs, 2–6 cm × 3–15 mm, lanceolate to ovate; pedicels < calyx or absent; flowers often subsessile in small cymes, whorled at nodes . . . . **S. scouleri** ssp. **scouleri** (in part)
          23' Cauline leaves in 2–4 pairs, 0.2–0.8 cm × 10–80 mm, narrowly oblanceolate to linear-lanceolate; pedicels > calyx, flowers in open cymes, or rarely solitary. . . . . . . **S. parryi** (in part)

14' Petal limbs 2-lobed, very rarely with a small, lateral tooth.
  24. Basal leaves densely tufted.
    25. Stems 3–15 cm; basal leaves fleshy; seeds winged . . . . **S. suksdorfii**
    25' Stems (10)20–60 cm; basal leaves not fleshy; seeds not winged.
      26. Veins in sepal lobes thickened and broader apically than commissural veins; petal limbs 2–10 mm.
        27. Sepal lobe tips acute to obtuse, scarious margins broad; petal limbs 2–5 mm . . . . **S. scaposa**
        27' Sepal lobe tips blunt to acute, scarious margins narrow; petal limbs 5–10 mm . . . . **S. pectinata**
      26' Veins in sepal lobes not thickened or broader apically than commissural veins; petal limbs 5–7 mm. . . . . **S. parryi** (in part)
  24' Basal leaves not densely tufted.
    28. Flowers nodding; stamens and styles long-exserted . . . . **S. bridgesii**
    28' Flowers spreading to erect; stamens and styles included, about equaling petal claws or somewhat exserted.
      29. Pedicels and calyx hairs short, grayish, retrorsely-puberulent, rarely also glandular . . . . **S. douglasii**
      29' Pedicels, and calyx hairs glandular-puberulent, neither grayish nor retrorse.
        30. Cauline leaves in 1–12 pairs, 2–6 cm × 3–15 mm, lanceolate to ovate; pedicels < calyx or absent; flowers often subsessile in small cymes, whorled at nodes . . . . **S. scouleri** ssp. **scouleri** (in part)
        30' Cauline leaves in 2–4 pairs, 2–8 cm × 2–8 mm, narrowly oblanceolate to linear-lanceolate; pedicels > calyx; flowers in open cymes or rarely solitary. . . . . **S. parryi** (in part)

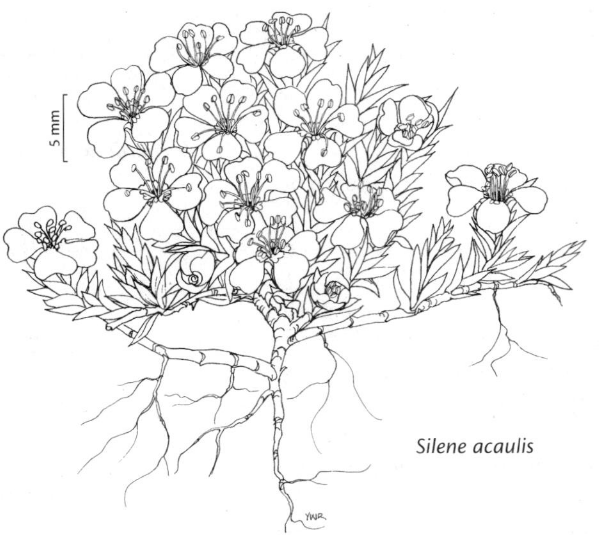

SILENE

Silene acaulis (Linnaeus) Jacquin (p. 568)

Moss campion

Silene acaulis (L.) Jacq. var. *exscapa* (All.) DC.

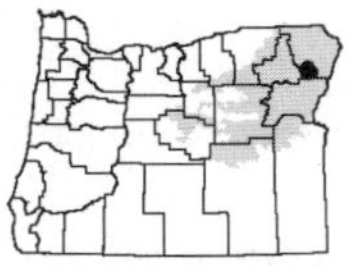

Plants perennial, mat-forming; caudices woody. **Stems** erect, 3–6(10) cm, branches many, glabrous. **Leaves** mostly basal, crowded, linear-subulate to lanceolate, 0.4–1 cm × 0.8–1.5 mm. **Inflorescences** flowers solitary; pedicels glabrous or glandular hairy. **Flowers** bisexual or unisexual, erect; calyces tubular to campanulate, 7–10 mm, glabrous, obscurely 10-veined, not netted above, pale commissures absent, not inflated in fruit, lobes 1–2 mm, tips obtuse; appendages vestigial; limbs 2.5–3.5 mm, bright pink, rarely white, entire or lobes 2; stamens exserted (staminate flowers) or not (pistillate flowers); styles 3, exserted. **Fruits** cylindric, teeth 6; stalks 1 mm. **Seeds** 0.8–1 mm, light brown, not winged. **2*n***=24.

Bare rocky slopes, alpine summits, gravel bars. Flowering Jul–Aug. 1600–3000 m. BW. ID, NV, WA; western and northern North America; eastern Asia, Europe. Native.

Silene acaulis is a variable, circumpolar species. Although our material is often cited as variety *exscapa*, we follow Morton (2005) in not recognizing infraspecific taxa.

Silene antirrhina Linnaeus (p. 569)

Sleepy catchfly

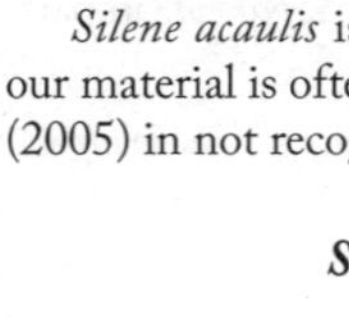

Plants annual; taproots slender. **Stems** erect, 12–80 cm, glabrous or retrorsely puberulent, upper internodes usually viscid. **Leaves** basal leaves oblanceolate to spatulate, 1–3(6) cm × 3–6(10) mm; upper cauline leaves linear to narrowly oblanceolate, gradually reduced upward. **Inflorescences** terminal, open cymes, not 1-sided; pedicels glabrous. **Flowers** bisexual, erect; calyces ovate to campanulate, not inflated in fruit, 4–9 mm, glabrous, 10-veined, not netted above, pale commissures present, lobes 1 mm, tips acute; petal claws glabrous (or rarely petals absent); appendages 0 or to 0.4 mm; limbs ~2.5 mm, white to red, lobes 2; stamens included; styles 3, included. **Fruits** ovoid, teeth 6; stalks 1 mm. **Seeds** 0.5–0.8 mm, gray-black, not winged. **2*n***=24.

Roadsides, meadows, slopes, open woods. Flowering May–Jul. 0–1200 m. BW, Casc, CR, ECas, Lava, Sisk, WV. CA, ID, NV, WA; throughout North America; Europe, South America. Native.

Silene bernardina S. Watson (p. 569)

Palmer's catchfly, mountain catchfly

Silene montana S. Watson

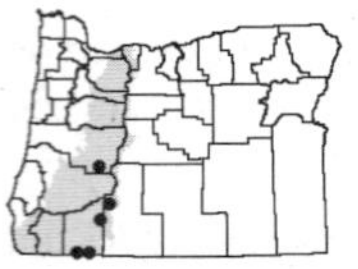

Plants perennial; caudices woody. **Stems** erect, 15–55 cm, branches generally few, puberulent to short-hairy, glandular above. **Leaves** basal leaves tufted or not, linear to oblanceolate, 2–8 cm × 2–6 mm; cauline leaves usually linear, 1–6 cm × 1–4 mm, gradually reduced upward, upper leaves not overlapping pedicels. **Inflorescences** terminal and axillary, open cymes, not 1-sided; pedicels glandular hairy. **Flowers** bisexual, erect; calyces tubular, not inflated in fruit, 12–15 mm, glandular-puberulent, 10-veined, not netted above, pale commissures present, commissural veins fork above and fuse with lobe veins, lobes 2–3.5 mm, tips acute; petal claws ciliate at base; appendages 2, deeply divided; limbs 4–6 mm, white, pink, or reddish, lobes 4, 1.5–3.5 mm; stamens slightly exserted; styles 3(4), equaling petal claws or slightly exserted. **Fruits** ellipsoid, teeth 6(8); stalks 2–5 mm. **Seeds** 1.5–2 mm, brown, not winged. **2*n***=48.

Dry meadows and woods. Flowering Jul–Aug. 1500–2300 m. Casc, Sisk. CA, ID, NV, WA; south to Mexico. Native.

Silene bernardina is a variable species and can be sometimes be difficult to distinguish from *S. oregana*. We follow Morton (2005) in not trying to apply infraspecific concepts.

Silene bridgesii Rohrbach

Bridges' catchfly

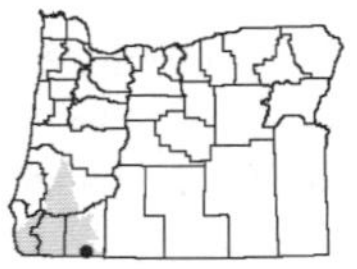

Plants perennial; caudices woody. **Stems** decumbent to erect, 16–50(65) cm, branches 1 or few, puberulent, viscid above. **Leaves** basal usually withering; middle cauline oblanceolate to elliptic, 2–6(8) cm × 6–15 mm; upper cauline elliptic, lanceolate, or oblong, 1–5 cm × 3–12 mm, gradually reduced above middle. **Inflorescences** terminal and axillary, open cymes, not 1-sided; pedicels bent distally, glandular hairy. **Flowers** bisexual, nodding; calyces tubular to campanulate, not inflated in fruit, 7–14 mm, glandular-puberulent, 10-veined, not netted above, pale commissures present, calyx lobes 2–3 mm, tips acute; petal claws generally glabrous; appendages 2; limbs 4–9 mm, white, lobes 2, 1–3 mm; stamens long-exserted; styles 3, long-exserted. **Fruits** ovoid, teeth 6; stalks 2–3 mm. **Seeds** 1–1.5 mm, brown, not winged. **2*n***=48.

Forests. Flowering Jun–Aug. 600–2300 m. Sisk. CA. Native.

Although sometimes listed as being endemic to California, *S. bridgesii* is known in Oregon from two collections from Ashland Butte made in 1886.

Silene campanulata S. Watson ssp. ***glandulosa*** C.L. Hitchcock & Maguire (p. 569)

Bell catchfly

Silene campanulata S. Watson var. *greenei* S. Watson

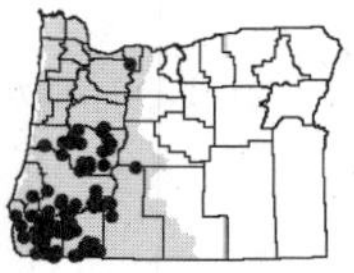

Plants perennial; caudices woody. **Stems** erect, 5–40 cm, branches generally many, hairy, glandular (especially above) or not, rarely glabrous. **Leaves** basal lanceolate to circular, 1–3(5) cm × 10–30 mm; upper cauline linear to broadly ovate, slightly reduced upward. **Inflorescences** terminal and axillary, open cymes, not 1-sided; pedicels deflexed in fruit, glandular-puberulent. **Flowers** bisexual, nodding to spreading; calyces broadly campanulate, enlarging in fruit, 6–10 mm, puberulent, glandular or not, obscurely 10-veined, not netted above, pale commissures absent, tips broadly acute to obtuse; petal claws glabrous to ciliate; appendages 2–4; limbs 5–7 mm, creamy white to greenish, lobes 4–6(8), 1–3 mm, quickly curling; stamens exserted; styles 3, exserted. **Fruits** ovoid, teeth 6; stalks 1–2.5 mm. **Seeds** 2–2.5 mm, brown, not winged. **2*n***=48.

Dry, open, rocky slopes, rock crevices, open coniferous-oak forests and forest margins. Flowering Apr–Aug. 100–2300 m. Casc, CR, ECas, Sisk, WV. CA. Native.

Plants with stems lacking glandular hairs may be treated as var. *greenei*. Subspecies *glandulosa* is more widespread than ssp. *campanulata*, which is known only from scattered sites in northern California.

Silene conoidea Linnaeus

Large sand catchfly

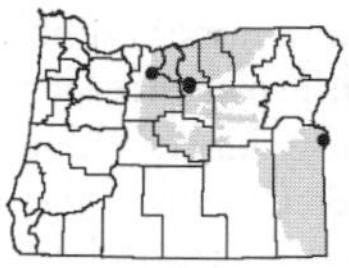

Plants annual; taproots slender. **Stems** erect, 50–80 cm, puberulent, viscid above. **Leaves** basal lanceolate to oblanceolate, 1–12 cm × 3–10 mm; upper cauline lanceolate, gradually reduced upward. **Inflorescences** terminal, open cymes, not 1-sided; pedicels densely glandular hairy. **Flowers** bisexual, erect; calyces narrowly conic, inflated in fruit, 18–26 mm, glandular-puberulent, (25)30-veined, not netted above, pale commissures present in fruiting calyx, lobes 5–10 mm, acuminate; petal claws glabrous; appendages 2; limbs 8–12 mm, white, deep pink, or purplish, entire or notched; stamens exserted; styles 3, slightly exserted. **Fruits** ovoid to conic, teeth 6; stalks 1–2.5 mm. **Seeds** 1–1.5 mm, gray-brown, not winged. **2*n***=20, 24.

Disturbed, open areas, riverbanks. Flowering May–Jul. 200–700 m. Col, Lava, Owy. CA, ID, NV, WA; western North America, MO, TX; Asia, Europe. Exotic.

An uncommon, introduced species, the earliest Oregon collection was made in Wasco County in 1950.

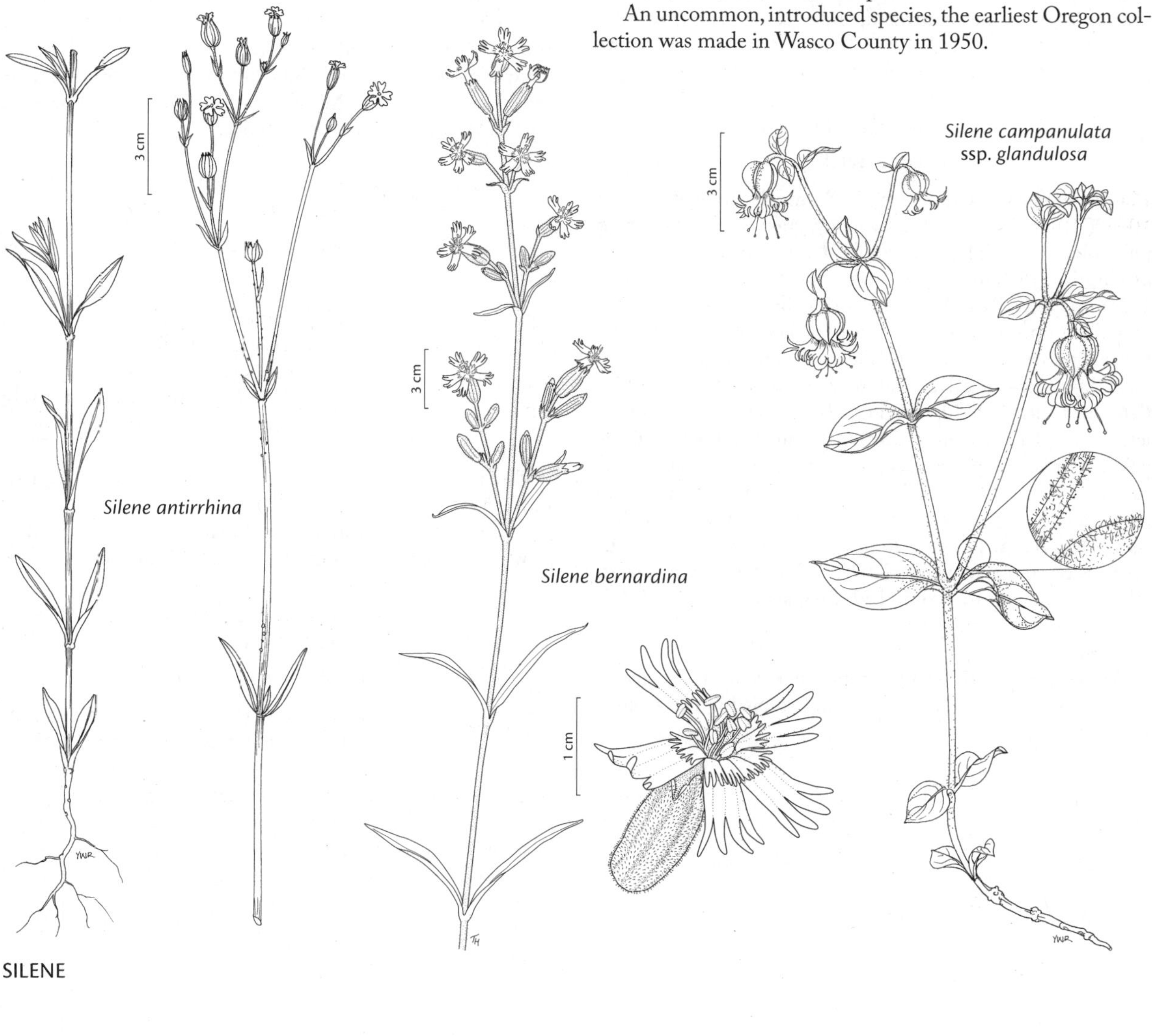

SILENE

Silene dioica (Linnaeus) Clairville

Red campion

Lychnis dioica L.

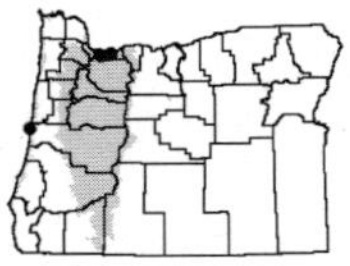

Plants perennial; taproots slender. **Stems** ascending, sometimes decumbent, 25–80 cm, branches few, hairs long-spreading below, ± glandular above. **Leaves** basal spatulate, 3–8.5 cm × 10–35 mm; upper cauline ovate to elliptic, gradually reduced upward. **Inflorescences** terminal, open cymes, not 1-sided; pedicels softly hairy, seldom glandular. **Flowers** unisexual, ascending to erect; calyces narrowly (staminate) or broadly (pistillate) campanulate, inflated in fruit, 10–15 mm, softly hairy, 8–12-veined, not netted above, pale commissures absent, lobes 2–3 mm, rounded to acute; petal appendages 4; limbs 8–12 mm, bright pink, entire or lobes 2, 1–2 mm; stamens equaling petal claws; styles 5, equaling petal claws. **Fruits** broadly ovoid to globose, teeth 5, splitting to 10; stalks 0. **Seeds** 1–1.6 mm, dark brown to black, not winged. **2*n***=24.

Roadside ditches, edge of woods. Flowering May–Jun. 0–200 m. Casc, Est, WV. ID, WA; scattered across northern US and southern Canada; Europe. Exotic.

McNeill (1979) noted that the earliest collection of *S. dioica* in the western United States was made in Portland in 1905. This species is sometimes cultivated, and it can escape from such settings. *Silene dioica* and *S. latifolia* are interfertile, and hybrid plants (*S.* ×*hampeana*) with pale pink petals can be expected.

Silene douglasii Hooker

Douglas' catchfly

Plants perennial; caudices woody. **Stems** decumbent to erect, 10–40(70) cm, branches few to many, grayish white retrorsely puberulent, rarely glandular above. **Leaves** basal oblanceolate or elliptic, 2–10 cm × 1.5–13 mm; cauline ± linear, 1–4 cm × 1–5 mm, gradually reduced upward. **Inflorescences** terminal or axillary, open cymes, not 1-sided; pedicels grayish white retrorsely puberulent, rarely glandular above. **Flowers** bisexual, ascending to erect; calyces campanulate, strongly (or not) inflated in fruit, 9–15(20) mm, grayish white retrorsely puberulent, rarely glandular, prominently 10-veined, not netted above, pale commissures present, commissural veins fork above and fuse with lobe veins, lobes 2–3 mm, tips blunt; petal claws ciliate at base; appendages 2; limbs 3–4(13) mm, white to pink or purplish, lobes 2, 1–2 mm, sometimes with two lateral teeth; stamens equaling petal claws; styles 3–5, not to slightly exserted. **Fruits** oblong to ovoid, teeth 3–5; stalks 3–4 mm. **Seeds** 1–2 mm, brown, not winged.

3 varieties.

While the grayish white retrorse pubescence is usually diagnostic, some plants may be difficult to distinguish from *S. parryi*.

Key to *Silene douglasii* varieties:

1. Leaves fleshy; calyces inflated, hairs not glandular; petals usually with small, lateral teeth . . . . ***S. d.*** var. ***oraria***

1' Leaves not fleshy; calyces not inflated, hairs often also glandular; petals without small, lateral teeth.

  2. Leaves oblanceolate or elliptic to broadly linear; pedicels and calyces rarely sparsely glandular; plants widespread . . . . ***S. d.*** var. ***douglasii***

  2' Leaves narrow, linear; pedicels and calyces often glandular; plants of the Columbia River Gorge . . . . ***S. d.*** var. ***rupinae***

Silene douglasii Hooker var. ***douglasii*** (p. 571)

Douglas' catchfly

Silene douglasii Hook. var. *monantha* (S. Watson) B.L. Rob.

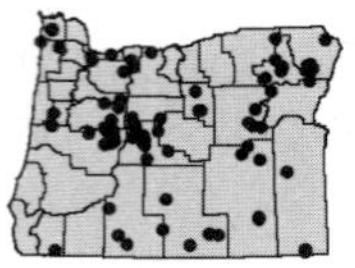

Leaves oblanceolate or elliptic to broadly linear, 2–10 cm × 2–9 mm, not fleshy. **Inflorescences** pedicels grayish white retrorsely puberulent, rarely sparsely glandular. **Flowers** calyces not strongly inflated, grayish white retrorsely puberulent, rarely sparsely glandular; petal limbs lacking lateral teeth. **2*n***=48.

Open rocky slopes and cliffs, dry woods, sagebrush, grasslands. Flowering May–Aug. 50–3000 m. All ecoregions. CA, ID, NV, WA; western North America. Native.

Silene douglasii Hooker var. ***oraria*** (M. Peck) C.L. Hitchcock & Maguire

Cascade Head catchfly

Silene oraria M. Peck

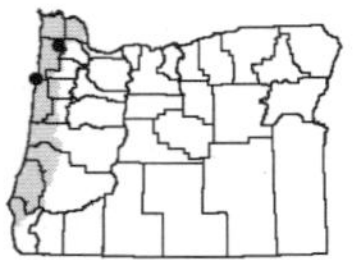

Leaves oblanceolate, 2–5(8) cm × 3–13 mm, fleshy. **Inflorescences** pedicels grayish white retrorsely puberulent. **Flowers** calyces strongly inflated, grayish white retrorsely puberulent, not glandular; petal limbs often with lateral teeth.

Coastal bluffs. Flowering May–Jul. 50–1000 m. CR, Est. Native.

This variety is considered as threatened in Oregon. It is known only from bluffs, primarily coastal, in Tillamook County.

Silene douglasii Hooker var. ***rupinae*** Kephart & Sturgeon

Cliff-dwelling campion

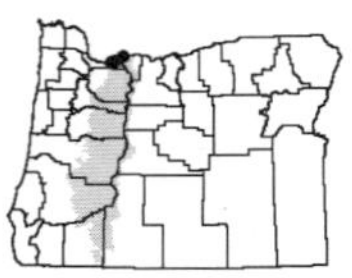

Leaves narrow, linear, 2–10 cm × 1.5–5 mm, not fleshy. **Inflorescences** pedicels grayish white retrorsely puberulent, often glandular. **Flowers** calyces not strongly inflated, grayish white retrorsely puberulent, often glandular; petal limbs lacking lateral teeth.

Rocky river slopes and bluffs, along trails. Flowering May–Jul. 0–200 m. Casc. WA. Native.

This variety is found only on rocky banks and slopes of the Columbia River Gorge.

Silene gallica Linnaeus

Small-flowered catchfly

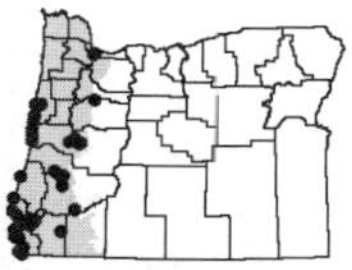

Plants annual; taproots slender. **Stems** erect, 10–40 cm, short-rough-hairy to minutely bristly, viscid above. **Leaves** basal oblanceolate or spatulate, 1–3.5 cm × 3–5 mm, often withering; cauline oblanceolate to oblong, 0.8–2.5 cm × 2–4 mm, gradually reduced upward. **Inflorescences** terminal, open cymes, 1-sided; pedicels 0 or short, densely glandular puberulent. **Flowers** bisexual, ascending to erect; calyces narrowly tubular-ovoid, enlarging slightly in fruit, 6–10 mm, long glandular-hairy especially on veins, prominently 10-veined, not netted above, pale commissures present, lobes 2–2.5 mm, acute; petal claws glabrous; appendages 2; limbs entire to notched, 3–6 mm, white, pink, or lavender, often with a central pink/dark red spot; stamens included; styles 3, included. **Fruits** ovoid, teeth 6; stalks < 1 mm. **Seeds** ~1 mm, red-brown, not winged. **2*n***=24.

Disturbed areas, roadsides, meadows, coastal bluffs. Flowering May–Aug. 0–700 m. CR, Est, Sisk, WV. CA, ID, WA; scattered in eastern North America; Europe. Exotic.

Common in disturbed areas, *Silene gallica* was first collected in Oregon in 1881.

Silene grayi S. Watson

Gray's catchfly

Plants perennial; caudices woody. **Stems** decumbent to erect, 10–20(30) cm, branches many, retrorsely puberulent below, glandular-viscid above. **Leaves** basal narrowly oblanceolate to spatulate, 1.5–4(6) cm × 1–5 mm, densely tufted, fleshy; cauline linear to oblanceolate, 0.4–1 cm × 1–3 mm, abruptly reduced upward, upper leaves not overlapping pedicels. **Inflorescences** terminal, open cymes, not 1-sided, or flowers solitary; pedicels glandular hairy. **Flowers** bisexual, erect or sometimes nodding; calyces broadly cylindric, enlarging and becoming campanulate in fruit, 8–10 mm, glandular-puberulent, prominently 10-veined, not netted above, pale commissures present, lobes 1–2 mm, apiculate; petal claws ciliate at base; appendages 2; limbs 3–5 mm, pink to purplish, lobes 4, 2 inner > outer; stamens equaling petal claws; styles 3(4), equaling petal claws. **Fruits** ovoid, teeth 6(8); stalks 1.5–3 mm. **Seeds** 2–3 mm, brown, winged. **2*n***=48.

Rocky, alpine slopes. Flowering Jun–Aug. 1000–2300 m. Sisk. CA, NV. Native.

Morton (2005) noted that small, alpine plants of *S. parryi* may be mistaken for *S. grayi*.

Silene hookeri Nuttall

Hooker's catchfly

Plants perennial; caudex woody. **Stems** decumbent to erect, 4–25 cm, branches many, hairs appressed below, often glandular-puberulent above. **Leaves** oblanceolate to spatulate, 2–8 cm × 5–15(25) mm, largest near middle of stem, upper leaves overlapping with or exceeding pedicels. **Inflorescences** terminal, open cymes, not 1-sided, or flowers solitary; pedicels short-hairy, glandular or not. **Flowers** bisexual, ascending to erect; calyces broadly tubular, enlarging in fruit, 12–20 mm, densely short-hairy or glandular-hairy, 10-veined, not netted above, commissures narrow, pale, lobes 4–7 mm, tips acute; petal claws glabrous or slightly ciliate at base; appendages 2; limbs 10–15 mm, pink or white, rarely bright red, lobes 4, 5–10 mm, of equal size or 2 inner > outer; stamens slightly exserted; styles 3, exserted. **Fruits** ovoid to oblong, teeth 6; stalks 0.5–5 mm. **Seeds** ~2 mm, dark-brown to black, not winged.

California and Oregon. 4 subspecies; 2 subspecies treated in *Flora*.

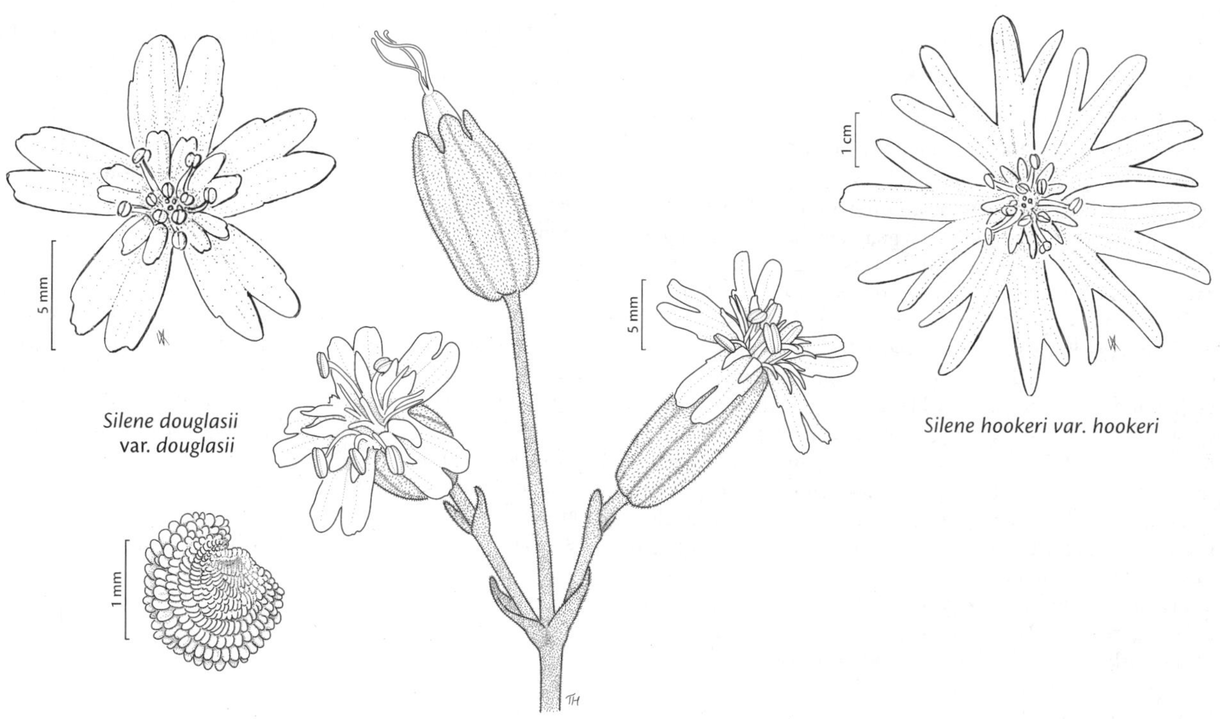

SILENE

Key to *Silene hookeri* subspecies:

1. Petal limbs pink or white . . . . ***S. h.*** ssp. ***hookeri***
1' Petal limbs bright red . . . .***S. h.*** ssp. ***serpentinicola***

Silene hookeri Nuttall ssp. ***hookeri*** (p. 571)

Hooker's catchfly

Silene pulverulenta M. Peck

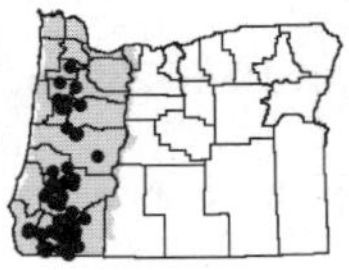

Flowers petal limbs pink or white. **Fruits** stalks 2–5 mm. ***2n***=72.

Dry rocky or sandy slopes, oak or coniferous woodlands, grassy meadows. Flowering Apr–Jun. 50–1300 m. Casc, CR, Sisk, WV. CA. Native.

We follow Morton (2005) in not recognizing plants with glandular hairs (known in Oregon from Jackson County) as ssp. *pulverulenta*.

Silene hookeri Nuttall ssp. ***serpentinicola*** (T.W. Nelson & J.P. Nelson) K.L. Chambers & S.C. Meyers

Serpentine Indian pink

Silene serpentinicola T.W. Nelson & J.P. Nelson

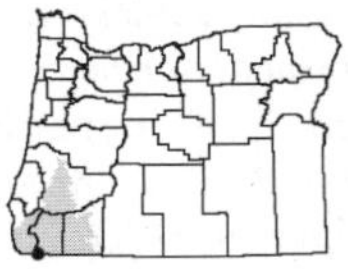

Flowers petal limbs bright red. **Fruits** stalks 0.5–1 mm. ***2n***=72.

Serpentine soils, coniferous woodlands. Flowering Apr–Jun. 100–800 m. Sisk. CA. Native.

We follow Chambers and Meyers (2011) in treating this taxon as a subspecies of *S. hookeri*; see Nelson and Nelson (2004) and Hartman et al. (2012) for treatments as a species. The only record for this taxon in Oregon is a collection made by Thomas Howell in 1884.

Silene latifolia Poiret (p. 573)

White campion

Lychnis alba Mill., *Silene latifolia* Poir. ssp. *alba* (Mill.) Greuter & Burdet

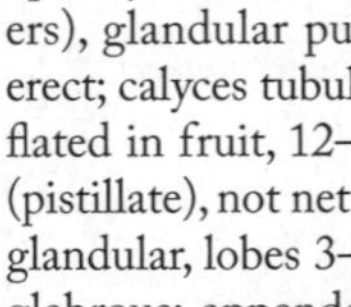

Plants annual, biennial or perennial; taproots and caudices woody. **Stems** erect or sometimes decumbent at base, 30–120 cm, branches few, generally hirsute, often glandular above. **Leaves** basal often withering; middle cauline oblanceolate to elliptic, 5–14.5 cm × 6–30(56) mm; upper cauline lanceolate to elliptic, 1.5–8 cm × 3–15 mm, gradually reduced upward. **Inflorescences** terminal, open cymes, not 1-sided; pedicels sometimes 0 (staminate flowers), glandular puberulent. **Flowers** unisexual; ascending to erect; calyces tubular (staminate) or ovoid (pistillate), much inflated in fruit, 12–20 mm, 10-veined (staminate) or 20-veined (pistillate), not netted above, pale commissures present, hirsute-glandular, lobes 3–6 mm, tips obtuse or acuminate; petal claws glabrous; appendages 2; limbs 7–9(12) mm, white, entire or lobes 2; stamens slightly exserted or equaling petal claws; styles 5, slightly exserted. **Fruits** ovoid, teeth 5, splitting to 10; stalks 1–2 mm. **Seeds** 1–2 mm, gray-brown, not winged. ***2n***=24.

Disturbed areas, field edges, shaded riverbanks. Flowering May–Sep. 0–1800 m. BW, Casc, CR, ECas, Sisk, WV. CA, ID, WA; throughout North America except far north and southeast; Europe. Exotic.

Silene latifolia is an introduced species that may have arrived via ballast; the first collection in Oregon was made in 1911.

Silene lemmonii S. Watson

Lemmon's catchfly

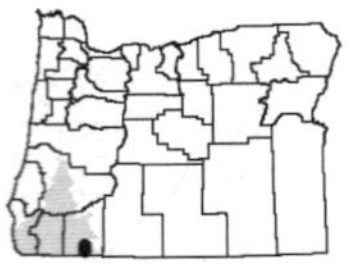

Plants perennial; caudices woody. **Stems** decumbent to erect, 15–45 cm, branches generally few, sparsely hairy below, glandular-viscid above. **Leaves** basal oblanceolate to elliptic, 2–3.5 cm × 8–10 mm, tufted but not densely; cauline in 2–3 pairs, linear-lanceolate to lanceolate, 0.5–2.5 cm × 3–5 mm, ± abruptly reduced upward, upper leaves not overlapping pedicels. **Inflorescences** terminal, open cymes, not 1-sided, rarely flowers solitary; pedicels glandular hairy. **Flowers** bisexual, nodding; calyces campanulate, enlarging in fruit, 6–10 mm, short-glandular-hairy, prominently 10-veined, not netted above, pale commissures present, lobes 1–2 mm, tips acute; petal claws short-woolly; appendages 2; limbs 8 mm, yellow-white to pink, lobes 4, linear, often curled, 4–8 mm; stamens exserted; styles 3, long-exserted. **Fruits** oblong to ovoid, teeth 6; stalks 2–3 mm. **Seeds** 1–1.5 mm, rusty brown, not winged. ***2n***=48.

Dry woods and slopes. Flowering Jun–Jul. 1000–2100 m. Sisk. CA. Native.

Much more widely distributed in California, *S. lemmonii* is known in Oregon only from a few collections in Jackson County.

Silene menziesii Hooker (p. 573)

Menzies' catchfly

Silene menziesii Hook. var. *viscosa* (Greene) C.L. Hitchc. & Maguire, *Silene menziesii* Hook. var. *menziesii*

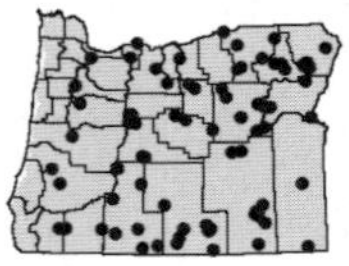

Plants perennial, sometimes matted; taproots slender. **Stems** decumbent to erect, 5–20(50) cm, branches many, puberulent to short, sometimes deflexed hairy, glandular-puberulent above. **Leaves** oblanceolate to elliptic, 2–6 cm × 3–20 mm, slightly reduced upward. **Inflorescences** terminal, open cymes, not 1-sided, or flowers solitary, axillary or terminal; pedicels glandular-hairy. **Flowers** usually bisexual, functionally unisexual, ascending to erect; calyces campanulate, not inflated in fruit, 5–7 mm, puberulent and glandular, obscurely 10-veined, not netted above, pale commissures absent, lobes 1–2 mm, tips subacute to acuminate; petal claws glabrous; appendages 2; limbs 1.5–3 mm, white, lobes 2; stamens included (pistillate flowers) or equaling petal claws (staminate flowers); styles 3(4), included (staminate flowers) or equaling petal claws (pistillate flowers). **Fruits** ovoid-ellipsoid, teeth 3(4), splitting to 6(8) teeth; stalks 1–2 mm. **Seeds** 0.5–1 mm, black, not winged. ***2n***=24, 48.

Coniferous or aspen woods, sage-juniper slopes, streambanks. Flowering Apr–Sep. 50–2600 m. All ecoregions except Est. CA, ID, NV, WA; western North America. Native.

Silene noctiflora Linnaeus

Night-flowering catchfly

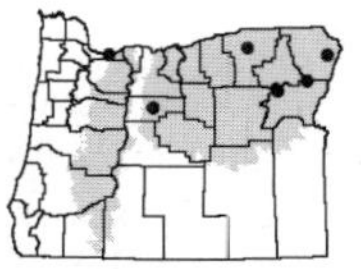

Plants annual; taproots slender. **Stems** generally erect, 20–60(80) cm, densely hairy, glandular especially above. **Leaves** basal elliptic to oblanceolate, 6–12(14) cm × 20–45 mm; upper cauline broadly elliptic to lanceolate, 1–7 cm × 3–12 mm, gradually reduced upward. **Inflorescences** terminal, open cymes, not 1-sided; pedicels covered with both short glandular and longer non-glandular hairs. **Flowers** bisexual, ascending to erect; calyces ovate-elliptic, inflated in fruit, 14–22 mm, hairs mostly glandular, prominently 10-veined, veins netted above middle of sepal, pale commissures present, lobes 6–13 mm, tips acuminate; petal claws glabrous; appendages 2; limbs 8–9 mm, white or yellowish white to pinkish, lobes 2, 4 mm; stamens included; styles 3, included. **Fruits** ovoid, teeth 6; stalks 1–3 mm. **Seeds** 1 mm, dark brown to black, not winged. ***2n***=24.

Open, disturbed areas, fields, lawns. Flowering May–Aug. 0–1400 m. BW, Casc, Col, Lava. CA, ID, WA; throughout North America except far north; Europe. Exotic.

First collected in Oregon from a grain field in 1907, *Silene noctiflora* appears to be far less common than *S. latifolia*, with which it is sometimes confused.

Silene nuda (S. Watson) C.L. Hitchcock & Maguire

Western fringed catchfly, sticky catchfly

Silene scaposa B.L. Rob. var. *lobata* C.L. Hitchc. & Maguire

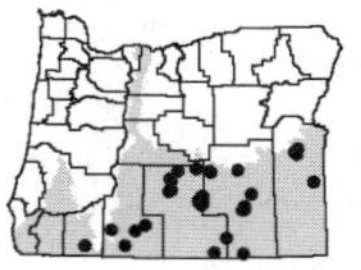

Plants perennial; caudices woody. **Stems** erect, 15–50 cm, branches few to many, puberulent, glandular-viscid above. **Leaves** basal narrowly oblanceolate, 2–10(20) cm × 2–12(20) mm, densely tufted, not fleshy; cauline in 1–3 pairs, linear-lanceolate, 1.2–4 cm × 1–4 mm, abruptly reduced upward, not overlapping pedicels. **Inflorescences** terminal and axillary open cymes, not 1-sided, or flowers solitary; pedicels glandular-puberulent. **Flowers** bisexual, erect; calyces campanulate, enlarging in fruit, 10–12 mm, glandular-viscid, prominently 10-veined, not netted above, pale commissures present, veins in lobes thickened and broader apically than commissural veins, commissural veins not forking/fusing with lobe veins, lobes 1.5–4 mm, tips acute to obtuse, with a broad scarious margin; petal claws ciliate at base; appendages 2–4; limbs 2.5–5.5 mm, off-white to dingy purple, lobes 4, 2–5 mm; stamens slightly exserted; styles 3–5, equaling petal claws. **Fruits** teeth 3–5; stalks 0.5–2 mm. **Seeds** 1.2–2 mm, brown, not winged. ***2n***=48.

Rocky slopes, outcrops, sage flats, lava craters. Flowering May–Jul. 900–2100 m. BR, ECas, Owy, Sisk. CA, ID, NV; CO. Native.

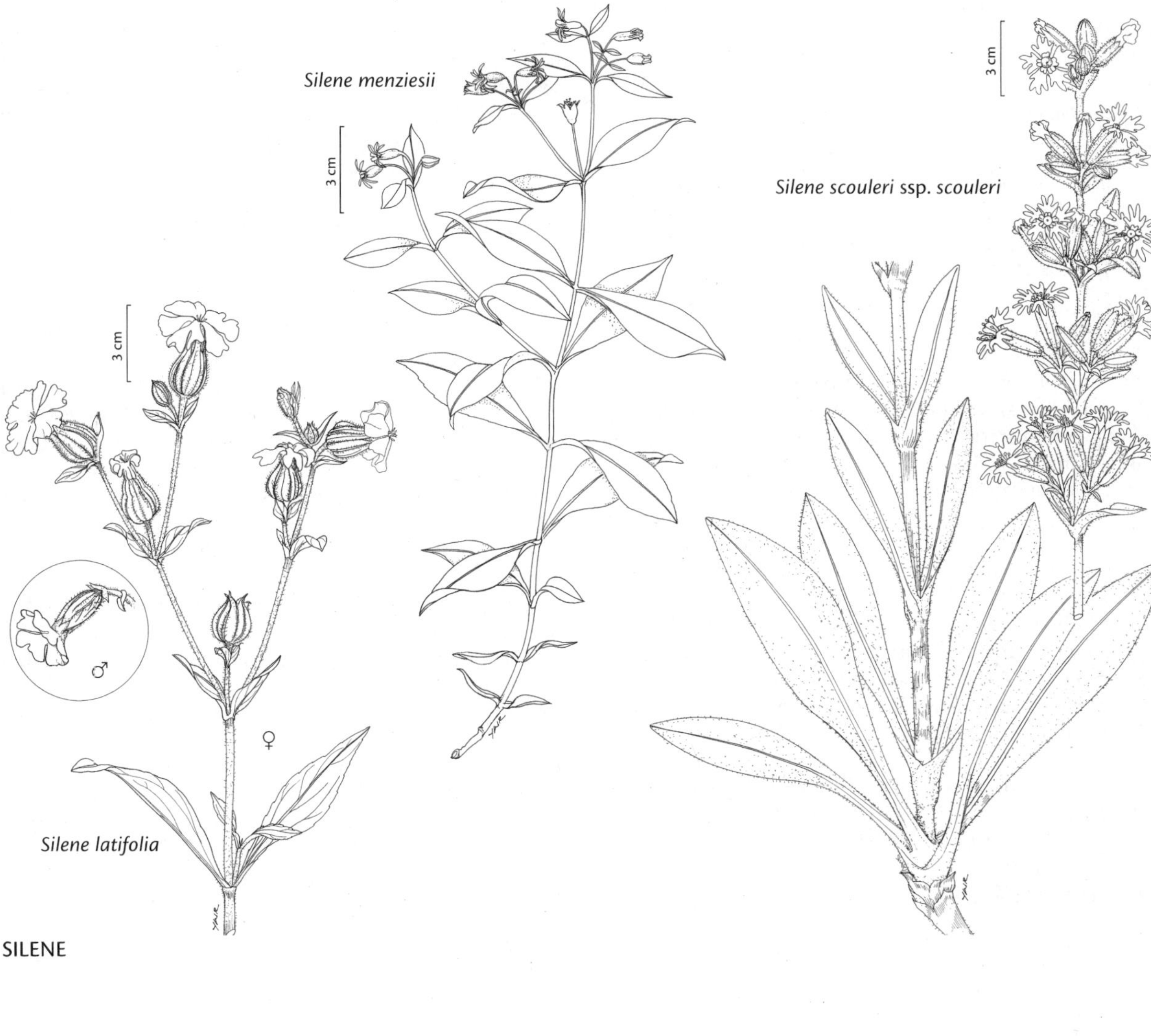

SILENE

We follow Holmgren and Holmgren (2012) in recognizing *S. nuda* as including *S. scaposa* var. *lobata*; see their discussion of the typification of these taxa for additional information. Although considered by Morton (2005) to have two-lobed petals, the original description clearly states "limbs of the petals 4-lobed" (Watson 1871).

Silene oregana S. Watson

Oregon catchfly

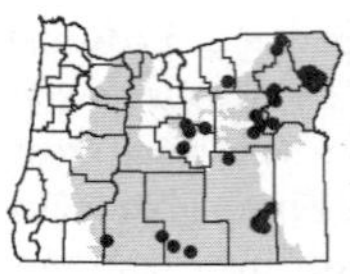

Plants perennial; caudices woody. **Stems** erect, 30–50(70) cm, branches 0–few, puberulent, glandular especially above. **Leaves** basal oblanceolate or spatulate, 5–8 cm × 7–12 mm, not tufted; upper lanceolate, 1–6(8) cm × 2–6 mm, gradually reduced upward, not overlapping pedicels. **Inflorescences** terminal and axillary, open cymes, not 1-sided, pedicels glandular hairy. **Flowers** bisexual, ascending; calyces narrowly campanulate, enlarging in fruit, 9–15 mm, glandular-puberulent, prominently 10-veined, not netted above, pale commissures present, lobes 2–3 mm, tips obtuse; petal claws glabrous; appendages 4–6; limbs 3–8 mm, white to pink, lobes 4–6, linear, some splitting further; stamens included to slightly exserted; styles 3(5), included to barely exserted. **Fruits** ellipsoid to ovoid, teeth 6(8, 10); stalks 3–4 mm. **Seeds** 1–2 mm, brown, not winged. ***2n***=48.

Dry slopes and cliffs, coniferous woods. Flowering Jun–Sep. 700–3000 m. BR, BW, Casc, ECas. CA, ID, NV, WA; northeast to MT, east to WY, southeast to UT. Native.

Silene parryi (S. Watson) C.L. Hitchcock & Maguire

Parry's catchfly

Silene douglasii Hook. var. *macounii* (S. Watson) B.L. Rob., *Silene macounii* S. Watson

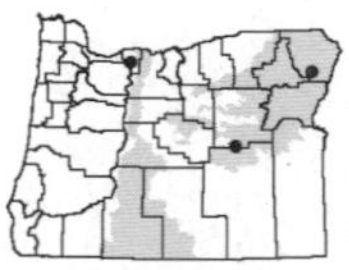

Plants perennial; caudices woody. **Stems** erect, (10)20–60 cm, branches few, puberulent, glandular-viscid above. **Leaves** basal spatulate or oblanceolate, 3–8 cm × 2–14 mm, densely tufted or not, not fleshy; cauline in 2–4 pairs, narrowly oblanceolate to linear-lanceolate, 2–8 cm × 2–8 mm, abruptly reduced upward, not overlapping pedicels. **Inflorescences** terminal, open cymes, not 1-sided, or flowers solitary; pedicels > calyx, glandular-viscid. **Flowers** bisexual, ascending to erect; calyces campanulate, inflated in fruit, 13–14 mm, glandular-hairy, prominently 10-veined, netted above, pale commissures present, veins in lobes not thickened or broader than commissural veins, commissural veins fork above and fuse with veins in lobes, lobes 2–3 mm, tips acute to rounded-obtuse; petal claws glabrous; appendages 2(4); limbs 5–7 mm, white, often green- or purple-tinged, lobes(2)4, 2 inner > outer; stamens equaling petal claws or barely exserted; styles 3(5), exserted. **Fruits** narrowly ovoid, teeth 3(5), splitting to 6, 8 or 10; stalks 2–3 mm. **Seeds** 1.5–2.5 mm, brown, not winged. ***2n***=48, 96.

Grassy meadows, rocky areas. Flowering Jun–Aug. 1400–2600 m. BW, ECas. ID, WA; north to AK, northeast to Alberta, east to WY, southeast to UT. Native.

Silene parryi can sometimes be confused with *S. douglasii*, *S. grayi*, and small plants of *S. scouleri*.

Silene pectinata S. Watson

Sticky catchfly

Silene insectivora L.F. Hend.

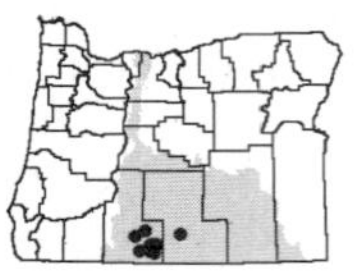

Plants perennial; caudices woody. **Stems** erect, 15–50 cm, branches 0–few, coarsely-hairy, glandular-viscid especially above. **Leaves** basal oblanceolate to elliptic, 6–15 cm × 10–30 mm, densely tufted, not fleshy; cauline in 2–3(4) pairs, linear to lanceolate, 0.8–4 cm × 3–8 mm, abruptly reduced above stem bases and much smaller than basal leaves. **Inflorescences** terminal, open cymes, not 1-sided, pedicels sometimes minute, glandular hairy. **Flowers** bisexual, erect or ascending; calyces tubular, enlarging in fruit, 12–16 mm, glandular hairy, prominently 10-veined, not netted above, pale commissures present, veins in lobes thickened and broader apically than commissural veins, commissural veins not forking/fusing with lobe veins, lobes 2–5 mm, margins narrow, scarious, tips blunt to acute; petal claws glabrous to ciliate at base; appendages 2; limbs 5–10 mm, pink, lobes 2, 1–3 mm; stamens included or equaling petal claws; styles 3–5, included or equaling petal claws. **Fruits** conic to ellipsoid, teeth 6(8, 10); stalks 1–2 mm. **Seeds** 1–1.5 mm, dark brown, not winged. ***2n***=48.

Dry meadows, alkali flats, rocky areas. Flowering Jun–Aug. 1100–1700 m. BR, ECas. CA, NV. Native.

We follow Holmgren and Holmgren (2012) in recognizing *S. pectinata* to include plants with two petal lobes that have most often been treated as *S. nuda*; see their discussion of the typification of *S. nuda* for additional information. Specimens originally identified as *S. nuda* ssp. *insectivora* would also key to *S. pectinata*. In Oregon, most populations of *S. pectinata* are found in Klamath County.

Silene scaposa B.L. Robinson

Robinson's catchfly

Silene scaposa B.L. Rob. var. *scaposa*

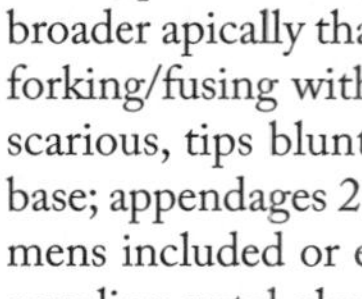

Plants perennial; caudices woody. **Stems** erect, 15–50 cm, branches few to many, puberulent, glandular-viscid above. **Leaves** basal narrowly oblanceolate, 2–10(20) cm × 2–12(20) mm, densely tufted, not fleshy; cauline in 1–3 pairs, linear-lanceolate, 1.2–4 cm × 1–4 mm, abruptly reduced upward. **Inflorescences** terminal and axillary, open cymes, not 1-sided, or flower ssolitary; pedicels glandular-puberulent. **Flowers** bisexual, erect; calyces campanulate, enlarging in fruit, 10–12 mm, glandular-viscid, prominently 10-veined, not netted above, pale commissures present, veins in lobes thickened and broader apically than commissural veins, commissural veins not forking/fusing with lobe veins, lobes 1.5–4 mm, with a broad scarious margin, tips acute to obtuse; petal claws ciliate below; appendages 2–4; limbs 2–5 mm, off-white to dingy purple, lobes 2, 1–2 mm; stamens slightly exserted; styles 3–5, equaling petal claws. **Fruits** teeth 3–5; stalks 1.5–2.5 mm. **Seeds** 1.2–2 mm, brown, not winged. ***2n***=48.

Moist hillsides, grassy slopes, coniferous woods. Flowering May–Jul. 700–2100 m. BW, Col, Lava, WV. Native.

Silene scaposa is defined here as having two petal lobes, with var. *lobata* (four petal lobes) placed into synonymy of *S. nuda*.

Silene scouleri Hooker ssp. ***scouleri*** (p. 573)

Scouler's catchfly

Silene grandis Eastw., *Silene pacifica* Eastw., *Silene scouleri* Hook. ssp. *grandis* (Eastw.) C.L. Hitchc. & Maguire, *Silene scouleri* Hook. var. *pacifica* (Eastw.) C.L. Hitchc., *Silene scouleri* Hook. var. *scouleri*

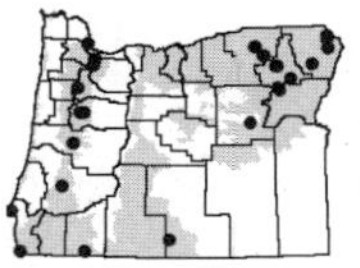

Plants perennial; caudices woody. **Stems** generally erect, 15–70 cm, generally unbranched, densely puberulent, glandular-viscid above. **Leaves** basal oblanceolate to elliptic, 5–10 cm × 12–20 mm, not tufted, not fleshy; cauline in 1–12 pairs, lanceolate to ovate, 2–6 cm × 3–15 mm, gradually reduced upward, not overlapping pedicels. **Inflorescences** terminal and axillary cymes, not 1-sided, often whorled at nodes; pedicels < calyx, glandular-puberulent. **Flowers** bisexual, spreading or ascending to erect; calyces tubular to narrowly campanulate, enlarging in fruit, 12–16 mm, glandular puberulent, prominently 10-veined, not netted above, pale commissures present, commissural veins fork above and fuse with lobe veins, lobes 2–4 mm, tips broadly obtuse; petal claws ciliate at base; appendages 2; limbs 3–5 mm, greenish white to rose, lobes 2–4(6), 2 inner > outer, larger lobes may be notched; stamens included or equaling petal claws; styles 3(4), equaling petal claws. **Fruits** ellipsoid to ovoid, teeth 6(8); stalks 3–6 mm. **Seeds** 1–1.5 mm, gray-brown, not winged. ***2n***=48.

Meadows, grassy and talus slopes, coastal bluffs. Flowering Jun–Aug. 0–2100 m. BW, Col, ECas, Est, Sisk, WV. CA, WA; north to British Columbia. Native.

Silene scouleri is a variable species, sometimes divided into three subspecies. Morton (2005) showed the presence of both ssp. *scouleri* and ssp. *hallii* in Oregon and annotated seven specimens at the Oregon State University Herbarium as "ssp. *hallii* varying to *S. oregana*." Morton noted that ssp. *hallii* is centered in Colorado, and most specimens identified to that taxon occur along the Rocky Mountains. Those seven specimens, most from northeastern Oregon, were annotated by S.C. Meyers as ssp. *scouleri*; we follow his annotations here.

Silene spaldingii S. Watson

Spalding's campion

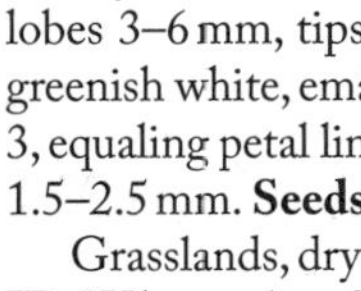

Plants perennial; caudices woody. **Stems** erect, 20–60 cm, branches few to many, villous and glandular-viscid. **Leaves** lanceolate, 3–7 cm × 5–15 mm, largest near middle of stem. **Inflorescences** terminal and axillary, open cymes, not 1-sided; pedicels glandular-viscid. **Flowers** bisexual, spreading or ascending to erect; calyces tubular-campanulate, enlarging in fruit, 10–15 mm, glandular-viscid, obscurely 10-veined, not netted above, pale commissures absent, lobes 3–6 mm, tips blunt; petal appendages 4(6); limbs 2 mm, greenish white, emarginate; stamens equaling petal limbs; styles 3, equaling petal limbs. **Fruits** ellipsoid to oblong, teeth 6; stalks 1.5–2.5 mm. **Seeds** ~2 mm, yellowish brown, winged. ***2n***=48.

Grasslands, dry areas. Flowering Jul–Sep. 600–1600 m. BW. ID, WA; north to British Columbia, northeast to MT. Native.

Silene spaldingii is a federally listed threatened species (listed endangered by the State of Oregon), and is known from only a few populations in Wallowa County.

Silene suksdorfii B.L. Robinson

Suksdorf's catchfly

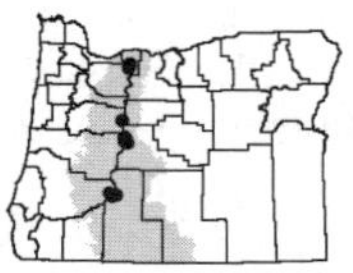

Plants perennial; caudices woody. **Stems** decumbent to erect, 3–15 cm, branches many, puberulent, glandular-viscid above. **Leaves** basal narrowly oblanceolate, 0.5–4.5 cm × 1.5–3.5 mm, densely tufted, fleshy; cauline in 1–3 pairs, linear-lanceolate, 0.5–1.5 cm × 1–2 mm, abruptly reduced upward. **Inflorescences** terminal, open cymes, not 1-sided, or flowers solitary; pedicels glandular-viscid. **Flowers** bisexual, erect; calyces campanulate, not inflated in fruit, 9–12 mm, short-glandular-hairy, prominently 10-veined, not netted above, pale commissures present, commissural veins fork above and fuse with lobe veins, becoming papery, cell walls of hairs purple, lobes 1–2 mm, tips acute to obtuse; petal claws ciliate at base; appendages 2; limbs 3–5 mm, white to purplish, lobes 2; stamens equaling petal claws; styles 3(4), equaling petal claws. **Fruits** ovoid, teeth 6(8); stalks 2–3.5 mm. **Seeds** 1–2 mm, brown, winged. ***2n***=48.

Rocky crevices, ridges, slopes. Flowering Jul–Sep. 1600–2800 m. Casc, ECas. CA, WA. Native.

Silene vulgaris (Moench) Garcke

Bladder campion

Silene cucubalus Wibel

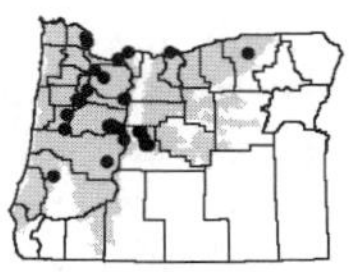

Plants perennial; caudices woody. **Stems** decumbent to erect, 20–80 cm, branches few, glabrous to glaucous. **Leaves** lower cauline lanceolate to oblanceolate, 4–8 cm × 5–20 mm; upper cauline lanceolate to ovate, 3–4.5 cm × 5–15 mm, gradually reduced upward. **Inflorescences** terminal, open cymes, not 1-sided; pedicels glabrous. **Flowers** bisexual or unisexual (pistillate), nodding to spreading; calyces campanulate, inflated and becoming papery in fruit, 7–10 mm, glabrous, obscurely 10–15-veined, netted throughout, pale commissures absent, lobes 2–3 mm, acute to mucronate; petal claws glabrous; appendages 0 or 2, minute; limbs 7 mm, white, emarginate or lobes 2, 3–4 mm; stamens exserted; styles 3, exserted. **Fruits** subspherical, teeth 6; stalks 2–3 mm. **Seeds** 1–1.5 mm, black, not winged. ***2n***=24.

Roadsides, fields, forest openings. Flowering Jun–Aug. 0–1300 m. Casc, Col, CR, Lava, WV. CA, ID, WA; throughout North America except far north and southeast; Europe. Exotic.

A widespread introduced species, *S. vulgaris* was first gathered in Oregon on ballast at Linnton (Portland) in 1911.

SPERGULA Spurrey

Plants annual. **Stems** ascending to erect. **Leaves** opposite, appearing whorled due to axillary clusters of 16–30 per node, linear; petioles 0; stipules 2 per node, ovate to triangular, scarious. **Inflorescences** terminal cymes, bracts present. **Flowers** usually bisexual, rarely pistillate via stamen abortion; sepals 5, usually free, glandular-hairy; petals 5, entire; stamens 10, arising on hypanthium rim, staminodes 0; styles and stigmas 5. **Fruits** capsules, opening by 5 spreading to recurved valves. **Seeds** 5–25, spherical, 1–1.5 mm, blackish, minutely roughened or white-papillate, winged.

Asia, Europe, North America, South America. 5 species; 1 species treated in *Flora*.

Spergula arvensis Linnaeus (p. 576)

Corn spurrey, stickwort

Spergula arvensis L. ssp. *arvensis*, *Spergula arvensis* L. var. *arvensis*, *Spergula arvensis* L. var. *sativa* (Boenn.) Rchb.

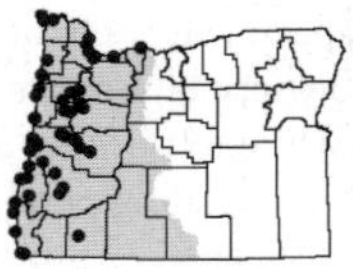

Stems 10–40+ cm, often branched at base, glabrous or usually glandular-hairy. **Leaves** 1–5 cm, margins often strongly revolute. **Inflorescences** bracts often purplish; pedicels erect to ascending in flower, spreading to reflexed in fruit. **Flowers** sepals elliptic to ovate, 2.5–5 mm; petals elliptic to ovate, 2.5–4 mm. **2*n***=18, 36.

Disturbed areas, cultivated fields, open fields, beaches. Flowering Apr–Jun. 0–400 m. Casc, CR, ECas, Est, Sisk, WV. CA, ID, WA; scattered throughout much of northern North America; worldwide. Exotic.

SPERGULARIA Sand-spurrey

Plants annual or perennial. **Stems** erect to sprawling. **Leaves** opposite, axillary clusters often present, 2–4+ per node, thread-like to linear; petioles 0; stipules 2 per node, lanceolate to broadly triangular, not partially fused to leaf bases, scarious. **Inflorescences** terminal cymes or flowers solitary and axillary, flowers few to many; bracts present. **Flowers** bisexual; sepals 5, basally united 20% of their length, glabrous to glandular-hairy, spurs 0; petals 5, entire; stamens 2–10, arising on hypanthium rim, staminodes 0; styles and stigmas 3. **Fruits** capsules, opening by 3 spreading valves. **Seeds** 30–150+, spherical to angular, 0.4–1.1 mm, brown, red-brown, or black, smooth, sculptured, papillate or not, often winged.

Central America, Mediterranean region, western North America, South America. ~60 species; 5 species treated in *Flora.*

Reports of two species of *Spergularia* from Oregon are based on collections from ballast near Portland between 1902 and 1916. *Spergularia bocconi* is most similar to *S. rubra* and is distinguished by having moderately fleshy leaves, dull, white to tan, deltate, somewhat inconspicuous stipules, and light brown seeds. *Spergularia villosa* is a woody-based perennial similar to *S. macrotheca*,

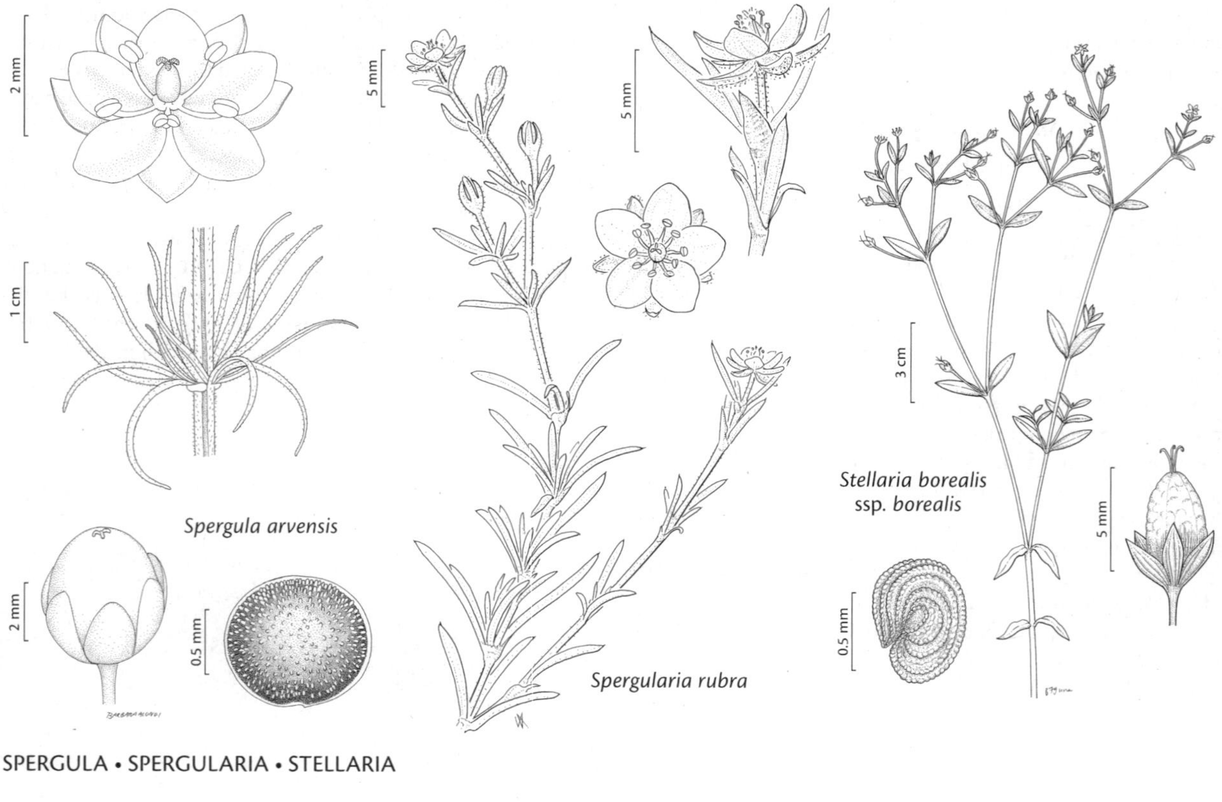

SPERGULA • SPERGULARIA • STELLARIA

differing in having shorter sepals and styles and smaller seeds. A third species, *Spergularia media*, was collected from ballast in 1902, and Rossbach (1940) cited a specimen (*Nelson 2355*, GH) from a salt marsh near Toledo in 1918. It would also key near *S. rubra* but has fleshy leaves and larger (0.8–1.1 mm) seeds, which are smooth and usually winged. Each of the species is known from California and might be expected to reappear in saline areas of Oregon.

Key to *Spergularia* species:

1. Plants perennial, with woody base; stamens 9–10 . . . . . ***S. macrotheca*** var. ***macrotheca***
1' Plants annual or short-lived perennial, lacking woody base; stamens 2–10.
  2. Seeds black. . . . . ***S. diandra***
  2' Seeds light to dark brown or red-brown.
    3. Stamens 6–10; seeds not winged. . . . . ***S. rubra***
    3' Stamens 2–5; seeds winged or not.
      4. Seeds 0.9–1.1 mm, wings often present . . . . . ***S. canadensis*** var. ***occidentalis***
      4' Seeds 0.5–0.7(0.8) mm, wings usually absent. . . . . ***S. marina***

Spergularia canadensis (Persoon) G. Don var. ***occidentalis*** R. Rossbach

Canadian sand-spurrey

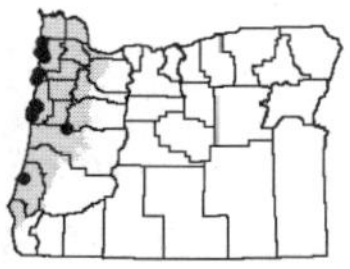

Plants annual; from slender taproots. **Stems** 7–25 cm. **Leaves** axillary leaf clusters usually absent; blades linear, 10–28 mm, fleshy; stipules inconspicuous, widely triangular, 1–2.7 mm, dull white, tips obtuse to acute. **Flowers** sepal lobes 2.5–3.5 mm, up to 4.5 mm in fruit; petals white or pink; stamens 2–4; styles 0.3–0.7 mm. **Seeds** 0.9–1.1 mm, red-brown, nearly smooth, minutely glandular-hairy, not papillate, often winged. **2*n***=36.

Salt marshes, tidal flats. Flowering Jun–Oct. 0–100 m. CR, Est, WV. CA, WA; north to AK. Native.

Spergularia diandra (Gussone) Heldreich

Alkali sand-spurrey

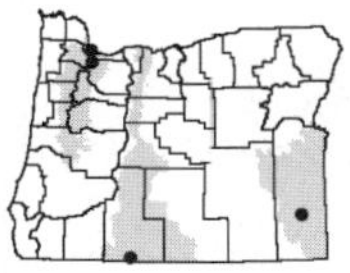

Plants annual; from slender taproots. **Stems** 5–15 cm. **Leaves** axillary leaf clusters usually absent; blades linear, 6–23 mm, slightly fleshy; stipules inconspicuous, widely triangular, 1.5–2 mm, dull to silvery tan, tips acuminate. **Flowers** sepal lobes 2.3–3.1 mm, barely longer in fruit; petals white; stamens 4–7; styles 0.4–0.6 mm. **Seeds** 0.4–0.6 mm, black, shiny, surface with low, elongate tubercles, not papillate, not winged. **2*n***=18.

Basalt scabland, sand and gravel bars. Flowering May–Sep. 0–700 m. ECas, Owy, WV. WA, ID; north to British Columbia, northeast to Saskatchewan; Mediterranean Africa and Europe, southwestern Asia. Exotic.

Spergularia macrotheca (Chamisso & Schlechtendal) Heynhold var. ***macrotheca***

Beach or sticky sand-spurrey

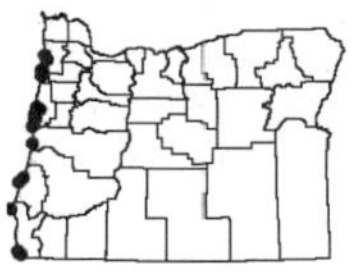

Plants perennial; from woody taproots. **Stems** 5–35 cm. **Leaves** 1 or 2+ leaves per axillary leaf cluster; blades linear, 10–40 mm, fleshy; stipules conspicuous, narrowly triangular, 4.5–11 mm, dull white to tan, tips long-acuminate. **Flowers** sepal lobes (4.5)5–7 mm, up to 8 mm in fruit; petals pink to rosy or blue; stamens 9–10; styles 0.5–3 mm. **Seeds** 0.6–0.8 mm, red-brown, smooth or surface tubercled, sculpted with lines, or of low rounded mounds, not papillate, often winged. **2*n***=36, 72.

Salt marshes and flats, sandy banks and cliffs. Flowering May–Aug. 0–50 m. Est. CA, WA; north to British Columbia, south to Mexico. Native.

Spergularia marina (Linnaeus) Grisebach

Salt-marsh sand-spurrey, salt sand-spurrey

Spergularia salina J. Presl & C. Presl

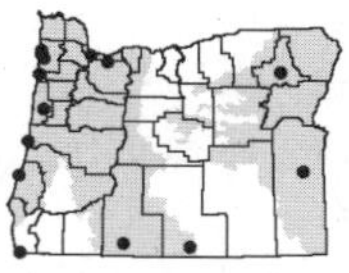

Plants annual; from slender taproots. **Stems** 4–25 cm. **Leaves** axillary leaf clusters absent; blades linear, 15–40 mm, fleshy; stipules inconspicuous, widely triangular, 1.2–3.5 mm, dull white, tips acute or short-acuminate. **Flowers** sepal lobes 2.5–4.5 mm, up to 4.8 mm in fruit; petals white or pink to rosy; stamens 2–5; styles 0.4–0.7 mm. **Seeds** 0.5–0.7(0.8) mm, light brown to red-brown, smooth or roughened, papillate or not, not usually winged. **2*n***=18?, 36.

Salt marshes, roadsides. Flowering May–Aug. 0–1400 m. BW, Casc, CR, ECas, Est, Owy, WV. CA, ID, NV, WA; most of North America; Asia, Europe. Native?

Formerly only found in salt marshes, *S. marina* has been collected twice along roadsides since 2005.

Spergularia rubra (Linnaeus) J. Presl & C. Presl (p. 576)

Red sand-spurrey

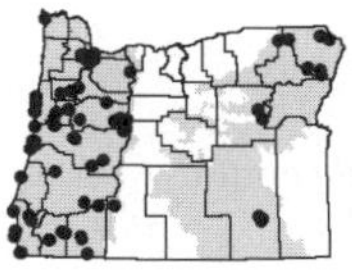

Plants annual or short-lived perennial; from slender to stout taproots. **Stems** 8–25 cm. **Leaves** 2–4+ leaves per axillary cluster; blades thread-like to linear, 4–15 mm, barely fleshy; stipules conspicuous, lanceolate, 3.5–5 mm, shiny, white, tips long-acuminate or apiculate. **Flowers** sepal lobes 2–3.2 mm, up to 4 mm in fruit; petals pink; stamens 6–10; styles 0.6–0.8 mm. **Seeds** 0.4–0.6 mm, red-brown to dark brown, surface sculpted with lines, minutely papillate, not winged. **2*n***=18, 27, 36, 54.

Roadsides, slopes, waste areas, riverbanks, lawns. Flowering Apr–Oct. 0–2300 m. BR, BW, Casc, CR, Est, Sisk, WV. CA, ID, NV, WA; north to AK, northeast to MT, east to CO, southeast to NM, eastern North America; Asia, Europe. Exotic.

STELLARIA Starwort, chickweed

Plants annual or perennial; rhizomes, when present, without tuber-like thickenings. **Stems** erect to prostrate. **Leaves** opposite, linear to ovate; petioles present (basal leaves) or 0; stipules 0. **Inflorescences** terminal or axillary cymes, or flowers solitary; bracts present; usually pedicellate. **Flowers** bisexual; sepals (4)5, free, glabrous to glandular-hairy; petals (0)1–5, white, tips 2-lobed > 20% length or often divided nearly to base; stamens (1)5 or 10, borne on nectariferous disk, staminodes 0; styles 3(4–5). **Fruits** capsules, ovoid or spherical to cylindric-oblong, opening by 6(8, 10) ascending to recurved valves. **Seeds** 3–20+, globose to compressed-ellipsoid, 0.5–1.3 mm, yellowish to dark brown, papillate or tuberculate, rarely smooth.

Worldwide, especially in north temperate regions. ~190 species; 11 species treated in *Flora.*

The hybrid *S. borealis* × *S. longifolia* has been collected once in Oregon along a stream in Klamath County in 1928. As expected, this sterile hybrid exhibits some features of both parents; the combination of lanceolate to lance-linear leaves, a lateral inflorescence, the lowermost flowers subtended by large, leafy bracts and all others subtended by much reduced scarious bracts, and an absence of capsules can be used to recognize this plant.

Two additional species closely resembling, and related to, *Stellaria media* are to be expected in Oregon. *Stellaria neglecta* (greater chickweed) has been collected in California and from Whidbey Island, Washington in 2004. It resembles a large *S. media* plant and is most consistently distinguished by having sepals 5–6 mm and seeds with conical, rather than rounded, tubercles, especially on the marginal face. *Stellaria pallida* (lesser chickweed) is also known from California and from several sites in Washington that have been made since 2000. This species resembles a smaller *S. media* with apetalous flowers with sepals 2–3 mm and seeds that are 0.7–0.8 mm and yellow-brown.

Key to *Stellaria* species:

1. Plants annual, from a taproot.
    2. Leaves crowded near base, upper linear-lanceolate, rarely ovate; internode hairs absent or scattered; bracts scarious . . . . ***S. nitens***
    2' Leaves evenly spaced, mostly ovate; internode hairs in a line; bracts leafy . . . . ***S. media***

1' Plants perennial, from slender rhizomes.
    3. Flowers solitary and axillary.
        4. Petals 5, ≥ sepals; leaves fleshy . . . . ***S. humifusa***
        4' Petals usually 0; leaves not fleshy.
            5. Sepals (4)5, acute or acuminate, veins in fruit 3, prominent; leaf blade margins generally wavy, glabrous . . . . ***S. crispa***
            5' Sepals usually 4(5), obtuse, veins in fruit 1, obscure; leaf blade margins generally flat, ciliate near base . ***S. obtusa***
    3' Flowers in terminal, axillary, or lateral cymes.

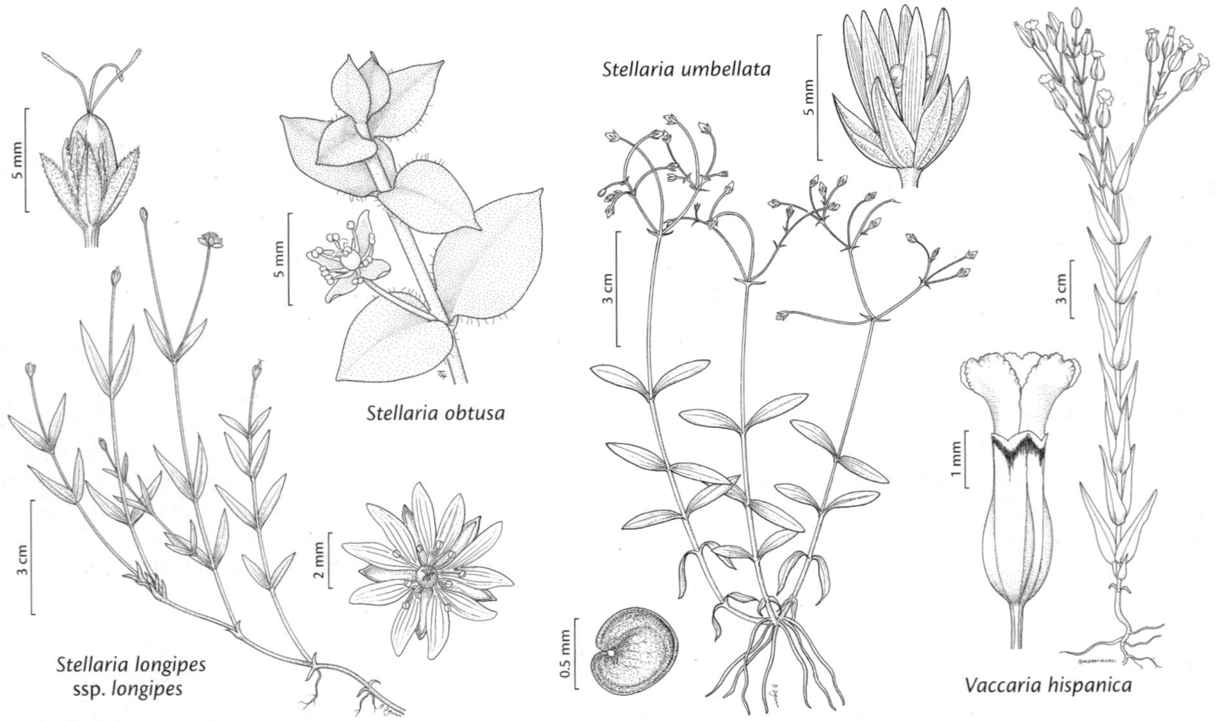

STELLARIA • VACCARIA

6. Petals > sepals.
   7. Inflorescences narrow; pedicels ascending to erect, flowers usually few (1–7) . . . . ***S. longipes*** ssp. ***longipes***
   7' Inflorescences open, branches widely spreading; pedicels spreading or reflexed, flowers generally many.
      8. Sepals 4–5.5 mm, margins densely ciliate or not; inflorescences terminal; seeds with prominent, elongate tubercles; leaf blade margins not papillate, shiny . . . . ***S. graminea***
      8' Sepals 3–4 mm, margins glabrous; inflorescences lateral; seeds minutely roughened; leaf blade margins papillate (at 20× magnification), dull . . . . ***S. longifolia***

6' Petals < 0.8 × length of sepals or 0.
   9. Flowers in an umbel-like cyme; bracts scarious; petals 0 or rarely minute. . . . . ***S. umbellata***
   9' Flowers in terminal cyme; bracts leafy (or uppermost rarely scarious); petals present or rarely absent.
      10. Leaves elliptic to ovate, primary blades mostly < 2.5 cm, margins not papillate, shiny; stem internodes not papillate; sepals 1.5–2.5 mm, ≤ 3 mm in fruit, with 1–3 obscure veins; styles 0.5–0.9 mm, yellowish or white, thick, curved outward . . . . ***S. calycantha***
      10' Leaves lanceolate to widely so, primary blades mostly > 2 cm, margins papillate (at 20× magnification) or not, dull; stem internodes finely papillate or not; sepals 1.5–3.5 mm, to 4.5 mm in fruit, with 1 or usually 3 prominent veins; styles generally 0.9–1.6 mm, white, slender, erect, later spreading . . . . ***S. borealis***

Stellaria borealis Bigelow

Boreal starwort

Plants perennial; from slender rhizomes. **Stems** sprawling to erect, 10–50 cm, internodes finely papillate or glabrous. **Leaves** evenly spaced, oblong to broadly lanceolate, 10–45 mm, not fleshy, margins papillate (at 20×) or not, sometimes ciliate near base. **Inflorescences** terminal or rarely axillary, flowers several to many; bracts leafy or upper scarious; pedicels erect to ascending in flower, erect, arching, or often sharply reflexed in fruit. **Flowers** sepals 5, lanceolate, 1.5–3.5 mm, to 4.5 mm in fruit, glabrous, acute, margins widely scarious, veins in fruit 1–3, visible near base, prominent; petals 5 or sometimes fewer or 0, 0.5–0.8 × length of sepals; styles erect or later spreading, slender, 0.9–1.6 mm, white. **Fruits** ovoid, brown or tan. **Seeds** 0.7–1.2 mm, red-brown, smooth or minutely roughened or tuberculate. ***2n***=52.

2 varieties.

Key to *Stellaria borealis* varieties:

1. Fruiting pedicels arching, often sharply reflexed; leaf blade margins papillate (at 20× magnification), dull; stem internodes finely papillate; sepals with 3 prominent veins in fruit; seeds roughened with low, elongate tubercles . . . . ***S. b.*** ssp. ***sitchana***

1' Fruiting pedicels erect or sometimes arching; leaf blade margins not papillate, often ciliate near the base; stem internodes glabrous; sepals with 1–3 veins visible near base in fruit; seeds smooth or minutely roughened . . . . ***S. b.*** ssp. ***borealis***

Stellaria borealis Bigelow ssp. ***borealis*** (p. 576)

Boreal starwort

Stellaria borealis Bigelow var. *bongardiana* Fernald, *Stellaria calycantha* (Ledeb.) Bong. var. *floribunda* Fernald

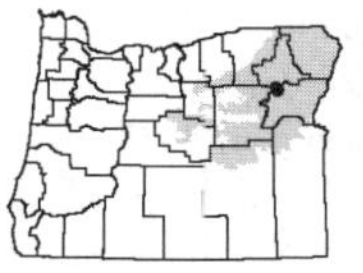

Stems erect, 10–40 cm, internodes glabrous. **Leaves** oblong to lanceolate, 10–30 mm, margins not papillate, ciliate near base. **Inflorescences** terminal or rarely axillary; bracts leafy or upper scarious; pedicels erect or sometimes arching. **Flowers** sepals 1.5–3 mm, to 3.5 mm in fruit, veins in fruit 1–3, visible near base; petals 5, sometimes fewer or 0. **Fruits** brown, opaque. **Seeds** 0.7–0.9 mm, red-brown, smooth or minutely roughened.

Creek banks. Flowering May–Sep. 0–3500 m. BW. CA, ID, NV, WA; north to AK, northeast to MT, east to CO, northeastern North America; circumboreal. Native.

Stellaria borealis ssp. *borealis* has only been collected in Oregon on one occasion in 1915 in Baker County.

Stellaria borealis Bigelow ssp. ***sitchana*** (Steudel) Piper & Beattie

Sitka starwort

Stellaria calycantha (Ledeb.) Bong. var. *sitchana* (Steud.) Fernald, *Stellaria borealis* Bigelow var. *sitchana* (Steud.) Fernald, *Stellaria calycantha* (Ledeb.) Bong. var. *bongardiana* Fernald

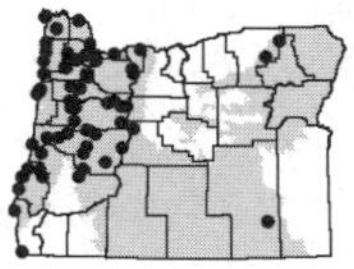

Stems sprawling to erect, 15–50 cm, internodes generally finely papillate. **Leaves** lanceolate to widely so, 15–45 mm, margins papillate (at 20× magnification), flat, dull, sometimes ciliate near base. **Inflorescences** terminal or seldom axillary; bracts leafy; pedicels arching, often sharply reflexed. **Flowers** sepals 3–3.7 mm, to 4.5 mm in fruit, veins in fruit 3, prominent; petals 5 or sometimes fewer. **Fruits** tan, translucent. **Seeds** 1–1.2 mm, dark red-brown; tubercles elongate, low, rough.

Grassy swales, swamps, wet meadows, streambanks, damp sands, moist coniferous forests. Flowering May–Sep. 0–2100 m. BR, BW, Casc, CR, ECas, Est, WV. CA, ID, WA; north to AK, northeast to MT, east to SD, southeast to UT. Native.

Stellaria calycantha (Ledebour) Bongard

Northern starwort

Stellaria borealis Bigelow var. *simcoei* (Howell) Fernald, *Stellaria calycantha* (Ledeb.) Bong. var. *calycantha*, *Stellaria calycantha* (Ledeb.) Bong. var. *simcoei* (Howell) Fernald, *Stellaria simcoei* (Howell) C.L. Hitchc.

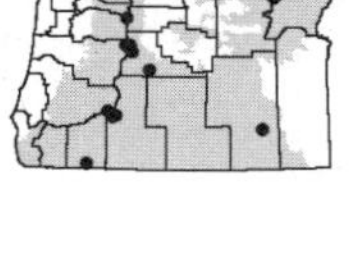

Plants perennial; from slender rhizomes. **Stems** prostrate to erect, 5–25 cm, internodes glabrous or with wavy scattered hairs, not papillate. **Leaves** evenly spaced, elliptic to ovate, 3–25 mm, not fleshy, margins not

papillate, flat to wavy, shiny, ciliate or not. **Inflorescences** terminal or axillary, flowers 1–10; bracts leafy; pedicels ascending to erect, sometimes curved in fruit. **Flowers** sepals 5, ovate to elliptic, 1.5–2.5 mm, to 3 mm in fruit, acute or rarely obtuse, glabrous, margins scarious, veins in fruit 1–3, obscure; petals 5, sometimes fewer or 0, 0.3–0.5 × length of sepals; styles curved outward, 0.5–0.9 mm, yellowish or white, thick. **Fruits** globose to globose-ovoid, greenish maturing to dark brown to black. **Seeds** 0.7–0.9 mm, red-brown, nearly smooth or minutely roughened. **2*n***=26.

Mossy banks, bogs, wet meadows, shaded areas. Flowering Jun–Aug. 1200–2600 m. BR, BW, Casc, ECas, Sisk. CA, ID, NV, WA; north to AK, northeast to MT, east to WY, southeast to NM; northeastern Asia. Native.

Stellaria calycantha and *S. crispa* have often been mistaken for each other. *S. calycantha* plants usually have a small terminal inflorescence, flowers usually with small petals, thick styles, and a globose capsule and are usually found at sites above 1700 m. *Stellaria crispa* flowers are strictly axillary, apetalous, with longer, thinner styles, and an ovoid capsule.

Stellaria crispa Chamisso & Schlechtendal

Crisp starwort

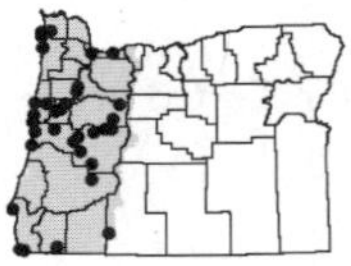

Plants perennial; from slender rhizomes. **Stems** prostrate to trailing, 10–40 cm, internodes glabrous. **Leaves** evenly spaced, broadly elliptic to ovate, 8–20 mm, not fleshy, margins not papillate, generally wavy, shiny, glabrous. **Inflorescences** flowers solitary in axils; pedicels ascending in flower, spreading to recurved in fruit. **Flowers** sepals (4)5, lanceolate, 2.5–4 mm, acute to acuminate, glabrous, margins scarious, veins in fruit 3, prominent; petals usually 0. **Fruits** ovoid to ovoid-ellipsoid, tan to brown. **Seeds** 0.8–1 mm, red-brown, tubercles low, elongate. **2*n***=26, 52.

Moist woods, boggy flats, moist shaded roadsides. Flowering Apr–Aug. 0–2100 m. Casc, CR, Est, Sisk, WV. CA, ID, WA; north to AK, northeast to Alberta. Native.

Stellaria graminea Linnaeus

Common starwort

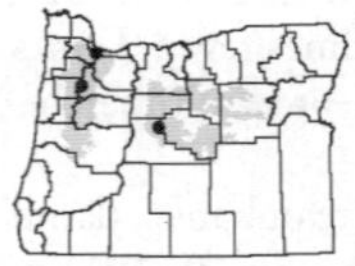

Plants perennial; from slender rhizomes. **Stems** sprawling to erect, 10–60 cm, internodes glabrous. **Leaves** evenly spaced, linear to lanceolate, 10–35 mm, not fleshy, margins not papillate, shiny, ciliate near base. **Inflorescences** terminal, flowers many; bracts scarious; pedicels spreading to erect, generally widely divergent in fruit. **Flowers** sepals 5, lanceolate or narrowly so, 4–5.5 mm, acute, margins widely scarious, densely ciliate or not, veins in fruit 3, prominent; petals 5, 1–1.4 × length of sepals. **Fruits** narrowly ovoid, greenish or tan. **Seeds** 0.9–1.1 mm, dark brown, tubercles prominent, elongate. **2*n***=39, 52.

Lawns, disturbed areas. Flowering Jun–Jul. 900–1700 m. Lava, WV. CA, ID, NV, WA; north to British Columbia, northeast to MT, eastern North America; China, Europe. Exotic.

Stellaria humifusa Rottbøll

Salt-marsh starwort

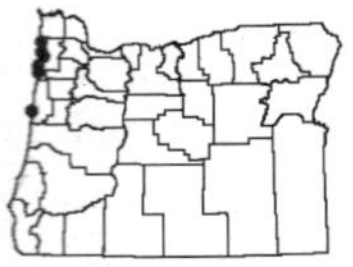

Plants perennial; from slender rhizomes. **Stems** reclining to ascending, 2–20 cm, internodes glabrous. **Leaves** evenly spaced, elliptic to elliptic-lanceolate, 4–15 mm, fleshy, margins not papillate, flat, shiny, glabrous or sparsely ciliate. **Inflorescences** flowers solitary in axils; pedicels ascending. **Flowers** sepals 5, lanceolate, 4–5 mm, acute, glabrous, margins narrowly scarious, veins in fruit 1–3, prominent; petals 5, 1 × length of sepals. **Fruits** ovoid, tan. **Seeds** 0.8–1 mm, pale brown, smooth or minutely roughened. **2*n***=26.

Marshland and tidal flats, beaches. Flowering Jul–Sep. 0–50 m. Est. WA; north to AK; circumboreal. Native.

This is a protected species in Oregon at the southern edge of its range in western North America.

Stellaria longifolia Willdenow

Long-leaved starwort

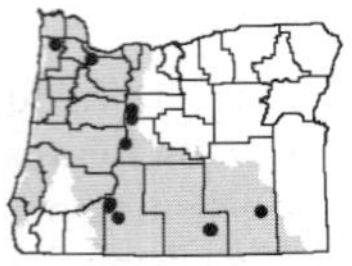

Plants perennial; from slender rhizomes. **Stems** sprawling to ascending, 15–40 cm, internodes minutely scabrous. **Leaves** evenly spaced, linear to linear-lanceolate, 15–35 cm, not fleshy, margins papillate (at 20× magnification), flat, dull, ciliate near base or not. **Inflorescences** lateral, flowers several or generally many; bracts scarious; pedicels spreading to erect in flower, widely divergent in fruit. **Flowers** sepals 5, lanceolate to narrowly elliptic, 3–4 mm, acute, glabrous, margins widely scarious, veins in fruit 3, visible near base; petals 5, 1–1.2 × length of sepals. **Fruits** ovoid-conic, dark purple or tan. **Seeds** 0.8–0.9 mm, red-brown, minutely roughened. **2*n***=26.

Swamps, damp thickets, creekbanks. Flowering Jun–Aug. 800–2500 m. BR, Casc, CR, ECas, WV. CA, ID, NV, WA; throughout Canada and US except southeastern US; circumboreal. Native.

Stellaria longipes Goldie ssp. *longipes* (p. 578)

Goldie's starwort

Stellaria longipes Goldie var. *altocaulis* (Hultén) C.L. Hitchc., *Stellaria longipes* Goldie var. *laeta* (Richardson) S. Watson, *Stellaria longipes* Goldie var. *longipes*

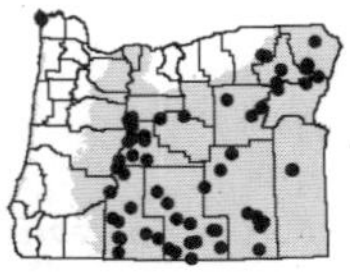

Plants perennial; from slender rhizomes. **Stems** ascending to erect, 5–35 cm, internodes glabrous or hairs scattered, wavy. **Leaves** evenly spaced, linear to linear-lanceolate, 10–40 cm, not fleshy, margins not papillate, flat, shiny, sometimes ciliate near base. **Inflorescences** terminal or axillary, flowers few (1–7) in a narrow cyme; bracts leafy or scarious; pedicels ascending to erect in flower, straight in fruit. **Flowers** sepals 5, lanceolate to ovate-lanceolate, 3–5.5 mm, acute, glabrous, margins widely scarious, veins in fruit 3, prominent; petals 5, 1–1.2 × length of sepals. **Fruits** ovoid to narrowly ovoid, dark purple. **Seeds** 0.7–1 mm, red-brown, minutely roughened. **2*n***=mostly 52, 78, 104.

Streambanks, moist or boggy meadows, seeps, moist forests. Flowering May–Aug. 0–2500 m. BR, BW, Casc, ECas, Est, Lava, Owy. CA, ID, WA; western and northern North America; circumboreal. Native.

Stellaria media (Linnaeus) Villars

Common chickweed

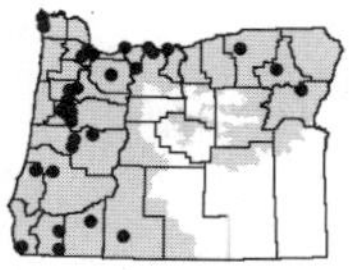

Plants annual but often over-wintering; from taproots. **Stems** prostrate to erect, 7–50 cm, internode hairs in a line. **Leaves** evenly spaced, ovate to broadly elliptic, 8–45 mm, not fleshy, margins not papillate, flat, shiny, often ciliate near base. **Inflorescences** terminal or axillary, flowers few to many; bracts leafy; pedicels spreading to erect in flower, curved to reflexed in fruit. **Flowers** sepals 5, lanceolate to ovate, 3–4.5 mm, to 6+ mm in fruit, acute to obtuse, glabrous or glandular-hairy, margins widely scarious, veins in fruit often 1 or 3 near base; petals 5, 0.7–0.9 × length of sepals. **Fruits** ovoid-oblong, green or tan. **Seeds** 0.9–1.3 mm, dark red-brown to brown, tubercles rounded. ***2n***=40, 42, 44.

Disturbed areas, cultivated fields, open woods, streambanks. Flowering Feb–Sep. 0–1500 m. BW, Casc, Col, CR, ECas, Est, Sisk, WV. CA, ID, NV, WA; cosmopolitan. Exotic.

Often a pernicious urban weed; this species is sometimes difficult to distinguish from *S. neglecta* and *S. pallida*.

Stellaria nitens Nuttall

Shining starwort

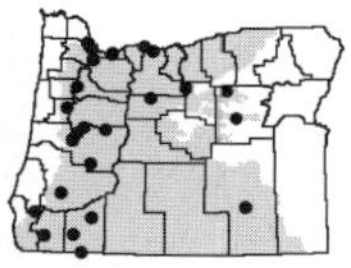

Plants annual; from taproots. **Stems** ascending to erect, 3–25 cm, internode hairs absent or scattered. **Leaves** crowded near base, 5–15 mm, lower oblanceolate to obovate; upper linear-lanceolate, rarely ovate, not fleshy, margins not papillate, shiny, often ciliate. **Inflorescences** terminal, flowers few to several; bracts scarious; pedicels ascending to erect in flower, generally straight in fruit. **Flowers** sepals 5, lanceolate to needle-like, 2.8–4.2 mm, long-acuminate, glabrous, margins widely scarious, veins in fruit 3, prominent; petals 5, 0.2–0.5 × length of sepals or 0. **Fruits** narrowly ovoid, green or tan. **Seeds** 0.5–0.7 mm, brown, tubercles minute. ***2n***=20, 40.

Dry bluffs, rocky slopes, meadows, roadsides, open woodlands. Flowering Mar–Jun. 0–1700 m. BR, Casc, Col, ECas, Lava, Sisk, WV. CA, ID, NV, WA; north to British Columbia, northeast to MT, southeast to AZ, south to Mexico. Native.

Stellaria obtusa Engelmann (p. 578)

Rocky Mountain starwort

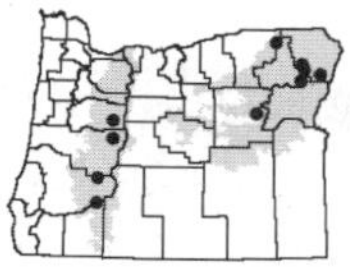

Plants perennial; from slender rhizomes. **Stems** generally prostrate, 4–20 cm, internodes glabrous. **Leaves** evenly spaced, ovate to elliptic, 5–12 mm, not fleshy, margins not papillate, generally flat, shiny, generally ciliate near base. **Inflorescences** flowers solitary in axils; pedicels ascending in flower, spreading in fruit. **Flowers** sepals 4(5), ovate, 1.5–3.5 mm, generally obtuse, glabrous, margins thinly scarious, veins in fruit 1, obscure; petals 0. **Fruits** globose to globose-ovoid, green or tan. **Seeds** 0.6–0.7 mm, grayish black, finely reticulate. ***2n***=26, 52, ~65, ~78.

Moist areas in meadows and woods, streambanks. Flowering May–Aug. 900–2000 m. BW, Casc. CA, ID, WA; north to British Columbia, northeast to Alberta, east to CO. Native.

Stellaria umbellata Karelin & Kirilov (p. 578)

Umbellate starwort

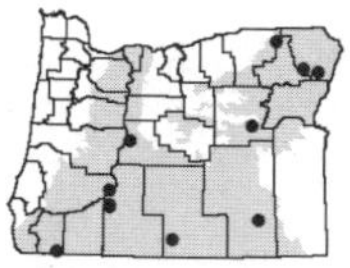

Plants perennial; from slender rhizomes. **Stems** prostrate to erect, 2–20 cm, internodes glabrous. **Leaves** evenly spaced, elliptic to ovate, 5–20 mm, not fleshy, margins not papillate, generally flat, shiny, glabrous. **Inflorescences** terminal, generally umbel-like, flowers few to generally many; bracts scarious; pedicels erect to spreading in flower, spreading to arching or reflexed in fruit. **Flowers** sepals 5, lanceolate to ovate, 1.5–2.5 mm, acute, margins widely scarious, veins in fruit 3, weak; petals 0 or rarely minute. **Fruits** narrowly ovoid, greenish to tan. **Seeds** 0.6–0.7 mm, brown, smooth. ***2n***=26.

Moist meadows and slopes, bogs, open woods. Flowering Jun–Aug. 1100–2900 m. BR, BW, Casc, ECas, Sisk. CA, ID, NV, WA; north to AK, northeast to Northwest Territories, east to CO, southeast to NM; northeastern Asia. Native.

VACCARIA Soapwort

Plants annual. **Stems** erect. **Leaves** opposite, lanceolate to ovate; basal leaves petiolate, cauline leaves sessile; stipules 0. **Inflorescences** terminal, open, flat-topped cyme; involucral bracts present; pedicels 5–40+ mm. **Flowers** bisexual; sepals 5, fused, glabrous, calyx tubes prominent, cylindric to ovoid, strongly 5-angled, 7.5–17 mm, whitish green, 5-veined, commissures between adjacent sepals absent, lobes < tube; petals 5, pink to reddish; appendages 0; stamens 10, borne at ovary base, staminodes 0; styles and stigmas 2. **Fruits** capsules, opening by 4 ascending to recurved teeth. **Seeds** ~10, subglobose, 1.6–1.8 mm, red-brown to black, tuberculate.

Asia, Australia, Europe, North America, South Africa, South America. 1 or 4 species; 1 species treated in *Flora*.

Vaccaria hispanica (Miller) Rauschert (p. 578)

Cow soapwort, cowcockle

Saponaria vaccaria L., *Vaccaria segetalis* (Neck.) Garcke, *Vaccaria vulgaris* Host

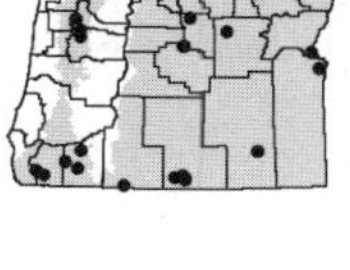

Stems (8)20–100 cm, glabrous, glaucous; branched above. **Leaves** 2–12 cm. **Flowers** calyx tubes cylindric to urceolate, 7.5–17 mm, veins 5, angled or keeled, each vein green-flanked, flanking < 1 mm wide, lobes ovate to triangular, 1.5–3 mm; petals 15–25 mm, limbs exserted, oblanceolate to obovate or obcordate, 3–8 mm. ***2n***=30.

Roadsides and disturbed areas, grain fields, woods. Flowering May–Jul. 0–1800 m. BR, BW, Col, ECas, Lava, Owy, Sisk, WV. CA, ID, NV, WA; widely scattered throughout North America; Asia, Europe. Exotic.

Vaccaria hispanica was first collected in Oregon in 1881. Once common in wheat fields, *V. hispanica* is far less common today; collections made since 2000 came from one population each in Crook and Wheeler counties.

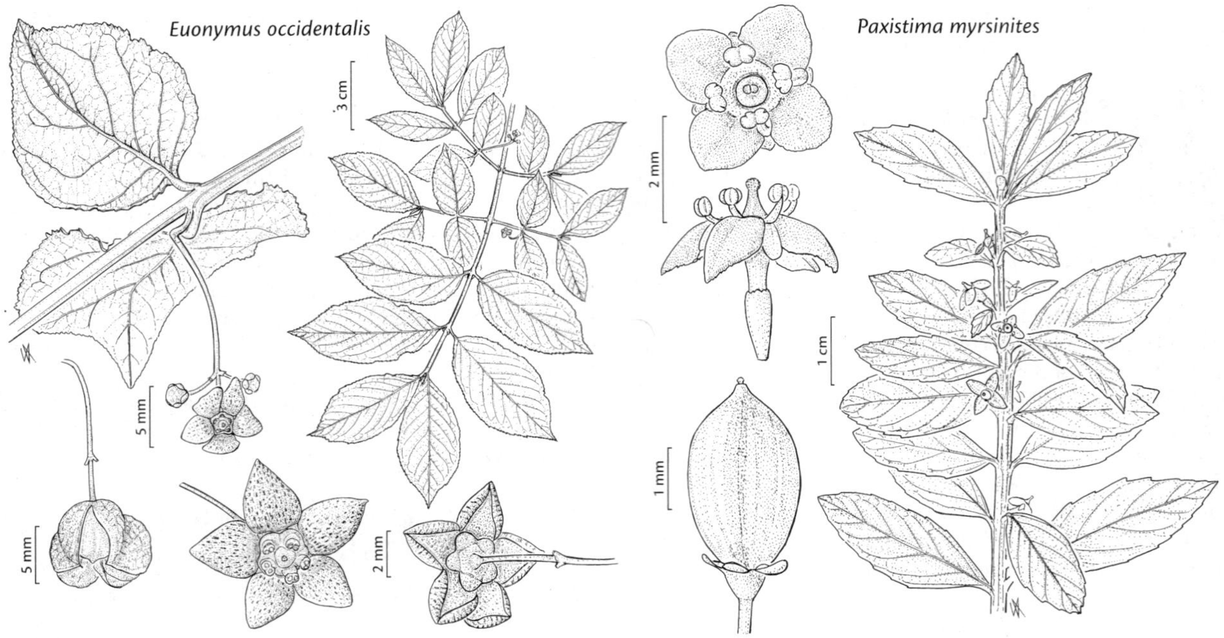

EUONYMUS • PAXISTIMA

CELASTRACEAE Staff-tree family

Stephen C. Meyers

Shrubs or **small trees** perennial, evergreen or deciduous, glabrous. **Stems** twigs 4-angled, with corky ridges. **Leaves** opposite, simple, margins scalloped or finely serrate, veins pinnate; stipules absent or small and shed early. **Inflorescences** axillary clusters; bracts present. **Flowers** bisexual, radially symmetrical, small; sepals 4 or 5; petals 4 or 5; disk fused to hypanthium; stamens 4 or 5; ovaries superior, styles 1, stigmas 2–5-lobed. **Fruits** capsules, dehiscence loculicidal. **Seeds** enclosed by arils.

Worldwide. 50 genera; 2 genera treated in *Flora*.

Contemporary authorities place the genus *Parnassia* within Celastraceae (Stevens 2012). This is based, however, on a study in which the molecular data was inconclusive (Simmons et al. 2008). As such, we are not including *Parnassia* within this family but rather maintaining it within Parnassiaceae.

Key to Celastraceae genera:

1. Shrubs or small trees deciduous, 2–5(6) m; flower parts 5-merous; leaves 3–13 cm . . . . **EUONYMUS**
1' Shrubs evergreen, up to 1 m; flower parts 4-merous; leaves 1–3 cm . . . . **PAXISTIMA**

EUONYMUS Burning bush

Shrubs or **small trees** deciduous. **Leaves** margins scalloped or finely serrate. **Flowers** parts 5-merous; ovaries sunken in disk, stigmas 3–5-lobed. **Fruits** 3–5-valved. **Seeds** brown, smooth; arils red to reddish orange.

Worldwide. 180 species; 1 species treated in *Flora*.

Euonymus europaeus has been recently reported from Eugene and Portland but does not yet appear naturalized. It is distinguished from our native *E. occidentalis* by its pink (versus red) capsules.

Euonymus occidentalis Torrey (p. 582)

Western burning bush, western wahoo

Euonymus occidentalis Torr. var. *occidentalis*

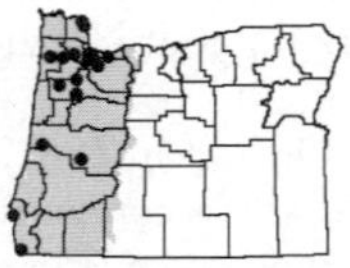

Plants 2–5(6) m. **Leaves** ovate to oblong-lanceolate, 3–13 cm, bases truncate to tapered, tips acuminate; petioles 2–14 mm. **Inflorescences** (1)3(5)-flowered. **Flowers** sepals rounded, 1–2 mm; petals 4–6 mm, greenish brown to purplish brown, mottled, margins transparent; stamens attached to disk margin; styles 3-lobed. **Fruits** red.

Moist areas and streambanks in forests. Flowering May–Jun. 0–1800 m. Casc, CR, Est, Sisk, WV. CA, WA; north to British Columbia. Native.

Two varieties of *E. occidentalis* are often recognized. Variety *occidentalis* has acuminate leaf tips, while those of variety *parishii* are rounded. Our plants are presumably variety *occidentalis*.

PAXISTIMA Mountain lover

Shrubs evergreen. **Leaves** margins finely serrate. **Flowers** parts 4-merous; disks square; ovaries sunken in disk, stigmas 2-lobed. **Fruits** 2-valved. **Seeds** brown to black, smooth, shiny; arils white.

North America. 2 species; 1 species treated in *Flora.*

Some references spell this genus as *Pachistima.* While *Pachistima* is correct etymologically, *Paxistima* is the name Rafinesque originally used when describing the genus. Thus, in accordance with the International Code of Botanical Nomenclature, we are unfortunately forced to use the etymologically incorrect name.

Paxistima myrsinites (Pursh) Rafinesque (p. 582)

Mountain lover, Oregon boxwood

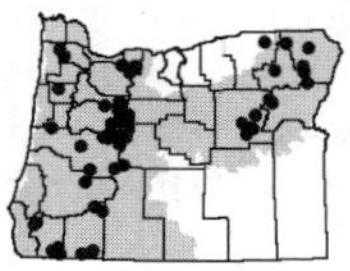

Plants 2–10 dm; stems branched in flat sprays. **Leaves** lanceolate to oblong-lanceolate, 1–3 cm, firm, bases attenuate, tips rounded to acute, surfaces glabrous, glossy; petioles 1–2 mm. **Inflorescences** 1–3-flowered. **Flowers** petals ovate, 3–4 × 1–3 mm, reddish orange to maroon; stamens alternating with petals. **Fruits** 4–5 mm. **Seeds** 1–2.

Shaded areas. Flowering Apr–Aug. 100–2400 m. BW, Casc, CR, ECas, Lava, Sisk, WV. CA, ID, WA; north to British Columbia, east to the Rocky Mountains, south to northern Mexico. Native.

CERATOPHYLLACEAE Hornwort family

Stephen C. Meyers

Herbs perennial, monoecious, aquatic; roots absent. **Stems** branched. **Leaves** whorled, dissected dichotomously, margins with minute teeth; sessile; stipules absent. **Inflorescences** axillary, solitary flowers. **Flowers** inconspicuous, sessile; stamens 3–many, filaments short, anthers with 2–3 teeth at tips; ovaries superior; styles 1, stigmas decurrent. **Fruits** achenes, indehiscent, beaked by persistent style. **Seeds** 1.

Worldwide. 1 genus.

CERATOPHYLLUM Hornwort

Family composed of one genus, *Ceratophyllum.*

6 species; 2 species treated in *Flora.*

Key to *Ceratophyllum* species:

1. Leaves forked 3–4 times, 5–15 mm . . . . *C. echinatum*

1' Leaves forked 1–2 (rarely 3) times, 15–25(30) mm . . . . *C. demersum*

Ceratophyllum demersum Linnaeus (p. 583)

Coontail

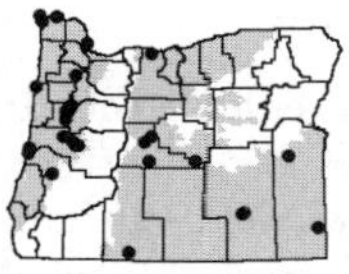

Plants submerged or floating. **Stems** 1–40 dm, flaccid. **Leaves** 5–12 per whorl, forked 1–2(3) times, 15–25(30) mm. **Flowers** subtending bracts 1.5–2 mm. **Fruits** flattened, ellipsoid, 4–6 mm, dark green to brown; persistent styles ~1–12 mm. **2*n***=24, 38, 40, 48.

Fresh water to brackish ponds, lakes, quiet streams and rivers. Flowering Jun–Aug. 0–1600 m. BR, Col, CR, ECas, Est, Lava, Owy, WV. CA, ID, NV, WA; throughout North America; worldwide. Native.

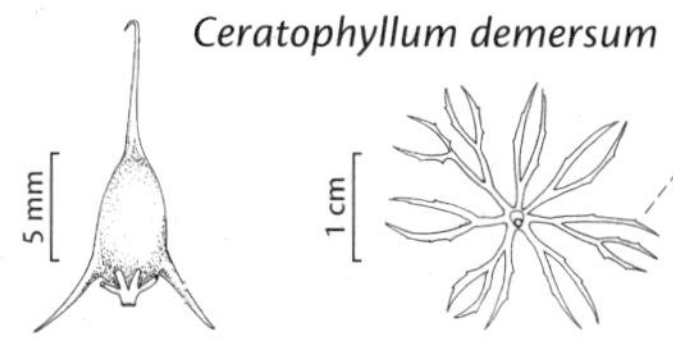

CERATOPHYLLUM

In Oregon, fertile plants are rarely encountered. Reproduction is primarily achieved by fragmentation of the branching stems.

Ceratophyllum echinatum A. Gray

Spineless hornwort

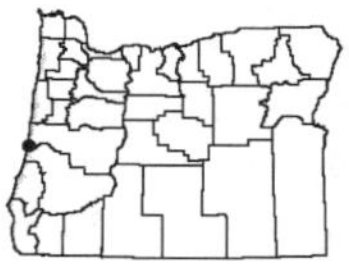

Plants submerged or floating. **Stems** 1–10 dm, flaccid. **Leaves** 5–12 per whorl, forked 3–4 times, 5–15 mm. **Flowers** subtending bracts 1–2 mm. **Fruits** flattened, ellipsoid, 4–6 mm, dark green to brown; persistent styles ~1–8 mm. **2*n***=24.

Fresh water ponds, lakes, marshes. Flowering Jun–Aug. 0–50 m. Est. WA; north to British Columbia; eastern North America. Native.

Herbarium records indicate a population of *C. echinatum* was present in Siltcoos Lake (Lane County) from 1937 until at least 1951. It has not been reported in Oregon since. Unlike *C. demersum,* this species is not tolerant of brackish waters and prefers cooler, clearer, and more acidic waters.

CHENOPODIACEAE

(see Amaranthaceae)

CLEOMACEAE Spiderflower family

Stephen C. Meyers

Herbs or **shrubs** annual or perennial, unpleasantly scented or not. **Leaves** alternate, palmately compound, rarely simple, entire or serrate; petiolate; stipules minute, usually scale- or hair-like. **Inflorescences** terminal or axillary racemes; bracts present; pedicels present. **Flowers** bisexual, radial or bilateral; sepals 4, distinct or partially fused, persistent; petals 4, equal or unequal, clawed; stamens 6 or 10–20; ovaries superior, styles 1, stigmas 1, unlobed, head-like. **Fruits** capsules, laterally dehiscent or in distal ½. **Seeds** few to many.

Nearly worldwide. 12–17 genera; 3 genera treated in *Flora*.

All genera currently placed within Cleomaceae were formerly included within Capparaceae. A molecular study, however, concluded that Capparaceae was paraphyletic, containing not only the genera now ascribed to Cleomaceae but also all Brassicaceae genera (Hall et al. 2002). In order to avoid the creation of one much larger family (Brassicaceae, due to priority), most authorities now accept three smaller, monophyletic groups (Capparaceae, Cleomaceae, and Brassicaceae).

Key to Cleomaceae genera:

1. Stamens 10–20; petals unequal . . . . **POLANISIA**
1' Stamens 6; petals equal.
  2. Leaflets not folded inward; fruits 2–10 mm . . . . **CLEOMELLA**
  2' Leaflets folded inward; fruits 10–60 mm . . . . **PERITOMA**

CLEOMELLA Stinkweed

Herbs annual, unpleasantly scented, glabrous; stems branched or not. **Leaves** palmately compound; leaflets 3(8), rarely unifoliolate, not folded inward; stipules 3–8-palmatifid, narrowly linear, 1–2 mm. **Inflorescences** terminal racemes, 3–15 cm, expanding up to 20 cm in fruit; bracts leaf-like, unifoliate. **Flowers** sepals equal, fused 25–50% of their length, green; petals equal; stamens 6, well exserted, anthers coiling with maturity. **Fruits** ovoid to rhomboidal. **2*n***=40.

North America. 10 species; 2 species treated in *Flora*.

Key to *Cleomella* species:

1. Leaflets ovate-elliptic to ovate 1.5–3 × as long as wide; petioles (1)2–6 cm . . . . ***C. hillmanii*** var. ***hillmanii***
1' Leaflets linear to linear-elliptic, (3)4–10 × as long as wide; petioles 0.5–2 cm . . . . ***C. plocasperma***

Cleomella hillmanii A. Nelson var. *hillmanii*

Hillman's stinkweed

Cleomella macbrideana Payson

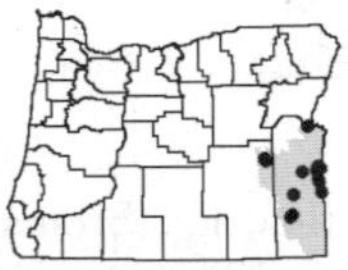

Plants 10–45(70) cm. **Leaves** leaflets ovate-elliptic to ovate, 10–25 × 4–14 mm, 1.5–3 × as long as wide; petioles (10)20–60 mm; stipules 3–5-palmatifid. **Pedicels** 7–17 mm. **Flowers** sepals 1–2 mm; petals oblong, 3–7 × 1–2.5 mm, yellow; stamens 5–12 mm; styles 1–2 mm. **Fruits** rhomboidal, 4–10 mm. **Seeds** 1–6, round, 2–3 mm, smooth.

Dry open areas and meadows. Flowering Apr–Jun. 800–1300 m. Owy. CA, ID, NV. Native.

The only other variety within *C. hillmanii*, var. *goodrichii*, is disjunct from the general range of *C. h.* var. *hillmanii* in Idaho and Nevada. It has shorter pedicels, usually less than 8 mm long.

Cleomella plocasperma S. Watson (p. 585)

Alkali cleomella, twisted cleomella

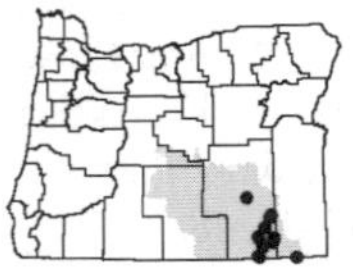

Plants 10–50(75) cm. **Leaves** leaflets linear to linear-elliptic, 15–40 × 2–7 mm, (3)4–10 × as long as wide; petioles 5–20 mm; stipules 3–8-palmatifid. **Pedicels** 4–12 mm. **Flowers** sepals 1–2 mm; petals oblong, 2.5–6 × 0.5–1.5 mm, yellow; stamens 6–12 mm; styles 1–2 mm. **Fruits** ovoid to rhomboidal, 2–7 mm. **Seeds** 2–4, round, 1–2.5 mm, smooth.

Wet meadows and near thermal springs. Flowering Jun–Sep. 1000–1500 m. BR. CA, ID, NV; UT. Native.

PERITOMA Bee plant

Herbs or **shrubs** unpleasantly scented when crushed, glabrous or glandular-hairy; stems branched. **Leaves** palmately compound; leaflets 3–5, folded inward; stipules minute or lacking. **Inflorescences** generally terminal racemes, sometimes axillary, 1–4 cm, expanding up to 35 cm in fruit; bracts unifoliate, obovate to spatulate, < leaves. **Flowers** sepals equal, distinct or fused 33–50% of their length, green, yellow or purple; petals equal; stamens 6, well exserted, anthers coiling with maturity. **Fruits** oblong.

North America. 6 species; 3 species treated in *Flora*.

Some floras include the species of *Peritoma* within the genus *Cleome*.

Key to *Peritoma* species:

1. Petals purple . . . . ***P. serrulata***
1' Petals yellow.
   2. Leaflets 5 (occasionally less), linear to elliptic; petals with a short claw . . . . ***P. lutea***
   2' Leaflets 3, ovate to obovate; petals without a claw. . . . . ***P. platycarpa***

Peritoma lutea (Hooker) Rafinesque

Yellow bee plant

Cleome lutea Hook.

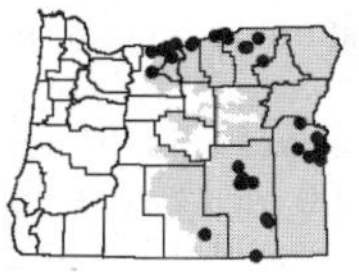

Herbs annual, 15–30 cm, glabrous. **Leaves** leaflets generally 5, occasionally < 5 distally, linear to elliptic, 15–60 × 5–15(20) mm, margins shallowly dentate to nearly entire. **Pedicels** 5–15 mm. **Flowers** sepals fused 50% of their length, 1.5–2.5 mm, yellow; petals oblong to ovate, 5–8 mm, yellow, with a short claw; stamens 10–15 mm, yellow; styles 0.5–1 mm. **Fruits** linear, 15–40 mm, striate. **Seeds** 10–20, triangular, 3–4 mm. ***2n***=34.

Scrubland, dry roadsides. Flowering May–Sep. 0–1700 m. BR, BW, Col, Owy. CA, ID, NV, WA; western US and Mexico. Native.

Peritoma platycarpa (Torrey) H.H. Iltis (p. 585)

Golden bee plant

Cleome platycarpa Torr.

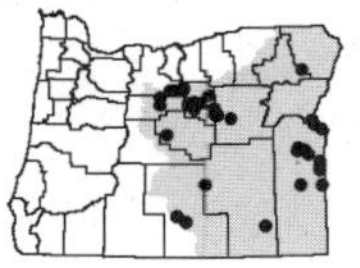

Herbs annual, 10–40(55) cm, glandular-hairy. **Leaves** leaflets 3, ovate to obovate, 15–50 × 5–15 mm, margins shallowly dentate to entire. **Pedicels** 5–18 mm. **Flowers** sepals distinct, 4–6 mm, yellow; petals oblong to ovate, 5–12 mm, yellow, without a claw; stamens 10–15 mm, yellow; styles 1–3 mm. **Fruits** ovate-oblong, 10–25 mm, striate. **Seeds** 10–20, elliptic to spherical, 3–4 mm. ***2n***=40.

Sagebrush scrub, fields, dry roadsides. Flowering May–Oct. 300–2000 m. BR, BW, Lava, Owy. CA, ID, NV. Native.

Peritoma serrulata (Pursh) de Candolle

Rocky Mountain bee plant

Cleome serrulata Pursh

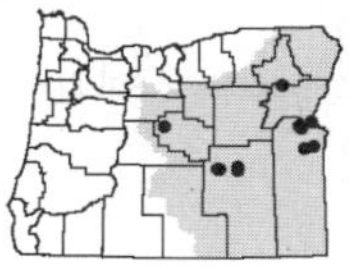

Herbs annual, 30–70 cm, glabrous to slightly puberulent. **Leaves** leaflets 3, lanceolate to obovate, 10–35 × 5–15 mm, margins shallowly dentate to entire. **Pedicels** 5–17 mm. **Flowers** sepals fused 33–50% of their length, 1.5–4 mm, green to purple; petals lanceolate to ovate, 6–12 mm, purple, with a short claw; stamens 15–25 mm, purple; styles 0.1–0.5 mm. **Fruits** linear, 25–60 mm, striate. **Seeds** 10–35, spherical, occasionally crescent-shaped, 3–4 mm. ***2n***=34, 60.

Fields, disturbed areas, roadsides. Flowering Jun–Sep. 600–1400 m. BR, BW, Lava, Owy. CA, ID, NV, WA. north to British Columbia, east across southern Canada and northern US, southeast to TX. Native.

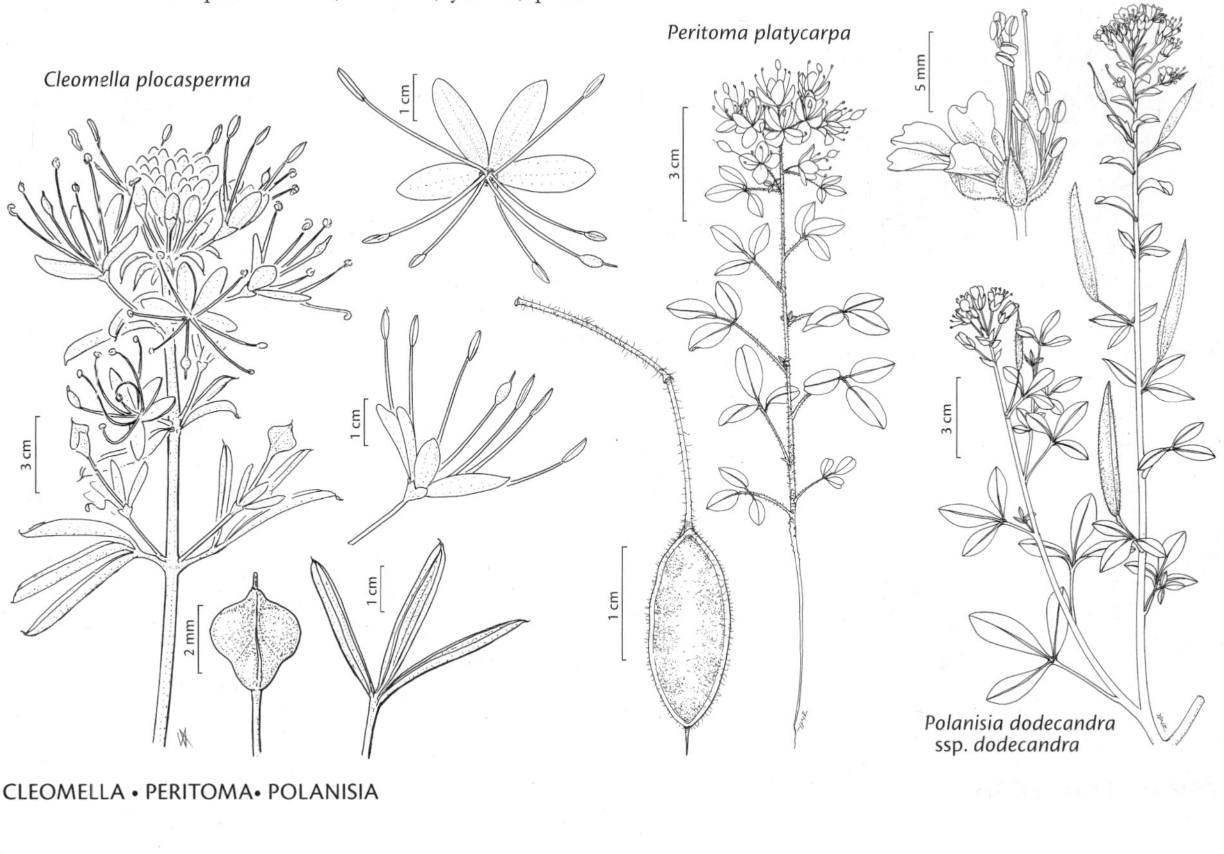

CLEOMELLA • PERITOMA• POLANISIA

POLANISIA Clammyweed

Herbs annual or perennial, unscented, glandular-pubescent, often sticky to touch; stems branched. **Leaves** palmately compound; leaflets 3 or rarely 5 distally, not folded inward; stipules minute or 0. **Inflorescences** terminal or axillary racemes, 1–15 cm, expanding up to 25 cm in fruit; bracts smaller than leaves, unifoliate. **Flowers** sepals equal, distinct; petals unequal; stamens 10–20, unequal, well exserted, anthers not coiling with maturity. **Fruits** oblong. ***2n***=20.

North America. 5 species; 1 species treated in *Flora*.

Polanisia dodecandra (Linnaeus) de Candolle

Red whisker clammyweed

Plants annual, rarely perennial, 10–50(75) cm. **Leaves** leaflets oblanceolate to ovate or broadly elliptic, 1.5–4.5 × 0.5–2 cm, margins entire to shallowly crenate, tips obtuse to acute. **Pedicels** 10–30 mm. **Flowers** sepals oblong, 3–7 mm, purple; petals oblong to ovate, 5–15 mm, adaxial pair > abaxial pair, white, with a long claw; stamens 5–30 mm, purple; styles deciduous, 5–15 mm, purple. **Fruits** 4–7 cm, green. **Seeds** 20–60, spherical to subspherical, 2–2.5 mm, brown to dark red.

Throughout most of US and Canada. 3 subspecies; 2 subspecies treated in *Flora*.

Key to *Polanisia dodecandra* subspecies:

1. Petals 4–8 mm; stamens slightly ≥ petals . . . . *P. d.* ssp. ***dodecandra***

1'. Petals 7–15 mm; stamens 1– 2 × length of petals . . . . *P. d.* ssp. ***trachysperma***

Polanisia dodecandra (Linnaeus) de Candolle ssp. ***dodecandra*** (p. 585)

Red whisker clammyweed

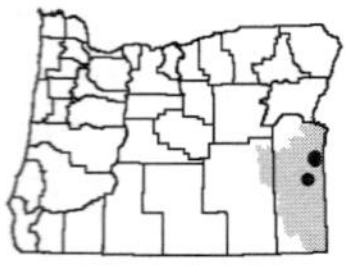

Flowers petals 4–8 mm; stamens 5–15 mm, slightly ≥ petals.

Fields, disturbed areas, roadsides. Flowering Jun–Sep. 600–1000 m. Owy. CA, ID, NV, WA; throughout most of US and southern Canada. Native.

Polanisia dodecandra (Linnaeus) de Candolle ssp. ***trachysperma*** (Torrey & A. Gray) H.H. Iltis

Sandyseed clammyweed

Polanisia trachysperma Torr. & A. Gray

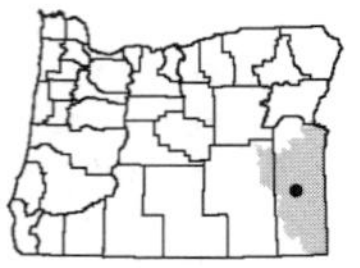

Flowers petals 7–15 mm; stamens 9–30 mm, 1–2 × length of petals.

Fields, disturbed areas, roadsides. Flowering Jun–Sep. 300–1100 m. Owy. CA, ID, NV, WA; most of US into southern Canada, widely scattered on east coast. Native.

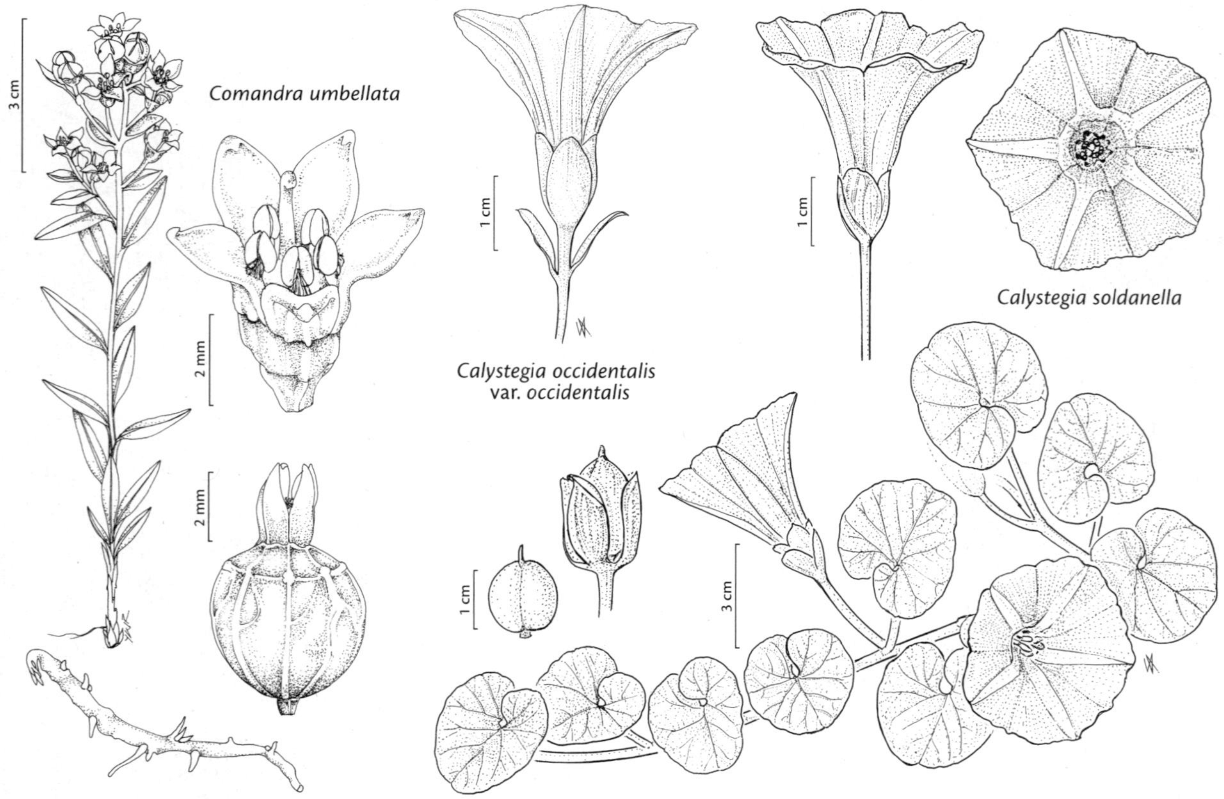

COMANDRA • CALYSTEGIA

COMANDRACEAE Bastard toadflax family

Stephen C. Meyers

Herbs or **subshrubs** perennial, glabrous, glaucous or not; facultative root parasite of neighboring green plants. **Leaves** alternate or scattered, simple, margins entire; stipules absent. **Inflorescences** simple or compound cymes; bracts present. **Flowers** bisexual; hypanthia elongated beyond ovary and resembling a green calyx tube, 5-lobed. **Seeds** 1, ovoid.

Worldwide. 2 genera; 1 genus treated in *Flora.*

Some references include the members of this family within Santalaceae. Here, we follow the classification proposed by Nickrent et al. (2010).

COMANDRA Bastard toadflax

Plants colonial by elongated, extensive rhizomes. **Leaves** sessile or subsessile. **Inflorescences** terminal; bracts present. **Flowers** hypanthium lobes white to pinkish; stamens 5, opposite lobes, bases with a tuft of hair; styles 1, stigma lobes 2–4. **Fruits** drupes.

Europe, North America. 2 species; 1 species treated in *Flora.*

Comandra umbellata (Linnaeus) Nuttall (p. 586)
Bastard toadflax

Comandra californica Eastw. ex Rydb., *Comandra pallida* A. DC., *Comandra umbellata* (L.) Nutt. ssp. *californica* (Eastw. ex Rydb.) Piehl, *Comandra umbellata* (L.) Nutt. var. *californica* (Eastw. ex Rydb.) C.L. Hitchc., *Comandra umbellata* (L.) Nutt. var. *pallida* (A. DC.) M.E. Jones

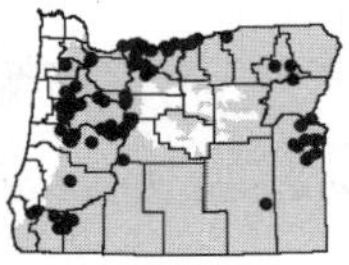

Herbs or **subshrubs** 1–3(4) dm. **Leaves** linear to lance-oblong, 1–5 × 0.3–1.4 cm, tips acute to obtuse. **Flowers** 2–7 mm; hypanthia persistent, hypanthium lobes lanceolate to triangular, 1–3 mm; stamens 0.5–1.5 mm. **Fruits** oblong-ovoid, 4–9 mm, blue to purple to brown.

Drier, rocky habitats. Flowering Apr–Aug. 0–2000 m. BR, BW, Casc, Col, ECas, Owy, Sisk, WV. CA, ID, NV, WA; throughout North America. Native.

Three poorly differentiated varieties of *C. umbellata* are sometimes recognized. Ours are either *C. u.* var. *californica* or *C. u.* var. *pallida*. The described differences between the varieties vary between authorities. Some mention subtle leaf shape differences and/or the amount of glaucousness, while others emphasize the shape of the miniscule hypanthium lobes. Geographically, variety *californica* will most likely be found from the Cascades westward, while variety *pallida* will be found in the eastern part of the state.

CONVOLVULACEAE Morning-glory family

Katie E. Mitchell

Vines, herbs, or **subshrubs** annual or perennial, deciduous. **Stems** trailing or climbing, glabrous or hairy, commonly with milky sap. **Leaves** alternate, simple, sometimes highly reduced, margins entire; petioles present; stipules absent. **Inflorescences** cymes, thyrses, or solitary flowers; bracts present, often leafy. **Flowers** bisexual, solitary or paired, radial; sepals (4)5, fused or distinct; corollas (4)5-parted, fused at least in part; stamens 5, epipetalous; pistils 1, ovaries 1- or 2-chambered, superior, styles 1 or 2, bifid to undivided, gynobasic to terminal, stigmas 1 or 2. **Fruits** capsules or utricles.

Worldwide. 50–60 genera; 5 genera treated in *Flora.*

Convolvulaceae is a diverse family closely related to Solanaceae. Its greatest diversity is found in the tropics where some are woody lianas and even trees, but most are herbaceous and twining or trailing, with strongly plicate corollas. *Cuscuta*, a distinctive genus of parasites, was historically placed within the monogeneric family, Cuscutaceae. Based on molecular evidence, *Cuscuta* is now placed in Convolvulaceae and is the only parasitic genus in the family (Stefanović & Olmstead 2004).

Key to Convolvulaceae genera:

1. Plants parasitic, non-photosynthetic, and not green. . . . . **CUSCUTA**

1' Plants photosynthetic and green.

   2. Flowers inconspicuous; corollas ≤ 6 mm, deeply 5-lobed, lobes > or ≈ to tubes; styles 2, united basally or distinct.

      3. Leaves elliptic; ovaries not lobed. . . . . **CRESSA**

      3' Leaves suborbicular to reniform; ovaries deeply 2-lobed. . . . . **DICHONDRA**

   2' Flowers showy; corollas ≥ 15 mm, entire or shallowly 5-lobed, if lobed, lobes < tubes; styles 1.

      4. Stigma lobes linear, usually flattened, tips acute; pair of bracts below flower attached well below calyx and not obscuring it . . . . **CONVOLVULUS**

      4' Stigma lobes oblong, usually cylindrical, tips obtuse; calyces often obscured by a pair of subtending bracts attached immediately below calyx, rarely with an obvious peduncle between bracts and calyx . . . . **CALYSTEGIA**

CALYSTEGIA Morning-glory, bindweed

Herbs or **vines** perennial; rhizomatous. **Stems** trailing or climbing, twisting to twining. **Leaves** reniform to sagittate to hastate. **Inflorescences** axillary, 1–2(4) flowers per peduncle; calyces often obscured by pair of subtending bracts attached immediately below calyx, rarely with an obvious peduncle between bracts and calyx. **Flowers** corollas funnel-shaped, large and showy, entire or shallowly 5-lobed, white, pink, or purple; ovaries 2-chambered; styles terminal; stigma lobes 2, oblong, usually cylindrical (flattened), tips obtuse. **Fruits** capsules, 1–4-seeded.

Worldwide in temperate to subtropical climates. ~25 species; 5 species treated in *Flora*.

The genus *Calystegia* is well supported as a monophyletic group nested within the more inclusive genus *Convolvulus* based on molecular data (Stefanović et al. 2002, 2003). Because this makes *Convolvulus s.s.* paraphyletic, some authors would have *Convolvulus* include *Calystegia*.

Calystegia is well defined morphologically by its oblong and more or less cylindrical stigma lobes with blunt tips, spheroidal pollen with about 20–40 apertures, and unilocular ovaries (often with an incomplete septum). In contrast, *Convolvulus* has linear to linear-spatulate, generally flattened stigma lobes with acute tips, ellipsoid to subspheroidal pollen with three to four apertures, and bilocular ovaries (Lewis & Oliver 1965).

Key to *Calystegia* species:

1. Bracts < 5 mm wide; flowers with an obvious (1)3–7 mm peduncle between bracts and calyx . . ***C. occidentalis*** ssp. ***occidentalis***

1' Bracts ≥ 5 mm wide; flowers without an obvious peduncle between bracts and calyx.

  2. Plants in sandy habitats, usually seashores; leaves reniform, thick and fleshy; flowers pink . . . . ***C. soldanella***

  2' Plants in various habitats, usually inland; leaves variously shaped but not thick and fleshy; flowers pink or white.

    3. Stems < 1 m, trailing or erect and not climbing . . . . ***C. atriplicifolia*** ssp. ***atriplicifolia***

    3' Stems ≥ 1.5 m, trailing and climbing.

      4. Bracts prominently keeled, not saccate at base, with margins not or only scarcely overlapping, tips acute to subobtuse. . . . . ***C. sepium*** ssp. ***angulata***

      4' Bracts slightly or not keeled, saccate at base, with overlapping margins, tips obtuse to truncate. . . . . ***C. silvatica*** ssp. ***disjuncta***

Calystegia atriplicifolia Hallier f. ssp. *atriplicifolia*

Night-blooming morning-glory

Convolvulus nyctagineus Greene

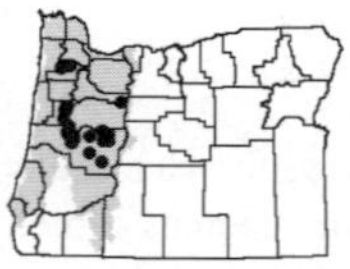

Herbs glabrous. **Stems** subacaulescent to suberect, to 35 cm. **Leaves** triangular-hastate to reniform, 3–5(7) cm, basal lobes parallel to widely spreading. **Bracts** attached immediately below and closely surrounding calyx, ovate, elliptic or suborbicular, 9–18 mm wide, not keeled, often saccate, bases cordate, tips rounded to emarginate. **Inflorescences** solitary flowers. **Flowers** corollas 40–73 mm, white.

Open, often rocky places, sometimes in open forests and disturbed sites. Flowering May–Jul. 50–1600 m. Casc, CR, WV. WA. Native.

Plants in Oregon are along a continuum of variation between typical *C. atriplicifolia* ssp. *atriplicifolia* found in Washington and *C. atriplicifolia* ssp. *buttensis* in California. The latter are more erect plants with smaller corollas 31–44 mm in length and narrower, 4–8 mm-wide bracts that more or less flat.

Calystegia occidentalis (A. Gray) Brummitt ssp. *occidentalis* (p. 586)

Pale morning-glory

Calystegia polymorpha (Greene) Munz, *Convolvulus occidentalis* A. Gray ssp. *fruticetorum* (Greene) Abrams, *Convolvulus polymorphus* Greene

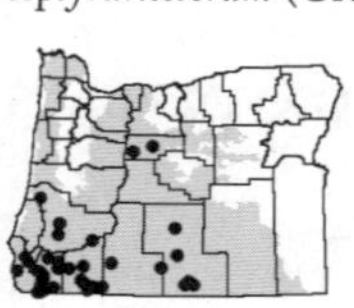

Herbs or **vines** with a woody rootstock or lower stems woody, all parts usually finely puberulent, or in part glabrous but with at least base of leaf blades and distal end of peduncles minutely pubescent. **Stems** climbing or trailing, to 4 m. **Leaves** highly variable in shape, generally 1.5–4 cm, basal lobes nearly absent and leaf bases almost cuneate or basal lobes short to almost equal to midrib, rounded to acutely 2-angled, sinuses rounded to nearly parallel-sided. **Bracts** inserted (1)3–7 mm below calyx, usually overlapping calyx by half its length or more, linear to oblanceolate, 1–4(5) mm wide, not keeled. **Inflorescences** solitary flowers or sometimes 2–4-flowered. **Flowers** corollas (20)25–42 mm, white or cream.

Grasslands, shrublands, open forests, disturbed habitats. Flowering May–Aug. 0–2000 m. BR, Casc, CR, ECas, Lava, Sisk. CA. Native.

Calystegia sepium (Linnaeus) R. Brown ssp. *angulata* Brummitt

Large bindweed, false hedge bindweed

Convolvulus sepium L., *Convolvulus sepium* L. var. *communis* R.M. Tryon

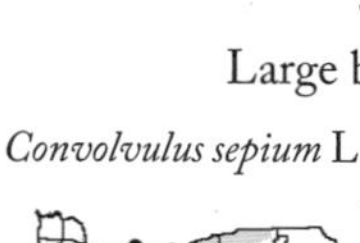

Vines glabrous. **Stems** twining, strongly climbing to several meters. **Leaves** sagittate, to 10 cm, basal lobes divergent, sinuses rounded. **Bracts** inserted immediately below and obscuring calyx, prominently keeled, not saccate at base, with margins not or only scarcely overlapping, 10–18 mm wide, tips acute to subobtuse. **Inflorescences** solitary flowers. **Flowers** corollas 48–60 mm, white or rarely white with pink on margin. $2n=22$.

Dry to moist, open areas. Flowering Jun–Sep. 50–200 m. Col, ECas, WV. ID, WA; most of US and southern Canada. Native.

This native species may be overlooked due to its similarity to *C. silvatica* ssp. *disjuncta*.

Calystegia sepium ssp. *erratica*, a native of eastern North America, was collected once by J.C. Nelson in a plowed field east of the State Prison in Salem in 1920 but does not appear to have naturalized. It is distinguished from ssp. *angulata* by its flowers having bracts and sepals not clearly distinguished from each other, pink corollas, and leaves with an almost closed sinus and lobes that are truncate (Brummitt 1980).

Calystegia silvatica (Kitaibel) Grisebach ssp. ***disjuncta*** Brummitt

Hedge bindweed, morning-glory

Convolvulus sepium L., *Convolvulus sepium* L. var. *communis* R.M. Tryon

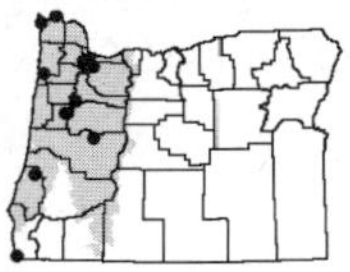

Vines glabrous. **Stems** twining, strongly climbing to several meters. **Leaves** sagittate to hastate to subreniform, generally 5–12 cm, basal lobes divergent, sinuses rounded. **Bracts** inserted immediately below and obscuring calyx, not or slightly keeled, saccate at base, margins strongly overlapping, tips obtuse to truncate. **Inflorescences** solitary flowers. **Flowers** corollas 48–65 mm, white, sometimes pink-tinged along mid-petal bands. **2*n***=22.

Disturbed sites, often overgrowing fences and shrubs. Flowering May–Sep. 0–1000 m. Casc, CR, Est, WV. CA, WA; north to British Columbia; Africa, Australia, Europe, New Zealand. Exotic.

This is the most common weedy bindweed in Oregon. Native to southwestern Europe, it is a serious weed along the west coast of North America from British Columbia to California. It is easily spread by rhizome-contaminated soil.

Calystegia silvatica is included in the *C. sepium* complex, characterized as vigorous, rhizomatous climbers with large, showy corollas (Brummitt 2002). *Calystegia silvatica* and *C. sepium* are very similar species with relatively minor distinguishing characters. Intermediate plants may be the result of hybridization.

Calystegia soldanella (Linnaeus) R. Brown (p. 586)

Coast morning-glory

Convolvulus soldanella L.

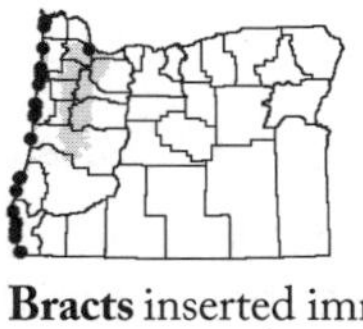

Herbs glabrous, not woody. **Stems** weakly or not twisting, mostly prostrate, up to 1(2.5) m. **Leaves** reniform, thick and fleshy, 1–3 cm, sinuses often with short parallel sides, tips broadly obtuse to emarginate. **Bracts** inserted immediately below calyx, ovate to suborbicular, 7–16 × 5–10 mm, flat or somewhat enfolding calyx. **Inflorescences** solitary flowers. **Flowers** corollas 32–52 mm, pink. **2*n***=22.

Sandy shores and beaches. Flowering May–Aug. 0–50 m. Est, WV. CA, WA; north to British Columbia, NC, VA; temperate regions nearly worldwide. Native.

CONVOLVULUS Morning-glory, bindweed

Herbs or **vines** usually glabrous. **Inflorescences** axillary, 1–2 flowers per peduncle; subtending bracts attached well below calyx, not obscuring it. **Flowers** corollas funnel-shaped, large and showy, entire or shallowly 5-lobed; stamens included, unequal; ovaries 2-celled, styles terminal, stigma lobes 2, linear, usually flattened, tips acute. **Fruits** capsules, 1–4-seeded.

Nearly worldwide. ~250 species; 1 species treated in *Flora*.

Convolvulus arvensis Linnaeus (p. 589)

Field morning-glory, field bindweed

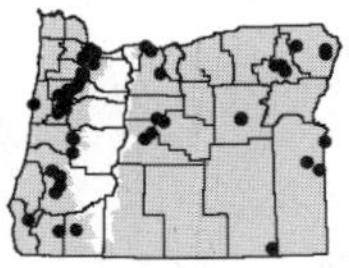

Herbs perennial, glabrous or short pubescent; deep-rhizomatous. **Stems** spreading and prostrate, or twining. **Leaves** highly variable, ovate to hastate or sagittate, 1–10 × 0.3–6 cm; petiolate. **Inflorescences** 1–2-flowered, pedunculate; bracts 2–4 mm, linear; pedicels reflexed in fruit, 5–18(35) mm. **Flowers** corollas 1.5–2.5 cm, white to pink, often white with pink. **Fruits** subglobose to ovoid, ~8 mm, glabrous.

Open places, disturbed sites. Flowering May–Sep. 0–1900 m. All ecoregions except Casc. CA, ID, NV, WA; widespread in North America; South America; Asia, Europe. Exotic.

This species is a troublesome weed with deep rhizomes that make it difficult to eradicate.

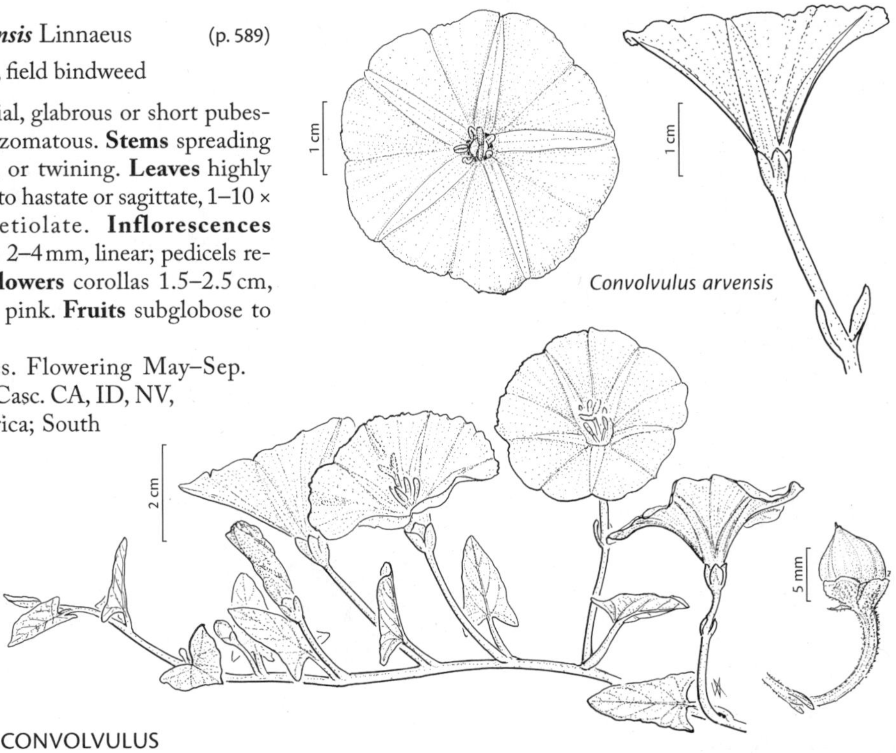

CONVOLVULUS

CRESSA Alkali weed, rosinweed

Herbs to **subshrubs** perennial, canescent. **Stems** not twining. **Inflorescences** axillary, solitary flowers; bracts present. **Flowers** sepals ovate to obovate; corollas salverform, deeply 5-lobed, lobes > or ≈ to tubes; stamens exserted, filaments filiform; styles 2, distinct to base, stigmas capitate. **Fruits** ovoid, dehiscent capsules, 1–4-seeded.

Worldwide. 4 species; 1 species treated in *Flora*.

Cressa truxillensis Kunth (p. 590)

Spreading alkali weed

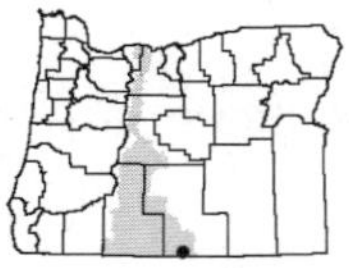

Herbs or **subshrubs** leafy throughout, silky canescent to woolly-villous. **Stems** mostly prostrate, 5–40 cm, usually branched. **Leaves** elliptic to lanceolate, 3–10 × 1–4 mm, bases cuneate to obtuse, tips usually acute; petioles 0.5–2 mm. **Inflorescences** concentrated on distal half of branches; peduncles 2–6 mm; bracts 2, 2–3 × 1 mm, foliaceous, subtending each flower. **Flowers** inconspicuous; calyces 3–4.5 mm, lobe tips acute; corollas ~6 mm, white, tubes 3 mm, lobes lanceolate; stamens ≈ corolla; ovaries ovoid, lanate apically; styles 2, unequal, 3–5 mm; stigmas smooth. **Seeds** 1, ovoid, 3–4 mm, brown, glabrous.

Shorelines and flats, on saline or alkaline soils. Flowering Jun–Jul. 1200–1500 m. ECas. CA, NV; southeast to OK, TX, and Mexico; South America. Native.

CUSCUTA Dodder

Herbs parasitic, usually glabrous; without roots. **Stems** twining, thread-like, attached to host by many haustoria, glabrous. **Leaves** alternate, rudimentary scales. **Inflorescences** thyrses or cymes, loose to dense. **Flowers** 4- or 5-parted, calyces and corollas fused at base; stamens alternate with corolla lobes, infrastaminal scales usually present; styles 1 or 2, terminal, stigmas capitate or elongate. **Fruits** capsules, 2-celled, 4-ovulate, 1–4-seeded.

Nearly worldwide. ~200 species; 10 species treated in *Flora.*

Cuscuta are stem parasites on a wide range of plants, attached to their hosts by numerous haustoria. The seeds germinate on the ground and grow toward a suitable host plant, attracted at least in part by chemical compounds released by the host. After attaching to the host, the lower portion of the plant dies. Haustoria of some species overwinter inside perennial host stems and generate new stems the following growth season. Some species are agricultural pests.

With flowers only a few millimeters in size, identification of *Cuscuta* species is best accomplished with a microscope and wetting solution for dried material. Measurements given are of rehydrated material. Infrastaminal scales are found along the attached portion of the filament within the corolla tube. Of the Oregon species that have infrastaminal scales, all are densely fimbriate except *C. suksdorfii*, which may only have denticulate wings or a few short fimbriae.

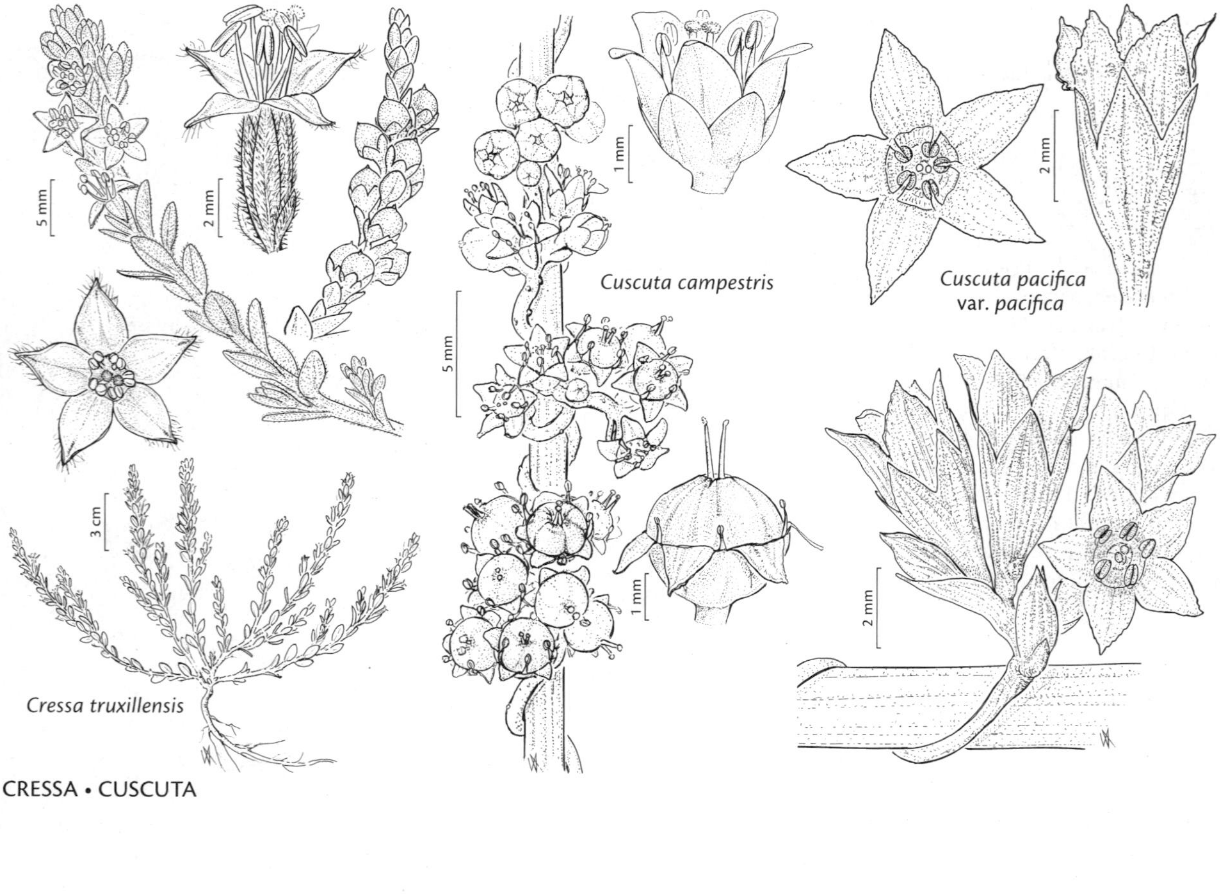

CRESSA • CUSCUTA

Key to *Cuscuta* species:

1. Stigmas elongate.
   2. Dried calyces golden, shiny, reticulate, ~100% length of corolla tube; calyx lobes tipped with a fleshy, short, obtuse appendage. . . . . ***C. approximata*** var. ***approximata***
   2' Dried calyces purplish or creamy, not shiny, not reticulate, ~50–70% length of corolla tube; calyx lobes without a fleshy appendage . . . . ***C. epithymum*** var. ***epithymum***

1' Stigmas capitate.
   3. Calyx lobe tips rounded or obtuse.
      4. Corolla lobe tips acute to acuminate; calyces ~100% length of corolla tube; flowers usually 5-parted . . ***C. campestris***
      4' Corolla lobe tips obtuse; calyces 50% length of corolla tube; flowers usually 4-parted . . . . ***C. cephalanthi***
   3' Calyx lobe tips acute to acuminate.
      5. Infrastaminal scales absent; flowers 5-parted.
         6. Capsules surrounded but not completely enclosed by corollas, capsule tips visible; anthers 0.2–0.5 mm; filaments 0.2–0.5 mm; styles 0.5–1(1.5) mm. . . . . ***C. occidentalis***
         6' Capsules enclosed by corollas, capsule tips not or barely visible; anthers 0.6–1.1 mm; filaments 0.3–1.1 mm; styles 1.2–3 mm.
            7. Calyces 75–~100% length of corolla tube; corolla lobes ≥ corolla tube; filaments 0.6–1.1 mm; styles 1.2–2.2 mm . . . . ***C. californica*** var. ***californica***
            7' Calyces 25–50% length of corolla tube; corolla lobes ≤ corolla tube; filaments 0.3–0.6 mm; styles 1.2–3 mm . . . . ***C. brachycalyx***
      5' Infrastaminal scales present, sometimes reduced to denticulate wings or a few short fimbriae (*C. suksdorfii*); flowers 4–5-parted (usually 4-parted in *C. suksdorfii*).
         8. Calyx lobe tips long-attenuate; corolla lobes straight, tips lanceolate-attenuate; infrastaminal scales reduced to denticulate wings or a few short fimbriae; plants of mountain meadows . . . . ***C. suksdorfii***
         8' Calyx lobe tips acute to acuminate; corolla lobes straight or inflexed, tips acute to cuspidate; infrastaminal scales well developed; habitats various.
            9. Corolla lobe tips acute to cuspidate, straight; plants of coastal salt marshes and tidal flats . . . . ***C. pacifica*** var. ***pacifica***
            9' Corolla lobe tips acute, inflexed; habitats various. . . . . ***C. indecora*** var. ***indecora***

Cuscuta approximata Babington var. ***approximata***

Clustered dodder, alfalfa dodder

Cuscuta approximata Bab. var. *urceolata* Yunck.

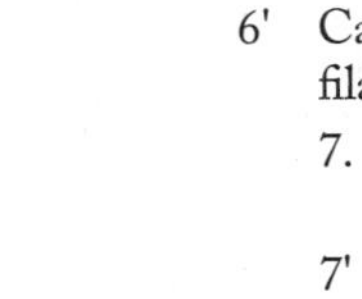

Stems 0.4–0.6 mm in diameter, yellow to orange. **Inflorescences** densely clustered; pedicels absent. **Flowers** 5-parted, 3–4.2 mm, fleshy; calyces campanulate, ≈ corolla tube, divided 33–50% of their length, golden yellow and shiny when dry, cellular-reticulate, lobes triangular to obovate, overlapping at base, margins entire, tips acute to acuminate, with a fleshy appendage distally; corollas campanulate with spreading lobes, 2.5–4 mm, white, creamy yellow when dry, tubes 1.5–2 mm, lobes ovate-orbiculate, ≤ tube, tips obtuse, straight; stamens exserted, < corolla lobes, filaments 0.3–0.6 mm, anthers 0.3–0.5 mm, infrastaminal scales ≤ corolla tube; stigmas elongate, terete, 0.4–0.7 mm. **Fruits** 1.5–2.3 × 1.6–2 mm, capped by withered corollas. **Seeds** 4. **2*n***=14, 28.

Often parasitic on species of Fabaceae, especially in disturbed sites and on cultivated alfalfa. Flowering Jun–Aug. 500–1000 m. BR, BW, Col, Lava, Owy, WV. CA, ID, NV, WA; north to British Columbia, east to ND, southeast to NM; Africa, Asia, Europe. Exotic.

Two additional varieties of this species are recognized in Europe, where this species is native.

Cuscuta brachycalyx (Yuncker) Yuncker

San Joaquin dodder

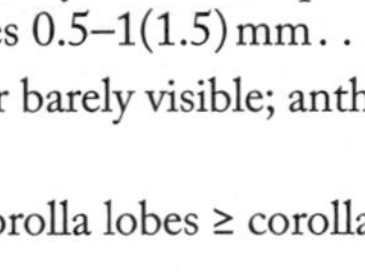

Stems 0.4–0.6 mm in diameter, yellow to orange. **Inflorescences** dense to loose; pedicels 1–6 mm. **Flowers** 5-parted, 4.5–6 mm, mostly membranous, fleshy at bases of calyces; calyces campanulate-cupulate, 25–50% as long as corolla tube, divided ≥ 67% of their length, brownish, not reticulate, not shiny, lobes triangular-ovate, slightly overlapping at base, margins entire, tips acute; corollas narrowly cupulate to campanulate, 4.4–5.8 mm, white, white to creamy brown when dry, tubes 2–3.5 mm, lobes lanceolate, reflexed, somewhat ≤ tube, tips acute, straight; stamens exserted, < corolla lobes, filaments 0.3–0.6 mm, anthers 0.7–1.1 mm, infrastaminal scales absent; styles 1.2–3 mm, evenly filiform, stigmas capitate, globose. **Fruits** 1.6–2.3 × 1.8–2.5 mm, enclosed by withered corollas. **Seeds** 1–4.

Parasitic on a wide variety of hosts. Flowering Jul–Sep. 200–1700 m. ECas, Sisk. CA. Native.

Cuscuta californica Hooker & Arnott var. ***californica***

California dodder, chaparral dodder

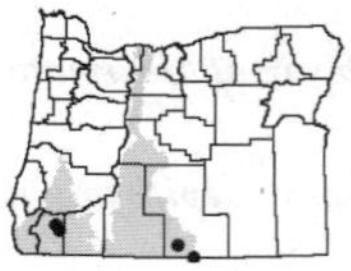

Stems 0.4–0.6 mm in diameter, yellow to orange. **Inflorescences** dense; pedicels (0.5)1–2.5 mm. **Flowers** 5-parted, 3–5(5.5) mm, membranous, sometimes basally fleshy; calyces turbinate-campanulate, 75%–~100% as long as corolla tube, divided 50–75% of their length, yellow, finely reticulate, shiny, lobes triangular to lanceolate, overlapping at base, margins entire, tips acute to acuminate; corollas cylindric-campanulate to obconic, 3–5 mm, white, creamy white when dry, tubes 1.6–2.4 mm, lobes narrowly lanceolate, ≥ tube, tips acute, straight; stamens exserted, < corolla lobes, reflexed, filaments 0.6–1.1 mm, anthers 0.6–1 mm, infrastaminal scales absent; styles 1.2–2.2 mm, evenly filiform, stigmas capitate, globose. **Fruits** 1.5–2.2 × 1.8–2.5 mm, enclosed by withered corollas. **Seeds** (1)2–4.

Parasitic on a wide variety of hosts. Flowering Jun–Sep. 200–2100 m. ECas, Sisk. CA, NV, WA; southeast to UT, AZ, and Mexico. Native.

Cuscuta campestris Yuncker (p. 590)

Field dodder

Cuscuta pentagona Engelm. var. *calycina* Engelm.

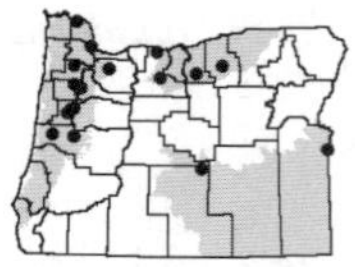

Stems 0.4–0.6 mm in diameter, yellow to orange. **Inflorescences** dense; pedicels 0.3–2.5(3.5) mm. **Flowers** (4)5-parted, 1.9–3.6 mm, membranous; calyces cup-shaped, ≈ corolla tube, divided 40–60% of their length, yellow, reticulate, shiny, lobes ovate-triangular, overlapping at base, margins entire, tips obtuse to rounded; corollas campanulate with spreading lobes, 2–3.5 mm, creamy white, yellow when dry, tubes (1.1)1.5–1.9 mm, lobes triangular to triangular-lanceolate, as long as tube, tips acute to acuminate, inflexed; stamens exserted, < corolla lobes, filaments 0.4–0.7 mm, anthers (0.3)0.4–0.5 mm, infrastaminal scales ≥ corolla tube; styles 0.5–1.6 mm, evenly filiform, stigmas capitate, globose. **Fruits** 1.3–2.8 × 1.9–3.8 mm, enveloped basally ≤ 33% by withered corollas. **Seeds** 4. **2*n***=56.

Parasitic on a wide variety of hosts. Flowering Jun–Sep. 0–1000 m. BR, Col, CR, Owy, WV. CA, ID, NV, WA; throughout North America; widespread worldwide. Native.

Some references synonymize *C. campestris* with *C. pentagona*, but *C. pentagona* is distinct, with a prominently angled calyx and somewhat smaller flowers (Costea et al. 2006). True *C. pentagona* has not been found in Oregon.

Cuscuta cephalanthi Engelmann

Buttonbush dodder

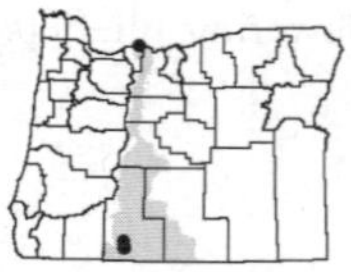

Stems 0.4–0.6 mm in diameter, yellow to orange. **Inflorescences** loose to dense; pedicels 0–1 mm. **Flowers** (3)4(5)-parted, 2–3 mm, membranous; calyces shallowly cup-shaped, 50% as long as corolla tube, divided ~67% of their length, yellow-brown, not reticulate, not shiny, lobes oblong-ovate, overlapping slightly at base, margins entire to serrulate, tips obtuse; corollas cylindric-campanulate to cylindric with reflexed lobes, 1.8–2.8 mm, white, creamy white when dry, tubes 1.1–2.2 mm, lobes ovate, 50–67% as long as tube, margins entire, lobe tips obtuse, straight; stamens enclosed to slightly exserted, filaments 0.2–0.4 mm, anthers 0.2–0.4 mm, infrastaminal scales slightly ≤ corolla tube; styles (0.6)1–2 mm, evenly terete, stigmas capitate, globose. **Fruits** 2.5–3.2(4) × 2–4 mm, capped by withered corollas. **Seeds** 1–2. **2*n***=60.

Parasitic on a wide variety of hosts. Flowering Jul–Oct. 100–1400 m. ECas. CA, ID, NV, WA; widespread in the US and Canada. Native.

Cuscuta epithymum (Linnaeus) Linnaeus var. ***epithymum***

Thyme dodder

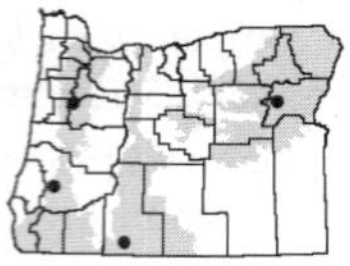

Stems 0.3–0.4 mm in diameter, reddish purple, sometimes yellow. **Inflorescences** densely clustered; pedicels usually absent, rarely up to 0.5 mm. **Flowers** 5-parted, 3–4(5) mm, fleshy; calyces cup-shaped, 50–70% as long as corolla tube, divided 50–67% of their length, creamy white or purplish, not reticulate, somewhat shiny, lobes ovate-triangular, overlapping at base, margins entire, tips acute to acuminate; corollas campanulate-cylindric with spreading lobes, 2.8–3.6 mm, white, pink or purplish, tubes 1.5–3 mm, lobes triangular to lanceolate, 50–75% as long as tube, tips acute to acuminate, straight; stamens exserted, < corolla lobes, filaments 0.4–0.7 mm, anthers 0.3–0.5 mm, infrastaminal scales ~80% as long as corolla tube; styles terete, stigmas elongate, terete, 0.6–1 mm. **Fruits** 1.6–2.2 × 1.6–2.3 mm, capped by withered corollas. **Seeds** 4. **2*n***=14.

Often parasitic on species of Fabaceae. Flowering Jul–Nov. 100–1400 m. BW, ECas, Sisk, WV. CA, NV, WA; scattered across US and southern Canada; nearly worldwide. Exotic.

Cuscuta indecora Choisy var. ***indecora***

Collared dodder

Cuscuta indecora Choisy var. *neuropetala* (Engelm.) Hitchc.

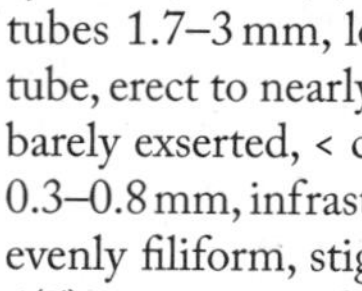

Stems 0.3–0.6 mm in diameter, yellow to orange. **Inflorescences** loose to dense; pedicels 0.5–6 mm. **Flowers** 5-parted, 3–4.5(5.3) mm, fleshy; calyces cup-shaped, 50–75% as long as corolla tube, divided 33–50% of their length, creamy yellow to brownish, not reticulate, not shiny, lobes triangular-ovate, overlapping at base, margins entire, tips acute; corollas campanulate to campanulate-cylindric, 2.5–4(5) mm, white, creamy white to brown when dry, tubes 1.7–3 mm, lobes triangular-ovate, 30–100% as long as tube, erect to nearly so, tips acute, inflexed; stamens enclosed to barely exserted, < corolla lobes, filaments 0.3–0.7 mm, anthers 0.3–0.8 mm, infrastaminal scales ≈ corolla tube in length; styles evenly filiform, stigmas capitate, globose. **Fruits** 2–3.5 × 1.9–4(5) mm, surrounded or capped by withered corollas. **Seeds** 2–4. **2*n***=30.

Parasitic on a wide variety of hosts. Flowering Jul–Dec. 0–900 m. Col, ECas, Owy, Sisk, WV. CA, ID, NV, WA; widespread in North America; South America. Native.

The common name, collared dodder, describes the thickened, elevated ring at the tip of the capsule surrounding the base of the styles, a feature unique to this species among Oregon *Cuscuta*.

Cuscuta occidentalis Millspaugh

Western dodder

Cuscuta californica Hook. & Arn. var. *breviflora* Engelm.

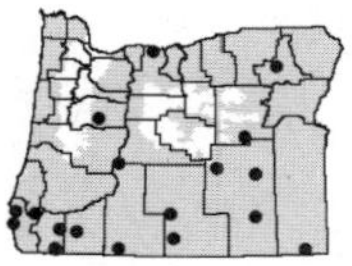

Stems 0.4–0.6 mm in diameter, yellow to orange. **Inflorescences** densely clustered; pedicels 0–0.5(1.5) mm. **Flowers** 5-parted, 2.7–3.4 mm, membranous; calyces campanulate, < corolla tube, divided 40–50% of their length, usually yellow, not reticulate, shiny, lobes ovate to lanceolate, not overlapping at base, margins entire, tips acuminate; corollas campanulate-cylindric with spreading (sometimes reflexed) lobes, 2.5–3.2 mm, white, creamy white to yellow when dry, tubes 1.4–2.1 mm, lobes lanceolate, slightly < tube, tips acuminate, straight; stamens ± exserted, < corolla lobes, filaments 0.2–0.5 mm, anthers 0.2–0.5 mm, infrastaminal scales absent; styles evenly filiform, stigmas capitate, globose. **Fruits** 1.8–2.2 × 2–2.6 mm, surrounded but not completely enclosed by withered corollas. **Seeds** 2–4.

Parasitic on a wide variety of hosts. Flowering Jun–Aug. 50–2300 m. BR, BW, Casc, Col, CR, ECas, Est, Owy, Sisk. CA, ID, NV, WA; southeast to CO, south to Mexico. Native.

Cuscuta pacifica Costea & M.A.R. Wright var. *pacifica* (p. 590)

Pacific salt marsh dodder

Cuscuta salina Engelm. var. *salina*

Stems 0.3–0.4 mm in diameter, orange. **Inflorescences** dense; pedicels 0.5–2 mm. **Flowers** 5-parted, 3.5–6 mm, membranous; calyces campanulate to cup-shaped, ≈ corolla tube, divided ~67% of their length, creamy yellow, brown or sometimes yellow when dry, not shiny, lobes ovate-triangular, slightly overlapping at base, margins entire, calyx tips acute to acuminate; corollas campanulate with erect to spreading lobes, 2.8–5.4 mm, white, dark brown or rarely creamy yellow when dry, tubes 1.5–2.6 mm, lobes ovate to rhombic-ovate, as long as tube, tips acute to cuspidate, straight; stamens included, filaments 0.3–0.6 mm, anthers 0.3–0.6 mm, infrastaminal scales 50–75% as long as corolla tube; styles evenly filiform, stigmas capitate. **Fruits** 2–3.6 × 1.4–2.1 mm, surrounded by withered corollas. **Seeds** 1–2.

Coastal salt marshes and tidal flats. Parasitic primarily on *Sarcocornia* and *Jaumea* but is also found on other hosts. Flowering Jun–Oct. 0–50 m. Est. CA, WA; north to British Columbia, south to Mexico. Native.

Cuscuta suksdorfii Yuncker

Mountain dodder

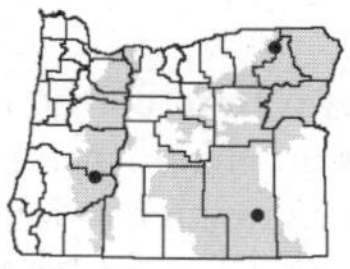

Stems 0.3–0.4 mm in diameter, yellow. **Inflorescences** loose; pedicels 0–2 mm. **Flowers** 4–5-parted, 2.8–3.3 mm, membranous; calyces slightly bilaterally symmetric, broadly campanulate, > corolla tube, divided 50–60% of their length, white to creamy yellow, not reticulate, not shiny, lobes ovate, not overlapping at base, margins entire, tips long-attenuate; corollas campanulate, 2.6–3 mm, white, creamy yellow when dry, tubes 1.2–1.5 mm, lobes triangular-ovate, nearly erect, > tube, tips lanceolate-attenuate, straight; stamens included to barely visible at corolla sinuses, filaments 0.2–0.5 mm, anthers 0.2–0.4 mm, infrastaminal scales 50–75% as long as corolla tube, often reduced to denticulate wings, sometimes with few, short fimbriae; styles slightly subulate to terete, stigmas capitate, globose. **Fruits** 2–3.2 × 2–3.6 mm, surrounded on basal half by withered corollas. **Seeds** 2–4.

Mountain meadows. Parasitic on a wide variety of hosts. Flowering May–Oct. 1100–2300 m. BR, BW, Casc. CA, WA. Native.

It is not known whether this species is rare in Oregon or just rarely collected.

DICHONDRA Ponysfoot

Herbs perennial; taprooted. **Stems** creeping or sprawling. **Leaves** suborbicular to reniform. **Inflorescences** axillary, solitary flowers; pedicellate. **Flowers** inconspicuous; sepals united only at base; corolla deeply 5-lobed, lobes > or ≈ to tubes; stamens with filiform filaments; ovaries 2-celled, each carpel with 2 ovules, styles 2, united basally but appearing distinct, filiform, stigmas capitate.

Temperate to tropical regions. ~15 species; 1 species treated in *Flora.*

Dichondra donelliana Tharp & M.C. Johnston (p. 593)

California ponysfoot

Plants pubescent with yellowish, appressed hairs. **Stems** rooting at nodes. **Leaves** (5)10–15(22) × (8)15–25(38) mm, tips rounded to emarginate; long-petiolate. **Inflorescences** pedicels erect, sharply recurved distally. **Flowers** sepals 2–3 mm at anthesis, 2.5–3.7 mm in fruit; corollas campanulate, 2–3 mm, white to greenish white, lobes broadly lanceolate, tips acute. **Fruits** deeply 2-lobed, separating into 2 nutlets or utricles, dark brown at maturity.

Open places near the ocean. Flowering Mar–Jun. 0–400 m. Est. CA. Native.

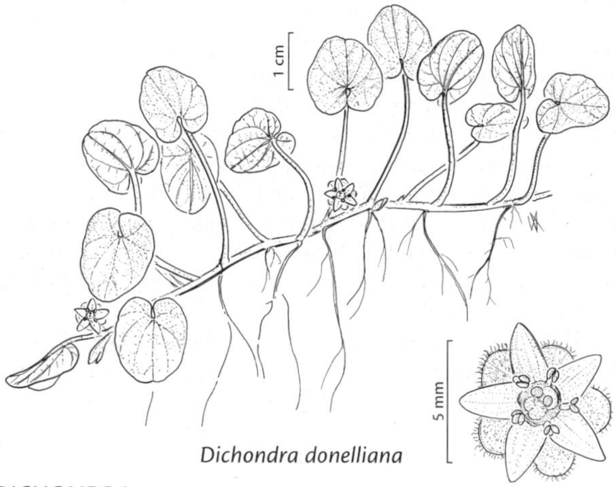

Dichondra donelliana

DICHONDRA

CORNACEAE Dogwood family

Alan T. Whittemore

Trees, **shrubs**, or **herbs** usually deciduous. **Leaves** simple, margins usually entire; stipules 0. **Inflorescences** clusters or cymes. **Flowers** bisexual; petals united; stamens alternate petals; styles simple. **Fruits** drupes with 2 stones.

Worldwide, but most species are in the north temperate zone. 2 genera; 1 genus treated in *Flora.*

CORNUS Dogwood

Leaves opposite, unlobed, veins arcuate. **Flowers** radially symmetrical; sepals united; petals 4, white, yellow, green, or purple; stamens 4; pistils compound, ovaries inferior, carpels 2, styles undivided.

North temperate zone and mountains in the tropics. ~65 species; 4 species treated in *Flora.*

Cornus amomum, native to eastern North America, escapes locally along the Willamette River in Eugene. It will key to *C. sericea* but differs in having blue (not white) fruit and in having brown, somewhat crisped hairs on the underside of the leaf (especially the veins).

The blackfruit dogwood, *C. sessilis*, is known from northern California, within 90 km of the Oregon border. *Cornus sessilis* is a shrub or small tree. Flowers and fruit grow on unbranched pedicels; the inflorescence is subtended by deciduous bracts, which are green or brown and not showy. Vegetative growth continues from the center of the umbel, so the fruits, which are black at maturity, are found on slender, unbranched pedicels at the base of the current year's growth on an ordinary leafy twig. Recently, a small stand (two trees) has been found in Jackson County; these may be waifs, descended from horticultural material, rather than a disjunct native stand.

Key to *Cornus* species:

1. Inflorescences subtended by 4–6 showy white bracts; flowers and fruits forming dense clusters 6–24 mm wide; fruits red.
    2. Trees to 25 m; bracts (4)6 per inflorescence, (3)5–5.9 cm, rounded or short-acuminate; fruits ellipsoidal. . . . . ***C. nuttallii***
    2' Rhizomatous herbs; bracts 4 per inflorescence, 1.2–2.1 cm, strongly acuminate; fruits spherical . . . . . . ***C. unalaschkensis***

1' Inflorescences not subtended by bracts; flowers and fruits in open flat-topped cymes 30–60 mm wide; fruits white, seldom bluish.
    3. Secondary veins 3–5 on each side of midrib, not projecting on underside of leaf; leaf bases acute (often narrowly so), seldom obtuse or rounded; twigs dark brownish purple . . . . ***C. glabrata***
    3' Secondary veins 4–7 on each side of midrib, strongly projecting on underside of leaf; leaf bases obtuse or rounded, rarely broadly acute; twigs purplish red, seldom red-brown, or green in shade . . . . ***C. sericea***

Cornus glabrata Bentham

Brown dogwood

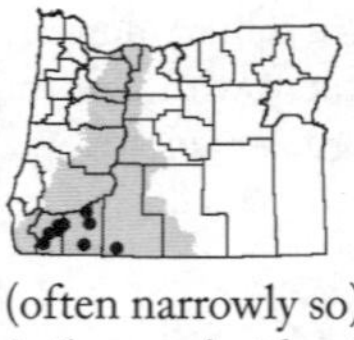

Shrubs 1.5–6 m. **Twigs** dark brownish purple, sparsely to moderately strigose with tightly appressed, parallel, 2-rayed hairs. **Leaves** elliptic to lance-elliptic, seldom ovate, 4–5.6(8) × 1.4–4.6 cm, bases acute (often narrowly so), seldom obtuse or rounded, secondary veins 3–5 on each side of blade, not projecting on leaf underside, midribs scarcely projecting, tips acuminate, generally abruptly acuminate, surfaces sparsely to moderately strigose with tightly appressed, parallel, 2-rayed hairs; petioles 5–11 mm. **Inflorescences** flat-topped, much-branched cymes, appearing with or after leaves, 3–5 cm wide, sparsely to moderately strigose with tightly appressed, parallel, 2-rayed hairs; bracts 0. **Flowers** corollas ~4 mm, lobes linear, white to cream. **Fruits** ellipsoidal, 8–9 mm, white, seldom bluish.

Thickets and streambanks. Flowering May–Jul. 300–1300 m. Casc, ECas, Sisk. CA. Native.

Cornus nuttallii Audubon ex Torrey & A. Gray (p. 595)

Pacific dogwood

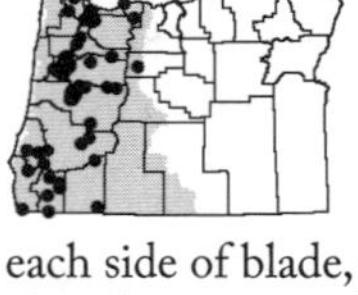

Trees to 25 m. **Twigs** purplish brown, covered with tightly appressed, 2-rayed hairs. **Leaves** usually elliptic, occasionally ovate or obovate, 8.2–11.6 × 4.1–6.8 cm, bases obtuse or rounded-obtuse, secondary veins 4–6 on each side of blade, projecting from leaf underside, tips acute or bluntly short-acuminate, undersides covered with appressed or spreading 2-rayed hairs, upper surfaces covered with tightly appressed 2-rayed hairs; petioles 7–17 mm. **Inflorescences** tight clusters of sessile flowers, 1.5–2.4 cm wide, buds developing in fall and blooming as leaves and bracts appear in spring; bracts (4)6, showy, plane, perpendicularly spreading, elliptic to obovate, (3)5–5.9 × 2.5–3.9 cm, white, narrowed to base, tips rounded or short-acuminate. **Flowers** corollas 1.5–2.5 mm, lobes triangular, cream to greenish white with purple or green tips. **Fruits** ellipsoidal, 10–15 mm, red. **2*n***=22.

Woods and thickets, often beneath taller trees in open conifer forests, but sometimes in deciduous forests and open habitats. Flowering Mar–Jul. 0–1600 m. Casc, CR, ECas, Sisk, WV. CA, ID, WA; north to British Columbia. Native.

Pacific dogwood is one of the most striking flowering trees of North America, but it is unfortunately difficult to cultivate. Two other flowering dogwoods, *C. florida* and *C. kousa*, often take its place in cultivation. Both are smaller trees. *Cornus florida*, native to eastern North America, has notched bract tips, and the bracts often have some pink pigmentation. *Cornus kousa*, native to East Asia, has the bracts widest below the middle and acute or acuminate, and the fruits are fused into a single compound fruit with many stones.

Cornus sericea Linnaeus
Red-osier dogwood

Cornus × *californica* C.A. Mey., *Cornus sericea* L. ssp. *occidentalis* (Torr. & A. Gray) Fosberg, *Cornus sericea* L. ssp. *sericea*, *Cornus stolonifera* Michx., *Cornus stolonifera* Michx. var. *occidentalis* (Torr. & A. Gray) C.L. Hitchc.

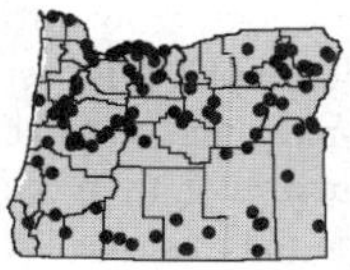

Shrubs 1.5–4 m. **Twigs** purplish red, seldom red-brown, or green in shade, strigose or villous, hairs either appressed or spreading and crisped. **Leaves** elliptic to ovate or lanceolate, 4.7–14.8 × 1.8–11.8 cm, bases obtuse or rounded, seldom broadly acute, secondary veins 4–7 on each side of blade, prominent beneath, tips usually gradually acuminate, surfaces abaxially pubescent with spreading, often somewhat crisped hairs, tightly appressed, parallel, 2-rayed hairs, or both hair types on different parts of leaf, adaxially strigose, hairs appressed (seldom, a few hairs spread slightly); petioles 6–30 mm. **Inflorescences** flat-topped, much-branched cymes, appearing with or after leaves, 3–6 cm wide, strigose or villous, hairs either appressed or spreading and crisped; bracts 0. **Flowers** corollas 2.5–4.5 mm, lobes linear or lanceolate, white to cream. **Fruits** ellipsoidal, 7–9 mm, white, seldom bluish.

Banks of rivers, creeks, ponds, and lakes, marshes and wet ground; Flowering May–Sep. 0–1900 m. All ecoregions except Est. CA, ID, NV, WA; north to AK, throughout Canada and US except TX and southeastern states, south to northern Mexico; Asia, Europe. Native.

Variation in *C. sericea* needs further study. In western North America, the species has been divided into several species or subspecies based on the size of the flowers, the shape and surface ridging of the stones, and the distribution of different hair types (appressed or spreading and crisped) on the twigs, leaves, and inflorescences (Rickett 1944; Fosberg 1942). In Oregon, there is no correlation between flower size and the indumentum characters. Unless other characters can be discovered, there is no justification for recognizing more than one taxon in Oregon.

Cornus unalaschkensis Ledebour (p. 595)
Western bunchberry

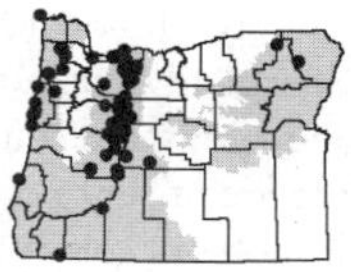

Herbs perennial, 10–22 cm; rhizomes woody, slender, 1–2(3) mm thick. **Stems** green or light brown, glabrous or sparsely appressed-pubescent with parallel tightly appressed 2-armed hairs. **Leaves** in terminal whorls of 4 on sterile stems, 6 on flowering stems; elliptic, 3.1–6 × 1.6–3.6 cm, bases acute or acuminate, secondary veins 2–3 on each side of blade, projecting (not strongly) on leaf underside, tips acute, obtuse, or acuminate, surfaces glabrous or sparsely appressed-pubescent with parallel tightly appressed 2-armed hairs; 1 or more pairs of bracts below, leaf-like distally to small and early deciduous proximally; petioles 1–5 mm. **Inflorescences** tight clusters of flowers, 1 per stem, 0.6–1.4 cm wide, glabrous or sparsely appressed-pubescent with parallel tightly appressed 2-armed hairs; bracts 4, showy, plane, perpendicularly spreading, broadly ovate to almost circular, 1.2–2.1 × 0.8–1.8 cm, white, bases broadly rounded, tips strongly acuminate. **Flowers** corollas 0.5–1 mm, lobes triangular, white; hypanthia deep purple. **Fruits** spherical, 6–8 mm, red. **2*n***=44.

Moist forests. Flowering May–Aug. 0–1800 m. BW, Casc, CR, ECas, Est, Sisk. CA, ID, WA; north to Alaska. Native.

Oregon bunchberries have often been called *C. canadensis*, but all seem to be referable to the allotetraploid species *C. unalaschkensis*.

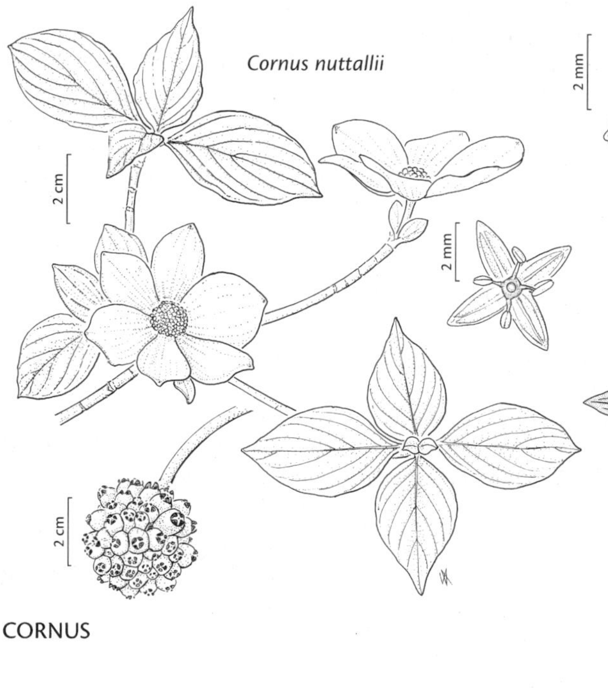

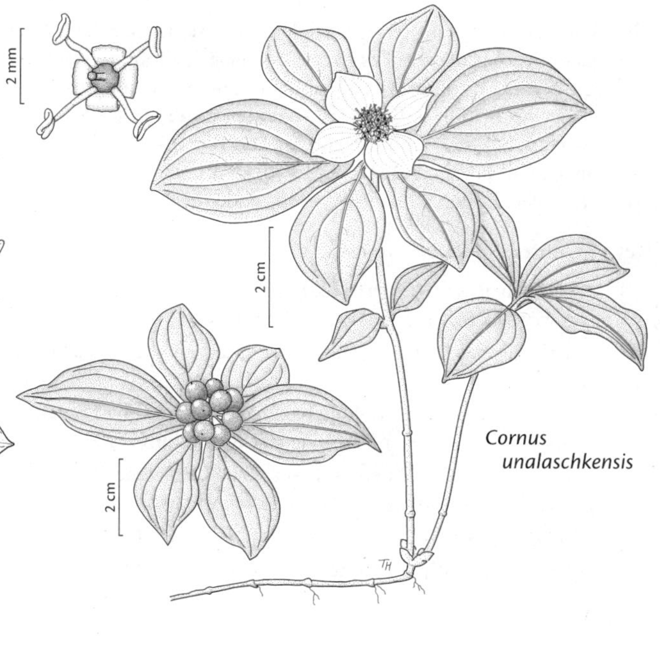

CORNUS

CRASSULACEAE Stonecrop family

Barbara L. Wilson, Richard E. Brainerd, and Nick Otting

Herbs or **rarely shrubs** annual, biennial, or perennial, succulent, leaves usually persistent, sometimes deciduous or easily detached. **Stems** alike, all potentially flowering, annual species axillary flowering stems overtopping primary stems. **Leaves** cauline or sometimes in dense basal rosettes, alternate, opposite or whorled, simple, margins usually entire, sometimes toothed or lobed; petioles usually absent; stipules absent. **Inflorescences** usually cymes, sometimes terminal or axillary spikes, thyrses or panicles, branches commonly scorpioid cymes and at first commonly coiled, rarely flowers solitary. **Flowers** bisexual (sometimes unisexual in *Rhodiola*), radially symmetric; hypanthia inconspicuous; sepals (3)4–5(30+), petals as many as sepals, distinct or connate basally or sometimes tubular, margins entire; nectaries as scales at bases of carpels; stamens as many as sepals or 2 times as many and in 2 series, free or adnate to corolla bases or tube; pistils usually as many as sepals, rarely fewer, ± distinct, ovaries superior or semi-inferior, 1-locular, placentation submarginal, styles 1 per pistil, terminal, distinct, stigmas 1, terminal, capitate. **Fruits** usually whorls of follicles, each opening along adaxial suture, rarely whorls of flimsy 1-seeded utricles (*Sedella*). **Seeds** 1–20+ per carpel, fusiform or ellipsoid, brownish; embryo straight; endosperm oily, scant.

Nearly worldwide, mostly north-temperate regions and southern Africa. ~35 genera; 5 genera treated in *Flora*.

Many herbarium specimens of Crassulaceae are of very poor quality because the plants dry slowly and often mold, grow into odd shapes, or lose parts. Collectors should take notes of colors and other traits on the live plants and/or take photographs. It is best to kill the plants quickly. A good way is to freeze the specimen solid, allow it to thaw, then puncture the larger parts and squeeze the juice from them. Papers and blotters in the press should be changed several times in short order to accelerate drying, then the specimen should be arranged carefully in a natural position before final drying. Alternatively, the plants can be microwaved for 15–60 seconds, arranged in a natural position, then pressed and dried. Because microwaving destroys DNA and other chemical traits, small, unmicrowaved pieces should be set aside to dry in packets.

Key to Crassulaceae genera:

1. Plants annual; leaves < 1 cm; petals 0.5–2(5) mm.
    2. All leaves opposite, bases fused around stem; petals < 2 mm . . . . **CRASSULA**
    2' Leaves alternate distally, bases free; petals 1.3–5 mm. . . . . **SEDELLA**

1' Plants perennial, biennial, or annual; longer leaves > 1 cm; petals > 4 mm (but 1.5–5 mm in *Rhodiola*).
    3. Flowering stems arising from axils of well-developed rosette leaves but not from tip of rosette. . . . . **DUDLEYA**
    3' Flowering stems arising from tip of rosette, or well-developed rosette leaves absent.
        4. Sepals and petals usually 4(5); horizontal stems with brown, scale-like leaves, not forming rosettes; flowering stems arising from axils of brown, scale-like leaves . . . . **RHODIOLA**
        4' Sepals and petals usually (4)5(6+); horizontal stems with well-developed, green to reddish, succulent leaves that often form rosettes; flowering stems arising from tips of rosettes . . . . **SEDUM**

CRASSULA Pygmyweed

Herbs annual, aquatic or terrestrial, glabrous. **Stems** succulent. **Leaves** opposite, connate basally, fleshy, bases fused around stem, margins entire. **Inflorescences** thyrses, panicles or solitary flowers in axils of leaves. **Flowers** 3–4(5)-parted; petals spreading or recurved, distinct or fused basally, whitish; stamens as many as sepals; pistils spreading to erect. **Fruits** follicles, recurved or erect.

Nearly worldwide. ~250 species; 4 species treated in *Flora*.

All species of *Crassula* in Oregon are annuals. It is possible that these will be split off as *Tillaea* again and indeed possible that the annuals will be divided into two genera: *Tillaea* with a flower in the axils of both leaves at a node and *Tillaeastrum* with a flower in the axil of only one leaf at each node.

Key to *Crassula* species:

1. Flowers 1 per node (or terminal); sepals ~½ as long as petals; seeds 3+ per follicle; plants generally aquatic or on wet substrates; stems generally rooting at nodes.
    2. Follicles erect, oblong, subtruncate; seeds with regular, continuous, longitudinal lines at 20× magnification, dull . . . . ***C. aquatica***
    2' Follicles ascending, obliquely lanceolate, distal half outcurved; seeds with irregular, interrupted, longitudinal lines at 20× magnification, shiny. . . . . ***C. solieri***

1' Flowers generally (1)2 per node; sepals ≥ petals; seeds (1)2 per follicle; plants generally terrestrial, often on seasonally wet substrates; stems not rooting at nodes.
    3. Sepals generally 3, attenuate; pedicels generally < 0.5 mm . . . . ***C. tillaea***
    3' Sepals generally 4 or 5, acute to acuminate; pedicels generally ≥ 1.5 mm. . . . . ***C. connata***

Crassula aquatica (Linnaeus) Schönland (p. 597)

Water pygmyweed

Tillaea aquatica L., *Tillaeastrum aquaticum* (L.) Britton

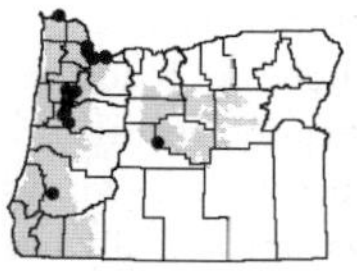

Plants aquatic but often stranded, reddish in age. **Stems** decumbent, erect if stranded, 2–14 cm, usually branched from base; rooting at nodes. **Leaves** oblanceolate to linear, 2–6 mm, tips acute. **Inflorescences** 1 flower per node; pedicels 0.5–20 mm. **Flowers** 4-parted; sepals ovate to oblong, 0.5–1.5 mm, tips obtuse to rounded; petals ovate to oblong, 1–2 mm. **Fruits** erect, oblong, subtruncate, old follicles spreading, flat. **Seeds** 6–17, 3+ per follicle, oblong-ellipsoid, (0.3)0.4–0.5(0.6) × 0.1–0.3 mm, with longitudinal lines (at 20× magnification), dull, not papillate, minutely rugulose.

Vernal pools, muddy edges of ponds or streams, coastal marshes, brackish mudflats. Flowering May–Jul. 0–1400 m. CR, Lava, Sisk, WV. CA, ID, NV, WA; north to AK, east to Labrador, south to Mexico; Asia, Europe. Native.

Crassula aquatica is a tiny plant of ephemeral wet places. Magnification reveals the single flower per node and the wide-spreading, flat, opened fruits. Pedicel length varies, from very short in northern salt marshes to as much as 20 mm long in fruit in some inland locations. Varieties have been named based on pedicel length, but intermediate plants are common and ranges overlap. It is best to treat this as variation within one species.

Crassula connata (Ruiz & Pavón) A. Berger (p. 597)

Sand pygmyweed

Tillaea erecta Hook. & Arn.

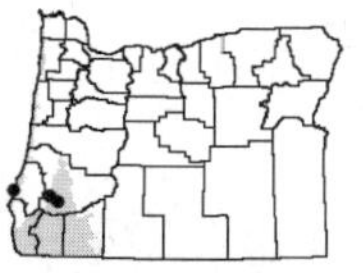

Plants terrestrial, green to red. **Stems** erect, 2–6(10) cm, simple or much branched; not rooting at nodes. **Leaves** ovate to oblong, 1–3(6) mm, tips acute to rounded. **Inflorescences** (1)2 flowers per node; pedicels (0.2)1.5–6 mm. **Flowers** (3)4–5-parted; sepals lanceolate, 0.5–2 mm, tips acute to acuminate; petals narrow-triangular, 0.6–1.5 mm. **Fruits** ascending, ovoid, with 1–2 seeds; old follicles ascending, boat-shaped. **Seeds** ellipsoid, 0.3–0.6 × 0.1–0.2 mm, not papillate, dull, rugulose.

Gravelly areas, mossy rocks, lawns, edges of paths, shallow wet areas that dry in spring. Flowering Feb–May. 0–400 m. Est, Sisk. CA, WA; north to British Columbia, southeast to TX, south to Argentina and Chile. Native.

Crassula connata is a small, uncommon, green to reddish plant of ephemerally wet areas. It may grow with the more common *C. tillaea*, and, like that species, it usually has two flowers per node. *Crassula tillaea* has flower parts in threes and usually has narrower leaves.

Crassula solieri (Gay) F. Meigen

Smooth-seed pygmyweed

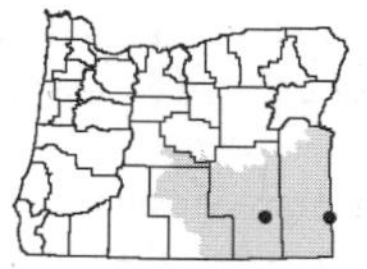

Plants aquatic or terrestrial, reddish purple with age. **Stems** erect when stranded, 2–10 cm, usually branched; rooting at nodes. **Leaves** oblong to linear, 1–5 mm, tips obtuse. **Inflorescences** 1 flower per node; pedicels 0.5–0.6(10) mm. **Flowers** 4-parted; sepals triangular-ovate to lanceolate, 0.4–1 mm, tips rounded to obtuse; petals lanceolate, 1–1.5 mm. **Fruits** ascending, obliquely lanceolate, with 3–14 seeds, distal half outcurved; old follicles spreading, flat. **Seeds** oblong-ellipsoid, 0.3–0.6 × 0.1–0.3 mm, with irregular interrupted longitudinal lines, not papillate, shiny, smooth.

Vernal pools, margins of lakes and streams. Flowering Mar–Jun. 1300–2600 m. BR, Owy. CA, NV; northeast to MT, southwest to TX, south to Mexico; Chile. Native.

Crassula solieri is a tiny, uncommon plant of ephemeral wet places. Its seeds are very shiny, looking wet even when they are dry. Aging plants turn a darker color than most other *Crassula* species.

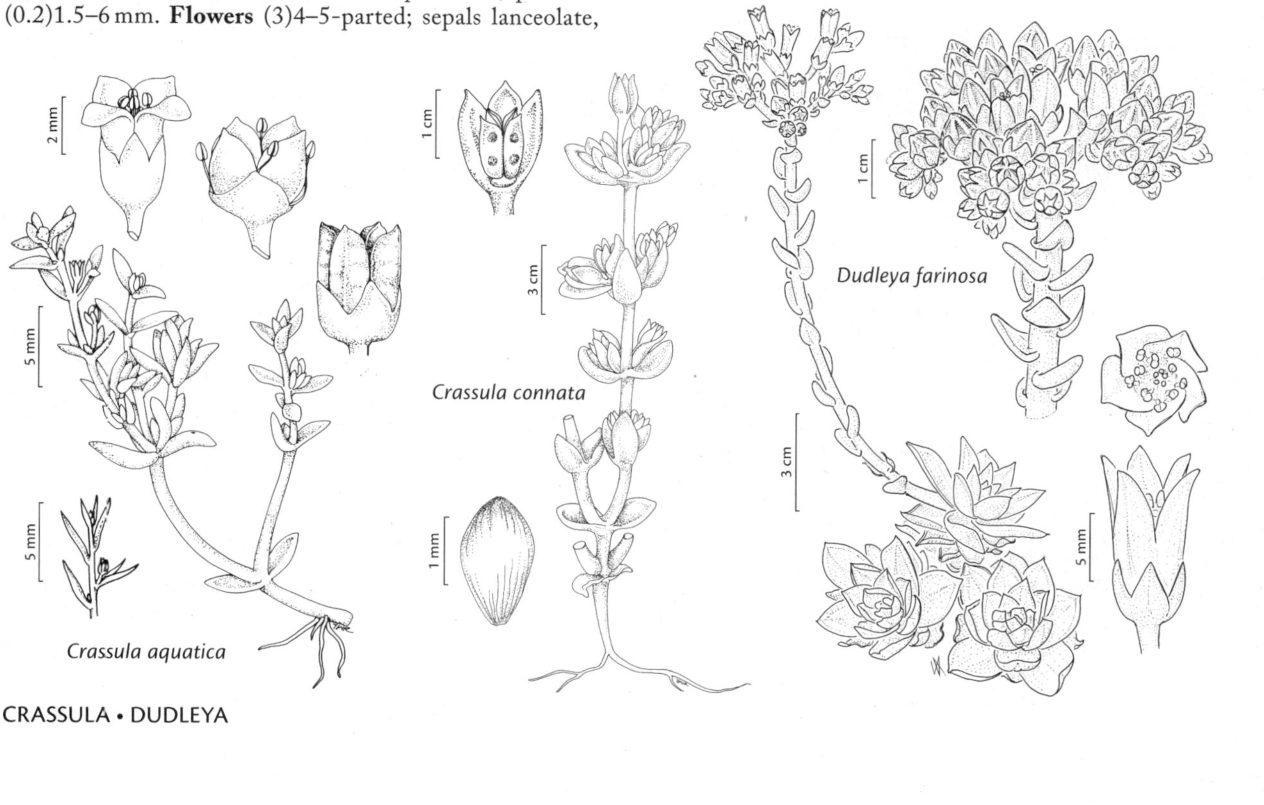

CRASSULA • DUDLEYA

Crassula tillaea Lester-Garland

Moss pygmyweed, tilliwort

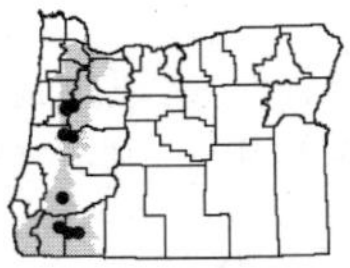

Plants terrestrial, reddish with age. **Stems** erect, 1–6 cm, simple or branched; not rooting at nodes. **Leaves** oblong, 1–3 mm, tips acute, apiculate or bristle-tipped. **Inflorescences** (1)2 flowers per node; pedicels 0.2–0.5(1.5) mm. **Flowers** 3(4)-parted; sepals lanceolate, 1–1.5 mm, tips attenuate, apiculate; petals narrow-lanceolate, 0.5–1 mm. **Fruits** ascending, with (1)2 seeds, ovoid; old follicles ascending, boat-shaped. **Seeds** ellipsoid, 0.2–0.3 mm, not papillate, dull, smooth to minutely rugulose.

Hard-packed, bare soil, sand, along trails and roads, lawns. Flowering Feb–May. 50–700 m. Sisk, WV. CA, WA; north to British Columbia; South America; Africa, Europe. Exotic.

This common, tiny plant can form red patches in roadsides or trampled lawns at low elevations. Magnification reveals the succulent leaves and three-parted flowers. It occasionally grows with *C. connata*, which has four-parted flowers and usually has broader leaves.

DUDLEYA Liveforever

Herbs perennial, 40–100 cm, glabrous; arising from usually fleshy, usually simple or 2-branched, modified stems forming above-ground caudices; leaves persistent or withering in early summer. **Leaves** alternate, densely crowded and forming a rosette, linear to orbicular, thickened, fleshy, clasping, margins entire; sessile. **Inflorescences** cymes arising from axils of well-developed rosette leaves, with 2–3(6) main branches (which may be branched again); pedicels erect. **Flowers** erect to ascending, 5-parted; sepals connate basally; petals erect and forming a tube, fused at least basally, white, yellow, orange, or red; stamens 10; pistils erect, fused only near base. **Fruits** follicles, erect.

Southwestern United States and northwestern Mexico. ~45 species; 1 species treated in *Flora*.

Dudleya are sometimes grown in rock gardens, but they have a reputation as being hard to maintain. In the wild, plants usually come up from seed, rather than the vegetative offsets so prolific in many *Sedum* species.

Dudleya is distinguished from *Sedum* by flowering shoots that arise from the axils of leaves at the side of the rosette. In *Sedum*, the flowering shoots arise as a continuation of the rosette stem, at the tips of the rosette.

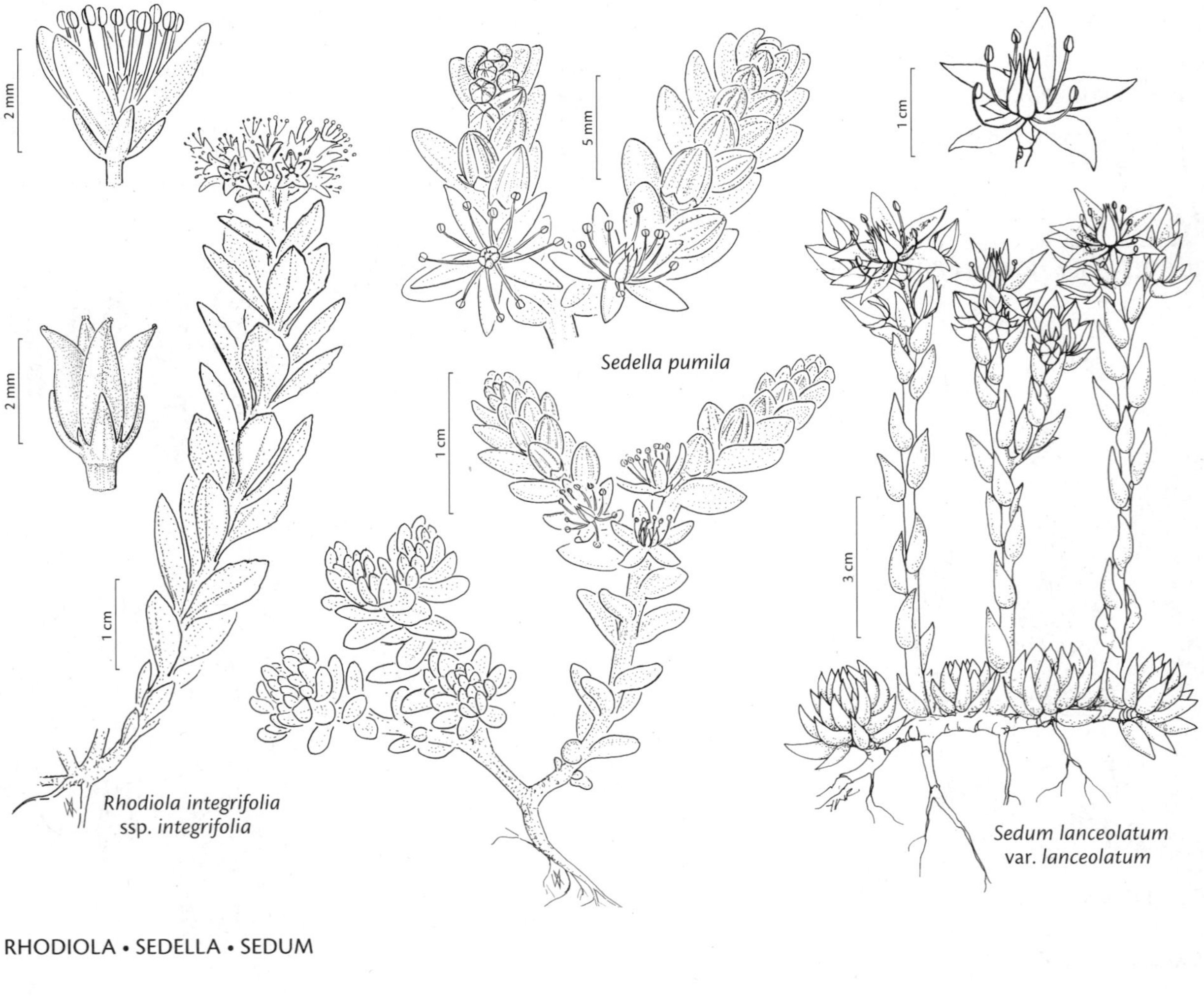

RHODIOLA • SEDELLA • SEDUM

Dudleya farinosa (Lindley) Britton & Rose (p. 597)

Powdery liveforever, bluff lettuce, powdery dudleya

Plants perennial. **Stems** thick, apically branching caudices, cespitose, becoming decumbent or hanging with age, 10–60 cm; rosettes (1)3–5, with 15–30 leaves, 4–10 cm in diameter. **Rosette leaves** persistent, ovate-oblong, 2.5–6 × 1–2.5 cm, 3–6 mm thick, usually whitish, occasionally gray-green or pale green, becoming reddish, bases 1–2.5 cm wide, tips acute, surfaces often farinose, not glaucous. **Flowering shoots** 10–35 cm, stem leaves 20–35, spreading, cordate-ovate, 10–25 × 10–20 mm, tips acute. **Inflorescences** with 3–5 branches, compact and flat-topped, 1–4 cm wide. **Flowers** sepals 5–8 mm; petals connate for 1–2 mm, 10–14 × 3–4 mm, pale yellow, tips acute to obtuse and often outcurved.

Cliffs and bluffs near the ocean. Flowering Jun–Sep. 0–100 m. Est. CA. Native.

Dudleya farinosa produces large, usually whitish, broad-leaved rosettes on cliffs near the ocean. Californian populations vary and have been treated as varieties or even species. It seems best, however, to treat all members as a single taxon, slightly differing because of isolated populations.

RHODIOLA Roseroot

Herbs perennial, dioecious or synoecious, glabrous. **Stems** forming rootstocks; horizontal stems spreading, decumbent or erect, usually much branched, with tips above ground. **Leaves** of flowering stems persistent, linear, oblanceolate, ovate or oblong, to ~1 mm thick, somewhat fleshy, sometimes clasping basally, margins entire or toothed. **Inflorescences** spike-like or head-like cymes, arising from tip of rosette. **Flowers** bisexual or unisexual; sepals connate basally; petals erect or spreading, distinct; stamens twice as many as sepals. **Fruits** follicles, erect.

Asia, Europe, North America. ~40 species; 1 species treated in *Flora.*

Phedimus spurius (Caucasian stonecrop) has crenate leaves, like some *Rhodiola* leaves. It is a garden plant that can persist long after a garden is abandoned or where yard waste has been discarded. Its leaves form sterile rosettes as well as flowering shoots.

Our species of *Rhodiola* is quite similar to eastern *R. rosea* and has sometimes been treated as a subspecies of that taxon. *Rhodiola rosea* has glaucous leaves and greenish yellow or yellow flowers, sometimes red at the tips. Another defining difference between the species is that *R. rosea* has 22 chromosomes and *R. integrifolia* has 36.

Our *Rhodiola integrifolia* and eastern *R. rosea* are called "roseroot" because of the rose-like odor of the rootstocks.

Rhodiola integrifolia Rafinesque ssp. ***integrifolia*** (p. 598)

Western roseroot, king's crown

Sedum rosea (L.) Scop. ssp. *integrifolium* (Raf.) Hultén

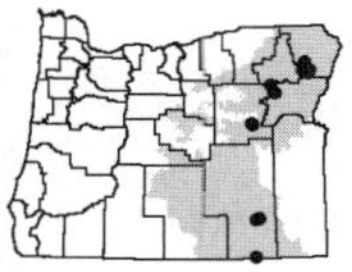

Stems rootstocks spreading or erect, 1–5 cm in diameter, scaly; flowering shoots deciduous, 3–50 cm. **Leaves** elliptic to ovate or oblanceolate to linear, 0.5–5.5 × 0.2–1.5(2) cm, green, rarely glaucous, tips acute to obtuse. **Inflorescences** with ≤250 flowers, ≤8 cm in diameter. **Flowers** mostly unisexual, 4(5)-parted; sepals lanceolate to ovate, 1.5–3 mm; petals elliptic-oblong, 1.5–5 mm, dark red, sometimes yellow basally or in lower 67%, spreading in staminate flowers, erect in pistillate flowers. **Fruits** 4–9 mm; beaks spreading. $2n$=36.

Rocky slopes, cliffs, alpine meadows. Flowering Jul–Aug. 2100–3000 m. BR, BW. CA, ID, NV, WA; north to AK, east to CO, southeast to NM; Asia. Native.

The other subspecies of *Rhodiola integrifolia* occur in isolated populations in New Mexico, Minnesota, and New York.

SEDELLA Mock stonecrop

Herbs annual, succulent, glabrous. **Leaves** opposite basally, alternate distally, falling by flowering time, succulent, bases free. **Inflorescences** terminal cymes with flowers ± sessile on branches. **Flowers** 5-parted; sepals fused basally; petals fused basally, yellow, straw-colored or greenish yellow, sometimes marked with red; stamens 5 or 10. **Fruits** utricles, 1-seeded, decomposing rather than splitting open. $2n$=18.

California and Oregon. 3 species; 1 species treated in *Flora.*

Sedella pumila (Bentham) Britton & Rose (p. 598)

Sierra mock stonecrop

Parvisedum pumilum (Benth.) R.T. Clausen

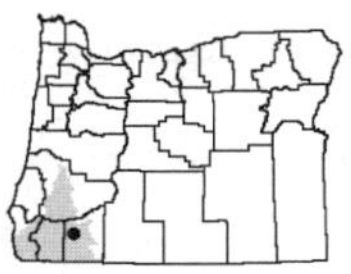

Plants 3–17 cm; branches usually erect, sometimes spreading, as long as main axis. **Stems** with 10–25 nodes. **Leaves** 4–7 × 1–3 mm, mostly falling by flowering time except for bracts. **Inflorescences** with 1–5 branches; branches 1–5 cm. **Flowers** with musty odor, 4–10 mm in diameter; petals widely spreading, erect in fruit, 1.3–5 × 0.3–1.2 mm, straw-colored to bright yellow, usually with red midrib, tips acute; stamens 10; pistils 1–2 mm, glandular on suture and angles, styles erect or recurved, 0.3–1.2 mm. **Fruits** erect or ascending, 1.2–1.5 mm. **Seeds** 0.8–1.5 mm.

Vernal pools, moss mats. Flowering Mar–May. 50–1500 m. Sisk. CA. Native.

Sedella pumila has widely spreading, yellow petals. Leaves usually fall off before flowering, except for the bracts in the inflorescence. Most populations are in the Great Valley of California.

SEDUM Stonecrop

Herbs or **rarely subshrubs** annual, biennial, or perennial, leaves deciduous or persistent, 2–10 dm, glabrous or hairy. **Stems** erect to creeping, usually much branched; horizontal stems well developed, green to reddish. **Leaves** alternate or opposite, sometimes forming dense rosettes, succulent, linear to orbicular or spatulate, terete to laminar, 0.1–8 cm, bases spurred or not, margins entire. **Inflorescences** terminal or axillary cymes, arising from tip of rosette. **Flowers** (4)5(6+)-parted; sepals distinct or basally connate; petals erect to spreading at anthesis, distinct or basally connate, yellow, white, pink, or red, often with red midribs or bases; stamens 2 × as many as sepals, rarely equal; pistils erect, distinct to basally fused. **Fruits** follicles, erect or spreading.

Africa, Asia, Central America, Europe, North America. ~450 species; 13 species treated in *Flora*.

Sedum, as defined here, is a diverse group, and not all the members are close relatives. The species now placed in *Sedella* and *Rhodiola* have already been removed from *Sedum*, and the genus may be split further. Recent research using DNA markers indicates that *Dudleya* arose within *Sedum*, as it is currently defined. Therefore, either *Dudleya* will be transferred into *Sedum* or *Sedum* will be broken up.

Use of the key may require fresh flowers from early in the flowering season or photographs or detailed descriptions of such flowers. "Stem leaves" are those of the flowering stems. Late in the season, stem leaves fall away. The last few flowers of the season of some taxa can be cleistogamous and are atypical in petal length and orientation. Anther color changes with age and may be hidden by yellow pollen.

Key to *Sedum* species:

1. Plants glandular-pubescent (sometimes minutely so) on lower stem, leaf margins, inflorescence branches, sepals, and/or petals.
   - 2. Basal leaves much wider than thick, in prominent basal rosettes; petals yellow; range canyons of southwestern Oregon . . . . . ***S. moranii***
   - 2' Basal leaves ~ as wide as thick, on ascending leafy shoots, basal rosettes lacking; petals white; range widespread, usually near human habitation . . . . . ***S. album*** (in part)

1' Plants not glandular-pubescent.
   - 3. Leaves of flowering shoots opposite or whorled.
     - 4. Dried flower stems ~1 mm thick; follicles erect; leaves pale green, speckled with pink, sometimes reddish, slightly glaucous, dull . . . . . ***S. debile***
     - 4' Dried flower stems 1.5–2 mm thick; follicles widely spreading; leaves bright green or red, not glaucous, shiny . . . . . ***S. divergens***
   - 3' Leaves of flowering shoots alternate.
     - 5. Most stem leaves broadest below middle or parallel-sided and similar to rosette leaves (if present).
       - 6. Petals white or pink.
         - 7. Leaf tips acute or acuminate; flowering stems usually bearing bulbils; leaf margins papillose-ciliate, papillae 0.1–0.15 mm . . . . . ***S. stenopetalum*** (in part)
         - 7' Leaf tips acute to obtuse; flowering stems without bulbils; leaf margins entire or low-papillose, papillae ≤ 0.1 mm.
           - 8. Anthers yellow, orange, or red; biennial or weak perennial with fibrous roots and widely scattered stems, not rhizomatous or stoloniferous; follicles strongly spreading in fruit; montane habitats in southwestern Oregon . . . . . ***S. radiatum*** ssp. ***depauperatum***
           - 8' Anthers dark red or dark purple; strong perennial with thick rhizomes or stolons, stems clustered and forming dense stands; follicles erect; urban weed . . . . . ***S. album*** (in part)
       - 6' Petals yellow.
         - 9. Leaves of vegetative shoots and rosettes flat, > 5 × as wide as thick; rosettes flat. . ***S. spathulifolium*** (in part)
         - 9' Leaves of vegetative shoots and rosettes narrow and often rounded in cross section, ≤ 2 × as wide as thick; rosettes not flat.
           - 10. Plants producing leafy bulbils in axils of stem leaves; longer leaves acuminate or attenuate; dried leaves strongly keeled . . . . . ***S. stenopetalum*** (in part)
           - 10' Plants not producing leafy bulbils in axils of stem leaves; leaves blunt or rounded, occasionally acute; dried leaves not keeled . . . . . ***S. lanceolatum*** var. ***lanceolatum***
     - 5' Most stem leaves broadest at or above middle or, if parallel-sided, then different from rosette leaves.
       - 11. Ovaries and fruits glandular-pustulose; mature follicles widely spreading; plants usually apparently biennial; rosette leaves narrowly oblanceolate; range eastern Oregon.
         - 12. Plants lacking bulbils along flowering stem; leaf tips blunt; leaves deciduous . . . . . ***S. leibergii***
         - 12' Plants with bulbils along flowering stem; leaf tips acute or acuminate; leaves persistent . . . . . ***S. stenopetalum*** (in part)

11' Ovaries and fruits not glandular-pustulose; mature follicles erect to spreading; plants perennial; rosette leaves various; range various.

13. Petals attenuate, gradually tapered to aristate tip, yellow; inflorescence with 3 primary branches that are dichotomously branched; leaves glossy, not glaucous . . . . **S. oreganum**

13' Petals blunt or acute to attenuate but never gradually tapered to a narrow aristate tip, white to yellow or pink; inflorescence with 3–5 primary branches or panicle-like; leaves glaucous, rarely glossy.

14. Leaves spreading, usually nearly parallel to ground, usually forming broad, flat rosettes; petals yellow; follicles erect until mature, then spreading; stolons slender, 0.4–1.3 mm in diameter when dried . . . . ***S. spathulifolium*** (in part)

14' Leaves not forming broad, flat rosettes; petals yellow, white, pink or red; mature follicles erect; stolons thicker, 1.4–5 mm in diameter when dried.

15. Fresh petals pink at or before anthesis, margins white, erect or spreading to 15–30° from axis of the flower; anthers dark red; petal tips acuminate . . . . **S. laxum**

15' Fresh petals white to yellow, sometimes dull pink basally at or before anthesis, erect to spreading to 45°; anthers yellow, orange, or red; petal tips obtuse to acute.

16. Rosettes loose with internodes easily visible; sepals, inflorescences, and leaves glaucous but without thick layer of granular white wax; rosette leaves broadly oblanceolate, often notched at tips; stem leaves round to oblong; range Cascade and Siskiyou mountains . . . . **S. oregonense**

16' Rosettes dense with internodes usually hidden; sepals, inflorescences, and leaves with thick layer of granular white wax leaving a residue on fingers when handled (although this weathers away so easiest to see on young flowering material); rosette and stem leaves narrowly oblanceolate; range southwestern Oregon . . . . **S. oblanceolatum**

Sedum album Linnaeus

White stonecrop

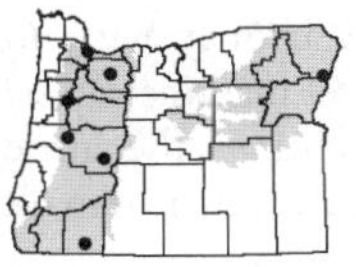

Plants perennial; thick rhizomes or stolons present. **Stems** creeping or growing upwards, clustered and forming dense stands, much branched, minutely puberulent and papillose, densely but inconspicuously glandular toward base. **Leaves** spreading to appressed, linear to ovate, 4–20(25) × 1–20 mm, green to reddish, thick, bases not or slightly spurred, margins entire or low-papillose, tips obtuse to rounded, papillae < 0.1 mm, surfaces glabrous or with sparse hairs generally < 0.1 mm; basal leaves approximately as wide as thick, on ascending leafy shoots; sessile. **Flowering shoots** erect, simple or branched, 5–18(30) cm, glabrous or sparsely minutely hairy; stem leaves alternate, linear to ovate, broadest below middle or parallel-sided, similar to other leaves. **Inflorescences** panicle-like cymes with 15–50+ flowers, 3–5-branched; bracts similar to leaves but smaller. **Flowers** 5-parted; sepals erect, 0.5–1.5 × 0.2–0.5 mm, green, tips acute; petals spreading, lanceolate, 2–4.5 mm, white or rarely pink, tips ± acute; filaments white, anthers dark red or dark purple. **Fruits** erect, white. **2*n***=34, 51, 68, 102, 136.

Disturbed places, trail sides, old house sites, sidewalks, gravel parking lots, coastal salt marshes, coastal rocks. Flowering Jul–Aug. 0–1400 m. BW, Casc, Sisk, WV. CA, WA; north to British Columbia, northeast to MT, southeast to UT, ME; Europe. Exotic.

Sedum album is a familiar cultivated succulent with white flowers. Many populations are stubbornly persistent in sites with sandy or gravelly soil where competition from other perennial plants is minimized. Populations in coastal marshes and rocks seem to be spreading.

Sedum debile S. Watson

Weakstem stonecrop

Gormania debilis (S. Watson) Britton

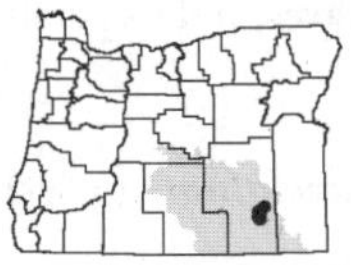

Plants perennial, tufted, glabrous. **Stems** decumbent, branched, with erect rosettes. **Leaves** opposite or rarely alternate, ascending, usually elliptic, oblanceolate or obovate, subterete, 4.2–7.2 × 2.8–4.3 mm, pale green, speckled with pink, sometimes reddish, bases not spurred, tips widely rounded, surfaces slightly glaucous, dull, minutely papillose. **Flowering shoots** erect, decumbent or ascending, simple, 3–12 cm; stem leaves opposite or whorled, ovate-elliptic; dried stems ~1 mm thick. **Inflorescences** cymes with 2–7 flowers, 2-branched; bracts similar to leaves but smaller. **Flowers** 5-parted; sepals erect, 2–4.2 × 1.3–2 mm, tips obtuse, surface glaucous; petals erect at bases but then widely spreading, elliptic-lanceolate, 6–9 mm, yellow, tips obtuse with apiculus; filaments yellow, anthers yellow. **Fruits** erect, connate basally, straw-colored with purple stripes. **2*n***=16.

Rock outcrops. Flowering Jun–Aug. 1500–2800 m. BR. ID, NV; northeast to MT, southeast to NM. Native.

In our area, the only similar *Sedum* growing with *S. debile* is *S. lanceolatum*. The two have similar flowers, but *S. lanceolatum* seems to bloom earlier. *Sedum debile* has paler, glaucous leaves with rounded tips, while *S. lanceolatum* has reddish leaves with pointed tips. The dense, rounded rosettes of *S. debile* break off and roll about in wind; they seem to be an important form of reproduction in this species.

Sedum divergens S. Watson

Cascade stonecrop, spreading stonecrop, Pacific stonecrop

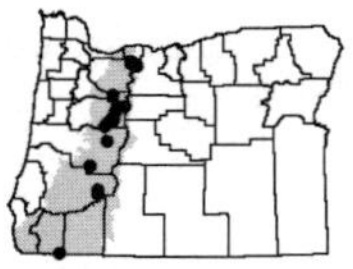

Plants perennial, somewhat tufted, glabrous. **Stems** decumbent, becoming erect or ascending, branched, without rosettes. **Leaves** opposite, rarely alternate, 2-ranked, suborbicular to obovate or oblong, subterete, ± globular, 4–9 × 4–6 mm, green or red, bases slightly clasping, tips rounded, surfaces not glaucous, shiny. **Flowering shoots** erect, rarely branched, 6–10 cm; stem leaves suborbicular to obovate or oblong; dried stems 1.5–2 mm thick. **Inflorescences** cymes with 5–25 flowers, 2–3-branched; branches recurved to erect; bracts similar to other leaves or elliptic to broadly lanceolate, 2–5 mm. **Flowers** 5(7)-parted; sepals ovate, 2–3 × 1.5–2 mm, green, tips obtuse; petals erect basally, widely spreading from about middle, usually distinct, narrow, ~5 mm, yellow, tips acute or obtuse and with an apiculus; filaments yellow, anthers yellow. **Fruits** widely spreading, fused at base, red or reddish brown to straw-colored. **2*n***=16.

Rocky slopes, talus, lava. Flowering Jun–Aug. 1200–2300 m. Casc, Sisk. CA, WA; north to AK, northeast to Alberta. Native.

Within this species, each pair of leaves is at a right angle to the pair below it, forming two columns of almost round, green to red leaves. This leaf arrangement is unique among our native *Sedum*. *Sedum debile* has opposite leaves, but they are glaucous and often whitish, and they form ball-like rosettes. The two species do not overlap in range. *Sedum divergens* may recall particularly thick-leaved individuals of *S. oreganum*, but the latter species does not have such neat columns of leaves, and its petals are much longer and nearly erect.

Sedum lanceolatum Torrey var. ***lanceolatum*** (p. 598)

Spearleaf stonecrop, lance-leaved stonecrop

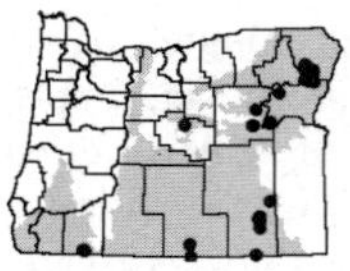

Plants perennial, tufted, glabrous. **Stems** decumbent and ascending, branched, sometimes papillose, bearing terminal rosettes and above-ground shoots. **Leaves** rosette leaves spreading to erect or ascending, lanceolate, elliptic-lanceolate or elliptic-ovate, narrow and often rounded in cross section, 4.2–9 × 1.5–2.5 mm, dull gray-green, bluish green, green, or dark reddish, bases with very short spurs, tips obtuse or obtusely apiculate, surfaces papillose; dried leaves not keeled. **Flowering shoots** erect, simple or branched, not forming rosettes in axils of leaves, 3–18 cm; stem leaves not easily detached, alternate, elliptic-lanceolate, broadest below middle, bases with short spur, not producing leafy bulbils in axils. **Inflorescences** cymes with 5–25 flowers, with (1)3(6) forked branches; bracts similar to leaves. **Flowers** 5-parted; sepals ovate or lanceolate, 2–4 × 1–2 mm, pale green to yellow-green, tips rounded to blunt or rarely acute, often papillose; petals widely spreading from erect base, distinct, lanceolate, elliptic-lanceolate or linear-lanceolate, keeled, 6–9 mm, yellow, tips acute to acuminate with minute apiculus; filaments yellow, anthers yellow, sometimes with red. **Fruits** erect, fused at base, not glandular. **2*n***=16, 32, 48.

Rock outcrops, cliffs. Flowering Jun–Sep. 1200–3000 m. BR, BW, ECas, Sisk. CA, ID, NV, WA; north to AK, east to SD, southeast to NM. Native.

Sedum lanceolatum is a common, yellow-flowered stonecrop, growing east of the Cascades. It produces tight rosettes of pointed, often reddish leaves. Similar *S. leibergii* has glandular-pustulose, glandular, horizontally spreading follicles. *Sedum stenopetalum* produces rosettes in the axils of the leaves of the flowering stems.

Sedum laxum (Britton) A. Berger

Roseflower stonecrop

Plants perennial, cespitose or not, glabrous; stolons 1.4–5 mm in diameter when dried. **Stems** ascending, much branched, bearing dense rosettes. **Leaves** rosette leaves obovate, spatulate or oblanceolate, subterete to somewhat flattened, 10–50 × (4.5)6–33 mm, green, bases not spurred, not scarious, tips truncate to rounded or obtuse, notched or not, surfaces glaucous or not, not pruinose. **Flowering shoots** erect or decumbent, 10–42 cm; stem leaves alternate, oblanceolate, elliptic, oblong or suborbicular, broadest near or above middle, auriculate or not, decurrent or not, different from rosette leaves. **Inflorescences** panicle-like cymes with 12–80 flowers, with 3+ branches; branches 2-forked; bracts similar to stem leaves but smaller. **Flowers** 5-parted; calyx lobes erect, closely appressed to corolla tube, ovate or lanceolate, (2)2.6–5.1 mm, pale green, tips acute or subacute; petals erect or spreading 15–30° from flower axis, 4–11 mm, pink with broad white margins, sometimes almost entirely white, aging red, tips acuminate, sometimes attenuate; anthers dark rose red, rarely dark red-orange, aging black. **Fruits** erect, brown.

Western North America. 4 subspecies; 2 subspecies treated in *Flora*.

Sedum laxum is recognized by its long, narrowly acute to acuminate, usually pink petals. The white petal margins can be broad, and, depending on environmental conditions, the entire petal may be white. Even when petals are white, anthers remain dark red, rarely dark red-orange.

Key to *Sedum laxum* subspecies:

1. Stem leaves usually suborbicular, rarely oblong, widely spreading to reflexed, rarely > 2 × as long as wide, bases clasping or auriculate . . . . ***S. l.*** ssp. ***heckneri***

1' Stem leaves elliptic to oblong, ascending to somewhat spreading, often > 2 × as long as wide, bases not clasping or auriculate . . . . ***S. l.*** ssp. ***laxum***

Sedum laxum (Britton) A. Berger ssp. ***heckneri*** (M. Peck) R.T. Clausen

Heckner's stonecrop

Sedum heckneri M. Peck, *Sedum laxum* (Britton) A. Berger var. *heckneri* (M. Peck) H. Ohba

Leaves rosette leaves mostly obovate or spatulate, 12–37 × (2)7–19 mm, 1–3(6) × as long as wide, tips often notched, surfaces glaucous; stem leaves widely spreading to reflexed, rarely > 2 × as long as wide, bases clasping or auriculate. **Flowering shoots** 8–30 cm; stem leaves usually suborbicular, rarely oblong, reflexed or saddle-shaped, 6–15(28) × 7–15(20) mm, 0.6–1.8(2+) × long as wide, bases clasping or auriculate. **Inflorescences** 2–12 cm. **Flowers** calyx lobes 2–5 mm, 15–45% as long as petals; petals 8–14 mm. **2*n***=30.

Cliffs and rock outcrops, road cuts, often on serpentine. Flowering Jun–Aug. 200–1900 m. Sisk. CA. Native.

Sedum laxum ssp. *heckneri* is recognized by its suborbicular, usually reflexed or saddle-shaped stem leaves. *Sedum laxum* ssp. *laxum* has elliptic to oblong leaves that are not auriculate, but may be decurrent. The two subspecies overlap only slightly in range where both leaf types may occur in a single population. *Sedum oregonense* can have suborbicular leaves but has shorter, blunter, white to yellowish petals, yellow anthers, and a more open growth form.

Sedum laxum (Britton) A. Berger ssp. ***laxum***

Roseflower stonecrop

Gormania laxa Britton, *Sedum laxum* (Britton) A. Berger ssp. *latifolium* R.T. Clausen, *Sedum laxum* (Britton) A. Berger var. *latifolium* (R.T. Clausen) H. Ohba, *Sedum laxum* (Britton) A. Berger var. *laxum*, *Sedum laxum* ssp. *perplexum* R.T. Clausen

Leaves rosette leaves flattened to unusually plump, 13–51 × 8–23 mm, 1–3 × as long as wide, tips rounded to notched, surfaces glaucous; stem leaves elliptic to oblong, ascending to somewhat spreading, often > 2 × as long as wide, bases not clasping or auriculate. **Flowering shoots** 10–42 cm; stem leaves ascending to somewhat spreading, elliptic to oblong, laminar to unusually plump, 9–22 × 4–13 mm, 1.2–3.5 × as long as wide, green or red, bases not clasping or auriculate, usually truncate, sometimes decurrent, tips rounded to obtuse, glaucous. **Inflorescences** 2.5–12(26) cm. **Flowers** calyx lobes 2.5–6(8.5) mm, 25–50(75)% as long as petals; petals 7–14 mm. **2*n***=30.

Cliffs, rock outcrops, boulder fields, roadcuts, usually on serpentine. Flowering May–Jul. 0–1400 m. CR, Est, Sisk. CA. Native.

Although flowers of *S. l.* ssp. *laxum* are essentially identical in different populations, stature and leaf shape vary, and the variations have some geographic coherence. As a result, this taxon has been divided into three subspecies. If one wants to split, plants of inland mountains that are tall, with flat, glaucous, often decurrent leaves have been called *S. l.* ssp. *laxum*. Plants along the main stem of the Smith River in California that are tall with very plump, green leaves that are usually not decurrent have been called *S. l.* ssp. *latifolium*. Coastal plants that are small, with somewhat plump, more or less green leaves that are usually not decurrent have been called *S. l.* ssp. *perplexum*. Studying morphological variation in populations across the Siskiyou Mountains would aid in understanding this group.

Sedum leibergii Britton

Leiberg's stonecrop

Sedum borschii (R.T. Clausen) R.T. Clausen

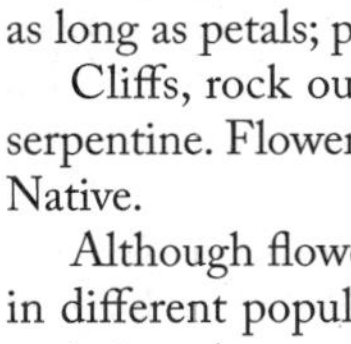

Plants biennial, glabrous. **Stems** horizontal, simple. **Leaves** rosette leaves deciduous, spreading, narrowly oblanceolate, 2–16 × 1.4–3.2 mm, green or greenish white, bases not spurred, tips blunt, surfaces papillose, not glaucous. **Flowering shoots** erect, simple, (5)11–18 cm; stem leaves alternate, not producing leafy bulbils in axils, ovate or elliptic, widest below middle, bases not spurred. **Inflorescences** cymes with 5–60 flowers, 3–6-branched; branches often strongly recurved, each 1–3 times dichotomously forked; bracts like stem leaves but smaller. **Flowers** (5)6(7)-parted; sepals ovate, 1.5–2 × 0.7–1.1 mm, green, tips acute; petals spreading, separate nearly to base, lanceolate to oblong, 4–6 mm, yellow, keel green or dark red, tips subobtuse to acute; filaments yellow, anthers yellow; ovaries glandular-pustulose. **Fruits** widely spreading when mature, fused near base, glandular-pustulose. **2*n***=16.

Cliffs, rocky slopes, on basalt or limestone. Flowering Apr–Jul. 50–1800 m. BW, Col, ECas, Lava, Owy. ID, WA. Native.

Sedum leibergii is a small, yellow-flowered *Sedum* like *S. lanceolatum*, *S. radiatum*, or *S. stenopetalum*. Unlike those species, *S. leibergii* has distinctly glandular-pustulose follicles. Additionally, *S. leibergii* does not produce offsets in the axils of stem leaves like *S. stenopetalum*.

Sedum moranii R.T. Clausen (p. 604)

Rogue River stonecrop

Gormania glandulifera (L.F. Hend.) Abrams, *Sedum glanduliferum* (L.F. Hend.) M. Peck

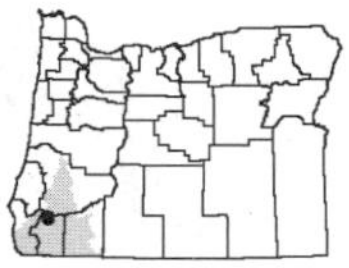

Plants perennial, tufted, glabrous proximally, glandular-pubescent distally. **Stems** vertical, branched, glandular, bearing dense terminal rosettes. **Leaves** rosette leaves ascending, oblong-spatulate, laminar, 14–40 × 9–32 mm, green, bases not spurred, tips widely rounded to obtuse or notched, surfaces with papillose-crenulate appendage, glaucous when young; basal leaves much wider than thick; proximal leaves glabrous; distal leaves glandular-hairy. **Flowering shoots** recurved when young, simple, 13–30 cm, glandular pubescent; stem leaves oblong-oblanceolate, bases not spurred. **Inflorescences** cymes with 20–30 flowers, (2)3-branched, glandular; branches not forked. **Flowers** 5-parted; sepals erect, ovate or lanceolate, 4–8.5 × 3–4.5 mm, 40% as long as petals, green, tips obtuse to acute, glandular-pubescent; petals erect, connate basally, oblong-lanceolate, 12–16 mm, yellow, tips aristate, glandular-pubescent; filaments greenish yellow, anthers yellow. **Fruits** somewhat divergent when mature, fused basally, brown. **2*n***=30.

Steep serpentine slopes and rock outcrops. Flowering May–Jul. 100–300 m. Sisk. Native.

Sedum moranii is distinctive with its usually three-branched inflorescence and densely glandular sepals, petals, and inflorescence. This species is extremely rare, known only from a short stretch of the Rogue River.

Sedum oblanceolatum R.T. Clausen

Oblongleaf stonecrop

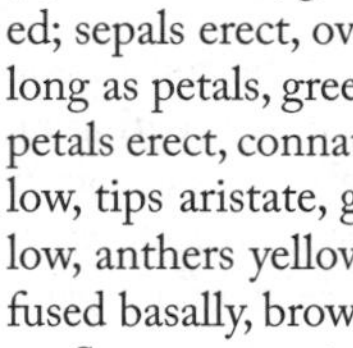

Plants perennial, glabrous; stolons 1.4–5 mm in diameter when dried. **Stems** erect, branched from base, bearing dense rosettes with hidden internodes. **Leaves** rosette leaves narrowly oblanceolate, 9–55 × 5–10 mm, 2.5–8 × as long as wide, with thick layer of granular white wax. **Flowering shoots** 10–22 cm; stem leaves ascending to somewhat spreading, alternate, narrowly oblanceolate, 15–30 × 3–10 mm, 2.2–5 × as long as wide, broadest above middle, bases not auriculate, surfaces strongly glaucous. **Inflorescences** panicle-like cymes, 2.5–8.5 cm, with thick layer of granular white wax, 3–5-branched. **Flowers** calyx lobes 4–8 mm, (30)40(80)% as long as petals, with thick layer of granular white wax, tips acute; petals ascending, 9–14 mm, white, sometimes

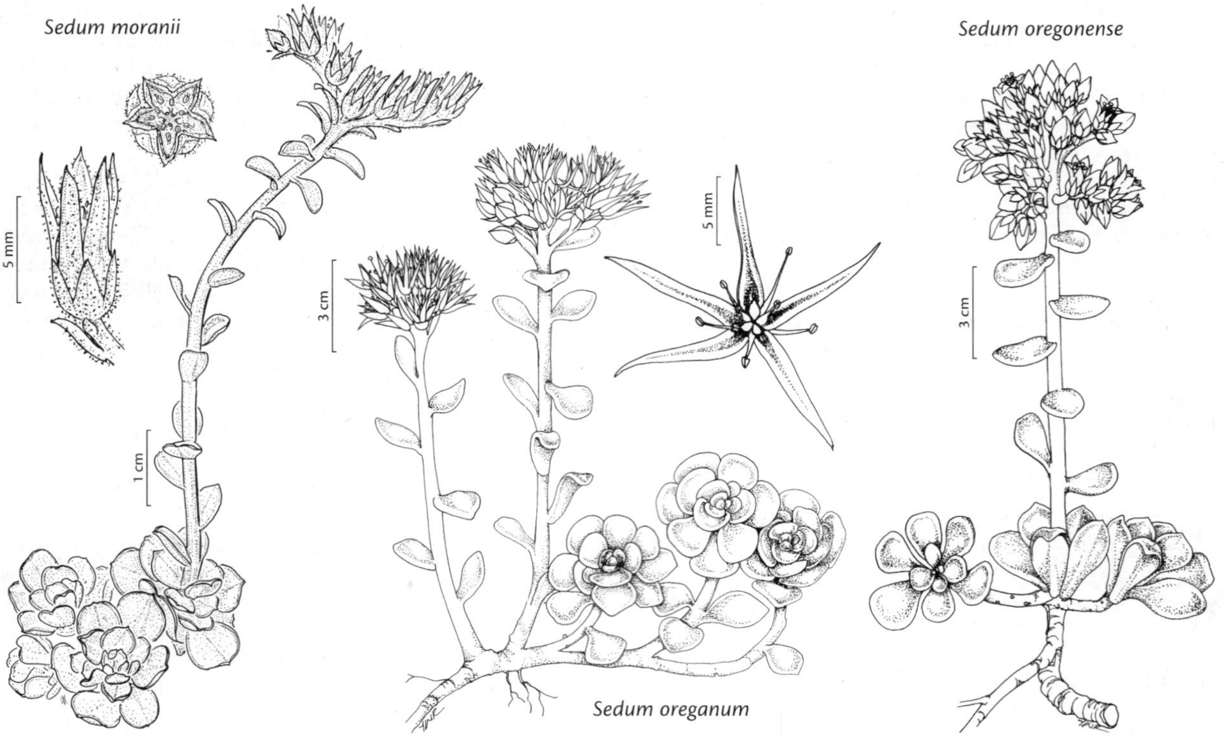

SEDUM

apparently yellow due to color of anthers, tips acute to attenuate; anthers yellow. **Fruits** erect, fused to slightly above base.

Rock outcrops, rocky slopes. Flowering May–Jul. 400–1800 m. Sisk. CA. Native.

Sedum oblanceolatum can be recognized by its narrow rosette and stem leaves. The dense, almost granular, waxy coating on the young leaves, panicle branches, and sepals imparts a waxy residue when handled. The distinctive glaucous coating wears off gradually as the season progresses. It also melts off in a plant press or microwave.

Sedum oreganum Nuttall (p. 604)

Oregon stonecrop

Gormania oregana (Nutt.) Britton, *Sedum oreganum* Nutt. var. *oreganum*, *Sedum oreganum* ssp. *tenue* R.T. Clausen, *Sedum oreganum* Nutt. var. *tenue* (R.T. Clausen) H. Ohba

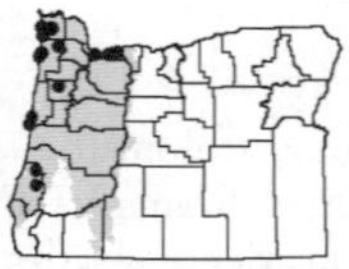

Plants perennial, glabrous. **Stems** rootstocks and decumbent branches with terminal rosettes. **Leaves** alternate, spreading, spatulate or cuneate, 7–12 × 4.4–9 mm, thickest near tip, green or red, bases with short spurs, tips broadly rounded or truncate, surfaces glossy, not glaucous. **Flowering shoots** erect, simple or branched, 8–13 cm; stem leaves alternate, broadest above middle, like those of sterile shoots. **Inflorescences** dense corymbs with 5–10 flowers, 3-branched; branches dichotomously forked; bracts spatulate or oblanceolate, 5–8 mm. **Flowers** erect, 5-parted; sepals erect, fused basally, ovate-lanceolate, ~3.5 × 2 mm, green, tips acuminate or acute; petals erect or distally somewhat spreading, connate basally, narrowly lanceolate, ~1 cm, yellow, tips attenuate to aristate; filaments yellowish or greenish, anthers yellow. **Fruits** erect, fused basally, light brown or yellow. **2*n***=24.

Rock outcrops, cliffs, steep rocky slopes. Flowering Jun–Aug. 0–1500 m. Casc, CR, Est, WV. CA, WA; north to AK. Native.

Sedum oreganum is recognized by its plump, shiny, crowded leaves and its long, narrow, erect, yellow flowers. *Sedum divergens* has more nearly spherical, opposite leaves.

Sedum oreganum is sometimes divided into coastal (*S. o.* ssp. *oreganum*) and Cascadian (*S. o.* ssp. *tenue*) subspecies. Because *Sedum* disperse poorly, a biologically significant difference may have developed between these disjunct populations. However, the reported morphological differences do not appear to differentiate the populations consistently.

Sedum oregonense (S. Watson) M. Peck (p. 604)

Creamy stonecrop

Gormania watsonii Britton

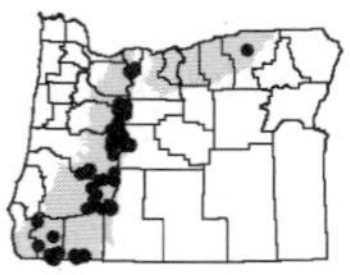

Plants perennial, glabrous; stolons 1.4–5 mm in diameter when dried. **Stems** much branched, bearing loose rosettes with internodes visible. **Leaves** rosette leaves broadly oblanceolate, 14–33 × 7–15 mm, 1.2–4.2 × as long as wide, tips rounded to truncate, often notched, surfaces glaucous. **Flowering shoots** 7–26 cm; stem leaves ascending to somewhat spreading, alternate, round to oblong, 6–17 × 3–13 mm, 1–5 × as long as wide, broadest near or above middle, auriculate or not. **Inflorescences** panicle-like cymes, columnar, 2.6–9 cm, glaucous. **Flowers** 5-parted; calyx lobes 1.5–6 mm, 20–55% as long as petals, tips acute, glaucous; petals ascending, 7–13 mm, white, cream, occasionally light yellow; anthers usually yellow, aging brown, rarely red or orange. **Fruits** erect, distinct, brown. **2*n***=30.

Rock outcrops, rocky slopes, lava fields. Flowering May–Sep. 500–2500 m. Casc, Col, Sisk. CA. Native.

Sedum oregonense has loose, spreading rosettes with visible internodes. Its flowers are white or cream-colored but may appear yellow due to the yellow anthers visible within. The stout stem leaves are round to oblong, differing in shape from the oblanceolate rosette leaves. No similar *Sedum* grow in the Cascades, but confusion may arise in southwestern Oregon. *Sedum oblanceolatum* has narrower leaves, denser rosettes, and a mealy waxiness that rubs off onto ones hands. *Sedum laxum* has dense rosettes and normally has pink flowers.

Sedum radiatum S. Watson ssp. ***depauperatum*** R.T. Clausen

Coast Range stonecrop

Sedum radiatum S. Watson var. *depauperatum* (R.T. Clausen) H. Ohba

Plants biennial or weak perennial, glabrous except pedicels and leaves of offsets sometimes ciliate; roots fibrous. **Stems** erect or horizontal near base; simple or branched, bearing rosettes. **Leaves** rosette leaves ascending, oblong-elliptic, oblong-lanceolate, lanceolate or ovate, subterete, 4.4–14 × 1.8–3 mm, yellow-green with 5 green to purple veins, bases with short spurs, scarious, sometimes low-papillose on margins, tips acuminate to acute, unlobed, surfaces not glaucous; papillae < 0.1 mm. **Flowering shoots** erect or slightly recurved, simple, 5.5–19 cm, sometimes ciliate, papillose; stem leaves alternate, lanceolate, often longer and wider than rosette leaves, broadest below middle, bases spurred, margins sometimes ciliate, tips unlobed or 3-lobed. **Inflorescences** cymes with 4–25 flowers, 3-branched, branches often dichotomously forked, 3–8 cm; bracts like other leaves but smaller. **Flowers** 5-parted; sepals distinct, lanceolate to ovate, 1.5–3.5 × 0.8–2 mm, green to yellow, bases broadly spurred, tips acuminate or acute, surface sometimes papillose; petals spreading, distinct, elliptic-lanceolate, 6–7 mm, white to cream-colored, sometimes drying yellowish, tips acute with apiculus; filaments white, anthers yellow, orange, or red. **Fruits** strongly spreading when mature, fused at base, straw-colored, streaked with reddish brown. **2*n***=16.

Rocky, sandy, gravelly slopes, rocks or talus along streams. Flowering Jun–Jul. 500–1700 m. Sisk. CA. Native.

Sedum radiatum ssp. *depauperatum* is a small, short-lived *Sedum* like *S. lanceolatum*, *S. leibergii*, and *S. stenopetalum*. However, flowers of *S. r.* ssp. *depauperatum* are white to cream-colored, thus differing from all except *S. stenopetalum* ssp. *ciliosum*, which nearly always has small rosettes in the axils of the leaves of the flowering stems.

Sedum spathulifolium

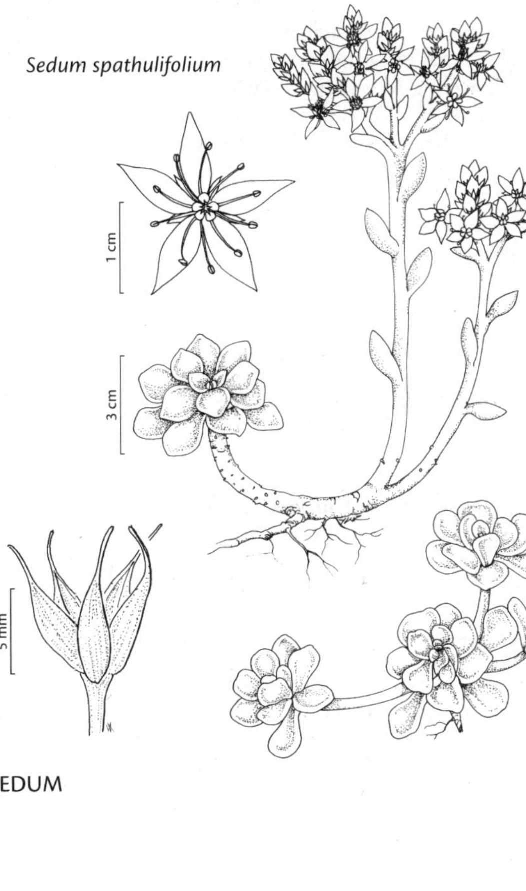

SEDUM

Sedum spathulifolium Hooker (p. 605)

Broadleaf stonecrop

Sedum spathulifolium var. *minus* Henderson, *Sedum spathulifolium* Hook. ssp. *pruinosum* (Britton) R.T. Clausen & C.H. Uhl, *Sedum spathulifolium* Hook. var. *pruinosum* (Britton) B. Boivin, *Sedum spathulifolium* ssp. *purdyi* (Jeps.) R.T. Clausen, *Sedum spathulifolium* Hook. ssp. *spathulifolium*, *Sedum spathulifolium* Hook. var. *spathulifolium*

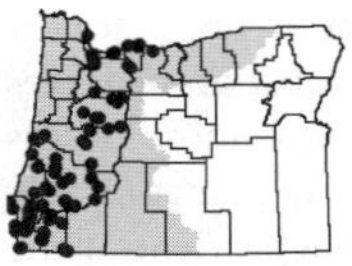

Plants perennial, forming open mats, glabrous; stolons slender, 0.4–1.3 mm when dried. **Stems** extensively stoloniferous, with dense terminal rosettes. **Leaves** rosette leaves spreading, usually nearly parallel to ground, usually forming flat rosettes, spatulate, narrowing to petioles, 7–19 × 4.5–10 mm, 5 × as wide as thick, green or pruinose and white, bases not spurred, not scarious, sometimes muricate or papillose near margins, tips rounded or truncate, surfaces usually glaucous. **Flowering shoots** erect, simple, 3–14 cm; stem leaves alternate, spatulate-oblong or elliptic-oblong, widest at or below middle, different from rosette leaves. **Inflorescences** cymes with ~30 flowers; 3-branched; branches forked; bracts oblong-spatulate or linear. **Flowers** 5-parted; sepals spreading to erect, lanceolate, linear-lanceolate, oblong-ovate or obovate, ~2.5 × 1.5 mm, green to yellow-green, tips acute or obtuse, surface glaucous or pruinose; petals strongly spreading above erect base, linear to oblanceolate, 4.5–9 mm, yellow, tips acute; filaments yellow, anthers yellow. **Fruits** erect until mature then spreading, fused basally, brown.

Rocks, cliffs and road cuts. Flowering Apr–Aug. 0–2400 m. Casc, Col, CR, ECas, Est, Sisk, WV. CA, WA; north to British Columbia. Native.

Sedum spathulifolium is highly variable. If subspecies are recognized, most of our plants are *S. spathulifolium* ssp. *spathulifolium*. A coastal form with thick, white-pruinose leaves, short, stout stolons, dense inflorescences and thick, crowded stem leaves can be called *S. s.* ssp. *pruinosum* (if it is considered a strictly coastal entity). If it is considered to include strongly glaucous or pruinose plants of the Coast Range that are less dense in growth form, the name *S. s.* var. *minus* would be applied. Plants with thin, green leaves, numerous long, slender stolons that tend to grow upwards before growing out, paler flowers, and more spreading follicles can be called *S. s.* ssp. *purdyi*, a rare form known only from southern Josephine County, and Del Norte and Siskiyou counties in California.

Sedum stenopetalum Pursh
Wormleaf stonecrop

Plants annual, biennial, or perennial, tufted to spreading, usually glabrous. **Stems** decumbent, branched, with terminal rosettes. **Leaves** rosette leaves persistent, spreading to erect, narrowly oblanceolate, narrow and often rounded in cross section, 4.3–13.8 × 1.4–2.7 mm, green to red, bases with small spur, margins sometimes papillose-ciliate, papillae 0.1–0.15 mm, tips acuminate or attenuate, surfaces not glaucous; dried leaves strongly keeled. **Flowering shoots** erect, branched, 10–43 cm; stem leaves alternate, linear, with basal scarious spurs and producing offset rosettes in their axils; bulbils present. **Inflorescences** solitary flowers or small cymes with 9–15(25) flowers, mostly 3-branched; primary branches sometimes dichotomously forked; bracts like other leaves but smaller. **Flowers** (3)5(8)-parted; sepals distinct, lanceolate to ovate, 2–3.7 × 0.8–1.7 mm, yellow, yellow-green, greenish, white, or cream-colored, tips acute or acuminate; petals spreading horizontally, distinct, lanceolate to elliptic, widest at or below middle, 5.4–11 mm, yellow or white with green or brown, keel white, tips obtuse to long-acuminate, sometimes apiculate; filaments yellow or white, anthers yellow; ovaries glandular-pustulose. **Fruits** widely spreading when mature, fused near base, light green to brown, papillose, rarely glandular-pustulose. **2*n*=64.**

Western North America. 2 subspecies.

Sedum stenopetalum is easily identified by its linear leaves, and by the small rosettes in the axils of all or most of the leaves on the flowering stems. These rosettes may bear more or less deformed, petal-less, terminal flowers. Because plants usually grow from overwintering rosettes that are independent of the parent plant, they usually appear biennial but can be considered clones of long-lived perennials.

Plants with a single, terminal flower (sometimes replaced by a rosette), have been segregated as *S. stenopetalum* ssp./var. *monanthum*. The range of this putative taxon is completely overlapped by the typical form, and the two are sometimes found in a single population.

Key to *Sedum stenopetalum* subspecies:

1. Petals white or cream-colored, often drying yellow, 7–11 mm; range southwestern Oregon . . . . ***S. s.*** ssp. ***ciliosum***

1' Petals yellow, yellow-green, or greenish, 5.4–8 mm; range widespread. . . . . . ***S. s.*** ssp. ***stenopetalum***

Sedum stenopetalum Pursh ssp. *ciliosum* (Howell) R.T. Clausen
White wormleaf stonecrop

Sedum ciliosum Howell, *Sedum douglasii* Hook. ssp. *ciliosum* (Howell) R.T. Clausen, *Sedum radiatum* S. Watson var. *ciliosum* (Howell) H. Ohba

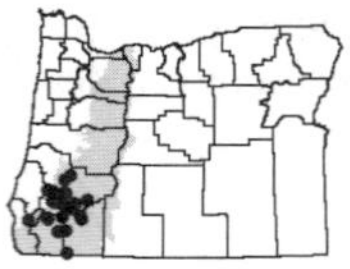

Flowering shoots erect, unbranched, 10–19 cm. **Inflorescences** branches sometimes dichotomously forked, 11–22 cm; pedicels to 1 mm, sometimes ciliate, papillose. **Flowers** 5-parted; sepals 1.5–3.5 × 0.8–2 mm; petals white or cream-colored, often drying yellowish, 7–11 mm, tips acute with mucronate appendage; filaments white, anthers yellow.

Talus, rocks along creeks, cliffs. Flowering May–Jul. 100–1200 m. Casc, Sisk. Native.

White to cream-colored flowers distinguish this species from the other small *Sedum* except for *S. radiatum* ssp. *depauperatum*, which lacks axillary rosettes on the flowering stems.

Sedum stenopetalum Pursh ssp. *stenopetalum*
Wormleaf stonecrop

Sedum douglasii Hook. var. *douglasii*, *Sedum stenopetalum* Pursh ssp. *monanthum* (Suksd.) R.T. Clausen, *Sedum stenopetalum* Pursh var. *monanthum* (Suksd.) H. Ohba, *Sedum stenopetalum* Pursh var. *stenopetalum*

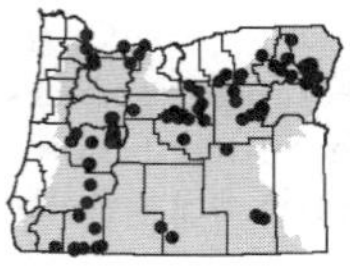

Flowering shoots branched or not, 10–43 cm; stem leaves of offsets glabrous. **Inflorescences** mostly 3-branched; bracts like other leaves but smaller; pedicels 0–0.5 mm, glabrous. **Flowers** (3)5(8)-parted; sepals 2–3.7 × 0.9–1.7 mm; petals 5.4–8 mm, yellow, yellow-green, or greenish, tips obtuse to long-acuminate, sometimes apiculate; filaments yellow, anthers yellow.

Rocky slopes, scablands, rocky meadows, open forests, talus slopes. Flowering May–Aug. 0–1900 m. BR, BW, Casc, ECas, Lava, Sisk, WV. CA, ID, WA; north to British Columbia, northeast to Alberta, east to WY. Native.

CROSSOSOMATACEAE Crossosoma family

Stephen C. Meyers

Shrubs deciduous, small, stems branched. **Leaves** alternate, simple, entire; stipules minute. **Flowers** bisexual, radial; sepals (3)5(6), persistent; petals (3)5(6), distinct; ovaries superior, styles 1, short and stout, stigmas head-like. **Fruits** follicles. **Seeds** with a thin aril.

Mexico and western United States. 4 genera; 1 genus treated in *Flora*.

GLOSSOPETALON Greasewood

Stems ribbed, bearing small spines at tips; bark green when young, yellowish brown to light orangish tan, light brown, or gray with age. **Leaves** petioles very short or absent. **Inflorescences** axillary, flowers 1. **Flowers** sepals persistent, ± equal, green; petals ± equal, white, short-clawed; stamens 6–10. **Fruits** ovoid, striate, beaked. **Seeds** 1, subglobose.

Northern Mexico and western United States. 5 species; 1 species treated in *Flora*.

Older floras often included *Glossopetalon* within the family Celastraceae.

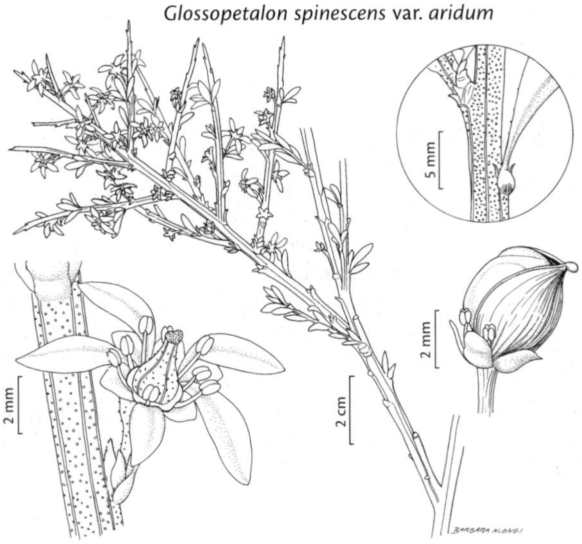

GLOSSOPETALON

Glossopetalon spinescens A. Gray var. ***aridum*** M.E. Jones (p. 607)

Nevada greasewood

Glossopetalon nevadense A. Gray, *Glossopetalon nevadense* A. Gray var. *stipuliferum* (H. St. John) C.L. Hitchc., *Glossopetalon stipulifera* H. St. John

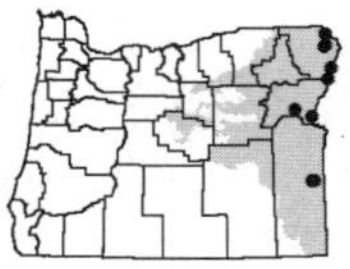

Plants 3–15 dm, glabrous to finely puberulent. **Leaves** oblanceolate to oblong, 3–15 × 1.5–4.5 mm, bases cuneate, tips acute or mucronate. **Inflorescences** outer bracts < inner bracts; pedicels 1–5 mm. **Flowers** sepals rounded, 1–2.5 mm; petals narrowly oblanceolate, 2–6 mm; stamens equal. **Fruits** 3–5 mm. **Seeds** 2–3 mm, smooth, shiny; arils white.

Rocky areas in sagebrush or juniper forests. Flowering Mar–Jun. 200–1400 m. BW, Owy. CA, ID, NV WA; northeast to MT, southeast to AZ, Mexico. Native.

Glossopetalon spinescens includes four other varieties. They are not likely to be encountered in Oregon and are currently known only from the southwestern United States and Mexico.

CUCURBITACEAE Cucumber family

Stephen C. Meyers

Vines perennial, monoecious. **Stems** climbing or trailing, 5-angled, with coiled tendrils. **Leaves** alternate, simple, palmately lobed; stipules lacking. **Staminate inflorescences** racemes or panicles. **Pistillate inflorescences** solitary flowers. **Flowers** radial; stamens 3–5(6), filaments and anthers twisted and fused; ovaries inferior. **Fruits** capsules, gourd- or melon-like. **Seeds** many.

Nearly worldwide. ~100 genera; 1 genus treated in *Flora*.

MARAH Bigroot, manroot, wild cucumber

Roots deep, large, fleshy. **Stems** ± scabrous or hairy, glabrous in age. **Leaves** round to cordate in outline, 3–7(9)-lobed. **Inflorescences** pistillate flowers 1 per axil. **Flowers** sepals 0; corollas cup-shaped, adaxially glandular-villous; stigmas 1. **Fruits** ellipsoid to ovate, tapered to a beak, prickly.

Western North America. 7 species; 1 species treated in *Flora*.

Meaning "bitter," *Marah* is one of only a few botanical names of Hebrew origin. The common names, manroot and bigroot, reflect the enormous size of the tubers, which can be as heavy and as large as a man, weighing over 100 kg.

Marah oregana (S. Watson) Howell (p. 607)

Oregon bigroot

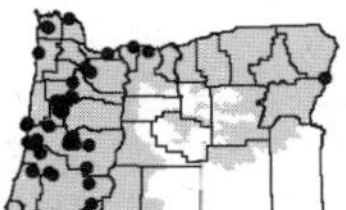

Stems 1–10 m. **Leaves** 8–15 × 8–15 cm. **Flowers** corollas 5–8-lobed, 3–15 mm wide, white. **Fruits** 3–8 cm. **Seeds** rounded, 15–25 mm, smooth. **2*n***=32.

Shrubby open areas, fields, meadows, rocky slopes, edges of forests. Flowering Mar–Jul. 0–1900 m. BW, Casc, Col, CR, ECas, Est, Sisk, WV. CA, WA; north to British Columbia. Native.

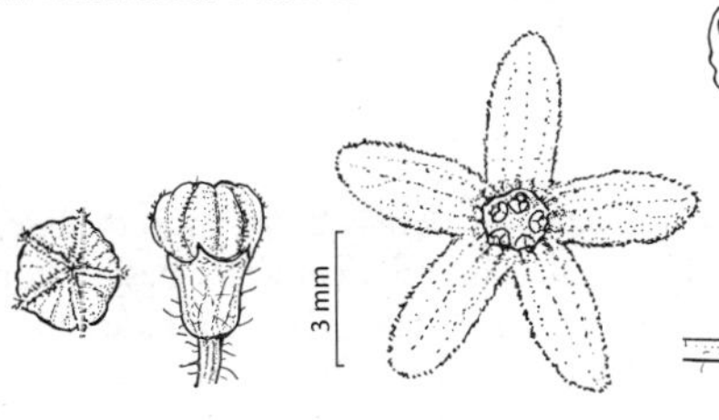

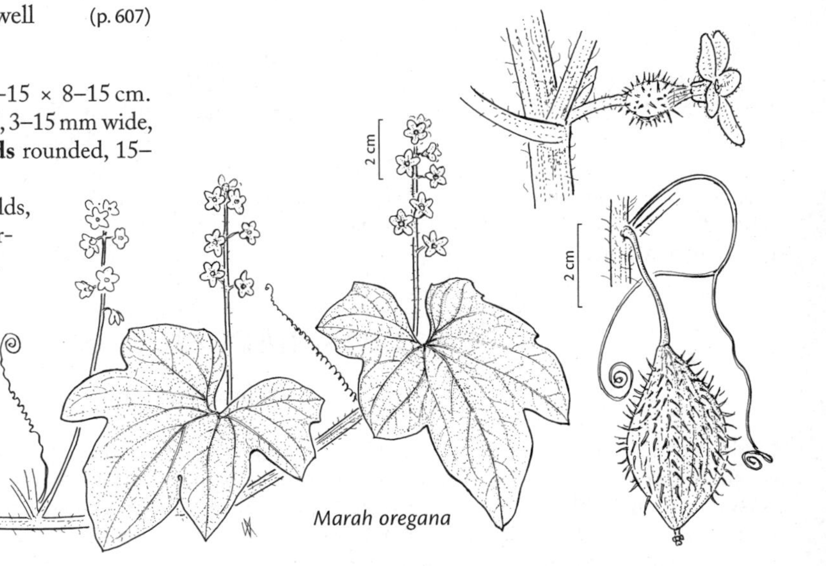

MARAH

CUSCUTACEAE

(see Convolvulaceae)

DIPSACACEAE Teasel family

Tamra Dickson Prior

Herbs biennial. **Leaves** opposite but appearing whorled, margins entire or toothed or pinnately lobed; stipules 0. **Inflorescences** dense, cymose, involucrate heads; involucel and receptacle bracts present. **Flowers** irregular, perfect, epigynous; calyces small, persistent, cup-shaped or cut into 4–5 irregular segments or into numerous teeth or hairs; corollas 4–5-lobed or with 2 lips; stamens distinct, 2 or 4, alternate with corolla lobes; pistils 1, ovaries inferior, 1-loculed, placentation apical, ovules 1, stigmas simple or 2-lobed. **Fruits** achenes, dry, indehiscent, surrounded by persistent calyces. **Seeds** 1 per fruit.

Africa, Asia, Europe, North America. ~10 genera; 1 genus treated in *Flora.*

Within the asterid clade, taxa within Dipsacaceae are very similar to those in Asteraceae except that they have unfused stamens.

Several species of this family are cultivated as ornamentals. *Knautia arvensis* is an exotic that has naturalized in the eastern United States. A single Oregon collection of this species was made in Wallowa County in 1985.

DIPSACUS Teasel

Stems erect, generally < 2 m, stout, prickly, few-branched. **Leaves** pairs often connate around stem; sessile. **Inflorescences** terminal on long, naked, prickly peduncles; receptacle bracts ovate or lanceolate, ending in a straight awn; involucel bracts linear, stiff. **Flowers** calyx limbs cup-shaped; corollas lavender or rarely white, tubes long; stamens 4; stigmas oblique, entire. **Fruits** 4-angled, appressed-hairy.

Africa, Asia, Europe, North America. ~12 species; 1 species treated in *Flora.*

The name *Dipsacus* likely derives from the Greek word *dipsa*, meaning thirst; a reference to how water is collected in the cup-like leaf base. *Dipsacus laciniatus* has been collected only once, in 2001, in Jackson County.

Dipsacus fullonum Linnaeus (p. 608)

Wild teasel, Fuller's teasel

Dipsacus sylvestris Huds.

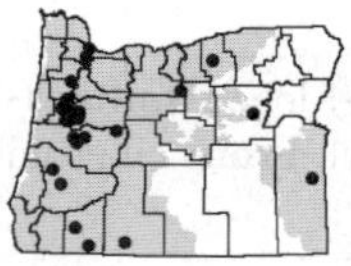

Plants taprooted. **Stems** striate angled; prickles denser toward top. **Leaves** ± prickly, particularly on lower midrib otherwise entirely or nearly glabrous; basal leaves oblanceolate, crenate, dying off second season; cauline leaves lanceolate, ≤ 3 dm, bases often connate forming a water collecting cup, margins entire. **Inflorescences** heads erect, ovoid-cylindric, 5–20 cm; receptacle bracts ≤ 20 cm. **Flowers** calyces ~1 mm, silky; corollas pubescent, tubes whitish, lobes white to purple. **Fruits** 6–8 mm. $2n$=16, 18.

Moist, disturbed areas, roadsides, pastures. Flowering Apr–Aug. 0–1600 m. Casc, Col, CR, ECas, Lava, Owy, Sisk, WV. CA, ID, NV, WA; throughout much of North America; Africa, Asia, Europe. Exotic.

Dipsacus fullonum is considered a noxious weed in most places it has been introduced. The heads of this plant were traditionally used for raising the surface on some kinds of cloth to give it more texture.

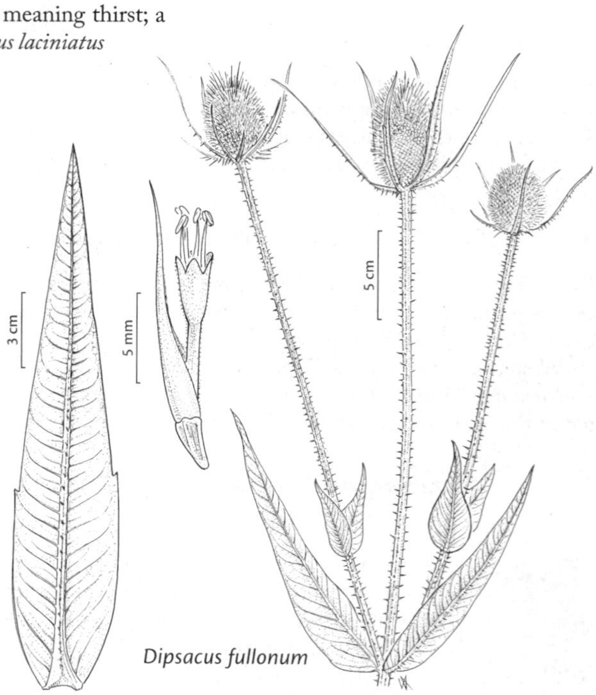

Dipsacus fullonum

DIPSACUS

DROSERACEAE Sundew family

Tamra Dickson Prior

Herbs perennial, carnivorous, glandular-pubescent, exuding viscid secretions; roots weak. **Leaves** generally basal, ± rosette-like; buds coiled. **Inflorescences** 1-sided raceme-like cymes of 1–10 flowers. **Flowers** bisexual, radial; sepals 5, distinct or basally connate, persistent; petals 5, free or slightly fused, persistent; stamens 5, distinct or connate basally, anthers 2-celled, dehiscent longitudinally; pistils 1, ovaries superior to slightly inferior, 1-celled with 3–5 parietal placentation or 3–5-loculed with axile placentation, styles 3–5, distinct, deeply bifid so that stigmas appear 6–10. **Fruits** capsules, valves generally 3–5. **Seeds** many, spindle-shaped, albuminous.

Nearly worldwide in temperate and tropical regions. 3 genera; 1 genus treated in *Flora.*

All members of this genus use stalked mucilaginous glands, found on their leaves, to snare and digest insects and arachnids. Animals are caught in the sticky fluid secreted by the glands then trapped by the leaf folding around them until they are digested by plant enzymes and bacteria. From these animals, the plants absorb minerals that are lacking or scarce in the soil of their habitats (bogs, swamps, marshes).

DROSERA Sundew

Herbs often reddish or brownish. **Leaves** adaxial surfaces and margins covered in gland-tipped hairs; long-petiolate. **Inflorescences** usually simple but may be forked. **Flowers** white to pink. **Seeds** usually with a loose seed coat that often forms a jointed to flattened wing at one or both ends.

Nearly worldwide. ~90 species; 2 species treated in *Flora.*

The flowers of many species, including ours, close at night and on overcast days.

Key to *Drosera* species:

1\. Leaves ascending to erect, at least 2 × as long as wide; testa not flattened . . . . ***D. anglica***

1' Leaves spreading, ~1 × as long as wide; testa flattened and wing-like . . . . ***D. rotundifolia***

Drosera anglica Hudson (p. 609)

English sundew, great sundew, line-leaved sundew

Drosera longifolia L.

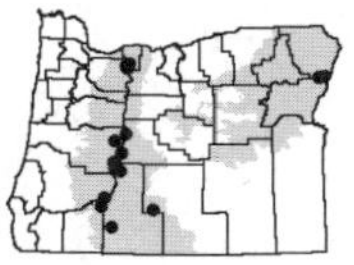

Leaves ascending to erect, narrowly oblong-oblanceolate to spatulate-oblong or cuneate-obovate, 10–30(50) × (2)3–5(7) mm, tapering gradually to petioles; petioles 1.5–8 cm, glabrous or sparsely glandular-hairy. **Scapes** 6–18 cm. **Inflorescences** usually 1, sometimes 2, 2–7-flowered. **Flowers** calyces (4)5–6 mm, connate about 33% of length; petals 8–12 mm, white; styles usually bifid ~67% from tips. **Seeds** 1–1.4 mm, black; seed coats longitudinally striate-netted, prolonged but not flattened at each end. **2*n*=40.**

Swamps and bogs, especially in sphagnum. Flowering Jul–Sep. 800–1900 m. BW, Casc, ECas. CA, ID, WA; north to AK, east to Newfoundland and ME; Asia, Europe. Native.

Drosera anglica has been known to hybridize with *D. rotundifolia* to form the named hybrid "species" *D.* × *obovata*, a sterile triploid. These hybrids are intermediate in nearly all morphological characters.

Drosera rotundifolia Linnaeus (p. 609)

Roundleaf sundew

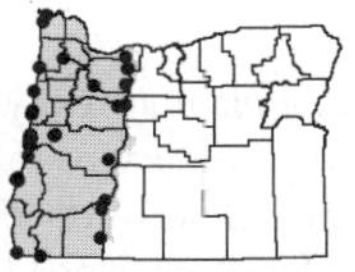

Leaves spreading, round to broadly obovate-cuneate, (4)6–12 × (4)6–12 mm, abruptly narrowed at base or ± cordate; petioles 2–9 cm. **Scapes** 7–25 cm. **Inflorescences** 1–several, (1)3–10-flowered. **Flowers** calyces fused at base, 4–6 mm; petals 3–6 mm, white to pink, styles bifid to base. **Seeds** 1–1.5 mm, light brown; seed coats finely and regularly longitudinally striate, prolonged and distinctly flattened at each end. **2*n*=20.**

Swamps, bogs, wet meadows, wet forests, especially in sphagnum. Flowering May–Sep. 0–2900 m. Casc, CR, Est, Sisk, WV. CA, ID, WA; northern North America and most states east of the Mississippi River; Asia, Europe. Native.

See *D. anglica* discussion.

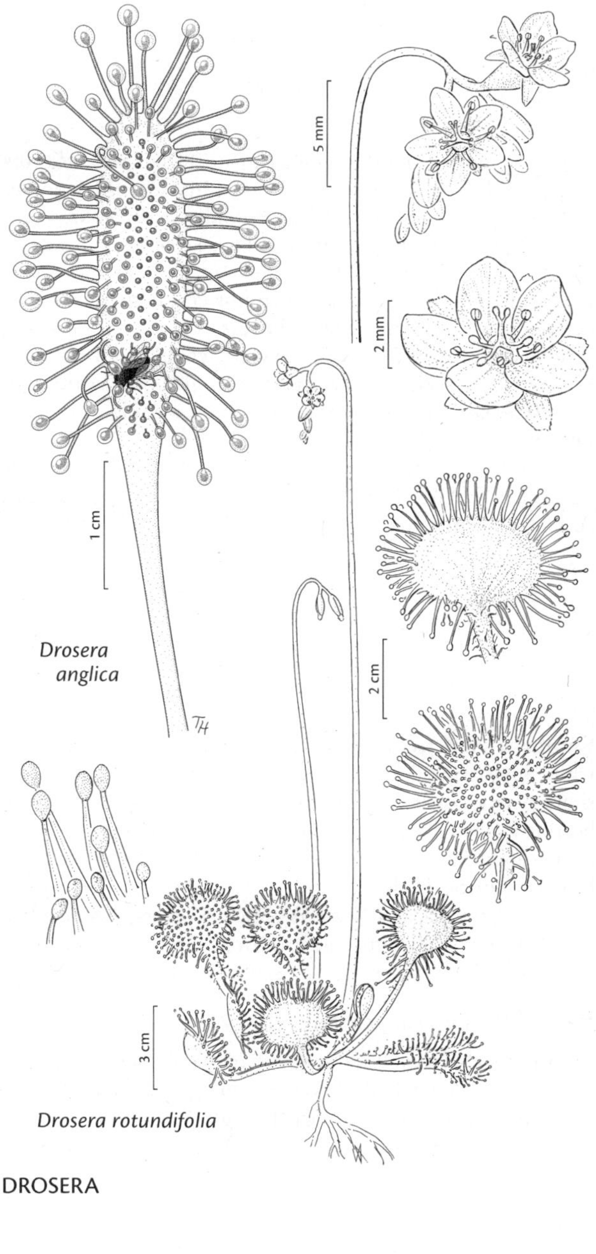

DROSERA

EHRETIACEAE Fukien tea family

Christian Feuillet

Trees, **shrubs**, or **herbs**. **Stems** appearing dichotomously branched. **Leaves** petiolate. **Inflorescences** bracts present, subtending flowers. **Flowers** (4)5-merous; styles terminal on ovary, once divided, falling away with pericarp in fruit, stigmas 2, capitate. **Fruits** schizocarps splitting into 4 nutlets; cotyledons not plicate.

Worldwide in warm regions. 9 genera; 1 genus treated in *Flora*.

Members of the Ehretiaceae have traditionally been included in Boraginaceae. Since the late 1990s, however, specialists of this group, and recently the Angiosperm Phylogeny Group (Stevens 2015), have recognized Ehretiaceae as a distinct family based on morphological and DNA data.

TIQUILIA Crinklemat

Herbs to **subshrubs** herbaceous from a woody caudex. **Stems** procumbent or spreading. **Leaves** in basal rosettes and in apical clusters on branches. **Inflorescences** axillary, flowers in clusters. **Flowers** calyces persistent, lobes linear to lanceolate; corollas deciduous, funnelform; stamens unequal, included, adnate to corolla tubes. **Fruits** dry.

Western Hemisphere, mostly in desert habitats. 28 species; 1 species treated in *Flora*.

Tiquilia nuttallii (Hooker) A.T. Richardson (p. 610)

Nuttall's crinklemat, Nuttall's sandmat

Coldenia nuttallii Hook.

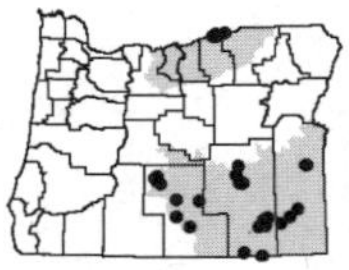

Herbs annual, up to 30 cm wide. **Stems** young stems with minute, stiff trichomes and a few bristles; older stems woody; caudices ≤ 5 mm wide. **Leaves** blades ovate to round, 4–10 × 2–8 mm, green or gray-green, with 2–3 pairs of lateral veins, margins entire, surfaces hairy; petioles filiform, ≤ 9 mm, hairy. **Flowers** calyces 3–5 mm, hairy; corollas 3–4 mm, limbs 1–2.5 mm wide, white to pink or violet, throats yellow, surfaces inner hairy, outer glabrous. **Fruits** 1.1–1.5 × 0.6–0.8 mm, gray, mottled with brown or black, smooth. **2*n***=16.

Deserts. Flowering May–Aug. 800–1400 m. BR, Col, Owy. CA, ID, NV, WA; east to WY, southeast to CO and AZ, MO; Argentina. Native.

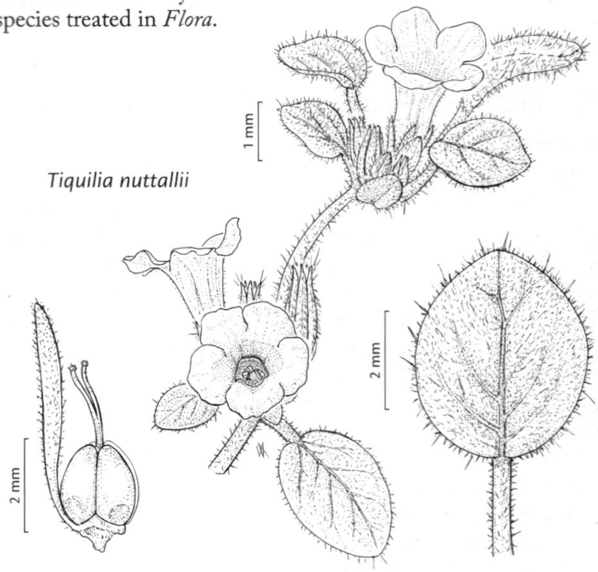

TIQUILIA

ELAEAGNACEAE Oleaster family

Alan T. Whittemore

Shrubs or **small trees** organs covered with circular, fimbriate scales, scales silvery or brown, often giving surface of plant a silvery white appearance. **Leaves** simple, unlobed, pinnately veined; petiolate; stipules 0. **Inflorescences** axillary, flowers solitary or in short racemes or sessile clusters; pedicels short. **Flowers** unisexual or bisexual, 4–9 mm; hypanthia well developed, inner surface bearing a lobed disk; sepals 4, united below; petals 0; stamens 4 or 8, free; pistils of 1 carpel, ovaries superior but closely enveloped by base of hypanthium, 1-locular with a solitary basal ovule, styles filiform, undivided, with 1 oblique stigma. **Fruits** drupes, tipped by persistent calyx lobes.

Asia, Australia, and North America. 3 genera; 2 genera treated in *Flora*.

Members of Elaeagnaceae fix nitrogen in root nodules that contain the filamentous bacterium *Frankia*. This allows the plants to invade habitats where the soil is deficient in nitrogen compounds.

Key to Elaeagnaceae genera:

1. Leaves alternate; flowering after leaves are developed; flowers perfect, or male and perfect flowers together in same inflorescences, emerging in axils of fully developed leaves; stamens 4, only tips of anthers projecting from hypanthium . . . . **ELAEAGNUS**

1' Leaves opposite; flowering before leaves expand; dioecious; stamens 8, fully exserted from hypanthium . . **SHEPHERDIA**

ELAEAGNUS Silverberry, oleaster

Shrubs or **small trees** 2–7 m, usually deciduous; older twigs often thorn tipped. **Leaves** alternate. **Inflorescences** solitary flowers or 2–8 flowers in sessile clusters, flowering after leaves are developed. **Flowers** hypanthia campanulate or tubular, throat open, similar in staminate and bisexual flowers; disks unlobed; stamens 4, alternating with sepals, tips of anthers projecting from hypanthium. **Fruits** yellow or red, with scattered silvery scales.

Asia, Australia, North America. ~60 species; 2 species treated in *Flora*.

Key to *Elaeagnus* species:

1. Leaves lanceolate or very narrowly oblong or elliptic, (3)4–8 × as long as wide; scales silver; hypanthia 2.5–3.5 mm wide, yellow inside; fruits yellow, mealy . . . . ***E. angustifolia***

1' Leaves elliptic or occasionally lance-ovate, 2–3.5(4.3) × as long as wide; scales silver or silver and brown; hypanthia 1–2 mm wide, white inside; fruits red, juicy. . . . . ***E. umbellata***

Elaeagnus angustifolia Linnaeus (p. 611)

Russian olive, oleaster

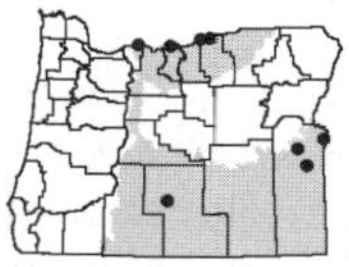

Small trees 5–7 m, scales silver; young twigs silvery white. **Leaves** lanceolate or very narrowly oblong or elliptic, 25–91 × 6–18 mm, (3)4–8 × as long as wide, bases rounded to acute, margins entire, tips bluntly acute, surfaces abaxially silvery white, adaxially dull green; petioles 6–10 mm. **Inflorescences** pedicels 2–4 mm. **Flowers** 1–3 per leaf axil; hypanthia 4–5 × 2.5–3.5 mm, uniformly silvery white outside, yellow inside; calyx lobes 2–4 mm. **Fruits** ellipsoidal, 8–12 × 7–8 mm, yellow, mealy. **2*n***=28.

Riverbanks, creeks, dry washes, roadsides, sagebrush areas, grasslands. Flowering May–Jun. 0–1500 m. BR, Col, ECas, Owy. CA, ID, NV, WA; north to British Columbia, east to Nova Scotia and VA, southeast to TX; Asia, Europe. Exotic.

Russian olive is often planted in arid areas, where few woody plants will survive. It is tolerant of saline or alkaline soil as well as heat and drought. It has escaped and become naturalized throughout the drier parts of North America. Sterile specimens resemble *Shepherdia argentea* but are easily distinguished by the alternate leaves.

Elaeagnus umbellata Thunberg

Autumn olive

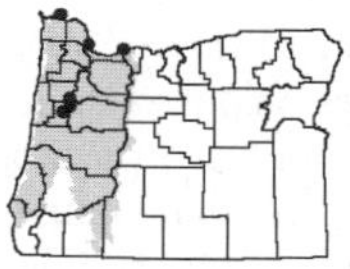

Shrubs 2.5–5 m, scales silver or mix of silver and brown; young twigs silvery white to brown. **Leaves** elliptic or occasionally lance-ovate, 32–71 × 11–24 mm, 2–3.5(4.3) × as long as wide, bases rounded to acute, margins entire, tips bluntly acute to rounded, surfaces abaxially silvery white, sometimes speckled with brown, adaxially dull green; petioles 4–6 mm. **Inflorescences** pedicels 6–7 mm. **Flowers** 1–8 per leaf axil; hypanthia 4–6 × 1–2 mm, silvery white and often speckled with brown outside, white inside; calyx lobes 2 mm. **Fruits** ellipsoidal, 8–10 × 7–8 mm, red, juicy. **2*n***=28.

Disturbed soil, fencerows, roadsides, dredge spoils, riparian areas. Flowering May–Jun. 0–200 m. Casc, CR, WV. WA; MT and eastern half of US; Asia. Exotic.

The brown coloration of some scales may deepen when the specimen is dry. The brown speckling that is often conspicuous in herbarium specimens of this species may be much less prominent in living plants.

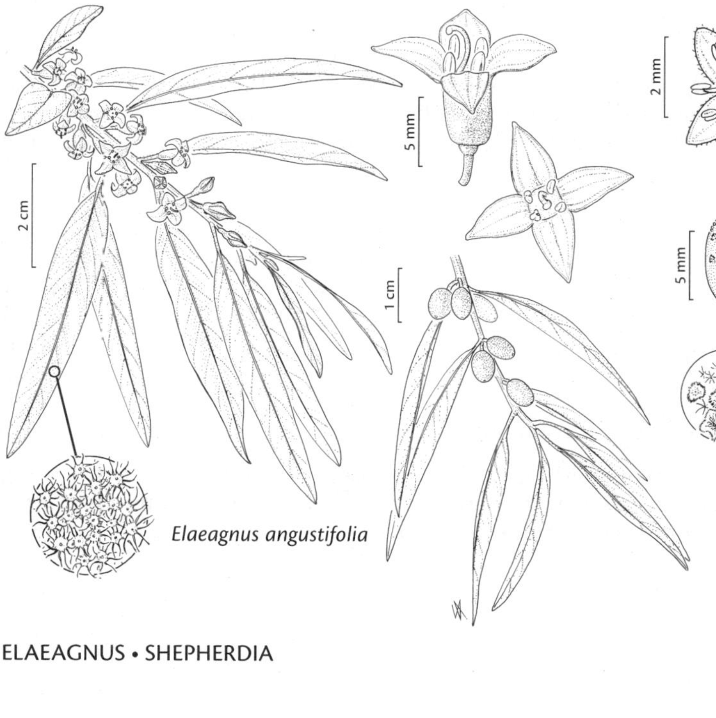

Elaeagnus angustifolia

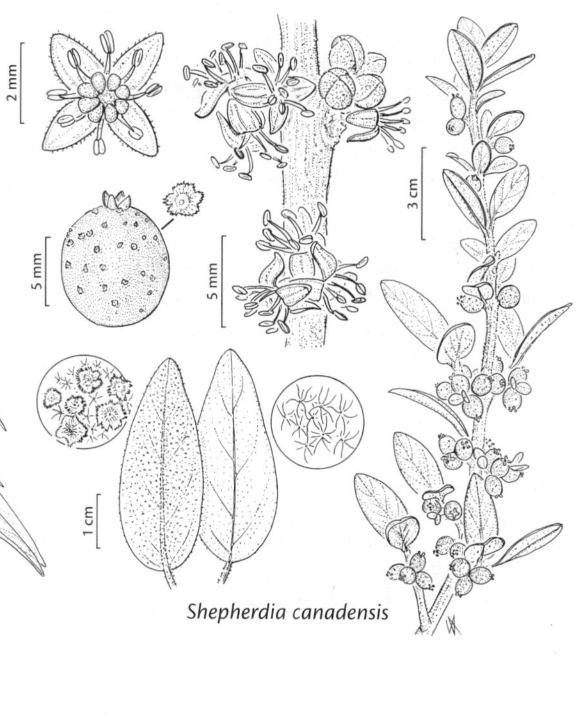

Shepherdia canadensis

SHEPHERDIA Buffaloberry, soapberry

Shrubs or **small trees** 0.7–7 m, dioecious. **Leaves** opposite. **Inflorescences** compact racemes, flowering before leaves expand; pedicels about 1 mm. **Flowers** throat of hypanthium ± closed by an 8-lobed pubescent disk. **Staminate flowers** hypanthia shallowly bowl-shaped; stamens 8, fully exserted from hypanthium; disk lobes alternate with stamens. **Pistillate flowers** hypanthia urn-shaped, nearly closed at top; disks deeply 8-lobed, lobes nearly meeting above ovary. **Fruits** dark red, with scattered silvery scales.

Northern and western North America. 3 species; 2 species treated in *Flora*.

Key to *Shepherdia* species:

1. Twigs silvery white, some branches thorn-tipped; leaves silvery white beneath . . . . ***S. argentea***

1' Twigs orange-brown, branches unarmed; leaves white with conspicuous brown speckles beneath . . . . ***S. canadensis***

Shepherdia argentea (Pursh) Nuttall

Thorny buffaloberry, silver buffaloberry

Elaeagnus utilis A. Nelson

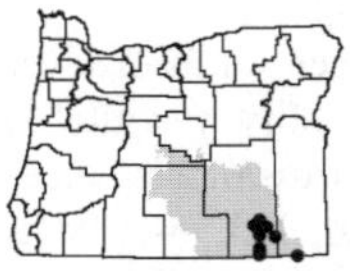

Shrubs or **small trees** deciduous, 2–7 m, some branches thorn-tipped (rarely not); twigs silvery white. **Leaves** oblong or elliptic, 1.3–6 × 1–2 cm, bases obtuse or acute, margins entire, tips rounded, seldom obtuse, surfaces abaxially silvery white, adaxially silvery green; petioles 3–7 mm. **Flowers** white. **Fruits** ellipsoidal, 5–7 mm, red. **2*n***=22.

Streambanks, slopes and dry flats, often in alkaline soil. Flowering Apr–Jun. 1200–1400 m. BR. CA, ID, NV; north to British Columbia, east to NY, southeast to NM. Native.

Shepherdia canadensis (Linnaeus) Nuttall (p. 611)

Canadian buffaloberry, russet buffaloberry, soapberry

Elaeagnus canadensis (L.) A. Nelson

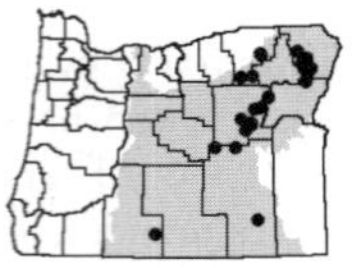

Shrubs deciduous, 0.7–4 m, unarmed; twigs orange-brown. **Leaves** oblong or ovate to triangular-ovate, 4–6 × 2–4 cm, bases rounded, margins entire, tips rounded-obtuse, surfaces abaxially white and conspicuously brown-dotted, adaxially green; petioles 4–9 mm. **Flowers** brown-scaly outside, yellow-green inside. **Fruits** ellipsoidal, 8–9 mm, red. **2*n***=22.

Streambanks, bottomlands, forests, meadows, shrublands. Flowering May–Jul. 900–2800 m. BR, BW, ECas, Lava. CA, ID, NV, WA; north to AK, east to Newfoundland, southeast to NM. Native.

ELATINACEAE Waterwort family

Hamid Razifard, Donald H. Les, and Gordon C. Tucker

Herbs or **subshrubs** annual or short-lived perennials. **Stems** erect to prostrate, green, reddish green, or red. **Leaves** opposite, simple, tips acute; petiolate; stipules 2. **Inflorescences** axillary, cymes or flowers solitary; bracts absent; pedicelled. **Flowers** bisexual, radial, bell- or urn-shaped; sepals 3–5, 1 smaller, free; petals 3–5, free or connate basally, nectaries 0; stamens 3–10, filaments free, anthers dehiscent by slits; pistils 1, ovaries superior, carpels 3–5, placentas axile, ovules 3–many per chamber, styles 3–5, stigmas 3–5. **Fruits** capsules. **Seeds** 9–36.

Nearly worldwide in aquatic habitats. 2 genera.

Elatinaceae is composed of aquatic species found near rivers as well as in wetlands, ponds, reservoirs, and lakes. Cleistogamous (i.e., non-opening, self-pollinating) flowers are very common in this family (Leach 1989; personal observation). A few species grow in rice fields in the United States and other countries (Rosman et al. 2016; DiTomaso & Healy 2007; Tucker 1986). Also, some species are popular aquarium plants (De Wit 1964).

Key to Elatinaceae genera:

1. Stems, leaves, and sepals covered with glandular hairs; average leaf length > 10 mm; sepals, petals, stamens, and carpels 5; sepal tips acute . . . . **BERGIA**

1' Stems, leaves, and sepals glabrous; average leaf length < 10 mm; sepals, petals, stamens, and carpels 3 or 4; sepal tips obtuse . . . . **ELATINE**

BERGIA Waterfire

Herbs or **subshrubs** annual or perennial; roots primary. **Stems** erect, with glandular hairs. **Leaves** narrowly elliptic, margins serrate, tips acute, surfaces glandular hairy. **Inflorescences** cymes; pedicels straight. **Flowers** sepals 5, with glandular hairs, tips acute; petals 5, pink; stamens 10, as long as petals; carpels 5. **Fruits** nearly globose. **Seeds** 0.4–0.5 mm.

Nearly worldwide. ~25 species; 1 species treated in *Flora*.

Bergia is a mostly Old World genus, with its greatest species diversity in southern Africa and Australia (Leach 1989; Tucker 1986).

Bergia texana (Hooker) Walpers (p. 614)

Texas waterfire, Texas bergia

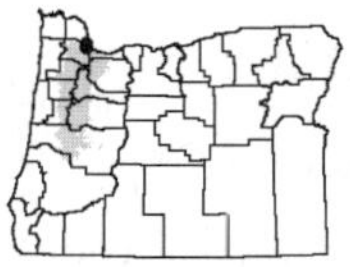

Stems 4–15 cm; average internode length ~12 mm. **Leaves** 10–30 mm; average petiole length > 2 mm. **Inflorescences** cymes, 2–5-flowered; pedicels > 0.5 mm. **Seeds** 30–50, nearly straight or slightly curved, surface pits ~15 per row.

Vernal pools, lakes, or mudflats near rivers. Flowering Jul–Aug. 0–1400 m. WV. CA, NV, WA; scattered in western, southern and central US, Mexico. Native.

Bergia texana is distinguished from all the New World species of Elatinaceae by the presence of glandular hairs on its stems, leaves, and sepals.

ELATINE Waterwort

Herbs annual; roots primary or adventitious. **Stems** erect or prostrate, glabrous. **Leaves** obovate, oblanceolate, or spatulate, usually < 10 mm, margins entire, tips obtuse, surfaces glabrous. **Inflorescences** flowers solitary; peduncles straight or recurved. **Flowers** sepals 3–4, tips obtuse, glabrous; petals 3–4, dull white to pink; stamens 3–8, as long as petals or slightly shorter; carpels 3–4. **Fruits** globose, compressed globose, or disk-shaped. **Seeds** 0.4–0.7 mm.

Nearly worldwide. ~25 species; 4 species treated in *Flora.*

Elatine has considerable species diversity both in the New World (approximately 15 species) and Old World (approximately 10 species). Although this genus has a nearly cosmopolitan range, most *Elatine* species are rare and grow in small patches in their native habitats. The majority of the native *Elatine* species growing in the United States belong to section *Crypta. Elatine californica* is the only representative of section *Elatine* (a mostly Old World group) in the United States.

Two *Elatine* species with a nearly cosmopolitan range and of Asian origin (*E. ambigua* and *E. triandra*) were introduced to the United States probably as a consequence of rice farming and/or fish stocking. Both species are popular aquarium plants, sold or traded through internet forums, and used for creating a "green mat" in fish tanks. Thus, aquarium disposal probably has contributed to their spread. In Oregon, *Elatine triandra* has been collected in Devils Lake, Lincoln County and Smith Lake, Clatsop County.

Many *Elatine* species have both submersed and emersed forms, which vary greatly in their morphology. For this treatment, we have selected morphological characters that are applicable to both forms. In their submersed form, *Elatine* species have adventitious roots originating from the stem nodes. The adventitious roots facilitate attachment of floating stem fragments (viable for more than a few weeks) to their substrate to generate new plants (asexual reproduction). However, sexual reproduction via seeds is the main reproductive strategy in this genus. The seeds released from the parent plants sometimes germinate immediately; thus, many *Elatine* species have more than one generation per season.

Key to *Elatine* species:

1. Seeds highly curved to nearly circular; sepals and petals 4; stamens 8; pedicels recurved, usually > 1.5 mm .... ***E. californica***

1' Seeds slightly curved to nearly straight; sepals and petals 3; stamens 3; pedicels straight, < 1.5 mm.

  2. Stems green; seed surface pits < 16 per row; carpels 2–4 .... ***E. brachysperma***

  2' Stems red or reddish green; seed surface pits 14–25 per row; carpels 3.

    3. Length of seed surface pits 3–5 × their width. .... ***E. chilensis***

    3' Length of seed surface pits 2 × their width. .... ***E. rubella***

Elatine brachysperma A. Gray

Short-seed waterwort

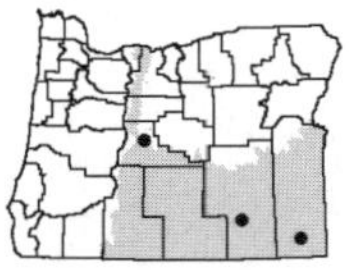

Plants 3–10 cm. **Stems** green; average internode length 6–15 mm. **Leaves** blades obovate or obovate-elliptic, 3–8 mm; petioles average length 0.5–1 mm. **Inflorescences** peduncles straight, ~0.1 mm. **Flowers** sepals 3; petals 3; stamens 3, as long as petals or slightly shorter; carpels 2–4. **Fruits** globose or pressed globose, 2- to 4-locular. **Seeds** 10–45, slightly curved to nearly straight, 0.4–0.5 mm; surface pits < 16 per row, length 0.5–nearly 1 × their width.

Mudflats near lakes and reservoirs. Flowering Jun–Jul. 1200–1700 m. BR, Owy, ECas. CA, NV, WA; throughout much of western, southern, and central US, Mexico; Argentina. Native.

Elatine brachysperma can be distinguished from other *Elatine* species growing in Oregon by the number of its seed surface pits, which is fewer than 16 per row. In Oregon, this species is critically imperiled, but it is secure in neighboring states (heritage rank: "G5, S1" [Oregon Biodiversity Information Center, 2016]), e.g., in California.

Elatine californica A. Gray

California waterwort

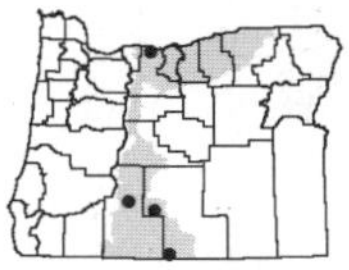

Plants 1.5–7 cm. **Stems** light green to green (reddish green); internodes 2–12 mm. **Leaves** blades spatulate or elliptic, 3–12 mm; petioles 0.5–6 mm. **Inflorescences** peduncles recurved, usually > 1.5 mm. **Flowers** (0.1)1.5–4 mm; sepals 4, sometimes 1 smaller; petals 4; stamens 8; carpels 4. **Fruits** disk-shaped to compressed globose, 4-locular. **Seeds** 16–24, highly curved to nearly circular, 0.5–0.7 mm; surface pits 20–27 per row, length 1.2–2 × their width.

In or near puddles, ponds, reservoirs, lakes, and marshes. Flowering Mar–Aug. 50–1600 m. Col, ECas. CA, ID, NV, WA; northeast to MT, southeast to NM, south to Mexico. Native.

Elatine californica can be distinguished from all other *Elatine* species in the Unites States by its four-merous flower parts and nearly circular seeds. Both morphological and molecular analyses (Razifard et al. 2017) support a close affinity between *E. californica* and *E. hydropiper*, which occurs in Central Asia, Europe, and North Africa.

Elatine chilensis Gay

Chilean waterwort

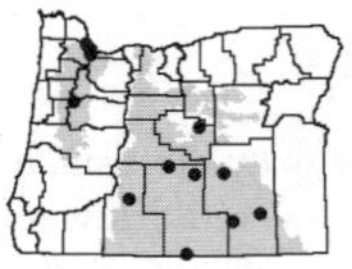

Plants 0.5–4 cm. **Stems** green to reddish green; internodes 2–7 mm. **Leaves** blades oblong to oblanceolate, 2–6 mm; petioles 0.4–1.2 mm. **Inflorescences** peduncles straight, < 1.5 mm. **Flowers** 0.1–0.5 mm; sepals 3, sometimes 1 smaller; petals 3; stamens 3; carpels 3. **Fruits** compressed globose to globose, 3-locular. **Seeds** 18–36, slightly curved to nearly straight, ~0.5 mm; surface pits 16–25 per row, length 3–5 × their width.

Emergent form on mudflats near lakes and reservoirs, submersed form in shallow lake waters. Flowering Apr–Sep. 0–2600 m. BR, ECas, Lava, WV. CA, NV, WA; western North America; South America. Native.

Elatine chilensis has an amphitropical disjunction in its distribution between southern South America and western North America. The exact cause of disjunction remains unclear, although birds might be responsible for dispersing the minute seeds of this species between the two regions (Carlquist 1983).

Elatine rubella Rydberg (p. 614)

Southwestern waterwort, red-stem waterwort

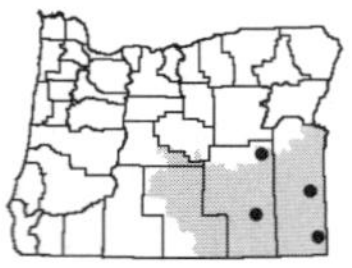

Plants 1.5–4 cm. **Stems** green to reddish green; internodes 1–7 mm. **Leaves** blades oblanceolate, 3–7 mm; petioles 0.5–1 mm. **Inflorescences** peduncles straight, < 1.5 mm. **Flowers** 0.1–0.5 mm; sepals 3, sometimes 1 smaller; petals 3; stamens 3; carpels 3. **Fruits** compressed globose to globose, 3-locular. **Seeds** 15–30, slightly curved to nearly straight, ~0.5 mm; surface pits 14–23 per row, length ~2 × their width.

Emergent form in mudflats near reservoirs, submersed form in marshes. Flowering Apr–Jul. 1100–1600 m. BR, Owy. CA, ID, NV, WA; western North America. Native.

Collections of *Elatine rubella* often are misidentified as *E. chilensis*, and vice versa, because of high similarity between the two species and the seed features not being readily visible. The emergent form of this species usually has more rigid stems and shorter internodes.

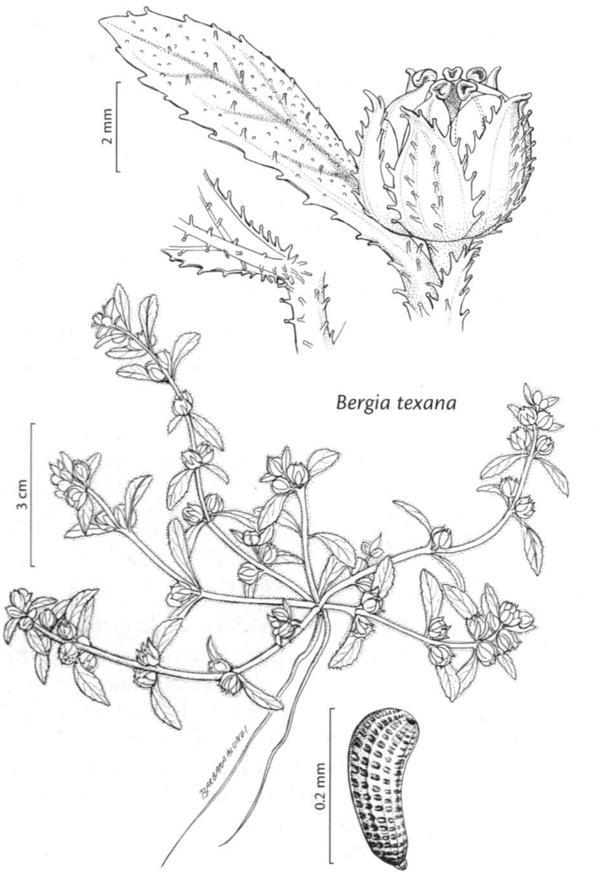

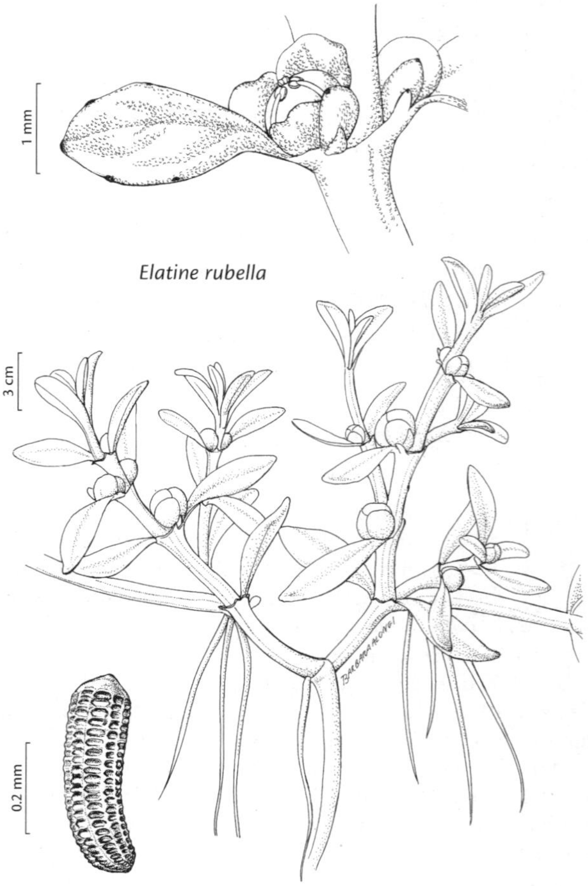

BERGIA • ELATINE

ERICACEAE Heath family

Stephen C. Meyers (except where noted)

Herbs, **shrubs**, or **trees** perennial, deciduous or evergreen, photosynthetic or non-photosynthetic. **Stems** erect to decumbent or prostrate, glabrous or hairy. **Leaves** normal, reduced, or absent, basal rosettes or cauline, usually alternate but sometimes opposite or whorled, simple, leathery or not, margins entire or toothed, plane or revolute; petioles present or absent; stipules absent. **Inflorescences** terminal or axillary racemes, panicles, cymes or flowers solitary; generally bracted. **Flowers** bisexual, radially or bilaterally symmetric; sepals 3–5, distinct or fused basally; petals 3–5, distinct or fused; stamens 5–10, anthers awned or not, dehiscent by pores or slits; styles 1, stigmas 1, generally 5-lobed. **Fruits** capsules, berries, or drupes. **Seeds** winged or not.

Nearly worldwide. ~120 genera; 26 genera treated in *Flora*.

Some authorities treat the subfamilies of Ericaceae as distinct families. Within Oregon, these would include Monotropaceae, Pyrolaceae, and Vacciniaceae. Conservatively, we have chosen not to recognize these families.

Non-photosynthetic ericaceous species are often referred to as saprophytes, plants that use decaying or non-living organic matter for nourishment. In reality, these species are true parasites, obtaining all of their nutrition from mycorrhizal fungi, which in turn parasitize photosynthetic plants, usually conifers. They are more appropriately referred to as epiparasites or mycotrophs.

Key to Ericaceae genera:

1. Plants non-photosynthetic (with no leaves or green parts).
  2. Plants bright red.
    3. Plants fleshy and succulent; flowers campanulate . . . . . **SARCODES**
    3' Plants not fleshy and succulent; flowers opening widely. . . . . **PYROLA** (in part)
  2' Plant stems red (pinkish) and white striped, reddish brown, brown, pink, yellow, cream-colored, or white.
    4. Stems with red (pinkish) and white stripes . . . . . **ALLOTROPA**
    4' Stems not striped.
      5. Petals fused.
        6. Corollas round-urceolate, narrowing near top, inner surface glabrous, < 1 cm . . . . . **PTEROSPORA**
        6' Corollas cylindric-campanulate, inner surface pubescent, > 1 cm. . . . . **HEMITOMES**
      5' Petals distinct.
        7. Inflorescences solitary flowers . . . . . **MONOTROPA** (in part)
        7' Inflorescences racemes.
          8. Corollas 5-petaled. . . . . **PYROLA** (in part)
          8' Corollas usually 4-petaled (sometimes 5-petaled in *Monotropa*).
            9. Inner faces of petals glabrous; anthers 2–4 mm . . . . . **PLEURICOSPORA**
            9' Inner faces of petals pubescent; anthers < 2 mm.
              10. Plants pinkish or yellowish; fruits capsules . . . . . **MONOTROPA** (in part)
              10' Plants whitish; fruits berries . . . . . **PITYOPUS**

1' Plants photosynthetic.
  11. Leaves scale-like and appressed. . . . . **CASSIOPE**
  11' Leaves not scale-like and appressed.
    12. Leaves broad, not linear or nearly needle-like.
      13. Petals ± distinct.
        14. Leaves prominently toothed . . . . . **CHIMAPHILA**
        14' Leaves entire or if toothed not prominently.
          15. Plants shrubs; mostly woody.
            16. Plants deciduous; inflorescences terminal solitary flowers, sometimes 2–3-flowered cymes; petals salmon to coppery-pink . . . . . **CLADOTHAMNUS**
            16' Plants evergreen; inflorescences terminal or axillary corymbs or rounded racemes; petals whitish yellow to white . . . . . **RHODODENDRON** (in part)
          15' Plants herbaceous.
            17. Flowers solitary and terminal. . . . . **MONESES**
            17' Flowers in racemes.
              18. Inflorescences 1-sided, often arching but erect in fruit . . . . . **ORTHILIA**
              18' Inflorescences symmetric, erect . . . . . **PYROLA** (in part)

13' Petals fused at least ⅓ of their length.
19. Corollas > 11 mm wide.
20. Corollas campanulate or funnelform . . . . **RHODODENDRON** (in part)
20' Corollas ± saucer-shaped.
21. Leaves with glandular dots; corollas without sac-like pockets. . . . . **KALMIOPSIS**
21' Leaves without glandular dots; corollas with sac-like pockets. . . . . **KALMIA**
19' Corollas < 11 mm wide (but ≤ 12 mm in *Vaccinium macrocarpon*).
22. Leaves > 7 cm; fruits warty; plants shrubs or medium-sized trees . . . . **ARBUTUS**
22' Leaves ≤ 7 cm (except in *Gaultheria shallon* leaves 4–10 cm); fruits not warty; plants erect, prostrate, or vine-like shrubs
23. Perianths 4-merous; stamens 8 . . . . **MENZIESIA**
23' Perianths 5-merous; stamens 10 (or if perianth 4-merous and stamens 8 then ovary truly inferior and leaves deciduous.)
24. Ovaries truly inferior; plants deciduous or evergreen . . . . **VACCINIUM**
24' Ovaries superior or appearing inferior because of fleshy calyx when ripe; plants evergreen.
25. Adaxial and abaxial leaf surfaces similar. . . . . **ARCTOSTAPHYLOS**
25' Adaxial and abaxial leaf surfaces not similar.
26. Adaxial leaf surfaces shiny; fruits drupe- or berry-like . . . . **GAULTHERIA**
26' Adaxial leaf surfaces not shiny; fruits capsules . . . . **LEUCOTHOË**
12. Leaves linear, nearly needle-like.
27. Leaves opposite . . . . **CALLUNA**
27' Leaves alternate or whorled.
28. Sepals 3; petals 3, < 2 mm . . . . **EMPETRUM**
28' Sepals 4–5; petals 4–5, > 3 mm.
29. Leaves < 1 mm wide, whorled; corollas persistent in fruit. . . . . **ERICA**
29' Leaves ≥ 1 mm wide, alternate; corollas deciduous . . . . **PHYLLODOCE**

ALLOTROPA Sugarstick, candystick

Genus composed of one species, *A. virgata.*

Allotropa virgata Torrey and A. Gray (p. 616)
Sugarstick, candystick

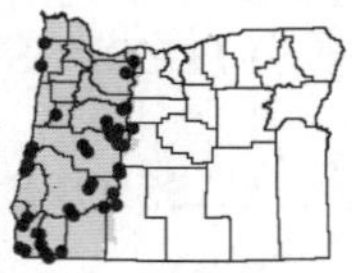

Herbs non-photosynthetic, 1–5 dm; rhizomatous. **Stems** erect, simple, reddish to pinkish with vertical white stripes, fleshy, glabrous. **Leaves** scale-like or absent, linear-lanceolate, pinkish to yellow-brown. **Inflorescences** terminal spike-like racemes, 5–20 cm; bracts 1–3 cm. **Flowers** tepals 5, distinct, triangular-ovate, 3–7 mm, white aging to reddish brown; corollas radially symmetric, cup-shaped; stamens 8 or 10, slightly exserted, purplish, anthers opening by basal pores; ovaries superior. **Fruits** capsules, flattened, 4–7 × 6–8 mm, loculicidal. **Seeds** many, minute, linear, narrowly winged. **2*n***=26.

Mixed or coniferous forests in humus. Flowering May–Aug. 0–2100 m. Casc, CR, Est, Sisk, WV. CA, ID, NV, WA; north to British Columbia, northeast to MT. Native.

The red-and-white-striped stems of *A. virgata* are unique among our non-photosynthetic ericaceous species.

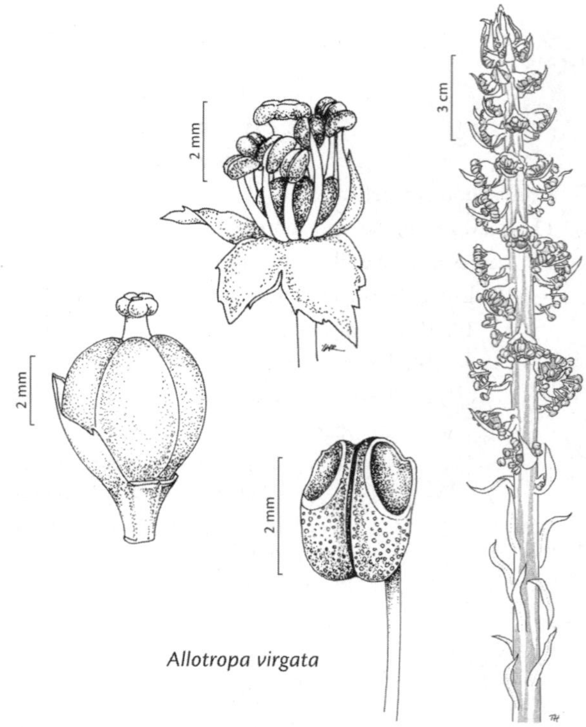

Allotropa virgata

ALLOTROPA

ARBUTUS Madrone

Shrubs to **medium trees** evergreen; bark smooth when young, fissured and exfoliating with age. **Leaves** alternate, leathery. **Inflorescences** terminal panicles. **Flowers** sepals 5 fused basally; petals 5, fused; stamens 10, included, anthers opening by slit-like pores; ovaries superior. **Fruits** berries.

Mediterranean regions in the Northern Hemisphere. ~20 species; 1 species treated in *Flora.*

Arbutus menziesii Pursh (p. 617)

Pacific madrone

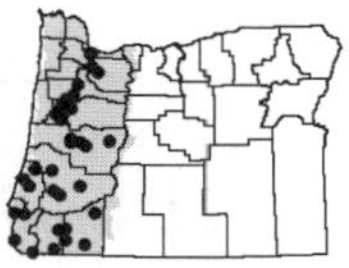

Shrubs and **trees** to 35 m; bark red and smooth to reddish brown and exfoliating. **Leaves** ovate to elliptic, 7–15 cm, pale green abaxially, dark green and shiny adaxially, margins entire to minutely serrate; petioles 1–2.5 cm. **Inflorescences** 5–15 cm, pubescent; bracts ovate, 4–6 mm, whitish. **Flowers** sepals ovate, 0.5–1.5 mm, thin; corollas urceolate, 5–8 mm, white to pinkish, lobes very short, reflexed. **Fruits** globose, 8–12 mm in diameter, orange to bright red, roughened by glandular-warty projections. **Seeds** 2–3 mm, light brown. **2*n*=26.**

Coniferous and mixed forests, rocky meadows. Flowering Mar–Jun. 0–1600 m. Casc, CR, Sisk, WV. CA, WA; north to British Columbia. Native.

The largest of the ericaceous species in Oregon, *Arbutus menziesii* is easily identifiable by its size and exfoliating bark. This tree is popular in cultivation throughout the Northwest and is often seen in parks and gardens. Native Americans found a variety of uses for the fruits, bark, and leaves. Its fragrant flowers and brightly colored fruit are attractive to wildlife as well.

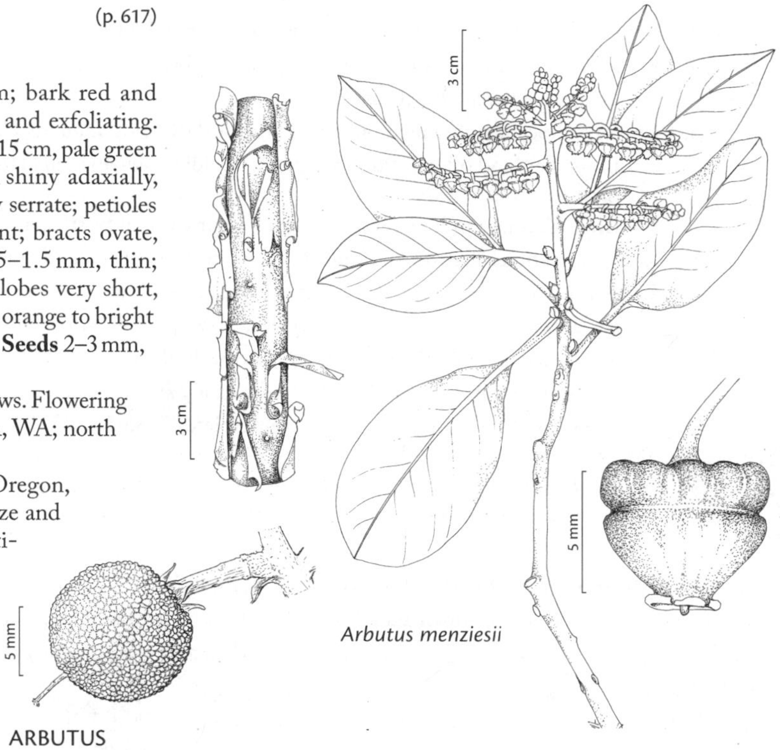

ARBUTUS

ARCTOSTAPHYLOS Manzanita

Kenton L. Chambers

Small trees or **shrubs** evergreen. **Stems** prostrate to erect, often contorted, much branched, rarely with a much enlarged root-crown (burl); bark shiny brown to dark red or blackish, smooth, often flaky on branches. **Leaves** alternate, simple, elliptic to lanceolate, ovate, or rounded, abaxial and adaxial surfaces similar, dull green or grayish, or adaxial surface brighter green, bases tapered to rounded or truncate, margins entire, tips acute, obtuse, or rounded; petioles short; stipules absent. **Inflorescences** terminal, racemes or panicles, condensed or open, few- to many-flowered; bracteate throughout. **Flowers** bisexual, regular, pedicelled, overwintering in bud; sepals (4)5, connate at base; corollas (4)5-merous, urn-shaped, < 11 mm wide, white or pink; petals connate nearly their entire length, with short distinct, spreading or recurved lobes; stamens (8)10, included in corolla, free, filaments broadened and usually hairy at base, anthers with curved awns, opening by terminal pores; ovaries superior, glabrous or pubescent, compound, 2–10 chambered; styles 1, stigmas capitate, hypogynous disk present. **Fruits** berry- or drupe-like, with 2–10 1-seeded nutlets, brown or red, fleshy, mealy, not warty.

Western North America (1 species circumpolar). ~60 species; 8 species treated in *Flora.*

This genus attains by far its greatest diversity in California, where 85 species and subspecies were recognized (Wells 1993) with even more added in recent years. This extreme proliferation of taxa is generally ascribed to interspecific hybridization followed by the stabilization of recombinant genetic types, often as small, morphologically distinct populations in unique habitats or isolated locations apart from the parental taxa. Our treatment follows a conservative approach, with several earlier recognized Oregon species being placed in synonymy or under various putative interspecific hybrid combinations. Included are two named hybrids of known parentage that, in our opinion, do not have distinct species status. Species identifications are well defined except in the Siskiyou Mountains region of southwestern Oregon, where co-occurring taxa frequently hybridize, especially in disturbed habitats, and key characteristics may recombine. Foliage pubescence traits are often best seen in young leaves, if available.

Key to *Arctostaphylos* species:

1. Plants prostrate, mat-forming shrubs.
   2. Leaf tips rounded, blunt or with a slight notch (rarely acuminate); stems lax; fruits bright red . . . . ***A. uva-ursi***
   2' Leaf tips acute or rounded with a sharp projection; stems stiff; fruits brownish red . . . . ***A. nevadensis***

1' Plants erect shrubs or small trees.
   3. Stems arising from an enlarged basal root-crown (burl).
      4. Branchlets glandular-bristly, tomentose; leaves often lightly villous . . . . ***A. glandulosa*** ssp. ***glandulosa***
      4' Branchlets and leaf bases minutely glandular, not bristly; leaves otherwise glabrous . . . . ***A. patula*** (in part)
   3' Stems lacking an enlarged basal root-crown.
      5. Leaves canescent-puberulent on one or both surfaces, not minutely glandular.
         6. Branchlets and often inflorescences with evident, few to many long, dense or scattered, bristly hairs . . . . ***A. columbiana***
         6' Branchlets and inflorescences tomentose but without bristly hairs . . . . ***A. canescens***
      5' Leaves sometimes lightly canescent, glabrous, or minutely glandular especially near bases of blades and on petioles.
         7. Leaves glabrous, lacking minute glandular hairs.
            8. Leaves whitish or grayish glaucous; floral pedicels with stalked, glandular hairs. . . . . ***A. viscida***
            8' Leaves green, not glaucous; floral pedicels glabrous. . . . . ***A. patula*** (in part)
         7' Leaves with minute, glandular hairs, especially at bases of blades and on petioles.
            9. Leaf blades narrowly elliptic to lanceolate, 1.5–3.5 × 0.5–1.5 cm, bases tapering or slightly rounded; petioles 3–6 mm . . . . ***A. hispidula***
            9' Leaf blades broadly lanceolate to ovate or rounded, (2)2.5–6 × 1.5–4 cm, bases usually rounded or truncate; petioles 6–15(20) mm. . . . . ***A. patula*** (in part)

Arctostaphylos canescens Eastwood (p. 619)

Hoary manzanita, downy manzanita

Arctostaphylos canescens Eastw. ssp. *canescens*, *Arctostaphylos* × *cinerea* Howell

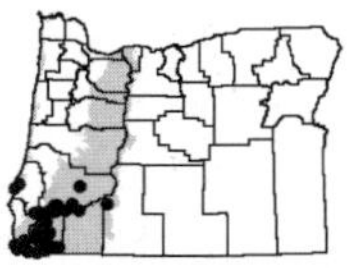

Shrubs erect, 0.3–3 m, much branched; bark dark red to blackish, smooth; upper branches and twigs tomentose. **Leaves** oblong-ovate to elliptic or rounded, 1.5–5.5 × 1–3 cm, green or gray-green, bases abruptly tapered, margins smooth or ciliate, tips obtuse, often with a sharp projecting points, surfaces both canescent-puberulent or strigose or adaxially sometimes nearly glabrous; petioles 3–10 mm, tomentose. **Inflorescences** compact panicles, axis tomentose but without bristly hairs; basal bracts triangular, tomentose but without bristly hairs. **Pedicels** 2–8 mm, tomentose, sometimes glabrous or with stalked glands. **Flowers** corollas white or pink; ovaries pubescent. **Fruits** brownish red. **2*n***=26.

Rocky slopes to brushy flats and mixed woods, usually non-serpentine soils, rarely coastal. Flowering Feb–May. 0–1700 m. Casc, Est, Sisk. CA. Native.

Arctostaphylos canescens, principally a species of non-serpentine soils, is frequently in contact with the serpentine-adapted *A. viscida*, especially in Josephine and Curry counties. Variable hybrid derivatives of these taxa are found in areas of mixed soil following disturbance or on roadsides and alluvial flats such as the upper Illinois River Valley. This hybridization pattern was studied by Gottlieb (1968) at the former mining town of Waldo, a site where Thomas Howell had described several new species. The hybrids usually have lightly canescent-puberulent foliage, more open inflorescences with shorter subtending bracts than *A. canescens*, and often have floral pedicels with mixed glandular and non-glandular hairs. Elsewhere in its range, *A. canescens* meets the similar but bristly-stemmed *A. columbiana*, and mixed populations of bristly and non-bristly plants may occur, which are difficult to classify as one or the other species.

Arctostaphylos columbiana Piper (p. 619)

Hairy manzanita, bristly manzanita, Columbia manzanita

Arctostaphylos nortensis P.V. Wells

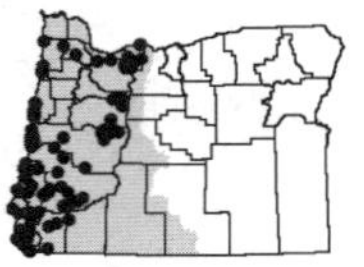

Small trees or **shrubs** erect, 1–10 m, much branched, glabrous on lower branches, tomentose above; bark reddish to blackish, smooth; branchlets hispid, with dense or scattered bristly hairs, their tips pointed, sometimes minutely glandular, or bristly hairs limited to inflorescence axis and bracts. **Leaves** ovate-elliptic to broadly lanceolate, 1–7 × 2–3 cm, green or grayish, bases tapering to rounded or truncate, margins sometimes ciliate, tips acute to obtuse, often with a sharp projecting point, surfaces one or both densely to minutely canescent-puberulent, sometimes hispid near base; petioles tomentose, 3–10 mm. **Inflorescences** compact racemes or panicles, axis tomentose, often hispid with bristly hairs; sometimes long-bracteate at base. **Pedicels** 1–4 mm, usually minutely tomentose, bracts at bases erect or reflexed, narrowly lanceolate to linear or triangular, tomentose, often hispid-margined (or lower bracts sometimes glabrous). **Flowers** corollas white, pink, or reddish orange; ovaries pubescent. **Fruits** brownish red. **2*n***=26.

Brushlands, forest edges, open woods, rocky slopes, roadsides, disturbed areas. Flowering Mar–May. 0–1600 m. Casc, CR, ECas, Est, Sisk, WV. CA, WA; north to British Columbia. Native.

Hybrids of *Arctostaphylos columbiana* with *A. nevadensis* and *A. uva-ursi* are discussed under those species.

It is commonly noted that populations of *A. columbiana*, even far from the range of *A. canescens*, may contain individuals completely lacking the expected bristly hairs on their twigs or inflorescences. This is evidently part of the normal genetic variation within the species. In the Siskiyous, where *A. columbiana* is sympatric with *A. canescens*, such pubescence variants are nearly identical with the latter species. It is likely that hybridization blurs distinctions between the two species, and that the identification one gives to plants lacking bristles in mixed populations will necessarily be arbitrary. The description above includes the putatively rare, serpentine-endemic taxon *A. nortensis*, of adjacent Del Norte County, California. The key traits of this species given by Wells (2000), including its compactly racemose inflorescences, recurved inflorescence bracts, puberulent pedicels, glabrous filaments, variably bristly to non-bristly populations, and serpentine habitat, have all been found to occur as variations within the widespread *A. columbiana* as it exists in Oregon.

Arctostaphylos glandulosa Eastwood ssp. ***glandulosa*** (p. 619)

Eastwood's manzanita

Arctostaphylos intricata Howell

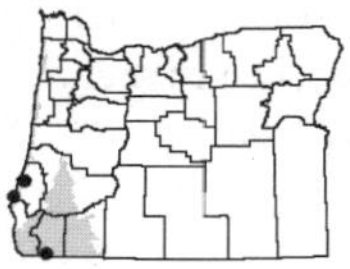

Shrubs erect, 1–2.5 m, spreading and much branched from an enlarged basal burl; bark reddish or blackish, smooth; branchlets glandular bristly. **Leaves** ovate to ovate-lanceolate or elliptic, 2–5 × 1–2.5 cm, green, rarely somewhat glaucous, bases tapering to obtuse, rounded, or truncate, margins smooth or ciliate, tips acute to obtuse or rounded, surfaces tomentose, often lightly villous; petioles 5–8 mm, tomentose, usually also glandular-bristly. **Inflorescences** panicles, axis tomentose, usually also glandular-bristly, long-bracteate at base. **Pedicels** 2–8 mm, tomentose, often glandular (glabrous); bracts at pedicel bases erect or reflexed, lanceolate, acuminate, glandular. **Flowers** corollas white or pink; ovaries tomentose or glandular. **Fruits** brown. **2*n***=52.

Forest edges and open sites, dry, rocky slopes and ridges. Flowering Feb–Apr. 0–400 m. Est, Sisk. CA. Native.

The form of this widespread Californian species found in Oregon, to which the name *A. intricata* also applies, is called ssp. *glandulosa* f. *glandulosa* by Wells (2000). We are provisionally using this subspecific designation, pending the collection of more Oregon specimens that include field notes on the presence of burls, for correlation with the alternative leaf and pubescence characteristics of other subspecies.

Arctostaphylos hispidula Howell

Howell's manzanita, Gasquet manzanita

Arctostaphylos hispidula Howell var. *hispidula*, *Arctostaphylos hispidula* Howell var. *viscosissima* (M. Peck) M. Peck, *Arctostaphylos stanfordiana* Parry

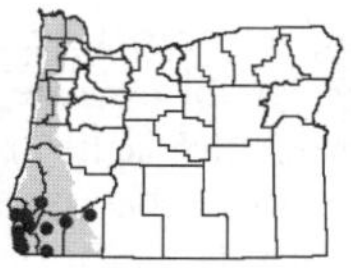

Shrubs erect, 1–2 m, much branched; bark reddish to blackish, smooth; upper branchlets minutely glandular and often lightly or densely hispidulous to tomentulose. **Leaves** narrowly elliptic or lanceolate, 1.5–3.5 × 0.5–1.5 cm, green, bases tapering or slightly rounded, margins smooth or finely ciliate near base, tips acute, surfaces canescent. **Inflorescences** compact racemes or panicles, axis densely glandular, often hispidulous (glabrous); ebracteate or short- to long-bracteate at base. **Pedicels** 2–5 mm, glabrous; bracts at pedicel bases erect, triangular, glandular. **Flowers** corollas white or pink; ovaries glabrous. **Fruits** brownish. **2*n***=26.

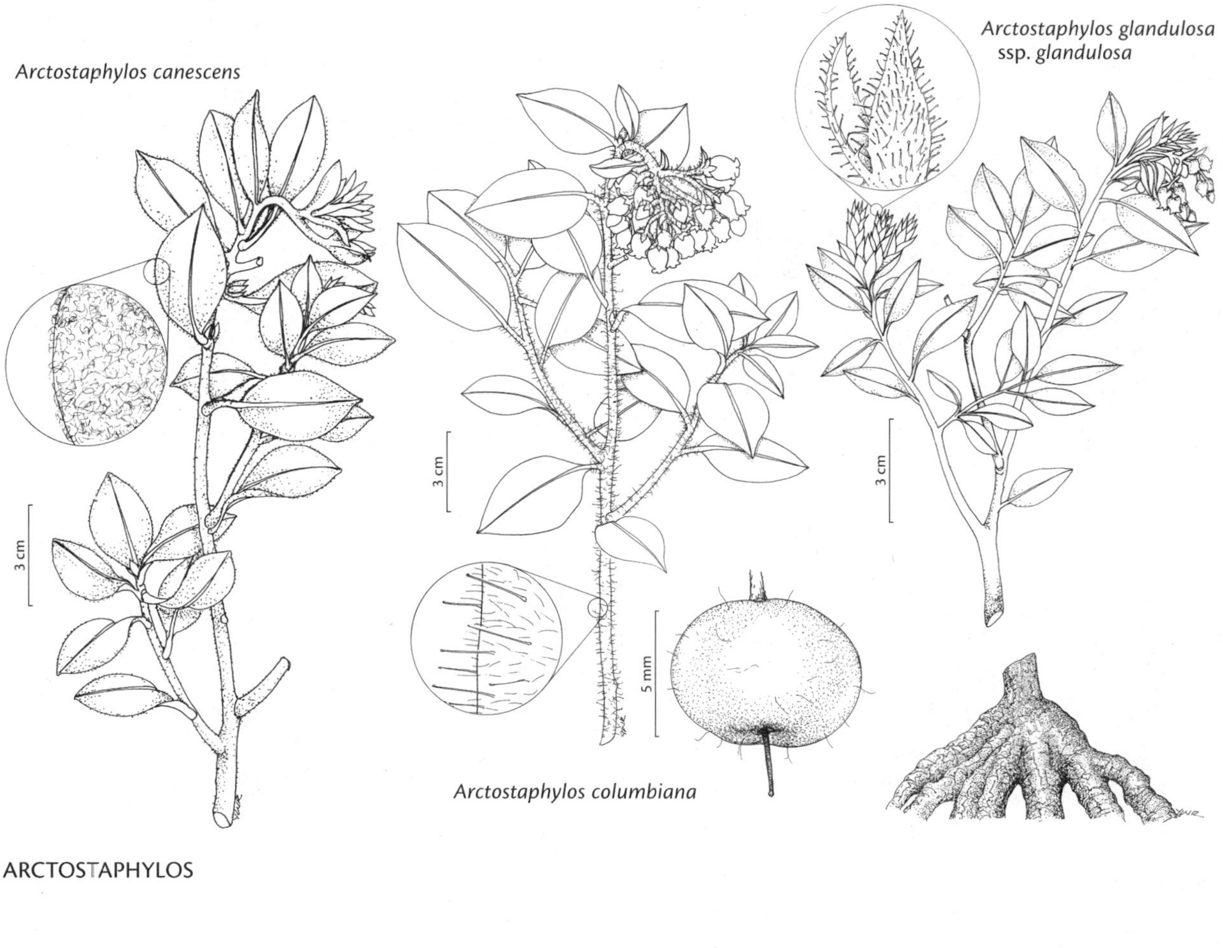

ARCTOSTAPHYLOS

Shrublands, grassy openings in woods, gravelly flats, slopes and ridges, often on serpentine. Flowering Mar–Jul. 100–1900 m. CR, Sisk. CA. Native.

Arctostaphylos hispidula most resembles *A. patula* in the fine glandular pubescence of its leaf bases, petioles, and branchlets. Its leaf blades are usually smaller, narrower, and more distinctly tapered to the petiole. The extent of its hybridization with co-occurring species is not known.

Arctostaphylos nevadensis A. Gray

Pinemat manzanita

Arctostaphylos nevadensis A. Gray ssp. *nevadensis*

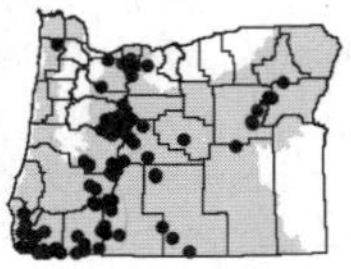

Shrubs stiff, mat-forming and ascending, usually < 4 dm; bark dark red to blackish, smooth and glabrous below, puberulent above, rough and flaking on branchlets. **Leaves** elliptic to oblanceolate or obovate, 1–3.5 × 0.5–1.5 cm, green, bases usually tapering, tips acute or rounded with sharp projection, surfaces glabrous or puberulent on midrib and margins, especially near base; petioles 3–7 mm, often puberulent. **Inflorescences** racemes or panicles, compact, short-bracteate at base, axis glabrous or finely puberulent. **Pedicels** 2–7 mm, usually glabrous; bracts at pedicel bases erect, triangular, glabrous or puberulent. **Flowers** corollas usually pink; ovaries glabrous. **Fruits** brownish red. ***2n***=52.

Montane coniferous forests, talus slopes, rock outcrops, roadsides, sandy, gravelly or loam soils. Flowering Mar–Jul. 300–2800 m. BR, BW, Casc, CR, ECas, Lava, Sisk. CA, NV, WA. Native.

Despite their similarity in prostrate habit, *A. nevadensis* and *A. uva-ursi* are adequately distinct, and no hybrids between them have been reported. Along the Oregon coast, *A. uva-ursi* is common, where it is known to hybridize with *A. columbiana*. Hybridization of *A. nevadensis* with *A. columbiana* was described by Kruckeberg (1977) from the west flank of Mt. Hood and from the Puget Trough region of Washington. The hybrids had abnormal meiosis and low pollen fertility, and they produced no viable seeds. Where *A. nevadensis* is sympatric with *A. patula* in the Cascade and Siskiyou mountains, plants have been collected that are suggestive of hybridization between these species. In addition, Wells (2000) stated that the type of *A. parvifolia*, described from the Illinois Valley, Josephine County, is similar to a hybrid of *A. nevadensis* with *A. glandulosa* in Del Norte County, California, named *A.* × *knightii*. However, *A. glandulosa*, a burl-producing species, is not known from the Illinois Valley, whereas *A. canescens* is common there. The type of *A.* × *parvifolia* is canescent on the leaves and inflorescence axes, identifying this hybrid as a likely cross of *A. nevadensis* with *A. canescens*.

Arctostaphylos patula Greene (p. 620)

Green-leaved manzanita

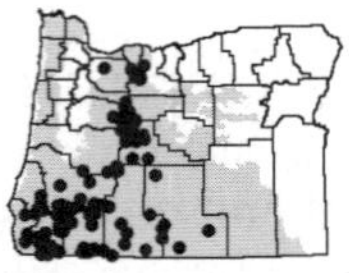

Shrubs erect, 1–2 m, rarely with an enlarged basal crown (burl) sprouting after fires, often clonal; bark reddish to blackish, smooth, often flaky on branches; lower branches prostrate and rooting, usually finely blackish or yellowish glandular on branchlets. **Leaves** blades broadly lanceolate to ovate or rounded, (2)2.5–6 × 1.5–4 cm, green, bases rounded or truncate, tips acute or obtuse, sometimes rounded, surfaces finely blackish or yellowish glandular hairy near base, becoming glabrous; petioles 6–15(20) mm, with usually minutely glandular hairs. **Inflorescences** panicles, rarely racemes, axis finely glandular, rarely glabrous; long-bracted at bases or

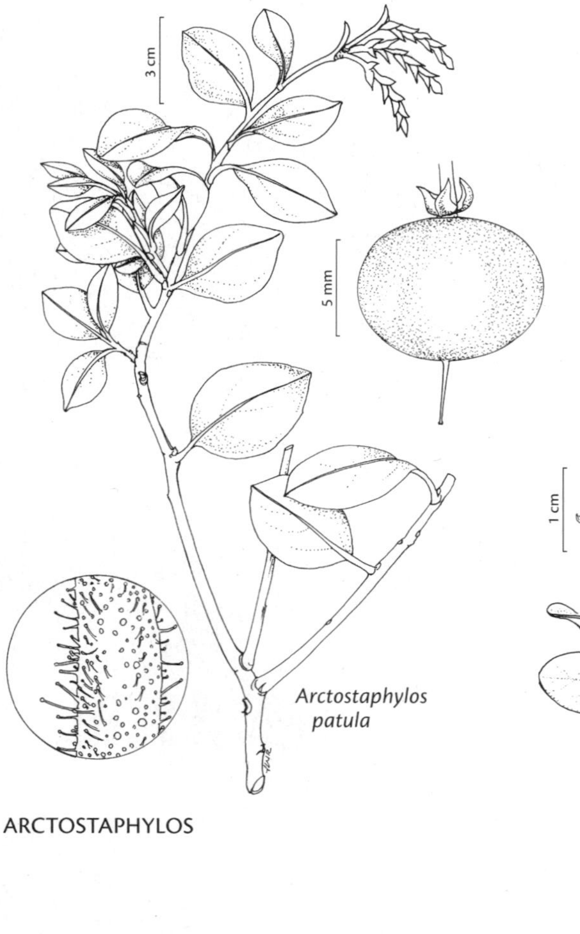

Arctostaphylos patula

Arctostaphylos uva-ursi

ARCTOSTAPHYLOS

not. **Pedicels** 2–8 mm, glabrous; bracts at pedicel bases erect, triangular, acute to acuminate, glabrous or minutely glandular. **Flowers** corollas pink; ovaries glabrous. **Fruits** dark brown to brownish red. **2*n***=26.

Shrublands, conifer forests, open slopes, flats on gravelly, sandy, or pumice soils, clear-cuts and other disturbed sites, rarely on serpentine. Flowering Mar–Jun. 100–2400 m. BR, Casc, CR, ECas, Lava, Sisk. CA, NV, WA; northeast to MT, southeast to NM, south to Mexico. Native.

This species often has a low, rounded growth form, spreading by rooting of the lower branches, and usually propagating by seeds following clear-cuts or fires. The burl-forming habit often present in *A. patula* in other states is uncommon in Oregon, found only in some populations in southwestern counties bordering California. No taxonomic distinction is made here between plants with and without burls. Leaf blades with acutely tapering bases are characteristic of related *A. hispidula,* but may also occur occasionally in *A. patula* well outside the range of this other species. Hybridization may occur with *A. uva-ursi* and *A. nevadensis.* In Curry County, some specimens having canescent puberulence on their inflorescences and floral pedicels and may represent hybrids of *A. patula* with *A. canescens.*

Arctostaphylos uva-ursi (Linnaeus) Sprengel (p. 620)

Kinnikinnick, bearberry

Arctostaphylos uva-ursi (L.) Spreng. var. *coactilis* Fernald & J.F. Macbr., *Arctostaphylos uva-ursi* (L.) Spreng. var. *uva-ursi*

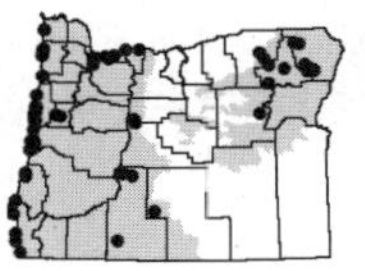

Shrubs prostrate, lax, mat-forming, often slightly ascending, usually < 3 dm; bark dark red to blackish, often flaky; branchlets glabrous to minutely puberulent or tomentose. **Leaves** oblanceolate or obovate, 1–2.5 × 0.5–1.5 cm, dull abaxially, shiny and green adaxially, bases tapering, tips rounded, blunt or notched, rarely acuminate, surfaces glabrous with ciliate margins; petioles 2–4 mm, tomentose or puberulent. **Inflorescences** compact racemes, axis puberulent; long- or short-bracteate at base. **Pedicels** 2–4 mm, usually glabrous; bracts at pedicel bases erect or spreading, triangular, glabrous or puberulent. **Flowers** corollas pink; ovaries glabrous. **Fruits** bright red. **2*n***=26, 52.

Ocean bluffs, sandy flats, rocky hillsides, mixed woods, coniferous forests. Flowering Feb–Jun. 0–2800 m. BW, Casc, CR, ECas, Est, Sisk, WV. CA, ID, NV, WA; north to AK, east to CO, Canada, northeastern US; Asia, Europe. Native.

The most common hybrid formed by *A. uva-ursi* in Oregon is with *A. columbiana,* with which it is commonly sympatric on the Oregon Coast and near Portland. The name *A.* × *media* applies to this hybrid, a taller, more upright plant with larger leaves than *A. uva-ursi* but lacking the bristly stems of the *A. columbiana* parent. Kruckeberg (1977) states that plants of *A.* × *media* have abnormal meiosis and produce no viable seeds. Therefore, it has no status as a hybrid-origin species.

A second hybrid type, found in Deschutes and Jefferson counties, is *A. uva-ursi* × *A. patula.* It tends to form mats, like *A. uva-ursi,* but is taller, more upright, and has larger leaves; its inflorescence has the fine glandular puberulence of *A. patula.* According to Wells (2000), the name *A.* × *coloradensis* is based on a local hybrid population of *A. uva-ursi* × *A. patula* in the Rocky Mountains, but we consider its application to hybrids in Oregon as uncertain.

Arctostaphylos viscida Parry (p. 621)

White-leaved manzanita

Arctostaphylos pulchella Howell, *Arctostaphylos viscida* Parry ssp. *pulchella* (Howell) P.V. Wells, *Arctostaphylos viscida* Parry ssp. *viscida*

Small trees or **shrubs** erect, 1–8 m, much branched, glabrous throughout or branchlets glandular-hispid, especially toward inflorescence; bark reddish to blackish, smooth. **Leaves** ovate to broadly elliptic or rounded, 2–6 × 1–2.5 cm, white or grayish glaucous, sometimes greenish, bases obtuse to truncate or slightly cordate, tips acute to obtuse or rounded, sometimes with small projecting point, surfaces glabrous; petioles glabrous or glandular, 5–15 mm. **Inflorescences** open panicles or rarely racemes, axis glabrous or glandular; not bracteate at base. **Pedicels** 6–10 mm, pubescent with long- or short-stalked glandular hairs; bracts at pedicel bases triangular, acute, glabrous or minutely glandular. **Flowers** corollas white or pink; ovaries glabrous or minutely glandular. **Fruits** brownish red. **2*n***=26.

Shrublands, open woods, often dominant on serpentine but also found in other habitats in various soils, on roadsides and in disturbed areas. Flowering Mar–Jul. 300–1100 m. Sisk. CA. Native.

The subspecies of *A. viscida* recognized by Wells (2000) differ principally in pubescence traits. He misapplied the epithet *pulchella* to one of these entities, since the type specimen of *A. pulchella* has glabrous ovaries, not glandular ones as he implies for "ssp. *pulchella.*" Presence of glandular hairs on twigs and fruits is variable in Oregon material, and no subspecies are recognized here.

Where *A. viscida* is sympatric with *A. canescens,* hybrid swarms occur, following habitat disturbance. These are discussed under the latter species. Also, introgression from *A. canescens* may lead to lightly canescent leaves and inflorescence axes in otherwise typical plants of *A. viscida.*

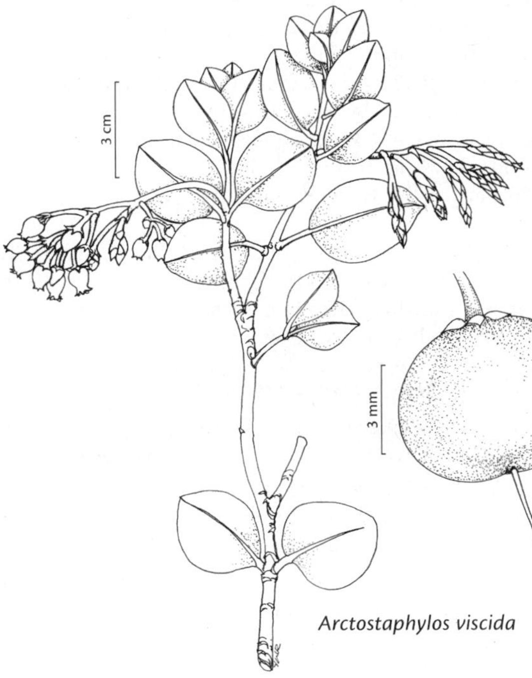

Arctostaphylos viscida

ARCTOSTAPHYLOS

CALLUNA Heather, Scotch heather

Genus composed of one species, *C. vulgaris*.

Calluna vulgaris (Linnaeus) Hull (p. 622)

Heather, Scotch heather

Shrubs evergreen, 15–80 cm; stems branched, pubescent to glabrate. **Leaves** opposite, nearly needle-like, keeled abaxially 2.5–3.5 mm, margins spinulose-ciliate; sessile. **Inflorescences** terminal panicles, racemes or spikes, often one-sided; bracts 4–8, linear, 0.7–1.5 mm. **Flowers** urceolate to slightly globose; sepals 4, oblong, 3–4 mm, pink to purplish, distinct; petals 4, 2–3 mm, pink to purplish, lobed nearly to base, persistent in fruit; stamens 8, included, anthers opening by slits along their whole length; ovaries superior, styles persistent. **Fruits** capsules, 1–2 mm wide, pubescent. **Seeds** ellipsoid, ~0.5 mm, brown to reddish brown. **2*n***=16.

Sandy soil. Flowering Jul–Sep. 0–50 m. Est. WA; north to British Columbia, northeastern North America; Europe. Exotic.

Long naturalized in northeastern North America, *Calluna vulgaris* has only recently been documented in Oregon (2006) near Yachats. This species is most easily recognized by its very small, near needle-like leaves.

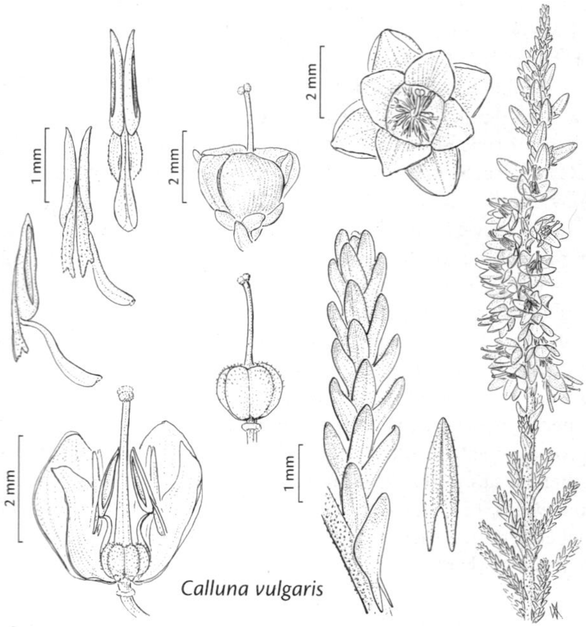

CALLUNA

CASSIOPE White heather, mountain heather, moss heather

Shrubs or **subshrubs** evergreen. **Stems** decumbent to prostrate, often rooting. **Leaves** opposite, 4-ranked, scale-like, keeled adaxially, ± appressed, surfaces glabrous, glandular, or hairy; sessile. **Inflorescences** axillary, solitary flowers; bracts 0, bractlets 4–6. **Flowers** sepals 4–5, ± distinct; petals 4–5, fused; stamens 8 or 10, anthers opening by terminal pores; ovaries superior. **Fruits** capsules.

Circumboreal. ~14 species; 1 species treated in *Flora*.

Cassiope mertensiana (Bongard) G. Don

White heather

Subshrubs 10–30 cm. **Stems** densely branched from base, 5–30 cm, glabrous, puberulent or glandular. **Leaves** ovate-lanceolate, 2–5 mm, margins entire, glabrous or ciliate with minute glandular hairs. **Pedicels** 5–30 mm, often nodding, glabrous or puberulent; subtended by bractlets. **Flowers** sepals ovate, 1.5–2.5 mm, reddish; corollas campanulate, petals 5–8 mm, white to pink, fused 67% of their length. **Fruits** round to ovoid, 1.5–4 mm, loculicidal. **Seeds** many, minute, winged. **2*n***=26.

2 varieties.

A high elevation alpine species, *C. mertensiana* is generally found only above 2000 meters on moist, rocky mountain slopes. This species is distinguished from other Oregon ericaceous species by its small, scale-like leaves.

Key to *Cassiope mertensiana* varieties:

1. Stems and pedicels puberulent; leaf margins glabrous. . . . . ***C. m.*** var. ***mertensiana***

1' Stems and pedicels glabrous, sometimes puberulent with age; leaf margins ciliate with minute glandular hairs. . . . . ***C. m.*** var. ***gracilis***

Cassiope mertensiana (Bongard) G. Don var. *gracilis* (Piper) C.L. Hitchcock

Slender white heather

Cassiope mertensiana (Bong.) G. Don ssp. *gracilis* Piper

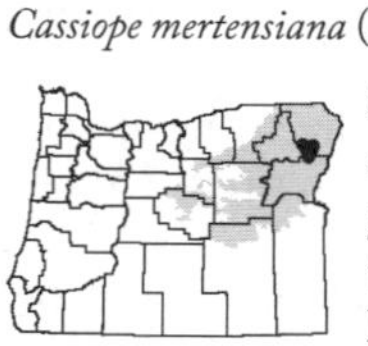

Stems glabrous, sometimes puberulent with age. **Leaves** margins ciliate with minute glandular hairs. **Pedicels** glabrous, sometimes puberulent with age. **Flowers** calyx lobes entire.

Rocky, high elevation montane areas. Flowering Jul–Aug. 2100–3000 m. BW. ID, WA; northeast to MT. Native.

Within Oregon, *C. mertensiana* var. *gracilis* has thus far only been collected in the Wallowa Mountains.

Cassiope mertensiana (Bongard) G. Don var. *mertensiana* (p. 623)

White heather

Cassiope mertensiana (Bong.) G. Don ssp. *mertensiana*

Stems puberulent. **Leaves** margins glabrous. **Pedicels** puberulent. **Flowers** calyx lobes often ciliate.

Rocky, high elevation montane areas. Flowering Jul–Sep. 1100–2800 m. Casc. CA, ID, WA; north to AK. Native.

Within Oregon, *C. mertensiana* var. *mertensiana* is found only on the crest of the Cascade Mountains.

CHIMAPHILA Prince's pine

Subshrubs evergreen, rhizomatous. **Stems** erect, rarely decumbent, simple or branched, slightly woody, usually glabrous, sometimes papillose or hispidulous. **Leaves** cauline, ± whorled, margins prominently toothed; petioles short. **Inflorescences** solitary flowers to umbel-like racemes, 1–10-flowered, rarely > 10. **Pedicels** erect in fruit, papillate to glandular-hairy; bracts ovate or lanceolate. **Flowers** radial, nodding; sepals 5, distinct almost to base, calyx lobes ovate; petals 5, white to red, distinct; stamens 10, included, anthers awnless; ovaries superior. **Fruits** capsules, erect, deeply furrowed.

Asia, Europe, and North America. 5 species; 2 species treated in *Flora*.

Key to *Chimaphila* species:

1. Flowers 1–3; sepals 4–7 mm; leaves lanceolate to elliptic; bracts ovate to obovate . . . . **C. menziesii**

1' Flowers 3–10; sepals 1–3 mm; leaves generally oblanceolate; bracts needle-like to lanceolate . . . . **C. umbellata**

Chimaphila menziesii (R. Brown) Sprengel

Little prince's pine

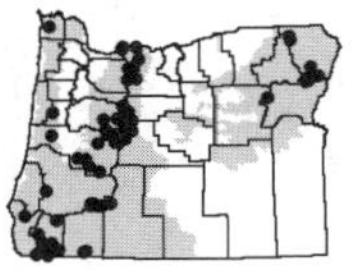

Plants 5–25 cm. **Stems** simple or sparingly branched, reddish. **Leaves** lanceolate to elliptic, occasionally ovate, 10–50 × 4–25 mm, light green abaxially, green and shiny adaxially, margins usually serrate, occasionally entire, tips acute, surfaces glabrous; petioles 4–8 mm. **Inflorescences** 1–3-flowered; bracts ovate to obovate, (2)3–7 mm, margins dentate. **Pedicels** 5–15 mm. **Flowers** sepals 4–7 mm, margins dentate, often membranous; petals concave, suborbicular, 5–7 × 5–8 mm, white, often tinged with pink, margins entire to dentate; filaments swollen near middle, ciliate. **Fruits** globose, 4–7 mm. **$2n$=26**.

Montane and subalpine coniferous forests. Flowering May–Aug. 100–2100 m. BW, Casc, CR, ECas, Sisk. CA, ID, NV, WA; north to British Columbia, northeast to MT. Native.

Throughout its range, *Chimaphila menziesii* shares the same habitats as *C. umbellata*.

Chimaphila umbellata (Linnaeus) W.P.C. Barton (p. 623)

Pipsissewa, prince's pine

Chimaphila umbellata (L.) W.P.C. Barton var. *occidentalis* (Rydb.) S.F. Blake, *Chimaphila umbellata* (L.) W.P.C. Barton ssp. *umbellata*

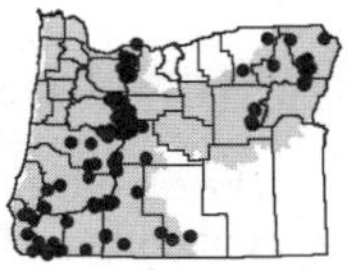

Plants 5–30 cm. **Stems** simple or sparingly branched, greenish to reddish. **Leaves** oblanceolate, occasionally ovate or elliptic, 20–80 × 7–22 mm, light green to brownish green abaxially, green and shiny adaxially, margins usually serrate, occasionally entire, tips acute, surfaces glabrous; petioles 5–12 mm. **Inflorescences** 3–10-flowered, rarely more; bracts needle-like to lanceolate, 3–8 mm, margins entire or dentate distally. **Pedicels** 5–30 mm. **Flowers** sepals 1–3 mm, margins dentate to erose, lobe margins often membranous; petals concave, ovate to obovate, 4–11 × 4–8 mm, white, often tinged with pink, margins dentate to erose; filaments swollen near base, occasionally toward middle, ciliate. **Fruits** depressed-globose, 4–9 mm. **$2n$=26**.

Montane and subalpine coniferous forests, occasionally at lower elevations. Flowering Jun–Sep. 100–2400 m. BW, Casc, CR, ECas, Lava, Sisk, WV. CA, ID, NV, WA; throughout much of North America except south-central states; Asia, Europe. Native.

Many authorities recognize six distinct subspecies of *C. umbellata* worldwide. With the exception of *C. umbellata* ssp. *domingensis* (geographically isolated on Hispaniola Island), the characters separating the subspecies are unreliable and inconsistent.

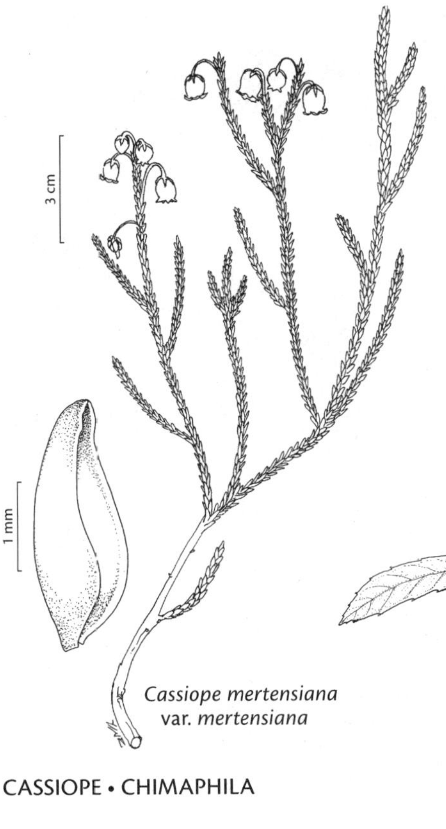

Cassiope mertensiana var. mertensiana

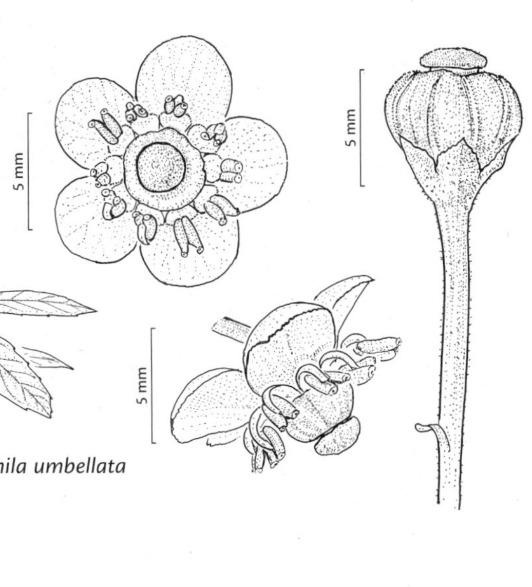

Chimaphila umbellata

CASSIOPE • CHIMAPHILA

CLADOTHAMNUS Copper bush

Genus composed of one species, *C. pyroliflorus.*

Various authors have treated *Cladothamnus*, *Elliottia*, and *Tripetaleia* as one genus (*Elliottia*). However, considering the marked morphological and geographic differences between these genera, we are treating *Cladothamnus* as a distinct and monotypic genus.

Cladothamnus pyroliflorus Bongard (p. 624)

Copper bush

Elliottia pyroliflora (Bong.) S.W. Brim & P.F. Stevens

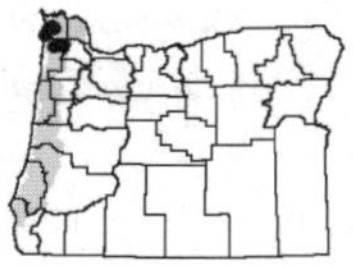

Shrubs deciduous, 0.5–2.5 m. **Stems** erect; twigs of current season green to greenish brown, glabrous to slightly pubescent, older twigs brown to copper with exfoliating bark. **Leaves** alternate or seemingly whorled, elliptic-oblanceolate to oblanceolate, 15–52 × 5–15 mm, light green abaxially, green adaxially, margins entire, tips mucronate, rarely rounded, surfaces glaucous; petioles 1–3 mm. **Inflorescences** terminal, solitary flowers, sometimes 2–3-flowered cymes; bracts leaf-like, lanceolate, 8–15 mm, margins entire. **Pedicels** 4–15 mm. **Flowers** perfect, radially symmetric; sepals 5, connate at base, lanceolate, 5–10 mm, often unequal, tips acute; corollas deciduous; petals 5, distinct, oblong-elliptic, 10–15 mm, salmon to coppery-pink; stamens 8–10, filaments glabrous, < petals, anthers 1–2 mm; ovaries superior, styles included, stigmas 5-lobed. **Fruits** depressed-globose, 5–8 mm, glabrous, pitted, 5–6-valved, opening septicidally. **2*n*=22.**

Montane coniferous forests. Flowering Jun–Jul. 500–1100 m. CR. WA; north to AK. Native.

Cladothamnus pyroliflorus is relatively rare in Oregon, which represents the southernmost extent of its range.

EMPETRUM Crowberry

Subshrubs or **shrubs** evergreen. **Stems** prostrate, low, trailing, woody; twigs glabrous to pubescent, sometimes glandular. **Leaves** alternate or whorled, margins entire, surfaces glabrous, often sparsely glandular; petioles short. **Inflorescences** axillary or terminal, flowers solitary or rarely more. **Flowers** unisexual or bisexual (generally dioecious but also synoecious or polygamous), radially symmetric; corollas deciduous; ovaries superior. **Fruits** drupes, juicy, falling with calyx.

Circumboreal, southern South America, and south Atlantic islands. 3–18 species; 1 species treated in *Flora.*

The current taxonomic understanding of *Empetrum* is incomplete. This is due, in part, to the complex mixture of ploidy levels, sexual systems, and morphology within the genus. Using various combinations of these characteristics, as many as 18 distinct taxa have been described. These combinations, however, do not seem to reflect evolutionary relationships within the genus. As such, most current authorities have reluctantly settled on dividing the genus into three species: *E. atropurpureum* and *E. eamesii*, which are restricted to northeastern North America, and *E. nigrum*, which contains the remaining populations in the world. Further study of this genus is needed.

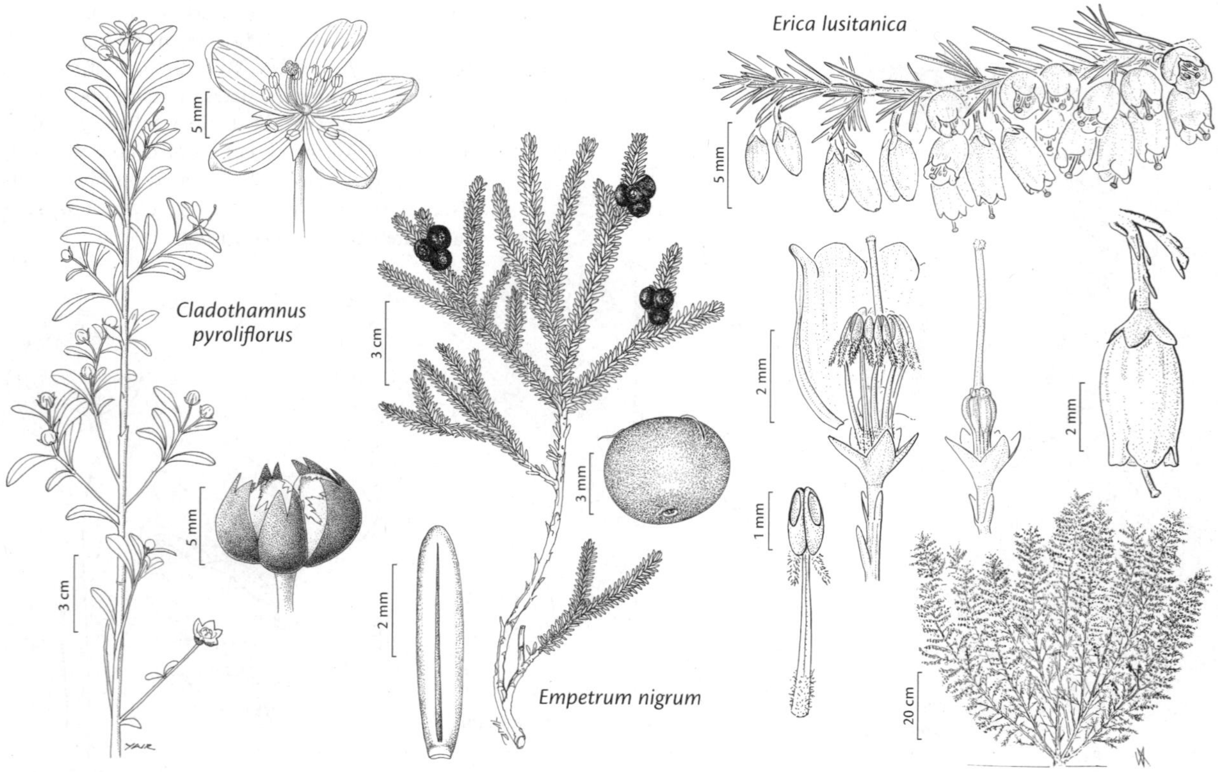

CLADOTHAMNUS • EMPETRUM • ERICA

Empetrum nigrum Linnaeus (p. 624)

Black crowberry

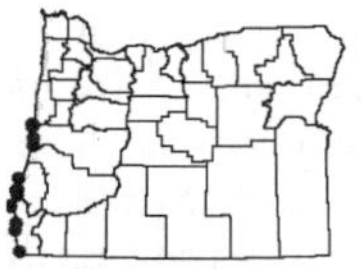

Stems slender, creeping and rooting, 15–100 cm, green to reddish. **Leaves** linear, nearly needle-like, deeply grooved abaxially, 3–7.7 × 0.5–2 mm, stiff, tips rounded to subacute, surfaces glabrous or sparsely glandular. **Flowers** sepals 3, green to red or purple; petals 3, 0.5–1.5 mm, white to purple; stamens (2)4, filaments < 4 mm, anthers without awns. **Fruits** 4–10 mm in diameter, black with a bloom. **Seeds** 1–3 mm, brown. ***2n***=26, 52.

Coastal bluffs. Flowering Apr–May. 0–50 m. Est. CA, WA; north to AK, east across Canada to northeastern states; circumboreal, southern South America, south Atlantic islands. Native.

Within Oregon this species is rare, but, when found, it is always growing in a coastal habitat. In other parts of the world, *E. nigrum* can be found inland in sphagnum bogs, subalpine forests, and on mountain summits. This may suggest that our plants are a distinct taxon. However, more data are needed to determine the range and relationships between our "taxon" and others within the genus.

ERICA Heath

Subshrubs or **shrubs** evergreen. **Stems** generally erect; twigs glabrate to pubescent. **Leaves** whorled, linear, leathery, margins entire or minutely dentate or sparsely ciliate; petioles present. **Inflorescences** axillary, flowers in small clusters; bractlets 2–3. **Flowers** bisexual, radially symmetric; sepals 4, distinct to fused at base; petals 4, fused, persistent in fruit; stamens 8(10), anthers awned, 2-lobed, dehiscent by pores; ovaries superior. **Fruits** capsules, loculicidal, dehiscent from tip to base. **Seeds** several.

Europe, North America. ~600 species; 1 species treated in *Flora.*

Thus far, the only *Erica* species known to have escaped and persisted in Oregon is *E. lusitanica. Erica carnea* has been collected twice, in Clatsop and Lincoln counties (1997 and 2006), but there is no indication these populations have become established.

Erica lusitanica Rudolphi (p. 624)

Spanish heath

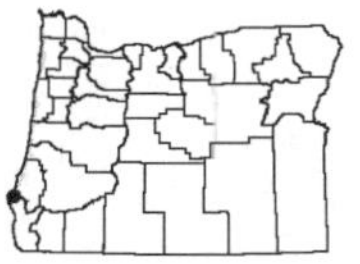

Plants 1–3 m, woody; twigs of current season green with course hairs, older twigs grayish brown, often with peeling bark and glabrate. **Leaves** in whorls of 3–4, 2–7 × 0.2–0.5 mm, bright green to grayish green, margins revolute; petioles 0.2–0.4 mm. **Inflorescences** axillary and terminal, 1–4-flowered clusters. **Pedicels** 1–3 mm. **Flowers** sepals deltate, lower 33% fused, margins entire, tips subacute; corollas campanulate, 3.5–5 mm, lobes deltate, ~1 mm; stigmas exserted. **Fruits** 1–4 mm. **Seeds** ~0.5 mm.

Sandy coastal habitats. Flowering Mar–Jul. 0–50 m. Est. CA; southwestern Europe. Exotic.

Although known from only one location in Oregon, this population has persisted and grown continually since 1998.

GAULTHERIA Wintergreen

Subshrubs or **shrubs** evergreen. **Stems** erect or prostrate and creeping; twigs glabrous or pubescent; rooting at nodes or not. **Leaves** alternate, shiny, leathery; blades ovate to elliptic or reniform, margins entire to serrate, glabrous to sparsely pubescent. **Inflorescences** axillary and solitary or in axillary or terminal racemes; bracts present. **Pedicels** bractlets present. **Flowers** sepals 5, connate at base to nearly entire length; corollas urceolate to campanulate, petals 5, connate 50% to nearly entire length, white to pink, lobes short; stamens 10, included with or without awns; ovaries superior or appearing inferior because of fleshy calyx when ripe. **Fruits** capsules surrounded by persistent, fleshy calyx and appearing berry-like.

Temperate and montane habitats in Asia and North America. 115–150 species; 3 species treated in *Flora.*

Key to *Gaultheria* species:

1. Leaves 4–10 cm; inflorescences racemes; plants erect or partially prostrate . . . . ***G. shallon***

1' Leaves 1–4 cm; inflorescences solitary flowers; plants creeping, mat-forming, or spreading.

  2. Sepals (and resulting fruit) glabrous; leaves 1–2(2.5) cm . . . . ***G. humifusa***

  2' Sepals (and resulting fruit) glandular pubescent; leaves 2–4 cm . . . . ***G. ovatifolia***

Gaultheria humifusa (Graham) Rydberg (p. 626)

Alpine wintergreen

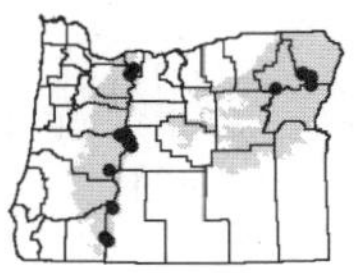

Subshrubs creeping, mat-forming; stoloniferous. **Stems** slender, 5–30 cm; twigs glabrous to pubescent, not glandular. **Leaves** ovate to elliptic, 1–2(2.5) cm, bases obtuse to rounded, margins entire or minutely dentate to serrate, teeth often bristle-tipped, tips generally subacute, sometimes rounded, surfaces abaxially sparsely pubescent along veins, adaxially glabrous. **Inflorescences** axillary, solitary flowers; bracts 3–5, 2–3 mm. **Pedicels** 0.5–2 mm; bractlets 1–3, ≤ 1 mm. **Flowers** sepals fused for ~33% of their length, 2–3 mm, pink to red, glabrous; corollas campanulate, petals 2.5–4 mm, connate at base, white to pinkish, glabrous, lobes 1.5–3 mm; filaments glabrous, anthers without awns. **Fruits** globose, 5–7 mm in diameter, red, glabrous.

Subalpine wetlands, meadows, forests, and slopes. Flowering May–Aug. 800–2700 m. BW, Casc. CA, ID, NV, WA; north to AK, east to CO. Native.

Morphologically, *G. humifusa* is very similar to *G. ovatifolia.* The most readily distinguishable differences between the two species are calyx (and resulting fruit) pubescence and leaf length.

Additionally, *G. humifusa* is generally found at higher elevations and in moister sites.

Gaultheria ovatifolia A. Gray (p. 626)

Slender wintergreen, western teaberry, Oregon wintergreen

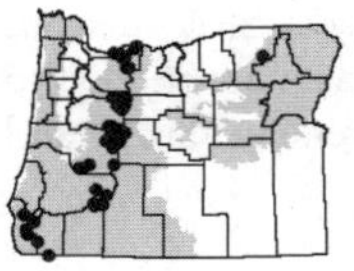

Subshrubs or occasionally **shrubs** spreading, generally not mat-forming; rhizomatous. **Stems** 10–35 cm; twigs glandular pubescent. **Leaves** ovate to subcordate, 2–4 cm, bases obtuse to rounded, margins dentate to serrate, teeth often bristle-tipped, tips subacute to acute, rarely rounded, surfaces abaxially occasionally sparsely pubescent along veins, adaxially glabrous. **Inflorescences** axillary, solitary flowers; bracts 3–5, 1–2.5 mm. **Pedicels** 0.5–2 mm; bractlets 2–6, ≤2 mm. **Flowers** sepals fused for ~33% of their length, 1.5–2.5 mm, pink to red, glandular pubescent; corollas campanulate, petals white to pinkish, 2.5–4 mm, glabrous, lobes 1–1.5 mm; filaments glabrous, anthers without awns. **Fruits** globose, 5–7 mm in diameter, red, glandular pubescent.

Montane mid-elevation forests. Flowering Jun–Jul. 400–1900 m. BW, Casc, CR, ECas, Sisk. CA, ID, WA; north to British Columbia, northeast to MT. Native.

See comments for *G. humifusa*.

Gaultheria shallon Pursh (p. 626)

Salal

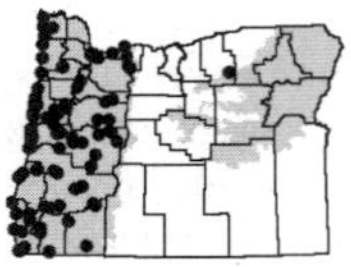

Shrubs erect or partially prostrate, not mat-forming, 5–25 dm; rhizomatous. **Stems** twigs glabrous to pubescent and glandular. **Leaves** ovate to elliptic, 4–10 cm, bases subcordate to rounded, margins serrate, teeth bristle-tipped, tips mucronate, sometimes acute, surfaces abaxially occasionally sparsely pubescent along veins, adaxially glabrous. **Inflorescences** axillary and terminal racemes; peduncles glandular pubescent; bracts 4–6 mm. **Pedicels** glandular pubescent; bractlets 1–5, 3–5 mm, pink, glandular pubescent. **Flowers** sepals 3–5 mm, pink to red, fused for ~50% of their length, glandular pubescent; corollas urceolate 6–10 mm, white to pink, glabrous, lobes 1–1.5 mm; filaments pubescent, anthers with awns. **Fruits** globose, 6–10 mm in diameter, purple to black with a bloom. **2*n***=22, 44.

Forests, cliffs, coastal bluffs. Flowering Mar–Jul. 0–1800 m. BW, Casc, CR, Est, Sisk, WV. CA, WA; north to AK. Native.

Gaultheria shallon is ubiquitous in lower elevation forests west of the crest of the Cascade Mountains. It has been collected once in the eastern portion of the state (Morrow County), approximately 250 km from the edge of its normal range.

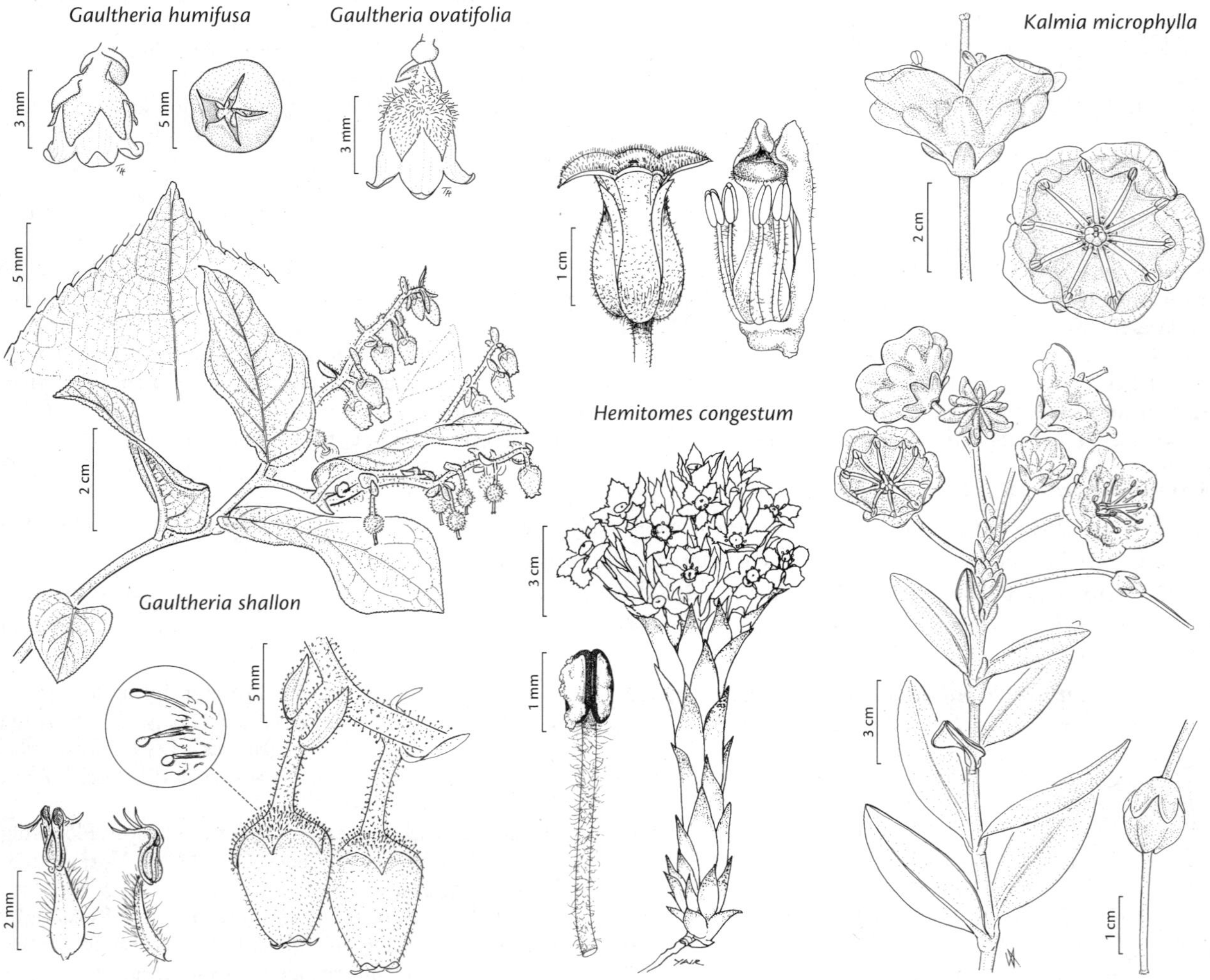

GAULTHERIA • HEMITOMES • KALMIA

HEMITOMES Gnome plant

Genus composed of one species, *H. congestum*.

Hemitomes congestum A. Gray (p. 626)

Gnome plant

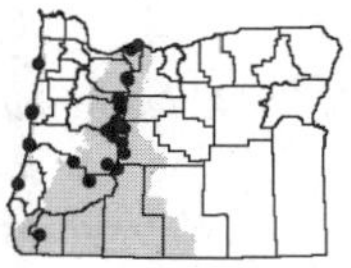

Herbs non-photosynthetic, heterotrophic, 2–15 cm, often waxy in appearance, pinkish when fresh, brownish with age and drying; rhizomatous. **Stems** (if present) short and fleshy, arising from short root-stocks. **Leaves** scale-like, numerous, overlapping, ovate to oblong, thin, margins irregularly toothed and ciliate. **Inflorescences** cymes, racemes, or solitary flowers, not persisting after seed dispersal; bracts 1–2 cm, cream to pink. **Flowers** radially symmetric, erect, developed as plant emerges from soil; sepals 2 in lateral flowers, 4 in terminal flowers, oblong, flattened or folded against corolla, 8–15 mm; corollas cylindric-campanulate, 10–20 mm, inner surface pubescent; petals 4, fused for over 50% of their length, unequally lobed; stamens 8, ~10 mm, included, filaments hairy, anthers without awns, elongate, dehiscent by slits; ovaries superior, styles straight, slightly lobed, stigmas yellow. **Fruits** capsules, berry-like, ~1 cm in diameter, red, pubescent. **Seeds** many. **2*n***=104.

Coastal and mid-elevation forests. Flowering May–Aug. 0–1800 m. Casc, ECas, Est, Sisk. CA, WA; north to British Columbia. Native.

The bimodal distribution of *Hemitomes congestum* is curious; this species is found only near the coast and in mid-elevation forests, mainly in the Western Cascades. Further study of this biogeographic pattern is warranted.

KALMIA Laurel

Shrubs evergreen. **Stems** ascending or erect. **Leaves** opposite or appearing whorled, occasionally alternate, leathery, margins entire. **Inflorescences** terminal racemes or solitary flowers in leaf axils; bracts present. **Flowers** bisexual, radially symmetric; sepals 5, distinct; corollas saucer-shaped, petals 5, fused for ~50% of their length, lobes containing sac-like pockets; stamens 10, included, anthers dehiscent by apical slits; ovaries superior, stigmas 5-lobed. **Fruits** capsules. **Seeds** many.

Asia, Caribbean, Europe, North America. 10 species; 1 species treated in *Flora*.

The flowers of most *Kalmia* species have a peculiar pollination mechanism. The anthers of these species are retained in sac-like pockets. When an insect lands upon a flower, the anthers snap forward, dousing the insect in pollen.

Kalmia microphylla (Hooker) A. Heller (p. 626)

Alpine laurel, western swamp laurel

Kalmia microphylla (Hook.) A. Heller var. *microphylla*, *Kalmia microphylla* (Hook.) A. Heller var. *occidentalis* (Small) Ebinger, *Kalmia occidentalis* Small, *Kalmia polifolia* Wangenh., *Kalmia polifolia* Wangenh. ssp. *microphylla* (Hook.) Calder & Roy L. Taylor, *Kalmia polifolia* Wangenh. var. *microphylla* (Hook.) Rehder, *Kalmia polifolia* Wangenh. ssp. *occidentalis* (Small) Abrams, *Kalmia polifolia* Wangenh. var. *polifolia*

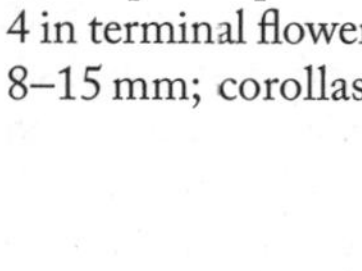

Shrubs 2–6 dm; rhizomatous. **Stems** glabrous; twigs round in cross section to slightly angled at nodes, viscid, glabrous, occasionally puberulent. **Leaves** ovate to oblong-elliptic, 1–3.5 cm, margins rolled under, tips obtuse to acute, surfaces white and granular puberulent abaxially, dark green and glabrous adaxially; petioles absent or very short. **Inflorescences** up to 12-flowered; bracts leaf-like. **Pedicels** 1–3.5 cm, slender. **Flowers** sepals 2.5–3.5 mm, light pink to reddish pink, glabrous, margins ciliate; petals 6–11 mm, pink to deep reddish pink, rarely white, glabrous, puberulent at bases near filaments; anthers without awns, filaments ≤5 mm; styles ≤7 mm. **Fruits** subglobose, 4–5 mm. **Seeds** ~1 mm; winged. **2*n***=24.

Bogs at mid-elevations, rarely coastal. Flowering May–Aug. 0–2900 m. BR, BW, Casc, ECas, Est, Sisk. CA, ID, NV, WA; north to AK, northeast to Ontario, east to CO. Native.

Some authorities recognize two varieties of this species. *Kalmia microphylla* var. *microphylla* grows at higher elevations and has short leaves, while the lower elevation *K. m.* var. *occidentalis* has longer leaves. We do not feel, however, that this minor morphological difference merits taxonomic recognition. Additionally, debate continues as to whether *K. microphylla* should be considered conspecific with the similar, mainly eastern North American species, *K. polifolia*.

KALMIOPSIS Kalmiopsis

Robert J. Meinke

Shrubs evergreen, sometimes rooting adventitiously and suckering. **Stems** erect or trailing, branching from base; twigs terete, puberulent and sparsely sessile- or stipitate-glandular, becoming glabrate, older twigs without peg-like projections. **Leaves** alternate, margins entire; buds with sessile glands; petioles present. **Inflorescences** terminal, corymbiform racemes, 2–15-flowered. **Flowers** bisexual, radially symmetric; sepals 5, nearly distinct, persistent; corollas deciduous, saucer-shaped, petals 5, connate ~33–50% of their length; stamens 10, exserted, anthers not awned, dehiscent by apical pores; ovaries superior, subglobose, stigmas capitate. **Fruits** capsules, subglobose, 5-valved, dehiscent. **Seeds** ~50–150, ovoid, not tailed, not winged.

Southwestern Oregon. 2 species.

Kalmiopsis is the only vascular plant genus endemic solely to the state of Oregon; it is cultivated for its showy flowers. The plants are long-lived and generally reproduce infrequently in nature. Flowers exhibit a pronounced style-height polymorphism; further study is needed to assess its breeding system significance. Distyly is otherwise unknown in the Ericaceae, with the possible exception of *Epigaea repens*, a dioecious species that exhibits a continuum of long- and short-styled flowers yet lacks other evidence of heterostyly.

Key to *Kalmiopsis* species:

1. Tufted erect shrubs, 2–4(8) dm; petals connate ~50% their length; inner surface of connate portion puberulent to glabrous; filaments 3–7(10) mm; endemic to Siskiyou Mountains. . . . . ***K. leachiana***

1' Loose, erect to trailing shrubs, usually to 12(30) dm; petals connate ~33% of their length; inner surface of connate portion with basal tufts of yellow cilia; filaments 7–16 mm; endemic to Cascade Mountains . . . . ***K. fragrans***

Kalmiopsis fragrans Meinke & Kaye (p. 628)
Umpqua kalmiopsis, fragrant kalmiopsis

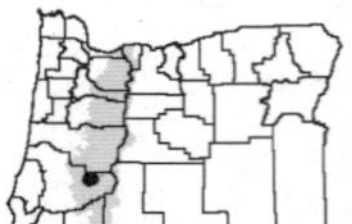

Plants erect, larger plants frequently trailing, loose, to 12(30) dm. **Twigs** reddish to purplish, becoming gray to dark gray, puberulent and sparsely sessile-glandular, becoming glabrate. **Leaves** sweetly aromatic, elliptic to elliptic-oblong, (5)8–30(45) × 4–10 mm, pale green abaxially, deep green adaxially, bases ± cuneate, margins entire, plane, tips obtuse, apiculate, surfaces abaxially moderately covered with sessile crystalline-punctate glands, adaxially glabrous or sparsely sessile dotted-glandular; petioles 1–4 mm, sparsely puberulent, glandular. **Inflorescences** erect, (2)4–8(12)-flowered; bracts leaf-like. **Pedicels** 0.5–2.5(3.3) cm, hairy, glandular. **Flowers** calyx lobes ovate, 3–5 mm, reddish pink to purple, margins ± glandular-ciliate; corollas 16–28(33) mm in diameter, pale reddish purple to deep pink, connate ~33% of their length, throats mostly glabrous, petal ridges connected within corolla lobes, abaxial surface ± puberulent and glandular toward tip; filaments 7–16 mm, with yellowish cilia densely tufted basally, anthers oblong, 0.7–1.8 mm, purple; ovaries puberulent, dotted-glandular, styles dimorphic, 11–15 mm (long form), 5–8 mm (short form). **Seeds** 0.3–0.7 mm.

Tuffaceous outcrops within shaded, mesic, coniferous forests, open ridges, bare rock or shallow soil at bases of cliffs or boulders. Flowering Apr–Jun. 400–1600 m. Casc. Native. ☙

In cultivation, *K. fragrans* has sometimes been sold as the "LePiniec" form of *Kalmiopsis*. In the wild, it has a narrow geographic range (restricted to Douglas County in the southern Cascades) and is significantly rarer than *K. leachiana*.

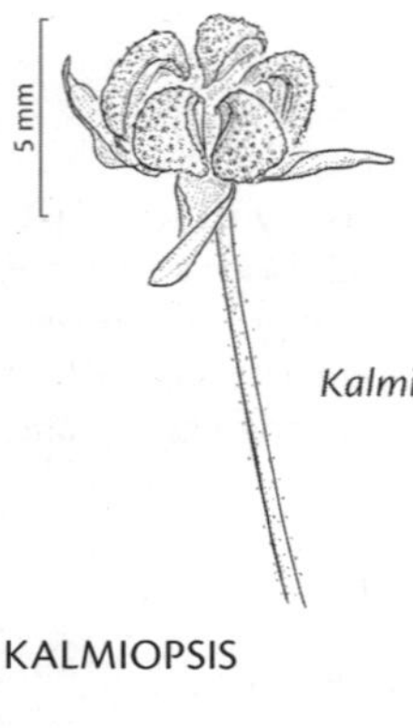

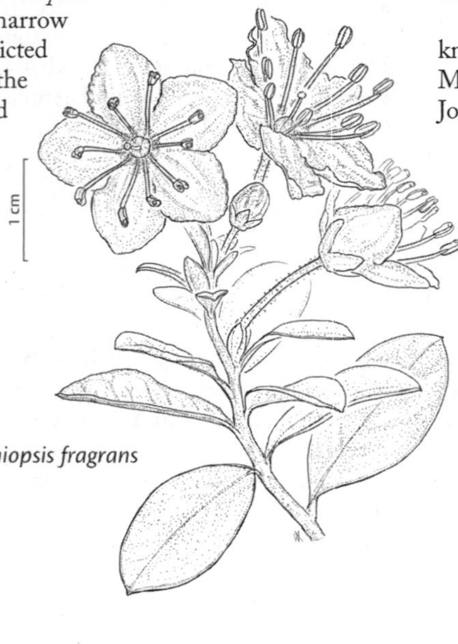

Kalmiopsis fragrans

Kalmiopsis leachiana (L.F. Henderson) Rehder (p. 628)
Siskiyou kalmiopsis

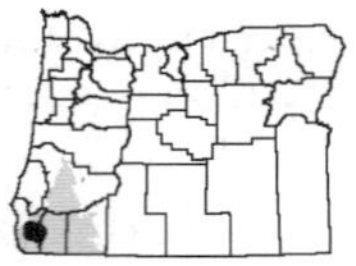

Plants erect, rarely trailing, usually tufted, 2–4(8) dm. **Twigs** pale reddish to purplish, becoming gray to dark gray, puberulent and sparsely stipitate-glandular, becoming glabrate. **Leaves** aromatic, elliptic to elliptic-oblong, 10–25 × 5–12 mm, light green abaxially, dark green adaxially, bases cuneate, margins entire, usually plane, tips obtuse, usually apiculate, surfaces abaxially densely covered with sessile crystalline-punctate glands, adaxially glabrous or sparsely sessile dotted-glandular; petioles 0.5–2 mm, sparsely puberulent and sessile-glandular. **Inflorescences** erect, (5)7–12(15)-flowered; bracts leaf-like. **Pedicels** 1–2 cm, hairy, glandular. **Flowers** calyx lobes ovate, 3–5 mm, light purple, margins ± ciliate-glandular; corollas (12)14–20 mm diameter, rose to deep pink, connate ~50% of their length, throats puberulent or glabrous, petal ridges connected between corolla lobes, abaxial surface ± puberulent and glandular apically; filaments 3–7(10) mm, rarely ciliate basally, anthers oblong, 0.7–1.8 mm, purple; ovaries puberulent, dotted-glandular, styles dimorphic, 7–10 mm (long form), 3–6 mm (short form). **Seeds** 0.3–0.7 mm. **2*n***=24.

Open sunny ridges in xeric shrub community or sparse woodland, rooted in shallow or, more often, deeper soils, on a range of substrates, often including ultramafics. Flowering Apr–Aug. 200–1900 m. Sisk. Native. ☙

Kalmiopsis leachiana is known only from the Siskiyou Mountains in Curry and Josephine counties.

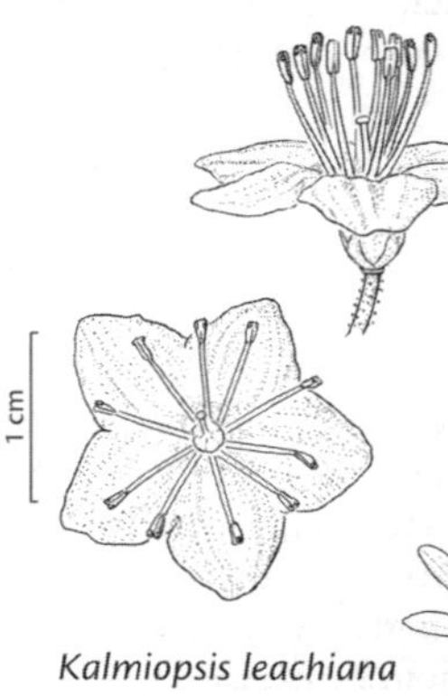

Kalmiopsis leachiana

LEUCOTHOË Hobblebush, dog-laurel

Shrubs evergreen. **Stems** erect, coarsely branched, glabrous to sparsely pubescent. **Leaves** alternate, flat, leathery, not glaucous; petioles present. **Inflorescences** terminal dense-flowered racemes; bracts present. **Flowers** sepals 5, nearly distinct; petals 5, fused entire length; stamens 8(10), included, filaments straight, flattened anthers opening by terminal pores; ovaries superior, styles slightly > stamens, stigmas 5-lobed. **Fruits** capsules, persistent. **Seeds** many, minute.

East Asia, North America. 5 species; 1 species treated in *Flora.*

Leucothoë davisiae Torrey (p. 629)

Sierra laurel, western leucothoë

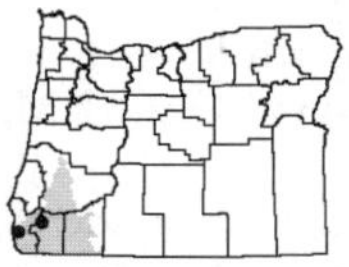

Plants 0.5–1.5 m. **Leaves** oblong to elliptic, 1–6 cm, light green abaxially, dark green adaxially, not shiny, glabrous, margins entire to serrate, tips rounded to acute, occasionally mucronate; petioles 1–5 mm. **Inflorescences** up to 50-flowered, 3–15 cm, dense; bracts ~2 mm. **Pedicels** 1–2 mm. **Flowers** sepals white, ~3 mm, tips rounded to acute; corollas urceolate, 5–8 mm; filaments glabrous, anthers 1–1.5 mm, with 0–2 awns. **Fruits** strongly depressed, thin-walled. **Seeds** winged. **2*n***=22.

Bogs. Flowering Jun–Aug. 800–1400 m. Sisk. CA. Native.

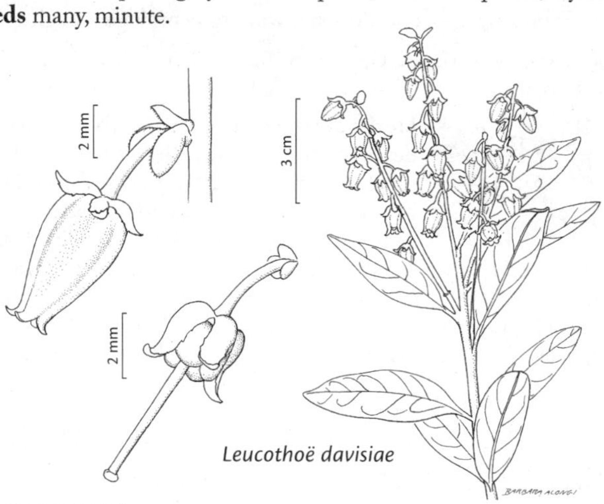

LEUCOTHOË

MENZIESIA False azalea

Shrubs deciduous. **Stems** erect or spreading; bark of older stems often finely shredding. **Leaves** alternate, thin, pubescent; petioles present. **Inflorescences** in terminal umbel-like clusters on previous year's growth; bracts present. **Flowers** bisexual, ± bilateral; sepals 4, rarely 5, fused for ~75% of their length; petals 4(5), fused for 67% of their length; stamens 8, included, anthers without awns, dehiscent by longitudinal terminal slits; ovaries superior, styles included, stigmas capitate.

Fruits capsules, dehiscent from tip to base. **Seeds** many.

Japan, North America. ~10 species; 1 species treated in *Flora.*

Menziesia ferruginea Smith (p. 629)

Fool's huckleberry, rusty menziesia

Menziesia ferruginea Sm. var. *ferruginea*, *Menziesia ferruginea* Sm. var. *glabella* (A. Gray) M. Peck

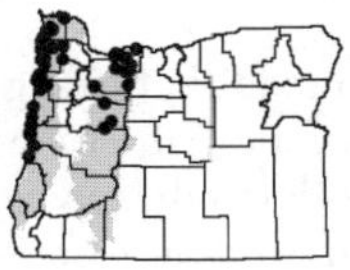

Plants 1–4 m. **Stems** younger twigs pubescent to glandular-pubescent, with peg-like projections; older twigs glabrous, without peg-like projections. **Leaves** lanceolate or elliptic to obovate, 1.5–5.5 cm, abaxially pale green, adaxially dark green, margins entire to serrate and ciliate, rolled under when young, tips rounded to acute with a mucronate tip, surfaces abaxially glabrous to pubescent or glandular-pubescent along veins, adaxially pubescent to glandular-pubescent; petioles 1–6 mm. **Inflorescences** 2–10-flowered. **Pedicels** 1–3 cm, glandular-pilose. **Flowers** sepals 0.5–1 mm, glandular-ciliate; corollas cylindric to urceolate, ≤ 1 cm, yellowish green to yellow tinged with red or bronze; filaments glabrous to pubescent near base, anthers linear. **Fruits** oblong, 5–8 mm. **Seeds** linear. **2*n***=26.

Coastal and mid-elevation forests. Flowering Apr–Aug. 0–1700 m. Casc, CR, Est. CA, ID, WA; north to AK, east to WY. Native.

Two varieties of *M. ferruginea* have been recognized (*M. f.* var. *ferruginea* and *M. f.* var. *glabella*), the latter having less pubescent leaves. The geographic boundaries of these varieties are highly overlapping, and intermediates are common.

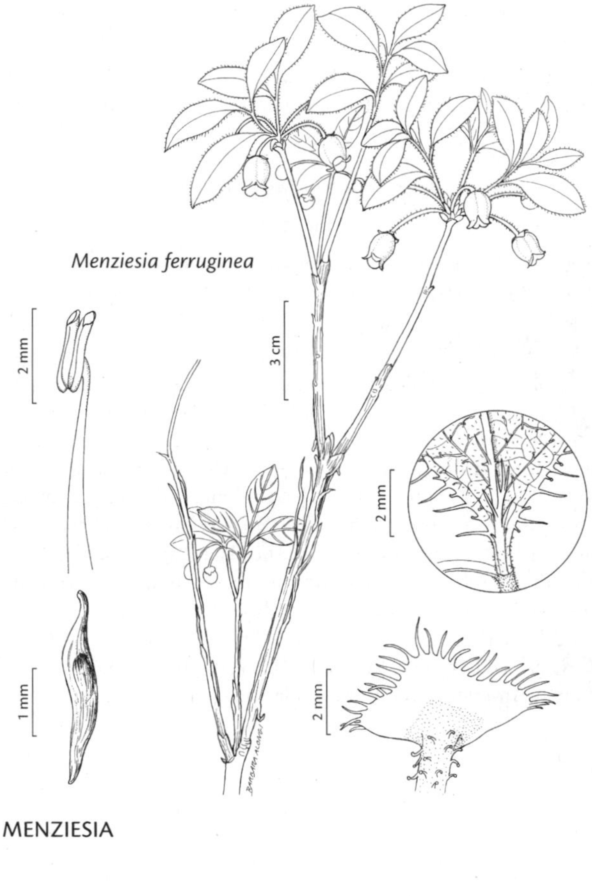

MENZIESIA

MONESES Wood-nymph

Genus composed of one species, *M. uniflora.*

Moneses uniflora (Linnaeus) A. Gray (p. 631)

Wood-nymph, one-flowered wintergreen, single delight

Moneses uniflora (L.) A. Gray var. *reticulata* (Nutt.) S.F. Blake, *Pyrola uniflora* L.

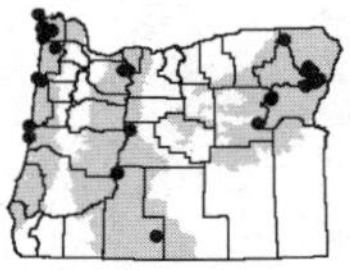

Herbs evergreen, to 15 cm; with very slender, rhizome-like roots. **Stems** simple, glabrous. **Leaves** opposite or in whorls, cauline but appearing basal, ovate to orbicular, 0.5–2 cm, light green and dull abaxially, dark green and shiny adaxially, thin, obscurely veined, bases often entire, margins crenulate to serrate, surfaces glabrous; petioles 2–10(15) mm. **Inflorescences** terminal, solitary flowers; peduncles 1.5–13 cm, glabrous, occasionally papillose above; bracts 0–2, ≤ 4.5 mm. **Flowers** radial, nodding, erect in fruit; sepals 5, 1–3 mm, whitish, thin, margins ciliate; petals 5, ovate, distinct, 0.5–1 cm, white to pinkish, margins denticulate; stamens 10, anthers without awns, with tubules, dehiscent by pores; ovaries superior, styles straight, stigmas 5-lobed. **Fruits** capsules, erect, globose to subglobose, 4–9 mm, dry, dehiscing from top to bottom. **Seeds** many, winged. ***2n***=26.

Moist woods, bogs. Flowering May–Aug. 0–2200 m. BW, Casc, CR, ECas, Est. CA, ID, NV, WA; north to AK, east across Canada to northeastern states, southeast to CO and NM; circumboreal. Native.

Morphologically, this monotypic genus can be considered intermediate to *Chimaphila* and *Pyrola*. Whether this genus is the result of a hybridization event is currently unknown.

MONOTROPA Indian pipe

Herbs non-photosynthetic, epiparasitic. **Stems** fleshy, simple, generally clustered from a clump of matted roots. **Leaves** scale-like. **Inflorescences** terminal, racemes or solitary flowers, initially nodding, becoming erect in fruit, persistent; bracts absent. **Flowers** radially symmetric; sepals absent or (3)4–5(6), distinct; petals (3)4–5(6), distinct, inner surface pubescent; stamens 8–10(14), anthers unawned, dehiscent by slits, < 2 mm; ovaries superior, styles straight, stigmas shallowly depressed or funnel-shaped, with or without a subtending ring of hairs. **Fruits** capsules. **Seeds** many, minute.

Asia, Europe, North America, South America. 2–3 species; 2 species treated in *Flora*.

Recent studies have suggested that *Monotropa* is not a monophyletic genus. These results are preliminary, however. Future studies may result in a realignment of this and other non-photosynthetic genera within Ericaceae.

Key to *Monotropa* species:

1. Inflorescences racemes, flowers rarely solitary; plants yellowish to red; stigmas shallowly depressed without a subtending ring of hairs; upper portion of stems sparsely pubescent to finely pubescent . . . . ***M. hypopitys***

1' Inflorescences solitary flowers; plants white, occasionally pinkish to reddish; stigmas funnel-shaped with subtending ring of hairs; upper portion of stems glabrous to sparsely pubescent. . . . . ***M. uniflora***

Monotropa hypopitys Linnaeus

Pinesap

Hypopitys fimbriata (A. Gray) Howell, *Hypopitys lanuginosa* (Michx.) Nutt., *Hypopitys latisquama* Rydb., *Hypopitys monotropa* Crantz

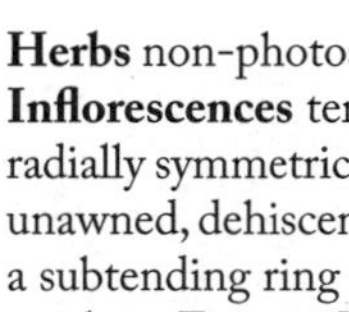

Stems 5–30 cm, yellowish to red, turning brown to black on drying, upper portion usually finely pubescent, occasionally sparsely pubescent. **Leaves** spatulate to ovate, occasionally linear or lanceolate, 2–10 mm. **Inflorescences** racemes. **Flowers** sepals absent or 4–5, elliptic, 6–11 mm; petals 4–5, oblong, 6–15 mm, yellowish to reddish, margins ciliate, often coarsely toothed; filaments sparsely pubescent, anthers horseshoe shaped; stigma shallowly depressed without a subtending ring of hairs. **Fruits** thin, 5–10 mm. **Seeds** < 1 mm, winged. ***2n***=48.

Moist forests. Flowering Jun–Sep. 0–2300 m. BW, Casc, CR, ECas, Est, Sisk, WV. CA, ID, WA; throughout much of North America; Asia, Europe. Native.

Monotropa uniflora Linnaeus (p. 631)

Indian pipe

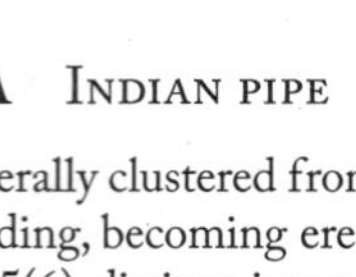

Stems 5–30 cm, white to reddish, turning black on drying, upper portion glabrous or sparsely pubescent. **Leaves** linear or lanceolate to ovate, 2–10 mm. **Inflorescences** solitary flowers. **Flowers** sepals (3)5(6), lanceolate to oblong, 6–10 mm; petals (3)5(6), oblong-spatulate, 9–20 mm, white to reddish, margins entire; filaments glabrous or sparsely pubescent; anthers ellipsoid to ovate; stigmas funnel-shaped with a subtending ring of hairs. **Fruits** stout, 5–12 mm. **Seeds** < 1 mm, winged. ***2n***=32, 48.

Moist forests. Flowering May–Sep. 0–1800 m. BW, Casc, CR, Est, WV. CA, ID, WA; throughout much of North America, south to Columbia; Asia, Europe. Native.

ORTHILIA One-sided wintergreen

Genus composed of one species, *O. secunda*.

Although past floras have traditionally placed *Orthilia* within *Pyrola*, the consensus of most modern botanists is to consider *Orthilia* a distinct genus.

Orthilia secunda (Linnaeus) House (p. 631)

One-sided wintergreen, one-sided pyrola, sidebells

Pyrola secunda L., *Pyrola secunda* L. var. *secunda*, *Remischia secunda* (L.) Garcke

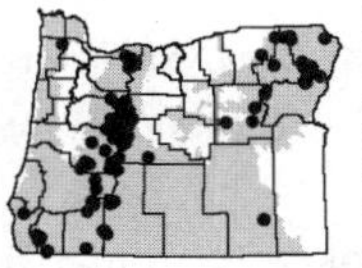

Herbs evergreen, 10–20 cm; rhizomes branched. **Stems** single, decumbent to erect, glabrous to papillose, often woody toward base. **Leaves** alternate or whorled, cauline, generally near base, ovate to orbiculate, 1.5–6 × 1–3 cm, dull and light green abaxially, dark green and shiny adaxially, bases rounded to acute, margins entire or serrulate, not prominently toothed, tips rounded to acute, surfaces glabrous; petioles 0.5–2 cm. **Inflorescences** 3–30-flowered racemes with flowers arranged on one side, arching in flower, erect in fruit; peduncles 10–20 cm; bracts ovate to lanceolate, ≤ 1 cm; pedicels ≤ 8 mm. **Flowers** radially symmetric; sepals 5, triangular, ~1 mm, green, often with white to pink hyaline margins, tips rounded; petals 5, distinct, broadly ovate, 4–6 mm, with two basal tubercles on inner surface, margins dentate or irregularly toothed; stamens 10, 5–8 mm, anthers included, ~1.5 mm, without awns, dehiscent by pores; ovaries superior, styles elongate, erect with maturity, straight, 3–6 mm, stigmas 5-lobed. **Fruits** capsules, depressed-globose, 3–6 mm in diameter, dehiscent by valves. **Seeds** many, winged. $2n$=38.

Middle to high elevation forests. Flowering Jun–Sep. 100–2600 m. BR, BW, Casc, CR, ECas, Sisk. CA, ID, NV, WA; circumboreal. Native.

Plants with a small growth form, light green coloring, and smaller orbicular leaves may be assigned to variety *obtusata*. Plants with mixed characters occur frequently, however, suggesting this variety does not warrant distinction.

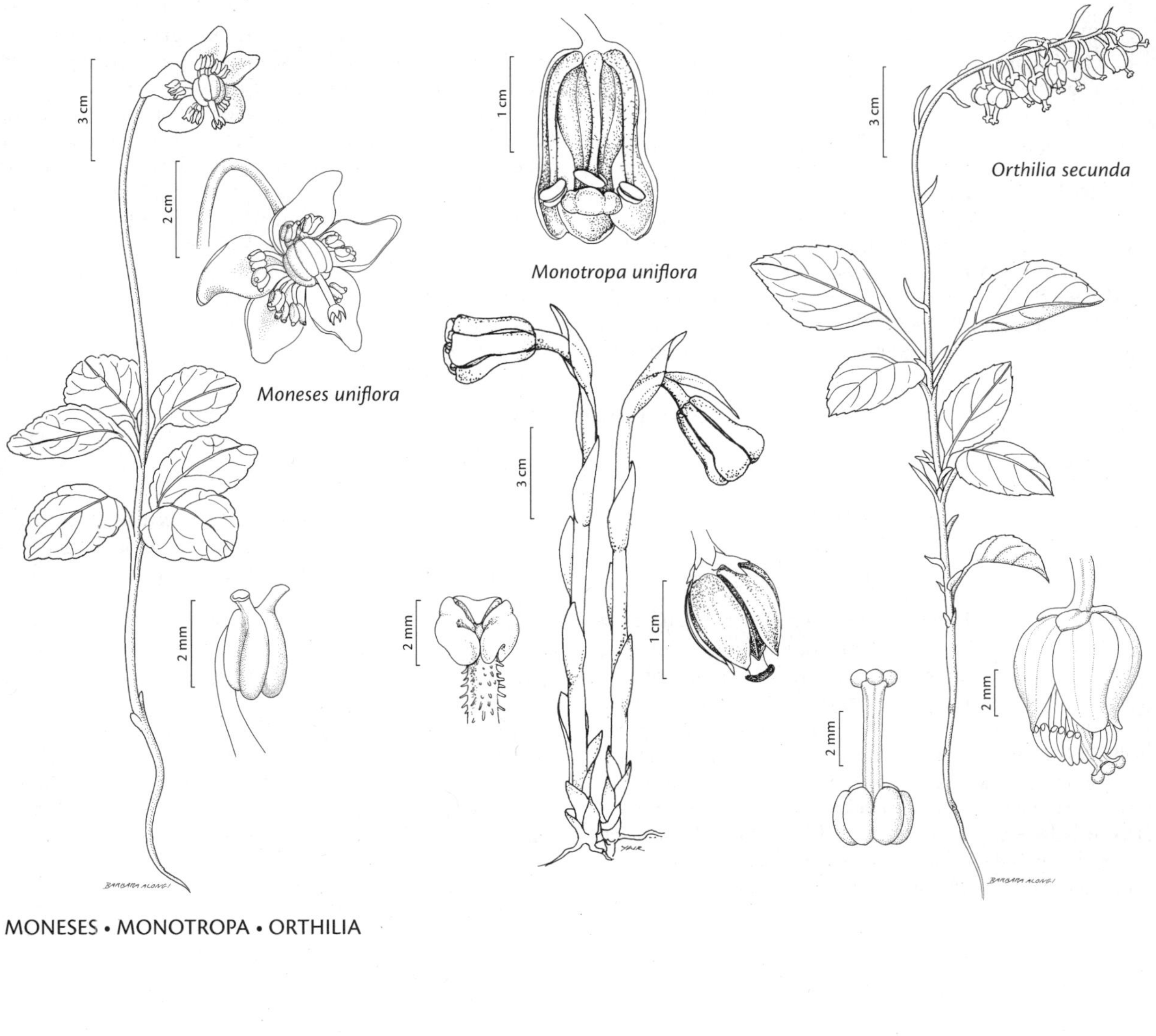

MONESES • MONOTROPA • ORTHILIA

PHYLLODOCE Mountain heather

Subshrubs to **shrubs** evergreen, low growing. **Stems** prostrate to ascending; younger branches glandular, older branches glabrous with rough peg-like projections. **Leaves** alternate, needle-like, curved; petioles short. **Inflorescences** umbel-like terminal clusters at end of branches; bracts present; pedicels glandular-pubescent. **Flowers** bisexual, radially symmetric; sepals 5, fused at base; petals 5, fused for at least 50% of their length, deciduous; stamens (8)10, anthers not awned, dehiscent by terminal slits; styles included or exserted. **Fruits** capsules. **Seeds** many, winged.

Circumboreal. 8 species; 2 species treated in *Flora.*

Phyllodoce breweri has been reported from Douglas County. More investigation is needed to confirm its presence. *Phyllodoce breweri* is similar to *P. empetriformis* but has exserted rather than included stamens.

Key to *Phyllodoce* species:

1. Corollas pinkish to rose; glabrous; campanulate . . . . ***P. empetriformis***
1' Corollas white to yellow or yellowish green; glandular-pubescent; narrowly urceolate . . . . ***P. glanduliflora***

Phyllodoce empetriformis (Smith) D. Don
Pink mountain heather

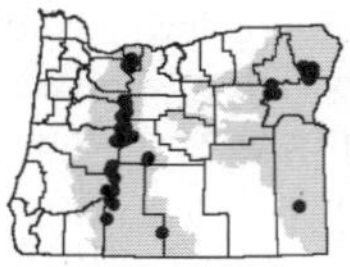

Plants erect to decumbent, 10–40 cm, much branched and often matted. **Stems** younger branches sparsely glandular, older branches glabrous. **Leaves** overlapping, generally curving upward, becoming straight with maturity, 5–14 × 1–1.5 mm, margins revolute and entire to minutely glandular, surfaces abaxially grooved and glandular, adaxially glabrous. **Inflorescences** 1–12-flowered; pedicels ≤2.5 cm, glandular. **Flowers** sepals ovate, 2–3 mm, lobed nearly to base, glabrous, margins ciliate; corollas campanulate, petals 5–8 mm, pinkish to rose, glabrous, lobes spreading, 1.5–2.5 mm; stamens included, filaments 1.5–3 mm, glabrous, anthers 1–2.5 mm; styles exserted, 5–7 mm. **Fruits** globose, ~3 mm in diameter, glandular. **2*n***=24.

Subalpine to alpine slopes, and occasionally level wetlands. Flowering May–Sep. 900–2900 m. BW, Casc, ECas, Owy. CA, ID, WA; north to Yukon, east to WY, southeast to AZ. Native.

Where *P. empetriformis* and *P. glanduliflora* are sympatric, hybrids are occasionally found (*P.* × *intermedia*). Whether or to what degree these hybrids are fertile has yet to be determined.

Phyllodoce glanduliflora (Hooker) Coville (p. 633)
Yellow mountain heather

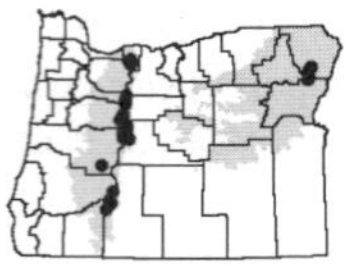

Plants erect to decumbent, 10–35 cm, much branched and often matted. **Stems** younger branches glandular-pubescent, older branches glabrous. **Leaves** overlapping, generally curving upward, becoming straight with maturity, 4–12 × 1–2 mm, margins revolute, glandular-serrulate, surfaces abaxially grooved and glandular, adaxially glabrous to glandular. **Inflorescences** 1–9-flowered; pedicels ≤3 cm, glandular. **Flowers** sepals narrowly ovate to lanceolate, 2.5–4 mm, lobed nearly to base, glandular-pubescent abaxially, margins not ciliate; corollas narrowly urceolate, petals 5–8 mm, white to yellow to yellowish green, glandular, lobes reflexed, 1–2 mm; stamens included, filaments 2–3 mm, pubescent, anthers 1–1.5 mm; styles included, 3–5 mm. **Fruits** globose, ~3 mm in diameter, glandular. **2*n***=24.

Cliffs and subalpine to alpine slopes. Flowering Jul–Aug. 300–3000 m. BW, Casc. WA; north to AK, east to WY. Native.

See comments for *Phyllodoce empetriformis.*

PITYOPUS Pinefoot

Genus composed of one species, *P. californicus.*

Pityopus californicus (Eastwood) H.F. Copeland (p. 633)
Pinefoot

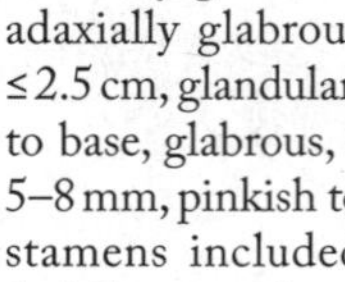

Herbs non-photosynthetic. **Stems** erect, simple or branched, sometimes largely underground, appearing 0–15 cm tall, white to cream-colored, drying to brownish, stout and fleshy. **Leaves** triangular-ovate proximally, lanceolate to linear distally, 1–2.5 cm, white, margins entire. **Inflorescences** spike-like racemes or flowers solitary, erect when emerging from soil, 1–6 cm, cream-colored, not persistent after seed dispersal; bracts 1–1.5 cm. **Flowers** radially symmetric, erect; sepals 4–5, distinct, oblong, 0.5–2 cm, flat against or clasping corolla, margins entire to jagged; petals 4(5), distinct, oblong, 1–1.5 cm, cream-colored to yellowish, outer surfaces glabrous, inner surfaces densely pubescent, margins jagged, often fringed, tips rounded to subacute; stamens 8, included, ≤1 cm, pubescent, anthers erect, horseshoe-shaped, < 2 mm, without awns, dehiscent by slits; ovaries superior, styles straight, 3–5 mm, pubescent, stigmas subtended by ring of hairs. **Fruits** berries, 5–9 mm in diameter, fleshy. **Seeds** ≤100, ovoid, < 0.5 mm in diameter, not winged.

Moist lower to mid-elevation forests. Flowering May–Jul. 0–2000 m. Casc, Est, Sisk. CA, WA. Native.

Pityopus californicus is distinguished from all other non-photosynthetic ericaceous species by its combination of white stems and distinct, pubescent petals.

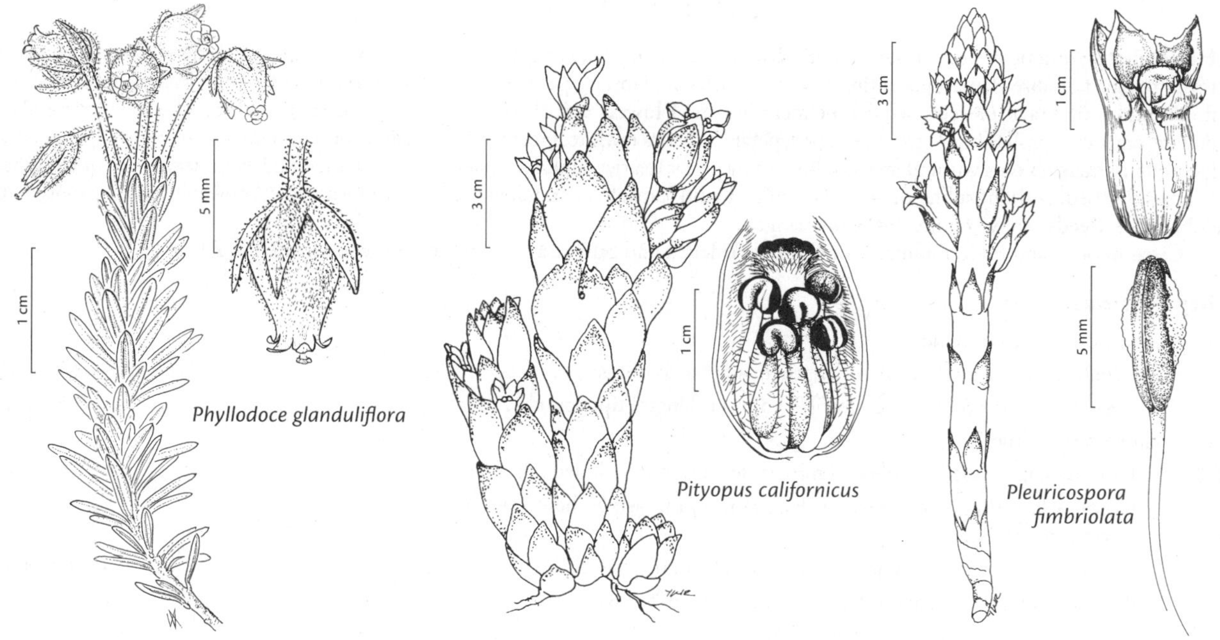

PHYLLODOCE • PITYOPUS • PLEURICOSPORA

PLEURICOSPORA Fringed pinesap

Genus composed of one species, *P. fimbriolata*.

Pleuricospora fimbriolata A. Gray (p. 633)

Fringed pinesap

Pleuricospora longipetala Howell

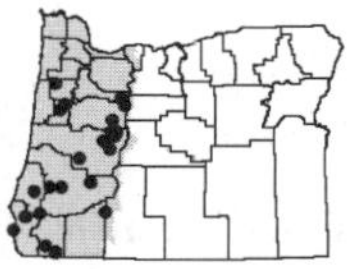

Herbs non-photosynthetic; roots brittle. **Stems** erect, simple, 5–20 cm, white to yellowish or brownish, drying to brownish or reddish, fleshy. **Leaves** ovate, 1–2.5 cm, white to yellowish, margins jagged. **Inflorescences** racemes, 1.5–8 cm, cream-colored, erect when emerging from soil, persistent after seed dispersal; bracts 1–1.5 cm. **Flowers** radially symmetric, erect; sepals 4, lanceolate to ovate, 0.5–1.5 cm, margins jagged; petals 4(5), distinct, oblong, 1–2 cm, cream-colored to yellowish, glabrous, margins jagged, tips rounded to subacute; stamens 8, ~1 cm, glabrous, anthers elongate, 2–4 mm, without awns, dehiscent by slits; ovaries superior, styles straight, slender, 1.5–2.5 mm, glabrous, stigmas funnel-shaped, not subtended by a ring of hairs. **Fruits** berry-like, 7–10 mm in diameter, initially white, turning darker with age, fleshy. **Seeds** ≤ 100, ovoid, < 0.5 mm in diameter, not winged. **2*n***=26.

Moist lower to mid-elevation forests. Flowering Jun–Aug. 100–1700 m. Casc, CR, Est, Sisk, WV. CA, WA; north to British Columbia. Native.

Pleuricospora fimbriolata is distinguished from all other non-photosynthetic ericaceous species by its combination of white to yellowish or brownish stems and distinct glabrous petals.

PTEROSPORA Pinedrops

Genus composed of one species, *P. andromedea*.

Pterospora andromedea Nuttall (p. 635)

Pinedrops

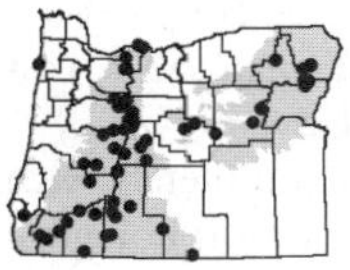

Herbs non-photosynthetic; roots in a thick ball-like mass. **Stems** erect, simple, 1.5–15 dm, pink to reddish brown, drying to reddish brown, fleshy, glandular-pubescent, sticky. **Leaves** crowded near bases of stem, sparse near top, awl-shaped to linear-lanceolate, 2–4 cm, pink to reddish brown, surfaces glandular-pubescent. **Inflorescences** racemes, 10–40 cm, pink to yellowish brown or reddish brown, erect when emerging from soil, persistent after seed dispersal; bracts 2–4 cm. **Flowers** radially symmetric, pendent; sepals 5, distinct, linear-lanceolate, 4–6 mm, densely glandular-pubescent; corollas round-urceolate; petals 5–9 mm, whitish to yellowish, fused 80% of their length, glabrous, lobes reflexed, tips rounded; stamens 10, included, ~3–5 mm, glabrous, anthers depressed-ovoid to ovoid, with awns, dehiscent by slits 1–3 mm; ovaries superior; styles straight, slender, stigmas disk-like, ~1.5 mm wide, not subtended by a ring of hairs. **Fruits** capsules, pendent, depressed-globose, 8–14 mm. **Seeds** over 100, < 1 mm, winged at one end. **2*n***=16.

Forests. Flowering Jun–Sep. 0–1900 m. BW, Casc, ECas, Est, Sisk. CA, ID, NV, WA; western and northern North America. Native.

PYROLA Wintergreen

Herbs photosynthetic or non-photosynthetic. **Stems** erect, bright red or green, glabrous. **Leaves** alternate, generally basal or sometimes reduced or absent, margins entire to serrate, surfaces glabrous; petioles present. **Inflorescences** symmetric racemes, erect in flower and fruit; bracts present; pedicels pendent in fruit. **Flowers** radially or bilaterally symmetric; sepals 5, calyx lobes triangular, deltate, or ovate; corollas bowl- to bell-shaped, petals 5, distinct, white, greenish, yellow, pink, or red; stamens 10, exserted, filaments glabrous, anthers oblong, without awns, dehiscent by pores; ovaries superior, styles exserted or included, bent downward or straight, stigmas 5-lobed, without subtending ring of hairs. **Fruits** capsules, pendulous, dehiscence loculicidal, cobwebby tissue exposed at dehiscence. **Seeds** many, minute, fusiform, winged.

Circumboreal and high mountains of Central America and Sumatra. ~30 species; 5 species treated in *Flora*.

Key to *Pyrola* species:

1. Plants non-photosynthetic.
   2. Sepals fused for ≥ 67% of their length, tips acute to obtuse . . . . *P. chlorantha* (in part)
   2' Sepal fused for < 67% (generally 50%) of their length, tips acute. . . . . ***P. aphylla***

1' Plants photosynthetic.
   3. Corollas cup-shaped; anthers < 1 mm; styles straight, included . . . . ***P. minor***
   3' Corollas spreading, not cup-shaped; anthers > 1 mm; styles bent sideways to downward, exserted.
      4. Most bracts > subtending pedicels; sepal length > width . . . . ***P. asarifolia***
      4' Most bracts ≤ subtending pedicels; sepal length ≈ width.
         5. Sepals fused ≥ 67% of their length, tips acute to obtuse . . . . ***P. chlorantha*** (in part)
         5' Sepal fused for < 67% (generally 50%) of their length, tips acute . . . . ***P. picta***

Pyrola aphylla Smith

Leafless wintergreen

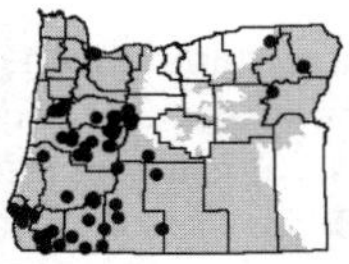

Plants non-photosynthetic, 5–40 cm; rhizomatous. **Leaves** reduced or absent. **Inflorescences** 1–2, 4–25-flowered; inflorescence bracts awl-shaped to ovate-lanceolate, 3.5–5 × 2–3 mm. **Pedicels** 3–16 mm. **Flowers** sepals deltate to ovate, 1–3 mm, usually fused 50% of their length, margins entire or denticulate, tips acute; corollas spreading, petals obovate, 4–10 × 3–6 mm, white to pink, margins entire to denticulate; stamens 5–9 mm, anthers 2–5 mm; styles exserted, bent sideways to downward, 4–9 mm, stigmas 1–1.5 mm. **Fruits** depressed-globose, 4–8 mm. **2*n*=46.**

Coniferous forests. Flowering May–Aug. 0–2100 m. BR, BW, Casc, CR, ECas, Sisk, WV. CA, WA; north to British Columbia, south to Mexico. Native.

Previously considered conspecific with *P. picta*, this species was recently shown to be phylogenetically differentiated by Jolles and Wolfe (2012).

Pyrola asarifolia Michaux

Liverleaf wintergreen, pink wintergreen

Plants 10–45 cm; rhizomatous. **Leaves** ovate, elliptic, round or reniform, 15–75 × 10–55 mm, not mottled, abaxially dull and light green to purplish, adaxially shiny and dark green, leathery, margins entire or crenulate or denticulate, tips obtuse to acute; petioles 10–80 mm, glabrous. **Inflorescences** 1 per stem, 4–29-flowered; bracts ovate to oblong-ovate, 3–15 × 1–3.5 mm. **Pedicels** 3–12 mm. **Flowers** sepals triangular to triangular-ovate, 1.5–5.5 × 1–3 mm, length > width, green or pinkish with thin white to pinkish margins, margins entire to denticulate, tips acute; corollas spreading, petals obovate to round, 4.5–9.0 × 3–6 mm, white to pink or purplish red, margins entire; stamens 4–7.5 mm, anthers 2–4 mm; styles exserted, bent sideways to downward, 7–10 mm, stigmas ~1 mm wide, without subtending ring of hairs. **Fruits** depressed-globose, 4–8 mm. **2*n*=46.**

Asia, North America. 3 subspecies; 2 subspecies treated in *Flora*.

Key to *Pyrola asarifolia* subspecies:

1. Leaf blade margins generally entire to crenulate; bracts ≤ 1.5 × length of pedicels . . . . *P. a.* ssp. ***asarifolia***

1' Leaf blade margins denticulate (rarely entire); bracts > 1.5 × length of pedicels *P. a.* ssp. ***bracteata***

Pyrola asarifolia Michaux ssp. *asarifolia*

Alpine pyrola, bog wintergreen

Pyrola asarifolia Michx. var. *asarifolia*, *Pyrola asarifolia* Michx. var. *incarnata* Fernald, *Pyrola uliginosa* Torr.

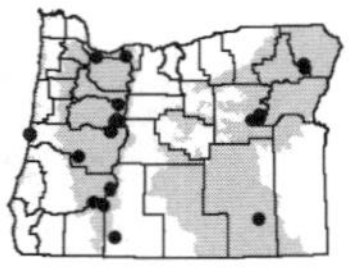

Leaves round, ovate, elliptic or reniform, margins entire to crenulate. **Inflorescences** bracts 1–1.5 × length of subtending pedicels. **Flowers** sepals 1.5–4 mm; anthers 2–3.4 mm.

Moist forests, bogs, streambanks. Flowering Jun–Aug. 0–1900 m. BR, BW, Casc, Est, WV. CA, ID, NV, WA; north to AK, east to Newfoundland, southeast to NM; Asia. Native.

The range of *P. a.* ssp. *bracteata* overlaps completely with that of *P. a.* ssp. *asarifolia*. Whether the slight morphological differences between these two subspecies are environmentally induced may be worthy of future research.

Pyrola asarifolia Michaux ssp. ***bracteata*** (Hooker) Haber

Long-bracted wintergreen, pink wintergreen

Pyrola bracteata Hook.

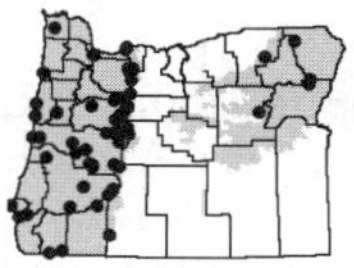

Leaves ovate or elliptic, margins denticulate (rarely entire). **Inflorescences** bracts > 1.5 × length of pedicels. **Flowers** sepals 3–5.5 mm; anthers 2.5–3.7 mm.

Moist or dry forests, occasionally in open areas. Flowering May–Aug. 0–1800 m. BW, Casc, CR, Est, Sisk, WV. CA, ID, WA; north to AK, northeast to MT. Native.

Pyrola chlorantha Swartz (p. 635)

Green-flowered wintergreen

Pyrola virens Schreb.

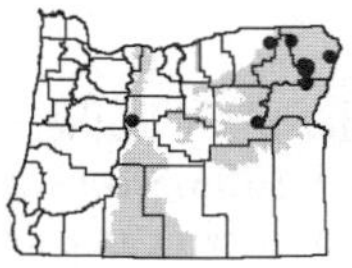

Plants photosynthetic or non-photosynthetic, 10–25 cm; rhizomatous. **Leaves** normal, but sometimes reduced or absent, ovate, elliptic, obovate, or round, 8–25 × 7–28 mm, not or occasionally slightly mottled along veins, abaxially dull and light green to purplish, adaxially shiny and dark green, leathery, margins entire or crenulate, tips obtuse to rounded; petioles 8–55 mm, glabrous. **Inflorescences** 1–2(3) per stem, 2–17-flowered; bracts awl-shaped to lanceolate, 2.5–7.0 × 0.5–0.8 mm. **Pedicels** 3–8 mm. **Flowers** sepals deltate to deltate-ovate, fused ≥ 67% of their length, 1–2 × 1–2 mm, green or pinkish with thin white margins, margins entire, erose or denticulate, tips acute to obtuse; corollas spreading, obovate, 4–9 × 3.3–5.5 mm, petals green, white or yellowish white, margins entire, erose or denticulate; stamens 4–7.5 mm, anthers 2–4 mm; styles exserted, bent sideways to downward, 4–7 mm, stigmas 1–1.5 mm wide. **Fruits** depressed-globose, 3–6.5 mm. ***2n***=46.

Montane forests, occasionally at lower elevations. Flowering Jun–Jul. 500–1900 m. BW, ECas. CA, ID, NV, WA; north to AK, east to Newfoundland, southeast to NM; Europe. Native.

While the photosynthetic form of this species is most common, non-photosynthetic plants are occasionally found. Given recent findings separating *P. aphylla* from *P. picta* (Jolles & Wolfe 2012), taxonomic recognition of non-photosynthetic plants may be warranted.

Pyrola chlorantha is reported to hybridize with *P. picta*. The morphological similarity between these two species is large. Research concerning their potential conspecificity is needed.

Pyrola minor Linnaeus (p. 635)

Lesser wintergreen, snowline wintergreen

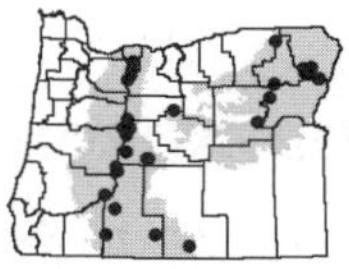

Plants 6–25 cm; rhizomatous. **Leaves** oblong-elliptic, ovate, obovate, or round, 10–37 × 8–25 mm, not mottled, abaxially dull and light green, adaxially dull to shiny and green to dark green, slightly leathery, margins crenulate, tips obtuse or rounded; petioles 2–30 mm, glabrous. **Inflorescences** 1 per stem, 3–15-flowered; bracts lanceolate to ovate-lanceolate, 3–5 × 0.5–1.5 mm. **Pedicels** 2.5–4.5 mm. **Flowers** radially symmetric; sepals deltate, ~1.5 mm, margins entire to denticulate, green or pink with thin white to pinkish margins, tips acute to obtuse; corollas cup-shaped, petals obovate to broadly elliptic, 3–5 × 2.5–4 mm, white or pink, margins entire; stamens 2–3.5 mm, anthers 0.3–0.9 mm; styles included, straight, 0.5–1.5 mm, stigmas ~1 mm wide. **Fruits** depressed-globose, 3–5.5 mm. ***2n***=46.

Moist montane forests and streambanks. Flowering Jun–Aug. 900–2200 m. BW, Casc, ECas. CA, ID, NV, WA; north to AK, east to Greenland, southeast to NM; Asia, Europe. Native.

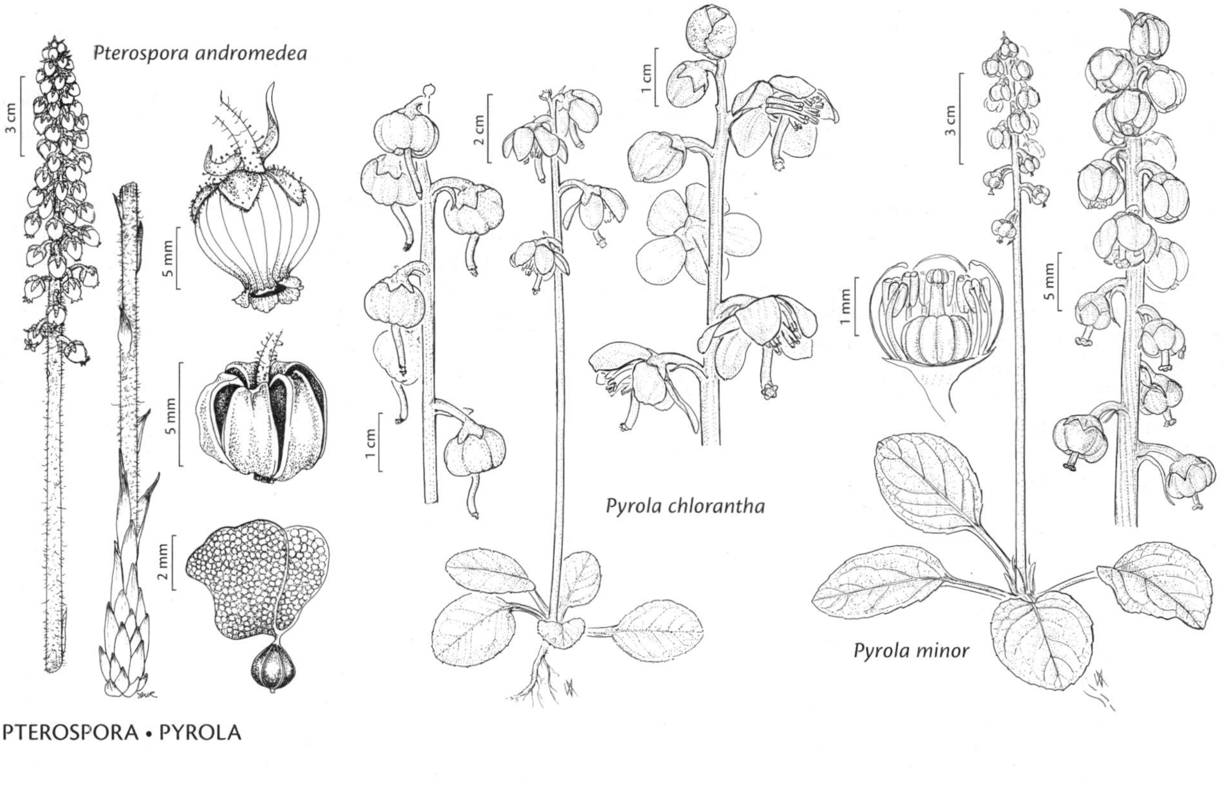

PTEROSPORA • PYROLA

Pyrola picta Smith (p. 637)

White-veined wintergreen, white-vein pyrola

Pyrola crypta Jolles, *Pyrola dentata* Sm., *Pyrola dentata* Sm. var. *dentata*, *Pyrola dentata* Sm. var. *integra* A. Gray

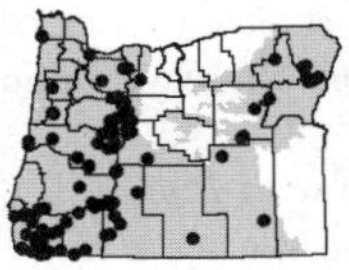

Plants 5–40 cm; rhizomatous. **Leaves** normal, but sometimes reduced or absent, ovate to spatulate, 5–75 × 4–45 mm, abaxially dull and light green to purplish, adaxially dull or shiny and green to dark green, veins white leathery, margins entire or denticulate, tips obtuse to acute; petioles 4–25 mm, glabrous. **Inflorescences** 1–2(3), 4–22-flowered; bracts awl-shaped to ovate-lanceolate, 3.5–6 × 2–3 mm. **Pedicels** 3–18 mm. **Flowers** sepals ovate to deltate, 1–4 × 1–4 mm, usually fused 50% of their length, green with thin white to pinkish margins, margins entire or denticulate, tips acute; corollas spreading, petals obovate, 4–11 × 3–7 mm, white, greenish white, pink, or reddish, margins entire to denticulate; stamens 5–9 mm, anthers 2–5 mm; styles exserted, bent sideways to downward, 4–9 mm, stigmas 1–1.5 mm wide. **Fruits** depressed-globose, 4–9 mm. **2*n***=46.

Coniferous forests. Flowering May–Sep. 0–2300 m. BR, BW, Casc, CR, ECas, Est, Sisk, WV. CA, ID, NV, WA; north to British Columbia, east to SD, southeast to NM. Native.

Plants with light green to purplish and dentate leaves are considered to be *P. dentata* by some authorities. Additionally, specimens with shorter sepals (less than two millimeters) and longer bracts (greater than four millimeters) have been described as *P. crypta*.

See also discussions for *P. aphylla* and *P. chlorantha*.

RHODODENDRON Rhododendron, azalea

Shrubs or **trees** deciduous or evergreen, rhizomatous or not. **Stems** erect, glabrous or pubescent. **Leaves** alternate, thin or leathery, margins entire to minutely serrate, plane to revolute; buds usually large with overlapping scales; petioles present. **Inflorescences** terminal or axillary, corymbs or rounded racemes. **Flowers** bisexual, bilaterally or radially symmetric; sepals 5; corollas campanulate or funnelform, petals 5; stamens 5–12, included or exserted, filaments unequal, anthers without awns, dehiscent by pores; ovaries superior, styles curved, stigmas capitate. **Fruits** capsules, elongate. **Seeds** many, minute, with or without tails, winged.

Worldwide, except Africa. ~1000 species; 4 species treated in *Flora*.

Although, for several decades, evidence has suggested that the genus *Ledum* should be placed within *Rhododendron* (Kron & Judd 1990), many contemporary floras have continued to treat the genera separately. Here, we have chosen to merge the genera.

Rhododendron groenlandicum has long been assumed to exist in Oregon. This is based on a lone specimen, presumably collected on "Russian Island" in the mid-19th century (the specimen is housed at the Wake Forest University Herbarium). Aside from this specimen, no evidence of this species' existence in Oregon has been documented.

Key to *Rhododendron* species:

1. Plants deciduous; leaves thin, not leathery, margins minutely serrate or entire.
   - 2. Inflorescences 1–2-flowered, axillary; flowers bowl- to bell-shaped, < 25 mm, without blotch or stripes on upper lobe (rarely with faint yellow coloring); leaf margins minutely serrate and glabrous . . . . ***R. albiflorum***
   - 2' Inflorescences 3–15-flowered, terminal; flowers funnel-shaped, > 25 mm, with yellow to orange blotch or stripes on upper lobe; leaf margins entire and ciliate . . . . ***R. occidentale***

1' Plants evergreen; leaves thick, leathery, margins entire.
   - 3. Most leaves > 8 cm; flowers > 20 mm, pink to rose (rarely white), with spots on upper lobe . . . . ***R. macrophyllum***
   - 3' Leaves ≤ 8 cm; flowers < 10 mm, white to cream, without spots on upper lobe . . . . ***R. columbianum***

Rhododendron albiflorum Hooker (p. 637)

White-flowered azalea, white rhododendron

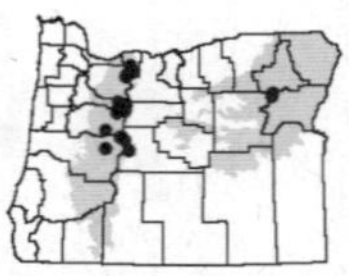

Shrubs deciduous, to 2.5 m; bark smooth to furrowed; rhizomatous. **Leaves** narrowly elliptic, ovate or obovate, 2.5–9 × 1–3 cm, abaxially light green, adaxially green, thin, not leathery, margins minutely serrate and glabrous, tips acute to rounded, surfaces sparsely pubescent. **Inflorescences** axillary, 1–2-flowered; bracts present and similar to bud scales. **Pedicels** 8–15 mm, hairy. **Flowers** opening soon after leaves, pendulous, very fragrant; sepals 5–17 mm, pubescent, margins glandular-hairy; corollas bowl- to bell-shaped, 9–22 mm, white, rarely with faint yellow coloring on upper lobe, minutely pubescent or glabrous on outer surfaces; petals connate, tubes 3–9 mm, lobes 6–15 mm; stamens 9(12), included, unequal, 5–12 mm. **Fruits** 5–8 mm. **Seeds** with tails. **2*n***=26.

Forests, streambanks, thickets. Flowering Jul–Aug. 800–2300 m. BW, Casc. ID, WA; north to British Columbia, northeast to MT. Native.

The white, pendulous, axillary flowers of *R. albiflorum* distinguish this species from all other *Rhododendron* species in Oregon.

Rhododendron columbianum (Piper) Harmaja

Western Labrador tea

Ledum ×columbianum Piper, *Ledum glandulosum* Nutt., *Ledum glandulosum* Nutt. var. *columbianum* (Piper) C.L. Hitchc., *Ledum glandulosum* Nutt. var. *glandulosum*

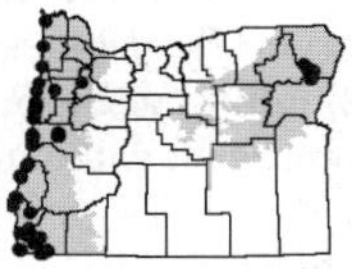

Shrubs or **small trees** evergreen, to 2 m; bark smooth, peeling with age; rhizomatous. **Leaves** ovate to lanceolate, 1–8 × 1.5–3 cm, abaxially light green to white, adaxially green, thick, leathery, margins entire, tips acute, surfaces abaxially glandular, usually velvety, occasionally with a few scattered hairs, adaxially glabrous, occasionally with a few scattered hairs. **Inflorescences** terminal, 10–35-flowered; bracts present. **Pedicels** 5–25 mm, short pubescent and glandular. **Flowers** opening after leaves, erect, somewhat fragrant; sepals 1–1.5 mm, pubescent, reflexed in fruit; corollas 5–9 mm, white to cream, without spots on upper lobe, inner surfaces pubescent, outer surfaces glabrous, petals distinct or only slightly fused at base, lobes 4–7 mm; stamens exserted, equal, (8)10(12), 4.5–8.5 mm. **Fruits** 2–6 mm. **Seeds** without tails. ***2n***=26.

Stream margins, bogs. Flowering May–Aug. 0–2500 m. BW, CR, Est, Sisk, WV. CA, ID, NV, WA; north to British Columbia, east to CO. Native.

Some authorities have segregated the coastal and various inland forms of this species into separate species or varieties. Further study and recognition of these taxa may be warranted.

Rhododendron macrophyllum D. Don

Pacific rhododendron, western rhododendron

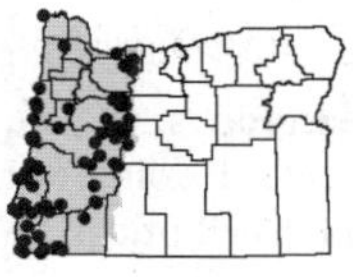

Shrubs or **small trees** evergreen, to 5 m; bark smooth to furrowed and shredding; occasionally rhizomatous. **Leaves** elliptic to obovate, (6)8.5–15(20) × 2.5–7.0 cm, abaxially light green to rusty colored, adaxially green, thick, leathery, margins entire, tips acute to obtuse, surfaces abaxially velvety, occasionally with a few scattered hairs, adaxially glabrous, occasionally with a few scattered hairs. **Inflorescences** terminal, 10–20-flowered; bracts present. **Pedicels** 30–60 mm, glabrous. **Flowers** opening after leaves, erect to horizontal, fragrant; sepals 1–1.5 mm, glabrous, margins pubescent; corollas broad, campanulate, 25–50 mm, white to pink or purple, yellow-green spots on upper lobe, outer surfaces glabrous, petals connate, lobes 15–30 mm; stamens 10, included, unequal, 15–35 mm. **Fruits** 10–25 mm. **Seeds** without tails. ***2n***=26.

Low to mid-elevation forests. Flowering Apr–Aug. 0–1900 m. Casc, CR, Est, Sisk, WV. CA, WA; north to British Columbia. Native.

Plants of this species are cultivated worldwide for their large showy flowers.

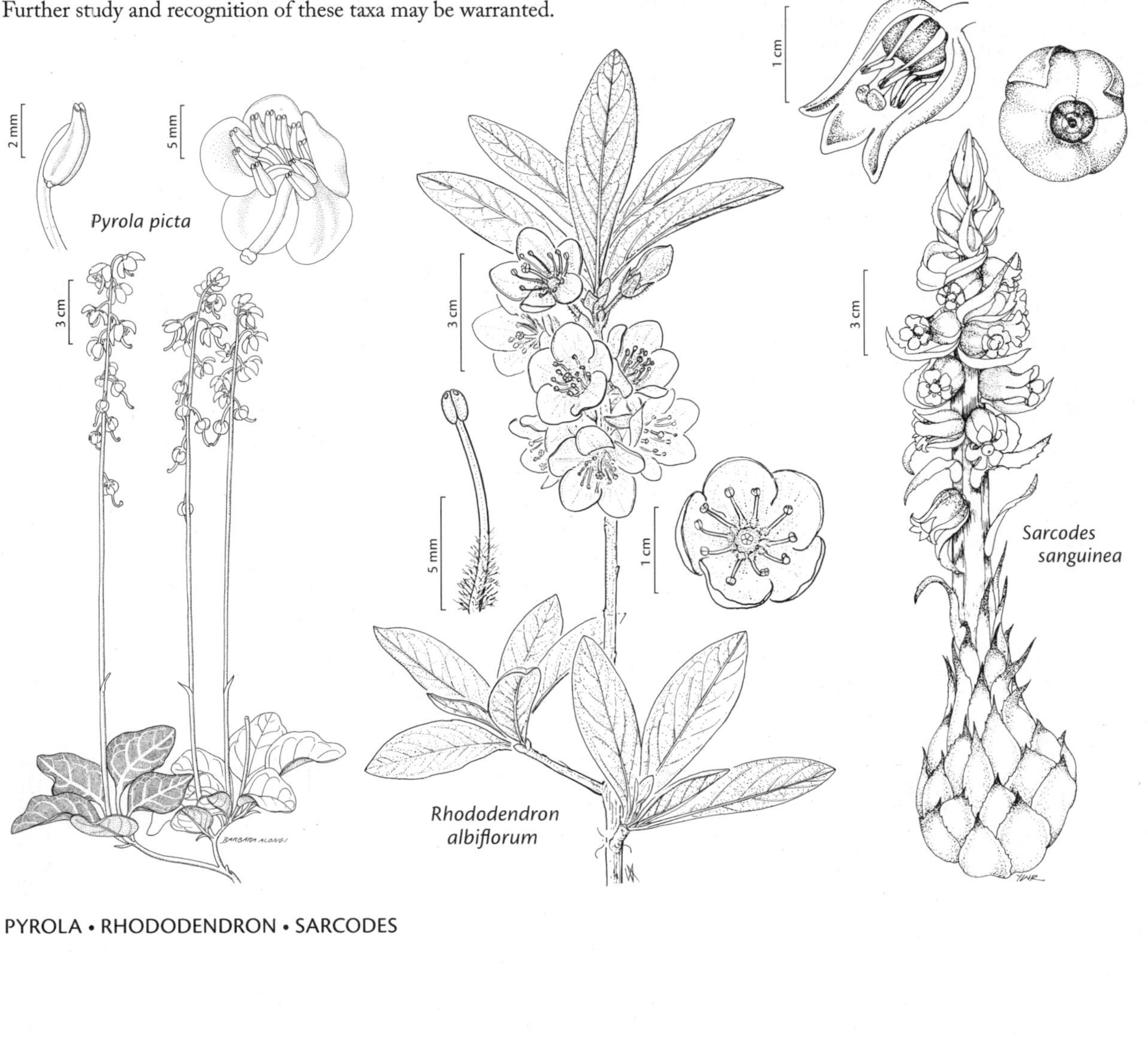

PYROLA • RHODODENDRON • SARCODES

Rhododendron occidentale (Torrey & A. Gray) A. Gray

Western azalea

Rhododendron occidentale (Torr. & A. Gray) A. Gray var. *paludosum* Jeps.

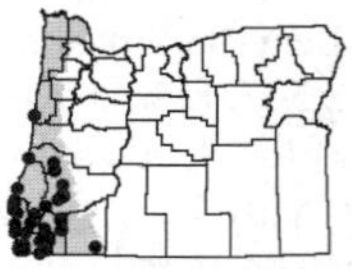

Shrubs or **small trees** deciduous, to 9 m; bark smooth to furrowed and shredding; not rhizomatous. **Leaves** ovate to elliptic, 2.5–10 × 1–3.5 cm, light green abaxially, green to dark green adaxially, thin, not leathery, margins entire and ciliate, tips acute or mucronate to obtuse, surdaces abaxially glabrous, often with a few scattered hairs, adaxially glabrous, occasionally with a few scattered hairs. **Inflorescences** terminal, 3–15-flowered; bracts present. **Pedicels** 10–25 mm, glandular-hairy. **Flowers** opening with or after leaves, erect to horizontal, fragrant or foul-scented; sepals 1–4(9) mm, usually glandular-hairy, margins usually glandular-hairy; corollas funnel-shaped, 30–55 mm, white to pink or salmon, with yellow to orange blotch or stripes on upper lobe, usually glandular-hairy on both surfaces, occasionally glabrous on outer surface, petals connate, lobes 14–27 mm; stamens 5, exserted, unequal, 40–70 mm. **Fruits** 10–20 mm. **Seeds** without tails. **2*n***=26.

Moist thickets, serpentine substrates, ocean bluffs. Flowering Apr–Aug. 0–1400 m. CR, Est, Sisk. CA. Native.

SARCODES Snow plant

Genus composed of one species, *S. sanguinea.*

Sarcodes sanguinea Torrey (p. 637)

Snow plant

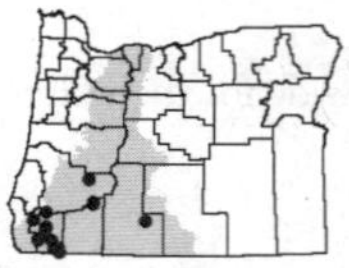

Herbs non-photosynthetic. **Stems** simple, 15–40 cm, bright red, drying to dark red or brown, fleshy, succulent, glandular-hairy. **Leaves** opposite or appearing whorled, often overlapping, 1–7 cm, red; lower leaves narrowly ovate; upper leaves broadly linear. **Inflorescences** racemes, erect at emergence from soil, fleshy, persistent after seed dispersal; bracts 0.5–1.5 cm, red. **Flowers** radially symmetric, horizontal; sepals 5, distinct, narrowly ovate, 0.5–1.5 cm, red; corollas campanulate, petals 5, 1–1.5(2) cm, red, lobes reflexed, rounded to blunt; stamens 10, included, 0.5–1.5 cm, anthers elongate, 3–4 mm, without awns, dehiscent by pore-like slits; ovaries superior, styles straight, 5–8 mm, stigmas reddish, without subtending ring of hairs. **Fruits** capsules, horizontal to erect, 1–2.5 cm in diameter, persistent to disintegrating, walls brittle. **Seeds** many, ovoid, 0.5–1 mm in diameter, unwinged. **2*n***=64.

Mid- to higher elevation forests. Flowering May–Jul. 400–1900 m. Casc, ECas, Sisk. CA, NV. Native.

Sarcodes sanguinea is the only non-photosynthetic, ericaceous, solid red species encountered in Oregon.

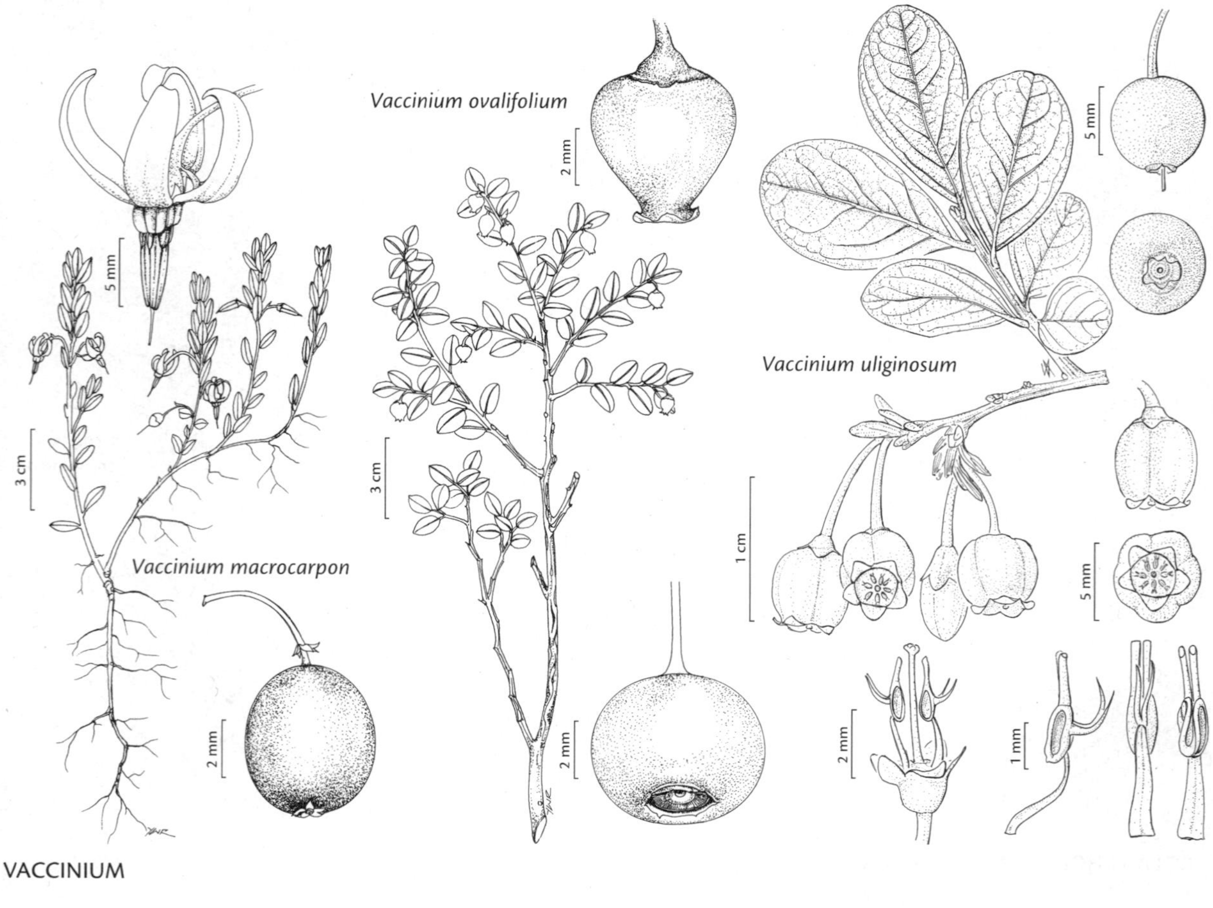

VACCINIUM

VACCINIUM Blueberry, huckleberry, bilberry, cranberry

Shrubs, **subshrubs**, or **vines** evergreen or deciduous. **Stems** erect, ascending, spreading or creeping, glabrous, puberulent or hairy. **Leaves** alternate, simple, margins entire or serrate, plane or revolute, surfaces glabrous to glaucous or puberulent; petioles absent or present. **Inflorescences** axillary, racemes or solitary, 1–10-flowered; bracteoles present or absent. **Flowers** sepals 4–5, 67% to fully fused; corollas globose to urceolate, petals 4–5, 67% to fully fused, white to red; stamens 8–10, included, filaments flat, straight, glabrous or hairy, anthers awned or awnless, dehiscent by pores; ovaries inferior, styles usually surpassing stamens. **Fruits** berries, ovoid to globose, red to black, fleshy. **Seeds** 2–40.

Temperate areas and tropical mountains in Asia, Europe, North America, and South America. ~450 species; 11 species treated in *Flora*.

Many *Vaccinium* keys rely, often heavily, on fruit characters. This unfortunately relegates the observer to identifying plants during one, often short, life stage. Furthermore, most herbarium specimens are lacking fruits. As such, presented here is a key based almost entirely on vegetative characters.

Vaccinium corymbosum (high-bush blueberry), a popular cultivar, has been reported to have escaped and established a small population in Eugene (Lane County). If persisting or found elsewhere, the species is distinguished from our other *Vaccinium* by a combination of its tall growth habit (1–5 m), large leaves (15–70 × 10–25 mm) and fruits with a bloom.

Key to *Vaccinium* species:

1. Leaf margins wholly or partially serrate.
    2. Plants evergreen; leaves thick, firm.
        3. Twigs round in cross section, reddish brown to grayish . . . . . ***V. ovatum***
        3' Twigs sharply angled, green to brown . . . . . ***V. parvifolium*** (in part)
    2' Plants deciduous; leaves thin, flexible.
        4. Leaf margins serrate only on proximal 50% . . . . . ***V. ovalifolium*** (in part)
        4' Leaf margins serrate on distal 50%
            5. Leaf tips usually rounded, sometimes subacute; fruits with a bloom.
                6. Leaves prominently reticulate-veined, green on both sides; filaments > anthers . . . . . ***V. caespitosum***
                6' Leaves weakly reticulate-veined, light green to gray abaxially, green adaxially; filaments < anthers . . . . . ***V. deliciosum***
            5' Leaf tips acute; fruits without a bloom.
                7. Leaves ≥ 3 cm . . . . . ***V. membranaceum***
                7' Leaves typically < 2 cm.
                    8. Branching habit crowded giving a "broom-like" appearance; twigs glabrous . . . . . ***V. scoparium***
                    8' Branching habit open, not "broom-like", or rarely weakly; twigs usually sparsely puberulent . . . . . ***V. myrtillus***

1' Leaf margins entire.
    9. Stems generally creeping and vine-like; plants trailing to ascending; flowers 4-merous, with reflexed petals.
        10. Leaves strongly glaucous abaxially; bracts < 1 mm wide; filaments 50% as long as anthers . . . . . ***V. oxycoccos***
        10' Leaves lightly glaucous abaxially; bracts 1–2 mm wide; filaments 25–33% as long as anthers . . . . . ***V. macrocarpon***
    9' Stems not creeping or vine-like; plants in dense mats, clumped, or erect; flowers 5-merous (sometimes 4-merous in *V. uliginosum*), urceolate.
        11. Stems and twigs sharply angled in cross section; mature berries red . . . . . ***V. parvifolium*** (in part)
        11' Stems and twigs round in cross section or only young twigs angled; mature berries blue to black.
            12. Inflorescences 2–4-flowered racemes (rarely 1-flowered); stems and twigs round in cross section . . . . . ***V. uliginosum***
            12' Flowers solitary; stems round in cross section; twigs angled or with sharp ridges . . . . . ***V. ovalifolium*** (in part)

Vaccinium caespitosum Michaux
Dwarf bilberry

Vaccinium caespitosum Michx. var. *caespitosum*

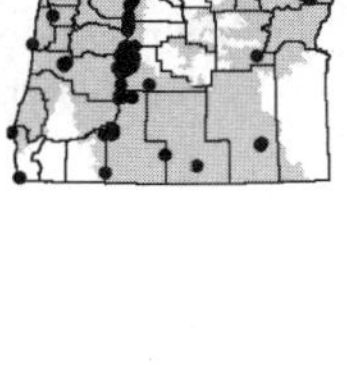

Shrubs deciduous, up to 6 dm; twigs yellow-green to reddish, round in cross section to angled, glabrous to puberulent; rhizomatous. **Leaves** obovate to oblanceolate, 10–30 × 3–12 mm, thin, flexible, green abaxially, margins serrate on distal 25–50%, prominently reticulate-veined, tips rounded, sometimes subacute, surfaces abaxially glabrous or glandular, adaxially glabrous. **Inflorescences** solitary flowers in axils. **Flowers** sepals 5, light green, calyx lobes vestigial; petals 5, corollas cylindric-urceolate to globose, 4–6 × 3–5 mm, white to pink, glaucous; filaments > anthers, glabrous. **Fruits** 5–8 mm in diameter, blue to bluish black, with a bloom. **2*n***=24.

Wet meadows to rocky mountain slopes. Flowering Apr–Aug. 50–2800 m. BR, BW, Casc, CR, ECas, Est, WV. CA, ID, WA; north to AK, east to Newfoundland, south to Guatemala. Native.

Morphologically, *V. caespitosum* is very similar to *V. deliciosum*, making identification of these species difficult, particularly at higher elevations where their ranges overlap. The chromosome numbers of *V. caespitosum* (2*n*=24) and *V. deliciosum* (2*n*=48) may indicate that *V. deliciosum* is an autopolyploid derivative of *V. caespitosum*.

Vaccinium deliciosum Piper

Cascade bilberry

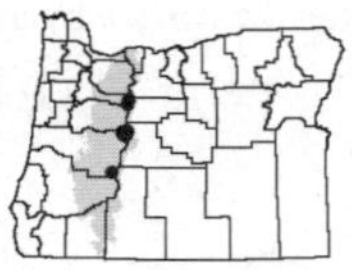

Shrubs deciduous, up to 4 dm; twigs round in cross section to angled, green to greenish brown, glabrous to puberulent; rhizomatous. **Leaves** obovate to oblanceolate, 10–35 × 5–17 mm, light green to gray abaxially, green adaxially, thin, flexible, margins serrate from tips to 50–75% of length of blade, weakly reticulate-veined, tips rounded, sometimes subacute, surfaces glabrous. **Inflorescences** solitary flowers in axils. **Flowers** sepals 5, calyx lobes indistinct or very shallow; petals 5, corollas globose-urceolate to globose, 4–6 × 4–7 mm, pinkish, glaucous; filaments < anthers, glabrous. **Fruits** 6–12 mm in diameter, red to bluish black, with a bloom. **2*n***=48.

Subalpine forests and slopes. Flowering Jul–Aug. 800–2100 m. Casc. CA, ID, WA; north to British Columbia. Native.

See also *V. caespitosum*.

Vaccinium macrocarpon Aiton (p. 638)

Cranberry

Shrubs evergreen, 3–15 dm; stems vine-like, trailing, sometimes ascending; twigs slender, round in cross section to angled, light red to brown, often peeling, glabrous to minutely puberulent; not rhizomatous. **Leaves** ovate to elliptic, 5–18 × 2–7 mm, lightly glaucous abaxially, green to dark green adaxially, stiff, margins entire, tips rounded to subacute, surfaces abaxially glaucous, adaxially glabrous. **Inflorescences** solitary flowers in axils; bracts 1–2 mm wide. **Flowers** sepals 4, calyces shallowly lobed, deltate, red, glaucous; petals 4, 5–12 mm, glabrous, whitish pink to pink, lobes deeply parted; filaments 25–33% as long as anthers, pubescent. **Fruits** 7–15 mm in diameter, pink to red, with or without a bloom. **2*n***=24.

Wet coastal habitats. Flowering May–Jul. 0–50 m. Est. WA; north to British Columbia, eastern North America; Europe. Exotic.

Native to eastern North America, *Vaccinium macrocarpon* has escaped cultivation and naturalized elsewhere. This species appears closely related to *V. oxycoccos*. Whether these species will hybridize in the wild remains to be investigated.

Vaccinium membranaceum Douglas ex Torrey

Thin-leaved huckleberry, black huckleberry

Vaccinium coccineum Piper, *Vaccinium globulare* Rydb.

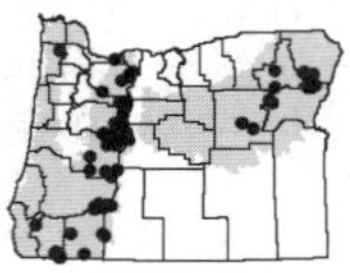

Shrubs deciduous, 5–20 dm; twigs round in cross section to angled, yellow-green to reddish green, thin, flexible, glabrous to puberulent; bark grayish on older branches; not rhizomatous. **Leaves** elliptic to ovate, 30–50 × 10–25 mm, light green abaxially, turning red to purple with age, thin, flexible, margins sharply serrate from tips to near base, tips acute, surfaces abaxially glandular, adaxially glabrous. **Flowers** sepals 5, calyces shallowly lobed; petals 5, corollas shallowly lobed, globose to urceolate, 3–6 × 5–7 mm, white to pink, glaucous; filaments glabrous. **Fruits** 6–12 mm in diameter, dark red to black, without a bloom. **2*n***=48.

Mid- to higher elevation coniferous forests (rarely at lower elevations). Flowering Apr–Aug. 1000–2100 m. BW, Casc, CR, Lava, Sisk. CA, ID, WA; north to British Columbia, east to Quebec, southeast to AZ. Native.

Renowned for the flavor of its berries, *V. membranaceum* is one of the most extensively harvested *Vaccinium* species in the wild. This species bears the state fruit of Idaho.

Vaccinium myrtillus Linnaeus

Whortleberry

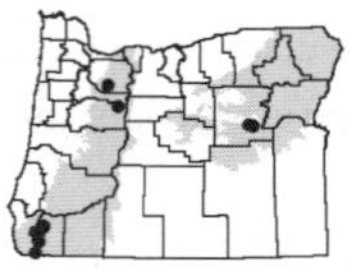

Shrubs deciduous, up to 4 dm; branching habit open, rarely weakly "broom-like"; twigs strongly angled, green to greenish brown, sparsely puberulent along grooves, sometimes glabrous; rhizomatous. **Leaves** elliptic to ovate, 11–20(30) × 5–10 mm, light green abaxially, green adaxially, thin, flexible, margins sharply serrate from tips to near base, prominently veined, tips acute, surfaces abaxially glandular, adaxially glabrous. **Inflorescences** solitary flowers in axils. **Flowers** sepals 5, calyces shallowly lobed, glabrous; petals 5, corollas ovoid-urceolate to globose, 3–7 × 4–7 mm, white to pink, glaucous; filaments glabrous. **Fruits** 4–9 mm in diameter, dark red to bluish black, without a bloom. **2*n***=24, 48.

Wet mid- to high elevation areas. Flowering Apr–Jul. 1100–2100 m. BW, Casc, Sisk. ID, NV, WA; northeast to Alberta and southeast to NM, Greenland; Asia, Europe. Native.

Morphologically, *Vaccinium myrtillus* is similar to *V. scoparium*, and at higher elevations their ranges overlap. *Vaccinium myrtillus* differs from *V. scoparium* by its generally larger leaves and flowers, often sparsely puberulent twigs, and its broom-like habit.

Vaccinium ovalifolium Smith (p. 638)

Early blueberry, oval-leaf blueberry

Vaccinium alaskaense Howell

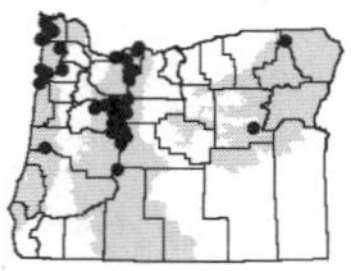

Shrubs deciduous, 3–35 dm, clumped; stems round in cross section; twigs slightly angled or with sharp ridges, yellow-green to brown, glabrous to minutely puberulent; not rhizomatous. **Leaves** ovate to elliptic, 25–40 × 15–20 mm, light green to bluish green, thin, flexible, margins entire to minutely serrate on lower 50%, tips usually rounded or sometimes acute, surfaces usually glaucous, adaxially occasionally glabrous. **Inflorescences** solitary flowers in axils. **Flowers** sepals 5, glabrous or sometimes glaucous, calyx lobes minute or absent; petals 5, corollas subglobose to urceolate, 4–7 × 3–5 mm, pink, glaucous; filaments < anthers, glabrous. **Fruits** 6–10 mm in diameter, blue to black, with a bloom. **2*n***=24.

Moist mid-elevations, occasionally coastal. Flowering Mar–Jul. 0–2100 m. BW, Casc, CR, ECas, Est. North to AK and east to Prince Edward Island; Asia. Native.

Vaccinium ovatum Pursh

Evergreen huckleberry, California huckleberry

Vaccinium ovatum Pursh var. *saporosum* Jeps.

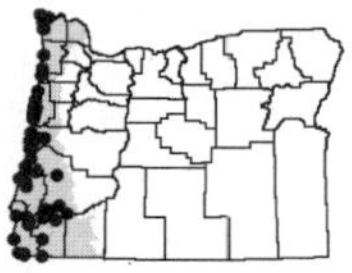

Shrubs evergreen, 4–35 dm, erect; twigs round in cross section, reddish brown to grayish, puberulent; rhizomatous. **Leaves** narrowly ovate, 15–40 × 8–20 mm, light green abaxially, green adaxially, thick, leathery, firm, margins sharply serrate from tips to 67% or near base, tips acute to acuminate, surfaces abaxially sparsely glandular with blackish hairs, adaxially glabrous. **Inflorescences** terminal racemes. **Flowers** sepals 5, calyx lobes fused at base, deltate, red, glaucous; petals 5, corollas campanulate to urceolate, 2.5–7 × 1.5–5 mm, pink, glabrous; filaments pubescent. **Fruits** 6–9 mm in diameter, blue to black, sometimes with a bloom. **2*n***=24.

Coastal and occasionally inland coniferous forests. Flowering Mar–Jun. 0–800 m. CR, Est, Sisk. CA, WA; north to British Columbia, south to Mexico. Native.

The distribution of *V. ovatum* populations is interesting from a biogeographic perspective. In British Columbia, Washington, northern Oregon, and southern California, *V. ovatum* is generally found within several kilometers of the coast. However, in northern California and Mexico, populations are found up to 50 kilometers inland and in southern Oregon up to 100 kilometers inland.

Vaccinium oxycoccos Linnaeus

Small cranberry, wild cranberry, bog cranberry, swamp cranberry

Oxycoccus palustris Pers. var. *intermedium* (A. Gray) Howell, *Vaccinium oxycoccos* L. var. *intermedium* A. Gray, *Vaccinium oxycoccos* L. var. *oxycoccos*

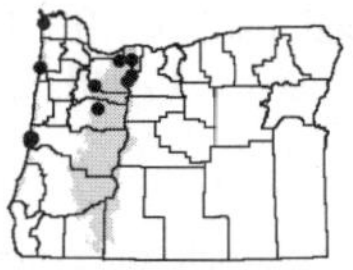

Vine-like creeping shrubs evergreen, 1–5 dm; twigs slender, round in cross section to angled, red to brown, puberulent or sometimes glabrous, often peeling; not rhizomatous. **Leaves** ovate to elliptic, 3–10 × 1–5 mm, strongly glaucous abaxially, green to dark green adaxially, stiff, margins entire, tips rounded to subacute, surfaces abaxially glaucous, adaxially glabrous. **Inflorescences** solitary flowers in axils; bracts < 1 mm wide. **Flowers** sepals 4, calyx lobes shallow, deltate, red, glaucous; petals 4, corollas 4.5–7.5 mm, whitish pink to pink, glabrous, lobes deeply parted and reflexed; filaments 50% as long as anthers, pubescent. **Fruits** 4–10 mm in diameter, dark red to bluish black, with a bloom. **2*n***=24, 48, 72.

Wet mid- to higher elevation bogs and wetlands, often in sphagnum, occasionally coastal. Flowering Apr–Jul. 0–1400 m. Casc, Est. CA, ID, WA; north to AK, east across Canada to northeastern states; circumboreal. Native.

The vine-like habit and tiny reflexed flowers of *V. oxycoccos* readily distinguish it from all other Oregon *Vaccinium* species, with the exception of escaped populations of cultivated cranberry (*V. macrocarpon*). The latter, however, has larger leaves, flowers, and fruits as well as lighter-colored leaves.

Vaccinium parvifolium Smith

Red huckleberry

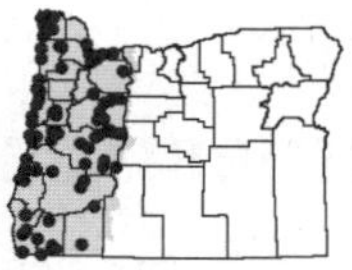

Shrubs deciduous to semi-evergreen, 10–50 dm, erect; stems sharply angled; twigs sharply angled, green to brown, glabrous to minutely puberulent; not rhizomatous. **Leaves** ovate to oblong-elliptic, 10–30 × 4–18 mm, thin, margins usually entire, rarely minutely serrate, tips usually rounded or sometimes acute, surfaces abaxially glabrous to somewhat puberulent, adaxially glabrous but sometimes minutely puberulent along midrib. **Inflorescences** solitary flowers in axils. **Flowers** sepals 5, calyx lobes green, glabrous; petals 5, corollas globose to urceolate, 4–6 × 3–5 mm, greenish yellow to pink, glaucous; filaments ≤ anthers, glabrous. **Fruits** 6–10 mm in diameter, red, with a faint bloom. **2*n***=24.

Coniferous forests, often on decaying wood. Flowering Mar–Jul. 0–1600 m. Casc, CR, Est, Sisk, WV. CA, WA; north to AK. Native.

Some old, semi-evergreen leaves of *V. parvifolium* may be minutely serrate and shiny, while most leaves are entire and dull. Thus, when using the above key, care must be taken to examine younger leaves for proper identification.

Vaccinium scoparium Leiberg ex Coville

Grouseberry, grouse whortleberry

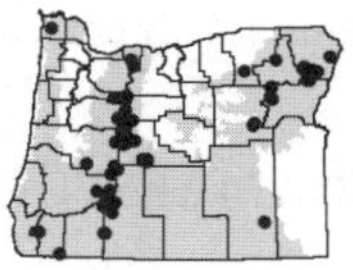

Shrubs deciduous, up to 4 dm; branching habit crowded, giving a "broom-like" appearance; twigs strongly angled, green to yellowish green, glabrous; rhizomatous. **Leaves** elliptic to ovate-lanceolate, 7–15 × 4–14 mm, light green abaxially, green adaxially, thin, flexible, margins finely serrate from tips to near base, prominently veined, tips acute, surfaces abaxially often glandular, adaxially glabrous. **Inflorescences** solitary flowers in axils. **Flowers** sepals 5, calyces very shallowly lobed, glabrous; petals 5, corollas urceolate to globose, 3–4 × 3–5 mm, pink, glaucous; filaments glabrous. **Fruits** 3–6 mm in diameter, bright red to bluish purple, without a bloom. **2*n***=24.

Subalpine and alpine meadows, slopes and pine forests. Flowering May–Aug. 1000–2700 m. BR, BW, Casc, CR, ECas, Sisk. CA, ID, NV, WA; northeast to Alberta, east to SD, southeast to NM. Native.

See also *V. myrtillus*.

Vaccinium uliginosum Linnaeus (p. 638)

Bog blueberry, bog bilberry

Vaccinium occidentale A. Gray, *Vaccinium uliginosum* L. var. *alpinum* Bigelow, *Vaccinium uliginosum* L. ssp. *occidentale* (A. Gray) Hultén, *Vaccinium uliginosum* L. var. *uliginosum*

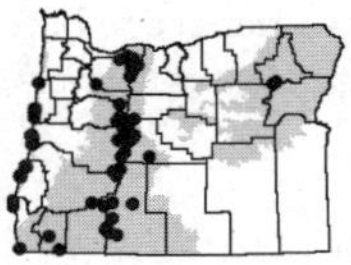

Shrubs deciduous, 0.2–0.7 m, in dense mats; stems round in cross section; twigs round in cross section to angled, green to grayish brown, glabrous to finely puberulent, older twigs slightly peeling; rhizomatous. **Leaves** ovate to elliptic, 9–29 × 4–18 mm, thin or thick at maturity, margins entire, tips rounded to acute, surfaces abaxially glabrous or glaucous, adaxially glabrous. **Inflorescences** racemes, 2–4-flowered, occasionally 1-flowered. **Flowers** distinct, sepals 4 or 5, usually glaucous; corollas urceolate, 4–6 × 2–5 mm, white to pink, glaucous; petals 4 or 5; filaments ≤ anthers, glabrous. **Fruits** 4–9 mm in diameter, blue to bluish black, with a faint bloom. **2*n***=24, 48, 72.

Wet mid-elevation montane and coastal habitats. Flowering May–Sep. 0–2400 m. BW, Casc, ECas, Est, Sisk. CA, ID, NV, WA; northern US and Canada; Asia, Europe. Native.

Some authorities consider the coastal form of *V. uliginosum* a distinct species (*V. occidentale*). If recognized, this species would be identified by its distinctly reticulated leaf veins, thin leaves, and smaller fruits.

EUPHORBIACEAE Spurge family

Stephen C. Meyers

Shrubs, **subshrubs**, or **herbs** annual, biennial, or perennial, monoecious or dioecious; stems branched. **Leaves** alternate, subopposite, opposite or whorled, simple, margins entire or toothed; petioles present or not; stipules present or not. **Inflorescences** various and diverse, in some genera individual flowers reduced and grouped into a bisexual inflorescence resembling a single flower (cyathium); bracts present or not, sometimes petaloid in appearance. **Flowers** unisexual, sometimes with a corolla-like involucre; sepals 0–5; petals usually 0; stamens 1–many; ovaries superior, styles 3. **Fruits** capsules, dry. **Seeds** 1 per chamber.

Worldwide. ~300 genera; 3 genera treated in *Flora*.

Mercurialis annua has been collected on ballast in Portland and rarely as a garden waif.

The plants and seeds of many species in Euphorbiaceae are toxic and highly poisonous to animals. The milky sap of *Chamaesyce* and *Euphorbia* species contains chemicals that help protect the plants from herbivory and can be very irritating to skin and eyes.

Key to Euphorbiaceae genera:

1. Leaves densely, silvery, stellate hairy; flowers with a perianth; inflorescenses terminal or axillary cymes; plants without milky latex .... **CROTON**

1' Leaves not densely, silvery, stellate-hairy; flowers without a perianth; inflorescenses cyathiums (multiple individual flowers grouped into a bisexual inflorescence resembling a single flower); plants with milky latex.

  2. Leaves opposite, 3–15 mm; stems prostrate (but sometimes ascending with *C. serpillifolia*) .... **CHAMAESYCE**

  2' Lower leaves alternate (upper and/or leaves near flowers decumbent to ascending in *E. myrsinites*), leaves 10–150 mm; stems usually erect (decumbent to ascending with *E. myrsinites*) .... **EUPHORBIA**

CHAMAESYCE Sandmat, spurge

Herbs with milky latex; stems usually prostrate, but occasionally ascending in some species. **Leaves** cauline, opposite, surfaces glabrous to hairy; short-petiolate or sessile; stipules present. **Inflorescences** axillary, multiple individual flowers grouped into a cyathium, generally in 5 groups of 3–many staminate flowers around a pistillate flower; involucres bell-shaped, usually with 4 prominent glands, glands often with petal-like appendages. **Flowers** sepals 0; petals 0. **Staminate flowers**; stamens 1. **Pistillate flowers** 1, central; styles separate or fused at base.

Worldwide. ~250 species; 3 species treated in *Flora*.

At various times, some authorities have relegated *Chamaesyce* to a subgenus of *Euphorbia*, while others have considered the genus paraphyletic and simply merged all taxa into *Euphorbia*. In view of the vacillating taxonomy within this group, we are recognizing both genera.

Single documentations of *Chamaesyce prostrata* occur from The Dalles, Wasco County, and Ashland, Jackson County, but the species does not appear established in Oregon. If encountered again, it can be distinguished from other Oregon members of this genus by its combination of hairy involucres and cross-ridged seeds.

Likewise, *Chamaesyce serpens* has been reported once from a housing development in Talent, Jackson County. It can be identified by its stipules, which are fused into wide membranous scales.

Key to *Chamaesyce* species:

1. Stems, involucres, and fruits hairy .... ***C. maculata***

1' Stems, involucres, and fruits glabrous.

  2. Seeds cross-ridged .... ***C. glyptosperma***

  2' Seeds smooth to wrinkled .... ***C. serpillifolia***

Chamaesyce glyptosperma (Engelmann) Small

Ridge-seeded spurge, corrugate-seeded spurge, ribseed sandmat

Euphorbia glyptosperma Engelm.

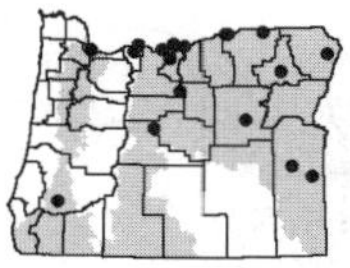

Plants annual; stems prostrate, 4–40 cm, glabrous; taprooted. **Leaves** oblanceolate to elliptic or ovate, 3–10 mm, occasionally with central red spot, margins entire to shallowly toothed, tips rounded, surfaces glabrous; petioles 0.3–1 mm; stipules linear, often dissected, 1–1.5 mm. **Inflorescences** involucres 0.5–1 mm, glabrous; glands elliptic, gland appendages white. **Fruits** ovoid, 1.5–2 mm, glabrous. **Seeds** ovoid, 1–1.5 mm, cross-ridged.

Dry areas, roadsides. Flowering Jun–Sep. 0–1600 m. BW, Col, ECas, Lava, Owy, Sisk, WV. CA, ID, NV, WA; southern Canada and most of US except southeastern states, Mexico. Native.

Chamaesyce maculata (Linnaeus) Small

Spotted spurge, spotted sandmat, prostrate spurge

Chamaesyce supina (Raf.) Moldenke, *Euphorbia maculata* L., *Euphorbia supina* Raf.

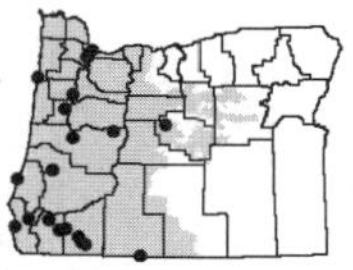

Plants annual; stems prostrate, 5–50 cm, hairy; taprooted. **Leaves** linear-oblong to ovate-oblong, 5–15 mm, often with a central red spot, margins entire to shallowly toothed, tips rounded to acute, surfaces glabrous or hairy; petioles 0.5–1.5 mm; stipules slender, fringed, 0.3–0.9 mm. **Inflorescences** involucres 0.5–1 mm, hairy; glands elliptic, gland appendages white to pink. **Fruits** ovoid, 1–1.5 mm, hairy. **Seeds** ovoid to oblong, 0.7–1.3 mm, usually wrinkled, occasionally smooth.

Sidewalks, cracks in pavement, roadsides, disturbed areas. Flowering Jun–Sep. 0–1300 m. Casc, CR, ECas, Est, Lava, Sisk, WV. CA, ID, NV, WA; throughout North America except northern Canada. Native.

Some sources speculate this species was originally native to eastern North America and was later introduced in other parts of the continent. Without definitive proof, however, we currently consider this species native.

Chamaesyce serpillifolia (Linnaeus) Small (p. 643)

Thyme-leaved spurge, thymeleaf sandmat

Chamaesyce serpillifolia (L.) Small ssp. *serpillifolia, Euphorbia serpillifolia* Pers.

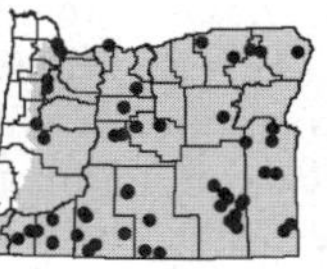

Plants annual; stems prostrate to ascending, 5–40 cm, glabrous, sometimes glaucous; taprooted. **Leaves** oblanceolate to oblong, 3–15 mm, occasionally with a central red spot, margins entire to shallowly toothed, tips rounded, surfaces glabrous; petioles 0.5–1.5 mm; stipules slender and entire to linear and dissected, 1–1.5 mm. **Inflorescences** involucres 0.5–1 mm, glabrous; glands elliptic, gland appendages white. **Fruits** spherical, 1–2 mm, glabrous. **Seeds** ovoid, 0.9–1.3 mm, smooth to wrinkled.

Usually dry places but sometimes in wet areas, roadsides, disturbed areas. Flowering Jun–Sep. 0–1900 m. BR, BW, Casc, Col, ECas, Lava, Owy, Sisk, WV. CA, ID, NV, WA; southern Canada, most of US except southeast and parts of northeast, Mexico. Native.

In California, some plants of this species have hairy stems, involucres, and fruits. These plants are accepted by some authorities as *C. s.* ssp. *hirtula* (ours, in turn, would be *C. s.* ssp. *serpillifolia*). Given the already small morphological differences between our species of *Chamaesyce*, we have chosen not to recognize the subspecies.

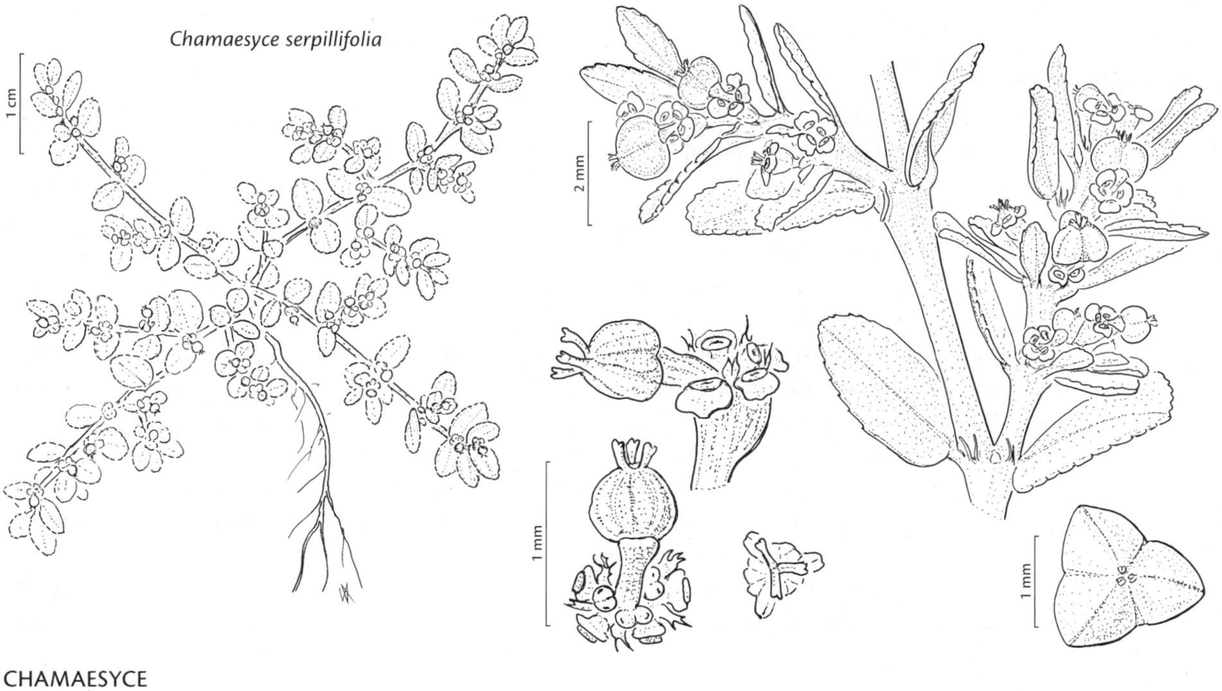

CHAMAESYCE

CROTON Croton

Herbs or **shrubs** without milky latex but often with colored or clear sap; stems spreading to erect. **Leaves** cauline, alternate or rarely subopposite, densely silvery, surfaces stellate-hairy; petiolate; stipules greatly reduced. **Flowers** perianth present; sepals 5; petals 0. **Staminate flowers** filaments free; pedicelled. **Pistillate flowers** short-pedicelled or 0; styles distinct, 2-lobed.

Worldwide. ~1000 species; 1 species treated in *Flora*.

Older floras may refer to this genus as *Eremocarpus*.

Croton setigerus Hooker (p. 644)

Turkey mullein, dove weed

Eremocarpus setigerus (Hook.) Benth.

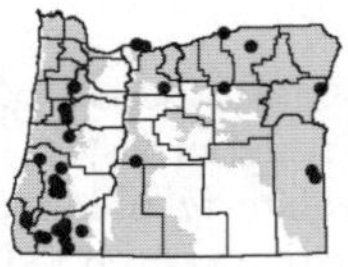

Herbs annual, 1–2 dm, forming mounds or tall mats up to 8 dm wide. **Leaves** strongly scented, ovate to elliptic, 1–5 cm; petioles 0.5–6 cm. **Inflorescences** terminal or axillary cymes. **Staminate flowers** terminal, 1–2 mm; stamens (5)6(9), exserted; pedicel 2–3 mm. **Pistillate flowers** axillary, 1–2 mm. **Fruits** 3–6 mm. **Seeds** 1, elliptic, 3–4 mm.

Dry open areas and meadows. Flowering Jun–Sep. 0–1300 m. BW, Col, CR, ECas, Owy, Sisk, WV. CA, ID, NV, WA; southeast to AZ. Native.

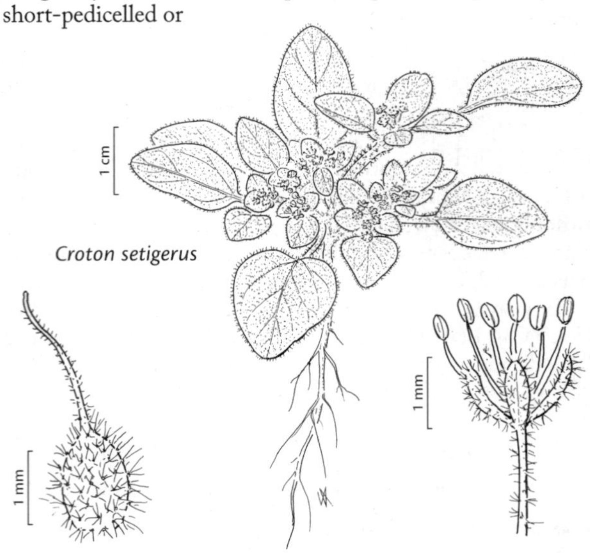

CROTON

EUPHORBIA Spurge

Herbs or **subshrubs** with milky latex; stems usually erect but occasionally prostrate in some species. **Leaves** cauline, alternate (upper leaves sometimes whorled) or occasionally opposite, 1–15 cm, surfaces usually glabrous; petiolate or sessile; stipules 0 or greatly reduced. **Inflorescences** multiple individual flowers grouped into a cyathium of 5 groups of 5–many staminate flowers around a pistillate flower; involucres usually with 4–5 prominent glands, glands often with horn-like projections. **Flowers** sepals 0; petals 0. **Staminate flowers**; stamens 1. **Pistillate flowers** 1, central; styles fused at base.

Worldwide. ~1500 species; 9 species treated in *Flora*.

Euphorbia dentata has been reported once in Sherman County (2008). If encountered again, it can be recognized by its dentate leaves.

Older floras may include current members of *Chamaesyce* in this genus.

Key to *Euphorbia* species:

1. All leaves opposite . . . . **E. lathyris**

1' Leaves alternate (at least toward base), leaves toward inflorescence may be opposite or whorled.

  2. Leaves densely crowded on stems . . . . **E. myrsinites**

  2' Leaves not densely crowded on stems.

    3. Leaves toothed or serrulate (at least in distal 25%).

      4. Stems moderately to densely hairy; plants perennial . . . . **E. oblongata**

      4' Stems glabrous to sparsely hairy; plants annual.

        5. Involucres 1–1.5 mm; styles divided less than halfway to base. . . . . **E. helioscopia**

        5' Involucres 2–3 mm; styles divided nearly to base. . . . . **E. spathulata**

    3' Leaves entire, wavy or slightly crenate.

      6. Leaves elliptic to spatulate; plants annual or biennial.

        7. Lower leaves petiolate; seeds 1–1.5 mm, with rows of large pits . . . . **E. peplus**

        7' Lower leaves sessile or with minute petioles; seeds 1.5–2.5 mm, with reticulate pitting . . . . **E. crenulata**

      6' Leaves linear to narrowly oblanceolate; plants perennial.

        8. Leaves linear, 1–3 mm wide; plants 1.5–3(4) dm . . . . **E. cyparissias**

        8' Leaves narrowly oblong to narrowly oblanceolate, 3–8(10) mm wide; plants 3–7 dm . . . . **E. esula**

Euphorbia crenulata Engelmann

Western wood spurge

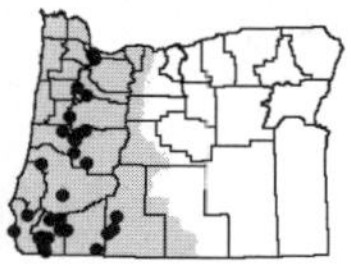

Herbs or **subshrubs** annual or biennial, 1.5–5 dm; stems erect, glabrous. **Leaves** elliptic to spatulate, 10–35 mm × 4–12 mm, margins entire to slightly crenate, tips obtuse to acuminate, surfaces glabrous; lower leaves alternate, sessile or with minute petioles; upper and floral leaves opposite or whorled, sessile. **Inflorescences** involucres bell-shaped, 2–3 mm, glands 2-horned. **Fruits** oblong, 3–3.5 mm. **Seeds** round to oblong, 1.5–2.5 mm, with reticulate pitting.

Dry areas. Flowering Apr–Jul. 0–1600 m. Casc, CR, ECas, Est, Sisk, WV. CA, NV; AZ, CO. Native.

Euphorbia crenulata greatly resembles the European native *E. peplus*. They differ only in seed pitting characteristics and lower leaf petioles. Their genetic affinity is worthy of study.

Euphorbia cyparissias Linnaeus

Cypress spurge

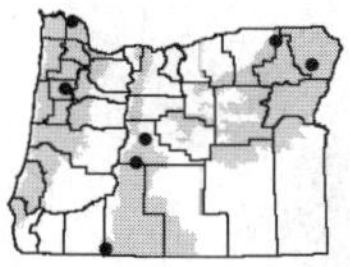

Herbs or **subshrubs** perennial, 1.5–3(4) dm; stems erect, glabrous. **Leaves** lower leaves alternate, sometimes appearing whorled, linear, 10–30 × 1–3 mm, margins entire, tips obtuse to acuminate, surfaces glabrous, sessile; floral leaves opposite or whorled, broadly ovate, 0.4–1 cm, margins entire to slightly crenate, sessile. **Inflorescences** involucres bell-shaped, 1–3 mm, glands 2-horned. **Fruits** oblong, 2.5–3 mm. **Seeds** ellipsoid, 1.5–2 mm, smooth. **2*n*** =20, 36, 42.

Disturbed areas. Flowering May–Aug. 0–1600 m. BW, CR, ECas, WV. CA, ID, WA; throughout North America; Asia, Europe. Exotic.

Euphorbia esula Linnaeus

Leafy spurge

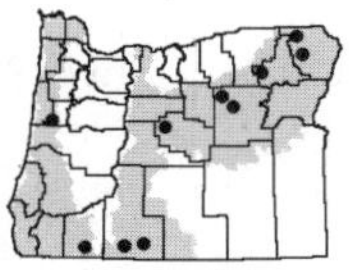

Herbs or **subshrubs** perennial, 3–7 dm; stems erect, glabrous and glaucous to sparsely pubescent. **Leaves** lower leaves alternate, narrowly oblong to narrowly oblanceolate, 30–90 × 3–8(10) mm, margins entire, tips obtuse to acuminate, surfaces glabrous, sessile or occasionally short-petiolate; floral leaves opposite or whorled, broadly ovate, 1–1.5 cm, margins entire to slightly crenate, sessile. **Inflorescences** involucres bell-shaped, 2–3 mm, glands 2-horned. **Fruits** oblong, 3–3.5 mm. **Seeds** ellipsoid, 2–2.5 mm, ± smooth. **2*n***=16, 20, 48, 60, 64.

Roadsides, disturbed areas. Flowering May–Jul. 0–1400 m. BW, CR, ECas, Lava, Sisk. CA, ID, NV, WA; throughout US except TX and southeastern states, Mexico; Asia, Europe. Exotic.

Euphorbia helioscopia Linnaeus

Wartweed, mad woman's milk

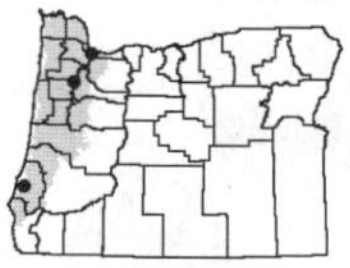

Herbs annual, 1–5 dm; stems,erect, glabrous to sparsely hairy. **Leaves** alternate (upper leaves may be opposite), obovate to spatulate, 10–30 × 5–15 mm, margins toothed in at least upper 25%, entire below, tips rounded, surfaces glabrous; sessile. **Inflorescences** involucres cone-shaped, 1–1.5 mm, glands not horned. **Flowers** styles divided less than halfway to base. **Fruits** round, 2.5–3 mm. **Seeds** ovoid, 1.5–2.5 mm, with reticulate pitting.

Disturbed areas. Flowering Apr–Jul. 0–200 m. CR, WV. CA, WA; scattered throughout North America; Asia, Europe. Exotic.

Euphorbia helioscopia is easily confused with *E. spathulata*. The former is an exotic species introduced from Eurasia, while *E. spathulata* is considered native. The most reliable characteristic separating these taxa is the size of the involucres.

Euphorbia lathyris Linnaeus

Caper spurge, mole plant, gopher plant

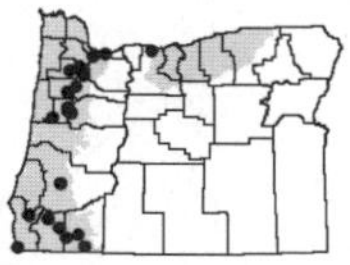

Herbs or **subshrubs** biennial, 1.5–12 dm; stems erect, glabrous, glaucous. **Leaves** opposite, narrowly oblong to ovate or subcordate, 30–150 × 5–30 mm, bases clasping, margins entire, tips acute, surfaces glabrous. **Inflorescences** involucres bell-shaped, 2.5–4 mm, glands 2-horned, horns rounded and short. **Fruits** spherical, 5–15 mm. **Seeds** round, 4–6 mm, wrinkled or with shallow reticulate markings.

Disturbed areas. Flowering May–Jul. 0–500 m. Col, CR, Est, Sisk, WV. CA, ID, WA; scattered across North America; Asia, Europe. Exotic.

Euphorbia lathyris is our only member of the genus to have only opposite leaves. It is sometimes planted in gardens based on the unsubstantiated belief it may deter moles and gophers.

Euphorbia myrsinites Linnaeus

Myrtle spurge

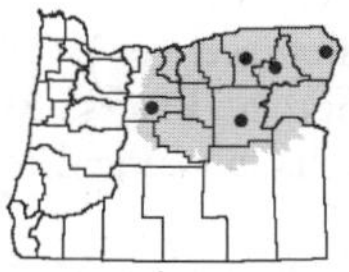

Herbs perennial, 1–4 dm; stems decumbent to ascending, glabrous. **Leaves** alternate (upper and floral leaves may be opposite or whorled), densely crowded on stem, obovate to obovate-oblong, 10–45 × 5–16 mm, margins entire to serrulate at least in upper 25%, tips mucronate, surfaces glabrous, glaucous; ± sessile. **Inflorescences** involucres cone- to bell-shaped, 1.5–3 mm, glands shortly 2-horned. **Fruits** round to oblong, 5–7mm. **Seeds** round to oblong, 3–4 mm, smooth or wrinkled.

Disturbed areas, escaped from nearby cultivation. Flowering Apr–Jun. 500–1000 m. BW, Col, Lava. CA, ID, WA; scattered in western US, MI, Ontario; Europe. Exotic.

The densely crowded leaves on the stems and decumbent to ascending habit easily distinguish *E. myrsinites* from all other members of this genus in Oregon.

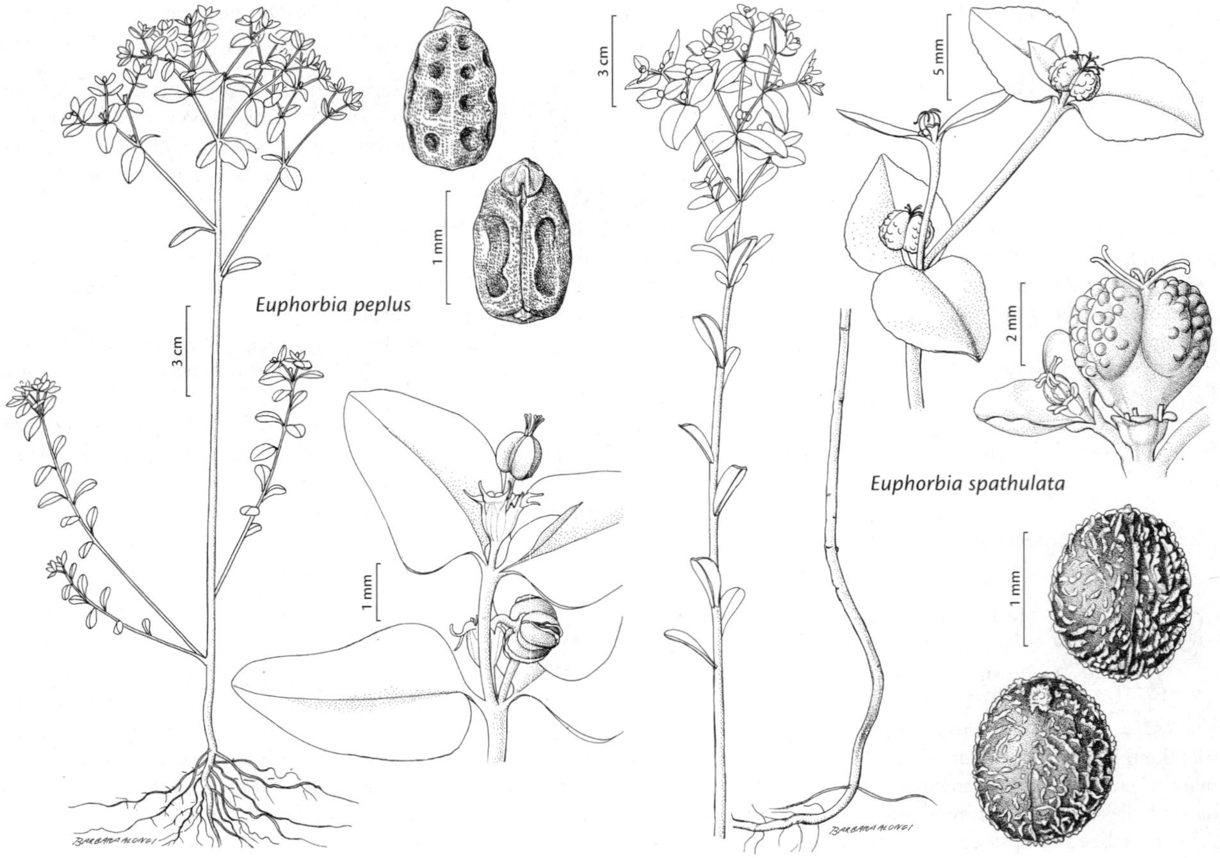

EUPHORBIA

Euphorbia oblongata Grisebach

Eggleaf spurge

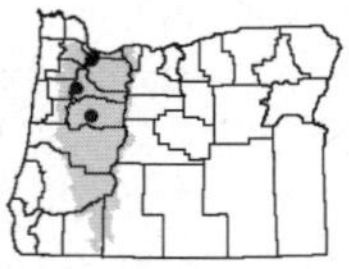

Herbs perennial, 5–8 dm; stems erect, moderate to densely hairy. **Leaves** alternate (upper and floral leaves generally opposite or whorled), oblong to narrowly obovate, 20–85 × 3–15 mm, margins serrulate at least in upper 33%, tips obtuse, surfaces glabrous. **Inflorescences** involucres bell-shaped, 1.5–3.5 mm, glands not horned. **Fruits** oblong, 3–5 mm. **Seeds** oblong, 2–3 mm, with shallow reticulate markings.

Disturbed areas. Flowering May–Sep. 50–300 m. Casc, WV. CA, WA; Europe. Exotic.

Euphorbia peplus Linnaeus (p. 646)

Petty spurge

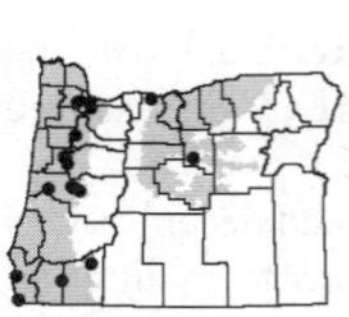

Herbs annual, 1–4(5) dm; stems erect, glabrous. **Leaves** lower leaves alternate, elliptic to obovate, 15–30 × 5–15 mm, margins entire, tips obtuse, surfaces glabrous, petiolate; upper and floral leaves opposite or whorled, round to rhombic-ovate, 0.5–1.5 cm, sessile or with short petioles. **Inflorescences** involucres bell-shaped, 1–1.5 mm, glands 2-horned. **Fruits** spherical, 1.5–3 mm. **Seeds** oblong, 1–1.5 mm, with rows of large pits.

Disturbed areas. Flowering Feb–Aug. 0–900 m. Col, CR, Est, Lava, Sisk, WV. CA, ID, NV, WA; scattered throughout North America; Europe. Exotic.

Euphorbia spathulata Lamarck (p. 646)

Prairie spurge

Euphorbia dictyosperma Fisch. & C.A. Mey.

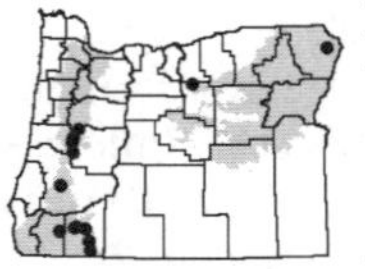

Herbs annual, 1–5(6) dm; stems, erect, glabrous. **Leaves** alternate (upper leaves may be opposite), obovate to spatulate, 10–30 × 5–15 mm, margins toothed at least in upper 25%, entire below, tips rounded, surfaces glabrous, sessile. **Inflorescences** involucres cone- to bell-shaped, 2–3 mm, glands not horned. **Flowers** styles divided nearly to base. **Fruits** round, 2–2.5 mm. **Seeds** ovoid to ellipsoid, 1–2 mm, with reticulate pitting.

Disturbed areas. Flowering Apr–Aug. 100–1000 m. BW, Sisk, WV. CA, ID, WA; throughout most of US, Ontario, Mexico. Native.

See also *E. helioscopia*.

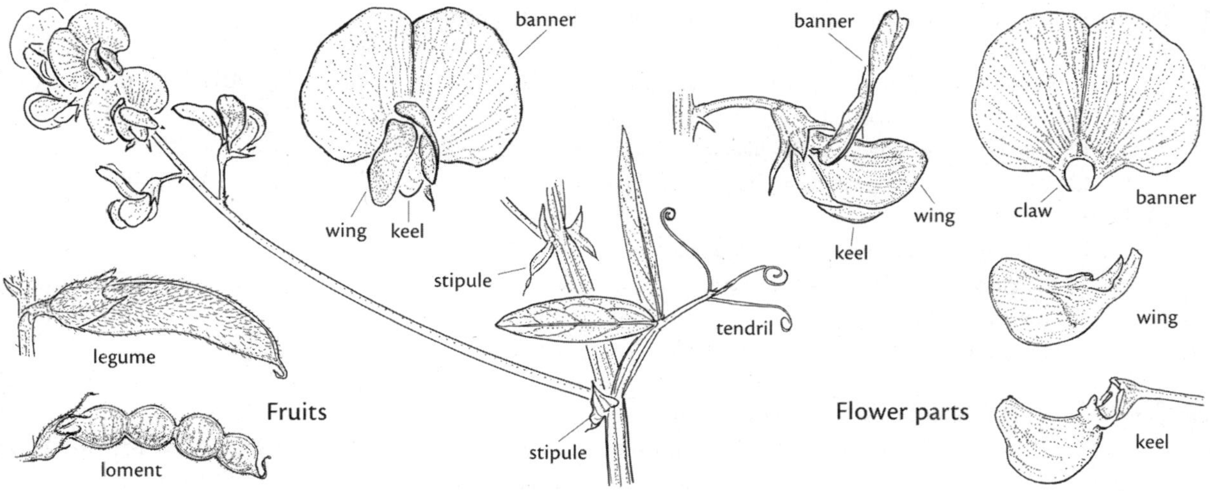

FABACEAE Pea family

Herbs, **shrubs**, or **trees** often with root nodules containing nitrogen-fixing bacteria. **Leaves** alternate; simple, 1(2)-pinnately compound to palmately compound or trifoliate; leaflets entire (toothed); stipules generally present. **Inflorescences** axillary or terminal, generally racemes, spikes, umbels or heads, or flowers 1–few in axils; bracts and bracteoles present or absent. **Flowers** bisexual; hypogynous or perigynous; (radially symmetric) bilaterally symmetric; sepals usually 5, distinct to connate, often bilabiate; petals (1)5, usually papilionaceous, with uppermost petal (banner) generally largest and reflexed, 2 lateral petals (wings) generally horizontal, generally enclosing lowermost and innermost, partially connate pair of petals (keel) which enclose stamens and pistil; stamens (5)10(many), distinct, or monadelphous (all 10 fused) or diadelphous (9 fused, 1 distinct); pistils 1, 1-carpellate, ovaries superior, placentation marginal, ovules 1–many, styles 1, stigmas 1. **Fruits** generally legumes (pods), dehiscent along 1 or both sutures, or loments and breaking crosswise into 1-seeded segments, or indehiscent. **Seeds** 1–many; often reniform, generally hard and smooth.

Worldwide. ~740 genera; 32 genera treated in *Flora*.

Fabaceae, formerly known as Leguminosae, is the third largest family of flowering plants. It is traditionally classified into three subfamilies (sometimes treated as separate families): Caesalpinioideae, Mimosoideae, and Papilionoideae (Faboideae). Caesalpinioideae has flowers with five or ten distinct stamens and five distinct petals; Oregon has only the genus *Cercis*. Mimosoideae has radially symmetric flowers with many stamens and five distinct petals; Oregon has only the genus *Acacia*. Papilionoideae has bilaterally symmetric (papilionaceous) flowers with typically ten stamens that are either monadelphous or diadelphous; this is the largest subfamily in Oregon.

Key to Fabaceae genera:

1. Plants woody, trees or shrubs.
    2. Leaves palmately compound . . . . **LUPINUS** (in part)
    2' Leaves simple, trifoliate, pinnately compound, or absent at flowering time.
        3. Leaves pinnately or 2-pinnately compound.
            4. Leaves 2-pinnately compound; stamens many . . . . **ACACIA**
            4' Leaves pinnately compound; stamens 10.
                5. Corollas reduced to single banner petal, < 7 mm; fruits 1(2)-seeded . . . . **AMORPHA**
                5' Corollas with all 5 petals, ≥ 15 mm; fruits with ≤ 10 seeds . . . . **ROBINIA**
        3' Leaves simple, trifoliate, or absent at flowering time.
            6. Plants armed, spiny . . . . **ULEX**
            6' Plants unarmed, not spiny.
                7. Corollas pink to purple; leaves cordate to reniform . . . . **CERCIS**
                7' Corollas yellow or white; leaves not cordate or reniform.
                    8. Calyces deeply split on upper side; stems ± leafless at flowering . . . . **SPARTIUM**
                    8' Calyces not deeply split on upper side; stems usually leafy at flowering.
                        9. Styles abruptly curved medially or gently curved throughout; calyces bilateral, upper lips minutely 2-lobed . . . . **CYTISUS**
                        9' Styles abruptly bent upwards below tip; calyces weakly if at all bilateral, upper lips strongly 2-lobed . . . . **GENISTA**

1' Plants herbaceous, sometimes suffrutescent but upper part of stems not woody.
  10. Leaves simple or 0.
    11. Inflorescences 7–35-flowered . . . . **ASTRAGALUS** (in part)
    11' Inflorescences 1(2)-flowered.
      12. Leaves absent; stipules leaflet-like; tendrils present . . . . **LATHYRUS** (in part)
      12' Leaves present; stipules gland-like or sometimes 0; tendrils absent . . . . **ACMISPON** (in part)
  10' Leaves compound.
    13. Leaflets 2 or trifoliate or palmately compound.
      14. Leaflets 2; tendrils generally present . . . . **LATHYRUS** (in part)
      14' Leaflets ≥ 3; tendrils absent.
        15. Leaflets toothed or wavy-margined.
          16. Fruits spirally coiled or reniform . . . . **MEDICAGO**
          16' Fruits not spirally coiled or reniform.
            17. Corollas persistent after anthesis; plants often > 5 dm . . . . **MELILOTUS**
            17' Corollas not persistent after anthesis; plants usually < 5 dm . . . . **TRIFOLIUM** (in part)
        15' Leaflets entire.
          18. Flowers in spikes, heads, umbels, 1–3-flowered clusters, or solitary in axils.
            19. Stipules conspicuous . . . . **TRIFOLIUM** (in part)
            19' Stipules gland-like or 0 (lowest pair of leaflets sometimes stipular in position).
              20. Leaflets usually 5, with lower 2 basal on petiole, others terminally ternate . . . **LOTUS** (in part)
              20' Leaflets 1–5, not arranged as above . . . . **ACMISPON** (in part)
          18' Flowers in racemes or pseudoracemes.
            21. Leaves palmately compound with > 3 leaflets . . . . **LUPINUS** (in part)
            21' Leaves usually trifoliate.
              22. Herbage not gland-dotted; stamens distinct; stipules leaf-like . . . . **THERMOPSIS**
              22' Herbage dotted with small, sessile glands; stamens monadelphous or diadelphous; stipules not leaf-like.
                23. Stamens monadelphous; corollas cream to yellowish . . . . **RUPERTIA**
                23' Stamens diadelphous; corollas blue to purple or white or bicolored . . . . **LADEANIA**
    13' Leaves pinnately compound.
      24. Leaves even-pinnate.
        25. Styles longitudinally folded; stipules generally > leaflets . . . . **PISUM**
        25' Styles not longitudinally folded; stipules sometimes ≈ leaflets.
          26. Styles terete, flattened abaxially or adaxially, not twisted, evenly pubescent just below stigma and/or abaxially tufted . . . . **VICIA**
          26' Styles abaxially compressed, sometimes twisted, adaxially pubescent . . . . **LATHYRUS** (in part)
      24' Leaves odd-pinnate.
        27. Leaves glandular.
          28. Flowers not papilionaceous; fruits without hooked bristles . . . . **DALEA**
          28' Flowers papilionaceous; fruits densely covered with hooked bristles, cocklebur-like . . . . **GLYCYRRHIZA**
        27' Leaves not glandular.
          29. Inflorescences umbels, 1–3-flowered clusters, solitary flowers or axillary heads.
            30. Stipules gland-like or 0; leaflets 1–5.
              31. Leaflets usually 5, with lower 2 basal on petiole, others terminally ternate . . . **LOTUS** (in part)
              31' Leaflets 1–5, not arranged as above . . . . **ACMISPON** (in part)
            30' Stipules not gland-like; leaflets 5–37.
              32. Fruits not loments; leaves subsessile to petiolate . . . . **HOSACKIA**
              32' Fruits loments; leaves sessile, rarely subsessile.
                33. Plants annual, pubescent; styles < 1 mm; corollas 3–8 mm . . . . **ORNITHOPUS**
                33' Plants perennial, glabrous or glabrate; styles 4–6 mm; corollas (8)10–15 mm . . . . **SECURIGERA**

29' Inflorescences racemes or spikes, occasionally subumbels.

34. Fruit margins with toothed wing and a few short, spiny teeth near tip, strongly reticulate-veined; flowers pinkish lavender . . . . **ONOBRYCHIS**

34' Fruit margins without toothed wing, not spiny; flower colors various.

35. Stamens not fused . . . . **SOPHORA**

35' Stamens monadelphous or diadelphous.

36. Fruits loments; corolla wings < keels . . . . **HEDYSARUM**

36' Fruits not loments; corolla wings > or subequal to keels.

37. Stamens monadelphous; fruits with oblique furrow on each face . . . . **GALEGA**

37' Stamens diadelphous; fruits without oblique furrow on each face.

38. Styles adaxially pubescent.

39. Flowers orange-red or brick red; pods bladdery-inflated, stipitate . . . . **SPHAEROPHYSA**

39' Flowers not orange-red or brick red; pods not bladdery-inflated, not stipitate . . . . **LATHYRUS** (in part)

38' Styles glabrous.

40. Keel petals abruptly narrowed to beak-like point; plants generally acaulescent; pods with ventral (seed-bearing) suture generally intruded forming a partial or complete septum . . . . **OXYTROPIS**

40' Keel petals not abruptly beaked; plants generally caulescent; pods variable but ventral suture not intruded, or less so than dorsal suture . . . **ASTRAGALUS** (in part)

ACACIA Wattle

Cameron J. Ross

Trees or **shrubs** evergreen. **Leaves** alternate, 2-pinnately compound, basal portion swollen; petioles present. **Inflorescences** axillary raceme- or panicle-like arrays of compact, globose heads with 13–42 flowers. **Flowers** small and inconspicuous; sepals 4–5; petals 4–5; stamens many, exserted; each flower subtended by an inconspicuous peltate bracteole. **Fruits** dehiscent, linear to oblong, flat to cylindrical, 2-valved; funicles enlarged to form fleshy arils under seeds. **Seeds** ovoid.

Tropical and subtropical areas. ~960 species; 1 species treated in *Flora.*

Acacia dealbata Link (p. 649)

Silver wattle, blue wattle

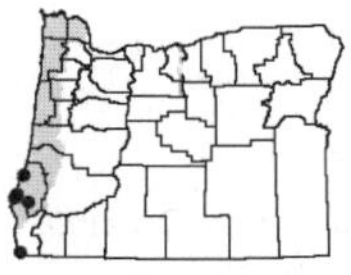

Plants up to 30 m. **Stems** canescent. **Leaves** evenly pinnate, up to 16 cm, margins pubescent; primary leaflets in 6–30 pairs with raised glands on axis at the base of each pair, secondary leaflets in 15–70 pairs, linear, 2–5 × 0.4–0.8 mm; petioles 7–20(25) mm, canescent. **Inflorescences** raceme- or panicle-like arrays of (5)10–30 heads; bracts 2, canescent. **Flowers** pale to bright yellow. **Fruits** 2–9(12) cm × 6–14 mm, flat, silver-blue, glabrous; arils pale yellow.

Disturbed areas. Flowering Jan–Apr. 0–50 m. CR, Est. CA; Australia. Exotic.

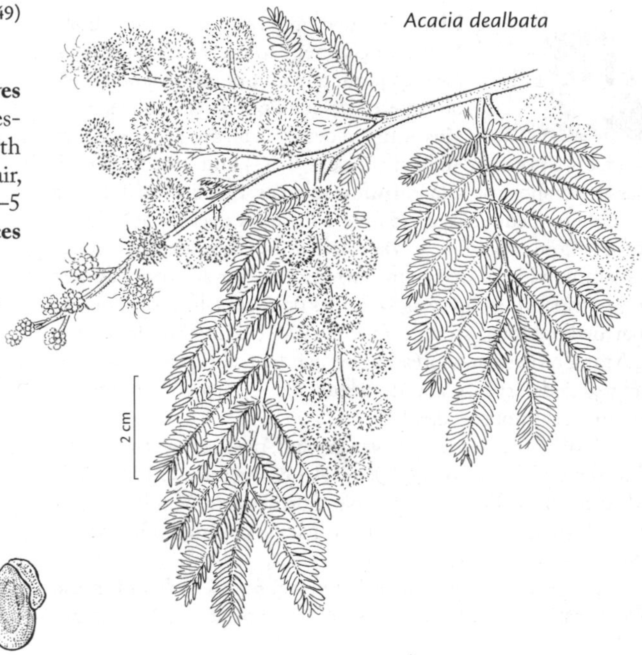

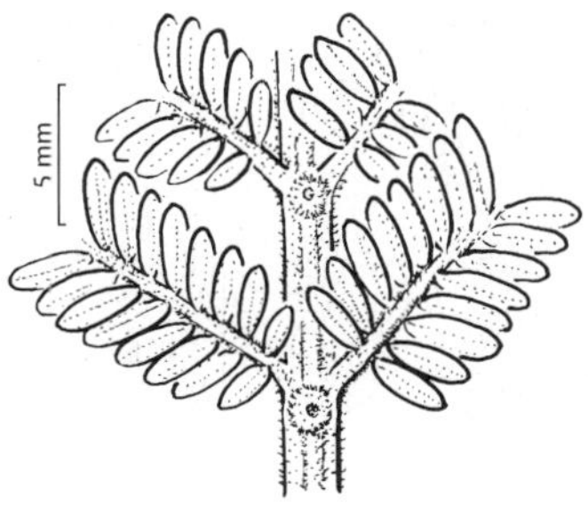

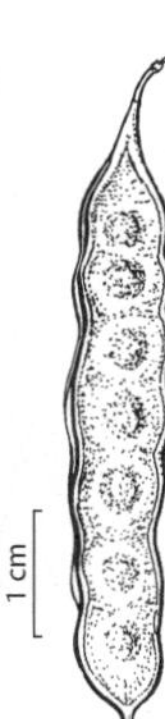

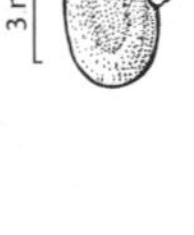

ACACIA

ACMISPON Deervetch

Richard R. Halse

Herbs annual or perennial, usually cespitose. **Stems** decumbent to ascending or erect. **Leaves** alternate; odd- or even-pinnately, palmately, or subpalmately compound, margins usually entire, sometimes denticulate, surfaces usually pubescent; stipules gland-like, sometimes 0. **Inflorescences** 1–12-flowered, axillary umbels or solitary flowers; bracts resembling leaflets, or 0. **Flowers** papilionaceous; calyces campanulate, lobes 5; corollas yellow, cream, greenish white, pinkish red, often marked with red, darkening with age; stamens 10, diadelphous, anthers basifixed; stigmas terminal. **Fruits** linear to oblong or ovoid, straight or curved, dehiscent or indehiscent, glabrous or pubescent. **Seeds** 1–10, oblong, ovoid, elliptic, lenticular, or reniform.

Central America, North America, South America. 34 species; 6 species treated in *Flora*.

Previous floras of Oregon and the Pacific Northwest have included *Acmispon* within the genus *Lotus*. However, recent studies have shown that it is distinct from the Eurasian *Lotus* and North American *Hosackia*.

Key to *Acmispon* species:

1. Plants perennial; inflorescences (3)5–12-flowered . . . . ***A. nevadensis*** var. ***nevadensis***

1' Plants annual; inflorescences 1–2-flowered.

  2. Bracts present; peduncles 3–25 mm.

    3. Calyx lobes 2–5 × tube length; leaf surfaces sparsely appressed-hairy . . . . ***A. americanus*** var. ***americanus***

    3' Calyx lobes ≈ tube length; leaf surfaces glabrous to glabrate. . . . . ***A. parviflorus***

  2' Bracts 0; peduncles 0–2 mm.

    4. Corollas cream, pink, or pale yellow . . . . ***A. denticulatus***

    4' Corollas yellow, turning red.

      5. Fruits 3–4 mm wide; plants conspicuously villous to glabrate . . . . ***A. brachycarpus***

      5' Fruits 2.2–3 mm wide; plants sparsely strigose to glabrate. . . . . ***A. wrangelianus***

Acmispon americanus (Nuttall) Rydberg var. *americanus* (p. 650)

Spanish clover, American deervetch

Hosackia americana (Nutt.) Piper, *Lotus americanus* (Nutt.) Bisch., *Lotus purshianus* Clem. & E.G. Clem. var. *purshianus*, *Lotus unifoliolatus* (Hook.) Benth. var. *unifoliolatus*

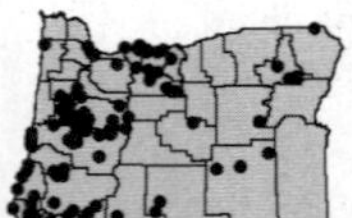

Plants annual, glabrate to pilose. **Stems** 1–10, procumbent to erect, 0.5–6(12) dm. **Leaves** pinnately compound; leaflets 1–5, obovate, ovate or elliptic to lanceolate, margins entire to denticulate, tips acute, surfaces sparsely appressed-hairy; rachises 0.5–4 mm, not flattened; sessile to short petiolate. **Inflorescences** 1–2-flowered; bracts unifoliate; peduncles 3–25 mm. **Flowers** calyces 2.5–6.5 mm, tubes pilose, lobes linear-lanceolate, 2–5 × tube length; corollas 4–9 mm, whitish or cream, turning pink or salmon, banners rose-striped, keel tips yellow. **Fruits** dehiscent, linear-oblong to narrowly oblong, straight or ± curved, (15)20–30(40) × 1.5–2.5 mm, brown, tips spreading or pendent, abruptly downward angled and curved, glabrous. **Seeds** (3)5–8, olive to light brown, often black mottled. **2*n***=14.

Grasslands, meadows, waste ground, riverbanks, sand dunes. Flowering May–Sep. 0–1800 m. All ecoregions. CA, ID, NV, WA; north to British Columbia, northeast to Manitoba, east to WI, southeast to LA, south to Mexico, scattered in eastern North America. Native.

A second variety, *A. americanus* var. *helleri*, is found in the southeastern United States, east of the Appalachian Mountains.

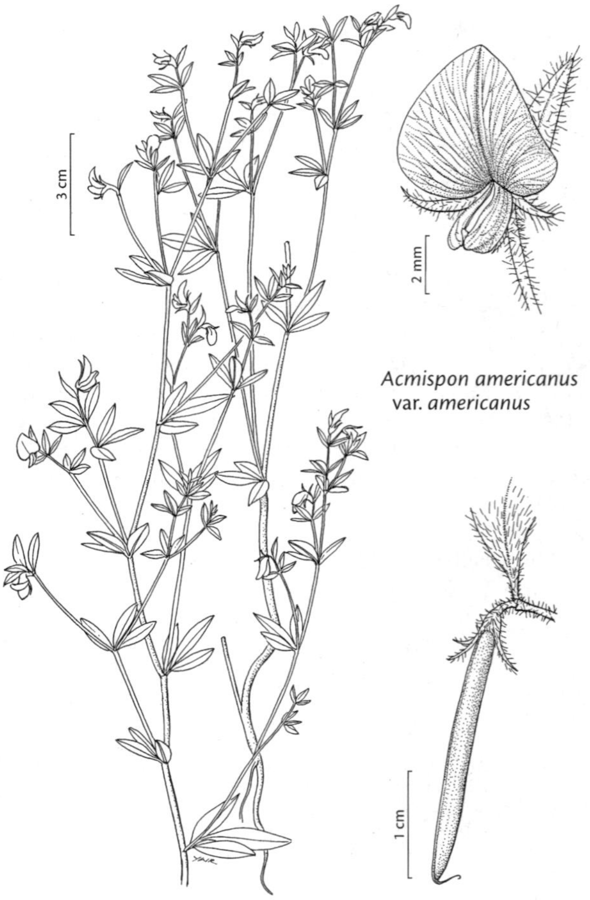

Acmispon americanus var. americanus

ACMISPON

Acmispon brachycarpus (Bentham) D.D. Sokoloff

Colchita, hill lotus

Lotus humistratus Greene

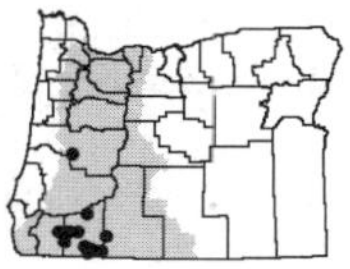

Plants annual, conspicuously villous to glabrate. **Stems** 1–20, procumbent to low-ascending, 0.5–4 dm. **Leaves** pinnately or palmately compound; leaflets 3–5, usually 2 on one side of the rachis and 2 terminal, elliptic to obovate, oblanceolate, tips usually obtuse, sometimes acute, surfaces villous to glabrate; rachises 4–10 mm, flattened; sessile or subsessile. **Inflorescences** 1-flowered; ± sessile; bracts 0. **Flowers** calyces 3–6 mm, tubes villous, lobes lanceolate, 1–2 × tube length; corollas 5–9 mm, yellow, reddish-tipped, turning red. **Fruits** dehiscent, oblong, straight, 6–14 × 3–4 mm, brown or tawny, tips erect, obtuse, villous. **Seeds** 2–5, tan to dark brown or blackish, faintly mottled. **2*n***=12.

Pastures, creek banks, roadsides, hillsides, woodlands. Flowering Apr–Jul. 100–1600 m. Casc, ECas, Sisk, WV. CA, ID, NV; southeast to NM, south to Mexico. Native.

Acmispon denticulatus (Drew) D.D. Sokoloff

Riverbar lotus, meadow lotus

Hosackia denticulata Drew, *Lotus denticulatus* (Drew) Greene

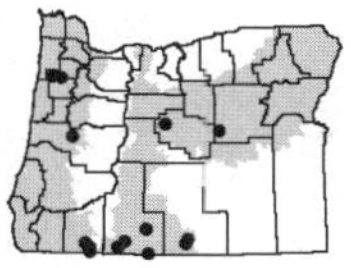

Plants annual, glabrous or hirsute, often glaucous. **Stems** 1–5, decumbent to erect, 0.5–4 dm. **Leaves** subpinnately, pinnately or palmately compound; leaflets 2–4, often 1 or 2 on one side of the rachis and 2 terminal, elliptic to obovate, margins entire or denticulate, tips acute to obtuse, surfaces hirsute; rachises 5–12 mm, flattened; petiolate. **Inflorescences** 1–2-flowered; bracts 0; ± sessile. **Flowers** calyces 3–5 mm, tubes hirsute or glabrous, lobes subulate, ~1.5 × tube length; corollas 5–8 mm, cream white, or pink to pale yellow, banners purple-tinged, keel tips yellowish. **Fruits** dehiscent, erect or spreading, straight, widely oblong, 8–20 × 1–4 mm, tawny, tips abruptly downward angled and curved, strigose or glabrous. **Seeds** 2–4, gray, faintly mottled. **2*n***=14.

Roadsides, fields, slopes, gravel bars, meadows. Flowering May–Aug. 50–1600 m. BW, CR, ECas, Lava, Sisk, WV. CA, ID, WA; north to British Columbia, southeast to UT. Native.

Acmispon nevadensis (S. Watson) Brouillet (p. 651) var. ***nevadensis***

Nevada deervetch, Nevada bird's-foot trefoil, Douglas' lotus

Lotus douglasii Greene, *Lotus nevadensis* (S. Watson) Greene var. *douglasii* (Greene) Ottley, *Lotus nevadensis* (S. Watson) Greene var. *nevadensis*

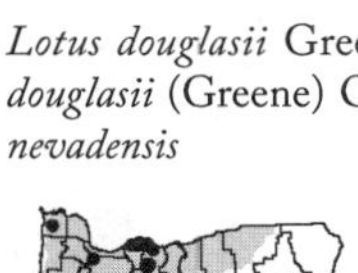

Plants perennial, mat-forming or cespitose, villous to glabrate. **Stems** 1–20, procumbent, 0.5–1(3–12) dm. **Leaves** irregularly pinnately compound to subpalmate; leaflets 3–5, obovate to elliptic, tips acute to obtuse, surfaces villous; rachises (3)5–10(12) mm, ± flattened or not; subsessile to petiolate. **Inflorescences** (3)5–12-flowered; bracts 1–3-foliate; peduncles 3–15(30) mm. **Flowers** calyces 3.5–7 mm, tubes sparsely villosulous, lobes subulate, ≤ tube length; corollas 4.5–8(10) mm, yellow, reddening with age. **Fruits** indehiscent, declined or deflexed, ovoid-ellipsoid, strongly arched, 6–10 × 1.8–2.2 mm, tawny, tips long-tapering to curved, slender beak, rugose, strigillose. **Seeds** 1–3, brown, not mottled. **2*n***=14.

Roadsides, scree, prairies, hillsides, ponderosa pine forests. Flowering May–Aug. 0–1800 m. BR, Casc, Col, CR, ECas, Sisk, WV. CA, ID, NV, WA; south to Mexico. Native.

A second variety, *A. nevadensis* var. *davidsonii,* occurs in California.

Acmispon parviflorus (Bentham) D.D. Sokoloff

Small-flowered lotus

Hosackia parviflora Benth., *Lotus micranthus* Benth.

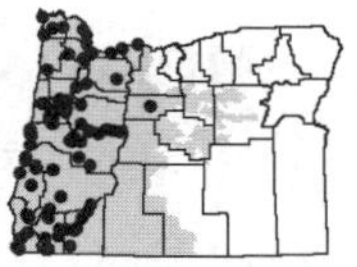

Plants annual, glabrous or sparsely strigillose. **Stems** 1–10, procumbent to erect, 0.5–4 dm. **Leaves** irregularly pinnately to palmately compound; leaflets 3–5, obovate to oblong-elliptic, tips obtuse, surfaces glabrous to glabrate; rachises 2–8 mm, sometimes flattened; subsessile to short-petiolate. **Inflorescences** 1-flowered; bracts 1-3-foliate; peduncles ascending, 3–15 mm. **Flowers** calyces 1–2.5 mm, tubes sparsely strigillose, lobes subulate, ≈ tube length; corollas (2.5)4–6 mm, pink or salmon, quickly fading, wings and keels yellowish. **Fruits** dehiscent, erect or spreading, curved or straight, narrowly oblong, 15–25 × 2 mm, tips short hook-beaked, smooth, glabrous or sparsely strigillose. **Seeds** 3–9, brown, not mottled. **2*n***=14.

Roadsides, meadows, openings in woodlands, riverbanks, coastal bluffs. Flowering Apr–Jul. 0–1500 m. Casc, CR, ECas, Est, Lava, Sisk, WV. CA, WA; north to British Columbia. Native.

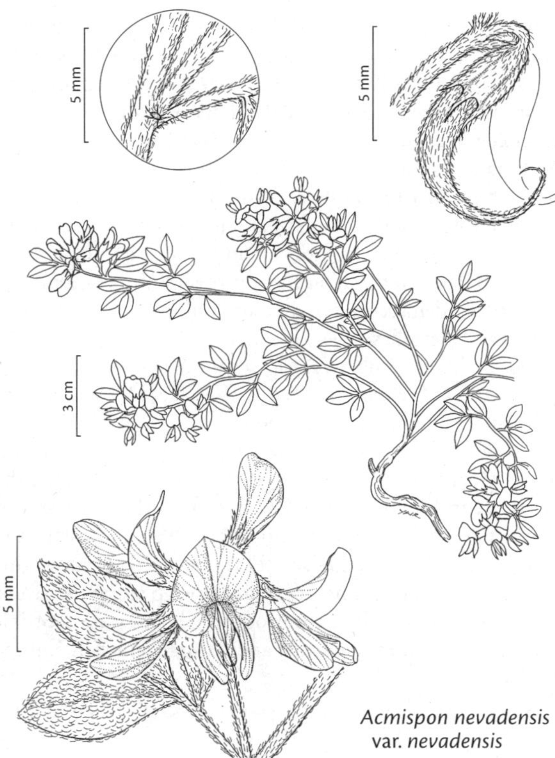

Acmispon nevadensis var. *nevadensis*

ACMISPON

Acmispon wrangelianus (Fisher & C.A. Meyer) D.D. Sokoloff
Hairy lotus

Lotus subpinnatus Lag., *Lotus wrangelianus* Fisch. & C.A. Mey.

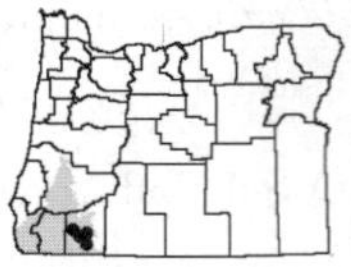

Plants annual, sparsely strigose to glabrate. **Stems** 1–20, usually procumbent, 0.5–3 dm. **Leaves** irregularly pinnately compound; leaflets 3–5, usually 2 on one side of rachis and 2 terminal, elliptic to obovate, tips usually obtuse, sometimes acute, surfaces sparsely long-ciliate, strigose; rachises 5–15 mm, flattened; petiolate. **Inflorescences** 1-flowered; bracts 0; peduncles 0–2 mm. **Flowers** calyces 2.5–5 mm, tubes pubescent to villous, lobes lanceolate, ≈ tube length; corollas 5–9 mm, yellow, turning red. **Fruits** dehiscent, erect or spreading, straight, oblong, 10–18 × 2.2–3 mm, tips abruptly downward angled and curved, pubescent to glabrate. **Seeds** 3–7, olive green to brown, not mottled. **2*n***=12.

Hillsides, open and rocky slopes, oak woodland and savannas. Flowering Apr–Jun. 600–900 m. Sisk. CA. Native.

Acmispon wrangelianus is common in California but in Oregon is currently known only from Jackson County.

AMORPHA False indigo

Richard R. Halse

Shrubs fragrant when bruised, often gland-dotted. **Stems** erect, glabrous or hairy. **Leaves** deciduous, alternate, odd-pinnately compound; leaflets opposite or subopposite, elliptic or oblong to ovate or nearly orbicular, margins entire or crenate; stipules setaceous to linear, usually caducous; petiolules present. **Inflorescences** terminal, spike-like racemes, either solitary or in clusters; bracts caducous. **Flowers** calyces slightly bilaterally symmetrical, greenish to reddish or purplish, persistent; corollas non-papilionaceous, reduced to single banner petal, banners clawed, entire or erose, glabrous; stamens 10, monadelphous, exserted; pistils sessile, styles slender, exserted, stigmas capitate. **Fruits** loments, oblong to obovoid, gland-dotted, dispersed with calyces. **Seeds** 1(2).

Asia, Europe, North America. 16 species; 1 species treated in *Flora.*

Amorpha fruticosa Linnaeus (p. 652)
Bastard indigo, tall indigo bush

Amorpha occidentalis Abrams

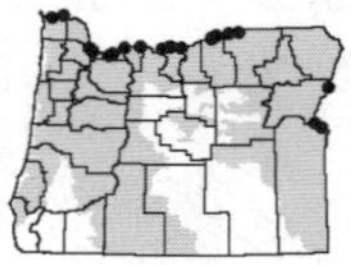

Plants 1–4 m. **Stems** puberulent to glabrate, eglandular to sparsely gland-dotted. **Leaves** 10–28 cm; leaflets (5)9–21(31), elliptic to oblong, (10)20–40 × 1–20 mm, margins ± entire, midveins terminated by slender mucros, mucros 0.5–2 mm, tips acute to round or emarginate, surfaces pilosulose or glabrate, eglandular or gland-dotted. **Inflorescences** racemes, erect, densely flowered, usually in clusters of 2–8 branches; pedicels 1–2.2 mm. **Flowers** calyces obconic to funnelform or campanulate, 2–4 mm, puberulent to pilosulose or glabrous, gland-dotted, ciliate marginally, teeth triangular-dentate, 0.2–1 mm; banners broadly obovate to obcordate, 5–6 × 3.5–4.2 mm, reddish purple, entire, slightly emarginate or erose. **Fruits** indehiscent, straight to curved, 5–9 × (2)3–4.5 mm, usually gland-dotted, sessile. **2*n***=40.

Riverbanks, streambanks. Flowering May–Aug. 0–700 m. BW, Casc, Col, CR, ECas, Owy, WV. CA, ID, WA; eastern North America; Asia, Europe. Exotic.

This species has been cultivated as an ornamental and used for erosion control. It is suggested that it was planted along the Boise River in Idaho by the CCC and from there has spread downriver to the Snake and Columbia rivers almost to the Pacific Ocean (Glad & Halse 1993). Dense thickets of this shrub can be found on sand, rock, and rip-rap along the rivers. Washington has listed it as a noxious weed. Cultivation of this species has resulted in its escape and naturalization in Europe and Asia.

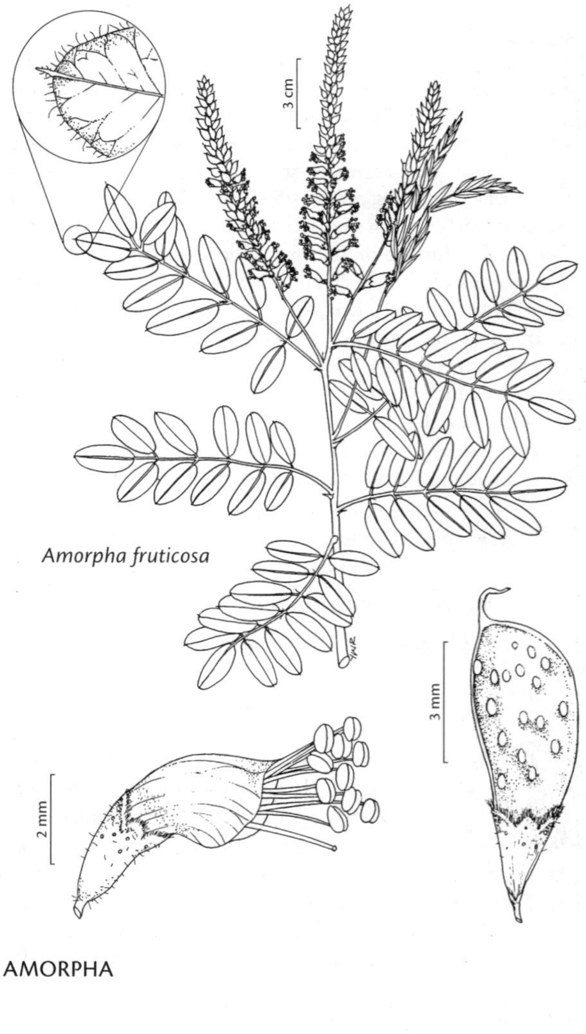

AMORPHA

ASTRAGALUS Milkvetch, locoweed

Richard R. Halse

Herbs annual, biennial, or perennial; acaulescent to caulescent, glabrous to pubescent, hairs basifixed or malpighian. **Stems** prostrate to decumbent to erect. **Leaves** alternate, unifoliate to odd-pinnately compound; leaflets entire, petiolulate or sessile or decurrent, terminal and lateral usually jointed to rachis; stipules free or connate-sheathing, especially on lowermost leaves. **Inflorescences** racemes, spikes, or subumbels; peduncles axillary or arising from the caudex; bracts subtending pedicels often small, membranous, early deciduous; bracteoles 0–2, small, at bases of calyces. **Flowers** papilionaceous, bisexual; calyces campanulate to cylindric, with 5 subequal teeth; banners erect, ovate-oblong, wings usually shorter than banners and longer than keels, keels suberect, obtuse to acute; stamens diadelphous; ovules 2–many, styles filiform, straight or incurved, glabrous, stigmas 1, terminal. **Fruits** dehiscent throughout or only at tips, erect to deflexed, straight to curved or coiled, linear to globose, turgid to inflated, glabrous to densely hairy, unilocular to partially or completely bilocular by intrusion of a longitudinal false septa arising from dorsal suture; valves thickened fleshy to membranous to coriaceous or woody; sessile or stipitate or borne on gynophores. **Seeds** 2–many, reniform.

Cosmopolitan in Northern Hemisphere, secondarily extending into Southern Hemisphere. ~1500–2000 species; 56 species treated in *Flora*.

Many species of *Oxytropis* can be mistaken for species of *Astragalus*; see the discussion under *Oxytropis* for differences between the genera.

Mature fruits are often required for the keys to work. Cross sections of the fruit body show if it is internally unilocular or bilocular through the intrusion of the dorsal suture. In some cases, the dorsal suture may not completely divide the locule. If the locule is almost but not quite divided, the pod can be said to be semibilocular; if the intrusion into the locule is present but there is an obvious space between the sutures, the pod can be said to be subunilocular. Gradations do exist between these forms. The stipe is derived from the ovary and is part of the pod; this stalk can elevate the pod some distance beyond the mouth of the calyx. The gynophore is of receptacular origin and not part of the fruit; it is a stalk that also elevates the pod.

Careful observation of the stipule is also important in species determination in *Astragalus*. Stipules may be free and monomorphic. Stipules may also be dimorphic, with the stipules at lowermost 1–3(4) nodes connate and sheathing, while the upper stipules are typically free.

Astragalus species are among the most important toxic plants in North America. Livestock losses due to ingestion of these plants are of major economic concern to ranchers. Locoism is caused by the presence of indolizidine alkaloids; cracker-heels disease is caused by the presence of toxic glycosides or nitrotoxins; selenosis is caused by the presence of high levels of selenium in the soil and its subsequent accumulation in the tissues of the plants growing in these soils. The toxicity of all species of *Astragalus* is unknown. In Oregon, species that cause problems include *Astragalus canadensis, A. curvicarpus, A. lentiginosus, A. reventus,* and *A. succumbens.*

Considering the large number of species of *Astragalus* worldwide and their ecological amplitude, it is surprising that there are essentially no exotic species found in Oregon. *Astragalus drummondii*, Drummond's milkvetch, has been collected in Oregon twice: once in Douglas County in 1887 and once in Wasco County in 1927. It has not persisted and is considered a waif. It is native from the east base of the Rocky Mountains from Montana to Utah, east to Colorado and the Dakotas. Drummond's milkvetch can be recognized by villous pubescence, large nodding yellow or white flowers and pendulous, stipitate, glabrous pods. *Astragalus cicer*, chickpea milkvetch, a native of Europe, has been introduced into North America as an experimental forage crop, ornamental novelty, and in reclamation plantings. It has escaped and become naturalized in California, Idaho, Nevada, Washington, and elsewhere, so it could appear in Oregon. It can be recognized by its sprawling growth form, yellowish flowers, and swollen, black-hairy, sessile, bilocular pods.

Key to *Astragalus* species:

1. Leaflets confluent with rachis, spine-tipped; plants somewhat prickly . . . . . ***A. kentrophyta***
1' Leaflets not or only terminal leaflet confluent with rachis, not spine-tipped; plants not prickly.
  2. Plants with malpighian hairs.
    3. Plants acaulescent . . . . . ***A. calycosus*** var. ***calycosus***
    3' Plants caulescent.
      4. Corollas 4.5–7 mm; fruits deflexed . . . . . ***A. caricinus***
      4' Corollas 11.3–17.5 mm; fruits erect . . . . . ***A. canadensis***
  2' Plants with basifixed hairs.
    5. Plants annual or biennial.
      6. Corollas 2.5–3.3 mm; fruits 2.8–4.2 mm . . . . . ***A. gambellianus***
      6' Corollas 5–9.1 mm; fruits 15–24 mm.
        7. Inflorescences (3)7–15-flowered; leaflets oblong-obovate, -ovate, or obovate-cuneate . . . . . ***A. diaphanus***
        7' Inflorescences 2–8-flowered; leaflets linear-oblong or linear-elliptic, oblanceolate . . . . . ***A. geyeri*** var. ***geyeri***
    5' Plants perennial.
      8. Fruits bladdery-inflated . . . . . **Subkey 1**
      8' Fruits not bladdery-inflated.

9. Fruits 4-winged when mature . . . . *A. tetrapterus*
9' Fruits not 4-winged when mature.
  10. Leaves all, or at least some of the upper ones, with terminal leaflet confluent with rachis, not jointed to it . . . . **Subkey 2**
  10' Leaves all with terminal leaflet jointed to rachis.
    11. Fruits canescent, tomentose, tomentulose, villous, or hirsute . . . . **Subkey 3**
    11' Fruits glabrous, strigillose, or villosulous.
      12. Stipes or gynophores absent . . . . **Subkey 4**
      12' Stipes or gynophores present.
        13. Fruits bilocular or semibilocular. . . . . **Subkey 5**
        13' Fruits unilocular or subunilocular . . . . **Subkey 6**

Subkey 1.

1. Fruits stipitate, stipes ≥ 2 mm.
  2. Leaflets oblong-oblanceolate, elliptic, or narrowly obovate . . . . *A. whitneyi*
  2' Leaflets linear to linear-filiform or linear-elliptic.
    3. Fruits mottled . . . . *A. sterilis*
    3' Fruits not mottled . . . . *A. cusickii* var. *cusickii*
1' Fruits sessile.
  4. Plants ± acaulescent . . . . *A. platytropis*
  4' Plants caulescent . . . . *A. lentiginosus*

Subkey 2.

1. Fruits stipitate, stipes 3–5 mm . . . . *A. mulfordiae*
1' Fruits sessile or subsessile, stipes 0–2 mm.
  2. Corollas 5–10 mm.
    3. Stipules free; fruits 14–20 mm . . . . *A. atratus* var. *owyheensis*
    3' Stipules connate-sheathing; fruits 5–8 mm. . . . . *A. peckii*
  2' Corollas 15–22(24) mm.
    4. Stipules free; fruits declined or pendulous . . . . *A. tetrapterus*
    4' Stipules, at least lowermost connate-sheathing; fruits erect . . . . *A. toanus* var. *toanus*

Subkey 3.

1. Inflorescences ovoid or subcapitate. . . . . *A. agrestis*
1' Inflorescences not ovoid or capitate.
  2. Fruits 3.5–9 mm; corollas 4.5–12 mm.
    3. Inflorescences lax, flowers not crowded. . . . . *A. caricinus*
    3' Inflorescences dense, flowers closely crowded.
      4. Corollas glabrous, whitish, often lavender-tinged . . . . *A. spaldingii*
      4' Corollas pubescent, pale yellow . . . . *A. tyghensis*
  2' Fruits (7)13–40(45) mm; corollas 12–32 mm.
    5. Fruits bilocular, triquetrously compressed. . . . . *A. malacus*
    5' Fruits unilocular, obcompressed.
      6. Plants acaulescent.
        7. Plants with thatch of persistent leaf bases . . . . *A. newberryi* var. *castoreus*
        7' Plants without thatch of persistent leaf bases . . . . *A. purshii* (in part)
      6' Plants caulescent.
        8. Fruits densely white or tawny tomentose or densely villous, nearly always concealing valve surfaces; leaflets elliptic to oblanceolate . . . . *A. purshii* (in part)
        8' Fruits villous-hirsute, hairs not or rarely concealing valve surfaces; leaflets obovate, ovate-cuneate or broadly oblanceolate.
          9. Leaflets (9)17–23(27); inflorescences (5)8–18-flowered; plants not of Malheur County. . . . *A. inflexus*
          9' Leaflets (7)11–17; inflorescences 2–8-flowered; plants of Malheur County. . . . . *A. nudisiliquus*

Subkey 4.

1. Fruits ascending to erect, not humistrate.
   2. Fruits erect, straight.
      3. Stipules, at least the lowermost, connate-sheathing.
         4. Fruits hirsutulous or villosulous . . . . *A. hoodianus*
         4' Fruits glabrous to strigillose or rugulose.
            5. Leaflets (9)13–25(31); calyces 7–12 mm, greenish. . . . . *A. conjunctus*
            5' Leaflets 3–11(13); calyces 4.6–8 mm, often purplish. . . . . *A. toanus* var. *toanus*
      3' Stipules free.
         6. Corollas 7–10.5 mm; fruits 2.4–4 mm wide; calyx tubes 2.3–4 mm; stipules 1.5–2.5 mm . . . . *A. obscurus*
         6' Corollas 15.2–34 mm; fruits 4–10 mm wide; calyx tubes 5.6–9.9 mm; stipules 3–10 mm.
            7. Fruits 7–10 mm wide, glabrous; calyx tubes 6.2–9.9 mm . . . . *A. reventus*
            7' Fruits 4–6.5 mm wide, glabrous or minutely strigillose; calyx tubes 5.6–7.5 mm. . . . . *A. sheldonii*
   2' Fruits ascending.
      8. Plants grayish hirsute; bracts 2.5–12 mm; corollas with some pink to purple. . . . . *A. succumbens*
      8' Plants strigillose, villosulous, pilose, or glabrate; bracts 1–2.5 mm; corollas whitish to yellowish . . . . *A. lentiginosus* (in part)
1' Fruits deflexed, declined, spreading, or pendulous, often humistrate.
   9. Stipules, at least the lowermost, connate-sheathing.
      10. Plants strigillose; fruits 3.5–4.5 mm . . . . *A. tegetarioides* var. *tegetarioides*
      10' Plants densely gray-villous or -pilose; fruits 6–9 mm . . . . *A. caricinus*
   9' Stipules free.
      11. Fruits 4–7 mm; corollas 4.8–6.1 mm . . . . *A. lemmonii*
      11' Fruits usually ≥ 10 mm; corollas 7.4–21.4 mm.
         12. Fruits turgid to some degree, straight to slightly incurved . . . . *A. lentiginosus* (in part)
         12' Fruits not turgid, usually curved through a quarter- to half-circle or more . . . . *A. iodanthus*

Subkey 5.

1. Fruits erect, rarely deflexed.
   2. Stipes 6–15(17) mm; corollas ochroleucous or purple; calyces 5.5–9.5 mm . . . . *A. eremiticus*
   2' Stipes to 1.5 mm; corollas white or cream; calyces 8.5–12.6 mm. . . . . *A. sheldonii*
1' Fruits not erect, or if ascending then curved.
   3. Fruits glabrous.
      4. Stipes 1–1.9 mm; fruits 2.5–3.6 mm wide. . . . . *A. umbraticus*
      4' Stipes 5–20 mm; fruits 4.5–19 mm wide.
         5. Plants thinly villosulous . . . . *A. accidens* (in part)
         5' Plants sparsely strigose to glabrous. . . . . *A. vallaris*
   3' Fruits pubescent.
      6. Stipules, at least the lowermost, connate-sheathing.
         7. Corollas pink-purple to purple, 7–12 mm . . . . *A. alpinus* var. *alpinus*
         7' Corollas white or whitish, sometimes lilac-veined or -tinged, 6–8.2 mm. . . . . *A. mulfordiae*
      6' Stipules free.
         8. Corollas 6.5–10 mm.
            9. Calyces 3.4–4.8 mm; leaflets 9–21; stipes 2.5–5 mm . . . . *A. misellus* var. *misellus*
            9' Calyces 5–8 mm; leaflets 17–33; stipes 4–14 mm. . . . . *A. howellii*
         8' Corollas 9.5–19.3 mm.
            10. Stipes 1.5–3.5 mm . . . . *A. salmonis*
            10' Stipes 5–15 mm.
               11. Fruits 10–25 mm . . . . *A. accidens* (in part)
               11' Fruits 25–40 mm . . . . *A. arthuri*

Subkey 6.

1. Calyces gibbous at base.
   2. Flowers erect to spreading at anthesis; fruits erect. . . . . ***A. tweedyi***
   2' Flowers nodding at anthesis; fruits pendulous.
      3. Fruits straight or gently incurved . . . . ***A. collinus***
      3' Fruits crescentic, hamate, falcate, or coiled . . . . ***A. curvicarpus***
1' Calyces not gibbous at base.
   4. Stipules, at least the lowermost, connate-sheathing.
      5. Stipes 6–16 mm; corollas yellow or white to ochroleucous.
         6. Plants villous-villosulous . . . . ***A. californicus***
         6' Plants densely strigillose to glabrate. . . . . ***A. filipes***
      5' Stipes 1–8 mm; corollas white to ochroleucous, pink-purple or purple.
         7. Fruits black hairy or with mixed black and white hairs; stipes 1.2–3 mm.
            8. Fruits sulcate dorsally. . . . . ***A. alpinus*** var. ***alpinus***
            8' Fruits not or scarcely sulcate dorsally. . . . . ***A. robbinsii*** var. ***alpiniformis***
         7' Fruits usually glabrous or with few scattered white hairs; stipes 1–8 mm.
            9. Stipules and ripe fruits turning dark-brown or black with maturity or drying; plants not of alpine habitats. . . . . ***A. multiflorus***
            9' Stipules and ripe fruits not blackening with maturity or drying; plants of alpine habitats . . . . ***A. australis*** var. ***glabriusculus*** (in part)
   4' Stipules free.
      10. Corollas 15.5–21.5 mm.
         11. Leaflets 7–17; stems prostrate to ascending; calyces nearly glabrous except sometimes with few black hairs on teeth; fruits sharply incurved or bent . . . . ***A. beckwithii*** var. ***weiserensis***
         11' Leaflets 17–39; stems ascending to erect; calyces pilosulose with black or mixed black and white hairs; fruits straight . . . . ***A. sheldonii***
      10' Corollas 5–16 mm.
         12. Fruits laterally compressed, bicarinate.
            13. Corollas white to purple; plants of alpine habitats. . . . . ***A. australis*** var. ***glabriusculus*** (in part)
            13' Corollas yellowish green, whitish or pale lilac, tinged lilac or purple or with yellowish veins; plants not of alpine habitats.
               14. Stipes 12–20 mm. . . . . ***A. sclerocarpus***
               14' Stipes 2–9 mm.
                  15. Fruits 12–25 mm; peduncles (3)5–15 cm . . . . ***A. solitarius***
                  15' Fruits 5–11 mm; peduncles 1–6 cm.
                     16. Peduncles 1–3.5 cm; stems erect; plants of Baker, Harney, and Malheur counties . . ***A. alvordensis***
                     16' Peduncles 3–6 cm; stems prostrate; plants of Klamath County. . . . . ***A. applegatei***
         12' Fruits obcompressed, not bicarinate.
            17. Inflorescences (2)5–12-flowered; pedicels 1.5–6.5 mm; stipes 1–2 mm; fruit valves papery . . . . ***A. atratus*** var. ***owyheensis***
            17' Inflorescences 2–8-flowered; pedicels 1–4 mm; stipes 1.5–3.5 mm; fruit valves coriaceous. . . . . . ***A. salmonis***

Astragalus accidens S. Watson

Rogue River milkvetch

Plants perennial, caulescent, thinly villosulous, hairs basifixed. **Stems** several–numerous, decumbent to ascending, 30–55 cm, basally leafless. **Leaves** (3)5–12 cm; leaflets (15)19–29, lance-oblong, oblong to oblong-elliptic, 6–22 × 3–9 mm, tips emarginate or retuse, surfaces abaxially pubescent, adaxially pubescent or glabrous; stipules 3–6 mm, free. **Inflorescences** racemes, 7–15-flowered; peduncles (4.5)7–15 cm; bracts 1.5–3.3 mm; pedicels 1.5–3.3 mm; bracteoles (0)2. **Flowers** horizontal or somewhat declined at anthesis; calyces 6–9.4 mm, villosulous with black, fuscous or white hairs, tubes 4.0–5.5 mm, teeth subulate, 1.7–4 mm; corollas 13.8–19.3 mm, whitish to greenish yellow or yellowish, drying ochroleucous; ovules 11–24. **Fruits** bilocular, pendulous, rarely humistrate, oblong-ovoid or half-ovoid, subterete, becoming ± laterally compressed and bicarinate, 10–25 × 4.5–12 mm, glabrous or loosely strigillose; valves subligneous; stipes 5–12 mm.

California and Oregon. 2 varieties.

Key to *Astragalus accidens* varieties:

1. Ovaries and fruits pubescent, fruits 10–16 mm. . . . . ***A. a.*** var. ***accidens***
1' Ovaries and fruits glabrous, fruits 16–25 mm. . . . . ***A. a.*** var. ***hendersonii***

Astragalus accidens S. Watson var. ***accidens***

Rogue River milkvetch

Leaflets 23–29, glabrous adaxially. **Ovaries** pubescent. **Fruits** 10–16 mm, loosely strigillose; stipes 5–8 mm.

Dry wooded and grassy slopes, roadside banks. Flowering Apr–Jun. 100–800 m. Sisk. Native.

This variety is endemic to southwestern Oregon.

Astragalus accidens S. Watson var. ***hendersonii*** (S. Watson) M.E. Jones

Henderson's milkvetch

Astragalus watsonii E. Sheld.

Leaflets (15)19–27, glabrous or thinly pubescent adaxially. **Ovaries** glabrous. **Fruits** 16–25 mm, glabrous; stipes 6–12 mm. $2n$=26.

Open and wooded slopes, roadside banks. Flowering Mar–Jun. 200–1500 m. Casc, Sisk. CA. Native.

Astragalus agrestis Douglas ex G. Don (p. 657)

Field milkvetch, purple milkvetch

Astragalus dasyglottis Fisch. ex DC., *Astragalus goniatus* Nutt.

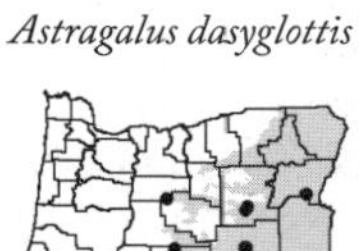

Plants perennial, caulescent, strigillose to villous-pilose, sometimes sparsely so, hairs basifixed. **Stems** few–numerous, decumbent to erect, 5–40 cm. **Leaves** 2–10 cm; leaflets (9)13–23, narrowly elliptic to lance-oblong, 4–18 × 2–5 mm, tips obtuse to retuse or acute, surfaces strigillose; stipules 2–11 mm, at least lowermost connate-sheathing. **Inflorescences** racemes ovoid or subcapitate, 5–15(20)-flowered; peduncles 1.5–15 cm; bracts 3–7 mm; pedicels 0.5–1.5 mm; bracteoles 0. **Flowers** ascending to erect at anthesis; calyces 7–12.5 mm, villous with black and white hairs, tubes 5–7.8 mm, teeth linear, 2.5–5.5 mm; corollas (14)17–24 mm, pink, purple, or blue-lavender, or ochroleucous suffused with lilac; ovules 14–26. **Fruits** bilocular, ascending to erect, oblong-elliptic, nearly straight, obcompressed to ± triquetrous, 7–10 × 2.8–4.5 mm, silky white-villous; valves subcoriaceous; stipes 0.3–1 mm.

Wet meadows, streambanks. Flowering Jun–Aug. 1000–1900 m. BR, BW, Owy. CA, ID, NV, WA; north to Yukon, northeast to Quebec, east to IL, southeast to NM; Asia. Native.

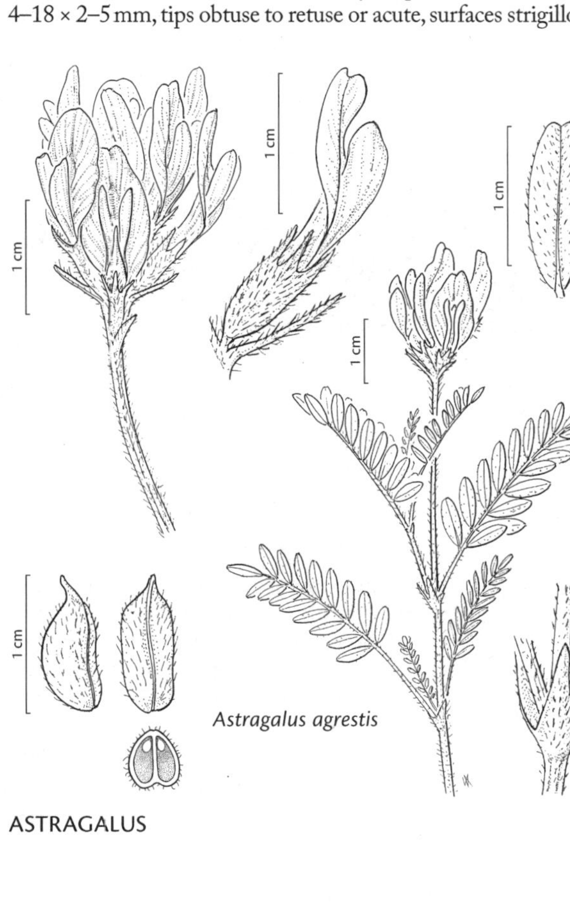

Astragalus agrestis

ASTRAGALUS

Astragalus alpinus Linnaeus var. ***alpinus*** (p. 659)

Alpine milkvetch

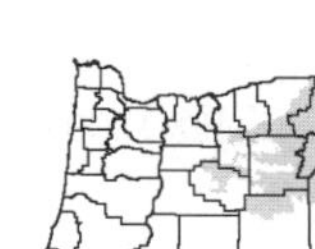

Plants perennial, caulescent (but may appear acaulescent), strigillose to villosulous, hairs basifixed. **Stems** several–numerous, decumbent to ascending, 1–15(25) cm. **Leaves** (2)3–15 cm; leaflets (11)15–25, ovate to elliptic-oblong, (2)6–20(24) × 2–10 mm, tips retuse to rounded, surfaces strigillose; stipules 1.5–8 mm, at least lowermost connate-sheathing. **Inflorescences** racemes elongate, 10–30-flowered; peduncles 3–15(17) cm; bracts 1–2.5 mm; pedicels 0.5–2.3 mm; bracteoles 0. **Flowers** erect to declined at anthesis; calyces 3.2–6.5 mm, strigillose to villosulous with black or some white hairs, tubes 2–4 mm, teeth subulate, 0.9–3 mm; corollas 7–12 mm, pink-purple to purple; ovules 5–11. **Fruits** semibilocular or subunilocular, pendulous, oblong-lanceolate, straight or slightly curved, cordate-triangular in cross section, sulcate dorsally to form a nearly complete partition, 7–14 × 2.5–4.2 mm, strigillose black-hairy or fuscous or mixed black and white hairs; valves papery-membranous; stipes 1.4–3 mm.

Alpine meadows and ridges, gravel creek banks. Flowering Jun–Sep. 1100–3000 m. BW. ID, NV, WA; north to AK, northeast to Greenland, east to Nova Scotia, southeast to NM; Asia, Europe. Native.

Alpine milkvetch can be confused with *A. robbinsii* var. *alpiniformis*. See also that discussion.

Astragalus alvordensis M.E. Jones

Alvord milkvetch

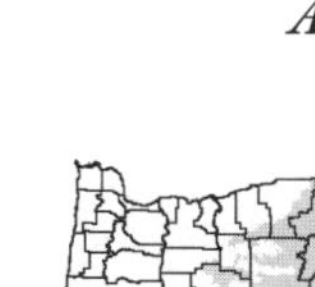

Plants perennial, caulescent, cinereous, hairs basifixed. **Stems** 1–few, erect, 15–30 cm, wiry. **Leaves** (1)1.5–5 cm; leaflets (7)11–21, narrowly or broadly obovate, 1–7 × 1–5 mm, tips retuse or obcordate, surfaces abaxially strigose, adaxially glabrescent or strigose; stipules 0.5–2.5 mm, free. **Inflorescences** racemes, loosely 5–14-flowered; peduncles 1–3.5 cm; bracts 0.5–1 mm; pedicels 1–2 mm; bracteoles 0. **Flowers** declined at anthesis; calyces 2.5–3 mm, villosulous with white and black hairs, tubes 2–2.5 mm, teeth triangular, 0.5–0.7 mm; corollas 7–8.2 mm, pale lilac or whitish to yellowish, veined and tinged with lilac; ovules 10–16. **Fruits** unilocular, pendulous, crescentic or hamate to coiled (1–2 spirals), strongly laterally compressed, bicarinate, 5–10 × 3–4 mm, villosulous; valves papery; stipes 2–9 mm.

Desert hillsides, sand dunes, sagebrush. Flowering May–Jun. 1000–1400 m. BR, BW, Owy. NV. Native.

Astragalus applegatei M. Peck

Applegate's milkvetch

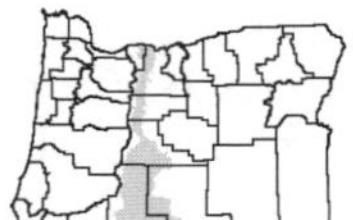

Plants perennial, caulescent, thinly strigillose, hairs basifixed. **Stems** 1–few, prostrate, 30–40 cm. **Leaves** 3.5–7 cm; leaflets 7–11, linear to linear-elliptic, (5)8–20 × 1–3 mm, tips obtuse or sometimes mucronulate, surfaces abaxially strigillose, adaxially glabrous; stipules 1.5–3 mm, free. **Inflorescences** racemes loosely 10–18-flowered; peduncles 3–6 cm; bracts 1–1.5 mm; pedicels 1–1.5 mm; bracteoles 0. **Flowers** spreading at anthesis, declined in age; calyces 3–3.3 mm, strigillose with white or a mixture of white and black hairs, tubes 2.3–2.5 mm, teeth triangular, 0.7–0.9 mm; corollas 6.5–7 mm, whitish, tinged with lilac; ovules 8–10. **Fruits** unilocular, spreading or declined, oblong-ellipsoid, ± straight, strongly laterally compressed, bicarinate, 8–11 × 2.4 × 3 mm, strigillose; valves thinly cartilaginous; stipes 4–5 mm.

Moist meadows, ditches, wet flats. Flowering Jun–Jul. 1200–1400 m. ECas. Native.

Applegate's milkvetch is currently known only from Klamath County, Oregon.

Astragalus arthuri M.E. Jones

Waha milkvetch, Arthur's milkvetch

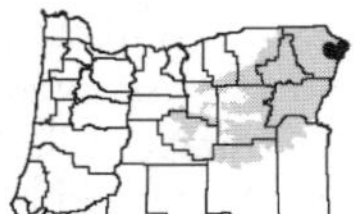

Plants perennial, subacaulescent to caulescent, strigillose, hairs basifixed. **Stems** several, ascending to erect, forming bushy clumps, 2–40 cm. **Leaves** 4–14 cm; leaflets (15)19–27, linear-elliptic, oval or oblanceolate, 2–20 × 2–4 mm, tips obtuse to subacute, truncate-mucronulate or shallowly retuse, surfaces abaxially strigillose, adaxially glabrous; stipules 2–7 mm, free. **Inflorescences** racemes loosely (5)8–20-flowered; peduncles (7)10–20 cm; bracts 1.5–3.5 mm; pedicels 1–4 mm; bracteoles usually 0. **Flowers** declined at anthesis; calyces 6.6–8.8 mm, strigose with mixed black and white hairs, tubes 4.5–6 mm, teeth lance-subulate 1.8–3.5 mm; corollas 11.7–15.2 mm, ochroleucous; ovules 16–30. **Fruits** bilocular, pendulous, linear-ellipsoid, compressed-triquetrous, tapering distally into a slender acuminate beak, nearly straight or gently incurved, deeply sulcate, (25)30–40 × 2.5–4 mm, strigillose; valves papery; stipes 6–15 mm. **2*n***=24.

Open coniferous forests, grassland slopes, meadows. Flowering May–Jul. 400–1400 m. BW. ID, WA. Native.

Early flowering plants appear subacaulescent as the leaves are in loose basal tufts, and the peduncles appear subscapose. Later in the season, the plants are more obviously caulescent, but even then the peduncles and racemes are longer than the stems. This feature plus the long slender linear pods help distinguish this species.

Astragalus atratus S. Watson var. *owyheensis* (A. Nelson & J.F. Macbride) M.E. Jones (p. 659)

Owyhee milkvetch, mourning milkvetch

Astragalus owyheensis A. Nelson & J.F. Macbr.

Plants perennial, caulescent, strigillose, hairs basifixed. **Stems** numerous, prostrate to ascending, 2–20(30) cm. **Leaves** (2)4–14 cm; leaflets 7–11(13), remote, scattered, linear-elliptic, 1.5–7(10) × 0.5–1 mm, tips obtuse to subacute, surfaces abaxially strigillose, adaxially glabrous, terminal leaflet generally confluent with slender or filiform rachis, or a small dilation of rachis tip; stipules 1–3.5 mm, free. **Inflorescences** racemes loosely (2)5–12-flowered; peduncles (1.5)3.5–12.5 cm; bracts 0.8–2.5 mm; pedicels 1.5–6.5 mm; bracteoles usually 0. **Flowers** nodding at anthesis; calyces 3–5.8 mm, strigillose with black, white, or mixed colored hairs, tubes 2.3–4 mm, teeth subulate, 0.7–2.1 mm; corollas 6.3–10 mm, whitish, faintly lavender-tinged; ovules 10–17. **Fruits** unilocular, pendulous, narrowly elliptic-lanceolate, obcompressed, nearly straight or slightly decurved, slightly sulcate, 14–20 × 3–4 mm, mottled or not, strigillose; valves papery; stipes 1–2 mm.

Dry hillsides, gravelly flats, sagebrush. Flowering May–Jun. 800–1300 m. BW, Owy. ID, NV. Native.

Astragalus australis (Linnaeus) Lamarck var. *glabriusculus* (Hooker) Isely

Aboriginal milkvetch

Astragalus aboriginum Richardson, *Astragalus forwoodii* S. Watson var. *wallowensis* (Rydb.) M. Peck

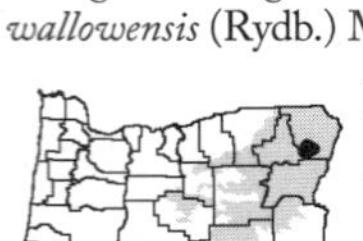

Plants perennial, caulescent, villous or villous-tomentose, hairs basifixed. **Stems** few–several, prostrate to ascending, 10–50 cm. **Leaves** 1–7(10) cm; leaflets 5–19, linear to oblong, lanceolate or elliptic, 3–16 × 1.5–7 mm, tips acute to obtuse, surfaces abaxially strigose or villous, adaxially strigose, villous, or glabrous; stipules 2–7(11) mm, lowermost shortly connate-sheathing or margins free but in contact and overlapping. **Inflorescences** racemes densely 6–40-flowered; peduncles (2)6.5–15 cm; bracts 1.2–5 mm; pedicels 0.8–3.5 mm; bracteoles 0. **Flowers** spreading to reflexed at anthesis; calyces 3.7–6.4 mm, villous with mainly black hairs, sometimes with white hairs, tubes 2.1–5 mm, teeth subulate, 1–3 mm; corollas 7–12.5 mm, white to purple; ovules 8–16. **Fruits** unilocular, spreading to pendulous or humistrate, narrowly elliptic, straight or slightly decurved, laterally compressed, bicarinate, 10–30 × 3–7 mm, usually glabrous or with few scattered white hairs; valves papery; stipes 2.5–8 mm. **2*n***=16.

Alpine, limestone slopes, fell fields. Flowering Jul–Aug. 1700–2900 m. BW. ID, NV; north to British Columbia, northeast to ND, east to NE, southeast to CO, Quebec. Native.

Astragalus australis, subarctic milkvetch, is a variable circumboreal species. The relationship between Eurasian and North American varieties is uncertain.

Astragalus beckwithii Torrey & A. Gray var. *weiserensis* M.E. Jones (p. 660)

Weiser's milkvetch

Astragalus weiserensis (M.E. Jones) Abrams

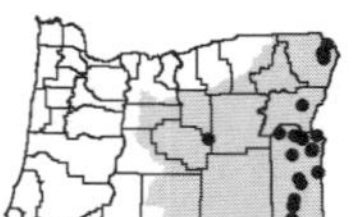

Plants perennial, caulescent, glabrous to sparsely strigose, hairs basifixed when present. **Stems** several, prostrate to ascending, 10–35(70) cm. **Leaves** 2–15 cm; leaflets (7)11–17, broadly ovoid to rhombic-elliptic, (6)10–25 × 5–13 mm, tips obtuse or sometimes mucronulate, surfaces glabrous to glabrate; stipules 3–10 mm, free. **Inflorescences** racemes loosely 7–16-flowered; peduncles 3–15 cm; bracts 3–7.5 mm; pedicels 1–2.5 mm; bracteoles usually 2.

Flowers ascending at anthesis; calyces (8.5)10.2–13.5 mm, nearly glabrous, sometimes with a few black hairs on teeth, tubes (4.6)5.1–6.3 mm, teeth lanceolate, 4.5–7 mm; corollas 16–20.5 mm, ochroleucous; ovules (18)25–41. **Fruits** unilocular, ascending to declined, oblong- or ovoid-ellipsoid, sharply incurved or bent, ± obcompressed and dorsally flattened or sulcate below the triangular beak, ventrally carinate, 15–30 × 6–12 mm, mottled, glabrous; valves stiffly coriaceous to subligneous; gynophores 1.5–5 mm. **2*n***=22.

Sagebrush-bunchgrass communities, ash-clay outcrops, dry hillsides, road cuts. Flowering Apr–Jun. 500–1600 m. BR, BW, Lava, Owy. ID, NV, WA; north to British Columbia. Native.

Astragalus californicus (A. Gray) Greene

California milkvetch, Klamath milkvetch

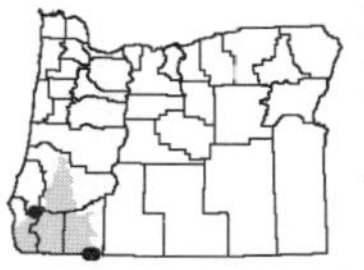

Plants perennial, caulescent, villous-villosulous, hairs basifixed. **Stems** several–numerous, decumbent to ascending, leafless at bases, 15–50 cm. **Leaves** 3–8.5 cm; leaflets 13–21, elliptic-oblong to oblanceolate, (4)6–20(23) × 2–4 mm, tips truncate-retuse to obtuse, surfaces usually strigillose; stipules 1.5–6.5 mm, at least lowermost connate-sheathing. **Inflorescences** racemes loosely 10–30-flowered; peduncles (3)6–14 cm; bracts (1.5)2–5 mm; pedicels 1.3–3.5 mm; bracteoles 0–2. **Flowers** declined or nodding at anthesis; calyces 6.4–9.7 mm, silky villous with white and some black hairs, tubes 5.2–7 mm, teeth triangular, 0.9–3 mm; corollas 11.5–17.4 mm, ochroleucous or yellowish; ovules (15)19–27. **Fruits** unilocular, pendulous, linear-oblong, strongly laterally compressed, straight or slightly decurved, bicarinate, 27–43 × 3.4–5 mm, sparsely to densely strigillose, rarely glabrous; valves papery; stipes 8–14 mm.

Oak woodlands, coniferous forests, chaparral. Flowering Mar–May. 600–1400 m. Sisk. CA. Native.

Astragalus calycosus Torrey ex S. Watson var. ***calycosus*** (p. 660)

Torrey's milkvetch

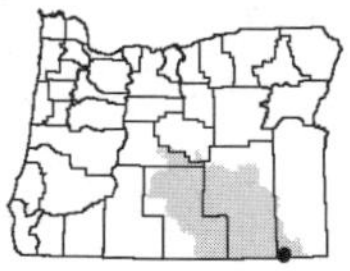

Plants perennial, acaulescent, tufted or matted, generally cinereous-silvery, hairs malpighian. **Stems** 0 or rarely to 2 cm, forming mounds 15–20 cm wide, often with persistent leaf bases. **Leaves** often palmately trifoliolate, 1–7 cm; leaflets 3–7, obovate, oblanceolate, or elliptic, 2–19 × 1–7 mm, tips obtuse to acute, surfaces silvery strigose; stipules 1.5–6 mm, free. **Inflorescences** 1–8-flowered; bracts 0.5–2 mm; pedicels 0.7–3 mm; bracteoles 0 or minute. **Flowers** ascending to spreading at anthesis; calyces 5.2–10.6 mm, white or partly black strigillose, tubes 4–6.4 mm, teeth subulate, 1.5–4.2 mm; corollas 10–16.5(20.8) mm, whitish to pink-purple; ovules 12–28. **Fruits** bilocular, ascending, usually humistrate, narrowly oblong, laterally compressed, usually curved, 8–25 × 3–4.5 mm, strigose; valves papery; stipes 0. **2*n***=22.

Mine tailings, ash deposits, sagebrush. Flowering May–Jul. 900–1700 m. BR. CA, ID, NV; southeast to NM. Native.

In Oregon, this taxon is currently known only from Malheur County.

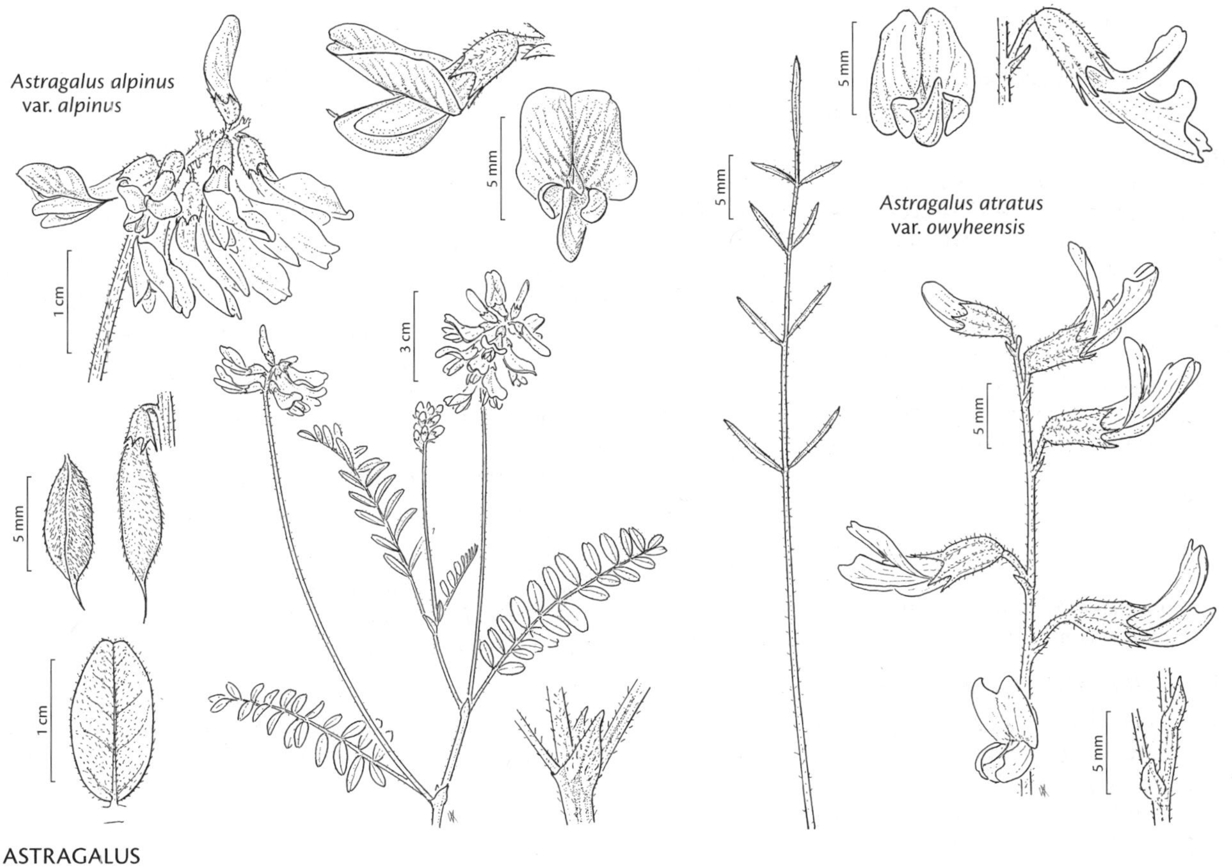

ASTRAGALUS

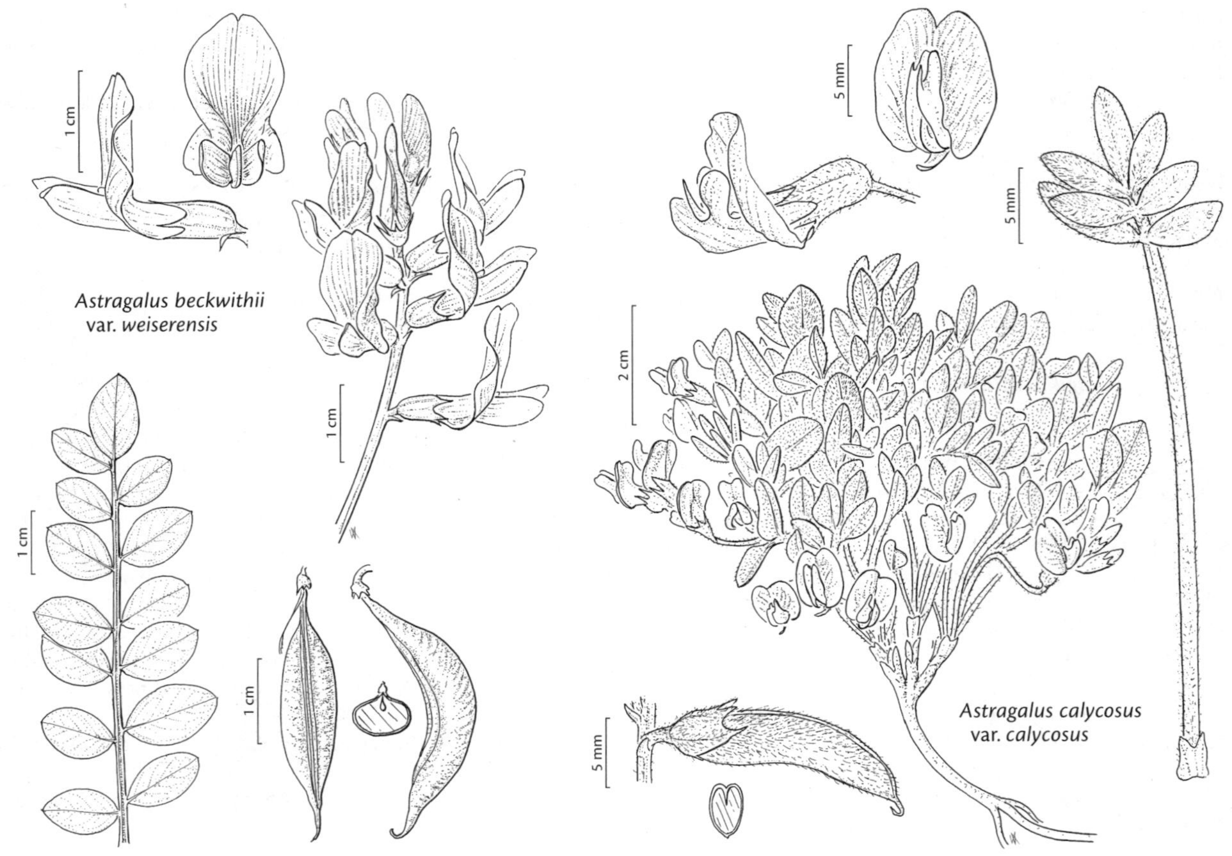

ASTRAGALUS

Astragalus canadensis Linnaeus
Canadian milkvetch

Plants perennial, caulescent, strigose, hairs malpighian. **Stems** 1–few, erect, rarely decumbent to ascending, 15–120 cm. **Leaves** 3–35 cm; leaflets (7)13–35, lanceolate, lance-oblong or elliptic, 10–52 × 6–16 mm, tips obtuse to emarginate, surfaces abaxially strigose, adaxially strigose or glabrous; stipules 3–14 mm, at least lowermost connate-sheathing. **Inflorescences** racemes, densely many-flowered; peduncles 4–22 cm; bracts 1.5–10 mm; pedicels 0.5–3.5 mm; bracteoles 0–2. **Flowers** spreading-declined at anthesis; calyces 4.6–10.5 mm, white-, black- or fuscous-strigillose or -pilosulose, tubes 4–8.5 mm, teeth subulate or triangular, 1.2–5 mm; corollas (11.3)13.2–17.5 mm, ochroleucous or greenish white; ovules 16–28. **Fruits** bilocular, erect, cylindroid, terete or grooved dorsally, 10–20 × 2.9–5.2 mm, strigose or glabrous; valves coriaceous or stiffly chartaceous; stipes 0. **2*n*=16.**

Canada and United States. 3 varieties; 2 varieties treated in *Flora*.

Canadian milkvetch is perhaps the most widely distributed *Astragalus* in North America, rivaled only by *Astragalus alpinus*.

Key to *Astragalus canadensis* varieties:

1. Plants ± robust; calyx teeth 1–3 mm; fruits densely strigose, beaks 1.5–3 mm . . . . . ***A. c.*** var. ***brevidens***

1' Plants relatively slender; calyx teeth 1.5–4.4 mm; fruits thinly strigose or glabrate, beaks 3–5 mm . . . . ***A. c.*** var. ***mortonii***

Astragalus canadensis Linnaeus var. *brevidens* (Gandoger) Barneby
Pasture milkvetch, short-tooth Canadian milkvetch

Astragalus brevidens Rydb.

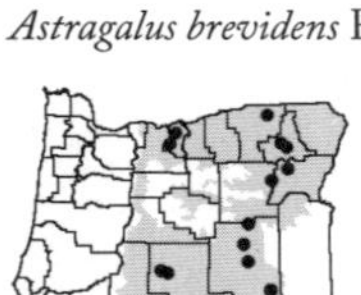

Stems ± robust, erect or sometimes decumbent to ascending, 10–55 cm. **Calyx teeth** 1–3 mm, ventral pair usually broader. **Fruits** densely strigose; beaks 1.5–3 mm. **2*n*=16.**

Alkaline meadows, lake shores, roadsides, sagebrush. Flowering May–Aug. 500–1800 m. BR, BW, Col, ECas. CA, ID, NV, WA; north to British Columbia, northeast to MT, east to WY, southeast to CO. Native.

Astragalus canadensis var. *mortonii* (Nuttall) S. Watson
Morton's milkvetch, Morton's Canadian milkvetch

Astragalus mortonii Nutt.

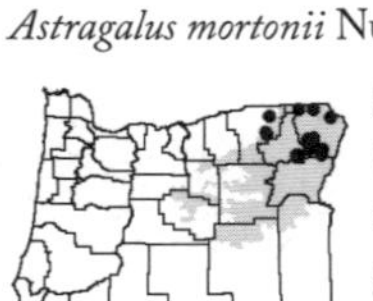

Stems relatively slender, erect, 25–70(90) cm. **Calyx teeth** 2–4.4 mm, ventral pair usually not much broader. **Fruits** thinly strigose or glabrate; beaks 3–5 mm. **2*n*=16.**

Coniferous forests, sagebrush hillsides. Flowering Jun–Aug. 800–1900 m. BW. ID, WA; north to British Columbia, northeast to MT. Native.

Astragalus caricinus (M.E. Jones) Barneby

Buckwheat milkvetch

Astragalus lyallii A. Gray, *Astragalus lyallii* A. Gray var. *caricinus* M.E. Jones

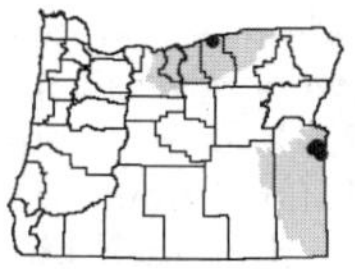

Plants perennial, caulescent, densely gray-villous or -pilose, hairs basifixed or malpighian. **Stems** numerous, ascending to erect, clustered, (10)15–30 cm. **Leaves** 3.5–9(10.5) cm; leaflets 11–19(23), narrowly elliptic to lance-elliptic to narrowly oblong, (3)5–18 × 1–4 mm, tips acute to obtuse or mucronulate, surfaces villous; stipules (2)3–8 mm, at least lowermost connate-sheathing. **Inflorescences** racemes lax, not crowded, (5)10–25-flowered; peduncles 0.5–5(7) cm; bracts 1–2.5 mm; pedicels 0.4–1.5 mm; bracteoles 0. **Flowers** spreading to declined at anthesis; calyces 3.5–5.5 mm, densely white- or partly black-villosulous, tubes 1.7–3 mm, teeth subulate-setaceous, 1.3–3 mm; corollas 4.5–7 mm, whitish or tinged lilac, drying yellowish; ovules 6–8. **Fruits** bilocular, deflexed, lance-elliptic, straight, compressed-triquetrous or laterally compressed, narrowly sulcate dorsally, 6–9 × 2–3 mm, white or gray tomentulose; valves papery; stipes 0.

Sand dunes. Flowering May–Jun. 100–1000 m. Col, Owy. ID, WA. Native.

Buckwheat milkvetch is distinguished by its dense gray pubescence, connate lower stipules, small pale flowers, and small, deflexed, densely hairy, bilocular fruits.

Astragalus collinus (Hooker) G. Don

Hillside milkvetch

Plants perennial, caulescent, villosulous or strigose throughout, hairs basifixed. **Stems** several–numerous, decumbent to ascending or erect, in clumps, 15–50 cm. **Leaves** 3–9(10.5) cm; leaflets (9)11–21(25), oblong-oblanceolate to linear-elliptic, 4–20(22) × 2–6 mm, tips truncate-retuse to obtuse, surfaces abaxially pubescent, adaxially pubescent, glabrescent, or glabrous; stipules 2–5(7) mm, free. **Inflorescences** racemes closely (10)15–40(55)-flowered; peduncles (4)5–16 cm; bracts 1.5–3.5 mm; pedicels 0.8–2.8 mm; bracteoles 2, rarely 0. **Flowers** nodding at anthesis; calyces gibbous-saccate at bases, 7–12.3 mm, villosulous with white or some black hairs, tubes 5.8–9.4 mm, teeth lanceolate to triangular, 0.8–2.7(3.5) mm; corollas (10.5)12–17.2 mm, white, cream, ochroleucous to lemon yellow; ovules 7–18. **Fruits** unilocular, pendulous, linear- or lance-oblong to lunate-oblong, straight or gently incurved, laterally compressed, bicarinate, 7–25 × 2.3–4.5 mm, villosulous or strigillose to glabrate; valves coriaceous; stipes (3.5)5–15 mm; **2*n***=24.

2 varieties.

Key to *Astragalus collinus* varieties:

1. Fruits linear-oblong, straight or rarely slightly incurved, 7–25 mm, villosulous or strigillose, hairs 0.25–0. 5 mm; ovules (10)12–18 . . . . ***A. c.*** var. ***collinus***

1' Fruits ovate-oblong, gently incurved, 8–15 mm, villosulous, rarely glabrous, hairs 0.5–1 mm; ovules (7)8–12 . . . . . . ***A. c.*** var. ***laurentii***

Astragalus collinus (Hooker) G. Don var. *collinus*

Hillside milkvetch

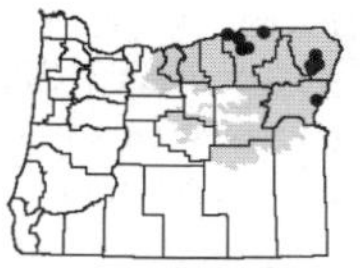

Ovules (10)12–18. **Fruits** linear-oblong, straight or rarely slightly incurved, 7–25 mm, villosulous or strigillose, hairs 0.25–0.5 mm; stipes 5–15 mm.

Grassy hillsides, bunchgrass prairies, sagebrush. Flowering May–Jul. 300–1600 m. BW, Col. ID, WA; north to British Columbia. Native.

Astragalus collinus (Hooker) G. Don var. *laurentii* (Rydberg) Barneby

Lawrence's milkvetch

Astragalus laurentii (Rydb.) M. Peck

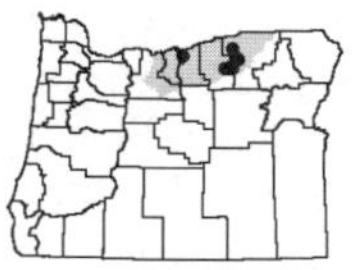

Ovules (7)8–12. **Fruits** ovate-oblong, gently incurved, 8–15 mm, villosulous, rarely glabrous, hairs 0.5–1 mm; stipes (3.5)5–10 mm.

Bunchgrass prairies, roadsides. Flowering May–Jul. 500–1100 m. Col. Native.

This variety is endemic to northeastern Oregon.

Astragalus conjunctus S. Watson

Stiff milkvetch, basalt milkvetch, Idaho milkvetch

Plants perennial, subacaulescent to shortly caulescent, strigillose, hairs basifixed. **Stems** several–numerous, ascending to erect, in clumps, 15–65 cm. **Leaves** leaflets (9)13–25(31), linear-oblong, linear-elliptic, lanceolate, or sub-filiform, 3–23 × 0.7–3(4) mm, tips obtuse, acute or retuse, surfaces abaxially pubescent, adaxially pubescent, glabrate, or glabrous, terminal leaflets sometimes confluent with rachis; stipules 3–11 mm, at least lowermost connate-sheathing. **Inflorescences** racemes loosely 7–17(20)-flowered; peduncles 10–33 cm; bracts 2–4.5 mm; pedicels 1–4.5 mm; bracteoles 0–2. **Flowers** ascending to spreading at anthesis; calyces (7)8.5–12 mm, greenish strigillose with black or mostly black hairs, tubes 4.9–9.2 mm, teeth subulate, 1.3–3(4) mm; corollas 15–26 mm, whitish, sometimes tipped with purple or rarely purple throughout; ovules 15–30. **Fruits** unilocular to semibilocular, erect, oblong-ellipsoid to oblong-ovoid, straight or slightly incurved, obcompressed, dorsal surface almost flat to shallowly sulcate, 12–25 × 4–8(10) mm, transversely rugulose; valves coriaceous to subligneous; stipes 0.

2 varieties.

Astragalus conjunctus is closely related to *A. hoodianus.* A third closely related species, *A. reventiformis*, has been reported from Sherman County, but no voucher has been found to confirm its presence in Oregon.

Key to *Astragalus conjunctus* varieties:

1. Fruits glabrous, 5–8 mm wide; calyces cylindric, (5.7)6–9.2 mm . . . . . ***A. c.*** var. ***conjunctus***

1' Fruits strigillose, 4.3–5 mm wide; calyces campanulate, 4.9–6.2 mm . . . . . ***A. c.*** var. ***rickardii***

Astragalus conjunctus S. Watson var. ***conjunctus***

Stiff milkvetch, basalt milkvetch, John Day Valley milkvetch

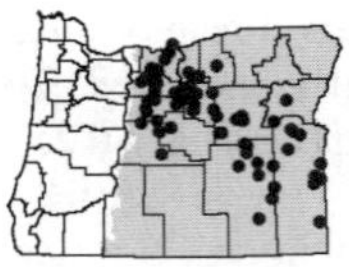

Flowers calyces cylindric, tubes (5.7)6–9.2 mm; corollas 16–25.5 mm; ovules 23–30. **Fruits** 12–25 × 5–8 mm, glabrous. **2*n***=24.

Dry rocky slopes, sagebrush and bunchgrass communities. Flowering Apr–Jul. 200–2200 m. BR, BW, Col, ECas, Lava, Owy. ID, WA. Native.

Astragalus conjunctus S. Watson var. ***rickardii*** S.L. Welsh, K. Beck & F. Caplow

Rickard's milkvetch

Astragalus reventus A. Gray var. *conjunctus* (S. Watson) M.E. Jones

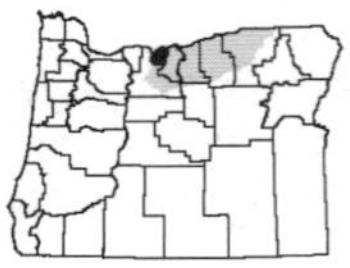

Flowers calyces campanulate, tubes 4.9–6.2 mm; corollas 15–18.5 mm; ovules 15–20. **Fruits** 13–20 × 4.3–5 mm, strigillose.

Roadsides, bunchgrass prairies. Flowering Apr–Jun. 100–1100 m. Col. WA. Native.

Astragalus curvicarpus (A. Heller) J.F. Macbride

Curvepod milkvetch

Plants perennial, caulescent, strigillose to villosulous or glabrate, hairs basifixed. **Stems** several–numerous, decumbent to ascending, 15–50 cm. **Leaves** 2.5–9 cm; leaflets (9)11–21(25), obovate-cuneate, oblong-ovate or elliptic to broadly oblanceolate, (7)9–19(21) × 2–9 mm, tips retuse or truncate to obtuse, surfaces strigose to glabrescent or glabrous; stipules 1–5 mm, free. **Inflorescences** racemes (5)10–25(35)-flowered; peduncles 4–15 cm; bracts 0.7–2.5 mm; pedicels 1–3.5 mm; bracteoles 0. **Flowers** nodding at anthesis; calyces gibbous-saccate or gibbous-truncate at bases, 6.1–13.6 mm, villosulous or strigose with white or black hairs or a mixture, tubes 5.4–12 mm, teeth triangular, 0.4–2.3 mm; corollas 13.6–21 mm, white, ochroleucous, or lemon yellow; ovules 14–25. **Fruits** unilocular, pendulous, narrowly oblong, crescentic, hamate, falcate or coiled through 1.5 spirals, strongly laterally compressed, bicarinate, 14–35 × 2.7–4.5 mm, cross reticulate, villosulous, strigillose, glabrate, or glabrous; valves stiffly papery or thinly coriaceous; stipes 6–20 mm.

3 varieties.

Key to *Astragalus curvicarpus* varieties:

1. Ovaries and fruits pubescent, rarely fruits glabrate or glabrous . . . . ***A. c.*** var. ***curvicarpus***

1' Ovaries and fruits glabrous.

  2. Calyces 6.1–8.5(9.3) mm; keels 9.4–11.6(12.3) mm; leaflets pubescent, sometimes adaxially thinly pubescent . . . . ***A. c.*** var. ***brachycodon***

  2' Calyces (8)9–13.6 mm; keels 11.5–14.5 mm; leaflets abaxially glabrous to thinly pubescent or pubescent along midrib only, adaxially glabrous, sometimes ciliate . . . . ***A. c.*** var. ***subglaber***

Astragalus curvicarpus (A. Heller) J.F. Macbride var. ***brachycodon*** (Barneby) Barneby

Curvepod milkvetch

Astragalus whitedii Piper var. *brachycodon* Barneby

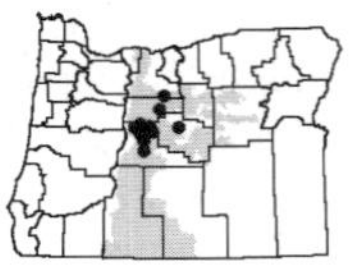

Leaflets pubescent, sometimes thinly pubescent adaxially. **Flowers** calyces 6.1–8.5(9.3) mm, tubes 5.4–7.6(8.6) mm, teeth 0.4–1 mm; corollas 13.6–15(16.8) mm, keels 9.4–11.6(12.3) mm; ovaries glabrous. **Fruits** glabrous; stipes 6–12(15) mm.

Juniper-sagebrush shrubland, sagebrush. Flowering May–Jul. 800–1400 m. ECas, Lava. Native.

This variety is endemic to central Oregon.

Astragalus curvicarpus (A. Heller) J.F. Macbride var. ***curvicarpus*** (p. 662)

Sickle milkvetch, curvepod milkvetch

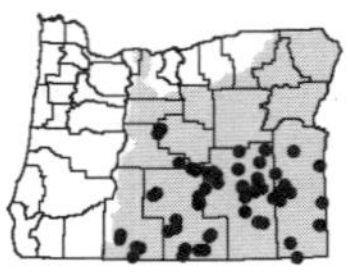

Leaflets pubescent. **Flowers** calyces (7.9)9.1–11.2 mm, tubes (6.9)7.5–9.4 mm, teeth (0.5)1–2.3 mm; corollas (15)16.4–21 mm, keels (11)11.7–15.2 mm; ovaries pubescent. **Fruits** villosulous or strigillose, rarely glabrate or glabrous; stipes (9)11–20 mm.

Sagebrush flats, juniper woodlands, roadsides, slopes. Flowering May–Aug. 800–2000 m. BR, BW, ECas, Lava, Owy. CA, ID, NV. Native.

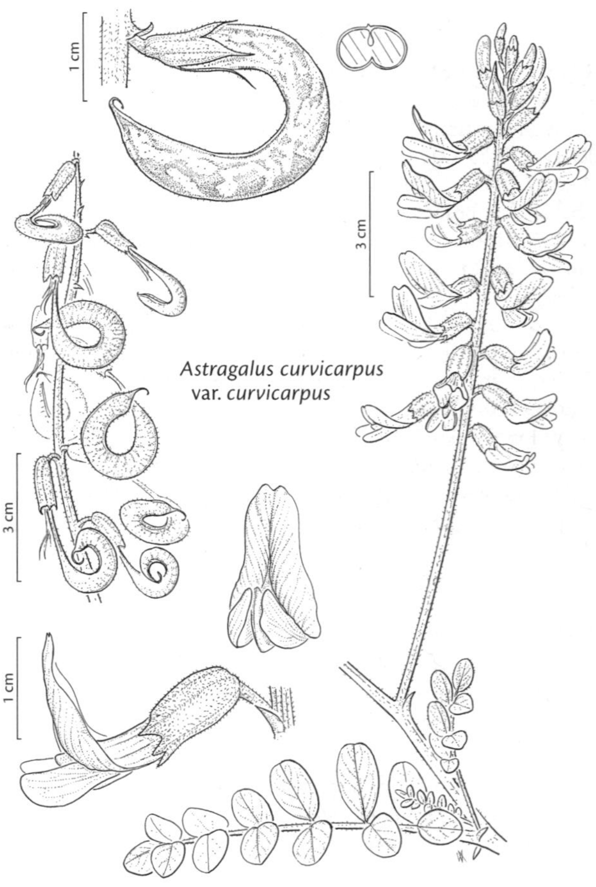

ASTRAGALUS

Astragalus curvicarpus (A. Heller) J.F. Macbride var. *subglaber* (Rydberg) Barneby

Smooth loco milkvetch, glabrous sickle milkvetch

Astragalus subglaber (Rydb.) M. Peck

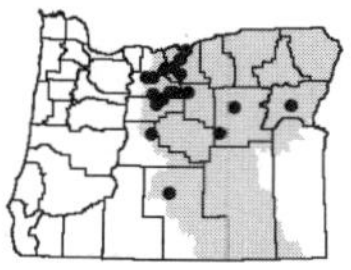

Leaflets abaxially glabrous to thinly pubescent or pubescent along midrib only, adaxially glabrous, sometimes ciliate. **Flowers** calyces (8)9–13.6 mm, tubes (7.5)8–11.9 mm, teeth 0.5–2.3 mm; corollas 13.5–19.5 mm, keels 11.5–14.5 mm; ovaries glabrous. **Fruits** glabrous; stipes 10–15 mm.

Roadsides, desert scabland, juniper grasslands, sagebrush. Flowering May–Jul. 200–1800 m. BR, BW, Col, Lava. Native.

This variety is endemic to eastern Oregon.

Astragalus cusickii A. Gray var. *cusickii*

Cusick's milkvetch

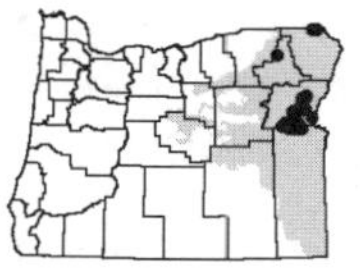

Plants perennial, caulescent, strigillose, hairs basifixed. **Stems** several–numerous, ascending to erect, 30–40(70) cm. **Leaves** 4–10(13) cm; leaflets 5–13, linear to linear-filiform, 2–22 × 1–2 mm, tips acute to obtuse, surfaces abaxially strigillose, adaxially strigillose or rarely glabrous, terminal leaflet jointed or confluent with rachis; stipules 0.5–4.5 mm, at least lowermost connate-sheathing. **Inflorescences** racemes loosely 3–14(18)-flowered; peduncles 5–15(22) cm; bracts 0.8–1.5 mm; pedicels 1.2–3.5 mm; bracteoles usually 0. **Flowers** declined at anthesis; calyces obliquely cuneate, 4–6.7 mm, strigillose to villous with black or mixture of black and white hairs, tubes 4.4–5.8 mm, teeth triangular, 0.5–1 mm; corollas 12.5–15.5 mm, white, becoming ochroleucous; ovules 10–20. **Fruits** unilocular, pendulous, bladdery inflated, obovoid, half-ellipsoid, 20–48 × 5–16 mm, not mottled, glabrous; valves membranous; stipes 3–6 mm. **2*n***=22.

Rocky and gravelly clay slopes, crevices of basalt cliffs, rangeland. Flowering May–Jul. 600–1200 m. BW, Owy. ID, WA. Native.

The stipitate, balloon-like pods are translucent at maturity so the seeds can be seen inside.

Astragalus diaphanus Douglas ex Hooker

Transparent milkvetch

Plants annual, rarely biennial, caulescent, strigillose to glabrate, hairs basifixed. **Stems** few–several, decumbent to ascending, 4–30(37) cm. **Leaves** 2–6 cm; leaflets (5)7–13, oblong-obovate, oblong-ovate or obovate-cuneate, 2–12 × 2–6 mm, tips obtuse, truncate or retuse, surfaces abaxially strigose, adaxially glabrous; stipules 1.5–4 mm, free. **Inflorescences** racemes loosely (3)7–15-flowered, peduncles 1–4(5.5) cm; bracts 0.8–1.8 mm; pedicels 0.6–1.8; bracteoles 0. **Flowers** declined at anthesis; calyces 3.5–4.5 mm, strigillose with all black or black and white hairs, tubes 2–2.6 mm, teeth lance-subulate, 1.3–1.9 mm; corollas 6.7–9.1 mm, whitish with lavender veins on banner, keels purple-tipped; ovules 6–14. **Fruits** unilocular or semibilocular, deflexed or declined, bladdery inflated or not, minutely strigillose; valves papery-membranous; stipes 0.

2 varieties.

Key to *Astragalus diaphanus* varieties:

1. Fruits linear-oblong to obliquely elliptic, not or slightly inflated, semibilocular. . ***A. d.*** var. ***diaphanus***

1' Fruits half-ovoid, strongly inflated or bladdery, unilocular. . . . . ***A. d.*** var. ***diurnus***

Astragalus diaphanus Douglas ex Hooker var. *diaphanus*

Transparent milkvetch

Astragalus drepanolobus A. Gray

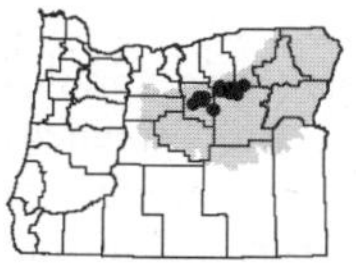

Fruits semibilocular, not or slightly inflated, linear-oblong to obliquely-elliptic, strongly arcuate when mature, obviously curved when young, 15–20 × 3–4 mm. **2*n***=28.

Juniper-ponderosa pine forest openings, roadbeds. Flowering May–Jul. 500–1200 m. BW, Lava. WA. Native.

Astragalus diaphanus Douglas ex Hooker var. *diurnus* (S. Watson) Barneby

John Day milkvetch

Astragalus diurnus S. Watson

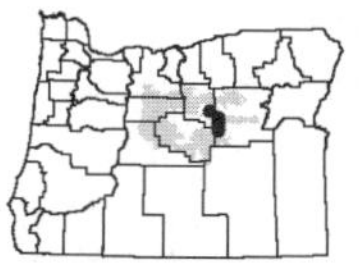

Fruits unilocular, strongly inflated or bladdery, half-ovoid, not arcuate at maturity, slightly curved when young; 15–20 × 6–12 mm.

Barren rocky slopes, sagebrush-juniper communities. Flowering May–Jun. 700–1400 m. Lava. Native.

This variety is currently known from only a few sites in Grant and Wheeler counties.

Astragalus eremiticus E. Sheldon (p. 664)

Hermit milkvetch

Astragalus eremiticus E. Sheld. var. *eremiticus*, *Astragalus eremiticus* E. Sheld. var. *malheurensis* (A. Heller) Barneby, *Astragalus eremiticus* E. Sheld. var. *spencianus* M.E. Jones, *Astragalus malheurensis* A. Heller

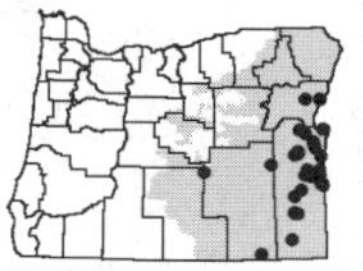

Plants perennial, caulescent, thinly to densely strigillose, hairs basifixed. **Stems** several, ascending to erect, 10–50 cm. **Leaves** (3)4.5–18 cm; leaflets ovate, elliptic to oblong, 5–26 × 1.5–6 mm, tips obtuse to retuse, surfaces abaxially strigillose, adaxially glabrate to glabrous; stipules 3–11 mm, free. **Inflorescences** racemes 10–25-flowered; peduncles (1.5)2.5–17 cm; bracts 1.5–4 mm; pedicels 0.7–3.5 mm; bracteoles 0–2. **Flowers** ascending, spreading, or declined at anthesis; calyces subsymmetric or strongly oblique at bases, 5.5–9.5 mm, strigillose with black or mixture of black and white hairs, tubes 4.5–8 mm, teeth triangular to subulate, 1–2.5 mm; corollas 12–20 mm, ochroleucous with keel tips usually purple, or petals purple with yellowish wing tips; ovules 17–32. **Fruits** bilocular, erect, ± straight, oblong to ellipsoid, trigonous, dorsally sulcate, 12–27(30) × 3.5–9 mm, glabrous; valves chartaceous or coriaceous; stipes 6–15(17) mm. **2*n***=24.

Ash-clay hillsides, grassy slopes, sagebrush-bunchgrass communities. Flowering Apr–Jul. 500–1800 m. BR, BW, Owy. ID, NV; southeast to AZ. Native.

Astragalus filipes Torrey ex A. Gray (p. 664)

Basalt milkvetch, threadstalk milkvetch

Astragalus stenophyllus Torr. & A. Gray, *Astragalus stenophyllus* Torr. & A. Gray var. *filipes* (Torr. ex A. Gray) Tidestr., *Astragalus stenophyllus* Torr. & A. Gray var. *stenophyllus*

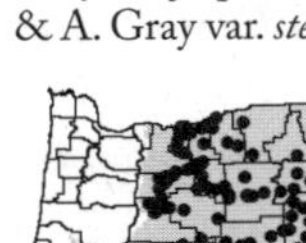

Plants perennial, caulescent, densely strigillose to glabrate, hairs basifixed. **Stems** several–numerous, erect or ascending, in large clumps, 20–90 cm. **Leaves** 2.5–12 cm; leaflets (5)9–23, linear, filiform or narrowly elliptic or oblong, 3–25(30) × 0.5–5 mm, tips obtuse to truncate, retuse or subacute, surfaces strigose or glabrous; stipules 2–5 mm, at least lowermost connate-sheathing. **Inflorescences** racemes loosely 4–30-flowered; peduncles 4.5–22 cm; bracts 1–3.5 mm; pedicels 1.5–6 mm; bracteoles usually 0. **Flowers** spreading to declined at anthesis; calyces 4–7.7 mm, strigillose with black and white hairs, tubes 3.3–6.4 mm, teeth triangular, 0.5–1.5 mm; corollas 10–15 mm, whitish to ochroleucous; ovules 11–22. **Fruits** unilocular, pendulous, oblong to elliptic, strongly laterally compressed, ± straight, bicarinate, 17–30(35) × 3–6.5 mm, glabrous to strigillose; valves papery; stipes 6–16 mm. **2*n***=22, 24.

Sagebrush, grasslands. Flowering May–Jul. 50–2400 m. BR, BW, Col, ECas, Lava, Owy, Sisk. CA, ID, NV, WA; north to British Columbia, southeast to UT, south Mexico. Native.

This species is one of the most abundant and widely dispersed milkvetches in Oregon, often forming large colonies. It is usually easily recognizable by its connate stipules, numerous erect stems in clumps, yellowish flowers and pendulous, stipitate, laterally compressed, straight pods.

Astragalus gambellianus Sheldon

Gambel's dwarf milkvetch, little blue loco

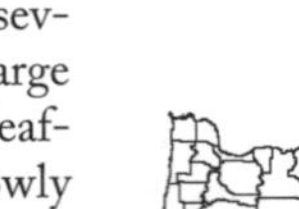

Plants annual, caulescent, pilosulose, hairs basifixed. **Stems** 1–several, erect to ascending, 3–30 cm. **Leaves** 1–4 cm; leaflets 7–15, oblong to oblanceolate, linear-oblanceolate, 1–9 × 1–2 mm, tips retuse, surfaces abaxially pubescent, adaxially pubescent or glabrescent; stipules 1–3 mm, free. **Inflorescences** racemes 4–15-flowered; peduncles (1)2–5(6) cm; bracts 0.5–1.3 mm; pedicels 0.2–0.9 mm; bracteoles 0. **Flowers** deflexed at anthesis; calyces 1.8–2.5 mm, black-pilosulose, tubes 1.1–1.7 mm, teeth subulate or triangular, 0.5–0.9 mm; corollas 2.5–3.3 mm, whitish-tinged, veined or margined with lavender or blue; ovules 2. **Fruits** bilocular, deflexed, broadly ovate or nearly circular in dorsal view, strongly obcompressed, 2.8–4.2 × 2.4–3.6 mm, hirsutulous to strigillose; valves papery; stipes 0. **2*n***=22.

Grassy slopes, clearings in chaparral. Flowering May–Jun. 600–1300 m. Sisk. CA; south to Mexico. Native.

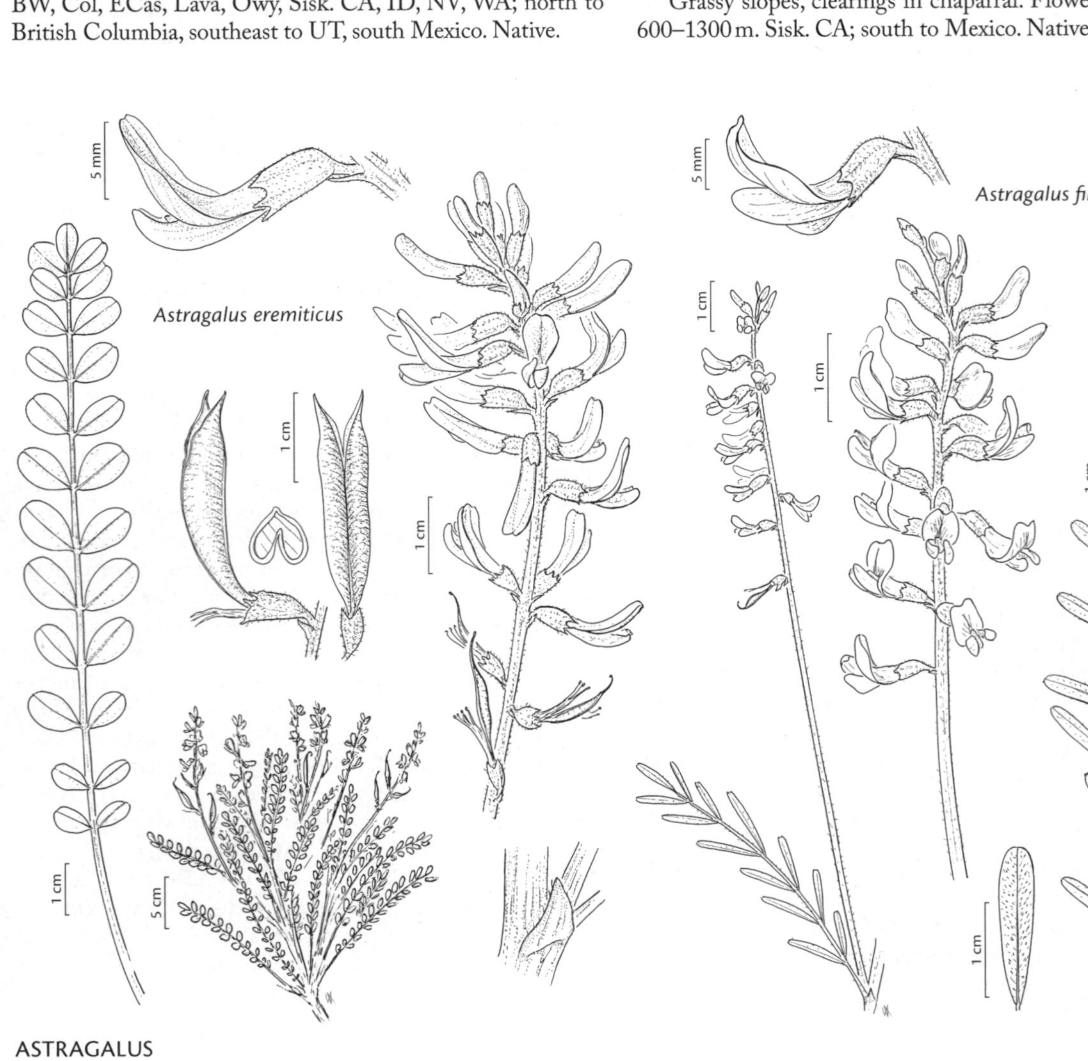

ASTRAGALUS

Astragalus geyeri A. Gray var. *geyeri* (p. 665)

Geyer's milkvetch

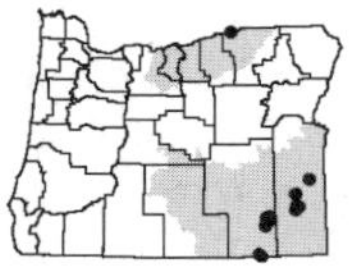

Plants annual, caulescent, strigose, hairs basifixed. **Stems** 1–many, decumbent to erect, 2–20(30) cm. **Leaves** (1.5)2–10.5(12) cm; leaflets 3–13, linear-oblong or -elliptic to oblanceolate, 3–18 × 1–5.2 mm, tips obtuse to retuse, surfaces abaxially strigose, adaxially strigose to glabrous; stipules 1.5–4 mm, free. **Inflorescences** racemes, loosely 2–8-flowered; peduncles 0.6–2.5 cm; bracts 0.7–2 mm; pedicels 0.6–1.5 mm; bracteoles 0–2. **Flowers** ascending at anthesis; calyces 2–4 mm, white- or fuscous-strigose, tubes 1.5–2.5 mm, teeth subulate to lanceolate, 0.6–1.5 mm; corollas 5–7.6 mm, whitish, suffused or veined with lilac; ovules 10–18. **Fruits** unilocular; spreading to declined, bladdery inflated, obliquely ovoid, 15–24 × 6–12 mm, strigose; valves papery-membranous; stipes 0.

Sand dunes, sandy alkaline flats, sagebrush. Flowering May–Jul. 800–1300 m. BR, Col, Owy. CA, ID, NV, WA; east to WY, southeast to UT. Native.

Astragalus hoodianus Howell

Hood River milkvetch

Astragalus conjunctus S. Watson var. *oxytropidoides* M.E. Jones, *Astragalus reventus* A. Gray var. *oxytropidoides* (M.E. Jones) C.L. Hitchc., *Cnemidophacos knowlesianus* Rydb.

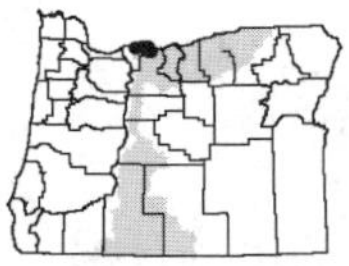

Plants perennial, subacaulescent or shortly caulescent, strigose, villosulous, or pilosulose, hairs basifixed. **Stems** several, ascending to erect, in clumps, 15–45 cm. **Leaves** (5)7–20 cm; leaflets 25–37, linear-elliptic, oblong, oblanceolate to ovate-oblong, 5–27 × 1.5–5 mm, tips obtuse, acute or retuse, surfaces pubescent, terminal leaflet often confluent with rachis; stipules (3)5–11 mm, at least lowermost connate-sheathing. **Inflorescences** racemes oval or oblong, soon elongating, densely (6)10–30-flowered; peduncles 12–30 cm; bracts 2–9.5 mm; pedicels 1–3.7 mm; bracteoles 0–2. **Flowers** spreading to ascending at anthesis; calyces 10–15 mm, strigillose-pilosulose with all black or mixture of black and white hairs, tubes 5.7–7.9 mm, teeth lanceolate or lance-caudate, (2.6)4.6–7 mm; corollas 18–23 mm, cream, unspotted; ovules 16–20. **Fruits** unilocular or incompletely bilocular, erect, broadly oblong-ovoid to narrowly oblong-ellipsoid, ± straight, obcompressed, 11–22 × 4.5–7.2 mm, hirsutulose or villosulous; valves coriaceous to subligneous; stipes 0.

Grassy, rocky or sandy slopes. Flowering Apr–Jun. 50–700 m. Col, ECas. WA. Native.

Astragalus howellii A. Gray

Howell's milkvetch

Astragalus howellii A. Gray var. *howellii*

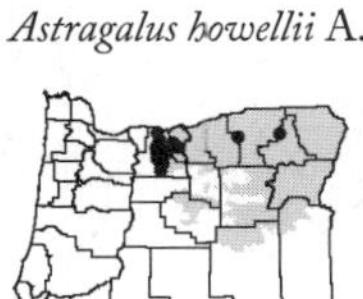

Plants perennial, caulescent, villosulous, hairs basifixed. **Stems** several, ascending to erect, forming bushy clumps, 5–40 cm. **Leaves** (4)5–14 cm; leaflets (17)21–27(33), oblong, obovate-cuneate or oblanceolate, 4–14 × 1–4 mm, tips obtuse to truncate-retuse, surfaces pubescent, adaxially sometimes glabrate to glabrous; stipules 3–5 mm, free. **Inflorescences** racemes loosely 10–25-flowered; peduncles 5.5–18 cm; bracts 1–2.5 mm; pedicels 1–4.5 mm; bracteoles 0.

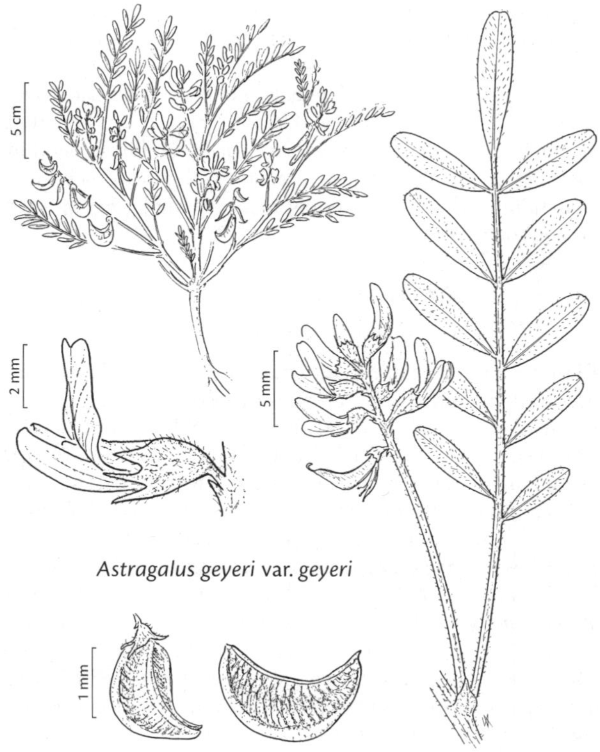

Astragalus geyeri var. *geyeri*

ASTRAGALUS

Flowers spreading or declined at anthesis; calyces 5–8 mm, strigose or villosulous with white or sometimes black hairs, tubes 4–5 mm, teeth subulate-triangular, 1–3 mm; corollas 6.5–8.5 mm, ochroleucous, unspotted; ovules 11–21. **Fruits** bilocular, pendulous, lunately lanceolate or linear-lanceolate, trigonously compressed, (16)20–30 × 3–5 mm, densely villosulous; valves stiffly papery; stipes (4)7–12(14) mm.

Scabland, dry rocky slopes and ridges, grassy hills. Flowering Apr–Jun. 50–1600 m. BW, Col. WA. Native.

Astragalus inflexus Douglas ex Hooker

Bent milkvetch

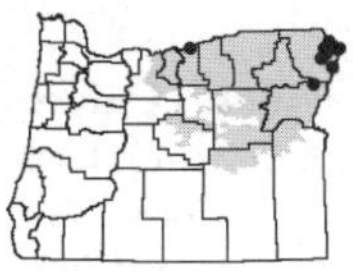

Plants perennial, caulescent, tufted, canescent, densely villous to tomentose, hairs basifixed. **Stems** several, prostrate and radiating with incurved-ascending tips, 10–50 cm. **Leaves** (3)4–12(16) cm; leaflets (9)17–23(27), obovate to broadly oblanceolate, (4)6–16(20) × 2–6 mm, tips acute to shortly acuminate, surfaces pubescent; stipules 5–12(16), free. **Inflorescences** racemes loosely (5)8–18-flowered; peduncles (2)3–8 cm; bracts 4.5–10 mm; pedicels 1.3–3.3 mm; bracteoles 0. **Flowers** ascending at anthesis; calyces (9.4)11.3–16.4 mm, usually purplish, villous with white hairs or rarely some black hairs, tubes (6.2)8.2–10.2 mm, teeth linear-lanceolate, 3.1–7 mm; corollas (16.5)19.5–23 mm, pink-purple; ovules 22–28. **Fruits** unilocular; ascending, obliquely ovoid, lance-ovoid or oblong-ellipsoid, ± straight proximally, incurved distally, obcompressed and sulcate at least ventrally, (13)15–25(30) × (5)7–9.5 mm, densely villous-hirsute, hairs rarely concealing valve surfaces; valves coriaceous; gynophores 0.5–1.8 mm. **2*n***=22.

Rocky slopes, dry grassy hillsides, river terraces. Flowering May–Jul. 400–900 m. BW, Col. ID, WA; northeast to MT. Native.

Astragalus inflexus is generally found along the Snake River and its tributaries in Idaho and Washington. It can be found on the banks of the Columbia River near The Great Bend in Washington, where the Snake enters the Columbia. A specimen from Gilliam County is probably an immigrant by way of seeds brought down by flood waters.

Astragalus iodanthus S. Watson

Humboldt River milkvetch

Plants perennial, caulescent, strigillose to villosulous, hairs basifixed. **Stems** several, prostrate to decumbent, 3–40 cm. **Leaves** 2–8 cm; leaflets 7–21, obovate to oblong or oblanceolate to elliptic, 3–18 × 2–12 mm, tips truncate to retuse, obtuse or mucronate, surfaces sparsely strigose to glabrous; stipules 2–6 mm, free. **Inflorescences** racemes 7–21-flowered; peduncles 1–4.5 cm; bracts 1–3 mm; pedicels 0.3–2 mm; bracteoles 0–2. **Flowers** widely ascending or spreading at anthesis; calyces 3.3–8 mm, strigillose or rarely villosulous with black or mixture of black and white hairs, tubes 2.6–5 mm, teeth subulate or triangular, 1–3 mm; corollas 9–15.5 mm, bright purple to lilac or whitish, sometimes lilac-tinged or -tipped; ovules 14–30. **Fruits** unilocular to semibilocular; deflexed, humistrately spreading, lanceolate, linear-lanceolate or lance-elliptic, curved through a quarter- to half-circle or more, dorsiventrally or trigonously compressed, ventrally sulcate, 15–40 × 2.5–8.5 mm, often mottled, strigillose or glabrous; valves stiffly papery or subcoriaceous; stipes 0.

2 varieties.

Key to *Astragalus iodanthus* varieties:

1\. Calyx tubes 2.6–3.3 mm; corollas pale lilac or whitish and sometimes lilac-tinged or -tipped, banners 7.5–11.5 mm. . . . . ***A. i.*** var. ***diaphanoides***

1' Calyx tubes (3.4)3.7–5 mm; corollas bright purple, banners 11–15 mm. . . . . ***A. i.*** var. ***iodanthus***

Astragalus iodanthus S. Watson var. *diaphanoides* Barneby

Snake milkvetch, violet milkvetch

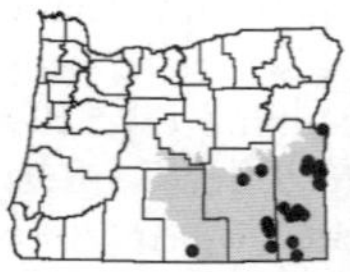

Stems usually finely strigillose. **Calyces** (3.3)3.7–5.3 mm, tubes 2.6–3.3 mm, teeth (0.6)1–2.2 mm; corollas pale lilac or whitish and sometimes lilac-tinged or -tipped, banners 7.5–11.5 mm.

Alkaline flats, ash-clay outcrops and slopes. Flowering Apr–Jun. 800–1500 m. BR, Owy. CA, ID, NV. Native.

Astragalus iodanthus S. Watson var. *iodanthus*

Humboldt River milkvetch

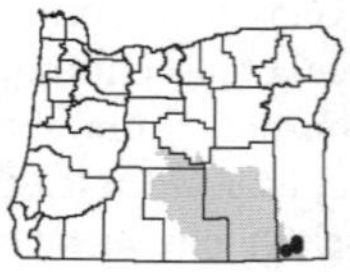

Stems glabrate. **Calyces** (5)5.5–8 mm, tubes (3.4)3.7–5 mm, teeth (1.3)1.5–3 mm; corollas bright purple, banners 11–15 mm. **2*n***=22.

Sagebrush-bunchgrass communities. Flowering May–Jul. 1300–2300 m. BR. CA, NV; southeast to UT. Native.

Astragalus kentrophyta A. Gray

Thistle milkvetch, spiny milkvetch, kentrophyta

Plants perennial, caulescent, mat-forming, somewhat prickly, strigillose, villosulous or hirsutulous, hairs basifixed or malpighian. **Stems** prostrate, much branched, mats 5–50 cm in diameter. **Leaves** 0.4–2.6 cm; leaflets 3–9, confluent with rachis, linear to narrowly elliptic or lanceolate, (1)3–13(17) × 0.5–1.5 mm, tips softly or stiffly spine-tipped, mucros or spinules, surfaces strigose; stipules 1.5–5 mm, at least lowermost connate-sheathing. **Inflorescences** racemes, 1–3-flowered, appearing axillary; peduncles 0–1.5(3) cm; bracts 0.8–3.5 mm; pedicels 0.5–2 mm; bracteoles 0. **Flowers** declined at anthesis; calyces 2.4–8.3 mm, strigose or hirsutulous with white, black or mixture of white and black hairs, tubes 1.3–3 mm, teeth subulate, 0.8–5 mm; corollas 3.9–10 mm, pink, purple or whitish to ochroleucous; ovules 2–8. **Fruits** unilocular, declined or spreading, elliptic to oblong, usually curved, laterally compressed, bicarinate, 3–10 × 1.3–4 mm, strigose; valves thick-papery; stipes 0. **2*n***=24.

Western North America. 9 varieties; 2 varieties treated in *Flora.*

This species of *Astragalus* is easily recognizable by its mat-forming growth habit, spinose, confluent leaflets, and tiny, subsessile to peduncled, few-flowered racemes. Thistle milkvetch is found in much of western North America from alpine to lowland habitats. This has led to infraspecific variation, which is complex and intergrading.

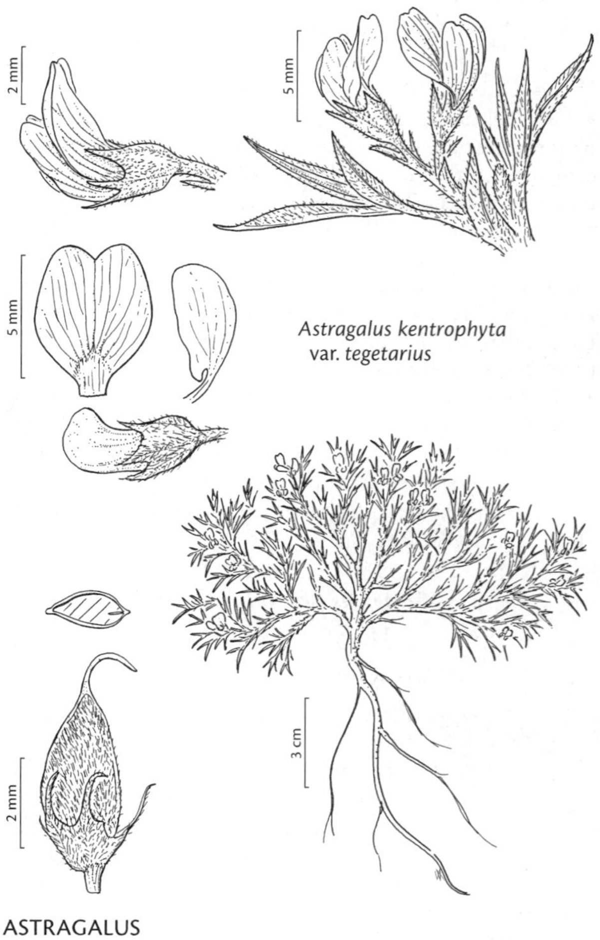

ASTRAGALUS

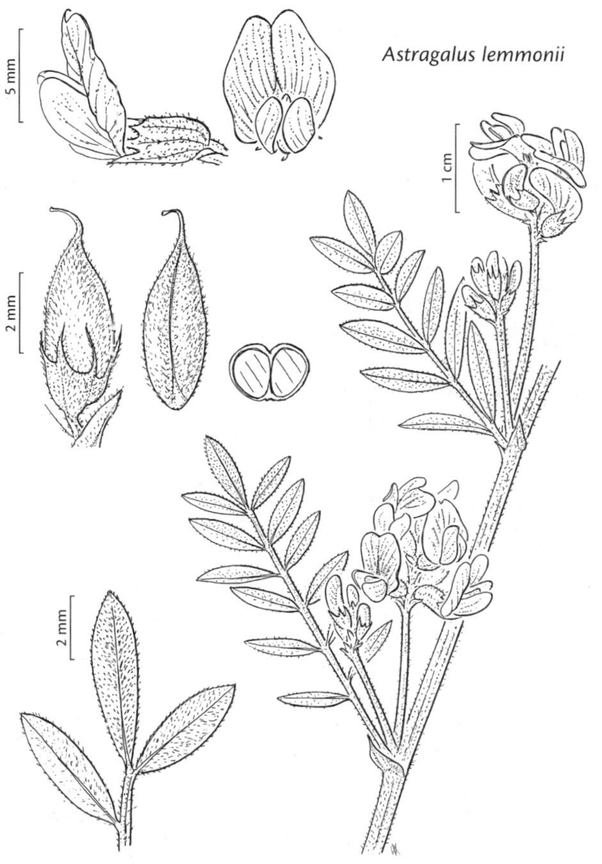

ASTRAGALUS

Key to *Astragalus kentrophyta* varieties:

1. Plant hairs mostly malpighian; corollas whitish, banners and keel tips sometimes faintly pink- or lilac-tinged. . . . . ***A. k.*** var. ***jessiae***

1' Plant hairs basifixed; corollas purple or purplish, sometimes white with pink or purplish keel tips. . . . . ***A. k.*** var. ***tegetarius***

Astragalus kentrophyta A. Gray var. ***jessiae*** (M. Peck) Barneby

Jessie's kentrophyta, spiny milkvetch

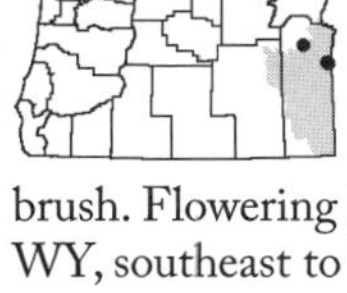

Plants strigose, hairs mostly malpighian. **Leaflets** 5. **Corollas** 5.2–6.5 mm, whitish, banners and keel tips sometimes faintly pink- or lilac-tinged. **Fruits** 3–4.5 mm.

Bluffs, dunes, dry, open areas in sagebrush. Flowering May–Jul. 800–1800 m. Owy. ID, NV; east to WY, southeast to UT. Native.

Douglas' kentrophyta, *Astragalus kentrophyta* var. *douglasii*, is a probably extinct taxon from the Walla Walla region of Washington, likely near The Great Bend of the Columbia River, and so could be in adjacent Oregon. It would resemble variety *jessiae* but has a definite beaked fruit whereas variety *jessiae* is beakless or nearly so.

Astragalus kentrophyta A. Gray var. ***tegetarius*** (S. Watson) Dorn (p. 666)

Mat milkvetch, mountain kentrophyta

Astragalus kentrophyta A. Gray var. *implexus* (Canby ex Porter & J.M. Coult.) Barneby, *Astragalus tegetarius* S. Watson

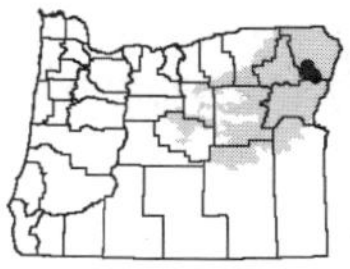

Plants densely to thinly strigose, villosulous or villous, hairs basifixed. **Leaflets** (3)5–9. **Corollas** (3.9)4.5–8(9.2) mm, purple or purplish, sometimes white with pink or purplish keel tips. **Fruits** (3)4–8(9) mm.

Near and above timberline, alpine ridges, gravel creek bars. Flowering Jun–Sep. 1700–3000 m. BW. CA, ID, NV; northeast to MT, east to WY, southeast to NM. Native.

Astragalus lemmonii A. Gray (p. 667)

Lemmon's milkvetch

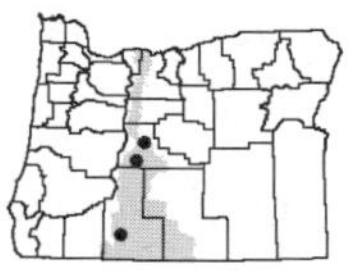

Plants perennial, caulescent, sparsely strigillose, hairs basifixed. **Stems** several–numerous, prostrate, loosely matted, 10–50 cm. **Leaves** (1)1.5–4.5 cm; leaflets (7)9–15, narrowly elliptic to elliptic-oblanceolate, 2–11 × 1–2 mm, tips obtuse to subacute, surfaces abaxially strigillose, adaxially glabrate or glabrous; stipules 2–5 mm, free. **Inflorescences** racemes, subcapitate, (2)5–13-flowered; peduncles 0.6–1.7 cm; bracts 0.8–2 mm; pedicels 0.7–2.5 mm; bracteoles 0. **Flowers** loosely ascending at anthesis; calyces 3–4 mm, strigillose with white or sometimes a few black hairs, tubes 1.7–2.2 mm, teeth subulate, 1.1–1.7 mm; corollas 4.8–6.1 mm, whitish or tinged with lilac, banners sometimes purple-veined; ovules 4–8. **Fruits** bilocular, spreading or somewhat declined, elliptic or oblong-elliptic, compressed trigonous, straight or somewhat incurved, ventrally carinate, 4–7 × 1.5–2.5 mm, strigillose; valves papery; stipes 0.

Meadows and flats along streams and lakeshores. Flowering May–Jul. 1200–2000 m. ECas. CA, NV. Native.

This species is distinguished by its racemes of tiny flowers borne on two or three short peduncles per node and its occupation of moist alkaline habitats on the shores of streams or lakes.

Astragalus lentiginosus Douglas ex Hooker

Freckled milkvetch

Plants perennial, caulescent, strigillose, villosulous, pilose, or glabrate, hairs basifixed. **Stems** few–numerous, prostrate to decumbent or erect, 10–100 cm. **Leaves** 1–17(24) cm; leaflets (5)11–29, ovate to narrowly elliptic-lanceolate, obovate, oblong-elliptic, 2–21(30) × 2–11 mm, tips obtuse or retuse; stipules 2–7 mm, free. **Inflorescences** subumbels or racemes, (3)5–35(48)-flowered; peduncles 1–7 cm; bracts 1–2.5 mm; pedicels 1–2.5 mm; bracteoles usually 0. **Flowers** ascending to spreading at anthesis; calyces 3.8–12.5 mm, strigillose or pilosulose with white or black hairs, tubes 2.8–9 mm, teeth subulate or triangular, 0.6–5 mm; corollas 7.4–21.4 mm, whitish to yellowish or faintly lilac, sometimes purple, keel tips usually spotted; ovules (7)10–42. **Fruits** bilocular or semibilocular, ascending or spreading, often humistrate, straight to incurved, bladdery inflated or not, turgid, often mottled, strigillose to glabrous; valves stiffly coriaceous to papery-membranous, opaque to translucent; stipes 0.

Western North America. ~40+ varieties; 4 varieties treated in *Flora*.

Astragalus lentiginosus is a polymorphic complex ranging from British Columbia to Baja California, west to the Rocky Mountains, east of the Rockies in Colorado, New Mexico, and Texas. This group has been treated in the past as up to 33 separate species but is now considered a single species with over 40 varieties. Many varieties are localized either geographically or ecologically; others are widespread.

Key to *Astragalus lentiginosus* varieties:

1. Corollas (12.6)14–21.4 mm . . . . . *A. l* var. ***platyphyllidius***
1' Corollas 7.4–11.5(13.3) mm.
  2. Fruit valves stiffly papery, ± opaque . . . . *A. l* var. ***lentiginosus***
  2' Fruit valves thinly papery, transparent or translucent.
    3. Racemes loosely flowered; fruits in loose heads. . . . . *A. l* var. ***salinus***
    3' Racemes densely flowered; fruits forming globose or oblong heads . . . . *A. l.* var. ***floribundus***

Astragalus lentiginosus Douglas ex Hooker var. ***floribundus*** A. Gray

Carson City milkvetch, floriferous freckled milkvetch

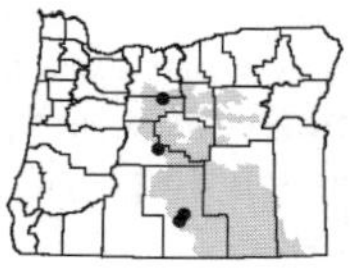

Plants glabrous or thinly strigillose. **Stems** prostrate to weakly ascending, 20–50 cm. **Leaves** 3–8(11) cm; leaflets 11–15(19), oblong-obovate, obovate, or oblanceolate, 5–15 × 5–8 mm, tips retuse, rarely subacute, surfaces glabrous, sometimes thinly strigillose adaxially. **Inflorescences** racemes, densely 10–30(40)-flowered; peduncles 1–4(7) cm. **Flowers** calyces 4–6.5 mm, strigillose, tubes 3–4 mm, teeth subulate, 0.6–2 mm; corollas 8.8–11 mm, whitish, keels and wings sometimes lavender-tipped; ovules 15–21(25). **Fruits** crowded into dense globose or oblong heads, humistrate, ovoid or subglobose, strongly bladdery inflated; (8)12–21 × (6)7–12 mm, faintly mottled, glabrous or thinly strigillose; valves thinly papery, transparent or translucent.

Sagebrush communities. Flowering May–Jul. 700–1700 m. BR, Lava. CA, NV. Native.

Astragalus lentiginosus Douglas ex Hooker var. ***lentiginosus***

Freckled milkvetch

Astragalus lentiginosus Douglas var. *carinatus* M.E. Jones

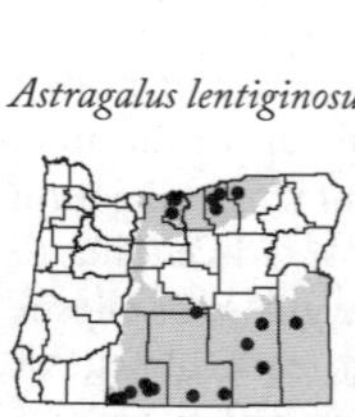

Plants thinly strigillose. **Stems** prostrate to ascending, 10–30(55) cm. **Leaves** 3–10 cm; leaflets (5)9–17(19), broadly obovate, obovate-cuneate, oblong-elliptic to suborbicular or oblanceolate, (3)5–15(18) × 2–11 mm, tips retuse or obtuse, sometimes truncate, surfaces thinly pubescent to glabrous. **Inflorescences** racemes 8–18(22)-flowered; peduncles 1–3.5 cm. **Flowers** calyces 4.1–6.4 mm, thinly strigillose with black, white or mixed hairs, tubes 2.8–4.2 mm, teeth subulate, 1–2.2 mm; corollas 7.4–11 mm, whitish or yellowish, sometimes faintly lilac; ovules 15–21. **Fruits** humistrate, obliquely ovate to lance-acuminate, incurved, strongly to scarcely inflated, 10–23 × (3)4–10 mm, sometimes green or red mottled, thinly strigillose or rarely glabrous; valves stiffly papery, opaque. **2*n***=22.

Fields, ponderosa pine forests, sagebrush communities. Flowering May–Jul. 200–1600 m. BR, Col, ECas, Owy. CA, ID, NV, WA; north to British Columbia. Native.

Astragalus lentiginosus Douglas ex Hooker var. ***platyphyllidius*** (Rydberg) M. Peck

Broad-leaved freckled milkvetch

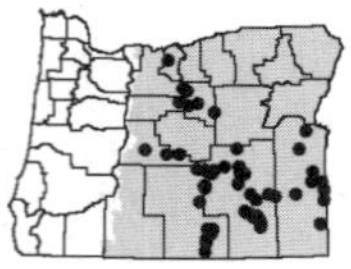

Plants subglabrous. **Stems** decumbent or ascending, (7)10–35 cm. **Leaves** (4)5–11 cm; leaflets (7)11–17(19), broadly obovate-cuneate, elliptic, or suborbicular, (4)7–20 × 3–11 mm, tips obtuse, retuse, truncate and apiculate, surfaces abaxially glabrous or thinly strigillose along the midrib, adaxially glabrous, sometimes ciliate. **Inflorescences** racemes loosely (5)7–15-flowered; peduncles 1–5 cm. **Flowers** calyces 8–12.5 mm, strigillose, tubes 5–9 mm, teeth 2.4–5 mm; corollas (12.6)14–21.4 mm, whitish, keel tips sometimes purplish; ovules (21)24–32. **Fruits** plumply ovoid or narrowly lance-ellipsoid, nearly straight or incurved, strongly or little inflated; (13)15–40(48) × 7–14 mm, minutely strigillose to glabrous; valves coriaceous or stiffly papery, opaque.

Sagebrush communities, flats, slopes. Flowering May–Jul. 500–2500 m. BR, BW, Col, ECas, Lava, Owy. CA, ID, NV; east to WY, southeast to CO. Native.

Astragalus lentiginosus Douglas ex Hooker var. ***salinus*** (Howell) Barneby

Harney milkvetch, salty freckled milkvetch

Astragalus lentiginosus Douglas var. *macrolobus* (Rydb.) Barneby, *Astragalus salinus* Howell

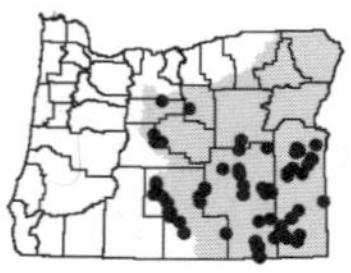

Plants subglabrous to strigillose or villosulous. **Stems** decumbent to ascending, 6–45 cm. **Leaves** 4–11 cm; leaflets 9–19, broadly obovate, obovate-cuneate, obcordate or oblong to oblanceolate, 4–20 × 2–9 mm, tips retuse or emarginate, rarely obtuse and apiculate or subacute, surfaces glabrous or pubescent. **Inflorescences** racemes loosely 10–30-flowered; peduncles 2–4.5(5) cm. **Flowers** calyces 5–6.4 mm, strigillose, tubes 3.6–4.6 mm, teeth subulate, 1–3 mm; corollas 9.5–11.5(13.3) mm, whitish or keels and wings sometimes lavender-tipped; ovules (7)16–25. **Fruits** in loose heads, humistrate, obliquely ovoid or subglobose, strongly bladdery inflated; 14–26(30) × (6)7–14 mm, green or mottled, glabrous or puberulent; valves thinly papery, transparent or translucent.

Saline flats, sand dunes, sagebrush, slopes. Flowering May–Aug. 700–1800 m. BR, BW, Lava, Owy. CA, ID, NV; northeast to MT, east to WY, southeast to UT. Native.

Astragalus malacus A. Gray

Shaggy milkvetch

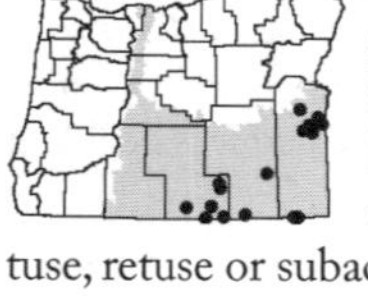

Plants perennial, caulescent, prominently grayish hirsute to pilose, hairs basifixed. **Stems** several, decumbent to erect, 2–30 cm. **Leaves** 4–15 cm; leaflets 7–21, elliptic to obovate, 5–20(22) × 4–8(13) mm, tips obtuse, retuse or subacute, sometimes minutely apiculate, surfaces

abaxially pubescent, adaxially pubescent or glabrescent; stipules 7–17 mm, free. **Inflorescences** racemes loosely 9–35-flowered; peduncles 4–11 cm; bracts 4–9 mm; pedicels 0.7–3 mm; bracteoles 2. **Flowers** declined at anthesis; calyces 8.6–14 mm, often purplish-tinged, hirsute with black or mixture of black and white hairs, tubes 6–10 mm, teeth subulate, 1.5–5 mm; corollas 12–19 mm, reddish purple; ovules 15–24. **Fruits** bilocular; declined, narrowly oblong or oblong-elliptic, shallowly to deeply incurved, dorsally deeply sulcate, triquetrously compressed, 18–38 × 4.5–6 mm, shaggy hirsute; valves stiffly papery; stipes 1–3 mm.

Sagebrush, rocky slopes. Flowering Apr–Jun. 700–1900 m. BR, ECas, Owy. CA, ID, NV. Native.

Astragalus misellus S. Watson var. *misellus*

Pauper milkvetch

Astragalus howellii A. Gray var. *aberrans* (M.E. Jones) C.L. Hitchc.

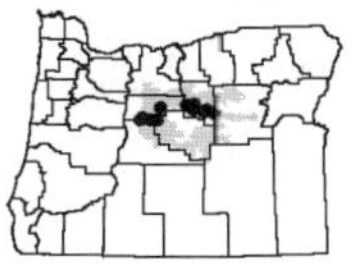

Plants perennial, caulescent, strigillose, hairs basifixed. **Stems** several–numerous, prostrate to ascending, 10–25 cm. **Leaves** 2.5–7 cm; leaflets (9)13–21, oblong-obovate to linear-oblong or -elliptic, 2–10 × 0.5–3 mm, tips subacute or retuse, surfaces abaxially strigillose, adaxially glabrate to glabrous; stipules 1.5–3 mm, free. **Inflorescences** racemes loosely 5–15-flowered; peduncles 1.5–8(10) cm; bracts 1–1.5 mm; pedicels 0.8–1.7 mm; bracteoles 0. **Flowers** horizontal, spreading, or declined at anthesis; calyces 3.4–4.8 mm, strigose with white or mixed white and black hairs, tubes 2.2–2.8 mm, teeth lance-subulate, 0.9–2.2 mm; corollas 6.5–10 mm, yellowish or greenish ochroleucous, sometimes slightly rose-tinged; ovules 11–17. **Fruits** bilocular, pendulous, lunately lance-elliptic, compressed triquetrous, 12–25 × 2.6–4 mm, strigillose-villosulous; valves papery; stipes 2.5–5 mm.

Scabland, rocky sagebrush slopes, clay hillsides. Flowering Apr–Jul. 600–1400 m. Lava. Native.

This variety is endemic to central Oregon. A second variety, var. *pauper*, is found only in Washington.

Astragalus mulfordiae M.E. Jones

Mulford's milkvetch

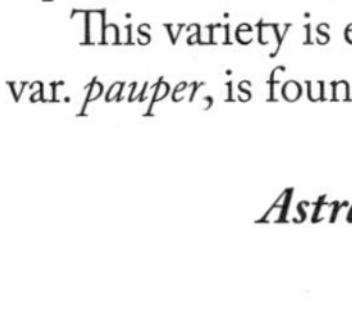

Plants perennial, caulescent, strigillose, hairs basifixed. **Stems** several–numerous, spreading to ascending, 10–30 cm. **Leaves** (2.5)4.5–10.5 cm; leaflets (7)15–25, linear-oblong, linear-oblanceolate, or filiform, terminal leaflet jointed to or confluent with rachis, 1–11 × 0.5–1.5 mm, tips obtuse or subacute, surfaces abaxially strigillose, adaxially glabrous; stipules 1.5–6 mm, at least lowermost connate-sheathing. **Inflorescences** racemes, loosely or remotely (5)8–20-flowered; peduncles (0.5)1–7 cm; bracts 0.7–2 mm; pedicels 0.7–2 mm; bracteoles 0–2. **Flowers** declined at anthesis; calyces 2.5–5 mm, usually white strigillose, tubes 1.5–3 mm, teeth subulate, 0.8–2 mm; corollas 6–8.2 mm, white or whitish, banners sometimes lilac-veined or -tinged; ovules 10–16. **Fruits** bilocular, pendulous, lunately half-ellipsoid, gently incurved, triquetrously compressed, 9–16 × 3–4.5 mm, finely strigillose; valves papery; stipes 3–5 mm.

Sagebrush, sandy bluffs. Flowering May–Jun. 600–1300 m. Owy. ID. Native.

Connate lower stipules, distant, narrow green leaflets, white or lilac-tinged flowers, and nodding, stipitate, trigonous, and bilocular fruits characterize this species.

Astragalus multiflorus (Pursh) A. Gray (p. 669)

Pulse milkvetch

Astragalus tenellus Pursh

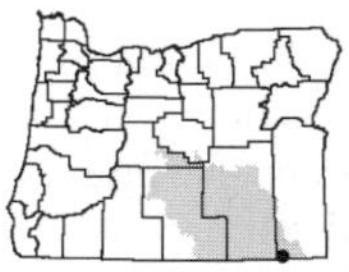

Plants perennial, caulescent, strigillose, hairs basifixed. **Stems** several–numerous, prostrate to ascending or erect, 10–50 cm. **Leaves** 2–9 cm; leaflets (7)11–21(25), narrowly oblong to elliptic, linear, oblanceolate or obovate, 3–24(27) × 0.4–6 mm, tips acute to obtuse, mucronate or emarginate, surfaces abaxially strigose, adaxially glabrous; stipules 1.5–8 mm, at least lowermost connate-sheathing. **Inflorescences** racemes loosely or remotely 3–23-flowered; peduncles 0.2–4 cm; bracts 0.5–3 mm; pedicels 0.7–3.2 mm; bracteoles 0–2. **Flowers** ascending at anthesis; calyces 2.5–6 mm, strigillose with black, white, or a mixture of black and white hairs, tubes 2–3 mm, teeth subulate or linear, 0.7–3 mm; corollas 6–9(11) mm, white to ochroleucous or pink-purple; ovules 3–9. **Fruits** unilocular, pendulous, elliptic to oblong, straight or curved, laterally flattened, bicarinate, 7–17 × 2.5–4.5 mm, dark brown to black at maturity, usually glabrous or with few scattered white hairs; valves papery; stipes 1–7 mm. ***2n***=24.

Sagebrush communities. Flowering May–Aug. 1200–1800 m. BR. ID, NV; north to Yukon Territory, northeast to Manitoba, east to MN, southeast to CO. Native.

This widely distributed species has only recently been found in Malheur County; it is probably native.

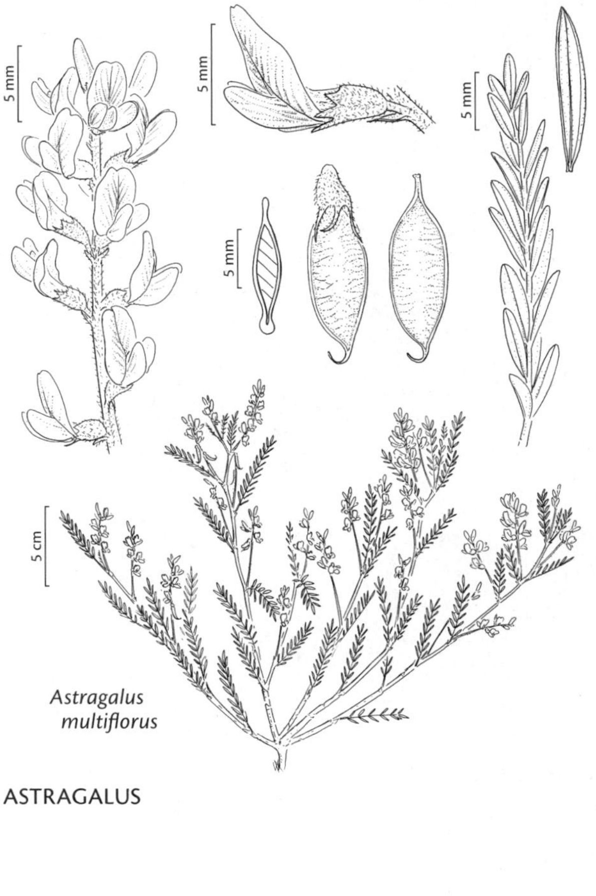

ASTRAGALUS

Astragalus newberryi A. Gray var. *castoreus* M.E. Jones

(p. 670)

Beaver Dam Mountains milkvetch

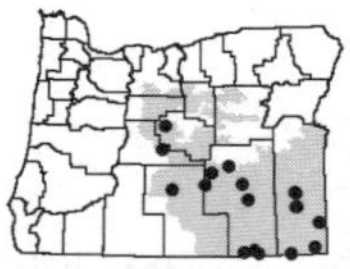

Plants perennial, acaulescent, usually with a thatch of persistent leaf bases, 2–12 cm tall, silky-pilose or -strigose and shortly tomentulose, hairs basifixed. **Stems** 0. **Leaves** 1.5–14(5) cm; leaflets 7–11(13), obovate to elliptic, oblanceolate, or orbicular, 3–20 × 2–14 mm, tips acute to obtuse or retuse, surfaces villous-tomentulose; stipules (2.5)4–14 mm, free. **Inflorescences** racemes subcapitate, 2–8-flowered; peduncles 0.5–11 cm; bracts 3.5–10 mm; pedicels 1.4–5 mm; bracteoles 0–2. **Flowers** ascending at anthesis; calyces 12–20 mm, often purplish-tinged, mainly white-pilose, tubes 10–16 mm, teeth subulate to lanceolate, 1.5–4 mm; corollas 21–32 mm, pink-purple; ovules 27–46. **Fruits** unilocular, ascending, humistrate, obliquely ovoid to ovoid-acuminate, obcompressed, curved, (13)17–28 × 7–13 mm, villous-hirsute and tomentulose; valves coriaceous; stipes 0.

Sagebrush, ridgetops, slopes. Flowering May–Jun. 900–1800 m. BR, Lava, Owy. CA, ID, NV; southeast to AZ. Native.

Astragalus newberryi var. *castoreus* may be misidentified as *A. purshii* var. *glareosus*, from which it can be distinguished by its strict acaulescent habit and persistent leaf bases.

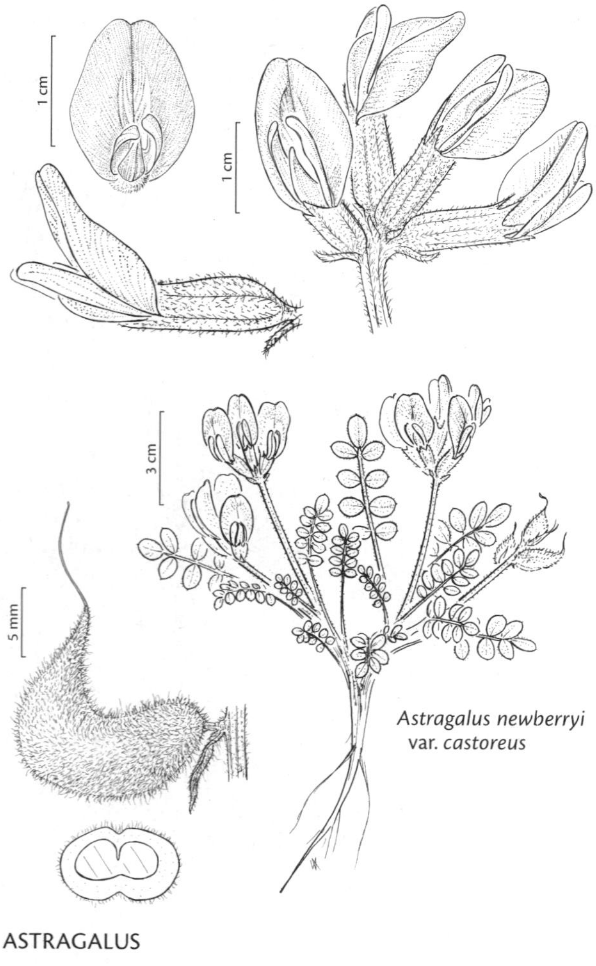

Astragalus newberryi var. castoreus

ASTRAGALUS

Astragalus nudisiliquus A. Nelson

Cobblestone milkvetch

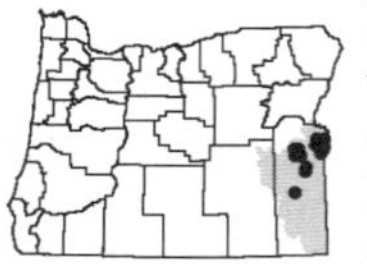

Plants perennial, caulescent, mat-forming, white tomentose, hairs basifixed. **Stems** several–numerous, prostrate, (2)5–25(30) cm. **Leaves** 3–11 cm; leaflets (7)11–17, broadly obovate, ovate-cuneate or broadly oblanceolate, 4–17 × 2.5–6.5 mm, tips obtuse, emarginate or subacute, surfaces abaxially white tomentose, adaxially tomentose, sometimes more thinly so and greenish; stipules (1.5)3–7 mm, free. **Inflorescences** racemes subcapitate, (2)4–8-flowered; peduncles 1–7 cm; bracts (1.5)3–6 mm; pedicels 1.4–3.4 mm; bracteoles 0–2. **Flowers** ascending at anthesis; calyces 12–17(19) mm, often purplish-tinged, thinly villosulous with mainly white hairs, tubes 10–14 mm, teeth subulate or subulate-setaceous, (2)3–5 mm; corollas 19–25.5 mm, pink-purple; ovules 30–41. **Fruits** unilocular, ascending, humistrate, obliquely-ovoid or oblong-ellipsoid, obcompressed, (15)20–40(45) × (8)9–13 mm, villous-hirsute, hairs not concealing valve surfaces; valves coriaceous; stipes 0.

Rocky slopes, sandy riverbanks, cobblestone alluvia, eroded slopes with sagebrush. Flowering May–Jun. 600–1400 m. Owy. ID. Native.

Astragalus nudisiliquus may resemble *A. purshii*, but the latter's valve surfaces are usually completely concealed by dense pubescence. In *A. nudisiliquus*, the valve surfaces are easily seen through the hairs.

Astragalus obscurus S. Watson

Arcane milkvetch

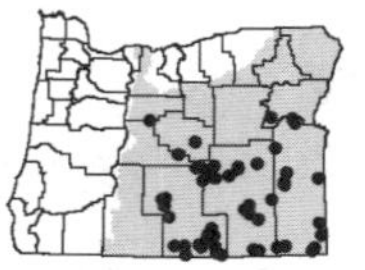

Plants perennial, caulescent, strigillose, hairs basifixed. **Stems** several–numerous, decumbent to ascending, 15–40 cm. **Leaves** (2.5)4–10(12) cm; leaflets 5–15, oblong-elliptic to lance-oblong or linear-elliptic, 2–15 × 0.5–2.5 mm, tips obtuse to truncate to shallowly retuse or subacute, surfaces abaxially strigillose, adaxially thinly pubescent to glabrous; stipules 1.5–2.5 mm, free. **Inflorescences** racemes compact in early anthesis, soon elongating, (3)6–15-flowered; peduncles 3–15 cm; bracts 0.8–2.5 mm; pedicels 0.5–2.2 mm; bracteoles 0–2. **Flowers** ascending at anthesis; calyces 3–5.5 mm, strigillose with black and white hairs, tubes 2.3–4 mm, teeth triangular to subulate, 0.4–1.6 mm; corollas 7–10.5 mm, ochroleucous or whitish, often tinged with dull purple; ovules 14–23. **Fruits** ± bilocular, erect, linear-oblong, straight, obtusely triquetrous, 10–25 × 2.4–4 mm, finely strigillose; valves coriaceous to cartilaginous; stipes 0.

Sagebrush, ridge tops, talus, hillsides, scabland. Flowering May–Jul. 700–2300 m. BR, BW, ECas, Lava, Owy. CA, ID, NV. Native.

Astragalus obscurus is recognizable by its few leaflets, small, dingy flowers ascending in a close raceme, and erect, sessile, linear-oblong, trigonously compressed fruits. It may be confused with the sympatric *A. atratus*, which is alike in general habit but has pendulous, decurved pods.

Astragalus peckii Piper

Peck's milkvetch

Astragalus lentiformis A. Gray

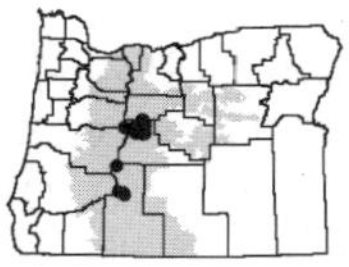

Plants perennial, caulescent, with a thatch of stiff persistent petioles, grayish strigose-pilosulose, hairs basifixed. **Stems** several–numerous, erect when young, prostrate with age, (1)5–20 cm. **Leaves** 1–4 cm; leaflets 6–14, lanceolate, oblanceolate or linear-elliptic, 1–6 × 0.5–2 mm, tips acute or subacute, surfaces abaxially pubescent, adaxially pubescent or glabrescent, terminal leaflet confluent with rachis, filiform or slightly expanded, sometimes as recurved hook; stipules 1–3 mm, connate-sheathing. **Inflorescences** racemes loosely 5–9-flowered; peduncles 0.7–2 cm; bracts 1–1.5 mm; pedicels 1–2.5 mm; bracteoles 0. **Flowers** declined at anthesis; calyces 3.3–4.4 mm, densely strigillose with white hairs, tubes 1.8–2.1 mm, teeth subulate, 1.4–2.4 mm; corollas 5–7.5 mm, ochroleucous, banners purple-veined; ovules 7–9. **Fruits** bilocular, deflexed, oblong-ellipsoid, ovate-elliptic to lenticular, compressed triquetrous, 5–8 × 1.9–3.3 mm, strigose to villosulous; valves papery; stipes 0–0.5 mm.

Pumice and sandy soils, sagebrush. Flowering May–Aug. 900–1500 m. Casc, ECas, Lava. Native. ✿

This endemic species is notable for its small, purple-veined flowers; small, compressed triquetrous fruits; prominent connate stipules; and thatch of persistent petioles.

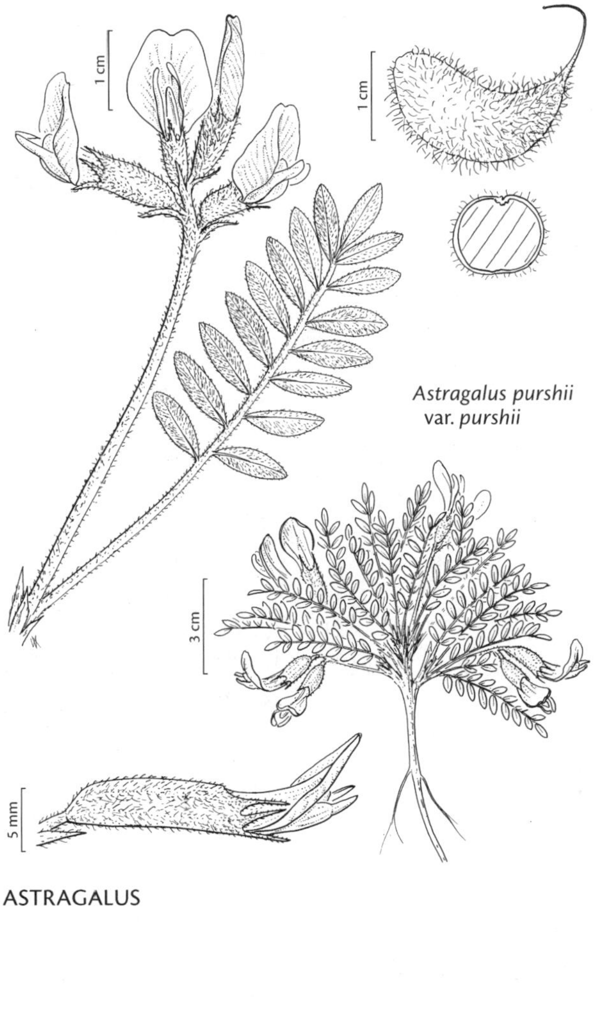

ASTRAGALUS

Astragalus platytropis A. Gray

Broad-keeled milkvetch

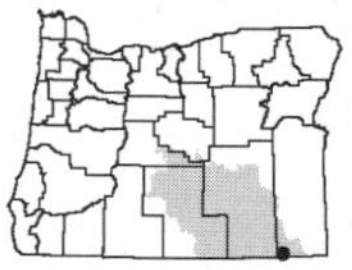

Plants perennial, ± acaulescent, densely tufted, with a thatch of persistent leaf bases, silvery- or gray-strigillose or sericeous, hairs basifixed. **Stems** prostrate or ascending to erect, often buried, 0–2 cm. **Leaves** 1–9 cm; leaflets 5–15, elliptic to obovate, oblong or oval, 2–11 × 1.5–7 mm, tips acute to obtuse or retuse, surfaces silvery strigose; stipules 1.5–5 mm, at least lowermost connate-sheathing. **Inflorescences** racemes or subumbels, 2–9-flowered; peduncles scapiform, 1.5–6.5 cm; bracts 0.6–2 mm; pedicels 0.7–1.9 mm; bracteoles 0–2. **Flowers** ascending at anthesis; calyces 2.5–4.5 mm, strigose with white hairs or a mixture of white and black hairs, tubes 2–3.4 mm, teeth subulate or triangular, 0.5–2 mm; corollas 6–9.5 mm, whitish or pink-purple, tinged distally with lilac; ovules 24–34. **Fruits** bilocular, ascending, humistrate, bladdery inflated, ovoid to subglobose, 15–33 × 10–22 mm, purple-mottled, strigillose; valves papery; stipes 0.

Ridge tops, screes, roadsides, talus. Flowering Jun–Aug. 1500–2000 m. BR. CA, ID, NV; northeast to MT, southeast to UT. Native.

In Oregon, *A. platytropis* is the only acaulescent, montane *Astragalus* with basifixed hairs and sessile, bladdery, bilocular fruits.

Astragalus purshii Douglas ex Hooker

Pursh's milkvetch, woollypod milkvetch

Plants perennial, acaulescent to shortly caulescent, densely villous to villous-tomentose, hairs basifixed. **Stems** prostrate, loosely to densely tufted, 0–20 cm. **Leaves** 1–12(17) cm; leaflets (3)5–17(21), elliptic to oblanceolate, 2–14(20) × 1–7 mm, tips obtuse to acute, surfaces densely villous; stipules 2.5–15 mm, free. **Inflorescences** racemes or subumbels, 1–12-flowered; peduncles 1–14 cm; bracts 4–9 mm; pedicels 2–4.3 mm; bracteoles 0–2. **Flowers** ascending at anthesis; calyces 5.5–16(19) mm, often purple, villous-pilose with white or mixed white and black hairs, tubes 8.5–12.5 mm, teeth subulate, 2.2–6 mm; corollas 19–27 mm, whitish to ochroleucous or pink-purple; ovules 14–40(46). **Fruits** unilocular, ascending, obliquely ovoid, usually curved, obcompressed, scarcely to deeply sulcate, (7)13–27(30) × 3.5–11 mm, densely white to tawny tomentose or densely villous, hairs nearly always concealing valve surfaces; valves coriaceous; sessile or on gynophores 0–1.6 mm.

Western North America. ~8 varieties; 4 varieties treated in *Flora*.

Throughout western North America, particularly in the Intermountain Region, this is a low, tufted milkvetch with white or gray villous hairs and pods resembling balls of cotton. Barneby (1964) stated, "Attempts to devise a practical key to the varieties of *A. purshii* are never wholly successful."

Variety *ophiogenes*, the Snake River milkvetch, a native of Idaho, has been reported from Malheur County, but this is apparently based on misidentifications of *A. purshii* var. *lagopinus*. Variety *ophiogenes* has 3–11-flowered racemes and 9–17 leaflets.

Key to *Astragalus purshii* varieties:

1. Calyces 5.5–9 mm; corollas 9–13.2 mm . . . . *A. p.* var. ***lagopinus***
1' Calyces (8)10–17(19) mm; corollas (14.6)15.4–25(27) mm.
    2. Corollas white or cream, keel tips spotted, wings and banners dull bluish purple-tipped. . . . . *A. p.* var. ***purshii***
    2' Corollas purple or pink purple.
        3. Fruits not, or only very shallowly and obscurely sulcate ventrally; leaflets (3)5–13(17).. . . . . *A. p.* var. ***tinctus***
        3' Fruits deeply sulcate ventrally on proximal ½; leaflets (7)9–17(21) . . . . *A. p.* var. ***glareosus***

Astragalus purshii Douglas ex Hooker var. ***glareosus*** (Douglas ex Hooker) Barneby

Gravel milkvetch

Astragalus glareosus Douglas ex Hook.

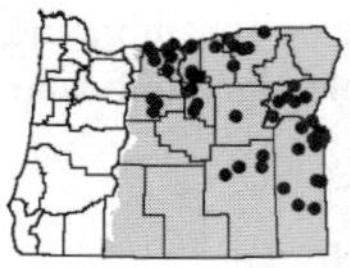

Stems 5–20 cm. **Leaves** (2)4–14(17) cm; leaflets (7)9–17(21). **Racemes** 2–5(10)-flowered. **Flowers** calyces 12–16(18.7) mm; corollas 19–25(26.5) mm, purple. **Fruits** deeply sulcate ventrally on proximal ½. **2*n***=22.

Sagebrush flats, dry hillsides, bunchgrass prairies, open ponderosa pine forests. Flowering Apr–Jul. 100–1700 m. BR, BW, Col, ECas, Lava, Owy. ID, NV, WA; southeast to UT. Native.

Astragalus purshii Douglas ex Hooker var. ***lagopinus*** (Rydberg) Barneby

Hare's-foot milkvetch

Astragalus lagopinus (Rydb.) M. Peck

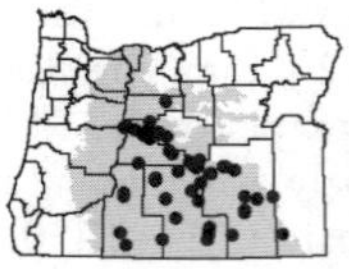

Stems 1–8 cm. **Leaves** 1–4.5(7) cm; leaflets (3)5–7(11). **Racemes** (2)3–5(7)-flowered. **Flowers** calyces (5.5)6–9 mm; corollas 9–13.2 mm, pink-purple. **Fruits** openly and shallowly sulcate dorsally on proximal ½.

Meadows, sagebrush communities, pumice and sandy areas. Flowering Apr–Jul. 800–2300 m. BR, Casc, ECas, Lava. CA, NV. Native.

Astragalus purshii Douglas ex Hooker var. ***purshii*** (p. 671)

Pursh's milkvetch

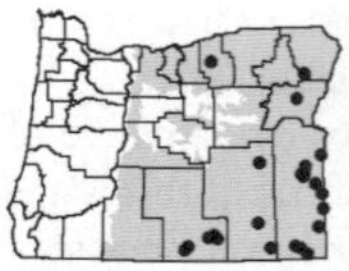

Stems 0–10 cm. **Leaves** (1.5)3–10(15) cm; leaflets (5)7–15(17). **Racemes** 1–6-flowered. **Flowers** calyces (12)13–16(19) mm; corollas 19–25(27) mm, white or cream, keel tips spotted, wings and banners dull bluish purple-tipped. **Fruits** not or very shallowly sulcate ventrally. **2*n***=22.

Dry hillsides and plains; sagebrush communities. Flowering May–Jun. 400–2200 m. BR, BW, Col, ECas, Owy. CA, ID, NV, WA; north to British Columbia, northeast to Saskatchewan, east to ND, southeast to UT, CO. Native.

Astragalus purshii Douglas ex Hooker var. ***tinctus*** M.E. Jones

Dyed milkvetch

Astragalus candelarius E. Sheld.

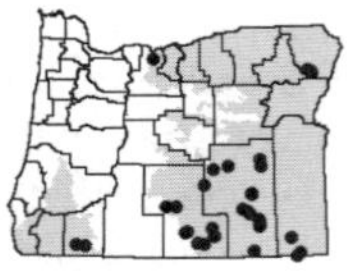

Stems 0–12 cm. **Leaves** 2–8(11) cm; leaflets (3)5–13(17). **Racemes** 3–11-flowered. **Flowers** calyces (8)10–14.2(15.8) mm; corollas (14.6)15.4–25 mm, pink-purple. **Fruits** not or only very shallowly sulcate ventrally.

Sagebrush flats, grasslands, alpine ridges. Flowering May–Jul. 400–3000 m. BR, BW, Col, Owy, Sisk. CA, NV. Native.

Astragalus reventus A. Gray

Blue Mountain milkvetch

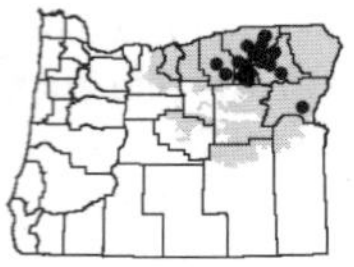

Plants perennial, caulescent, strigillose to villosulous, hairs basifixed. **Stems** several–numerous, ascending to erect, 15–40 cm. **Leaves** (6)8–18(21) cm; leaflets (17)23–41, lance-oblong or -ovate, oblanceolate or narrowly elliptic to linear, 5–21 × 2–6 mm, tips obtuse, retuse or acute, surfaces abaxially pubescent, adaxially glabrate to glabrous; stipules 3.5–9 mm, free. **Inflorescences** racemes, at first dense but elongating and becoming looser, 8–21-flowered; peduncles 10–20 cm; bracts 2.7–5.5 mm; pedicels 1.4–5.4 mm; bracteoles (0)2. **Flowers** spreading to nodding at anthesis;

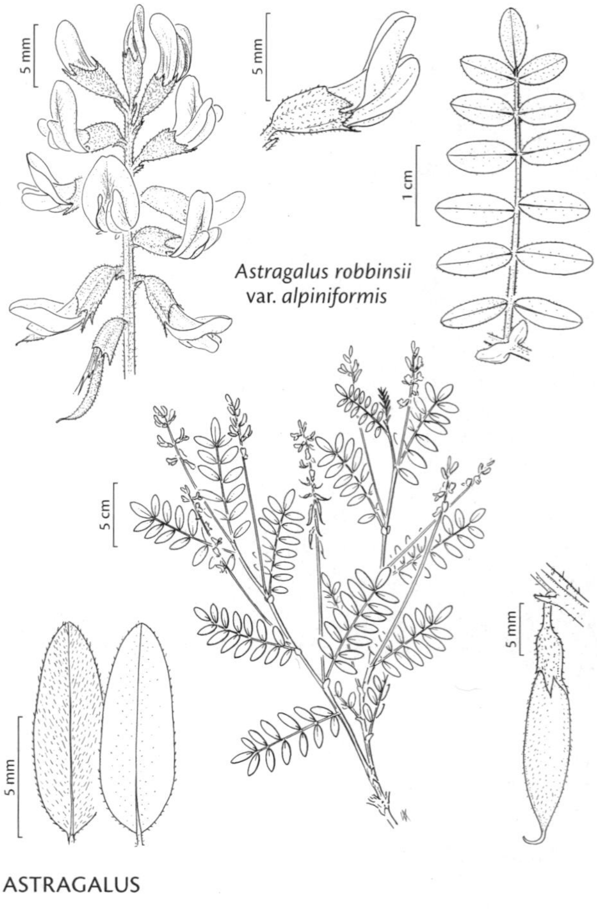

ASTRAGALUS

calyces (8.9)10.7–14.8 mm, strigillose to villosulous with black or mixed black and white hairs, tubes 6.2–9.9 mm, teeth subulate, (2)3–5 mm; corollas (15.2)16.2–25(34) mm, white or cream, keel tips sometimes purplish, fading yellowish; ovules 25–36. **Fruits** unilocular or incompletely bilocular, erect, ovoid-acuminate to broadly lance- or oblong-ellipsoid, subterete, somewhat obcompressed, straight, carinate, (15)17–30 × 7–10 mm, glabrous; valves coriaceous to woody; stipes 0.

Openings in pine forests, rocky hillsides. Flowering May–Jul. 300–1800 m. BW, Col. WA. Native.

The growth habit and form of the fruits of this species resemble *Astragalus conjunctus*, but the latter species is distinguished by its scarious, connate lower stipules.

Astragalus robbinsii (Oakes) A. Gray var. *alpiniformis* (Rydberg) Barneby (p. 672)

Wallowa elegant milkvetch

Astragalus macounii Rydb., *Atelophragma alpiniforme* Rydb.

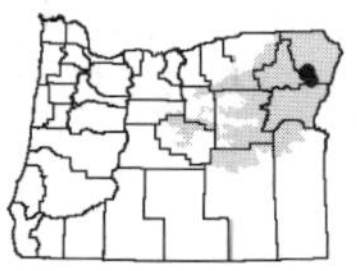

Plants perennial, caulescent, thinly to densely strigillose to pilose-villosulous, hairs basifixed. **Stems** several, ascending, (10)15–45 cm. **Leaves** 4–9.5 cm; leaflets 9–17, ovate to obovate or lance-elliptic, (3)5–12 × 4–8 mm, tips rounded, surfaces abaxially strigillose, adaxially glabrous; stipules 1.5–6 mm, at least lowermost connate-sheathing. **Inflorescences** racemes (3)6–12-flowered; peduncles (6)8–21 cm; bracts 1–3.5 mm; pedicels 0.5–3 mm; bracteoles usually 0. **Flowers** declined at anthesis; calyces 4–6.8 mm, strigillose with black hairs, tubes 3.2–4.5 mm, teeth subulate, 0.7–2.3 mm; corollas 6–9 mm, purplish; ovules 3–6. **Fruits** unilocular, pendulous, broadly ellipsoid, flattened, not or scarcely sulcate dorsally, 8–13 × 3.7–5 mm, strigose with mostly black hairs; valves papery; stipes 1.2–2 mm.

Alpine streambanks and gravel bars. Flowering Jun–Aug. 1200–2700 m. BW. Native.

Astragalus robbinsii is widespread in western North America and disjunctly eastward to Labrador, Newfoundland, Nova Scotia, and New England. Eight varieties have been described; variety *alpiniformis* is the only one that occurs in Oregon, and it is also endemic to the state.

Astragalus robbinsii var. *alpiniformis* can be confused with *Astragalus alpinus* var. *alpinus* as the two species are alike in habitat, growth form, flowers, and fruits, and also occur sympatrically. In *A. robbinsii* var. *alpiniformis*, the keel is shorter than the wings and much shorter than the banner, leaflet number is 9–17, and the pod is not or scarcely sulcate dorsally, whereas in *A. alpinus* var. *alpinus*, the keel is as long or longer than the wings and equals or is a little shorter than the banner, leaflet number is usually 15–25, and the pod is strongly sulcate dorsally.

Astragalus salmonis M.E. Jones

Trout Creek milkvetch

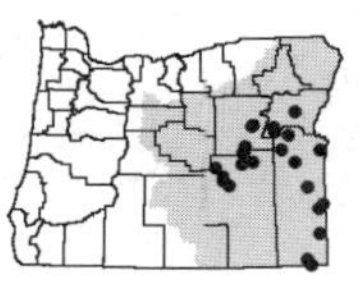

Plants perennial, shortly caulescent or subacaulescent, strigillose, hairs basifixed. **Stems** several–numerous, ascending-erect, 1–5 cm. **Leaves** mostly clustered at bases, 2.5–10 cm; leaflets 9–15(19), narrowly oblong-elliptic, linear-oblong or oblong-obovate, (1)2–10 × 1–3 mm, tips obtuse to subacute, surfaces abaxially pubescent, adaxially glabrous to glabrate; stipules 2–5 mm, free. **Inflorescences** racemes loosely 2–8-flowered; peduncles 3–12(15) cm; bracts 1–2 mm; pedicels 1–4 mm; bracteoles 0. **Flowers** horizontal or nodding at anthesis; calyces 5–9 mm, strigillose with black or partly black hairs, tubes 3.5–6 mm, teeth subulate, 1.4–3 mm; corollas 9.5–13.2 mm, whitish, sometimes lavender-tinged; ovules (14)17–29. **Fruits** unilocular to partially bilocular, irregularly pendulous, humistrate, obliquely oblong-ellipsoid, somewhat arcuate, strongly obcompressed, 12–22 × 4–6.5 mm, generally red-mottled, strigillose; valves coriaceous, woody; stipes 1.5–3.5 mm.

Dry sites, sagebrush, stony flats and hillsides. Flowering Apr–Jun. 800–2000 m. BR, BW, Lava, Owy. ID, NV. Native.

Astragalus sclerocarpus A. Gray

The Dalles milkvetch

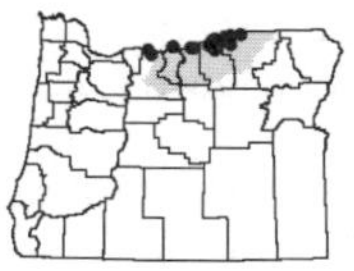

Plants perennial, caulescent, silvery strigillose, hairs basifixed. **Stems** several, spreading or decumbent to ascending, clump-forming, 20–50 cm. **Leaves** (3)5–13 cm; leaflets (5)7–17(21), linear, linear-oblanceolate or narrowly oblong, 3–23 × 1–4 mm, tips obtuse or retuse, surfaces abaxially pubescent, adaxially pubescent, glabrescent, or glabrous; stipules 1.5–5 mm, free. **Inflorescences** racemes loosely (3)7–21(30)-flowered; peduncles 4–9.5 cm; bracts 1–2 mm; pedicels 1–2.5 mm; bracteoles usually 2. **Flowers** spreading-ascending at anthesis; calyces 5.9–8 mm, finely strigillose with black or largely black hairs, tubes 5.3–7 mm, teeth triangular, 0.5–1.3 mm; corollas 13–16 mm, whitish, banners and wings lilac-tinged, keels spotted with purple; ovules 30–36. **Fruits** unilocular, pendulous, crescentic or falcate, laterally compressed, sigmoidally incurved distally, bicarinate, 20–35 × (5.5)6.5–9 mm, densely to thinly strigillose; valves stiffly coriaceous or subligneous; stipes 12–20 mm.

Sandy barrens, dunes, roadsides, sagebrush. Flowering Apr–Jun. 50–900 m. Col. WA; north to British Columbia. Native.

Astragalus sheldonii (Rydberg) Barneby

Sheldon's milkvetch

Astragalus reventus A. Gray var. *sheldonii* (Rydb.) C.L. Hitchc.

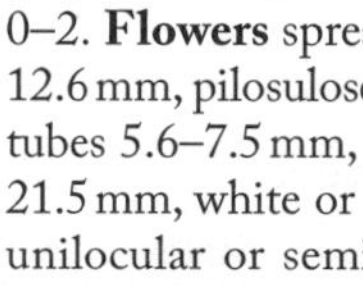

Plants perennial, caulescent, strigillose-pilosulose, hairs basifixed. **Stems** several–numerous, ascending to erect, in clumps, 20–50 cm. **Leaves** 7–19 cm; leaflets (17)23–35(39), oblong-obovate, elliptic or oblanceolate, 5–22 × 1–4 mm, tips retuse, obtuse or subacute, surfaces abaxially pubescent, adaxially glabrous; stipules 3–10 mm, free. **Inflorescences** racemes loosely 10–30(35)-flowered; peduncles (11)16–30 cm; bracts 2.5–9 mm; pedicels 1.3–6 mm; bracteoles 0–2. **Flowers** spreading to nodding at anthesis; calyces 8.5–12.6 mm, pilosulose with black or mixed black and white hairs, tubes 5.6–7.5 mm, teeth subulate, 2–5.2(6) mm; corollas 15.5–21.5 mm, white or cream, unspotted; ovules (20)24–31. **Fruits** unilocular or semibilocular, erect, narrowly oblong-elliptic, straight, obcompressed, (15)17–23 × 4–6.5 mm, glabrous or minutely strigillose; valves coriaceous; stipes 0–1.5 mm. ***2n*=24.**

Prairies, ridges, meadows. Flowering May–Jul. 1100–1700 m. BW. ID, WA. Native.

Astragalus sheldonii is closely related to *A. reventus* from which it differs by its narrower and more nearly bilocular fruits, shorter calyx tube, and usually the presence of a small stipe.

Astragalus solitarius M. Peck

Lonesome milkvetch, weak milkvetch

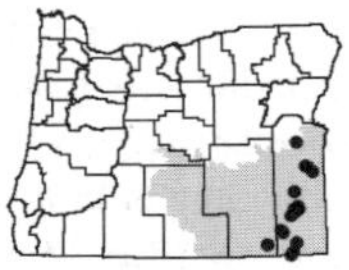

Plants perennial, caulescent, cinereous-strigillose, hairs basifixed. **Stems** solitary or few, erect, (15)20–50 cm. **Leaves** 2–7 cm; leaflets 5–9(13), linear to linear-oblong, (1.5)3–15 × 0.5–3 mm, tips obtuse or subacute, surfaces abaxially strigose, adaxially strigose or glabrescent; stipules 1–3 mm, free. **Inflorescences** racemes, loosely or remotely (3)7–30-flowered; peduncles (3)5–15 cm; bracts 0.6–1.5 mm; pedicels 1–2 mm; bracteoles 0. **Flowers** spreading at anthesis; calyces 2.3–3.4 mm, strigose, hairs white or partly fuscous or black, tubes 2.1–2.8 mm, teeth deltoid or subulate, 0.3–0.8 mm; corollas (5.5)6–8 mm, yellowish green, purple-tinged; ovules (6)8–15. **Fruits** unilocular, pendulous, narrowly oblong or lance-oblong, straight or falcately decurved, strongly compressed, bicarinate, 12–25 × 3–5.5 mm, strigillose; valves papery to thinly cartilaginous; stipes 4–8 mm.

Slopes, flats, sagebrush. Flowering May–Jul. 900–1600 m. BR, Owy. NV. Native.

Astragalus spaldingii A. Gray

Spalding's milkvetch

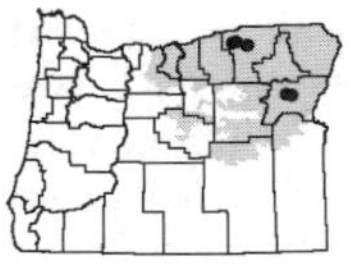

Plants perennial, caulescent, canescent-villous, hairs basifixed. **Stems** numerous, prostrate, decumbent, or ascending, 10–50 cm. **Leaves** 2.5–13 cm; leaflets (9)15–25(29), narrowly elliptic to oblanceolate, 4–18 × 1–4.5 mm, tips obtuse to acute, surfaces villous or villous-pilose; stipules 2.5–8 mm, free. **Inflorescences** racemes dense, closely crowded, elongating and spiciform with age, (6)12–30-flowered; peduncles 3–10 cm; bracts 1.5–4.5 mm; pedicels 0.2–0.8 mm; bracteoles 0. **Flowers** spreading at anthesis; calyces 5–8.3 mm, white villous or villous-tomentose, tubes (2.5)3.1–4.5 mm, teeth narrowly subulate, 2–4.2 mm; corollas 7–11.7 mm, whitish, often lavender-tinged, glabrous; ovules 4–10. **Fruits** bilocular, enclosed by persistent calyces, spreading or ascending, obliquely ovoid, straight, somewhat compressed, (3.5)4–6 × 2.5–3.5 mm, densely villous-tomentulose; valves thickly papery; stipes 0–0.3 mm. **2*n***=24.

Slopes and roadside banks, sagebrush and bunchgrass communities. Flowering Apr–Jul. 300–1300 m. BW, Col. ID, WA. Native.

This species is readily distinguished by its spike-like racemes of small, crowded, pallid flowers and small, densely pubescent ovoid fruits enclosed for most of their length by the persistent calyces.

Astragalus sterilis Barneby

Barren milkvetch, sterile milkvetch

Astragalus cusickii A. Gray var. *sterilis* (Barneby) Barneby

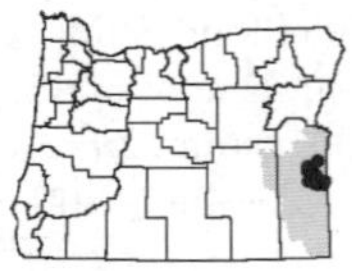

Plants perennial, caulescent, strigillose, hairs basifixed. **Stems** single–few, ascending to erect, 6–16 cm. **Leaves** 2–9 cm; leaflets 6–12, linear-elliptic, obscurely jointed, terminal one represented by a dilation of rachis, 1–7 × 0.5–1.5 mm, tips obtuse, surfaces strigillose; stipules 1.5–5 mm, at least lowermost connate-sheathing. **Inflorescences** racemes loosely 1–5-flowered; peduncles (1)2–7 cm; bracts 0.7–1.3 mm; pedicels 1–2 mm; bracteoles 0. **Flowers** declined at anthesis; calyces 3.5–4 mm, thinly strigillose with white and a few black hairs, tubes 3–3.6 mm, teeth triangular, 0.4–0.6 mm; corollas 9–10 mm, ochroleucous, fading yellowish; ovules 17–20. **Fruits** unilocular, pendulous, bladdery-inflated, obovoid, 20–25 × 9–11 mm, purple-mottled, glabrous; valves membranous; stipes 2–4 mm.

Barren clay or ash soils, talus slopes, gravel bars. Flowering May–Jun. 700–2000 m. Owy. ID. Native.

Astragalus succumbens Douglas ex Hooker

Crouching milkvetch, Columbia milkvetch

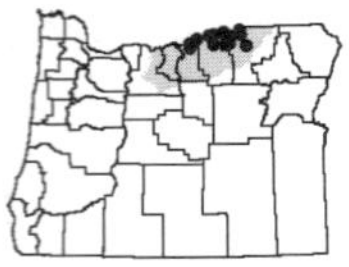

Plants perennial, caulescent, densely to thinly grayish hirsute, hairs basifixed. **Stems** several, decumbent to erect, 2–40(50) cm. **Leaves** 3–10(11) cm; leaflets (7)9–19, oblong to oblong-ovate, obovate to elliptic, 5–19 × 3–9 mm, tips acute to obtuse or retuse, surfaces abaxially hirsute, adaxially hirsute to glabrescent; stipules (1.5)2.5–8(9) mm, free. **Inflorescences** racemes densely 10–25-flowered; peduncles 1.5–5.5 cm; bracts (2.5)4–12 mm; pedicels 0.5–2.3 mm; bracteoles 0–2. **Flowers** spreading-ascending at anthesis; calyces 9.5–13(15) mm, hirsute to hirsutulous with all white or white with some black hairs, tubes 7–8.6 mm, teeth subulate to linear-lanceolate, 2.5–5.5 mm; corollas 14–20(23) mm, bicolored, banners pink to white, wings white, keels pink to purple, tips spotted; ovules 27–38. **Fruits** bilocular, ascending, lunately linear-lanceolate or -oblong, arcuate to straight, triquetrous, strongly compressed, deeply sulcate dorsally, (20)25–40 × 4–8 mm, glabrous; valves cartilaginous; stipes 0. **2*n***=24.

Sandy hills and dunes, roadbanks, sagebrush. Flowering Apr–Jul. 50–900 m. Col. WA. Native.

Astragalus tegetarioides M.E. Jones var. *tegetarioides*

Deschutes milkvetch, bastard kentrophyta

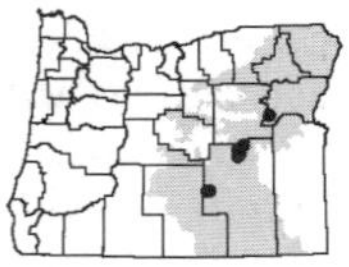

Plants perennial, caulescent, strigillose, hairs basifixed. **Stems** numerous, prostrate, forming mats or cushions, 5–15 × 10–40 cm. **Leaves** 1–4(6) cm; leaflets (5)7–11, obovate-cuneate, 1.5–5.5(7) × 2–4 mm, tips obtuse, truncate or emarginate or minutely apiculate, surfaces pubescent; stipules 0.8–3(5) mm, at least lowermost connate-sheathing. **Inflorescences** racemes loosely 2–8-flowered; peduncles 0.3–2.5 cm; bracts 1.2–2.7 mm; pedicels 0.4–1.3 mm; bracteoles 0. **Flowers** spreading to declined at anthesis; calyces (2.2)2.6–4 mm, thinly white-villosulous, tubes 1.1–2 mm, teeth subulate, 1–1.9 mm; corollas 4.4–7 mm, whitish, banners pale lavender-veined; ovules 2–4. **Fruits** unilocular, spreading to declined, often humistrate, ovoid-lenticular, obscurely trigonous, 3.5–4.5 × 1.5–2.8 mm, minutely strigose to silky villous; valves papery; stipes 0. **2*n***=24.

Juniper-sagebrush communities, dry pine forests. Flowering May–Aug. 1300–1700 m. BR, BW. Native.

This variety is endemic to eastern Oregon. A second variety, var. *anxius*, is found only in California.

Astragalus tetrapterus A. Gray (p. 675)

Four-wing milkvetch

Astragalus cinerascens (Rydb.) Tidestr., *Astragalus tetrapterus* A. Gray var. *cinerascens* (Rydb.) Barneby, *Astragalus tetrapterus* A. Gray var. *tetrapterus*

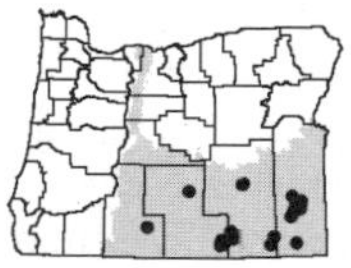

Plants perennial, caulescent, strigillose, hairs basifixed. **Stems** several, ascending to erect, becoming decumbent in fruit, 10–35(45) cm. **Leaves** 1.5–9 cm; leaflets 9–23, linear, narrowly oblong or elliptic, 1–33 × 0.3–3.2 mm, tips obtuse to acute, surfaces glabrous to strigose, terminal leaflets confluent with rachis in the uppermost leaves; stipules 2–5.5 mm, free. **Inflorescences** racemes 6–15-flowered; peduncles 1–6.5 cm; bracts 1.5–3.5 mm; pedicels 1.4–4.3 mm; bracteoles 0–2. **Flowers** ascending at anthesis; calyces (4)5.5–8.7 mm, strigose with black or mixture of black and white hairs, tubes 3.5–7 mm, teeth subulate or triangular, 0.5–2.8 mm; corollas 15–19 mm, white to yellowish and pink or bright pink-purple-tinged, keel faintly to darkly purple-tipped; ovules 28–42. **Fruits** unilocular, declined or pendulous, obliquely oblong or clavate-oblong, curved or coiled, strongly carinate, obcompressed early, becoming sharply tetragonal and 4-winged, wings up to 2.5 mm wide at maturity, (16)20–40 × (4)6–10 mm, glabrous to strigose; valves stiffly papery to subcoriaceous; stipes 0. **2*n***=22.

Hillsides of ash-clay-tuff, sandy soils, sagebrush. Flowering May–Jun. 900–1700 m. BR, ECas, Owy. ID, NV; southeast to AZ. Native.

The four-winged fruits of this species are unique among Oregon milkvetches.

Astragalus toanus M.E. Jones var. ***toanus***

Toano milkvetch

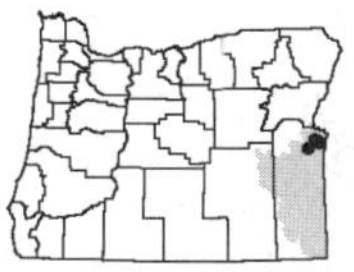

Plants perennial, caulescent, rush-like, robust, sparsely leafy, strigillose, hairs basifixed. **Stems** numerous, ascending to erect, in bushy clumps, 15–50 cm. **Leaves** 2–10 cm; leaflets 3–11(13), irregularly spaced, linear-filiform to oblong, 3–30 × 1.4–2.5 mm, tips obtuse to acute, surfaces abaxially strigillose, adaxially strigillose or glabrous, uppermost simple or with terminal leaflet confluent with rachis and appearing as a linear or recurved prolongation of rachis; stipules 4–13 mm, at least lowermost connate-sheathing. **Inflorescences** racemes loosely 7–35-flowered; peduncles 6–25 cm; bracts 1–3 mm; pedicels 0.8–3.5 mm; bracteoles 0–2. **Flowers** ascending at anthesis; calyces 4.6–8 mm, often purplish, strigose with white or dark hairs, tubes 4–6.5 mm, teeth subulate, 0.5–2 mm; corollas 15–22(24) mm, pink-purple; ovules 14–26. **Fruits** unilocular, erect, oblong or lance- to linear-elliptic, straight or a little incurved, slightly compressed laterally, bicarinate, 13–25 × 3.7–7 mm, glabrous to strigose; valves coriaceous or subligneous; stipes 0. **2*n***=22.

Barren sandy slopes, sagebrush and salt-desert communities. Flowering May–Jul. 700–1000 m. Owy. ID, NV; southeast to UT. Native.

This species is disagreeably selenium-scented and characterized by its rush-like growth form; reduced foliage; abundant purple-pink flowers; and erect, sessile, laterally compressed fruits.

Astragalus tweedyi Canby

Tweedy's milkvetch

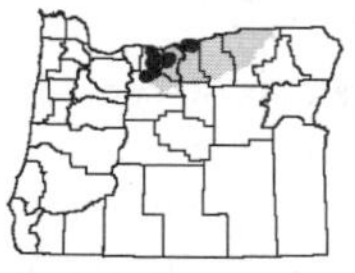

Plants perennial, caulescent, loosely strigillose to villosulous, hairs basifixed. **Stems** few, ascending to erect, 20–80 cm. **Leaves** 3–10(13) cm; leaflets (7)11–23, narrowly oblong-oblanceolate to linear-oblong, 5–22(25) × 1–2 mm, tips retuse to obtuse, surfaces abaxially strigillose, adaxially strigillose to glabrous; stipules 1–7 mm, free. **Inflorescences** racemes dense but elongating and becoming looser, 10–35(50)-flowered; peduncles (5)7–15 cm; bracts 1.5–6 mm; pedicels 0.8–2.5 mm; bracteoles 0. **Flowers** erect to spreading at anthesis; calyces gibbous-convex at bases, 8–10.5 mm, villous or villous-tomentulose with mostly white hairs, tubes 7.5–9 mm, teeth triangular, 1–2.1 mm; corollas 15–18.6 mm, ochroleucous, unspotted; ovules 12–19. **Fruits** unilocular, erect, oblong-ellipsoid, straight or slightly arcuate, laterally compressed, bicarinate, 12–15 × 3–4(5.4) mm, rugulose, sparsely pilosulose or glabrous; valves coriaceous; stipes 6–10 mm.

Dry hillsides, ridges, roadside banks, sagebrush. Flowering May–Jul. 100–1000 m. Col. WA. Native.

In Oregon, there are three species of *Astragalus* that have calyces that are obviously gibbous at the base so that the pedicel is attached obliquely: *A. tweedyi*, *A. collinus*, and *A. curvicarpus*; the three species are closely related. *Astragalus tweedyi* can be distinguished by its erect fruits, while the other species have pendulous fruits.

ASTRAGALUS

Astragalus tyghensis M. Peck (p. 676)

Tygh Valley milkvetch

Astragalus spaldingii A. Gray var. *tyghensis* (M. Peck) C.L. Hitchc.

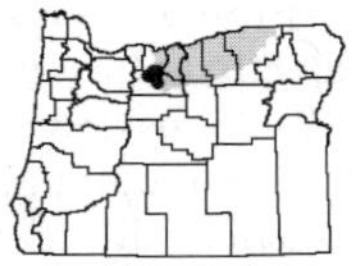

Plants perennial, caulescent, canescent villous-tomentose, hairs basifixed. **Stems** several or numerous, prostrate or decumbent to ascending, forming loose mats, 10–55 cm. **Leaves** 5–14 cm; leaflets (7)15–21, oval-obovate to elliptic, 6–17 × 2–7 mm, tips obtuse to subacute, surfaces densely villous-tomentose; stipules 3–6 mm, free. **Inflorescences** racemes, dense, closely crowded, ovoid to oblong, later elongating and becoming cylindrical, (10)20–40-flowered; peduncles 5–12 cm; bracts 2–5 mm; pedicels 0.5–1 mm; bracteoles 0. **Flowers** spreading to ascending at anthesis; calyces 6.6–7.8 mm, densely white, silky villous-tomentose, tubes 3.9–4.5 mm, teeth narrowly subulate, 2.7–3.5 mm; corollas 9–12 mm, pale yellow, pubescent, marcescent; ovules 6–8. **Fruits** bilocular, mostly enclosed by persistent calyces, spreading to somewhat declined, obliquely ovoid, straight, obscurely trigonous, 4.5–6 × 3 mm, canescently tomentulose; valves stiffly papery; body contracted basally into stipe-like necks, 0.4–0.6 mm.

Dry hillsides and valleys, open pine woodlands, sagebrush communities. Flowering May–Jun. 300–1000 m. Col. Native. ✤

The pubescent petals of this narrow endemic species distinguish it from all other Oregon milkvetches.

Astragalus umbraticus Sheldon (p. 676)

Bald Mountain milkvetch, sylvan milkvetch

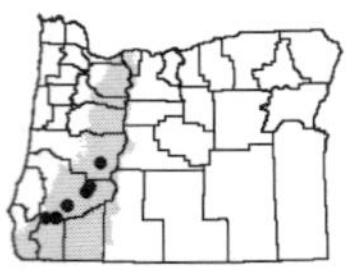

Plants perennial, caulescent, thinly villous, glabrate, hairs basifixed. **Stems** several, spreading to ascending, 20–50 cm. **Leaves** 4–12 cm; leaflets 11–23, broadly oblong, ovate, or obovate to suborbicular, 4–20 × 2.5–8 mm, tips retuse or obtuse, surfaces glabrate to ciliate; stipules 3.5–9.5 mm, free. **Inflorescences** racemes loosely but shortly 10–25-flowered; peduncles 5–12 cm; bracts 1.7–4 mm; pedicels 0.6–1.2 mm; bracteoles 0. **Flowers** spreading to declined at anthesis; calyces 5.2–7 mm, thinly strigillose with dark hairs, tubes 3.1–4 mm, teeth subulate, 2.1–3.2 mm; corollas 10–14 mm, cream, greenish white, pale yellow, or ochroleucous, unspotted; ovules 10–15. **Fruits** bilocular, spreading or declined, lunately or falcately linear-lanceolate or linear, incurved to a half-circle, compressed triquetrous, 14–24 × 2.5–3.6 mm, glabrous; valves papery to thinly coriaceous; stipes 1–1.9 mm.

Oak and pine woodlands, ridges, burned areas. Flowering May–Jul. 400–1300 m. Casc, Sisk. Native. ✤

This species is endemic to southwestern Oregon.

Astragalus vallaris M.E. Jones

Snake Canyon milkvetch

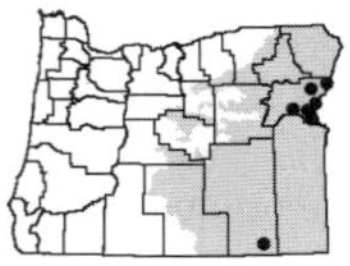

Plants perennial, caulescent, sparsely strigose to glabrous, hairs basifixed. **Stems** numerous, decumbent, 8–40 cm. **Leaves** 3–10 cm; leaflets 13–25(29), oblanceolate or obovate-cuneate, 6–17 × 3–8 mm, tips

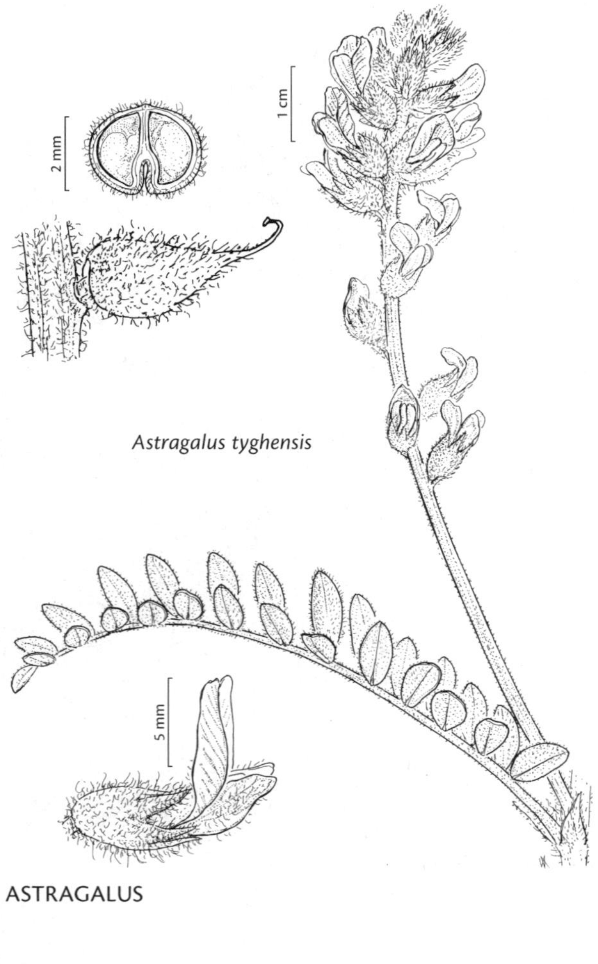

Astragalus tyghensis

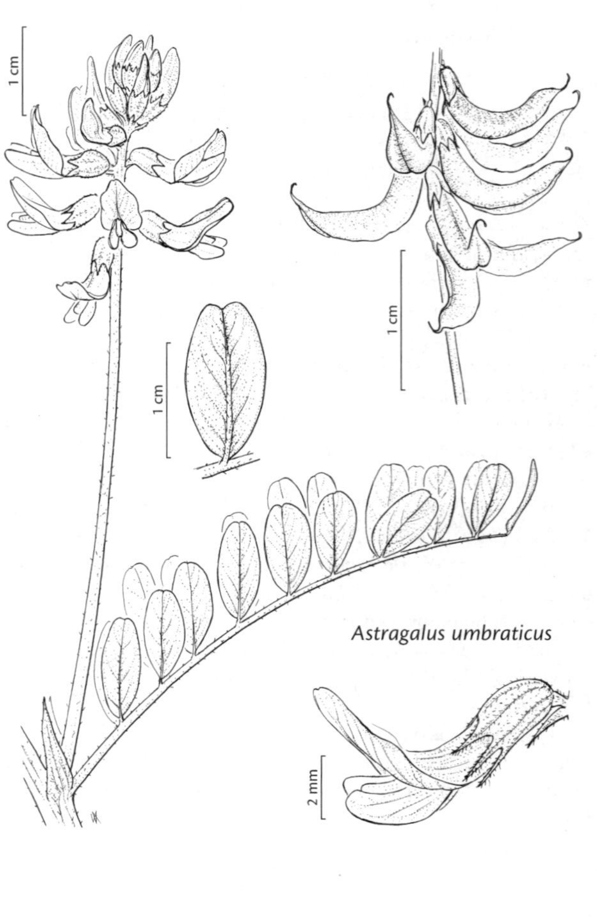

Astragalus umbraticus

ASTRAGALUS

obtuse or retuse or mucronulate, surfaces abaxially sparsely strigose on the margins and midribs, adaxially glabrous; stipules 2.5–5 mm, free. **Inflorescences** racemes shortly but loosely 5–10-flowered; peduncles 2–6 cm; bracts 2–4 mm; pedicels 1.5–5 mm; bracteoles 2. **Flowers** ascending at anthesis; calyces (8)9.5–11 mm, strigillose with black hairs, tubes 5.5–6.5 mm, teeth linear-subulate, 3.5–4.5 mm; corollas 16–23 mm, whitish; ovules 26–28. **Fruits** bilocular, ascending, humistrate, obliquely oblong-ovoid, slightly incurved, obcompressed, shallowly sulcate on dorsal and ventral surfaces, 20–40 × (9)11–19 mm, rugose, glabrous; valves woody; stipes 5–20 mm.

Roadsides, hillsides, sagebrush and bunchgrass communities. Flowering Apr–Jun. 600–1300 m. BR, BW, Owy. ID. Native.

Astragalus whitneyi A. Gray

Balloon milkvetch, balloonpod milkvetch

Plants perennial, caulescent, strigillose, hairs basifixed. **Stems** several–numerous, decumbent to erect, 2–40 cm. **Leaves** 1–10 cm; leaflets (5)9–21, oblong-oblanceolate, elliptic, or narrowly obovate, 2–21 × 1–5 mm, tips obtuse to acute, surfaces abaxially strigillose, adaxially strigose, glabrate, or glabrous; stipules 1–5 mm, at least lowermost connate-sheathing. **Inflorescences** racemes loosely 3–16-flowered; peduncles 1–9.5(12) cm; bracts 1–3 mm; pedicels 0.5–2.5 mm; bracteoles usually 0. **Flowers** declined at anthesis; calyces 4–9.3 mm, strigillose with white and black hairs, tubes 3.5–5.9 mm, teeth triangular, 0.5–3.6 mm; corollas 8.3–17.2 mm, ochroleucous, pinkish white, lilac, or pink-purple; ovules (13)15–30(32). **Fruits** unilocular, pendulous, bladdery inflated, obovoid to ellipsoid, 15–60(75) × (8)10–25(35) mm, usually red- or purple-mottled, glabrous or strigillose; valves membranous; stipes 2–9 mm.

Western United States. 5 varieties; 3 varieties treated in *Flora*.

Key to *Astragalus whitneyi* varieties:

1. Ovaries and fruits glabrous. . . . . . . *A. w.* var. ***siskiyouensis***

1' Ovaries and fruits strigillose.

- 2. Calyces 4–6 mm; corollas 9–12.8 mm; stems prostrate to weakly ascending, 2–12(17) cm. . . *A. w.* var. ***sonneanus***
- 2' Calyces (5.5)6–8 mm; corollas (12.8)13.5–17.2 mm; stems decumbent to erect, 5–30 cm . . . . . *A. w.* var. ***confusus***

Astragalus whitneyi A. Gray var. *confusus* Barneby

Alturas balloon milkvetch, cone-like milkvetch

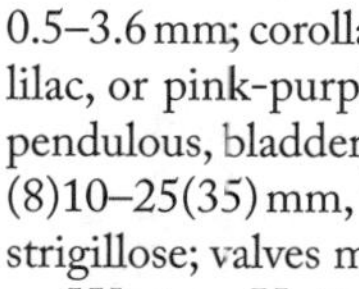

Stems robust, decumbent to erect, 5–30 cm. **Leaves** (2)3–10 cm; leaflets (7)1–21, (3)5–20 mm. **Inflorescences** (5)8–16-flowered. **Flowers** calyces (5.5)6–8 mm, tubes (4)4.6–5.9 mm, teeth (1.5)1.8–3.6 mm; corollas (12.8)13.5–17.2 mm, whitish, tinged with pink or lavender; ovaries strigillose. **Fruits** (17)20–60 × (10)13–25 mm, strigillose; stipes 4–9 mm.

Scabland, sagebrush communities. Flowering Jun–Aug. 1300–3000 m. BR, BW, ECas. CA, ID, NV. Native.

Intermediates may be found between var. *confusus* and var. *siskiyouensis*.

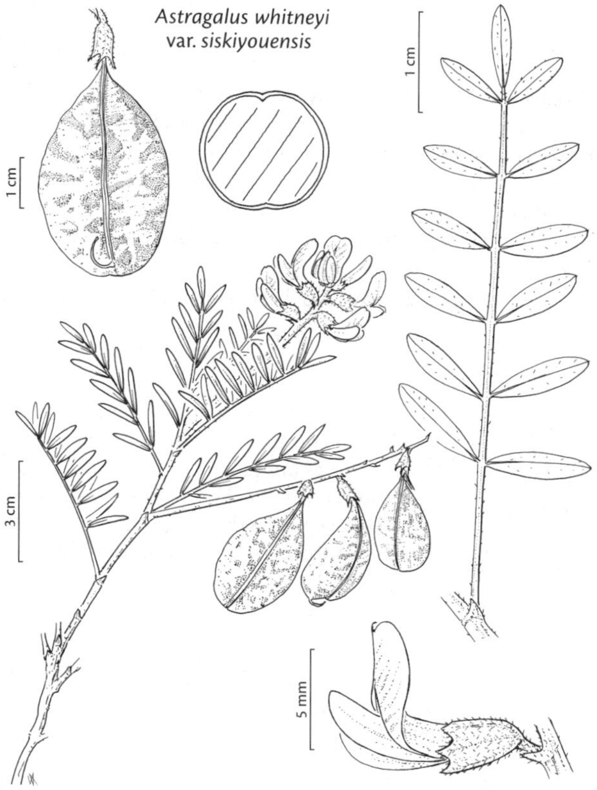

ASTRAGALUS

Astragalus whitneyi A. Gray var. *siskiyouensis* (Rydberg) Barneby

(p. 677)

Siskiyou balloon milkvetch

Astragalus whitneyi A. Gray ssp. *siskiyouensis* (Rydb.) Abrams

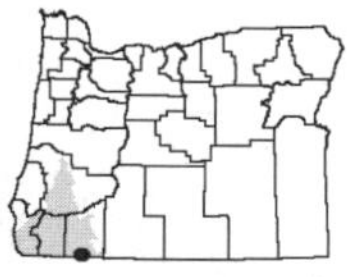

Stems slender, decumbent to ascending, 10–40 cm. **Leaves** 3–9 cm; leaflets 9–19, 4–18(21) mm. **Inflorescences** 4–16-flowered. **Flowers** calyces 4.6–5.6(6) mm, tubes 3.8–4.7 mm, teeth 0.5–1.7 mm; corollas 9.5–13.5 mm, ochroleucous; ovaries glabrous. **Fruits** (15)22–60(75) × 10–25(35) mm, glabrous; stipes 3–6 mm.

Pine forests, rocky slopes, ridge crests. Flowering Jun–Aug. 1200–2200 m. Sisk. Native.

Astragalus whitneyi A. Gray var. *sonneanus* (Greene) Jepson

Sonne's balloon milkvetch

Astragalus whitneyi A. Gray ssp. *hookerianus* (Torr. & A. Gray) Abrams

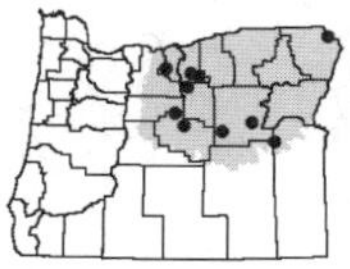

Stems slender, prostrate to weakly ascending, 2–12(17) cm. **Leaves** 1–4 cm; leaflets (5)11–17, 2–10(12) mm. **Inflorescences** 4–9(11)-flowered. **Flowers** calyces 4–6 mm, tubes 3.5–4.3 mm, teeth 1–1.7 mm; corollas 9–12.8 mm, whitish or cream, sometimes lilac-tinged; ovaries strigillose. **Fruits** (15)20–20(55) × (8)10–22(26) mm, loosely and sparsely strigillose; stipes 2.5–6 mm.

Rocky crests and summits, ridges near timberline. Flowering May–Jul. 800–2800 m. BW, Col, Lava. ID, WA. Native.

CERCIS Redbud

Christopher A. Royce

Trees or **shrubs** deciduous. **Stems** 1–several trunks, slender, branched, glabrous. **Leaves** alternate, cordate to reniform, bases cordate to rounded, margins entire, surfaces glabrous or pubescent; petiolate. **Inflorescences** short corymbs or racemes; bracts present. **Flowers** blooming before leaves unfurl; calyces short-campanulate, pink to purple, inconspicuously 5-toothed, glabrous; corollas pink to purple, banners reduced and interior, reflexed, keels free, wings reflexed; stamens 10, persistent at dehiscence. **Fruits** dehiscent, flat, papery, red, copper, or pale to dark brown, tips conspicuously beaked, glabrous to puberulent.

Asia, Europe, North America. 10 species; 1 species treated in *Flora.*

Cercis occidentalis Torrey ex A.Gray (p. 678)

Western redbud

Cercis canadensis L.

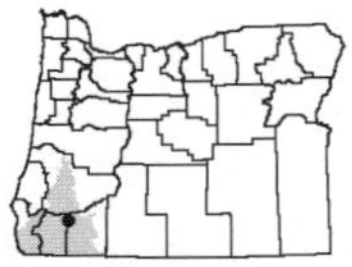

Plants 2–7 m. **Stems** gray; youngest twigs black or conspicuously darker. **Leaves** 14–53 × 26–74 mm, glossy, tips ± retuse or inconspicuously notched; petioles 11–29 mm. **Inflorescences** pedicels 7–14 mm, orange to copper, glabrous distally, proximally pilosulose to tomentose. **Flowers** (3)4–12(14); calyces 3–5 mm; corollas 5–10 mm. **Fruits** 35–92 mm. **Seeds** round. $2n$=14.

Forests, dry slopes, streamsides. Flowering Mar–May. 100–1500 m. Sisk. CA; southeast to AZ. Native.

This species is rare in Oregon and not usually observed north of the Klamath National Forest, California.

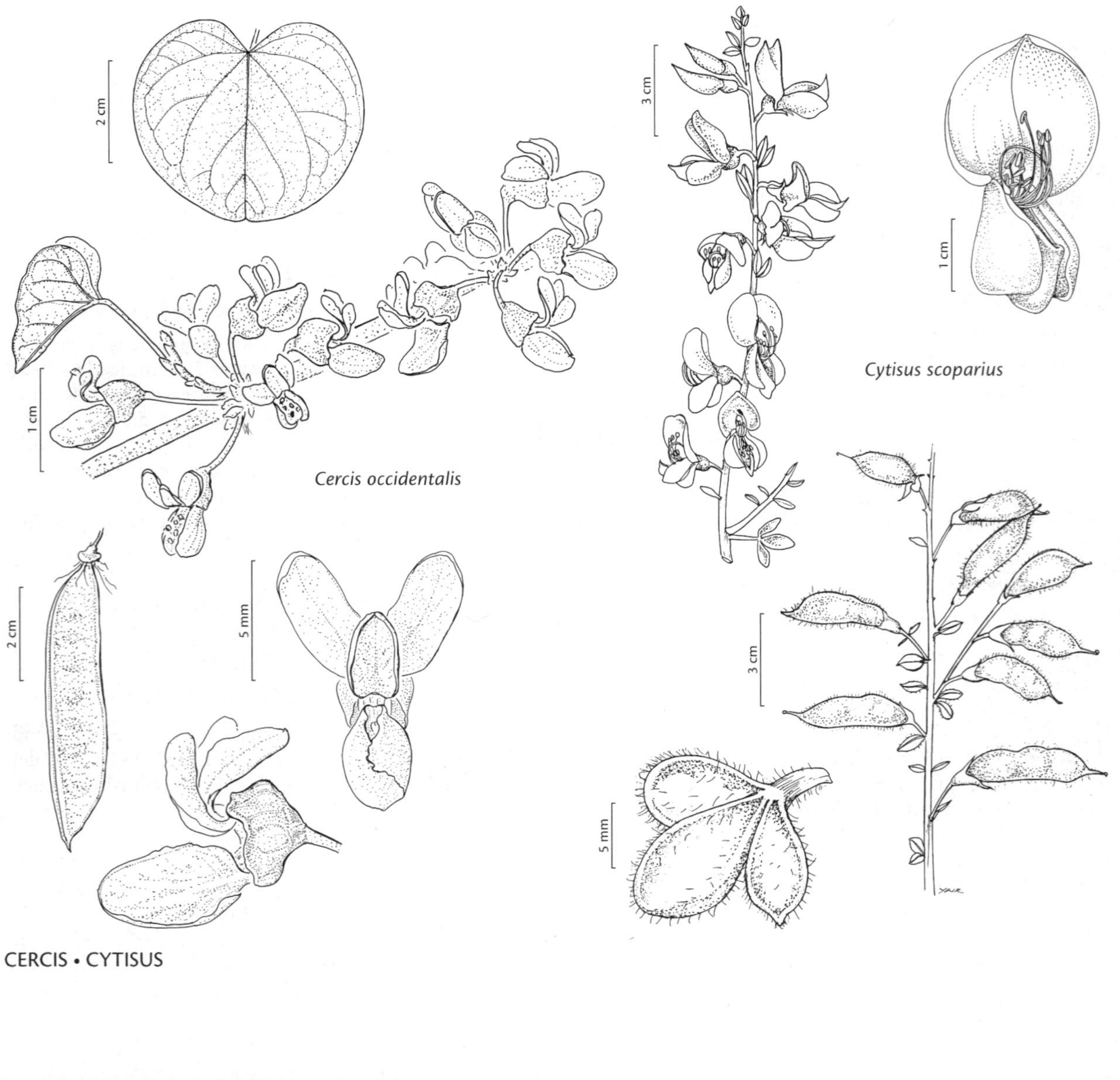

CERCIS • CYTISUS

CYTISUS Broom

Melanie A. Link-Pérez

Shrubs usually deciduous. **Stems** unarmed, ribbed, striate, or flanged, often green, glabrous or pubescent. **Leaves** alternate, generally ternate, reduced upwards; leaflets 3 proximally, 1 distally, margins entire, surfaces glabrous or pubescent; petioles present or absent; stipules absent or inconspicuous. **Inflorescences** axillary 1–3-flowered clusters; pedicels short. **Flowers** papilionaceous; calyces bilateral, fused, campanulate, 2-lipped, upper lips minutely 2-lobed, lower lips minutely 3-lobed; corollas generally yellow or white, petals 5, keels upcurved; stamens 10, monadelphous; styles abruptly curved medially or gently curved throughout, villous proximally, glabrous distally. **Fruits** dehiscent, oblong, papery to leathery, pubescent. **Seeds** several, reniform, ovoid, or rounded, carunculate.

Africa, Asia, Europe, North America. ~65 species; 3 species treated in *Flora*.

Key to *Cytisus* species:

1. Corollas white, 9–13 mm; calyces 2–3 mm; fruits beige . . . . **C. multiflorus**

1' Corollas yellow, 15–25 mm; calyces 3–7 mm; fruits brown to reddish brown or black.

  2. Calyces 5–7 mm, glabrous; fruits flat, glabrous but margins pilose; leaflets 2.5–6 mm wide, 1–2 × longer than wide . . . . **C. scoparius**

  2' Calyces 3–5 mm, appressed-hairy; fruits ± inflated, densely white-hairy; leaflets 1.5–2.5 mm wide, ≥ 2 × longer than wide . . . . **C. striatus**

Cytisus multiflorus (L'Héritier) Sweet

White Spanish broom, Portuguese broom

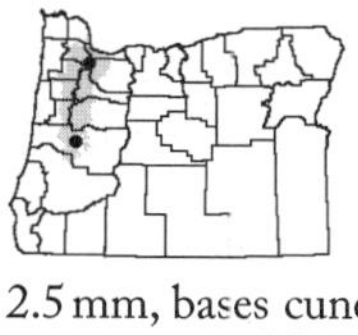

Plants 1–4 m, erect. **Stems** branches many, 5-angled, flexible, silvery-sericeous, becoming glabrous, upper often leafless at flowering. **Leaves** unifoliate to palmate; leaflets obovate-oblong to lanceolate, up to 10 × 2.5 mm, bases cuneate, tips obtuse to acute, surfaces sericeous; petioles 0–5(9) mm. **Inflorescences** axillary, 1–3-flowered; pedicels 3–7 mm. **Flowers** calyces 2–3 mm, anthocyanic, silky-hairy; corollas 9–13 mm, white, banners generally reflexed, lower center dark-lined; styles bent abruptly near tips, villous proximally. **Fruits** narrowly oblong, 15–30 mm, beige, appressed-hairy. **Seeds** 4–6, reniform, 2.5–3 mm, olive to brown. $2n$=46.

Roadsides, pastures. Flowering Apr–May. 0–200 m. WV. CA, WA; Australia, Europe, New Zealand. Exotic.

Cytisus multiflorus is native to Spain and Portugal.

Cytisus scoparius (Linnaeus) Link (p. 678)

Scots broom, Scotch broom

Cytisus scoparius (L.) Link var. *andreanus* (Puiss.) Dippel

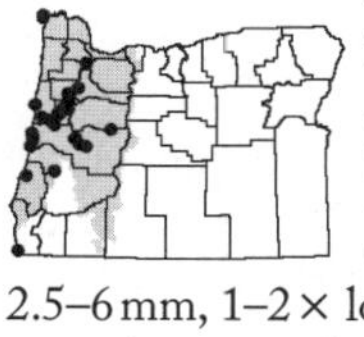

Plants 1–3 m, erect, stiffly ascending. **Stems** branches many, generally 5-angled or sharply flanged, green, glabrous, sparsely foliose. **Leaves** unifoliate to palmate; leaflets obovate to elliptic or elliptic-oblong, 6–15 × 2.5–6 mm, 1–2 × longer than wide, bases cuneate, tips obtuse, emarginate, or apiculate, surfaces with hairs appressed or absent; petioles 0–5 mm. **Inflorescences** axillary, 1–3-flowered; pedicels 4–10 mm. **Flowers** calyces 5–7 mm, anthocyanic, glabrous; corollas 15–23 mm, yellow, banners reflexed or not; styles long and incurved, often looping, villous proximally. **Fruits** oblong, flat, 25–45(70) mm, brown or black at maturity, glabrous but margins pilose. **Seeds** 5–12, reniform, 2–3 mm, brown. $2n$=46,48.

Roadsides, clearings, thickets, sand dunes. Flowering Mar–Jun. 0–1400 m. Casc, CR, Est, WV. CA, WA; north to AK, southeast to UT, scattered in eastern North America; Africa, Australia, Europe, Japan, New Zealand, South America. Exotic.

Cytisus scoparius is native to Europe and has been introduced to the United States and adjacent Canada. In Oregon, this species can be a problem invasive and is a "B" designated weed. The plants are densely branched with sparse foliage and large, yellow flowers that form showy inflorescences. The plants lose their leaves and fruits at the end of the growing season and have the appearance of highly branched green-black sticks during the winter months. There are horticultural forms (Bailey & Bailey 1976), and some plants with reddish wing petals are recognized as var. *andreanus* in some floras.

Cytisus striatus (Hill) Rothmaler

Striated broom, hairy-fruited broom

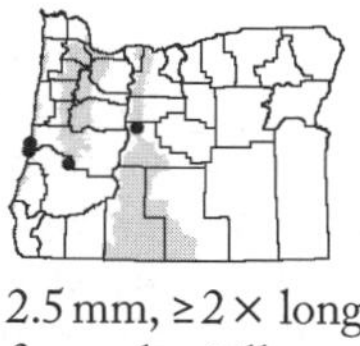

Plants 2–3 m, erect. **Stems** branches many, striate, generally 8–10-angled, green to greenish brown, sericeous, soon becoming glabrous. **Leaves** unifoliate to palmate; leaflets obovate or oblanceolate, 4–10 × 1.5–2.5 mm, ≥ 2 × longer than wide, bases cuneate, tips acute, surfaces abaxially strigose with stiff appressed hairs, adaxially hirsute to glabrous; petioles 1–9 mm. **Inflorescences** axillary, 1–2-flowered; pedicels 5–15 mm. **Flowers** calyces 3–5 mm, appressed-hairy; corollas 15–25 mm, yellow, banners not reflexed; styles long, curved throughout and often looping, hairy. **Fruits** obovoid to ellipsoid, ± inflated, 20–30(35) mm, dark brown, densely white-hairy. **Seeds** 3–6, reniform, 2–4 mm, reddish brown to brown. $2n$=46,48.

Sand dunes, roadsides, disturbed areas. Flowering Apr–Jul. 0–1000 m. ECas, Est, WV. CA; Europe, South America. Exotic.

Cytisus striatus is native to Spain and Portugal. The first record for Oregon was collected in 1982 from sand dunes southwest of Florence.

DALEA Prairie clover

Rory Nichols

Herbs glandular, aromatic when crushed; taprooted. **Stems** branches several to many. **Leaves** odd-1-pinnate, glandular; stipules inconspicuous. **Inflorescences** many-flowered spikes. **Flowers** calyces campanulate or turbinate, tubes glandular, 10-ribbed; banners attached to receptacles, other petals attached to fused filament tube alternating with 5 stamens; ovules 2. **Fruits** indehiscent, obovate to deltate. **Seeds** 1.

North America, South America. ~165 species; 1 species treated in *Flora*.

The center of diversity of *Dalea* is in Mexico with about 115 species.

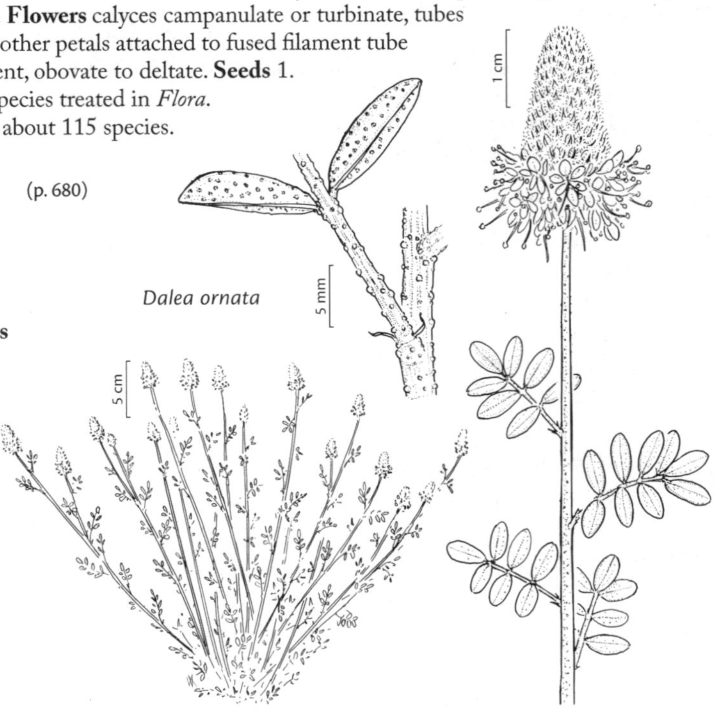

DALEA

Dalea ornata (Douglas ex Hooker) Eaton & Wright (p. 680)

Western prairie clover, ornate dalea

Petalostemon ornatum Douglas ex Hook.

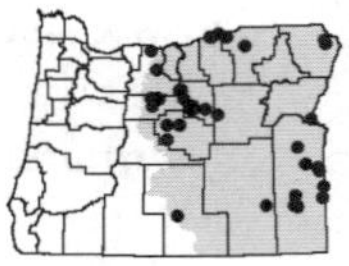

Plants perennial, 30–60 cm. **Stems** ascending. **Leaves** leaflets 5–9, ovate to elliptic, 10–20 mm. **Inflorescences** dense, ovoid to cylindric, 10–60 × 8–15 mm. **Flowers** calyces 4–6 mm, papery, hairs ascending, long, silky; corollas pink or lavender to rose-purple, banners 6–8 mm, other petals 4–6 mm.

Sagebrush-steppe in open, dry, clay, rocky or sandy slopes. Flowering May–Jul. 50–1600 m. BR, BW, Col, Lava, Owy. CA, ID, NV, WA. Native.

Dalea ornata is used in sagebrush steppe restoration and is important for pollinators.

GALEGA Goat's rue

Dominic Maze

Herbs perennial, glabrous to pubescent. **Stems** much branched. **Leaves** alternate, odd-pinnate; leaflet bases rounded to cuneate, rarely oblique, margins entire; stipules herbaceous, sagittate to semi-sagittate to ovate or ovate-orbicular; petiolate. **Inflorescences** terminal or axillary, many-flowered racemes; pedunculate; bracts linear, 3–5 mm. **Flowers** papilionaceous, bisexual; calyces campanulate, sepals with 5 subequal teeth; petals subequal; stamens 10, monadelphous, anthers uniform or alternately larger and smaller, elliptic, dorsifixed; ovaries unilocular, many-ovuled, styles incurved, stigmas terminal, small. **Fruits** dehiscent, flattened to linear-cylindric, 2-valved. **Seeds** several to many.

Nearly worldwide. 3–6 species; 1 species treated in *Flora*.

The generic epithet is derived from the Greek "gala" (milk) in reference to the genus' galactagogue properties (a substance that promotes lactation).

Galega officinalis Linnaeus (p. 681)

Goat's rue, professor-weed

Plants ascending, 4–18(20) dm; with woody root crowns. **Stems** hollow. **Leaves** leaflets (8)12–20, oblong, elliptic, lanceolate, or ovate-oblong, 15–50 × 4–15 mm, tips mostly acute and mucronate, terminal leaflet often emarginate; stipules semi-sagittate to sagittate. **Inflorescences** 25–50-flowered. **Flowers** calyces sparsely pubescent, teeth ≈ length of tubes; corollas 10–15 mm, white to lilac, wings ≥ keels. **Fruits** semi-erect, linear-cylindric, torulose, 20–50 × 2–3 mm, tipped with remnant styles. **Seeds** 2–10, oblong to reniform, 3–4.5 × 1–2 mm, beige to reddish brown, with oblique furrow on each face.

Roadsides, waterways, croplands, seasonally inundated grasslands. Flowering May–Sep. 50–200 m. WV. WA; widely scattered in North America; Africa, Asia, Europe, New Zealand, South America. Exotic.

Native to southeastern Europe and Asia Minor, *Galega officinalis* has been introduced to temperate regions worldwide as a medicinal and ornamental. It was additionally introduced by researchers at the Utah Agricultural Experiment Station in 1891 as a potential forage plant (which accounts for common name "professor-weed"). The plant contains the alkaloid galegine, making it unpalatable to livestock and toxic in larger quantities. It is a serious weed where naturalized.

GENISTA Broom

Melanie A. Link-Pérez

Shrubs perennial. **Stems** erect, angled or flanged. **Leaves** alternate, palmate; leaflets 1 or 3; petiolate. **Inflorescences** axillary. **Flowers** papilionaceous; sepals 5, calyces campanulate, fused at bases, weakly if at all bilateral; petals 5, yellow, keels obtuse or nearly straight; stamens 10, monadelphous, filaments of alternating lengths, anthers dimorphic; styles abruptly bent upwards below tip. **Fruits** dehiscent or indehiscent, oblong, generally flat; sessile or stalked. **Seeds** several, round to oval, generally carunculate.

Worldwide. ~80 species; 1 species treated in *Flora*.

Genista monspessulana (Linnaeus) L. Johnson (p. 681)

French broom

Cytisus monspessulanus L.

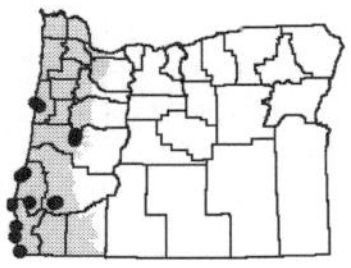

Plants to 3 m, pubescent, densely branched. **Stems** flanged, silvery-sericeous when young. **Leaves** alternate; leaflets 3, cuneate-obovate, rarely lanceolate, 10–25 × 5–12 mm, 1.5–2.5 × longer than wide, bases cuneate, midribs raised and hairy, margins entire, involute, tips mucronate, surfaces abaxially pubescent; petioles < 5 mm; stipules lanceolate, ≤ 3 mm, margins entire. **Inflorescences** 4–10-flowered, clustered or racemose, 15–25 × 15–22 mm; pedicels 1–4 mm. **Flowers** calyces 5–7 mm, sericeous to pilose, upper lips strongly 2-lobed; corollas 10–14 mm, glabrous, keels nearly straight to slightly curved distally; styles exserted. **Fruits** dehiscent, 15–30 × 4–6 mm, brown to black, villous or sericeous; stalked. **Seeds** generally 3–8, quadrangular to round, flattened, 2.5–3 × 2.2–2.5 mm, brown or black, lustrous. ***2n***=44, 46, 48.

Roadsides, riverbanks, and sandy areas, dunes. Flowering Mar–Jun. 0–300 m. CR, Est, Sisk, WV. CA, WA; Africa, Australia, Europe, South America. Exotic.

Genista monspessulana is the only species in the genus that is widely established along the Pacific Coast, where it contributes as a parent in most alleged hybrids. In Oregon, the putative hybrid *Genista monspessulana* × *G. stenopetala* has been observed only in Curry County where it has been collected from two populations; these plants have larger leaflets (up to 3 cm long and 1 cm wide) that are three times as long as wide and elliptic in shape and flowers that are more numerous, larger, and fragrant.

Genista monspessulana is most likely to be confused with *Cytisus scoparius* based on overall appearance, but upon closer inspection the presence of a style that is curved only near the tip and abundant trifoliate leaves present throughout the plant will confirm the specimen as *G. monspessulana* and not *C. scoparius* (which has strongly curved styles and much smaller leaves that are commonly unifoliate on new growth).

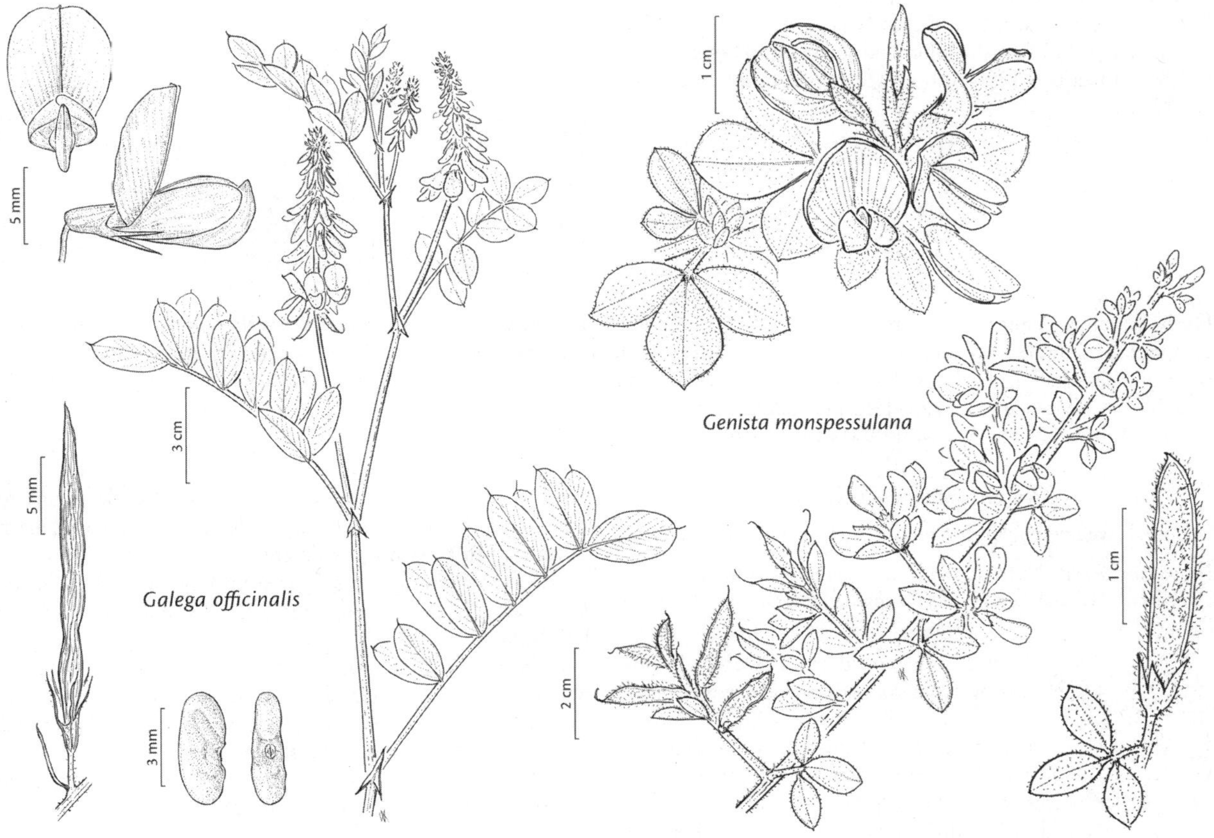

GALEGA • GENISTA

GLYCYRRHIZA Licorice

Richard R. Halse

Herbs perennial, glabrous to glandular-pubescent; from creeping rootstalks. **Stems** solitary to many. **Leaves** alternate, odd-1-pinnately compound; leaflets ovate, lanceolate, or elliptic, margins entire; petiolate; stipules caducous, linear, small. **Inflorescences** axillary racemes; peduncled. **Flowers** papilionaceous; calyces campanulate, slightly bilabiate with 2 partially connate adaxial teeth; corollas yellow-white or purple-tinged to bluish; stamens diadelphous; ovaries glabrous or glandular, styles incurved, glabrous. **Fruits** indehiscent, compressed, turgid, ellipsoid to oblong, brown or purplish. **Seeds** 2–4, brown.

Asia, Australia, Europe, North Africa, North America, South America.

~18 species; 1 species treated in *Flora*.

Most species of *Glycyrrhiza* have a sweetish, licorice-flavored root. *Glycyrrhiza glabra*, European licorice, has been cultivated for at least 2000 years and is the source for licorice flavoring.

Glycyrrhiza lepidota Pursh (p. 682)

American licorice

Glycyrrhiza lepidota Pursh var. *glutinosa* (Nutt.) S. Watson, *Glycyrrhiza lepidota* Pursh var. *lepidota*

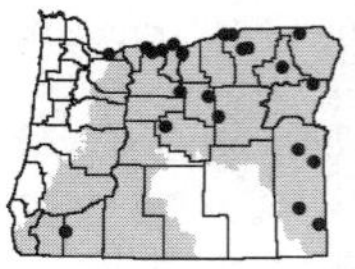

Plants 3–12 dm, finely to densely glandular-pubescent. **Stems** with stipitate or punctate glands. **Leaves** leaflets 7–19, 2.5–5 × 0.5–1.5 cm, tips mucronate, surfaces glabrous to sparsely pubescent, densely glandular-punctate. **Inflorescences** racemes dense, flowers numerous, 3–7 cm; peduncles 1.5–9 cm; pedicels 0.5–1 mm. **Flowers** calyces 8–10 mm, pubescent, glandular, teeth subulate; corollas 9–15 mm; ovaries glandular-bumpy, quickly becoming bristly. **Fruits** 12–15 × 8–10 mm, densely covered with hooked bristles, appearing cocklebur-like. **2*n***=16.

Riverbanks, roadsides, floodplains. Flowering May–Jul. 0–1600 m. BW, Casc, Col, ECas, Lava, Owy, Sisk. CA, ID, NV, WA; north to British Columbia, northeast to Quebec, east to IN, southeast to TX, south to Mexico, scattered in eastern US. Native.

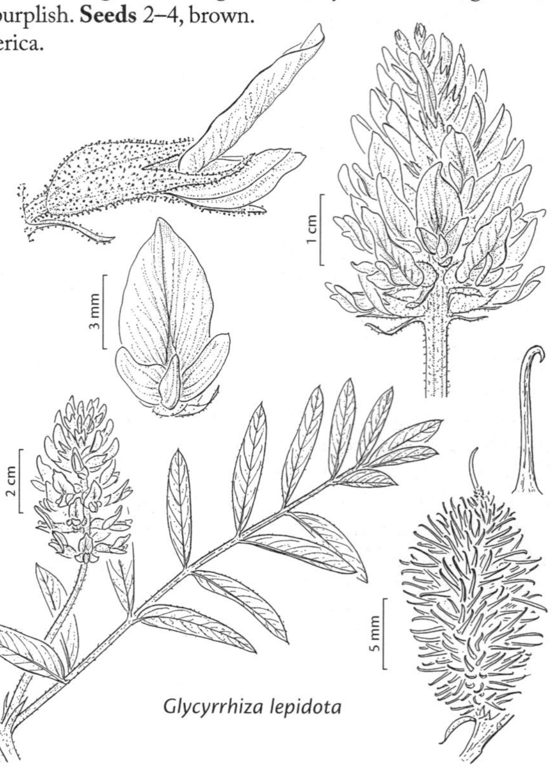

GLYCYRRHIZA

HEDYSARUM Sweetvetch

Cameron J. Ross

Herbs perennial; taprooted. **Stems** erect, branched or unbranched, glabrous to strigose. **Leaves** usually opposite, odd-pinnately compound, minutely punctate; stipules persistent, scarious. **Inflorescences** showy, secund racemes; bracts present. **Flowers** calyces campanulate, sepals 5, equal; corollas papilionaceous, white to purple, keels exserted beyond wings and longer than banners; stamens diadelphous; styles curved. **Fruits** loments, indehiscent, flat.

Asia, Europe, North America. ~100 species; 1 species treated in *Flora*.

Hedysarum boreale Nuttall (p. 984)

Boreal sweetvetch, Utah sweetvetch, northern sweetvetch

Hedysarum boreale Nutt. var. *boreale*, *Hedysarum boreale* Nutt. var. *mackenziei* (Richardson) C.L. Hitchc., *Hedysarum mackenziei* Richardson

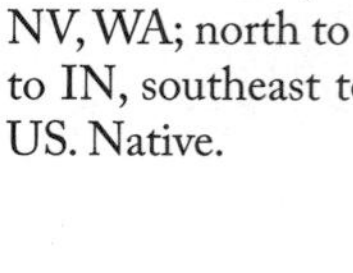

Plants 15–60 cm; taproots woody. **Stems** arising from robust crowns. **Leaves** up to 15 cm; leaflets 7–15, linear-elliptic to broadly obovate, 10–30 mm, bases rounded, tips acuminate, surfaces minutely pubescent to sericeous. **Inflorescences** 5–30(50)-flowered, elongated or compact, 2–15 cm; peduncles 10–25 cm; pedicels 1–3 mm. **Flowers** calyces 5–8 mm, > corolla tubes; corollas light to deep pink to dark purple, banners 10–16 mm, keels 10–20 mm, ≤ 3 × length of wings. **Fruits** with 2–8 segments.

Rocky or gravelly areas. Flowering Jun–Aug. 900–2500 m. BW, Lava. ID, NV; throughout most of Canada and western US. Native.

Some references assign some plants from the Wallowa Mountains to *H. boreale* var. *mackenzii* (versus the nominate variety elsewhere in the state) based on pubescence, flower size, and fruit segment number. Most plants in the Wallowa Mountains, however, appear morphologically intermediate. Thus, here we conservatively recognize only to the species level.

HOSACKIA Lotus, deervetch

Gerald D. Carr

Herbs perennial, low to robust; sometimes stoloniferous or rhizomatous. **Stems** prostrate, decumbent, ascending, sprawling or erect, generally 10–150 cm. **Leaves** alternate, odd-pinnate; leaflets 3–17, ± opposite, margins entire, tips usually terminating in small mucros, surfaces glabrous to pubescent; subsessile to long-petiolate; stipules fragile or rarely leaf-like, membranous or scarious. **Inflorescences** ± umbels, (1)2–25+-flowered, frequently ± secund. **Peduncles** axillary; bracts 1–5(7)-foliolate, usually leaf-like, sometimes scale-like, sometimes absent, ± medial to somewhat distal or directly subtending umbels. **Flowers** bisexual, papilionaceous; calyces weakly bilaterally symmetric, cylindric to obconic-cylindric, lobes 5, posterior pair usually broader, sometimes differentiated or ± similar; corollas combinations of white, cream, yellow, pink, purple, or red, banners ± reflexed, wings ≥ keels; stamens 10, 9 filaments fused, 1 free. **Fruits** dehiscent, linear to oblong, cylindric or slightly compressed, glabrous, ± beaked. **Seeds** mostly 4–20, ± oblong or subreniform, olive-brown to black, often mottled.

Central America, western North America. 14 species; 6 treated in *Flora*.

Prior to recent molecular studies, e.g., Allan & Porter (2000), species that are here treated in the genus *Hosackia* were commonly included in a broader circumscription of the genus *Lotus*. In accord with molecular data, *Hosackia* is now recognized as belonging to a unique New World clade that sets it apart from the Old World genus *Lotus*.

Key to *Hosackia* species:

1. Plants robust, 70–150 cm; inflorescences mostly 10–25+-flowered; flowers nodding, banners and wings inflexed . . . . *H. crassifolia* var. *crassifolia*

1' Plants low to robust, 5–70 cm; inflorescences mostly 2–15(20)-flowered; flowers not or only slightly nodding, banners and wings spreading.

  2. Stipules broad, some leaf-like, asymmetric, auriculate; stems glabrous, red- to yellow-pustulose, somewhat resinous . . . . *H. stipularis* var. *ottleyi*

  2' Stipules narrow, membranous or scarious, ± symmetric, not auriculate; stems usually with some pubescence, not pustulose or resinous.

    3. Leaflets 7–17; bracts usually well separated proximally from umbels, rarely absent; pedicels 2–5 mm; calyx tubes cylindric; banners white to rose or red-purple . . . . *H. rosea*

    3' Leaflets 3–11; bracts directly subtending umbels, or absent; pedicels 0.5–1.5 mm; calyx tubes obconic-cylindric; banners yellow to yellow-orange, red-orange, or reddish.

      4. Leaves hirsute-strigillose; calyces hispid-strigillose, lobes similar in size, shape, and spacing . . . . *H. oblongifolia* var. *oblongifolia*

      4' Leaves glabrous or with minute appressed trichomes; calyces glabrous, posterior pair of lobes and anterior 3 dissimilar in size, shape, and spacing.

        5. Peduncles 1.5–8 cm, bracts leaf-like, 1–3(5)-foliolate; inflorescences 2–5(6)-flowered; wings pale pink to lavender or deep red-violet, keel tips purple . . . . *H. gracilis*

        5' Peduncles 2–17 cm, bracts absent or scale-like, scarious, or very rarely leaf-like, simple; inflorescences 3–10(11)-flowered; wings pale yellowish, white, or slightly pink, keel tips yellow . . *H. pinnata*

Hosackia crassifolia Bentham var. *crassifolia*

Big deervetch

Lotus crassifolius (Benth.) Greene var. *crassifolius*

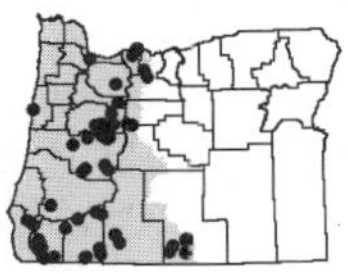

Plants robust, usually 70–150 cm. **Stems** (1)5–15+, sprawling or spreading to ascending, proximally fistulose, sparingly branched distally, often sparsely strigillose distally. **Leaves** mostly 5–17 cm; leaflets 7–19, oval to elliptic-oblong or obovate, 6–20 × 14–35 mm, margins generally ciliate, tips obtuse, surfaces glabrate to sparsely or moderately strigillose; petiolules 1.5–3 mm. **Stipules** broadly ovate, oblong, or narrowly lanceolate, 3–8(9) × 1–5 mm, membranous, margins entire or sometimes toothed or lobed, tips rounded to acute-acuminate. **Inflorescences** mostly 10–25+-flowered, with flowers generally facing away from stem axis; pedicels ± strongly unilaterally deflexed, 3–6 mm. **Peduncles** mostly horizontal to inclined, 3–11 cm, generally < subtending leaf, little elongating in fruit, glabrous or distally strigillose, bracts present or occasionally lacking, ± medial or rarely sub-approximate to umbel, sometimes scale-like, inconspicuous, more often leaf-like, 1–5-foliolate. **Flowers** nodding, usually 12–14 mm; calyces 5–8 mm, green to pink or purple, glabrous, tubes cylindric, 4.5–6.5 mm, lobes ± deltoid, 0.5–1.5 mm, sparsely ciliolate or eciliate; corollas narrow, not showy, claws > calyx tubes, banners narrow, somewhat boat-shaped, greenish or whitish with bold red to purple medial blotch, margins strongly inflexed, wings not spreading laterally, proximally greenish or whitish, distally red to purple, margins inflexed, ± embracing keel, keels slightly < wings, distally pinkish to purplish. **Fruits** ± pendulous, linear, 20–65 × 3–5 mm, reddish brown to brown, glabrous. **Seeds** 7–12, oblong, ~3 mm, olive or reddish to dark brown, sometimes black-mottled. $2n$=14.

Conifer or mixed woodlands, forest edges and openings, shrublands, burned areas, roadsides. Flowering May–Aug. 0–1900 m. Casc, CR, ECas, Est, Sisk, WV. CA, NV, WA; south to Mexico. Native.

Hosackia crassifolia var. *crassifolia* is the most robust Oregon member of the genus. It is easily recognized by its size and nodding inflorescences with up to 25 or more strongly deflexed, narrow flowers.

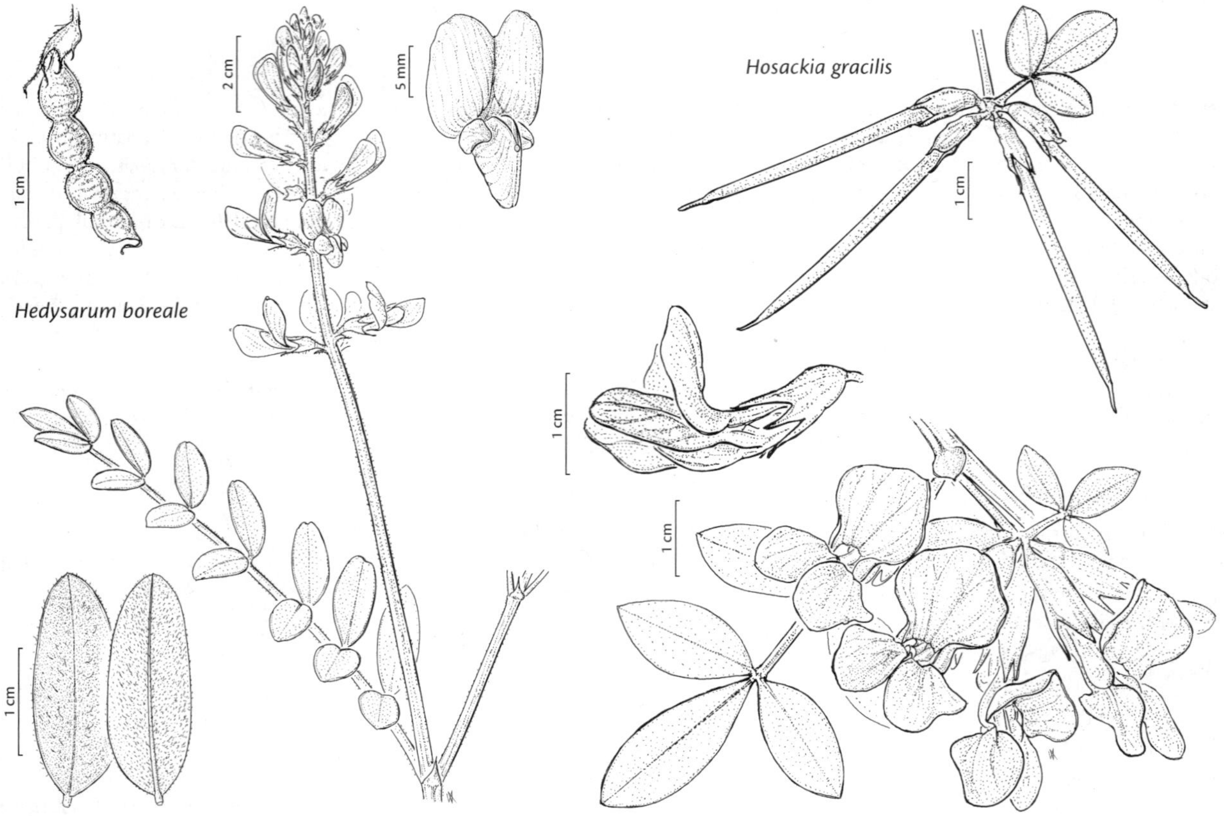

HEDYSARUM • HOSACKIA

Hosackia gracilis Bentham (p. 684)

Seaside lotus

Lotus formosissimus Greene

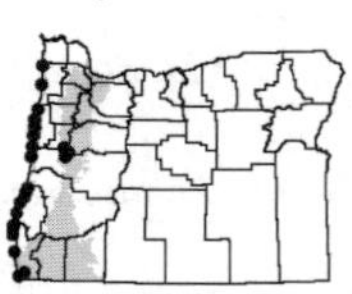

Plants usually low, 5–50 cm; stoloniferous or rhizomatous. **Stems** 1–20+, prostrate or decumbent to ascending, slender, sometimes sparsely branched, distally usually with scattered or numerous closely appressed, flattened trichomes. **Leaves** 1–6 cm; leaflets 3–7, elliptic or oblanceolate to obovate, 3.5–19 × 1.5–9 mm, margins eciliate or rarely sparsely ciliate, tips obtuse to emarginate, surfaces glabrous; petiolules 0.3–1 mm. **Stipules** narrow, deltoid or ovate to lanceolate, 2–6 × 1–3 mm, ± symmetric, scarious, margins entire or occasionally toothed, tips acute or acuminate, sometimes bifid. **Inflorescences** 2–5(6)-flowered, with flowers ± on one side; pedicels 0.5–1.5 mm. **Peduncles** spreading to ascending or erect, 1.5–8 cm, often much > subtending leaf, elongating in fruit, often bearing a few scattered or sometimes copious closely appressed, flattened trichomes, bracts directly subtending umbel, leaf-like, 1–3(5)-foliolate. **Flowers** 10–14(15) mm; calyces 4.5–6.5 mm, green, often reddish or with reddish highlights, glabrous, tubes obconic-cylindric, 2.5–4 mm, lobes subulate to narrowly triangular, 1.5–2.5 mm, sparsely ciliolate or eciliolate, posterior pair of lobes and anterior 3 dissimilar in size, shape, and spacing; corollas broad, showy, claws conspicuously > calyx tubes, banners large and spreading, yellow to sometimes yellow-orange or red-orange, wings spreading widely, pale pink to lavender or deep red-violet, inner margins sub-approximate to commonly overlapping, keels proximally yellow, tips purple, ~1.5–2 mm < wings. **Fruits** inclined to declined, linear, 20–40 × 1.5–2.5(3) mm, brown, glabrous. **Seeds** usually 10–14, subreniform, nearly round in outline, 1–1.5 mm, olive to brown. **2*n***=14.

Coastal bluffs and dunes, bogs, grasslands, grassy glades. Flowering Apr–Aug. 0–700 m. Est, Sisk, WV. CA, WA; north to British Columbia. Native.

Hosackia gracilis is primarily a coastal species that is most similar to the inland species, *H. pinnata.* Plants from a few inland populations, especially in Lane County, seem to be somewhat intermediate and may be difficult to place based on corolla color patterns alone. This treatment emphasizes the nature of the peduncle bract in determining the taxonomic placement of questionable material of these two species.

Hosackia oblongifolia Bentham var. *oblongifolia* (p. 685)

Streambank bird's-foot trefoil

Hosackia torreyi A. Gray, *Lotus oblongifolius* (Benth.) Greene var. *oblongifolius*, *Lotus torreyi* (A. Gray) Greene

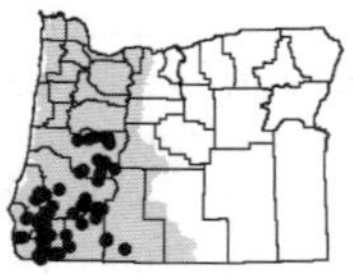

Plants low to robust, 10–65 cm. **Stems** 1–15+, decumbent to ascending, slender or sometimes proximally fistulose, rarely 2–3-branched, glabrous or glabrate proximally, usually copiously hirsute-strigillose distally. **Leaves** 2–8 cm; leaflets 3–11, usually elliptic to obovate, sometimes lanceolate, 5–36 × 2–14 mm, margins usually ciliate, tips acute or obtuse, surfaces usually sparsely to copiously hirsute-strigillose, sometimes ± glabrous especially adaxially; petiolules 0.5–1 mm. **Stipules** narrow, triangular to ovate or lanceolate, 2–4(5) × 1.5–2.7 mm, scarious, margins entire, tips obtuse to attenuate. **Inflorescences** (1)2–10(20)-flowered, with flowers on one side or ± symmetric; pedicels 0.5–1 mm. **Peduncles** mostly ascending to erect, 2–15(17) cm, > subtending leaf, much elongating in fruit, ± hirsute-strigillose, bracts directly subtending umbels, leaf-like, 1–3(5)-foliolate, rarely absent. **Flowers** 8–13 mm; calyces 3.5–6.5 mm, green, often suffused with red, hispid-strigillose, tubes obconic-cylindric, 2.5–3.5 mm, lobes similar, subulate, 1–3 mm, ciliolate; corollas broad, showy, claws ≥ calyx tubes, banners large and spreading, pale to bright yellow or orange to reddish, veins often prominent, faintly reddish to dark red, wings spreading laterally away from and exposing keel, proximally yellowish, distally white, keels ~1.5 mm < wings, pale yellowish. **Fruits** inclined to declined or pendulous, ± linear, 20–45 × 1.5–2 mm, brown to blackish, glabrous. **Seeds** 8–15, oblong, 1.7–2.2 mm, olive to black mottled with orange. **2*n***=14.

Swamps, bogs, marshes, wet meadows, moist banks, wet ditches. Flowering May–Sep. 100–1800 m. Casc, CR, ECas, Sisk, WV. CA, NV; southeast to AZ, south to Mexico. Native.

Hosackia oblongifolia var. *oblongifolia* is an attractive variable species with a wide distribution beyond southwestern Oregon. Its flowers are unique among Oregon species in having wing petals that flare outward, often completely exposing the keel. This taxon is similar to *H. gracilis* and *H. pinnata* and is sometimes confused with them. However, among these taxa, the conspicuous occurrence of somewhat spreading, often curved trichomes on the calyx and elsewhere in *H. oblongifolia* var. *oblongifolia* is diagnostic. Some Oregon populations of *H. oblongifolia* var. *oblongifolia* have umbels with considerably more flowers than generally reported for this taxon.

Hosackia pinnata (Hooker) Abrams
Bog deervetch

Lotus pinnatus Hook.

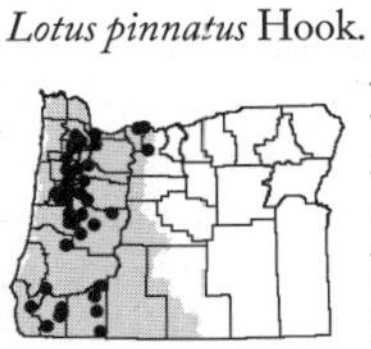

Plants low to moderately robust, 7–70 cm; rhizomatous. **Stems** 1–10+, ascending or decumbent, often fistulose, occasionally branched near bases, distally usually with few or more numerous closely appressed, flattened trichomes. **Leaves** 2–11 cm; leaflets 5–9, elliptic to obovate or oblanceolate, sometimes oval, 8–39 × 3–16 mm, margins eciliate or ciliate, tips emarginate or rounded to obtuse or acute, surfaces glabrous or occasionally with minute trichomes appressed to abaxial midribs, young leaflets sometimes heavily invested with minute flattened trichomes; petiolules 0.6–1.2 mm. **Stipules** narrow, deltoid or ovate to ovate-lanceolate, 2.5–5 × 1.7–4, ± symmetric, scarious, margins entire, tips usually obtuse or acute. **Inflorescences** 3–10(11)-flowered, with flowers on one side or occasionally ± radially disposed; pedicels 0.5–1.5 mm. **Peduncles** ascending, 2–17 cm, usually much ≥ subtending leaf, much elongating in fruit, often bearing a few scattered or sometimes copious closely appressed trichomes, bracts directly subtending umbels, simple, generally scale-like, scarious, rarely leaf-like, 2–3(4) mm, sometimes absent. **Flowers** slightly nodding, (12)13–16 mm; calyces 6–8 mm, green or reddish, glabrous, tubes obconic-cylindric, 4–5 mm, lobes subulate to triangular, 1–3 mm, eciliate or sometimes sparsely ciliate, posterior pair of lobes and anterior 3 dissimilar in size, shape, and spacing; corollas broad, showy, claws conspicuously > calyx tubes, banners large and spreading, yellow to yellow-orange, sometimes with reddish blush or reddish veins, wings spreading widely, proximally pale yellowish, distally white, sometimes slightly pink, inner margins ± approximate, sometimes overlapping, keels up to 3 mm < wings, bright yellow throughout or proximally paler with brighter yellow tips. **Fruits** ± inclined to declined, ± linear, 30–70 × 1.5–3 mm, dark brown to reddish. **Seeds** few, oblong, 2–2.7 × 1–1.5 mm. **2*n***=14.

Wetlands, meadows, stream margins, bogs. Flowering Apr–Jul. 0–1200 m. Casc, CR, ECas, Sisk, WV. CA, ID, WA; north to British Columbia. Native.

The similarity of *H. pinnata* to *H. gracilis* and their possible confusion is discussed under the latter species. The two species share an unusual type of minute trichome not seen among the other Oregon species of the genus. These trichomes (0.2–0.5 mm), which may be found on distal portions of stems and peduncles and on unexpanded leaflets, are strongly appressed, completely flat, and pointed at both ends.

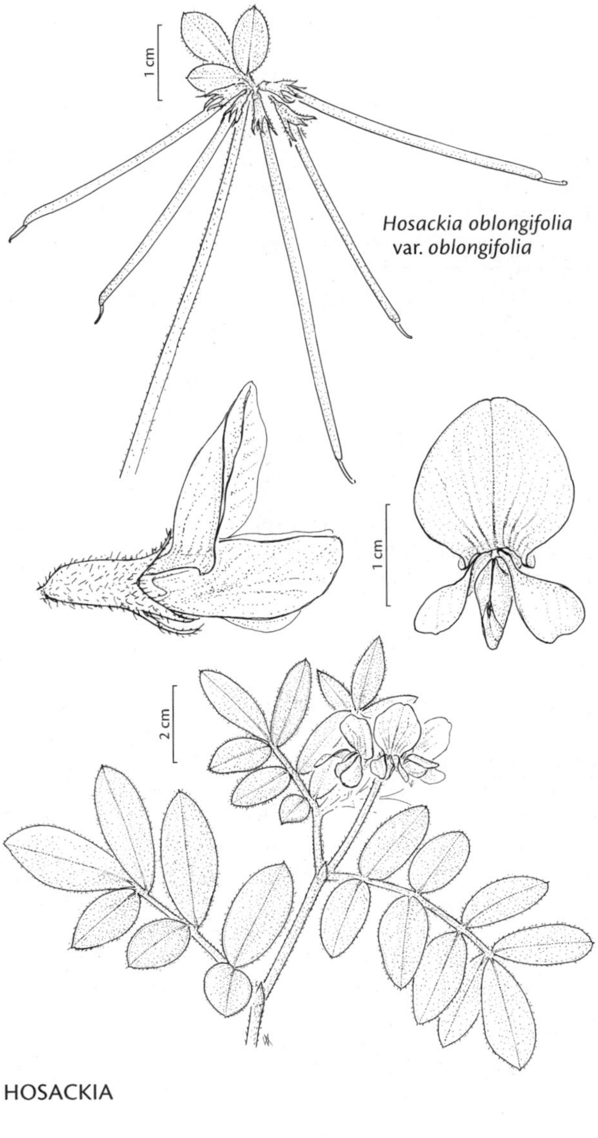

HOSACKIA

Hosackia rosea Eastwood

Rosy bird's-foot trefoil

Lotus aboriginus Jeps., *Lotus crassifolius* (Benth.) Greene var. *subglaber* (Ottley) C.L. Hitchc.

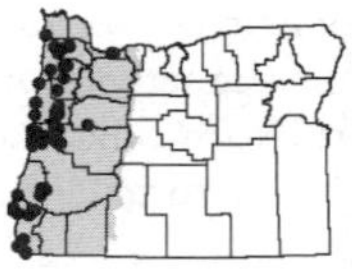

Plants usually robust, 10–70 cm. **Stems** 1–10+, inclined or spreading to erect, simple or sparsely branched, glabrous throughout or sparsely puberulent or strigillose distally. **Leaves** 3–15 cm; leaflets 7–17, elliptic or oblong to obovate or oblanceolate, 8–40 × 9–18 mm, margins usually ciliate, tips obtuse or rounded to truncate or retuse, surfaces moderately strigillose, less so adaxially, glabrate to glabrous with age; petiolules 0.5–1.5 mm. **Stipules** narrow, triangular or ovate-lanceolate to lanceolate, 3–7 × 1.5–3 mm, ± symmetric, scarious, margins entire or occasionally toothed or lobed, tips usually acute to acuminate. **Inflorescences** 5–15-flowered, with flowers usually facing in same direction; pedicels 2–5 mm. **Peduncles** mostly inclined to erect, 3.5–13(15) cm, often ≥ subtending leaf, little elongating in fruit, strigillose to glabrate; bracts usually well separated proximally from umbels, usually somewhat leaf-like, 1–5(7)-foliolate, occasionally scale-like, inconspicuous, rarely absent. **Flowers** slightly or not nodding, 9–13 mm; calyces 4–6 mm, green or suffused with red or purple, glabrous, tubes cylindric, 3.5–4 mm, lobes triangular to triangular-subulate, 0.5–2 mm, ciliolate; corollas broad, showy, claws > calyx tubes, banners large, spreading widely, white to rose or red-purple, generally becoming darker, often purple especially along veins, margins slightly inflexed, wings widely spreading laterally, colored similar to banner or only veins pinkish, inner margins approximate or overlapping, keels ~2.5 mm < wings, distally pink or purple. **Fruits** mostly declined to inclined, ± linear, 20–43 × 3–4 mm, reddish brown, glabrous. **Seeds** 4–10, oblong, 2–3.5 mm, olive or reddish brown to nearly black.

Coastal areas, Douglas-fir forests, logged areas, roadsides. Flowering Apr–Aug. 0–1000 m. Casc, CR, Est, Sisk, WV. CA, WA. Native.

Robust plants of *H. rosea* may approach the stature of *H. crassifolia* var. *crassifolia.* As a result, vegetative material of the two species could be rather easily confused. However, the showy flowers of *H. rosea* with heavily veined, widely spreading petals, contrast sharply with the inconspicuously veined, narrow, inflexed petals of *H. crassifolia* var. *crassifolia.*

Hosackia stipularis Bentham var. *ottleyi* (Isely) Brouillet

Thicket lotus, pine lotus

Lotus stipularis (Benth.) Greene var. *ottleyi* Isely

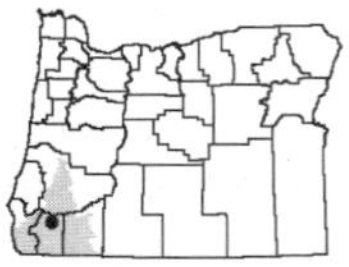

Plants ± robust, 15–30(50) cm, rhizomatous. **Stems** erect to ascending, wiry, unbranched or sparsely branched, glabrous, yellow- to red-pustulose, somewhat resinous, often appearing ± varnished. **Leaves** 1.5–7 cm; leaflets 9–15, oval to elliptic or obovate to oblanceolate, 5–12 × 3–6 mm, margins eciliate, tips obtuse to acute, surfaces glabrous; petiolules 0.3–0.6 mm. **Stipules** broad, widely ovate or deltoid to ovate-lanceolate, 2–7 × 1.3–7 mm, proximal and medial leaf-like, somewhat firm and green, strongly asymmetric, generally auriculate at base on one side, tips displaced laterally toward opposite side, usually acute or abruptly attenuate, distal generally small, scarious, ± symmetric. **Inflorescences** mostly 4–8-flowered, with flowers usually on one side; pedicels straight to arcuate, ± unilaterally deflexed, 1–2.5 mm. **Peduncles** ascending then spreading, 1–4(6) cm, sub-equal to subtending leaf, little elongating in fruit, glabrous, sometimes yellowish pustulose and varnished, bract ± medial, leaf-like, 1–5-foliolate, rarely absent. **Flowers** sometimes slightly nodding, 10–12 mm; calyces 5–6 mm, green, purplish, or pink, glabrous, tubes obconic-cylindric, ~4 mm, lobes subulate or short-triangular, 1–1.5 mm, eciliate; corollas broad, showy, claws ≥ calyx tubes, banners large and spreading, proximally pink to dark purplish red, distally white or whitish with reddish or purplish veins, wings spreading laterally, white, sometimes with pale pinkish veins, inner margins approximate or overlapping, keels ~2.5 mm < wings, distally purplish or whitish with purplish veins. **Fruits** ± declined or pendulous, oblong, 15–30 × 2.5–4 mm, olive brown, glabrous, ± pustulose, resinous. **Seeds** 5–7, oblong, 1.5–2.5 mm, olive-brown.

Disturbed clearcut or logged forests. Flowering Apr–Jun. 300–1300 m. Sisk. CA. Native.

The generally well-developed, asymmetric, leaf-like stipules of *H. stipularis* var. *ottleyi* are unique among Oregon members of the genus. In Oregon, this species is documented by only three Josephine County collections from two locations separated by less than six miles. Our limited material represents only a portion of the variation seen in California populations of this taxon.

LADEANIA Scurfpea

Ashley Egan

Herbs perennial, glandular. **Stems** smooth or lightly striate, sometimes ribbed apically, lower portions often with tan to brown modified leaves, usually branched basally, sparingly branched above. **Leaves** persistent or deciduous and sometimes lacking at maturity, alternate, palmately compound or pseudopalmate, (1)3–5-foliate; leaflets linear to oblanceolate; petioles not jointed to stems; stipules quickly to tardily deciduous. **Inflorescences** indeterminate pseudo-racemes with 3–22 nodes. **Flowers** bisexual, papilionaceous; calyces campanulate, glandular, usually pubescent, sometimes glabrate to sericeous, teeth 5, triangular, ≈ in length, usually < ½ calyx tube, abaxial tooth sometimes slightly longer, flaring back and tearing along lateral sinus in fruit; corollas cream or violet to blue-purple; stamens diadelphous; ovaries 1-ovulate. **Fruits** indehiscent, deciduous above receptacles, globose, pilose; beaks reduced. **Seeds** round to broadly elliptic, somewhat compressed but not flattened, brown. **2*n***=22.

Western North America. 2 species; 1 species treated in *Flora.*

Historically, this genus has been described as having as many as 14 species (Rydberg 1919), based on morphometric characters. However, many specimens are intermediate with each other, leaving no clear break in morphological characters to delineate separate species (Isely 1998; Grimes 1990). Thus, only the widespread species, *L. lanceolata*, along with the narrowly endemic *L. juncea*, merit recognition, with only *L. lanceolata* found in Oregon.

Ladeania lanceolata (Pursh) A.N. Egan & Reveal (p. 687)

Lanceleaf scurfpea, lemon scurfpea, dune scurfpea

Psoralea lanceolata Pursh, *Psoralea lanceolata* Pursh ssp. *lanceolata*, *Psoralea lanceolata* Pursh ssp. *scabra* (Nutt.) Piper, *Psoralidium lanceolatum* (Pursh) Rydb.

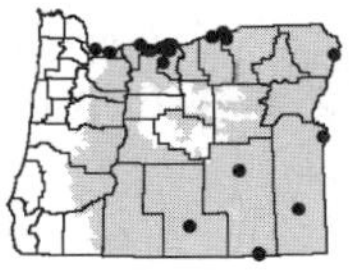

Plants 2–7 dm. **Stems** branching basally and distally. **Leaves** persistent through flowering, palmate, rarely pseudopalmate, 3(5)-foliate; leaflets obovate or oblanceolate, 10–26 × 4–10 mm, surfaces glabrate to sparsely pubescent, gland-dotted; petioles (5)9–29 mm. **Inflorescences** 1–10 cm. **Flowers** calyces 1.5–2.5 mm; corollas 4–7 mm, blue to purple or white or bicolored. **Fruits** 4–6 mm, hirsute to long-villous, conspicuously glandular. **2*n***=22.

Dry sandy areas, sagebrush. Flowering May–Jul. 0–1300 m. BR, BW, Casc, Col, ECas, Owy. CA, ID, NV, WA; northeast to Saskatchewan, east to IA, southeast to TX. Native.

Ladeania lanceolata is highly variable in morphology, which has resulted in the recognition of up to five taxa (Rydberg 1919). The length of the inflorescence, pod vestiture, and leaflet shape all vary along geographic trend lines throughout its range but with no clear break, making species distinctions based on morphology alone untenable. Thus, a single widespread species is recognized here.

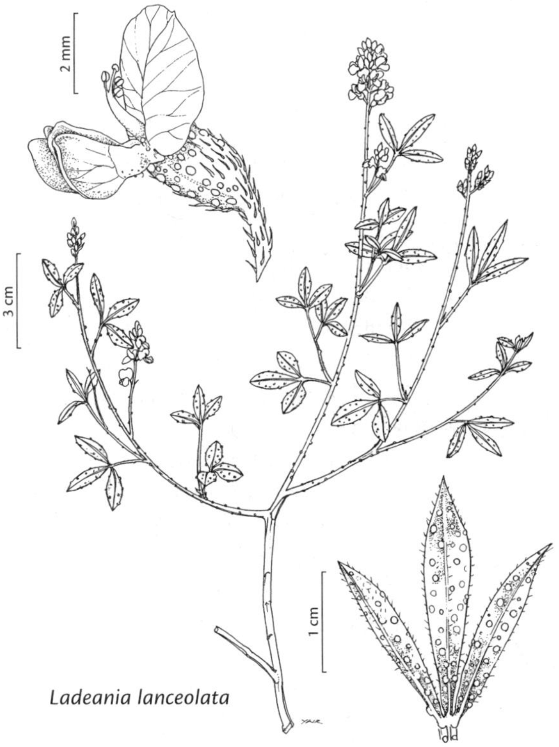

Ladeania lanceolata

LADEANIA

LATHYRUS Sweet pea, pea, vetchling

Steven L. Broich

Herbs annual or perennial, glabrous or pubescent, with unicellular, eglandular and/or multicellular, glandular trichomes; taprooted or rhizomatous. **Stems** usually angled, sometimes winged. **Leaves** even-pinnate, rachises sometimes winged, usually terminated in a simple or branched, prehensile tendril, sometimes reduced to bristle or absent; leaflets 0–16, paired or unpaired, ovate to linear, usually decreasing in size from base of rachis, margins entire; stipules foliose, semi-sagittate, rarely sagittate or hastate. **Inflorescences** erect or lax, axillary racemes, with 1–many flowers; bracts absent; bracteoles absent. **Flowers** calyx tubes symmetrically or obliquely campanulate with 5 equal or unequal lobes, dorsal lobes, if unequal, short and triangular, lateral lobes and keel lobes longer and lanceolate or linear; corollas white, pink, lavender, rose, red, blue, or purple, wings and keels usually lighter than banners; banner claws convolute around other petal claws, banner blades usually folded perpendicular to axis of flowers, occasionally reflexed against calyx tubes, wings adherent to keels with ball and socket mechanism, usually subequal to keels, keels incurved, attenuate, lower and distal margins fused; stamens diadelphous, anthers uniform; ovaries glabrous or pubescent, styles abaxially compressed, sometimes twisted, adaxially pubescent. **Fruits** linear, flattened or terete, glabrous or pubescent, non-septate, usually with obscure margins, splitting along both margins, valves twisted after dehiscence; sessile, rarely stipitate. **Seeds** 3–15, spherical, sometimes angular, usually dull brown, black, tan, or dark green, often mottled. **2*n***=14.

Asia, Europe, North America, South America. ~150 species; 22 species treated in *Flora*.

Within some polymorphic species, morphological variation is such that distinct varieties can be delimited. Within other species, however distinctive some populations may be, intergradation among forms is such that well marked taxa cannot be reliably described.

Unless otherwise noted in the taxon description: vestiture, if reported present, consists of unicellular, eglandular trichomes; the number of leaflets and tendril morphology are for upper stem leaves bearing inflorescences; leaflet dimensions given are for basal leaflets on the rachis of leaves subtending inflorescences; flower lengths reported are the distance from the base of the calyx tube to the distal keel petal margin; references to the relationship between calyx lobe length and calyx tube length always take into account the longest lobe present; wing and keel petals are almost always lighter in color than the banner petal; and colors given in descriptions are for the most pigmented part of the corolla.

Lathyrus odoratus has been reported once in Oregon. It is recognized by its hairy appearance, rose-purple flowers, pustulose-hirsute fruits, and leaves with two leaflets. Similarly, *L. nissolia* has been sporadically reported in Oregon but is not yet established. It can be recognized by its red flowers, simple, grass-like leaves, and lack of tendrils.

Key to *Lathyrus* species:

1. Leaflets ≥ 4.
  2. Tendrils not well developed.
    3. Leaflets usually paired.
      4. Stipules ≥ terminal leaflets.
        5. Leaflets glabrous; plants of eastern Oregon . . . . . ***L. rigidus***
        5' Leaflets densely villous; plants coastal . . . . . ***L. littoralis***
      4' Stipules much < terminal leaflets . . . . . ***L. tracyi*** (in part)
    3' Leaflets usually unpaired.
      6. Leaflets 8–16, villous, 5–15 mm . . . . . ***L. torreyi***
      6' Leaflets 4–10, glabrous to sparsely or finely pubescent, 20–110 mm.
        7. Flowers 15–22 mm, pink- to blue-purple or white . . . . . ***L. nevadensis*** (in part)
        7' Flowers 7–15 mm, white . . . . . ***L. lanszwertii*** (in part)
  2' Tendrils well developed.
    8. Stipules sagittate; plants coastal . . . . . ***L. japonicus*** var. ***glaber***
    8' Stipules semi-sagittate; plants usually not coastal.
      9. Wings exceeding keels by 2–4 mm.
        10. Stipules 5–15 mm . . . . . ***L. brownii***
        10' Stipules 15–25 mm . . . . . ***L. pauciflorus***
      9' Wings ≈ keels in length.
        11. Stems distinctly winged.
          12. Leaflets 4–8, paired . . . . . ***L. palustris*** (in part)
          12' Leaflets 9–12, unpaired . . . . . ***L. delnorticus***
        11' Stems ribbed, angled, or narrowly winged.
          13. Corollas white to yellow, tan, or orange-tinged.
            14. Corollas yellow, tan, or orange-tinged . . . . . ***L. sulphureus***
            14' Corollas white.
              15. Flowers 7–13 mm . . . . . ***L. tracyi*** (in part)
              15' Flowers generally 12–18 mm.
                16. Stems basally branched; lateral calyx lobes generally lanceolate . . . . . ***L. vestitus*** (in part)
                16' Stems often branched mid-stem; lateral calyx lobes narrowly triangular . . . . ***L. holochlorus***
          13' Corollas (at least in part) rose, deep red, blue-purple, or purple.
            17. Leaflets 10–16; stipules approaching leaflets in size . . . . . ***L. polyphyllus***
            17' Leaflets 4–12; stipules much < leaflets.
              18. Inflorescences 5–20-flowered; lateral calyx lobes ≥ tubes, lobes lanceolate . . ***L. vestitus*** (in part)
              18' Inflorescences 2–8-flowered; lateral calyx lobes ≤ tubes, lobes deltoid.
                19. Lateral calyx lobes ≈ tubes; plants coastal . . . . . ***L. palustris*** (in part)
                19' Lateral calyx lobes < tubes; plants usually inland.
                  20. Flowers 10–20 mm, banner blades > claws . . . . . ***L. nevadensis*** (in part)
                  20' Flowers 7–15 mm, banner blades ≈ claws . . . . . ***L. lanszwertii*** (in part)

1' Leaflets 0 or 2.
  21. Leaflets 0 (but stipules may appear as 2 leaflets); corollas yellow . . . . . ***L. aphaca***
  21' Leaflets 2; corollas blue- or rose-purple, red, reddish orange, magenta, pink, or white.
    22. Plants perennial; inflorescences 4–15-flowered.
      23. Leaflets 7–50 mm wide, > ½ as wide as stems . . . . . ***L. latifolius***
      23' Leaflets 4–10 mm wide, < ½ as wide as stems . . . . . ***L. sylvestris***
    22' Plants annual; inflorescences 1–3-flowered.
      24. Stems not winged; corollas red to reddish orange . . . . . ***L. sphaericus***
      24' Stems winged; corollas blue- to rose-purple.
        25. Flowers 20–30 mm . . . . . ***L. tingitanus***
        25' Flowers 8–10 mm.
          26. Fruits pustulose-hirsute . . . . . ***L. hirsutus***
          26' Fruits glabrous . . . . . ***L. angulatus***

Lathyrus angulatus Linnaeus (p. 689)

Angled pea

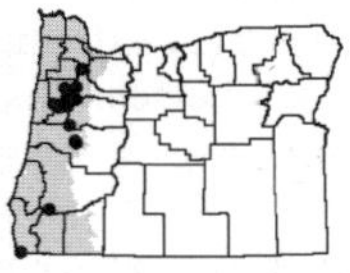

Plants annual. **Stems** erect to clambering, 1–5 dm, winged, basally branched. **Leaves** 3–5 cm; leaflets 2, paired, linear, 20–70 × 1–4 mm, surfaces glabrous; tendrils well developed on upper leaves; stipules lanceolate, as wide as leaflets, bases semi-sagittate. **Inflorescences** flowers solitary, 2–7 cm. **Flowers** 8–10 mm; corollas rose-purple to blue-purple. **Fruits** glabrous, pods not distinctly veined.

Disturbed areas. Flowering May–Jul. 50–500 m. CR, Est, Sisk, WV. CA. Exotic.

Lathyrus angulatus can be distinguished from the closely related *L. sphaericus* by the absence of distinct venation on its mature fruits.

Lathyrus aphaca Linnaeus

Yellow pea, yellow vetchling

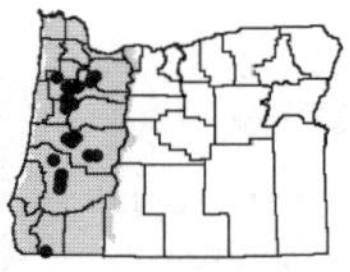

Plants annual. **Stems** erect to clambering, 2–6 dm, basally branched. **Leaves** 1–4 cm; leaflets 0; tendrils well developed; stipules leaf-like, ovate, bases hastate, surfaces glabrous. **Inflorescences** flowers solitary, 2–4 cm. **Flowers** 6–10 mm; corollas yellow.

Disturbed areas. Flowering May–Jun. 50–400 m. Casc, CR, Sisk, WV. CA, WA; widely scattered in North America; Asia, Europe. Exotic.

Lathyrus angulatus

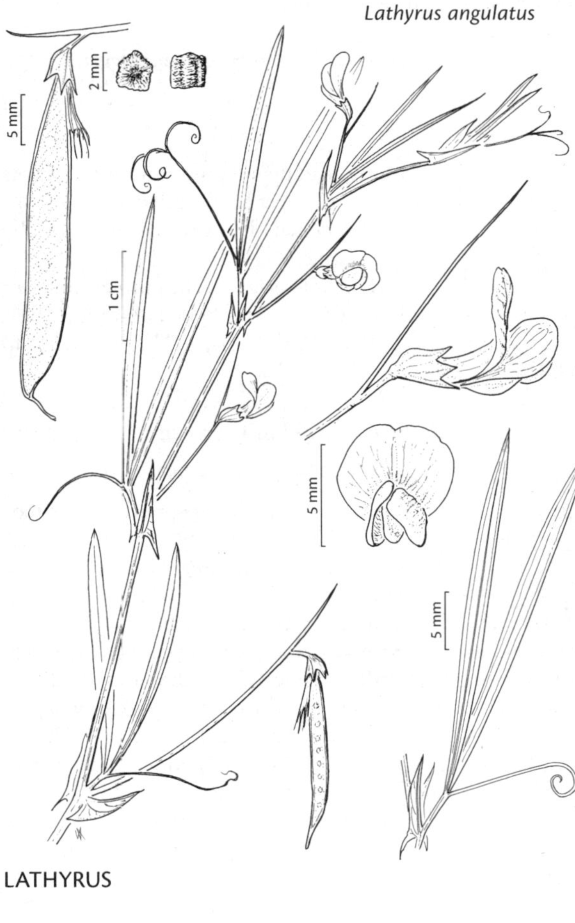

Lathyrus brownii Eastwood

Brown's pea

Lathyrus lanszwertii Kellogg var. *brownii* (Eastw.) Jeps., *Lathyrus pauciflorus* Fernald ssp. *brownii* (Eastw.) Piper, *Lathyrus schaffneri* Rydb.

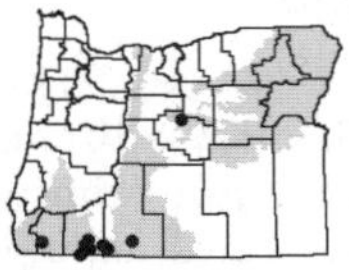

Plants perennial. **Stems** sprawling, 2–5 dm, basally branched 0–3 times. **Leaves** 4–6 cm; leaflets 8–10, unpaired, ovate to linear, 15–40 × 1–10 mm, surfaces glabrous; tendrils well developed; stipules linear, 5–15 mm, smaller than leaflets, bases semi-sagittate. **Inflorescences** 2–4-flowered, 3–4 cm. **Flowers** 12–16 mm; corollas purple, wings exceeding keels by 2–4 mm.

Streambanks, open ponderosa pine forests. Flowering Apr–Jul. 600–1900 m. BW, ECas, Sisk. CA, NV. Native.

Lathyrus brownii has been variously allied with *L. pauciflorus* or with *L. lanszwertii*. It is distinct from both: from the former by its smaller leaflets and flowers and by range; from the latter by its larger flowers, smaller leaflets, general lack of pubescence, and more typically prostrate, sprawling habit.

Lathyrus delnorticus C.L. Hitchcock

Del Norte pea

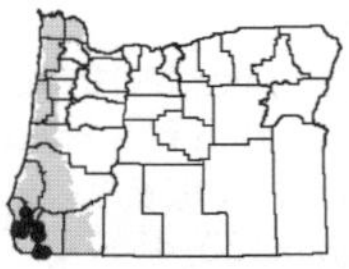

Plants perennial. **Stems** erect to ascending, 2–8 dm, winged, basally branched 0–several times. **Leaves** 9–13 cm; leaflets 9–12, unpaired, lanceolate, 30–50 × 5–15 mm, surfaces glabrous; tendrils well developed; stipules lanceolate, sometimes as large as leaflets, bases semi-sagittate. **Inflorescences** 8–12-flowered, 7–10 cm. **Flowers** 9–12 mm; corollas cream to white with purple striations, wings ≈ keels.

Serpentine near rivers and creeks in coniferous and mixed evergreen forests. Flowering May–Jul. 50–1400 m. CR, Sisk. CA. Native.

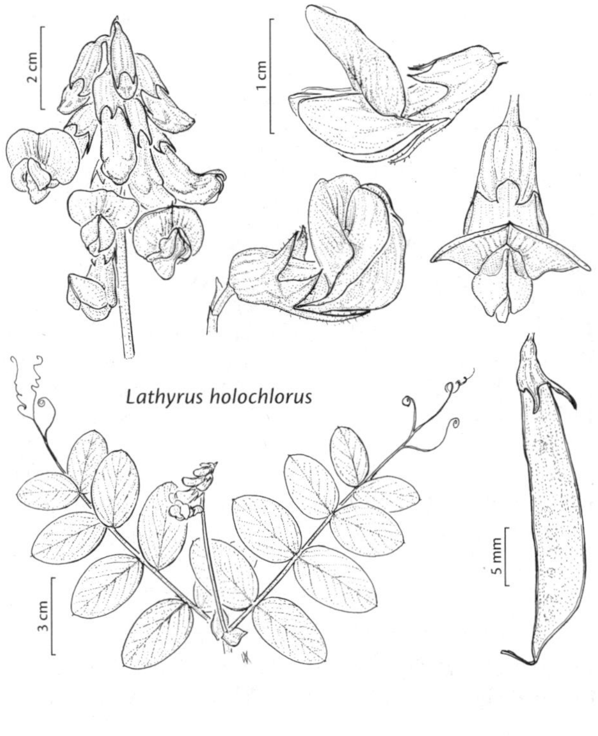

Lathyrus holochlorus

Lathyrus delnorticus is endemic to Coos, Curry, and Josephine counties in Oregon and Del Norte and western Siskiyou counties in California.

Lathyrus hirsutus Linnaeus

Rough pea, hairy vetch, Caley pea

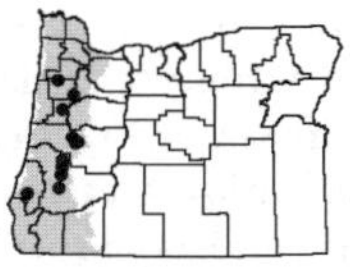

Plants annual. **Stems** climbing, 2–10 dm, winged, basally branched 0–many times. **Leaves** 8–30 cm; leaflets 2, paired, lanceolate to linear, 25–75 × 5–12 mm, surfaces glabrous; tendrils well developed above; stipules linear, much < leaflets, bases semi-sagittate. **Inflorescences** 1–2-flowered, 4–15 cm. **Flowers** 8–10 mm; corollas rose-purple with paler wings and keels. **Fruits** pods pustulose-hirsute.

Roadsides, pastures, cultivated fields, disturbed areas. Flowering Jun–Aug. 0–400 m. CR, Sisk, WV. CA; scattered in southeastern US; Europe. Exotic.

Lathyrus hirsutus is occasionally cultivated as a forage crop.

Lathyrus holochlorus (Piper) C.L. Hitchcock (p. 689)

Thin-leaved pea

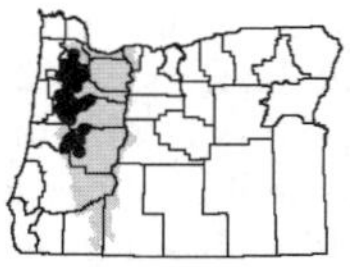

Plants perennial. **Stems** climbing, 3–10 dm, angled to narrowly winged, often branched mid-stem. **Leaves** 6–10 cm; leaflets 8–12, unpaired, ovate to lanceolate, 20–45 × 15–25 mm, surfaces glabrous; tendrils well developed; stipules linear to lanceolate, much < leaflets, bases semi-sagittate. **Inflorescences** 7–15-flowered, 5–15 cm. **Flowers** 12–14 mm; calyx lateral lobes narrowly triangular; corollas white to cream, wings ≈ keels in length.

Roadside fencerows, streambanks, grasslands, open oak woodlands. Flowering Apr–Jun. 0–400 m. Casc, WV. Native.

Lathyrus holochlorus is endemic to the Willamette Valley of western Oregon.

Lathyrus japonicus Willdenow var. *glaber* (Seringe) Fernald

Beach pea, maritime pea

Lathyrus maritimus (L.) Bigelow

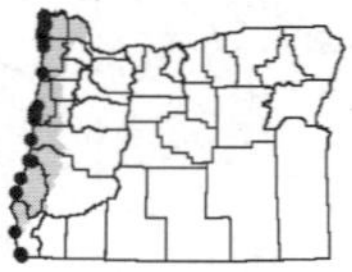

Plants perennial. **Stems** erect to sprawling or climbing, 1–3.5 dm, basally branched 0–4 times. **Leaves** 3–7 cm; leaflets 6–8, unpaired, ovate to lanceolate, 15–55 × 5–35 mm, surfaces glabrous; tendrils usually well developed; stipules ovate, as large as leaflets, bases sagittate. **Inflorescences** 4–8-flowered, 4–15 cm. **Flowers** 12–20 mm; corollas blue to purple.

Sandy or gravelly coastal areas. Flowering Apr–Aug. 0–50 m. CR, Est. CA, WA; northeastern and northwestern North America. Native.

Lathyrus lanszwertii Kellogg

Nevada pea, thick-leaved pea, Lanszwert's pea

Plants perennial. **Stems** sprawling or climbing, 2–8 dm, angled to narrowly winged, basally branched 0–4 times. **Leaves** 5–8 cm; leaflets 4–10, unpaired, lanceolate, rarely linear, 30–70 × 5–20 mm, glabrous to sparsely pubescent; tendrils well developed or not; stipules linear to lanceolate, much < leaflets, bases semi-sagittate. **Inflorescences** 2–6-flowered, 3–10 cm. **Flowers** 7–15 mm; calyx lateral lobes deltoid, < tubes; corollas blue-purple to white, banners ≈ claws in length, wings ≈ keels in length.

Western United States. 5 varieties; 2 varieties treated in *Flora*.

Key to *Lathyrus lanszwertii* varieties:

1. Stems usually > 3 dm, climbing; upper leaf tendrils well developed; corollas usually blue-purple . . . . *L. l.* var. ***lanszwertii***

1' Stems often < 3 dm, erect; upper leaf tendrils reduced to simple bristles; corollas white . . . . *L. l.* var. ***aridus***

Lathyrus lanszwertii Kellogg var. *aridus* (Piper) Jepson

Dryland Nevada pea

Lathyrus lanszwertii Kellogg ssp. *aridus* (Piper) Bradshaw

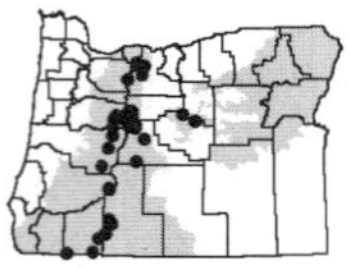

Stems often < 3 dm, erect. **Leaves** upper leaf tendrils reduced to simple bristles, not branched. **Flowers** corollas white.

Open ponderosa pine woodlands, sagebrush steppe. Flowering May–Jul. 700–1800 m. BW, Casc, ECas, Sisk. CA, ID, WA. Native.

Variety *aridus* appears to intergrade with var. *lanszwertii* in central Washington and Oregon.

Lathyrus lanszwertii Kellogg var. *lanszwertii*

Lanszwert's pea

Lathyrus lanszwertii Kellogg ssp. *lanszwertii*

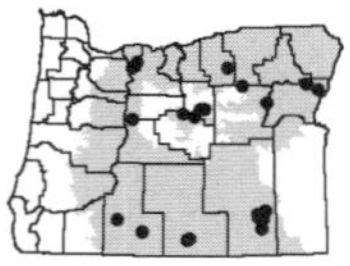

Stems usually > 3 dm, climbing. **Leaves** tendrils of upper leaves well developed, usually branched. **Flowers** corollas usually blue-purple.

Thickets, sagebrush, juniper, pine, or aspen woodlands. Flowering Jun–Aug. 800–2500 m. BR, BW, Casc, Col, ECas. CA, ID, NV, WA; east to WY, southeast to CO. Native.

Lathyrus latifolius Linnaeus (p. 691)

Everlasting pea, perennial pea

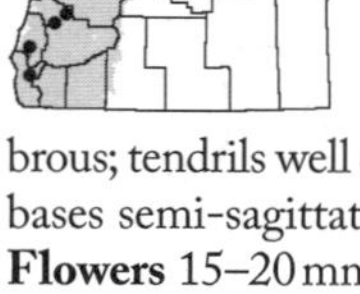

Plants perennial. **Stems** sprawling or climbing, 8–20 dm, broadly winged, basally branched 1–many times. **Leaves** 2–5 cm; leaflets 2, paired, lanceolate, 30–120 × 7–50 mm, > ½ as wide as stems, surfaces glabrous; tendrils well developed; stipules lanceolate, much < leaflets, bases semi-sagittate. **Inflorescences** 5–15-flowered, 10–25 cm. **Flowers** 15–20 mm; corollas purple, magenta, pink, or white.

Roadsides, disturbed areas. Flowering May–Aug. 0–1300 m. Casc, CR, Sisk, WV. CA, ID, NV, WA; widely scattered in Canada, throughout US; Europe. Exotic.

Lathyrus latifolius is widely naturalized in North America and forms large thickets along roadsides, especially in the Pacific Coast states.

Lathyrus littoralis (Nuttall) Endlicher (p. 691)

Gray beach pea, silky beach pea

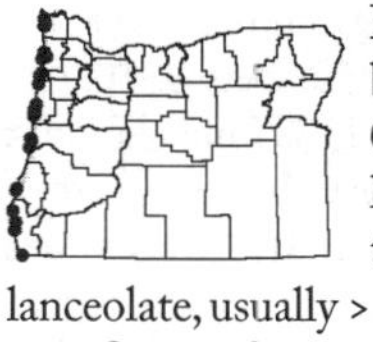

Plants perennial. **Stems** sprawling, 1–3 dm, basally branched 0–several times. **Leaves** 0.5–3 cm, leaflets 4–6, paired, obovate to oblanceolate, 10–20 × 4–8 mm, densely villous; tendrils flattened bristles; stipules ovate to lanceolate, usually > leaflets, bases semi-sagittate. **Inflorescences** 4–6-flowered, 3–9 cm. **Flowers** 13–15 mm; corollas 2-colored, dark purple to pink-purple and white.

Coastal sand dunes, deflation plains. Flowering Apr–Jul. 0–100 m. Est. CA, WA; north to British Columbia. Native.

Lathyrus nevadensis S. Watson

Sierra Nevada pea

Plants perennial. **Stems** erect, sprawling or climbing, 1–6 dm, ribbed or angled, basally branched 0–several times. **Leaves** 2–10 cm; leaflets 4–10, usually unpaired, ovate to lanceolate, 20–110 × 2–30 mm, surfaces glabrous, rarely finely pubescent; tendrils well developed or reduced to bristles; stipules linear, much < leaflets, bases semi-sagittate. **Inflorescences** 2–8-flowered, 4–15 cm. **Flowers** 10–22 mm; calyx lateral lobes deltoid, < tubes; corollas pink-purple, blue-purple, or white, banners > claws, wings ≈ keels.

Western United States and Canada. 3 varieties; 2 varieties treated in *Flora*.

Key to *Lathyrus nevadensis* varieties:

1. Corollas pink- to blue-purple, rarely white; leaflets 4–10, tendrils well developed or reduced to bristles; inflorescences 2–8-flowered, 10–20 mm . . . . ***L. n.*** var. ***nevadensis***

1' Corollas white; leaflets 4–6, tendrils reduced to bristles; inflorescences 2–4-flowered 2–4, 18–22 mm . . . . ***L. n.*** var. ***cusickii***

Lathyrus nevadensis S. Watson var. *cusickii* (S. Watson) Broich

Sierra Nevada pea

Lathyrus cusickii S. Watson, *Lathyrus nevadensis* S. Watson ssp. *cusickii* (S. Watson) C.L. Hitchc.

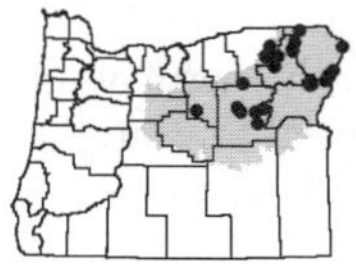

Leaflets 4–6; tendrils bristle-like. **Inflorescences** 2–4-flowered. **Flowers** 18–22 mm; corollas white.

Open coniferous forests. Flowering May–Jun. 700–1800 m. BW, Lava. ID, NV, WA. Native.

Variety *cusickii* includes two distinct races, one with ovate to lanceolate leaflets, the other with linear leaflets. Plants from an intermediate population have been reported in Umatilla County.

Lathyrus nevadensis S. Watson var. *nevadensis*

Sierra Nevada pea

Lathyrus nevadensis S. Watson ssp. *lanceolatus* (Howell) C.L. Hitchc., *Lathyrus nevadensis* S. Watson ssp. *nevadensis*, *Lathyrus nevadensis* S. Watson var. *nuttallii* (S. Watson) C.L. Hitchc., *Lathyrus nevadensis* S. Watson var. *pilosellus* (M. Peck) C.L. Hitchc., *Lathyrus nuttallii* S. Watson

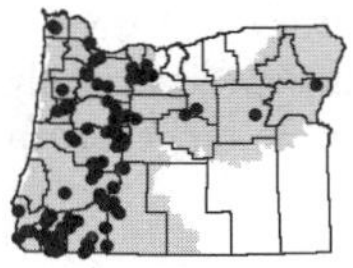

Leaflets 4–10; tendrils well developed or reduced to bristles. **Inflorescences** 2–8-flowered. **Flowers** 10–20 mm; corollas pink- to blue-purple, rarely white.

Open coniferous and mixed forests. Flowering Apr–Jul. 0–1900 m. BW, Casc, CR, ECas, Lava, Sisk, WV. CA, ID, WA. Native.

Lathyrus nevadensis var. *nevadensis* includes two races. One race consists of short (less than 3 dm), erect plants bearing 4–6 leaflets on leaf rachises ending in, at most, a short (less than 1 cm) bristle rather than a tendril, and with 2–4(6)-flowered racemes

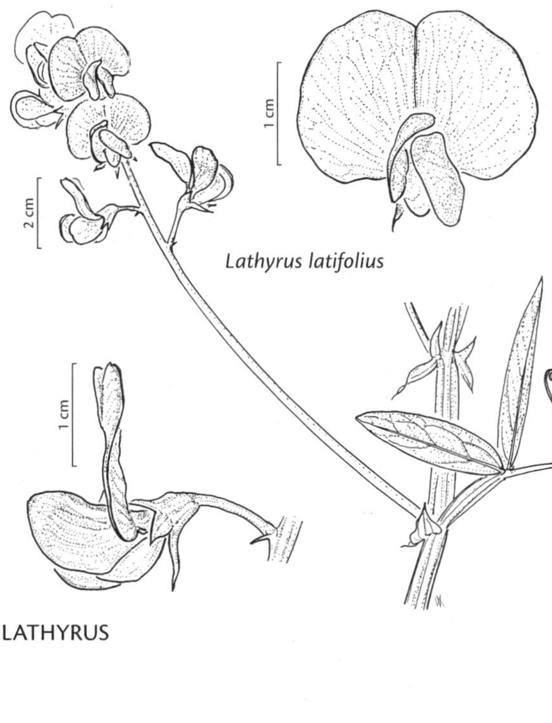

Lathyrus latifolius

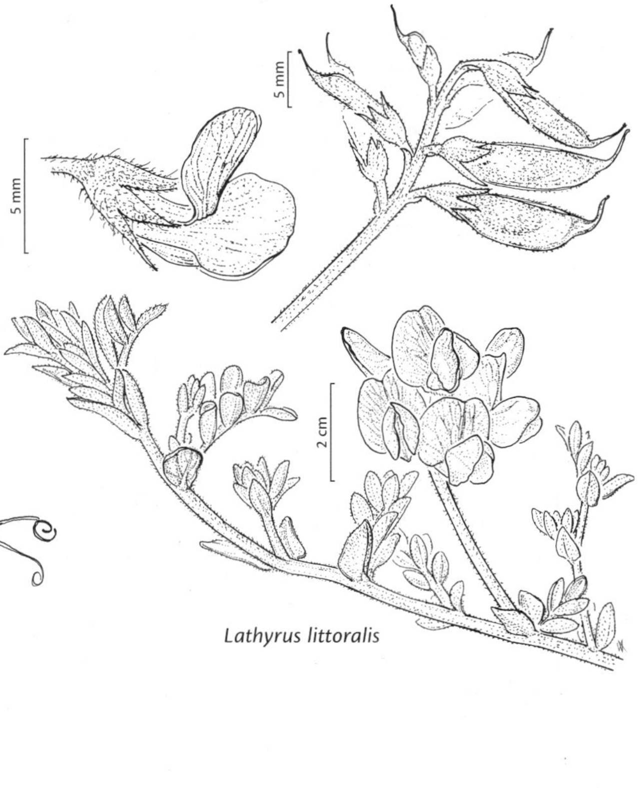

Lathyrus littoralis

LATHYRUS

of rather large (15–20 mm) flowers (*L. nevadensis s.s.*). A second race consists of sprawling or clambering plants up to 6 dm tall bearing leaves of 6–8(10) leaflets on rachises which end in a long (> 2 cm), often branched and prehensile tendril, and with racemes of 4–8 smaller (10–17 mm) flowers (*L. lanceolatus*). The *nevadensis* race is more common in the southwestern part of the range; the *lanceolatus* race is found throughout the range of the variety but is more common to the north and in the eastern foothills of the Cascades in central Washington. In northwestern California and southwestern Oregon, the two races intergrade extensively.

Lathyrus palustris Linnaeus

Marsh pea

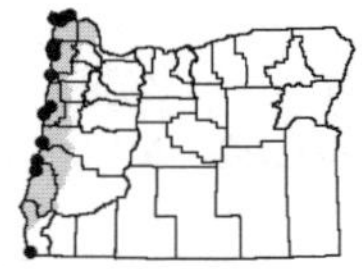

Plants perennial. **Stems** sprawling or climbing, 2–10 dm, angled to winged, basally branched 0–4 times. **Leaves** 1–6 cm; leaflets 4–8, paired, lanceolate to linear, 15–60 × 2–20 mm, surfaces glabrous, glabrate or pubescent; tendrils well developed; stipules lanceolate to linear, much < leaflets, bases semi-sagittate. **Inflorescences** 2–7-flowered, 3–14 cm. **Flowers** 10–15 mm; calyx lateral lobes deltoid, ≈ tubes, petals blue-purple, wings ≈ keels.

Coastal wetlands. Flowering Jun–Aug. 0–50 m. CR, Est. CA, WA; scattered in North America; Asia, Europe. Native.

Although in Oregon (and California) this species is only found in coastal wetlands, in all other areas of its range, it can also be found far inland.

Lathyrus pauciflorus Fernald

Few-flowered pea

Plants perennial. **Stems** erect, sprawling or climbing, 3–8 dm, basally branched 1–5 times. **Leaves** 6–9 cm; leaflets 8–10, unpaired, lanceolate to linear or ovate, 20–80 × 1–25 mm, glabrous; tendrils well developed; stipules ovate to lanceolate, 15–25 mm, smaller than leaflets, bases semi-sagittate. **Inflorescences** 4–5-flowered, 5–18 cm. **Flowers** 12–20 mm; corollas purple to pale blue-lilac or nearly white, wings exceeding keels by 2–4 mm.

2 varieties.

While the range of leaflet shape (lanceolate to linear) in *L. pauciflorus* is striking, there are a complete series of intermediate forms.

Key to *Lathyrus pauciflorus* varieties:

1. Leaflets lanceolate to linear, width < ½ length . . . . ***L. p.*** var. ***pauciflorus***

1' Leaflets ovate, width ≥ ½ length . . . . . . ***L. p.*** var. ***utahensis***

Lathyrus pauciflorus Fernald var. *pauciflorus* (p. 692)

Few-flowered pea

Lathyrus pauciflorus Fernald ssp. *pauciflorus*, *Lathyrus pauciflorus* Fernald ssp. *tenuior* Piper

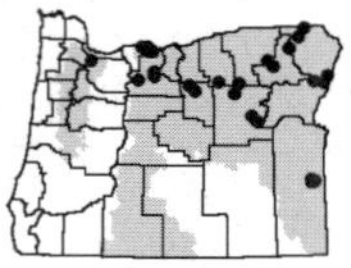

Leaflets lanceolate to linear, width < ½ length.

Sagebrush steppe, ponderosa pine and pine-juniper woodlands. Flowering Apr–Jul. 0–2000 m. BW, Col, ECas, Lava, Owy, WV. ID, WA. Native.

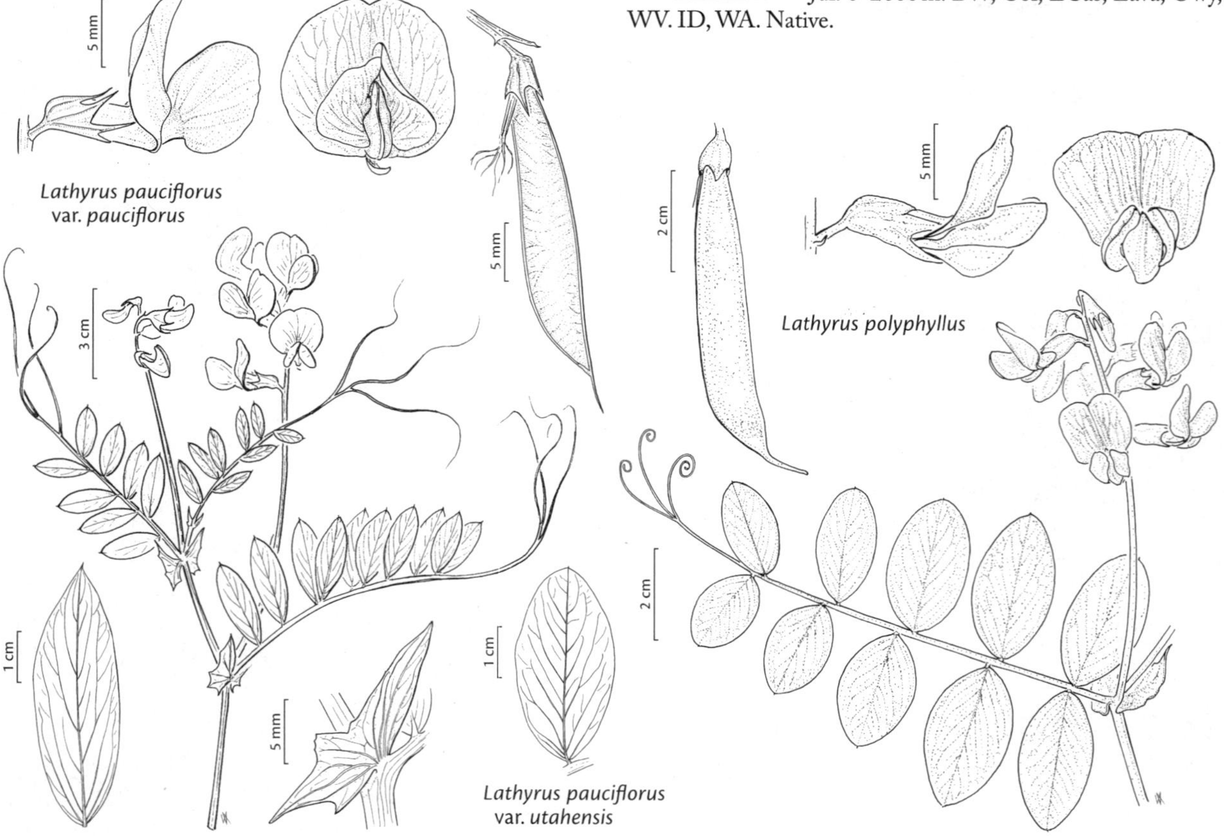

LATHYRUS

Lathyrus pauciflorus Fernald var. ***utahensis*** (M.E. Jones) R.J. Davis (p. 692)

Utah pea

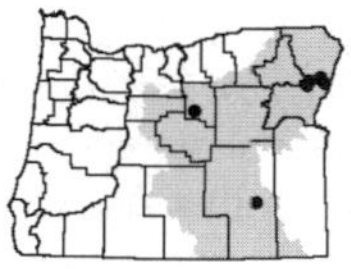

Leaflets ovate, width ≥ ½ length.

Sagebrush steppe, pine and pine-juniper woodlands. Flowering May–Jul. 800–1900 m. BR, BW, Lava. ID, WA; southeast to UT, AZ. Native.

Lathyrus polyphyllus Nuttall (p. 692)

Leafy pea

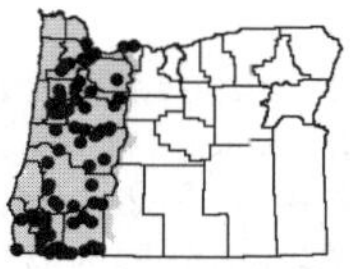

Plants perennial. **Stems** erect, climbing, 4–8 dm, ribbed or angled, basally branched 0–2 times. **Leaves** 11–16 cm; leaflets 10–16, unpaired, ovate to lanceolate, 30–50 × 10–30 mm, surfaces glabrous; tendrils well developed; stipules ovate to lanceolate, approaching leaflets in size, bases semi-sagittate. **Inflorescences** 8–12-flowered, 9–14 cm. **Flowers** 15–18 mm; corollas blue-purple, wings ≈ keels.

Openings in coniferous forests. Flowering Apr–Jul. 0–1600 m. Casc, CR, Est, Sisk, WV. CA, WA. Native.

Lathyrus rigidus White (p. 693)

Rigid pea

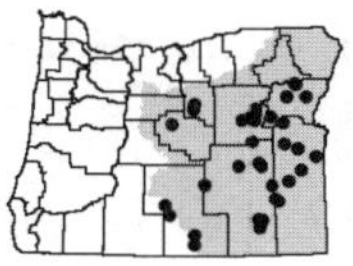

Plants perennial. **Stems** erect, 1–3 dm, basally branched 5–10 times. **Leaves** 1–3 cm; leaflets 4–8, paired or not, lanceolate, 15–30 × 5–10 mm, surfaces glabrous; tendrils bristle-like, usually < 1 cm; stipules lanceolate, as large as terminal leaflets, bases semi-sagittate. **Inflorescences** 2–3-flowered, 4–9 cm. **Flowers** petals lilac to white; 12–18 mm.

Basalt scablands, sagebrush. Flowering Apr–Jun. 800–1600 m. BR, BW, Lava, Owy. CA, ID, NV. Native.

Lathyrus sphaericus Retzius

Grass pea

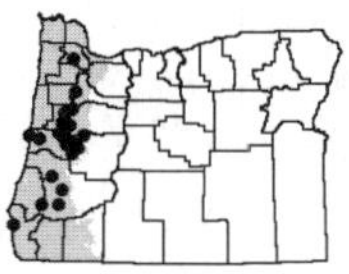

Plants annual. **Stems** erect or sprawling, 2–5 dm, not angled, basally branched. **Leaves** 0.5–1.5 cm; leaflets 2, paired, linear, 20–90 × 2–10 mm, glabrous; tendrils to 4 cm and often unbranched; stipules linear, much < leaflets, bases semi-sagittate. **Inflorescences** flowers solitary, 1–2 cm. **Flowers** 6–10 mm; corollas red to reddish orange. **Fruits** pods distinctly veined.

Roadsides, pastures, disturbed areas. Flowering Apr–Jun. 0–600 m. CR, Est, Sisk, WV. CA, WA; north to British Columbia; Africa, Asia, Europe. Exotic.

Lathyrus sphaericus can be distinguished from the closely related *L. angulatus* by the presence of distinct venation on the mature fruit.

Lathyrus sulphureus W.H. Brewer ex A. Gray

Brewer's sweet pea, snub pea, sulfur pea

Lathyrus nevadensis S. Watson ssp. *stipulaceus* (T.G. White) Bradshaw, *Lathyrus sulphureus* W.H. Brewer ex A. Gray var. *sulphureus*

Plants perennial. **Stems** climbing, 5–15 dm, ribbed or angled, branched 0–3 times at the base. **Leaves** 9–12 cm; leaflets 8–12, unpaired, ovate to lanceolate, 25–50 × 10–20 mm, surfaces glabrous; tendrils well developed; stipules lanceolate, sometimes as large as terminal leaflets; bases semi-sagittate. **Inflorescences** 9–15-flowered, 6–10 cm. **Flowers** 11–13 mm; corollas yellow, tan, or orange-tinged, wings ≈ keels.

Mixed oak-pine woodlands. Flowering Apr–Jun. 0–700 m. Est, Sisk. CA. Native.

Lathyrus sylvestris Linnaeus

Flat pea

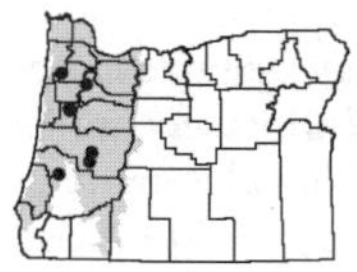

Plants perennial. **Stems** climbing, 6–20 dm, broadly winged, basally branched 1–3 times. **Leaves** 2–4 cm; leaflets 2, paired, linear, 20–150 × 4–10 mm, < ½ as wide as stems, surfaces glabrous; tendrils well developed; stipules linear, much < leaflets, bases semi-sagittate. **Inflorescences** 4–6-flowered, 9–15 cm. **Flowers** 12–17 mm; corollas red-purple.

Roadsides, fencerows, disturbed areas. Flowering Jun–Aug. 50–500 m. Casc, CR, WV. ID, WA; scattered in North America; Africa, Asia, Europe. Exotic.

Lathyrus sylvestris is similar to *L. latifolius*, and intermediate populations are known to occur in Europe. In North America, *L. sylvestris* is recognizable by its distinctly narrow stipules.

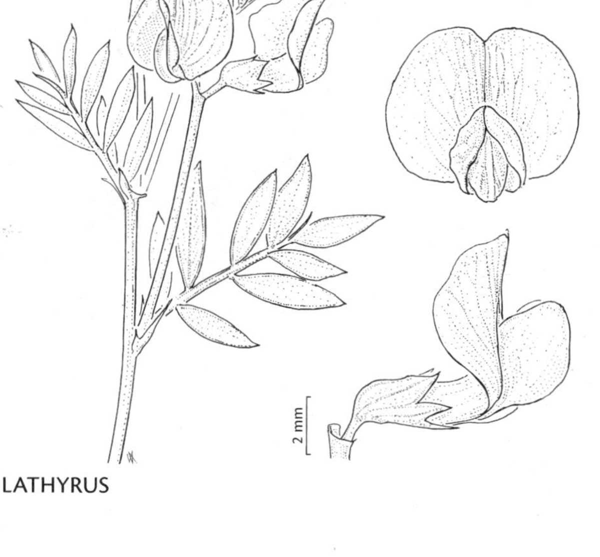

LATHYRUS

Lathyrus tingitanus Linnaeus

Tangier pea

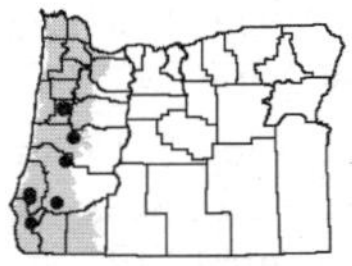

Plants annual. **Stems** climbing, 8–20 dm, broadly winged, basally branched 1–3 times. **Leaves** 2–4 cm; leaflets 2, paired, ovate/obovate to lanceolate, 25–60 × 5–35 mm, surfaces glabrous; tendrils well developed; stipules lanceolate, < leaflets, bases semi-sagittate. **Inflorescences** 1–3-flowered; 4–20 cm. **Flowers** 20–30 mm; corollas rose-purple.

Disturbed areas. Flowering May–Jun. 0–300 m. CR, Sisk, WV. CA, ID, WA; UT; Africa, Europe, Hawaii. Exotic.

Lathyrus torreyi A. Gray (p. 694)

Torrey's pea

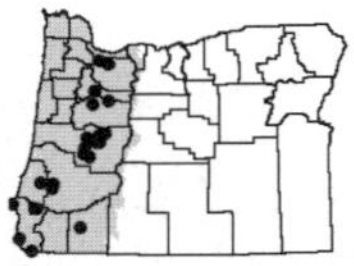

Plants perennial. **Stems** erect to sprawling, 1–4 dm, basally branched 0–3 times. **Leaves** 2–7 cm; leaflets 8–16, unpaired, ovate to lanceolate, 5–15 × 3–10 mm, surfaces villous; tendrils absent or mucronate; stipules lanceolate, much < leaflets, bases semi-sagittate. **Inflorescences** 1–2-flowered, 0.5–3 cm. **Flowers** 8–12 mm; corollas blue-purple.

Open coniferous forests. Flowering May–Jul. 0–800 m. Casc, CR, Est, Sisk, WV. CA, WA. Native.

Lathyrus tracyi Bradshaw

Tracy's pea

Lathyrus lanszwertii Kellogg var. *tracyi* (Bradshaw) Isley

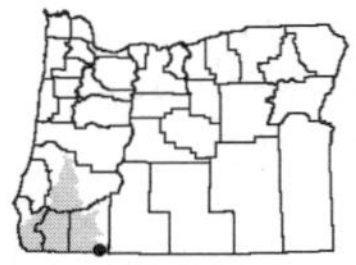

Plants perennial. **Stems** erect, sprawling or climbing, 2–6 dm, ribbed or angled, often branched at mid-stem 1–4 times. **Leaves** 2–6 cm; leaflets 4–8, usually paired, ovate to linear, 15–30 × 5–20 mm or 30–70 × 1–10 mm, surfaces glabrous or pubescent; tendrils well developed or not; tendrils well developed or not; stipules linear, < leaflets, bases semi-sagittate. **Inflorescences** 5–15-flowered; 2–7 cm. **Flowers** 7–13 mm; corollas white, wings ≈ keels.

Roadsides, open coniferous and mixed evergreen forests. Flowering May–Jul. 1400–1800 m. Sisk. CA. Native.

Lathyrus tracyi includes both ovate- and linear-leafleted forms. Linear-leafleted populations certainly suggest a relationship to *L. lanszwertii*, but ovate-leafleted forms, the presence of mid-stem branching, and floral structure suggest that *L. tracyi* may be better allied to *L. holochlorus* found to the north in the Willamette Valley.

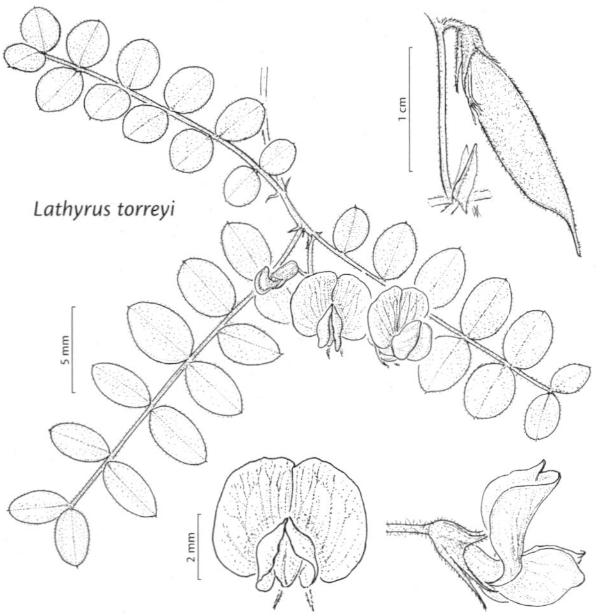

LATHYRUS

Lathyrus vestitus Nuttall

Pacific pea

Plants perennial. **Stems** sprawling, climbing, 2–20 dm, angled to narrowly winged, basally branched 0–4 times. **Leaves** 4–14 cm; leaflets 6–12, unpaired, ovate to linear, 12–40 × 5–25 mm, surfaces sparsely to densely pubescent, sometimes glabrous abaxially; tendrils well developed; stipules lanceolate to linear, much < leaflets, bases semi-sagittate. **Inflorescences** 5–20-flowered, 5–25 cm. **Flowers** 12–18 mm; calyx lateral lobes lanceolate, ≥ tubes; corollas blue-purple to white, wings ≈ keels.

Pacific states and northwestern Mexico. 3 varieties; 2 varieties treated in *Flora*.

Key to *Lathyrus vestitus* varieties:

1. Corollas cream to white; plants glabrous. . . . . *L. v.* var. ***ochropetalus***

1' Corollas blue-purple, lavender, pink, or white; plants sparsely to densely pubescent . . . . *L. v.* var. ***vestitus***

Lathyrus vestitus Nuttall var. *ochropetalus* (Piper) Isely

Pacific pea

Lathyrus bolanderi S. Watson, *Lathyrus vestitus* Nutt. ssp. *bolanderi* (S. Watson) C.L. Hitchc., *Lathyrus vestitus* Nutt. ssp. *ochropetalus* (Piper) C.L. Hitchc.

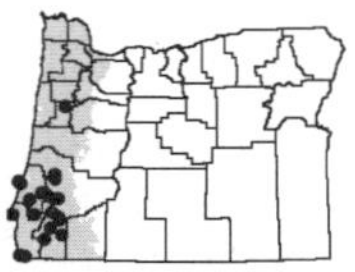

Plants glabrous. **Flowers** 12–15 mm; corollas cream to white.

Roadside fencerows, grasslands, openings in coastal coniferous forests. Flowering Apr–Jul. 0–900 m. CR, Est, Sisk, WV. CA, WA. Native.

Lathyrus vestitus Nuttall var. *vestitus*

Pacific pea

Lathyrus vestitus Nutt. ssp. *puberulus* (White ex Greene) C.L. Hitchc., *Lathyrus vestitus* Nutt. ssp. *vestitus*

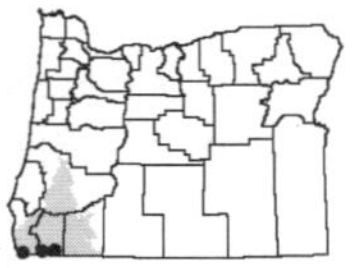

Plants sparsely to densely pubescent. **Flowers** 12–18 mm; corollas blue-purple, lavender, pink, or white.

Roadside fencerows, grasslands, oak or conifer woodlands along creeks, chaparral. Flowering Apr–Jul. 100–1200 m. Sisk. CA. Native.

Variety *vestitus* includes populations found in habitats ranging from moist coastal coniferous forests to chaparral. Numerous attempts to describe distinct taxa within this series of populations have been undermined by extensive clinal variation north to south and intergradation among the populations.

LOTUS Bird's-foot trefoil

Cameron J. Ross

Herbs annual or perennial; with or without rhizomes. **Stems** decumbent to ascending, glabrous, pilose to pubescent, bases sometimes woody. **Leaves** odd-pinnately compound; leaflets usually 5, with lower 2 basal on petiole, others terminally ternate, tips obovate to acuminate; sessile or short-petiolate; stipules reduced, gland-like. **Inflorescences** umbels or 1–3-flowered clusters, axillary; bracts generally ternate. **Peduncles** 0–15 cm. **Flowers** bilaterally symmetric, papilionaceous; calyces campanulate, bases connate, lobes outcurved or not; corollas yellow with red-tinged striations; stamens 10, filiform, diadelphous. **Fruits** dehiscent, laterally compressed, spiraling, linear to oblong, extending beyond calyces, with or without beaks. **Seeds** 1–30(+), globose to ellipsoid, 1–2 mm, brown, usually with black markings.

Worldwide. 125 species; 2 species treated in *Flora.*

Native taxa, formerly placed in this genus, have been moved to either *Acmispon* or *Hosackia.*

Lotus subbiflorus var. *subbiflorus* has been collected in Curry County. It is not yet naturalized, however.

Key to *Lotus* species:

1. Inflorescences (1)3–6(10)-flowered; calyx lobes usually ≤ tubes, not reflexed in bud; rhizomes absent; stems solid, pithy . . . . **L. corniculatus**

1' Inflorescences (5)7–12(15)-flowered; calyx lobes usually ≥ tubes, slightly reflexed in bud; rhizomes present; stems hollow . . . . **L. uliginosus**

Lotus corniculatus Linnaeus (p. 695)

Common bird's-foot trefoil

Lotus tenuis Waldst. & Kit. ex Willd.

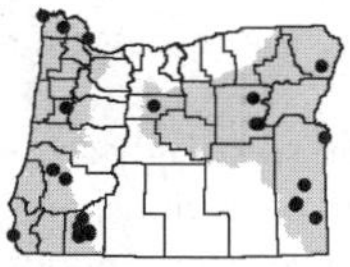

Plants perennial, to 1–8 dm, glabrous to ciliate; taprooted, rhizomes absent. **Stems** decumbent to ascending, branched, solid, pithy, sometimes trailing and rooting at nodes. **Leaves** leaflets obovate to oblanceolate, (4)6–22(25) × 2–10(12) mm, tips acuminate or obtuse. **Inflorescences** (1)3–6(10)-flowered, with tufted hairs at bases. **Peduncles** 1.5–12 cm. **Flowers** calyces 5–7(8) mm, lobes usually ≤ tubes, not reflexed in bud; corollas 10–16 mm, yellow. **Fruits** 1.5–3 cm, glabrous. **2*n*=24.**

Disturbed areas, roadsides. Flowering May–Aug. 0–1900 m. BW, CR, Est, Lava, Owy, Sisk, WV. CA, ID, NV, WA; throughout North America; Africa, Asia, Europe. Exotic.

Lotus uliginosus Schkuhr

Big trefoil, greater bird's-foot trefoil

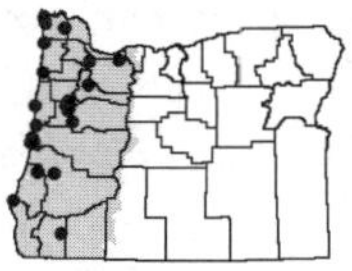

Plants perennial, 2–10 dm, glabrous to pilose and pubescent; rhizomes present. **Stems** ascending to erect, branched, hollow. **Leaves** leaflets obovate, 8–25(33) × 3–15 mm, tips acute to obtuse. **Inflorescences** (5)7–12(15)-flowered, with tufted hairs at bases. **Peduncles** 8–15 cm. **Flowers** calyces 7–9 mm, lobes usually ≥ tubes, slightly reflexed in bud; corollas 10–18 mm, yellow. **Fruits** 2–3.5 cm, glabrous. **2*n*=24.**

Wetlands, disturbed areas. Flowering Jun–Sep. 0–400 m. Casc, CR, Est, Sisk, WV. CA, ID, WA; widely scattered in North America; Africa, Asia, Europe. Exotic.

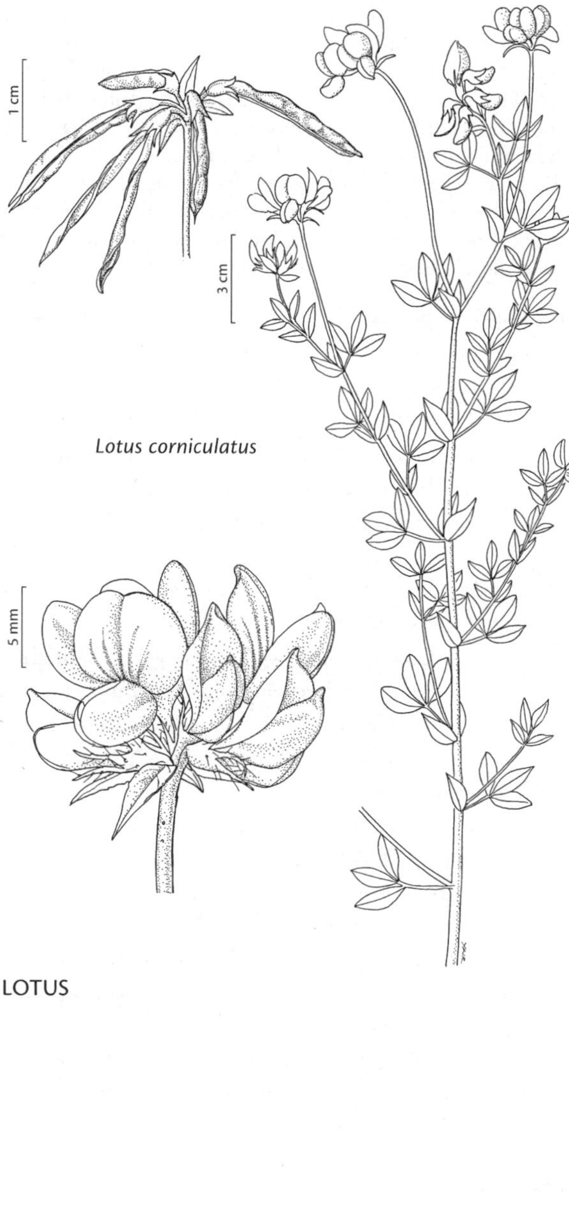

LOTUS

LUPINUS Lupine

Paul M. Severns, Stephen C. Meyers, and Christopher A. Royce

Herbs or **shrubs** annual or perennial; caulescent to subacaulescent or acaulescent; taprooted or rhizomatous. **Stems** erect, ascending, decumbent, spreading or prostrate, glabrous to tomentose, branched or unbranched. **Leaves** basal and/or cauline, alternate, palmately compound; leaflets 3–15, linear to ovate, margins entire, tips rounded to acute, sometimes apiculate, surfaces glabrous to pubescent; petiolate. **Inflorescences** racemes 1–several, usually terminal, sometimes axillary; pedicellate bracts usually deciduous. **Flowers** calyces 2-lipped, lips equal, subequal, or unequal, pubescent; corollas white, cream, yellow, pink, purple, blue, or bicolored; banners centrally grooved, frequently with spot of contrasting color near base, lateral petals reflexed to varying degrees, keels generally pointed; stamens 10, filaments of alternating lengths (5 long with short anthers, 5 short with long anthers); styles glabrous; stigmas with collar of hairs. **Fruits** legumes, dehiscent, oblong to elliptic. **Seeds** 2–12, smooth.

Worldwide. ~220 species; 26 species treated in *Flora.*

The identification of *Lupinus* species can lead to frustration as the combination of vegetative and floral characters occurring in keys or the species descriptions infrequently matches the specimen in hand to satisfaction. Much of this ambiguity arises from porous species barriers that allow frequent hybridization and introgression, producing plants with trait combinations not addressed in keys and treatments. Those hoping to identify *Lupinus* should expect morphological evidence of hybridization when perennial lupine taxa co-occur. It seems to be a near certainty that even the most well-behaved lupine will eventually find another congener with which it will hybridize. *Lupinus* species generally have a chromosome number of $2n$=32 to $2n$=52. As a result of widespread hybridization, we cautiously do not list chromosome numbers for individual species.

The key presented here attempts to minimize the influence of allometric measurements (such as the banner index) and often refers to three-dimensional flower forms (with particular attention given to the banner—the upper two fused petals). The three-dimensional flower forms we use are: 1. The "typical" type has a strongly reflexed banner with the upper fused petals arranged vertically and more or less parallel to each other (usually no rolling). 2. The "cup" type has a banner that spreads widely to form a bowl or cup shape. 3. The "pinched" type has a banner that is cupped by the sides and pinched inward (other authors refer to this morphology as a "ruffled banner"), causing the distal tip of the banner to be reflexed. In practice, these floral types are easier to diagnose on live plants in full anthesis. However, the banner types are also present on pressed specimens when an inflorescence contains the full range of flower development.

There are several *Lupinus* species that are problematic for identification because an arguably unnecessary, and seemingly arbitrary, number of varieties have been described. While the varieties may appear in discrete morphological states in collections, they often show no pattern of geographical segregation that are commonly found in varietal taxa and in the field display a continuous range of morphological intermediates.

Key to *Lupinus* species:

1. Plants annual.
  2. Plants usually ≤ 1 dm.
    3. Plants 0.1–0.2(0.3) dm; corollas white, sometimes blue- or purple-tinged . . . . ***L. uncialis***
    3' Plants 0.5–1(1.2) dm; corollas blue to purple.
      4. Inflorescences 0.8–2(3) cm; peduncles 1–6 cm . . . . ***L. brevicaulis***
      4' Inflorescences 2–4 cm; peduncles 0–1 cm . . . . ***L. pusillus*** var. ***intermontanus***
  2' Plants 1–10 dm.
    5. Leaves mostly sub-basal . . . . ***L. microcarpus***
    5' Leaves cauline.
      6. Corollas usually yellow, rarely pink to bright blue; seeds (1)2 . . . . ***L. luteolus***
      6' Corollas white, pink, blue, or purple, sometimes with patches of yellow; seeds (3)4–8.
        7. Corollas 9–13 mm; pedicels (3)4–8 mm . . . . ***L. affinis***
        7' Corollas 5–9 mm; pedicels 1–3 mm . . . . ***L. bicolor***

1' Plants perennial.
  8. Leaves basal, sub-basal, or basally clustered.
    9. Pedicels 1–3(4) mm; fruits 0.9–2 cm . . . . ***L. lepidus***
    9' Pedicels 4–8 mm; fruits 2–5 cm.
      10. Stems villous . . . . ***L. nevadensis***
      10' Stems glabrous to inconspicuously pubescent . . . . ***L. onustus***
  8' Leaves cauline, mostly cauline, or cauline and basal.
    11. Calyces spurred.
      12. Wings usually with tuft of hair near tips . . . . ***L. arbustus***
      12' Wings glabrous or sometimes midveins pubescent . . . . ***L. argenteus*** (in part)
    11' Calyces not spurred (but may be slightly saccate in *L. leucophyllus, L. polyphyllus,* and *L. sericeus*).

Banner Types

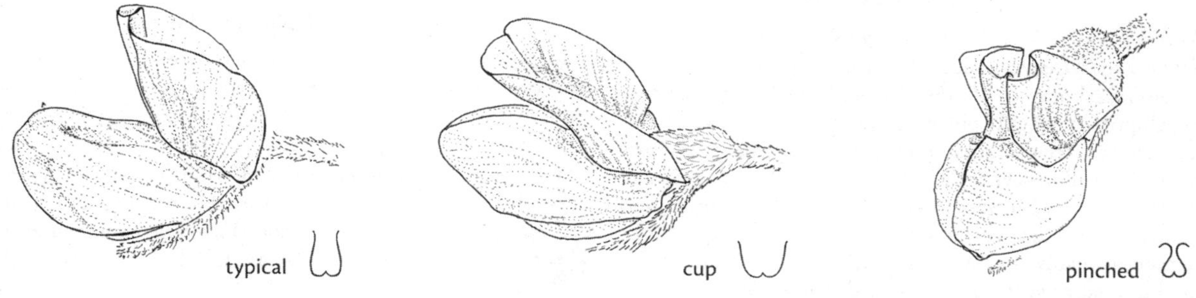

13. Banners pinched type.
  14. Banner margins inrolled. . . . . **L. albicaulis**
  14' Banner margins not inrolled.
    15. Calyces sericeous . . . . **L. oreganus**
    15' Calyces glabrous . . . . **L. tracyi**
13' Banners cup or typical type.
  16. Corollas yellow or partly yellow.
    17. Leaflets (40)50–120(150) mm. . . . . **L. sabinianus**
    17' Leaflets 15–60 mm.
      18. Banners typical type; stems woody; plants 10–20 dm, usually coastal . . . . **L. arboreus**
      18' Banners cup type; stems not woody; plants 4–8 dm, of northeastern Oregon . . . . . **L. sulphureus** (in part)
  16' Corollas white, pink, blue, or purple.
    19. Stems prostrate to decumbent; plants usually cushion- or mat-forming.
      20. Leaflet surfaces abaxially glabrous to strigose, adaxially strigose; plants coastal. . . . . **L. littoralis** var. **littoralis**
      20' Leaflet surfaces tomentose; plants inland. . . . . **L. breweri**
    19' Stems ascending to erect; plants not cushion- or mat-forming.
      21. Banners cup type.
        22. Basal leaves usually not withered by flowering; plants of northern Oregon . . . . **L. sulphureus** (in part)
        22' Basal leaves usually withered by flowering; plants of central and southern Oregon . . . **L. argenteus** (in part)
      21' Banners typical type (with a slight cup in *L. albifrons* var. *albifrons*).
        23. Banners abaxially pubescent.
          24. Corollas violet to lavender; banners with a slight cup; plants 1–4(5) dm . . . . . **L. albifrons** var. **albifrons**
          24' Corollas white to blue; banners without a slight cup; plants 2–10(14) dm.
            25. Pedicels 1–3 mm; often with some leaves clustered near bases. . . . . **L. leucophyllus**
            25' Pedicels 3–5(7) mm; rarely with some leaves clustered near bases . . . . **L. sericeus**
        23' Banners abaxially glabrous.
          26. Leaves basal and cauline . . . . **L. polyphyllus**
          26' Leaves cauline.
            27. Leaflets (20)30–100 × 5–30 mm; inflorescences usually dense. . . . . **L. latifolius**
            27' Leaflets 15–35(50) × 5–15 mm; inflorescences usually open . . . . **L. rivularis**

Lupinus affinis J. Agardh

Fleshy lupine

Lupinus nanus Douglas ex Benth., *Lupinus nanus* Douglas ex Benth. var. *carnosulus* (Greene) C.P. Sm.

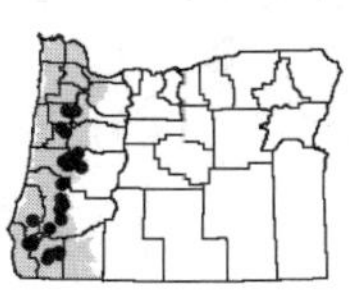

Plants annual, 2–5(6) dm. **Stems** erect, hirsute to villous. **Leaves** cauline; leaflets 5–8, 20–350 × 4–15 mm, oblanceolate, tips acute, often apiculate, surfaces hirsute to minutely papillose-hispid, denser abaxially; petioles 30–100 mm. **Inflorescences** (3.6)5–16(20) cm; peduncles 5–20 cm; pedicels (3)4–8 mm. **Flowers** calyces 5–7 mm, not spurred, lips equal to subequal; corollas 9–13 mm, blue to purple, banners typical type, spot white. **Fruits** 2–4.5 cm, pubescent. **Seeds** 5–8.

Openings in forests, meadows. Flowering Mar–Jun. 0–400 m. CR, Sisk, WV. CA. Native.

Lupinus albicaulis Douglas ex Hooker

Sickle-keeled lupine, pine lupine

Plants perennial, 3–9(12) dm. **Stems** erect, not woody, strigose to hirsutulous. **Leaves** cauline, lower leaves generally withered by flowering; leaflets 5–10, 15–70 × 2–23 mm, linear-oblanceolate, elliptic, or broadly oblanceolate, tips acute, often apiculate, surfaces strigose, sometimes glabrate adaxially; petioles 14–70 mm. **Inflorescences** 9.3–15(25) cm; peduncles 2–12 cm; pedicels 2–7 mm. **Flowers** calyces 5–15 mm, not spurred, lips unequal to subequal; corollas 7–15 mm, light purple to light blue, occasionally cream or yellow, banners pinched type, margins usually tightly inrolled, spot absent. **Fruits** 20–50 mm, villous. **Seeds** 3–8.

2 varieties.

This species may hybridize with *L. arbustus*, *L. latifolius*, *L. oreganus*, and *L. tracyi*.

Key to *Lupinus albicaulis* varieties:

1. Calyces 8–15 mm; corollas 10–15 mm . ***L. a.*** var. ***albicaulis***
1' Calyces 5–8 mm; corollas 7–11 mm . . . ***L. a.*** var. ***shastensis***

Lupinus albicaulis Douglas ex Hooker var. *albicaulis* (p. 698)

Sickle-keeled lupine, pine lupine

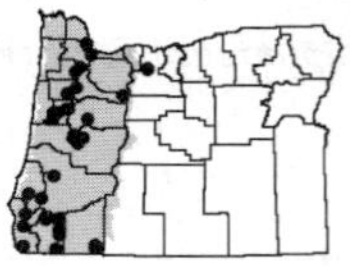

Flowers calyces 8–15 mm; corollas 10–15 mm.

Conifer forests, dry meadows, streamsides. Flowering Apr–Aug. 0–1800 m. Casc, CR, Sisk, WV. CA, WA. Native.

Lupinus albicaulis Douglas ex Hooker var. *shastensis* (A. Heller) C.P. Smith

Sickle-keeled lupine, pine lupine

Lupinus adsurgens Drew, *Lupinus adsurgens* Drew var. *adsurgens*, *Lupinus andersonii* S. Watson, *Lupinus formosus* Greene, *Lupinus formosus* Greene var. *clemensiae* C.P. Sm., *Lupinus klamathensis* Eastw., *Lupinus pumicola* A. Heller

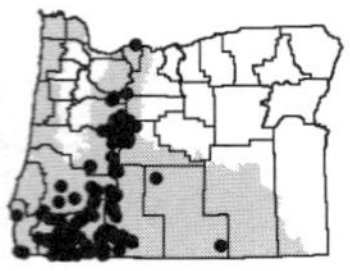

Flowers calyces 5–8 mm; corollas 7–11 mm.

Conifer forests, dry meadows and openings, streamsides. Flowering Apr–Sep. 200–2500 m. BR, Casc, CR, ECas, Est, Sisk. CA. Native.

Lupinus albifrons Bentham var. *albifrons*

Silver bush lupine, silver lupine, white-leaf bush lupine

Lupinus albifrons Benth. var. *eminens* C.P. Sm., *Lupinus albifrons* Benth. var. *fissicalyx* (A. Heller) C.P. Sm., *Lupinus albifrons* Benth. var. *flumineus* C.P. Sm.

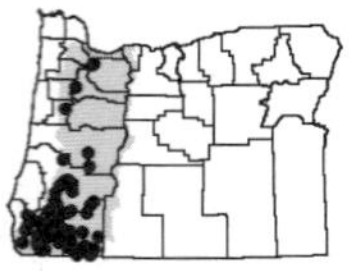

Plants perennial, 1–4(5) dm. **Stems** ascending or erect, woody, canescent to pubescent. **Leaves** cauline; leaflets 6–10, 10–45 × 3–10 mm, linear-elliptic to oblong-oblanceolate, occasionally spatulate, tips acute, apiculate, surfaces sericeous; petioles (15)23–96 mm. **Inflorescences** 3.5–24 cm; peduncles 2.5–20.5 cm; pedicels 3–8 mm. **Flowers** calyces 5–8 mm, not spurred, lips subequal; corollas 9–15 mm, violet to lavender, banners typical type with a slight cup, abaxially pubescent, spot white to yellow, turning purple. **Fruits** 25–50 mm, pubescent. **Seeds** 4–9.

Dry meadows, roadsides. Flowering Apr–Jul. 0–2000 m. Casc, Sisk, WV. CA. Native.

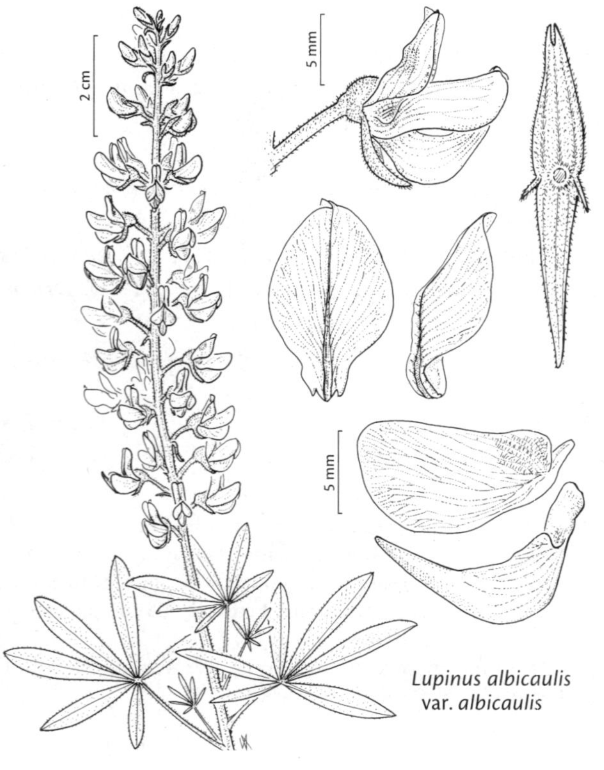

Lupinus albicaulis var. *albicaulis*

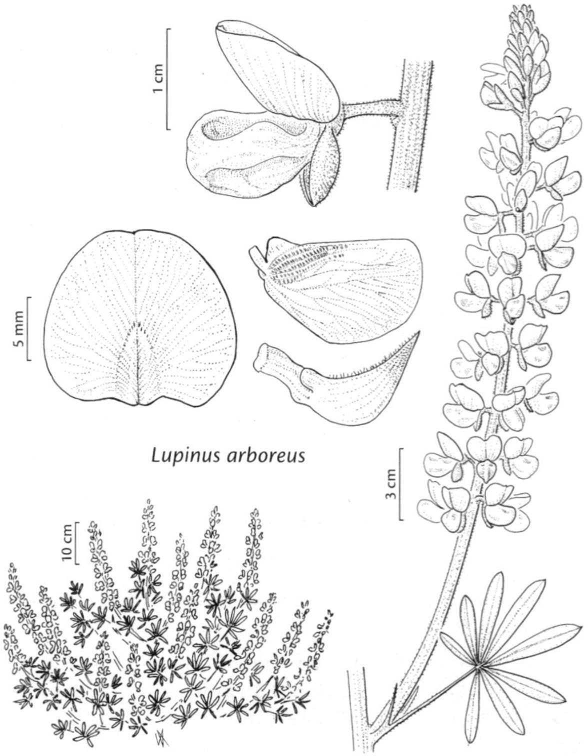

Lupinus arboreus

LUPINUS

Lupinus arboreus Sims

(p. 698)

Tree lupine, yellow bush lupine

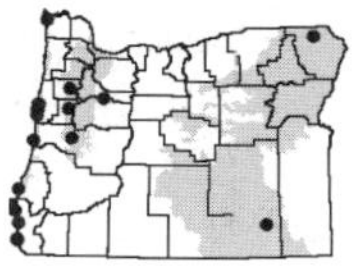

Plants perennial, 10–20 dm. **Stems** ascending to erect, woody, obscurely to slightly pubescent. **Leaves** cauline; leaflets 5–10(12), 15–60 × 4–10 mm, linear-elliptic to oblanceolate, tips acute, usually apiculate, surfaces abaxially pubescent to sericeous, adaxially glabrate to pubescent; petioles (15)20–40 mm. **Inflorescences** (4.5)7.5–21 cm; peduncles (3)5–11.5 cm; pedicels 5–11 mm. **Flowers** calyces 8–10 mm, not spurred, lips equal to subequal; corollas (12)14–18 mm, usually yellow, rarely yellow and blue or purple; banners typical type, spot absent. **Fruits** (20)30–70 mm, villous. **Seeds** 2–8.

Dunes, sandy areas. Flowering Apr–Jul. 0–300 m. Est, WV. CA, WA; north to British Columbia. Native and exotic.

While usually occurring near the ocean, *Lupinus arboreus* is occasionally found further inland, often as an ornamental plant in parking lot vegetation strips. This species appears to be expanding its range northward. It may occasionally hybridize with *L. littoralis*.

Lupinus arbustus Douglas ex Lindley

(p. 700)

Longspur lupine, spur lupine

Lupinus caudatus Kellogg, *Lupinus caudatus* Kellogg var. *submanes* C.P. Sm., *Lupinus caudatus* Kellogg var. *subtenellus* C.P. Sm., *Lupinus laxiflorus* Douglas ex Lindl. var. *calcaratus* (Kellogg) C.P. Sm., *Lupinus laxiflorus* Douglas ex Lindl. var. *cognatus* C.P. Sm., *Lupinus laxiflorus* Douglas ex Lindl. var. *elmerianus* C.P. Sm., *Lupinus laxiflorus* Douglas ex Lindl. var. *laxiflorus*, *Lupinus laxiflorus* Douglas ex Lindl. var. *lyleianus* C.P. Sm., *Lupinus laxiflorus* Douglas ex Lindl. var. *pseudoparviflorus* (Rydb.) C.P. Sm. & H. St. John, *Lupinus laxiflorus* Douglas ex Lindl. var. *silvicola* (A. Heller) C.P. Sm., *Lupinus lyleianus* C.P. Sm., *Lupinus mucronulatus* Howell var. *umatillensis* C.P. Sm., *Lupinus sulphureus* Douglas ex Hook. var. *applegatianus* C.P. Sm.

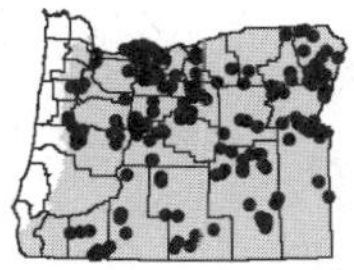

Plants perennial, 2–8 dm. **Stems** ascending to erect, sometimes decumbent, not woody, pubescent to sericeous. **Leaves** mostly cauline; leaflets 6–10(11), 15–60 × 2–11 mm, linear-oblanceolate to oblanceolate, tips acute, often apiculate, surfaces usually pubescent, sometimes glabrate adaxially; petioles 22–93(166) mm. **Inflorescences** (6.4)7–16.5(21.9) cm; peduncles (1.5)2–8.4 cm; pedicels (2)3–6 mm. **Flowers** calyces 3–6 mm, spurred, lips unequal; corollas 7–14 mm, blue, purple, yellow, or white; banners cup type, spot absent or white to yellow, wings usually with tuft of hair near tips. **Fruits** 15–35 mm, pilose. **Seeds** 3–6.

Rocky areas, meadows, open slopes, sagebrush, open forests. Flowering Apr–Sep. 0–2600 m. BR, BW, Casc, Col, ECas, Lava, Owy, Sisk, WV. CA, ID, NV, WA; north to British Columbia, northeast to MT, southeast to UT. Native.

This species hybridizes with *L. oreganus*, *L. albicaulis*, and potentially other broadly sympatric, perennial lupines.

Lupinus argenteus Pursh

Silvery lupine

Plants perennial, (1)1.5–7 dm. **Stems** ascending to erect, not woody, canescent to sericeous. **Leaves** cauline or basal and cauline, basal usually withered by flowering; leaflets 5–10, 10–60 × 2–12 mm, linear-oblanceolate or linear-elliptic to broadly oblanceolate, tips obtuse to acute, apiculate, surfaces glabrate to canescent; petioles 12–180 mm. **Inflorescences** 4–23.5 cm; peduncles 1–10.4 cm; pedicels 1–6 mm. **Flowers** calyces 2–7 mm, spurred, oblique, or gibbous, lips subequal to unequal; corollas 5–14 mm, blue, or purple, rarely white, banners cup type, spot absent or white to yellow, wings glabrous or sometimes midveins pubescent. **Fruits** 11–26 mm, pilose. **Seeds** 2–6.

Western North America. ~12 varieties; 4 varieties treated in *Flora*.

Key to *Lupinus argenteus* varieties:

1. Calyces spurred.
   2. Flowers in profile appearing open . . . . *L. a.* var. ***heteranthus***
   2' Flowers in profile appearing closed . . . . *L. a.* var. ***utahensis***

1' Calyces oblique or gibbous.
   3. Corollas 8–12 mm . . . . *L. a.* var. ***argenteus***
   3' Corollas 5–8 mm . . . . *L. a.* var. ***holosericeus***

Lupinus argenteus Pursh var. *argenteus*

Silvery lupine

Lupinus corymbosus A. Heller, *Lupinus johannis-howellii* C.P. Sm.

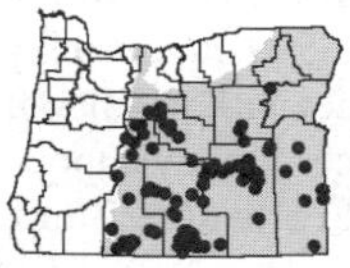

Flowers in profile appearing open; calyces 4–7 mm, oblique or gibbous; corollas 8–12 mm.

Dry meadows, sagebrush, roadsides, open forests. Flowering May–Aug. 700–1800 m. BR, BW, ECas, Lava, Owy. CA, ID, NV, WA; north to British Columbia, northeast to Saskatchewan, east to SD, southeast to NM. Native.

Lupinus argenteus Pursh var. *heteranthus* (S. Watson) Barneby

Tailcup lupine, Kellogg's spurred lupine

Lupinus caudatus Kellogg, *Lupinus caudatus* Kellogg var. *caudatus*, *Lupinus hendersonii* Eastw.

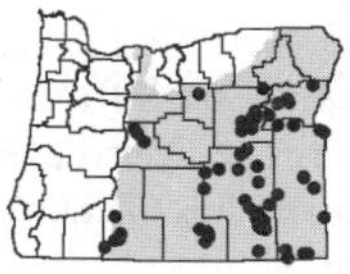

Flowers in profile appearing open; calyces 5–6(7) mm, spurred; corollas 5–8 mm.

Rocky areas, dry meadows, sagebrush, open forests. Flowering Apr–Aug. 900–2800 m. BR, BW, ECas, Lava, Owy. CA, ID, NV. Native.

Lupinus argenteus Pursh var. *holosericeus* (Nuttall) Barneby

Little-flowered lupine, silky lupine, holo lupine

Lupinus holosericeus Nutt.

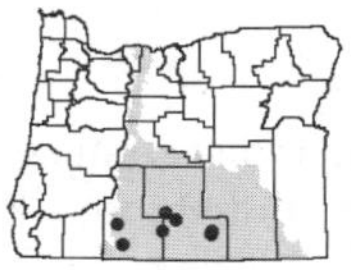

Flowers in profile appearing open; calyces 3–4 mm, slightly oblique; corollas 5–8 mm.

Gravelly areas, sagebrush, meadows. Flowering May–Jul. 900–1900 m. BR, ECas. ID, NV, WA. Native.

Lupinus argenteus Pursh var. ***utahensis*** (S. Watson) Barneby

Kellogg's spurred lupine

Lupinus caudatus Kellogg var. *rubricaulis* (Greene) C.P. Sm.

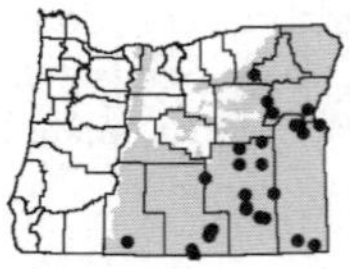

Flowers in profile appearing closed; calyces 2–5 mm, spurred; corollas 5–8 mm.

Rocky areas, dry meadows, open forests, sagebrush. Flowering May–Jul. 900–2700 m. BR, BW, ECas, Owy. ID, NV; southeast to CO. Native.

Lupinus bicolor Lindley

Two-color lupine, miniature lupine

Lupinus bicolor Lindl. var. *bicolor*, *Lupinus bicolor* Lindl. var. *microphyllus* (S. Watson) C.P. Sm., *Lupinus bicolor* Lindl. var. *tridentatus* Eastw. ex C.P. Sm., *Lupinus micranthus* Douglas, *Lupinus nanus* Douglas ex Benth., *Lupinus nanus* Douglas ex Benth. var. *apricus* (Greene) C.P. Sm., *Lupinus polycarpus* Greene

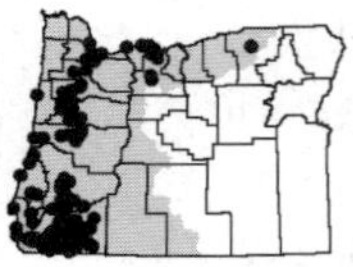

Plants annual, 1–5 dm. **Stems** ascending or erect, pubescent. **Leaves** cauline; leaflets 5–8, 10–40 × 1–5 mm, linear to oblanceolate, tips acute, surfaces pubescent to sericeous; petioles 12–55 mm. **Inflorescences** 1–9 cm; peduncles 2–6.5 cm; pedicels 1–3 mm. **Flowers** calyces 4–8 mm, not spurred, lips equal to subequal; corollas 5–9 mm, usually blue, rarely light blue, pink, or white, sometimes with patches of yellow; banners usually typical type, sometimes weakly cupped or pinched, spot white, purple, or absent. **Fruits** 10–25 mm, villous. **Seeds** 3–6.

Pastures, meadows, roadsides, open forests. Flowering Mar–Jul. 0–1000 m. Casc, Col, CR, ECas, Est, Sisk, WV. CA, WA; north to British Columbia, south to Mexico. Native.

Lupinus brevicaulis S. Watson

Short-stemmed lupine, sand lupine

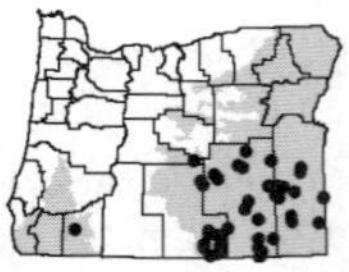

Plants annual, acaulescent, 0.5–1 dm. **Leaves** basal; leaflets 5–7(9), 8–16 × 3–7 mm, broadly oblanceolate to obovate, tips rounded to acute, surfaces abaxially pilose, adaxially glabrate; petioles 16–60(80) mm. **Inflorescences** 0.8–2(3) cm; peduncles 1–6 cm; pedicels 0.2–1.3 mm. **Flowers** calyces 5–9(10) mm, not spurred, lips unequal; corollas 3–8 mm, blue, banners cup type, spot white or yellow. **Fruits** 7–12 mm, villous. **Seeds** 2–3(4).

Sandy or gravelly areas. Flowering Apr–Jul. 600–1900 m. BR, BW, Owy, Sisk. CA, NV; east to WY, southeast to NM, south to Mexico. Native.

Lupinus breweri A. Gray

Brewer's lupine

Lupinus breweri A. Gray var. *breweri*

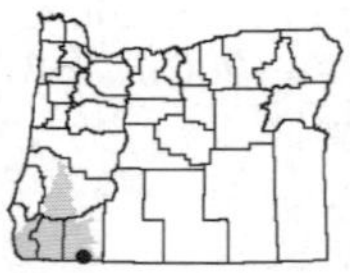

Plants perennial, cushion or mat-forming, 0.2–1.5 dm. **Stems** usually prostrate, woody, pubescent. **Leaves** cauline; leaflets 5–10, 4–15 × 2–5 mm, usually linear-elliptic, rarely oblanceolate, often silvery, tips rounded to acute, surfaces tomentose; petioles 9–25 mm. **Inflorescences** 0.8–2(3) cm; peduncles 1–6 cm; pedicels (1.5)3–4 mm. **Flowers** calyces 4–6.5 mm, not spurred, lips subequal; corollas 4–11 mm, violet to blue, banners cup type, spot white or yellow. **Fruits** 8–20 mm, villous to tomentose. **Seeds** 2–5.

Meadows, open forests. Flowering Jul. 1200–2300 m. Sisk. CA, NV. Native.

Lupinus latifolius Lindley ex J. Agardh

Broad-leaved lupine

Plants perennial, 1–10(15) dm. **Stems** erect, not woody, glabrate to inconspicuously strigillose or villous. **Leaves** cauline; leaflets 5–11, (20)30–100 × 5–30 mm, linear-oblanceolate to broadly oblanceolate, tips acute, sometimes apiculate, surfaces abaxially strigose to villous, adaxially glabrous to strigose; petioles (17)22–133(180) mm. **Inflorescences** (2.9)5–34.2 cm, usually dense; peduncles 2.2–13.5; pedicels 1–10 mm. **Flowers** calyces 5–13 mm, not spurred, lips unequal to subequal; corollas 5–16 mm, blue, purple, or pink, banners typical type, glabrous, spot white or yellow. **Fruits** 12–38 mm, densely pubescent. **Seeds** 6–10.

Western United States and Mexico. 6–9 varieties; 3 varieties treated in *Flora*.

This species is known to hybridize with *Lupinus arbustus*, *L. albicaulis*, *L. littoralis*, and *L. lepidus*.

Key to *Lupinus latifolius* varieties:

1. Corollas 8–11 mm. . . . . *L. l.* var. ***viridifolius***

1' Corollas 10–15 mm.

  2. Plants (2)3–10(15) dm . . . . *L. l.* var. ***latifolius***

  2' Plants 1–3 dm . . . . *L. l.* var. ***subalpinus***

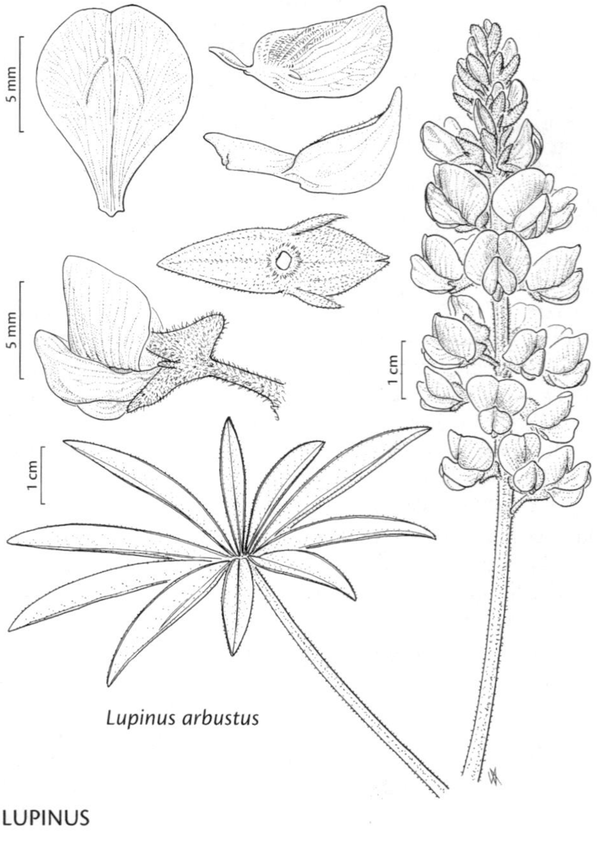

Lupinus arbustus

LUPINUS

Lupinus latifolius Lindley ex J. Agardh var. *latifolius*

Broad-leaved lupine

Lupinus latifolius Lindl. ex J. Agardh var. *columbianus* (A. Heller) C.P. Sm., *Lupinus latifolius* Lindl. ex J. Agardh var. *ligulatus* (Greene) C.P. Sm., *Lupinus latifolius* Lindl. ex J. Agardh var. *longipes* (Greene) C.P. Sm.

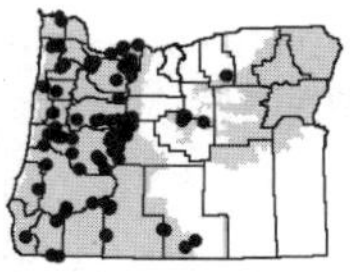

Plants (2)3–10(15) dm. **Flowers** corollas 10–15 mm.

Moist meadows, roadsides, open forests, rocky areas. Flowering May–Aug. 0–2100 m. BW, Casc, CR, ECas, Sisk, WV. CA, WA; south to Mexico. Native.

Lupinus latifolius Lindley ex J. Agardh var. *subalpinus* (Piper & B.L. Robinson) C.P. Smith

Cascade lupine

Lupinus arcticus S. Watson var. *subalpinus* (Piper & B.L. Rob.) C.P. Sm.

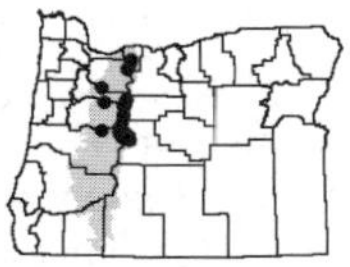

Plants 1–3 dm. **Flowers** corollas 10–15 mm.

Mountain meadows, streamsides, open forests. Flowering Jun–Sep. 1000–2300 m. Casc. WA. Native.

Lupinus latifolius Lindley ex J. Agardh var. *viridifolius* (A. Heller) C.P. Smith

Broad-leaved lupine

Lupinus latifolius Lindl. ex J. Agardh var. *barbatus* (L.F. Hend.) Munz

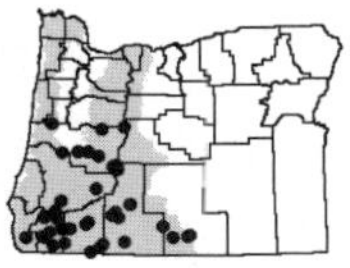

Plants 2–9 dm. **Flowers** corollas 8–11 mm.

Moist meadows, streamsides, roadsides, forests. Flowering May–Aug. 100–1700 m. Casc, CR, ECas, Sisk. CA. Native.

Lupinus lepidus Douglas ex Lindley

Pacific lupine, prairie lupine

Plants perennial, acaulescent, subacaulescent, shortly-caulescent, or caulescent, sometimes mat-forming, 0.5–5(6) dm. **Stems** prostrate to erect, not woody, pubescent. **Leaves** mostly basal; leaflets 5–8, 5–40 × 2–10 mm, oblanceolate to narrowly obovate, surfaces sericeous to strigose; petioles 10–120(150) mm. **Inflorescences** (1)1.5–20(25) cm; peduncles 0–15(20) cm; pedicels 1–3(4) mm. **Flowers** calyces 4–8 mm, not spurred, lips equal to subequal; corollas 6–11 mm, blue, purple, or pink, banners cup type, spot white or yellow. **Fruits** 9–20 mm, pubescent. **Seeds** 2–5.

Western United States. 11 varieties; 7 varieties treated in *Flora*.

The varietal taxa within this species are arguably arbitrary and extraneous. However, since *L. l.* var. *cusickii* is a listed federal and state taxon, we include the varieties here.

Key to *Lupinus lepidus* varieties:

1. Inflorescences usually not exserted from leaves.
   2. Pedicels 2–4 mm . . . . *L. l.* var. ***cusickii***
   2' Pedicels 0.5–2 mm . . . . *L. l.* var. ***utahensis***

1' Inflorescences usually exserted from leaves.
   3. Plants strongly caulescent, 1–4(5) dm. . . . . *L. l.* var. ***lepidus***
   3' Plants acaulescent, subacaulescent, or shortly-caulescent, 0.1–2.5 dm.
      4. Peduncles usually ≤ leaves . . . . *L. l.* var. ***aridus***
      4' Peduncles usually > leaves.
         5. Inflorescences oblong-globose to globose . . . . *L. l.* var. ***lobbii***
         5' Inflorescences cylindric to oblong or oblong-globose.
            6. Plants of Mount Ashland, Jackson County. . . . . . . *L. l.* var. ***ashlandensis***
            6' Plants not of Mount Ashland, Jackson County . . . . *L. l.* var. ***sellulus***

Lupinus lepidus Douglas ex Lindley var. *aridus* (Lindley) Jepson

Dry ground lupine

Lupinus aridus Douglas var. *aridus*

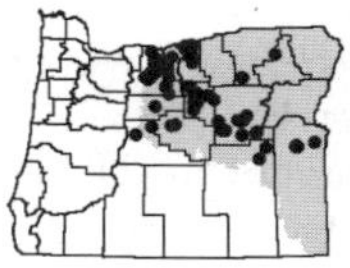

Plants subacaulescent or shortly-caulescent, 0.4–2.5 dm. **Inflorescences** cylindric to oblong-globose, exserted from leaves; peduncles usually ≤ leaves; pedicels 1–3 mm.

Sagebrush, rocky, sandy area. Flowering Apr–Jul. 50–1800 m. BW, Col, Lava, Owy. CA, ID, NV, WA; northeast to MT, southeast to AZ. Native.

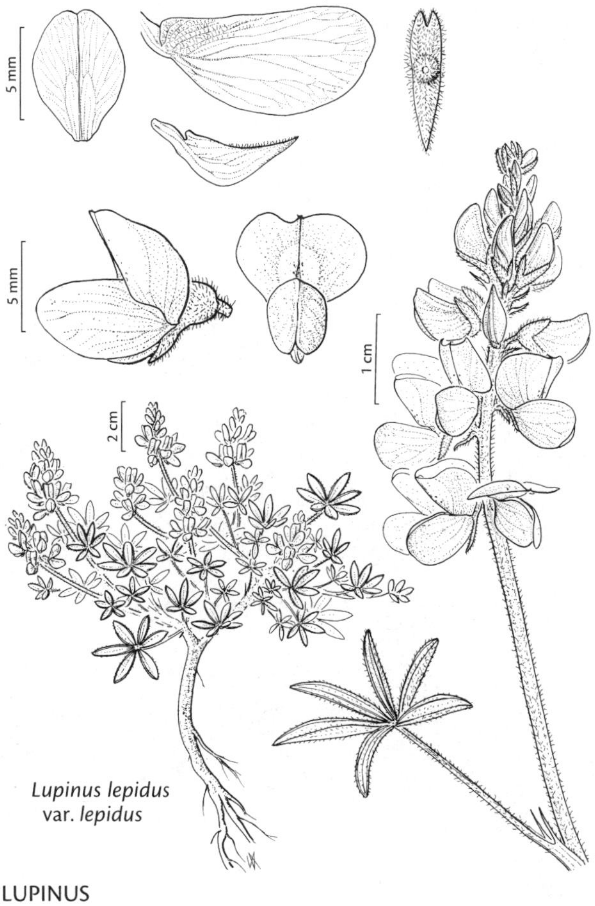

Lupinus lepidus var. *lepidus*

LUPINUS

Lupinus lepidus Douglas ex Lindley var. ***ashlandensis*** (B.J. Cox) Isey

Ashland lupine

Lupinus aridus Douglas ssp. *ashlandensis* B.J. Cox

Plants acaulescent, 0.1–1 dm. **Inflorescences** cylindric to oblong, exserted from leaves; peduncles > leaves; pedicels 1–4 mm.

Alpine slopes. Flowering Jul–Aug. 2000–2300 m. Sisk. Native.

Lupinus lepidus Douglas ex Lindley var. ***cusickii*** (S. Watson) C.L. Hitchcock

Cusick's lupine

Lupinus aridus Douglas var. *abortivus* (Greene) C.P. Sm., *Lupinus aridus* Douglas var. *cusickii* (S. Watson) C.P. Sm.

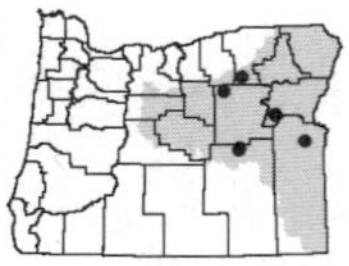

Plants acaulescent or subacaulescent, 0.1–0.8 dm. **Inflorescences** cylindric to oblong-globose, usually not exserted from leaves; peduncles usually < leaves; pedicels 2–4 mm.

Sagebrush, gravelly areas. Flowering Jun–Jul. 600–1600 m. BW, Lava, Owy. Native.

This taxon is listed as a Species of Concern by the United States Fish and Wildlife Service, and as endangered by the state of Oregon.

Lupinus lepidus Douglas ex Lindley var. ***lepidus*** (p. 701)

Pacific lupine, dwarf lupine

Lupinus minimus Douglas ex Hook.

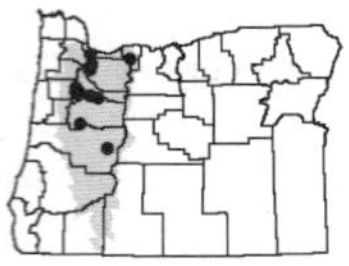

Plants strongly caulescent, 1–4(5) dm. **Inflorescences** cylindric to oblong, usually exserted from leaves; peduncles > leaves; pedicels 2–4 mm.

Dry meadows, gravelly areas, roadsides, streamsides. Flowering May–Jul. 0–1500 m. Casc, WV. WA; north to British Columbia. Native.

Lupinus lepidus Douglas ex Lindley var. ***lobbii*** (A. Gray) C.L. Hitchcock

Prostrate lupine

Lupinus lyallii A. Gray var. *fruticulosus* C.P. Sm., *Lupinus lyallii* A. Gray var. *lobbii* (A. Gray ex S. Watson) C.P. Sm., *Lupinus lyallii* A. Gray var. *lyallii*, *Lupinus minutifolius* Eastw.

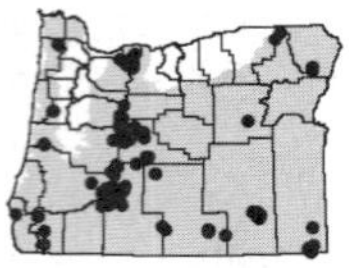

Plants subacaulescent or short-caulescent, 0.1–0.9 dm. **Inflorescences** oblong-globose to globose, exserted from leaves; peduncles usually > leaves; pedicels 1–3 mm.

Rocky areas, subalpine meadows. Flowering May–Sep. 400–2900 m. BR, BW, Casc, CR, ECas, Lava, Owy, Sisk. CA, ID, NV, WA. Native.

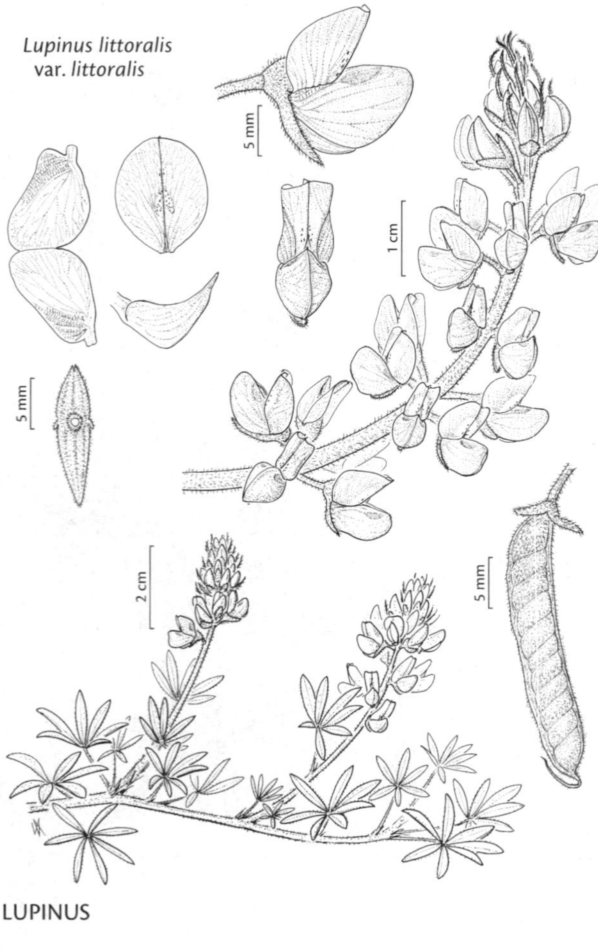

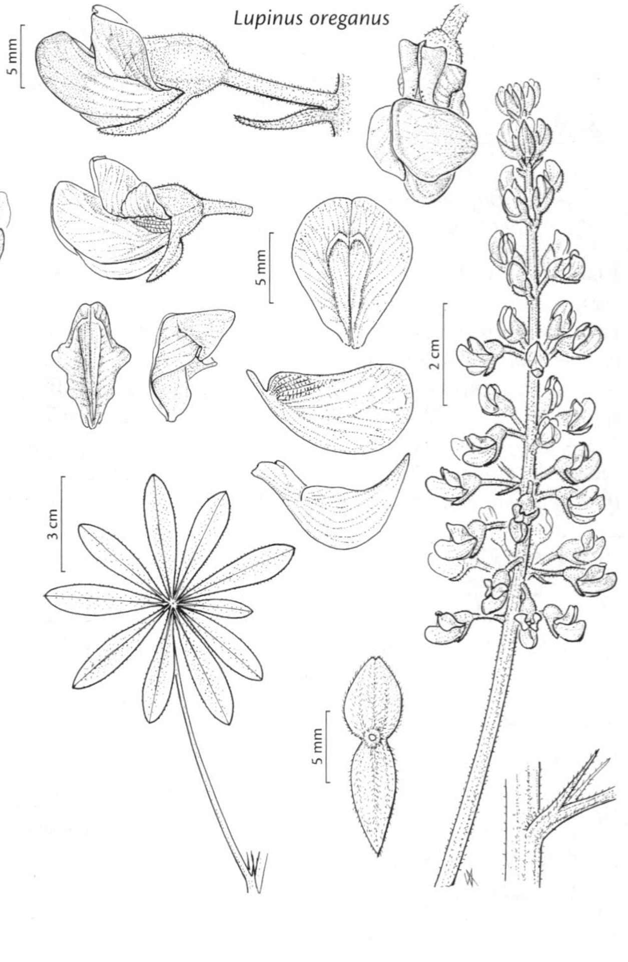

LUPINUS

Lupinus lepidus Douglas ex Lindley var. ***sellulus*** (Kellogg) Barneby

Donner Lake lupine

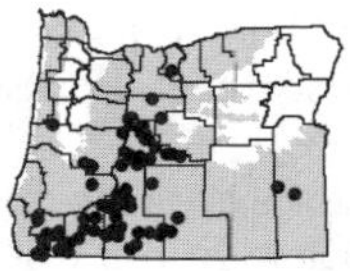

Plants acaulescent or subacaulescent, 0.4–1.5 dm. **Inflorescences** cylindric to oblong-globose, usually exserted from leaves; peduncles usually > leaves; pedicels 1–3 mm.

Sagebrush, rocky, sandy area. Flowering May–Sep. 100–1900 m. BR, Casc, Col, CR, ECas, Lava, Owy, Sisk. CA, NV. Native.

Lupinus lepidus Douglas ex Lindley var. ***utahensis*** (S. Watson) C.L. Hitchcock

Utah lupine

Lupinus caespitosus Nutt.

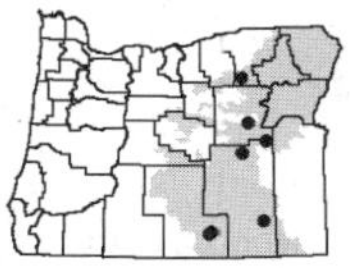

Plants acaulescent or subacaulescent, 0.1–0.7 dm. **Inflorescences** oblong to sub-globose, usually not exserted from leaves; peduncles usually < leaves; pedicels 0.5–2 mm.

Sagebrush, juniper-ponderosa pine forests. Flowering Jun–Jul. 1000–2100 m. BR, BW. CA, ID, NV; northeast to MT, southeast to NM. Native.

Lupinus leucophyllus Douglas ex Lindley

Velvet lupine, woolly-leaved lupine

Lupinus leucophyllus Douglas ex Lindl. var. *belliae* C.P. Sm., *Lupinus leucophyllus* Douglas ex Lindl. var. *canescens* (Howell) C.P. Sm., *Lupinus leucophyllus* Douglas ex Lindl. var. *leucophyllus*, *Lupinus leucophyllus* Douglas ex Lindl. var. *tenuispicus* (A. Nelson) C.P. Sm.

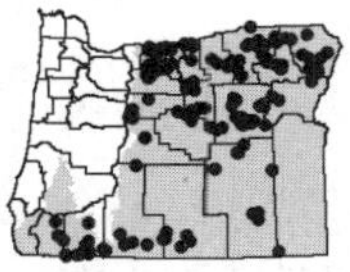

Plants perennial, 2.5–9 dm. **Stems** erect, not woody, villous to tomentose. **Leaves** cauline, often with some leaves clustered near bases; leaflets 6–10, 25–65 × 4–16 mm, linear-oblanceolate to narrowly oblanceolate, tips usually acute, sometimes apiculate, surfaces pilose to villous; petioles (18)22–124(160) mm. **Inflorescences** 5.5–20.5(37) cm; peduncles 1.9–8.4 cm; pedicels generally 1–3 mm. **Flowers** calyces 5–10 mm, not spurred, slightly saccate, lips equal to unequal; corollas 5–11 mm, blue, purple or whitish, banners typical type, abaxially pubescent, spots yellow to dark brown. **Fruits** 13–26 mm, pubescent. **Seeds** (2)3–6.

Sagebrush, meadows, roadsides. Flowering May–Sep. 50–2600 m. BR, BW, Col, ECas, Lava, Owy, Sisk. CA, ID, NV, WA; north to British Columbia, northeast to MT, east to WY, southeast to CO. Native.

Lupinus littoralis Douglas var. ***littoralis*** (p. 702)

Seashore lupine

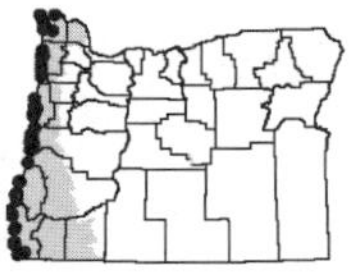

Plants perennial, 0.5–3 dm, often cushion- or mat-forming. **Stems** prostrate to decumbent, not woody, pilose to strigose. **Leaves** cauline, leaflets 5–9, 10–40 × 2–10 mm, surfaces abaxially glabrous to strigose, adaxially strigose, tips acute, apiculate; petioles 20–60 mm. **Inflorescences** 10–15 cm; peduncles 3–9 cm; pedicels 3–5 mm. **Flowers** calyces 9–13 mm, lips subequal to unequal, not spurred; corollas 8–13 mm, pink to purple or blue, banners typical type, spots white. **Fruits** 30–40 mm, pubescent. **Seeds** 8–12.

Coastal dunes, sandy areas, grassy areas. Flowering Apr–Aug. 0–400 m. CR, Est, Sisk. CA, WA; north to British Columbia. Native.

Although predominately a coastal taxon, *Lupinus littoralis* var. *littoralis* is occasionally found inland. This species is known to hybridize with *L. latifolius*.

Lupinus luteolus Kellogg

Bushy yellow lupine, butter lupine

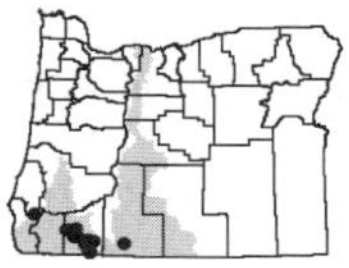

Plants annual, 3–10 dm. **Stems** erect, strigillose. **Leaves** cauline; leaflets (5)6–8(9), 8–40 × 4–11 mm, linear to oblanceolate, tips rounded, surfaces abaxially strigose, adaxially glabrate to strigose; petioles 11–64 mm. **Inflorescences** (2.5)3.3–10.4(13.5) cm; peduncles 2.7–10.6 cm; pedicels 1–3 mm. **Flowers** calyces not spurred, lips unequal; corollas 1–1.4(1.6) cm, usually yellow, rarely pink to bright blue, banners cup type, spot absent. **Fruits** (10)12–15 mm, strigose to sericeous. **Seeds** (1)2.

Meadows, roadsides, openings in forests. Flowering Jun–Aug. 300–1600 m. ECas, Sisk. CA. Native.

Lupinus microcarpus Sims

Chick lupine

Lupinus densiflorus Benth. var. *densiflorus*, *Lupinus microcarpus* Sims var. *microcarpus*, *Lupinus subvexus* C.P. Sm. var. *transmontanus* C.P. Sm.

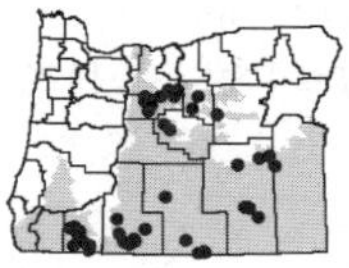

Plants annual, 1–5(7) dm. **Stems** decumbent to erect, pubescent. **Leaves** mostly subbasal; leaflets (5)7–11, 10–60 × 2–12 mm, linear-elliptic to oblanceolate or obovate, tips rounded to acute, often apiculate, surfaces abaxially sparsely villous, adaxially glabrous; petioles 25–119 mm. **Inflorescences** (2)4–12(20) cm; peduncles 4–10.5 cm; pedicels 1–3 mm. **Flowers** calyces (6.5)7.5–17 mm, not spurred, lips unequal; corollas 0.9–1.7 cm, purple to lilac, sometimes pink or pink-purple, banners cup type, spot absent. **Fruits** 10–16(20) mm, sericeous. **Seeds** 2–3.

Disturbed areas, roadsides, meadows, chaparral, sagebrush. Flowering May–Jul. 300–1700 m. BR, ECas, Lava, Owy, Sisk. CA, WA; north to British Columbia; Chile. Native.

Lupinus nevadensis A. Heller

Nevada lupine

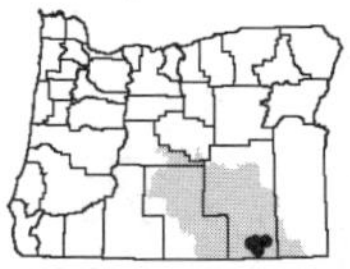

Plants perennial, 1–4 dm. **Stems** erect, not woody, villous. **Leaves** basally clustered; leaflets 6–10, 20–45 × 2–7 mm, elliptic to oblanceolate, tips acute, surfaces abaxially villosulous, adaxially villous; petioles 35–120 mm. **Inflorescences** (4.5)6.8–15 cm; peduncles 2.9–3.5 cm; pedicels 4–8 mm. **Flowers** calyces 3–5 mm, not spurred, lips equal to subequal; corollas 6–12 mm, blue, banners cup type, margins often inrolled, spot white to yellow. **Fruits** 25–30 mm, pilose. **Seeds** 3–4.

Sagebrush. Flowering May–Sep. 1000–2000 m. BR. CA, NV. Native.

Lupinus onustus S. Watson

Plumas lupine

Lupinus mucronulatus Howell var. *mucronulatus*, *Lupinus oreganus* A. Heller, *Lupinus oreganus* A. Heller var. *pusillulus* C.P. Sm.

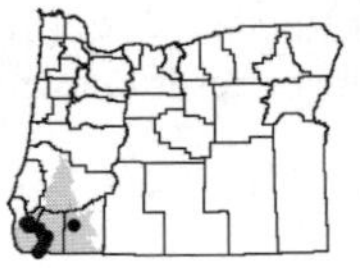

Plants perennial, 1–3 dm. **Stems** decumbent to ascending, not woody, glabrous to inconspicuously pubescent. **Leaves** mostly subbasal; leaflets 5–9, 15–50 × 7–19 mm, obovate to oblanceolate, tips acute, apiculate, surfaces abaxially strigose, adaxially glabrate; petioles (4.5)7–16 cm. **Inflorescences** (6)7.1–19.6(36) cm; peduncles (3.5)5–9.7 cm; pedicels 4–8 mm. **Flowers** calyces 5–8 mm, not spurred, lips subequal; corollas 6–12 mm, blue to violet, banners cup type, spot absent. **Fruits** 20–50 mm, pilose. **Seeds** 2–4.

Pine forests, serpentine. Flowering Apr–Jul. 50–1600 m. Sisk. CA. Native.

This species may hybridize with other co-occurring perennial *Lupinus* species.

Lupinus oreganus A. Heller (p. 702)

Kincaid's lupine

Lupinus leucopsis J. Agardh, *Lupinus leucopsis* J. Agardh var. *hendersonianus* C.P. Sm., *Lupinus oreganus* A. Heller var. *kincaidii* C.P. Sm., *Lupinus oreganus* A. Heller var. *oreganus*, *Lupinus sulphureus* Douglas ex Hook. ssp. *kincaidii* (C.P. Sm.) L.Ll. Phillips

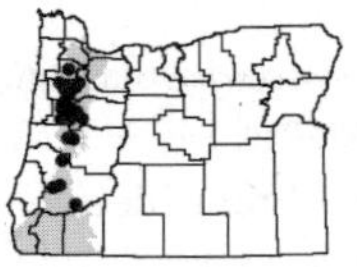

Plants perennial, 4–8 dm. **Stems** erect, not woody, clustered, slightly pubescent. **Leaves** mostly cauline; leaflets 6–11, (20)25–40(50) × 6–11 mm, elliptic to oblanceolate, tips acute, sometimes apiculate, surfaces abaxially sparsely pubescent to sericeous, adaxially glabrous to sericeous; petioles (40)70–140(165) mm. **Inflorescences** (1)1.3–2.5 dm; peduncles 4.5–12.5(14.5) cm; pedicels 3–8 mm. **Flowers** calyces 6–10 mm, not spurred, asymmetrical, sericeous, lips unequal to subequal; corollas 8–12 mm, blue to purple, occasionally white, banners pinched type, spot absent. **Fruits** 20–30 mm, pilose. **Seeds** 2–3.

Remnant prairies, edges and opening in oak forests. Flowering May–Jun. 50–900 m. Sisk, WV. WA; north to British Columbia. Native.

Lupinus oreganus has been treated in the past as *L. sulphureus* var. *kincaidii*. It is not, however, closely related to *L. sulphureus* (Severns, Meyers & Tran 2012). It is listed as Threatened by the United States Fish and Wildlife Service. This species is an important larval host for the endangered Fender's blue butterfly (*Plebejus icarioides* ssp. *fenderi*). This species hybridizes with *L. albicaulis* and *L. arbustus*.

Lupinus polyphyllus Lindley

Bigleaf lupine

Plants perennial, 1.5–15 dm. **Stems** erect, not woody, glabrous to sparsely pubescent. **Leaves** basal and cauline; leaflets 5–17, (30)40–100(150) × 3–20(30) mm, tips acute, sometimes apiculate, surfaces abaxially glabrate to strongly hirsute, adaxially glabrous to glabrate; petioles 3–35 cm. **Inflorescences** 5–35 cm; peduncles 3–13 cm; pedicels 4–10 mm. **Flowers** calyces 4–8 mm, not spurred, sometimes slightly saccate, lips equal to subequal; corollas 5–8 mm, blue to violet, sometimes white, banners typical type, glabrous, margins sometimes inrolled, spot white or occasionally purple. **Fruits** 25–40 mm, pilose. **Seeds** 3–9.

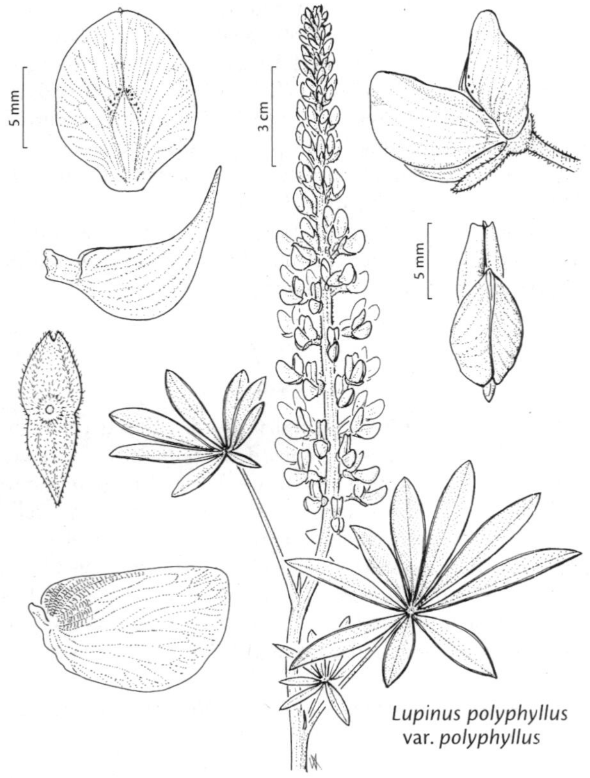

Lupinus polyphyllus var. *polyphyllus*

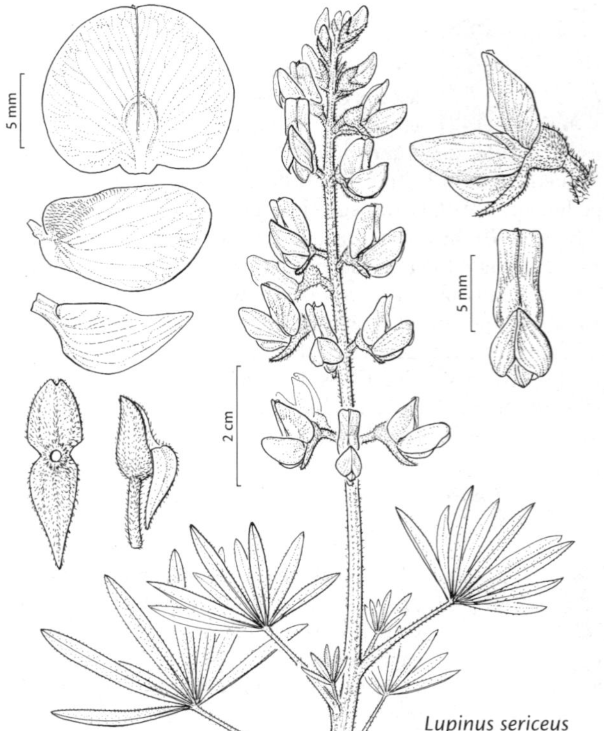

Lupinus sericeus

LUPINUS

Western and northeastern North America. 6–9 varieties; 4 varieties treated in *Flora*.

Colorful ornamental cultivars of this species are found frequently in roadside ditches. The cultivars are usually larger and have flowers with a greater color range than our native *L. polyphyllus* taxa.

Key to *Lupinus polyphyllus* varieties:

1. Plants (5)10–15 dm; stems strongly fistulose; leaflets 9–17 . . . . ***L. p.*** var. ***polyphyllus***
1' Plants 1.5–10 dm; stems weakly fistulose; leaflets 5–11.
  2. Plants of wet meadows and forests or streamsides; pedicels 2–5 mm . . . . ***L. p.*** var. ***burkei***
  2' Plants of sagebrush, dry grasslands, juniper forests; pedicels 4–10 mm.
    3. Plants (3)3.5–7 dm; stems often in clumps . . . . ***L. p.*** var. ***prunophilus***
    3. Plants 1.5–4 dm; stems usually not in clumps. . . . . ***L. p.*** var. ***saxosus***

Lupinus polyphyllus Lindley var. *burkei* (S. Watson) C.L. Hitchcock

Burke's lupine

Lupinus burkei S. Watson

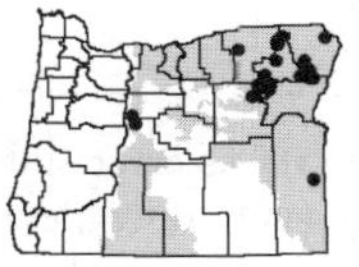

Plants 3–10 dm. **Stems** weakly fistulose. **Leaflets** 5–11. **Inflorescences** pedicels 2–5 mm.

Wet meadows or forests, streamsides. Flowering Jun–Aug. 900–2400 m. BW, Col, ECas, Owy. ID, NV, WA; north to British Columbia, northeast to MT, east to WY. Native.

Lupinus polyphyllus Lindley var. *polyphyllus* (p. 704)

Bigleaf lupine

Lupinus polyphyllus Lindl. var. *pallidipes* (A. Heller) C.P. Sm., *Lupinus superbus* A. Heller var. *bernardinus* Abrams, *Lupinus superbus* A. Heller var. *elongatus* (Greene) C.P. Sm.

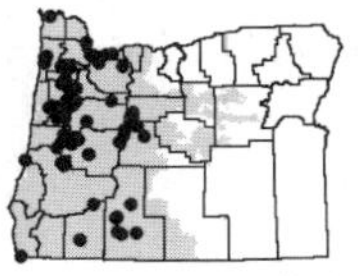

Plants (5)10–15 dm. **Stems** strongly fistulose. **Leaflets** 9–17. **Inflorescences** pedicels 5–10 mm.

Meadows, streamsides, forests. Flowering May–Sep. 0–1900 m. Casc, CR, ECas, Est, Lava, Sisk, WV. CA, NV, WA; north to British Columbia, northeast to Alberta. Native.

Lupinus polyphyllus Lindley var. *prunophilus* (M.E. Jones) L. Phillips

Hairy bigleaf lupine

Lupinus biddlei L.F. Hend. ex C.P. Sm.

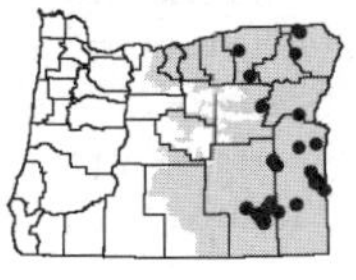

Plants (3)3.5–7 dm. **Stems** often in clumps, weakly fistulose. **Leaflets** 5–11. **Inflorescences** pedicels 4–10 mm.

Sagebrush, dry grassland. Flowering Apr.–Aug. 500–2000 m. BR, BW, Col, Owy. CA, ID, NV, WA; east to WY, southeast to NM. Native.

Plants with white corollas are treated as *L. biddlei* by some floras.

Lupinus polyphyllus Lindley var. *saxosus* (Howell) Barneby

Rock lupine

Lupinus saxosus Howell, *Lupinus saxosus* Howell var. *saxosus*, *Lupinus saxosus* Howell var. *subsericeus* (B.L. Rob. ex Piper) C.P. Sm.

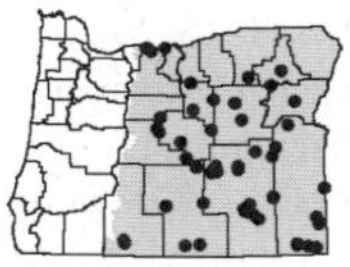

Plants 1.5–4 dm. **Stems** usually not in clumps, weakly fistulose. **Leaflets** 5–11. **Inflorescences** pedicels 4–10 mm.

Sagebrush, dry grasslands, juniper forests. Flowering Apr–Jun. 100–2200 m. BR, BW, Col, ECas, Lava, Owy. CA, ID, NV, WA. Native.

Lupinus pusillus Pursh var. *intermontanus* (Heller) Dunn

Intermountain lupine

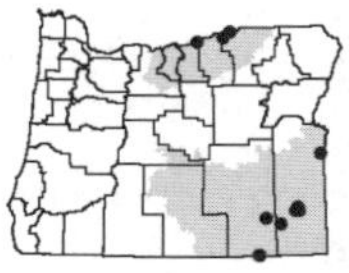

Plants annual, often subacaulescent, 0.5–1(1.2) dm. **Stems** decumbent to erect, pilose. **Leaves** basal; leaflets 5–7, 10–25 × 4–7 mm, tips acute, often apiculate, surfaces abaxially setose to pilose, adaxially glabrous; petioles 30–60 mm. **Inflorescences** 2–4 cm; peduncles 0–1 cm; pedicels 0.2–1 mm. **Flowers** calyces 4–5 mm, not spurred, lips unequal; corollas 5–8(10) mm, blue to purple, banners cup type, spot white, yellow, or absent. **Fruits** 10–20 mm, hirsute. **Seeds** 2–3.

Sandy or gravelly areas. Flowering May–Jun. 900–1600 m. BR, Col, Owy. CA, ID, NV, WA; northeast to MT, east to WY, southeast to AZ. Native.

Lupinus rivularis Douglas ex Lindley

Riverbank lupine

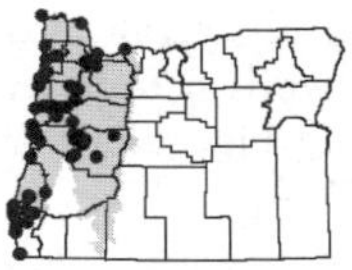

Plants perennial, 4–10(12) dm. **Stems** ascending to erect, woody or not, inconspicuously pubescent to villous. **Leaves** cauline; leaflets 5–10, 15–35(50) × 5–15 mm, linear to oblanceolate, tips acute, often apiculate, surfaces abaxially strigillose, adaxially glabrous to sparsely pubescent; petioles 5–12 mm. **Inflorescences** 10–15(20) cm, usually open; peduncles 4–9 cm; pedicels 4–8 mm. **Flowers** calyces 8–9 mm, not spurred, lips subequal to unequal; corollas 12–17 mm, blue, or blue and white, banners typical type, abaxially glabrous, margins sometimes inrolled, spots white or absent. **Fruits** 16–55 mm, villous. **Seeds** 6–8.

Gravelly or sandy riparian areas, roadsides, grasslands, wetlands. Flowering Mar–Aug. 0–700 m. Casc, CR, Est, WV. CA, WA; north to British Columbia. Native.

True *L. rivularis* (with a solid blue banner) is restricted to riparian habitats often on gravel and sandbars. A cultivar of *L. rivularis* was bred and released by the NRCS (Natural Resources Conservation Service) in the mid to late 1990s and widely planted on roadsides. The cultivar has a banner that is obviously two-toned, with a white to pale blue center and surrounding blue banner. All roadside populations of *L. rivularis* should be assumed to be the released cultivar. Seeds from the roadside cultivar have been collected and widely planted in restored western Oregon grasslands and wetlands.

Lupinus sabinianus Douglas ex Lindley

Sabin's lupine

Lupinus sabinii Douglas ex Lindl.

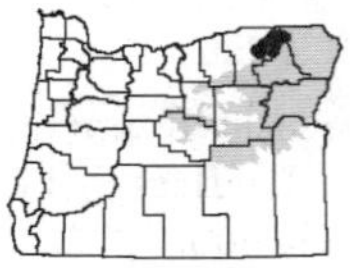

Plants perennial, 6–12 dm. **Stems** erect, not woody, pubescent. **Leaves** cauline; leaflets (5)7–11, (40)50–120(150) × 8–15(20) mm, oblanceolate, tips acute, often apiculate, surfaces sericeous to strigose; petioles 30–126 mm. **Inflorescences** 9–19.5 cm; peduncles 4.9–8.8 cm; pedicels 7–11 mm. **Flowers** calyces 5–6 mm, not spurred, slightly oblique, lips unequal; corollas 10–14(18) mm, yellow, banners typical to cup type, spot absent. **Fruits** 30–45 mm, sericeous. **Seeds** 4–7.

Openings in forests, grassy areas, roadsides. Flowering May–Jul. 500–1700 m. BW. WA. Native.

Lupinus sericeus Pursh (p. 704)

Silky lupine

Lupinus leucopsis J. Agardh, *Lupinus leucopsis* J. Agardh var. *mollis* (A. Heller) C.P. Sm., *Lupinus leucopsis* J. Agardh var. *shermanensis* C.P. Sm., *Lupinus ornatus* Douglas ex Lindl., *Lupinus sericeus* Pursh var. *fikeranus* (Sm.) C.L. Hitchc., *Lupinus sericeus* Pursh var. *flexuosus* (Lindl.) C.P. Sm., *Lupinus sericeus* Pursh var. *sericeus*, *Lupinus sericeus* Pursh var. *wallowensis* C.P. Sm.

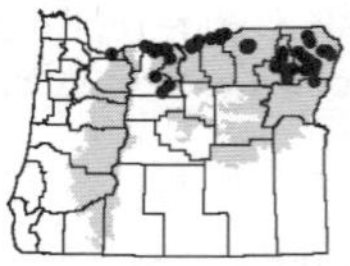

Plants perennial, 2–10(14) dm. **Stems** erect, not woody, pubescent. **Leaves** cauline, rarely with some leaves clustered near bases; leaflets 6–9, 20–70 × 4–12 mm, linear-elliptic to oblanceolate, tips acute, often apiculate, surfaces strigillose; petioles (18)23–90 mm. **Inflorescences** 9–17(24) cm; peduncles 3–8.7 cm; pedicels 3–5(7) mm. **Flowers** calyces 4–10 mm, not spurred, sometimes slightly saccate, lips subequal to unequal; corollas 12–16 mm, whitish to blue; banners typical type, abaxially pubescent, spot white, yellow or brown. **Fruits** 16–35 mm, pilosulose. **Seeds** 2–6.

Disturbed areas, roadsides, meadows, sagebrush open forests. Flowering May–Aug. 50–2200 m. BW, Casc, Col. ID, NV, WA; north to British Columbia, northeast to Alberta, east to WY, southeast to NM. Native.

This species may be locally abundant in recently burned areas.

Lupinus sulphureus Douglas ex Hooker

Sulphur lupine

Plants perennial, 4–8 dm. **Stems** erect, not woody, finely pubescent. **Leaves** basal and cauline, basal usually present at flowering; leaflets 6–11, (20)25–40(50) × 3–9 mm, tips acute, often apiculate, surfaces abaxially sparsely pubescent to sericeous, adaxially glabrous to sericeous; petioles 40–150 mm. **Inflorescences** 5–20 cm; peduncles 4–12 cm; pedicels 3–8 mm. **Flowers** calyces 6–9 mm, not spurred, asymmetrical, lips subequal to unequal; corollas 8–12 mm, yellow or blue to purple, sometimes white; banners cup type, spot absent. **Fruits** 20–30 mm, pilose. **Seeds** 4–5.

2 subspecies.

Key to *Lupinus sulphureus* subspecies:

1. Corollas blue to purple, sometimes white . . . . *L. s.* ssp. ***subsaccatus***

1' Corollas yellow . . . . *L. s.* ssp. ***sulphureus***

Lupinus sulphureus Douglas ex Hooker ssp. ***subsaccatus*** (Suksdorf) L.Ll. Phillips

Whitish lupine, Bingen lupine

Lupinus bingenensis Suksd. var. *subsaccatus* Suksd., *Lupinus leucopsis* J. Agardh, *Lupinus leucopsis* J. Agardh var. *bingenensis* (Suksd.) C.P. Sm., *Lupinus sulphureus* Douglas ex Hook. var. *subsaccatus* (Suksd.) C.L. Hitchc.

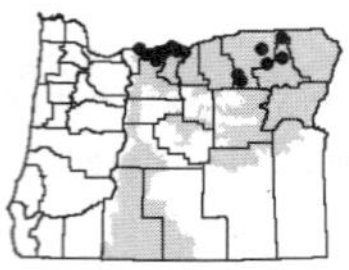

Flowers corollas blue to purple, sometimes white.

Sagebrush, rocky or gravelly areas, dry meadows. Flowering Apr–Jun. 0–1500 m. BW, Col, ECas. ID, WA; north to British Columbia, northeast to MT. Native.

Lupinus sulphureus Douglas ex Hooker ssp. ***sulphureus***

Sulphur lupine

Lupinus sulphureus Douglas ex Hook. var. *sulphureus*

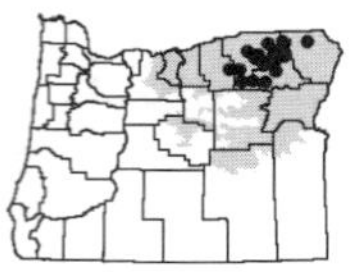

Flowers corollas yellow.

Sagebrush, rocky or gravelly areas, dry meadows. Flowering May–Jul. 700–1700 m. BW, Col. WA; north to British Columbia. Native.

Lupinus tracyi Eastwood

Tracy's lupine

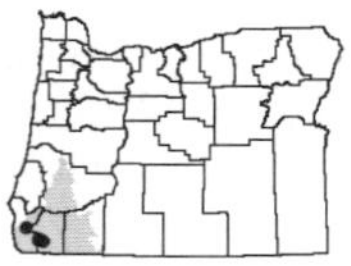

Plants perennial, 4–7 dm. **Stems** erect, not woody, glabrous. **Leaves** cauline; leaflets 5–7, 7–40 × 2–10 mm, broadly oblanceolate, tips rounded to acute, often apiculate, surfaces glabrous, rugulose; petioles 9–22(30) mm. **Inflorescences** 4–10(20) cm; peduncles 1.9–5.5 cm; pedicels 3–6 mm. **Flowers** calyces 3–7 mm, not spurred, glabrous, lips subequal to unequal; corollas 7–11 mm, whitish to blue, banners pinched type, spot absent. **Fruits** 15–35 mm, pubescent. **Seeds** 3–4.

Forests, disturbed areas, serpentine. Flowering May–Jun. 300–1400 m. Sisk. CA. Native.

This species may hybridize with *L. onustus* and *L. albicaulis*.

Lupinus uncialis S. Watson

Inch high lupine

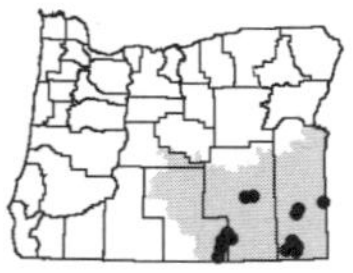

Plants annual, subacaulescent, usually forming tufts, 0.1–0.2(0.3) dm. **Stems** prostrate to decumbent, sericeous to pilose. **Leaves** basal; leaflets 3–5, (3)4–7 × 1–3 mm, oblanceolate, tips rounded to acute, surfaces densely pilose to tomentose; petioles 8–18 mm. **Inflorescences** 0.5–1.5 cm; peduncles 0.3–1.5 cm; pedicels 3–6 mm. **Flowers** calyces 3–5 mm, not spurred, 2-lipped, equal to unequal; corollas 3–6 mm, white, sometimes blue- or purple-tinged, banners cup type, spot absent. **Fruits** 5–9 mm, pubescent. **Seeds** 1–2.

Rocky or gravelly areas, sagebrush, juniper forests. Flowering May–Jun. 900–2000 m. BR, Owy. CA, ID, NV. Native.

MEDICAGO Alfalfa, medic, burclover

Nurul Ain Khalib and Melanie A. Link-Pérez

Herbs annual or perennial. **Stems** prostrate to erect, generally pubescent. **Leaves** alternate, pinnate or sub-palmate; leaflets 3, margins toothed distally, tips mucronate or mucronulate, surfaces glabrous or pubescent; petioles present; stipules present, adnate to petioles, entire to lacerate. **Inflorescences** 1–30-flowered axillary racemes, sometimes umbellate or sub-capitate. **Flowers** papilionaceous; calyces radial or bilateral, campanulate, fused, lobes 5; corollas petals 5, often ephemeral, free, yellow, violet, or variegated yellow-violet, keels obtuse or rounded; stamens 10, diadelphous; styles short, glabrous. **Fruits** indehiscent, becoming exserted, generally spiraled in 2–6 turns and forming burs, sometimes reniform or rarely sickle-shaped, spiny or unarmed. **Seeds** 1–several, reniform or sometimes oval.

Worldwide. ~87 species; 6 species treated in *Flora*.

All of Oregon's *Medicago* species are introduced. These include several widely established species that are cultivated for forage or used for soil improvement due to their symbiotic associations with nitrogen-fixing bacteria. Alfalfa (*M. sativa*) is perhaps the most important forage plant in the temperate world.

Two taxa (*M. falcata* and *M. turbinata*) have been collected only once or twice and do not appear to be established in our flora. *Medicago falcata*, yellow-flowered alfalfa, is a recent waif with yellow corollas and straight or falcate fruit, which distinguish this species from all other *Medicago* found in Oregon. *Medicago turbinata*, southern medic, has been collected only twice over a century ago on ballast.

Fruit measurements include the length of the spines.

Key to *Medicago* species:

1. Plants perennial; flowers violet or variegated yellow-violet; calyces ≥ 4 mm; fruits both coiled and unarmed, with doughnut-like hole visible in center of coils . . . . *M. sativa*

1' Plants annual; flowers yellow; calyces ≤ 3 mm; fruits not as above.

  2. Fruits 1-seeded, oval to reniform, unarmed; inflorescences 10–20-flowered . . . . *M. lupulina*

  2' Fruits ≥ 2-seeded, coiled, generally armed with spines; inflorescences 1–6-flowered.

    3. Leaflets generally with central dark anthocyanic blotch, length ≈ width; fruits grooved along edges . . . . *M. arabica*

    3' Leaflets without central dark blotch, length > width; fruit edges not grooved.

      4. Stipules entire or slightly serrate; stems densely pubescent . . . . *M. minima*

      4' Stipules laciniate; stems glabrous to sparsely pubescent.

        5. Leaflets 8–22 mm; stipules 4–12 mm; corollas 3.5–6 mm . . . . *M. polymorpha*

        5' Leaflets 3–8 mm; stipules 2–4 mm; corollas 2.5–4 mm . . . . *M. praecox*

Medicago arabica (Linnaeus) Hudson

Spotted medic, spotted burclover

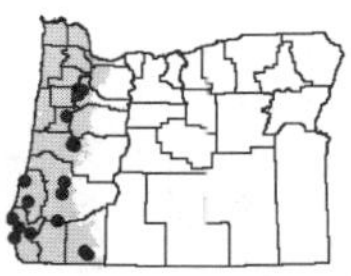

Plants annual, 3–70 cm. **Stems** decumbent to erect, glabrate or sparsely pubescent with septate hairs. **Leaves** pinnate to sub-palmate; leaflets obcordate to broadly cuneate-obovate, length ≈ width, 8–20(25) mm, generally with central dark anthocyanic blotch, bases cuneate, margins distally serrate, veins raised or slightly thickened, tips usually emarginate, surfaces glabrous to slightly pubescent, more so abaxially; petioles 10–60 mm; petiolules 0.1–5 mm; stipules 4–8 mm, margins shallowly lacerate. **Inflorescences** 2–4-flowered, globose, 6–10 × 6–10 mm; peduncles 10–20 mm; pedicels 1–2 mm. **Flowers** calyces 2–2.5 mm; corollas 4–5(6) mm, yellow. **Fruits** spirally coiled 4–6 turns, subglobose or shortly cylindric, 8–10 mm, glabrous, edges grooved, dorsal sutures spiny; spines curved, 2–3 mm. **Seeds** several, reniform, 2–3.5 × 1.2–1.5 mm, yellow or yellow-brown. **2*n***=16.

Grassy slopes, roadsides, open woodlands, disturbed areas. Flowering Mar–Jun. 0–1600 m. CR, Est, Sisk, WV. CA, WA; scattered in North America; Asia, Australia, Europe, South America. Exotic.

Medicago arabica can be distinguished easily from other species of *Medicago* by the dark spots on its leaflets.

Medicago lupulina Linnaeus (p. 708)

Black medic, hop clover

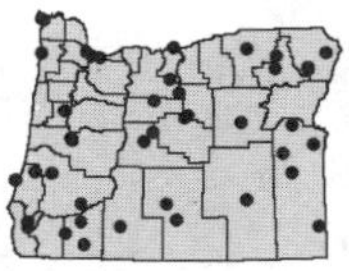

Plants annual, 10–45(80) cm. **Stems** pubescent. **Leaves** sub-palmate; leaflets obovate, length ≥ width, 7–13(20) mm, bases cuneate, margins distally serrate, veins prominent, tips obtuse to rounded, surfaces slightly pubescent, more so abaxially; petioles 2–10(15) mm; petiolules 0.5–4 mm; stipules 4–8 mm, margins entire or toothed. **Inflorescences** 10–20-flowered, ovoid, 7–20 × 5–8 mm; peduncles 10–30 mm; pedicels 0.8–1 mm. **Flowers** calyces 1–1.5 mm; corollas 2–3 mm, yellow. **Fruits** reniform, 2.5–3 × 2–2.5 mm, becoming black with age, glabrous to pubescent, strongly concentric-veined, unarmed. **Seeds** 1, oval to reniform, 1.5–2 × 1–1.2 mm, yellow to olive-green. **2*n***= 16,32.

Riparian areas, meadows, lawns, roadsides, disturbed areas. Flowering May–Jul. 0–1600 m. All ecoregions. CA, ID, NV, WA; worldwide. Exotic.

Medicago lupulina is characterized by dense, ovoid inflorescences consisting of ten or more tiny yellow flowers. This is the only species of *Medicago* in Oregon with one-seeded fruits.

Medicago minima (Linnaeus) Bartalini

Little burclover

Medicago minima (L.) Bartal. var. *pubescens* Webb

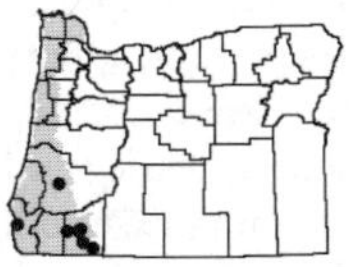

Plants annual, 3–30 cm. **Stems** decumbent or ascending, densely pubescent. **Leaves** sub-palmate; leaflets obovate-obelliptic, length > width, 4–10 mm, bases cuneate, margins distally denticulate or serrulate, veins prominent, tips obtuse to rounded, surfaces densely silky; petioles 2–20 mm; petiolules 0.5–3 mm; stipules 3–6 mm, margins entire or slightly serrate. **Inflorescences** 3–6-flowered, subcapitate; peduncles 2.5–10 mm; pedicels 0.5–1 mm. **Flowers** calyces 2–2.5(3) mm; corollas 3–5 mm, yellow. **Fruits** spirally coiled 3–4 turns, spherical to shortly cylindric, 4.5–9 mm, sparsely pubescent, dorsal sutures spiny, rarely tuberculate; spines hooked, 1–2 mm. **Seeds** 3–10, slightly reniform, 1.5–2.5 × 0.9–1.3 mm, yellow to light brown. **2*n***=16.

Roadsides, riparian areas, disturbed areas. Flowering Apr–Jul. 0–1200 m. CR, Sisk. CA, WA; worldwide. Exotic.

Medicago polymorpha Linnaeus (p. 708)

California burclover, toothed medic, bur medic

Medicago hispida Gaertn., *Medicago hispida* Gaertn. var. *apiculata* (Willd.) Urb., *Medicago hispida* Gaertn. var. *hispida*

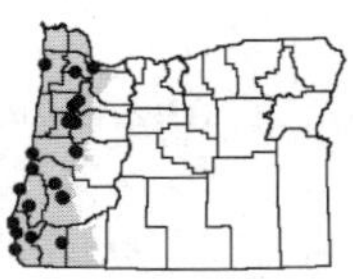

Plants annual, commonly mat-forming, 10–80(100) cm. **Stems** prostrate to ascending, generally glabrous. **Leaves** pinnate; leaflets wedge-shaped to obovate, rarely obcordate, length > width, 8–22 mm, bases cuneate, margins distally serrate, veins prominent, tips obtuse to rounded or emarginate, surfaces glabrate, rarely puberulent; petioles 5–40(60) mm; petiolules 0.5–6 mm; stipules 4–12 mm, margins deeply laciniate. **Inflorescences** 2–6-flowered; peduncles 4–30 mm; pedicels straight, 0.5–1 mm. **Flowers** 4–6(8) mm; calyces 2.5–3 mm; corollas 3.5–6 mm, yellow. **Fruits** spirally coiled 2–6 turns, ovoid to shortly cylindric, 5–10 mm, glabrous, dorsal sutures spiny or unarmed; spines hooked, 1–2 mm. **Seeds** 2–12, reniform, 2–4 × 1.5–2.2 mm, light yellow to brownish. **2*n***=14,16.

Open grassy areas, fields, disturbed areas, roadsides. Flowering Mar–Jun. 0–1000 m. CR, Est, Sisk, WV. CA, ID, NV, WA; worldwide. Exotic.

Medicago polymorpha can be distinguished from most other *Medicago* species by its combination of deeply laciniate stipules and lack of pubescence.

Medicago praecox de Candolle

Mediterranean medic, early medic

Plants annual, 10–20 cm. **Stems** decumbent, glabrous to sparsely pubescent. **Leaves** pinnate; leaflets wedge-shaped or obovate, length > width, 3–8 mm, bases cuneate, margins distally denticulate, veins raised, tips emarginate, surfaces abaxially slightly pubescent, adaxially glabrous; petioles 2–5 mm; petiolules 0.5–2 mm; stipules 2–4 mm, margins strongly lacerate. **Inflorescences** 1–3-flowered; peduncles 1.5–3 mm; pedicels 0.5–1 mm. **Flowers** calyces 1.5–2 mm; corollas 2.5–4 mm, yellow or variegated yellow-purple, ephemeral. **Fruits** spirally coiled 3–4 turns, spherical or

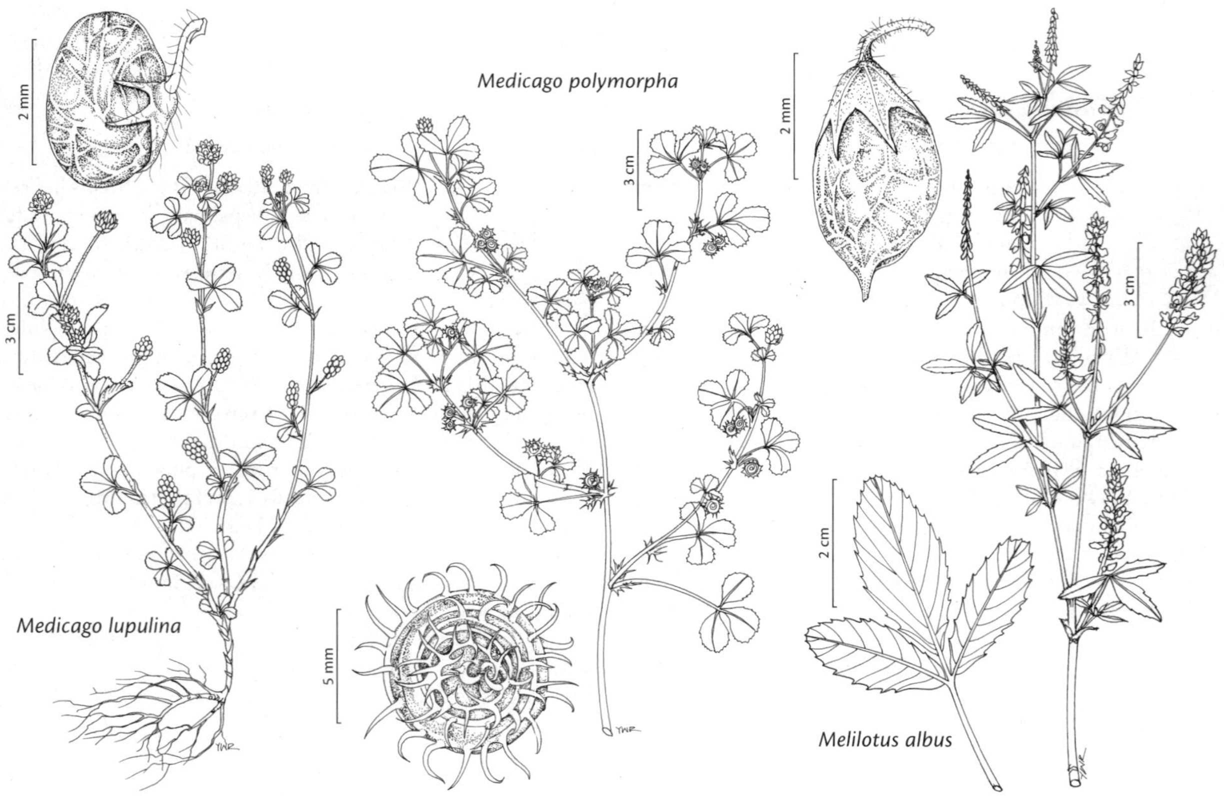

MEDICAGO • MELILOTUS

disk-like, 5–10 mm, sparsely hairy, dorsal sutures spiny; spines hooked, 2–3 mm. **Seeds** several, reniform, 1.7–2.4 × 0.9–1.3 mm, yellow to brownish yellow. **2*n***=14.

Gravelly and grassy areas, roadsides, near beaches. Flowering Apr–May. 0–300 m. Est. CA; Australia, Europe. Exotic.

Medicago praecox has tiny flowers that appear similar to *M. polymorpha*, but the leaflets of the latter are usually glabrous and generally twice as long as those of *M. praecox*. *Medicago praecox* also has fewer flowers per inflorescence.

Medicago sativa Linnaeus

Alfalfa, lucerne

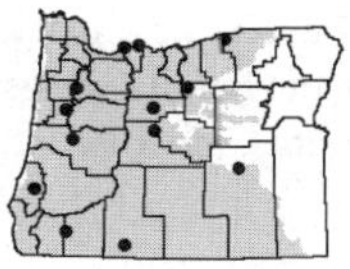

Plants perennial, 20–80 cm, glabrous to puberulent. **Stems** decumbent to erect, glabrous to puberulent. **Leaves** sub-palmate; leaflets narrowly lanceolate to obovate, length > width, 8–29 mm, bases cuneate, margins distally serrate, veins prominent, tips truncate to rounded, surfaces pubescent; petioles 3–15 mm; petiolules 1–6 mm; stipules 6–14 mm, margins entire to sharply toothed. **Inflorescences** 8–30-flowered, ovoid to oblong; peduncles 8–40 mm; pedicels 1–3 mm. **Flowers** calyces 4–5.5 mm; corollas 8–10 mm, violet or variegated yellow-violet. **Fruits** spirally coiled 2–3 turns, rarely sickle-shaped, with doughnut-like hole visible in center of coils, 3–5 mm, villous, unarmed. **Seeds** several, reniform, 1–2.5 × 1–1.5 mm, yellow, brownish, greenish yellow or violet-brown. **2*n***=16, 32.

Sandy or rocky areas, riparian areas, fields, roadsides, disturbed areas. Flowering Jun–Aug. 0–1500 m. BR, Casc, Col, CR, ECas, Lava, Sisk, WV. CA, ID, NV, WA; worldwide. Exotic.

Medicago sativa is the only Oregon species with purple or variegated flowers and with coiled, non-spiny fruits containing several seeds. It also has the largest flowers (approaching 1 cm in length) and more flowers per inflorescence than most of our other species. *Medicago sativa* is commonly used as a cover crop or green manure.

MELILOTUS Sweetclover

Richard R. Halse

Herbs annual or biennial, strigillose to glabrate; taprooted. **Stems** erect or ascending. **Leaves** cauline, alternate, pinnately trifoliolate; leaflets 3, margins entire to toothed; stipules narrow, bases partially adnate to petioles. **Inflorescences** narrow, elongate, loosely-flowered, spike-like, axillary racemes; bracts minute; bracteoles 0. **Flowers** papilionaceous, usually pendulous; calyces campanulate, 5-toothed; corollas yellow or white; stamens 10, diadelphous, anthers basifixed, monomorphic; ovules 1–3, styles glabrous. **Fruits** indehiscent to tardily dehiscent, straight, usually globose to ovoid, reticulate-veined or rugose, glabrous. **Seeds** 1–3, ovoid to subglobose or elliptic.

Asia, Europe, North Africa, North America. 18 species; 3 species treated in *Flora.*

The sweetclovers are recognizable by their slender, typically elongate racemes of numerous small yellow or white flowers and indehiscent 1–3-seeded fruits. They are used as soil builders (green manure), forage crops, and as honey plants. The presence of coumarins gives them a characteristic vanilla or sweet odor, especially when cut. All of our species have become naturalized, widespread weeds.

Key to *Melilotus* species:

1. Corollas white, ochroleucous when dried; banners > wings . . . . ***M. albus***

1' Corollas yellow, white or cream when dried; banners ≈ wings.

  2. Corollas 4–7 mm; pedicels 1.2–2.5 mm, fruits transversely rugose . . . . ***M. officinalis***

  2' Corollas 1.5–3 mm; pedicels 0.5–0.8 mm, fruits reticulate-veined. . . . . ***M. indicus***

Melilotus albus Medikus (p. 708)

White sweetclover

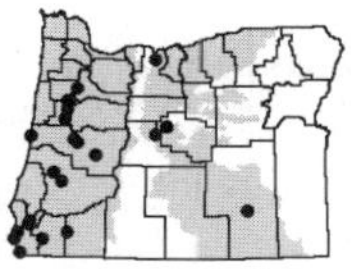

Plants usually biennial, rarely annual, 5–20 dm. **Leaves** leaflets narrowly oblong-obovate to suborbicular-oblong, 10–24(50) × 5–15 mm; stipules lanceolate-subulate to setaceous, margins usually entire, sometimes toothed at bases. **Inflorescences** 40–120-flowered; pedicels 1–2 mm. **Flowers** calyx tubes 1.2–1.8 mm, teeth 1–1.5 mm, acuminate; corollas white, ochroleucous when dried, 3.5–6 mm, banners distinctly > wings. **Fruits** 2.5–6 mm, reticulate-veined. **2*n***=16.

Roadsides, riverbanks, disturbed areas, fields. Flowering May–Sep. 0–1600 m. BR, Casc, Col, CR, Est, Lava, Sisk, WV. CA, ID, NV, WA; nearly worldwide. Exotic.

Melilotus indicus (Linnaeus) Allioni

Annual yellow sweetclover, Indian sweetclover, sourclover

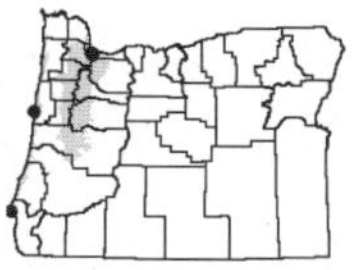

Plants annual, 1–8 dm. **Leaves** leaflets lanceolate-oblong, oblanceolate, or obovate, 7–30 × 2–14 mm; stipules lanceolate-subulate to setaceous, margins subentire, rarely entire or toothed at bases. **Inflorescences** 10–50-flowered; pedicels 0.5–0.8 mm. **Flowers** calyx tubes 0.6–0.8 mm, teeth 0.5–0.7 mm, triangular; corollas yellow, white when dried, 1.5–3 mm, banners ≈ wings. **Fruits** 1.5–4 mm, reticulate-veined. **2*n***=16.

Ballast, waste ground, roadsides. Flowering May–Sep. 0–1300 m. Est, WV. CA, ID, NV, WA; nearly worldwide. Exotic.

Melilotus officinalis (Linnaeus) Pallas

Yellow sweetclover

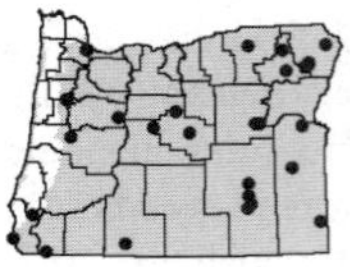

Plants biennial, rarely annual, 5–25 dm. **Leaves** leaflets obovate, obovate-oblong, ovate or ovate-lanceolate, 8–38 × 3–16 mm; stipules lanceolate or subulate, margins entire. **Inflorescences** 20–80-flowered; pedicels 1.2–2.5 mm. **Flowers** calyx tubes 1–1.8 mm, teeth 1–1.5 mm, acuminate; corollas yellow, cream when dried, 4–7 mm, banners ≈ wings. **Fruits** 2.5–5 mm, transversely rugose. **2*n***=16.

Roadsides, ballast, riverbanks, disturbed areas. Flowering May–Sep. 0–2200 m. All ecoregions except CR. CA, ID, NV, WA; nearly worldwide. Exotic.

ONOBRYCHIS Sainfoin

Richard R. Halse

Herbs perennial. **Leaves** alternate, odd-pinnately compound; leaflets entire; stipules amplexicaul, scarious. **Inflorescences** axillary, spike-like racemes; bracts present, bracteoles minute. **Flowers** papilionaceous; calyces campanulate, teeth subulate; corolla wings < keels, keels prominent, obliquely truncate; stamens diadelphous proximally, monadelphous distally. **Fruits** loments, indehiscent, reduced to 1 segment, flattened to rounded, margins warty to spinescent. **Seeds** 1, reniform.

Asia, Europe, North America. ~130 species; 1 species treated in *Flora.*

Onobrychis viciifolia Scopoli (p. 710)

Common sainfoin

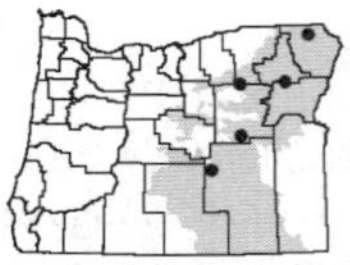

Plants inconspicuously strigose to glabrate. **Stems** erect, clustered, 3–8 dm. **Leaves** 15–20 cm; leaflets 9–21, obovate to narrowly elliptic, 10–25 × 3–7 mm, tips apiculate, surfaces abaxially appressed-pubescent, adaxially red-dotted and glabrous. **Inflorescences** 10–50-flowered; peduncles 10–20 cm; bracts 2–4 mm; pedicels 1–2.5 mm. **Flowers** ascending-spreading; calyces 5–7.5 mm, tubes 1.5–2 mm, teeth 4–6 mm; corollas 8–15 mm, pink-purple, red-veined. **Fruits** ascending, obovate to ovate, flattened, exserted from persistent calyces; 5–8 mm, coarsely reticulate, coriaceous, densely appressed-pubescent, margins with toothed wing. **2*n***=8.

Roadsides, disturbed areas, farmland. Flowering Jun–Aug. 800–1500 m. BR, BW. CA, ID, NV, WA; widely scattered in Canada and US; Asia, Europe. Exotic.

Common sainfoin is cultivated for forage and soil improvement. It is recognizable by its long, spike-like racemes of pink-purple flowers with a large blunt keel and small wings, and single-seeded, indehiscent fruits. It could be mistaken for a species of *Hedysarum*, but the fruits in that genus are several-seeded.

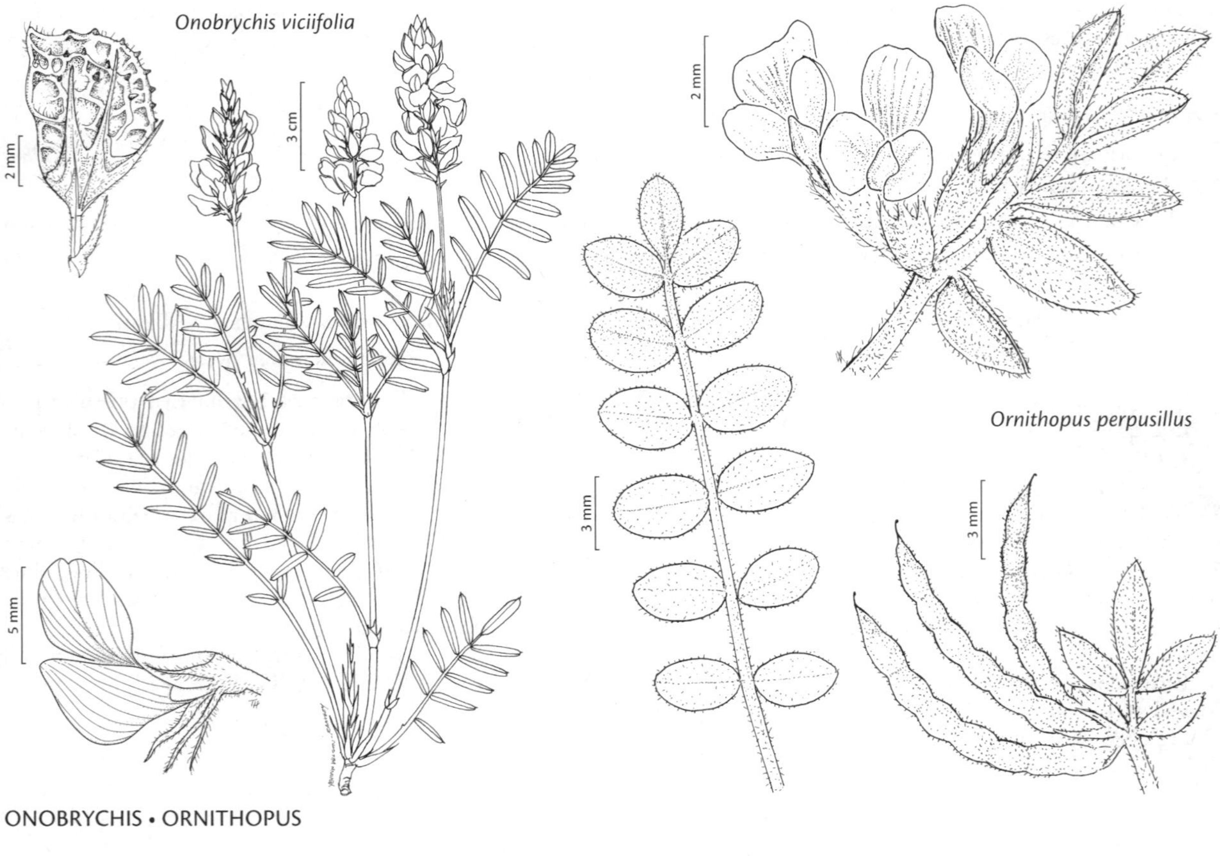

ONOBRYCHIS • ORNITHOPUS

ORNITHOPUS Birdsfoot, serradela

Richard R. Halse

Herbs annual, caulescent. **Stems** prostrate to ascending. **Leaves** alternate, odd-pinnately compound; leaflets entire, pubescent; sessile, rarely subsessile; stipules free, membranous, dark-tipped. **Inflorescences** axillary, capitate; pedunculate. **Heads** 1–5-flowered; pedicels present or absent. **Flowers** papilionaceous, bisexual; calyces tubular to campanulate; corollas white, pink, purple, or yellow, keels obtuse; stamens diadelphous; styles terminal. **Fruits** loments, usually constricted between seeds, curved, laterally compressed, reticulate-veined. **Seeds** 3–9.

Asia, Europe, North America. 6 species; 2 species treated in *Flora.*

Key to *Ornithopus* species:

1. Corollas yellow, 5–8 mm; fruits 20–50 × 2–3 mm. . . . . ***O. compressus***
1' Corollas white or pink, 3–5 mm; fruits 10–18(25) × 1.5–2 mm . . . . ***O. perpusillus***

Ornithopus compressus Linnaeus

Yellow serradela

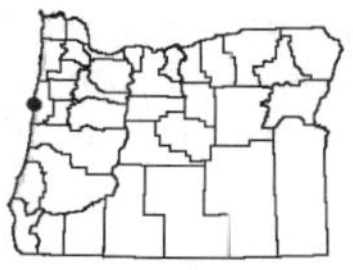

Plants 1–5 dm, pubescent. **Leaves** 3–6 × 1–1.5 cm; leaflets 15–37, elliptic to oblong-lanceolate. **Heads** 7–10 mm in diameter; subtended by leafy pinnate bract. **Flowers** calyces 3.5–4.3 mm, teeth 1.2–1.8 mm; corollas 5–8 mm, yellow. **Fruits** generally slightly constricted between seeds, 20–50 × 2–3 mm. 2n=14.

Disturbed areas, sandy soils. Flowering Apr–Jun. 0–50 m. Est. Europe. Exotic.

Ornithopus perpusillus Linnaeus (p. 710)

Little white birdsfoot

Plants 1–3 dm, pubescent. **Leaves** 1–4 × 0.5–1 cm; leaflets 15–27, ovate, elliptic, or oblong. **Heads** ~5 mm in diameter. **Flowers** calyces 2–2.5 mm, teeth 0.4–0.7 mm; corollas 3–5 mm, white or pink. **Fruits** constricted between seeds, 10–18(25) × 1.5–2 mm. ***2n***=14.

Disturbed areas, sandy soils. Flowering May–Jul. 0–50 m. Est. British Columbia, PA; Europe. Exotic.

OXYTROPIS Crazyweed, locoweed, oxytrope

Richard R. Halse

Herbs perennial, cespitose, with branched caudices, generally acaulescent, rarely with short leafy stems, hairs basifixed, rarely malpighian. **Leaves** usually basally clustered, rarely alternate, odd-pinnately compound; stipules free or adnate to petioles, often connate around stems. **Inflorescences** racemes or spikes, usually scapose, showy; bracts usually persistent. **Flowers** papilionaceous, bisexual; calyx tubes cylindric or campanulate, 5-lobed, lobes unequal; corolla petals 5, reddish, purple, white, or pale cream, keels abruptly narrowed into points or teeth, with straight or curved beaks, < wings; stamens 10, diadelphous, anthers monomorphic; styles glabrous, stigmas capitate. **Fruits** unilocular, bilocular, or semibilocular, ascending or rarely deflexed, usually persistent, strongly exserted or enclosed in persistent calyces, body ovoid to oblong or lanceolate, turgid to inflated, usually ventrally sulcate; valves papery to coriaceous to woody; sessile to shortly stipitate.

Temperate and arctic Northern Hemisphere. ~300 species; 4 species treated in *Flora.*

Many species of *Oxytropis* and *Astragalus* look alike and often share a common name—locoweed. *Oxytropis* can usually be distinguished from *Astragalus* by its strongly beaked keel and its typically acaulescent and scapose growth form. In addition, the fruits of *Oxytropis* tend to be partly or mostly divided into two locules by the intrusion of a false septum arising from the ventral suture compared with the false septum (when present) arising from the dorsal suture in *Astragalus.*

Three species of oxytrope (*O. campestris, O. deflexa, O. sericea*) have been reported to cause locoism—a poisoning of livestock due to the presence of indolizidine alkaloids. *Oxytropis sericea* also causes right-heart failure or brisket disease in cattle. In Oregon, oxytrope tends to grow in areas not subject to grazing or else is rare with small population sizes so that locoism due to these plants is uncommon.

Key to *Oxytropis* species:

1. Fruits deflexed; flowers ascending, becoming deflexed . . . . ***O. deflexa*** var. ***sericea***
1' Fruits erect; flowers ascending, not deflexed.
   2. Plants glandular . . . . ***O. borealis*** var. ***viscida***
   2' Plants not glandular.
      3. Fruits fleshy, becoming hardened and bony; scapes 7–35 cm; plants of southeastern Oregon . . ***O. sericea*** var. ***sericea***
      3' Fruits membranous; scapes 2.5–10 cm; plants of northeastern Oregon . . . . ***O. campestris*** var. ***cusickii***

Oxytropis borealis de Candolle var. ***viscida*** (Nuttall) S. Welsh

Nuttall's oyxtrope, sticky crazyweed, sticky oxytrope

Oxytropis viscida Nutt., *Oxytropis viscida* Nutt. var. *viscida*

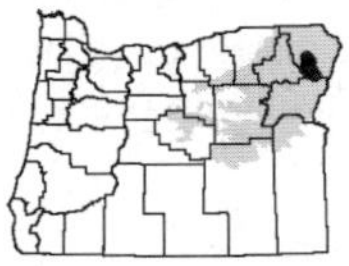

Plants acaulescent, 4–18 cm, pilose-hirsute, with wart-like glands. **Leaves** 2–20 cm; leaflets 15–45, oblong to lanceolate or elliptic, 1.5–25 × 1–5 mm, surfaces glabrate to glabrous, often glandular. **Inflorescences** racemes 7–30-flowered; scapes 2–28 cm, spreading hairy, generally glandular. **Flowers** ascending, not deflexed; calyces 5–11 mm, tubes cylindric, teeth 1.5–3.5 mm; corollas 10–19 mm, purplish, yellowish, or whitish tinged with blue. **Fruits** bilocular, erect, 8–20 × 4–6 mm, glandular.

Alpine ridges, open meadows, gravel bars of creeks. Flowering Jun–Aug. 1200–3000 m. BW. CA, ID, NV, WA; north to AK, northeast to Nunavut, east to Quebec, southeast to CO. Native.

This species is unique in the genus for being glandular. The degree of glandularity is quite variable over its geographic range; in Oregon, the scapes usually have obvious wart-like glands.

Oxytropis campestris (Linnaeus) de Candolle var. ***cusickii*** (Greenman) Barneby (p. 712)

Field locoweed, Cusick's oxytrope, yellow locoweed

Oxytropis cusickii Greenm.

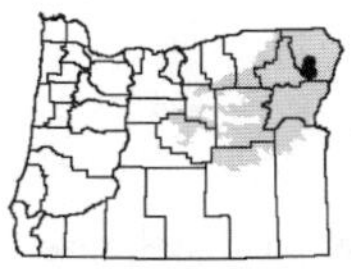

Plants acaulescent, 4–18 cm; pilose. **Leaves** 1.5–12 cm; leaflets 7–17, oblong to lanceolate or obovate, 3–23 × 1–6 mm, surfaces silky villous. **Inflorescences** racemes 4–20-flowered; scapes 2.5–10 cm, pilose to glabrate. **Flowers** ascending, not deflexed; calyces 7–10 mm, tubes cylindric, teeth 1.5–3 mm; corollas 10–20 mm, yellowish white to creamy yellow. **Fruits** semibilocular, erect, 8–18 mm, membranous, pilose. **2*n***=48.

Cliffs, meadows, streamsides. Flowering Jun–Aug. 1100–3000 m. BW. ID, WA; north to British Columbia, northeast to Alberta, east to WY, southeast to CO. Native.

Oxytropis campestris var. *spicata* has been reported from Oregon, but no specimens have been seen to confirm this. Variety *spicata* can be distinguished from var. *cusickii* by having more than 17 leaflets and strongly hairy stipules.

Oxytropis deflexa (Pallas) de Candolle var. ***sericea*** Torrey & A. Gray (p. 712)

Shortstem oxytrope, nodding locoweed

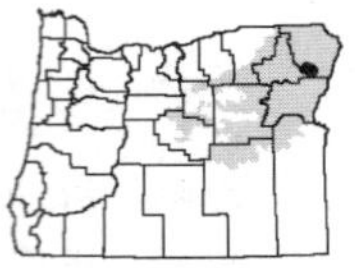

Plants caulescent to acaulescent, 5–50 cm, villous-pilose. **Leaves** 2–20 cm; leaflets 23–41, ± lanceolate, 3–24 × 1–7 mm, surfaces strigillose to villous-pilose. **Inflorescences** racemes 3–25-flowered; scapes 3.5–30 cm, villous-pilose. **Flowers** ascending, becoming deflexed; calyces 4–8 mm, tubes cylindric, teeth 1.5–5 mm; corollas 6–8 mm, white to pale blue or purple. **Fruits** deflexed, ± unilocular, 8–18 × 3–5 mm, pilose. **2*n***=16.

Ridgetops, streamsides. Flowering Jun–Aug. 2400–2800 m. BW. CA, ID, WA; north to AK, northeast to Ontario, east to WY, southeast to NM. Native.

In the few collections of this species in Oregon, the plants appear to be acaulescent, but the deflexed fruits easily identify it.

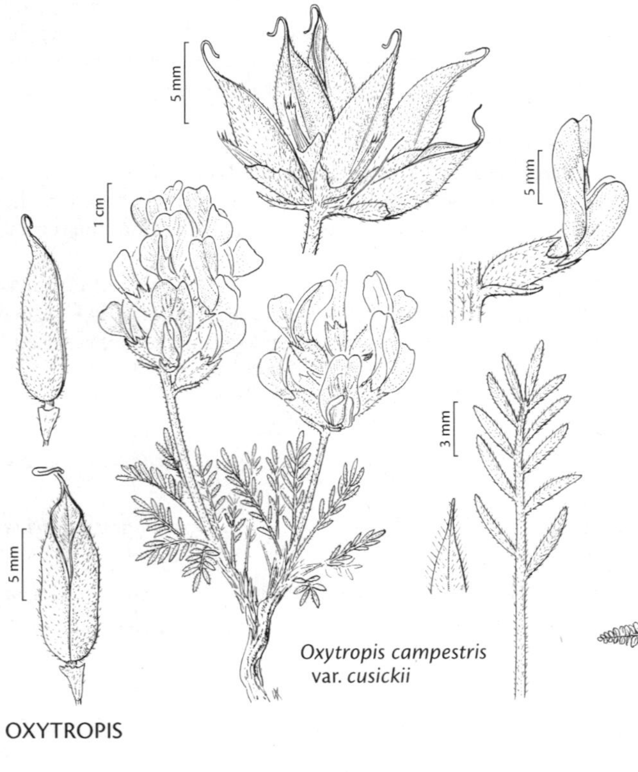

OXYTROPIS

Oxytropis sericea Nuttall ex Torrey & A. Gray var. ***sericea***

White locoweed, silky locoweed, silky crazyweed

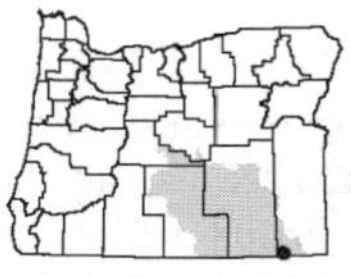

Plants acaulescent, 13–32 cm, silky pilose. **Leaves** 3.5–30 cm; leaflets 9–23, elliptic to ovate, 4–40 × 1.5–10 mm, surfaces canescent. **Inflorescences** racemes 6–30-flowered; scapes 7–35 cm, pilose. **Flowers** ascending, not deflexed; calyces 8–13 mm, tubes cylindric, teeth 1.5–4.5 mm; corollas 15–26 mm, white or tinged with purple, particularly on keels. **Fruits** ± bilocular, erect, 10–25 × 4–7.5 mm, fleshy, becoming hardened and bony, strigose. **2*n***=48.

Ash deposits, gravelly areas. Flowering May–Jul. 1400–1800 m. BR. ID, NV; northeast to MT, east to SD, southeast to TX. Native.

PISUM Pea

Richard R. Halse

Herbs annual, glabrous. **Stems** erect or sprawling, ribbed, branched. **Leaves** alternate, even-pinnately compound, rachises winged, terminating in branched tendrils; leaflets opposite, ovate to lanceolate, margins entire or dentate; stipules present. **Inflorescences** axillary racemes, erect or lax; bracts caducous. **Flowers** papilionaceous; calyces campanulate, teeth 5, equal, lanceolate, > tubes; corollas white or bicolored; stamens 10, diadelphous; ovaries glabrous or pubescent, styles longitudinally folded, pubescent on inner sides. **Fruits** oblong-linear, terete. **Seeds** spherical or angular, usually yellow or green, sometimes mottled.

Worldwide. ~5 species; 1 species treated in *Flora.*

Pisum sativum Linnaeus var. ***arvense*** (Linnaeus) Poiret (p. 713)

Field pea

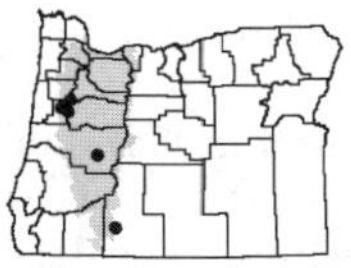

Stems 1–20 dm. **Leaves** 5–10 cm; leaflets (2)4–6(8), (9)20–50(60) × (6)15–30(40) mm; stipules ovate to lanceolate, semi-sagittate, 15–80 × 10–40 mm, usually > leaflets. **Inflorescences** 1–3(4)-flowered; peduncles 3–10 cm; pedicels 7–20 mm. **Flowers** calyces 10–13 mm, tubes 4–5 mm, teeth 6–9 mm; corollas 15–30 mm, bicolored, banners lilac, wings dark reddish purple. **Fruits** 50–90 × 15–20 mm. **Seeds** angular, 4–8 mm. **2*n***=14.

Roadsides, fields. Flowering May–Jun. 50–400 m. Casc, WV. CA, ID, WA; worldwide. Exotic.

Field pea is cultivated as a cover crop for soil improvement or as a seed crop. It can be found throughout the United States in cultivation and as a recurrent waif.

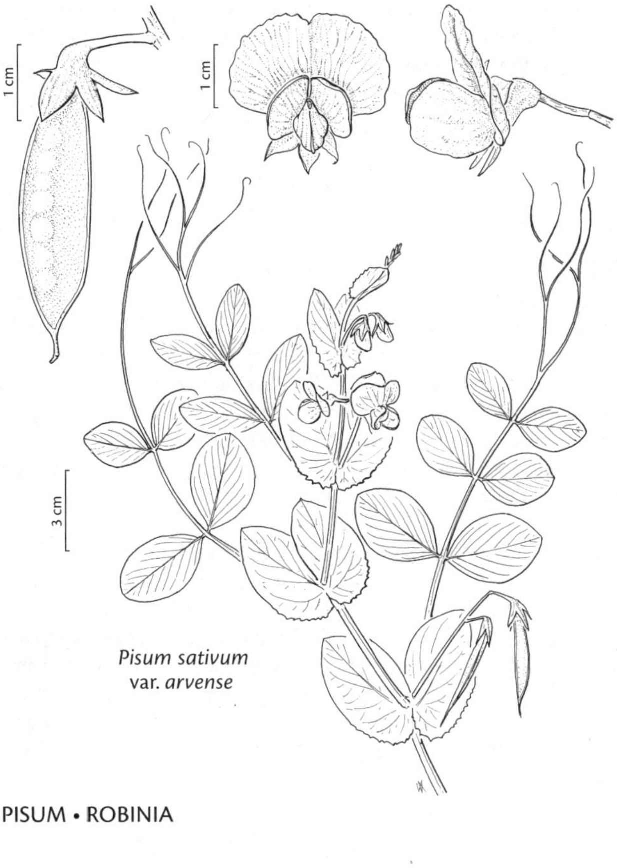

Pisum sativum var. *arvense*

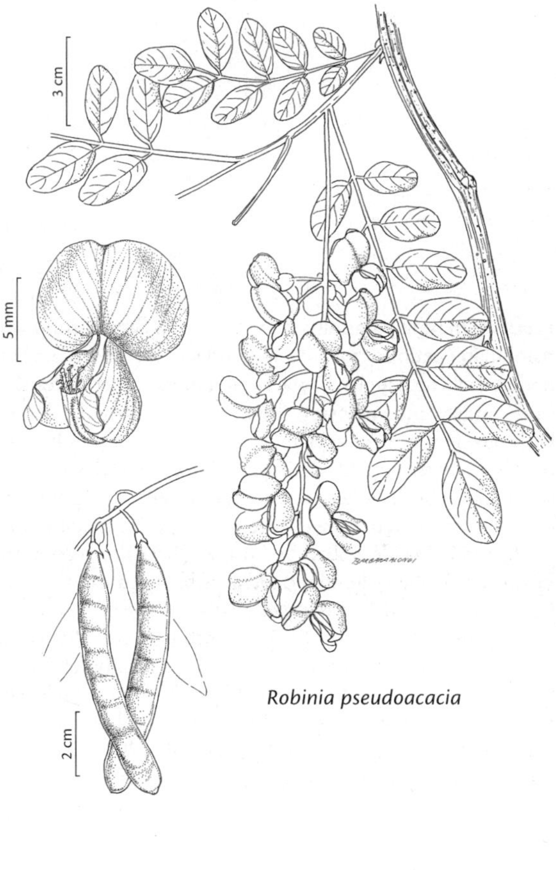

Robinia pseudoacacia

PISUM • ROBINIA

ROBINIA Locust

Cameron J. Ross

Trees or **shrubs** generally spreading, glabrous or hispid; rhizomatous or not. **Stems** ascending, branched, usually armed. **Leaves** deciduous, alternate, odd-pinnately compound; petioles present; stipules setaceous and/or caducous, or modified into spines. **Inflorescences** terminal or axillary pendulous racemes; bracts present, sometimes deciduous. **Flowers** fragrant, papilionaceous; calyces campanulate, lobes 5; petals 5, white or pink to purple or red, banners reflexed; ovary placentation marginal. **Fruits** dehiscent, oblong, laterally compressed, 4–12 × 0.5–1.5 cm, 2-valved, stipitate. **Seeds** 3–10, reniform, 5.5–6.5 × 3.5–4 mm.

Nearly worldwide. 4 species; 2 species treated in *Flora.*

This genus was named after French botanists Jean and Vespasien Robin, who introduced the genus to Europe in 1601 from plants collected in eastern North America.

Key to *Robinia* species:

1. Trees or shrubs up to 10 m tall, glandular; inflorescences 4–11-flowered; corollas purple to pink-red . . . . ***R. hispida***

1' Trees up to 25 m tall, not glandular; inflorescences 12–30-flowered, corollas white to light pink . . . . ***R. pseudoacacia***

Robinia hispida Linnaeus

Bristly locust

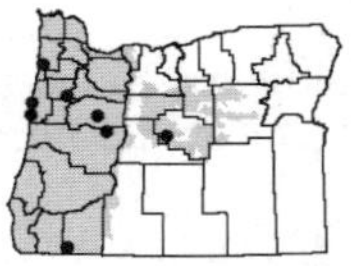

Trees or **shrubs** to 10 m, glabrous to hispid, glandular; rhizomatous. **Stems** armed. **Leaves** leaflets 9–15, 20–60 × 15–35 mm; stipules modified into spines. **Inflorescences** 4–11-flowered; pedicels 4–16 mm. **Flowers** calyces 10–16 mm, tubes 4–10 mm, 3–8 mm; corollas 2–3 cm, purple to pink-red.

Roadsides, pastures, near cultivation. Flowering May–Jun. 0–1100 m. Casc, CR, Est, Lava, Sisk, WV. CA, WA; scattered in US and Mexico; Asia, Europe. Exotic.

Robinia pseudoacacia Linnaeus (p. 713)

Black locust

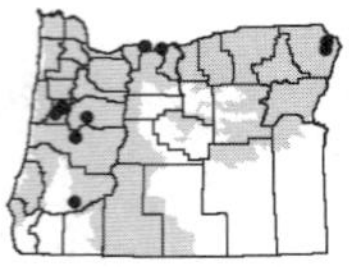

Trees to 25 m, glabrous to puberulent, not glandular; rhizomatous. **Stems** armed. **Leaves** leaflets 11–21(25), 15–60 × 12–25 mm; stipules modified into spines. **Inflorescences** 12–30-flowered; pedicels 6–13 mm. **Flowers** calyces 5–7 mm, tubes 5–7 mm, 3–5 mm; corollas 1.5–2 cm, white to light pink. **2*n***=20.

Roadsides, pastures, near cultivation. Flowering May–Jun. 0–1000 m. BW, Casc, Col, CR, ECas, WV. CA, ID, NV, WA; throughout US, scattered in Canada and Mexico; Africa, Asia, Europe. Exotic.

This plant has been known to poison and kill livestock that digest its foliage. It is native to the eastern United States but has been widely cultivated around the world for gardens, street trees, firewood, and honey. It has escaped cultivation and become invasive in some places, including the Pacific states.

RUPERTIA Scurfpea

Ashley Egan

Herbs perennial, glandular; roots woody. **Stems** somewhat decumbent to erect, 0–few-branched. **Leaves** alternate, pinnately trifoliolate; petioles ribbed; stipules deciduous, reflexed. **Inflorescences** axillary pseudoracemes, ovoid to ellipsoid, 2–3 flowers per node; bracts deciduous. **Flowers** papilionaceous; calyces tubular-campanulate in bud, gibbous in flower, swelling to become broadly campanulate and enclosing fruits, teeth 5, triangular, lowermost more lanceolate, glandular; corollas cream to yellow, occasionally with purple spot near keel and wing tips; stamens initially monadelphous, upper stamen becoming free; anthers in 2 series, upper basifixed, lower dorsifixed; ovaries 1-ovulate, styles reflexed. **Fruits** indehiscent, elliptic or depressed-obovate, apiculate or with a short, broad beak, at least sparingly glandular and pubescent. **Seeds** 1, reniform, smooth. **2*n***=22.

Western United States. 3 species; 1 species treated in *Flora.*

Rupertia was described by Grimes (1990) to accommodate three species previously aligned within *Hoita* by Rydberg (1919). However, Rydberg described *Hoita* as having a calyx that is not gibbous at the base with flowers purple or cream-colored. Grimes (1990) recognized that three of Rydberg's species—those with cream-colored flowers and a gibbous calyx that elongates through fruiting—were erroneously allied within *Hoita. Rupertia* is distinct within the Psoraleeae tribe by having a gibbous calyx that elongates through fruiting coupled with strongly reflexed stipules, indehiscent fruits, and white, yellow, or cream-colored flowers. Grimes (1990) conjectured that *Rupertia* was most closely related to *Pediomelum* due to calyx morphology and flower color, but his cladistic analysis of several morphological characters placed *Rupertia* as sister to *Ladeania* (Grimes' *Psoralidium*), a finding congruent with that of Egan & Crandall (2008a, 2008b) based on molecular data. The genus is named for Rupert C. Barneby, a noted botanist who specialized in Fabaceae.

Rupertia physodes (Douglas ex Hooker) J. Grimes (p. 715)

Forest scurfpea, California tea

Psoralea physodes Douglas ex Hook.

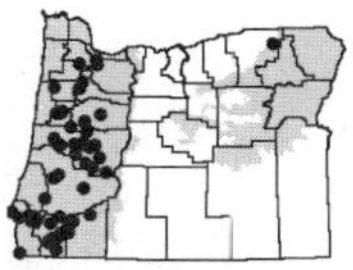

Plants to 75 cm, white- to black-pubescent. **Stems** sometimes stoloniferous at bases. **Leaves** leaflets lanceolate to triangular, or broadly elliptic to ovate, 2–6.5 × 1.5–5 cm, bases widely attenuate to somewhat cordate, often differently textured or colored, tips acute, surfaces glabrate to sparsely pubescent, dotted with blonde glands; lowermost nodes with brown, modified, sometimes bifid leaves, 3–7 mm; petioles ribbed; stipules ± linear, 4–10 mm. **Inflorescences** 2–12 cm; bracts elliptic or lanceolate to caudate; pedicels 1.5–2.5 mm, subtended by subulate, recurved, 1–2 mm bractlets. **Flowers** 9–14 mm; calyces 5–8 mm, usually black- and/or rarely white-pubescent, tubes 4–5.5 mm, to 15 mm in fruit, pubescent internally, upper teeth deltate, to 1 mm, lateral teeth triangular, lower tooth triangular, 2–3 mm; corolla banners oblanceolate to rhombic, 10–14 × 5–7 mm, wings and keels 10–13 × 2–3 mm, keels slightly < wings. **Fruits** depressed-obovate, 4–7 mm, apiculate, golden red, with sparse red-brown hairs. **Seeds** 5–6.5 × 3–4 mm, dark brick red. **2*n*=22.**

Slopes, canyons, woodlands. Flowering Apr–Aug. 0–1500 m. BW, Casc, CR, Est, Sisk, WV. CA, ID, WA; north to British Columbia. Native.

Rupertia physodes is the most widely distributed species of its genus and the only one to occur outside of California.

Rupertia physodes

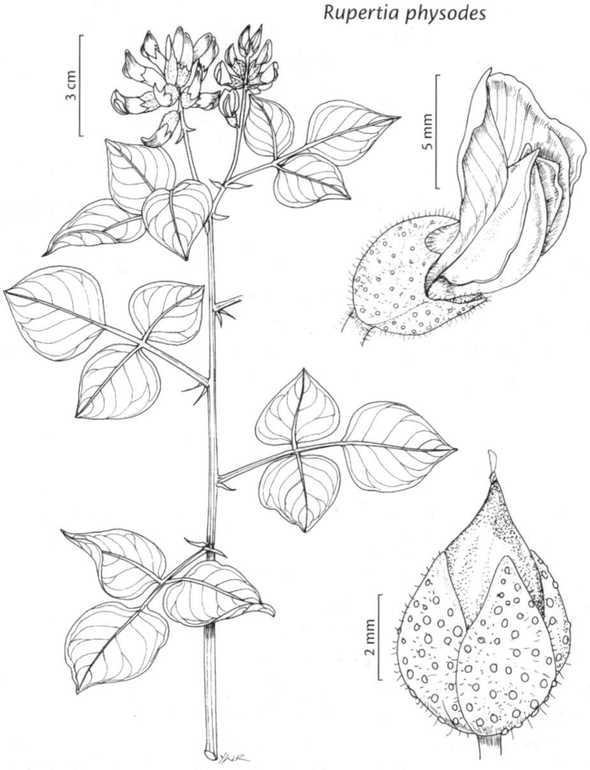

RUPERTIA

SECURIGERA Crown vetch

Richard R. Halse

Herbs perennial. **Stems** furrowed or ridged. **Leaves** alternate, odd-pinnately compound; leaflets opposite, entire, stipels 0; sessile; stipules membranous, dark-tipped. **Inflorescences** axillary umbels; peduncles long; bracts present. **Flowers** papilionaceous; calyces campanulate, weakly bilaterally symmetric; corollas pink, purple, wings almost completely concealing keels, keels acute; stamens diadelphous; styles glabrous. **Fruits** loments, linear, cylindrical or angled in cross section. **Seeds** 2–12, oblong.

Asia, Europe, North America. 12 species; 1 species treated in *Flora.*

Securigera varia (Linnaeus) Lassen (p. 715)

Crown vetch, purple crown vetch

Coronilla varia L.

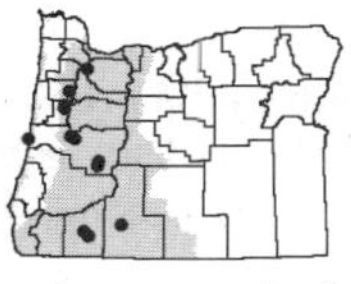

Plants glabrous or glabrate. **Stems** decumbent to sprawling, vine-like, 2–12 dm. **Leaves** 5–16 cm; leaflets (9)11–25, oblong, obovate, or elliptic, 6–25 × 2.5–12 mm, red-dotted above, tips obtuse, truncate, or mucronate; stipules linear to oblong, 1–6 mm. **Inflorescences** 6–25-flowered; peduncles 4–20 cm; pedicels 2–7 mm. **Flowers** spreading to deflexed; calyces 2–3 mm, tubes 1–1.5 mm, teeth 0.5–1 mm, 2 ventral teeth partly connate into bidenticulate tips, lateral 3 teeth separated by wide shallow sinuses; corollas (8)10–15 mm, banners pink to purplish, wings white, keels white with purple tips; styles 4–6 mm. **Fruits** ascending to erect, clustered, linear, torulose, longitudinally ridged, 20–60(80) × 1.5–2.2 mm, breaking up into (1)2–12 bluntly quadrangular joints. **Seeds** 3.5–4 × 1–1.5 mm. **2*n*=24.**

Securigera varia

SECURIGERA

Roadsides, disturbed areas, fields. Flowering Jun–Jul. 0–1300 m. Casc, ECas, Est, Sisk, WV. CA, ID, NV, WA; scattered throughout North America; Africa, Asia, Europe. Exotic.

Crown vetch has been cultivated as an ornamental, for fodder, and to control erosion. The long-peduncled umbels of *Securigera* may resemble some species of *Acmispon, Hosackia,* or *Lotus*, but in those genera, the fruits are not lomentaceous.

SOPHORA Necklacepod

Richard R. Halse

Herbs perennial. **Stems** erect. **Leaves** alternate, odd-pinnately compound; leaflets alternate to sub-opposite, margins entire; stipels absent or minute; stipules usually present, caducous. **Inflorescences** terminal or axillary racemes; bracts caducous; pedicels present; bracteoles caducous or absent. **Flowers** papilionaceous; calyces campanulate, gibbous at bases, teeth 5; corollas white, creamy white, or yellow; stamens 10; ovules (1)2–18. **Fruits** cylindric, fusiform, or torulose.

Worldwide. ~50 species; 1 species treated in *Flora*.

Sophora leachiana M. Peck (p. 716)

Western sophora, western necklacepod

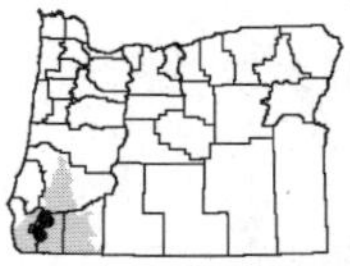

Plants finely gray-tomentose; rhizomatous. **Stems** 2–4 dm. **Leaves** 5–18 cm; leaflets 15–33, obovate to oblong, 15–38 × 7–15 mm, tips obtuse, rounded to truncate, surfaces abaxially villous-tomentose, adaxially appressed-pubescent. **Inflorescences** 20–50-flowered, 5–14 cm; pedicels 2–4 mm; bracteoles 1–2. **Flowers** divergent or declined; calyces tubular-campanulate, 6–11 mm, tubes 5–8 mm, teeth broadly triangular, 2–3 mm; corollas 9–16 mm, creamy or greenish white; stamens distinct. **Fruits** indehiscent, torulose, 30–40 × 3–5 mm, light brown, leathery, densely short villous-tomentose; stipes 3–4 mm. **Seeds** 1–2, 4–5 mm. **2*n***=36, 54.

Woodlands, roadsides. Flowering May–Jun. 100–800 m. Sisk. Native.

Sophora leachiana is found only in the Siskiyou Mountains of southwestern Oregon. It was discovered by and named after Oregon botanist Lilla Leach. Flowering depends on high levels of light, and when the tree canopy is established plants cease flowering but persist vegetatively. Plants do have vigorous asexual reproduction by rhizomes, but even these patches will die out as the forest matures. Even in populations with high light levels, there are low levels of fruit/seed development.

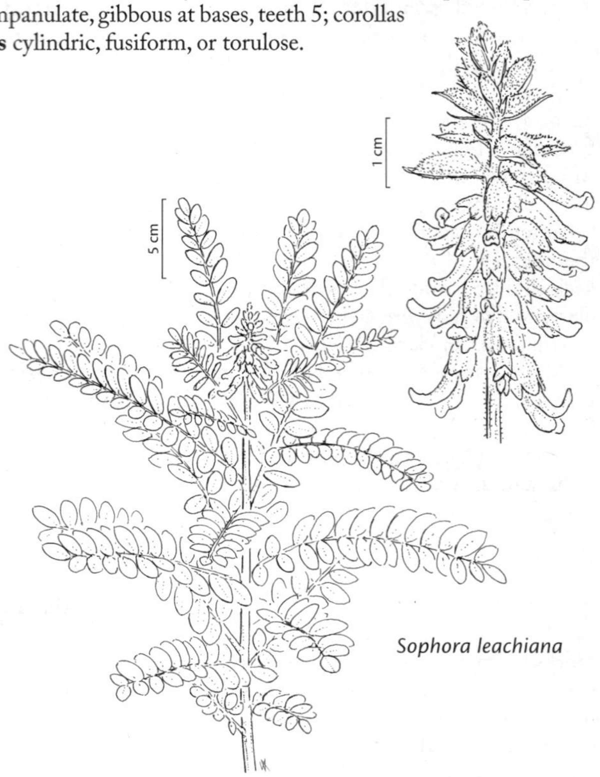

SOPHORA

SPARTIUM Spanish broom

Richard R. Halse

Genus composed of one species, *S. junceum*.

Spartium junceum Linnaeus (p. 717)

Spanish broom

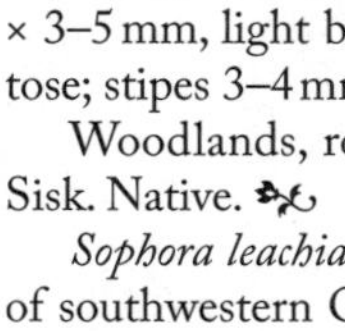

Shrubs 1–4 m. **Stems** erect, usually leafless at flowering, rush-like, cylindrical, glaucous green, striate, glabrous, branches many. **Leaves** deciduous, alternate to subopposite, simple or lowermost trifoliate, oblong-linear to lanceolate, 10–30 × 2–5 mm, tips acute to obtuse, surfaces abaxially appressed-sericeous, adaxially glabrous; subsessile. **Inflorescences** open, terminal racemes, 5–20-flowered; bracts present, pedicels 1–6 mm, bracteoles present. **Flowers** papilionaceous; calyces deeply split above, with 5-toothed lip below; corollas fragrant, 20–25 mm, bright yellow, keels ciliate along lower edges; stamens 10, monadelphous, anthers dimorphic. **Fruits** dehiscent, spreading-ascending, linear-oblong, flattened, 30–100 × 5–9 mm, dark brown to black, sericeous, becoming glabrous. **Seeds** 4–18.

Roadsides, disturbed areas. Flowering May–Jul. 50–600 m. Casc, Sisk, WV. CA, WA; TX, HI; Mediterranean region. Exotic.

Spanish broom is closely related to other brooms in the genera *Cytisus* and *Genista*. It was introduced into North America as an ornamental. In Oregon, it is a B-listed weed; Washington lists it as a Class A noxious weed; California also considers it a noxious weed.

SPHAEROPHYSA Austrian peaweed

Richard R. Halse

Herbs perennial. **Stems** erect, strigillose. **Leaves** alternate, odd-pinnately compound; stipules small. **Inflorescences** axillary racemes; bracts present; bracteoles present. **Flowers** papilionaceous; calyces campanulate, 5-lobed, lobes subequal, < tubes; petals 5, reddish, rarely blue; stamens 10, diadelphous; pistils stipitate, incurved, ovaries 1-chambered, styles adaxially pubescent. **Fruits** unilocular, indehiscent, inflated; valves papery, membranous, semi-transparent. **Seeds** reniform. **2*n***=16.

Central and eastern Asia, western North America. 2 species; 1 treated in *Flora.*

Sphaerophysa salsula (Pallus) A.P. de Candolle (p. 717)

Austrian peaweed, alkali swainsonpea, swainsona

Swainsona salsula (Pall.) Taub.

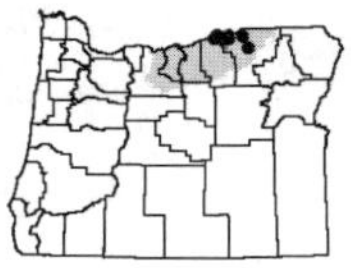

Plants green to greenish gray, strigillose; roots creeping, horizontal, woody. **Stems** simple or sparsely branched, 4–15 dm. **Leaves** 3–11 cm; leaflets (9)15–25, oblong-obovate to narrowly oblong to narrowly elliptic, 3–18 × 3–9 mm, tips retuse to obtuse, apiculate, or acute; petioles 3–20 mm. **Pedicels** persistent, 2.5–8 mm, recurved in fruit. **Inflorescences** 5–17-flowered, ascending; peduncles 3–9 cm in flower, 5–18 cm in fruit. **Flowers** calyces 5–6 mm; corollas 10–15 mm, dull brick or orange-red, pinkish brown, drying lavender or brownish, wings ≥ keels; ovaries densely strigillose. **Fruits** opening by weathering or decay, bladdery inflated, ovoid or globose, 1.3–3.5 cm, green to purplish-tinged, strigillose to glabrate, sulcate below; stipes 4–8(12) mm. **Seeds** ~40, brown to greenish or reddish brown, smooth, dull.

Disturbed areas, roadsides, sagebrush, salt flats. Flowering Jun–Aug. 50–400 m. Col. CA, ID, WA; northeast to Saskatchewan, east to WY, southeast to TX; Asia. Exotic.

Sphaerophysa salsula is a state-listed noxious weed in California, Nevada, Oregon, and Washington. Plants are generally considered to be halophytes, occurring in saline, sandy, river, or lakeside soils.

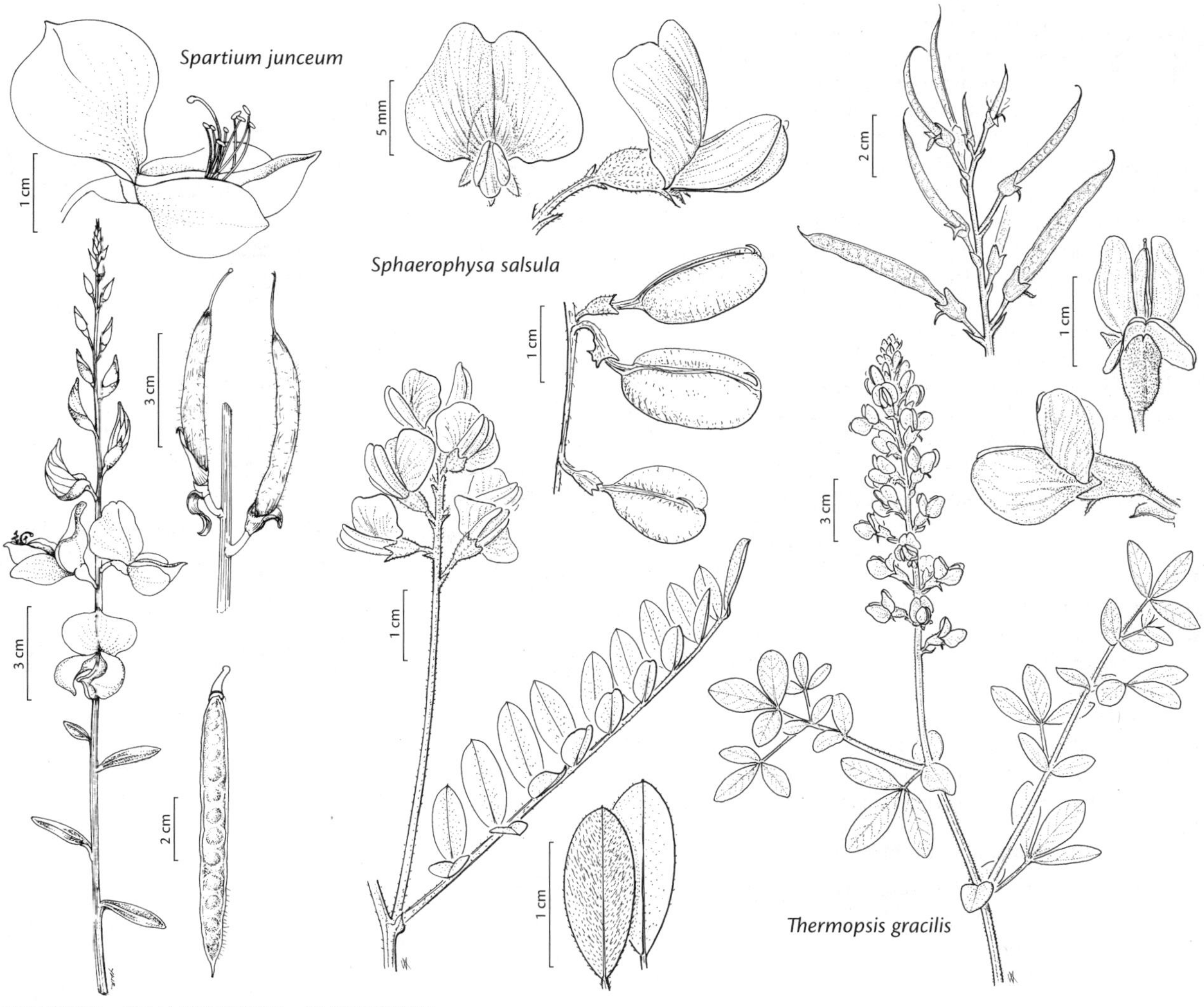

SPARTIUM • SPHAEROPHYSA • THERMOPSIS

THERMOPSIS Golden pea, goldenbanner, thermopsis

Rory Nichols

Herbs perennial; rhizomatous or with woody roots. **Stems** spreading or erect, branches few to many. **Leaves** cauline, palmately compound; leaflets 3, elliptic to rhombic, surfaces glabrous to densely pubescent; stipules leaf-like. **Inflorescences** racemes, 5–25(40)-flowered. **Flowers** 1–3 per node, 16–25 mm; calyx lobes 5, upper 2 ± fused; corollas bright yellow; stamens 10, distinct. **Fruits** dehiscent, spreading to erect, ± straight, flat, glabrous to tomentose. **Seeds** 7–15.

Central and east Asia, temperate North America. ~23 species; 2 species treated in *Flora*.

Some plants, particularly in Curry County, are very hairy throughout; this may represent introgression between *T. gracilis* and *T. californica*. *Thermopsis californica* appears hairy at arm's length and has 2–5 flowers per node with ascending fruits. It is reasonable to speculate that the northern extent of *T. californica* may reach into southwestern Oregon; however, little evidence exists. More study is needed.

Key to *Thermopsis* species:

1. Fruits spreading, occasionally nearly ascending; 1–2 flowers per node. . . . . ***T. gracilis***

1' Fruits erect; 2–3 flowers per node. . . . . ***T. montana***

Thermopsis gracilis Howell (p. 717)

Slender goldenbanner, slender thermopsis

Thermopsis macrophylla Hook. & Arn., *Thermopsis macrophylla* Hook. & Arn. var. *venosa* (Eastw.) Isely, *Thermopsis montana* Nutt. var. *venosa* (Eastw.) Jeps., *Thermopsis subglabra* L.F. Hend.

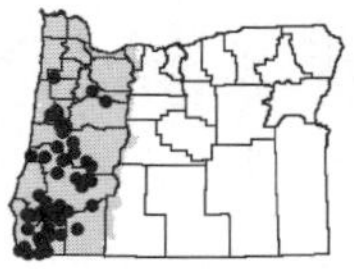

Plants to 10 dm, glabrous to sparsely pubescent. **Stems** branches many. **Leaflets** elliptic to obovate or rhombic, 28–80 × 15–40 mm, mostly < 4 × as long as wide; surfaces abaxially glabrous to sparsely appressed-pubescent, adaxially glabrous. **Inflorescences** bracts elliptic or ovate. **Flowers** 1–2 per node. **Fruits** spreading, occasionally nearly ascending, straight, sometimes slightly curved, 35–65 mm, glabrous to appressed-pubescent. **2*n***=36.

Roadsides, open forests, meadows; moist or dry soil. Flowering Apr–Jul. 0–1800 m. Casc, CR, Est, Sisk, WV. CA. Native.

In western Oregon, the ranges of *T. gracilis* and *T. montana* var. *ovata* overlap; they are difficult to tell apart without mature pods. *Thermopsis gracilis* is more common and typically has ovate floral bracts versus elliptic to more commonly narrowly elliptic in *T. montana* var. *ovata*.

Thermopsis montana Nuttall

Mountain goldenbanner, mountain golden pea

Plants to 9(10) dm, ± pubescent. **Stems** branches few to several. **Leaflets** narrowly elliptic to obovate, 25–85 × 7–35 mm, surfaces glabrate to pubescent. **Inflorescences** bracts narrowly elliptic to widely ovate. **Flowers** 2–3 per node. **Fruits** erect, straight, rarely slightly curved, appressed-pubescent to tomentose. **2*n***=18.

2 varieties.

Some plants from southeastern Oregon cannot be assigned to either variety, suggesting some intergradation occurs.

Key to *Thermopsis montana* varieties:

1. Leaflets narrow, 4–6 × as long as wide; leaflets 7–12 mm wide. . . . . ***T. m.*** var. ***montana***

1' Leaflets wide, mostly < 4 × as long as wide; leaflets 15–35 mm wide. . . . . ***T. m.*** var. ***ovata***

Thermopsis montana Nuttall var. ***montana***

Mountain goldenbanner, mountain golden pea

Thermopsis rhombifolia (Nutt. ex Pursh) Nutt. ex Richardson var. *montana* (Nutt.) Isely

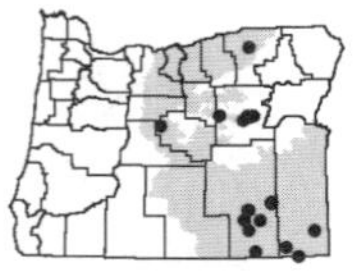

Plants 3.5–9 dm. **Leaflets** narrowly elliptic to sometimes narrowly oblanceolate, 25–60 × 7–12 mm, 4–6 × as long as wide; surfaces abaxially densely appressed-pubescent, adaxially glabrate to appressed-pubescent. **Inflorescences** bracts elliptic to widely ovate. **Fruits** 47–72 mm, tomentose.

Streambanks, wet meadows, open, moist areas. Flowering Apr–Jul. 300–2300 m. BR, Col, Lava, Owy. ID, NV, WA; northeast to MT, east to WY, southeast to NM. Native.

Thermopsis montana var. *montana* is our most distinct golden pea with its narrowly elliptic leaflets and overall slender form.

Thermopsis montana Nuttall var. ***ovata*** (B.L. Rob. ex Piper) Isely

Common thermopsis, slender goldenbanner

Thermopsis rhombifolia (Nutt. ex Pursh) Nutt. ex Richardson var. *ovata* (B.L. Rob. ex Piper) Isely

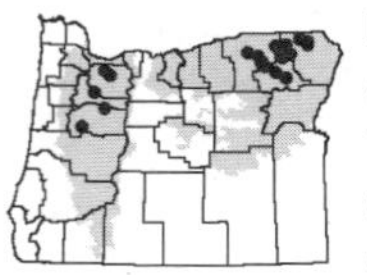

Plants 4–7(10) dm. **Leaflets** widely elliptic or widely oblanceolate to obovate, 45–85 × 15–35 mm, mostly < 4 × as long as wide; surfaces abaxially sparsely appressed-pubescent, adaxially generally glabrous. **Inflorescences** bracts narrowly elliptic, sometimes elliptic. **Fruits** 48–80 mm, appressed-pubescent.

Streambanks, rocky, sandy areas, open conifer forests. Flowering May–Jul. 100–1700 m. BW, Casc, Col, WV. ID, WA; northeast to MT. Native.

TRIFOLIUM Clover

Michael A. Vincent

Herbs annual or perennial. **Stems** prostrate to erect, glabrous or pubescent. **Leaves** alternate, odd-pinnate or palmate; leaflets 3(5–9), margins entire or toothed, surfaces glabrous or pubescent; petioles present or absent; stipules present, adnate to petioles. **Inflorescences** axillary or terminal racemes, umbels, spikes, or heads; involucral bracts present or absent, fused or distinct; bracteoles present or absent. **Pedicels** present or absent. **Flowers** usually fertile; papilionaceous; sepals fused, calyces radially or bilaterally symmetric, tubular or campanulate, sometimes inflated in fruit, lobes 5; petals 5, persistent after anthesis, free; stamens 10, diadelphous. **Fruits** legumes, cylindrical or laterally compressed, dehiscent or indehiscent; sessile or stipitate. **Seeds** globose, ovoid-oblong, mitten-shaped, or reniform.

Africa, Asia, Central America, Europe, North America, South America. ~280 species; 48 species treated in *Flora*.

Trifolium differs from *Medicago*, *Melilotus*, and *Trigonella* by persistence of the corolla after anthesis, a legume that is included in or only slightly exserted from the persistent corolla, and filaments that are dilated distally (Wei & Vincent 2010; Gleason & Cronquist 1991).

Trifolium has a major center of diversity in the California region in North America and another in the eastern African highlands (Zohary & Heller 1984). *Trifolium* consists of two subgenera, subg. *Chronosemium* (represented in *Flora* by *T. aureum*, *T. campestre*, and *T. dubium*) and subg. *Trifolium* (all remaining species in *Flora*) (Ellison et al. 2006).

Species of *Trifolium* have been cultivated for centuries in parts of Europe as forage crops, and several species were introduced into the flora area for that purpose, including *T. hirtum*, *T. hybridum*, *T. incarnatum*, *T. pratense*, *T. repens*, *T. subterraneum*, and *T. vesiculosum*.

Trifolium retusum (teasel clover), which is native to Eurasia, is a very recent introduction in the western United States; it is known from only a few counties in California, Oregon, and Washington.

Trifolium echinatum (prickly clover) and *T. tomentosum* (woolly clover) have been found only as waifs on ballast. *Trifolium gracilentum* (pinpoint clover) is known only from three historical collections from the 1870s and 1880s and is likely extirpated. *Trifolium andersonii* (five-leaf clover) was reported for Oregon by Peck (1961) and Barneby (1989, as var. *andersonii*), but no specimen verifying its presence in the state has been located, even though it is documented within 5 miles of the state. In *T. andersonii*, the leaflet margins are entire or very slightly dentate distally, the leaflet adaxial surfaces are densely pubescent, and the calyx lobes are subulate and plumose.

Key to *Trifolium* species:

1. Inflorescences involucrate; involucral bracts ± connate, variously incised or dentate; involucres > pedicels, sometimes > calyces.
    2. Plants perennial.
        3. Abaxial calyx lobes ≈ tube, rarely longer; involucres lobed ⅓–½ their length, lobes sharply and acutely serrate; leaflets obovate to elliptic; roots not tuberous . . . . ***T. wormskioldii***
        3' Abaxial calyx lobes > tube, other lobes often also as long; involucres deeply lobed ≥ ⅔ their length, lobes bifurcate or irregular; leaflets oblanceolate, obovate, or elliptic; roots tuberous . . . . ***T. siskiyouense***
    2' Plants annual.
        4. Calyces and/or banners inflated in fruit.
            5. Banners inflated entire length in fruit, not distally twisted; flowers 10–27 mm; involucral bract tips acuminate, sometimes 2- or 3-fid . . . . ***T. fucatum***
            5' Banners proximally inflated in fruit, distally narrowed into twisted tip; flowers 4–15 mm; involucral bract tips acute, toothed, not 2- or 3-fid.
                6. Tips of calyx lobes unbranched, setaceous; calyces > corollas . . . . ***T. barbigerum***
                6' Tips of at least some calyx lobes branched, flattened or setaceous; calyces ≤ corollas . . . . ***T. cyathiferum***
        4' Neither calyces nor corollas inflated in fruit.
            7. Involucres flattened or cup- or bowl-shaped, when flattened laterally, nearly completely hiding calyces.
                8. Involucral lobes entire or slightly toothed proximally; involucres conspicuously villous; corollas ≈ calyces . . . . ***T. microcephalum***
                8' Involucral lobes conspicuously toothed; involucres glabrous or inconspicuously hairy; corollas > calyces . . . . ***T. microdon***
            7' Involucres flattened to bowl-shaped, when flattened laterally, not hiding flowers except proximally.
                9. Calyx tubes not slit between adaxial lobes; leaflets elliptic, obcordate, obovate, or oblanceolate.
                    10. Banner petals 6.5–16 mm, exserted 1.5–8.4 mm beyond tips of calyx lobes, keel petals beaked or not; mature seeds 1.9–2.8 mm . . . . ***T. appendiculatum***
                    10' Banner petals 3.2–9.7 mm, included in or exserted 0.1–5.7 mm beyond tips of calyx lobes, keel petals not beaked; mature seeds 1.1–1.6 mm. . . . . ***T. variegatum***

9' Calyx tubes slit between adaxial lobes; leaflets linear, oblong, lanceolate, oblanceolate, elliptic, rhombic, or obovate.

11. Flowers 11–16 mm; involucres 1–2 mm, divided 4/5–9/10 their length, lobes lanceolate-subulate; inflorescences 3–8 mm wide, 3–15-flowered. . . . . ***T. oliganthum***

11' Flowers 12–20 mm; involucres 3–9 mm, divided 1/4–1/3 their length, lobes deeply lacerate; inflorescences 10–30 mm wide, 10–50-flowered.

12. Peduncles and calyces glabrous; calyx lobes 3-fid or shouldered; leaflet margins serrate . . . . ***T. willdenovii***

12' Peduncles and calyces glandular; calyx lobes usually entire, rarely 3-fid or shouldered; leaflet margins coarsely spinulose-serrate . . . . ***T. obtusiflorum***

1' Inflorescences without involucres, or involucral bracts short, connate or distinct, and pedicels much > involucres, or calyces inflated in fruit, or distal-most stipules and/or leaves appearing to form involucres.

13. Calyces and/or corollas inflated in fruit.

14. Fruiting calyces symmetrical, veins 5 or 20–36; calyx lobes subequal or unequal.

15. Calyces inflated in fruit, but not corollas; calyx veins 20–36, lobes reflexed, subequal; tubes of fruiting calyces with transversely interconnecting veins; involucres absent; bracteoles present, nearly as long as calyces; plants 15–70 cm . . . . ***T. vesiculosum***

15' Corollas inflated in fruit, but not calyces; calyx veins 5, lobes not reflexed, unequal; tubes of fruiting calyces without transversely interconnecting veins; involucres vestigial, reduced to narrow ring; bracteoles absent; plants 1–30 cm. . . . . ***T. depauperatum*** var. ***depauperatum***

14' Fruiting calyces markedly asymmetric-bilabiate, veins 5–10; calyx lobes unequal.

16. Plants perennial, prostrate or creeping; flowers not resupinate; inflorescences distinctly involucrate by bracts of proximal flowers . . . . ***T. fragiferum***

16' Plants annual, prostrate, procumbent, ascending, or erect; flowers resupinate; inflorescences inconspicuously involucrate by a narrow rim . . . . ***T. resupinatum***

13' Neither calyces nor corollas inflated in fruit.

17. Corollas yellow; fruits stipitate, seeds 1–2.

18. Terminal leaflets sessile or subsessile; stipules 9–15 mm; banners emarginate . . . . ***T. aureum***

18' Terminal leaflets petiolulate; stipules 3–8 mm; banners rounded or broadly acute.

19. Stipules 3–5 mm; flowers 2.8–3.5 mm; leaflets obovate to elliptic-lanceolate; corollas not or slightly ribbed . . . . ***T. dubium***

19' Stipules 5–8 mm; flowers 4.3–5 mm; leaflets rhombic to obovate; corollas obviously ribbed. . . . ***T. campestre***

17' Corollas not yellow or greenish yellow; fruits sessile or stipitate, seeds 1–9.

20. Inflorescences with 2–8 fertile flowers and many sterile flowers without corollas; inflorescences forming burs, pushing into substrate in fruit . . . . ***T. subterraneum***

20' Inflorescences with all or most flowers fertile and with corollas; inflorescences not forming burs, not pushing into substrate in fruit.

21. Flowers sessile or subsessile; calyx orifices usually thickened, narrowly opening or closed by callosity or ring of hairs (open in *T. arvense*, *T. hirtum*, and *T. striatum*, slightly closed in *T. pratense*); fruits leathery distally, transversely dehiscent; seeds 1–2.

22. Calyx veins 20; inflorescences sessile, appearing involucrate, disarticulating at maturity into fruiting calyces. . . . . ***T. hirtum***

22' Calyx veins 10; inflorescences sessile or pedunculate, not involucrate (appearing so in *T. pratense* because of leaves and expanded stipules subtending inflorescence), disarticulating or not.

23. Plants perennial; inflorescences subtended by involucre-like leaves and expanded stipules . . . . ***T. pratense***

23' Plants usually annual, rarely biennial; inflorescences not subtended by involucre-like leaves and stipules (except *T. striatum* subtended by stipular pseudo-involucre).

24. Leaflets broadly ovate or obovate to oblong, < 1.5(2) × as long as wide.

25. Peduncles 20–100 mm; inflorescences 20–70 mm; stipules ovate, tips blunt or somewhat tapering distally . . . . ***T. incarnatum***

25' Peduncles 0–10 mm; inflorescences 6–10 mm; stipules lanceolate-ovate, tips acute-acuminate . . . . ***T. striatum***

24' Leaflets linear-lanceolate, or linear-oblong to narrowly elliptic, usually > (2.5)3 × as long as wide.

26. Inflorescences 30–80 mm; calyces 8–13 mm; calyx orifices closed by bilabiate callosities . . . ***T. angustifolium***

26' Inflorescences 10–23 mm; calyces 5–8 mm; calyx orifices open, hairy . . . . ***T. arvense***

21' Flowers usually pedicellate, sometimes sessile; calyx orifices open; fruits not leathery distally, longitudinally dehiscent; seeds 1–9.

27. Calyces glabrous throughout.

28. Plants perennial, creeping, rooting at nodes; peduncles erect, from prostrate stems; pedicels strongly reflexed in fruit; corollas 1–2 × as long as calyces . . . . ***T. repens***

28' Plants annual or perennial, usually erect, ascending, rarely prostrate, not rooting at nodes; peduncles erect or not, from prostrate or upright stems; pedicels reflexed or not; corollas < to > calyces.

29. Corollas ≥ 2 × as long as calyces.

30. Pedicels erect, straight, or slightly reflexed.

31. Plants annual, procumbent, decumbent, or ascending; inflorescences 8–10 mm wide, peduncles 0–1 mm . . . . ***T. glomeratum***

31' Plants perennial, erect; inflorescences 20–35 mm wide, peduncles 30–120 mm . . . ***T. douglasii*** (in part)

30' Pedicels strongly reflexed (at least in fruit).

32. Rachises not prolonged beyond flowers; inflorescences without sterile flowers . . . ***T. hybridum*** (in part)

32' Rachises extending beyond flowers; inflorescences often bearing a cluster of sterile flower buds distally.

33. Plants 30–100 cm, fistulose; leaflets 23–50 mm wide . . . . ***T. howellii***

33' Plants 5–50 cm, not fistulose; leaflets 3–20 mm wide.

34. Peduncles straight; inflorescences 30–70+-flowered; leaflets 25–50 mm, tips obtuse or retuse; calyces 4–6 mm, veins 5 . . . . ***T. beckwithii***

34' Peduncles slightly twisted distally; inflorescences 15–30-flowered; leaflets 5–20 mm, tips acute to acuminate; calyces 3–3.5 mm, veins 10, 5 sometimes faint. . . . . ***T. productum***

29' Corollas < 2 × as long as calyces.

35. Calyx lobe margins hyaline, markedly ciliate, dentate or pectinate . . . . ***T. ciliolatum*** (in part)

35' Calyx lobe margins green, purple, or pink, not ciliate, not dentate or pectinate.

36. Leaflet margins serrate distally or entire; peduncles 30–80 mm; flowers 6–8 mm; calyx lobes straight or slightly curved in fruit . . . . . . ***T. bifidum*** var. ***decipiens*** (in part)

36' Leaflet margins coarsely dentate; peduncles 6–15 mm; flowers 4–5.5 mm; calyx lobes markedly recurved in fruit. . . . . ***T. cernuum***

27' Calyces entirely or partly pubescent, sometimes sparsely so.

37. At least some leaves 5–9-foliolate.

38. Inflorescences very large, 25–80 × 30–70 mm; calyx lobes 2–4 × tube length; leaflets (5)7–9 . . ***T. macrocephalum***

38' Inflorescences smaller, 10–20 × 13–24 mm; calyx lobes ≤ 1.5 × tube length; leaflets 3–5. . . . . . . ***T. gymnocarpon***

37' All leaves 3-foliolate.

39. Pedicels, or at least distal, not reflexed in fruit, or absent.

40. Leaflets ≥ 3 × as long as wide.

41. Calyx veins 15–20; leaflets 30–110 mm; inflorescences 15–60 × 20–35 mm, 30–80-flowered.

42. Leaflets sometimes folded lengthwise, falcate; leaflets adaxially villous; inflorescences spicate, ovoid to cylindrical . . . . ***T. plumosum*** ssp. ***plumosum***

42' Leaflets flat, not falcate; leaflets adaxially glaucous or glabrous; inflorescences umbellate, globose to ovoid . . . . ***T. douglasii*** (in part)

41' Calyx veins 5–10; leaflets 8–60 mm; inflorescences 10–35 × 10–32 mm, 10–40-flowered . . . . ***T. longipes*** (in part)

40' Leaflets ≤ 2 × as long as wide.

43. Plants perennial, cespitose, stems little branched distally.

44. Stipules fused at bases, broadly obovate, margins slightly lobed; leaflets slightly overlapping, ovate, obovate, or orbiculate, tips emarginate . . . . ***T. owyheense***

44' Stipules distinct at bases, ovate to lanceolate, margins entire, sometimes slightly serrate proximally; leaflets not overlapping, linear, lanceolate to elliptic, or oblong, cauline sometimes ovate, tips acute . . . . ***T. longipes*** (in part)

43' Plants annual, not cespitose, stems branched along their length or from bases, or unbranched.
45. Inflorescences usually paired, sessile or subsessile, sometimes 1 head slightly stalked, partially hidden by terminal leaves and stipules . . . . ***T. macraei***
45' Inflorescences single, pedunculate, not hidden by terminal leaves or stipules.
46. Calyces campanulate, veins 10 (additional faint veins sometimes present), sinuses rounded, lobes divergent; corollas < calyces; bracteoles absent. . . . . ***T. albopurpureum***
46' Calyces tubular, veins 20–30, sinuses acute, lobes appearing rigidly erect; corollas < or > calyces; bracteoles cup-like. . . . . ***T. dichotomum***
39' Pedicels all reflexed in fruit.
47. Leaflets ≥ 3 × as long as wide.
48. Plants annual; leaflet tips rounded or truncate, shallowly to deeply retuse . . . . ***T. bifidum*** var. ***decipiens*** (in part)
48' Plants perennial; leaflet tips acute or acuminate, not usually retuse.
49. Peduncles bent or curved proximal to inflorescences, inflorescences inverted or horizontal; calyces curved in fruit, tubes 1.5–2.5 mm, lobes often strongly curved or twisted, plumose. . . . . ***T. eriocephalum***
49' Peduncles straight, inflorescences erect; calyces not curved in fruit, tubes 2.5–3 mm, lobes straight or lax, glabrous or hairy . . . . ***T. longipes*** (in part)
47' Leaflets usually ≤ 3 × as long as wide.
50. Plants annual; calyx lobe margins dentate or pectinate. . . . . ***T. ciliolatum*** (in part)
50' Plants perennial; calyx lobe margins entire.
51. Plants canescent; margins of stipules and leaflets coarsely spinulose-dentate . . . . ***T. leibergii***
51' Plants glabrous or pubescent, not canescent; margins of stipules and leaflets entire or toothed, not spinulose.
52. Rachises not prolonged beyond flowers; banners ovate-oblong, tips rounded-denticulate . . . . ***T. hybridum*** (in part)
52' Rachises sometimes prolonged beyond flowers; banners elliptic, obovate, ovate-oblong, or lanceolate-ovate, tips not denticulate.
53. Stipule tips long-acuminate, margins entire; corollas 6–10 mm; banners as long as wings and keel petals. . . . . ***T. breweri***
53' Stipule tips acuminate or acute, margins entire proximally, 1–3-toothed distally; corollas 12–15 mm; banners > wing and keel petals . . . . ***T. latifolium***

Trifolium albopurpureum Torrey & A. Gray
Rancheria clover

Trifolium albopurpureum Torr. & A. Gray var. *albopurpureum*, *Trifolium macraei* Hook. & Arn. var. *albopurpureum* (Torr. & A. Gray) Greene

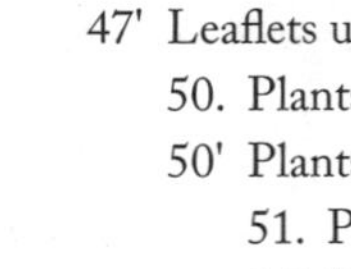

Plants annual, erect, 5–40 cm, pubescent; branched from bases. **Leaves** palmate; leaflets 3, obovate or elliptic, 5–20 × 2–10 mm, bases cuneate, margins serrate, veins obscure or slightly thickened, tips acute or obtuse, often retuse, surfaces pubescent; petioles 5–70 mm; petiolules 1 mm; stipules elliptic-ovate, 6–10 mm, margins entire, toothed, or remotely lobed, ciliate, tips acuminate. **Inflorescences** axillary or terminal, single, 25–50-flowered, ovoid-ellipsoid, 10–18 × 8–15 mm; involucres absent; bracteoles absent. **Peduncles** 5–100 mm. **Pedicels** absent. **Flowers** 6–9 mm; calyces campanulate, 6–8 mm, densely pubescent, veins 10 (additional faint veins sometimes present), tubes 1–2 mm, lobes narrowly triangular, subulate, unequal, sinuses rounded, lobes divergent, orifices open; corollas 4–6 mm, < calyces, white with purple tips, banners narrowly elliptic, 6–7 × 2 mm, tips acute, slightly erose. **Fruits** longitudinally dehiscent, broadly ellipsoid, 2.8–3.2 mm, ≥ calyces. **Seeds** 1(2), ellipsoid, 2–2.6 mm, yellow or brown, often red-mottled, smooth, glossy. **2*n***=16.

Grassy meadows and valleys. Flowering Apr–Jun. 0–1200 m. Est, Sisk, WV. CA, WA; southeast to AZ, south to Mexico. Native.

Trifolium albopurpureum is a widespread member of a closely allied group of species that includes *T. dichotomum*.

Trifolium angustifolium Linnaeus
Narrowleaf crimson clover

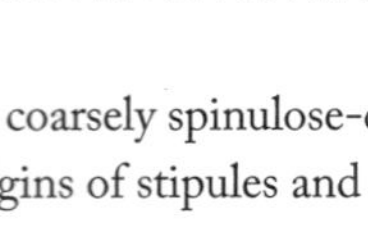

Plants annual, erect, 10–30 cm, appressed-pubescent; unbranched to sparsely branched. **Leaves** pinnate; leaflets 3, linear-lanceolate, 30–50 × 2–4 mm, bases cuneate, margins entire, veins fine, tips acute, surfaces hairy; petioles 20–30 mm; petiolules 1+ mm; stipules lanceolate to linear, 5–22 mm, margins entire, tips acute. **Inflorescences** terminal, 25–75-flowered, spiciform, cylindrical or conical, 30–80 × 15–20 mm; involucres absent; bracteoles absent. **Peduncles** 10–22 mm. **Pedicels** straight, 0.5 mm. **Flowers** 10–13 mm; calyces tubular, 8–13 mm, tuberculate with appressed, stiff hairs, veins 10, tubes 3–5 mm, lobes unequal, abaxial lobes subulate-setaceous, longest spreading stellate in fruit, orifices closed by bilabiate callosities, hairy; corollas 9–12 mm, banners ovate-elliptic, 9–12 × 2–2.5 mm, pale pink, purple, or white, tips acute. **Fruits** transversely dehiscent, ovoid, 2.2–2.5 mm, ≥ calyces, leathery distally. **Seeds** 1, ovoid, 1.6–2 mm, light brown or yellow, smooth, glossy. **2*n***=14, 16.

Disturbed areas. Flowering May–Jul. 0–700 m. CR, Est, Sisk, WV. CA; southeastern US; worldwide. Exotic.

Trifolium angustifolium is native to Eurasia and Africa and is widely distributed as an agricultural weed in many regions globally (Randall 2002).

Trifolium appendiculatum Lojacono

Large white-tip clover

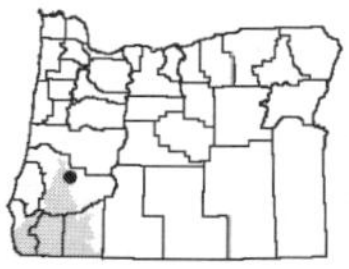

Plants annual, ascending, 7–50 cm, glabrous; branched. **Leaves** palmate, dimorphic with gradual transition between proximal and distal leaves; leaflets 3, obovate-obcordate to elliptic-obovate, 3–26 × 1–11 mm, bases cuneate, margins serrate, veins fine, tips emarginate-mucronate or rounded-mucronate, surfaces glabrous; petioles 2–73 mm; petiolules 0.3–0.9 mm; stipules broadly triangular, 3–13 mm, margins laciniate, tips acute. **Inflorescences** axillary, 4–25-flowered, subglobose, 8–31 mm, not elongating in fruit; involucres present, flat, laciniate, 6–12 mm, 4–7-lobed, lobes toothed, longest tooth 1.1–3 mm; bracteoles absent. **Peduncles** 17–73 mm. **Pedicels** straight 0.3–0.5 mm. **Flowers** 6.5–16 mm; calyces tubular, 3.5–7.4 mm, glabrous, veins 10–20, tubes 1.5–3.3 mm, lobes awl-like to shouldered, subequal, tips and shoulders purple, orifices open, glabrous; corollas 6.5–16 mm, exserted beyond tips of calyces 1.5–8.4 mm, deep purple, white-tipped, banners obovate, 6.5–16 × 1.5–4 mm, tips emarginate, keel petals beaked or not. **Fruits** laterally compressed, ellipsoid, 3.6–3.8 mm, < calyces, chartaceous. **Seeds** 2, globular to mitten-shaped, 1.9–2.8 mm, blackish brown, black-speckled.

Meadows. Flowering May–Jun. 200–2500 m. Sisk. CA. Native.

See also *T. variegatum* discussion.

Trifolium arvense Linnaeus

Rabbitfoot clover

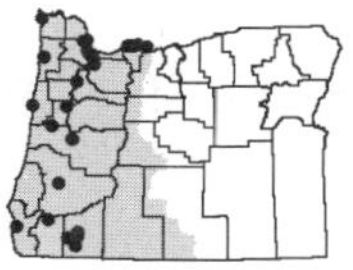

Plants annual, erect, 5–30 cm, appressed-villous; branched distally. **Leaves** pinnate; leaflets 3, linear-oblong to narrowly elliptic, 10–20 × 2–5 mm, bases cuneate, margins denticulate distally, veins fine, tips mucronate, surfaces hairy; petioles 4–45 mm; petiolules 1+ mm; stipules linear-lanceolate to ovate-oblong, 5–8 mm, margins entire, tips filiform. **Inflorescences** axillary or terminal, 30–100-flowered, ovoid or cylindrical, 10–23 × 9–12 mm, elongate in fruit; involucres absent; bracteoles obtuse, ≤ 0.1 mm. **Peduncles** 5–25 mm. **Pedicels** straight, ≤ 0.5 mm. **Flowers** 5–8 mm; calyces campanulate, 5–8 mm, villous, veins 10, tubes 1.2–1.8 mm, lobes setacous, subequal, purple or pink, orifices open, hairy; corollas 3–6 mm, much < calyx, white to pink, banners narrowly ovate-elliptic, 5–5.5 × 1–1.5 mm, tips obtuse. **Fruits** transversely dehiscent, ovoid, 1.5–2 mm, ≥ calyces, leathery distally. **Seeds** 1, globose, ~1 mm, yellow, smooth, glossy. **2*n***=14.

Waste places, roadsides, fields. Flowering May–Jul. 0–1300 m. Casc, CR, ECas, Est, Sisk, WV. CA, WA; throughout North America, except rarely in Great Basin and Great Plains; worldwide. Exotic.

Trifolium arvense is native to Eurasia and northern Africa and is widely distributed worldwide as a weed. It is listed among cultivated clovers by Hermann (1953), though Gillett (1985) expresses doubt that the species has been cultivated.

Trifolium aureum Pollich

Golden hop clover, golden clover, yellow clover

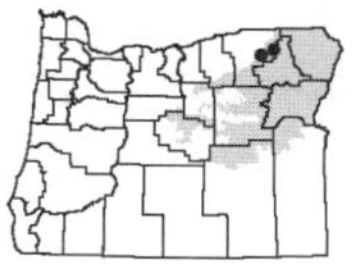

Plants annual, erect, 20–60 cm, glabrous or hirsute; unbranched or profusely branched. **Leaves** palmate; leaflets 3, obovate to elliptic-lanceolate, 22–28 × 6–8 mm, bases cuneate, margins denticulate, lateral veins prominent, ± parallel, tips acute, truncate, or retuse, mucronate, surfaces glabrous or with scattered hairs; petioles 8–10 mm; petiolules ≤ 1 mm; stipules oblanceolate, 9–15 mm, margins entire, tips acuminate or cuspidate. **Inflorescences** axillary or terminal, 20–80-flowered, usually ovoid-cylindrical, sometimes globose, 12–20 × ~13 mm; involucres absent; bracteoles reduced to fringe of red setae. **Peduncles** 20–50 mm. **Pedicels** reflexed, 0.3–0.6 mm. **Flowers** 7–8 mm; calyces campanulate, 2–2.5 mm, glabrous, veins 5, tubes 0.8–1 mm, lobes deltate, unequal, 0.6–0.9 mm, abaxial linear, adaxial reduced, ½ length of abaxial, orifices open; corollas 5–6 mm, golden yellow becoming brown, banners persistent, obovate or boat-shaped, 4–4.5 × 2.4–2.6 mm, tips emarginate. **Fruits** oblong, 3–3.5 mm, ≥ calyces; stipitate. **Seeds** 1, ovoid, 1–1.2 mm, pale yellow-green, smooth, glossy. **2*n***=14, 16.

Roadsides, railroads, fields, disturbed woodland glades. Flowering Jun–Jul. 1000–1500 m. BW. CA, ID, WA; eastern and western North America; worldwide. Exotic.

Trifolium agrarium is a confused name that has been historically used for *T. aureum* (Dandy 1958) but has been rejected under Article 56 of the *International Code of Nomenclature for algae, fungi, and plants* (Turland et al. 1996). Native in Eurasia, *T. aureum* is weedy worldwide.

Trifolium barbigerum Torrey

Bearded clover

Trifolium barbigerum Torr. var. *barbigerum*

Plants annual, decumbent or erect, 7–20 cm, puberulent or glabrous; branched. **Leaves** palmate; leaflets 3, oblanceolate or obovate, 15–25 × 4–7 mm, bases cuneate, margins serrate to ± entire, veins fine, thickened distally, tips rounded or retuse, surfaces pubescent or glabrous; petioles 10–100 mm; petiolules ≤ 0.5 mm; stipules ovate-lanceolate, 5–11 mm, margins toothed or lacerate, tips acute-acuminate. **Inflorescences** axillary or terminal, 5–20-flowered, subglobose or globose, 6–15 × 10–15 mm; involucres bowl-shaped, bracts ± connate, 5–15 mm, lobes 6–10, shallow, sharply setaceous-toothed, tips acute; bracteoles absent. **Peduncles** 50–100 mm. **Pedicels** straight, 0.2 mm. **Flowers** 4–6 mm; calyces campanulate, 5–9 mm, > corollas, usually pubescent, rarely glabrous, veins 5, lobes ± equal, often exceeding banner, tips unbranched, setaceous, tubes 2–3 mm, orifices open; corollas 5–8 mm, usually lavender to purple, rarely white, usually with white tips, banners broadly oblong, proximally inflated in fruit, distally narrowed into twisted tip, 6–8 × 6–8 mm, tips broadly acute. **Fruits** ovoid-ellipsoid, 2–3 mm. **Seeds** 1–2, ellipsoid to subglobose, 1–1.5 mm, brown, slightly roughened. **2*n***=16.

Vernal pools, streambanks, meadows, lawns. Flowering May–Jul. 0–100 m. Est. CA. Native.

Trifolium barbigerum is morphologically similar to the Chilean species *T. physanthum*, to which it is a sister species in phylogenetic studies (Ellison et al. 2006).

Trifolium beckwithii Brewer

Beckwith's clover

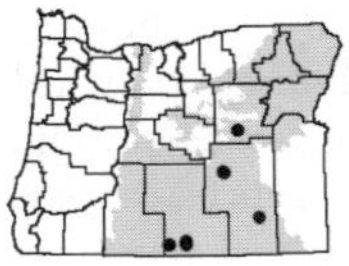

Plants perennial, erect or ascending, 5–50 cm, glabrous; unbranched or clumped from crown. **Leaves** palmate; leaflets 3, usually rhombic, oblong, or elliptic, sometimes ovate, 25–50 × 6–20 mm, bases cuneate, margins setose-serrulate, veins slightly thickened, tips obtuse or retuse, surfaces glabrous, abaxially glaucous; petioles 5–200 mm; petiolules 1 mm; stipules narrowly ovate, lanceolate, or linear-lanceolate, 10–20 mm, margins entire or toothed, tips acute. **Inflorescences** terminal or axillary, 30–70+-flowered, globose or ovoid becoming subglobose, 15–40 × 18–33 mm; rachises extending beyond flowers, undivided or forked, often bearing cluster of sterile flower buds distally; involucres absent; bracteoles minute. **Peduncles** straight, thick, 50–250 mm. **Pedicels** strongly reflexed in fruit, 1–1.5 mm. **Flowers** sweetly fragrant, 11–14 mm; calyces campanulate, gibbous, 4–6 mm, glabrous, veins 5, tubes 2–3 mm, lobes narrowly triangular or subulate, unequal, orifices open; corollas 10–13 mm, light purplish, sometimes with pink tips, banners straight, obovate to broadly elliptic, 13–18 × 6–8 mm, tips rounded or retuse, apiculate. **Fruits** oblong, ~5 mm, ≥ calyces. **Seeds** 2–4, flattened globose, 1.5–2 mm, reddish brown, smooth. ***2n***=48.

Moist grassy meadows along streams. Flowering Jun–Jul. 1200–1800 m. BR, BW, ECas. CA, ID, NV, WA; northeast to MT, southeast to UT, SD. Native.

Disjunct populations of *T. beckwithii* in South Dakota are over 1200 km east of the nearest populations in Montana and appear to have been long-distance introductions from populations in northern California, based on molecular data (Duvall et al. 1999).

Trifolium bifidum A. Gray var. ***decipiens*** Greene

Pinole clover, notchleaf clover

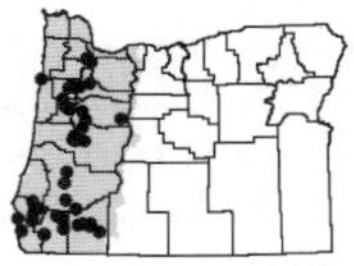

Plants annual, erect, 5–55 cm, sparingly pubescent or glabrous; branched. **Leaves** palmate; leaflets 3, narrowly obcordate, obovate, oblanceolate, or linear, 10–25 × 3–7 mm, bases cuneate, margins serrate distally or entire, veins moderately thickened, tips rounded or truncate, shallowly to deeply retuse, surfaces glabrous, sometimes hairy abaxially along midvein; petioles 10–70 mm; petiolules ≤ 0.5 mm; stipules ovate to lanceolate, 8–15 mm, margins entire or slightly serrate, tips acicular. **Inflorescences** axillary or terminal, 5–30-flowered, globose becoming subglobose, 8–15 × 8–15 mm; rachises prolonged beyond flowers; involucres with very narrow rim, ≤ 0.5 mm; bracteoles broadly triangular, membranous, ≤ 0.5 mm. **Peduncles** 30–80 mm. **Pedicels** reflexed in fruit, 1–3 mm. **Flowers** 6–8 mm; calyces campanulate, 3–3.6 mm, slightly hairy or glabrous, veins 10, tubes 1–1.5 mm, lobes subulate, straight or slightly curved in fruit, unequal, margins green or purple, orifices open; corollas 5–7 mm, < 2× as long as calyces, pink or purple, banners elliptic to oblong, 5–7 × 3–4 mm, tips rounded, apiculate. **Fruits** longitudinally dehiscent, obovoid, 3–4 mm, ≥ calyces; stipitate. **Seeds** 1–2, oblong, 2–3 mm, brown, mottled, smooth. ***2n***=16.

Open woodlands, fields, roadsides, slopes, streambanks, meadows. Flowering May–Jun. 50–1200 m. Casc, CR, Est, Sisk, WV. CA, WA; north to British Columbia, south to Mexico. Native.

Trifolium bifidum consists of two somewhat poorly differentiated varieties that may co-occur in some populations and may intergrade; it is possible that these varieties should not be recognized as distinct.

Trifolium breweri S. Watson

Brewer's clover

Plants perennial, erect or ascending, 5–50 cm, glaucous, sparsely appressed-pubescent; branched. **Leaves** pinnate; leaflets 3, obovate to obcordate, 3–15 × 3–12 mm, bases cuneate, margins dentate, veins thickened, recurved, tips obtuse or emarginate, surfaces abaxially sparsely hairy, adaxially glabrous; petioles 3–40 mm; petiolules ~0.5 mm; stipules lanceolate to narrowly ovate, 3–10 mm, margins entire, tips long-acuminate. **Inflorescences** axillary, 5–25-flowered, globose, 13–15 × 13–15 mm; rachises sometimes prolonged beyond flowers; involucres absent; bracteoles minute. **Peduncles** proximal to inflorescence, 30–50 mm, flowers appearing inverted. **Pedicels** strongly reflexed in fruit, 2–3 mm. **Flowers** 11–13 mm; calyces tubular, 6–8 mm, pilose or glabrous, veins 10, tubes 1.5–2 mm, lobes subulate, subequal, margins entire, orifices open; corollas 6–10 mm, rose-pink or creamy white, banners ovate-oblong, 6–10 × 3 mm, as long as wing and keel petals, tips rounded or emarginate. **Fruits** longitudinally dehiscent, ovoid-ellipsoid, 7–10 mm, ≥ calyces; short-stipitate. **Seeds** 1–2, ovoid-ellipsoid, 2.5–3 mm, brown, smooth. ***2n***=16.

Wooded areas, roadsides, mixed evergreen or pine forests. Flowering Jun–Jul. 500–1000 m. Sisk. CA; south to Mexico. Native.

Trifolium breweri was considered by Martin (1943) as a remarkably morphologically uniform species; it was considered by Clements (1920) as a foundational species in the Sierran Montane Forest *Pinus* Association climax formation.

Trifolium campestre Schreber

Hop clover

Trifolium procumbens L.

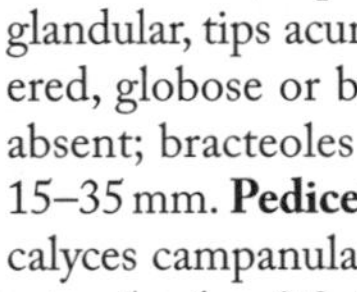

Plants annual, erect to ascending, 5–40 cm, slightly villous or glabrous; branched. **Leaves** pinnate; leaflets 3, rhombic to obovate, 4–15 × 4–8 mm, bases cuneate, margins denticulate distally, lateral veins prominent, ± parallel, ascending, tips truncate or retuse, surfaces adaxially glabrate to strigose; petioles ≤ 1–1.5 times leaflet length; petiolules 0.5–0.6 mm in lateral leaflets, 3–5 mm in terminal leaflets; stipules ovate-lanceolate, 5–8 mm, margins entire, glandular, tips acuminate. **Inflorescences** axillary, 30–60-flowered, globose or broadly ovoid, 8–15 × 7–10 mm; involucres absent; bracteoles reduced to fringe of red setae. **Peduncles** 15–35 mm. **Pedicels** reflexed, 0.7–1.2 mm. **Flowers** 4.3–5 mm; calyces campanulate, strongly bilabiate, 1.3–1.5 mm, glabrous, veins 5, tubes 0.3–0.6 mm, lobes unequal, adaxial < abaxial and deltate, lateral and abaxial linear subulate, each tipped with 1–2 stiff hairs, orifices open; corollas 3.5–6 mm, yellow becoming brown, obviously ribbed, banners persistent, enveloping other petals, boat-shaped, obovate, 4.3–5 × 2.8–3.2 mm, tips broadly acute. **Fruits** oblong, 2–2.5 mm, ≥ calyces; stipitate. **Seeds** 1–2, ellipsoid, 1–1.5 mm, yellow, smooth, lustrous. ***2n***=14.

Roadsides, fields. Flowering May–Jul. 0–1900 m. BR, BW, Casc, Col, CR, WV. CA, ID, WA; throughout most of North America; worldwide. Exotic.

Trifolium campestre is native to Eurasia and northern Africa and is found worldwide as a weed. *Trifolium procumbens* is a name that has been historically used for *T. campestre* (Dandy 1958) but has been rejected under Article 56 of the *International Code of Nomenclature for Algae, Fungi, and Plants* (Turland et al. 1996).

Trifolium cernuum Brotero

Nodding clover

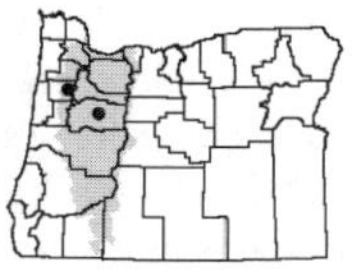

Plants annual, prostrate, ascending or erect, 5–40 cm, glabrous; branched. **Leaves** pinnate; leaflets 3, obovate or obcordate, 4–15 × 4–10 mm, bases cuneate, margins coarsely dentate, veins prominent, recurved, tips rounded, truncate, or emarginate, surfaces glabrous; petioles 10–100 mm; petiolules ≤ 1 mm; stipules triangular-lanceolate, 9–10 mm, margins entire, tips long-acuminate, reflexed. **Inflorescences** axillary or terminal, 8–20-flowered, depressed-globose, 9–11 × 8–10 mm; involucres absent; bracteoles lanceolate, ~1 mm. **Peduncles** 6–15 mm, becoming shorter distally. **Pedicels** reflexed in fruit, 0.5–2 mm. **Flowers** 4–5.5 mm; calyces tubular-campanulate, 4 mm, glabrous, veins 10, tubes 2–2.2 mm, lobes triangular-subulate, markedly recurved in fruit, subequal, margins green, pink, or purple, orifices open; corollas 4–5 mm, < 2× as long as calyces, pink, banners obovate, 4–5 × 1–2 mm, tips deeply emarginate. **Fruits** longitudinally dehiscent, ovoid-ellipsoid, ~4 mm, ≥ calyces. **Seeds** 1–4, ovoid, 0.8–1 mm, yellow, minutely papillate. ***2n***=16.

Roadsides, lawns. Flowering May–Jun. 50–400 m. Casc, WV. CA; Africa, Asia. Exotic.

Trifolium cernuum, which is native to western Asia and northern Africa, is a relatively recent introduction. Other than a record from wool mill waste in North Carolina in 1932, the earliest records in the United States are from the 1990s in California.

Trifolium ciliolatum Bentham (p. 725)

Foothill clover

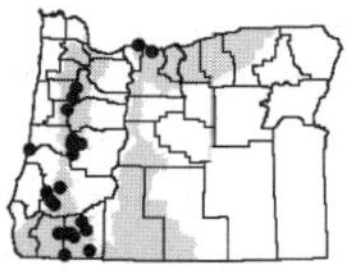

Plants annual, erect, 5–50 cm, glabrous or sparsely hairy; branched. **Leaves** palmate; leaflets 3, elliptic to oblong or obovate, 8–35 × 5–15 mm, bases cuneate, margins serrate proximally, obscurely denticulate distally, veins thickened, tips usually rounded or retuse, rarely acute, surfaces glabrous; petioles 10–130 mm; petiolules ~0.5 mm; stipules ovate-lanceolate, 10–15 mm, margins entire, sometimes ciliate, tips acuminate. **Inflorescences** axillary or terminal, 10–30-flowered, ovoid becoming subglobose, 7–22 × 5–20 mm; involucres a narrow, membranous, dentate rim, ~0.5 mm; bracteoles linear or cup-shaped, ≤ 1 mm. **Peduncles** 25–120 mm. **Pedicels** erect but becoming reflexed, 0.5–6 mm. **Flowers** 6–13 mm; calyces broadly campanulate, 5–11 mm, glabrous, veins 10, tubes 1–5 mm, lobes elliptic to linear, unequal, margins hyaline, markedly ciliate, dentate or pectinate, sinuses narrow, orifices open; corollas 5–13 mm, white, pink, or purple, banners broadly ovate, 6–13 × 4–7 mm, tips rounded, apiculate. **Fruits** longitudinally dehiscent, ovoid, 5–10 mm, < 2× as long as calyces; short-stipitate. **Seeds** 1–2, ovoid, 2.5–3 mm, brown, mottled, smooth. ***2n***=16.

Oak-pine chaparral, meadows, roadsides. Flowering May–Jun. 50–800 m. Col, ECas, Sisk, WV. CA, WA; south to Mexico. Native.

Trifolium cyathiferum Lindley (p. 725)

Cup clover, bowl clover

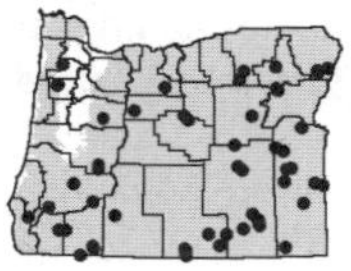

Plants annual, erect, 5–50 cm, glabrous; few to many branched. **Leaves** palmate; leaflets 3, obovate, elliptic, or oblanceolate, 5–25 × 3–15 mm, bases cuneate, margins finely toothed, veins prominent, tips rounded, blunt, or retuse, surfaces glabrous; petioles 10–100 mm; petiolules ≤ 0.5 mm; stipules ovate to lanceolate, 6–12 mm, margins entire, toothed, or lacerate, tips acute or aristate. **Inflorescences** axillary or terminal, 1–35-flowered, subglobose or ovoid, 5–18 × 5–20 mm; involucres broadly bowl-shaped, bracts ± connate,

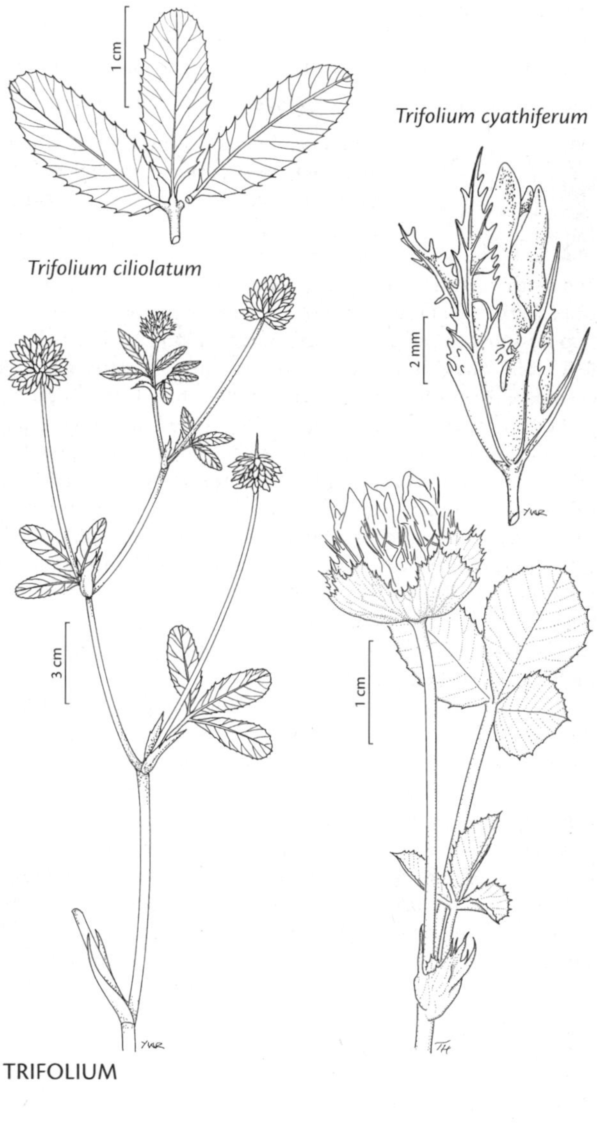

TRIFOLIUM

4–22 mm, shallowly lobed, lobes 3–15, toothed, tips broadly acute, veins prominent, green; bracteoles absent. **Peduncles** 10–80 mm. **Pedicels** straight, 0.1–0.2 mm. **Flowers** 7–15 mm; calyces campanulate, inflated in fruit, 6–13 mm, ≤ corollas, glabrous, veins 13–20, lobes strongly oblique, broadly triangular, unequal, tips setaceous, (adaxial) unbranched or (abaxial and lateral) conspicuously 2- or 3-fid, segments often flattened, glabrous, tubes 3–7 mm, orifices open; corollas 6–13 mm, usually creamy white or rose to pink, sometimes whitish with pinkish tips, banners obovate or elliptic, distally narrowed into twisted tip, proximally inflated in fruit, 6–13 × 3–6 mm, tips rounded to broadly acute. **Fruits** ellipsoid, 2.5–3 mm. **Seeds** 1–2, ovoid, 0.4–0.6 mm, yellow-brown, smooth. ***2n***=16.

Wet meadows, seeps, roadsides, fields. Flowering May–Aug. 100–2300 m. BR, BW, Casc, Col, CR, ECas, Lava, Owy, Sisk. CA, ID, NV, WA; north to AK, northeast to MT, southeast to UT. Native.

Trifolium cyathiferum may be of hybrid origin; phylogenetic studies indicate it has the same *trn*L intron sequence as *T. buckwestiorum*, *T. polyodon*, and *T. variegatum*, while *nr*DNA and combined analyses place the species in a clade with *T. barbigerum*, *T. fucatum*, *T. jokerstii*, and *T. physanthum*, all with inflated fruiting corollas (Ellison et al. 2006).

Trifolium depauperatum Desvaux var. ***depauperatum***

Poverty clover, cowbag clover, dwarf sack clover

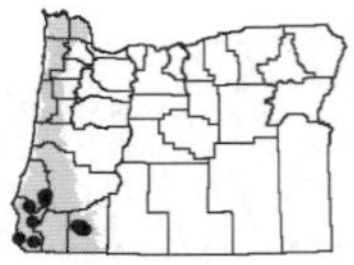

Plants annual, erect or decumbent, 1–30 cm, glabrous; branched. **Leaves** palmate; leaflets 3, lanceolate, linear, obovate, or oblanceolate, 4–25 × 1–7 mm, bases cuneate, margins entire, serrate, or pinnately lobed, veins moderately thickened abaxially, tips rounded, truncate, acute, or retuse, surfaces glabrous; petioles 10–50 mm; petiolules ≤ 1 mm; stipules sheathing proximally, ovate-oblong to lanceolate, 4–13 mm, margins entire or serrulate, tips short triangular-cuspidate. **Inflorescences** axillary or terminal, 3–15-flowered, globose or depressed-globose, 10–15 × 10–15 mm; involucres vestigial, reduced to a narrow ring; bracteoles absent. **Peduncles** 20–60 mm. **Pedicels** straight, 0–0.5 mm. **Flowers** 7–10 mm; calyces campanulate, symmetrical, 2–5 mm, glabrous, veins 5, tubes 1–2.5 mm, lobes triangular-lanceolate or subulate, unequal, orifices open; corollas inflated in fruit, 3–11 mm, white or pink, banners broadly ovate, 4–7 × 3–8 mm, tips rounded, broadly acute or retuse. **Fruits** ovoid, 3–4 mm, < calyces; ± sessile. **Seeds** 1–6, mitten-shaped, 1.2–1.6 mm, gray or tan, mottled, irregularly bumpy. ***2n***=16.

Meadows, grassy slopes, lawns. Flowering Apr–May. 100–900 m. CR, Sisk. CA, ID, WA; north to British Columbia, south to Mexico; northwestern South America. Native.

Two of the three varieties of *T. depauperatum* (var. *amplectens* and var. *truncatum*) are restricted to California and easily distinguished from var. *depauperatum* by their expanded involucres. In the phylogenetic treatment of Ellison et al. (2006), *Trifolium depauperatum* is not allied with *T. fucatum* and other clovers with inflated corollas but rather with *T. willdenovii*, *T. obtusiflorum*, and *T. trichocalyx* (though with weak support).

Trifolium dichotomum Hooker & Arnott

Branched Indian clover

Trifolium albopurpureum Torr. & A. Gray var. *dichotomum* (Hook. & Arn.) Isely, *Trifolium macraei* Hook. & Arn. var. *dichotomum* (Hook. & Arn.) Brewer ex S. Watson

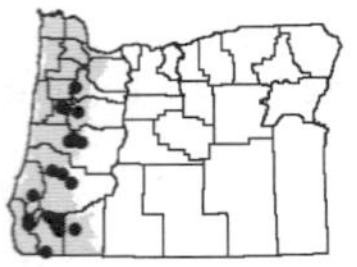

Plants annual, erect, 15–30 cm, canescent; branched from bases and distally, or unbranched. **Leaves** palmate; leaflets 3, obovate or oblanceolate, 5–15 × 2–11 mm, bases cuneate, margins dentate, veins obscure or slightly thickened, tips rounded or emarginate, surfaces pilose or glabrescent; petioles 5–65 mm; petiolules 0.5–0.7 mm; stipules ovate or elliptic, 3–10 mm, margins entire, ciliate, tips acuminate. **Inflorescences** terminal or axillary, single, 10–35-flowered, ovoid-ellipsoid, 11–22 × 10–15 mm; involucres absent; bracteoles cup-like, ≤ 0.5 mm. **Peduncles** 25–100 mm. **Pedicels** absent. **Flowers** 10–12 mm; calyces tubular, 4–8 mm, densely pubescent, veins 20–30, tubes 2–4 mm, lobes appearing rigidly erect, linear-setaceous, nearly equal, plumose, sinuses acute, orifices open; corollas 7–12 mm, red-violet to pink, with pink or white tips, banners ovate-oblong, 9–11 × 3–4 mm, tips broadly rounded, emarginate. **Fruits** longitudinally dehiscent, obovoid, 3–3.5 mm, ≥ calyces. **Seeds** 1, ellipsoid, 2–2.5 mm, yellow-brown, smooth. ***2n***=32.

Open slopes, serpentine soil, oak woodlands, disturbed areas. Flowering Apr–Jun. 0–1000 m. CR, Sisk, WV. CA, WA; north to British Columbia. Native.

Trifolium dichotomum reportedly is a tetraploid species (Wexelsen 1928), closely related to the diploid *T. albopurpureum*.

Trifolium douglasii House

Douglas' clover

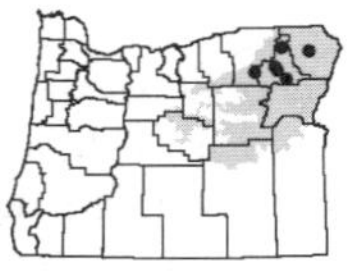

Plants perennial, erect, 20–75 cm, glabrous or slightly pilose; unbranched or sparsely branched. **Leaves** palmate; leaflets 3, elliptic-oblanceolate to linear-elliptic, 30–95 × 5–16 mm, flat, not falcate, bases cuneate, margins setose-denticulate to serrate, lateral veins prominent, recurved, dichotomously forked near margin, tips rounded or acute, surfaces glaucous, glabrous, sometimes abaxially sparsely pubescent; petioles 15–150 mm; petiolules ~1 mm; stipules lanceolate to ovate, 15–65 mm, margins setose-serrulate or entire, tips acuminate. **Inflorescences** terminal or axillary, 30–50+-flowered, umbellate, globose to ovoid, 15–50 × 20–35 mm; involucres absent; bracteoles minute. **Peduncles** 30–120 mm. **Pedicels** erect or slightly reflexed, ≤ 0.2 mm. **Flowers** 14–16 mm; calyces campanulate, 6–9 mm, glabrous or sparingly pilose, veins 15–20, tubes 1.5–3 mm, lobes narrow, triangular or subulate, unequal, abaxial lobes straight, lateral and adaxial curved downwards and inwards around corolla, tortuous, orifices open; corollas 12–16 mm, ≥ 2 × as long as calyces, magenta or pink, banners oblong, 12–16 × 4–6 mm, tips flared, acute or obtuse; ovaries glabrous or pubescent distally. **Fruits** longitudinally dehiscent, ovoid, 3–4.5 mm **Seeds** 1–2, ovoid, 1–1.5 mm, tan to brown, slightly roughened. ***2n***=16.

Moist meadows, rich soil in prairies, creek bottoms, streambanks, openings in pine forests. Flowering Jun–Jul. 800–1600 m. BW. ID, WA. Native.

Trifolium douglasii is rare throughout its range and has been impacted by agricultural practices (Camp et al. 2011).

Trifolium dubium Sibthorp

Little hop clover, suckling clover

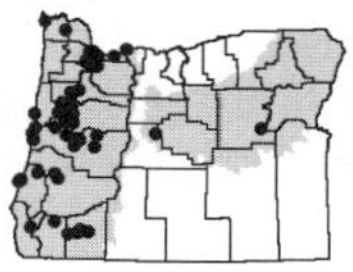

Plants annual, erect to prostrate, 20–40 cm, glabrous or sparingly hairy; branched from bases. **Leaves** pinnate; leaflets 3, obovate to elliptic-lanceolate, 10–20 × 5–15 mm, bases cuneate, margins dentate distally, lateral veins prominent, ± parallel, ascending, tips rounded or retuse, surfaces glabrous or with scattered hairs; petioles ≤ 15 mm; petiolules ≤ 0.5 mm in lateral leaflets, 1–1.5 mm in terminal leaflets; stipules ovate, 3–5 mm, margins entire, tips acute. **Inflorescences** axillary or terminal, 5–20-flowered, ovoid or globose, 5–9 × ~6 mm; involucres absent; bracteoles reduced to fringe of red setae. **Peduncles** 10–15 mm. **Pedicels** reflexed, 0.2–0.5 mm. **Flowers** 2.8–3.5 mm; calyces campanulate, 1.5–2 mm, glabrous, veins 5, tubes 0.8–1 mm, lobes unequal, > tube, adaxial 2 shorter, orifices open; corollas 2.6–3.2 mm, pale yellow becoming brown, not or slightly ribbed, banners persistent, spatulate, 2.6–3.2 × 2 mm, tips rounded to broadly acute. **Fruits** ellipsoidal, 1.5–2 mm, ≥ calyces, short-beaked; stipitate. **Seeds** 1, ellipsoid, 0.9–1 mm, yellow or pale brown, surfaces smooth, glossy. ***2n***=16, 28, 32.

Disturbed ground, fields. Flowering Apr–Jul. 0–1700 m. BW, Casc, CR, Est, Lava, Sisk, WV. CA, ID, NV, WA; throughout North America but absent from central Canada; worldwide. Exotic.

Trifolium dubium is native to Eurasia and is widely introduced as a weed worldwide. It may be confused with *Medicago lupulina*, but the latter can be distinguished by its obviously toothed stipules, deciduous corollas, and shiny, black fruits. Little hop clover may be the "shamrock" of Irish folklore (Nelson 1991), though other candidates include other species of *Trifolium* or species of *Medicago* or *Oxalis*.

Trifolium eriocephalum Nuttall

Woolly-head clover

Plants perennial, erect or decumbent, 12–46 cm, sparsely hairy to densely villous, appressed-pilose, or glabrate; branched. **Leaves** palmate; leaflets 3, linear, linear-elliptic, elliptic, lanceolate, lanceolate-oblong, or obovate, 10–50 × 4–12 mm, bases cuneate, margins denticulate or serrulate, veins fine or somewhat thickened, tips acuminate or acute, surfaces abaxially appressed-pilose, adaxially glabrous, densely villous, or sparsely or densely pilose; petioles 10–150 mm; petiolules 1 mm; stipules lanceolate or ovate, 10–50 mm, margins entire or denticulate, tips acuminate or acute. **Inflorescences** terminal or axillary, inverted or horizontal, 25–70-flowered, ellipsoidal or globose becoming subglobose, 20–30 × 10–30 mm; rachises prolonged beyond flowers; involucres absent; bracteoles absent. **Peduncles** bent just proximal to inflorescence, 30–180 mm. **Pedicels** reflexed, ≤ 0.3 mm. **Flowers** 9–18 mm; calyces campanulate, curved in fruit, 4–11 mm, villous, veins 10, tubes 1.5–2.5 mm, lobes subulate, often strongly curved and twisted, unequal, plumose, orifices open; corollas 8–16 mm, white, cream, or purple, banners oblanceolate, 8–16 × 3–5 mm, tips rounded or retuse, recurved; ovaries pubescent distally. **Fruits** longitudinally dehiscent, ovoid, 2–3.5 mm. **Seeds** (1)2–4, mitten-shaped or subglobose, 1.2–2 mm, yellow-brown to brown, smooth. ***2n***=16.

Western United States. 6 subspecies; 5 subspecies treated in *Flora*.

Trifolium eriocephalum was revised by Gillett (1971). The subspecies exhibit unusually complex distributional patterns and overlapping morphological features that sometimes make identifications problematic (Zohary & Heller 1984).

Key to *Trifolium eriocephalum* subspecies:

1. Ovules usually 2, rarely 1 or 3; calyx lobes usually green, rarely purple, straight, subequal, 3–4 × tube length.
   2. Plants densely villous distally; basal and median leaflets elliptic or lanceolate-oblong; roots cylindrically thickened, not predominantly fusiform-tuberous . . . . *T. e.* ssp. ***eriocephalum***
   2' Plants slightly villous, appressed-pilose, or glabrate distally; basal and median leaflets usually narrowly lanceolate or linear, rarely narrowly elliptic; roots fusiform-tuberous. . .*T. e.* ssp. ***cascadense***

1' Ovules 3 or 4; calyx lobes usually purple, contorted, abaxial lobes longer, lateral lobes 1–1.5 × tube length.
   3. Basal leaflets linear or linear-elliptic, tips acute, surfaces sparsely hairy adaxially; roots cylindrically thickened, branched, often rhizome-like, not predominantly fusiform-tuberous . . . . *T. e.* ssp. ***cusickii***
   3' Basal leaflets obovate, tips rounded to retuse, surfaces usually glabrous adaxially, sometimes sparsely hairy; roots fusiform-tuberous, without rhizomes (except by burial of crowns).
      4. Inflorescences ~25 mm wide, significantly overtopping distal leaves; median leaflets broadly lanceolate, surfaces adaxially sparsely hairy or glabrous . . . . *T. e.* ssp. ***arcuatum***
      4' Inflorescences ~10 mm wide, scarcely overtopping distal leaves; median leaflets narrowly lanceolate or linear, surfaces adaxially densely villous. . . . . . . *T. e.* ssp. ***martinii***

Trifolium eriocephalum Nuttall ssp. *arcuatum* (Piper) J.M. Gillett

Arched clover

Trifolium eriocephalum Nutt. var. *arcuatum* McDermott, *Trifolium eriocephalum* Nutt. var. *piperi* J.S. Martin

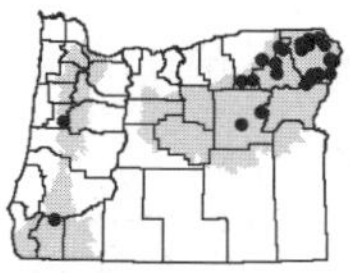

Plants sparsely hairy or glabrous; roots fusiform-tuberous, not rhizome-like (except by burial of crowns). **Leaflets** basal obovate, median broadly lanceolate, tips rounded to retuse, surfaces adaxially sparsely hairy or glabrous. **Inflorescences** significantly overtopping distal leaves, ~25 mm wide. **Flowers** calyx lobes contorted, usually purple, abaxial lobes longer, lateral lobes 1–1.5 × tube length; corollas white, cream, or purple; ovules 3 or 4.

Open pine forests, riverbanks, clearings, wet meadows. Flowering May–Aug. 900–2300 m. BW, Lava, Sisk, WV. ID, WA; northeast to MT. Native.

Trifolium eriocephalum Nuttall ssp. *cascadense* J.M. Gillett

Cascade clover

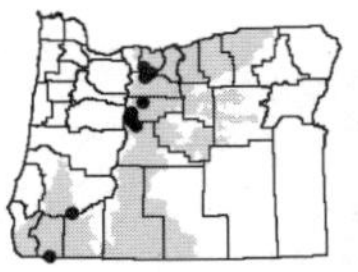

Plants slightly villous, appressed-pilose, or glabrate distally; roots fusiform-tuberous. **Leaflets** basal and median usually narrowly lanceolate or linear, rarely narrowly elliptic, tips rounded to acute, surfaces adaxially sparsely hairy. **Flowers** calyx lobes straight, subequal, 3–4 × tube length, green; corollas white; ovules (1)2(3).

Open pine forests, riverbanks, clearings, wet meadows. Flowering May–Jul. 700–1500 m. Col, ECas, Lava, Sisk. Native.

Trifolium eriocephalum Nuttall ssp. *cusickii* (Piper) J.M. Gillett

Cusick's clover

Trifolium eriocephalum Nutt. var. *cusickii* (Piper) J.S. Martin, *Trifolium eriocephalum* Nutt. var. *harneyense* (Howell) McDermott

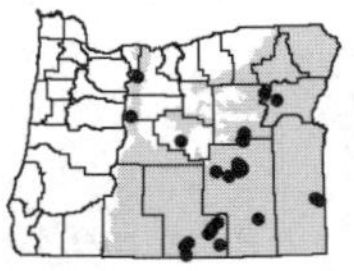

Plants slightly villous; roots cylindrically thickened, branched, often rhizome-like, not predominantly fusiform-tuberous. **Leaflets** basal and median linear or linear-elliptic, tips acute, surfaces adaxially sparsely hairy. **Flowers** calyx lobes contorted, usually purple, abaxial lobes longer, lateral lobes 1–1.5 × tube length; corollas white, cream, or purple; ovules 3 or 4.

Open pine forests, riverbanks, clearings, wet meadows. Flowering Jun–Jul. 1200–2000 m. BR, BW, ECas, Owy. ID, NV. Native.

Trifolium eriocephalum Nuttall ssp. *eriocephalum* (p. 728)

Woolly-head clover

Trifolium eriocephalum Nutt. var. *eriocephalum*

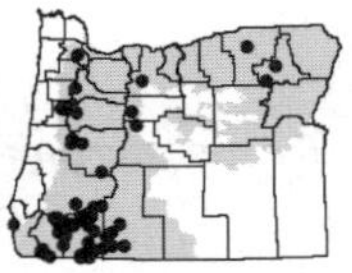

Plants densely villous distally; roots cylindrically thickened, not predominantly fusiform-tuberous. **Leaflets** basal and median elliptic or lanceolate-oblong, tips rounded to acute, surfaces adaxially densely villous. **Flowers** calyx lobes straight, subequal, 3–4 × tube length, usually green, rarely purple; corollas white; ovules (1)2(3).

Open pine forests, riverbanks, clearings, wet meadows. Flowering Apr–Aug. 50–1500 m. BW, Casc, Col, ECas, Sisk, WV. CA. Native.

Trifolium eriocephalum Nuttall ssp. *martinii* J.M. Gillett

Martin's clover

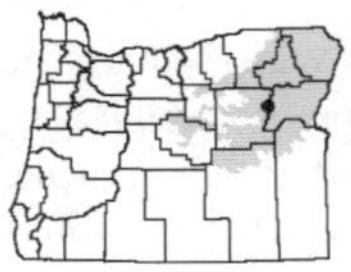

Plants densely villous; roots fusiform-tuberous, not rhizome-like except by burial of crown. **Leaflets** tips rounded to retuse; basal obovate, surfaces adaxially glabrous; median narrowly lanceolate or linear, surfaces densely villous adaxially. **Inflorescences** scarcely overtopping distal leaves, ~10 mm wide. **Flowers** calyx lobes contorted, usually purple, abaxial lobes longer, lateral lobes 1–1.5 × tube length; corollas white, cream, or purple; ovules 3 or 4.

Open pine forests, riverbanks, clearings, wet meadows. Flowering May–Jun. 0–2000 m. BW. ID. Native.

Trifolium fragiferum Linnaeus

Strawberry clover

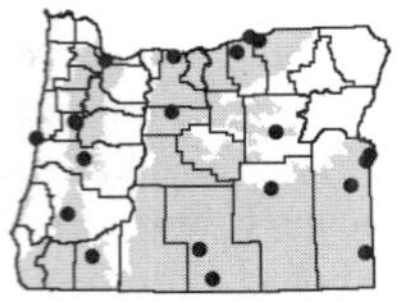

Plants perennial, prostrate or creeping, 5–50 cm, pubescent or glabrous; branched; rooting at nodes. **Leaves** pinnate; leaflets 3, ovate, obovate, or elliptic, 5–30 × 3–15 mm, bases cuneate, margins spinulose-denticulate, veins prominent, thickened, curved, tips obtuse, often retuse, surfaces abaxially hairy, adaxially glabrous; petioles 10–90 mm; petiolules ~1 mm; stipules linear or lanceolate, dilated proximally, 13–20 mm, margins entire, tips subulate or acuminate. **Inflorescences** axillary, erect or ascending, 10–30-flowered, globose, 8–20 × 8–20 mm, distinctly involucrate by bracts of proximal flowers; involucres cup-shaped, bracts distinct or connate proximally, lanceolate-oblong, 3–6 mm; bracteoles linear or lanceolate, 2–3 mm, acuminate or 2-fid. **Peduncles** 30–170 mm. **Pedicels** straight, ≤ 0.5 mm. **Flowers** 6–8 mm; calyces tubular, inflated in fruit, markedly asymmetric-bilabiate, 2.5–7 mm, usually pilose to woolly, sometimes glabrescent, veins 5–10, connected by reticulating lateral veins, tubes 2–7 mm, lobes unequal, abaxial straight, ≥ tube, very unequal in fruit, adaxial spreading in fruit, < tube, orifices open; corollas 5–8 mm, white to pink, banners oblong-lanceolate, 5–8 × 2 mm, tips emarginate to crenulate. **Fruits** ovoid, 2.5–3 mm, < calyces, long-beaked. **Seeds** 1–2, globose-reniform, 1–1.4 mm, tan or brown-spotted, smooth, dull. **2*n***=16.

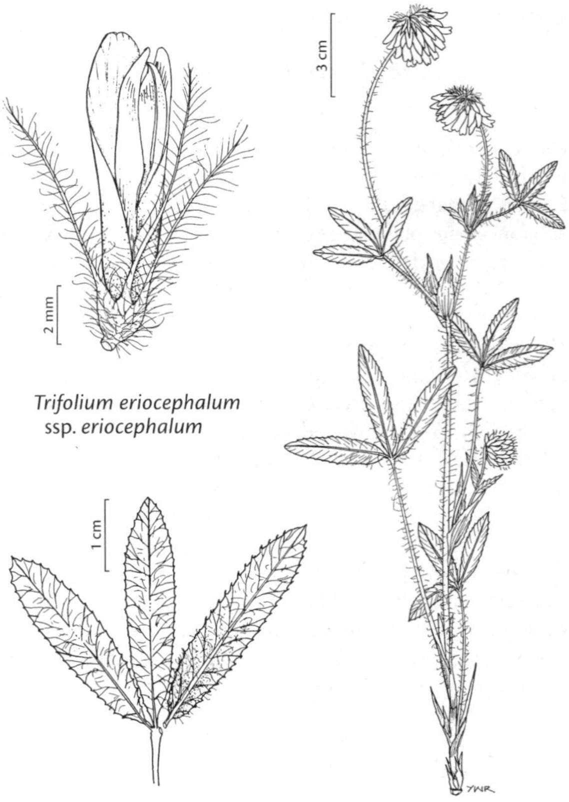

Trifolium eriocephalum ssp. *eriocephalum*

TRIFOLIUM

Meadows, fields, roadsides, sandy and saline soil. Flowering Jul–Sep. 0–1500 m. BR, Col, ECas, Est, Lava, Owy, Sisk, WV. CA, ID, NV, WA; scattered in North America; Africa, Asia, Europe. Exotic.

Trifolium fragiferum is native to Eurasia. It is cultivated for pastures and in mixtures with grass for hay and silage and to a lesser extent as groundcover in orchards and vineyards and as a green manure cover crop (St. John et al. 2010).

Trifolium fucatum Lindley (p. 729)

Bull clover, sour clover

Trifolium flavulum Greene

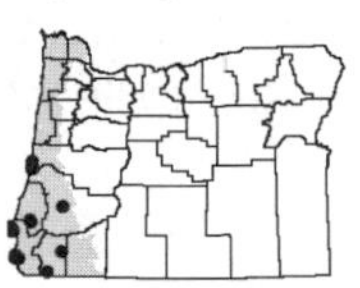

Plants annual, erect or ascending, 10–80 cm, glabrous or glabrescent; unbranched to numerous dichotomous branches. **Leaves** palmate; leaflets 3, broadly obovate, orbiculate, or rhombic-obovate, 8–40 × 7–30 mm, bases broadly cuneate, margins remotely dentate to densely serrulate-dentate, veins obscure, thickened near leaflet margin, tips rounded or slightly retuse, surfaces glabrous or glabrate; petioles 30–150 mm; petiolules 1–1.5 mm; stipules ovate or lanceolate, 10–30 mm, margins entire or toothed, tips usually acuminate, sometimes 2-fid. **Inflorescences** terminal or axillary, 10–30-flowered, subglobose or globose, 10–40 × 10–40 mm; involucres broadly bowl-shaped, involucral bracts ± connate, 4–15 mm, lobes 3–8, lanceolate, tips undivided or sometimes 2- or 3-fid, acuminate; bracteoles distinct or connate, broadly ovate, ~1 mm. **Peduncles** 30–130 mm. **Pedicels** straight, ~1 mm. **Flowers** 10–27 mm; calyces campanulate, 3–8 mm, glabrous, veins 10, tubes 1.5–2.5 mm, lobes 5–10, unequal, undivided or 3-fid, long-acuminate, orifices open; corollas 10–27 mm, creamy white to yellow, becoming pink to purple in age, banners broadly ovate, inflated entire length in fruit, not distally twisted, 10–27 × 6–15 mm, tips rounded, erose, keel petals rarely dark purple. **Fruits** linear to ovoid, 7–8 mm; stipitate. **Seeds** 3–8, globose, 1.6–2 mm, gray, mottled, reticulate. **2*n***=16.

Moist places, meadows, roadsides. Flowering May–Jun. 0–700 m. CR, Est, Sisk. CA, WA. Native.

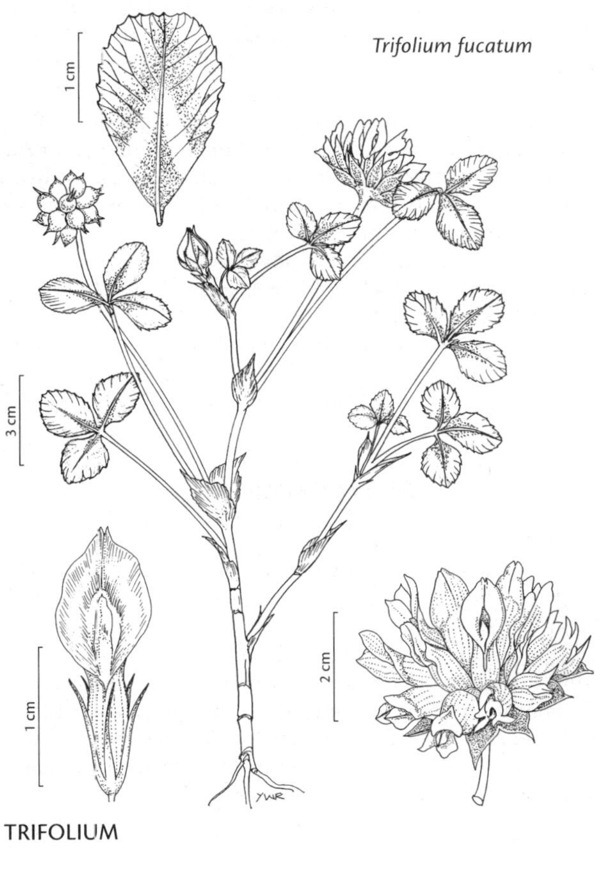

TRIFOLIUM

Trifolium glomeratum Linnaeus

Clustered clover

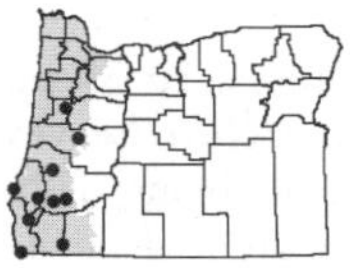

Plants annual, procumbent, decumbent, or ascending, 10–30 cm, glabrous; branched. **Leaves** pinnate; leaflets 3, obovate or obcordate, 6–15 × 4–10 mm, bases cuneate, margins spinulose-serrate, veins thickened distally, tips retuse or rounded, surfaces glabrous; petioles 0–70 mm; petiolules ≤ 0.5 mm; stipules ovate, 10–15 mm, margins entire, tips subulate-setaceous occasionally acute or apiculate. **Inflorescences** axillary, 30+-flowered, globose, 8–10 × 8–10 mm; involucres absent; bracteoles linear, ≤ 0.5 mm. **Peduncles** 0–1 mm. **Pedicels** straight, ≤ 0.2 mm. **Flowers** 6–8.5 mm; calyces tubular-obconic, 3–4 mm, glabrous, veins 10–12, tubes 1.5–2 mm, lobes triangular-ovate, spreading to recurved in fruit, equal, orifices open; corollas 6–8 mm, ≥ 2 × as long as calyces, pink, occasionally white, banners obovate, 6–8 × 1–2 mm, tips acute. **Fruits** longitudinally dehiscent, obovoid, 2–3 mm. **Seeds** 2, reniform, 1 mm, brown, tuberculate. **2*n***=14, 16.

Roadsides, lawns. Flowering May–Jun. 0–600 m. CR, Est, Sisk, WV. CA; scattered in southeastern US; Africa, Europe. Exotic.

Trifolium glomeratum is native to Europe and northern Africa. It is occasionally cultivated as a forage crop (Hermann 1953) and sometimes spreads.

Trifolium gymnocarpon Nuttall (p. 730)

Nuttall's clover

Trifolium gymnocarpon Nutt. ssp. *plummerae* (S. Watson) J.M Gillett, *Trifolium gymnocarpon* Nutt. var. *plummerae* (S. Watson) J.S. Martin

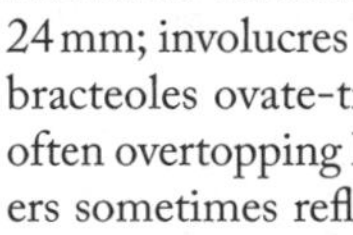

Plants perennial, cespitose, 2–15 cm, pubescent; numerous short stems branched from woody crown. **Leaves** palmate; leaflets 3–5, ovate, obovate, oblong, or elliptic, 8–30 × 2–18 mm, bases cuneate, margins spinose-dentate, veins thickened distally, tips rounded or acute, surfaces pubescent, adaxially sometimes glabrous; petioles 15–100 mm; petiolules ≤ 0.5 mm; stipules ovate to lanceolate, 5–15 mm, margins entire or irregularly dentate, tips acute to acuminate. **Inflorescences** terminal, 6–15-flowered, subglobose, 10–20 × 13–24 mm; involucres narrow rims, 0.5 mm, membranous, dentate; bracteoles ovate-triangular, ≤ 0.8 mm. **Peduncles** 10–65 mm, often overtopping leaves. **Pedicels** erect, those of proximal flowers sometimes reflexed, 1–4 mm. **Flowers** 7.5–13 mm; calyces tubular-campanulate, 3–7.5 mm, strigose, veins 10, tubes 2–3 mm, lobes narrowly triangular, subequal, ≤ 1.5 × tube length, orifices open; corollas 7–12 mm, pink-purple, often with whitish tips, sometimes nearly entirely whitish, banners oblong, 7–14 × 5–7 mm, tips rounded, usually retuse. **Fruits** longitudinally dehiscent, ovoid, 4–5 mm. **Seeds** 1, mitten-shaped, 3–4.5 mm, tan to brown, roughened.

Rocky slopes, clay or gumbo soil on plains. Flowering May–Jun. 1200–2700 m. BR, BW, Owy. CA, ID, NV; northeast to MT, east to WY, southeast to NM. Native.

Two loosely differentiated varieties or subspecies of *T. gymnocarpon* have sometimes been recognized: *plummerae* with 3–5 adaxially pilose leaflets, and *gymnocarpon* with 3 adaxially glabrous leaflets (Isely 1998; Gillett 1972). Considerable overlap among characters has led other authors to reject the distinction (Barneby 1989; Zohary & Heller 1984).

Trifolium hirtum Allioni

Rose clover

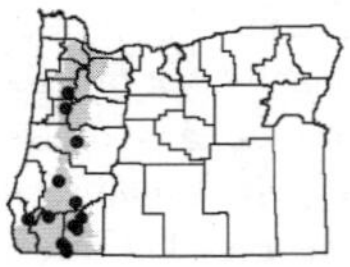

Plants annual, curved-ascending, 10–35 cm, densely spreading-hairy; branched. **Leaves** pinnate; leaflets 3, obovate to oblong, 8–25 × 5–13 mm, bases cuneate, margins denticulate distally, veins prominent, closely-spaced, tips rounded, surfaces densely spreading-hairy; petioles 5–50 mm; petiolules ~0.5 mm; stipules lanceolate-ovate, 8–18 mm, margins entire, tips long-setaceous. **Inflorescences** terminal on branches, 10–50-flowered, globose or ovoid, 15–25 × 15–25 mm, disarticulating at maturity into fruiting calyces; involucres enlarged stipules; bracteoles absent. **Peduncles** absent. **Pedicels** absent. **Flowers** 10–17 mm; calyces campanulate, 7–11 mm, pilose, veins 20, tubes 2–5 mm, lobes subequal, abaxial slightly longer, orifices open, pilose; corollas 10–14 mm, pink to purplish red, banners lanceolate, 10–14 × 1–2 mm, tips acute-acuminate. **Fruits** transversely dehiscent, ovoid, 2–3 mm, ≥ calyces, leathery distally. **Seeds** 1, globose-ellipsoid, 1.5–2 mm, tan or brown, smooth, glossy. $2n=10$.

Fields, roadsides. Flowering Apr–Jun. 50–800 m. Sisk, WV. CA; southeastern US; Africa, Asia, Europe. Exotic.

Trifolium hirtum is native to the Mediterranean region. It was first cultivated in California in the 1940s and is used as a forage plant and as a nitrogen source in roadside grass plantings (Love 1985).

Trifolium howellii S. Watson

Howell's clover, canyon clover

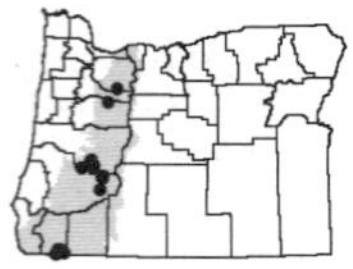

Plants perennial, erect, 30–100 cm, glabrous; stems unbranched or branched distally, fistulose. **Leaves** palmate; leaflets 3, ovate, elliptic, or rhombic, 35–95 × 23–50 mm, bases cuneate, margins serrate to nearly entire, veins obscure, tips acute, rounded, or obtuse, surfaces glabrous; petioles 5–200 mm; petiolules 1–1.5 mm; stipules ovate, 15–45 mm, margins entire, sometimes lobed, tips acuminate. **Inflorescences** terminal or axillary, 20–70-flowered, globose or ellipsoidal, 25–40 × 20–25 mm; rachises prolonged beyond flowers, undivided or forked, often bearing cluster of sterile flower buds distally; involucres absent; bracteoles minute, membranous, blunt or acute. **Peduncles** 30–110 mm. **Pedicels** erect in bud, reflexed in flower and fruit, ~1 mm. **Flowers** 10–14 mm; calyces campanulate, gibbous, 4–5.5 mm, glabrous, veins 10, tubes 2–2.5 mm, lobes linear-lanceolate, subequal, orifices open; corollas 11–12 mm, white to greenish white or greenish yellow, banners elliptic-oblong, 11–12 × 4–5 mm, tips obtuse. **Fruits** longitudinally dehiscent, oblong or clavate, 4–5 mm, ≥ calyces. **Seeds** 1–3, angular, mitten-shaped, 2.5 mm, reddish black or brown, smooth or slightly roughened, dull.

Wet streambanks and meadows, river floodplains, shady woodlands, springs, thickets. Flowering Jun–Aug. 400–1900 m. Casc, Sisk. CA. Native.

Trifolium howellii is of conservation concern and is very narrowly distributed in northern California and Oregon (Gillett 1972). It has some of the largest leaves of any *Trifolium* species in North America.

Trifolium hybridum Linnaeus

Alsike clover

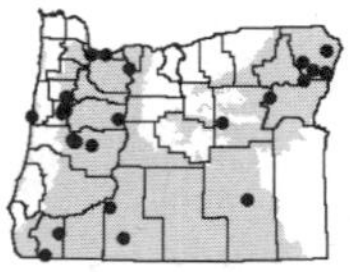

Plants perennial, erect or ascending, 10–70 cm, glabrous or sparsely hairy; branched. **Leaves** pinnate; leaflets 3, usually obovate or ovate, rarely rhombic, 10–40 × 10–25 mm, bases cuneate, margins sharply serrate, veins prominent, tips broadly acute, obtuse, or emarginate, surfaces glabrous; petioles 10–210 mm; petiolules 1 mm; stipules obovate or lanceolate, 10–25 mm, margins entire or toothed, tips long-acuminate. **Inflorescences** terminal or axillary, 10–70+-flowered, globose becoming subglobose; rachises not prolonged

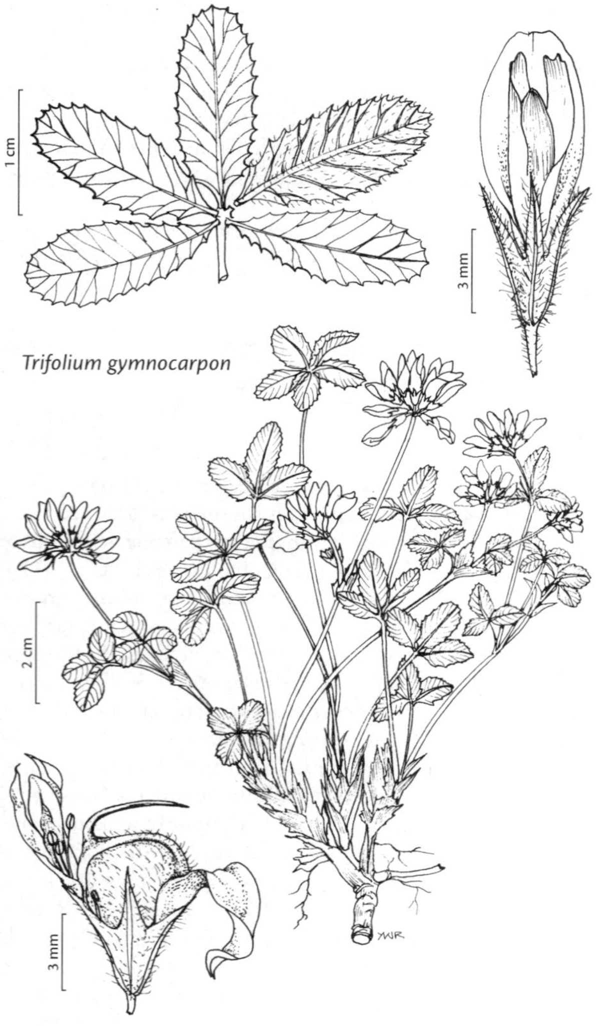

TRIFOLIUM

beyond flowers, 10–25 × 10–25 mm; involucres absent; bracteoles lanceolate, ~1 mm. **Peduncles** 20–100 mm. **Pedicels** strongly reflexed, 2–3.5 mm. **Flowers** 8–12 mm; calyces campanulate, 3.5–4 mm, glabrous or sparsely hairy, veins 10, tubes 1.5–2 mm, lobes lanceolate-subulate, unequal, margins entire, orifices open; corollas 7–10 mm, ≥ 2 × as long as calyces, pale to dark pink, banners ovate-oblong, 6.5–10.5 × 4–6 mm, tips rounded-denticulate. **Fruits** longitudinally dehiscent, ellipsoid, 3–7 mm, ≥ calyces; stipitate. **Seeds** 2–4, mitten-shaped, 1–1.4 mm, tan, brown, or red-brown, smooth, dull. ***2n***=16.

Roadsides, fields, meadows, waste places. Flowering May–Aug. 0–1900 m. BR, BW, Casc, ECas, Est, Sisk, WV. CA, ID, NV, WA; throughout North America; Asia, Europe. Exotic.

Trifolium hybridum is native to Eurasia, is widely cultivated as a forage crop, and is weedy worldwide. It may cause dermatitis in humans (Hardin & Arena 1974) and is implicated as a cause of dermatitis and biliary fibrosis in horses (Fisher 1995), though the connection between the clover and the diseases is inconclusive (Nation 1989).

Trifolium incarnatum Linnaeus

Crimson clover

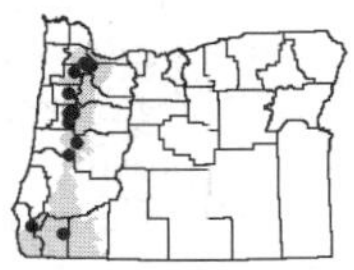

Plants usually annual, rarely biennial, erect, 20–60 cm, short-villous; unbranched or sparsely branched. **Leaves** pinnate; leaflets 3, broadly ovate, 10–30 × 10–15 mm, bases cuneate, margins denticulate, veins fine, tips emarginate or retuse, surfaces with spreading pustule-based hairs; petioles 10–80 mm; petiolules 1 mm; stipules ovate, 10–20 mm, margins wavy or toothed, tips blunt or somewhat tapering distally. **Inflorescences** terminal, 25–100-flowered, oblong, 20–70 × 10–25 mm; involucres absent; bracteoles absent. **Peduncles** 20–100 mm. **Pedicels** straight, ~0.5 mm. **Flowers** 10–15 mm; calyces tubular-campanulate, ~10 mm, villous, veins 10, tubes 3–4 mm, lobes spreading in fruit, equal, > tube, orifices narrowly opening; corollas 11–17 mm, usually scarlet to red, rarely pink or white, banners oblong-elliptic, much > wings and keel petals, 10–16 × 2 mm, tips acute. **Fruits** transversely dehiscent, ovoid, 2.5–3 mm, leathery distally. **Seeds** 1, ellipsoid, 2–2.5 mm, reddish, smooth, glossy. ***2n***=14.

Meadows, roadsides, especially in sandy soil. Flowering May–Jun. 0–200 m. Sisk, WV. CA, ID, WA; scattered throughout North America; worldwide. Exotic.

Trifolium incarnatum is native to Europe and was introduced to the United States in 1818 as a forage and green manure crop; it is commonly used as a winter grazing crop and in roadside grass plantings as a nitrogen source, especially in the southeastern states (Knight 1985). It is commonly found as a weed worldwide.

Trifolium latifolium (Hooker) Greene

Broad-leaved clover, twin clover

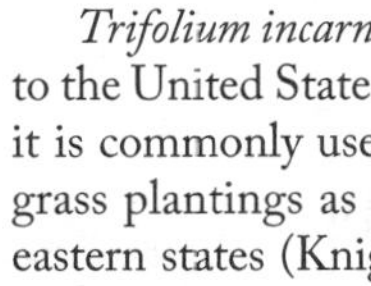

Plants perennial, ascending, 4–40 cm, appressed-pubescent; branched from slender, rhizomatous crown. **Leaves** palmate; leaflets 3, obovate or elliptic, 5–42 × 3–25 mm, bases cuneate, margins serrulate, veins fine or slightly thickened, tips acute, rounded, or retuse, apiculate, surfaces appressed-pubescent; petioles 15–100 mm; petiolules ≤ 1 mm; stipules ovate or lanceolate, 5–20 mm, margins entire proximally, 1–3-toothed distally, tips acuminate or acute. **Inflorescences** terminal, 6–30-flowered, globose, 23–31 × 20–30 mm; rachises prolonged beyond flowers; involucres absent; bracteoles minute, scale-like, membranous, ≤ 0.3 mm. **Peduncles** bent proximal to inflorescence, flowers appearing inverted, 25–120 mm. **Pedicels** reflexed in fruit, 1.5–2 mm. **Flowers** sweetly fragrant, 13–17 mm; calyces campanulate, 4.5–5 mm, appressed-pubescent, veins 5, tubes ~2 mm, lobes linear, subequal, margins entire, orifices open; corollas 12–15 mm, white, sometimes with purplish or buff-pink veins, banners elliptic, 12–15 × 4 mm, > wing and keel petals, tips tapered, sometimes retuse. **Fruits** longitudinally dehiscent, oblong to subglobose, ~5 mm, ≥ calyces. **Seeds** 1–2, flattened globose, 1.2 mm, dark brown, smooth. ***2n***=16, 32.

Clearings in conifer forests, moist gravelly, rocky, or clay soil, grassy hillsides and gullies, prairies. Flowering May–Jul. 900–1800 m. BW. ID, WA; northeast to MT. Native.

Trifolium latifolium is morphologically allied to *T. longipes* (Gillett 1969), and molecular phylogenetic studies place it as sister to that species (Ellison et al. 2006).

Trifolium leibergii A. Nelson & J.F. Macbride

Leiberg's clover

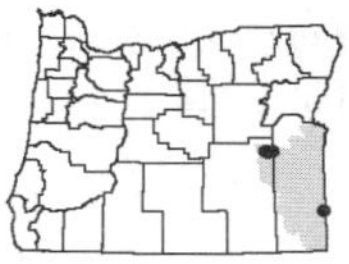

Plants perennial, erect, 10–15 cm, canescent; aerial stems sparsely branched. **Leaves** palmate; leaflets 3, obovate, rhombic, or elliptic, sometimes folded, 9–15 × 5–13 mm, bases cuneate, margins coarsely spinulose-dentate, veins very thick, tips rounded, long-apiculate, surfaces canescent; petioles 10–30 mm; petiolules ≤ 0.5 mm; stipules lanceolate, 7–15 mm, margins coarsely spinulose-dentate, tips acuminate. **Inflorescences** axillary, 5–25-flowered, globose, 22–25 × 23–27 mm; involucres a narrow, chartaceous, irregular rim, ≤ 0.5 mm; bracteoles triangular, ≤ 0.2 mm. **Peduncles** 20–45 mm. **Pedicels** reflexed in fruit, 1.5–4 mm. **Flowers** 11–15 mm; calyces campanulate, 5–11 mm, pilose, veins 10, tubes 1.8–2.2 mm, lobes lanceolate-setiform, subequal, margins entire, orifices open; corollas 10.5–13 mm, creamy white with purplish veins, becoming pinkish purple-tinged, banners broadly rhombic-ovate or obovate, 10.5–13 × 5–7 mm, tips rounded, slightly emarginate. **Fruits** longitudinally dehiscent, ovoid, 3–4 mm. **Seeds** 1–2, mitten-shaped, 1.5–2 mm, brown, smooth.

Open, rocky slopes and ridges. Flowering May–Jul. 1000–1600 m. Owy. NV. Native.

Trifolium leibergii is known from only a few populations in Oregon and Nevada. It is in the Center for Plant Conservation's National Collection of Endangered Plants.

Trifolium longipes Nuttall

Longstalk clover

Plants perennial, erect or ascending, cespitose, 5–46 cm, slightly pilose; unbranched to little branched distally. **Leaves** palmate; leaflets 3, not overlapping, linear, lanceolate to elliptic, or oblong, cauline sometimes ovate, 8–60 × 2–16 mm, bases cuneate, margins entire or serrate, veins slightly to very thickened, curving, tips acute, surfaces abaxially glabrous, adaxially usually appressed-pubescent; petioles 10–130 mm; petiolules ≤ 1 mm; stipules ovate to lanceolate, 10–50 mm, distinct at bases, margins usually entire, sometimes slightly serrate proximally, tips acute or acuminate. **Inflorescences** terminal, erect, 10–40-flowered, globose, ellipsoidal, obconic, or subglobose, 10–35 × 10–32 mm; involucres absent; bracteoles scale-like, ~0.5 mm. **Peduncles** straight, 10–300 mm. **Pedicels** erect or reflexed in fruit, 0.5–2 mm. **Flowers** 10–19 mm; calyces tubular, 4–10 mm, hairy or glabrous, veins 5–10, tubes 2.5–3 mm, lobes staight or lax, linear, unequal, ≤ 6 × tube length, orifices open; corollas 10–18 mm, white, cream, pink, lavender, or purple, sometimes pale with purple tips, banners ovate to oblanceolate, 10–18 × 3–4 mm, tips usually not beaked (except ssp. *multipedunculatum*), usually obtuse, acute, or acuminate, rarely emarginate; ovaries pubescent distally. **Fruits** longitudinally dehiscent, oblong to ellipsoid, 4–5 mm. **Seeds** 1–4, flattened globose, 2–3 mm, dark brown, smooth.

Western US. 12 subspecies; 7 subspecies treated in *Flora*.

Trifolium longipes encompasses a morphologically diverse and taxonomically difficult suite of populations. Gillett (1969) united *T. longipes* with *T. rusbyi* and differentiated ten subspecies. Zohary and Heller (1984) maintained *T. rusbyi* (with 7 subspecies) as distinct from *T. longipes* (with 5 subspecies). Molecular work supports the union of *T. rusbyi* and *T. longipes* (Ellison et al. 2006).

Key to *Trifolium longipes* subspecies:

1. Banners, wings, and often keel petals beaked. . . . . *T. l.* ssp. ***multipedunculatum***

1' Petals without beaks, tips acute to attenuate.

- 2. Pedicels erect in fruit, or proximal mostly horizontal or only slightly reflexed in fruit.
  - 3. Plants rhizomatous, roots not fusiform, sometimes thickened.
    - 4. Calyx lobes lax, usually only slightly > tube length, tube veins not prominent; leaflets lanceolate, elliptic to oblong, or ovate . . . . *T. l.* ssp. ***hansenii***
    - 4' Calyx lobes often somewhat rigid, ~6 × tube length, tube veins usually prominent; leaflets linear to elliptic-linear or broadly elliptic . . . . . *T. l.* ssp. ***elmeri***
  - 3' Plants not rhizomatous (stems ascending from crowns), roots often fusiform, thickened . . . . . *T. l.* ssp. ***longipes***
- 2' Pedicels of at least proximal flowers, usually all, strongly reflexed in fruit.
  - 5. Pedicels slender, 1.5–2 mm; plants with slender taproots. . . *T. l.* ssp. ***oreganum***
  - 5' Pedicels stout, ≤ 1 mm; plants rhizomatous.
    - 6. Inflorescences globose or ellipsoidal; proximal flowers with more strongly reflexed pedicels than distal ones . . *T. l.* ssp. ***caurinum***
    - 6' Inflorescences hemispheric, obconic, or subglobose; all flowers with strongly reflexed pedicels . . . . *T. l.* ssp. ***reflexum***

Trifolium longipes Nuttall ssp. *caurinum* (Piper) J.M. Gillett

Beaked longstalk clover

Trifolium longipes Nutt. var. *multiovulatum* (L.F. Hend.) C.L. Hitchc.

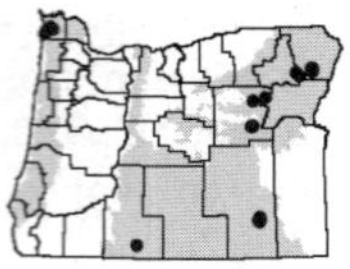

Plants 10–20 cm; rhizomatous. **Inflorescences** 28–32-flowered, globose or ellipsoidal, 30–32 mm wide. **Pedicels** at least proximal, but often all, strongly reflexed, proximal more strongly reflexed than distal, stout, ≤ 1 mm. **Flowers** 14–16 mm; calyces usually glabrous, rarely very sparsely hairy; corollas lavender, banner tips acute or acuminate. **2*n***=48.

Montane forests. Flowering May–Aug. 700–2900 m. BR, BW, CR, ECas. WA. Native.

Trifolium longipes Nuttall ssp. *elmeri* (Greene) J.M. Gillett

Elmer's clover

Trifolium longipes Nutt. var. *elmeri* (Greene) McDermott

Plants 20–40 cm; rhizomatous, roots not fusiform, sometimes thickened. **Leaflets** linear to elliptic-linear to broadly elliptic. **Inflorescences** 10–30-flowered, subglobose, 15–20 mm wide. **Pedicels** erect, proximal-most horizontal or only slightly reflexed, stout, ≤ 1 mm. **Flowers** 10–14 mm; calyces sparsely hairy, lobes ~6 × tube length, often somewhat straight and rigid, tube veins usually prominent; corollas white, lavender, or pink, banner tips acute or acuminate. **2*n***=16.

Meadows. Flowering Jun–Jul. 0–2000 m. Est, Sisk. CA. Native.

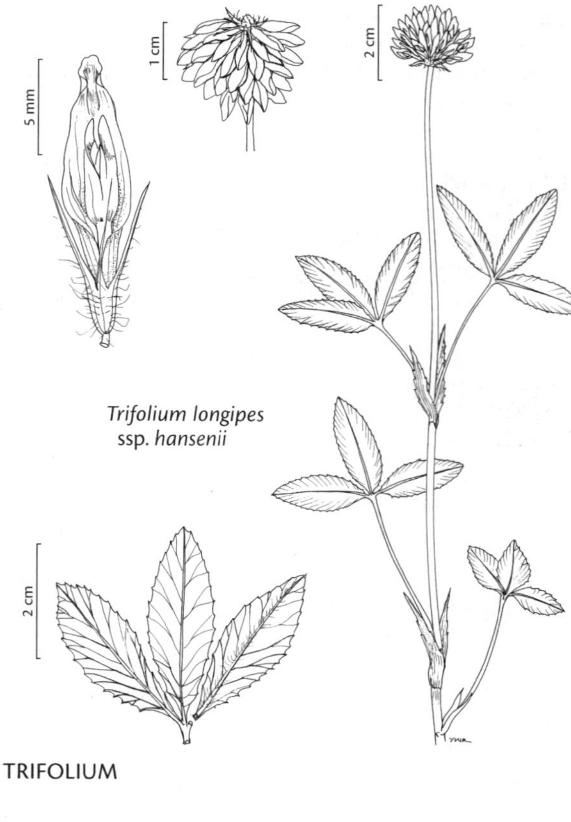

TRIFOLIUM

Trifolium longipes Nuttall (p. 732)
ssp. ***hansenii*** (Greene) J.M. Gillett

Hansen's clover

Trifolium longipes Nutt. var. *hansenii* (Greene) Jeps., *Trifolium longipes* Nutt. var. *nevadense* Jeps.

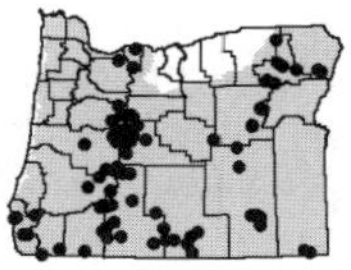

Plants 15–40 cm; rhizomatous, roots not fusiform, sometimes thickened. **Leaflets** lanceolate, elliptic to oblong, or ovate. **Inflorescences** 15–25-flowered, subglobose, 15–30 mm wide. **Pedicels** erect, proximal-most horizontal or only slightly reflexed, stout, ≤ 1 mm. **Flowers** 13–15 mm; calyces sparsely hairy, lobes lax, usually only slightly > tube length, tube veins not prominent; corollas white, lavender, or pink, banner and wing tips acute or acuminate. **2*n***=48.

Moist meadows, pine forests. Flowering Jun–Aug. 900–3000 m. All ecoregions except Col, Est. CA. Native.

Trifolium longipes Nuttall ssp. ***longipes***

Longstalk clover

Trifolium longipes Nutt. var. *longipes*

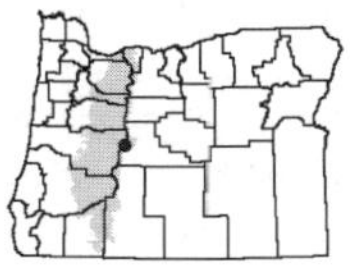

Plants 15–30 cm; roots thickened, often fusiform. **Inflorescences** 15–40-flowered, subglobose, 10–20 mm wide. **Pedicels** erect, proximal-most horizontal or only slightly reflexed, stout, ≤ 1 mm. **Flowers** 10–12 mm; calyces sparsely hairy or glabrous; corollas white, banner and wing tips slender-acuminate. **2*n***=16.

Boggy meadows. Flowering May–Jul. 1200–2300 m. Casc. WA. Native.

Trifolium longipes Nuttall
ssp. ***multipedunculatum*** (P.B. Kennedy) J.M. Gillett

Multistalk clover

Trifolium longipes Nutt. var. *multipedunculatum* (P.B. Kenn.) J.S. Martin, *Trifolium multipedunculatum* P.B. Kenn.

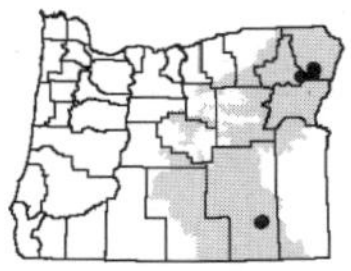

Plants 10–17 cm; roots slender, or taproots bearing a cluster of rhizomes. **Inflorescences** 10–25-flowered, globose, 20–30 mm wide. **Pedicels** at least proximal but often all strongly reflexed, slender, 1.5–2 mm. **Flowers** 12–14 mm; calyces curly-pubescent at sinuses; corollas purple, sometimes keel petals white, banners, wings, and often keel petal tips beaked. **2*n***=32, 48.

Open pine or spruce-fir forest to alpine, on talus, rocky places, meadows. Flowering Jul–Aug. 2200–2900 m. BR, BW. CA, WA. Native.

Trifolium longipes Nuttall
ssp. ***oreganum*** (Howell) J.M. Gillett

Oregon clover

Trifolium longipes Nutt. var. *oreganum* (Howell) Isely, *Trifolium oreganum* Howell var. *oreganum*

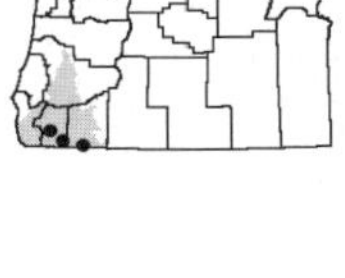

Plants 7–25 cm; bearing a crown of pilose wiry stems; taproots slender. **Inflorescences** 10–25-flowered, globose or obconic, 20–28 mm wide. **Pedicels** at least proximal but often all strongly reflexed, slender, 1.5–2 mm. **Flowers** 11–13 mm; calyces sparsely hairy; corollas pink or white, often bicolored, banner tips acute or acuminate.

Forested slopes. Flowering Apr–Aug. 400–2000 m. Sisk. CA. Native.

Trifolium longipes Nuttall
ssp. ***reflexum*** (A. Nelson) J.M. Gillett

Reflexed longstalk clover

Trifolium longipes Nutt. var. *reflexum* A. Nelson, *Trifolium oreganum* Howell var. *rydbergii* (Greene) McDermott

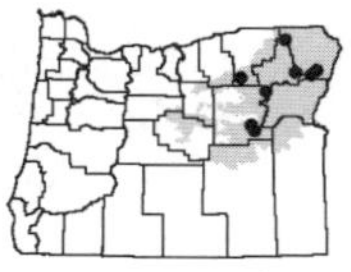

Plants 15–37 cm; rhizomatous. **Inflorescences** 10–20-flowered, hemispheric, obconic, or subglobose, 20–30 mm wide. **Pedicels** strongly reflexed, stout, ≤ 1 mm. **Flowers** 13–15 mm; calyces slightly pilose or curly-pilose at sinuses, often glabrous proximally; corollas white, cream, or violet, banner tips acute or acuminate. **2*n***=24, 48.

Moist meadows. Flowering Jun–Aug. 1200–2800 m. BW. WA; northeast to MT, east to WY, southeast to NM. Native.

Trifolium macraei Hooker & Arnott

Macrae's clover

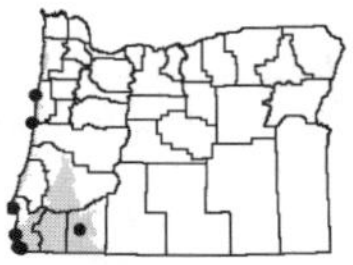

Plants annual, erect, ascending, or prostrate, 3–30 cm, pubescent; branched. **Leaves** palmate; leaflets 3, obovate, oblanceolate, or elliptic, 6–16 × 3–12 mm, bases cuneate, margins subentire to serrate, veins obscure or slightly thickened, tips rounded, retuse, surfaces pubescent; petioles 5–60 mm; petiolules ≤ 0.5 mm; stipules ovate to oblong, 6–10 mm, margins entire, tips acuminate to cuspidate. **Inflorescences** terminal or subterminal, usually paired, sometimes 1 head slightly stalked and partially hidden by terminal leaves and stipules, 10–40-flowered, ovoid or subglobose, 8–25 × 5–20 mm; involucres formed from terminal leaves and stipules; bracteoles broadly ovate to linear, 1–2 mm. **Peduncles** 0–2 mm. **Pedicels** absent. **Flowers** 5.5–7.5 mm; calyces tubular, 4–5 mm, pubescent, veins 5–10, tubes 1.5–2 mm, lobes subulate, subequal, orifices open; corollas 5–7 mm, usually purple or pink, rarely white, banners obovate-oblong, 6–7 × 2–3 mm, tips rounded, slightly denticulate. **Fruits** longitudinally dehiscent, oblong, 2.5–3 mm. **Seeds** 1, ellipsoid, 1.6–2 mm, yellow, mottled, smooth. **2*n***=16.

Grassy fields, sandy ocean bluffs. Flowering May–Jul. 0–400 m. Est, Sisk. CA; South America. Native.

Trifolium macraei was described from specimens collected in Chile, and the South American material has broader banner petals, smaller auricles on the wing petals, and a style that is barely curved upwards distally (Isely 1998). Further examination of the disjunct populations may reveal other differences and prompt reconsideration of the identity of North American specimens.

Trifolium macrocephalum (Pursh) Poiret (p. 734)

Largehead clover, big-headed clover

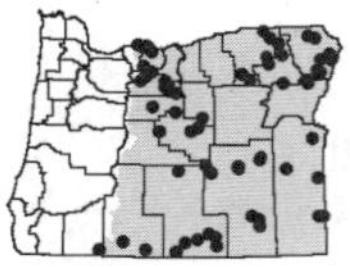

Plants perennial, erect or ascending, 5–25 cm, villous; branched. **Leaves** palmate; leaflets (5)7–9, broadly to narrowly obovate, often folded, 10–27 × 4–11 mm, bases cuneate, margins serrulate or irregularly lobed, veins thickened, especially distally, tips rounded or truncate, apiculate, surfaces abaxially villous, adaxially sparsely villous to glabrate; petioles 10–140 mm; petiolules 0.9–1.2 mm; stipules ovate, obovate, or oblong, 10–30 mm, margins entire. **Inflorescences** terminal, 20–32-flowered, globose or ovoid-ellipsoidal, 25–80 × 30–70 mm; involucres absent; bracteoles broadly ovate, membranous, ≤ 0.5 mm, truncate. **Peduncles** 20–80 mm. **Pedicels** erect, 1–1.5 mm. **Flowers** 20–30 mm; calyces campanulate, 10–22 mm, villous, veins 10–15, tubes 2.5–4 mm, lobes subulate, subequal, 2–4 × tube length, plumose, orifices open; corollas 20–28 mm, white, creamy white, or pinkish, banners ovate or oblong, 20–28 × 10–13 mm, tips rounded or slightly emarginate, keel petals deep pink. **Fruits** longitudinally dehiscent, ovoid, 4–5 mm, ≥ calyces. **Seeds** 1–2, mitten-shaped to ellipsoid, 2.5–3 mm, yellow or reddish, smooth. ***2n***=32, 48.

Rocky places, hard, compacted clay gumbo, lava beds, sage-covered slopes. Flowering Apr–Jul. 0–2200 m. BR, BW, Col, ECas, Lava, Owy. CA, ID, NV, WA. Native.

Trifolium macrocephalum has the largest inflorescences of any species in the genus. The name *T. megacephalum* is an illegitimate replacement name for *Lupinaster macrocephalum*.

Trifolium microcephalum Pursh

Smallhead clover

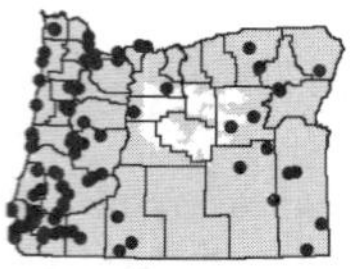

Plants annual, erect to ascending, 3–55 cm, densely to sparsely pubescent; branched. **Leaves** palmate; leaflets 3, oblanceolate, obovate, or obcordate, 4–17 × 3–11 mm, bases cuneate, margins setose, often dentate distally, veins fine or slightly thickened, tips usually retuse, rarely rounded, surfaces villous; petioles 10–50 mm; petiolules 0.5 mm; stipules obliquely ovate, 4–12 mm, margins entire or slightly serrate, tips acuminate. **Inflorescences** axillary or terminal, 10–40-flowered, globose, 5–13 × 4–12 mm; involucres flattened or bowl-shaped, bracts ± connate, 4–10 mm, conspicuously villous, lobes 5–12, divided ½ their length, lanceolate-ovate, acuminate, entire or slightly toothed proximally, when flattened laterally nearly completely hiding calyces; bracteoles absent. **Peduncles** 10–80 mm. **Pedicels** absent. **Flowers** 4–6 mm; calyces campanulate-tubular, 3–5.5 mm, pubescent, veins 10, tubes 1.8–2.5 mm, lobes triangular-aristate, ± equal, margins hyaline, wavy, orifices open; corollas 4–6 mm, ≈ calyces, white or pink to lavender, banners oblong, 3–6 × 1–2 mm, tips emarginate. **Fruits** broadly ellipsoid, 1.5–2 mm. **Seeds** 1, oblong, 1.5 mm, yellow, reddish mottled, smooth. ***2n***=16.

Meadows, roadsides, streambanks, forest clearings, grassy slopes. Flowering Apr–Aug. 0–2000 m. All ecoregions except Lava. CA, ID, NV, WA; north to AK, northeast to MT, southeast to AZ, south to Mexico. Native.

Trifolium microcephalum is common in much of its range in California, north into Oregon, and is rare in much of the rest of its range. It is closely allied with and greatly resembles the Chilean species *T. vernum* (Ellison et al. 2006), from which it is distinguished by its shorter calyx lobes (Zohary & Heller 1984).

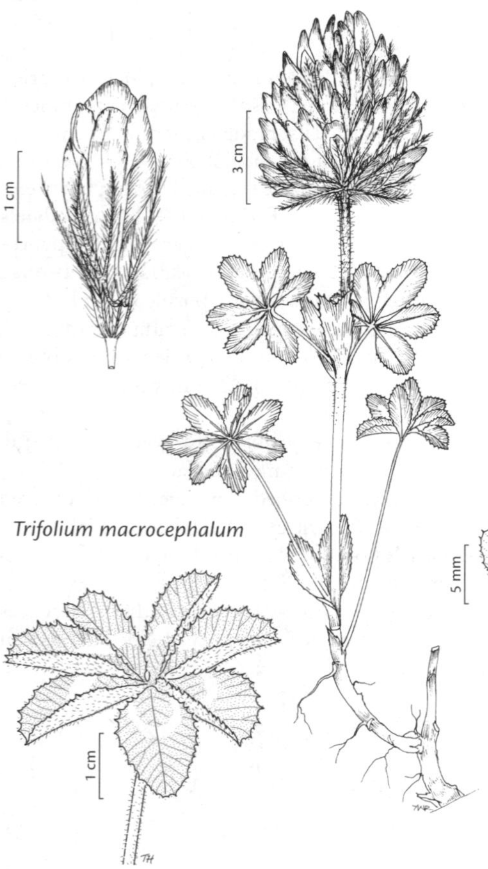

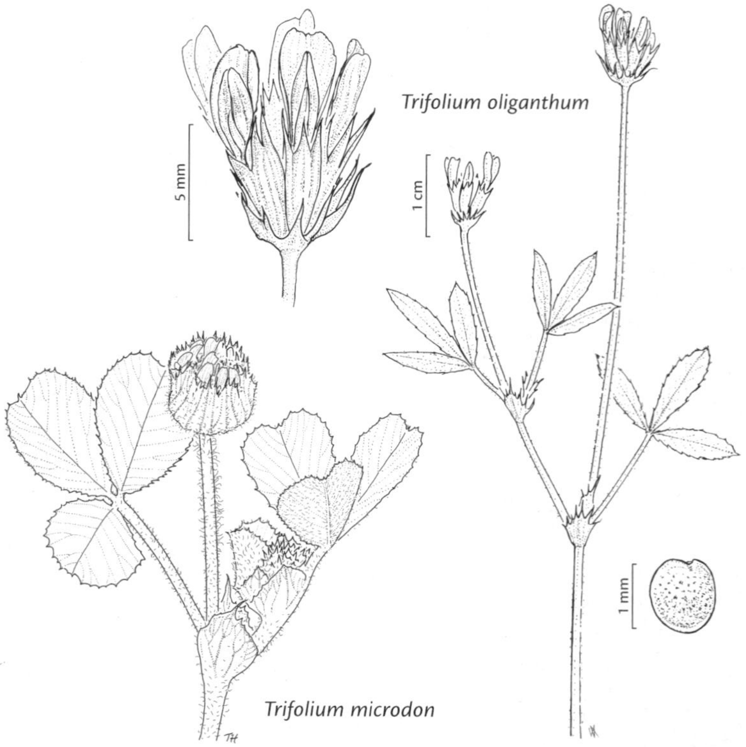

TRIFOLIUM

Trifolium microdon Hooker & Arnott (p. 734)

Thimble clover

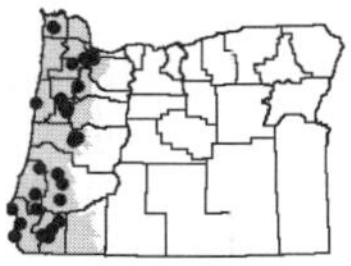

Plants annual, erect or ascending, 6–35 cm, villous or glabrous; branched. **Leaves** palmate; leaflets 3, obovate or obcordate, 4–14 × 3–12 mm, bases cuneate, margins serrate, veins fine, thickened distally, tips rounded or emarginate, surfaces villous or glabrous; petioles 10–80 mm; petiolules ~0.5 mm; stipules obliquely ovate, 4–12 mm, margins entire, toothed, or lacerate, tips acuminate. **Inflorescences** axillary or terminal, 10–17-flowered, subglobose, 8–17 × 5–15 mm; involucres cup-shaped, bracts ± connate, 5–15 mm, glabrous or inconspicuously hairy, lobes 8–12, ovate, conspicuously sharply toothed; bracteoles absent. **Peduncles** 15–74 mm. **Pedicels** absent. **Flowers** 6–7 mm; calyces tubular-campanulate, 3–4 mm, glabrous, veins 10, tubes 2.5–3.5 mm, lobes triangular, ± equal, not or minutely aristate, conspicuously toothed, margins membranous, orifices open; corollas 6–9 mm, pale pink or white, banners oblong, 6–9 × 2–3 mm, tips narrowly rounded or emarginate-mucronate. **Fruits** ovoid, 2–3 mm. **Seeds** 1–2, oblong, 1.5–2 mm, greenish, sometimes mottled, smooth. ***2n***=16.

Meadows, roadsides, dry slopes, fields, open oak or pine forests. Flowering Apr–Jul. 0–1000 m. CR, Est, Sisk, WV. CA, ID, WA; north to British Columbia, south to Mexico; South America. Native.

Trifolium microdon is one of several taxa in the genus that have an apparently natural disjunct distribution between the western coast of South America (Chile) and western North America. Others include *T. depauperatum* var. *depauperatum* and *T. macraei*.

Trifolium obtusiflorum Hooker

Clammy clover

Trifolium tridentatum Lindl.

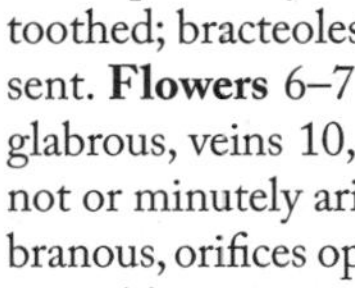

Plants annual, erect or ascending, 2–100 cm, resinous stipitate-glandular; branched. **Leaves** palmate; leaflets 3, elliptic, lanceolate, oblanceolate, rhombic, or obovate, 15–40 × 3–17 mm, bases cuneate, margins coarsely spinulose-serrate, veins thickened, tips acute, mucronate, surfaces glandular; petioles 15–100 mm; petiolules ≤ 0.5 mm; stipules ovate, 10–15 mm, sheathing, margins deeply lacerate, tips acuminate. **Inflorescences** axillary or terminal, 10–50-flowered, globose or ovoid, 10–35 × 10–30 mm; involucres flattened or bowl-shaped, divided $1/4$–$1/3$ their length, when flattened laterally not hiding flowers except proximally, 3–8 mm, deeply lacerate; bracteoles absent. **Peduncles** 30–150 mm, glandular. **Pedicels** erect, ~1 mm. **Flowers** 13–20 mm; calyces tubular-campanulate, slit between adaxial lobes, 10–13 mm, glandular, veins 20+, tubes 5–7 mm, lobes narrowly triangular or lanceolate-subulate, unequal, ≤ tube, usually entire, rarely 3-fid or shouldered, orifices open; corollas 10–18 mm, white or pale pinkish with dark purple spot, banners broadly elliptic, 10–18 × 2–4 mm, tips blunt. **Fruits** obovoid, 3.5–4 mm. **Seeds** 1–2, ellipsoid or mitten-shaped, 2.5 mm, brown, mottled, smooth. ***2n***=16.

Moist swales, creek bottoms. Flowering May–Jun. 0–100 m. Est, Sisk. CA; south to Mexico. Native.

Trifolium obtusiflorum is much less common than the similar *T. willdenovii*, occurring in moist areas in cismontane California and north into Oregon (Jepson 1925). It is easy to distinguish from *T. willdenovii* by its glandularity, which in fresh specimens causes them to be sticky to the touch.

Trifolium oliganthum Steudel (p. 734)

Fewflower clover

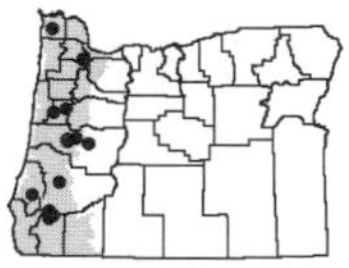

Plants annual, erect, 10–50 cm, glabrous or sparsely glandular; dichotomously branched. **Leaves** palmate; leaflets 3, linear, oblong, oblanceolate, or elliptic, 5–25 × 1–5 mm, bases cuneate, margins entire, spinulose, or dentate, veins fine or moderately thickened, tips rounded or truncate, mucronate, surfaces glabrous; petioles 5–40 mm; petiolules ≤ 0.5 mm; stipules lanceolate, 4–12 mm, margins lacerate, tips subulate. **Inflorescences** axillary or terminal, 3–15-flowered, obconic, 3–12 × 3–8 mm; involucres flattened to bowl-shaped, bracts ± connate, divided $4/5$–$9/10$ their length, 1–2 mm, lobes lanceolate-subulate, when flattened laterally not hiding flowers except proximally; bracteoles absent. **Peduncles** 25–75 mm. **Pedicels** erect, 0.5–2 mm. **Flowers** 11–16 mm; calyces campanulate, slit between adaxial lobes, 4–5 mm, glabrous, veins 10, tubes 2.5–5.2 mm, lobes triangular-subulate, ≤ tubes, 3-fid or shouldered, orifices open; corollas 6–8 mm, lavender to violet with white tips, banners narrowly ovate-oblong, 6–8 × 2 mm, tips narrowly rounded, erose, keel petals purple. **Fruits** oblong, 2.1–3.2 mm; sessile. **Seeds** 1–2(3), lenticular or reniform, 1.2–1.3 mm, reddish brown, mottled, smooth. ***2n***=16.

Riverbanks, grassy, rocky slopes, meadows, fields. Flowering Apr–Jun. 0–1000 m. CR, Sisk, WV. CA, WA; north to British Columbia. Native.

Trifolium oliganthum is relatively common throughout much of its range. Even though it closely resembles *T. variegatum*, it is placed near *T. bifidum* in molecular studies (Ellison et al. 2006).

Trifolium owyheense Gilkey

Owyhee clover

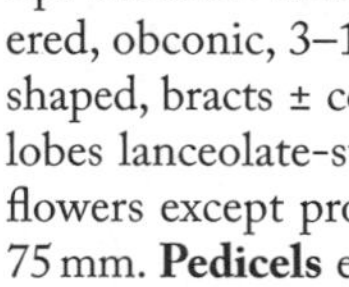

Plants perennial, cespitose, spreading, 10–20 cm, glabrous, glaucous; branched proximally, little branched distally. **Leaves** palmate; leaflets 3, slightly overlapping, ovate, obovate, or orbiculate, 10–20 × 7–23 mm, bases truncate to rounded, margins sparsely dentate, veins prominent, tips rounded, emarginate, surfaces glabrous; petioles 20–60 mm; petiolules 1–1.5 mm; stipules broadly obovate, 10–20 mm, margins slightly lobed, tips acute, fused at bases. **Inflorescences** terminal, 20–30-flowered, usually formed of 2 sessile heads, globose, 25–50 × 25–50 mm; involucres absent; bracteoles cup-like, ~0.5 mm. **Peduncles** 30–70 mm, > subtending leaves. **Pedicels** erect, reflexed in only proximal-most flowers, ~1 mm. **Flowers** 15–21 mm; calyces tubular-campanulate, 9–12 mm, pilose, veins 10, tubes 4.5–6 mm, lobes subequal, abaxial most subulate, longest, orifices open; corollas 20–23 mm, deep pink or magenta, banners tubular for most of length, 18–22 × 5–7 mm, tips flared. **Fruits** longitudinally dehiscent, ellipsoid, 4–5 mm. **Seeds** 1–2, mitten-shaped, 2–3 mm, yellow-mottled, smooth.

Dry shale hillsides on diatomaceous earth. Flowering May–Jun. 800–1700 m. Owy. ID. Native.

Trifolium owyheense is a rare endemic that is restricted to approximately 40 populations in a small portion of east-central Malheur County, Oregon, and immediately adjacent Owyhee County, Idaho (Mancuso 2001).

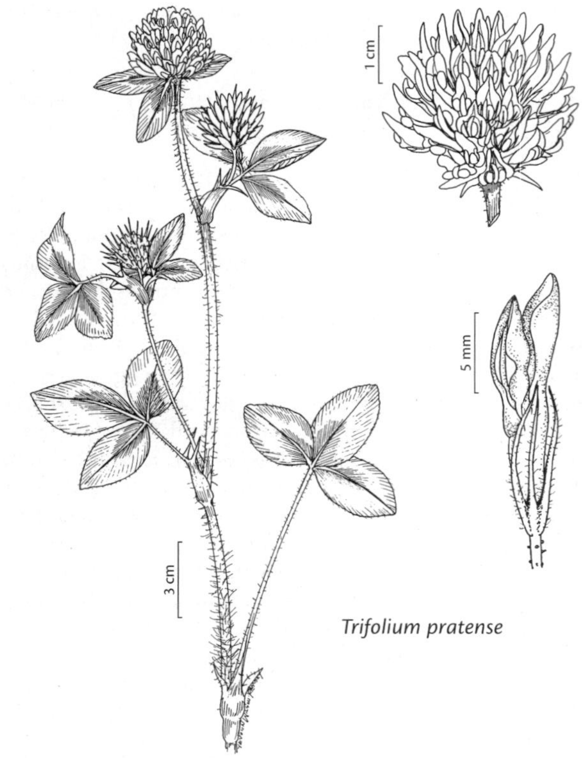

Trifolium pratense

TRIFOLIUM

Trifolium plumosum Douglas ssp. ***plumosum***

Plumed clover

Trifolium plumosum Douglas var. *plumosum*

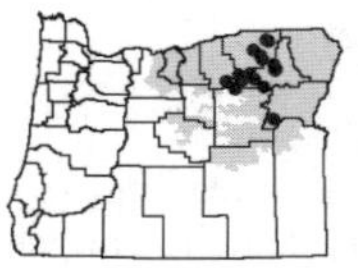

Plants perennial, erect, 25–35 cm, villous; branched or unbranched. **Leaves** palmate; leaflets 3, linear, falcate, sometimes folded lengthwise, 40–110 × 1–7 mm, bases cuneate, margins denticulate, veins thickened, tips acuminate, surfaces villous; petioles 10–200 mm; petiolules ≤ 0.5 mm; stipules linear, oblong, or ovate, 15–25 mm, margins entire or slightly serrulate, tips acute to acuminate. **Inflorescences** terminal, 70–80-flowered, spicate, ovoid to cylindrical, 30–60 × 25–30 mm; involucres absent; bracteoles absent. **Peduncles** 20–60 mm. **Pedicels** erect or absent, ≤ 0.5 mm. **Flowers** 17–18 mm; calyces campanulate, ~5 mm, villous, veins 20, tubes 2.5 mm, lobes subulate, subequal, orifices open; corollas 12–22 mm, creamy white, banners oblong, keel petals crimson, 12–22 × 3–5 mm, tips acute; ovaries glabrous or pubescent distally. **Fruits** longitudinally dehiscent, obovoid, 4–5 mm, ≥ calyces. **Seeds** 1–3, ellipsoid, 1.8–2 mm, reddish brown or tan, smooth, glossy. **2*n***=32.

Dry rocky hillsides, clay roadsides, fields, open pine forests. Flowering Jun–Jul. 700–1500 m. BW, Col. WA. Native.

The two subspecies of *T. plumosum* are relatively easy to distinguish. Gillett (1972) stated that the basal leaflet width is the best distinction between the two subspecies, which are also geographically isolated. In subspecies *plumosum*, basal leaflets are 2–5(7) mm wide, with acuminate tips, and the plants are 20–35 cm tall, while in ssp. *amplifolium* (which is endemic to Idaho), basal leaflets range from (8)9–16 mm wide, with acute tips, and the plants are 10–60 cm tall.

Trifolium pratense Linnaeus (p. 736)

Red clover

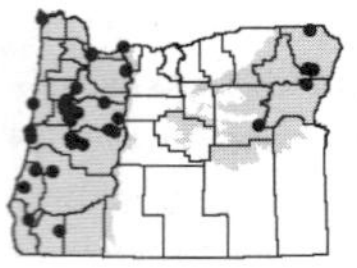

Plants perennial, erect, ascending, or decumbent, 20–70 cm, pilose or glabrous; branched. **Leaves** pinnate; leaflets 3, elliptic to obovate, 15–50 × 7–15 mm, bases obtuse to cuneate, margins subentire, veins prominent, tips obtuse, acute, or retuse, surfaces appressed-pubescent; petioles 10–90 mm; petiolules ~1 mm; stipules broadly triangular, 10–20 mm, margins entire, sometimes ciliate, tips mucronate or setaceous. **Inflorescences** terminal, solitary or paired, 75–100+-flowered, globose or ovoid, 12–77 × 7–22 mm; involucres formed by stipules of distal-most leaves; bracteoles absent. **Peduncles** essentially absent, subtended by stipules of distal leaves. **Pedicels** straight, 0–1 mm. **Flowers** 15–18 mm; calyces tubular-campanulate, 8–11 mm, hairy, veins 10, tubes 3–4 mm, lateral and abaxial lobes 2 × tube, adaxial lobes ≈ tube, orifices slightly closed, hairy; corollas 13–18 mm, usually rose-purple, rarely pink or white, banners elliptic-ovate, 10–13 × 3–5 mm, tips narrowly rounded. **Fruits** transversely dehiscent, oblong, ~3 mm, leathery distally. **Seeds** 1–2, yellow, yellow-brown, or purple, ovoid, 1.6–2 mm, smooth, dull. **2*n***=14.

Fields, prairies, roadsides, disturbed areas. Flowering May–Sep. 0–2000 m. BW, Casc, CR, Est, Sisk, WV. CA, ID, NV, WA; throughout North America; worldwide. Exotic.

Trifolium pratense is native to Eurasia. It is morphologically variable and numerous varieties have been recognized (Zohary & Heller 1984). The species is commonly cultivated as a forage crop, and has become weedy worldwide.

Trifolium productum Greene (p. 737)

Shasta clover

Trifolium kingii S. Watson, *Trifolium kingii* S. Watson var. *productum* (Greene) Jeps.

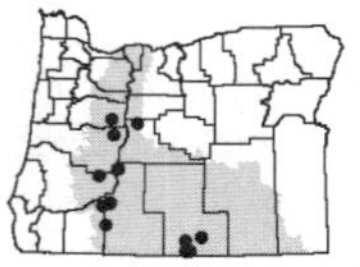

Plants perennial, leaves basal and cauline, ± erect, loosely cespitose, 15–45 cm, glabrous; branched. **Leaves** palmate; leaflets 3, lanceolate or elliptic, 5–20 × 3–13 mm, bases cuneate, margins serrate, veins prominent, tips acute, apiculate, surfaces glabrous; petioles 5–120 mm; petiolules ~1 mm; stipules ovate-lanceolate, 5–20 mm, margins entire, sometimes lobed, tips acute to acuminate. **Inflorescences** terminal or axillary, 15–30-flowered, ellipsoidal or conical, 15–20 × 10–30 mm; rachises prolonged beyond flowers, undivided or forked, often bearing cluster of sterile flower buds distally; involucres absent; bracteoles minute, lanceolate. **Peduncles** 20–130 mm, slightly twisted distally. **Pedicels** strongly reflexed, ~0.5 mm. **Flowers** 12–14 mm; calyces campanulate, 3–3.5 mm, pink to purple, glabrous, veins 10 (5 sometimes faint), tubes 1.5–1.7 mm, lobes triangular-subulate, subequal, orifices open; corollas 12–14 mm, pink to deep purple, banners oblong, 12–14 × 4–5 mm, tips rounded or retuse. **Fruits** longitudinally dehiscent, obliquely ellipsoid, ~5 mm, ≥ calyces. **Seeds** 1–2, flattened-ovoid, 2–3 mm, brown, often purple-mottled, smooth. **2*n***=16.

Open coniferous woods, streambanks, grassy meadows, rocky areas. Flowering May–Aug. 1200–2400 m. BR, Casc, ECas. CA, NV. Native.

Trifolium productum is most morphologically similar to *T. kingii* but differs from the latter by its glabrous calyx and a rachis that is elongated beyond the flowers ≤ 15 mm and is apically forked (Zohary & Heller 1984). *Trifolium productum* is geographically isolated from *T. kingii*. The former is restricted to northern California, western Oregon, and western Nevada, while the latter is restricted to Utah and eastern Nevada (Gillett 1972).

Trifolium repens Linnaeus

White clover, Dutch clover

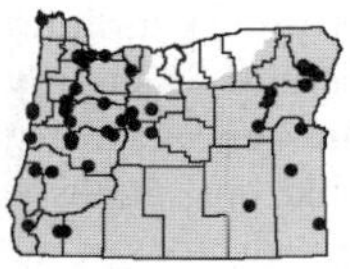

Plants perennial, creeping, 10–40 cm, glabrous or glabrescent; branched; rooting at nodes. **Leaves** pinnate; leaflets 3, obovate, obcordate, or orbiculate, 6–40 × 4–25 mm, often with light markings, bases cuneate, margins serrulate distally, veins moderately prominent, tips rounded, emarginate, or retuse, surfaces glabrous; petioles 50–200 mm; petiolules ≤ 1 mm; stipules lanceolate, 9–13 mm, margins entire, tips short-subulate. **Inflorescences** axillary, 20–40+-flowered, globose, 15–35 × 15–35 mm; involucres absent; bracteoles lanceolate, 1–2 mm, white. **Peduncles** erect from prostrate stems, 15–300 mm. **Pedicels** elongate, strongly reflexed in fruit, 3–5 mm. **Flowers** 8–13 mm; calyces campanulate, 3–5 mm, glabrous, veins 6–10, tubes 1.5–2.5 mm, lobes unequal to subequal, adaxial triangular-lanceolate, < tube, orifices open; corollas 4–12 mm, 1–2 × calyces, white, often becoming pinkish in age, banners ovate-lanceolate or oblong, 4–12 × 1–4 mm, tips rounded. **Fruits** longitudinally dehiscent, linear-oblong, 4–5 mm, ≥ calyces. **Seeds** 3–4, ovoid-reniform, 1 mm, yellow, reddish brown, or light brown, smooth, glossy. ***2n***=16, 28, 32, 48, 64.

Fields, lawns, roadsides, waste places. Flowering Apr–Jul. 0–2200 m. All ecoregions except Col. CA, ID, NV, WA; throughout North America; worldwide. Exotic.

Trifolium repens is native to Eurasia. It may very well be the most important temperate pasture plant (Baker & Williams 1987) and has been considered the most important perennial pasture plant in North America (Piper 1924). It was introduced at least as early as the mid-1800s (Mack 2003) and spread so rapidly that it became known to Native Americans as "White Man's Foot Grass" (Strickland 1801). The species is morphologically diverse; most material from North America represents var. *repens*, but some specimens fit within the circumscriptions given by Zohary and Heller (1984) of var. *giganteum*, with inflorescences to 35 mm wide and leaflets nearly 40 mm, while others have smaller, pale pink petals with hairy petioles and pedicels, and approach var. *biasolettii*.

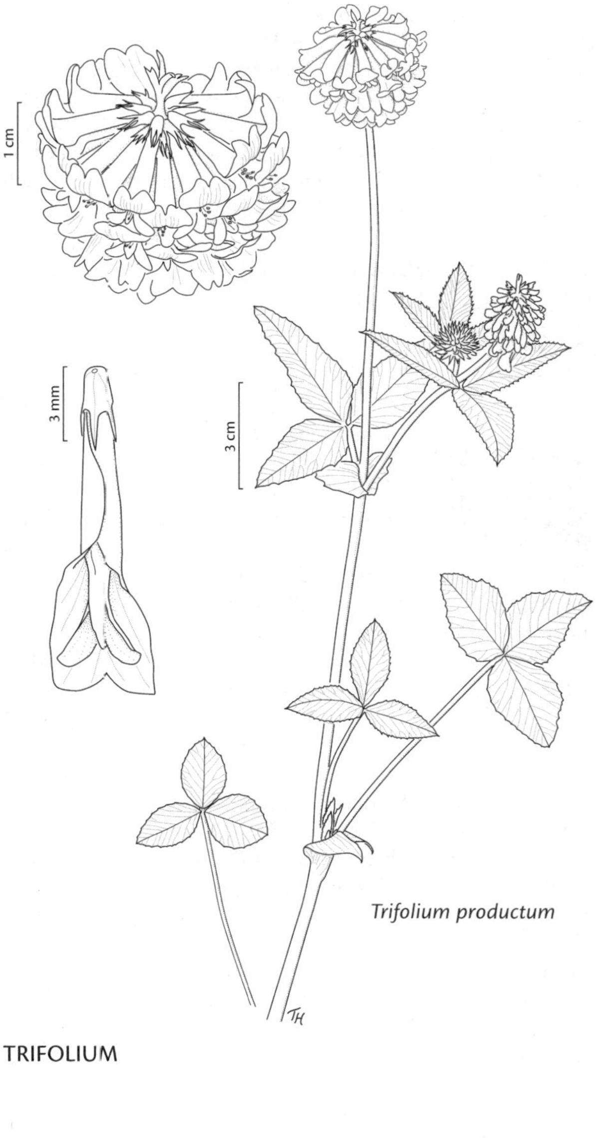

Trifolium productum

TRIFOLIUM

Trifolium resupinatum Linnaeus

Reversed clover, Persian clover

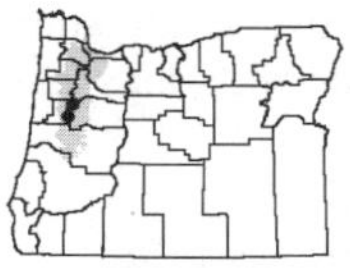

Plants annual, ascending, erect, prostrate, or procumbent, 20–60 cm, glabrous or glabrescent; branched. **Leaves** pinnate; leaflets 3, obovate, ovate, elliptic, or rhombic, 10–30 × 5–30 mm, bases cuneate, margins spinulose-dentate, veins moderately prominent, tips rounded or broadly acute, surfaces abaxially hairy, adaxially glabrous; petioles 5–200 mm; petiolules ≤ 1 mm; stipules lanceolate or ovate, 5–18 mm, margins entire, tips acuminate-filiform. **Inflorescences** axillary, 15–30-flowered, subglobose becoming globose, forming stellate-spreading clusters of calyces in fruit, 5–16 × 8–15 mm; inconspicuously involucrate by a narrow rim, rim ~0.2 mm; bracteoles minute or absent. **Peduncles** 10–60 mm. **Pedicels** slightly reflexed, ≤ 0.5 mm. **Flowers** resupinate, 5–6 mm; calyces tubular-campanulate, inflated in fruit, markedly asymmetric-bilabiate, slit between adaxial lobes, 2.5–3 mm, becoming 8–11 mm in fruit, hairy, short-hairy, or glabrescent in fruit, veins 10, connected by lateral veins, tubes 2 mm, becoming 7–9 mm in fruit, lobes erect, triangular, becoming divergent and linear-lanceolate, unequal, green, orifices open; corollas 5–8 mm, pink to purple, banners oblong, 5–8 × 2 mm, tips emarginate to crenulate. **Fruits** lenticular, 1.8–2.2 mm, < calyces. **Seeds** 1, ovoid, 1.5–1.9 mm, dark purple, olive-green, yellow, or reddish brown, smooth, dull. ***2n***=14, 16, 32.

Wet meadows, lawns, roadsides, fields, waste places. Flowering Jun–Jul. 50–200 m. WV. CA, WA; eastern and midwestern North America; worldwide. Exotic.

Trifolium resupinatum is native to Eurasia and northern Africa and has become weedy throughout the world. Widespread use of *T. resupinatum* as a forage crop in the United States began in the late 1920s.

Trifolium siskiyouense J.M. Gillett

Siskiyou clover

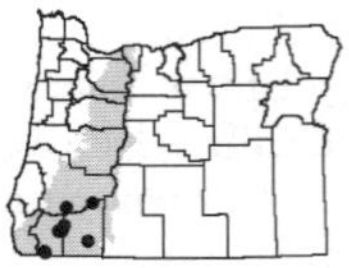

Plants perennial, erect 20–50 cm, glabrous; sparsely branched; roots tuberous, rhizomes elongate, slender. **Leaves** palmate; leaflets 3, obovate, oblanceolate, or elliptic, 8–30 × 5–7 mm, bases cuneate, margins finely serrulate, veins fine or thickened distally, tips acute, mucronate, or retuse, surfaces glabrous; petioles 16–24 mm; petiolules ~0.5 mm; stipules lanceolate, 8–15 mm, margins entire to slightly lacerate, tips acuminate. **Inflorescences** mostly terminal, sometimes axillary, 30–50-flowered, subglobose, 12–18 × 15–20 mm; involucres shallowly bowl-shaped, deeply lobed ≥ ⅔ of length, lobes irregular to bifurcate, 18–22 mm; bracteoles linear, 1–2 mm. **Peduncles** 28–34 mm. **Pedicels** straight, 0.5–0.8 mm. **Flowers** 10–13 mm; calyces tubular, 0.5–0.6 mm, glabrous, veins 10, tubes ~2 mm, lobes ± equal, 3–4 mm, abaxial lobes > tube, other lobes often also as long, orifices open; corollas 9–12 mm, white to purple, banners oblong, 0.7–1.2 × 2–3 mm, tips retuse. **Fruits** oblong, 2–4 mm. **Seeds** 2–4, subglobose or mitten-shaped, ~1.5 mm, brown, smooth.

Wet meadows, grassy hillsides. Flowering May–Aug. 400–1500 m. Casc, Sisk. CA. Native.

Trifolium siskiyouense is restricted to a small geographical region in Shasta and Siskiyou counties, California, and Josephine County, Oregon. It is difficult to distinguish from *Trifolium wormskioldii* without underground parts (showing the tubers) but can be separated from the latter by deeply cut involucral segments, long-lobed, slender calyces, and the combination of long wing petals and short keel petals, with the wings 0.1–0.2 mm > the keel (Gillett 1980).

Trifolium striatum Linnaeus

Knotted clover

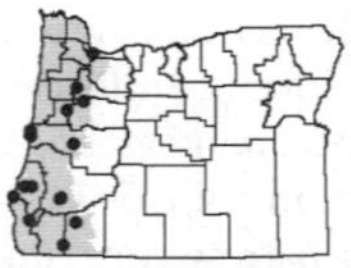

Plants usually annual, rarely biennial, erect, ascending, or decumbent, 10–50 cm, patent-villous; branched from bases. **Leaves** pinnate; leaflets 3, obovate to oblong, 9–16 × 4–10 mm, bases cuneate, margins denticulate, veins fine, tips obtuse, obcordate, or emarginate, surfaces hairy; petioles 5–50 mm; petiolules ~0.5 mm; stipules lanceolate-ovate, 9–11 mm, margins entire, tips acute-acuminate, setaceous, ciliate. **Inflorescences** axillary or terminal, solitary or paired, 20–60-flowered, ovoid or oblong, 6–10 × 8–16 mm; involucres absent but subtended by stipular pseudo-involucre; bracteoles absent. **Peduncles** 0–10 mm. **Pedicels** absent. **Flowers** 5–7 mm; calyces ellipsoidal to urceolate, 3–4 mm, hairy, veins 10, tubes 2–3 mm, lobes erect or spreading, unequal, abaxial lobes subulate, longest equal to tube, orifices open; corollas 25–35 mm, pink, banners oblong, 2.5–3.5 × 1.5–2 mm, tips retuse. **Fruits** transversely dehiscent, ovoid, 2–2.5 mm, ≥ calyces, leathery distally. **Seeds** 1, globose to ovoid, 1–1.5 mm, tan or reddish brown, smooth, glossy. **2*n***=14.

Disturbed areas. Flowering May–Jul. 0–700 m. CR, Est, Sisk, WV. CA; scattered in eastern US, AR, OK, MO; Asia, Europe. Exotic.

This species is native to Eurasia and has been recorded as an occasional weed in the United States, having been first collected on ballast in New Jersey in 1880. It is found sporadically as a weed of disturbed habitats and appears to be spreading rapidly.

Trifolium subterraneum Linnaeus (p. 739)

Subterranean clover, subclover

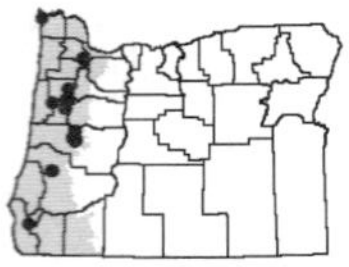

Plants annual, prostrate to ascending, 10–80 cm, glabrous or appressed-soft-pubescent; branched. **Leaves** palmate; leaflets 3, broadly obcordate, 8–28 × 10–30 mm, bases cuneate, widely spaced, margins mostly entire, slightly dentate distally, veins delicate, tips emarginate, surfaces appressed-sericeous; petioles 10–200 mm; petiolules ~1 mm; stipules ovate, 5–30 mm, margins entire or slightly toothed, ciliate, tips acute to acuminate. **Inflorescences** axillary, globose or cylindrical, 5–15 × 8–15 mm, bur-like in fruit; involucres cup-shaped, bracts ± connate, 5–15 mm, glabrous or inconspicuously hairy, lobes 8–12, ovate, sharply toothed; bracteoles absent. **Peduncles** 15–74 mm, in fruit curving, growing into ground. **Pedicels** absent. **Flowers** outer 2–8, fertile, calyx tubes often with purple band at top; corollas 7–10 mm, white, pink, or pink-striped, banners ovate-elliptic, 7–10 × 1.5–2 mm, tips rounded; inner many, sterile, calyces stalk-like, with 4–5 bristles at tip; corollas absent, recurving and forming burs to surround fruits, delivered to ground by growing peduncles. **Fruits** obovoid, 3–4 mm, ≥ calyces. **Seeds** 1, ellipsoid, 2.6–3 mm, purplish black, smooth, dull. **2*n***=16.

Meadows, roadsides, disturbed areas. Flowering Apr–Jun. 0–900 m. CR, Est, Sisk, WV. CA, WA; north to British Columbia, southeastern US; Africa, Asia, Australia, Europe. Exotic.

Trifolium subterraneum is native to Eurasia and northern Africa and has become weedy in areas where it has been cultivated, such as South Africa and Australia. It was first introduced by the USDA in about 1921 as a pasture crop; it is utilized as such in the western and southern United States (McGuire 1985).

Trifolium variegatum Nuttall (p. 739)

White-tip clover

Trifolium geminiflorum Greene, *Trifolium melananthum* Hook. & Arn., *Trifolium variegatum* Nutt. var. *geminiflorum* (Greene) Vincent, *Trifolium variegatum* Nutt. var. *major* Lojac.

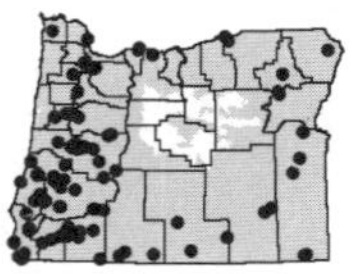

Plants annual, prostrate to erect, 2.5–50 cm, glabrous; unbranched or branched. **Leaves** palmate, dimorphic, with gradual or abrupt transition between proximal and distal leaves; leaflets 3, obovate-obcordate to elliptic-obovate, 2.9–11.4 × 1.9–5 mm, bases cuneate, margins serrate, veins fine, tips emarginate-mucronate or rounded-mucronate, surfaces glabrous; petioles 1–95 mm; petiolules 0.1–0.8 mm; stipules broadly triangular to ovate, recurved or not, 2–10 mm, margins laciniate, tips acute. **Inflorescences** axillary, 1–25-flowered, racemose-umbellate, 4–12 × 1–10 mm, not elongate in fruit; involucres flat, laciniate, 3–12 mm, 2–7-lobed, lobes toothed, longest tooth 0.6–3 mm; bracteoles absent. **Peduncles** 4–6 mm. **Pedicels** straight, 0.1–0.5 mm. **Flowers** 3–10 mm; calyces tubular, 2.1–6.8 mm, glabrous, veins 10–20, tubes 1.9–2.4 mm, lobes awl-like to shouldered, subequal, tips purple or stramineus, orifices open; corollas 3–10 mm, included in or exserted beyond tips of calyces 0.1–5.7 mm, purple, usually white-tipped, banners obovate, 3.2–9.7 × 0.6–2.6 mm, tips emarginate, keel petals not beaked. **Fruits** laterally compressed, ellipsoid, 2.2–4 mm, < calyces, chartaceous. Seeds 1–2, globular to mitten-shaped, 1.1–1.6 mm, blackish brown, dark-speckled or not. **2*n***=16.

Wet places, meadows. Flowering Apr–Aug. 0–2200 m. All ecoregions except Lava. CA, ID, NV, WA; north to AK, east to WY, south to Mexico. Native.

Trifolium variegatum is a widespread and morphologically variable species. Because of this variability, *T. variegatum* and the related species *T. appendiculatum* and *T. polyodon* were lumped into seven "phases" by Isely (1998), who described these as "ecotypic groups," some of which might deserve varietal or specific rank. A recent revision of *T. variegatum* (Foster 2015) resulted in recognition of two species, a widespread *T. variegatum* (including *T. melananthum*, *T. geminiflorum*, and *T. subsalinum*), and *T. appendiculatum*, which is restricted to California, except for reports from Grants Pass (Josephine County) from 1910, and Sutherlin (Douglas County) from 1932. Cooper (1957) considered this species a good choice for cultivation in wet areas for grazing and hay production.

Trifolium vesiculosum Savi (p. 739)

Arrowleaf clover

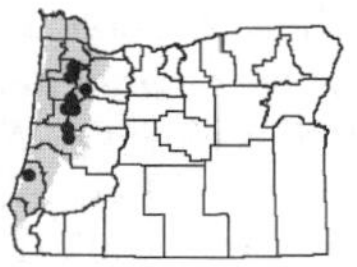

Plants annual, erect or ascending, 15–70 cm, glabrous; branched. **Leaves** pinnate; leaflets 3, obovate to oblong, elliptic or lanceolate, 5–40 × 5–15 mm, bases cuneate, margins spinulose-denticulate, veins prominent, thickened, tips apiculate, surfaces glabrous; petioles 5–100 mm; petiolules ~1 mm; stipules linear-lanceolate, 10–35 mm, margins entire, tips subulate or setaceous. **Inflorescences** terminal or axillary, 50–100+-flowered, globose, ovoid, or oblong, 30–60 × 20–35 mm; involucres absent; bracteoles lanceolate, 6–7 mm, nearly as long as calyces, acuminate. **Peduncles** 10–120 mm. **Pedicels** absent. **Flowers** 12–16 mm; calyces urceolate, symmetrical, inflated in fruit, 6–10 mm, glabrous, veins 20–36, connected transversely in fruit, tubes 3–5 mm, lobes reflexed, subulate, subequal, as long as tube, orifices constricted; corollas 12–15 mm, white becoming pink, banners ovate, broadly clawed, 12–15 × 2–4 mm, striate, tips acute-acuminate. **Fruits** ellipsoid, 2.5–3.5 mm, < calyces. **Seeds** 2–3, ovoid, 1–1.5 mm, brown, roughened. **2*n***=16.

Fields, roadsides, forest openings. Flowering Jun–Aug. 0–300 m. CR, WV. CA, WA; southeastern US; Asia, Europe. Exotic.

Trifolium vesiculosum is native to Eurasia and is a recently introduced forage crop, the first cultivar of which was introduced in 1963; it is cultivated in the southern and western United States.

Trifolium willdenovii Sprengel

Tomcat clover

Trifolium tridentatum Lindl., *Trifolium tridentatum* Lindl. var. *tridentatum*

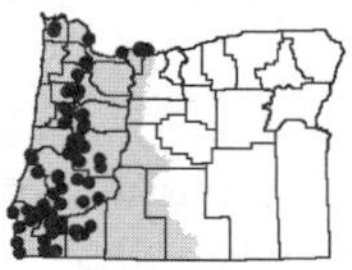

Plants annual, erect, 10–60 cm, glabrous; branched. **Leaves** palmate; leaflets 3, linear, lanceolate, or elliptic, 10–50 × 2–15 mm, bases cuneate, margins serrate, veins fine, tips acute, blunt, or retuse, surfaces glabrous; petioles 10–80 mm; petiolules ~0.5 mm; stipules lanceolate or ovate, 10–20 mm, margins dentate or lacerate, tips acuminate. **Inflorescences** axillary or terminal, 10–50-flowered, globose or ovoid, 10–35 × 10–30 mm; involucres flattened or bowl-shaped, divided ¼–⅓ their length, when flattened laterally not hiding

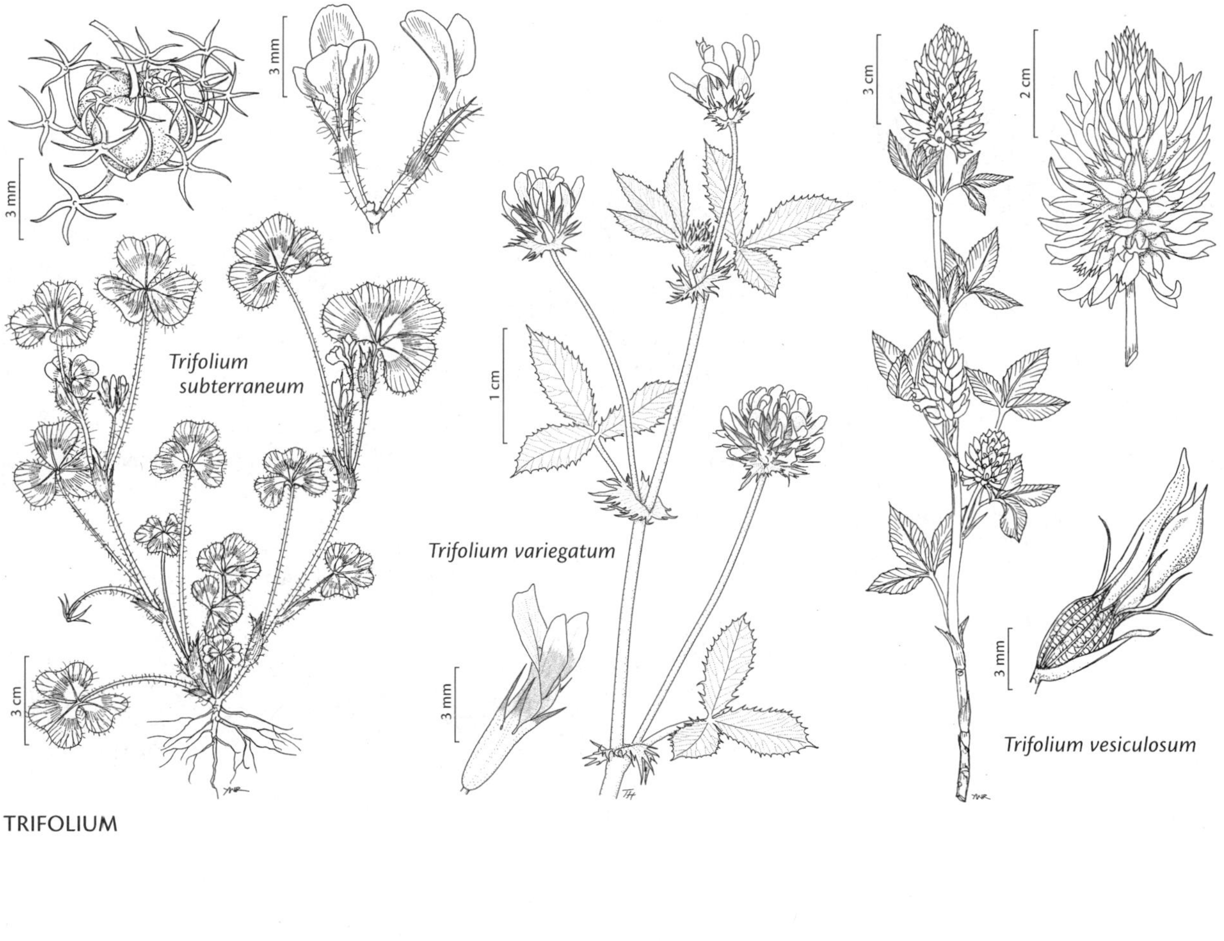

TRIFOLIUM

flowers except proximally, 3–9 mm, deeply lacerate; bracteoles absent. **Peduncles** 20–150 mm, glabrous. **Pedicels** erect, ~0.5 mm. **Flowers** 12–20 mm; calyces tubular-campanulate, slit between adaxial lobes, 6–9 mm, glabrous, veins 10–15, tubes 3–7 mm, lobes triangular or subulate, 3-fid or shouldered, unequal, ≤ tube, orifices open; corollas 10–18 mm, usually white with purple spot, sometimes lavender or pink or all white, banners narrowly oblong, 12–20 × 3–4 mm, tips obtuse or retuse. **Fruits** ellipsoid, 3 mm. **Seeds** 1–2, globose-ellipsoid, 1.5–2.5 mm, yellow, mottled, smooth. ***2n***=16.

Roadsides, meadows, lawns, hillsides, stream margins, open oak chaparral. Flowering Apr–Jul. 0–1500 m. Casc, CR, ECas, Est, Sisk, WV. CA, ID, WA; north to British Columbia, southeast to AZ, south to Mexico. Native.

Trifolium willdenovii was considered by Zohary and Heller (1984) to be one of the most variable clover species in North America.

Trifolium wormskioldii Lehmann (p. 740)

Springbank clover

Trifolium fimbriatum Lindl., *Trifolium wormskioldii* Lehm. var. *wormskioldii*

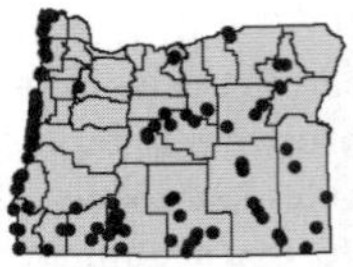

Plants perennial, erect, ascending, or decumbent, 10–40 cm, glabrous; branched; roots not tuberous, rhizomes elongate. **Leaves** palmate; leaflets 3, obovate to elliptic, 4–42 × 2–13 mm, bases cuneate or rounded, margins spinulose-serrate, veins fine or thickened distally, tips usually acute to rounded, sometimes retuse or mucronate, surfaces glabrous; petioles 10–110 mm; petiolules ~0.5 mm; stipules ovate to lanceolate, 15–20 mm, margins entire (proximal stipules), lacerate (distal stipules), tips acute to acuminate. **Inflorescences** axillary or terminal, 20–50-flowered, subglobose or globose, 18–25 × 20–25 mm; involucres broadly bowl-shaped, bracts ± connate, divided ⅓–½ their length, 12–20 mm, lobes > 10, sharply and acutely serrate, veins prominent; bracteoles linear or ovate, 1–2 mm. **Peduncles** 15–110 mm. **Pedicels** straight, 1.5–2 mm. **Flowers** odor somewhat unpleasant, 12–14 mm; calyces tubular, 5–6 mm, whitish, glabrous, veins 10, tubes 2–3 mm, lobes subequal, ≈ tube, rarely longer, green to purple, orifices open; corollas 10–12 mm, usually magenta to purple, rarely white or bicolored, banners ovate-oblong, 12–14 × 0.2–0.3 mm, tips truncate or rounded, slightly retuse. **Fruits** oblong, 2–4 mm. **Seeds** 3–4, subglobose or mitten-shaped, ~1.5 mm, brown, sometimes mottled, smooth. ***2n***=16, 32.

Saline flats, beaches, meadows, grassy areas, alluvial soil. Flowering May–Oct. 0–2300 m. All ecoregions. CA, ID, NV, WA; north to British Columbia, east to WY, southeast to CO and NM, south to Mexico. Native.

Trifolium wormskioldii, a largely tetraploid species, is widespread in western North America. Distinguishing *T. wormskioldii* from closely allied species (especially *T. mucronatum*) is sometimes difficult, which led Barneby (1989) to synonymize *T. mucronatum* and *T. pinetorum* under the former name. However, *Trifolium wormskioldii* produces long white rhizomes, while *T. mucronatum* produces only short rhizomes at most, often only small, fibrous roots or taproots (Gillett 1980). While the distinction of *T. pinetorum* from *T. wormskioldii* was supported by Ellison et al. (2006), material of *T. mucronatum* was not included in their DNA analyses, leaving the relationships murky.

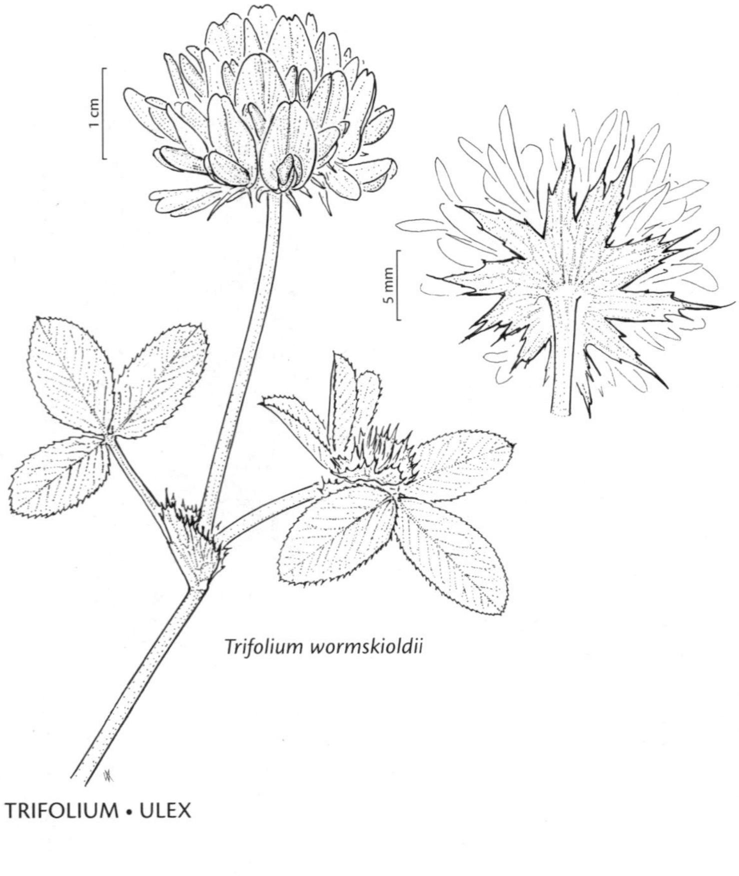

Trifolium wormskioldii

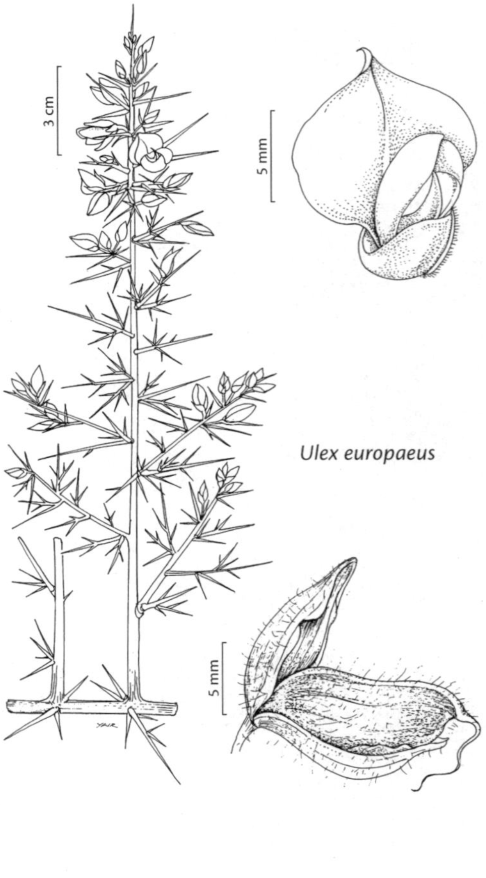

Ulex europaeus

TRIFOLIUM • ULEX

ULEX Gorse

Cameron J. Ross

Shrubs armed, evergreen. **Stems** strongly angled, spinescent, branched. **Leaves** alternate, simple, linear, becoming spines with age. **Inflorescences** axillary clusters, sometimes racemes or flowers solitary. **Flowers** calyces yellow, 2-lipped, persistent, upper lips 2-toothed, lower lips 3-toothed; corollas yellow, petals equal; stamens 10, monadelphous; styles 1, stigmas 1; bracteoles present. **Fruits** dehiscent, protruding from calyces, ovate to linear-oblong, pubescent. **Seeds** 1–6(8).

Worldwide. ~20 species; 1 species treated in *Flora.*

Ulex europaeus Linnaeus (p. 740)

Common gorse

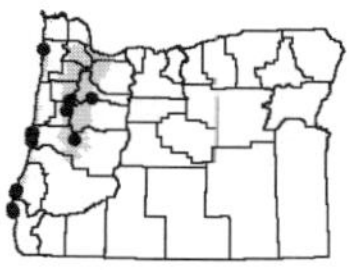

Plants 5–25(30) dm. **Leaves** margins entire. **Inflorescences** in 1–3-flowered clusters, rarely racemes; bracteoles 2, persistent; pedicels (3)5(6) mm, pubescent. **Flowers** fragrant, papilionaceous; calyces 10–16 mm, ⅔ corolla length; corollas 15–18(20) mm, banners ovate, wings oblong to obtuse, keels oblong to obtuse. **Fruits** 1–2 cm, dark brown to black, villous.

Coastal and disturbed areas. Flowering Apr–Jul. 0–500 m. Est, WV. CA, WA; north to British Columbia, mid-Atlantic and northeastern US; Africa, Europe. Exotic.

This species was likely introduced to the Pacific Northwest in Bandon, Oregon and has since spread to California, Washington, and British Columbia.

VICIA Vetch

Steven L. Broich

Herbs perennial or annual, glabrous, glabrate or pubescent with unicellular trichomes; from rhizomes or taproots. **Stems** angled or ridged. **Leaves** even-pinnate; leaflets 2–20, alternate or opposite; rachises usually terminated in a simple or branched, sometimes mucronate tendril; stipules inconspicuous or foliose, semisagittate or ligulate, with or without nectariferous patch. **Inflorescences** erect or lax, 1–many-flowered axillary racemes; bracts 0. **Flowers** chasmogamous or cleistogamous; calyx bases sometimes gibbous, tubes symmetrically or obliquely campanulate, with 5 equal or unequal lobes, when unequal dorsal lobes usually short, triangular, lateral lobes longer and linear; corollas white, pink or lavender to rose, blue to purple, or yellow, wings and keels often lighter than banners, banner claws convolute around other petal claws, banner blades usually folded perpendicular to axis of flower, wings adherent to keels with a ball and socket mechanism, usually > keels, keels incurved, attenuate, lower and distal margins fused; stamens diadelphous, anthers uniform; ovaries glabrous or pubescent, styles terete, flattened abaxially or adaxially, evenly pubescent just below stigma and/or abaxially tufted. **Fruits** dehiscent, terete or flattened, elliptic to linear, glabrous or pubescent, margins usually obscure, splitting along both margins; valves twisted after dehiscence; sessile or stipitate. **Seeds** 3–15, spherical, sometimes angular, usually brown or black.

Africa, Asia, Europe, North America. ~160 species; 10 species treated in *Flora.*

Vicia faba and *V. lathyroides* have been collected in Oregon, but neither has become established.

Key to *Vicia* species:

1. Stipules lacking nectariferous patch; inflorescences usually > subtending leaves, 1–50-flowered.
   2. Leaflets 16–30; banners oblong, not constricted between blades and claws . . . . ***V. nigricans*** var. ***gigantea***
   2' Leaflets 4–24; banners ovate, oblong, or fiddle-shaped, constricted between blades and claws.
      3. Banners fiddle-shaped, blades and claws of equal width; calyx bases gibbous or symmetrical; styles laterally compressed.
         4. Flowers 8–18 mm; inflorescences ≥ 10-flowered.
            5. Plants perennial; calyx bases symmetrical or slightly gibbous; banner blades ≈ claws; fruits 4–8 mm wide . . . . ***V. cracca***
            5' Plants annual or rarely biennial; calyx bases distinctly gibbous; banner blades ½ claws; fruits 6–10 mm wide . . . . ***V. villosa***
         4' Flowers 2–9 mm; inflorescences 1–7-flowered.
            6. Leaflets 4–8; fruits 22–38 mm, stipitate. . . . . ***V. hassei***
            6' Leaflets 8–16; fruits 6–10 mm, sessile . . . . ***V. hirsuta***
      3' Banners ovate or oblong, blades wider than claws; calyx bases not gibbous; styles terete or abaxially compressed.
         7. Plants perennial; inflorescences 5–9-flowered; flowers 12–25 mm; fruits distally asymmetrical, upcurved . . . . ***V. americana***
         7' Plants annual; inflorescences 1–3-flowered; flowers 3–6 mm; fruits distally rounded . . . . ***V. tetrasperma***

1' Stipules with abaxial nectariferous patch; inflorescences usually < subtending leaves, 1–4-flowered.
  8. Calyx lobes subequal; corollas violet-purple, lavender, or whitish; fruits glabrous to pubescent . . . . ***V. sativa***
  8' Calyx lobes unequal; corollas yellow or cream, sometimes purplish-tinged; fruits villous or pilose.
    9. Banners glabrous; corollas yellow . . . . ***V. lutea***
    9' Banners adaxially pubescent; corollas cream, sometimes purplish-tinged . . . . ***V. pannonica***

Vicia americana Muhlenberg (p. 742)

American vetch

Vicia americana Muhl. ssp. *americana*, *Vicia americana* Muhl. var. *americana*, *Vicia americana* Muhl. var. *linearis* (Nutt.) S. Watson, *Vicia americana* Muhl. ssp. *oregana* (Nutt.) Abrams, *Vicia americana* Muhl. var. *oregana* (Nutt.) Nels., *Vicia americana* Muhl. var. *truncata* (Nutt.) Brewer, *Vicia americana* Muhl. var. *villosa* (Kellogg) F.J. Herm., *Vicia californica* Greene

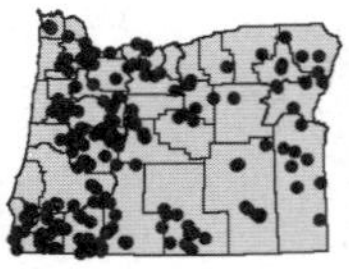

Plants perennial, 1–10 dm. **Leaves** leaflets 8–18, ovate or elliptic to linear, 15–35 × 1–14 mm, surfaces glabrous to finely pubescent; rachises 2–8 cm; stipules inconspicuous. **Inflorescences** 5–9-flowered, 2–8 cm, usually > subtending leaves. **Flowers** 12–25 mm; calyx bases symmetrical, lobes unequal, < tubes; corollas bluish purple, rarely white, glabrous, banners oblong, constricted between blades and claws, blades wider than claws; styles abaxially compressed. **Fruits** distally asymmetrical, upcurved, 25–35 × 5–6 mm, tawny to brown, glabrous or pubescent; stipes to 5 mm.

Open moist woods, thickets, meadows, tall grass prairies, sagebrush steppes, roadsides, disturbed areas. Flowering Mar–Sep. 0–2300 m. All ecoregions. CA, ID, NV, WA; north to AK, northeast to Quebec, east to NY, southeast to TX, south to Mexico. Native.

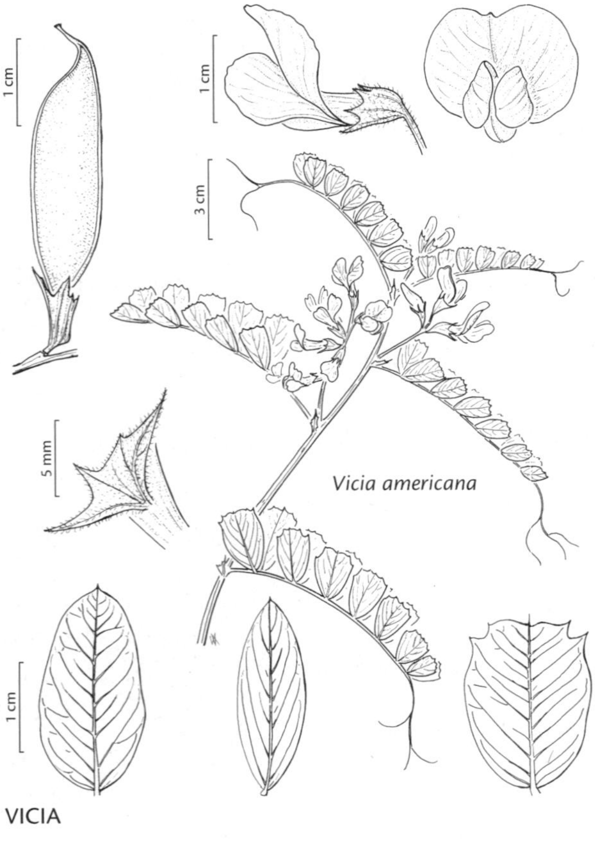

VICIA

Vicia americana intergradation is widespread. Gunn (1968) described two relatively distinct entities (varieties *americana* and *minor*) that may represent the extremes of a continuum of variation in some regions. If these taxa were accepted, only *V. americana* var. *americana* would occur in Oregon.

Vicia cracca Linnaeus

Bird vetch, tufted vetch, cat peas, tinegrass

Vicia semicincta Greene

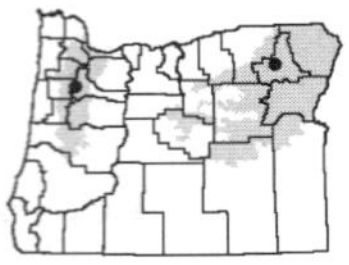

Plants perennial, 5–20 dm. **Leaves** leaflets 10–24, narrowly oblong or elliptic to linear, 10–25 × 5–8 mm, surfaces glabrate to appressed-pilose; rachises 2–6 cm; stipules foliose. **Inflorescences** 20–50-flowered, 4–15 cm, > subtending leaves. **Flowers** 8–12 mm; calyx bases symmetrical or slightly gibbous, lobes unequal, longer ones ≈ tubes; corollas blue-violet or purple, sometimes lilac or whitish, glabrous, banners fiddle-shaped, constricted between blades and claws, blades and claws ≈ length and width; styles laterally compressed. **Fruits** 15–30 × 4–8 mm, tawny, glabrous; stipes to 1.5 mm.

Open woodlands, old fields, thickets, roadsides. Flowering Jun–Aug. 0–900 m. BW, WV. CA, ID, WA; north to AK, northeast to Greenland, east to Nova Scotia, southeast to GA; Asia, Europe. Exotic.

Vicia hassei S. Watson

Hasse's vetch, slender vetch

Vicia exigua Nutt. var. *hassei* (S. Watson) Jeps.

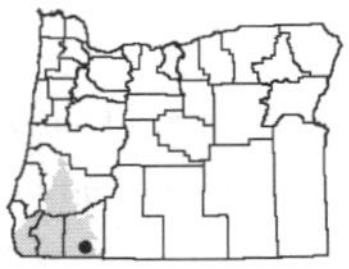

Plants annual, to 10 dm. **Leaves** leaflets 4–8, elliptic to lanceolate-linear, 8–40 × 1–9 mm, surfaces glabrous to sparsely pubescent; rachises 2–4 cm; stipules inconspicuous. **Inflorescences** 1–2-flowered, 1–3 cm, > subtending leaves. **Flowers** 6–9 mm; calyx bases symmetrical, lobes subequal, ½ tubes; corollas white to faint bluish or lavender, glabrous, banners fiddle-shaped, blades and claws of equal width; styles laterally compressed. **Fruits** 22–38 × 5–7 mm, tawny, glabrous or finely pubescent; stipes to 3 mm.

Grassy or brush-covered slopes, streamsides, floodplains, forest margins. Flowering Mar–May. 400–600 m. Sisk. CA; south to Mexico. Native.

Vicia hirsuta (Linnaeus) S.F. Gray

Tiny vetch, hairy vetch

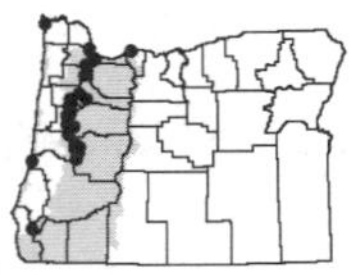

Plants annual, 2–8 dm. **Leaves** leaflets 8–16, narrowly oblong-elliptic to linear, 5–15 × 1–4 mm, surfaces glabrous; rachises 2–8 cm; stipules inconspicuous. **Inflorescences** 2–7-flowered, 1–5 cm, usually > subtending leaves. **Flowers** 2–4 mm; calyx bases symmetrical, lobes

subequal, ≥ tubes; corollas white to purplish blue, glabrous, banners fiddle-shaped, blades and claws of equal width; styles laterally compressed. **Fruits** 6–10 × 3–4 mm, black, hirsute; sessile.

Roadsides, fields, open woodlands, streamsides, disturbed areas. Flowering May–Jul. 0–800 m. Casc, Est, Sisk, WV. CA, WA; widely scattered in North America; Asia, Europe. Exotic.

Vicia lutea Linnaeus

Yellow vetch

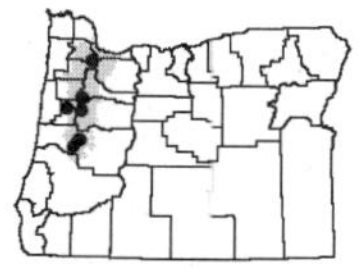

Plants annual, 2–6 dm. **Leaves** leaflets 6–16, oblong to lanceolate-linear, 8–20 × 1–5 mm, surfaces abaxially sparsely pubescent with long, pustule-based trichomes, adaxially glabrous; rachises 2–6 cm; stipules inconspicuous, with abaxial nectariferous patch. **Inflorescences** 3-flowered, < subtending leaves. **Flowers** 18–25 mm; calyx bases symmetrical, lobes unequal, ≈ tubes; corollas yellow, glabrous, banners oblong or blades slightly wider than claws; styles abaxially compressed. **Fruits** 25–35 × 7–14 mm, reddish black, pilose; stipes to 3 mm.

Roadsides and adjacent fields. Flowering May–Jun. 0–200 m. WV. CA; AL, LA, TX, VA; Africa, Asia, Europe. Exotic.

Vicia nigricans Hooker & Arnott var. ***gigantea*** (Hooker) Broich

Giant vetch

Vicia gigantea Hook.

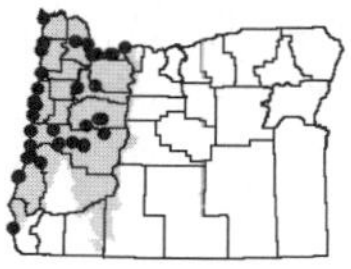

Plants perennial, 6–20 dm. **Leaves** leaflets 16–30, ovate to lanceolate, 20–50 × 5–20 mm, surfaces glabrous; rachises 9–23 cm; stipules foliose. **Inflorescences** 6–19-flowered, 5–25 cm, usually > subtending leaves. **Flowers** 15–25 mm; calyx bases symmetrical, lobes unequal, ≈ tubes; corollas reddish purple to yellowish, glabrous, banners oblong, not constricted between blades and claws; styles abaxially compressed. **Fruits** 33–55 × 10–20 mm, black, glabrous; stipes to 4 mm.

Upper margins of beaches, coastal forests. Flowering Mar–Jul. 0–700 m. Casc, CR, Est, WV. CA, WA; north to AK; South America. Native.

Lassetter and Gunn (1979) made the case for combining the common North American Pacific Coast taxon, *Vicia gigantea,* with *V. nigricans* of the South American coast.

Vicia pannonica Crantz

Hungarian vetch

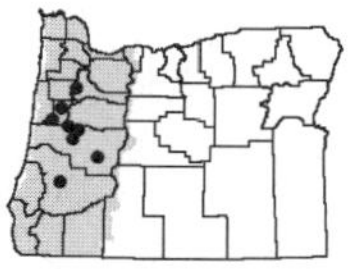

Plants annual, 3–8 dm. **Leaves** leaflets 10–20, obovate or oblong-elliptic to linear, 5–20 × 2–6 mm, surfaces pilose; rachises 4–8 cm; stipules inconspicuous, with abaxial nectariferous patch. **Inflorescences** 2–4-flowered, < subtending leaves. **Flowers** 14–20 mm; calyx bases symmetrical, lobes unequal, ≈ tubes; corollas cream, sometimes purplish-tinged, banners oblong or abruptly contracted into narrow claws, adaxially pubescent; styles abaxially compressed. **Fruits** 20–30 × 7–11 mm, purple-black, villose; stipes to 2 mm.

Pastures, meadows, roadsides. Flowering May–Jun. 50–400 m. Casc, CR, Sisk, WV. CA, ID, WA; AR, CT, NC, GA; Asia, Europe. Exotic.

Vicia pannonica is cultivated for forage and as a seed crop and has become locally established, particularly in Pacific Coast states.

Vicia sativa Linnaeus

Common vetch, spring vetch, tare

Plants annual, 3–10 dm. **Leaves** leaflets 8–14, ovate-oblong to linear, 15–30 × 5–15 mm, surfaces abaxially sparsely pubescent, adaxially glabrous; rachises 3–8 cm; stipules foliose, with abaxial nectariferous patch. **Inflorescences** 1–4-flowered, < subtending leaves. **Flowers** 10–30 mm, calyx bases symmetrical, lobes subequal, ≤ tubes; corollas violet-purple, lavender, or whitish, glabrous, banners abruptly contracted into narrow claws; styles abaxially compressed. **Fruits** 25–60 × 3–11 mm, reddish brown to black, glabrous or pubescent; sessile.

Worldwide. 7 varieties; 2 varieties treated in *Flora.*

Key to *Vicia sativa* varieties:

1. Flowers 10–20 mm; calyx lobes usually < tubes; fruits reddish brown to black, glabrous. . . . . *V. s.* var. ***angustifolia***

1' Flowers 20–30 mm; calyx lobes usually ≈ tubes; fruits yellow to brown, pubescent. . . *V. s.* var. ***sativa***

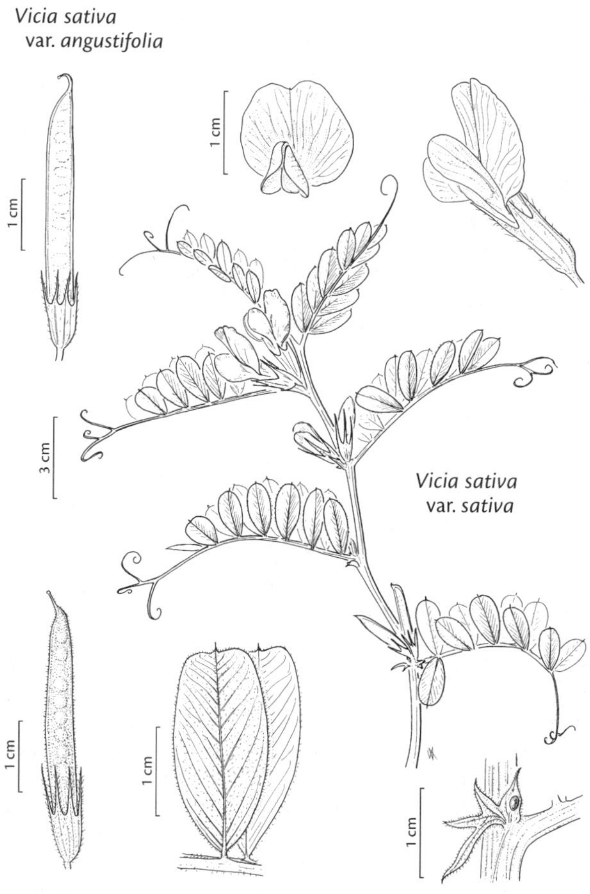

VICIA

Vicia sativa Linnaeus var. **_angustifolia_** (Linnaeus ex Reichard) Wahlenberg (p. 743)

Common vetch

Vicia angustifolia (L.) Rchb., *Vicia sativa* L. ssp. *nigra* (L.) Ehrh.

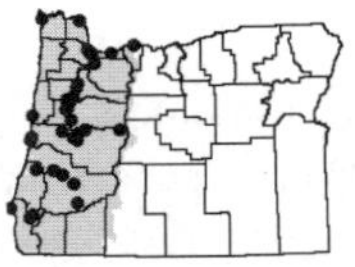

Leaflets narrowly elliptic to linear. **Flowers** 10–20 mm; calyx lobes usually < tubes. **Fruits** reddish brown to black, glabrous. Fields, roadsides, waste areas. Flowering Apr–Jul. 0–1400 m. Casc, CR, Est, Sisk, WV. CA, ID, WA; throughout most of North America; Asia, Europe. Exotic.

Vicia sativa Linnaeus var. **_sativa_** (p. 743)

Common vetch, spring vetch, tare

Vicia sativa L. ssp. *sativa*

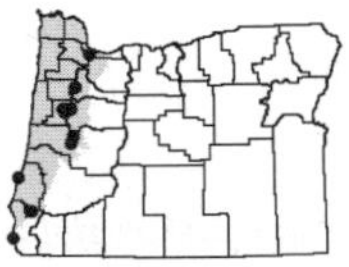

Leaflets usually ovate-oblong to narrowly elliptic, sometimes linear. **Flowers** 20–30 mm; calyx lobes usually ≈ tubes. **Fruits** yellow to brown, pubescent.

Fields, roadsides, waste areas. Flowering May–Jun. 0–1200 m. CR, Est, WV. CA, ID, NV, WA; worldwide. Exotic.

Vicia tetrasperma Linnaeus

Lentil tare, slender vetch, sparrow vetch

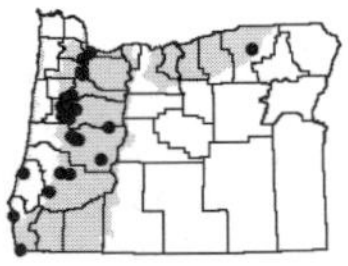

Plants annual, 1–5 dm. **Leaves** leaflets 4–12, elliptic to linear, 6–20 × 1–3 mm, surfaces glabrous; rachises 1–4 cm; stipules inconspicuous. **Inflorescences** 1–3-flowered, 1–3 cm, > subtending leaves. **Flowers** 3–6 mm; calyx bases symmetrical, lobes unequal, longer ones ≈ tubes; corollas light purple to pale lavender, glabrous, banners ovate to oblong, constricted between blades and claws, blades wider than claws; styles terete. **Fruits** distally rounded, 10–15 × 3–4 mm, yellow to light brown, glabrous; stipes to 1 mm.

Fields, roadsides, woodlands, disturbed areas. Flowering May–Jun. 0–800 m. Casc, Col, Est, Sisk, WV. CA, ID, WA; scattered in North America; Africa, Asia, Europe. Exotic.

Vicia villosa Roth

Hairy vetch, winter vetch, fodder vetch

Plants annual or rarely biennial, 2–13 dm. **Leaves** leaflets 10–18, narrowly oblong to linear, 10–35 × 1–10 mm, surfaces glabrous to villous; rachises 4–9 cm; stipules inconspicuous. **Inflorescences** 10–20+-flowered, 3–12 cm, > subtending leaves. **Flowers** 14–18 mm; calyx bases gibbous, lobes unequal, usually > tubes, corollas violet to rose and white, glabrous, banners fiddle-shaped, constricted between blades and claws, blades ½ claw length and ≈ width; styles laterally compressed. **Fruits** 15–40 × 6–10 mm, tawny, glabrous; stipes to 3 mm.

Worldwide. 5 varieties; 2 varieties treated in *Flora*.

Introduced as a forage crop, *V. villosa* has become widely established in North America.

Vicia benghalensis, a closely related species, is cultivated as a cover crop along the Pacific Coast. It has become locally established in California and perhaps elsewhere where it is cultivated. *Vicia benghalensis* can be distinguished from *V. villosa* by its pubescent fruits.

Key to _Vicia villosa_ varieties:

1. Plants glabrate with inconspicuous, subappressed to spreading pubescence; inflorescences usually 10–20-flowered. . . . . *V. v.* var. ***glabrescens***

1' Plants villous; inflorescences usually ≥ 20-flowered . . . . *V. v.* var. ***villosa***

Vicia villosa Roth var. **_glabrescens_** W.D.J. Koch

Hairy vetch, winter vetch, fodder vetch

Vicia villosa Roth ssp. *varia* (Host) Corb.

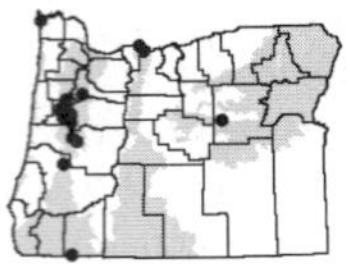

Plants glabrate with inconspicuous, subappressed to spreading pubescence. **Inflorescences** usually 10–20-flowered. **Flowers** 12–16 mm.

Fields, roadsides, waste areas. Flowering May–Aug. 0–1900 m. BW, ECas, Est, Sisk, WV. CA, ID, WA; throughout most of North America; Europe. Exotic.

Vicia villosa Roth var. **_villosa_** (p. 744)

Hairy vetch, winter vetch, fodder vetch

Vicia villosa Roth ssp. *villosa*

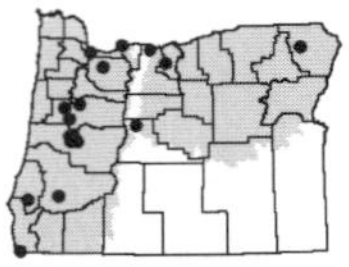

Plants villous. **Inflorescences** usually ≥ 20-flowered. **Flowers** 14–18 mm.

Fields, roadsides, waste areas. Flowering May–Jul. 50–1300 m. BW, Casc, Col, CR, Est, Lava, Sisk, WV. CA, ID, NV, WA; throughout most of North America; Europe. Exotic.

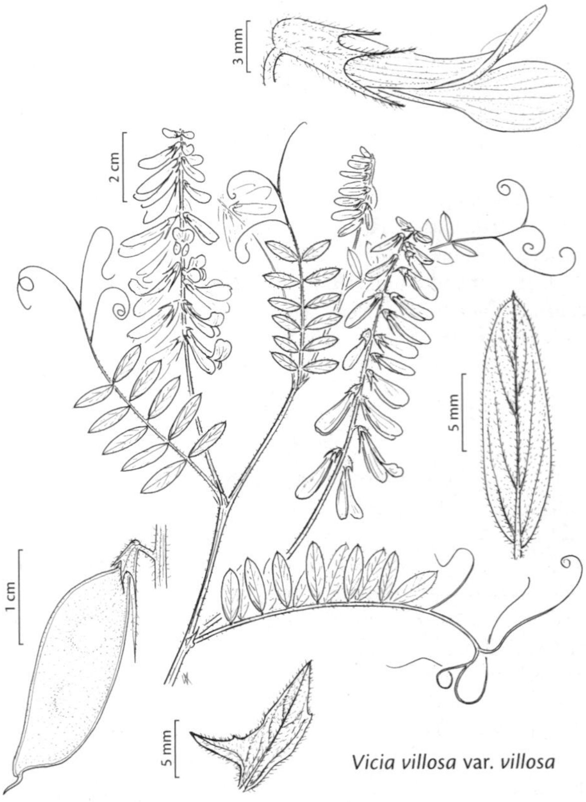

Vicia villosa var. *villosa*

VICIA

FAGACEAE Beech family

Alan T. Whittemore

Trees or **shrubs** evergreen or deciduous. **Buds** with several imbricate scales; terminal buds present. **Leaves** alternate, simple, pinnately veined, sometimes pinnately lobed; stipules present. **Inflorescences** spike-like catkins in leaf axils of current year, pistillate flowers near bases and staminate flowers above, or unisexual with staminate and pistillate inflorescences on same plant, or pistillate flowers solitary. **Flowers** radially symmetrical; sepals 4–6, free or basally fused; corollas absent. **Staminate flowers** stamens 6–many, free, pistillodes present. **Pistillate flowers** ovaries inferior, 3-locular, placentation axile, styles 3, linear or broadened distally; staminodes absent. **Fruits** large nuts, partly or completely enclosed in woody cups, with a spiny or scaly outer surface.

Throughout temperate northern hemisphere and mountains in northern tropics. 10 genera; 3 genera treated in *Flora.*

Members of Fagaceae are dominant trees in parts of western Oregon. The nuts are large, rich in oils, and are an important food source for wildlife. The quality of the timber is high, but these hardwoods have never been important in the Oregon timber industry.

Key to Fagaceae genera:

1. Plants deciduous; leaves thin and flexible, lobed or margins regularly toothed .... **QUERCUS** (in part)
1' Plants evergreen; leaves thick and rigid, not lobed, margins entire or spinose-toothed.
  2. Fruits usually 3 per cup, cups deeply divided into several irregular woody valves, completely enclosing fruits, margins and surfaces of valves with spines; leaves abaxially with dense bright yellow scales, margins entire .... **CHRYSOLEPIS**
  2' Fruits 1 per cup, cup undivided, enclosing up to half the fruit, surfaces with scales; leaves abaxially green or glaucous (but yellow hairs in *Quercus chrysolepis*), not scaly; margins entire or toothed.
    3. Cup scales spreading or reflexed; staminate and bisexual inflorescences rigid, ascending or erect; pistillate flowers at base of bisexual inflorescences; leaves abaxially densely tomentose, becoming glabrescent with age .... **NOTHOLITHOCARPUS**
    3' Cup scales appressed or slightly spreading; staminate inflorescences flaccid, pendent; pistillate flowers in separate inflorescences; leaves abaxially glabrous (but yellow hairs in *Quercus chrysolepis*) .... **QUERCUS** (in part)

CHRYSOLEPIS Chinquapin

Trees or **shrubs** evergreen; bark split into parallel ridges. **Leaves** thick and rigid, margins entire, not lobed, secondary veins turning aside before reaching margins, abaxially with dense bright yellow scales. **Inflorescences** staminate and bisexual on the same plant, ascending or erect, elongate, rigid, bisexual inflorescences with few pistillate flowers at bases, staminate flowers on remainder of inflorescences. **Fruits** usually 3 per cup, fewer if some not pollinated, ripening 2nd year after flowering, round or 3-angled in cross section; cups completely enclosing fruits, deeply divided into several irregular woody valves, margins and outer surfaces of valves covered with sharp, slender, interlocking branched spines; spines 1–2 cm.

Western North America. 2 species.

Key to *Chrysolepis* species:

1. Shrubs or trees 1–40 m; leaf tips acute or acuminate; bark thick, fissured .... ***C. chrysophylla***
1' Shrubs 0.3–2.5 m; leaf tips rounded, rarely rounded and apiculate; bark thin, smooth .... ***C. sempervirens***

Chrysolepis chrysophylla (Douglas ex Hooker) Hjelmquist
Giant chinquapin, golden chinquapin

Shrubs or **trees** 1–40 m; bark thick, fissured with age. **Terminal buds** ovoid to globose, 3–7 mm, puberulent; lateral buds smaller. **Leaves** plane or folded along midribs, narrowly elliptic to ovate or lanceolate, 4.6–13.1 × 1.5–3.1(3.8) cm, bases acute, acuminate, obtuse, or rounded-obtuse, secondary veins 6–13 on each side, tips acute or acuminate; petioles 3–16 mm.

2 varieties.

Key to *Chrysolepis chrysophylla* varieties:

1. Trees or occasionally shrubs 2–40 m; leaves plane, 6.6–13.1 cm .... ***C. c.* var. *chrysophylla***
1' Shrubs 1–4(10) m; leaves folded along midribs, 4.6–6.8 cm .... ***C. c.* var. *minor***

Chrysolepis chrysophylla (Douglas ex Hooker) Hjelmquist var. ***chrysophylla***

Giant chinquapin, golden chinquapin

Castanopsis chrysophylla (Douglas ex Hook.) A. DC. var. *chrysophylla*

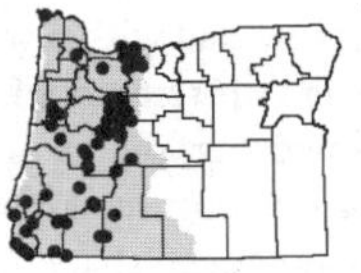

Trees or occasionally **shrubs** 2–40 m. **Terminal buds** ovoid, ~7 mm. **Leaves** plane, narrowly elliptic to lanceolate, 6.6–13.1 × 1.7–3.1(3.8) cm, bases acute or acuminate, tips slenderly acuminate; petioles 8–16 mm.

Dense or open forests. Flowering Jun–Sep. 0–1900 m. Casc, CR, ECas, Sisk, WV. CA, WA. Native.

The two varieties of *C. chrysophylla* are difficult to distinguish, and some botanists prefer not to separate them. Plants of variety *chrysophylla* are typically trees, but populations growing in exposed sites in the mountains may be dwarfed and shrubby.

Chrysolepis chrysophylla (Douglas ex Hooker) Hjelmquist var. ***minor*** (Bentham) Munz

Dwarf chinquapin

Castanopsis chrysophylla (Douglas ex Hook.) A. DC. var. *minor* (Benth.) A. DC.

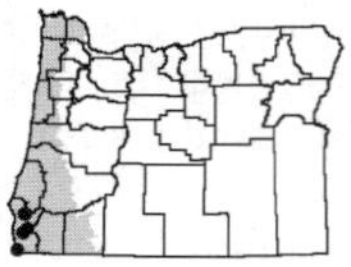

Shrubs 1–4(10) m. **Terminal buds** broadly ovoid to globose, 3–6 mm. **Leaves** concave and folded along midrib, at least in basal half, narrowly elliptic to ovate or lanceolate, 4.6–6.8 × 1.5–2.6 cm, bases acute to obtuse or rounded-obtuse, tips acute or acuminate; petioles 3–9 mm.

Openings in forest, open woodlands. Flowering Jun–Sep. 0–1900 m. CR, Est, Sisk. CA. Native.

Chrysolepis sempervirens (Kellogg) Hjelmquist (p. 746)

Bush chinquapin

Castanopsis sempervirens (Kellogg) Dudley ex Merriam

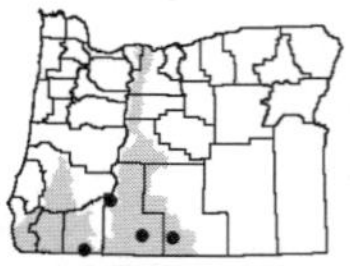

Shrubs 0.3–2.5 m; rhizomatous; bark thin, smooth. **Terminal buds** ovoid, to 4 mm, puberulent; lateral buds smaller. **Leaves** plane, oblong-elliptic, 3.2–6 × 1.3–2.2 cm, bases usually obtuse, occasionally rounded or acute, secondary veins 6–11 on each side, tips rounded, rarely rounded and apiculate; petioles 4–7 mm.

Chaparral, open woodland on dry slopes, sandy or rocky areas. Flowering Jul–Aug. 500–2300 m. ECas, Sisk. CA, NV. Native.

This species is often confused with *C. chrysophylla* var. *minor*, but *C. chrysophylla* does not show the rhizomatous, colonial growth form of *C. sempervirens*. In the herbarium, the shape of the leaf tip is the best character for distinguishing them.

CHRYSOLEPIS

NOTHOLITHOCARPUS Tanoak, tanbark-oak

Genus composed of one species, *N. densiflorus.*

Notholithocarpus densiflorus (Hooker & Arnott) Manos, Cannon, & S. Oh

Tanoak, tanbark-oak

Trees or **shrubs** evergreen, 1–40 m; bark thick, split into parallel ridges, fissured with age. **Terminal buds** ovoid, 2.5–7 mm, densely pubescent. **Leaves** plane or rugose, oblong or elliptic, 4.1–13 × 1.5–5.2 cm, thick, leathery, margins plane to deflexed, entire or toothed, secondary veins 5–14 on each side, surfaces abaxially densely tomentose, becoming glabrescent with age; petioles 4–18 mm. **Inflorescences** staminate and bisexual on same plant, ascending or erect, elongate, rigid; bisexual with few pistillate flowers at bases and staminate flowers on remainder. **Fruits** 1 per cup, ripening 2nd year after flowering, ovoid or cylindrical, 15–26 × 14–20 mm, tips rounded or obtuse; cups covering only base of nut, undivided, shallowly saucer-shaped, 5–11 × 15–22 mm, scales spreading or reflexed, 2–6 mm, overlapping, covering whole surface.

Western North America. 2 varieties.

Tanoak is rich in tannins, and it was important for tanning when cattle ranching was widely practiced in areas where the tree is native. Leather is no longer an important product of the Pacific states, and the economic importance of tanoak is much reduced. The North American *Notholithocarpus* superficially resembles the Asian genus *Lithocarpus*, and until recently it was included in that genus. It definitely does not belong to *Lithocarpus* but represents a distinct genus that is only distantly related to *Lithocarpus* and *Quercus* (Manos et al. 2009).

Tanoak has suffered heavy mortality due to sudden oak death (*Phytophthora ramorum*). While many genera of trees and shrubs are liable to infection and can serve as hosts for sudden oak death, only species of *Notholithocarpus* and some species of *Quercus* are normally killed by it (Kliejunas 2010).

Key to *Notholithocarpus densiflorus* varieties:

1. Trees; petioles (8)10–18 mm; leaves (4.9)6.4–13 cm, rugose and strongly impressed over secondary veins, margins with sharp teeth, tips acute, obtuse, or short-acuminate . . . . ***N. d.*** var. ***densiflorus***

1' Shrubs; petioles 4–9 mm; leaves 4.1–7.7 cm, plane or weakly impressed over secondary veins, margins entire or with crenations or weak teeth, tips rounded or obtuse . . . . ***N. d.*** var. ***echinoides***

Notholithocarpus densiflorus var. *densiflorus*

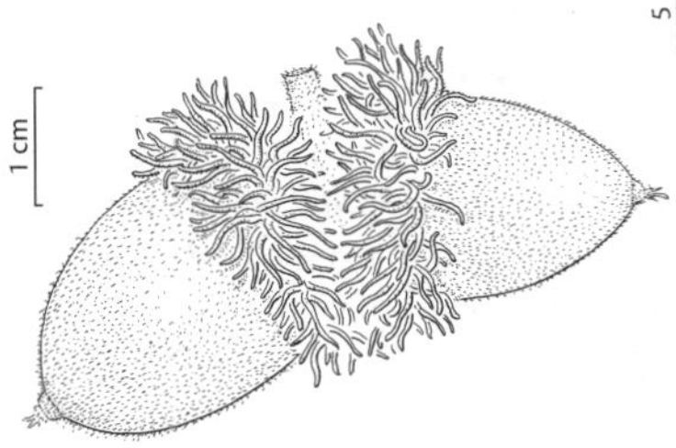

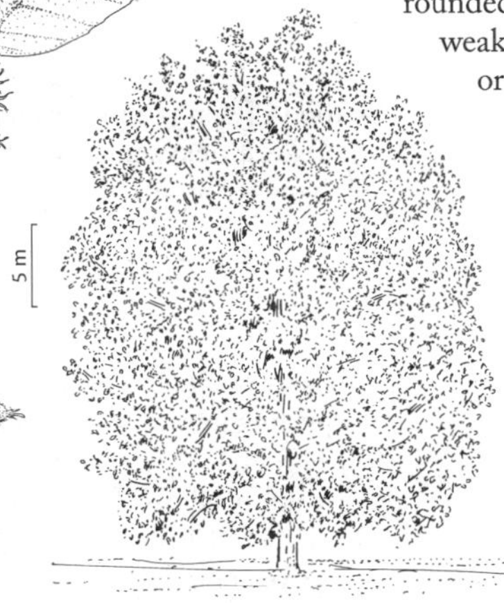

NOTHOLITHOCARPUS

Notholithocarpus densiflorus (Hooker & Arnott) Manos, Cannon, & S. Oh var. ***densiflorus***

(p. 747)

Tanoak

Lithocarpus densiflorus (Hook. & Arn.) Rehder var. *densiflorus*

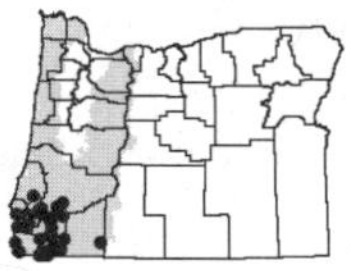

Trees to 40 m, with long shaggy hairs; trunks 1(3). **Terminal buds** 4–7 mm, usually enveloped in persistent stipules. **Leaves** rugose and strongly impressed over secondary veins, (4.9)6.4–13 × (2.1)2.3–5.2 cm, bases rounded, rarely rounded-obtuse or truncate, margins with sharp teeth, secondary veins 9–14 on each side, tips acute, obtuse, or short-acuminate, teeth 0.7–1 mm; petioles (8)10–18 mm.

Forests, edges of clearings. Flowering Jun–Oct. 0–1400 m. Casc, CR, Est, Sisk. CA. Native.

Notholithocarpus densiflorus (Hooker & Arnott) Manos, Cannon, & S. Oh var. ***echinoides*** (R. Brown) Manos, Cannon, & S. Oh

(p. 747)

Shrub tanoak

Lithocarpus densiflorus (Hook. & Arn.) Rehder var. *echinoides* (R. Br.) Abrams

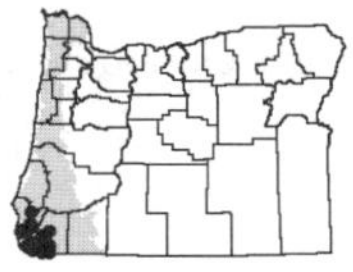

Shrubs forming colonies, ~1 m, with short hairs. **Terminal buds** 2.5–3 mm, stipules usually not persistent. **Leaves** plane or weakly impressed over secondary veins, 4.1–7.7 × 1.5–3.1 cm, bases usually obtuse or narrowly rounded, seldom acute, margins entire or with crenations or weak teeth, secondary veins 5–11 on each side, tips rounded or obtuse, crenations or teeth to 0.5 mm; petioles 4–9 mm.

Rocky serpentine slopes, chaparral, open woodlands. Flowering Jun–Aug. 300–1800 m. CR, Sisk. CA. Native.

Notholithocarpus densiflorus var. *echinoides*

QUERCUS Oak

Trees or **shrubs** evergreen or deciduous; bark split into parallel ridges or loose plates or strips. **Leaves** thick and leathery or thin and flexible, margins entire, toothed, crenate, or lobed, secondary veins reaching margins or turning aside, surfaces abaxially glabrous or with appressed yellow hairs. **Inflorescences** unisexual; staminate catkins flaccid, usually pendent, elongate spikes; pistillate flowers 1–few. **Fruits** (acorns) 1 per cup, round in cross section, ripening 1st or 2nd year after flowering; cups undivided, enclosing only lower part of fruits, scaly, scales appressed or slightly spreading, overlapping and covering surface of cups, 2–7 mm.

Temperate area and tropical mountains of Northern Hemisphere. ~400 species; 5 treated in *Flora*.

Several exotic species of oak are widely planted as shade trees, and seedlings of two of these species are sometimes found in the vicinity of cultivated trees. Red oak (*Quercus rubra*), of the eastern United States, will key to *Q. kelloggii*. It can be recognized by having completely glabrous leaves except for small tufts of hairs in the vein axils on the leaf underside and the upper leaf surface is dull. Additionally, its cups are saucer-shaped and cover only ⅓ or less of the nuts, and its scales are distinctly convex-thickened at their bases. English oak (*Q. robur*) will key to *Q. garryana*, but it can be recognized by short (2–8 mm) petioles, narrowly auriculate leaf bases, and fruits on peduncles that are 2–8 cm.

Key to *Quercus* species:

1. Leaves evergreen, margins entire or spine-tipped, not lobed.
   2. Shrubs; fruit attachment scars 2–4 mm wide . . . . *Q. vacciniifolia*
   2' Trees or sometimes shrubs; fruit attachment scars 5–6 mm wide . . . . *Q. chrysolepis*

1' Leaves deciduous, margins lobed or toothed.
   3. Secondary veins (extending from midrib to margin) usually 9–15 on each side of midrib, leaf margins toothed, unlobed . . . . *Q. sadleriana*
   3' Secondary veins (extending from midrib to margin) usually 3–5 on each side of midrib, leaf margins entire or with bristles, lobed.
      4. Leaf lobes with bristles . . . . *Q. kelloggii*
      4' Leaf lobes without bristles . . . . *Q. garryana*

Quercus chrysolepis Liebmann (p. 748)

Canyon live oak

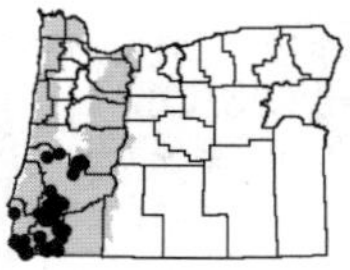

Trees or sometimes **shrubs** evergreen, 1–25 m, solitary; bark splitting into narrow plates and shed as thin flakes, pale to medium gray; twigs with spreading, multiradiate hairs. **Buds** ovoid or ellipsoid, tan to red-brown, tips acute or rounded, glabrous below, puberulent above; terminal buds 2–4 mm. **Leaves** oblong to elliptic or lanceolate, (2.2)2.5–4.7 × (1.2)1.5–2.4 cm, thick and rigid, bases rounded, occasionally truncate, margins entire or spine-toothed, not lobed, secondary veins 5–9 on each side, unlobed, tips obtuse, acute, acuminate, or apiculate, teeth 1–8 on each side, triangular, 1–2 mm, spines 0.5–1 mm, surfaces abaxially with appressed yellow hairs and sometimes multiradiate hairs on midribs and bases early, strongly glaucous with age, adaxially glabrous or with multiradiate hairs at base of midribs; petioles 2–8 mm. **Fruits** ripening 2nd year; nuts ovoid or cylindrical, 1.5–3 × 1–1.8 cm, tips obtuse or rounded; cups discoid or shallowly bowl-shaped, thin, 3–9 × 11–19 mm, covering only bases of nuts, scales plane, ~2 mm; attachment scars 5–6 mm wide.

Mixed evergreen woodlands, edges of grasslands, chaparral, canyons, ridges, rocky areas. Flowering Apr–May. 0–1600 m. Casc, CR, Sisk. CA, NV; southeast to NM, south to Mexico. Native.

This species is extremely variable. Shrubs and large trees may be found together at the same site, and entire leaves and strongly toothed leaves sometimes occur on the same branch. Acorns from Oregon populations of *Q. chrysolepis* have thin cups, not too different from the cups of *Q. vaccinifolia*, but examination of additional acorns is needed to assess the variability in this character properly. See also *Q. vacciniifolia* discussion.

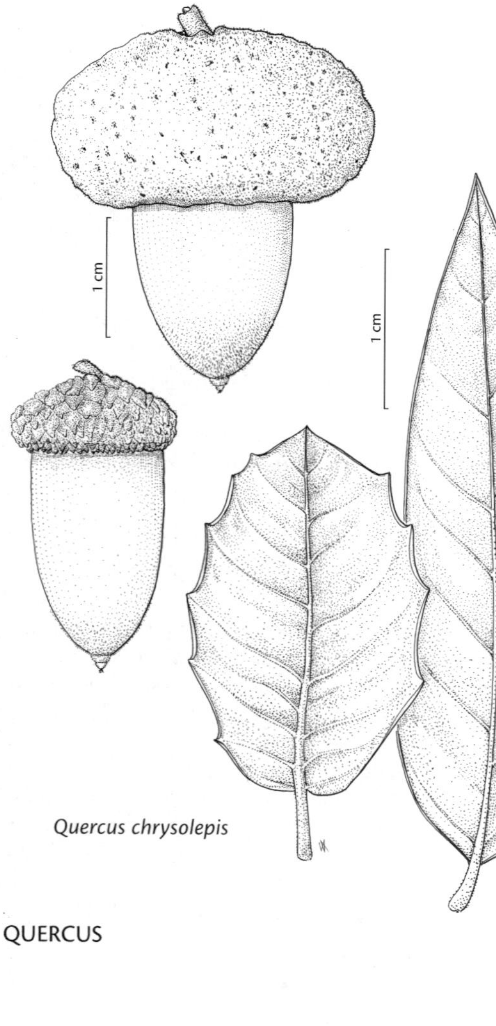

Quercus chrysolepis

Quercus garryana Douglas ex Hooker

Oregon white oak, Garry oak

Trees or **shrubs** deciduous, 1–20 m, solitary or colonial; bark splitting into irregular plates and shed as thin flakes, pale to medium gray; twigs glabrous to densely pubescent. **Buds** narrowly ovoid or ellipsoidal, tan or red-brown, tips acute or rounded, surfaces glabrous to densely pubescent; terminal buds 2–8 mm. **Leaves** elliptic to obovate, 3.6–12.6 × 2–8.8 cm, bases obtuse or truncate, seldom rounded or shallowly cordate, lobes 3–5 on each side, margins entire, secondary veins usually 3–5 on each side, tips apiculate, rounded or rounded-obtuse, surfaces abaxially pale green with appressed or spreading stellate hairs and often unbranched hairs on midribs, adaxially dark, ± shiny, glabrous or with scattered stellate hairs; petioles 3–18 mm. **Fruits** ripening 1st year; nuts cylindrical, 1.8–3.3 × 1–2 cm, tips narrowly rounded; cups bowl-shaped, thick, 6–11 × 15–20 mm, covering only bases of nuts, scales plane, keeled, or with strongly convex corky thickening; attachment scars 6–7 mm wide.

3 varieties.

QUERCUS

Key to *Quercus garryana* varieties:

1. Trees, (5)8–20 m; trunks 1–few; twigs pubescent; terminal buds 5–8 mm, tan, pubescent . . . . *Q. g.* var. ***garryana***

1' Small trees or shrubs, 1–4 m; trunks multiple; twigs glabrous or pubescent; terminal buds 2–5 mm, red-brown, glabrous to sparsely pubescent.

   2. Abaxial stellate leaf hairs with 2–6(8) spreading, rarely weakly appressed rays, rays 0.25–0.4(0.5) mm . . . . *Q. g.* var. ***breweri***

   2' Abaxial stellate leaf hairs with 4–10 appressed rays, rays ~0.1–0.15 mm . . . . *Q. g.* var. ***semota***

Quercus garryana Douglas ex Hooker
var. ***breweri*** (Engelmann) Jepson
Brewer's oak

Quercus breweri Engelm.

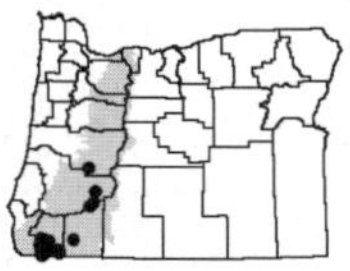

Small trees or **shrubs** 1–4 m, colonial; trunks few–many. **Twigs** red-brown or purplish brown, seldom gray-brown, sparsely to densely pubescent. **Terminal buds** ellipsoidal or ovoid, 2–5 mm, red-brown, tips rounded, glabrous to sparsely pubescent. **Leaves** 3.6–8.3 × 2–5.9 cm, bases obtuse, margins lobed 56–82% of width, adaxial veins not or scarcely impressed, surfaces abaxially with stellate hairs with 2–6(8) spreading or rarely weakly appressed rays, 0.25–0.4(0.5) mm.

Rocky ridges and slopes, in chaparral. Flowering Apr–Jun. 400–1900 m. Casc, Sisk. CA. Native.

Quercus garryana var. *breweri* and var. *semota* differ in little except the hairs of the leaf underside, which are relatively large and spreading away from the leaf surface in var. *breweri*, and small and tightly appressed in var. *semota*. The two pubescence forms are scattered over the range of the species in Oregon and south into the Klamath region of California, but only var. *semota* is found farther south, in the Sierra Nevada. Many authors have combined the two as a single variety, var. *breweri*.

Melanerpes formicivorus
Acorn woodpecker

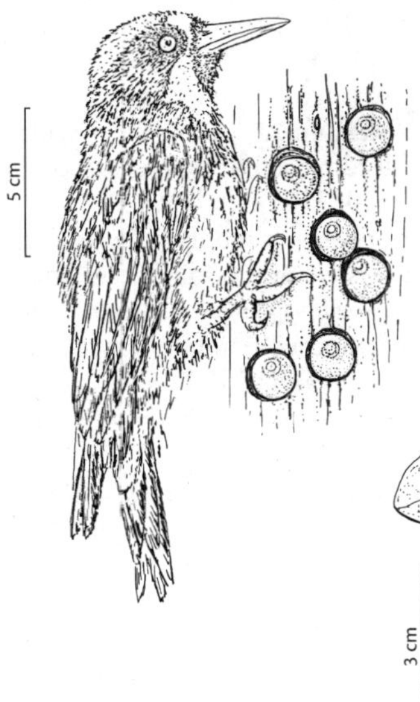

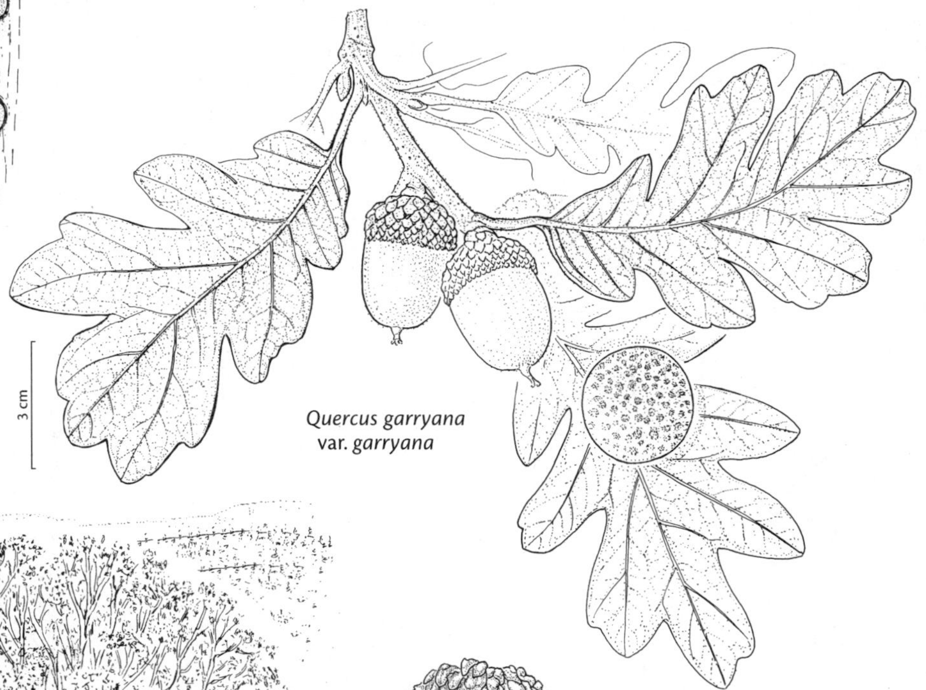

Quercus garryana
var. *garryana*

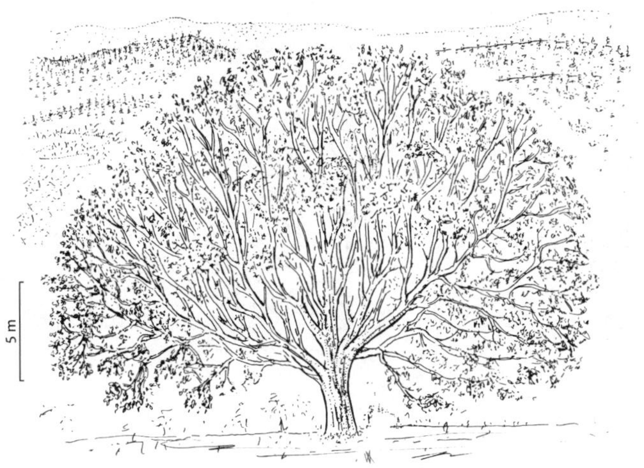

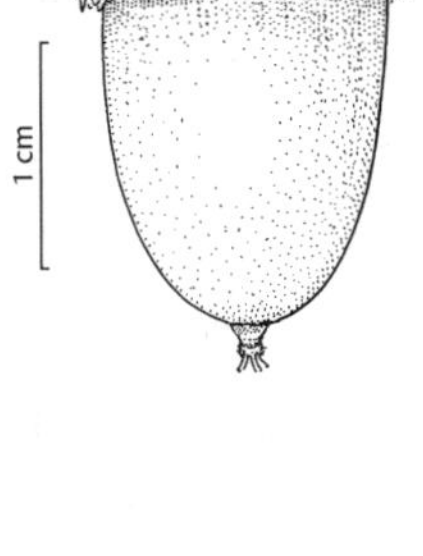

QUERCUS

Quercus garryana Douglas ex Hooker var. ***garryana*** (p. 749)

Garry oak, Oregon white oak

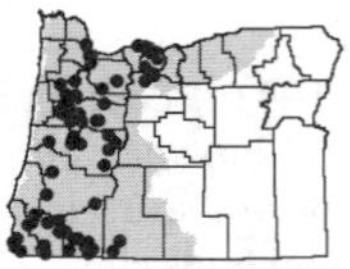

Trees to (5)8–20 m; trunks 1–few. **Twigs** gray-brown to dull purplish brown, pubescent. **Terminal buds** narrowly ovoid, 5–8 mm, tan, tips acute, pubescent. **Leaves** 5.7–12.6 × 4.4–8.8 cm, bases obtuse, truncate, or shallowly cordate, margins lobed (45)52–86% of width, adaxial veins slightly impressed, surfaces abaxially with stellate hairs with (2)4(6) spreading rays, 0.3–0.5 mm.

Deciduous woodland in bottomlands and slopes. Flowering Apr–Jun. 0–1700 m. Casc, Col, CR, ECas, Sisk, WV. CA, WA; north to British Columbia. Native.

Quercus garryana var. *garryana* is normally quite different from the two shrubby varieties, var. *semota* and var. *breweri*. Puzzling forms, suggesting significant levels of introgression, are seen where the range of var. *garryana* overlaps with the ranges of var. *semota* and var. *breweri*. Study of this variation at a population level is needed to resolve the status of these taxa.

Quercus garryana Douglas ex Hooker var. ***semota*** Jepson

Jepson's Oregon white oak

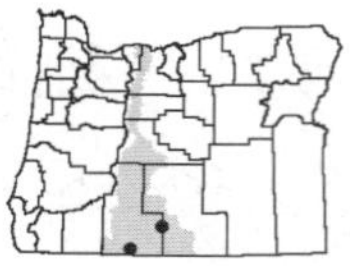

Shrubs 1–3 m, colonial; trunks multiple. **Twigs** red-brown or purplish brown, glabrous or sometimes pubescent. **Terminal buds** ellipsoidal to ovoid, 2–5 mm, red-brown, tips rounded, glabrous to sparsely pubescent. **Leaves** 3.7–8.5 × 2–5.5 cm, bases obtuse to truncate, seldom retuse, margins lobed (18)37–75% of width, adaxial veins not or scarcely impressed, surfaces abaxially with stellate hairs with 4–10 appressed rays, 0.1–0.15 mm.

Dry ridges and slopes. Flowering Apr–Jun. 800–1800 m. ECas. CA. Native.

Quercus kelloggii Newberry (p. 750)

California black oak

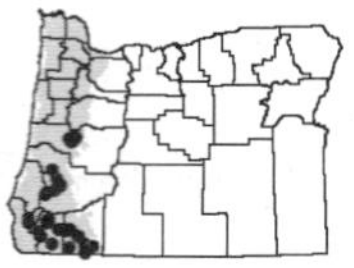

Trees deciduous, to 25 m, solitary; bark splitting into narrow plates, hard and not flaking, medium to dark gray; twigs with scattered, small, multiradiate hairs, becoming glabrescent. **Buds** broadly to narrowly ovoid, brown, tips obtuse or acute, with long-ciliate scale margins, backs of scales glabrous or puberulent near tips of buds; terminal buds 5–7 mm. **Leaves** elliptic or obovate, (8.2)10.5–15.5 × (6.3)8.8–14.7 cm, surfaces abaxially bright green, bases broadly obtuse or truncate, lobes 3–4 on each side, each with (3)4 marginal bristles, secondary veins 4 on each side, tips acute, surfaces abaxially with scattered, small, multiradiate hairs and long, multiradiate hairs tufted in axils of veins, becoming glabrescent, adaxially with scattered, small, multiradiate hairs, becoming glabrescent; petioles (8)23–29 mm. **Fruits** ripening 2nd year; nuts ovoid to cylindrical, 19–34 × 16–18 mm, tips obtuse; cups cylindrical, thin, 16–20 × 20–24 mm, covering ~ ½ of nuts, scales loosely appressed; attachment scars 7–8 mm wide.

Open woodlands on dry hillsides and bottomlands. Flowering Apr–May. 0–1400 m. CR, Sisk, WV. CA. Native.

In California, *Q. kelloggii* grows commonly with an evergreen oak, *Q. wislizenii*. The two species hybridize occasionally. The hybrid is known as *Q.* × *morehus*. *Quercus wislizenii* is not known from Oregon, but populations with morphology resembling *Quercus* × *morehus* are known from localities in Jackson and Douglas counties. It is presumed that *Q. wislizenii* was once present at these sites and hybridized with *Q. kelloggii*, and populations of hybrid origin have persisted after *Q. wislizenii* died out.

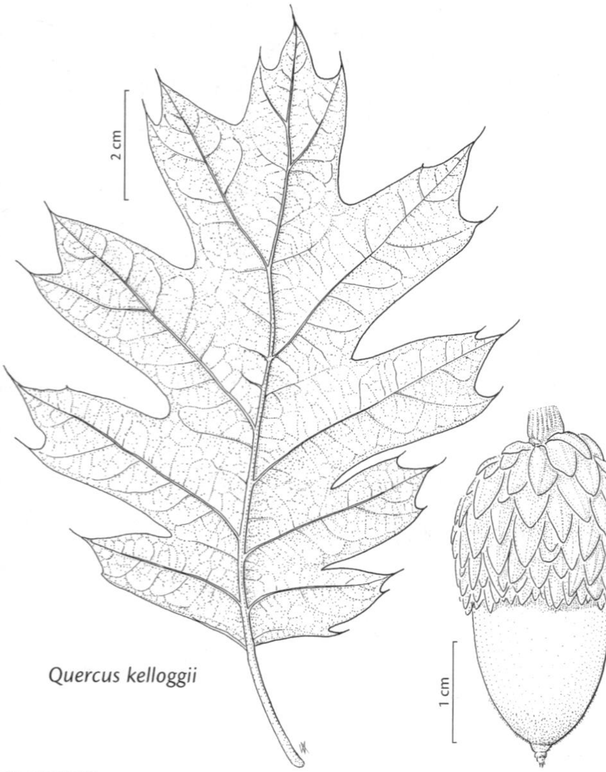

Quercus kelloggii

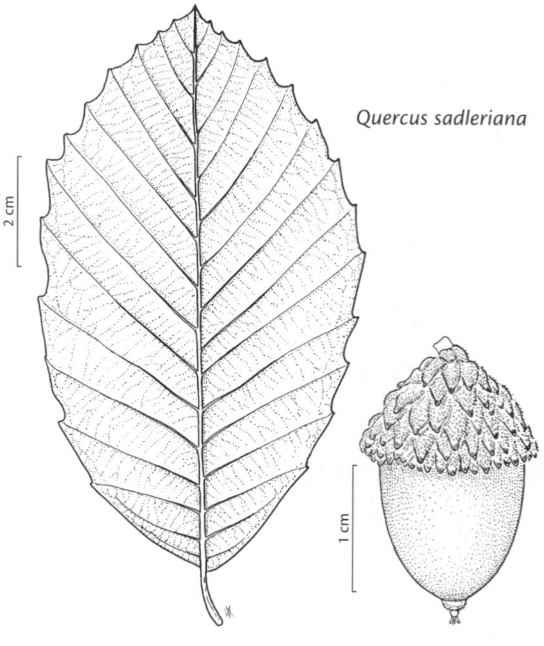

Quercus sadleriana

Quercus sadleriana R. Brown (p. 750)

Deer oak

Shrubs deciduous, 1–2 m, colonial; bark usually smooth, pale to medium gray; twigs glabrous or with few, scattered hairs. **Buds** broadly ovoid, gray-pubescent, tips rounded; terminal buds 5–7 mm. **Leaves** elliptic to obovate, 7.3–13.2 × 3.9–6.7 cm, bases rounded or obtuse, margins toothed, not lobed, teeth acuminate or rounded and mucronate, secondary veins usually 9–15 on each side, tips obtuse, surfaces abaxially glaucous, with small, appressed, stellate hairs when young, becoming glabrescent, adaxially glabrous; petioles 12–16 mm. **Fruits** ripening 1st year; nuts ovoid, 13–22 × 7–13 mm, tips rounded, cups cup-shaped or conical, thin, 6–13 × 14–22 mm, covering ~½ of nuts, scales appressed; attachment scars 4–5 mm wide.

Rocky openings in conifer forests, ridges, slopes. Flowering Apr–Jun. 1000–2000 m. CR, Sisk. CA. Native.

This species is endemic to the Klamath region of southwestern Oregon and adjacent California. It is known to hybridize with *Q. garryana* var. *breweri* at a few sites where the two oaks grow together. Leaves of the hybrids are lobed for 20–30% of their width, with 6–7 shallow obtuse lobes on each side.

Quercus vacciniifolia Kellogg (p. 751)

Huckleberry oak

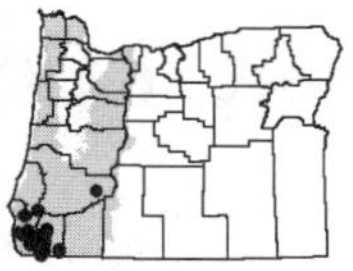

Shrubs evergreen, to 1.5 m, solitary; bark usually smooth, pale to medium gray; twigs glabrous or puberulent. **Buds** ovoid, red-brown, tips rounded, glabrous or puberulent; terminal buds 2–2.5 mm. **Leaves** oblong to oblong-ovate or elliptic, 1.9–3.6 × 0.7–1.9 cm, bases rounded to truncate, margins entire or rarely spine-toothed, unlobed, secondary veins 6–8 on each side, tips acute or obtuse, sometimes rounded or apiculate, surfaces abaxially glaucous, glabrous, rarely with scattered, appressed, yellow hairs, adaxially glabrous or with appressed, yellow hairs; petioles 2–5 mm. **Fruits** ripening 2nd year; nuts ovoid to ellipsoidal, 11–22 × 9–14 mm, tips acute to rounded; cups shallowly bowl-shaped, thin, 5–10 × 10–19 mm, covering only base of nuts, scales plane, appressed; attachment scars 2–4 mm wide.

Rocky places in chaparral or open woodlands. Flowering May–Jul. 300–1600 m. Casc, CR, Sisk. CA, NV. Native.

Huckleberry oak is closely related to *Q. chrysolepis*, and shrubby forms of the latter species can sometimes cause confusion. The habit of the shrub is different; *Q. vacciniifolia* tends to form low shrubs around rock outcrops in the mountains, usually with a dense crown having many crowded, parallel, slender, pliable stems. *Quercus chrysolepis* has a more open crown, and the branches are stiffer, more divergent, and not so much branched. These characters show some variation, and the two species may be difficult to distinguish from isolated twigs in the herbarium. Hybrids are sometimes found where they grow together, usually around 1000 m elevation. The name was originally spelled "*Quercus vaccinifolia*," but a recent change in the International Code of Botanical Nomenclature (Art. 60.8, Vienna code) requires the addition of a second "i."

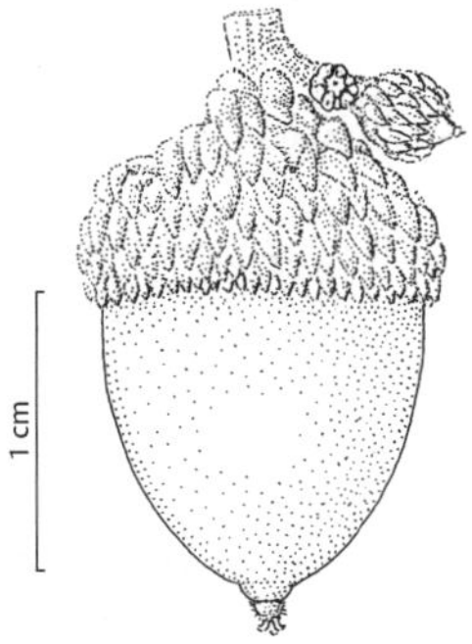

Quercus vacciniifolia

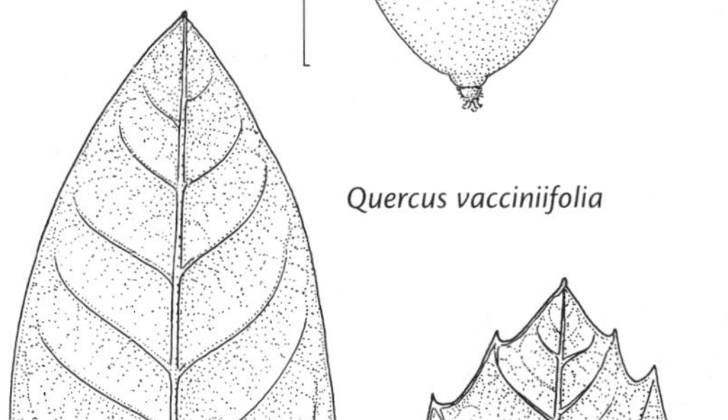

QUERCUS

FUMARIACEAE

(see Papaveraceae, Volume 3)

GLOSSARY

For terms that are illustrated, the page numbers of the illustration can be found at the end of the entry. The references *Plant identification terminology: An illustrated glossary* by Harris and Harris (2001) and *Merriam-Webster's collegiate dictionary* (2000) were consulted in the preparation of these definitions.

A

abaxial The side facing away from the axis or stem (compare to adaxial). (p. 754)

abdomen The third and posterior major division of an insect body.

abort (adj. **abortive**) To fail to develop fully.

abrupt Terminating suddenly. (p. 757)

acaulescent Lacking a distinct stem.

accrescent Becoming larger with age.

accumbent Of cotyledons: lying against the radicle on one edge.

achene Dry, indehiscent fruit containing one seed attached to the ovary wall at a single point. (p. 765)

acicular Needle-shaped, round or grooved in cross section.

acid (adj. **acidic**) Having a pH less than 7 (compare to alkaline).

acorn Hard, dry, indehiscent fruit of oaks, with a single seed and a cup-like base.

actinomycete One of a diverse order of bacteria common in soils and other habitats; some form symbiotic nitrogen fixing associations with plants (see *Landscaping* chapter).

acuminate Gradually tapering to a narrow point with concave sides. (p. 757)

acute Tapering to a pointed apex with more or less straight sides. (p. 757)

adaxial The side facing the axis or stem (compare to abaxial). (p. 754)

adherent Of dissimilar organs or parts sticking together but easily separated and not fused.

adnate Fused to unlike parts (compare to connate).

adventitious Developing or arising in an unusual position, usually said of roots on a stem.

adventive Not native.

aerial Occurring in the air, above ground or water.

agamospermy Formation of seed without fertilization.

aggregate To cluster densely.

albuminous Resembling or containing albumen, the nutritive tissue in a seed.

alkaline Basic; having a pH greater than 7 (compare to acidic).

allelochemical A chemical produced by an organism that exerts a detrimental effect on other species (see *Insect* chapter).

allopatric Of two or more taxa with geographic ranges that do not overlap.

allopolyploid Having more than two complete sets of chromosomes, from two or more different species.

allotetraploid Having four complete sets of chromosomes, from two or more different species.

alluvial Composed of unconsolidated sediment which has been eroded and deposited by water.

alpine Above timberline.

alternate Borne singly at a node; as leaves on a stem; borne between other organs (compare to opposite). (p. 754)

alveolate With pits separated by thin ridges; honey-combed.

amplexicaul Clasping the stem. (p. 757)

androgynous With staminate flowers above the pistillate flowers.

angustiseptate With a septum crossing the narrowest part of a fruit.

annual A plant that germinates from seed, flowers, sets seed, and dies in one year or less (compare to biennial, perennial).

annular In the form of a ring.

antenna (pl. **antennae**) Paired, segmented sensory appendages on the head of an insect.

anterior Toward the front; on the side away from the main axis.

anther The expanded, apical, pollen-bearing portion of the stamen. (p. 762, 769)

anthesis The flowering period, when a flower is fully expanded and functioning.

anthocyanin Water-soluble plant pigment, ranging in color from blue to purple to red.

antrorse Directed forward or toward the apex or tip (compare to retrorse).

aperture Openings or thin spots on a pollen grain.

apetalous Without petals or corolla.

apex (pl. **apices**) The tip; point farthest from the point of attachment.

apical At or toward the apex or tip (compare to basal). (p. 766)

apiculate Ending abruptly with a short, slender point. (p. 757)

apiculus A small, slender point.

apomixis Reproduction by seed without fertilization.

aposematic Of an insect that displays a warning of unpalatability, usually with color.

appendage An attached, supplementary or secondary part.

appressed Pressed flat against another organ.

aquatic Growing in water.

arachnoid Cobweb-like; formed of tangled hairs or fibers. (p. 758)

arcuate Moderately curved like a bow.

areole A small, well-defined area on a surface, such as the angular space between leaf veins, or the area on a cactus bearing the flowers and/or spines.

aril Fleshy, corky, or hard appendage arising at or near the point of seed attachment, sometimes covering the entire seed.

aristate Awn- or bristle-tipped.

armed Bearing prickles, thorns, barbs, or spines.

arthropod An animal with an exoskeleton, segmented body, and paired, jointed appendages, such as insects, spiders, and crustaceans.

articulate Jointed; separating freely along a well-defined line.

ascending Rising obliquely or curving upwards.

aseriate Not arranged in a row or rows.

asexual Reproducing without fertilization.

asymmetric Irregular in shape, not divisible into identical mirror-image halves.

asynchrony Not happening at related times (see *Insect* chapter) (compare to synchrony).

attenuate Gradually narrowed or tapered, more gradual than acuminate. (p. 757)

auricle Ear-shaped appendage or lobe.

auriculate With auricles. (p. 757)

awl-like Narrow throughout but with a broader base tapering to sharp tip.

awn A stiff, narrow, bristle-like appendage from the tip or dorsal surface.

axil The upper angle between a stem and a leaf or branch of the stem.

axile On or pertaining to the axis.

axillary Pertaining to the axil; in the axil.

axis (pl. **axes**) The main or central line or stem about which the lateral organs or parts are arranged.

B

backcross Cross of a hybrid with one of the parental species.

bald Lacking the usual covering; usually an area without trees.

ballast Heavy material placed low in a vessel such as a ship to improve stability.

band Stripe.

banner Upper, and usually largest, petal of a papilionaceous flower (Fabaceae). (p. 770)

barb Short, rigid, reflexed point.

barbed Bearing sharp, rigid, reflexed points like barbs of a fishhook.

barbellate With short, stiff hairs or barbs. (p. 758)

bark The outermost layers of a woody stem.

basal Related to or situated at the base (compare to apical).

basalt A dark gray to black dense to fine-grained igneous rock.

basic Alkaline, with a pH greater than 7 (compare to acid).

basifixed Attached at the base.

basin A depression; watershed.

Batesian mimicry A type of mimicry in which a harmless prey animal is avoided by predators because of its resemblance to a harmful prey animal.

beak (adj. **beaked**) A prolonged, firm tip. (p. 769)

bearded Bearing one or more tufts of long hair.

berry A fleshy fruit with more than one seed, developing from a single pistil. (p. 765)

bicarinate Having two keels.

bicolored Of two distinct colors.

bidenticulate With two small teeth.

biennial A plant which lives for more than one but less than two years after germination (compare to annual, perennial).

bifid Forked or divided for about half its length.

bifurcate Divided into two forks or branches.

bilabiate Two-lipped.

Leaf Types & Parts

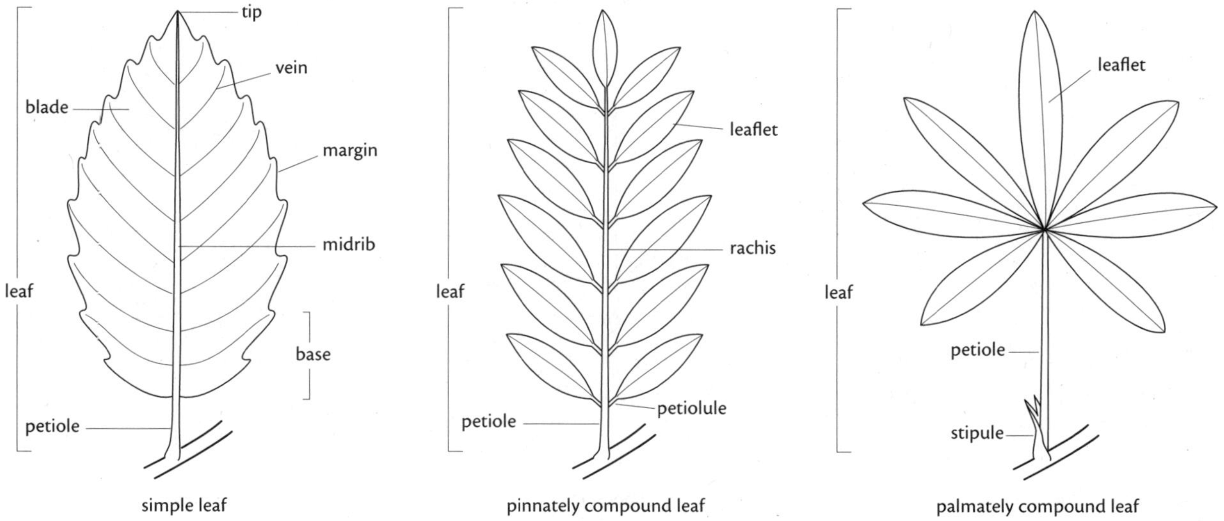

bilateral Arranged on opposite sides.

bilaterally symmetric Divided by a single plane into two sides that are mirror images of each other.

bilobed Divided into two lobes.

bilocular With two locules, as in some ovaries.

biomass The total mass of organisms in a given area or volume.

bioswale A storm water runoff conveyance system using plants and soil to absorb low flows, treat runoff, and improve water quality (see *Landscaping* chapter).

biota The array of organisms found in an area at a specific time.

biotic Relating to or resulting from living things.

bipinnate Twice or doubly pinnate.

biseriate Arranged in two rows.

bisexual Of a flower with both male and female reproductive organs; perfect (compare to unisexual).

biternate Divided into three, each third again divided into three.

bladder A thin-walled and inflated structure.

blade The broad part of a leaf or petal. (p. 753)

bloom A whitish or glaucous powdery coating of a surface; a flower; to produce flowers.

bog A wetland with spongy peat deposits, acidic water (pH less than 5.5), and low nutrient levels.

boreal Of, relating to, or located in northern regions.

bottomland Low lying, alluvial land by a river or stream.

brackish Water that is saltier than fresh water, but not as salty as seawater.

bract A modified leaf or leaf-like structure at the base of a flower or inflorescence. (p. 766)

bracteate With bracts.

bracteole A small bract.

bractlet A small bract. (p. 766)

branchlet A small branch growing from a larger branch.

bristle A short, stiff, hair-like structure. (p. 769)

brood A clutch of individuals that hatch at the same time from eggs produced by one set of parents.

bud An undeveloped shoot or flower, often covered with protective scales.

bulb An underground bud with thickened fleshy scales, such as an onion.

bulbil A small blub or bulb-like structure.

bur A structure armed with hooks or barbs.

burl Outgrowth, usually rounded, of chaotically-grained wood on a tree trunk or limb.

C

caducous Falling off early compared to similar structures in other plants (compare to persistent).

calcareous Derived from limestone; containing lime.

callous Hardened or thickened.

calyculus (pl. **calyculi**) Small bract below the involucre; false sepal.

calyx (pl. **calyces**) The outer whorl of the perianth; collective term for all of the sepals of a flower. (p. 762)

campanulate Bell-shaped.

canescent Gray to white in color due to a fine pubescence. (p. 758)

capillary Slender and fine; hair-like.

capitate Head-like or in a rounded cluster. (p. 760)

capsule A dry, dehiscent fruit composed of more than one carpel. (p. 765)

carinate Longitudinally keeled.

carnivorous A plant that derives some or most of its nutrients by trapping and digesting animals (usually insects and arachnids).

carnivory The eating of animal flesh (compare to herbivory).

carpel A simple pistil or one unit of a compound pistil.

carpophore A slender floral axis extension between adjacent carpels. (p. 765)

cartilaginous Tough and firm yet flexible, like cartilage.

caruncle A protuberance at or surrounding the hilum (scar at point of attachment) of a seed.

Leaf Arrangement & Orientation

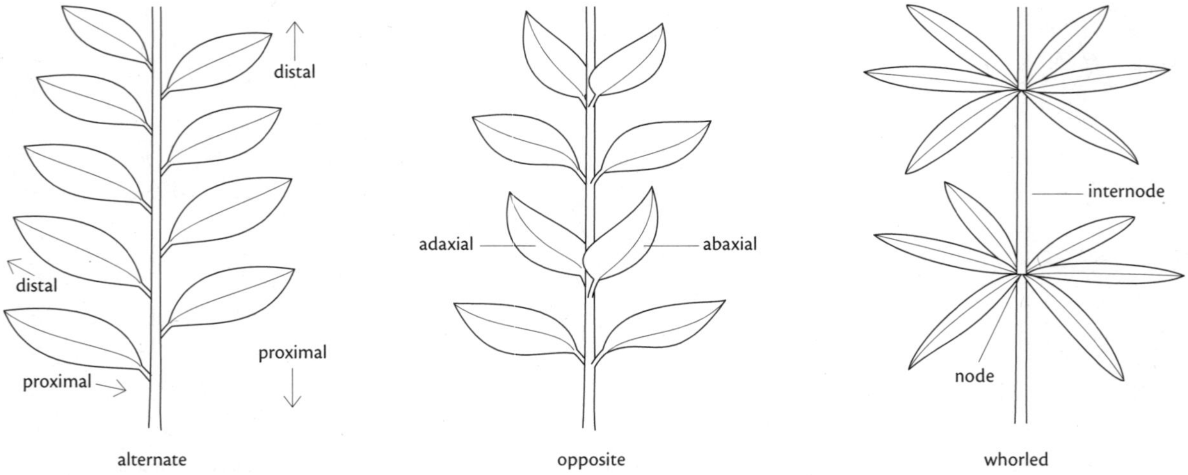

caterpillar The larva of a butterfly, skipper, or moth, with an elongate segmented body, three pairs of true legs, and several pairs of leg-like appendages.

catkin An unbranched inflorescence densely covered with many apetalous, unisexual flowers that are sessile or nearly so. (p. 760)

caudate Having a tail-like appendage.

caudex (pl. **caudices**) Persistent, often woody base of an herbaceous perennial; compact stem or trunk of a fern.

caulescent With an obvious leafy stem above the ground.

cauline Of or relating to the stem.

cell A hollow cavity or compartment within a structure; the basic structural unit of which plant tissue is composed.

cellular Made of small cavities or compartments.

cespitose Growing in dense tufts.

chaffy Bearing thin, dry scales or bracts.

chamaephyte A woody plant whose perennating buds are just above the ground.

chamber A hollow space.

channeled With a deep groove.

chaparral A vegetation type consisting of dense thickets of evergreen shrubs.

chartaceous With a thin, papery texture, usually not green.

chasmogamous Flowers that open before fertilization and are usually cross-pollinated (compare to cleistogamous).

chlorophyll The green pigment in plants.

chlorosis Abnormal loss of green color in plant leaves (see *Landscaping* chapter).

chrysalis The immobile, pupal stage between caterpillar and adult in butterflies.

ciliate With a fringe of hairs along the margin. (p. 756)

ciliolate With a fringe of minute hairs along the margin.

cilium (pl. **cilia**) Small hair, usually along a margin.

cinereous Light, ashy gray, often due to a covering of hairs.

circinate Coiled from the tip downward.

circumboreal Throughout the boreal (northern) regions of North America and Eurasia.

circumpolar Throughout the polar regions of North America and Eurasia.

circumscissile Splitting circumferentially, with the top coming off like a lid.

clade A group composed of an ancestor taxon and all of its descendents (a monophyletic group).

clambering Climbing weakly.

clasp To surround, entirely or in part.

clavate Club-shaped, widening gradually toward the apex.

claw The narrow base of some petals or sepals. (p. 770)

cleft Cut or split about halfway to the middle or base.

cleistogamous Flowers that self-fertilize without opening (compare to chasmogamous).

cliff A steep, vertical or over-hanging face of rock or earth.

climbing Growing more or less erect by leaning or twining on a support.

clones A set of genetically identical individuals that are produced by vegetative reproduction from a single parent.

coat The outer covering of a structure, usually a seed.

coevolve For two organisms to influence each other in the process of evolution, such as plants and pollinators, or hosts and parasites.

collar The area on the outside of a grass leaf at the junction of the sheath and blade; a ring or semi-circular band like the collar of a shirt.

colleter A group or tuft of hairs that secrete mucilage.

colony A group of plants, usually connected underground by organs.

column A structure formed by united filaments or by united style and filaments.

columnar Shaped like a column.

Leaf Shapes

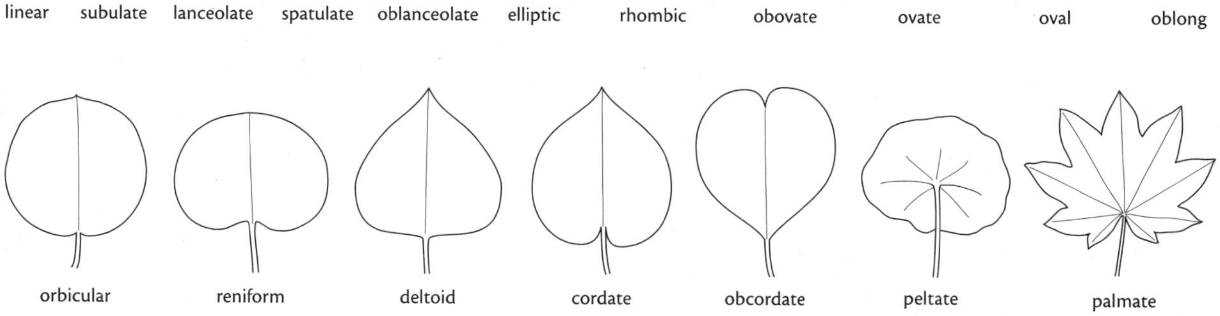

coma A tuft of hairs, especially on the tip of a seed.

commissure Joint or seam; the face along which the carpels adjoin each other.

comose With a tuft or tufts of hair.

compound Composed of two or more like parts.

compressed Flattened.

concave Hollowed out or curved inward (compare to convex).

concentric Arcs or rings sharing a common center.

conduplicate Folded lengthwise, with the upper surface within.

cone A dense series of spirally arranged woody bracts or sporophylls.

confluent Blending or coming together.

congener (adj. **congeneric**) A species or subordinate taxon belonging to the same genus.

congested Densely crowded.

conic Cone-shaped, with the point of attachment at the broad end.

conifer A plant that bears cones.

connate Of like parts fused together (compare to adnate, distinct).

connivent Converging, but not fused or united, usually with bases separate and apices approaching each other.

conspecific Belonging to the same species.

constricted Narrowed.

contorted Twisted or bent.

contracted Narrowed; shortened.

contractile Having the ability to contract.

convex Rounded and curved outward on the surface (compare to concave).

convolute Rolled together longitudinally.

copious Large in number or quantity; abundant.

cordate Heart-shaped, with the notch at the base. (pp. 755, 757)

cordiform Heart-shaped.

coriaceous Of a leathery texture.

corm A short, solid, vertical, underground stem with thin, papery leaves.

corolla Collective term for all of the petals of a flower. (pp. 762, 769)

corona The set of petal- or crown-like appendages between the corolla and stamens in some flowers.

coroniform Crown-shaped.

corrugated Wrinkled or folded into alternating ridges and furrows.

corymb A flat- or round-topped raceme with lower pedicels longer than the upper. (p. 760)

corymbiform Appearing like a corymb.

corymbose Having flowers in corymbs; having the general appearance of a corymb.

cosmopolitan Distributed worldwide.

cotyledon Primary leaf of an embryo; a seed leaf.

creeping Growing along or just beneath the ground surface and producing roots.

crenate With rounded teeth along the margin. (p. 756)

crenulate With very small, rounded teeth along the margin.

crescentic Crescent-shaped.

crest An elevated rib or ridge on a surface.

crevice An opening or crack in a rocky substrate.

crinkled Flattened and somewhat twisted, kinked, or curled.

crisp (alt. **crisped**) Curly, wavy, twisted or wrinkled.

crown The top part of a tree or shrub; the persistent base of an herbaceous perennial; a corona.

cruciform Cross-shaped.

cryptically In a way that is hidden, camouflaged or concealed.

cryptophyte Having perennating buds hidden under ground or water.

MARGINS

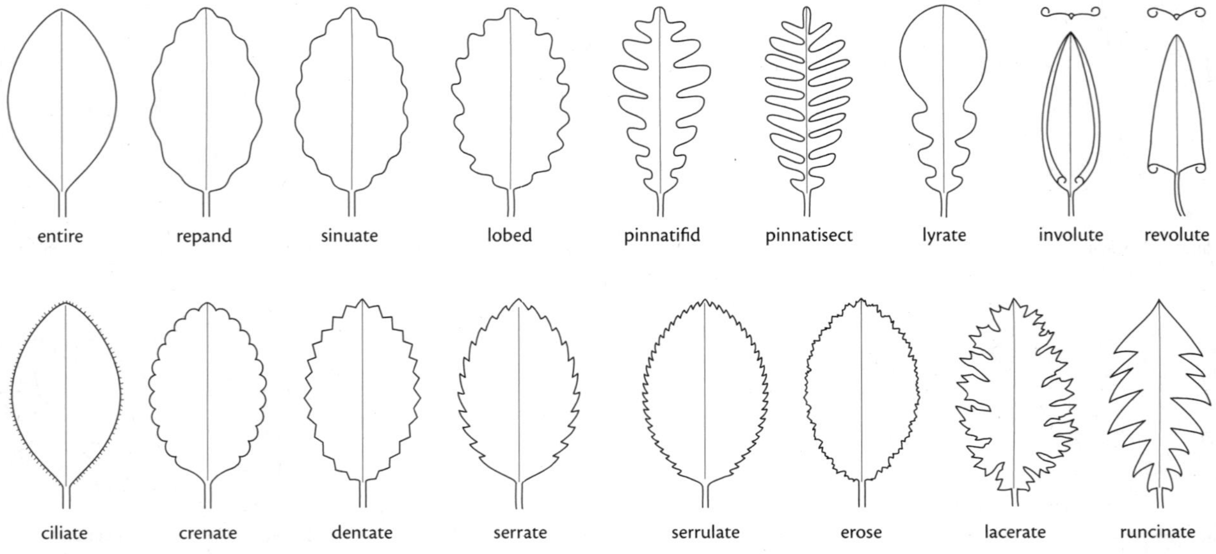

cuckoo A species that lays its eggs amongst another species' eggs; a nest parasite.

cultivar A form of plant originating under cultivation.

cuneate Wedge-shaped, narrowly triangular and attached at the narrow end; with the sides coming together at an acute angle at the base. (p. 757)

cupulate Cup-shaped; having a cup-like structure.

cuspidate Tipped with a short, sharp, abrupt point. (p. 757)

cyanic Blue or bluish; with blue pigments.

cylindric Cylinder-shaped; elongate and round in cross section.

cylindroid Cylindrical but elliptic in cross section.

cyathium (pl. **cyathia**) The inflorescence in the genus *Euphorbia*, consisting of a cup-like involucre containing a single pistil and male flowers with a single stamen.

cyme A branched, flat- or round-topped, determinate inflorescence in which the terminal (often central) flower matures first. (p. 760)

cymose With flowers arranged in a cyme.

cymule Small cyme, or section of a compound cyme.

D

deciduous Falling off; not persistent (compare to evergreen).

decompound More than once divided or compound.

decumbent Lying on the ground but with the tip ascending.

decurrent Extending downward from the point of insertion. (p. 757)

decurved Curved downwards.

decussate Four-ranked in opposite pairs, with each pair at right angles to the next pair.

deflation plain Part of a sand dune landscape immediately behind the foredune where the sand has been wind-scoured down to the summer water table level.

deflexed Bent abruptly downward.

dehisce (adj. **dehiscent**) To split open, generally at maturity.

deltate See deltoid.

deltoid Triangular, with the point of attachment on a side. (p. 755)

dendritic Having a tree-like branching pattern. (p. 758)

dentate Sharp-toothed along the margin, with the teeth directed outward, perpendicular to the margin. (p. 756)

denticle A small tooth.

denticulate Dentate, but with very small teeth.

depauperate Impoverished, stunted or poorly developed.

depressed Flattened from above.

descending Directed downward at a moderate angle.

determinate Of definite or limited growth; of an inflorescence in which the terminal flower opens first, halting further development on the main axis (compare to indeterminate).

detoxify To make a harmful substance harmless.

diadelphous Stamens united in two often unequal sets by their filaments.

diapause Suspended development controlled by particular environmental conditions.

dichotomous With the axis forked into two equal or nearly equal divisions or branches.

dicot Plant with two cotyledons (seed leaves) (compare to monocot).

didymous Twinned, the two parts similar and attached by a short portion of their inner surface.

diffuse Widely or loosely spreading, usually with many branches.

dilated Flattened or expanded.

dimorphic In two forms.

dioecious Having staminate (male) and pistillate (female) flowers on separate plants (compare to monoecious).

diploid Having two sets of chromosomes.

disarticulate To separate at a joint or articulation.

disc In Asteraceae, the central portion of the head, bearing tubular (disc) florets. (p. 769)

Tip and Base Shapes

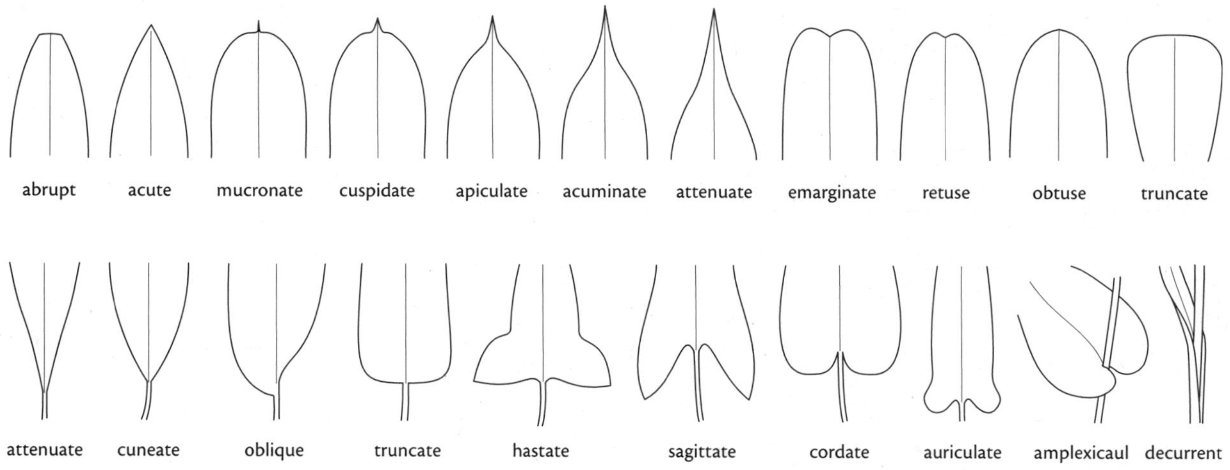

disciform Of a head with two kinds of disc florets and ray florets absent, usually with outer pistillate and inner bisexual florets (Asteraceae); in the form of a disk.

discoid Of a head with all disc florets of one kind and ray florets absent (Asteraceae) (p. 769); resembling a disk; orbicular with convex faces.

discrete Clearly seperate from each other.

disjunct Occurring in widely separated geographic areas.

dissected Deeply divided or cut into many segments.

distal Toward the tip, or the end of the organ opposite the end of attachment (compare to proximal). (p. 754)

distinct Separate, not attached to like parts (compare to connate).

distyly Having flowers with styles of two different lengths within a population.

dithecal Having two pollen sacs or cells.

diurnal Opening or occurring in the daytime (compare to nocturnal).

divaricate Spreading at a very wide angle, extremely divergent.

diverge Spread away from one another, usually at a wide angle.

divided Cut or lobed to the base or to the midrib.

dominant Of an organism or taxon that by its size, abundance, or coverage exerts considerable influence on a community's biotic and abiotic conditions.

dormancy A period when normal physical functions are suspended or slowed in order to minimize metabolic activity in unfavorable environmental conditions.

dorsal Pertaining to the back or outward surface of an organ in relation to the axis, as in the lower surface of a leaf; abaxial (compare to ventral).

dorsifixed Attached at or by the back.

dorsiventral With an upper and lower surface; from top to bottom or back to front (not side to side).

downy Covered with soft, fine hairs.

drought A period of abnormally dry weather.

drupe A fleshy indehiscent fruit with a stone or pit surrounding usually a single seed, as in a peach or cherry. (p. 765)

duff Organic matter in various stages of decomposition on the floor of a forest.

dune (adj. **dunal**) A mound or ridge of sand or other loose sediment, formed by wind.

Hairs and Surfaces

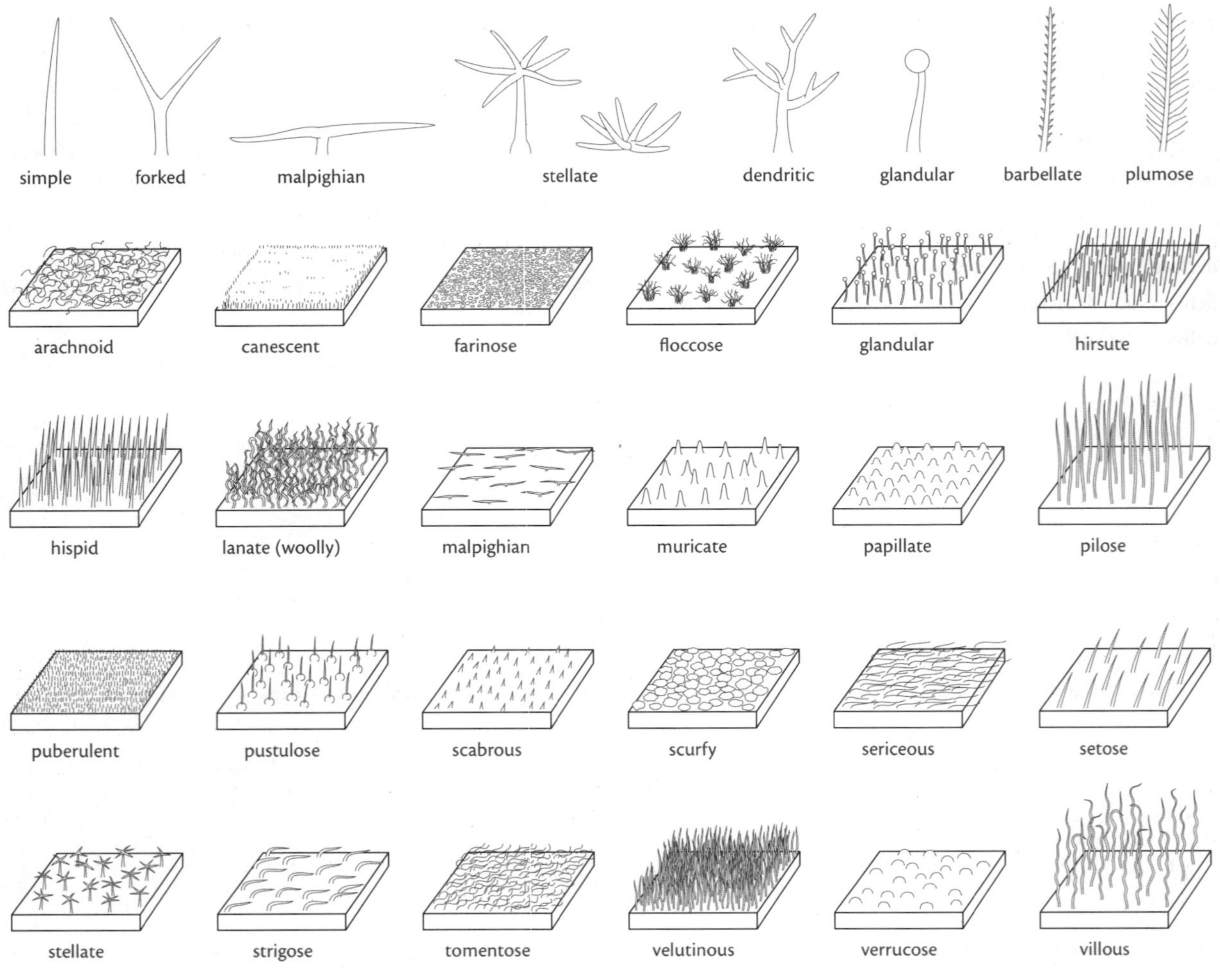

E

ecoregion A geographically defined area that has similar climate, soils and geology, and a distinctive composition of species.

ecosystem The system of all living organisms and the physical environment interacting in a particular place (see *Landscaping* chapter).

ecotone A transition area between two biomes or plant communities.

ecotype A genetically distinct geographic variety adapted to a certain environment.

ectomycorrhiza (pl. **ectomycorrhizae**) An association involving a fungus and plant roots, characterized by the formation of a fungal sheath around fine root tips with hyphae growing between root cells and outwards in the soil (see *Landscaping* chapter).

eglandular Lacking glands.

elaiosome A lipid- and protein-rich structure on the seeds of some plants that is attractive to ants which disperse the seeds.

ellipsoid Elliptic in long section or outline and circular in cross section.

elliptic Oval-shaped; of a plane with the shape of an ellipse, longer than wide and rounded at both ends, the widest part near the middle. (p. 755)

elongate Drawn out in length.

emarginate Notched at the apex. (p. 757)

emasculated With stamens removed or made ineffective.

embryo The new plant in a seed or archegonium.

emergent Emersed.

emersed Rising above the surface of the water.

endemic Confined to a given region.

endocarp Inner layer of pericarp of a fruit.

endogenous Originating from within.

endomycorrhiza (pl. **endomycorrhizae**) An association involving a fungus and plant roots, characterized by hyphae growing into root cells and outwards in the soil (see *Landscaping* chapter).

endophytic Of an organism that lives between living plant cells.

endosperm The nutritive tissue surrounding the embryo of a seed.

enfolding Wrapping up or enveloping.

entire With a margin that is not toothed, notched, or divided. (p. 756)

ephemeral Lasting a very short time.

epidermis Outermost layer of cells covering the plant, beneath the cuticle.

epigynous Of an inferior ovary with petals, sepals, and stamens attached to the top.

epiparasite A parasite that feeds on another parasite.

epipetalous Attached to the petals.

erect Upright.

erose With irregular teeth on the margin. (p. 756)

evapotranspiration The movement of water from land to atmosphere by evaporation from soil and other surfaces and by transpiration from plants (see *Landscaping* chapter).

evergreen Having leaves throughout the year (compare to deciduous).

eviscerate To remove the internal organs.

excurrent Extending beyond what is typical, such as a vein that proceeds beyond the margin.

exfoliate To peel off in scales or flakes.

exotic Not native; post-colonial arrivals.

exserted Projecting out or beyond surrounding parts (compare to included).

exudate A liquid, resinous, or gelatinous substance secreted by organs or parts of the plant, or yielded when the plant is damaged.

F

facultative Capable of, but not restricted to a particular mode of life.

facultatively In a way that is optional or discretionary; not by necessity.

falcate Strongly curved, sickle-like.

falciform Sickle-shaped; shaped like the beak of a falcon.

family A taxonomic grouping of closely related genera.

farinose Bearing a mealy or powdery coating. (p. 758)

fascicle A close cluster or bundle.

fen A wetland fed by groundwater and dominated by mosses, herbs, and shrubs, with a pH generally higher than 5.5.

fenestrate With window-like openings or translucent areas.

fertile Producing seeds, spores, or pollen capable of germination; of a plant with reproductive organs.

filament Any thread-like body, especially the stalk of stamens. (p. 762)

filiform Thread-like.

fimbria (pl. **fimbriae**) A single, slender projection; a single unit of a fringe.

fimbriate Fringed, with slender processes along the margin.

fimbrillate Fringed with very fine hairs or hair-like structures.

fissured Cracked or fractured.

fistulose Tubular; hollow and cylindrical.

flaccid Limp, not rigid or turgid.

flag The distal, scarious portion of a leaf.

flange A projecting rim or edge.

flexuous With curves, bends or zigzags.

floccose With tufts of long, soft, tangled hairs. (p. 758)

floodplain A low-lying area near a river or stream subject to flooding.

floret A small flower. (p. 769)

foliaceous Leaf-like; bearing leaves.

foliose Leafy or leaf-like.

follicle A dry, dehiscent fruit of one carpel opening along one side. (p. 765)

forb A non-grass-like herb.

forked Divided into two more or less equal branches. (p. 758)

fornix (pl. **fornices**) A small crest or scale in the throat of a corolla as in some Boraginaceae.

frass The solid excreta of insects, usually of the larvae.

free Not fused or united.

freshwater Water with a low salt concentration.

fringe (adj. **fringed**) Hairs along a margin.

fruit A mature ovary along with any other structures that are attached and ripen with it as a unit. (pp. 765, 766, 770)

funicle Stalk connecting the ovule to the placenta.

funnelform Funnel-shaped.

furrow A groove or depression in a surface.

fuscous Of a dark, gray-brown color.

fused Joined and growing together.

fusiform Spindle-shaped, widest in the middle and tapering at both ends.

G

gall An abnormal growth of plant tissue in response to the activities of another organism, often an insect.

generalist An adaptable species that is able to thrive in a variety of enviromental conditions or make use of a variety of different resources (see *Insect* chapter) (compare to specialist).

geniculate With a knee-like bend.

genus (pl. **genera**) A taxonomic grouping of closely related species.

geophysical Related to the earth's physical properties and the physical processes acting upon, above, and within the earth.

gibbous Somewhat swollen on one side.

glabrate Nearly glabrous; becoming glabrous.

glabrescent Nearly glabrous; becoming glabrous.

glabrous Naked, lacking hairs or scales.

glade A clearing in a woodland.

gland A structure that secretes oily or sticky substances.

glandular Bearing glands. (p. 758)

glaucous With a distinct whitish or bluish, waxy bloom or powder on the surface.

globose Almost spherical in shape.

globular Globe-shaped.

glochid (pl. **glochids**) A barbed hair or bristle. (p. 765)

glochidiate Barbed at the tip; bearing barbed hairs or bristles.

glomerate Densely clustered.

glomerule A head or dense cluster of flowers.

glutinous Gummy; covered with sticky exudate.

graduated Arranged in a series.

granular With a sand-like texture.

grassland A community dominated by grasses.

guild A group of species that exploit the same class of environmental resources in a similar way (see *Insect* chapter).

gynobase Elongation or enlargement of the receptacle.

gynobasic Attached to the gynobase.

gynoecious With pistillate flowers only.

gynomonoecious Having both pistillate and bisexual flowers on the same plant.

gynophore An elongated stalk bearing the pistil.

gynostegium (pl. **gynostegia**) A structure formed by the fusion of the anthers with the stigmatic region of the gynoecium.

H

habit The general appearance or characteristic mode of growth of a plant.

habitat The kind of place in which a plant grows.

hair A fine epidermal outgrowth, usually elongate.

halophyte A plant that grows in salty environments.

hamate Hook-shaped, especially at the tip.

hastate Arrowhead-shaped with the basal lobes pointed, narrow, and spreading at a wide angle. (p. 757)

haustoria Root-like organs of parasitic plants that penetrate host tissues.

head An inflorescence of densely clustered, sessile, or nearly sessile flowers. (p. 760)

helicoid Coiled like a spiral or helix.

hemispheric In the shape of half of a sphere.

herb A vascular plant which is not woody.

herbaceous Not woody; leaf-like in appearance or texture.

herbivory The eating of plants (compare to carnivory).

heteromorphic Existing in two or more easily recognizable forms (compare to homomorphic).

heterostylous Having styles of different lengths in flowers of the same species.

heterotrophic Dependent on external sources for organic nutrients.

hilum (pl. **hila**) Attachment scar of a seed.

hirsute With distinct hairs, often rather stiff or bristly. (p. 758)

Inflorescence Types

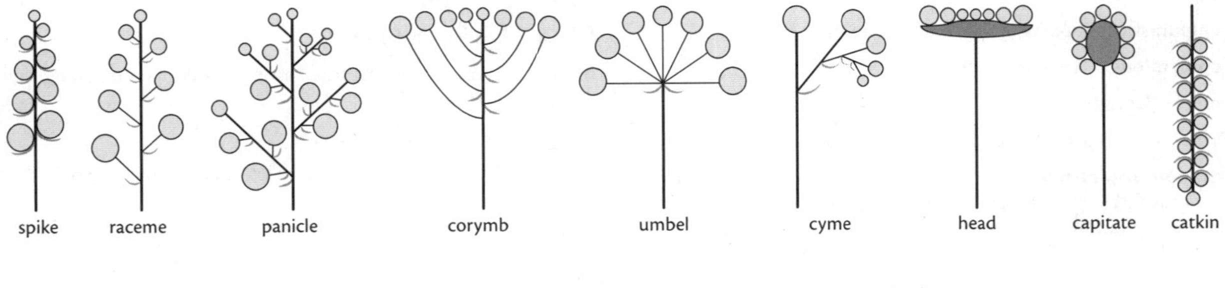

hirsutulous Pubescent with very small, coarse, stiff hairs.

hispid With stiff, spreading hairs or bristles. (p. 758)

hispidulous With very small, stiff, spreading hairs or bristles.

homomorphic All the same kind or form (compare to heteromorphic).

homostylous Styles of more or less consistent length.

hood A hollow, arched covering; the arched upper petals or sepals in some species.

host The plant on which a parasite lives and from which it derives its nourishment; the plant on which an epiphyte grows.

humistrate Lying on the ground.

hummock A hill or mound.

humus The dark-colored, organic material in soil derived from the decomposition of biomass.

hyaline Thin, membranous, and translucent or transparent.

hybrid The progeny from a cross between different taxa (genera, species, or subspecific taxa).

hydrologic Related to the distribution, circulation, or properties of water on the earth or in the atmosphere.

hygroscopic Absorbing moisture and changing shape as a result.

hypanthium (pl. **hypanthia**) The fused or coalesced basal portion of floral parts (sepals, petals, stamens) around the ovary.

hypogynous Having floral parts (stamens, petals, or sepals) borne on the receptacle beneath the pistil and free of the ovary.

I

igneous Of rock from lava or magma.

imbricate With overlapping edges.

imperfect Flower having either stamens or pistils, but not both (compare to perfect).

impressed Sunk below the surface.

incised Cut sharply and deeply.

inclined Rising upward at a moderate angle.

included Enclosed, not protruding (compare to exserted).

incomplete Lacking an expected part.

incumbent Resting or leaning upon; bent backwards with sides of cotyledons against hypocotyl.

incurved Curved inward.

indehiscent Not opening at maturity along definite lines or by pores.

indeterminate An inflorescence in which the lower flowers bloom first, permitting indefinite growth of the main axis (compare to determinate).

indigenous Native; of people, the earliest known inhabitants of an area.

indumentum The covering of a plant surface, especially hairs and scales.

inferior Situated below another organ or part; an inferior ovary has other floral parts attached above its point of attachment (compare to superior).

inflated Swollen, expanded or distended.

inflexed Bent abruptly inward.

inflorescence The flowering part of the plant. (p. 760)

infraspecific Of taxonomic ranks below species, such as subspecies and variety.

infrastaminal Below the stamens; in *Cuscuta*, infrastaminal scales are digitate or lobed protrusions at the base of the epipetalous stamens.

infructescence Fruiting stage of an inflorescence.

inoculant In gardening or agriculture, a microbe or suite of microbes introduced to improve plant growth (see *Landscaping* chapter).

inrolled Rolled inwards; involute or revolute.

insectivorous Of an organism that eats insects.

inserted Attached to or growing out of.

instar The growth stage between two successive molts in the development of an invertebrate animal, usually an insect larva.

intercostal Between the ribs or nerves.

intergrade To merge into one another gradually.

internode Portion of stem between two nodes. (p. 754)

interpetiolar Between petioles.

interrupted Discontinuous.

interspace The space between ridges (*Plagiobothrys*).

intertidal Of the area of seashore that is covered with water at high tide and exposed at low tide.

interwebbed Interwoven, as with branches of a pappus.

introgression The modification of species by the repeated back-crossing of hybrids with the parent species.

intruded Protruded into.

intrusion Protrusion into.

inverted Positioned opposite the typical direction.

invested Covered with.

involucel A small or secondary involucre. (p. 766)

involucrate With an involucre.

involucre (adj. **involucral**) One or more whorls of bracts subtending an inflorescence. (pp. 766, 769)

involute Rolled inwards or to the adaxial (upper) side (compare to revolute). (p. 756)

irregular Not similar in size or arrangement.

J

joint A node; the section of the stem from which a leaf or branch arises.

jointed Having one or more nodes or points of articulation.

juvenile Young, immature.

K

keel A ridge, usually on the back, like the keel of a boat; a boat-shaped structure formed by fusion along the basal margin of the two lower petals in a papillonaceous flower (Fabaceae). (p. 770)

L

lacerate With an irregular, ragged margin. (p. 756)

laciniate Irregularly divided into very narrow, pointed segments.

lamina Expanded portion of an organ such as leaf or petal; a thin, flat part.

lanate Covered with long, intertwined hairs; cottony, woolly. (p. 758)

lanceolate Lance-shaped, several times longer than wide, widest below the middle and tapering toward the ends, applied to flat structures. (p. 755)

lanceoloid Lance-shaped; lanceolate in long section or outline and circular in cross section, thickest below the middle and tapering toward the ends, applied to three-dimensional structures.

larva (pl. **larvae**) An immature insect after emerging from the egg, usually used for insects that undergo a complete metamorphosis.

lateral To the side. (p. 766)

latex A milky sap or exudate, drying to a rubbery consistency.

latiseptate Having broad partitions or septae.

lax Loosely arranged; open and spreading.

leader The highest terminal shoot(s) of a plant.

leaf An expanded, usually photosynthetic organ of a plant. (pp. 753, 766)

leaflet A segment of a compound leaf. (pp. 753, 766)

legume A dry dehiscent fruit derived from a single carpel and usually opening along two lines of dehiscence. (pp. 765, 770)

lenticel A slightly raised, somewhat corky, often lens-shaped area on the surface of a stem.

lenticular Shaped like a biconvex lens.

lepidote Clad in small, scurfy scales.

ligneous Woody.

ligular Tongue- or strap-shaped, flattened.

ligulate With a ligule; strap-shaped; of a head with only ligulate florets (Asteraceae). (p. 769)

ligule The flattened, elongate portion of a ligulate floret (Asteraceae); an appendage on the adaxial side of the leaf at the junction of the sheath and the blade. (p. 769)

limb The expanded part of a petal, sepal, or leaf; the expanded portion of a sympetalous corolla (p. 762); a tree branch.

limestone A type of sedimentary rock composed mainly of calcium carbonate, usually white or light gray in color.

linear Narrow with parallel sides. (p. 755)

lineate Marked with lines.

lip The usually larger and distinct lower petal or group of fused petals in some bilaterally symmetric flowers.

loam Soil of clay, silt, and sand containing humus.

lobe (adj. **lobed**) A rounded division or segment of an organ. (pp. 756, 762, 766)

lobulate With small lobes.

locule (adj. **locular**) The cavity or compartment of an organ. (p. 762)

loculicidal Dehiscing through the locules, usually near the center of the locule on the ovary wall (compare to septicidal).

loment A legume that is constricted between the seeds. (pp. 765, 770)

lomentaceous Loment-like; having loments.

lowland Low elevation areas; not montane.

lunate Crescent-shaped.

lyrate Pinnatifid with the terminal lobe rounded and much larger than the lower lobes. (p. 756)

M

maggot A legless larval phase of an insect, frequently a fly.

malpighiaceous See malpighian.

malpighian A short-stalked hair that is straight, tapering to both ends, with the point of attachment near the middle. (p. 758)

marcescent Withering but persistent.

margin (adj. **marginal**) The edge. (p. 753)

maritime Coastal; near the sea.

marsh A wetland that is dominated by herbaceous vegetation and flooded most or all of the year (compare to swamp).

Flower Parts

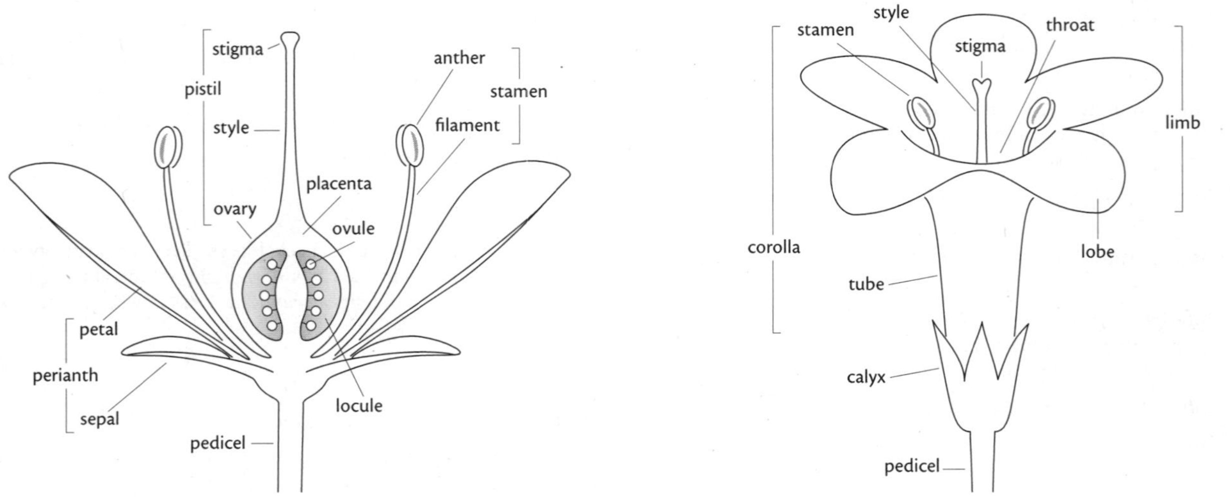

meadow A vegetation type dominated by grasses and forbs.

mealy With the consistency of coarse granules; powdery, dry, and crumbly.

medial Of the middle.

membrane (adj. **membranous**) A thin, flexible, often translucent piece of tissue.

mericarp A section of a schizocarp; one of the two halves of the fruit in Apiaceae.

mesic Moist (compare to xeric).

metamorphosis The relatively abrupt change in body form between the end of the immature development phase and the onset of the adult phase.

microclimate The climate of a small area that differs from that of the surrounding area (see *Landscaping* chapter).

midrib The central vein of a leaf or other organ. (p. 753)

midvein The main or central vein of a leaf.

mimic To resemble a model, usually another species, as a means of protection or advantage.

monadelphous Stamens united by the filaments and forming a tube around the gynoecium.

moniliform Cylindrical and constricted at regular intervals, giving a beaded necklace-like appearance.

monocarpic Flowering and bearing fruit only once, then dying.

monocot Plant with one cotyledon (compare to dicot).

monoecious Having staminate (male) and pistillate (female) flowers on the same plant (compare to dioecious).

monomorphic Having only one form.

monophyletic Composed of members that descended from a single ancestral line; usually, a group composed of all of the descendants of the most recent common ancestor (compare to polyphyletic).

monotypic Having only one representative (e.g., a genus or family with a single species).

montane In the mountains.

moraine A deposit of rock and sediment left by a glacier.

morphology Form and structure; the study of form and structure.

mucilage A viscous secretion of slimy consistency.

mucilaginous Slimy.

mucro A short, sharp tip.

mucronate With a mucro. (p. 757)

mucronulate Tipped with a small, sharp point.

multicipital A root or rhizome producing multiple crowns; having multiple heads.

multiradiate With many rays.

multiseriate Arranged in many rows.

muricate Covered with short, rough protuberances. (p. 758)

muriculate With fine, small, sharp projections.

mycorrhizal Pertaining to the symbiotic relationship between a fungus and the root system of a vascular plant.

mycotrophic Having a symbiotic or parasitic relationship with a fungus, obtaining some or all of its water or food from fungi.

myrmecochory The collection and dispersal of seeds by ants.

N

naked Lacking appendages.

nativar A cultivated selection of a native species chosen for its horticultural qualities.

native Of a species of a region that established in the landscape prior to European settlement, reproduces, and persists without human influence.

nectar a sweet fluid, commonly found in insect-pollinated flowers.

nectariferous Bearing nectar.

nectary Gland that secretes nectar.

needle-like Linear and usually stiff.

nematode One of a diverse phylum of tiny, cylindrical animals that are free-living in soil or water or parasitic in animals or plants (see *Landscaping* chapter).

nerve A strand of vascular and supporting tissue.

nocturnal Opening or occurring at night (compare to diurnal).

node A point on the stem where leaves or branches are inserted. (p. 754)

nodule A swelling or a knob-shaped protuberance.

nut Dry fruit with a hard exterior that typically contains a single seed. (p. 765)

nutlet A small, dry, indehiscent fruit; one of the sections of a mature ovary (Boraginaceae, Lamiaceae, Verbenaceae). (p. 765)

O

ob- Prefix meaning inversion, in the reverse direction.

obcompressed Flattened vertically or anteriorly instead of laterally.

obconic Cone-shaped, with the narrow end at the point of attachment.

obcordate Heart-shaped, with the pointed end at the point of attachment. (p. 755)

obdeltoid Deltoid, with the attachment at the pointed end.

obelliptic Almost elliptical, with the distal end larger than the proximal end.

oblanceolate Inversely lanceolate, with attachment at narrow end. (p. 755)

oblanceoloid Oblanceolate in outline or long section and circular in cross section.

oblique With unequal sides; slanting. (p. 757)

oblong Two to four times longer than wide with the sides mostly parallel. (p. 755)

obovate Of a flat structure that is egg-shaped in outline and attached at the narrower end. (p. 755)

obovoid Of a three-dimensional structure that is egg-shaped and attached at the narrower end.

obpyramidal Pyramidal with the attachment at the narrower end.

obsolete Lacking or rudimentary and not likely functional.

obtriangular Triangular with an angle at the point of attachment or base.

obtuse Blunt or rounded, with the sides coming together at an angle wider than 90°.

ochroleucous Yellowish white or cream-colored.

offset A short, often prostrate shoot originating at the base of another shoot.

opposite Having two leaves attached at each node, directly across from each other; borne on the same radius as another part (compare to alternate). (p. 754)

orbicular (alt. **orbiculate**) Rounded or circular in outline. (p. 755)

organism An individual animal, plant, or other life form.

outcrop A rock formation that is visible on the surface.

outcross To pollinate a plant with pollen from a different plant of the same species.

oval Broadly elliptic, narrowing somewhat from the middle to rounded ends. (p. 755)

ovary The basal portion of the pistil that develops into fruit after fertilization and contains the seeds. (p. 762, 769)

ovate Of a flat structure with an egg-shaped outline, attached at the broad end. (p. 755)

oviposit To lay an egg or eggs, especially of an insect.

ovoid Of a three-dimensional structure that is egg-shaped, with the point of attachment at the broad end.

ovule An immature seed. (p. 762)

P

palea (pl. **paleae, paleas**) A chaffy scale or bract; bracts on the receptacle between the florets (Asteraceae); the uppermost bract subtending a grass floret. (p. 769)

paleate Having paleae.

palmate Radiating from a common point, like the fingers of a hand. (p. 755)

palmatifid Palmately cleft or lobed.

panicle A branched inflorescence maturing from the bottom upwards. (p. 760)

papilionaceous Of an irregular flower with a banner, two wings, and a keel.

papilla (pl. **papillae**) A small, rounded, superficial projection.

papillate With papillae. (p. 758)

pappus (pl. **pappi**) A modified calyx of some Asteraceae florets, consisting of awns, bristles, and/or scales at the summit of the achene. (pp. 765, 769)

parasite An organism that obtains its food or water, at least partly, from a host organism.

parasitoid A parasite that kills its host.

parietal Pertaining to the wall of an organ.

peat Soft, soil-like, acidic material composed of partially decomposed plant matter.

pectinate Comb-like; deeply divided with narrow segments regularly spaced.

pedate Palmately divided with the lateral lobes two-cleft.

pedicel The stalk of an individual flower in an inflorescence. (pp. 762, 766)

pedicellate With a pedicel.

peduncle Stalk of a solitary flower or an inflorescence. (pp. 766, 769)

peduncular Relating to the peduncle.

pedunculate Having a peduncle.

peltate Shield-shaped; of leaves, with the stalk attached on the lower surface, not at the margin. (p. 755)

pendent Drooping or hanging down.

pendulous See pendent.

perennial With a life cycle lasting more than two years (compare to annual, biennial).

perfect With both male and female parts in the same flower; bisexual (compare to imperfect).

perforate With holes or perforations.

perianth The sepals and petals of a flower, collectively. (p. 762)

pericarp The wall of a fruit.

peridotite Coarse-grained igneous rock.

perisperm Food storage tissue in some seeds, arising from the nucellus.

persistent Remaining attached after similar structures in other plants normally fall off, after the function has been completed (compare to caducous).

petal Each of the segments of the corolla of a flower. (p. 762)

petaliferous Bearing petals.

petaloid Petal-like.

petiolar Pertaining to the petiole.

petiolate Having a petiole.

petiole Stalk at base of the leaf blade that attaches to the stem. (p. 753)

petiolulate With a petiolule.

petiolule The stalk of a leaflet. (p. 753)

phagostimulant A substance that promotes feeding.

phenotype The observable characteristics of an organism produced by the interaction of its genotype and the environment.

phyletic Relating to the evolutionary development of a group, such as a species.

phyllary An involucral bract. (p. 769)

phylogeny The evolutionary history of a group of organisms and the relationships among its members.

physiochemical Pertaining to both physical and chemical properties.

physiognomy (adj. **physiognomic**) The overall size and shape; the overall shape and structure of a plant community (see *Insect* chapter).

physiology The branch of biology that deals with the normal functions of living organisms.

phytochemical A chemical compound produced in plants that confers competitive advantage, deters herbivory, or thwarts pathogens.

phytochemistry The branch of chemistry concerned with plants and plant-produced chemicals.

phytogeography The study of the geographical distribution of plants.

pilose With long, soft, straight, distinct hairs. (p. 758)

pilosulose Bearing minute, long, soft, straight hairs.

pinna (pl. **pinnae**) The leaflet of a pinnate leaf, or the primary division of a bi- or tripinnate leaf.

pinnate (adv. **pinnately**) Resembling a feather, as in a compound leaf with leaflets arranged on opposite sides. (p. 753)

pinnatifid Pinnately lobed half the distance, or more, to the midrib (but not reaching the midrib). (p. 756)

pinnatisect Pinnately lobed almost to the midrib. (p. 756)

pinnule The second or third order divisions of a bi- or tripinnate leaf.

pistil The female reproductive organ of a flower, consisting of a stigma, style, and ovary. (p. 762)

pistillate Having pistils and lacking stamens (compare to staminate).

pistillode A sterile, vestigial pistil.

pith Cork-like tissue.

pitted With small pits or depressions.

placenta (pl. **placentae**) The portion of the ovary bearing ovules. (p. 762)

placentation The arrangement of the placenta or placentae in an ovary.

plains An area of flat land with few trees.

plane With a flat surface.

plano-convex Flat on one side, convex on the other.

plicate Folded into pleats.

plumose Feather-like. (p. 758)

pod A dry fruit that opens to release seeds.

pollen The microspores or the dust-like fertilizing cells of gymnosperms and angiosperms.

pollinate To transfer pollen from anthers to stigmas.

pollinator An animal that transfers pollen from anther to stigma, effecting pollination.

pollinium (pl. **pollinia**) A waxy cluster of pollen grains transported as a unit during pollination.

polygamodioecious Mostly dioecious, but with some perfect flowers.

polygamous With unisexual and bisexual flowers on the same plant.

polygonal With three or more sides, but typically used when there are five or more sides.

polymorphic Existing in many forms.

polyphyletic Composed of members that descended independently from two or more ancestral lines (compare to monophyletic).

polyploid With more than two sets of the basic chromosome number.

poricidal Opening by pores.

posterior Toward the back; on the side toward the axis.

prairie A large open area of grassland.

predator (adj. **predatory, predaceous**) An animal that kills and eats other animals.

prehensile Adapted for grasping.

prickle A sharp outgrowth of epidermis or bark.

FRUITS

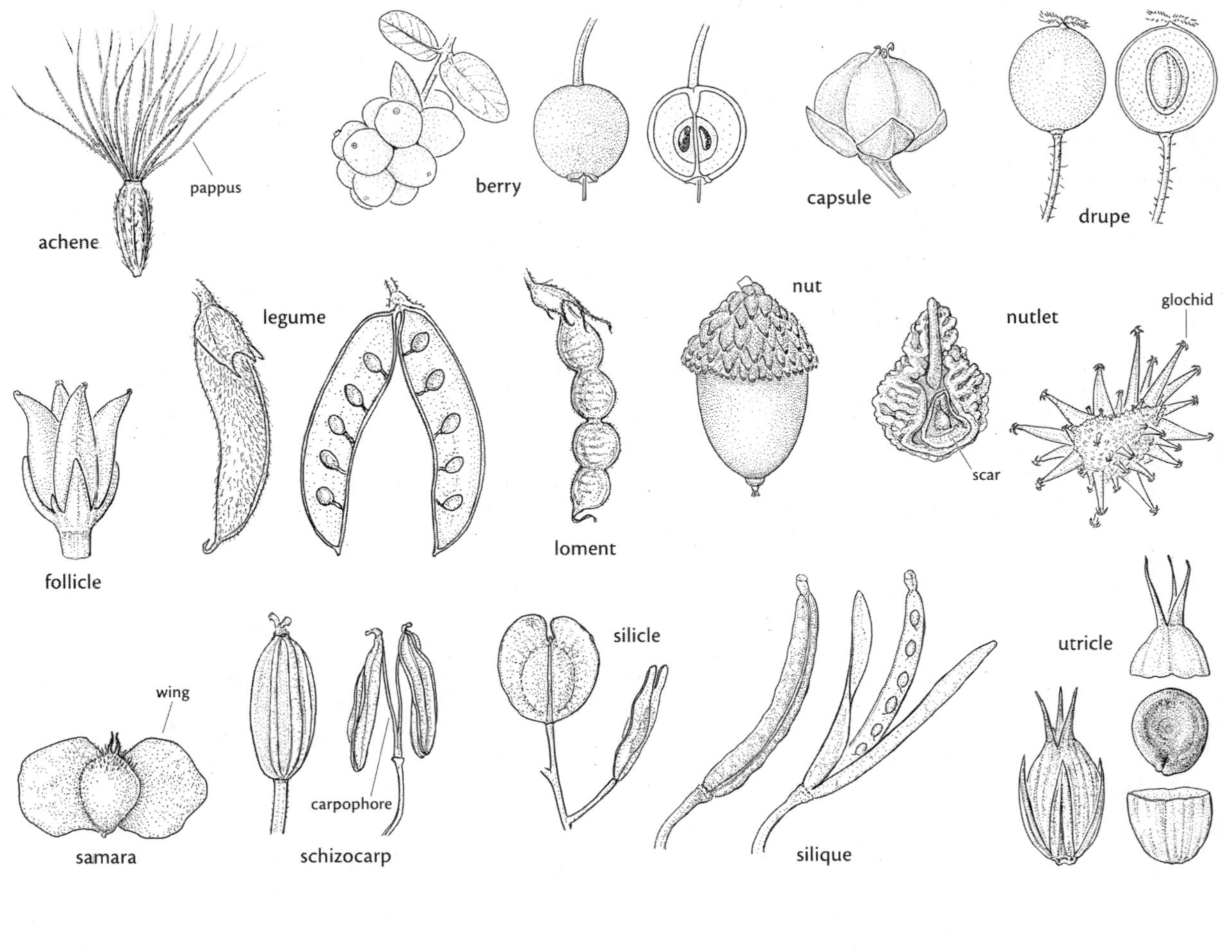

primary The first division of a leaf or inflorescence. (p. 766)

process A projecting appendage.

procumbent Trailing or lying flat, not rooting at the nodes.

proliferous Producing plantlets or bulblets, usually on the leaves.

prostrate Lying flat along the ground.

protuberance A rounded swelling or bump on the surface.

proximal Toward the base or attached end of an organ (compare to distal). (p. 754)

pruinose With a waxy, powdery coating.

pseudoscape A false scape, where the leaves appear to be basal but are not truly so; a slender, leafless, underground stalk arising from a corm and bearing at ground level the leaves and flowering stem (Apiaceae).

pseudoverticillate Not actually whorled but appearing so.

puberulent Minutely pubescent. (p. 758)

pubescence Hairiness; with short, soft hairs.

pubescent Covered with hairs, especially short, soft hairs.

pulvinate Cushion-like or mat-like.

pumice Volcanic rock.

punctate Dotted with pits, dots, or with translucent, sunken glands.

pungent Ending in a rigid or sharp point; having strong, sharp taste or smell.

punky Dry and crumbly or soft.

pupa The intermediate, immobile stage of life of many insects in which a larva transforms into an adult.

pupate (of a larva) to become a pupa.

pustule Small, blister-like or pimple-like projection.

pustulose With small blisters or pustules. (p. 758)

pyramidal Pyramid-shaped.

pyriform Pear-shaped.

Q

quadrangular Four-angled.

quadrate Square or rectangular.

R

raceme An inflorescence with pedicellate flowers on an elongate, unbrached axis maturing from the bottom upwards. (p. 760)

racemiform Of an infloresence that has the appearance of a raceme, but not necessarily the structure of a raceme.

racemose Having flowers in a raceme.

rachilla (pl. **rachillae**) A small rachis; the axis of a spikelet.

rachis (pl. **rachises**) The axis (excluding petiole) of a pinnately compound leaf or inflorescence. (pp. 753, 766)

radially symmetric With identical parts radiating from the center like spokes on a wheel.

radiate Of a head with both ray florets and disc florets (Asteraceae). (769)

rangeland Open, uncultivated areas capable of producing forage for grazing animals.

rank A (vertical or lengthwise) row.

ranked Arranged into (vertical or lengthwise) rows; one-ranked: pointing up like brush bristles; two-ranked: in flat sprays.

ray The flattened, elongate portion of a ray floret (Asteraceae) (p. 769); a branch of an umbel. (p. 766)

receptacle (adj. **receptacular**) The often enlarged end of the stalk on which flower parts are attached. (p. 769)

reclined Curved downward; lying on a support.

recurved Curved backward (compare to incurved).

reflexed Bent backward.

regular Uniform or symmetrical in shape and structure.

remote Spaced far apart.

reniform Kidney-shaped. (p. 755)

Apiaceae

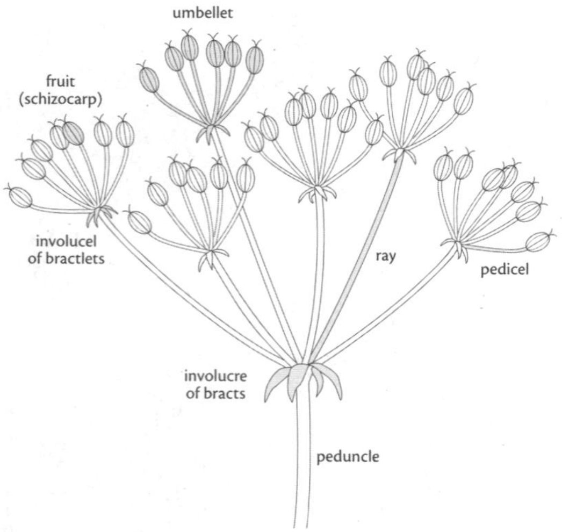

Compound Umbel

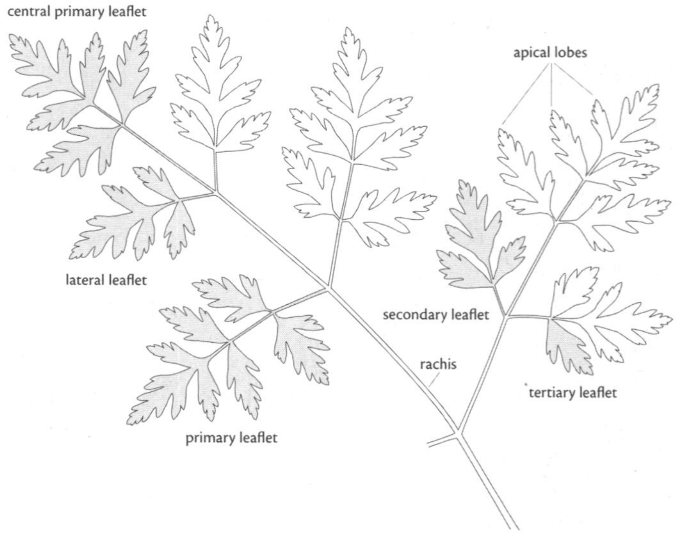

Typical Apicaeae Leaf

repand Shallowly sinuate or slightly wavy. (p. 756)

replum Partition between the two compartments of silicles or siliques; persistent septum after dehiscence of a fruit.

resin A thick, sticky liquid that hardens into a solid.

resupinate Upside down due to twisting of the pedicel.

reticulate In the form of a network; net-veined.

retrorse Directed downwards or backwards, away from the tip (compare to antrorse).

retuse A rounded or blunt tip with a small notch. (p. 757)

revolute With margins rolled downwards or to the abaxial side (compare to involute). (p. 756)

rhizomatous With rhizomes.

rhizome A modified underground stem, usually growing horizontally.

rhizomal Related to the rhizome; attached to or arising from the rhizome.

rhombic Diamond-shaped. (p. 755)

rhomboidal A diamond-shaped plane.

ribbed With prominent veins or raised bands.

riparian Relating to areas adjacent to rivers and streams.

root The typically underground extension of the plant axis that tends to grow downwards and is without leaves or nodes.

rootstock A rhizome; a swollen, very short, erect, underground stem.

rosette A group of organs radiating from the center, usually refering to leaves at ground level.

rosulate With leaves in a rosette.

rotate Flat and circular or disc-shaped.

rudimentary Not fully developed.

rugose Wrinkled.

rugulose Slightly or finely wrinkled.

runcinate Pinnately and rather sharply incised with the lobes directed away from the tip. (p. 756)

S

sac A pouch-like structure.

saccate Sac-shaped; with a sac.

sagittate Arrowhead-shaped. (p. 757)

saline Salty.

salverform With a slender tube and an abruptly spreading, flattened limb.

samara A winged, dry fruit. (p. 765)

samaroid Like a samara.

sap The fluids that circulate through a plant.

savanna A grassland region with scattered trees.

scaberulose Slightly rough to the touch due to minute, hard projections or to the presence of short, stiff hairs; diminutive of scabrous.

scabland Barren, rocky plains with little or no soil.

scabrid Roughened.

scabridulous Minutely roughened.

scabrous Rough to the touch due to minute, hard projections or to the presence of short, stiff hairs. (p. 758)

scale A small, often leaf-like organ, often dry and membranous. (p. 769)

scape A leafless flowering stalk rising from ground level.

scapiform Scape-like, but not completely leafless.

scapose With flowers on a scape; scape-like.

scar A mark left by the separation of two structures, as a leaf scar on a stem or on a seed after detachment from the placenta. (p. 765)

scarious Thin, dry, and translucent.

schizocarp A dry, indehiscent fruit that separates into one-seeded pieces at maturity. (p. 765, 766)

scorpioid Coiled like the tail of a scorpion; an inflorescence with a zigzag rachis.

scree An accumulation of loose stones or rock, typically at the base of a slope or cliff.

scrub A plant community dominated by shrubs or stunted trees.

sculptured With smooth curves.

scurfy Covered with small, coarse scales. (p. 758)

secund With parts to one side of the axis only.

seed A ripened ovule.

seep (alt. **seepage**) A place where ground water oozes up to the surface.

segment Each free part of a divided whole.

semiannular A shape resembling part of a ring.

semibilocular An ovary partly divided into two locules.

senesce (adj. **senescent**) To deteriorate or die.

sepal The outermost whorl of parts in a flower, often green and leaf-like. (p. 762)

septate Internally divided by one or more partitions.

septicidal Dehiscing through the septa (compare to loculicidal).

septum (pl. **septa**) A partition or cross-wall.

sequester To accumulate a compound.

sericeous Silky with soft, straight, appressed hairs. (p. 758)

series Rows.

serpentine Usually describes soil containing serpentinite, a dark green or brown mineral that inhibits the growth of many plants.

serrate With sharp teeth along the margin pointing toward the tips. (p. 756)

serration One of the teeth along a serrated margin.

serrulate Serrate, but with very small teeth. (p. 756)

sessile Without a stalk.

seta (pl. **setae**) Fine bristle.

setaceous Bristle-like.

setose Covered with bristles. (p. 758)

setulose Covered with minute bristles.

sheath A tubular structure surrounding an organ or part.

sheathe To form a sheath.

shoot A young branch or stem.

shrub A woody plant, typically branched from base.

sigmoid S-shaped.

silicified Impregnated with silica.

silicle A dry, dehiscent fruit, less than twice as long as wide, with two valves separating from the persistent placentae and septum. (p. 765)

silique A dry, dehiscent fruit, more than twice as long as wide, with two valves separating from the persistent placentae and septum. (p. 765)

silky Covered with very fine, more or less appressed, lustrous hairs.

simple Composed of one part, not divided or branched. (p. 753)

sinuate Having a wavy or curvy margin, with alternating rounded lobes and notches. (p. 756)

sinuous Shallowly wavy, with many curves.

sinus The gap or recess between two lobes or segments.

slough Soft, muddy ground; backwater.

solitary Borne singly.

spadix (pl. **spadices**) A spike with small, sessile flowers crowded along a fleshy axis.

spatulate Shaped like a spatula or spoon, rounded at the tip and gradually tapering to the base. (p. 755)

specialist A species that is able to thrive only in a narrow range of environmental conditions or make use of limited resources, often dietary (see *Insect* chapter) (compare to generalist).

species A group of plants of the similar morphology, capable of interbreeding and producing fertile offspring.

spherical Round or globe-shaped.

spheroid Nearly spherical.

spicate Arranged in a spike.

spiciform An inflorescence with a spike-like appearance, but without the structure of a true spike.

spiculate Bearing spicules; spicule-like.

spicule A short, pointed, epidermal projection.

spike An elongate, unbranched, indeterminate inflorescence with sessile flowers or spikelets. (p. 760)

spinal Along midline of the back surface.

spine A sharp and pointed modified leaf or stipule.

spinescent Spine-tipped; bearing spines.

spinose With spines.

spinule A small spine.

spinulose With small spines.

spore A vegetative, reproductive cell; the first cell of the gametophyte generation.

spray A slender shoot or branch with leaves, flowers, or fruits.

spreading Nearly horizontal.

spur A hollow, sac-like projection.

stalk A secondary stem, usually the supporting structure of an organ like a leaf, flower, or fruit.

stamen Male organ of the flower that bears pollen, composed of a filament and anther. (p. 762)

staminate Having only stamens and lacking pistils (compare to pistillate).

staminode (pl. **staminodia**) A modified stamen that is sterile and doesn't produce pollen.

stellate Star-shaped; with several to many branches radiating from a central point. (p. 758)

stem Central support of a plant.

steppe A large, semi-arid, treeless area of flat land.

sterile Not producing seeds, pollen, or spores.

stigma (adj. **stigmatic**) The apical part of the pistil that is receptive to pollen. (p. 762)

stipe A stalk or support.

stipel A small, stipule-like structure at the base of a leaflet.

stipitate With a stipe.

stipular Pertaining to the stipule.

stipulate Bearing stipules.

stipule One of a pair of leaf- or scale-like appendages inserted at or near the base of a petiole. (pp. 753, 770)

stolon A horizontal, elongate stem along the ground that can produce roots, stems, and leaves.

stoloniferous Bearing stolons.

stomata Small pores or apertures on the epidermis, allowing gas exchange between the tissue and outside air.

stomatal Relating to the stomata.

stone The hard, woody endocarp enclosing the seed of a drupe.

stramineous Straw-colored.

strap-shaped Straight, flat, and narrow.

striate With fine parallel lines or grooves.

strigillose Minutely strigose.

strigose With pointed, appressed and rigid hairs, scales or bristles. (p. 758)

style The elongate part of the pistil between the ovary and the stigma. (pp. 762, 769)

styliferous Bearing a style.

sub- Prefix meaning slightly, somewhat, under, nearly, or almost.

subalpine Vegetative zone immediately below timberline and above the montane zone.

subarctic The area of the Northern Hemisphere south of the Arctic Zone where winters are long and cold and the growing season is short, falling between approximately 50°N and 70°N latitude.

submersed (alt. **submerged**) Under water.

subshrub A small shrub, often woody only at the base.

subspecies A grouping within a species above the level of variety; usually a geographically isolated variant of a species.

subtend To be immediately below.

subterranean Underground.

subulate Awl-shaped, narrow and gradually tapering to a fine point. (p. 755)

succulent Juicy or fleshy.

sucker A shoot arising from underground parent tissues.

suffrutescent Somewhat shrubby, with base of stems slightly woody.

suffused Tinted.

sulcate With longitudinal grooves or furrows.

superior Situated above another organ or part; a superior ovary has other floral parts attached below its point of attachment (compare to inferior).

surculose Producing suckers or runners from the base or from rootstocks.

suture A seam or line formed at the junction of two margins.

swale A low tract of land that is usually moist or marshy.

swamp A wetland with some woody vegetation that is flooded most or all of the year (compare to marsh).

symbiosis Biological interaction between two different organisms living in close physical association.

symbiotic Pertaining to a close relationship between species in which each species benefits.

sympatric Of two or more species occupying the same geographic area.

sympetalous With united petals.

synchrony (adj. **synchronous**) The state of simultaneous action, development, or occurrence (see *Insect* chapter) (compare to asynchrony).

synoecious With staminate and pistillate flowers in the same head.

synsepalous With united sepals.

T

talus An accumulation of rock debris forming a slope, usually at the base of a cliff.

taproot The main root axis from which smaller root branches arise; a type of root with a main root axis and smaller branches.

tawny Tan in color.

taxon (pl. **taxa**) A unit of biological classification, such as a particular species, genus, or family.

temperate The area between the tropics and the Arctic in the Northern Hemisphere, and tropics and the Antarctic in the Southern Hemisphere.

ASTERACEAE

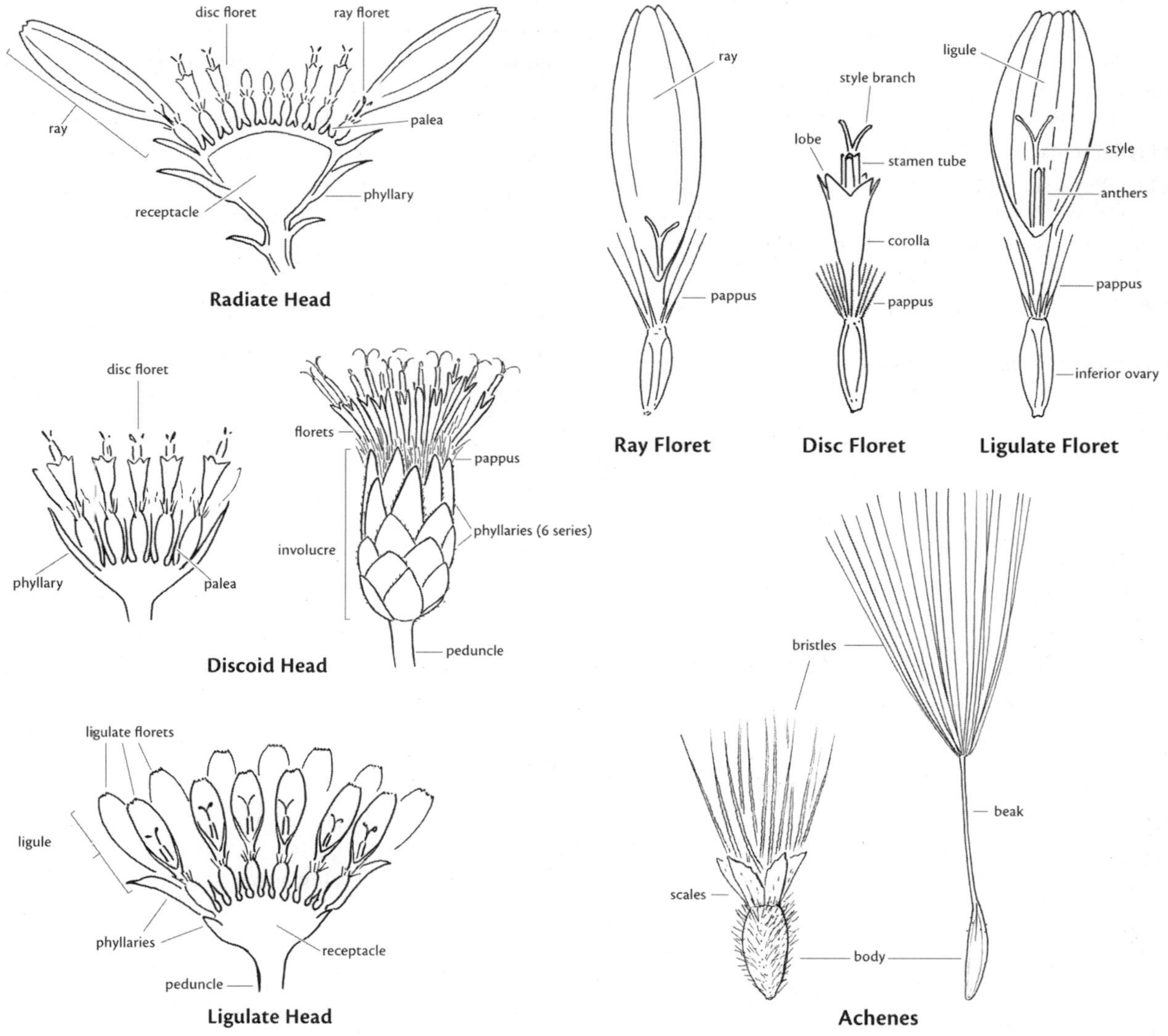

tendril A slender, twining structure by which a climbing plant may attach itself. (p. 770)

tepal A segment of a perianth in which the sepals and petals are not well differentiated.

terete Cylindrical, round in cross section.

terminal At the end of a stem or axis.

ternate Arranged or divided into threes.

terrestrial Growing on the ground.

tertiary Third in order or level. (p. 766)

testa The seed coat.

tetradynamous Having four long and two short stamens, as in most Brassicaceae.

tetragonal Four-angled.

tetrahedral Having four sides.

tetraploid Having four sets of chromosomes.

thorax The middle of the three major divisions of an insect body which may bear legs and wings.

thorn A stiff, woody modified stem with a sharp point.

throat The orifice or expanded end of the tube in a flower with fused sepals or petals; the upper region of the grass leaf sheath, where it begins to spread apart and expose the stem; in orchids, the expanded portion of the corolla at the base of the column. (p. 762)

thyrse Compact, cylindrical or ovate panicle with an indeterminate main axis and cymose sub-axes.

tibia (adj. **tibial**) The fourth leg segment (lower leg) of an insect.

tilth Physical condition of the soil, especially with respect to its suitability for growing plants (see *Landscaping* chapter).

timberline Elevation above which conditions prevent tree growth.

tomentose With a covering of short, matted or tangled, soft, woolly hairs. (p. 758)

tomentulose Slightly tomentose.

toothed With teeth; with small points or lobes along the margin.

topography The shape and physical features of an area.

tortuous Twisted or bent.

torulose Cylindrical with alternate slight swellings and contractions; of a fruit constricted slightly between the seeds.

trailing Growing along the ground but not rooting.

translucent Letting light through, almost transparent.

transverse Lying across, set crosswise.

tree A large, woody plant with usually one main stem (trunk).

tribe A taxonomic grouping of closely related genera, a division of the family below the level of subfamily.

trichome A hair or hair-like outgrowth of the epidermis.

trifoliate Having three leaflets (or leaves).

trifoliolate Of a compound leaf with three leaflets.

trigonous With three angles, triangular in cross section.

tripinnate With three orders of leaflets, each pinnately compound.

triploid Having three sets of chromosomes.

triquetrous Three-edged, with three protruding angled ridges.

triternate Thrice divided into threes; triply ternate.

truncate With the base or tip flattened abruptly as though cut off. (p. 757)

trunk The erect, unbranched portion of a tree or tree-like plant.

tube A hollow, cylindrical structure; the constricted, basal portion of a sympetalous corolla. (p. 762)

tuber A swollen, usually underground, part of a stem or root.

tubercle A small, wart-like swelling or projection.

tuberculate Bearing tubercles.

tubercule A nodule, as on the roots of some legumes.

tuberous Producing tubers; tuber-like.

Fabaceae

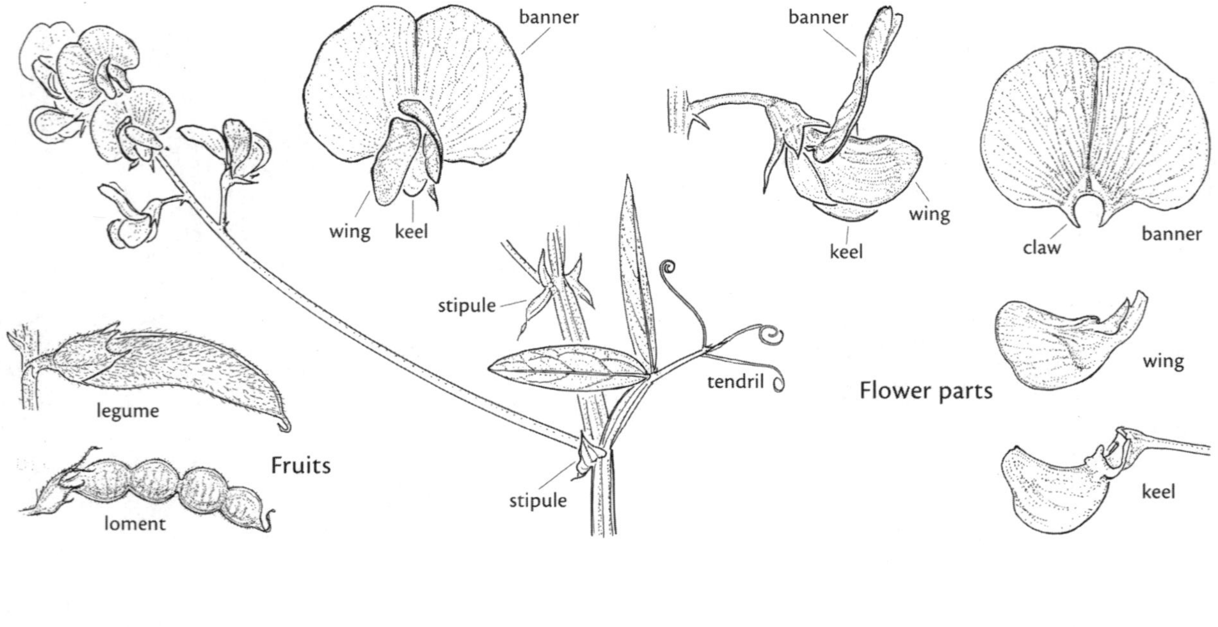

tuff A sedimentary rock formed from an accumulation of volcanic ash or dust and igneous rock.

tufted In a small, dense cluster.

turbinate Top-shaped.

turgid Swollen, rigid and stiff by means of internal pressure.

twining Coiling or spiraling around a support for climbing.

U

ultimate The final section or division of a structure.

ultramafic Substrates derived from igneous rock containing high amounts of magnesium and iron.

umbel A flat- or convex-topped inflorescence with the pedicels arising from a common point. (pp. 760, 766)

umbellate In umbels.

umbellet An ultimate umbellate cluster of a compound umbel. (p. 766)

umbo A protuberance on the end or side, usually of a seed cone scale.

unarmed Without prickles, thorns, barbs, or spines.

uncinate Hooked at the tip.

undulate Corrugated or wavy.

unicellular Of a single cell.

unifoliate With a single leaf; not divided into leaflets.

unifoliolate A leaf theoretically counpound, though expressing as a single leaf and appearing simple.

unilaterally One-sided or to one side only.

unilocular With a single cavity or chamber.

uniseriate Arranged in a single row.

unisexual With either female or male reproductive parts, but not both (compare to bisexual).

united Fused and growing together.

unpaired Not in pairs; one or more leaflets not exactly across from a leaflet on the opposite side of the rachis (*Lathyrus*).

upland Land elevated above the alluvial plain or fluvial terrace.

urceolate Pitcher-like; hollow and contracted at the mouth like an urn or pitcher.

utricle A small, thin-walled, one-seeded, more or less bladder-like fruit. (p. 765)

V

valvate Opening by valves; with margins touching along their entire length but not overlapping.

valve One of the segments of a dehiscent fruit that separates from the other segments at maturity.

valvular Relating to, having, or acting as a valve.

variegated Irregularly colored in patches.

variety A grouping within a species, below the level of subspecies, with a recognizable appearance but not usually genetically isolated from other varieties.

vascular Of tissue containing xylem and phloem, the conducting elements of the plant.

vegetative Without floral parts and/or sex organs.

vein Vessel by which water and nutrients are transported; often most easily seen in leaves. (p. 753)

velutinous Velvety; covered with fine, soft, spreading hairs. (p. 758)

venation The arrangement of the veins, especially in leaves.

ventral On or pertaining to upper or inward surface in relation to the axis; adaxial (compare to dorsal).

ventricose Inflated or swollen on one side only, as in some corollas.

vernal In the spring.

vernicose Having a shiny surface, as though polished.

verrucose Warty. (p. 758)

vesicule (adj. **vesicular**) A small, bladder-like structure.

vestigial Reduced to a rudiment and not functional.

vestiture The epidermal coverings of a plant, collectively.

villosulous Bearing minute, long, soft, shaggy but unmatted hairs.

villous With long, soft, shaggy but unmatted hairs. (p. 758)

vine Trailing or climbing plant with a long, flexible stem, often supporting itself using tendrils.

viscid Sticky or gummy.

viviparous Germinating or sprouting from a seed or bud while attached to the parent plant.

W

waif An unusual species found in the wild that is not native and persists only one or a few generations.

wetland An area of land where the seasonal or constant saturation with water determines the nature of plant and animal communities living there.

whorl (adj. **whorled)** A ring-like arrangement of three or more similar parts at the same node. (p. 754)

wing A flat, thin margin extending from or bordering a structure (p. 765); one of the lateral petals of a papilionaceous corolla (Fabaceae). (p. 770)

woodland An area covered with trees.

woolly Long, soft, usually entangled hairs. (p. 758)

X

xeric Dry (compare to mesic).

REFERENCES

Abrams, L. & R.S. Ferris. 1923–1960. Illustrated flora of the Pacific states. Vols. 1–4. Stanford University Press, Stanford, CA.

Alexander, P.J., M.D. Windham, R. Govindarajulu, I.A. Al-Shehbaz & C.D. Bailey. 2010. Molecular phylogenetics and taxonomy of the genus *Thysanocarpus* (Brassicaceae). Systematic Botany 35:559–577.

Allaby, R. 2010. Integrating the processes in the evolutionary system of domestication. Journal of Experimental Botany 61(4):935–944.

Allan, G.J. & J.M. Porter. 2000. Tribal delimitation and phylogenetic relationships of Loteae and Coronilleae (Faboideae: Fabaceae) with special reference to *Lotus*: Evidence from nuclear ribosomal ITS sequences. American Journal of Botany 87:1871–1881.

Al-Shehbaz I.A., K. Mummenhoff & O. Appel. 2002. *Cardaria*, *Coronopus*, and *Stroganowia* are united with *Lepidium* (Brassicaceae). Novon 12:5–11.

Al-Shehbaz, I.A. & K. Mummenhoff. 2010. *Stubendorffia* and *Winklera* belong to the expanded *Lepidium* (Brassicaceae). Edinburgh Journal of Botany 68:165–171.

Al-Shehbaz, I.A. 2012. A generic and tribal synopsis of the Brassicaceae (Cruciferae). Taxon 61:931–954.

American Standard for Nursery Stock. 2014. Available at https://www.americanhort.org/page/standards. Accessed March 2019.

Anderson, J.T., J.H. Willis & T. Mitchell-Olds. 2011. Evolutionary genetics of plant adaptation. Trends in Genetics 27(7):258–266.

Arias, T. & J.C. Pires. 2012. A fully resolved chloroplast phylogeny of *Brassica* crops and wild relatives (Brassicaceae: Brassiceae): Novel clades and potential taxonomic implications. Taxon 61:980–988.

Baad, M.F. 1969. Biosystematic studies of the North American species of *Arenaria* subgenus *Eremogone* (Caryophyllaceae). Ph.D. dissertation. University of Washington, Seattle, WA.

Babcock, E.B. & G.L. Stebbins, Jr. 1938. The American species of *Crepis*. Carnegie Institution of Washington 504:1–199.

Babcock, E.B. 1947. The genus *Crepis*. Part 2. University of California Publications in Botany 22:199–1030.

Bailey, L.H. & E.Z. Bailey. 1976. Hortus Third: A concise dictionary of plants cultivated in the United States and Canada, 1st edition. Macmillan, New York, NY.

Baker, M.J. & M.W. Williams. 1987. White clover. Centre for Agriculture and Bioscience International, Wallingford, UK.

Baldwin, B.G. 1999. New combinations and new genera in the North American tarweeds (Compositae-Madiinae). Novon 9:462–471.

______. 2003a. Characteristics and diversity of Madiinae. In: Carlquist, S., B.G. Baldwin & G.D. Carr, eds. Tarweeds & silverswords: Evolution of the Madiinae (Asteraceae). Missouri Botanical Garden Press, St. Louis, MO.

______. 2003b. A phylogenetic perspective on the origin and evolution of Madiinae. In: Carlquist, S., B.G. Baldwin & G.D. Carr, eds. Tarweeds & silverswords: Evolution of the Madiinae (Asteraceae). Missouri Botanical Garden Press, St. Louis, MO.

______. 2005. Origin of the serpentine-endemic herb *Layia discoidea* from the widespread *L. glandulosa* (Compositae). Evolution 59:2473–2479.

______. 2009. Heliantheae alliance. In: Funk, V.A., A. Susanna, T.F. Stuessy & R.J. Bayer, eds. Systematics, evolution, and biogeography of Compositae. International Association for Plant Taxonomy, Vienna, Austria.

Baldwin, B.G. & B.L. Wessa. 2000. Phylogenetic placement of *Pelucha* and new subtribes in Helenieae sensu stricto (Compositae). Systematic Botany 25:522–538.

Baldwin, B.G., B.L. Wessa & J.L. Panero. 2002. Nuclear rDNA evidence for major lineages of Helenioid Heliantheae (Compositae). Systematic Botany 27:161–198.

Bandick, A.K. & R.P. Dick. 1999. Field management effects on soil enzyme activities. Soil Biology and Biochemistry 31(11):1471–1479.

Barkley, T.M. 1978. *Senecio*. In: Britton, N.L. 1905+. North American flora. 34 vols. New York, NY. Vol 2 Part 10.

Barneby, R.C. 1964. Atlas of North American *Astragalus*. Memoirs of the New York Botanical Garden 13:1–1192.

______. 1989. Fabales. In: Cronquist, A., A.H. Holmgren, N.H. Holmgren, J.L. Reveal, P.K. Holmgren & R.C. Barneby. 1972–2013. Intermountain flora: Vascular plants of the Intermountain West, U.S.A. Vols. 1–6. Bronx, NY. Vol 3 Part B.

Bayer, R.J. & G.L. Stebbins. 1987. Evolution and phylogenetic relationships of the *Antennaria* (Asteraceae: Inuleae) polyploid agamic complexes. Biologisches Zentralblatt. 106:683–698.

Beard, J.S. 1978. The physiognomic approach. In: Whittaker, R.H., ed. Classification of plant communities. Dr. W. Junk bv Publishers, The Hague, Netherlands.

Bearden, B.N. & L. Petersen. 2000. Influence of arbuscular mycorrhizal fungi on soil structure and aggregate stability of a vertisol. Plant and Soil 218(1–2):173–183.

Beatley, T. 1995. Planning and sustainability: The elements of a new (improved?) paradigm. Journal of Planning Literature 9(4):383–395.

Belkin, D. 2018. "A rose by any other name might just be a rhododendron." Wall Street Journal, 15 August, p. 1.

Bogler, D.J. 2006. *Crepis*. In: Flora of North America Editorial Committee, eds. 1993+. Flora of North America north of Mexico. 18+ vols. New York and Oxford. Vol 19.

Boldt, P.E. 1989. *Baccharis* (Asteraceae): A review of its taxonomy, phytochemistry, ecology, economic status, natural enemies and the potential for its biological control in the United States. Texas Agricultural Experiment Station, Temple, TX.

Bolund, P. & S. Hunhammar. 1999. Ecosystem services in urban areas. Ecological Economics 29(2):293–301.

Bowman, J.L., H. Brüggemann, J.-Y. Lee & K. Mummenhoff. 1999. Evolutionary changes in floral architecture within the genus *Lepidium* (Brassicaceae). International Journal of Plant Sciences 160:917–929.

Brouillet, L. 2006. *Aster*. In: Flora of North America, Editorial Committee, eds. 1993+. Flora of North America north of Mexico. 18+ vols. New York and Oxford. Vol 20.

Brown, T.C., J.C. Bergstrom & J.B. Loomis. 2007. Defining, valuing, and providing ecosystem goods and services. Natural Resources Journal 47(2):329–379.

Brummitt, R.K. 1980. Further new names in the genus *Calystegia* (Convolvulaceae). Kew Bulletin 35:330–331.

______. 2002. *Calystegia sylvatica* (Convolvulaceae) in western North America. Madroño 49:130–131.

Brundrett, M.C. 2002. Coevolution of roots and mycorrhizas of land plants. New Phytologist 154(2):275–304.

Burns, T.D., M.I. Bidartondo & D.L. Taylor. 2002. Host specificity in ectomycorrhizal communities: What do the exceptions tell us? Integrative & Comparative Biology 42(2):352–359.

Buslaff, J. & L. Johnson, eds. 2004. Landscaping with native plants. 4th edition. Wild Ones, Appleton, WI.

Butterflies and moths of North America. www.butterfliesandmoths.org.

Cacho, N.I., A.M. Burrell, A.E. Pepper & S.Y. Strauss. 2014. Novel nuclear markers inform the systematics and the evolution of serpentine use in *Streptanthus* and allies (Thelypodieae, Brassicaceae). Molecular Phylogenetics and Evolution 72:71–81.

Carlquist, S., B.G. Baldwin & G.D. Carr, eds. 2003. Tarweeds & silverswords: Evolution of the Madiinae (Asteraceae). Missouri Botanical Garden Press, St. Louis, MO.

Carlquist, S. 1959. Studies on Madiinae: Anatomy, cytology, and evolutionary relationships. Aliso 4:171–236.

______. 1983. Intercontinental dispersal. Sonderbdaeni De des Naturwissenschaftlichen Vereins in Hamburg 7:37–47.

______. 2003. Diversity in trichomes and glandular structures of Madiinae. In: Carlquist, S., B.G. Baldwin & G.D. Carr, eds. Tarweeds & silverswords: Evolution of the Madiinae (Asteraceae). Missouri Botanical Garden Press, St. Louis, MO.

Carr, G.D. 2003. Chromosome evolution in Madiinae. In: Carlquist, S., B.G. Baldwin & G.D. Carr, eds. Tarweeds & silverswords: Evolution of the Madiinae (Asteraceae). Missouri Botanical Garden Press, St. Louis, MO.

Carr, R.L. & G.D. Carr. 1983. Chromosome races and structural heterozygosity in *Calycadenia ciliosa* Greene (Asteraceae). American Journal of Botany 70:744–755.

Cellinese, N. In press. Campanuloideae. In: Queiroz, K. de., P.D. Cantino & J.A. Gauthier, eds. Phylonyms: A companion to the PhyloCode. CRC Press, Boca Raton, FL.

Chambers, K.L. & S.C. Meyers. 2011. Nomenclatural changes for some taxa in the Oregon flora. Journal of the Botanical Research Institute of Texas 5:619–623.

Chambers, K.L. 1955. A biosystematic study of the annual species of *Microseris*. Contributions of the Dudley Herbarium 4:207–312.

______. 2017. *Artemisia estesii*, a new diploid species of the *Artemisia ludoviciana* complex in the Pacific Northwest. Journal of the Botanical Research Institute of Texas 11:1–9.

Chandler, G.T. & G.M. Plunkett. 2004. Evolution in Apiales: Nuclear and chloroplast markers together in (almost) perfect harmony. Botanical Journal of the Linnean Society 144:123–147.

Chan, R. & R. Ornduff. 2006. *Lasthenia*. In: Flora of North America Editorial Committee, eds. 1993+. Flora of North America north of Mexico. 18+ vols. New York and Oxford. Vol 21.

Chan, R., B.G. Baldwin & R. Ornduff. 2002. Cryptic goldfields: A molecular phylogenetic reinvestigation of *Lasthenia californica* sensu lato and close relatives (Compositae: Heliantheae sensu lato). American Journal of Botany 89:1103–1112.

Chmielewski, J.G. 1993. *Antennaria pulvinata* Greene: The legitimate name for *A. aromatica* Evert (Asteraceae: Inuleae). Rhodora 95:261–276.

Christy, J.A. & E.R. Alverson. 2011. Historical vegetation of the Willamette Valley, Oregon, circa 1850. Northwest Science 85(2):93–107.

Christy, J.A., A. Kimpo, V. Marttala, P.K. Gaddis & N.L. Christy. 2009. Urbanizing flora of Portland, Oregon, 1806–2008. Native Plant Society of Oregon Occasional Paper 3:1–319.

Clark, W.D. 1979. The taxonomy of *Hazardia* (Compositae: Astereae). Madroño 26:105–127.

______. 2006. *Hazardia*. In: Flora of North America Editorial Committee, eds. 1993+. Flora of North America north of Mexico. 18+ vols. New York and Oxford. Vol 20.

Clausen, J., D.D. Keck & W.M. Hiesey. 1948. Experimental studies on the nature of species. 3rd edition. Environmental responses of climatic races of *Achillea*. Carnegie Institution of Washington, Washington DC.

Clausen, J. 1951. Stages in the evolution of plant species. Cornell University Press, Ithaca, NY.

Clements, F.E. 1920. Plant indicators: The relation of plant communities to process and practice. Carnegie Institution, Washington, DC.

Cooper, C.S. 1957. A legume for native flood meadows: I. Establishment and maintenance of stands of white-tip clover (*Trifolium variegatum*) in native flood meadows and its effect upon yields, vegetative and chemical composition of hay. Agronomy Journal 49:473–477.

Costea, M., G.L. Nesom & S. Stefanović. 2006. Taxonomy of the *Cuscuta pentagona* complex (Convolvulaceae) in North America. Sida 22:151–175.

Crawford, C.S. 1981. Temporary vegetation community: Emphasis on herbivores. In: Biology of desert invertebrates. Springer, Berlin, Germany.

Cronquist, A. 1981. An integrated system of classification of flowering plants. Columbia University Press, New York, NY.

______. 1994. *Chrysothamnus*. In: Cronquist, A., A.H. Holmgren, N.H. Holmgren, J.L. Reveal, P.K. Holmgren & R.C. Barneby. 1972–2013. Intermountain flora: Vascular plants of the Intermountain West, U.S.A. Vols. 1–6. Bronx, NY. Vol 5.

______. 1994. *Rudbeckia*. In: Cronquist, A., A.H. Holmgren, N.H. Holmgren, J.L. Reveal, P.K. Holmgren & R.C. Barneby. 1972–2013. Intermountain flora: Vascular plants of the Intermountain West, U.S.A. Vols. 1–6. Bronx, NY. Vol 5.

______. 1997. Apiaceae. In: Cronquist, A., A.H. Holmgren, N.H. Holmgren, J.L. Reveal, P.K. Holmgren & R.C. Barneby. 1972–2013. Intermountain flora: Vascular plants of the Intermountain West, U.S.A. Vols. 1–6. Bronx, NY. Vol 3 Part A.

Cronquist, A. & A.H. Holmgren. 1984. *Sambucus*. In: Cronquist, A., A.H. Holmgren, N.H. Holmgren, J.L. Reveal, P.K. Holmgren & R.C. Barneby. 1972–2013. Intermountain flora: Vascular plants of the Intermountain West, U.S.A. Vols. 1–6. Bronx, NY. Vol 4.

Crowl, A.A., E. Mavrodiev, G. Mansion, R. Haberle, A. Pistarino, G. Kamari, D. Philtos, T. Borsch & N. Cellinese. 2014. Phylogeny of Campanuloideae (Campanulaceae) with emphasis on the utility of nuclear pentatricopeptide repeat (PPR) Genes. PLoS One 9(4):e94199.

Dandy, J.E. 1958. List of British vascular plants. British Museum, London, England.

Danforth, B.N., R.L. Minckley & J.L. Neff. 2019. The solitary bees: Biology, evolution, conservation. Princeton University Press, Princeton, NJ.

Darrach, M.E. & C.E. Hinchliff. 2014. *Lomatium tarantuloides* (Apiaceae), a new narrowly endemic species from northeast Oregon. Phytoneuron 2014-27:1–8.

Darrach, M.E. & D.H. Wagner. 2011. *Lomatium pastoralis* (Apiaceae), a new narrow endemic from Northeastern Oregon. Journal of the Botanical Research Institute of Texas 5:427–435.

Dasmann, R.F. 1991. The importance of cultural and biological diversity. In: Oldfield, M.L. & J.B. Alcorn, eds. Biodiversity: Culture, conservation, and ecodevelopment. Westview Press, Boulder, Colorado.

De Wit, H.C.D. 1964. Aquarium plants. Blandford Press Ltd., London, England.

Dillenberger, M.S. & J.W. Kadereit. 2014. Maximum polyphyly: Multiple origins and delimitation with plesiomorphic characters require a new circumscription of *Minuartia* (Caryophyllaceae). Taxon 63:64–88.

DiTomaso, J.M. & E.A. Healy. 2007. Weeds of California and other Western States. Vol 1. Aizoaceae-Fabaceae. University of California Division of Agriculture and Natural Resources. Berkley, CA.

Dugle, J.R. 1966. A taxonomic study of western Canadian species in the genus *Betula*. Canadian Journal of Botany 44:929–1007.

Duvall, M.R., J.D. Noll & G.E. Larson. 1999. Disjunct populations of *Trifolium beckwithii* (Fabaceae) in eastern South Dakota: Vicariance or recent long-distance dispersal? A preliminary analysis. Madroño 46:199–204.

Dwyer, J.F., E.G. McPherson, H.W. Schroeder & R.A. Rowntree. 1992. Assessing the benefits and costs of the urban forest. Journal of Arboriculture 18(5):227–227.

Ellison, N.W., A. Liston, J.J. Steiner, W.M. Williams & N.L. Taylor. 2006. Molecular phylogenetics of the clover genus (*Trifolium*–Leguminosae). Molecular Phylogenetics and Evolution 39:688–705.

Emmel, T.C., ed. 1998. Systematics of Western American butterflies. Mariposa Press, Gainesville, FL.

Enke, N. & B. Gemeinholzer. 2008. Babcock revisited: New insights into generic delimitation and character evolution in *Crepis* L. (Compositae: Cichorieae) from ITS and *matK* sequence data. Taxon 57:756–768.

Excoffier, L., T. Hofer & M. Foll. 2009. Detecting loci under selection in a hierarchically structured population. Heredity 103:285–298.

Ferris, R.S. 1960. *Rudbeckia*. In: Abrams, L. & R.S. Ferris, eds. 1940. Illustrated flora of the Pacific States. Stanford University Press, Stanford, CA. Vol 4.

Fisher, C. 1995. Horse care: Perilous pasture plants. Rural Heritage 20:44–45.

Fosberg, F.R. 1942. *Cornus sericea* L. (*C. stolonifera* Michx.). Bulletin of the Torrey Botanical Club 69:583–589.

Foster, B.G. 2015. A taxonomic reevaluation of *Trifolium variegatum* Nutt. based on morphology. Master of Science Thesis, Miami University, Oxford, OH.

Fukuda, I. 1967. The biosystematics of *Achlys*. Taxon 16:308–316.

Fukuda, I. & H. Baker. 1970. *Achlys californica* (Berberidaceae) a new species. Taxon 19:341–344.

Furlow, J.J. 1997. Betulaceae. In: Flora of North America Editorial Committee, eds. 1993+. Flora of North America north of Mexico. 18+ vols. New York and Oxford. Vol 3.

Gaskin, J.F. & L.M. Wilson. 2007. Phylogenetic relationships among native and naturalized *Hieracium* (Asteraceae) in Canada and the United States based on plastid DNA sequences. Systematic Botany 32:478–485.

Gillett, J.B. 1985. Taxonomy and morphology. In: Taylor, N.L., ed. 1985. Clover Science and Technology. American Society of Agronomy 25:7–47.

Gillett, J.M. 1969. Taxonomy of *Trifolium* (Leguminosae). II. The *T. longipes* complex in North America. Canadian Journal of Botany 47:93–113.

______. 1971. Taxonomy of *Trifolium* (Leguminosae). III. *T. eriocephalum*. Canadian Journal of Botany 49:395–405.

______. 1972. Taxonomy of *Trifolium* (Leguminosae). IV. The American species of section *Lupinaster* (Adanson) Seringe. Canadian Journal of Botany 50:1975–2007.

______. 1980. Taxonomy of *Trifolium* (Leguminosae). V. The perennial species of section *Involucrarium*. Canadian Journal of Botany 58:1425–1448.

Glad, J.B. & R.R. Halse. 1993. Invasion of *Amorpha fruticosa* L. (Leguminosae) along the Columbia and Snake rivers in Oregon and Washington. Madroño 40:62–63.

Gleason, H.A. & A. Cronquist. 1991. Manual of vascular plants of northeastern United States and adjacent Canada. New York Botanical Garden, Bronx, NY.

Gómez-Baggethun, E. & D.N. Barton. 2013. Classifying and valuing ecosystem services for urban planning. Ecological Economics 86:235–245.

Gottlieb, L.D. 1968. Hybridization between *Arctostaphylos viscida* and *A. canescens* in Oregon. Brittonia 20:83–93.

Greenberg, A.K. & M.J. Donoghue. 2011. Molecular systematics and character evolution in Caryophyllaceae. Taxon 60:1637–1652.

Grimes, J.W. 1990. A revision of the New World species of Psoraleeae (Leguminosae: Papilionoideae). Memoirs of the New York Botanical Garden 61:1–113.

Guilliams, C.M. 2015. Diversification, biogeography, and classification of Amsinckiinae (Boraginaceae), with an emphasis on the popcornflowers (*Plagiobothrys*). Ph.D. dissertation. University of California, Berkeley, CA.

Guilliams, C.M., K.E. Hasenstab-Lehman & B.G. Baldwin. In preparation. Nomenclatural changes in Boraginaceae subtribe Amsinckiinae.

Guppy, G.A. 1978. Species relationships of *Hieracium* (Asteraceae) in British Columbia. Canadian Journal of Botany 56:3008–3019.

Hall, J.C., K.J. Sytsma & H.H. Iltis. 2002. Phylogeny of Capparaceae and Brassicaceae based on chloroplast sequence data. American Journal of Botany 89:1826–1842.

Hall, J.C., T.E. Tisdale, K. Donohue, A. Wheeler, M.A. Al-Yahya & E.M. Kramer. 2011. Convergent evolution of a complex fruit structure in the tribe Brassiceae (Brassicaceae). American Journal of Botany 98:1989–2003.

Hanson, T. 2018. Buzz. The nature and necessity of bees. Basic Books, New York, NY.

Harbaugh, D.T., M. Nepokroeff, R.K. Rabeler, J. McNeill, E.A. Zimmer & W.L. Wagner. 2010. A new lineage-based tribal classification of the family Caryophyllaceae. International Journal of Plant Sciences 171:185–198.

Hardin, J.W. & J.M. Arena. 1974. Human poisoning from native and cultivated plants. Duke University Press, Durham, NC.

Hartman, R.L. & R.K. Rabeler. 2005. *Loeflingia*. In: Flora of North America Editorial Committee, eds. 1993+. Flora of North America north of Mexico. 18+ vols. New York and Oxford. Vol 5.

Hartman, R.L., R.K. Rabeler & D.H. Wilken. 2012. *Silene*. In: Baldwin, B.G., D. Goldman, D.J. Keil, R. Patterson, T.J. Rosatti & D. Wilken. 2012. The Jepson manual: Vascular plants of California. 2nd edition. University of California Press, Berkeley, CA.

Hasenstab-Lehman, K.E. & M.G. Simpson. 2012. Cat's eyes and popcorn flowers: Phylogenetic systematics of the genus *Cryptantha* s.l. (Boraginaceae). Systematic Botany 37:738–757.

Hermann, F.J. 1953. A botanical synopsis of the cultivated clovers (*Trifolium*). U.S. Department of Agriculture, Washington, DC.

Hinchliff, J. 1994. An atlas of Oregon butterflies. The Oregon State University Bookstore, Inc., Corvallis, OR.

Hitchcock, C.L. & A. Cronquist. 1964. *Betula pumila*. In: Hitchcock, C.L., A. Cronquist, M. Ownbey & J.W. Thompson. 1955–1969. Vascular plants of the Pacific Northwest. Vols. 1–5. Seattle, WA. Vol 2.

______. 2018. Flora of the Pacific Northwest: An illustrated manual. 2nd edition. Giblin, D.E., B.S. Legler, P.F. Zika & R.G. Olmstead, eds. University of Washington Press, Seattle, WA.

Hitchcock, C.L., A. Cronquist & M. Ownbey. 1959. *Sambucus*. In: Hitchcock, C.L., A. Cronquist, M. Ownbey & J.W. Thompson. 1955–1969. Vascular plants of the Pacific Northwest. Vols. 1–5. Seattle, WA. Vol 4.

Hitchcock, C.L., A. Cronquist, M. Ownbey & J.W. Thompson. 1955–1969. Vascular plants of the Pacific Northwest. Vols. 1–5. University of Washington Press, Seattle, WA.

Holmgren, N.H. & P.K. Holmgren. 2012. *Silene*. In: Cronquist, A., A.H. Holmgren, N.H. Holmgren, J.L. Reveal, P.K. Holmgren & R.C. Barneby. 1972–2013. Intermountain flora: Vascular plants of the Intermountain West, U.S.A. Vols. 1–6. Bronx, NY. Vol 2 Part A.

Hopkins, M. 1937. *Arabis* in eastern and central North America. Journal of the New England Botanical Club 39:63–188.

Isely, D. 1998. Native and naturalized Leguminosae (Fabaceae) of the United States. Monte L. Bean Life Science Museum, Brigham Young University, Provo, UT.

James, D.G. & D. Nunnallee. 2011. Life histories of Cascadia butterflies. Oregon State University Press, Corvallis, OR.

Jansen, R.K., R.S. Wallace, K. Kim & K.L. Chambers. 1991. Systematic implications of chloroplast DNA variation in the subtribe Microseridinae (Asteraceae: Lactuceae). American Journal of Botany 78:1015–1027.

Jepson, W.L. 1925. A manual of the flowering plants of California. University of California Press, Berkeley, CA.

Jeyanny, V., M.A. Husni, B.S. Sumar, K.W. Rasidah & W. Suhaimi. 2016. Contribution of forest floor fractions to carbon storage and abundance patterns of arbuscular mycorrhizal fungal colonisation in a tropical montane forest. Southern Forests: A Journal of Forest Science 78(4):315–319.

Johnston, I.M. 1924. Studies in the Boraginaceae II. A synopsis of the American native and immigrant Borages of the subfamily Boraginoideae. Contributions from the Gray Herbarium of Harvard University 70:3–61.

Jolles, D.D. & A.D. Wolfe. 2012. Genetic differentiation and crypsis among members of the myco-heterotrophic *Pyrola picta* species complex (Pyroleae: Monotropoideae: Ericaceae). Systematic Botany 37:468–477.

Jones, J.A., R. Hutchinson, A. Moldenke, V. Pfeiffer, E. Helderop, E. Thomas, J. Griffin & A. Reinholtz. 2018. Landscape patterns and diversity of meadow plants and flower-visitors in a mountain landscape. Landscape Ecology 34(5):997–1014.

Jones, M.E. 1902. Contributions to western botany, No. 10. Mammoth Record Print, Robinson, UT.

Jordon-Thaden, I., I. Hase, I.A. Al-Shehbaz & M. Koch. 2010. Molecular phylogeny and systematics of the genus *Draba* s.l. (Brassicaceae) and identification of its closest related genera. Molecular Phylogenetics and Evolution 55:524–540.

Kadereit, G. & H. Freitag. 2011. Molecular phylogeny of Camphorosmeae (Camphorosmoideae, Chenopodiaceae): Implications for biogeography, evolution of C4-photosynthesis and taxonomy. Taxon 60:51–78.

Keck, D.D. 1959. *Madia*. In: Munz, P.A. A California flora. University of California Press, Berkeley, CA.

Keil, D.J. 2006. *Carthamus*. In: Flora of North America Editorial Committee, eds. 1993+. Flora of North America north of Mexico. 18+ vols. New York and Oxford. Vol 19.

Kemper, T. & R.P. McNeill. 2012. *Lomatium brunsfeldianum*: A new species *Lomatium* (Umbelliferae) from northern Idaho. Journal of the Botanical Research Institute of Texas 6:29–36.

King, R.M. & H. Robinson. 1987. The genera of the Eupatorieae (Asteraceae). Monographs in Systematic Botany. Vol 22. Missouri Botanical Garden Press, St. Louis, MO.

Kliejunas, J.T. 2010. Sudden oak death and *Phytophthora ramorum*: A summary of the literature. United States Department of Agriculture, Pacific Southwest Research Station, Albany, CA.

Knight, W.E. 1985. Crimson clover. In: Taylor, N.L., ed. 1985. Clover science and technology. American Society of Agronomy 25:491–502.

Kramer, A.T., B. Crane, J. Downing, J.L. Hamrick, K. Havens, A. Highland, S.K. Jacobi, T.N. Kaye, E. Lonsdorf, J. Ramp Neale, A. Novy, D.W. Tallamy, A. White & J. Zeldin. 2019. Sourcing native plants to support ecosystem function in different planting contexts. Restoration Ecology 27(3):470–476.

Kron, K.A. & W.S. Judd. 1990. Phylogenetic relationships within the Rhodoreae (Ericaceae) with specific comments on the placement of *Ledum*. Systematic Botany 15:57–68.

Kruckeberg, A.R. 1977. Manzanita (*Arctostaphylos*) hybrids in the Pacific Northwest: Effects of human and natural disturbance. Systematic Botany 2:233–250.

Leach, G.J. 1989. Taxonomic revision of *Bergia* (Elatinaceae) in Australia. Journal of Adelaide Botanic Garden 11:75–100.

Lee, C.-S. & S.R. Downie. 2006. Phylogenetic relationships within *Cicuta* (Apiaceae tribe Oenantheae) inferred from nuclear rDNA ITS and cpDNA sequence data. Canadian Journal of Botany 84:453–468.

Lee, J., B.G. Baldwin & L.D. Gottlieb. 2002. Phylogeny of *Stephanomeria* and related genera (Compositae–Lactuceae) based on 18S–26S nuclear rDNA ITS and ETS sequences. American Journal of Botany 89:160–168.

______. 2003. Phylogenetic relationships among the primarily North American genera of Cichorieae (Compositae) based on analysis of 18S–26S nuclear rDNA ITS and ETS sequences. Systematic Botany 28:616–626.

Lesica, P. & P.M. Kittelson. 2013. Morphological and ecological segregation of two sympatric *Lomatium triternatum* (Apiaceae) varieties in Montana. Madroño 60:211–216.

Lesica, P., R.R. Leary, F.W. Allendorf & D.E. Bilderback. 1988. Lack of genic diversity within and among populations of an endangered plant, *Howellia aquatilis*. Conservation Biology 2:275–282.

Lewis, W.H. & R.L. Oliver. 1965. Realignment of *Calystegia* and *Convolvulus* (Convolvulaceae). Annals of the Missouri Botanical Garden 52:217–222.

Li, W.L., G.P. Berlyn & P.M.S. Ashton. 1996. Polyploids and their structural and physiological characteristics relative to water deficit in *Betula papyrifera* (Betulaceae). American Journal of Botany 83:15–20.

Linderman, R.G. 1991 Mycorrhizal interactions in the rhizosphere. In: Keister, D.L. & P.B. Cregan, eds. The rhizosphere and plant growth. Beltsville Symposia in Agricultural Research.

Linderman, R.G. & E.A. Davis. 2004. Evaluation of commercial inorganic and organic fertilizer effects on arbuscular mycorrhizae formed by *Glomus intraradices*. HortTechnology 14(2):196–202.

Liu, Y., Y-S. Chen & Q-E. Yang. 2013. Generic status, circumscription, and allopolyploid origin of *Faberia* (Asteraceae: Cichorieae) as revealed by ITS and chloroplast DNA sequence data. Taxon 62:1235–1247.

Love, R.M. 1985. Rose clover. In: Taylor, N.L., ed. 1985. Clover science and technology. American Society of Agronomy 25:535–546.

Mabry, M.E., R.A. Dowdy, L.M. Simpson, J.P. Rebman & M.G. Simpson. 2016. Taxonomy of the winged popcorn flower: *Cryptantha pterocarya* (Boraginaceae). Phytotaxa 253:97–130.

Mabry, M.E. & M.G. Simpson. 2018. Evaluating the monophyly and biogeography of *Cryptantha* (Boraginaceae). Systematic Botany 43:53–76.

Macbride, J.F. 1916. The true mertensias of western Northern America. Contributions from the Gray Herbarium of Harvard University 48:1–20.

Mack, R.N. 2003. Plant naturalizations and invasions in the eastern United States: 1634–1860. Annals of the Missouri Botanical Garden 90:77–90.

Mancuso, M. 2001. The status of *Trifolium owyheense* (Owyhee clover) in Idaho. Idaho Department of Fish and Game, Natural Resource Policy Bureau, Boise, ID.

Manos, P.S., C.H. Cannon & S.-H. Oh. 2009. Phylogenetic relationships and taxonomic status of the paleoendemic Fagaceae of western North America: Recognition of a new genus, *Notholithocarpus*. Madroño 55:181–190.

Mansion, G., G. Parolly, A.A. Crowl, E. Mavrodiev, N. Cellinese, M. Oganesian, K. Fraunhofer, G. Kamari, D. Phitos, R. Haberle, G. Akaydin, N. Ikinci, T. Raus & T. Borsch. 2012. How to handle speciose clades? Mass taxon-sampling as a strategy towards illuminating the natural history of *Campanula* (Campanuloideae). PLoS One 7(11):e50076. doi:10.1371/journal.pone.0050076.

Markos, S. & J.L Strother. 2012. *Corethrogyne*. In: Flora of North America Editorial Committee, eds. 1993+. Flora of North America north of Mexico. 18+ vols. New York and Oxford. Vol 20.

Martin, J.S. 1943. A revision of the native clovers of the United States. Ph.D. dissertation, University of Washington, Seattle, WA.

Marschner, P. 2007. Plant-microbe interactions in the rhizosphere and nutrient cycling. In: Marschner, P. & Z. Rengel, eds. Nutrient cycling in terrestrial ecosystems. Soil Biology, Springer, Berlin, Heidelberg, Germany.

Martinson, R., J. Lambrinos & R. Mata-Gonzalez. 2019. Water stress patterns of xerophytic plants in an urban landscape. Hort Science 54(5):812–823.

Martinson, R., K. Theodore & A. Truax. 2017. Native plant landscaping. A primer for Central Oregon. WinterCreek Restoration, Bend, OR.

Mason, G. 1975. Guide to the plants of the Wallowa Mountains of northeastern Oregon. Museum of Natural History, University of Oregon, Eugene, OR.

Mata-González, R., T.L. Evans, D.W. Martin, T. McLendon, J.S. Noller, C. Wan & R.E. Sosebee. 2014. Patterns of water use by great basin plant species under summer watering. Arid Land Research and Management 28(4):428–446.

Mata-González, R., T. McLendon & D.W. Martin. 2005. The inappropriate use of crop transpiration coefficients (Kc) to estimate evapotranspiration in arid ecosystems: A review. Arid Land Research and Management 19(3):285–295.

McGuire, W.S. 1985. Subterranean clover. In: Taylor, N.L., ed. 1985. Clover science and technology. American Society of Agronomy 25:515–534.

McNeill, J. 1962. Taxonomic studies in the Alsinoideae: I. Generic and infra-generic groups. Notes from the Royal Botanic Garden 24:79–155.

______. 1979. *Silene alba* and *S. dioica* in North America and the generic delimitation of *Lychnis*, *Melandrium*, and *Silene* (Caryophyllaceae). Canadian Journal of Botany 56:297–308.

Meyers, S.C., T. Jaster, K.E. Mitchell & L.K. Hardison, eds. 2015. Flora of Oregon. Volume 1: Pteridophytes, gymnosperms, and monocots. Botanical Research Institute of Texas, Fort Worth, TX.

Miller, J.C. & P.C. Hammond. 2003. Caterpillars and adult Lepidoptera of Northwest forests and woodlands. USDA, USFS, FHTET-2003-03.

______. 2007. Butterflies and moths of pacific northwest forests and woodlands: Rare, endangered, and management sensitive species. USDA, USFS, FHTET-2006-07.

Moldenke, A.R. 1971. Studies of the species diversity of California plant communities. Ph.D. dissertation. Stanford University, Stanford, CA.

______. 1975. Niche specialization and species diversity along a California transect. Oecologia 21(3):219–242.

______. 1976. California pollination ecology and vegetationtypes. Phytologia 34:305–361.

Moore, A.J. & M.S Dillenberger. 2017. A conspectus of the genus *Cherleria* (*Minuartia* s.l., Caryophyllaceae). Willdenowia 47:5–14.

Mooring, J.S. 1980. A cytogeographic study of *Chaenactis douglasii* (Compositae, Helenieae). American Journal of Botany 81:919–926.

Morin, N.R. 2012. Campanulaceae (except *Downingia*, *Lobelia*). In: Baldwin, B.G., D. Goldman, D.J. Keil, R. Patterson, T.J. Rosatti & D. Wilken. 2012. The Jepson manual: Vascular plants of California. 2nd edition. University of California Press, Berkeley, CA.

Morton, J.K. 2005. *Silene*. In: Flora of North America Editorial Committee, eds. 1993+. Flora of North America north of Mexico. 18+ vols. New York and Oxford. Vol 5.

Mulligan, G.A. 1980. The genus *Cicuta* in North America. Canadian Journal of Botany 58:1755–1767.

______. 1996. Synopsis of the genus *Arabis* (Brassicaceae) in Canada, Alaska and Greenland. Rhodora 97:109–163.

Mummenhoff, K., H. Brüggemann & J.L. Bowman. 2001. Chloroplast DNA phylogeny and biogeography of *Lepidium* (Brassicaceae). American Journal of Botany 88:2051–2063.

Mummenhoff, K., A. Polster, A. Mühlhausen & G. Theißen. 2009. *Lepidium* as a model system for studying the evolution of fruit development in Brassicaceae. Journal of Experimental Botany 60:1503–1513.

Nation, P.N. 1989. Alsike clover poisoning: A review. The Canadian Veterinary Journal 30:410–415.

Nazaire, M. & L. Hufford. 2014. Phylogenetic systematics of the genus *Mertensia* (Boraginaceae). Systematic Botany 39:268–303.

Nelson, A. 1904. New plants from Nevada. Proceedings of the Biological Society of Washington 17:91–98.

Nelson, E., G. Mendoza, J. Regetz, S. Polasky, H. Tallis, D.R. Cameron, K.M.A. Chan, G.C. Daily, J. Goldstein, P.M. Kareiva, E. Lonsdorf, R. Naidoo, T.H. Ricketts & M.R. Shaw. 2009. Modeling multiple ecosystem services, biodiversity conservation, commodity production, and tradeoffs at landscape scales. Frontiers in Ecology and the Environment 7(1):4–11.

Nelson, E.C. 1991. Shamrock: Botany and history of an Irish myth. Boethius Press, Aberystwyth, Wales.

Nelson, L.S., R.D. Shih & M.J. Balick. 2007. Handbook of poisonous and injurious plants. Springer, Boston, MA

Nelson, T.W. & J.P. Nelson. 2004. A new species of *Silene* (Caryophyllaceae) from the serpentines of Del Norte County, California. Madroño 51:384–386.

Nesom, G.L. & T.J. Leary. 1992. A new species of *Ionactis* (Asteraceae: Astereae) and an overview of the genus. Brittonia 44:247–252.

Nesom, G.L. & H. Robinson. 2007. Tribe Astereae: 284–342. In: Kadereit, J.W. & C. Jeffrey, eds. 1990–2004. The families and genera of vascular plants. Vols. 1–8. New York, NY. Vol 8.

Noyes, R.D. 2000. Biogeographical and evolutionary insights on *Erigeron* and allies (Asteraceae) from ITS sequence data. Plant Systematics and Evolution 220:93–114.

Noyes, R.D. & L.H. Rieseberg. 1999. ITS sequence data support a single origin for North American Astereae (Asteraceae) and reflect deep geographical division in *Aster* s.l. American Journal of Botany 86:398–412.

Oberprieler, C. 2002. A phylogenetic analysis of *Chamaemelum* Miller (Compositae, Anthemideae) and related genera based upon nrDNA, ITS, and cpDNA *trnL/trnF* IGS sequence variation. Botanical Journal of the Linnean Society 138:255–273.

O'Donnell Young, B. 2011. The naturescaping workbook. Timber Press, Portland, OR.

Omernik, J.M. & G.E. Griffith. 2014. Ecoregions of the conterminous United States: Evolution of a hierarchical spatial framework. Environmental Management 54(6):1249–1266.

Oregon Bee Project. 2020. Oregon Department of Agriculture, Oregon State University Extension, and Oregon Department of Forestry. Available at https://www.oregonbeeproject.org. Accessed December 2019.

Oregon Department of Fish and Wildlife. 2015. Naturescaping. Available at https://www.dfw.state.or.us/conservationstrategy/naturescaping/index.asp. Accessed March 2019.

Ornduff, R. 1963. Experimental studies in two genera of Helenieae (Compositae): *Blennosperma* and *Lasthenia*. The Quarterly Review of Biology 38:141–150.

______. 1966. A biosystematic survey of the goldfield genus *Lasthenia*. University of California Publications in Botany. 40:1–92.

Oxelman, B., M. Líden, R.K. Rabeler & M. Popp. 2000. A revised generic classification of the tribe Sileneae (Caryophyllaceae). Nordic Journal of Botany 20:743–748.

Panero, J.L. 2007. Tageteae. In: Kubitzki, K., ed. The families and genera of vascular plants. Springer–Verlag, Berlin, Germany.

Peck, M.E. 1961. A manual of higher plants of Oregon. 2nd edition. Binfords & Mort Publishers, Metropolitan Printing Company, Portland, OR.

Pelham, J.P. 2008. A catalogue of the butterflies of the United States and Canada. Journal of Research on the Lepidoptera. 40:1–652. Revised online: www.butterfliesofamerica.com/US-Can-Cat.htm.

Perring, M.P., P. Audet & D. Lamb. 2014. Novel ecosystems in ecological restoration and rehabilitation: Innovative planning or lowering the bar? Ecological Processes 3(8).

Piper, C.V. 1906. Flora of the state of Washington. Contributions from the United States National Herbarium 11:1–637.

______. 1924. Native American forage plants. Agronomy Journal 6:682–682.

Ponte, J.-P.E. 2018. Jeweled spider flies (*Eulonchus tristis*) are important pollinators of *Iris bracteata*, a rare Siskiyou Mountain endemic. Theses and projects. 240. https://digitalcommons.humboldt.edu/etd/240

Pozo, M.J. & A.-A. Concepcion. 2007. Unraviling mycorrhiza-induced resistance. Current Opinion in Plant Biology 10(4):393–398.

Price, R.A. 1993. *Erysimum*. In: Hickman, J.C., ed. The Jepson manual: Higher plants of California. 1993. University of California Press, Berkeley, CA.

Proctor, M., P. Yeo & A. Lack. 1996. The natural history of pollination. Timber Press, Portland, OR.

Pujadas Savalà, A.J. 2002. El complejo *Daucus carota* L. (Apiaceae) en la flora Ibérica. Anales del Jardín Botánico de Madrid 59:368–375.

Pyke, D.A., J.E. Herrick, P. Shaver & M. Pellant. 2002. Rangeland health attributes and indicators for qualitative assessment. Journal of Range Management 55(6):584–597.

Pyle, R.M. 2002. The butterflies of Cascadia: A field guide to all the species of Washington, Oregon, and surrounding territories. Seattle Audubon Society, Seattle, WA.

Pyle, R.M. & C.C. LaBar. 2018. Butterflies of the Pacific Northwest. Timber Press, Portland, OR.

Randall, R.P. 2002. A global compendium of weeds. R.G. & F.J. Richardson, Melbourne, Australia.

Razifard, H., A. Rosman, G.C. Tucker & D.H. Les. 2017. Systematics of the cosmopolitan aquatic genus *Elatine*. Systematic Botany 42:73–86.

Ricker, J.G. 2019. Suitability of cultivated forms of native shrubs to support pollinators. Unpublished thesis (M.S.), University of Connecticut.

Rickett, H.W. 1944. *Cornus stolonifera* and *Cornus occidentalis*. Brittonia 5:149–159.

Roberts, R.P. & L.E. Urbatsch. 2003. Molecular phylogeny of *Ericameria* (Asteraceae, Astereae) based on nuclear ribosomal 3' ETS and ITS sequence data. Taxon 52:209–228.

______. 2004. Molecular phylogeny of *Chrysothamnus* and related genera (Asteraceae, Astereae) based on nuclear ribosomal 3' ETS and ITS sequence data. Systematic Botany 29:199–215.

Roberts, R.P., L.E. Urbatsch & K.M. Neubig. 2005. *Nestotus* and *Toiyabea*, two new genera of Asteraceae: Astereae from the western United States and Canada. Sida 21:1647–1655.

Rolfsmeier, S. 2013. Taxonomy and phylogeny of the genus *Lappula* Moench (Boraginaceae) in North America. Ph.D. dissertation. Kansas State University, Manhattan, KS.

Rollins, R.C. 1993. The Cruciferae of continental North America: Systematics of the mustard family from the Artic to Panama. Stanford University Press, Stanford, CA.

Rosman, A.J., H. Razifard, G.C. Tucker & D.H. Les. 2016. New records of *Elatine ambigua* (Elatinaceae), a nonindigenous North American species. Rhodora 118:235–242.

Rossbach, R.P. 1940. *Spergularia* in North and South America. Rhodora 42:105–143.

Rydberg, P.A. 1919. Fabaceae, tribe Psoraleeae. North American Flora 24:1–64.

Sadeghian S., S. Zarre, R.K. Rabeler & G. Heubl. 2015. Molecular phylogenetic analysis of *Arenaria* (Caryophyllaceae: tribe Arenarieae) and its allies inferred from nuclear DNA ITS and plastid DNA rps16 sequences. Botanical Journal of the Linnean Society 178:648–669.

Saroyan, J.P., D.R. Parnell & J.L. Strother. 2000. Revision of *Corethrogyne* (Compositae: Asteraceae). Madroño 47:89–96.

Scheen, A.C., C. Brochmann, A.K. Brysting, R. Elven, A. Morris, D.E. Soltis, P.S. Soltis & V.A. Albert. 2004. Northern hemisphere biogeography of *Cerastium* (Caryophyllaceae): Insights from phylogenetic analysis of noncoding plastidnucleotide sequences. American Journal of Botany 91:943–952.

Schierenbeck, K.A. & F. Phipps. 2010. Population genetics of *Howellia aquatilis* (Campanulaceae) in disjunct locations throughout the Pacific Northwest. Genetica 138:1161–1169.

Schultheis, L.M. 2010. Morphologically cryptic species within *Downingia yina* (Campanulaceae). Madroño 57:20–41.

Sell, P.D. 1976. *Crepis*. In: Tutin, T.G., V.H. Heywood, N.A. Burges, D.M. Moore, D.H. Valentine, S.M. Walters & D.A. Webb, eds. Flora Europaea. Cambridge University Press, Cambridge, UK. Vol 4.

Semple, J.C., S.B. Heard & L. Brouillet. 2002. Cultivated and native asters of Ontario (Compositae: Astereae). University of Waterloo Biological Series 41, Waterloo, Canada.

Severns, P. & A. Moldenke. 2010. Management tradeoffs between focal species and biodiversity: Endemic plant conservation and solitary bee extinction. Biodiversity and Conservation 19(12):3605–3609.

Shanahan, D.F., R.A. Fuller, R. Bush, B.B. Lin & K.J. Gaston. 2015. The health benefits of urban nature: How much do we need? Bioscience 65(5):476–485.

Shepherd, L.D. 2017. A non-destructive DNA sampling technique for herbarium specimens. PloS One 12(8):e0183555.

Simmons, M.P., J.J. Cappa, R.H. Archer, A.J. Ford, D. Eichstedt & C.C. Clevinger. 2008. Phylogeny of Celastreae (Celastraceae) and the relationships of *Catha edulis* inferred from morphological characters and nuclear and plastid genes. Molecular Phylogenetics and Evolution 48:745–757.

Simmons, M.T., H.C. Venhaus & S. Windhager. 2007. Exploiting the attributes of regional ecosystems for landscape design: The role of ecological restoration in ecological engineering. Ecological Engineering 30(3):201–205.

Simpson, M.G., R. Dowdy, J.P. Rebman, R.B. Kelley & L.M. Simpson. 2014. Recognition of two species in *Eremocarya* (Boraginaceae): Evidence from fornix bodies, nutlets, corolla size, and biogeography. Madroño 61:259–275.

Simpson, M.G., C.M. Guilliams, K.E. Hasenstab-Lehman, M.E. Mabry & L. Ripma. 2017. Phylogeny of the popcorn flowers: Use of genome skimming to evaluate monophyly and interrelationships in subtribe Amsinckiinae (Boraginaceae). Taxon 66:1406–1420.

Simpson, M.G., L.M. Simpson & J.P. Rebman. 2016. A new, large-flowered variety of *Eremocarya micrantha* (Boraginaceae). Madroño 63:39–54.

Smith, S.E. & F.A. Smith. 2012. Fresh perspectives on the roles of arbuscular mycorrhizal fungi in plant nutrition and growth. Mycologia 104(1):1–13.

Soltis, D.E., V. Evgeny, S.C. Meyers, P.M. Severns, L. Zhang, M.A. Gitzendanner, T. Ayers, M. Chester & P.S. Soltis. 2012. Additional origins of Ownbey's *Tragopogon mirus*. Botanical Journal of the Linnean Society 169:297–311.

Spalik, K., J.P. Reduron & S.R. Downie. 2004. The phylogenetic position of *Peucedanum* sensu lato and allied genera and their placement in tribe Selineae (Apiaceae, subfamily Apioideae). Plant Systematics and Evolution 243:189–210.

Standish, R., R. Hobbs & J. Miller. 2013. Improving city life: Options for ecological restoration in urban landscapes and how these might influence interactions between people and nature. Landscape Ecology 28(6):1213–1221.

Stefanović, S., D.F. Austin & R.G. Olmstead. 2003. Classification of Convolvulaceae: A phylogenetic approach. Systematic Botany 24:791–806.

Stefanović, S., L. Krueger & R.G. Olmstead. 2002. Monophyly of the Convolvulaceae and circumscription of their major lineages based on DNA sequences of multiple chloroplast loci. American Journal of Botany 89:1510–1522.

Stefanović, S. & R.G. Olmstead. 2004. Testing the phylogenetic position of a parasitic plant (*Cuscuta*, Convolvulaceae, asteridae): Bayesian inference and the parametric bootstrap on data drawn from three genomes. Systematic Biology 53:384–399.

Stevens, P. F. (2001 onwards). Angiosperm Phylogeny Website. Version 14, July 2017. http://www.mobot.org/MOBOT/research/APweb/.

St. John, L., D.G. Ogle & D. Tilley. 2010. Plant guide for strawberry clover (*Trifolium fragiferum*). USDA Natural Resources Conservation Service, Aberdeen, ID Plant Materials Center.

Strickland, W. 1801. Observations on the agriculture of the United States of America. W. Buhner, London, U.K.

Strother, J.L. 2006. *Hieracium*. In: Flora of North America Editorial Committee, eds. 1993+. Flora of North America north of Mexico. 18+ vols. New York and Oxford. Vol 19.

Stuckey, R.L. 1972. Taxonomy and distribution of the genus *Rorippa* (Cruciferae) in North America. Sida 4:279–430.

Sundberg, S. 2003. New Oregon records. Oregon Flora Newsletter 9:5.

Thaman, R.R. 2013. Ethno-biodiversity, taxonomy and bioinformatics for all ages: Engaging and educating the next generation of taxonomists as a foundation for sustainable living on Planet Earth—Challenges and opportunities. In: Brooks, L.A. & S. Aricò, eds. Tracking key trends in biodiversity science and policy: Based on the proceedings of a UNESCO International Conference on Biodiversity Science and Policy. Pp. 23–25.

Thiers, B. [continuously updated]. Index herbariorum: A global directory of public herbaria and associated staff. New York Botanical Garden's Virtual Herbarium. http://sweetgum.nybg.org/science/ih/.

Thompson, W.C. 1983. A biosystematic study of *Lagophylla* (Compositae: Heliantheae) and related taxa. Ph.D. dissertation. University of California, Davis, CA.

Torchio, P.F., G.E. Trostle & D.J. Burdick. 1988. The nesting biology of *Colletes kincaidii* Cockerell (Hymenoptera: Colletidae) and development of its immature forms. Annals of the Entomological Society of America 81(4):605–625.

Tucker, G.C. 1986. The genera of Elatinaceae in the southeastern United States. Journal of the Arnold Arboretum 67:471–483.

Turland, N., J. Kirschner & J. Štěpánek. 1996. Proposals to reject the names *Trifolium agrarium*, *T. filiforme* and *T. procumbens* (Leguminosae). Taxon 45:549–551.

Ungerer, M.C., L.C. Johnson & M.A. Herman. 2008. Ecological genomics: Understanding gene and genome function in the natural environment. Heredity 100:178–183.

Urbatsch, L.E. & P.B. Cox. 2006. *Rudbeckia*. In: Flora of North America Editorial Committee, eds. 1993+. Flora of North America north of Mexico. 18+ vols. New York and Oxford. Vol 21.

Van Auken, O.W. 2009. Causes and consequences of woody plant encroachment into western North American grasslands. Journal of Environmental Management 90(10):2931–2942.

Van Breemen, N., R. Finlay, U. Lundstrom, A.G. Jonsmans, R. Giesler & M. Olsson. 2000. Mycorrhizal weathering: A true case of mineral plant nutrition? Biogeochemistry 49:63–67.

Venkatesh, C.S. 1958. A cyto-genetic and evolutionary study of *Hemizonia* section *Centromadia* (Compositae). Evolution 39:1236–1241.

Wang, B. & Y.-L. Qiu. 2006. Phylogenetic distribution and evolution of mycorrhizas in land plants. Mycorrhiza 16(5):299–363.

Warren, A.D. 2005. Butterflies of Oregon: Their taxonomy, distribution, and biology. Contributions of the C.P. Gillette Museum of Arthropod Diversity, Colorado State University, Fort Collins, CO.

Warren, A.D., K.J. Davis, E.M. Stangeland, J.P. Pelham, K.R. Willmott & N.V. Grishin. 2016. Illustrated lists of American butterflies. www.butterfliesofamerica.com/intro.htm.

Warwick, S.I. & C. Sauder. 2005. Phylogeny of tribe Brassiceae (Brassicaceae) based on chloroplast restriction site polymorphisms and nuclear ribosomal internal transcribed spacer and chloroplast trnL intron sequences. Canadian Journal of Botany 83:467–483.

Watson, L.E., T.M. Evans & T. Boluarte. 2000. Molecular phylogeny and biogeography of tribe Anthemideae (Asteraceae), based on chloroplast gene ndhF. Molecular Phylogenetics and Evolution 15:59–69.

Watson, S. 1871. United States Geological Expolration [sic] of the Fortieth Parallel. Vol. 5. Government Printing Office, Washington, DC.

Weber, W.A. 1984. New names and combinations, principally in the Rocky Mountain flora IV. Phytologia 55:1–11.

Wei, Z. & M.A. Vincent. 2010. Fabaceae. 28. Tribe Trifolieae. In: Wu, Z.-Y. & P.H. Raven, eds. 2010. Flora of China. Missouri Botanical Garden, St. Loius, MO.

Weiss, M.R. 1995. Floral color change: A widespread functional convergence. American Journal of Botany 82:167–185.

Wells, P.V. 2000. The manzanitas of California, also of Mexico and the world. Self published. Lawrence, KS.

Wendling, B.M., K.E. Galbreath & E.G. DeChaine. 2011. Resolving the evolutionary history of *Campanula* (Campanulaceae) in western North America. PLoS One 6(9):e23559.

Wexelsen, H. 1928. Chromosome numbers and morphology in *Trifolium*. University of California Publications in Agricultural Sciences Cytologia 2:256–276.

Williams, J.H., Jr. & M.L. Arnold. 2001. Sources of genetic structure in the woody perennial *Betula occidentalis*. International Journal of Plant Sciences 162:1097–1109.

Williams, L.O. 1937. A monograph of the genus *Mertensia* in North America. Annals of the Missouri Botanical Garden 24:17–159.

Wilson, J.S. & O. M. Carril. 2015. The bees in your backyard: A guide to North America's bees. Princeton University Press, Princeton, NJ.

Windham, M.D. & I.A. Al-Shehbaz. 2007. New and noteworthy species of *Boechera* (Brassicaceae) II: Apomictic hybrids. Harvard Papers in Botany 11:257–274.

______. 2007b. New and noteworthy species of *Boechera* (Brassicaceae) III: Additional sexual diploids and apomictic hybrids. Harvard Papers in Botany 12:235–257.

Wolters, E.A. & M.L. Hubbard. 2014. Oregon water: Assessing differences between the Old and New Wests. The Social Science Journal 51(2):260–267.

Wright, D.P., J.D. Scholes & D.J. Read. 1998. Effects of VA mycorrhizal colonization on photosynthesis and biomass production of *Trifolium repens* L. Plant, Cell & Environment 21(2):209–216.

Xiang, W.F. & J.C. Semple. 1996. Molecular systematic study of *Aster* sensu lato and related genera (Asteraceae: Astereae) based on chloroplast DNA restriction site analyses and mainly North American taxa. In: Hind, D.J.N., et al., eds. Proceedings of the International Compositae Conference, Kew, England.

Young-Cheol, K., B.R. Glick, Y. Bashan & R. Choong-Min. 2012. Enhancement of plant drought tolerance by microbes. In: Arcola, R., ed. Plant responses to drought stress. Springer, Berlin, Heidelberg, Germany.

Zacharias, E.H. & B.G. Baldwin. 2010. A molecular phylogeny of North American Atripliceae (Chenopodiaceae), with implications for floral and photosynthetic pathway evolution. Systematic Botany 35:839–857.

Zardini, E. 1992. *Madia sativa* Mol. (Asteraceae-Heliantheae-Madiinae): An ethnobotanical and geographical disjunct. Economic Botany 46:34–44.

Zhang, J.-W., Z.-L. Nie, J. Wen & H. Sun. 2011. Molecular phylogeny and biogeography of three closely related genera, *Soroseris*, *Stebbinsia*, and *Syncalathium* (Asteraceae, Cichorieae), endemic to the Tibetan Plateau, SW China. Taxon 60:15–26.

Zohary, M. & D. Heller. 1984. The genus *Trifolium*. Israel Academy of Sciences and Humanities, Jerusalem, Israel.

Appendix 1

Misapplied and In Part Names

This table lists taxa not treated because they represent names misapplied to Oregon taxonomic concepts. A *misapplied* name is one that is accepted in a key reference but does not truly refer to the taxonomic description with which it is associated. This happens for two reasons. Either the name was applied in error or a name is used in a flora to describe a taxonomic concept that is subsequently determined to be new to science. Once the new concept is published with a new name, the initial name—although correctly applied to a different taxon—is henceforth considered *misapplied* to the new concept. Misapplied names are not true synonyms.

A few of the names listed are either misapplied to more than one Oregon taxon or are good synonyms that refer, *in part*, to more than one accepted name in this flora.

The name under the column heading **Old Name** shows the name used in one of the key local floras whereas the name(s) in the column **Accepted Name(s)** shows the name(s) recognized and described in this volume.

Old Name	Accepted Name(s)
MISAPPLIED	
AMARANTHACEAE	
Amaranthus graecizans L.	*Amaranthus blitoides*
Chenopodium hybridum L.	*Chenopodium simplex*
Corispermum hyssopifolium L.	*Corispermum americanum* var. *americanum*, *C. pacificum*
Salsola kali L. var. *tenuifolia* Moq.	*Salsola tragus*
ANACARDIACEAE	
Rhus radicans L.	*Toxicodendron rydbergii*
Rhus toxicodendron L.	*Toxicodendron rydbergii*
APIACEAE	
Lomatium grayi (J.M. Coult & Rose) J.M. Coult & Rose	*Lomatium papilioniferum*
Lomatium grayi (J.M. Coult & Rose) J.M. Coult & Rose var. *grayi*	*Lomatium papilioniferum*
ARALIACEAE	
Hydrocotyle umbellata L.	*Hydrocotyle verticillata*
ASTERACEAE	
Agoseris apargioides (Less.) Cronquist var. *eastwoodiae* (Fedde) Munz	*Agoseris apargioides* var. *maritima*
Agoseris heterophylla (Nutt.) Greene ssp. *californica* (Nutt.) Piper	*Agoseris heterophylla* var. *heterophylla*
Antennaria alpina (L.) Gaertn.	*Antennaria media*
Aster falcatus Lindl. var. *crassulus* (Rydb.) Cronquist	*Symphyotrichum ericoides* var. *pansum*
Aster foliaceus Lindl. *ex* DC. var. *canbyi* A. Gray	*Symphyotrichum foliaceum* var. *parryi*
Aster junciformis Rydb.	*Symphyotrichum spathulatum*
Baccharis douglasii DC.	*Baccharis glutinosa*
Baeria macrantha (A. Gray) A. Gray var. *pauciaristata* A. Gray	*Lasthenia ornduffii*
Balsamorhiza macrolepis W.M. Sharp var. *platylepis* (W.M. Sharp) Ferris	*Balsamorhiza hookeri*, *B. sericea*
Balsamorhiza platylepis W.M. Sharp	*Balsamorhiza hookeri*, *B. sericea*
Chaenactis douglasii (Hook.) Hook. & Arn. var. *montana* M.E. Jones	*Chaenactis douglasii*
Chaenactis rubricaulis Rydb.	*Chaenactis douglasii*

Old Name	Accepted Name(s)
Chrysopsis villosa (Pursh) Nutt. ex DC. var. *foliosa* (Nutt.) D.C. Eaton	*Heterotheca villosa* var. *villosa*
Cirsium foliosum (Hook.) DC.	*Cirsium scariosum*
Cirsium utahense Petr.	*Cirsium inamoenum* var. *inamoenum*
Erechtites prenanthoides (A. Rich.) DC.	*Erechtites minimus*
Erigeron acris L. var. *asteroides* (Andrz. ex Besser) DC.	*Erigeron acris* var. *kamtschaticus*
Erigeron decumbens Nutt. ssp. *robustior* Cronquist	*Erigeron eatonii* var. *plantagineus*
Erigeron decumbens Nutt. var. *robustior* (Cronquist) Cronquist	*Erigeron eatonii* var. *plantagineus*
Erigeron foliosus Nutt. var. *hartwegii* (Greene) Jeps.	*Erigeron foliosus* var. *confinis*
Erigeron petrophilus Greene var. *viscidulus* (A. Gray) G.L. Nesom	*Erigeron petrophilus* var. *petrophilus*
Erigeron speciosus (Lindl.) DC. var. *macranthus* (Nutt.) Cronquist	*Erigeron speciosus*
Erigeron subtrinervis Rydb.	*Erigeron speciosus*
Erigeron ursinus D.C. Eaton	*Erigeron glacialis* var. *glacialis*
Gnaphalium japonicum Thunb.	*Euchiton sphaericus*
Gnaphalium viscosum Kunth	*Pseudognaphalium macounii*
Grindelia camporum Greene	*Grindelia hirsutula*
Grindelia nana Nutt. var. *integrifolia* Nutt.	*Grindelia nana* var. *nana*, *G. integrifolia*
Hymenoxys lemmonii (Greene) Cockerell	*Hymenoxys cooperi*
Leontodon hastilis L. var. *vulgaris* Koch	*Leontodon saxatilis*
Pseudognaphalium viscosum (Kunth) W.A. Weber	*Pseudognaphalium macounii*
Rudbeckia occidentalis Nutt. var. *montana* (A. Gray) Perdue	*Rudbeckia occidentalis*

BORAGINACEAE

Amsinckia vernicosa Hook. & Arn.	*Amsinckia carinata*
Mertensia oblongifolia (Nutt.) G. Don	*Mertensia foliosa*, *M. umbratilis*
Mertensia oblongifolia (Nutt.) G. Don var. *amoena* (A. Nelson) L.O. Williams	*Mertensia foliosa*
Mertensia oblongifolia (Nutt.) G. Don var. *nevadensis* (A. Nelson) L.O. Williams	*Mertensia foliosa*
Mertensia oblongifolia (Nutt.) G. Don var. *oblongifolia*	*Mertensia foliosa*
Mertensia viridis A. Nelson	*Mertensia cusickii*
Myosotis macrosperma Engelm.	*Myosotis verna*
Myosotis virginica (L.) Britton, Sterns & Poggenb.	*Myosotis verna*

BRASSICACEAE

Arabis hirsuta (L.) Scop.	*Arabis eschscholtziana*, *A. pycnocarpa*
Arabis holboellii Hornem.	*Boechera pinetorum*, *B. pauciflora*, *B. pendulocarpa*, *B. retrofracta*
Arabis holboellii Hornem. var. *pinetorum* (Tidestr.) Rollins	*Boechera pinetorum*, *B. pauciflora*
Arabis lyallii S. Watson var. *lyallii*	*Boechera lyallii*, *B. calderi*
Dentaria californica Nutt. var. *californica*	*Cardamine californica*
Descurainia pinnata (Walter) Britton ssp. *halictorum* (Cockerell) Detling	*Descurainia pinnata* ssp. *brachycarpa*
Descurainia pinnata (Walter) Britton var. *osmiarum* (Cockerell) Shinners	*Descurainia pinnata* ssp. *brachycarpa*
Lepidium pubescens Desv.	*Lepidium strictum*

Old Name	Accepted Name(s)
CACTACEAE	
Coryphantha vivipara (Nutt.) Britton & A. Br.	*Pediocactus nigrispinus*
Pediocactus simpsonii (Engelm.) Britton & Rose	*Pediocactus nigrispinus*
Pediocactus simpsonii (Engelm.) Britton & Rose var. *robustior* J.M. Coult.	*Pediocactus nigrispinus*
CAPRIFOLIACEAE	
Lonicera caerulea L.	*Lonicera cauriana*
Lonicera hispidula (Lindl.) Douglas ex Torr. & A. Gray var. *vacillans* (Benth.) A. Gray	*Lonicera hispidula*
Symphoricarpos albus (L.) S.F. Blake var. *albus*	*Symphoricarpos albus* var. *laevigatus*
Symphoricarpos rotundifolius A. Gray var. *oreophilus* (A. Gray) M.E. Jones	*Symphoricarpos rotundifolius*
CARYOPHYLLACEAE	
Cerastium viscosum L.	*Cerastium glomeratum*
Minuartia elegans (Cham. & Schlecht.) Schischkin	*Sabulina austromontana*
CONVOLVULACEAE	
Cuscuta pentagona Engelm.	*Cuscuta campestris*
Cuscuta pentagona Engelm. var. *pentagona*	*Cuscuta campestris*
Cuscuta salina Engelm.	*Cuscuta pacifica* var. *pacifica*
Cuscuta salina Engelm. var. *major* Yunck.	*Cuscuta pacifica* var. *pacifica*
Dichondra occidentalis House	*Dichondra donelliana*
CORNACEAE	
Cornus canadensis L.	*Cornus unalaschkensis*
CRASSULACEAE	
Sedum douglasii Hook.	*Sedum stenopetalum*, *S. radiatum* ssp. *depauperatum*
Sedum douglasii Hook. ssp. *radiatum* (S. Watson) R.T. Clausen	*Sedum radiatum* ssp. *depauperatum*
Sedum lanceolatum Torr. var. *rupicola* (G.N. Jones) C.L. Hitchc.	*Sedum lanceolatum* var. *lanceolatum*
ELATINACEAE	
Elatine americana (Pursh) Arn.	*Elatine chilensis*
Elatine triandra Schkuhr	*Elatine chilensis*
ERICACEAE	
Arctostaphylos × *cinerea* Howell	*Arctostaphylos canescens*, *A.* × *oblongifolia*
Kalmia polifolia Wangenh. ssp. *polifolia*	*Kalmia microphylla*
FABACEAE	
Astragalus alpiniformis (Rydb.) M. Peck	*Astragalus alpinus* var. *alpinus*
Astragalus gibbsii Kellogg	*Astragalus curvicarpus* var. *curvicarpus*
Astragalus kentrophyta A. Gray var. *elatius* S. Watson	*Astragalus kentrophyta* var. *tegetarius*
Astragalus lectulus S. Watson	*Astragalus purshii* var. *tinctus*
Astragalus lentiginosus Douglas var. *chartaceus* M.E. Jones	*Astragalus lentiginosus* var. *platyphyllidius*

Old Name	Accepted Name(s)
Astragalus miser Douglas	*Astragalus obscurus*
Astragalus newberryi A. Gray var. *newberryi*	*Astragalus newberryi* var. *castoreus*
Astragalus occidentalis (S. Watson) M.E. Jones	*Astragalus robbinsii* var. *alpiniformis*
Astragalus purshii Douglas var. *lectulus* (S. Watson) M.E. Jones	*Astragalus purshii* var. *tinctus*
Astragalus purshii Douglas var. *ophiogenes* (Barneby) Barneby	*Astragalus purshii*
Astragalus speirocarpus A. Gray	*Astragalus curvicarpus* var. *curvicarpus*
Astragalus whitedii Piper	*Astragalus curvicarpus* var. *curvicarpus*
Astragalus whitedii Piper var. *whitedii*	*Astragalus curvicarpus* var. *curvicarpus*
Lupinus dalesiae Eastw.	*Lupinus albicaulis* var. *shastensis*
Lupinus wyethii S. Watson	*Lupinus polyphyllus* var. *prunophilus*
Trifolium longipes Nutt. ssp. *shastense* (House) J.M. Gillett	*Trifolium longipes* ssp. *caurinum*
Trifolium longipes Nutt. var. *shastense* (House) Jeps.	*Trifolium longipes* ssp. *caurinum*

IN PART

ASTERACEAE

Antennaria neglecta Greene var. *attenuata* (Fernald) Cronquist	*Antennaria howellii* ssp. *neodioica*, *A. howellii* ssp. *petaloidea*
Aster chilensis Nees	*Symphyotrichum ascendens*, *S. chilense*, *S. hallii*
Aster foliaceus Lindl. ex DC.	*Symphyotrichum cusickii*, *S. foliaceum*, *S. hendersonii*
Aster foliaceus Lindl. ex DC. var. *lyallii* (A. Gray) Cronquist	*Symphyotrichum cusickii*, *S. hendersonii*
Haplopappus bloomeri A. Gray	*Ericameria bloomeri*, *E. greenei*

BORAGINACEAE

Cryptantha sobolifera Payson	*Oreocarya schoolcraftii*, *O. sobolifera*, *O. subretusa*

BRASSICACEAE

Arabis × *divaricarpa* A. Nelson	*Boechera acutina*, *B. divaricarpa*, *B. pratincola*
Arabis lemmonii S. Watson	*Boechera lemmonii*, *B. paddoensis*
Arabis platysperma A. Gray	*Boechera covillei*, *B. howellii*, *B. platysperma*
Arabis platysperma A. Gray var. *howellii* (S. Watson) Jeps.	*Boechera covillei*, *B. howellii*
Arabis sparsiflora Nutt.	*Boechera atrorubens*, *B. pauciflora*, *B. sparsiflora*
Arabis suffrutescens S. Watson	*Boechera horizontalis*, *B. suffrutescens*
Boechera holboellii (Hornem.) Á. Löve & D. Löve	*Boechera pinetorum*, *B. retrofracta*
Lesquerella occidentalis (S. Watson) S. Watson	*Physaria kingii* ssp. *diversifolia*, *P. occidentalis* ssp. *occidentalis*

CONVOLVULACEAE

Convolvulus sepium L.	*Calystegia sepium* ssp. *angulata*, *C. silvatica* ssp. *disjuncta*
Convolvulus sepium L. var. *communis* R.M. Tryon	*Calystegia sepium* ssp. *angulata*, *C. silvatica* ssp. *disjuncta*

FABACEAE

Lupinus caudatus Kellogg	*Lupinus argenteus* var. *heteranthus*, *L. arbustus*
Lupinus leucopsis J. Agardh	*Lupinus oreganus*, *L. sericeus*, *L. sulphureus* ssp. *subsaccatus*
Lupinus nanus Douglas ex Benth.	*Lupinus affinis*, *L. bicolor*
Trifolium tridentatum Lindl.	*Trifolium obtusiflorum*, *T. willdenovii*

Appendix 2

Select Species for Native Plant Gardens

The following list of Oregon native plant species is chosen based on commercial availability, ease of growth, suitablility for the home garden, and general aesthetic qualities such as showy flowers and attractive foliage. Good species to grow for insect pollinators and as caterpillar host foods are in appendices 4 and 5. Some species have varieties or subspecies whose characteristics may vary slightly from the average values presented for that species. For additional plants and more in-depth information about those listed here, as well as a list of native plant vendors in Oregon, visit the OregonFlora website (https://oregonflora.org) or consult with the vendor.

Region: northwest (NW), northeast (NE), southwest (SW), southeast (SE). The Cascades mark the east-west boundary; southern borders of Lane and Deschutes counties and eastward mark the the north-south boundary.

Type: deciduous (d), evergreen (e), tree (T), shrub (S), vine (V), perennial (P), annual (A)

Light tolerance: sun (sun), part shade (part shade), shade (shade)

Bloom: period in the region(s) where a species occurs: early (E), middle (M), late (L)

Water tolerance: dry (dry), moderate (moderate), wet (wet)

Landscape use: groundcover (C), erosion control (E), green roof (G), hedgerow (H), rock garden (R), water margin (W)

Species	Common Name	Region	Type	Light	Bloom	Water	Use
PTERIDOPHYTES							
ATHYRIACEAE							
Athyrium filix-femina	lady fern	NW, NE, SW, SE	dP	part shade, shade	–	moderate, wet	W
BLECHNACEAE							
Struthiopteris (Blechnum) spicant	deer fern	NW, SW	eP	part shade, shade	–	moderate, wet	C, W
DRYOPTERIDACEAE							
Dryopteris arguta	coastal wood fern	NW, SW	eP	part shade, shade	–	moderate	E, W
Dryopteris expansa	spreading wood fern	NW, SW	dP	part shade, shade	–	moderate, wet	E, W
Dryopteris filix-mas	male fern	NE, SE	eP	part shade	–	moderate	E, W
Polystichum munitum	sword fern	NW, NE, SW, SE	eP	sun, part shade, shade	–	moderate	E, G, W
POLYPODIACEAE							
Polypodium glycyrrhiza	licorice fern	NW, SW	dP	part shade, shade	–	dry, moderate	C, G, R
PTERIDACEAE							
Adiantum aleuticum	maidenhair fern	NW, NE, SW	dP	part shade, shade	–	moderate, wet	C, R
Pentagramma triangularis	goldback fern	NW, NE, SW	eP	sun, part shade	–	dry, moderate	R
MONOCOTS							
ALISMATACEAE							
Alisma triviale	northern water plantain	NW, NE, SW, SE	dP	sun, part shade	M	wet	W
Sagittaria latifolia	wapato	NW, NE, SW, SE	dP	sun, part shade	M	wet	W
AMARYLLIDACEAE							
Allium acuminatum	tapertip onion	NW, NE, SW, SE	dP	sun	M	dry, moderate	G, R
Allium amplectens	slim-leaf onion	NW, SW, SE	dP	sun, part shade	M	moderate, wet	R, W
Allium cernuum	nodding onion	NW, NE	dP	sun, part shade, shade	E–M	dry, moderate	G, R, W
Allium validum	Pacific swamp onion	NE, SW, SE	dP	sun, part shade	M	wet	W

Species	Common Name	Region	Type	Light	Bloom	Water	Use
ARACEAE							
Lysichiton americanus	skunk cabbage		dP	◐ ●	E–M	●	W
ASPARAGACEAE							
Brodiaea coronaria	crown brodiaea		dP	☼ ◐	M	○ ◐	R
Brodiaea elegans	elegant brodiaea		dP	☼ ◐	M	○ ◐	R
Camassia leichtlinii	great camas		dP	☼ ◐	E–M	○ ◐ ●	W
Camassia quamash	common camas		dP	☼ ◐	E–M	○ ◐ ●	G, W
Dichelostemma congestum	ookow		dP	☼ ◐	M	○ ◐	R
Dichelostemma ida-maia	firecracker flower		dP	☼ ◐	M	○	R
Hastingsia alba	white-flowered rush-lily		dP	☼	M	●	W
Leucocrinum montanum	sand lily		dP	☼	E	○	C, R
Maianthemum dilatatum	false lily-of-the-valley		dP	◐ ●	M	◐ ●	C
Maianthemum racemosum	large false Soloman's seal		dP	☼ ◐ ●	M	◐	C
Maianthemum stellatum	starry false Soloman's seal		dP	☼ ◐ ●	M	◐	C
Triteleia grandiflora	large-flowered triteleia		dP	☼ ◐	E–M	○ ◐	R
Triteleia hyacinthina	hyacinth triteleia		dP	☼	M	◐	R
CYPERACEAE							
Carex densa	dense sedge		eP	☼ ◐	M	●	E, W
Carex leptopoda	slender-foot sedge		dP	◐ ●	M	◐	E, W
Carex lyngbyei	Lyngbye's sedge		dP	☼	M	●	W
Carex mertensii	Mertens' sedge		dP	☼ ◐	M	◐ ●	E, W
Carex obnupta	slough sedge		eP	☼ ◐ ●	M	◐ ●	C, E, H, W
Carex rossii	Ross' sedge		dP	☼ ◐	M	○ ◐	C, E, G, R
Carex stipata	awl-fruit sedge		dP	☼ ◐	E–M	◐ ●	E
Carex tumulicola	foothill sedge		dP	☼ ◐ ●	M	◐	C, E
Carex unilateralis	one-sided sedge		dP	☼	E–M	●	E
Eleocharis palustris	common spikerush		eP	☼ ◐	M–L	◐ ●	E, W
Schoenoplectus tabernaemontani	tule		eP	☼ ◐	M–L	●	E
IRIDACEAE							
Iris bracteata	Siskiyou iris		dP	☼ ◐ ●	E–M	○ ◐	R
Iris chrysophylla	yellow-leaf iris		eP	☼ ◐ ●	E–M	○ ◐	R
Iris douglasiana	Douglas iris		eP	☼ ◐	E–M	○ ◐	C, E, H, R
Iris innominata	golden iris		eP	☼ ◐ ●	E–M	○ ◐	R
Iris missouriensis	western blue flag		dP	☼ ◐	E–M	◐ ●	E, W
Iris tenax	tough-leaf iris		dP	☼ ◐ ●	E–M	○ ◐	C, E, R
Olsynium douglasii	grass widows		dP	☼	E	○ ◐ ●	R
Sisyrinchium bellum	western blue-eyed grass		dP	☼ ◐	M	◐ ●	R
Sisyrinchium californicum	golden-eyed grass		dP	☼ ◐	M	◐ ●	G, R
Sisyrinchium idahoense	Idaho blue-eyed grass		d/eP	☼ ◐	M	◐ ●	E, G, R, W

Species	Common Name	Region	Type	Light	Bloom	Water	Use
JUNCACEAE							
Juncus balticus	Baltic rush		eP	☼	M–L	◒ ●	C, E, H, W
Juncus bolanderi	Bolander's rush		eP	☼	M–L	◒ ●	W
Juncus ensifolius	dagger-leaved rush		eP	☼ ◐	M–L	◒ ●	C, E, H, W
Juncus patens	gray rush		eP	☼ ◐	M–L	◒ ●	C, E, H
Juncus tenuis	path rush		eP	☼ ◐	M	◒ ●	C, E, R
LILIACEAE							
Calochortus macrocarpus	sagebrush mariposa lily		dP	☼ ◐	M–L	○	R
Calochortus tolmiei	Tolmie's cat's ear		dP	☼ ◐ ●	E–M	○ ◒	R
Clintonia uniflora	queen's cup		dP	◐ ●	M	◒	C
Erythronium hendersonii	Henderson's fawn-lily		dP	◐	E	◒	R
Erythronium oregonum	Oregon fawn-lily		dP	☼ ◐	E	◒	C, R
Erythronium revolutum	pink fawn-lily		dP	☼ ◐ ●	E	◒	W
Fritillaria affinis	checker-lily		dP	☼ ◐ ●	E–M	○ ◒	R
Fritillaria pudica	yellow bells		dP	☼ ◐	E	○	R
Lilium columbianum	Columbia lily		dP	☼ ◐	M–L	◒	H, W
Lilium pardalinum	leopard lily		dP	☼ ◐	M–L	◒ ●	W
Lilium washingtonianum	Cascade lily		dP	◐	M–L	○	H, R
Prosartes hookeri	Hooker's fairybells		dP	◐ ●	E–M	○ ◒	E
Prosartes smithii	fairy lanterns		dP	◐ ●	E–M	◒	E
Streptopus amplexifolius	claspleaf twisted-stalk		dP	◐ ●	E–M	◒ ●	W
MELANTHIACEAE							
Trillium albidum	giant trillium		dP	◐ ●	E–M	◒ ●	W
Trillium ovatum	western trillium		dP	◐ ●	E–M	○ ◒	W
Trillium rivale	brook trillium		dP	☼ ◐ ●	E–M	○ ◒	R, W
Veratrum californicum	false-hellebore		dP	☼ ◐	M	◒ ●	W
Xerophyllum tenax	bear-grass		eP	☼ ◐ ●	M–L	◒	E, R
POACEAE							
Beckmannia syzigachne	American sloughgrass		A	☼ ◐	M–L	●	W
Danthonia californica	California oatgrass		dP	☼ ◐ ●	M	○ ◒	E, R
Deschampsia cespitosa	tufted hairgrass		dP	☼	M	◒ ●	E, W
Festuca californica	California fescue		dP	☼ ◐	E–M	○ ◒	E
Festuca idahoensis	Idaho fescue		dP	☼	M	○	C, E, G, R
Festuca occidentalis	western fescue		eP	◐	E–M	○	C, G, R
Festuca roemeri	Roemer's fescue		dP	☼ ◐	M	○ ◒	C, E, H, R
Koeleria macrantha	junegrass		dP	☼	E–M	○	C, G, R
DICOTS							
AMARANTHACEAE							
Krascheninnikovia lanata	winter fat		eS	☼ ◐	M	○	H, R

Species	Common Name	Region	Type	Light	Bloom	Water	Use
ANACARDIACEAE							
Rhus aromatica	fragrant sumac		dS	☼	E–M	○ ◒	E, H
Rhus glabra	smooth sumac		dS/T	☼ ◐	M	○ ◒	E, H, R, W
APIACEAE							
Lomatium columbianum	Columbia desert parsley		dP	☼ ◐	E–M	○	R
Lomatium dissectum	fernleaf biscuitroot		dP	☼ ◐	E–M	○ ◒	R
Lomatium macrocarpum	large-fruited desert parsley		dP	☼ ◐	E	○	R
Lomatium nudicaule	barestem lomatium		dP	☼	E–M	○ ◒	H
Lomatium triternatum	nineleaf lomatium		dP	☼ ◐	E	◒ ●	R
Lomatium utriculatum	spring gold		dP	☼ ◐	E	○ ◒	R
APOCYNACEAE							
Apocynum androsaemifolium	spreading dogbane		dP	☼ ◐	M	○ ◒	E, R
Asclepias cordifolia	purple milkweed		dP	☼ ◐	M	○ ◒	R
Asclepias fascicularis	narrowleaf milkweed		dP	☼	M	○ ◒	C, E, R, W
Asclepias speciosa	showy milkweed		dP	☼	M	○ ◒	H, W
ARISTOLOCHIACEAE							
Asarum caudatum	western wild ginger		eP	◐ ●	E–M	◒ ●	C
Asarum marmoratum	marbled wild ginger		eP	◐ ●	E–M	○ ◒	R
ASTERACEAE							
Achillea millefolium	common yarrow		dP	☼ ◐ ●	M–L	○ ◒	C, G, H, R
Ageratina occidentalis	western boneset		dP	☼	M–L	○ ◒	R
Anaphalis margaritacea	pearly everlasting		dP	☼ ◐	M–L	○ ◒	E, G, R
Anisocarpus madioides	woodland tarweed		dP	◐ ●	M	◒	–
Antennaria microphylla	littleleaf pussytoes		dP	☼ ◐	M	○	C, E, G, R
Artemisia ludoviciana	western wormwood		dP	☼ ◐	M–L	◒	E, G, R, W
Artemisia tridentata	big sagebrush		eS	☼	M	○	H, R
Baccharis pilularis	chaparral broom		eS	☼ ◐	L	○ ◒	C, E, H, R
Balsamorhiza deltoidea	deltoid balsamroot		dP	☼ ◐	E–M	○ ◒	E
Balsamorhiza sagittata	arrowleaf balsamroot		dP	☼ ◐	E–M	○ ◒	E, R
Canadanthus modestus	great northern aster		dP	☼ ◐	M–L	◒ ●	W
Chaenactis douglasii	Douglas' dusty maidens		dP	☼ ◐	M	○ ◒	R
Chrysothamnus viscidiflorus	sticky-flowered rabbitbrush		dS	☼	L	○	H, R
Coreopsis tinctoria	Columbia coreopsis		A	☼ ◐	M–L	◒ ●	E, G, R
Ericameria nauseosa	rubber rabbitbrush		dS	☼ ◐	L	○	E, H, R
Erigeron compositus	fernleaf fleabane		dP	☼ ◐	M	○	G, R
Erigeron linearis	desert yellow fleabane		dP	☼ ◐	E–M	○	G, R
Erigeron poliospermus	purple-cushion fleabane		eP	☼ ◐	E–M	○	G, R
Erigeron speciosus	showy fleabane		dP	☼ ◐ ●	M–L	○ ◒	G, R
Eriophyllum lanatum	woolly sunflower		dP/S	☼ ◐	M	○ ◒	E, G, R

Species	Common Name	Region	Type	Light	Bloom	Water	Use
Gaillardia aristata	blanket flower		dP	☼	M	○ ◒	G, H, R
Grindelia integrifolia	Puget Sound gumplant		dP	☼ ◐	M–L	◒ ●	H
Helenium autumnale	common sneezeweed		dP	☼ ◐	L	◒ ●	W
Helianthella uniflora	Rocky Mountain helianthella		dP	☼ ◐	M	○	H, R
Hymenoxys hoopesii	owl's claw		dP	☼	M–L	◒	H
Madia elegans	common madia		A	☼	M–L	○ ◒	G
Petasites frigidus	coltsfoot		dP	☼ ◐ ●	E	◒ ●	C, W
Rudbeckia occidentalis	western coneflower		dP	☼ ◐	M–L	◒ ●	H, W
Solidago elongata	West Coast goldenrod		dP	☼ ◐	M–L	○ ◒	E, H, W
Solidago multiradiata	northern goldenrod		dP	☼ ◐	L	○ ◒	E, H
Symphyotrichum hallii	Hall's aster		dP	☼	L	◒ ●	–
Symphyotrichum subspicatum	Douglas' aster		dP	☼ ◐	L	◒ ●	C, E, H, W
Townsendia florifer	showy townsendia		dP	☼	M	○	C, R
Wyethia angustifolia	narrow leaved mule's ears		dP	☼ ◐	M	◒ ●	W
BERBERIDACEAE							
Achlys triphylla	vanillaleaf		dP	◐ ●	M	○ ◒	C
Berberis aquifolium	tall Oregon-grape		eS	☼ ◐ ●	E	○ ◒	E, H
Berberis nervosa	Cascade Oregon-grape		eS	☼ ◐ ●	E	○ ◒	C, E, H
Berberis repens	creeping Oregon-grape		eS	☼ ◐ ●	E–M	○ ◒	C, E, G, R
Vancouveria hexandra	northern inside-out flower		dP	◐ ●	M	○ ◒	C
BETULACEAE							
Betula occidentalis	water birch		dS/T	☼ ◐	M	◒ ●	H, W
Corylus cornuta	California hazel		dS/T	☼ ◐ ●	E	○ ◒	H
BORAGINACEAE							
Adelinia grandis	Pacific hound's tongue		dP	☼ ◐ ●	E	○ ◒	R
Lithospermum ruderale	western stoneseed		dP	☼	M	○	R
Mertensia subcordata	broadleaf bluebells		dP	◐	M	◒	W
Plagiobothrys figuratus	fragrant popcornflower		A	☼	M	◒ ●	C, W
BRASSICACEAE							
Cardamine nuttallii	Nuttall's toothwort		dP	☼ ◐ ●	E	◒	R, W
Erysimum capitatum	western wallflower		dP	☼ ◐	M	○ ◒	G, R
CAMPANULACEAE							
Campanula rotundifolia	American harebell		dP	☼ ◐ ●	L	○ ◒	C, G, R
CAPRIFOLIACEAE							
Lonicera ciliosa	orange honeysuckle		dV	☼ ◐ ●	M	○ ◒	H
Lonicera involucrata	twinberry		dS	☼ ◐ ●	E–M	○ ◒ ●	E, H, W
Symphoricarpos albus	common snowberry		dS	◐ ●	M	○ ◒	E, H
Symphoricarpos mollis	creeping snowberry		dS	☼ ◐	M	◒	C, E, H
Symphoricarpos rotundifolius	roundleaf snowberry		dS	☼ ◐	M–L	○	E, R

Species	Common Name	Region	Type	Light	Bloom	Water	Use
CELASTRACEAE							
Euonymus occidentalis	western burning bush		dS/T	◐ ●	M	◐ ●	E, H, W
Paxistima myrsinites	mountain lover		eS	○ ◐ ●	M	○ ◐	E, H, R
CORNACEAE							
Cornus nuttallii	Pacific dogwood		dT	◐ ●	M	○ ◐	E, H
Cornus sericea	red-osier dogwood		dS	○ ◐ ●	E–M	◐ ●	E, H, W
Cornus unalaschkensis	western bunchberry		dP	◐ ●	M–L	◐	C, W
CRASSULACEAE							
Sedum divergens	Cascade stonecrop		eP	○ ◐	M	○ ◐	C, G, R
Sedum oreganum	Oregon stonecrop		eP	○ ◐	M–L	○ ◐	C, G, R
Sedum oregonense	creamy stonecrop		eP	○ ◐	M–L	○	C, G, R
Sedum spathulifolium	broadleaf stonecrop		eP	○ ◐	M	○ ◐	C, G, R
ERICACEAE							
Arbutus menziesii	Pacific madrone		eT	○ ◐	E–M	○ ◐	E, H
Arctostaphylos columbiana	hairy manzanita		eS/T	○ ◐	E	○ ◐	E, R
Arctostaphylos nevadensis	pinemat manzanita		eS	◐	E–M	◐	E, G, R
Arctostaphylos patula	green-leaved manzanita		eS	○	E–M	○ ◐	E, G, H, R
Arctostaphylos uva-ursi	kinnikinnick		eS	○ ◐	E–M	○ ◐	C, E, G, R
Gaultheria shallon	salal		eS	◐ ●	E–M	○ ◐ ●	C, E, H
Rhododendron macrophyllum	Pacific rhododendron		eS/T	◐ ●	M	○ ◐	E, H
Rhododendron occidentale	western azalea		dS	○ ◐	M	◐	E, H, W
Vaccinium ovatum	evergreen huckleberry		eS	○ ◐ ●	E–M	○ ◐	E, H
Vaccinium parvifolium	red huckleberry		dS	◐ ●	E–M	○ ◐	E, H
FABACEAE							
Cercis occidentalis	western redbud		dS/T	◐ ●	E–M	◐	H
Dalea ornata	western prairie clover		dP	○	M	○	G, R
Lupinus albifrons	silver bush lupine		eP/S	○	E–M	○	R
Lupinus argenteus	silvery lupine		dP	○	M	○	G, H
Lupinus latifolius	broad-leaved lupine		dP	○ ◐ ●	E–M	○ ◐	E, H, R
Lupinus polyphyllus	bigleaf lupine		dP	○ ◐ ●	M	○ ◐ ●	E, H, R, W
Lupinus rivularis	riverbank lupine		dP	○	E–M	○ ◐ ●	E, R, W
Lupinus sericeus	silky lupine		dP	○ ◐	M	○ ◐	G, H
Thermopsis gracilis	slender goldenbanner		dP	○ ◐	E–M	◐ ●	H, W
FAGACEAE							
Quercus garryana	Oregon white oak		dS/T	○	E	○ ◐	E, H, R
Quercus kelloggii	California black oak		dS/T	○ ◐	E	○ ◐	E, H
GARRYACEAE							
Garrya elliptica	wavyleaf silktassel		eS/T	○ ◐ ●	E	○ ◐	E, H

Species	Common Name	Region	Type	Light	Bloom	Water	Use
GERANIACEAE							
Geranium oreganum	western geranium		dP	☼ ◐	M	○ ◒	E
Geranium viscosissimum	sticky purple geranium		dP	☼ ◐	M–L	○ ◒	E, G, H, W
GROSSULARIACEAE							
Ribes aureum	golden currant		dS	☼ ◐	E	○ ◒	H, W
Ribes cereum	wax currant		dS	☼ ◐	E–M	○ ◒	H, R
Ribes divaricatum	spreading gooseberry		dS	◐	E–M	◒	E, H
Ribes lobbii	gummy gooseberry		dS	☼ ◐	E–M	○ ◒	E, H
Ribes sanguineum	red-flowering currant		dS	☼ ◐	E	○ ◒	E, H, R
HYDRANGEACEAE							
Philadelphus lewisii	Lewis' mockorange		dS	☼ ◐ ●	M	○ ◒	E, H, R, W
HYDROPHYLLACEAE							
Hydrophyllum tenuipes	Pacific waterleaf		dP	◐ ●	M	◒	C
LAMIACEAE							
Agastache urticifolia	nettle-leaved horsemint		dP	☼ ◐	M–L	◒	E, W
Clinopodium douglasii	yerba buena		dP	◐ ●	M	◒	C, R
Monardella odoratissima	mountain monardella		eP	☼	M–L	○ ◒	G, H, R
Prunella vulgaris	common selfheal		dP	☼ ◐ ●	M	◒	C, G
Salvia dorrii	purple sage		eS	☼ ◐	E–M	○	G, H, R
Scutellaria angustifolia	narrowleaf skullcap		dP	☼	E–M	○ ◒	C, R, W
LAURACEAE							
Umbellularia californica	Oregon myrtle		eS/T	☼ ◐ ●	E–M	○ ◒	E, H
LINACEAE							
Linum lewisii	wild blue flax		dP	☼	M	○ ◒	E, G, R
LINNAEACEAE							
Linnaea borealis	western twinflower		eP	◐ ●	M	◒	C, E
MALVACEAE							
Iliamna rivularis	streambank globemallow		dP	☼ ◐	M–L	◒ ●	H, W
Sidalcea asprella (malviflora)	rose checkermallow		dP	☼ ◐	M	○ ◒	H, W
Sidalcea campestris	meadow checkermallow		dP	☼ ◐	M	◒ ●	W
Sidalcea cusickii	Cusick's checkermallow		dP	☼ ◐	M	◒	W
Sidalcea oregana	Oregon checkerbloom		dP	☼ ◐	M–L	◒	W
Sphaeralcea munroana	Munro's globemallow		dP	☼	E–M	○	G, R
MONTIACEAE							
Lewisia columbiana	Columbian lewisia		eP	◐ ●	M–L	○ ◒	R
Lewisia cotyledon	Siskiyou lewisia		eP	◐ ●	M–L	○ ◒	G, R
Lewisia rediviva	bitterroot		dP	☼	E–M	○	R

Species	Common Name	Region	Type	Light	Bloom	Water	Use
MYRICACEAE							
Myrica californica	Pacific wax myrtle		eS/T		E–M		H
OLEACEAE							
Fraxinus latifolia	Oregon ash		dT		M		E, H
ONAGRACEAE							
Chamaenerion angustifolium	fireweed		dP		M–L		E, G, W
Clarkia amoena	farewell to spring		A		M		G, R
Clarkia pulchella	ragged robin		A		M		G, R
Epilobium canum	California fuchsia		dP		L		E, R
Epilobium densiflorum	denseflower spikeprimrose		A		L		W
OROBANCHACEAE							
Castilleja hispida	harsh Indian paintbrush		dP		M		R
Castilleja miniata	scarlet paintbrush		dP		M–L		R
OXALIDACEAE							
Oxalis oregana	Oregon wood sorrel		dP		E–M		C, E
PAEONIACEAE							
Paeonia brownii	western peony		dP		M		H, R
PAPAVERACEAE							
Corydalis scouleri	Scouler's corydalis		dP		M		C, W
Dicentra cucullaria	western dutchman's breeches		dP		E–M		W
Dicentra formosa	Pacific bleeding heart		dP		E–M		C, W
Eschscholzia californica	California poppy		A/dP		M–L		E, R
PHRYMACEAE							
Diplacus aurantiacus	bush monkeyflower		eP		M		R
Erythranthe cardinalis	scarlet monkeyflower		dP		M–L		W
Erythranthe guttata	common monkeyflower		A/dP		M		W
Erythranthe lewisii	great purple monkeyflower		dP		L		W
PLANTAGINACEAE							
Penstemon cardwellii	Cardwell's penstemon		dS		M–L		C, R
Penstemon deustus	hotrock penstemon		dP		M–L		R
Penstemon euglaucus	glaucous penstemon		dP		M–L		C, R
Penstemon humilis	lowly penstemon		eP		M		C, G, R
Penstemon procerus	littleflower penstemon		dP		M		C, E
Penstemon rydbergii	Rydberg's penstemon		dP		M		R
Penstemon serrulatus	coast penstemon		dP		M–L		R, W
Penstemon speciosus	royal penstemon		dP		M–L		G, R
Penstemon venustus	Venus penstemon		dS		M		E, G, H
Veronica regina-nivalis	snow queen		eP		E		C

Species	Common Name	Region	Type	Light	Bloom	Water	Use
POLEMONIACEAE							
Gilia capitata	bluehead gilia		A	☼ ◐ ●	M	○ ◒	C, G, R
Ipomopsis aggregata	scarlet gilia		dP	☼	M–L	○	G, R
Phlox adsurgens	woodland phlox		dP	◐	M	○ ◒	C, R
Phlox diffusa	spreading phlox		dP	☼ ◐	E	○	C, G, R
Phlox hoodii	Hood's phlox		eP	☼ ◐	E	○	C, G, R
Phlox longifolia	long-leaved phlox		dP	☼ ◐	E–M	○	C, G, R
Phlox speciosa	showy phlox		dP	☼ ◐ ●	E–M	○ ◒	C, G, R
Polemonium carneum	great polemonium		dP	◐	M	◒	W
Polemonium occidentale	western Jacob's-ladder		dP	☼ ◐ ●	M	◒ ●	C, W
POLYGONACEAE							
Eriogonum compositum	arrowleaf buckwheat		dP	☼ ◐	M	○ ◒	C, G, R
Eriogonum heracleoides	parsnipflower buckwheat		dP	☼ ◐	M	○	C, G, R
Eriogonum niveum	snow buckwheat		dP	☼ ◐	M–L	○	C, G, R
Eriogonum nudum	naked buckwheat		dP	☼	M–L	○ ◒	G, R
Eriogonum ovalifolium	cushion buckwheat		dP	☼	M–L	○	C, G, R
Eriogonum sphaeracephalum	rock buckwheat		dS	☼ ◐	M	○	C, G, R
Eriogonum strictum	Blue Mountain buckwheat		dP	☼ ◐	M	○	G, R
Eriogonum umbellatum	sulphur-flower buckwheat		eP/S	☼ ◐	M	○	C, E, G, R
PRIMULACEAE							
Dodecatheon dentatum	white shooting star		dP	◐	M	◒ ●	W
Dodecatheon hendersonii	Henderson's shooting star		dP	☼ ◐	E	○ ◒	R
Dodecatheon poeticum	poet's shooting star		dP	◐	M	○ ◒ ●	W
Dodecatheon pulchellum	darkthroat shooting star		dP	☼ ◐	E–M	◒ ●	W
Lysimachia latifolia	western starflower		dP	◐ ●	M	○ ◒	C
RANUNCULACEAE							
Actaea rubra	western red baneberry		dP	◐ ●	E–M	◒	W
Aquilegia formosa	western columbine		dP	☼ ◐	M	○ ◒	R, W
Clematis ligusticifolia	western clematis		dV	●	M–L	◒	E, H, W
Delphinium menziesii	Menzies' larkspur		dP	☼ ◐	E–M	○ ◒ ●	W
Delphinium nuttallianum	upland larkspur		dP	☼ ◐	E–M	○ ◒	R
Delphinium trolliifolium	poison larkspur		dP	◐ ●	E–M	◒ ●	H
Ranunculus occidentalis	western buttercup		dP	☼ ◐	E–M	◒	C, R
Ranunculus orthorhynchus	straightbeak buttercup		dP	☼ ◐	E–M	●	C, W
Thalictrum occidentale	western meadowrue		dP	☼ ◐ ●	E–M	◒	E, W
Thalictrum polycarpum	tall western meadowrue		dP	☼ ◐	E–M	○ ◒	W
RHAMNACEAE							
Ceanothus cuneatus	buckbrush		eS	☼ ◐	E–M	○ ◒	E, H, R
Ceanothus integerrimus	deerbrush		dS	◐ ●	M	○ ◒	H

Species	Common Name	Region	Type	Light	Bloom	Water	Use
Ceanothus prostratus	Mahala mat		eS	☼ ◐	E–M	◊	C, E, G, R
Ceanothus sanguineus	redstem ceanothus		dS	☼ ◐	E–M	◊ ⬙	E, H, R
Ceanothus thyrsiflorus	blue blossom		eS/T	☼ ◐	M	◊	E, H, R
Ceanothus velutinus	tobacco brush		eS	☼ ◐	M	◊ ⬙	E, H, R
Rhamnus purshiana	cascara		dS/T	☼ ◐ ●	E–M	⬙	E, H
ROSACEAE							
Amelanchier alnifolia	Saskatoon serviceberry		dS/T	☼ ◐ ●	E–M	⬙ ♦	E, H, W
Amelanchier utahensis	Utah serviceberry		dS/T	☼	E–M	◊ ⬙	E, H, R
Aruncus dioicus	sylvan goatsbeard		dP	◐ ●	M	⬙	R, W
Cercocarpus ledifolius	curlleaf mountain mahogany		eS/T	☼ ◐	E–M	◊	H, R
Chamaebatiaria millefolium	desert sweet		d/eS	☼	M–L	◊	H, R
Crataegus douglasii	Douglas' hawthorn		dS/T	☼ ◐ ●	E–M	◊ ⬙	E, H, W
Dasiphora fruticosa	shrubby cinquefoil		dS	☼ ◐	M–L	◊	H, R
Drymocallis glandulosa	sticky cinquefoil		dP	☼ ◐	M	◊ ⬙	C
Fragaria chiloensis	beach strawberry		eP	☼ ◐	E–M	◊ ⬙	C, E, G, R
Fragaria vesca	woodland strawberry		dP	☼ ◐ ●	E–M	⬙	C, E, R
Fragaria virginiana	Virginia strawberry		dP	☼ ◐ ●	E–M	◊ ⬙	C, E, G, R
Geum macrophyllum	largeleaf avens		dP	☼ ◐ ●	M	⬙ ♦	G, W
Geum triflorum	old man's whiskers		dP	☼ ◐	M	◊ ⬙	E, G, R
Holodiscus discolor	oceanspray		dS	☼ ◐ ●	M	◊ ⬙	E, H, R
Holodiscus microphyllus	rock spiraea		dS	◐	M–L	◊ ⬙	E, G, H, R
Malus fusca	western crabapple		dS/T	☼ ◐ ●	E–M	⬙ ♦	E, H, W
Oemleria cerasiformis	osoberry		dS/T	☼ ◐ ●	E	◊ ⬙	E, H
Physocarpus capitatus	Pacific ninebark		dS	☼ ◐ ●	M	⬙ ♦	E, H, W
Physocarpus malvaceus	mallow ninebark		dS	☼ ◐	M	⬙	E, H, W
Potentilla anserina	silverweed cinquefoil		dP	☼ ◐ ●	M	⬙ ♦	C, E, W
Potentilla gracilis	slender cinquefoil		dP	☼ ◐	M	⬙	E, H, W
Prunus emarginata	bitter cherry		dS/T	☼ ◐ ●	M	◊ ⬙	E, H
Prunus subcordata	Klamath plum		dS	☼ ◐	E–M	◊ ⬙	E, H
Prunus virginiana	chokecherry		dS/T	☼	E–M	⬙	E, H, R, W
Purshia tridentata	bitterbrush		dS	☼	E–M	◊	H, R
Rosa gymnocarpa	baldhip rose		dS	☼ ◐ ●	M	◊ ⬙	E, H, W
Rosa nutkana	Nootka rose		dS	☼ ◐	M	◊ ⬙ ♦	E, H, W
Rosa pisocarpa	cluster rose		dS	☼ ◐	M	◊ ⬙ ♦	E, H
Rosa woodsii	Woods' rose		dS	☼ ◐	M–L	◊ ⬙	E, H, R, W
Rubus leucodermis	blackcap		dS	☼ ◐	M	◊ ⬙	E, H
Rubus nutkanus	thimbleberry		dS	☼ ◐ ●	M	◊ ⬙	E, H
Rubus spectabilis	salmonberry		dS	☼ ◐ ●	E	◊ ⬙ ♦	E, H
Sorbus scopulina	Greene's mountain-ash		dS	☼ ◐	M	◊ ⬙	H, W
Sorbus sitchensis	Sitka mountain-ash		dS	◐ ●	M	⬙	H, W
Spiraea douglasii	hardhack		dS	☼ ◐	M–L	⬙ ♦	E, H, W

Species	Common Name	Region	Type	Light	Bloom	Water	Use
Spiraea lucida	shinyleaf spiraea		dS	○ ◐ ●	M–L	○ ◐	H, R
Spiraea splendens	subalpine spiraea		dS	○ ◐	M–L	◐	H, R, W
SALICACEAE							
Populus tremuloides	quaking aspen		dT	○	E–M	◐ ●	E, H, W
Salix geyeriana	Geyer's willow		dS/T	○	E	○ ◐ ●	E, H, W
Salix hookeriana	coastal willow		dS/T	○ ◐	E	◐ ●	E, H, W
Salix scouleriana	Scouler's willow		dS/T	○ ◐ ●	E	○ ◐ ●	E, H, W
Salix sitchensis	Sitka willow		dS/T	○ ◐	E	◐ ●	E, H, W
SAPINDACEAE							
Acer circinatum	vine maple		dS/T	○ ◐ ●	E–M	○ ◐	E, H, W
Acer glabrum	Rocky Mountain maple		dS/T	○ ◐	E–M	◐	H, R, W
SAXIFRAGACEAE							
Boykina major	large boykinia		dP	◐ ●	M	◐ ●	C, W
Darmera peltata	umbrella plant		dP	◐ ●	E–M	◐ ●	W
Heuchera chlorantha	meadow alumroot		dP	○ ◐	M	◐	C
Heuchera cylindrica	roundleaf alumroot		eP	○ ◐	M	○ ◐	C, E, G, R
Heuchera glabra	smooth alumroot		dP	○ ◐ ●	M–L	◐	E, G, R
Heuchera micrantha	small-flowered alumroot		eP	○ ◐ ●	M	◐	C, E, R, W
Micranthes oregana	Oregon saxifrage		dP	○ ◐	E–M	◐ ●	W
Tellima grandiflora	fringecup		dP	◐ ●	M	○ ◐	C, E
Tiarella trifoliata	three-leaf foamflower		dP	◐ ●	M	○ ◐	C
Tolmiea menziesii	piggyback plant		dP	◐ ●	M	◐	C, W
VALERIANACEAE							
Plectritis congesta	rosy plectritis		A	○ ◐	E–M	◐	G, R
VERBENACEAE							
Verbena lasiostachys	western verbena		dP	○	M–L	○ ◐ ●	R
VIBURNACEAE							
Sambucus mexicana	blue elderberry		dS/T	○ ◐	M	○ ◐	H, W
Sambucus racemosa	red elderberry		dS/T	◐ ●	E–M	○ ◐ ●	E, H, W
Viburnum edule	highbush cranberry		dS	○ ◐	M	◐ ●	E, H
Viburnum ellipticum	Oregon viburnum		dS	○ ◐	M	○ ◐	E, H, W
VIOLACEAE							
Viola adunca	hookedspur violet		dP	○ ◐ ●	E–M	○ ◐	C
Viola glabella	stream violet		dP	◐ ●	E–M	◐ ●	C
Viola praemorsa	upland yellow violet		dP	○ ◐	E	○ ◐	C
Viola sempervirens	evergreen violet		eP	◐ ●	E–M	◐	C, R

Appendix 3

Native Plants for Insects

The plant genera listed here are suggested for inclusion in gardens and landscapes because of their ability to provide the wildlife support indicated. Certain species of these genera are cultivated; since not all are necessarily native to Oregon, check for its origin status within the *Flora* species descriptions, on the OregonFlora website (https://oregonflora.org), or with the vendor. Ecoregions these genera inhabit in the wild are listed (see front endpaper). These serve as guidelines for regions that are appropriate for inclusion in planted settings.

Btrfly Nectar: a nectar source for adult butterflies and moths
Bees & Other: supports bees and other pollinating insects
Benef. Insect: provides food, shelter, or habitat for beneficial insects
Food Plant: a food source for the caterpillar stage of butterflies and moths

Genus	Common Name	Ecoregion	Btrfly Nectar	Bees & Other	Benef. Insect	Food Plant
GYMNOSPERMS						
CUPRESSACEAE						
Calocedrus	incense-cedar	BW, Casc, Col, ECas, Sisk, WV			x	x
Chamaecyparis	white-cedar	CR, Est, Sisk				x
Thuja	arborvitae	BW, Casc, CR, ECas, Est, Sisk, WV				x
PINACEAE						
Abies	fir	BR, BW, Casc, CR, ECas, Sisk, WV			x	x
Picea	spruce	BW, Casc, CR, ECas, Est, Sisk			x	x
Pinus	pine	BR, BW, Casc, Col, CR, ECas, Est, Sisk, WV			x	x
Pseudotsuga	Douglas-fir	BW, Casc, Col, CR, ECas, Est, Sisk, WV				x
Tsuga	hemlock	BR, BW, Casc, CR, ECas, Est, Sisk, WV			x	x
MONOCOTS						
ALISMATACEAE						
Sagittaria	arrowhead	BR, BW, Casc, Col, CR, ECas, Lava, Owy, WV	x	x	x	x
AMARYLLIDACEAE						
Allium	onion	All	x	x	x	
ARACEAE						
Lysichiton	skunk cabbage	Casc, CR, ECas, Est, Sisk, WV		x	x	
ASPARAGACEAE						
Brodiaea	brodiaea	Casc, CR, ECas, Est, Sisk, WV	x	x		
Camassia	camas	All except Lava	x	x	x	
Dichelostemma	snake-lily	Casc, CR, ECas, Est, Sisk, WV	x	x		
Hastingsia	rush-lily	Sisk	x	x		
Leucocrinum	sand-lily	Br, BW, ECas, Lava, Owy	x	x		
Maianthemum	false Solomon's seal	All		x	x	
Triteleia	triteleia	All	x	x		

Genus	Common Name	Ecoregion	Btrfly Nectar	Bees & Other	Benef. Insect	Food Plant
CYPERACEAE						
Carex	sedge	All				x
IRIDACEAE						
Iris	iris	All	x	x	x	
Sisyrinchium	blue-eyed grass	All except Col	x	x	x	
JUNCACEAE						
Juncus	rush	All		x	x	x
LILIACEAE						
Calochortus	mariposa lily	All	x	x		
Erythronium	fawn-lily	All except BR	x	x		
Fritillaria	fritillary	All	x	x	x	
Lilium	lily	Casc, CR, ECas, Est, Sisk, WV	x	x		x
Prosartes	fairybells	BR, BW, Casc, CR, ECas, Est, Lava, Sisk, WV		x		
MELANTHIACEAE						
Trillium	trillium	BW, Casc, Col, CR, ECas, Est, Lava, Sisk, WV		x	x	
Veratrum	false hellebore	BR, BW, Casc, CR, ECas, Est, Sisk, WV	x	x		
Xerophyllum	beargrass	Casc, CR, ECas, Est, Sisk	x	x		
POACEAE						
Danthonia	oatgrass	All		x	x	x
Deschampsia	hairgrass	All				x
Festuca	fescue	All		x		x
Koeleria	junegrass	All		x		x
DICOTS						
ANACARDIACEAE						
Rhus	sumac	BW, Casc, Col, ECas, Lava, Owy, Sisk		x		
APIACEAE						
Angelica	angelica	All	x	x		x
Heracleum	cow parsnip	All	x	x		x
Lomatium	desert parsley	All	x	x		x
APOCYNACEAE						
Asclepias	milkweed	All except Est	x	x	x	x
ASTERACEAE						
Achillea	yarrow	All	x	x	x	x
Anaphalis	everlasting	BW, Casc, Col, CR, ECas, Est, Lava, Sisk, WV	x	x	x	x
Baccharis	baccharis	CR, Est, Sisk, WV	x	x	x	
Balsamorhiza	balsamroot	BR, BW, Casc, Col, ECas, Lava, Owy, Sisk, WV	x	x		x

Genus	Common Name	Ecoregion	Btrfly Nectar	Bees & Other	Benef. Insect	Food Plant
Canadanthus	aster	BW, Casc, CR, ECas, WV	x	x		
Ericameria	rabbitbrush	BR, BW, Casc, Col, ECas, Lava, Owy, Sisk	x	x	x	x
Erigeron	fleabane	All	x	x	x	x
Eriophyllum	woolly sunflower	All	x	x	x	x
Eucephalus	aster	BW, Casc, CR, ECas, Sisk, WV	x	x		x
Gaillardia	blanket flower	BW, Casc, Col, ECas, Lava		x		
Madia	tarweed	All	x	x		
Rudbeckia	coneflower	BR, BW, Casc, CR, Lava, Sisk, WV	x	x		x
Solidago	goldenrod	All	x	x	x	x
Symphyotrichum	aster	All	x	x	x	x
Wyethia	mule's ears	All except CR	x	x	x	x
BERBERIDACEAE						
Achlys	vanilla-leaf	Casc, CR, Sisk, WV		x		x
Berberis	Oregon-grape	All	x	x	x	
Vancouveria	inside-out flower	Casc, CR, ECas, Est, Sisk, WV		x	x	x
BETULACEAE						
Alnus	alder	All				x
Corylus	hazel	Casc, CR, ECas, Est, Sisk, WV	x		x	x
BRASSICACEAE						
Cardamine	bittercress	All	x	x		x
Erysimum	wallflower	All	x	x		x
CAPRIFOLIACEAE						
Lonicera	honeysuckle	BR, BW, Casc, CR, ECas, Est, Sisk, WV	x	x	x	x
Symphoricarpos	snowberry	All	x	x	x	x
CLEOMACEAE						
Peritoma	bee plant	BR, BW, Lava, Owy		x	x	x
CORNACEAE						
Cornus	dogwood	All	x	x		x
CRASSULACEAE						
Sedum	stonecrop	All	x	x	x	x
ERICACEAE						
Arbutus	madrone	Casc, CR, Est, Sisk, WV	x	x	x	x
Arctostaphylos	manzanita	BR, BW, Casc, CR, ECas, Est, Lava, Sisk, WV	x	x	x	x
Gaultheria	wintergreen	BW, Casc, CR, ECas, Est, Sisk, WV	x	x	x	
Rhododendron	rhododendron	BW, Casc, CR, Est, Sisk, WV	x	x		
Vaccinium	blueberry	BR, BW, Casc, CR, ECas, Est, Lava, Sisk, WV	x	x		x

Genus	Common Name	Ecoregion	Btrfly Nectar	Bees & Other	Benef. Insect	Food Plant
FABACEAE						
Astragalus	milkvetch	BR, BW, Casc, Col, ECas, Lava, Owy, Sisk		x		x
Cercis	redbud	Sisk		x	x	
Dalea	prairie clover	BR, BW, Col, Lava, Owy	x	x	x	x
Lupinus	lupine	All	x	x	x	x
Trifolium	clover	All	x	x	x	x
FAGACEAE						
Chrysolepis	chinquapin	Casc, CR, ECas, Est, Sisk, WV	x	x	x	x
Quercus	oak	Casc, Col, CR, ECas, Sisk, WV	x	x	x	x
GERANIACEAE						
Geranium	cranesbill	BR, BW, Casc, Col, ECas, Lava, Owy, Sisk, WV	x	x		x
GROSSULARIACEAE						
Ribes	currant, gooseberry	All	x	x	x	x
HYDRANGEACEAE						
Philadelphus	mock orange	All	x	x	x	x
HYDROPHYLLACEAE						
Hydrophyllum	waterleaf	All	x	x		
Phacelia	phacelia	All	x	x		
LAMIACEAE						
Clinopodium	yerba buena	Casc, CR, ECas, Est, Sisk, WV		x		
Monardella	coyote mint	BR, BW, Casc, Col, ECas, Lava, Owy, Sisk	x	x	x	
Prunella	self heal	BR, BW, Casc, CR, ECas, Est, Lava, Sisk, WV	x	x	x	
Salvia	sage	BR, BW, Col, ECas, Lava, Owy		x	x	
Scutellaria	skullcap	All except Est		x	x	
Stachys	hedge nettle	All except BW, Col		x	x	
LAURACEAE						
Umbellularia	California laurel	CR, Est, Sisk, WV		x		
LIMNANTHACEAE						
Limnanthes	meadowfoam	CR, Sisk, WV		x		
LINNAEACEAE						
Linnaea	twin flower	BW, Casc, CR, ECas, Est, Sisk, WV	x	x		
MALVACEAE						
Sidalcea	checkermallow	All	x	x	x	x
Sphaeralcea	globemallow	BR, BW, Col, Lava, Owy	x	x	x	x

Genus	Common Name	Ecoregion	Btrfly Nectar	Bees & Other	Benef. Insect	Food Plant
MONTIACEAE						
Claytonia	spring beauty	All		x		
OLEACEAE						
Fraxinus	ash	Col, CR, ECas, Est, Sisk, WV				x
ONAGRACEAE						
Chamaenerion	fireweed	BR, BW, Casc, CR, ECas, Est, Lava, Sisk, WV	x	x	x	x
Clarkia	clarkia	All	x	x		x
OXALIDACEAE						
Oxalis	wood sorrel	Casc, CR, Est, Sisk, WV	x	x		x
PAPAVERACEAE						
Corydalis	fumewort	BR, BW, Casc, CR, ECas, Est, Lava, Sisk, WV	x	x		x
Dicentra	bleeding heart	All except Lava	x	x	x	x
Eschscholzia	California poppy	Casc, Col, CR, ECas, Sisk, WV	x	x	x	
PHRYMACEAE						
Erythranthe	monkey flower	All	x	x	x	x
PLANTAGINACEAE						
Penstemon	penstemon	All	x	x	x	x
Synthyris	kittentails	BW, Casc, CR, ECas, Est, Lava, Sisk, WV	x	x	x	
Veronica	speedwell	All	x	x		
PLUMBAGINACEAE						
Armeria	armeria	Est	x	x	x	
POLEMONIACEAE						
Gilia	gilia	All	x	x	x	x
Ipomopsis	gilia	BR, BW, Casc, Col, ECas, Lava, Owy, Sisk	x	x	x	x
Phlox	phlox	BR, BW, Casc, Col, CR, ECas, Lava, Owy, Sisk	x	x		x
POLYGONACEAE						
Eriogonum	buckwheat	All	x	x	x	x
PRIMULACEAE						
Dodecatheon	shooting star	All except Est		x	x	x
RANUNCULACEAE						
Aconitum	monkshood	BR, BW, Casc, ECas, Lava, Owy, Sisk		x		
Actaea	baneberry	BR, BW, Casc, CR, ECas, Est, Sisk, WV		x		
Aquilegia	columbine	All except Owy	x	x	x	x
Delphinium	larkspur	All	x	x		x
Ranunculus	buttercup	All	x	x		

Genus	Common Name	Ecoregion	Btrfly Nectar	Bees & Other	Benef. Insect	Food Plant
RHAMNACEAE						
Ceanothus	ceanothus	All	x	x	x	x
Rhamnus	buckthorn	BR, BW, Casc, CR, ECas, Est, Sisk, WV	x	x	x	x
ROSACEAE						
Aruncus	goatsbeard	Casc, CR, Est, Sisk, WV	x	x	x	x
Chamaebatiaria	fernbush	BR, ECas		x		x
Crataegus	hawthorn	All	x	x		x
Drymocallis	cinquefoil	Casc, CR, ECas, Sisk, WV	x	x	x	x
Fragaria	strawberry	BW, Casc, Col, CR, ECas, Est, Lava, Sisk, WV	x	x	x	x
Geum	avens	All	x	x		x
Holodiscus	oceanspray	All	x	x	x	x
Malus	apple	Casc, CR, ECas, Est, Sisk, WV	x	x	x	x
Oemleria	osoberry	Casc, CR, Est, Sisk, WV	x	x		x
Physocarpus	ninebark	BW, Casc, Col, CR, ECas, Est, Lava, Sisk, WV	x	x	x	x
Potentilla	cinquefoil	All	x	x		x
Prunus	plum	All	x	x	x	x
Purshia	bitterbrush	BR, BW, Casc, Col, ECas, Lava, Owy, Sisk	x	x	x	x
Rosa	rose	All	x	x	x	x
Rubus	blackberry	BW, Casc, Col, CR, ECas, Est, Owy, Sisk, WV	x	x	x	x
Spiraea	spirea	All except Owy	x	x	x	x
SALICACEAE						
Populus	poplar	All				x
Salix	willow	All	x	x	x	x
SAPINDACEAE						
Acer	maple	All except BR		x	x	x
SAXIFRAGACEAE						
Heuchera	alumroot	All	x	x		x
Tellima	fringecup	Casc, CR, ECas, Est, Sisk, WV		x	x	
Tiarella	foamflower	BW, Casc, CR, ECas, Est, Sisk, WV		x		
Tolmiea	piggyback plant	Casc, CR, Est, Sisk, WV	x	x	x	
VALERIANACEAE						
Plectritis	plectritis	All	x	x	x	x
VIBURNACEAE						
Viburnum	viburnum	BW, Casc, CR, ECas, Sisk, WV	x	x	x	x
VIOLACEAE						
Viola	violet	All	x	x	x	x

Appendix 4

Butterfly Foodplants

This appendix lists the caterpillar foodplants of known, documented species of Oregon butterflies, organized by plant family. Several subtaxa that are of conservation concern are also included. Known plant genera used by the butterfly species are listed, but these may be from observations in other states. Pages listed refer to those where the butterfly is mentioned in the Insects as Plant Taxonomists chapter. Listed also are the allelochemicals present in the plants that are relevant to the plant-caterpillar interaction. Symbols used after butterfly species indicate the following:

- * specialist using only one family
- ° generalist using three or more families
- ^ species of conservation concern

Butterfly Species	Common Name	Plant Genera	Page	Chemicals
GYMNOSPERMS				
CUPRESSACEAE				
*Callophrys gryneus**	cedar hairstreak	*Calocedrus, Juniperus, Thuja*		
PINACEAE				
*Callophrys eryphon**	western pine elfin	*Pinus*	29	terpenes
*Neophasia menapia**	pine white	*Abies, Cedrus, Picea, Pinus, Pseudotsuga, Tsuga*	29	terpenes
MONOCOTS				
CYPERACEAE				
*Euphyes vestris**	dun skipper	*Carex*	43, 53	
Oarisma garita	Garita skipperling	*Carex*		
*Polites mardon^*	Mardon skipper	*Carex*	53	
POACEAE				
*Amblyscirtes vialis**	common roadside skipper	*Agrostis, Avena, Bromus, Cynodon, Elymus, Phleum, Poa*	53	
*Atalopedes campestris**	sachem	*Cynodon, Digitalis, Eleusine, Festuca, Hordeum, Poa*		
*Carterocephalus palaemon**	Arctic skipper	*Bromus, Calamogrostis*		
*Cercyonis oetus**	dark wood nymph	*Festuca, Poa*		
*Cercyonis pegala**	common wood nymph	*Andropogon, Avena, Danthonia, Festuca, Poa, Stipa*		
*Cercyonis sthenele**	Great Basin wood nymph	*Andropogon, Avena, Danthonia, Poa*		
*Coenonympha tullia**	common ringlet	*Carex, Festuca, Poa*		
*Erebia epipsodea**	Butler's alpine	*Poa, Setaria*		
*Hesperia colorado**	western branded skipper	*Achnatherum, Andropogon, Bromus, Festuca, Lolium*		
*Hesperia columbia**	Columbian skipper	*Danthonia, Koeleria*	54	
*Hesperia juba**	Juba skipper	*Bromus, Deschampsia, Poa, Stipa*		
*Hesperia lindseyi**	Lindsey's skipper	*Danthonia, Festuca, Phalaris*	54	
*Hesperia nevada**	Nevada skipper	*Achnatherum, Elymus, Festuca, Koeleria*		
*Hesperia uncas**	Uncas skipper	*Achnatherum*		

Butterfly Species	Common Name	Plant Genera	Page	Chemicals
*Neominois ridingsii**	Ridings' satyr	*Hesperostipa*		
Oarisma garita	Garita skipperling	*Agrostis, Bromus, Elymus, Koeleria, Muhlenbergia, Poa, Stipa*		
*Ochlodes agricola**	rural skipper	*Pseudoroegneria*		
*Ochlodes sylvanoides**	woodland skipper	*Agropyron, Agrostis, Avena, Cynodon, Danthonia, Elymus, Leymus*		
*Ochlodes yuma**	Yuma skipper	*Phragmites*		
*Oeneis nevadensis**	great arctic	no field records, captivity reared on grasses		
*Polites mardon^*	Mardon skipper	*Danthonia, Festuca, Poa*	53	
*Polites peckius**	Peck's skipper	*Bromus, Distichlis, Poa*		
*Polites sabuleti**	sandhill skipper	*Agrostis, Bromus, Cynodon, Distichlis, Festuca, Hordeum, Poa*		
*Polites sonora**	Sonoran skipper	*Festuca, Poa*		
DICOTS				
AMARANTHACEAE				
*Brephidium exilis**	western pygmy blue	*Atriplex, Chenopodium, Salicornia, Salsola, Suaeda*		
*Hesperopsis libya**	Mojave sootywing	*Atriplex, Chenopodium, Salsola*		
*Pholisora catullus**	common sootywing	*Amaranthus, Atriplex, Chenopodium, Salsola*		
APIACEAE				
*Papilio indra**	Indra swallowtail	*Cymopterus, Lomatium*	46, 47	furanocoumarins
*Papilio zelicaon**	anise swallowtail	*Angelica, Foeniculum, Heracleum, Lomatium, Sphenosciadium*	47	furanocoumarins
APOCYNACEAE				
*Danaus plexippus**	monarch	*Asclepias*	42, 48, 53	cardiac glycosides
ARISTOLOCHIACEAE				
*Battus philenor**	pipevine swallowtail	*Aristolochia*	53	aristolochic acids
ASTERACEAE				
*Chlosyne acastus**	sagebrush checkerspot	*Erigeron, Chrysothamnus, Dieteria*		
*Chlosyne hoffmanni**	Hoffmann's checkerspot	*Eucephalus, Eurybia*	31	
*Chlosyne palla**	northern checkerspot	*Erigeron, Solidago, Symphyotrichum*		
*Papilio machaon oregonia**	Oregon swallowtail	*Artemisia dracunculus*	47, 48	anisic aldehyde
*Phyciodes cocyta**	northern crescent	*Eucephalus, Eurybia, Symphyotrichum*		
*Phyciodes mylitta**	Mylitta crescent	*Centaurea, Cirsium*		
*Phyciodes orseis**	California crescent	*Cirsium*		
*Phyciodes pallida**	pale crescent	*Cirsium*		
*Phyciodes pulchella**	field crescent	*Eucephalus, Eurybia*		
Vanessa cardui°	painted lady	*Carduus, Centaurea, Cirsium*	54	
Vanessa virginiensis	American lady	*Anaphalis, Antennaria, Gnaphalium*		

Butterfly Species	Common Name	Plant Genera	Page	Chemicals
BERBERIDACEAE				
Callophrys augustinus°	brown elfin	*Berberis*	35	
BETULACEAE				
Nymphalis antiopa°	mourning cloak	*Alnus*		
Nymphalis l-album	Compton tortoiseshell	*Betula*		
Papilio eurymedon°	pale swallowtail	*Alnus*		
Polygonia faunus	green comma	*Alnus, Betula*		
BRASSICACEAE				
*Anthocharis lanceolata**	gray marble	*Boechera, Streptanthus, Turritis*		glucosinolates
*Anthocharis julia**	Julia's orangetip	*Arabis, Boechera, Cardamine, Streptanthus, Turritis*		glucosinolates
*Euchloe ausonides**	large marble	*Arabis, Boechera, Descurainia, Erysimum, Sisymbrium, Turritis*	50	glucosinolates
*Euchloe hyantis**	California marble	*Arabis, Strepanthus*		glucosinolates
*Euchloe lotta**	desert marble	*Arabis, Boechera, Caulanthus, Descurainia, Sisymbrium*		glucosinolates
*Pieris marginalis**	margined white	*Arabis, Boechera, Cardamine, Erysimum, Lepidium, Nasturtium*	38	glucosinolates
*Pieris rapae**	cabbage white	*Barbarea, Boechera, Brassica, Lepidium, Nasturtium, Sisymbrium*	50	glucosinolates
*Pontia beckerii**	Becker's white	*Brassica, Descurainia, Lepidium, Sisymbrium, Thelypodium*		glucosinolates
*Pontia occidentalis**	western white	*Boechera, Lepidium, Rorippa, Stanleya, Sisymbrium, Thelypodium, Thlaspi*		glucosinolates
*Pontia protodice**	checkered white	*Arabis, Boechera, Descurainia, Lepidium, Rorippa, Sisymbrium, Thylepodium*		glucosinolates
*Pontia sisymbrii**	spring white	*Arabis, Boechera, Descurainia, Sisymbrium, Streptanthus*	50	glucosinolates
CAPRIFOLIACEAE				
Euphydryas anicia	Anicia checkerspot	*Symphoricarpos*		iridoid glycosides
Euphydryas chalcedona	chalcedona checkerspot	*Symphoricarpos*		iridoid glycosides
Euphydryas colon	snowberry checkerspot	*Lonicera, Symphoricarpos*		iridoid glycosides
*Euphydryas gillettii^*	Gillett's checkerspot	*Lonicera, Symphoricarpos*		iridoid glycosides
CORNACEAE				
Celastrina echo°	echo azure	*Cornus sericea*	41	
CRASSULACEAE				
*Callophrys mossii**	Moss's elfin	*Sedum*		
*Parnassius smintheus**	mountain parnassian	*Sedum*	38	
ERICACEAE				
Callophrys augustinus°	brown elfin	*Arbutus, Arctostaphylos, Gaultheria, Rhododendron, Vaccinium*	35	
*Callophrys polios**	hoary elfin	*Arctostaphylos uva-ursi*	35	

Butterfly Species	Common Name	Plant Genera	Page	Chemicals
*Callophrys polios maritima**^	seaside hoary elfin	*Arctostaphylos uva-ursi*		
Celastrina echo°	echo azure	*Arbutus, Arctostaphylos, Vaccinium*	41	
*Colias interior**	pink-edged sulphur	*Vaccinium*		
*Colias skinneri**	Skinner's sulphur	*Gaultheria, Vaccinium*		
Lycaena mariposa	mariposa copper	*Vaccinium*		
FABACEAE				
*Callophrys dumetorum**	bramble hairstreak	*Acmispon, Hosackia*	39	
*Colias alexandra**	Queen Alexandra's sulphur	*Astragalus, Lathyrus, Medicago*	51	
*Colias christina**	Christina's sulphur	*Hedysarum, Lathyrus, Thermopsis*		
*Colias eurytheme**	orange sulphur	*Astragalus, Lathyrus, Lotus, Lupinus, Medicago, Melilotus, Thermopsis, Trifolium, Vicia*	51	
*Colias occidentalis**	western sulphur	*Lathyrus, Lupinus, Melilotus, Vicia*		
*Colias philodice**	clouded sulphur	*Astragalus, Lathyrus, Lupinus, Medicago, Melilotus, Thermopsis, Trifolium, Vicia*		
*Cupido amyntula**	western tailed blue	*Astragalus, Lathyrus, Oxytropis, Vicia*		
*Cupido comyntas**	eastern tailed blue	*Acmispon, Astragalus, Lathyrus, Lotus, Lupinus, Trifolium, Vicia,* others		
*Epargyreus clarus**	silver-spotted skipper	*Glycyrrhiza, Hosackia, Robinia*		
*Erynnis persius**	Persius duskywing	*Acmispon, Astragalus, Lotus, Lupinus, Thermopsis*		
*Glaucopsyche lygdamus**	silvery blue	*Astragalus, Hosackia, Lathyrus, Lupinus, Vicia*		
*Glaucopsyche piasus**	arrowhead blue	*Lupinus*		
*Icaricia acmon**	acmon blue	*Acmispon, Lotus, Lupinus, Melilotus*		
*Icaricia icarioides**	Boisduval's blue	*Lupinus*	44	
*Icaricia icarioides fenderi**^	Fender's blue	*Lupinus oreganus*	35	
*Icaricia saepiolus**	greenish blue	*Trifolium*		
*Icaricia saepiolus littoralis**^	greenish blue	*Trifolium*	35	
*Icaricia shasta**	Shasta blue	*Astragalus, Lupinus, Oxytropis, Trifolium*		
Leptotes marina	marine blue	*Astragalus, Lathyrus, Trifolium, Vicia*		
*Plebejus anna**	Anna's blue	*Astragalus, Hosackia, Lathyrus, Lupinus, Vicia*		
*Plebejus idas**	northern blue	*Astragalus*		
*Plebejus melissa**	Melissa blue	*Acmispon, Astragalus, Glycyhirriza, Hosackia, Lupinus, Medicago, Sphaerophysa, Vicia*		
*Satyrium fuliginosa**	sooty hairstreak	*Lupinus*		
*Satyrium semiluna**	halfmoon hairstreak	*Lupinus*		
Strymon melinus°	gray hairstreak	*Hosackia, Lotus, Trifolium, Vicia*		
*Thorybes diversus**	western cloudywing	*Trifolium*		
*Thorybes mexicana**	Mexican cloudywing	*Astragalus, Trifolium, Vicia*		
*Thorybes pylades**	northern cloudywing	*Acmispon, Astragalus, Lathyrus, Hosackia, Trifolium, Vicia*		
Vanessa cardui°	painted lady	*Lupinus*	54	

Butterfly Species	Common Name	Plant Genera	Page	Chemicals
FAGACEAE				
*Adelpha californica**	California sister	*Chrysolepis, Quercus*	44	
*Erynnis propertius**	Propertius duskywing	*Quercus*	44	
*Habrodais grunus**	golden hairstreak	*Chrysolepis, Notholithocarpus, Quercus*	44	
*Satyrium auretorum**	gold-hunter's hairstreak	*Quercus*		
GROSSULARIACEAE				
*Lycaena arota**	tailed copper	*Ribes*		
*Polygonia gracilis**	hoary comma	*Ribes*		
*Polygonia oreas**	oreas anglewing	*Ribes*	32	
MALVACEAE				
*Heliopetes ericetorum**	northern white skipper	*Iliamna, Malva, Malvella, Sidalcea, Sphaeralcea*		
*Pyrgus communis**	common checkered skipper	*Iliamna, Malva, Malvella, Sidalcea, Sphaeralcea*		
Pyrgus ruralis	two-banded checkered skipper	*Sidalcea*		
Strymon melinus°	gray hairstreak	*Sidalcea*		
Vanessa annabella	West Coast lady	*Alcea, Iliamna, Malva, Malvella, Sidalcea, Sphaeraclea*		
Vanessa cardui°	painted lady	*Alcea, Malva*	54	
OLEACEAE				
Euphydryas colon	snowberry checkerspot	*Fraxinus*		
Papilio multicaudata	two-tailed swallowtail	*Fraxinus*	46	
OROBANCHACEAE				
Chlosyne leanira	Leanira checkerspot	*Castilleja, Cordylanthus*		iridoid glycosides
Euphydryas anicia	Anicia checkerspot	*Castilleja*		iridoid glycosides
Euphydryas chalcedona	chalcedona checkerspot	*Castilleja*		iridoid glycosides
Euphydryas colon	snowberry checkerspot	*Castilleja*		iridoid glycosides
Euphydryas editha	Edith's checkerspot	*Castilleja*		iridoid glycosides
Euphydryas editha taylori^*	Taylor's checkerspot	*Castilleja*	36	iridoid glycosides
*Euphydryas gillettii^*	Gillett's checkerspot	*Pedicularis*		iridoid glycosides
PAPAVERACEAE				
*Parnassius clodius**	clodius parnassian	*Dicentra*	32	
PHRYMACEAE				
Euphydryas chalcedona	chalcedona checkerspot	*Erythranthe*		
Euphydryas colon	snowberry checkerspot	*Erythranthe*		
Euphydryas editha	Edith's checkerspot	*Erythranthe*		
Junonia coenia	common buckeye	*Erythranthe*		

Butterfly Species	Common Name	Plant Genera	Page	Chemicals
PLANTAGINACEAE				
Euphydryas anicia	Anicia checkerspot	*Collinsia, Penstemon*		iridoid glycosides
Euphydryas chalcedona	snowberry checkerspot	*Penstemon*		iridoid glycosides
Euphydryas colon	snowberry checkerspot	*Penstemon, Plantago*		iridoid glycosides
Euphydryas editha	Edith's checkerspot	*Penstemon*		iridoid glycosides
Euphydryas editha taylori	Taylor's checkerspot	*Plantago*	36	iridoid glycosides
Euptoieta claudia	variegated fritillary	*Plantago*		iridoid glycosides
Junonia coenia	common buckeye	*Antirrhinum, Penstemon, Plantago*		iridoid glycosides
POLYGONACEAE				
*Apodemia mormo**	Mormon metalmark	*Eriogonum*	54, 55	
*Callophrys affinis**	western green hairstreak	*Eriogonum*		
Callophrys augustinus°	brown elfin	*Eriogonum*	35	
*Callophrys sheridanii**	Sheridan's green hairstreak	*Eriogonum*	39	
Euphilotes unnamed*	Cascadia blue	*Eriogonum heracleoides, E. douglasii*		
Euphilotes unnamed*	pumice blue	*Eriogonum marifolium*		
*Euphilotes ancilla**	Rocky Mountain dotted blue	*Eriogonum sphaerocephalum*		
*Euphilotes baueri**	Bauer's blue	*Eriogonum ovalifolium*		
*Euphilotes columbiae**	Columbia dotted blue	*Eriogonum*		
*Euphilotes enoptes**	Pacific dotted blue	*Eriogonum*	39	
*Euphilotes glaucon**	summit blue	*Eriogonum umbellatum*	54	
Icaricia unnamed*	volcano blue	*Eriogonum pyrolifolium* var. *coryphaeum*		
Icaricia acmon	acmon blue	*Eriogonum*		
*Icaricia lupini**	lupine blue	*Eriogonum*		
*Lycaena cupreus**	lustrous copper	*Oxyria, Rumex*		
*Lycaena editha**	Edith's copper	*Rumex*	49	
*Lycaena gorgon**	gorgon copper	*Eriogonum*		
*Lycaena helloides**	purplish copper	*Polygonum, Rumex*		
Lycaena heteronea	blue copper	*Eriogonum*		
*Lycaena nivalis**	lilac-bordered copper	*Polygonum douglasii*		
*Lycaena phlaeas**	American copper	*Oxyria, Rumex*		
*Lycaena rubidus**	ruddy copper	*Oxyria, Rumex*		
*Lycaena xanthoides**	great copper	*Rumex*		
*Philotiella leona^*	Leona's little blue	*Eriogonum spergulinum*	54	
PRIMULACEAE				
Agriades glandon	Arctic blue	*Douglasia*		
Agriades podarce^*	Sierra Nevada blue	*Dodecatheon*	39	

Butterfly Species	Common Name	Plant Genera	Page	Chemicals
RHAMNACEAE				
Callophrys augustinus°	brown elfin	*Ceanothus, Rhamnus*	35	
Celastrina echo°	echo azure	*Ceanothus*	41	
*Erynnis pacuvius**	Pacuvius duskywing	*Ceanothus*		
Limenitis lorquini	Lorquin's admiral	*Ceanothus*		
*Nymphalis californica**	California tortoiseshell	*Ceanothus*	33	
Papilio eurymedon°	pale swallowtail	*Ceanothus, Rhamnus*		
Satyrium californica	California hairstreak	*Ceanothus*		
*Satyrium saepium**	hedgerow hairstreak	*Ceanothus*	34	
ROSACEAE				
Callophrys augustinus°	brown elfin	*Amelanchier, Prunus, Purshia, Pyrus*	35	
Celastrina echo°	echo azure	*Holodiscus, Physocarpus, Prunus, Purshia, Spiraea*	41	
Limenitis archippus	viceroy	*Holodiscus, Malus, Prunus, Spiraea*	42	
Limenitis lorquini	Lorquin's admiral	*Amelanchier, Holodiscus, Malus, Prunus, Spiraea*		
Limenitis weidemeyerii	Weidemeyer's admiral	*Amelanchier, Holodiscus, Physocarpus, Prunus*		
*Lycaena helloides**	purplish copper	*Potentilla*		
Nymphalis antiopa°	mourning cloak	*Prunus, Rosa, Spiraea*		
Papilio eurymedon°	pale swallowtail	*Amelanchier, Crataegus, Holodiscus, Prunus, Spiraea*		
Papilio multicaudata	two-tailed swallowtail	*Amelanchier, Prunus*	46	
Pyrgus ruralis	two-banded checkered skipper	*Fragaria, Potentilla*		
*Satyrium behrii**	Behr's hairstreak	*Purshia*		
Satyrium californica	California hairstreak	*Purshia*		
*Satyrium tetra**	mountain mahogany hairstreak	*Cercocarpus*		
*Satyrium titus**	coral hairstreak	*Prunus, Rosa*		
Strymon melinus°	gray hairstreak	*Fragaria*		
SALICACEAE				
Erynnis icelus	dreamy duskywing	*Populus, Salix*		
Limenitis archippus	viceroy	*Populus, Salix*	42	
Limenitis lorquini	Lorquin's admiral	*Populus, Salix*		
Limenitis weidemeyerii	Weidemeyer's admiral	*Populus, Salix*		
Nymphalis antiopa°	mourning cloak	*Populus, Salix*		
Nymphalis l-album	Compton tortoiseshell	*Populus, Salix*		
Papilio rutulus°	western tiger swallowtail	*Populus, Salix*		
Polygonia faunus	green comma	*Salix*		
Satyrium sylvinus	sylvan hairstreak	*Salix*		

Butterfly Species	Common Name	Plant Genera	Page	Chemicals
SAPINDACEAE				
Nymphalis antiopa°	mourning cloak	*Acer*		
Papilio rutulus°	western tiger swallowtail	*Acer*		
SAXIFRAGACEAE				
Agriades glandon	Arctic blue	*Saxifraga*		
SCROPHULARIACEAE				
Euphydryas chalcedona	chalcedona checkerspot	*Verbascum*		
URTICACEAE				
*Aglais milbertii**	Milbert's tortoiseshell	*Urtica*		
*Polygonia satyrus**	satyr comma	*Urtica*		
Vanessa annabella	West Coast lady	*Urtica*		
*Vanessa atalanta**	red admirable (red admiral)	*Parietaria, Urtica*		
VERBENACEAE				
Junonia coenia	common buckeye	*Verbena*		iridoid glycosides
VIOLACEAE				
*Boloria bellona**	meadow fritillary	*Viola*		
*Boloria epithore**	western meadow fritillary	*Viola*		
*Boloria selene**	silver-bordered fritillary	*Viola*		
*Speyeria callippe**	callippe fritillary	*Viola*		
*Speyeria coronis**	coronis fritillary	*Viola*		
*Speyeria cybele**	great-spangled fritillary	*Viola*		
*Speyeria egleis**	Great Basin fritillary	*Viola*		
*Speyeria hesperis**	northwestern fritillary	*Viola*		
*Speyeria hydaspe**	hydaspe fritillary	*Viola*		
*Speyeria mormonia**	Mormon fritillary	*Viola*		
*Speyeria zerene**	zerene fritillary	*Viola*		
Speyeria zerene hippolyta^*	Oregon silverspot	*Viola*	35	
VIBURNACEAE				
Celastrina echo°	echo azure	*Sambucus*	41	
VISCACEAE				
*Atlides halesus**	great purple hairstreak	*Phorodendron serotinum*	46	thionins
Callophrys johnsoni^*	Johnson's hairstreak	*Arceuthobium*	30	
*Callophrys spinetorum**	thicket hairstreak	*Arceuthobium*		

Appendix 5

Specialist Solitary Bee Species

This table represents known and potential specialized pollinators in the state of Oregon and the plant species or genera on which they are specialized. This represents approximately one third of the known bee taxa within the Pacific Northwest, and certainly within Oregon. Bees are the insurance policy for plants in natural circumstances. Even if visited by generalists, having a specialist pollinator around will result in cross-pollination. These common names will be applied in the developing Oregon Bee Atlas. Symbols used after bee species indicate the following:

- ? tentative assignment of specialist
- * known polylege (collector of pollen from multiple unrelated flowers) and its emphasized plant genus
- ^ not recorded but likely to occur in Oregon
- ( ) plant genera used less frequently, may be in other than preferred plant family

Bee Genera	Common Name	Plant Genera	Page
MONOCOTS			
LILIACEAE			
Calliopsis cincta	mariposa-lily black-&-yellow gnat bee	*Calochortus*	
Dufourea calochorti	mariposa-lily low-antenna bee	*Calochortus*	52
Perdita californica^	California black-&-yellow gnat bee	*Calochortus*	
Perdita tularensis	mariposa-lily black-&-yellow gnat bee	*Calochortus*	
Perdita wyomingensis	Wyoming black-&-yellow gnat bee	*Calochortus*	
DICOTS			
APIACEAE			
*Andrena chlorogaster**	interrupted-banding tiny spring bee	*Lomatium* (*Sanicula*)	
*Andrena merriami**	Merriam's broad-tibia spring bee	various	
Andrena microchlora	complete-banding tiny spring bee	*Lomatium* (*Sanicula*)	
Andrena pallidiscopa	lomatium wild-celery spring bee	*Lomatium*	47
Perdita nevadensis	montane black-&-yellow gnat bee	*Perideridia*	
ASTERACEAE			
Andrena ablegata	white-capped two-celled grassland spring bee	*Agoseris* (*Microseris, Nothocalais*)	
Andrena accepta	yellow-face Mexican summer bee	*Helianthus* (+)	
Andrena apacheorum^	Apache autumn bee	*Chrysothamnus* (*Ericameria, Gutierrezia*)	
*Andrena bendensis**	Bend autumn bee	*Chrysothamnus*	
Andrena bocensis	shiny-back autumn bee	*Chrysothamnus* (*Ericameria, Gutierrezia*)	
Andrena canadensis^	Canadian autumn bee	fall-blooming genera	
Andrena chalybioides	shiny-blue two-celled grassland spring bee	*Agoseris* (*Microseris, Nothocalais*)	
Andrena chromotricha^	white-haired autumn bee	fall-blooming genera	
Andrena citrinihirta	lemon-haired autumn bee	fall-blooming genera	
Andrena colletina	transparent-wing autumn bee	*Chrysothamnus* (*Ericameria, Gutierrezia*)	
Andrena columbiana	Columbian autumn bee	*Erigeron* (*Senecio*)	
Andrena costillensis	costilla autumn bee	various	
Andrena duboisi	goldfields western spring bee	*Lasthenia* (*Layia*)	
Andrena evoluta	shiny-green two-celled grassland spring bee	*Agoseris* (*Microseris, Nothocalais*)	

Bee Genera	Common Name	Plant Genera	Page
Andrena helianthi	sunflower Mexican summer bee	*Helianthus* (+)	
*Andrena hirticincta^*	greenish yellow autumn bee	fall-blooming genera	
*Andrena luteihirta^*	yellow-haired spring bee	fall-blooming genera	
Andrena nothocalaidis	dandelion two-celled grassland spring bee	*Agoseris* (*Microseris, Nothocalais*)	
Andrena nigripes	red-tailed blue-eyes spring bee	asters (*Solidago*)	36
Andrena nubecula	missing-fascia autumn bee	asters (*Solidago*)	
*Andrena pallidifovea**	fasciate flat-lipped spring bee	*Lasthenia* (*Layia*)	
Andrena pensilis	tidytips flat-lipped spring bee	*Lasthenia* (*Layia*)	
Andrena puthua	goldfields two-celled grassland spring bee	*Lasthenia* (*Layia*)	
Andrena scutellinitens	protruding-clypeus autumn bee	*Chrysothamnus* (*Ericameria, Gutierrezia*)	
Andrena subchalybea	strong-shouldered two-celled grassland spring bee	*Agoseris* (*Microseris, Nothocalais*)	
Andrena submoesta	golden-backed two-celled grassland spring bee	*Lasthenia* (*Layia*)	
Andrena sulcata	grooved-terga autumn bee	*Chrysothamnus* (*Ericameria, Gutierrezia*)	
Andrena surda	black-winged autumn bee	*Chrysothamnus* (*Solidago*)	
Andrena vulpicolor	fox-colored Mexican summer bee	*Chrysothamnus* (*Ericameria, Gutierrezia*)	
*Anthophora urbana**	common digger bee	various	37, 49
Ashmeadiella bucconis	low-ocellus pre-existing cavity-nesting bee	various	
*Ashmeadiella californica**	narrow-cheek pre-existing cavity-nesting bee	fall-blooming genera	
*Ashmeadiella cubiceps**	cube-headed pre-existing cavity-nesting bee	various	
Ashmeadiella difugita	black-midfemur pre-existing cavity-nesting bee	various	
*Calliopsis coloradensis**	fasciate black gnat bee	*Solidago* (*Rudbeckia*)	
*Colletes compactus hesperius**	elongate-malar silk-lining bee	asters (*Solidago*)	
*Colletes fulgidus**	ochre-backed silk-lining bee	fall-blooming genera	
*Colletes gypsicolens**	gray-backed silk-lining bee	*Chrysothamnus* (*Ericameria, Gutierrezia*)	
*Colletes hyalinus oregonensis**	variable silk-lining bee	various	
*Colletes lutzi monticola**	dense-fascia silk-lining bee	various	
Colletes simulans	transcontinental silk-lining bee	various	
Diadasia enavata	sunflower turret-building fuzzy bee	*Helianthus* (+)	
*Dianthidium curvatum sayi**	spiny-forecoxa resin-collecting bee	various	
Dianthidium heterulkei	acute-midtail resin-collecting bee	fall-blooming genera	
Dianthidium parvum	rounded-midtail resin-collecting bee	various	
*Dianthidium pudicum consimile**	elongate-midtail resin-collecting bee	various	
*Dianthidium pudicum pudicum**	elongate-midtail resin-collecting bee	various	
Dianthidium singulare	inflated-terga resin-collecting bee	various	
Dianthidium subparvum	truncate-midtail resin-collecting bee	various	
Dianthidium ulkei	no-midtail resin-collecting bee	various	
Dufourea marginata	sunflower low-antenna bee	*Helianthus* (+)	
Heriades cressoni	Cresson's pre-existing cavity-nesting bee	summer-blooming genera	
Hesperapis carinata	chaenactis melittid bee	spring-blooming genera	
*Megachile brevis**	small non-fisted leaf-cutting bee	various	

Bee Genera	Common Name	Plant Genera	Page
Megachile fidelis	three-horned cylindrical leaf-cutting bee	various	
Megachile inimica	unmodified-clypeus cylindrical leaf-cutting bee	various	
Megachile mellitarsis	rusty-tarsi cylindrical leaf-cutting bee	various	
Megachile parallela	large non-fisted leaf-cutting bee	various	
Megachile pugnata	cheek-spined cylindrical leaf-cutting bee	various	
Megachile latimanus	small-fisted leaf-cutting bee	various	
Megachile nevadensis	yellow-scopa leaf-cutting bee	fall-blooming genera	
Megachile perihirta	small-fisted leaf-cutting bee	various	
*Megachile pseudonigra**	melanic leaf-cutting bee	various	
*Megachile subnigra**	black-scopa leaf-cutting bee	various	
Megachile wheeleri	white-backed leaf-cutting bee	various	
Melissodes agilis	protruding-clypeus long-horned bee	various	
Melissodes appressa	yellow-sternum long-horned bee	*Chrysothamnus* (*Ericameria, Gutierrezia*)	
Melissodes bimatris	ochre-scopa long-horned bee	*Chrysothamnus* (*Ericameria, Gutierrezia*)	
Melissodes confusa	confusing long-horned bee	various	
Melissodes glenwoodensis	red-backed flanged long-horned bee	*Chrysothamnus* (*Ericameria, Gutierrezia*)	
*Melissodes lupina**	v-tailed flanged long-horned bee	various	
Melissodes lustra	u-tailed flanged long-horned bee	*Chrysothamnus* (*Ericameria, Gutierrezia*)	
Melissodes lutulenta	black-mandible long-horned bee	various	
*Melissodes menuachus**	white-labrum long-horned bee	*Grindelia* (*Solidago*)	
*Melissodes metenua**	black-forehead flanged long-horned bee	various	
Melissodes microsticta	thin-moustache long-horned bee	various	
Melissodes moorei	coastal long-horned bee	summer-blooming genera	
Melissodes pallidisignata	red-sternum long-horned bee	summer-blooming genera	
Melissodes paulula	half-moustache long-horned bee	summer-blooming genera	
Melissodes plumosa	pale-forehead flanged long-horned bee	summer-blooming genera	
Melissodes pullatella	chocolate long-horned bee	*Grindelia*	41
Melissodes rivalis	thistle long-horned bee	*Cirsium*	
Melissodes robustior	pale-sternum long-horned bee	summer-blooming genera	
Melissodes saponellus	white-backed long-horned bee	summer-blooming genera	
Melissodes semilupina	depressed-flagellum long-horned bee	*Chrysothamnus* (*Ericameria, Gutierrezia*)	
Melissodes stearnsi	hooked-tongue flanged long-horned bee	summer-blooming genera	
Melissodes subagilis	white-scopa long-horned bee	summer-blooming genera	
Melissodes verbesinarum	black-faced long-horned bee	various	
Melissodes vernalis	shiny-backed long-horned bee	various	
Osmia californica	thistle iridescent mason bee	*Cirsium* (*Wyethia*)	
Osmia coloradensis	Colorado iridescent mason bee	various	
Osmia grinnelli	bare-wing iridescent mason bee	various	
Osmia marginipennis	chaenactis iridescent mason bee	various	
Osmia montana montana	smooth-faced mason bee	various	
Osmia montana quadriceps	smooth-faced mason bee	various	
Osmia subaustralis	widespread iridescent mason bee	various	

Bee Genera	Common Name	Plant Genera	Page
*Osmia texana**	Texas iridescent mason bee	*Cirsium*	
Perdita adjuncta	rabbitbrush black-&-yellow gnat bee	*Chrysothamnus* (*Ericameria, Gutierrezia*)	
Perdita aemula	goldenweed composite black-&-yellow gnat bee	*Chrysothamnus* (*Ericameria, Gutierrezia*)	
Perdita albescens^	Canadian composite black-&-yellow gnat bee	fall-blooming genera	
Perdita ciliata	hairy composite black-&-yellow gnat bee	fall-blooming genera	
Perdita idahoensis^	Idaho composite black-&-yellow gnat bee	*Chrysothamnus* (*Ericameria*)	
Perdita lepidosparti	broomsage composite black-&-yellow gnat bee	fall-blooming genera	
Perdita melanostoma^	black-mouthed composite black-&-yellow gnat bee	*Chrysothamnus* (*Ericameria*)	
Perdita nuda^	bare composite black-&-yellow gnat bee	fall-blooming genera	
Perdita oregonensis	Oregon composite black-&-yellow gnat bee	*Chrysothamnus* (*Ericameria, Gutierrezia*)	
Perdita rivalis	aster composite black-&-yellow gnat bee	asters (*Erigeron*)	
Perdita similis	similar composite black-&-yellow gnat bee	fall-blooming genera	
Perdita stottleri	Stottler's composite black-&-yellow gnat bee	fall-blooming genera	
Perdita zonalis aequalis	equal composite black-&-yellow gnat bee	fall-blooming genera	
Svastra obliqua expurgata	sunflower long-horned bee	*Helianthus* (+)	
BERBERIDACEAE			
Andrena berberidis?	Oregon-grape hip-finger spring bee	*Berberis*	
BORAGINACEAE			
Andrena chapmanae	Chapman's broad-tibia spring bee	*Cryptantha*	
Andrena cryptanthae	cryptantha spear-tongued spring bee	*Cryptantha*	
Andrena osmioides	metallic cryptantha spear-tongued spring bee	*Cryptantha*	
Colletes paniscus sculleni	bluebells silk-lining bee	*Mertensia*	
Eucera amsinckiae	fiddleneck spring long-horned bee	*Amsinckia*	
*Hoplitis grinnellii**	Grinnell's pre-existing cavity-nesting bee	(*Cryptantha*)	
Hoplitis jacintana	San Jacinto pre-existing cavity-nesting bee	*Cryptantha*	43
*Osmia mertensiae**	bluebells iridescent mason bee	*Mertensia* (*Lappula*)	
Osmia sculleni	spiny-tongued iridescent mason bee	*Hackelia*	
BRASSICACEAE			
Andrena gordoni?	Gordon's broad-tibia spring bee	various	
Andrena hallii?	Hall's earlike-pronotum spring bee	*Stanleya*	
Andrena piperi	bladderpod tiny spring bee	various	
Andrena prima	bladderpod radish spring bee	*Physaria*	
Andrena scurra	non-metallic broad-tibia spring bee	*Lomatium*	
CACTACEAE			
Diadasia australis^	cactus turret-building fuzzy bee	various	
CAMPANULACEAE			
Dufourea campanulae	bellflower low-antenna bee	*Campanula*	

Bee Genera	Common Name	Plant Genera	Page
CAPRIFOLIACEAE			
Andrena cheyennorum	Cheyenne cone-faced spring bee	*Symphoricarpos*	
Andrena trevoris?	unique cheek-ridged spring bee	*Symphoricarpos*	
Dufourea holocyanea	snowberry low-antenna bee	*Symphoricarpos*	
CLEOMACEAE			
Perdita zebrata	bee-flower black-&-yellow gnat bee	*Cleome*	
CONVOLVULACEAE			
Diadasia bituberculata	morning-glory turret-building fuzzy bee	*Convolvulus*	
CORNACEAE			
*Andrena flocculosa**	rudimentary-banding osier spring bee	*Cornus*	
Andrena persimulata	complete-banding osier spring bee	*Cornus*	
CUCURBITACEAE			
Peponapis pruinosa	light-colored gourd long-horned bee	various	
Xenoglossa angustior	dark-colored gourd long-horned bee	various	
ERICACEAE			
Andrena arctostaphylae	manzanita flat-necked spring bee	*Arctostaphylos*	
Andrena cristata	labrum-crista bearded spring bee	*Arctostaphylos*	
Andrena monogonoparia?	metallic Arctic spring bee	*Arctostaphylos*	
Andrena obscuripostica	manzanita bearded spring bee	*Arctostaphylos*	
Andrena principalis?	earliest long-tongued spring bee	*Arctostaphylos*	
Andrena vandykei	Vandyke's flat-necked spring bee	*Arctostaphylos*	
Eucera acerba	manzanita spring long-horned bee	*Arctostaphylos*	
Habropoda cineraria	manzanita digger bee	*Arctostaphylos*	
Osmia inermis	ericaceous iridescent mason bee	*Kalmia* (*Phyllodoce, Rhododendron*)	
*Osmia ribifloris**	western iridescent mason bee	*Arctostaphylos*	
FABACEAE			
*Andrena buckeli**	Buckell's Arctic spring bee	*Lupinus*	
*Andrena lupinorum**	pea-preferring black-haired spring bee	various	
Andrena plana	clover broad-tibia spring bee	*Trifolium*	
Anthidium clypeodentatum	four-toothed hornet-mimic bee	various	
Anthidium collectum	black-brushed hornet-mimic bee	various	
Anthidium edwardsii	v-clypeus hornet-mimic bee	various	
Anthidium emarginatum	variable hornet-mimic bee	(*Acmispon, Astragalus, Lupinus*)	
*Anthidium formosum**	high-mountain hornet-mimic bee	various	
Anthidium illustre	large hornet-mimic bee	various	
*Anthidium maculosum**	acute-tailspines hornet-mimic bee	various	
Anthidium mormonum	two-toothed hornet-mimic bee	various	

Bee Genera	Common Name	Plant Genera	Page
Anthidium palliventre	sand-nesting hornet-mimic bee	(*Acmispon, Astragalus, Lupinus*)	
Anthidium tenuiflorae	alpine hornet-mimic bee	various	
Anthidium utahense	montane hornet-mimic bee	*Acmispon* (*Astragalus, Lupinus*)	41
*Ashmeadiella aridula**	deervetch pre-existing cavity-nesting bee	*Acmispon*	
Atoposmia anthodyta	flower-visiting pre-existing cavity-nesting bee	(various)	
Calliopsis anthidius	clover black-&-yellow gnat bee	*Trifolium*	
Calliopsis micheneri	Michener's black-&-yellow gnat bee	*Trifolium*	
Calliopsis trifolii	two-brush black-&-yellow gnat bee	*Trifolium*	
Dufourea afasciata	clover low-antenna bee	*Trifolium*	
Dufourea spinifera	alpine-clover low-antenna bee	*Trifolium*	
*Eucera edwardsii**	red-backed spring long-horned bee	various	
*Eucera frater albopilosa**	black-scopa spring long-horned bee	various	
*Eucera frater frater**	red-scopa spring long-horned bee	various	
*Hoplitis albifrons**	white-faced pre-existing cavity-nesting bee	*Acmispon*	
*Hoplitis fulgida**	bright-green iridescent mason bee	(*Acmispon, Lupinus*)	
*Hoplitis grinnellii**	Grinnell's pre-existing cavity-nesting bee	*Acmispon*	
Hoplitis hypocrita	Cockerell's pre-existing cavity-nesting bee	*Acmispon* (*Astragalus, Lupinus*)	
Hylaeus bisinuatus?	introduced pollen-regurgitating bee	*Melilotus*	
*Megachile angelarum**	Los Angeles petal-plucking bee	*Acmispon*	
*Megachile brevis**	small non-fisted leaf-cutting bee	various	
*Megachile frigida**	frigid non-fasciate leaf-cutting bee	various	
Megachile melanophaea	black-bearded non-fasciate leaf-cutting bee	various	
*Megachile rotundata**	alfalfa non-fisted leaf-cutting bee	*Medicago*	
*Nomia melanderi**	commercial alkalai bee	various	51
*Nomia tetrazonata tetrazonata**	red-tailed alkalai bee	various	
*Osmia albolateralis**	white-sided iridescent mason bee	various	
*Osmia atrocyanea**	reticulate-propoduem iridescent mason bee	various	
Osmia calla	gorgeous iridescent mason bee	various	
*Osmia densa**	confluently-punctate iridescent mason bee	various	
*Osmia gabrielis**	San Gabriel iridescent mason bee	various	
Osmia integra	red-backed iridescent mason bee	various	
Osmia longula	long-spined iridescent mason bee	*Astragalus*	
*Osmia malina**	depressed-clypeus iridescent mason bee	various	
Osmia nifoata	Cockerell's iridescent mason bee	various	
Osmia nigrifrons	flat-chinned iridescent mason bee	various	
Osmia nigrobarbata	confluent-punctation iridescent mason bee	various	
Osmia obliqua	acute-basitarsus iridescent mason bee	various	
Osmia regulina	clover iridescent mason bee	various	
Perdita trisignata^	lotus black-&-yellow gnat bee	*Acmispon*	

Bee Genera	Common Name	Plant Genera	Page
GROSSULARIACEAE			
Andrena hemileuca	white-scopa Arctic spring bee	*Ribes*	
Andrena latifrons?	wide-faced bearded spring bee	*Ribes*	
Andrena porterae	Porter's hip-finger spring bee	*Ribes*	
Andrena submaura	currant hip-finger spring bee	*Ribes*	
Osmia pikei	currant iridescent mason bee	*Ribes*	
HYDROPHYLLACEAE			
Andrena macrocephala	red-tailed baby-blue-eyes spring bee	*Nemophila*	
Andrena nemophilae	baby-blue-eyes broad-headed spring bee	*Nemophila*	
*Andrena nigroclypeata^*	black-faced baby-blue-eyes spring bee	*Nemophila*	
Andrena torulosa	baby-blue-eyes nemophila spring bee	*Nemophila*	
Anthidium banningense	narrow cross-stripe hornet-mimic bee	*Phacelia*	
Anthidium collectum	black-brushed hornet-mimic bee	(*Phacelia*)	
Anthidium emarginatum	variable hornet-mimic bee	*Phacelia*	
Anthidium illustre	large hornet-mimic bee	(*Phacelia*)	
Anthidium mormonum	two-toothed hornet-mimic bee	(*Phacelia*)	
Anthidium palliventre	sand-nesting hornet-mimic bee	*Phacelia*	
Anthidium tenuiflorae	alpine hornet-mimic bee	(*Phacelia*)	
*Anthophora centriformis**	phacelia digger bee	*Phacelia*	
Atoposmia copelandica	metallic pre-existing cavity-nesting bee	*Phacelia* (*Nemophila*)	
Calliopsis xenus	phacelia black-&-yellow gnat bee	*Phacelia*	
Chelostoma californicum	California pre-existing cavity-nesting bee	*Phacelia*	
Chelostoma minutum	tiny pre-existing cavity-nesting bee	*Phacelia*	
Chelostoma phaceliae	heliotrope pre-existing cavity-nesting bee	*Phacelia*	
Colletes consors pascoensis?	phacelia silk-lining bee	*Phacelia*	
Dufourea trochantera	heliotrope low-antenna bee	*Phacelia*	
Dufourea virgata	widely-nectaring low-antenna bee	(*Phacelia*)	
*Hoplitis albifrons**	white-faced pre-existing cavity-nesting bee	(*Phacelia*)	
*Hoplitis fulgida**	bright-green iridescent mason bee	*Phacelia*	
*Hoplitis grinnellii**	Grinnell's pre-existing cavity-nesting bee	(*Phacelia*)	
Hoplitis laevibullata?	inflated-tarsus iridescent mason bee	*Nemophila*	
Hoplitis louisae?	Louisa's iridescent mason bee	*Phacelia*	
Panurginus atriceps	baby-blue-eyes black gnat bee	*Nemophila*	
LAMIACEAE			
Anthidium edwardsii	v-clypeus hornet-mimic bee	(*Trichostema*)	
*Anthidium maculosum**	acute-tailspines hornet-mimic bee	(*Stachys*)	
Calliopsis zonalis sierrae	monardella black-&-yellow gnat bee	*Monardella*	
*Osmia exigua**	mint iridescent mason bee	various	

Bee Genera	Common Name	Plant Genera	Page
LIMNANTHACEAE			
Andrena pulverea	meadowfoam western spring bee	*Limnanthes*	
Panurginus occidentalis	meadowfoam black gnat bee	*Limnanthes*	
MALVACEAE			
Diadasia lutzi lutzi	globemallow turret-building fuzzy bee	*Sphaeralcea*	
Diadasia nigrifrons	checkermallow turret-building fuzzy bee	*Sidalcea* (*Iliamna*)	
Diadasia nitidifrons	mallow turret-building fuzzy bee	*Sidalcea* (*Iliamna*)	
Diadasia olivacea	bushmallow turret-building fuzzy bee	*Sidalcea* (*Iliamna*)	
Macrotera arcuata^	big-headed gnat bee	*Sphaeralcea*	
MELANTHIACEAE			
Andrena astragali	zigadenus paleo spring bee	*Toxicoscordion*	
NAMACEAE			
Calliopsis fracta	yerba santa black-&-yellow gnat bee	*Eriodictyon*	
Calliopsis linsleyi	Linsley's black-&-yellow gnat bee	*Eriodictyon*	
ONAGRACEAE			
Andrena anograe^	red-backed evening-primrose spring bee	*Oenothera*	
Andrena chylismiae	suncup evening-primrose spring bee	*Camissonia*	
Andrena lewisorum	Lewis' two-celled grassland spring bee	*Clarkia*	
Andrena linsleyana^	Linsley's evening-primrose spring bee	*Oenothera*	
Andrena nevadae?	Nevada evening-primrose spring bee	*Camissonia*	
Andrena omninigra^	crew-cut evening-primrose spring bee	*Clarkia*	
Andrena raveni	Raven's evening-primrose spring bee	*Camissonia*	
Andrena rozeni^	Rozen's evening-primrose spring bee	*Camissonia*	
Andrena vanduzeei^	gayophytum evening-primrose spring bee	*Gayophytum*	
Ceratina sequoiae	clarkia micro-carpenter bee	*Clarkia*	
Diadasia angusticeps	clarkia turret-building fuzzy bee	*Clarkia*	
Dufourea orovada	evening-primrose low-antenna bee	*Camissonia*	
Dufourea scabricornis	gayophytum low-antenna bee	*Gayophytum*	
Dufourea spilura	spine-bearing low-antenna bee	*Gayophytum*	
Hesperapis regularis^	clarkia melittid bee	*Clarkia*	
Lasioglossum aberrans	evening-primrose sociality-adaptive bee	*Oenothera*	
Megachile gravita	bare-belly leaf-cutting bee	*Clarkia*	
Megachile pascoensis	spotted-scutum leaf-cutting bee	*Clarkia*	
Melissodes clarkiae	clarkia flanged long-horned bee	*Clarkia*	
OROBANCHACEAE			
Anthidium edwardsii	v-clypeus hornet-mimic bee	(*Cordylanthus*)	
*Anthidium placitum**	cordylanthus hornet-mimic bee	*Cordylanthus*	
Dufourea versatilis rubriventer	monkeyflower low-antenna bee	(*Castilleja*)	
*Megachile angelarum**	Los Angeles petal-plucking bee	(*Cordylanthus*)	

Bee Genera	Common Name	Plant Genera	Page
PAPAVERACEAE			
Andrena cercocarpi	poppy concave-lipped spring bee	*Eschscholzia*	
*Calliopsis obscurella**	poppy black-&-yellow gnat bee	*Eschscholzia*	
*Perdita quadrisignata^*	poppy black-&-yellow gnat bee	*Eschscholzia*	
PHRYMACEAE			
Calliopsis trifolii	two-brush black-&-yellow gnat bee	(*Erythranthe*)	
Dufourea versatilis rubriventer	monkeyflower low-antenna bee	*Erythranthe*	
PLANTAGINACEAE			
Ashmeadiella australis	penstemon pre-existing cavity-nesting bee	*Penstemon*	
Atoposmia abjecta	displaced pre-existing cavity-nesting bee	*Penstemon*	
Atoposmia anthodyta	flower-visiting pre-existing cavity-nesting bee	*Penstemon* (*Collinsia*)	
Atoposmia elongata	elongate pre-existing cavity-nesting bee	*Penstemon*	
*Atoposmia oregona**	Oregon pre-existing cavity-nesting bee	*Penstemon*	
Atoposmia triodonta shastensis	three-toothed pre-existing cavity-nesting bee	*Penstemon*	
*Osmia bakeri**	Chinese-houses iridescent mason bee	*Collinsia*	
*Osmia brevis**	penstemon iridescent mason bee	*Penstemon* (*Collinsia*)	
POLEMONIACEAE			
*Andrena levipes**	gilia Mexican summer bee	*Gilia* (*Linanthus*)	
Andrena ribblei?	Ribble's paleo spring bee	*Polemonium*	
Andrena segregans	Jacob's ladder paleo spring bee	*Polemonium*	
Dufourea bernardina	San Bernardino low-antenna bee	*Gilia*	
Dufourea femorata	gilia low-antenna bee	*Gilia*	
Dufourea virgata	widely-nectaring low-antenna bee	*Gilia*	
POLYGONACEAE			
Perdita claypolei limulata	buckwheat black-&-yellow gnat bee	*Eriogonum*	32
PRIMULACEAE			
*Macropis nuda^*	loosestrife oil-collecting bee	*Lysimachia*	41
RANUNCULACEAE			
Andrena caerulea	iridescent-blue paleo spring bee	*Ranunculus*	
Andrena cuneilabris	buttercup two-celled grassland spring bee	*Ranunculus*	
Andrena suavis	buttercup paleo spring bee	*Ranunculus*	
*Eucera delphinii**	larkspur spring long-horned bee	*Delphinium*	
Panurginus nigrihirtus	buttercup black gnat bee	*Ranunculus*	
RHAMNACEAE			
Andrena candidiformis	mountain-lilac tiny spring bee	*Ceanothus* (*Rhamnus*)	
Andrena ceanothifloris	melanic Arctic spring bee	*Ceanothus*	
Andrena cleodora	mountain-lilac moonscape-waist spring bee	*Ceanothus*	

Bee Genera	Common Name	Plant Genera	Page
Andrena edwardsii?	Edward's Arctic spring bee	*Ceanothus*	
*Andrena pertristis carliniformis**	lilac-preferring black-haired spring bee	*Ceanothus*	
Panurginus atriceps	baby-blue-eyes black gnat bee	(*Ceanothus*)	
ROSACEAE			
Andrena birtwelli	Birtwell's Arctic spring bee	*Potentilla* (*Drymocallis*)	
Andrena forbesii?	Forbes' moonscape-waist spring bee	woody genera	
*Andrena hippotes**	red-legged moonscape-waist spring bee	woody genera	
*Andrena medionitens**	crew-cut scoop-jawed spring bee	woody genera	
Andrena melanochroa	non-metallic tiny spring bee	*Fragaria* (*Potentilla*)	
*Andrena prunorum prunorum**	red-legged twisted-spur spring bee	woody genera	
Andrena topazana?	yellow-haired spring bee	*Potentilla* (*Prunus*)	
Andrena washingtoni?	Washington Arctic spring bee	*Rosa* (*Rubus*)	
Calliopsis edwardsii	cinquefoil black-&-yellow gnat bee	*Potentilla* (*Rosaceae*)	
Colletes nigrifrons?	cinquefoil silk-lining bee	*Potentilla* (*Drymocallis*)	
Dufourea fimbriata	cinquefoil low-antenna bee	*Drymocallis, Potentilla*	
Hylaeus basalis	big-scape pollen-regurgitating bee	*Potentilla* (*Rosaceae*)	
Hylaeus nunenmacheri	Nunenmacher's pollen-regurgitating bee	*Potentilla* (*Drymocallis*)	
SALICACEAE			
*Andrena andrenoides^*	red-tailed two-celled willow spring bee	*Salix*	
Andrena bucculenta	crew-cut two-celled willow spring bee	*Salix*	
*Andrena clarkella^*	Clark's Arctic spring bee	*Salix*	
Andrena concinnula	appressed-hair two-celled willow spring bee	*Salix*	
Andrena forbesii?	Forbes' moonscape-waist spring bee	(*Salix*)	
Andrena frigida	frigid Arctic spring bee	*Salix*	
Andrena gibberis?	hump-backed two-celled willow spring bee	*Salix*	
Andrena illinoiensis?	iridescent-blue tiny spring bee	*Salix*	
*Andrena ishii?^*	Ishi's tiny spring bee	*Salix*	
Andrena macoupinensis	willow Arctic spring bee	*Salix*	
Andrena mariae	red-tailed moonscape-waist spring bee	*Salix*	
Andrena nevadensis	mountain two-celled willow spring bee	*Salix*	
Andrena nigrae	red-tailed tiny spring bee	*Salix*	
Andrena perarmata	mandible-spur Arctic spring bee	*Salix*	
Andrena salictaria	willow tiny spring bee	*Salix*	
Andrena semipunctata	impunctate-tergum moonscape-waist spring bee	*Salix*	
Andrena sigmundi	Sigmund's moonscape-waist spring bee	*Salix*	
Andrena striatifrons	willow moonscape-waist spring bee	*Salix*	
Andrena subaustralis	willow earlike-pronotum spring bee	*Salix*	
Andrena trizonata	willow scoop-jawed spring bee	*Salix*	
*Andrena wellesleyana^*	Wellesley two-celled willow spring bee	*Salix*	
Perdita salicis imperialis	willow black-&-yellow gnat bee	*Salix*	

INDEX

Scientific and common names for plants described in this volume and scientific names in Appendix 1 are indexed. Accepted family names of plants in the flora are in **ALL CAPS BOLD**, accepted and treated generic and lower taxa names are in **bold**. Taxonomic synonyms, misapplied names, and taxa not fully treated in the flora are in *italics*. Common names are in SMALL CAPS, and common family names are in ALL CAPS. Page numbers of the main descriptions are in **bold**; illustrations are in regular; other entries such as appearances in keys or discussion paragraphs are in light.

Concepts that are more narrowly circumscribed are indented and listed below their inclusive name. Page numbers are listed for names (scientific or common) as they appear in correspondence to a description. An inclusive name may occur only in the context of one of these narrower concepts, and therefore page numbers are given only for the narrower concept, whether an accepted name, a synonym, or a common name.

A

C

D

E

F

H

I

M

N

O

Q

T

U

V

X

Y

Z

BIOGRAPHIES

LINDA K. HARDISON grew up in West Texas. Gardening, camping, and backpacking were some of her early botanical inspirations. She discovered her passion for biological research studying seagrass ecology at the University of Texas at Austin, where she earned bachelor's degrees in botany and marine biology. Laboratory work and graduate studies shifted her focus to progressively smaller elements of botany: she studied cell wall formation, photosynthesis, plant-microbe interactions, and completed a Ph.D. in molecular genetics. OregonFlora brought her happily back to the study and appreciation of whole organisms and ecosystems. Linda has volunteered for OregonFlora since its inception and has been its director since 2005.

TANYA HARVEY grew up in New England. She fell in love with plants while still a small child and spent her childhood drawing her animals and her family's garden. Moving to the country while in high school gave her the opportunity to hike, draw, raise butterflies, and learn about wildflowers. Tanya earned a bachelor's degree in Math at Dartmouth College, but with the advent of home computers, she taught herself graphic design and found her way back to art. When not doing art and design, she is usually out botanizing in the Western Cascades or doing restoration work on her rural property. Working on the *Flora of Oregon* since 2012 has truly been the ideal job as it allows her to combine her skills in graphic design, drawing, and photography with her lifelong passion for plants.

THEA JASTER joined OregonFlora in 2002. She works primarily as a data manager and flora editor and also enjoys working with undergraduates and volunteers. Thea graduated from the University of New Hampshire ('90) with a B.A. in Botany and has worked in various roles involving her botanical training ever since. Her main interests are in plant conservation, biodiversity, and plant taxonomy. Thea's understanding and interest in database management (originally sourcing from her role at The Nature Conservancy in Arlington, Virginia in the '90s) have also served her well in her role with OregonFlora. She enjoys knowing her work (on the *Flora* and website) will not only provide important knowledge for fellow botanists but will also engage and inspire those new to the delights of learning Oregon's diverse flora.

STEPHEN C. MEYERS was born and raised in rural upstate New York. Stephen earned a bachelor's degree in aeronautical science on Long Island and worked for several years as an air traffic controller. In his mid-20s, he switched careers and inscribed cemetery monuments for the next decade. During this time, an interest in alpine mountaineering introduced Stephen to the mountains of the Northwest and, in turn, to its botanical wonders. Lured by the latter, he moved to Oregon and obtained a master's degree in genetics and a doctorate in botany from Oregon State University. Employed by OregonFlora since 2010, Stephen currently resides in Corvallis, Oregon.

KATIE E. MITCHELL was born and raised in western Washington where camping, hiking, and other outdoor pursuits sparked her interest in the natural world. After earning a B.S. in botany and biology from the University of Washington, she got her first job in the field as a weed warrior at Olympic National Park. She later conducted plant surveys and studies for several years in different areas of Washington. She felt honored to be invited to join the team working on this flora and has been part of the project since 2004.

JOHN MYERS began illustrating plants at the Missouri Botanical Garden and was staff illustrator during its most bodacious years of tropical plant exploration. There he was introduced to plant taxonomy, field collecting and true weirdness by some of the best botanists of that era. It was in this collegial environment, where everyone held a fascination for plant forms and relationships, felt a joy for the natural world, and witnessed the destruction of the world's great forests that he developed a grotesque but accurate world view. His work is found in various journals, field guides, and floras, and his favorite, the *Flora of Oregon*. Myers came from Ohio, and has a bachelor's degree in Zoology from Miami University. He now lives in Oregon's Willamette Valley AVA with his dog and curses whenever he hears the whine of a chainsaw.

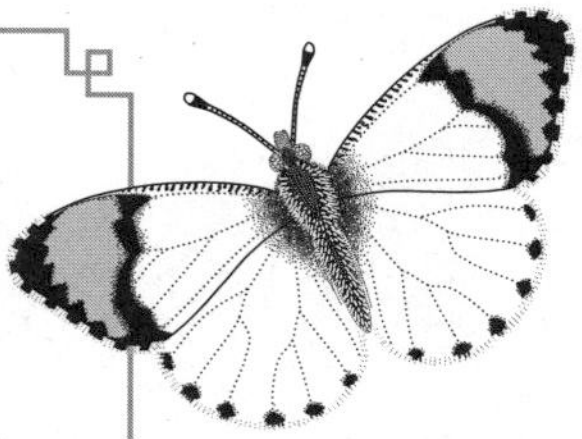

Flora of Oregon
Volume 2

The main typeface is Adobe Caslon Pro. Cronos Pro is used for captions, appendices, and the index. The symbol ❦, used to indicate endemics, is an ornament from Minion Pro.

Text pages are acid-free 70 gsm New Classic White paper.

Color pages are acid-free 100 gsm matte art paper.

End papers are 180 gsm woodfree paper.

The dust jacket is 150 gsm gloss art paper with UV coating.

The binding cloth for the cover is BK-3 from Dong-A Co. with Murata Gleam metallic foil stamping.

The book was designed and produced on a Mac in Adobe InDesign®.

Pen & ink illustrations created for this volume were scanned and arranged in Adobe Photoshop®. Other illustrations were created in Adobe Illustrator®.

Maps were created in ArcGIS.

Printed and bound in Korea by ArtPac.

Book design is by Tanya Harvey Design.

Oregon Counties

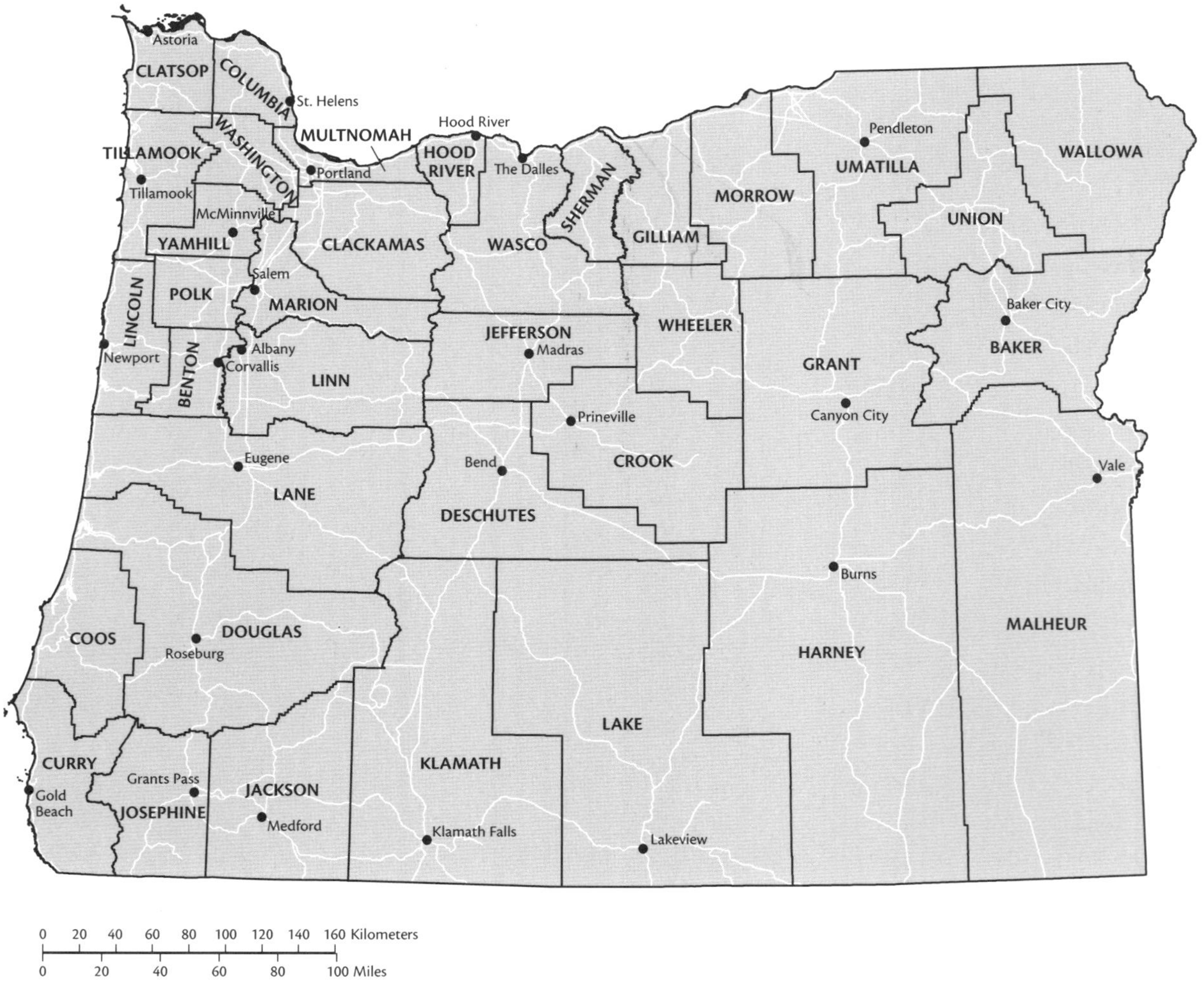